Proceedings

29th ACM/IEEE Design Automation Conference

29th ACM/IEEE DESIGN AUTOMATION CONFERENCE®

JUNE 8 - 12, 1992
ANAHEIM, CALIFORNIA
ANAHEIM CONVENTION CENTER

PROCEEDINGS 1992

sponsored by ACM SIGDA® IEEE COMPUTER SOCIETY DATC® IEEE CIRCUITS & SYSTEMS SOCIETY® *in cooperation with* EDAC Electronic Design Automation Companies

IEEE Computer Society Press The Institute of Electrical and Electronics Engineers, Inc.

Proceedings

29th ACM/IEEE Design Automation Conference

June 8 - 12, 1992
Anaheim, California

Sponsored by
IEEE Computer Society Technical Committee on Design Automation
IEEE Circuits & Systems Society
ACM SIGDA

In cooperation with
EDAC (Electronic Design Automation Companies)

IEEE Computer Society Press
Los Alamitos, California

Washington • Brussels • Tokyo

The papers in this book comprise the proceedings of the meeting mentioned on the cover and title page. They reflect the authors' opinions and, in the interests of timely dissemination, are published as presented and without change. Their inclusion in this publication does not necessarily constitute endorsement by the editors, the IEEE Computer Society Press, or the Institute of Electrical and Electronics Engineers, Inc.

Published by the
IEEE Computer Society Press
10662 Los Vaqueros Circle
PO Box 3014
Los Alamitos, CA 90720-1264

IEEE Computer Society Press Order Number 2822
Library of Congress Number 85-644294
IEEE Catalog Number 92CH3144-3
ACM Order Number 477920
ISBN 0-8186-2821-9 (microfiche)
ISBN 0-8186-2822-7 (case)
ISBN 0-89791-515-1(ACM)
ISBN 0-89791-516-X (ACM Library Series)
ISSN 0738-100X

Additional copies can be ordered from

IEEE Computer Society Press
Customer Service Center
10662 Los Vaqueros Circle
PO Box 3014
Los Alamitos, CA 90720-1264

IEEE Service Center
445 Hoes Lane
PO Box 1331
Piscataway, NJ 08855-1331

IEEE Computer Society
13, avenue de l'Aquilon
B-1200 Brussels
BELGIUM

IEEE Computer Society
Ooshima Building
2-19-1 Minami-Aoyama
Minato-ku, Tokyo 107
JAPAN

ACM Order Department, PO Box 64145, Baltimore, MD 21264, 1-800-342-6626, 1-410-528-4261 (outside US, MD, and AK)

Production Editor: Robert Werner
Printed in the United States of America by Braun-Brumfield, Inc.

The Institute of Electrical and Electronics Engineers, Inc.

EXECUTIVE COMMITTEE

GENERAL CHAIR
Daniel G. Schweikert
Cadence Design Systems, Inc.
555 River Oaks Pkwy., Bldg. 4
San Jose, CA 95134
(408) 944-7297

VICE CHAIR
Alfred E. Dunlop
AT&T Bell Labs.
600 Mountain Ave., Rm. 3D-484
Murray Hill, NJ 07974-2070
(908) 582-6579

FINANCE CHAIR
Thomas Minot
Sun Microsystems, Inc.
2550 Garcia Ave.
Mountain View, CA 94043-1100
(415) 336-2055

ARRANGEMENTS CHAIR
Joseph D. Mastroianni
Cadence Design Systems, Inc.
2475 Augustine Dr.
Santa Clara, CA 95054-3082
(408) 987-5317

TUTORIAL CHAIR
Mary Jane Irwin
Penn State University
Dept. of CS, 333 Whitmore Lab.
University Park, PA 16802
(814) 865-9505

AUDIO-VISUAL CHAIR
Charles A. Shaw
Cadence Design Systems, Inc.
2475 Augustine Dr.
Santa Clara, CA 95054-3082
(408) 987-5025

EUROPEAN/MIDDLE EAST REP.
Egon Horbst
Siemens AG
Zentrale Labs.
Otto-Hahn-Ring 6
Munchen 83, D-8000 Germany
(49) 89-636-3354

ACM REPRESENTATIVE
Michael Lorenzetti
Mentor Graphics Corp.
8005 SW Boeckman Rd.
Wilsonville, OR 97070-7777
(503) 685-7000

IEEE/CAS REPRESENTATIVE
Ian Getreu
Analogy, Inc.
P.O. Box 1669
Beaverton, OR 97075-1669
(503) 626-9700

EXHIBIT MANAGER
Marie R. Pistilli
MP Associates, Inc.
7490 Clubhouse Rd., #102
Boulder, CO 80301
(303) 530-4333

TECHNICAL PROGRAM CHAIR
Bryan T. Preas
Xerox Parc
3333 Coyote Hill Rd.
Palo Alto, CA 94304
(415) 812-4845

PUBLICATIONS CHAIR
Basant R. Chawla
AT&T Bell Labs.
1247 S. Cedar Crest Blvd.
Rm. 2E-246
Allentown, PA 18103
(215) 770-3484

INDUSTRY PROGRAM CHAIR
Richard C. Smith
Texas Instruments
6550 Chase Oaks Blvd.
Plano, TX 75086
(214) 575-3673

PUBLICITY CHAIR
Dennis W. Shaklee
AT&T Bell Labs.
11900 N. Pecos St., Rm. 1F58
Denver, CO 80234
(303) 538-4280

PAST CHAIR
A. Richard Newton
University of California
Dept. of EECS, Rm. 512 Cory Hall
Berkeley, CA 94720
(510) 642-2967

ASIA/INDIA/S. PACIFIC REP.
Tokinori Kozawa
Hitachi Ltd.
Semi-Cond. Des. & Dev. Ctr.
5-20-1, Josuihoncho, Kodaira
Tokyo 187, Japan
(81) 42-321-4146

IEEE REPRESENTATIVE
Ronald Waxman
University of Virginia
Thornton Hall, Dept. of EE
Charlottesville, VA 22903
(804) 924-6086

CONFERENCE MANAGER
P.O. Pistilli
MP Associates, Inc.
7490 Clubhouse Rd., #102
Boulder, CO 80301
(303) 530-4333

General Chair's Message

It is with great pleasure that I take this opportunity to introduce you to the ACM/IEEE Design Automation Conference. This is our 29th year of operation. By the conventional statistics, the Conference continues to be very successful: starting with 137 attendees in 1963, attendance has risen almost every year to 17,000 attendees last year in San Francisco. The electronic design automation industry has grown even more incredibly, with annual software sales of $1.3 billion. In the face of this rapid change, the DAC Executive Committee is newly challenged every year to make the Conference valuable, interesting, and exciting to an ever increasing number of attendees with an increasingly diverse range of interests.

This year, we have chosen to increase our emphasis on presentations which we think will be of interest to the design automation *user* community. Every time slot offers something in this area: short tutorials, panels, and selected papers. The Technical Program Committee and the Industry Program Committee has cooperated closely to provide a better balance between theoretical/algorithmic papers and presentations which dealt with the equally challenging task of building/using a complete DA system in a creative, production design environment.

Overall, the Technical Program has attracted the usual flood of excellent contributions. Bryan Preas, our Technical Program Chair received over 440 papers submissions. The Technical Program Committee selected the best 125 papers, organizing them into 44 sessions.

Having attended over 20 DACs in the past, I have personally found the tutorial offerings to have been an outstanding benefit of the Conference. Many attendees have come to the same conclusion. Our full-day tutorial program has been extremely popular, with many of the tutorials selling out well before the start of the Conference. Our Tutorial Chair, Mary Jane Irwin, has tried to satisfy the demand and has been able to expand the program to nine tutorials.

A major benefit for many attendees has been the design automation industry Exhibits. DAC offers you a chance to stay abreast of the rapid advances available in the EDA marketplace. Almost all of the major EDA companies are represented; but DAC also offers the chance to get information on a steady stream of small newcomers.

The Executive Committee and I welcome you to DAC and to Anaheim. We hope that both the Conference and the venue meet your expectations. We invite your comments and your personal participation in making DAC even better in the future.

Daniel G. Schweikert
General Chair

TECHNICAL PROGRAM COMMITTEE

CHAIR
Bryan T. Preas
Xerox Parc
3333 Coyote Hill Rd.
Palo Alto, CA 94304
(415) 812-4845

Robert K. Brayton
Univ. of California
Dept. of EECS
Rm. 519 Cory Hall
Berkeley, CA 94720
(510) 643-9801

Randal E. Bryant
Carnegie Mellon Univ.
School of Computer Science
Rm. 7128 Wean Hall
Pittsburgh, PA 15213-3890
(412) 268-8821

Giovanni De Micheli
Stanford Univ.
Ctr. for Integrated Systems
Stanford, CA 94305-4070
(415) 725-3632

Satoshi Goto
NEC Corp.
Daito Bldg. 7F
14-22, Shibaura, 4-chome
Minato-ku, Tokyo 108 Japan
+81-3-3456-6348

Dwight D. Hill
AT&T Bell Labs.
600 Mountain Ave., Rm. 3D-446
Murray Hill, NJ 07974
(908) 582-7766

Min-Yu Hsueh
Cadence Design Systems, Inc.
555 River Oaks Pkwy., Bldg. 3
San Jose, CA 95134
(408) 944-7571

Kurt Keutzer
Synopsys, Inc.
700 E. Middlefield Rd.
Mountain View, CA 94043
(415) 962-5000

Ulrich Lauther
Siemens AG
Otto-Hahn-Ring 6
Postfach 83-09-55
D-8000 Munich 83 Germany
(49) 89-636-48834

Janak H. Patel
Univ. of Illinois
Coordinated Science Lab.
1101 W. Springfield Ave.
Urbana, IL 61801
(217) 333-6201

Massoud Pedram
Univ. of S. California
Dept. of EE-Systems
Los Angeles, CA 90089
(213) 740-4458

Thomas P. Pennino
AT&T Bell Labs.
Rm. 1M-415
101 Crawfords Corner Rd.
Holmdel, NJ 07733
(908) 949-7340

Rob A. Rutenbar
Carnegie Mellon Univ.
Dept. of ECE
Rm. 3105 Hamerschlag Hall
5000 Forbes Ave.
Pittsburgh, PA 15213-3890
(412) 268-3334

Resve A. Saleh
Univ. of Illinois
Rm. 4-129, CSL
1101 W. Springfield Ave.
Urbana, IL 61801
(217) 244-6038

Alberto Sangiovanni-Vincentelli
Univ. of California
Dept. of EECS
Rm. 515 Cory Hall
Berkeley, CA 94720
(510) 642-4882

Andrzej J. Strojwas
Carnegie Mellon Univ.
Dept. of ECE
Rm. 2106 Hamerschlag Hall
Pittsburgh, PA 15213-3890
(412) 268-3530

Donald E. Thomas
Carnegie Mellon Univ.
Dept. of ECE
5000 Forbes Ave.
Pittsburgh, PA 15213-3890
(412) 268-3545

Steve Trimberger
Xilinx Inc.
2100 Logic Dr.
San Jose, CA 95124
(408) 559-7778

Thomas. W. Williams
IBM Corp.
Dept. 56B/O25R
6300 Diagonal Hwy., Box 1900
Boulder, CO 80301-9191
(303) 924-7692

Ellen J. Yoffa
IBM Corp.
Bldg. 4, Rm. B-01
P.O. Box 100, Route 100
Somers, NY 10589
(914) 766-4654

1991 ACM/IEEE Best Paper Award

Each year, awards are made for the best papers in three categories. Winners are determined from detailed reviews of the accepted papers in the technical sessions. Each award is accompanied by a plaque and a cash award of $400. The awards are given by the ACM Special Interest Group in Design Automation and the IEEE Computer Society Design Automation Technical Committee.

SIMULATION, VERIFICATION AND TEST

Paper 2.2: "ADAPTS: A Digital Transient Simulation Strategy for Integrated Circuits"
Authors: Alexander D. Stein, Binay J. George, Tuyen V. Nguyen, Ronald A. Rohrer
Affiliation: Carnegie Mellon University, Pittsburgh, PA

Paper 5.4: "The Interdependence Between Delay Optimization of Synthesized Networks and Testing"
Authors: T.W. Williams; M. Ray Mercer; B. Underwood
Affiliation: IBM Corporation, Boulder, CO; University of Texas, Austin, TX; MCC, Austin, TX

PHYSICAL DESIGN, DATABASES, FRAMEWORKS AND SYSTEMS

Paper 13.3: "SIDECAR Design Support for Reliability"
Authors: Charles R. Yount, Daniel P. Siewiorek
Affiliation: Carnegie Mellon University, Pittsburgh, PA

BEHAVIORAL, REGISTER-LEVEL AND LOGIC SYNTHESIS

Paper 19.2 "Algorithms for Synthesis of Hazard-Free Asynchronous Circuits"
Authors: Luciano Lavagno, Alberto Sangiovanni-Vincentelli; Kurt Keutzer
Affiliation: University of California, Berkeley, CA; Synopsys, Inc., Mountain View, CA

IEEE Computer Society Awards

Meritorious Service Certificate

Richard Newton
University of California
Berkeley, CA

1991 ACM/SIGDA Awards

For Contributions to SIGDA

James P. Cohoon
University of Virginia
Charlottesville, VA

Patrick M. Hefferan
SIGDA Newsletter Editor
St. Paul, MN

1991 SIGDA Travel Grant Awards

85 Awards to attend DAC '91, including, from Eastern Europe:

A. L. Stempkovsky - USSR
I. P. Norenkov - USSR
A. N. Solovyov - USSR
V. N. Gridin - USSR
A. Poppe - Hungary
G. G. Kazyonhov - USSR
J. J. Petrichkovich - USSR
E. Piworarska - Poland
S. L. Ulanov - USSR
I. Kovacic - Czechoslavokia
V. N. Loshakov - USSR
V. A. Mamatov - USSR
L. B. Osipov - USSR
T. Kruse - Germany
M. Niewczas - Poland

1992 ACM/IEEE Best Paper Award

This year, awards are made for the best papers in four categories. Winners are determined from detailed reviews of the accepted papers in the technical sessions. Each award is accompanied by a plaque and a cash award of $400. The awards are given by the ACM Special Interest Group in Design Automation and the IEEE Computer Society Design Automation Technical Committee.

SIMULATION, VERIFICATION, TEST

Session 26.1: "Validating Discrete Event Simulations Using Event Pattern Mappings"

Authors: Benoit A. Gennart; David C. Luckham
Affiliation: NTT LSI Lab., Kanagawa, Japan; Stanford University, Stanford, CA

PHYSICAL DESIGN

Session 21.3: "Hierarchical Pitchmatching Compaction Using Minimum Design"

Authors: Cyrus S. Bamji, Ravi Varadarajan
Affiliation: Cadence Design Systems, Inc.
San Jose, CA

BEHAVORIAL, REGISTER TRANSFER AND LOGIC SYNTHESIS

Session 3.2: "Implicit and Incremental Computation of Prime and Essential Prime Implicants of Boolean Functions"

Authors: O. Coudert, Jean Christophe Madre
Affiliation: Bull Corporate Research Center
Les Clayes-sous-bois, France

USE OF CAD, FRAMEWORKS, DATABASES, SYSTEMS

Session 32.1: "Design Process Management for CAD Frameworks"

Authors: Margarida F. Jacome, Stephen W. Director
Affiliation: Carnegie Mellon University
Pittsburgh, PA

SIGDA-DATC DESIGN AUTOMATION GRADUATE SCHOLARSHIPS

Each year the ACM SIGDA and its colleagues in the IEEE DATC are sponsoring scholarships of $12,000 each to support graduate research and study in Design Automation, with emphasis in "electronic and computer design and test automation". These scholarships are awarded at the Design Automation Conference directly to a university for the Faculty Investigator to expend in accordance with the proposal.

Scholarships can be awarded in support of new projects or for renewal of any of last year's SIGDA/DATC grants.

The university receiving such a scholarship is then free to use this money in direct support of one or more of its Design Automation graduate students in the manner outlined in its proposal, except that the scholarship funds shall NOT be used to support indirect costs or overhead.

The Faculty Investigator of the project receiving a scholarship award will be expected to submit a brief report of the year's activities supported by the scholarship for publication in the Sponsors Newsletters. This report should be submitted at the conclusion of the year supported.

Applications for the scholarship, either new or renewal, should be submitted by the Faculty Investigator or Department Chair and should include:

- A brief biography and transcript of the student(s) proposal for the scholarship;
- A paragraph for each student with an outline proposing his or her goals and objectives;
- A brief proposal concerning the research to be conducted and the way in which the scholarship is to be expended;
- A brief statement of the impact of the scholarship on the Design Automation Program at his or her institution.

Scholarships will be awarded based on the following criteria:

- Academic credentials of the student;
- Quality of the proposed research;
- Impact of the award on the DA program at the institution. Preference will be given to institutions which are trying to establish new DA research programs.

Beginning in 1992 these scholarships will be funded directly by the Design Automation Conference.

Applications for the 1993 awards should be received no later than April 1, 1993 by:

H.H. Loomis, Jr.
Department of ECE, Code EC/Lm
Naval Postgraduate School
Monterey, CA 93943-5000

For further information contact: Herschel Loomis at (408) 646-3214 or on MILNET at loomis@ece.nps.navy.MIL

Notification of the awards will be made on May 1, 1993.

1991Recipients SIGDA-DATC Scholarships

North Carolina State University
Advisor: Wentai Liu
Student: C. Thomas Gray

University of Florida
Advisor: Sartaj Sahni
Student: Andrew Lim

1991 SIGDA Scholarships

University of Arizona
Advisor: Sy-Yen Kuo
Student: Ing-Yi Chen

Portland State University
Advisor: Marek Perkowski
Student: Edmond Pierzchala

University of Washington
Advisor: Andrew Yang
Student: Julie Chen

University of Cincinnati
Advisor: Ranga Vemuri
Student: Ram Vemuri

University of Virginia
Advisor: James Cohoon
Students: Shekhar Bapat/Patrick Heck

Western Michigan University
Advisor: Naveed Sherwani
Students: Nancy Holmes/Moazzem Hos

University of California, Irvine
Advisor: Fadi J. Kurdahi
Student: Champaka Ramachandran

University of Texas, Austin
Advisor: Martin D. F. Wong
Students: Yang Cai/Ting-Chi Wang

1991 SIGDA $1000 Library Grants

University of Massachusetts
Ohio State University
Cornell University
Syracuse University
State University of New York

Purdue University
University of North Carolina
State University of New York, Buffalo
University of Nebraska

Washington State University
Princeton University
Washington University
University of Utah

1991 SIGDA ACSEE Undergraduate Scholarships

San Francisco Bay Area, California

Ms. Maria Guadalupe Ferhandez Rocha
Mr. Howell Martinez
Ms. Rosella M. Papa
Ms. Anne Jin Lee

CALL FOR PAPERS

30TH ACM/IEEE DESIGN AUTOMATION CONFERENCE®

DALLAS CONVENTION CENTER • JUNE 14 -18, 1993

DAC is the premier conference devoted solely to the field of Design Automation. All aspects of the use of computers as aids to the design process are included, from conceptual design through manufacturing. Four types of sessions are included: regular paper sessions, short paper sessions, panels, and tutorials.

REQUIREMENTS FOR SUBMISSION OF PAPERS

Authors should submit their papers to the Program Chair postmarked no later than October 23, 1992. Each submission should include one cover page and eight stapled copies of the complete manuscript.

The one cover page should include:

- Name, affiliation, and complete address for each author
- A designated contact person including his/her telephone number, fax number, and e-mail address
- A designated presenter, should the paper be accepted
- A list of topic numbers, **ordered by relevancy**, most clearly matching the content of the paper. It is critical that the topic number(s) accurately reflect the content of the paper. If your paper spans more than one topic, please indicate this.
- The following signed statement: "All appropriate organizational approvals for the publication of this paper have been obtained. If accepted, the author(s) will prepare the final manuscript in time for inclusion in the Conference proceedings and will present the paper at the Conference."

The eight copies of the complete manuscript should include the following items. **To permit a blind review, do not include name(s) or affiliation(s) of the author(s) on the manuscript.**

- Title of paper
- 60-word abstract indicating significance of contribution
- The complete text of the paper in English, including all illustrations and references, not exceeding 5000 words. Overly long papers will not be considered. The papers will be reviewed as finished papers. Preliminary submissions will be at a disadvantage.

Notice of acceptance will be mailed to the contact person by February 20, 1993. Authors of accepted papers must sign a copyright release form.

TOPICS OF INTEREST

Authors are invited to submit original technical papers describing recent and novel research or engineering developments in all areas of design automation. Topics include, but are not limited to:

1.1 Electrical Simulation
1.2 Discrete Simulation
1.3 Timing Verification
2.1 Testing, Fault Simulation, Test Pattern Generation, Test Validation and Design-for-Testability
2.2 Formal Hardware Verification
3.1 Floorplanning and Placement
3.2 Global and Detailed Routing
3.3 Physical Module Generation
3.4 Symbolic Layout and Compaction
3.5 Complete Layout Systems
3.6 Layout Verification (DRC, ERC)
4.1 Technology-independent, Combinational Logic Synthesis and Optimization
4.2 Technology Mapping and the Interaction Between Logic Synthesis and Layout
4.3 Sequential Synthesis and Optimization
4.4 High-level Synthesis and System-level Design Aids
4.5 Asynchronous Synthesis
5.1 Hardware Description Languages
5.2 Design Systems and Databases
6.1 DA for IC Fabrication
6.2 Computer Aids for Manufacturing
7.1 DA for Analog Circuits
7.2 High-speed Systems and Microwave DA
7.3 DA for Electronic Packaging
7.4 Technology CAD
8.1 Human Factors in DA
8.2 Frameworks in DA
8.3 Software Engineering in DA
8.4 Hardware/Software Codesign
9.1 Complete DA Systems
9.2 User Experience with DA Systems
9.3 Electronic Design Using DA, Design Methodology

PROGRAM CHAIR

MP Associates, Inc.
ATTN: Bryan Preas
Program Chair, 30th DAC
7490 Clubhouse Rd., Suite 102
Boulder, Colorado 80301
For information call: (303) 530-4333

sponsored by ACM SIGDA® IEEE COMPUTER SOCIETY DATC® IEEE CIRCUITS & SYSTEMS SOCIETY® *in cooperation with* EDAC Electronic Design Automation Companies

DALLAS, TEXAS
JUNE 14-18, 1993

sponsored by

IEEE COMPUTER SOCIETY DATC® | acm SIGDA® | IEEE CIRCUITS & SYSTEMS SOCIETY

in cooperation with

EDAC
Electronic Design Automation Companies

PANELS, TUTORIALS, SPECIAL TOPIC SESSIONS

Proposals for topics of Panels, Tutorial Sessions, and Full-day Tutorials should be submitted to the Program Chair no later than October 23, 1992. Each proposal should not exceed two pages in length and should include a description of the topic, structure of the session or tutorial, and a list of participants. You should make contact with all participants in advance.

Special Topic Sessions may be either independent papers with a common theme or a set of closely related papers describing an overall system. In both cases, independent reviews of each paper and evaluation of the session as a whole will be used to select sessions. Proposals for Special Topic Sessions should be submitted along with the papers to be included in the session and should describe the theme of the session. These proposals and paper submissions must be postmarked no later then October 23, 1992.

USE OF DA TOOLS

The Design Automation Conference is broadening its scope of interest to include the use of automation design process. We are soliciting papers and proposals for panels and tutorials of interest to system and circuit designers, design managers, and DA support engineers in the areas of design process, design methodology, results of actual DA usage, and comparison of DA tools and systems. Areas include, but are not limited to:

- Use of automation in the design of state-of-the-art systems
- Design methodology and design process to take advantage of DA systems
- Technology-independent design: ASICs vs. FPGAs across vendors
- Requirements of a DA environment that minimize the challenges in using the tools
- Strategies for evaluating DA tools and for choosing DA vendors
- Comparative results of using multiple DA systems on specific circuits
- Complete DA systems; tools built on top of frameworks; integrating vendor tools within your system

REVIEWERS

A total of 442 manuscripts were submitted to the 1992 DAC. The Conference Executive and Program Committees wish to acknowledge the time and effort spent by the following people who reviewed these manuscripts and returned the review forms completed. Our thanks to all of those who participated and contributed to the success of the Conference.

ELFRIEDE ABEL
MIRON ABRAMOVICI
CHARLOTTE ACKEN
JOHN M. ACKEN
BRYAN ACKLAND
GORDON ADSHEAD
PRATHIMA AGRAWAL
VISHWANI AGRAWAL
ZUBAIR AHMED
ROBERT AITKEN
HUSSEIN K. AL-HUSSEIN
KATHLEEN ALBERTINI
DAVID ALLEN
HAL ALLES
HIROSHI ANDOU
FRANK ANGELOTTI
KURT ANTREICH
GINETTI ARNOLD
DAVID ARTZ
SUNIL ASUTAPUTRE
VAHE AVEDISSIAN
JAMES AYLOR
SCOTT BAEDER
MARK BALES
PORAS T. BALSARA
BEHROOZ BAMDALI
SHEKHAR BAPAT
KAREN BARTLETT
HERBERT BAUER
DEREK BEATTY
BERND BECKER
PHILIP M. BENJAMIN
NACQUES BENKOSA
WILLIAM BERG
REINALDO BERGAMASCHI
MIKE BEUNDER
BHARAT BHUVA
PAUL BLACK
ROBERT BLACKBURN
JOHN P. BLANKS
VIJAY BOBBA
RICHARD BONNEAU
MEHDI BONYADI
AHSAN BOOTEHSAZ
JOHN BORKENHAGEN
GAETANO BORRIELLO
JONATHAN BOWEN
DAVID BOYER
SAIED BOZORGUI-NESBAT
DANIEL BRAND
JAMES BRANDT
MELVIN BREUER
FORREST BREWER
JACK BRINER
RAINER BRUECK
KEVIN BUCHS
JERRY BURCH
HOLGER BUSCH
WILLIAM BUSH
MICHAEL BUSHNELL
KENNETH M. BUTLER
YANG CAI
DANIEL CAMPBELL
RAUL CAMPOSANO
PAOLO E. CAMURATI
ROBERT CARDEN
ANDY CARPENTER
GIORGIO CASINOVI
ALAN CAVE
RALPH CAVIN
RAKESH CHADHA
VIRAPHOL CHAIYAKUL
TAPAN CHAKRABORTY
ROGER CHAMBERLAIN
HEMING H. CHAN
M. CHANDRASEKHAR
K.C. CHANG
MI-CHANG CHANG
PAUL CHANG
CHIEN-IN CHEN
DE-SHENG CHEN
HOWARD CHEN
HSI-CHUAN CHEN
KUANGCHIEN CHEN
SAO-JIE CHEN
KWANG-TING CHENG
JUE-HSIEN CHERN
MARKO CHEW
GEORGE CHI
KUANG-WEI CHIANG
VIVEK CHICKERMANE
SCOTT CHIU
JUN-DONG CHO
K-C CHU
MICHAEL CILETTI
LUC CLAESEN
RICHARD CLOUTIER
JOHN COHN
JAMES COHOON
JASON CONG
DAVID COOKE
LORNE COOPER
HENRY COX
SCOTT CRAVENS
FRANCESCO CURATELLI
ANTON DAHBURA
SUMIT DASGUPTA
SCOTT DAVIDSON
ALBERT T. DAVIS
JOANNE E. DEGROAT
EDWARD DEJESUS
MARTIN J. DEVILLY
ALLEN DEWEY
DAVID DILL
FAVSTO DISTANTE
DILEEP DIVEKAR
ANTUN DOMIC
MICHEL T. DOREAU
JEAN-CLAUD DUFOURD
JAMES R. DULEY
NIKIL DUTT
ROBI DUTTA
EARL ECKLUND, JR.
MARTYN EDWARDS
AVI EFRATI
NORMAN ELIAS
RON ELLICKSON
DIANE ELMER
RICHARD ENBODY
BERN ESCHERMANN
RICHARD EVERSOLE
FRANCINE FERGUSON
JOEL FERGUSON
KEN FIDUK
DENNIS C. FOGG
THOMAS FOX
MARK FREDRICKSON
MATTHIAS FREI
ERNEST FREY
STEPHEN FREZZA
EBY G. FRIEDMAN

SHAODI GAO
HANS GETHOFFER
INDRANEEL GHOSH
EMIL GIRCZYC
CLINTON GLOVER
JAMES GODA
JIE GONG
GERT GOOSSENS
NOBUYUKI GOTO
ROBERT GRAFTON
JOHN GRAHAM
JOHN GRANACKI, JR.
W. GRASS
LAURENCE GRODD
PATRICK GROENEVELD
JOHN S. GROUT
ANURAG GUPTA
RAJESH GUPTA
PAUL GUTWIN
TEDD S. HADLEY
LOU HAFER
CARL HAGE
JOHN HAGERMAN
J. HANCOCK
MICHAEL HANEY
IKUO HARADA
JUSTIN HARLOW
HAIDAR HARMANANI
MARK R. HARTOOG
PATRICK L. HECK
RICHARD HEIDENREICH
SYBILLE K. HELLEBRAND
JURGEN HERRMANN
DWIGHT HILL
SYLVIA HILL
JOHN HILLAWI
KANJI HIRABAYSHI
TAMOTSU HIWATASHI
JAN-MING HO
MARK HOFMANN
SHERVIN HOJAT
ERNST HORNEBER
PAUL W. HORSTMANN
SUSAN M. HORSTMANN
ARDING HSU
YU-CHIN HSU
XIAOBO HU
RAGHU V. HUDLI
JOSEPH HUGHES
GURDEEP S. HURA
KATHERINE HUTCHISON
CHI-YI HWANG
TINGTING HWANG
KAZUHIKO IIJIMA
ELAINE IODICE
MARY JANE IRWIN
MASAKI ISHIKAWA
NAGISA ISHIURA
JAMES JACOB
NARENDRA JAIN
PREM JAIN
DHANENDRA JANI
CHRISTIAN JAY
RAJEEV JAYARAMAN
ANURA JAYASUMANA
KJELL JEPPSON
J. JESS
PRADIP K. JHA
FRANK JOHANNES
I. JOHNSTONE
WEN-BEN JONE
LARRY JONES
JING-YANG JOU
SUNGTAE JUNG
KNUT JUST
SUDHIR KADKADE
ANDREW KAHNG
LALGUD KANNAN
ROHIT KAPUR
OSAMU KARATSU
PHILIP KAUFMAN
NOBUAKI KAWATO
MICHAEL KEARNEY
BRION KELLER
PAWEL KERNTOPF
KEI-YONG KHOO
CHOON KIM
CHRIS KINGSLEY
JAMES R. KIPPS
ROBERT KIRK
HITOSHI KITAZAWA
DAVID KNAPP
MING-TAT KO
KOLAR KODANDAPANI
MATTHIAS M KOEHLER
GEORGE KONSTANTINOW
THOMAS E. KOPEC
THADDEUS KOWALSKI
HEINRICH KRAMER
ANDRZEJ KRASNIEWSKI
DAVID KU
KEN E. KUBIAK
WILLIAM KUBITZ
PHILIP KUEKES
NAND KUMAR
RAMACHANDR KUNDA
SY-YEN KUO
FADI KURDAHI
DON LACY
MAUREEN LADD
HOWARD K. LANE
DAVID LAPOTIN
ROBERT LAPP
TRACY LARRABEE
ROBERT LARSEN
RICHARD LAZANSKY
CHIN-HWA LEE
JAE-HWANG LEE
JIN-FUW LEE
GUANG-TSAI LEI
ROBERT LEMBACH
REGIS LEUVEUGLE
STEVEN LEVIJAN
DAVID LEWIS
J.T. LI
WING-NING LI
MICHIEL LIGTHART
JOHNSON LIMQUECO
LI-SHIN LIN
WEI-LIANG LIN
YOUN-LONG LIN
ANTONIO LIOY
JOSEPH LIS
DICK LIU
EDWARD LIU
LUNG-CHUN LIU
SALLY LIU
WENTAI LIU
CHI-YUAN LO
MICHAEL LORENZETTI
THOMAS LOVETT
JOHN K. LOWELL
JOHN LUDWIG
CHIDCHANOK LURSINSAP
STEPHEN LUSKY
HI-KEUNG MA
JEAN-CHRISTOPHE MADRE
MOSSADDEQ MAHMOOD
AMITAVA MAJUMDAR
RAFIC Z. MAKKI
YASHWANT MALAIYA
RICHARD MALERO
ABDUL MALIK
SOUMMYA MALLICK
GIOVANNI MANCINI
ALAN MANTOOTH
ALAN MARTELLO
GRANT MARTIN
PETER MARWEDEL
TSUNEO MATSUDA
HEINZ MATTES
PETER MAURER

RONALD MAXWELL
KART MAYARAM
ROBERT MAYO
PANKAJ MAYOR
KENNETH MCELVAIN
MICHAEL MCFARLAND
RICHARD MCKNIGHT
BRUCE MCMILLIN
KRISTEN MCNALL
ZAFER MEHMOOD
STEPHEN MEIER
PAUL MENCHINI
TERESA H. MENG
KRISHNA MENON
STEVE J. MEYER
SCOTT MIDKIFF
MITCHELL MILNAR
PETER MOCEYUNAS
HOSSEIN MODARRES
MOSUR K. MOHAN
PAUL MOLITOR
JAMES MONZEL
CHO MOON
HAJIMU MORI
KLAUS MUELLER-GLASER
ROBERT MUELLER-THUNS
ASHUTOSH S. MUJUMDAR
WILLIAM MULLEN
PERRY MULLINIX
MICHIO MURAKAMI
SABURO MUROGA
GERALD MUSGRAVE
FARID NAJM
RANJANI NARAYAN
SANJIV NARAYAN
VINOD NARAYANAN
RAMESH NARAYANASWAMY
WILLIAM NEEDLES
JOHN NESTOR
A. RICHARD NEWTON
THERESA NIX
CHARLES NJINDA
STEFAAN NOTE
RAMAKRISHA NUNNA
WILLIAM NYE
HIDETOSHI ONODERA
AKIRA ONOZAWA
ALEX ORAILOGLU
PAUL PAINTER
JENGWEI PAN
BARRY PANGRLE
C. PAPACHRISTOU
GEORGINA PAPP
KESHAB PARHI

JANAK PATEL
SRINIVAS PATIL
SATHYAMURT PATTANAM
PIERRE PAULIN
WILLIAM PAULSEN
MASSOUD PEDRAM
PAUL PENFIELD
RICHARD PEREZ
MAREK PERKOWSKI
GEORGE PERSKY
JEN-I PI
EDWARD PITTY
CARL PIXLEY
FRANCK J. POIROT
EDWIN PORTER
MIODRAG POTKONJAK
SHIV PRAKASH
PAOLO PRINETTO
MICHAEL QUAYLE
JAN RABAEY
DAVID RABINOWITZ
SRINIVAS RAGHVENDRA
ROCHIT RAJSUMAN
KUNDA RAMACHANDRA
CHAMPAKA RAMACHANDRAN
LOGANATH RAMACHANDRAN
TULSI RAMCHANDANI
F.J. RAMMIG
VASANT B. RAO
SHISHPAL RAWAT
WILLIAM READ
SUDHAKAR REDDY
DOUGLAS REEVES
RICARDO REIS
MARK RENCHER
MINJOONG RIM
KENNETH ROBERTS
MARTIN ROBERTS
GABRIEL ROBINS
MARKUS ROBINSON
WOLFGANG ROSENSTIEL
DON ROSS
JACQUES ROUILLARD
JAMES ROWSON
JAYANTA ROY
RABINDRA ROY
RICHARD L. RUDELL
MARTIN RUDOLPH
JEFFREY RUPLEY
ROB A. RUTENBAR
DANIEL SAAB
KAREM SAKALLAH
KEWAL SALUJA
JERE SANBORN

MAJID SARRAFZADEH
MASAO SATO
GABRIELE SAUCIER
HAMID SAVOJ
PRASHANT SAWKAR
JAMES SCHARF
LOUIS SCHEFFER
WERNER SCHIELE
H. DANIEL SCHNURMANN
BERND SCHUERMANN
KENNETH SCOTT
CARL SECHEN
RAJEEV SEHGAL
JOEL SEIDMAN
MASATOSHI SEKINE
ALOK SHARMA
BETTY C. SHEN
STEVEN SHERMAN
NAVEED SHERWANI
KATHERINE SHERWOOD
BING SHEU
JOHN SHIELDS
TSUGUO SHIMIZU
HYUNCHUL SHIN
HIROSHI SHIRAISHI
NOBORN SHIRAKI
SAJJAN SHIVA
ERNST SIEPMANN
ANOOP SINGHAL
ERIC SKULDT
DAVID SMART
J.W. SMITH
JEFFREY SMITH
SKIP SMITH
GERALD SOBELMAN
ARUN SOMANI
FABIO SOMENZI
MATTEO SONZA REORDA
LAMBERT SPAANENBURG
BRUCE K. SPEYER
RICK L. SPICKELMIER
GOPAL SRINATH
GANAPAT SRINIVASA
ROBERT STANION
BALSHA R. STANISIC
JORGEN STAUNSTRUP
VICTORIA STAVRIDOU
HELMUT STECKENBILLER
UDO STEINKOPF
NOEL STRADER
PAUL STRAVERS
DAVID W. STRINGFELLOW
ALBRECHT STROELE
CHARLES STROUD

CARSON STUART
STEPHEN SUGIYAMA
MARK SUMNER
HARI SURAPANENI
KEI SUSUKI
MARK SWANSON
FRANK SZORC
THOMAS SZYMANSKI
MICHAEL TEBO
KARL-HEINZ TEMME
C.J. TERMAN
KIT THAM
DONALD THOMAS
JEFFREY THOMAS
CIHAN TINAZTEPE
JAMES TOLAR
QIAO TONG
LOUISE TREVILLYAN
STEPHEN TRIMBERGER
ERWIN TRISCHLER
CARY TRLICA
CHARLES TRULLEMANS
RAYMOND TSUI
AKIRA TSUKIZOE
TAI-CHING TUAN
JON TURINO
KAZUHIRO UEDA
CARY USSERY
MATTHIAS UTESCH
FRANK N. VAHID
RENE VAN LEUKEN
KENNETH A. VANBREE
WILLIAM VANCLEEMPUT
LUKAS VANGINNEUEN
IAN VANHOOF
PRAB VARMA
JAMES VELLENGA
RANGANADHA VEMURI
KUMAR VENKATRAMAN
KOFI E. VIDA-TORKU
ARNALDO VIEGAS DE LIMA
V. VISVANATHAN
RAGHUNATH VUTUKURU
RONALD WADSACK
MICHAEL WAHL
KAZUTOSHI WAKABAYASHI
DUNCAN WALKER
ROBERT WALKER
DAVID WALLACE
ALBERT WANG
JEN-PIN WENG
NEIL H.E. WESTE
JACOB WHITE
RICHARD P. WILEY
JOHN WILLIS
JOHN WILLOUGHBY
WARREN WIEMANN
SHMUEL WIMER
RUDI WINKELMANN
MARIANNE WINSLETT
JOHN WISNIEWSKI
ANTHONY WOJCIK
STEPHEN WOLFE
KEN WONG
MARTIN D.F WONG
STANLEY WONG
NAM-SUNG WOO
GRAHAM WOOD
CHUNG-HAO WU
DAVID M. WU
KWOK WU
ZEEV WURMAN
AKIHIKO YAMADA
MASAAKI YAMADA
ANDREW YANG
CHYAN YANG
MEHMET YANILMAZ
HIROTO YASUURA
KOK-HOO YEAP
DAVID YEH
KEN YEH
PING YEUNG
ELLEN YOFFA
GARY YORK
TAKESHI YOSHIMURA
ALEXANDRU ZAMFIRESCU
DAVID ZEEK
DAVID ZEIN

Table of Contents

* *Best Paper Award Candidates*

Session 5: Partitioning and Floorplanning
Chair: M. Pedram; Organizers: M. Pedram and E. Yoffa

Session 6: Interconnect Simulation
Chair: R.A. Saleh; Organizers: R. Saleh and A. Strojwas

Session 7: Scheduling and Allocation
Chair: P. Paulin; Organizers: G. De Micheli and D. Thomas

* *Best Paper Award Candidates*

** Best Paper Award Candidates*

Session 13: Asymptotic Waveform Evaluation
Chair: J. White; Organizers: R. Saleh and A. Strojwas

Session 14: System-Level Synthesis
Chair: D.E. Thomas; Organizers: G. De Micheli and D. Thomas

Session 15: Performance Issues in Logic Synthesis
Chair: P. McGeer; Organizers: R. Brayton and A. Sangiovanni-Vincentelli

Session 16: Panel — Why Data Models Will Become the Fastest Growing Segment of the EDA Market 260
Chair: B. Lattin; Organizer: M. Middlewood

Panel Members:

J. Hoover, H. Jones, G. Langeler, B. Lattin, A.R. Newton, and C. Widdoes

Session 17: High-Level Test Generation
Chair: M.A. Bruer; Organizers: J. Patel and T. Williams

** Best Paper Award Candidates*

Session 18: Allocation and Binding
Chair: M. McFarland; Organizers: G. De Micheli and D. Thomas

Session 19: Panel — CASE -- Why It Doesn't Work for CAD 297
Chair: R. Potter

Panel Members:
S. Baeder, E. Boerger, J. Sullivan, G. Symons, and A. Wasserman

Session 20: Tutorial — Directions to Watch in Design Technology 298
Chair: G. Langeler; Organizer: T. Pennino

Speaker:
A.R. Newton

Session 21: Design Verification and Compaction
Chair: C.-M. Kyung; Organizers: D. Hill and E. Yoffa

Session 22: Fault Simulation and Fault Diagnosis
Chair: J.A. Abraham; Organizers: J. Patel and T. Williams

** Best Paper Award Candidates*

Session 23: FPGA Synthesis
Chair: E. Detjens; Organizers: R. Brayton and A. Sangiovanni-Vincentelli

Session 24: Tutorial — Acquiring and Maintaining State-of-the-Art DA Systems
Chair: B. Hauser; Organizer: T. Pennino

Session 25: Timing Optimization and Verification
Chair: K.A. Sakallah; Organizers: R. Rutenbar and A. Strojwas

Session 26: Discrete Simulation
Chair: P. Moorby; Organizers: R. Bryant and R. Saleh

** Best Paper Award Candidates*

** Best Paper Award Candidates*

Session 32: Frameworks
Chair: T. Miller; Organizer: T. Pennino

Session 33: Global Issues in Routing
Chair: U. Lauther; Organizers: S. Goto and U. Lauther

Session 34: Path Delay Analysis
Chair: J. Benkoski; Organizers: R. Rutenbar and A. Strojwas

Session 35: Sequential Logic Synthesis
Chair: G. Saucier; Organizers: R. Brayton and A. Sangiovanni-Vincentelli

Chair: L. Lanzo; Organizer: J. Bhat

Panel Members:

T. Hiwatashi, R. Kling, N. Nettleton, G. Porter, B. Taylor, J. Toler, Y. Sugiura, and J. Whitmore

** Best Paper Award Candidates*

Session 37: Multi-Layer Channel and Over-the-Cell Routing
Chair: S. Goto; Organizers: S. Goto and U. Lauther

Session 38: Automated Approaches to Formal Verification of Hardware
Chair: J.C. Madre; Organizers: R. Bryant and K. Keutzer

Session 39: Advances in High-Level Synthesis
Chair: A.C. Parker; Organizers: G. De Micheli and D. Thomas

Session 40: Tutorial — EDIF/CFI - A User's Perspective
Chair: N. English; Organizer: T. Pennino

* *Best Paper Award Candidates*

* *Best Paper Award Candidates*

1992 KEYNOTE ADDRESS

Prof. Michael A. Cusumano
Sloan School of Management
Massachusetts Inst. of Tech.
Cambridge, MA

Many companies that have traditionally been producers of hardware equipment with only small amounts of programming now find that they are rapidly becoming software producers. This means that they now have to manage large-scale programming projects that consume enormous amounts of engineering effort but are necessary to add functionality to their hardware. Their software skills, increasingly, are becoming either a source of competitive advantage or disadvantage. But while many traditional hardware firms have not built up much expertise in developing large-scale software systems, there is considerable knowledge in the world industry. In particular, the problems that many hardware vendors face today have been faced already by computer manufacturers during the past two decades ago, as software programs have taken on greater importance in commercial and industrial computer systems. Several Japanese firms have been especially strong in producing routine but complex software on time and with high levels of reliability and predictability.

This keynote address focuses on the management of large-scale software development and software organizations in a systematic or "factory-like" way, through a "process" rather than "product" focus and the use of common development methods and tools, historical project and quality data, planned reusability, incremental improvement, and other measures. The major proponents of this notion of the "software factory" have been Japanese firms -- Hitachi, Toshiba, Fujitsu, and NEC -- building on ideas that came orginally from U.S. firms such as GE, AT&T, SDC and IBM. The Japanese philosophies for software development and organizational innovations are discussed as solutions to the problems that frequently occur in software development.

Michael A. Cusumano, 37, is an Associate Professor at the Massachusetts Institute of Technology's Sloan School of Management, where he teaches strategy and technology management. Fluent in Japanese and an expert on Japanese engineering and production management, he is the author of two widely acclaimed books on Japan: Japan's Software Factories: A Challenge to U.S. Management (Oxford University Press, 1991) and The Japanese Automobile Industry (Harvard University Press, 1985, 1989).

Dr. Cusumano holds a Ph.D from Harvard University in Japanese studies and a B.A. degree from Princeton University in the history of ideas. He also spent two years as a postdoctoral fellow in production/operations management at Harvard Business School and lived in Japan for five years, including two terms as a Fulbright Fellow at the University of Tokyo.

MAXIMUM CURRENT ESTIMATION IN CMOS CIRCUITS

Harish Kriplani†, *Farid Najm*†† and *Ibrahim Hajj*†

†Coordinated Science Laboratory, and
Dept. of Electrical and Computer Engineering
University of Illinois at Urbana-Champaign
1101 W. Springfield Avenue
Urbana, IL 61801.

††Semiconductor Process & Design Center
Texas Instruments Inc.
Dallas, TX 75265.

Abstract: *Excessive power supply and ground currents in integrated circuits can severely affect circuit reliability and performance. Some of the problems arising from excessive current flow are : (1) excessive voltage drop (glitches) on the power/ground lines, which can lead to soft errors, and (2) large instantaneous power dissipation, which causes overheating and ultimately leads to performance degradation. Maximum current estimates are, therefore, needed in the supply lines in order to determine the severity of these problems. These currents, however, depend on the specific input patterns that are applied to the circuit. Most previous work in this area has focused on search techniques that attempt to locate the worst case current by searching for the corresponding worst case input patterns. However, since the input space is huge, search-based algorithm for this problem can take an exponential amount of time, in the worst case. In this paper, we propose a pattern-independent, linear-time algorithm that estimates an upper bound for the Maximum Envelope Current (MEC) waveform. The MEC waveform is a point-wise maximum on all the possible waveforms that the circuit can draw. Experimental results on several benchmark circuits are provided to establish the usefulness of this approach.*

1 Introduction

A major concern in present day VLSI circuits is the design of supply and ground lines in a way that ensures design reliability and performance. Excessive supply currents can severely affect both circuit lifetime and performance. Some of the problems arising from excessive current flow are : (1) excessive voltage drops in the power/ground lines, and : (2) large instantaneous power dissipation. Excessive voltage drop manifests itself as *glitches* on the supply lines, and causes erroneous logic signals and soft errors. Large instantaneous power dissipation causes overheating of the devices and ultimately degrades the circuit performance. Maximum current estimates are needed in the supply lines in order to determine the severity of these problems.

Power supply and ground lines deliver power to all the gates in a circuit. The points at which the individual gates or cells are tied to the bus are called *contact points*. We need to estimate the maximum currents drawn by the circuit at every contact point in order to properly design the supply lines. A design method that uses these current waveforms to redesign supply lines for acceptable voltage drop can be found in [4].

A CMOS gate draws a pulse of current from the supply lines only when its output changes state, i.e., when it switches either low → high or high → low. For this reason, CMOS circuits pose severe difficulties for supply current estimation. We will, therefore, focus on CMOS technology. The set of excitations that are possible at each input of a gate at a particular time instant are low, high, low → high and high → low. The gate will draw different (or no) current pulses for each of these input excitations. Furthermore, the current drawn by the circuit depends also upon the time instants at which these excitations are applied. For example, if a circuit has two inputs, then the current drawn from the supply lines depends not only upon the specific excitations applied at each input but also upon the delay (or *skew*) between them.

As is clear from the above discussion, the current drawn by a CMOS circuit is a complex function of the input excitations and timing. Therefore, we need to define very carefully what we mean by the maximum current waveform at a contact point. Chowdhury and Barkatullah in [2] find the maximum of the peaks of various contact currents for all possible input patterns. In the final analysis of the supply lines, they then assume that these constant peak values are applied at the contact points for all time, i.e., they have dc currents flowing in the lines. This assumption, however, gives pessimistic results since separate sections in the circuit rarely draw their maximum currents simultaneously. In this paper, we propose a better measure of the maximum current waveform called *maximum envelope current waveform*.

The *Maximum Envelope Current (MEC) waveform* is a waveform whose value at any time t is the maximum of all the current values that the circuit can draw at that time. There is a unique MEC waveform at every contact point. Let $U = \{u_1(t_1), u_2(t_2), \ldots, u_n(t_n)\}$ be an input vector that is applied to the circuit C, where $u_i(t_i)$ is the input excitation that is applied to the ith input at time instant t_i, and n is the number of circuit inputs. Thus, the ith input changes state at time instant t_i and t_i is irrelevant if the input does not change state. Furthermore, if $[0, T]$ is the time interval of interest over which the

0738-100X/92 $3.00

circuit is being analyzed, then :

$$I_{MEC}(t) = \max_{\substack{u_i(t_i) \in \{low, high, lh, hl\} \\ t_i \in [0,T] \\ 1 \leq i \leq n}} I(t)$$

Here, $I_{MEC}(t)$ is the value of the MEC waveform at time t and $I(t)$ is the value of the current that the circuit C draws from the supply lines at time t when the input vector U is applied to it. The abbreviations lh and hl stand for low $\rightarrow$ high and high $\rightarrow$ low transitions, respectively. Thus, $I_{MEC}(t)$ is the maximum possible current value that could be drawn from the supply lines at time t, given that each input can switch at anytime during the interval $[0, T]$. We are interested in algorithms that efficiently estimate the MEC waveform.

Accurate estimation of the MEC waveform at every contact point is extremely difficult since we need to determine current waveforms corresponding to all possible input patterns. If the circuit has n primary inputs then we need to explore the set of 4^n input patterns to calculate the MEC waveforms. Moreover, as was pointed out before, each component of the input vector can be applied to the circuit at any time during the interval $[0, T]$. This compounds the problem and makes it practically impossible to handle by any of the known search procedures. As will be shown in the next section, most previous work in this area has been based on search techniques. In this paper, we propose linear time (in the number of gates) algorithms that provide tight upper bounds on MEC waveforms. The proposed approach represents a trade-off between execution speed and tightness of these bounds.

This paper is organized as follows. In the next section, we briefly discuss previous and related work in this area. In section 3, we discuss various assumptions on which our approach is based. In section 4, we present the algorithm in detail. Experimental results are presented in section 5. In order to maintain reasonable execution times, our algorithm neglects signal correlations that exist inside the circuit. As mentioned above, the price one has to pay for this is looser upper bounds on the MEC. The signal correlation problem is described in section 6. In section 7, we present a heuristic technique that partially accounts for signal correlations in order to obtain tighter bounds. Finally, in section 8, conclusions and some guidelines for future work are presented.

2 Previous Work

Chowdhury and Barkatullah have addressed the problem of maximum current estimation in [2]. In their methodology, they divide the circuit into a set of macros, each of which consists of a combinational interconnection of logic gates. Considering each macro separately, they use either a (branch and bound) search technique or a heuristic technique to find the maximum of its transient current, assuming its inputs switch simultaneously. However, they do not discuss the effect of the various macro interconnections on the overall maximum current. In addition, they assume that internal nodes make single transitions. Our experience with various benchmark circuits indicates that the contribution of multiple node transitions to the supply currents can be significant. Due to the huge size of the input space, their branch and bound search technique is slow on large circuits. Furthermore, their heuristic approach does not guarantee an upper bound on the maximum currents.

Devadas et. al. have addressed a similar problem in [3]. They consider the estimation of worst case power dissipation in CMOS combinational circuits. They reduce this problem to a weighted max-satisfiability problem, on a set of multi-output Boolean functions obtained from the circuit logic description. These functions are appropriately weighted to account for the different load capacitances. A branch and bound algorithm is then used to solve the ($\mathcal{NP}$-complete) max-satisfiability problem. They are able to account for multiple node transitions. However, for a multilevel logic circuit, even under a unit gate delay assumption, the functions generated by their algorithm are fairly complex. Consequently, even for small circuits, their analysis is slow. Analysis of multi-level circuits under a general delay model was not attempted.

From this brief survey, it is clear that existing methods for the calculation of maximum current are computationally too expensive to handle large VLSI circuits. In order to be able to handle present and next generation of VLSI circuits, linear time algorithms are necessary. In section 4, we present a linear time algorithm for the MEC waveform estimation. But first, we discuss various simplifying assumptions that will be used in that algorithm.

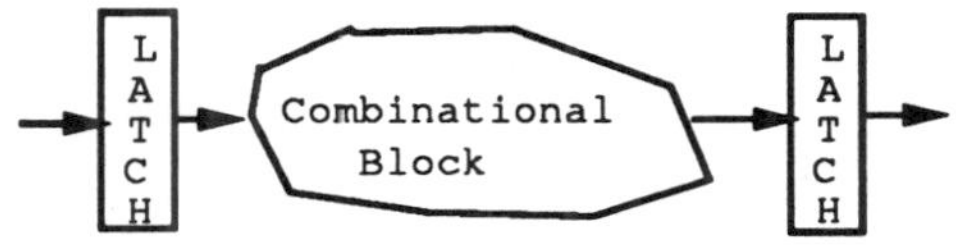

Figure 1. An example of a latch controlled synchronous digital circuit.

3 Assumptions

In order to reduce the complexity of the problem, we focus on a specific, but very common design style, namely (edge-triggered) latch-controlled synchronous digital circuits. These circuits consist of combinational blocks separated by latches (see Figure 1) such that all inputs to each combinational block switch simultaneously. As a result, we will be able to focus the analysis in the next section on a single combinational block whose inputs all switch at time 0. This effectively eliminates the time domain uncertainty about the input transitions, and significantly simplifies the problem. This assumption has also been used by both previous approaches surveyed above, although it has not always been clearly mentioned.

We assume that the delay of each gate in the circuit is fixed and is specified ahead of time. But different gates can have different delays. Further, we assume that each time the output of a gate switches, a triangular pulse of current is drawn from the supply lines, as shown in Figure 2. The duration of this pulse is

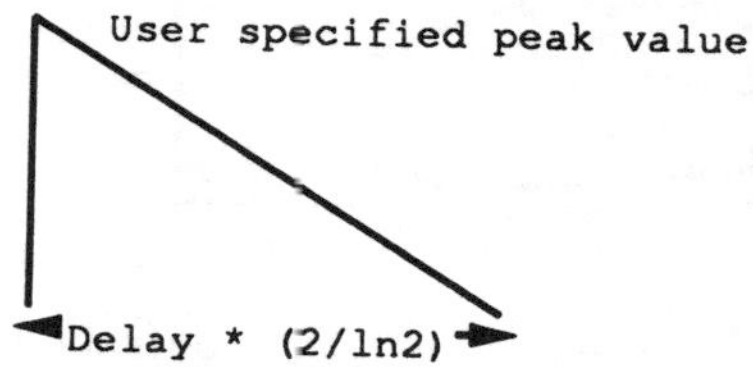

Figure 2. Model of a gate current pulse.

computed from the delay of the gate (by charge conservation), and its height (peak value) is a user-specified value. Two separate values for the peak current are allowed, corresponding to *lh* and *hl* transitions at the gate output.

Given the specific clocking scheme of the synchronous circuit, the MEC waveforms of combinational blocks driven by the same clock can be grouped together and used to find the bus branch currents. For purposes of this paper, we will focus (in the next section) on the analysis of a single combinational block, tied to the bus at a single contact point.

4 Description of the Proposed Algorithm (iMax)

Given a gate level description, we assume that little or nothing is known about the values at the primary inputs, except that they may only transition at time zero. We call this an *uncertainty* about these input signals. The basic idea is to propagate this uncertainty into the circuit, so that, at every logic gate, we know the range of possible values that its inputs can take. From this, the worst case gate current is computed, as explained below.

Perhaps the first question to come to mind is as to the kind of information that one maintains in order to represent the uncertainty about internal circuit nodes. Ideally, one would like to compute the set of *all* possible transitions (along with their timing information) that occur at every gate in the circuit. That would certainly be enough to estimate the MEC waveform. However, given the uncertainty at the inputs, the number of possible transitions at internal gates grows exponentially, and quickly becomes a bottleneck. To avoid this problem we maintain information, not about individual transitions, but about *intervals* during which the output of the gate *might* switch. Thus, for each of the excitations *low*, *high*, *hl* and *lh*, we store a list of intervals during which a node might carry that excitation. These intervals, which might overlap, serve to describe the signal uncertainty.

An example of this signal representation is given in Figure 3. In that figure, we show an uncertain signal $X(t)$ represented as four sets of intervals along the time axis. Thus, if $x(t)$ is a logic signal that belongs to the family $X(t)$, i.e., $x(t) \in X(t)$, then $x(t)$ is low up to t_1, switches from low to high sometime between t_1 and t_2, is then high up to t_3, etc. Thus at any time between t_1 and t_2 the signal can be either high or low. Notice that between t_6 and t_7 the signal may make any number of low to high and high

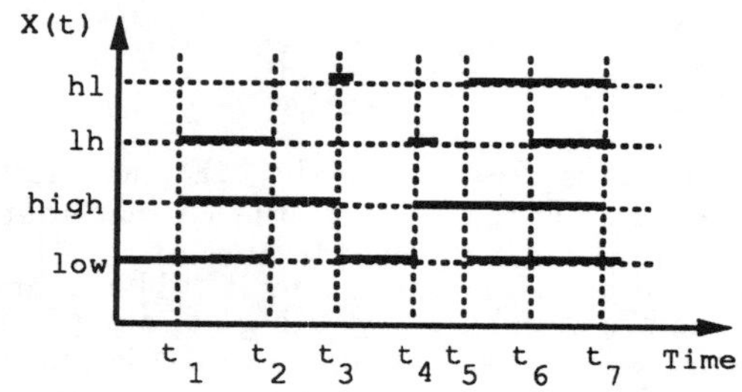

Figure 3. An illustration of signal uncertainty at a gate output.

to low transitions. At the primary inputs, the signals are represented by such waveforms with a single point of *possible* transition at time 0. As internal signals are generated, the number of possible transition points increases. In order to contain the complexity, we then start to merge neighboring transition points into intervals. In general, this strategy can be stated as follows : when the number of intervals associated with a gate corresponding to any excitation exceeds a certain user-specified threshold (Max_No_Hops), we repeatedly merge closest-neighbor intervals, so as to keep their number down.

Given such waveform information at the inputs of a logic gate, we would like to derive the corresponding information about its output. One cannot do this accurately, though, without knowing how some of these inputs, if any, are *correlated*. For instance, certain *combinations* of the gate input values may not be possible. Unfortunately, maintaining information on the correlation between various circuit nodes is too expensive. We, therefore, use a *conservative* approximation, one that does *not* underestimate the MEC waveform, as follows. If we assume that *all* combinations of the gate input values are possible, i.e., that the gate inputs are *independent*, then the worst case current in that case *is* an upper-bound on the gate current for that range of inputs. In other words, the worst case current over *all* combinations of inputs is certainly an upper bound on the worst case current over *some*.

Given the type of a Boolean gate, and using the independence assumption, it is a simple matter to compute the intervals at the gate output from those at its inputs (see Figure 4). From this, the supply current contribution of the gate is also calculated. For instance, if a gate can switch at any time during a certain interval, then the waveform shown in Figure 5 is considered to be its contribution to the supply current for that interval. This waveform is an envelope of all possible currents that the gate can draw due to a transition occurring at any time during that interval. Thus, it represents the worst case scenario. At every gate, there are two possible current waveforms, one due to low → high transitions (LHCurrent) and the other due to high → low transitions (HLCurrent). The maximum of the two current waveforms at every time point is considered to be the worst case supply current. Once all the gate currents have been computed, the current waveform at the contact point is determined by adding together the individual gate contributions.

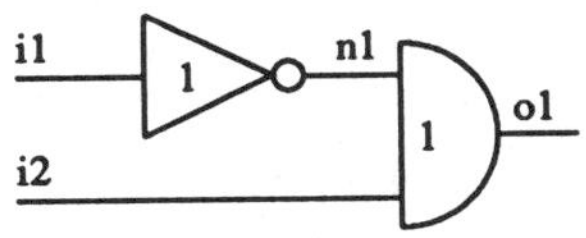

Input Description : i1, i2 $\in \{low, high, hl, lh\}$ at time 0.
Uncertainty Intervals :
i1, i2: $lh[0, 0]$, $hl[0, 0]$, $low[0, \infty)$, $high[0, \infty)$
n1: $lh[1, 1]$, $hl[1, 1]$, $low[0, \infty)$, $high[0, \infty)$
o1: $lh[1, 1][2, 2]$, $hl[1, 1][2, 2]$, $low[0, \infty)$, $high[0, \infty)$
if MAX_NO_HOPS = 1 then
o1: $lh[1, 2]$, $hl[1, 2]$, $low[0, \infty)$, $high[0, \infty)$

Key : Excitation[Interval Begin, Interval End]
Unit Gate Delays

Figure 4. An example of output intervals computation.

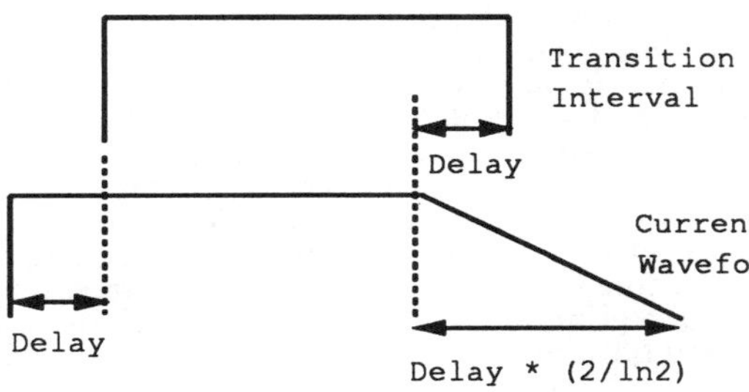

Figure 5. Current waveform due to a transition interval.

The above approach has been implemented in a program called **iMax**. In the program, the circuit is first *levelized* so that a gate at level j does not feed any other gate at a level less than or equal to j. Any user-specified restrictions on certain inputs are then imposed, while all other inputs are assumed to take all possible values from the set $\{$ *low*, *high*, *lh*, *hl* $\}$. After this, the circuit is analyzed in a level by level fashion, starting from the lowest level, propagating the uncertainty at the inputs of every gate to its output. As a result, we get a current waveform that is a point-wise upper bound on the MEC waveform.

In order to assess the quality of the solution, we need to determine how tight the upper bound obtained by iMax is. One way of doing this would be to perform an exhaustive enumeration over all possible input patterns and actually calculate the MEC. However, this would be very expensive and practically impossible for circuits with more than 9-10 inputs (Note: $4^{10} = 1,048,576$). We, therefore, resort to the following random enumeration approach. We repeatedly apply input patterns to the circuit, randomly selected from the set of all possible patterns, and use an event-driven logic simulator to calculate the possible excitations at the outputs of various gates. From these excitations, supply current waveforms are easily calculated. By maintaining the upper-bound envelope of all these waveforms, we basically get a lower-bound on the MEC. Naturally, the more patterns are simulated the closer this waveform will be to the MEC. Ideally, one would like to see the upper-bound obtained from iMax come as close to this lower-bound as possible.

Table1. iMax and iLogSim results for 9 small circuits.

Circuit	No. Gates	No Inputs	iMax10	iLogSim	Ratio
BCD Decoder	18	4	32.41	32.41	1.00
Comparator A	31	11	54.47	54.47	1.00
Comparator B	33	11	54.00	54.00	1.00
Decoder	16	6	26.75	26.75	1.00
P. Decoder A	29	9	44.11	44.11	1.00
P. Decoder B	31	9	49.35	49.35	1.00
Full Adder	36	9	58.00	55.44	1.05
Parity	46	9	47.57	47.57	1.00
Alu	63	14	78.89	70.87	1.11

Table 2. iMax and iLogSim results for 10 ISCAS-85 circuits.

	No.	No.	Peak Currents			CPU Times (sec)	
Circuit	Gates	Inputs	iMax10	iLogSim	Ratio	iMax	iLogSim
c432	160	36	182.63	148.00	1.23	1.92	1677.63
c499	202	41	247.97	144.00	1.72	1.65	655.61
c880	383	60	420.27	267.88	1.57	4.72	2842.85
c1355	546	41	635.01	413.69	1.53	7.22	4459.27
c1908	880	33	734.62	439.68	1.67	12.96	23446.32
c2670	1193	233	1177.27	684.48	1.72	12.49	18935.79
c3540	1669	50	1751.05	785.61	2.23	20.57	76788.50
c5315	2307	178	2327.27	1296.19	1.80	27.39	110218.80
c6288	2406	32	4097.67	2985.89	1.37	61.72	2107521.87
c7552	3512	207	4365.80	2260.47	1.93	46.40	191569.24

The program that implements this random enumeration technique is called **iLogSim** (Current Logic Simulator).

5 Experimental Results

In all the circuit examples considered below, two assumptions are made. First, a constant number is assigned to every gate as its delay value. This delay value can be different for different gates. Second, the peak of the transition current for every gate for both *lh* and *hl* transitions is (arbitrarily) taken to be 2 units of current. The results of iLogSim on various circuits were collected after running the logic simulator 10,000 times with random input patterns. As mentioned earlier, we also assume that all the gates are connected to a single contact point, and we report the peak values of the current waveforms obtained from iMax and iLogSim.

Table 1 lists the results of running iMax and iLogSim on nine small CMOS circuits. The number (10) next to iMax in the table indicates the value of Max_No_Hops. In most of the cases, the results of iMax are in perfect agreement with the iLogSim results. In Table 2, we report the results of running the algorithms on the ten ISCAS-85 benchmark circuits [1].

We observe that for all the circuits, the linear time iMax algorithm took only a few seconds of cpu time on a sun SPARC station 2. Further, for most of these circuits, the upper bound obtained from iMax is within 80% of the lower bound obtained from iLogSim. There are two possible reasons for this mismatch. Firstly, it

Table 3. iMax results vs. Max_No_Hops				
Circuit	iMax: Max_No_Hops			
	1	2	5	10
c432	236.82	190.35	185.68	182.63
c499	292.10	255.37	247.97	247.97
c880	552.52	443.06	425.66	420.27
c1355	751.04	654.54	647.68	635.01
c1908	910.29	761.79	742.80	734.62
c2670	1478.11	1293.78	1196.07	1177.27
c3540	2093.29	1802.01	1762.63	1751.05
c5315	2638.15	2468.82	2350.40	2327.27
c6288	4135.69	4111.81	4099.86	4097.67
c7552	5170.48	4717.09	4424.91	4365.80

is quite possible that the lower bound obtained from iLogSim is not very close to the MEC waveform. Since all the circuits have at least 32 inputs, the space of possible input patterns is huge, and one should not expect that the waveform based on 10,000 input patterns will be very close to the MEC. For smaller circuits (Table 1) where the input space is not so huge, 10,000 input patterns were enough to get a lower bound estimate of MEC waveform that is quite close to the actual waveform, as is reflected by the results. The second possible source of mismatch is our conservative independence assumption for the signals at the gate inputs. One can improve on this assumption by attempting to resolve the signal correlations, as discussed in the following sections.

We next discuss the effect of varying the Max_No_Hops parameter on the performance of iMax. Table 3 lists the results obtained from iMax on ISCAS-85 circuits.

From the table, it can be seen that the peak of the total current waveform improves as Max_No_Hops increases, though it does not improve significantly beyond Max_No_Hops = 5. As the value of this parameter is increased, the amount of memory being used by the program increases, and the number of intervals being merged decreases. A value between 5 and 10 seems to be a good choice for this parameter.

6 Signal Correlations

In general, signal values at internal nodes are correlated, which limits the number of transitions that can occur at the output of a gate. The source of the correlation problem is situations where the output of a gate fans out to several other gates. Such gates are called ***multiple fan-out gates.*** While analyzing each of the gates that are directly connected to a multiple fan-out gate, we should assign the same value to all the lines connected to its output. Thus, the outputs of each of the gates that are directly connected to a multiple fan-out gate are correlated. In other words, even though each of these gates can assume all possible transition values calculated by iMax, they may not ***simultaneously*** carry their worst case excitations. As one goes deeper into the circuit, where these correlated outputs reconverge and feed the same gate, the inputs of that gate become correlated. Such gates are called ***reconvergent fan-out gates.*** With correlated inputs, the number of transitions that can possibly occur at the output of the gate is reduced. In iMax, while analyzing a gate, we assume that each of its inputs can take all possible values from the set of possible values as calculated by iMax, i.e., we assume that the gate input lines are independent. Thus, the basic iMax algorithm described above completely ignores all signal correlations and, therefore, overestimates the supply currents.

The advantage of ignoring correlations in the basic iMax algorithm is its, very desirable, linear time performance. In the next section, we propose a heuristic that accounts for signal correlations, at the expense of increased execution time.

7 Multi-Cone Analysis (MCA)

One can always compute the MEC waveform by brute-force input enumeration. By the same token, we can *improve* on the result of the basic iMax algorithm by doing *partial enumeration.* By enumerating the possible signal values at only a small subset of the circuit nodes, we achieve a trade-off between accuracy and speed. Because of their influence on signal correlation, it turns out that the best candidates for partial enumeration are the multiple fan-out nodes. By enumerating all possible values at a fan-out node and estimating the total circuit current corresponding to each, one eliminates the signal correlations arising from it and obtains a better upper bound on the MEC.

To illustrate these ideas, let N be a multiple fan-out node at the primary inputs. Let $X(t)$ be an *ambiguous* or uncertain waveform at N, as described in section 4. Thus $X(t)$ is a set of possible waveforms $x(t) \in X(t)$, each of which is a possible logic signal at that node. Let t_0 be a fixed time instant and define $X_l(t)$ to be the set of all $x(t) \in X(t)$ for which $x(t_0)$ is low. Likewise, define $X_h(t)$, $X_{lh}(t)$, and $X_{hl}(t)$, based on $x(t_0)$. These sets form a disjoint partition of $X(t)$. Let $Y_i, i = 1, \ldots, m$, be the uncertain waveforms at all *other* primary inputs. If $I_{MEC}(t)$ is the MEC waveform and $I(t)$ is the current corresponding to a specific selection of $x(t) \in X(t)$ and $y_i(t) \in Y_i(t)$, then :

$$I_{MEC}(t) \leq \max_{\substack{x(t) \in X(t) \\ y_i(t) \in Y_i(t)}} I(t) = \max \left\{ \max_{\substack{x(t) \in X_l(t) \\ y_i(t) \in Y_i(t)}} I(t), \max_{\substack{x(t) \in X_h(t) \\ y_i(t) \in Y_i(t)}} I(t), \max_{\substack{x(t) \in X_{lh}(t) \\ y_i(t) \in Y_i(t)}} I(t), \max_{\substack{x(t) \in X_{hl}(t) \\ y_i(t) \in Y_i(t)}} I(t) \right\}$$

Here equality holds when all $Y_i(t)$'s are uncorrelated.

By restricting $X(t)$ so that $x(t_0)$ is low, we can use iMax to obtain an upper bound on $\max_{\substack{x(t) \in X_l(t) \\ y_i(t) \in Y_i(t)}} I(t)$, call it $I_l(t)$. We can likewise compute the upper bounds $I_h(t)$, $I_{lh}(t)$, and $I_{hl}(t)$. From the above, it's easy to see that :

$$I_{MEC}(t) \leq \max\{I_l(t), I_h(t), I_{lh}(t), I_{hl}(t)\}$$

Thus, by enumerating the possible values of $x(t_0)$, running iMax for each case and taking the maximum, we obtain an upper bound on the MEC waveform. Since in each such run of iMax the fanout branches of N take *specific* known values, correlation between them at t_0 is no longer an issue, and the resulting upper bound on MEC should be an improvement on the original upper bound. This analysis is true for any t_0. Therefore, if we select a number of candidate t_0 values and do this enumeration at each, the *minimum* of the upper bounds obtained in each case can be chosen to be the best available upper bound on the MEC waveform. In our implementation, we choose t_0 to be at the beginning and at the end of each of the intervals in $X(t)$.

Ideally, one would like to enumerate all possible *combinations* of values at the multiple fan-out nodes. However, this becomes too expensive since there are typically many multiple fan-out lines in a circuit. We, therefore, enumerate every fan-out node on its own. The overall procedure is as follows. We first run the basic iMax algorithm on the circuit and determine the set of (unconstrained) transitions that can occur at every gate. In a second pass, we separately examine each multiple fan-out line and enumerate its values at a number of discrete time points t_0.

While enumerating a particular multiple fan-out line, we do not need to recompute the signal transitions at *all* the gates in the circuit. It is enough to process only those gates that lie in the COne of INfluence (COIN) of the fan-out line. The COIN of a line consists of all the gates whose outputs are affected by a change in excitation at the line. Thus, a gate is in the COIN of a line if it is either directly connected to it or connected to the output of a gate which is in the COIN. The COIN can be constructed by a simple depth-first traversal starting at the fan-out line. The *distance* of a gate from a line is defined as the minimum number of gates present on any path between the gate and the line. A fan-out node has little direct influence on gates that are very far removed from it. It, therefore, makes little sense to maintain very deep COINs. Based on this, in a COIN, we only include those gates whose distance from the parent fan-out line is less than or equal to a user-specified parameter Depth_FO_Cone.

While a COIN is being analyzed, we consider that the remainder of the circuit is unconstrained, and draws its full-range of possible currents as initially computed by iMax. As a result, irrespective of the choice of COINs and of the order in which they are enumerated, our partial enumeration algorithm always gives an upper bound on the MEC.

In table 4, we report the results of applying the above algorithm (MCA) to the ISCAS benchmark circuits. For most of the circuits, this leads to about 7–8% improvement in results at a small increase in execution time.

Table 4. Results of MCA algorithm for 10 ISCAS-85 benchmark circuits.

Circuit	Peak Currents					CPU Time
	iMax10	MCA	Best iLogSim	Old Ratio	NewRatio	MCA (sec)
c432	182.63	182.38	148.00	1.23	1.23	59.71
c499	247.97	223.75	144.00	1.72	1.55	11.95
c880	420.27	403.65	267.88	1.57	1.51	96.76
c1355	635.01	630.43	413.69	1.53	1.52	185.84
c1908	734.62	689.05	439.68	1.67	1.57	436.25
c2670	1177.27	1168.48	684.48	1.72	1.71	370.24
c3540	1751.05	1692.45	785.61	2.23	2.15	939.67
c5315	2327.27	2251.22	1296.19	1.80	1.74	616.76
c6288	4097.67	4097.56	2985.89	1.37	1.37	2916.37
c7552	4365.80	4317.48	2260.47	1.93	1.91	1329.93

8 Conclusions

We have proposed a linear time algorithm that computes the maximum currents in the supply lines. Most of the previous algorithms on maximum current estimation suffer from exponential complexity and are not adequate for large circuits. Our approach avoids the exponential complexity by adopting a pattern independent approach. The algorithm presented is able to find an upper bound on MEC waveform in circuits with 100's of inputs and 1000's of gates in a few minutes of cpu time on a sparc SUN workstation. For circuits with few inputs, where it is possible to explore the input search space and determine what the actual MEC waveform is, our algorithm produced extremely good results. For most of the larger circuits, the upper bounds produced by our algorithm are within 60% of the lower bounds obtained by exploring a very small portion of the input space. We are currently conducting further research on the signal correlation problem.

Acknowledgements

A major part of this research was performed while the first author was at Texas Instruments for the Summer of 1991. The authors are thankful to the technical staff members of the Semiconductor Process and Design Center at TI Dallas, especially Dr. Ping Yang and Dr. Jue-Hsien Chern, for providing valuable discussions and guidance. This research is supported by Texas Instruments Inc. and the Semiconductor Research Corporation.

References

[1] F. Brglez and H. Fujiwara, "A Neutral Netlist of 10 Combinational Benchmark Circuits and a Target Translator in Fortran," *Proc. of International Symposium on Circuits and Systems*, 663–698, June 1985.

[2] S. Chowdhury and J. S. Barkatullah, "Estimation of Maximum Currents in MOS IC Logic Circuits," *IEEE Trans. on Computer-Aided Design*, Vol. 9, No. 6, pp. 642–654, June 1990.

[3] S. Devadas, K. Keutzer and J. White, "Estimation of Power Dissipation in CMOS Combinational Circuits," *IEEE Custom Integrated Circuits Conference*, pp. 19.7.1–19.7.6, 1990.

[4] J. E. Hall, D. E. Hocevar, P. Yang and M. J. McGraw, "SPIDER – A CAD System for Modeling VLSI Metallization Patterns," *IEEE Trans. on Computer-Aided Design*, Vol. CAD-6, No. 6, pp. 1023-1031, November 1987.

Incremental Circuit Simulation Using Waveform Relaxation

Yun-Cheng Ju and *Resve A. Saleh*
Coordinated Science Laboratory
University of Illinois at Urbana-Champaign
Urbana, IL 61801

ABSTRACT

In this paper, two algorithms are developed using waveform relaxation for the rapid re-simulation of circuits that have been modified slightly compared to a previous simulation run. Both *local* and *global* changes can be handled so long as the changes are relatively small. In this approach, the window sizes, step sizes, and final waveforms from a previous simulation are used to drive the incremental simulation. In addition, a selective storage scheme is used to reduce the overhead associated with saving the waveforms from a previous run. Experimental results indicate that the run-time using the incremental approach can be reduced by an order of magnitude compared to a complete re-simulation.

1. INTRODUCTION

Incremental logic simulation algorithms [1,2] for digital circuits have been developed over the past few years and shown to provide a substantial savings in CPU time. It is based on the observation that when a circuit is slightly modified, only a small portion of the circuit behavior will be affected in some short intervals. Typical changes include modifications of gate delays, localized modifications of the logic design, and variations in the input waveforms to the logic circuit. In principle, only those portions of the circuit affected by the changes need to be recomputed in a subsequent simulation when the behavior differs from the previous run. This is the basic strategy behind incremental simulators. Waveforms of a previous run are stored and re-used in simulating a slightly modified circuit to reduce the overall run-time of the new simulation. As a result, incremental simulators run in a time that is proportional to the size and implication of design changes, rather than the size and the amount of activity of the entire circuit under simulation.

The incremental simulation algorithms can be categorized into two types: *incremental-in-space* and *incremental-in-time*. Based on an assumption of unidirectionality of the influence of a design change, incremental-in-space algorithms usually employ a network complier to detect the boundary of the worst case influence of the design change before the new simulation run and re-simulate the parts of circuit which may be affected throughout the new simulation time frames. The nodes on the boundary of the influence of the design change are treated as external nodes and their waveforms from the previous run are fed into the new simulation as input signals. On the other hand, incremental-in-time algorithms maintain the exact influence of a design change dynamically during the new simulation run. They maximally utilize the waveforms of the last run to reduce the number of re-evaluations to a minimum by re-simulating a component only for the intervals over which different simulation results could occur.

In our previous work [3], we applied the incremental methods to the electrical level using the event-driven simulation techniques of the iSPLICE3 program [4] which uses Iterated Timing Analysis (ITA)[4]. This implementation was based on the *incremental-in-space* approach. Events were initiated by the subcircuit containing the local modifications and propagated outward to fanout subcircuits. A boundary for event scheduling was defined at nodes for which the new solution and the previous solution did not differ. However, once a component was identified as being in the cone of influence of the local change, it was re-simulated for the rest of the new simulation run. Therefore, the overall speedup using this approach depended on the location of the design change and how early in the simulation period it began to affect neighboring subcircuits. We obtained speedups in the range of 1 to 10, depending on the nature of the change, with a typical value of around 4.0.

2. MOTIVATION FOR A NEW APPROACH

The next step in our work was to extend iSPLICE3 by implementing the incremental-in-time algorithm to improve the performance. In gate-level logic simulation, the assumption of unidirectionality of signal flow is valid and this greatly simplifies the development of incremental-in-time techniques. On the other hand, in electrical-level circuit simulation, any change at a given node may result in changes at its feedforward and feedback nodes due to local bidirectional coupling through floating capacitances or resistors. Therefore, the boundary of influence must be determined dynamically during the new simulation run in both the forward and backward directions, and involves iterating waveform fragments to produce the solution. An initial attempt to introduce the

0738-100X/92 $3.00 © 1992 IEEE

incremental-in-time algorithm into the iSPLICE3 program was unsuccessful, primarily due to the complexity and overhead of the implementation. It became clear that a new approach was needed to develop an efficient incremental-in-time simulator for local changes in the design.

Another motivation for abandoning the approach based on event-driven ITA was to handle global changes. Consider the re-simulation of circuits where all subcircuits have been modified slightly. An example of this is a decrease in the channel length or the threshold voltage of all transistors by 10%. In such cases, all of the subcircuits must be completely re-simulated. Event-driven incremental techniques are clearly not appropriate for this purpose, even though the waveforms of the previous run and the new run may be very similar.

An alternative circuit simulation algorithm, *waveform relaxation* (WR) [5,6], is better suited to addressing both of the situations described above. In the WR algorithm, the speed of convergence strongly depends on how close the initial solutions are to the final solutions. For incremental simulation, the waveforms from the previous run are provided as initial guesses, so that most of the subcircuits are expected to take fewer iterations to converge. In addition, it is well-known that subdividing the entire simulation period into smaller *windows* [6] can greatly improve the efficiency of WR algorithms in handling circuits with feedbacks loops (eg., finite state machines, ring oscillators, etc.). However, the optimal window sizes are difficult to determine *a priori*. We observed that when a design change is small, the previous window sizes can provide valuable information to reduce the amount of CPU-time spent on rejected windows.

In this paper, we explore *incremental-in-iteration* and *incremental-in-time* algorithms using waveform relaxation techniques for rapid re-simulation of circuits with small *global* or *local* changes. These techniques have been implemented in the *RELAX2.3* program [6].

3. INCREMENTAL-IN-ITERATION SCHEME FOR GLOBAL CHANGES

To determine the effectiveness of WR for this application, we first implemented a version of Relax2.3 that performs an incremental-in-iteration scheme on circuits with small global changes. The procedure begins by simulating a given circuit using the non-incremental version of Relax2.3. Next, all of the final waveforms and the final window sizes are stored on disk. The circuit is modified slightly by changing some global parameter, such as the threshold voltage or the oxide thickness. Before the re-simulation of the modified design, the window sizes from the previous run, which will be referred to as *super-windows*, are used to define simulation intervals for the incremental run. Each super-window is treated as a physically separate simulation in the incremental simulator. If a subcircuit can not be re-simulated during the incremental run within one super-window, then it is divided into a number of *sub-window*. The last sub-window is forced to align with the corresponding super-window boundary as shown in Figure 1.

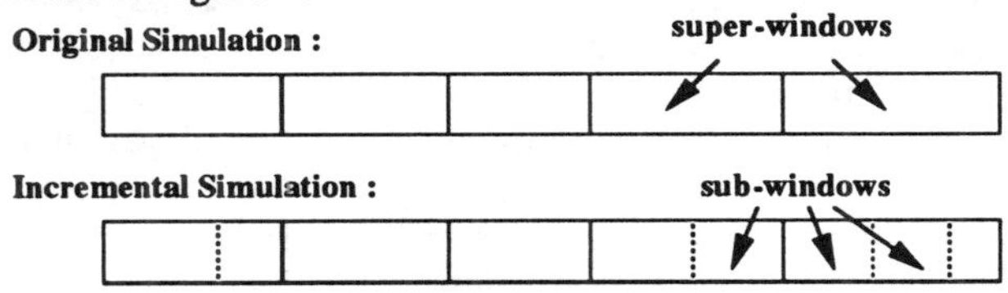

Figure 1: Window Definitions

At the beginning of each super-window, all the associated waveforms from the previous run are read into memory from disk. A waveform is identified as being *invalid* and discarded if there is a discontinuity between the previous waveform and the new result produced in the last sub-window. All the other waveforms left are treated as *valid* waveforms and are used as initial guess waveforms. To simplify the waveform comparison procedure, a subcircuit is forced to take the same step sizes as in the previous run. However, this requirement is removed if the subcircuit is identified as being affected in the new run. A side benefit of this policy is to reduce the number rejected time points in the new run.

The incremental-in-iteration scheme is based on partial waveform convergence [6]. If a subcircuit converges in a sub-window, it is no longer processed unless it is affected by another subcircuit in the same sub-window. Since the initial guess waveforms are derived from the previous run, the new waveforms usually converge in one or two iterations. If the super-window is completed using one sub-window, the next super-window is selected and the procedure repeats. Otherwise after simulating the first reduced and accepted sub-window, all of the waveforms from the previous run are ignored and a standard WR algorithm is used to simulate the rest of the super-window.

Experimental Results

An industrial circuit, a portion of a 64K CMOS static RAM with 277 transistors, was used to assess the performance of the incremental-in-iteration scheme. Since the incremental methods all produce the same waveforms as the non-incremental approach, no waveform comparisons are presented. In this example, the zero-bias threshold voltage V_{TO} associated with the four MOS models of transistors (two for N-channel, two for P-channel) are modified with a ΔV_{TO} from –0.1 to 0.1 volt, respectively, which produces ±10% variation. The waveforms and the window sizes produced by simulating the original configuration are used as initial solutions for the re-simulation of all modified designs. The overhead associated with reading and writing previous waveforms is reduced to the minimum by saving and restoring the whole memory image of the waveform array in binary form using low-level I/O calls. Figure 2 shows the speedups of our incremental-in-iteration scheme against the standard scheme, including the CPU times spent in reading previous waveforms. The experimental results show that the speedup depends on the effect of changing a parameter on the circuit behavior. In the case of pmos1, the speedup remains at around 8 as the threshold voltage is varied. However, the speedup drops of noticeably for nmos1 and significantly for pmos2 and nmos2. Hence, this circuit behavior is sensitive to the *nmos2* and *pmos2* parameter set.

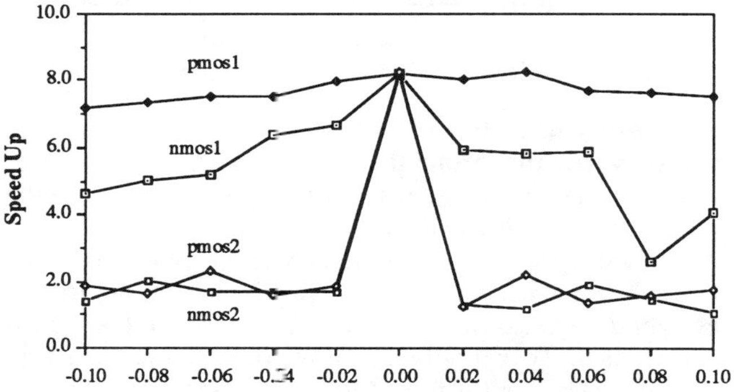

Figure 2: Performance of Global Changes

4. AN INCREMENTAL-IN-TIME SCHEME FOR LOCAL CHANGES

The results in the previous section indicate that some speedup can be derived from WR even if the whole circuit is re-simulated during the incremental run. Clearly, the speedup can be improved substantially if the changes are local in nature. In this section, we present an incremental-in-time scheme for local changes that combines the waveform relaxation and event-driven techniques. Incremental-in-time algorithms attempt to locate the true influence of a design change dynamically during the new simulation run. Unfortunately, the high costs associated with restarting a transient analysis and waveform checking needed for suspending the analysis make it very difficult to develop a true incremental-in-time technique at the electrical level. Instead, we use incremental-in-time methods at the super-window level. To be more specific, a subcircuit is re-simulated for an entire super-window if any different simulation results may occur at this particular subcircuit in a given super-window. It then schedules its fanout subcircuits, and they are re-evaluated in this super-window. If the fanout subcircuits do not change relative to the previous waveforms, they do not schedule their fanouts. If a fanout subcircuit produces a different solution compared to the previous run, it will reschedule its fanouts, and the process continues in this manner.

The major difference between this method and the previous scheme is that only modified subcircuits and subcircuits with invalid waveforms are marked to be processed at the beginning of each super-window. As more and more subcircuits are affected, they are scheduled on an event-driven basis in the same super-window. However, if the implication of the design change stops, the associated subcircuits are dropped from the re-simulation list at the beginning of the next super-window. Finally, if a sub-window is shorter than the corresponding super-window, a standard WR algorithm is used to simulate the rest of the super-window.

Experimental Results

In this section, the static RAM circuit (with 277 transistors) is used to illustrate the performance of the WR-based incremental method. The first experiment was to perform 250 runs of the incremental simulator after changing the W/L values of 250 different transistors. The total CPU-time for 250 non-incremental runs was 18,277 seconds. The total CPU-time for the 250 incremental runs was 1055 seconds. The time spent in disk I/O was negligible. The maximum and minimum speedups were 272 and 1.8 respectively, and the overall speedup was approximately 17.

In the second experiment, the WR method was compared to the ITA method for incremental simulation. Two different simulations were performed using non-incremental methods and the two incremental methods. The CPU-times and speed improvements are listed in Table 1 for two local changes in the circuit. In these two representative cases, and just about all other cases, the WR-based method was superior to the ITA method.

		Incremental-in-space (ITA)	Incremental-in-time (WR)
Mod. #1	Standard	68.54 Sec.	75.52 Sec.
	Incremental	12.97 Sec.	6.74 Sec.
	Speed Up	5.28	11.20
Mod. #2	Standard	65.62 Sec.	73.59 Sec.
	Incremental	73.50 Sec.	7.57 Sec.
	Speed Up	0.89	9.72

Table 1: Performance of Local Changes

There are two reasons why the *incremental-in-time* scheme outperforms the *incremental-in-space* scheme presented in [3] in most cases. The primary reason is that the incremental-in-time scheme exploits more latency in the simulation domain. In the incremental-in-space scheme, all affected subcircuits will be re-simulated to the end of the new run even if the implication of the design change stops. This is clearly demonstrated in Figure 3 which is an activity map of the re-evaluations performed in the incremental run. Each subcircuit is numbered along the y-axis and the simulation period is shown along the x-axis. The dark regions indicate that evaluations of a given subcircuit were performed in the incremental run. Clearly the WR-based method is superior to the ITA-based method. The two vertical stripes in the WR activity map are points at which all the subcircuits were re-evaluated for the entire super-window. This was because more than one sub-window was needed to complete the super-window, so the program reverted back to the standard WR

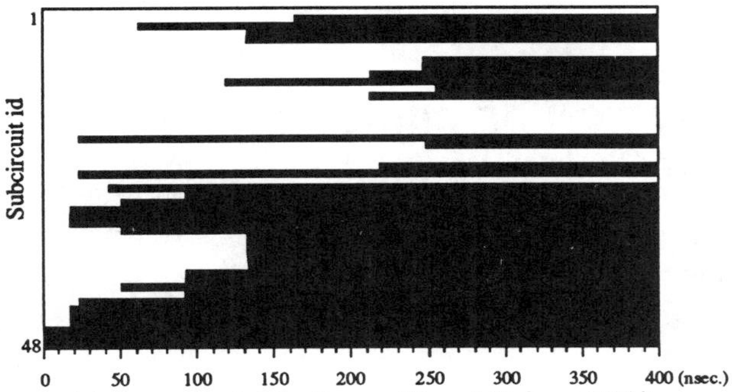

(a) Incremental-in-Space (Event-Driven ITA)

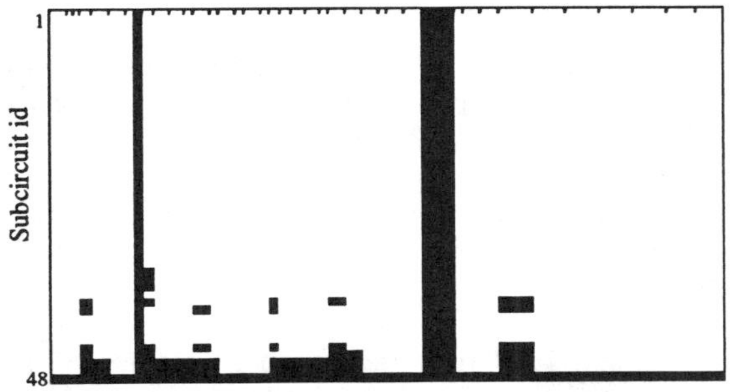

(b) Incremental-in-Time (WR)

Figure 3: Activity Maps for ITA and WR

method.

The other major reason is that the waveform comparison is easier and more accurate in the modified incremental-in-time scheme using WR. In WR, each subcircuit is evaluated separately so it can take the same time steps as it did in the previous run to avoid the need for interpolation. However, interpolation is virtually unavoidable in the ITA approach since the time steps taken in the new run will usually be different from the previous run. As a result, some subcircuits may be incorrectly identified as being affected by a change. This is also the cause of the significant amount of re-evaluation shown in the ITA activity map.

5. A SELECTIVE STORAGE SCHEME

As in other incremental algorithms, there is a tradeoff between memory usage and performance. Saving thousands of waveforms containing thousands of points may require several megabytes of disk space. Fortunately, there are effective ways of reducing the disk storage at the expense of some additional CPU-time. For example, since most of the speedups come from the windows that required more iterations to converge in the first run, storing only the waveforms associated with those windows can still achieve good speedups. In other words, only those critical super-windows are saved and re-used in the incremental simulation. In addition, re-using only the previous window sizes can achieve large savings in the disk space with a further increase the CPU-time for the incremental run.

Table 2 shows the disk storage, I/O overhead, and execution times associated with different waveform storage strategies. If all the waveforms are saved, the speedup is approximately 7 but the memory requirements are 351Kb for 400ns simulation of a circuit with 150 nodes. Here, 4 bytes are needed per node per time-point of simulation. However, saving only the nodes that required more than a certain number of iterations to converge reduced the disk space significantly with a corresponding reduction in the performance. When only the super-window sizes were saved, there was a negligible amount of disk space used but the speedup was reduced to roughly 2.0. Therefore, under normal circumstances, all the waveforms should be saved, but when disk space is limited, the partial waveform storage scheme provides an effective way to trade off memory overhead and re-simulation speed.

Different Partial Storage Scheme	Total Execution Time (Secs.)	Disk I/O Overhead (Secs.)	Disk Storage Taken (Kbytes)
Non-incremental	114.34	-	-
All Windows	16.52	0.30	351
Iterations > 3	16.93	0.21	301
Iterations > 4	21.33	0.16	256
Iterations > 5	26.90	0.08	142
Iterations > 6	39.82	0.03	66
No Waveforms (Window Sizes Only)	55.43	0	0

Table 2: Performance of Selective Waveform Storage

In conclusion, the waveform relaxation method provides an effective means to perform incremental simulation of VLSI circuits. Financial support of this work was provided by the Semiconductor Research Corporation under Contract No. 91-DP-109.

6. REFERENCES

[1] K. Choi, S. Y. Hwang, and T. Blank, "Incremental-in-Time Algorithm for Digital Simulation," *Proceedings of the 25th ACM/IEEE DAC*, pp. 501-505, June 1988.
[2] S. Y. Hwang, T. Blank, and K. Choi, "Fast Functional Simulation: An Incremental Approach," *IEEE Transaction on CAD*, vol. 7, no. 7, pp. 765-774, 1988.
[3] Y. C. Ju, F. L. Yang, and R. A. Saleh, "Mixed-Mode Incremental Simulation and Concurrent Fault Simulation," *Proceedings of the ICCAD*, pp. 158-161, Nov. 1990.
[4] R. Saleh, A. R. Newton, **Mixed-Mode Simulation** Kluwer Academic Publishers, Norwell, Massachusetts, 1990.
[5] E. Lelarasmee, A. E. Ruehli, and A. L. Sangiovanni-Vincentelli, "The Waveform Relaxation Method for Time Domain Analysis of Large Scale Integrated Circuits," *IEEE Transactions on CAD*, vol. 1, no. 3, pp. 131-145, July 1982.
[6] J. White and A. Sangiovanni-Vincentelli, **Relaxation Techniques for the Simulation of VLSI Circuits** Kluwer Academic Publishers, Norwell, Massachusetts, 1986.

PARALLEL WAVEFORM RELAXATION OF CIRCUITS WITH GLOBAL FEEDBACK LOOPS

T.A. Johnson
A.E. Ruehli

IBM Thomas J. Watson Research Center
Yorktown Heights, N.Y.

Abstract

Feedback loops often severely degrade the performance of waveform relaxation techniques in solving large circuit analysis problems. Several new approaches have been studied in this investigation to provide greater parallelism and faster convergence for such circuits. WR_V256, an experimental waveform-relaxation-based parallel circuit simulator for the Victor family of distributed memory parallel machines, was used to study performance tradeoffs of partitioning and scheduling algorithms for circuits containing global feedback loops. This investigation includes circuits ranging from less than 300 to over 93,000 transistors. Several of the circuits were extracted directly from a 16 Mbit DRAM design.

1. Introduction

A barrier exists today in the simulation of VLSI circuits by conventional general purpose circuit programs like SPICE or ASTAP. Generally, the polynomial run time limits an analysis to circuits with fewer than 5,000 to 10,000 transistors, depending on the complexity of the transistor models employed. Meanwhile, the number of transistors on VLSI chips is increasing beyond one million. The result is an ever-widening gap between circuit simulation verification needs and capability.

Three different circuit simulation approaches exist for the transient analysis of VLSI circuits. In the first approach, accuracy is traded for speed e.g. SPECS [1]. The second approach, waveform relaxation (WR) [2-4], aims at SPICE level accuracy. Both approaches lead to an approximately linear relationship between circuit size and run time. The third approach, WR applied to parallel processors, is the topic of this paper. The utility of the parallel WR (pWR) method for VLSI circuit simulation has been demonstrated by several researchers [5-9]. Although, the algorithms applied in these programs have been different, results obtained show consistent trends. Early parallel processors were not very powerful. Hence, early pWR research concentrated on techniques to apply more than one processor to solve a single circuit. Today, with faster parallel processors, many subcircuits are assigned to each processor.

The primary challenge facing an implementation of parallel WR is to consistently achieve high processor utilization and parallel efficiency. The efficiency and speed of parallel waveform relaxation circuit analysis strongly depends on the ability to partition a large circuit into many smaller subcircuits so that workload can be distributed well among the processors. Circuits which partition well generally can be solved much faster using waveform relaxation than with conventional techniques. However, the presence of feedback signals provides global coupling among the otherwise weakly coupled subcircuits, and inhibits partitioning. In general, most previous work to improve the performance of waveform relaxation when analyzing circuits containing global feedback loops has been targeted to sequential machines. This paper describes a new approach that has been developed specifically for pWR. These new techniques provide greater parallelism and faster convergence for circuits with global feedback loops than is achievable with other published methods.

2. Victor and the WR_V256 Simulator

TOGGLE, a serial waveform relaxation circuit simulator [3], served as the starting point for WR_V256. This program was ported to the Victor family of distributed memory MIMD machines. Briefly, the Victor parallel processors are built around the INMOS transputer, a microprocessor that integrates serial communication links (i.e. channels) into its architecture. Each Victor processor node consists of one transputer with 4 Mbytes of

DRAM. Each Victor disk node consists of one transputer with 1 MByte of SRAM and a SCSI interface to either a 300 or a 600 Mbyte disk. The largest member of the Victor family interconnects 256 processor nodes in a 16X16 mesh and connects the top and bottom rows of the mesh with a row of 16 disks, as shown in Figure 1. A total of 1 GByte of main memory and 10 GBytes of disk storage are available.

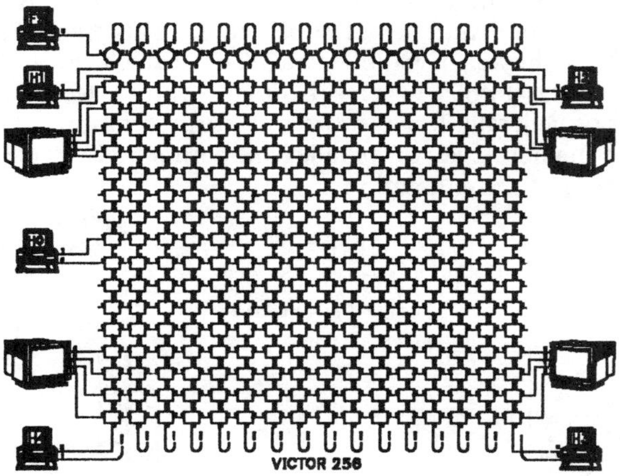

Figure 1. VICTOR V256 topology.

TOGGLE has been optimized for a serial environment. A Gauss-Seidel (G-S) relaxation algorithm is used that takes advantage of subcircuit dormancy to reduce the amount of work per iteration by skipping the analysis of subcircuits whose inputs have not changed. For a sequential machine, the G-S relaxation algorithm is enforced by leveling the subcircuits and ordering them, by level, on a centralized dispatch queue. One complication of porting to a distributed memory machine is that the centralized dispatch queue must be split into many independent queues. Hence, the data dependencies that were enforced by the single queue in the serial program are lost. Therefore, we added a waveform data-driven approach that ensures a subcircuit is not analyzed until all of its inputs have been updated.

Distribution of subcircuits among the Victor nodes is done statically. Sequential chains of subcircuits are generated based on subcircuit interconnections. Several chains may be grouped together based on the FET count per chain, and each group is then allocated to a separate processor. Note that a uniform number of FETs per group does not guarantee that all groups will require the same CPU time. This is a first order approximation based on observed behavior that indicates model evaluation dominates solution time for small subcircuits. For large subcircuits the effectiveness of FET count decreases since analysis time may be dominated by matrix solution.

3. Waveform Relaxation Algorithms

Waveform-relaxation-based circuit simulators are generally implemented using either a Gauss-Seidel (G-S) or a Gauss-Jacobi (G-J) algorithm. A properly implemented G-S waveform relaxation with signal flow ordering and scheduling will, in general, out-perform G-J in a sequential machine environment. In both approaches, subcircuits are formed such that the local WR iterations converges. The approximate contraction factor must be $\gamma = g_f g_b < 1$, where g_f is the worst case forward gain and g_b is the worst case backward gain [10]. The requirement for G-J is that both gains are small, e.g. $g_f \cong g_b < 0.3$. In G-S, with proper signal flow scheduling, it is desirable that $g_b << g_f$. The ultimate example of this is the one-way circuit for which $g_b = 0$ and g_f can be arbitrary large [11] and convergence is obtained in one iteration.

Signal flow (data flow) is used in the G-S algorithm to order computations, and generally results in fewer waveform iterations than a G-J implementation. Although improving the rate of convergence, this ordering of computations may reduce parallelism. The key to proper ordering and scheduling for a G-S WR program is precise knowledge of the logic signal flow within a circuit. For circuits with global feedback loops, it is often difficult to correctly determine signal propagation. Sequential implementations of G-S algorithms generally characterize feedback loops according to their estimated delay, and process them differently depending upon that delay [11]. Feedback loops that are determined to be "long" are generally broken. For each WR iteration, the waveform information used for the feedback signals is the waveforms obtained from the previous WR iteration. This essentially introduces G-J type behavior into a G-S analysis, and may slow down convergence due to the additional iterations required. Feedback loops characterized as "short", are merged into a single subcircuit. In essence, the global feedback loop is removed by localizing it. For sequential machine implementations, these approaches work quite well, but improvements are possible for pWR.

4. Parallel Processing of Feedback Loops

Broken feedback loops force portions of a circuit to be solved with at least some input waveforms not precisely known for the current iteration. This results in more waveform iterations than is the case in the absence of such broken loops. Feedback loops which are merged into a single subcircuit maintain strict G-S ordering, but they create larger subcircuits. This has a negative impact on load balancing, matrix size and multi-rate properties. Most G-S algorithms apply a *basic* scheduling technique where the WR relaxation spans all non-converged subcircuits. This can result in unnecessary iterations and propagation of signals which may be in error. The impact of this is more severe in early iterations, where few, if any, of the subcircuits can be skipped in the analysis. The above observations as well as extensive experimentations lead us to the following algorithm for pWR partitioning and scheduling.

1. Break all feedback loops, independent of delay.
2. Break feedback loops such that the workload can be distributed as uniformly as possible among the processors. Referring to Figure 2, breaking the loop at "A" will create two chains of subcircuits with nearly the same workload. (Assuming each subcircuit contains approximately the same number of FETs.) If each chain is assigned to a different processor, they can be solved concurrently.
3. Locally schedule subcircuits associated with feedback loops multiple times for each WR iteration. Multiple scheduling of feedback loop subcircuits is most effective during the first 1 or 2 WR iterations. Therefore, the number of local iterations can be reduced as the number of WR iterations increases. Additionally, the Newton convergence criteria can be loosened during the early local iterations, and tightened as the iterations increase.

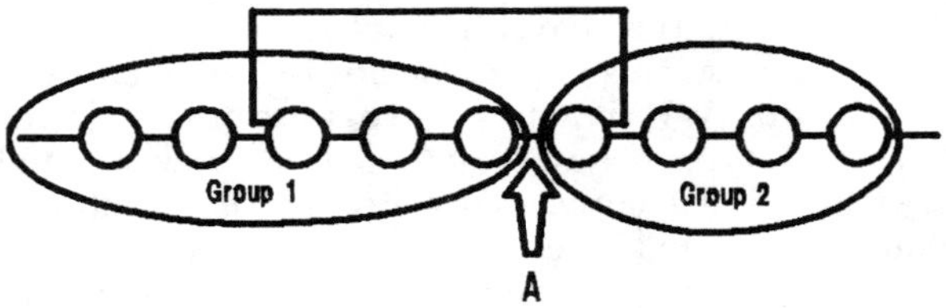

Figure 2. Feedback loop cutting.

5. Experiments/Results

A suite of circuits ranging in size from under 300 to over 93,000 FETs was simulated using both conventional partitioning/scheduling techniques and those described in Section 4. The larger circuits were extracted from a 16 Mbit DRAM design. These circuits were simulated using the Victor, V256 processor, described in Section 2 with the larger circuits utilizing all 256 processors. Figure 3 compares run times for each circuit when feedback loops are cut regardless of length, versus a conventional approach which cuts only "long" feedback loops. In all cases, the increased parallelism resulting from cutting all feedback loops produces a substantially faster run time. Cutting feedback loops allows the partitioner to create fewer large subcircuits and a greater total number of subcircuits. This translates into a higher level of parallelism, which in turn translates into a faster run time. Cutting feedback loops generally forces more waveform iterations, but the higher level of parallelism permits faster execution of those iterations.

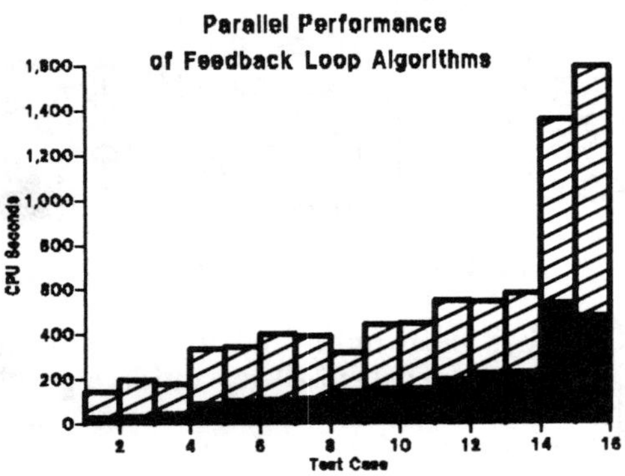

Figure 3. . Run time comparison of new feedback loop algorithm (solid), to conventional (crossed).

The next step toward even faster run times is made by implementing multiple local iterations among subcircuits forming feedback loops during a single WR iteration. Table 1 shows performance results for a small circuit of 640 transistors. In Table 1, "X_Y" means that "X" local feedback iterations were performed during the first WR iteration, and "Y" local iterations were performed during each successive WR iteration.

Approach	4 nodes	16 nodes	64 nodes
Merged	183	81	50
1_1	177	53	40
2_2	165	51	43
4_2	125	42	33
6_2	151	50	39

Table 1. Run Time comparison for different processor configurations.

As the number of local iterations is increased, the run time reaches a minimum, after which it increases again. This increase is caused by the overhead of performing additional local iterations. Once a subcircuit has converged, it will not be simulated again unless its inputs change. However, there still is a performance penalty associated with checking to determine if a subcircuit should be re-simulated. As more processors are used, more parallelism can be exploited by cutting feedback loops. The ratio of run time for merged feedback loops to the "best" cut loop algorithm improves from 4 processors to 16 processors. Eventually however, most available parallelism is exploited, and increasing from 16 to 64 processors shows a smaller percentage improvement. Table 2 shows run time results for several other circuits with similar results. This data suggests that 2 to 4 local feedback iterations in WR iteration 1, and 2 feedback iterations in subsequent WR iterations is a good choice on average.

Approach	Ckt1	Ckt2	Ckt3	Ckt4	Ckt5
Merged	351	385	629	114	242
2_2	248	304	591	107	124
4_2	262	284	556	89	129
6_2	293	312	728	49	138
9_2	287	325	783	50	146

Table 2. Run time for different feedback loop algorithms (seconds)

6. Conclusions

Parallel waveform relaxation is clearly a viable approach for future large circuit simulation. However, techniques like the ones introduced in this paper are needed to achieve a uniformly high efficiency for a large class of circuits. An exciting result obtained by the techniques given in this paper is the analysis of an 18,000 transistor circuit in just 9 minutes.

References

[1] C. Visweswariah and R. Rohrer, "Piecewise Approximate Circuit Simulation," *IEEE Trans. on CAD of ICs and Systems*, pp. 861-870, 1991.

[2] J. White and A. Sangiovanni-Vincentelli, "RELAX2: A New Waveform Relaxation Approach for the Analysis of LSI MOS Circuits," *Proc. International Symposium on Circuits and Systems*, pp. 756-759, 1983.

[3] T.J. LeBlanc, T.J. Cockerill, P.J. Ledak, H.Y. Hsieh and A.E. Ruehli, "Recent Advances in Waveform Relaxation Based Circuit Simulation," *IEEE International Conference on Computer Design: VLSI in Computers*, pp. 594-596, 1985.

[4] C.H. Carlin and A. Vachoux, "MOSART: A Large Scale Time-Domain Simulator Based on Waveform Relaxation," *Proceedings of the European Conference on Circuits Theory and Design ECCTD 87*, pp. 227-233, 1987.

[5] L. Peterson and S. Mattisson, "Circuit Simulation on a Hypercube," *IEEE International Symposium on Circuits and Systems*, pp. 1119-1122, 1988.

[6] R. Saleh, K. Gallivan, M. Chang, I. Hajj, D. Smart, and T. Trick, "Parallel Circuit Simulation on Supercomputers," *Proceedings of the IEEE*, vol. 77, no. 12, pp. 1915-1913, 1989.

[7] T.A. Johnson and D.J. Zukowski, "Waveform Relaxation Based Circuit Simulation on the VICTOR (V256) Parallel Processor," *to appear in: IBM Journal of Research and Development*, November 1991.

[8] P. Odent, L Claesen, H. De Man, "Acceleration of Relaxation-Based Circuit Simulation Using a Multiprocessor System," *IEEE Trans. on Computer-Aided Design*, no. V 9 No. 10, pp. 1063-1072, 1990.

[9] D.W. Smart and T.N. Trick, "Waveform Relaxation on Parallel Processors," *Int. Journ. of Circuit Theory and Appl.*, no. V 16, pp. 447-456, 1988.

[10] J. White and A. Sangiovanni-Vincentelli, "Partitioning Algorithm and Parallel Implementations of Waveform Relaxation Algorithms for Circuit Simulation," *Proc. International Symposium on Circuits and Systems*, pp. 1069-1072, 1985.

[11] A. Ruehli, Ed., *Circuit Analysis, Simulation and Design*, North-Holland, 1987.

ON THE OVER-SPECIFICATION PROBLEM IN SEQUENTIAL ATPG ALGORITHMS

Kwang-Ting Cheng
AT&T Bell Laboratories
Murray Hill, NJ 07974

Hi-Keung Tony Ma
Synopsys Inc.
Mountain View, CA 94043

Abstract - Most sequential ATPG (Automatic Test Pattern Generation) programs employ the time-frame expansion technique. Within a time-frame, combinational test generation algorithms that are variations of D-algorithm or PODEM are used. In this paper, we show that some ATPG programs may err in identifying untestable faults. In other words, these test generators may not be able to find the test sequence for a testable fault, even allowed infinite run time, and furthermore may mistakenly claim it as untestable. The main problem of these programs is that the underlying combinational test generation algorithm may over-specify the requirements at the present state lines. We show a necessary condition that the underlying combinational test generation algorithm must satisfy to ensure a correct sequential ATPG program. It is shown that the simple D-algorithm satisfies this condition while PODEM and the enhanced D-algorithm do not. We also study the impact of over-specification on the length of the generated test sequence. Over-specification causes a longer test sequence. Experimental results are presented.

1. Introduction

Test generation for sequential circuits has been studied widely. The sequential circuit test generation problem can be stated as following: Given a circuit and a stuck-at fault, find an input sequence that produces an erroneous response at some primary output in the presence of the fault. The erroneous response is defined in general by the results of fault simulation, i.e. the input sequence produces opposite values (0/1 or 1/0) at a primary output for the good and the faulty circuits. The time-frame expansion technique [1] is usually used and combinational test generation algorithms like D-algorithm [2] and PODEM [3], are extended to handle the iterative array model of the sequential circuits. If the entire problem space is searched and no solution is found, the fault is reported as *untestable*. In contrast to combinational circuits, an untestable fault in a sequential circuit is not necessarily a redundant fault [4 ,5]. A fault is called a redundant fault if the presence of the fault does not change the output responses to any input sequence.

Almost all test generation approaches use three-valued logic (0, 1, x) to handle the unknown initial state of the circuit. It is well-known that three-valued simulation produces pessimistic results. Given a partially specified input vector (some of the inputs have values x's), unknown values x's may be produced by a three-valued logic simulator at some primary outputs that would have logic 0 or 1 values if evaluated by a symbolic simulator. When three-valued logic is used in fault simulation for sequential circuits, a similar problem occurs. A fault detected by a given test sequence, regardless of what the initial state is, may be reported pessimistically as undetected by a three-valued fault simulator. Similarly, a testable fault may be claimed as untestable by the three-valued-logic-based test generator due to the limitation of the three-valued logic. We call those faults that are untestable by a three-valued fault simulator/test generator *three-valued untestable* faults. Recent research reported in [6] has been targeted to generate test sequences for testable but three-valued untestable faults.

Existing sequential ATPG algorithms are designed to solve the following problem: Given a circuit and a stuck-at fault, assuming all memory elements initially have unknown values x's and assuming the use of three-valued logic for computing the signal responses, find an input sequence that would produce a D or $\overline{D}$ at some primary output in the presence of the fault. In this paper, we show that most existing sequential test generation programs do not correctly solve the above problem. In other words, these algorithms may not be able to find a test sequence, even if infinite run time is allowed, for a fault that has a test (that can be verified by a three-valued fault simulator) and may mistakenly claim such fault as three-valued untestable. The problem originates from the fact that the underlying combinational test generation algorithm used by these programs may over-specify requirements at the present state lines while processing a time-frame. PODEM [3] is a typical combinational test generation algorithm that cannot produce a *minimally-specified* test. PODEM frequently generates a test that specifies values at some primary inputs that are x's for detecting the target fault.

During test generation, PODEM searches a test for a fault by continuously dividing the current search space into two disjoint sub-spaces. The complete search space is defined by the set of primary inputs. By searching through disjoint sub-spaces, PODEM avoids exploring any sub-spaces that contain fully-specified test vectors that were declared invalid earlier in the search process. However, this strategy also prevents the algorithm from always producing a minimally-specified test. Furthermore, if the complete space needs to be searched exhaustively (and implicitly) to generate all tests for a fault, the set of tests generated by PODEM will not always contain all minimally-specified test vectors. Similarly, the enhanced D-algorithm also utilizes the concept of disjoint search space where the complete search space is defined by the set of all nodes in a circuit. The over-specification phenomenon does not cause a problem for the correctness of combinational test generation programs since all primary inputs can be set to any known values. This is not so in sequential test generation problem as shown in the next section. The over-specification problem of PODEM-based combinational test generation programs has been studied in [7] for the purpose of vector compaction for combinational circuits.

In this paper, a necessary condition for the underlying combinational test generation algorithm to ensure a complete sequential algorithm is formulated. The condition, in short, is that the underlying combinational algorithm must be able to generate all minimally-specified combinational tests for a fault. PODEM and enhanced D-algorithm do not satisfy this condition, but the D-algorithm does. A 4-valued PODEM that satisfies the condition is suggested. The authors further show a sufficient condition on the circuit and the target fault for the PODEM-based or enhanced-D-algorithm-based methods to be complete. This condition is that both fault-free and faulty circuits are initializable from the unknown state and the initialization sequence is given. Over-specifying values at the present state lines during test generation may also result in longer test sequences. Experimental results are presented in Section 7 to illustrate the impact of over-specification on the length of test sequences.

2. Impact Of Using A Disjoint-Search-Space Algorithm

PODEM and enhanced D-algorithm both utilize the concept of disjoint search space for combinational test generation. When these algorithms are used within a time frame of a time-frame-expansion-based sequential test generator, it may give rise to incorrect results. In the following, we show an example of a testable fault that is mistakenly classified as untestable by a PODEM-based sequential test generator using the time-frame expansion technique.

Example 1: Consider the circuit given in Fig. 1 that has three flip-flops, one input (I) and one output (O). The next state lines are A, B and C and the corresponding present state lines are a, b, and c. Note that flip-flop A in the circuit is not initializable.

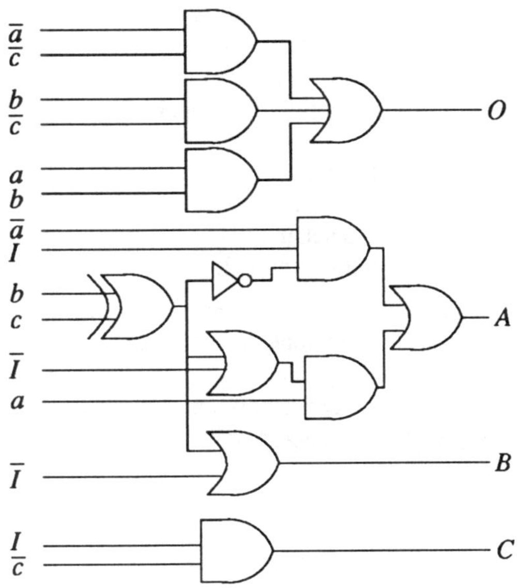

Fig. 1: An example

Let us try to generate a test for the stuck-at-0 fault at output O using the reverse time processing technique [8 ,9] and iterative array model. Combinational test generation algorithm PODEM is used within each time frame. Since the fault is at the primary output, no fault propagation is required and only backward justification of state is needed. Assuming that the order of assigning values at the present state lines during the search is c followed by a, followed by b, the set of combinational tests found by PODEM for the last time frame is:

	I	a	b	c
T_1	x	0	x	0
T_2	x	1	1	0
T_3	x	1	1	1

Because flip-flop A is not initializable, none of the three required states (abc = $0x0$, 110, 111) specified in the combinational tests can be justified. The test generator concludes that the fault is untestable.

However, vector T_4 listed below, that is jointly covered by tests T_1 and T_2,

	I	a	b	c
T_4	x	x	1	0

is also a combinational test that is not generated by PODEM. The required state of T_4 ($x10$) has an x value at present state line a and can be justified by a single input vector I = 0. Thus, the stuck-at-0 fault at output O can be detected by the test sequence $\{I = 0, I = x\}$. □

It may be intuitive to resolve the above problem by expanding the combinational tests to convert the over-specified present-state inputs into x's. The simple solution is to choose a present-state input i whose value is specified (i.e. set to 0 or 1) and change it to an unknown value x. If the new vector is still a combinational test, a less specified vector is found. This process is continued until all present-state inputs are processed once. This simple heuristic may help but still does not guarantee to find all minimally-specified tests and hence a complete solution, as shown by the following example.

Example 2: Consider the same circuit given in Example 1. Assuming that the order of the present state lines considered for expansion is c followed by b, followed by a. The resulting combinational tests after the simple expansion process are listed below:

	I	a	b	c
T_1'	x	0	x	0
T_2'	x	1	1	x
T_3'	x	1	1	x

All three vectors still have specified values at present state line a and, thus, are not justifiable. □

Although a more sophisticated strategy can be used to resolve the problem shown in this example (e.g. using different order of present state lines for expanding different vectors), even exhaustive expansion on each test cannot always guarantee to produce all minimally-specified combinational tests if a PODEM-based test generator is used. This is illustrated in the following example.

Example 3: Consider a single-output combinational circuit with three inputs a, b and c. The ON-set of the circuit consists of three input cubes -00, $10-$ and $1-1$. The circuit is implemented in two-level AND-OR logic with each AND gate corresponding to an ON-set cube. If a PODEM-based test generator is used to generate all tests for the output stuck-at-0 fault and the first assignment is made to input c, the set of tests produced will be $\{x00, 1x1\}$. Note that the test $10x$ is missing in the test set. Even with exhaustive expansion on each test in the test set, test $10x$ cannot be produced. The only way to produce test $10x$ from the test set is to use a 2-level minimization program like [10] to generate all the prime cubes from the test set. However, this could be a very expensive procedure for sequential circuits with large number of flip-flops. □

3. A Necessary Condition for Complete Algorithms

In this section, we present a necessary condition for a three-valued sequential ATPG algorithm using the time-frame expansion technique to be complete.

Given a fault f and a time-frame tf during the test generation process for a sequential circuit, the underlying combinational test generator is used to achieve one of the following three objectives: (1) Activation: find a combinational vector (consisting of primary input and present-state parts) to propagate the fault to a primary output or a next state line. (2) Justification: given a state, find a combinational vector that produces the given state at the next state line. (3) Propagation: given a partially specified present state containing some D or $\bar{D}$, find a combinational vector, with its present-state part covered by the given partially specified present state, that propagates the fault effects to a primary output or a next state line. Different sequential ATPG algorithms may assign these three objectives to the underlying combinational ATPG program in different order and combine them with different backtracking mechanisms. We use $F_{tf}(I, PS)$ to denote the function of the given objective.

Definition: Given a function F_{tf}, a **three-valued prime cube** p of F is a combinational vector for the given time-frame tf that will achieve the given objective and has the property that it is minimally specified in the present-state part of the vector based on three-valued logic simulation, i.e. any present-state line values cannot be further expanded.

Note that the **three-valued prime cube** p is defined in terms of three-valued logic and may be further expanded in the present-state part and still be a valid vector for F_{tf} in the Boolean sense in achieving the objective.

Example 4: Let $O = Iab + I\bar{b}c$ be the output function of a sequential circuit where I is the primary input and a, b and c are the present-state inputs. The output function O is implemented in two-level AND-OR logic with two AND gates, each corresponding to one of the components in the equation. If the objective is setting the output O to 0 for the output stuck-at-1 fault, then the three-valued prime cubes of the funtion are $Iabc = 111-$ and $Iabc = 1-01$. Note that the cube $Iabc = 11-1$ is a Boolean prime cube but not a three-valued prime cube, i.e. simulating $11-1$ on the circuit using three-valued logic simulation will produce an unknown value at the output O. □

A necessary condition: If a sequential ATPG algorithm using time-frame expansion technique is (three-valued) complete, the underlying combinational ATPG algorithm must be able to generate all three-valued prime cubes for the function F_{tf} of any time frame.

The well-known D-algorithm can generate all three-valued prime-cubes for a given objective. The D-algorithm never over-specifies values at signal lines during test generation. If a logic 0 is specified at the output of a two-input AND gate, the D-algorithm assigns a logic 0 at one of the inputs and a logic x at the other input. If this decision eventually leads to a conflict, it backtracks and unspecifies the first input back to x and assigns a logic 0 to the second input. Thus, every value assignment is essential.

An extension of the D-algorithm, called enhanced D-algorithm, is commonly used in sequential ATPG algorithms. The main difference between the D-algorithm and the enhanced D-algorithm is the strategy of backtracking. If the assignment $0x$ to the two inputs of an AND gate for justifying a logic 0 at the gate's output eventually leads to a conflict, instead of changing it to $x0$, enhanced D-algorithm changes the input assignment to 10. This strategy is similar to that used in PODEM. Whenever a backtrack occurs, it searches for a solution in a sub-space that is disjoint from the sub-spaces that have been explored. The advantage of this strategy is the reduction of the computational complexity of search. However, this strategy may cause value assignment at signals that are not essential to achieve the given objective. Therefore, enhanced D-algorithm does not satisfy the condition given above.

As shown by Example 2 and 3, PODEM or enhanced D-algorithm followed by simple expansion cannot generate all three-valued prime cubes for a given function either.

The recently developed Boolean Difference method [11] does not produce minimally specified combinational tests either. The heuristics used for solving the 3SAT problem may make some value assignments that are not essential for satisfying the Boolean Difference formula.

4. 4-Valued PODEM Algorithm

The PODEM algorithm can be modified to generate all three-valued prime cubes. The modified PODEM algorithm, called 4-valued PODEM, requires one more logic value and results in the use of 4-valued logic (0, 1, d, x). Values d has the logic meaning don't care. Initially, all inputs are assigned x values. In 4-valued PODEM, the process of choosing a primary input and a logic value (0 or 1) to achieve a certain objective is exactly the same as that of PODEM. However, at most 2 backtrackings are necessary for an input. When a conflict occurs, instead of complementing the value assigned at the latest chosen primary input, 4-valued PODEM assigns logic d at that primary input. This is basically unspecifying the value at the latest chosen primary input. After the subspace under this assignment is fully exploited, the binary value (0 or 1) that is not yet tried at this primary input is then assigned. The 4-valued PODEM creates a decision tree as shown in Fig. 2 where each node has three children corresponding to specifying 0, d, and 1 at a primary input.

Some of the combinational tests generated by 4-valued PODEM are still over-specified. However, it can be

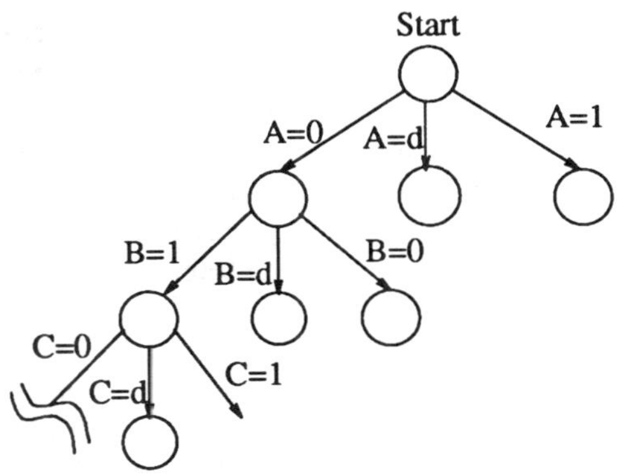

Fig. 2: Decision tree created by 4-valued PODEM.

shown easily that the set of generated combinational tests includes all 3-valued prime-cubes.

Note that value d, like value x, need not be justified if assigned to a present state line. Also, to search the vector space more effectively, we can use only three values (0, 1 and x) for primary inputs and use four values (0, 1, x and d) for present state lines. This change will not sacrifice the accuracy because over-specifying requirements at primary inputs does not cause incorrectness to test generation. If this strategy is adopted, a node in the decision tree corresponding to a primary input will have two children, corresponding to specifying 0 and 1 at the input, and a node corresponding to a present state line will have three children, corresponding to specifying 0, d, and 1 at the line.

5. Initializable Circuits

Even though three-valued PODEM-based and enhanced D-algorithm-based algorithms may produce incorrect results for sequential circuits with an unknown initial state as shown in Example 1, they are correct if the initial states of fault-free and faulty circuits are both known (no flip-flop has a logic value x). Over-specifying requirements at the present state lines will not cause problems to test generation for circuits and faults with known fault-free and faulty initial states. Furthermore, if an initialization sequence that initializes both fault-free and faulty circuits to known states S_g and S_b is given, existing sequential test generators can correctly generate a test for the fault or identify it as three-valued untestable by using S_g and S_b as the initial states for test generation.

Theorem: If a fault is found untestable by a PODEM-based (or enhanced D-algorithm-based) test generation algorithm with given known fault-free and faulty initial states (no flip-flop has a logic value x), S_g and S_b, and there exists a sequence T that can initialize the fault-free and faulty circuits, using three-valued logic, from the unknown state (every flip-flop has a logic value x) to S_g and S_b, the fault is three-valued untestable.

Proof: If a fault is three-valued testable, there exists a sequence T' that detects the fault for unknown fault-free and faulty initial states. By definition, T' also detects the fault for any known initial states including S_g and S_b. Note that if the initial states are known, the resultant fault-free and faulty states for any fully-specified input sequence will be fully specified too. Therefore, from S_g and S_b, there exists a test sequence for the fault such that the required states for each time frame are all fully-specified and over-specification does not cause problem. Thus, a PODEM-based (or enhanced D-algorithm-based) test generator would be able to generate a test, given enough time, for the fault. □

In existing test generation systems, once a test for the target fault is generated, a fault simulator is used to identify those faults in the fault list that are also detected by the generated test sequence. The fault simulator is also used to compute the final states of the fault-free and faulty circuits for each of the undetected faults. These states are recorded and used by the test generator. When a target fault is selected, the test generator assumes the recorded final states of the fault-free and target faulty circuits as the initial states for test generation. The newly generated sequence for the target fault is then attached to the end of the previously generated sequence to form the final test sequence. Because almost all fault-free circuits are initializable in practice and, for most faults in an initializable fault-free circuit, the faulty circuits are also initializable, the test generators have known initial states for test generation for most of the faults. Only the first few faults in the fault list and those faults that prevent the circuit from initialization may have an unknown or partially known initial states when they are selected as target fault for test generation.

6. The Impact of Over-Specification Problem On Test Length

Over-specifying values at the present state lines has great impact on the length of the generated test sequence. Consider a circuit that contains an 8-bit counter. If the combinational test generator over-specifies a logic 1 at the most significant bit of the counter during test generation, hundreds of additional vectors may be needed to detect the fault.

7. Experimental Results

To study the impact of over-specifying values at the present state lines on test length and on the accuracy of identifying three-valued untestable faults, we implemented two sequential circuit test generators. Both generators use the BACK algorithm [9]. The only difference between the two generators is in the underlying combinational algorithm. One, named D_STG, uses the simple D-algorithm and the other, named ED_STG, uses the enhanced D-algorithm. As mentioned in Section 3, the enhanced D-algorithm may over-specify values at the present state lines. Thus, ED_STG may incorrectly claim a fault, that prevents the circuit from initialization, as three-valued untestable. For s298 of ISCAS-89 sequential benchmark circuits [12], both D_STG and ED_STG are allowed 400 seconds for processing a fault such that no fault is aborted and every fault is reported either as detected or three-valued untestable. D_STG reported that 43 faults out of 308 faults are three-valued untestable while ED_STG reported 44 faults are three-valued untestable. There is a fault incorrectly reported as three-value untestable by ED_STG. After careful examination, it is found that this fault prevents a flip-flop from initialization in the faulty circuit. The detection of this fault does not require specification of any value at that flip-flop in the faulty circuit. This is verified by examining the fault-free and faulty states produced by the test sequence generated by D_STG.

Both D_STG and ED_STG were run on a set of ISCAS-89 benchmark circuits. A reasonable time limit for processing a fault is set in this experiment. The D_STG and ED_STG time limits, computed automatically according to circuit characteristics, are identical for the same circuit but vary from circuit to circuit. Experiments were also made on some partial scan circuits, where a set of flip-flops were scanned and made controllable and observable to break all or most cycles for easier test generation [13]. The results are listed in Table 1. The partial scan circuits are named as *cktname-ps* where *cktname* is the name of the corresponding non-scanned benchmark circuit. Circuit s9234.1 is a recently released benchmark circuit from MCNC [14] that is modified and corrected from ISCAS-89 circuit s9234. There are two numbers listed under the column titled fault coverage/test efficiency. The first number is the fault coverage which is defined as the number of detected faults divided by the total number of faults. The second number (in parentheses) is the test efficiency which is defined as the sum of the number of detected faults and the number of identified three-valued untestable faults divided by the number of total faults. Since the program was not given a very large time limit, in many cases, less than 100% test efficiency was obtained. In most cases, D_STG produces shorter test sequences while requiring longer CPU times. ED_STG requires smaller CPU time than D_STG because enhanced D-algorithm exploits the vector space more effectively than D-algorithm as explained in Section 3. The last column shows the percentage of test length reduction which is defined as the test length of ED_STG minus the test length of D_STG divided by the test length of ED_STG. Test length reduction varies from 0% to 27.6%.

Table 1 - Results of D_STG and ED_STG

Ckt	CPU seconds (Sparc 2)		fault coverage (test efficiency)		test length		
	ED_STG □	D_STG †	ED_STG	D_STG	ED_STG	D_STG	reduction
s298	1547	2084	85.7(96.4)	86.0(96.8)	211	163	22.7%
s344	131	132	99.1(99.2)	99.1(99.2)	358	358	0.0%
s349	287	295	95.7(99.1)	95.7(99.1)	111	104	6.3%
s386	681	1207	81.3(100.0)	81.3(100.0)	232	174	25.0%
s400	7002	6947	90.1(93.6)	90.1(93.9)	2291	2036	11.1%
s1196	37	38	99.8(100.0)	99.8(100.0)	362	362	0.0%
s1488	5302	41796	92.1(93.9)	92.1(93.9)	449	325	27.6%
s400-ps	8	8	97.2(100.0)	97.2(100.0)	95	90	5.3%
s5378-ps	2367	4614	93.6(99.6)	93.7(99.4)	1452	1410	2.9%
s9234.1-ps	82086	91300	75.5(89.9)	75.6(89.9)	30094	29692	1.3%

□ ED_STG - using enhanced D-algorithm † D_STG - using simple D-algorithm

8. Conclusion

In this paper we investigated the impact of over-specifying values at the present state lines by combinational ATPG algorithms on the accuracy of the sequential ATPG algorithm and on the length of the generated test sequences. An example was presented to show that over-specification may cause the sequential ATPG to be inaccurate. It was shown that using an underlying combinational ATPG algorithm capable of generating all three-valued prime cubes for a given function is a necessary requirement for an accurate sequential ATPG algorithm based on the time-frame expansion technique. Some of the existing combinational algorithms do not satisfy this condition. Modifications of those algorithms to meet this condition were suggested and discussed. It was also shown that if both fault-free and faulty circuits are initializable and the initialization sequence is given, over-specification at present state lines does not pose problems to test generation. Over-specification causes the test generator to produce longer test sequences as indicated in our experimental data. However, as also shown by our experimental data, techniques that are accurate and capable of generating shorter sequences may also require longer run time. Further research in finding an algorithm that solves correctly the over-specification problem while retaining the efficiency of disjoint-search-space algorithm is worth pursuing.

REFERENCES

1. H. Kubo, "A Procedure for Generating Test Sequences to Detect Sequential Circuit Failur es," *NEC Res. & Dev.*, pp. 69-78 (October 1968).
2. J. P. Roth, "Diagnosis of Automata Failures: A Calculus and A Method," *IBM J. Res. Develop.* **10**, pp. 278-291 (July 1966).
3. P. Goel, "An Implicit Enumeration Algorithm to Generate Tests for Combinational Logic Circuits," *IEEE Trans. on Compters* **C-30**, pp. 215-222 (March 1981).
4. M. Abramovici and M. A. Breuer, "On Redundancy and Fault Detection in Sequential Circuits," *IEEE Trans. on Computers* **C-28**, pp. 864-865 (November 1979).
5. K.-T. Cheng, "On Removing Redundancy in Sequential Circuits," *Proc. 28th Design Automation Conf.*, pp. 164-169 (June 1991).
6. I. Pomeranz and S. M. Reddy, "Test Generation for Synchronous Sequential Circuits Using Multiple Observation Times," *Fault-Tolerant Computing Symp.* (June 1991).
7. I. Pomeranz, L. N. Reddy, and S. M. Reddy, "COMPACTEST: A Method to Generate Compact Test Sets for Combinational Circuits," *Proc. Int'l Test Conference* (Oct. 1991).
8. R. Marlett, "EBT: a comprehensive test generation technique for highly sequential circuits," *Proc. 15th Design Automation Conf.*, pp. 332-339 (June 1978).
9. W. T. Cheng, "The BACK Algorithm for Sequential Test Generation," *Proc. Int. Conf. Computer Design (ICCD-88)*, Rye Brook, NY, pp. 66-69 (October 1988).
10. R. K. Brayton, C. T. McMullen, G. D. Hachtel, and A. L. Sangiovanni-Vincentelli, "Logic Minimization Algorithms for VLSI Synthesis," *Kluwer Academic Publishers* (1984).
11. T. Larrabee, "Efficient Generation of Test Patterns Using Boolean Difference," *Proc. International Test Conference*, pp. 795-801 (August 1989).
12. F. Brglez, D. Bryan, and K. Kozminski, "Combinational Profiles of Sequential Benchmark Circuits," *Int'l Symposium of Circuits and Systems*, pp. 1929-1934 (May 1989).
13. K. -T. Cheng and V. D. Agrawal, "A Partial Scan Method for Sequential Circuits with Feedback," *IEEE Trans. Computers* **39-4**, pp. 544-548 (April 1990).
14. S. Yang, "Logic Synthesis and Optimization Benchmarks User Guide," *MCNC Technical Report* (Jan. 1991).

FREEZE!: A New Approach for Testing Sequential Circuits

Miron Abramovici
AT&T Bell Laboratories
1100 E. Warrenville Rd.
Naperville, IL 60566

Krishna B. Rajan[†]
Illinois Institute of Technology
Dept. Electrical & Computer Engineering
Chicago, IL 60616

David T. Miller
AT&T Bell Laboratories
1100 E. Warrenville Rd.
Naperville, IL 60566

ABSTRACT

In this paper we introduce a new approach for testing sequential circuits. The classical approach applies only one vector in every state. In contrast, the new approach temporarily disables the sequential behavior of the circuit by holding the clock inactive ("freezing" the clock) and applies a group of vectors in every state. In this way many faults can be combinationally detected. We also present the results obtained by the first sequential test generator that employs the new approach, a fault-independent algorithm called FIRST. FIRST detects a large percentage of the faults detected by a conventional fault-oriented sequential test generator in less CPU time and with shorter sequences.

1. Introduction

Semiconductor product quality is strongly related to the fault coverage of the test sequence used for acceptance testing [13]. Even for a mature manufacturing process with yield greater than 0.8, a fault coverage greater than 99.6% is needed to achieve a defect level of less than 200 defective parts per million [8]. Efficient test generation for sequential circuits is badly needed today, because even state-of-the-art sequential test generators, such as GENTEST [6] and HITEC [10], cannot achieve satisfactory fault coverage for complex sequential circuits within reasonable time limits. The full-scan design for testability technique simplifies the problem of testing a sequential circuit by converting it into a combinational one during testing, so that more efficient combinational test generation (CTG) algorithms can be applied. Although the testing problem is simplified significantly, this technique has its own drawbacks. All flip-flops are modified to accommodate a scan-chain, thus adversely affecting the area and the performance of the circuit. If there are tight restrictions on area and/or performance, full scan is not possible.

Partial-scan techniques, in which only a subset of the flip-flops are scanned, can reduce the costs associated with full-scan design [12]. A class of partial-scan techniques, such as [5], try to simplify the cyclic structure of the circuit and hence reduces its sequential complexity. These techniques, however, still rely on conventional sequential test generation (STG) algorithms for which the partial-scan circuit may be too difficult to process efficiently [3]. Partial-scan techniques that use CTG algorithms may require too many flip-flops to be scanned [2], [7].

Because of the complexity of the STG problem, we believe that any approach that maintains the classical framework of targeting one fault at a time using an iterative array model of the circuit, can offer only marginal improvements over existing techniques. It is also likely that the much-needed breakthrough cannot be achieved by only one algorithm, but by a combination of different techniques. In [11], an increased utilization of the internal states of the circuit is achieved by targeting faults that are easily detectable in the current state. However, there is more room for exploiting each state as will be shown by the new approach.

This paper presents a new approach for testing sequential circuits which extends the classical concept of a test sequence. By temporarily disabling the sequential behavior of the circuit by holding the clock inactive ("freezing" the clock), we can apply a group of vectors (instead of only one) in every state. This allows many faults to be combinationally detected and enables the use of much faster CTG algorithms (with simple extensions).

[†] The research of this author was supported by a grant from AT&T Bell Laboratories.

We have developed a fault-independent STG algorithm called FIRST (Fault-Independent Rapid Sequential Test Generator) based on the new approach. The preliminary results obtained so far are encouraging. On the average, FIRST is able to detect nearly 84 percent of the faults detected by a conventional fault-oriented algorithm (GENTEST). In a separate run done for the purpose of comparison, GENTEST was asked to target only the faults detected by FIRST. For most circuits, GENTEST spent two to three times more CPU time to detect the same faults, and generated about twice as many vectors as FIRST.

The rest of the paper is organized as follows. Section 2 outlines the new approach for testing sequential circuits and contrasts it with the classical one. Section 3 discusses the implications of the new approach on test generation. Section 4 presents the results of FIRST and Section 5 concludes the paper.

2. A New Approach for Testing Sequential Circuits

2.1 The Classical Approach

The classical testing method for a synchronous sequential circuit relies on a *test sequence*, where every step in the sequence consists of applying one vector at the primary inputs (PIs), followed by a CLOCK pulse (see Figure 1). The state of the circuit $S(y)$ is determined by the values of the state variables y corresponding to the outputs of the flip-flops (F/Fs). A conventional STG algorithm deals with one *target fault* at a time, and searches for a test sequence to activate the target fault and propagate its effect to a primary output (PO). This search is two-dimensional, as it proceeds both in space (through the circuit) and in time. A test sequence takes the circuit through a sequence of states. The temporal behavior of the sequential circuit is modeled by a spatial sequence of *time frames*, where each time frame is a replica of the circuit C loaded with the values corresponding to a state in the state sequence.

2.2 The New Approach

The main idea of the new approach is that *once the circuit enters a new state S(y), this state is maintained* (by freezing CLOCK) *while a group of several vectors are applied* at the PIs (see Figure 1). In this way we can fully exploit state *S(y)*, by detecting many faults whose detection is conditioned by the circuit being in state *S(y)*. In contrast, since the classical approach applies only one vector in state *S(y)*, every time state *S(y)* is needed to detect a fault, a test sequence may be required to bring the circuit again to state *S(y)*.

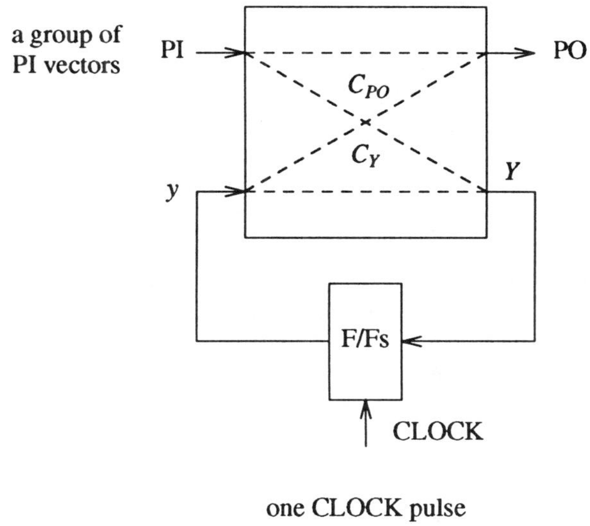

Figure 1 The STG model

The group of vectors applied while the circuit remains in state *S(y)* consists of $n_y + 1$ vectors. The first n_y vectors try to detect faults at the POs, and they can be applied in any order. The last vector tries to propagate fault effects toward the F/Fs inputs Y. A CLOCK pulse is applied after the last vector in the group. As the new approach applies a sequence of groups of vectors, it represents an extension of the classical one, which can be obtained as a degenerate case for n_y=0 for every state *S(y)*. *Therefore the new approach can never be worse than the classical one.*

3. Implications on Test Generation

Several STG algorithms can be developed based on the new method. The most important advantage in the new approach is that *most of the faults detected while the circuit is held in state S(y) are combinationally detected,* because the sequential behavior of the circuit is disabled by holding CLOCK inactive. Hence we can use CTG algorithms, which are much more efficient than STG algorithms. Because the F/F values do not change during this period, the values of the state variables y are "frozen" in the circuit C, that is, the test generator is not allowed to change them.

The only faults that *may* not be combinationally testable in the current state *S(y)* are those that have been activated in previous time frames, and whose effects are stored in F/Fs (that is, the value of some F/F changes in the presence of the fault). These faults can be handled by simple extensions to a CTG algorithm to take into account possible interactions between their effects propagating through F/Fs and their current effects in the circuit C. (These interactions have been analyzed in [9]).

For the first n_y vectors applied in state $S(y)$, the algorithm deals only with the subcircuit C_{PO}, containing all the logic that feeds the POs. While the classical approach concentrates only on one of the stored fault effects, the new approach can deal with many more, because it can apply several vectors so that each vector tries to propagate different subsets of the stored fault effects to the POs. (The fault effects remain stored in F/Fs as long as the circuit remains in the same state $S(y)$).

After generating n_y vectors to detect as many faults as possible while the circuit is in state $S(y)$, one more vector is generated with the goal of propagating effects of the remaining undetected faults to the F/F inputs Y. Here only the subcircuit C_Y, containing all the logic that feeds Y, needs to be processed.

The first STG algorithm based on the new testing method is a fault-independent algorithm called FIRST, whose goal is to quickly generate an initial sequence of groups of vectors that detects a large fraction of the detectable faults, without explicitly targeting any fault. FIRST uses and extends the concepts and techniques employed by the fault-independent CTG algorithm SMART (Sensitizing Method for Algorithmic Random Testing) [1]. Because of space limitations, we shall not discuss the details of FIRST and its implementation.

4. Results

A prototype implementation FIRST was run on the ISCAS-89 sequential benchmark circuits [4]. Since the fault simulator currently used by FIRST is approximate, the test vectors generated by FIRST were simulated by a separate simulator. Table 1 compares the fault coverage obtained by FIRST(F) with the one obtained by GENTEST(G). Being fault-independent, FIRST cannot prove that a fault is undetectable. This is why the results for the higly redundant circuits are not included here. FIRST detected between 58.59% and 100.00% (on the average 83.54%) of the faults detected by GENTEST. Although a fault-independent algorithm is expected to achieve a lower fault coverage than a fault-oriented algorithm, for three circuits - s953, s1423, and s9234 - FIRST detected more faults than GENTEST (4.6 times more on s1423).

Both FIRST and GENTEST were run on a SUN SPARC2 workstation. For comparison, we asked GENTEST to target only the faults detected by FIRST. These results are provided in Table 2. On the average, GENTEST generated nearly two times more vectors than FIRST. Although on several circuits GENTEST was faster than FIRST, the total CPU time spent by GENTEST (on all circuits) was 18.3 times larger than the time used by FIRST. (The CPU time showed for FIRST includes the separate fault simulation time.)

Table 3 illustrates the relation between the vectors and clocks generated by FIRST. FIRST generated between 1.43 and 5.31 vectors for every clock (on the

TABLE 1. Results of FIRST

Circuit	Fault Coverage(%)		
	F	G	F/G
s208	58.53	63.13	92.71
s298	65.38	85.58	76.40
s344	65.00	95.83	67.83
s349	57.34	95.38	60.12
s420	38.43	41.44	92.74
s641	60.79	85.15	71.39
s713	70.05	80.52	87.00
s820	53.64	91.55	58.59
s832	53.10	88.42	60.05
s953	8.87	8.78	101.03
s1196	99.13	99.61	99.52
s1238	91.23	94.63	96.41
s1423	53.67	11.62	461.88
s1488	79.37	93.88	84.54
s5378	59.47	71.28	83.43
s9234	0.26	0.25	104.00
s13207	5.49	6.22	88.26
s35932	85.32	89.77	95.04
Average	55.84	66.84	83.54

TABLE 2. Results for the same target faults: FIRST & GENTEST

Circuit	Vectors			CPU Time(secs)		
	F	G	G/F	F	G	G/F
s208	54	135	2.50	11.8	3.4	0.29
s298	44	162	3.68	11.5	22.6	1.97
s344	12	51	4.25	4.8	24.6	5.14
s349	12	44	3.67	4.6	4.0	0.87
s420	34	113	3.32	14.8	7.6	0.51
s641	70	89	1.27	24.7	7.2	0.29
s713	63	154	2.44	32.9	345.7	10.51
s820	85	225	2.65	42.1	2685.2	63.75
s832	109	281	2.58	50.5	2052.1	40.65
s953	20	19	0.95	26.1	1.5	0.06
s1196	276	338	1.22	177.9	41.5	0.23
s1238	306	306	1.00	179.8	42.5	0.24
s1423	94	21	0.22	725.2	111870.1	154.25
s1488	117	527	4.50	107.1	6081.8	56.80
s5378	252	454	1.80	1082.5	437.5	0.40
s9234	3	5	1.67	64.5	3.2	0.05
s13207	103	58	0.56	5866.1	1797.6	0.31
s35932*	148	548	3.70	3702.0	96582.0	26.09
Total	1802	3530	1.96	12129	222010	18.30

* GENTEST run on Convex C-200

TABLE 3. Relation between vectors and clocks

Circuit	Vectors	Clocks	Vects./Clks.
s208	54	32	1.69
s298	44	30	1.47
s344	12	6	2.00
s349	12	6	2.00
s420	34	17	2.00
s641	70	22	3.18
s713	63	14	4.50
s820	85	37	2.30
s832	109	52	2.10
s953	20	14	1.43
s1196	276	52	5.31
s1238	306	72	4.25
s1423	94	56	1.68
s1488	117	50	2.34
s5378	252	110	2.29
s9234	3	0	-
s13207	103	70	1.47
s35932	148	54	2.74
Total	1802	694	2.56

average 2.56 vectors/clock).

5. Conclusions

This paper has presented a new approach for testing synchronous sequential circuits.* By applying a group of vectors in every state, the new approach exploits many fault detection opportunities ignored by the classical approach in its exclusive pursuit of one target fault at a time. The sequence of groups of tests is a generalization of the classical test sequence where each group consists of only one vector. By temporarily disabling the sequential behavior of the circuit, the new approach can employ CTG algorithms (or simple extensions thereof) to detect most faults. Several STG algorithms can be developed based on the new method. The first such algorithm is a fault-independent algorithm called FIRST. The results of our prototype implementation are encouraging in terms of fault coverage, vectors and CPU time, and we expect the final implementation to achieve better results. Eventually FIRST will be used as a first phase of an STG system whose second phase will be a fault-oriented algorithm also based on the new approach.

Acknowledgement

The authors would like to thank T.J. Chakraborty, K.T. Cheng, and F. Maamari for useful discussions. The support provided by J. Bierbauer, J. Dussault, and H. Nham is gratefully acknowledged. Our appreciation to R. Bencivenga for helping with GENTEST package.

*Patent applied for

References

[1] M. Abramovici, J. J. Kulikowski, P. R. Menon, and D. T. Miller, "SMART and FAST: Test Generation for VLSI Scan-Design Circuits," *IEEE Design & Test of Computers*, Vol. 3, No. 4, pp. 43-54, August, 1986.

[2] M. Abramovici, J. J. Kulikowski, and R. K. Roy, "The Best Flip-Flops to Scan," *Proc. Intn'l Test Conf.*, pp. 166-173, October, 1991.

[3] S. Bhawmik, C. J. Lin, K. T. Cheng, and V. D. Agrawal, "PASCANT: A partial scan and test generation system," *Proc. Custom Integrated Circuits Conf.*, pp. 17.3.1-17.3.4, May 1991.

[4] F. Brglez, D. Bryan, and K. Kozminski, "Combinational Profiles of Sequential Benchmark Circuits," *Proc. Int'l Symp. Circuits and Systems*, IEEE Press, New York, 1989, pp. 1929-1934.

[5] K.-T. Cheng and V. D. Agrawal, "An Economical Scan Design for Sequential Logic Test Generation," *Digest of Papers 19th Intn'l. Symp. on Fault-Tolerant Computing*, pp. 28-35, June, 1989.

[6] W. T. Cheng and T. J. Chakraborty, "Gentest - An Automatic Test-Generation System for Sequential Circuits," *IEEE Computer*, pp. 43-49, April, 1989.

[7] R. Gupta, R. Gupta, and M. A. Breuer, "BALLAST: A methodology for partial scan design," *Proc. 19th Intn'l. Symp. on Fault-Tolerant Computing*, pp. 118-125, 1989.

[8] E. J. McCluskey and F. Buelow, "IC Quality and Test Transparency," *Proc. Intn'l. Test Conf.*, pp. 295-301, September, 1988.

[9] P. R. Menon, Y. H. Levendel, and M. Abramovici, "SCRIPT: A Critical Path Tracing Algorithm for Synchronous Sequential Circuits," *IEEE Trans. on Computer-Aided Design*, pp. 738-747, June, 1991.

[10] T. Niermann and J. H. Patel, "HITEC: A Test Generation Package for Sequential Circuits," *IEEE EDAC*, pp. 214-218, 1991.

[11] T. Ono and M. Yoshida, "A Test Generation Method for Sequential Circuits based on Maximum Utilization of Internal States," *Proc. Intn'l Test Conf.*, pp. 75-82, October, 1991.

[12] E. Trischler, "Incomplete scan path with an automatic test generation methodology," *Proc. Test Conf.*, pp. 153-162, Nov. 1980.

[13] T. W. Williams and N. C. Brown, "Defect Level as a Function of Fault Coverage," *IEEE Trans. on Computers*, Vol. C-30, No. 12, pp. 987-988, December, 1981.

SWiTEST: A Switch Level Test Generation System for CMOS Combinational Circuits*

K.J. Lee
Dept. of Electrical Engineering
National Cheng-Kung University
Tainan 70101, Taiwan
R.O.C.

C.A. Njinda, M.A. Breuer
Dept. of EE–Systems
University of Southern California
Los Angeles, CA 90089-2560
U.S.A.

Abstract

For CMOS circuits, switch level test generation (SLTG) is potentially more powerful than conventional gate level test generation (GLTG). In this paper we present a SLTG system called SWiTEST. SWiTEST deals with bridging, breaking, stuck-open/on and stuck-at faults. It employs both logic and current (IDDQ testing) monitoring and takes into account the invalidation problem associated with stuck-open tests. The framework for SWiTEST is based on the PODEM algorithm.

1 Introduction

This paper describes SWiTEST, a system for automatically generating tests for CMOS combinational circuits. Conventional test generation (TG) algorithms usually deal with the detection of line stuck-at faults at the gate-level [1, 2, 3]. These algorithms have been applied to benchmark circuits containing more than ten thousand primitive gates, and have been shown to be quite efficient in generating tests for line stuck-at faults.

However, there exists several unique problems when testing CMOS circuits [4, 5]. These problems point out the inadequacy of gate level modeling and therefore a different form of representation is required. Judging from previous work [6, 7, 8, 9], switch level modeling [6] appears to be a more appropriate model to use for test generation of CMOS circuits. In this model the basic unit of a CMOS circuit is a transistor and each transistor is considered as an ON/OFF switch whose status depends on the logic value at its gate terminal. Most of the structural information of a circuit can be captured by this model. Realistic CMOS defects can be modeled as not only stuck-at faults, but also *transistor stuck-on/open*, *bridging* and *breaking faults*.

The inadequacy of gate level test generation (GLTG) in detecting CMOS faults does not imply that the concepts developed in GLTG should be discarded altogether. Test generation is basically a search problem where an appropriate test is sought. Many concepts used to search for tests at the gate level are also applicable at the switch level. Therefore we have based the implementation of SWiTEST on a well-known GLTG algorithm, namely PODEM [1]. To apply the concepts in PODEM at the switch level we divide the switch level test generation (SLTG) task into three subtasks: circuit manipulation, fault analysis and a test generation framework. Circuit manipulation attempts to preprocess a CMOS circuit so that useful information for test generation can be obtained. Fault analysis aims at identifying what fault models are most realistic and how to efficiently deal with each of these faults. The test generation framework consists of the major components that can be shared by the test generators for each fault model. To improve fault detectability SWiTEST employs IDDQ testing [10, 11]. By monitoring the steady state current, this technique can detect many CMOS defects such as shorts between two metal lines, latch-up, gate-oxide short, and a floating node with a voltage that turns on both P- and N-type transistors. More information about IDDQ testing can be found in [12, 13].

The rest of this paper is structured as follows. Section 2 provides an overview of the SWiTEST system. In Section 3 some experimental results are presented and discussed. We conclude in Section 4.

*This work was supported by the Defense Advanced Research Projects Agency and monitored by the Office of Naval Research under Contract No. N00014-87-K-0861, and by the Federal Bureau of Investigation under Contract No. JFBI90092.

2 System Overview

A block diagram of the major components of SWiTEST is presented in Figure 1. Next these components are described. Due to space limitation the reader should refer to [14] for more detailed information.

Circuit Partitioning: At the gate level the I/O relations between two gates are well defined, while at the switch level each transistor is a bidirectional device. To make use of PODEM concepts such as efficient backtracing, transistors must be clustered into groups so that during test generation each group can be considered as a basic unit and the I/O relations among groups can be defined. SWiTEST adopts a commonly-used partitioning scheme for switch level circuits which clusters a MOS circuit into a number of disjoint "channel-connected" components called *transistor groups* (*TGs*) [15]. The relationship between two *TGs* can be defined by the interconnections through the gate terminals of transistors in the original circuit. Each interconnection between two *TGs* can be considered as an input or output of the corresponding *TGs*.

Complex Gate Identification: Most switch level circuits consist of complex gates and pass transistors, where a complex gate is made up of a P-network and an N-network and contains only one output node. The output can never be treated as an "input" to the gate. Also to have a logic 1(0) at the output of a complex gate, a conducting path of P(N)-type transistors from VDD(GND) to the output must be turned on. Such knowledge can aid test

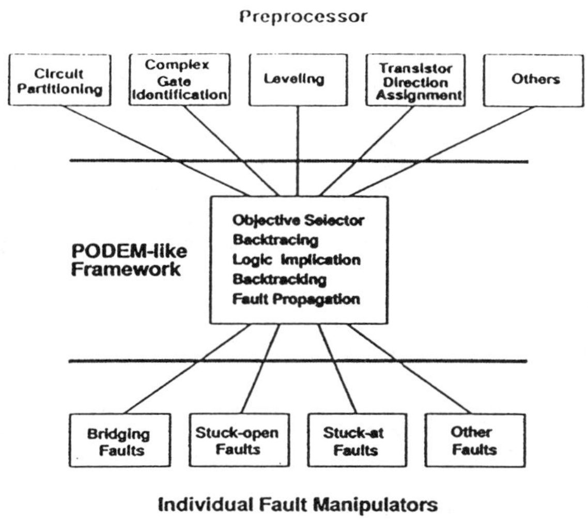

Figure 1: Organization of SWiTEST

generation processes such as backtracing and fault propagation. Moreover, many complex gates are in fact primitive gates (NAND, NOR or inverters) which have well defined I/O relationship. Thus further identifying primitive gates can potentially reduce the switch level test generation complexity to an extent that is close to that of gate level test generation.

Transistor Group Leveling: To determine the next objective to be achieved for backtracing and fault propagation, many gate level PODEM implementations use the "level" information, i.e., "how far" a gate is from its closest primary input or output. Similar level information based on transistor groups is employed in SWiTEST.

Transistor Direction Assignment: The partitioning or gate identification scheme only identifies the I/O relation among partitions or complex gates. In SWiTEST the signal flow directions of transistors [16] inside a partition or complex gate are determined and used to aid test generation. More information about this technique can be found in [14].

Faults Considered: SWiTEST can generate tests for stuck-open, stuck-on, bridging, breaking faults as well as stuck-at faults. A transistor stuck-on fault is considered as a bridging fault between the drain and source of the faulty transistor. For a line breaking fault a transistor is used to replace the line such that the line breaking fault can be modeled as a stuck-open fault of the transistor. It is assumed that 2-pattern tests are required for stuck-open faults while single pattern tests are enough for other faults. IDDQ testing is employed for bridging faults and for the invalidation problem of stuck-open fault testing due to charge sharing [14]. An efficient algorithm to generate stuck-open tests that cannot be invalidated by any circuit delay has been developed. The algorithm is based on deriving an intermediate vector, and then generating a two vector test set T_1 and T_2. To guarantee that a test is not invalidated due to circuit delays, an algorithm that enumerates all possible cutsets of an undirected graph is employed.

Test Generation Framework: Based on the gate level PODEM algorithm, the SWiTEST framework contains five major components as shown in Figure 1.

Objective selection: In the original PODEM algorithm only one objective is to be achieved at a time. When dealing with switch level faults it may be necessary to achieve more than one objective simultaneously. For example the initial objectives for detecting a bridging fault are to set the two shorted nodes to complementary logic values. Also when generating a test for a stuck-open fault it is necessary to turn off all but the faulty transistor in a cutset of transistors that separate a power source and an output node of a transistor group. SWiTEST uses an *objective array* to store the set of objectives that are achieved by each primary input assignment. Thus if backtracking is necessary, the objectives achieved by the latest PI assignment can be easily identified. Also to detect a bridging fault between two nodes x and y, two sets of initial objectives are possible, either $x = 1$ and $y = 0$, or $x = 0$ and $y = 1$. The correct selection of the set of objectives may have a large impact on the test generation performance since it is quite possible that only one set of objectives can be achieved. A set of guidelines is used in SWiTEST to determine the best objective for each fault.

Backtracing: Backtracing always succeeds in the original PODEM algorithm, i.e., there always exist some unspecified primary input(s) so that by assigning logic values to these PIs the status of the objective line (or node) can be changed. This, however, is not always true at the switch level. One such example is shown in Figure 2(a) where node x cannot be assigned a logic value unless a backtracking process is performed. Also in the gate level PODEM algorithm it is assumed that loops do not exists in the circuit. Thus backtracing can be a *memoryless* process, i.e., when a new temporary objective is determined, the old objective can be simply discarded. However at the switch level the old objective must be stored because backtracing might lead to a *switch level loop* as illustrated in Figure 2(b) (y to x to z to y). Another consideration is that the direction of backtracing at the switch level is not as clear as that at the gate level. We have developed a *hybrid, search-based* method to deal with these problems. This method makes use of the information about complex gates and transistor signal flow directions during backtracing. When possible, backtracing is performed at the gate level. Otherwise signal flow direction information is used to guide switch level backtracing. The problem of discovering that a floating node cannot be assigned a logic value is solved by a search-based mechanism.

Logic implication: The output of a transistor group can be set to 1 (0) only when a conducting path from *VDD*(*GND*) or a primary input with logic 1(0) to that node is established. Based on this observation, an efficient recursive, event-driven algorithm has been developed. This algorithm is basically a "depth-first search" starting with the node N that is to be assigned a logic value V. Any node that is connected to N through conducting transistors is assigned the logic value V. When the value of a

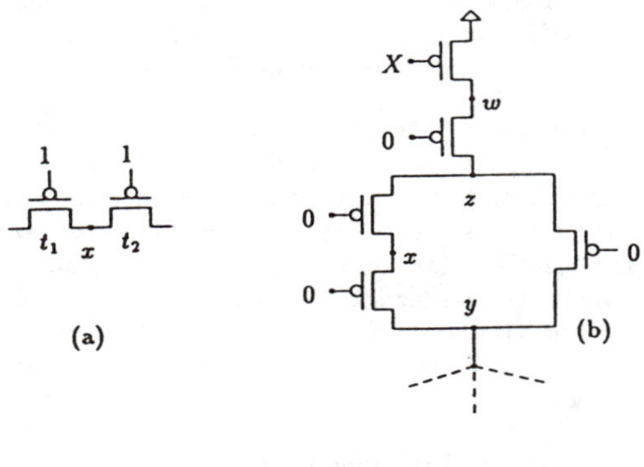

Figure 2: Problems at the switch level backtracing

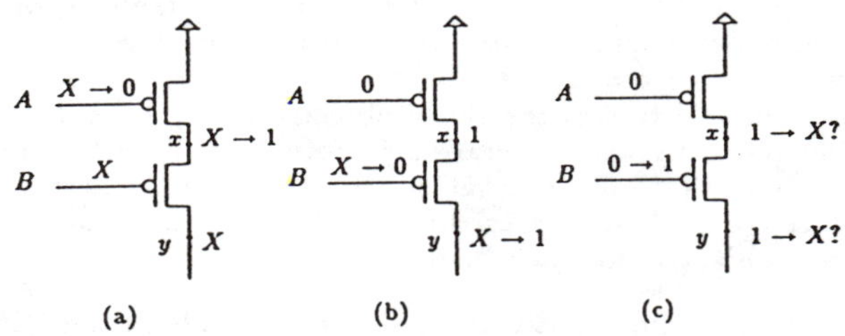

Figure 3: Difficulty in switch level backtracking

node is changed, the transistors that are controlled by the node are also processed so that the effect of assigning a logic value to a primary input can be propagated throughout the circuit in an event-driven manner.

Backtracking: Backtracking in the gate level PODEM algorithm is simply a forward implication starting with the primary input whose value has just been changed. It is not necessary to keep track of the information about which primary input assignment causes the status change of any internal line (or node) of the circuit. At the switch level this information is necessary as explained in the following example. In Figure 3(a) A is assigned 0. This results in x being assigned a value of 1. Then B is assigned a value of 0 as shown in Figure 3(b), which results in the assignment of value 1 to y. If later it is found that $AB = 00$ cannot lead to a test, then backtracking must be performed and a value of 1 must be assigned to B, as shown in Figure 3(c). If the information that x drives y to 1 due to $B = 0$ was not recorded, then unless a new simulation process on the entire transistor group containing x and y is done, it is impossible to go back to the status before $B = 0$ was assigned.

Fault propagation: Fault propagation in SWiTEST can be divided into two stages: to propagate the fault effect from the fault site to an output of the faulty transistor group, and to propagate the fault effect from an output of the faulty transistor group to some primary output. Both stages can be done by setting objectives that will give a transistor group output different logic values in the fault-free and the faulty circuits, and then trying to backtrace and assign some PI values to achieve these objectives. The fault propagation mechanism used in SWiTEST is similar to that used in the PODEM algorithm. The difference is that in the PODEM algorithm both the fault-free and faulty circuits are processed concurrently, while in SWiTEST, they are processed separately. During fault propagation, if a node has different values in the fault-free and faulty circuits (including 1 and X, and 0 and X), then it is considered to be in the D-frontier [1]. When one primary input is assigned a new value the simulation must be performed on the fault-free and faulty circuits separately, thereby apparently requiring twice the computation time when compared to the original PODEM algorithm. However, at the switch level if a node has complementary logic values in the fault free and faulty circuits, then the node turns on different type of transistors. Thus only half of the circuit will be processed. This implies that at the switch level, dealing with both circuits concurrently is not more efficient than dealing with them separately.

3 Experimental results

Nine circuits consisting of a 4-bit parity generator (parity), a combinational adder (cadder) and seven other circuits including six ISCAS-85 benchmark circuits have been used for experiments on a SUN Sparc 1 Workstation. The gates in a ISCAS-85 circuit are flatted into fully complementary CMOS switch level implementation. All transistor stuck-open and stuck-on faults in each circuit are processed. For stuck-on faults (which are considered as bridging faults), it is assumed that IDDQ testing is used. For stuck-open faults, test invalidation due to circuit delay is considered. Charge sharing is not explicitly considered because it is assumed that IDDQ testing is used. Tests for all node stuck-at faults can also be generated. To compare SWiTEST to existing gate level generators, experimental results on stuck-at faults at the output of each transistor group are given. Each stuck-on, stuck-open or stuck-at fault in each circuit is processed independently. No fault collapsing or fault simulation is used. The experimental results are shown in Table 1. The maximum number of backtrackings allowed per fault is set to 1000. Each columns of this table is defined as follows: [ntran] – number of transistors in the circuit, which is also the number of transistor faults; [mem] – size of memory required in 10^3 bytes; [nd] – number of faults detected; [nu] – number of faults identified as undetectable; [%fd] – percentage of detected faults among all faults; [s/f] – average time (seconds) spent on each fault, which does not include the relatively small preprocessing time; and [df] – number of stuck-at faults under consideration.

All faults in the first three circuits are either detected or identified as undetectable. For other circuits there exists faults that cannot be detected or identified as undetectable within the limited backtrackings. This is expected since it is well known that the PODEM algorithm cannot efficiently identify all undetectable faults even at the gate level.

The total test generation time is dependent on circuit size. However it appears that the average test generation time for each type of fault does not totally depend on circuit size. This is due to the fact that some circuits (e.g., c3540) have a larger percentage of aborted faults than oth-

circuit	ntran	mem	s-on				s-op				s-at				
		(K)	nd	nu	%fd	s/f	nd	nu	%fd	s/f	nf	nd	nu	%fd	s/f
cadder	28	128	28	0	100.	0.001	28	0	100.	0.002	10	10	0	100.	0.001
parity	32	132	32	0	100.	0.001	32	0	100.	0.002	16	16	0	100.	0.002
c60	110	140	108	2	98.18	0.002	102	8	92.73	0.004	86	84	2	97.67	0.003
c880	1802	400	1751	0	97.17	0.070	1802	0	100.	0.014	1178	1178	0	100.	0.010
c1355	2308	468	2126	0	92.11	0.180	2068	72	89.60	0.248	1290	1290	0	100.	0.062
c1908	3446	656	3132	0	90.89	0.275	3393	0	98.46	0.162	2226	2204	0	99.01	0.138
c2670	5668	1044	5555	29	98.01	0.085	5381	48	94.94	0.257	4088	3955	16	96.75	0.219
c3540	7504	1300	7325	12	97.61	0.257	6888	0	91.79	1.176	5020	4697	0	93.57	1.167
c7552	15396	2556	15100	26	98.08	0.173	14762	0	95.88	0.472	10330	10061	0	97.40	0.334

Table 1: Experimental results

ers.

The memory requirement for each circuit is clearly linear with circuit size. For the largest circuit which contains 15,396 transistors, the memory requirement is less than 2.6 Mbytes, which is significantly less than that reported in [7], which requires 15 Mbytes for a circuit containing 770 transistors.

As expected, the percentage of detected stuck-open faults is in general smaller than the other two faults. However the difference is not significant. This result illustrates that the robust detection of stuck-open faults using a PODEM based algorithm is achievable.

The average time for SWiTEST to process a stuck-at fault for c7552 is 0.334 seconds. In [3], which is probably the fastest gate level test generation system currently available, it takes approximately 41.5 seconds to generate 380 test vectors. This also includes the time for fault simulation. From experimental data reported recently, fault simulation and test generation time are approximately the same [2, 3]. Therefore the average time to generate a test for a given fault is roughly $(41.5/2)/380 = 0.0546$ seconds, which is within an order of magnitude of 0.334 seconds. This comparison is not very precise due to different hardware resources and program implementation, but it does illustrate that SLTG can be done in time compatible to that required for GLTG. Note that in [3] the number of backtrackings is close to zero due to the efficient heuristics employed. If similar heuristics are employed in SWiTEST, the performance of SLTG can be further improved.

4 Conclusion

Automatic test generation for CMOS circuits at the switch level is an important and complex problem. In this paper we have presented a switch level test generation system for CMOS combinational circuits. The experimental results on SWiTEST imply that switch level test generation can be done in CPU time that is within an order of magnitude of that required for gate level test generation. This is true even for stuck-open faults which have previously been considered as much more difficult to detect due to the test invalidation problem.

References

[1] P. Goel. An implicit enumeration algorithm to generate tests for combinational logic circuits. *IEEE Trans. on Computers*, C-30(3):215–222, Mar. 1981.

[2] M.H. Schulz and E. Auth. Advanced automatic test pattern generation and redundancy identification techniques. In *Proc. Int'l. Symp. on Fault Tolerant Computing*, pages 30–35, 1988.

[3] J.A. Waicukauski, P.A. Shupe, D.J. Giramma, and A. Matin. ATPG for ultra-large structured designs. In *Proc. Int'l. Test Conf.*, pages 44–51, Sept. 1990.

[4] Y.K. Malaiya and S.Y. Su. A new fault model and testing technique for CMOS devices. In *Proc. Int'l. Test Conf.*, pages 25–34, 1982.

[5] H.H. Chen, R.G. Mathews, and J.A. Newkirk. An algorithm to generate tests for MOS circuits at the switch level. In *Proc. Int'l. Test Conf.*, pages 304–312, 1985.

[6] R.E. Bryant. A switch-level model and simulator for MOS digital systems. *IEEE Trans. on Computers*, C-33(2):160–177, Feb. 1984.

[7] K. Cho and R.E. Bryant. Test pattern generation for sequential MOS circuits by symbolic fault simulation. In *Proc. Design Automation Conf.*, pages 418–428, 1989.

[8] C.H. Chen and J.A. Abraham. Mixed-level sequential test generation using a nine-valued relaxation algorithm. In *Int'l Conf. on Computer-Aided Design*, pages 230–233, Nov. 1990.

[9] M.K. Reddy. *Testable CMOS digital design and switch-level test generation for MOS digital circuits.* PhD thesis, Univ. of Iowa, Iowa City, 1991.

[10] J.M. Soden and C.F. Hawkins. Test considerations for gate oxide shorts in CMOS ICs. *IEEE Design & Test of Computers*, 3(4):56–64, Aug. 1986.

[11] W. Maly (editor). Built-in current testing — Part I. Technical Report CMUCAD-88-27, Carnegie-Mellon Univ., June 1988.

[12] T.M. Storey and W. Maly. CMOS bridging faults detection. In *Proc. Int'l. Test Conf.*, pages 842–851, Sept. 1990.

[13] F.J. Ferguson and T. Larrabee. Test pattern generation for realistic bridging faults in CMOS ICs. In *Proc. Int'l. Test Conf.*, pages 492–499, 1991.

[14] K.J. Lee. *Switch level test generation for CMOS circuits.* PhD thesis, Univ. of Southern California, Los Angeles, August 1991.

[15] R.E. Bryant. An algorithm for MOS LSI logic simulation. *LAMBDA*, pages 46–53, Fourth Quarter 1980.

[16] N.P. Jouppi. Derivation of signal flow direction in MOS VLSI. *IEEE Trans. on Computer-Aided Design of Integrated Circuits and Systems*, CAD-6(3):480–490, May 1987.

Fast Exact and Quasi-Minimal Minimization of Highly Testable Fixed-Polarity AND/XOR Canonical Networks

Andisheh Sarabi

Department of Electrical Engineering
Portland State University
Portland, OR 97223

Marek A. Perkowski

Department of Electrical Engineering
Portland State University
Portland, OR 97223

Abstract

The high testability of AND/XOR networks and new technologies that make their use more possible call for new minimization and synthesis tools. This paper introduces the fast Exact and Quasi-minimal algorithms for minimal fixed polarity AND/XOR canonical representation of Boolean functions. The method uses features of array of disjoint cubes representation of functions to identify the minimal networks. These features can drastically reduce the search space and provide high quality heuristics for quasi-minimal representations. Experimental results show that these special AND/XOR networks, on the average, have similar number of terms to Boolean AND/OR networks while there are functions for which AND/XOR circuits are much smaller. The circuits generated here are much more testable.

1 Introduction

AND/XOR realizations of switching functions have certain advantages over traditional AND/OR realizations but, due to the lack of circuit technologies and necessary synthesis tools, they have not been used extensively.

In many applications the AND/XOR realizations of the circuits require less layout area than their AND/OR counterparts [1]. The AND/XOR PLAs often require fewer products than AND/OR PLAs [2]. These networks are also highly testable and many of their applications can be found in arithmetic, encoding, telecommunication, and linear systems.

Although XOR gates already exist in EPLDs (Signetics LHS501, arithmetic PALs, etc.), and standard cell libraries, the full potential of this logic has not been realized in the mainstream logic design and synthesis. One disadvantage, in terms of realization, has been the slow speed and area constraints of the XOR gate. Recent advances in FPGA technology have made significant progress in this regard. FPGAs such as the Xilinx 3000 family, which incorporate a high-speed lookup table approach to implementing any function of up to five input variables, make an XOR gate with 5 inputs as fast as any other gate of the same number of inputs. Another example is that of the 1010/1020 family of devices from Actel. The cascaded multiplexer structure of Actel makes it possible to implement two- and most three-input functions, that include XOR operators. Of special interest here are new programmable technologies that include XOR as one of few logic gates available in them. For instance, the CLi6006 Field Programmable Gate Array from Concurrent Logic, Inc. includes a two-input XOR gate in a small granularity block [3].

Currently, few, if any, minimization and synthesis methods exist which use AND/XOR-based methods for these devices. Two-level minimizers such as ESPRESSO and multi-level minimizers such as MISII, BOLD, or FPGA specific ones such as mis-pga do not include XOR synthesis schemes.

In this paper methods for fast minimization of functions with large numbers of variables and fixed-polarity AND/XOR networks will be presented. These special AND/XOR networks are not only the most easily testable of all general purpose networks, but also have other merits of their own. These canonical forms which are known as Consistent Generalized Reed-Muller *CGRM* [4] forms, are used for general Boolean function representations in such applications as the classification of functions, Ashenhurst and other decomposition methods, and multi-level design. In addition, these forms have been studied for applications in image processing and used as the basis for minimization of other AND/XOR networks [5].

In the following sections after a brief introduction of *CGRM* forms, the testability of AND/XOR networks is discussed. Next, fast realization of *CGRM* expansion for a given polarity is introduced along with several new cube operations. In section 5, an Exact method to find the minimal polarity *CGRM* is presented together with characteristics of arrays of disjoint cubes which can reduce the required search space. In the following section a Quasi-Minimal approach is described and in section 7 some experimental results are presented and discussed.

†This research was partially supported by the NSF grant MIP-9110772.

2 Background

While there are only two canonical forms for the Boolean AND/OR networks, the number of possible canonical representations of a function for the AND/XOR networks is very large [4]. One class of AND/XOR networks, the first to be noted in the literature, is that of the *Reed-Muller canonical (RMC)* form [9, 10] and its superset, the *Consistent Generalized Reed-Muller (CGRM)* canonical forms [4].

The *RMC* representation consists of only positive product terms and is given as:

$$f(x_1, x_2, \ldots, x_n) = \bigoplus_{i=0}^{2^n-1} a_i \mu_i \quad (1)$$

where $a_i \in \{0,1\}$ and $\mu_i = x_n{}^{e_n} x_{n-1}{}^{e_{n-1}} \ldots x_2{}^{e_2} x_1{}^{e_1} = \prod_{j=1}^{n} x_j{}^{e_j}$ where $e_j \in \{0,1\}$ such that $e_n e_{n-1} \ldots e_2 e_1$ is a binary number which equals i. Moreover $x_i{}^0 = 1$ and $x_i{}^1 = x_i$. $\oplus$ denotes the summation over GF(2), the Galois field of two elements.

If the restriction that all the variables should take positive polarity is removed and they are also allowed to take negative polarities, one can have a *Generalized Reed-Muller (GRM)* canonical form. If the variables are, however, restricted to retain the same polarity, either positive or negative, in all product terms, the canonical form will be that of the *Consistent Generalized Reed-Muller (CGRM)* form.

For n variables, there are 2^n possible arrangements of polarities; hence, there are 2^n possible *CGRM* forms for a function. Depending on the polarity of the variables, the number of terms in the expansion varies. As an example for three variables, the function $\bar{x}_1\bar{x}_2\bar{x}_3$ is represented in *RMC* as:

$$\bar{x}_1\bar{x}_2\bar{x}_3 = 1 \oplus x_1 \oplus x_2 \oplus x_1x_2 \oplus x_3 \oplus x_1x_3 \oplus x_2x_3 \oplus x_1x_2x_3. \quad (2)$$

The same function will just be represented as $\bar{x}_1\bar{x}_2\bar{x}_3$ if negative polarities are chosen for all the three variables. This *CGRM* obviously has less number of terms. The problem of finding those polarities of the variables which result in a *CGRM* with a least number of terms is essentially the minimization problem of interest here.

In order to differentiate between the Boolean product terms and the terms in Reed-Muller forms, the term "monoterm" will be used for the latter.

Definition 1 *A* monoterm *is a product term in Reed-Muller canonical forms.*

The minimization problem can be formulated in terms of the minimal polarity. In this paper, the binary number $\mathbf{p} = p_1 p_2 \ldots p_n$ where $p_i \in \{0,1\}$ such that $\dot{x}_i = x_i{}^{p_i}$ is called the *polarity*. If $p_i = 0$, $\dot{x}_i = \bar{x}_i$ and if $p_i = 1$, $\dot{x}_i = x_i$. For all possible CGRM forms of a switching function, the *minimal polarity* is the polarity which results in a CGRM form with the least number of monoterms. The polarity of each literal in this polarity is called the *polarity literal.*

3 Testability of AND/XOR Networks

The inherent high testability of AND/XOR networks has long been known [11, 12]. It has been shown [11] that all single stuck-at-faults of the *RMC* network, assuming the primary inputs are fault free, can be detected by the set of $(n+4)$ tests which are independent of the function. With faulty primary inputs, the same number was shown to be sufficient to detect stuck-at-faults provided that an extra observable AND gate is added to the network. With addition of extra observable outputs, the same $n+4$ independent tests can detect all single and multiple stuck-at-faults. It has been shown that the tests for detecting bridging faults of the *RMC* network are also independent of the function. It has to be noted that the test set for the *CGRM* networks is of the same cardinality as one for the *RMC* network. The only difference is that the test vector bits for the variables with negative polarities are the opposites of the ones in the *RMC* network. The distinction between *CGRM* networks and other AND/XOR networks was made by Pradhan [12]. Although the test set for the other AND/XOR networks is also independent of the function, its cardinality is higher than *CGRM* networks.

4 Fast Realization of the CGRM

Following Fisher [6], the problem of *CGRM* minimization of a switching function can be divided into two steps. The first step is to identify the minimal polarity and the second is to realize the *CGRM* expansion of the function with this polarity. In this section a fast method dealing with the second step will be presented. The identification of the minimal polarity is the subject of sections 5 and 6.

One efficient method for realizing a *CGRM* expansion of switching functions is by operating on the set of disjoint cubes which represent the function [6, 7, 8]. In this method, the function is represented by disjoint cubes rather than minterms to reduce the memory requirements. Monoterms representing each cube are expanded and those occurring in an odd number of cubes are retained as the ones representing the function. The fast method introduced here uses the new operations of cube *commonality*, *difference*, and *symmetric difference* together with a fast Gray-code approach to realize a *CGRM* expansion. Before introducing the method, the monoterms representing each cube, originally reported by Fisher [6] for the case of *CGRM*, are given by Theorem 1:

Theorem 1 *The monoterms originating from a cube for the* RMC *expansion are all the cubes that have their 1s in the same literal positions as the 1s of the original cube and either "0" or "-" in the 0-literal positions of the original cube.*

These monoterms will be referred to as monoterms *representing* the cube from now on.

≡	0	1	–
0	1	0	–
1	0	1	–
–	–	–	–

Table 1: Equivalence Operator for a Simple Bit

Γ	0	1	-
0	0	1	-
1	1	1	∅
-	-	∅	-

Table 2: Cube Commonality Operator for a Single Bit

Observation 1 *The monoterms are generated by the ones of the cube and zeros being replaced by DCs and then alternately replacing the positions of zeros with ones and DCs going through all the possible combinations. If the number of zeros in a cube is denoted by N_0, the total number of monoterms representing that cube will be 2^{N_0} as this is the total possible number of combinations. This can be proven based on the fact that $\bar{x}$ can be represented by $1 \oplus x$. Any negative variable then contributes two terms to the expansion.*

Theorem 1 gives the monoterms representing the cube at polarity 11...1 or the *RMC*. In order to obtain the representation for any other polarity, first a matching operation of the original cube and the polarity cube is performed to find the equivalent *RMC* representation. Then Theorem 1 is used to generate the monoterms of this equivalent cube. Finally the matching operation is performed on each monoterm generated to get *CGRM* form of interest. The matching operation here is the bitwise equivalence operation. Table 1 shows this operation for a single bit.

When a function is comprised of more than one cube, the monoterms that represent this function are not the simple union set of the monoterms of each cube but only those that occur in an even number of cubes. These will be referred to as *expansion* monoterms from now on.

To formulate the process of *CGRM* generation, analogies with set theory will be used. For two cubes, the expansion monoterms are those that are not common in the two cubes. For this purpose the *commonality* of two cubes needs to be identified. The cube commonality is a cube representing the monoterms that are common in two cubes and is given below:

Definition 2 *Let C_1 and C_2 be two cubes. The* cube commonality *operator on C_1 and C_2 is defined as follows:*

$$C_1 \Gamma C_2 = \begin{cases} \emptyset & \text{if } \exists i \text{ such that } C_{1i} \Gamma C_{2i} = \emptyset \\ C_3 & \text{otherwise, where } C_{3i} = C_{1i} \Gamma C_{2i} \end{cases} \tag{3}$$

where C_{ki} represents the i-th *literal of the cube C_k, and the commonality operator for a single bit is defined in Table 2.*

By the same analogy with set theory, the expansion monoterms of two disjoint cubes are the result of the *symmetric difference* of the two. The *difference* of two cubes is the set of all the monoterms which represent the first cube and do not represent the second. This operation which is the counterpart of the *sharp* operation for the case of monoterms is given below:

-	0	1	-
0	ϵ	-	1
1	ϵ	ϵ	∅
-	ϵ	∅	ϵ

Table 3: Cube Difference Operator for a Single Bit

Definition 3 *Let C_1 and C_2 be two cubes. The* cube difference *operator of $C_1 and C_2$, $C_1 - C_2$, is:*

$$\begin{cases} C_1 & \text{if } C_{1i} - C_{2i} = \emptyset \text{ for some } i; \\ \emptyset & \text{if } C_{1i} - C_{2i} = \epsilon \text{ for all } i = 1, 2, \ldots, n \\ \bigcup_i & \text{if } (C_{11}, C_{12}, \ldots, \alpha_i, \ldots, C_{1n}) \text{ otherwise,} \\ & \text{where the } \bigcup \text{ is for all those } i \text{ for which} \\ & C_{1i} - C_{2i} = \alpha_i \in \{-, 1\} \end{cases} \tag{4}$$

where C_{ki} is the i-th *literal of the cube C_k and the literal difference operator is defined in Table 3.*

Definition 4 *Let C_1 and C_2 be two cubes. The* symmetric difference *of the two cubes is:*

$$C_1 \oplus C_2 = (C_1 - C_2) \bigcup (C_2 - C_1). \tag{5}$$

Difference and symmetric difference follow the same properties as in set theory. As an example, the symmetric difference is commutative and associative:

$$C_1 \oplus C_2 = C_2 \oplus C_1; \tag{6}$$

$$(C_1 \oplus C_2) \oplus C_3 = C_1 \oplus (C_2 \oplus C_3) = C_1 \oplus C_2 \oplus C_3. \tag{7}$$

The associativity of symmetric difference, given in Equation (7), makes it possible to extend this operation to more than two cubes. That is, the expansion monoterms of a switching function, given as a set of disjoint cubes, are obtained from the n-argument symmetric difference of all the cubes, i.e. $C_1 \oplus C_2 \oplus \cdots \oplus C_n$.

The process of realization of *CGRM* expansion for a given polarity can now be outlined. First, the function is represented as a set of disjoint cubes. The cubes are then operated by the equivalence operation with the polarity of the *CGRM*. The symmetric difference of all these cubes is found and monoterms representing each of the resulting cubes are generated in a Gray-code order. Finally, the Equivalence operation with the polarity cube is performed on each of these monoterms to give the *CGRM* expansion.

The number of the resulting monoterms can be found from the inclusion-exclusion principle. This number is given by the following theorem:

Theorem 2 *Let $C_1, C_2, \ldots, C_n$ be a set of* n *disjoint cubes. Let S_k denote the sum of the number of all monoterms common in all possible k cubes. The number of monoterms representing the set of cubes is:*

$$S_1 - 2S_2 + 4S_3 - 8S_4 + \cdots + (-2)^{k-1}S_k + \cdots + (-2)^{n-1}S_n. \quad (8)$$

In the above equation, S_1 denotes the number of all monoterms that are only in one cube, S_2 denotes the number of all monoterms that are common in any two cubes, etc.

5 Exact Minimization

Although identification of the minimal polarity is an NP-hard problem, certain features of the array of disjoint cubes can be used to reduce the required search space. Some of these features, if present in an array, can be used to find the minimal polarity without any search. Others can just reduce the amount of search needed to find the exact solution. In the case that none of the features exist, the whole exhaustive search needs to be performed.

The number of expansion monoterms for a set of disjoint cubes is the difference between the total sum of the number of monoterms representing each cube and the number of monoterms that are subtracted because they occur in an even number of cubes. Both of these numbers change with different polarities. The minimal polarity is the one which results in the optimum balance between these two numbers resulting in the least number of expansion monoterms.

There are certain features of the function that can be used to reduce the search space for identifying the minimal polarity. For the case of functions that are comprised of only one cube, the minimal polarity can be found directly without any search.

Theorem 3 *The minimal polarities for a single cube are the polarities which match all the literals in the cube. The number of such polarities is equal to $2^{N_{DC}}$ where N_{DC} is the number of DC-literals in the cube.*

When a function is comprised of more than one cube, there are other features that if they exist, lead to the search space reduction. Theorem 4 provides one criteria for identifying a minimal polarity literal based on the columns of an array of disjoint cubes, using Theorem 2.

Theorem 4 *Let ${S^1}_i$ denote the sum of S_is of the cubes which have a value of 1 in a given column. Let ${S^0}_i$ denote the sum of S_is of the cubes which have a value of 0 in that column. Let ${S^{1-DC*}}_i$ denote the sum of S_is of the cubes that have both 1s and DCs in the column, assuming the 1 has been changed to a 0. Let ${S^{0-DC}}_i$ denote the sum of S_is of the cubes that have both 0s and DCs in the column. The corresponding minimal polarity literal for a column in the array of disjoint cubes should be changed when*

$$\sum_{k=1}^{n}(-2)^{k-1}S_k^1 + \sum_{k=2}^{n}(-2)^{k-1}S_k^{1-DC*} \quad (9)$$

$$< 1/2\sum_{k=1}^{n}(-2)^{k-1}S_k^0 + \sum_{k=2}^{n}(-2)^{k-1}S_k^{0-DC}.$$

From Theorem 4, it is possible to infer the following theorem:

Theorem 5 *For a column comprised of all 0s or all 1s, the corresponding minimal polarity literal is the same as the value in the column. (If the opposite is chosen, the number of monoterms representing the cubes would be doubled.) For a column comprised of all DC values, either 0 or 1 will be the minimal literal value.*

In the Exact method of minimization, first it is checked if the function is only comprised of one cube or two. Direct solution for these cases is found using Theorems 3, 4, and 5. If there are more cubes involved, first Theorem 5 is used to identify any columns in the array of disjoint cubes for which minimal polarity literal can be found readily. All the other columns are set to the zero polarity and a search for minimal polarity is performed in Gray-code order, changing one column at a time. This method is given formally below:

```
The exact algorithm
CGRMIN_EXACT
{
  If ( the function is comprised of one or two cubes )
  {
     Preset the columns to the minimal polarity;
     Generate the CGRM of the array; }
  else {
     Preset the columns
        to minimal polarity literals if possible, or
        to polarity 0 otherwise;
     for ( all combinations of the columns with
              undetermined polarities )
     {
        Set one column to its opposite polarity using
        Gray-code order;
        Generate the CGRM of the array;
        if ( No. of monoterms decreases )
              min_polarity = polarity_of_current_CGRM;
     }
  Generate the CGRM of min_polarity; }
}
```

6 Heuristic Minimization

In this section, a fast heuristic approach to the minimization problem is introduced. The corresponding heuristics combine the characteristics of the the overall number of monoterms and the ones subtracting, in order to identify the minimal *CGRM* polarity for a given array of disjoint cubes. Based on these heuristics, a priority of search for different polarities is devised and a minimization algorithm is introduced.

Here, similar to the Exact method, the columns minimal polarities of which can be identified without any search are first pre-set to their minimal polarities. Next, for all the remaining columns in the array of disjoint cubes, the polarity literals are chosen such that there will be the least number of 0s in each column. Referring back to Table 1, the least number of 0s in a column results from a polarity which is equivalent to the value that occurs the most in that column. That is, there will be the least number of non-matching values in that column with its corresponding polarity literal. Since the overall number of monoterms is determined by the number of zeros in the cubes, this polarity is used as a starting point for the minimization. However, the minimum number of 0s in the array does not guarantee the minimum number of expansion monoterms. For many functions it was shown experimentally that this first choice gives the minimal solution. For other cases, however, a search scheme is required. The general heuristics below can be used in any minimization scheme:

- *As the number of monoterms for a cube is 2^{N_0}, where N_0 is the number of 0s in the cube, if a cube has a relatively large number of 0s, some polarity literals should be changed to reduce the contribution of this cube to the number of expansion monoterms.*
- *From (9), if the number of occurrences of a 1 or a 0 in a column with no DC values is much higher than its opposite value, it is more likely that this value will be the minimal polarity literal for that column. A ratio of 4 to 1 for the most occurring and least occurring values almost guarantees this likelihood.*
- *When there are DC values involved in a column, the greater the number of the DC values, the less will be the overall number of monoterms affected if the polarity of that column is changed.*
- *For functions with large number of cubes, the monoterms which are common in two or three cubes are the main subtracting monoterms.*

Here, based on the above heuristics, a priority is calculated for every column minimal polarity of which can not be found without search. This priority is used to guide the search towards the minimal polarity.

Let the *priority* of the column i be represented by Γ_i. Let the number of the 1s in that column be represented by $N^1{}_i$, and the number of 0s by $N^0{}_i$. Then

$$\Gamma_i = \frac{(N^1{}_i)^2}{N^1{}_i + N^0{}_i} \tag{10}$$

The Quasi-minimal method can now be presented formally as follows.

```
The quasi-minimal algorithm
CGRMIN
{
  Preset the columns
     to minimal polarity literals if possible, else
     to most occurring values in the columns;
        Find the values of column priorities, Γs ;
  Sort Γs in descending order;
  for (k = 1; k <No_of_undetermined_polarities; k++)
  {
     Set the column with priority k to its opposite
        polarity;
     Generate the CGRM of the array;
     if ( No. of monoterms representing the array or
        the overall number of monoterms decreases );
     {
        min_polarity = polarity_of_current_CGRM;
        continue;
     }
     else
        change the column back to its previous polarity;
  }
  Generate the CGRM of min_polarity;
}
```

In this algorithm, the number of searches required for the worst case will be of order $2n$, for a function of n variables which significantly contributes to its speed.

7 Experimental Results

The Exact and Quasi-Minimal methods were tested for 112 MCNC benchmark functions. As most of the benchmarks are multi-output, the BLIF format of the functions was used to generate single output components of these functions for testing. Several of these functions are shown for comparison in Table 4. In this table n stands for the number of variables in the functions and No. for the number of terms.

The times given in the table are the CPU times in seconds on a Sequent S27 - two 386 processors - machine. The table includes the time comparisons of the functions for MINGRM [7], an existing exact *CGRM* minimizer, ESPRESSO, and CGRMIN, the *CGRM* minimizer developed based on the algorithms described earlier.

For the functions tested, the two level AND/OR and *CGRM* were found to be relatively comparable. For the 112 functions overall, while ESPRESSO found 1193 terms, this number for exact *CGRM* was found to be 1382 and the quasi-minimal program gave 1557. Depending on the functions, AND/OR can be a better choice for realization while for others it is the *CGRM* that certainly gives the better alternative. The functions rd532, rd732, and rd844 in Table 4 are examples of the latter; mis20, mis59, and sao22 are examples of the former. EXORCISM [1] gives 102 terms for 9sym, 5 for con12 and rd532, 9 for f51m4, 22 for sao23, and 16 for misex20. It is obvious that classifying functions according to their linearity and using multi-level AND/OR/XOR minimizers will definitely

Name	n	ESPRESSO		MINGRM		CGRMIN	
		No.	sec	No.	sec	No.	sec
5xp11	7	7	0.1	12	5.6	12	0.0
9sym	9	85	3.9	173	1851.3	173	10.3
bw7	5	6	0.0	8	0.4	12	0.0
con12	7	5	0.0	8	1.2	8	0.0
duke8	22	5	0.0	4	*	6	0.0
f51m4	8	10	0.1	7	4.5	9	0.1
rd532	5	16	0.1	5	1.1	5	0.1
rd732	5	64	0.9	7	49.6	7	1.5
rd842	8	128	3.1	8	364.2	8	5.7
sao22	10	20	0.2	52	642.1	61	0.4
sao23	10	22	0.9	47	905.9	59	0.5
mis20	10	7	0.0	62	309.6	66	0.2
mis59	12	4	0.0	26	1166.1	26	0.0
vg28	25	5	0.0	13	*	13	0.0
z42	7	28	0.4	9	12.3	13	0.4

Table 4: Two Level AND/OR Compared to Two Level *CGRM*

lead to more economical realizations of switching functions.

The differences between the minimal and quasi-minimal results demonstrate the quality of heuristics used. For majority of the functions, 66 out of 112 to be precise, the heuristic program found the exact minimum solutions. There were only 6 functions that differed by more than 10 terms; the largest being 12 terms for the function sao23. As indicated in the table, CGRMIN is extremely fast and overall with the results presented, it demonstrates the high quality of the approach.

8 Conclusions

In this paper fast Exact and Quasi-Minimal methods for minimization of fixed polarity AND/XOR canonical forms were presented. First, a fast cube-based method for the generation of these forms was introduced which can handle functions with large numbers of variables. In addition, certain features of switching functions represented as arrays of disjoint cubes were presented. It was shown that if these features are present in the function, they can significantly reduce the minimization task. Based on these features, exact and heuristic algorithms were introduced, implemented, and tested with 112 single output functions created from the MCNC benchmark functions. These methods give faster results than any method published previously. The Quasi-Minimal method proved to be based on high quality heuristics. Out of 112 functions, it gives the exact solution for 66 while only 6 deviate by 10 or more terms, the highest being 12 terms.

The results presented confirm also the merits of the AND/XOR networks. By identifying XOR contents of the Boolean functions, one can create more compact realizations of these functions. This study can be used for classification of functions and multi-level AND/OR/XOR and other XOR realizations. The fast programs introduced can be used to bring more XOR utilization into the realm of logic synthesis. This research will also be expanded into multi-output, incompletely specified, multi-valued functions.

* stands for unavailable data.

References

[1] M. Helliwell and M. A. Perkowski, "A Fast Algorithm to Minimize Multi-Output Mixed-Polarity Generalized Reed-Muller Forms", *Proc. of the 25th ACM/IEEE Design Automation Conference*, pp. 4427-432, 1988.

[2] T. Sasao, and Ph. Besslich, "On the Complexity of MOD-2 Sum PLAs", *IEEE Trans. on Computers*, Vol. 39, No. 2, pp. 262-266, 1990.

[3] "Concurrent Logic, Inc.", *CLi 6000 Series Field Programmable Gate Arrays*, Preliminary Information, Dec 1991, Rev 1.3.

[4] M. Davio, J. P. Deschamps, and A. Thayse, *Discrete and Switching Functions*, Mc-Graw-Hill, 1978.

[5] Ph. W. Besslich and M. W. Riege, "An Efficient Program for Logic Synthesis of Mod-2 Sum Expressions" *Euro ASIC'91*, pp. 136-141, Paris, France, 1991.

[6] L. T. Fisher, "Unateness Properties of AND Exclusive OR Logic Circuits", *IEEE Trans. on Computers*, Vol. C-23, No. 2 pp. 166-172, 1974.

[7] I. Schäfer and M. A. Perkowski, "Multiple-Valued Input Generalized Reed-Muller Forms", *Proc. of ISMVL'91*, pp. 40-48, May 1991.

[8] D. Varma and E. A. Trachtenberg, "Computation of Reed-Muller Expansions of Incompletely Specified Boolean Functions From Reduced Representations", *Proc. of IEE*, Vol. 138, Part E, No. 2, pp. 85-92, 1991.

[9] I. S. Reed, "A Class of Multiple-Error-Correcting Codes and Their Decoding Scheme", *IRE Trans. on Inf. Theory*, Vol. PGIT-4, pp. 38-49, 1954.

[10] D. E. Mueller, "Application of Boolean Algebra to Switching Circuit Design and to Error Detection", *IRE Trans. on Elec. Computers*, Vol. EC-3, pp. 6-12, September 1954.

[11] S. M. Reddy, "Easily Testable Realization for Logic Functions", *IEEE Trans. on Computers*, Vol. C-21, No. 11, pp. 1183-1188, 1972.

[12] D. K. Pradhan, "Universal Test Sets for Multiple Fault Detection in AND_EXOR Arrays", *IEEE Trans. on Computers*, Vol. C-27, No. 2, pp. 181-187, 1978.

Implicit and Incremental Computation of Primes and Essential Primes of Boolean functions

O. Coudert and **J. C. Madre**
Bull Corporate Research Center
Rue Jean Jaurès
78340 Les Clayes-sous-bois, FRANCE

Abstract

Recently introduced *implicit* set manipulation techniques have made it possible to formally verify finite state machines with state graphs too large to be built. This paper shows that these techniques can also be used with success to compute and manipulate implicitly extremely large sets of prime and of essential prime implicants of incompletely specified Boolean functions. These sets are denoted by *meta-products* that are represented with binary decision diagrams. This paper describes two procedures. One is based on the standard BDD operators, and the other, more efficient, takes advantage of the structural properties of BDDs and of meta-products to handle a larger class of functions than the former one.

1 Introduction

We have recently introduced a technique [4] for verifying finite state machines that can deal with machines with state graphs too large to be built. The key concepts that make this verification possible are to denote subsets of $\{0,1\}^{\aleph}$ with their characteristic functions, and represent these Boolean functions with binary decision diagrams (BDDs) that are a very compact graph representation of such functions [3]. Since there is no relation between the size of a set and the size of the BDD that denotes it, the computation cost of this technique is completely independent from the number of states of the machines [4].

In this paper we show that these concepts can be used with success to implement *implicit* computation procedures of the sets of prime and of essential prime implicants of incompletely specified Boolean functions. These procedures have costs that are independent of the sizes of these sets, and thus they overcome the limitations of methods based on explicit prime manipulations [2, 8, 11, 12].

This paper is divided in 6 parts. Section 2 presents the elementary concepts that will be used to solve the problem treated here. Section 3 introduces the canonical *meta-product* representation of sets of products built out of finite sets of Boolean variables. Section 4 gives the expressions that define the sets of prime and of essential prime implicants of incompletely specified Boolean functions, and shows that these equations can be evaluated with the standard BDD operators. Section 5 describes how this standard procedure can be transformed into a more efficient incremental procedure. Section 6 gives experimental results obtained with these procedures and discusses them.

2 Definitions

A Boolean function f from $\{0,1\}^n$ into $\{0,1\}$ denotes a *unique* subset S_f of $\{0,1\}^n$, made of all the elements x of $\{0,1\}^n$ such that $f(x)=1$. Conversely any subset S of $\{0,1\}^n$ can be represented by a unique Boolean function χ_S from $\{0,1\}^n$ into $\{0,1\}$, called its *characteristic function*, such that $\chi_S(x) = 1$ if and only if $x \in S$. Characteristic functions are a very interesting representation of Boolean sets because Boolean operators correspond with set operators [4].

We note P_n the set of products that can be built out of the set of variables $\{x_1, \ldots, x_n\}$. A *product* is a formula of the form $(l_1 \ldots l_k)$, $k \leq n$, whose literals l_j are built out of distinct variables. The product p contains the product p' ($p \succeq p'$) if and only if $S_p \supseteq S_{p'}$. The relation "$\succeq$" is a partial order on P_n.

An *incompletely specified* Boolean function f_c, also called a function with a *care set*, is defined by a couple (C, f), where f is a function from $\{0,1\}^n$ into $\{0,1\}$, and the care set C is a subset of $\{0,1\}^n$. This function is defined by $f_c(x) = f(x)$ if $x \in C$, and $f_c(x) = *$ if $x \notin C$, where the symbol "$*$" can be either 0 or 1 [2].

The product p of P_n is a *cube* of the function $f_c = (C, f)$ if and only if the set $S_p \cap C$ is not empty, and for any element of this set, the function f evaluates to 1. The cube p of f_c is a *prime implicant* or *prime* of f_c if and only if p is a maximal element of the set of cubes of f_c with respect to the partial order "$\succeq$". Finally the prime p of f_c is an *essential prime implicant* or *essential prime* of f_c if and only if the function denoted by the sum of all the other primes of f_c is not equal to f on the care set C [2].

Quantified propositional formulas are formulas of the form $(Q_1x_1 \ldots Q_kx_k\ f)$, where $Q_1, \ldots, Q_k$ are either the universal or the existential quantifier, f is a propositional formula, and $x_1, \ldots, x_k$ are variables occurring in f [4]. Any quantified propositional formula can be rewritten into a propositional formula, using the identities $(\forall x\ f) = (f_{\overline{x}} \wedge f_x)$, and $(\exists x\ f) = (f_{\overline{x}} \vee f_x)$, where x is free in f, and $(f_{\overline{x}}, f_x)$ is Shannon's expansion of f with respect to x [4]. Adding quantifiers to Propositional Logic does not increase its expressive power, but it allows us to write very concisely expressions whose quantifier-free forms have exponential sizes.

3 Meta-Products

The set of products P_n has 3^n elements, so $\lceil \log_2(3^n) \rceil$ Boolean variables are sufficient to represent any element and any subset of P_n. However, using this optimal encoding, there is not direct link between an element of P_n and its representation. In order to establish such a direct link, we use $n + n$ variables, the first n noted $\{o_1, \ldots, o_n\}$, and called the *occurrence* variables, and the last n noted $\{s_1, \ldots, s_n\}$, and called the *sign* variables. We define the mapping σ from the set $\{0,1\}^n \times \{0,1\}^n$ into the set P_n in the following way:

$$\sigma(o, s) = l_1 \ldots l_n,$$

where l_k is the empty string if $o_k = 0$, l_k is $\overline{x_k}$ if $o_k = 1$, and $s_k = 0$, and finally l_k is x_k if $o_k = 1$, and $s_k = 1$. For instance, the couple ([0111], [1101]) denotes the product $(x_2\overline{x_3}x_4)$. The *meta-product* $\mathcal{P}$ of the subset $P = \{p_1, \ldots, p_m\}$ of P_n is the characteristic function of the subset $(\cup_{k=1}^{m} \sigma^{-1}(p_k))$ of $\{0,1\}^n \times \{0,1\}^n$. It is a canonical functional representation [7].

3.1 Properties of Meta-Products

A couple (o, s) of $\{0,1\}^n \times \{0,1\}^n$ denotes a unique product of P_n, which itself denotes a unique subset of $\{0,1\}^n$, whose characteristic function, noted $\chi(o, s)$, is defined by the equation

$$\chi(o, s) = \lambda x.(\bigwedge_{k=1}^{n} (o_k \Rightarrow (x_k \Leftrightarrow s_k))).$$

The predicate "$\doteq$" expresses that the products p and p' represented by the couples $c = (o, s)$, and $c' = (o', s')$ respectively are identical, which is the case if and only if they are made of the same literals:

$$c \doteq c' = (\bigwedge_{k=1}^{n} (o_k \Leftrightarrow o'_k) \wedge (o_k \Rightarrow (s_k \Leftrightarrow s'_k))).$$

The predicate "$\dot{\succeq}$" expresses that the product p denoted by the couple $c = (o, s)$ contains the product p' denoted by the couple $c' = (o', s')$, which is the case if and only if all the literals of p also occur in p':

$$c \dot{\succeq} c' = (\bigwedge_{k=1}^{n} (o_k \Rightarrow (o'_k \wedge (s_k \Leftrightarrow s'_k)))).$$

3.2 Representing Meta-Products with BDDs

The implicit prime and essential prime computation procedures that will be presented in the next sections make an intensive use of the predicates $\doteq$, $\dot{\succeq}$, and of the function $\chi(o, s)$. The variable orderings that minimize the BDDs of these predicates are all compatible with the relation

$$\begin{array}{lll} o_{\pi(1)} < o'_{\pi(1)} & < & \{s_{\pi(1)}, s'_{\pi(1)}, x_{\pi(1)}\} < \\ & \cdots & \\ o_{\pi(n)} < o'_{\pi(n)} & < & \{s_{\pi(n)}, s'_{\pi(n)}, x_{\pi(n)}\}, \end{array}$$

where π is a permutation of the integers $\{1, \ldots, n\}$. These variable orderings are good candidates for building BDDs of meta-products. Among them, we have chosen the ordering that minimizes the computational cost of the routines manipulating meta-products [7]:

$$o_1 < o'_1 < s_1 < s'_1 < x_1 < \cdots < o_n < o'_n < s_n < s'_n < x_n$$

4 Implicit Prime and Essential Prime Computation

The theorems given in this section show that the usual Boolean operators provided by available BDD packages, for instance [1, 10], are sufficient to compute in an implicit way the sets of primes and of essential primes of incompletely specified Boolean functions.

4.1 Implicit Cube Computation

The meta-product $Cube(f_c)$ of the set of cubes of the function (C, f) evaluates to 1 for the couple $c = (o, s)$ if and only if the product p denoted by c is such that the set $p \cap C$ is not empty, and is included in S_f. This is expressed by the following theorem [7].

Theorem 1 *The meta-product $Cube(f_c)$ of the set of products that are cubes of the function f_c is*

$$\lambda o.\lambda s.((\exists x\ \chi_C(x) \wedge \chi(o, s)(x)) \wedge (\forall x\ (\chi_C(x) \wedge \chi(o, s)(x)) \Rightarrow f(x))).$$

4.2 Implicit Prime Computation

By definition the prime implicants of the function (C, f) are cubes of this function that are maximal with respect to the partial order "$\dot{\succeq}$" defined in Section 2 [7].

Theorem 2 *The meta-product $\mathcal{P}rime(f_c)$ of the set of products that are primes of the function f_c is:*

$$\lambda o.\lambda s.(Cube(f_c)(o, s) \wedge (\forall (o', s')\ (Cube(f_c)(o', s') \wedge (o', s') \dot{\succeq} (o, s)) \Rightarrow (o', s') \doteq (o, s))).$$

4.3 Implicit Essential Prime Computation

By definition, the meta-product $\mathcal{E}ss(f_c)$ evaluates to 1 for the couple (o, s) if and only if the subset of $\{0,1\}^n$ that this couple denotes contains an element of the care set C that is not contained in any of the subsets of $\{0,1\}^n$ denoted by the other primes of f_c [2, 7].

Theorem 3 *The meta-product $\mathcal{E}ss(f_c)$ of the set of essential primes of the partial function f_c is equal to*

$$\lambda o.\lambda s.(\exists s'\ \chi_C(s') \wedge f(s') \wedge (\forall o'\ \mathcal{P}rime(f_c)(o',s') \Rightarrow (o',s') \doteq (o,s))).$$

It is immediate, using the preceding theorems, to compute in a implicit way the BDDs of the meta-products $\mathcal{P}rime(f_c)$ and $\mathcal{E}ss(f_c)$ from the BDDs of f and χ_C. These computations, referred to as **SP** and **SE** in Section 6, that are shown by experience to be fairly efficient, have nevertheless a major drawback. It is necessary in both cases to compute the BDD of *Cube*(f_c), that is shown by experience to be in most cases much larger than the resulting BDDs.

5 Incremental Prime and Essential Prime Computation

It has been shown [2] that the primes of the function $f = (\overline{x_k} f_{\overline{x_k}} + x_k f_{x_k})$ can be obtained by combining the primes of the functions $f_{\overline{x_k}}$ and f_{x_k} respectively. This process gives a way to compute the primes of f without computing its cubes. Since each vertex of the BDD of f represents one of the functions that can be obtained by applying, with a given order, Shannon expansion on f, the number of vertices in this BDD is the number of sets of primes and of combinations that are necessary to get the primes of f.

We show here that the procedure given in [11], which implements this computation scheme with explicit prime manipulations, can be generalized to obtain a combination technique that manipulates sets of primes implicitly. This new procedure relies on implicit prime set manipulations that are presented in [5]. We will consider here that the function to be treated is completely defined, the general case is handled in [5].

5.1 Incremental Prime Computation

The following theorem explains how the meta-products $\mathcal{P}rime(f_{\overline{x_k}})$ and $\mathcal{P}rime(f_{x_k})$ can be combined to get $\mathcal{P}rime(\overline{x_k} f_{\overline{x_k}} + x_k f_{x_k})$.

Theorem 4 *The meta-product $\mathcal{P}rime(f)$ of the set of primes of the function $f = (\overline{x_k} f_{\overline{x_k}} + x_k f_{x_k})$ is*

$$(\overline{o_k} \wedge \mathcal{P}rime(f_{\overline{x_k}} \wedge f_{x_k})) \vee$$
$$(o_k \wedge \overline{s_k} \wedge \mathcal{P}rime(f_{\overline{x_k}}) \wedge \neg\mathcal{P}rime(f_{\overline{x_k}} \wedge f_{x_k})) \vee$$
$$(o_k \wedge s_k \wedge \mathcal{P}rime(f_{x_k}) \wedge \neg\mathcal{P}rime(f_{\overline{x_k}} \wedge f_{x_k}))$$

Computing the meta-product $\mathcal{P}rime(f_{\overline{x_k}} \wedge f_{x_k})$ can be done in two ways. The first and simplest way, used in the method referred to as **IP1** in Section 6, consists in building the BDD of the function $(f_{\overline{x_k}} \wedge f_{x_k})$ and in applying the incremental procedure on this BDD. Since this new BDD can contain vertices that are in none of the BDDs of the functions $f_{\overline{x_k}}$ and f_{x_k}, this method can induce a non polynomial number of recursions with respect to the size of the BDD of the function f. The second method [5], used in the procedure referred to as **IP2** in Section 6, consists in building the meta-product *Prod* of all the possible conjunctions of one prime of $f_{\overline{x_k}}$ and one prime of f_{x_k}, and then in extracting the required meta-product from *Prod*. Using this method, each vertex of the BDD of f is treated once.

5.2 Incremental Essential Prime Computation

The problem addressed here is to combine the meta-products $\mathcal{P}rime(f_{\overline{x_k}})$, $\mathcal{E}ss(f_{\overline{x_k}})$, $\mathcal{P}rime(f_{x_k})$, and $\mathcal{E}ss(f_{x_k})$ in order to get $\mathcal{E}ss(\overline{x_k} f_{\overline{x_k}} + x_k f_{x_k})$.

This combination process relies on 3 set operators that are evaluated implicitly. The operators $\mathcal{FC}$ and $\mathcal{FNC}$ are closely related. By definition, for any meta-product $\mathcal{P}$ and any function f, $\mathcal{FC}(\mathcal{P}, f)$ is the meta-product of the subset of products denoted by $\mathcal{P}$ that are cubes of the function f, and $\mathcal{FNC}(\mathcal{P}, f)$ is the meta-product of the subset of products denoted by $\mathcal{P}$ that are not cubes of the function f. Both operators have computational costs in $O(|\mathcal{P}| \times |f|)$, where $|\mathcal{P}|$ and $|f|$ are the sizes of the BDDs of $\mathcal{P}$ and f respectively. The operator $\mathcal{FAE}$ computes, in a non-polynomial cost in the worst case, for any meta-products $\mathcal{P}$ and $\mathcal{P}'$, the meta-product of the subset of products of $\mathcal{P}$ that contain an element that is not covered by any of the products denoted by $\mathcal{P}'$. The implicit essential combination method is based on the following theorems.

Theorem 5 *The meta-product of the set of essential primes of the function $(\overline{x_k} f_{\overline{x_k}} + x_k f_{x_k})$, in which the literal $(\overline{x_k})$ occurs, is*

$$o_k \wedge \overline{s_k} \wedge (\mathcal{FC}(\mathcal{E}ss(f_{\overline{x_k}}), \neg f_{x_k}) \vee \mathcal{E}ss^r(f_{\overline{x_k}})).$$

In this equation, $\mathcal{E}ss^r(f_{\overline{x_k}})$ is defined by

$$\mathcal{E}ss^e(f_{\overline{x_k}}) = \mathcal{FNC}(\mathcal{E}ss(f_{\overline{x_k}}), \neg f_{x_k}) \wedge \mathcal{FNC}(\mathcal{E}ss(f_{\overline{x_k}}), f_{x_k})$$
$$\mathcal{E}ss^r(f_{\overline{x_k}}) = \mathcal{FAE}(\mathcal{E}ss^e(f_{\overline{x_k}}), \mathcal{P}rime(f_{\overline{x_k}}) \vee \mathcal{P}rime(f_{\overline{x_k}} \wedge f_{x_k})).$$

Theorem 6 *The meta-product of the set of essential primes of the function $(\overline{x_k} f_{\overline{x_k}} + x_k f_{x_k})$, in which neither the literal $(\overline{x_k})$ nor the literal (x_k) occur, is*

$$\overline{o_k} \wedge (\mathcal{FC}(\mathcal{E}ss(f_{\overline{x_k}}), f_{x_k}) \vee \mathcal{FC}(\mathcal{E}ss(f_{x_k}), f_{\overline{x_k}})).$$

We will compare two implicit essential prime computation procedures in Section 6. Procedure **IE1**, based on Theorem 3, computes meta-products of essential primes from meta-products of primes computed using Procedure **IP1**. Procedure **IE1** is based on the preceding theorems.

6 Experimental Results

Table 1 gives the characteristics of each treated circuit : **#in** and **#fun** are the numbers of inputs and of outputs respectively ; **#Prime** and **#Ess** are the cumulated sums of the numbers of primes and of essential primes respectively obtained for each output ; **CS** is "yes" if output functions have a care set. The entries *indust*, *math*, and *random* correspond to the directories of the MCNC 2-level minimization benchmark. For these entries, column **#file** is the number of circuits in this directory.

Name	#in #file	#fun	CS	#Prime	#Ess
math	23	128	*yes*	8616	1937
indust	111	2412	*yes*	87797	17284
random	11	189	*yes*	118141	169
dsip	452	421	*no*	22850	2223
*mul*07	14	14	*no*	28972	1551
*mul*08	16	16	*no*	152051	3879
addsub	31	15	*no*	408190	102413
*s*1423	91	79	*no*	469307	36226
*add*4	29	12	*no*	688167	247
*cbp*32	65	33	*no*	4.3*e*10	4.3*e*10

Table 1. Characteristics of Examples.

Each file has first been translated from its original format into an intermediate form in which Boolean functions are represented with a textual form of BDDs. The CPU times needed to perform these translations for all files of *indust*, *math*, and *random* directories is 160 seconds on a Sun Sparc2 machine. The variable ordering used to build these BDDs is the same for all functions of a circuit, and the occurrence and sign variables are ordered according to this input variable ordering.

Name	SP	IP1	IP2	SE	IE1	IE2
math	7	7	16	32	22	115
indust	258	83	364	577	424	–
random	495	211	–	873	636	–
dsip	24	17	34	63	49	343
*mul*07	137	57	345	272	199	–
*mul*08	1002	377	–	–	1429	–
addsub	3	1	29	8	5	32
*s*1423	153	36	102	347	231	–
*add*4	18	7	35	43	33	329
*cbp*32	13	2	5	31	16	175

Table 2. CPU times in seconds on Sun Sparc2.

The experimental results given in Table 2 show that these procedures can handle Boolean functions with very large numbers of primes and of essential primes. It appears clearly that the incremental prime computation method **IP1** is much more efficient that the standard method **SP**. Surprisingly, though the computational cost of Procedure **IP2** is theoretically lower than the cost of **IP1**, the former is much less efficient than the latter on these examples. In the same way, Procedure **IE1** is much more efficient than Procedure **IE2**. In this table, symbol "-" means that CPU time is larger than 1500 seconds.

7 Conclusion

In this paper we have presented procedures for computing and manipulating in an implicit way the sets of primes, and of essential primes of incompletely specified Boolean functions. These methods allow us to deal with functions for which these sets are too large to be handled by enumeration based procedures [2, 8, 11, 12].

It has been shown [12] that computing the primes and essential primes of a multiple-valued input and output function comes down to computing the primes and essential primes of a multiple-valued inputs Boolean function. There are several ways to extend the methods presented here in order to deal with such Boolean functions, that correspond to different ways to encode the multiple-valued input variables of the function into Boolean variables [9]. A procedure dedicated to multiple output Boolean functions has also been developed [6].

References

[1] K. S. Brace, R. L. Rudell, and R. E. Bryant, "Efficient Implementation of a BDD package", in *Proc. of 27th ACM/IEEE DAC*, June 1990.

[2] R. E. Brayton, G. D. Hachtel, C. T. McMullen, A. L. Sangiovanni-Vincentelli, *Logic Minimization Algorithms for VLSI Synthesis*, Kluwer Academic Publishers, 1984.

[3] R. E. Bryant, "Graph-based Algorithms for Boolean Functions Manipulation", *IEEE Trans. on Computers*, Vol C35, 1986.

[4] O. Coudert, J. C. Madre, "A Unified Framework for the Formal Verification of Sequential Circuits", in *Proc. of ICCAD'90*, Santa-Clara, November 1990.

[5] O. Coudert, J. C. Madre, "Implicit and Incremental Computation of Primes and Essential Primes of Boolean functions", *BULL Research Report* N°91033, November 1991.

[6] O. Coudert, J. C. Madre, "Symbolic Prime Computation of Multiple Output Boolean functions", *BULL Research Report* N°91034, November 1991.

[7] O. Coudert, J. C. Madre, "A New Method to Compute Prime and Essential Prime Implicants of Boolean Funtions", in *Proc. of MIT VLSI Conference*, March 1992.

[8] S. J. Hong, S. Muroga, "Absolute Minimization of Completely Specified Switching Functions", *IEEE Trans. on Computers*, Vol 40, pp. 53–65, 1991.

[9] B. Lin, O. Coudert, J. C. Madre, "Symbolic Prime Generation for Multiple-Valued Functions", in *Proc. of 29th ACM/IEEE DAC*, June 1992.

[10] J. C. Madre, "Benchmarks for Tautology Checking - Experimental Results", in *Formal VLSI Correctness Verification*, L. J. M. Claesen Editor, North-Holland, 1990.

[11] J. C. Madre, O. Coudert, "A Logically Complete Reasoning Maintenance System Based on Logical Constrain Solver", in *Proc. of IJCAI'91*, Sydney, Australia, August 1991.

[12] T. Sasao, "An Application of Multiple-Valued Logic to a Design of Programmable Logic Arrays", *Proc. of 8th Int'l Symposium on Multiple Valued Logic*, 1978.

Symbolic Prime Generation for Multiple-Valued Functions

Bill Lin[1,*] Olivier Coudert[2] Jean Christophe Madre[2]

[1]Department of EECS, University of California, Berkeley, CA 94720
[2]Bull Corporate Research Center, Rue Jean Jaurès, 78340 Les Clayes-sous-bois, France

Abstract

The minimization of multiple-valued functions has many important applications in logic synthesis. In particular, the problems of two-level multiple-output minimization, symbolic input encoding and state assignment rely on efficient solutions to the multiple-valued minimization problem. Previously, exact solution to this problem can only be afforded for moderate size problems. A fundamental bottleneck in existing state-of-the-art exact minimizers is the inability to efficiently generate and represent all prime implicants for large functions. This paper presents new techniques based on the implicit representation and generation of primes and essential primes that can tackle multiple-valued functions with sets of primes several orders of magnitude larger than existing methods. Functions with over 10^{10} primes have been successfully generated using our proposed method.

1 Introduction

Two-level logic minimization is a key problem that arises in many contexts in the domain of logic synthesis. For over 20 years, the subject has been under intensive investigation. Initial research was directed towards developing techniques for minimizing binary-valued logic functions [1]. These techniques have been generalized for the minimization of multiple-valued (MV) functions [6,5], in which variables can assume more than two discrete values. More specifically, a multiple-valued function $f : D \times B^m$ is a function mapping the set D to the set B^m. The output set B^m is the m dimensional Boolean space and the input set D is the Cartesian product of n sets $D_1 \times \cdots \times D_n$, where D_i is a set of P_i values, $D_i = \{0, 1, \ldots, P_i - 1\}$, and P_i is a positive integer. Binary-valued input variables are a special case of multiple-valued variables. The significance of the multiple-valued minimization problem is in its wide applications in such areas as PLA optimization, multi-level logic synthesis symbolic input encoding, and state assignment. Multiple-valued minimization can be used even in the context of minimizing multiple-output binary-valued functions. Sasao [6] showed that the minimization of multiple-output functions can be reduced to an equivalent single-output minimization problem with an added multiple-valued variable. Thus, binary-valued, multiple-valued, and multiple-output functions can be treated uniformly.

Heuristic minimization techniques have been developed for minimizing both binary-valued and multi-valued functions [1,5]. These techniques have been successful in tackling relatively large size problems. Exact minimization techniques, on the other hand, have only been affordable in relatively moderate size problems [5]. A fundamental bottleneck in existing state-of-the-art minimizers is their inability to efficiently generate and represent all prime implicants for large functions. This limitation is due to the fact that a function can have in the worst case exponential number of prime implicants with respect to the number of variables. Thus, exact minimization techniques that require the explicit generation and representation of primes are inherently limited to functions with relatively limited number of variables and primes [1,5].

In this paper, we present new techniques based on the implicit representation and generation of primes for multiple-valued functions. Solving this problem is fundamental in developing more powerful two-level minimization algorithms. Our proposed techniques can generate and represent all primes for functions with prime sets several orders of magnitude larger than existing methods. The key idea that makes this computation possible is the symbolic representation of multiple-valued cubes in a characteristic function form called the **characteristic-cube function**. This symbolic representation can be efficiently denoted using a binary decision diagrams (BDD's) [2], which is known to be a very compact representation for Boolean formulas. Since there is no direct correspondence between the number of elements in a characteristic function and the size of the BDD representation that denotes it, very large sets of primes may be captured symbolically using the characteristic-cube function representation. Using this representation, we have developed a **symbolic prime generation** procedure based on implicit set operations. The proposed method requires only basic Boolean operators that are available in any BDD package. For a 32-bit adder, we are able to symbolically generate the complete set of over 10^{10} primes in only 8.5 seconds of CPU time. The characteristic-cube function that denotes this set has less than 8500 BDD nodes.

The rest of this paper is organized as follows. In Section 2, basic definitions and terminology are provided. In Section 3, we describe our symbolic representation for multiple-valued cubes and covers. In our approach, both binary-valued and multiple-valued variables can be treated uniformly. In Section 4, we present a simple but very efficient symbolic prime generation method based on the characteristic-cube representation for both completely specified and incompletely specified multiple-valued functions. In Section 5, the problem of extracting essential primes is considered. In Section 6, strategies for handling multiple-output functions are described. In Section 7, experimental results are presented on a range of benchmarks. Finally, conclusions are drawn in Section 8.

2 Definitions and Notations

We use the standard definitions and notations widely used in the areas of logic synthesis and verification. Only a subset of these will be summarized for the sake of brevity.

2.1 Multiple-Valued Logic

A **multiple-valued function** f is a mapping $f : D \rightarrow B^m$. $D = D_1 \times \cdots \times D_n$ is called the **domain** of f, where $D_i =$

*The author is currently at the IMEC Laboratory in Leuven, Belgium.

0738-100X/92 $3.00 © 1992 IEEE

$\{0, \ldots, P_i - 1\}$ and P_i is a positive integer. B^m is the **co-domain** of f. Associated with each input variable X_i is a finite set of values $D_i = \{0, \ldots, P_i - 1\}$, where P_i is some positive integer, that provides the set of admissible values that X_i can assume. X_i is referred to as a **multiple-valued variable**. A binary-valued variable is a special case where $P_i = 2$. Let X_i be a variable taking a value from the set D_i. Let $S_i \subseteq D_i$ be a subset of D_i. Then $X_i^{S_i}$ represents the characteristic function

$$X_i^{S_i} = \begin{cases} 0 & \text{if } X_i \notin S_i \\ 1 & \text{if } X_i \in S_i \end{cases}$$

$X_i^{S_i}$ is called a **literal** of variable X_i. A **product term** S is a Boolean product (AND) of literals of all the variables of D. Thus, S represents a subset of D. If $|S_i| = 1$ for all X_i, then the product term is also a **minterm**. If a product term evaluates to 1 for a given minterm, the product term is said to **contain** the minterm.

2.2 BDD's and Characteristic Functions

A binary decision diagram (BDD) [2] is a very compact canonical graph representation of Boolean functions. The reader is referred to [2] for details. BDD's hold key properties that make them a better representation for reasoning about Boolean functions than previously used representations, e.g. the sum-of-products representation. In addition to the canonical property, BDD's are in most cases much more compact than any sum-of-products that denotes the same function. BDD's can also be used to represent finite sets efficiently. Any subset A of B^n can be denoted by a unique Boolean function $\chi_A : B^n \rightarrow B$. This function is called the **characteristic function** of A and is defined in the following way: $\chi_A(\mathbf{x}) = 1$ if and only if $\mathbf{x} \in A$. Characteristic function is a very interesting implicit representation of Boolean sets because there is a direct correspondence between the set operators and the logical operators. Since a characteristic function is a Boolean function, it can be represented with a binary decision diagram and basic set operations can be performed directly on the graph representation.

2.3 Boolean Operations

In addition to the basic propositional logic operations, the existential operator ($\exists$) and the universal operator ($\forall$) are required for the implicit manipulation of sets. These Boolean quantifiers can also be computed directly on a BDD representation as follows. The existential quantification of a set of Boolean variables $X = \{x_1, \ldots, x_n\}$ with respect to the Boolean formula f can be evaluated as

$$\begin{aligned} \exists X(f) &= \exists x_1(\exists x_2(\ldots \exists x_{n-1}(\exists x_n(f)) \ldots)), \\ \exists x_i(f) &= f_{x_i} + f_{\overline{x_i}}, \end{aligned}$$

where $f_{x_i}(f_{\overline{x_i}})$ denotes the **cofactor** of formula f with respect to a literal $x_i(\overline{x_i})$. Likewise, the universal quantifier can be evaluated as

$$\begin{aligned} \forall X(f) &= \forall x_1(\forall x_2(\ldots \forall x_{n-1}(\forall x_n(f)) \ldots)), \\ \forall x_i(f) &= f_{x_i} \cdot f_{\overline{x_i}}. \end{aligned}$$

In this paper, we will use the notations $f \Rightarrow g$ to designate a logical implication from f to g (evaluated as $\overline{f} + g$), $f \equiv g$ to designate an XNOR operation, and $f - g$ to designate a sharp product operation (evaluated as $f \cdot \overline{g}$). These alternative notations are used to improve the readability of the manuscript.

3 Characteristic-Cube Function

3.1 Representation of Multiple-Valued Cubes

A multiple-valued function $f : D \rightarrow B$ can be represented efficiently as a characteristic function in BDD form by first encoding the multi-valued variables into binary vectors. This problem was addressed in [7]. Product terms and primes, however, have been represented explicitly using two-level cover concepts. This is a fundamental limitation of previous prime generation algorithms. Here, we propose to alleviate this bottleneck by using a symbolic representation based on binary decision diagrams to implicitly represent and manipulate multiple-valued product terms and primes. Given a function $f : D \rightarrow B$, where $D = D_1 \times \cdots \times D_n$, and each D_i is some finite set of P_i ($P_i \geq 0$) values $\{0, \ldots, P_i - 1\}$, there can be $\prod_{i=1}^{n} 2^{P_i}$ possible product terms. This is because a literal $X_i^{S_i}$ of the *i-th* variable can be any subset of D_i. In order to represent in an efficient way any possible set of product terms with a characteristic function in BDD form, an efficient encoding of all possible product terms is required. For this purpose, many encoding schemes are possible. Effectively, we need some function

$$\zeta : B^l \rightarrow \mathcal{P}(D_i),$$

where $\mathcal{P}(D_i)$ is the powerset of D_i, to encode each variable X_i. Clearly, at least P_i unique binary variables are required to encode $\mathcal{P}(D_i)$. Here, we present an efficient symbolic representation based on the positional cube notation.

Definition 3.1 *Let $X_1^{S_1} X_2^{S_2} \cdots X_n^{S_n}$ be a product term. It can be represented in positional cube notation form as*

$$c_{1_0} c_{1_1} \cdots c_{1_{P_1 - 1}} - c_{2_0} c_{2_1} \cdots c_{2_{P_2 - 1}} - \cdots - c_{n_0} c_{n_1} \cdots c_{n_{P_n - 1}}$$

where $c_{i_j} = 0$ if $j \notin S_i$ and $c_{i_j} = 1$ if $j \in S_i$.

Definition 3.2 *Let $\mathcal{C} = \{T_1, \ldots, T_m\}$ be a set of product terms. A* **characteristic-cube function** *is simply a characteristic function*

$$\chi_{cube} : B^{\mathcal{L}} \rightarrow B$$

defined as follows: (1) $\mathcal{L} = \sum_{i=1}^{n} P_i$. (2)

$$\chi_{cube}(\mathbf{c}) = \begin{cases} 1 & \text{if and only if } \exists T \in \mathcal{C} \text{ such that} \\ & \forall i, \forall j = \{0, \ldots, P_i - 1\}, \quad c_{i_j} \equiv (j \in S_i), \\ 0 & \text{otherwise} \end{cases}$$

where $\mathbf{c} = c_{1_0} c_{1_1} \cdots c_{1_{P_1 - 1}} - \cdots - c_{n_0} c_{n_1} \cdots c_{n_{P_n - 1}}$ and $T = X_1^{S_1} X_2^{S_2} \cdots X_n^{S_n}$.

Each component $c_{i_0} c_{i_1} \cdots c_{i_{P_i - 1}}$ represents a set of values contained in the *i-th* literal of $\mathbf{c}$, and consists of P_i binary bits. Each bit is called a part. We denote $I(\mathbf{c})$ as the set of minterms of D contained in $\mathbf{c}$.

The key to the characteristic-cube form is the symbolic representation of product terms with binary decision diagrams. There is no direct correspondence between the number of product terms in the set and the size of the BDD that denotes the characteristic-cube function. Depending on the structural regularity of the set under consideration, very large sets of product terms may be symbolically represented in a compact form.

3.2 All-Zero Code Simplification

In the characteristic-cube representation, each literal $X_i^{S_i}$ is represented by its corresponding component $c_{i_0}c_{i_1}\cdots c_{i_{P_i-1}}$. With P_i bits, all possible subsets $S_i \subseteq D_i$ of P_i values can be represented. However, in the two-level minimization problem (and most other problems), we do not need to consider the case where a literal contains no values (*i.e.* $X_i^{S_i}$ where $S_i = \{\}$) since it has no real meaning. Hence, we do not need to consider product terms with any empty literals either. The collection of these "non-real" product terms can be used implicitly as "don't care" conditions to simplify the BDD representation for representing multiple-valued product terms. More specifically, the possible don't care conditions can be succinctly captured in the expression

$$\sum_{i=1}^{n} \prod_{j=0}^{P_i-1} \overline{c_{i_j}} \qquad (1)$$

The term $\prod_{j=0}^{P_i-1} \overline{c_{i_j}}$ expresses the condition for an empty literal and the outer summation is used to check if any literal violates this condition.

3.3 Relations and Properties

Characteristic-cube representation has a number of key properties that follow directly from the fact that it is a characteristic function representation. Therefore, set operations corresponds directly to logical operations. The characteristic-cube representation also has the interesting property that many useful relations can be easily defined and computed. We develop in this section several of these relations that will be used for the symbolic prime generation procedure. For example, the equivalence of two product terms can be determined simply by checking if their characteristic-cube representations are identical.

Definition 3.3 *Let $f : D \rightarrow B$ be a multiple-valued function, where $D = D_1 \times \cdots \times D_n$ and $D_i = \{0, \ldots, P_i - 1\}$. The* **equality relation** *of f is a characteristic function $\chi_{equal} : B^{\mathcal{L}} \times B^{\mathcal{L}} \rightarrow B$, where $\mathcal{L} = \sum_{i=1}^{n} P_i$, such that $\chi_{equal}(\mathbf{c}, \hat{\mathbf{c}}) = 1$ if and only if the multiple-valued product terms $\mathbf{c}$ and $\hat{\mathbf{c}}$ are identical (contain exactly the same minterms).*

Theorem 3.1 *Let $f : D \rightarrow B$ be a multiple-valued function, where $D = D_1 \times \cdots \times D_n$, and $D_i = \{0, \ldots, P_i - 1\}$. The expression*

$$\chi_{equal}(\mathbf{c}, \hat{\mathbf{c}}) = \prod_{i=1}^{n} \prod_{j=0}^{P_i-1} c_{i_j} \equiv \hat{c_{i_j}} \qquad (2)$$

computes the equality relation.

Proof: Two product terms $\mathbf{c}$ and $\hat{\mathbf{c}}$ are identical if and only if they are products of the same literals. The expression $\prod_{j=0}^{P_i-1} c_{i_j} \equiv \hat{c_{i_j}}$ ensures that the *i-th* literal of both product terms contain exactly the same subset of values from D_i. The outer product $\prod_{i=1}^{n}$ ensures that this holds for all variables. ■

A relation expressing the containment of two product terms can also be easily computed. A product term $S = X_1^{S_1} X_2^{S_2} \cdots X_n^{S_n}$ contains another product term $T = X_1^{T_1} X_2^{T_2} \cdots X_n^{T_n}$ if and only if $S_i \supseteq T_i$, for all i. The pairwise containment relation can be defined as follows.

Definition 3.4 *The* **containment relation** *of f is a characteristic function $\chi_{contain} : B^{\mathcal{L}} \times B^{\mathcal{L}} \rightarrow B$, $\mathcal{L} = \sum_{i=1}^{n} P_i$, such that $\chi_{contain}(\mathbf{c}, \hat{\mathbf{c}}) = 1$ if and only if the multiple-valued product term $\mathbf{c}$ contains $\hat{\mathbf{c}}$.*

Theorem 3.2 *The expression*

$$\chi_{contain}(\mathbf{c}, \hat{\mathbf{c}}) = \prod_{i=1}^{n} \prod_{j=0}^{P_i-1} \hat{c_{i_j}} \Rightarrow c_{i_j} \qquad (3)$$

computes the containment relation.

Proof: $\prod_{j=0}^{P_i-1} \hat{c_{i_j}} \Rightarrow c_{i_j}$ ensures that the *i-th* literal of the product term $\mathbf{c}$ contains all the values contained in the same literal of the product term $\hat{\mathbf{c}}$. The outer product $\prod_{i=1}^{n}$ ensures that this holds for all variables. ■

In the above expression, $\chi_{contain}(\mathbf{c}, \hat{\mathbf{c}}) = 1$ if and only if $\mathbf{c} \supseteq \hat{\mathbf{c}}$.

Definition 3.5 *The* **strict containment relation** *of f is a characteristic function $\chi_{strict\_contain} : B^{\mathcal{L}} \times B^{\mathcal{L}} \rightarrow B$, $\mathcal{L} = \sum_{i=1}^{n} P_i$, such that $\chi_{strict\_contain}(\mathbf{c}, \hat{\mathbf{c}}) = 1$ if and only if the multiple-valued product term $\mathbf{c}$ strictly contains $\hat{\mathbf{c}}$.*

The characteristic function representing pairwise strict containment ($\mathbf{c} \supset \hat{\mathbf{c}}$) can be computed as follows.

Theorem 3.3 *The expression*

$$\chi_{strict\_contain}(\mathbf{c}, \hat{\mathbf{c}}) = \chi_{contain}(\mathbf{c}, \hat{\mathbf{c}}) - \chi_{equal}(\mathbf{c}, \hat{\mathbf{c}}) \qquad (4)$$

computes the strict containment relation.

Proof: Follows from the definitions. ■

4 Symbolic Prime Generation

In this section, techniques are given for implicit generation and representation of implicants and prime implicants. These techniques are simple but very efficient. They require only basic Boolean operations available in any BDD package. Given a function $f : D \rightarrow B$, $D = D_1 \times \cdots \times D_n$, the set of possible implicants and prime implicants can be efficiently generated using the characteristic-cube representation and the global relations defined in the previous section. We first consider the case for completely specified functions. Then we will consider the case for incompletely specified functions.

4.1 Completely Specified Functions

We first need to establish the relationship between the characteristic-cube representation and the multiple-valued function.

Definition 4.1 *Let $f : D \rightarrow B$ be a multiple-valued function, where $D = D_1 \times \cdots \times D_n$ and $D_i = \{0, \ldots, P_i - 1\}$. The* **mapping relation** *for the domain $D = D_1 \times \cdots \times D_n$ is the characteristic-cube function*

$$\chi_{mapping}(\mathbf{c}, \mathbf{x}) = \prod_{i=1}^{n} \prod_{j=0}^{P_i-1} X_i^{\{j\}} \Rightarrow c_{i_j} \qquad (5)$$

where $X_i^{\{j\}}$ corresponds to the j-th value of the i-th variable of D.

This relation provides the correspondence between a product term $\mathbf{c}$ and the minterms that it contains.

Theorem 4.1 *Let $f : D \to B$ be a multiple-valued function, where $D = D_1 \times \cdots \times D_n$ and $D_i = \{0, \ldots, P_i - 1\}$. $\chi_{mapping}(\mathbf{c}, \mathbf{x}) = 1$ if and only if $\mathbf{x} \in I(\mathbf{c})$.*

Proof: If the *i-th* variable of the minterm $\mathbf{x}$ has the value $X_i^{\{j\}}$, then the corresponding characteristic-cube representation of $\mathbf{c}$ should have a 1 in the *j-th* part of the *i-th* component (*i.e.* $c_{i_j} = 1$). This is precisely given by the expression $X_i^{\{j\}} \Rightarrow c_{i_j}$. ■

Using the mapping relation, we can next derive a characteristic-cube function corresponding to the set of all possible implicants for the function f. This can be simply expressed as follows.

Theorem 4.2 *Let $f : D \to B$ be a completely specified multiple-valued function, where $D = D_1 \times \cdots \times D_n$, and $D_i = \{0, \ldots, P_i - 1\}$, and let $\chi_{mapping}(\mathbf{c}, \mathbf{x})$ be the mapping relation for the domain D. The characteristic-cube function for the set of product terms that form the set of all possible implicants of f is given by the expression*

$$\chi_{implicants}(\mathbf{c}) = \forall \mathbf{x} \, (\chi_{mapping}(\mathbf{c}, \mathbf{x}) \Rightarrow f(\mathbf{x})) \quad (6)$$

Proof: A product term $\mathbf{c}$ is an implicant of f if and only if all the minterms that it covers (*i.e.* $I(\mathbf{c})$) belong to the *ON-set* of f. That is, $\forall \mathbf{x} \in I(\mathbf{c}), f(\mathbf{x}) = 1$. This is precisely given by the expression $\forall \mathbf{x} \, (\chi_{mapping}(\mathbf{c}, \mathbf{x}) \Rightarrow f(\mathbf{x}))$. ■

The computation only makes use of standard BDD operations described in Section 2.

From the set of implicants f, we can extract out symbolically the subset of implicants that are primes of f using the global relations developed in the previous section. Precisely, we remove from the characteristic function $\chi_{implicants}(\mathbf{c})$ those implicants $\mathbf{c}$ that are fully contained (single-cube containment) by some other implicant $\hat{\mathbf{c}} \in \chi_{implicants}$. The remaining set of implicants are guaranteed to be primes. This is achieved as follows.

Theorem 4.3 *Let $\chi_{implicant}(\mathbf{c})$ be the complete set of implicants for the multiple-valued function f defined above. Then the characteristic-cube function for the complete set of primes is given by the expression*

$$\chi_{prime}(\mathbf{c}) = \chi_{implicant}(\mathbf{c}) - (\exists \hat{\mathbf{c}} \, (\chi_{strict\_contain}(\hat{\mathbf{c}}, \mathbf{c}) \chi_{implicant}(\hat{\mathbf{c}}))) \quad (7)$$

Proof: $(\exists \hat{\mathbf{c}} \, (\chi_{strict\_contain}(\hat{\mathbf{c}}, \mathbf{c}) \chi_{implicant}(\hat{\mathbf{c}})))$ computes the set of product terms where each product term $\mathbf{c}$ in the set is fully contained by some implicant $\hat{\mathbf{c}}$ in $\chi_{implicant}(\hat{\mathbf{c}})$. These product terms can be removed from $\chi_{implicant}(\mathbf{c})$, as indicated by the sharp operation. ■

4.2 Incompletely Specified Functions

We now consider the case of incompletely specified function. We will assume an incompletely specified function $\mathcal{F}$ is given by a couple (f, C), where $f : D \to B$ is a function and $C : D \to B$ is the **care** set. Minterms $\mathbf{x} \in D$ outside the care set C are considered don't care points. Any function $\hat{f} : D \to B$ such that $\hat{f}(\mathbf{x}) = f(\mathbf{x})$ when $C(\mathbf{x}) = 1$ is a possible implementation of $\mathcal{F}$.

In the case of an incompletely specified function, the set of all possible implicants is computed differently. A product term $\mathbf{c}$ is an implicant of $\mathcal{F}$ if it does not contain any minterm outside of $f + \overline{C}$, and it contains at least one minterm inside of C. The computation of all implicants of an incompletely specified function is stated by the following theorem.

Theorem 4.4 *Let $\mathcal{F} = (f, C)$ be an incompletely specified multiple-valued function, where f is a multiple-valued function, and C is the care set. Let $\chi_{mapping}(\mathbf{c}, \mathbf{x})$ be the mapping relation. Then, the characteristic-cube function for the set of all possible implicants is given by the expression*

$$\chi_{implicants}(\mathbf{c}) = (\exists \mathbf{x} (\chi_{mapping}(\mathbf{c}, \mathbf{x}) C(\mathbf{x})))(\forall \mathbf{x} (C(\mathbf{x}) \Rightarrow (\chi_{mapping}(\mathbf{c}, \mathbf{x}) \Rightarrow f(\mathbf{x})))) \quad (8)$$

Proof: The first component $(\exists \mathbf{x} \, (\chi_{mapping}(\mathbf{c}, \mathbf{x}) C(\mathbf{x})))$ ensures that the cube $\mathbf{c}$ covers at least one point in the care set. The second component expresses the condition that every minterm in the care set covered by $\mathbf{c}$ must also belong to the *ON-set* of f. ■

The set of primes of $\mathcal{F}$ can then be extracted from the characteristic-cube function $\chi_{implicant}(\mathbf{c})$ by applying Theorem 4.3 above.

5 Extraction of Essential Primes

We now consider the problem of computing essential primes. This can be achieved by extracting symbolically from the set of primes those that are essential. Recall that a prime is deemed essential if it covers at least one minterm in the *ON-set* that is not covered by any other prime. The basic strategy here is to compute implicitly for each prime $\mathbf{c}$ those minterms that can still be covered if it is not considered. Precisely, the characteristic cube function corresponding to the set of all essential primes can be computed as follows. Here, we consider incompletely specified functions directly.

Theorem 5.1 *Let $\mathcal{F} = (f, C)$ be an incompletely specified multiple-valued function, where f is a multiple-valued function, and C is the care set. Let $\chi_{prime}(\mathbf{c})$ be the set of prime implicants for $\mathcal{F}$. The characteristic-cube function for the set of essential prime implicants is given by the expressions*

$$\chi_{still\_cover}(\mathbf{c}, \mathbf{x}) = \exists \hat{\mathbf{c}} ((\chi_{prime}(\hat{\mathbf{c}}) - \chi_{equal}(\mathbf{c}, \hat{\mathbf{c}})) \chi_{mapping}(\hat{\mathbf{c}}, \mathbf{x})) \quad (9)$$

$$\chi_{essential}(\mathbf{c}) = \exists \mathbf{x} (C(\mathbf{x})(f(\mathbf{x}) \Rightarrow \chi_{still\_cover}(\mathbf{c}, \mathbf{x})))$$

Proof: The characteristic function $\chi_{still\_cover}(\mathbf{c}, \mathbf{x})$ represents for each possible $\mathbf{c}$ (not necessarily prime), the set of $\mathbf{x}$'s still covered by $\chi_{prime}(\mathbf{c}) - \mathbf{c}$. The expression $\exists \mathbf{x} \, (C(\mathbf{x}) \, (f(\mathbf{x}) \Rightarrow \chi_{still\_cover}(\mathbf{c}, \mathbf{x})))$ imposes the condition that every care set minterm in the *ON-set* of f must still be covered. If there exists one $\mathbf{x}$ not covered, then $\mathbf{c}$ is essential. Note that we don't need to further check if $\mathbf{c}$ is a prime because by definition the set of primes can cover f. ■

The above computation again requires only standard Boolean operations.

6 Multiple-Output Functions

In the previous sections, single-output multi-valued functions were considered. As shown in [6], a m-output n-input multiple-valued function can be transformed into a *n+1*-input single-output function where the *n+1-th* variable (X_{n+1}) can take on m values. That is, given

$$f : D \to B^m,$$

where $D = D_1 \times \cdots \times D_n$, we can transform it into another function of the form

$$\hat{f} : D \times D_{n+1} \to B,$$

where $D_{n+1} = \{0, \ldots, m-1\}$. The techniques described in the previous sections are directly applicable.

However, more efficient computation can be achieved by deriving the mapping relation $\chi_{mapping}$ and the set of all possible implicants $\chi_{implicants}$ directly from the multiple-output function form. This avoids computing the characteristic function for $\hat{f} : D \times D_{n+1} \rightarrow B$, which can be time consuming if the number of outputs is large. This is accomplished by introducing m new variables $y = \{y_0, \ldots, y_{m-1}\}$ for representing the outputs. Each y_j corresponds uniquely an individual output $f^{(j)}$ of f. To represent the literals corresponding to the output parts, exactly m unique variables $\{c_{n+1_0}, \ldots, c_{n+1_{m-1}}\}$ are needed. There is also the one-to-one correspondence between c_{n+1_j} and y_j. The computation of the mapping relation $\chi_{mapping}$ is modified as

$$\chi_{mapping}(\mathbf{c}, \mathbf{x}, \mathbf{y}) = \left(\prod_{i=1}^{n} \prod_{j=0}^{P_i - 1} X_i^{\{j\}} \Rightarrow c_{i_j} \right) \left(\prod_{j=0}^{m-1} c_{n+1_j} \Rightarrow y_j \right) \qquad (10)$$

The characteristic functions χ_{equal}, $\chi_{contain}$, $\chi_{strict\_contain}$, and χ_{prime} are computed the same way as in previous sections since these characteristic functions do not depend on the variables in $\mathbf{x}$ or $\mathbf{y}$. The set of all possible implicants needed for prime computation needs to be modified as

$$\chi_{implicants}(\mathbf{c}) = \forall \mathbf{x}, \mathbf{y} \left(\chi_{mapping}(\mathbf{c}, \mathbf{x}, \mathbf{y}) \Rightarrow \left(\prod_{j=0}^{m-1} f^{(j)}(\mathbf{x}) \equiv y_j \right) \right) \qquad (11)$$

Incompletely specified multiple-output functions can be handled in a similar fashion. Also, the computation for detecting essential primes can be extended as well.

7 Implementation and Results

We have implemented the techniques described in this paper and have tested them on a range of large benchmarks. The experimental results are provided in Table 1. The CPU times are reported in seconds on a DECstation 5000 machine. In our experiments, the characteristic-cube representation was used to represent and generate primes symbolically. We used the all-zero code concept described in Section 3.2 to simplify the BDD sizes and improve the computation times. The experimental results show that the techniques presented in this paper can indeed be applied to functions with very large number of primes.

The first set of benchmarks is the MCNC FSM benchmark set. A total of 43 examples were considered. For these state machine examples, the state variable must be treated as a multiple-valued variable. The total number of primes and CPU time for all the examples combined are reported under the columns labeled **#primes** and **time**, respectively. The examples **ex4**, **ibm**, and **misg** are large PLA examples from the ESPRESSO benchmark set. For each of these examples, the techniques of Section 6 was applied to treat a multiple-output function as an equivalent single-output multiple-valued function. The number of primes can be quite significant for these examples when multiple-output primes are considered. Again, the number of primes and CPU times are reported appropriately. The third set of examples is a collection of multi-level logic benchmarks. The examples cbp32 and cbp16 are 32-bit and 16-bit adders, respectively. The example **clm** is a memory controller. The example **s420** is from the ISCAS sequential test set. The examples **mm10** and **mm20** are min max circuits.

name	#files	#primes	time
mcnc.fsm	43	8.34e+03	34.76
name	**#I/O**	**#primes**	**time**
ex4	128/28	2.04e+09	16.37
ibm	47/17	1.05e+09	29.95
misg	56/23	2.20e+09	10.95
cbp32	65/33	4.29e+10	8.37
cbp16	33/17	6.55e+05	1.95
clm	46/34	1.51e+04	42.24
s420	35/18	6.56e+04	15.70
mm10	43/30	5.14e+03	3.34
mm20	83/60	5.24e+06	14.80

Table 1: Experimental results.

As can be seen from the table, large problem instances with prime sets well beyond the size of traditional prime generation methods can be efficiently handled using our proposed techniques with only modest CPU times. More extensive experimental results can be found in [4].

8 Conclusions

In this paper, a novel approach for generating primes and essential primes implicitly for multiple-valued functions has been presented. The approach is based on a symbolic representation of multiple-valued cubes called the characteristic-cube function that can efficiently denote very large sets of prime implicants with compact data structures like BDD's. The symbolic generation algorithms make use of only simple logic operators that are available in any BDD package. Alternative techniques based on a different symbolic representation have also been developed for Boolean functions [3]. However, the framework developed in this paper is much more general in the sense that binary-valued, multiple-valued, and multiple-output functions can all be treated uniformly.

References

[1] R. K. Brayton, G. D. Hachtel, C. T. McMullen, and A. Sangiovanni-Vincentelli. Logic Minimization Algorithms for VLSI Synthesis. Kluwer Publisher, 1984.

[2] R. E. Bryant. Graph-based algorithms for Boolean function manipulation. IEEE Transactions on Computers, C-35(8):677–691, August 1986.

[3] O. Coudert and J. C. Madre. Implicit and incremental computation of primes and essential primes of boolean functions. In Proceedings of the ACM/IEEE Design Automation Conference, June 1992.

[4] B. Lin, O. Coudert, and J. C. Madre. A novel framework for the symbolic generation of primes and essential primes of multiple-valued functions. In IMEC internal report. Available upon request, February 1992.

[5] R. Rudell and A. Sangiovanni Vincentelli. Multiple valued minimization for PLA optimization. IEEE Transactions on Computer-Aided Design, CAD-6(5):727–750, September 1987.

[6] T. Sasao. An application of multiple-valued logic to a design of programmable logic arrays. In Proceedings of the 8th International Symposium on Multiple-Valued Logic, 1978.

[7] A. Srinivasan, T. Kam, S. Malik, and R. K. Brayton. Algorithms for discrete function manipulation. In Proceedings of the IEEE International Conference on Computer-Aided Design, November 1990.

FPGA Design Principles (a tutorial)

Dwight D. Hill
AT&T Bell Labs

Ewald Detjens
Exemplar

Abstract

This tutorial describes current Field Programmable Gate Array (FPGA) technology from three directions: programming technology, functional block capabilities, and routing architectures. It then sketches FPGA CAD issues: logic synthesis for FPGA's, placement and routing, and issues in the design of new FPGA's.

1 Technology Taxonomy

FPGA's constitute a radical new direction in microelectronics. This is because they significantly reduce the overhead involved in designing and producing new logic circuits. Today's FPGA's can be viewed as an evolution of PAL's where size is increased by an order of magnitude, or a refinement of mask-programmed gate arrays, where the reprogramming time and cost are drastically reduced. To the CAD professional, FPGA's represent a new challenge: what is the best way to exploit this new technology?

Users choose a part based on external metrics such as speed, density, cost and CAD effectiveness for their particular application. Suppliers try to optimize these across many applications by making technology and architectural tradeoffs. The result is a wide spectrum of features: anti-fuse versus reprogrammable configuration, block-structured versus channel-structured routing, and lookup table (LUT) versus multiplexor versus sum-of-products logic.

Generally speaking, the technology and architecture of the routing fabric is the most important factor in determining the effectiveness of an FPGA for a particular application. FPGA routing enables (almost) arbitrary connection among logic blocks, but at the cost of tying up more area and incurring more delay than present in a mask programmed part. Likewise, the logic architectures of FPGA's are larger and slower than mask defined gates, since their functionality must be programmable. But in comparison to the routing fabric, their delays tend to be more predictable and less of a limiting factor.

2 Basic Technology Issues

Anti-fuse parts, typified by the Actel Act-1 and Act2 architectures [2], use special processing steps to fabricate insulating layers between the routing wires. These anti-fuses have relatively low resistance when turned on and are relatively compact. Unfortunately, they require high voltages to activate, and large isolation transistors are usually required to protect logic from the programming voltages. Their channelled architecture is relatively friendly to CAD tools and has advantages for predicting routing delays, since almost all designs that can be fit onto the chip can be routed within a limit of 2 or 3 anti-fuses in series. However, the lower resistance of the anti-fuses comes at the cost of reconfigurability.

One alternative to anti-fuses is SRAM-based programmable interconnection points (PIP's^{tm}). These consist of an SRAM cell controlling one or more FET's. When the SRAM is set, the FET is turned on, and the application signal can propagate from one node to another. Reconfigurable parts are attractive for many reasons. Most obviously, they can be used over and over again on the lab bench to debug and/or refine a design, (which is important since many projects die before even the first copy is produced). Also, even after the design is working, many low volume products choose to include reconfigurable FPGA's in their production versions. This is because features can be added, errors corrected, and interfaces modernized with little or no cost in the field. Such changes can be implemented by shipping a new ROM, floppy, or even by modem, depending on the application.

Reconfigurable parts are typified the Xilinx 3000 and 4000 series [13]. These architectures consist of a set of Configurable Logic Blocks (CLB's^{tm}) connected to routing wires, which are in turn connected to other routing wires by "switch boxes." Typical parts would have from 60 to 320 CLB's, and an equal number of switch boxes. A two-dimensional array of CLB's and switch-box blocks define the routing architecture. Because the routing is SRAM based, each PIP is several times larger than an anti-fuse. It is therefore essential to get the most routing functionality out of each PIP. This is done by increasing the complexity of the routing fabric, and by tolerating a greater number of PIP's in series. (On these chips, "maze routers" are often used, since they can handle arbitrary node-to-node interconnections.) The result is a system that can have high speed for some application nets, but not all. A complex, timing-directed CAD system is needed to make best use of these architectures.

Reconfigurable parts can also be based on EPROM technology, as typified by the Lattice ispLSI parts [8]. In some ways, these parts resemble a collection of PAL's connected with a pool of on-chip routing. In a PAL, the logic functionality is intermingled with the routing in an array of minterms.

3 CAD Issues

Today's system designers make wide use of Programmable Logic Devices (PLD's) and mask-programmed application specific integrated circuits (MP-ASIC's), a category which includes gate-arrays, standard-cell and full-custom parts. FPGA architectures differ from PAL's and MP-ASIC's by varying degrees. The more differences there are, the more nontraditional optimization techniques are necessary, whether performed by hand or by a program.

3.1 Design Capture and Environment

PLD's are typically designed with a low-level logic specification language. Placement, routing, and testing are simplified and hidden from the user. By contrast, MP-ASIC's are usually specified with schematics, and recently, via hardware description languages (HDL's) such as VHDL. Today's larger FPGA's tend to have a design flow closer to that of MP-ASIC's rather than PLD's. For example, place-and-route is an important and sometimes painful step, not guaranteed to complete. Also, since on-chip delays can vary by several orders of magnitude, back annotation of routing delays may be needed, just as in an MP-ASIC. But there are two ways that FPGA CAD tends to differ from ASIC CAD: the architecture of the parts tends to eliminate the need for test vectors, and the majority of users today still use PC environments, rather than UNIX ^{tm} workstations.

3.2 Synthesis issues

Synthesis begins with logic optimization, which is the process of restructuring logic to minimize delay or area. Traditionally, PAL's and PLD's had sum-of-products architectures that required two-level minimization. Programs like MINI and Espresso were written to address this problem.

But most MP-ASIC's and FPGA's support multiple levels of logic. Multi-level minimization involves decomposing complex logic functions and finding common subfunctions [10], [12]. MP-ASIC's benefit from algebraic techniques, where logic functions are treated simply as algebraic formulas that can be decomposed by factoring. The MIS program from UC Berkeley uses this type of technique to minimize the number of logic literals that appear in the logic. These "literal count" metrics are well suited to mask programmed CMOS parts, but are less effective for FPGA's.

For example, LUT's have the same size and delay no matter what function they implement. The best fit for them is usually complex functions with a constant number of inputs (typically 4 or 5). Similarly, in Actel's logic modules, multiplexors are the dominant element. Optimization based on Binary Decision Diagrams (BDD's) or "If-Then-Else" graphs is appropriate.

Ideally, one would like to synthesize logic directly into the available hardware primitives. However, this is not yet generally done. Rather, most systems use a generic optimization package followed by a technology binding stage. Technology mapping is the process of binding logic to specific cells or structures on the target device. For an ASIC, this means selecting cells from a library. For LUT based FPGA's, this means collapsing AND/OR/INV gates into as few LUT's as possible [4]. For a multiplexor architecture it means graph matching, combined with hardware specific facilities such as bridging inputs or tying them to power and ground [3] [9] [6]. Two other FPGA specific problems are *packing* which is putting more than one LUT and/or flip-flop in each CLB, and *partitioning* which is dividing a large circuit across several FPGA's.

3.3 Integration and Performance

CAD systems strive to extract the highest possible performance from the available hardware. Unfortunately, this is difficult in logic synthesis for FPGA's because most of the delay is in the routing, and the routing cannot usually be characterized until after synthesis is complete. (One area that CAD can influence is state assignment. For MP-ASIC's, the best state encoding is usually a minimum-length encoding. For FPGA's with large register resources one-hot encoding is often better.)

At the present time, human intervention in logic assignment, placement and routing is sometimes needed to achieve the full potential of the silicon, especially in the SRAM parts. Future systems will probably have to integrate the physical design into the logic synthesis. This may mean that the physical place-and-route software will be integrated with the optimize-map-pack software, to acheive a total solution that can intelligently trade off gate-counts against routing delays.

4 Evaluation of New Architectures

Another area of CAD support for FPGA's is the evaluation of new FPGA architectures. This is of greatest interest to the FPGA's makers, but users may find the discussions interesting. Some of the work has concentrated on finding "optimal" values of architecture parameters such as the number of routing tracks [1] [11], or the type of lookup tables [7]. One limitation of such studies is that real FPGA architectures are usually very complex, and some information is lost if the study is based on a simplified model, such a routing fabric without "long-lines".

Other CAD research has considered the question of software structures to describe and manipulate FPGA primitives so that complex architectures can be accurately modeled [5]. But such studies tend to simplify the CAD algorithms involved, choosing more general but less effective techniques. And all paper studies lose accuracy from the absence of accurate electrical information, such as routing parasitics. In the end, it is difficult to predict the overall effectiveness of an architecture without actually laying it out and optimizing the CAD system around the measured values.

5 Closing Thoughts

At the present time, FPGA sales are increasing dramatically, as is the number of FPGA companies. In this environment each vendor tries to provide some distinguishing features that improve performance and reduce cost. But users may want a portable environment that allows them to migrate freely among vendors, and between FPGA's and mask programmed parts. This schism is partially resolved when the design can be processed automatically via synthesis from a common HDL, perhaps even using a common intermediate form such as the Library of Parameterized Macros (LPM). But other issues, such as investments in fuse-programmers, software, and engineer training, tend to lock users into one supplier. In many ways, the flurry of FPGA activity is similar to the melee that took place during the early days of microprocessors; its resolution will strongly influence the direction of microelectronics and of CAD in the 1990's.

References

[1] S. Brown, J. Rose, and Z. Vranesic. A detailed router for field-programmable gate arrays. Proc ICCAD, pages 382–385, Nov. 1990.

[2] A. ElGamal, J. Green, J. Reyneri, E. Rogoyski, K. El-Ayat, and A. Mohsen. An architecture for electrically configurable gate arrays. IEEE J. Solid State Circuits, 24(2):394–398, apr 1989.

[3] S. Ercolani and G. D. Micheli. Technology mapping for electrically programmable gate arrays. 28th ACM/IEEE Design Automation Conference, pages 234–239, 1991.

[4] R. Francis, J. Rose, and K. Chung. Chortle: A technology mapping program for lookup table-based field programmable gate arrays. Proc. Design Automation Conference, pages 613–691, 1990.

[5] D. Hill. A cad system for the design of field programmable gate arrays. Design Automation Conference, 1991.

[6] K. Karplus. Amap: a technology mapper for selector-based field-programmable gate arrays. 28th ACM/IEEE Design Automation Conference, pages 244–247, 1991.

[7] J. L. Kouloheris and A. ElGamal. Fpga area versus cell granulatity - lookup tables and pla cells. in the book FPGA's by Abindon Books, 1991.

[8] Lattice Corporation. pLSI and ispLSI Data Book and Handbook. Lattice, 1992.

[9] F. Mailhot and G. D. Micheli. Technology mapping with boolean matching. European Design Automation Conference, Glasgow, Scotland, pages 212–216, March 1990.

[10] R. Murgai, Y. Nishizaki, N. Shenoy, R. Brayton, and A. Sangiovanni-Vincentelli. Logic synthesis for programmable gate arrays. Proc. Design Automation Conference, pages 620–625, 1990.

[11] J. Rose and S. Brown. Flexibility of interconnection structures in field-programmable gate arrays. IEEE Journal of Solid State Circuits, 26(3):277–282, March 1990.

[12] G. Saucier, P. Sicard, and L. Bouchet. Multi-level synthesis on pals. EDAC, pages 542–546, 1990.

[13] Xilinx Corporation. Xilinx Programmable Gate Array User's Guide. Xilinx, 1988.

Net Partitions Yield Better Module Partitions

Jason Cong, Lars Hagen and Andrew Kahng

UCLA Department of Computer Science
Los Angeles, CA 90024-1596

Abstract

In this paper, we demonstrate that the "dual" intersection graph of the netlist strongly captures circuit properties relevant to partitioning. We apply this transformation within an existing testbed that uses an eigenvector computation to derive a linear ordering of nets, rather than modules [12]. We then find a good module partition with respect to the ratio cut metric [23] via a sequence of incremental independent-set computations in bipartite graphs derived from the net ordering. An efficient matching-based algorithm called IG-Match was tested on MCNC benchmark circuits as well as additional industry examples. Results are very encouraging: the algorithm yields an average of 28.8% improvement over the results of [23]. The intersection graph representation also yields speedups over, e.g., the method of [11], due to additional sparsity in the netlist representation.[1]

1 Preliminaries

A standard model for VLSI layout associates a graph $G = (V, E)$ with the circuit netlist; vertices in V represent modules and edges in E represent signal nets. The vertices and edges of G may be weighted to reflect module area and the multiplicity or importance of a wiring connection. Because nets often have more than two pins, the netlist is more generally represented by a *hypergraph* $H = (V, E')$, where hyperedges in E' are the subsets of V contained by each net.

Two basic partitioning formulations are:

- **Minimum-Width Bisection:** Given $G = (V, E)$, find the partition of V into disjoint U and W, with $|U| = |W|$, such that $e(U, W)$ i.e., the number of edges in $\{(u, w) \in E \mid u \in U$ and $w \in W\}$, is minimized.
- **Minimum Ratio Cut:** Given $G = (V, E)$, find the partition of V into disjoint U and W such that $\frac{e(U,W)}{|U|\cdot|W|}$ is minimized.

Minimum-width bisection has been a popular objective, particularly within hierarchical approaches. However, the area bisection requirement is unnecessarily restrictive, and various ad hoc thresholds and penalty functions (e.g., the r-bipartition formulation of Fiduccia and Mattheyses [8]) have had varying degrees of success in relaxing this constraint. With this in mind, the recent *ratio cut* metric of Wei and Cheng [23] has proved to be a highly successful objective function for many applications, ranging from layout to hardware simulation. The ratio cut metric intuitively allows freedom to find "natural" partitions: the numerator captures the minimum-cut criterion, while the denominator favors an even partition. Unfortunately, finding either a minimum-width bisection or a minimum ratio cut is NP-complete, so heuristic methods must be used. Previous approaches to min-width bisection, most of which can also be applied to the minimum ratio cut objective fall into several classes, as surveyed in [6] [12] [18].

In practice, the class of iterative methods are popular either as stand-alone strategies or as a postprocessing refinement to other methods. Iterative methods are based on local perturbation of a current solution and can be greedy or stochastic. Practical implementations will use a number of random starting configurations and return the best result [18] [23] in order to adequately search the solution space and give predictable performance, or "stability". For example, Wei and Cheng [23] use an adaptation of the shifting and group swapping methods in [8], and take the best result of 10 runs, in order to achieve the results cited in Section 4.

With respect to the present work, the most important class of partitioning algorithms consists of "spectral" methods which use eigenvalues and eigenvectors of matrices derived from the netlist graph. Recall that the circuit netlist may be represented by the undirected graph $G = (V, E)$ with $|V| = n$ vertices $v_1, \ldots, v_n$. Often, we use the $n \times n$ *adjacency matrix* $A = A(G)$, where A_{ij} is equal to the weight of (v_i, v_j) if $(v_i, v_j) \in E$ and $A_{ij} = 0$ otherwise. By convention $A_{ii} = 0$ for all $i = 1, \ldots, n$. If we let $d(v_i)$ denote the degree of node v_i (i.e., the sum of the weights of all edges incident to v_i), we obtain the $n \times n$ *diagonal degree matrix* D defined by $D_{ii} = d(v_i)$.

Early theoretical work connecting graph spectra and partitioning is due to Barnes, Donath and Hoffman [1] [6] [7]. Most recent eigenvector and eigenvalue methods have dealt with both module placement (Frankle and Karp [9], Kleinhans et al. [17] and Tsay and Kuh [21]) and graph min-cut bisection (Blanks

[1] Research supported in part by NSF grants MIP-9110511 and MIP-9110696.

[2] and Boppana [3]). In general, these previous works formulate the partitioning problem as the assignment or placement of nodes into bounded-size clusters, i.e., chip locations. The problem is then transformed into a quadratic optimization, and Lagrangian relaxation is used to derive an eigenvector formulation [13]. In [11], Hagen and Kahng established a close relationship between the optimal ratio cut cost and the second smallest eigenvalue of the matrix $Q = D - A$, where D and A are as defined above:

Theorem 1 (Hagen-Kahng): Given a netlist graph $G = (V, E)$ with adjacency matrix A, diagonal degree matrix D, and $|V| = n$, the second smallest eigenvalue λ of $Q = D - A$ yields a lower bound on the cost c of the optimal ratio cut partition, with $c \geq \frac{\lambda}{n}$. □

This result suggests that the eigenvector x corresponding to λ, i.e., the solution of the matrix equation $Qx = \lambda x$, be used to guide the partitioning. In [11], x was used to induce a linear ordering of the modules, and the best "split" in terms of ratio cut cost was returned. To be more specific, the n components x_i of the eigenvector were sorted, yielding an ordering $v = v_1, \ldots, v_n$ of the modules. The splitting index r, $1 \leq r \leq n-1$, was then found which gave the best ratio cut-cost when modules with index $> r$ were placed in U and modules with index $\leq r$ were placed in W.

The main contribution of the present work follows [12] in highlighting advantages to using the *dual* representation of the logic design, and confirming that *net* structure and interrelationships, rather than *module* adjacencies, should constitute the primary descriptors of a circuit. In particular, the dual *intersection graph* representation of the netlist hypergraph yields much more natural circuit partitioning formulations, since it inherently emphasizes relationships between signal nets. Moreover, the intersection graph yields a sparser circuit representation than traditional net models.[2]

When we use the intersection graph representation of the netlist, we may formulate the minimum-cost module partitioning as a two-stage process. In the first stage, we partition the *nets* of the design. Some modules will belong only to nets on one side of the partition; these modules can be unambiguously assigned to that side. However, other modules may belong to nets on both sides of the partition. Thus, the second stage of the module partitioning involves finding the best *completion* of the net partition, i.e., an assignment of each shared module to one side or the other such that the partition cost is minimized. We propose an efficient algorithm, called IG-Match, for completing the net partition; the algorithm is so named because it is based on a matching computation in a special bipartite graph. IG-Match affords a tight bound on the number of nets cut in completing any given partition; this bound is essentially best-possible. When we move from one net partition to another based on a shift in the splitting index of the sorted eigenvector, the corresponding change in the bipartite graph is very small. Therefore, an incremental strategy is possible, with the computational burden of examining all splits of the eigenvector effectively amortized. Empirical results are very encouraging. The IG-Match method yields significant improvements over the previous ratio-cut partitioning methods: results are an average of 28.8% better than those of Wei and Cheng [22] [23], and this also represents a 6% average improvement over the recent "voting" scheme of Hagen and Kahng [12].

The remainder of this paper is organized as follows. In Section 2, we discuss netlist representations and define the netlist intersection graph. In Section 3, we formulate the problem of completing the module partition from a given net partition as a maximum independent set instance in a bipartite graph, and then present the IG-Match algorithm along with analyses of its performance and complexity. Section 4 gives performance results on benchmarks from MCNC and industry sources; we also give comparisons with the RCut1.0 program of Wei and Cheng [23]. Finally, Section 5 gives conclusions and directions for future work.

2 The Netlist Intersection Graph

The transformation of the netlist hypergraph to a graph is accomplished via a *net model* which determines the A_{ij} values in the weighted graph representation of the design. The most common net model is that of a weighted clique, where a k-pin net will induce $C(k, 2)$ edges among its k modules. Recent work has widely adopted a "standard" weighted clique model [18], wherein a k-pin net contributes $\frac{1}{k-1}$ to each of $C(k, 2)$ A_{ij} values. Advantages of the clique model stem chiefly from its "fairness"; however, the model also has a number of disadvantages. For circuit netlists with large nets the clique model will induce dense adjacency matrices poorly suited for the numerical methods used to calculate the eigenvector.

If we consider the partitioning problem from a slightly different perspective, we realize that the minimum (ratio) cut metric is not only asking for an assignment of *modules* to the two sides of the partition, but is equivalently asking us to assign *nets* to the two sides of the partition, with the objective of maximizing the number of nets that are *not* cut by the partition. In other words, we want to assign the greatest possible number of nets *completely* to one side or the other of the partition. A central observation [12] is that such an objective can be captured using the graph dual of the netlist hypergraph, also known as the *intersection graph* of the hypergraph.

The dualization of the problem is as follows. Given a netlist hypergraph $H = (V, E')$ with $|V| = n$ and $|E'| = m$, we consider the graph $G' = (V', E_{G'})$ which has $|V'| = m$, i.e., G' has m vertices, one for each hy-

[2] For example, as reported in [12], the intersection graph of the MCNC Test05 benchmark has an adjacency matrix that is over *ten times* sparser than the adjacency matrix created using the standard clique model (19935 nonzeros versus 219811 nonzeros). This allows significant speedup of numerical computations: the eigenvector computation for Test05 using the intersection graph representation requires 48 CPU seconds on a Sun 4/60, versus 619 seconds with the standard clique model.

peredge of H (that is to say, each signal in the netlist). Two vertices of G' are adjacent if and only if the corresponding hyperedges in H have at least one module in common. G' is called the ***intersection graph*** of the hypergraph H. For any given H, the intersection graph G' is uniquely determined; however, there is no unique reverse construction. An example of the intersection graph is shown in Figure 1.

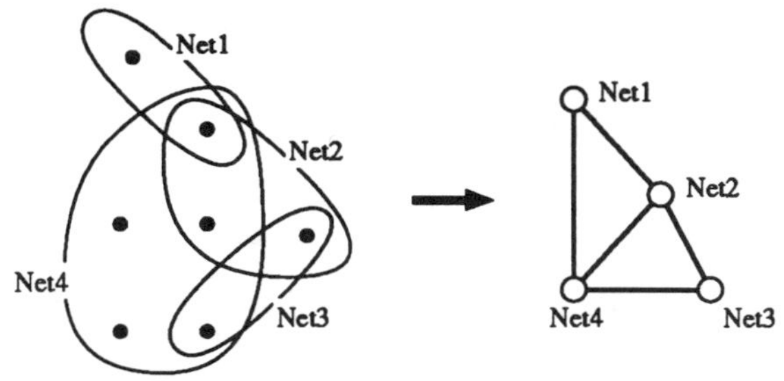

Figure 1: Left: the hypergraph for a netlist with four signal nets (each node represents a module). Right: the intersection graph of the hypergraph (each node represents a signal net).

Given this definition, the adjacency matrix A' of the intersection graph G' will have a nonzero element A'_{ab} when signal nets s_a and s_b share at least one module. As with the usual mapping of the netlist hypergraph to a graph via the weighted clique net model, there are a number of possible heuristic edge weighting methods for the intersection graph. We have tried several approaches, most of which lead to extremely similar, high-quality partitioning results; this indicates that the intersection graph is indeed a robust, natural representation. In the discussion below, we use the following weighting in the intersection graph construction.

For each pair of signal nets s_a and s_b with $q \geq 1$ nodes $v_1, \ldots, v_q$ in common, let $|s_a|$ and $|s_b|$ be the number of nodes in s_a and s_b respectively. The element A'_{ab} is then given by

$$A'_{ab} = \sum_{k=1}^{q} \frac{1}{(d_k - 1)} \left(\frac{1}{|s_a|} + \frac{1}{|s_b|}\right)$$

where d_k is again equal to the degree of the kth common node v_k, i.e., the number of nets incident to module k.[3] The diagonal degree matrix D' is constructed analogously to the matrix D described in Section 1 above, with the D'_{jj} entry equal to the sum of the entries in the jth row of A', i.e., $D'_{jj} = \sum_{i=1}^{m} A'_{ij}$.

Given A' and D', we then find the eigenvector x' corresponding to the second smallest eigenvalue λ' of $Q' = D' - A'$, using the same Lanczos code as in [11] [12]. As described above, the sorted eigenvector yields an ordering v' of the ***net*** indices, and we use this ordering to derive a heuristic ***module*** partition. Before presenting our new partitioning algorithm, we note that the intersection graph has had only limited previous application in the CAD literature. Pillage and Rohrer [20] applied the "nets-as-points metric" to module placement, the idea being that a heuristic 2-D placement of nets would establish preferred regions for each module – i.e., a module would wish to lie somewhere within the convex hull of the locations of nets to which it belonged. For partitioning, Kahng [16] used diameters of the intersection graph to yield an approximate hypergraph bisection heuristic; more recently, Yeh et al. [24] proposed to compute gains from a "net perspective" in an iterative multiway partitioning approach. Finally, Hagen and Kahng [12] have recently given a spectral approach which also involves the netlist intersection graph.

3 Module Partitions Based on Net Partitions

Recall that sorting the eigenvector of the second smallest eigenvalue of the netlist intersection graph yields a linear ordering v' for the signal nets of the original netlist. Nets at one end of the sorted eigenvector will usually have very weak connections to nets at the other end, so our basic strategy is to test all possible splitting points of the linear ordering, in order to see which splits might lead to a good module partition. Although such an approach seems computationally expensive, an efficient incremental method can be applied.

Consider what happens when we split the vertices of the intersection graph into two sets L and R. A net $l_i \in L$ might share one or more modules with a net $r_j \in R$. If we draw an edge $\{l_i, r_j\}$ between all pairs of signal nets (l_i, r_j) which are on opposite sides of the split and which have at least one module in common, we induce a bipartite graph $B(L, R, E_B)$ (see Figure 2). Note that according to such results as Theorem 1 [11], use of the eigenvector-based ordering suggests that $|E_B|$, i.e., the number of edges in $\{(l_i, r_j) | l_i \in L, r_j \in R\}$, will be small.

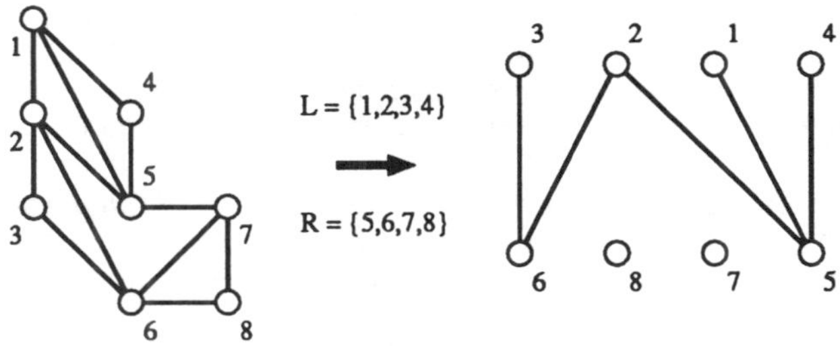

Figure 2: Inducing a bipartite graph from the intersection graph.

If a given edge $\{l_i, r_j\}$ is present in B, the key observation is that for any module partition, at most

[3] This net weighting scheme is designed so that overlaps between large nets are accorded somewhat lower significance than overlaps between small nets.

one of the following can be true: (i) l_i has all of its modules on the L-side of the partition, or (ii) r_j has all of its modules on the R-side of the partition. This follows immediately from the fact that l_i and r_j have some module in common. In [16], nets left uncut by the final module partition were called *winners*, and those cut were called *losers*. Adopting this terminology, we see that the min-cut objective is to maximize the number of winner nets, or equivalently, to minimize the number of loser nets.

In formalizing this optimization, the following graph-theoretic terms are useful.

Definition: Given a graph $G = (V, E)$, an *independent set* in G is a subset $V' \subseteq V$ such that no two nodes of V' are connected by an edge. A *maximum independent set* (MIS) is an independent set with largest possible cardinality.

Definition: Given a graph $G = (V, E)$, a *vertex cover* (VC) of G is a subset $V' \subseteq V$ such that for every edge $\{v_i, v_j\} \in E$, either $v_i \in V'$ or $v_j \in V'$. A *minimum vertex cover* (MVC) is a vertex cover with smallest possible cardinality.

Definition: Given a graph $G = (V, E)$, a *matching* in G is a set of k edges in E, no two of which have a vertex in common; we say that k is the *size* of the matching. A *maximum matching* (MM) is a matching with largest possible size.

Using these terms, the problem of maximizing the number of winners (minimizing the number of cut nets) is equivalent to finding a maximum independent set in B. While the MIS problem is NP-complete in general, it is efficiently solved for bipartite graphs. The following two standard results (see, e.g., Theorems 10.1 and 10.2 in [14]) motivate our algorithmic approach.

Theorem 2: For a bipartite graph $B = (L, R, E_B)$ with $|L| + |R| = n$, the sizes of any minimum vertex cover and any maximum independent set sum to n. Moreover, the complement of any MIS will be a MVC, i.e., $(L \cup R) - MIS = MVC$. □

Theorem 3: For a bipartite graph $B = (L, R, E_B)$, the size of a minimum vertex cover of B is equal to the size of a maximum matching in B. □

Given a minimum vertex cover, by Theorem 2 we may simply take its complement to obtain a maximum independent set, and vice versa. (Our plan will be to derive an MIS and make all of the corresponding signal nets winners.) Theorem 3 provides a lower bound for the size of the vertex cover (i.e., the set of loser nets).

Our high-level strategy is as follows. Given the intersection graph $G' = (V', E_{G'})$, we will split the eigenvector-based net ordering v' at some index r. Placing v'_i, $i \leq r$, in L, and v'_j, $j > r$, in R, induces a bipartite subgraph $B = (L, R, E_B)$ of G'. The algorithm will try all splitting indices $r = 1, \ldots, n-1$, where $n = |V|$.

For each bipartite subgraph B, Phase I of the IG-Match main loop (the reader is referred to Figure 4) finds a maximum independent set in B. We first find a maximum matching in B using the standard augmenting-path technique [19] and breadth-first search. The size of the MM gives the size of the MVC, which is the number of cut nets that we hope to attain. From the maximum matching, we construct a maximum independent set in B (i.e., the set of winner nets), as follows. Any unmatched vertex of B is a winner (Figure 3 shows the unmatched vertices on each side of the partition as U_L and U_R). Starting from each vertex in U_L or U_R, we trace alternating paths; none of these paths will be an augmenting path since the matching was maximum. We mark the second, fourth, etc. vertices in each of these alternating paths as losers; the (first,) third, fifth, etc. vertices in each path are marked as winners. We do this because the vertex cover (i.e., the set of losers) must contain at least one vertex from every edge, and in particular, it must contain at least one node from every edge in the set of alternating paths. (If we make nodes of the opposite parity into losers, i.e., the first, third, etc., the resulting vertex cover would not necessarily be minimum.) In Figure 3, winners on paths starting at vertices of U_L are denoted as the set $Even(L)$ since they are at even distance from U_L, and losers on these paths are denoted as $Odd(L)$; $Even(R)$ and $Odd(R)$ are similarly denoted. Note that $U_L \subseteq Even(L)$ and $U_R \subseteq Even(R)$.[4]

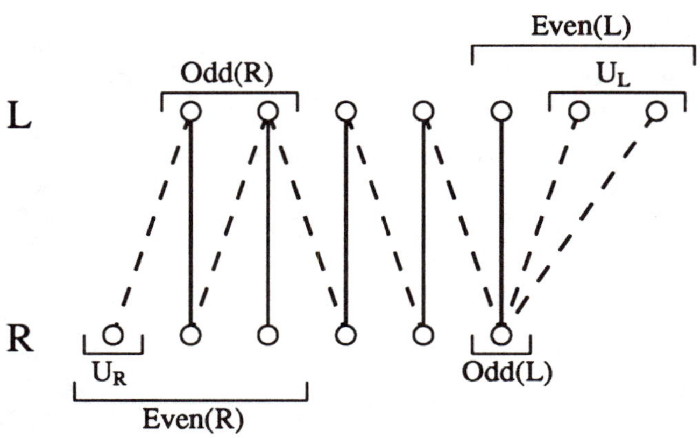

Figure 3: An example bipartite graph showing the set M of matching edges (solid lines), the edges not in the matching (dashed lines), and the sets U_L, U_R, $Even(L)$, $Even(R)$, $Odd(L)$, and $Odd(R)$.

It is possible that after Phase I has been executed, some edges in the matching M remain whose ver-

[4] It is worth noting that the set of loser nets computed during Phase I of the IG-Match main loop is the so-called *critical set* described by Hasan and Liu in [15]. The critical set $C = Odd(L) \cup Odd(R)$ is the unique subset of vertices in B such that every minimum vertex cover of B contains C. Note that even though we start with an arbitrary maximum matching M in B, a result of [15] shows that we will always end up with the same *Odd* and *Even* sets, independent of which maximum matching we use.

tices do not belong to any one of the sets $Even(L)$, $Even(R)$, $Odd(L)$, or $Odd(R)$. These remaining vertices and their induced bipartite subgraph are denoted as $B' = (L', R', E_{B'})$. Phase II of the IG-Match main loop will consider all ***modules*** which have not been assigned to a side of the partition via any of the winners determined in Phase I. Essentially, Phase II will put all of these unassigned modules first on one side, then the other, and determine which option yields the better ratio cut cost. Note that this will force all nets in L' to be winners and all nets in R' to be losers, or vice versa.

The IG-Match algorithm is optimal in that the number of nets cut by the module partition will never exceed the size of the maximum matching in B (see [5] for all theorem proofs).

Theorem 4: The set of loser nets output by Algorithm IG-Match is an MVC in B. □

In practice, the number of nets cut by the completed module partition can be less than the size of the MM. This is because a loser net v in $Odd(L)$ may in some instances only have modules in common with nets in $Even(R)$. When Phase II assigns all the modules of nets in $Even(R)$ to the R-side of the partition, the net v will end up with all its modules on the R-side and none of its modules on the L-side, i.e., net v will actually ***not*** be cut by the partition, even though it is a loser.

With each test of a split of the sorted eigenvector, we move one node from L to R. Since the change in the bipartite graph B is very small, we may retain information between the successive maximum matching computations, as well as between the successive MIS constructions. This allows an efficient implementation of the IG-Match algorithm which has small amortized complexity.

Theorem 6: Given the intersection graph $G' = (V', E_{G'})$ of the netlist hypergraph, The IG-Match algorithm requires $O(|V'| * (|V'| + |E_{G'}|))$ time to complete the module partition for each of the $|V'| - 1$ net partitions derived by splitting the sorted eigenvector v' of $Q'(G')$. □

4 Experimental Results

Table 1 shows computational results for the IG-Match algorithm on benchmark circuits from the MCNC layout test suite, as well as two additional industry circuits analyzed in [23]. The experiments were conducted using uniform module areas as in [12]; thus, the values L and R sum to the total number of modules in each benchmark circuit. For each benchmark, we compare our results with the best result over 10 runs of the RCut1.0 program, following the experimental procedure in [23].[5] Overall, our results are an average of 28.8% better than those of RCut1.0. IG-Match also uniformly dominates results from the precursor intersection graph based algorithm in [12], averaging almost 6% improvement.

The CPU times required by our numerical algorithms are very competitive with those cited in [23] for Fiduccia-Mattheyses optimization: for example, the eigenvector computation for PrimSC2 requires 83 seconds of CPU time on a Sun4/60, versus 204 seconds of CPU for 10 runs of RCut1.0.

[5] The results reported in [23] are already an average of 39% better than Fiduccia-Mattheyses output in terms of the ratio cut metric; in obtaining this assessment, the authors of [23] compared the best of 10 Rcut1.0 runs to the best of 20 F-M runs, all with random starting seeds.

```
IG-Match for B = (L, R, E_B).

L, R : sets of net-vertices in left, right partitions
E_G' : set of edges in the intersection graph G'
E_B : set of edges between L and R
M : set of edges in maximum matching between L and R
N : set of net-vertices to examine in breadth first search
W_L, W_R : sets of winner net-vertices in L, R resp.
V_L, V_R : sets of modules contained by nets in W_L, W_R resp.
V_N : set of modules not contained by nets in W_L or W_R

Main Loop Phase I: Selecting Winner Nets
  v ∈ L    /* v is the next net in the sorted IG eigenvector */
  L := L − {v}
  for all edges (v, y) ∈ E_B do
      E_B := E_B − {(v, y)}
  R := R ∪ {v}
  for all edges (x, v) ∈ E_G' do
      if x ∈ L then
          E_B := E_B ∪ {(x, v)}

  M := a maximum matching of B

  /* Construct a maximum independent set */
  W_L := set of unmatched net-vertices of L, i.e., U_L
  N := W_L
  while N ≠ ∅ do
      let x ∈ N
      N := N − {x}
      for all edges (x, y) ∈ E_B do
          if (x', y) ∈ M and x' ∉ W_L then
              W_L := W_L + {x'}
              N := N + {x'}
  endwhile /* W_L = Even(L) */
  W_R := set of unmatched net-vertices of R, i.e., U_R
  N := W_R
  while N ≠ ∅ do
      let y ∈ N
      N := N − {y}
      for all edges (x, y) ∈ E_B do
          if (x, y') ∈ M and y' ∉ W_R then
              W_L  = W_L + {y'}
              N := N + {y'}
  endwhile /* W_R = Even(R) */

Main Loop Phase II: Module Assignment
  V_L := ∅
  V_R := ∅
  for all nets x ∈ W_L do
      V_L := V_L ∪ { modules in net x}
  for all nets y ∈ W_R do
      V_R := V_R ∪ { modules in net y}
  V_N := V − (V_L ∪ V_R)
  calculate ratio-cut with partition (V_L ∪ V_N | V_R)
  calculate ratio-cut with partition (V_L | V_R ∪ V_N)
```

Figure 4: Algorithm IG-Match.

Test problem	Wei-Cheng RCut1.0 L:R:cutsize	IG-Match L:R:cutsize	Gain (%)
bm1	9:873:1	21:861:1	57
19ks	1011:1833:109	650:2194:85	-1
Prim1	152:681:14	154:679:14	1
Prim2	1132:1882:123	740:2274:77	21
Test02	372:1291:95	211:1452:38	37
Test03	147:1460:31	803:804:58	38
Test04	401:1114:51	73:1442:6	50
Test05	1294:1391:110	105:2490:8	53
Test06	145:1607:18	141:1611:17	3

Table 1: Output from IG-Match algorithm, compared with results from the RCut1.0 program of Wei and Cheng.

5 Conclusions

We have presented a new approach to module partitioning, based on a combination of spectral techniques and the intersection graph representation G' for the circuit hypergraph. Following [12], we use a sparse Lanczos code to induce a linear ordering of *nets* via the sorted second eigenvector of $Q'(G')$. We then formulate the optimal *completion* of the module partition as a maximum independent set computation in a bipartite graph. Our IG-Match algorithm is guaranteed to complete a module partition without cutting more nets than the size of a maximum matching in the bipartite graph; this bound is tight. Furthermore, the computation is efficient in an amortized sense when we test *all* possible partitions ("splits") of the sorted eigenvector to see which leads to the best module partition: IG-Match tests all splits in $O(|V'| * (|V'| + |E|)$ time. Since the computational complexity of the Lanczos implementation scales well with increasing problem sizes [10], we believe that this overall methodology will continue to be useful even when problem sizes grow very large.

A number of interesting open issues remain. The eigenvector computation can be sped up further by additionally sparsifying the input through thresholding, or by relaxation of the numerical convergence criteria. A hybrid algorithm which uses clustering to condense the input before applying the partitioning algorithm (such an approach is discussed by Bui et al. [4] and by Lengauer [18]) is also promising. Parallel speedups of the Lanczos code are also possible. With any of these heuristics, the ratio cuts so obtained may optionally be improved by using standard iterative techniques. Finally, following the successes reported by Wei and Cheng [22] [23], the intersection graph based ratio cut partitioning should be applied to ratio cut partitioning for other CAD applications, particularly test and the mapping of logic for hardware simulation.

References

[1] E. R. Barnes, "An Algorithm for Partitioning the Nodes of a Graph", *SIAM J. Alg. Disc. Meth.* 3(4) (1982), pp. 541-550.

[2] J. Blanks, "Partitioning by Probability Condensation", *Proc. ACM/IEEE Design Automation Conf.*, 1989, pp. 758-761.

[3] R.B. Boppana, "Eigenvalues and Graph Bisection: An Average-Case Analysis", *IEEE Symp. on Foundations of Computer Science*, 1987, pp. 280-285.

[4] T. N. Bui, S. Chaudhuri, F. T. Leighton and M. Sipser, "Graph Bisection Algorithms with Good Average Case Behavior", *Combinatorica* 7(2) (1987), pp. 171-191.

[5] J. Cong, L. Hagen and A. B. Kahng, "Net Partitions Yield Better Module Partitions", UCLA CS Dept. TR-910075, November 1991.

[6] W.E. Donath, "Logic Partitioning", in *Physical Design Automation of VLSI Systems*, B. Preas and M. Lorenzetti, eds., Benjamin/Cummings, 1988, pp. 65-86.

[7] W.E. Donath and A.J. Hoffman, "Lower Bounds for the Partitioning of Graphs", *IBM J. Res. Dev.* (1973), pp. 420-425.

[8] C.M Fiduccia and R.M. Mattheyses, "A Linear Time Heuristic for Improving Network Partitions", *ACM/IEEE Design Automation Conf.*, 1982, pp. 175-181.

[9] J. Frankle and R.M. Karp, "Circuit Placement and Cost Bounds by Eigenvector Decomposition", *IEEE Intl. Conf. on Computer-Aided Design*, 1986, pp. 414-417.

[10] G. Golub and C. Van Loan, *Matrix Computations*, Baltimore, Johns Hopkins University Press, 1983.

[11] L. Hagen and A. B. Kahng, "Fast Spectral Methods for Ratio Cut Partitioning and Clustering", *Proc. IEEE Intl. Conf. on Computer-Aided Design*, Santa Clara, 1991, pp. 10-13.

[12] L. Hagen and A. B. Kahng, "New Spectral Methods for Ratio Cut Partitioning and Clustering", to appear in *IEEE Transactions on CAD*; also available as UCLA CSD TR-910073.

[13] K. M. Hall, "An r-dimensional Quadratic Placement Algorithm", *Management Science* 17(1970), pp. 219-229.

[14] F. Harary, *Graph Theory*, Addison-Wesley, 1969.

[15] N. Hasan and C. L. Liu, "Minimum Fault Coverage in Reconfigurable Arrays", *Proc. 18th IEEE Intl. Symp. on Fault-Tolerant Computing Systems*, 1988, pp. 348-353.

[16] A. B. Kahng, "Fast Hypergraph Partition", *Proc. ACM/IEEE Design Automation Conf.*, 1989, pp. 762-766.

[17] J. M. Kleinhans, G. Sigl, F. M. Johannes and K. J. Antreich, "GORDIAN: VLSI Placement by Quadratic Programming and Slicing Optimization", *IEEE Trans. on CAD* 10(3) (1991), pp. 356-365.

[18] T. Lengauer, *Combinatorial Algorithms for Integrated Circuit Layout*, Wiley-Teubner, 1990.

[19] C. L. Liu, *Introduction to Combinatorial Mathematics*, McGraw-Hill, 1968.

[20] L. T. Pillage and R. A. Rohrer, "A Quadratic Metric with a Simple Solution Scheme for Initial Placement", *Proc. ACM/IEEE Design Automation Conf.*, 1988, pp. 324-329.

[21] R.S. Tsay and E.S. Kuh, "A Unified Approach to Partitioning and Placement", *Princeton Conf. on Inf. and Comp.*, 1986.

[22] Y. C. Wei, "Circuit Partitioning and Its Applications to VLSI Designs", Ph.D. Thesis, UCSD CSE Dept., September 1990.

[23] Y. C. Wei and C. K. Cheng, "Ratio Cut Partitioning for Hierarchical Designs", *IEEE Trans. on CAD*, July 1991, pp. 911-921.

[24] C. W. Yeh, C. K. Cheng and T. T. Lin, "A General Purpose Multiple Way Partitioning Algorithm", *Proc. ACM/IEEE Design Automation Conf.*, June 1991, pp. 421-426.

Performance-Driven System Partitioning on Multi-Chip Modules

Minshine Shih, Ernest S. Kuh
Department of EECS, UC Berkeley
Berkeley, CA 94720

Ren-Song Tsay
IBM T.J. Watson Research Center
Yorktown Heights, NY 10598

Abstract

We propose an efficient and effective algorithm for system partitioning under timing and capacity constraints. We consider the problem of assigning functional blocks into slots on Multi-Chip Modules during high level design in order to have fast feedback on impact of high level design decisions. We use a clustering step to ensure timing correctness, followed by packing and K&L algorithm to satisfy capacity constraints while minimizing net crossings. Our method is unique in that we minimize net crossings while satisfying timing and capacity constraints. Test results showed that our method eliminated timing violations and obtained comparable number of net crossings to that of K&L algorithm using similar CPU time. Our method can be extended to use partitioning algorithms other than K&L's.

1 Introduction

The goal of performance-driven system partitioning is to allow simultaneous high level functional (or logic) design and low level layout design of large digital systems (such as mainframe computers). For example, system partitioning and mapping onto MCM (Multi-Chip Module) has profound impact on the performance. As interconnection delay becomes more and more dominating in overall delay, the feasibility of system cycle time can only be determined with confidence after mapping onto physical packaging. An automatic partitioning/mapping tool can reduce the number of turn-arounds between high level design and physical design.

We consider a special type of MCM, namely, TCM (Thermal Conduction Module)[6]. We developed a new partitioning algorithm to assign functional blocks into TCM slots while meeting the cycle time constraint and slot (size) capacity constraint.

Kernighan and Lin proposed a well-known heuristic for improving graph partitioning[1]. Fiduccia and Mattheyses proposed an efficient algorithm to handle hypergraphs[2]. Recently Wei and Cheng proposed ratio cut approach in order to extract "natural clusters" in the original network[3]. They further apply Fiduccia-Mattheyses algorithm to the contracted network (from ratio cut) to obtain better and more stable results[4]. All these previous approaches address only the number of cuts (i.e. global nets crossing partitions) together with size constraints. The timing constraints are not taken into account. We propose a simple algorithm to consider cycle time constraints as well as size constraints while minimizing the total number of cuts.

2 System Modeling

2.1 High Level Design and Chip Carrier

In system design phase, designers only have a rough sketch of the system architecture, usually a set of main functional blocks and their interconnections. Most of these blocks lack of details or are not fixed at this time. Only *estimates* of delays, areas, power consumptions, and I/O pins usage are given. As the design evolves, there will be minor changes/updates of these estimates constantly.

The TCM carrier contains a regular two-dimensional (e.g., 12 × 12) array of chip slots. Let S_{TCM} represent the set of all slots. Each slot $s \in S_{TCM}$ has constraints on area capacity A_s, thermal capacity P_s, and maximum I/O pin count IO_s.

2.2 System Graph

A synchronous digital system consists of registers and blocks of combinational logic. For simplicity, all clock generation and distribution circuits are ignored. Then the system can be represented by a *System Graph*, which is a directed graph, $G(V, E) \equiv G(R \bigcup C, E)$, where R is the set of register nodes, and C is the set of combinational block nodes. E is the set of all *directed edges*, which correspond to signal flow in the system.

In this paper, we make a simplified assumption that each combinational block node has exactly one input edge and one output edge, and each edge connects to a register node. Each edge $e_{ij} \in E$ is associated with a parameter w_{ij}, which represents the number of wires. This reflects the fact that each edge may represent a bundle of physical wires (such as data buses). Thus the combinational block node can have multiple inputs and multiple outputs. We use rectangles to represent register nodes and circles to represent combinational block nodes (Figure 1).

Each node $v_i \in V$ has three attributes, i.e., circuit area a_i, power consumption p_i, and internal delay d_i. If v_i is a combinational block, then d_i represents the intrinsic delay from input to output of the combinational block plus any on-chip wiring delay. Note that each register node may represent an array of physical registers, and its delay is assumed to be zero.

2.3 Wiring Delay Models

There are two ways to model inter-chip TCM wiring delay.

1). Constant Delay Model:

0738-100X/92 $3.00 © 1992 IEEE

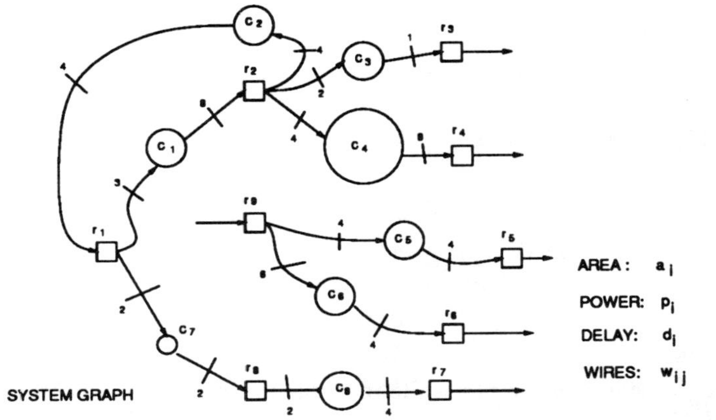

Figure 1:

$D_{TCM}(s_{source}, s_{sink}) = D_{constant}, \forall\ s_{source} \neq s_{sink}$
Due to the very different scale of chip v.s. TCM carrier, the delay between any two different TCM slots is *approximated* as a constant, independent of distance.

2). Transmission Line Delay Model:

$D_{TCM}(s_{source}, s_{sink}) \quad = \quad D_{unit\_length} \times length$
Where D_{unit_length} is the per unit length wiring delay which depends on the material (dielectric constant ϵ_r). *length* is the manhattan distance between s_{source} and s_{sink}.

3 Problem Formulation

Given a set of slots S_{TCM} on a TCM, and a design represented by a system graph $G(R \bigcup C, E)$, we want to find an optimum mapping $\mathcal{T} : R \bigcup C \longrightarrow S_{TCM}$ that minimizes the total inter-chip wire crossings while satisfying:

C1: (Timing Constraints)
$d_j + D_{TCM}(\mathcal{T}(r_i), \mathcal{T}(c_j)) + D_{TCM}(\mathcal{T}(c_j), \mathcal{T}(r_k)) \leq T_{cycle}$, $\forall c_j \in C$ and assuming $r_i, r_k \in R$; $c_j \in C$; $e_{ij}, e_{jk} \in E$ and T_{cycle} is the given cycle time.
C2: (Area Constraints)
$\sum_{\forall i, \mathcal{T}(v_i)=s} a_i \leq A_s, \quad \forall s \in S_{TCM}$
C3: (Thermal Constraints)
$\sum_{\forall i, \mathcal{T}(v_i)=s} p_i \leq P_s, \quad \forall s \in S_{TCM}$
C4: (I/O Pin Count Constraints)
$\sum_{\forall i,j, \mathcal{T}(v_i)=s, \mathcal{T}(v_j)\neq s} w_{ij} \leq IO_s, \quad \forall s \in S_{TCM}$

The objective function is the number of total inter-chip wire crossings

$$W = \sum_{\forall i,j,\ \mathcal{T}(v_i)\neq \mathcal{T}(v_j)} w_{ij},$$

and the problem is to

$$minimize\ W, \quad subject\ to\ C1,\ C2,\ C3,\ and\ C4.$$

Since the thermal and I/O constraints can be treated in a similar fashion as area constraints, we shall focus on timing and area constraints for simplicity. Thus in the following sections, we actually consider the following simplified problem

$$minimize\ W, \quad subject\ to\ C1\ and\ C2.$$

One possible way of minimizing cycle time is to probe different cycle time by using our algorithm in a fashion similar to binary search in order to achieve the minimum possible cycle time.

4 Basic Operations

4.1 K-Way Packing

Let $S = \{s_1, \ldots, s_k, \ldots, s_K\}$ be a set of K slots and each slot s_k has an area capacity A_k. We are given a set of N objects $O = \{o_1, \ldots, o_n, \ldots, o_N\}$, each object o_n has size a_n. The problem is to find a mapping $\mathcal{M} : O \longrightarrow S$ such that $\sum_{\forall n, \mathcal{M}(o_n)=s_k} a_n \leq A_k, \forall s_k \in S$.

Our simple *first-fit* heuristic is to first sort N objects in non-increasing order according to their sizes a_n. The second step is to assign sequentially (in that order) each object o_n to the *first* slot s_k such that $a_n + \sum_{\forall i, \mathcal{M}(o_i)=s_k} o_i \leq A_k$.

4.2 K-Way K&L Under Area Constraints

Here we only describe our modifications to the basic K&L algorithm (using Fiduccia & Mattheyses' data structure).

Given a set of K slots S and a set of N objects O as specified in the previous section. Suppose we have an initial assignment $\mathcal{M} : O \longrightarrow S$. Now we can construct a *gain table* according to this initial assignment. In the gain table each object o_n has $K - 1$ entries. Each entry $g_n(k)$ represents the gain (reduction of wire crossings) of object o_n being moved to slot s_k. Starting from the entry $g_n(k)$ with the highest gain, an object o_n is moved to slot s_k if $a_{o_n} + \sum_{\forall i, \mathcal{M}(o_i)=s_k} a_i \leq A_k$. This is equivalent to testing if there is enough room in slot s_k to accommodate object o_n. If the inequality does not hold, the move is skipped and the next highest gain entry is tried.

5 Main Algorithm

5.1 Overview and Strategy

We propose a Divide-and-Conquer strategy, i.e.,
1). we first ensure that the timing constraints are satisfied by merging nodes in the system graph into Super-Nodes,
2). pack the Super-Nodes into TCM slots to satisfy area constraints while maintaining correct timing,
3). apply K&L algorithm iteratively on Super-Nodes to reduce wire crossing while maintaining correct timing and satisfying area constraints.

5.2 The Algorithm

A *Super-Node Graph* is constructed from a system graph $G(V, E) = G(R \bigcup C, E)$ by the following steps:

Algorithm: Super-Node Graph Construction
Label each $v_i \in V$ as a Super-Node
for each $c_j \in C$\{
(assuming $v_i, v_k \in V$ are super-nodes connected to the input and output of $c_j \in C$, respectively)
    if ($d_j > T_{cycle}$) \{
        no solution exit;
    \} else if ($d_j + D_{constant} > T_{cycle}$) \{
        merge c_j, v_i and v_k into a super-node v_l;

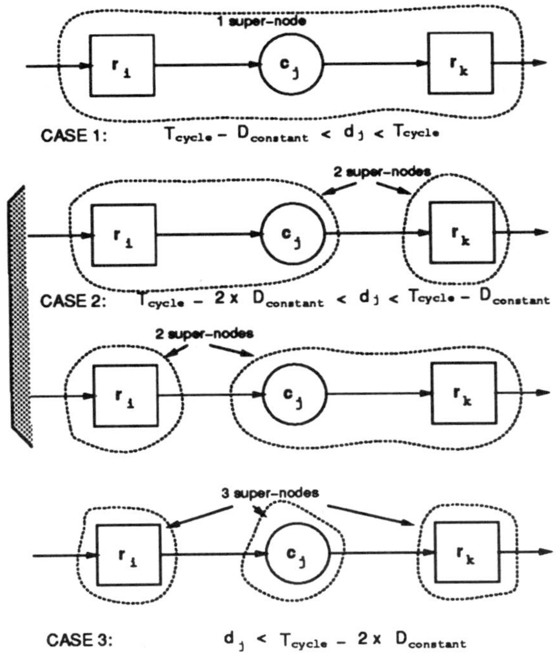

Figure 2:

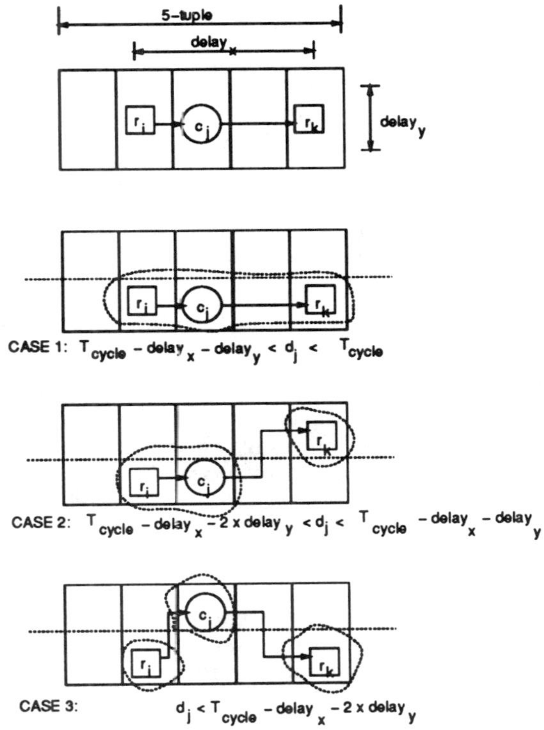

Figure 3:

```
            update edges to connect to super-node v_l;
            a_l = a_i + a_j + a_k;
      } else if (d_j + 2 × D_constant > T_cycle) {
            randomly merge c_j with (either) v_i (or
            v_k) into super-node v_l;
            update edges to connect to super-node v_l;
            a_l = a_i + a_j;
            record c_j as "semi-free";
      } else {
            do nothing;
      }
}
```

Figure 2 shows all the cases of super-node merging. Once Super-Node Graph is constructed we use K-Way Packing algorithm mentioned earlier to obtain slot assignments for Super-Nodes. Each Super-Node corresponds to an object in the packing algorithm.

Then we apply K-Way K&L algorithm mentioned earlier to move super-nodes around among different TCM slots in order to minimize wire crossings. One deviation from standard K&L approach is that we also allow the extra freedom of migrating combinational nodes (already merged into super-nodes) which had been labeled "semi-free" during Super-Node graph construction. These "semi-free" blocks provide more flexibility to satisfy area constraints and thus enable some K&L moves which are not possible when there are not enough room in the destination slot to accommodate the incoming Super-Node.

A key observation is that after the Super-Node merging loop in the algorithm, any arbitrary slot assignment on the super-nodes will not introduce timing constraint violation, as long as all original functional blocks merged into the same Super-Node are assigned to the same slot. Notice that this is a necessary and sufficient condition for satisfying timing constraint.

5.3 Extension

Our algorithm can be extended to handle Transmission Line Delay Model. We only give the main ideas here (see [7] for details). The algorithm proceeds in a top-down bi-partitioning style by recursively dividing a rectangular area into two subareas with a cutline. Similar to Min-Cut algorithm, whenever a cutline is applied, all Super-Nodes in the rectangle are assigned to either side of the cutline and are not allowed to cross in later steps. For example, in figure 3, the rectangle was devided into 5 regions by previous vertical cutlines and the current horizontal cutline (dotted) divides the whole rectangle into top and bottom subareas, each subarea contains 5 regions. Now our task is to assign Super-Nodes into either top or bottom subarea. Here the notion of Super-Node is generalized to be a 5-tuple (corresponding to 5 regions), it may contain some register or combinational block nodes in any of the 5 tuples due to merging. The top and bottom subareas correspond to 2 slots, each slot is a 5-tuple. The assignment of a Super-Node to either slot reduces the 5-tuple slot capacity in each tuple respectively. The Super-Nodes are merged according to $delay_x$ (x-direction delay) and $delay_y$ (y-direction delay), which are proportional to distances. The Packing and K&L algorithms are modified to handle 5-tuple objects and are very similar to what described earlier.

6 Implementation and Test Results

Our first implementation is using Constant Delay Model. The Super-Node merging is implemented using UNION-FIND algorithm with path compression[5]. It is known to have effectively linear run time for all practical cases. Since each Super-Node has $K-1$ gain entries it requires more sophisticated gain updating method to obtain efficiency.

All 6 test cases are part of real designs from industry. We compare test results among three approaches. The first one is random partitioning, the best results from 1000 runs are chosen and listed. Notice that for all 6 cases the minimum number of net crossings and minimum number of timing violations are not obtained from the same partitioning.

The second approach is the basic K-Way K&L. Since timing constraints are completely ignored, the number of timing violations is not predictable. In one circuit (sys4 in 2-slot and 4-slot) the number of timing violations were reduced to 0 by K-way K&L algorithm. However for all other cases the timing violation cannot be eliminated by K-Way K&L algorithm.

The third approach is our method. Because of our Super-Node merging step, there is no timing violation in any case. But this step produces larger Super-Nodes and they have less freedom to move (due to slot capacity constraints) during K&L optimization latter. Therefore when the number of slots increases (and consequently slot capacity decreases) our method produces more wire crossings than the basic K-Way K&L algorithm. Comparing results from basic K&L algorithm for 2, 4 and 8 slots, we find that CPU run time (in seconds on DEC 5000) grows sub-linearly as the number of slots increases, although the number of gain-entries increases linearly with the number of slots. This implies that run time is dominated by updating gain entries during a move. The amount of computations for gain updating is directly related to the density of the interconnections of the graph. This helps explaining why our method is sometimes slower than basic K-Way K&L algorithm.

Acknowledgements

The authors wish to thank K.S. Huang at IBM Research for valuable discussions and suggestions. This research was sponsored by NSF Grant No. MIP 88-03711 and SRC grant No. 91-DC-008.

References

[1] B.W. Kernighan and S. Lin, "An Efficient Heuristic Procedure for Partitioning Graphs," *The Bell System Technical Journal*, 49(2). Feb. 1970. pp. 291-307

[2] C.M. Fiduccia and R.M. Mattheyses, "An Linear Time Heuristic for Improving Network Partitions," *Proc. 19th Design Automation Conference*, 1982, pp. 175-181

[3] Y. Wei and C. Cheng, "Toward Efficient Hierarchical Designs by Ratio Cut Partitioning," *Proc. IEEE Int. Conf. on Computer-Aided Design*, 1989, pp. 298-301

[4] Y. Wei and C. Cheng, "A Two-Level Two-Way Partitioning Algorithm," *Proc. IEEE Int. Conf. on Computer-Aided Design*, 1990, pp. 516-519

[5] R. Tarjan, "Data Structures and Network Algorithms," Society for Industrial and Applied Mathematics, 1983

[6] R.R. Tummala and E.J. Rymaszewski, "Microelectronics Packaging Handbook, Chap. 16" Van Nostrand Reinhold, 1989

[7] M. Shih, E.S. Kuh and R. Tsay, "Performance-Driven System Partitioning On Multi-Chip Modules," *IBM Research Report RC 17315(#76556)*, Oct. 1991

			Best Random			Kernighan & Lin			Our Algorithm		
name	# of Reg.	# of Comb.	# of net cross.	# of Viol.	run time	# of net cross.	# of Viol.	run time	# of net cross.(+-%)	# of Viol.	run time
System Partitioning onto 2 Slots											
sys1	62	518	472	18	2.0	276	8	0.2	251 (-9.0%)	0	0.2
sys2	101	1862	1782	56	8.5	933	9	1.2	916 (-1.8%)	0	1.7
sys3	457	2630	2533	116	13.3	1126	27	2.9	1067 (-5.2%)	0	3.5
sys4	467	5814	5675	141	30.1	2368	0	16.0	2332 (-1.5%)	0	22.2
sys5	472	12811	12590	455	67.0	5683	65	92.7	5552 (-2.3%)	0	129.3
sys6	782	17567	17327	651	93.4	5655	60	227.0	6633 (+17.3%)	0	159.8
System Partitioning onto 4 Slots											
sys1	62	518	739	44	2.0	398	10	0.3	395 (-0.7%)	0	0.4
sys2	101	1862	2726	122	8.6	1337	16	2.1	1344 (+0.5%)	0	3.4
sys3	457	2630	3845	231	13.5	1468	41	6.4	1645 (+12.0%)	0	7.6
sys4	467	5814	8580	334	30.4	3547	0	25.1	3268 (-7.8%)	0	27.6
sys5	472	12811	19055	944	67.9	8513	97	198.0	8747 (+2.7%)	0	155.8
sys6	782	17567	26090	1367	94.7	8699	104	239.0	9786 (+12.4%)	0	279.0
System Partitioning onto 8 Slots											
sys1	62	518	882	61	2.0	450	10	0.8	477 (+6.0%)	0	0.4
sys2	101	1862	3213	172	8.6	1569	15	3.6	1627 (+3.7%)	0	5.3
sys3	457	2630	4543	308	13.6	1921	46	11.6	1987 (+3.4%)	0	12.2
sys4	467	5814	10071	466	30.6	4418	2	43.7	4301 (-2.7%)	0	43.8
sys5	472	12811	22311	1306	68.4	9957	117	135.8	10258 (+3.0%)	0	290.2
sys6	782	17567	30566	1835	94.8	11516	127	356.9	11558 (+0.4%)	0	613.0

A Wire Length Estimation Technique Utilizing Neighborhood Density Equations

Takeo Hamada, Chung-Kuan Cheng, and Paul M. Chau

University of California at San Diego, La Jolla, CA 92093-0407

Abstract

A new wire length estimation technique is presented. Wire length distribution is modeled by wire density on a 2-dimensional lattice. Assuming a pointwise independent branching process, the wire length distribution is found by solving the neighborhood density equations. For several industrial circuits tested, this technique gives an average 9.0% of estimation accuracy, which compares favorably with other techniques recently proposed.

1 Introduction

To obtain the best placement results for current very complex circuits, it may take several days of computer time even on the fastest workstation available. Therefore it is quite profitable to invest a small percentage of CPU time to perform wire length estimation. An accurate wire length estimation is critically important in deriving proper timing models for performance driven designs.

When the design style is uniform in spatial distribution, as in row-based design, cell positions can be approximated by 2-dimensional lattice points. Since we are only interested in wire length calculations, we can simplify computations by assuming cell positions that are probabilistically distributed over the lattice rather than a deterministic one. This leads to the approximation of wire length distribution by a random field [5] over a 2-dimensional lattice.

The local topological structure of the network is often reflected in the local structure of the placement. Two components which are topologically close are likely to be physically close in the optimized placement. We present a technique of topological analysis of the network, in which the local structure of the network is characterized by a growing sequence of multi-level neighborhoods. By assuming a pointwise independent branching process, the neighborhood density equations are derived, which describes the distribution of multi-level neighborhoods. By minimizing the wire length subject to the neighborhood constraints, the wire length distribution of an optimized placement can be found.

This paper is organized as follows. We discuss previously proposed wire length estimation techniques in the following section. The wire length distribution model of a multi-pin net is discussed in section 3, and a technique for multi-level neighborhood analysis is presented in section 4. The neighborhood density equations and a proposed wire length estimation technique is presented in section 5, followed by experimental results in section 6.

2 Previous Works

2.1 Estimation Using Rent's Rule

In the domain of row-based design style, e.g. standard cells and gate arrays, various wire length estimation techniques have been proposed. Donath [1] and Feuer [2] showed that the average interconnection wire length can be calculated based upon Rent's constant. Although the results of these studies are of theoretical importance, they are often difficult to apply in practice, since estimation is sensitive to Rent's constant, which varies in different test cases.

2.2 Local Neighborhood Analysis

Sechen [9] presented an interconnection estimation technique. He proposed a *checker-board* model which assumes that square cells are placed on a two-dimensional square grid. In this model, the pins on a net are randomly placed among the set of cells incident to the neighboring nets. The set of distinct cells C_1 which comprise all the neighbors of net n_0 are called the *neighborhood* of n_0. Pins of n_0 are assumed to be distributed randomly on the square lattice of $\sqrt{|C_1|+1}$ grids in width and height. He enumerated states in the checker-board model, thereby establishing a size for a bounding box, to calculate the average wire length of the optimized placement.

Pedram and Preas [6,7] presented the *net neighborhood population* technique. Following the same assumption as Sechen, they assumed that pins are randomly distributed on a rectangular grid which has the size of the neighborhood population of the net. The two methods have the following common features:

(i) The average wire length is estimated from a local analysis of the neighborhood.

(ii) Both methods extract the interconnection structure directly from the circuit.

(iii) Rent's rule is not employed.

This local neighborhood analysis technique was successful in predicting the average wire length for small to medium sized circuits. In calculation of neighborhood population, to remove the impact of global nets

0738-100X/92 $3.00 © 1992 IEEE

i.e. clock and ground, Pedram and Preas [6] used a cut-off criterion, which ignores nets that connect more than 25% of the cells in the circuit. They [7] report that the estimation accuracy of this technique is within 10% of the actual wire length, for the test cases with a size ranging from 95 to 2907 cells.

3 Wire Length Distribution Model

Several probabilistic models have been proposed to account for the wire length distribution of the optimized placement. Sastry and Parker [8] found that the actual wire length distribution fits a Weibull distribution well.

A simple test to characterize probability distribution is to take the ratio between the average μ and the standard deviation σ. The wire length distribution, especially that of small pin count nets is characterized by large secondary moments. This fact is reflected by the small μ/σ ratio for small pin count nets. From the measurement on the wire length distribution, we observe the following.

(i) The wire length distribution can be identified as a Weibull distribution.

(ii) The average wire length of multi-pin net follows a simple relationship.

(iii) The μ/σ ratio for each pin count is relatively a constant across different circuits, suggesting that they follow the same wire length distribution with different scale parameters.

The Weibull distribution is a 2-parameter probability distribution defined by

$$f(a,b;r) = g \cdot r^{b-1} exp(-ar^b). \quad (1)$$

The following two subsections discuss the details of (ii) and (iii). Once the relationships are established, the Weibull distribution can be expanded to nets with various pin counts.

3.1 Average Wire Length of Multi-Pin Net

When the design style is uniform in space, as is the case with row-based design, cell positions can be approximated by 2-dimensional lattice points. The observed wire length distribution is a result of placement optimization. Suppose we sample a cell, and fix it on a lattice point. We assume that the objective of placement optimization is minimum wire length. The cells neighboring the sampled cell in the netlist can be seen competing against each other to obtain lattice points closer to the sampled cell. On the other hand, each neighboring cell of the sampled cell is perturbed by its own neighboring cells. We assume that these two opposite forces reach an equilibrium at the optimized placement. In other words, at equilibrium, a small perturbation to the positions of neighboring cells does not change the wire length significantly. When this condition stands, we can show the existence of a global parameter, which is equivalent to the temperature in thermodynamics.

Table I shows average wire length of multi-pin net. In measuring the wire length, multi-pin net is expanded into a clique, and rectilinear distance of each pin pair on the clique is measured. Each wire length is normalized by its corresponding 2-pin net wire length.

We observe in the table that the smaller the pin count the smaller the average wire length. Intuitively, the small pin count net has better chance to minimize its wire length than the large pin count net. We can formally state this fact by using the concept of potential as follows; let E be the objective function to be minimized. Suppose we choose a sample point and one of its neighboring cells, say cell A. Imagine that a small perturbation is given to the position of A. The potential of A w.r.t. the sampled cell at lattice point (x, y), $e_A(x,y)$, is defined by

$$\Delta E = \frac{\partial e_A(x,y)}{\partial x} \cdot \delta x + \frac{\partial e_A(x,y)}{\partial y} \cdot \delta y$$

where δx and δy are perturbation and ΔE is the measured difference in the objective function. When half perimeter wire length is used as the objective function, we show [4] that the potential of m-pin net is given by

$$e_A(x,y) = \frac{3m}{m^2 + 2m - 2} \cdot (|x| + |y|). \quad (2)$$

In terms of potential, the small pin count net has larger potential than the large pin count net. Through similar arguments as in statistical mechanics, we can conclude that μ_m, the average wire length of m-pin net, is given by

$$\mu_m = \frac{2(m^2 + 2m - 2)}{3m} \cdot \frac{1}{\beta}$$

where β is the equilibrium parameter. Table I shows average wire length predicted by the above formula under the column $\mu_{pred.}$. It shows good correspondence with experimental data, especially with the largest test case 26k.

3.2 The μ/σ Ratio Relationship

On the μ/σ relationship, we made the following observations;

(i) μ/σ ratio monotonically decreases w.r.t. the number of pin count. This is due to the fact that the smaller the pin count the larger its secondary moment of wire length distribution.

(ii)μ/σ ratio converges to $\sqrt{2}$ as pin count becomes large.

We use clique expansion to translate each multi-pin net into equivalent 2-pin net pairs. Based on these observations, we derived the following empirical formula from the 7 test cases in Table II. We used minimum squared error criteria to identify the parameters.

$$\mu_m/\sigma_m = 1.41 \cdot (1 - 0.5 \cdot exp(-0.261 \cdot (m-2)))$$

where m is pin count, μ_m and σ_m are its average and standard deviation, respectively. As a remark, note that the expression of potential depends on the objective function.

4 Analysis of Multi-Level Neighborhood

We decompose the circuit into a growing sequence of multi-level neighborhoods which is characterized by *local dimension*. We utilize a *concentric presentation* to describe the relationship between each multi-pin net and its equivalent 2-pin net pairs.

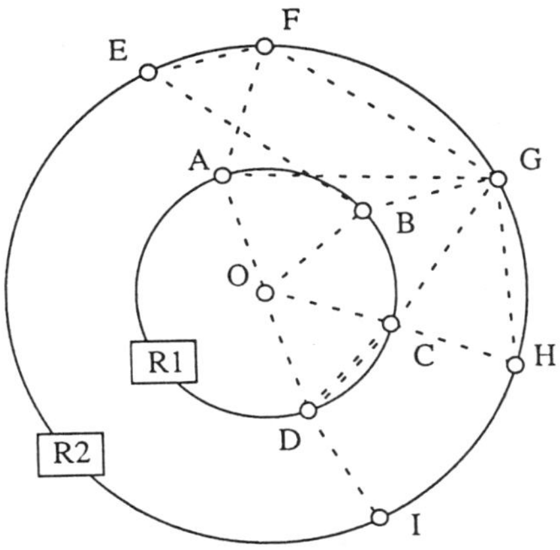

Fig. 1. Concentric Presentation of Network

4.1 Local Dimension

The local structure of a network can be characterized by a growing sequence of multi-level neighborhoods. A cell is called a neighbor of another if there is a net connecting them. We define the topological distance between any two arbitrary cells by the number of nets on the shortest path between the cells. We define the d-ball of O, $\varepsilon_d(O)$, as a distinct set of cells whose topological distance from O is less than or equal to d. The *local dimension* referenced to cell O, is a sequence $\{\nu_i(O)\}$ defined by the following for each i,

$$\nu_i(O) = |\varepsilon_{i+1}(O) - \varepsilon_i(O)|.$$

The local dimension reflects the local structure of placement around O. We calculate the local dimension of a circuit by taking the average of local dimensions at its cells, i.e.

$$\{\overline{\nu_i}\} = \frac{1}{n}\sum_{j=1}^{n}\{\nu_i(C_j)\}$$

where n is the number of the cells in the circuit.

4.2 Clique Expansion on Concentric Presentation

By applying clique expansion, we translate each multi-pin net into equivalent 2-pin pairs. Then we can look at the distribution of wires on the lattice as a random field, where wires are independently branching from each lattice point. The local dimension of a circuit is understood as a result of independent branching process from one generation to the next of multi-level neighborhoods. Fig. 1 illustrates a result of clique expansion and its concentric presentation around a cell O, up to the second level neighbors.

We categorize the multi-level neighbors of O by its topological distance. On the ring $R1$ are the first level neighbors of O, and on the ring $R2$ are the second level neighbors. We view that the edges are independently branching out of $R1$ to form $R2$. To pursue this point of view further, we examine the dependency among branching edges. We classify edges in Fig. 1 into the following four categories:

(i) Δ^{01}: edges branching from O to $R1$.
(ii) Δ^{11}: edges branching from $R1$ to $R1$.
(iii) Δ^{12}: edges branching from $R1$ to $R2$.
(iv) Δ^{22}: edges branching from $R2$ to $R2$.

We distinguish pin count in each category by subscript. For example, Δ^{01}_m is a set of m-pin net edges in the category of Δ^{01}. Based on this classification, we define the ***edge branching factor*** at each level of neighborhood as follows; We assume that edges branching from O to $R1$ are independent. Thus the total branching factor of m-pin net edge from O to $R1$, κ^0_m, is given by

$$\kappa^0_m = |\Delta^{01}_m|.$$

Multi-pin net edges in Δ^{11} and Δ^{12} may fall upon the same cell, thus their distribution may not be independent. We assign weights to these dependent edges so that they can be treated as if they were independently branching edges. Fig. 1 shows that three edges branching from $R1$, $[AG]$, $[BG]$, and $[CG]$, fall upon the same cell G on $R2$. We assign a weight of 1/3 to each of the three edges. In general, let e be an edge, and let $w(e)$ be its weight. We suppose that e is connecting node x on R_i and node y on R_j. We define the head of e, h(e), and the tail of e, $t(e)$, as follows; $h(e) = x$ and $t(e) = y$ if $i \le j$, or $h(e) = y$ and $t(e) = x$ otherwise. The weight $w(e)$ of an edge e in Δ^{ij} is given by

$$w(e) = \frac{1}{|\{e'|h(e') \in Ri \text{ and } t(e') = t(e) \in Rj\}|}.$$

Using this edge weight, the contributions to the m-pin net edge branching factor at $R1$, λ^{11}_m and λ^{12}_m, are given by

$$\lambda^{ij}_m = \frac{1}{|R1|}\sum_{e \in \Delta^{ij}_m} w(e).$$

We assume that the total branching factor of m-pin net edge at $R1$, κ^1_m, is approximated by a linear combination of the two,

$$\kappa^1_m = \alpha \cdot \lambda^{11}_m + \lambda^{12}_m$$

where α is a constant of $0 < \alpha < 1$. By calculating neighborhood density, we found empirically that $\alpha = 0.6$ gives a good approximation. The branching factors of a given circuit is the average of ones calculated at its cells.

5 The Neighborhood Density Equations

Given a sample point O on a 2-dimensional lattice, we derive equations to describe the relationship between local dimension and wire length distribution. Without loss of generality, we assume O is at the origin of the lattice. We define the wire density $w_m(x, y)$ of an m-pin net edge as the probability that an m-pin net edge exist between O and (x, y). In practice, we assume that $w_m(x, y)$ follows the distribution of (1), where $r = |x| + |y|$.

Let $\varphi^k(x, y)$ be the probability that location (x, y) belongs to the k-th neighbor of O. The condition that location (x, y) belongs to the 1st neighbor of O is that at least one edge branching from O falls upon the (x, y) location. Therefore, $\varphi^1(x, y)$ is given by

$$\varphi^1(x, y) = 1 - \prod_m (1 - w_m(x, y))^{\kappa_m^0}.$$

We show [4] that the 2nd neighborhood density, $\varphi^2(x, y)$ is given by

$$\varphi^2(x, y) \approx (1 - \varphi^1(x, y)) \cdot [1 - \prod_{(s,t) \neq (x,y)} (1 - \varphi^1(s, t) \cdot (\sum_m \kappa_m^1 \cdot w_m(x - s, y - t)))].$$

where κ_m^1 is the total branching factor of m-pin net edge at $R1$.

We calculate the amount of the increase in the objective function caused by the neighborhood of O, which is naturally represented as the energy of the system around O. When placement is optimized, the energy E is minimized. E is defined by

$$E = \sum_{x,y} \sum_m \kappa_m^0 \cdot w_m(x, y) \cdot e_m(x, y).$$

where $e_m(x, y)$ is the potential of an m-pin edge branching from O to (x, y), and is given by (2). We assume that the optimized placement attains the minimum energy, whereas it satisfies topological constraints imposed by local dimension. Therefore we find the wire density $w_m(x, y)$ of the optimized placement by solving the following constrained optimization problem:

$$min\ E$$
$$\sum_{x,y} \varphi^i(x, y) \geq \overline{\nu_i}. \quad (i = 1, 2, ...k) \tag{3}$$

which is called the ***neighborhood density equations***. In practice, we solve the case $k = 2$. Note that the equilibrium parameter β is the only unknown in the equations. We show [4] that the energy E in the equation (3) is inversely related to β. Using this fact, we solve (3) by Algorithm 1. This algorithm searches the largest feasible β, which satisfies constraints of (3), using the bi-section search. When $\beta_{opt.}$ is found, we have the wire density $w_m(x, y)$ of the optimized placement. The average wire length of m-pin net is known from $w_m(x, y)$, from which the half perimeter wire length is calculated.

In applying this method, we considered up to 12-pin nets. Adding very large pin count nets (>30) to the equation (3) tends to make it unstable. The vertical grid size is set to the average row interval, and the horizontal grid size is set to the average cell width. When the gate utilization is not 100%, the horizontal grid size needs to be adjusted proportional to the gate utilization factor. A lattice of up to 100×100 grids are used to calculate wire density $w_m(x, y)$. Considering that $w_m(x, y)$ is symmetric, we can save the grid size up to 75%. Since this technique uses only the local information of the circuit, its convergence behavior is relatively independent of the circuit size. It takes only a few minutes including multi-level neighborhood analysis, for small to medium test cases on Sun 4/280. For the largest test case 26k, it takes less than 25 min. in total.

Algorithm 1

```
Search_Beta() {
    /* find initial value of β. */
    β = very large;
    while (β is in infeasible region) {
        β = β ÷ 2;
        calculate wire density w_m(x, y)
        calculate multi-level neighborhoods;
    }
    /* bi-section search on β_opt. */
    β_0: smallest infeasible β;
    β_1: largest feasible β;
    while (|β_1 − β_0| > ε) {
        β_2 = (β_0 + β_1)/2;
        calculate wire density w_m(x, y)
        calculate multi-level neighborhoods;
        if (β_2 is feasible)
            β_1 = β_2;
        else /* β_2 is infeasible. */
            β_0 = β_2;
    }
    β_opt. = largest feasible β.

}   /* end of algorithm. */
```

6 Experimental Results

The proposed estimation technique is implemented in approximately 3,000 lines of C.

Table II summarizes statistics of each test case. The first four test cases, PrimSC1, PrimSC2, Test06, and Test08, are standard cell test cases taken from the MCNC benchmark set. The next three test cases, 19ks, 19kstw, and 26k, are from Hughes Aircraft. We removed the effect of I/O pads in calculating the wire length throughout the following experiments.

Table II: Statistics on Test Cases

circuit	cell	pad	net	pin
PrimSC1	752	81	904	5888
PrimSC2	2907	107	3029	19195
Test06	1691	61	2048	7696
Test08	3744	60	4266	14119
19ks	2684	172	3343	11578
19kstw	3079	172	3731	12660
26k	25917	360	29151	92530

Table III summarizes our experimental results. The first column show estimation results by the proposed technique. The next two columns show estimation results by the technique proposed by Pedram and Preas [7]. Two different cut-off point for large pin count nets are examined. For Pedram_25%, the 25% criteria [6] is used, and nets connecting more than 25% of the cells are ignored. For Pedram_40pins, nets larger than 40-pin are ignored. Each experimental result shows average neighborhood population size, wire length, and relative error from measured wire length. The last column shows the measured wire length of an optimized placement. All test cases were placed using a hierarchical simulated annealing placement program PRC [3] with aspect ratio 1:1.

For Pedram_25%, Good estimation results were observed for test cases PrimSC1, PrimSC2, and 26k, which gave small average neighborhood population sizes. On the other hand, large estimation errors were observed for test cases Test06, Test08, 19ks and 19kstw, which gave average neighborhood population sizes larger than 100. Average error was 92.3% with a maximum deviation of +273.6% (test case 19kstw).

For Pedram_40pins, average neighborhood population sizes become normal, and better estimation results were obtained. We found that 40-pin is the best cut-off criteria of this technique for these test cases. Average error was 15.7% with a maximum deviation of +26.6% (test case 26k).

Our proposed technique achieved **more stable results** than the method by Pedram and Preas in all 7 test cases. On average it achieved estimation error of 9.0% with a maximum deviation of +16.3% (test case 19kstw).

Acknowledgment

The authors acknowledge helpful discussions with Mr. Michael Alston and Ms. Karen Stephen. This work was supported in part by NSF MIP-9009260, by NSF Initiation Grant MIP-99-10437, and by the NSF/IUCRC program at the University of California at San Diego, Amdahl, AT & T, Cadence, Elite, Hughes, and Pie under MICRO.

References

[1] W. E. Donath, "Placement and Average Interconnection Lengths of Computer Logic," *IEEE Trans. on Circuits and Systems*, vol. CAS-26, pp. 272-277, Apr. 1979.

[2] M. Feuer, "Connectivity of Random Logic," *IEEE Trans. on Computers*, vol. C-31, no. 1, pp. 29-33, Jan. 1982.

[3] T. Hamada, C. K. Cheng, and P. M. Chau, "An Efficient Multi-Level Placement Technique Using Hierarchical Partitioning," submitted to *IEEE Trans. on Circuits and Systems*,

[4] T. Hamada, C. K. Cheng, and P. M. Chau, "A Wire Length Estimation Technique Utilizing Neighborhood Density Equations on 2-Dimensional Random Field," submitted to *IEEE Trans. on Computer-Aided Design*,

[5] C. Itzykson and J-M. Drouffe, ***Statistical field theory***, vols. 1 and 2, Cambridge, UK: Cambridge University Press, 1989.

[6] M. Pedram and B. Preas, "Accurate Prediction of Physical Design Characteristics for Random Logic," *Proc. of Int. Conf. on Computer Design*, pp. 390-393, Nov. 1989.

[7] M. Pedram and B. Preas, "Interconnection Length Estimation for Optimized Standard Cell Layouts," *Proc. of Int. Conf. on Computer-Aided Design*, pp. 100-108, Oct. 1989.

[8] S. Sastry and A. C. Parker, "Stochastic Models for Wireability Analysis of Gate Arrays," *IEEE Trans. on Computer-Aided Design*, vol. CAD-5, no. 1, Jan. 1986.

[9] C. Sechen, "Average Interconnection Length Estimation for Random and Optimized Placements," *Proc. of Int. Conf. on Computer-Aided Design*, pp. 190-193, Nov. 1988.

Table I: Average Wire Length of Multi-Pin Net

pin count	$\mu_{exp.}$ PrimSC1	PrimSC2	Test06	Test08	19ks	19kstw	26k	$\mu_{pred.}$
2	1.00	1.00	1.00	1.00	1.00	1.00	1.00	1.00
3	1.15	1.50	1.29	1.56	1.74	1.55	1.64	1.44
4	1.30	1.98	1.45	2.05	2.27	2.56	1.65	1.83
5	1.32	1.92	3.85	2.36	2.64	2.71	2.24	2.20
6	1.69	2.25	5.52	2.83	2.65	3.74	2.44	2.56
7	1.52	3.30	2.91	3.44	2.73	5.82	2.77	2.91
8	2.32	2.45	2.07	3.20	1.81	4.53	3.40	3.25

Table III: Experimental Results

circuit	Proposed Method		Pedram_25%			Pedram_40pins			measd.
	w.r.l.[mm]	err.[%]	nbhd.	w.r.l.[mm]	err.[%]	nbhd.	w.r.l.[mm]	err.[%]	w.r.l.[mm]
PrimSC1	736.0	+0.3	18.3	706.3	-3.8	18.3	706.3	-3.8	733.9
PrimSC2	2940.2	-11.8	36.6	2875.2	-13.8	36.6	2875.2	-13.8	3335.0
Test06	2323.9	+5.5	89.8	4169.0	+89.4	35.6	2450.9	+11.3	2202.8
Test08	4983.8	-7.5	138.9	9951.6	+84.8	34.8	4261.1	-20.9	5385.1
19ks	4342.4	-6.5	205.1	11788.6	+153.9	25.1	4140.5	-10.8	4643.8
19kstw	5331.1	+16.3	340.4	17128.6	+273.6	16.6	3544.6	-22.7	4585.0
26k	26275.8	-14.9	93.0	39076.9	+26.6	93.0	39076.9	+26.6	30877.2

A Graph Theoretic Technique to Speed up Floorplan Area Optimization*

Ting-Chi Wang and D. F. Wong
Department of Computer Sciences
University of Texas at Austin
Austin, Texas 78712

Abstract

A well known approach for the floorplan area optimization problem is to first determine a list of all non-redundant implementations of the entire floorplan and then select an optimal floorplan from the list [4,6,9,10,12]. For large floorplans, this approach may fail due to insufficient memory space available to store the implementations of sub-floorplans generated during the computation. To effectively reduce both memory usage and running time, we present in this paper two algorithms to optimally reduce the number of implementations for rectangular and L-shaped sub-floorplans. The common key idea of our two algorithms is to reduce the problem to a constrained shortest path problem, which we can solve optimally in polynomial time. Our algorithms are designed specifically for [10] but they can also be applied to other algorithms such as [4,6,12] as well. The experimental results are very encouraging.

1 Introduction

A general approach to floorplan design is to first determine the topology of the floorplan primarily using the interconnection information among the modules [2,3,5,8]. The floorplan topology specifies the relative positions of the modules and is usually represented by a floorplan tree [5,10]. If each module is rectangular and has a finite number of implementations, the *floorplan area optimization problem* is to select an appropriate implementation for each module such that, without changing the given floorplan topology, the area of the entire floorplan is minimized.

A well known approach for the floorplan area optimization problem is to first determine a list of all non-redundant implementations of the entire floorplan and then select an optimal floorplan from the list [4,6,9,10,12]. The computation is done in a bottom-up fashion starting at the leaves of the floorplan tree (representing the modules) and ending at the root (representing the entire floorplan). For large floorplans, this approach may fail due to insufficient memory space available to store the implementations of sub-floorplans generated during the computation. An effective method to reduce memory usage is as follows: During the computation, whenever the set of non-redundant implementations of a sub-floorplan exceeds a predefined size, we only retain a subset of the implementations that can best approximate the original set. In this paper, we present two algorithms to optimally select implementations for rectangular and L-shaped sub-floorplans. Our algorithms are designed specifically for the floorplan optimization algorithm in [10] but they can also be applied to other algorithms such as [4,6,12] as well.

The common key idea of our two algorithms is to reduce the problem of optimally selecting a subset of implementations to a constrained shortest path problem, which we can solve optimally in polynomial time. We have incorporated the two algorithms into [10], and experimental results indicate that for the test cases where [10] was able to run alone, with the incorporation of the two algorithms, both memory usage and running time were reduced significantly while the solutions remained of comparable quality. Furthermore, for the test cases where [10] failed to run, the two algorithms helped to produce satisfactory solutions.

Due to space limitation, the proofs of theorems and lemmas in this paper are omitted, but they can be found in [11].

2 Preliminaries

A *floorplan* for m modules consists of an enveloping rectangle subdivided by horizontal and vertical line segments into m non-overlapping rectangles called *basic rectangles*. Each basic rectangle must be large enough to accommodate the module assigned to it. A *floorplan tree* is a hierarchical description of a floorplan. It specifies how the floorplan is obtained by recursively partitioning a rectangle into a finite number of parts. Figure 1 shows a floorplan and its corresponding floorplan tree. Each leaf in the floorplan tree corresponds to a basic rectangle and each internal node corresponds to a composite rectangle in the floorplan.

A *block* in a floorplan is defined as a *connected* set of basic rectangles. Two types of blocks will be considered in this paper: *L-shaped* and *rectangular* blocks. Figure 2 shows an L-shaped block and a rectangular block, where the L-shaped block consists of three basic rectangles and the rectangular block consists of four basic rectangles. An *implementation* of an L-shaped block is represented by a

*This work was partially supported by the National Science Foundation under grant MIP-8909586, by an IBM Faculty Development Award, and by an ACM SIGDA scholarship.

0738-100X/92 $3.00 © 1992 IEEE

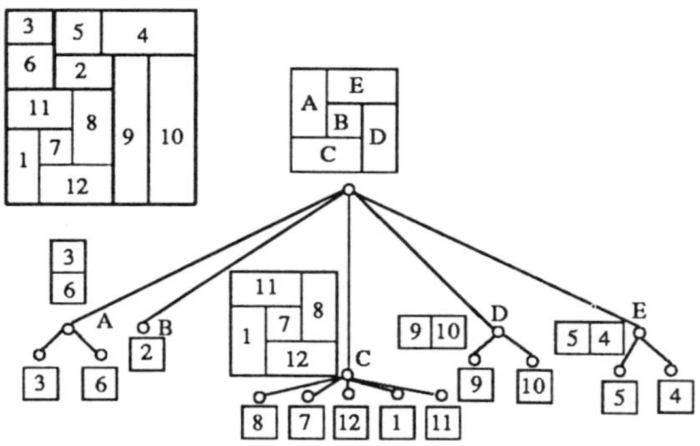

Figure 1: A floorplan and its floorplan tree.

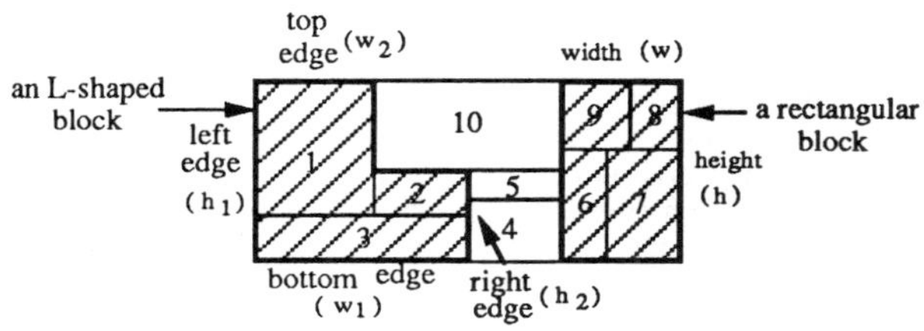

Figure 2: An L-shaped and a rectangular blocks.

4-tuple (w_1, w_2, h_1, h_2) with $w_1 \geq w_2$ and $h_1 \geq h_2$, where w_1 and w_2 represent the widths of the bottom edge and the top edge, respectively, and h_1 and h_2 represent the heights of the left edge and the right edge, respectively. (See Figure 2.) An *implementation* of a rectangular block is represented by a pair (w, h), where w is the width and h is the height. (See Figure 2.) We now describe below some definitions relevant to this paper.

Definition 1 Let $I_1 = (w_1, w_2, h_1, h_2)$ and $I_2 = (w_1', w_2', h_1', h_2')$ be two implementations of an L-shaped block. We say that I_1 *dominates* I_2 if the following four inequalities hold:

$$(1)\ w_1 \geq w_1'\ (2)\ w_2 \geq w_2'\ (3)\ h_1 \geq h_1'\ (4)\ h_2 \geq h_2'.$$

Similarly, if $I_1 = (w, h)$ and $I_2 = (w', h')$ are two implementations of a rectangular block, we say that I_1 *dominates* I_2 if the following two inequalities hold:

$$(1)\ w \geq w'\ (2)\ h \geq h'.$$

Definition 2 Let I_1 and I_2 be two implementations of a block (L-shaped or rectangular). We say that I_1 is a *redundant implementation* if I_1 dominates I_2.

Definition 3 A list $\{(w_{1,1}, w_{2,1}, h_{1,1}, h_{2,1}), (w_{1,2}, w_{2,2}, h_{1,2}, h_{2,2}), ..., (w_{1,n}, w_{2,n}, h_{1,n}, h_{2,n})\}$ is an *L-list* if the following four inequalities hold for all i and j, $1 \leq i < j \leq n$:

$$(1)\ w_{1,i} \geq w_{1,j}\ (2)\ w_{2,i} = w_{2,j}\ (3)\ h_{1,i} \leq h_{1,j}\ (4)\ h_{2,i} \leq h_{2,j}.$$

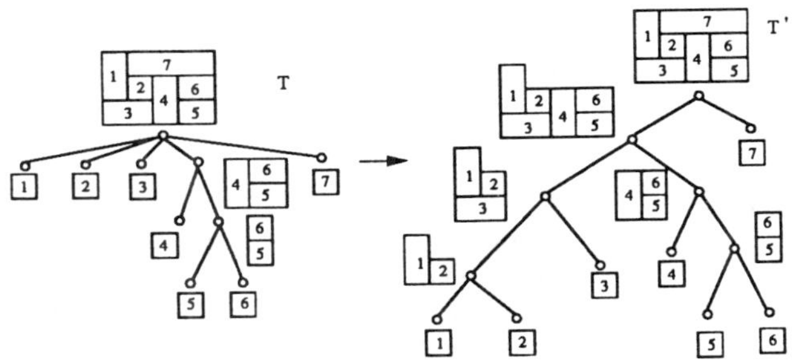

Figure 3: Restructuring T into a binary tree T'.

Definition 4 A list $\{(w_1, h_1), (w_2, h_2), ..., (w_n, h_n)\}$ is an *R-list* if the following two inequalities hold for all i and j, $1 \leq i < j \leq n$:

$$(1)\ w_i \geq w_j\ (2)\ h_i \leq h_j.$$

Definition 5 An *irreducible L-list* is an L-list in which there are no redundant implementations. Similarly, an *irreducible R-list* is an R-list in which there are no redundant implementations.

3 Our Approach

In this section, we present our approach to reducing memory usage by [10]. We first briefly review the algorithm in [10]. The inputs to this algorithm consist of a floorplan tree specifying the topology of the floorplan, and a finite set of non-redundant implementations for each module.

Given a floorplan tree T, it is first restructured into a binary tree T' such that each internal node in T' either corresponds to a rectangular or an L-shaped block. Figure 3 shows an example of how to restructure T into T'. At the beginning, the algorithm uses an irreducible R-list to store all non-redundant implementations for each leaf in T'. (Note that all non-redundant implementations of a leaf are the same as those of its corresponding module.) Next, the algorithm computes all non-redundant implementations for each internal node in T' in a bottom-up fashion. Let u be an internal node in T'. If u represents a rectangular block, the algorithm constructs an irreducible R-list to store all the non-redundant implementations for u. If u represents an L-shaped block, the algorithm constructs a set of irreducible L-lists to store all the non-redundant implementations for u. During computation, the algorithm only considers the *candidates* which are very likely to be non-redundant, and then eliminates the redundant ones from the candidates. The details of the algorithm can be found in [10].

A key observation regarding the memory space requirement of [10] is that every time when the algorithm generates all non-redundant implementations for an internal node in T', all of them have to be stored as each such implementation is possible to be part of an optimal solution. Hence, for large floorplans[1], the numbers of non-

[1] In this paper a large floorplan refers to a floorplan either having a large number of modules or having a large number of non-redundant implementations for each module.

redundant implementations for higher level nodes could be very large and consequently the algorithm may fail due to insufficient memory space. To reduce memory usage, we describe below an effective and efficient approach to controlling the number of non-redundant implementations for each internal node in T'.

Every time after the algorithm generates all non-redundant implementations for an internal node, we check the number of non-redundant implementations of that internal node. If it exceeds a predefined limit, we reduce it to the limit by applying an implementation selection algorithm. Since each internal node represents either a rectangular or an L-shaped block, two different implementation selection algorithms are needed. For a rectangular block, we design an algorithm to optimally select a subset of non-redundant implementations from its irreducible R-list. For an L-shaped block, since its non-redundant implementations are stored in a set of irreducible L-lists, we design an algorithm to optimally select a subset of non-redundant implementations from an irreducible L-list, and sequentially apply the algorithm to each of its irreducible L-lists. The two algorithms run in polynomial time and are described in the next section.

4 Optimal Selection of Implementations

In this section, we first introduce the *Constrained Shortest Path Problem (CSPP)* and present a polynomial time optimal algorithm for this problem. Then, our two implementation selection algorithms based on reducing the problem of optimally selecting a subset of implementations from an irreducible R-list or L-list to the CSPP are presented, respectively.

4.1 Constrained Shortest Path Problem

Let $G = (V, E)$ be a weighted *directed acyclic graph (DAG)*, where $V = \{v_1, v_2, ..., v_n\}$ and E are the vertex set and the edge set, respectively. For each edge $(v_i, v_j) \in E$, let $w(v_i, v_j)$ be its weight and we assume $w(v_i, v_j) > 0$. Given two vertices $s, t \in V$ and a positive integer $k \leq n$, the solution to the CSPP in G is either a simple path of minimum total weight from s to t with exactly k vertices or an indication that no such path exists. Note that the CSPP is different from the classical shortest path problem in the sense that the number of vertices on the path is restricted to be exactly k. Figure 4 shows a weighted DAG in which the number on each edge represents the weight. It is easy to see that the shortest path from v_1 to v_6 is the path $v_1 \rightarrow v_2 \rightarrow v_3 \rightarrow v_4 \rightarrow v_5 \rightarrow v_6$ with total weight of 8. However, if we consider the CSPP with $k = 4$, the constrained shortest path from v_1 to v_6 should be the path $v_1 \rightarrow v_2 \rightarrow v_4 \rightarrow v_6$ with total weight of 11.

Let $W(s, v_i, l)$ be the minimum total weight of a constrained shortest path in G from s to v_i with exactly l vertices. If there exists no such path from s to v_i with l vertices, we simply let $W(s, v_i, l) = \infty$. We present below a polynomial time optimal algorithm called Constrained_Shortest_Path to solve the CSPP. The algorithm Constrained_Shortest_Path is based on the dynamic programming technique [1].

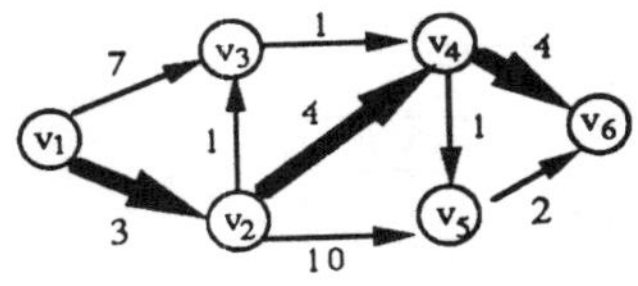

Figure 4: A constrained shortest path with $k = 4$.

```
Algorithm: Constrained_Shortest_Path
 Begin
   /* Initialization */
   W(s, s, 1) := 0;
   for each vertex v_i ∈ V − {s} do
     W(s, v_i, 1) := ∞;
   end for
   /* Calculate W(s, t, k) */
   for l := 2 to k do
     W(s, s, l) := ∞;
     for each vertex v_i ∈ V − {s} do
       /* Calculate W(s, v_i, l) */
       W(s, v_i, l) := ∞;
       for each edge (v_j, v_i) ∈ E do
         if W(s, v_j, l − 1) + w(v_j, v_i) < W(s, v_i, l) then
           W(s, v_i, l) := W(s, v_j, l − 1) + w(v_j, v_i);
           update the vertex, i.e., v_j, which gives rise to the
             current W(s, v_i, l);
         end if
       end for
     end for
   end for
   if W(s, t, k) = ∞ then
     return("Can not find such a path.");
   else /* Determine the constrained shortest path. */
     P := ∅; v := t; l := k;
     while l > 2 do
       P := P ⋃ {v_j}, where v_j is associated with W(s, v, l);
       v := v_j; l := l − 1;
     end while
     if k ≥ 2
       P := {s} ⋃ P ⋃ {t};
     else /* k = 1 and s = t */
       P := {s};
     end if
     return(P);
   end if
 End.
```

Theorem 1 *Algorithm Constrained_Shortest_Path correctly solves the CSPP in $O(k(|V| + |E|))$ time, where $k, |V|$, and $|E|$ are the number of vertices on the constrained shortest path, the number of vertices, and the number of edges, in G, respectively.* □

4.2 Rectangular Blocks

In this subsection, we present an algorithm to optimally select a subset of non-redundant implementations from an irreducible R-list. That is, given an irreducible R-list R storing n non-redundant implementations of a rectangular block B, and a positive integer $k < n$, the algorithm selects from R a subset of k implementations which can best approximate R.

We denote R by $\{r_1, r_2, ..., r_n\}$ with each r_i equal to (w_i, h_i). Let $R' = \{r_{d_1}, r_{d_2}, ..., r_{d_k}\}$ be a subset of k im-

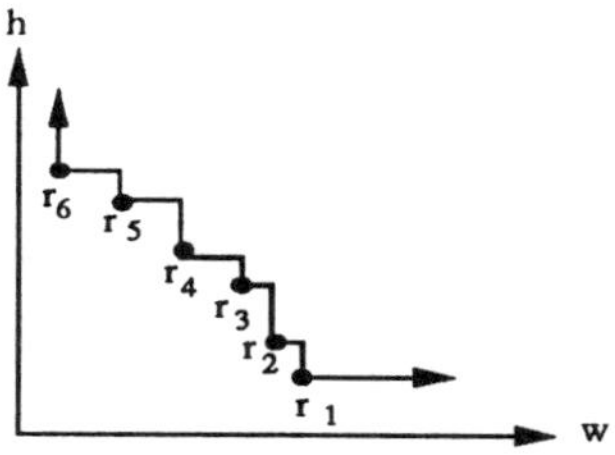

Figure 5: A staircase curve.

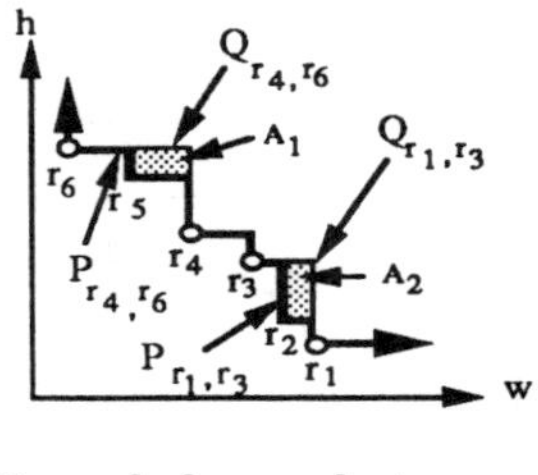

Figure 6: Bounded area between two curves.

plementations selected from R with $1 = d_1 < d_2 < ... < d_k = n$. We may assume $2 \le k < n$. We can use a *staircase curve* C_R to represent R such that R constitutes the set of *corners* on C_R with r_1 and r_n being the rightmost and the leftmost corners, respectively. (Figure 5 shows the staircase curve C_R for $R = \{r_1, r_2, ..., r_6\}$.) Note that any point on or above C_R represents a *feasible* (either redundant or non-redundant) implementation of B, but only the corners are non-redundant. Similarly, we can use another staircase curve $C_{R'}$ to represent R', and any point on or above $C_{R'}$ is also a feasible implementation of B. Intuitively, we can consider the bounded area between C_R and $C_{R'}$ as the amount of feasible implementations discarded due to selecting R' from R. Hence, we define the cost $ERROR(R, R')$ of selecting R' from R as the bounded area between C_R and $C_{R'}$, and our objective is to select R' such that $ERROR(R, R')$ is minimized. Figure 6 shows an example in which $R = \{r_1, r_2, r_3, r_4, r_5, r_6\}$, $R' = \{r_1, r_3, r_4, r_6\}$, and by definition, $ERROR(R, R')$ is the sum of the areas of A_1 and A_2.

Clearly, $C_{R'}$ can be partitioned into a set of subcurves $\{Q_{r_{d_1}, r_{d_2}}, Q_{r_{d_2}, r_{d_3}}, ..., Q_{r_{d_{k-1}}, r_{d_k}}\}$, where for all q, $1 \le q < k$, subcurve $Q_{r_{d_q}, r_{d_{q+1}}}$ has r_{d_q} and $r_{d_{q+1}}$ as its two endpoints. Similarly, for all q, $1 \le q < k$, let $P_{r_{d_q}, r_{d_{q+1}}}$ be the subcurve of C_R with r_{d_q} and $r_{d_{q+1}}$ as its two endpoints. For example, in Figure 6, A_1 corresponds to the bounded area between Q_{r_4, r_6} and P_{r_4, r_6}, and A_2 corresponds to the bounded area between Q_{r_1, r_3} and P_{r_1, r_3}. If we define $error(r_{d_q}, r_{d_{q+1}})$ as the bounded area between $Q_{r_{d_q}, r_{d_{q+1}}}$ and $P_{r_{d_q}, r_{d_{q+1}}}$, we have

$$ERROR(R, R') = \sum_{q=1}^{k-1} error(r_{d_q}, r_{d_{q+1}}).$$

That is, $ERROR(R, R')$ can be calculated in terms of $error(r_{d_q}, r_{d_{q+1}})$ for $q = 1, 2, ..., k-1$. We present below

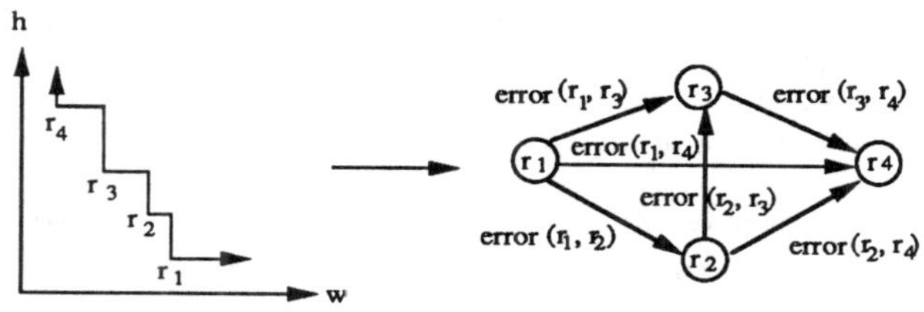

Figure 7: Constructing a weighted DAG.

an $O(n^2)$ algorithm called Compute_R_Error to calculate $error(r_i, r_j)$ for all i and j, $1 \le i < j \le n$.

```
Algorithm: Compute_R_Error
  Begin
    for i := 1 to n − 1 do
      error(r_i, r_{i+1}) := 0;
    end for
    for l := 2 to n − 1 do
      for i := 1 to n − l do
        error(r_i, r_{i+l}) := error(r_i, r_{i+l−1})+
                        (w_i − w_{i+l−1})(h_{i+l} − h_{i+l−1});
      end for
    end for
  End.
```

We now describe how to reduce the problem of selecting an optimal R' from R to an instance of the CSPP. Given R, we construct a weighted DAG $G = (V, E)$ with $V = R$ and $E = \{(r_i, r_j) | \forall i, j, 1 \le i < j \le n\}$. Each edge (r_i, r_j) has its weight equal to $error(r_i, r_j)$. (See Figure 7 for an illustration of the construction.) The construction can be done in $O(n^2)$ time. We now have the following lemma.

Lemma 1 *Let R be an irreducible R-list, and let G be the weighted DAG constructed from R. We have, each path P in G from r_1 to r_n with k vertices corresponds to a subset of k implementations of R. Moreover, $ERROR(R, P)$ is equal to the total weight of P.* □

It follows from Lemma 1 that the problem of selecting an optimal R' from R is equivalent to the CSPP in G with $s = r_1$ and $t = r_n$. The following algorithm R_Selection optimally selects a subset R' of k implementations from R such that $ERROR(R, R')$ is minimized.

```
Algorithm: R_Selection
  Begin
    1. Apply the algorithm Compute_R_Error to calculate
       error(r_i, r_j) for all i and j, 1 ≤ i < j ≤ n;
    2. Construct the corresponding weighted DAG G from R;
    3. Apply the algorithm Constrained_Shortest_Path to G
       with s = r_1 and t = r_n;
  End.
```

Theorem 2 *Algorithm R_Selection optimally selects a subset of k implementations from an irreducible R-list of n implementations in $O(kn^2)$ time.* □

4.3 L-shaped Blocks

In this subsection, we present an algorithm to optimally select a subset of non-redundant implementations from an irreducible L-list. Since each L-shaped block has a set of irreducible L-lists to store its non-redundant implementations, we will sequentially apply the algorithm to reduce the size of each of its irreducible L-lists.

Let $L = \{l_1, l_2, ..., l_n\}$ be an irreducible L-list storing n implementations of an L-shaped block B, where $l_i=(w_{1,i}, w_{2,i}, h_{1,i}, h_{2,i})$ for all i, $1 \le i \le n$. Let $L' = \{l_{d_1}, l_{d_2}, ..., l_{d_k}\}$ be a subset of k implementations selected from L with $1 = d_1 < d_2 < ... < d_k = n$. Again, we may assume $2 \le k < n$. Since each $l_i \in L$ can be viewed as a point in the 4-dimensional real space R^4, we will use $dist(l_i, l_j)$ to denote the Manhattan distance[2] between l_i and l_j, for all $l_i, l_j \in L$. Hence, we have

$$dist(l_i, l_j) = |w_{1,i} - w_{1,j}| + |w_{2,i} - w_{2,j}| + |h_{1,i} - h_{1,j}| + |h_{2,i} - h_{2,j}|.$$

Since $w_{2,i} = w_{2,j}$, we have

$$dist(l_i, l_j) = |w_{1,i} - w_{1,j}| + |h_{1,i} - h_{1,j}| + |h_{2,i} - h_{2,j}|.$$

In general, we expect the less distance between l_i and l_j, the more similar in shape they are. Hence, $dist(l_i, l_j)$ measures the amount of difference in shape between l_i and l_j. For each discarded implementation $l_i \in L - L'$, we define $cost(l_i)$ as the distance between l_i and the implementation in L' which is closet to l_i. That is, for any $l_i \in L - L'$, we have

$$cost(l_i) = \min_{l_j \in L'} dist(l_i, l_j).$$

Intuitively, we can consider $cost(l_i)$ as the cost of not selecting l_i from L, and hence the implementation in L' which has the minimum distance from l_i is responsible for replacing l_i. Therefore, we define the total cost $ERROR(L, L')$ of selecting L' from L as the sum of the costs of the implementations not selected into L'. That is,

$$ERROR(L, L') = \sum_{l_i \in L - L'} cost(l_i).$$

Now, our goal is to select an optimal L' such that $ERROR(L, L')$ is minimized. We first state below some important lemmas.

Lemma 2 *Let L be an irreducible L-list, and let $l_{i'}, l_i, l_j, l_{j'} \in L$ with $i' < i < j < j'$. We have $dist(l_i, l_j) \le dist(l_{i'}, l_j)$ and $dist(l_i, l_j) \le dist(l_i, l_{j'})$.* □

For any discarded implementation $l_i \in L - L'$, there exist exactly two consecutive implementations $l_{d_q}, l_{d_{q+1}} \in L'$ with $d_q < i < d_{q+1}$. We define such l_{d_q} and $l_{d_{q+1}}$ as the *left neighbor* and the *right neighbor* of l_i, respectively. Based on Lemma 2, we can deduce Lemma 3 which states that $cost(l_i)$ can be computed only by considering the distance between l_i and its two neighbors.

[2]In fact, we can use any L_p metric [7] to measure the distance. It is well known that the Manhattan distance is the L_1 metric. Note that all the lemmas and theorem presented in this subsection remain correct for any L_p metric.

Lemma 3 *Let L and $L' \subseteq L$ be two irreducible L-lists. For any $l_i \in L - L'$, if l_{d_q} and $l_{d_{q+1}}$ are its left and right neighbors, then $cost(l_i) = \min(dist(l_{d_q}, l_i), dist(l_i, l_{d_{q+1}}))$.* □

For any two consecutive implementations $l_{d_q}, l_{d_{q+1}} \in L'$, let $X = \{l_i, l_{i+1}, ..., l_j\} \subseteq L - L'$ with $i = d_q + 1$ and $j = d_{q+1} - 1$. We define the cost of discarding X from L as

$$error(l_{d_q}, l_{d_{q+1}}) = \sum_{l_i \in X} cost(l_i),$$

where each $cost(l_i)$ can be computed just based on Lemma 3. Now, it is clear that $ERROR(L, L')$ can be computed in terms of $error(l_{d_q}, l_{d_{q+1}})$ for $q = 1, 2, ..., k-1$, i.e.,

$$ERROR(L, L') = \sum_{q=1}^{k-1} error(l_{d_q}, l_{d_{q+1}}).$$

We present below an $O(n^3)$ algorithm called Compute_L_Error to compute $error(l_i, l_j)$ for all i and j, $1 \le i < j \le n$.

Algorithm: Compute_L_Error

```
Begin
  for i := 1 to n - 1 do
    for j := i + 1 to n do
      error(l_i, l_j) = 0;
      for q := i + 1 to j - 1 do
        error(l_i, l_j) := error(l_i, l_j)+
                    min(dist(l_i, l_q), dist(l_q, l_j));
      end for
    end for
  end for
End.
```

We now describe how to reduce the problem of determining an optimal L' from L into an instance of the CSPP. Given L, we construct a weighted DAG $G = (V, E)$ with $V = L$ and $E = \{(l_i, l_j) | \forall i, j, 1 \le i < j \le n\}$. Each edge (l_i, l_j) has its weight equal to $error(l_i, l_j)$. The construction is very similar to that described in Section 4.2. We now have the following lemma similar to Lemma 1.

Lemma 4 *Let L be an irreducible L-list, and let G be the weighted DAG constructed from L. We have, each path P in G from l_1 to l_n with k vertices corresponds to a subset of k implementations of L. Moreover, $ERROR(L, P)$ is equal to the total weight of P.* □

It follows from Lemma 4 that the problem of determining an optimal L' from L is equivalent to the CSPP in G with $s = l_1$ and $t = l_n$. The following algorithm L_Selection optimally selects a subset L' of k implementations from L such that $ERROR(L, L')$ is minimized.

Algorithm: L_Selection

```
Begin
  1. Apply the algorithm Compute_L_Error to compute
     error(l_i, l_j) for all i and j, 1 <= i < j <= n;
  2. Construct the corresponding weighted DAG G from L;
  3. Apply the algorithm Constrained_Shortest_Path to G
     with s = l_1 and t = l_n;
End.
```

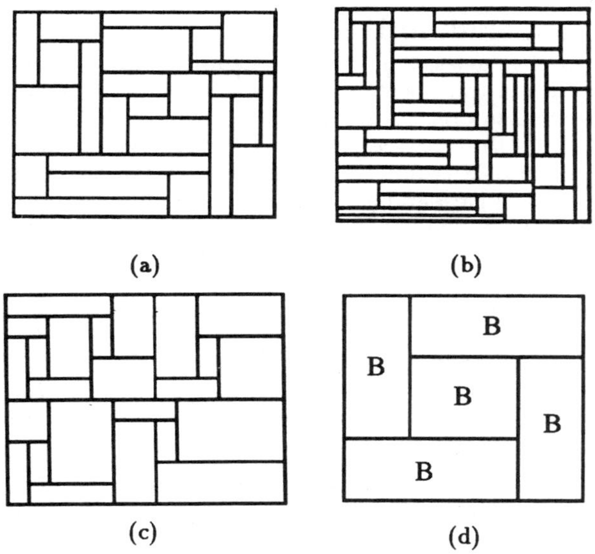

Figure 8: Test floorplans.

Theorem 3 *Algorithm L_Selection optimally selects a subset of k implementations from an irreducible L-list of n implementations in $O(n^3)$ time.* □

Finally, we explain how to apply the algorithm L_Selection to each of the irreducible L-lists of an L-shaped block B. Suppose B has p irreducible L-lists storing a total of N implementations. Also suppose we want to reduce the number of implementations of B from N to K. Let L be one of the p L-lists, and $|L|$ be the number of implementations stored in L. When we apply the algorithm L_Selection to L, the limit on the number of implementations for L is set to $\lfloor \frac{K|L|}{N} \rfloor$. Thus, the limit on the number of implementations for each of the p L-lists is dynamically adjusted.

5 Experimental Results

We have implemented the algorithms R_Selection and L_Selection in C language on a Sun SPARC station running Unix operating system, and also incorporated them into [10]. We used 16 test cases based on the following 4 floorplans FP1–FP4 to test the two algorithms. (For each floorplan, we had 4 test cases corresponding to 4 different sets of modules.)

- FP1: the 25-module floorplan as shown in Figure 8(a);
- FP2: the 49-module floorplan as shown in Figure 8(b);
- FP3: the 120-module floorplan as shown in Figure 8(d) in which each rectangular block B denotes the 24-module floorplan in Figure 8(c);
- FP4: the 245-module floorplan as shown in Figure 8(d) in which each rectangular block B denotes the 49-module floorplan in Figure 8(b).

We first used the test cases of FP1–FP3 to measure the performance of the algorithm R_Selection. The results are reported in Tables 1–3, where N is the number of non-redundant implementations of each module, K_1 is the user specified limit on the number of implementations for each rectangular block, M is the maximum number of implementations ever stored in memory during the computation[3], A_{OPT} is the optimal area of the floorplan produced by [10], and A_R is the area of the floorplan produced by [10] with R_Selection. It is clear that M and $\frac{A_R - A_{OPT}}{A_{OPT}}$ measure the memory usage and the quality of solutions produced by using R_Selection, respectively. As can be seen from Tables 1–3, the algorithm R_Selection helped to significantly reduce both memory usage and running times while its solutions were very close to optimal. In particular, [10] failed to run for the last three test cases of FP3 due to insufficient memory space, and hence incorporating the algorithm R_Selection into [10] became necessary.

Test Case #	N	[10] M	[10] CPU (sec)	[10] + R_Selection K_1	[10] + R_Selection M	[10] + R_Selection CPU (sec)	$\frac{A_R - A_{OPT}}{A_{OPT}}$
1	20	67871	16.2	20	15834	5.3	1.21%
				30	19607	7.4	0.99%
				40	26656	10.5	0.04%
2	20	139580	40.4	20	18143	5.7	0.46%
				30	25375	9.2	0.09%
				40	35950	14.3	0.09%
3	40	257944	181.0	40	52891	33.6	0.18%
				50	59142	38.8	0.18%
				60	72721	46.8	0.0%
4	40	402398	997.8	40	79810	50.6	0.05%
				50	84928	61.3	0.23%
				60	96672	75.6	0.23%

Table 1: The experimental results of FP1 (25 modules).

Test Case #	N	[10] M	[10] CPU (sec)	[10] + R_Selection K_1	[10] + R_Selection M	[10] + R_Selection CPU (sec)	$\frac{A_R - A_{OPT}}{A_{OPT}}$
1	20	343603	76.8	20	45474	10.3	1.99%
				30	63025	15.3	0.28%
				40	90199	23.8	0.03%
2	20	495836	117.7	20	46541	11.0	0.27%
				30	59299	15.2	0.76%
				40	90548	23.0	0.19%
3	40	608891	187.8	40	230002	73.0	0.13%
				50	239653	81.4	0.1%
				60	252735	90.3	0.1%
4	40	473772	407.7	40	131751	51.7	0.15%
				50	142940	57.8	0.15%
				60	158288	65.5	0.0%

Table 2: The experimental results of FP2 (49 modules).

[3] Note that M is not equal to the total number of implementations ever generated because as soon as R_Selection eliminates some implementations, we no longer need to store them.

Test Case #	N	[10] M	[10] CPU (sec)	[10] + R_Selection K_1	[10] + R_Selection M	[10] + R_Selection CPU (sec)	$\frac{A_R - A_{OPT}}{A_{OPT}}$
1	20	439234	90.3	20	50841	15.6	1.29%
				30	53514	20.8	0.4%
				40	57947	27.6	0.16%
2	20	> 806553	-	20	61566	37.0	-
				30	67717	51.1	-
				40	74211	65.7	-
3	40	> 798494	-	40	230891	170.5	-
				50	236818	192.1	-
				60	244656	221.3	-
4	40	> 800552	-	40	269127	210.8	-
				50	282360	241.3	-
				60	299268	284.9	-

Table 3: The experimental results of FP3 (120 modules).

Finally, we used the test cases of FP4 to measure the performance of R_Selection and L_Selection. This time, [10] again failed to run for each of the 4 test cases. Table 4 reports the results produced by using R_Selection with and without L_Selection. In Table 4, K_2 is the user specified limit on the number of implementations for each L-shaped block, and A_{R+L} is the area of the floorplan produced by [10] with R_Selection and L_Selection. Clearly, $\frac{A_{R+L} - A_R}{A_R}$ measures the effectiveness of L_Selection. As shown in Table 4, the algorithm L_Selection also helped to significantly reduce memory usage while its running times and solution qualities still remained comparable. In particular, for the last two test cases, using L_Selection became necessary because using R_Selection alone was unable to prevent the failure caused by insufficient memory space. Since we observed that the number of implementations of an L-shaped block in general is much larger than that of a rectangular block, it is much more expensive to use L_Selection than R_Selection. Therefore in our current implementation we employed two special techniques to speed up the computation. Firstly, instead of using L_Selection whenever the number X of implementations of an L-shaped block exceeds K_2, we use L_Selection only when $\frac{K_2}{X} < \theta$, for some user specified parameter θ, $0 < \theta \leq 1$. Secondly, since L_Selection is extremely slow when X is very large, whenever $X > S$ for some user specified parameter S, we first use a heuristic version of L_Selection to reduce the number of implementations to S, and then use L_Selection to further reduce it to K_2 implementations.

Test Case #	N	[10] + R_Selection K_1	[10] + R_Selection M	[10] + R_Selection CPU (sec)	[10] + R_Selection + L_Selection K_2	[10] + R_Selection + L_Selection M	[10] + R_Selection + L_Selection CPU (sec)	$\frac{A_{R+L} - A_R}{A_R}$
1	20	40	379943	106.8	1000	232666	101.3	6.43%
					1500	239972	109.9	4.2%
					2000	292279	94.7	3.15%
2	20	40	462441	145.4	1000	209744	136.1	6.25%
					1500	255946	133.3	4.02%
					2000	268903	171.1	2.44%
3	40	40	> 782146	-	1000	341626	338.0	-
					1500	507107	455.5	-
					2000	600522	532.8	-
4	40	40	> 775390	-	1000	346208	341.7	-
					1500	458501	494.7	-
					2000	589706	552.2	-

Table 4: The experimental results of FP4 (245 modules).

6 Concluding Remarks

The two implementation selection algorithms presented in this paper in fact has other applications. Firstly, they can be applied to other existing floorplan design algorithms (e.g., [4,6,12]) to reduce both memory usage and computation time. Secondly, if each module in the floorplan is allowed to have an infinite number of implementations specified by a continuous shape curve, we still can solve this case efficiently by first approximating each such shape curve by a large number of points and then applying [10] together with the two algorithms.

References

[1] A. V. Aho, J. E. Hopocroft and J. D. Ullman, *The Design and Analysis of Computer Algorithms,* Addison Wesley, 1974.

[2] D. P. Lapotin and S. W. Director, "Mason: A Global Floor-planning Tool," in *Proc. IEEE. Intl. Conf. on Computer-Aided Design,* 1985, pp. 143-145.

[3] U. Lauther, "A Min-Cut Placement Algorithm for General Cell Assemblies Based on a Graph Representation," *Journal of Digital Systems,* Vol. IV, Issue 1, pp. 21-34, 1980.

[4] T. Lengauer and R. Müller, "A Robust Framework for Hierarchical Floorplanning with Integrated Global Wiring," in *Proc. IEEE International Conf. on Computer-Aided Design,* 1990, pp. 148-151.

[5] R. H. J. M. Otten, "Automatic Floorplan Design," in *Proc. 19th ACM/IEEE Design Automation Conf.,* 1982, pp. 261-267.

[6] M. Pedram and B. Preas, "A Hierarchical Floorplanning Approach," in *Proc. IEEE International Conf. on Computer Design,* 1990, pp. 332-338.

[7] F. P. Preparata and M. I. Shamos, *Computational Geometry: An Introduction,* Springer-Verlag, 1985.

[8] N. R. Quin and M. A. Breur, "A Force Directed Component Placement Procedure for Printed Circuit Boards," *IEEE Trans. on Circuits and Systems,* Vol. CAS-26, No. 6, pp. 377-388, 1979.

[9] L. Stockmeyer, "Optimal Orientations of Cells in Slicing Floorplan Designs," *Information and Control,* Vol. 59, pp. 91-101, 1983.

[10] T.-C. Wang and D. F. Wong, "An Optimal Algorithm for Floorplan Area Optimization", in *Proc. 27th ACM/IEEE Design Automation Conf.,* 1990, pp. 180-186.

[11] T.-C. Wang and D. F. Wong, "A Graph Theoretic Technique to Speed up Floorplan Area Optimization," Tech. Rep. TR-91-26, Dept. of Comp. Sci., Univ. of Texas at Austin, 1991.

[12] G. Zimmerman, "A New Area and Shape Function Estimation Technique for VLSI Layouts," in *Proc. 25th ACM/IEEE Design Automation Conf.,* 1988 pp. 60-65.

Canonical Embedding of Rectangular Duals with Applications to VLSI Floorplanning

Susmita Sur-Kolay and Bhargab B. Bhattacharya
Electronics Unit
Indian Statistical Institute
Calcutta - 700 035, India
email: shakti!ncst.ernet.in!isical!bhargab

Abstract: *In this paper, the notion of equivalent embedding of rectangular duals (floorplans) is introduced which leads to a new concept of canonical embedding of a rectangular dual; this is a floorplan corresponding to a given neighborhood graph such that the number of directed cycles in its channel digraph is minimum. Strongly maximal rectangular hierarchy (sMRH) in nonslicible floorplans is then defined. The canonical form of any arbitrary floorplan consists of at most one nonslicing core for each member of sMRH. Such an embedding therefore, represents a floorplan with minimum deviations from a slicing structure. An* $O(n^2)$ *algorithm for realizing a canonical embedding is also presented. Canonical embedding lends deep insight to the yet unsolved problem of characterizing inherent nonslicibility and motivates design for slicibility. It also makes determination of safe routing order amazingly simple.*

1. Introduction

Rectangular dualization is a well-known method of VLSI floorplanning. This produces a floorplan by rectilinearly embedding the geometric dual of a given neighborhood graph [BS86]. In a valid embedding, the boundary and all the modules of the floorplan should be of rectangular shape. Although much emphasis has been given on a special type of floorplans known as slicings or slicible floorplans [Ot83, LSW87], general floorplans i.e. nonslicings appear very often [WKC 88], [LK89]. A related concept of inherent nonslicibility of neighborhood graphs appeared in [SB88]; there exists an infinite family of neighborhood graphs which do not have any corresponding slicible floorplans [SB91a]. Complete characterization of inherent nonslicibility is yet to be established.

2. Main results

In this paper, first a new method of **re-embedding** a nonslicible floorplan to a slicible equivalent, if one exists, is described. The *equivalence* is in the sense of keeping the associated neighborhood graph invariant.

Secondly, the novel concept of **canonical embedding** of a rectangular dual is introduced. This is defined to be a valid, equivalent rectilinear embedding which has the minimum number of 4-cycles and no longer cycles in the corresponding channel digraph [SS83], among all possible valid floorplans corresponding to the given rectangular graph. The maximal rectangular hierarchy (MRH), given in [SB91a, SB91b] is further classified as strong and weak. Strong MRH is utilized for obtaining a canonical embedding.

Next we expose a very interesting feature of canonical embeddings. We show that for any floorplan F, its canonical embedding **has either 0 or 1 nonslice core** for every member of strong MRH in F. This implies that all slicible rectangular graphs have canonical embeddings with no nonslice cores and all *inherently nonslicible rectangular graphs* have an equivalent embedding with at most one nonslice core at each level of strong MRH. Intuitively, it also gives a common base for assessing the degree of inherent nonslicibility and a new measure of shape complexity of floorplans.

The immediate consequences of these results are as follows. From a theoretician's viewpoint, this lends handles towards recognition of inherently nonslicible floorplans. From an engineer's viewpoint, a canonical embedding makes the NP-complete problem of deriving the optimal feasible (safe) routing order [DAK85, GW88, SB91b] amazingly simple. Such an embedding produces an equivalent floorplan with minimum number of nonslicing cores at the cost of some changes in aspect ratios of modules. Furthermore, the proposed method for transformation of a nonslicible floorplan (perhaps obtained by linear time algorithms such as in [BS86]) to an equivalent slicible one, if possible, is very useful because efficient algorithms for wiring, orientation [LSW87, St83] in slicings can then be utilized.

3. Background

A **floorplan** is a rectangle dissection of an enveloping rectangle by isothetic line segments, called **cuts or channels**, into a finite number of indivisible non-overlapping rectangles known as modules. Conventionally, it is assumed that two perpendicular cuts can meet but not cross, forming T-junctions only. A **node** in a floorplan is either a corner or a T-junction. A **channel graph** of a floorplan is a digraph where there is a distinct vertex for each channel and there is an arc (v,w) if the corresponding channels v and w are respectively the base and crosspiece of a T-junction in the floorplan (Fig. 1e). It has directed cycles iff the floorplan is nonslicible [SS83]. In nonslicible floorplans, directed cycles existing in the channel graph give cyclic dependencies which need to be resolved efficiently for feasible channel routing [SB91b].

A **rectangular graph** (Fig. 1a) is a neighborhood graph for which a rectangular dual i.e. a valid rectangular floorplan (Fig. 1d) exists [SB88]. The **inner dual** (Fig. 1c) of a rectangular graph G is an isothetic embedding of the geometric dual [Ha69] of G, excluding the vertex corresponding to the exterior face (Fig. 1b) of G. Each internal face of the inner dual is bounded by four or more edges and embedded as a rectangle. Efficient algorithms for constructing a rectangular dual have been developed in [KK85], [LL84], [BS86], [LL88], but these do not pay attention to producing slicible floorplans when possible.

Non-uniqueness of rectangular duals

Given a rectangular graph G, its rectangular dual may not be unique and both nonslicible and slicible duals may exist for G (Figs. 2a & 2b). A rectangular graph is **inherently nonslicible** if there exists no corresponding slicible floorplan. The resulting floorplan is also called inherently nonslicible. A minimum example [SB88] is shown in Fig. 1.

Converting nonslicing to equivalent slicing

In order to gain insight to the characteristics of nonslicibility, an interactive tool was developed [SB91a]. It essentially consists of a graphics screen editor for drawing floorplans, and a slicer routine for transforming a given floorplan to an equivalent one by straightening a user-specified jog path in the existing floorplan. A **jog path** is a sequence of consecutive nodes in a floorplan which specify portions of cuts alternately horizontal and vertical, the first and the last portions being of the same orientation. The equivalence is in the sense of *preserving adjacency between modules*. The ultimate goal is to convert a given nonslicible floorplan to an equivalent slicing, if possible, by straightening a series of jog paths. The equivalence is achieved by *keeping the cyclic order of nodes around each module invariant*; however with respect to a module, the nodes defining its four corners may differ. A jog path P is said to be **valid** if P can be straightened to yield an equivalent floorplan whose boundary and all modules remain rectangular. The necessary and sufficient conditions for validity of a jog path appeared in [SB91a].

Example: A jog path in the nonslicible floorplan shown in Fig. 2a is {18, 7, 15, 6, 16, 22, 5}. After straightening it, an equivalent slicible floorplan (Fig. 2b) is found. In contrast the jog path {18, 7, 14, 8, 9, 23, 22, 5} in Fig. 2a is invalid.

The core section of the slicer is the jog path straightener. The slicer asks the user to specify a jog path. Given a path, the slicer checks its validity and reconstruct an equivalent floorplan with all the nodes on the jog path lying along a *single cut*.

4. Canonical embedding of a rectangular dual

The method outlined above provides a transformation operation from a valid floorplan to another valid one; the invariants of the operation are the number of modules and the adjacency relation between modules i.e., the corresponding rectangular graphs are isomorphic. But floorplans, equivalent under above transformation, are not isomorphic in terms of definition of the channels. Nonisomorphic floorplans correspond to different rectangular dissections, so the associated cutlines are different. Therefore, the transformation operation on a floorplan changes its channel graph. The number of vertices and arcs in the two channel graphs are equal, but the set of arcs are different. The primary intention of canonical embedding is to reduce the number of directed cycles in the channel graph.

The keystone in the above method is finding a valid *cross-country* (from one side of the bounding rectangle to the opposite) jog path. In the search involved, one can start at one of the boundary nodes (excluding corners); from each node i.e. T-junction there are two possible successive nodes. Since the validity criterion is global, selecting a valid jog path may turn out to be very expensive. Besides, a series of valid jog paths are necessary in most cases to obtain equivalent slicing. But choice of one valid path over another may necessitate a much longer series of subsequent jog paths. There exist floorplans with exponential number of valid jog paths (Fig.3). Moreover, we shall discuss an example shortly which has exactly one valid jog path and thereby search may need backtracking several times.

In this scenario, we turn to local (bottom-up) search for valid jog paths, which will ultimately lead to an efficient polynomial time transformation algorithm. The basic idea is to start from the core of nonslicibility and proceed outwards to the boundary instead of beginning from a boundary T-junction for cross-country jog path. A handful of primitive transformation operations on the core (highlighted with heavy lines) can be defined; these have rotation symmetry

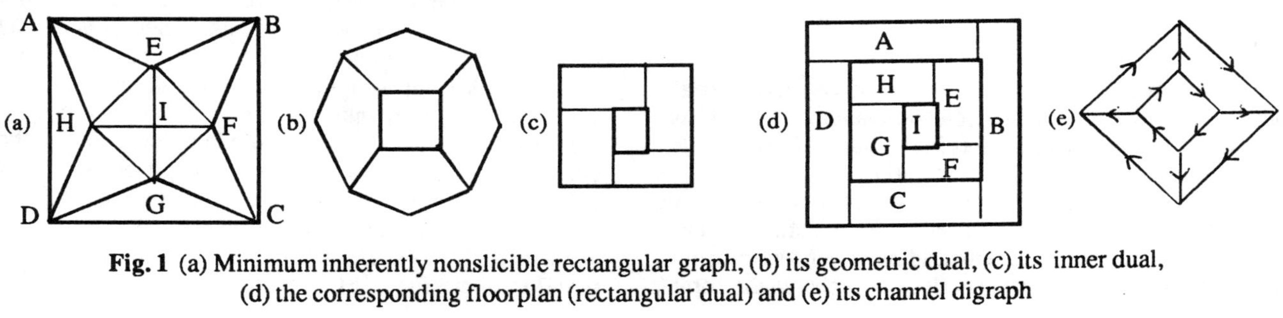

Fig. 1 (a) Minimum inherently nonslicible rectangular graph, (b) its geometric dual, (c) its inner dual, (d) the corresponding floorplan (rectangular dual) and (e) its channel digraph

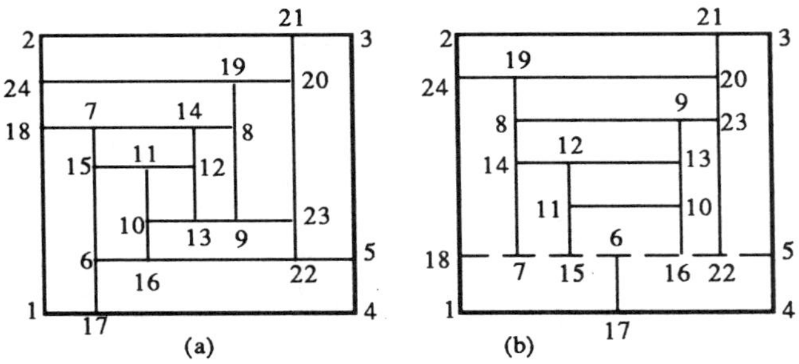

Fig. 2 Converting nonslicing to slicing by straightening jog path

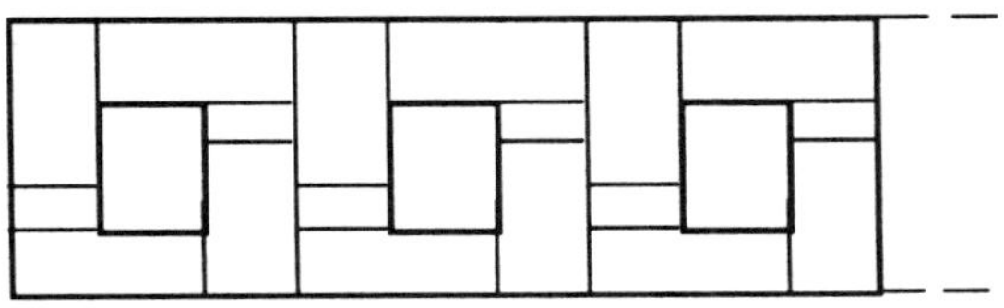

Fig. 3 A floorplan with n modules and $O(2^n)$ valid jog paths

over the four quadrants. The first two primitives are illustrated in Figs. 4a & 4b. Each of these nonslicing patterns has exactly one valid jog path ({5,4,3,6,7} and {6,7,3,2,10} respectively), which after straightening yields equivalent slicing patterns. The third possible primitive arises in the case shown in Fig. 4c, where there are three valid jog paths ({5,4,3,6,7}, {9,1,4,3,8} and {5,4,1,2,10}). Straightening any of them produces a slicing pattern which is specific to the jog path chosen. Fig. 4d depicts another possibility where there are three valid jog paths ({14,10,2,3,8}, {13,9, 14, 10,6,7} and {13, 5, 12, 8, 11}). The latter two paths do not touch the nonslice core {1,2,3,4}. Slicing patterns can be obtained by straightening any of these three. The validity of some these jog paths may be violated depending on the surrounding topology. So the primitive to be chosen for application must be decided accordingly.

We digress to mention that in rectangular dualization method, the boundary of an inner dual may or may not have a rectangular embedding depending on k, the number of boundary vertices having degree 2. If k = 4 (the minimum value is 4 from rectangularity), then the embedding is necessarily rectangular. For k > 4 (as in Fig. 4d), both rectangular and (isothetic) polygonal embeddings may be possible. However, if the inner dual has p pendant vertices (degree 1) on the boundary, then for p > 4, only polygonal embedding is possible.

Application of a primitive operator either dissolves the core entirely or *inflates* it towards the boundary. Inflation of nonslice core is said to occur if the number of modules enclosed by the core increases. These jog paths are exterior to the core. For example, in Fig 5a there are two nonslice cores {20, 21, 25, 26} and {22, 23, 24, 25} sharing a cut. The jog path {10, 19, 20, 21, 22, 23, 14} is chosen by applying primitive operations of Figs. 4b and 4c. After straightening, there is a single inflated nonslice core, shown in Fig. 5b. It should be noted that after inflation, the altered channel graph has smaller number of longer (length $\geq$ 6) directed cycles. By repeatedly inflating the core (Fig. 5d), it is possible to do away with all the longer directed cycles.

In some rare cases, although the pattern appears to match Fig. 4c, none of the three corresponding jog paths on the exterior of the core are valid. Only one valid jog path {10,19,20,22,13,14,5} through the interior of the core (Fig. 6) exists. This is a minimum example of an inner dual which exhibits such a property. Straightening this path results in a slicing. Next, *deflation* of the nonslice core i.e., decrease in the number of modules in the core, may occur in some cases. In the example shown in Fig. 7, straightening any of the two valid jog paths ({2,18,19,10,11,7} or {9, 10, 11, 12, 13, 5}) causes deflation. In the next phase of transformation, primitive operation of Fig. 4b produces slicing. So, oscillation between alternate inflation and deflation is precluded and termination is always guaranteed.

Definition: A canonical embedding of a rectangular dual is a rectilinear embedding which
(a) is a valid equivalent floorplan, and
(b) the corresponding channel digraph has the minimum number of directed 4-cycles.

Fig. 5d shows a canonical embedding of the nonslicible floorplans in Figs. 5a, 5b & 5c.

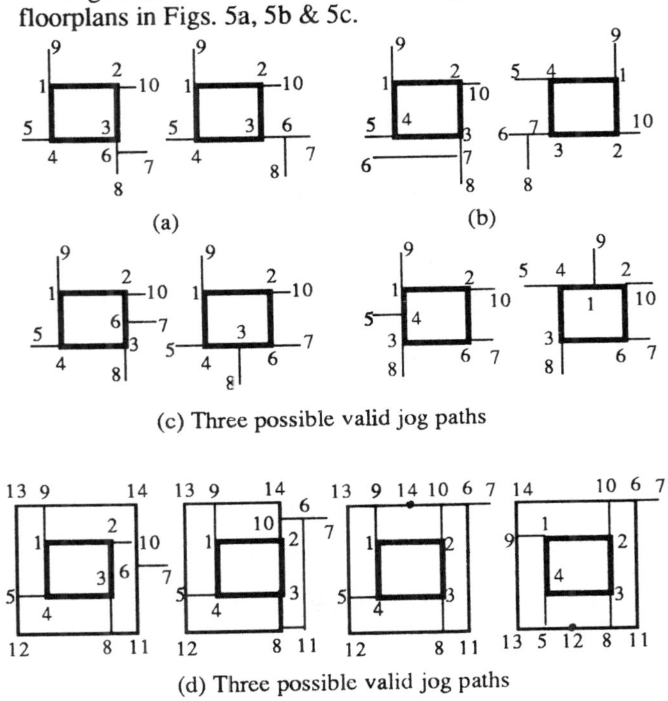

Fig. 4 Primitive operations in jog-path straightening

5. Generation of canonical embedding

Next, we discuss canonical embedding in general, including the family of inherently nonslicible floorplans. It is observed that the inherently nonslicible ones either do not have any valid jog path or straightening one produces another equivalent nonslicible floorplan and so on.

Fig. 8 shows a typical example of inherently nonslicible (INS) floorplans. In addition to this, numerous other nonisomorphic members occur [SB88, SB91a].

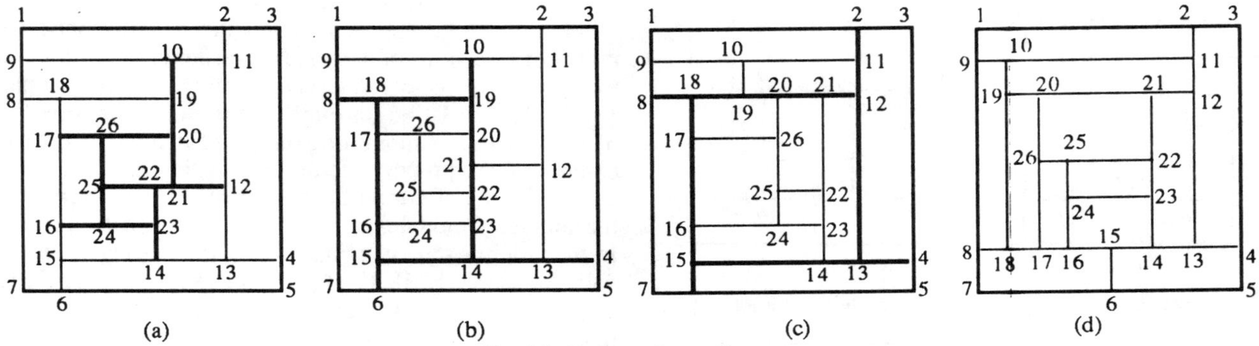

Fig. 5 Inflation of nonslice core

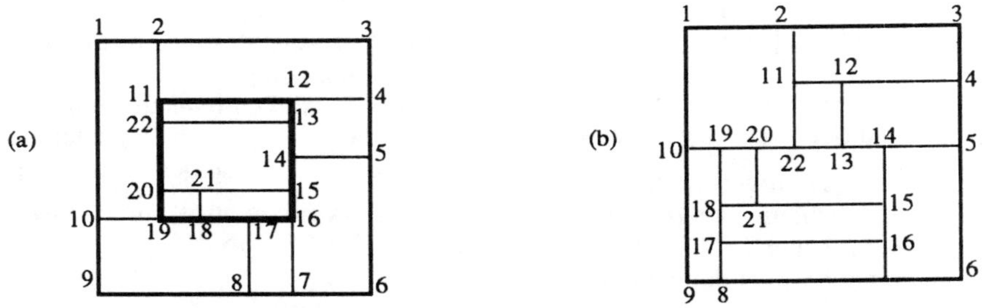

Fig. 6 : (a) A nonslicing inner dual where no valid jog path exists through the exterior of the nonslicing core {11, 12, 16, 19}; (b) Straightening the only valid jog path {10, 19, 20, 22, 13, 14, 5} passing through the *interior* of the core yields a slicible equivalent.

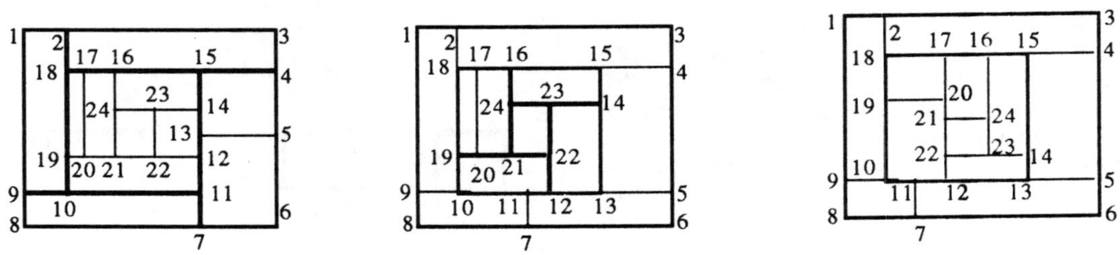

Fig. 7 Deflation of nonslice core

5.1 Maximal rectangular hierarchy

At this juncture, we revisit the concept of maximal rectangular hierarchy (MRH) [SB91a, SB91b, TW91]. In a floorplan, a **maximal rectangle** can be defined as one which is not contained in any rectangle in the floorplan other than the enveloping one. A nonslicible floorplan can be decomposed into a nonempty set of mutually exclusive (non-overlapping) and collectively exhaustive maximal rectangles.

Maximal rectangles can be defined recursively, as illustrated by heavy lines in Fig. 9a, to produce a hierarchy of maximal rectangles. A maximal rectangle may contain a slicible or a nonslicible pattern of rectangles. At the top level, we have just the bounding rectangle of the floorplan.

Definition: A maximal rectangle at any level of the MRH is said to be **strongly maximal** (smr) if it has exactly four T-junctions around it externally (e.g. the boundary, {8,9,10,11} {12,13,14,15} in Fig. 9a).

A *weakly maximal rectangle* (wmr) is therefore one which has more than four T-junctions around it externally (e.g. maximal rectangle {1,2,3,4,5} in Fig. 9a). In the case of a wmr, the next level within it cannot be ignored any longer while transforming nonslicible floorplan to its equivalent slicing because the cuts present in the inner level may provide a valid jog path for an adjacent nonslicing pattern and produce an equivalent slicing. For example, the inner dual in Fig. 6a can be transformed to an equivalent slicing since the cut (14,5) through a wmr {12,7,6,4} acts a handle [SB91a]. Furtermore, it is the wmr which has handles on the exterior. The nonslice cores in Figs. 4a , 4b & 4d are smrs wheras the one in Fig. 4c is wmr and so is the rectangle {11,12,13,14} in Fig. 4d.

In order to obtain a canonical embedding, we need to consider a strong MRH for a floorplan which focusses on the smrs only and their rectangular hierarchy. Moreover, it may be concluded that the recognition problem of INS can be simplified by utilizing the strong MRH decomposition.

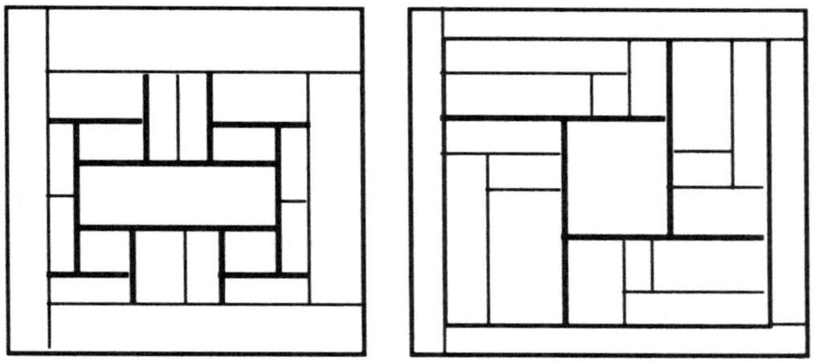

(a) (b)

Fig. 8 A member of INS

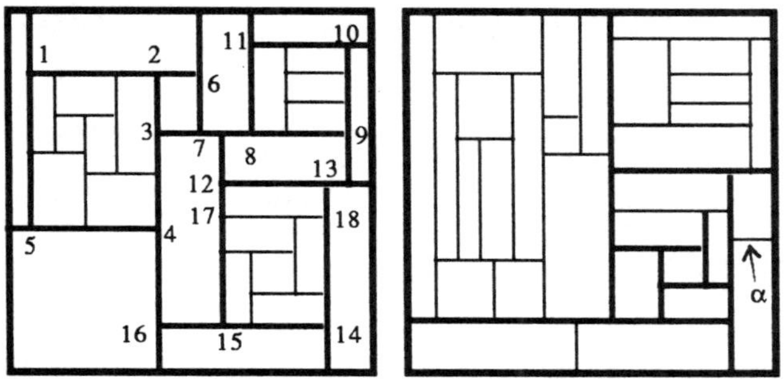

(a) (b)

Fig. 9 Maximal Rectangular Hierarchy

5.2 Examples and properties

Primitive operations for obtaining canonical embedding cause inflation or deflation of only those rectangles which are wmr's. Consider the INS floorplan of Fig. 8a which has four nonslice cores. Applying primitive operations of Fig. 4, the floorplan of Fig. 8b is obtained which has one nonslice core; the corresponding cut lines touch a bounding smr. No more reduction of the nonslice core is possible and this is a canonical embedding.

From the definition of primitives and the strong connectedness [SB91b] of channel graphs at any MRH level, it can be shown that the canonical embedding of a member of INS contains a smr having exactly one nonslice core.

A floorplan of n modules having an iterative structure as in Fig. 10a will have $O(\mathrm{n})$ directed 4-cycles and $O(2^{\mathrm{n}})$ longer directed cycles in its channel graph [SB91b]. Application of appropriate primitive operations produces an equivalent slicing, as shown in Fig. 10b. The canonical embedding of this floorplan has no nonslice cores.

Fig. 11a shows another interesting nonslicing floorplan. The channel graph of such an example with n modules has exactly one directed 4-cycle but $O(\mathrm{n})$ longer directed cycles [SB91b]. However, selecting the jog path (marked with thick lines in Fig. 11a), for transformation gives us a slicible equivalent, i.e., a canonical embedding with no nonslice cores as in Fig. 11b.

A canonical embedding of a rectangular dual may not be unique but the number of nonslice cores in it is minimum. If a canonical embedding has a nonslice core, then the smallest rectangle enclosing it must be an smr. All wmr's

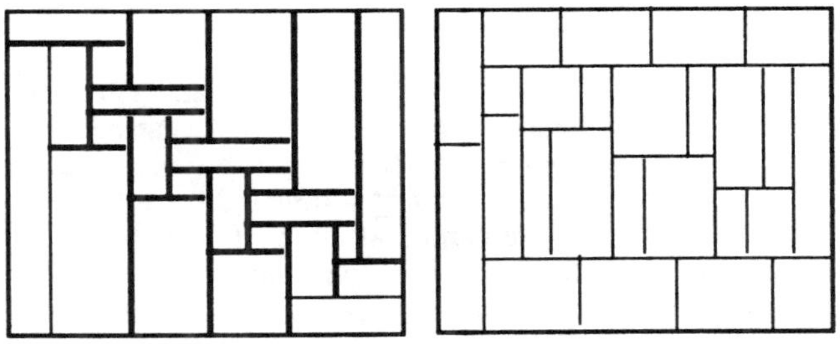

(a) (b)

Fig. 10 Canonical embedding of a typical example

can be made slicible. There *cannot* be any *longer cycle* (length >4) in the channel digraph of canonical embedding and each of the four cuts defining the corresponding core meets the bounding smr of that core. These observations lead to the following theorem.

Theorem: A canonical embedding of a rectangular dual has either 0 or 1 nonslice core at any level of the strong MRH.

Corollary: If the inner dual of a floorplan has σ smr's with nonempty interior, then there exists a canonical embedding with at most σ directed 4-cycles in its channel digraph.

Example: Consider a floorplan whose inner dual is shown in Fig. 9a. It has 4 smr's with nonempty interior e.g., the bounding one, {8,9,10,11}, {12,13,14,15} and {14,15,17,18}. Its canonical embedding consists of 2 nonslice cores (4-cycles) as shown in Fig. 9b (without the cut α). Thus it is a member of INS.

5.3 Algorithm for obtaining canonical embedding

Algorithm: Canonical-embedding
Input: A nonslicible floorplan with n modules
Output: Canonical embedding of the floorplan

Step 1. Extract the corresponding channel graph.
Step 2. Using depth-first-search method, find strongly connected components of channel graph to get MRH.
Step 3. Refine MRH to sMRH using strongly connected components (an smr corresponds to a directed 4-cycle).
Step 4. Process each level of sMRH independently (directed 4-cycles in the corresponding strongly connected component have already been identi fied in Step 3).

repeat
  if appropriate primitive operation can yield a valid jog path
  then straighten; either inflation or deflation occurs;
  else search for valid jog path through the interior of core;
until slicing is found or no further transformation is possible.

The channel graph changes after each operation of straightening a jog path, so updating is implied. The final output is a canonical embedding of the given floorplan.

Complexity

Step 1 requires linear time because the channel graph is planar. The depth-first search involved in step 2 can be done in linear time. In step 3, the above algorithm again takes $O(n)$ time since identification of a directed 4-cycle takes constant time and there can be at most $O(n)$ such cycles. Finally, in step 4, there can be linear number of iterations, since there can be those many inflations in the worst case. If there is deflation in any step, then it can be shown that in the next step, the concerned maximal rectangle becomes slicible. While choosing primitives, backtrack is bounded because the degree of search tree is bounded i.e. choice of jog paths during local inflation/deflation is bounded by a constant. Straightening a jog path including updation, which is the heart of the iterative step 4, can take $O(n)$ in the worst case. Hence the overall complexity of the above technique for obtaining a canonical embedding of a rectangular dual is $O(n^2)$. This method is extremely useful in both deciding membership in INS and finding feasible routing order from the canonical embedding.

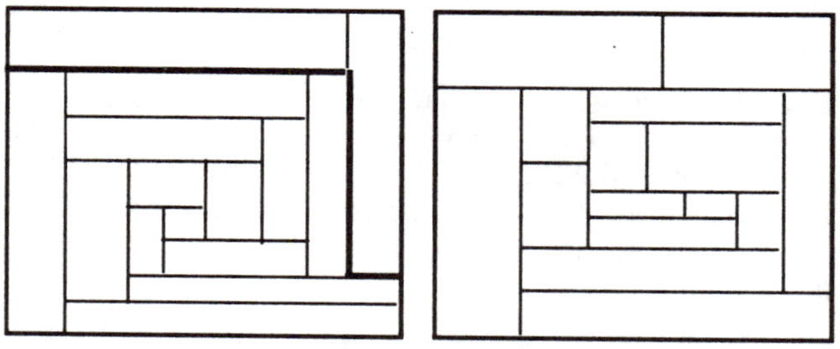

(a) (b)

Fig. 11 Canonical embedding of a nonslicing floorplan

6. Applications and Concluding Remarks

The concept of canonical embedding introduced in this paper, is significant for both the floorplanner and the channel router. It takes us a step further towards the open problem of completely characterizing the INS family. Next, the canonical embedding has the minimum number of directed cycles thereby simplifying the problem of determining the feasible order of routing channels immensely. At each level of the strong MRH of the canonical embedding, there remains at most one directed cycle.

Another important application of this concept is *design for slicibility*. A canonical embedding of a member of INS provides a floorplan closest to a slicible pattern. Having identified such a member of INS from its canonical embedding, one can proceed to propose floorplan design for slicibility by introducing appropriately extra cuts. For example, in Fig. 9b one needs to add simply the cut α to obtain an equivalent slicible inner dual. The physical implication of this special cut added is that one block is split into two.

The significant open problem related to design for slicibility is to minimize the number of extra cuts to be introduced. Naively, one may infer that one cut per level of sMRH is required. But as in the above example, there are two levels of strongly maximal rectangles whereas one cut is sufficient. There is scope for optimization in terms of nesting in the sMRH. Such a technique is currently being investigated. Furthermore, in the algorithm sketched here, aspect ratios of modules and in some cases orientation may change. These issues have also to be taken care of.

References

[BS86] Bhasker, J. & S. Sahni, "A Linear Algorithm to find a Rectangular Dual of a Planar Triangulated Graph," *Proc. 23rd DAC*, June 1986, pp. 108-114.

[DAK85] Dai, W. M., T. Asano & E. S. Kuh, " Routing Region Definition and Ordering Scheme for Building-Block Layout," *IEEE TCAD*, July 1985, pp. 189-197.

[GW88] Guruswamy, M. & D. F. Wong, " Channel Routing Order for Building-Block Layout with Rectilinear Modules," *Proc. ICCAD (IEEE)*, 1988, pp. 184-187.

[Ha69] Harary, F., Graph Theory, *Addison-Wesley Publishing Co.*, Reading, 1969.

[HSM82] Heller, W. R., G. Sorkin & K. Maling, "The Planar Package Planner for System Designers," *Proc. 19th DAC*, 1982, pp. 253-260.

[KK85] Kozminski, K. & E. Kinnen, "Rectangular Dual of Planar Graphs," *Networks*, Vol. 15, 1985, pp. 145-157.

[LL84] Leinwand, S. M. & Y.T. Lai, " A n Algorithm for Building Rectangular Floor-plans," *Proc. 21st DAC*, 1984, pp. 663-664.

[LL88] Leinwand, S.M. & Y.T. Lai, "Algorithm for Floorplan Design Via Rectangular Dualization," *IEEE TCAD*, Vol. CAD-7, No. 12, Dec. 1988, pp. 1278 - 1289.

[LK89] Lokanathan B. & E. Kinnen, " Performance Optimized Floor Planning by Graph Planarization," *Proc. 26th DAC, 1989*, pp. 116-121.

[LSW87] Luk, W. K., P. Sipala & C.K. Wong, "Minimum-area Wiring for Slicing Structures," *IEEE Trans. on Computers*, Vol. C-36, No.6, June 1987, pp. 745-760.

[Ot83] Otten, R.H.J.M., "Efficient Floorplan Optimization," *Proc. ICCD*, 1983, pp. 499-502.

[St83] Stockmeyer, L.J., "Optimal Orientations of Cells in Slicing Floorplan designs," *Information and Control*, Vol. 57, 1983, pp. 91-101.

[SB88] Sur-Kolay, S. & B. B. Bhattacharya, "Inherent Nonslicibility of Rectangular Duals in VLSI Floorplanning," *Lecture Notes in Computer Science*, No. 338, Springer Verlag (Berlin), 1988, pp. 88-107.

[SB91a] Sur-Kolay, S. & B. B. Bhattacharya, "On the Family of Inherently Nonslicible Floorplans in VLSI Layout Design," *Proc. ISCAS* (IEEE), pp. 2850-2853, Singapore, June 1991.

[SB91b] Sur-Kolay, S. & B. B. Bhattacharya, "The Cycle Structure of Channel Graphs in Nonslicible Floorplans and A Unified Algorithm for Feasible Routing Order," *Proc. ICCD*, pp. 524 - 527, October 1991.

[SS83] Supowit, K.J. & E.F. Slutz, "Placement Algorithms for Custom VLSI," *Proc. 20th DAC*, 1983, pp. 164-170.

[TW91] The, K. S. & D. F. Wong, "Area Optimization for Higher Order Hierarchical Floorplans," *Proc. ICCD*, pp. 520-523, October 1991.

[WKC88] Wimer, S., I. Koren & I. Cederbaum, "Floorplans, Planar Graphs and Layouts," *IEEE Trans. Circuits and Systems*, Vol. CAS-35, No. 3, March 1988, pp. 267-278.

Simulating Lossy Interconnect with High Frequency Nonidealities in Linear Time*

Jaijeet S. Roychowdhury   A. Richard Newton   Donald O. Pederson

Department of Electrical Engineering and Computer Sciences
University of California, Berkeley

Abstract

Transient simulation of lossy lines with high frequency nonidealities (such as skin effect) is usually performed by convolution, which involves computation quadratic in the number of time-points in the simulation. In this paper, a linear time simulation technique is described in which each impulse response is decomposed into two terms, with widely separated rates of decay. One term corresponds to a "simple lossy" line (a line with frequency-independent electrical parameters), while the other represents the contribution of the high frequency nonideality. The latter term can decay to insignificant levels in an interval much shorter than simulation times of interest, making truncation possible and leading to computation in linear time. The former term usually decays over much longer intervals, so truncation is not generally feasible. However, this term is also computed in linear time by a recursive procedure that involves no approximation of impulse responses. Experimental results verify the linear complexity of the new technique and demonstrate computational advantage for simulations of 5 or more cycles, with speedups of more than 10 times for long simulations.

1 Introduction

Efficient transient simulation of lossy transmission lines within nonlinear circuits has recently become a prominent problem. A major application is in the simulation of multi-chip modules (MCMs), high-performance packages that contain long thin interconnections connecting chips. At the high operating speeds of MCMs, interconnections exhibit transmission line effects like reflection, attenuation and dispersion, which can seriously degrade signal quality.

For many applications, transmission lines characterized by constant resistance, inductance, and capacitance (R_{lf}, L_{lf} and C_{lf}) per unit length ("simple lossy lines") are adequate for accurate simulation. At very fast rise times, however, it can be necessary to model these parameters as varying with frequency to account for nonideal effects ("frequency-varying models"). For several nonideal effects, the parameters differ from R_{lf}, L_{lf} and C_{lf} only at high frequencies (hence the term "high frequency nonidealities"). The most important phenomenon requiring frequency-varying models is skin effect, a high frequency nonideality that causes current to concentrate near the surface of the interconnect at high frequencies, thereby increasing the effective resistance. Skin effect can become significant in typical MCM interconnect at rise times less than 1ns[1], rounding off rising and falling edges (*skin effect induced dispersion* [1]).

Many techniques exist for the simulation of transmission line models within nonlinear circuits (see Section 2 for a brief review). The most accurate time domain technique for the simulation of lines with high frequency nonidealities is direct convolution, but this requires computation that rises quadratically with the number of time-points in the simulation. Existing schemes that reduce the complexity to linear rely on assuming specific forms for impulse responses (such as sums of exponentials in the time domain, or rational functions in the frequency domain).

In this work, the impulse responses of frequency-varying lines are first decomposed into two parts, one the response of a simple lossy line, the other the contribution of the nonideality. Second, the fact that the physical mechanisms of *high frequency* nonidealities operate in time scales that are short relative to time scales of interest in practical MCM applications is exploited. This translates to quickly decaying responses for the nonideal part, making truncation, and hence computation in linear time, possible. Truncation is not generally feasible for the responses of the simple lossy line part which can decay slowly. It is shown that the convolution operations involved in the simple lossy part of the computation can be avoided, replaced by a linear time recursive computation which involves no approximation of impulse responses. The validity of the recursive procedure does not follow immediately from our earlier work [2] (the state-based method)[2]. The addition of nonideal terms changes relationships between waveforms that were part of the basis of the recursive formulation for the simple lossy case[3].

Apart from linear complexity, this technique has the advantage of not assuming specific forms for the nonideal responses. Models of differing levels of sophistication exist for nonideal behaviour [3, 1, 4, 5], with responses that differ in the details but share the same time scale. The approach presented here is capable of simulating any such model, including empirically measured data characterising nonidealities. Detailed knowledge of internal physical mechanisms is not required; phenomena such as skin effect that arise from non-uniform current or voltage distributions along the transverse dimensions of the line can be simulated without keeping track of the transverse variation. The price paid for this generality is convolution with the nonideal response; however, exploiting the short time scale of the phenomena generating this response makes computation in linear time possible. For the impulse responses of the simple lossy line, the time scale is not short; for this part, knowledge of the internal mechanism is exploited for efficient simulation, through the recursive state-based technique. This is not restrictive because this internal mechanism is fundamental to all models of transmission lines. All features of the state-based technique for simple lossy lines, like dynamic allocation of sample points for the state (see [2]), apply fully to the the recursive computation part of the nonideal line.

In Section 2, previous approaches to lossy line simulation are reviewed. Section 3 contains the mathematical details of the decomposition and recursive computation. Experimental results are presented in Section 4.

*This research was supported by AT&T Bell Laboratories, IBM, Raytheon and the California State MICRO program.

[1] Other high frequency nonidealities such as dielectric loss [1] can usually be ignored for typical MCM packages.

[2] The state-based method is a linear time recursive procedure for simple lossy lines, i.e., lines *without* nonidealities.

[3] If the transmission line equations are expressed in the admittance, impedance or hybrid representations (i.e., two of the port variables can be clearly identified as "outputs" and the other two as "inputs"), the validity of the recursive procedure follows immediately from our previous work. Using these representations, however, results in impulse responses for the nonideal term that are not eventually monotonically decreasing, and truncation is not possible. Another seemingly workable approach is to precondition the port variables of a simple lossy line such that the preconditioned system is equivalent to a frequency-varying line. This approach is not feasible in a time-stepping simulator because the preconditioning operation can involve convolution with non-causal responses. This work (in Section 3) justifies recursion in the scattering-parameter representation (in which the impulse responses do decay monotonically) with nonidealities.

0738-100X/92 $3.00 © 1992 IEEE

2 Previous Work

The convolution approach is the most accurate and general approach for the time domain simulation of nonideal lossy lines within nonlinear circuits. ([6, 7] contain descriptions of other methods for lossy line simulation.) Liu et al [8] were apparently first to use convolution to simulate devices similar to lines, i.e., antennas, in conjunction with nonlinear loads. Djordjević et al [9] applied the technique to simulate lossy lines. That the lossy line is a linear two-port was exploited by expressing the "outputs" of the two-port as a convolution of the "inputs" with impulse responses (or Green's functions) characteristic of the line. In a comparison of techniques for lossy line simulation [6], convolution was found slow and only slightly more accurate than the approaches involving lumped approximations. In the early formulation, the duration and nature of the impulse responses was a limitation: for lossless lines, the impulse responses were infinite in duration and consisted of periodic sharp spikes; those of lossy lines diminished gradually while being otherwise similar. Impulse responses that fall to negligible levels after a short time are desirable for computational efficiency[4]. Augmenting the line with a quasi-matched load was attempted in [9] to shorten the responses. Schutt-Aine et al [10, 11] reformulated the convolution equations using a scattering parameter approach that led to "well behaved" impulse responses[5]. Even though these responses decreased monotonically, their effective duration (depending on the values of R, L, G, and C) could be greater than the total simulation time, leading to quadratic complexity. Approaches reducing this complexity to linear have relied mainly on assuming specific functional forms for the impulse responses. Semlyen et al [12] have used sums of exponentials for time domain approximation, while frequency domain simplifications using rational function and Padé approximations [13, 14], essentially translating to exponentials in the time domain, are available. A different approach to linear complexity applicable only to simple lossy lines (the state-based method, [2]) was proposed recently by the present authors. This technique is equivalent to convolution in accuracy and does not involve any approximation of impulse responses.

3 State-Based Formulation with Nonideal Convolution Terms

For a frequency-varying transmission line with parameters $R(s)$, $L(s)$, $C(s)$ and $G(s)$[6], the constitutive equations are[15]:

$$\left[Y_0 V_2(s) H_Y(s) - I_2(s)\right] - \left[Y_0 V_1(s) H_{\gamma Y}(s) + I_1(s) H_\gamma(s)\right] = 0 \quad (1)$$

$$\left[Y_0 V_2(s) H_{\gamma Y}(s) + I_2(s) H_\gamma(s)\right] - \left[Y_0 V_1(s) H_Y(s) - I_1(s)\right] = 0 \quad (2)$$

where $V_1(s)$, $V_2(s)$, $I_1(s)$ and $I_2(s)$ are the Laplace transforms of the voltages and currents at the two ends of the line, which is of length l, and

$$Y_0 = \sqrt{\frac{C_{lf}}{L_{lf}}}, \quad H_Y(s) = \frac{Y(s)}{Y_0}, \quad H_\gamma(s) = e^{-\lambda(s) l} \quad (3)$$

$$H_{\gamma Y}(s) = H_Y(s) H_\gamma(s), \quad Y(s) = \sqrt{\frac{sC(s)+G(s)}{sL(s)+R(s)}} \quad (4)$$

$$\lambda(s) = \sqrt{\left(sC(s)+G(s)\right)\left(sL(s)+R(s)\right)}, \quad (5)$$

[4]If the responses remain significant for a period greater than the total simulation time, convolution from time $t_0 = 0$ to $t = t_n$ is necessary at every time-point t_n. This operation requires computation proportional to n, resulting in total computation proportional to N^2 (quadratic time complexity) in a simulation with N time-points.

[5]The "well behaved" impulse response for a lossless line is a single Dirac delta function at the line delay; for lossy lines, the impulse responses decrease asymptotically and monotonically to zero.

[6]s denotes the Laplace variable ($= j2\pi f$ for the Fourier transform).

$$R_{lf} = R(0), \quad L_{lf} = L(0) \quad (6)$$

$$C_{lf} = C(0), \quad G_{lf} = G(0) \quad (7)$$

Inverse Laplace transformation yields the time domain constitutive equations:

$$\left[Y_0 v_2(t) * h_Y(t) - i_2(t)\right] - \left[Y_0 v_1(t) * h_{\gamma Y}(t) + i_1(t) * h_\gamma(t)\right] = 0 \quad (8)$$

$$\left[Y_0 v_2(t) * h_{\gamma Y}(t) + i_2(t) * h_\gamma(t)\right] - \left[Y_0 v_1(t) * h_Y(t) - i_1(t)\right] = 0 \quad (9)$$

where $*$ is the convolution operator[7], $v_1(t)$, $v_2(t)$, $i_1(t)$, $i_2(t)$ are the port variables of the line, and $h_Y(t)$, $h_\gamma(t)$ and $h_{\gamma Y}(t)$ are the inverse Laplace transforms of $H_Y(s)$, $H_\gamma(s)$ and $H_{\gamma Y}(s)$, respectively.

The above convolution formulation can be used directly to simulate the frequency-varying line. However, it is convenient to rewrite Equations 1 and 2 in the following equivalent form for the purposes of the decomposition mentioned in Section 1:

$$\left[Y_0 V_2(s) H_Y^{lf}(s) - I_2(s)\right] - \left[Y_0 V_1(s) H_{\gamma Y}^{lf}(l,s) + I_1(s) H_\gamma^{lf}(l,s)\right] + U_1(s) = 0 \quad (10)$$

$$\left[Y_0 V_2(s) H_{\gamma Y}^{lf}(l,s) + I_2(s) H_\gamma^{lf}(l,s)\right] - \left[Y_0 V_1(s) H_Y^{lf}(s) - I_1(s)\right] + U_2(s) = 0 \quad (11)$$

where:

$$H_Y^{lf}(s) = \frac{Y^{lf}(s)}{Y_0}, \quad H_\gamma^{lf}(x,s) = e^{-\lambda^{lf}(s)x} \quad (12)$$

$$H_{\gamma Y}^{lf}(x,s) = H_Y^{lf}(s) H_\gamma^{lf}(x,s), \quad Y^{lf}(s) = \sqrt{\frac{sC_{lf}+G_{lf}}{sL_{lf}+R_{lf}}} \quad (13)$$

$$\lambda^{lf}(s) = \sqrt{\left(sC_{lf}+G_{lf}\right)\left(sL_{lf}+R_{lf}\right)} \quad (14)$$

$U_1(s)$ and $U_2(s)$ represent the contribution of the frequency-varying part of the line:

$$U_1(s) = \left[Y_0 V_2(s) \Delta H_Y(s)\right] - \left[Y_0 V_1(s) \Delta H_{\gamma Y}(s) + I_1(s) \Delta H_\gamma(s)\right] \quad (15)$$

$$U_2(s) = \left[Y_0 V_2(s) \Delta H_{\gamma Y}(s) + I_2(s) \Delta H_\gamma(s)\right] - \left[Y_0 V_1(s) \Delta H_Y(s)\right] \quad (16)$$

where:

$$\Delta H_Y(s) = H_Y(s) - H_Y^{lf}(s), \quad \Delta H_\gamma(s) = H_\gamma(s) - H_\gamma^{lf}(l,s) \quad (17)$$

$$\Delta H_{\gamma Y}(s) = H_{\gamma Y}(s) - H_{\gamma Y}^{lf}(l,s) \quad (18)$$

Laplace inversion of Equations 10 and 11 yields the time domain formulation:

$$\left[Y_0 v_2(t) * h_Y^{lf}(t) - i_2(t)\right] - \left[Y_0 v_1(t) * h_{\gamma Y}^{lf}(l,t) + i_1(t) * h_\gamma^{lf}(l,t)\right] + u_1(t) = 0 \quad (19)$$

$$\left[Y_0 v_2(t) * h_{\gamma Y}^{lf}(l,t) + i_2(t) * h_\gamma^{lf}(l,t)\right] - \left[Y_0 v_1(t) * h_Y^{lf}(t) - i_1(t)\right] + u_2(t) = 0 \quad (20)$$

where $h_Y^{lf}(t)$, $h_\gamma^{lf}(x,t)$ and $h_{\gamma Y}^{lf}(x,t)$[8] are the inverse Laplace transforms of $H_Y^{lf}(s)$, $H_\gamma^{lf}(x,s)$ and $H_{\gamma Y}^{lf}(x,s)$ respectively. $u_1(t)$ and $u_2(t)$, the inverse transforms of $U_1(s)$ and $U_2(s)$, are given by:

[7]$a(t) * b(t) = \int_0^t a(\tau) b(t-\tau)\, d\tau$.

[8]Analytic expressions for these are given later in this section.

$$u_1(t) = \left[Y_0\, v_2(t) * \triangle h_Y(t)\right] - \left[Y_0\, v_1(t) * \triangle h_{\gamma Y}(t) + i_1(t) * \triangle h_\gamma(t)\right] \quad (21)$$

$$u_2(t) = \left[Y_0\, v_2(t) * \triangle h_{\gamma Y}(t) + i_2(t) * \triangle h_\gamma(t)\right] - \left[Y_0\, v_1(t) * \triangle h_Y(t)\right] \quad (22)$$

where

$$\triangle h_Y(t) = h_Y(t) - h_Y^{lf}(t), \qquad \triangle h_\gamma(t) = h_\gamma(t) - h_\gamma^{lf}(l,t) \quad (23)$$

$$\triangle h_{\gamma Y}(t) = h_{\gamma Y}(t) - h_{\gamma Y}^{lf}(l,t) \quad (24)$$

Equations 19 and 20 are equivalent to Equations 8 and 9. The motivation behind rewriting the latter equations as the former stems from that all terms except u_1 and u_2 can be computed recursively, as shown later in this section. Calculating $u_1(t)$ and $u_2(t)$ (according to Equations 21 and 22) requires convolution; if, however, the responses $\triangle h_Y$, $\triangle h_\gamma$ and $\triangle h_{\gamma Y}$ can be truncated with insignificant loss of accuracy (as in the high frequency nonidealities being considered), convolution transforms into an integral with fixed limits.

The existing formulation of the state-based method [2] is applicable only to the dc-parameter case, for which the constitutive equations are (Equations 19 and 20 with $u_1(t) = u_2(t) \equiv 0$):

$$\left[Y_0\, v_2(t) * h_Y^{lf}(t) - i_2(t)\right] - \left[Y_0\, v_1(t) * h_{\gamma Y}^{lf}(l,t) + i_1(t) * h_\gamma^{lf}(l,t)\right] = 0 \quad (25)$$

$$\left[Y_0\, v_2(t) * h_{\gamma Y}^{lf}(l,t) + i_2(t) * h_\gamma^{lf}(l,t)\right] - \left[Y_0\, v_1(t) * h_Y^{lf}(t) - i_1(t)\right] = 0 \quad (26)$$

The state-based method, as derived in [2], asserts that Equations 25 and 26 are equivalent to a different set of equations that do not use convolution but involve an internal state term. Thus the expressions on the LHS of Equations 25 and 26 are in essence computed in a recursive manner, without using past values of v_1, v_2, i_1 and i_2. *Implicit in this procedure is the assumption that $v_1(t)$, $v_2(t)$, $i_1(t)$ and $i_2(t)$ are constrained to obey Equations 25 and 26 at any t. If this assumption does not hold (as in Equations 19 and 20 with nonzero $u_1(t)$ and $u_2(t)$), then it does not automatically follow from [2] that the recursive procedure is still equivalent to the LHS of Equations 25 and 26.*

In this work, it is proved that the recursive computation of the state-based procedure is always equivalent to the LHS of Equations 25 and 26, i.e., for *arbitrary* $v_1(t)$, $v_2(t)$, $i_1(t)$ and $i_2(t)$. A set of partial differential equations (PDEs) are constructed such that their solution yields Equations 19 and 20. An alternative solution procedure that replaces all terms except $u_1(t)$ and $u_2(t)$ by a recursive computation is developed. Then the fact that the PDEs must have a unique solution given a unique set of external influences is used to assert the equivalence of the recursive procedure and the LHS of Equations 25 and 26. The key difference between this procedure and the one in [2] is the addition of terms w_1 and w_2 (see Equations 27 and 28 in the next paragraph) that are contributed by the nonideality. While these terms do not admit of any obvious physical interpretation, their inclusion leads to both the nonideal convolution equations (Equations 19 and 20) and the recursive technique for the simple lossy line part. As mentioned earlier, attempts to introduce the nonideal term through the port variables and formulations other than the scattering-parameter one lead to noncausal situations or destroy the quickly decaying property of the purely nonideal responses.

Let $w_1(x,t)$ and $w_2(x,t)$ be two arbitrary functions defined on $[0,l] \times [0,\infty)$. Consider the following (the familiar Telegrapher Equations [11] with "inputs"[9] w_1 and w_2):

$$\frac{\partial v}{\partial x} = -\left(L_{lf}\,\frac{\partial i}{\partial t} + R_{lf}\, i\right) + w_1 \quad (27)$$

$$\frac{\partial i}{\partial x} = -\left(C_{lf}\,\frac{\partial v}{\partial t} - G_{lf}\, v\right) + w_2 \quad (28)$$

The above equations hold for x varying between 0 and l, the length of the transmission line. $v(x,t)$ and $i(x,t)$ are the voltage and current at the point x in the line at time t, respectively. It is assumed that the simulation starts from time 0.

The circuit interacts with the transmission line through the port variables $v_1(t) = v(0,t)$, $i_1(t) = i(0,t)$, $v_2(t) = v(l,t)$ and $i_2(t) = -i(l,t)$. These four port variables specify the *boundary conditions* of Equations 27 and 28.

In addition to w_1, w_2, and the port variables, the internal state of the transmission line also determines the future behaviour of the line. This internal state is stored in the energy-storing distributed inductance and capacitance and is specified by the voltages and currents in the line's interior at time 0, $v_0(x) = v(x,0)$ and $i_0(x) = i(x,0)$. $v_0(x)$ and $i_0(x)$ are the *initial conditions* for Equations 27 and 28. The combination of the Telegrapher Equations, the boundary and initial conditions and the "inputs" w_1 and w_2 specify the future behaviour of the line uniquely.

Equations 27 and 28 are solved to obtain time domain relations between w_1, w_2, $v(x,t)$, $i(x,t)$ and the internal state $v_0(x)$, $i_0(x)$. (The derivation is omitted for brevity; see [16] for the details.) In the process, w_1 and w_2 are chosen as functions of u_1 and u_2 (of Equations 21 and 22). Moreover, the state $v_0(x)$, $i_0(x)$ is represented by samples v_{0j}, i_{0j} at points x_j, $j = 1, \ldots, n_l$, inside the line. n_l is the total number of state samples; the index of a sample point at x is denoted by n_x, with $n_0 = 1$. The time domain equations finally obtained for the nonideal line, valid at any $x \in [0,l]$, are:

$$\left[Y_0\, v(x,t) * h_Y^{lf}(t) + i(x,t)\right] - \left[Y_0\, v(0,t) * h_{\gamma Y}^{lf}(x,t) + i(0,t) * h_\gamma^{lf}(x,t)\right] + u_1(t)\,\chi(x-l) = \sum_{j=1}^{n_x} \left\{ \begin{array}{l} i_{0j}\left[h_{S\gamma Y}(x-x_j,t) - h_{S\gamma Y}(x-x_{j-1},t)\right] + \\ Y_0\, v_{0j}\left[h_{S\gamma}(x-x_j,t) - h_{S\gamma}(x-x_{j-1},t)\right] \end{array} \right\} \quad (29)$$

$$\left[Y_0\, v(l,t) * h_{\gamma Y}^{lf}(l-x,t) - i(l,t) * h_\gamma^{lf}(l-x,t)\right] - \left[Y_0\, v(x,t) * h_Y^{lf}(t) - i(x,t)\right] + u_2(t)\,\chi(x) = \sum_{j=n_x}^{n_l} \left\{ \begin{array}{l} i_{0j}\left[h_{S\gamma Y}(x_{j-1}-x,t) - h_{S\gamma Y}(x_j-x,t)\right] \\ -Y_0\, v_{0j}\left[h_{S\gamma}(x_{j-1}-x,t) - h_{S\gamma}(x_j-x,t)\right] \end{array} \right\} \quad (30)$$

where:

$$h_Y(t) = \left[\delta(t) + \alpha\{I_1(\alpha t) - I_0(\alpha t)\}\right] e^{-\beta t} \quad (31)$$

$$h_\gamma(x,t) = \left[\delta(t - \gamma_0 x) + u(t-\gamma_0 x)\,\frac{\alpha \gamma_0 x\, I_1\left(\alpha\sqrt{t^2 - (\gamma_0 x)^2}\right)}{\sqrt{t^2 - (\gamma_0 x)^2}}\right] e^{-\beta t} \quad (32)$$

$$h_{\gamma Y}(x,t) = \left[\delta(t-\gamma_0 x) + u(t-\gamma_0 x)\,\alpha \times \left\{\frac{t I_1\left(\alpha\sqrt{t^2-(\gamma_0 x)^2}\right)}{\sqrt{t^2-(\gamma_0 x)^2}} - I_0\left(\alpha\sqrt{t^2-(\gamma_0 x)^2}\right)\right\}\right] e^{-\beta t} \quad (33)$$

$$h_{S\gamma}(x,t) = u(t-\gamma_0 x)\, I_0\left(\alpha\sqrt{t^2-(\gamma_0 x)^2}\right) e^{-\beta t} \quad (34)$$

[9] Not to be confused with the port variables.

$$h_{S\gamma Y}(x,t) = u(t-\gamma_0 x)\left[e^{-\alpha(t-\gamma_0 x)} + \alpha\gamma_0 x\int_{\gamma_0 x}^{t} e^{-\alpha(t-\tau)}\frac{I_1\left(\alpha\sqrt{\tau^2-(\gamma_0 x)^2}\right)}{\sqrt{\tau^2-(\gamma_0 x)^2}}\,d\tau\right] e^{-\beta t} \quad (35)$$

In the above, $u(\cdot)$ and $\delta(\cdot)$ denote the unit step and delta functions, respectively. $\chi(x)$ is defined to be 1 if $x=0$, 0 otherwise.

The equations obtained by substituting $x=l$ in Equation 29 and $x=0$ in Equation 30 are of special interest:

$$\begin{aligned}&\left[Y_0\, v_2(t) * h_Y^{lf}(t) - i_2(t)\right] \\ &\quad - \left[Y_0\, v_1(t) * h_{\gamma Y}^{lf}(l,t) + i_1(t) * h_\gamma^{lf}(l,t)\right] + u_1(t) \\ &\quad = \sum_{j=1}^{n_l}\left\{\begin{array}{c} i_{0j}\left[h_{S\gamma Y}(l-x_j,t) - h_{S\gamma Y}(l-x_{j-1},t)\right] + \\ Y_0\, v_{0j}\left[h_{S\gamma}(l-x_j,t) - h_{S\gamma}(l-x_{j-1},t)\right]\end{array}\right\}\end{aligned} \quad (36)$$

$$\begin{aligned}&\left[Y_0\, v_2(t) * h_{\gamma Y}^{lf}(l,t) + i_2(t) * h_\gamma^{lf}(l,t)\right] \\ &\quad - \left[Y_0\, v_1(t) * h_Y^{lf}(t) - i_1(t)\right] + u_2(t) \\ &\quad = \sum_{j=1}^{n_l}\left\{\begin{array}{c} i_{0j}\left[h_{S\gamma Y}(x_{j-1},t) - h_{S\gamma Y}(x_j,t)\right] \\ -Y_0\, v_{0j}\left[h_{S\gamma}(x_{j-1},t) - h_{S\gamma}(x_j,t)\right]\end{array}\right\}\end{aligned} \quad (37)$$

The above are the equations of the convolution technique with an additional RHS term from the initial state. At the start of the simulation, the initial state is assumed zero[10]. The RHS then becomes zero, and the two equations become identical to Equations 19 and 20, which, by construction, are equivalent to Equations 8 and 9. These equations may be used directly for simulation by convolution: at time-point t_i, each convolution on the LHS is carried out from $t_0 = 0$ to t_i, using stored values of v_1, v_2, i_1 and i_2 at t_j for all j, $0 \le j < i$. The computation required at each t_i is proportional to i, leading to the quadratic complexity property.

The fact that Equations 36 and 37 are derived from a system of PDEs makes it possible to solve them in another way: at time t_1 (the first time-point after $t_0 = 0$), the procedure is identical to the convolution method, with convolution over the interval $[t_0, t_1]$, and initial state zero. The new state of the line at time t_1 is calculated next, using Equations 29 and 30 for each x chosen to sample the internal state. At the next time-point t_2, the newly calculated, nonzero internal state at t_1 is used as $v_0(x)$ and $i_0(x)$, instead of the zero internal state at time t_0, in Equations 29, 30, 36 and 37. In other words, the time-invariance property of the Telegrapher Equations is used. Thus the convolution operation at time t_2 starts from the previous time-point t_1, i.e., the time at which the internal state is used in the above equations, instead of from t_0 as in the convolution method. The internal state at t_2 is then calculated, and this internal state is used as the initial state at time t_3, and so on. This procedure is the basis of the recursive computation; information from the port variables' history, which keeps increasing in size as the simulation progresses, is condensed into the state, which is of constant size. The key feature leading to linear complexity in this technique is that the computation at time t_i is independent of i, since convolution is always performed over only one interval, $[t_{i-1}, t_i]$.

This recursive procedure is valid for arbitrary $u_1(t)$ and $u_2(t)$, not just the particular choices of Equations 21 and 22. The validity of the recursion in computing the LHS of Equations 25 and 26 for arbitrary $v_1(t)$, $v_2(t)$, $i_1(t)$ and $i_2(t)$ follows immediately.

4 Experimental Results

The techniques outlined in the previous section were implemented in the circuit simulator SPICE3 version 3e1 [17]. Results on an example circuit with typical MCM interconnect are presented in this section.

The physical parameters of the aluminium interconnect (taken from [18]) are: width = 11μm, thickness = 2μm, height above ground plane = 10μm, SiO_2 dielectric, length of line = 10cm, R_{lf} = 12 ohms/cm, L_{lf} = 8.79 nH/cm, C_{lf} = 1 pF/cm. Dielectric loss (parallel conductance G) is assumed to be zero at all frequencies. Skin effect is accounted for with the following model [1]:

$$R(f) = \begin{cases} R_0\sqrt{f} & f \ge f_T \\ R_{lf} & f < f_T \end{cases} \quad (38)$$

$$L(f) = \begin{cases} L_{ext} + \frac{L_0}{\sqrt{f}} & f \ge f_T \\ L_{lf} & f < f_T \end{cases} \quad (39)$$

$C(f)$ is assumed constant at C_{lf}; f_T, the transition frequency after which skin effect becomes significant, is taken to be 1 GHz from [1]; L_{ext}, the inductance external to the wire and less than L_{lf}, is assumed to be 4 nH/cm. From these, R_0 and L_0 are calculated to be 3.7947×10^{-4} and 1.5147×10^{-4} respectively. The ideal delay $T = l\sqrt{L_{lf} C_{lf}}$ is 0.9375 ns.

The impulse responses of the pure nonideality Δh_Y, Δh_γ and $\Delta h_{\gamma Y}$ contain impulse (delta function) components which are first separated out. $\Delta h'_Y$, $\Delta h'_\gamma$ and $\Delta h'_{\gamma Y}$, the components without impulses, were calculated by numerical inversion of the Fourier transform, and are shown in Figs. 1, 2 and 3. The following relations hold:

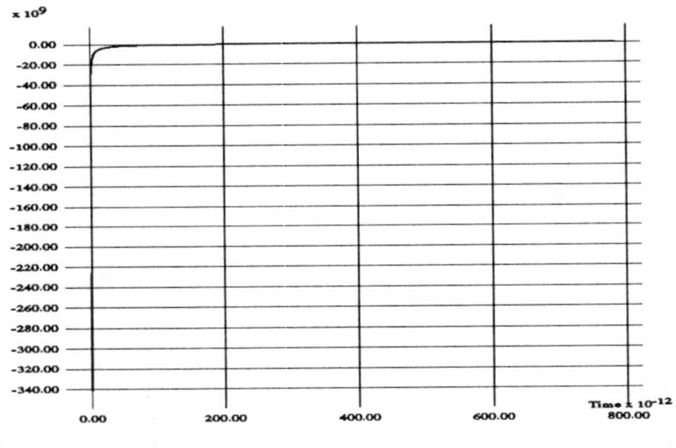

Figure 1: $\Delta h'_Y(t)$

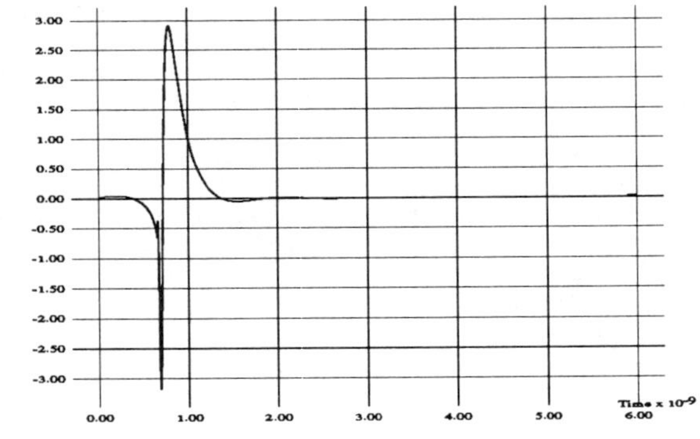

Figure 2: $\Delta h'_\gamma(t)$

[10] If the circuit is at a nonzero DC initial state, a simple reformulation exploiting the linearity of the transmission line yields a system with zero initial state.

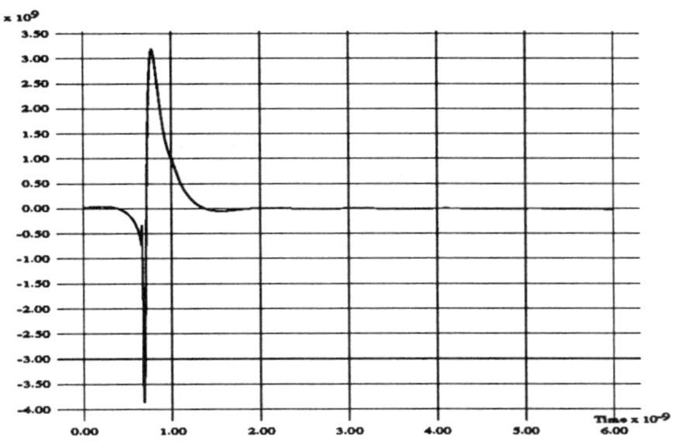

Figure 3: $\triangle h'_{\gamma Y}(t)$

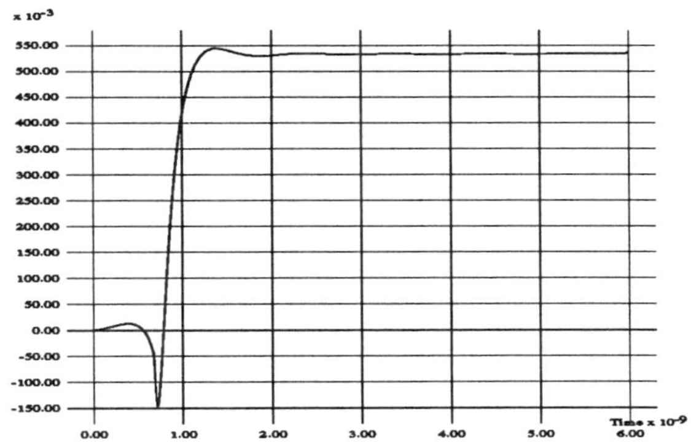

Figure 5: $\int_0^t \triangle h'_{\gamma}(\tau)\,d\tau$

$$\triangle h_Y(t) = M_Y\,\delta(t) + \triangle h'_Y(t) \tag{40}$$

$$\triangle h_{\gamma}(t) = M_{\gamma}\,\delta(t-T) + \triangle h'_{\gamma}(t) \tag{41}$$

$$\triangle h_{\gamma Y}(t) = M_{\gamma Y}\,\delta(t-T) + \triangle h'_{\gamma Y}(t) \tag{42}$$

For this example, $M_Y = 0.420210266, M_{\gamma} = -0.5326360212$, and $M_{\gamma Y} = -0.5160118569$. In order to obtain a quantitative estimate of the error caused by truncation, the functions $\int_0^t \triangle h'_Y(\tau)\,d\tau$, $\int_0^t \triangle h'_{\gamma}(\tau)\,d\tau$, and $\int_0^t \triangle h'_{\gamma Y}(\tau)\,d\tau$ are useful. Plots of these functions for this example are shown in Figs. 4, 5 and 6; the point after which these functions achieve a steady value is the effective length of the response.

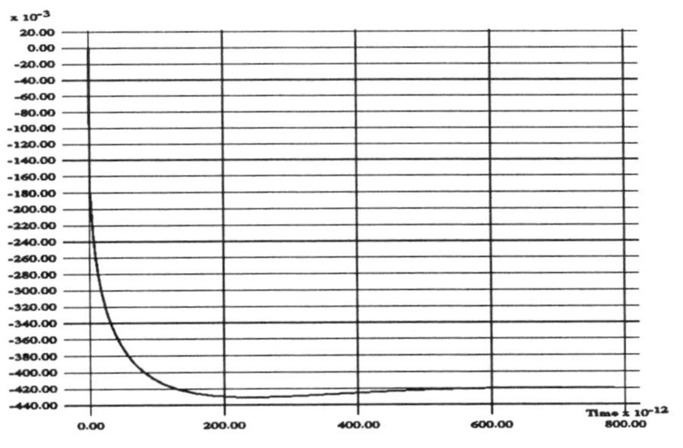

Figure 4: $\int_0^t \triangle h'_Y(\tau)\,d\tau$

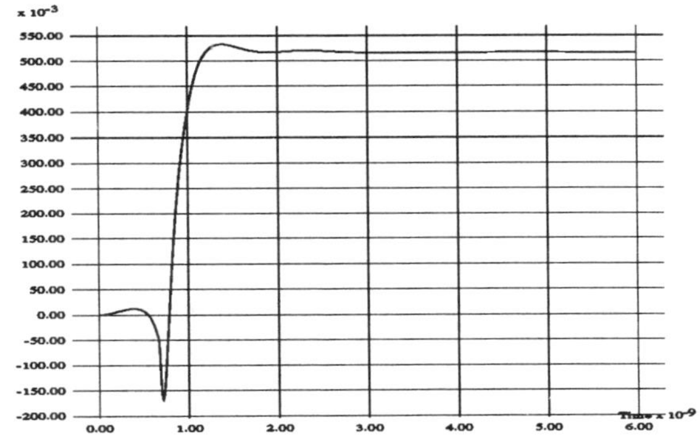

Figure 6: $\int_0^t \triangle h'_{\gamma Y}(\tau)\,d\tau$

For this example, the effective durations of $\triangle h'_Y$, $\triangle h'_{\gamma}$ and $\triangle h'_{\gamma Y}$ were 600ps, 3ns and 3ns respectively, as seen from the plots of their integrals. Convolution was performed using the numerical formula in [15].

The circuit consisted of a nonlinear load of two clamping diodes driven by a voltage source with a series resistance through the lossy line. Rise times of 10ps and 1ns were used.

Fig. 7 illustrates the voltage at the load end for the 10ps rise-time case. Simulation by pure convolution and the state-based/conv. technique of this paper are seen to yield identical results. Also shown is the result of simulation without skin effect, i.e., with only DC losses considered. The rise time degradation caused by skin effect, and the rounding of edges, can be seen clearly.

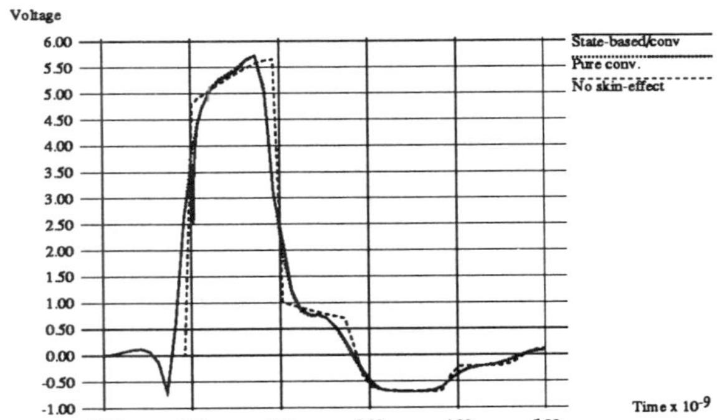

Figure 7: Load voltage, 10ps rise time

Fig. 8 shows the load voltage for a rise time of 1ns. It is seen that skin effect makes virtually no difference to the waveform in this case.

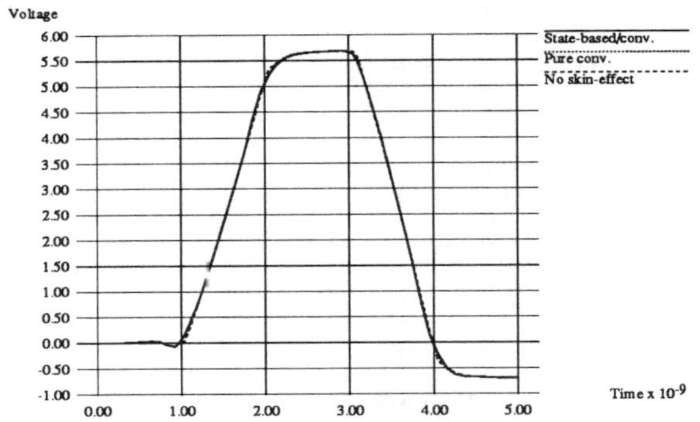

Figure 8: Load voltage, 1ns rise time

Table 1 and Fig. 9 show computation time as a function of total simulation length for the state-based and the pure convolution techniques. The linear and quadratic complexities of the two methods are apparent from the figure.

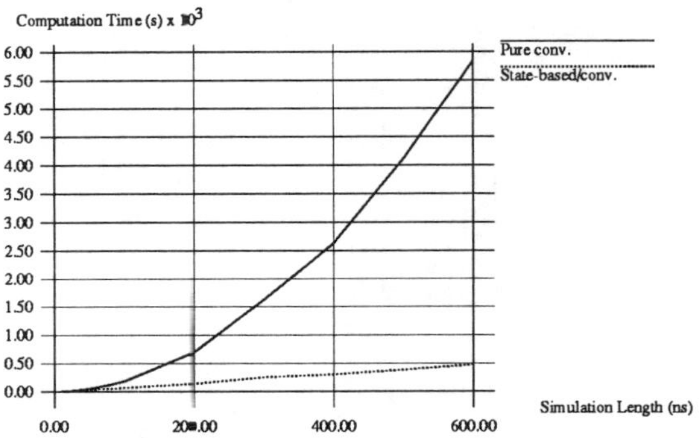

Figure 9: Execution time vs. simulation length

Acknowledgments

Gerry Marino's encouragement and support of this work is greatly appreciated. Technical discussions with Albert Ruehli and Jacob White were very helpful. Thanks are due to Sally Liu and Ta-Fang Fang for encouragement and to Chris Lennard for assistance during implementation and for suggesting the title of this paper. Administrative help from Kia Cooper is very much appreciated.

Simulation Length	Execution Time[a]	
	State-based/conv.	Convolution only
10 ns	6.82 s	4.90 s
20 ns	13.4 s	12.83 s
40 ns	26.72 s	38.26 s
60 ns	40.35 s	75.8 s
80 ns	53.59 s	126 s
100 ns	70.16 s	186.2 s
200 ns	140.78 s	695 s
300 ns	253.8 s	1619.5 s
400 ns	302.4 s	2635.1 s
500 ns	383.5 s	4110.6 s
600 ns	473.2 s	5839.9 s

[a]CPU times on a DEC 5000/200 running Ultrix 4.2.

Table 1: Comparison of Execution Times

References

[1] A. Deutsch et al High-speed signal propagation on lossy transmission lines. *IBM J. Res. Dev.*, 34(4):601–615, July 1990.

[2] J.S. Roychowdhury et al. An Impulse-Response Based Linear Time-Complexity Algorithm for Lossy Interconnect Simulation. In *Proc. IEEE International Conference on Computer Aided Design*, pages 62–65, Santa Clara, CA, November 1991.

[3] H.A. Wheeler. Formulas for the skin effect. *Proc. IRE*, 30:412–424, 1942.

[4] W. Wlodarczyk and V. Besch. Skin-Effect Losses of Interconnect Lines in Frequency and Time-Domain. *Elec. Let.*, 26(16):1237–1233, 2nd August 1990.

[5] R.A. Sainati and T.J. Moravec. Estimating High Speed Circuit Interconnect Performance. *IEEE Trans. Ckts. Sys.*, 36(4):533–541, April 1990.

[6] A.R. Djordjević et al. Time-Domain Response of Multiconductor Transmission Lines. *Proc. IEEE*, 75(6):743, June 1987.

[7] J.S. Roychowdhury, A. R. Newton, and D.O. Pederson. Algorithms for the Transient Simulation of Lossy Interconnect. Submitted to IEEE Trans. CAD.

[8] T. K. Liu and F. M. Tesche. Analysis of antennas and scatterers with nonlinear loads. *IEEE Trans. Antennas Propagat.*, AP-24:131, March 1976.

[9] A.R. Djordjević et al. Analysis of Lossy Transmission Lines with Arbitrary Nonlinear Terminal Networks. *IEEE Trans. Microwave Th. Tech.*, MTT-34(6):660, June 1986.

[10] J.E. Schutt-Aine and R. Mittra. Scattering Parameter Transient Analysis of Transmission Lines loaded with Nonlinear Terminations. *IEEE Trans. Microwave Th. Tech.*, MTT-36:529–536, 1988.

[11] J.E. Schutt-Aine and R. Mittra. Nonlinear Transient Analysis of Coupled Transmission Lines. *IEEE Trans. Ckts. Sys.*, 36(7), July 1989.

[12] A. Semlyen and A. Dabuleanu. Fast and Accurate Switching Transient Calculations on Transmission Lines with Ground Return Using Recursive Convolution. *IEEE Trans. Power App. Sys.*, PAS-94:561–571, 1975.

[13] F.Y. Chang. Waveform Relaxation Analysis of RLCG Transmission Lines. *IEEE Trans. Ckts. Sys.*, 37(11):1394–1415, November 1990.

[14] S. Lin and E.S. Kuh. Padé Approximation Applied to Transient Simulation of Lossy Coupled Transmission Lines. In *IEEE Multi-Chip Module Conference*, Santa Cruz, CA, March 1992. To appear.

[15] J.S. Roychowdhury and D.O. Pederson. Efficient Transient Simulation of Lossy Interconnect. In *Proc. IEEE Design Automation Conference*, pages 740–745, San Francisco, CA, June 1991.

[16] J.S. Roychowdhury, A. R. Newton, and D.O. Pederson. Simulating Lossy Interconnect with High Frequency Nonidealities in Linear Time. To be submitted to IEEE Trans. CAD.

[17] T.L. Quarles. *Analysis of Performance and Convergence Issues for Circuit Simulation.* PhD thesis, EECS Dept., Univ. Calif., Berkeley, Elec. Res. Lab., April 1989. Memorandum no. UCB/ERL M89/42.

[18] C.A. Neugebauer et al. High Performance Interconnections between VLSI Chips. *Solid State Tech.*, June 1988.

Transient Simulation of Lossy Interconnect *

Shen Lin and Ernest S. Kuh
Department of Electrical Engineering and Computer Sciences
University of California, Berkeley, CA 94720

Abstract

A new approach for transient simulation of lossy transmission lines terminated in arbitrary nonlinear elements is presented. The approach is based on convolution simulation. By making use of the Pade approximations of each line's characteristic admittance function and exponential propagation function, we derive a recursive convolution formulation, which greatly reduces the computation used to perform convolutions. In this paper, we analyze the errors introduced by Pade approximations and develop a scheme to determine the necessary order for an approximation. We incorporate the proposed approach into our Stepwise Equivalent Conductance timing simulator, SWEC [1] [2]; therefore, no Newton Raphson iteration is needed to simulate lossy lines with nonlinear terminals. The comparisons with SPICE3.e [5] indicate that SWEC, which gives accurate results, can be one to two order-of-magnitudes faster.

1 Introduction.

With higher signal speeds, electrical length of interconnects can become significant fraction of the signal wavelength. Consequently, the conventional lumped-impedance interconnect model is no longer adequate. Instead, a distributed transmission line model should be used.

The fundamental difficulty encountered in integrating transmission line simulation into a transient circuit simulator arises because circuits containing nonlinear devices or time-dependent characteristics must be characterized in the time domain while transmission lines with loss, dispersion, or discontinuities are best characterized in the frequency domain. To cope with this difficulty, two types of approaches have been proposed in the previous works. One uses a network of lumped elements and segments of ideal transmission line to approximate the frequency response of each lossy or dispersive line [6]. The approximated circuit models are suitable for existing general-purpose circuit simulators. However, the drawback of this type of approach is that the amount of computation increases for the simulation because a large number of extra nodes and elements are introduced [5]. The other adopts the convolution technique. For each integration, the output of a linear transmission line is the convolution of the input with its impulse response (Green's function). The difficulty of this type of approach lies in how to determine the impulse response of an arbitrary transmission line. People used the inverse Fast Fourier transformation technique [3], the numerical inverse Laplace transformation technique [4], inverse Fourier transformation of frequency-domain scattering parameters [10] [11], and even the explicit analytical approach[5] to determine the impulse response. However, the convolution simulations suffer from a common drawback: the convolution operation needs to extend over the entire past history. The computation time required at any time point t is then proportional to t; therefore, the convolutions at large time points will be very time-consuming. Furthermore, the inverse Fourier transformations of [3], [10], and [11] will suffer from either the aliasing effects or that too many frequency points are needed for the transformations to avoid aliasing.

In this paper, by making use of the Pade approximation (or the reduced moment method) we develop a recursive convolution formulation, which greatly reduces the computations. We give mathematical analyses of the errors introduced by the Pade approximation and propose a scheme to determine the necessary order of the approximation. The approach is implemented in the Stepwise Equivalent Conductance MOS timing simulator, SWEC [1] [2]. The key feature of SWEC is that Newton Raphson iteration is not needed for the implicit integration of a circuit even with lossy lines terminated in nonlinear elements.

The Pade approximation has been used in the literatures [7], [8], and [9] for the delay estimation of linear RLC networks or linear RLC distributed networks. The transfer function of the output (or outputs) of the network is expanded into a Maclaurin series of s, the Laplace transform variable, around $s = 0$. The series is then truncated to a necessary order. The first order Elmore delay will be the first moment of the series. An approximated time domain output waveform can also be determined by matching the truncated transfer function. However, the previous approaches can not be directly applied to the lossy line simulation due to the following two reasons:

1. The transfer function should be expanded into a series of $\frac{1}{s}$ around $s = \infty$ because high frequency responses are more important.
2. For distributed networks, the transfer function of the output will have infinite number of poles, which makes it impossible to be approximated by a finite degree Pade rational function.

*This work was supported by SRC under Grant 91-DC-008.

We will give more explanation to the first point when we analyze the errors introduced by the Pade approximation in section 4. The second point will be discussed in section 2.

2 Convolution Simulation of $RLGC$-lines.

The Telegrapher Equations for a $RLGC$-line are:

$$\frac{\partial v}{\partial x} = -(L\frac{\partial i}{\partial t} + Ri) \quad (1)$$

$$\frac{\partial i}{\partial x} = -(C\frac{\partial v}{\partial t} + Gv) \quad (2)$$

The transmission line stretches from 0 to l on the x coordinate, where l is the length of the line; $v(x,t)$ is the voltage at point x at time t; $i(x,t)$ is the current in the $+x$ direction at point x at time t. The boundary conditions for Eq.(1) and Eq.(2) are:

$$v(0,t) = v_1(t), \qquad v(l,t) = v_2(t) \quad (3)$$

$$i(0,t) = i_1(t), \qquad i(l,t) = -i_2(t) \quad (4)$$

V_1, V_2, I_1, and I_2, the Laplace transforms of v_1, v_2, i_1, and i_2, can be solved on the frequency domain to get the relation

$$\begin{bmatrix} I_1 \\ I_2 \end{bmatrix} = Y_0(s) \begin{bmatrix} \frac{e^{\lambda l}+e^{-\lambda l}}{e^{\lambda l}-e^{-\lambda l}} & -2 \\ -2 & \frac{e^{\lambda l}+e^{-\lambda l}}{e^{\lambda l}-e^{-\lambda l}} \end{bmatrix} \begin{bmatrix} V_1 \\ V_2 \end{bmatrix}, \quad (5)$$

where

$$\lambda = \lambda(s) \equiv \sqrt{(sL+R)(sC+G)} \quad (6)$$

$$Y_0(s) \equiv \sqrt{\frac{sC+G}{sL+R}}. \quad (7)$$

Each entry of the 2x2 admittance matrix on the right hand side of Eq.(5) has an infinite number of poles, the s such that $e^{\lambda(s)l} = e^{-\lambda(s)l}$, which makes it impossible to approximate any entry of that matrix by a finite degree Pade rational function. Papers [8] and [9] expand the entries around $s = 0$ to perform the delay and waveform estimation. We believe this will incur inaccuracy in their results.

To overcome this difficulty, Eq.(5) is rearranged [5] to get the following form:

$$Y_0V_1 - I_1 = e^{-\lambda l}(Y_0V_2 + I_2) \quad (8)$$

$$Y_0V_2 - I_2 = e^{-\lambda l}(Y_0V_1 + I_1), \quad (9)$$

where the Y_0 and the $e^{-\lambda l}$ have no pole on the right half of the s plane; therefore, we can apply the Pade approximation to these two frequency-domain functions, which will be introduced in the next section.

The inverse Laplace transforms of Eq.(8) and Eq.(9) lead to the following equations for the transmission line in the time domain:

$$v_1(t) * h_1(t) - i_1(t) = v_2(t) * h_3(t) + i_2(t) * h_2(t) \quad (10)$$

$$v_2(t) * h_1(t) - i_2(t) = v_1(t) * h_3(t) + i_1(t) * h_2(t), \quad (11)$$

where $*$ stands for the convolution operator, and the impulse response functions

$$h_1(t) \equiv \mathcal{L}^{-1}\{Y_0(s)\} \quad (12)$$

$$h_2(t) \equiv \mathcal{L}^{-1}\{e^{-\lambda(s)l}\} \quad (13)$$

$$h_3(t) \equiv \mathcal{L}^{-1}\{Y_0(s)e^{-\lambda(s)l}\} \quad (14)$$

In [5], the authors derived the analytical solutions of $h_1(t)$, $h_2(t)$, and $h_3(t)$ explicitly[1].

3 Recursive Convolution Simulation.

3.1 The Simulation Algorithm.

In [1] and [2], we have discussed that nonlinear circuits can be treated as linear circuits composed of time-varying conductors, and for the integration of a time step, say from t_n to t_{n+1}, the time-varying conductances can be replaced by their middle values within (t_n, t_{n+1}) interval to achieve a second order of accuracy. The middle conductance is equal to the conductance at t_n plus $\frac{t_{n+1}-t_n}{2}$ times the time derivative of the conductance at t_n. After the substitutions of the effective constant conductances, the original nonlinear circuit is transformed into a linear time-invariant circuit; therefore, the implicit integration does not involve solving any nonlinear equation. This integration technique is called the Stepwise Equivalent Conductance Integration. A timing simulator for digital CMOS circuits, SWEC, was implemented based on the technique.

When simulating nonlinear circuits with transmission lines, we employ the same technique. The nonlinear circuit is treated as a linear time-varying circuit. For each implicit integration, we replace every time-varying conductor by a time-invariant conductor of its effective conductance; therefore, we end up integrating a linear circuit with transmission lines. A modified nodal analysis is used to solve the voltage at each node and the current of each line. We incorporate the transmission line simulation into SWEC.

3.2 The Pade approximations of $Y_0(s)$ and $e^{-\lambda(s)l}$.

$$Y_0(s) = \sqrt{\frac{sC+G}{sL+R}} = \sqrt{\frac{C}{L}}\sqrt{\frac{1+\frac{G}{C}y}{1+\frac{R}{L}y}} \quad (15)$$

where y is used to denote $\frac{1}{s}$. $\sqrt{\frac{1+\frac{G}{C}y}{1+\frac{R}{L}y}}$ can be expanded into a Maclaurin series of y around $y = 0$, or $s = \infty$. Therefore,

$$\sqrt{\frac{1+\frac{G}{C}y}{1+\frac{R}{L}y}} = 1 + m_1y + m_2y^2 + \cdots + m_ky^k + \cdots, \quad (16)$$

[1]Please note that $Y_0(s)$ stretches to $s = \infty$ and is not periodic; therefore, it is difficult to solve $h_1(t)$ by using inverse Fourier transform.

where the k-th moment is

$$m_k = \frac{\frac{d^k\sqrt{\frac{1+\frac{G}{C}y}{1+\frac{R}{L}y}}}{dy^k}}{k!}. \quad (17)$$

The differentiation of Eq.(17) can be evaluated exactly by using symbolic differentiations.

Let us focus for now on the case that the first $2n-1$ moments of $\sqrt{\frac{1+\frac{G}{C}y}{1+\frac{R}{L}y}}$ are matched to a Pade approximation with both the numerator polynomial and the denominator polynomial of the same degree n, that is

$$\sqrt{\frac{1+\frac{G}{C}y}{1+\frac{R}{L}y}} \approx \frac{a_ny^n + a_{n-1}y^{n-1} + \cdots + a_1y + 1}{b_ny^n + b_{n-1}y^{n-1} + \cdots + b_1y + 1} \quad (18)$$

The reason for that the numerator polynomial and the denominator polynomial are of the same degree is as $y \longrightarrow \infty$, the value of $\sqrt{\frac{1+\frac{G}{C}y}{1+\frac{R}{L}y}}$ approaches to a finite nonzero number, $\sqrt{\frac{LG}{RC}}$, which also leads to the constraint that $\frac{a_n}{b_n} = \sqrt{\frac{LG}{RC}}$.

After considering $y = \frac{1}{s}$, we have

$$\begin{aligned} Y_0(s) &= \sqrt{\frac{sC+G}{sL+R}} \\ &\approx \sqrt{\frac{C}{L}}\frac{a_ny^n + a_{n-1}y^{n-1} + \cdots + a_1y + 1}{b_ny^n + b_{n-1}y^{n-1} + \cdots + b_1y + 1} \\ &= \sqrt{\frac{C}{L}}\frac{s^n + a_1s^{n-1} + a_2s^{n-2}\cdots + a_n}{s^n + b_1s^{n-1} + b_2s^{n-2}\cdots + b_n} \quad (19) \end{aligned}$$

Let $P(s)$ denote $s^n + b_1s^{n-1} + b_2s^{n-2}\cdots + b_n$ and $Q(s)$ denote $s^n + a_1s^{n-1} + a_2s^{n-2}\cdots + a_n$. Eq.(19) is then decomposed by applying the Heaviside's theorem,

$$\begin{aligned} Y_0(s) &\approx \sqrt{\frac{C}{L}}\frac{Q(s)}{P(s)} \\ &= \sqrt{\frac{C}{L}}(1 + \sum_{i=1}^{n}\frac{q_i}{s-p_i}), \quad (20) \end{aligned}$$

where $p_i's$ are the roots of $P(s) = 0$, and

$$q_i = \frac{Q(p_i) - P(p_i)}{\dot{P}(p_i)} \quad (21)$$

The Pade approximation helps us to find the inverse Laplace transform of $Y_0(s)$,

$$h_1(t) = \mathcal{L}^{-1}\{Y_0(s)\} \approx \sqrt{\frac{C}{L}}(\delta(t) + \sum_{i=1}^{n} q_ie^{p_it}), \quad (22)$$

where $\delta(t)$ is the Dirac function.

The Pade approximation of $e^{-\lambda(s)l}$ is similar.

$$\begin{aligned} \lambda(s) &= \sqrt{(sL+R)(sC+G)} \\ &= s\sqrt{LC} + s\sqrt{LC}[\sqrt{(1+\frac{R}{sL})(1+\frac{G}{sC})} - 1] \\ &= s\sqrt{LC} + s\sqrt{LC}[1 + \sum_{i=1}^{\infty}\frac{m_i}{s^i} - 1] \quad (23) \end{aligned}$$

The moments m_i can be found out by iteratively applying symbolic differentiations on $\sqrt{(1+\frac{R}{L}y)(1+\frac{G}{C}y)}$ with respect to y, that is

$$m_k = \frac{\frac{d^k\sqrt{(1+\frac{G}{C}y)(1+\frac{R}{L}y)}}{dy^k}}{k!}. \quad (24)$$

$$e^{-\lambda(s)l} = e^{-s\sqrt{LC}l}e^{-m_1\sqrt{LC}l}e^{\sum_{i=1}^{\infty} -\sqrt{LC}l\frac{m_{i+1}}{s^i}} \quad (25)$$

$e^{\sum_{i=1}^{\infty} -\sqrt{LC}l\frac{m_{i+1}}{s^i}}$ is, then, approximated by a Pade approximation,

$$\begin{aligned} e^{\sum_{i=1}^{\infty} -\sqrt{LC}l\frac{m_{i+1}}{s^i}} &\approx \frac{s^n + a_1s^{n-1} + a_2s^{n-2} + \cdots + a_n}{s^n + b_1s^{n-1} + b_2s^{n-2} + \cdots + b_n} \\ &= 1 + \sum_{i=1}^{n}\frac{q_i}{s-p_i}. \quad (26) \end{aligned}$$

Here, $\frac{a_n}{b_n}$ is equal to $e^{m_1\sqrt{LC}l}e^{-\sqrt{GR}l}$ by matching the point at $s = 0$[2].

Therefore,

$$e^{-\lambda(s)l} = e^{-s\sqrt{LC}l}e^{-m_1\sqrt{LC}l}(1 + \sum_{i=1}^{n}\frac{q_i}{s-p_i}). \quad (27)$$

The time domain effect of $e^{-s\sqrt{LC}l}$ is the constant delay of $\sqrt{LC}l$. The inverse Laplace transform of $1 + \sum_{i=1}^{n}\frac{q_i}{s-p_i}$ is the form that we mentioned, a Dirac function plus the sum of n exponential functions of t, based on which we derive our recursive convolution formulation.

The Pade approximation of $Y_0(s)e^{-\lambda(s)l}$ can be computed by multiplying Eq.(20) and Eq.(27). Since $\frac{q_i}{s-p_i}\frac{q_i'}{s-p_i'} = \frac{q_iq_i'}{p_i-p_i'}(\frac{1}{s-p_i} - \frac{1}{s-p_i'})$, the $p_i's$ of the multiplication result are composed of the $p_i's$ for $Y_0(s)$ and those for $e^{-\lambda(s)l}$. The leading delay term is still $e^{-s\sqrt{LC}l}$.

3.3 Recursive Convolution Formulation.

From the discussion of the previous subsection, we know $h_1(t)$ is a Dirac function plus n exponential functions and $h_2(t)$ and $h_3(t)$ are of the same form except they have the $\sqrt{LC}l$ delay. The convolution of any function, say $v_1(t)$,

[2]The n here may be different from that of Eq.(18) for $Y_0(s)$.

with a Dirac function, $v_1 * \delta(t_{n+1})$ is $v_1(t_{n+1})$. And, the convolution of $v_1(t)$ with the exponential function of $e^{p_i t}$ at time t_{n+1} will be

$$\begin{aligned}
&\int_0^{t_{n+1}} v_1(\tau)e^{p_i(t_{n+1}-\tau)}d\tau \\
= &\int_0^{t_n} v_1(\tau)e^{p_i(t_{n+1}-\tau)}d\tau + \int_{t_n}^{t_{n+1}} v_1(\tau)e^{p_i(t_{n+1}-\tau)}d\tau \\
= &e^{p_i(t_{n+1}-t_n)} \int_0^{t_n} v_1(\tau)e^{p_i(t_n-\tau)}d\tau \\
+ &\int_{t_n}^{t_{n+1}} v_1(\tau)e^{p_i(t_{n+1}-\tau)}d\tau \qquad (28)
\end{aligned}$$

we have the recursive relation to obtain $\int_0^{t_{n+1}} v(\tau)e^{p_i(t_{n+1}-\tau)}d\tau$ if we have computed $\int_0^{t_n} v_1(\tau)e^{p_i(t_n-\tau)}d\tau$, which is true because the convolution is evaluated before t_{n+1}. Therefore, the convolution integration $h_1(t) * v_1(t)$ can be computed recursively.

The recursive formulation to compute $i_2 * h_2$ follows a similar procedure as that of computing $v_1 * h_1(t_{n+1})$, where $v_1(t)$ is replaced by $i_2(t - \sqrt{LC}l)$. $i_2(t_n - \sqrt{LC}l)$ and $i_2(t_{n+1} - \sqrt{LC}l)$ are determined by employing interpolations between previous i_2 points. If $t_{n+1} - \sqrt{LC}l$ is less than t_n, then both $i_2(t_n - \sqrt{LC}l)$ and $i_2(t_{n+1} - \sqrt{LC}l)$ can be determined before the integration from t_n to t_{n+1}; otherwise, $i_2(t_{n+1} - \sqrt{LC}l)$ is represented as an interpolating point between $i_2(t_n)$ and $i_2(t_{n+1})$, and $i_2(t_{n+1})$ is left as an unknown variable to be determined through the integration. A queue is implemented to store the previous i_2 data. The data points preceding $t_n - \sqrt{LC}l$ can be deleted from the queue.

As we mentioned in the last subsection that $Y_0(s)e^{-\lambda(s)l}$ has the same form as $e^{-\lambda(s)l}$. The convolution of $v_2 * h_3$ can be computed similarly as we compute $i_2 * h_2$, where i_2 is replaced by v_2.

Therefore, we can compute the three convolutions of Eq.(10) recursively. According to the same discussion, the recursive convolutions of Eq.(11) follows.

3.4 Circuit Partition.

In previous simulation works, very often transmission lines were neglected or were modeled as lump RLC elements when accuracy was traded for efficiency. By using our approach, which considers the transmission line, we gain another advantage in circuit partitioning. Furthermore, we can achieve a considerable speedup on the overall simulation. A circuit can be partitioned at the transmission lines; therefore, a large circuit can be decomposed into small subcircuits.

Let us consider the situation that the node 1 and the node 2 of Eq.(10) or Eq.(11) belong to different subcircuits. Then, Eq.(10) is the convolution equation which will be considered when we integrate the subcircuit containing node 1, and Eq.(11) will be considered when the subcircuit containing node 2 is integrated. Because of the $\sqrt{LC}l$ delay introduced by $h_2(t)$ and $h_3(t)$, the subcircuit containing node 1 can be integrated, say from its t_n to its t_{n+1}, independently from the subcircuit containing node 2 if the subcircuit containing node 2 has already been evaluated beyond $t_{n+1} - \sqrt{LC}l$, which means $v_2(t_{n+1} - \sqrt{LC}l)$ and $i_2(t_{n+1} - \sqrt{LC}l)$ were available. It is similar to the subcircuit containing node 2. In order to do the integration on each subcircuit independently, we need to place the upper bound, $\sqrt{LC}l$, on the time steps of the two subcircuits. Therefore, whenever we make a cut on a transmission line, we place the upper bound of the line's $\sqrt{LC}l$ value on the time steps of the two subcircuits connected by the line. This technique is applicable to both digital and analog circuits, including MOS and bipolar circuits. Since some lines would have very small $\sqrt{LC}l$ values, causing very small time steps and degrading the efficiency, we should avoid making cuts on those lines.

4 Error Analysis of the Pade Approximation.

In this section, we are going to analyze the errors introduced by these approximations and to determine the necessary order for the approximations.

Let us consider a frequency domain function $H(s)$ which has no poles on the right half plane or at $s = 0$. $H(s)$ represents any transfer function, such as $Y_0(s)$. It can be expanded into a series of $\frac{1}{s}$ around $s = \infty$, which is

$$\begin{aligned}
H(s) &= m_0 + m_1\frac{1}{s} + m_2\frac{1}{s^2} + \cdots \\
&+ m_k\frac{1}{s^k} + m_{k+1}\frac{1}{s^{k+1}} + m_{k+2}\frac{1}{s^{k+2}} + \cdots \qquad (29)
\end{aligned}$$

Let $H_P(s)$ be the Pade approximation of $H(s)$ which, when expanded into a series of $\frac{1}{s}$, matches the $m_i's$ of $H(s)$ up to m_k. This means

$$\begin{aligned}
H_P(s) &= m_0 + m_1\frac{1}{s} + m_2\frac{1}{s^2} + \cdots \\
&+ m_k\frac{1}{s^k} + m'_{k+1}\frac{1}{s^{k+1}} + m'_{k+2}\frac{1}{s^{k+2}} + \cdots \qquad (30)
\end{aligned}$$

We say $H_P(s)$ is a k-th order approximation of $H(s)$ at $s = \infty$[3]. Let $E(s)$ be the difference of the two, which is the frequency domain approximation error. So

$$\begin{aligned}
E(s) &= H(s) - H_P(s) \\
&= (m_{k+1} - m'_{k+1})\frac{1}{s^{k+1}} + (m_{k+2} - m'_{k+2})\frac{1}{s^{k+2}} + \cdots \qquad (31)
\end{aligned}$$

Let

$$h(t) \equiv \mathcal{L}^{-1}\{H(s)\} \qquad (32)$$

$$h_P(t) \equiv \mathcal{L}^{-1}\{H_P(s)\} \qquad (33)$$

$$e(t) \equiv \mathcal{L}^{-1}\{E(s)\} = h(t) - h_P(t) \qquad (34)$$

[3]Please note that for our convolution approach, k should be an odd number. The degree of the numerator polynomial and that of the denominator polynomial are both equal to $\frac{k+1}{2}$.

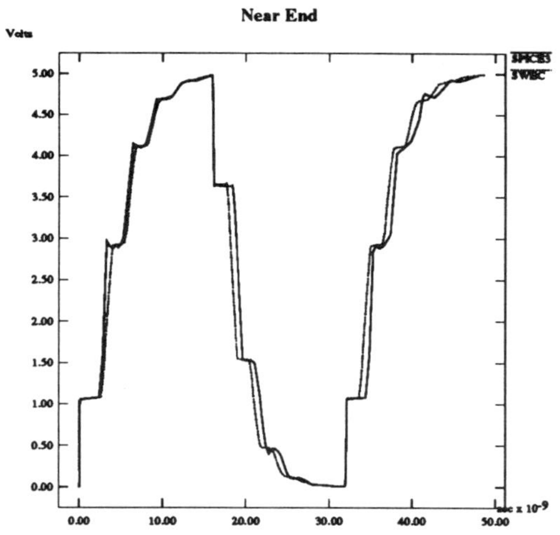

Figure 1: A line driven by an inverter (3 input vectors).

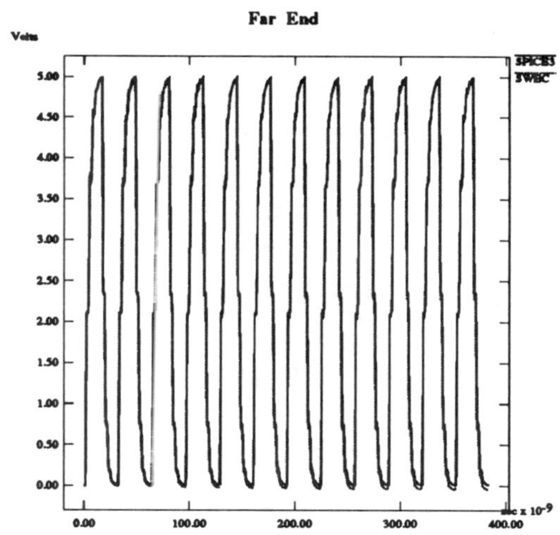

Figure 2: A line driven by an inverter (24 input vectors).

From the initial value theorem and Eq.(31),

$$\lim_{t\to 0} e(t) = \lim_{s\to\infty} E(s) = 0 \tag{35}$$

Furthermore, since the Laplace transform of the i-th derivative of $e(t)$, $e^{(i)}(t)$, is $s^i E(s)$, we have

$$\lim_{t\to 0} e^{(i)}(t) = \lim_{s\to\infty} s^i E(s) = 0 \qquad \forall\, i = 1..k. \tag{36}$$

Therefore, around $t = 0$, $h_P(t)$ matches $h(t)$ up to the k-th derivative. This property is important to assure the accuracy when $h_P(t)$ is used to evaluate the convolution integration which originally involves $h(t)$. Let us consider how we compute the convolution integration $h * v(t_n)$: (1) $h(t)$ is flipped around $t = 0$ and, then, shifted to t_n, and (2) we integrate from $t = 0$ to $t = t_n$ the values of $v(t)$ times the flipped and shifted h. Hence, the values of h around $t = 0$ are crucial; therefore, $h_P(t)$ should match $h(t)$ around $t = 0$ inorder to achieve accurate convolution results. Since $e(t)$ has its first k derivatives vanished at $t = 0$, we can expect that $h_P(t)$ matches $h(t)$ for an interval beyond $t = 0$. As k increases, the interval will increase. If the simulation time is small, a small k will guarantee the accuracy when Pade approximations are applied to the transfer function. Hence, the necessary order for the approximation depends on the simulation time.

The above discussion justifies our approach to expand $H(s)$ into a series of $\frac{1}{s}$ around $s = \infty$ instead of a series of s around $s = 0$. The Pade approximation $H_P(s)$ should match the $\frac{1}{s}$ series in order to achieve accurate convolution results. This is different from the work in [7], [8], and [9], which is based on the expansion at $s = 0$. In order to hightlight this point further, in the next section, we will also show the simulation results based on the Pade approximations of the s expansions of $Y_0(s)$ and $e^{-\lambda(s)l}$ in the convolutions.

Even though $H(s)$ does not have poles in the right half plane or the origin, the Pade approximation may lead to right half plane poles. In that case we need to increase the order of the approximation. As the approximation becomes accurate, the poles will be pushed to the left half plane. The other fact to determine the necessary order for a Pade approximation is that the maximum errors in the time domain, i.e. $h(t) - h_P(t)$, should be less than a pre-specified error bound, which is as follows. After neglecting higher order terms, we have

$$E(s) \approx (m_{k+1} - m'_{k+1})\frac{1}{s^{k+1}}, \tag{37}$$

and,

$$e(t) \approx (m_{k+1} - m'_{k+1})\frac{t^k}{k!}. \tag{38}$$

Therefore, the absolute value of $e(t)$ is a monotonic function of t, which will have its maximum $|(m_{k+1} - m'_{k+1})\frac{T_s^k}{k!}|$ at T_s, the ending simulation time. We want it to be less than an upper bound, say ϵ, that is

$$(m_{k+1} - m'_{k+1})\frac{T_s^k}{k!} \le \epsilon \tag{39}$$

We need to increase the order k until the condition of Eq.(39) is satisfied.

For our convolution approach, there are two $\epsilon's$, one for the approximation of $Y_0(s)$ and one for that of $e^{-\lambda(s)l}$. Those two $\epsilon's$ can be tuned to achieve the best trade-off between accuracy and efficiency.

5 Experimental Results.

We compared the simulation results of SWEC with the results of SPICE3.e, which implemented the work of [5]. Since SPICE3.e can not allow leakage resistances, G is set to zero for all the examples. The first test circuit is a $RLGC$-line driven by a CMOS inverter and terminated in a capacitor. Fig.1 shows the result at the near end. The first two rows of Table 1 summarize the CPU times for the case with 3 input

Experiment	SPICE3.e	SWEC
INV with 3 vec.	2.7 s	0.3 s
INV with 24 vec.	79.5 s	1.2 s
NAND	4.0 s	0.4 s

Table 1: CPU times.

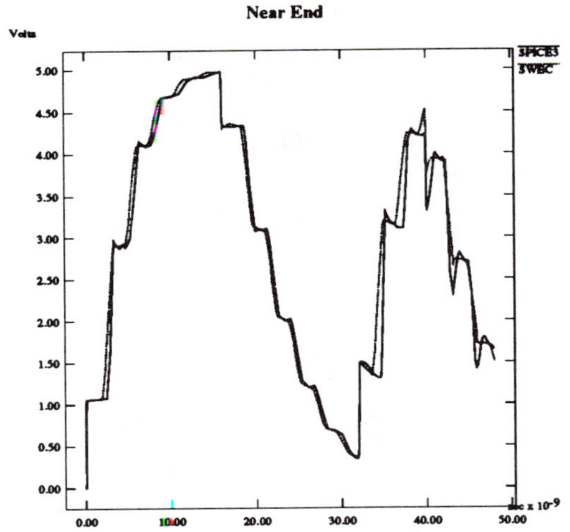

Figure 3: A line driven by a NAND.

vectors and that with 24 input vectors. The 24-vector case simply repeats the 3-vector case eight times. The CPU time increase for SWEC to finish the two was four times, while 30 times for SPICE3.e. Fig. 2 shows the far end result of the 24-vector case. We did not observe any accuracy degradation although the simulation time increased dramatically. A fifth order Pade approximation was used for the simulation.

The second circuit was a line driven by a CMOS 2 input NAND gate. The inputs to the NAND gate were used to create a glitch on purpose at the output; therefore, we could observe more transient information. Fig. 3 shows the result at the near end. The third row of Table 1 summarizes the CPU times.

To highlight that the expansions of $Y_0(s)$ and $e^{-\lambda(s)l}$ into series of s are not appropriate for simulation, we also implement a similar simulation program as SWEC except it uses the s expansions rather than $\frac{1}{s}$. We simulated the inverter circuit using this simulator. Fig.4 shows the near end result. As shown, the high frequency information is all lost.

6 Conclusion.

We have presented a new approach for transient simulation of lossy transmission lines terminated in arbitrary nonlinear elements. The approach employs the recursive convolution and the stepwise equivalent conductance integration. We also present a new circuit partition technique when the simulation of transmission is considered.

Experimental results are very encouraging. We can achieve one to two order-of-magnitude speed gain compared with SPICE3.e. We believe the proposed approach can make the multi-chip module simulation feasible.

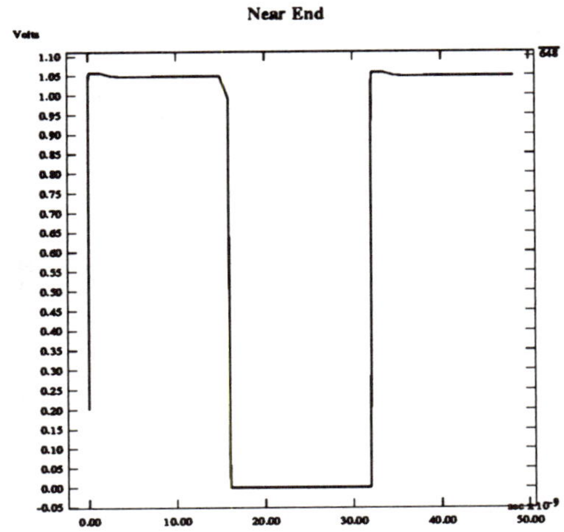

Figure 4: Expansion in s.

References

[1] Shen Lin, M. Marek-Sadowska, Ernest Kuh, "SWEC: A Step-Wise Equivalent Conductance Timing Simulator for CMOS VLSI Circuits," *Proc. EDAC*, pp. 142-148, Feb. 1991.

[2] Shen Lin and Ernest Kuh, "A New Approach to Circuit Simulator," *Proc. 10th European Conf. Circuit Theory and Design*, pp. 264-273, July. 1991.

[3] A. R. Djordjevi'c, T. K. Sarkar, and R. F. Harrington, "Analysis of lossy transmission lines with arbitrary nonlinear terminal networks," *IEEE Trans. Microwave Theory Tech.*, vol. MTT-34, pp. 660-666, June 1986.

[4] J. R. Griffith and M. S. Nakhla, "Time-Domain Analysis of Lossy Coupled Transmission Lines," *IEEE Trans. Microwave Theory Tech.*, vol. MTT-38, pp. 1480-1487, Oct. 1990.

[5] J. S. Roychowdhury and D. O. Pederson, "Efficient Transient Simulation of Lossy Interconnect," *Proc. DAC*, pp. 740-745, June 1991.

[6] F. Y. Chang, "Waveform Relaxation Analysis of RLGC Transmission Lines", *IEEE Trans. on Circuits and Systems*, Vol. 37 , pp. 1394-1415, November 1990.

[7] L. T. Pillage and R. A. Rohrer, "Asymptotic Waveform Evaluation for Timing Analysis", *IEEE Trans. on CAD*, Vol. 9 , pp. 352-366, April 1990.

[8] S. P. McCormick and J. Allen, "Waveform Moment Methods for Improved Interconnection Analysis", *Proc. DAC*, Orlando, FL, pp. 406-412, June 1990.

[9] T. K. Tang and M. S. Nakhla, "Analysis of High-Speed VLSI Interconnects Using the Asymptotic Waveform Evaluation Technique", *Proc. ICCAD*, pp. 542-545, November 1990.

[10] J. E. Schutt-Aine and R. Mittra, "Scattering Parameter Transient Analysis of Transmission Lines loaded with Nonlinear Terminations", *IEEE Tran. Microwave Tech.*, MTT-36:529-536, 1988.

[11] D. Winklestein, M. B. Steer, and R. Pomerleau, "Simulation of Arbitrary Transmission Line Network with Nonlinear Terminations", *IEEE Tran. on Circuits and Systems*, Vol. 38, pp. 418-412, 1991.

AWESpice: A General Tool for the Accurate and Efficient Simulation of Interconnect Problems

Vivek Raghavan, J. Eric Bracken and Ronald A. Rohrer
SRC-CMU Research Center for CAD, Department of Electrical and Computer Engineering
Carnegie Mellon University, Pittsburgh PA 15213

Abstract

AWESpice is a tool for the efficient and accurate simulation of circuits dominated by interconnect. It accepts a general nonlinear circuit and uses the AWE algorithm to efficiently convert its linear interconnect portion into multiport admittance macromodels. The macromodels are then simulated in conjunction with the nonlinear devices using a modified version of the SPICE algorithm. This technique leads to a significant reduction in CPU time while retaining the benefits of a conventional circuit simulator.

I. Introduction

In this age of shrinking device sizes, increasing system complexity and faster signal speeds, interconnections are rapidly becoming the dominant determinant of electronic circuit and system performance. In many of today's high performance digital electronic system designs, interconnect delay dominates system delay. In addition, coupling, ringing and transmission line effects in interconnections have begun to increasingly affect signal quality. These problems manifest themselves not only at the chip level but also at the package, multi-chip module, board and backplane levels. Thus there exists a need for finding efficient yet accurate simulation techniques to analyze circuits dominated by interconnect.

There have been many previous attempts to address this problem. Penfield et al. [18] pioneered an efficient technique to determine the delays through RC tree interconnect structures. While adequate for a broad class of useful and interesting circuits, the delays found using this technique were not always accurate and the method was restricted severely with respect to the interconnect topologies allowed. Much effort [21] [20] [2] has been made to improve and bound the accuracy of these methods as well as to broaden the class of topologies addressed. However, with increasing system frequency, interconnect models are becoming too complex to be handled with such techniques.

At the printed circuit board level, interconnect simulation has been approached in terms of the transmission line problem. Branin developed the method of characteristics to accurately and efficiently simulate a single lossless transmission line [1] . This method was extended to handle multiple coupled lossless transmission lines using modal decomposition techniques [5] . A number of techniques have been proposed to extend this technique to general lossy coupled transmission lines [3] [8] [7] [16] [6] .

While these methods provide reasonably accurate results, the costs involved preclude the solution of large practical interconnect problems. In addition, only a subset of interconnect problems consist solely of transmission lines. General interconnect problems consist of arbitrary combinations of lumped and distributed elements with general nonlinearities. This is especially true in today's technologies when the same net may go from on-chip interconnect through nonlinear buffers, through chip packaging to traces on a multi-chip module (MCM) or a printed circuit board (PCB) and back again onto the IC through packaging interconnect.

II. The AWE Approach

Asymptotic Waveform Evaluation (AWE) is a technique to efficiently evaluate the time domain response of a linear circuit in terms of a few dominant poles and residues [14] [22] . AWE is general in that it can handle all linear(ized) elements connected in any topology. The results produced by AWE are comparable in accuracy to those obtained using conventional circuit simulation techniques. The AWE technique involves the computation of the circuit moments in an efficient and recursive manner. A reduced order pole-residue model is determined from the moments using a form of Padé approximation. The AWE algorithm has also been extended to handle general lossy coupled distributed lines [4] [10] .

However, the AWE technique can handle only linear(-ized) circuits, while the general interconnect analysis problem is nonlinear due to the presence of nonlinear drivers, loads and sometimes nonlinear transmission elements such as pass transistor logic. So while the interconnect problem is largely linear, a full fledged nonlinear transient analysis is necessary to obtain accurate results. Previous attempts to extend AWE to solve this problem [13] [17] addressed only the nonlinear driver problem. The solutions presented were also technology specific.

This work was supported by SRC (Contract 91-DC-068) and Intel corporation

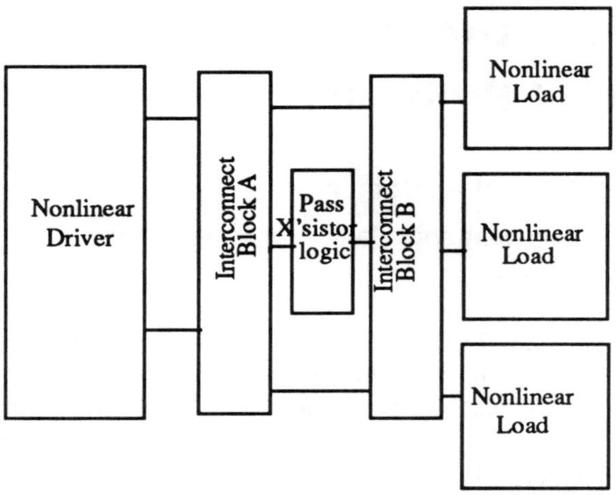

Figure 1: The generalized interconnect problem

This paper attempts to solve the above problem in a general fashion.The philosophy is to exploit the efficiency and accuracy of AWE for the linear interconnect portion of the circuit while using the tried and tested SPICE [9] simulator for the nonlinear portions of the circuit. This approach enables the use of all existing SPICE nonlinear device models.

The linear interconnect portions of the circuit are identified and preprocessed using AWE to produce compact multiport macromodels. These macromodels can be simulated along with the nonlinear portions of the circuit in a modified SPICE like transient simulator, AWESpice. AWESpice has been implemented as a modification of SPICE3C1 [15] . The major efficiency gain is that the size of the overall circuit to be simulated is greatly reduced, resulting in a significant speed-up in the resulting SPICE run. In this sense, AWESpice can be said to be a SPICE accelerator in software. An alternative technique which solves the problem of handling nonlinear terminations using AWE but outside the context of SPICE has been described in [11] .

III. Finding AWE Macromodels

In order to combine the AWE and SPICE algorithms it is desired to macromodel the linear portions of the circuit with multiport admittance macromodels. In previous publications [22] [12] , AWE approximations were found for voltage and current transfer functions. However, AWE approximations can be determined for multiport admittance parameters in an equally simple manner. The admittance parameter matrix *Y(s)* of a multiport linear circuit can be expressed using the following notation:

$$\begin{bmatrix} y_{11}(s) & \dots & \dots & y_{1n}(s) \\ \dots & \dots & \dots & \dots \\ \dots & \dots & \dots & \dots \\ y_{n1}(s) & \dots & \dots & y_{nn}(s) \end{bmatrix} \begin{bmatrix} v_1(s) \\ \dots \\ \dots \\ v_n(s) \end{bmatrix} = \begin{bmatrix} i_1(s) \\ \dots \\ \dots \\ i_n(s) \end{bmatrix} \qquad (1)$$

where $V(s)$ and $I(s)$ correspond to the port voltage and current vectors respectively. Admittance parameters are chosen because they fit naturally into the Modified Nodal Analysis (MNA) [19] formulation employed by SPICE. Alternatively, a hybrid or a scattering formulation may also be used to macromodel the multiport partition.

Moments for the entire admittance matrix of a linear partition can be determined by changing only the independent source values. Thus we can efficiently determine AWE approximations for the entire admittance parameter matrix using a single LU factorization. The cost of finding the moment of the entire admittance parameter matrix is similar to finding the moments of any circuit output.

AWE approximations for the admittance parameters can then be determined through the Padé approximation technique [22] . Each AWE approximation $y_{ij}(s)$ consists of a partial fraction expansion. It should be noted that the poles in each approximation are different and even the number of poles in each approximation need not be the same. This observation is important as a different number of poles may be required to approximate different system functions accurately. The frequency domain approximation can be easily transformed into a time domain impulse response:

$$\sum_{j=1}^{q_{ij}} \frac{k_{rij}}{s - p_{rij}} + d_{ij} \Leftrightarrow \sum_{rij=1}^{q_{ij}} k_{rij} e^{p_{rij} t} + d_{ij} \qquad (2)$$

where p_{rij} and k_{rij} are possibly complex poles and residues and d_{ij} represents any direct coupling between the input and output. Direct coupling terms are usually found in the diagonal terms of the admittance matrix.

In some cases the multiport is augmented with resistances before determining the AWE approximations of its admittance parameters. This approach is used when the multiport has extremely long impulse responses. The augmenting resistances are compensated for in the time domain simulation by adding equivalent negative resistances. It can be shown that this approach is analogous to using a scattering parameter formulation.

IV.The AWESpice Algorithm

In order to solve the combined AWESpice circuit both the port current and port voltage vectors must be determined for the duration of the transient simulation. At each time point, t_k, in the simulation the voltage vectors for all previous time points are known while the voltage vector for the present time point, $V(t_k)$, is treated as unknown. The a_{ij}'s and b_j's in the circuit MNA matrix (3) correspond to conductance and current stencils of all linear and nonlinear not modeled using the admittance macromodels. The $y_{ij}(t_k)$'s and the $i_j(t_k)$'s, which will be evaluated in following paragraphs, correspond to the admittance macromodel stencils to the global MNA matrix.The equation assumes that the common node of the multiport is ground; thus all port voltages are also node voltages. In the general case the port voltage is the difference between the positive and common node voltages of the port. The admittance macromodel is thus stenciled into the MNA circuit matrix at each time point.The following paragraphs describe the methodology to find the admittance macromodel stencils.

$$\begin{bmatrix} y_{11}(t_k)+a_{11} & \cdots & \cdots & \cdots & y_{1n}(t_k)+a_{1n} & \cdots & a_{1N} \\ \cdots & \cdots & \cdots & \cdots & \cdots & \cdots & \cdots \\ \cdots & \cdots & y_{ij}(t_k)+a_{ij} & \cdots & \cdots & \cdots & \cdots \\ \cdots & \cdots & \cdots & \cdots & \cdots & \cdots & \cdots \\ y_{n1}(t_k)+a_{n1} & \cdots & \cdots & \cdots & y_{nn}(t_k)+a_{nn} & \cdots & \cdots \\ \cdots & \cdots & \cdots & \cdots & \cdots & \cdots & \cdots \\ a_{N1} & \cdots & \cdots & \cdots & \cdots & \cdots & a_{NN} \end{bmatrix} \begin{bmatrix} v_1(t_k) \\ \cdots \\ v_i(t_k) \\ \cdots \\ v_n(t_k) \\ \cdots \\ v_N(t_k) \end{bmatrix} = \begin{bmatrix} i_1(t_k)+b_1 \\ \cdots \\ i_j(t_k)+b_j \\ \cdots \\ i_n(t_k)+b_n \\ \cdots \\ b_N \end{bmatrix} \qquad (3)$$

There is a simple and direct relationship between the time and frequency domain representations of the admittance matrix. The frequency domain product corresponds in general to a time domain convolution integral to express the port current vector $I(t)$ in terms of the port voltage vector and the admittance matrix. This relationship forms the basis upon which stencils for the admittance macromodels are determined at each time point.

$$I(t) = \int_0^t Y(t-\tau)\, V(\tau)\, d\tau \qquad (4)$$

In general the evaluation of a convolution integral becomes progressively more expensive as the simulation time increases. *However, the AWESpice algorithm reduces the convolution to an operation with a constant cost at each time point.*

In a circuit simulation context the port voltages can be assumed to be linear functions of time between successive time steps. Thus the port voltage waveforms [Figure 2] are piecewise linear functions of time and can be decomposed into a set of ramps [Figure 3]. Also each term in the admittance matrix, $y_{ij}(s)$, is an AWE approximation and thus is in partial fraction form. So, in order to evaluate the time domain port currents, it is possible to find the frequency domain product of each ramp with each pole-residue pair in

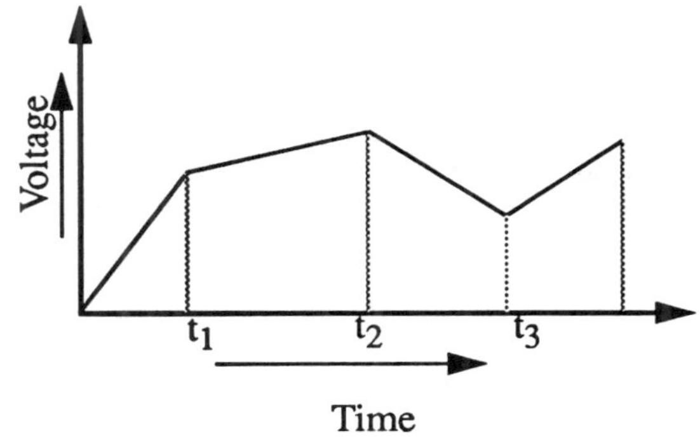

Figure 2: Port Voltage Waveforms

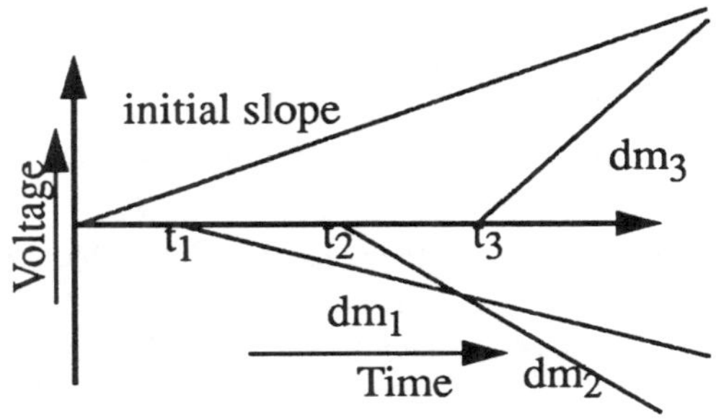

Figure 3: Decomposition of the port voltage waveform

$y_{ij}(s)$ and then find the inverse Laplace transform symbolically on a term by term basis. The symbolic expression for the time domain current when a real pole-residue pair is excited with a single ramp is derived below.

Let m be the slope of the input voltage and p_r and k_r form a real pole-residue pair. The subscripts i and j corresponding to the indices of the admittance matrix have been dropped. Then the current contribution $i(s)$ of the pole-residue pair is given in (5). Expressing (5) in partial fraction form, its inverse Laplace transform is trivially given by (6) where $u(t)$ is the unit step function.

$$i(s) = \frac{m}{s^2} \times \frac{k_r}{s - p_r} \qquad (5)$$

$$i(t) = m\left(\frac{-k_r}{p_r^2} + \frac{(-k_r)\,t}{p_r} + \frac{k_r}{p_r^2} e^{p_r t}\right) u(t) \qquad (6)$$

The current due to a single element of the macromodel admittance matrix excited by a single ramp is the sum of the currents due to each individual pole-residue pair. Complex conjugate pole pairs are handled together to obtain a real time domain current. The direct coupling component of the current can also be computed in a similar manner.

The port voltage can be decomposed into a set of ramps as illustrated in Figure 3. A new ramp with a slope equal to the change in the slope at that time point originates at each time step. Using arguments similar to those above for the single ramp case we can compute the change in current due to a slope change *dm* at time t_k for a single real pole-residue pair to be:

$$di(t) = dm\left(\frac{-k_r}{p_r^2} + \frac{(-k_r)(t-t_k)}{p_r} + \frac{k_r}{p_r^2}e^{p_r(t-t_k)}\right)u(t-t_k) \qquad (7)$$

As the port voltage consists of a large number of such ramps, both storage and computation requirements increase greatly as a function of simulation time. However, as shall be shown, we can reduce this problem to evaluating three terms at each time point: A constant term, a term linear in time, and a $e^{p(t-t_k)}$ term. The coefficients of these terms are updated at each time point.

Let d_k be the coefficient of the constant term, e_k the coefficient of the linear term and f_k the coefficient of the $e^{p(t-t_k)}$ term at time t_k. Each new change in slope will adjust the coefficients of these terms. From (7), adjustments in the terms for a change in slope *dm* at time t_k are given by:

$$d_k = d_{k-1} + dm\left(\frac{-k_r}{p_r^2} + \frac{k_r t_k}{p_r}\right) \qquad (8)$$

$$e_k = e_{k-1} + dm\left(\frac{-k_r}{p_r}\right) \qquad (9)$$

$$f_k = f_{k-1}e^{p(t_k - t_{k-1})} + dm_k\frac{k_r}{p_r^2} \qquad (10)$$

The above technique can be used to determine the time domain current if the entire voltage waveform at each port is known. However at any time point t_k the voltage at time t_k, $v(t_k)$, is *unknown*. Hence the unknown change in slope at time t_{k-1}, dm_{k-1} is given by

$$dm_{k-1} = \frac{v(t_k) - v(t_{k-1})}{t_k - t_{k-1}} - \frac{v(t_{k-1}) - v(t_{k-2})}{t_{k-1} - t_{k-2}} \qquad (11)$$

This change of slope can be divided into the known and unknown parts. The current contribution of the known part can be computed using the earlier formulas. The unknown part is the coefficient of $v(t_k)$, which forms part of the $y_{ij}(t_k)$'s stenciled into the left hand side of the circuit MNA matrix. The known current contribution, corresponding to all known slopes, for each pole in the admittance parameter $y_{ij}(s)$ will form the $i_j(t_k)$'s which are stencilled into the right hand side of the circuit simulation equations at each time point t_k. We assume a trapezoidal integration technique for all non macromodel circuit elements.

The technique described above calculates the exact instantaneous current due to the multiport macromodel at each time point assuming a correct port voltage waveform. However as we know the impulse response of the admittance parameters for all time, we can find the integral of the current, ensuring that the charge transferred at each time interval is correct given the port voltage waveform. The equations for this alternate approach can also be stenciled into the MNA matrix in a similar manner.

V. Results

Some examples have been run to show the accuracy and efficiency of AWESpice on a variety of interconnect problems. The first example is an MOS gate driving through RLC interconnect into a nonlinear transmission gate which drives another lossy transmission line into an MOS load. The RLC interconnect was modeled using 50 lumped RLC sections.- Figure 4 shows the comparison between the waveforms

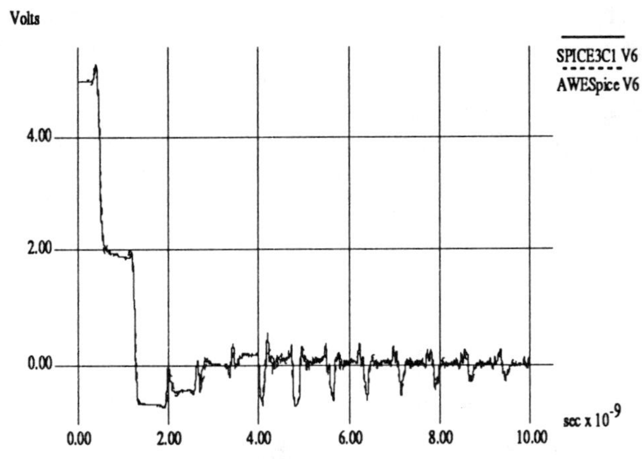

Figure 4: MOS gate driving interconnect through a pass gate which drives the MOS load through interconnect

obtained with AWESpice and SPICE3C1 at one end of the transmission gate. The waveforms are almost identical. The AWESpice simulation took 8.2 CPU-sec while the SPICE3C1 simulation took 418.2 CPU-sec. The time required to create the AWE macromodel for the interconnect circuit was 0.7 CPU-sec. A new version of SPICE3 [7] which incorporates a single lossy transmission line model required 53.63 CPU-sec to simulate the same circuit. It must also be noted that not all interconnect can be modeled using a lossy transmission line model. The second example has a high speed ECL gate driving RLC interconnect having 50

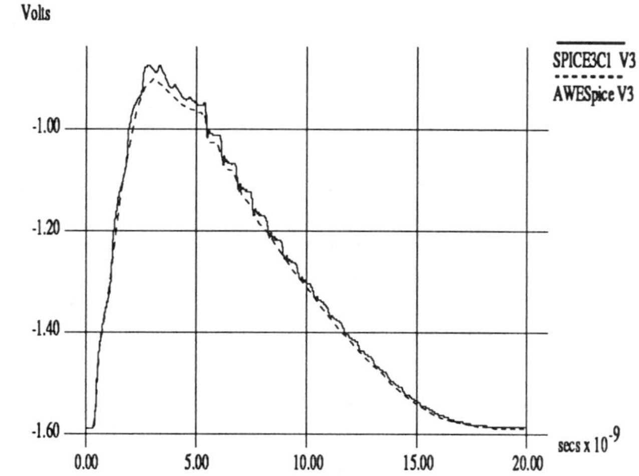

Figure 5: Waveform of an ECL gate driving through RLC interconnect through to a ECL gate load.

elements which drives another ECL gate. A comparison of the SPICE3C1 and the AWESpice waveforms is shown in Figure 5 at the receiver end of the RLC interconnect. In this case too the waveforms agree closely, while AWESpice took 3.1 CPU-sec and SPICE3C1 took 41.3 CPU-sec. When interconnect was replaced by another interconnect having 1000 linear elements the AWESpice time was almost unchanged while the SPICE3C1 time went up to 158.2 CPU-sec. The macromodel time was 0.3 CPU-sec in the first case and 3.1 CPU-sec in the latter case.

AWE has been extended to model lossy coupled distributed lines [10] [4]. To demonstrate the use of AWESpice to simulate these elements, it was applied to a circuit containing two systems of coupled lossy lines, lumped elements and nonlinear 54F240 line drivers. The circuit is shown below in Figure 6. Note that one distributed element consists of four coupled lossy lines, and the other of two. Each 54F240 driver model contains 7 transistors and 14 diodes.

The moment computations for the distributed elements required a total of 0.6 CPU-sec. The transient analysis of the circuit using the AWESpice program consumed 16.6 more CPU-sec. (All CPU times are for a DECstation 3100). The same simulation—using a four-segment approximation to each of the systems of lines—required 23.4 CPU-sec. The resulting waveforms at the source and load points are shown in Figure 7 below. An example of a crosstalk waveform from the same circuit is shown in Figure 8.

The speed up obtained with AWESpice depends on the ratio of the number of transistors to the number of linear elements. For linear interconnect dominated circuits, AWESpice can be expected to show a significant speedup. We also

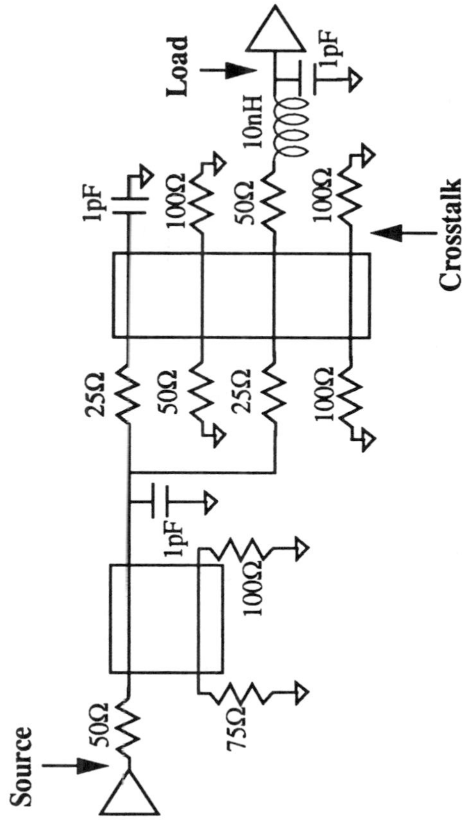

Figure 6:Example circuit with lossy coupled lines

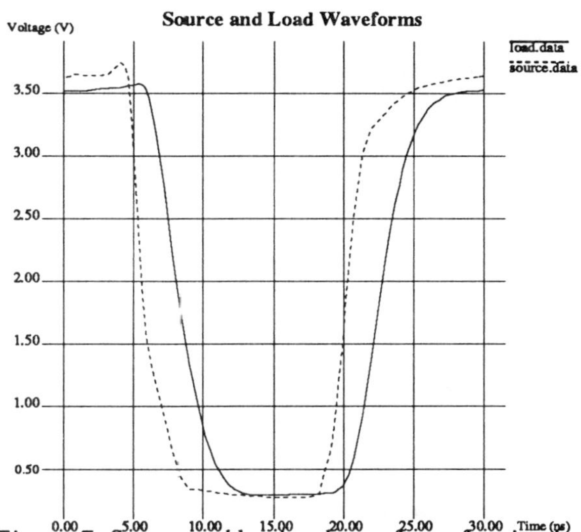

Figure 7: Source and load end waveforms for circuit in Figure 6

note that while the macromodel creation time goes up with the size of the linear circuit, it is insignificant compared to the SPICE3C1 runtimes. The use of distributed elements in AWE further reduces the macromodel time.

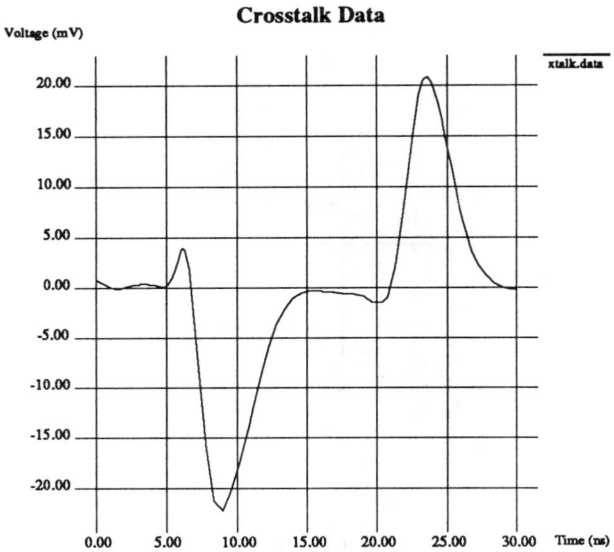

Figure 8:Crosstalk waveform for circuit in Figure 6

VI. Conclusions

In this paper we have developed a technique to bring together the speed and accuracy of the AWE method for linear circuits and the generality of a SPICE based nonlinear transient simulator. The technique is general and can be used for all kinds of interconnect problems. The results demonstrate both the accuracy and efficiency of this technique. Thus AWESpice shows the potential to be an extremely useful tool for interconnect analysis.

VII. Bibliography

[1]F.H.Branin Jr. Transient analysis of lossless transmission lines. *Proc. IEEE. 55* Nov, 1967.

[2]Chorng-Yeong Chu and Mark Alan Horowitz. Charge-Sharing Models for Switch-Level Simulation. *IEEE Trans. Computer-Aided Design.* CAD-6(6):1053-1061, November, 1987.

[3]A.R.Djordjevic, T.P. Sarkar and R.F. Harrington. Time Domain Response of Multiconductor Transmission Lines. *Proc. IEEE. 75*(6), June, 1987.

[4]Eric Bracken, Vivek Raghavan and Ronald A. Rohrer. Simulating Distributed Elements with Asymptotic Waveform Evaluation (AWE). 1992. Accepted for Publication at IEEE International Microwave Symposium.

[5]F.Y.Chang. Transient analysis of lossless coupled transmission lines in a nonhomogenous dielectric medium. *IEEE Trans. Microwave Theory and Techniques.* MTT-18616-626, Sept., 1970.

[6]F.Y.Chang. The generalized method of characteristics for waveform relaxation analysis of lossy coupled transmission lines. *IEEE Trans. Microwave Theory and Techniques.* MTT-37(12):2028-2038, Dec, 1989.

[7]Jaijeet S. Roychowdhury and Donald O. Pederson. Efficient Transient Simulation of Lossy interconnect. *28th ACM/IEEE Design Automation Conference Proceedings,* pages 740-745. 1991.

[8]J.E.Schutt-Aine and R.Mittra. Nonlinear Transient Analysis of Coupled Transmission Lines. *IEEE Trans. Circuits and Systems. 36*(7):959-967, July, 1987.

[9]L.W. Nagel. *SPICE2, A Computer Program to Simulate Semiconductor Circuits.* Technical Report ERL-M520, UC-Berkeley, May, 1975.

[10]Tak K. Tang and Michel Nakhla. Analysis of High Speed VLSI Interconnects Using The Asymptotic Waveform Evaluation Technique. *Technical Digest of the IEEE International Conference on Computer-Aided Design,* pages 542-545. 1990.

[11]Dong H. Xie and Michel Nakhla. Delay and Crosstalk Simulation of High-Speed Interconnects with Nonlinear Terminations. *Technical Digest of the IEEE International Conference on Computer-Aided Design,* pages 66-69. 1991.

[12]Curtis L. Ratzlaff, Nanda Gopal and Lawrence T. Pillage. RICE: Rapid Interconnect Circuit Evaluator. *28th ACM/IEEE Design Automation Conference Proceedings.* 1991.

[13]Douglas R. Holberg, Santanu Dutta and Lawrence Pillage. DC Parameterized Piecewise-Function Transistor Models for Bipolar and MOS Logic Stage Delay Evaluation. *Technical Digest of the IEEE International Conference on Computer-Aided Design,* pages 546-549. 1990.

[14]Lawrence T. Pillage and Ronald A. Rohrer. Asymptotic Waveform Evaluation for Timing Analysis. *IEEE Trans. Computer-Aided Design. 9*(4):352-366, April, 1990.

[15]T.L.Quarles. *The SPICE3 Implementation Guide.* Technical Report ERL-M89/44, UC-Berkeley, April, 1989.

[16]R.Wang and O.Wing. Analysis of VLSI Multiconductor Systems by Bi-Level Waveform Relaxation. *Technical Digest of the IEEE International Conference on Computer-Aided Design,* pages 166-169. November, 1990.

[17]Vivek Raghavan and Ronald A. Rohrer. A New Nonlinear Driver Model for Interconnect Analysis. *28th ACM/IEEE Design Automation Conference Proceedings,* pages 561-566. 1991.

[18]Jorge Rubinstein, Paul Penfield Jr. and Mark Horowitz. Signal Delay in RC Tree Networks. *IEEE Trans. Computer-Aided Design.* CAD-2(3):202-211, July, 1983.

[19]J. Vlach and K. Singhal. *Computer Methods for Circuit Analysis and Design.* Van Nostrand Reinhold Company, New York, New York, 1983.

[20]John L. Wyatt Jr. Signal Delay in RC Mesh Networks. *IEEE Trans. Circuits and Systems.* CAS-23(5):507-510, May, 1985.

[21]J. L. Wyatt Jr. C.A. Zukowski and P. Penfield Jr. Step Response Bounds for Systems Described by M-Matrices, with Application to Timing Analysis of Digital MOS Circuits. *Proceedings of 24th Conference on Decision and Control,* pages 1552-1557. Ft. Lauderdale, FL., December, 1985.

[22]Xiaoli Huang, Vivek Raghavan and Ronald A. Rohrer. AWEsim: A Program for the Efficient Analysis of Linear(ized) circuits. *Technical Digest of the IEEE International Conference on Computer-Aided Design,* pages 534-537. November, 1990.

A Boundary-Element approach to Transient Simulation of Three-dimensional Integrated Circuit Interconnect

David D. Ling
I.B.M. T. J. Watson Research Center
Yorktown Heights, NY

S. Kim, J. White
Department of EECS
MIT, Cambridge, MA

Abstract

It has been recently suggested that sufficiently accurate integrated circuit cross-talk simulations can be performed by computing the time evolution of electric fields both inside and outside three-dimensional integrated circuit conductors via a finite-difference discretization of Laplace's equation. In this paper the same calculation is performed, but the volume mesh associated with finite-difference methods is avoided through the use of a boundary-element method in which only conductor surfaces are discretized. Two boundary-element approaches are investigated, and it is shown that the straight-forward approach leads to unacceptable discretization errors, and a less intuitive second approach yields good results even with coarse surface meshes. Finally, numerical experiments demonstrating the effectiveness of the second approach in calculating cross-talk are presented.

1 Introduction

A simplified approach to analyzing the parasitic coupling between nearby lines of interconnect on an integrated circuit, referred to as cross-talk, is to compute a lumped resistor-capacitor model for the interconnect structure. Such an approach provides a certain amount of insight, but doesn't model the distributed effects accurately enough to make aggressive design decisions. Recently it was suggested that sufficiently accurate cross-talk simulations could be performed by computing the time evolution of the electric fields both inside and outside the integrated circuit conductors via a finite-difference discretization of Laplace's equation[4]. In that approach, the simplified equations were derived by assuming the magnetic fields and volume charges were negligible.

In this paper, we perform this same calculation but avoid the volume mesh associated finite-difference methods. To see why this is feasible, consider that the absence of volume charge and magnetic fields imply that currents which flow through conductors only serve to increase or decrease conductor surface charge with time. This suggests that a boundary-element method, in which only conductor surface quantities are computed, can capture the transient phenomenon. In this paper, two boundary-element approaches are investigated, and it is shown that the most straight-forward approach leads to unacceptable discretization errors, and a less intuitive second approach yields good results even with coarse surface meshes. Finally, numerical experiments demonstrating the effectiveness of the second approach in calculating cross-talk are presented.

In the next section, our assumptions about the cross-talk problem are given, and several basic results derived. In section 3, we derive a source formulation for the transient interconnect problem, and demonstrate the difficulties with discretizations of that approach. A Green's theorem based formulation is derived in section 4, and its advantages compared to the source formulation are made clear. In section 5, we present some application results. Finally, we give our conclusions and acknowledgements in Section 6

[0]This work was supported by the Defense Advanced Research Projects Agency contract N00014-91-J-1698, the National Science Foundation contract (MIP-8858764 A02) and grants from I. B. M.

2 Basic Assumptions and Consequences

For homogeneous dielectric media, the electric field, E satisfies

$$\nabla \cdot E = \frac{\rho}{\epsilon}, \tag{1}$$

where ρ is the volume charge density and ϵ is the dielectric permitivity. Inside a conductor, the current density, J, is given by $J = \sigma E$ where σ is the conductivity of the material. Conservation of charge implies that

$$\nabla \cdot J = -\frac{\partial \rho}{\partial t} \tag{2}$$

and substituting for J in terms of the electric field yeilds

$$\nabla \cdot E = -\frac{1}{\sigma}\frac{\partial \rho}{\partial t}. \tag{3}$$

Combining (1) and (3) leads to an equation for the time evolution of the volume charge inside a conductor, as in

$$\frac{\partial \rho}{\partial t} = -\frac{\rho}{\tau} \tag{4}$$

where $\tau = \frac{\epsilon}{\sigma}$ is the dielectric relaxation time, and is of the order of hundreds of femtoseconds. Note that at such time scales, the assumed constitutive relations between J and E are not likely to still apply, so (refeq:qdecay) should only be used as an indicator of longer time behavior.

From (4), it follows that any initial charge in the interior of a conductor must rapidly decay, and this volume charge *never* increases, regardless of what happens on the periphery of the conductor. For this reason, volume charge inside the conductor can *always* be neglected. The computationally important implication of zero conductor volume charge is that all currents flow through the conductor to the conductor surface, where they produce a build-up of surface charge. Note that this surface charge may be "bled off" by external circuitry at points where contact is made to the conductor.

It is generally assumed that for integrated circuit interconnect, length scales are such that magnetic effects can be ignored, though this may not continue to be the case as integrated circuit speeds increase. This assumption implies that the electric field, E, is given by $E = \nabla\psi$, where ψ is a scalar potential. Also, there is zero volume charge in the free space outside the conductor, and as shown above, zero volume charge inside the conductor. Together, these several observations imply that (1) can be simplified to

$$\nabla^2\psi = 0 \tag{5}$$

everywhere except on the conductor surface. Therefore, ψ can be related to the conductor surface charge, ρ_s, through the superposition integral

$$\psi(x) = \int_S \frac{\rho_s(x')}{4\pi\epsilon\|x - x'\|} da' \tag{6}$$

where S is the union of all the conductor surfaces, and $\|x - x'\|$ is the Euclidean distance between x and x'.

As conductor surface charge is now the only charge in the problem, we have from conservation of charge that at any point x on the conductor surface which is not an external contact point,

$$-\frac{\partial\rho_s(x)}{\partial t} = J_{normal}(x) = -\sigma\frac{\partial\psi(x)}{\partial n}, \tag{7}$$

and on any point x on the conductor surface which is a contact point

$$\begin{aligned} -\frac{\partial\rho_s(x)}{\partial t} &= J_{normal}(x) - J_{external}(x) \\ &= -\sigma\frac{\partial\psi(x)}{\partial n} - J_{external}(x) \end{aligned} \tag{8}$$

where $J_{normal}(x)$ is the current density along the inward directed normal to the conductor surface, $J_{external}(x)$ is the current density supplied to the conductor through the contact, ρ_s is the conductor surface charge, and $\frac{\partial\psi(x)}{\partial n}$ is the spatial derivative of ψ along the inward directed normal to the conductor surface.

3 Source Formulation

In this section, a physically simple boundary integral formulation for the transient interconnect problem is presented, along with a discretization procedure. An example is given to demonstrate that the solutions obtained by discretizing this simple formulation have unacceptable errors. In the next section, a less intuitive formulation is given which allieviates these difficulties.

3.1 Integral Formulation

A simple boundary formulation for the transient interconnect problem can be derived by eliminating the normal derivative of the potential from (7) and (8) by differentiating (6) and substituting. For non-contact points on the conductor surface this yeilds

$$-\frac{\partial\rho_s(x)}{\partial t} == -\sigma \int_S \frac{\partial}{\partial n}\frac{\rho_s(x')}{4\pi\epsilon\|x - x'\|} da', \tag{9}$$

and for external contact points,

$$-\frac{\partial\rho_s(x)}{\partial t} = -\sigma \int_S \frac{\partial}{\partial n}\frac{\rho_s(x')}{4\pi\epsilon\|x - x'\|} da' - J_{external}(x). \tag{10}$$

3.2 Charge Discretization

To numerically solve (10) for ρ_s at non-contact conductor surfaces, and both $J_{external}$ and ρ_s at contact surfaces, the surfaces of all the conductors are broken into small panels or tiles. It is then assumed that on each panel l, there is a constant surface charge q_l. In addition, it is assumed that for each panel of a contact surface, l_c, there is a constant current density J_c.

If there are n total panels, m of which belong to contact surfaces, then there are n unknown panel charges, and m unknown contact panel currents, for a total of $m + n$ unknowns. To generate a system of $n + m$ equations from which these $n + m$ unknowns can be computed, a collocation scheme is used[2][5]. That is, (9) is enforced at the center point in each of $n - m$ non-contact panels and (10) is enforced at the center point of each of the m contact panels. The result is a linear system of the form

$$4\pi\tau\frac{\partial}{\partial t}q(t) = -Dq(t) + AJ_c(t). \tag{11}$$

where $q(t) \in \Re^n$ is the vector of time-varying panel charges, $J_c(t) \in \Re^m$ is the vector of time-varying contact panel currents, $A \in \Re^{n\times m}$ is a matrix of nearly all zeros, except $A_{l,l-(n-m)} = 1$ if l is a contact panel, and $D \in \Re^{n\times n}$ is given by

$$D_{kl} = \frac{\partial}{\partial n'} \int_{panel_l} \frac{1}{\|x' - x_k\|} da'; \tag{12}$$

where x_k is the center of panel k. Note that in defining A, it is assumed the elements of J_c are appropriately normalized.

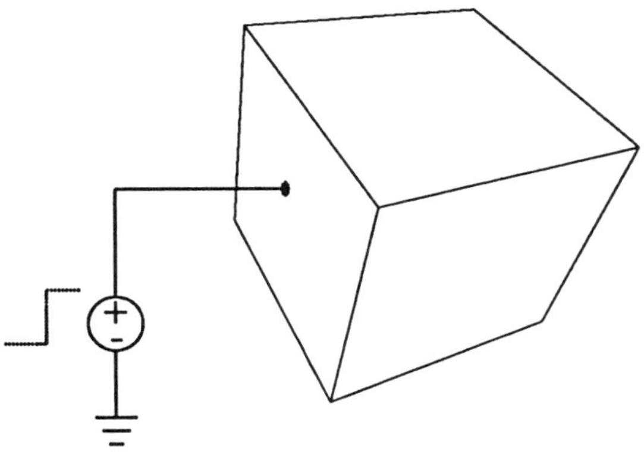

Figure 1: Conductive cube with one side driven by a voltage step.

Given that charge density is assumed constant on each panel, from (6) it follows that the m contact panel potentials, which are assumed constant over the panel, are related to the n panel charges by

$$\Psi_c = Pq \tag{13}$$

where $\Psi_c \in \Re^m$ is the vector of contact panel potentials and $P \in \Re^{m \times n}$ is given by

$$P_{kl} = \frac{1}{a_l} \int_{contact\ panel_l} \frac{1}{\|x' - x_k\|} da' \tag{14}$$

where a_l is the area of the lth panel, and x_k is the center of the kth panel.

3.3 Numerical Difficulties

The combination of (11) and (13) yeilds a differential-algebraic system of $n+m$ equations in $n+m$ unknowns. It is straight-forward to solve this system numerically for the time evolution of the panel charges, and to use (6) to compute potentials at any conductor surface point of interest. Unfortunately, the results from such a calculation contain spatial discretization errors that preclude the scheme's use as part of a circuit simulation program.

This numerical difficulty can best be demonstrated by examining a simple example, such as the conductive cube shown in Figure (1). One face of the cube is assumed to be driven by an ideal voltage source which makes a unit step transition, and therefore this driven cube face is an isopotential surface whose potential changes, at time zero, from zero to one volt. The correct behavior of the potentials at points on non-driven cube surfaces is obvious, they should change, over time, from zero to one volt.

Figure (2) is a plot of the time behavoir of the potential at an observation point on the cube face opposite the driven face, for three sucessively finer surface discretizations. As even the plot based on discretizing the surface into 1944 panels shows, the discretization error is such that the computed equilibrium, or final

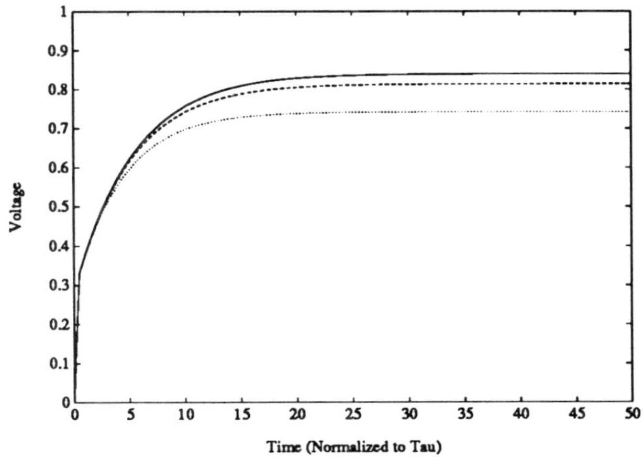

Figure 2: Potential computed at the observation point using three different uniform surface discretizations of the source formulation. The dotted line is the result of using 54 panels, the dashed line the result of using 486 panels, and the solid line is the result of using 1944 panels.

value, for the potential is only about 0.85 volts. Such an error is clearly unacceptable if the intention is to use the calculation procedure as part of a circuit simulator.

4 Green's Formulation

The nonphysical nature of the discretization error in the source formulation stems primarily from the fact that charge is discretized, rather than potential. This can be remedied by switching to a formulation based on Green's identity. Using this second formulation allows the potential to be discretized directly, and insures that isopotential equilibrium behavior is retained *regardless* of the spatial discretization. In the three subsections below we derive this Green's identity based formulation, describe the numerical discretization procedure, and compare computed results with those produced using the source formulation to demonstrate the Green's formulation's advantages.

4.1 Formulation Derivation

As ψ satisfies (5), it is easily derived from Green's theorem that[3]

$$\frac{\psi(x)}{\epsilon} = \int_{S_i} \psi(x') \frac{\partial}{\partial n'} \frac{1}{4\pi\epsilon\|x - x'\|} da' - \int_{S_i} \frac{\partial \psi(x')}{\partial n'} \frac{1}{4\pi\epsilon\|x - x'\|} da', \tag{15}$$

where S_i is the closed surface of the ith conductor, and x is some point on S_i. It also follows from Green's theorem that if x is outside conductor i, then

$$\int_{S_i} \psi(x') \frac{\partial}{\partial n'} \frac{1}{4\pi\epsilon\|x - x'\|} da' - \int_{S_i} \frac{\partial \psi(x')}{\partial n'} \frac{1}{4\pi\epsilon\|x - x'\|} da' = 0. \quad (16)$$

Summing appropriately (15) or (16) over all the conductor surfaces yields

$$\frac{\psi(x)}{\epsilon} = \int_S \psi(x') \frac{\partial}{\partial n'} \frac{1}{4\pi\epsilon\|x - x'\|} da' - \int_S \frac{\partial \psi(x')}{\partial n'} \frac{1}{4\pi\epsilon\|x - x'\|} da'. \quad (17)$$

where S is the union of the conductor surfaces S_i, and x is some point in S.

Decomposing the second integral in (17) into contact and non-contact surfaces, and then substituting using (7) and (8) leads to

$$\frac{\psi(x)}{\epsilon} = \int_S \psi(x') \frac{\partial}{\partial n'} \frac{1}{4\pi\epsilon\|x - x'\|} da' - \int_{S_{noncontact}} \frac{1}{\sigma} \frac{\partial \rho(x')}{\partial t} \frac{1}{4\pi\epsilon\|x - x'\|} da' - \int_{S_{contact}} \frac{1}{\sigma} \frac{\partial \rho(x')}{\partial t} \frac{1}{4\pi\epsilon\|x - x'\|} da' + \int_{S_{contact}} \frac{1}{\sigma} J_{external}(x') \frac{1}{4\pi\epsilon\|x - x'\|} da'. \quad (18)$$

Two of the integrals in (18) can be eliminated by noticing that the time-derivative of (6) yields

$$\frac{\partial \psi(x)}{\partial t} = \int_S \frac{\partial \rho_s(x')}{\partial t} \frac{1}{4\pi\epsilon\|x - x'\|} da'. \quad (19)$$

Combining (18) with (19), and then substantially reorganizing, produces

$$4\pi\tau \frac{\partial \psi(x)}{\partial t} = -4\pi\psi(x) + \int_S \psi(x') \frac{\partial}{\partial n'} \frac{1}{\|x - x'\|} da' + \frac{1}{\sigma} \int_{S_{contact}} \frac{J_{external}(x')}{\|x - x'\|} da' \quad (20)$$

where again note that

$$J_{external}(x) = \frac{\partial \rho_s(x)}{\partial t} - \sigma \frac{\partial \psi(x)}{\partial n'} \quad (21)$$

is, not surprisingly, the sum of displacement and resistive current at the contact point x.

Although perhaps tedious to derive, (20) has the advantage of including as unknowns only the variables of interest, that is, only the surface potentials and the contact surface currents.

4.2 Discretization

To numerically solve (20) for ψ at non-contact conductor surfaces, and for J at contact surfaces (as defined in (21)), the surfaces of all the conductors are broken into small panels or tiles. It is then assumed that on each panel l, there is a constant surface potential Ψ_l. In addition, it is assumed that for each panel of a contact surface, l_c, there is a constant current density J_c.

If there are n total panels, m of which belong to contact surfaces for which the potential is known a-priori, the vector of constant panel potentials, $\Psi \in \Re^n$, has $n - m$ unknown entries. As none of the entries of the vector of m constant contact panel currents, $J_c \in \Re^m$, is known a-proiri, a total of n unknowns must be determined. To generate a system of n equations from which the unknown elements from the vectors Ψ and J_c can be computed, as with the source formulation a collocation scheme is used[2][5]. That is, (20) is enforced at the center point in each of n panels. The result is a dense linear system of the form

$$4\pi\tau \frac{\partial}{\partial t} \Psi = (D - 4\pi)\Psi + PJ \quad (22)$$

where $D \in \Re^{n \times n}$ whose elements are

$$D_{kl} = \int_{panel_l} \frac{\partial}{\partial n'} \frac{1}{\|x' - x_k\|} da'; \quad (23)$$

and where $P \in \Re^{n \times m}$ whose elements are

$$P_{kl} = \frac{1}{a_l} \int_{contact\ panel_l} \frac{1}{\|x' - x_k\|} da' \quad (24)$$

where x_k is the center of the kth panel. Note here that the elements of J here are normalized by σ.

4.3 Comparison to the Source Formulation

The Green's identity based formulation has several advantages over the source formulation. The discretization of the source formulation produces a system of $n+m$ equations in $n+m$ unknowns, and the discretization of the Green's formulation produces an system of only n equations in n unknowns. In addition, all the surface potentials are calculated in the case of the discretized Green's formulation, but in the case of the discretized source formulation the surface potentials at points of interest must be derived from integrals of computed surface charges.

To show the superior accuracy in the surface potential produced by the discretized Green's formulation, again consider the simple conductive cube example shown in Figure (1). Figure (3) is a plot of the time behavior of the potential at an observation point on the cube face opposite the driven face, for three sucessively finer surface discretizations. As comparing the plots in Figures (2) and (3) make clear, the discretized Green's formulation converges with panel refinement

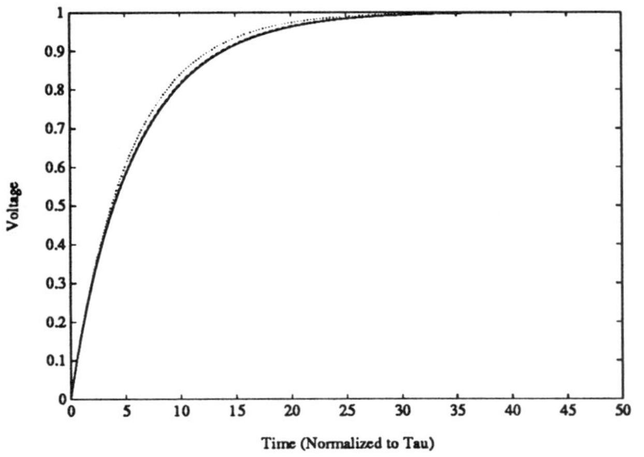

Figure 3: Potential computed at observation point in Figure 1, using three different uniform surface discretizations of the Green's formulation. The dotted line is the result of using 54 panels, the dashed line the result of using 486 panels, and the solid line is the result of using 1944 panels.

much more rapidly than the discretized source formulation, and in addition, the discretized Green's formulation computes the correct equilibrium, or final value, of the potential *regardless of the coarseness of the discretization.*

5 Application Experiments

The most obvious application for the above technique is determining how much changes in voltages on a given conductor are capacitively transmitted to nearby conductors. In particular, we consider the scenario depicted in Figure (4), in which a voltage step is applied to the near end of one conductor, and its effect is monitored at the far end of a parallel conductor which has its near end grounded. The intent is to model realistic interconnect, in which typically one end of every conductor is driven and the other end is connected to a high-impedence input. Clearly it is difficult to approximate how much the voltage will rise at the far end of the grounded conductor.

To compute this voltage rise on the far end of the grounded conductor as a function of the length of the parallel pair, we used the discretized Green's formulation given in (22) combined with a fixed-timestep backward-Euler algorithm, and used 200 fixed timesteps. The resulting matrix problems were solved using one L-U factorization, and 200 forward-elimination/backward-substitutions. To examine the length effects, the conductors were each assumed to be of unit square cross-section and a unit distance apart. Transient solutions were computed for conductors of 2, 4, 6, and 8 units. The surface discretizations for the length 2 and 8 computations are given in Figures (5) and (4) respectively, and the transient behaviors for several lengths are plotted in Figure (6).

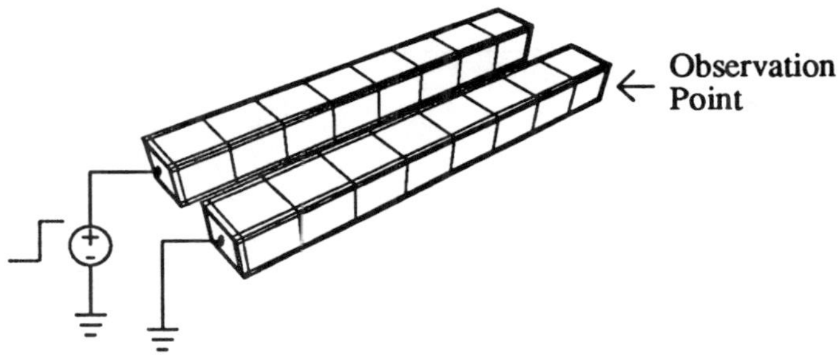

Figure 4: Cross-Talk Simulation Scenario

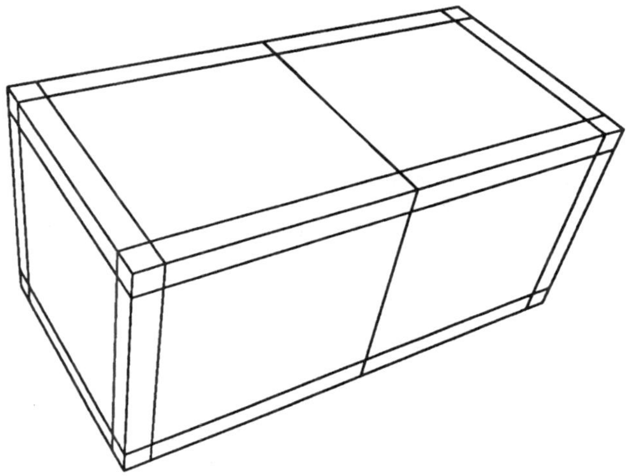

Figure 5: Two-unit length discretization

Note that the time-axis in all these plots is normalized to $\tau = \frac{\epsilon}{\sigma}$. This makes a point that was not initially obvious to the authors, that the *peak magnitude* of the cross-talk pulse is uneffected by changes in either the conductivity, σ, or the dielectric permittivity ϵ. Though, of course, the period of the pulse is directly proportional to $\tau = \frac{\epsilon}{\sigma}$.

Finally, the key point of this work is demonstrated in Table 1, in which the number of panels, or unknowns, and the CPU time required for the transient calculation on an IBM RS6000 model 540 is presented. Note that the most expensive calculation requires less than half a minute, which compares very favorably to computation times required by finite-difference based techniques[4]

length	unknowns	CPU Time
2	132	4.7s
4	180	8.9s
6	228	14.2s
8	276	30.8s

Table 1: CPU times for Transient Calculation

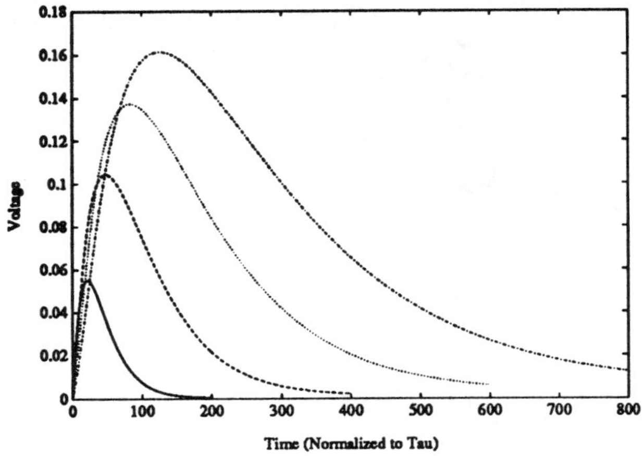

Figure 6: Cross-Talk waveforms for several interconnect lengths. The solid line is for a 2-unit long interconnect, the dashed line for a 4-unit long interconnect, the dotted line for a 6-unit long interconnect, and the dash-dotted line for a 8-unit long interconnect.

6 Conclusions and Acknowledgements

In this paper it is demonstrated that boundary-element techniques can be used to perform very efficient transient simulation of three-dimensional interconnect structures, fast enough to easily be included in a circuit simulator. It was first shown that reasonable discretizations of the most obvious integral formulation do not produce results of acceptable accuracy, and a less intuitive formulation was derived and shown to insure reasonable results even for coarse discretizations.

Note that for the experiments presented in this paper, the differential-algebraic system in (22) was solved using a simple backward-Euler scheme. However, a variety of techniques such as backward-difference methods[1] or moment-matching algorithms[4] can also be used. Also, although the system in (22) is dense, the formulation has advantages over using finite-difference or finite-element techniques applied to solving (5) directly. Since only the two-dimensional conductor surfaces are discretized, rather than the three-dimensional volume containing the problem, as would be the case with finite-difference methods, the discretization is easier to generate and the resulting system contains many fewer unknowns. Unfortunately, the resulting matrix problem is dense, and therefore the comparison to finite-difference methods is not clear.

Note that the direct factorization algorithm used in the numerical experiments presented above limits the size of problems which can be analyzed to relatively simple, but still interesting, structures. In addition, homogeneous dielectric media was assumed, and this may not be realistic. To address these short-comings, we are investigating using iterative techniques to perform the matrix solution, as well as trying to improve the efficency of the matrix coefficient calculation using multipole algorithms. Also, we are examining how to include dielectric interfaces in our formulation.

The authors would like to thank Dr. Ruehli of the I.B.M. T. J. Watson Research Center, and Dr. Ali of the Research Laboratory of Electronics at M.I.T, for the several helpful discussions. In addition, we would like to thank Prof. Newman and Dr. Korsmeyer of the Ocean Engineering department at M.I.T. for their help in understanding several aspects of panel methods. Finally, we would like to acknowledge the assistance of the many members of the M.I.T. custom integrated circuits group.

References

[1] *Numerical Initial Value Problems in Ordinary Differential Equations*, Prentice-Hall, Inc., Englewood Cliffs, New Jersey, 1971.

[2] R. F. Harrington, *Field Computation by Moment Methods*, Macmillan, New York, 1968.

[3] J. D. Jackson, *Classical Electrodynamics*, J. Wiley and Sons, New York, 1975.

[4] S. Kumashiro, R. Rohrer and A. Strojwas, "A New Efficient Method for the Transient Simulation of Three Dimensional Interconnect Structures," *Proc. Int. Electron Devices Meeting*, San Francisco, December 1990.

[5] A. Ruehli and P. A. Brennen, "Efficient capacitance calculations for three-dimensional multiconductor systems," *IEEE Transactions on Microwave Theory and Techniques*, vol. MTT-21, no. 2, pp. 76–82, February 1973.

Move Frame Scheduling and Mixed Scheduling-Allocation for the Automated Synthesis of Digital Systems †

Mehrdad Nourani and Christos Papachristou

Department of Computer Engineering
Case Western Reserve University
Cleveland, OH 44106

Abstract

This paper presents new algorithms for the scheduling and allocation phases in high-level synthesis under time and resource constraints. This is achieved by formulating these problems in terms of Liapunov's stability theorem using a transformation technique between the design space and the dynamic system space.

These algorithms are based on moves in the design space, which correspond to the moves towards the equilibrium point in the dynamic system space. The scheduling algorithm (MFS) takes care of mutually exclusive operations, loop folding, multi-cycle operations, chained operations and pipelining (structural and functional). The mixed scheduling-allocation algorithm (MFSA) can handle all of the above scheduling applications as well as simultaneously performing allocation of functional units, registers and interconnects while minimizing the overall cost.

1 Introduction

High-level synthesis deals with the automatic design of register transfer level (RTL) implementations of digital systems from behavioral descriptions. A behavior may represent a general purpose computer, an application specific system (ASIC), or a combination of them. Behavioral synthesis is normally achieved in two basic steps: 1) Data path synthesis (operation scheduling and hardware allocation), and 2) Control path design. In the scheduling phase, operations are assigned to the appropriate control steps. The allocation phase specifies the design at a register transfer level by assigning operations into the hardware components.

Operation scheduling determines the hardware cost-speed tradeoffs of a design. The simplest way to perform scheduling is to relegate the task to the user to explicitly define the parallelism of the design such as in the *SLICE* system [1]. The *FACET* system, presented in [2], used ASAP schedule for finding near optimal solution. Conditional deferment methods, such as in [3], were based on the fact that if the operation concurrency is higher than the number of available units, those operations should be postponed. List scheduling techniques, such as in [4], sort and then schedule operations in topological order based on priorities (or mobilities), are more complex class of algorithms.

The fact that decisions at higher levels of synthesis (i.e. allocation) may dominate the results produced by an independent scheduling phase was a motivation to develop another class of methods in which these two phases are interdependent. *MAHA* [5] attempts to perform scheduling and allocation in parallel by allocating hypothetical average functional units during the list scheduling phase. *HAL* [6] uses a stepwise refinement technique which uses force-directed scheduling assuming single-function units. A simulated annealing based algorithm was used to perform scheduling and allocation together in one process [8]. Some researchers [3][11][9][10] use linear programming formulation to develop scheduling and allocation algorithms. Neural network-based models such as [12]have been recently attempted.

†This work was partially supported by the Semiconductor Research Corporation (SRC) under contract No. 91-DJ-147

This paper presents new algorithms for scheduling and mixed scheduling-allocation process in high level synthesis. The main contribution of our method is in the following:

1. A new fast scheduling method, Move Frame Scheduling (MFS) which performs guided search in the huge state space of schedules, driven by Liapunov's stability theorem.
2. A new fast scheduling-allocation method, Move Frame Scheduling-Allocation (MFSA), which extends the guiding mechanism of MFS to search the scheduling and allocation spaces simultaneously.
3. The methods can be applied to different synthesis aspects under time and resource constraints.

The main advantage of our methods over existing scheduling and allocation algorithms is in running time. Two specific characteristics provide this important advantage. First, the main decisions to schedule or allocate an operation in data flow graph (DFG) are made within a relatively small 2-dimensional frame. Second, our methods do not suffer from the *size explosion* problem for variables and/or constraints as some methods (e.g. linear programming-based) do. Furthermore, based on the Liapunov's theorem, we use the Liapunov (energy) function as the guiding mechanism to approach optimal solution while avoiding the *probablistic exploration* and *tunning* problems in some energy-based approaches such as annealing and neural networks.

This paper is organized as follows: Section 2 reviews the transformation technique and stability problem. Sections 3 and 4 describe the MFS and MFSA methods, respectively. Section 5 extends MFS and MFSA to include a number of important synthesis aspects. The experimental results are shown in section 6. Finally, concluding remarks are in section 7.

2 The Synthesis Problem Transformed to a Stability Problem

2.1 Transformation Technique

The basic idea is to use the similarities between the *synthesis problem* in computer engineering and the *stability problem* in control theory. In this analogy, *system*, *stability* and *equilibrium point* have the same notion as *algorithm*, *convergence* and *optimal solution*, respectively. The mathematical proof for correctness of this transformation is beyond the scope of this article. The details can be found in [13] and [14].

2.2 Stability Concept and Liapunov Theorem

Based on [15] a system is asymptotically stable at the equilibrium point if and only if a trajectory beginning in a suitable neighborhood of the equilibrium point tends to the equilibrium point as the time passes.

The problem of determining whether or not a system is stable is fundamental in all control system design. For non-linear systems the Liapunov's theorem has the greatest potential for investigating stability in such systems. This theorem is more a principle than a technique. Intuitively speaking, if the time rate of change of the energy E of an isolated system is negative for every possible state X of the system, except for a single equilibrium state X_e, then the

0738-100X/92 $3.00 © 1992 IEEE

energy will continually decrease until it finally assumes its minimum. The generalization of this concept was considered by *M. A. Liapunov.* He expressed his idea by the following theorem:

A dynamic system described by the discrete time vector equation $X(k+1) = A(k) * X(k)$ $(k = 0, 1, 2, \cdots)$ is stable if and only if there exists a scalar Liapunov function $V(X) : R^n \rightarrow R^+$ with the following properties:

1. $V(X(k)) > 0$ *when* $X(k) \neq X_e$.
2. $V(X(k+1)) - V(X(k)) < 0$ *when* $X(k) \neq X_e$.
3. $V(X(k)) = 0$ *when* $X(k) = X_e$.
4. $\lim_{\|X\| \rightarrow \infty} V(X) = \infty$.

The main practical point of this theorem is that if such a function $V(X)$ exists, then it guarantees that the system is stable and if we let time grow, by selecting an appropriate movement mechanism (i. e. make $V(X)$ decreasing monotonically) the system tends towards, and finally reaches, the equilibrium point. Complete proof of this theorem for different cases (discrete or continuous time) can be found in many of control theory books such as [15].

2.3 New Formulation of Synthesis Problem

To use Liapunov's theorem, we must first formulate the synthesis problem as a system whose behavior can be expressed by the vector equation $X(k+1) = A(k) * X(k)$.

Given a DFG for each operation O_i $(1 \leq i \leq l)$, consider a 2-dimensional placement (grid) table shown in Figure 1. One dimension represents the number of control steps and the other shows the instances of functional units (hardware modules). In a scheduling algorithm, the functional units are assumed to be single function operators, while in a simultaneous scheduling-allocation algorithm they are also allowed to be multifunctional units (ALU's). This assumption is due to the fact that in the scheduling phase the concurrency of similar operations is an important factor, while in the allocation phase merging different operations can significantly decrease the overall ALU cost. In Figure 1, the horizontal coordinate represents *the number of functional units (FU)* of a specific type, and the vertical coordinate shows *the number of control steps*. For example, the *present position* of operation O_i, denoted by O_{ip}, shows that O_i is performed by the fourth functional unit of that type in the sixth control step. Note that because we normally have more than one type of functional units, the complete space will be a 3-dimensional space where the third dimension represents the type of FU's. Note that Figure 1 does not show this but Figure 2 shows a 2-dimensional representation of this 3-dimensional space for a typical example.

We define $X_e = \vec{0}$ as the equilibrium point (which is a dummy one in the synthesis problem because obviously $\vec{0} = (0,0)$ is not accessible). Let us assume that the system is stable, and let time grow. Then in a trajectory towards the equilibrium point, the present position of O_i, denoted by O_{ip}, at time k (k-th step of the movement) is: $X_i(k) = (x_{i,k}, y_{i,k})^T$. Similarly, the next position of O_i, denoted by O_{in}, at time $k+1$ is: $X_i(k+1) = (x_{i,k+1}, y_{i,k+1})^T$. Assuming: $\begin{pmatrix} x_{i,k+1} \\ y_{i,k+1} \end{pmatrix} = \begin{pmatrix} a_{i,k} & 0 \\ 0 & b_{i,k} \end{pmatrix} \begin{pmatrix} x_{i,k} \\ y_{i,k} \end{pmatrix}$ where $a_{i,k}$ and $b_{i,k}$ are two positive real numbers, we have: $X_i(k+1) = A_i(k) * X_i(k)$. If we join all of similar elements corresponding to all operations O_i $(1 \leq i \leq l)$ in DFG, we will totally get: $X(k+1) = A(k) * X(k)$ where:

$$X(k+1) = \begin{pmatrix} X_1(k+1) \\ \vdots \\ X_l(k+1) \end{pmatrix}, X(k) = \begin{pmatrix} X_1(k) \\ \vdots \\ X_l(k) \end{pmatrix},$$

$$A(k) = \begin{pmatrix} A_1(k) & \cdots & \mathbf{0} \\ \vdots & \cdots & \vdots \\ \mathbf{0} & \cdots & A_l(k) \end{pmatrix} \text{ and } \mathbf{0} = \begin{pmatrix} 0 & 0 \\ 0 & 0 \end{pmatrix}.$$

2.4 Liapunov Function in Synthesis Problems

So far we have formulated the general synthesis problem so as to apply Liapunov's theorem. Application of this theorem involves finding a function satisfying those basic properties mentioned in section 2.2. Unfortunately, there is no standard method for finding such a function. Moreover, a Liapunov function may not be unique. Different Liapunov functions will produce different performances, and consequently different results. The general role of the Liapunov function in MFS and MFSA is the same as the energy function in annealing and neural network based approaches. In fact, some researchers in the latter area call their energy function *Liapunov function*.

Scheduling (and allocation) in MFS (and MFSA) is accomplished by *moves* in the 2-dimensional grid space. The direction of moves and the moves themselves should satisfy the Liapunov theorem criteria. The Liapunov function is used not only to guarantee approaching to the equilibrium point, but also to specify the direction towards it. The first role is guaranteed if the energy function satisfies those four properties. However, the problem constraints and specifications determine which direction is more desirable.

3 Move Frame Scheduling Algorithm (MFS)

Our goal, in introducing MFS as an independent scheduling algorithm, is twofold. First, the scheduling process is less complicated and easier to describe. Moreover, in many practical applications we need to do the scheduling first. Second, we want to show that MFS itself has enough credibility as an independent scheduling algorithm for producing a balanced schedule for any desirable allocation process that follows.

3.1 Liapunov Function Used in MFS

When we schedule an operation under time constraints, we should not waste any control steps. In other words, the Liapunov (energy) function should make sure that control step t is selected before $t+1$, if possible. Assume that max_j number of functional units of type j (denoted by $FU_{j,1}, FU_{j,2}, \cdots, FU_{j,max_j}$) are available. Then, to select control step t before $t+1$, it is sufficient to assign a lower energy value for position (FU_{j,max_j}, t) than position $(FU_{j,1}, t+1)$. This relation should be valid for any FU and any control step. Based on these requirements, consider this function: $V(X(k)) = \sum_{i=1}^{l}(x_{i,k} + n * y_{i,k})$ where $n = Max\{max_j\}, 1 \leq j \leq$ *number_of_types* of FU's. max_j may be specified by the user as a resource constraint or can be set to a presummed big number as the upper bound.

By a similar argument, we can easily show that the function: $V(X(k)) = \sum_{i=1}^{l}(cs * x_{i,k} + y_{i,k})$ where cs is an upper bound for total number of control steps, is a Liapunov function which can be applied to scheduling under resource constraints. This function selects a position in control step $t+1$ performed by an existing FU instead of adding a new FU in control step t.

For each operation $O_i (1 \leq i \leq l)$, the next position must be closer to the equilibrium point. The closeness will be interpreted by the Liapunov function. It is trivial to show that properties (1), (3) and (4) are held. To satisfy property (2) we must have : $x_{i,k+1} < x_{i,k}$ and $y_{i,k+1} < y_{i,k}$. This means that the next position of each operation should be to the *left* and/or *above* of its present position. Moreover, the next position should satisfy the data dependency constraints.

3.2 Scheme and Example

To generate a balanced schedule (minimum concurrency) under fixed time constraint, MFS proceeds in the following four steps:

- *Step 1:* Find ASAP (As Soon As Possible) and ALAP (As Late As Possible) schedules, within the given number of control steps, to specify the time frame for each operation. Figure 2(a) shows these schedules.

- *Step 2:* Select an arbitrary ordering of different types of FU's and do all the subsequent phases in this order.
  - Set the maximum number of different types of FU's, $max_j (1 \leq j \leq number\_of\_types)$ to the numbers given by the user as the hardware constraints. For example, if the user requires up to 4 adders then $max_+ = 4$. If they are not given by the user, find the maximum number of FU's in ASAP and ALAP schedules and select them as the upper bound.
  - Calculate the mobility * for each operation: $mob[O_i] = ALAP_{c-step}[O_i] - ASAP_{c-step}[O_i]$
  - Determine the priorities of operations in ALAP schedule based on their mobilities. The rule is very simple: If $mob[p] < mob[q]$ then p has more priority than q. Priority determination starts from the first control step and will cover all control steps in ALAP. Ties will be broken arbitrarily.
- *Step 3:* Construct ASNAP (As Soon and Near As Possible) and ALFAP (As Late and Far As Possible) tables. Briefly, the ASNAP and ALFAP tables, specify a 2-dimensional frame (rectangle) for moving each operation individually and independently. For an operation of type j, this frame is specified by $[1, cs]$ and $[1, max_j]$.
- *Step 4:* Schedule each operation in the order of its priority by the following process:
  - Determine the *Primary Frame* (PF): This is a 2-dimensional frame for each operation restricted by its place in ASNAP and ALFAP tables.
  - Determine the *Redundant Frame* (RF): Since we do not know how many of them we really need, we consider a variable to show the current number of FU's ($current_j$). We set this variable to $\lceil N_j/cs \rceil$ where N_j is the total number of operations of type j in DFG and cs is the total number of control steps. At each step the frame defined by $[ASAP_{c-step}, ALAP_{c-step}]$ and $[current_j+1, max_j]$ represents the *redundant frame.*
  - Determine the *Forbidden Frame* (FF): We must exclude those positions whose control steps are less than or equal to the predecessors (of O_i) control step, in order to satisfy the data dependency constraints. These excluded positions form the forbidden frame.
  - Determine the *Move Frame* (MF): This is the valid frame in which an operation can be scheduled. For each operation we have: $MF_{O_i} = PF_{O_i} - (RF_{O_i} + FF_{O_i})$. Note that this equation represents a ***set relation*** between the frames because each frame is really a set of positions. Figure 2, (a) and (b), illustrates the above frames for a typical operation r of type j. In this figure, we assume that operation r has two predecessors, namely K_1 and K_2, and the current number of functional units of type j at that step is two ($current_j = 2$). These operations have already been assigned to positions K_1^* and K_2^* in previous steps (X shows other occupied positions).
  - Select one of the available positions in MF (move frame) which has the smallest Liapunov value and assign operation O_i in it. If there is no valid position for O_i (i.e. MF_{O_i} is empty or fully occupied), increase $current_j$ by one and then do a local rescheduling by going back to step 3. For the example shown in Figure 2, the best position in MF to assign operation r will be r^*.

*The mobility is the difference in control steps between the scheduled places of an operation in ALAP and ASAP schedules, respectively. Many researchers such as [4] and [6] have already used this notion.

Analysis of MFS shows that the algorithm runs in $O(l^3)$ in the worst case, where l is total number of operations.

4 Move Frame Scheduling-Allocation Algorithm (MFSA)

Many researchers [7-9] believe that because of the tight interaction between scheduling and allocation, a simultaneous algorithm has a better chance to capture the global optimal solution. However, such algorithms may be slower and more complex. They usually suffer from size explosion (e.g.linear programming based), probablistic nature (e.g. simulated annealing) or unreliable approximations for nonlinear objective functions.

MFSA is an algorithm in which the scheduling, allocation and data path optimization processes interact with each other by means of a complex Liapunov function. The Liapunov function in MFSA is capable of selecting and optimizing different types of functional units (ALU's), optimizing multiplexers (or buses), registers and data path connections. As we will show shortly, although MFSA is a bit slower than MFS, the complexity of MFSA is the same as MFS because the 2-dimensional tables (search space) and general movement mechanism within the move frame for operations remain unchanged. The Liapunov function in MFSA is nonlinear and complex but we are not going to optimize it. In fact, similar to MFS, this function is used as an evaluation tool to guide the movement in the search space.

4.1 Liapunov Function Used in MFSA

MFSA uses the same 3-dimensional space (combination of 2-dimensional tables) as MFS explained in section 3. So, the behavior of the system can still be represented by the vector equation: $X(k+1) = A(k) * X(k)$. To have the necessary information for multiplexer (MUX) and register (REG) optimization, we should annotate the input signals (input variables) of each operation, together with its name in the DFG and other tables. As we will show shortly, these signals play a major role in the Liapunov function and MFSA algorithm. The following important rules will be considered in the new Liapunov function:

- In time constrained problems, control step t should be selected before $t+1$, as explained in MFS.
- When a new functional unit (called "ALU" in MFSA) is used for performing an operation, its cost should be added to the incremental ALU cost up to that iteration. The cost of an existing ALU, previously used to perform other operations, is zero.
- When the Liapunov function evaluates an ALU for assigning an operation, the input signals associated with all operations in that ALU should maximally share MUX inputs to minimize MUX cost. If we assume that each operation has at most two input signals, after assigning a new operation to an ALU it may share zero, one or two signals with other operations.
- When the Liapunov function evaluates a specific control step to assign an operation, the life span of its input signals (coming from its predecessors scheduled and allocated at previous iterations) can be determined. Thus, we can evaluate the effect of this selection to the cost of registers up to that point.

Assume that the Liapunov function finds the best position for O_i in the 2-dimensional valid frame (MF_{O_i} frame) in kth iteration (i.e. kth time evaluation of appropriate time slots and ALU's). Let $y_{i,j,k}$ and $x_{i,j,k}$ denote the time slot (expressed in control steps) and ALU index of type j (index of ALU of type j when many ALU's of this type are available) chosen in kth iteration by the Liapunov function, respectively. Based on the above points and notations, consider this Liapunov function:

$V(X(k)) = \sum_{i=1}^{l}(f_{i,j,k}^{TIME} + f_{i,j,k}^{ALU} + f_{i,j,k}^{MUX} + f_{i,j,k}^{REG})$. $f_{i,j,k}$ represents the contribution of O_i in the Liapunov function when it is evaluated in the kth iteration. Recall that in MFS, each operation can be only assigned to a functional unit of its type, e.g. an addition is always assigned to an adder (+), but in MFSA each operation can be assigned to different functional units, e.g. an addition may be assigned to single or multifunction ALU's such as (+), (+−), (+ >) or (+− >) based on the cell library given by the user. Note also that all other operations scheduled and allocated in the previous iterations keep their contributions in the Liapunov function unchanged. Now, let's explain each term in $V(X(k))$ for a specific operation O_i. For brevity, we will show the four terms involved in the function without their subscripts, i.e. f^{TIME}, f^{ALU}, f^{MUX} and f^{REG}:

- $f^{TIME} = f^{TIME}(y_{i,j,k}) = C * y_{i,j,k}$
  C is a constant that guarantees control step t is selected before $t+1$ if it is possible. To guarantee this in iteration k (in which control step $y_{i,j,k}$ has been selected) we must satisfy:
  $C * y_{i,j,k} + f_{max}^{ALU} + f_{max}^{MUX} + f_{max}^{REG} < C * (y_{i,j,k} + 1) + f_{min}^{ALU} + f_{min}^{MUX} + f_{min}^{REG}$
  or: $C > [f_{max}^{ALU} + f_{max}^{MUX} + f_{max}^{REG}] - [f_{min}^{ALU} + f_{min}^{MUX} + f_{min}^{REG}]$
  The maximum and minimum values for f^{ALU}, f^{MUX} and f^{REG}, i.e. f_{max}^{ALU}, f_{min}^{ALU}, etc., will be determined shortly when we describe each term.
- $f^{ALU} = f^{ALU}(x_{i,j,k})$
  When we use a new ALU, its cost should be added to the overall ALU cost. So, $f^{ALU}(x_{i,j,k}) = Cost(ALU_j)$ if ALU specified by $x_{i,j,k}$ has not already been used. Otherwise the cost of using an existing ALU is zero. Clearly, $f_{max}^{ALU} = Cost(ALU_j)$ and $f_{min}^{ALU} = 0$.
- $f^{MUX} = f^{MUX}(x_{i,j,k}, S_{i,j,k})$
  $S_{i,j,k}$ is the set of input signals of operations in the ALU specified by $x_{i,j,k}$ ($x_{i,j,k}$th ALU of type j). The term f^{MUX} represents the possibility of signal sharing after adding the new operation in this ALU. In a straight forward design in which each ALU has two multiplexers (MUX^1 and MUX^2) feeding the ALU input signals, we have: $f^{MUX} = f^{MUX}(x_{i,j,k}, S_{i,j,k}) = [Cost(MUX_{after}^1) + Cost(MUX_{after}^2)] - [Cost(MUX_{before}^1) + Cost(MUX_{before}^2)]$, where subscripts *before* and *after* denote the multiplexer configurations before and after adding operation O_i, respectively. Practically, the cost of a multiplexer with r data inputs and 1 output, denoted by $Cost(MUX_r)$, is not a linear function of r. So, depending on the cell library used in the design :
  $f_{max}^{MUX} = 2 * max\{Cost(MUX_{r+1}) - Cost(MUX_r)\}$ where $r = 1, 2, \cdots$ and $f_{min}^{MUX} = 0$.
- $f^{REG} = f^{REG}(y_{i,j,k}, P_{i,j,k})$
  $P_{i,j,k}$ is the set of predecessors of operation O_i which have been scheduled and allocated in the previous iterations. When O_i is assigned to control step $y_{i,j,k}$, the consuming time of signals generated by those predecessors (up to iteration k) will be $y_{i,j,k}$. In other words, in a backward look at the partially constructed schedule (up to iteration k) the life spans of input signals of O_i (coming from predecessor nodes or registers) are determined. Assuming that O_i has at most two inputs, then zero, one or two registers are needed to store those signals until they are consumed by O_i. Obviously, $f_{max}^{REG} = 2 * Cost(REG)$ and $f_{min}^{REG} = 0$.

If the user wishes to put more emphasis on optimizing one of the four factors participating in the Liapunov function (TIME, ALU, MUX and REG), he/she may consider a weighted Liapunov function as: $f_{i,j,k} = (w^{TIME} * f^{TIME}) + (w^{ALU} * f^{ALU}) + (w^{MUX} * f^{MUX}) + (w^{REG} * f^{REG})$, where w^{TIME}, w^{ALU}, w^{MUX} and w^{REG} are weights based on the relative importance of each factor. $w^{TIME} = w^{ALU} = w^{MUX} = w^{REG} = 1$ gives an overall optimizer without emphasising any particular factor. Similar to MFS, properties (1),(3) and (4) are clearly held. Property (2) is also satisfied if we consider a movement mechanism that moves O_i from $(x_{i,j,k}, y_{i,j,k})$ to $(x_{i,j,k+1}, y_{i,j,k+1})$ if and only if $f_{i,j,k+1} < f_{i,j,k}$.

4.2 Scheme and Example

The MFSA generates an RTL (register transfer level) structure in the following steps:

- *Step 1, Step 2,* and *Step 3* are exactly the same as the first three steps in MFS algorithm. As we have already pointed out, the only difference is that we record the input signals together with the operations when different tables and frames are constructed in the process.
- *Step 4:* Consider each operation in order of priorities as in MFS by the following process:
  - Determine all ALU's (i.e single or multifunction ALU of type j) capable of performing operation O_i.
  - Determine the frames $PF_{O_i,j}$, $RF_{O_i,j}$, $FF_{O_i,j}$ and $MF_{O_i,j}$ for those ALU's found in the previous step.
  - Compute the Liapunov function values for each empty position in $MF_{O_i,j}$. Then, select a position in this frame with the smallest associated Liapunov value and assign O_i to that position ($x_{i,j,k}$th control step and $y_{i,j,k}$th ALU) and update the function and tables.

It is important to note that the Liapunov function in MFS is a *static* function. It assigns a fixed value to each position in the tables or frames in the entire process. In MFSA, however, the Liapunov function is a *dynamic* function which is updated in each iteration based on the best configuration of the table (operations, input signals, multiplexers and registers) found in the previous iterations. The running time is also $O(l^3)$, the same order as in MFS. To show how other specifications (other than those considered directly in the Liapunov function) can be handled in MFSA, we consider two design styles:

1. Conventional data path design style (unrestricted RTL structure).
2. Restricted design style which gives an RTL structure without a self loop around ALU's. This has been proven to be useful for having a self testable structure [18]. So, in this case no operation is allowed to be with its successors or predecessors within the same ALU.

5 Applications to Synthesis Problems

5.1 Conditional Statements

If-then-else and *case* statements cause branches in the DFG. Operations in different branches are mutually exclusive, thus they can be executed on the same type of FU and scheduled into the same control step without increasing the required number of FU's. To take advantage of this observation, we remove all of the operations which are shared between branches except one of them. Obviously, those shared operations can be executed by the same FU.

5.2 Loops

The user should specify a constraint on the loop iteration time. This can be done by adding two more operations (addition and comparison or increment and comparison) into the DFG corresponding to the body of the loop. Then the operations bounded by the new DFG are considered within the constraint using MFS or MFSA.

For nested loops, the operations of the inner most loop are scheduled and allocated first, relative to the local time constraint. When this is done, the entire loop is treated as a single operation with an execution time that is equal to the loop's local time constraint. This process is repeated for all loops until the outer most loop is scheduled and allocated. Loop unfolding by using *functional pipelining* is another way to handle loops. We will comment on this issue shortly.

5.3 Multi-Cycle Operations

The fact that different operations have different execution time, has been of concern in the literature. MFS and MFSA consider k-cycle operations as k consecutive single-cycle operations of different types. To be able to run our methods on multi-cycle operations, we should consider the following modifications:

- All assumed k single-cycle operations must be scheduled in consecutive control steps because no delay is allowed during the execution of an operation.
- Two k-cycle operations are concurrent if at least one of their k consecutive stages is concurrent.
- If the difference of mobilities between two k-cycle operations is less than k, we will reverse the previous rule for priority determination because, in this special case, the operation with more mobility has always a better chance to use the empty positions.
- As a rule for tie breaking, the operation with earlier predecessors (in terms of control steps), will get higher priority.

5.4 Chained Operations

Scheduling of consecutive data dependent operations in a single control step is called *chaining*. This feature is supported by our methods, MFS and MFSA, by considering the following extensions:

- ASAP and ALAP schedules (and consequently the mobilities and priorities) are determined based on the given execution time of operations and the length of control step clock (T).
- The Forbidden Frame (FF) which originally excludes the control steps of predecessors, should be changed to allow chaining.

5.5 Pipelining

The optimization of scheduling when we have pipelined behavior in synthesis is a very important issue according to recent developments in this area. Two simple and efficient pipelining techniques (*structural* and *functional*) have been presented in the literature [6][16][17]. Both of these pipelining techniques are supported by our methods using simple modifications.

5.5.1 Structural Pipelining

In structural pipelining, a modified schedule is obtained through the use of pipelined FU's, e.g. a two-stage pipelined multiplier, so that the operation instances are executed in an overlapping fashion. The extension of MFS and MFSA for structural pipelining is based on the fact that once any stage of a pipelined FU is empty, it is considered available. We then consider these modifications:

- Change multi-cycle operations (for which a pipelined FU's are available) to single-cycle operations of different types. After this modification, different operations represent different stages of a multi-stage pipelined functional unit.
- Run the algorithm on the new DFG considering the fact that different stages of pipelined operations can be concurrent but must be scheduled in consecutive control steps.

5.5.2 Functional Pipelining

In functional pipelining we try to divide the DFG description (operations) into some partitions that can be performed concurrently. Successive partitions are streamed into the pipe so that different instances are executed in an overlapping fashion. In loop unfolding application, each instance corresponds to a new initiation of the loop. For a given latency L, the operations scheduled into control step $t + k * L$ ($k = 1, 2, 3, \cdots$) run concurrently, so we must balance the distribution of operations across all individual control steps. To consider functional pipelining we need the following process:

1. Given the DFG description (i.e. body of a loop), the time constraint for DFG (cs control steps) and latency L, consider a new DFG consisting of two *instances* with delay of L cycles in between.
2. Divide the new DFG into two partitions DFG_{p1} and DFG_{p2}. DFG_{p1} consists of all operations between control step 1 and $\lceil \frac{cs+L}{2} \rceil$. Similarly, DFG_{p2} consists of all operations between control step $\lceil \frac{cs+L}{2} \rceil + 1$ and $cs + L$. Clearly $DFG_{double} = DFG_{p1} \cup DFG_{p2}$.
3. Run the algorithm on DFG_{p1}, taking into account the time frame for operations belonging to *Instance 2* by adding dummy operations if it is necessary.
4. Adjust the result of the previous step in order to get identical instances. To do this, we should add all operations which are in $(Instance1) \cap DFG_{p1}$ but not in $(Instance2) \cap DFG_{p1}$.
5. Eliminate adjusted operations (added in previous step) from DFG_{p2} and run the algorithm on the remaining operations in DFG_{p2}.

5.6 Multiplexer Optimization

Term f^{MUX} in the Liapunov function, explained in section 4, evaluates the contribution of operation O_i to the multiplexer cost. To consider multiplexer optimization, two terms in f^{MUX}, i.e. $Cost(MUX^1_{after})$ and $Cost(MUX^2_{after})$ should be evaluated under the *best case* of sharing for the input signals. Assuming that each operation has at most two inputs, MFSA uses a constructive algorithm which reads the set of operations assigned to a specific ALU and their corresponding inputs and constructs two lists of input signals L_1 and L_2 such that $|L_1| + |L_2|$ is minimum.

Briefly, the algorithm first assigns the non-commutative operations to the appropriate MUX's of an ALU and then checks two possibilities for arranging input signals for each commutative operation in L_1 and L_2.

5.7 Interconnect Optimization

When some of the operations performed by ALU_r are successors (or predecessors) of operations in ALU_s, the connection lines carrying the data between ALU_r and $ALUs$, and consequently the input lines to the multiplexers of these ALU's, can be shared unless there is a conflict between them (e.g. both predecessors of an operation are located in the same ALU). Line sharing for data transfer between ALU's has a secondary effect on $Cost(MUX^1_{after})$ and $Cost(MUX^2_{after})$ terms in f^{MUX} before the Liapunov function makes its final decision.

5.8 Register Optimization

Term f^{REG} itself, considers register optimization by finding the least number of registers saving signals required to perform operations within an ALU at that iteration. To find this number, we use an expanded version of the *activity selection algorithm*†. This is a greedy algorithm capable

†This algorithm is an extended version of the *left edge algorithm* used by many researchers such as [19].

of finding the best solution for one register in $\Theta(m)$ where m is the number of signals some of which are selected to be saved in that REG. Briefly, the signal with the smallest death time is selected and if it is compatible (no time conflict) with other signals in the register it will be assigned to that register.

6 Experimental Results

The MFS and MFSA algorithms described in the previous sections have been implemented in C on a SUN SPARC-SLC workstation. These methods can easily be integrated with any synthesis tool. Currently, they are some of the options for scheduling and allocation in our high level synthesis tool called SYNTEST [20] which produces a self-testable RTL structure from a behavioral description. If MFS is selected, it reads the initial behavior (DFG) and the user has to specify the constraints and specifications (including total control steps, execution time for each type of operations, pipeline options and so on). Then MFS finds a balanced schedule under design constraints. If MFSA is selected, in addition to the above constraints and specifications, the cell library (which may be restricted to some specific types) and other limitation imposed to RTL structure (i.e. design styles explained before) should be prepared for the system. Then MFSA generates a schedule and its corresponding RTL structure while optimizing the overall cost.

This section presents brief results for six design examples from the literature. Table 1 shows a summary of the design results produced by MFS for the above six examples. In the second column, "1" means all operations take one cycle and "2" shows only multiplication consumes two cycles. Also "*C*", "*F*" and "*S*" in this column show the special feature of that design, i.e *chaining, functional pipelining and structural pipelining*, respectively. Note, the CPU time for all examples is less than 0.2 seconds on a SPARC station.

The result of MFSA which is a simultaneous scheduling-allocation algorithm are tabulated in Table 2. Style 1 is unrestricted RTL structure while style 2 is RTL structure without self loop around ALU's. Overall cost of RTL designs (in micron square) is based on a NCR library [21]. The CPU time for running MFSA is less than 0.4 seconds for these examples. The wall clock time is about three seconds. For comparison, we have estimated the overall RTL structure cost (in micron square) based on the NCR library [21]. Comparison between the presented methods in the literature [6][5][11] and the first design style obtained by MFSA shows an improvement of $1-5\%$ for examples 3, 5 and 6, while for examples 1 it shows 4% overhead. However, direct comparisons with other methods are not possible because of the difference in cell libraries and specification of the design and unavailability of their complete RTL structure. On the other hand, design style 2 shows $2-11\%$ overhead compared to design style 1. This is due to the fact that in style 2, the method imposes more restrictions for merging operations and sharing input signals within ALU's [18][20].

7 Conclusion

In this paper, we have presented a new approach to data path synthesis under time or resource constraints which is suitable for computationally dense behaviors such as digital signal processors and other ASIC systems. An important feature of our method is that by applying the global stability concept of dynamic systems to the transformed problem of scheduling and allocation, the MFS and MFSA algorithms explore the search space in a guided and global fashion and produce optimal or near-optimal results for all of the examples attempted to date.

Furthermore, the flexibility of Liapunov's stability approach as a guiding mechanism of the MFS and MFSA algorithms was highlighted by supporting different practical aspects of synthesis including mutually exclusive operations, loop folding, chaining, multi-cycle operations, pipelining and multiplexer and register optimization.

References

[1] T. Blackman et al., "The Slice compiler: Language and features," *Proc. 22nd Design Automat. Conf.*, June 1985.

[2] C. Tseng and D. P. Siewiorek, "FACET: A Procedure for the Automated Synthesis of Digital Systems," *Proc. 20th Design Automat. Conf.*, June 1983.

[3] P. Marwedel, "A new synthesis algorithm for the MIMOLA software system," *Proc. 23rd Design Automat. Conf.*, July 1986.

[4] B. M. Pangrel and D. D. Gajski, "Slicer: A state synthesizer for intelligent silicon compilation," *Proc. IEEE Int. Conf. Computer Design (ICCD-87)*, Oct. 1987.

[5] A. C. Parker et al., " MAHA: A program for data path synthesis," *Proc. 23rd Design Automat. Conf.*, July 1986.

[6] P. G. Paulin and J. P. Knight, "Forced-Directed Scheduling for the Behavioral Synthesis of ASIC's," *IEEE Trans. Computer-Aided Design*, June 1989.

[7] R. J. Clotier and D. E. Thomas, "The Combination of Scheduling, Allocation and Mapping in a Single Algorithm," *Proc. 27th Design Automat. Conf.*, June 1990.

[8] S. Devadas and A. R. Newton, "Algorithms for Hardware Allocation in Data Path Synthesis," *IEEE Trans. Computer-Aided Design*, July 1989.

[9] C. H. Gebotys and M. I. Elmasry, "Simultaneous Scheduling and Allocation for Cost Constrained Optimal Architectural Synthesis," *Proc. 28th Design Automat. Conf.*, June 1991.

[10] C. A. Papachristou and H. Konuk, "A Linear Programming Driven Scheduling and Allocation Method Followed by an Interconnect Optimization Algorithm," *Proc. 27th Design Automat. Conf.*, June 1990.

[11] C. T. Hwang Y. C. Hsu and Y. L. Lin, "Optimum and Heuristic Data Path Scheduling Under Resource Constraints," *Proc. 27th Design Automat. Conf.*, June 1990.

[12] A. Hemani, "NISCHE: A neural net inspired scheduling algorithm," *High-Level Synthesis Workshop*, 1989.

[13] L. Ljung, "On Positive Real Transfer Functions and the Convergence of Some Recursive Schemes," *Trans. Automatic Control*, Aug. 1977.

[14] M. Nourani and C. Papachristou "Move Frame Scheduling Based on Liapunov Stability Theorem for the Automated Synthesis of Digital Systems," Technical Report CES-91-05, Dept. of Computer Eng., Case Western Reserve Univ., 1991.

[15] S. P. Banks, *Control system engineering: Modeling and simulation*, Prentice-Hall, 1986.

[16] N. Park and A. C. Parker, "SEHWA: A program for synthesis of pipelines," *Proc. 23rd Design Automat. Conf.*, July 1986.

[17] C. T. Hwang, Y. C. Hsu and Y. L. Lin, "Scheduling for Functional Pipelining and Loop Winding," *Proc. 28th Design Automat. Conf.*, June 1991.

[18] C. A. Papachristou, S. Chiu and H. Harmanani "A Data Path Synthesis Method for Self-Testable Designs," *Proc. 28th Design Automat. Conf.*, June 1991.

[19] F. J. Kurdahi and A. C. Parker, "REAL: A Program for REgister ALlocation," *Proc. 24rd Design Automat. Conf.*, July 1987.

[20] C. A. Papachristou, S. Chiu, H. Harmanani, "SYNTEST: a method for high-level SYNthesis with self-TESTability," *Proc. of the Internat. Conf. on Computer Design (ICCD-91)*, Oct. 1991.

[21] *NCR ASIC Data Book*, NCR Corporation, 1989.

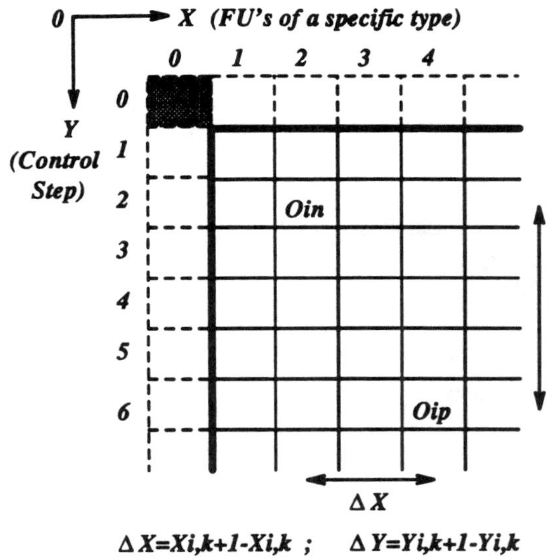

Figure 1: Present (O_{ip}) and next (O_{in}) position of an operation in the placement table

Table 1: The MFS result for six examples

Ex.	Special feature	Total number of control steps		
#1		T=4	T=5	
	1	*,++,−, /, &, |	*,+,−, /, &, |	
#2		T=4		
	C	+,−		
#3		T=4	T=6	T=8
	1	**,+,−,>		
	F		* * *, +, >	
	S			*,+,−,>
#4		T=8	T=9	T=13
	1	**, ++, −−	**, ++, −	*,+,−
#5		T=9	T=10	T=13
	2	*****, ++,−−	***,++, −−	**,+,−
	S	***,++ −−	**,++, −−	*,+,−
#6		T=17	T=19	T=21
	2	***,+++	**,++	*,++
	S	**,+++	*,++	

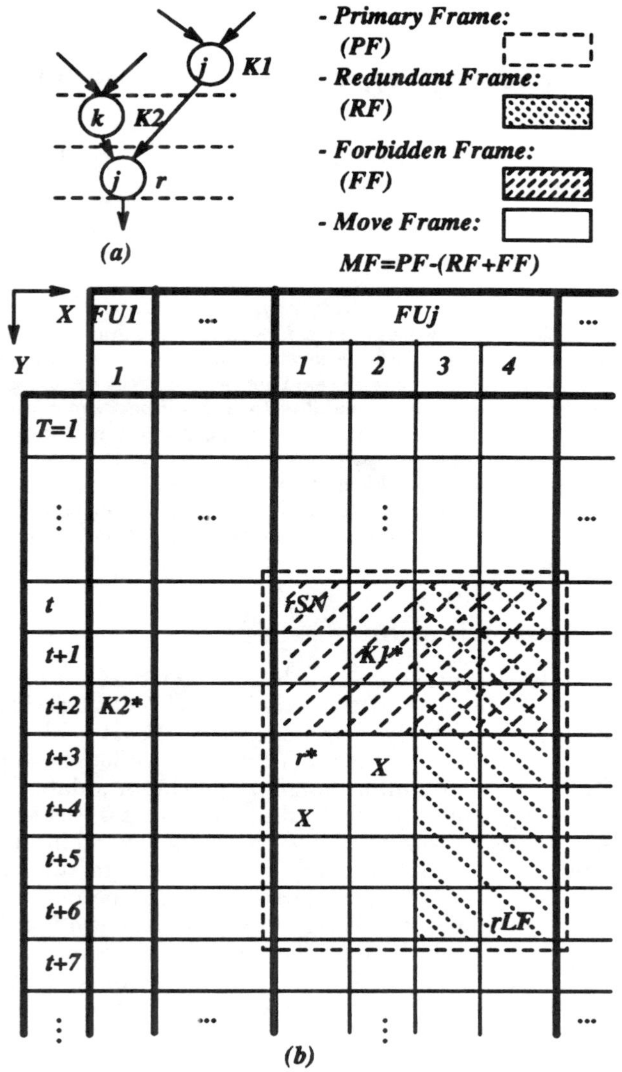

Figure 2: Different frames in MFS and process

Table 2: The result of MFSA algorithm

Ex.	T	Style	ALU's	Cost	REG	MUX	MUX_{in}
#1	4	1	$(*+\|), (+-), (\&/)$	52530	8	4	9
		2	$(/\|), (*+), (\&-), (+)$	56536	9	5	10
#2	7	1	$(+- >!), (\& + /), 2(- >)$	40706	5	7	8
		2	$(+/ >!), (+- >), (\&-)$	41485	5	6	12
#3	4	1	$(+- >), (*+), (*-)$	55356	10	4	12
		2	$(* + -), (*- >), (*+)$	62122	11	4	11
#4	8	1	$2(*+), 2(*-)$	88216	8	8	29
		2	$4(* + -), (+)$	88954	7	8	25
	9	1	$3(* + -), 2(+)$	76656	8	6	24
		2	$4(* + -)$	83680	8	8	28
#5	9	1	$3(* + -), (*+), (*-)$	101420	6	10	30
		2	$2(* + -), 2(*+), (*-)$	100436	6	9	29
	10	1	$3(* + -)$	70344	8	6	21
		2	$(* + -), 2(*+), (-)$	74756	8	8	28
#6	17	1	$3(*), 3(+)$	82352	12	8	27
		2	$3(*), 4(+)$	86270	13	8	26

Representing Conditional Branches for High-Level Synthesis Applications

Minjoong Rim and Rajiv Jain
Department of Electrical and Computer Engineering
University of Wisconsin
Madison, WI 53706

Abstract

In this paper we outline a new representation of behavioral specification for high-level synthesis applications. The main features of our representation are: correct handling of conditional branches, capability to tradeoff between control-select and data-select forms, keeping minimum necessary precedence relationships, correct representation of all conditional actions, simplified mutual exclusion testing and correct determination of bit-widths and value transfers. The representation is simple and can be easily generated automatically.

1 Introduction

High-level synthesis research is targeted at automatically producing register-transfer level (RTL) designs. High-level synthesis programs accept a behavioral specification either in a hardware descriptive language [1] [13], or a programming language [3] [15] or a data flow graph [5] [8] [9] [10] [11]. In the first two cases, the input specification is compiled into one of the several internal representations [9] [12] [18]. An RTL design can only be as good as its input specification can permit it to be, and a fundamental improvement in representing the specification which allows better exploitation of parallelism and pipelining can have a major impact on the final design. The basic issue addressed in this paper is the representation of conditional branches. Despite the wide use of data flow graphs (DFGs) there is no consensus on representation of conditional branches. In this paper we focus on such a data flow graph representation which can be used by next generation high-level synthesis tools to implement better and more efficient RTL designs.

To provide good resource sharing among operations in data flow graphs with conditional branches, mutual exclusion testing must be done. Several algorithms which perform mutual-exclusion test have been reported in literature [9] [17]. In this paper, we will see that these algorithms do not always provide a correct solution, and a new algorithm which overcomes the deficiencies of these algorithms is presented.

The paper is organized as follows. In Section 2 we address the problem of representing conditional branches. The existing methods and their shortcomings are mentioned. Next, we present a new representation which overcomes these shortcomings. In Section 3 some problems with conventional methods for mutual exclusion testing are given and a new algorithm is proposed. The paper concludes with a brief discussion on research in progress.

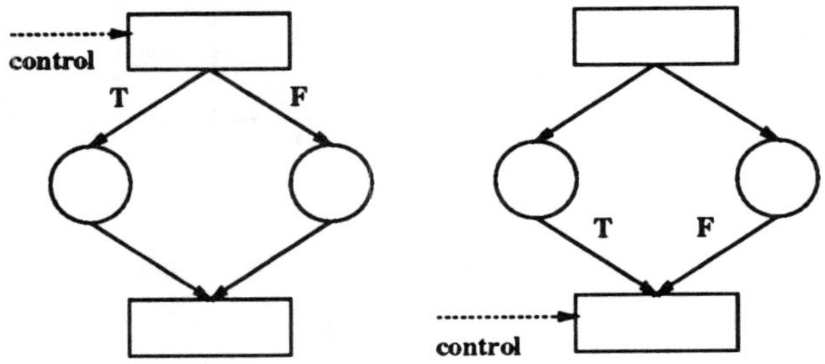

(a) C-select representation (b) D-select representation

Figure 1: Conditional representations

2 Representing Conditional Branches

The following terms are used in this paper. Two nodes[1] in a DFG are *mutually exclusive* if during any single execution at most one of these nodes is necessarily performed. A node is *unconditional* if it does not belong to any conditional branch else it is *conditional*. The *true conditional path* of a conditional branch refers to path(s) which are activated with true condition. The *false conditional path* of a conditional branch is defined similarly. The assumption of a conditional branch having two-way branches is used for clarity of presentation, and our results can be easily extended to multi-way branches.

2.1 Existing Conditional Branch Representations

Existing representations for conditional branches can be broadly classified into two types: control select (C-select) and data select (D-select) [5]. In C-select representation (Figure 1a) condition selection is performed before any conditional operation is executed and a conditional branch selected. Since alternate branches of a conditional branch are mutually exclusive, there is no resource conflict between operations in different branches and all mutually exclusive operations of the same type can share resources. In Figure 1a, one module can perform the two operations without any resource conflict (assuming, of course, that these operations are of the correct type). In C-select representation the decision as to which operations share resources is postponed until scheduling or binding is done. Due to the absence of information on sharing resources a

[1] The words node and operation are used interchangeably.

priori, the precedence relationships between condition selection and execution paths must exist in the representation. As we will see later, some of these precedence relationships may be unnecessary and result in inferior schedules.

In D-select representation (Figure 1b) all conditional branches are executed separately in parallel and correct data values are selected at the end of the conditional branch [12]. This representation may result in fast execution since conditional operations can be executed before the condition selection is made. One drawback with this representation is that mutually exclusive operations cannot share resources, resulting in expensive designs. The use of D-select representation, however, may be preferable to C-select in certain cases, for example, when conditional branches do not contain any mutually exclusive operations due to type conflict. Another drawback with the D-select representation is that some conditional actions cannot be represented [5]. This is because the correct conditional path is chosen by the selection of output values and some conditional actions do not have an output (Figure 5a, 5c).

We say that a node is *scheduled in a C-select manner* with respect to a conditional branch if it is scheduled after the condition selection for the conditional branch is scheduled. Otherwise, it is *scheduled in a D-select manner* implying that it cannot share resources with any of conditional nodes in the branch. In order to produce good designs, the scheduler must decide which operations must be scheduled in a C-select manner and which operations are to be scheduled in a D-select manner. Suppose a conditional branch has a pair of mutually exclusive operations. If they are scheduled in the same time-step, a D-select representation will prevent them from sharing resources. This may produce an inferior schedule as compared to a schedule produced using the C-select representation. However, when they are scheduled in different time-steps, C-select representation may delay scheduling of operations until the condition selection is made, resulting in an slower schedule. Additional advantages of tradeoff between C-select and D-select forms by the scheduler are illustrated in [17].

Select representation [12] of *value trace* [12] of CMU-DA system [14] was originally developed to support the D-select. Later, the select representation was implemented as a C-select form [18] and transformations between C-select and D-select forms are allowed [2]. However, these transformations have to be applied prior to scheduling and dynamic tradeoffs are not possible.

Another widely used representation is the *distribute-join* [9] [10] of USC ADAM system [4] [6] (Figure 2a), which is a C-select type and does not permit dynamic tradeoff between C-select and D-select forms. One problem with the distribute-join representation occurs while representing precedence relationships. The distribute-join representation enforces unnecessary precedence relationships which may mislead the scheduler into producing inferior designs. In distribute-join representation, all out-going edges must go through the join node. Suppose, in Figure 2a, b and

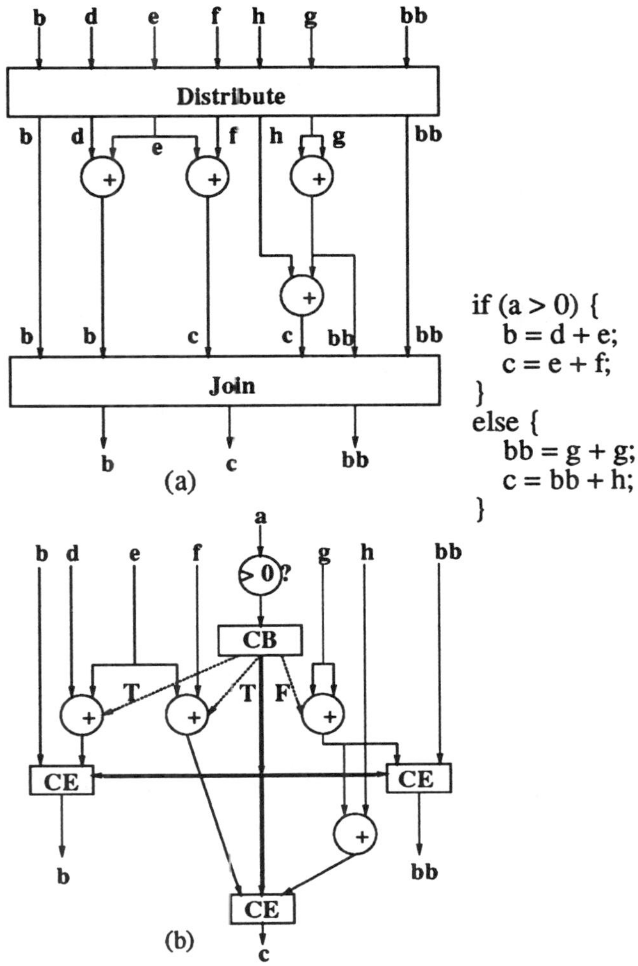

Figure 2: Distribute-join and our representations

bb values are available before c is computed, thus enabling operations requiring b and/or bb value(s) to be ready for scheduling. However, these ready operations must wait until the join operation is scheduled, which in turn, can only be scheduled after c becomes available. This prevents scheduling of the ready operations in parallel or before c value is evaluated. Thus, the use of distribute-join representation may delay execution of ready operations which may result in an inferior schedule.

This disadvantage can be easily overcome by splitting distribute and join nodes. The representation in [17] uses *fork-join* representation which is similar to the distribute-join representation with split distribute and join nodes resulting in too many special nodes. It retains the fork node which is similar to the distribute node. The one drawback of the use of fork or distribute nodes is that some conditional actions which do not have inputs (Figure 5b, 5c) cannot be represented without adding dummy inputs.

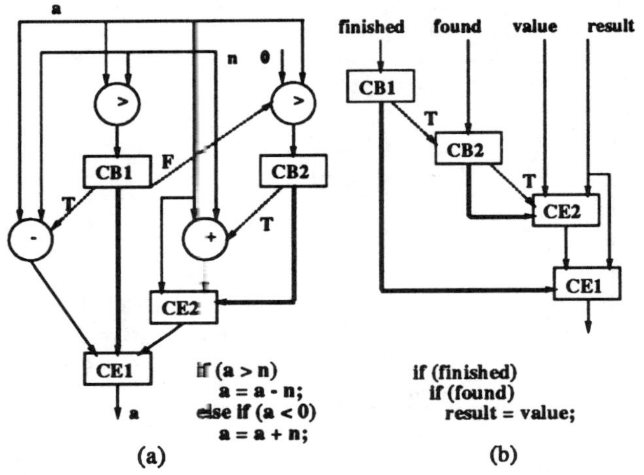

Figure 3: Nested conditionals in our representation

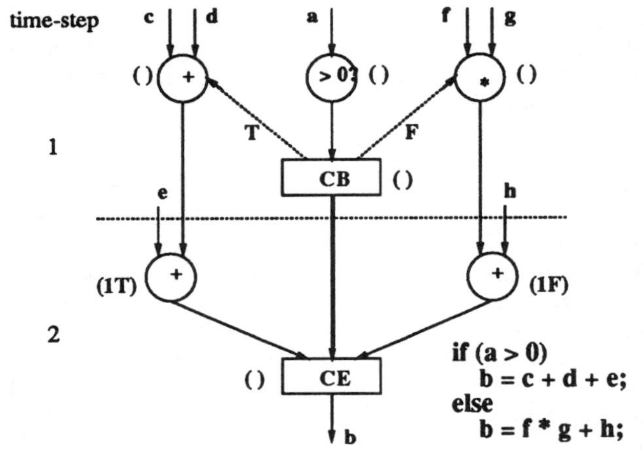

Figure 4: A scheduled DFG with operations in D-select and C-select forms

2.2 A New Representation

We now present a new representation which overcomes the above mentioned disadvantages. Examples of our representation are given in Figures 2b, 3 and 4. In our representation we have three types of edges which represent different precedence relationships. A value transfer between two operations implies that the value producing (predecessor) operation should be completed before the value consuming (successor) operations start execution. A precedence relationship due to a value transfer is named *strong precedence* relationship, and is depicted by an arrow called an *S-edge*. Some precedence relationships which do not carry values and are required for correct execution of the specification are called *pure precedence* relationships. Pure precedence relationships are used for special cases illustrated later, and can also be used by the designer to impose constraints, such as timing constraints or loop dependencies. The edge representing pure precedence relationship is called a *P-edge* and is denoted by a bold arrow. The third type of relationship is not a precedence relationship in a strict sense in that it can be violated for correct execution. We call this a *weak precedence* relationship and its sole use is to identify correct conditional paths. The edge representing a weak precedence relationship is called a *W-edge* and is denoted by a broken arrow. The W-edge carries an attribute T or F denoting whether this edge is in the true conditional path or the false conditional path respectively. The relationships between a node and its predecessors is determined by the type of edges connecting the node and a predecessor. Thus, *S-predecessors*, *P-predecessors*, and *W-predecessors* are immediate predecessors connected by S-, P-, and W-edges respectively.

The reason for distinguishing the three types of edges is as follows. During mutual exclusion testing, P-edges are not required since they do not relate to control sequence. During scheduling, W-edges are not required since they do not represent real precedence relationships. During binding, we only use S-edges, which represents value transfers. Most conventional representations intermingle these different edges which restricts the design of efficient and/or correct algorithms for mutual exclusion testing, scheduling or binding.

A conditional branch is represented by one CB (conditional branch begin) node and several CE (conditional branch end) nodes, one for each value being transferred out of the conditional branch. A CB node selects and activates the correct conditional execution path in the conditional branch according to the condition selection. Boolean operations which determine control flow are explicitly represented. A CB node represents the control for the conditional branch. Edges emanating from a CB node represent weak precedence. No value transfer occurs between a CB node and any of its successor nodes. All S-edges incident to operations in a conditional branch go directly to the operations without going through the CB node.

A CE node marks the termination of a path in the conditional branch. Each value being transferred out of a conditional branch must go through a CE node. This allows the scheduler to schedule the CE nodes in different time-steps eliminating unnecessary precedence relationships mentioned earlier. In Figure 2b operations requiring *b* or *bb* can be scheduled independent of when *c* is produced. Also value transfers are clearly represented since inputs and output of each CE node represent the same variable. This also allows the correct computation of bit-width requirements. The number of inputs to a CE node is always two since there are two input alternatives (true or false) to choose from.

Our representation allows tradeoff between C-select and D-select during scheduling. In C-select representation condition selection precedes initiation of the condition execution paths. This, however, is not essential to the correct execution of the program, since initiation of the conditional execution path without condition selection only eliminates possibilities of combining some

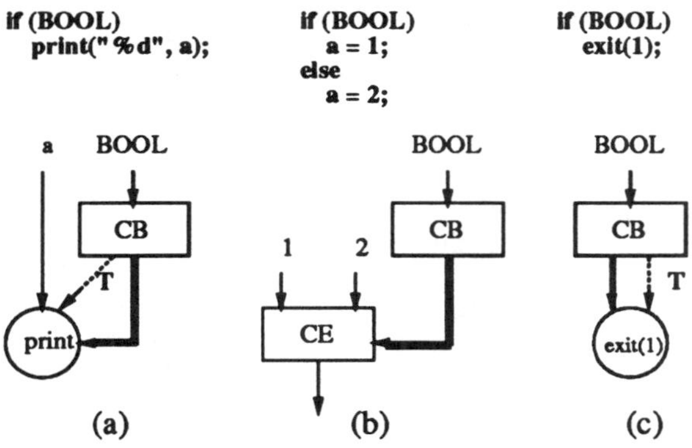

Figure 5: Some special conditional actions

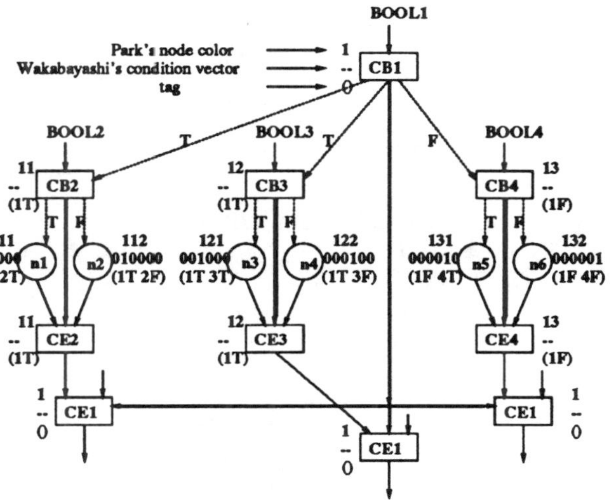


```
if (BOOL1) {
  if (BOOL2)
    ...;      : n1
  else
    ...;      : n2
  if (BOOL3)
    ...;      : n3
  else
    ...;      : n4
}
else {
  if (BOOL4)
    ...;      : n5
  else
    ...;      : n6
}
```

Figure 6: Node color, condition vector, and tag

mutually exclusive operations. In our representation, W-edges emanating from a CB node represents a weak precedence which can be violated, permitting operations in a conditional branch to be scheduled before the CB node. Operations scheduled before or in parallel with the CB node are executed in a D-select manner (time-step 1 of Figure 4). Operations scheduled after the CB node are treated in C-select manner (time-step 2 of Figure 4). In any case, a CE node cannot be scheduled before the corresponding CB node which is taken care of by a pure precedence relationship between the CB and the CE nodes. Our representation can handle special conditional actions shown in Figure 5. We summarize our representation with the construction rules.

1. Construct nodes representing operations.
2. For each conditional branch, construct a CB node.
3. For each output value of a conditional branch, construct a CE node. Since there are always two possibilities (true or false) according to condition selection, the CE node has two input values and one output value.
4. All value transfers are represented by S-edges.
5. If a node in a conditional branch has no S-predecessor in the same innermost conditional branch, then add a W-edge from the CB node to the node in question. The W-edge has an attribute true or false indicating the true or false conditional path.
6. P-edges are inserted between a CB node and all its corresponding CE nodes. P-edges may also be added for special actions such as those illustrated in Figures 5a and 5c.

3 Mutual Exclusion Testing

Having discussed our representation we now address the issue of determining mutually exclusive operations. This is done in two steps. First, the nodes are tagged and next, mutually exclusive operations are determined. Since resource sharing among mutually exclusive operations can only be determined during scheduling, the latter step is deferred until scheduling is in progress. Details on how to tag the nodes and modify the tags during scheduling to reflect resource sharing are discussed in this section. We begin by surveying existing algorithms for determining mutual exclusivity and their shortcomings.

3.1 Existing Algorithms

Two well-known algorithms for mutual exclusion testing are the node coloring algorithm by Park [9] and the condition vector method by Wakabayashi [17]. These algorithms, however, cannot handle certain types of conditional branches correctly. In Figure 6, node $n1$ is mutually exclusive to node $n5$ but is not mutually exclusive to node $n3$. The algorithms of Park and Wakabayashi incorrectly indicate that nodes $n1$ and $n3$ are mutually exclusive and can share resources. This error can result in a design which will not meet the input specification. If BOOL3 was replaced by BOOL2 in CB3, then $n1$ becomes mutually exclusive to $n4$ as well. This shows *that mutual exclusion depends upon the boolean variable selecting the conditional branch.* For mutual exclusion testing we have to consider condition selections which activate the conditional branches and not the conditional branches themselves [16]. Further, *the test for mutual exclusion should not depend on the order of nested conditional branches.*

As mentioned before our representation allows tradeoff between C-select and D-select forms during or prior to scheduling. This tradeoff can complicate mutual exclusion testing. In Figure 7, since conditional branch CB1 is not initiated in time-step 1, node $n1$ is not mutually exclusive to nodes $n3$ or $n4$. Such a case can eas-

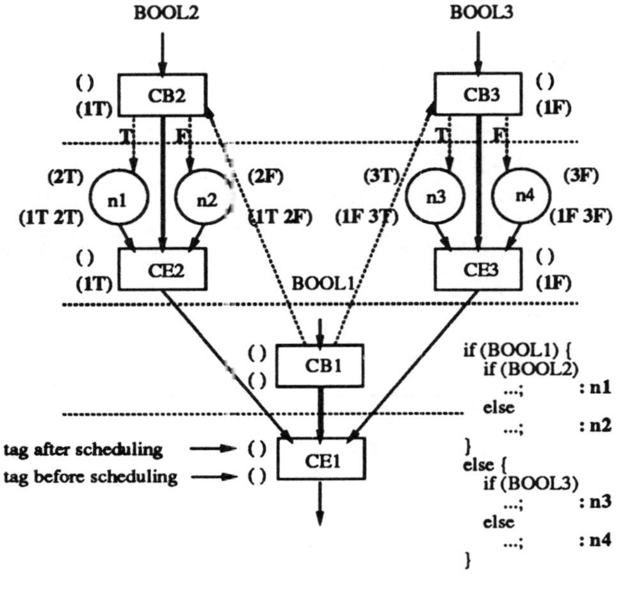

Figure 7: Handling tradeoffs between C-select and D-select forms

ily occur if BOOL2 and BOOL3 are evaluated before BOOL1. Wakabayashi's representation cannot handle such cases even though tradeoff between C-select and D-select forms is allowed in their representation. This is because Wakabayashi implicitly assumes that for any two nested conditional branches the condition selection of the inner conditional branch is always done after the outer one.

3.2 A New Algorithm

We now present a new and simple algorithm for mutual exclusion detection which overcomes the above problems. The tag of a conditional operation is different if it is executed in a C-select or D-select manner for some fixed conditional branch. Before scheduling, the tag of an operation is determined assuming that the operation is executed in a C-select manner for every conditional branch, and the final tag is determined during scheduling. Examples of tags are given in Figures 4, 6 and 7.

Our algorithm assigns a tag [2] to each operation in the data flow graph. The tag of a node represents the condition selections which activates the node. A tag is an unordered set of elements representing condition selection of the conditional branches. Each element contains two fields: the *condition* field and the *selection* field. The condition field represents the boolean values of the conditional branch. The selection field represents either true or false decision which activates the node. In this paper, we use an integer for the condition field and T and F for the selection field. A unique integer is used for each distinct boolean value incident to a CB node.

[2] A tag is similar to *condition cube* in [7] and [16].

For example, a tag of (1T 2F) means that the node is activated if the condition 1 is true and condition 2 is false. Since unconditional operations are activated regardless of any condition selection, they have an empty tag denoted by ( ).

If a node is activated with the true selection of a certain condition and another node is activated with the false selection of the same condition, they are mutually exclusive. Thus, two nodes with tags (1T 2F) and (2T 3T) are mutually exclusive. For correct design we use a pessimistic approach, that is, if there is no evidence that two nodes are mutually exclusive, they are assumed to be not mutually exclusive. This guarantees us of a correct albeit a possibly inferior design. For example, nodes with tags (1F) and (2T) respectively are not mutually exclusive without additional information. Of course, there exist cases in which they could be mutually exclusive. If condition 1 is (a + b == 0) and condition 2 is ((a == 0) && (b == 0)), they become mutually exclusive. If these types of cases can be detected during data flow graph generation stage, then we can take advantage of them. To summarize, the mutual exclusion test is:

- If the tags of two nodes have an element with same condition fields and different selection fields, then they are mutually exclusive.

We now present a tagging algorithm which will tag the nodes such that the mutual exclusion test can be correctly applied. [3]

Node Tagging Algorithm (before scheduling)

1. A unique integer is assigned to each distinct boolean value incident to a CB node.
2. The tag tag_i of a node $node_i$ is determined as follows:
   (a) if $node_i$ does not have S- or W-predecessors, tag_i is empty (node is unconditional).
   (b) if $node_i$ is a CE node, tag_i is the same as its corresponding CB node.
   (c) if $node_i$ does not have a W-predecessor its tag_i is the longest tag of its S-predecessors.
   (d) tag_i is the same as its W-predecessor with one more element, whose condition field is determined by the boolean value incident to the CB node and whose selection field is determined by the true or false attribute of the W-edge.

When a node is scheduled before or in parallel with a CB node containing this node (that is, in a D-select fashion), its tag must be changed. In this case, the node is not dependent on the boolean value incident to the CB node any more, and the element in the tag

[3] After nodes are tagged, we may perform backward traversing to correct bad programming style using the fact that a node is necessary to be executed only if its outputs are necessary.

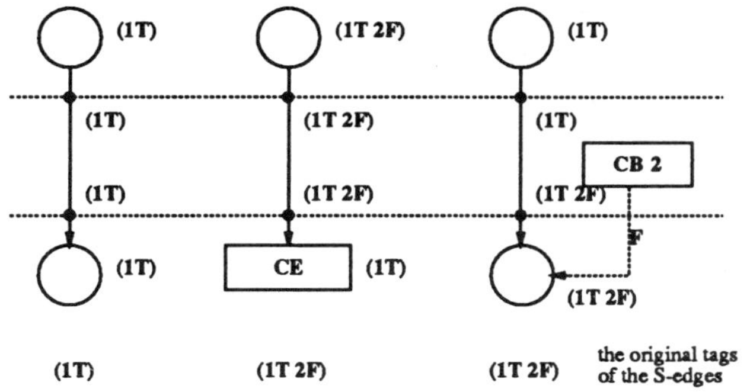

Figure 8: Tagged R-values for mutually exclusive values after scheduling

corresponding to the CB node is eliminated. Examples are given in Figure 7. Notice that the set of all pairs of mutually exclusive operations after scheduling is a subset of that before scheduling.

Having considered mutual exclusion for operations, we now turn our attention to determining mutually exclusive values which can share registers. A value produced in one time-step and used in another time-step must be stored in a register. We define a *R(egister)-value* to be a value which needs to be stored in a register over a time-step. Note that if an S-edge crosses several time-steps, then it has several R-values, one for each time-step. If two R-values are mutually exclusive, then only one register is required to store them. If an S-edge is connected to only one node, it is either an input or an output edge in the data flow graph and the corresponding R-values cannot be mutually exclusive to any other R-value. Thus, our interest lies in the S-edges connecting two nodes. Before scheduling, the source and the sink node are tagged assuming they are C-select. A value corresponding to an S-edge is required if and only if both the sink and the source nodes are activated. The condition for which both nodes are activated can be easily obtained. Of the sink and source nodes, let $node_1$ be the node with longer tag and the other node be $node_2$. Then, we can easily show that the tag of $node_1$ must be a super set of the tag of $node_2$, and if $node_1$ is activated, then $node_2$ must also be activated and the value corresponding to the S-edge is required.

Edge Tagging Rule (before scheduling)

- The tag of an S-edge is the longer tag of its source and sink nodes.

During or after scheduling the tags of R-values are determined using the S-edges and scheduling information in a similar way as the nodes. Of course, tags of different R-values of the same S-edge can differ indicating that a certain part of the S-edge can be shared and the other part cannot be shared. The test for mutual exclusion is the same as for operation. Examples of R-value tags after scheduling are shown in Figure 8.

4 Conclusion and Future Work

In this paper we have presented a new representation for conditional branches and a new algorithm for mutual exclusion testing. The new representation handles conditional branches correctly and satisfies scheduling as well as binding requirements. The new algorithm for mutual exclusion testing is simple and powerful, and has been coded and verified with several conditional data flow graphs. Currently, we are developing scheduling algorithms which can take advantage of this representation. We have extended the ideas presented in this paper to representing loops as well.

References

[1] R. K. Brayton, et al. The Yorktown Silicon Compiler System. *Silicon Compilation*. Addison-Wesley, 1988.

[2] T. E. Fuhrman. Industrial Extensions to University High Level Synthesis Tools: Making It Work in the Real World. *28th DAC*, 1991.

[3] E. Girczyc, et al. Applicability of a Subset of Ada as an Algorithmic Hardware Description Language for Graph-Based Hardware Compilation. *IEEE Trans. CAD*, April 1985.

[4] J. J. Granacki, et al. The ADAM Advanced Design Automation System: Overview, Planner and Natural Language Interface. *22th DAC*, 1985.

[5] L. Hafer and A. Parker. A Formal Method for the Specification, Analysis, and Design of Register-Transfer Level Digital Logic. *IEEE Trans. CAD*, Jan. 1983.

[6] R. Jain, et al. Experience with the ADAM Synthesis System. *26th DAC*, 1989.

[7] J. A. Nestor and G. Krishnamoorthy. SALSA: A New Approach to Scheduling with Timing Constraints. *ICCAD*, 1990.

[8] B. M. Pangrle and D. D. Gajski. Design Tools for Intelligent Silicon Compilation. *IEEE Trans. CAD*, Nov. 1987.

[9] N. Park and A. C. Parker. Sehwa: A Software Package for Synthesis of Pipelines from Behavioral Specifications. *IEEE Trans. CAD*, March 1988.

[10] A. C. Parker, et al. MAHA: A Program for Datapath Synthesis. *23th DAC*, 1986.

[11] P. G. Paulin and J. P. Knight. Force-Directed Scheduling for the Behavioral Synthesis of ASIC's. *IEEE Trans. CAD*, June 1989.

[12] E. A. Snow, et al. A Technology-Relative Computer-Aided Design System: Abstract Representations, Transformations, and Design Tradeoffs. *15th DAC*, 1978.

[13] D. E. Thomas, et al. The System Architect's Workbench. *25th DAC*, 1988.

[14] D. E. Thomas, et al. Automatic Data Path Synthesis. *IEEE Computer*, Dec. 1983.

[15] H. Trickey. Flamel: A High-Level Hardware Compiler. *IEEE Trans. CAD*, March 1987.

[16] C.-J. Tseng, et al. Bridge: A Versatile Behavioral Synthesis System. *25th DAC*, 1988.

[17] K. Wakabayashi and T. Yoshimura. A Resource Sharing and Control Synthesis Method for Conditional Branches. *ICCAD*, 1989.

[18] R. A. Walker and D. E. Thomas. Behavioral Transformation for Algorithmic Level IC Design. *IEEE Trans. CAD*, Oct. 1989.

Global Scheduling Independent of Control Dependencies Based on Condition Vectors

Kazutoshi WAKABAYASHI
C&C Systems Research Laboratories, NEC Corp.
4-1-1 Miyazaki Miyamae-ku Kawasaki, Japan
wakaba%btl.cl.nec.co.jp@sj.nec.com

Hirohito TANAKA
NEC Scientific Information System Development Ltd.
R&D 637, KSP, 3-2-1 Sakado, Takatsu-ku
Kawasaki, Japan

Abstract

This paper presents a global scheduling method based on *condition vectors*. The proposed method exploits a more "global parallelism". That is, it can parallelize multiple nests of conditional branches and optimize across the boundaries of basic blocks thoroughly, and so on. Moreover, it can optimize all possible execution paths. Also proposed is an algorithm which generates a single finite state machine controller from parallel individual control sequences derived in the global parallelization process. Experimental results proves that the global parallelization is markedly effective.

1 Introduction

In order to minimize the number of states in the final schedule for a given behavior, it is very important to exploit both parallelism and alternation among operations. Some methods such as the Force-directed [2], the Sehwa [3], the Bridge [4] and Kim's methods [5], are able to accomplish conditional resource sharing for mutually exclusive operations. However, they cannot exploit sufficient potential parallelism, because they cannot move an operation inside a fork-join pair to a control step outside of it, even if there are available resources left.

On the other hand, path-based scheduling [6] and percolation scheduling[7] can somewhat exploit parallelism across basic blocks boundaries. Though path-based scheduling finds all possible execution paths and optimizes each of them, the fundamental order of operations for a given path is chosen in advance, then the scheduling result may not be sufficiently optimized across the basic limit boundaries. "Percolation" scheduling starts with optimal schedule and applies semantics-preserving transformations. It can extract parallelism beyond the basic block limits. However, it does not consider conditional resource sharing well, and it optimizes only the total number of control steps, not all possible paths.

Moreover, no conventional method can exploit "global parallelism"; no method deals with parallelization of top level conditional branches.

We previously proposed a List Scheduling method based on Condition Vectors (CVLS) [9]. This method can deal with single nested conditional branches systematically. It can accomplish conditional resource sharing thoroughly, and can optimize all possible execution paths. However, this method did not overcome all of the above problems.

In order to overcome all of them, we have developed some essential new improvements to the CVLS method. These new techniques can schedule operations independent of control dependencies. In other words, they transform control structure of the given behavior drastically, while preserving semantics, in order to minimize the number of states in final schedule.

2 Preliminaries

2.1 Condition Vectors and Varieties of Conditional Mutual Exclusiveness

We introduced the **Condition Vector (CV)** concept in [9]. CVs show the execution conditions of an operation in a vector form, which is a 1-hot encoding for all leaf branches in the behavior description, the value of CV for a particular operation is the OR-ed result of all leaf branches reachable from that operation. Mutual exclusiveness is derived by means of a CV associated with each operation. The CVs are defined for each top level nest of conditional branches, called a **"conditional tree"**. Examples of CVs for a conditional tree are shown in Fig.1. The "basic" column shows basic CVs for each operation, which expresses explicit mutual exclusiveness. If CVs have no common '1' bit, they are mutually exclusive, such as (3) and (4).

The scheduling method proposed in this paper allows operations inside a conditional branch to be executed before the condition test operation itself, in order to save resources and states. In that case, the operations are *always* executed and afterward the result of operations will be used or abandoned according to whether the conditional is true or false. Therefore, the execution condition of operations in a branch changes according to whether it is executed before the conditional operation or after. CVs can express such a dynamic execution condition. In the Figure, the execution condition of operations in a branch will change on the basis of whether or not condition test operations in (2) and (4) are previously executed. Columns (a) (b) and (c) show three possible CV values for each operation. Such a *dynamic* CV can be obtained as the bitwise OR-ed result between a operation and the "not done" condition test operation. In column (c), if conditional (2) is "not done", the values of CV for all operation becomes [111], regardless of whether conditional (4) is 'done' or 'not'.

Note that the value of operation of (1) in column (a) is [1,0,1], while that in 'basic' column is [1,1,1]. The variable 'x' is not used under the condition for executing branch (5)(CV=[0,1,0]), therefore it is not necessary to calculate the value of 'x'. Thus, the actual execution condition is expressed as [1,0,1]. This shows the operations in (1) and (5) are actually mutually exclusive when conditionals (2) and (4) are "done", which is called *data-dependent mutual exclusiveness.*

Similar solutions for detecting mutual exclusiveness in vector form are adopted in Sehwa[3] and R^2S[8]. However,

29th ACM/IEEE Design Automation Conference®

0738-100X/92 $3.00 © 1992 IEEE

they can detect neither data-dependent mutual exclusiveness nor dynamic mutual exclusiveness. Moreover, for a more significant difference, CVs can be used for optimizing individual execution paths, as well as resource sharing as in the following section, while vectors in Sehwa and R^2S are used only for resource sharing.

		basic	(a)	(b)	(c)
$x = a+b-c+d;$	-(1)	[111]	[101]	[111]	[111]
if $(a \neq 0)$	-(2)	[111]	[111]	[111]	[111]
$y = x + c;$	-(3)	[100]	[100]	[100]	[111]
else if $(a + b < c)$	-(4)	[011]	[011]	[011]	[111]
$y = c + d;$	-(5)	[010]	[010]	[011]	[111]
else $y = x + d;$	-(6)	[001]	[001]	[011]	[111]
Conditionals:	$a \neq 0$ -(2)		done	done	not
	$a+b<c$ -(4)		done	not	don't care

Figure 1: Example description and its CVs

2.2 Scheduling and control synthesis for a conditional tree

The List Scheduling can be extended to deal with nested conditional branches through use of the CV concept. Such extension is called **CV based list scheduling (CVLS)**. The CVLS can achieve both "conditional sharing" and "optimization of all possible paths". Scheduling and control synthesis techniques are both based on CVs.

In order to to keep track of the usage for various types of FUs, we introduced Function unit Utilization Vector : $\mathbf{FUV}_{\mathrm{FU}\,f}$(c-step k), each component of which represents the number of FU f used in the c-step k under each CV condition. $\mathbf{FUV}_f(k)$ is the vector sum of CVs for all operations in the c-step k which are assigned to the FU f in the scheduling. The value of the largest component of $\mathbf{FUV}_f(k)$ shows the necessary FU f number in c-step k.

In addition, FUV is used for optimizing all possible paths. The number of possible paths for a conditional tree is equal to the dimensions of its CVs. Here, **FUVALL**(c-step k) is the vector sum of the CVs of any kinds of operations scheduled in a c-step k; i.e. $\mathbf{FUVALL}(k) = \sum_{f \in allFUs} \mathbf{FUV}_f(k)$. If any components of a FUVALL is '0', the c-step k can be skipped under '0' component conditions, because the '0' components show that no operation is performed under those conditions. The control sequences for all possible paths are generated by using such characteristics of FUVALL. (See [9] for a more detailed discussion.)

Fig.4 demonstrates the above discussions. There are two conditional trees CT1 and CT2, and an unconditional operation (3) in (a). The scheduling result for the CT1 and the unconditional operation is shown in (CT1) of Fig.4 (b). The maximum components of FUV+ and FUV- can be seen to lie within the constraint (two adder and one subtractor) shown in "R0" columns. Any c-step whose FUVALL has a '0' component can be skipped under the condition for '0' components. Thus, the generated control sequence becomes "FSM for CT1" in Fig.4(c).

3 Scheduling independent of control dependencies

3.1 Operation Node Dividing Technique

This section presents a "node dividing technique", by which an unconditional operation can be duplicated and moved inside conditional branches, if extra resource sharing

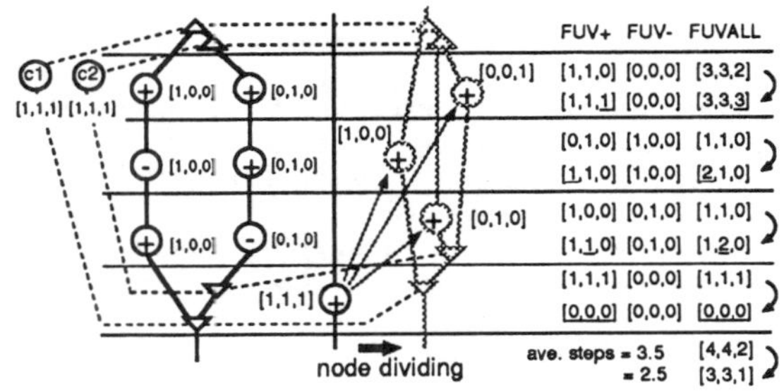

Figure 2: Node dividing technique

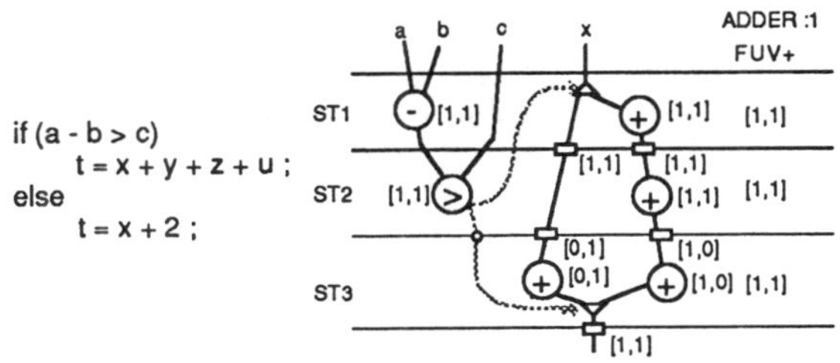

Figure 3: Pre-execution of an operation in a branch

is possible. This technique minimizes the number of c-steps.

Fig.2 shows a typical example. The scheduling result without node dividing, requires [4,4,2] c-step for each path. If the unconditional node, whose CV is [111], is divided into tree nodes whose CVs are [100], [010] and [001], the scheduling can be improved into [3,3,1] c-steps. This example shows that *an unconditional operation and an operation inside conditional branches can be mutually exclusive*, even if they have no relation with each other.

Such node dividing can be applied for any operation node whose CV has plural '1' components. If a node dividing makes length of some paths shorter, the dividing will be adopted. Node dividing is applied when a node becomes a candidate for allocation in List Scheduling as in the Section 4. Note that unconditional operation can be moved into any conditional trees by node dividing, while operations inside a conditional tree can be moved into only inner branches of the tree. In addition, one operation will be executed in different c-steps for each execution instance, when the node is divided.

3.2 Execution of operations inside a branch before the condition test

This section presents a scheduling technique which allows operations inside a conditional branch to be executed before the conditional test operation itself, in order to save resources and the number of c-steps. Fig.3 is an example showing scheduling under one adder and one subtractor. CVs of operators and registers are shown beside them. The data lines for y, z, u and 2 are omitted. Only three c-steps are required to execute the entire behavior, while, if the scheduler didn't allocate operations in a branch before the conditional operation, five c-steps are required.

As explained in Section 2, CVs express dynamic mutual exclusiveness. Consequently, the scheduler can neglect the control dependencies. The semantic of the given behavior is preserved by means of dynamic CVs. For example, the '+'[1,1] operations in ST1 and ST2 are always executed. On the other hand, the result of '+' in ST1 is always registered (which CV is [1,1]), but the result of '+' in ST2 is registered only when the branching condition is true because the registers CV is [1,0].

Note that an operation executed before a conditional test, can be handled as unconditional operation. Hence, the operation moved outside the conditional tree can be divided and moved into another conditional tree by node dividing. This means that *operations in different conditional trees can be mutually exclusive, and can share resources.*

3.3 Conditional tree duplication

At the first stage of the scheduling, "conditional tree duplication" is optionally performed, which duplicates a tree and move them into the previous tree as long as the number of leaves of the combined tree is less than 32, because a CV is implemented as one word. This could parallelize conditional trees. The following shows an example of the duplication: CT1 and CT2 are combined into CT1-2.

```
CT1: if (c1)  [10]     CT1-2: if (c1) if (c2)  [1000]
     else     [01]  →                else      [0100]
CT2: if (c2)  [10]            else   if (c2)   [0010]
     else     [01]                   else      [0001]
```

3.4 Parallelization of multiple conditional trees

Previous sections presented how to exploit mutual exclusiveness among conditional trees and unconditional operations. This section presents how to exploit potential parallelism among multiple conditional trees. The proposed scheduling method parallelizes conditional trees, independent of control dependencies, while preserving data dependencies and the semantics of the original design. For instance, in Fig.4(a), some plus operations in CT1 and CT2 can be executed concurrently without violating the data dependencies between them; 'x' in this case. ([1] denotes a vector whose components are all '1'.)

Conditional trees are scheduled in the order of the control flow of the given behavior. Operations in the first conditional tree are scheduled first, under the given resources constraints. When the first tree didn't use all FUs in any c-steps, some FUs remain in the c-steps. Thus, the remaining FUs can be available for the next conditional tree. Then, operations of the next tree are scheduled under the remaining FUs. Namely, operations of each conditional tree are scheduled under various resource constraints in each c-step.

Fig.4(b) shows a scheduling result for (a) under two adders and one subtractor. (In the figure, FUVALL in ST1 contains CVs for $c1, c2, c3$.) First, operations in the conditional tree CT1 are scheduled. The number of remaining FUs in each c-step are shown in column 'R1'. Next, operations in the second tree CT2 are scheduled, using the remaining FUs: R1. Unconditional operations are schedules with any conditional trees. They are scheduled according to the priority function, and when all operations in a conditional tree are scheduled, the remaining unconditional operations will be scheduled with the next branch. (See Section 4.)

3.5 Control sequence synthesis for globally parallelized branches

This section presents how to synthesize and how to optimize the control sequence for the parallelized conditional trees as in the previous section. Each conditional tree has an active execution path, therefore, parallel conditional

```
CT1: if (c1)   x = a + b;                 -(1) [1,0]
     else      x = a + b − c;             -(2) [0,1]
     z = a − b;                           -(3) [1]
CT2: if (c2)   y = a + c + x;             -(4) [1,0,0]
     else {    y = a + c;                 -(5) [0,1,1]
           if (c3) y = y − c − d; }       -(6) [0,1,0]
```

(a) Multiple conditional trees

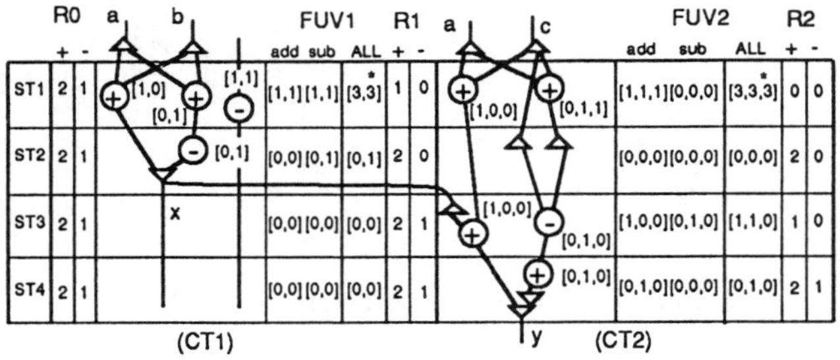

(b) Parallelization of multiple conditional trees in (a)

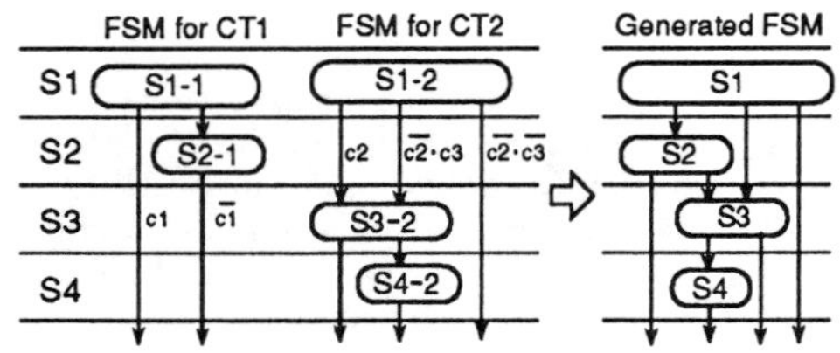

(c) Control sequence for the scheduling result shown in (b)

Figure 4: Scheduling of multiple conditional trees

trees requires the simultaneous activation of as many paths as the number of them. "FSM for CT1" and "CT2" in Fig.4 (c) shows control sequences for the scheduling of CT1 and CT2 in (b). Note that these two control sequences are dependent, since there are some data dependencies and some resources conflicts between them. Hence, the control for the parallelized conditional trees cannot be realized in two independent FSMs. A FSM shown in right-hand in (c) shows the control sequence for the parallelized trees. The FSM is obtained by combining the two FSMs expressing sequences of individual conditional trees so that they could move synchronously. That is, states in the same c-step for the two FSMs should be executed at the same time.

Such combined FSM can be generated by determining the transition destination and its condition from each c-step. They are simply determined by enumerating all possible combinations of paths, but the number of the combination is equal to the product of the number of each path and optimizing them. Then, we propose better generation method. The method determines the nearest transition destination and its condition first. In the figure example, they are 'S2 ' and $\overline{c1}$. Under *other* conditions, it can transit to a farer state. Second nearest destination is 'S3', and the condition is '$c2 + \overline{c2} \cdot c3$'. By this way, the state transition from ST1 (*DS1*) becomes the left-hand in the following:

$$\begin{aligned} DS1 = \ & \text{if } (\overline{c1}) && \rightarrow S2; && \text{if } (\overline{c1}) \rightarrow S2; \\ & \underline{\text{else}} \text{ if } (c2 + \overline{c2} \cdot c3) && \rightarrow S3; \ \rightarrow && \text{else} \ \rightarrow S5; \\ & \underline{\text{else}} && \rightarrow S5; && \end{aligned}$$

The destinations for the other states can be determined in the same way. If there are some relations between conditional test operation $c1, c2, c3$, the FSM can be optimized more. For example, if a relation such as '$c1 \cdot c2 = 0$' and '$c1 \cdot \overline{c3} = 0$'holds, then state transition *DS1* is optimized as the right-hand in the previous example. In this way, the proposed method can deal with the "false path" problem described in [6] by using such boolean relations among conditional test operations.

4 Entire Scheduling Algorithm

```
conditional tree duplication; /* option */
R_FU_f = number of FU f;
for CT = the first conditional tree to the last do {
   Irest = operations inside CT;
   Orest = unconditional operations;
   Cstep = 0;
   while (Irest ≠ φ) {
      Cstep = Cstep + 1;
      compile ReadyList for Cstep from Irest and Orest;
      Irest, Orest = the remainder operations;
      while (ReadyList ≠ φ) {
         n = operation with the largest priority;
         calculate dynamic CV of n in Cstep;
         if n is dividable (CV has plural '1') then
            divide n;
            if the schedule becomes better then
               allocate the divided nodes in the former c-step
               goto L1;
         if max(FUV_f) ≤ R_f then  allocate n to Cstep
         else Irest or Orest = n;
      L1:
         /* for operation chaining */
         if n is allocated then renew Readylist, Irest,Orest;
      }
   }
   R_FU_f = R_FU_f - number of used FU f for CT;
}
/* node re-allocation */
for Cstep = the last c-step -1 downto the first do
   re-allocate n to the later c-step if '0' of FUVALLL increases;
Control sequence synthesis for the scheduling result;
```

5 Experimental results and remarks

The proposed method is implemented in *Cyber* system [10]. Table.1 shows comparisons with "MAHA" in [1], "PATH" in [6], "KIM" in [5], "R^2S" in [8] and our method "CVLS". In the table, "maha", "kim" are examples in [1] and [5], and "parker' is parker1986 in the HLS benchmark1991. "Wmaha" and "25maha" are obtained by duplicating "maha" twice and twenty five times respectively, and connecting them in series (we put one data dependency among each "maha" data). "Paths" shows the length of longest and shortest path and the average where all conditionals have equal probabilities. "Full" shows results for the proposed method. "Not par" shows results without parallelization of multiple conditional trees and node dividing, and "not cnd" shows results also without branch operation pre-execution.

Examples "maha" and "parker" have only two conditional trees, but our method gives significantly better results than other methods. Note that our method produce good results without chaining, this shows our method has high ability to exploit potential parallelism. Our method gives a better result also for "Kim". "Kim" has only one conditional trees, therefor the result shows the ability of node dividing and pre-execution of branch operations are effective. The results for "Wmaha" and "25maha" show the powerful ability to exploit potential parallelism of CVLS. The "full" results are much better than "not par" and "not cnd". In addition, the path length for "25maha" is far less than 25 times of maha. Moreover, the path length in "full" drastically decreases according to the increase of FU numbers, on the other hand, that in "not cnd" decrease slightly. These results prove that the CVLS scheduling can utilize FUs sufficiently.

The CVLS algorithm is so simple that it runs fast. The behavior containing more than one thousand operations can be scheduled in about a minute on 28.5 MIPS work station.

Table 1: Scheduling Results

Data	Method	Constraint			Results: Paths max/min(ave.)			
		add	sub	Ch	St	full	not par	not cnd
maha	KIM	1	1	1	8	8/3		
	CVLS	1	1	**1**	**5**	**5/2**(3.31)	6/3(4.12)	8/3(4.62)
	MAHA	1	1	2	8	8		
	PATH	1	1	2	9	5/2		
	KIM	1	1	2	6	5/2		
	MAHA	2	3	3	4	4		
	KIM	2	3	3	3	3/2		
	CVLS	2	3	3	3	3/1(1.56)	3/3(3.00)	4/3(3.25)
	CVLS	2	3	**1**	4	**4/1**(2.38)	5/3(3.75)	8/3(4.62)
	R^2S	2	1	2	4	4/4		
	CVLS	2	1	2	4	4/1(2.38)	4/3(3.38)	5/3(4.00)
parker	CVLS	1	1	1	5	5/2(3.31)	6/3(4.12)	8/3(4.62)
	CVLS	2	3	1	4	4/1(2.38)	5/3(3.75)	8/3(4.62)
kim	KIM	2	1	1	7	7/6(6.25)		
	CVLS	2	1	1	**6**	**6/5**(5.75)	6/6(6.00)	8/7(7.25)
*w*maha	CVLS	2	3	1	7	7/2(4.80)	10/5(6.38)	14/5(7.75)
	CVLS	2	3	3	5	5/2(3.75)	6/5(5.38)	6/5(5.38)
	CVLS	1	1	1	10	10/3(6.77)	13/6(8.25)	16/6(9.25)
25maha	CVLS	5	5	1	32	**32**(31.96)	102(66.25)	153(83.00)
	CVLS	2	3	1	65	**65**(63.95)	107(71.75)	157(87.12)
	CVLS	1	1	1	125	**125**(94.88)	156(88.5)	180(96.38)

add/sub: No. of adder/subtractor; Ch: No. of chaining; St: No. of state;

6 Conclusions

This paper proposed a novel scheduling method based on condition vectors which can schedule operations independent of the control dependencies, such as node dividing, pre-execution of branch operation and parallelization of multiple conditional trees. Also, presented is a FSM synthesis algorithm for the parallelized multiple conditional trees. The experimental results show that our method can produce remarkably better scheduling and it runs efficiently.

References

[1] A.C.Parker, et al, "MAHA:A Program for Datapath Synthesis", 23rd DAC, pp416-424, 1986.

[2] P.G.Paulin, J.P.Knight, "Force-Directed Scheduling in Automatic Data Path Synthesis", 24th DAC, pp195-202, 1987.

[3] N. Park and A.C.Parker, "SEHWA:A Software Package for Synthesis of Pipelines from Behavioral Specifications", IEEE trans. on CAD, vol.7, no.3, March 1988.

[4] C.J.Tseng, et al, "Bridge:A Versatile Behavioral Synthesis System", Proc. 25th DAC, pp415-420, 1988.

[5] T.Kim, et al., "A Scheduling Algorithm For Conditional Resource Sharing", ICCAD'91, pp84-87. 1991.

[6] R.Camposano, "Path-Based Scheduling for Synthesis",IEEE trans. on CAD, Vol.10,No.1, pp85-93,1991.

[7] R. Potasman, et al., "Percolation Based Synthesis", 27th DAC, pp444-449, 1990.

[8] P.F. Yeung and D.J.Rees, "Resources Restricted Global Scheduling", VLSI'91, 7.2, 1991.

[9] K.Wakabayashi and T.Yoshimura, "A Resource sharing and Control Synthesis Method for Conditional Branches", ICCAD'89,62-65,Nov 1989

[10] K.Wakabayashi, "Cyber: High Level Synthesis System from Software into ASIC", High-level VLSI Synthesis, Kulwer Academic Publishers, June 1991.

Optimal Scheduling and Allocation of Embedded VLSI Chips

Catherine H. Gebotys

Department of Electrical and Computer Engineering,
University of Waterloo, Waterloo, Ontario. N2L 3G1 Canada

A new integer programming (IP) model is presented for synthesizing high speed embedded VLSI chips. For the first time a model is presented for simultaneous scheduling, selecting types of functional units, allocation and determining the clock period. Functional units can be chained, multicycled, or pipelined, with different speed and area characteristics. This research is important for industry since, a larger number of optimal architectures, satisfying area, speed, and interface constraints, can be synthesized. Simultaneous optimization of the clock period is extremely important for high speed architectures. The synthesis problem is transformed into a tight IP model based on polyhedral theory. A branch and bound algorithm produces globally optimal architectures in practical cpu execution times. This research breaks new ground by 1. simultaneously scheduling and allocating with complex and asynchronous interface constraints, to minimize both the average execution time and the area, 2. automatically selecting the clock period and types of functional units (including chained operations), and 3. synthesizing globally optimal architectures of embedded VLSI chips in practical cpu execution times.

1. Introduction

An architectural synthesis tool that supports design exploration for embedded VLSI chips with complex and asynchronous interface constraints is key to the acceptance of high level synthesis tools in industry in the 1990's. Secondly a tool which provides exploration of speed optimal architectures by exploring different clock periods and automatically selecting different types of functional units such as chained operations is also a necessity for high speed applications. Although research in architectural synthesis has been active and developing over the last 10 years, few researchers have addressed these two important problems. Embedded VLSI chips, or those that communicate with their surrounding system components, are the prime application area for VLSI synthesizers. Interface constraints are very important for the introduction of architectural synthesis tools in industry because frequently the performance bottleneck of digital signal processing applications and other control dominated applications is at the interface. For example the ability of the embedded chip to continue processing internal data (independent of the interface) simultaneously with asynchronous data (data synchronized with the clock but indeterminate with respect to state) received from external system components, has a large impact on the overall performance of the system. Furthermore the synthesizer should be able to select from a library of functional units, and support chaining operations to meet high speed requirements. The later tradesoff clock speed for algorithm latency (or delay).

The objective of architectural synthesizers is to transform an input algorithm (or behavior) into a hardware architecture that satisfies all interface constraints and minimizes some cost. Synthesizers must minimize the average execution time given area constraints or minimize the chip area given constraints on the execution time. Area should be modeled as a piecewise linear cost function[1] of the number of functional units, registers, and busses. The architectural synthesis problem is believed to be NP-hard, since many of its subtasks have been defined as NP-complete.

2. Problem Description and Previous Research

The following problem, describes an important part of the architectural synthesis problem that will be examined in this paper. Problem 1. *By mapping each code operation of the input algorithm to a time (or cstep) and a type of functional unit, produce a schedule, allocate hardware (functional units, registers, busses) and determine the optimal clock period such that area and timing constraints are met.* The timing constraints include latency, throughput, complex, and asynchronous constraints (determined by the communication with other system components). The problem of minimizing area (or delay, defined as the clock period multiplied by the number of csteps), given constraints on delay (or area), will be addressed. Extensions of problem 1 include supporting conditional code, loops, and functional pipelining. Functional units may be multicycled, pipelined or chained. For example a chained functional unit is one which performs two consecutive operations without using a register to hold the data output from the first operation as input to the second operation.

Previous research in architectural synthesis has separated the selection of a clock period from the scheduling and allocation tasks. However this is unsuitable for speed critical applications where the clock period has a significant impact on satisfying timing constraints and selecting functional units from a library. A large MILP model[2] selected functional units from a library and simultaneously chained operations, however only very small examples could be solved. Recently an algorithm[3] extending the force-directed scheduling algorithm to real time was developed to address choosing functional units from a realistic library. Unfortunately chaining of operations was not discussed and a fixed clock period was used in the results section. Aside from minimum and maximum timing constraints, few synthesizers support asynchronous interfaces. A relative scheduling algorithm[4] was developed to support minimum, maximum, fixed, and asynchronous timing constraints. However it is assumed that module binding and hardware allocation has already been done before the scheduling, even though these three tasks are highly interdependent.

A new model for synthesizing embedded VLSI chips, OAS (for optimal architectural synthesis), is introduced below. The basis of this model is an extension of the previously researched IP approach[5] to simultaneous scheduling and resource allocation. Using polyhedral theory[6], constraints were formulated to reduce the size of the search space and the problem was solved using the branch and bound algorithm. The new model differs from previous research[5,10,4] by 1. modeling the clock period as a variable (whereas previously it was assumed fixed), 2. formulation of timing constraints as a function of the clock period, and 3. presenting a new formulation for asynchronous interfaces that models average execution time. This research breaks new ground by (1) providing industry with synthesized high performance architectures of embedded VLSI chips in practical cpu execution times (thus supporting complex and asynchronous interfaces), and (2) automatically optimizing clock period and selecting types of functional units, including chained operations, simultaneously with scheduling and allocation tasks.

The following terminology will be used in this paper: k = a code operation of the input algorithm. A partial order between k_1 and k_2 is represented by $k_1 <\bullet\, k_2$, or in other words code operation k_1 outputs data that is used by code operation k_2. The main variables of the model are p_c, and $x_{j,k,t}$. The selection of a clock period is modeled with p_c=1 if clock period c is selected, where $p(c)$= the clock period in nanoseconds. When $x_{j,k,t} = 1$, code operation k is assigned to cstep j and to type of functional unit t. $t_z \varepsilon T(k_z)$ means that code operation k_z can be implemented by functional unit type t_z. For example a multiplication operation, k, can be implemented by t=1 a two-cycle multiplier and t=2 a pipelined multipler or $T(k)=\{1,2\}$ (further details will be provided later in the paper concerning the use of t for chained operations). $j_z \varepsilon R(k_z)$ means that $asap(k_z) \leq j_z \leq alap(k_z)$ (where $asap/alap(k_z)$ is the as soon/late as possible csteps). For example $asap(k)=min_c\ asap_c(k)$, where $asap_c(k)=asap(k)|p_c=1$. The variable I_t represents the number of functional units of type t. For simplicity we will assume that $t_z \varepsilon op(C_z,L_z)$ which means that the type of functional unit t_z requires C_z csteps to produce output data and can accept new input data every L_z csteps. For example a two cycle type of functional unit is $t\varepsilon op(2,2)$. We use the notation $time(k_1,k_2) \leq$ or $\geq$ or $=T$ ($time(k_1,k_2)=\sum_j (j)x_{j,k_2}-\sum_j (j)x_{j,k_1}$, however not formulated as such in the model) to represent the maximum, minimum or fixed time constraint between the two operations. From now on it will be assumed that $x_{j,k,t}$ implies $j\varepsilon R(k)$ *and* $t\varepsilon T(k)$.

3. The IP Formulation

Consider the functional units in table 1 (columns 1 through 3) from a library[7] where registers also occupy 500 mil^2 (A) and have 5ns delay (D). The possible clock periods and types of functional units are given in columns 4 through 8 (generated for the elliptical wave filter, EWF). For example the fifth column of table 1 uses a 156ns clock period and the types of functional units are a single cycle adder (C_+=1 in row 4) of type t=2 (row 5) and a three cycle multiplier of type t=3. In column 7, the clock period is 307ns and there are two types of functional units, the adder-adder (++) and the multiplier-adder (*+). The adder-adder type of functional unit performs two additions $+_{t=3} <\bullet +_{t=2}$ in one cstep (row 4). The multiplier-adder performs the multiplication operation followed by the addition operation, $* <\bullet +$, in two csteps (C_{*+}=2). Note that this type of functional unit can also perform only one multiplication (C_*=2) or only one addition (C_+=1). The clock periods and different types of functional units in table 1 will be used to illustrate how the OAS model works. Unfortunately there is no room in this paper to provide the complete model, however some inequalities will be given.

Table 1. Types of Functional Units vs Clock Periods for EWF

Library			c	1	2	3	4
k	A	D	p_cp(c)	156	190	307	380
	(mil^2)	(ns)	k	+,*	+,*	++,*+	++,*
+	1200	151	C_k	1,3	1,2	1,2(1)	1,1
*	49000	375	t_k	2,3	2,2	3/2,2	3/2,1

The first constraint ensures that only one clock period will be chosen: $\sum_c p_c=1$ (1). The code operation assignment constraint, $\sum_{t\varepsilon T(k)} \sum_{j\varepsilon R(k)} x_{j,k,t} = 1, \forall k$ (2), ensures that each code operation of the input algorithm will be assigned to only one cstep and one type of functional unit.

Let us consider formulation of the partial order constraints. Consider a multiplication operation that precedes an addition operation and can be implemented by four types of functional units (see row 3 and 4 of table 1). The following is an example of one inequality that models the partial order: $\sum_{t=1}^{t=2} \sum_{j2 \geq j} x_{j2,*,t} + \sum_{j3 \geq j-2} x_{j3,*,3} + \sum_{j \geq j1} x_{j1,+,2} \leq 1, \forall * <\bullet +$. This is a new facet of the underlying scheduling problem.

The partial order constraints for the addition operations that map to the adder-adder type of functional unit can be modelled using fixed timing constraints[5]. These constraints are formulated between an addition operation with t=3 (the first addition) and the second addition operation with t=2.

The number of multipliers can be formulated as: $\sum_{*}(\sum_{t\varepsilon T(*)} x_{j,*,t} + \sum_{t=2,3} x_{j+1,*,t} + x_{j+2,*,3}) \leq I_*, \forall j\varepsilon R(*)$. We will now present inequalities to model the relationship between selecting a type of functional unit and the clock period.

The selection of a clock period can be modeled using the following inequalities: $\sum_j x_{j,*,3} \leq p_1$, $\sum_j x_{j,*,2} \leq p_2+p_3$, $x_{j,*,2}+x_{j+1,+,2} \leq 1+p_3$, $\sum_j x_{j,+,3} \leq p_3+p_4$, and $\sum_j x_{j,*,1} \leq p_4$. For example the last inequality forces the clock period to be 380ns if multipliers of type 1 are selected and allocated.

The cost function can be formulated as a peicewise linear area cost or as the delay in nanoseconds. The delay is $(\sum_c p(c)p_c)$ $(\sum_j (j-1)x_{j,e,0})$,where operation e (preceded by all operations) is used to mark the completion of the input algorithm. Although this cost function is nonlinear, a solution strategy that transforms the problem into a linear IP problem will be discussed in section 5. The area cost function and register allocation constraints can be modelled as described in [5] by substituting $\sum_{t \in T(k)} x_{j,k,t}$ for $x_{j,k}$. A more detailed modification is required for the bus allocation constraints, to be described in a longer paper.

4. Interface Constraints

Since the new IP formulation models the clock period as a variable the minimum, maximum, fixed and asynchronous timing constraints must be formulated as a function of the clock period, unlike previous models. For example assume that a maximum timing constraint $time(+25,+11) \leq 770ns$ is given and a subset of the partially ordered list of operations is $+25 <\cdot *21 <\cdot +19 <\cdot +11$. Given the clock periods to be selected from table 1 we can rewrite the timing constraint as $time(+25,+11) \leq 4p_1+4p_2+2p_3+2p_4$. This timing constraint is represented by the following two partial order constraints: $\sum_{k=+25,+11} \sum_{j} x_{j,k,3} + \sum_{j1>j} x_{j1,+11,2} + \sum_{j2 \leq j-4} x_{j2,+25,2} \leq 2-p_1-p_2$ and $\sum_{t=2,3} (\sum_{j1>j} x_{j1,+11,t} + \sum_{j2 \leq j-2} x_{j2,+25,t}) \leq 2-p_3-p_4$. Note that a similar formulation of these timing constraints can be used to model algorithm latency or minimum timing constraints.

A new approach is presented in this paper for modeling bounded or unbounded asynchronous interfaces that minimizes an average execution time, $\overline{T_e}$. We assume that each possible arrival time for input data has an associated probability, obtained from previous simulation studies. For example the probability of data being accessed after one cycle (ie. located in cache) could be 0.85 otherwise a delay of 10 cycles is required. By modeling only selected high probability arrival times using a conditional code construct in the model, the average execution time of the input algorithm can be optimized. The cost function is *Minimize* $\overline{T_e}$ where $\overline{T_e} = \sum_i P_i T_{e_i}$, $T_{e_i} = \sum_t \sum_{j \in R(e_i)} (j-1) x_{j,e_i,t}$ (the execution time assuming data arrives during the ith selected time) and P_i is the probability that the input data arrives at the ith selected time. The code operation e_i marks the completion of the algorithm assuming data from the asynchronous operation arrives during the ith selected time. When the input data arrives at a (lower probability) time that was not selected, it is stored in a register allocated by the IP model until the controller reaches the next selected time at which point it can process the data.

5. Synthesizer Results

The OAS architectural synthesizer was used to synthesize optimal high speed architectures for the well known EWF example[8]. Problem 1 was solved using the data in table 1, which is easily generated during the preprocessing stages of OAS. Finally results on minimizing average execution time given constraints on area and asynchronous interfaces are presented using a modified EWF example. The IP solver was GAMS/ZOOM which uses a branch and bound algorithm running on a MIPS RC6280 workstation.

Optimal schedules, selection of types of functional units, allocation and selection of clock period was solved for the EWF (using table 1) in two stages. In stage 1 the number of csteps is fixed and the clock period is minimized given constraints on the area. This model is solved as a relaxed LP to obtain a lower bound on the clock period (rounded up to the nearest clock period of table 1), shown in table 2. Since the model is very tight, good bounds on the clock period have been found. The second stage then minimizes the area of the architecture. A reduced model (obtained by deleting p_c), which uses the clock period from stage 1, is solved to produce an architecture. Results on the second stage are also shown in table 2. Figures 1,2, and 3 show the OAS schedules corresponding to rows 8,6, and 4 of table 2 respectively (Var,Eqn is number of variables,equations of IP model).

To give an example of the speed advantage OAS provides one can map the solutions to the library described in [3] where areas of *, + are 17000, 1500 units and delays of *, + are 80ns, 55 ns (and registers have delay of 2ns). The minimum delay for the EWF in [3] (using a fixed clock period of 100ns) was 1800ns with an area of approximately 21000 units. The OAS minimum delay, which uses c=4 or 112ns clock period (in last row of table 2 except mapped to a different library), was 1456ns with an area of 21500 units.

Table 2. OAS synthesized architectures vs CPU seconds

Var	Eqn	CPU	$I_{+,*}$	p_c	T_e	Delay	Area
Stage 1: Minimize $\sum p_c p(c)$							
850	1921	0.29	-	156	21	3276	≤1.518
554	1301	0.25	-	190	17	3230	≤1.518
184	526	0.12	-	307	12	3684	≤1.518
Stage 2: Minimize $0.49*I_* + 0.012*I_+$ ($10^5 mil^2$)							
384	435	3.22	2,3	156	21	3276	1.494
252	304	0.58	3,3	190	17	3230	1.506
150	349	1.01	3,3	307	12	3684	1.506
208	488	1.28	3,2	307	13	3991	1.016
150	316	0.89	3,2	380	12	4560	1.016
208	447	8.15	3,1	380	13	4940	0.526

To illustrate the asynchronous interface for the modified EWF an area constraint of 2 adders and 1 two-cycle multiplier was used and the average execution time was minimized. The asynchronous operation, a, can at the earliest produce data after two cycles (with probability 0.6) and at the lastest after 4 cycles (with probability 0.4). The optimal schedule, shown in figure 4, used 425 variables, 610 equations, and was synthesized in 190 cpu seconds. OAS also minimized chip area, (area = $f(I_{t_+}, I_{t_*}, R, B)$ where R, B are the number of registers and busses), with constraints on the average execution time of $T_{e_1}=21$ and $T_{e_3}=26$. OAS required 254 variables, 592 inequalities, and synthesized a globally optimal architecture in 8.3 cpu seconds. To illustrate the impact of the model formulation on the efficiency of solving the IP, the data precedence constraints were replaced by a previously researched constraint[9,10]. This modified IP model had the same number of variables, only 469 inequalities, however it required 274 cpu seconds to solve (using the same IP solver on the same cpu).

6. Conclusions

In this paper we have defined a new model, OAS, that supports simultaneous scheduling, selection (of functional unit types and clock periods), and allocation, and support for complex timing constraints and asynchronous interfaces. This is in contrast to previous research [5] that could not simultaneously select types of functional units and clock periods nor model average execution times. Globally optimal schedules and allocations which minimize an area cost function, an average execution time, or delay have been synthesized in practical cpu execution times. It has been shown that by simultaneously selecting clock periods and chaining operations, the architectures, synthesized by OAS, are up to 23% faster than previously published architectures. Secondly up to 30 times improvement in cpu execution times were found attributable to the (polyhedral) formulation of the data precedence constraint. Although the worst case complexity is exponential (since we are solving an IP), we have found that many problems can be solved extremely quickly to global optimums. As input algorithms become larger we can partition the behavior or take advantage of the mathematical flexibility of OAS to model systems with hierarchy and regularity.

In summary this research for the first time synthesizes architectures in practical cpu execution times, supports complex constraints including asynchronous interfaces and automatically selects the clock period and chained types of functional units. This is important for industry since its has been estimated it would take an effort larger than the synthesis itself to modify a self contained synthesized design to interface with other external processes in a system. Also these early decisions made during high level synthesis have the greatest impact on the final design. In contrast to other synthesizers we can obtain exact globally optimal schedules, allocations and selection (of types of functional units and clock periods), that support complex interface constraints. Previous synthesizers could at best guarantee a locally optimal solution, which may not satisfy all timing constraints and could not explore strategies for asynchronous interfaces (simultaneously with scheduling and allocation) nor simultaneously select chained operations. The mathematical basis of the approach provides the flexibility of modeling complex interface constraints which are critical to achieving high performance. The polyhedral theory behind this mathematical model explains why we can solve many problems in practical cpu execution times. Finally the methodology forms a DA tool that can be brought to market quickly since it based on highly reliable mathematical software. For the first time this research 1) provides industry with a DA tool for synthesizing high performance embedded VLSI architectures, and 2) a new methodology that can support architectural design exploration of interface strategies, to minimize area-delay costs. Future research involves synthesis of multichip systems, models for simultaneous scheduling, allocation and binding, and optimization model for mixed analog/digital synthesis. This research is supported in part by University of Waterloo Start Up Grant, NSERC, and MICRONET. The author would like to thank Dr.R.J.Gebotys for his helpful discussions.

References

[1] S.Devadas,A.R.Newton,"Algorithms for Hardware Allocation in Data Path Synthesis",*IEEE Trans. on CAD*, 1989.

[2] L.Hafer,A.Parker,"A formal Method for the Specification, Analysis and Design of Register-Transfer-Level Digital Logic",*IEEE Trans. on CAD*,1983.

[3] L.Ramachandran,D.D.Gajski,"An Algorithm for Component Selection in Performance Optimized Scheduling",*ICCAD*,1991.

[4] D.C.Ku,G.DeMicheli,"Relative Scheduling Under Timing Constraints",DAC,1990.

[5] C.H.Gebotys,M.I.Elmasry,*Optimal VLSI Architectural Synthesis: Area, Performance and Testability*,Kluwer,1992.

[6] G.L.Nemhauser,L.A.Wolsey,*Integer and Combinatorial Optimization*, Wiley Interscience,1988.

[7] A.Parker,"Tutorial on High Level Synthesis", Can. Conf. on VLSI,1990.

[8] High-Level Synthesis Workshop Clearinghouse, email: hlsw-request@decwrl.dec.com,1988.

[9] K.R.Baker,*Introduction to Sequencing and Scheduling*,John Wiley & Sons,1974.

[10] C-T.Hwang,J-H.Lee,Y-C.Hsu,"A Formal Approach to the Scheduling Problem in High-Level Synthesis",*IEEE Trans. on CAD*,1991.

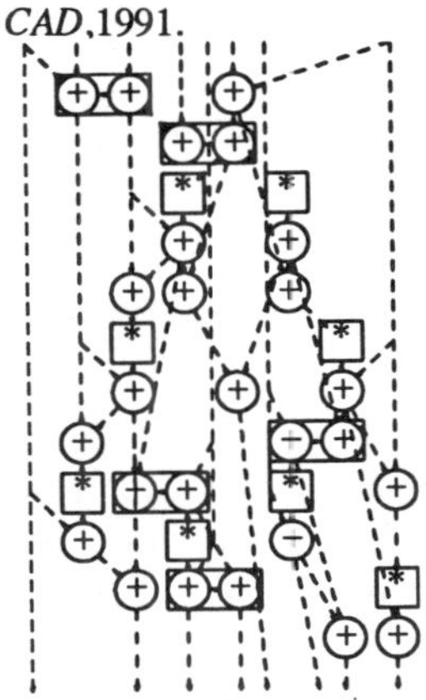

Figure 1. Schedule for 380ns clock period.

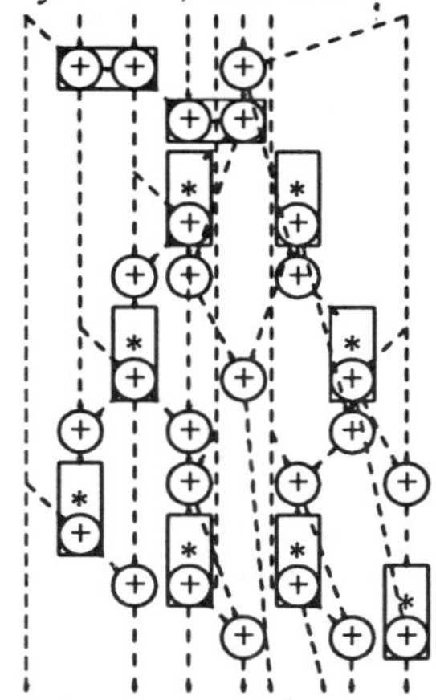

Figure 2. Schedule for 307ns clock period.

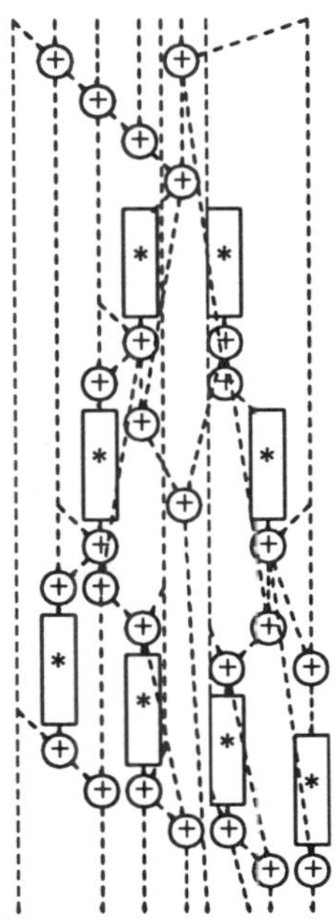

Figure 3. Schedule for 156ns clock period.

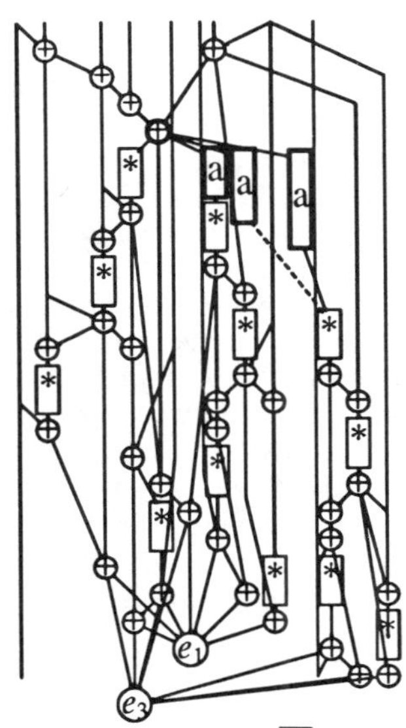

Figure 4. Schedule with $\overline{T_e}=23.8$.

Optimal Allocation and Binding in High-Level Synthesis

Minjoong Rim[1], Rajiv Jain[1] and Renato De Leone[2]
[1]Department of Electrical and Computer Engineering
[2]Department of Computer Sciences
University of Wisconsin
Madison, WI 53706

Abstract

In this paper, we present an integer linear program (ILP) formulation for the allocation and binding problem in high-level synthesis. Given a behavioral specification and a time-step schedule of operations, the formulation minimizes wiring and multiplexer areas. This is the first time that an ILP model for minimizing multiplexer and wiring areas has been mathematically formulated and optimally solved. The model handles chaining, multi-cycle operations, pipelined modules, conditional branches and trades off wiring area with resource area.

1 Introduction

Automatic generation of RTL (register-transfer level) designs from a behavioral description is known as high-level synthesis. To synthesize a desired RTL design several conflicting and interacting sub-problems have to be solved. These sub-problems are scheduling, module allocation, module selection, module binding, and multiplexer and register allocation. In this paper, we present a new technique for allocation and binding with an objective of minimizing wiring and multiplexing areas required for interconnecting the design. *Allocation* is the task of ascertaining the number of modules of each type, and the number of registers and multiplexers required for the design. *Binding* is a mapping of operations to modules and values to registers. The binding and allocation tasks are inter-dependent and also depend upon scheduling, interconnect costs and delays [10]. The input to our allocation and binding formulation is a scheduled data flow graph. First, we formulate the problem as a zero-one integer linear programming (ILP) problem and then use existing ILP solution techniques to solve the formulation.

The synthesis problem was first formulated as a mixed ILP problem by Hafer and Parker [4]. Hafer's formulation, however, does not minimize the number of wires or multiplexers used in the design with the result that the final design may actually require large multiplexer and wiring areas. Furthermore, this formulation entails enormous runtime and is not practical. By decomposing synthesis into two sub-problems, namely, (i) scheduling, and (ii) allocation and binding, we can approach each part independently and quickly solve them optimally. Solving each individual sub-problem independently and combining the solutions, however, does not necessarily lead to a globally optimal solution. The assumption of breaking synthesis into these two sub-problems has been used by almost all synthesis systems found in literature [10]. The notable exceptions are [4] in which all sub-components are solved concurrently, and [1] in which the synthesis problem is solved by partitioning it into two sub-problems different from the above breakdown: the scheduling, module allocation and module binding problem and, the register and multiplexer allocation problem. Recently, an ILP model for performing scheduling, allocation and binding concurrently has been proposed by Gebotys [2]. This model, however, like the work by Hafer and Parker [4] ignores multiplexer and wiring area and the solution can thus give an inferior design. Hafer has recently proposed solution to the high-level synthesis problem with an objective of minimizing storage, without considering interconnect and multiplexer area [3]. Hsu, et al. [5] schedule and allocate but do not bind. We believe the formulation presented in this paper to be the first attempt at solving the allocation and binding problem with the objective of minimizing wiring and multiplexer area. Wiring comprises of a substantial part of chip real estate and one cannot over-emphasize the importance of minimizing it.

2 Problem Formulation

We assume point-to-point interconnection style. After binding, we can get a partial floorplan and global routing information and wires can be combined to form buses using techniques like those in [9]. Even when other bus structures are used, our method is expected to result in shorter buses and less interconnections between buses and resources. The model is initially developed for a non-pipelined design style. Chaining, multi-cycle operations, multi-cycle values, and pipelined modules are allowed. An important advantage of our formulation is that extensions for handling these features do *not* require additional constraints or variables. Thus, the formulation complexity does not grow with these generalizations. Extensions for functional pipelining are elementary and do *not* require any additional constraints or variables, as we will see later. We assume that all operations produce only one output. This assumption is made for simplicity of explanation and is relaxed later. Handling of DFGs with conditional branches and modules with functional pipelining have been successfully modeled and have been omitted from the paper due to the lack of space. Details of these are given in [13]. The following terms will aid in the discussion. In the formulation, there is no difference between operations and values or between modules and registers. In the remainder of the paper operations and values are collectively called ***tasks*** and modules and registers are collectively called ***agents***. *TaskType* is a set of task types. *AgentType* is a set of agent types. *Time* is a set of time-steps. I is a set of tasks. J is a set of agents. $I_t \equiv \{i \in I \mid$ task i is active in time-step $t\}$ where $t \in Time$. A task is said to be *active* if it is scheduled in that time step. A unit-cycle operation is active during only one time-step whereas a multi-cycle operation is active over several time-steps. A pipelined task with initiation interval l is active in all the first l time-steps starting from the time-step it is first scheduled in. The output of a pipelined task, however, is available only after the total delay of the task. *Interconnection* is a set of ordered triples (i_1, i_2, l) each of which represents interconnection between the output of task i_1 and the lth input of task i_2. If tasks i_1 and i_2 are operations, it represents chaining of operations. Notice that multi-cycle operations, chaining, and pipelined modules are considered by extracting proper input data. Even though in the formulation we do not explicitly discuss multi-cycle operations, chaining of operations or pipelined modules, they are taken care of. We write $type_{ta}(i)$ to denote the type of task i, and $type_{ag}(j)$ to denote the type of an agent j.

The objective of allocation and binding is to minimize the overall cost (area) of an RTL design. That is,

$$\text{minimize} \quad \textit{module area} + \textit{register area} + \textit{multiplexer area} + \textit{wiring area}$$

A minimum number of modules of each type and registers required by the implementation are determined by the schedule and the max-cut of the input description respectively. While solving the allocation and binding problem, module and register areas are assumed to be constants [1], and the objective function minimizes wiring and multiplexer areas. In general, a large number of connections between resources (register and modules) requires large quantities of multiplexers and wires. Hence, first, we assume that the number of multiplexers can be reduced by minimizing the number of wires and use this as an objective function. A similar objective has been adopted in [11]. Minimizing wiring and multiplexer areas concurrently is described later. We now formulate our problem with an objective of minimizing wiring area.

Module and Register Binding: Let $x_{i,j} = 1$ if task i is assigned to agent j and zero otherwise. If task i cannot be performed by agent j due to type conflict, we set

$$x_{i,j} = 0 \qquad (1)$$

Each task should be performed by at least one agent. This is enforced by,

$$\sum_{j \in J} x_{i,j} \geq 1, \quad i \in I$$

Wiring area may be reduced by having more than one agent perform the same task but this is very unlikely. Hence, we require each task be assigned to exactly one agent, and

$$\sum_{j \in J} x_{i,j} = 1, \quad i \in I \qquad (2)$$

Since an agent can perform at most one task in a time-step, we have

$$\sum_{i \in I_t} x_{i,j} \leq 1, \quad j \in J, \ t \in Time \qquad (3)$$

Constraints 2 and 3 represent a simple assignment problem.

Wiring Area: A wire is defined to be an interconnection between two different agents. It excludes multiplexer area and split wires at the inputs of the multiplexers. Let $w_{j_1,j_2} = 1$ if there exists a value transfer from agent j_1 to agent j_2 in any time-step. If $(i_1, i_2, l) \in$ *Interconnection* there is a data transfer from task i_1 to task i_2, and therefore there must be a wire from the agent executing task i_1 to the agent executing task i_2. This can be modeled by,

$$w_{j_1,j_2} \geq x_{i_1,j_1} + x_{i_2,j_2} - 1, \quad j_1, j_2 \in J \qquad (4)$$

Wiring area can now be written as

$$\sum_{j_1 \in J} \sum_{j_2 \in J} c_{j_1,j_2} w_{j_1,j_2} \qquad (5)$$

where c_{j_1,j_2} is the cost of connecting agent j_1 to agent j_2, and is the area of the wire. If preliminary floorplanning has already been done, then these values may be estimated from the floorplan and used in the equation. In the absence of a floorplan, an estimate of the average wire length can be obtained using techniques given in [8]. In this case, all c_{j_1,j_2}'s are equal and can be eliminated from the objective function.

The number of variables in the formulation can be reduced by taking into account type conflicts. Let G_k be the number of tasks of type k ($k \in TaskType$). Let F be the total number of agents, and F_k be the number of agents which can perform a task of type k. The number of $x_{i,j}$ decision variables is $\sum_{k \in TaskType} G_k F_k$, and the number of w_{j_1,j_2} decision variables is at most F^2. [2] Thus, at most $(F^2 + \sum_{k \in TaskType} G_k F_k)$ binary decision variables are needed in the model. Note that the value of w_{j_1,j_2} is completely determined by the variables $x_{i,j}$'s (assuming c_{j_1,j_2}'s are positive). The complexity of the problem can be reduced by declaring variables w_{j_1,j_2} as non-negative real variables, and using mixed integer programming techniques with at most $\sum_{k \in TaskType} G_k F_k$ binary decision variables.

Functional Pipelining: The formulation presented so far is aimed at non-pipelined designs but it can be

[1] Tradeoffs between active, wiring and multiplexer areas are discussed in [13].

[2] In practice many w_{j_1,j_2} variables can be set to zero since certain agents cannot be connected.

easily extended to the pipelined design style as well. Let l be the initiation interval of the pipelined data path. Then, the operations in time steps $t+n\times l$ $(n = 1,2,\ldots)$ are executed simultaneously and cannot share modules. Consequently, a set of active tasks in time-step t is defined to be,

- $I_t \equiv \{i \in I|$ task i is active in time-step $t+n\times l$ $(n=1,2,\ldots)\}$ $\; t = 1,2,\ldots,l.$

The problem formulation itself remains unaltered.

Multi-Output Operations: So far we have assumed that an operation generates only one unique value. Many operations generate several values which are fed to more than one resources, for example, an addition operation generates a carry value and a sum of its two inputs. Such operations can be accommodated in the following simple manner. We redefine *Interconnection* to be a set of ordered quadruples (i_1, m, i_2, l) each of which represents interconnection between the mth output of task i_1 and the lth input of task i_2. If task i produces g outputs, then $M_{ta}(i) = \{1,2,\ldots,g\}$. If task i is of type value, $M_{ta}(i) = \{1\}$. If agent j produces g outputs, then $M_{ag}(j) = \{1,2,\ldots,g\}$. If agent j is of type register, $M_{ag}(j) = \{1\}$.

Wiring area is now modeled as follows. Let $w_{j_1,m,j_2} = 1$ if there exists a value transfer from the mth output of agent j_1 to agent j_2 in any time-step. If $(i_1, m, i_2, l) \in$ *Interconnection*, there is a data transfer from the mth output of task i_1 to task i_2, and there must be a wire from the mth output of the agent executing task i_1 to the agent executing task i_2. This is given by,

$$w_{j_1,m,j_2} \geq x_{i_1,j_1} + x_{i_2,j_2} - 1, \quad j_1, j_2 \in J \qquad (6)$$

Wiring area can now be written as

$$\sum_{j_1\in J}\sum_{m\in M_{ag}(j_1)}\sum_{j_2\in J} c_{j_1,m,j_2} w_{j_1,m,j_2} \qquad (7)$$

where c_{j_1,m,j_2} is the cost of connecting the mth output of agent j_1 to agent j_2.

Multiplexer Area Minimization: One assumption made in the research literature is that minimizing multiplexer area results in minimum wiring area as well (or vice-versa) [11]. One can easily construct an example where minimizing one does not necessarily minimize the other. In order to achieve an optimal design we need to minimize both multiplexer and wiring areas. We now establish constraints with the objective function for minimizing multiplexer area. These constraints, when combined with the constraints and objective function for wiring area, minimize both multiplexer and wiring areas.

First, we define a few terms. If task i is a g-ary operation, then $L_{ta}(i) = \{1,2,\ldots,g\}$. For a task i of type value, $L_{ta}(i) = \{1\}$. If an agent j is a g-input module, then $L_{ag}(j) = \{1,2,\ldots,g\}$. If agent j is of type register, then $L_{ag}(j) = \{1\}$.

Multiplexer area at an input port of an agent is assumed to increase linearly in the number of distinct inputs to the input port. Let $z_{j_1,j_2,l} = 1$ if there exists a transfer of a value from agent j_1 to the lth input port of agent j_2 and zero otherwise. Suppose that task i_2 is a register or a non-commutative operation. If $(i_1, i_2, l) \in$ *Interconnection*, [3] then there must be a connection from the agent executing task i_1 to the lth input port of the agent executing task i_2.

$$z_{j_1,j_2,l} \geq x_{i_1,j_1} + x_{i_2,j_2} - 1, \quad j_1, j_2 \in J \qquad (8)$$

For commutative operations we need a slightly more complicated model. The difficulty with a commutative operation stems from the fact that a value can go to any input port of the corresponding agent. We introduce a concept of ***pseudo input port***. Each input value arrives at a pseudo input port which is connected to each functionally equivalent input port of the agent. The mapping from pseudo-input ports to input ports is an assignment problem. Let $s_{l,l',i} = 1$ if there exists a connection between pseudo-input port l and input port l' of commutative task i. The constraints can be written as

$$\sum_{l\in L_{ta}(i)} s_{l,l',i} = 1, \quad l' \in L_{ta}(i) \qquad (9)$$

$$\sum_{l'\in L_{ta}(i)} s_{l,l',i} = 1, \quad l \in L_{ta}(i) \qquad (10)$$

for each commutative operation. Then,

$$z_{j_1,j_2,l'} \geq x_{i_1,j_1} + x_{i_2,j_2} + s_{l,l',i_2} - 2, \quad l' \in L_{ta}(i_2),\; j_1, j_2 \in J \qquad (11)$$

The multiplexer area is given by

$$\sum_{j_1\in J}\sum_{j_2\in J}\sum_{l\in L_{ag}(j_2)} d_{j_1,j_2,l}\; z_{j_1,j_2,l} \qquad (12)$$

where $d_{j_1,j_2,l}$ is the multiplexer cost corresponding to $z_{j_1,j_2,l}$. The objective is to minimize a linear combination of expressions 5 and 12.

For some cases, the number of multiplexers for each port at an agent may be limited. For example, total delay of multiplexers at a module should not exceed some value due to timing constraints. This is modeled as

$$\sum_{j_1\in J} e_{j_1,j_2,l} z_{j_1,j_2,l} \leq f_{j_2,l} \qquad (13)$$

3 Examples and Results

By convention adopted in literature [6] [11] [12] input registers are read-only buffers and cannot be reused.[4] Also, a multi-cycle value is restricted to be stored in one register over several time-steps [7]. This restriction can

[3] If *Interconnection* is a set of 4-tuples, then the constraints are modified accordingly.

[4] This convention can be easily overruled in our formulation.

DFG	# of		Time	DFG, # of			RTL Design, # of			
	x	w	(secs.)	+	×	values †	+	×	reg.	wires
complex	13	18	0.25	2	4	6	1	2	3	6
diff. eqn.	42	32	25.9	5	6	11	2	2	4	10
AR filter ‡	156	50	190.84	12	16	24	2	2	6	16
EW filter	223	48	3137	26	8	35	2	1 ⋆	10 *	30

†: number of values to be stored in registers
‡: chaining of operations
⋆: pipelined multiplier
*: input registers reused

Table 1: Summary of results

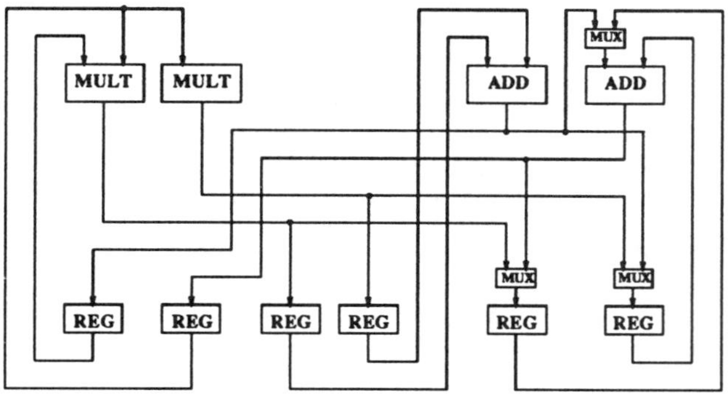

Figure 1: Synthesized RTL design for AR filter

result in inferior designs, specially in the pipelined design style where multi-cycle values are generally stored in several registers. This restriction can be easily eliminated without any modification to our formulation by simply breaking a multi-cycle value into several single cycle values.

To verify the formulation, we synthesized several RTL designs. We chose the complex multiplication, differential equation, AR filter and EW filter from high-level synthesis literature. The results are summarized in Table 1. All solutions are optimal with respect to the objective function of minimizing number of wires in the design. These examples illustrate chaining of operations, use of pipelined modules and the re-use of input buffer registers and were executed with an objective of minimizing wiring area alone. For the first three examples, input registers and wires from input registers are omitted whereas the last result includes both assuming that input registers can be reused. We observed that there are too many optimal solutions for EW filter. This implies that even a non-judicious method may be able to get an optimal solution. In contrast, AR filter had a relatively few optimal solutions, and thus we recommend AR filter as a benchmark for heuristics for binding and allocation problem. A synthesized design for AR filter is shown in Figure 1 (connections with the input registers have not been shown for clarity). The scheduled DFGs and other synthesized designs are given in [13]. We are currently experimenting with both wiring and multiplexer area minimization, and the preliminary results indicate there is no improvement in cost. These problems were formulated and solved using XPRESS on a SUN SPARCstation 1 with 16 Mb main memory.

4 Conclusion

In this paper we have formulated the resource allocation and binding problem in high-level synthesis. The formulation handles chaining, multi-cycle operations, pipelined modules, conditional branches and makes tradeoffs between wiring area and module and register area. Most of these generalizations are achieved without an increase in the number of variables or constraints. We have also synthesized several examples from literature to demonstrate the practicality of the approach. All solutions were optimal.

References

[1] R. Cloutier and D. Thomas. The Combination of Scheduling, Allocation and Mapping in a Single Algorithm. *27th DAC*, 1990.

[2] C. H. Gebotys and M. I. Elmasry. A Global Optimization Approach for Architectural Synthesis. *ICCAD*, 1990.

[3] L. Hafer. Constraint Improvements for MILP-Based Hardware Synthesis. *28th DAC*, 1991.

[4] L. Hafer and A. Parker. A Formal Method for the Specification, Analysis, and Design of Register-Transfer Level Digital Logic. *IEEE Trans. CAD*, January 1983.

[5] C.-T. Hwang, et al. Optimum and Heuristic Data Path Scheduling Under Resource Constraints. *27th DAC*, 1990.

[6] K. Kucukcakar and A. C. Parker. Data Path Tradeoffs Using MABAL. *27th DAC*, 1990.

[7] F. J. Kurdahi and A. C. Parker. REAL: A Program for REgister ALLocation. *24th DAC*, 1987.

[8] F. J. Kurdahi and A. C. Parker. Techniques for Area Estimation of VLSI Layouts. *IEEE Trans. CAD*, January 1989.

[9] T. A. Ly, et al. A Generalized Interconnect Model for Data Path Synthesis. *27th DAC*, 1990.

[10] M. C. McFarland, et al. The High-Level Synthesis of Digital Systems. *Proceedings of the IEEE*, February 1990.

[11] N. Park and F. J. Kurdahi. Module Assignment and Interconnect Sharing in Register-Transfer Synthesis of Pipelined Designs. *ICCAD*, 1989.

[12] P. G. Paulin and J. P. Knight. Force-Directed Scheduling for the Behavioral Synthesis of ASIC's. *IEEE Trans. CAD*, June 1989.

[13] M. Rim, et al. Optimal Allocation and Binding in High-Level Synthesis of VLSI Digital Systems. Tech. report, Dept. of Electrical and Computer Engineering, University of Wisconsin - Madison, Sept. 1991.

Time Constrained Allocation and Assignment Techniques for High Throughput Signal Processing*

Werner Geurts Francky Catthoor Hugo De Man†

IMEC, Kapeldreef 75, B-3001 Leuven, Belgium

Abstract

In this paper, a technique for the allocation of complex application specific datapaths will be presented. The technique is especially suited for the synthesis of application specific architectures for high-throughput signal processing applications. Such applications comprise hierarchical compositions of nested loops and condition blocks. A minimum area set of datapaths is allocated and the available cycle budget is automatically distributed over the different blocks of the hierarchy in one global optimization process. Results for a number of real life examples are presented.

1 Introduction

In this paper we will focus on architectural synthesis techniques for irregular high-throughput signal processing applications. The signal flow graphs of such applications exhibit a high degree of repetitivity (FOR and WHILE loops), combined with decision making (IF blocks). Typical application examples can be found in the image processing, medium-level video processing and front-end telecommunication domains [1].

While the number of primitive operations to be performed is high, the number of machine cycles that are available is limited due to the high sample rate of the application. This low opportunity for multiplexing will lead to a large number of primitive functional units (FUs) such as adders, multipliers and comparators which will have to be interconnected and supplied with input data. It is important to optimize not only the cost of FUs but also the significant cost of the interconnection network and of the storage elements.

Traditional techniques [2, 3] which work at the level of primitive operations and primitive FUs fail in this objective (see [4]). Most often the optimum number of FUs is found, but the interconnect and register cost is prohibitive. Even powerful datapath synthesis techniques such as the ones proposed in e.g. [5, 6] are incapable of building reasonable datapaths for a graph scheduled with an inadequate objective.

The problem of high interconnect and register cost can also be approached from an architectural standpoint [1, 7]. Instead of primitive functional units, more complex execution units (EXUs), composed of a few functional units with fixed interconnections, are then defined before scheduling. In the HYPER system [7], these complex combinatorial modules are not programmable. They correspond to chains of operations that have a high degree of occurrence. In the CATHEDRAL-2ND system [1], the highly multiplexed programmable execution units are composed by the user and a chaining algorithm assigns chains of primitive operations to the execution units.

The approach we present here is a part of the CATHEDRAL-III design system [1, 4]. In the architectural style of CATHEDRAL-III, the datapaths are so called *application specific units* (ASUs). Such units are composed of FUs in such a way that their functionality is highly tailored towards the needs of a specific application. It is the aim of the synthesis process to allocate a minimum area set of ASUs and to find the most suitable composition for each of these ASUs. The main constraint for this optimization process is the *hardware sharing factor*: $HSF = \lceil T_{SAMPLE}/T_{CLOCK} \rceil$, where T_{SAMPLE} is the sample period at which the application has to run and T_{CLOCK} is the required clock period.

2 The ASU Allocation Task

As mentioned in the introduction, a CATHEDRAL-III architecture consists of ASUs (see figure 2 for an example) whose composition is tailored to the application at hand. One approach to perform the mapping of an application on such a target architecture could be to select the most suitable ASUs from a predefined library. However, in view of the size this library (all possible interconnection schemes of various sized set consisting of adders, subtractors, comparators, multipliers, shifters, etc... would have to be included), we have rejected this approach. Instead we have adopted an approach in which the ASUs are synthesized, starting from the signal flow graph of the application. A global description of all the steps involved in this process can be found in [4]. We will now briefly sketch the three steps directly relevant to the work presented in this paper:

1 In the **Operation Clustering** step, the atomic operations of the signal flow graph are grouped into clusters (see figure 1). A cluster c of operations - represented by the graph $G_c(V_c, E_c)$ - is a connected subgraph of the signal flow graph ($G_{SFG}(V, E)$). The vertices $v \in V_c$ represent arithmetic, relational and logical operations. The edges $u \rightarrow v \in E_c$ represent scalar signals and are characterized by a signal type. A cluster can have a duration (or initiation interval) of one or

*This research was sponsored by the ESPRIT Basic Research Action 3281 "ASCIS" of the E.C.

†Professor at the Katholieke Universiteit Leuven

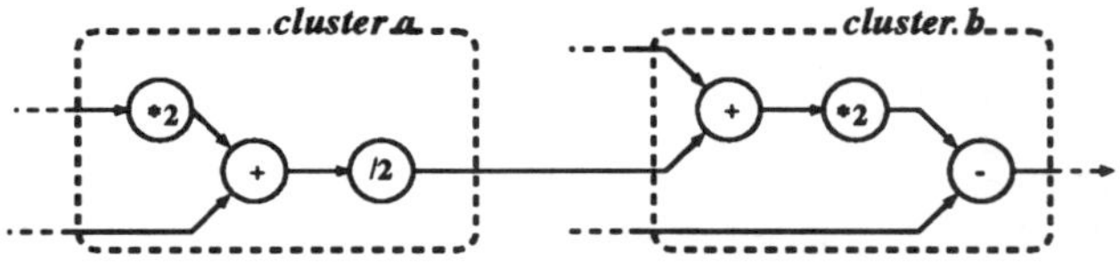

Figure 1: *Part of a signal flow graph in which the operations have been grouped into clusters.*

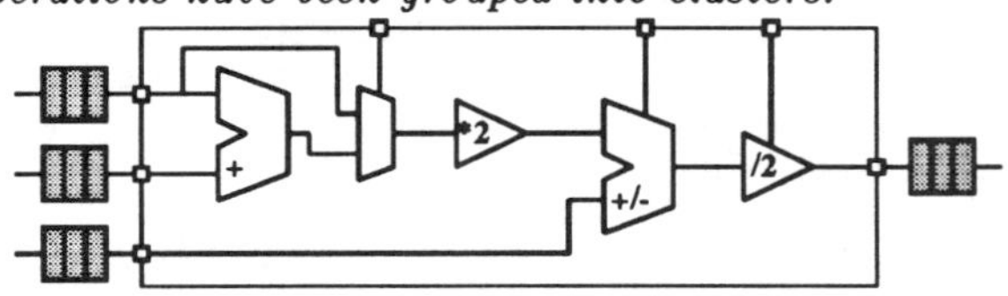

Figure 2: *Application specific unit for the clusters of figure 1. The inputs and outputs are connected to memories. The control signals for the multiplexor, adder/subtractor and second shifter impose the correct mode.*

more clock cycles ($d_c \geq 1$) and a latency of zero or more pipeline sections ($l_c \geq 0$). We are currently investigating CAD techniques for this step.

2 In the **ASU Assignment** step, the clusters are assigned to ASUs in such a way that the total ASU area is minimized under the constraint of a given global cycle budget and I/O constraints. This problem will be addressed in detail in this paper.

3 In the **ASU Definition** step, the internal composition of each ASU is constructed, so that it can execute all clusters assigned to it. For this purpose, the *constructive matching* technique of [4], is used.

3 The ASU Assignment Problem

The goal of the ASU assignment problem is to assign each of the N_{cl} clusters to exactly one of the N_{ASU} ASUs. A naive search in which the area of each of the $N_{ASU}^{N_{cl}}$ possible assignments is evaluated is clearly out of the question. We will therefore break the problem up into two steps. In a first step, a heuristic objective function based on the *pairwise compatibility* of clusters is constructed. In the second step, this objective function is optimized by means of a global optimization technique.

A first version of this technique has already been published [8]. This paper contains significant extensions to the cost functions and to the modeling of block hierarchy. Moreover, some experimental results will be presented.

3.1 The Objective Function

The first step for the modeling of the cost is the construction of a *compatibility graph* (figure 3). This graph is complete, its vertices represent clusters and the edges are weighted with a measure $C_M(i,j)$ which indicates the pairwise compatibility of the incident vertices i and j. This compatibility measure expresses the hardware area *overhead* that is encountered when two clusters are assigned to the same ASU. It is zero if the two clusters are identical (i.e. there is no overhead), and becomes larger when the clusters are more dissimilar. Both the internal composition of the clusters and their embedding in the signal flow graph is taken into account.

Internal compatibility measure: The internal compatibility, $C_{MI}(i,j) = 2 \times C_{ij} - C_i - C_j$, of two clusters i and j is defined as the difference between the area C_{ij} of an ASU capable of executing both clusters i and j, and the area of two dedicated ASUs for i and j. This measure determines the area of the ASU the two clusters are assigned to. The area C_i (C_j) of the dedicated ASUs is the area of the FUs needed to perform the operations of cluster i (j). In order to compute C_{ij}, we must find the cheapest set of resources that can execute both clusters i and j. A detailed description of the computation of C_{ij} is given in [9].

Embedding compatibility measure: The embedding compatibility $C_{ME}(i,j)$ determines the area of the inter-ASU interconnections. The underlying observation for the modeling of interconnect area is that it is more opportune to assign two clusters to the same ASU if:

- Some signals produced or consumed by the clusters have the same source (e.g. if they come from the same memory port). This decreases the required number of multiplexor inputs.
- There is communication between the two clusters. This changes a global (chip level) interconnection into an interconnection local to the ASU.

More details on how to compute $C_{ME}(i,j)$ can be found in [9].

The total compatibility measure is given by: $C_M(i,j) = C_{MI}(i,j) + C_{ME}(i,j)$.

In the second step, we will now formulate the ASU assignment problem as a partitioning problem. The N_{cl} vertices of the compatibility graph must be partioned into N_{ASU} disjoint subsets. The assignment of cluster i to ASU (subset) s is modeled by a boolean variable $x_{i,s}$.

$$x_{i,s} = \begin{cases} 1 & \text{if } i \text{ is assigned to ASU } s \\ 0 & \text{otherwise} \end{cases}$$

The objective function which has to be minimized is the sum of the weights of all edges that do not cut across the boundaries of a partition:

$$\min \sum_{s=1}^{N_{ASU}} \sum_{\forall i-j} C_M(i,j) \cdot x_{i,s} \cdot x_{j,s} \qquad (1)$$

This function is an upper bound for the overhead involved in multiplexing N_{cl} clusters on N_{ASU} ASUs [9].

We will now present the constraints for the optimization process. First, we will present the basic constraints that apply to each block of the hierarchy. Then, the necessary constraints for the modeling of the block hierarchy will be given.

3.2 Basic Constraints for a Single Block

Model without precedence constraints: Two types of constraints are needed for a single block: first, each cluster should be uniquely assigned to a single ASU; and second, the number of clusters belonging to a block that are assigned to the same ASU should not

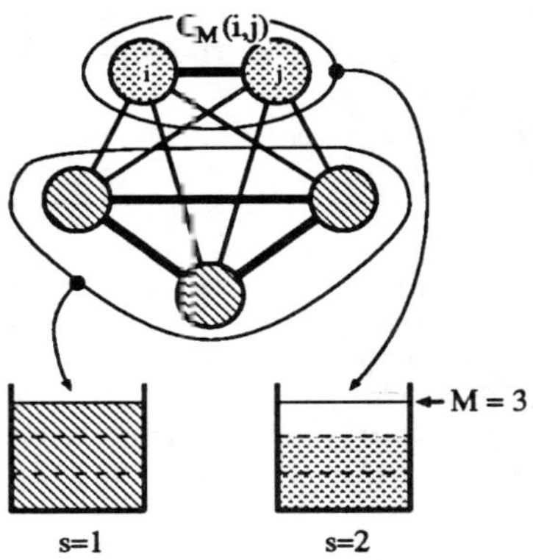

Figure 3: *Model for the cluster to ASU assignment based on pairwise compatibility of clusters.*

exceed the available cycle budget M_B for the block.

$$\sum_{s=1}^{N_{ASU}} x_{i,s} = 1, \quad i = 1 \cdots N_{cl} \tag{2}$$

$$\sum_{\forall i \in B} d_i \cdot x_{i,s} \leq M_B, \quad s = 1 \cdots N_{ASU} \tag{3}$$

Constraint model with precedences: In order to take precedences into account, a time index has to be added to the assignment variable: $x_{i,s,p} = 1$ means that cluster i will be executed by ASU s at time potential p. Each cluster is given an earliest and latest potential S_c and L_c, relative to the beginning of the block they are nested in. With these basic ingredients the modeling of precedences can be done in a way similar to the traditional precedence constraints used in ILP scheduling formulations [3].

We believe though that the extra dimension added to the problem will make the approach intractable for industrial size problems. Furthermore, the modeling of precedences is not strictly required for the ASU assignment since scheduling will be performed later [4]. If it is discovered during scheduling that no feasible solution exists within the specified cycle budget, an additional design iteration is needed in which more ASUs are allocated or a different operation clustering is considered. It is our belief that such an iterative process will sooner lead to good results than the CPU intensive alternative of taking precedences into account.

3.3 Modeling of the Block Hierarchy

The hierarchy in the signal flow graph is modeled as a block structure where a block B contains (i) nested blocks $b \in B$ and (ii) nested clusters $c \in B$. Two types of blocks exist:

- **Conditions:** Clusters which are executed under the same condition are combined in one condition block. Different condition blocks which are mutually exclusive are encapsulated in a *condition merge block*. A condition block can contain loops, condition merge blocks and clusters.
- **Loops:** These can be either manifest (FOR) loops or data dependent (WHILE) loops. The body of a loop can contain other loops, condition merge blocks and clusters.

The blocks are executed one at a time. This corresponds to a single thread of control implementation. The relative ordering of the basic blocks (but not of the clusters) is done during the high level memory management step [4].

In order to distribute the user specified cycle budget for the complete application over the nested blocks, each block is characterized by two values: M_B is the number of cycles available for the clusters of block B; N_B is the number of cycles available for the complete block, i.e. for the clusters $c \in B$ and for the nested blocks $b \in B$. The number of cycles consumed by a block should not exceed the available number of cycles:

$$\sum_{b \in B} n_b \cdot N_b + M_B \leq N_B \ , \ \forall B \tag{4}$$

where n_b is the number of times a nested block is executed. Equation (4) holds for loops and condition blocks. For a conditions merge block K, equation (4) must be substituted by the following constraint, stating that the cycle budget for each of the condition blocks $k \in K$ should not exceed the cycle budget of the merge block:

$$N_k \leq N_K \ , \ \forall k \in K \tag{5}$$

Based on the properties of the signal flow graph and on decisions made during previous synthesis steps, tight minimum and maximum bounds for the N and M variables can be derived. The lower bound N^{MIN} for the cycle budget for a block may be given by:

- *The structure of the algorithm.* In case of a condition block, the *critical path* is a lower bound for N. For loops , the lower bound is given by the *iteration period bound* of the loop.
- *Allocation and assignment decisions that were taken during previous synthesis steps.* Decisions made at a higher level in the synthesis script (e.g. during memory management) may impose a constraint on the minimum number of cycles needed for a block.

The upper bound for the available cycle budget of a certain block N^{MAX} can be computed under the assumption that all other blocks take minimum time to execute. The smaller the slack, $N^{MAX} - N^{MIN}$, of a block; the smaller the search space for the global optimization algorithm becomes. Especially due to the many constraints from the preceding memory synthesis task in our practical applications, we believe that the search space is always severely restricted.

Based on the minimum and maximum bounds, two additional constraints can be added to the problem:

$$N_B^{MIN} \leq N_B \leq N_B^{MAX} \tag{6}$$

$$M_B^{MIN} \leq M_B \leq M_B^{MAX} \tag{7}$$

4 Complexity Analysis

The basis of our synthesis approach is a global optimization technique. In section 5 we will show that it is possible to apply this technique to industrial-size examples. This is true because the grouping of primitive operations into clusters gives a significant reduction in problem size. We will now indicate the complexity of

Applic.	N_{cl}	N_{bl}	N_{ASU}	HSF	CPU time st.1	CPU time st.2
VLSI	5	3	2	32	8	12
EX1	8	3	2	64	12	103
EX2	8	3	4	32	12	1400
FLAT	8	1	3	3	17	345
TREE	12	7	2	220	28	573
RD	8	3	2	138	43	135

Table 1: *Statistics of the examples. N_{bl} is the number of blocks in the application. The other entries are explained in the text. CPU times (in seconds) were obtained for a DECStation 3100.*

the two steps introduced in section 3 and discuss the possibilities to reduce the required CPU time.

Step 1 (Pre-processing): The construction of the compatibility graph involves the computation of $\mathcal{O}(N_{cl}^2)$ compatibility measures. This problem can be solved in $\mathcal{O}(N_{op/cl}^3)$ time [9], where $N_{op/cl}$ is the average number of operations per cluster.

Step 2 (ILP): The optimization problem composed of equations (1), (2), (3), (4), (5), (6), and (7) has a quadratic cost function, it contains zero-one variables as well as general integers and constraints of both of equality and inequality type. This problem is transformed into a pure zero-one linear program. The number of zero-one variables required to model the cluster to ASU assignment is of the order $\mathcal{O}(N_{ASU} \cdot N_{cl}^2)$. The modeling of the block hierarchy requires an additional number of zero-one variables which is linearly dependent on the number of blocks N_{bl} and on the number of bits required to encode the slack of each block: $\mathcal{O}(N_{bl} \cdot \log_2(N^{MAX} - N^{MIN}))$.

Pruning possibilities: From this complexity analysis we can conclude that the size of the search space is dominantly determined by the N_{cl}^2 factor which stems from the number of edges in the complete compatibility graph. The complexity can be arbitrarily decreased by the pruning of this graph. At the end of step 1, the edges which have a high relative weight (low degree of compatibility), i.e. those for which $C_M(i,j) > 1.5 \cdot \max(C_i, C_j)$, can be deleted. This directly influences the number of zero-one variables. The pruning is stopped when (i) an acceptable number of zero-one variables is obtained or (ii) when the risk of excluding good solutions arises. As a terminating heuristic we state that the degree of a vertex in the compatibility graph may not become less than N_{ASU}. The pruning should also be done evenly over all vertices of the compatibility graph, so that the degree of all vertices is approximately equal. An unbalanced graph would deteriorate the quality of the pairwise compatibility heuristic.

5 Experimental Results

The described technique, without the pruning strategy, has been implemented in a tool called CLASH. The statistics for a number of examples are shown in table 1. The VSLI example is taken from [8]. The EX1, EX2 and FLAT are relatively complex artificial examples which are shown to illustrate the complexity behavior. The TREE example is a quadtree composition algorithm, which is a submodule of a video codec. It takes as input an 8 by 8 pixel block of an image and encodes this in a quadtree. The RD example is an algorithm which detects regularity between two contour segments of an image.

6 Conclusions

In this paper, a technique has been presented for the allocation of dedicated lowly multiplexed datapaths in the architectural synthesis for high-throughput signal processing applications. The technique - based on zero-one quadratic programming - minimizes the total area consumed by the datapaths and performs this optimization across the boundaries of the blocks of a hierarchical signal flow graph. We have shown the feasibility of this global approach by means of several industrial examples. We have also indicated the opportunity for the pruning of the search space if CPU time becomes a problem.

Acknowledgements We would like to thank the members - present and past - on the CATHEDRAL research group for the many stimulating discussions, and the reviewers for their constructive comments.

References

[1] D. Lanneer e.a. "Architectural synthesis for medium and high throughput signal processing with the new CATHEDRAL environment" in *Trends in high-level synthesis*, editors R.Camposano and W.Wolf, Kluwer, 1991.

[2] N. Park, A. Parker "Sehwa: A Software System for the Synthesis of Pipelines from Behavioral Specifications" is *IEEE Trans on CAD* vol. 7 no.3, March 1988, pp. 356-370.

[3] J.H Lee, Y.C. Hsu and Y.L. Lin, "A New Integer Linear Programming Formulation For The Scheduling Problem in Data Path Synthesis" in *Proc. ICCAD-89*, Santa Clara, Calif., Nov. 1989, pp.20-23.

[4] S. Note, W. Geurts, F. Catthoor, H. De Man "CATHEDRAL-III: Architecture-Driven High-Level Synthesis for High Throughput DSP Applications" in *Proc. DAC-91*, San Francisco, Calif., 1991.

[5] T. A. Ly, W. L. Elwood and E. F. Girczyc "A Generalized Interconnect Model for Data Path Synthesis." in *Proc. DAC-90*, June 1990, pp. 168-173.

[6] N. Park and F. Kurdahi "Module Assignment and Interconnect Sharing in Register Transfer Synthesis of Pipelined Data Paths" in *Proc. ICCAD-89*, Santa Clara, Calif., Nov. 1989, pp. 16-19.

[7] J.M. Rabaey, C. Chu, P. Hoang and M. Potkonjak "Fast prototyping of Datapath-Intensive Architectures." in *IEEE Design and Test of Computers*, June 1991, pp.40-51.

[8] W. Geurts, S. Note, F. Catthoor, H. De Man "Partitioning-Based Allocation of Dedicated Data-Paths in the Architectural Synthesis for High-Throughput Applications." in *Proc. VLSI91 Conference*, Edinburgh, Scotland, Aug. 1991.

[9] W. Geurts *Synthesis techniques for lowly multiplexed datapaths.* IMEC internal report ASCIS/IMEC/m30/C3/4, available upon request.

SESSION 8

PANEL: IS TECHNOLOGY INDEPENDENT DESIGN REALLY PRACTICAL?

Chair: Peter Hillen - Congruity, Palo Alto, CA
Organizer: Tom Pennino - AT&T Bell Labs., Holmdel, NJ

We all dream of the day when a design can be done at an abstract architectural level and then with the push of a button it is converted to the real world silicon implementation of our choice. Today this is not just a pipe dream for designers, it is a real world imperative, demanded by the pressures of shrinking time to market, increased end product functionality and more complex silicon architectures. This panel will examine the practicality of technology independent design using currently available technology (tools and ICs) and what the future should and will hold from both a user and vendor perspective.

Questions about technology independence necessarily focus around logic synthesis. Gate arrays, PLDs and now FPGAs provide design engineers a broad selection of logic design realizations. This has resulted in a proliferation of solution-specific design processes and tool sets supplied by semiconductor and third party vendors. Each of them have their own unique algorithmic approach to solving the synthesis problem. PLD synthesis has taken advantage of the PLD's relatively simple structure and has used exhaustive fitting of the optimized netlist to the available devices. ASIC synthesis (primarily Gate Array) has taken advantage of the large number and regular layout of gates.

FPGAs present a whole new synthesis challenge. They offer the enticement of field programmability to system designers but are burdened with exotic architectures which seem more a convenience to silicon circuit designers. To date there are three popular and uniquely different FPGA architectures, so a designer can't just specify a generic FPGA. Re-targeting a design either among PLDs, FPGAs or Gate Arrays or from one vendor to another has meant re-synthesis of the design, often on a different set of tools, a major endeavor.

The newest logic synthesis tools promise to relieve the designer from the burden of efficient technology implementation through automatic synthesis and optimization of the design followed by automatically mapping it to the targeted technology. They are the "switching yard" for taking an architectural concept and creating a physical realization. Their targeting ability also introduces technology and vendor independence and re-targetability of designs.

Synthesis also facilitates top-down design by being the CAD link from higher level descriptive abstractions like VHDL and the corresponding physical realization. However, given the radical differences in existing targeted synthesis, it remains to be seen whether these new "universal" or "modular backend" synthesis tools can deliver what they promise.

Even though the panel will focus on synthesis, other levels of the design process are also important for technology independent design. They are fair game for discussion too. Can a HDL description really be blind to the target technology? Is there an "art" to the way a design is coded into a HDL? Will the designer always be involved at the conceptual level in directing his design to the target technology? How can libraries help? Do they need to be universal across all physical implications in order to be effective?

This panel is made up of users, CAD vendors and semiconductor manufactures. Against the background of a promising but young targetable synthesis technology, we will have a spirited discussion of what is reality today and what might become reality in the future.

Panel Members:

Brian Box - Lockheed Sanders, Hudson, NH
Rob Mendes da Costa - Mentor Graphics Corp., Wilsonville, OR
Roger Neiderland - McDonnel Douglas, Santa Ana, CA
Ravi Ravakumar - Cadence Design Systems, Inc., Santa Clara, CA
Tim Teckman - Intel Corp., Hillsboro, OR
Steve Trimberger - Xilinx Inc., San Jose, CA

An Engineering Environment for Hardware/Software Co-Simulation

David Becker, Raj K. Singh, Stephen G. Tell

Department of Computer Science
University of North Carolina at Chapel Hill

Abstract

We describe an environment supporting concurrent hardware and software engineering for high performance systems. In place of a conventional bread-boarded prototype, we used distributed communicating processes to allow software and simulated hardware to interact. We extended the Cadence Verilog-XL simulator to enable software debugging and testing using hardware simulation. The environment was proven during a successful system design.

1 Introduction

Many applications demand system designs with high performance packed in a small space. Prototyping practices such as wire-wrapping or bread-boarding cannot handle the speed and density of these high performance systems. Even the first prototype must be a printed circuit board. Debugging an assembled board is time consuming and difficult. Modifications are often impossible without building a new board. Hardware simulation handles these challenges with tools that analyze the design at speed and account for the effects of physical layout on signal integrity. A designer can expect the resulting system to be as good as the simulation of that system.

An accurate simulation cannot ignore the software components of a system. A design may require firmware for an embedded microprocessor or might interact with software on existing systems. Thorough hardware simulation needs the developing software to exercise all parts of the design and, ideally, the system software should be developed in parallel with the hardware. Developing software using a hardware prototype fails on both counts so a method is needed to exercise the simulated hardware with the real software.

We created a co-simulation engineering environment that links the software components of a system to the simulation of the hardware components. Using Unix networking facilities and the Verilog-XL simulator from Cadence Design Systems, the software components can control and react to events within the hardware simulation. This facility was applied to development of software for a high-speed custom communication interface, called a Network Interface Unit (NIU)[1].

The hardware and software components of the NIU were developed concurrently. The NIU hardware design was simulated with a Verilog model and exercised by the evolving software components, which in turn were tested by running them on the simulated hardware. The resulting circuit board worked at its initial power-up and the software performed as expected.

Before describing the simulation interface, a brief description of the NIU system is presented in section 2. The concurrent engineering environment is discussed in section 3 and two implementations are discussed in section 4. We conclude with the results of our effort and discuss possible directions for future work.

2 NIU system overview

The NIU is a custom high speed intelligent communication interface for a gigabit B-ISDN network[2]. The system interfaces a 800 Mbit/s HIPPI[3] link to the backplane ring network of the Pixel Planes 5 (PXPL5) heterogeneous graphics multicomputer[4]. The ring backplane is connected to the NIU and other boards through a "ring board" providing two 640 Mbit/s message-passing ports. A Cypress CY7C611 SPARC integer processing unit supervises the inbound and outbound hardware data pipelines and handles all protocol processing for the HIPPI packets.

Figure 1 shows the NIU software components and how they fit into the hardware design. The processor firmware responds to interrupts from the data pipelines and supervises the data flow with control and status registers. Monitor software, executing on the PXPL5 host workstation, communicates with the NIU processor by passing messages over the ring backplane.

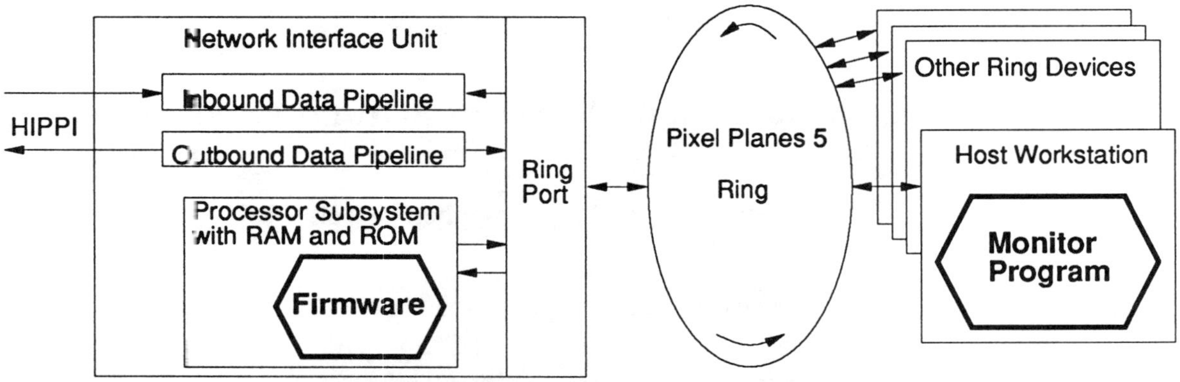

Figure 1. The NIU software components and their relation to the hardware.

The completed NIU circuit board consists of 619 parts on a 10-layer board connected by 1351 signals. The hardware model was written in 4000 lines of Verilog code. An additional 6000 lines of Verilog code modeled the logic devices including 60 PLDs. The two software components totalled 2000 lines of C++ code.

3 Simulation architecture

The goals we had for co-simulation were to be able to write the majority of the NIU software, minimize the changes needed when the real hardware arrived and use a minimum of project resources. The first solution considered was integrating our software into the simulator as user extensions to the Verilog simulator. Preventing the software components and the simulator proper from interfering with each other would have been very difficult. Another approach considered was purchasing or writing a complete model of the SPARC processor for the hardware simulation. Verilog could then simulate all the firmware but such models are expensive. This solution would not address development of the host monitoring code.

Our solution implements the software components as separate programs that use Unix interprocess communication (IPC) mechanisms to interact with the hardware simulation. The co-simulation environment is a layer of software below the system's software and hardware simulation components that encapsulates the interprocess communication links among the components. This solution does not accurately simulate the relative speeds of the hardware and software components, but we deemed this limitation acceptable in our application.

One key to creating a link between a software component and the hardware is selecting the appropriate level of abstraction for the interprocess communication. Our environment places no constraint on IPC message formats, but we find it best to model them after low-level software/hardware interaction, e.g. "put value A in register X." A hardware level model with messages like "signals x1, x3, x4 are asserted this clock cycle," would require significant simulation-specific software. Similarly, a higher level of abstraction with messages like "shut down pipeline", would increase the amount of software changes needed to switch to real hardware.

This link between the software component and hardware component has three parts. First, the *hardware interface functions* of the software component are rewritten to communicate with the simulation. Second, a *simulation module* must be modified to handle the communications with its corresponding software component. These first two parts are the end points of a link and must be customized for each link. The final part are the *simulator extensions* which connect the end points of the link. These extensions allow simulator modules to communicate over a TCP connection to software components.

The NIU co-simulation contains two instances of this technique: one link for the firmware and another for the host monitor software components. Figure 2 illustrates our environment as applied to the NIU co-simulation. The next section will describes how all parts of the software/hardware links were implemented in the NIU co-simulation.

4 Interface implementation

The NIU host monitor link has simple examples of the two parts of the co-simulation environment specific to each link: the hardware interface functions and the simulation module. The NIU processor firmware is a more complex example. The next two subsections will describe these two co-simulation links and examine the considerations in designing both the communication model and the interface at each endpoint.

Later, we describe the implementation of the IPC simulator extensions, which are largely independent of the interface issues handled in the other two components. Our original implementation of this environment required an IPC extension customized for each co-simulation link. The

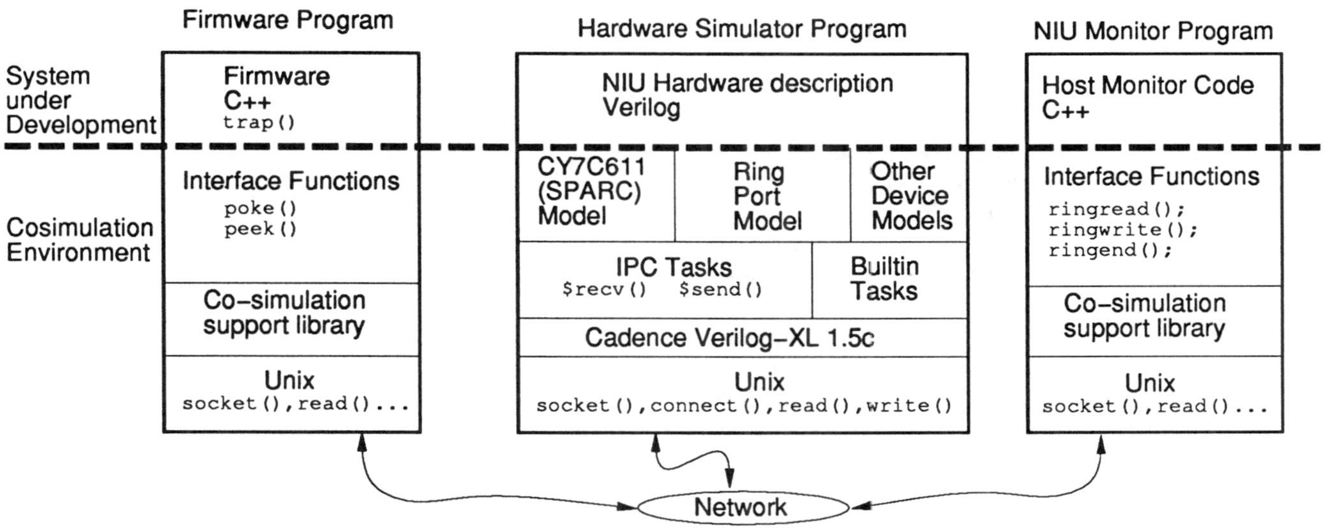

Figure 2. Components of the NIU co-simulation environment.

current environment employs a general-purpose simulator extension that is reused for all co-simulation links.

4.1 NIU host monitor code

The NIU and the other boards in Pixel Planes 5 send messages to each over a high speed ring. The ring packets are of unlimited length, sent a word at a time with a data valid signal and completed by an end of message signal. The IPC message format used by the host monitor co-simulation link reflects how the real ring packets function. Messages to the simulator are "put X in the write register and assert data-valid" and "assert end of message signal". Return messages are "X was written to the read register" and "end of message".

The format for a host monitor message is two words long where the first word is the message type identifier and the second is a data word. A simulated ring packet is sent by successively sending messages of type WRITEDATA for each data word in the ring packet. The hardware interface function that writes these data words to a register is replaced with a function that sends WRITEDATA messages to the simulation followed by a WRITEEND message. When the host expects a ring packet from the NIU, it waits for READDATA messages followed by a READEND message from the simulator that together form the simulated ring packet.

IPC messages are sent and received by a Verilog simulation module that represents the ring board connector in the hardware description. This module checks each clock cycle for an incoming IPC message and when one arrives, it asserts the appropriate signals on the simulated NIU ring connector pins. If the module sees outgoing data on the connector, it sends a message containing that data back to the host monitor program. The Verilog module is responsible for the correct signal timing on its input and output wires.

```
module ringconnector (/*...*/);
reg [31:0] ringserv_fd, msg_type, msg_data;
initial $makeserver(ringserv_fd, 7000);
always @(posedge CLOCK) begin
   $recv(ringserv_fd, msg_type, msg_data);
   if (msg_type==WRITEDATA)
     /* put signals for msg_data on connec-
     tor wires */
   else if (msg_type == WRITEEND)
     /*raise end of packet connector wire*/
   else if (msg_type == 0)
     /* no message; set connector write
     wires accordingly */
   if /* outgoing dataword valid */
     $send(ringserv_fd,READDATA,dataword)
   else if /* outgoing end wire raised */
     $send(ringserv_fd, READEND);
end
```

Figure 3. Ring connector module managing IPC messages

The host monitor software component of the NIU system was written using this co-simulation interface to exchange messages with simulated hardware. When the real hardware arrived, the functions to read and write words of a ring packet were rewritten to use the real hardware registers rather than send IPC messages. The rest of the code remained the same and ran correctly since it had already be tested.

4.2 NIU processor firmware

The processor firmware controls the NIU data pipelines through several control, status, and data registers mapped into the processor address space. Also, several data pipeline events interrupt the processor. In co-simulation, the firmware sends messages to the simulator of the form "poke X into address A", "peek what is in address A" or "trap handler done". The simulator sends messages of the form "address A holds X" or "interrupt X has occurred".

The firmware program uses the hardware interface functions `poke()` and `peek()` for all operations on memory-mapped I/O registers. These two functions are written to send IPC messages during co-simulation and are replaced with simple macros when using real hardware. The processor firmware has a function called `trap()` for trap handling, where the traps of interest are the hardware interrupts. In the real system, `trap()` is called from assembly language via the trap vector table. In co-simulation, an interface function checks for incoming messages and calls `trap()` asynchronously.

The simulator module of this link is a behavioral model of the SPARC processor's internal pipeline which generates accurate control signals for testing the memory and I/O devices on our board. Our model normally fetches instructions from memory, but treats all instructions as no-operation (NOP) instead of decoding fetched instructions. The decode stage of the pipeline checks for IPC messages from the firmware. If a POKE or PEEK message is received, a store or load cycle is executed instead of NOP. When a load cycle completes, a PEEKREPLY message with the result of the load is sent back to the firmware program. When external hardware asserts the interrupt lines of the processor model, an INTERRUPT message is sent to the program indicating which interrupt occurred. When the firmware returns a TRAPDONE message, a return from trap cycle is simulated.

An INTERRUPT message needs to interrupt the firmware asynchronously to allow trap handling to be realistically simulated. We used the Unix signal mechanism to duplicate this asynchronous behavior. The co-simulation support library linked with the firmware program requests Unix to send a signal when a message arrives from the simulator. This signal interrupts the firmware program and moves execution to a signal handling function which is part of the support code. This handler reads an INTERRUPT message and calls the firmware's trap handling function. If a PEEK is in progress, the signal handler waits for the PEEKREPLY before calling `trap()` because PEEK and PEEKREPLY comprise a single indivisible instruction. When the trap handler completes, a TRAPDONE message is sent to the simulator and Unix moves firmware execution back to where it was interrupted.

The most complex part of the firmware co-simulation was writing the processor pipeline model. The pipeline model, however, would also have to be written for other simulation strategies and only a small part of it is concerned with sending and receiving IPC messages. The bootstrap code and assembly language parts of trap handling were written after the board arrived. The firmware was linked to this assembly and ran on the new hardware just as it did on the simulation, only a bit faster.

4.3 Communication through the Verilog simulator

The Verilog hardware description language does not incorporate any interprocess communication facilities. It does have a Programming Language Interface (PLI) that allows user written C or C++ code to be called from within the simulation[5]. We used this PLI facility to add extensions to the simulator so the remote software programs could make a TCP connection to the simulator program and communicate with modules within the simulation. The first strategy we used involved specialized extensions for each co-simulation link. From that experience, we devised a general mechanism that we feel is much easier to use and describe.

The Verilog PLI facility associates a user written subroutine with a Verilog task name. Invoking a user written task in a Verilog program will cause the simulation to call the C++ function associated with that task name. These functions are called under several circumstances. One possible configuration is to have the C++ function called whenever an argument to the Verilog task changes during simulation. Our first communication mechanism is based on this form of user function.

Two tasks were added to Verilog in the original solution: `$sparc()` and `$ring()`. In the processor module, the pipeline executes NOP instructions unless the `$sparc()` task signals through one of its arguments that the firmware requests execution of an LD, ST or RETT instruction. The Verilog module and `$sparc()` communicate through a small set of arguments, shown in Figure 4.

```
initial begin
   // start up the SPARC interface task
$sparc(data, address, read_inst,
   write_inst, rett_inst, ResetL,
   cycle_done, PSR, traptype, Clock);
end
```

Figure 4. Parameters passed to `$sparc()` by the `cy7c611()` module.

Since the clock signal is an argument, the C++ code associated with `$sparc()` is called every clock cycle.

During each cycle the C++ code can read and write the signal lines connected to it. On the initial cycle, it connects to the firmware process. When the reset signal is negated, a RESET message is sent to the firmware process. On subsequent cycles, `$sparc()` checks for messages from the firmware process. The C++ code sets the `$sparc()` parameters to the instruction, address and data as needed for that cycle. When the cycle completes, `$sparc()` is signalled with the results so the C++ code can send a message back to the firmware process. With the exception of the clock, these arguments do not represent any actual electrical signals, but are only a communication mechanism between the simulation module and C++ code.

The ring port simulation module functions in a similar fashion, associating `$ring()` with its C++ code. When `$ring()` is initially called, it creates a socket to which the monitoring program can connect. Thereafter, on every clock cycle `$ring()` checks to see if the command program has connected. Once connected, `$ring()` relays messages between the simulated hardware port and the monitor program. When a WRITE message is received, it begins clocking the message into the port one word per cycle. When the hardware sends data to the port, `$ring()` stores it until the end of message signal is asserted. At the end of the message, a READ message is sent over the socket to the monitoring process.

A custom addition to the Verilog simulator for each link had several problems. One was writing a third piece of custom code for each link and another was the difficulty of modifying the simulator program. Further, the original C++ code performed some link-specific operations on the messages which were easily moved into Verilog or deemed unnecessary. Verilog is more suited for checking signals each clock cycle and sending acknowledgment signals. C++ is suited for making the networking system calls, so our new implementation extends Verilog with tasks only for generalized interprocess communication (IPC). The form of task/user function association used with these tasks arranges for the C++ function to be called every time its task is invoked during simulation.

The IPC facility added to Verilog allows modules to create TCP connections to remote Unix programs during the simulation. The `$makeserver()` task creates a Unix socket at the specified TCP port on the machine running the simulation. The software component of the co-simulation can connect to the simulation by using the IP address of the computer running Verilog and the TCP port number being served.

The `$send()` and `$recv()` tasks communicate with programs that connect to the TCP ports set up by `$makeserver()`. The send task takes a variable number of arguments, all of which are put into a packet and sent to the remote program. If the server is not connected to a remote program, the `$send()` call is ignored. Each call of the `$recv()` task checks for an arriving message. If a message has arrived, the arguments to `$recv()` are filled with the message data. When no remote program is connected or no incoming data is waiting, `$recv()` sets its first data parameter, conventionally used for a message type identifier, to zero and ignores the other parameters.

Some software components need to know the state of the simulation when they connect. The `$connect()` task returns true if the associated server is connected to a remote program. Start-up messages can be sent when the connection is established. The `$disconnect()` task terminates the current connection of the server and can be preceded by shutdown messages. Only one remote program can connect to a server at a time. After a server disconnects, a new connection can be polled for with the connect call. This mechanism is used by our processor co-simulation to start the firmware process when the processor reset signal is negated.

The software developed for the co-simulation environment includes Verilog simulator extensions, Verilog simulation modules and modified versions of the hardware interface functions. The Verilog modules for the processor and ring port were moderately complex, requiring about 700 lines of commented code for behavioral simulation. Only a small portion of this code is concerned with interprocess communications; the rest would have to be written for other simulation strategies as well. The IPC tasks added to the interpreter were written in about 200 lines of commented C++ code and the simulation interface functions for the software components required 300 lines of commented C++ code.

5 Conclusions and future work

We have developed an environment for the concurrent design and implementation of software controlled hardware systems. Independent processes executing the controlling software use interface function to connect to Verilog models of the hardware via custom code linked into the simulator. With this connection, the Verilog models of the software controlled devices can exhibit their high level behavior during simulation. Concurrent development allowed hardware and software division of labor to happen on the fly as the trade-offs became apparent. This should be especially useful for developing systems with embedded software components.

This environment was an important factor in the successful development of a high speed network interface circuit board, its processor firmware, and the command software running on a host computer. The final simulations of the hardware design were driven by completed software components. To prepare for the actual hardware, the software interface functions were recast from simulated form to hardware form. After these had been debugged with the delivered hardware, the entire system performed as expected.

Other models of communication may prove useful. Perhaps all the signal changes on the interface could be sent over the network for handling by other types of programs or other Verilog interpreters. Another potential direction is to add a thread package to the simulator and run the software processes as separate threads in the same process instead of running them as independent processes. This could simplify the communication and make the communication timing more predictable. It may also be possible to direct unmodified hardware interface functions to unmapped memory and handle the resulting page faults by emulating the I/O instruction with co-simulation messages. Another suitable application of the co-simulation would be to link a software microprocessor emulation to the board level Verilog simulation.

Acknowledgments

We would like to acknowledge the support of Cadence Design Systems, Inc. for providing us with the Verilog-XL software under their university program. The work reported in this paper is supported, in part, by the National Science Foundation and the Defense Advanced Research Projects Agency under cooperative agreement NCR 8919038 with the Corporation for National Research Initiatives. Support is also provided by Bell South and GTE Corp.

References

1. Raj K. Singh, S.G. Tell, and D. Becker, "VISTAnet Network Interface Unit: Prototype System Specifications," TR91-017, Department of Computer Science, University of North Carolina at Chapel Hill, 1991.

2. B. E. Basch, "VISTAnet: A B-ISDN Field Trial," *IEEE-LTS*, pp. 22-30, Aug 1991.

3. H. Fuchs, "Pixel Planes 5: A Heterogeneous Multiprocessor Graphics System Using Processor Enhanced Memories," *Computer Graphics*, vol. 23, no. 3, pp. 79-88, 1989.

4. X3T9.3 Task Group, "HIPPI Framing Protocol Specification (HIPPI-FP)," Revision 2.8, American National Standard for Information Systems, 24 June 1991.

5. Verilog-XL Version 1.5c, *Reference Manual*, Cadence Design Systems, June 1990.

High Level Synthesis of Pipelined Instruction Set Processors and Back-End Compilers

Ing-Jer Huang and Alvin M. Despain
Department of Electrical Engineering - Systems
University of Southern California
Los Angeles, CA 90089-2561
ijhuang@usc.edu, despain@usc.edu

Abstract

Designing instruction set processors and constructing their compilers are mutually dependent tasks. Piper is a high level synthesis tool of ADAS which controls the hardware-software interactions at the micro-architecture level. The key function of Piper is to perform pipeline scheduling with any fixed instruction-latency, and generate a reorder table for a back-end compiler to resolve potential pipeline hazards existing in the design. Models for performance and cost of both hardware and software are developed in order to characterize the design space. A synthesis example of a simple instruction set processor illustrates Piper's synthesis capabilities and how the performance and cost of hardware and software are estimated.

1. Introduction

Pipelining is a technique to improve the throughput of digital systems. Throughput is improved by overlapping the execution of successive iterations. The time delay between the issues of consecutive iterations is called *latency*. However, two major problems limit the power of pipelining. First, *loop carried dependencies* (LCD's) [7], the data dependencies between different iterations, prevent the next iteration from being fired right after the issue of the current iteration (latency of 1). In instruction set processor specifications, LCD appears as the inter-instruction dependency. A *pipeline hazard* is introduced when the LCD prevents a successive instruction from being executed at its designated cycle. A design automation (DA) system targeted at pipelined architectures has to construct a proper data path structure and a control scheme so as to ensure the satisfaction of LCD's. Secondly, *conditional branching*, the conditional execution of operations within the loop body, further complicates the detection of existences and lengths of LCD's. General approaches to these problems are either to find out the *minimal achievable latency* (MAL) [7] and issue iterations with latency no less than MAL, or to issue an iteration at a latency of one, but stall the hardware as soon as the pipeline hazards are detected. Some DA systems choose to restrict their applications to algorithms without conditional branches, whereas some other DA systems perform the worst case analysis by expanding traces with all possible branches to find the LCD's with longest temporal distance which possibly happen. Such solutions can be found in DA systems for DSP applications [2][5][6][7]. The basic limitations of these approaches are that the idle cycles (bubbles) are wasted, and that the synthesized hardware is tuned to worst cases which may rarely happen. We will discuss these systems in Section 2.

One area which has not received much attention from researchers in design automation for instruction set processors is to construct the back-end compilers for the machines generated. The decisions made in the micro-architecture implementations may affect the implementations of some phases in back-end compilers. For example, the typical 'delayed load' and 'delayed branch' are caused by the pipeline hazards in the pipeline structure, and require a phase *reorderer* in the back-end compiler to reorder the instructions or insert *nop*'s to ensure the correctness of the instruction behavior and improve the performance.

In this paper we propose a synthesis methodology for high-performance pipelined instruction set (IS) processors which out performs the current DA systems in that it is capable of making the most use of the idle cycles with the presence of LCD's and conditional branching in the instruction set architecture (ISA) specification. Piper also distinguishes itself from the other DA systems in that it bridges the semantic gaps between hardware and software at the micro-architecture level.

The performance (execution time) and cost (memory usage) of the reorderer depend on the characteristics of the reorder table which is subject to the pipeline synthesis decision. Therefore, we are able to add an additional dimension to the design space such that designers can explore instruction-level parallelism between hardware (synthesized IS processors) and software (reorderer). We have developed a model for determining the performance and cost of both hardware and software to encapsulate the qualitative interaction between them. Instruction benchmarks are used to evaluate the run time performance of pipeline.

This synthesis methodology and performance/cost model have been implemented in a high level synthesis system Piper. It is the behavioral domain component of the Advanced Design Automation System (ADAS) which transforms an abstract description of instruction set architecture into a mask layout [1] [13].

The rest of this paper is organized as follows. Section 2 reviews the related work. Our approach to pipeline synthesis with any fixed instruction-latency is described in Section 3. An algorithm to construct the reorder table for the back-end compiler (reorder) is presented in Section 4. In Section 5 the model of the performance and cost for the pipelined instruction set processors and their reorders is described. Our techniques are illustrated by a synthesis example of a simple instruction set processor *sml* and compared to other DA systems in Section 6. Conclusion and future work are presented in Section 7.

2. Related work

In the pursuit of high performance in digital circuit designs, many high level synthesis systems focus on automating the design of pipeline structures. Several pipeline scheduling algorithms have been developed. These algorithms differ in their application scope and achieved performance. SEHWA uses two algorithms: feasible scheduling and maximal scheduling [2]. HAL applies a modified force-directed scheduling algorithm to achieve better quality of pipeline scheduling and allocation at the cost of higher complexity [5]. These two systems do not work with applications which exhibit loop carried dependencies. PLS and CATHEDRAL II are capable of working on applications with LCD's by means of an iterative folding algorithm [7] and loop folding algorithm [4], respectively. Both systems adopt an iterative approach to find the MAL. This MAL then serves as the upper bound of the performance of the synthesized designs. The major application domain of the above synthesis systems is DSP.

ASPD further improves the pipeline scheduling performance by applying an enhanced percolation scheduling algorithm and using a wide-instruction-word (WIW) architecture as a template [6]. ASPD is capable of synthesizing WIW-based processors with instruction latency less than MAL at the cost of control for flushing pipeline when the pipeline hazards are detected. However, the practicality of ASPD is limited by the facts of microcode explosion and the complexity of pattern merging.

The System Architect's Workbench consists of two synthesis methodologies: one is tuned specifically to microprocessor designs, and the other supports general design style [8]. In the latter approach, pipelined designs are derived by interactively performing transformations at the system level. However, it does not attempt to solve the pipeline hazard automatically.

The difference between Piper and these systems is that Piper focuses on instruction set processor designs, and further improves the performance of the pipeline structures by considering simultaneously the synthesis of hardwares and the construction of back-end compilers.

In compiler designs, there has been some work in machine-independent code generation and optimization to automate the generation of code-generators and -optimizers by separating the machine-dependent information from the machine-independent algorithms. The machine-dependent information is derived from a machine description. In most systems the machine-dependent information is tabularized. Cattell's and Graham's works focus on the generation of code-generators [10][11], while Giegerich's work is on the derivation of machine-specific optimizers [12]. They derive their machine-dependent tables from instruction set semantics which describes the instruction formats, addressing modes, and operations. However, the micro-architectural features such as pipeline configuration are not considered in these systems. Piper is different from these works in that it focuses on the construction of back-end compiler phases which are related to the micro-architecture. Currently, the focus is restricted to generating the reorderers for instruction set processors with various pipeline configurations.

3. Pipeline synthesis with any fixed instruction latency

It is commonly found in the designs of modern pipelined microprocessors that a microprocessor operates with the instruction latency shorter than the MAL. For example, in the MIPS R2000/R3000 microprocessor which operates with instruction latency of one, the minimal achievable latency found in the data/control flow is two (read after load from memory, for example) [9]. The processor does not flush its pipeline, nor does it insert *nop*'s itself in order to resolve pipeline hazard. The optimizing compiler is thus responsible for inserting *nop*'s or reordering the instructions when such patterns of instructions are found. Compared with previous approaches in Section 2, this approach has higher performance and lower cost because of the shorter instruction latency and the absence of the flush/stall circuitry. However, a phase in the compiler has to be constructed to properly reorder instructions or insert *nop*'s, and, therefore, the time and space complexities are increased in the compilation process.

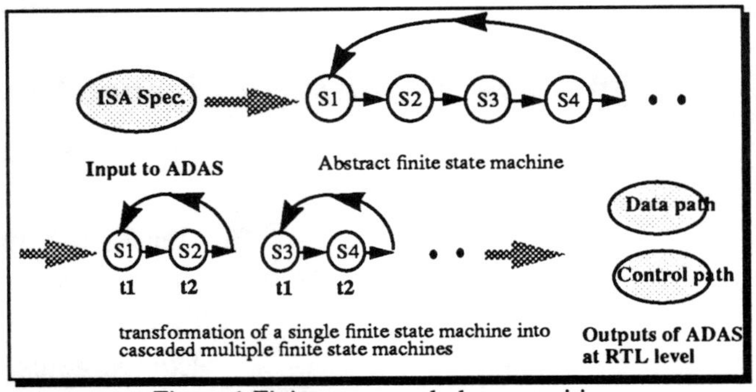

Figure 1 Finite state graph decomposition

We adopt this approach in Piper to achieve high performance comparable to the manually-designed pipelined processors. This approach is generalized in Piper such that scheduling with any fixed instruction latency is supported, regardless of the MAL. This allows designers to explore design spaces ranging from a non-pipelined design to a maximal throughput pipeline (instruction latency of one). The scheduling process of Piper begins with a global data-flow analysis for intra-instruction and inter-instruction dependencies. It tags the conditional branches and associates the tags to the dependencies. The scheduler of Piper then schedules the micro-operations according to the desired instruction latency given by the designers. After the micro-operations are scheduled, the data path and control path can be constructed.

Note that this approach is independent of the scheduling algorithm used by the synthesis system. It lifts the MAL constraint such that scheduling algorithms are free to schedule with any fixed instruction-latency. This enhances the capability of the scheduling algorithms, and provides more flexibility for the designers.

In the current implementation of Piper, the pipeline scheduling is done in the stage partitioning phase by decomposing an abstract finite state graph generated by Fiper, the preprocessor of Piper, into several smaller finite state graphs, according to the desired instruction latency specified by designers. Each smaller finite state graph represents a stage in the pipeline. Figure 1 illustrates the concept of stage partitioning. An ISA specification is transformed into an abstract finite state machine consisting of four states which is then decomposed into two cascading finite state machines. Each smaller finite state machine iterates between its two states, and is synchronized with each other by a system counter counting between *t1* and *t2*. The result is a two-stage machine with instruction latency of two. Piper is able to deal with conditional branches inside the loop body. In the case where non-deterministic inner loops exist in the loop body, the current implementation of Piper does not attempt to pipeline the design.

Figure 4 shows the inter-instruction dependencies after the stage partitioning for an instruction set processor with three instructions `foo`, `bar`, and `goo`. In this example the processor is partitioned into three stages with the stage time of two cycles. This processor consists of fourteen states, `S1` to `S14` shown as bubbles in the figure. For each state there is a state label shown at the left side of the bubble, and a conditional tag shown at the right side. A conditional tag with non-empty string indicates that the state is conditional executed. For example, state `S1` with tag of `[]` is non-conditionally executed, whereas `S3` with tag of `[foo]` is conditionally executed when the opcode is `foo`. The tags are derived from the conditional branches such as opcode decoding and the values of status registers. The state transitions are represented by light arcs. The body of the state represents the active register transfers. For simplicity, we only show the register transfers of interest in this discussion. For example, in state `S1`, `PC<-` indicates a write to register `PC`, and in state `S7`, `<-X` indicates a read from register `X`. The heavy bi-directional arcs indicate the inter-instruction dependencies. The longest inter-instruction dependency spans over four cycles. However, the machine is implemented with latency of two. Therefore this is not a hazard-free pipeline implementation.

Note that better resource utilization may be obtained if rescheduling after the stage partitioning is performed. For example, assume that there is an addition operation in states `S11` and `S13` (in Figure 2), respectively. Then it would require that two adders be allocated since these two additions may be executed concurrently at cycle 2. When the mobility allows, the former addition can be rescheduled to state `S12` or the latter to state `S14`, then only one adder will be sufficient since these two additions do not execute concurrently. However, while this rescheduling may improve the resource utilization, it may degrade the performance. For example, by delaying the addition in `S11` to `S12` (with tag `goo`) may introduce a pipeline hazard between an instruction which follows `goo` and uses the result of that addition. The rescheduling and its impact on the pipeline hazards will be integrated into the future version of Piper.

An algorithm described in Section 4 is then invoked to solve the pipeline hazards by extracting a reorder table for the reorderer.

4. Construction of the reorder table

To solve the pipeline hazards existing in the high performance implementations of pipelined micro-architectures, Piper assumes that there is a retargetable reorderer to reorder the instructions. The reorderer takes in the instruction format specification and a reorder table to properly reorder instructions in order to generate hazard-free and efficient target codes. An exemplary entry of the table is `reorder(foo-after-bar-by-3-nop)`. This

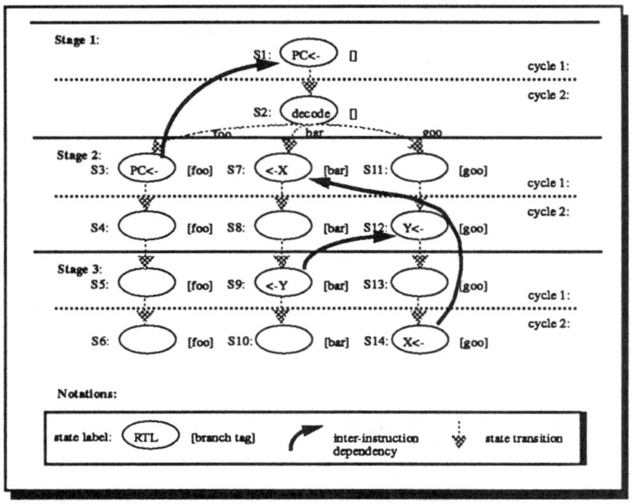

Figure 2 global data-flow analysis

```
1:   for each upward dependency pair Si-Sj do
2:          % Si is deeper than Sj in the pipe
3:   {
4:       if either Si or Sj is in an inner loop then
              terminate;% Not pipelinable
5:       if Si and Sj are not of the same stage then
6:       {
7:            if cycle index of Sj > cycle index of Si then notInTime=1
8:            else notInTime = 0;

9:            Displacement = stage index of Si - stage index of Sj + notInTime;
10:           Nops = Displacement - 1,

11:           if tag of Si contains opcode then Inst2= this opcode
12:           else Inst2= allInstructions;
13:           if tag of Sj contains opcode then Inst1= this opcode
14:           else Inst1= allInstructions;

15:           report table entry: reorder(Inst1 after Inst2 by Nops nop)
16:           }
17:  }
18:  delete entries which are subsets of others
```

Figure 3 Reorder table generation

means that there should be at least three *nop*'s or other independent instructions (*safe instruction distance*=3) in between the instruction pattern of `foo` following `bar`.

An algorithm which generates the reorder table is described in Figure 3. The algorithm takes as inputs the pipeline schedule and the dependencies from the global data-flow analysis, and generates as output a reorder table. The details of the algorithm is described as follows.

The algorithm first collects all the upward dependencies (from deeper pipeline stage to shallower one). The downward dependencies do not cause pipeline hazards. For example, in Figure 2, the dependency due to a read of register `X` in state `S7` and a write of register `X` in state `S14` causes a hazard for `bar`-after-`goo` but not for `goo`-after-`bar` since the potential hazard for the latter one is automatically solved by the structure of the pipeline: `bar` reads `X` at stage 2, and at that time `goo` which follows `bar` has not arrived at the stage 3 where it writes to `X`.

The next step is to generate the reorder entry for each upward dependency. This task consists of three sub-tasks. First, we check that whether any one of the involving states *Si* and *Sj* is in an internal loop. If true, pipelining is not a suitable implementation for this processor, and thus we terminate this algorithm. Second, the safe instruction distance for the dependent instruction pair has to be determined. Third, the names of the instructions which involve in this dependency have to be identified. Line 7 to 10 calculate the safe instruction distance. Line 11 to 14 determine the names of the instructions from the tags of the involving states. In the case where the opcode can not be inferred from the tags, we unify the name of the instruction to *allInstructions* which means this reorder case apply to all instructions. In the example of Figure 2, the `S14-S7`, and `S3-S1` dependencies result in the reorder entries `reorder(bar-after-goo-by-1-nop)` and `reorder(allInstruction-after-foo-by-1-nop)`, respectively. The last step deletes the reorder entries which are subsets of others.

In Figure 4 we show the reorder tables for instruction latencies of 1, 2, 3, and 4 for the example in Figure 2. As we expect, the larger the instruction latency, the less number of instruction pairs need reordering. The instruction latency of 1 results in three reorder pairs while the instruction latency of 3 resulting in only one reorder pair. Also the safe instruction distances decrease as the instruction latency getting larger. The longest safe instruction distance is four for instruction latency 1, while being one for instruction latency 3. Note that the reorder table for instruction latency 4 has zeros for all entries which means that this processor can be

Reorder patterns	Safe instruction distance			
Instruction latency	1	2	3	4
bar after *goo*	4	1	1	0
all Instructions after *foo*	2	1	0	0
goo after *bar*	1	1	0	0

Figure 4 Reorder tables

pipelined with instruction latency of 4 (2 stages) with no pipeline hazards for all instruction patterns. Therefore, this implementation makes the pipelined micro-architecture transparent to the compiler.

The generated reorder table affects the execution time and memory space of the reorderer. The more entries in the reorder table and the larger the safe instruction distance, the longer time the reorder takes to execute and the more memory space it consumes. We will discuss these effect in the following section.

5. Performance/cost evaluation

We characterize the synthesized design by the performance and cost of the hardware and its supporting software (the reorderer). The important metrics are peak/run-time performance and cost of hardware, and time/space complexities of compilers. With these metrics we are able to provide the designers with a broader and finer design space which spans both the hardware and the software. This capability of modeling the software and hardware components of complete systems is essential in extending the application scope of the design automation systems into embedded systems design.

5.1. The performance estimation of hardware

Two performance metrics are used to evaluate designs: maximal and run-time performances. The maximal performance can be obtained when the instruction latency and the clock rate are known. It does not vary with the applications. However, the run-time performance depends on the characteristics of the applications. The instruction trace expansion due to inter-instruction dependencies and pipeline hazards degrades the run-time performance of a deeply pipelined processor from its maximal performance. Consider a fragment of instructions compiled for non-pipelined machine in case (a) of Figure 5. Suppose that there is a synthesized pipelined machine which requires two delay slots between branch `brn` and its successive instruction of any type.

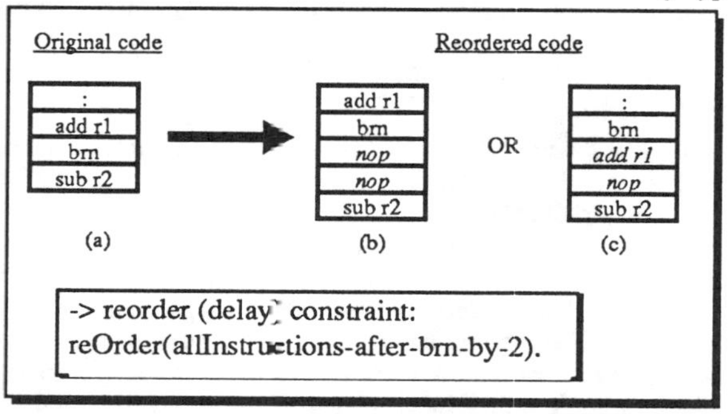

Figure 5 Instruction reordering

In case (b) two `nop`'s are inserted between `brn` and the successive `sub r2`. The instruction trace is expanded by two `nop` slots which degrade the pipeline performance by two pipeline cycles. In case (c) the size of the `nop` slots is compressed from two to one by reordering an independent instruction `add r1` into the `nop` slots. Piper uses a set of benchmarks (instruction set programs) to estimate the run-time performance. Instead of repeating the phase of compiling and simulating the benchmarks for each pipeline implementation, Piper applies the characteristics of simulation traces of benchmarks compiled for non-pipelined processors, and the characteristics of the reorder table to predict the run-time performance. With this approach, the run-time performance of different synthesized pipelined processors can be quickly obtained.

The instruction set simulator of ADAS provides a set of instruction benchmark characteristics. One of the characteristics used by Piper is the pattern frequency of instruction pairs which is the frequency of the consecutive instruction pairs in the instruction trace. Shown in Figure 3 is the pattern frequency of a small instruction benchmark for the instruction set in Figure 7. For example, the first entry represents that `load` immediately followed by `add` (the `load-add` pair) happens twelve times in the trace which has a frequency of 14.5%.

For each pipelined processor p, the run time performance metrics with respect to a benchmark can be approximated by the following equations:

$$Exp = \sum_j Dist_{p_j} \cdot (1 - C_j) \cdot R_j \cdot A + A$$

$$S_p = \frac{A \cdot M}{L_p \cdot Exp}$$

$$T_p = \frac{A}{L_p \cdot Exp}$$

where E_{xp} is the trace expansion due to the insertion of *nop's*, S_p is the speedup of pipelined processor p with respect to non-pipelined processor, and T_p is the throughput of pipelined processor p.

$Dist_{pj}$ is the number of *nop*'s (in the worst case) to be inserted between a consecutive instruction pattern j for pipelined processor p, R_j is frequency of the consecutive instruction pattern j, L_p is the instruction latency of pipelined processor p, M is the instruction cycle time of non-pipelined processor, A is the total number of instructions in the simulation trace, and C_j is the compression factor for consecutive instruction pattern j which means that, empirically, the percentage of the number of *nop* slots which can be filled by reordering instructions into the slots. C_j can be set to zero for the worst case analysis.

The product AM is the total number of cycles when the benchmark is executed on the non-pipelined processor while the product L_pE_{xp} is the total number of cycles executed on the pipelined processor p.

5.2. The cost estimation of hardware

We represent the cost of the hardware by the sizes of the data path and control path. The size of the data path is obtained by estimating the number of transistors in the data path. The wiring cost

```
instruction_pattern_count([load,add],12,0.144578).
instruction_pattern_count([store,add],11,0.13253).
instruction_pattern_count([load,and],1,0.0120482).
instruction_pattern_count([add,brn],12,0.144578).
```

Figure 6 Pattern frequency of instruction pairs

is not considered in our current version of Piper. The control model adopted by Piper is data stationary model. Each stage has its own controller. We implement these controllers as PLAs. Therefore, we derive the control path cost by summing up the estimated PLA size of each stage.

5.3. The time/space complexities of instruction set compilers

When comparing the time and space complexities of instruction set compilers for different pipeline implementations of the same instruction set architecture, we focus on the reorder phase of the compiler since the other phases of the compiler remain constant. We apply the characteristics of the reorder table to the equations of time and space complexities to derived the relative performance and cost of the reorderer, respectively. The possible characteristics of reorder table include the number of reorder entries (consecutive instruction patterns) E, the maximal safe instruction distance D_{max}, the minimal safe instruction distance D_{min}, and the average safe instruction distance D_{avg}.

The time/space complexities of the reorderer can be considered as the constraint imposed on the synthesis system by the software (the reorderer). For example, a reorderer with time complexity of $E^3 D_{max}$ implies that it is not feasible for a synthesis system to synthesize a design which produces a larger reorder table; however, a design with large safe instruction distance but a smaller reorder table may still be an acceptable design.

To demonstrate the idea, let's consider a simple reorderer which reads in the reorder table of size E, and then sweeps a window of size $2*D_{max}$ through the instructions to look for the target patterns, and then searches within the window for independent instructions to be inserted into the *nop* slots. Assume that the table is kept in an array. Then for each iteration, the reorderer takes $O(E*D_{max}+D_{max}^2)$ time. At least, memory of size $O(E+D_{max})$ is necessary. Its time and space complexities may be expressed as:

$$\text{time complexity} = L \cdot D_{max} \cdot (E + D_{max})$$

$$\text{space complexity} = E + D_{max}$$

where L is the benchmark length. By substituting the variables for values, we can obtain the relative complexities. Similarly, we can derive the time/space complexities in terms of reorder table parameters for any reordering algorithms. This set of equations is kept in a system configuration file in Piper and should be modified if the reordering technique used in the compiler is changed.

6. Synthesis example and comparison

In this section we present a synthesis example for an instruction set processor *sm1*. Its ISA specification in Prolog [14] is shown in Figure 7. It is an accumulator-based processor with eight instructions. Piper returns six pipelined designs with instruction latency ranging from one to six. The instruction latency of six is the non-pipelined design. In Figure 8 to Figure 10 we show the performance/cost evaluations for these designs. In these figures, we characterize the designs by their instruction latencies. From the dependency analysis, the MAL is five for *sm1*. Therefore, the designs with latency from one to four are not hazard-free, and require the reorderer to resolve the pipeline hazards.

In Figure 8 are the costs of hardware including data path and control path. Figure 9 shows the maximal and average (run-time) performance curves and the performance/cost curves. In this example, we use the speed-up to represent the performance. The average performance is obtained by applying the benchmark characteristics from a small straight-line instruction benchmark

```
class([register,master_sllave,r1,16]).
class([counter,programable,c1,16]).
class([memory,ram,m1,[16,64]]). % 16bit x 64K
attributed([r1,ac,0,fsm,1]).
attributed([c1,pc,0,fsm,2]).
attributed([m1,m,_,fsm,3]).
opcode(execute,1).
fsm( AC, PC , Memory):-
fetch( PC, Op, Adr, Memory ),
 P1 is PC + 1,
execute( [Op], Adr, AC, P1, Memory, AC_next,
PC_next, Memory_next ),
fsm( AC_next, PC_next, Memory_next ).
fetch( PC, Op, Adr, Memory ):-
memory_read( PC, Memory, [Op,Adr] ).
execute( [add], X, AC, PC, Memory, AC_next, PC,
Memory ) :-
memory_read( X, Memory, Data ),
AC_next is AC + Data.
execute( [and], X, AC, PC, Memory, AC_next, PC,
Memory ) :-
memory_read( X, Memory, Data ),
AC_next is AC ∧.
execute( [sub], X, AC, PC, Memory, AC_next, PC, Mem-
ory ) :-
 memory_read( X, Memory, Data ),
 AC_next is AC - Data.
execute( [shr], X, AC, PC, Memory, AC_next, PC, Mem-
ory ) :-
 AC_next is AC >> 1.
execute( [store], X, AC, PC, Memory, AC, PC, Memory_n-
ext ):-
memory_w irite( X, AC, Memory, Memory_next ).

execute( [load], X, _, PC, Memory, AC_next, PC, Mem-
ory):-
memory_read( X, Memory, Data ),
AC_next = Data.
execute( [jump], X, AC, _, Memory, AC, PC_next, Mem-
ory ):-
 PC_next = X.
execute( [bm], X, AC, PC, Memory , AC, PC_next, Mem-
ory ) :-
(AC < 0 -> PC_next = X;
 PC_next = PC).
```

Figure 7 ISA specification in Prolog: sm1

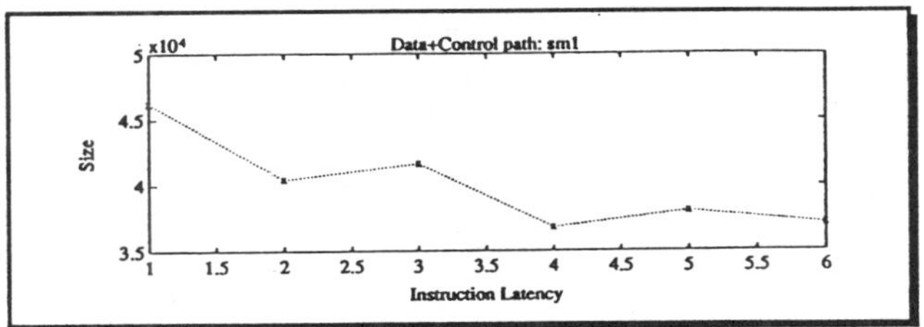

Figure 8 costs of data path and control path for *sml*

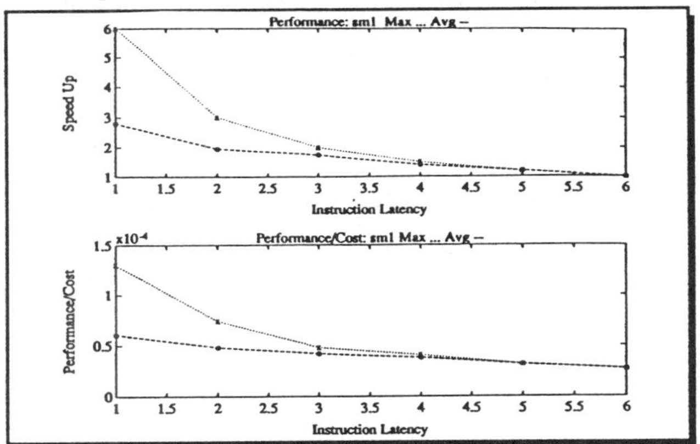

Figure 9 maximal/average performance for *sml*

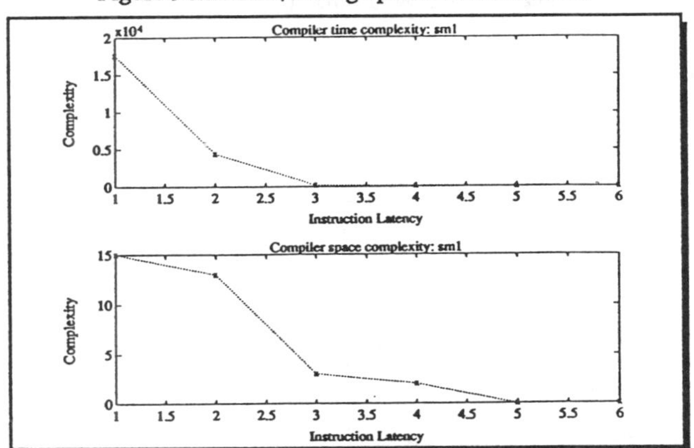

Figure 10 compiler time/space complexities for *sml*

System	Latency = 1[a]	Latency = 2[a]	Latency = 3[a]	Latency = 4[a]	Latency = 5	Latency = 6
SEHWA	n/a	n/a	n/a	n/a	Yes	Yes
PLS, HAL, CATHEDRALII	no	no	no	no	Yes	Yes
ASPD	Yes[b]	Yes[b]	Yes[b]	Yes[b]	Yes	Yes
Piper	Yes	Yes	Yes	Yes	Yes	Yes

a. These designs are not hazard-free.
b. The pipeline is stalled whenever an instruction which may cause pipeline hazard is found in decode stage.

Figure 10 Comparison of synthesis example: sml

with conditional branches to the speed-up equation listed in Section 5. Note that the smaller the instruction latency, the greater the average performance degrades from the maximal performance due to the insertions of nop's. The performance/cost curves indicate that pipelining, with higher utilization of resources, dramatically improves the performance at slightly extra cost. This is also observed in [3] and [4].

The relative time/space complexities of reordering phase of the compilers are illustrated in Figure 10. As we expect, the complexities increase dramatically as the instruction latency decreases. These curves show that the maximal throughput design (latency of one) has a relatively large compiler complexity, even though it offers the largest performance/cost ratio. Therefore it might not be a feasible design from a compiler's point of view. Note that the relative complexities of time and space are zeros for designs of instruction latency 5 and 6. This means that no reordering is required for these two designs. This result is interesting in that if the reordering is not supported in a computing environment, then a speed-up of 1.2 (instruction latency 5) is still achievable at an additional 5.7% hardware cost of the non-pipelined design.

In Figure 11 we show a comparison for the synthesis capability of several DA systems with our *sml* example. PLS, HAL, and CATHEDRAL II perform pipeline synthesis under the constraint of MAL and thus are not able to produce the designs with latency less than five. SEHWA performs pipeline synthesis for data path only. Therefore, SEHWA is unavailable for designs with pipeline hazards since they require special cares in the control path design. ASPD is able to synthesize designs with pipeline hazards. The control path generated by ASPD stalls the pipeline as soon as a leading instruction which may cause hazards is decoded. ASPD assumes the worst cases. Suppose that the only instruction-pair with pipeline hazard is `loadR1-addR1`. ASPD will always stall the pipeline as soon as `loadR1` is decoded, no matter which instruction follows. This strategy leaves compilers no room to make use of the stalled cycles. Therefore, the performance of the synthesized designs by ASPD is inferior to Piper's designs.

7. Conclusions and future work

We have presented a synthesis model which encapsulates the interactions between the hardware and software at the micro-architecture level. We have shown that even higher performance of pipelined instruction set processor designs can be achieved by our design automation system Piper with a scheduling technique which allows pipeline scheduling with any fixed instruction-latency, and with the help of the reorderer in the back-end compiler. We also have presented the performance and cost models for both hardware and software. This capability of modeling the aspects of the synthesis, performance, and cost for the software and hardware components of complete systems makes Piper a promising design automation system for embedded systems design.

In addition to the improvement of the automatically generated pipeline structures, Piper contributes to the knowledge of automating the generation of back-end compilers, especially the micro-architecture dependent phases, which has been lack of profound success during the past decade. We have presented an algorithm to extract the reorder table for the reorderer from a description of the pipelined micro-architecture.

The rescheduling guided by the characteristics of instruction benchmarks after stage partitioning in order to improve the resource utilization will be implemented into the future version of Piper. Piper will be able to deal with the non-deterministic inner loops by means of inserting internal opcodes. We will develop and implement further synthesis techniques to fine tune the micro-architecture to take advantages of some advanced compiling techniques such as delay slots with annulling or branch prediction.

Acknowledgments

The authors would like to thank Gino Cheng, Barton Sano, Apoorv Srivastava, Chi-Ying Tsui and anonymous reviewers for their suggestions and comments. We also appreciate the research supports from Yong-Seon Koh, Shih-Ming Liu, Ricky Pan, Iksoo Pyo, Ching-Long Su, and Shiqun Wu.

Reference

[1] William Bush, et al., "An Advanced Silicon Compiler in Prolog," *Proc. ICCAD*, 1987

[2] Nohbyung Park and Alice C. Parker, "Sehwa: A Software Package for Synthesis of Pipelines from Behavioral Specifications," *Trans. on CAD, Vol 7. No. 3*, March 1988

[3] Michael C. McFarland, Alice C. Parker and Raul Camponsano, "The High-Level Synthesis of Digital Systems," *Proc. of the IEEE, Vol. 78, No. 2*, February 1990

[4] Gert Goossens, Jan Rabaey, Joos Vandewalle and Hugo De Man, "An Efficient Microcode Compiler for Application Specific DSP Processors," *IEEE Trans. on CAD*, September 1990

[5] Pierre B. Paulin and John P. Knight, "Force-Directed Scheduling for the Behavioral Synthesis of ASIC's," *IEEE Trans. on CAD*, June 1989

[6] Mauricio Breternitz Jr. and John Paul Shen, "Architecture Synthesis of High-Performance Application-Specific Processors", *Proc. of the 27th DAC*, 1990

[7] Cheng-Tsung Hwang, et al., "Scheduling for Functional Pipelining and Loop Winding", *Proc. of the 28th CAD*, 1991

[8] D. E. Thomas, et al., "The System Architect's Workbench," *Proc. of the 25th DAC*, 1988

[9] Gerry Kane, *MIPS RISC architecture*, Prentice Hall, 1989

[10] R. G. G. Cattell, "Automatic Derivation of Code Generators from Machine Description," *ACM Trans. on Programming Languages and Systems*, Vol. 2, No. 2, April 1980

[11] S. L. Graham, "Table-Driven Code Generation," *IEEE Computer*, August 1980

[12] Robert Giegerich, "A Formal Framework for the Derivation of Machine-Specific Optimizers," *ACM Trans. on Programming Languages and Systems*, Vol. 25 No. 3, July 1983

[13] Iksoo Pyo, et al., "Application Driven Design Automation for Microprocessor Design," *Proc. of the 29th DAC*, 1992

[14] Gino Cheng, et al., "A Full-Range Design Automation for Instruction Set Processors," *First International Conference on the Practical Application of Prolog*, April 1992

APT: AN AREA-PERFORMANCE-TESTABILITY DRIVEN PLACEMENT ALGORITHM

Sungho Kim Prithviraj Banerjee Vivek Chickermane Janak H. Patel
Center for Reliable and High-Performance Computing
Coordinated Science Laboratory
University of Illinois
Urbana, IL 61801

ABSTRACT

Conventional placement algorithms have been designed to minimize the area of circuit layout. Recently researchers have proposed performance-driven placement algorithms. In this paper, in addition to a performance-driven placement, we present a testability driven placement algorithm using the notion of partial scan. The cost of including a flip-flop cell in a scan path is obtained in the placement process and guides the selection of scan flip-flops. This is then followed by a cell resizing step in which logic cells change their widths and timing characteristics using the templates provided by a cell library. This reduces delays in the critical timing paths and delays caused by inserting scan flip-flops. Experimental results on the ISCAS benchmark show that combining the performance and testability optimizations in the placement level results in better solutions in terms of area, timing and fault coverage.

1. INTRODUCTION

As the size of modern VLSI circuits keeps increasing, meeting the timing and testability requirements of complex chips is becoming more difficult. Conventional physical design CAD tools, however, have mostly tried to minimize the area of the layout only.

There has been extensive research in performance driven placement recently [1-5]. In this paper, we extend these efforts in developing a better placement tool (1) by adding a more accurate timing model, (2) by allowing cell resizing, and (3) by making it testability driven.

As most standard cell libraries provide multiple templates with different driving capability for each gate, we can leave the problem of choosing the proper sizes of cells to the standard-cell placement tool. Previously, this important decision which influences the area and the timing of the circuit was made only by the technology mapping step in the logic synthesis [6]. The disadvantage of making the decision at this level is that the interconnect capacitances and resistances are not available at that stage. The templates selected optimally at the logic synthesis stage will no longer be optimal later in the physical synthesis stage where large interconnect capacitances are introduced to the circuit. To best deal with this problem, the cells must be resized at the physical level where more accurate wire delay estimates are available.

† This research was supported by the Semiconductor Research Corporation under Contract SRC 91-DP-109.

Also, in our approach, the testability of the circuit is addressed at the placement level. For many years, full scan design (where all of the flip-flops in the circuit are chained in a scan path) was the most commonly used method to test complex sequential VLSI circuits. With advances in automated sequential test generators, the partial scan design methodology, in which only a subset of flip-flops are selected to be scanned, has been proposed as a cost-efficient alternative to complete scan design. The problem facing the partial scan design is in choosing which flip-flops to scan. If the partial scan decision is made before the placement step, each candidate flip-flop is weighted equally in terms of the area and testing penalty. But, if the decision is made in the placement step, each flip-flop can be assigned with a more realistic cost.

The example in Figure 1 illustrates our approach. An intermediate step in the standard cell placement optimization process is shown in Figure 1(a). Each rectangle represents a standard cell and some cells overlap their neighboring cells. Cells A and B are flip-flops. A critical path that terminates at the input of flip-flop A is also shown in the figure. In Figure 1(b), flip-flops A and B are replaced by scan flip-flops, which are about 40% wider, and put in a scan chain.

The increased area of a scan flip-flop is largely due to a multiplexer inside. Because of this multiplexer, the input signal coming into the flip-flop is delayed by τ_{mux}. The system performance is degraded if the scan flip-flop is put in critical paths as seen in the figure. To compensate for the extra delay τ_{mux} introduced by a scan flip-flop, cells D and F in the critical path are replaced by cells of higher drive, hence larger templates, as seen in Figure 1(c). As seen in the figure, only the cells with enough free space around them are replaced with a larger cell. On the contrary, cells C and E are replaced with smaller cells to reduce the cell overlaps and the overall placement area. This reduction in cell widths does not affect the performance because the cells are not in critical timing paths.

Recently, we have proposed a related approach of layout driven design for testability [7], where observable

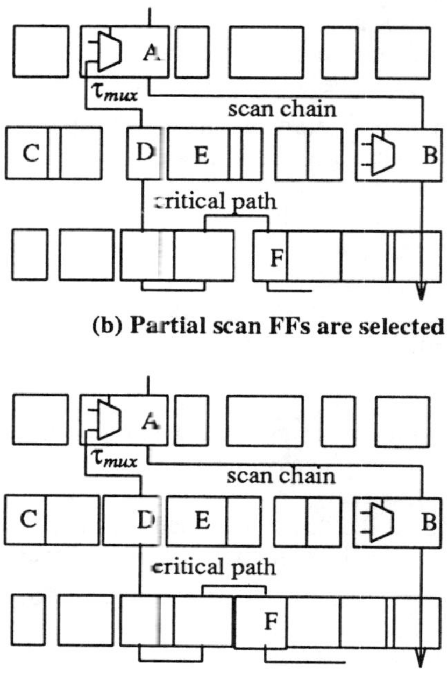

Figure 1. Illustration of APT approach

test points are added using transistors in the channel region of standard cell layouts after the placement and routing is completed. Our work in this paper deals with the addition of scan flip-flops during the placement process. Also our work explicitly address the performance driven placement issues.

The remainder of this paper is organized as follows. Section 2 presents an overall view of our placement strategy. In Section 3, we discuss the problem of timing driven placement and introduce the delay model. Section 4 describes the cell resizing problem. Section 5 discusses the partial scan selection problem. Section 6 presents some experimental results on ISCAS benchmark circuits.

2. OVERVIEW OF APT PLACEMENT ALGORITHM

The area-performance-testability driven placement tool (called APT) that we have developed was built around the Timberwolf5.4 placement package which uses the simulated annealing algorithm [8]. The cost, C, that the algorithm tries to minimize, consists of four independent terms as follows:

$$C = W + \mu P_o + \lambda P_R + \sigma P_T$$

The first term is the total wire length, represented by W. The second term, P_o, is the overlap penalty function. The third term, P_R, is the row length control penalty function. The fourth term, P_T, is the net delay penalty function which is defined in the next section.

Figure 2 shows the flow chart of APT. At the end of each outer iteration I of the simulated annealing algorithm, it enters the performance and testability optimization phase as seen in the right half of Figure 2. In this phase, slacks on timing requirements are calculated and partial scan flip-flops are selected using OPUS [9], an optimization based partial scan selection program. OPUS uses the latest slack value and placement configuration obtained in the previous iteration to select partial scan flip-flops. After the selection, the cell resizing step takes place where cells in the most critical paths are upgraded with larger cells to reduce the delay and those in the least critical paths are replaced with smaller ones to reduce the area and cell overlaps. If needed, the delay caused by inserting a scan flip-flop is compensated in this cell resizing step. After the cell resizing, timing bounds are calculated for each net using a timing analysis algorithm similar to the zero slack algorithm [5]. After the bound calculation, the next iteration of simulated annealing starts. In each iteration of the annealing step, cells switch their

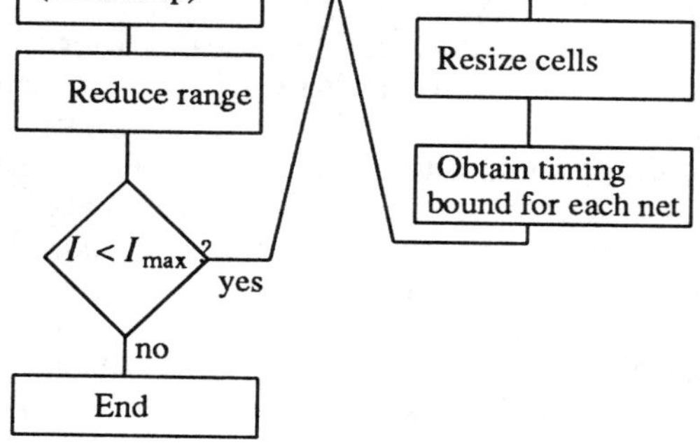

Figure 2. Overview of APT placement algorithm

positions attempting to reduce the total wire lengths and also to meet the timing bound given for each net. In doing so, it uses the timing model described in the next section.

3. PERFORMANCE DRIVEN PLACEMENT AND AN ACCURATE TIMING MODEL

In performance driven placement, "slack" and critical path information are incorporated into the objective function of the optimization. This can be done effectively by assigning higher weights to the nets in the critical timing paths [1,2]. However, these higher weights do not specify the amount by which it should shorten the critical paths. In the timing constraints approach [5], an upper bound is obtained for each net delay such that if each local delay on the net stays below the bound, then the timing requirement is met. In APT, we use the combined approach similar to [10] to result in a better solution.

In this section, we improve the timing performance by introducing a better timing model. Figure 3 shows the timing model we propose. The linear RC model was an acceptable approach for modeling digital MOS circuits in the past. In particular, Elmore's model [11] was used widely to approximate the time taken for a signal to start from an initial value and reach half of its final value. In general, calculating Elmore's delay involves a matrix inversion which is time-consuming. However, if the RC network is a tree, then R can be determined by inspection, and τ_d can be computed in linear time [12]. The delay at node e, $\tau_{De} = \sum_k r_{ek} c_k$, where r_{ek} is the resistance between node e and the source that is common with the resistance between node k and the source. In order to simplify the computation even further, we assume a star tree as seen in Figure 3. The delay at the input of gate A, τ_A, is then $R_s(C_s + C_a + C_b + C_c) + R_A C_A$. The branch resistance, R_A, is estimated by α·(distance in x-direction between the source and the sink A) + β·(distance in y-direction between the source and the sink A). Here, α and β are per unit length resistances in the interconnects in the x-direction (e.g. metal 1) and in interconnects in the y-direction (e.g. metal 2) respectively. C_A is the sum of the input capacitance of gate A and γ·(the distance in the x-direction between the source and the sink A) and ω · (the distance in y-direction between the source and sink A), where γ = per unit length capacitance of interconnect in x-direction (metal 1), and ω = per unit capacitance in y-direction (metal 2).

Having the delay estimate for each source-to-sink segment, we can use the following cost function to influence the placement to meet the timing requirement similar to [10]:

$$P_T = \sum_{net} \sum_{sink} W_{sink}\ \delta_{RC}(\tau_{sink} - \overline{\tau_{sink}})$$

$W_{sink} = \dfrac{k}{slack_{sink}}$ and $\delta_{RC}()$ is a function to penalize the objective cost when the estimated delay, τ, exceeds its delay upper bound, $\bar{\tau}$. The upper bound on delay is obtained by an algorithm similar to the zero slack algorithm [5]. In our algorithm, delay bounds are obtained for each source-to-sink segment unlike the zero slack algorithm which treats each net as one delay element.

4. CELL RESIZING

The timing weight and bound assigned to each net as seen previously will control the wire delay to meet the timing requirement. But, if excessive timing weights are assigned, the wirability may suffer. By introducing cell resizing to the placement process, the conflict may be solved easily. Figure 4(a) shows a net with an excessive delay. In conventional timing-driven placement, all of the pins connected to the net must move closer to each other to decrease the delay as seen in Figure 4(b). This may increase the wire lengths of other nets attached to the gates. But, in our method, we have the choice of replacing the source cell with one with a higher driving capability as seen in Figure 4(c) to reduce the delay at the net. In this way, the wirelengths do not have to increase to meet the

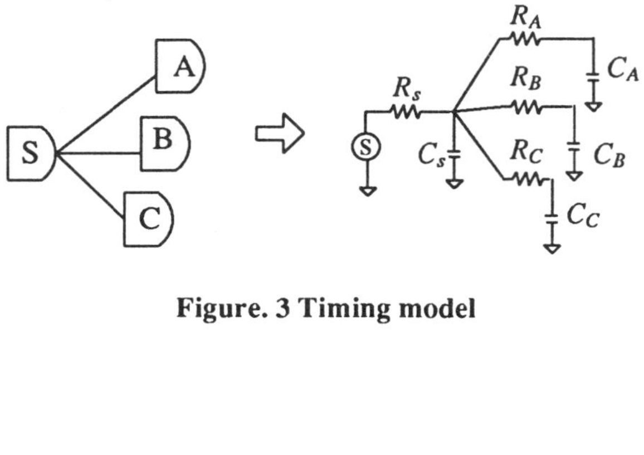

Figure. 3 Timing model

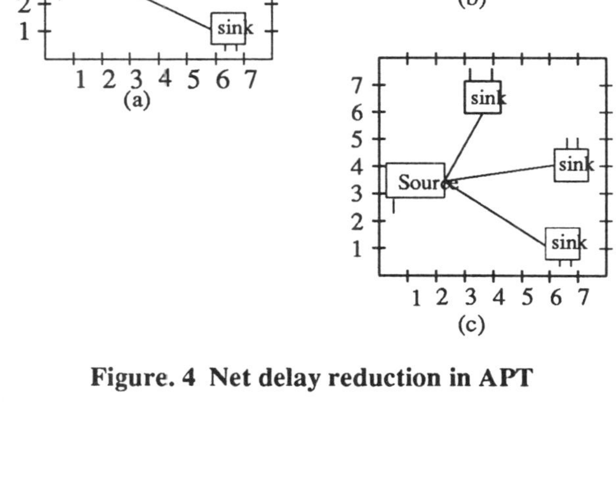

Figure. 4 Net delay reduction in APT

timing requirement. Also, as seen in Figure 5, a cell may be replaced with a cell with a larger delay if the delay at the output net is smaller than necessary. By switching to a smaller cell, the area is saved. The area saved can be used by other cells to reduce the delay as seen in Figure 1(c). Therefore, in our method, a net delay can be reduced in two ways; cell displacement, or cell resizing.

Another benefit of resizing cells in the placement is that it helps reducing cell overlaps. Because standard cells have different widths, the problem of cell overlap occurs in placement. Often two cells are selected for interchange which differ in width. Therefore, an exchange of locations of these two cells often results in some overlap with the other cells. To solve this problem, placement algorithms such as [8] add a penalty term $\mu \cdot P_{overlap}$ to its cost function to penalize movement of cells with large overlaps as seen in Figure 2. In APT, as seen in Figure 6, the cell resizing algorithm also tries to reduce the amount of the overlap if such a resizing does not violate the timing constraints. By reducing the amount of overlap in this way, the placement algorithm can concentrate more on minimizing the wirelengths instead of minimizing the cell overlaps.

The cell resizing problem is itself an NP-hard problem [6, 13]. Therefore, we use an $O(n^2)$ heuristic algorithm similar to [6] in this research. The algorithm, which optimizes the area/performance iteratively, consists of an increasing phase and a decreasing phase. In the increasing phase, a number of cells (set to parameter *Quantum*) of the largest weights are replaced by larger templates to meet the timing requirement. When the delay requirement is met, the process is reversed. The weight W_{time} used for each gate is defined as

$$W_{time} = Sensitivity - Criticality + Availability$$
$$= \frac{\Delta delay}{\Delta area} + \frac{R_{out} \cdot C_L}{slack} + \Delta Gap + \Delta Row$$

The *sensitivity* and *criticality* measures in the weight function were first used in [6]. The *sensitivity* measure indicates how much the local delay per unit area can be decreased if another template is selected for this cell. It is defined as $\frac{\Delta\, delay}{\Delta\, area}$, where $\Delta\, delay$ is the difference between the new local delay of this cell after we substitute a new template and the original delay. The $\Delta\, area$ is equal to the area of the new template minus the area of the original one. The *criticality* measure is defined as $\frac{R_{out} \cdot C_L}{slack}$ and tells us whether a cell needs to be replaced by a larger template to meet the delay constraints of the circuit. For each cell N, the slack is defined to be:

$$slack(N) = Req(N) - Act(N)$$

$Req(N)$, in the above equation, is the last time a signal has to arrive at the output pin of cell N in order to make it on time to the output pad of the circuit. $Act(N)$ is the actual signal arrival time at the output pin in the worst case. A slack tells us how much, at most, a cell delay can increase so that the delay constraints are not violated. The smaller the slack of a cell, the more critical this cell is. The *availability* consists of the sum of the $\Delta\, Gap$ and $\Delta\, Row$ measures and indicates whether there is enough space available for a cell width increase. $\Delta\, Gap$ is the amount of empty region around the cell and $\Delta\, Row$ is $Rowlength_{desired} - Rowlength_{actual}$.

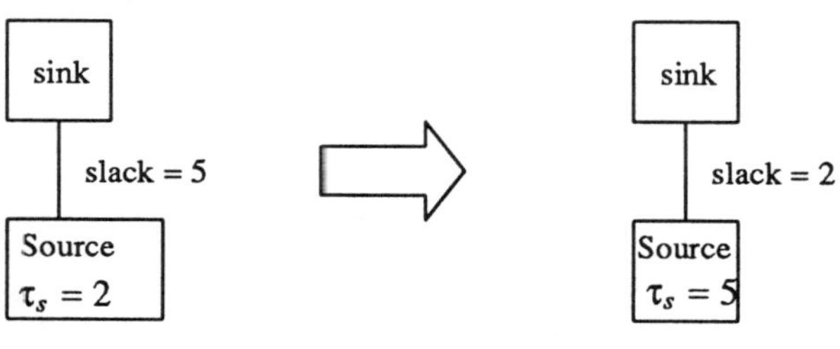

Figure. 5 Cell resizing to reduce the area

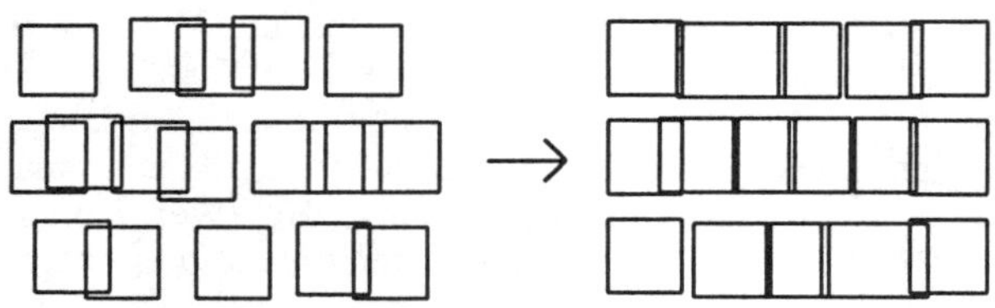

Figure. 6 Cell resizing to reduce cell overlaps

5. PARTIAL SCAN FLIP-FLOP SELECTION

Sequential circuits are in general very difficult to test. A well-known design for testability technique, the full scan design, connects all flip-flops to a scan chain. In partial scan, one connects only a subset of flip-flops in a chain to make the ATPG process easier. Because choosing the right set of flip-flops is a hard problem, researchers have proposed many heuristics to select the partial scan flip-flops [9, 14]. In this paper, we use only one algorithm for selecting partial scan flip-flops. In this algorithm [9], flip-flops are selected to break all of the cycles in the reduced graph of the circuit. Because a scan flip-flop is about 40% larger in size and slows down the signal latched into the memory element, it must be carefully selected so that the timing and area requirements of the circuit are not violated.

This can be done effectively by formulating the problem as an optimization problem such as OPUS [9]. Given a bound on *cost*, the OPUS program tries to select scan flip-flops such that the testability profit, the number of cycles broken in the circuit, is maximized. In our method, we provide an accurate cost to OPUS, which uses it to select a set of flip-flops to break the cycles. The cost used for each flip-flop i, $W(i)$, has been improved over the previous measure [9] and defined as:

$$w(i) = w_c(i) + w_s(i) + w_d(j)$$

$$w_c(i) = \frac{1}{|\,x(i) - x_{center}\,| + k}$$

$$w_s(i) = \begin{cases} 0 & \text{if } \omega + \varepsilon < desired; \\ \dfrac{\omega + \varepsilon - desired}{r}, & otherwise \end{cases}$$

$$w_d(i) = \begin{cases} 0 & \text{if } slack(i) - \tau_{mux} > 0; \\ -\,slack(i) + \tau_{mux} & otherwise \end{cases}$$

In the above equation, W_c is a weight that reflects the channel density. Because the middle region of a standard cell placement is usually the most dense, we want to avoid putting a flip-flop on the scan path in the central region by giving a higher weight. $x(i)$ is the x-coordinate of flip-flop i. x_{center} is the x-coordinate of the center of the placement. The constraint on scan design is tighter for flip-flop cells in rows with wider dimensions than for those in smaller rows. This constraint can be modeled by a cost w_s as shown above, where ω is the current length of the row in which i resides, ε is the extension in the row width if all of the flip-flops in the row are put in the scan chain, *desired* is the desired row width, and r is the number of flip-flops in the row. Finally, one needs to estimate the cost of the scan when a flip-flop lies in the critical timing path. When $slack(i) > \tau_{mux}$, the delay added by a scan flip-flop, we let the delay weight, w_d, to be 0, because the increased delay does not degrade the circuit timing even after inserting a scan flip-flop. Otherwise, we increase the weight as shown.

The cell resizing step is followed by the partial scan selection step as seen in Figure 2. If there is timing degradation due to a scan flip-flop, it is compensated in the cell resizing step. In each iteration, scan flip-flops selected in the previous iteration are changed back to non-scan flip-flops and the partial scan selection algorithm starts all over again. The flip-flops selected in the previous iteration are likely to be selected again. However, if scan flip-flops have moved to less favored locations, they may not be selected again because of the increased weight. On the other hand, a flip-flop not scanned before may be selected to be a scan flip-flop if it has moved to a better location in the previous iteration. Therefore, as the placement of the cells keeps improving in each iteration, so is the choice of the scan flip-flops.

6. RESULTS

The APT placement algorithm is built around the Timberwolf 5.4 placement package [8] which is written in C. The program was run on a SUN SPARC IPC. Table 1 shows the results of APT with no timing and testability optimization on some of the ISCAS sequential benchmark circuits. The ISCAS benchmark circuits were mapped onto MSU (Mississippi State University) standard cells using the technology mapper in the OCT/VEM tools [15].

Table 1. Placement with no optimization

Ckts.	Width	Delay (*nsec*)	Fault Cov. (Eff) in %	CPU (min)
s298	973	42.23	86.0(94.4)	6
s953	1820	74.51	99.1(99.4)	18
s1238	1914	100.16	94.6(100.0)	29
s5378	5115	126.68	68.4(72.2)	156

Because only one template was provided for each cell in the MSU library, we made additional templates using the dimensions and parameters in the MCNC scalable CMOS library.

The second and third columns in the table show the width of the longest row in λ, and the delay in the most critical path. The next column provides the fault coverage and efficiency in percentage which were obtained by the HITEC [16] sequential circuit test generator. The last column shows the CPU time in minutes.

Table 2 shows the results of *APT* with the timing optimization only. The timing requirement was that the most critical path should have 90% of the delays obtained in Table 1.

Table 3 shows *APT* results with both timing and testability optimizations. In spite of the inserted scan flip-flops, the timing requirements were met for all of the circuits. Compared to Table 2 in which only timing optimization was done, the placement widths in Table 3 are larger because of the additional area due to scan flip-flops. Here, one can make several tradeoffs. If there

Table 2. Placement w/ timing optimization

Ckts	Area O. H.(%)	Timing Improv(%)	Fault Cov (Eff) in %	CPU (min)
s298	3	10.2	86.0(94.4)	8
s953	2	11.2	99.1(99.4)	24
s1238	4	10.6	94.6(100.0)	39
s5378	2	10.5	68.4(72.2)	217

Table 3. Placement w/ timing & test. optimization

Ckts	Area O. H.(%)	Timing Improv(%)	Fault Cov. (Eff) in %	CPU (min)
s298	10	10.2	98.9(100)	9
s953	8	10.3	100.0(100)	26
s1238	11	10.5	94.9(100)	43
s5378	12	10.1	97.2(100)	235

Table 4. Placement w/ test. optimization

Ckts	Area O.H.(%)	Delay O. H.(%)	Fault Cov (Eff) in %	CPU (min)
s298	7	1.9	99.4(100)	7
s953	6	0.0	100.0(100)	20
s1238	7	0.2	94.9(100)	33
s5378	10	0.0	97.2(100)	174

exists unused space at the end of the rows, one may reduce $\tau_{required}$ and rerun the *APT* to obtain a faster circuit. If more fault coverage and fault efficiency are desired, one may increase $cost_{max}$ of OPUS and run *APT* again. The tradeoffs between the performance and the testability are done by adjusting both $\tau_{required}$ and $cost_{max}$. Table 4 shows the results of *APT* with testability optimization only. As seen in the table, the performance degradation due to the scan flip-flops is little or none in all cases. This is because any performance degradation due to insertion of scan flip-flop in critical paths is countered by a modified placement of cells.

7. CONCLUSIONS

In this paper, we have incorporated the timing and the testability optimizations into the placement algorithm *APT* using

(1) an accurate timing model and timing bounds
(2) cell resizing
(3) partial scan flip-flop selection and insertion.

APT uses a more accurate cost available in the placement process to determine cell size and scan flip-flops. By addressing the testability and the performance optimization at the physical level, this approach enables the designer to make effective tradeoffs among area, timing and testability of a circuit.

REFERENCES

[1] A. E. Dunlop and V. D. Agrawal et. al., "Chip layout optimization using critical path weighting," *Proc. 21th Design Automation Conf.*, pp. 133-136, 1984.

[2] M. Burstein and M. Youssef, "Timing influenced layout design," *Proc. 22nd Design Automation Conf.*, pp. 124-130, 1985.

[3] M. Marek-Sadowska and S. P. Lin, "Timing driven placement," *Proc. of Int. Conf. on Computer-aided Design*, pp. 94-97, 1989.

[4] M. A. Jackson and E. S. Kuh, "Performance-driven placement of cell based IC's," *IEEE Design Automation Conf.*, pp. 370-375, 1989.

[5] P. S. Hauge, R. Nair, and E. J. Yoffa, "Circuit placement for predictable performance," *Proc. of International Conf. on Computer Aided Design*, pp. 88-91, 1987.

[6] S. Lin, M. Marek-Sadowska, and E. Kuh, "Delay and area optimization in standard-cell design," *27th IEEE Design Automation Conf.*, pp. 349-352, 1990.

[7] S. Kim, P. Banerjee, and S. Patil, "A layout driven design for testability technique for MOS VLSI circuits," *Proc. Int. Test Conference*, pp. 157-165, October, 1991.

[8] C. Sechen and et al., *The TimberWolfSC Distribution 5.4.* Yale University, July, 1989.

[9] V. Chickermane and J. H. Patel, "An optimization based approach to the partial scan design problem," *Proc. International Test Conference*, pp. 377-386, September, 1990.

[10] W. K. Luk, "A fast physical constraint generator for timing driven layout," in *28th Design Automation Conference*, San Francisco, pp. 626-631, June 1991.

[11] W. C. Elmore, "The transient response of damped linear networks with particular regard to wideband amplifiers," *J. Appl. Phys.*, vol. 19, pp. 55-63, Jan. 1948.

[12] P. Penfield, Jr., J. Rubinstein, and M. A. Horowitz, "Signal Delay in RC Tree Networks," *IEEE Trans. Computer Aided Design*, vol. CAD-2, no. 3, pp. 202-211, July 1983.

[13] P. K. Chan, "Algorithm for library-specific sizing of combinational logic," *Proc. 27th Design Automation Conference*, pp. 353-356, 1990.

[14] K. T. Cheng and V. D. Agrawal, "An economical scan design for sequential logic test generation," *Proc. 19th Int'l. Symp. on Fault-Tolerant Computing*, pp. 28-35, 1989.

[15] R. L. Spickelmier et al, *Octtools Distribution 3.0.* Electronics Research Lab, UC Berkeley, August, 1989.

[16] T. M. Niermann and J. H. Patel, "HITEC: a test generation package for sequential circuits," *Proc. European Conference on Design Automation*, pp. 214-218, February 1991.

A Performance Driven Macro–Cell Placement Algorithm†

Tong Gao, P. M. Vaidya, C. L. Liu

Department of Computer Science
University of Illinois at Urbana–Champaign

Abstract

In this paper, we present a new performance driven macro–cell placement algorithm. Placement of modules is guided by a set of upper– and lower–bounds on the net wire lengths. A convex programming algorithm is used to compute a set of upper–bounds on the net wire lengths which will ensure that timing requirements between input and output signals are satisfied. A set of lower–bounds is also computed to control signal skews at intermediate points of the circuit. Artificial nets are introduced between all pairs of modules. Lower–bounds on the lengths of the artificial nets are computed to avoid module overlaps in the placement. A modified min–cut placement algorithm is then used to generate a placement that satisfies the upper– and lower–bounds. We then employ an iterative procedure to modify the set of upper– and lower–bounds to improve the quality of the placement result.

1. Introduction

As technology advances, the size of modules in integrated circuits becomes smaller, while more modules are included in the circuits. Consequently, while the effect of intra-module delay becomes less significant, the effect of inter-module wire delay becomes more prominent. Thus, minimizing interconnecting wire delay becomes an important issue in improving VLSI circuit performance. Also, for a module to operate properly, input signals to the module must arrive more or less at the same time. The time difference between the latest arriving signal and the earliest arriving signal at a module is called the ***signal skew*** at the module. Both wire delays and signal skews depend on net wire lengths which in turn depend heavily on the physical placement of the modules. It is, therefore, important to study ***performance driven placement algorithms*** which can produce placements that satisfy the timing requirements of a design. Performance driven placement problems are in general difficult because of the incompatibility between timing requirements and physical requirements. Timing requirements are path oriented since delay between two points in the circuit depends on all the paths between them, while physical placement is net oriented since net wire lengths are determined by the placement of the modules.

The performance driven placement problem has been studied extensively in the past [2–13]. Most previous work can be grouped into two major categories, namely, net–based approach and path–based approach. In a path–based algorithm [3,4,5], timing requirements are considered explicitly and the algorithm tries to satisfy both timing requirements and physical requirements simultaneously during placement. In a net–based algorithm [6–11], timing requirements are first translated into physical requirements, such as net constraints. The placement algorithm then tries to generate a placement that satisfies the timing requirements under the guidance of the net constraints. One typical net–based approach is to first translate timing requirements into a set of upper–bounds on net wire lengths [6,9,10]. The upper–bounds are then used to guide the placement process. Utilization of upper–bounds solves the incompatibility problem between timing requirements and physical requirements. However, since timing requirements are path oriented, representing timing requirements by net constraints usually over–constrains the placement problem which in turn, might exclude some feasible placement solutions. In [11], an iterative approach was introduced to find a feasible placement solution that might have been excluded by a particular choice of upper–bounds. However, the algorithm in [11] considers only timing delays between input and output signals. Also, it only deals with gate–array technology.

In this paper, we present a new macro–cell placement algorithm which besides considering timing delay between input and output pins, also includes signal skew as part of the timing requirements. In our algorithm, upper–bounds on wire lengths of signal nets are computed to ensure that timing delay requirements are satisfied, lower–bounds on wire lengths of signal nets are computed to ensure that signal skew requirements at the modules are satisfied. Furthermore, artificial nets are introduced between all pairs of modules. Lower–bounds on the wire lengths of these nets based on the dimensions of the macro–cells are introduced to control module overlaps during placement. A new modified min–cut algorithm is then used to generate placements that satisfy the upper– and lower–bounds on the net wire lengths. By introducing bounds on net wire lengths, we are able to incorporate all the timing requirements and physical requirements in an unique way. In our algorithm, we also employ an iterative approach to modify the upper– and lower–bounds according to the previously generated placement in order to improve the quality of the placement result.

2. Problem formulation

Input to the placement algorithm consists of circuit specification and timing specification. A circuit consists of a set of m modules $\{M_1, M_2, \cdots, M_m\}$, a set of n nets $N = \{N_1, N_2, ..., N_n\}$, and a set of p primary input pins (PIs) and primary output pins (POs) $\{R_1, R_2, ..., R_p\}$. Module M_i has dimensions $w_i \times h_i$, and is to be placed anywhere on the chip as long as it does not overlap with another module. The placement area has dimensions $W \times H$. A net is an interconnection of a set of pins on the boundaries of the modules, and net wire length is estimated by the half perimeter of the bounding rectangle of the pins in the net. For simplicity in computation, it is assumed that all pins are located at the centers of the modules and a net is viewed as an interconnection of the modules where the pins reside. Both primary input pins and primary out-

† This work was partially supported by the National Science Foundation under grants MIP 89–06932, CCR–9057481, and CCR–9007195.

put pins have predetermined positions on the boundary of the placement area. The jth pin of module M_i is denoted P_{ij}. Timing specification of the circuit includes signal arrival times at the *PIs* (the signal arrival time at *PI* R_i is denoted a_i), the required signal arrival times at the *POs* (the required signal arrival time at *PO* R_k is denoted r_k), and the maximum allowable signal skew C_i at module M_i. Internal delays of the modules are given constants. We use d_i to denote the internal delay of module M_i. Delay in a net is proportional to the lumped capacitance of the net, which in turn is proportional to the net wire length. To simplify our notation, we consider net wire length and net delay as equivalent throughout this paper. The net wire length of net N_i is denoted z_i, and the signal skew at module M_i is denoted c_i.

The timing problem has been studied in [1]. For a combinational circuit without feedback loop, the arrival time a_{ij} of the signal at pin P_{ij} and the arrival time a_k at primary output pin R_k is computed by a forward traversal of the circuit starting from the *PIs*. It is assumed that the early arriving signals wait for the latest arriving signal at the modules. The signal skew c_i at module M_i can be computed by taking the time difference between the latest arriving signal and the earliest arriving signal at module M_i. Obviously, a given placement satisfies the timing requirements if $a_k \leq r_k$ for all *POs* R_k and $c_i \leq C_i$ for all modules M_i.

It was shown in [11] that the timing delay requirements can be expressed as a set of linear constraints

$$a_i + \sum_{all\ M_u} d_u - \sum_{all\ N_v} z_v \leq r_k \tag{2.1}$$

for all possible paths from any primary input pin R_i to any primary output pin R_k where the M_u's and N_v's are the modules and nets on the path from R_i to R_k, respectively.

If module M_i has only one input signal, the signal skew c_i is equal to 0. If M_i has $q > 1$ input pins $\{P_{i1}, P_{i2}, ..., P_{iq}\}$, the signal skew requirement $c_i \leq C_i$ can be expressed as a set of linear constraints

$$a_{ij} - a_{il} \leq C_i \tag{2.2}$$

for all input pins P_{ij} and P_{il}, $j \neq l$, which in turn, are equivalent to

$$(a_s + \sum_{all\ M_u} d_u + \sum_{all\ N_v} z_v) - (a_t + \sum_{all\ M_w} d_w + \sum_{all\ N_y} z_y) \leq C_i \tag{2.3}$$

for all the possible paths from any primary input pin R_s to pin P_{ij} and all possible paths from any primary input pin R_t to pin P_{il}. M_u's and N_v's are the modules and nets on the path from R_s to P_{ij}, respectively. M_w's and N_y's are the modules and nets on the path from R_t to P_{il}, respectively.

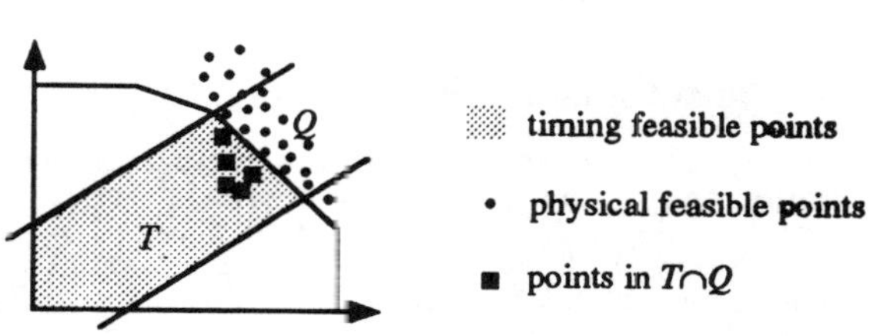

Figure 1. Relation between T and Q.

It is, therefore, possible to express the timing requirements as a set of linear constraints on the net wire lengths. A set of *positive* wire lengths $\hat{X} = \{\hat{z}_1, \hat{z}_2, ..., \hat{z}_n\}$ for the nets in $N = \{N_1, N_2, .., N_n\}$ is said to be *timing feasible* if it satisfies the linear constraints in (2.1) and (2.3), and is said to be *physical feasible* if there is a physical placement with net wire lengths equal to that in $\hat{X}$. A set of wire lengths $\hat{X}$ can be represented by a point in n dimensional space. Let T denote the set of all timing feasible points which will be referred to as the *timing feasible region* (Figure 1), and Q denote the set of all physical feasible points which will be referred to as the *physical feasible region* (Figure 1). A solution to the placement problem should have its net wire lengths corresponding to some point in $T \cap Q$ (Figure 1). Note that T is a convex polytope bounded by linear constraints whereas Q can be either a continuous region or a set of discrete points depending on the physical constraints in the placement problem.

3. Computing the bounds

Let $\hat{X} = \{\hat{z}_1, \hat{z}_2, ..., \hat{z}_n\}$ be a timing feasible point. Consider the box bounded by the following hyperplanes which we shall denote $B_{\hat{X}}$:

$$z_1 = \hat{z}_1,\ z_2 = \hat{z}_2,\ ...,\ z_n = \hat{z}_n$$

$$z_1 = 0,\ z_2 = 0,\ ...,\ z_n = 0.$$

Since the linear constraints in (2.1) have only positive coefficients, any point in $B_{\hat{X}}$ satisfies the constraints in (2.1) (Figure 2). On the other hand, since there are negative coefficients in (2.3), some of the points in $B_{\hat{X}}$ might not satisfy the constraints in (2.3) (e.g., point Y in Figure 2). Therefore, if $\hat{X}$ is used as a set of upper-bounds, a physical placement with net wire lengths satisfying the upper-bounds is guaranteed to satisfy the timing delay requirements, but is not guaranteed to satisfy the signal skew requirements. To enforce the signal skew requirements, we also compute a set of lower-bounds on the net wire length. Initially, the lower-bounds for all signal nets are set to 0. As will be discussed in section 6, they will be modified iteratively to control the signal skews in the circuit.

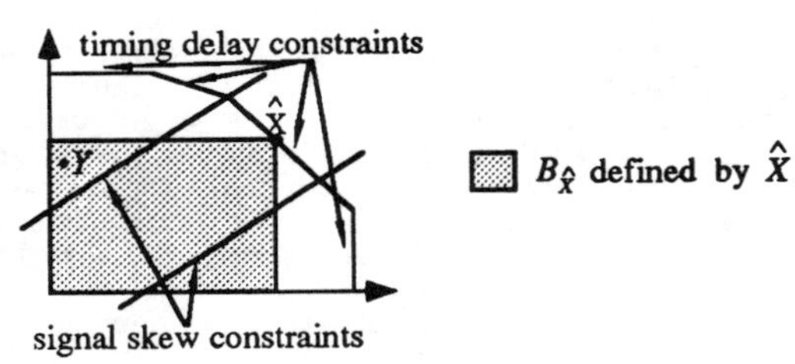

Figure 2. Upper-bounds and timing constraints.

As was suggested in [11], among the many possible choices of a timing feasible point $\hat{X}$ as a set of upper-bounds, we choose the one that maximizes the volume of $B_{\hat{X}}$ which shall allow maximum flexibility in physical placement. Consequently, we want to choose a point that is an optimal solution to the problem of maximizing the product $z_1 z_2 \cdots z_n$ subject to the linear constraints in (2.1) and (2.3). We use, instead, the objective function

$$\alpha_1 \log z_1 + \alpha_2 \log z_2 + \cdots + \alpha_n \log z_n$$

where $\alpha_1, \alpha_2, ..., \alpha_n$ are constants determined by the algorithm and are modified iteratively during successive iterations. The constants $\alpha_1, \alpha_2, ..., \alpha_n$ give us the flexibility to choose different sets of upper-bounds for different values of these constants. When all the α_i's are set to 1, maximizing the objective function

$$\alpha_1 \log z_1 + \alpha_2 \log z_2 + \cdots + \alpha_n \log z_n$$

is equivalent to maximizing the function $z_1 z_2 \cdots z_n$.

To solve the convex programming problem, we used a powerful convex programming algorithm described in [14]. The details of the algorithm can be found in [11,14], here we shall give only a brief summary of the algorithm.

The convex programming algorithm starts with an initial polytope β which encloses the optimal solution (Figure 3a), and iteratively generates cutting planes which cut off parts of β that do not contain the optimal solution (Figures 3b, 3c). In each iteration, a cutting plane is generated as follows: First, the center (see [14] for definition) of the polytope β is computed and its feasibility is checked by running a timing analysis for the circuit using the coordinates of the point as net wire lengths. If the center is feasible (e.g., all the timing requirements are satisfied), a cutting plane tangent to the surface of the objective function at the center is generated (Figure 3b). If the center is not feasible, a violated constraint in (2.1) or (2.3) is found and a cutting plane passing through the center and parallel to the constraint is generated (Figure 3c). The cutting plane divides the polytope β into two parts, and the cutting process continues for the part of the polytope that contains the optimal solution until the polytope β is small enough that a good estimation of the optimal solution can be made.

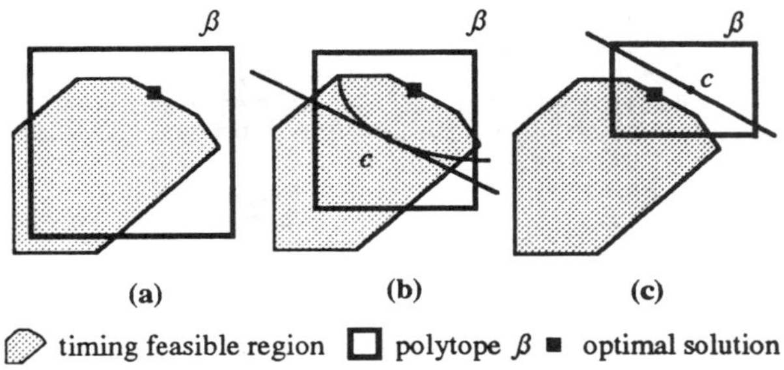

Figure 3. Computing a set of upper–bounds.

4. Minimizing overlaps between macro–cells

To control module overlaps during placement, we specify a minimum separation between each pair of modules. The minimum separation is expressed by the Manhattan distance between the centers of the two modules. Because the placement algorithm deals only with net wire lengths rather than distances between modules, an artificial net is introduced between each pair of modules. Denote the kth artificial net by N_{n+k}. (Nets N_1, N_2, ..., N_n are signal nets in the circuit.) The minimum separation requirement between a pair of modules is expressed as a lower–bound on the length of the corresponding artificial net. Note that there is no upper–bound on the length of an artificial net.

To guarantee that M_i and M_j will not overlap, it is sufficient to have $x_{n+k} \geq (w_i + w_j + h_i + h_j)/2$ where x_{n+k} is the wire length of the artificial net between M_i and M_j. Clearly, excessively stringent requirements might cause the placement algorithm to fail even though a feasible placement does exist.

To avoid over–constraining the placement problem, five different lower–bounds are computed for each artificial net according to the sizes and shapes of the modules the net connects. The five lower–bounds for net N_{n+k} are:

$$l_{k1} = (w_i + w_j)/2, \quad l_{k2} = (h_i + h_j)/2$$

$$l_{k3} = (w_i + h_j)/2, \quad l_{k4} = (h_i + w_j)/2$$

$$l_{k5} = (w_i + w_j + h_i + h_j)/2.$$

Note that each of the first four lower–bounds corresponds to a non–overlapping arrangement of M_i and M_j. In the placement phase, the placement algorithm tries to find a placement that satisfies as many of the lower–bounds as possible for each artificial net. The reason to have the lower–bounds l_{k3} and l_{k4} is that the placement algorithm allows a module to be rotated by 90 degrees. Obviously, the more lower–bounds that are satisfied, the better the chance that the modules will not overlap. Note that for the artificial nets it is not necessary to satisfy all the lower–bounds in order to have a non–overlapping placement.

5. The placement algorithm

After all the upper– and lower–bounds are computed, a placement algorithm is used to generate a placement that satisfies these bounds. A modified min–cut algorithm which tries to satisfy as many bounds as possible is used. In the placement process, the macro–cells are treated as points. The lower–bounds on net wire lengths of the artificial nets provide an effective way to handle variable sized modules as well as the case of non–gridded placement.

Basically, our placement algorithm is a modified min–cut algorithm. At each level of the partitioning hierarchy, net wire lengths are estimated and the estimated values are compared with their corresponding upper– and lower–bounds. Based on the result of the comparison, different weights are assigned to different nets. The weights are then used to compute the gains of the modules. Here we define the gain of a module to be the sum of the weights of the nets that are removed from the cut–set minus the sum of the weights of the nets that are added to the cut–set when the module is moved to its complementary partition.

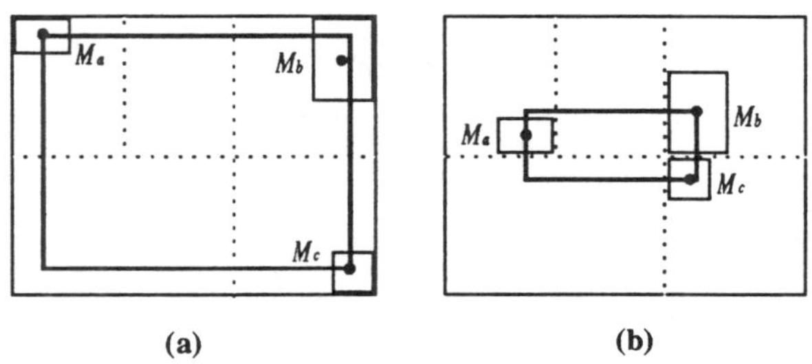

Figure 4. $\max_i$ and $\min_i$ for N_i connecting M_a, M_b, and M_c.

At any intermediate level of the partitioning hierarchy, the maximum and minimum possible wire lengths for each net are computed. Because a module can eventually be placed anywhere within its partition in the chip, the maximum wire length $\max_i$ for net N_i is the half perimeter of the smallest bounding rectangle that contains all the points at which the centers of the modules in the net may locate (Figure 4a), and the minimum possible wire length $\min_i$ is estimated as the half perimeter of the smallest bounding rectangle that contains at least one possible location for the center of each module in the net (Figure 4b).

The weight f_i for signal net N_i is the sum of f_{i1} and f_{i2} where f_{i1} and f_{i2} are computed as follows: Recall that $\hat{x}_i$ denotes the upper–bound on the net wire length of N_i. If $\max_i \leq \hat{x}_i$, f_{i1} is set to 0 because no matter how the partitions are further refined, the upper–bound $\hat{x}_i$ will be satisfied. If $\min_i > \hat{x}_i$, f_{i1} is set to a predetermined constant *maxcrit*1 because no matter how the partitions are further refined, the wire length will exceed $\hat{x}_i$. If $\max_i \geq \hat{x}_i \geq \min_i$, f_{i1} is computed according to the formula

$$round\,(\frac{\max_i - \hat{x}_i}{\max_i - \min_i}\times maxcrit1)^{\dagger}.$$

Similarly, for the lower-bound l_i on the net wire length of N_i, if $\min_i \geq l_i$, then f_{i2} is set to 0. If $\max_i \leq l_i$, then f_{i2} is set to $-maxcrit1$. If $\min_i \leq l_i \leq max_i$, f_{i2} is computed according to the formula

$$-round\,(\frac{l_i - \min_i}{\max_i - \min_i}\times maxcrit1).$$

Next, we discuss how the weight f_k is computed for artificial net N_{n+k}. Because the artificial nets are introduced to ensure minimum separation between modules, it is desirable to have them included in the cut-set. Thus, the weight for an artificial net should always be non-positive. The value of f_k lies within a predetermined range, from $-maxcrit2$ to 0. It is computed as follows: If the two modules M_i and M_j connected by N_{n+k} belong to different partitions, f_k is set to 0 because these two modules will not overlap. If these two modules are in the same partition of dimensions $width \times length$, f_k is computed as

$$round(-maxcrit2\times\frac{9-u}{9})$$

where u is, among the 9 inequalities listed below, the number of inequalities that are satisfied.

$$width - w_i/2 - w_j/2 \geq l_{k1}, \quad length - w_i/2 - w_j/2 \geq l_{k1}$$

$$width - h_i/2 - h_j/2 \geq l_{k2}, \quad length - h_i/2 - h_j/2 \geq l_{k2}$$

$$width - w_i/2 - h_j/2 \geq l_{k3}, \quad length - w_i/2 - h_j/2 \geq l_{k3}$$

$$width - h_i/2 - w_j/2 \geq l_{k4}, \quad length - h_i/2 - w_j/2 \geq l_{k4}$$

$$length + width - w_i/2 - h_i/2 - w_j/2 - h_j/2 \geq l_{k5}$$

Because a module can be rotated by 90 degrees, each of the inequalities corresponds to a possible non-overlapping placement of the two modules. Thus, if more of these inequalities are satisfied, the algorithm will have a better chance of finding a non-overlapping placement of the two modules.

After the gains are computed, Fiduccia's min-cut algorithms is called to divide each partition, A, into two sub-partitions, B and C. Since we are partitioning a collection of macro-cells, some area balancing rule is needed. Experimental results suggested that

$$area_A/2 - maxmod \leq area_B \leq area_A/2 + maxmod$$

is a reasonable balancing rule where $maxmod$ is the area of the largest module in partition A. After A is partitioned into B and C, the cut line is shifted to further balance the areas of the partitions according to the areas of the module in B and C. Suppose that the coordinates of the lower left corner of partition A are $(llx,\ lly)$ and the coordinates of the upper right corner of partition A are $(urx,\ ury)$. If the cut line is vertical, the new cut line will be positioned at

$$llx + \frac{area_B(urx - llx)}{area_B + area_C}.$$

If the cut line is horizontal, the new cut line will be positioned at

$$lly + \frac{area_B(ury - lly)}{area_B + area_C}.$$

Note if the shapes of the modules are flexible, by shifting the cut line, it is guaranteed that each partition will have enough area for all the modules in the partition.

The partitioning process continues on the sub-partitions until each sub-partition contains at most one module and the area of the sub-partition is small enough so that a good estimation of the module position can be made.

Even though the modified min-cut placement algorithm tries to satisfy as many of the upper- and lower-bounds as possible, it does not guarantee that all of them will be satisfied. Therefore, if the shape of the macro-cells are rigid, there might be some overlaps in the placement generated. These overlaps can be removed by local movement and rotation of the modules. Because the modules are moved locally, the timing performance of the placement will not be changed substantially.

6. Iterative modification of the bounds

Even though the placement algorithm tries to satisfy as many of the upper- and lower-bounds as possible, it does not guarantee that all the timing requirements will be satisfied. Therefore, after a placement is generated, a timing analysis is carried out for the placement obtained. If the placement satisfies all the timing requirements, then we are done. Otherwise, a violated timing requirement is found. Depending on whether it is a timing delay requirement or a signal skew requirement, either the set of upper-bounds or the set of lower-bounds will be modified for the next iteration of the placement algorithm. By adjusting the bounds in accordance with information obtained from previous placements, we are able to improve the quality of placement iteratively.

As was pointed out in [11], physical feasible points are clustered in groups. If the placement algorithm fails to generate a physical placement that satisfies the timing delay requirements, the set of upper-bounds should be moved closer to the physical feasible point corresponding to the physical placement generated. This is achieved in our algorithm by adjusting the weights α_i's in the objective function $\alpha_1 \log x_1 + \alpha_2 \log x_2 + \cdots + \alpha_n \log x_n$. The new α_i's are computed as

$$\alpha_i = \alpha_i' \times (1 - (\hat{x}_i - \hat{y}_i)/q)$$

where α_i' is the old weight for x_i in the objective function, $\hat{y}_i$ is the wire length of net N_i in the placement, $\hat{x}_i$ is the old upper-bound for net N_i, and $q = \max_{i=1 \text{ to } n} |\hat{x}_i - \hat{y}_i|$. Since in the new objective function the coefficients are larger for nets the wire lengths of which exceeded the corresponding upper-bounds, the convex programming algorithm will try to increase the values of the upper-bounds for these nets.

On the other hand, if a signal skew requirement is violated (Figure 5a,b), in order to force the point corresponding to the physical placement (e.g., Y in Figure 5a,b) into the timing feasible region, we can either decrease the value of the upper-bounds as shown in Figure 5a (changing the upper-bounds from $\hat{X}$ to $\hat{X}'$) or increase the value of the lower-bounds as shown in Figure 5b. If we decrease the value of the upper-bounds on wire lengths for some nets, the set of upper-bounds for all the nets need to be recomputed (Figure 5a). Therefore, we choose to increase the values of the lower-bounds in the case that a signal skew requirement is violated. If the signal skew requirement at M_i is violated, by backtracking from the net with the earliest arriving signal at M_i, we can find a path p from some primary input pin R_j to M_i the length of which

---

† $round(x)$ represents the smallest integer greater than x if the fractional part of x is greater than 0.5, and represents the largest integer less than or equal to x otherwise.

should be increased in order to satisfy the constraint $c_i \leq C_i$. Consequently, the lower-bounds for the nets on path p should be increased so that

$$a_j + \sum_{all\ N_k \in p} l_k + \sum_{all\ M_k \in p} d_k = a_{il} - C_i$$

where a_j is the signal arrival time at R_j and a_{il} is the arrival time of the latest arriving signal at M_i. In order to give the placement algorithm maximum flexibility during placement, the quantity $q = a_{il} - C_i - a_j - \sum_{all\ M_k \in p} d_k$ should be distributed over the lower-bounds of the nets on p such that $\prod_{all\ N_a \in p} (\hat{x}_a - l_a)$ is maximized. It can be shown that the solution to this optimization problem is to set the value of the lower-bounds for each net N_a on p to be

$$l_a = \frac{q\ \hat{x}_a}{\sum_{all\ N_k \in p} \hat{x}_k}.$$

We omit the proof of this result here.

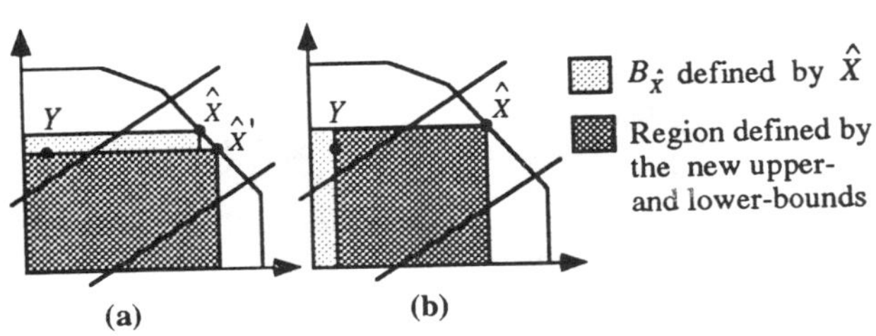

Figure 5. Modifying the upper- or lower-bounds.

7. Experimental result

Our algorithm was implemented in C and executed on a Sparc 2 workstation. We present the results on eight test examples the specification of which are summarized in Table 1. The columns "req." and "skew" are the required arrival time at the *POs* and the maximum allowable signal skew at the modules, respectively. Although our program is capable of handling different arrival times and required times at the *PIs* and *POs*, for simplicity of presentation of our results, we set the arrival times at all *PIs* to 0 and required time at all *POs* to be the same in our test runs. Also, even though our program is capable of handling different maximum allowable signal skew constraints at different modules, we set them to the same value.

circuit no.	no. of modules	no. of nets	W	H	req.	skew
1	72	145	16	16	100	20
2	83	172	16	16	165	20
3	97	197	16	16	110	22
4	97	222	16	16	110	19
5	102	213	16	16	110	22
6	102	236	16	16	110	22
7	104	219	16	16	115	22
8	102	229	16	16	115	22

Table 1. Circuit specifications.

Table 2 lists the results of the test runs. The result produced by Fiduccia's min-cut algorithm [15] is also listed for comparison. In Table 2, the column "arrive" is the latest arrival time at the primary outputs, the column "skew" is the maximum signal skew at all the modules in the circuit, and the column "wire" is the total wire length of the placement. On the average, our algorithms has a 26.0 percents improvement on input-output delays and a 42.0 percents improvement on the maximum signal skews when compared with the min-cut algorithm. Note that there is usually a trade off between timing performance and total wire length of the placement. (Since we are concerned with both timing delays and signal skews, improving timing performance does not always reduce total wire length.)

circuit	min-cut algorithm			our algorithm		
no	arrive	skew	wire	arrive	skew	wire
1	88.1	26.6	910.5	68.6	19.5	882.6
2	171.5	48.8	1057.3	107.9	19.6	1079.3
3	105.2	34.4	917.3	85.3	21.0	1043.2
4	136.1	42.3	1045.2	84.9	16.7	1117.6
5	96.2	31.1	1043.8	85.0	21.9	1242.5
6	132.7	47.9	1334.9	86.6	20.5	1333.6
7	94.3	25.9	1057.2	83.2	20.6	1235.5
8	116.2	39.3	1165.4	78.0	19.8	1213.6

Table 2. Results.

To illustrate the iterative procedure presented in section 6, intermediate results of each iteration when circuit 8 was placed are listed in Table 3. The row "spec." is the specified required time at the primary output pins and maximum allowable signal skew of the circuit. After the first iteration, both the timing delay and the signal skew requirements are not satisfied. Therefore, the set of upper-bounds was modified. After the second iteration, the timing delay requirements are satisfied but the signal skew requirements are still not satisfied. Therefore, the set of lower-bounds was modified. In the third iteration, the placement generated satisfies all the timing requirements. The placement generated by our algorithm is shown in Figure 6.

circuit 8	arrive	skew
spec.	115.0	22.0
iter. 1	120.6	44.2
iter. 2	82.5	31.2
iter. 3	78.0	19.8

Table 3. Intermediate results of placing circuit 8.

The average running time for computing the upper-bounds and generating a placement for each circuit are listed in Table 4. Note that the computation of the upper-bounds consumes a large protion of the total computation time.

circuit no.	1	2	3	4	5	6	7	8
upper-bounds	3.5	4.9	6.2	7.5	7.2	8.5	7.5	8.0
placement	0.3	0.4	0.5	0.4	0.5	0.6	0.5	0.5

Table 4. Running time in minutes.

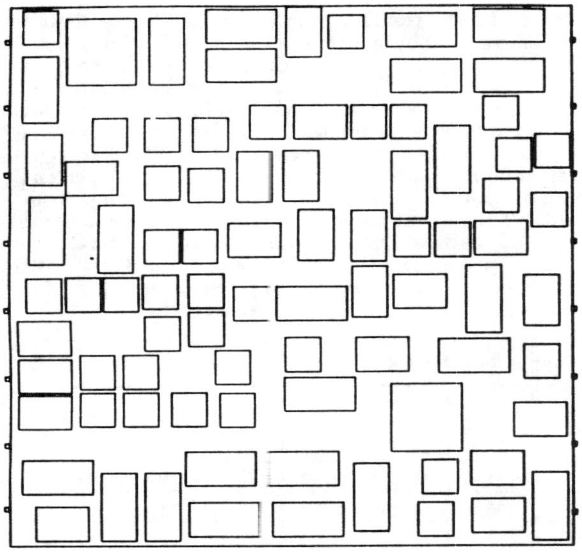

Figure 6. Placement of circuit 8.

8. Conclusion

In this paper, we presented a unique and effective approach to the performance driven macro–cell placement problem by translating both the timing and physical requirements into a set of upper– and lower–bounds on net wire lengths. We then presented a modified min–cut algorithm which tries to generate placements that satisfy as many of the upper– and lower–bounds as possible. We then discussed how the upper– and lower– bounds are to be modified to improve the timing performance of the placement. Experimental result are very encouraging.

References

[1] R. B. Hitchcock, G. L. Smith, and D. D. Cheng, "Timing Analysis of Computer Hardware," *IBM Journal of Research and Development*, 1982, vol. 26, no. 1, pp. 100–105.

[2] M. Burstein and M. N. Youssef, "Timing Influenced Layout Design," *Proc. 22th Design Automation Conference*, 1985, pp. 124–130.

[3] W. E. Donath et. al., "Timing Driven Placement Using Complete Path Delays," *Proc. 27th Design Automation Conference*, 1990, pp. 84–89.

[4] M. A. B. Jackson and E. S. Kuh, "Performance–Driven Placement of Cell Based IC's," *Proc. 26th Design Automation Conference*, 1989, pp. 370–375.

[5] A. Srinivasan, K. Chaudhary, and E. S. Kuh, "Ritual: A Performance Driven Placement Algorithm for Small Cell ICs," *Proc. ICCAD*, 1991, pp. 48–51.

[6] P. S. Hauge, R. Nair, and E. J. Yoffa, "Circuit Placement for Predictable Performance," *Proc. ICCAD*, 1987, pp. 88–91.

[7] A. E. Dunlop et. al., "Chip Layout Optimization Using Critical Path Weighting," *Proc. 21st Design Automation conference*, 1984, pp. 133–136.

[8] M. Marek–Sadowska and S. P. Lin, "Timing Driven Placement," *Proc. ICCAD*, 1989, pp. 94–97.

[9] Y. Ogawa et. al., "Efficient Placement Algorithms Optimizing Delay for High–Speed ECL Masterslice LSI's", *Proc 23rd Design Automation Conference*, 1986, pp. 404–410.

[10] W. M. Dai et. at., "BEAR: A New Building–Block Layout System," *Proc. ICCAD*, 1987, pp. 34–38.

[11] T. Gao, P. M. Vaidya, and C. L. Liu, "A New Performance Driven Placement Algorithm," *Proc. ICCAD*, 1991, pp. 44–47.

[12] S. Teig, R. L. Smith, and J. Seaton, "Timing Driven Layout of Cell–Based IC's," *VLSI System's Design*, pp 63–73, May 1986.

[13] M. Terai, K. Takahashi, and K. Sato, "A New Min–Cut Placement Algorithm for Timing Assurance Layout Design Meeting Net Length Constraint," *Proc. Design Automation Conference*, 1990, pp. 96–102.

[14] P. M. Vaidya, A New Algorithm for Minimizing Convex Functions Over Convex Sets," *Proc. 30th Annual IEEE Symposium on Foundations of Computer Science*, 1989, pp. 332–337.

[15] C. M. Fiduccia and R. R. Mattheyses, "A Linear–Time Heuristic for Improving Network Partitions," *Proc. 19th Design Automation Conference*, 1982, pp. 175–181.

Fuzzy Logic Approach to Placement Problem

Rung-Bin Lin and Eugene Shragowitz

Computer Science Department
University of Minnesota, Minneapolis, MN 55455

Abstract

A contemporary definition of the placement problem is characterized by multiple objectives. These objectives are : minimal area, routability, timing and possibly some others. This paper contains a description of the placement system based on the fuzzy logic approach. Linguistic variables, their linguistic values and membership functions are defined. Fuzzy logic rules govern the placement process. Details of implementation and experimental results are provided.

1. Introduction

Placement is one of the important steps in VLSI design. Many placement algorithms [1-7] have been introduced in the past few years. All these algorithms do not explicitly address two important aspects of the placement problem : presence of multiple conflicting objectives and utilization of expert knowledge in decision making. These aspects of the placement problem can be addressed by the technique based on the fuzzy set theory [9-11]. One attempt to apply the fuzzy set theory to problems of physical design was made in [8]. In this work the connectivity matrix was modified and then the fuzzy set theory was used to construct clusters for placement. The aforementioned work does not address the most interesting multi-objective aspect of the design problem, which is in the center of this paper. In this paper, we apply fuzzy reasoning to the placement of sea-of-gate arrays. We compare results obtained with and without fuzzy logic and draw some conclusions from this comparison.

2. Basics of Fuzzy Set Theory

A fuzzy set A, as defined by L.A. Zadeh, is a class of objects with continuum of grades of membership. An element may partially belong to a fuzzy set. This is contrary to the ordinary set theory, where an element is either in a set or not in a set. Formally, a fuzzy set is defined as follows [9] :

Definition 1 : *A fuzzy subset A of a universe of discourse X is defined as $A = \{ (x,\mu_A(x)) \mid all\ x \in X \}$, where X is a space of points and $\mu_A(x)$ is a membership function of $x \in X$ being an element of A.*

In general, a membership function $\mu(x)$ is mapping from X to the interval $[0,1]$. If $\mu_A(x)$=1 or 0, the fuzzy set A becomes an ordinary set. One possible membership function for the fuzzy set A is illustrated by Fig. 1. On this figure point x_1 belongs to the fuzzy set A with degree of membership $\mu_A(x_1)$. An example of the fuzzy set A is *a person's age near 50*. In this case, the space X is defined as the interval $[0,t]$, where t is the maximum age humans have ever had. A person of 45 has a degree of membership in the fuzzy set *a person's age near 50*. We will return to this example later in the paper.

Set operations such as union, intersection, and complementation, etc., are naturally introduced into the theory of fuzzy sets. Due to the nature of the membership function, results of fuzzy set operations are also fuzzy sets. The membership function of the resulting fuzzy set for basic operations on two fuzzy sets A and B is usually defined in one of the following forms :

$$\mu_{A \cap B}(x) = min\,(\mu_A(x),\mu_B(x)). \quad (1)$$

$$\mu_{A \cup B}(x) = max\,(\mu_A(x),\mu_B(x)). \quad (2)$$

$$\mu_{\neg A}(x) = 1.0 - \mu_A(x). \quad (3)$$

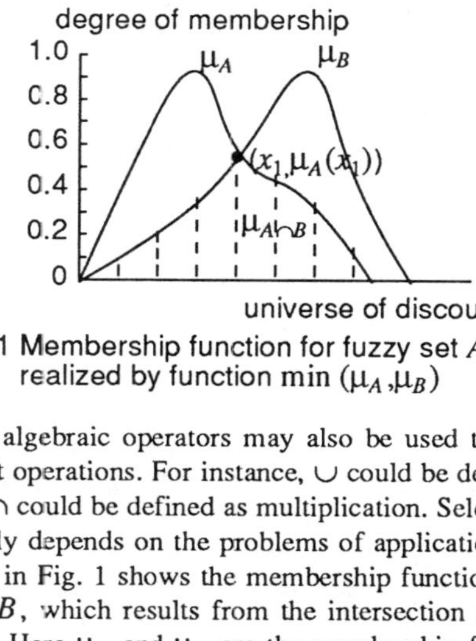

Fig. 1 Membership function for fuzzy set $A \cap B$ realized by function min (μ_A,μ_B)

Some other algebraic operators may also be used to define these fuzzy set operations. For instance, $\cup$ could be defined as addition and $\cap$ could be defined as multiplication. Selection of operators solely depends on the problems of applications. The shaded region in Fig. 1 shows the membership function of the fuzzy set $A \cap B$, which results from the intersection of fuzzy sets A and B. Here μ_A and μ_B are the membership functions for the fuzzy sets A and B respectively. For example, μ_A could be a membership function for the fuzzy set of *a person's age*

*This work is supported in part by the Mentor Graphics Corporation.

0738-100X/92 $3.00

near 50 and μ_B could be a membership function for the fuzzy set of *a person's height around 6 feet*.

3. Approximate Reasoning by Linguistic Variables

In the Webster's dictionary logic is defined as a science of correct reasoning. In the domain of problem solving, imprecise information always prevents us from making precise reasoning. Reasoning by a human is seldom exact. Therefore, the traditional binary logic based on Boolean theory of ordinary sets is not adequate to model human reasoning. Fuzzy logic is defined as "the art of fuzzy reasoning". Whenever "art" is mentioned, it involves more or less subjective judges and empirical studies about underlying domains. Reasoning made on imprecise domain is called fuzzy reasoning. A concept of fuzzy logic can be better understood if a "degree of membership" in a fuzzy set is interpreted as a "degree of truth". For example, equation (1) can be interpreted as that an element $x \in X$ is in the fuzzy set A and in the fuzzy set B with the degree of truth $\mu_{A \cap B}(x)$. That is, A *and* $B = \{(x, \mu_{A \cap B}(x)) \,|\, all\ x \in X\}$.

As defined by L.A. Zadeh, a linguistic variable is a variable whose values are words or sentences in natural or artificial language. The formal definition is referred to [10]. Informally, the concept of a linguistic variable can be explained by an example.

Example 1: As it was mentioned earlier, the *age near 50* is a fuzzy set. Here, *age* is a linguistic variable. The linguistic variable *age* may have many linguistic values. For instance, the set of linguistic values of *age* may be defined as {very young, somewhat young, young, somewhat old, old, very old, near 50, not close to 60 and not too old, ... }. Each linguistic value defines a fuzzy set on X, where X is a possible range of *age* measured in years. For instance, X can be the interval [0,200].

By using hedges such as *some, more, very, quite, fair, extremely*, etc., and connectives such as *and* and *or*, we can make approximate reasoning based on linguistic variables and their values. For instance, suppose we have another linguistic variable called *health* defined for each person. Linguistic values of *health* are defined as *very good, good, bad*, and *very bad*. Then we can make some reasoning based on values of these variables. For example,

> If a person's *age* is *young* and his *health* is *very good*, he can donate *some blood*.

Here *donate blood* is another linguistic variable with values *donate no blood, donate some blood, donate more blood, donate a lot of blood*. As it can be observed from above descriptions, fuzzy sets based on linguistic variables are used to model the imprecise information. Fuzzy logic is then used to make reasoning about the correct actions from the inputs. In the same manner, we will use this approach for the physical layout problems.

4. Multi-objective Optimization by Fuzzy Logic Approach

Consider a multi-objective optimization problem [11]

$$Maximize\ F(x), \tag{4}$$

where $F(x)=(f_1(x), f_2(x), \dots, f_m(x))$ is a vector-function, X is a solution space, and $f_i(x)$ is mapping from X to the set of the real number for all $i \in I=\{1,2,3,\dots,m\}$. Unlike the single objective optimization problem, no concept of optimal solution is universally accepted for the multi-objective optimization. In practical cases, the rating of an individual objective reflects the preference of decision makers. At best, a compromise between competing objectives can be expected. Fuzzy approach has some properties useful for this situation.

Let $g_i(\bullet)$ be mapping from the domain Y_i to the interval [0, 1] for all $i \in I$, where $Y_i \subset R$ is the range of $f_i(\bullet)$. $g_i(\bullet)$ is a measurement of the objective $f_i(\bullet)$ according to the degree of satisfaction of decision makers. $g_i(y_i)=1$, $y_i \in Y_i$ means complete satisfaction of the i-th objective, while $g_i(y_i)=0$ could mean 100% dissatisfaction. Therefore, a good compromise for multiple objectives will be the solution, which is the most satisfactory to decision makers. Therefore, the multi-objective optimization problem can be presented as follows :

$$Maximize\ G(x), \tag{5}$$

where $G(x)=G(g_1(f_1(x)), g_2(f_2(x)), \dots, g_m(f_m(x)))$ is a function of membership functions, $f_i(x) \in Y_i$ for all $i \in I$. The same definition can be applied to minimization. From now on, the function (5) is a target for optimization.

5. Fuzzy Logic Approach to Constructive Placement

When placement is performed manually by experts, they should select a cell and a position for it. Experts have to use their visual perceptions of partial placements to make the best selections. It is quite often that qualitative characteristics are available instead of quantitative measurements. Layout experts usually reason using qualitative characteristics and often are able to obtain good layouts. For example, one line of reasoning based on qualitative characteristics could be : *If a cell has strong connectivity to the partial placement and it has critical timing delays on its nets, it is a good candidate to be selected for placement.* The reasoning statement is called a "rule" in a fuzzy logic system. Three linguistic variables are employed here to make an approximate reasoning. They are *connectivity, timing delay*, and *candidate. The strong connectivity, the critical timing delay*, and *the good candidate* are respectively their linguistic values. Each linguistic value defines a fuzzy set. It is likely that more than one rule is necessary to make the best selections. For example, another rule could be : *If a cell has weak connectivity to the partial placement or it has small timing delays on its nets, it is a bad candidate.*

It seems, that the negation of the first rule is equivalent to the second rule. It is actually so in the binary logic reasoning. However, in fuzzy logic it is not. The reason is that *does not have strong connectivity* is not equivalent to *has weak connectivity* in terms of linguistic values. For the same reason the statements from the following pairs *not critical timing delay, good timing delay*, and *not bad candidate, good candidate* are not equivalent to each other. In case when rules are used to make a decision, we need to get a *crisp value* based on evaluations of these rules. A *crisp value* is a precise value in fuzzy logic system. In case of the placement problem, it is an evaluation of the degree of urgency for candidate cells and evaluation of the degree of quality for candidate positions. One can see that the given above rules of selection of candidate cell describe the multi-criteria decision making based on the fuzzy information (sets). In the following text, we will use the fuzzy approach to solve these problems.

5.1 Constructive Strategy of Placement

Two basic strategies of placement are known : constructive and iterative. Sometimes combinations of these two are used when the initial solution is obtained by the constructive placement and improved by the iterative technique.

The constructive placement consists of two major tasks (1) *select a cell,* and (2) *select a position for placing the selected cell.* These two tasks are repetitively executed until all cells are placed. The placement process can be defined as a quadruple $S(P,H,L,F)$, where

(1). P is a partial placement characterized by the set of already placed cells, area used, number of used and remaining routing tracks, amount of remaining space and other values;

(2). H is the set of remaining unplaced cells;

(3). $L = L^{\alpha} \cup L^{\beta}$ is a set of rules and functions that are used to generate a new state. Here L^{α} is a set of rules for decision making, and L^{β} is a set of objective functions involved in decision making;

(4). F is the set of final states.

Each instance of (P,H) defines a state of the placement process. A state is called the final state when $H = \varnothing$ or no space is available for placement. The placement process stops when it reaches the final state. A final state is called feasible, if all cells are successfully placed and no violations of timing constraints are reported. The purpose of the placement process is to reach a feasible final state in a reasonable amount of time.

Let a state of the constructive placement process at time t be described by (P_t,H_t), where P_t and H_t are respectively partial placement and set of remaining unplaced cells at time t. A state transition from (P_t,H_t) to (P_{t+1},H_{t+1}) is the result of solving the two major tasks. Selection of a cell and selection of a position for placement are governed by the set L of rules and functions. Multi-objective optimization problems arise in the process of solving these tasks.

In general, timing performance and usage of silicon area are the most important objectives for optimization in design of VLSI circuits . In order to optimize these two goals, the constructive placement must consider all the factors influencing the course of the process.

5.2 Criteria Related to Selecting a Cell

Placement is constructed by adding one cell at a time to the partial placement starting from the left side of the chip and moving to the right. Incremental global routing is performed after each cell is placed. For a partial placement P_t, some nets are completely routed, some nets are partially routed and the remaining nets are not routed. To select a cell η from the set H_t of remaining unplaced cells, we consider several criteria :

(1). connectivity $f_1((P_t,H_t),\eta)$ of selected cell η to partial placement P_t,

(2). connectivity ratio $f_2((P_t,H_t),\eta) = f_1((P_t,H_t),\eta)/f_\eta$, where f_η is the total number of pins for cell η,

(3). approximate average timing delay $f_3((P_t,H_t),\eta)$ on nets computed for the cell η.

Timing delays in this work were computed according to the methodology described in [6]. As one can see multiple criteria are important for selecting a candidate cell. By simple reasoning we can explain that a good layout is hard to obtain if a single criterion is used for selection of cells. For example, suppose criterion (1) is used alone. The cells with the largest numbers of pins are going to be selected first. In general, such cells are larger. Therefore, the larger cells will very likely concentrate on the left side of the chip. This phenomenon was observed in the solutions described in [6]. If criterion (1) is used alone, cells with $f_2((P_t,H_t),\eta) = 1$ will be delayed from being selected, even though these cells could be perfect candidates to be placed next. As a result, some interconnections will become very long. But, if the criterion (2) is used alone, the larger cells will be delayed from being selected because of the large value of f_η. This explanation illustrates why multiple criteria should be taken into consideration.

5.3 Applications of Fuzzy Logic to Selection of Cells

As it was defined earlier, the placement process is characterized by a quadruple $S(P,H,L,F)$. For each new state, a cell is selected first and then a position is found for it. The state transition is accompanied by updating information on resource usage in partial placement P. Suppose, the current state of a placement process is (P_t,H_t), and the space of points for selecting a cell η is defined as $X_t = \{all\ \eta \in H_t\}$. To select a cell is to find a point $x_\eta \in X_t$ such that it is the most satisfactory to the layout experts.

A set of key parameters that influence the placement process must be identified. For selecting a cell, these parameters are identified in the previous section. Three objective functions $f_1((P_t,H_t),\eta)$, $f_2((P_t,H_t),\eta)$, and $f_3((P_t,H_t),\eta)$ were defined there. Three linguistic variables *connectivity, connectivity ratio,* and *timing delay* are defined for these functions. Then the problem of selecting a cell may be formulated as follows :

$$Maximize\ F_1(x), \tag{6}$$

where $F_1(x)=(f_1(x_\eta),f_2(x_\eta),f_3(x_\eta))$ is a vector-function. A set of linguistic values is defined for each component of the objective function. In our case, only one linguistic value is defined for each variable. That is *strong connectivity, large ratio,* and *critical timing delay,* respectively.

As it was described in section 4, instead of finding maximum of (6), we are solving another problem :

$$Maximize\ G_1(x), \tag{7}$$

where $G_1(x)=G_1(g_1(f_1(x_\eta)),g_2(f_1(x_\eta)),g_3(f_3(x_\eta)))$ is a function. $g_i(\bullet)$ is a degree of satisfaction of layout experts with objectives $f_i(\bullet)$ for $i \in \{1,2,3\}$. These degrees of satisfaction are described by membership functions on fuzzy sets of linguistic values. That is to say, $g_1(\bullet)$, $g_2(\bullet)$, and $g_3(\bullet)$ are respectively the membership functions for *strong connectivity, large ratio, and critical timing delay.*

Membership functions are very important for successful applications of fuzzy logic. They quantify the qualitative characteristics of parameters. Empirical data, if they are available, or general knowledge are used to build the membership functions. Otherwise, the subjective judgement of decision makers is employed to define them. In our case, membership functions are easy to build. They are assumed to be non-decreasing functions, because each function $g_i(\bullet)$ measures degree of satisfaction of objective $f_i(\bullet)$ for all $i \in \{1,2,3\}$. For instance, the larger is the connectivity $f_i(\bullet)$ to the partial placement, the higher is the degree of membership in the fuzzy set *strong connectivity*. Specific instances of these membership functions for linguistic values of the linguistic variables *connectivity, connectivity ratio*, and *timing delay*, are given in Fig. 3, Fig. 4 and Fig. 5. These functions reflect some level of knowledge about cell library available to an expert prior to layout. The membership functions may be adjusted according to the physical and electrical characteristics of the circuits. For example, the membership function for *strong connectivity* may be adjusted with respect to the average connectivity for different designs. To simplify possible changes in the shapes of membership functions they are constructed as piece-linear functions.

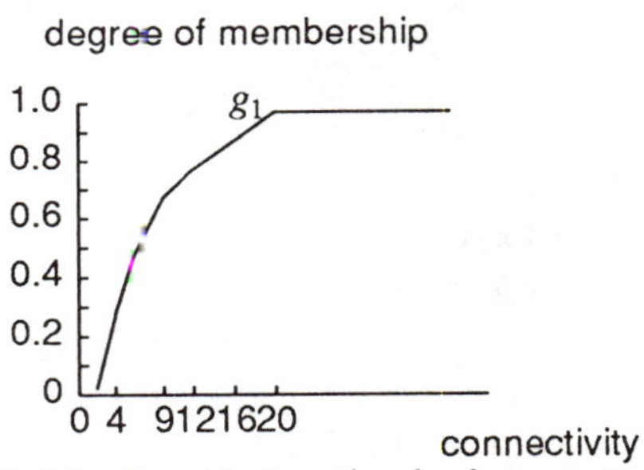

Fig. 3 Membership function for fuzzy set *strong connectivity*

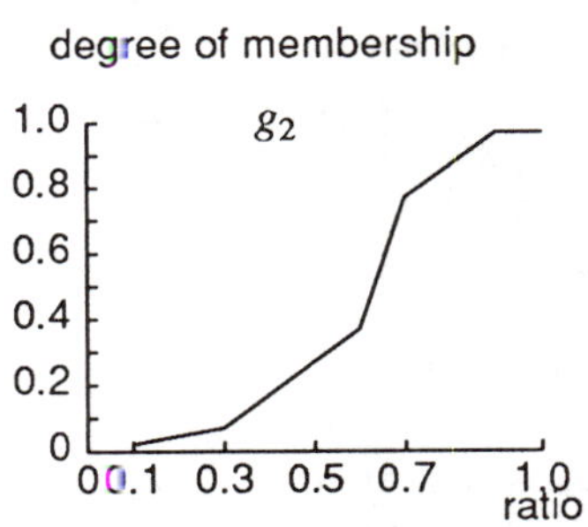

Fig. 4 Membership function for fuzzy set *large ratio*

In fuzzy logic maximization (or minimization) of the function $G_1(x)$ is formulated in terms of the values of linguistic variables as a rule-based algorithm. In our case, different rules represent alternative evaluations of the objectives. It is quite similar to the case when objectives are evaluated from different viewpoints. The results of evaluations of different rules are taken together to produce a value for making decisions. Two decision rules are formulated for this problem.

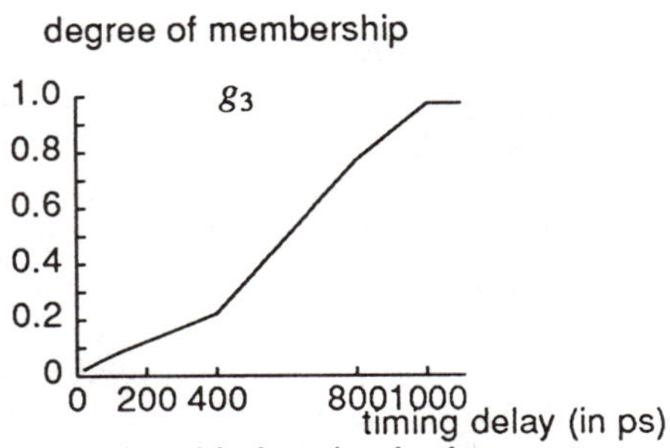

Fig. 5 Membership function for fuzzy set *critical timing delay*

Rule 1 : If a cell η has (*strong connectivity to the partial placement* or *large ratio*) and it produces *critical timing delay*, it is a *good candidate* to be placed next.

Rule 2 : If a cell η has *strong connectivity* and *large ratio* and it produces *critical timing delay*, it is a good candidate.

Here *a candidate* is a linguistic variable, and *a good candidate* is its linguistic value. The membership function of *a good candidate* is implicitly defined as the function $g(x_\eta)$ where $g(x_\eta)$ is a value resulting from an evaluation of a rule. At the first glance, it seems that Rule 2 is contained in Rule 1. But this is only true when these two rules are evaluated according to binary logic. It will be demonstrated below that in the fuzzy logic domain these two rules produce different results.

Finally, for each point $x_\eta \in X_t$, these rules are evaluated together to get a crisp value for decision making.

Evaluation of Rule 1 : It is easy to see that Rule 1 allows us to get a value of function $G_1(x)$ for any x_η in consideration in terms of the membership functions of linguistic values. Formally, it can be written according to laws of fuzzy logic in the following form :

$$G'_1(x_\eta) = max\ [\ g_1(f_1(x_\eta)), g_2(f_2(x_\eta))] \times g_3(f_3(x_\eta)).$$

In fuzzy logic "or" is realized by the function *max*. Fuzzy logic "and" is realized by algebraic *multiplication*. These two operators are mentioned in section 2. Notice, that *max* is not a compensatory operator. That is, $max\ (\ g_1(f_1(x_\alpha)),\ g_2(f_2(x_\alpha)))$ $> max\ (\ g_1(f_1(x_\beta)),\ g_2(f_2(x_\beta)))$ even if $min\ (\ g_1(f_1(x_\beta)),$ $g_2(f_2(x_\beta))) >> min\ (\ g_1(f_1(x_\alpha)),\ g_2(f_2(x_\alpha)))$. For example, if cell β has a fairly large connectivity and a fairly large ratio, it is still ranked inferior to cell α, which has a very large ratio, but a very small connectivity. Obviously, it is not wise to make a decision solely based on Rule 1. The same argument can be applied to the situation when, instead of fuzzy "or", fuzzy "and" is used to combine the first two functions $g_1(\bullet)$, $g_2(\bullet)$ and function *min* is used to realize Rule 1.

Evaluation of Rule 2 : Rule 2 proposes an another formulation of the function $G_1(x)$:

$$G''_1(x_\eta) = g_1(f_1(x_\eta)) \times g_2(f_2(x_\eta)) \times g_3(f_3(x_\eta)). \quad (8)$$

In fuzzy logic "and" operator is realized by algebraic *multiplication*, which is a compensatory operator. As a result, cell β may be ranked superior to cell α in the aforementioned case. However, if decision is made only based on Rule 2, it will not provide the best selection, either. Notice, that $g_3(\bullet)$ always participates in making decisions to account for timing factor. Evaluations for Rule 1 and Rule 2 are weighed according to following rule to produce a crisp value $G_1(x_\eta) = \lambda \times G'_1(x_\eta) + (1.0 - \lambda) \times G''_1(x_\eta)$, where $0 < \lambda < 1.0$ is the weighing factor. In our case, $\lambda = 0.5$, because it is assumed that these two rules are of equal importance. A function $G_1(x_\eta)$ gives a crisp value for the function $G_1(x)$ from formula (7). We compute for each cell η a value $G_1(x_\eta)$ and choose the cell with the largest value as the next cell for placement.

5.4 Numerical Example of Selecting a Cell

Let us assume, that *connectivity* f_1=12 for a candidate cell η, that corresponds to $g_1(12)$=0.78 degree of participation in the set *strong connectivity* from Fig. 3. Let *connectivity ratio* f_2=0.7, that corresponds to $g_2(0.7)$ =0.8 degree of membership in the set *large ratio* according to Fig. 4. Let the average *timing delay* $f_3 = 800$ picoseconds, i.e., the degree of membership in the fuzzy set *critical timing delay* is $g_3(800)$=0.75 according to Fig. 5. Then, evaluation of the Rule 1 for these data gives $G'_1(x_\eta)$= $max(0.78,0.8) \times 0.75$ $=0.8 \times 0.75=0.6$. Evaluation of the Rule 2 gives $G''_1(x_\eta)$= $0.78 \times 0.8 \times 0.75=0.468$. And, finally, the crisp value for the cell is $G_1(x_\eta)=0.5 \times 0.6+0.5 \times 0.468=0.534$. This value is be used to compare the given cell with other cells.

5.5 Criteria for Selecting a Position

After a cell has been chosen, a position must be selected for this cell. The decision-making process involves analysis of the current status of the partial placement. The following criteria could be considered when a new position α is selected for a cell η added to the partial placement P_t :

(1). timing delay $f_4((P_t,H_t),\eta,\alpha)$,

(2). wasted area $f_5((P_t,H_t),\eta,\alpha)$ = *wasted area / (cell area + wasted area)*,

(3). effect on future utilization of chip area $f_6((P_t,H_t),\eta,\alpha)$.

A feasible position α for the selected cell η must first satisfy constraints on geometric fit and routability. It is clear that the set of feasible positions varies for different cells due to these two constraints. As the partial placement grows, its right boundary will be indented. Effect of this phenomenon is registered by criterion (3). The unevenness of the border in y-dimension may result in wasting a substantial chip area if large cells are placed next to the indented boundary of the partial placement. As one can see, selection of a position is a multi-criteria decision problem. Optimization based on any single objective seldom creates good layouts.

5.6 Applications of Fuzzy Logic to Selection of Positions

The approach similar to the one described in the section on cell selection can be used to find positions for cells. A position should be considered as the most satisfactory one by the layout experts.

Three objective functions $f_4((P_t,H_t),\eta,\alpha)$, $f_5((P_t,H_t),\eta,\alpha)$, and $f_6((P_t,H_t),\eta,\alpha)$ were defined for this purpose in the previous section. The problem of deciding a position α for the selected cell η becomes

$$Minimize\ F_2(x), \quad (9)$$

where $F_2(x)=(f_4(x_\alpha),f_5(x_\alpha),f_6(x_\alpha))$ is a vector-function. There is no unique way to introduce a norm for a vector-function. In classical norms such as Eucleadian norm and others are counterintuitive in such situations and produce bad results in applications. Fuzzy logic provides solutions for such situations.

Three linguistic variables *timing delay, wasted area,* and *border evenness* are respectively defined for the above three functions. A set of linguistic values is defined for each linguistic variable. In our case, two linguistic values *small timing delay* and *large timing delay* are defined for *timing delay*. Two linguistic values *large wasted area* and *small wasted area* are defined for *wasted area*. Two linguistic values *good border evenness* and *bad border evenness* are also defined for *border evenness*.

Instead of finding an optimum of (9), we are solving the following problem :

$$Maximize\ G_2(x), \quad (10)$$

where $G_2(x)=G_2(g_4(f_4(x_\alpha)),g_5(f_5(x_\alpha)),g_6(f_6(x_\alpha)))$ is a function of membership functions. $g_i(\bullet)$ is a degree of satisfaction of layout experts with value of objective $f_i(\bullet)$ for $i \in \{4,5,6\}$. Simultaneously, we minimize a function $H_2(x) = H_2$ $(h_4(f_4(x)), h_5(f_5(x)), h_6(f_6(x)))$ which characterizes a degree of dissatisfaction of layout expert with values of objectives $f_i(\bullet)$ for $i \in \{4,5,6\}$.

Membership functions for these linguistic values are omitted here due to the lack of space. These membership functions are piece-linear and could be easily changed during tuning. They are constructed based on analysis of available data from the underlying domain. A clock period is an important element to shape membership functions for the *critical timing delay*. In our case, empirical data about average timing delay and average wiring length are available. They become great assets in building membership functions. If these data are not immediately available, the subjective judgement of decision makers will play a major role. In this case, after some trial runs, the shapes of membership functions can be easily adjusted.

Two rules have been introduced for selection of positions :

Rule 1 : If a position produces *small timing delay* and *small wasted area* and *good border evenness*, it is a *good position.*

Rule 2 : If a position does *not* produce ((*large timing delay* or *large wasted area*) and *bad border evenness*), it is a *good* *position.*

Here *position* is a linguistic variable, and *good position* is its linguistic value. These two rules are then evaluated to get a crisp value for each position α. Our approach works as follows :

Evaluation of Rule 1 : Rule 1 is based on the evaluation of the function $G_2(x)$. According to the laws of fuzzy logic, Rule 1 can be realized in the following form :

$$G'_2(x_\alpha) = min[g_4(f_4(x_\alpha)), g_5(f_5(x_\alpha)), g_6(f_6(x_\alpha))]$$

Evaluation of Rule 2 : Rule 2 is based on the evaluation of the function $H_2(x)$. Respectively, Rule 2 has the following formulation :

$$G''_2(x_\alpha) = 1.0 - H''_2(x_\alpha), \quad \text{where}$$

$$H''_2(x_\alpha) = max[h_4(f_4(x_\alpha)), h_5(f_5(x_\alpha))] \times h_6(f_6(x_\alpha)).$$

Values obtained from evaluation of Rule 1 and Rule 2 are used to obtain a crisp value for the position in consideration, i.e. $G'_2(x_\alpha)$ and $G''_2(x_\alpha)$ are combined to get a crisp value $G_2(x_\alpha) = \lambda \times G'_2(x_\alpha) + (1.0 - \lambda) \times G''_2(x_\alpha)$, where $0 < \lambda < 1.0$ is the weighing factor. In our case, we take $\lambda = 0.5$. Crisp values $G_2(x_\alpha)$ are compared for all possible positions α and a position with the largest value of $G_2(x_\alpha)$ is selected for placement. In the case of cell selection we used two different ways to characterize satisfaction with each possible solution. In this case the crisp value $G_2(x_\alpha)$ is obtained by maximizing the degree of satisfaction with the solution and by minimizing the degree of dissatisfaction with the same solution.

6. Experimental Results

The approach described above is tested on three industrial chips provided by Control Data Corporation. The implementation is based on the system developed in [6]. Only the decision-making procedures for selection of cells and positions are modified to implement the fuzzy logic approach. Table 1 lists the main characteristics of these chips. Table 2 compares our results with the results from [6]. It shows a dramatic improvement in timing performance even though the average connection length does not improve too much. It is worth to mention that we always obtained good timing for these three circuits for different shapes of membership functions. Moreover, no unrouted connections are produced in ETA test case. This is a very important characteristic of the solution because the unrouted connections could produce many long paths when they are routed. The increase in running time for all three chips is small, since the membership functions are relatively easy to evaluate.

	clock period	# of cells	# of nets	# of pins	utilization of cells
CDC-A	120 ns	1156	2515	7484	66%
CDC-B	120 ns	1705	2537	7330	61%
ETA	21 ns	[illegible]092	4983	13781	76%

Table 1. Three industrial sea-of-gates designs

	JUNE3	Fuzzy Approach
slack of the longest path (CDC-A)	+19 ns	+31 ns
# of long paths	0	0
average connection length	19	17
# of unrouted connections	0	0
slack of the longest path (CDC-B)	+8 ns	+65 ns
# of long paths	0	0
average connection length	17	21
# of unrouted connections	0	0
slack of the longest path ((ETA)	0 ns	+2 ns
# of long paths	0	0
average connection length	15	15
# of unrouted connections	3	0

Table 2. Placement results

7. Conclusion

We successfully applied fuzzy logic to optimize a process of decision making in physical design. Multiple objectives such as utilization of area, routability, and timing were considered simultaneously and balanced by fuzzy logic algorithms. Our experiments demonstrated that solutions obtained by fuzzy logic were of much better quality than those achieved by standard techniques.

8. References

[1] Michael Burstein and Mary N. Youssef, "Timing Influenced Layout Design", in Proc. of ACM/IEEE 22nd Design Automation Conference, pp. 124-130, 1985.

[2] A.E. Dunlop, V.D. Agrawal, and et. al., "Chip Layout Optimization Using Critical Path Weighting", Proc. 21st Design Automation Conf., pp. 142-146, 1987.

[3] M. Igusa, M. Beardslee, and A. Sangiovanni-Vincentelli, "ORCA : A Sea-of-Gates Place and Route System", Proc. 26th Design Automation Conf., pp. 122-127, 1989.

[4] M. Jackson, E.S. Kuh, "Performance-Driven Placement of Cell Based IC's", Proc. 26th Design Automation Conf., pp. 370-375, 1989.

[5] M. Marek-Sadowska, S.P. Lin, "Timing Driven Placement", in Proc. ICCAD'89, pp. 94-97.

[6] Suphachai Sutanthavibul and Eugene Shragowitz, "JUNE: An adaptive Timing-Driven Layout System", in Proc. of ACM/IEEE 27th Design Automation Conference, pp. 90-95, 1990.

[7] Suphachai Sutanthavibul and Eugene Shragowitz, "Dynamic Prediction of Critical Paths and Nets for Constructive Timing-Driven Placement", in Proc. of ACM/IEEE 28th Design Automation Conference, pp. 632-635, 1991

[8] M. Razaz and J. Gan, "Fuzzy Set Based Initial Placement for IC Layout", In Proc. European DAC, pp. 655-659, 1990.

[9] L.A. Zadeh, "Fuzzy Sets", Information and Control 8, 338-353 (1965).

[10] L.A. Zadeh, "The Concept of a Linguistic Variable and its Application to Approximate Reasoning-I", Information Science 8, pp. 199-249 (1975).

[11] Hans J. Zimmermann, "Fuzzy Sets , Decision Making, and Expert Systems", Kluwer Academic Publishers, Boston, 1987.

Delay Fault Test Generation for Scan/Hold Circuits using Boolean Expressions[1]

Debashis Bhattacharya[2]
Yale University
P.O. Box 2157, Yale Station
New Haven, CT 06518

Prathima Agrawal and Vishwani D. Agrawal
AT&T Bell Laboratories
600 Mountain Avenue
Murray Hill, NJ 07974

Abstract

A new test generation technique for path delay faults in scan/hold type circuits is presented. It uses reduced ordered binary decision diagrams (ROBDDs) to represent Boolean functions implemented by the subcircuits in a circuit, as well as to represent the constraints to be satisfied by the delay fault test. Two faults are considered for each path in the circuit under test and a pair of constraint functions, corresponding to the two time frames that constitute a transition, is evaluated for each fault. These constraints are then manipulated to obtain robust tests, if they exist; otherwise, non-robust tests are obtained from the constraint functions, if they exist. An implementation of this technique is used to analyze delay fault testability of several ISCAS '89 benchmark circuits. For circuits amenable to analysis using BDDs, the Boolean algebraic technique is seen to be much faster than existing branch-and-bound algorithms. For example, test generation for all path delay faults in s38584, without the aid of any path delay fault simulator, was completed in about 1.25 hours on a DECstation 5000/200.

1. INTRODUCTION

Most of previously reported research in test generation for delay faults in digital circuits is based on conventional branch-and-bound algorithms like PODEM, originally developed for single stuck-line (SSL) fault detection, modified to handle delay faults [1, 6, 12, 7]. However, automatic generation of delay tests for large circuits has still not become a practice and there is need for more efficient methods of generating such tests.

In this paper, we present a new technique for generation of tests for delay faults in sequential circuits with special scan features, viz., scan/hold type latches [13], using Boolean algebraic expressions represented as reduced ordered binary decision diagrams (ROBDDs). We show that this approach is significantly faster than existing techniques, and hence is very attractive for the development of a fast general purpose delay fault test generator program. Extensions of the method to conventional scan and non-scan sequential circuits, though not discussed, are possible and are currently under investigation. We use a *path delay* fault model [13], where a path starts at a primary input or a flip-flop output, and ends at a primary output or a flip-flop input. A fault causes the time of propagation of a transition through this path to exceed the clock period of the circuit. Assuming scan/hold type latches, we focus our attention only on the two time frames during which the transition in question occurs. Thus, the present consideration only involves *path activation*; the issues of *initialization* and *error propagation*, important for general non-scan sequential circuits, remain topics of on-going research.

Path activation requires a pair of vectors that causes a transition at the origin of the path, and propagates the transition to the end of the path. In the fully-transitional path (FTP) approach [9], the entire path is *sensitized* by each of the two vectors. In an alternative approach, proposed by Lin and Reddy [12], the entire path is sensitized in the second time frame, while in the first time frame it is only sensitized through the gates propagating non-controlling values. Note however, that merely satisfying the necessary constraints in the two time frames does not guarantee the absence of hazards on lines that are required to remain stable across the two vectors, and some post-processing may be required to ensure the absence of hazards.

Our algorithm uses a judicious mix of the these approaches to path activation. First, we attempt the generation of FTP and robust delay fault tests. If that fails, robust test generation using the criteria of Lin and Reddy (these are designated SIT tests, and are discussed in detail in Section 3), is attempted. If no robust test exists, we may generate non-robust tests also using the criteria of Lin and Reddy. The merits of this three-tier approach to delay fault test generation will be shown later in the paper.

Section 2 summarizes some important facts about ROBDDs. Details of our delay fault test generation algorithm are presented in Section 3. Section 4 contains experimental results from a prototype implementation of the algorithm, using several circuits from the ISCAS '89 benchmark suite [4].

[1] This research was supported in part by NSF grants MIP-8909658 and MIP-9101966.

[2] Consultant, AT&T Bell Laboratories.

0738-100X/92 $3.00 © 1992 IEEE

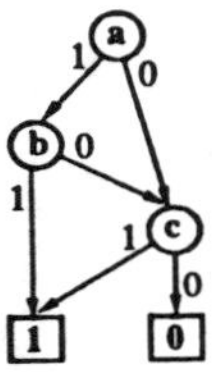

Fig. 1: BDD for the function $F(a,b,c) = a.b + c$

2. OVERVIEW OF BINARY DECISION DIAGRAMS

The principal data structure used to represent the behavior of various subcircuits in the circuit under test as well as to represent various fault detection constraints, is the reduced ordered binary decision diagram or ROBDD. This is a graphical representation of the Shannon decomposition of a Boolean function. A binary decision diagram (BDD) is a directed acyclic graph in which every node represents some Boolean function. It was originally described by Lee [11] and Akers [2]. In the BDD graph, every non-terminal node has associated with it one input variable. There are also exactly two outgoing edges for every non-terminal node. If a node represents the function F and is associated with the variable x, then one edge points to the node representing the function $F(\cdots, x = 1, \cdots)$ which is the Shannon cofactor with respect to x. Similarly, the other edge points to the node that represents the function $F(\cdots, x = 0, \cdots)$ which is the Shannon cofactor with respect to $\bar{x}$. There are only two terminal nodes representing the constant 1 and 0 functions. These obviously have no outgoing edges. An example of a BDD representing the function $F(a,b,c) = a.b + c$ is shown in Fig. 1.

BDDs have recently gained popularity as a method to represent and manipulate Boolean functions as a result of extensions due to Bryant [5]. He showed that if a fixed variable ordering is used and the graphs are reduced to their minimal form, then they form canonical representations, i.e., two ROBDDs with the same variable ordering represent the same function if and only if the two graphs are isomorphic. A recent computer implementation [3] makes this test even simpler by guaranteeing that every unique ROBDD will have a unique pointer associated with it. As a result, the check for equivalence of two functions generated using two different Boolean expressions is a constant time comparison operation. Bryant also gave polynomial time algorithms (in the size of the BDDs) for manipulating ROBDDs using unary and binary Boolean operators, and for performing composition of ROBDDs. In particular, if we have the ROBDDs for functions F and G and a binary Boolean operator OP, then we may obtain the ROBDD for the function (F OP G) in polynomial time.

We make extensive use of these two properties in our algorithm. For example, we use the first property to quickly determine when the test set is null; if the test set is null, it will always be represented by the constant 0 function. We use the second property to efficiently calculate the constraints that must be satisfied in order for a given path delay fault to be testable, as well as to obtain the actual test sets from the constraint functions.

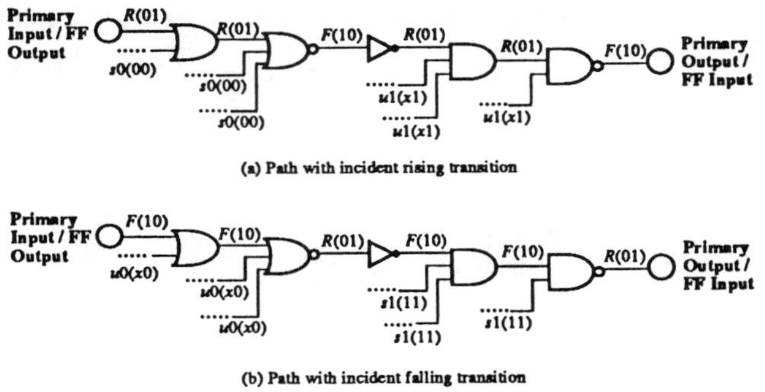

Fig. 2: Propagation of rising and falling transitions along a sample path.

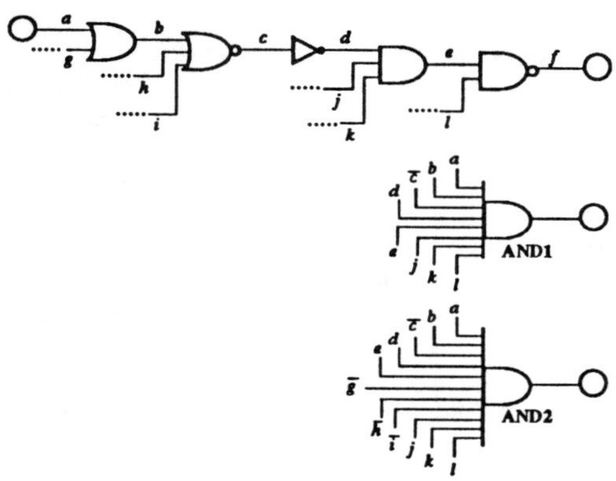

Fig. 3: Modified circuit with AND gates corresponding to constraints in t_1 and t_2 (incident falling transition).

3. TEST GENERATION

Our delay fault test generation mechanism is derived from the idea of satisfying constraints on off-path values specified according to Lin and Reddy [12]; the key concepts of this technique are illustrated in Figure 2. As explained in Section 2, we focus only on the two time frames that are denoted as t_1 and t_2. In Fig. 2, we show the propagation of a rising and falling transition along a sample path. The interpretation of the various signals is given in Table 1 [12]. The lines assigned the signals $s0$ and $s1$ are designated as *stable lines*. Obviously, the off-path signals in t_1 and t_2 rep-

Signal name	Value in t1	Value in t2	Hazard Requirement
$u0$	x	0	None
$u1$	x	1	None
$s0$	0	0	No static hazard
$s1$	1	1	No Static hazard
R	0	1	No dynamic hazard
F	1	0	No dynamic hazard
X	x	x	None

Table 1: Signal values used by Lin and Reddy for delay fault test generation.

resent two sets of constraints that must be satisfied in order for the transition to propagate along the given path. Moreover, these constraints depend not only on the path being considered, but also on the transition being considered.

In a recent paper [1], this concept was developed with a different view of the test generation problem. In that formulation, two extra AND gates designated, AND1 and AND2, were introduced for each combination of path and transition: output of AND1 (AND2) corresponds to the constraints in t_1 (t_2) being satisfied. For the path of Fig. 2 and the rising transition, the modified circuits with AND1 and AND2 gates are shown in Fig. 3. The path activation problem now becomes simply one of identifying a pair of vectors so that the first vector sets the output of AND1 to 1 and second vector sets the output of AND2 to 1. Note, however, this approach requires post processing steps for hazard elimination.

We adopt an approach somewhat similar to that of [1]. However, instead of introducing the AND1 and AND2 gates, we compute two constraint functions designated C_1 and C_2, so that the set of inputs satisfying C_1, (C_2) has a one-to-one correspondence to the set of inputs to the circuit under test which satisfies all the off-path constraints in time frame t_1 (t_2). In the discussion that follows, the path input is always designated a. We also assume that for any given function f, $\mathcal{S}(f)$ denotes the ON-set of f. Further, for a function F, $F|_x$ and $F|_{\bar{x}}$ denote the Shannon co-factors of F with respect to variable x and $\bar{x}$, respectively. Finally, given a cube q of a function F, $q|_{x=0}$, $q|_{x=1}$ and $q|_{x=X}$ represent the three cubes obtained by substituting 0, 1, and X, respectively, for variable x in q. The following theorems (proofs omitted due to page limitation) form the basis of our path delay fault test generator.

Theorem 1 For a given path fault with path input a, an FTP test exists *iff* the function $\mathcal{S}(R)$ is non-null, where R is a function calculated as follows:

$$R = \forall a(C_2) = C_2|_a \, . \, C_2|_{\bar{a}} \qquad (1)$$

Let q be a cube of the function R computed using equation 1. Then, the test vector pair ($q|_{a=1}$, $q|_{a=0}$) represents a robust test for the delay fault with incident falling transition, with only one possible exception, as stated next.

Theorem 2 For a given path fault with path input a, incident falling transition, and a pair of vectors ($q|_{a=1}$, $q|_{a=0}$), where q is a cube of the constraint function R from equation (1), an off-path line L cannot have a hazard *unless* the value of L is solely dependent on the convergence of a and $\bar{a}$ via multiple sensitized paths.

Theorem 2 implies that the vector pair ($q|_{a=1}$, $q|_{a=0}$) can be easily checked for robustness using following procedure. First, we assign values to primary inputs from the cube q in terms of 0, 1, and X (for don't care). Next, we compute all the implications resulting from these assignments. If any off-path line that contributes to C_1 is X, a static hazard is possible on this line, and hence the given vector pair is not robust.

If a robust FTP test cannot be found for the fault in question, we attempt to generate a pair of test vectors in which only the path input a changes, and which satisfies the general activation conditions given by Lin and Reddy [12]. Such tests are designated Single Input Transition or SIT tests.

Theorem 3 For a given path fault with input a and incident falling transition, a SIT test exists *iff* $\mathcal{S}(S)$ is non-null, where S is a function calculated as follows:

$$S = C_1|_a \, . \, C_2|_{\bar{a}} \qquad (2)$$

Modifications to Theorems 2 and 3 when the incident transition is rising instead of falling, follow obviously. Moreover, the following interesting necessary condition is seen to be true:

Theorem 4 A path delay fault is not robustly testable if it does not have a SIT test.

The converse is not true, since a theorem similar to Theorem 2 holds for SIT tests as well. Hence, the existence of a SIT test does not guarantee robustness, which has to be checked separately.

If $\mathcal{S}(S) = \Phi$, or if $\mathcal{S}(S) \neq \Phi$, but a robust test cannot be found, we may look for a non-robust test. The fault is non-robustly testable if

$$\mathcal{S}(N_1) = \mathcal{S}(C_1|_a) \neq \Phi \text{ and } \mathcal{S}(N_2) = \mathcal{S}(C_2|_{\bar{a}}) \neq \Phi$$

In such cases, the test consists of a pair of cubes, one each from N_1 and N_2, respectively. As in the case of SIT tests, this test pair truly uses the minimal path activation condition presented by Lin and Reddy. If either $\mathcal{S}(N_1) = \Phi$ or $\mathcal{S}(N_2) = \Phi$, path activation is impossible for this fault, i.e., the fault is redundant.

A test generation algorithm based on the above Boolean algebraic manipulations and checks is presented in pseudo-code format in Fig. 4. Note that currently only two cubes are extracted from the constraint functions R and S and checked for robustness. If the check fails, the fault is declared to be robustly untestable.

```
rdtests(path) {
  C2 = build_constr(FRAME2,path);
  if (C2 == Null_set)
    both fauls redundant;
  else {
    R = forall(head(path),C2);
    if (R == Null_set)
      for two incident transitions
        gen_SIT_test(path,C2,transition);
    else {
      extract_two_cubes(R);
      for two incident transitions {
        if ((robust(cube[0],path,transition) or
            (robust(cube[1],path,transition))
          fault tested robustly;
        else
          gen_SIT_test(path,C2,transition);
      }
    }
  }
}

gen_SIT_test(path,C2,transition) {
  C1 = build_constr(FRAME1,path,transition);
  if (C1 == Null_set)
    fault is redundant;
  else {
    S = SIT_constr(C1,C2,path,transition);
    if (S == Null_set)
      gen_NRD_test(path,C1,C2,transition);
    else {
      extract_two_cubes(S);
      if ((robust(cube[0],path,transition) or
          (robust(cube[1],path,transition))
        fault tested robustly;
      else
        gen_NRD_test(path,C1,C2,transition);
    }
  }
}

gen_NRD_test(path,C1,C2,transition) {
  (N1,N2) = NRD_constrs(path,C1,C2,transition);
  if ((N1 == Null_set) or (N2 == Null_set))
    fault is redundant;
  else {
    extract_cubes(N1,N2);
    fault is tested non-robustly;
  }
}

BOOLEAN robust(cube,path,transition) {
  assign_inputs(cube);
  for_all_stable_lines(path,transition) {
    if (find_value(line) == X)
      return(FALSE);
  }
  return(TRUE);
}

nrdtests(path) {
  C2 = build_constr(FRAME2,path);
  if (C2 == Null_set)
    fault is redundant;
  else {
    for two incident transitions {
      C1 = build_constr(FRAME1,path,transition);
      if (C1 == Null_set)
        fault is redundant;
      else
        gen_NRD_test(path,C1,C2,transition);
    }
  }
}
```

Fig. 4: Pseudo code for our BDD-based delay fault test generation algorithm.

```
extract_paths(netlist) {
  compute MIN_LENGTH;
  for each primary input I {
    if (I->min_dist_to_PO >= MIN_LENGTH)
      find_paths(I,MIN_LENGTH);
  }
}

find_paths(G,trheshold) {
  add G to path traced so far;
  if (G is a primary output) {
    if (generate_path_file)
      print path traced;
    else
      rdtests(path traced);
  }
  else {
    threshold--;
    for each output B of G {
      if (B->min_dist_to_PO >= threshold)
        find_paths(B, threshold);
    }
  }
  delete G from path traced so far;
}
```

Fig. 5: Pseudo code for path-extraction algorithm.

This obviously provides a conservative estimate of the robust detectability of path faults in the circuits considered. However, judgement about this heuristic should be reserved until experimental results are discussed in Section 4. A refinement to the algorithm that permits extraction of multiple cubes from R and S and checks them all for robustness, is currently under way.

4. IMPLEMENTATIONAL DETAILS AND EXPERIMENTAL RESULTS

The algorithm of Fig. 4 was implemented in a C language program, TSUNAMI-D, and run on a DECstation 5000 /200. It contains a path extractor to extract all paths from the given circuit having path lengths greater than MIN_LENGTH, where *path length* is measured by the number of on-path lines, treating fanout stems and fanout branches as separate lines. MIN_LENGTH is computed as follows:

$$\text{MIN\_LENGTH} = \max\{\,2,\ \text{MIN\_P\_LENGTH},\ \text{FRAC*MAX\_P\_LENGTH}\,\}$$

where MIN_P_LENGTH and MAX_P_LENGTH represent the shortest and longest path lengths in the circuit, and FRAC (default zero) is an optional user supplied fraction ($0 < \text{FRAC} \leq 1$) that allows only those paths whose lengths fall within the given fraction of the longest path in the circuit, to be extracted; paths of length smaller than two correspond to direct connections between inputs and outputs, and such paths are not considered explicitly. This path extractor is simply a modified depth-first search of the circuit as outlined in Fig. 5. In our implementation, one has the option to either print out all paths in a file, or to generate tests directly as paths are extracted.

This implementation was used to perform experi-

Circuit	Total faults	Non-robustly detected	%Cov.	Total time (secs.)	BDD Memory (Mb)
s208	290	290	100.0	0.1	0.10
s298	462	352	76.2	0.1	0.06
s344	710	643	90.6	0.3	0.17
s382	800	704	88.0	0.3	0.14
s386	414	414	100.0	0.2	0.12
s420	738	738	100.0	1.0	0.62
s510	738	738	100.0	0.5	0.26
s526n	816	706	86.5	0.3	0.14
s641	3,488	2,096	60.1	6.7	1.31
s820	984	984	100.0	1.0	0.51
s838	2,018	2,018	100.0	8.3	1.31
s953	2,266	2,266	100.0	2.3	0.87
s1196	6,194	3,708	59.8	8.7	1.31
s1488	1,924	1,916	99.6	2.4	0.96
s27	56	50	89.3	0.0	0.00
s349	730	643	88.1	0.3	0.16
s400	898	725	80.8	0.4	0.14
s444	1,070	723	67.5	0.4	0.15
s526	820	708	86.4	0.3	0.14
s713	43,624	2,066	4.7	15.1	0.54
s832	1,012	996	98.4	1.0	0.53
s1238	7,116	3,663	51.4	9.0	1.31
s1423	89,452	33,981	38.0	878.5	1.31
s1494	1,952	1,926	98.7	2.4	0.96
s5378	27,084	19,413	71.7	284.1	1.31
s13207	2,690,638	162,798	6.1	1,566.9	1.31
s35932	394,282	38,372	9.7	127.4	1.31
s38584	2,161,442	170,291	7.9	1,934.7	5.24

Table 2: Non-robust test generation for all delay faults.

ments on several circuits from the ISCAS '89 suite of benchmark circuits [4]. First, results from non-robust test generation using TSUNAMI-D are presented in Table 2. In order to compare these results to previously reported delay fault test generation algorithms, e.g., those in [1, 6, 7], all CPU times need to be normalized with respect to the corresponding machine speeds. The data in Table 2 clearly show that for circuits amenable to analysis using BDDs, TSUNAMI-D is significantly faster than branch-and-bound based delay fault test generators. This is especially noticeable for circuits with large number of paths like s1423, s38584, etc., for which, TSUNAMI-D is faster than the other techniques by nearly two orders of magnitude.

Next, tests were generated using the FTP and SIT strategies explained in Section 3, and outlined in Fig. 4. A summary of the experimental results are presented in Table 3. Again, for circuits with large number of paths, TSUNAMI-D is faster than the other methods by two orders of magnitude or more. Moreover, memory requirement of TSUNAMI-D is seen to be quite small, ranging from less than 100Kbytes to about 5.24Mbytes, for the circuits considered.

As discussed in Section 3, the limitation to the extraction of only two cubes from functions R and S provides us with a *conservative estimate* of the actual robust delay fault coverage. However, from Theorem 4, the number of faults for which SIT tests can be found provides a simple upper bound on the robust delay fault testability for the circuit under test. An inspection of Table 3 shows that the percentage of faults which are SIT testable, but not robustly testable, is less than 7 percent in all the cases. For most of the circuits considered, the difference is very small at less than 4 percent. Thus, the heuristic of extracting only two cubes for each fault, although simple, is extremely useful in practice.

The run times for robust and non-robust test generation, are not easily comparable, since the time required to manipulate ROBDDs is very sensitive to number of common subgraphs met in the process. For circuits with large number of paths, the run times do exhibit the expected behavior, i.e., non-robust test generation is significantly faster than robust test generation, although the same does not hold for the smaller circuits. Interestingly, the run time per fault, for robust or non-robust test generation, is remarkably close for a wide range of circuit sizes with widely varying number of paths.

Finally, the data in Tables 2 and 3 reveal an interesting relationship between the delay fault testability and the single stuck-line (SSL) fault testability of a circuit. In both tables, the benchmark circuits have been separated into two groups, the first contains no undetectable SSL fault while the second group contains undetectable SSL faults. The group with no undetectable SSL fault is clearly seen to have significantly higher delay fault coverage than the other group, both for robust and non-robust test generation. We believe this is explained by the fact that a redundant SSL fault, especially on a line shared between multiple paths, leads to a large number of undetectable path delay faults. Implementations of better path selection criteria are also currently under way.

5. CONCLUSIONS AND FURTHER WORK

We have presented a new ROBDD-based technique to generate tests for path delay faults in scan/hold circuits. We employ the conventional FTP approach to path activation, as well as a new approach, designated SIT, for robust (hazard-free) test generation. If this attempt fails, we try to generate a non-robust test. Both SIT tests and non-robust tests are generated using the general path activation conditions presented by Lin and Reddy [12]. Unlike other published procedures, every major step in the test generation process presented consists of manipulation of Boolean functions, and does not require the enumeration of input patterns or covers. This algorithm has been implemented in C, and experimental results presented using several benchmark circuits from the ISCAS '89 suite show that the algebraic approach could be faster than existing methods by as much as two to three orders of magnitude for large circuits.

Circuit	Total faults	FTP tested	%Cov.	FTP+ robustly tested	%Cov.	Total time (secs.)	FTP+SIT tested	%Cov.	FTP+SIT +robustly tested	%Cov.	Total time (secs.)	BDD Memory (Mb)
s208	290	290	100.0	290	100.0	0.1	290	100.0	290	100.0	0.1	0.07
s298	462	336	72.7	353	72.1	0.1	345	74.7	342	74.0	0.2	0.05
s344	710	602	84.8	578	81.4	0.5	623	87.7	599	84.4	0.5	0.11
s382	800	658	82.3	643	80.4	0.4	680	85.0	665	83.1	0.4	0.11
s386	414	412	99.5	412	99.5	0.2	413	99.8	413	99.8	0.2	0.09
s420	738	738	100.0	738	100.0	1.1	738	100.0	738	100.0	1.1	0.37
s510	738	720	97.6	720	97.6	0.7	729	98.8	729	98.8	0.7	0.14
s526n	816	686	84.1	683	83.7	0.4	697	85.4	694	85.1	0.4	0.10
s641	3,488	1,682	48.2	1,627	46.6	6.7	2,044	58.6	1,979	56.7	8.2	1.21
s820	984	974	98.0	972	98.8	1.1	982	99.8	980	99.6	1.2	0.29
s838	2,018	2,018	100.0	2,018	100.0	9.2	2,018	100.0	2,018	100.0	9.2	1.31
s953	2,266	2,246	99.1	2,246	99.1	3.5	2,256	100.0	2,256	100.0	3.5	0.49
s1196	6,194	3,140	50.7	3,083	49.8	11.7	3,618	58.4	3,562	57.6	15.0	0.98
s1488	1,924	1,832	95.2	1,832	95.2	2.8	1,875	97.5	1,875	97.5	3.0	0.59
s27	56	48	85.7	48	85.7	0.0	50	89.3	50	89.3	0.0	0.00
s349	730	600	82.2	576	78.9	0.4	623	85.3	599	82.1	0.5	0.11
s400	898	626	69.8	626	69.8	0.4	696	77.5	664	74.0	0.5	0.12
s444	1,070	624	58.3	564	52.7	0.5	694	64.9	585	54.7	0.7	0.14
s526	820	686	83.7	682	83.2	0.4	697	85.1	693	84.5	0.4	0.10
s713	43,624	1,652	3.8	895	2.1	16.2	2,009	4.6	1184	2.7	20.6	0.49
s832	1,012	974	96.2	968	95.7	1.2	991	97.9	984	97.2	1.2	0.32
s1238	7,116	2,874	40.4	2,834	39.8	11.6	3,602	50.6	3,654	49.9	16.6	0.94
s1423	89,452	24,880	27.8	20,057	22.4	762.1	29,883	33.4	23,220	26.0	926.5	1.31
s1494	1,952	1,826	93.5	1,826	93.5	2.9	1,882	96.4	1,882	96.4	3.0	0.59
s5378	27,084	17,638	65.1	17,196	63.5	263.9	18,886	69.7	18,248	67.3	296.9	1.31
s13207	2,690,638	81,540	3.0	25,453	0.9	2,619.4	83,588	3.1	27,484	1.0	4,286.0	1.31
s35932	394,282	35,172	8.9	21,367	5.4	292.9	38,052	9.6	21,655	5.5	439.8	1.31
s38584	2,161,442	118,262	5.5	79,135	3.7	2,786.3	138,780	6.4	90,146	4.2	4,790.2	5.24

Table 3: Robust test generation for all delay faults using FTP and SIT strategies.

We are currently working on extending the method presented in several directions including extraction of multiple cubes from the constraint functions R and S, and generalization to handle conventional-scan, partial-scan, and non-scan circuits.

REFERENCES

[1] Agrawal, P., Agrawal, V., and Seth, S.C., "A new method for generating tests for delay faults in non-scan circuits," *Proc. 5th Int'l Conf. on VLSI Des.*, Bangalore, India, Jan. 1992, pp. 4-11.

[2] Akers, S.B., "Binary decision diagrams", *IEEE Trans. Comp.*, Vol. C-27, June 1978, pp. 509-516.

[3] Brace, K.S., Rudell, R.L., and Bryant, R.E., "Efficient implementation of a BDD package," *Proc. 27th ACM/IEEE Des. Auto. Conf.*, June 1990, pp. 40-45.

[4] Brglez, F., Bryan, D., and Kozminski, K., "Combinational profiles of sequential benchmark circuits," *Proc. Int'l. Symp. on Circuits and Systems*, May 1989, pp. 1929-1934.

[5] Bryant, R.E., "Graph-based algorithms for boolean function manipulation," *IEEE Trans. Comp.*, Vol. C-35, August 1986, pp. 677-691.

[6] Cheng, K.T., Devadas, S., and Keutzer, K., "Robust delay-fault test generation and synthesis for testability under a standard scan design methodology," *Proc. 28th ACM/IEEE Des. Auto. Conf.*, 1991, pp. 80-86.

[7] Fuchs, K., Fink, F., and M.H. Schulz, "DYNAMITE: An efficient automatic test pattern generation system for path delay faults," *IEEE Trans. CAD*, Vol. 10, October 1991, pp. 1323-1335.

[8] Ghosh, A., Devadas. S., and Newton, A.R., "Test generation and verification for highly sequential circuits," *IEEE Trans. on CAD*, Vol. 10, May 1991, pp. 652-667.

[9] Glover, C.T., and Mercer, M.R., "A method of delay fault test generation," *Proc. 25th ACM/IEEE Des. Auto. Conf.*, June 1988, pp. 90-95.

[10] Goel, P., "An implicit enumeration algorithm to generate tests for combinational logic," *IEEE Trans. Comp.*, Vol C-30, March 1981, pp. 215-222.

[11] Lee, C.Y., "Representation of switching circuits by binary-decision programs," *Bell Sys. Tech. J.*, Vol. 38, July 1959, pp. 985-999.

[12] Lin, C.J., and Reddy, S., "On delay fault testing in logic circuits," *IEEE Trans. CAD*, Vol. CAD-6, No. 5, Sept. 1987, pp. 694-703.

[13] Malaiya, Y.K., and Narayanaswamy, R., "Modeling and testing for timing faults in synchronous sequential circuits," *IEEE Des. & Test of Computers*, Vol. 1, Nov. 1984, pp. 62-74.

[14] Stanion, T., and Bhattacharya, D., "TSUNAMI: A path-oriented scheme for algebraic test generation," *Proc. 21st Symp. on Fault-Tol. Comp.*, June 1991, pp. 36-43.

Delay Fault Models and Test Generation for Random Logic Sequential Circuits

Tapan J. Chakraborty
AT&T Bell Laboratories
P. O. Box 900,
Princeton, NJ 08540

Vishwani D. Agrawal
AT&T Bell Laboratories
600 Mountain Avenue.
Murray Hill, NJ 07974

Michael L. Bushnell
Dept. of Electrical & Computer Eng.
Rutgers University,
New Brunswick, NJ 08855

Abstract **- In this paper, we study delay fault models and the corresponding test generation methodology for general sequential circuits. Our test generation method is based on *transition* and *hazard* states of signals and is applicable to any sequential circuit irrespective of its structure, including the circuits where the flipflops can form a scan register. No hold requirement need be imposed on the flipflops. To test a combinational path between two flipflops of the sequential circuit a signal transition is placed at the source flipflop and is propagated through the path. The final logic value is latched in the destination flipflop. An incorrect logic value will be captured in the destination flipflop if the path delay exceeds the clock period. The logic value captured in the destination flipflop is observed at a primary output through path sensitization. The test thus contains two vector sequences. The first sequence initiates and propagates a signal transition from the source flipflop to the destination flipflop; the second vector sequence propagates the logic value captured in the destination flipflop to a primary output. We propose three fault models based on specific assumptions about the initial states of flipflops for the vector sequence that propagates the captured logic value to a primary output. To ensure that the tests are independent of gate delays, we introduce an additional *hazard* condition that identifies possible multiple transitions between vectors. Both vector sequences are generated by time-frame expansion procedure. We present the results of our test generation method for various fault models on the ISCAS benchmark circuits.**

1. Introduction

The timing behavior of a synchronously clocked digital design is often verified by static path analysis [1]. This analysis can identify all potentially critical paths. It, however, makes no attempt to propagate signals through these paths. In reality, some paths may be nonfunctional or *false*. The current state of the art allows analysis of false paths in combinational or scan type of sequential circuits [2]. For device testing, where the stuck-at fault model still dominates the field, the designer must manually augment the tests for critical path testing.

Timing faults are often referred to as *delay faults*. Their testing is important in the production of today's VLSI circuits that are increasingly becoming faster and more complex. In a synchronous design, the delays of combinational paths must not exceed the clock period. Testing of delay faults checks if the propagation time of a signal transition through a combinational path exceeds the clock period. The test should, (1) produce a signal transition at the beginning of the path, (2) propagate the transition through the designated path, and (3) allow observation of the value latched at the destination flipflop of the path. In addition, a test must measure the delay of the specified path irrespective of all other delays in the circuit. Such tests are known as *robust* [3,4].

It is possible to select a minimal set of paths for delay testing [5]. A large amount of recent literature [3,4,6,7,8,9,10,11,12,13,14,15,16,17] has been devoted to the generation and evaluation of delay tests. However, as we will explain, most of this work applies to circuits having a rather restricted architecture.

Smith [16] proposed a model for delay faults based on paths. A test for a path fault is generated using six propagation values and the test is independent of individual gate delay values. The basic assumption in Smith's proposal is that flipflops or latches feeding the combinational circuit are directly controllable from primary inputs and the flipflops receiving the outputs from the combinational circuit are directly observable at the primary outputs. A schematic of this architecture is shown in Fig. 1.

A delay test for the combinational logic in a synchronous sequential circuit of the above type is defined as a two-vector sequence that correctly propagates a signal transition (0 to 1 or 1 to 0) from a set of latches to the next set of latches.

Unfortunately, many practical circuits do not conform to the architecture of Fig. 1. It is possible to apply the two-vector tests to *scan* circuits provided the scan flipflops are either designed as *double strobe flipflops* [18] or contain additional *hold* latches controlled by a separate hold clock. The hardware overhead in such a design may be high. Also, the hold latch imposes performance penalty. In spite of the fact that a vast majority of work on delay testing has only limited applicability recent attempts of delay testing in production environment have been reported [19].

0738-100X/92 $3.00 © 1992 IEEE

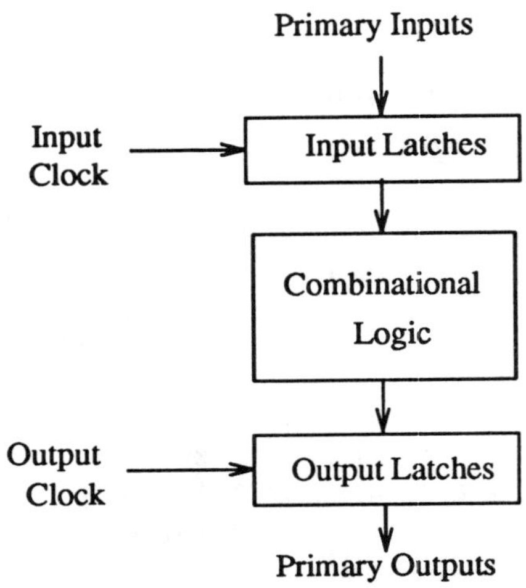

Fig. 1: A constrained hardware model for delay test.

Two-vector testing of delay faults in normal scan (without hold latch) circuits requires special consideration. The two patterns should be separated by a single clock application which can be either in normal or in test mode. Generation of such vectors has been studied in the literature [9]. For tests generated for combinational logic alone, it has been shown [12] that the restriction of single bit shift for the two-vector tests significantly reduces the fault coverage.

Test generation methods for delay fault based on stuck-at fault test generation for non-scan synchronous circuits have also been presented [20,21]. In one approach [21], tests are generated for *fully transitional paths* (FTP) for possible delay faults. Unfortunately, for many practical circuits delay fault testability using *fully transitional path* model is rather low. Hence, with this method many paths in the circuit will be declared untestable, while they may be testable using a different method of signal transition propagation through a combinational path. The method of Agrawal *et al* [20] does not use FTP. However, it uses a fault model that corresponds to our most optimistic assumption. Also, our present methodology is easier to extend to fault simulation.

Fault simulation for path based delay faults has received very little attention from researchers. Smith [16] proposed a six-value algebra for path based delay fault simulation and Schulz *et al* [15] modified it to a four-value algebra to identify strictly delay tests using parallel pattern simulation technique. Both of these methods are applicable to essentially combinational circuits or scan based designs.

In this paper, we study the problem of delay fault modeling and test generation for any random logic sequential circuit. The proposed test generation method, based on *transition* and *hazard* states of signals, is applicable to any sequential circuit of either non-scan, scan or scan-hold type of design. Three fault models based on different initial state assumption during the propagation of the fault effect to a primary output are proposed and analyzed using the proposed delay fault test generation method. We also propose a novel thirteen-value algebra to simplify the analysis of robust and nonrobust tests during fault simulation of path delay faults.

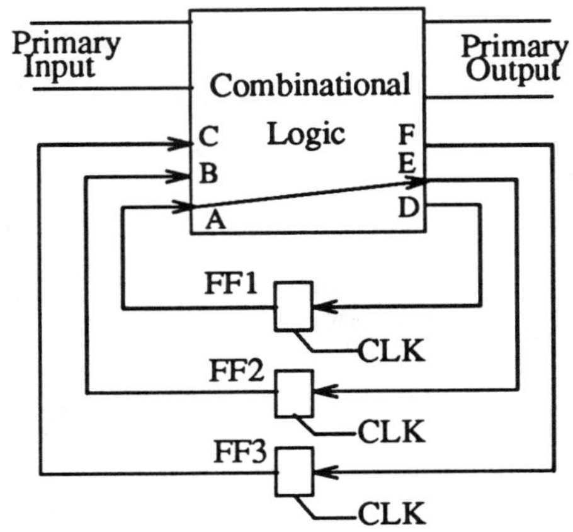

Fig. 2: Circuit Model for Test Generation

2. Hardware Model for Delay Fault Testing

Figure 2 shows the hardware model we use to study the delay test problem of a general sequential circuit. Consider a path between flipflops FF1 and FF2. A delay test involves the propagation of a signal transition from FF1 to FF2 through a designated combinational path between them. It may require a number of vectors at the primary input of the circuit to initiate a signal transition at the output of FF1 and propagate it to FF2. If the delay fault exists in the circuit, an incorrect logic value will be latched in FF2. This captured logic value must be observed at the primary output by path sensitization. Therefore, the test for delay fault contains two sequences. The first vector sequence V1 initiates a signal transition at FF1 and propagates it to FF2, and the second vector sequence V2 propagates the captured logic state of FF2 to the primary output. These two vector sequences V1 and V2 together form the complete test. This method could be used to test any combinational path between primary inputs, outputs and flipflops in the circuit.

3. Test Generation Model

Figure 2 is a model of a synchronous sequential circuit. The logic values at the primary output of the circuit are determined by the current logic values of the primary input and the state values at the previous time frame. The previous state values of the flipflops FF1, FF2 and FF3 are fed into the combinational network at A, B and C and the current state values appear at D, E and F to be fed to the input of the flipflops FF1, FF2 and FF3. The flipflops are clocked in the following time frame and the new values appear at A, B and C inputs of the combinational network. The line between A and E represents a combinational path between flipflops FF1 and FF2 selected for delay fault test generation. The test generation method consists of two steps. In the first step a signal transition initiated at the output of FF1 is propagated through the path from A to E and the final logic value at E is captured in the flipflop FF2. The second step is to propagate this value to a primary output using path sensitization method.

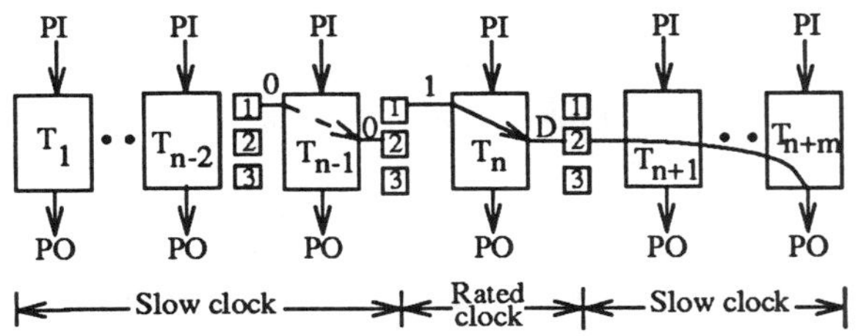

Fig. 3: Iterative logic array model for delay test.

The *iterative logic array model* for delay fault test generation is shown in Fig. 3. This model is similar to that used for sequential circuit test generation except that no stuck fault is assumed present [22]. Each copy of the circuit represents the behavior of the circuit during one time frame. A time frame is one clock period. Thus, the sequential circuit is represented as an equivalent combinational circuit. In Fig. 3 the time frame for each copy is represented by T_i where *i* represents the time frame number. The top inputs PI and outputs PO represent the primary inputs and outputs at each copy; the flipflops FF1, FF2 and FF3 of Fig. 3 are represented by boxes marked 1, 2 and 3, respectively. The state values of these flipflops are fed into the combinational copy at the subsequent time frame.

Let us assume that our test measures the propagation time of a rising transition (0 to 1). This transition is propagated in the time frame T_n. This requires that FF1 should produce a 0 in T_{n-1} and then a 1 in T_n. We assume that the 0 propagates to FF2 and initializes it at the end of T_{n-1}. In fact, it is not absolutely necessary to sensitize the entire path in T_{n-1} [3]. Since the path delay is measured only during T_n and we perform initialization in timeframe T_{n-1} and earlier and fault propagation (in time frames T_{n+1} and beyond) assuming fault-free operation, it implies that the clock is run at the rated speed only during T_n. This is illustrated in Fig. 3.

It is clear that the signal values in the two time frames, T_{n-1} and T_n, are jointly responsible for hazards during path activation. Similarly, the signal values in T_{n-1} and T_{n-2} are related. We, therefore, consider pairs of time frames. A signal can assume nine values (0,1, or X in each time frame). Such signal representation has been used by others [9].

Because of the dual time frame procedure, we always consider two time frames at a time. As we will explain later, this allows us to account for hazard conditions that may occur between two time frames. At the end of T_n, the state of FF2 will be 1 if the path delay did not exceed the clock period. Otherwise, its state will be 0. This condition is represented by *D* which is finally propagated to a primary output.

4. A Thirteen-Value Algebra

Hazards are timing anomalies in digital circuits whose precise behavior may depend on the analog operation of circuit elements. A detailed study of hazards was conducted by Eichelberger [23]. It is also recognized that the hazards play an important role in two-vector tests [13,24]. The presence of hazard between the two vectors can destroy the initialization produced by the first vector, thus invalidating the test.

In this paper, we propose a new thirteen-value algebra that enables us to compute hazard free (or robust) test sequences. In this value system the logic values of two consecutive clock periods are described by a *transition* state. With three (0, 1 and X) signal states there will be nine transition states. Manipulation of transition states requires simultaneous consideration of two time frames of the clock period. Since we neglect the actual delays of gates, an additional *hazard* condition is introduced to identify multiple transitions. With the presence of *hazard* condition the nine transition states are represented by eighteen-value transitions (with hazard and no_hazard condition) and these eighteen-values eventually collapse to unique thirteen transition values. Each of these thirteen-values is represented by a triplet, e.g., $1|h|0$ represents a logic 1 to 0 transition with the presence of hazard, in between, while $1|nh|0$ represents the same transition with no hazard. Similarly, $0|h|0$ and $0|nh|0$ represents a steady logic zero value with hazard and without hazard, respectively. Each triplet represents the values for two consecutive time frames; while the first and last value represents the logic values for the first and second time frames, the middle *h* or *nh* value represent the presence or absence of hazard. If the logic value is X (don't care) in the first and/or second time frame then a don't care hazard condition is specified by ~ in between the timeframes indicating an unstable transition value. As an example, the truth table of a two-input AND gate is given in TABLE I. For simplicity, a numerical encoding for signals as defined in TABLE II is used.

It can be seen from the truth table for the AND gate that if two opposite transitions without hazard appear at

TABLE-I: Truth Table For a Two-Input AND Gate

	1	2	3	4	5	6	7	8	9	10	11	12	13
1	1	1	1	1	1	1	1	1	1	10	1	1	1
2	1	2	3	1	2	3	1	2	3	10	2	2	1
3	1	3	3	1	3	3	1	3	3	10	3	3	1
4	1	1	1	4	4	4	7	7	7	10	1	4	4
5	1	2	3	4	5	6	7	8	9	10	2	5	4
6	1	3	3	4	6	6	7	9	9	10	3	6	4
7	1	1	1	7	7	7	7	7	7	10	1	7	7
8	1	2	3	7	8	9	7	8	9	10	2	8	7
9	1	3	3	7	9	9	7	9	9	10	3	9	7
10	10	10	10	10	10	10	10	10	10	10	10	10	10
11	1	2	3	1	2	3	1	2	3	10	11	11	1
12	1	2	3	4	5	6	7	8	9	10	11	12	13
13	1	1	1	4	4	4	7	7	7	10	1	13	13

TABLE-II: Code Definition

Values	Codes	Values	Codes
0|*h*|0	1	*X*|~|1	8
0|*h*|1	2	*X*|~|*X*	9
0|~|*X*	3	0|*nh*|0	10
1|*h*|0	4	0|*nh*|1	11
1|*h*|1	5	1|*nh*|1	12
1|~|*X*	6	1|*nh*|0	13
X|~|0	7		

the input, then the output must contain a hazard value. If one of the inputs has no-hazard controlling value (0|*nh*|0 for AND gate) then the output contains no-hazard controlling value. In all other cases, the output signal contains hazard (*h* or ~) if any input signal has hazard. The proposed thirteen-value system for hazard analysis during the test generation of delay faults is an extension of the ten value algebra proposed for delay fault test generation by Schulz *et al* [14]. Although both the ten and thirteen-value algebra are complete value systems for hazard analysis, the resolution of the proposed thirteen-value system is higher. Thus it is easier to distinguish between a robust and non-robust test for delay faults using the thirteen-value algebra. The next example illustrates this difference.

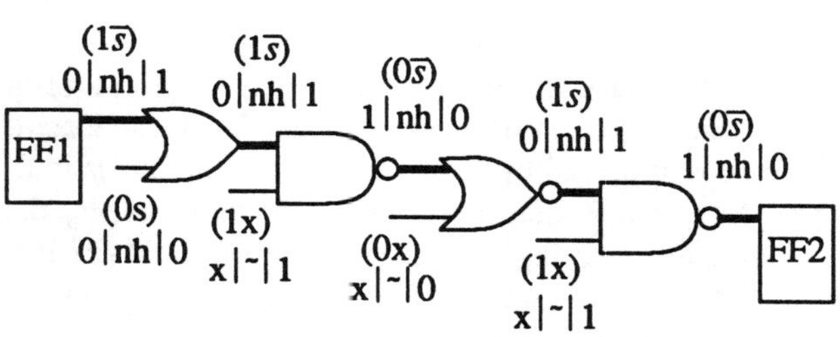

Fig. 4: Path with incident rising transition.

In Fig. 4, we consider a path (shown in bold lines) between two flipflops FF1 and FF2 for delay fault test generation. We have shown the values for propagation of rising transition (0 to 1) from the output of FF1 through the path for capture at FF2. The off-path sensitization values at various combinational elements are also shown. At each line, the logic values are shown using our proposed thirteen-value system and the values in parenthesis are from the ten-value system given by Schulz *et al* [14]. In that system, signals with hazards (static or dynamic) and signal transitions with or without hazards are not distinguished and thus are represented by the same value. Rising and falling transitions with or without hazard are represented by $1\bar{s}$ and $0\bar{s}$, respectively. Similarly, steady values with logic 0 or 1 are represented by 0s and 1s, respectively. It is clear from Fig. 4 that the input value of the flipflop FF2 is a falling transition with no hazard in the thirteen-value system, whereas in the ten-value notation, the $0\bar{s}$ means a falling transition with or without hazard. In order to determine, the possibility of hazard in this signal, all off path sensitization values must be examined. Thus, at the end of simulation of a vector, all off-path sensitization values are reexamined for every fault in the circuit. On the other hand, with thirteen-value algebra, at the end of each vector simulation one has to examine only the input of the destination flipflop of each path fault in the circuit to determine if it is a robust or non robust fault detection. Schulz *et al* have also proposed a delay fault simulation method [15] to determine robust test using six values or nonrobust test using four values. This also requires checking all off-path sensitization values for each fault after simulation of each vector to distinguish between robust and nonrobust test.

Hence during test generation, all decisions should be made to generate a robust test, which will guarantee the detection of the target fault. If, however, all choices are exhausted and no tests are found with no_hazard conditions, then the fault is untestable by a robust test. A non-robust test can be generated although it will not guarantee the detection of the target fault. In the next section, we illustrate the test generation algorithm by an example.

5. Test Generation Algorithm

The proposed test generation algorithm is shown in Fig. 5. It consists of two parts. The complete test generation algorithm with both the Parts (I & II) has the following steps:

(1) Select a combinational path between the source flipflop FF1 and the destination flipflop FF2. The path can be selected based on the criterion of longest time delay, or any other criterion [5].

(2) Select a signal transition (0 to 1 or 1 to 0) at the output of the source flipflop.

(3) Mark the logic values using the thirteen-value

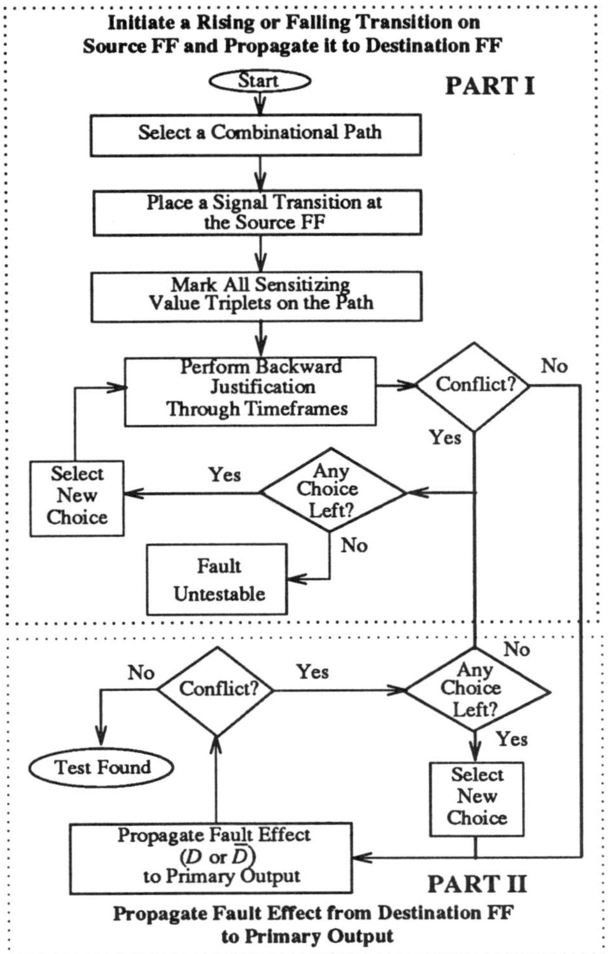

Fig. 5: Test Generation Algorithm For Delay Faults.

algebra at the input of the logic gates on the selected combinational path as required for the propagation of the signal transition. These values initiate and propagate the signal transition through the selected path from the source flipflop FF1 to the destination flipflop FF2.

(4) Perform backward justification of these logical assignments to the primary input of the circuit to generate the input vector sequence V1. Because of the specific signal representation, two time frames are simultaneously considered.

(5) The logic value latched at FF2 can be considered as D or $\overline{D}$, where the good and faulty machine values are logic 1 and 0 or vice-versa, respectively. Next, the logic value captured in FF2 is propagated to a primary output by path sensitization technique using vector sequence V2.

(6) The vector sequence V1 and V2 together form the delay fault test for the path between FF1 and FF2.

(7) If conflicts arise during backward justification in step four or during path sensitization in step five, then these conflicts are resolved using all available choices for robust and nonrobust test. If, however, no choices are left and requirements of step four or five are not met, then the delay fault is declared untestable.

Here steps 1 through 5 belong to Part I and the remaining steps 6 and 7 belong to Part II of the algorithm. Due to conflicting requirements, the algorithm may switch back and forth between Part I and Part II many times before declaring the delay fault to be *untestable* as shown in Fig 5.

6. Path Delay Fault Models

In this section, we propose three different fault models and analyze their effect on the proposed test generation algorithm. Let us refer to Fig. 3 that shows the iterative logic array model of delay fault test generation. These fault models are based on different assumptions of the initial state values of the flipflops at time frame T_{n+1} during the propagation of the fault effect from the destination flipflop to a primary output. It is possible that states of the flipflops FF1 and/or FF3 may be affected by the delayed propagation of signal transition in T_n, since there can be combinational paths to them through any of the elements on the selected path between FF1 and FF2. Hence, some flipflops other than FF2 may also latch a D or $\overline{D}$ at the end of T_n. Consequently, during the propagation of D or $\overline{D}$ at the output of FF2 to a primary output, the state of other flipflops could also be D or $\overline{D}$ if affected by the delay fault or else may be the correct good machine state 0 or 1 or X. A complete test generation algorithm should propagate the D or $\overline{D}$ from the output of FF2 to a primary output without being masked by the states of other flipflops.

Fault Model A. In this fault model we assume that the states of all flipflops other than the destination flipflop are not affected by excessive delay of the selected, as well as that of any other, path. The initial states of all these flipflops are assumed to be the same as the good machine state values after the transition has been propagated from the source flipflop to the destination flipflop. This is an optimistic model and the fault coverage can be expected to be higher than the other two fault models. The tests are not *robust* and can be invalidated by specific delay patterns in the circuit.

Fault Model B. A conservative approach is to assume that the initial states of all flipflops except the destination (FF2) to be X (don't care) at the T_{n+1} time frame while the state of FF2 is D. Then this D is propagated to the primary output assuming the initial state at all other flipflops to be in don't care state. The fault coverage now is most pessimistic and will often be lower than the fault coverage with fault model A. However, the vectors generated using this model are guaranteed to be correct even if other flipflop states are affected by delays in the circuit. The tests are *robust*.

Fault Model C. In this fault model all flipflop states other than the destination flipflop are held unchanged without hazard during the last two time frames of the transition propagation phase. This ensures that the delay fault or any other hazard condition does not affect the states of all flipflops other than the destination flipflop. This assumption may restrict the successful propagation of the signal transition in some cases and the overall fault coverage now will be lower than that for model A. The tests are *robust*.

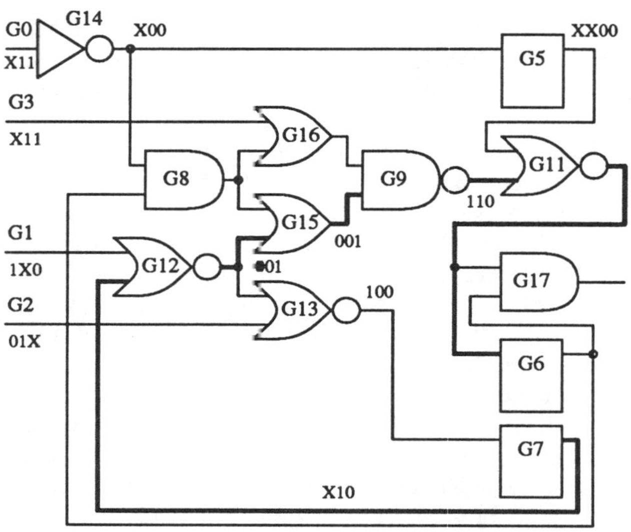

Fig. 6: Example Circuit (Modified s27).

Figure 6 shows an example that is a modified ISCAS '89 benchmark circuit. We will illustrate the above fault models using the proposed delay test generation method applied to this circuit. Here G5, G6 and G7 are flipflops. Since this is synchronous circuit, clock signals are implied and not shown in the diagram. This circuit has four primary inputs (G0, G1, G2, and G3) and one primary output. We have selected G7 and G6 as the source and destination flipflops. The combinational path between them is selected for delay fault test generation. This path is shown in the Fig. 6 by bold lines. We initiate a signal transition 1 to 0 at the input of the flipflop G7 and it is propagated to the flipflop G6 thorough the combinational gates G12, G15, G9, G11 and G6. All values produced during test generation are written on the figure. The test takes three vectors. It can be verified that there is no occurrence of hazard value anywhere on the selected path. Hence it is a robust test for this delay fault. The captured value at G6 can be considered as $\bar{D}$ since the good value is 0 and the faulty value is 1. Using path sensitization this $\bar{D}$ has been propagated to the primary output through the NAND gate G17. First, using the *fault model A* we determine the input vector sequence for initialization and propagation of the signal transition form G7 to G6 is X10X, 1X11 and 10X1 and since there are no other flipflop is between G6 and the primary output, the last vector also propagates the fault effect ($\bar{D}$) to the primary output. Here, the state of the flipflop G5 is X00 during these three time frames and using the fault model A and C the fault effect latched at flipflop G6 is successfully propagated to primary output. But under fault model B the state of G5 would be X0X and the fault effect propagation is masked by the output of G11 which also becomes X and the fault becomes *untestable*. Thus fault model B is a pessimistic model compared to model A and C. Similarly, a delay test can also be generated for the 0 to 1 transition.

7. Results

In this section, we discuss the implementation strategy for the presented algorithms and evaluate the performance. We have implemented the proposed delay fault test generation algorithm by modifying an existing sequential circuit test generator, Gentest [25]. The system is divided into three parts.

The first part deals with path selection. In the present prototype implementation, a single combinational path between each *source* and *destination* pair is selected for test generation. The *sources* are all primary input and flipflops and *destinations* are all flipflops and primary outputs. Although in the current implementation no specific criterion is used to select a combinational path except that it must be a physical path between the *source* and *destination*, any definition of *longest path* or *critical path* could be used as a valid path selection criterion [5].

The second part deals with the propagation of fault effect from the *destination* to a primary output. If the selected path terminates at a primary output then this part is not necessary and the test generation process proceeds directly to the final part for propagation of signal transition. If the fault effect propagation is impossible then the delay fault for this path is declared *untestable*, but the test generation still proceeds to check whether or not the transition propagation is *untestable*. Although this strategy increases the total CPU time for test generation it gives more useful and detailed information about testability of the selected path. Of the three fault models given in the previous section, only A and B are currently implemented.

In the final part, a signal transition is initiated at the

source and propagated to the *destination*. Basically, the logic values of signals for the gates on the targeted path during the two time frames between which the transition occurs are justified backwards using the time frame expansion method. The last two vectors obtained in this phase are responsible for *path activation* and all other vectors prior to these two vectors are responsible for *initialization* of the circuit. It should be noted that during this phase there is no notion of fault in the circuit due to the test procedure of slowing down the clock as mentioned earlier. Thus, the generation of of this part of the test sequence deals with *fault free* machine only.

For each selected path, two kinds of signal transitions are considered, *rising transition* and *falling transition*. The rising and falling transitions initiate logic 0 to 1 and 1 to 0 changes, respectively, at the *source*. A transition is propagated through the path and the final logic value is captured at the *destination*. If the transition propagation (a *necessary* condition) is not possible then the fault is declared *untestable*.

TABLE-III: Test Generation With Fault Model A

Circ. Name	Faults Tot.	Det.	Untest	Unact	Cov. %	CPU Sec.	Tot. Vec.
S27	36	25	11	11	69.4	0.2	122
S208	290	27	263	239	9.3	48	512
S298	144	29	115	101	20.1	831	1179
S344	212	51	161	158	24.0	1353	472
S349	212	52	160	158	24.5	1400	459
S820	416	139	277	276	33.1	19025	1113
S953	1120	40	1080	1075	3.6	16336	120

TABLE-IV: Test Generation With Fault Model B

Circ. Name	Faults Tot.	Det.	Untest	Unact	Cov. %	CPU Sec.	Tot. Vec.
S27	36	19	17	11	52.7	0.2	88
S208	290	22	268	239	7.2	47	492
S298	144	21	123	101	14.6	513	786
S344	212	48	164	158	22.6	1106	429
S349	212	47	165	158	22.6	1117	450
S820	416	136	280	276	32.6	18165	1095
S953	1120	38	1082	1075	3.4	15820	114

Tables III and IV present the results of delay fault test generation for ISCAS sequential benchmark circuits. Except for s298, only a partial set of paths, that includes one path per source-destination pair, is selected for test generation. For the circuit s298, all combinational paths were considered. For each path two faults were considered (signal transition 0 to 1 and 1 to 0, respectively, at the *source* that is either a flipflop or a primary input). The test generation begins with an *unknown* initial state since no initial state is assumed. For the circuits that are not initializable, the delay fault coverage and, also the stuck-at fault coverage will be low. All experiments were carried out on a SUN Sparc2 workstation.

Table III gives the results for fault model A. The reported CPU time includes the path search time. The test generation algorithm is *complete*. Hence, given enough time it will find the test if one exists. However, if test is not found for a fault, then no test exists for this fault and therefore, the fault is declared *untestable* (shown as "Untest" in the tables). As shown in Table III the delay fault coverage ranges from a low of 3.6% for s953 to 69% for s27. All undetected faults are declared *untestable* and a majority of them are untestable due to the inability to propagate a signal transition between *source* and *destination*. The reasons is that either the path is combinationally nonfunctional, i.e., *combinational false path*, or there exists no state pair for the sequential circuit which enables the propagation of signal transition through the path. This number is shown in the table as "Unact." which is close to that of the untestable faults. Based upon this observation, we selected model B for robust tests. Notice that model C favors conditions for fault propagation and, as a result, may place additional constraints on path activation. It is observed that a fault is detected rather quickly, but when a fault is not detectable it takes a long time to prove the fault to be *untestable* as the complete search space must be exhausted. When neither transition is testable, the path can be considered as a *sequentially redundant path*.

The results for fault model B are given in Table IV. The fault coverage is somewhat lower than that for *fault model A*. This is expected since the fault effect propagation condition is quite restrictive for *fault model B* as compared to *fault model A*. However, the test vectors are robust. The difference in the fault coverage is about 17% for s27 and 6% for s298. We must point out that while the tests generated using *fault models B* and *C* need not be validated, the tests generated for *fault model A* must be validated by simulation using hazard states.

We performed fault simulation for stuck-at faults for circuits s27, s820 and s953 using the vectors generated by the delay test generator. The resulting fault coverages were 100%, 75.52% and 7.97%, respectively. For comparison, we also generated tests for stuck-at faults using Gentest [25]. The coverages for these tests were 100%, 95.15% and 8.18%, respectively. The reason the fault coverage for the circuit s953 for stuck-at and path delay fault is so poor because the circuit is not initializable. Even this limited experience indicates that the circuits with poor testability (with respect to stuck-at faults) may, in fact, also have many untestable paths. Also, the tests for path delay faults may have to be augmented to have an adequate stuck-at fault coverage if these tests were to be used for device testing. The consequences arising from low path delay fault coverage, e.g., test quality, design for testability, etc., require detailed investigation.

8. Conclusion

Path delay fault models for nonscan sequential circuits is one of the main contributions of this paper. Further, we have proposed a novel method for test generation for delay faults in random logic sequential circuit. Our method can generate tests for any synchronous sequential circuit irrespective of its structure. The presence of scan register or hardware reset is not required. When a scan register is available, the initialization and the final fault propagation phases will become trivial. The new thirteen-value algebra represents logic values at two consecutive time frames and the hazard between them. Using this algebra the detection of robust and nonrobust test becomes much easier, hence it would be extremely useful for path based delay fault simulation. This algebra also has possible applications in generating hazard free tests for stuck faults in asynchronous sequential circuits. With the present work, it is now possible to study the coverage, quality and design for testability issues related to delay faults in nonscan circuits. The method of path delay testing in which the clock is run at the rated speed only during the path activation phase simplifies test generation. It may also help in diagnosing the failed paths. However, the slowing of clock must be carefully examined for dynamic circuits.

References

[1] V.D. Agrawal, "Synchronous Path Analysis in MOS Circuit Simulator," *Proc. 19th Design Autom. Conf.*, pp. 629-635, June 1982.

[2] P.C. McGeer and R.K. Brayton, *Integrating Functional and Temporal Domains in Logic Design,* Kluwer Academic Publishers, Boston, 1991.

[3] C.J. Lin and S.M. Reddy, "On Delay Fault Testing in Logic Circuits," *IEEE Trans. CAD*, Vol. CAD-6, pp. 694-701, September 1987.

[4] S. Patil and S.M. Reddy, "A Test Generation System for Path Delay Faults," *Proc. Int'l Conf. Comput. Design*, pp. 40-43, 1989.

[5] W.-N. Li, S.M. Reddy, and S. Sahni, "On Path Selection in Combinational Logic Circuits," *IEEE Trans. CAD*, Vol. 8, pp. 56-63, January 1989.

[6] Z. Barzilai and B.K. Rosen, "Comparison of AC Self-Testing Procedures," *Proc. Int'l Test Conf.*, pp. 89-94, 1983.

[7] M. Breuer, "The Effects of Races, Delays and Delay Faults on Test Generation," *IEEE Trans. Comput.*, Vol. C-23, pp. 1078-1092, October 1974.

[8] T. Hayashi et al, "A Delay Test Generator for Logic LSI," *Proc. 14th Fault-Tolerant Comput. Symp.*, pp. 146-149, June 1984.

[9] V.S. Iyengar, B.K. Rosen, and I. Spillinger, "Delay Test Generation 2 -- Algebra and Algorithms," *Proc. Int'l Test Conf.*, pp. 867-876, 1988.

[10] J.D. Lesser and J.J. Shedletsky, "An Experimental Delay Test Generator for LSI Logic," *IEEE Trans. Comput.*, Vol. C-29, pp. 235-248, March 1980.

[11] Y.K. Malaiya and R. Narayanaswamy, "Modeling and Testing for Timing Faults in synchronous Sequential circuits," *IEEE Design & Test of Comput.*, Vol. 1, pp. 62-74, November 1984.

[12] W. Mao and M. D. Ciletti, "Arrangement of Latches in Scan-path Design to Improve Delay Fault Coverage," *Proc. Int'l Test Conf.*, pp. 387-393, 1990.

[13] A.K. Pramanick and S.M. Reddy, "On the Fault Coverage of Delay Fault Detecting Tests," *Proc. European Des. Autom. Conf.*, pp. 334-338, March 1990.

[14] K. Fuchs, F. Fink, and M. Schulz, "DYNAMITE: An Efficient Automatic Test Pattern Generation System for Path Delay Faults," *IEEE Trans. CAD*, Vol. 10, pp. 1323-1335, October 1991.

[15] M. Schulz, K. Fuchs, and F. Fink, "Parallel Pattern Fault Simulation Of Path Delay Faults," *Proc. 26th Des. Autom. Conf.*, pp. 357-363, 1989.

[16] G.L. Smith, "Model For Delay Faults Based Upon Paths," *Proc. Int'l Test Conf.*, pp. 342-349, 1985.

[17] T. M. Storey and J. W. Barry, "Delay Test Simulation," *Proc. 14th Des. Autom. Conf.*, pp. 492-494, June 1977.

[18] B.I. Dervisoglu and G.E. Strong, "Design for Testability: Using Scanpath Techniques for Path-Delay Test and Measurement," *Proc. Int'l Test Conf.*, pp. 365-374, Oct. 1991.

[19] I. Deol, C. Mallipeddi, and T. Ramakrishnan, "Amdahl Chip Delay Test System," *Proc. Int'l Conf. Comput. Design*, pp. 200-205, October 1991.

[20] P. Agrawal, V.D. Agrawal, and S.C. Seth, "A New Method for Generating Tests for Delay Faults in Nonscan Circuits," *Proc. 5th Int'l Conf. VLSI Design*, pp. 4-11, January 1992.

[21] S. Devadas, "Delay Test Generation for Synchronous Sequential Circuits," *Proc. Int'l Test Conf.*, pp. 144-152, Sept. 1989.

[22] P. Muth, "A Nine-Valued Circuit Model for Test Generation," *IEEE Trans. Comput.*, Vol. C-25, pp. 630-636, June 1976.

[23] E.B. Eichelberger, "Hazard Detection in Combinational and sequential Switching Circuits," *IBM Journal*, pp. 90-99, March 1965.

[24] S.M. Reddy, M.K. Reddy, and V.D. Agrawal, "Robust Tests For Stuck-Open Faults in CMOS Combinational Logic Circuits," *Proc. 14th Fault-Tolerant Comput. Symp.*, pp. 44-49, June 1984.

[25] W.-T. Cheng and T.J. Chakraborty, "Gentest: An Automatic Test Generation System for Sequential Circuits," *Computer*, Vol. 22, pp. 43-49, April 1989.

Equivalence of Robust Delay-Fault and Single Stuck-Fault Test Generation *

Alexander Saldanha   Robert K. Brayton   Alberto L. Sangiovanni-Vincentelli

University of California - Berkeley CA

Abstract

A link between the problems of robust delay-fault and single stuck-fault test generation is established. In particular, it is proved that all the robust test vector pairs for any path delay-fault in a network are directly obtained by all the test vectors for a corresponding single stuck-fault in a modified network. Since single stuck-fault test generation is a well solved problem, this result yields an efficient algorithm for robust delay-fault test generation.

1 Introduction

Due to the increasing need for reliability in high-performance VLSI circuits, static testing which only ensures logical correctness is becoming insufficient. In particular, defects and random variations in process parameters often cause delays to fall outside the specifications. Static testing may not detect such defects since there may be no impact on the logical (static) behavior of the circuit. However, since most circuits operate in a clocked environment, it is essential that the final logic value on the circuit be asserted before the clock arrives. This need for ensuring the temporal (dynamic) correctness of circuits, given delay specifications, has led to the development of delay-testing for chips.

The problem of delay-fault test generation has been studied in [1, 6, 2]. Lin and Reddy [1] and Schulz *et al.* [6] use multiple-valued calculi to generate robust delay-fault tests. The former uses a 5-value calculus, whereas the latter employs a 10-value calculus. Besides apparently increasing the complexity of the problem over stuck-fault test generation, both approaches require the development of new (implicit search) algorithms combined with modification of existing stuck-fault test generation programs. A recent approach of McGeer *et al.* [2] is restricted to test generation for the strongest model of robust delay-fault testing, namely hazard-free single path propagation delay-fault testing. Although this approach is easily extended to the general model, there is a significant increase in size of the problem. In short, a satisfactory solution for delay-fault test generation (under the general model) has been lacking up to now. There are several proposals made for approximate delay-fault test generation in the literature(*e.g.*, [4]). None guarantee completeness of the test generation process, and hence are excluded from consideration here.

The contribution of this paper is the establishment of a link between robust delay-fault test generation to single stuck-fault test generation. A robust test for any path delay-fault in a network is obtained by a single stuck-fault test generation on a slightly modified network. This allows the *unaltered* application of mature and efficient static testing techniques [5] to delay-fault test generation.

2 Delay-faults

A few basic definitions are reviewed, before a formal definition of delay-fault testing conditions are provided. A **controlling value** for a gate f is the value at its input that determines the value at the output independent of the other inputs, and is denoted as $A(f)$. For example, *0* is a controlling value for an AND gate. A **non-controlling value** for a gate f is the value at its input which is not a controlling value for the gate, and is denoted as $I(f)$. For example, *1* is a non-controlling value for an AND gate. A **simple gate** is any one of AND, OR, NAND, NOR, and NOT. All the results described here apply only to simple gates. Only these gates have controlling vs. non-controlling values for each input. However, using the macro-expansion operator [2], any complex gate can be converted to an equivalent connection of simple gates.

Let $P = \{f_0, f_1, ..., f_m\}$ be a path. The inputs of f_i other than f_{i-1} are called **side-inputs** of f_i along P and denoted as $S(f_i, P)$. A path that starts at a primary input and ends at a side-input of P is a **side-path** of P. A **path delay-fault** is said to occur along path P if the delay along P falls outside its specified limits.

Definition 2.1 (RPDFT)*: A path $P = \{f_0, f_1, ..., f_m\}$ is said to be* **robust delay-fault testable** *for the rising (falling) transition at f_m by the vector pair $< v_1, v_2 >$ if at each node f_i, $f_i(v_1) \neq f_i(v_2)$ yields the desired transition being tested, and for each $g_j \in S(f_i, P)$:*

1. *$g_j(v_2) = I(f_i)$, and*
2. *if $f_{i-1}(v_1) = I(f_i)$, then there is no transition on g_j.*

The vector v_2 is assumed to be applied after v_1, delayed by an amount greater than the delay of the circuit.

This general robust test may involve multiple path propagation with hazards. This definition does not specify the value on side-inputs for v_1 when a path transition from a controlling value is being tested. This is because even in the presence of hazards on side inputs, a delay on this path will appear as a delay on the transition at the output of the gate. However, for the opposite transition (from non-controlling to controlling value) on the path input to the gate, no transitions are allowed on the side-inputs to this gate. Possible advantages of using delay-tests with hazards are a reduction in the size of the test set and a higher degree of fault coverage. This is because more paths can potentially be tested in the case when hazards are allowed compared to the hazard-free case; circuits are known with paths for which no robust test exists under the stronger delay-fault model, *viz.* hazard-free single-path propagation, while a general robust test does exist [3].

3 RPDFT and single stuck-fault test generation

In this section we present the main results of the paper, namely, a relationship between robust delay-fault and single stuck-fault test generation. For ease of exposition and proofs, we first consider only a particular circuit structure, functionally equivalent to any given circuit; this structure is termed a *leaf-dag*. A **leaf-dag** is a circuit composed of AND and OR gates with fanout and inverters only permitted at the inputs. An inverter is not allowed fanout. Every circuit (composed of simple gates,

* Research support by Fujitsu Laboratories Ltd.

0738-100X/92 $3.00 © 1992 IEEE

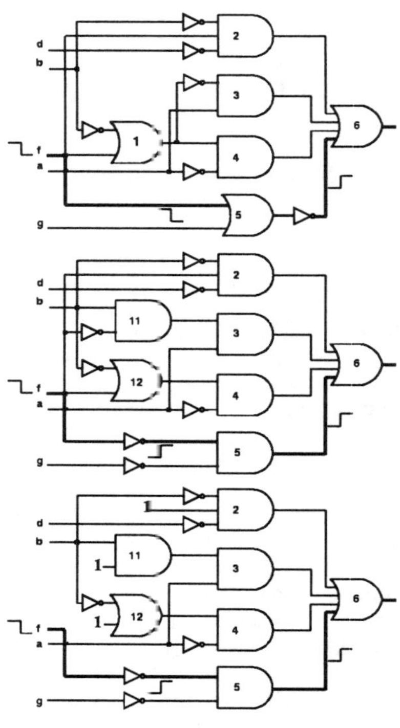

Figure 1: Example: Leaf-dag and rising-smooth network

AND, OR, NAND, NOR, and NOT gates) can be converted to a leaf-dag, albeit with possibly an exponential number of gate duplications.

Definition 3.1 *The I-**edge** of path P with input i in a leaf-dag refers to the first connection of P after the inverter, if it exists. The I-edge is said to be **associated** with input i.*

Definition 3.2 *Let P be a path starting with primary input i in leaf-dag η. Let P_R be the set of side-paths to all the OR gates along P, such that the I-edge of each $Q \in P_R$ is associated with i. The **rising-smooth-circuit** for path P is obtained by replacing the I-edge of each $Q \in P_R$ by 1.*

A **falling-smooth-circuit** is defined analogously, by considering AND gates instead of OR gates, and *0* instead of *1*.

Figure 1 illustrates the terms defined above. The initial circuit is shown at the top. A leaf-dag (shown in the middle) is derived by pushing all inverters to primary input leads. Note that gate 1 is first duplicated to gates 11 and 12; gate 12 becomes an AND gate when the inverter is pushed from the output to its inputs (using DeMorgan's law). Similarly, the OR gate 5 becomes an AND gate. The rising-smooth network for the bold path P is shown at the bottom of the figure. (Note the I-edge of P is the connection from the inverter to the AND gate 5.) This is obtained from the leaf-dag by replacing those I-edges associated with f, of side-paths to all the OR gates along P by *1*. Thus, the I-edges of the paths starting at f through 11, 12, and 2 are replaced.

Theorem 3.1 *Let P be a path in leaf-dag η. $< v_1, v_2 >$ is a robust delay-fault test for the rising transition along P in the rising-smooth-circuit η_R for P, if and only if $< v_1, v_2 >$ is a robust delay-fault test for the rising transition along P in η.*

Proof Let i be the input of P. Let P_I denote the side-paths to OR gates along P in η, such that the I-edge of each $Q \in P_I$ is associated with input i.

If part: $< v_1, v_2 >$ is a robust delay-fault test for the rising transition along P in η. By condition 2 of Definition 2.1, each side-input along path $Q \in P_I$ to an OR gate along P in η is at a steady non-controlling (*0*) value with no transitions on the application of v_2 after v_1. This implies that these side-inputs are insensitive to a change on i under $< v_1, v_2 >$, *i.e.* asserting any constant value on the I-edge of each $Q \in P_I$ leaves the value on these side-inputs unchanged under $< v_1, v_2 >$. Hence $< v_1, v_2 >$ remains a test for P in η_R.

Only if part: Only the I-edges of paths in P_I are set to *1* in obtaining η_R from η. By condition 1 and 2 of Definition 2.1, since $< v_1, v_2 >$ is a robust delay-fault test for P in η_R, each side-input to OR gates along P is at a steady non-controlling (*0*) value. This implies (since there are no inverters along the way) that the constant *1* asserted on the I-edge of each path in P_I does not propagate to the side-inputs of the these OR gates. This in turn implies that on application of the test $< v_1, v_2 >$ in η, none of the transitions on these I-edges propagates to the side-inputs of OR gates along P. Since the side-inputs of all other gates along P remain unchanged in η and η_R, $< v_1, v_2 >$ is a robust delay-fault test for P in η. ■

v_i denotes the value of primary input i in v, where v is a vector of the primary inputs of a network.

Theorem 3.2 *Let P be a path with input i in leaf-dag η. v is a test for the stuck-0 fault on the I-edge of P in the rising-smooth-circuit η_R for P if and only if $< \tilde{v}, v >$, where $\tilde{v}_i = \overline{v}_i$, $\tilde{v}_j = v_j$, for $j \neq i$, is a robust delay-fault test for the rising transition along P in η.*

Proof **If part:** Let $< \tilde{v}, v >$ be a robust delay-fault test for the rising transition along P in η. Then by Theorem 3.1 it is a robust delay-fault test for the rising transition along P in η_R. The output of P is *1* when v is applied to the rising-smooth-circuit η_R for P. Consider what happens in the presence of the stuck-*0* fault on the I-edge of P in η_R under vector v. By condition 1 of Definition 2.1 each side-input to P is at a non-controlling value under v. The I-edge of P is *1* (in the absence of the fault) under vector v in η_R, and the stuck-*0* fault effect thus propagates to the output in η_R. Hence v is a test for the stuck-*0* fault on the I-edge of P in η_R.

Only if part: Let v be a test vector for the stuck-*0* fault on the I-edge of P in the rising-smooth-network η_R for P. i is the only input in η_R that changes under the vector pair $< \tilde{v}, v >$. Let P_S denote the side-paths to P whose I-edges are not associated with i. Each side-input to P that belongs to a side-path in P_S must be at a non-controlling value under v; this is due to the test condition for the stuck-fault. Of the remaining side-inputs, let I_{and} denote the side inputs that meet P at an AND gate, and let I_{or} denote the side-inputs that meet P at an OR gate. I_{or} is empty by construction of the rising-smooth-network η_R. Since a stuck-*0* fault is being tested, each side-input to P in I_{and} is at value *1* in η_R under v (in the absence of the fault). Apply the vector $\tilde{v}$ to the circuit. The side-inputs to P in I_{and} may be at either value *0* or *1*. All other side-inputs to P are at non-controlling values. At some time greater than the delay of the network, apply vector v. The side-inputs to P in I_{and} may change to *1*, or they may remain at *1*, possibly with dynamic or static hazards, respectively. All the other side-inputs to P remain at hazard-free non-controlling values, since the transition on i under the vector pair cannot propagate to these side-inputs in the rising-smooth-network η_R. The transition that propagates along P is a *0* -to-*1* transition, which propagates a delay-fault at each AND gate along P irrespective of the presence of hazards on the side-inputs, provided the side-inputs have a final value of *1*. Since this condition holds, $< \tilde{v}, v >$ is a robust delay-fault for the rising transition along P in η_R, and hence in η by Theorem 3.1. ■

```
/* P is a path with transition t at input i
   in network η.  */
compute_rpdft_test(η, P, t) {
   Make P fanout free.
   For given t,
      Q_or = OR gates in P with rising transition.
      Q_and = AND gates in P with falling transition.
   Q = transitive fanout of i in reverse topological order.
   compute_path_parity(P, Q, Q_or, Q_and, i, parity).
   For each (n ∈ Q) {
      If n has any fanout f with parity(f) = {} {
         n' = duplicate(n).
         For each fanout f of n with parity(f) == {}
            Replace connection n to f by n' to f.
      }
      If (parity(n) contains both 0 and 1) {
         n' = duplicate(n).
         parity(n') = {1}.
         For each fanout f of n
            If (parity(f) == {1} )
               Replace connection n to f by n' to f.
         parity(n) = {0}.
      }
   }
   For each gate g ∈ Q with input i
      If (parity(g) == {0}) Set g to 0.
      Else If (parity(g) == {1}) Set g to 1.
   If (t is rising transition at i)
      v = test for stuck-0 fault on input of P.
   Else
      v = test for stuck-1 fault on input of P.
   RPDFT test for t is < v̄,v >.
}
```

Figure 2: Robust delay-fault test generation

A similar equivalence between RPDFT testing of the falling transition and single stuck-fault testing exists. The results stated apply to leaf-dags. Since most large Boolean networks cannot be converted to leaf-dags without an exponential (worst-case) increase in circuit size, an algorithm is presented in the next section that avoids this problem. Briefly summarized, the algorithm converts a Boolean network to another network using a sequence of duplications and transfer of connections, on which stuck-fault test generation is performed. As shown later, the size of this resulting network is at most four times the size of the initial network; in actual practice, the increase is a few gates.

4 RPDFT test generation

In the previous section robust delay-fault testing is shown to be equivalent to single stuck-fault testing on leaf-dag networks. Here we extend of the equivalence to arbitrary Boolean networks and develop an efficient process for generating the delay-fault tests.

Let P be a path in a Boolean network η for which a robust delay-fault test is required for transition t at the input i of P. η is composed of AND, OR, and NOT gates, where all inversions are explicitly represented. For ease of exposition, assume P is fanout-free; this is easily achieved by duplicating each gate along P at most once. For each gate along P, determine whether the transition along P is a rising or falling transition. Let Q_{or} denote the OR gates along P that have a rising transition along P, and Q_{and} the AND gates along P that have a falling transition.

Perform a reverse topological traversal of the gates Q in the transitive fanout of i, to determine the *parity* of all gates along side-paths to P. The algorithm for this is shown in Figure 3. A gate may have no parity, *0*, *1*, or both parities. For example,

```
compute_path_parity(P, Q, Q_or, Q_and, i, parity) {
   For each (n ∈ Q that is not a gate along P) {
      parity(n) = {}.
      If n terminates in gate ∈ Q_or
         parity(n) = {1}.
      If n terminates in gate ∈ Q_and
         parity(n) = {0}.
      For each fanout f of n that is also in Q
         If f is a NOT gate
            parity(n) = parity(n) ∪ {1 − v|v ∈ parity(f)}.
         Else
            parity(n) = parity(n) ∪ parity(f).
   }
}
```

Figure 3: Parity calculation

a side-input to an OR gate in Q_{or} has parity 1 and a side-input to an AND gate in Q_{and} has parity 0. If some fanouts of a gate have parity 1 and others have parity 0, the gate is assigned both parities. The parity is complemented across a NOT gate. Informally, the parity computation at g indicates which constant (*0* or *1*), if any, is to be asserted on the inputs of some paths through g.

In the second phase (shown in Figure 2), the algorithm performs at most two duplications of each gate to ensure that the parity of each resulting gate is either nothing, 0, or 1, but not both. The gates in Q are traversed in reverse topological order. When a node n is encountered that has both parities, it is duplicated. Since its fanouts have already been processed, each fanout itself has single parity, either 0 or 1. Thus, by transferring some fanout connections from n to its duplicate n', a single parity can be ensured for both n and n'.

In the final phase of the algorithm (Figure 2) a network $\tilde{\eta}$ is obtained where any gate with input i and parity 0 is set to *0* and any gate with input i and parity 1 is set to *1*. By performing stuck-fault test generation on $\tilde{\eta}$ on the appropriate fault on input i of P, a robust delay-fault test is obtained for the transition t.

Figure 4 illustrates the working of the algorithm on an example circuit. The bold path in the figure is to be tested for a delay-fault for the transition shown. Gate 6 has a rising transition along the path under test and is added to the set Q_{or}; Q_{and} is empty for this example. Gates 2, 3 and 4 all have parity 1. However, since gate 3 is connected to the inverted output of gate 1, while gates 2 and 4 are not, gate 1 has both parity 0 and 1. Thus, gate 1 is duplicated to gate 11 and 12 as shown, and gate 2 is connected to 11 while gate 4 is connected to 12. Now the inputs f of gates 11, 12, and 2 are set to 0, 1, and 1, respectively. Finally, any test for the stuck-*1* fault on input f of gate 5 on this circuit (shown on the bottom of the figure) yields the desired delay-fault test vector. Since $ab'df'g'$ is a test for this fault, $< ab'dfg', ab'df'g' >$ is a robust delay-fault test for the bold path in the original circuit.

It is apparent from the algorithm of Figure 2 that each node is duplicated at most twice. An additional duplication may be done for gates along the path being tested to make it fanout-free. Thus the area of the circuit on which stuck-fault testing is done is at most four times the area of the initial circuit.

The correctness of this procedure is proved using recourse to Theorem 3.2.

Theorem 4.1 *Let P be a path with input i in network η and let $\tilde{\eta}$ be the modified network obtained using the procedure of Figures 2 and 3. v is a test for the stuck-0 fault on the first edge of P in $\tilde{\eta}$ if and only if $< \tilde{v}, v >$, where $\tilde{v}_i = \overline{v}_i, \tilde{v}_j = v_j, j \neq i$, is a robust delay-fault test for the rising transition on input i of P in η.*

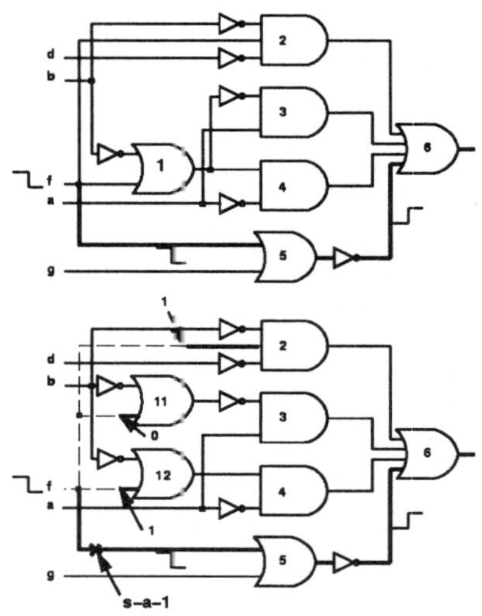

Figure 4: Example: Delay-fault test generation

Name	# Faults	Single-path		Multiple-path		
		Tested	CPU	Tested	CPU	SAT
apex7	2076	1654	55.0	1771	46.0	17.0
apex6	8984	3374	440.0	4120	398.0	147.0
rot	26048†	6340	1205.0	8202	1787.0	716.0
seq	21626†	2100	1989.0	4637	2914.0	962.0
des	89064†	5936	7040.0	7734	13953.0	5176.0
C880	12784†	11242	1797.0	12118	1883.0	722.0
s1196	6196	3088	1081.0	3587	619.0	267.0
s1238	7118	2852	1319.0	3590	840.0	350.0
s1438	32356†	13372	3431.0	15190	5133.0	1975.0
s1494	1952	1826	182.0	1833	138.0	45.0
s1488	1924	1832	176.0	1875	135.0	45.0
s5378	26754†	16960	4301.0	18309	4004.0	2210.0

\# Faults: Number of path delay-faults attempted
Tested: Number of testable delay-faults. Remainder proved untestable (no aborted delay-faults)
Single-path: Hazard-free single-path propagation RPDFT [2]
Multiple-path: Multiple-path RPDFT
†: Limited to 1000 longest paths from each primary input
CPU : Time including path tracing, network manipulation and Boolean satisfiability
SAT : Time for Boolean satisfiability

Table 1: Robust delay-fault test generation results

Sketch of proof: Since each Boolean network has a unique equivalent leaf-dag circuit that is derived from it, let η_L and $\tilde{\eta}_L$ denote the leaf-dag's derived from η and $\tilde{\eta}$, respectively. Assume the rising transition on input i (of P) in η corresponds to a rising transition along P in η_L. Then it can be shown that the leaf-dag $\tilde{\eta}_L$ is (structurally and logically) identical to the rising-smooth circuit for P in η_L. The result of the theorem follows from Theorem 3.2. Similar reasoning applies to the case when a rising transition on i in η corresponds to a falling transition along P in η_L. Here the leaf-dag $\tilde{\eta}_L$ is identical to the falling-smooth circuit for P in η_L. ■

An anaogous statement holds for robust delay-fault test generation of falling transitions.

5 RPDFT test generation results

Table 1 gives the results of robust delay-fault test generation using equivalent single stuck-fault testing for some of the larger optimized circuits from the MCNC and ISCAS benchmark suite. All CPU times are on a DEC 5000. Circuits with very low testability have been excluded from the table, since they do not help in discriminating between the use of single-path or multiple-path propagation delay-fault tests. The first column is the name of the example, and the second is the number of path delay-faults for which test generation is attempted. The third and fourth column give the testability and CPU time (including path-tracing and Boolean satisfiability) using the test generation algorithm for hazard-free single-path propagation delay-faults [2]. The fifth column gives the testability (using the general RPDFT model) obtained using the approach described in this paper. The sixth column lists the total CPU time consumed by this approach, including path-tracing, network manipulations (duplications and connection movement) inside the SIS system, and Boolean satisfiability that is invoked to check single stuck-fault testability. The actual number of satisfiability calls made by the single-path approach is half the number made by the multiple-path approach. This is because both transitions are tested by the same vector pair along a path in the former case, if a test exists. The last column shows the CPU time spent on stuck-fault test generation after the circuit $\tilde{\eta}$ is constructed.

The following observations can be made from the table. Multiple-path propagation tests provide up to 10% additional coverage over single-path propagation tests. Further test set compaction has not been explored in this experiment. Unlike all other reported approaches for delay-fault test generation, notably [1, 6], as with [2], *there are no aborted faults*. Overall, the new proposal provides a very efficient and rugged technique for delay-fault test generation.

6 Conclusions

We have established a link between the problems of robust delay-fault and single stuck-fault test generation. It is proved that the robust test vector pairs for any path delay-fault in a network can be generated by all the test vectors for a single stuck-fault in a modified network. Since single stuck-fault test generation is a well solved problem, this yields an efficient algorithm for robust delay-fault test generation. Experimental results demonstrate the efficiency of the proposed technique. Future work includes delay-fault test set compaction and synthesis for delay-fault testability in large Boolean circuits.

References

[1] C. Lin and S. Reddy. On delay fault testing in logic circuits. *IEEE Transactions on Computer-Aided Design*, C-6(5):694–703, September 1987.

[2] P. McGeer, A. Saldanha, P. Stephan, R. Brayton, and A. Sangiovanni-Vincentelli. Timing analysis and delay-fault test generation using path recursive functions. In *The Proceedings of the International Conference on Computer-Aided Design*, pages 180–183, November 1991.

[3] A. Pramanick and S. Reddy. On the design of path delay fault testable combinational circuits. In *The Proceedings of the International Fault Tolerant Computing Symposium*, pages 374–381, June 1990.

[4] J. Savir and W. McAnney. Random pattern testability of delay faults. In *The Proceedings of the International Test Conference*, pages 263–273, September 1986.

[5] M. Schulz and E. Auth. Advanced automatic test pattern generation and redundancy identification techniques. In *The Proceedings of the International Fault Tolerant Computing Symposium*, pages 30–35, June 1988.

[6] M. Schulz, K. Fuchs, and F. Fink. Advanced automatic test pattern generation techniques for path delay faults. In *The Proceedings of the International Fault Tolerant Computing Symposium*, pages 44–51, June 1989.

At-Speed Delay Testing of Synchronous Sequential Circuits

Irith Pomeranz and Sudhakar M. Reddy+
Electrical and Computer Engineering Department
University of Iowa
Iowa City, IA 52242

Abstract

Methods to test sequential circuits for delay faults are discussed. A method called at-speed testing is proposed for simplifying test application and reducing test length. A value system to allow at-speed testing is developed, and a test generation procedure is presented. The effect of at-speed test application on the path delay fault model is described. Experimental results are presented, demonstrating the applicability of at-speed testing and its effect on test length.

1. Introduction

Test generation for delay faults in combinational circuits has received wide attention [1-7]. With the exception of [8,9], delay faults in sequential circuits have mostly been considered under full-scan that allows the circuit to be tested similar to a combinational circuit [1-7]. The latter approaches may in general require a three-latch design for flip-flops to hold the first pattern of a test, while the second pattern is being shifted in (the need for a three latch design can be eliminated or reduced in some cases by proper ordering of flip-flops [10,11]).

It is well-known that to test delay faults in combinational circuits, two-pattern tests are required. The first pattern, T_1, of a two pattern test $<T_1,T_2>$, initializes the circuit. After the application of T_1, "sufficient" time is allowed to insure that all signal values have stabilized, before the second input pattern, T_2, is applied. The signal changes throughout the circuit, initiated by changing the inputs from T_1 to T_2, activate the fault under consideration and exhibit the fault effect at a primary output. The activation requirements depend on the fault model used and the type of tests desired. Two fault models have been proposed, gate delay faults [4] and path delay faults [14]. Two types of tests have been proposed: robust [1,14] and non-robust [3,4]. Several classes of robust [15] and non-robust tests can be defined.

The method described above to test delay faults in combinational logic circuits can be directly extended to test delay faults in sequential circuits. For example, the following testing method was proposed in [8,9]. Considering a sequential circuit that does not use scan-design, initialization may require several input patterns to bring the circuit to the appropriate state, all applied at a clock speed that allows all signals to stabilize before the next input is applied. In this paper, we call a clock that allows logic values to stabilize a *slow clock*. After initialization, a single input pattern is applied to activate the fault and exhibit the fault effect at a primary output or next-state variable line. The effect of this input pattern is sampled at the end of the normal clock period, called in this paper the *fast clock*. The fault effect is either propagated to a primary output or is latched into a flip-flop (for a path delay fault, the fault effect can reach only a single flip-flop). If the fault effect is latched into a flip-flop (rather than reaching a primary output), it has to be propagated. Propagation is achieved by applying additional input patterns using, again, a slow clock, whose period allows all signals to stabilize before the next input pattern is applied. The signal values latched into flip-flops at the end of the fast clock cycle must be determined using transient analysis (matched to the fault model used), since signal transitions may not allow circuit outputs to stabilize within the fast clock cycle. A value system suitable for such analysis is the ternary logic from [12,13], which we adopt in this work.

The method described above has several shortcomings, which we address in this paper. First, it allows only one fast clock cycle in a test sequence, precluding activation of faults in consecutive clock cycles. Thus, large numbers of test sequences may result. The second issue is that the method assumes the applicability of fast and slow clocks, which may not be feasible in the case of some circuits that do not use scan designs. This suggests the need to investigate the use of only the fast clock, or the normal system clock, in the application of tests to detect delay faults. We call the proposed test method, that uses consecutive fast clocks, *at−speed* testing. We discuss the assumption one has to make for at-apeed testing to be applicable and develop methods to mechanize the derivation of at-speed tests.

At-speed testing is investigated in this work in conjunction with conventional test generation approaches, through the use of consecutive fast (at-speed) clock cycles embedded in sequences of slow clock cycles. In addition, at speed testing is investigated in conjunction with scan designs, where initialization and propagation are achieved by scanning in and scanning out state-variable values and do not require a slow clock. At-speed testing where only the normal clock is used during test application is also important as a test strategy in itself, under the following design methodologies. (1) For Built-In Self Test, where pure or weighted random patterns are used as test patterns, and the incorporation of a slow clock complicates the test generation process and involves a hardware overhead. (2) For fault-tolerant systems where testing is performed by observing the behavior of the circuit under normal, at-speed, operation. In both cases, the value system developed for at-speed test generation is useful for at-speed fault simulation, to obtain the fault coverage of the sequences applied. It is further hoped that the proposed method would stimulate research to design circuits

+ Research supported in part by NSF Grant No. MIP-9109568 (IP) and the SDIO/IST Contract No. N00014-90-J-1793 managed by US Office of Naval Research (SMR).

0738-100X/92 $3.00 © 1992 IEEE

which are at-speed robust testable.

Robust tests for path delay faults are considered in this work. Application to non-robust tests is possible. However, it is conjectured (and supported by existing experimental evidence) that in most cases, when a non-robust test exists, a robust one also exists.

The paper is organized as follows. In Section 2, we introduce at-speed testing and the fault model to which it applies. Rules for computing next state values, that allow correct propagation of values from one time unit to the other, are presented in Section 3. In Section 4, we describe an at-speed test generation procedure for path delay faults. Experimental results are given in Section 5. Concluding remarks are given in Section 6.

2. Preliminaries

The use of a slow clock allows values in the circuit-under-test to stabilize before the next input is applied. If the values are not guaranteed to stabilize, as is the case when at-speed testing is used, the memory elements may latch unexpected values. As we propose to use multiple consecutive cycles of the normal system clock during a test sequence, we cannot guarantee correct activation and propagation using the value system used in [9]. In Section 3 we develop rules for robustly propagating values through the circuit, which are valid under *normal* (or fast) clock speed. This value system allows propagation of faults using the normal clock cycle, thus allowing additional faults to be activated and finally detected by the same test. The number of tests and the total test length are thus reduced.

The assumption underlying the use of consecutive fast clocks is that a transition that started one time unit earlier disappears and does not affect the next state or primary outputs at the current time unit. In other words, we assume that all transitions finish propagating within one clock cycle after they were initiated (this assumption is necessary independent of whether or not robust tests are used). This allows for the delay defects to increase the delay at a gate (for gate delay models) or along a path (for path delay models) by an amount equal to a clock period, and still fall within the assumptions made here. Hence most defects causing delay faults are expected to fall within the assumptions made. It should also be pointed out that even when "slow clocks" are used to activate and propagate a delay fault effect as in [9], one is assuming that the magnitudes of the delays in the potentially faulty circuit are bounded by some maximum corresponding to the period of the slow clock.

3. A Value System

In this section, we develop a value system suitable for conventional and at-speed testing. Note that whenever a slow clock is used, it can be assumed that the circuit under test is free of delay faults, as the slow clock cycle is selected to guarantee stabilization of all transients, even in the faulty circuit. Thus the logic values in the circuit under test can be computed using fault-free logic simulation. However, when a fast clock is applied, the changes in signal values from the previous time unit to the current time unit must be considered, as these values may or may not (depending on whether or not the circuit contained delay faults) reach primary outputs and next-state variable lines before the end of the fast clock cycle. For determining the logic values observed at the outputs or latched into flip-flops when a fast clock is used, we use the values from [12,13], that include triples of the form $(\alpha_1\alpha_2\alpha_3)$ on every line, where $\alpha_k \in \{0,1,x\}$, x stands for an unknown value and $k \in \{1,2,3\}$. At a given time unit i that starts when the clock becomes active for the ith time and ends before the clock becomes active for the $i+1$st time, the triple $(\alpha_1\alpha_2\alpha_3)$ has the following meaning. α_1 is the value at the beginning of time unit i, α_3 is the expected value at the end of time unit i if no delay faults occur, and α_2 is the intermediate value during the transition (if one occurs), e.g., $0x1$ is a $0 \rightarrow 1$ transition, $0x0$ is a hazardous 0 and 000 is a stable 0. At a given time unit, to compute an output value of a gate given its input values, the rules for computing the output value are applied to each one of the members of the input triples separately.

Rules for computing the values of the flip-flops at the end of a fast clock are considered next. We first consider the case where a single fast clock cycle is used. We then consider the case where more than one fast clock cycle is used. The following example illustrates the values for a single fast clock.

Example: Consider the circuit of Figure 1 and the test sequence (01011), where the fast clock is applied at time unit 3, when the second 0 is applied. We assume that the circuit is reset to state (0,0) before the test sequence is applied. Let us trace the values in the circuit for each time unit. The first two time units are applied using a slow clock, that allows all values to stabilize, and we obtain that the circuit reaches present state (0,1) and that the next state is (0,0). At time unit 3, the primary input (circuit line number 1) changes from 1 (its value at time unit 2) to 0 (its value at time unit 3). Line 2 (the output of the first flip-flop) remains at 0 and line 3 (the output of the second flip-flop) changes from 1 to 0. This creates the transitions $1 \rightarrow 0$ on line 1 and $1 \rightarrow 0$ on line 3. Since a fast clock is used at this time unit, we cannot assume that all the changes occurring in the circuit due to these transitions settle before the next time unit starts. We must therefore evaluate the values in the circuit under these transitions. We assign to the inputs the following values. Input 1 is assigned $1x0$, pseudo-input 2 is assigned 000 and pseudo-input 3 is assigned $1x0$. Propagating these values through the circuit, we obtain the following values. On line 11, we have the value $1x0$, and we have robust activation of the slow-to-fall fault on path (3-8-11). On line 12, we have the value $0x1$, and the slow-to-fall fault on path (1-4-12) is activated. On the output of gate 13, we have a hazard 0, or the value $0x0$. At the end of time unit 3, the following values are ready to be latched in the flip-flops.

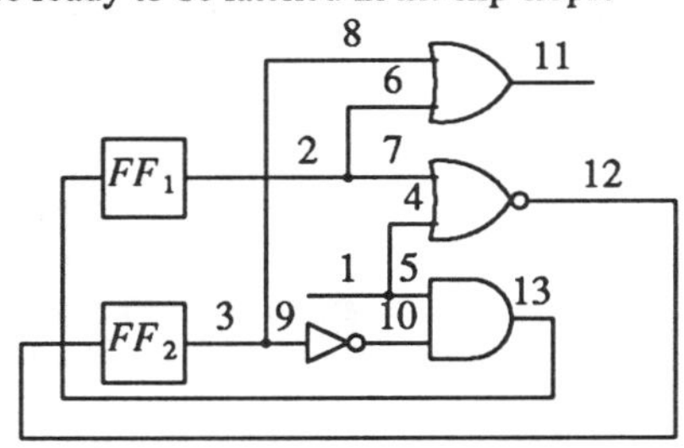

Figure 1: An example of test application

• Flip-flop 1 has the value $0x0$ on its input. If the circuit is fault free and all transitions propagate within the specified time, the value of flip-flop 1 is 0. Otherwise, its value is unknown. To capture this information for the purposes of fault propagation in the following slow clock cycles, we use the notation α/β, where α is the value in the fault free circuit and β is the value in the faulty circuit. Using this notation, the value latched in flip-flop 1 is $0/x$, where x stands for an unknown.

• Flip-flop 2 has the value $0x1$ on its input. If the circuit is fault free, the value of flip-flop 2 is 1. Otherwise, the value of flip-flop 2 is 0. Its value is therefore captured as 1/0.

During time units 4 and 5, propagation is performed, using a slow clock, resulting in the value 0/1 on the primary output at the end of time unit 5. The slow-to-fall fault activated along path (1-4-12) at time unit 3 is thus detected at time unit 5. □

The previous example illustrated the case where a single fault exhibited at a flip-flop was activated (the other fault activated was exhibited at a primary output, and was immediately detected). The treatment is similar if multiple faults are activated, which are all exhibited at a single flip-flop. However, if multiple paths are activated and exhibited at different flip-flops, the values latched in the flip-flops for the purposes of fault propagation are determined as follows. Let the path delay fault(s) in the set P_1 reach flip-flop 1 with the value $\alpha x \bar{\alpha}$ and let the path delay fault(s) in the set P_2 reach flip-flop 2 with the value $\beta x \bar{\beta}$ at the end of a fast clock cycle. We divide the testing process into two subprocesses, as follows. To test the faults in P_1, we can assign flip-flop 1 the value $\bar{\alpha}/\alpha$, in accordance with the example above. However, when the faults in P_1 are tested, it is not known whether or not the faults in P_2 are present or not. If at least one fault of the set P_2 is present, the value of flip-flop 2 should be set to β, otherwise, the value of flip-flop 2 should be set to $\bar{\beta}$. The value of flip-flop 2 is therefore set to x. Similarly, to test the faults in P_2, the value of flip-flop 1 is set to x and the value of flip-flop 2 is set to $\bar{\beta}/\beta$.

Next we consider the value system for at-speed testing, where multiple fast clock cycles are used. Consider the values at time unit i on a present state variable line y and on the next state variable line, denoted Y, corresponding to y (the value of Y is obtained by computing line values through the combinational logic of the circuit for a given input combination). Let us compute the value of y at the next time unit, $i+1$, based on the values of y and Y at time unit i. This value is the *effective* value of Y at time unit i (we call it effective since it may be different from the actual value of Y at time unit i, as subsequently shown). The following example illustrates this computation.

Example: Consider the circuit given in Figure 2. We assume that the circuit is first initialized to the all-zero state, that the primary inputs are set to zero, and that all transitions are allowed to settle at this state. We next consider the circuit at three consecutive time units at which fast clock cycles are used. The values assigned to the primary inputs are the following. The circuit is allowed to stabilize under input (1246) = (0000). The following sequence is then applied (1246) = (1110,1110,1111). At time unit 1, inputs 1 and 2 change from 0 to 1. Using the value system above, the values of inputs 1 and 2 are $0x1$, and therefore the value of line 7 is $0x1$. In this case, transitions are robustly propagated along paths (1,7) and (2,7) [1,14]. If paths (1,7) and (2,7) are fault free, the value latched in FF_1 is 1, otherwise, if at least one of the paths is faulty, the value latched in FF_1 is 0. In addition, at time unit 1, input 4 changes from 0 to 1, however, since line 3 is stable at 0 the transition is not propagated. If the clock at time unit 2 were a slow clock, then the contents of flip-flop 1 would be 1/0, according to the computations discussed above. However, given that the clock at time unit 2 is a fast clock we need to develop the following procedure. To compute the value of line 3 at time unit 2, we observe that if paths (1,7) and (2,7) are fault free, line 3 changes from 0 to 1 at time 2, otherwise, line 3 remains at 0. We therefore assign line 3 the value $0x1$. This can be seen to be a correct assignment, from the following arguments. If the transition at time unit 1 does not occur on line 7 by the time it is sampled at time unit 2, then the expected $0 \rightarrow 1$ transition on line 3 also does not occur at time unit 2. As long as we insure that the signal value on line 3 is propagated in subsequent clock cycles to an output, we can claim that paths (1,7) and (2,7) activated for slow-to-rise faults are detected, as follows. At time unit 2, input 4 is a stable 1, therefore, the $0x1$ on line 3 is propagated to line 8, activating the slow-to-rise fault on path (3,8). If paths (1,7), (2,7) and (3,8) are fault free, the value reaching line 8 at the end of time unit 2 is 1. Otherwise, if any of these paths if faulty, the value reaching line 8 is 0. The assignment of $0x1$ to line 3 allows us to chain paths, continuing the propagation of paths (1,7) and (2,7) by activating path (3,8). Note that we assign line 3 the value $0x1$ although it may never change to 1, if paths (1,7) or (2,7) are slow-to-rise. Although the value latched in FF_2 at the end of time unit 2 cannot identify which fault occurred, it does allow detection of each one of the three faults considered. At time unit 3, we assign line 5 the value $0x1$ in accordance with the discussion above, and line 6 assumes the value $0x1$. The primary output value is $0x1$. If the observed output value is 0, one (or more) of the following paths is slow-to-rise: (1,7), (2,7), (3,8), (5,9) or (6,9). Note that under the conventional testing scheme, at most two path delay faults can be detected by each test, whereas using at-speed testing and three consecutive fast clock cycles, we can detect five faults by a single test. It is also interesting to note that the path delay faults detected in this circuit are detected even if they are larger than a fast clock cycle, since different combinational subcircuits propagate the fault effect in different time units. □

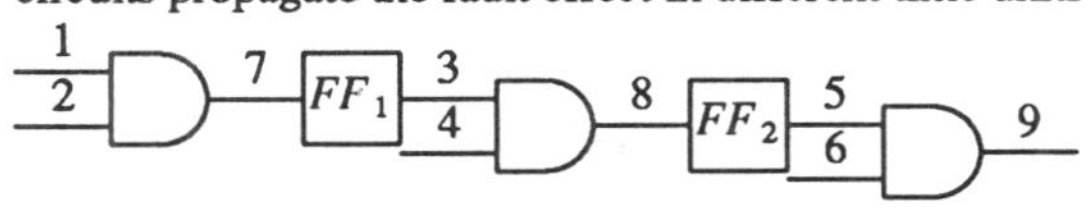

Figure 2: An example of next state computation

Note the difference between the value $0x1$ on line 1 at time unit 1 and the value $0x1$ on line 3 at time unit 2 in the previous example. The $0x1$ on line 1 is created by an input change from 0 to 1. This implies that the value of line 1 is guaranteed to change to 1 by the end of time unit 1. Therefore, if the value of line 1 in the succeeding time unit is 0, we can assign line 1 the value $1x0$ at that time unit, knowing that the initial value of line 1 is 1. In contrast, line 3 at time unit 2 may never assume the value 1, since this value may never reach FF_1 at time unit 1 (if path (1,7) or (2,7) is faulty). Therefore, the value of line 3 at the end of time unit 2 is unknown. Therefore, its value at the next time unit must be of the form $xx\alpha_3$. To take this into account, we add to each transition on a present state variable the qualifier L or R. R (*real*) stands for the case where the final value of the line can be determined, whereas L (*logic*) stands for the case where a transition allows fault propagation, however, the final value under the transition is unknown (the qualifiers are not propagated through the circuit to the next state lines, they are only used for computing the next value of a present state variable).

To derive the complete set of rules for determining present state values, let the values on y and Y at time unit i be $(\alpha_1\alpha_2\alpha_3)$ and $(\beta_1\beta_2\beta_3)$, respectively. We next consider in detail the case where $\alpha_3 = 0$ and if $\alpha_1\alpha_2\alpha_3$ is a transition, it is marked by R. In this case, at the end of time unit i, y assumes the value α_3, regardless of α_1 and α_2. Only α_3 is therefore relevant to the computation of y at time unit $i+1$. When the clock becomes active at the beginning of time unit $i+1$, either β_1 or β_3 propagate from Y to y, depending on whether or not there was a transition on Y that propagated within the required time frame. Let us now consider some combinations of $\alpha_3 = 0$ and $(\beta_1\beta_2\beta_3)$ (the case of $\alpha_3 = 1$ is similar).

$\alpha_3 = 0$, $(\beta_1\beta_2\beta_3) = (0x1)$: If no delay fault exists that delays the transition on Y, 1 appears on Y at the end of time unit i and the transition $(0x1)$ occurs on y at time $i+1$. Otherwise, if the transition on Y is delayed because of a fault, 0 appears on Y at the end of time unit i and the value of y remains constant, (000), at time unit $i+1$. This situation provides the conditions required to activate a delay fault on a path starting from y. The value of y at time unit $i+1$ is set to ($0x1$), and the effective value of Y at time i is (0x1). The qualifier L is given to the value of Y to indicate that the value of the flip-flop is actually unknown.

$\alpha_3 = 0$, $(\beta_1 \beta_2 \beta_3) = (1x0)$: If no delay fault exists that delays the transition on Y, 0 appears on Y at the end of time unit i. Therefore, the value of y at time $i+1$ is constant, (000). Otherwise, if a fault exists, 1 appears on Y at the end of time unit i and a transition ($0x1$) occurs on y at time unit $i+1$. If there is another delay fault on a path from y, that delay fault may invalidate a test that propagates ($0x1$) from y at time unit $i+1$, masking the delay fault leading to the transition on Y. Consequently, we cannot activate a new fault at time $i+1$ based on the transition on Y. Since the real final value of Y is unknown, the effective value of Y at time unit i is set to $0xx$ (in practice, this case can only occur at the first time unit).

$\alpha_3 = 0$, $(\beta_1 \beta_2 \beta_3) = (111)$ At the beginning of time unit $i+1$, y changes from 0 to 1. The value of y is therefore ($0x1$), and the effective value of Y at time i is (0x1). This is a real transition, and the qualifier R is added.

$\alpha_3 = 0$, $(\beta_1 \beta_2 \beta_3) = (1x1)$: The value of y at $i+1$ is unknown and no transition is robustly propagated to Y. The effective value of y is therefore $0xx$.

Other rules are obtained in a similar way. The complete set of rules for computing the effective value of Y at time unit i (or the value of y at time $i+1$), given the values of y and Y at time unit i, are summarized in Table 1.

Table 1: Computing the next value of a present state variable

		Y					
		000	$0x0$	$0x1$	$1x0$	111	$u \in U_1$
y	$v_1 \in V_1$	000	$0xx$	$0x1L$	$0xx$	$0x1R$	$0xx$
	$v_2 \in V_2$	$1x0R$	$1xx$	$1xx$	$1x0L$	111	$1xx$
	$v_3 \in V_3$	$xx0$	xxx	xxx	xxx	$xx1$	xxx

$V_1 = \{000, xx0, 1x0R\}$ $V_2 = \{111, xx1, 0x1R\}$ $V_3 = \{0xx, 1xx, xxx, 1x0L, 0x1L\}$ $U_1 = \{1x1, 0xx, 1xx, xx0, xx1, xxx\}$

4. Test Generation

In this section, we describe a test generation procedure for generating tests containing multiple consecutive normal clock cycles. Test generation using either single or multiple fast clocks require that at some fast clock cycle, the path to be tested be activated (the activation criteria depending on whether robust or non-robust tests are designed). If the path of interest terminates at a primary output, test generation is complete. If the path terminates at a flip-flop, the latched values must be propagated, again robustly or non-robustly, to a primary output in subsequent clock cycles.

We start by illustrating the test generation process when multiple fast clock cycles are used, using the circuit of Figure 2. The circuit has six paths, and therefore 12 path delay faults. In Table 2 we list some input combinations and the paths that they activate robustly. We refer to the entries of Table 2 as *state−transitions*. Test generation consists of chaining state-transitions such that new faults would be activated and faults activated during previous clock cycles would be propagated. The following conditions must be satisfied when chaining state-transitions to obtain a test sequence.

If input I assumes the combination $(\alpha_1 \alpha_2 \alpha_3)$ at time unit i, then at time unit $i+1$, input I must assume a combination $(\beta_1 \beta_2 \beta_3)$ such that $\beta_1 = \alpha_3$ (there cannot be a discontinuity in the value of I). Similarly, if the effective value of next state variable Y at time unit i is $(\alpha_1 \alpha_2 \alpha_3)$, then at time unit $i+1$, y must assume the same combination $(\alpha_1 \alpha_2 \alpha_3)$.

To ensure correct propagation of faults, an additional condition on the chaining of state-transitions to obtain a test sequence is set. Let t_1, t_2 be two state-transitions applied consecutively under a fast clock at times i and $i+1$, respectively. Let t_1 activate a fault of a path p, such that p ends at a next state variable Y. To propagate the fault on p further, t_2 must activate a path that starts from y (the present state variable of Y) and reaches a next state variable or primary output, as illustrated by the following example.

Table 2: Activating transitions for Figure 2

	pi 1 2 4 6	ps 3 5	ns 7 8	po 9	paths (all 0→1)
1	$0x1$ $0x1$ 000 000	000 000	$0x1R$ 000	000	(1,7) (2,7)
2	$1x0$ $1x0$ $0x1$ 000	$0x1R$ 000	xxx $0x1R$	000	(3,8) (4,8)
3	000 000 $1x0$ $0x1$	xxx $0x1R$	$xx0$ $1xx$	$0x1$	(5,9) (6,9)

Example: All state-transitions given in Table 2 can be chained together in order of appearance, as follows. (1) The primary input values for every pair of consecutive state-transitions are compatible. (2) The next state under every transition is equal to the present state under the following transition. (3) State-transition 1 in Table 2 activates the slow-to-rise faults of paths (1,7) and (2,7). State-transition 2 activates a fault along the path from line 3 to line 8, thus propagating the fault effect of the faults activated by state-transition 1, which reached line 7. State-transition 3 propagates all faults to primary output 9, by activating the additional path (5,9). □

In the example above, the test was terminated when a transition reached a primary output. Instead of terminating the test at that point, it is also possible to continue activating new faults, using transitions created on primary inputs or present state variables.

Propagation can also be performed using a slow clock. When an additional transition using a fast clock cannot be found that propagates the already activated faults further, an attempt is made to propagate the values latched in the flip-flops using a slow clock. In such a case, the translation described in Section 3 is required from the fast clock value system to the conventional value system used for slow clock propagation.

Chaining of state-transitions applied under a fast clock and input vectors applied under a slow clock, once the above value systems and chaining rules are established, is similar to test generation for stuck-at faults in sequential circuits. The details of the test generation procedure are therefore omitted.

5. Experimental Results

Experimental results of at-speed test generation are reported in this section. For the purposes of test generation, we extracted complete state tables from the gate level description of the circuit under test, containing all possible state-transitions. This allowed us to obtain the fault coverage using a 3-latch full-scan design, where the circuit can be tested as a combinational circuit, for comparison with the fault coverage obtained in the sequential circuit. It also allowed us to obtain a complete test set in some cases where the computation time for complete test generation was not too high.

The circuits were first assumed to have hardware reset to the all-zero state. Sequences of at-speed clock cycles were embedded in initialization and propagation sequences using slow clocks. Both slow-clock and fast-clock transitions were chained in all possible ways according to the rules of Section 4. To restrict the computation time, a limit was set on the maximum number of chains explored without finding a test. The test generation process was terminated when that limit was reached.

To study the effect of at-speed testing on test length and number of tests in a controlled way, three experiments were performed. First, only a single fast clock was allowed, and slow clock cycles were used for initialization and propagation. Second, up to two consecutive fast clock cycles were embedded in sequences of slow clock cycles. Third, up to three consecutive fast clock cycles were used. We did not go beyond three clock cycles, since the effect on test length diminishes as more fast clock cycles are used, partly due to the introduction of unknown values due to transitions marked L. Fault simulation was used to drop all faults detected once a test was generated. No overlapping of tests was attempted. Results are reported in Table 3. The reduction in test length can be seen to be more significant when going from one fast clock to two consecutive ones, than when going from two fast clocks to three. A similar trend exists for the number of tests.

A reduction in the number of tests is especially important for full-scan designs. For such designs, the following testing approach was investigated. Two input sequences are first shifted in to create a transition on the state variables. Simultaneously, transitions are created on primary inputs. The circuit is then allowed to run for one, two or three clock cycles at its normal clock speed. The flip-flop values latched at the end of all fast clock cycles are then shifted out. The number of sequences to be shifted in is reduced as more fast clock cycles are allowed, reducing test application time (primary input values need to be set at every clock cycle and primary outputs are observed at every clock cycle). Results are reported in Table 4. All detectable path delay faults are detected. The number of tests is reduced as more fast clock cycles are used. Test length increases, since state variables are not observed after every fast clock cycle, but only at the end of the test sequence.

6. Concluding Remarks

A method for at-speed test generation for delay faults in sequential circuits was proposed. At-speed testing reduces test application time by allowing more faults to be detected by a single test than in the conventional approach, and can help remove the requirement for a slow clock. It is especially important for random pattern testing and for fault-tolerant circuits, tested during normal operation. A value system to allow at-speed testing was developed and a test generation procedure was presented. Experimental results were given for the case where consecutive fast clock cycles are embedded within sequences of slow clocks and for full-scan designs, demonstrating the reduction in the number of tests and in the total test length. Reduction in test lengths appears to be moderate. However, further investigation of larger circuits, using test overlapping may reveal further improvement. It is also hoped that the at-speed testing approach proposed and the logic value system developed would stimulate further research into designing at-speed testable circuits, and would find use in fault simulation, especially using functional tests. Arbitrary combinations of fast and slow clocks also remain to be investigated.

References

[1] C. J. Lin and S. M. Reddy, "On Delay Fault Testing in Logic Circuits", IEEE Trans. CAD, Sept. 1987, pp. 694-703.

[2] S. M. Reddy, C. J. Lin and S. Patil, "An Automatic Test Pattern Generator for the Detection of Path Delay Faults", ICCAD, 1987, pp. 284-287.

[3] T. Glover and M. R. Mercer, "A Method of Delay Fault Test Generation, 25th DAC, 1988, pp. 90-95.

[4] A. K. Pramanick and S. M. Reddy, "On the Detection of Delay Faults", 1988 ITC, pp. 845-856.

[5] W.-W. Mao and M. D. Ciletti, "A Simplified Six-Waveform Type Method for Delay Fault Testing", 26th DAC, 1989, pp. 730-733.

[6] V. S. Iyengar, B. K. Rosen and J. A. Waicukauski, "On Computing the Sizes of Detected Delay Faults", IEEE Trans. Comput., March 1990, pp. 299-312.

[7] K. Fuchs, F. Fink and M. H. Schultz, "DYNAMITE: An Efficient Automatic Test Pattern Generation System for Path Delay Faults", IEEE Trans. CAD, Oct. 1991, pp. 1323-1335.

[8] Y. K. Malaiya and R. Narayanaswamy, "Modeling and Testing for Timing Faults in Synchronous Sequential Circuits", IEEE Design&Test, Nov. 1984, pp. 62-74.

[9] S. Devadas, "Delay Test Generation for Synchronous Sequential Circuits", 1989 ITC, pp. 144-152.

[10] W. Mao and M. D. Ciletti, "Correlation-Reduced Scan-path Design To Improve Delay Fault Coverage", 28th DAC, pp. 73-79, June 1991.

[11] K. T. Cheng, S. Devadas, and K. Keutzer, "Robust Delay Fault Test Generation and Synthesis for Testability Under a Standard Scan Design Methodology", 28th DAC, pp. 80-86, June 1991.

[12] M. Yoeli and S. Rinon, "Applications of Ternary Algebra to the Study of Static Hazards", JACM, pp. 84-97, Jan. 1964.

[13] E. B. Eichelberger, "Hazard Detection in Combinational and Sequential Circuits", IBM J. of R&D, pp. 90-99, March 1965.

[14] G. L. Smith, "Model for Delay Faults Based on Paths", 1985 ITC, pp. 342-349.

[15] A. K. Pramanick and S. M. Reddy, "On the Design of Path Delay Fault Testable Combinational Circuits", 20th FTCS., June 1990, pp. 374-381.

Table 3: Test generation results - reset

circ	tot./comb flts/det.	F tsts/len	F flts	FF tsts/len	FF flts	FFF tsts/len	FFF flts
bbara	140/140	41/240	63-	27/165	63-	25/186	63-
dk14	302/297	95/344	135-	54/247	135-	48/233	135-
dk15	202/200	70/234	105-	45/171	105-	44/160	105-
dk16	790/784	56/366	80-	42/305	82-	40/295	82-
dk17	184/181	52/222	71+	27/130	71+	23/116	71+
ex2	816/670	30/173	52-	23/144	52-	23/144	52-
ex3	270/239	62/236	86-	34/205	86-	35/193	86-
s27	56/ 50	18/ 89	19-	10/ 57	19-	8/ 51	19-

tot. flts - number of logical paths, equal to number of faults; **comb. det** - number of faults detectable using a 3-latch full-scan design; **tsts** - number of tests; **len** - total test length; **flts** - number of detected faults; **F, FF, FFF** - one, two and three fast clock cycles per test; +/- - the limit on the number of tests explored is not / is reached during the test generation process.

Table 4: Test generation results - 3-latch scan-design

circ	F tsts/len	F flts	FF tsts/len	FF flts	FFF tsts/len	FFF flts
bbara	74/ 74	140+	74/ 88	140+	69/ 87	140+
dk14	144/144	297+	132/180	297+	124/177	297+
dk15	82/ 82	200+	75/110	200+	69/112	200+
dk16	501/501	784+	473/520	784+	472/520	784+
dk17	101/101	181+	93/114	181+	89/111	181+
ex2	391/391	670+	385/409	670+	384/408	670+
ex3	102/102	239+	103/128	239+	96/120	239+
s27	13/ 13	50+	17/ 24	50+	14/ 19	50+

The Princeton University Behavioral Synthesis System

Wayne Wolf Andrés Takach Chun-Yao Huang Richard Manno
Ephrem Wu
Department of Electrical Engineering
Princeton University
Princeton NJ 08544

Abstract

The Princeton University Behavioral Synthesis System (PUBSS) is a high-level synthesis system targeted to control-dominated machines. PUBSS accepts a VHDL subset, in which the design can be described as multiple communicating processes plus registers, and generates a register-transfer implementation. This paper describes the compiler with emphasis on the automaton-based algorithms which allow efficient specification and manipulation of control-dominated systems.

1 Introduction

The Princeton University Behavioral Synthesis System (PUBSS) is designed for the efficient specification and optimization of control-dominated architectures. PUBSS compiles a VHDL behavior model, written as multiple communicating processes and data registers, into an efficient register-transfer implementation. Control-dominated architectures are important because many application-specific ICs (ASICs) are primarily control [1] and because ASICs, with their intense time-to-market and design cost pressures, are ideal candidates for high-level synthesis.

The key to synthesis of control-dominated architectures is *communication*—control is often much easier to design as concurrent processes which communicate between each other. As an example, consider a multiple queue machine [2]—the machine accepts enqueue and dequeue commands for multiple queues whose data is stored in a shared memory. The natural architectural description of the machine has three processes: enqueue, dequeue, and memory management. Coordination of access to shared memory is arbitrated by a single machine, the memory manager; at other times, the enqueue and dequeue processes can run independently. Describing the system as one large process is much more difficult and error-prone.

Compiling concurrent architectures requires new ways to represent and manipulate scheduling information alongside control:

- The designer must be able to easily specify both the component machines in the architecture and the constraints on communication between them. Control-dominated architectures do not have a standard concurrent architecture—the architectural decomposition is highly problem-dependent. Previous work in high-level synthesis has concentrated on the compilation of large uniprocessors or, in the case of DSPs, on compilation into highly specialized processing networks.

- Control structures specified as multiple processes are harder to schedule because communication between the processes must be coordinated. While some communication may require handshaking, much interprocess communication in control structures can be statically scheduled—the times of operations can be determined during chip design. Previous work has concentrated on scheduling single processes.

PUBSS is based on the *behavior finite-state machine* (BFSM) model [3], which captures scheduling constraints in a finite-state model. Compiling VHDL into a network of communicating BFSMs gives us powerful algorithmic tools for manipulation of the design's control structure: events can be moved through control structures and systems built from communicating components can be combined and repartitioned.

The greatest progress in high-level synthesis has been made by developing compilation algorithms for a target domain and/or target architecture. The SAW system from CMU [4] and the ADAM system from USC [5] are both aimed at datapath-controller architectures, often found in CPUs and other data-intensive processors; many other datapath-controller synthesis systems have been built [6, 7, 8]. Successful synthesis systems for digital signal processing include the Cathedral systems [9] and a family of compilers built at GE [10].

Compared to datapath and DSP synthesis, relatively little is understood about the synthesis of control-dominated architectures. Existing scheduling techniques [8, 11] work on a single machine but are not easily extended to communicating machines. Several groups [12, 13] have identified specification and scheduling of distributed control as a critical research problem.

The Value Trace (VT) proposed by Snow [14] and formalized by McFarland [15], is the prototypical data/control flow graph representation. A number of design representations have since been developed [16, 17, 13, 18, 19]. Several projects have represented system control as a state transition graph [18, 19], but they have not used the state machine to represent scheduling constraints on the control operations themselves.

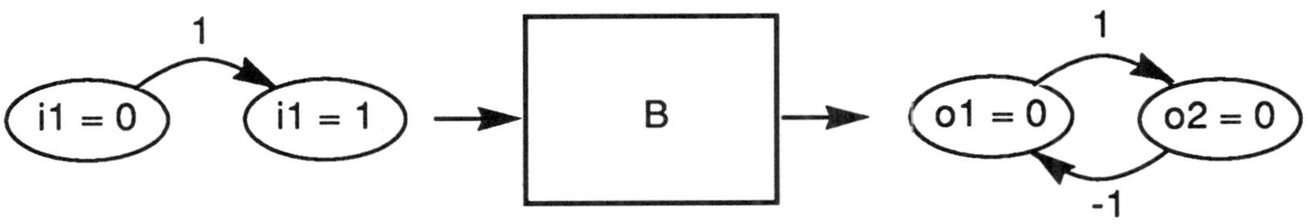

Figure 1: The input and output events of a behavior FSM are partially ordered in time.

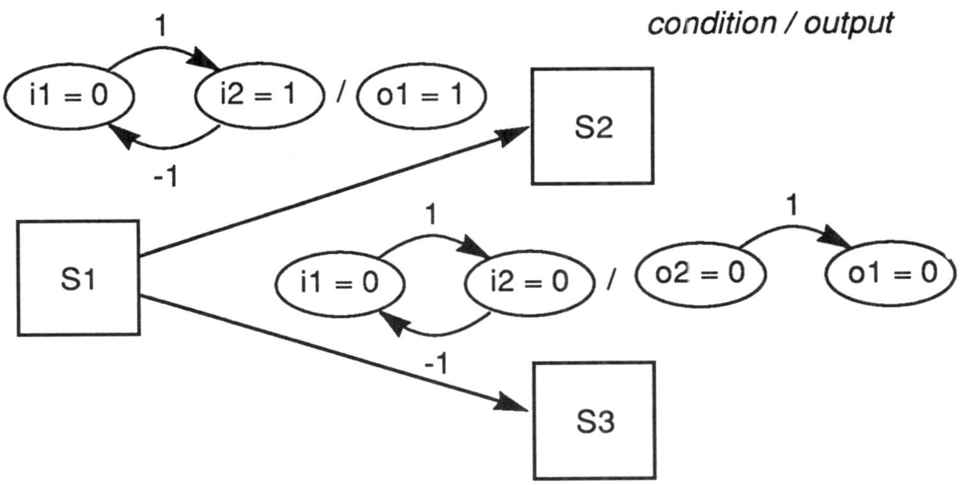

Figure 2: A behavior state transition graph.

2 Behavior FSMs

The behavior FSM model embeds scheduling information in a finite-state model. PUBSS models a design's control structure as a set of communicating behavior FSMs. A BFSM is a state machine whose inputs and outputs are *partially ordered* in time; in contrast, a standard register-transfer machine's I/O behavior is totally ordered in time because we know its possible input and output values on every clock cycle. A BFSM's I/O behavior is specified as *events* whose times are constrained by linear inequalities The inequalities can be represented by weighted, directed graph edges: edge direction shows the direction of the inequality and edge weight gives the minimum time between the events. In Figure 1, the machine will looks for the input sequence $i_1 = 0, i_1 = 1$, with at least one cycle between the two events. The machine then produces the output event sequence $o_1 = 0, o_2 = 0$, with exactly one cycle separating the outputs.

As with register-transfer FSMs, we prefer to write the system behavior as a state transition graph. Figure 2 shows a state transition graph fragment for a BFSM. The conditions and outputs are specified as *I/O event constraint graphs*, consisting of I/O events and the linear constraints on their times. Most constraints are within a transition, but constraints across states are allowed. Execution of a transition may require more than one clock cycle: the transition to S3 requires exactly two cycles to recognize the input condition and at least two cycles to generate the output.

Multiple communicating machines are easily specified as a network of BFSMs. Scheduling constraints of one component in the network may affect the feasible schedules of other components in the network. Scheduling constraints of the components interact when the components must communicate. Such synchronous transfer of information is essential to avoid the overhead associated with handshake communication. By scheduling most communication to be synchronous we can specify or partition designs into smaller components without the overhead incurred by handshaking.

Both Probst [20] and Devadas and Keutzer [21] have proposed alternatives to the BFSM model. We find Probst's model cumbersome, and it has a less-direct relationship to register-transfer implementations than does the BFSM model. Devadas and Keutzer use a nondeterministic FSM to specify scheduling requirements. Nondeterminism is best at abstracting function but is more limited in the abstraction of time.

3 System modeling

Compilation starts with a *behavior model* written by the user in VHDL. The VHDL subset understood by PUBSS allows description of architectures as multiple communicating machines. That subset is mapped onto a reasonably unrestrictive architecture of behavior FSMs and registers.

Figure 3 illustrates PUBSS's VHDL subset and the architectural schema onto which the VHDL is mapped. The VHDL model is described as one or more processes. Processes may communicate in either of two ways. Commu-

```
– signals connect processes
signal full_in, . . . : data := '0';
– registers store data
head : unsigned_register
   generic map (Addrwidth)
   port map (head_in,
      head_out, head_readwrite, head_ack);
– processes describe behavior
main: process
   begin
      wait on enq'transaction, . . .;
      if reset = '1' then
         head_in <= unsigned(0, Addrwidth);
```

our generic VHDL model

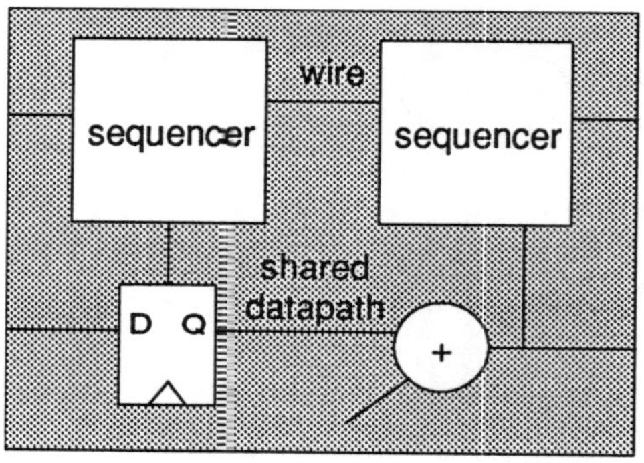

our architectural schema

Figure 3: The VHDL subset and the target architecture.

nication which does not require state may be performed through signals such as *full_in*. If the reader and writer are unsynchronized, values can be stored in registers such as *head*. The process and register structure is natural for hardware description—declared registers correspond to architected registers which the designer usually wants to control.

Compilation starts with a direct translation of the VHDL into a BFSM network. Elements in the VHDL structure are mapped onto components in the BFSM network: a process becomes a component BFSM, a register becomes a register component, and a signal becomes a wire connecting components and/or network primary inputs and outputs. We model registers separately from BFSMs to avoid state explosion. Registers and combinational logic together form a datapath shared by all the component BFSMs.

To make PUBSS's compilation technology accessible, we had to develop a VHDL subset that was both simulatable and synthesizable, with the simulation results matching as closely as possible the behavior of the synthesized system. (Some projects have used VHDL as a syntax which means one thing for simulation and something else for the special application.) We found VHDL poorly suited to our style of behavioral description—the VHDL model is not a behavior specification, but an implementation tuned for

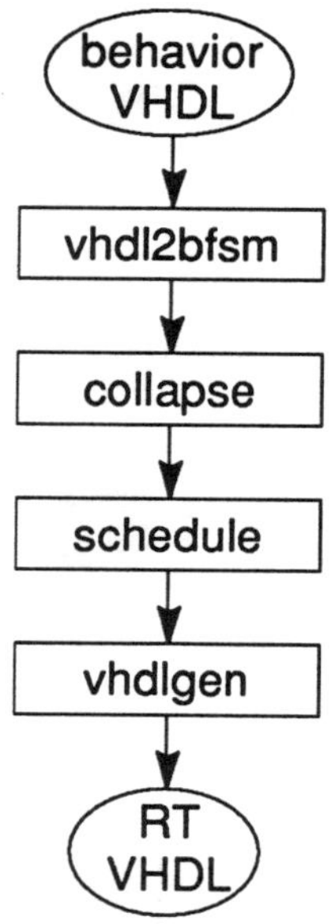

Figure 4: The structure of PUBSS.

execution on general-purpose CPUs. Many statements are required for simulation which have no synthesis meaning: the *wait on* in the example, assignments to the unknown value, *etc.* The registers used in the descriptions are library components developed for PUBSS; simulation models for the various register types are kept in a separate library. The compiler is made more complex by these problems but the result is a description format natural to the VHDL programmer.

4 The compilation process

The sequence of operations presently performed by PUBSS is shown in Figure 4. Compilation separates the operations specified in the VHDL into a datapath, which holds the shared registers, plus sequencers.

- *Vhdl2bfsm* generates a BFSM network description from the VHDL source. It performs simple source-level optimizations such as constant folding. It also algorithms based on the program dependence graph [22] to move statements to increase parallelism [23].
- *Collapse* forms the product of the component BFSMs, transforming the architecture to have one master BFSM and any declared registers. Collapsing the component BFSMs together statically schedules direct communication between them.
- *Schedule* finds a schedule for the system BFSM. It uses constraint-solving algorithms [24] to assign times to input and output events along paths in the behavior state transition graph.
- *Vhdlgen* generates a register-transfer style VHDL description of the scheduled system.

The BFSM network model is critical to successful compilation of control. The state machine semantics of BFSM let us both analyze the communication between system

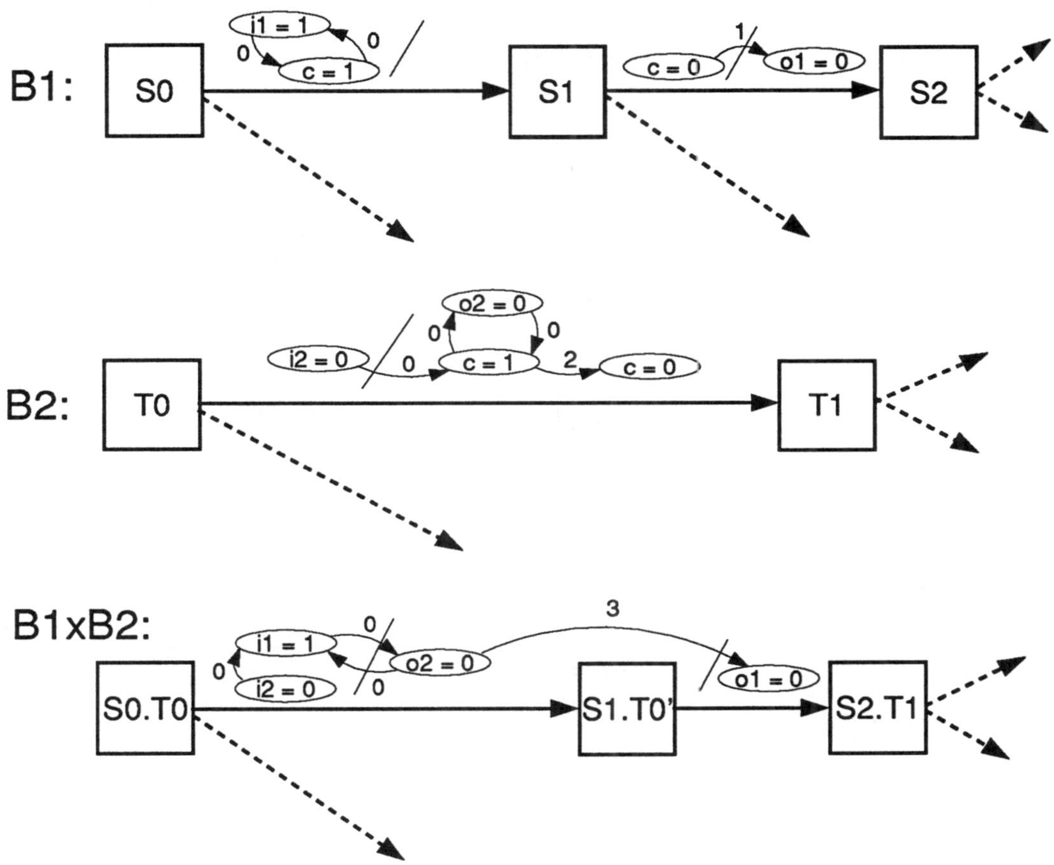

Figure 5: An example of BFSM network collapsing.

components and change the system's partitioning. *Collapsing* a BFSM network, or forming the product of the network's components, is a key operation. Collapsing a network into a single BFSM lets us analyze the system's communication—generating the product machine computes all possible communications between the component BFSMs while eliminating combinations of actions that cannot occur.

Collapsing a BFSM network requires much more care than generating the product of a system of register-transfer FSMs. Register-transfer communication is implicitly synchronized, since all I/O is fully scheduled; in contrast, input and output events in communicating BFSMs must be paired together. Register-transfer transitions are also guaranteed to take exactly one cycle, while BFSM transitions may take zero or more cycles; as a result, one component BFSM may have completed a transition and moved to a state while another component is still in the middle of its transition's events.

The collapsing algorithm symbolically executes the components in unison to determine the system behavior. For any of the component BFSMs to change state, the component must receive the set of input events which trigger a transition. Input events of a component BFSM are *matched* with corresponding output events that are supplied either by the network environment, through the network's primary inputs, or by any of the other BFSMs, through the nets that connect the BFSMs. Each output event communicates a Boolean *value* to the input events that is matched with. Matching events are constrained to occur at the same time, *i.e.* they inherit each other constraints.

Collapsing starts out with all component BFSMs in their behavior reset state. We let the environment to the network supply primary input events with events that exercise every possible transition of each of the component BFSMs. As soon as one of the component BFSM changes state we form a new composite state. Primary input/output events that where consumed/produced are kept in the product machine while the rest of the events are thrown away (their constraints are propagated back to primary IO events). We find all next states from the current state and procceed in breath-first search fashion until all states in the network are found.

Figure 5 gives an example of BFSM product formation. A wire c connects one an output of B2 to an input of B1. The output events generated by B2 on wire c must be matched to input events expected by B1; constraints follow the matched events from one component to another.

BFSM network collapsing has a variety of uses during compilation. One important use is modular constraint specification [25], which lets us separate internal and external scheduling constraints. Most hardware description systems require the designer to fold constraints imposed by external interfaces into the description of the system under design. Using BFSM networks, the external system's interface is described as one or more BFSMs. The system under design deals with the external interface through a simplified protocol. The external constraints are imposed on the system under design by collapsing the constraint machine into the system machines. Collapsing uniformly handles both simple and data-dependent scheduling constraints.

Forming the product of BFSMs is also the key to scheduling systems specified internally as communicating machines. Scheduling communicating machines requires imposing the scheduling constraints from the communicating machines upon each other. Communication may be scheduled either *dynamically* or *statically*: communication times are static when they can be determined when the times of the communication events can be determined at compile time; communication scheduling is dynamic when the communication time is data-dependent and must be coordinated by handshaking during chip execution.

Though some inter-component communication must be dynamic, a majority of communication in typical designs can be scheduled statically. Statically-scheduled communication is faster because the data is transferred in a single cycle; dynamically-scheduled communication adds handshaking overhead. However, the scheduling algorithm must work from the constraints of both machines—the schedule is limited by the union of the components' constraints. Traditional scheduling methods are not designed to simultaneously schedule multiple components [12]: communication must be scheduled manually—the designer must manually determine when each component can send/receive data and add constraints to each component to enforce the communication times—or all communication must be performed by handshaking.

By collapsing all the component machines into a single BFSM, we can combine all the constraints which limit internal communication. Scheduling the system when it is modeled as a network of communicating machines would require constantly comparing candidate schedules to be sure that they do not violate any component's constraints; scheduling the system as a single large BFSM is conceptually simpler and easier to program. State explosion has not been a serious problem for two reasons: data is segregated from sequencers and kept in registers, which are not collapsed into the BFSMs; and since much timing information is kept as scheduling constraints, BFSMs have fewer states than their register-transfer implementations.

5 Conclusions

Control-dominated machines are an important domain for high-level synthesis. PUBSS introduces new methods for architectural optimization of control-dominated systems: modular constraint generation and joint scheduling of communicating components are both made possible by the BFSM network model used to represent the design.

Acknowledgements

Miriam Leeser of Cornell University collaborated with us on the development of the BFSM model. Daaven Messenger of AT&T designed the original version of the multiple queue machine. This work was supported by the Semiconductor Research Corporation under contract 91-DJ-179 and by the National Science Foundation under contract MIP-9009960 (including a Research Opportunity for Undergraduates award).

References

[1] Kurt Keutzer. Three competing design methodologies for ASICs: architectural synthesis, logic synthesis and module generation. In *Proceedings, 26th Design Automation Conference*, pages 308–313. ACM/IEEE, June 1989.

[2] Wayne Wolf, Andrés Takach, and Tien-Chien Lee. Architectural optimization methods for control-dominated machines. In Raul Camposano and Wayne Wolf, editors, *High-Level VLSI Synthesis*. Kluwer Academic Publishers, 1991.

[3] Andrés Takach, Wayne Wolf, and Miriam Leeser. An automaton model for scheduling constraints. Technical Report CE-W92-16, Princeton University, February 1992. Submitted to *IEEE Transactions on Computers*.

[4] D. E. Thomas, E. D. Lagnese, R. A. Walker, J. A. Nestor, J. V. Rajan, and R. L. Blackburn. *Algorithmic and Regiter-Transfer Level Synthesis: The System Architect's Workbench*. Kluwer Academic Publishers, Boston, 1990.

[5] Alice C. Parker, Jorge T. Pizarro, and Mitch Mlinar. MAHA: a program for datapath synthesis. In *Proceedings, 23th Design Automation Conference*, pages 461–466. ACM/IEEE, June 1986.

[6] Wolfgang Rosentiel and Raul Camposano. Synthesizing circuits from behavioral level specifications. In C. J. Koomen and T. Moto-oka, editors, *Computer Hardware Description Languages and their Applications*, pages 391–403. Elsevier Science Publishers B. V., 1985.

[7] R. K. Brayton, R. Camposano, G. De Micheli, R. H. J. M. Otten, and J. van Eijndhoven. The Yorktown Silicon Compiler. In Daniel D. Gajski, editor, *Silicon Compilation*, pages 204–310. Addison-Wesley, 1988.

[8] P. G. Paulin and J. P. Knight. Force-directed scheduling for the behavioral synthesis of asics. *IEEE Transactions on Computer-Aided Design of Integrated Circuits and Systems*, 8(6):661–679, June 1989.

[9] Dirk Lanneer, Stefaan Note, Francis Depuydt, Marc Pauwels, Francky Catthoor, Gert Goossens, and Hugo De Man. Architectural synthesis for medium and high throughput signal processing with the new CATHEDRAL environment. In Raul Camposano and Wayne Wolf, editors, *High-Level VLSI Synthesis*. Kluwer Academic Publishers, Norwell MA, 1991.

[10] Fathy F. Yassa, Jefferey R. Jasica, Richard I. Hartley, and Sharbel E. Noujaim. A silicon compiler for digital signal processing: methodology, implementation, and applications. *Proceedings of the IEEE*, 75(9):1272–1282, September 1987.

[11] Raul Camposano. Path scheduling for synthesis. *IEEE Transactions on CAD/ICAS*, 10(1):85–93, January 1991.

[12] D. E. Thomas and T. E. Fuhrman. Industrial uses of the System Architect's Workbench. In Raul Camposano and Wayne Wolf, editors, *High-Level VLSI Synthesis*. Kluwer Academic Publishers, Norwell MA, 1991.

[13] G. De Micheli, D. Ku, F. Mailhot, and T. Truong. The Olympus synthesis system. *IEEE Design & Test of Computers*, pages 37–53, October 1990.

[14] Edward A. Snow. *Automation of Module Set Independent Register-Transfer Level Design*. PhD thesis, Carnegie-Mellon University, April 1978.

[15] Michael C. McFarland, S. J. The Value Trace: A database for automated digital design. Master's thesis, Carnegie-Mellon University, December 1978.

[16] David W. Knapp and Alice C. Parker. A data structure for VLSI synthesis and verification. Technical report, Computer Research Institute, University of Southern California, August 1985.

[17] Alex Orailoglu and Daniel D. Gajski. Flow graph representation. In *Proceedings, 23th Design Automation Conference*, pages 503–509. ACM/IEEE, June 1986.

[18] R. Camposano and W. Rosentiel. Synthesizing circuits from behavioral descriptions. *IEEE Transactions on Computer-Aided Design of Integrated Circuits and Systems*, CAD-8(2):171–180, February 1989.

[19] Leon Stok. *Architectural Synthesis and Optimization of Digital Systems*. PhD thesis, Eindhoven University of Technology, July 1991.

[20] David K. Probst and Hon F. Li. Using partial-order semantics to avoid the state explosion problem in asynchronous systems. In Robert Kurshan and E. M. Clarke, editors, *Workshop on Computer-Aided Verification*. DIMACS, 1990.

[21] Srinivas Devadas and Kurt Keutzer. An automata-theoretic approach to behavioral equivalence. In *Proceedings, ICCAD-90*, pages 30–33. IEEE Computer Society Press, 1990.

[22] Jeanne Ferrante, Karl J. Ottenstein, and Joe D. Warren. The program dependence graph and its use in optimization. *ACM Transactions on Programming Languages and Systems*, 9(3):319–349, July 1987.

[23] Ephrem Wu and Wayne Wolf. Fine-grained parallelism extraction for high-level synthesis. Submitted to *Jounal of Computer and Software Engineering*, February 1992.

[24] Wayne H. Wolf and Alfred E. Dunlop. Symbolic layout and compaction. In Bryan T. Preas and Michael J. Lorenzetti, editors, *Physical Design Automation of VLSI Systems*, chapter 6, pages 211–281. Benjamin-Cummings, 1988.

[25] Andrés R. Takach and Wayne Wolf. Modular scheduling constraint specification. In *Proceedings, ISCAS '92*. IEEE Circuits and Systems Society, 1992.

High-Level Synthesis from VHDL with Exact Timing Constraints

A. Stoll P. Duzy

Siemens AG, Corporate Research and Development
Otto-Hahn-Ring 6, W-8000 München 83, Germany

Abstract

We present a solution to the interface timing problem in high-level synthesis by requiring that the algorithmic specification must completely determine the interface timing on the basis of cycles. We show how this approach has been integrated into the high-level synthesis system CALLAS.

1 Introduction

High-level synthesis of digital systems from an algorithmic specification of the behavior [24, 7] has already been tested in industry [23]. Naturally, partial systems have been synthesized first. Experience has shown that correct timing of the interface is often a critical point in realistic designs. Typical reasons for unacceptable timing are implicit assumptions of the designer that are not expressed explicitly in the algorithmic specification, if this is possible at all. High-level synthesis is only practical, if it offers a methodology for handling the interface timing problem efficiently.

Therefore, we implemented a new approach to this problem in the Siemens high-level synthesis system CALLAS [9]. Our approach is motivated by the question how to partition between determination and incompleteness in the behavioral specification. Although incompleteness cannot be handled very well by conventional simulators, it is necessary for having optimization potential in synthesis. This trade-off is especially important for the timing behavior of the circuit interface. Looking for a consequent solution, we examined the rigorous requirement that the algorithmic specification must completely determine the interface timing on the basis of cycles. This concept proved to be feasible and will be described in the following.

This paper is organized as follows: first, in Section 2 we explain the timing problem and discuss our solution, which is closely related to a specific subset of VHDL described in Section 3. Then we show how our timing concept is implemented in the synthesis system CALLAS; in Section 4 we give an overview of the synthesis techniques of CALLAS, in Section 5 we detail controller reduction, which is the key transformation for satisfying the timing constraints, and in Section 6 we describe one of our algorithms more formally. Finally, we compare our timing concept to other approaches in Section 7, give experimental results in Section 8, and draw a conclusion in Section 9.

2 Timing concept

2.1 Preliminaries

CALLAS does not necessarily aim at the synthesis of a complete system. On the contrary, a circuit to be synthesized with CALLAS will be part of a bigger system in general. As VHDL has become a standard now, we assume that the complete system is specified in VHDL at any stage of the design process. This system description is hierarchical and multi-level in nature. Some modules are predefined and fixed, like a microprocessor, RAM, etc.; some other modules may already be given as a structural description, either obtained by synthesis or another method; finally, some modules are given as an algorithmic description and have to be synthesized. In particular, flexible means must be provided for describing the communication of a module to be synthesized with its given environment at the algorithmic level.

In our opinion, this philosophy is crucial for the acceptance of high-level synthesis, because in practice designers often want to apply high-level synthesis to some parts of a system only. Moreover, incremental synthesis becomes possible.

CALLAS targets synchronous digital circuits with one global *clock* signal; since we shall not consider combinational circuits in the following, there is always a global *reset* signal. Because the clock frequency is given, the cycle length is not a variable to be minimized, but a constraint for the optimization of other variables such as estimated area or number of cycles.

2.2 The timing problem

With earlier versions of CALLAS we gained experience when trying some realistic examples from Siemens divisions. With few exceptions [14], these circuits can be characterized as control-dominated circuits with small computational part, as described in [13]. For these applications, the bottleneck was neither area overhead nor exceeding cycle length, but the difficulty to control the schedule of the interface operations by the algorithmic specification.

Therefore we did not integrate the latest scheduling and allocation algorithms (e.g., [21]) in CALLAS; actually most synthesis techniques of CALLAS have already been used in early systems such as YSC [3], CATHEDRAL II [20], or SAW [22]. Instead we investigated a completely new approach to the interface timing problem, which can be stated as follows:

On the one hand, the timing constraints on the I/O operations in the algorithmic description must be sufficient such that the communication with the environment works for every implementation of the algorithm satisfying these constraints; on the other hand, the synthesis system must be able to satisfy the I/O constraints and must have enough optimization potential.

2.3 Our solution

If intervals are allowed for specifying timing constraints on I/O operations, e.g., as in ISYN [18], then a validation problem will arise: as VHDL simulators are commercially

0738-100X/92 $3.00 © 1992 IEEE

available now, the most common and attractive method for validating the algorithmic specification is simulation (*"Is what I asked for really what I want?"*). But it is hard to imagine how standard simulation techniques can be used for checking the sufficiency of I/O constraints that are time intervals in general.

This simulation problem motivated our timing concept. We experimented with the rather radical requirement that the algorithmic description must completely determine the interface timing on the basis of cycles. This approach turned out to be feasible in CALLAS, although obviously more effort must be put into the algorithmic description.

So, we adopted the philosophy that the interface timing is an intrinsic property of the algorithmic specification. We consider read/write operations at different cycles as different behavior, because it depends on the external environment whether the input signals do not change or the output signals are not used in between. Such external properties could be treated as sequential don't cares on the I/O operations, although they may be difficult to simulate.

Notice that powerful language constructs such as nested loops and branches can express very complicated I/O behavior also in the time dimension, i.e., the output waveforms are a functional of the input waveforms, both given in terms of cycles. In our approach, this functional must be invariant under synthesis, i.e., algorithmic simulation before synthesis and register-transfer simulation after synthesis must coincide. This principle can be expressed as *"What you simulate is what you synthesize"*.

2.4 Benefits

Obviously it can happen that the synthesis system cannot satisfy a timing constraint given in the algorithmic specification either because it is impossible in principle, e.g., a data-dependent loop cannot be executed in zero cycles, or because the techniques implemented in CALLAS cannot find the solution. However, the designer gets a quick feedback whether the specified timing can be satisfied. Then a second simulation at register-transfer level is superfluous, because CALLAS checks the consistency with the algorithmic simulation. Therefore the communication with the environment can be exactly simulated already at the algorithmic level. This feature is crucial for the iterative synthesis of several communicating modules. At any stage a mixed-level simulation is possible, where some modules are already synthesized and others not yet.

Although, in our concept, the timing behavior of the interface only is determined by the algorithmic specification, whereas scheduling of the internal operations is left to the synthesis system, this approach may appear to restrict the scheduler too much and to force the user to anticipate synthesis results. However, in contrast to the algorithm proposed in [11], our method produces a regular interface timing even if the interface timing constraints cannot be satisfied. The interface timing that could be achieved is backannotated into the original algorithmic description; therefore the user can apply the system also for exploring the optimal interface timing by specifying a minimal default timing.

3 VHDL subset

Sequential statements inside a VHDL process are a natural starting point for high-level synthesis. For realizing our timing concept in the VHDL synthesis subset, we rely on a one-to-one correspondence between the signals read or written by a process and the interface of the resulting circuit, whereas the inner life of the circuit has to be described with variables. Time can only be advanced by calling the procedure *cycles(n)*, which essentially waits n times on the rising edge of the *clock*. Since this affects only signals, operations on variables can be executed at any time that fits the data dependencies. For more details about the CALLAS synthesis subset of VHDL, we refer to [11]. As an illustration, our VHDL formulation of the prefetch example [4] is given in Figure 1. For matching the RT-simulation exactly, we describe the *reset* signal and the corresponding infinite overall loop explicitly, using the C preprocessor for simplicity.

```
library callas; use callas.callas.all ;
procedure cycl (signal reset,clk : in bit; n : in natural) is
begin
  for i in 1 to n loop
    wait until (clk = '1'); exit when (reset = '1');
  end loop;
end;

#define inf_begin(n)  cycl(reset, clk, n);\
 if (reset = '0') then outest_loop: loop
#define inf_end       end loop outest_loop; end if;
#define cycles(n)     cycl(reset, clk, n);\
 exit outest_loop when (reset = '1');

entity prefetch is
  port ( reset, clk        : in  bit;
         branchpc, ibus    : in  bit_vector (31 downto 0);
         branch, ire       : in  bit;
         ppc, popc, obus   : out bit_vector (31 downto 0)  );
end prefetch;
architecture behavior of prefetch is
begin
  process
    variable pc, oldpc : bit_vector (31 downto 0);
  begin
    pc := conv_bit_vector (32, 0); oldpc := pc;
    ppc <= pc; popc <= oldpc;
    inf_begin(1)
      while (ire = '0') loop cycles(1) end loop;
      oldpc := pc; pc := pc + 4;
      ppc <= pc; popc <= oldpc; obus <= ibus + 4;
      if (branch = '1') then  pc := branchpc; ppc <= pc; end if;
      cycles(1)
    inf_end
  end process;
end behavior;
```

Figure 1: Prefetch example (VHDL synthesis subset)

In our paradigm of a synchronous Moore machine, the input signal *ire* can be evaluated only at the rising edge of the *clock*. If $ire = 0$, the values of the output signals are kept constant according to the specification; if $ire = 1$, all computations that are necessary to produce the new output values can be done without loosing any cycle. In particular, if additionally $branch = 1$, then *branchpc* overwrites *pc* and immediately arrives at the output *ppc*; in this case, there seems to be a difference to [4], because in our version neither an obsolete value of *pc* appears at the port *ppc* nor an extra cycle is needed until *ire* can be evaluated again. So we take the view that the specification must determine the temporal interface access; leaving this to the scheduler as in [4] only works under additional assumptions on the environment, i.e., requires special communication models.

4 Synthesis system

The overall structure of the CALLAS system is shown in Figure 2. The VHDL compiler translates the algorithmic specification into a language-independent intermediate format, a flow graph representation, on which already certain behavioral transformations such as loop unrolling can be done. Then the CALLAS synthesizer, which originated in CADDY [6], converts the flow graph into an initial hardware structure, represented by a data and a control graph. This is basically a direct compilation as described in [9]. The resulting initial hardware structure is characterized by maximal parallelism and chaining of operations. Registers and multiplexors have been created for resolving data collisions. Their control is specified in the initial control graph, which reflects the loops and branches in the flow graph. Unnecessary registers are avoided by applying a data flow analysis, which includes the task of variable disambiguation (cf. [16]).

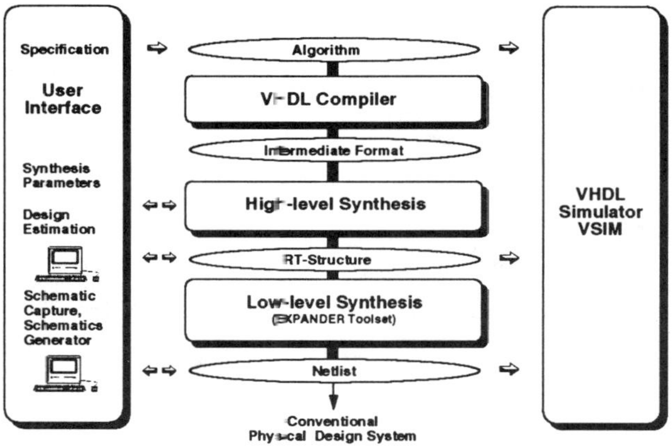

Figure 2: CALLAS system architecture

On the internal data and control graph, various transformations are performed, e.g., register folding (cf. [3]), multiplexor minimization [25], arithmetic and logic transformations [8], controller reduction (see next section), and operator folding, exploiting mutual exclusion (cf. [16]). These transformations are not commutative; therefore some experience is necessary in choosing the synthesis script. The final RT-structure is generated as a net list whose instances are abstract components to be mapped onto a real library by the EXPANDER tool set (cf. [9]). One of these components represents the controller and therefore is described by a symbolic flow table extracted from the control graph.

Contrary to common high-level synthesis systems (cf. [16]), which generally determine a favorable scheduling and allocation scheme first and then build data and control part accordingly, in CALLAS a reverse synthesis strategy is pursued. Since design space is explored by transforming an initial implementation, the information about scheduling and allocation is contained only implicitly in the actual implementation. The execution time of the functional units is determined by the control of the surrounding registers and multiplexors.

The crucial transformation for matching the timing of the interface operations with the algorithmic specification is controller reduction; it transforms the control part only, preserving the semantics specified by the data part. If resources are shared in the data part after appropriate transformations, the necessary data shifting will result in semantic constraints for the associated control in the control part. In this case, controller reduction may produce paths with more cycles than for the maximally parallel implementation. Similar constraints emerge, if violations of the maximal cycle length are repaired before controller reduction.

After controller reduction, all transformations of the control and data part must also preserve the achieved interface timing; nevertheless, if there are some idle control steps, rescheduling and reallocation is possible in principle for improving cycle length and area (cf. [17]).

5 Controller Reduction

5.1 Target architecture

Since our timing concept requires reading of input signals at specified cycles, we use input registers for all interface input signals. Similarly, we use transparent output registers for all interface output signals such that the values of these output signals are constant between subsequent assignments. This initial implementation appears somewhat redundant, but the register overhead can be reduced by appropriate transformations such as register folding.

Reading of an input signal is synthesized as follows: for picking up the value at the specified cycle, the controller sends the enable signal of the corresponding input register in the preceding cycle, since the input register causes a delay of one cycle. Consequently, all synthesized sequential circuits are Moore machines. Notice that in our VHDL subset it is impossible to describe a Mealy machine by a single process. Because of the input registers we realize the controller as a Mealy machine. Therefore our target architecture guarantees that there is no combinational feedback in the data path from the controller outputs to the controller inputs.

5.2 Data structure of the control part

The internal controller representation is a directed graph (V, E), where the vertices V represent the states of the FSM and the edges in E correspond to the state transitions. Each edge is attributed with a condition expression of the FSM inputs, which determines whether the edge is valid, and a list of control values for the FSM outputs, which have to be asserted in this case. Such a representation can obviously be translated into a symbolic flow table.

Among the control signals, which determine loading of the registers and selection of the multiplexors, those controls are marked that control reading of the circuit inputs or writing of the circuit outputs. The timing requested by the user is expressed by the *cycles* call in the algorithmic description; this information is represented by an edge in the control graph carrying the specified number of cycles as time information. The other edges initially carry the time information *"0 cycles"*. An edge is marked as **"time-relevant"**, if it carries a positive time information or if it has any I/O control attached.

5.3 Reduction technique

During controller reduction, also edges with I/O controls may get a positive time information. This means in our semantics that the I/O controls have to be followed by the specified number of wait cycles. Due to the presented data structure it is possible to reproduce the requested interface timing by a controller emulation spending time according to this time information. Of course, a real controller will take exactly one cycle for each edge. Thus the task of controller reduction is to match both timings.

To this end, the control graph is compacted first; the goal is to minimize the number of cycles for all paths, including conditional branches and loops, without changing the circuit semantics. The same objective has the AFAP scheduling described in [4]. In CALLAS, however, a hardware structure already exists when controller compaction is done; therefore, not the operations in the flow graph are scheduled, but the generated register-load and multiplexor-select operations. For determining whether the circuit semantics are preserved, the dependencies in the data part are checked.

Our technique of controller compaction can be characterized as a transformational approach based on "code" (i.e., control) motion similar to [19]; however, also the branching structure of the control graph is changed. We do not give the details of our compaction algorithm here, because the underlying problem is well known from microcode compaction, and an optimal solution in the context of high-level synthesis has already been published [4]. But we shall describe the extensions that are necessary for meeting the timing constraints. Besides the data dependencies, the additional constraint has to be observed that no I/O control is moved into a preceding edge that contains a positive time information. Also the marking of the time-relevant edges as well as their time information are updated for staying consistent with the requested timing.

Making all paths AFAP, however, is not really what is desired, because the number of vertices between two subsequent time-relevant edges should not depend on the path, according to our requirement that the user can predict and control also the temporal I/O behavior. A control graph having this property is called **"well-timed"**. In the next step, which will be described formally in the next section, the compacted graph is again expanded by adding wait states such that it becomes well-timed. Moreover, the time information of each time-relevant edge is supplemented with the actual timing besides the requested timing.

If the achieved actual timing coincides with the requested timing, controller reduction was successful. Otherwise, the achieved actual timing could already be backannotated for the user, but controls that caused a violation of the requested timing can potentially be moved into wait states which were inserted in the previous step. Therefore, another transformation algorithm is executed on the control graph. The difference to the first step is that now all transformations must preserve the "well-timed" property.

Finally, the user is informed whether the requested timing could be satisfied. In the positive case, RT simulation after synthesis will coincide with algorithmic simulation before synthesis. If the requested timing cannot be met, the system will nevertheless generate a controller with a well-defined timing, which is partly slower than requested, and will backannotate this timing in the original VHDL description.

6 Algorithm for insertion of wait states

In our controller model, the following **assumptions** hold (an example is shown in Figure 3):

- The control part is represented by a directed graph (V, E), and there is a subset F of E containing the time-relevant edges (non time-relevant edges are dotted arrows in Figure 3). Denominate $G = (V, E, F)$.
- For each vertex v in V, there is an edge e in E that starts at v, i.e., $e = (v, w)$. Moreover, there is a distinct vertex v_{reset} in V (not shown in Figure 3); for every other vertex v in V, there is a path in (V, E) from v_{reset} to v.
- The **requested timing** is given by a function $t_r : F \to \mathcal{Z}_+$ from F into the set $\mathcal{Z}_+$ of nonnegative integers with the interpretation that $t_r(f)$ cycles must be spent after the controls of an edge f in F have been asserted, assuming that all edges in $E \setminus F$ are traversed in zero time. Therefore such a function completely determines the temporal I/O behavior of the circuit.

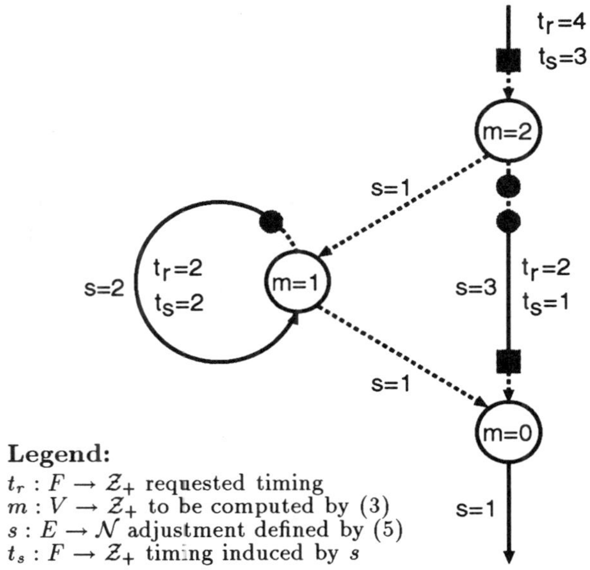

Figure 3: Insertion of wait states

Our intention is to extend the controller by inserting wait states (filled squares and circles in Figure 3) such that the actual timing, which is defined by counting edges, coincides with the requested timing. A necessary condition is that the actual timing does not depend on the path. That means, if one starts at a given edge $e_0 \in F$ and then goes through $E \setminus F$, one must finally come up again at F; every such path must have the same length depending only on e_0. More formally:

Definition: *A* **bridge** *p in G is a path $(e_0, \ldots, e_n)$ in (V, E) such that $e_0, e_n \in F$ and $e_1, \ldots, e_{n-1} \in E \setminus F$ and $n \geq 1$. G is called* **well-timed** *, iff each e in $E \setminus F$ can be extended to a path ending in F, and there is a function t on F such that $t(e_0) = n$ holds for every bridge $(e_0, \ldots, e_n)$ in G. Then the function $t : F \to \mathcal{Z}_+$ is called* **actual timing** *of G.*

If we assume that G is already well-timed with actual timing t (in Figure 3, the nodes represented by filled circles make the control graph well-timed with actual timing t_s), then the following is an obvious observation:

- There is no edge f in F with $t(f) = 0$.
- There is no cycle in the subgraph $(V, E \setminus F)$.
- A controller with actual timing $t_0 := \max(t, t_r)$ can be obtained by inserting $t_0(f) - t(f)$ wait states (filled squares in Figure 3) after every edge $f \in F$. Then the requested timing is matched, iff $t \leq t_r$.

Now the problem is left to obtain a well-timed controller by inserting wait states.

Definition: *A function $s : E \to \mathcal{N}$ from E into the set $\mathcal{N}$ of positive integers is an* **adjustment** *of G, iff there is a function t_s on F such that*

$$t_s(e_0) = s(e_1) + \cdots + s(e_n) \tag{1}$$

holds for every bridge $(e_0, \ldots, e_n)$ in G. Then the function $t_s : F \to \mathcal{Z}_+$ is called **timing induced by** *s .*

Note that an adjustment s of G induces a well-timed control graph with actual timing t_s, if one inserts $s(e) - 1$

wait states (filled circles in Figure 3) before every edge $e \in E$. For every edge f in F, the value $t_s(f) - 1$ is obviously bounded from below by the length of any path in $E \setminus F$ starting at the end vertex of f. The next theorem shows that this bound suffices.

Theorem: *If the subgraph $(V, E \setminus F)$ is acyclic, then there is an adjustment s of G with induced timing t_s such that*

$$t_s(f) = \max \{ \lg(p) \mid \text{path } p \text{ continues } f \text{ in } E \setminus F \} \qquad (2)$$

holds for all edges f in F, where $\lg(p)$ denotes the length of the path p.

Proof: Let $m : V \to \mathcal{Z}_+$ be the minimal function having the property

$$m(v) = \max \{ m(w) + 1 \mid e = (v, w) \in E \setminus F \} \qquad (3)$$

for all v in V. Because $(V, E \setminus F)$ is acyclic, such a function exists and can be computed recursively by (3). In fact, it can be shown by induction that the function m has the property

$$m(v) = \max \{ \lg(p) \mid p \text{ path in } E \setminus F \text{ starting at } v \} \qquad (4)$$

for all v in V. Now define the function s on E by

$$s(e) := \begin{cases} m(v) - m(w) \,, \text{ if } e = (v, w) \in E \setminus F \\ m(v) + 1 \,, \text{ if } e = (v, w) \in F \end{cases} \qquad (5)$$

for all e in E. Then the function s has positive values. Moreover, it follows from (4) and (5) that the function t_s given by (2) has property (1), i.e., s is an adjustment of G. □

Now it is straightforward to design an algorithm that first checks by depth-first search whether the subgraph $(V, E \setminus F)$ is acyclic and then exploits equation (3) for the recursive computation of the function m. Both requires no more than $|V|$ iterations of at most $|E|$ steps, i.e., the algorithm is of complexity $O(|V||E|)$. In the acyclic case, the controller is made well-timed by inserting wait states according to the adjustment s given by (5). Finally, the timing $\max(t_s, t_r)$ is obtained by inserting additional wait states.

7 Comparison to other approaches

7.1 CYCLE-BLOCK construct

The V compiler is able to synthesize I/O protocols at the cycle level specified by the CYCLE-BLOCK construct [1] in the V language. Such a construct encloses a region of code, where STEP statements are used for determining the relative cycle sequence for all operations involving signals in a specified list; the scheduling of other operations in such a region is only restricted by data dependencies. In an algorithmic description, several CYCLE-BLOCK constructs may be used with different signal lists. In terms of V, our timing concept corresponds to the extreme case that the complete algorithmic description is embedded in one CYCLE-BLOCK construct, where the list of referenced signals consists of all interface signals. Inside a CYCLE-BLOCK, however, complex loops and branches do not seem to be allowed; only a GOTO statement can be used for interrupting the sequential code. It is obvious that the V compiler gives more freedom to the synthesis system, because only relative schedules are fixed for certain signals. The price for this freedom is to give up our basic principle "What you simulate is what you synthesize". That makes the validation of the algorithmic description more difficult and restricts the style of communication with the environment that can be handled.

7.2 Local min/max timing constraints

In the HERCULES system, the method of relative scheduling [15] allows to simultaneously meet minimum and maximum timing constraints between arbitrary operations, if these constraints are well-posed. All operations must have a fixed or unbounded delay. The circuit is represented by a polar hierarchical acyclic graph. E.g., a vertex may expand to a data-dependent loop in the lower hierarchy; then it has unbounded delay.

In particular, it is possible to fix the interface timing completely by imposing a complete set of identical minimum and maximum timing constraints between all interface read/write and loop begin/end vertices. But this does not cover the general case of timing constraints that result from a VHDL description in our subset, because there may also be timing constraints across loop boundaries. Generally, our principle "What you simulate is what you synthesize" cannot be realized in HERCULES, although HERCULES allows more flexible timing constraints.

The kind of timing constraints used in HERCULES has already been proposed by Nestor [18] as an extension of ISPS. For satisfying these constraints, ISYN [18] uses a different approach based on a suitable list scheduling. The simulated annealing technique of SALSA [17] allows to improve a schedule that already satisfies the timing constraints for reducing the estimated design costs. In principle, such an improvement would also be possible and profitable in CALLAS.

7.3 Different concepts

Originally, timing constraints on sequences of statements were considered, e.g., in the ELF system [10]. A method for checking the consistency of such constraints was described in [5]. But this kind of constraints does not really meet the application needs and is too restrictive. The JANUS system [2] is able to synthesize an interface consisting of asynchronous logic circuitry between the internal data flow and the external timing requirements, given by an event graph. In the ADAM system, the tool CONSPEC [12] allows flexible control specifications, including asynchronous protocols and local min/max timing constraints. But CONSPEC assumes that the data path has already been scheduled, i.e., that there is a binding between data and control part, which assigns data operations to clock periods.

8 Results

In Table 1, we list results obtained by CALLAS with the new interface concept. Some of the examples originated in the 1989 Workshop on High-Level Synthesis, but the determination of the temporal interface behavior requires more careful modeling as we have shown in Section 3. The advantage of this effort is that high-level simulation before synthesis already coincides with RT-level simulation after synthesis; this has been checked for most of the examples because we were interested in validating our synthesis algorithms.

For every example, the number of lines in the VHDL specification is reported in column "loc". The next two columns contain data on the resulting control part, namely the number of states and the number of product terms in the FSM realization, whereas the column "regbits" shows the total number of register bits in the data part implementation. Finally, in the last two columns the size of the synthesized circuit is given in terms of equivalent gates and active cell area in *square mm* after mapping onto a Siemens 1.5μm standard cell library.

9 Conclusion

We have presented a new timing concept for high-level synthesis that subjects the interface timing to the principle

Design	loc	sta-tes	prod terms	reg bits	equiv gates	kernel area
adc	35	3	4	14	232	0.29
traffic	73	7	24	17	406	0.53
prefetch	33	2	3	162	2303	2.76
gcd	33	3	5	48	1054	1.42
display	112	0	0	0	287	0.41
serpar	77	3	5	56	849	1.19
rrq	84	7	15	75	1004	1.26
adap2	170	14	20	170	5229	7.59
watch	313	3	4	117	2138	2.64
biomed	429	13	24	559	9392	11.11

Table 1: CALLAS synthesis examples

"What you synthesize is what you simulate". This concept provides a clear separation between specification and implementation in the design process. On the one hand, the designer must determine the complete observable behavior already in the algorithmic specification, on the other hand, the synthesized circuit must exactly meet this behavior.

It may appear as a disadvantage of our method that it sometimes requires more accuracy in the algorithmic specification than necessary. Therefore, a useful extension will be to allow sequential don't cares on specific I/O operations to increase the optimization potential for scheduling and allocation. However, then it is doubtful that simulation is applicable for validating the specification, especially the interaction with the environment. Therefore, such an extension must go along with a new validation concept, which will probably restrict the kind of communication. Thus an area of future work is the automatic support of special communication protocols, also including pipelining.

References

[1] V. Berstis, "Cycle-level timing constraints in behavioral hardware descriptions," in *CHDL, Proceedings of the Ninth IFIP Symposium* (J. A. Darringer and F. J. Rammig, eds.), pp. 197–205, 1989.

[2] G. Borriello and R. Katz, "Synthesis and optimization of interface transducer logic," in *Proc. of IEEE ICCAD-87*, pp. 274–277, 1987.

[3] R. Brayton, R. Camposano, G. D. Micheli, R. Otten, and J. van Eijndhoven, "The Yorktown silicon compiler," in *Silicon Compilation* (D. Gajski, ed.), pp. 204–310, Addison-Wesley, 1988.

[4] R. Camposano, "Path-based scheduling for synthesis," *IEEE Transactions on Computer Aided Design*, vol. 10, pp. 85–93, Jan. 1991.

[5] R. Camposano and A. Kunzmann, "Considering timing constraints in synthesis from a behavioral description," in *Proc. of ICCD 1986*, pp. 6–9, 1986.

[6] R. Camposano and W. Rosenstiel, "Synthesizing circuits from behavioral descriptions," *IEEE Transactions on Computer Aided Design*, vol. 8, pp. 171–180, Feb. 1989.

[7] R. Camposano and W. Wolf, *Trends in High-Level Synthesis*. Kluwer Academic Publishers, 1991.

[8] J. Darringer and W. Joyner, "A new look at logic synthesis," in *Proc. of 17th Design Automation Conference*, pp. 543–549, 1980.

[9] P. Duzy, H. Krämer, M. Neher, M. Pilsl, W. Rosenstiel, and T. Wecker, "CALLAS - conversion of algorithms to library adaptable structures," in *Proc. of IFIP VLSI Conference 1989*, pp. 197–208, 1989.

[10] E. Girczyc and J. Knight, "An ADA to standard cell hardware compiler based on graph grammars and scheduling," in *Proc. of ICCD 1984*, pp. 726–731, 1984.

[11] W. Glunz and G. Umbreit, "VHDL for high-level synthesis of digital systems," in *Proc. of 1st European Conference on VHDL Methods*, 1990.

[12] S. Hayati and A. Parker, "Automatic production of controller specifications from control and timing behavioral descriptions," in *Proc. of 26th Design Automation Conference*, pp. 75–80, 1989.

[13] K. Keutzer, "Three competing design methodologies for ASIC's: Architectural synthesis, logic synthesis and module generation," in *Proc. of 26th Design Automation Conference*, pp. 308–313, 1989.

[14] M. Koster, M. Geiger, and P. Duzy, "ASIC design using the high-level synthesis system CALLAS: A case study," in *Proc. of ICCD 1990*, pp. 141–146, 1990.

[15] D. Ku and G. D. Micheli, "Relative scheduling under timing constraints," in *Proc. of 27th Design Automation Conference*, pp. 59–64, 1990.

[16] M. McFarland, A. Parker, and R. Camposano, "The high-level synthesis of digital systems," in *Proceedings of the IEEE*, vol. 78, pp. 301–318, Feb. 1990.

[17] J. Nestor and G. Krishnamoorthy, "SALSA: A new approach to scheduling with timing constraints," in *Proc. of IEEE ICCAD-90*, pp. 262–265, 1990.

[18] J. Nestor and D. Thomas, "Behavioral synthesis with interfaces," in *Proc. of IEEE ICCAD-86*, pp. 112–115, 1986.

[19] R. Potasman, J. Lis, A. Nicolau, and D. Gajski, "Percolation based synthesis," in *Proc. of 27th Design Automation Conference*, pp. 444–449, 1990.

[20] J. Rabaey, H. D. Man, J. Vanhoof, G. Goossens, and F. Catthoor, "CATHEDRAL II: A synthesis system for multiprocessor DSP," in *Silicon Compilation* (D. Gajski, ed.), pp. 311–360, Addison-Wesley, 1988.

[21] J. Scheichenzuber, W. Grass, U. Lauther, and S. März, "Global hardware synthesis from behavioral dataflow descriptions," in *Proc. of 27th Design Automation Conference*, pp. 456–461, 1990.

[22] D. Thomas, E. Lagnese, R. Walker, J. Nestor, J. Rajan, and R. Blackburn, *Algorithmic and Register-Transfer Level Synthesis: The System Architect's Workbench*. Kluwer Academic Publishers, 1990.

[23] J. Van Meerbergen and H. D. Man, "A true silicon compiler for the design of complex ICs for digital signal processing," *Philips Technical Review*, vol. 44, No. 7 1989.

[24] R. Walker and R. Camposano, *A Survey of High-Level Synthesis Systems*. Kluwer Academic Publishers, 1991.

[25] N. Wehn, J. Biesenack, and M. Pilsl, "A new approach to multiplexer minimization in the CALLAS synthesis environment," in *Proc. of IFIP VLSI Conference 1991*, pp. 203–214, 1991.

Synthesis from Production-Based Specifications

Andrew Seawright and Forrest Brewer
Department of Electrical and Computer Engineering
University of California, Santa Barbara, CA 93106

Abstract

This paper describes a model for, and an implementation of, production-based synthesis of hardware description language (HDL) code in which the overall structure of the resultant machine is derived from a hierarchy of sub-machine descriptions, each represented by a production. The production-based specification (PBS) consists of productions annotated with HDL action code, and forms the input to a design tool which outputs procedural HDL tailored for hardware synthesis. Due to the concise nature of this form of specification, for certain machines such as protocol controllers, the technique can save enormous labor in the construction of procedural specifications for these machines. Novel aspects of this research include the compilation of a PBS with HDL action clauses into synthesizable procedural HDL and the approach to specification of machine behavior in the event of exceptional conditions.

1.0 Introduction

A common way to describe the behavior of a machine is in a procedural form using a hardware description language (HDL) such as VHDL [13]. This type of description is procedural in the sense that the designer describes the behavior in a step by step sequence. An example procedural specification is shown in Figure 1, which describes a

```
process
begin
wait until CLK'event and CLK = '1';
if(SEEN_TRAILING and DATA = '0') then
   IS_LEGAL <= '0';
   COUNT <= 0;
elsif(SEEN_ZERO and DATA = '1') then
   SEEN_TRAILING := TRUE;
elsif(DATA = '0') then
   SEEN_ZERO <= '1';
   COUNT := COUNT + 1;
end if;
end process;
```

Figure 1

machine that counts consecutive zero bits in a sequential single bit input stream. Simulation and direct hardware synthesis is commonly performed from procedural descriptions. An alternative description is shown in Figure 2.

```
Count -> Valid | Invalid;
Valid -> ONE* Low* ONE*;
{
     IS_LEGAL <= '1';
}
Invalid ->ONE* ZERO+ ONE+ ZERO;
{
     IS_LEGAL <= '0';
     COUNT := 0;
}
Low -> ZERO; { COUNT := COUNT + 1;}
```

Figure 2

This second type of description is based on productions. A production is a rule describing certain compositions of symbols as a single symbol of higher abstraction. Symbols at the lowest level of abstraction (tokens) describe atomic sets of external signal transitions, while those at higher levels describe arbitrarily complex sequences of these tokens. In the example above, ONE and ZERO are tokens. Note that some productions initiate actions which are fragments of HDL code enclosed in braces.

For many types of complex behavior, production based specifications (PBS) are superior to procedural specifications, in that they can be more concise, and easier to debug and understand. Examples include communications protocols, cache protocols, bus controlers, adaptive coding, and translators. The simplicity of interpretation stems from the local nature of each production in the specification. Each production is simultaneously active for all input transitions so that the designer need not worry about the explicit construction of the global control flow, which is necessary in designing a procedural specification.

The synthesis procedure described in this work converts a production specification of machine behavior into a procedural description encoded in a specified hardware description language and tailored explicitly for hardware synthesis. This output description contains explicitly the minimal controlling FSM described implicitly by the productions. The advantage of the HDL output description is that it enables fast turnaround simulation or synthesis of the machine using state-of-the-art HDL tools. Since for many applications the production hierarchy is very concise, the resulting HDL has a greater chance of representing the designer's intent than a hand coded procedural description. In this respect, the user model for this technique is a hardware analog of the very successful compiler generation tools prevalent in software design [3] [4] [8].

2.0 Related Work

The related software tools compile grammar productions and pattern recognition expressions into parsers, and lexical scanner routines. Many of the advantages present in using these software tools *must be* applicable to the specification of control dominated behavior. However, since the trade-offs involved in designing a hardware machine are different than for generating a software program, the models, metrics, and techniques used in compiling these specifications differ substantially. With respect to hardware, the most closely related work is due to Ullman et. al. [1] [2] in their study of the compilation of regular expressions into PLA implementations. In this work, a single regular expression is compiled into a non-deterministic finite automata (NFA). This NFA was directly implemented as a PLA, with combinational feedback terms. The non-determinism of the current state in the machine was represented by these feedback terms, each representing a single active state. Verification of equivalent behavior of differing implementations of machines was done using automata based models [5]. Their model for representing classes of equivalent behavior is in some respects similar to the model presented in this paper representing the target machine's behavior after the production description is compiled.

Due to the fact that the PBS describes a hardware entity, a multitude of CAD techniques and technologies are available to create an efficient implementation. These techniques include: 1) The synthesis of logic networks from procedural hardware description languages such as implemented in the BDSYN tool [6] and industrial CAD packages [14]. Of particular importance to this technique is the notion of resource allocation and sharing among the several possibly mutually exclusive operations implied in the action clauses in the PBS. 2) The optimization of logic networks to meet timing and area constraints offered by logic synthesis [6] [7]. 3) The techniques of state assignment [11] [12] and minimization [10] of finite state machines.

3.0 Methodology and Advantages

In use, the tool provides an HDL generator. The output VHDL can be both simulated and synthesized. Desired changes in the specification are made by constructively modifying the PBS description and re-compiling. This technique has several advantages: 1) Modification of a design can be readily made, as the productions form a natural partitioning of the machine's behavior. 2) The compilation and optimizations of the productions are inherently symbolic, enabling the designer to specify behavior at many levels of abstraction, from single bit events to valid protocol recognition. 3) Powerful control and data flow analysis may be possible from the hierarchical PBS description. 4) After synthesis the structural architecture can be simulated and compared with the procedural simulation in order to track design integrity. 5) The procedural HDL output can be tailored and annotated by the production based specification compiler to match a synthesis policy. This allows customization of the output for direct logic synthesis using currently available tools.

4.0 Model

A Mealy [9] machine is defined as a 6-Tuple (Q, S, D, d, l, q0), where Q is a finite set of states, S the input alphabet, and D the output alphabet. The transition function d maps Q x S to Q, while l maps Q x S to D. The initial state of the machine is q0. Our goal is to design an entity M, which is essentially an FSM when viewed at the interface level, but could in fact be a complex controller with arbitrary datapath or other interior extensions. We model M as an 8-Tuple (Q, S, A, d, L, q0, I, X) where Q, S, d, and q0 are the same as in the Mealy machine above. L is a function of domain Q x S mapping to a range consisting of arbitrary ordered subsets of A. The elements of A are "actions" performed by the FSM under certain state transitions. These actions are segments of procedural HDL code. Unlike the Mealy machine's output function, L maps to ordered sets of these arbitrary high-level segments. I is a vector of boolean signals providing the totality of the machine's external interface. We define X as an function that maps the interface, I, to elements of the alphabet S. X provides a level of abstraction between the interface signal vector and the internal input language for the FSM and other possible uses for the interface signals (such as operands to/from the actions).

A production-based specification describing M is compiled using the PBS compiler to obtain complete register transfer level procedural description of the machine M. This RTL description consists of, among other things, the implementations of the functions X, d, and L. The PBS input consists of a set of productions, annotated with associated actions forming the set A, and auxiliary HDL code. The productions specify the behavior of the machine through the actions, as a function of internal state and the input stimulus. The terminal symbols of the productions form the set S. The auxiliary HDL code defines the interface I, the mapping function X, initialization code for the state q0, and other user defined HDL code. The user, or another tool, is free to define the specifics of the clocking strategy and the mapping X in the specification, as long as it can be synthesized. For the rest of this paper we assume a synchronous framework.

The following statements codify our model with respect to the specification of the behavior of the machine through the action clauses:

- The behavior of the machine with respect to the external world is defined by the behavior of a machine, derived from the production specification, in which the actions are combinationally executed in the designated points of the protocol and within a single clock transition.
- When there are several actions defined during a state transition of M, the actions of more primitive productions have higher precedence and conceptually execute before the actions of more advanced productions. To clarify, we define the behavior of the machine with several simultaneous actions to be the behavior of the machine if the actions were procedurally executed in step by step order even though they may be resolved by a single combinational logic function.
- A production is more primitive if it is called by the more advanced production.

Given the above points concerning the *behavior* of the machine as defined by the specification and viewed by the designer, any transformation of the machine is certainly allowed that results in the same external behavior. Such transformations would be made to improve performance or minimize cost.

4.1 Hardware vs. Software

There are different trade-offs and implicit assumptions in specifying and constructing a finite state machine to perform some task as a hardware device, rather than as a software program, using productions. In compiling programs in software, we often have the luxury of lookahead, e.g. looking at the future input tokens to disambiguate the meaning of the input. In hardware, lookahead may be difficult if not impossible. For example, in this model, input is defined through the X function as the state of the interface I during a clock transition. In this case, lookahead is simply *not* possible (lookahead can be simulated using an X function with state). Another difference in hardware is due to the presence of rigid timing constraints. For example, between successive cycles of a clock a FSM must compute its next state based on the current input. A controller may be required to respond to an input event in the next clock cycle. Although software compilers are designed to be efficient it may be of little concern that several "cycles" of the parsing loop are required to interpret the meaning of an input token before the next token is read. This is true in the classic shift/reduce parser [8]. A further distinction is found in the meaning and mechanism for producing "tokens". In hardware, we have a wide variety of possible candidates for tokens including operands on busses and transitions of input signals. Finally, the issue of error recovery is important in both hardware and software. In program compilers, the emphasis is usually placed on identifying errors in the input stream, and terminating the compilation. On the other hand, in hardware the emphasis is on conveniently specifying well-defined behavior of the machine with respect to abnormal or exceptional input. For example an exception may require the machine to perform additional processing, resynchronization, and recovery procedures.

5.0 Implementation

A PBS compiler has been implemented, in approximately 4600 lines of C++, which compiles input specifications containing productions, VHDL actions, and VHDL code segments. The productions are composed of symbols and operators. The symbols represent sub-machines ("calls" to other productions), or terminal symbols. In the current implementation, recursion in the productions is illegal. Each symbol in a production represents an instantiation of a sub-machine that performs some task. Terminal symbols represent machines that recognize a single input symbol. During the compilation process, any symbols in the productions that are not calls to other productions, i.e. are not non-terminals, are considered terminals forming the set S. The user explicitly specifies in the PBS description explicitly how the tokens are generated; i.e. the X function.

The operators in the productions combine the sub-machines into more complex machines through compositions of the sub-machines. These operators consist of the standard regular expression operators [8] [9] [10] plus several special operators. The special '^' operator is for convenience and is used as a shorthand for representing the concatenation of several identical sub-machines by the specified number of instantiation. For example the production below illustrates its use:

```
block -> byte^4;
```

The production defines `block` to be a sequence of four `bytes`.

The exception operator, '!', is used to specify conveniently the behavior of a sub-machine in the circumstance of non-recognition. For example in the production:

```
data -> block!resync;
```

The exception operator states that if the sub-machine `block` recognizes events that can not be `block` *or any other valid production*, then the state is transferred immediately to the sub-machine represented by the `resync` production. When then `resync` production accepts, the effect is as if the `block` machine accepts. Several levels of exception handling are possible, with well defined scoping rules and precedences. The '!!' operator is another exception operator. When used, the annotated sub-machine is constructed with exceptional behavior causing a "reset" of the sub-machine.

The exception operators allow the user to specify concisely behavior in the case of events not explicitly described in the ordinary productions, by giving access to the complement space of the production. The complement space for the production is defined with respect to all of the ensemble productions. This allows a convenient mechanism for specifying re-synchronization. Of course the user is also free to explicitly add specific error productions in the specification as well.

The productions can be annotated with VHDL code segments as "actions". These actions may be associated with any sub-machine symbol in the production or with the entire production as a whole. In either case, the action is assumed to take effect on valid recognition of the input conditions as described in the model section. Consider the example machine introduced earlier. When the machine detects an invalid input frame, the last token will be a ZERO. Two actions are indicated: the action attached to the ZERO token in the `low` production, and the action associated with the `invalid` production. Due to precedence the more primitive action "computes" first. In this case the overall effect is the `COUNT` variable reset to zero.

5.1 Compilation

The productions are collapsed and compiled into an intermediate tree representation. The interior nodes of the tree represent the various composition operators, and the leaves of the tree represent the terminal token symbols. To be efficient, common sub-trees are shared when there are multiple references to a production from other productions in the PBS. Since each node of the tree represents a particular point in the protocol, action objects are attached to lists provided at the nodes. In the current implementation, the action objects consist of wrappers around VHDL code.

A non-deterministic finite automata (NFA) is constructed from this tree using a process based on Thompson's Construction [9], however suitably modified to implement the exception handling mechanism and other extensions. Actions are propagated from the tree to the NFA while maintaining their precedence. Next the NFA is converted to a DFA. This is done using the well-known subset construction algorithm [9], again with extensions for implementing the exception handling mechanism, particularly in the method in which the 'death' state is handled. The death state is a DFA state generated in the construction process when a group of NFA states do not have successor(s) for some input symbol. All of the transitions leaving the DFA death state return to itself, thus the machine can never leave this state. Exception handling is described in more detail in the next section.

The next step in the compilation process is the optimization and encoding of the of the DFA. The number of states in the DFA is minimized using a partition refinement technique [10], and the states are encoded using a heuristic based on [12]. Finally synthesizable VHDL code is generated.

5.2 Exceptions and Compilation

When a PBS containing exception operators is compiled, the exception operators mark sub-trees in the intermediate representation with a specific exception "scope". These scope regimes may be hierarchically nested and associate non-recognition in the indicated sub-tree with a specific error sub-machine (represented as another sub-tree). Each exception scope is represented by a "color". The *natural* color annotates sub-trees not contained in any exception scope; e.g. no exception was specified. When the NFA is generated, all of the NFA nodes are colored in accordance to their relationship to the tree representation.

Recall, when the DFA ia constructed, a "death" state is generated when a group of NFA states have no successor state(s) for a transition on a particular input symbol. To implement the exception scoping, *several* different death states are generated. Each death state is marked by a unique *combination* of exception colors collected from the NFA nodes implying the transition. Due to the potential for non-determinism expressed in the NFA, a DFA death state may be marked by several exception colors. In the most typical cases of exception operator use, this is not manifest. However, complex production expressions containing exception operators are certainly possible. In this case, one of the exception colors dominates. This dominating color is determined by a simple precedence rule. In the final steps, all of the colored death states, except the one annotated by the natural color, are removed from the DFA and the transitions to these states are redirected to the DFA equivalents of the exception machines (note that no actions are lost in this process since the DFA is a mealy machine and actions are associated with the transitions, not the states). The exception behavior due to the '!!' operation is handled in a similar way.

5.3 VHDL output

The PBS compiler's output is two files: a VHDL package, and a VHDL entity/architecture body. The package file contains definitions, declarations, token and state encodings, and a state transition table. The entity/architecture body consists of the machine body with action code, and other user defined VHDL code segments "dropped" into the VHDL entity after compilation. The skeleton format of the VHDL entity body is shown in Figure 3. The main body of the machine is contained in the "machine core" section which contains the action code as well. The gray regions are locations in the skeleton where the user code segments are inserted using special directives. The directives, and the corresponding code enclosed by braces, is placed in the beginning of the PBS description, and can occur in any order. Several directives are mandatory such as "port{}" since a useful VHDL entity must have an interface. Most are optional. Figure 4. illustrates the directives required by the example.

5.4 Descriptive Partitioning Issue

We introduce this idea by noting the existence of input specifications, even quite small ones, that result in the generation of an enormous number of states in the DFA during compilation. In worst case the growth of the number of DFA states is an exponential function of the growth of the number of NFA states [8]. This arises essentially when the DFA states are used as "memory" as opposed to strictly "control state" due to the manner of specification. An example is the storage and later comparison of permutations of large numbers of past input symbols. Although extremely bad state explosion behavior certainly exists for pathological specifications, it does not detract from the utility of this specification technique. We have found that, for most types of machine behavior suited to PBS style specification, a natural partitioning of design specification is typical. This partitioning is between the description of the control aspects of the design through the productions, and the specification of data-paths, memory storage, and semantics through the HDL action clauses, and other procedural or structural entities. Storage-in-mass of input data can also be achieved via interactive behavior specified by productions, with memory storage entities.

6.0 Experiments

Several example production base specifications have been compiled and synthesized. Table 1 illustrates the results of the experiments[1] to date. The cache example is a simple cache coherency protocol from [15]. Parity is a sequential machine that recognizes only even parity bytes. It is an interesting example because it is a good benchmark for the optimization algorithms. The bounce example behaves as a bit sequential low pass filter. This example is from [1]. The count0 example is the complete description of the machine introduced earlier, and is a PBS implementation of the procedural design in [14]. The pager2 example is a large example adapted from a protocol for radio pagers, requiring complex control. The VHDL actions in this specification perform such functions as assembling data words and implementing linear feedback shift registers for error detection. The PBS and VHDL size columns are the total number of lines in the input descriptions and output descriptions respectively. Note that the VHDL output makes use of very concise jump transition tables and hand coded procedural versions of these files would probably be much larger. It should be noted that the smaller examples have fixed overhead in the form of reset conditions and interface descriptions which dominate the size of the smaller PBS descriptions. This can be seen by comparing the number of lines of productions and actions versus the total PBS description size. In the larger example (pager2), a hand coded procedural description would have to represent at least 536 distinct states, and 39 distinct actions, while this behavior can be concisely represented in 139 lines of productions. The NFA state column contains data for the total number of NFA states before conversion to a DFA. The compilation times are CPU seconds running on a Sun Sparcstation 1® workstation for compiling the PBS to procedural VHDL. This time includes state minimization, but excludes state assignment. The last two rows of the table show the results of further compilation and logic synthesis of the designs using the Synopsys® synthesis tools.

1. The PBS descriptions for these examples may be obtained via anonymous ftp from bears.ece.ucsb.edu.

7.0 Conclusion and Acknowledgment

A model and an implementation for hardware specification, simulation, and synthesis from production based specifications has been described. For many types of complex behavior, production based specifications have advantages over procedural specifications. We believe this specification technique to be useful for the ASIC designer, as well as having potential for further research in the field of system level design. Currently we are studying optimization of the data flows implied by the HDL actions in the framework of the model presented in this work.

The authors wish to acknowledge Synopsys Inc. for valuable discussion and for the use of their VHDL compiler and logic synthesis tools in obtaining the data for this paper. This research was supported through grants from Synopsys and the California MICRO program #90-195.

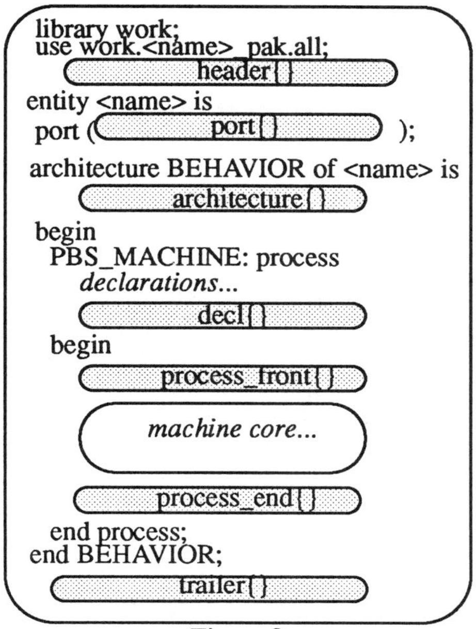

Figure 3

```
port { CLK, INPUT: in BIT;
       CNT: out INTEGER range 0 to 7;
       IS_LEGAL: out BIT}

process_front {
       wait until CLK'event and CLK = '1';
       if (input = '1')
           then PBS_TOKEN := ONE;
           else PBS_TOKEN := ZERO;
       end if;}

decl { variable COUNT: INTEGER range 0 to 7;}
```

Figure 4

Table 1: Experiments

metric	cache	parity	bounce	count0	pager2
number of productions	5	17	5	4	21
number of actions	2	2	2	3	39
lines of productions and actions	11	21	9	4	139
PBS size (lines)	41	48	36	41	187
procedural VHDL output	83	120	96	108	1269
number of NFA states	18	1020	13	30	1688
final number of states (DFA)	3	16	5	4	536
number of transitions with actions	3	4	2	6	548
compilation time	0.1	2.0	0.1	0.1	18.9
number of standard cells	9	26	14	30	***[a]
relative area	23	35	45	102	***[a]

a. hardware synthesis results not available

References

1. A. R. Karlin, H. W. Trickey, and J. D. Ullman, "Experience With A Regular Expression Compiler," Proceedings of the ICCD: VLSI in Computers, 1983.
2. R. W. Floyd, and J. D. Ullman, "The Compilation of Regular Expressions into Integrated Circuits," Journal of the Association for Computing Machinery. 29:2, 1982.
3. M. E. Lesk, "Lex - a lexical analyzer generator," Computing Science Technical Report 39, AT&T Bell Laboratories, Murray Hill, N. J. 1975.
4. S. C. Johnson, "Yacc: Yet Another Compiler Compiler," Computing Science Technical Report 32, AT&T Bell Laboratories, Murray, Hill, N. J. 1975.
5. S. Devadas, and K. Keutzer, "An Automata-Theoretic Approach to Behavioral Equivalence," proc. ICCAD 1990.
6. [BRSW87] R. K. Brayton, R. Rudell, A. Sangiovanni-Vincentelli, and A. R. Wang, "MIS: A Multiple-Level Logic Optimization System," IEEE T-CAD, Vol. 6, No. 6, November 1987.
7. A. J. de Geus, and D. J. Gregory, "The Socrates Logic Synthesis and Optimization System," in Design Systems for VLSI Circuits Logic Synthesis and Silicon Compilation. Dordrecht: Martinus Nijhoff Publishers, 1987.
8. A. V. Aho, R. Sethi, and J. D. Ullman, *Compilers Principles, Techniques, and Tools.* Reading, MA: Addison-Wesley, 1986.
9. J. E. Hopcroft, and J. D. Ullman, *Introduction to Automata Theory, Languages, and Computation.* Reading MA: Addison-Wesley, 1979.
10. Z. Kohavi, *Switching and Finite Automata Theory,* New York: McGraw-Hill, 1978.
11. G. De Micheli, R. K. Brayton, and A. Sangiovanni-Vincentelli, "Optimal State Assignment for Finite State Machines," IEEE T-CAD, Vol. 4, No. 3, July 1985.
12. S. Devadas, H-K. Ma, A. R. Newton, and A. Sangiovanni-Vincentelli, "MUSTANG: State Assignment of Finite State Machines Targeting Multilevel Logic Implementations," IEEE T-CAD, Vol. 7 No. 12. December 1988.
13. *IEEE Standard VHDL Language Reference Manual,* ANSI/IEEE 1076-198, New York: IEEE, 1988.
14. S. Carlson, *Introduction to HDL-Based Design Using VHDL,* Mountain View: Synopsys, 1990.
15. J. L. Hennessy, and D. L. Patterson, *Computer Architecture A Quantitative Approach,* pp 469-471. Morgan Kaufmann Publishers: San Mateo 1990.

SESSION 12

PANEL: WHICH ASIC TECHNOLOGY WILL DOMINATE THE 1990s

Chair: Ron Collett - Dataquest, Inc., San Jose, CA
Organizer: Steve Trimberger - Xilinx Inc., San Jose, CA

What does the future hold for ASIC technology? The question has wide implications for the direction that system design will take this decade. For instance, what kind of performance, cost and gate densities can designers expect from field programmable gate arrays (FPGAs) in the future? Will FPGAs become competitive enough to replace mask programmable solutions for high volume applications? If so, when? Will system performance requirements demand greater use of cell-based IC solutions and less use of gate arrays? What role will multichip modules play in system design? What kind of design automation tools will be needed to accommodate the next generation of ASIC technologies? These are just a few of the issues confronting electronic systems manufacturers.

Today, design engineers, managers and electronics systems manufacturers have an unprecedented breath of application specific IC (ASIC) technologies to select from when implementing systems. The range of options stretches from gate arrays, cell-based ICs and full custom ICs to field programmable gate arrays, complex programmable logic devices and simple programmable logic devices. Selecting the optimal technology for a particular design project is becoming a daunting task, especially given the performance, cost, functionality and time-to-market tradeoffs that define product development requirements.

Further clouding the decision-making process are the overlapping capabilities of different fabrication process technologies such as CMOS, BiCMOS and gallium arsenide. Packaging options including TAB, flip-chip, COB and multichip modules are also becoming significant issues within the technology selection equation. The challenge facing systems manufacturers is to assess the future directions that these technologies will follow and to select the right technologies for next generation projects.

Panelists will discuss their vision of the path that these technologies will follow over the next five years, debating issues such as cost, performance, design methodology and time-to-market advantage of the various alternatives. Panelists will also discuss the implications that these next generation semiconductor technologies will have on the electronic design automation tools.

Panel Members:

Bob Blair - LSI Logic Corp., Milpitas, CA
Dave Chavoustie - VLSI Tech., San Jose, CA
Werner Koepf - European Silicon Structures, Munich, Germany
Larry Palley - Intel Corp., Folsom, CA
Wes Patterson - Xilinx Inc., San Jose, CA
Kenji Yoshida - Toshiba Corp., Kawasaki, Japan

Generalized Moment-matching Methods for Transient Analysis of Interconnect Networks

Eli Chiprout and Michel Nakhla

Dept. of Electronics, Carleton University
Ottawa, Ontario, Canada K1S 5B6

Abstract

An approach is introduced which improves published moment matching methods used in transient waveform estimation of large linear networks including lossy, coupled transmission lines. The method, which selects from a general set of moment-matching approximations, ensures stability while increasing the accuracy of the transient response. The developed technique is useful for analysis of high-speed interconnects including lumped and distributed linear components with non-linear terminations. Examples are presented which demonstrate the stability and accuracy of the new method.

1. Introduction

Asymptotic Waveform Evaluation (AWE) has been proposed as a method to obtain the approximate transient response and system delay of linear(ized) lumped-element networks, while employing considerably less CPU time than conventional circuit simulators such as SPICE [1]. The AWE has been shown to give an improved delay estimation compared to an RC-tree interconnect model and has been generalized to apply to distributed-element networks which may include lossy coupled transmission lines [2],[3] and non-linear terminations [4]-[6]. As such, it is a useful tool for delay and crosstalk estimation in high-speed VLSI interconnects.

The AWE method approximates the network poles and the required output's residues using a moment-matching technique and the Maclaurin series expansion of the output frequency response function. However, the published methods, while being an approximation, can result in a large variation of error in different networks and at different outputs. By increasing the approximation order of stable networks, right-half plane poles are generated, and the error in the response depends critically on choice of network and output at which the method is applied. Sometimes only two or three left-half plane poles are obtained and all may be inaccurate approximations of the dominant stable network poles. Since the approximate residues of each output are derived from the approximate poles, and the transient response is derived from both poles and residues, applying the AWE can sometimes give large errors in the approximate system response [7].

It has been noted in [7] that, as with RC tree methods, the error in the approximation of the AWE is large closest to the driven end of the line. Depending on the network, we have observed that large approximation error can result at additional outputs other than those close to the driven end and, generally, different magnitudes of error result when the AWE is applied to different outputs of a network [8]. This poses a problem since it is not known *a priori* which outputs will yield small error.

The drawbacks of the moment matching approximations are these problems of stability and sensitivity [9].

In this paper, we introduce a generalized approach to moment matching based on the properties of Padé approximations. Rather than using one moment matching approximation, several are generated from the Padé Table at a minimal additional expense for a large network. Then, using two selection criteria based on the poles of smallest modulus and the total moment skew, an optimal moment matching set is selected.

For each entry in the Table, filtering the parasitic and/or right-half plane poles while leaving the remaining poles unaffected, leads to even better accuracy. Several examples are presented which show the stability and visible improvement in the approximate transient response over the published methods.

2. Outline of the AWE

Details of the AWE method are described in [1],[7]. A summary of the basic principles as applied to the impulse response is reviewed in this section.

Consider a linear network π which contains linear lumped components. Without loss of generality, the Modified Nodal Admittance (MNA) matrix equations [10] of the network π with an impulse excitation can be written as,

$$Gx(t) + Cx'(t) - E\delta(t) = 0 \qquad \textbf{(EQ.1)}$$

where,

$x(t) \in \Re^N$ is the vector of node voltage waveforms ap-

0738-100X/92 $3.00 © 1992 IEEE

Paper 13.1

pended by independent voltage source currents and linear inductor current waveforms,

$C \in \Re^{N \times N}$ and $G \in \Re^{N \times N}$ are constant matrices with entries determined by the lumped, linear components,

$E \in \Re^{N}$ is a constant vector with entries determined by the independent voltage and current sources,

$\delta(t)$ is the unit impulse function.

The Laplace transform of (1) assuming the initial conditions are all zero is,

$$(G + sC)X(s) - E = 0 \qquad \text{(EQ.2)}$$

The actual frequency and transient responses at a particular i-th output are given respectively by,

$$X_i(s) = c + \sum_{j=1}^{N} \frac{k_j}{s - p_j} \qquad \text{(EQ.3)}$$

$$x_i(t) = c\delta(t) + \sum_{j=1}^{N} k_j e^{p_j t} \qquad \text{(EQ.4)}$$

where N is the number of actual poles p_j of the circuit and the number of corresponding residues k_j of the output. In practice it is complicated and expensive to obtain the actual set of network poles and output residues. In order to extract an approximate set for a given output, (2) is expanded in a Maclaurin series in the form,

$$X(s) = \sum_{n=0}^{\infty} M_n s^n \qquad \text{(EQ.5)}$$

where,

$$M_n = (-G^{-1}C)^n G^{-1}E \qquad \text{(EQ.6)}$$

All moment vectors may be calculated recursively from M_0 by,

$$M_{n+1} = (-G^{-1}C)M_n \qquad \text{(EQ.7)}$$

The moments, m_n, $n = 0, 1, 2, \ldots$, of a specific output i, form the Maclaurin series of that output,

$$X_i(s) = \sum_{n=0}^{\infty} m_n s^n \qquad \text{(EQ.8)}$$

For a selected the output, the first $2q - 1$ moments of the series are then matched to an approximating lower order frequency domain function in the form,

$$X_i(s) \approx \hat{X}_i(s) = \hat{c} + \sum_{j=1}^{q} \frac{\hat{k}_j}{s - \hat{p}_j} \qquad \text{(EQ.9)}$$

where $\hat{p}_1$ through $\hat{p}_q$ are the approximate network poles, $\hat{k}_1$ through $\hat{k}_q$ their corresponding approximating residues, and $\hat{c}$ is an approximation to c in (3). Usually, $q \ll n$. Consequently, the impulse response is approximated by,

$$x_i(t) \approx \hat{x}_i(t) = \hat{c}\delta(t) + \sum_{j=1}^{q} \hat{k}_j e^{\hat{p}_j t} \qquad \text{(EQ.10)}$$

The reduced model characteristic polynomial is,

$$a_0 + a_1 s + a_2 s^2 + \ldots + a_{q-1} s^{q-1} + s^q \qquad \text{(EQ.11)}$$

where the coefficients of (11) are related to the moments by,

$$\begin{bmatrix} m_{-1} & m_0 & \cdots & m_{q-2} \\ m_0 & m_1 & \cdots & m_{q-1} \\ \vdots & \vdots & & \vdots \\ m_{q-2} & m_{q-1} & \cdots & m_{2q-3} \end{bmatrix} \begin{bmatrix} -a_0 \\ -a_1 \\ \vdots \\ -a_{q-1} \end{bmatrix} = \begin{bmatrix} m_{q-1} \\ m_q \\ \vdots \\ m_{2q-2} \end{bmatrix} \qquad \text{(EQ.12)}$$

and the roots of (11) are the reciprocal of the approximate poles $\hat{p}_j$. The moment m_{-1} represents the first moment of the expansion about $s = \infty$ of the system equations and may be obtained as described in [11]. The approximate residues $\hat{k}_j$, and $\hat{c}$ are related to the approximate poles and the moments by,

$$\begin{bmatrix} 1 & 1 & \cdots & 1 & 0 \\ \hat{p}_1^{-1} & \hat{p}_2^{-1} & \cdots & \hat{p}_q^{-1} & 1 \\ \hat{p}_1^{-2} & \hat{p}_2^{-2} & \cdots & \hat{p}_q^{-2} & 0 \\ \vdots & \vdots & & \vdots & \vdots \\ \hat{p}_1^{-q+1} & \hat{p}_2^{-q+1} & \cdots & \hat{p}_q^{-q+1} & 0 \end{bmatrix} \begin{bmatrix} \hat{k}_1 \\ \hat{k}_2 \\ \vdots \\ \hat{k}_q \\ \hat{c} \end{bmatrix} = \begin{bmatrix} -m_{-1} \\ -m_0 \\ -m_1 \\ \vdots \\ -m_{q-2} \end{bmatrix} \qquad \text{(EQ.13)}$$

It is characteristic of the problem, that the matrix on the left of (12) quickly becomes ill-conditioned as q becomes larger, eventually producing a polynomial that yields right-half plane poles for a stable network. These poles result in an unstable approximation because (10) grows without bound. This property of the moment-matching approach typically permits only certain low orders of approximation resulting in a small number of approximate poles. Even a well-conditioned matrix can result in a polynomial that yields right-half plane poles for a stable network or approximate poles with relatively large error compared to the real dominant poles. Methods have been attempted to minimize the problem by constraining the poles to the left-hand plane using non-linear optimization [12].

It has also been noted that large approximation errors can result when the method is applied at any arbitrary output. This causes a difficulty since it is not know *a priori* which outputs will give small error magnitudes.

3. Padé Table and selection criteria

The present method differs from the published AWE technique in that we note that (12) is a specific case of a general class of Padé approximations where the moments of a Maclaurin expansion of a function are matched to the moments of the rational approximation,

$$X_i(s) \approx \frac{P_L(s)}{Q_M(s)} \qquad \text{(EQ.14)}$$

where both numerator and denominator polynomials are of order L and M respectively [13]. If the denominator polynomial is described by $b_0 + b_1 s + b_2 s^2 + \ldots + b_M s^M$, the moment matching equations are given by:

$$\begin{bmatrix} m_{L-M+1} & m_{L-M+2} & \cdots & m_L \\ m_{L-M+2} & m_{L-M+3} & \cdots & m_{L+1} \\ \vdots & \vdots & & \vdots \\ m_L & m_{L+1} & \cdots & m_{L+M-1} \end{bmatrix} \begin{bmatrix} b_M \\ b_{M-1} \\ \vdots \\ b_1 \end{bmatrix} = - \begin{bmatrix} m_{L+1} \\ m_{L+2} \\ \vdots \\ m_{L+M} \end{bmatrix} \qquad \text{(EQ.15)}$$

in which $m_j = 0$ if $j < 0$. A unique solution for $Q_M(s)$ is generated when $b_0 = 1$. Roots of $Q_M(s)$ yield approximate poles of the output response. The coefficients of the numerator polynomial, $P_L(s)$, follow from the solution of the denominator coefficients but will not be needed in this approach, (approximate residues of each pole will be calculated directly). The resulting approximation is called an [L/M] Padé approximation. The denominator polynomials will be designated as $Q_{[L/M]}$ to indicate the order of the approximation. By inspection, if $m_{-1} = 0$, then (12) can be mapped to a $[L = q-2/M = q]$ Padé approximation with a reversed order of coefficients for the denominator polynomial. (The roots of a polynomial B with coefficients reversed from polynomial A, and the roots of polynomial A are reciprocals of each other).

With different possibilities for L and M, it is customary to build a subset of an infinite table where each entry of the table consists of one Padé approximation corresponding to a specific choice of L and M [13]. This array is referred to as a Padé Table. For a large network, the greatest CPU expense is in obtaining the moments, rather than forming a subset of the Padé Table for a small range of L and M [15]. And even here, the greatest cost is in the original LU decomposition, not in the generation of several moments. The approach proposed in this paper requires more moments than the published AWE but is outweighed by the accuracy achieved.

Our goal is to choose one entry, one selection of L and M, of the Padé Table that yields the best set of approximate poles to the dominant output poles. Once these poles are obtained, the residues may be calculated using (13). Therefore, our Table entries need only contain the $Q_{[L/M]}$. Baker and Graves-Morris [15] suggest searching as broad band as possible about the $[L=M/M]$ diagonal entries in the Padé Table in order to search for the poles of the function. However, no clear indication of a search algorithm is given. Chan has remarked [9] on the lack of means of determining the best moment matching approximation. We develop in this paper two criteria for selection of the best entry in the Padé Table.

Definition:
Given the network poles $\{p_j,\ j = 1, 2, \ldots, N\}$, *the pole(s) of smallest modulus are defined as,*
$\{p_s \in \{p_j\} \mid |p_s| = min(|p_1|, |p_2|, \ldots, |p_N|)\}$.

The number of these poles, n_s, is usually 1 or 2; there being either a smallest pole or a pair of smallest complex conjugate ones. These poles are simply referred to as the smallest pole(s) or $\{p_s\}$. The $\{p_s\}$ and their residues, $\{k_s\}$, may be calculated accurately and independently of any of the other poles. They may therefore be used to gauge the accuracy of an approximation.

To determine the smallest poles, we use a property of the Padé Table. It can be shown [14],[15] that in the Padé Table for a rational function,

$$\lim_{L \to \infty} [Q_{[L/M]}(s)] \to \prod_{j=1}^{M} \left(1 - \frac{s}{p_j}\right) \qquad \text{(EQ.16)}$$

where p_j are the actual poles of the network arranged in order of increasing modulus: $|p_1| \le |p_2| \le \ldots \le |p_M| < \ldots$ and $|p_M| \ne |p_{M+1}|$.

In other words, the entries of the Padé Table for a given M will converge to a polynomial with the smallest M poles as its roots, provided those poles are all smaller than any other network pole. The smallest pole(s) may be obtained with a high degree of accuracy by taking L large and solving for the roots of the resulting $Q_{[L/M]}$. For instance, if there is one smallest pole p_s, $(n_s = 1)$, (16) can be rewritten as,

$$\lim_{L \to \infty} [Q_{[L/1]}(s)] = 1 - s\left(\lim_{L \to \infty} \frac{m_{L+1}}{m_L}\right) \to 1 - \frac{s}{p_s} \qquad \text{(EQ.17)}$$

The residues of the smallest poles may be also be obtained by using the higher order moments. Equation (16) generally demand a more moments than required in the published AWE. The additional moments needed will depend on convergence to the smallest poles.

Once the $\{p_s\}$ are calculated, we determine a large band of entries in the Padé Table and calculate the denominator polynomial for each. Two selection criteria allow the successive filtering out of inaccurate Padé entries which will allow extraction of an accurate set of poles and subsequent residues.

Criterion 1:

A table entry is eliminated if its roots do not include the smallest poles, $\{p_s\}$.

A Maclaurin expansion is only valid within its radius of convergence R, which among rational system response functions is equal to the modulus of the smallest pole(s). Padé approximations perform a type of analytic continuation of the function described by the Maclaurin series to give information about the function in the region beyond the radius of convergence. However, as with the Maclaurin series themselves, a Padé approximation decreases in accuracy as the distance is increased from the origin [15]. If an approximation does not contain the smallest pole(s) as its roots, it is safe to assume that it is not an accurate approximation. Several Table entries may be eliminated in this way without further processing. This is a quick way to zero in on the accurate approximations.

This criterion is easily checked by either extracting the roots of each polynomial in the Table, or with less CPU, by simply substituting the $\{p_s\}$ into the denominator polynomials and verifying that the result is indeed numerically zero. Criterion 1 however, is not a sufficient condition for accuracy.

3.1 Moment Skew

Expanding (9) in a Maclaurin series gives,

$$\hat{X}_i(s) = \left(\hat{c} + \sum_{j=1}^{q} \frac{-\hat{k}_j}{\hat{p}_j}\right) + \sum_{n=1}^{\infty}\left(\sum_{j=1}^{q} \frac{-\hat{k}_j}{\hat{p}_j^{n+1}}\right)s^n \quad \text{(EQ.18)}$$

in which the approximate moments of the response are,

$$\hat{m}_0 = c + \sum_{i=1}^{q} \frac{-\hat{k}_i}{\hat{p}_i} \qquad \hat{m}_j = \sum_{i=1}^{q} \frac{-\hat{k}_i}{\hat{p}_i^{j+1}} \quad \text{(EQ.19)}$$

Approximate poles and residues satisfying (13) implies that,

$$m_n = \hat{m}_n \qquad n = 0, 1, \ldots, q-2. \quad \text{(EQ.20)}$$

An error criterion for a set of poles and residues is possible by comparing the next moments unused in (13) with the corresponding moments of the approximate response. In other words, a comparision is made between m_n and $\hat{m}_n$ for $n > q-2$.

For a given moment m_n and respective approximate moment $\hat{m}_n$, the n-th moment skew, ε_n, is defined as the scaled difference between the two values,

$$\varepsilon_n = \frac{m_n - \hat{m}_n}{m_n} \quad \text{(EQ.21)}$$

There are an infinite number of these moment skews but the sequence of them converges to zero for any set of poles that passes criterion 1 because the smallest pole(s) dominate both the system moments and the approximate moments. The total moment skew, ε, for a particular set of poles and residues is given by,

$$\varepsilon = \left(\sum_{i=q-1}^{\infty} |\varepsilon_n|^a\right)^{1/a} \qquad a \geq 1 \quad \text{(EQ.22)}$$

A large error in approximation implies a large moment skew. We now introduce the second criterion.

Criterion 2:

Of the entries remaining from criterion 1, the set of filtered (non-parasitic, non-positive) poles and subsequent residues of each entry that minimizes the total moment skew, ε, *is the one selected.*

A set of approximate poles is extracted from the each $Q_{[L/M]}$ of the Padé Table entries not rejected by the first criterion. Each set of approximate poles however, is not the final set that will be used for error comparison, because it may contain largely inaccurate poles. Poles known to be inaccurate may be removed from a set without reference to the other possibly accurate poles in the set. The nature of Padé approximations is that they may produce both inaccurate and accurate poles in the same approximation. In effect, the known inaccurate poles are "filtered" from each set of approximate poles. This procedure may be carried out because criterion 2 based on the unused moments will determine if a set of poles is collectively accurate.

Inaccurate poles may be in the form of "positive" poles or parasitic ones. Positive poles refer to those poles with positive real parts which are known to be inaccurate when the approximation is of a stable network. These poles often result when actual left-half plane network poles are large relative to their residues and to the smallest pole(s) which causes a large error in their approximation. They often appear large in the approximation but on the positive side of imaginary axis. Rather than rejecting an entire approximation, positive poles are removed from the set of approximate poles. This implies that all generated approximations will be guaranteed stability.

Parasitic poles are defined to be spurious poles generated by a Padé approximation which are relatively close to the origin and are associated with relatively small residues [15]. These poles are caused by numerical error and do not reflect actual network poles. They can usually be detected. Any poles clearly smaller than the known $\{p_s\}$ are deemed to be spurious. Other parasitic poles must be detected by calculating the approximate residues using (13) and the remaining poles. An error margin must be set for determination of small residues which will disqualify remaining parasitic poles. This successive filtering of inaccurate poles allows the examination of many approximations not considered in the published AWE.

Criterion 2 implies that approximate q poles and q residues of the selected Table entry not only match a q-th order approximation, they also minimize the error of higher order approximations. Since the moment matching is done in the frequency domain, any added accuracy this method provides compared to the AWE can mean a large improvement in the transient response approximation. This is clearly visible in the examples outlined.

4. Implementation and CPU considerations

Difficulties in implementation arise due to numerical error and the setting of margins. The moments will allow for quick converges to the smallest pole(s) if there are no significant poles nearby, otherwise it may take many moments. A tolerance limit is required typically to recognize convergence or a maximum number of moments. Each additional moment will cost a forward backward substitution from the LU decomposed matrix.

The limit to the accuracy of Padé approximations hinges critically on the number of accurate digits in the moments. It is necessary for any implementation to use as much accuracy as possible on the particular hardware running the approximations.

The Padé Table calculation of denominator polynomials may be done on an entry by entry basis using (15). However, [15] lists several published algorithms to calculate the polynomials recursively for each entry without needing matrix solutions. For a large number of Table entries, this approach can prove to be less costly. A polynomial root finding algorithm must be applied only to the few entries left over from criterion 1. Typically, roots of polynomial no greater than degree 10-12 are needed.

5. Examples

Example 1:

Consider the six-pole circuit shown in Figure 1. The output response is sought at node 7. The Padé Table row M=1 converges to the smallest pole, $p_s = -1.283258442e8$, as L⟶8. Having obtained the smallest pole, the Padé Table is filled for L=1⟶6 and M=2⟶6 with the various denominator polynomials. The p_s is then substituted into each polynomial and the error is tabulated. Following criterion 1, we eliminate all but the smallest results. Several of the approximations generated by the AWE methods are eliminated.

Subsequently criterion 2 is applied, (using $a = 1$ in (24)). The [0/6] approximation yields the lowest moment skew and is the one chosen.

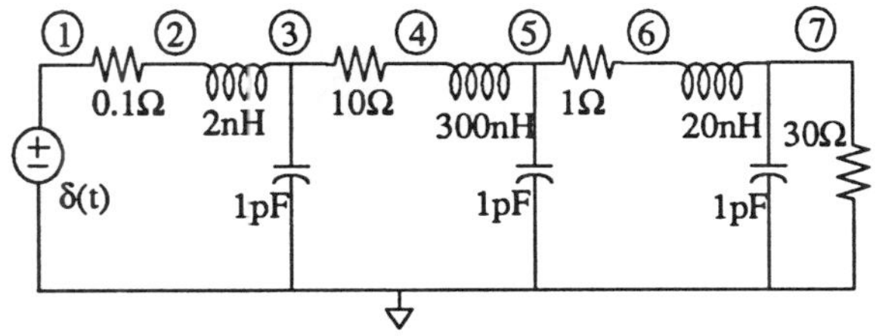

Figure 1. Six pole circuit

The actual network poles and output residues were obtained using eigenvalue methods. The 5th order (and higher) AWE approximation is unstable due to the right-half plane poles and the best stable AWE approximation which we were able to obtain is 4th order. The transient impulse response of the network is compared to the transient approximation of the 4th order AWE and that of the proposed method in Figure 2. The true response was obtained using (4).

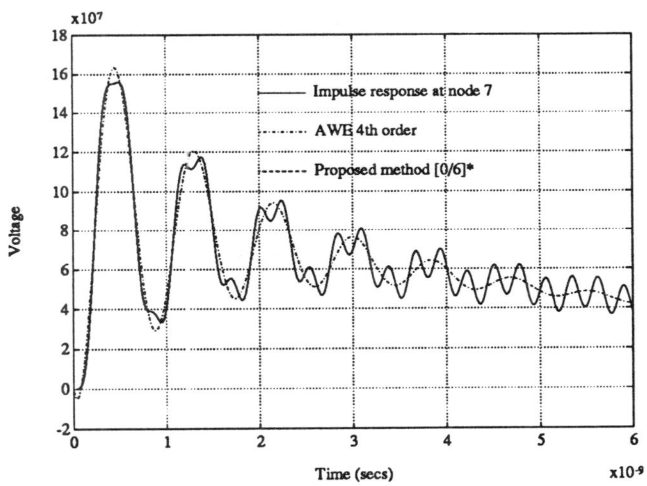

Figure 2. Impulse response and approximations*

Example 2:

Consider the filter shown in Figure 3. The output response is sought at V_{out}. The AWE method yields order 6 as the highest stable approximation. The proposed approach finds a minimum moment skew at [L=2/M=10] which yields 8 filtered poles. A comparison of the approximate transient impulse response of the two methods is given in Figure 4. The actual transient impulse response was obtained using eigenvalue analysis.

*The approximate responses of the proposed method and the true responses in figures 2 and 4 are almost indistinguishable.

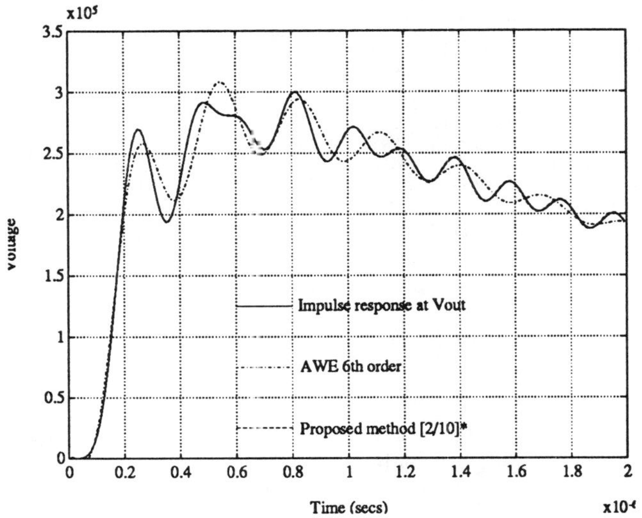

Figure 4. Impulse response and approximations*

6. Conclusion

An enhanced moment matching technique has been introduced for large network transient response estimation. The method, consisting of generating several sets of approximate poles from the Padé Table, calculating the poles of smallest modulus and filtering the sets of poles for positive and parasitic poles, and finally, choosing the entry with the least total moment skew, generalizes on the published moment matching techniques.

The examples presented and many other examples worked out by us show the advantages of the proposed method over the published moment matching methods both in producing only stable approximations and in improving the transient waveform approximation. The method presented can be equally extended to cover distributed networks with high-speed interconnects as in [2],[3] and to those containing nonlinear elements as in [4]-[6].

7. References

[1] L. T. Pillage and R. A. Rohrer, "Asymptotic waveform evaluation for timing analysis" *IEEE Trans. Computer-aided Design*, vol. 9, pp.352-366, April 1990.

[2] T. K. Tang and M. Nakhla, "Analysis of high-speed VLSI interconnects using the Asymptotic Waveform Evaluation technique" *IEEE Transactions on Microwave Theory and Techniques*, vol. MTT-39, Dec. 1991.

[3] T. K. Tang, M. Nakhla and R. Griffith, "Analysis of lossy multiconductor transmission lines using the Asymptotic Waveform Evaluation technique", *IEEE Transactions on Computer-aided Design of Integrated Circuits and Systems*, vol. 10, Nov. 1991.

[4] T. K. Tang, M. Nakhla and R. Griffith, "Analysis of VLSI interconnects with nonlinear Terminations", *IEEE International Symposium on Circuits and Systems (ISCAS)*, pp. 2339-2342, June 1991.

[5] D. Xie and M. Nakhla, "Delay and crosstalk simulation of high-speed VLSI interconnects with nonlinear terminations", *IEEE International Conference on Computer-Aided Design (ICCAD)*, pp. 66-69, Nov. 1991.

[6] E. Chiprout and M. Nakhla, "Fast nonlinear waveform estimation of large distributed networks" *Proceedings, IEEE International Microwave Symposium (MTT-S)*, June 1992.

[7] L. T. Pillage, "Asymptotic Waveform Evaluation for timing analysis", *Carnegie-Mellon University*, Research Report No. CMUCAD-89-34.

[8] E. Chiprout and M. Nakhla, "Optimal pole selection in asymptotic waveform evaluation" Proceedings, *IEEE International Symposium on Circuits and Systems (ISCAS)*, May 1992.

[9] P. K. Chan, "Comments on 'Asymptotic Waveform Evaluation for timing analysis'", *IEEE Transactions on Computer-Aided Design*, pp. 1078-1079, Aug. 1991.

[10] C.W. Ho, A.E. Ruehli and P.A. Brennan, "The modified nodal approach to network analysis," IEEE Trans. Circuits and Systems, vol. CAS-22, pp. 504-509, June 1975.

[11] X. Huang, V. Rahavan and R. A. Rohrer, "AWEsim: a program for the efficient analysis of linear(ized) circuits" *IEEE International Conference on Computer-Aided Design (ICCAD)*, Nov. 1990.

[12] C. L. Ratzlaff, N. Gopal, L. T. Pillage, "RICE: rapid interconnect circuit evaluator" *ACM/IEEE Design Automation Conference*, pp. 555-560, June 1991.

[13] J. L. Morris, *Computational Methods in Elementary Numerical Analysis*. New York: Wiley, 1983.

[14] G. A. Baker Jr., *Essential of Padé Approximants*. New York: Academic, 1975.

[15] G. A. Baker Jr. and P. Graves-Morris, *Padé Approximants - Encyclopedia of Mathematics and its Applications: Vols. 13 & 14, G. C. Rota Ed.*, Reading, Mass.: Addison-Wesley, 1981.

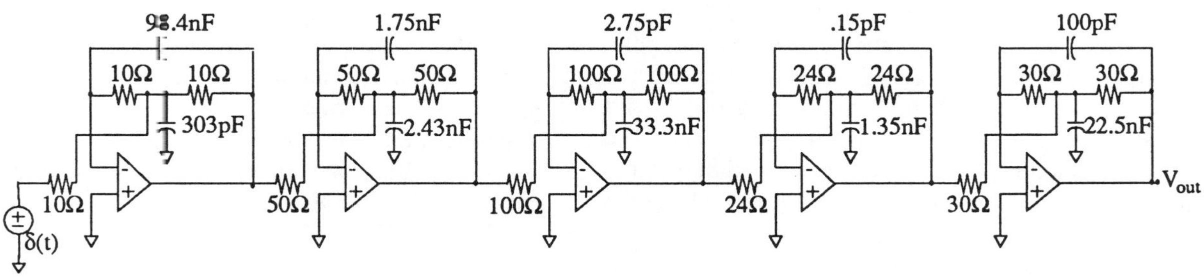

Figure 3. Sample filter circuit

On the Stability of Moment-Matching Approximations in Asymptotic Waveform Evaluation*

D. F. Anastasakis, N. Gopal, S. Y. Kim and L. T. Pillage
Computer Engineering Research Center
The University of Texas at Austin
Austin, Texas 78758

Abstract

Asymptotic Waveform Evaluation (AWE) has been demonstrated as an efficient approach for CAD circuit simulation/analysis. However, since it is based upon moment-matching, it is prone to yielding unstable approximations for stable circuits. This paper describes an approach for overcoming the inherent instability associated with AWE and moment-matching methods as they apply to circuit analysis problems. The efficiency and accuracy of this algorithm are demonstrated on several large, RLC interconnect-circuit analysis problems.

1 Introduction

Due to its efficiency, Asymptotic Waveform Evaluation (AWE) [15] has been applied to a myriad of CAD problems: RLC interconnect evaluation [10, 16], transmission line simulation [18], pole/zero analysis [9, 10], RC extraction [11], timing analysis [15] and piecewise linear circuit simulation [5]. By matching the first $2q$ moments of a linear(ized) circuit to those of a reduced order response model, AWE yields accurate frequency- and time-domain approximations in terms of the circuit's q most-dominant poles. Moments are easily obtained from a dc-equivalent circuit model as evidenced by the transient analysis of large, linear RLC circuits at speeds 10,000 times faster than a circuit-level simulation [10, 16]. Calculating the moments by other means can result in equally impressive efficiency [4].

However, AWE, and moment-matching approximations in general, suffer from two main problems: (1) it is difficult to select the required order of approximation *a priori* since it is not known how many poles are dominant, and (2) it is not always possible to select a particular order of approximation due to instability problems.

This paper describes an approach for overcoming the instability associated with AWE by considering a *partial* Padé approximation. It is shown that since the system poles are invariant to changes in the input forcing-function, applying input excitation forms of decreasing high-frequency energy isolates the dominant low-frequency poles more accurately while ignoring the high-frequency noise effects that cause instability. Further, since the poles are the same at all nodes of the circuit, a single set of dominant-pole approximations computed at an appropriately selected node can be used as a *common denominator* for the response approximation at any node [18]. The term "partial Padé" refers to the computation of the corresponding residues at any node by using the common set of dominant-pole approximations and matching only the first q moments of the impulse- or step-response *at that node*. Such an approach also facilitates the selection of an appropriate order of AWE approximation based on the maximum frequency of interest, as described in [7] and in a later section.

These methods for ensuring the stability of the AWE models, and for obtaining appropriate orders of moment-matching approximation have been implemented into existing versions of AWE. We demonstrate the utility of these techniques in the analysis of large, lumped RLC-interconnect models of off-chip and PCB interconnections.

*This work was supported in part by the National Science Foundation under MIP #9157263, the Semiconductor Research Corporation under contract #91-DP-142, and International Business Machines Corporation.

2 Background

Asymptotic Waveform Evaluation, a qth-order extension of the Elmore approximation [6] for RC trees, is based on a moment-matching technique recognized to be a Padé approximation [1, 9]. Using AWE, it is possible to accurately approximate the time- or frequency-domain response of large, linear RLC circuits in terms of a reduced, q^{th}-order, or q-pole model of the form:

$$\begin{aligned} \tilde{h}_q(t) &= \tilde{k}_1 e^{\tilde{p}_1 t} + \tilde{k}_2 e^{\tilde{p}_2 t} + \cdots + \tilde{k}_q e^{\tilde{p}_q t} \\ \tilde{H}_q(s) &= \frac{\tilde{k}_1}{s - \tilde{p}_1} + \frac{\tilde{k}_2}{s - \tilde{p}_2} + \cdots + \frac{\tilde{k}_q}{s - \tilde{p}_q} \end{aligned} \qquad (1)$$

where $\tilde{p}_1, \tilde{p}_2, \ldots, \tilde{p}_q$ are the dominant-pole approximations, and $\tilde{k}_1, \tilde{k}_2, \ldots, \tilde{k}_q$ are the corresponding residues. Such an approximation is possible since a response for a large, linear RLC circuit with thousands of poles, tends to be dominated by only a few of the poles [14]. This permits excellent reduced-order approximations in terms of only these *dominant* poles.

As derived in [15], the q dominant-poles in Eq.(1) are approximated by matching the moments m_0 through m_{2q-1} of the exact impulse response with those of the qth-order model. This yields the following set of linear equations [15]:

$$\begin{bmatrix} m_0 & m_1 \ldots m_{q-1} \\ m_1 & m_2 \ldots \; m_q \\ \vdots & \vdots \ddots \; \vdots \\ m_{q-1} & m_q \ldots m_{2q-2} \end{bmatrix} \begin{bmatrix} a_q \\ a_{q-1} \\ \vdots \\ a_1 \end{bmatrix} = - \begin{bmatrix} m_q \\ m_{q+1} \\ \vdots \\ m_{2q-1} \end{bmatrix}. \qquad (2)$$

The roots of the characteristic polynomial formed by the solution of Eq.(2):

$$1 + a_1 s + a_2 s^2 + \cdots + a_{q-1} s^{q-1} + a_q s^q = 0 \qquad (3)$$

yield the dominant pole approximations.

With a set of dominant-pole approximations $(\tilde{p}_1, \tilde{p}_2, \ldots, \tilde{p}_q)$, the corresponding q residues are obtained by matching the first q moments, resulting in the linear system:

$$-\left(\frac{\tilde{k}_1}{\tilde{p}_1} + \frac{\tilde{k}_2}{\tilde{p}_2} + \ldots + \frac{\tilde{k}_q}{\tilde{p}_q}\right) = m_0$$

$$-\left(\frac{\tilde{k}_1}{\tilde{p}_1^2} + \frac{\tilde{k}_2}{\tilde{p}_2^2} + \ldots + \frac{\tilde{k}_q}{\tilde{p}_q^2}\right) = m_1$$

$$\vdots$$

$$-\left(\frac{\tilde{k}_1}{\tilde{p}_1^q} + \frac{\tilde{k}_2}{\tilde{p}_2^q} + \ldots + \frac{\tilde{k}_q}{\tilde{p}_q^q}\right) = m_{q-1}. \qquad (4)$$

The inherent efficiency of AWE stems from the ease with which the moments of the exact response are calculated. These moments are obtained by the recursive analysis of an equivalent dc circuit, for which capacitors are replaced by current sources, and inductors by voltage sources [15]. In the case of RC- and RLC-interconnect circuit models, the tree-like topologies can be exploited to obtain the moments with even greater efficiency [16].

3 Frequency-Domain Considerations

For digital circuit analysis, most of the signal energy is low-frequency in nature due to the large dc component of the signal. Consequently, the dominant poles tend to be the lower-frequency poles, i.e., closest to the origin. By dominant poles, we refer to those system poles that are significantly stimulated by the input-signal frequency components, and hence, influence the response relatively more than the other poles. This is reflected by the relatively large magnitudes of the corresponding residues for these dominant poles.

To demonstrate this pole dominance, consider a typical, linear(ized) n^{th}-order circuit with poles $|p_1| \leq |p_2| \leq \cdots \leq |p_n|$. Let the impulse response be denoted by:

$$H(s) = m_0 + m_1 s + m_2 s^2 + m_3 s^3 + \cdots, \qquad (5)$$

where $[m_0, m_1, m_2, \ldots]$ represent the moment series, corresponding to the coefficients of $s^0, s^1, s^2, \ldots$ respectively. With the impulse response in Eq.(5), the response to any excitation can be found in the Laplace domain by multiplying the transform of the input signal with the impulse response. Thus, for a step input which has a Laplace transform of s^{-1}, the series expansion of the response is given by:

$$Y_{step}(s) = \frac{m_0}{s} + m_1 + m_2 s + m_3 s^2 + \cdots \qquad (6)$$

The *transient* moment set is now given by $[m_1, m_2, m_3, \ldots]$, with the value m_0 representing the initial conditions of the circuit. Similarly, for a ramp input, with a Laplace transform of s^{-2}, the series expansion of the response is:

$$Y_{ramp}(s) = \frac{m_0}{s^2} + \frac{m_1}{s} + m_2 + m_3 s + m_4 s^2 + \cdots \qquad (7)$$

and the transient moment set is given by $[m_2, m_3, m_4, \ldots]$, with initial conditions specified by m_1 and m_0 . Thus, it is observed that shifting the transient moment set is equivalent to changing the form of the input excitation. In other words, since the moment sets for each input excitation are identical to an appropriately shifted set of impulse-response moments, it is not necessary to explicitly compute the moments for each input excitation.

From the Cayley-Hamilton theorem [8], we know that the system can be fully characterized by the first $2n$ moments (coefficients in the series expansion of its transfer function about s=0). In other words, the exact transfer function can be reconstructed from the first $2n$ moments, $[m_0, m_1, \ldots, m_{2n-1}]$. Using a higher-order set of $2n$ moments, say $[m_j, m_{j+1}, \ldots, m_{j+2n-1}]$, is equivalent to solving for the exact response due to a lower-frequency input, namely the input-forcing function with transform s^{-j}. For this input, the exact response is obtained from the following $2n$ moment equations:

$$-\left(\frac{k_1}{p_1^{i+j+1}} + \frac{k_2}{p_2^{i+j+1}} + \cdots + \frac{k_n}{p_n^{i+j+1}}\right) = m_{i+j}, \quad i = 0, 1, \ldots, 2n-1, \qquad (8)$$

which may be rewritten as

$$-\left(\frac{k_1'}{p_1^{i+1}} + \frac{k_2'}{p_2^{i+1}} + \cdots + \frac{k_n'}{p_n^{i+1}}\right) = m_{i+j}, \quad i = 0, 1, \ldots, 2n-1, \qquad (9)$$

where

$$k_r' = \frac{k_r}{p_r^j}, r = 1, 2, \ldots, n. \qquad (10)$$

Clearly, the response obtained for an input forcing-function with decreased high-frequency energy has the same poles as the transfer function, but with different residues. Since, by assumption, $|p_1| \leq |p_2| \leq \cdots \leq |p_n|$, it is apparent from Eq.(10) that the residues of the high-frequency poles are scaled relatively more than those of the low-frequency poles. Thus, decreasing the high-frequency energy of the input signal serves to increase the relative dominance of the low-frequency poles, which are indeed the poles of interest.

For illustration, consider the 6-pole circuit in Fig.1, with the poles as listed. When successively applying

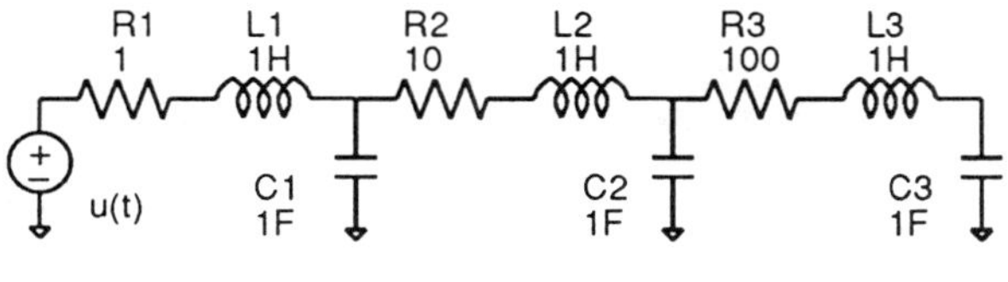

p_1	-8.9246e-3
p_2	-1.0295e-1
$p_{3,4}$	-5.5561e-1±j8.9648e-1
p_5	-9.7969
p_6	-9.998e1

Figure 1: Linear, 6-pole RLC circuit.

signals of decreasing high-frequency energy[1], the shift in the relative pole dominance at the capacitor C_3, can be easily observed, from the results in Table 1.

Input	$\delta(t)$	$u(t)$	$t.u(t)$	$t^2.u(t)/2!$
$\|k_1\|$	1.00	1.00	1.00	1.00
$\|k_2\|$	1.10	9.57e-2	8.28e-3	7.17e-4
$\|k_{3,4}\|$	5.81e-2	4.91e-4	4.15e-6	3.51e-8
$\|k_5\|$	1.38e-4	1.25e-7	1.14e-10	1.04e-13
$\|k_6\|$	1.14e-8	1.02e-12	9.06e-17	8.08e-21

Table 1: Illustration of the decrease in the relative dominance of high-frequency poles with decrease in input high-frequency energy for the circuit in Fig.1 (residues normalized by $|k_1|$).

4 Moment-Matching Instability

The choice of an appropriate order of approximation is typically based on the maximum frequency of interest. In the case of non-periodic digital signals, the maximum frequency is taken to be a measure of the highest frequency component of the signal spectrum with significant magnitude [7]. As shown in [7], approximating all the dominant circuit poles below this measure of maximum signal frequency results in a model with a frequency response that matches that of the exact circuit over the frequency range of interest.

The dominant-pole approximations obtained in AWE have been shown to converge to the exact circuit poles as the order of approximation is increased [9]. Using this result, a given AWE approximation is sufficient for a specified maximum signal frequency if all dominant-pole approximations within this frequency range, have converged.

[1]An impulse has more high-frequency energy than a step; a step has more high-frequency energy than a ramp; etc.

In the case of very high-frequency input signals (such as impulses and steps), however, satisfying this maximum-frequency criterion may lead to excessively high orders of moment-matching approximation, thus requiring a large number of moments. In the case of MOS, on-chip interconnects, the RC-circuit model poles are usually in the megahertz range, which causes the moment values to diverge very quickly (Refer Eq.(4)). This, in turn, may cause the matrix in Eq.(2)– recognized to be the Hankel matrix representing the series expansion of the response function– to become ill-conditioned or singular [3], resulting in an unstable approximation. This form of instability, alluded to as numerical instability in earlier sections, can be alleviated to some degree by frequency scaling, as described in [15]. The highest possible order of moment-matching approximation however, is limited by finite-precision computers.

A more serious problem is the inherent instability problem associated with the Padé approximation and moment-matching methods in general [2, 13, 17]. Some pole-zero configurations of response functions cannot be modeled at certain moment-matching approximation orders [9]. This is mainly due to the effects of high-frequency poles that are either outside the range of the significant frequencies of the propagating signal and are hence, not strongly stimulated, or due to the close presence of masking high-frequency zeros. For instance, Table 2 shows the location of the zeros of the impulse response across the capacitor C_1 of the circuit in Fig.1. Relating the location of the zeros with those of the poles in Fig.1, it is observed that z_4 masks the effect of p_6 almost exactly. Hence, it is expected that an AWE approximation of the response across C_1 would not reveal p_6, and this is indeed the case.

z_1	-9.01e-03
z_2	-1.12e-01
z_3	-9.90e+00
z_4	-9.998e+01

Table 2: Zeros of the impulse response across capacitor C_1 of the circuit in Fig.1.

Similarly, due to the low-pass nature of the circuit in Fig.1, there is insufficient high-frequency energy at response nodes far from the driver, e.g. across C_3, to stimulate the high-frequency poles. As such, an AWE approximation of the response across C_3 would once again, fail to reveal the high-frequency pole locations.

However, these high-frequency poles, while having an insignificant effect on the response, do influence the low-order moments very minimally. This minimal effect is interpreted as noise in the low-order moment values, which can, at times, result in unstable AWE approximations. This form of inherent instability is present at even low orders of approximation and occurs fairly frequently in the analysis of passive, linear RLC circuits.

Both forms of instability, numerical and inherent,

are demonstrated by the ladder model of an MCM interconnect segment, shown in Fig.2. Using AWE to

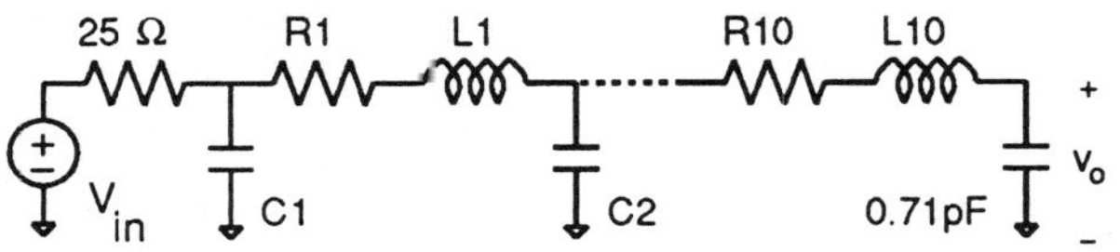

Figure 2: Uniform lumped model of interconnection segment (R_i=0.02Ω, C_i=0.21pF, L_i=1nH)

model the step response at every capacitor, it was observed that instability occurred fairly frequently (≈ 21% of the time) for second- through eighth-order approximations, as illustrated in Table 3. Also note

	Nodes									
	2	4	6	8	10	12	14	16	18	20
O(2)	-	s	s	s	s	s	s	s	s	s
O(3)	s	s	s	s	s	s	s	s	s	s
O(4)	-	s	s	s	s	s	s	-	s	s
O(5)	s	s	s	s	s	s	s	s	s	s
O(6)	-	s	s	-	s	s	s	-	-	s
O(7)	s	s	-	s	-	s	s	s	s	s
O(8)	-	s	s	s	-	-	s	-	-	-

Table 3: Illustration of instability in modeling the step response at several nodes of the circuit in Fig.2 for various orders of AWE approximation. (A "-" indicates that the AWE model was unstable, while "s" indicates a stable model.)

that instability is much more evident at high-order AWE models (≥ 6) than at lower-orders. These high-order unstable models can be attributed to the finite numerical precision of the moments and the resultant truncation/rounding errors that are present. Instability at lower-orders are due to the inherent instability problems caused by high-frequency pole influences.

Classical approaches for moment-matching suggest overcoming this inherent instability problem by increasing the order of approximation [2, 19]. While this approach may lead to stable, higher-order approximations, as shown in Table 3, it does not guarantee a solution in many cases. And we should point out that low-order approximations are insufficient to accurately model the underdamped responses at these nodes.

5 Stabilizing Moment-Matching Approximations

As explained in the previous section, instability in an AWE approximation can be attributed to the misinterpretation of the minimal effects of masked high-frequency poles on the lower-order moments. This can be avoided by exciting the circuit by signals with decreasing high-frequency energy, thus decreasing the relative dominance of the high-frequency poles and their effect on the moment values until they do not interfere in the approximation process. The omission of these high-frequency poles from the reduced-order approximation does not affect the accuracy since they influence relatively insignificant signal energy.

In addition, the decrease in input high-frequency energy results in increasingly accurate low-frequency pole approximations, due to their increase both in dominance and effect on the moments. This convergence of the dominant pole approximations to the actual low-frequency poles of the circuit, is similar to that obtained by a scan of the horizontal entries in the Padé table of the response approximation, as reported in [9]. Knowing the exact low-frequency poles of the circuit (within the available precision), facilitates the use of a common denominator, comprising these pole values, to approximate the response at all nodes of the circuit, to any form of input excitation. Such an approach would not only avoid the computational expense of obtaining dominant-pole approximations at every response node, but also the problem of worrying about possible instability at every node. The effectiveness of such a technique depends greatly on the selection of the node at which the dominant poles are computed.

In the case of lossy interconnect circuit models, which are predominantly low-pass in nature, there is attenuation of high-frequency energy along the line. Hence, the maximum bandwidth and high-frequency energy of the signal is found at the driving-point. This results in a stronger stimulation of the poles, and over a larger frequency range, at the driving-point than at any other response node. As a result, the highest order of stable dominant-pole approximation is found at the driving-point than at any other response node. Since this order of approximation is assumed at all response nodes, more accurate results are obtained than if the lower-order dominant-pole approximations computed at the response node were used. This establishes the driving-point as a good choice for the node at which the low-frequency poles are computed.

In terms of implementation, the only moments that need to be calculated are the step-response moments at the driving-point and at each of the desired response nodes. This can be achieved very efficiently using the algorithms outlined in [15, 16]. The dominant poles are then computed using Eq.(2) and the moments of the response at the driving-point. Convergence to the actual low-frequency poles is then observed by re-solving Eq.(2) using the moments for an input signal with reduced high-frequency energy. As shown in Eq.(5)-(7), this does not require any additional computation of moments; rather, the effect of altering the input signal specification is achieved by simply shifting the existing moment set.

This shifting procedure is repeated until the low-frequency pole approximations at the driving-point are observed to have converged. The only limitations on the number of shifts is that at some point, numerical noise in the high-order moments is encountered, thus precluding further shifting and eventually suppressing even some of the dominant low-frequency poles of interest (For example, one cannot decrease input signal frequency to such an extent that only two poles are stimulated, and then expect to obtain a third-order approximation!). If it is found imperative to obtain pole approximations of higher frequency than possible with the available precision, other transformations such as frequency shifting may be employed [9, 10]. However, for the interconnect problems we have encountered, the effects of such high-frequency poles are not in evidence.

The highest order of accurate, stable, dominant-pole approximation achieved at the driving-point is then assumed at each of the response nodes of the circuit, since these same pole values are used as a common denominator. This is validated by the argument that in the case of low-pass interconnect circuit models, any pole whose influence is not felt at the driving-point will not significantly influence the response at any other node and can be ignored for all practical purposes.

The decoupling of the solutions for the poles and residues, i.e., Eq.(2) and Eq.(4), allows the alteration in input signal specification to affect only the computation of the poles, which are invariant to the change, as stated earlier. The succeeding computation of the residues using Eq.(4) employs the moments of the response to the *original* excitation, thus ensuring that the correct response approximation is obtained. For instance, as in Eq.(5)-(7), if $[m_1, m_2, m_3, \ldots]$ are the step-response moments at a given response node, the moments of the response to a quadratic input excitation form are obtained by simply shifting the step-response moments twice to the left, i.e., $[m_3, m_4, m_5, \ldots]$. Thus, while it may be necessary to use the moments of the response to a quadratic input excitation in Eq.(2) to obtain stable dominant-pole approximations, a step-response approximation can be computed by substituting the values of these dominant-pole approximations and the *step-response* moments into Eq.(4) to solve for the corresponding residues.

6 Results

The proposed algorithms were tested on the RLC circuit in Fig.2, driven by a 25Ω source and terminated by a 1 pF capacitor. The oscillatory response waveforms of such underdamped circuits require high orders of AWE approximation to obtain results of sufficient accuracy. From Table 3, it is noted that such high orders of approximation are not realizable at many response nodes due to the instability problems discussed earlier.

Employing the proposed "moment-shifting" algorithm, it was possible to stabilize all the AWE approximations in Table 3. However, the stabilized dominant-pole approximations at the response nodes far from the driver yielded step responses that were slightly inaccurate when compared to the results from a conventional circuit simulation of the same circuit, as shown in Fig.3(a). This inaccuracy was overcome by using the stabilized dominant-pole approximations computed at the driving-point, as shown in Fig.3(b).

It is observed that the higher-frequency poles which added subtle waveform details were not observable in the moment calculations at the output, even though

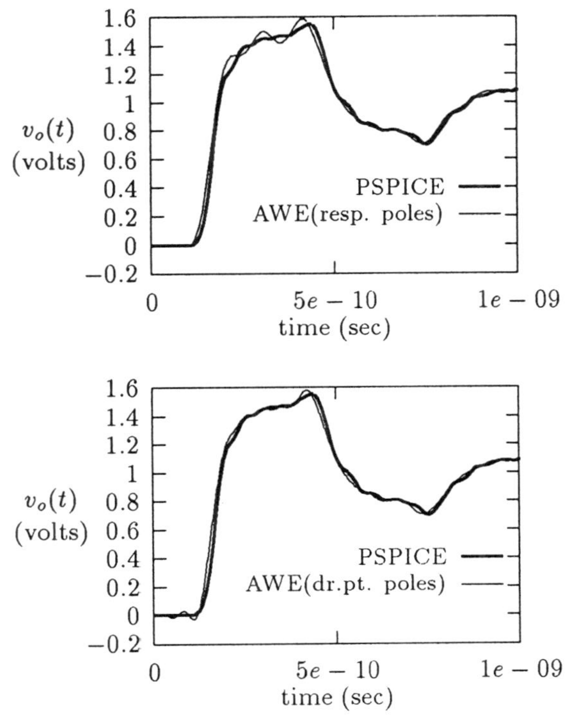

Figure 3: Comparison of the step response of the circuit in Fig.2 as obtained using from a circuit simulator with (a) a stabilized $8th$-order AWE approximation using poles computed at the response node, and (b) a $12th$-order partial Padé approximation using poles computed at the driving point.

the waveforms clearly show their effect. This is due to the greater dominance of the low-frequency poles in influencing the moment values at the output. At the driving-point however, these high-frequency poles are more strongly stimulated and influence the driving-point moments sufficiently to be detected by the AWE algorithm. This also leads to the earlier argument that high-frequency poles that are not stimulated and do not sufficiently influence the moments at the driving-point, will not influence the response moments at points farther from the driver.

Using the driving-point poles is even more accurate in eliminating spurious oscillations that may result from an AWE approximation of the response when the input signal rise-time is finite, as in a finite ramp. Figure 4 compares the actual response of a 10 cm. long low-loss interconnection (R=10.5Ω/cm, L=2.9nH/cm, C=1.17pF/cm) calculated using conventional FFT techniques, with the responses of a lumped, ladder model evaluated using moment-matching techniques. A conventional AWE approximation using the dominant-pole approximations computed at the response node shows spurious oscillations, as shown in Fig.4(a). Employing a partial-Padé approach us-

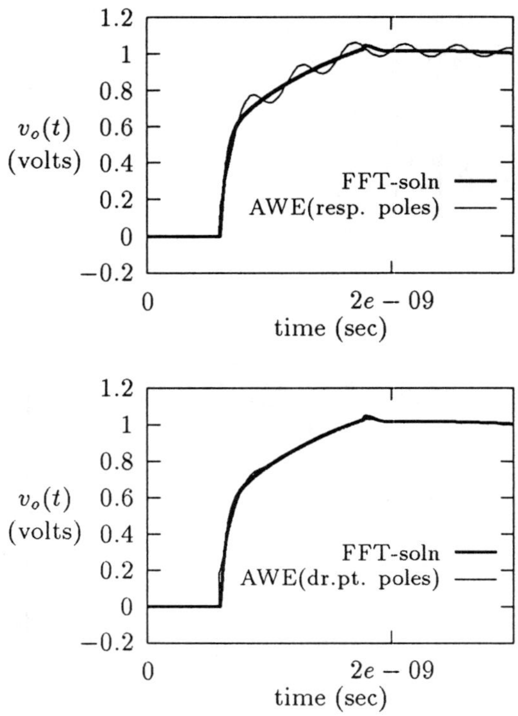

Figure 4: Comparison of the ramp response (t_rise = 0.1ns.) of a low-loss interconnect segment compared with (a) a stabilized $8th$-order AWE approximation using poles computed at the response node, and (b) a $10th$-order partial Padé approximation using poles computed at the driving point.

ing the dominant-pole approximations computed at the driving-point yielded a waveform that is nearly indistinguishable from the FFT result, as shown in Fig.4(b).

7 Conclusion

This paper has introduced an effective method of ensuring the stability moment-matching approximations in Asymptotic Waveform Evaluation. This greatly enhances the reliability of AWE for the analysis of linear(ized) RLC circuits. The minimal constraints used in the development of the stabilization algorithm might also allow its application to other model-reduction techniques based on moment-matching [4, 12]. The introduction of partial Padé techniques to model the response at all nodes greatly enhances the accuracy and efficiency of AWE for VLSI interconnect evaluation.

References

[1] G. A. Baker, Jr. *Essentials of Pade Approximants.* Academic Press, 1975.

[2] R. F. Brown. Model stability in use of moments to estimate pulse transfer functions. *Electron. Lett.*, 7, 1971.

[3] P. K. Chan. Comments on asymptotic waveform evaluation for timing analysis. *IEEE Trans. Computer-Aided Des.*, 10(8), Aug 1991.

[4] C. Chu and M. Horowitz. Charge sharing models for switch-level simulation. *IEEE Trans. Comp. Aided Design*, 6, 1987.

[5] C. T. Dikmen, M. M. Alaybeyi, S. Topcu, A. Atalar, E. Sezer, M. A. Tan, and R. A. Rohrer. Piecewise linear Asymptotic Waveform Evaluation for transient simulation of electronic circuits. In *Proc. IEEE Int'l. Symp. Ckt. Sys.*, Jun 1991.

[6] W. C. Elmore. The transient response of damped linear networks with particular regard to wideband amplifiers. *J. Applied Physics*, 19(1), 1948.

[7] N. Gopal, D. P. Neikirk, and L. T. Pillage. Evaluating on-chip RC-interconnect using moment-matching approximations. In *Proc. IEEE Int'l. Conf. Computer-Aided Des.*, Nov 1991.

[8] L. Ho and R. E. Kalman. Effective construction of linear state-variable models from input/output functions. In *Proc. 3rd Allerton Conf.*, 1966.

[9] X. Huang. *Pade approximation of linear(ized) circuit responses.* PhD thesis, Carnegie Mellon Univ., Nov 1990.

[10] X. Huang, V. Raghavan, and R. A. Rohrer. AWEsim: A program for the efficient analysis of linear(ized) circuits. In *Proc. IEEE Int'l. Conf. Computer-Aided Des.*, Nov 1990.

[11] S. Kumashiro, R. Rohrer, and A. Strojwas. A new efficient method for the transient simulation of 3D interconnect structures. In *IEDM Tech. Dig.*, 1990.

[12] S. P. McCormick and J. Allen. Waveform moment methods for improved interconnection analysis. In *Proc. 27th ACM/IEEE Des. Auto. Conf.*, 1990.

[13] J. Pal. Stable reduced-order Pade approximants using the Routh-Hurwitz array. *Electron. Lett.*, 15, 1979.

[14] D. O. Pederson and G. Wilson. Dominant zero and excess phase of a Hurwitz polynomial. *IEEE Trans. Ckt. Th.*, 11, Mar 1964.

[15] L. T. Pillage and R. A. Rohrer. Asymptotic waveform evaluation for timing analysis. *IEEE Trans. Comp. Aided Design*, 9, 1990.

[16] C. L. Ratzlaff, N. Gopal, and L. T. Pillage. RICE: Rapid Interconnect Circuit Evaluator. In *Proc. 28th ACM/IEEE Des. Auto. Conf.*, Jun 1991.

[17] Y. Shamash. Stable reduced-order models using Pade-type approximations. *IEEE Trans. Auto. Control*, 19, 1974.

[18] T. K. Tang, M. S. Nakhla, and R. Griffith. Analysis of lossy multiconductor transmission lines using the Asymptotic Waveform Evaluation technique. *IEEE Trans. Microwave Th. Tech.*, 39(12), Dec 1991.

[19] V. Zakian. Simplification of linear time-invariant systems by moment approximants. *Int'l. J. Control*, 18, 1973.

AWEsymbolic: Compiled Analysis of Linear(ized) Circuits using Asymptotic Waveform Evaluation

John Y. Lee and Ronald A. Rohrer

SRC-CMU Center for Computer Aided Design
Department of Electrical and Computer Engineering
Carnegie Mellon University
Pittsburgh, Pennsylvania 15213

Abstract

Asymptotic Waveform Evaluation (AWE) is shown to be effective in the symbolic analysis of linear(ized) circuits. AWEsymbolic efficiently and accurately produces reduced order models of both frequency and time domain behavior of linear(ized) circuits. By use of moment level partitioning, the numeric and symbolic computations are substantially decoupled, hence increasing computation speed. Results for AWEsymbolic show a marked decrease in execution time as compared to a full AWE analysis, while producing identical results. This portends the use of AWEsymbolic as an efficient modeling methodology, where AWE approximations are compiled into a reduced set of operations.

1. Introduction

THE goal of symbolic analysis tools, as opposed to traditional circuit simulators [5], is to obtain symbolic forms for network functions in the frequency domain. Some or all of the elements in a circuit are treated as symbols that have no assigned numerical values. The resulting network functions are pure symbolic, or mixed symbolic-numeric expressions.

Such symbolic forms have served many roles [10], including modeling, sensitivity calculation and pedagogy. Most efforts, including [12], [9], [10], [2], [8] compute an exact form of the network functions in the frequency domain. For high order systems, this can lead to complex symbolic forms, even when the number of symbols is low. In order to simplify these expressions, heuristic pruning can be employed [8].

These pruning mechanisms can be unreliable, because suppressed terms may play an important role in the response of a circuit. Determining which terms will be significant and which not, over a range of element values, is difficult. In particular, pruning at the level of the coefficients of the numerator and denominator polynomials of the impulse response is dangerous as the effects on pole-zero locations are difficult to discern.

This paper proposes the use of Asymptotic Waveform Evaluation (AWE) [7] for the symbolic analysis of linear and linear(ized) circuits. AWE affords the ability to reliably and efficiently compute reduced order symbolic expressions, without the use of unreliable pruning methods. Efficient pole-zero sensitivity analysis in AWE provides a mechanism to identify elements that have a significant effect on the dominant poles and zeros or residues of the circuit. Such sensitivities may serve as an automatic means for identifying suitable symbolic elements.

Through the use of moment level circuit partitioning [1], AWEsymbolic is able to significantly decouple the numeric and symbolic computations. The advantage of this approach is decreased computation time. In essence, the symbolic form provides a *compiled* set of operations which can quickly produce a final AWE approximation, where the operands are the values of the symbols. Examples later in the paper will show that the cost of evaluating such a model is much lower than that of a full AWE analysis, and hence can be useful in highly iterative applications.

Since the order of a reasonably accurate AWE approximation is typically low, often less than five, it is possible to factor the symbolic forms obtained. Not only can the symbolic forms for the dominant poles and zeros of a circuit be computed, but the transient response of a circuit can be expressed symbolically as well. Given the low iterative computational cost, and the ability to model in the time domain, AWEsymbolic should serve as a useful mechanism for modeling interconnect delay in physical CAD design tools.

2. Methodology

This section covers the methodology of AWEsymbolic. Notation is introduced below, and AWE is discussed briefly (a reader unfamiliar with Asymptotic Waveform Evaluation is encouraged to consult [7]). AWEsensitivity [4] and AWE partitioning [1], as they apply to AWEsymbolic are then discussed.

2.1. Definitions

Consider a single input, single output, linear time invari-

0738-100X/92 $3.00 © 1992 IEEE

ant system, represented in state variable [6] form:

$$\dot{x} = Ax + bu \quad (1)$$

$$y = c^T x + du \quad (2)$$

The frequency domain system function (the Laplace transform of the impulse response) is expressed as follows:

$$H(s) = c^T (sI - A)^{-1} b + d \quad (3)$$

Given a set of symbolic elements e:

$$e = \{e_1, e_2, \ldots, e_n\} \quad (4)$$

the goal of symbolic analysis is to compute the transfer function as $H(s, e)$. Given that not all variables of $H(s)$ are treated as symbols, the final result will be in mixed numeric-symbolic form.

It is easily shown that the coefficients of the numerator and denominator polynomials of $H(s)$ are multi-linear in the symbolic elements [10]. For example, consider the simple circuit shown in Figure 1. The full symbolic system function (where all elements are treated as symbols) is:

$$H(s) = \frac{G_1 G_2}{C_1 C_2 s^2 + (G_2 C_1 + G_2 C_1 + G_1 C_2) s + G_1 G_2} \quad (5)$$

If the value of G1 is set to 5, the mixed numeric-symbolic expression is:

$$H(s) = \frac{5 G_2}{C_1 C_2 s^2 + (G_2 C_1 + G_2 C_1 + 5 C_2) s + 5 G_2} \quad (6)$$

The coefficients of the numerators and denominators of (5) and (6) are linear in each of the elements. Often times, it is not necessary or desirable to compute an exact solution, whether the answer be numeric or symbolic. AWE addresses this issue.

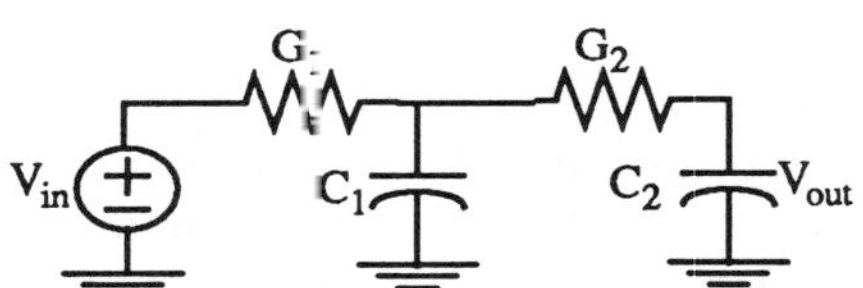

Figure 1: Sample RC circuit for symbolic analysis

2.2. AWE Approximation

In lieu of computing an exact symbolic solution for $H(s)$, AWE constructs an approximation [7] by matching moments of the impulse response, where the moments are the coefficients of the Maclaurin series expansion of $H(s)$ about $s = 0$:

$$H(s) = m_0 + m_1 s^1 + m_2 s^2 + \ldots + m_i s^i + \ldots \quad (7)$$

The moments can be computed efficiently from a DC circuit related simply to the original system. The moments are used to form a rational Padé approximation to $H(s)$. AWE is typically more than an order of magnitude or more efficient than traditional circuit simulation techniques, and its results can be applied to the analysis of time and frequency domain problems [7]. AWE analysis serves as the cornerstone in AWEsymbolic.

2.3. AWESensitivity

Efficient pole-zero sensitivity, using the adjoint method, has been implemented in AWE [4]. Because the poles and zeros completely define the behavior of a system, their sensitivities can serve as a robust mechanism to identify significant elements.

If a choice of symbolic elements has not been made, a pole-zero sensitivity analysis is performed using AWE. Elements with large normalized sensitivities are pruned as symbolic elements. Since it is possible to express all behavior of a linear system in terms of the poles and zeros, the pruning mechanism is easily extended to performance measures such as gain, ringing, phase margin, etc...

Given that the sensitivities computed by AWE provide only local information, it may be necessary to validate the choice of symbolic elements over the range spanned by the symbolic elements. Once the symbolic functions have been compiled (as will be shown later in this paper) though, the cost of validation is low.

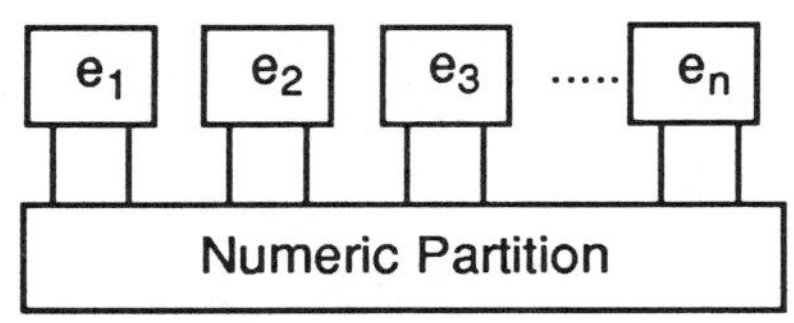

Figure 2: Partitioning of symbolic and numeric blocks

2.4. Partitioning

AWEsymbolic uses moment level partitioning to separate the numeric and symbolic computations are far as possible. The circuit is split into blocks that contain only symbols (symbolic partitions), and partitions that contain only numeric elements (numeric partitions), as shown in Figure 2. The moments are computed for each partition.

In a typical AWE analysis, the time needed to compute the moments far outweighs the time used to form the Padé approximation from the moments [13]. Since a mixed numeric-symbolic analysis is inevitably slower than a numeric simulation, separating the symbolic and numeric moment calculation provides the bulk of the execution time improvement.

To compute the moments of the overall (composite) circuit, the multi-port admittance parameters are computed for each partition. For a partition with n-ports, the admittance parameters are represented as:

$$Y_{nxn} = \begin{bmatrix} y_{11} & y_{12} & \cdots & y_{1n} \\ y_{21} & y_{22} & \cdots & y_{2n} \\ \cdots & \cdots & \cdots & \cdots \\ y_{n1} & \cdots & \cdots & y_{nn} \end{bmatrix} \quad (8)$$

Each admittance function representation is expanded into a Maclaurin series in s:

$$Y_{nxn} = Y_{nxn_0} + Y_{nxn_1} s + Y_{nxn_2} s^2 + \ldots \quad (9)$$

Given that only one symbolic element is encapsulated in any given symbolic block, its port representation is finite. In fact, when a general formulation technique such as modified nodal analysis (MNA) [3] is used, the expansion has only one term per element. With MNA, impedance elements are handled through auxiliary equations, so the symbolic admittance parameters for resistances and capacitances are stencilled using admittance forms, while inductances are stenciled as impedances. Together, we see that their moment expansion is finite:

$$Y_{symbolic} = G + s\,(C + L) \quad (10)$$

where G, C and L are matrices of appropriate dimension.

The port parameters from each partition (numeric or symbolic) are stenciled into a global admittance matrix Y_{global}. Its dimensions are proportional to the number of ports, which is generally proportional to the number of symbolic elements. Thus the size of Y_{global} will generally be much smaller than size of the original circuit, especially when the ratio of numeric elements to symbolic elements is high. The cost of solving this symbolic circuit is much less than that involved in solving a non-partitioned circuit.

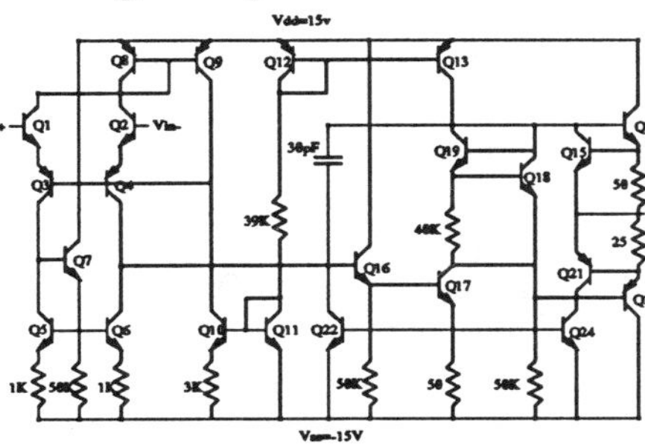

Figure 3: 741 Operational Amplifier

The MNA equations for the composite circuit are expressed as:

$$Y_{global}(s) V(s) = I(s) \quad (11)$$

where $V(s)$ are the outputs of the composite circuit, and I is the vector of independent sources. Each constituent of (11) is expanded into a Maclaurin series:

$$(Y_{global0} + Y_{global1} s + \ldots)\,(V_0 + V_1 s + \ldots) = (I_0 + I_1 s + \ldots) \quad (12)$$

Matching powers of s from the above equation yields:

$$\left.\begin{aligned} V_0 &= (Y_{global0})^{-1} I_0 \\ V_1 &= (Y_{global0})^{-1} (I_1 - Y_{global1} V_0) \\ V_2 &= (Y_{global0})^{-1} (I_2 - Y_{global2} V_0 - Y_{global1} V_1) \\ &\cdots \\ V_i &= (Y_{global0})^{-1} \left(I_i - \sum_{j=1}^{i} Y_{globalj} V_{(i-1-j)} \right) \\ &\cdots \end{aligned}\right\} \quad (13)$$

Each composite moment $\{V_0, V_1, \ldots\}$ can be computed by recursive solution of the resistive $Y_{global0}$ circuit. Since in general the impulse response is sought, all terms except I_0 of the Taylor series expansion of I are 0.

The moments $\{V_0, V_1, \ldots\}$ are then used to form an AWE approximation. Since the moments are symbolic functions, the result is a *reduced order symbolic* approximation, where the order is controlled by the number of moments matched [7]. The cost for finding the approximation is small —though somewhat hindered due to the mixed numeric-symbolic computation demands. In general, if an nth order approximation is sought, an nxn full matrix is LU factored, and a separate nxn forward substitution performed.

3. Examples

This section will illustrate the use of AWE symbolic in the frequency domain, and in the time domain. For each case, it will be seen that a first order analysis can provide a simple intuitive result. Higher order approximations are not as simple, yet are algebraically compact. It is shown that they can be evaluated at a fraction of the cost of a full AWE analysis. Thus, they can be used in applications where a low incremental cost of analysis is required.

3.1. Frequency Domain Symbolic Analysis

The 741 operational amplifier, shown in Figure 3, is analyzed using AWE symbolic. After linearization, the small signal circuit contains 170 linear elements, 62 of which are energy storage elements. Based on an AWEsensitivity analysis, two most significant elements, g_{outQ14} and C_{comp}, are chosen to be treated as symbols (due to their contribution to gain and phase of the amplifier's frequency response). A five port representation is used for the numeric partition.This includes the input and output ports, which must be preserved.

The first order AWEsymbolic approximation is shown below:

$$p_1 = \frac{-(8.75\times10^{-15} g_{outQ14} + 8.88\times10^{-20})}{(0.28C_{comp} + 1.27\times10^{-9} g_{outQ14} + 27.7C_{comp} g_{outQ14} + 9.9\times10^{-15})}$$

$$A_o = -\frac{(1.13\times10^{-6} g_{outQ14} + 1.51\times10^{-8})}{(7.74\times10^{-9} g_{outQ14} - 7.87\times10^{-14})} \qquad (14)$$

$$H(s, g_{outQ14}, C_{comp}) = \left(\frac{A_o}{s - p_1}\right)$$

The symbolic form is stable for all values of g_{outQ14} and C_{comp}, as is the case with the real circuit. Since only two moments are used in the approximation, the symbolic forms are multi-linear in terms of the symbolic elements. This is the case in general for first order AWEsymbolic approximations.

The symbolic expression in (14) is used to generate 3 dimensional plots of performance versus values of the symbolic elements. Figure 4 plots the first pole as a function of

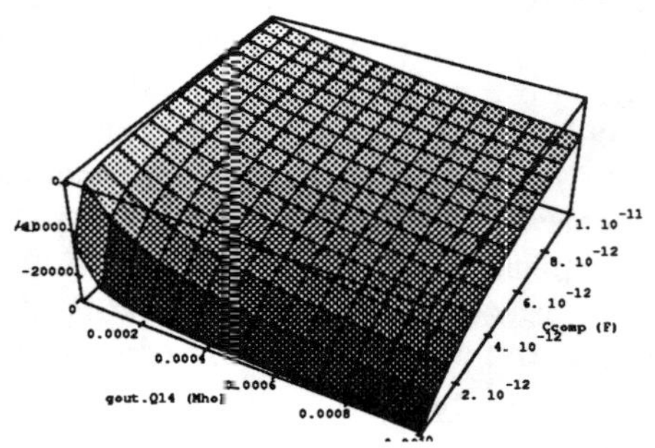

Figure 4: Plot of p_1 from first order symbolic form

the symbolic parameters. Figure 5 plots the DC gain of the op-amp as a function of the symbolic parameters.

The symbolic forms are computed in 3.03 seconds on a DECstation 5000, using Mathematica [11] for the symbolic analysis. The time to evaluate the symbolic forms, given numeric assignments for each symbol, is 0.37 microseconds. In contrast, the time needed to do an AWE analysis is 80.4 milliseconds, of which 25 milliseconds is overhead due to parsing and setup. Ignoring the overhead in both scenarios, the per iteration evaluation cost of AWEsymbolic is roughly *five orders of magnitude* faster.

The second order symbolic form is more complex and of course more accurate. In general, four moments are required for a second order approximation, equation (13) is recursively solved four times. However, since the complexity of each successive moment can increase dramatically, a partial Pade approximation, using derivatives is used. This decreases the number of moments needed in the approximation, with little sacrifice in accuracy.

The second order approximation is shown below. The notation $P(x^i, y^j)$ has been adopted to indicate a polynomial term of *ith* degree in the symbol *x*, and *jth* degree in the symbol *y*:

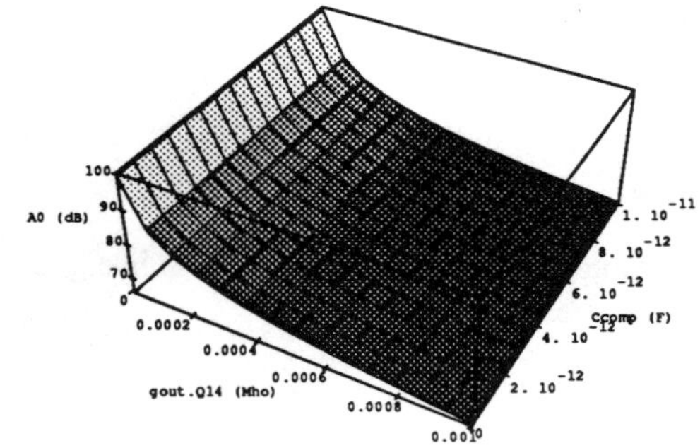

Figure 5: Plot of DC gain from first order symbolic form

$$H(s) = \frac{(b_0 s + b_1)}{(a_0 s^2 + a_1 s + a_2)} k$$

$$b_0 = P(C_{comp}, g^4_{outQ14})$$
$$b_1 = P(g^5_{outQ14})$$
$$a_0 = P(C_{comp}, g^3_{outQ14}) \qquad (15)$$
$$a_1 = P(C_{comp}, g^2_{outQ14})$$
$$a_2 = P(g^2_{outQ14})$$
$$k = P(g^3_{outQ14})$$

The symbolic form is not in multi-linear form—that is, our symbolic elements do not have a physical representation in the symbolic form. However, each term can be evaluated very quickly, and the results are identical to those obtained by a numeric AWE analysis (and as accurate), but with a much quicker iterative evaluation time.

Table 1 cites run times needed to generate the data directly from AWE, without any symbolic analysis. The run times were generated on a DECstation 5000, and do not include common overhead such as parsing and plotting. As is seen, the symbolic analysis is considerably faster than a pure numeric analysis for multiple evaluations of the system function at multiple symbol values.

Per iteration, the symbolic analysis cost is roughly 330 times less than for a numeric AWE analysis. The incremental cost for AWEsymbolic is 0.16 milliseconds, while it is 53.2 milliseconds for AWE. Recall that AWE has also been benchmarked to be at least an order of magnitude faster than SPICE [5] for this class of problem, so AWEsymbolic's speedup over traditional techniques may be quite high.

Figure 6 plots the unity gain frequency of the operational amplifier, as a function of the symbolic parameters. Similarly, Figure 7 plots the phase margin of the operational amplifier. The DC gain plot from the second order form is identical to that of the first order form shown above - since the first moment computed by AWE is always an exact form

Data Points	AWE	AWESymbolic
1	0.079s	2.27s
100	5.35s	2.29s
1000	53.2s	2.43s

Table1: AWEsymbolic run times

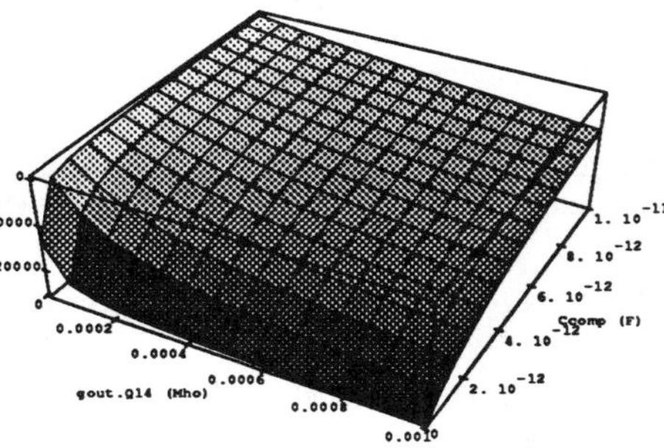

Figure 6: Plot of ϕ_m from 2nd order symbolic form

of the DC gain. Note that each plot was generated by use of the symbolic forms for the poles and zeros. The data is identical to that obtained from a pure numerical AWE analysis.

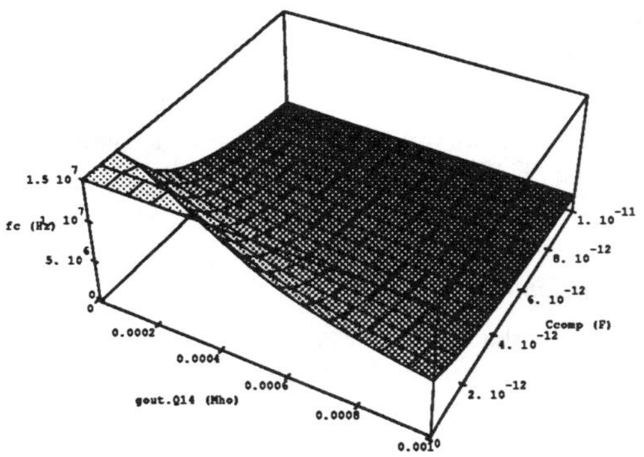

Figure 7: Plot of f_c from 2nd order symbolic form

3.2. Time Domain Symbolic Analysis

A timing model for the circuit shown in Figure 8 is to be

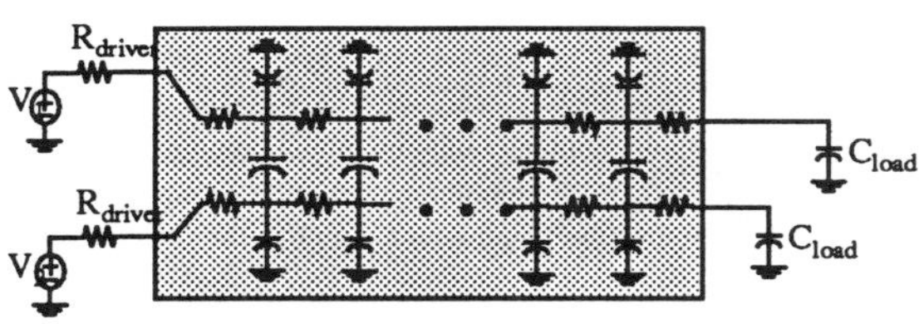

Figure 8: Lumped model for coupled lines.

constructed. Two coupled lines have been modelled using a lumped RC equivalent, with capacitive coupling along the length of the line. The lines are symmetric. The driver at each line is modeled by a linearized Thevenin equivalent, and the loading is assumed to be purely capacitive. Each line has been approximated with a 1000 segment model.

Using AWEsymbolic, a timing model for this circuit is developed. In this case, the driver resistance and the load capacitance are treated as symbols, and the time domain behavior computed. In order to model the non-monotonic nature of the cross coupling response, a second order AWE approximation is used to insure accuracy in the cross talk analysis from one line to the other. A first order approximation suffices to model the direct transmission down each line:

$$V_{out1}(t) = \frac{k_1}{p_1}\left(e^{p_1 t} - 1\right)$$

$$p_1 = \frac{(-1.875)\,G_{driver}}{4.14\times10^{-11} + 1.8C_{load} + 4.16\times10^{-9}G_{driver} + 375C_{load}G_{driver}} \qquad (16)$$

$$k_1 = \frac{1.875G_{driver}}{4.14\times10^{-11} + 1.8C_{load} + 4.16\times10^{-9}G_{driver} + 375C_{load}G_{driver}}$$

Using the notation from (15), the cross coupling is expressed as:

$$H(s) = \frac{b_0 s}{a_0 s^2 + a_1 s + P(3)a_2}$$
$$b_0 = P(G^2_{driver})$$
$$a_0 = P(C^2_{load}, G^2_{driver}) \qquad (17)$$
$$a_1 = P(C_{load}, G^2_{driver})$$
$$a_2 = P(G^2_{driver})$$

The results of the symbolic analysis coincide exactly with those of a numeric AWE analysis. The difference lies in the run time. A single AWE analysis for this circuit requires on average 1.12 seconds on a DECStation5000, while the AWEsymbolic analysis requires 5.41 seconds (using Mathematica for the symbolic computation) [11]. However, the incremental cost, which is crucial in iterative applications, is 0.11 milliseconds for AWE symbolic. This is four orders of magnitude faster than a numeric analysis with AWE.

Illustrations of the second order symbolic form are shown in Figures 9 and 10. The step response cross talk is plotted from the second order symbolic form, as a function of the symbolic parameters.

4. Conclusion

A new method for the symbolic analysis of linear(ized) circuits has been presented. The methodology uses Asymptotic Waveform Evaluation to produce a low order symbolic approximation to the system response of the circuit. The computational cost of evaluating such symbolic forms is orders of magnitude less than that of multiple numeric AWE simulations, as much as five orders of magnitude quicker in one example shown. This marked increase in performance

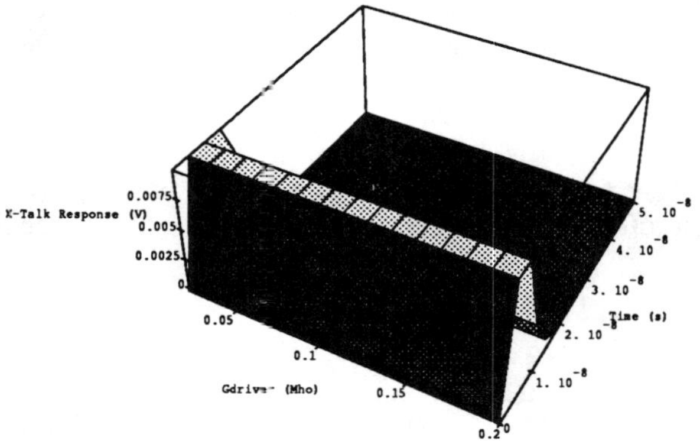

Figure 9: Cross-talk transient response as G_{driver} is varied

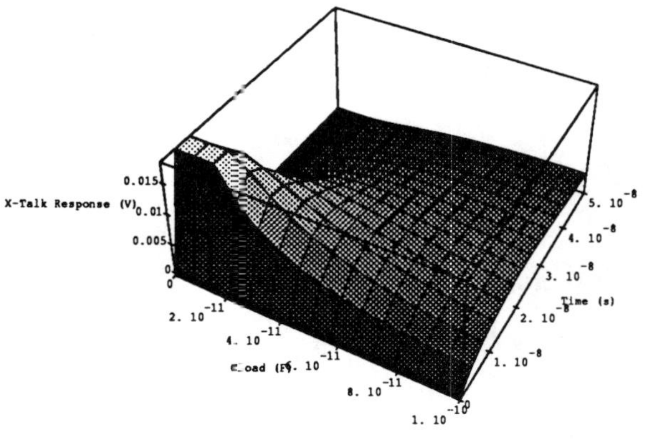

Figure 10: Cross-talk transient response, as C_{load} is varied

results from the compilation of an AWE analysis into a reduced set of operations, which are parameterized in terms of the symbolic elements. In addition, efficient pole-zero sensitivity in AWE allows significant elements to be identified with little additional cost. The encouraging results for AWEsymbolic portend its use for modeling in highly iterative applications, where low incremental cost, robustness and accuracy (in either the time or frequency domains) are required.

5. Acknowledgments

The authors thank Eric Bracken and Vivek Raghavan for their strong contributions, especially with AWE partitioning. Financial support for this research was provided by the Semiconductor Research Corporation (Contract #DC-91-068) and the Digital Equipment Corporation.

6. References

[1] M.A. Alaybeyi, J.E. Bracken, J.Y. Lee, V. Raghavan, R.J. Trihy and Ronald A. Rohrer. Exploiting Partitioning in Asymptotic Waveform Evaluation (AWE). *To appear in Proceedings IEEE Custom Integrated Circuits Conference*, 1992.

[2] G.M. Wierzba and A. Srivastava and K.V. Noren and J.A. Svoboda. Sspice-A Symbolic Spice Program for Linear Active Circuits. *32nd Midwest Symposium on Circuits and Systems*, pages 1197-1201. 1990.

[3] C.W. Ho and A.E. Ruehli and P.A. Brennan. The Modified Nodal Approach to Network Analysis. *IEEE Trans. Circuit Theory.* CAS-22504-509, June, 1975.

[4] John Y. Lee and Xiaoli Huang and Ronald Rohrer. Efficient Pole Zero Sensitivity Calculation in AWE. *Technical Digest of the IEEE International Conference on Computer-Aided Design*, pages 538-541. 1990.

[5] L.W. Nagel. *SPICE2, A Computer Program to Simulate Semiconductor Circuits*. Technical Report ERL-M520, UC-Berkeley, May, 1975.

[6] E. S. Kuh and R. A. Rohrer. The State-Variable Approach to Network Analysis. *Proceedings of the IEEE*. 53(7), July, 1965.

[7] Lawrence T. Pillage and Ronald A. Rohrer. Asymptotic Waveform Evaluation for Timing Analysis. *IEEE Trans. Computer-Aided Design*. 9(4):352-366, April, 1990.

[8] Georges G.E. Gielen and Herman C.C. Walsharts and Willy M.C. Sansen. ISAAC: A Symbolic Simulator for Analog Integrated Circuits. *IEEE Journal of Solid-State Circuits*. 24(6), December, 1989.

[9] Gary E. Alderson and P.M. Lin. Computer Generation of Symbolic Network Functions-A New Theory and Implementation. *IEEE Transactions on Circuit Theory*. CT-20(1), January, 1973.

[10] P.M. Lin. A Survey of Applications of Symbolic Network Functions. *IEEE Transactions on Circuit Theory*. CT-20(6), November, 1973.

[11] Stephen Wolfram. *Mathematica: A System for Doing Mathematics by Computer*. Addison Wesley, 1991.

[12] Kishore Singhal and Jiri Vlach. Symbolic Analysis of Analog and Digital Circuits. *IEEE Transaction on Circuits and Systems*. CAS-24(11), November, 1977.

[13] Xiaoli Huang and Vivek Raghavan and Ronald A. Rohrer. AWEsim: A program for the efficient Analysis of Linear(ized) circuits. *Technical Digest of the IEEE International Conference on Computer-Aided Design*, pages 534-537. November, 1990.

Specification Partitioning for System Design

Frank Vahid and Daniel D. Gajski
Department of Information and Computer Science
University of California, Irvine, CA, 92717

Abstract

Behavioral partitioning can be applied to attain various goals, one of which is the satisfaction of chip-packaging constraints. Such partitioning heavily influences decisions made in subsequent structural design, and may therefore lead to higher performance designs and more efficient use of area and pins than possible when structure is designed before partitioning. Current behavioral partitioning approaches are limited in that they partition at the level of control/dataflow graph operations. We introduce a new approach which partitions entire computations of a behavioral specification, such as processes and procedures, into chip behavioral specifications. We demonstrate the approach's usefulness and highlight results of several examples.

1 Introduction

Given a behavioral specification, our overall aim is to implement it as custom hardware and/or as software. Behavioral partitioning can be used to achieve one or more of the following subgoals in which each partition is implemented:

- using a specific class of hardware technology, such as custom layout, ASICs, or off-the-shelf processor chips. The essential tradeoff is between design/manufacturing costs and system performance, both being heavily influenced by the technologies used. Note that hardware/software partitioning is encompassed by this goal.
- as a chip that satisfies chip-capacity constraints. Here performance is affected by the amount of interchip communication, while cost depends on the chip packages chosen within the technology being used. Note that different chip technologies will have differing size constraints in terms of the silicon-area, gates, or number of instruction-words, as well as differing pin and power constraints.
- on a single sequential processor, where there may be one or more processors per chip. In the context of behavioral partitioning, a processor is either a custom-designed controller and datpath, or an off-the-shelf component such as a microcontroller. Here fewer processors mean less cost and less interprocessor communication time, but more processors can yield more parallel execution.
- by a synthesis tool, separately from other partitions. If the partitions are chosen well, results obtained by the synthesis tool may be superior to those obtained by the tool when applied to the entire unpartitioned behavior. Also, the unpartitioned behavior might exceed the tool's runtime or memory capacity; partitioning can create smaller behaviors that can be handled by the tool.

These goals are closely related and may overlap. Various combinations of these goals result in partitioning problems which differ in the types and weights of constraints, in the techniques and accuracy of estimations made of the constrained parameters, and in the objective functions which evaluate partitionings based on the estimations and constraints. In this paper, we focus on the goal of partitioning a behavior to satisfy chip-capacity constraints while considering system-performance constraints. We assume a hardware (as opposed to software) implementation with a uniform chip technology.

The paper is organized as follows. In Section 2 we discuss the need for a new partitioning approach, outline our specification partitioning approach, and highlight its advantages. In Section 3 we describe details of our approach. In Section 4 we provide the results of partitioning several examples using the specification partitioning tool we are developing. In Section 5 we discuss future work.

2 Specification partitioning overview

Structural partitioning can be used to attempt to satisfy chip-capacity and system-performance constraints, but there are drawbacks when compared to behavioral partitioning. In a structural partitioning approach, a behavior is first converted to structure, and then the structure is partitioned among chips. While yielding accurate size and pin estimates, the approach ignores the fact that the design of the structure can be heavily influenced by the chip-partitioning. For example, two sequential procedures that each perform an addition can share an adder if both procedures are implemented on the same chip. If the procedures are separated among two chips, the same sharing results in poor performance due to interchip communication time; a better design would include two adders, one on each chip.

The effect of partitioning on the structural design supports the need for behavioral partitioning. There are essentially two levels at which such behavioral partitioning can be performed. At the *operation* level, dataflow-level operations such as addition and subtraction are grouped. At the *algorithmic* level, entire program-grained computations such as processes and procedures, making up a set of sequential and concurrent behaviors, are grouped. Figure 1 shows goals that are achievable at each level. Most goals involve tradeoffs between cost and performance. In general, algorithmic-level goals relate to larger systems than

0738-100X/92 $3.00 © 1992 IEEE

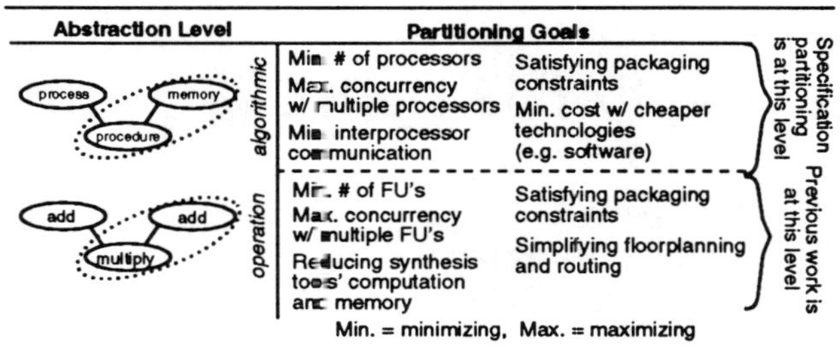

Figure 1: Abstraction levels vs. goals

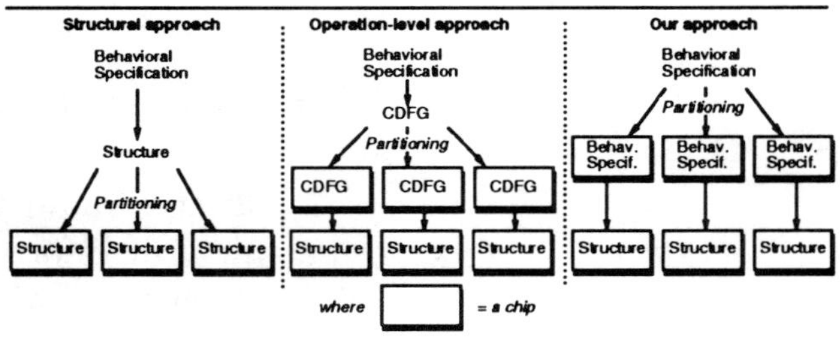

Figure 2: Specification partitioning vs. other methods

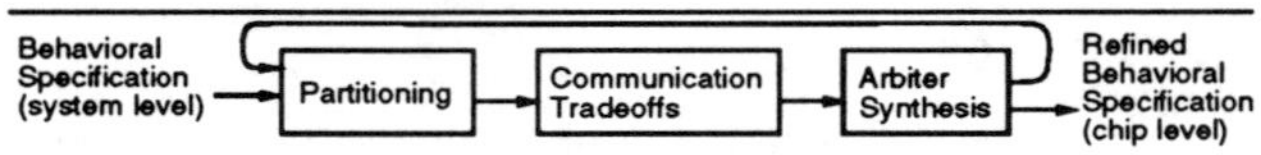

Figure 3: Iterative specification-refinement methodology

do operation-level goals.

Previous approaches to behavioral partitioning have been at the operation level; see [1] for a detailed survey of these approaches. In discussing these approaches, we shall use the term "CDFG" to denote any graph-type representation of behavior consisting of control and dataflow operations. Such representations are used internally by most synthesis tools.

YSC [2, 3] partitions logic containing behavioral operations, such as addition and subtraction, so that logic-synthesis runtime and memory-use constraints are not exceeded and so that follow-up floorplanning quality is improved. BUD [4] partitions the dataflow-graph operations of a CDFG in a manner that encapsulates scheduling and allocation decisions. Various partitionings are evaluated with respect to estimations of area and time. A similar approach is found in [5]. In Aparty [6, 7], the partitioning algorithm is extended to allow for the use of differing closeness functions and evaluation criteria at each one of multiple partitioning stages. CHOP [8] provides area and time estimates of an acyclic dataflow graph's operations for a given chip partitioning. Vulcan [9] partitions CDFG vertices, each of which represents a finite-state machine or combinational block, among chips.

As chip capacities as well as behavior-description sizes increase, we feel that algorithmic-level rather than operation-level partitioning approaches will be necessary. In this paper, we introduce an algorithmic-level approach (see Figure 2). **The main contribution of this research is the elevation of partitioning to a higher level of abstraction, i.e. grouping entire computations as opposed to merely operations.** In our specification partitioning approach, a behavioral description is viewed as a set of behaviors, such as processes, procedures, substates, and other code groupings imposed by the language, and a set of storage elements, including registers, memories, stacks, and queues. These behaviors and storage elements are then partitioned such that each partition represents a chip. To find a partitioning that satisfies constraints, we must be able to estimate values for pins, area, and performance for a given partitioning, thus requiring estimation models. The results of partitioning are reflected in a refined specification consisting of a set of interconnected chips, each possessing behaviors and storage elements.

Specification partitioning has several advantages. There are far fewer algorithmic-level objects than operation-level objects. In addition, because partitioning is performed on the specification itself, most of the objects are familiar to the designer. The first fact permits the application of more thorough partitioning algorithms. The two facts combined enable *designer control* over the partitioning decisions. Also, *functional test patterns* can be written for each chip to test intended behavior. If the most abstract description of a chip consists of thousands of gates or low-level control/data operations, such test patterns are difficult to obtain.

Another advantage is the localization of the effects of *late specification changes.* Each chip specification consists of entire portions of the original specification behaviors (e.g. processes), with little or no modification. If a process needs to be internally changed after structural design of the chips, only the affected chip need be changed, leaving the rest of the multi-chip system intact. In CDFG approaches, a single simple behavior may be spread over multiple chips.

Now let us consider the most important advantage, which relates to the ability to modify the partitioned specification. To discuss this, we must first describe the overall design task of which behavioral partitioning is actually a subtask. We distinguish between two specification levels: the *system specification*, which denotes desired system behavior, and *chip specifications*, which denote the division of the system into interconnected chip specifications. Conversion from the first to the second level is part of what we call **System Design**, and often results in modified or added behaviors. We view system design as the refinement of a specification through iterations of partitioning, communication tradeoffs (e.g. bus merging), and arbiter synthesis, among other tasks (see Figure 3).

In specification partitioning, it is *easy to modify or add to the specification* when chip information is obtained, since the specification is maintained near its original abstraction level. For example, a 16-bit parallel data transfer between two behaviors on separate chips could easily be converted to two 8-bit transfers or even to a serial transfer in order to reduce pins, simply by changing the internal details of the two behaviors involved. As another example, multiple data buses can be merged into a single bus, which may then require the addition of a bus arbiter process to resolve contending bus-transfers.

In contrast, CDFG approaches usually view a specification as unchangeable, thus limiting synthesis to tasks such as scheduling, binding and logic synthesis. For example, multiple data transfers can use a single bus only if there would not be contention; adding an arbiter is not consid-

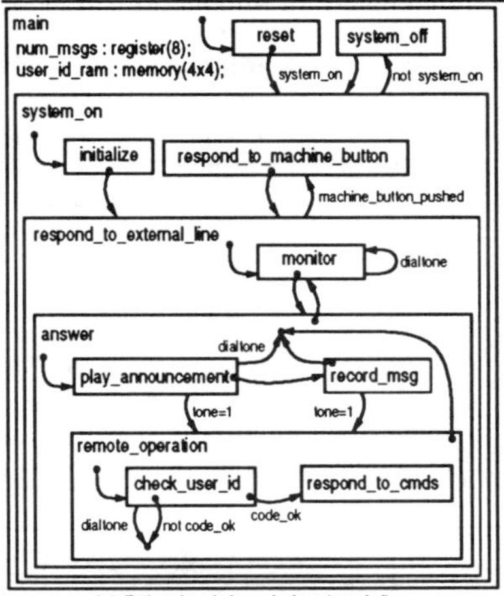

(a) Behavioral description (partial)

```
check_user_id
    entered_code : memory(4x4);
    i : integer range 1 to 5;
  i := 1;
  while i <= 4 loop
    wait until button_tone;
    entered_code[i] := button;
    i := i + 1;
  end loop;

  if (entered_code[1]=user_id_ram[1])
   and
     ...
    (entered_code[4]=user_id_ram[4])
  then
    code_ok <= true;
  else
    code_ok <= false;
  end if;
```

(b) a leaf behavior

Figure 4: Answering machine example.

ered. The arbiter would have had to exist in the original specification, even though its necessity may only arise when trying to satisfy pin constraints *after* partitioning.

In the cases when the specification partitioning approach does not decompose a behavior into fine enough granularity to satisfy chip-capacity constraints, it can be followed by an operation-level or structural-level partitioning approach. However, we believe such cases will become less frequent as chip capacities continue to increase.

3 Method

3.1 The behavioral specification

We specify behavior with the SpecCharts language [10], which allows behavior to be decomposed into concurrent sub-behaviors, sequential sub-behaviors sequenced by arcs, or VHDL sequential statements. SpecCharts is intended as a system specification language rather than a hardware description language (HDL); its main advantages over HDLs are the support of behavioral hierarchy and a unified representation of control, function, and structure. These features ease accurate capture and aid comprehension of the refined specification. Since SpecCharts subsumes behavioral VHDL, our approach is also applicable to VHDL.

An example of a behavioral specification is the telephone answering machine shown in Figure 4(a). It consists of a hierarchy of behaviors which respond to a phone line such that the machine can answer the phone, play an announcement, record a message, or be programmed remotely after a user identification number is checked. The code for the *check_user_id* behavior is shown in Figure 4(b).

We now describe the main steps of specification partitioning.

3.2 Object determination

Behavioral objects are all SpecChart behaviors and procedures (for VHDL, they would simply be all processes and procedures). Using VHDL-like declarations, memory storage is any variable or signal of array type where the elements are scalars or bit-vectors. Register storage is any other VHDL signal of register-kind or any global variable.

The example in Figure 4(a) shows 14 behaviors (e.g. *main, reset, system_on, system_off*, etc.) and 3 storage elements (*num_msgs, user_id_ram* and *entered_code*).

3.3 Estimation model creation

Estimation models are needed in order to estimate pins, area, and performance for a given partitioning of objects. The estimates are used by the partitioning algorithm to evaluate various partitionings.

Pin model: The model for pins is the simplest of the three. We use a hypergraph model where each vertex represents an object and each hyperedge represents communication between objects. There are several types of hyperedges:

- *storage access*: A hyperedge connects each storage with all behaviors which globally access (read and/or write) it. The weight of the hyperedge equals the register bitwidth or memory address/data bitwidth plus some control lines depending on the nature of the access.
- *behavior control*: A hyperedge connects each behavior with the parent behaviors that control it (e.g. calling procedures). Its weight is the sum of the bitwidths of any parameters, plus two control lines representing activation/completion control.
- *global-wire/external-port access*: For each non-storage signal or external port, a hyperedge connects it to all accessing behaviors. Its weight is the signal/port bitwidth.

Given a hypergraph, the number of pins for a partition is calculated as the sum of the weights of all hyperedges which cross the partition's boundary (i.e. the partition's cutsize).

Area model: Given a partition containing behavior and storage objects, we must estimate an area for the partition. This task is far more difficult than that for pins. An accurate area-estimation approach would perform synthesis for each partition, mapping sequential behaviors to a single control-unit and datapath (CU/DP). Unfortunately, such synthesis is computationally expensive, so performing it for each of the thousands of possible partitionings explored by a partitioning algorithm is infeasible.

An alternative is to assign a fixed area to each behavior and storage object, and then estimate a partition's area as the sum of its objects' areas. For each storage, area is determined by querying a component library. For each behavior, we synthesize a CU/DP and then use the resulting area. The advantage of this approach is that it only performs the computationally-expensive synthesis once per behavior, before partitioning begins. However, estimated partition area may be inaccurate since eventual implementation may not actually use a separate CU/DP for each behavior. The inaccuracy of this estimation approach is small if most behaviors are concurrent (i.e. processes) and thus will actually be implemented as separate CU/DPs, or if functional-unit sharing between sequential behaviors (which will occur if the behaviors use the same CU/DP) does not substantially affect overall area for a partition.

To minimize the error related to functional-unit sharing as well as reducing partitioning computation-time, only a subset of the specification's behaviors can be selected for

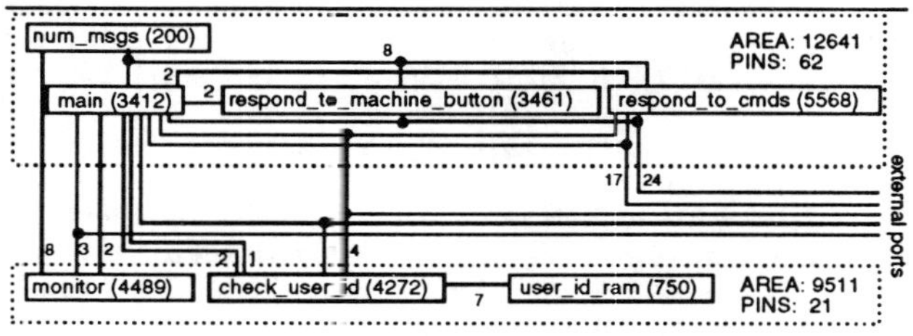

Figure 5: Partitioned hypergraph for answering machine

partitioning. By default, each remaining behavior is then encompassed by its parent behavior in the hierarchy.

We shall now discuss the synthesis technique used to create a CU/DP and estimate its area, although the technique is somewhat independent of our partitioning approach. We use the tool described in [11] to perform scheduling, allocation, and binding. We provide a basic set of functional units (e.g. an adder, a subtractor, and so on) and restrict synthesis to use only one functional unit of each type when building a datapath. This restriction could easily be changed to allow multiple small functional units but only one of each large functional-unit type (e.g. a multiplier) if such a change would more closely match the structural design strategy that will eventually be used. Once the structure is synthesized, the datapath area is determined by creating a bit-slice style floorplan of the muxes, registers, and functional units comprising the datapath, estimating wire-routing area, and then computing a bounding box. The control-unit area is determined from the state-register size and the number of gates.

Once each object has been assigned an area value, we map these values to the vertices of the hypergraph model already created for the pin model, so that standard hypergraph partitioning algorithms can be used. Figure 5 shows the hypergraph created for the answering machine example, along with a sample partitioning showing pin and area estimates. Areas are shown with each object, and interconnect-widths with each hyperedge. Note that many objects in the specification are not in the hypergraph; those objects are encompassed through hierarchy by objects in the hypergraph.

Performance model: Any behavior can have its average start-to-finish execution-time constrained. Execution time for a behavior can be modeled as the sum of two parts: (1) time spent communicating with other behaviors or storage over communication channels, and (2) time spent performing computations. The computation time is estimated by simulating the scheduled behavior with a representative data set, with all communication times over channels set to zero. This simulation also provides the average number of times that the behavior communicates over each channel during a single pass through the behavior; this number is multiplied by the latency of a channel's communication protocol to obtain total communication time over each channel. A communication is either a storage access or a sub-behavior access (activation/completion) plus the sub-behavior's execution-time. Default access protocols with given latencies are used for each type of access. Each channel actually has two protocol-latency values, one for accessing the object when it is on-chip, and one for the case when it is off-chip.

A given partitioning determines whether the on-chip or off-chip latency is to be used for each channel. The expected execution-time for a behavior is then the computation time plus the total communication time of its channels, which in turn are calculated using the appropriate latencies. We extend the hypergraph such that each vertex has a computation time. We add edges to represent channels between vertices, along with associated on-chip/off-chip communication times.

3.4 Partitioning

Let each hypergraph vertex be denoted as v_j, each channel edge by $c_{j,k}$ and each partition as V_i. We currently use the following straightforward objective function, which attempts to minimize constraint violations, to evaluate a given partitioning:

$$OBJFCT = k_1 \sum_i \left(100 \times \frac{excessarea(V_i)}{maxarea(V_i)}\right)^2$$

$$+k_2 \sum_i \left(100 \times \frac{excesspins(V_i)}{maxpins(V_i)}\right)^2 + k_3 \left(100 \times \frac{excesschips}{maxchips}\right)^2$$

An excess value (e.g. $excessarea(V_i)$) is equal to the estimated value (e.g. $area(V_i)$) minus the imposed constraint (e.g. $maxarea(V_i)$), with negative values being treated as 0. Thus, any partitionings that meet all constraints are considered equal since $OBJFCT$ will evaluate to 0. The k's are user-defined constants which indicate the relative importance of each metric. To favor balanced over unbalanced excesses, we multiply each term by 100 and then square. Other objective functions can also be used.

To extend $OBJFCT$ to consider performance, we add a fourth term:

$$OBJFCT = ... + k_4 \sum_j \left(100 \times \frac{excessexectime(v_j)}{maxexectime(v_j)}\right)^2$$

$excessexectime(v_j)$	$= exectime(v_j) - maxexectime(v_j)$, or 0 if negative,
$exectime(v_j)$	$= commtime(v_j) + comptime(v_j)$,
$comptime(v_j)$	the expected time to execute a single pass of the behavior associated with v_j assuming communication times are 0,
$commtime(v_j)$	$= \sum_k commdelay(c_{j,k})$ for all v_k with which v_j communicates
$commdelay(c_{j,k})$	$=$ $offchipdelay(c_{j,k})$ if v_j and v_k are not in the same partition V_i, $onchipdelay(c_{j,k})$ otherwise (recall that these delays incorporate expected frequency of channel use). If v_k is a behavior, it's execution-time equation is added.

Note that this term involves v_j, not V_i. Specifically, excess execution time is determined per vertex, not per partition. In other words, although execution time is affected by the chip partitioning, time-constraints are specified for behaviors (v_j), not partitions (V_i).

Well-known partitioning algorithms such as clustering and group migration are then applied. We currently use a clustering algorithm similar to that in [4], modified to use a closeness function defined for behaviors rather than operations. Evaluation of clusterings is done with the above objective function, rather than an area/time function. Our group migration algorithm is similar to the Kernighan/Lin algorithm discussed in [12], except that each move involves one object, not two. Also, an object can be moved to any one of multiple partitions, as opposed to just the opposite partition in bipartitioning. The objective function is that given above.

3.5 Refining the specification

The partitioning results are used to refine the original specification with chip information. Each chip will contain the behaviors and storage partitioned into it. If a storage is accessed by a behavior on a separate chip, that storage is converted to a process and all accesses are made via ports. Likewise, if a behavior is controlled by another behavior on a separate chip, control must be added between these behaviors. To aid designer comprehension, refinement should make minimal changes from the original specification. Each chip specification will serve as the behavioral input to a subsequent structural design phase, which may be done by a high-level synthesis tool.

Assume a function $anc\_obj(v_i)$ returns the vertex whose behavior is the closest ancestor to the behavior of v_i in the specification hierarchy. The function $accessing\_objs(v_i)$ returns the set of vertices which access the storage associated with v_i. We use the following algorithm to refine the original specification with the partitioning information:

Algorithm 3.1 : Specification refinement

for each vertex v_i representing a behavior
    if $anc\_obj(v_i)$ is on a different chip than v_i
        convert v_i's behavior to a concurrent behavior b_i
for each vertex v_i representing storage
    if any of $accessing\_objs(v_i)$ is on a different chip
        convert v_i's storage to a concurrent behavior b_i
for each chip-partition V_j
    create a concurrent behavior B_j representing a chip
for each conc. beh. b_i (created from a v_i above)
    move b_i to a conc. subbehavior of B_j such that $v_i \in V_j$

As an example, let us assume that the best partitioning in terms of area and pins has *check_user_id*, *monitor*, and *user_id_ram* on one chip, and all other objects on another chip, The refined specification resulting from this partitioning is shown in Figure 6.

Using ports to model storage access or subbehavior control unnecessarily models communication at a wire level. Communication should be modeled at a higher level in order to maintain information regarding the port groupings and the protocols over those ports. Therefore, whenever possible, we use the constructs of *protocols and channels* (see [10]) rather than ports. A protocol is defined as a set of ports (e.g. address and data) and a behavior over those ports. A channel is simply an instantiation of an associated protocol's ports. A channel is connected to other channels just as ports are, and the associated behavior is activated similar to a procedure call. These constructs separate communication from computation, making specifications readable and modifiable. Communication tradeoffs can be made later by merging channels, merging ports of a single channel, or associating a different protocol with a channel. All storage access and sub-behavior control in the refined specification uses channels with default protocols.

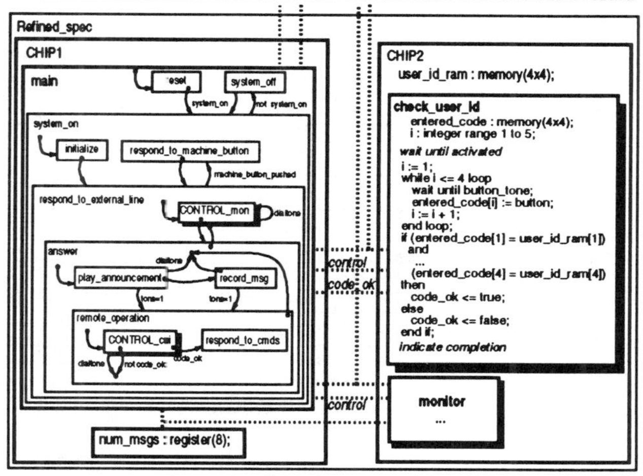

Figure 6: Refined specification

4 Results

SpecPart is a tool we are developing for specification partitioning. Its input is a specification in the SpecCharts language. Object selection, manual and automated partitioning, and partitioning evaluation are all aided by an X-based graphical interface. SpecPart currently consists of approximately 16,000 lines of C code.

Figure 7 summarizes partitioning results on several examples. All area units are generic; a particular technology would result in values in square mils or equivalent gates/transistors. The second column provides information on the size and characteristics of the specification, i.e. the number of specification lines, the number of behavioral hierarchical levels, and the number of behaviors and storage elements that could be selected for partitioning. Refined specifications were created for all partitionings.

We briefly discuss each example; for more details, see [13]. Rockwell's DRACO interfaces 16 I/O ports to a microprocessor's 8-bit multiplexed address/data bus, with extensive error and security checking. Two partitionings are shown with differing pin constraints. In the latter case, communication tradeoffs multiplex communication with 11 registers over a single bus, reducing pin counts from (134,124) to (58,48). A different merging resulted in (79,69) pins but had less effect on performance. The refined specification for one partitioning had 460 lines.

The telephone answering machine is the complete model from which the example in this paper was derived. By selecting several approximately equal-sized behaviors from the hierarchy, balanced two-way and three-way partitions are obtainable, as shown. The refined specification for the two-way partitioning had 1200 lines.

The Intel 8237 is a DMA controller. We tried three different selections of behaviors to obtain differing granularities for partitioning, as shown in the table. For the third case, pins can be reduced from (163,153) to (81,71) by merging the buses used by a particular behavior to access several registers on the opposite chip. Since the particular behavior (the program mode of the 8237) is only performed occasionally, the decrease in speed caused by these merges is of little consequence. For the fourth case, we imposed

Example	# spec. lines, hier. levels, poss. objects	# chips	area const.	pin const.	Area	Pins	comments
DRACO	302,4,19	1	--	--	15200	55	
		2	8000	100	5267 9933	107 108	
		2	8000	150	6677 8523	134 124	comm. tradeoffs reduce pins to 58 and 48.
Answering machine	726,10,59	1	--	--	29002	39	
		2	15000	50	14649 14254	52 46	
		2	15000	30	11076 17296	25 43	
		3	10000	40	10313 10459 8131	26 45 46	
I8237	697,5,54	1	--	--	38980	51	
		2	15000	80	9425 31804	67 114	coarse objects were chosen
		2	15000	80	26781 10907	94 47	area balance sacrificed for performance
		2	15000	150	19105 18029	163 153	comm. tradeoffs reduce pins to 81 and 71.
		2	15000	50	28535 8599	76 29	
AST-like bus controller	464,3,10	1	--	--	2455	130	
		2	5000	100	1620 803	88 58	model was respecified into control and data

Figure 7: Results of partitioning several examples

a time-constraint on a particular behavior (the transfer mode). In this case, several register files which were previously placed on a separate chip, were grouped with the constrained behavior to reduce communication time.

We also modeled a subset of protocols handled by an AST bus controller which interfaces 32-bit address and data buses with two 8-bit buses. To obtain a partitioning which met pin constraints, the original model had to be rewritten as two concurrent behaviors, one implementing the controlling state machine, the other the bus routing.

The above examples represent only the lower end of the domain for which SpecPart is intended. We are currently working on a much larger industrial example, consisting of a RISC signal processor, timers, several memories, and multiple I/O behaviors. We believe that the fact that such extensive functionality will be implemented on a single chip (with approximately one million transistors) lends support to the need for algorithmic-level partitioning approaches. We plan to examine not only multichip implementations, but also to extract partitions which lead to better single chip designs by using multiple custom processors, as discussed in the introduction.

5 Future work

We are currently developing a method for estimating area far more accurately than discussed in this paper, but in a manner which still permits constant-time updates of the estimate when an object is moved among partitions.

Currently, our refinement algorithm converts sequential behaviors which are separated from their parent into concurrent processes. However, if the behaviors on a chip are sequential, they can be merged into a single process in order to encourage more efficient implementation as a single CU/DP.

As mentioned in the introduction, attacking other problems such as hardware/software partitioning, or partitioning among multiple processors to extract concurrency (as discussed for the RISC example above) involves changes in the constraints, estimations, and objective functions. We plan to make these changes to broaden the scope of the specification partitioning approach.

If a satisfactory partitioning is not found, respecifying the behavior using more procedures can provide finer granularity. Automated techniques to isolate and convert candidate code portions should be developed.

We also plan to develop tools for generating and matching communication protocols, for performing communication tradeoffs, for binding memories, and for synthesizing arbiters to resolve bus contention.

6 Conclusion

We have introduced a partitioning approach that is the first to address behavioral partitioning at a process/procedural level as opposed to a dataflow-operation level. We have shown how the approach can be used to solve the specific problem of obtaining multichip partitions. As behavioral description sizes continue to increase, there are many other behavioral partitioning problems that must be addressed. We feel that building on the concepts outlined in this paper should lead to promising solutions to these problems.

7 Acknowledgements

We would like thank Sanjiv Narayan for his work on estimation which enabled implementation of our partitioner, and for his numerous other contributions. This research was supported by the National Science Foundation (grant #MIP-8922851) and the Semiconductor Research Corporation (grant #90-DJ-146); we are grateful for their support.

References

[1] F. Vahid, "A Survey of Behavioral-Level Partitioning Systems." UC Irvine, TR ICS 91-71, 1991.

[2] R. Camposano and R. Brayton, "Partitioning Before Logic Synthesis," in *Proc. ICCAD*, 1987.

[3] R. Camposano and J. van Eijndhoven, "Partitioning a Design in Structural Synthesis," in *Proc. ICCD*, 1987.

[4] M. McFarland and T. Kowalski, "Incorporating Bottom-Up Design into Hardware Synthesis," *IEEE Transactions on Computer-Aided Design*, September 1990.

[5] J. Rajan and D. Thomas, "Synthesis by Delayed Binding of Decisions," in *Proc. 22nd DAC*, 1985.

[6] E. Lagnese and D. Thomas, "Architectural Partitioning for System Level Synthesis of Integrated Circuits," *IEEE Transactions on Computer-Aided Design*, July 1991.

[7] E. Lagnese, *Architectural Partitioning for System Level Design of Integrated Circuits.* PhD thesis, Carnegie Mellon University., March 1989.

[8] K. Kucukcakar and A. Parker, "CHOP: A Constraint-Driven System-Level Partitioner," in *Proc. 28th DAC*, 1991.

[9] R. Gupta and G. Micheli, "Partitioning of Functional Models of Synchronous Digital Systems," in *Proc. ICCAD*, 1990.

[10] S. Narayan, F. Vahid, and D. Gajski, "System Specification and Synthesis with the SpecCharts Language," in *Proc. ICCAD*, 1991.

[11] S. Narayan and D. Gajski, "Area and Performance Estimation from System-Level Specifications." UC Irvine, TR ICS 92-16, 1992.

[12] B. Preas and M. Lorenzetti, *Physical Design Automation of VLSI Systems.* California: Benjamin/Cummings, 1988.

[13] S. Narayan, F. Vahid, and D. Gajski, "Modeling with SpecCharts." UC Irvine, TR ICS 90-20, 1990.

Synthesis and Simulation of Digital Systems Containing Interacting Hardware and Software Components

Rajesh K. Gupta   Claudionor Nunes Coelho, Jr.   Giovanni De Micheli

Center for Integrated Systems
Stanford University, Stanford, CA94305.

Abstract

Synthesis of systems containing application-specific as well as reprogrammable components, such as off-the-shelf microprocessors, provides a promising approach to realization of complex systems using a minimal amount of application-specific hardware while still meeting the required performance constraints. We describe an approach to synthesis of such hardware-software *systems starting from a behavioral description as input. The input system model is partitioned into hardware and software components based on imposed performance constraints. Synchronization between various elements of a mixed system design is one of the key issues that any synthesis system must address. In this paper, we consider software and interface synchronization schemes that facilitate communication between system components.*

We present tools to perform synthesis and simulation of a system description into hardware and software components. In particular, we describe a program, Poseidon, *that performs concurrent event-driven simulation of multiple functional modules implemented either as a program or as behavioral or structural hardware models. Input to* Poseidon *consists of description of interacting functional models with their respective clock cycle times and the interface synchronization scheme chosen by the partitioner. The resulting software component is assumed to be implemented for the DLX machine, a load/store microprocessor. We present simulation examples and design of a graphics controller demonstrate the feasibility of mixed system synthesis.*

1 Introduction

In this paper we consider an approach to extend the high-level synthesis techniques to synthesize system designs using application-specific and reprogrammable components. We refer to the application-specific component as the *hardware* component, while the program running on the reprogrammable component as the *software* component. Our approach to system synthesis using hardware and software components is inspired by the fact that in practice most systems used in embedded control and telecommunication applications consist of application-specific hardware components as well as reprogrammable components. While most digital functions can be implemented by software programs, a major reason for building dedicated ASIC hardware is satisfaction of performance constraints. The performance constraints can be on the overall time (latency) to perform a given task or on the input/output data rates. Typically, the pure software implementations of a system design are often too slow to meet the imposed performance constraints. Therefore, specialized hardware chips are often needed to complement or assist the reprogrammable component on certain performance-critical tasks. Further, *mixed* system designs reduce the size of the synthesis task by reducing the number of application-specific chips required while at the same time achieving the flexibility of software reprogramming to alter system behavior. Such a flexibility is also important in achieving rapid prototyping of complex system designs where non performance-critical and unconstrained operations are shifted into a program running on an off-the-shelf microprocessor.

Whereas the focus of high-level synthesis techniques thus far has been to generate a purely hardware implementation of a system design either as a single chip or as an interconnection of multiple chips, each of which is individually synthesized [1] [2] [3] [4], attempts at system synthesis using both hardware and software components have been rare and limited to developing frameworks for facilitating the design process [5].

The problem of synthesis of mixed systems is fairly complex. There are many subproblems that must be solved before an effective synthesis system can be developed. Among the important issues are the problems of modeling of system functionality and constraints, determination of the boundary between hardware and software components in the system model, specification and synthesis of the hardware-software interface, and implementation of hardware and software components. In this paper, we summarize a systematic approach to automatic synthesis of mixed systems and focus on the issue of interface design and the synchronization mechanisms that are used to facilitate communication between system models and ensure correctness of system functionality.

Figure 1 shows organization of the CAD design system used for synthesis of mixed system designs. The input to our synthesis system is an algorithmic description of system functionality. We model system behavior using the *HardwareC* [6] language that has a C-like syntax and supports timing and resource constraints. *HardwareC* supports specification of unknown delay operations that can arise from data dependent decisions and external synchronizations. The *HardwareC* description is compiled into a **system graph model** based on data-flow graphs[6]. The system graph model consists of vertices representing operations, and edges which represent either a data dependency or a control dependency. Overall the system graph model is composed of concurrent data-flow sections which are ordered by the system control flow. The data-flow sections simplify the tasks of identification of concurrency in the system model, while use of control constructs such

0738-100X/92 $3.00 © 1992 IEEE

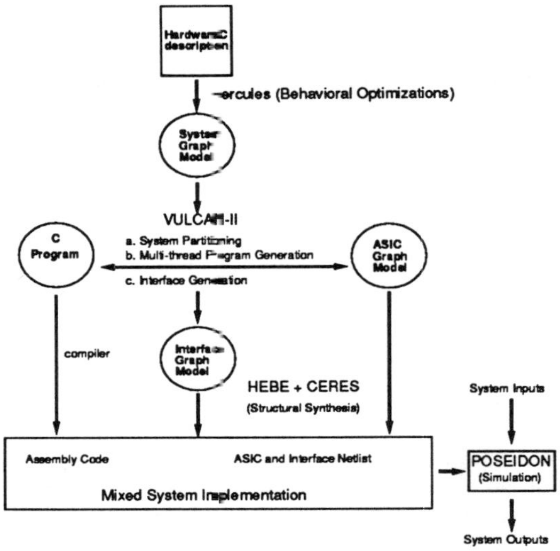

Figure 1: *System Synthesis Procedure*

as conditionals and loops obviate the need for a separate description of control flow. Associated with input/output statements, we specify corresponding constraints on input/output data rates. The input (output) rate constraints refer to the rates at which the data is required to be consumed (produced). The system graph model is input to *Vulcan-II* which partitions the system graph model into portions to be implemented either as dedicated hardware modules or as a sequence of instructions on a reprogrammable processor based on feasibility of satisfaction of externally imposed data-rate constraints. System partitioning constitutes an important phase of the system synthesis process. However, it is not the intent of this paper to delve into the partitioning issues. For an approach to system partitioning the reader is referred to [7]. Hardware synthesis of the models identified by *Vulcan-II* is performed by program *Hebe* [6]. For synthesis of the software component, we generate a corresponding C-description after generating a total order of operations in accordance with the partial order imposed by the graph model. The C-code is then compiled into assembly code for the target processor using existing software compilers. The interface synthesis is performed by *Vulcan-II* under timing constraints imposed on the system model. At the present time, *Vulcan-II* is a framework to carry out various synthesis tasks, and it is not yet an automated tool.

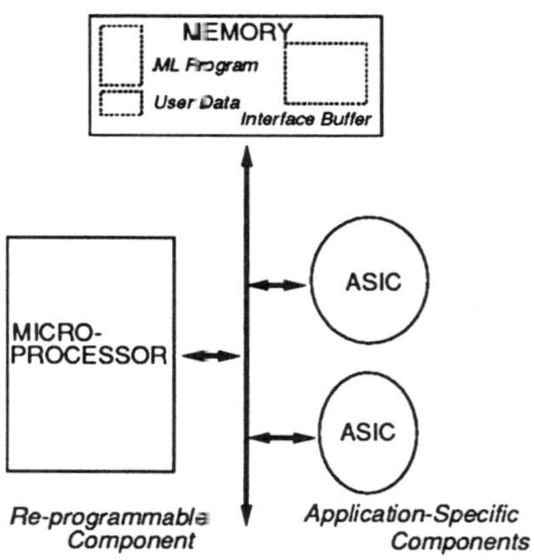

Figure 2: *Target System Architecture*

Target System Architecture

Figure 2 illustrates the *broad* features of the system architecture that is the target of our system synthesis approach. The target architecture consists of a general-purpose processor assisted by application-specific hardware components. The memory used for program and data-storage may be on-board the processor. However, the interface buffer memory needs to be accessible to the hardware modules directly. Because of the complexities associated with modeling hierarchical memory design, in this paper we consider the case where all memory accesses are to a single level memory, i.e., outside the reprogrammable component. The hardware modules are connected to the system address and data busses. Thus all the communication between the processor and different hardware modules takes place over a shared medium. Further, the mechanisms of data transfer between processor and application-specific components are constrained to those supported by the processor.

Concurrently executing hardware modules generate data that is consumed by the program(s) running on the processor. Due to the inherent serialization of operations implemented on the processor, any implementation must ensure that the data transfer across components takes place in an efficient manner that reduces the associated area and time overheads. The data transfer between the program(s) and the hardware modules is facilitated by the hardware and software synchronization mechanisms used. The choice of a data transfer scheme is determined by the individual execution rates of different models and control model used for scheduling various components. For example, a blocking transfer protocol may ensure correctness of the data-transfer but it may impose undue overheads on speed of execution of hardware modules. Whereas a non-blocking transfer may starve or overrun the interface buffers.

In the following sections, we outline our approach to system synthesis, and discuss different synchronization mechanisms used for synthesis for software and hardware components. The issue of synchronization between operations in a system model is closely related to the issue of communication between operations. For this reason, a choice for synchronization scheme is influenced by the selection of the communication scheme between system components. We first present the communication model supported in our synthesis system and then describe the synchronization schemes in context. We then address issues related to design of the hardware-software interface.

2 System Synthesis

Synthesis of application-specific hardware components under timing and resource constraints requires generation of a schedule of operations which satisfies the imposed timing constraints, an allocation of hardware resources that satisfies the resource constraints and finally the construction of a suitable control to facilitate hardware execution. Our model of hardware supports specification of data-dependent operations. Since data-dependent operations may offer unbounded delays it becomes necessary to schedule these operations dynamically. Therefore, we refer to data-dependent delay operations as points of synchronization in the system model. Our approach to synthesis of hardware under relative scheduling formulation has been described in detail elsewhere [6]. Briefly, the relative scheduling formulation makes it possible to achieve a data-driven dynamic schedule of operations with respect to a set of synchronization points (also referred to as *anchors* in [6]).

Here we focus on the problem of synthesis of the software component of the target system design. The software component is implemented as a program running on the onboard processor, i.e., the reprogrammable component. We assume that this program is small enough that it can be mapped to real memory, so that the issues related to virtual memory management can be ignored. As indicated in Figure 1, we start with a partition of the system graph model. System partitioning into hardware and software components is performed under the constraint that specified system input/output data rates can be supported by the final system implementation. One such partitioning approach relies on identifying and partitioning unbounded delay operations [7]. As a result of system partitioning we essentially have a set of concurrently executing hardware and software models. The software component consists of a set of concurrently executing routines, called *threads*. All threads begin with a point of synchronization and as such these are scheduled dynamically. However, within each thread of execution all the operations are statically scheduled. Therefore, for a given reprogrammable component the latency of each thread is known statically. As an example, data-dependent loops in software are implemented as a single thread with a data-dependent repeat count.

The problem of concurrent multi-thread implementation is well known [8]. In general, the program threads may be implemented either as a subroutines to a global task scheduler or as coroutines. However, in the context of mixed system designs where the processor is completely dedicated to the implementation of the system model and all software tasks are known statically, it is possible to use simpler and more relevant schemes to implement the software component. In the following, we present two schemes for implementation of the software component.

Software Implementation as Coroutines

Coroutines provide an attractive means of achieving concurrency between various program threads by reducing the cost of switching execution from one thread to another [9]. In this scheme, the reprogrammable component runs a task scheduler based on a priority assigned to various routines which are maintained in a co-operative, rather than hierarchical, relationship to each other. Each coroutine maintains a local state and *willingly* relinquishes control of the execution machine at points of synchronization. Coroutines provide a limited form of message passing via following two primitive operations: *resume* and *detach*. A coroutine switch consists in saving the current machine status and restoring the machine status of the next process to be executed. In the most general case, where any interruptions or exceptions may cause a context switch, all machine registers and flags should be saved. In case of an R/M processor, that is a processor that provides instructions with a register and memory operands such as 8086, the code for a coroutine based scheduler amounts to 34 instructions taking about 100 bytes. The coroutine switch takes 364 cycles when implemented for 8086 processor. By contrast, implementation of a global task scheduler using subroutines takes 728 clock cycles for the 8086 processor [10].

It is possible to reduce the overhead due to context switch if all the coroutine switches are explicit and known at the compile time. By making sure that during code optimization, variable lifetimes do not cross the coroutine boundaries, then the only register that needs to be saved is the *program counter* of the current coroutine and also only register that should be restored is the *program counter* of the next coroutine to be executed. The code for a such a scheduler on 8086 processor takes 103 cycles for each context switch. By comparison, on an load/store (L/S) machine, such as DLX [11], the code for task scheduler is reduced to 17 instructions (19 machine cycles), as opposed to the general case when all 64 registers would have to be saved requiring 192 instructions.

Software Implementation using Case Descriptions

In this approach, we merge different routines and describe all operations in a single routine using a method of description by cases [12]. This scheme is simpler than the coroutine scheme presented above. Here we construct a single program which has a unique case assignment for each point of synchronization. Thus each thread now corresponds to a case description of a rather large conditional in the final program. A global state register is used to store the state of execution of a thread. This method is restrictive since it precludes use of nested routines and requires description as a single switch statement, which in cases of particularly large software descriptions, may be too cumbersome. Overhead due to state save and restore amounts to 85 clock cycles for every point of synchronization when implemented on a 8086 processor. Consequently this scheme entails smaller overheads when compared to the general coroutine scheme described earlier. Corresponding overheads for the DLX processor amounts to 35 clock cycles for every point of synchronization.

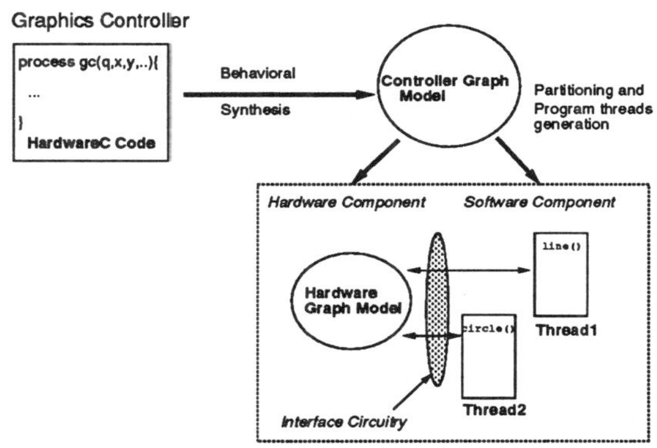

Figure 3: *System Synthesis Example*

In order to illustrate our system synthesis approach we consider synthesis of a graphics controller that provides for drawing of lines and circles given the end coordinates (and radius in case of a circle). Figure 3 illustrates some of the steps in synthesis of the graphics controller. The *HardwareC* description consisting of 457 lines of code is input to the behavioral synthesis phase. The resulting system graph model is input to *Vulcan-II*. As a result of system partitioning and program threads generation in *Vulcan-II*, the system design at this stage consists of interacting hardware modules modeled by the hardware graph models and a software component modeled by program threads. Next step is to synthesize the interface circuitry that would facilitate synchronization and communication between heterogeneous system components. Synthesis of interface circuitry is driven by the requirements imposed by system synchronization. We shall revisit this example in Section 4 to show how multiple program threads are synchronized with the concurrently operating hardware portions.

3 System Synchronization

A system design consists of various components which carry out operations in response to input data. An event refers to the execution of a data input/output operation. Synchronization in a system design refers to constraints on system design that ensure the partial ordering of events and operation executions imposed by the system model must be observed in *any* execution trace of the system model. Some synchronization constraints are needed to ensure correctness of the execution model, for example, all the data generated within the system model must be consumed in the time order in which it was generated. Typically this is guaranteed by appropriate choice of the execution semantics for the system model. Additional constraints may be needed to ensure correctness of a set of concurrently executing models. Further, some synchronization conditions may be externally imposed. For example, a certain *precedence* or *simultaneity* condition between execution of two operations imposed by the system control flow.

Communication Model

In the system graph model, communication between two operations is indicated by presence of an edge between respective operation vertices. When considering hardware synthesis, an edge between two operations may translate into either a physical wire connection, or it may be buffered and/or blocked to facilitate asynchronous communication. Final selection of data-transfer mechanism is made based on the data transfer requirement and how individual communicating models are implemented. However, note that in a mixed system implementation, due to inherently different rates of computation between hardware and software modules, it is necessary to allow multiple executions of individual models in order to achieve high system throughput. However, in presence of variation in rates of communication across different models appropriate buffering and handshake mechanisms may be required.

3.1 Software Synchronization

Our model of software component relies on the sequential execution of different threads of execution. Due to this serialization of the input system model, software synchronization is needed to ensure correct ordering of operations within the program threads and between different threads. A thread of execution already maintains an order of execution of its instructions, so a schedule of the operations is implicit to the sequential execution model of the instructions in a reprogrammable component. This solves the problem when a single thread of execution can be found for an entire description or among operations implemented in software belonging to the same thread - synchronization is only needed in points of synchronization and where the control is transferred between software and hardware. When data-dependent loops, and asynchronous message passing are present in the code, it may not always be possible to find a static schedule of the operations. If the order of execution can still be found, a single thread of execution could be determined that preserves the order in which the operations are executed. In case no such thread of execution can be determined, multiple threads of execution are required. In presence of multiple threads of executions (whether implemented as multiple programs or a single program using case descriptions described before) software synchronization consists of a mechanism to transfer control from one thread to another. In case of small number of threads, such a transfer can be done based on a statically defined priority of threads. For example, in case of two threads, control would simply switch from one thread to the other. In the general case, however, due to unbounded delay operations, we look for a *dynamic* scheduling of different threads of execution. Such a scheduling is done based on availability of data. Suppose we were to time stamp each data generated and also for each data request. Then the next thread of execution to be scheduled would be the one with the smallest request time stamp. Further, in order to maintain the correct order of data production and consumption, at any time the data being consumed is the one with the smallest time stamp. Such an scheme is implemented using a control FIFO that contains pointer to the next thread to be scheduled for execution [7]. Data transfer between two threads of execution can be implemented with shared memory or message passing. Shared memory can be facilitated by maintaining read and write pointers on each data-transfer. Such an scheme would add the overhead of maintaining and updating the read and write pointer for each data transfer across the program threads. Non-register based data-transfers (or data transfers which could culminate in control transfer) are well suited to be implemented as a queue connected with the control FIFO. On the other hand, register based transfers have the characteristic that once something is written, the data may be read many times. It is possible to use processor registers to transfer information between threads. However, such a scheme requires global register assignments which are not available for reassignment by the compiler. A limited form of message passing can be achieved by using co-routine model of implementation described before.

3.2 Hardware-Software Synchronization

Synchronization between hardware and software components is determined by the data transfer requirements between the sender and the receiver. A data transfer between two models can be either blocking or non-blocking. A blocking transfer protocol requires the sender(receiver) to block transfer until the corresponding receiver(sender) is ready to receive(send) data. Blocking can also be made conditional so as to reduce the associated timing penalties due to blocking [13]. With respect to their overheads, a non-blocking transfer consumes the system bus bandwidth, whereas a blocking transfer costs not only system bus bandwidth but also additional control lines that are needed to implement the required handshake. Therefore, for lower system costs, it is necessary to implement blocking only when absolutely necessary. A blocking transfer protocol can be thought of as a non-blocking transfer with an infinitely deep queue buffer. The queue size may be bounded by addition of handshake signals that treat queue as the sender or receiver of data. Alternatively, in presence of specific constraints on rates of data transfer, the queues can be sized algorithmically [14].

For a given data-transfer edge in the system graph model, we first attempt to determine the rates of data production and consumption associated with the sender and receiver models. Such a rate determination requires specification of data rates for external inputs and outputs. In case of unknown or varying data rates, a blocking protocol for both sending and receiving ends is selected. Either sender or receiver end of a transfer can be made non-blocking if it can be determined that the corresponding operations are always slower. In case of perfectly matched data-rates a synchronous non-blocking protocol is selected. After selecting transfer protocols for different data-transfers across the hardware and software models, the interface circuitry can be synthesized using asynchronous and synchronous logic synthesis techniques [15] [16]. For a description of the interface architecture the reader is referred to [7].

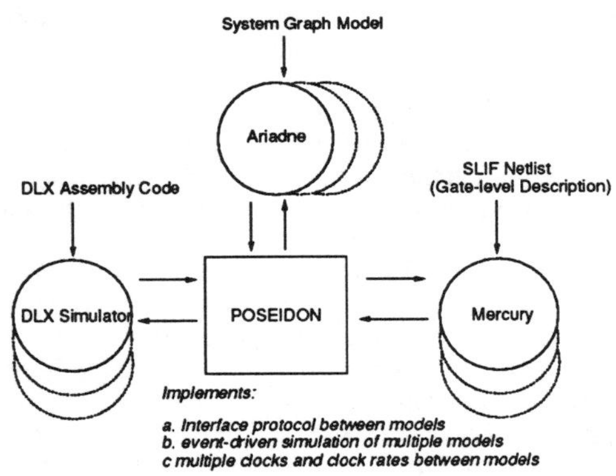

Figure 4: *Event-driven simulation of a mixed system design*

4 Simulation of Hardware-Software Systems

We have developed an event-driven simulator, named *Poseidon*, that performs concurrent simulation of multiple functional models

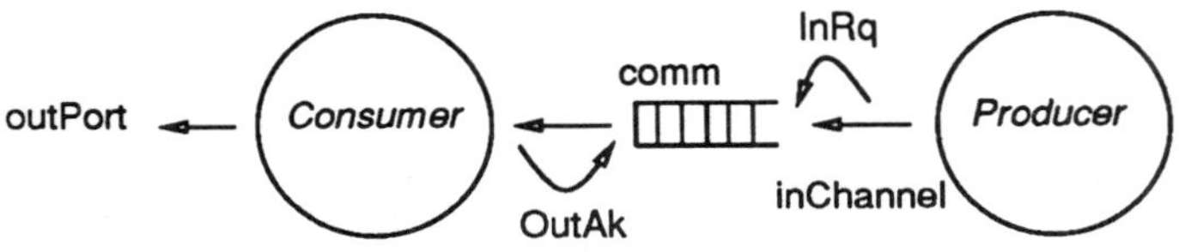

Figure 5: *Simulation Example 1*

implemented either as a program or as application-specific hardware. The software component is compiled into the assembly code of the target microprocessor. *Poseidon* currently supports simulation of assembly code for the DLX microprocessor, a RISC oriented load/store processor [11]. The hardware component of system design can be simulated either before or after the structural synthesis phase. The graph model before structural synthesis is simulated using program *Ariadne*. A gate-level description of the hardware component of system design is generated using structural synthesis techniques in program *Hebe* and simulated using program *Mercury*. Thus, *Poseidon* supports simulation of partially synthesized hardware modules along with the software component of the system design. *Poseidon* maintains an *event queue* which stores all simulation models sorted by their activation times. After simulating an event, the event is enqueued in the event queue. A system specification in *Poseidon* consists of following parts:

1. **Model declarations**: consists of declarations of the concurrently executing simulation models. Models can be either software or hardware models. Each model has an associated clock signal and clock cycle-time used for its simulation. It is assumed that the clock cycle-times are a rational multiple of each other. Further it is assumed that different models supply (latch) data at the interface using flip-flops at the interface edge-triggered by their respective clock signals.
2. **Model interconnections**: The interface between different system components is specified by *connections* among models. A connection between two models may be either a direct connection through a wire, or a port connection through a register or a queue. Queues can have multiple fanins and fanouts. Signal assignments indicate direct connections between respective models. For connections such as queues that require existence of additional control signals for synchronization, it is possible to group signals having identical synchronization requirements together for a given set of synchronization signals.
3. **Communication protocols**: Interface protocol for data-transfer is specified via *guarded* commands [17]. A guarded command is executed only when some precondition is true. Each precondition is specified as a logic equation of signal values and transitions. There are four commands recognized by the connection types. *Enqueue* and *dequeue* are used for queues port connections and *load* and *store* are used for register port connections.
4. **System outputs**: Outputs to be observed during simulation runs may be indicated by direct connections to the internal signals in the system model.

For illustration purposes, we consider a simple example of two models, `Producer` and `Consumer` connected by means of a finitely sized queue as shown in Figure 5. We consider two cases: one in which the producer model is implemented in software and consumer in hardware and the other in which producer, consumer implementations are reversed. Example 1 shows system specification for this example for the first case. The three first lines of the specification declare the models to be simulated. Model `io` models the external system inputs and outputs. The following parameter specifies the clock period of the clock signal associated with the respective model. A value of 3.0 for the consumer model

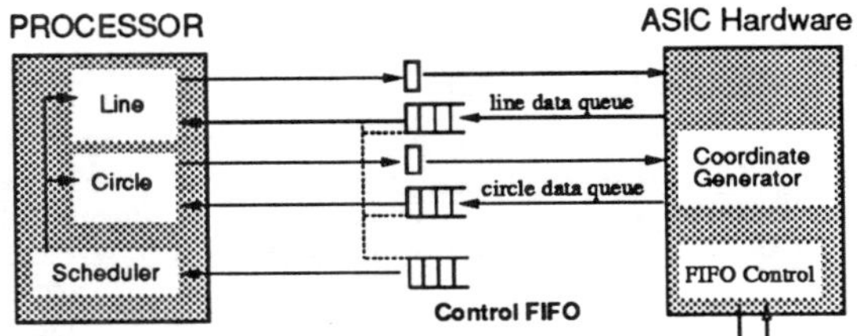

Software Component: ***main program***

```
int lastPC[MAXCRS]={scheduler,circle,line,main);
int current=MAIN;
int *controlFIFO = (int *) 0xaa0000;
int *controlFIFO_rq = (int *) 0xaa0004;

main(){
      resume(SCHEDULER);
};

int nextCoroutine;

scheduler(){
      resume(LINE);
      resume(CIRCLE);
      while (!RESET){
         do {
            nextCcroutine = *controlFIFO;
         } while ((nextCoroutine & 0x4) != 0x4);
       resume(nextCoroutine & 0x3); }
}
```

Figure 6: *Example 2: Graphics Controller Design*

indicates that consumer is implemented in an ASIC technology that uses a clock signal that is three times slower than the clock used by the reprogrammable component, which is usually a custom designed component. The system input/outputs are sampled here at the same rate as the consumer. The last two parameters specify the directory location where the model description can be found and the model name. The `queue` statement declares a queue named, `comm`, which is 4 bits wide and 3 words deep. We use `rq` and `ak` signals to implement a blocking communication protocol as indicated by the guarded commands. A '+' suffix indicates rising edge transition of the corresponding signal. A '-' suffix indicates falling edge transition. Symbols '&' and '!' indicate the boolean *and* and *not* operations.

Example 1: Specification of a producer-consumer pair (Figure 5).

```
# Models
model IO io 1.0 /local/ioDir IO;
model P dlx 1.0 /local/ProducerDir Producer;
model C mercury 3.0 /local/ConsumerDir Consumer;

# Connections
queue [4] comm[3];
C.RESET = IO.RESET;
C.r[0:0] = IO.r[0:0];

# Communication protocol
P.0xff004[0:0] = !comm.full;
C.b_rq = !comm.empty;
when (P.0xff000_wr+ & ! comm.full) do comm[0:3] enqueue P.0xff000[0:3];
when (C.b_ak+ & ! comm.empty) do comm[0:3] dequeue C.b[0:3];

# Outputs
IO.inChannel[0:3] = P.0xff000[0:3];
IO.outPort[0:3] = C.c[0:3];
IO.InRq = P.0xff000_wr;
IO.OutAk = C.b_ak;
```

In order to illustrate the effect of software and hardware-software synchronization mechanisms we now consider the design of the graphics controller introduced in Figure 3. Figure 6 shows the final implementation of the system design. The design consists of application-specific portions containing initial coordinate generators and control logic for controlFIFO and a software portion implemented on the DLX processor. The software component consists two threads of execution corresponding to the line and circle drawing routines. Both program threads generate coordinates that are used by the dedicated hardware. Input to *Poseidon* consists of gate-level description of the ASIC hardware, assembly code of the coroutines, and a description of the interface. Example 2 shows the *Poseidon* interface specification of the graphics controller design.

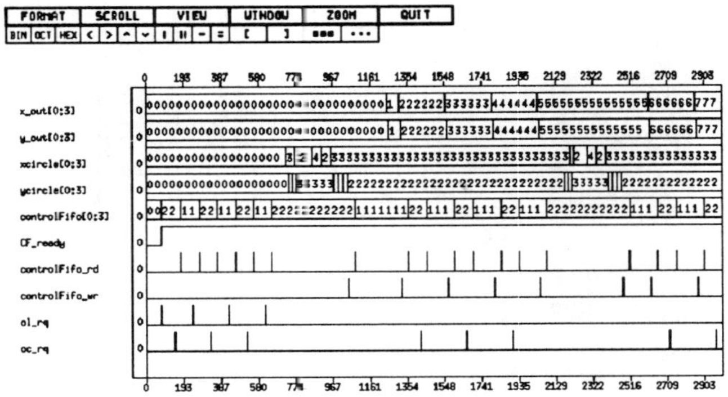

Figure 7: *Example 2: Simulation of Graphics Controller*

Example 2: Specification of the graphics controller interface (Figure 6).

```
model gc  io  1.0  DIR  GraphicsController;
model ccoord mercury 1.0  DIR  gcircle;
model lcoord mercury 1.0  DIR  gline;
model mp  dlx  1.0  DIR main;
model CF mercury 1.0 DIR control;

queue [1] lqueue[16], cqueue[16];
queue [3] controlFifo[2];

CF.r[0:0] = lcoord.run[0:0] = ccoord.run[0:0] = gc.run[0:0];
CF.RESET = lcoord.RESET = ccoord.RESET = gc.RESET;
CF.lrq[0:0] = !lqueue.empty;
CF.lak[0:0] = mp.0xff004_rd;
CF.crq[0:0] = !cqueue.empty;
CF.cak[0:0] = mp.0xee004_rd;
mp.0xee004[0:0] = !cqueue.empty;
mp.0xff004[0:0] = !lqueue.empty;

# Lqueue
when (lcoord.queue_rq+ & !lqueue.full) do lqueue[15:0] enqueue
lcoord.queue[15:0];
lcoord.queue_ak = !lqueue.full;

when (mp.0xff000_rd+ & !lqueue.empty) do lqueue[15:0] dequeue
mp.0xff000[15:0];
mp.0xff000[16:16] = !lqueue.empty;
...

# ControlFifo
when (CF.outline_rq+ & !controlFifo.full) do controlFifo[1:0] enqueue
outline[1:0];
CF.outline_ak = !controlFifo.full;
...

# Output specification
gc.x_out[7:0] = mp.0xff100[7:0];
gc.y_out[7:0] = mp.0xff104[7:0];
gc.controlFifo[1:0] = controlFifo[1:0];
gc.CF_ready = !controlFifo.empty;
...
```

Figure 7 shows some results of *Poseidon* simulations of the graphics controller. The hardware-software synchronization specified in Example 2 follows the scheme described in Section 3.2. The data-driven dynamic scheduling of program threads is achieved through the use of a 3-deep controlFIFO. In Figure 7, the circle and line drawing program threads are identified by id numbers 1 and 2 respectively. The program threads are implemented using the coroutine scheme described in Section 2. Signal `ol_rq` and `oc_rq` in Figure 7 indicate when the line and circle thread id's are being enqueued on data request from the software component. Variable time distance between `oc_rq` requests is due to data-dependent delay offered by the circle drawing routine in software. `controlFifo_wr` represents when a thread of execution enqueues its thread id in the control FIFO and then yields the execution to another thread. `x_out` and `y_out` are the coordinates generated by the line routine and `xcircle` and `ycircle` are the coordinates generated by the circle routine. `CF_ready` signals when the control FIFO is not empty and `controlFifo_rd` shows when the scheduler checks to see if the control FIFO is not empty (and eventually read it).

5 Conclusions

Synthesis using application-specific as well as reprogrammable components provides a promising extension of high-level synthesis approaches to realize complex system designs without corresponding increase in the magnitude of the synthesis tasks. Use of a reprogrammable component, however, poses interesting problems due to inherently serial nature of program execution that must interact with concurrently operating hardware portions. Thus synchronization between various components constitutes one of the most important issues in system synthesis.

We have presented an approach to synthesis of systems containing both application-specific and reprogrammable components and synchronization schemes that are used to facilitate data-transfer across concurrently executing system models. The resulting hardware and software components are simulated using an event-driven simulator, *Poseidon* which provides cycle-by-cycle simulation results. Since the selection of a synchronization scheme is driven by requirements of data-transfer rates, automatic selection of interface protocol can be made based on the data-rate constraints imposed on the system model. Work is underway to develop an automated tool to generate *Poseidon* interface specification and to synthesize appropriate interface circuitry once such a selection is made of all data transfers.

6 Acknowledgments

This research was sponsored by NSF-ARPA, under grant No. MIP 8719546 and, by DEC jointly with NSF, under a PYI Award program, and by a fellowship provided by Philips/Signetics. We acknowledge also support from ARPA, under contract No. J-FBI-89-101. The second author was partially supported by CNPq-Brazil under contract 200212/90.7.

References

[1] G. D. Micheli, D. C. Ku, F. Mailhot, and T. Truong, "The Olympus Synthesis System for Digital Design," *IEEE Design and Test Magazine*, pp. 37–53, Oct. 1990.

[2] J. Rabaey, H. D. Man, and *et. al.*, "Cathedral II: A Synthesis System for Multiprocessor DSP Systems", in *Silicon Compilation*, editor: D. Gajski , pp. 311–360. Addison Wesley, 1988.

[3] D. Thomas, E. Lagnese, R. Walker, J. Nestor, J. Rajan, and R. Blackburn, *Algorithmic and Register-Transfer Level: The System Architect's Workbench*. Kluwer Academic Publishers, 1990.

[4] R. Camposano and W. Rosenstiel, "Synthesizing Circuits from Behavioral Descriptions," *IEEE Transactions on CAD/ICAS*, vol. 8, no. 2, pp. 171–180, Feb. 1989.

[5] M. B. Srivastava and R. W. Broderson, "Rapid-Prototyping of Hardware and Software in a Unified Framework," in *Proceedings of the International Conference on Computer-Aided Design*, (Santa Clara), pp. 152–155, 1991.

[6] D. C. Ku and G. D. Micheli, "Synthesis of ASICs with Hercules and Hebe", in *High-level VLSI Synthesis*, editors: Raul Camposano, Wayne Wolf, pp. 177–203. Kluwer Academic Publishers, 1991.

[7] R. K. Gupta and G. D. Micheli, "System-level Synthesis Using Re-programmable Components," in *Proceedings of the European Design Automation Conference*, Mar. 1992.

[8] G. R. Andrews and F. Schneider, "Concepts and Notations for Concurrent Programming," *ACM Computing Surveys*, vol. 15, no. 1, pp. 3–44, Mar. 1983.

[9] M. E. Conway, "Design of a Separate Transition-Diagram Compiler," *Comm. of the ACM*, vol. 6, pp. 396–408, 1963.

[10] R. K. Gupta and G. D. Micheli, "System Synthesis via Hardware-Software Co-design," CSL Technical Report CSL-TR, Stanford University, 1992.

[11] J. L. Hennessey and D. A. Patterson, *Computer Architecture: A Quantitative Approach*, ch. 3. Morgan-Kaufmann, 1990.

[12] P. J. H. King, "Decision Tables," *The Computer Journal*, vol. 10, no. 2, Aug. 1967.

[13] I. C. Wand and A. J. Wellings, *Distributed Computing, F. B. Chambers et. al. editors*, ch. 14: Programming Languages, pp. 201–215. Academic Press, 1984.

[14] T. Amon and G. Borriello, "Sizing Synchronization Queues: A Case Study in Higher Level Synthesis," in *Proceedings of the 28th Design Automation Conference*, June 1991.

[15] T. H. Meng, *Synchronization Design for Digital Systems*, ch. Synthesis of Self-Timed Circuits, pp. 23–63. Kluwer Academic Publishers, 1991.

[16] G. Borriello and R. Katz, "Synthesis and Optimization of Interface Transducer Logic," in *Proceedings of the IEEE Transactions on CAD/ICAS*, Nov. 1987.

[17] E. W. Dijkstra, "Guarded Commands, Nondeterminacy, and Formal Derivation of Programs," *CACM*, vol. 18, no. 8, pp. 453–457, Aug. 1975.

High-Level Synthesis with Pin Constraints for Multiple-Chip Designs

Yung-Hua Hung[†] Alice C. Parker

Department of Electrical Engineering - Systems
University of Southern California
Los Angeles, CA 90089-0781

Abstract

In this paper, we describe an approach to data-path synthesis from a partitioned behavioral-level description into a multiple-chip pipelined RTL design. The problem is divided into *interchip connection determination* and *scheduling*. A heuristic search technique is described for interchip connection determination. Synthesis results are presented.

1 Introduction

Most modern digital systems do not fit into a single chip. Interchip communication is important and should be considered at an early stage.

Pin constraints should be taken into account during the process of high-level synthesis for multiple-chip designs. However, many current data path synthesis tools make an implicit assumption that every input value is ready whenever it is required. In general, inputs cannot be assured to be available at any time due to the limited number of I/O pins. Some tools can take pin limitations into account implicitly by the use of time constraints on inputs and outputs.

We describe an approach to multi-chip data path synthesis, given a behavioral description which has already been partitioned into a number of clusters, with the feasibility of clusters determined. A pipelined RTL design consisting of multiple chips is produced by our software. Also, a set of communication buses among the chips is determined, and the values to be transferred are scheduled on the pins and buses. The RTL design produced satisfies user-supplied constraints, including the number of I/O pins on individual chips.

2 Related Research

Most current high-level synthesis techniques, e.g. [6, 5, 2], do not consider communication across chips. Only one research project explicitly focusing on high-level synthesis of multiple-chip designs can be found in the literature. De Micheli et al. [3, 1] applied an iterative approach consisting of resource binding, non-pipelined scheduling, and partitioning. In their approach, resource binding is performed before scheduling. In the binding stage, operations are bound to specific resources. In each iteration, one of the binding configurations is selected, either by an exhaustive search or a heuristic search with some cost criteria. Once a resource binding is selected, the operations bound to the same resource are serialized to resolve resource conflicts. Branch-and-bound is used to explore the alternatives. Then, a variation of a control data flow graph (CDFG) containing binding and serialization information is partitioned into multiple chips. Pin sharing among I/O operations is not reported, as the pin cost, which is used as one of the constraints in the process of partitioning, is computed by adding the costs of all I/O operations in the partition.

[†]Sponsored by Computer & Communication Laboratories, Industrial Technology Research Institute

3 Problem Statement

The problem inputs include (1) a CDFG which has been partitioned into a number of clusters, each of which will be implemented on a separate chip; (2) a set of hardware modules, with cost and delay parameters, which will be used to implement the operations in the CDFG; (3) constraints on cost and pin count for each individual chip, or a performance requirement of the whole design; and (4) a specified goal of maximizing performance of the whole design, or minimizing the total cost of the design, depending on what constraints are given.

The problem outputs include (1) a schedule for the functional operations and I/O transfers; (2) a set of interchip connections; and (3) an assignment of I/O operations to communication buses.

A global clock is used to synchronize all functional operations and I/O transfers. Two minor clocks, one for data manipulation and one for I/O transfers, can also be handled. Both clocks must be derived from the global clock, and the initiation interval must be a multiple of the period of the global clock.

I/O transfers are activated at the beginning of a clock cycle, and must be completed in a single cycle. When two minor clocks are used, I/O transfers can be activated at the beginning of a minor clock cycle and must be completed in the same minor cycle.

A distributed control mechanism is assumed. Otherwise, a large number of I/O pins would have to be reserved for control signals. Also, control signals traveling off-chip would cause delays in the control path.

0738-100X/92 $3.00 © 1992 IEEE

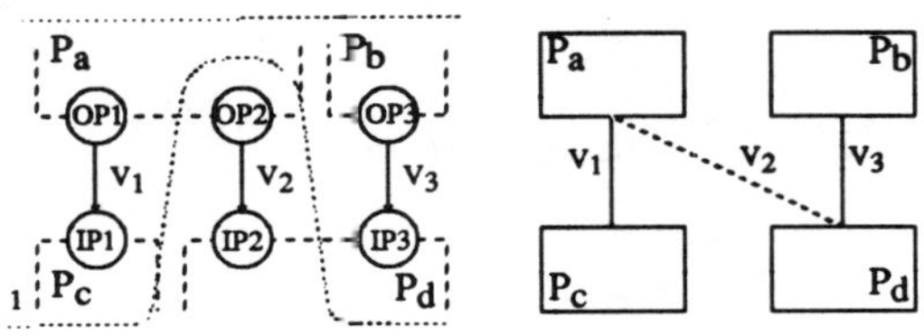

(a) Partially scheduled CDFG (b) Partial interchip connection

Figure 1: A partial schedule and interchip connection

4 Scheduling Difficulties

Scheduling I/O operations is quite different from and harder than scheduling functional operations because no switching elements are allowed on interchip connections. A CDFG with I/O operations is shown in Figure 1(a). We assume that each of the chips P_a, and P_b has 1 output pin, and each of the chips P_c, and P_d has 1 input pin. If I/O operations *OP1, IP1, OP3*, and *IP3* are scheduled in control step 1, as shown in Figure 1(a), I/O operations *OP2* and *IP2* are unable to be scheduled in any one control step. This can be illustrated by the partial RTL design as shown in Figure 1(b) corresponding to the partial schedule. Transferring value V_2 requires a connection between the output pin of partition P_a, and the input pin of partition P_d, as indicated by a dotted line in Figure 1(b). However, the connection would cause the transfers of the values V_1 and V_3, which occur in the same control step, to conflict even though partitions P_a, and P_d have 1 output pin, and 1 input pin, respectively, unallocated in control step 2. For the given constraints, the design requires 3 control steps. Scheduling an I/O operation in a control step in which there are unallocated I/O pins does not guarantee that there will be no communication conflict in the schedule. Moreover, scheduling an I/O operation in a control step, even with no communication conflict, can make achieving a completed schedule impossible.

Values can have a variety of bit widths. The manner in which I/O pins are allocated for transferring these values efficiently is an even harder problem.

5 Problem Approach

An I/O operation is modeled as a single I/O operation node which consists of both one output operation for a partition and one input operation for another partition. For a value required by more than one partition, there are several I/O operation nodes, one for each requiring partition, since a value is allowed to be output several times if necessary.

A cost is associated with each I/O operation to indicate the number of input or output pins required to transfer a value. There is a delay associated with each I/O operation, defined as the duration from the beginning of the clock cycle where the I/O operation is activated to the time the value becomes valid at the destination. An estimated delay is assumed for all I/O operations, because the actual delays of output drivers and interchip connections are unknown *a priori*.

Every communication bus can only be used to transfer at most one value in one control step. Values with smaller bit width are not grouped and transferred on a single communication bus. Also, values have to be transferred as a whole[1]. The problem has been approached in two phases:

1. determining the interchip connections before scheduling, and

2. scheduling given the interchip connections.

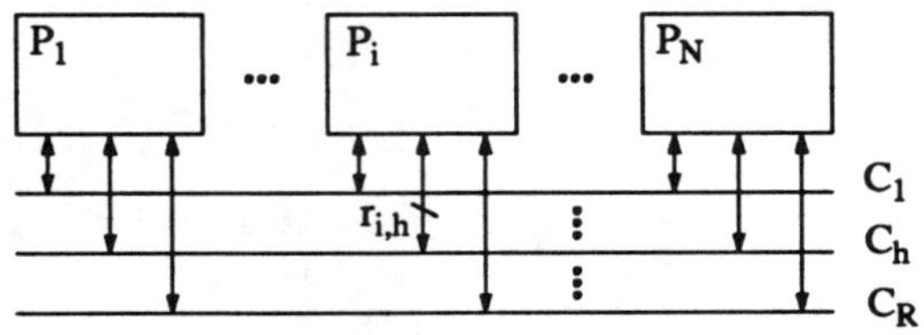

Figure 2: Interchip connection model

```
determine_interchip_connection
L = a sorted list of I/O's in descending order of bit width
C = assign_nodes ( L )

assign_nodes ( L )
if ( L = φ ) then return φ
w = first element of L
L' = L - w
select some of communication buses for w having best gain g
for ( each communication bus C_h ) do
   if ( there are I/O pins available to assign w to C_h ) then
      assign w to C_h tentatively and update connection table
      C' = assign_nodes ( L' )
      if ( C' ≠ FAIL ) then return C_h + C'
      restore connection table
   endif
endfor
return FAIL
```

Figure 3: Procedure to find an interchip connection

5.1 Determining Interchip Connections

The interchip connection model has the bus structure shown in Figure 2. In the figure, $r_{i,h}$ is the number of pins of partition P_i connected to bus C_h. In our model, each bus can be connected to more than one partition. The number of pins of a partition connected to a bus is not necessarily the same as the number of pins of other partitions connected to the bus.

The problem was formulated as an ILP and solved. However, for practical size problems, the size of the formulation is too large to obtain a solution within a reasonable time. So, a heuristic search technique has been developed as shown in Figure 3.

At the beginning, the I/O operations are sorted. An I/O operation with a higher bit width requires more I/O pins, and is assigned before the others. At each level of the search tree, only a small number of communication buses having the best gain are explored. For

[1] This assumption is being relaxed in ongoing research.

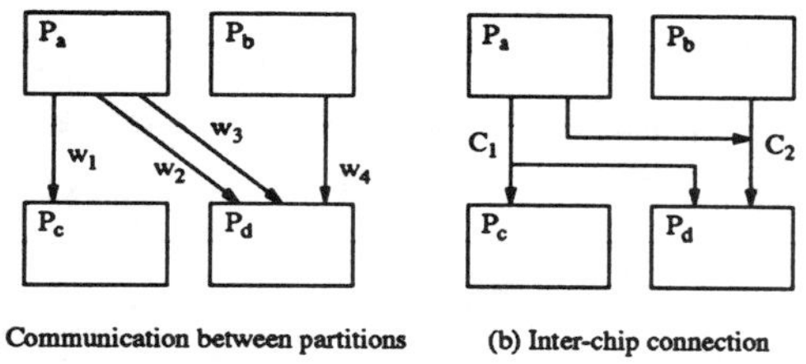

Figure 4: An example of an interchip connection

run-time efficiency, the communication buses to be explored do not have the same topology. The number of branches to be explored at each node is controlled probabilistically by an average branching factor, which is given by users to trade off the execution time and the chance of finding a solution. The search time is still exponential in term of the number of I/O operations in the worst case if the average branching factor is greater than 1.

The gain, g, of assigning I/O operation $w = (P_i, P_j)$ to communication bus C_h is

$$g = 10000 * g_1 + 100 * g_2 + g_3.$$

g_1 is used to make the I/O operation try to utilize an already existing communication path in order to save pin resources, and is given as

$$g_1 = \begin{cases} 0 & \text{if } r_{i,h} = 0 \wedge r_{j,h} = 0 \\ wf_i & \text{if } r_{i,h} \neq 0 \wedge r_{j,h} = 0 \\ wf_j & \text{if } r_{i,h} = 0 \wedge r_{j,h} \neq 0 \\ wf_i + wf_j & \text{otherwise} \end{cases}$$

where

$$wf_i = \begin{cases} ratio_i & \text{if } ratio_i > 1 \\ 0 & \text{otherwise} \end{cases}$$

$$ratio_i = \frac{\#\text{bits of unassigned I/O operations in } P_i}{\#\text{unallocated pins in } P_i}$$

Priority is given to the partition with a tight pin resource. This is reflected by the weighting factor wf_i.

g_2 is used to force the I/O operations transferring the same value to be assigned to a common communication bus.

$$g_2 = \begin{cases} 1 & w \text{ and } w' \text{ assigned to } C_h \text{ transfer same value} \\ 0 & \text{otherwise} \end{cases}$$

g_3 tries to balance the use of communication buses, and is the number of unassigned slots in C_h.

Every I/O operation is tentatively assigned to a communication bus in the procedure of determining interchip connection.

5.2 Scheduling given Connections

After the interchip connection has been determined, data path and I/O scheduling can be started. A list scheduling technique is used. Before scheduling each I/O operation w in control step s, any communication bus available in the control step for this I/O operation is checked. A direct way of checking availability of communication buses for the I/O operation is to check whether the communication bus previously assigned to the I/O operation is free in the control step. However, the resulting schedule using the static assignment can be too pessimistic. Therefore, it is of benefit to reassign communication buses dynamically as the scheduling proceeds. The problem can be modeled as a matching problem of a bipartite graph $G = (V, U, E)$, in which V is a set of nodes representing I/O operations, U is a set of nodes representing communication slots, which are 2-tuples (C_h, s_k), where C_h is a communication bus, and s_k is a group of control steps s satisfying $k = s \bmod L$. E is a set of edges representing the capability of a communication bus to be used to perform an I/O operation. The process of checking for a free communication bus to assign an I/O operation to at a control step and reassigning I/O operations to communication buses if necessary is similar to the one of finding an augmentation path in the matching problem. The reassignment of I/O operations to communication slots within a communication bus can be done by simply switching the assignment. So, the communication slots within a communication bus are grouped to eliminate unnecessary search. The reassignment of I/O operations for the example in Figure 4 is shown in Figure 5. A thin line indicates that the communication bus is capable of performing the I/O operation. A heavy line indicates that the I/O operation is currently assigned to the communication bus. A shaded line indicates that the assignment has been fixed because the communication bus has been allocated for an I/O operation at a control step. Circles in ellipses show the status of communication buses at certain control steps. An empty (filled) one indicates that the bus is (not) free at the control step. The dotted line in Figure 5(b) corresponds to an augmentation path in the matching problem.

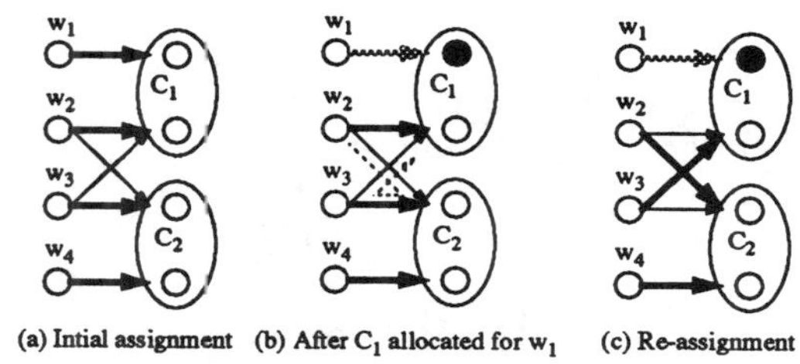

Figure 5: Snapshots of communication bus allocation

6 Results

The AR lattice filter element [4] is partitioned as shown in Figure 6, and a variety of bit widths is assumed. The number next to the I/O operation is the bit width. The I/O operations without numbers next to them have 8 bits. Chip 1 has 100 pins, and chips 2 and 3 have 90 pins. The stage time is 250 ns. Inputs arrive every 3 cycles. I/O operations take 10 ns. Additions are performed by adders with 30 ns delay. Multiplications are performed by multipliers with 210 ns delay. A minimum number of functional units are used. Chained operations are allowed.

The required functional operators for all partitions are 2 adders and 2 multipliers. The partitioned AR

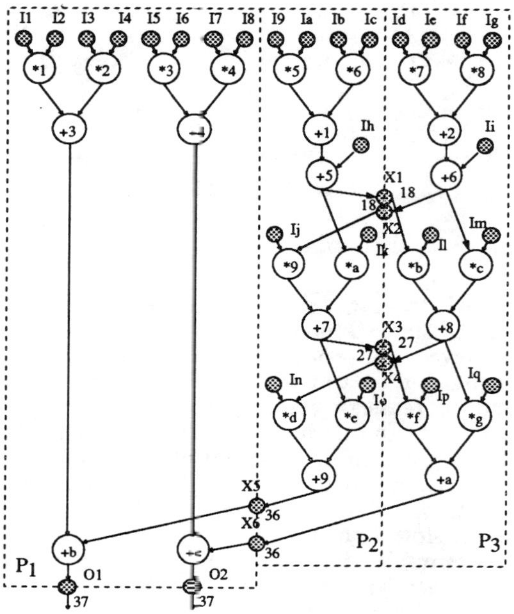

Figure 6: A partitioned AR filter

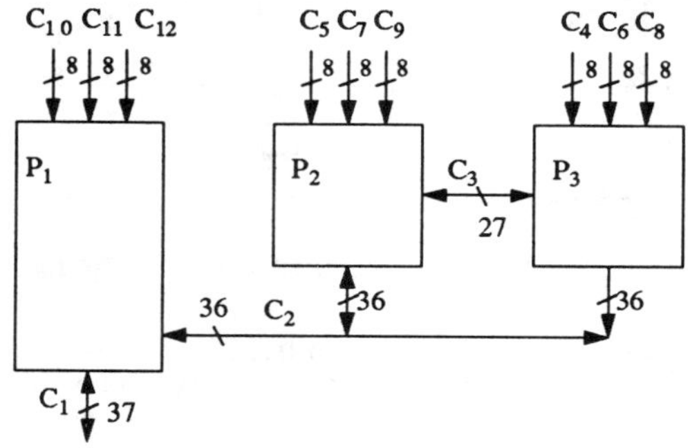

Figure 7: Interchip connection for partitioned AR filter

Control step	Partition P_1	Partition P_2	Partition P_3
0	I5 I6 I7 I8 *3 *4 +4	Ia Ib Ic *6	Ie If Ig *8
1	I2 I3 I4 *2	I9 Ih Ik *5 +1	Id Ii Im *7 +2
2	I1 *1 +3	Ij In Io *a +5	Il Ip Iq *c +6
3		X1 X2 *9 +7	X1 X2 *b +8
4		X4 *d	X4 *g
5		X3 *e +9	X3 *f +a
7	X5 +b	X5	
8	O1 X6 +c		X6
10	O2		

Table 1: Scheduling of partitioned AR filter

Comm. Bus	Initial Assignment	Final Assignment
C_1	I8 O1 O2	I8 O1 O2
C_2	X3 X5 X6	X1 X5 X6
C_3	X1 X2 X4	X2 X3 X4
C_4	Im Ip Iq	Ie Im Iq
C_5	Ik In Io	Ia Ik Io
C_6	Ig Ii Il	Ig Ii Il
C_7	Ic Ih Ij	Ic Ih Ij
C_8	Id Ie If	Id If Ip
C_9	I9 Ia Ib	I9 Ib In
C_{10}	I5 I6 I7	I2 I7
C_{11}	I2 I3 I4	I1 I4 I5
C_{12}	I1	I3 I6

Table 2: I/O operation to bus assignment

filter is scheduled as shown in Table 1. The interchip connection is shown in Figure 7. The actual numbers of pins required are 97, 87, and 87, respectively. The pin allocation is optimal in term of total number of I/O pins. The initial and final assignments of I/O operations to communication buses for the schedule are shown in Table 2. I/O operations have been reassigned to improve the performance. The number of control steps would be 12 instead of 11 otherwise. The CPU time is 0.2 sec. on a Sun 3/280.

References

[1] G. De Micheli and *et al*, "Partitioning of Functional Models of Synchronous Digital Systems," *Proc. of ICCAD*, Nov. 1990.

[2] C. Hwang, Y. Hsu, and Y. Lin, "Scheduling for Functional Pipelining and Loop Winding," *Proc. of the 28th DAC*, pp. 764-769, June 1991.

[3] D. Ku and G. De Micheli, "High-level Synthesis and Optimization Strategies in Hercules and Hebe," *EURASIC, Proc. of the European Conference on ASIC design*, May 1990.

[4] S. Y. Kung, "On Supercomputing with Systolic/Wavefront Array Processor," *Proc. of the IEEE*, pp. 867-884, July 1984.

[5] P. G. Paulin and J. P. Knight, "Force-Directed Scheduling for the Behavioral Synthesis of ASIC's," *IEEE Trans. on CAD*, vol. 8, pp. 661-679, June 1989.

[6] N. Park and A. C. Parker, "Sehwa: A Software Package for Synthesis of Pipelines from Behavioral Specifications," *IEEE Trans. on CAD*, vol. 7, pp. 356-370, March 1988.

Partitioning by Regularity Extraction*

D. Sreenivasa Rao and Fadi J. Kurdahi
Department of Electrical & Computer Engineering,
University of California, Irvine, CA 92717.

Abstract

It is well known that proper usage of regularity in digital systems leads to efficient as well as economic designs. We examine this important question of regularity extraction, and present a general methodology for doing so. We consider system-level partitioning to demonstrate how regularity can lead to reduced design costs. A prototype system based on these ideas has been built; and some real-life examples are considered for testing. Research is underway to explore some of the other promising applications that such a system could have.

1 PRELIMINARIES

1.1 Introduction

An important feature of practical digital system designs that a variety of CAD tools try to exploit is *regularity*. Several computationally powerful circuits (like systolic arrays, multipliers, etc.) that are being increasingly used in VLSI are characterized by inherent regularity. Further, since regularity is not an apparent feature of typical digital system descriptions, a tool that can extract regularity could be very useful in various areas of CAD. In the current paper, we present a general methodology for extracting regularity at any level of hierarchy, and examine the problem of system partitioning with respect to this concept. If we have a large digital system to be designed as a collection of various chips, the design cost would be greatly reduced by reducing the number of *distinct* chips to be designed, and constructing the whole system by repeating these few chips as many times as possible. Therefore, the designer is faced with the task of selecting the *components* of a digital system (by which we mean a collection of hardware units) in such a way that as few *different types* of them as possible are used, and for obvious reasons, as few *total number* of them as possible. Since this process takes into account factors like area bound, pin bound, critical path delay, etc., the designer will have effectively partitioned his system to optimize costs.

1.2 Previous Related Work

The problem of identifying regularity has not been extensively researched in the past, but there has been some previous related work that needs to be mentioned. One of the earliest works that addressed this problem was Kodres [1]. A sub-optimal greedy matching procedure for silicon compilers is described in [2]. A later effort was by Keutzer [3], who modeled the system as a directed, acyclic graph (DAG) and utilized compiler design techniques to match trees (rather than graphs), thus achieving linear time operation at the cost of optimality. More recently, [4] reported a rule based approach to construct a digital system totally from off-shelf components, while minimizing the *part count*. [5] utilizes this concept for logic synthesis, by extracting *factors* to decompose sequential FSM's, and replacing the original FSM with calls to these factors.

Partitioning has been extensively used for placement, routing and related physical design issues [6]. With the growing interest in high level synthesis, researchers have considered partitioning at functional and structural levels. One of the earliest works in the area of behavioral partitioning was [7]. Other important contributions to partitioning methods were made in [8], [9], [10], [11].

2 PROBLEM FORMULATION

A system architecture typically consists of hardware modules and their interconnections. We shall refer to a small collection of these modules as *templates*, indicating the fact that similar ones are to be identified in the circuit. Thus, there are three distinct steps in the process : *(a)* a suitable set of templates have to be *identified*; *(b)* the designer needs to minimize the number of such *distinct* templates used in the circuit; and *(c)* he also needs to minimize the *total number* of all templates used. We model this problem as a directed graph, where each node represents a hardware module, with an attached *label* to indicate its type. An important contribution of this work is that it allows *cyclic* directed graphs for modelling, unlike some previous works which allowed only DAG's [3]. We state the problems formally as follows:

Problem 1 : *"Given a directed, labelled graph G(N,A); a library of templates, each of which is a directed labelled graph $T_i(V,E)$; find all those subgraphs of G that are isomorphic to T_i."* This problem is essentially the subgraph isomorphism problem, with the simplification that we have a directed, labelled graph.

Problem 2 : *"Given an infinite number of each of a set of templates O = $T_1,\ldots, T_k$ and an overlapping set of subgraphs of the given graph G(N,A) which are isomorphic to some member of O; minimize k (ie., the number of distinct templates used) as well as $\sum x_i$ (ie., the total number of*

*This work was supported by NSF Grants # MIP-8909677

0738-100X/92 $3.00 © 1992 IEEE

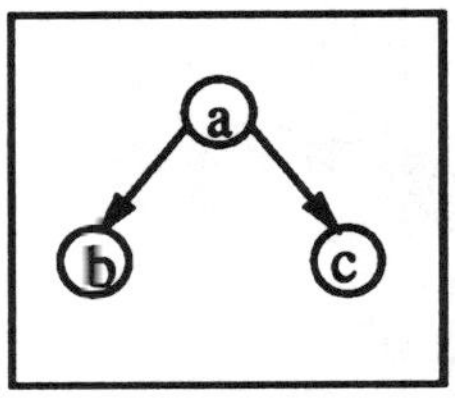

Figure 1: Digraph with K-formula **abc*

templates used, x_i being the number of times the template T_i is used) such that the number of nodes left uncovered is the minimum." This problem, although deceptively similar to the well known graph cover problem and the coin-changing problem, is actually much more complicated. In the following sections, we propose heuristics to efficiently solve the above problems.

3 SINGLE TEMPLATE MATCHING

3.1 Identification of Subgraphs

The first step in trying to efficiently determine the subgraphs of the given graph G(N,A) isomorphic to the reference template is to overcome the need for brute-force search of the whole graph in a way so that the optimality is not totally lost. An intuitive way to do this is to reduce the search domain by efficiently identifying non-overlapping subgraphs of G which have a high likelihood of having the patterns of the template under consideration. We start with the directed, labelled graph of the template, and designate one of the nodes as a *lead node*, which works as a "seed" for subgraph identification. The salient features of **Algorithm A.1**, *identify_subgraphs()* are described here [12]: We start with the lead node *v* and compute the *eccentricity(v)*. Any node *n* such that *distance(v,n)* $\leq \delta$ ($\delta >$ *eccentricity(v)*) is included in the subgraph. This, when repeated for all nodes in the system graph of the same type as *v* gives the required subgraphs.

It can be shown that the above algorithm has a linear time complexity in the number of nodes of the system graph. We will now discuss an efficient representation of directed graphs, and then present a heuristic to perform the template matching.

3.2 Linear Representation of Graphs

We present in this subsection a brief overview of a scheme proposed by Berztiss [13] that facilitates a linear representation of directed graphs. Berztiss proposed that directed graphs can be represented by a prefix notation as linear strings which he calls *K-formulas*. Informally, a K-formula is a combination of operator symbols called *K-operators* (represented by '*') and the node names. The simple graph shown in Fig. 1 would have the K-formula **abc, meaning that there are arcs originating from node *a* and terminating on nodes *b* and *c*.

Definition : Consider a formula $s_1 s_2 .. s_i .. s_m$. Let k_i and n_i denote the number of K-operators and the node symbols, respectively, in the sub-formula $s_1 s_2 .. s_i$. The formula is a K-formula *iff* the following conditions are satisfied:

$n_i \leq k_i$, i=1,2,....m-1.

$n_m = k_m + 1$.

We will now need a method to construct the K-formula of any given arbitrary digraph. The algorithm used for this relies on two key rules: The *substitution rule*, which allows us to combine two different K-formulas; and the *check step*, which ensures that the final set of K-formulas obtained is a minimal set. The algorithm [12] first constructs K-formulas for every pair of nodes (every arc $< a, b >$ has the K-formula *ab), and applies the substitution rule to these until it can no longer be applied. The check step ensures the minimality of the K-formula generated.

3.3 Practical Considerations

As can be easily seen, the linear representation of digraphs has the potential problem that it could lead to an exponential number of K-formulas. For example, the K-formula for the graph in Fig. 1 could as well have been written **acb instead of **abc. To overcome this, we follow a certain *pin ordering* for each hardware module in the design. For example, if in Fig. 1, module *a* has two output pins ordered as (p_0, p_1) then the K-formula of the digraph has to be **abc rather than **acb by virtue of the pin ordering. We also adhere to the convention of numbering the nodes in a breadth-first traversal (for the subgraphs as well as the reference template). This way, the number of alternate K-formulas is greatly reduced. The current implementation of the algorithm also allows for *partially configured templates*, so that templates with an almost match are accepted subject to certain criteria [12].

3.4 String Matching Algorithm

From the K-formulas of the system subgraphs and reference templates as obtained above, we need to determine if the pattern of the template occurs in the system subgraphs. This problem is one of string matching, although standard string matching algorithms cannot be applied here. This is because a K-formula is a string with two different types of characters. We use an $O(n^2)$ algorithm for this purpose, where n is the quantity $|V| + |E|$, or the total number of nodes and arcs in the graph. We note that, in general, a K-formula can be expressed as $* * * \cdots * a\alpha_1 \cdots \alpha_k$. Here, there are k leading K-operators, a leading node a, followed by k subformulas represented by α_1, α_2,....., α_k. As a first step, we need to identify the subformulas of a given K-formula, a process we refer to as *parenthesizing*.

Based on these ideas, we sketch **Algorithm A.2**, *string_match()*, as follows : We start by parenthesizing the K-formula of the subgraph and of the reference template. We then "slide" the template K-formula over that of the subgraph and look for a match of the leading node of template with a leading node of one of $\alpha_1 .... \alpha_k$, if the original leading node a does not match. If the leading node of say α_i has matched, we go to the sub-formulas of α_i, and match with subsequent nodes. If none of the leading nodes of $\alpha_1 .... \alpha_k$ match, then we go on to the next level of hierarchy and repeat the procedure.

4 MULTIPLE TEMPLATE MATCHING

The single template matching procedure can be repeated to obtain matches for all the templates that the designer has. Thus the trade-off involves choosing the templates in such a way as to minimize *(a)* the nodes left uncovered, or the *residue nodes*; *(b)* the number of distinct templates chosen; and *(c)* the total number of templates required to produce this minimum residue cover. Further, since performance is an important consideration, the choice of the templates has to

Nodes	T_1			T_2		T_3		
	T_1^1	T_1^2	T_1^3	T_2^1	T_2^2	T_3^1	T_3^2	T_3^3
v_1		1		1		1		
v_2		1		1				
v_3	1			1				1
v_4			1		1			1
v_5	1				1		1	
v_6			1		1		1	
v_7						1		

(a)

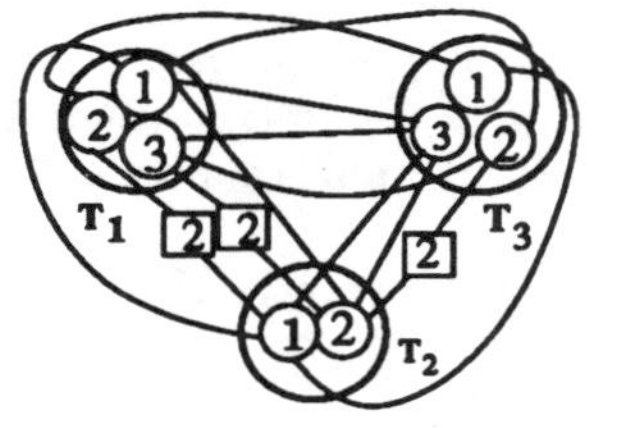

(b)

Figure 2: (a) Node cover table; and (b) its template constraint graph

be such that the delay along the critical path is minimized. The current implementation provides for user control of the relative importance of the performance vs. the above three considerations.

We propose in this section an essentially greedy approach to solving the complex problem described above. We differentiate between a template type T_i and its instantiation, which we denote by an additional superscript as T_i^m. We start by constructing a *node cover table*, as shown in Fig. 2(a). We have a '1' entry if a particular template member T_i^m covers a node v_j, and a '0' if not. We also construct the *template constraint graph* by considering each template instance as a node, and joining two templates with an n-node overlap by an edge of weight n. (Fig. 2(b))

We briefly describe here **Algorithm A.3**, *multiple_template_match()*, details can be found in [12]. If performance is given a high priority, we mark all the templates on the critical path and choose those that minimize the path length. Otherwise, we start by choosing the template type that is repeated the maximum number of times, and cover the graph with all its instances that do not overlap any other templates. We do this for all template types, considering them in decreasing order of their cost. We then move on to those template instances that have an overlap of 1, 2, etc. nodes with others, again considering the template types in decreasing order of cost. It can be shown that the formation of the template constraint graph takes time $O(m^2.n)$, where m is the number of distinct templates. Apart from the sorting procedures (that take $O(p \log p)$ time), the rest of Algorithm A.3 takes approximately linear time in p, where p is the number of nodes in the template constraint graph.

5 TEMPLATE GENERATION

Hitherto, the heuristics developed have assumed that the templates are made available by the user for the algorithm to work on. In a general case, an automatic partitioning algorithm must determine the best possible templates by itself before coming up with the matchings. We model this problem as a bin-packing problem and solve it using the first-fit approach. We assign each node we come across in a specific search order to the first bin, and when the bin is full we consider this collection of nodes and the inclusive edges to be our first template. We characterize the bin not only in terms of its area but also its external pin count. We further allow a margin for the area and pin attributes of the bin, ie., we accept the contents of the bin as a valid template if the area and pin count match the bin's capacity to within a certain threshold.

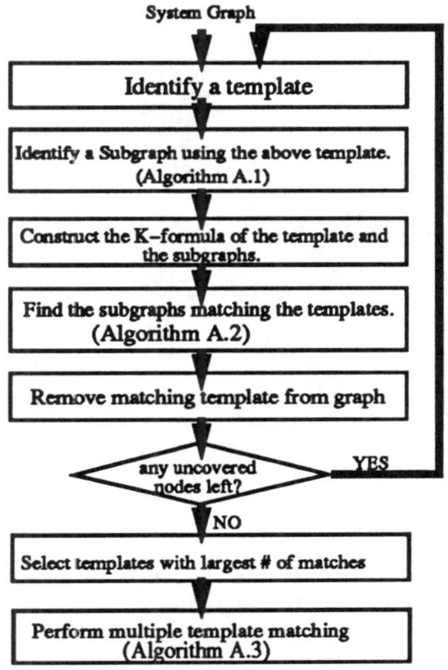

Figure 3: Overall flow of the partitioning system

6 RESULTS AND DISCUSSIONS

The various components described above can be pieced together to give the overall flow of the partitioner (Fig. 3). The results presented here are based on a prototype implementation of the algorithm on SUN/Sparc workstation. The system accepts VHDL structural description, which is then parsed to construct the directed graph needed for further processing. Let us first discuss the Elliptic filter example in detail. We assume that each operation is to be implemented by a dedicated 16-bit functional unit. Such a situation would arise in cases where the system is pipelined with an initiation interval of 1 for maximum throughput. Thus we are operating on a *structural* rather than *behavioral* representation. The template generation procedure comes up with three templates that are repeated more than once; these are shown in Fig. 4(b).

The single template matching procedure when repeated for the three templates, yields the results presented in Table I. It is not immediately apparent what combination of these templates should be chosen to obtain the minimal cost cover. Table II shows that solutions 1 and 2 (generated manually) are inferior to the solution produced by Algorithm A.3 - it chooses 2 of T_1 and 7 of T_2, leaving 10 nodes uncovered, and having 5 off-chip jumps along the critical path. The 10 residue nodes could easily fit on three chips, so we have a total of 12 templates, which are of three different types. This solution is shown pictorially in Fig. 4(a).

A second example we present is the AR filter. The template generation algorithm gives three different types of templates, shown in Fig. 5(b). The data path bit width here is assumed to be 8 bits, so it is possible to fit more operators on a template (or chip in this case). Table I shows that we have 8 occurrences of T_1, 2 of T_2, and 2 of T_3. The possible solutions

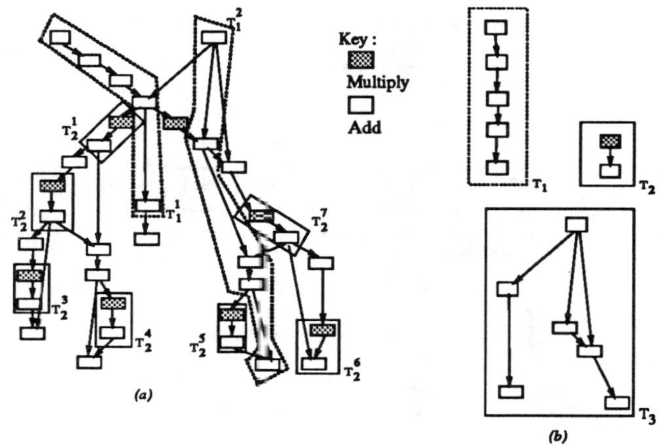

Figure 4: (a) Solution to the elliptic filter example; (b) The templates generated by the algorithm

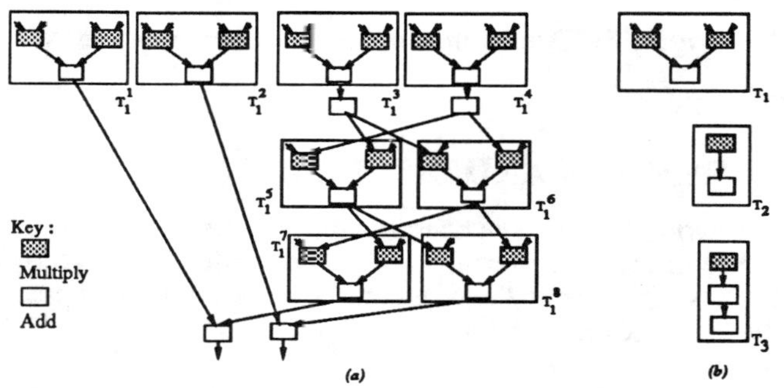

Figure 5: (a) Solution to the AR filter example; (b) The templates generated

Template Type	# Times Used	runtime
T_1	2	2.5 sec
T_2	8	
T_3	3	

(a)

Template Type	# Times Used	runtime
T_1	8	2.7sec
T_2	2	
T_3	2	

(b)

Table I : Results of Single Template Match. (a) Elliptic Filter; (b) AR Filter.

Solution	# templates				# residue nodes	# extra templates	Final Solution		# off-chip jumps	run-time
	T_1	T_2	T_3	Total			Types	# Parts		
1	0	4	3	7	11	5	5	12	6	–
2	1	5	2	8	9	4	5	13	6	–
AlgA.3 [Fig. 4(a)]	2	7	0	9	10	3	3	12	5	3.1sec

(a)

Solution	# templates				# residue nodes	# extra templates	Final Solution		# off-chip jumps	run-time
	T_1	T_2	T_3	Total			Types	# Parts		
1	4	2	0	6	8	3	3	9	4	–
2	2	2	2	6	8	3	3	9	4	–
AlgA.3 [Fig. 5(a)]	8	0	0	8	4	1	2	9	4	2.8sec

(b)

Table II : Results of Multiple Template Match. (a) Elliptic Filter; (b) AR Filter.

and their relative tradeoffs are shown in Table II. It is clear that the solution generated by Algorithm A.3 - 8 of T_1 and none of the others - is superior to the other solutions shown. This solution is shown in Fig. 5(a).

7 CONCLUSIONS AND FUTURE WORK

We have explored the problem of digital system partitioning by extraction of regularity. We have demonstrated how the total system design cost can be minimized by maximally 'cloning' certain components. The problem we have tackled represents an unexplored domain in system partitioning, and a lot more research needs to be done. The concept of regularity extraction opens up a whole range of applications that can be extensively simplified. Research is underway to investigate these and other outstanding issues.

References

[1] U. Kodres. Partitioning and Card Selection. In M. A. Breuer, editor, *Design Automation of Digital Systems*, Prentice-Hall, Englewood Cliffs, 1972.

[2] M. Kahrs. Matching a Parts Library in a Silicon Compiler. In *Proc. ICCAD-86*, pages 169–172, 1986.

[3] K. Keutzer. DAGON: Technology Binding and Local Optimization by DAG Matching. In *Proc. 24th DAC*, pages 617–623, 1987.

[4] W. P. Birmingham, A. Gupta and D. P. Siewiorek. The MICON System for Computer Design. In *Proc. 26th DAC*, pages 135–140, 1989.

[5] S. Devadas and A. R. Newton. Decomposition and Factorization of Sequential FSM's. In *Proc. ICCAD-88*, 1988.

[6] W. E. Donath. Logic Partitioning. In B. Preas and M. Lorenzetti, editors, *Physical Design Automation of VLSI Systems*, Benjamin Cummings, Menlo Park, 1988.

[7] M. McFarland. Computer-Aided Partitioning of Behavioral Hardware Descriptions. In *Proc. 20th DAC*, 1983.

[8] R. Camposano and R. K. Brayton. Partitioning Before Logic Synthesis. In *Proc. ICCAD-87*, pages 324–326, 1987.

[9] E. Dirkes-Lagnese and D. Thomas. Architectural Partitioning for System Level Design. In *Proc. 26th DAC*, 1989.

[10] R. Gupta and G. DeMicheli. Partitioning of Functional Models of Synchronous Digital System. In *Proc. ICCAD-90*, pages 216–219, 1990.

[11] K. Kucukcakar and A. C. Parker. *CHOP : A Constraint Driven System Level Partitioner*. Technical Report CEng. 90-26, Dept. of EE-Sys., USC, November 1990.

[12] D. Sreenivasa Rao and F. J. Kurdahi. *Digital System Partitioning by Efficient Extraction of Regularity*. Technical Report 91-14, Dept. of ECE, Univ. of Calif., Irvine, October 1991.

[13] A. T. Berztiss. *Data Structures : Theory and Practice*. Academic Press, New York, 1971.

A Path-Oriented Approach for Reducing Hazards in Asynchronous Designs

Meng-Lin Yu and P. A. Subrahmanyam
AT&T Bell Laboratories, Holmdel, N.J. 07733

Abstract

We describe a path-oriented approach for analyzing hazards in an asynchronous circuit that is synthesized from a specification in the form of a signal transition graph . We also describe a technique to avoid hazards by adjusting the delays in a circuit; rather than doing this piecewise for each path, we show how improved results can be obtained by solving an appropriately formulated linear programming problem.

1 Introduction and Motivation

A hazard in an asynchronous circuit implementation refers to an unexpected "glitch" in a signal value, i.e., a transition that is not allowed by the circuit specification. Hazards can lead to malfunctions in a system by triggering an erroneous sequence of transitions (possibly leading to erroneous states). It is therefore important to avoid hazards in an asynchronous circuit implementation. In so far as is feasible, it is desirable to detect the potential for hazards at the level of a circuit "specification" rather than analyzing the synthesized circuit implementation.

A convenient way to specify an asynchronous circuit behavior is via a Signal Transition Graph (STG) in which vertices represent signal transitions and arcs represent causal dependencies between signal transitions[1]. Synthesis is then performed by mapping the STG into a state graph, and eventually into an asynchronous circuit.

In this paper, we describe a path-oriented approach for analyzing hazards in an asynchronous circuit that is synthesized from an STG specification. For many designs, such an approach provides a good visual way of determining potential hazards, and assists designers in assessing a design and making appropriate changes. We also describe a technique for avoiding hazards by adjusting the delays in a circuit; rather than doing this piecewise for each path, we show how improved results can be obtained by solving an appropriately formulated linear programming problem. In addition to the benefits of working at a higher level of abstraction, this method handles some classes of dynamic hazards in addition to static hazards.

The paper is organized as follows: Section 2 summarizes some of the relevant concepts and definitions from literature. Section 3 discusses a path-oriented approach for analyzing hazards. Section 4 describes how this technique can be used to avoid hazards by formulating and solving an appropriate linear programming problem. Section 5 contains preliminary experimental results, while Section 6 draws some conclusions.

2 Preliminaries and Background

This section summarizes some of the relevant concepts relating to Petri nets and signal transition graphs.

Petri Nets. A Petri net is a triple consisting of a finite set of *transitions* T, a finite set of *places* P and a binary relation between transitions and places called a *flow relation* $F \subseteq T \times P \cup P \times T$.

A Petri net thus defines a bipartite graph, and is commonly pictorially depicted as such. The nodes are either places or transitions, and arcs between the nodes in P and T represent the relation F. If $(p, t) \in F$, an arc is drawn from place p to transition t. If $(t, p) \in F$, an arc is drawn from transition t to place p. If $(v_1, v_2) \in F$ then v_1 is called a *predecessor* of v_2, and v_2 is called a *successor* of v_1.

A *free-choice net* (FC net) is a Petri net, where, if a place p has more than one successor, say t_1, , ... t_n, then p is restricted to be the *only* predecessor of t_1, , ... t_n.

Signal Transition Graphs. A *signal transition graph* is an *interpreted free-choice Petri Net.* Transitions of the petri net are *interpreted* as signal transitions, i.e., changes of values on input, output or internal signals. Places are interpreted as conditions on the signal states.

A *positive* or *rising* transition of a signal s is denoted s^+; a *negative* or *falling* transition is denoted s^-. Henceforth, we use s^* to denote any transition (either s^+ or s^-) of the signal s. In such a case, $s^{*'}$ denotes the transition complementary to s^*, i.e., either s^- or s^+.

Dynamic Behavior of a Petri Net/STG: Marking and Firing. A *token marking* of a petri net is an assignment of a non-negative integer to each place in the net. Tokens are typically depicted by means of dots in the place, or along the edges leading to a place/transition.

A transition is *enabled*, implying that the corresponding event can actually occur in the system, when there is at least one token on each place (equivalently, the fanin arc) leading into the transition.

Once a transition is enabled, it must eventually *fire.* When a transition fires, one token is removed from each predecessor of the transition, and one token is placed in each successor of the transition.

A set of transitions is said to be *concurrent* if all the transitions in the set are simultaneously enabled in some marking reachable from the initial marking. If t_2 is constrained to occur after t_1, we write $t_1 \rightarrow t_2$; t_1 and t_2 are said to be *(sequentially) ordered*. Transitions that are neither concurrent nor sequential are said to be in *conflict*.

Live and Safe Nets. A free-choice net is said to be *live* if it is always the case that each of its transitions is enabled in some marking that is reachable from its initial marking, i.e., no transition is permanently disabled in a live net. A free-choice net is said to be *safe* if no place is

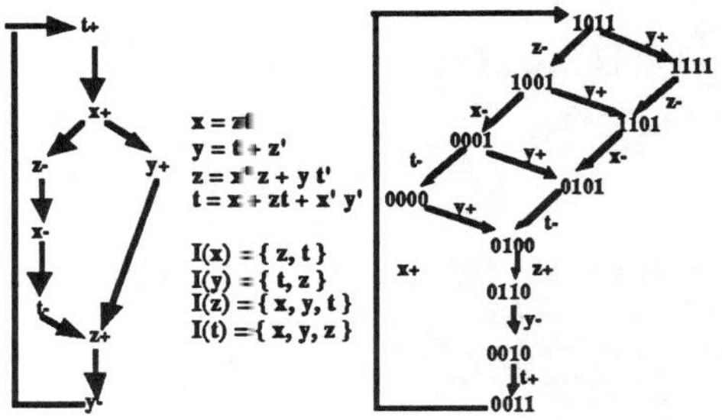

Figure 1: An STG (DH), logic equations and input sets (I), and its state graph.

ever assigned more than one token after any sequence of firings from the initial marking. All of the STGs considered in this paper will be assumed to be live and safe.

State Graphs. A state graph is an alternative representation of the behavior of an asynchronous circuit, where all of the concurrency in the transitions is resolved by representing it as a set of interleaved sequential transitions. The state graph corresponding to an STG can be computed by starting with the initial marking of the STG and exhaustively simulating the dynamic behavior of the STG[1]. Figure 1 shows an example STG and its state graph.

Unique State Coding. The state graph SG corresponding to an STG is used in the synthesis technique described in [1], and in most related techniques. Further, the signals in the STG are used as state variables in the synthesized circuit. Clearly, since the nodes in SG represent distinct states, enough "state information" must be contained in the state encoding in the synthesized circuit to be able to disambiguate distinct markings (states in SG). An STG has the *unique state coding* (USC) property if two distinct states in the state graph of the STG do not have identical binary codes.

Static and Dynamic Hazards. A *static hazard* is a 0→1→0 transition or a 1→0→1 transition in a signal where no transition for that signal should occur according to the specification.

A *dynamic hazard* is a 0→1→0→1 (or 1→0→1→0) transition in a signal where a single positive (or negative) transition for that signal should occur according to the specification.

Valid Pairs. In analyzing the hazards in a circuit, it is typical to analyze the behavior when the inputs of the circuit change from an initial vector v_1 to a final vector v_2. Each such vector pair yields a *transition cube* where all the signals that can change value along a path from v_1 to v_2 in the state graph have an undetermined value "-", whereas the invariant signal components have the value in v_1.

Given an STG specification, the synthesis process yields a network for each logic function f_i. In analyzing static hazards, the value of the i-th component is expected to be constant, hence the only vector pairs that need to be considered are those wherein $f_i(v_1) = f_i(v_2)$. Each such pair is called a *valid pair* for signal i.

3 Hazard Analysis and Avoidance

3.1 Previous Work

In the synthesis algorithm discussed in [3],[2], STGs are in essence converted into state graphs for finding a prime cover for a given output x. Candidate state pairs, or valid pairs, with the same x values and the same values of the next state function f_x, are formed. A transition cube is then determined for each valid pair and a check is made to see if this cube can only be partially covered by an on-set prime cover. A hazard is found if the transition cube is strictly partially covered as there exists a path in the transition cube that consists of subpaths of 1→0→1 or 0→1→0 values for f_x. If $t(i)$ denotes the time at which transition i occurs, the hazard condition is then identified as $t(j) - t(i) < \text{delay}(i \to x) - \text{delay}(j \to x)$, where j and i are transitions causing the value of output x to be incorrect and correct respectively. The quantity $t(j) - t(i)$ is named $d3$ and delay values are added independently (actually sequentially) to the value $d3$ associated with each hazard condition so as to preclude their occurrence. It is shown that such a delay insertion method is always feasible and satisfies all the hazard avoidance conditions.

Such methods do not operate directly at the STG level, and thus do not directly provide an intuitive explanation of the causes of hazards at the behavioral/STG level. They do not directly address dynamic hazards, as it is not easy to extend the basic concept to including transition cubes containing both on-set and off-set cubes for even hazard-free state pairs. In addition, the hazard avoidance procedure treats each hazard condition independently, since each condition is individually ensured by inserting a delay. If no attempt is made to use the slacks among conditions, the results from the procedure can be suboptimal. The linkage between the hazard-avoidance conditions and the system timing specifications was also not directly addressed.

3.2 Hazard Analysis: Overview

Since neither does a speed-independent design (a description at the STG level) guarantee a speed-independent implementation (at the gate level), nor are we interested in pure speed-independent designs and applications, hazard analysis has to be performed in order to have correct working asynchronous circuits. Further, it is desirable to perform such an analysis at the highest possible level.

We now delineate a method that directly operates on the STG, or the specification level. It uses a hazard analyzer for identifying the conditions that can potentially cause hazards, and produces a set of path delay (hazard avoidance) conditions under which the implementation will be hazard-free. The hazard analysis is based on a path-oriented approach for generating *potential* hazard candidates, both dynamic and static. We then simulate (evaluate) each candidate and compare its results with the correct results according to the specification. In addition to the benefits of working at a higher level of abstraction, this method handles some classes of dynamic hazards.

The hazard-avoidance conditions generated by the analysis phase are then merged with the initial timing constraints in the specification to form a system of linear inequalities. An appropriate cost function is then defined and the entire linear inequality system is solved to yield an optimal solution using a linear programming procedure. A typical cost function is, for example, the total number of delay elements inserted or the total delay along a specific path. The solutions are delays that each logic function has to have along a path; this can be implemented either by choosing an appropriate implementation from a library, for example, or by inserting delay elements.

For many designs, this path-oriented approach provides a good visual way of determining potential hazards. By assisting in the identification of where and how hazards are possible at the STG level, it provides an easy visual technique for designers to assess a design and to make changes accordingly.

3.3 Generating Hazard Candidates

Given an STG, we next discuss a path-oriented technique that identifies a set of potential hazard conditions. This set of condition is pruned by a subsequent simulation phase that is detailed later.

Given an STG and a signal x in question, we first check to see if the STG has the USC property. If it does not, then the USC property is ensured by either inserting internal variables automatically or by asking the designer to modify the specification. Once an STG has the USC property, we can then simplify the graph by contracting the graph [1] such that the contracted/simplified graph consists of only the signal x and signals in the input set of x. A signal y is in the *input set* of x if the signal y is a direct input to the logic generating module for the signal x in the final implementation. Techniques for determining the input set of a signal can be found in [7]. This simplification is useful, since the complexity of the following hazard checking procedure usually decreases significantly after the simplification.

For graphs where x only goes up and down once each, the contracted graph is then divided into 2 groups, group $x^+ \ldots x^-$ and group $x^- \ldots x^+$. Group $x^+ \ldots x^-$ consists of all signal transitions, together with their ordering relations, that can occur after x^+ and before x^- in the specification. Group $x^- \ldots x^+$ consists of all signal transitions that can occur after x^- and before x^+. For signals that go up and down more than once, one group is created for each adjacent $x^*, x^{*'}$ pair. The following discussions will assume that signals only go up and down once. Multiple up-and-down transitions of a signal can be handled similarly.

All transitions in these two groups maintain the same order as in the contracted graph. For signals transitions ordered with respect to x^* and $x^{*'}$, it is trivial to determine which group they belong to. The signals that are potentially concurrent with x^* belong to both groups.

Consider, for example, the STG in Figure 1. For signal x, group $x^+ \ldots x^-$ includes z^-, and group $x^- \ldots x^+$ includes t^-, z^+, t^+, where "," indicates sequential transitions. For signal y, group $y^+ \ldots y^-$ includes z^-, t^-, z^+ and group $y^- \ldots y^+$ includes t^+, z^-, t^-. For signal z, group $z^+ \ldots z^-$ includes y^-, t^+, x^+, y^+ and group $z^- \ldots z^+$ includes $(x^-, t^-) || y^+$, where $||$ indicates concurrent transitions. For signal t, group $t^+ \ldots t^-$ includes $x^+, ((z^-, x^-) || y^+)$, and group $t^- \ldots t^+$ includes y^+, z^+, y^-.

Given the above groups, we can now motivate the conditions that can lead to hazards.

Definition 1 (Inverted Signal Ordering) *An ordering of the signals in either of the groups* $x^+ \ldots x^-$ *or* $x^- \ldots x^+$ *that is different from the specified order is said to be* inverted.

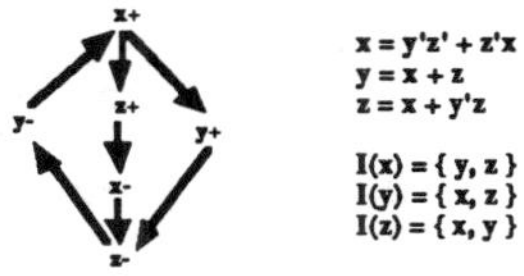

Figure 2: An Example STG (XYZ), associated logic equations and input sets (I).

Observe that a hazard can potentially occur at signal x only if some signal transitions in either the group $x^+ \ldots x^-$ or the group $x^- \ldots x^+$ are out of order in time when they reach the logic element that computes x, i.e., if ordering of the signals is inverted. A *potential hazard condition* is said to occur when an inverted ordering can lead to a hazard for output x.

The hazard conditions in [2] are related to the notion of valid pair of input vectors. Recall that a valid pair in the context of static hazards is a pair of states reachable from each other (i.e., having a path between them in the state graph) and having the same x and f_x value. In comparison, in the definition of a hazard condition given above, the "corresponding" notion of valid pairs is a pair of states reachable from each other but that do not necessarily need to have the same f_x (i.e., next-state) value. It is easy to see that this hazard condition is weaker than the condition that just detects static hazards.

Consider the example shown in Figure 2. Here, $x^- \ldots x^+ = \{y^+, z^-, y^-\}$ and $x^+ \ldots x^- = \{z^+ \mid\mid y^+\}$. Since y^+, z^- are in group $x^- \ldots x^+$, an inverted order for y^+ and z^- produces a hazard for x. Similarly, since z^+, x^- are in both $y^+ \ldots y^-$ and $y^- \ldots y^+$ if the transitions z^+ and x^- are inverted, they produce two hazards for y, one when $y = 0$, and one when $y = 1$. There are no hazards for z. This is consistent with the analysis of static hazards in [3] for this example.

This perspective on the origin of hazards is intuitively easy to understand. A synthesized circuit is designed to achieve the behavior of the "correct" STG description. If signals a and b are inputs to (the logic implementing) signal x, with transition a^* preceding transition b^* in the specification, the circuit will perform correctly if the assumption holds in the actual implementation. However, since the delays of each gate and wire are different, it is possible that a^* traverses a slower route than b^* to reach x, thus violating the assumption in the specification. This situation causes hazards if the sequences of transitions a^*, b^* and b^*, a^* generate different x output values. If $t(a^*)$ denotes the time at which the transition a^* occurs, and $d(a^* \rightarrow x^*)$ denotes the delay between a^* and x^*, then $t(a^*) + d(a^* \rightarrow x^*)$ must be less than $t(b^*) + d(b^* \rightarrow x^*)$. Thus, the condition required to prevent a hazard from occurring is $t(b^*) - t(a^*) > d(a^* \rightarrow x^*) - d(b^* \rightarrow x^{*'})$. The task of implementing a hazard free circuit from an STG specification then involves solving the conjunctions of all such conditions.

3.4 Hazard Analysis By Simulation

The conditions that leads to an inverted signal ordering is not a sufficient condition for the presence of hazards. An inverted signal ordering will generate a hazard only if the logic implementation produces differing results for the original and inverted orderings.

In order to determine precisely when hazards can occur, the circuit is first synthesized from the given STG specification. This process yields a boolean equation for each output variable, and an initial circuit implementation. The inverted orderings are then evaluated to detect potential hazard conditions. Two variations in the evaluation technique can be used.

The first method is a straightforward approach. When an input transition y^* to x arrives, that new y value is used in the boolean evaluation and a new x value is computed. When the next input transition z^* arrives, the new x value and the new z value are used for computing the next x output value. This evaluation process then repeats until all input transitions in question are evaluated in sequence.

The second method does the same except that the original x value at the beginning of computation is always used for computing any next x values. This method detects all sequences that may potentially switch x twice, assuming that the memory device for implementing x does not switch fast enough, e.g., like an ideal flip-flop; such behavior leads to static hazards. This second method, together with the restriction that at least one of the transitions immediately preceding x is always the last transition in a inverted signal ordering, makes this approach to be equivalent to the static hazard analysis in [3].

Note that in general the feedback loop for a variable can have an arbitrary delay in the range $[0, \infty]$. This entire range of delays needs to be considered to detect all of the possible hazards. However, in practice, the delay is usually in a bounded range [*min, max*] and this can be used to significantly reduce the computation complexity for the full blown analysis. Also note that the second method is weaker than the first one for detecting static hazards, as the first method may record the switch and detect no further change of value of x. The second approach thus depends on the design of the flip-flops used in the circuit. If the flip-flop internal feedback time is negligible through proper buffering before the final output, then the first method is a reasonably good method of detecting any hazard. Otherwise, both methods should be used together and the union of the detected hazards are reported. For simplicity, only the first method will be used in the discussion of the remainder of this paper.

The method can be summarized as follows.

0. For each variable x, contract the STG.

1. Synthesize the logic for x: this yields a boolean equation and an associated two-level implementation for x.

2. Construct $x^+ \ldots x^-$ and $x^- \ldots x^+$.

3. Evaluate the value of the output x for each inverted signal ordering, and compare the result with the correct output. If the results differ in number or parity of signal transitions, a hazard is detected.

For *a posteriori* design analysis, the number of inverted signal orderings can be large in the worst case ($O(n!)$). In practice, the number of input signals involved for a given signal output is usually quite small, and this computation may not pose a problem in typical control circuits. For synthesis, very few inverted signal orderings can lead to hazards; further, these cases can be identified during synthesis, as discussed in a forthcoming report.

Figure 3: Delay model.

3.5 An example

The following example illustrates the detection of dynamic hazards by this procedure. In Figure 1, if we restrict ourselves to static hazards for the signal z, when $z = 1$ (i.e., the states after z^+), the only valid pair is 0110 and 0011, with the second and fourth bit representing y and t respectively. The transition cube $x' - z-$ is covered by a prime implicant z and therefore there is no hazard. Another way of verifying this is to note that for both orderings y^-, t^+ and t^+, y^-, the output of z stays at 1 and there is thus no static hazard. However, if t^+, x^+, and y^+ are out of order, and the order becomes y^-, x^+, y^+, t^+ when they reach z, the transitions associated with z become $1 \to 0 \to 1 \to 0$ as opposed to the expected sequence $1 \to 1 \to 0 \to 0$. This indicates a dynamic hazard, which is associated with the hazard condition $t(y^+)-t(t^+) < d(t^+ \to z^+) - d(y^+ \to z^-)$.

4 The Hazard-Avoidance Procedure

In a design specification, timing information can be incorporated by describing the constraints on the delays between various pairs of transitions for correct asynchronous operations. Such timing constraints are usually guaranteed either by (1) the environment that a circuit operates in or by (2) the synthesizer, that views the timing information as being mandated by the environment and then ensures that they are met by the synthesized circuit.

The hazard-avoidance conditions and the original timing constraints form a system of inequalities involving delays between transitions. Linear programming techniques can be used to solve the system, once a linear objective function is defined.

In the most general case, each wire of a electrical net can have a different delay and each I/O path through a simple/complex gate can have a different delay. We can bundle the delays of a wire and I/O path together. The resultant circuit model we use has delays for each wire and zero delay for gates, as shown in Figure 3. The hazard analysis method discussed here can be intuitively explained in the context of such a delay model. Since the internal gate delays are zero, the only way hazards to occur is for the input signals to arrive in an order different from the specification. In such cases, delays have to be added to ensure that the correct order is restored.

Given an initial circuit implementation, the procedure below yields a set of inequalities embodying the hazard free conditions. The function `lca`(s_i, s_j, s_k) returns the set of least common ancestors of s_i and s_j that lead to s_k.

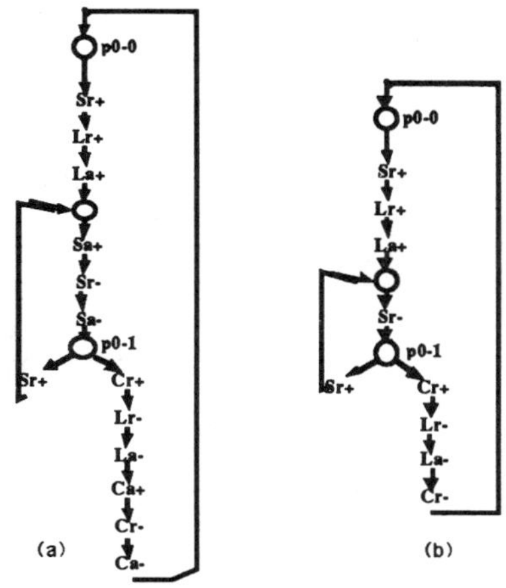

Figure 4: (a) Example STG (b) contracted STG for Lr.

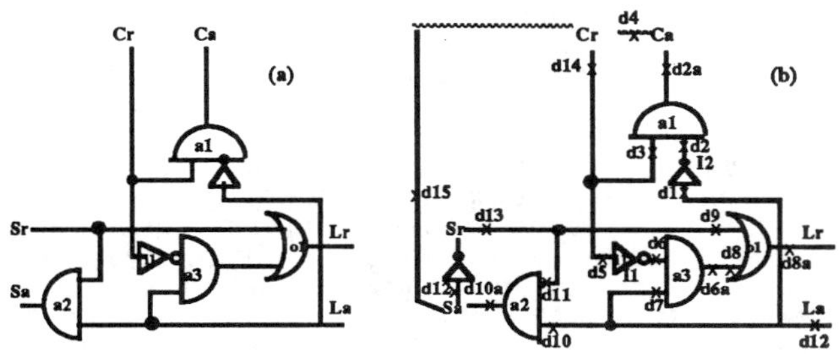

Figure 5: (a) An initial implementation (b) timed.

```
Hazard-free-conditions ← ∅
for each output s_k
  for each s_i, s_j causing hazards
    for each node_x in lca(s_i, s_j, s_k)
      Hazard-free-conditions ←
      Hazard-free-conditions ∪
      t(s_j) − t(s_i) > d(s_i→node_x) - d(s_j→node_x)
```

4.1 An Example

We will illustrate the hazard avoidance procedure with an example used in [1] and [2]. Figure 4 shows the STG description; a circuit initially synthesized from this STG is shown in Figure 5(a). We will assume that each inverter has a delay in the range [2,5], i.e., an inverter has a delay that is a minimum of 2 units and maximum of 5 units, and that all other gates have a delay in the range [3,7].

In order to enable a comparison with earlier methods e.g., in [2], we will here consider only static hazards. Note that a dynamic hazard is possible for Lr (Figure 4(b)), because the loop corresponding to $(Sr^+,Sr^-)^*$ can be traversed any number of times before Cr^+ is enabled. If the path from Sr to Lr involves an extraordinary amount of delay and therefore Cr^+ causes Lr^- before all the $Sr^-,Sr^+,Sr^-,Sr^+,\ldots,Sr^-$ transitions reach Lr, Lr will undergo the transitions $0\rightarrow1\rightarrow0\rightarrow1\rightarrow\ldots$ as many times as the signal Sr.

By contracting the STG and synthesizing the logic, we get $Lr=Sr + Cr'*La$, $Ca=Cr*La'$ and $Sa=Sr'*La$. In the group $Lr^-\ldots Lr^+$ for signal Lr, a hazard occurs when La^- and Cr^- are out of order when they reach Lr, *i.e.*, Cr^- effects Lr before La^-. The condition for avoiding such a hazard is

$$t(Cr^-) - t(La^-) > d(La^- \rightarrow a3^-) - d(Cr^-\rightarrow a3^+). \quad \ldots(0)$$

Note that the two paths join at $a3$; at this stage, one of the two signal will dominate and propagate further. Since La^- precedes Cr^- (Figure 5), following the path from La to Ca through the logic, we obtain $t(Cr^-) = t(La^-) + d(i2^+) + d(i2^+ \rightarrow Ca^+) + d(Ca^+ \rightarrow Cr^-)$, where $d(i2^+)$ denotes $d(i2^-\rightarrow i2^+)$. We can then rewrite (0) as

$$d(i2^+) + d(i2^+ \rightarrow Ca^+) + d(Ca^+ \rightarrow Cr^-) > d(La^- \rightarrow a3^-) - d(i1^+) - d(i1^+ \rightarrow a3^+) \quad \ldots(1)$$

In the group $Lr^+\ldots Lr^-$, a hazard occurs when La^+ and Sr^- are out of order. The hazard avoidance condition is

$$t(Sr^-) - t(La^+) > d(La^+\rightarrow Lr^-) - d(Sr^-\rightarrow Lr^-)$$

and it can be similarly expanded into

$$d(La^+\rightarrow Sa^+) + d(Sa^+\rightarrow Sr^-) > d(La^+\rightarrow a3^+) + d(a3^+\rightarrow Lr^+) - d(Sr^-\rightarrow Lr^-) \quad \ldots..(2)$$

For signal Ca, a hazard occurs when transitions La^+ and Cr^+ reach Ca out of order. The hazard avoidance condition is

$$d(La^+\rightarrow Sa^+) + d(Sa^+\rightarrow Sr^-) + d(Sr^-\rightarrow Sa^-) + d(Sa^-\rightarrow Cr^+) > d(i2^-) + d(i2^-\rightarrow Ca^-) - d(Cr^+\rightarrow Ca^+) \quad \ldots(3)$$

There is no hazard for signal Sa.

In order to make the system solvable, we insert delays for each wire and each gate output before fanout if the number of either fanin or fanout is greater than one. We then have,

$$D1 + d(i2^+) + D2 + d(i2^+\rightarrow Ca^+) + D2a + d(Ca^+\rightarrow Cr^-) + D4 > D7 + d(La^-\rightarrow a3\text{-}) - D14 - D5 - d(i1+) - D6 - d(i1+\rightarrow a3+), \quad \ldots..(4)$$

$$D10 + d(La^+\rightarrow Sa^+) + D10a + D12 + d(Sa^+\rightarrow Sr^-) > D7 + d(La^+\rightarrow a3^+) + D6a + D8 + d(a3^+\rightarrow Lr^+) - D13 - D9 - d(Sr^-\rightarrow Lr^-), \text{ and } \ldots\ldots\ldots(5).$$

$$D10 + d(La^+\rightarrow Sa^+) + D10a + D12 + d(Sa^+\rightarrow Sr^-) + D13 + D11 + d(Sr^-\rightarrow Sa^-) + D10a + D15 + d(Sa^-\rightarrow Cr^+) > D1 + d(i2^-) + D2 + d(i2^-\rightarrow Ca^-) - D14 - D3 - d(Cr^+\rightarrow Ca^+). \quad \ldots..(6)$$

Note that delays have been added to both the output side and the input side of a gate. Strategies can be developed such that fewer variables are used if the cost function used can be defined on only those variables.

If system timing constraints on the external signals exist, they are treated the same way as the above inequalities. There is no external system timing specified for this example. We will assume that d($Ca^+ \rightarrow Cr^-$)=0, d($Sa^+ \rightarrow Sr^-$)=0, d($Sa^- \rightarrow Cr^+$)=0. Further, we use the lower bound values for the left-hand side terms and negative right hand side terms, and upper bound values for positive right hand side terms and negative left hand side terms. From (4) (5) (6), we have,

D1 + D2 + D2a + D4 > D7 - D14 - D5 - D6 - 3

D10 + D10a + D12 > D7 + D6a + D8 - D13 - D9 + 8

D10 + 2*D10a + D12 + D13 + D11 + D15 > D1 + D2 - D14 - D3+ 3.

A simple cost function (to be minimized) can be the sum of total delay value, i.e., sum of Di's, as this may approximate the additional area needed and the overall slow-down once the delay elements are inserted. A cost function may also consider the total number of delays inserted, the slacks (the safety margin) and/or the delay along a certain path.

For this example, we use $\sum_i D_i$ as the cost function. A possible solution is D10a=9, which says that if gate $a2$ is slowed down by 9 units, the circuit is hazard-free. Note that this implementation happens to eliminate the dynamic hazard, caused by Cr^+ reaching Lr earlier than Sr^-, in this particular example. In comparison, [2] requires a delay to be added to each inequality *independently*. Thus their solution is D13=9 and D14=4. The total delay is 13 and consists of two delays, both of which have to be delay buffers. Moreover, their solution increases the delay along the longest path $Cr \rightarrow Lr$.

Note that it is usually more costly to use buffers as delays than to slow down a gate. For circuit implementations that employ a cell library, several implementations having different delays for the same functionality are available, and it may not be necessary to use delay elements for gate output delays. We can use the fastest available gates as the initial solution to such a system, and then obtain needed delays by solving the system. Slower gates approximating the solution can then be used instead of inserting delay buffers, as long as the approximation still satisfies the inequalities.

Further, if certain specific design styles or constraints, for example, similar I/O path delays for a single gate, isochronic forks, etc., are known, these constraints can be added to the system to simplify the problem and to restrict the solution domain. Tools can be written to provide designers options to tailor the inequality system to their specific design environments.

5 Experimental Results

The appropriateness of a particular implementation is strongly dependent on the constraints imposed by the environment that a circuit is designed to operate in. This information is captured by the initial timing specifications and restrictions. Note that, unlike synchronous designs, the longest delay between flip-flops may be on a path that is non-critical for asynchronous operations. In the experiments below, a simple criterion, consisting of the sum of total inserted delays, is used. This provides a measure of how much the circuit will be slowed down overall and how much additional area is needed, in the absence of environmental timing specifications. The results for a set of STG examples in the literature are shown below. The results from the method used in [2] are also included for comparison. "No. of dynamic hazards" is the number of hazards detected by the current method. "No. of delays" is the number of delay elements or delay adjustment for each gate. Only static hazards are considered here for the purpose of comparison. "Total delay" is the sum of all delay values needed to avoid hazards.

Ckt. name	No. of dynamic hazards	No. of delays [2]	No. of delays	Total delay [2]	Total delay
C133	0	4	3	15	13
C150	0	2	2	4	4
C172	1	3	2	13	9
xyz1	0	0	0	0	0
xyz2	0	0	0	0	0
xyzt	3	1	1	3	3

6 Conclusions

The primary objective of this paper was to explore a technique that provides a designer an intuitive (visual) correlation between specifications at the level of signal transition graphs and conditions that lead to hazards. The motivation was to provide insight into how the STG specifications should be altered, if necessary. This led to a path-oriented approach for hazard analysis. In essence, inverted signal orderings in an implementation can potentially lead to hazards; this set is pruned by simulation in order to detect those conditions that can actually lead to hazards. We have also described a technique for avoiding hazards by adjusting the delays in a circuit; rather than doing this piecewise for each path, we have shown how improved results can be obtained by solving an appropriately formulated linear programming problem. In addition to the benefits of working at a higher level of abstraction, this method handles some classes of dynamic hazards (under bounded delay assumptions) in addition to static hazards.

References

[1] T.A. Chu. Synthesis of self-timed VLSI circuits from Graph-theoretic Specifications. *MIT/LCS/TR-393*, June 1987.

[2] L.Lavagno et al. Algorithms for synthesis of hazard-free circuits. *DAC*, pp. 302-308, June 1991.

[3] L.Lavagno et al. Synthesis of verifiably hazard-free asynchronous control circuits. Technical Report, UCB/ERL M90/99, U.C. Berkeley, 1990.

[4] T. H.-Y.Meng. Asynchronous design for digital signal processing architectures. Ph.D.Thesis, UC Berkeley, Nov. '88.

[5] P.Vanbekbergen et al. Time and Area Performance Synthesis of Asynchronous Control Circuits. *IMEC Technical Report*, IMEC, Belgium, 1990.

[6] P.Vanbekbergen et al. Optimized Synthesis of Asynchronous Control Circuits from Graph-theoretic Specifications. *Proc. ICCAD*, pp. 184-187, Nov.1990.

[7] M.-L. Yu and P. A. Subrahmanyam. A New Approach for Checking the Unique State Coding Property of Signal Transition Graphs. *Proc. EDAC*, March 1992.

Circuit Structure Relations to Redundancy and Delay: The KMS Algorithm Revisited *

Alexander Saldanha Robert K. Brayton Alberto L. Sangiovanni-Vincentelli

University of California - Berkeley CA

Abstract

Keutzer, Malik, and Saldanha (KMS) [1] have presented an algorithm that derives an equivalent irredundant circuit implementation from a given redundant high-performance circuit, with no increase in delay (measured using viability analysis). In this paper we resolve the main bottlenecks in the KMS algorithm, arising due to an iterative loop of timing analysis, gate duplications, and redundancy removal. A circuit structure property based on path lengths is related to testability (redundancy) and delay. Based on this relationship, an efficient (single-pass) implementation of the KMS algorithm is presented. It consists of the transformation of any Boolean network to an equivalent circuit structure on which a single redundancy removal achieves the same effect as the original KMS algorithm.

1 Introduction

The relationship between the optimization criteria of area, performance, and testability in combinational circuits has been explored recently [1, 5]. While the relationship between the area and testability of a circuit has been known for some time, the relationship between the performance and testability of a circuit was first studied in [1]. There it is proved that redundancy is not necessary to reduce the delay of any circuit. An algorithm is provided that guarantees an irredundant implementation of any given redundant circuit with no increase in the delay. A principal drawback of the algorithm is the number of iterations of timing analysis, duplications, and redundancy removal that are performed. The algorithm operates on one longest false path at a time. Hence, it does not complete on circuits with large numbers of false paths.

Motivated by this observation, the relationship between circuit structure and path lengths versus testability and delay is explored. In particular, a new circuit structure property, termed L-path-disjoint, is defined that allows the simultaneous elimination of all (long) false paths via a single redundant multiple stuck-fault removal. An operation that transforms any circuit to possess this desired property is developed. The new single-pass KMS algorithm (referred to as the sKMS algorithm) takes time proportional to the size of the circuit to perform these transformations. With efficient timing analysis [4] and redundancy removal techniques, the running time of the complete sKMS algorithm is very small. The area (and fanout) increase of the algorithm is empirically shown to be very small. Thus, the sKMS algorithm shows promise in eliminating the false path problem (and the accompanying reliability problems [3, 1]) in a typical synthesis scenario.

2 The KMS algorithm

This section is a review of the algorithm that derives an equivalent irredundant circuit at least as fast as a given redundant circuit [1]. First, a few basic definitions are reviewed.

* Research support by Fujitsu Laboratories Ltd.

Definition 2.1 *A* **non-controlling value** *for a gate f is the value at its input which is not a controlling value for the gate. It is denoted as $I(f)$. For example, 1 is a non-controlling value for an* AND *gate.*

Definition 2.2 *Let $P = \{f_0, f_1, ..., f_m\}$ be a path. If P includes both a primary input and a primary output, P is an* **IO-path**. *The inputs of f_i other than f_{i-1} are called* **side-inputs** *of f_i along P and denoted as $S(f_i, P)$. A path that starts at a primary input and ends at a side-input of P is a* **side-path** *of P.*

Definition 2.3 *A path is* **statically sensitizable** *if there exists an input vector which sets all the side-inputs to the path to non-controlling values. The condition for* **static sensitization** *of a path $P = \{f_0, f_1, ..., f_m\}$ composed of simple gates is $\prod_{i=0}^{m} \prod_{g \in S(f_i,P)} (g = I(f_i))$.*

Similar to the approach of [1], we use the notion of **viability** [2] in determining the conditions under which a path may contribute to the delay of a circuit. Refer to [2, 4] for the formal definition of viability. It is mentioned here that if a path is statically sensitizable then it is viable, though the converse is not true.

If any longest path in the circuit is (statically) sensitizable, the KMS algorithm invokes standard redundancy removal to derive an irredundant circuit. This follows simply because redundancy removal cannot increase the longest (and true) delay of the circuit. If all of the longest paths are not sensitizable, the KMS algorithm chooses one path, say P, and attempts to remove a redundant fault on the first edge of P. If P is fanout-free, *i.e.* each gate along P has fanout $= 1$, then such a redundant connection always exists. If P is not fanout-free, duplication of some of the gates along P (gates duplicated include those between the first edge of P and the last gate along P with fanout > 1) is done to obtain a corresponding path P' that is fanout-free. The first edge of P' is now redundant, and is removed. It is proved in [1] that redundancy removal on the first edge of a longest path does not increase the delay of the circuit. Hence, by iterating the KMS algorithm, one path at a time in decreasing path length, an irredundant circuit, guaranteed to be no slower than the original is obtained.

An implementation of the original KMS algorithm is reported in [5]. The principal bottleneck of the algorithm is the number of iterations required. An iteration includes selecting a longest path, determining whether it is sensitizable or not; possible duplication and redundancy removal on the first edge of the path is performed if it is false. This algorithm fails to complete on all but small circuits or circuits with very few or no long false paths.

3 A single-pass algorithm

Due to the large number of false long paths in some circuits, it is imperative that any efficient algorithm must not explicitly

0738-100X/92 $3.00 © 1992 IEEE

enumerate each false path while performing the KMS transformation. This section develops such an algorithm and proves its correctness.

Definition 3.1 *The set of all the paths beginning at connection* c *and terminating at a primary output is called the* **path-set** *of the connection* c*, and is denoted* $\mathcal{PS}_c$.

Note that the paths in the path-set of a connection are IO-paths only if the connection is from a primary input.

Consider a connection c from a primary input in a circuit η. Additionally, assume the computed delay of η is $< L$ and let every path in $\mathcal{PS}_c$ be of length $\geq L$. Note that every path in $\mathcal{PS}_c$ is an IO-path. On completion of the KMS algorithm, all the paths in the resulting circuit η', are of length $< L$. The KMS algorithm removes connections between primary inputs and some gates, *i.e.* only the first edge of any path is removed. Thus, c cannot exist in η', since every IO-path through c is of length $\geq L$. This notion is captured formally by the next definition and theorems.

Definition 3.2 *A* L**-path-disjoint circuit** *is one where the paths in* $\mathcal{PS}_c$*, for any primary input connection* c*, are either all of length* $\geq L$ *or all of length* $< L$.

Theorem 3.1 *Let* η *be a circuit whose longest viable path is of length* $< L$*. Let* $C = \{c\}$ *be the set of all primary input connections such that each path in* $\mathcal{PS}_c$ *is of length* $\geq L$ *(and hence non-viable). Then any multiple stuck-fault composed of any combination of single stuck-0 or stuck-1 faults on each* $c \in C$ *is redundant.*

Proof Assume that some multiple stuck-fault F_C on C is irredundant. Let a vector v be a test for the fault. Consider the set of IO-paths P_{F_C} that propagate the fault effect to a primary output under vector v. The length of each $Q \in P_{F_C}$ is $\geq L$ by assumption. Pick a path $Q = \{f_0, f_1, ..., f_k\} \in P_{F_C}$ with the property that for each $f_i \in Q$, f_{i-1} under v is the earliest arriving input of f_i that propagates the fault effect. Such a path Q exists since the fault effect is propagated under v along some path in P_{F_C}. Consider each gate f_i along Q in circuit η when vector v is applied. All the side-inputs to f_i that do not propagate the fault effect are at non-controlling values. The remaining side-inputs each propagate the fault effect but each of them is available only at or after the time input f_{i-1} arrives at f_i. Hence these side-input are smoothed out (*i.e.* set to the non-controlling value) when considering the viability of the sub-path Q from f_0 up to f_i. Hence the sub-path of Q up to f_i is viable. Since this is true at f_k, Q is viable, contradicting that no path through c is viable. ■

Theorem 3.1 proves that the logical behavior of a circuit that meets the conditions of the theorem remains unchanged by a multi-fault removal. The next theorem proves that the delay, measured using viability analysis, remains unchanged under this kind of redundancy removal.

Theorem 3.2 *Let* η *be a circuit such that all IO-paths in* η *of length* $\geq L$ *are non-viable. Let* $C = \{c\}$ *be the set of all connections from a primary input such that each path in* $\mathcal{PS}_c$ *is of length* $\geq L$ *(and hence non-viable). Let* η' *be the circuit resulting after asserting a multiple stuck-fault composed of any combination of the single stuck-0 or stuck-1 faults on each* $c \in C$*. For any viable path* π' *in* η'*, the corresponding path* π *is viable in* η.

Proof Similar to the proof of Theorem 7.2 of [1]. The difference is that each gate may have several late-arriving side-inputs that are set to constant non-controlling values by the redundancy removal. However, these side-inputs always get smoothed out (*i.e.* set to the non-controlling value) and do not change the viability conditions of any path π' in η' from the viability of the corresponding path π in η. ■

```
/* Derive an irredundant network no slower than η.
η has no sensitizable paths of length ≥ L.  */
single_pass_kms(η, L) {
   /* Circuit η has only simple gates.
   A_f and R_f is arrival time and required time at f.
   atimes(f) are distinct path lengths from PI's to f.
   etimes(f) are distinct path lengths from f to PO's.  */

   /* Compute atimes(f) and etimes(f) for each f.  */
   kms_setup_times(η, atimes, etimes).

   /* Duplicate gates */
   node_list = gates in η in reverse topological order.
   Foreach node f in node_list {
      kms_duplicate_gate(f, η, L, atimes, etimes).
   }

   /* Set constants on first edge of paths ≥ L.  */
   kms_set_constant(η, L, atimes, etimes).
   Propagate constants as far as possible.
   Remove remaining redundancy in any order.
}
```

Figure 1: Single-pass (SKMS) algorithm

With Theorems 3.1 and 3.2 it is apparent that a multi-fault redundancy removal of the type specified can be done without increasing the computed delay (length of the longest viable path) or changing the logical behavior of the circuit. Moreover, no duplication is performed. Of course, not all circuits have such connections. A procedure is now described that transforms every circuit to a functionally equivalent L-path-disjoint circuit. In this case, all paths of length $\geq L$ will be removed by the application of the above theorems. The only operation used is the duplication of gates and the transfer of some fanouts of a gate to its duplicate.

Definition 3.3 *The set of distinct path lengths from primary inputs to a gate* f *is denoted by* **atimes(f)**. *The distinct path lengths from each gate* f *to the primary outputs is denoted by* **etimes(f)**.

If f is a primary input, *atimes(f)* is the single arrival time specified for f. If f is a primary output, *etimes(f)* is *0* . However, the algorithm could be generalized to take account of required times at the output; just take the maximum required time, R_{max}, and think of a buffer on each output f with delay $R_{max} - R_f$, where R_f is the required time for f. The algorithm to derive an L-path-disjoint network for a given network η is described in Figures 1 through 4. The main procedure, shown in Figure 1, consists of three phases.

The first phase computes the distinct paths lengths from primary inputs to each gate f, and the distinct path lengths from each gate f to the primary outputs. This is done by simply performing a topological traversal from the inputs to the outputs for computing *atimes*, and a reverse topological traversal for computing *etimes* (Figure 2).

The second phase consists of gate duplication and the transfer of some fanouts of a gate f to one or more duplicates of f, and is shown in Figure 3. The essential operation performed on a gate with multiple fanout is the transfer of a set of fanout connections of the gate to a duplicate gate. The gates are traversed in reverse topological order from primary outputs to primary inputs. Let f be a gate that is to be processed by the algorithm. Let P_f be any path from a primary input up to f. A duplication is only performed if there are at least two paths, $Q1_f$ and $Q2_f$

```
kms_setup_times(η, atimes, etimes) {
   Perform a delay trace on the network.
   Foreach node f of η {
      atimes(f) = {}.
      etimes(f) = {}.
   }
   /* For each node f, compute atimes(f) */
   node_list = gates in η in topological order.
   Foreach node f in node_list {
      If f is a primary input
         atimes(f) = A_f.
      Else Foreach fanin g of f
         atimes(f) = {u + d(f, g)|u ∈ atimes(g)}.
   }
   /* For each node f, compute etimes(f) */
   node_list = gates in η in reverse topological order.
   Foreach node f in node_list {
      If f is a primary output
         etimes(f) = 0.
      Else Foreach fanout g of f
         etimes(f) = {u + d(f, g)|u ∈ etimes(g)}.
   }
}
```

Figure 2: Path length calculations

```
kms_duplicate_gate(f, η, L, atimes, etimes) {
   g = f.
   Foreach time t ∈ atimes(f) in ascending order {
      If (t + min(etimes(g)) < L && t + max(etimes(g)) ≥ L) {
         /* some fanout of gate must be split */
         g' = duplicate_gate(g).
         atimes(g') = atimes(g).
         etimes(g') = etimes(g) − {t_e|t_e ∈ etimes(g), t + t_e ≥ L}.
         etimes(g) = etimes(g) − {t_e|t_e ∈ etimes(g), t + t_e < L}.
         Foreach fanout h of g {
            If (t + min(etimes(h)) + d(g, h) < L)
               Replace connection from g to h by g' to h.
         }
         /* repeat on duplicate gate of g */
         g = g'.
      }
   }
}
```

Figure 3: Gate duplication

from f to the primary outputs such that[1] $|P_f| + |Q1_f| < L$ and $|P_f| + |Q2_f| \geq L$. Following the duplication, one or more fanout connections are transferred from f to its duplicate f_{dup}, according to the following rule. Let Q_f represent any path from f to the primary outputs after some fanout connections are transferred to f_{dup}. Then $|P_f| + |Q_f| \geq L$ for any P_f and Q_f. Lemma 3.1 below ensures that this condition can always be satisfied. It is important to note that *etimes(f)* is updated whenever a fanout connection is moved to f_{dup}. Similarly, *etimes*(f_{dup}) is created to reflect the paths through the fanout connections transferred from f. Duplication is not done on any node with only one path to any output. Hence primary outputs are not duplicated.

This transformation is repeated on f_{dup} (the assignment $g = g'$ in Figure 3) until no further duplication and transfer of fanout connections is required. It is now shown that the resulting network after all the gates are processed by the algorithm is a L-path-disjoint network.

[1] $|P|$ denotes the length of path P.

```
kms_set_constant(η, L, atimes, etimes) {
   /* η is a L-path-disjoint network */
   Foreach primary input f of η {
      /* atimes(f) has exactly one entry */
      t = atimes(f).
      Foreach fanout g of f {
         If (t + d(f, g) + min(etimes(g)) ≥ L)
            Replace g by constant 0 or 1.
      }
   }
}
```

Figure 4: Setting constants on false paths

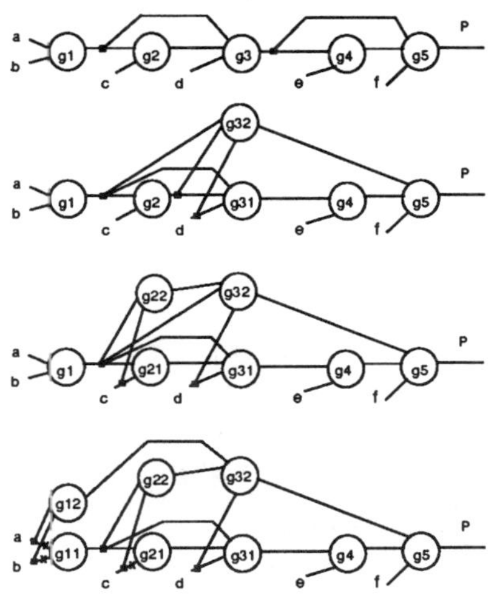

Figure 5: Example: Construction of an L-path-disjoint network

Lemma 3.1 *Consider gate f, in network η, that is processed by the algorithm of Figure 3. Assume that the lists* atimes(f) *and* etimes(f) *are computed using the procedure of Figure 2. Let f' refer to gate f or any of its duplicates*[2]*. Then the following invariant is true for each f': for each $f'_{atime} \in$* atimes(f'), *f'_{atime} +* etimes(f') *$< L$, or f'_{atime} +* etimes(f') *$\geq L$*[3].

Proof By induction on the level of a node. The level of a node is the maximum number of nodes along any path from the node to the primary outputs. See [5] for details. ∎

Since the invariant stated in Lemma 3.1 holds for each primary input, the final network is an L-path-disjoint network. Thus on completion of the second phase, the path-set of each first edge of any path of length $\geq L$ contains only paths of length $\geq L$. Hence, all such edges are set to constant *0* or *1* in the final phase of the single-pass algorithm (Figure 4).

Figure 5 illustrates the working of the algorithm on an example network. Each gate has unit delay and all primary inputs arrive at $t = 0$. Assume that the initial network (top of the figure) has a longest viable path of length 3. Hence an L-path-disjoint network for $L = 4$ is required. Following the first phase of the algorithm, the variables $atimes$ and $etimes$ are as follows:

$$atimes(g1) = \{1\} \qquad etimes(g1) = \{2,3,4\}$$
$$atimes(g2) = \{1,2\} \qquad etimes(g2) = \{2,3\}$$
$$atimes(g3) = \{1,2,3\} \qquad etimes(g3) = \{1,2\}$$

[2] There is no loss in generality in referring to the gates by a single representative f', since, for each path from the primary inputs to f, there exists a corresponding path to some duplicate gate of f.

[3] The notation $x + S < L$, for x a scalar and S a set, means that $x + s_i < L$ for all $s_i \in S$. $x + S \geq L$ has an analogous meaning.

Name	# Red.	Initial delay		Final delay	# Gates	
		Long.	True	True	Init.	sKMS
csa 8.2	8	36.8	24.0	20.8	97	100
csa 16.2	16	69.6	36.0	30.4	193	212
csa 16.4	8	59.2	35.2	32.0	185	202
csa 32.4	16	114.4	47.2	42.0	369	400
csa 32.8	8	104.0	57.6	54.4	361	386
csa 64.8	16	204.0	69.6	64.8	721	818
5xp1	1	30.8	25.2	24.0	118	116
rot	48	57.8	52.6	46.0	1004	857
clip	4	23.4	22.4	20.8	150	132
des	17	90.6	90.0	87.0	2007	2001
f51m	39	27.0	26.0	23.8	173	138
C1908	7	36.2	32.6	32.6	325	369
C6288	2	173.8	172.2	172.0	2353	2362
s713	38	43.6	40.0	32.0	154	131
s1238	67	34.4	32.0	30.6	429	409
s9234	224	40.8	38.8	33.0	1735	1535
s15850	310	88.8	87.4	78.2	3268	3013

sKMS : single-pass KMS algorithm
Final circuit verified against initial circuit

Table 1: sKMS algorithm on MCNC and ISCAS circuits

$$atimes(g4) = \{1,2,3,4\} \quad etimes(g4) = \{1\}$$
$$atimes(g5) = \{1,2,3,4,5\} \; etimes(g5) = \{0\}$$

The L-path-disjoint network is obtained by a reverse topological traversal. Since both $g5$ and $g4$ have single fanout, no duplication is done on these gates. $g3$ is duplicated to obtain gates $g31$ and $g32$ (second network from the top of the figure). $g31$ is connected to $g4$ while $g32$ is connected to $g5$. Now, $etimes(g31) = \{2\}$, and $etimes(g32) = \{1\}$, and the invariant of Lemma 3.1 is now satisfied for $g31$ and $g32$. Similarly, on duplicating $g2$ and $g1$ as shown in the figure, an L-path-disjoint network for $L = 4$ is obtained (bottom of the figure). Any multiple stuck-fault on the inputs a and b of $g11$, and c of gate $g21$ can now be removed. By the theorems discussed earlier in this section, the computed delay and logical behavior of the resulting circuit remains unchanged.

3.1 Results using the single-pass algorithm

Results of an implementation of the single-pass algorithm for redundancy removal, guaranteeing the delay does not increase are shown in Table 1. Each circuit is composed of two-input gates and the delay of a gate is computed using a block delay of 1.0 plus a drive delay of 0.2 for each fanout connection. For each circuit the length of the longest sensitizable path, T, is first determined using an efficient timing analysis algorithm [4]. The smallest distinct path length $L > T$ is also known. Using the transformation of Figure 3, an L-path-disjoint circuit is derived from the original circuit. Finally, using Theorem 3.1 and 3.2, the first edge of every path of length $\geq L$ is set to a constant. The topologically longest path in the resulting circuit is now of length $< L$ ($\leq T$). Finally, standard redundancy removal is used to derive an irredundant circuit that is no slower than the original circuit. The algorithm provides the same effect as the KMS algorithm in a single-pass following the timing analysis phase. Often the resulting circuits are identical. The number of operations performed by the algorithm is linear in the number of connections of the original circuit and the number of distinct path lengths $\geq T$ in η.

The first column is the name of the circuit and the second indicates the initial number of redundancies. The third and fourth columns give the longest path delay (or topological delay) and the computed delay (using viability analysis) of the initial circuit, respectively. The fifth column gives the computed delay of the irredundant circuit obtained after application of the sKMS algorithm. The last two columns compare the size of the initial redundant circuit against the irredundant circuit obtained by applying the sKMS algorithm. The first set of examples in the table are different sized carry-skip adders [1]. The second set are optimized MCNC and ISCAS benchmark examples. Although not shown in the table, direct redundancy removal on the initial circuit leads to an increase in delay in several circuits.

The single-pass algorithm completes on all the circuits experimented with. The CPU time on a DEC 5000, not including the timing analysis phase and the final redundancy removal, is only a few seconds even for the largest example. Similar results have been observed on *mapped* circuits that use technology dependent delay information, rather than the unit-fanout delay model of these experiments.

4 Conclusions

In this paper we resolve an important drawback of the KMS [1] algorithm. On circuits with large numbers of false paths, the iterative loop of duplication and redundancy removal in the original algorithm leads to huge amounts of running time. Based on the notion of an L-path-disjoint network, where the paths of an input connection of the network are either all of length $\geq L$ or all of length $< L$, for some selected L, an efficient single pass algorithm is developed. The testability of a multiple stuck-fault on the L-path-disjoint network is related to the viability of paths (of length $\geq L$) in the original network. Thus, by a single multiple stuck-fault redundancy removal, any given redundant circuit with false paths is converted to a circuit with a longest sensitizable path. The time to derive an L-path-disjoint network is proportional to the size of the original network. Although a tight area bound by the KMS algorithm is still open[4], empirically, the area penalty incurred is very small.

In summary, application of the single-pass KMS algorithm performs the efficient removal of long false paths while guaranteeing that the delay does not increase, achieves full (single stuck-fault or multiple stuck-fault) testability, and increases the reliability of the circuit at a small expense in area. These results are a step towards resolving the question of whether it is ever necessary to synthesize circuits with long false paths.

References

[1] K. Keutzer, S. Malik, and A. Saldanha. Is redundancy necessary to reduce delay? *IEEE Transactions on Computer-Aided Design*, C-10(4):427–435, April 1991.

[2] P. McGeer and R. Brayton. Provably correct critical paths. In *The Proceedings of the Decennial Caltech VLSI Conference*, 1989.

[3] P. McGeer, R. Brayton, R. Rudell, and A. Sangiovanni-Vincentelli. Extended stuck-fault testability for combinational networks. In *The Proceedings of the 6th MIT Conference on Advanced Research in VLSI*, pages 239–259. MIT Press, April 1990.

[4] P. McGeer, A. Saldanha, P. Stephan, R. Brayton, and A. Sangiovanni-Vincentelli. Timing analysis and delay-fault test generation using path recursive functions. In *The Proceedings of the International Conference on Computer-Aided Design*, pages 180–183, November 1991.

[5] A. Saldanha. Performance and testability interactions in logic synthesis. *Ph.D. Thesis, University of California - Berkeley*, October 1991.

[4] A weak bound of nG for the area increase has been proved where G is the number of gates in the circuit and n is the number of distinct path lengths in the circuit between the longest viable path and the longest path length.

Circuit Enhancement by Eliminating Long False Paths

Hsi-chuan Chen*† David H.C. Du* Siu Wing Cheng‡
Department of Computer Science
University of Minnesota

Abstract

One way to improve circuit performance is to introduce redundancies into the circuit. However, this results in higher number of long false paths in the circuit and the testability of the circuit is reduced. Keutzer et al [4] developed a long false path elimination algorithm to remove the redundancies without sacrificing its performance. We propose a more efficient algorithm and two very effective heuristics to remove long false paths. Our experimental results show that the proposed algorithm and heuristics are feasible for large designs and the increase of circuit area is significantly reduced.

1 Introduction

Due to the fast increasing demands for high-speed computation and highly reliable designs, the quality of a circuit is now evaluated by three criteria: area, performance, and testability. Therefore, in addition to minimizing circuit area, current synthesis systems also attempt to speed up circuit performance and to enhance circuit testability. However, these three objectives are usually conflicting to one another. The performance of the circuit is defined to be the delay of the *longest sensitizable* paths in the circuit [2]. Therefore, one way to enhance the performance of a circuit is to add stuck-at redundancies to make the current longest paths false [4, 6]. However, the existence of stuck-at redundancies is detrimental to circuit testability, because it makes test pattern generation difficult. Furthermore, a side-effect of this performance optimization approach is that many long false paths are created. As indicated in [7], the existence of long false paths in a circuit usually results in low delay-fault detection coverage.

Keutzer, Malik, and Saldanha [4] are the first to propose an algorithm (denoted as KMS algorithm) which not only enhances circuit testability but also retains circuit performance. KMS algorithm is used to eliminate all long false paths in a circuit [4]. When there is no long false path, the delay-fault detection coverage of the circuit increases. Also, most stuck-at redundancies are removed because of the elimination of long false paths. So, the circuit testability is improved.

KMS algorithm is based on the following observation on a long false path in a circuit: if each gate of the false path

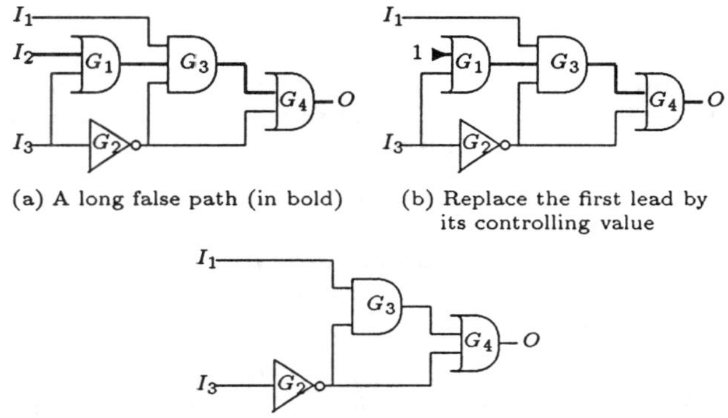

(a) A long false path (in bold) (b) Replace the first lead by its controlling value

(c) New implementation

Figure 1. KMS algorithm

has **single fanout**, then the **first** lead of the false path is both stuck-at-0 and stuck-at-1 redundant. Therefore, it can be replaced by either logic 0 or logic 1. By replacing the first lead of the path by the controlling value to its succeeding gate, the value at the output of the succeeding gate is fixed. Therefore, the succeeding gate as well as all its input leads can be removed. Since the value at the output of the succeeding gate is fixed, all leads originating from the gate need to be replaced by the fixed value. These induced replacements can further remove more gates and leads in the circuit. As a result, the long false path is shortened and circuit area decreases. A long false path is considered to be *eliminated* if it becomes not longer than the longest sensitizable paths. Figure 1 is an example of the logic value replacement. A long false path may contain gates with multiple fanouts. For such a long false path, the KMS algorithm first duplicates gates and leads (path isolation) so that each gate of the path in the modified circuit has single fanout before shortening it. The KMS algorithm repeats the long false path elimination process until the longest path in the final circuit is sensitizable.

The KMS algorithm has been shown to successfully eliminate all long false paths in small circuits. However, it may not be able to handle large circuits due to excessive gate/lead duplications in a path isolation and limited gate/lead removals in a logic value replacement. The reasons are as follows. First, each long false path in a large circuit may contain many gates. Very likely, most gates have multiple fanouts. Therefore, in order to isolate a complete long false path, the KMS algorithm may need to duplicate quite a lot of gates and leads, and consequently

*Research supported in part by NSF grant MIP-9007168.
†Current address: AT&T Bell Labs., Murray Hill, NJ 07974.
‡Current address: Dept of Computer Science, HKUST, HK.

0738-100X/92 $3.00 © 1992 IEEE

the area of the circuit increases dramatically.

In this paper, we propose to identify and perform logic replacement for a portion, called **segment**, of a long path instead of the path itself. This results in less gate and lead duplication and hence reduction in the increase of circuit area. We give sufficient conditions for the choice of segments that can retain the performance of the circuit. If we relax the restriction on the choice of segments, then we obtain two very effective heuristics for removing long false paths. Our experimental results show that these two heuristics are much more superior in reducing the increase in circuit area. Moreover, the performances of all the circuits tested are retained.

2 Definitions

Definition 1 [1] *A path is said to be* statically sensitizable *if there exists at least one primary input vector that stabilizes all side-inputs of P at their non-controlling values. A path is considered to be a* statically false *path if it is not statically sensitizable.*

According to [1, 2], a long false path must be statically false, but the critical path may or may not be statically sensitizable.

Definition 2 (SFS) *Partial path Q is called a* Statically False Segment *if there exists no primary input vector that stabilizes all side-inputs of Q at their non-controlling values.*

It is important to note that any partial path including an SFS segment is also an SFS segment. A statically false path itself is the largest SFS segment of the path. Thus, the KMS algorithm only deals with SFS segments that are paths by themselves.

Definition 3 (SZES (SOES)) *Partial path Q is called a* Statically Zero-Enforced (One-Enforced) Segment *if each primary input vector which stabilizes all side-inputs of Q at their non-controlling values also stabilizes the first lead of Q at logic* 0 (1).

We will use *segment* as a collective name for the SFS segment, the SZES segment, and the SOES segment.

3 Selecting A Segment

The first lead of an SFS segment is replaceable by logic 0 or logic 1 without altering the functionality of the circuit, if each gate of the segment has single fanout. Similarly, the first lead of an SZES (SOES) segment is replaceable by logic 0 (1) if each gate of the segment has single fanout. Figure 2 is an example of performing logic replacement for an SFS segment.

When any gate in a segment has multiple fanouts, we need to isolate the segment. To isolate a segment, all the gates (and so their associated input leads) between the last gate with multiple fanouts in the segment and the first gate of the segment need to be duplicated. We then replace the first lead of the segment by an appropriate logic value.

However, performing logic value replacement for an arbitrary segment may not retain the circuit performance.

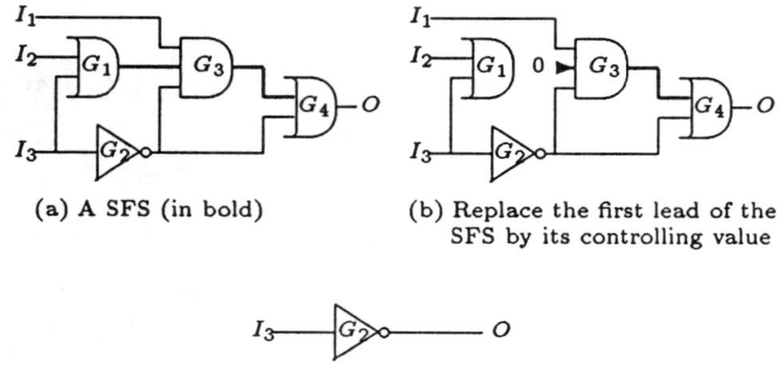

(a) A SFS (in bold)

(b) Replace the first lead of the SFS by its controlling value

(c) New implementation

Figure 2. SFS elimination

Suppose that τ is the circuit delay of the original circuit, then we select the *time-constrained segments* to guarantee that the delay of the modified circuit will not be greater than τ. We first define time-constrained SFS segments as follows.

Definition 4 *A* **Time-Constrained SFS Segment** $S = (f_s, G_{s+1}, \cdots, f_{t-1}, G_t)$ *is an SFS segment that satisfies the conditions C1 or C2 for all* $s+2 \le k \le t$:

C1: $Min\_in(f_s) + \sum_{i=s+1}^{k-1} d(G_i) \ge Max\_in(FI(G_k) - \{f_{k-1}\})$

C2: $Max\_in(FI(G_k) - \{f_{k-1}\}) + d(G_k) + Max\_out(G_k) \le \tau$

where $Min\_in(f_s)$ *is the minimum delay from the primary inputs to* f_s*, and* $d(G_i)$ *is the gate delay of* G_i*, and* $Max\_in(FI(G_k) - \{f_{k-1}\})$ *is the maximum delay from the primary inputs to the inputs of* G_k *other than* f_{k-1}*, and* $Max\_out(G_k)$ *is the maximum delay from* G_k *to the primary outputs.*

The time-constrained SZES and SOES can be defined similarly. However, if logic 0 (resp. logic 1) is non-controlling to the first gate of an SZES (resp. SOES), then k should vary from $s+1$ to t instead of from $s+2$ to t in the above.

Using time-constrained segments, the following result can be established. We omit the proof and the details can be found in [3].

Theorem 1 *The circuit delay will not increase after performing isolation and logic value replacement for a time-constrained segment.* □

The additional time constraint has the effect of pulling the beginning of a segment towards the primary inputs. Thus, a better lower bound for $Min\_in(\cdot)$ and better upper bounds for $Max\_in(\cdot)$ and $Max\_out(\cdot)$ should produce time-constrained segments of shorter lengths.

Remark : *If the delay* τ *of the original circuit is not known, then the alternative condition C2 can be dropped without affecting the correctness of the algorithm. Then the algorithm just runs until no time-constrained segment can be found along the longest paths of a circuit. This restriction has the advantage of avoiding timing verification as a preliminary step (which is often quite time-consuming) as well as allowing for possible improvement in circuit performance. But this restriction also results in a stricter time constraint which implies that we may work with longer segments.*

4 Long False Paths Elimination

Before we describe our algorithm, we first elaborate on how to determine whether a partial path is a segment. Let $Q = (f_0, G_1, f_1, \cdots, f_{t-1}, G_t)$ be a partial path. The initial condition set for Q is composed of the conditions requiring all side-inputs of Q at their non-controlling values. Using the implication and justification procedures in D-algorithm [1], these conditions are propagated both forward and backward and hence new conditions are generated and are added into the condition set. This condition propagation process continues until either a conflict is detected in the final condition set or no more new conditions can be added into the final condition set. For the case when a conflict is detected in the final condition set, Q is determined to be an SFS segment. For the case when there is no conflict in the final condition set: if f_s is assigned a logic 0 (resp. logic 1) in the final condition set, Q is determined as an SZES (resp. SOES) segment; otherwise, Q is not a segment.

For comparison (with the KMS algorithm), we define our long false paths elimination algorithm to be repeated iterations (runs) of path-tracing, time-constrained segment identification, segment isolation, and logic value replacement. The algorithm continues until the longest path in the final circuit contains no time-constrained segment. A longest path is traced by Best First Search. Then we identify a time-constrained segment along this path as follows.

Let $P = (I, f_0, G_1, f_1, G_2, \cdots, G_m, f_m, O)$ be the path traced. We divide the identification process into two phases. In Phase 1, we first locate gate G_t in P closest to I such that partial path $Q = (f_0, G_1, \cdots, f_{t-1}, G_t)$ is a segment. Then, we locate gate G_{s+1} in Q closest to G_t in Phase 2 such that partial path $S = (f_s, G_{s+1}, f_{s+1}, \cdots, f_{t-1}, G_t)$ is a time-constrained segment. In Phase 1, partial path Q is initialized to be (f_0, G_1). If Q is determined to be a segment, Phase 1 stops. Otherwise, we expand Q forward to include the next pair of lead and gate. Then, the expanded Q is under determination again. This expansion-and-determination process continues until Q is determined to be a segment. Let the final Q of Phase 1 be $(f_0, G_1, f_1, G_2, \cdots, f_{t-1}, G_t)$. Phase 2 is similar to Phase 1. The partial path S is initialized to be (f_{t-1}, G_t) and we expand S backward until S is determined to be a segment. Then we verify the conditions C1 or C2. If the conditions C1 or C2 are satisfied, then S is a time-constrained segment and we can stop. Otherwise, we expand S backward by one lead and one gate and perform the verification again. In our implementation, we use the topological shortest delay (resp. longest delay) to approximate $Min\_in(\cdot)$ (resp. $Max\_in(\cdot)$ and $Max\_out(\cdot)$).

After a time-constrained segment is identified, we duplicate gates and leads if necessary to isolate the segment. Then, the first lead of the segment is replaced by an appropriate logic value according to the type of the segment. The effect of the replacement will be fully propagated so as to remove as many gates and leads as possible.

5 Two Very Effective Heuristics

If the segments could be chosen freely without worrying about the additional time constraint, then much less increase in circuit area is expected. The reasons are as follows.

First, much shorter segments may possibly be obtained if the additional time constraint is disregarded. Much fewer gates and leads would be duplicated during isolation. Second, gate/lead removals in a logic value replacement depends on the extent of its propagation. The extent of the forward propagation of a logic value replacement is often limited, because AND gates and OR gates of a path usually interleave in reality. Replacing a lead of a path by the controlling value usually fixes the next lead at the non-controlling value, and consequently the forward propagation stops. Most gate/lead removals corresponding to a logic value replacement are due to the backward propagation in the input cone. When a lead is removed, if it is the only output lead of its originating gate, the gate and all its input leads are useless and can be removed too. Hence, if a segment whose first lead is far from the primary inputs could be chosen, then many gates and lead may possibly be removed in a logic value replacement. Furthermore, all the paths passing through that segment are eliminated in one run.

Unfortunately, such an approach cannot, in theory, retain the performance of a circuit because a long false partial path that merges with the segment at some point may become sensitizable. Thus, theoretically speaking, a long false path may become sensitizable and the circuit delay may then increase (a more detailed explanation is offered in [3]). However, our experimental results for the two heuristics actually show the opposite. Indeed, the performance of all the circuits in our experiments are retained. It is probably due to the fact that in circuits, which are speeded up by introducing redundancies to bypass long paths, short bypasses are likely to meet a long path close to the primary outputs. Therefore, the long partial path that becomes sensitizable will be blocked by some (earlier) controlling inputs later. Hence, the circuit performance will most probably not be affected.

The two heuristics differ from the algorithm in Section 4 only in the identification of segments. Since it is important to identify segments that are as short as possible, we give the definition of a prime segment as follows.

Definition 5 *A segment is a prime segment if no proper partial path of the segment is a segment.*

It is very time consuming to identify the best prime segment. Instead, we propose two prime segment identification procedures: one is to identify the prime segment closest to the primary input of a path (front segment identification procedure, denoted as FSIP), and the other is to identify the prime segment closest to the primary output of the path (rear segment identification procedure, denoted as RSIP).

We first trace the longest path P. In FSIP, we locate the shortest segment on P that begins at the primary inputs. Let it be $(f_0, G_1, \cdots, f_{t-1}, G_t)$. Then we initialize a partial path S to be (f_{t-1}, G_t) and we expand S backward until S is determined to be a prime segment. The determination process is implemented as before using the D-algorithm. In RSIP, we first locate the shortest segment on P that ends at the primary outputs. Let it be $(f_s, G_{s+1}, \cdots, G_m, f_m, O)$. Then starting with S equal to (f_s, G_{s+1}), we expand S forward until S is determined to be a prime segment.

Circuit Name	No. of Gates	Topological Delay (ns.)	Critical Delay (ns.)
C1908	913	76.60	66.70
C2670	1426	86.90	85.10
C3540	1719	98.70	95.20
C5315	2485	99.30	95.70
C6288	2448	319.90	316.30
C7552	3719	85.30	83.80

Table 1. Characteristics of ISCAS85 benchmark circuits

We denote the heuristics that uses FSIP and RSIP to identify prime segments by FLPEP and RLPEP, respectively.

6 Experimental Results

We have implemented our long false path elimination algorithm (denoted by LPEA) and FLPEP and RLPEP in C on a SUN Sparcstation. Due to the concern on computational efficiency, we use a simplified D-algorithm as in [1], which implements logic implication only, in identifying segments. We have experimented all of them on some ISCAS85 circuits. Table 1 shows the characteristics of each ISCAS85 circuit tested.

Table 2 shows our experimental results. For comparison, we also implemented the KMS algorithm. In Table 2, τ represents the circuit delay after eliminating all long false paths. $\Delta(G)$ denotes the net gate increase and $\#(R)$ the number of runs needed to eliminate all the long paths. Notation "-" inside Table 2 denotes the failure of completion due to $\Delta(G) > 20,000$. As expected, LPEA outperforms the KMS algorithm. LPEA results in much smaller increase in circuit area than the KMS algorithm does. For circuits that both LPEA and the KMS algorithms successfully terminate, the average area increase ratio of LPEA to the KMS algorithm is less than 17%. The two heuristics FLPEP and RLPEP show much superior ability in reducing the increase in circuit area. Also, the performances of all the circuits tested are retained. Both FLPEP and RLPEP successfully terminates even for circuits like C1908 and C3540, and RLPEP even terminates for C6288. The average area increase ratio of FLPEP and RLPEP to the KMS algorithm is less than 3%.

According to our experimental results, RLPEP outperforms FLPEP in most circuits (i.e., C2670, C3540, and C6288). Possible reasons are as follows. Since a RSIP prime segment is closer to the primary output, performing logic replacement for a RSIP prime segment will give us a larger reduction in path length. Therefore, RLPEP tends to use fewer runs than FLPEP to eliminate all long false paths. Furthermore, a RSIP prime segment is closer to the primary outputs and hence the input cone of the succeeding gate of its first lead may be large. As a consequence, RLPEP very likely removes more gates and leads than FLPEP in a logic value replacement.

Circuit C6288 is the most surprising example. While FLPEP fails to complete, RLPEP efficiently eliminates all long false paths with a negligible area increase as well as an improvement in performance. This can be explained by the fact that performing logic replacement for segments may also reduce the delay of the longest sensitizable paths. The best known estimation on its critical delay before elimination is 316.30 nano seconds [5].

However, as shown in our experiments there are some circuits (e.g., C5315 and C7552) on which FLPEP outperforms RLPEP. That might be because the number of long false paths sharing an FSIP prime segment is larger than that sharing the corresponding RSIP prime segment. In this case, FLPEP can eliminate more long false paths in one run than RLPEP does.

Acknowledgement

We would like to thank an anonymous referee for the comments that lead to the discovery of an error in an earlier version of this paper.

References

[1] J. Benkoski and et al. Timing Verification Using Statically Sensitizable paths. *IEEE TCAD*, cad-9(10):1073–1084, 1990.

[2] H. Chen and David Du. Path Sensitization in Critical Path Problem. In *ICCAD*, pages 208–211, 1991.

[3] Hsi chuan Chen, David H.C. Du, and Siu Wing Cheng. Circuit enhancement by eliminating long false paths. Technical Report TR 91–45, University of Minnesota, Department of Computer Science, 1991.

[4] K. Keutzer, S. Malik, and A. Saldanha. Is Redundancy Necessary to Reduce Delay? In *27th DAC*, pages 228–234, 1990.

[5] L. Liu, D. Du, and H. Chen. An efficient parallel critical path algorithm. In *28th DAC*, pages 535–540, 1991.

[6] P. McGeer, R. Brayton, S. Sahni, and A. Sangiovanni-Vincentelli. Performance Enhancement through the Generalized Bypass Transform. In *ICCAD*, pages 184–187, 1991.

[7] S. Patil and S. Reddy. A Test Generation System for Path Delay Faults. In *ICCAD*, pages 40–43, 1989.

Circuit	τ (ns.)	*KMS*		*LPEA*		*FLPEP*		*RLPEP*	
		Δ(G)	#(R)	Δ(G)	#(R)	Δ(G)	#(R)	Δ(G)	#(R)
C1908	66.70	-	-	-	-	784	354	784	354
C2670	85.10	502	162	79	12	39	10	14	7
C3540	95.20	-	-	2624	689	1580	472	356	74
C5315	95.70	5558	684	441	69	82	19	266	49
C6288	313.70	-	-	-	-	-	-	114	425
C7552	83.80	123	7	33	7	11	5	39	3

Table 2. Experimental results

Estimation of Average Switching Activity in Combinational and Sequential Circuits

Abhijit Ghosh
Mitsubishi Electronics America

Srinivas Devadas
MIT, Cambridge

Kurt Keutzer
Synopsys

Jacob White
MIT, Cambridge

Abstract

We address the problem of estimating the average power dissipated in VLSI combinational and sequential circuits, under random input sequences. Switching activity is strongly affected by gate delays and for this reason we use a general delay model in estimating switching activity. Our method takes into account correlation caused at internal gates in the circuit due to reconvergence of input signals. In sequential circuits, the input sequence applied to the combinational portion of the circuit is highly correlated because some of the inputs to the combinational logic are flip-flop outputs representing the state of the circuit. We present methods to probabilistically estimate switching activity in sequential circuits. These methods automatically compute the switching rates and correlations between flip-flop outputs.

1 Introduction

We address the problem of estimating the average power dissipated in combinational and sequential VLSI circuits given random input sequences. This measure can be used to make architectural or design-style decisions during the VLSI synthesis process.

Probabilistic methods for power or current estimation are attractive because statistical estimates can be obtained without recourse to time-consuming exhaustive simulation. In the past, probabilistic peak current estimation methods (*e.g.*, [5]) that compute probabilistic current waveforms for combinational circuits have been developed. Estimation of worst-case power dissipation (*e.g.*, [10], [7]) is a difficult problem requiring a branch-and-bound search and these methods have been applied to small to moderate sized circuits.

The problem of estimating average power in combinational circuits can be reduced to one of computing signal probabilities [1] of a multilevel circuit derived from the original circuit by a process of symbolic simulation. The work closest to our own is the work on transition density calculation by Najm [11]. Transition densities correspond to average switching rates for gates in the circuit. In [11], an interconnection of combinational logic modules, each with a certain delay, makes up a circuit. Transition densities are propagated through combinational logic modules without regard to their structure. Correlations between internal lines due to reconvergence are ignored during propagation. It is possible to take into account correlations by lumping all the modules into one large module, but in this case the information regarding the delay of the individual modules is lost. Cirit in [8] gives methods to calculate dynamic power dissipation based on approximate signal probability evaluation procedures.

Our work improves upon the state-of-the-art in several ways. We use a general delay model for combinational logic in our symbolic simulation method, which correctly computes the Boolean conditions that cause *glitching* (multiple transitions at a gate) in the circuit. In some cases, glitching may account for a significant percentage of the dissipated power or switching activity. Symbolic simulation produces a set of Boolean functions that represent the conditions for switching at different time points, for each gate in the circuit. Given input switching rates, we can use various exact or approximate methods to compute the probability of each gate switching at any particular time point. We then sum these probabilities over all the gates to obtain the expected switching activity in the entire circuit over all the time points corresponding to a clock cycle. Our method takes into account correlation caused at internal gates in the circuit due to reconvergence of input signals (reconvergent fanout).

Most VLSI circuits are sequential. We make an important extension to our estimation algorithms to handle sequential circuits. The combinational part of the circuit receives primary inputs as well as inputs from flip-flop outputs. Given the pin restrictions on VLSI chips, the number of flip-flops in large sequential circuits can be many times the number of primary inputs. Previous methods for power estimation assume uniform switching rates for all the inputs to the combinational logic or expect that switching rates are pre-computed using methods like random simulation. Further, the correlation [1] between the flip-flop outputs between one time frame and the next is ignored. Using inaccurate switching rates and ignoring this correlation can cause substantial errors in power estimation, especially for datapath circuits. We develop a method that ***automatically computes the switching rates*** for each of the flip-flop inputs, ***taking into account this correlation between the flip-flops.***

[1] For example, flip-flop A may only switch from $0 \rightarrow 1$ when flip-flop B switches from $1 \rightarrow 0$.

0738-100X/92 $3.00 © 1992 IEEE

2 Preliminaries

2.1 A Power Dissipation Model

The amount of energy dissipated by a CMOS logic gate each time its output changes is roughly equal to the change in energy stored in the gate's output capacitance. If the gate is part of a synchronous digital system controlled by a global clock, it follows that the average power dissipated by the gate is given by:

$$P_{avg} = 0.5 \times C_{load} \times (V_{dd}^2 / T_{cyc}) \times E(transitions) \quad (1)$$

where P_{avg} denotes the average power, C_{load} is the load capacitance, V_{dd} is the supply voltage, T_{cyc} is the global clock period, and $E(transitions)$ is the *expected value* of the number of gate output transitions per global clock cycle[11], or equivalently the average number of gate output transitions per clock cycle. All of the parameters in (1) can be determined from technology or circuit layout information except $E(transitions)$, which depends on both the logic function being performed and the statistical properties of the primary inputs.

2.2 Static Probabilities

Consider the case of dynamic CMOS logic (e.g. Domino). At the beginning of each clock cycle, all the gates are precharged, and gates make transitions only if their associated Boolean functions are satisfied. For example, a three-input AND-OR gate's Boolean function might be

$$(i_1 \cdot i_2) \vee (i_2 \cdot i_3), \quad (2)$$

where i_1, i_2, and i_3 are primary inputs. In this case, the expected value of the number of transitions at this gate's output is

$$E(transitions) = 2 \times P((i_1 \cdot i_2) \vee (i_2 \cdot i_3) = 1) \quad (3)$$

where $P(x)$ is defined as probability that x is true, and the factor of two in the equation accounts for the reset transition during precharge.

To evaluate (3), it is necessary to determine the primary input probabilities. We assume that *primary inputs are uncorrelated*, and that each is a waveform in time whose value is either zero or one, changing instantaneously at global clock edges. Assuming ergodicity without further comment, the probability of a particular input i_j being one at a given point in time, denoted $p_j{}^{one}$, is given by

$$p_j{}^{one} = \lim_{N \to \infty} \frac{\sum_{k=1}^{N} i_j(k)}{N} \quad (4)$$

where N is the total number of global clock cycles and $i_j(k)$ is the value of input i_j during clock cycle k. Clearly, the probability that i_j is zero at a given point in time, denoted $p_j{}^{zero}$ is

$$p_j{}^{zero} = 1 - p_j{}^{one}$$

We refer to $p_j{}^{zero}$ and $p_j{}^{one}$ as static probabilities. Note that

$$P((i_1 \cdot i_2) \vee (i_2 \cdot i_3) = 1) \neq p_1^{one} p_2^{one} + p_2^{one} p_3^{one}$$

because the first and second product terms are not independent. Rather,

$$P((i_1 \cdot i_2) \vee (i_2 \cdot i_3) = 1) =$$

$$P((i_1 \cdot i_2) \vee (\overline{i_1} \cdot i_2 \cdot i_3) = 1) = p_1^{one} p_2^{one} + p_1^{zero} p_2^{one} p_3^{one}$$

where the second equality holds because $i_1 \cdot i_2$ is disjoint from $\overline{i_1} \cdot i_2 \cdot i_3$. In this example, $(i_1 \cdot i_2) \vee (\overline{i_1} \cdot i_2 \cdot i_3)$ represents a disjoint cover for the logic function, and the terms $i_1 \cdot i_2$ and $\overline{i_1} \cdot i_2 \cdot i_3$ are referred to as *cubes* in the cover[2]. The equivalent logical expression, $(i_1 \cdot i_2) \vee (i_2 \cdot i_3)$, does not represent a disjoint cover because $i_1 \cdot i_2 \cdot i_3$ is contained in both cubes $i_1 \cdot i_2$ and $i_2 \cdot i_3$.

In general, given a disjoint cover for a Boolean function of uncorrelated inputs described by static probabilities, it is easy to determine the probability of the function evaluating to a 1. The procedure is given in the following two theorems whose proof follows directly from elementary probability[12].

Theorem 2.1 : *Given any disjoint cover for a Boolean function, the probability of the function evaluating to a 1 is equal to the sum of the probabilities of each of the cubes in the cover evaluating to a 1.*

Theorem 2.2 : *Given a logical function of uncorrelated inputs in the form*

$$f = i_{\alpha_1} \cdot i_{\alpha_2} \cdot \dots \cdot i_{\alpha_M} \cdot \overline{i_{\beta_1}} \cdot \overline{i_{\beta_2}} \cdot \dots \cdot \overline{i_{\beta_N}}$$

where the i_{α_j}*'s are nonnegated inputs and the* $\overline{i_{\beta_k}}$*'s are negated inputs, then*

$$P(f = 1) = p_{\alpha_1}^{one} \cdot p_{\alpha_2}^{one} \cdot \dots \cdot p_{\alpha_M}^{one} \cdot p_{\beta_1}^{zero} \cdot p_{\beta_2}^{zero} \cdot \dots \cdot p_{\beta_N}^{zero}$$

2.3 Transition Probabilities

For static combinational CMOS logic, a gate output can only change when its inputs change, and then only if the Boolean function describing the gate evaluates differently. For example, a 2-input AND gate's output will change between clock cycle t and $t+1$ if

$$(i_1(t) \cdot i_2(t)) \oplus (i_1(t+1) \cdot i_2(t+1)) \quad (5)$$

evaluates to 1, where $i_1(t)$, $i_2(t)$ and $i_1(t+1)$, $i_2(t+1)$ are the inputs to the gate at clock cycle t and $t+1$ respectively. The disjoint cover for (5) is

$$(i_1(t) \cdot i_2(t)) \cdot \overline{(i_1(t+1))} \vee (i_1(t) \cdot i_2(t)) \cdot (i_1(t+1) \cdot \overline{i_2(t+1)}) \vee$$

$$\overline{i_1(t)} \cdot (i_1(t+1) \cdot i_2(t+1)) \cdot \vee (i_1(t) \cdot \overline{i_2(t)})(i_1(t+1) \cdot i_2(t+1)). \quad (6)$$

It is not possible to use Theorem (2.2) to evaluate the probability of (6), because an input at time $t+1$ is correlated to its behavior at time t (as in a sequential circuit). Instead, transition probabilities for the transitions $0 \rightarrow 0$, $0 \rightarrow 1$, $1 \rightarrow 0$ and $1 \rightarrow 1$ must be used. We denote these transition probabilities by ${p_i}^{00}$, ${p_i}^{01}$, ${p_i}^{10}$, and ${p_i}^{11}$, respectively, where for example ${p_i}^{10}$ is defined by:

$$ {p_j}^{10} = \lim_{N\rightarrow\infty} \frac{\sum_{k=1}^{N} i_j(k)\overline{i_j(k+1)}}{N} \tag{7} $$

The other transition probabilities follow similarly.

Static probabilities can be computed from transition probabilities, but *not* vice-versa, because of correlation between one time frame and the next. So in general it is necessary to specify transition probabilities. The relations between static probabilities and transition probabilities follow directly from the definitions in (4) and (7), specifically,

$$ \begin{aligned} {p_j}^{zero} &= {p_j}^{00} + {p_j}^{01} \\ {p_j}^{one} &= {p_j}^{11} + {p_j}^{10} \end{aligned} \tag{8} $$

Both static and transition probabilities are used to compute $E(transitions)$ for static logic circuits, as can be seen from the expression for the probability that (6) evaluates to a 1,

$$ p_1^{10}\cdot p_2^{one} + p_1^{11}\cdot p_2^{10} + p_1^{01}\cdot p_2^{one} + p_1^{11}\cdot p_2^{01} \tag{9} $$

which for this example is also $E(transitions)$. For all primary inputs, it may be assumed that successive input vectors are uncorrelated and a 1 or a 0 are equally likely. Therefore, all transition probabilities may be be assumed to be 0.25, and all static probabilities to be 0.5.

2.4 General Combinational Networks

The algorithm for computing the average power dissipated in a dynamic CMOS combinational network follows directly from the approach in Section (2.2). That is, for each gate g_i in a logical network, we first determine the Boolean function of the gate, f_i, in terms of the network's primary inputs. Then we find a disjoint cover for f_i, and use Theorem (2.1) and (2.2) to evaluate $P(f_i = 1)$.

The average power dissipation for the network is then

$$ \begin{aligned} P_{avg} &= (V_{dd}^2/T_{cyc}) \times \sum C_i \times E(transitions\ of\ g_i) \\ &= (V_{dd}^2/T_{cyc}) \times \sum C_i \times P(f_i = 1) \end{aligned} \tag{10} $$

where C_i is the load capacitance of the ith gate and the summation is over all gates in the circuit. Note that (10) follows from the fact that the average value of a sum is equal to the sum of the average values, regardless of correlation[12].

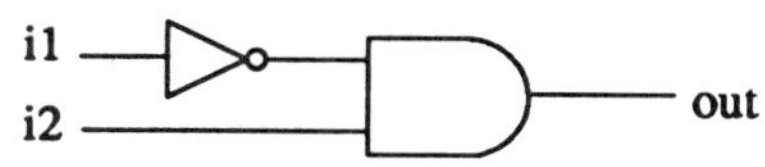

Figure 1: Glitching in a Static CMOS Circuit

A similar overall approach can be used to compute the average power dissipated in a static CMOS combinational network. However, as described in Section (2.3), a two-vector input sequence is required to stimulate activity in static gates. Therefore, the Boolean function for static gate output activity is different than that for a dynamic gate, and the probability of the function being satisfied requires transition probabilities for evaluation. In particular, assuming negligible gate delays, a static CMOS combinational logic gate's output will switch with a change in the primary input vector from $V0$ to Vt if:

$$ \begin{aligned} f_i = &((h_i(V0) = 0) \wedge (h_i(Vt) = 1)) \bigvee \\ &((h_i(V0) = 1) \wedge (h_i(Vt) = 0)) \end{aligned} \tag{11} $$

is satisfied, where h_i is the logic function corresponding to gate g_i's output.

The average power dissipated in the static network is then computed by using (10), with f_i being given by (11), and $P(f_i = 1)$ is evaluated using both transition and static probabilities, as described in Section (2.3).

Instead of enumerating disjoint covers, Binary Decision Diagrams (BDDs) [3] can be used for the calculation of signal probability. It has been shown in [6] and [11] that exact signal probability calculation for a given function can be performed by a linear traversal of a BDD representation of a logic function. We have implemented methods for signal probability calculation using BDDs.

3 Gate Delay Effects

As mentioned above, for static CMOS circuits, switching activity must be analyzed based on considering an input vector pair, denoted $< V0, Vt >$. If the gates have appreciable delays, there may be output glitches that can contribute significantly to dissipated power.

For instance, consider the circuit of Figure 1. Assuming that the delays of the inverter and the AND gate are both 1 time unit, if we apply the vector $i_1 = 0$, $i_2 = 0$, followed by $i_1 = 1$, $i_2 = 1$, we will obtain a glitch at the output *out*, which could cause power dissipation. Below, we present a symbolic simulation method that can be used to generate a multiple-output function that represents total switching activity over any possible input vector pair, assuming a general delay model for the gates in the circuit.

3.1 Symbolic Simulation

3.1.1 An Example

Consider the multilevel combinational circuit shown in Figure 2.

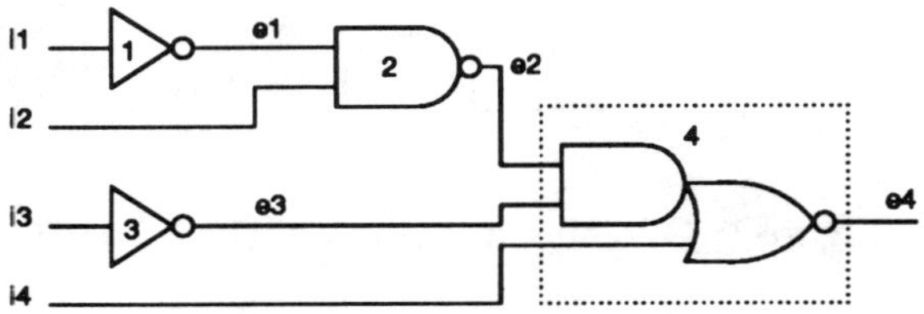

Figure 2: A multilevel combinational circuit

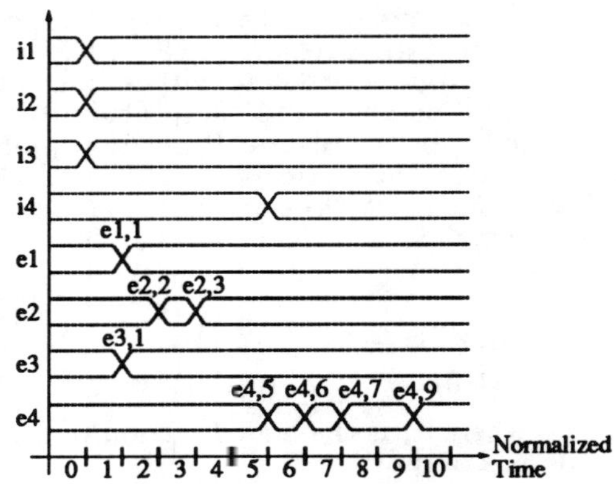

Figure 3: Signal waveforms for the primary inputs and the outputs of the logic gates

It has four primary inputs and one output and consists of four CMOS gates. Figure 3 shows the signal transitions at the primary input nodes as well as the possible transitions at the internal nodes of the network. The time points are in normalized units. The first three inputs, i_1, i_2, and i_3, switch simultaneously between time periods 0 and 1. The fourth input, i_4, is a late arriving signal that switches between the time points 5 and 6. In this example, the delays of both Gate 1 and Gate 3 are one time unit. Gate 2 has a delay of two units while Gate 4 has a delay of four units.

Also shown in Figure 3 are the waveforms, e_i's, representing the signals at the outputs of ith logic gates. Each of the possible transitions, $e_{i,j}$, represents either a low-to-high or high-to-low signal transition between $[j]^{th}$ and $[j+1]^{th}$ time points. The number of all possible transitions at a gate output may equal the sum of all possible transitions at the gate inputs. These transitions are delayed by the gate's propagation delay.

3.1.2 Unit-Delay Model

Even under the idealization of a unit-delay model, the gate output nodes of a multilevel network can have multiple transitions in response to a two-vector input sequence. In fact, it is possible for a gate output to have as many transitions as levels in the network.

We construct the Boolean functions describing the gate outputs at the discrete points in time implied by the unit-delay model. That is, we consider only discrete times $t,\ t+1,\ \ldots,\ t+l$, where t is the time when the inputs change from $V0$ to Vt, and l is the number of levels in the network. For each gate output i, we use symbolic simulation to construct the $l+1$ Boolean functions $f_i(t+j)$, $j \in 0,\ \ldots,\ l$ which evaluate to 1 if the gate's output is 1 at time $t+j$. Note that as we assume no gate has zero delay and that the network has settled before the inputs are changed from $V0$ to Vt, $f_i(t)$ is the logic function performed for $V0$ at the ith gate output. Finally, we can determine whether a transition occurs at a boundary between discrete time intervals $t+j$ and $t+j+1$ by exclusive-OR'ing $f_i(t+j)$ with $f_i(t+j+1)$.

3.1.3 General Delay Model

A gate with a large fanin may have several times the delay of an inverter. If one uses normalized time units, one can always introduce unit-delay buffers at the output of gates in a circuit, which have a delay greater than unity, in order to model differing delays among logic gates.

Further, all gates have *inertial delays*, *i.e.*, all gates require energy to switch their output nodes. For a transition at an input terminal to propagate through the gate, the state of the input signal before and after the transition must persist for a minimum duration equal to the delay of the gate. Any high frequency pulse at the inputs whose pulse duration is shorter than the gate delay (or some fraction of the gate delay) will be filtered out.

Our general symbolic simulator is able to simulate circuits with arbitrary gate transport and inertial delays (without introducing any unit delay buffers in the circuit). The simulator processes one gate at a time, moving from the primary inputs to the primary outputs of the circuit. For each gate, the possible transition times of its inputs are first obtained. Then, possible transitions at the output of the gate is derived, taking into account transport and inertial delays. If the inertial delay of a gate is non-zero, then all inputs to the gate must remain the same for a period equal to the inertial delay (or a fraction of it) for the output of the gate to make a transition.

4 Sequential Circuits

4.1 Introduction and Motivation

VLSI circuits are sequential, *i.e.*, they contain memory elements or flip-flops. This sequential nature of VLSI circuits makes the estimation of power dissipation more complicated than for combinational circuits.

A common model for a sequential circuit is shown in Figure 4. We will denote the vector pair applied to the combinational logic as $< V0,\ Vt >$. $V0$ and Vt have a primary input part and a present-state part. $V0$ is denoted $I0@P0$ and Vt is denoted $It@Pt$, where $I0$ and It correspond to the primary input parts, and $P0$ and Pt correspond to the present-state parts.

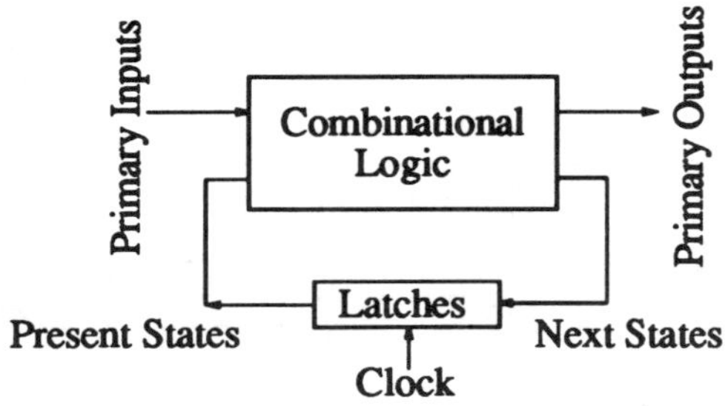

Figure 4: A General Synchronous Sequential Circuit

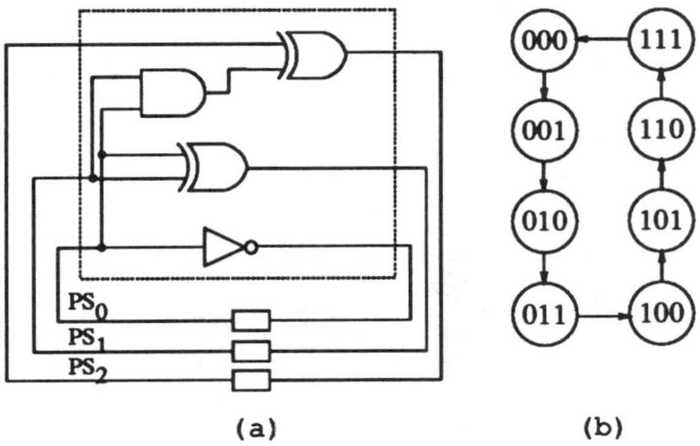

Figure 5: A 8-State Counter

One can ignore the feedback corresponding to the next-state lines and present-state lines, and estimate the power dissipated by the combinational logic using the methods presented in the previous sections. This strategy is a relatively crude approximation for two reasons as we will illustrate with a simple example.

Consider the STG of Figure 5(b). It represents the behavior of an 8-state autonomous counter, whose logic implementation is shown in Figure 5(a). Assuming that the counter can begin from any state with uniform probability, the probability of line PS_2 making a $1 \rightarrow 0$ transition is 0.125, and probabilities for $0 \rightarrow 1$, $0 \rightarrow 0$ and $1 \rightarrow 0$ transitions are 0.125, 0.375 and 0.375, respectively. On the other hand, the probability of the line PS_0 making a $1 \rightarrow 0$ transition is 0.5, and the probabilities for a $0 \rightarrow 1$, $0 \rightarrow 0$ and $1 \rightarrow 1$ are 0.5, 0, and 0, respectively. The transition probabilities for the different present-state lines are different.

To make matters worse, the transitions on one present-state line are correlated to the transitions on the other present-state lines. For instance, in Figure 5, it is easy to see that PS_2 makes a $0 \rightarrow 1$ transition only when PS_0 makes a $1 \rightarrow 0$ transition. Simply computing the correct switching rates does not necessarily result in exact power dissipation estimation if we ignore this correlation of the transitions. Note however, that in this example, the present-state lines are uncorrelated, *i.e.*, given the static probability of each present-state line, the probability of a particular state can be computed correctly.

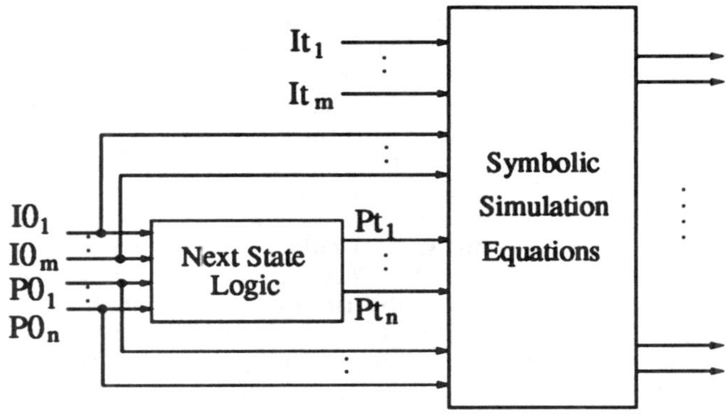

Figure 6: Taking Correlation Into Account

4.2 Power Estimation Method

We will assume that the circuit, upon power-up, can begin in any of the 2^n states, where n is the number of flip-flops. Further, we assume that the STG of the machine is strongly connected, and that an arbitrary number of cycles after power-up, the probability that the machine is in any of its 2^n states is uniform.

Consider the circuit of Figure 6. It has two blocks: the first is the combinational logic implementing the next-state function of the given machine, and is used to derive the next state from the present state. The first block feeds a second block, which represents the Boolean function obtained by symbolic simulation of the combinational logic of the machine.

The decomposition of Figure 6 implies that the gate output switching activity can be determined given the vector pair $< I0, It >$ for the primary inputs, but only $P0$ for the states. Therefore, to compute gate output transition probabilities, we require the transition probabilities for the primary inputs, but only require the static probabilities for the present state. This use of the next-state logic generates Boolean equations which model the correlation between the present and the next states, whereby the transition probabilities are automatically computed, taking into account the correlation between transitions. However, to use the average power estimation techniques described in Section 2, we must assume that the present-state lines are uncorrelated. We discuss the validity of this assumption in the next section.

4.3 STG Characteristics

The state bits in a synchronous sequential circuit are uncorrelated if the machine is equally likely to be in any of its 2^n states an arbitrary number of cycles after power-up (in which case the static probability of each present-state line is 0.5). There are many reasons why this might not be true. In particular if the entire state space is not strongly connected, the machine might operate only in a strongly connected portion of the state space. In general, after an arbitrarily long period of time, the machine will be in a strongly connected portion of its STG, which might contain just a single state. If the assumption above is not valid, then static probability of 0.5 cannot be assumed for the present-state lines. Moreover, because of the correlation between the present-state lines, the power estimate obtained is approximate.

Given a set of states in a strongly connected portion of the STG in which the machine can be, the probability of being in any particular state within the set is not necessarily uniform. In particular, it might depend on the *in-degree* [2] of a state. States with a large in-degree may have greater probabilities of being entered. The exact probabilities of being in particular states can be derived from the Chapman-Kolmogorov equations for discrete-time Markov Chains [12]. However, the number of equations will be exponential in the number of latches and inputs in the circuit, and therefore cannot be solved for most practical circuits.

In practice, large datapath circuits satisfy the assumptions because of their highly connected STGs that include almost all of 2^n states, and states with similar in-degrees. Thus, the assumption made regarding the structure of the STG is quite realistic for such circuits. However, controllers might violate these assumptions. In that case, because of the correlation between present-state lines, we will only be able to obtain an approximate power estimate. We can increase the accuracy of the power estimate by calculating the static probabilities of the present-state lines more accurately. Techniques to compute the reachable set of states, or the strongly connected portion of the STG of a sequential machine, can be developed, based on the strategies of [9, 4]. These techniques are viable for machines with up to approximately 50 flip-flops. Thereafter, the static probability of each present-state line can be calculated. For example, if a machine has states 00, 01, and 10, and it is in these states with proabilities 0.1, 0.5, 0.4, respectively, then the probability of the first present-state line being 1 is 0.4 and that of the second is 0.5. Finally, during signal probability calculation, we have to exclude the set of unreachable states. This is performed by computing the logical AND of the BDD representing the set of reachable states with the BDD representing the switching function.

5 Experimental Results

Throughout this section, we will be measuring the average power dissipation of the circuit by using (1) summed over all the gates in the circuit. The N_G values are computed for the gates in the circuit under different delay models. The capacitance values of the gates are assumed to be given, or can be computed using the fanin and fanout numbers and the size (in terms of the number of transistors) for each gate, or can be obtained after mapping.

The statistics of the examples used are shown in Table 1. All of the examples except the last two belong to the ISCAS-89 Sequential Benchmark set. Example **add16** is a 16-bit adder and accumulator, where the output of the adder is fed back to one set of inputs. Example **max16** is similar, with the adder replaced by maximum finder. Some of these circuits have over 200 inputs and outputs (*e.g.*, **s5378**). All the circuits considered are technology mapped circuits.

[2] The number of edges in the STG that enter this state.

CKT	Inputs	Outputs	Latches	Gates
s27	4	1	3	10
s298	3	6	14	119
s349	9	11	15	150
s386	7	7	6	159
s420	19	2	16	196
s510	19	7	6	211
s641	35	24	19	379
s713	35	23	19	393
s838	35	2	32	390
s1238	14	14	18	508
s1494	8	19	6	647
s5378	35	49	179	2779
add16	33	17	16	288
max16	33	16	16	154

Table 1: Statistics of examples

CKT	Combinational		Sequential	
	Power	Time	Power	Time
s27	14	0.1	15	0.1
s298	212	0.9	259	1.8
s349	280	1.6	258	2.9
s386	264	1.5	428	2.1
s420	287	1.8	336	2.8
s510	425	2.2	425	3.7
s641	582	3.5	580	6.7
s713	650	4.3	649	6.3
s838	568	4.4	672	4.6
s1238	914	6.2	904	6.4
s1494	1002	8.3	1185	13.8
s5378	4752	11.0	4823	45.1
add16	58	0.4	58	0.6
max16	56	0.4	73	0.7

Table 2: Power estimation for dynamic circuits

If the circuits are treated as purely combinational dynamic circuits, then the average-case power dissipation is shown in the second column of Table 2, while the time required to obtain the power dissipation is shown in the third column. These times have been obtained on a DEC-station 5900 with 440Mb of memory, and are in seconds. The power estimates, in microWatts, have been obtained assuming a clock frequency of 20MHz. Treating the circuit as a dynamic sequential circuit, the average power dissipation and the time required to derive the power dissipation are shown in columns four and five respectively of Table 2. Note that taking correlations of state transitions into account, as described in Section 4.2, produces significant differences in the power estimate.

For the remainder of this section, we consider the circuits to be static CMOS circuits. In Table 3, we treat the circuits as purely combinational and show the effects of various delay models on the power estimate. In the zero delay model, all gates have zero delay and therefore they switch instantaneously. In the unit delay model, all gates have one unit delay. Using the zero delay model ignores glitches in the circuit, and therefore power dissipation due to glitches are not

CKT	Zero Delay	Unit Delay	Variable Delay	
	Power	Power	Power	Time
s27	15	19	17	0.1
s298	190	215	198	1.6
s349	238	358	308	4.7
s386	252	288	271	3.8
s420	108	137	128	8.2
s510	271	346	304	3.1
s641	481	665	596	41.0
s713	513	669	637	45.1
s838	184	253	214	49.3
s1238	622	790	716	31.8
s1494	1032	1280	1194	20.0
s5378	3677	4513	4099	2022.2
add16	58	102	102	8.1
max16	56	85	85	6.4

Table 3: Power estimation for combinational circuits

CKT	Zero Delay	Unit Delay	Variable Delay	
	Power	Power	Power	Time
s27	11	15	14	0.2
s298	168	193	173	5.2
s349	225	275	262	6.5
s386	300	352	333	5.4
s420	126	152	150	10.5
s510	211	265	237	7.8
s641	399	539	493	40.4
s713	429	537	528	50.1
s838	230	278	275	62.4
s1238	613	781	706	45.0
s1494	1072	1498	1303	45.1
s5378	3644	4620	3978	226.8
add16	71	129	129	11.3
max16	46	75	75	8.5

Table 4: Power estimation for sequential circuits

taken into account. The unit delay model takes into account glitches, but does not take the inertial delay of gates into account and therefore may overestimate the power dissipation. The variable delay model with inertial gate delays is the most realistic one, and its estimates, as expected, lie between the zero delay and the unit delay model. Only the times required to obtain the power estimate for the variable delay model is shown in the last column.

The results obtained by treating the circuits as sequential, using the method of Section 4.2, are shown in Table 4. While considering these circuits as sequential, we have taken into account the correlation between two successive values on a present-state line using the method of Section 4.2. This again results in significant differences in the power estimate, as can be seen by comparing Tables 3 and 4.

For each circuit, we compared the power estimates obtained by our method to that obtained by simulating the circuit. The power estimate was obtained by logic simulation, until convergence, with randomly generated input sequence applied to the sequential circuit. The same power dissipation model was used. We obtained excellent agreement (to within 2%) between the power estimate using our method and the random simulation estimate. However, power estimation by simulation took anywhere between 2 to 100 times longer.

References

[1] P. H. Bardell, W. H. McAnney, and J. Savir. *Built-In Test for VLSI: Pseudorandom Techniques.* John Wiley and Sons, New York, New York, 1987.

[2] R. Brayton, G. Hachtel, C. McMullen, and A. Sangiovanni-Vincentelli. *Logic Minimization Algorithms for VLSI Synthesis.* Kluwer Academic Publishers, 1984.

[3] R. Bryant. Graph-Based Algorithms for Boolean Function Manipulation. In *IEEE Transactions on Computers*, volume C-35, pages 677–691, August 1986.

[4] J. Burch, E. Clarke, K. McMillan, and D. Dill. Sequential Circuit Verification Using Symbolic Model Checking. In *Proceedings of the* 27th *Design Automation Conference*, pages 46–51, June 1990.

[5] R. Burch, F. Najm, P. Yang, and D. Hocevar. Pattern-independent Current Estimation for Reliability Analysis of CMOS Circuits. In *Proceedings of the* 25th *Design Automation Conference*, pages 294–299, June 1988.

[6] S. Chakravarty. On the Complexity of Using BDDs for the Synthesis and Analysis of Boolean Circuits. In *Proceedings of the* 27th *Annual Allerton Conference on Communication, Control, and Computing*, pages 730–739, September 1989.

[7] S. Chowdhury and J. S. Barkatullah. Estimation of Maximum Currents in MOS IC Logic Circuits. In *IEEE Transactions on Computer-Aided Design*, volume 9, pages 642–654, June 1990.

[8] M. A. Cirit. Estimating Dynamic Power Consumption of CMOS Circuits. In *Proceedings of the Int'l Conference on Computer-Aided Design*, pages 534–537, November 1987.

[9] O. Coudert, C. Berthet, and J. C. Madre. Verification of Sequential Machines Using Boolean Functional Vectors. In *IMEC-IFIP Int'l Workshop on Applied Formal Methods for Correct VLSI Design*, pages 111–128, November 1989.

[10] S. Devadas, K. Keutzer, and J. White. Estimation of Power Dissipation in CMOS Combinational Circuits. In *Proceedings of the Custom Integrated Circuits Conference*, pages 19.7.1–19.7.6, May 1990.

[11] F. Najm. Transition Density, A Stochastic Measure of Activity in Digital Circuits. In *Proceedings of the* 28th *Design Automation Conference*, pages 644–649, June 1991.

[12] A. Papoulis. *Probability, Random Variables and Stochastic Processes.* McGraw Hill, 1984.

SESSION 16

PANEL: WHY DATA MODELS WILL BECOME THE FASTEST GROWING SEGMENT OF THE EDA MARKET

Chair: Bill Lattin - Logic Automation, Inc., Beaverton, OR
Organizer: M. Middlewood - Waggener Edstrom, Portland, OR

The panel will focus on a fundamental shift occurring in the EDA industry. No longer is the EDA industry's emphasis on powerful tools (work stations, simulators, synthesizers, and graphical environments) and the issues surrounding their integration. Today the industry is facing a explosive need for "data" libraries and models to drive these tools. Once found in data sheets and specification books, increasingly a variety of different data types are needed in electronic form. The panel will explore the implications of this shift upon various segments of the EDA industry.

Panel Members:

John Hoover - IBM Corp., Poughkeepsie, NY
Harvey Jones - Synopsys, Inc., Mountain View, CA
Gerry Langeler - Mentor Graphics Corp., Wilsonville, OR
Bill Lattin - Logic Automation, Inc., Beaverton, OR
A. Richard Newton - Univ. of California, Berkeley, CA
Curt Widdoes - Logic Modeling Systems, Inc., Milpitas, CA

Hierarchical Test Generation under Intensive Global Functional Constraints

Jaushin Lee and *Janak H. Patel*

Center for Reliable and High-Performance Computing
University of Illinois at Urbana-Champaign
Urbana, IL 61801, U.S.A.

Abstract

In hierarchical test generation, the test vectors for the low level structure of the module under test are computed and then justified at a high level. In the module test computation procedure, a low level ATPG tool is conventionally applied to the complete structure of that particular module without adding extra information. Due to the global circuit functional constraints applied to the inputs of that module, many of the test vectors being computed are not justifiable at the high level. Therefore high efficiency cannot be achieved without managing the functional constraint problem in the hierarchical ATPG process. In this paper, two schemes, *valid control code abstraction* and *test cube justification*, are proposed to overcome the global functional constraint problem and achieve very high efficiency in test computation. The proposed algorithms have been implemented in our hierarchical ATPG package and promising experimental results have been derived. We conclude that global functional constraints can be efficiently avoided through these techniques.

1. Introduction

Many novel approaches have been proposed for gate level sequential circuit test generation [1-4]. These approaches remain at gate level without looking at any of the inherent functional information in the sea of logic gates. The increase in VLSI circuit size will soon degrade the efficiency of the gate level ATPG approaches. The hierarchical circuit design style has been widely adopted in the design of modern VLSI systems. A number of approaches have been proposed to generate tests hierarchically and exploit high level functional information in circuits to speedup the test generation process [5-8]. However, the ATPG algorithms used in these approaches are extended from D-algorithm [10] or PODEM [11], so that the module partition cannot go much further than gate level. In [6], the module partition could be at a higher level, but it should be flattened to gate level if any interior stuck-at fault is considered. Therefore, the speedup of these approaches over gate level ones is limited. The problem of system level functional constraints for these approaches does not exist because the module under test either contains very few logic gates, or is flattened to gate level. The functional constraint problem will be explained in more detail in the next section. Sarfert, et. al. updated SOCRATES[9] by including higher level modules which implement non-trivial functions, like multiplexer, decoder, etc [10]. However, their approach still adopted single *bit-level* signals at the high level and the speedup is limited. Kunda, et. al. take advantage of both higher level primitives and *bit-vector* representation of signals to speedup the test generation process [11]. Murray and Hayes[12] assume the test set for module under test is precomputed, and again the global functional constraint problem has not been addressed in any of them. Anirudhan and Menon[13] have addressed this problem in their paper. Symbolic constraints are derived in the *top-down* approach and the module tests are computed such that these constraints could be avoided. However, for complex circuits the symbolic constraint representation could be computational prohibitive, and the algorithm of computing module tests under such constraints is non-trivial. Lee and Patel have proposed a relaxation-based test generation algorithm at an architectural level . The global functional constraints were manually derived and put into the module test computation process [14].

In this paper, we propose two approaches to overcome architectural level functional constraint problem in hierarchical test generation. We separate these constraints into **control constraints** and **bus constraints**. For each module, the control input lines can be unambiguously separated from the bus input lines based on high level functions. Bus lines carry input data for either transferring or computation, and control inputs determine the transferring paths or computation functions. For control constraints, valid control codes occurring at the control lines of the modules of interest are determined by searching through control words of machine instructions. These valid control codes are stored in a table which are then intersected in the module test computation process. For bus constraints, we observed in some circuits that the relations among input buses might be very complex, and it is not practical to derive these exact relations given reasonable amount of time. Instead of deriving the exact bus constraints, we try to justify the partially specified test vectors at the high level. A partially specified vector has several of its bits set to value X (don't care). Due to the fact that the bus constraints are less likely to be violated in most partially specified vectors, this approach is expected to be effective in avoiding bus constraints.

This research was supported by the Semiconductor Research Corporation under Contract SRC 91-DP-109.

In this paper, we assume that circuits are represented in two levels of hierarchy, that is, the architectural level and gate level. The relaxation-based architectural level test generator, **ARTEST** [14], and a gate level ATPG are used at these two levels. The contents of this paper are outlined as follows. Some preliminary definitions are given in Section 2. Section 3 introduces the derivation of control constraints and the constant value bus constraints, and the modification of gate level test generation algorithms. The approach of justifying partially assigned test vectors will be described in Section 4. Some experimental results will be shown in Section 5 to compare different approaches, followed by conclusions.

2. Definitions

In a hierarchical circuit design environment, some information barriers exist between levels of hierarchy. A phenomenon termed *hierarchy blindness* has been introduced by Lee and Patel [15], which illustrates two different kinds of information barriers. The first type of hierarchy blindness occurs when the low level circuit structure is not available to the high level. In this paper, it is assumed that the low level structure of the module under test is available. *However, the complete low level structure of the circuit is never required in our approach.* We are concentrating on the second type of hierarchy blindness, in which the high level architectural constraints are not visible to low level test computations. Figure 1 is an example of the second type of hierarchy blindness where only the four functions of an ALU are valid, when it is used as a module in a large circuit. In this case, test vectors for the ALU requiring functions other than these four ones will never be justified at the system level. These vectors are sometimes called "can't happen don't cares" in the logic synthesis literature. The high level architectural constraints applied to the inputs of the module under test are termed **global functional constraints** for that particular module. The global functional constraints for a module can be subdivided into control constraints and bus constraints for control lines and bus lines. For the ALU in Figure 1, any module test requiring a function other than ADD, SUB, AND, and OR, or requiring a value at bus A not equal to 15 plus value at bus B can never be justified at the system level. Therefore, a gate level test generator must take into account these constraints when generating test vectors for the

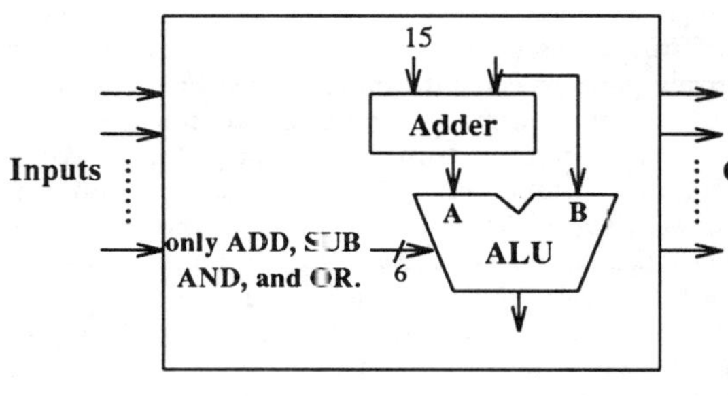

ALU : control constraints : 4 functions out of 64.
bus constraints : A = B + 15.

Figure 1. An example for global functional constraints.

ALU. In this paper, only data path faults are considered. All control constraints derived are from the valid behaviors of the control machine.

An n-dimensional **vector** can be represented as $V = <v_1, v_2, ..., v_n>$. Each $v_i \in$ **R**, and **R** is the range of data space at certain level of hierarchy. If considering gate level, **R** = {0,1}. In logic simulation and test generation at gate level, another value X, **unassigned don't care**, has been frequently used. So **R** can be updated to {0,1,X}. An n-dimensional Boolean **cube** is therefore defined as $C = <c_1, c_2,..., c_n>$ and $c_i \in$ {0,1,X}. A cube can be *expanded* to specific vectors in the dimensions with the data type X. In the extreme case that no X appears at any dimension of a cube, then this cube will also be a vector. An n-dimensional **test vector** at the gate level is defined as an input vector with all bits assigned values 0 or 1 for a circuit with n inputs. An n-dimensional **test cube** is a *partially assigned test vector*, if some input values remain X. For high level modules, we can also define a **control test subcube** and **bus test subcube** for each module, which is simply the test cube for the circuit divided into two disjoint parts for control input lines and bus input lines.

The **distance** between two vectors or cubes means the number of dimensions in these two vectors or cubes having different values. The **constraining weight** of the test cube is number of bits having 0 or 1 in the test cube. **Relative constraining weight** of a test cube is the ratio of the weight to the dimension of the test cube. For test vectors, the relative constraining weight is 1. For test subcubes, the relative constraining weight can also be calculated as the ratio of the weight of the subcube to the number of bits in either bus or control. A **test cover** is a collection of cubes that covers a specified set of test vectors. A cover is a **tautology** if expanding all its cubes, the data space covered is all 2^n vectors.

Some terms need to be defined in the context of the *discrete-relaxation algorithm* in ARTEST at the high level. A **fixed value** at a high level is a fixed constraint implied by the injected test vector or the control words of the instruction being used. A **value conflict** happens when the current assigned values at some internal lines contradict some fixed values. A **symbol** is a variable attached to a control line or a data bus, and its value is bounded by the bus size. If the value assigned to a symbol is out of the predefined range, it will be considered as a value conflict also. A **fixed symbol** is defined as a symbol with a fixed value.

3. Functional Constraint Abstraction

For a high level module in data path, its control input lines usually come from the control unit so that the valid codes on these lines depend on the control words or micro orders of machine instructions. However, bus input lines would strongly depend on data path configurations which implicitly determine relations among buses. Based on this observation, the strategies used for global functional constraint abstraction should be different for control lines and bus lines.

3.1. Control Line Constraints

Following the assumptions we have been using in our previous work, we again assume that the high level circuit under test can be represented as an interconnection of a data path portion and a control machine. The instruction set of the control machine and its corresponding control word sequences for each instruction should be available. For each module of interest, we can simply traverse the high level module diagram to ensure the connection of control input lines with the control unit. If they are connected, record the valid control codes from the control words of each instruction, and put those codes into the **Boolean cover** of that module. If any of the control input lines is not connected to the control unit, the corresponding bits in the code for that particular input line will be all Xs'. In the preprocessing phase of the high level ATPG algorithm we are using, a symbolic simulation is applied to the module diagram to derive behavior information and a system of equations for each instruction. When each valid control code is found, it has to be approved by checking the observability of the output symbol of the module at that cycle of the instruction. If the symbol is not observable, then the derived control code will be judged as invalid.

After all valid control codes are derived, a minimal Boolean cover is formed on the control vectors, using any of the available minimization procedures. If the Boolean cover is found to be a tautology, then no control constraint exists at that particular module, i.e., all binary vectors on the control lines are valid control codes.

Invalid control codes are caused by global constraints of the circuit and sometimes termed "can't happen don't cares" in logic synthesis literature. It should be noted that it is very difficult to derive these "can't happen don't cares" in large circuits. These information is also very difficult to derive by simulating the complete gate level structure of the circuit. Our approach exploits the high level behavior information of the circuit so that the valid control codes can be easily derived.

In this paper, four high level circuits will be used for all experiments. The first one is a microprogram sequence controller with the same functional implementation as that of Am2910 [16]. The second one is a 16-bit general purpose microprocessor with nine instructions (MP2). The third one is an 8-bit 2's complement multiplier (MP1), and the last one is a digital signal filter (PCONT2). The control constraints have been derived for some modules in these circuits, and the information is shown in Table 1. CSP represents Cover Space Percentage, which means the percentage of space covered by all expanded cubes to 2^n, where n is the number of control input lines. From Table 1, we know that Am2910 and PCONT2 have moderate control constraints, and MP1 has no control constraints. MP2 has intensive control

Control Constraints				
circuits	modules	# c_bits†	# cubes	CSP(%)
Am2910	4-to-1 Mux	3	3	62.5
	Incrementer	1	T‡	100.0
	RegCnt_CL2	3	2	75.0
MP2	16-bit ALU	6	4	6.3
MP1	8-bit Adder	1	T	100.0
	8-bit XOR	1	T	100.0
	2-to-1 Mux	1	T	100.0
PCONT2	8-bit Adder	2	1	25.0

† # c_bits : number of control bits.
‡ T means the cover is Tautology.

Table 1. Control constraint information for modules.

constraints at the ALU, where only 4 control codes out of 64 are valid. This information will be used in the module test computation procedure which will be illustrated in section 3.3.

3.2. Bus Line Constraints

Bus constraints have significantly different characteristics than control constraints. Most of the bus constraints are contributed by the complex data path configuration. Take the multiplier MP1 as an example, the data path module diagram performs iterative additions of the partial sums for multiplication. There is an Adder embedded in the circuit performing the addition function. The values of the two input bus A and B of the Adder have a very complex relation during each cycle. There is only one instruction, Multiply, in MP1. Through symbolic simulation in the preprocessing phase, the bus constraints at the last cycle can be **exactly** expressed by about **50** module equations. These module equations look like those shown in Figure 2. They can also be expressed in a RTL format which clearly illustrates the implicit functions. These module equations in a RTL format are shown in Figure 3. The inter-dependent relation between bus A and B is very intensive and can be expressed by these 50 equations. The bus constraint in the example shown in Figure 1 is relatively much less intensive since only 1 equation is involved. These 50 module equations fully characterize the circuit behavior at the high level. Injecting the required values at A and B into the equations, a solution satisfying these equations justifies the values at bus A and B. So the problem of avoiding this bus constraint, when tests of the Adder is computed, is as complex as the complete high level ATPG problem [14]. Those vectors which do not satisfy these 50 equations will be classified as "can't happen don't cares". It is very difficult to derive all those "can't happen don't cares" for either control covers or bus constraints. Based on this observation, we conclude that it will not be efficient to check for contradictions using these bus constraints, when each bit assignment is made at gate level ATPG algorithm.

Instead of dumping the exact bus constraint information to the gate level test generator, we propose a bottom-up approach to

$$\begin{cases} \ldots \\ 41\ (MUX2)(v_3, 0, v_{56}) = (v_{57}) \\ 42\ (ADDER)(v_{57}, v_{51}, 0) = (v_{58}) \\ 43\ (BusSplit\_LSB)(v_{58}) = (v_{59}, v_{60}) \\ 44\ (BusJoin\_MSB)(v_{60}, v_{55}) = (v_{61}) \\ \ldots \end{cases}$$

Figure 2. An example of system of module equations expressing bus constraints.

$$\begin{cases} \ldots \\ v_{57} \leftarrow v_3 \text{ if } v_{56} = 1, \textit{ else } v_{57} \leftarrow 0; \\ v_{58} \leftarrow v_{57} + v_{51}; \\ v_{59} \leftarrow v_{58}[8:1];\ v_{60} \leftarrow v_{58}[0:0]; \\ v_{61}[7:7] \leftarrow v_{60};\ v_{61}[6:0] \leftarrow v_{55}; \\ \ldots \end{cases}$$

Figure 3. Module equations in a RTL format.

avoid bus constraints. This approach will be addressed in the next section. However, sometimes some bus constraints are easily specified, e.g. fixed values on the data lines. Such constraints we do record and use in the gate level test generation.

3.3. Module Test Computation

Given the derived control constraints and constant value bus constraints, the gate level PODEM algorithm can be modified without making a major change. It is explained as follows:

1. Consider *n* input bits as an *n*-dimensional cube. When any assignment is made at an input bit, the cube is updated and intersected with the control cover of the corresponding module under test. If the intersection is null, then do backtracking.
2. If there exists any fixed value at an input bus, convert the fixed value to a vector and intersect with the cube when it is updated.
3. The Xs' in the control test subcube need to be assigned with 0 or 1 according to the control cover such that it is easier to derive an instruction sequence at the high level. A decision stack is kept for these assignments. A new test vector can be derived by doing boundary backtracking on the decision stack. If constant value happens at a bus line, the Xs' in the corresponding bus test subcube also need to be assigned with 0 or 1 according to the constant value.

The rules of bit-wise intersection of two cubes is shown in Table 2. The intersection of a test cube with the control cover is ∅ if the intersection of each control cube in the cover with the test cube results in ∅. If any bit in the resulting cube ends with ∅, it means conflict. The techniques for both control constraint and constant value bus constraint abstractions have been implemented in the preprocessing phase of ARTEST. A modified PODEM algorithm has also been implemented for the module test computation.

4. High Level Test Cube Justification

4.1. Important Observations

Some important observations have been obtained from the discussions in previous sections. Based on which the test cube justification technique is developed.

1. Observing those test cubes generated by PODEM, we found that *most of the control test subcubes have very high relative constraining weights, and contrarily most of the bus test subcubes have very low relative constraining weights*. In other words, given a single stuck-at fault in the module, most of the time very few bits corresponding to input buses will be assigned to a 0 or 1 by a PODEM like test generator.
2. Most of the modern high level test generators need to justify complete test vectors since *integer* representation for bus signals are used to enlarge the module partitions and speedup the test generation process at a high level. Therefore, test cubes generated by PODEM are fully specified to obtain integer values. However, this cube expanding procedure overspecifies many Xs which would very likely cause conflicts with various implicit constraints at the high level. If a test cube can be justified through the high level algorithm without being expanded, most bus constraints could be avoided.

bit-wise intersection			
∩	**0**	**1**	**X**
0	0	∅	0
1	∅	1	1
X	0	1	X

Table 2. Bit-wise intersection rules for cubes.

When the module tests are computed, only those Xs' corresponding to the control lines are expanded according to the control cover. The bus test subcubes derived by PODEM will not be expanded and injected directly to the high level. The modified high level ATPG algorithm which can justify the test cubes at a high level is presented in the next section.

4.2. Modification of the Discrete-Relaxation Algorithm at High Level

In the relaxation-based ARTEST, the high level test generation problem has been transferred to the problem of solving a system of equations. The injected test vector for the module under test will be translated to values corresponding to each input bus and control line. These fixed values will be assigned to symbols in the system of equations, and the *discrete-relaxation algorithm* will be applied to solve this system of equations. Those corresponding symbols with fixed values will be marked as fixed symbols. During the equation solving process, if there is no conflict at any of the fixed symbols in the system of equations, the solution is derived.

When a test cube is injected at the high level, it is partitioned into test subcubes corresponding to each input line. If any test subcube has relative constraining weight 0, i.e. all bits are X, this subcube will not be assigned to any symbol. If any test subcube is also a test subvector, i.e. all bits fully specified, the corresponding value will be calculated and assigned to a symbol as the previous approach. Except for the above two conditions, the test subcube in which some bits are specified and some are Xs, will be assigned to the corresponding symbol and a special data structure will be maintained to store all those **partially assigned symbols**. Those partially assigned symbols will not be marked as fixed symbols, since multiple values can be applied to them. In each iteration of the discrete-relaxation algorithm, when the value of a symbol is updated, the symbol has to be checked whether it is a partially assigned symbol. If it is, the updated value needs to be translated to a subvector and then intersected with the subcube. If there is a conflict, a new updated value will be calculated based on some rules. The iteration can therefore continue until convergence.

When a new updated value for a partially assigned symbol conflicts with the corresponding test subcube, the algorithm for calculating a new updated value can be illustrated by Figure 4. The conflict value is 236, in which the bits conflicting with the assigned ones in the test subcube need to be updated based on the values at the same dimensions of the test subcube. The other bits corresponding to Xs' are not updated. The new value is therefore 174. The concept is simple. We only update those dimensions having a conflict with the test subcube, and keep all other dimensions unchanged. From our experience, the exact new updated value does not seem to have a significant effect on the convergence of the algorithm. This heuristic has been proved to be effective in our experiments.

Test subcube : x x x 0 1 x x x 1 0

Conflict value : 0 0 1 1 1 0 1 1 0 0 = 236

↓ ↓

New updated value : 0 0 1 0 1 0 1 1 1 0 = 174

Figure 4. An example of conflict handling for test cubes.

5. Results and Comparisons

ARTEST has been continuously updated with the testcube justification technique to overcome intensive bus constraints. A highly efficient fault simulator, PROOFS [17], has been linked with this package to collapse module stuck-at faults. The four circuits introduced in the previous section are used and the test results for these techniques are shown in Table 3. All experiments were run on a SUN 4 machine. The first approach does not apply any technique to avoid the global functional constraints. The second approach exploits control cover, constant value bus constraints, and the test cube justification technique to overcome both control and bus constraints. It has been pointed out that MP1 has intensive bus constraints. PCONT2 is a digital signal filter and it also contains intensive bus constraints. MP2 has intensive control constraints at the ALU. Am2910 is a moderate circuit for either of the two types of functional constraints. The results show that these proposed techniques are very effective in avoiding global functional constraints. For the control constraint at the ALU in MP2, the speedup achieved by the valid control code abstraction technique is approaching 3.34 with much higher fault coverage. Considering the bus constraints at the Adder of MP1, the speedup achieved by the test cube justification technique is almost 66 times. The speedup for the Adder in PCONT2 is also extremely high. For the easy circuit, Am2910, the Approach 2 also derives speedup of 4.9 for RegCnt_CL2.

To compare the complexity of these approaches, the statistics of total number of boundary backtracks and retrials of PODEM are recorded and shown in Table 4. When an injected test vector or cube fails at the high level, the gate level test generator will be retried until the implicit limit is exceeded. For this experiment, the maximum limits for boundary backtrack and retrial of the gate level ATPG are 50 and 20. Considering the total number of boundary backtracks and retrials of the gate level ATPG as an index of algorithm complexity, these data demonstrate the efficiency of Approach 2. For the Adder in MP1, the total number of boundary backtracks consumed by Approach 2 is 254 times less than that of Approach 1. In computing tests for RegCnt_CL2 in Am2910 and Adders in PCONT2, Approach 2 also shows significant superiority over other approaches. If the ALU of MP2 or Adder of MP1 were embedded in other larger VLSI circuits, those amount of boundary backtracks in Approach 1 would not be affordable. Based on these experimental results, we can conclude that the test cube justification technique using discrete relaxation method is very promising in avoiding bus constraints. Combining the control cover abstraction technique for control constraints together with the test cube justification technique for bus constraints gives a complete solution to hierarchical test generation under intensive global functional constraints. This approach not only supports a higher level of module partitions for speedup, but is also unaffected by the drawbacks of overspecifying unassigned values at the module boundary for satisfying the requirements of high level ATPG algorithms.

circuit	primitive	# modules	# faults	Approach 1				Approach 2			
				# drop	# red	eff.(%)	CPU	# drop	# red	eff.(%)	CPU
Am2910	4-to-1 MUX	1	294	0	0	100.00	44 sec	0	0	100.00	30 sec
	Incrementer	1	184	0	0	100.00	4 sec	0	0	100.00	3 sec
	RegCnt_CL2	1	367	5	40	98.64	9.9 min	2	43	99.46	2.0 min
MP2	16-bit ALU	1	930	104	16	88.82	1.78 hr	0	16	100.00	34.0 min
MP1	8-bit XOR	1	84	0	0	100.00	5 sec	0	0	100.00	2 sec
	2-to-1 Mux	6	408	83	0	79.66	1.33 hr	80	0	80.39	1.9 min
	Adder	1	332	21	32	93.67	5.05 hr	20	32	93.98	4.6 min
PCONT2	8-bit Adder	8	2240	345	254	84.60	29.72 hr	166	355	93.20	56.9 min

Approach 1 : without any functional constraint information.
Approach 2 : with control cover, constant value bus constraint abstraction and test cube justification techniques.
drop = number of faults dropped.
red = number of faults identified as redundant.
eff.(%) = test generation efficiency = (#faults - #drop)/#faults.
RegCnt_CL2 : combinational part of the Register Counter.

Table 3. Fault efficiency and run time comparison of different approaches.

circuits	primitive	Approach 1		Approach 2	
		bd. back†	retry‡	bd. back	retry
Am2910	4-to-1 MUX	0	0	0	0
	Incrementer	0	0	0	0
	RegCnt_CL2	1597	36	42	24
MP2	16-bit ALU	99664	2022	0	0
MP1	8-bit XOR	0	0	0	0
	2-to-1 Mux	6739	181	171	171
	Adder	2540	51	10	10
PCONT2	16-bit Adder	99035	3329	1010	844

† Total number of module input vector backtracks.
‡ Total number of retrials of gate level ATPG.

Table 4. Algorithm complexity comparisons.

6. Conclusions

In this paper, we have formally addressed the system level functional constraint problem for hierarchical test generation. We have also proposed several approaches to solve both control constraints and bus constraints. For control constraints, circuit behavior information is exploited to derive valid control Boolean covers for different modules. For bus constraints, we propose a constant value bus constraint abstraction technique and a test cube justification technique. These proposed algorithms have been implemented in our hierarchical test generation package, ARTEST, and four high level circuits with different constraint characteristics have been tried in the experiments. The experimental results show the effectiveness of combining the control cover abstraction technique and the test cube justification technique as a complete solution to the global functional constraint problem.

References

[1] H-K. T. Ma, S. Devadas, A.R. Newton, and A. Sangiovanni-Vincentelli, "Test Generation for Sequential Circuits," *IEEE Trans. on Computer-Aided Design*, pp. 1081-1093, Oct. 1988.

[2] W.T. Cheng, "The Back Algorithm for Sequential Test Generation," *Proc. IEEE Int. Conf. on Computer Design*, pp. 66-69, Oct. 1988.

[3] V.D. Agrawal, K.T. Cheng, and P. Agrawal, "CONTEST: A Concurrent Test Generator for Sequential Circuits," *Proc. 25th IEEE/ACM Design Automation Conference*, June 1988.

[4] T. Niermann and J.H. Patel, "HITEC: A Test Generation Package for Sequential Circuits," *Proc. European Design Automation Conference*, Feb. 1991.

[5] S.J. Chandra and J.H. Patel, "A Hierarchical Approach to Test Vector Generation," *Proc. 24th IEEE/ACM Design Automation Conf.*, pp. 495-501, June 1987.

[6] B. Krishnamurthy, "Hierarchical Test Generation: Can AI Help?," *Internation Test Conf.*, pp. 694-700, Aug. 1987.

[7] J.D. Calhoun and F. Brglez, "A Framework and Method for Hierarchical Test Generation," *Proc. International Test Conf.*, pp. 480-490, Sept. 1989.

[8] D. Bhattacharya and J.P. Hayes, "A Hierarchical Test Generation Methodology for Digital Circuits," *Journal of Electronic Testing: Theory and Application (JETTA)*, vol. Vol. 1, pp. 103-123, 1990.

[9] M.H. Schulz, E. Trishler, and T.M. Sarfert, "SOCRATES: A highly Efficientc Automatic Test Pattern Generation System," *International Test Conf.*, pp. 1016-1026, Sept. 1987.

[10] T.M. Sarfert, R. Markgraf, E. Trischler, and M.H. Schulz, "Hierarchical Test Pattern Generation Based on High-level Primitives," *Proc. International Test Conf.*, pp. 470-479, Sept. 1989.

[11] R.P. Kunda, P. Narain, J.A. Abraham, and B.D. Rathi, "Speedup of Test Generation Using High-Level Primitive," *Proc. 27th ACM/IEEE Design Automation Conf.*, pp. 580-586, June 1990.

[12] B.T. Murray and J.P. Hayes, "Hierarchical Test Generation Using Precomputed Tests for Modules," *Proc. International Test Conf.*, pp. 221-229, 1988.

[13] P.N. Anirudhan and P.R. Menon, "Symbolic Test Generation for Hierarchically Modeled Digital Systems," *International Test Conf.*, pp. 461-469, Sept. 1989.

[14] Jaushin Lee and Janak H. Patel, "A Signal-driven Discrete Relaxation Technique for Architectural Level Test Generation," *Proc. IEEE Int. Conf. on Computer-Aided Design*, pp. 458-461, Nov. 1991.

[15] Jaushin Lee and Janak H. Patel, "An Architectural Level Test Generator for a Hierarchical Design Environment," *Proc. 21th Symposium on Fault-Tolerant Computing*, pp. 44-51, June 1991.

[16] Advanced Micro Devices, in *The Am2910, A Complete 12-bit Microprogram Sequence Controller.*

[17] T. Niermann, W.T. Cheng, and J.H. Patel, "PROOFS: A Fast, Memory-Efficient Sequential Circuit Fault Simulator," *IEEE Trans. on Computer-Aided Design*, vol. Vol. 11, No. 2, pp. 198-207, Feb. 1992.

A Methodology to Reduce the Computational Cost of Behavioral Test Pattern Generation

Jean-François Santucci Gérard Dray Norbert Giambiasi Marc Boumédine

L.E.R.I.
Laboratoire d'Etudes et de Recherche en Informatique
30000 NIMES FRANCE

Abstract

This paper presents different ways of computing testability measures for behavioral descriptions of digital circuits. These testability measures must be used in order to accelerate behavioral deterministic test pattern generation processes. The purpose of this paper is also to validate the proposed testability measures and to carry out a comparative study between them.

1. Introduction

Recently, new automatic test generation methods for digital circuits described according to a behavioral view have been developed [1, 2, 3, 4, 5]. This work is useful for three main reasons :

-the structural description of the circuit is not always available,

-the circuit complexity may not allow us to use conventional test pattern generation tools developed for structural descriptions,

-the test generation process is an integral part of the design process and one must start implementing it during the behavioral design phase.

The test generation process concerning this work is based on a fault detection technique and a path sensitization principle. This implies the definition of a behavioral fault modeling scheme [6] and the resolution of several well-known problems : manifestation of a fault, sensitization, propagation and justification. All the work mentioned previously, point out the necessity of defining testability measures associated with elements of the behavioral description in order to reduce the computational cost of the test generation process. Any test generation algorithm presented in the previous work builds a decision tree and applies a backtracking search procedure in order to find a solution. Behavioral testability measures will be helpful in speeding up the search process by guiding the choices involved by the decision tree.

In this paper, we propose to define testability measures for behavioral descriptions and study the effectiveness of these measures in the test generation process[1].

We are interested in behavioral descriptions in HDL. These descriptions can be divided into two categories, namely non-procedural descriptions and procedural descriptions. In this paper, we are only interested in procedural ones. They are characterized by a control flow and a data flow. The control flow corresponds to the sequence of operations organized according to control structures such as sequential, parallel, iterative and selective. The data flow corresponds to the objects handled in the description and the operations which manipulate these objects.

In order to facilitate the definition of testability measures, it is essential to explain in detail the basic concepts linked to the behavioral descriptions. We have therefore defined an internal model which highlights, on the one hand, the sequential and concurrent aspects and, on the other hand, the separation and interaction between the control flow and the data flow [7].

As for most testability analysis approaches for structural descriptions at gate-level and register-transfer-level [8, 9, 10], we use the concepts of controllability and observability to define the testability of an element of the internal model of a behavioral description [11]. This definition involves two steps :

- the definition of the notion of controllability and observability for the basic elements of the internal model.
- the definition of different ways to compute the testability measures according to the authors' [12] suggestions in order to compensate the fact that testability measures are not totally accurate. They suggest switching among the testability measures during the test generation process.

We therefore propose several ways to compute behavioral testability measures based upon the computations of testability measures defined for structural descriptions [13].

The computation of testability measures based on the work described in this paper has been implemented and integrated into a behavioral deterministic test pattern generator in order to study the effectiveness of the proposed testability measures.

The internal model derived from behavioral descriptions is presented in Section 2. In Section 3, we give the main features of a behavioral test pattern generation algorithm we have implemented. In Section 4, we give the definition of controllability and observability of the basic elements of the internal model. The different ways of computing these measures are presented in Section 5. In Section 6, experiments performed on test generation are described. The results are summarized and discussed in Section 7.

---

[1] This work is supported by the Esprit Everest project 2318.

0738-100X/92 $3.00 © 1992 IEEE

2. The internal model

In order to describe the circuit behavior, specific languages called HDL (Hardware Description Languages) are used, for example DACAPO [14] and VHDL [15]. A behavioral view allows a circuit to be described independently of its internal structure by defining how its outputs react when values are applied to its inputs. The internal model allows the main features of procedural behavioral descriptions to be highlighted.

An internal model is associated with each behavioral description. This model must satisfy two requirements : it should be easy to handle and able to highlight the concepts linked to the behavioral descriptions for a high abstraction level. In order to satisfy the first requirement, the definition of the internal model is based on a hierarchical multi-view modeling based on graph concepts. In order to satisfy the second requirement, there are two levels of internal modeling : an external view and an internal view. The external view represents the inputs/outputs of the description. The internal view is made up of two parts, the separation and the interaction of which are explicit :

- A control model represented by a bipartite graph involving two kinds of nodes : transitions and places. This graph has the following two characteristics :
  - the graph is represented in a hierarchical way insofar as a sub-graph can be associated with a transition. A transition with which a sub-graph is associated is called "decomposable". The upper hierarchical level transition is called the initial transition.
  - the graph is structured. Four types of sub-graphs allow control structures to be modeled : sequential, selective, repetitive and parallel.
- A data model represented by a graph which has two types of nodes : operation and data.
  - The data nodes which represent variables, signals and constants.
  - The operation nodes which are of two types : assignment and decision. A sub-graph is associated with each operation node. This sub-graph represents the expression involved in the modeling operation. It is composed of nodes representing the basic operators involved in the language.
- The interaction between these models takes place thanks to the association of an operation node with a transition of the control model. Such a transition is called a non-decomposable transition and can therefore represent the assignment of a variable or a decision point for the selection of a branch. A transition with which an assignment node is associated is called operative. A transition with which a decision node is associated is called a control transition.

3. Brief overview of the implemented deterministic test pattern generator

In this section, we outline the main features of a deterministic behavioral test pattern generator [5] that we have implemented in Common LISP in an object-oriented environment.

The deterministic test pattern generation algorithm for behavioral descriptions we have defined comes from the framework of studies carried out by [1, 2, 3, 4]. Since it is a fault-oriented automatic test pattern generator, an exhaustive fault model has been defined on the basic elements of the internal model [16]. A sub-set of conventional faults [6] has been selected for test generation. These fault hypotheses are classified as follows :

- on the elements of the data model : stuck-at fault hypotheses of an element of the data model,
- on the elements of the control model : hypotheses of a bad selection of a path in a selective control structure,
- on the elements of interaction between control/data : hypotheses of operation skipping.

Searching for a test pattern consists in solving three kinds of problems : problems concerning local fault effect manifestation problems (first to be solved), constraint justification and fault effect propagation. The search for a solution involves a decision process. Whenever there are several alternatives to solve a justification problem or to propagate a fault effect, we choose one of them. But by doing so, we may select a decision that leads to an inconsistency (conflict). Therefore, in our search for a test, we use a backtracking algorithm that allows a systematic exploration of the complete space of possible solutions.

In order to minimize the number of incorrect decisions (and, as a matter of fact, the number of backtracks), testability measures have to be defined.

4. Notion of controllability and observability of basic elements of the internal model

In this section, we define the cost functions used to measure "difficulty" in solving a justification or a propagation problem.

Since elements of the control model and elements of the data model are involved in the solving of a justification or a fault propagation problem, we define testability measures in terms of controllability and observability measures for each basic element of the internal model.

- For the basic elements of the data model involved in the generation process :
  - -The controllability measures associated with an operation node indicate the difficulty in setting the output of the operation node at a given value from the primary inputs, and the observability measures indicate the difficulty in propagating a fault effect from the output of the operation node to a primary output.
  - -The controllability measures associated with a data node indicate the difficulty in setting the data node to a given value from the primary inputs, and the observability measures indicate the difficulty in propagating a fault effect from the data node to a primary output.
- For the basic elements of the control model involved in the generation process :
  - -The controllability measures for a transition T indicate the difficulty in reaching the transition T from the initial transition of the control model.
  - -The observability measures indicate the difficulty in reaching an observable transition from the transition T. An observable transition is a transition associated with an operative node allowing a primary output to be assigned.

These measures will be used to choose from among several ways of solving a constraint or locally propagating a fault effect. From among different solutions of a problem, we first have to try the easiest one. The controllability and observability measures will be used as selection criteria.

5. Computations of controllability and observability measures of the basic modeling elements

Controllability and observability measures associated with elements of the data model (or the control model) are computed by taking into account only information given by the data model (or the control model). We have defined three different ways to compute the measures associated with the basic elements of the data model (distance-based measures, recursive formulae measures and fanout-based measures) while only two different ways are defined on the control model (distance-based and recursive formulae measures).

5.1. Controllability and observability measures based on distance

Any controllability and observability measure should show that the input nodes of the model are the easiest to be controlled and the output nodes are the easiest to be observed. From this principle, we can consider that the controllability measure of a node decreases with its distance from input nodes and its observability measure decreases with its distance from output nodes.

5.1.1.Computation of controllability and observability measures for the data model. The data model is transformed into an acyclic graph by opening all the cycles; the opened arcs are considered as primary inputs and primary outputs. A topological sorting of the data model is then performed. The controllability measures of a target node are given by the minimum number of nodes to be crossed from a primary input to the target node. The observability measures of a target node are given by the minimum number of nodes to be crossed from the target node to a primary output.

5.1.2.Computation of controllability and observability measures for the control model. Since the control model is represented by a structured graph, the computations of the distance-based controllability and observability measures are easy to perform. The controllability measures of a target transition are given by the minimum number of transitions to be crossed to reach the target transition from the initial transition. The observability measures of a target transition are given by the minimum number of transitions to be crossed to reach an observable transition from the target transition.

Although these measures are crude, we will observe, in Section 7, that using them still gives better results than not using any controllability or observability measures.

5.2. Recursive controllability and observability measures

The main drawback of the distance-based controllability and observability measures is that they do not take into account the semantics associated with the basic elements of the model. In this sub-section, we present a more complex way to compute controllability and observability measures. These measures are based on a fine analysis of the basic elements of the model.

5.2.1.Computation of controllability and observability measures for the data model. Since test information must be propagated through the data model, we define transfer formulae of controllability and observability for the basic elements of the data model. These formulae express the difficulties in propagating the test data through each basic element of the model.

The four basic elements of the data model are represented in Figure 1.

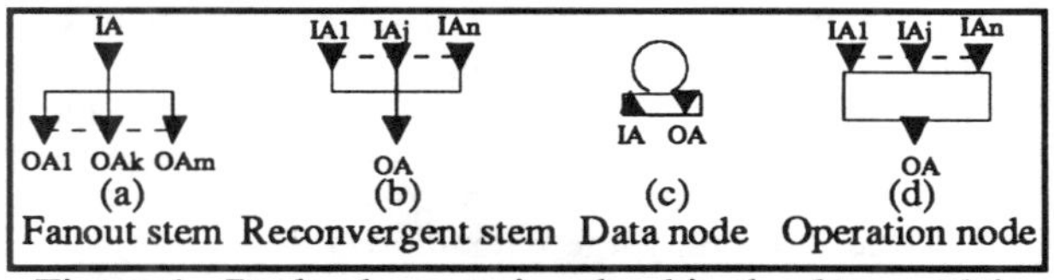

Figure 1 : Basic elements involved in the data model.

Subsequently, we will use the following notation : IAj (resp. OAk) represents the input (resp. output) arc of a basic element. CO(A) (resp. OB(A)) represents the controllability (resp. observability) value of the arc A. We have defined transfer formulae of controllability and observability for each basic element of the data model. These formulae allow the computing of the controllability (resp. observability) value at the output (resp. inputs) of a data model element, knowing the controllability (resp. observability) values at the inputs (resp. output) of this element. In Figure 2, we give the transfer formulae associated with three of the basic elements which are easily understood.

Controllability propagation formula through a fanout stem :
CO(OAk) = CO(IA), $(1 \leq k \leq m)$.
Observability propagation formula through a fanout stem :
OB(IA) = min (OB(OAk)), $(1 \leq k \leq m)$.
Controllability propagation formula through a reconvergent stem : CO(OA) = min (CO(IAj)), $(1 \leq j \leq n)$.
Observability propagation formula through a reconvergent stem : OB(IAk) = OB(OA), $(1 \leq k \leq m)$.
Controllability propagation formula through a data node :
CO(OA) = CO(IA).
Observability propagation formula through a data node :
OB(IA) = OB(OA).

Figure 2 : Recursive formulae associated with fanout-stems, reconvergent stems and data nodes.

The controllability and observability propagation formulae through an operation node are derived from recursive transfer formulae associated with the basic operators involved in the operation node. Each operation node is composed of an interconnection of basic operators defined in the language. For each basic operator, we have defined a controllability transfer coefficient CTC and an observability transfer coefficient OTC. In Figure 3, we give the recursive controllability and observability formulae associated with a basic operator.

Controllability propagation formula through a basic operator
$CO(OA) = CTC \times \min (CO(IAj)), (1 \leq j \leq n).$
Observability propagation formula through a basic operator
$OB(IAj) = OTC \times OB(OA) \times \min (CO(IAl)),$
$(1 \leq l \leq n, 1 \leq j \leq n, j \neq l).$

Figure 3 : Recursive formulae associated with basic operators.

The transfer formulae of controllability and observability measures for an operation node are expressed using a recursive algorithm and the previous formulae.

As for distance-based measures, the computation of controllability measures proceeds level by level. First, the controllability values of every primary input are set to 1. The controllability measures are then computed for every node at level 1, then for every node at level 2, and so on. If the data model has no cycle then the computation is achieved, otherwise the computation is repeated until the stabilization of the values is reached. The correct design of the behavioral description guarantees the convergence of the cost computation. The measures usually reach their stable values after only a few iterations.

The computation of the observability measures is achieved in the same way as for the controllability measures; only the initial nodes and the direction of the propagation are inverted.

5.2.2.Computation of controllability and observability measures for the control model. These measures are computed by means of six parameters which are propagated through the basic elements of the control model. Therefore, we have defined the transfer formulae of these parameters for each basic element of the control model. We will denote the six parameters involved in the measure computation by : NC(A) the number of paths from the initial transition to the arc A, LC(A) the sum of all the transitions on the paths between the initial transition to the arc A, PC(A) the probability of reaching the arc A from the initial transition, NO(A) the number of paths between the observable transitions and the arc A, LO(A) the sum of all the transitions on the paths between the observable transitions and the arc A, and PO(A) the sum of the probabilities of reaching observable transitions from the arc A. The computation of controllability measures proceeds level by level. First the controllability parameters for the initial transition are set at NC=1, LC=1, PC=1. The controllability parameters are then computed for every node at level 1, then for every node at level 2, and so on. The controllability of a transition is expressed by the following formula given in Figure 4. Ntrans represents the total number of transitions involved in the control model and OA represents the output arc of the transition.

$$CO(\text{transition}) = \left[1 - \frac{1}{\text{Ntrans}} \times \frac{LC(OA)}{NC(OA)}\right] \times PC(OA)$$

Figure 4 : Controllability formula for a transition.

The computation of observability measures proceeds level by level. First, the observability parameters for the final transition are set at NO=0, LO=0, PO=0. The observability parameters are then computed for every node at the preceding level, and so on. The observability of a transition node is expressed by the formula given in figure 5. Not represents the number of observable transitions, and IA represents the input arc of the transition.

$$OB(\text{transition}) = \left[1 - \frac{1}{\text{Ntrans}} \times \frac{LO(IA)}{NO(IA)}\right] \times \left[\frac{PO(IA)}{\text{Not}}\right]$$

Figure 5 : Observability formula for a transition.

5.3. Fanout-based controllability/observability measures for data model elements

On the data model, information may be propagated from one data node (called A) to more than one operation node by means of a fanout stem. In this case, we say that the data node is a potential fanout. Similarly, data can be propagated from more than one data node to an operation node (called B); we say that the operation node is a potential reconvergent node. If the node A and the node B are linked by more than one path, they form a fanout/reconvergence structure [cf. Figure 6]. The fanout /reconvergence structures may produce inconsistent data node assignments and thus increase the difficulty of the test generation process. In order to take this testing problem into account, we have studied the influence of fanout/reconvergence structures on the controllability and observability measures of elements of the data model.

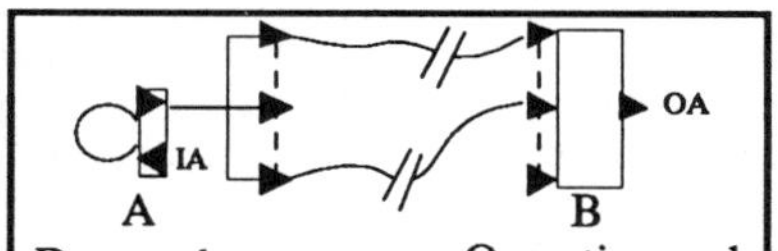

Figure 6 : Fanout/reconvergence structure on the data model.

We have defined a correction term of the controllability at the output node of a fanout/reconvergence structure. This correction term takes into account the dependences between the inputs of the node B. The more correlated the inputs are, the smaller the correction term will be. This correction term is inserted in the previous formulae defined in 5.2.1 for elements of the data model in order to reflect the influence of fanout/reconvergence.

6. Experiments in Test Generation

6.1. Methods issued from the previous computations and used for guiding the test pattern generation algorithm

In this sub-section, seven methods which have been used for experiments are described.

Method 1 : The test pattern generation algorithm is performed without any testability measure.

Method 2 : This method uses distance-based controllability and observability measures on both the data and control models.

Method 3 : Distance-based controllability and observability are used on the control model while recursive controllability and observability formulae are used on the data model.

Method 4 : Distance-based controllability and observability measures are used on the control model while fanout-based controllability and observability measures are used on the data model.
Method 5 : Recursive controllability and observability formulae are used on the control model while distance-based controllability and observability measures are used on the data model.
Method 6 : This method uses recursive controllability and observability formulae measures on both the data and control models.
Method 7 : Recursive controllability and observability formulae are used on the control model while fanout-based controllability and observability measures are used on the data model.

6.2. Description of the experiments

The approach used for the experiments stems from the work of Patel [12]. The behavioral test pattern generation method described in Section 3 was coded in Common LISP on an Apollo work station DN3500. Five separate programs, corresponding to the methodologies described in Section 5, were written to compute the controllability and observability measures. The seven methods described in Section 6.1 were performed on three behavioral descriptions of circuits (C1 : ADD_BCD, C2 : ALU8, C3 : ALU64) and on three experimental descriptions (C4, C5, C6). We consider the fault hypothesis defined in Section 3. No more than 50 faults were randomly selected for test generation to reduce the total execution time. Two types of experiments were conducted.

- The first experiment was conducted to propose a classification of the seven methods. To reach this goal, four numbers were computed for each behavioral description and for each method :
  - -the number of detected faults
  - -the number of undetectable faults
  - -the number of dropped faults (a fault is considered as dropped if the total number of backtracks exceeds a given limit)
  - -the number of backtracks per fault, obtained by dividing the total number of backtracks by the sum of the detected and undetectable faults
- The second type of experiment was conducted to determine if most of the dropped faults for all the previously described methods were similar. Such an experiment was performed in order to see how often a fault was dropped by the different methods.

The two types of experiments were performed with two backtrack limit values : 30 and 100.

7. Experiment results

7.1. First type of experiment

- Backtrack limit of 30.

The results of this experiment are summarized in Table 1. For most behavioral descriptions, if the number of backtracks per fault of a method is smaller compared with other methods, then the total number of faults detected is larger compared with other methods. Method 1, which uses no testability measures, detects the least number of faults. It points out that the use of behavioral testability measures improves the performance of the test pattern generation algorithm. The results using method 2 are not as good as other methods. For behavioral description C2, the number of backtracks per fault is very high compared with other methods. The best results are obtained by methods 7 and 6, followed by methods 4 and 3. The results obtained by methods 5 and 2 are not as good as those obtained by the four previous ones.

		Methods						
		1	2	3	4	5	6	7
C1	Backtracks/fault	8.36	7.78	5.14	4.82	7.32	4.96	4.62
	Detected+Undetectable	45	46	48	48	46	48	48
	Dropped	5	4	2	2	4	2	2
	Rank	7	6	4	3	5	2	1
C2	Backtracks/fault	21.58	19.1	14.38	13.12	15.02	12.56	11.22
	Detected+Undetectable	27	32	38	40	34	40	42
	Dropped	23	18	12	10	16	10	8
	Rank	7	6	4	3	5	2	1
C3	Backtracks/fault	24.42	19.36	18.22	17.46	20.48	18.6	15.3
	Detected+Undetectable	14	28	31	33	28	30	35
	Dropped	36	22	19	17	22	20	15
	Rank	7	5	3	2	6	4	1
C4	Backtracks/fault	15.32	15.35	11.61	7.58	15.41	11.51	7.9
	Detected+Undetectable	22	21	23	26	21	26	30
	Dropped	9	10	8	5	10	5	1
	Rank	5	6	4	2	7	3	1
C5	Backtracks/fault	9.52	9.04	9.19	11.47	8.66	9.57	13.85
	Detected+Undetectable	21	19	19	17	19	18	16
	Dropped	0	2	2	4	2	3	5
	Rank	1	3	4	6	2	5	7
C6	Backtracks/fault	8.96	4.52	4.56	4.16	2.96	3.64	4.96
	Detected+Undetectable	20	22	22	23	24	22	21
	Dropped	5	3	3	2	1	3	4
	Rank	7	4	5	2	1	3	6

Table 1 : Results of the first type of experiment. (Backtrack limit of 30)

- Backtrack limit of 100

The first experiment was performed with a backtrack limit of 100. The backtrack limit was increased from 30 to 100 to see if dropped faults were difficult to detect. In most cases, the total number of dropped faults is reduced by a very small number. As expected, the number of backtracks per fault is higher compared with those for a backtrack limit of 30. The interesting conclusion that results from this experiment is that if a fault is not detected within 30 backtracks, there is a high probability that it will not be detected within a larger backtrack limit.

7.2. Second type of experiment

- Backtrack limit of 30.

In Table 2, the results of this experiment are described. For each behavioral description, we give the number of dropped faults obtained by one, two ... or seven methods within a backtrack limit of 30. For example, for behavioral description C2, the entry under column 3 shows that out of a total of 34 distinct dropped faults, the same eight faults were dropped by three methods.

	1	2	3	4	5	6	7	Total
C1	0	2	2	0	1	1	0	6
C2	13	5	8	1	1	2	4	34
C3	10	3	7	5	1	3	10	39
C4	1	2	2	8	1	0	0	14
C5	3	2	1	2	0	0	0	8
C6	4	1	1	2	0	0	0	8

Table 2 : Results of the second type of experiment. (Backtrack limit of 30)

The results point out the fact that with a high probability, a detectable fault will not be dropped by all methods. Some faults are difficult to detect for one method, but easier to detect for others.

- Backtrack limit of 100.

The backtrack limit was increased to 100. The results issued from this experiment show that a dropped fault under the backtrack limit of 30 will most likely be dropped even with a backtrack limit of 100.

7.3. Discussion

The previous results necessitate some remarks. The first point is that among all proposed testability measures none is found to be consistently superior in all cases. The second point is that increasing the backtrack limit does not allow us to obtain noticeably better results. This suggests that it will be more efficient to keep the backtrack limit low and to switch between the different controllability and observability measures. The same conclusion was obtained by Patel [12]. A similar idea is presented by Min and Rogers [17] incorporating multiple search strategies to speed test generation with a technique for selecting which strategy to use and for switching strategies. In our case, we have proposed numerous ways to compute behavioral testability measures. According to the previous results, we propose using method 7 with a lower backtrack limit in the test generation process first. If the fault is not detected or is not determined to be undetectable, method 6 can be used. Method 4 and method 3 can then be used. Finally methods 5 and 2 should be used.

8. Conclusions and future work

We have presented different ways of computing testability measures for behavioral descriptions of digital circuits. These testability measures must be used in order to accelerate the behavioral deterministic test pattern generation process [1, 2, 3, 4, 5]. The purpose of this paper is also to validate the proposed testability measures and to carry out a comparative study between them. It has been observed that the use of the measures presented here improves the performance of the test generation process. As a result of this study, we have proposed a strategy of test generation. Finally, it should be explicitly stated that most of the experiments performed are based upon the ideas found in Patel's paper [12].

Our future work will concentrate on a number of investigations which are briefly described below :

- all testability measures implemented to speed up behavioral deterministic test pattern generation are static in nature. An interesting question appears to be whether benefits can be gained from the application of dynamic testability measures [18].

- the techniques issued from testability measures in order to improve and accelerate behavioral ATPG are heuristic. However, it would be interesting to develop algorithmic techniques that definitely lead to improvements of behavioral ATPG [19]. By using heuristic techniques, the ATPG process provides only average benefits.

References :

[1] D.S. Barclay, J.R. Armstrong, "A heuristic chip-level test generation algorithm", 24th Design Automation Conference, pp. 257-261, 1987.

[2] U.H. Levendel, P.R. Menon, "Test generation algorithms for computer hardware description languages", IEEE Trans. on computers, Vol. C-31, N° 7, pp. 577-588, July 1982.

[3] M.D. O'Neill, D.D. Jani,C.H. Cho, J.R. Armstrong, "BTG : a behavioral test generator", CHDL and their applications, IFIP, pp. 347-361, 1990.

[4] F. E. Norrod, "An automatic test generation algorithm for hardware description languages", 26th ACM/IEEE Design Automation Conference, pp. 429-434, 1989.

[5] N. Giambiasi, J.F. Santucci, A.L. Courbis, V. Pla, "Test pattern generation for behavioral descriptions in VHDL", Euro-VHDL 91, Stockholm, pp. 228-235 September 1991.

[6] S. Ghosh, T.J. Chakraborty, "On behavior fault modeling for digital designs", Journal of electronic testing : theory and applications, N°2, pp. 135-151, 1991.

[7] N. Giambiasi, J.F. Santucci, A.L. Courbis, M. Boumédine, R. Alali, K. Djafari, "General behavioural modeling framework for hardware systems" Poster session, EDAC, March 1990.

[8] L. H. Goldstein, E. L. Thigpen, "SCOAP: Sandia Controllability/Observability Analysis Program", 17th Design Automation Conf., Minneapolis, pp. 190-196, June 1980.

[9] J. E. Stephenson, J. Grason, "A testability measure for register transfer level digital circuits", International Symposium on Fault Tolerant Computing, Pittsburg, pp. 101-107, June 1976.

[10] K.Thearling and J.Abraham, "An Easily Computed Functional Level Testability Measure", Proc. Intn'l. Test Conf.., pp. 381-390, 1989.

[11] N. Giambiasi, J.F. Santucci, M. Boumédine, G. Dray, "Methods to reduce computational cost of behavioural test pattern generation", Poster session, Intn'l. Test Conf.., 1991.

[12] S. Patel and J. Patel, "Effectiveness of heuristic measures for automatic test pattern generation", 23rd Design Automation Conference, pp. 547-552, 1986

[13] M. Abramovici, M.A. Breuer, A.D. Friedman, Digital Systems Testing And Testable Design, Ch. 9, Computer Science Press, New York 1990.

[14] "DACAPO III", user manual Version 1.0, Dosis GmbH, September 1988.

[15] "IEEE standard VHDL language reference manual", IEEE standard 1076, 1987.

[16] N. Giambiasi, J.F. Santucci, M. Boumédine, "Definition and study of a general fault model on the control and data graphs", Doc.N°UN0002CRXX, LERI, Nîmes, June 1989.

[17] H.B. Min, W.A. Rogers, "Search strategy switching : a cost model and an analysis of backtracking", Journal of electronic testing : theory and applications, N°1, pp. 125-137, 1990.

[18] A. Ivanov, V.K. Agarwal, "Dynamic testability measures for ATPG", IEEE Trans. on computer-aided design, Vol. 7, N° 5, pp. 598-608, May 1988.

[19] M. Schulz, E. Trischler, T.M. Sarfert, "Socrates : A Highly Efficient Automatic Test Pattern Generation System", Proc. Intn'l. Test Conf.., pp. 1016-1026, September 1987.

Automatic Test Knowledge Extraction From VHDL (ATKET*)

Praveen Vishakantaiah, Jacob Abraham

Computer Engineering Research Center
University of Texas at Austin
Austin, TX 78758

Magdy Abadir

Microelectronics and
Computer Technology Corporation
Austin, TX 78759

Abstract

Behavioral information about modules in a design is exploited by test engineers to handle the complex problem of test generation for large designs. However, automation of the process of exploiting behavioral information to facilitate test generation is in its infancy. This paper introduces ATKET, an Automatic Test Knowledge Extraction Tool which automatically generates pieces of test knowledge by using structural and behavioral information in the VHDL description of a design. Results obtained from ATKET for a circuit which is difficult to test are presented.

1 Introduction

The emergence of Computer Aided Design technology in the past decade has facilitated the design of highly dense and complex integrated circuits. There has been a continued effort to introduce new test methodologies which not only would refrain from increasing product cost or degrading performance, but also keep pace with the design technology and reduce the burden of testing on the design engineer. Present-day test engineers exploit the behavior of modules in the design while deriving tests manually for large circuits using tools capable of generating tests for small circuits.

Sarfert et al. [9], Roy et al. [8] and Kunda et al. [6] propose methodologies to speed up the test generation process by using high level primitives. These techniques lack intelligence to combine module knowledge at a global level to generate a priori knowledge which can guide test generation process.

Work by Bending [3] requires the user to provide knowledge to guide test generation while the work by Crastes de Paulet [5], Shirley [10] and Thearling [11] use some form of knowledge extraction. Work by Crastes de Paulet is restricted to generating accessibility modes for each chip by using structural descriptions. Shirley's work is restricted to data paths. Thearling's work is aimed at designs developed using a silicon compiler. Also, none of the above techniques exploit VHDL behavioral descriptions.

ExperTest [7] uses a subset of VHDL for its work. Details of techniques used by ExperTest to generate test knowledge are not available.

*Automatic Test Knowledge Extraction Tool – pronounced as etiquette.

2 Research motivation

Automatic test generation for large sequential circuits targeted at gate or switch level faults is very expensive. This is because of the large search space that needs to be considered during test generation process. Knowledge about the behavior of modules in the design can be exploited to prune the search space [4] so as to make test generation feasible while still targeting faults at any desired level of abstraction.

Consider the hierarchical module example shown in Figure 1. While targeting a fault inside mod-

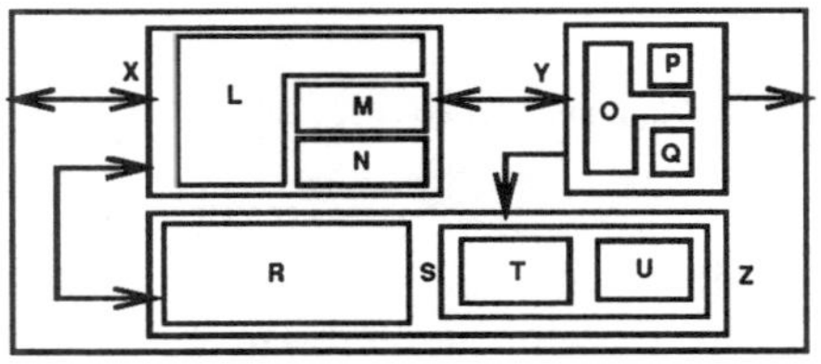

Figure 1: Hierarchical module example

ule O, a typical automatic test generator does not take into consideration the constraints due to modules X, Z, P and Q. Instead, justification is done through these modules whenever choices are available on the lines connecting module O with X, Z, P and Q. This is a very expensive procedure and can be

0738-100X/92 $3.00 © 1992 IEEE

eliminated by identifying justification and propagation paths through these modules in a preprocessing step and storing the knowledge.

This paper describes ATKET, an Automatic Test Knowledge Extraction Tool which synthesizes test knowledge using structural and behavioral information available in the VHDL description of a design. ATKET can be used in conjunction with tools like TIGER [1] and thereby eliminate the need for the designer to provide transparent modes of operation for modules.

A VHDL analyzer (available from Microelectronics and Computer Technology Corporation [MCC]) produces an intermediate representation of the information contained in a VHDL design. ATKET interfaces to this intermediate representation to access structural and behavioral information in the design and stores it in suitable data structures. A convenient representation called the Module Operation Tree (MOT) is used to capture the behavior of modules in the design. Information stored in the MOT along with structural information describing connections between modules in the design is used to generate test knowledge.

Section 3 describes extraction of information from the intermediate representation and introduces the concept of a MOT. Section 4 deals with the types of test knowledge and techniques used to generate them. Section 5 discusses an example circuit and results obtained using ATKET. The concluding remarks are given in Section 6.

3 Information extraction

Structural and behavioral information of the circuit described in VHDL is obtained by accessing the intermediate representation produced by MCC's VHDL analyzer.

3.1 Module operation tree

A Module Operation Tree (MOT) is used to represent behavioral information of a module internally. The MOT is a convenient representation which facilitates generation of pieces of test knowledge for the modules in the design. It ensures that only the designer specified set of operations are used during high-level justification and propagation. It also captures the sequentiality of statements within the VHDL *process* statement.

The MOT is a binary tree in which the nodes correspond to VHDL conditional statements. Each node has a left child which corresponds to the conditional statement at the node being false and a right child which corresponds to the conditional statement at the node being true. The nodes in the MOT are categorized into two types depending upon the conditional statement in the conditional constructs. The first type is one in which the conditional statement involves a port/signal/variable of width 1 (number of bits is 1) whose binary value decides module behavior. The C_RST = 1 node in Figure 2 is an example of this type of node. The second type is one in which the conditional statement involves a port/signal/variable of width greater than 1 and validity of the statement decides module behavior. The node containing the statement COUT = 500 in Figure 2 is an example of this type of node.

The nodes of MOT may contain a list of statements which correspond to statements at the same level as the conditional statement. Storage of such a list at MOT nodes is necessary since these are applicable to each of the subsequent nodes that are generated. For example, in Figure 3, the statement, $C \Leftarrow A + B$ applies to both leaf nodes which means that if the condition at the node is false, $C \Leftarrow A + B$ as well as Ovf is set to '0'. If the condition is true, $C \Leftarrow A + B$ as well as Ovf is set to '1'.

```
entity COUNTER is
  port (CLK : in BIT;
        C_RST : in BIT;
        LD : in BIT;
        CIN : in INTEGER range 0 to 4095;
        COUT : out INTEGER range 0 to 4095 );
end COUNTER;
architecture behavior of COUNTER is
 begin
   process
    variable C_REG : INTEGER range 0 to 4095;
    begin
      if (C_RST = '1') then
        C_REG := 0;
      elsif (CLK'Last_Value = '0') then
        if (CLK = '0') then
            C_REG := C_REG;
        elsif (LD = 1) then
            C_REG := CIN;
        elsif (C_REG = 500) then
            C_REG := 0;
        elsif (C_REG <= 3000) then
            C_REG := C_REG + 12;
        else
            C_REG := C_REG - 2999;
        end if;
      else
        C_REG := C_REG;
      end if;
      COUT <= C_REG;
   end process;
end behavior;
```

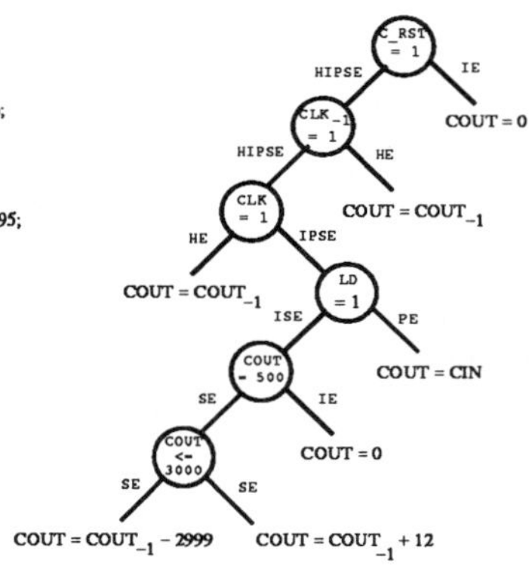

I: Initialization ; P: Propagation ; H: Hold ; S: State ; E:Edge

Figure 2: Description of a Counter and its MOT

Leaf nodes in the MOT contain a list of statements which correspond to the input-output relationships defined for the module by the designer. The list may contain only one statement as is the case in Figures 2 and 3.

Depending on the module input-output relationship, edges in the MOT are categorized so that information about leaf nodes is readily available in the MOT. The basic categories are:

1. Initialization Edge (IE), which will lead to a leaf corresponding to a module output being initialized to a constant value. The right edge of node C_RST = 1 in Figure 2 is an example.
2. Propagation Edge (PE), which will lead to a leaf corresponding to an identity mode [2] of the module. The right edge of node LD = 1 in Figure 2 is an example.
3. Hold Edge (HE), which will lead to a leaf corresponding to a module output that holds the previous value. The left edge of node CLK = 1 in Figure 2 is an example.
4. State Edge (SE), which will lead to a leaf corresponding to a module output which changes its value depending on its previous value. The right edge of node $COUT <= 3000$ in Figure 2 is an example.

```
entity ADDER is
  port(A, B : in Integer range 0 to 65535;
       Ovf  : out Bit;
       C    : out Integer range 0 to 65535);
end ADDER;

architecture behavior of ADDER is
begin
  RUN_ADDER : process(A, B)
  variable Temp : Integer range 0 to 131070;
  begin
     Temp := A + B;
     if (Temp >= 65536) then
        Ovf <= '1';
     else
        Ovf <= '0';
     end if;
     C <= Temp;
  end process RUN_ADDER;
end behavior;
```

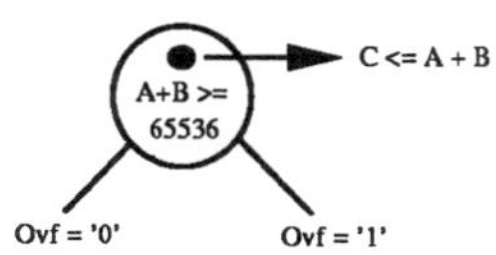

Figure 3: Description of an Adder and its MOT

A combination of these basic categories can also be used to categorize edges in the MOT. Each statement in the leaf node is analyzed using the rules shown in Figure 4 to decide the type of an edge pointing to the leaf node. The type of an edge which does not point to a leaf node is the union of edge types of the children edges of the node to which the edge points. For example, the type of left edge of node LD = 1 in Figure 2 is ISE (which means Initialization and State Edge) because the node COUT = 500 to which it is connected has a left edge whose type is SE and a right edge whose type is IE.

The MOT is built for every *process* statement in the behavioral description of a module. This preserves the concurrency of the *process* statement. The VHDL conditional constructs, *if-then-else-if* and *case* statements within the *process* statement are analyzed to generate the MOT. Nested conditional constructs are handled.

After extracting all of the necessary information from the intermediate representation, compiler optimization techniques, namely copy and constant propagation and dead code elimination, are used to compact information stored in the MOT.

RHS of Statement			E_b	E_a
Constant value (zero or any specific constant)			X	IE
			Q	Q + IE
Single port and not an expression		Port is same as LHS	X	HE
			Q	Q + HE
		Port is a module I/P port	X	PE
			Q	Q + PE
Expression	Atleast one port	Atleast one port is same as LHS	X	SE
			Q	Q + SE
		All ports are module input ports	X	PIE
			Q	Q + PIE
	Does not contain any port		X	IE
			Q	Q + IE

E_a -- Edge type after analyzing statement
E_b -- Edge type before analyzing statement
X -- Edge with no type
Q -- Edge is of some type 'Q'

Figure 4: Rules to categorize edges pointing to leaf nodes

Signal/variable assignment statements (referred to as statements from now on) at MOT nodes may contain internal signals (signals and variables that are not module ports). Such signals are eliminated to ensure that test knowledge is in terms of module ports and does not contain any information related to temporary signals or variables used in the design description. Relevant signal relationships between module ports and internal signals in the MOT nodes are used for the elimination process.

4 Test knowledge generation

Test generation for sequential circuits (especially targeted at switch level faults) can be very complex, time consuming and even impractical for large circuits due to line justification and error propagation involving multiple time frames. Test knowledge which guides line justification and error propagation by pruning the search space is obtained using information stored in the MOT in conjunction with structural connections between various modules in the design. The different pieces of test knowledge generated using the MOT for each module in the design are :

1. *Initialization Modes* which enable initialization of any module output to zero or to constant values.

2. *Propagation Modes* which correspond to identity modes of the module.
3. *Hold Modes* which correspond to any module output retaining its previous state.
4. *State Modes* in which the new state of any module output depends on its previous state.

The information corresponding to each mode generated contains a set of constraints on the module inputs which needs to be satisfied in each time frame and the action of the mode.

Modes are directly extracted from the MOT using the algorithm shown in Figure 5. If the modes contain constraints on module outputs, they are eliminated by using the algorithm shown in Figure 6. This ensures that the constraints are restricted to module inputs.

```
find_fund_modes(MOT, constraint_list) {
  if (statement list at the node is not empty)
    for (each statement in the statement list)
      Add modes, if any to the relevant mode list
    endfor
  endif
  if (left child is a leaf node)
    Generate modes using constraint_list
  else
    constraint_list.append(node_constraint)
    find_fund_modes(MOT_left, constraint_list)
  endif
  if (right child is a leaf node)
    Generate modes using constraint_list
  else
    constraint_list.append(node_constraint)
    find_fund_modes(MOT_right, constraint_list)
  endif
} end find_fund_modes()
```

Figure 5: Algorithm to find fundamental modes

4.1 Default modes for VHDL operators

The behavior of modules in the design can be specified using VHDL expressions which involve operators like '+', '-', and so on. Default modes are associated with the operators and stored as part of the test knowledge. For example, the VHDL expression, $C \Leftarrow A+B$ has a default initialization mode that corresponds to A and B being zero and a default propagation mode that

```
find_advanced_modes(MOT, fund_modes) {
  for (each fund_mode, m_i)
    if (constraints involve module outputs)
      Search MOT to eliminate constraint
      on module output
      if (such a mode, m_j is found)
        Combine m_i and m_j to update m_i
      else
        Unrealizable mode
      endif
    endif
  endfor
} end find_advanced_modes()
```

Figure 6: Algorithm to find advanced modes

corresponds to A being zero and B being a symbolic value.

4.2 Recognition of Shift operation

VHDL does not provide an operator to specify the Shift operation of modules and hence cannot be directly recognized. ATKET identifies the Shift operation (serial-serial, serial-parallel, and parallel-serial) automatically by using statements in each block of the VHDL process statement. For example, the technique used to identify the serial-parallel Shift operation is described below.

Let I_M be an input vector of module M and O_M be an output vector of M. Let 'b' and 'd' represent two specific bit positions of O_M and 'c' represent a specific bit position of I_M. If there is a statement $O_M[d] \Leftarrow O_M[b]$ followed by a statement, $O_M[b] \Leftarrow I_M[c]$ in the same block (indicates same time frame), a serial-parallel Shift operation is recognized.

The information extracted about the Shift operation is whether the Shift is to the left or to the right and the number of bits which can be shifted in from the module input without losing any bits.

4.3 Combining module modes

For each module M in the design, the modes of each of the subcomponents in M are combined to generate useful test knowledge relevant to module M.

The information used to combine the modes of modules in the design include the structural connectivity between modules, existence of feedback loops in the circuit, constraints on module inputs due to some signals being permanently tied to a fixed value, and constraints on outputs of modules connected to inputs of other modules.

5 Test knowledge example

Test knowledge for the example circuit [7] shown in Figure 7 was automatically generated in about 1.0 CPU second on a SUN 4/110. The research prototype of ATKET has been implemented in C++ (about 7000 lines of source code) using object-oriented techniques.

The behavior of the Counter and Adder shown in Figure 7 are described in Figures 2 and 3. The Reset of the Counter is connected to the Overflow of the Adder. REG_SLT controls the output of the Register on the positive edge of CLK. A '0' value of REG_SLT results in an hold operation for the Register while a '1' value results in a latch operation. RST initializes the output of the Register to zero.

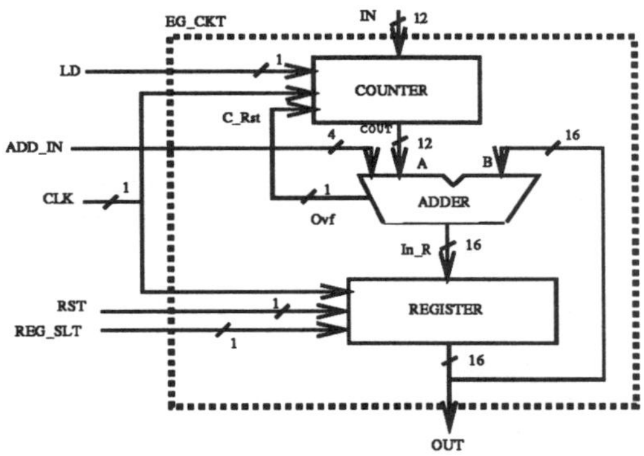

Figure 7: Test knowledge example

The modes generated for the Counter and Adder are shown in Figures 8 and 9. The modes generated for EG_CKT are shown in Figure 10. Constraints shown between two bold horizontal lines correspond to a mode. For example, in Figure 8, there are two initialization modes each spanning six time frames. The first initialization mode requires that the constraint CLK = 0 be satisfied in the first time frame, set of constraints CLK = 1, LD = 1, IN = 0, $ADD\_IN = 0$ and RST = 1 be satisfied in the second time frame and so on.

It should be noted that each of the modes in Figure 8 correspond to the seven leaf nodes in the MOT shown in Figure 2. In Figure 9, the first two initialization modes correspond to the two leaf nodes in the MOT shown in Figure 3 while the propagation mode

Mode	Time Frame	Constraints	Action
Required to set C_RST = 0	1	(CLK 0)	
	2	(CLK 1) (LD 1) (IN 0) (ADD_IN 0) (RST 1)	
Initialization	3	(CLK 0)	COUT = 0
	4	(CLK 1) (LD 1) (IN 0) (ADD_IN > 0) (RST 1)	
	5	(CLK 0) (RST 0)	
	6	(CLK 1) (LD 1) (IN x) (ADD_IN y) (REG_SLT 1) (RST 0)	
	3	(CLK 0)	COUT = 0
	4	(CLK 1) (LD 1) (IN 500) (ADD_IN 0) (RST 1)	
	5	(CLK 0)	
	6	(CLK 1) (LD 0)	
Propagation	3	(CLK 0)	COUT = S_V
	4	(CLK 1) (LD 1) (IN S_V)	
Hold	3	(CLK 1)	None
	4	None	
	3	(CLK 0)	None
	4	(CLK 0)	
State	3	(CLK 0)	COUT = $COUT_{-1} - 2999$
	4	(CLK 1) (LD 1) (IN <= 3000)	
	5	(CLK 0)	
	6	(CLK 1) (LD 0)	
	3	(CLK 0)	COUT = $COUT_{-1} + 12$
	4	(CLK 1) (LD 1) (IN > 3000)	
	5	(CLK 0)	
	6	(CLK 1) (LD 0)	

S_V — Symbolic Value x = 4095; y = 15

Figure 8: Modes generated for the Counter

corresponds to the default mode associated with the '+' operator in the statement at the node of the MOT in Figure 3. The state and hold modes of EG_CKT are not shown in Figure 10 due to space considerations.

For another example, the EG_CKT in Figure 7 was modified such that the 4 bit wide ADD_IN port is permanently tied to GND. Only one interesting mode which is automatically generated for this case is shown in Figure 11. This mode is generated as a result of identifying that the realization of constraint, $ADD\_IN > 0$ in the second initialization mode of the

Mode	Time Frame	Constraints	Action
Initialization	1	(CLK 0)	In_R = 0, Ovf = 0
	2	(CLK 1) (LD 1) (IN 0) (ADD_IN 0) (RST 1)	
	1	(CLK 0)	Ovf = 1
	2	(CLK 1) (LD 1) (IN 0) (ADD_IN > 0) (RST 1)	
	3	(CLK 0)	
	4	(CLK 1) (LD 1) (IN x) (ADD_IN y) (RST 0) (REG_SLT 1)	
Propagation	1	(CLK 0)	In_R = S_V
	2	(CLK 1) (LD 1) (IN 0) (ADD_IN 0) (RST 1)	
	3	(CLK 0)	
	4	(CLK 1) (LD 1) (IN S_V) (ADD_IN S_V) (RST 1)	

x = 4095; y = 15

Figure 9: Modes generated for the Adder

Mode	Constraints			Action
Initialization			(RST 1)	OUT = 0
	Time Frame	1	(CLK 0)	COUT = 0 In_R = 0 Ovf = 0 OUT = 0
		2	(CLK 1) (LD 1) (IN 0) (ADD_IN 0) (RST 1)	
Propagation	Time Frame	1	(CLK 0)	OUT = S_V
		2	(CLK 1) (LD 1) (IN 0) (ADD_IN 0) (RST 1)	
		3	(CLK 0)	
		4	(CLK 1) (LD 1) (IN S_V) (ADD_IN S_V) (RST 1)	
		5	(CLK 0)	
		6	(CLK 1) (RST 0) (REG_SLT 1)	

Figure 10: Modes generated for EG_CKT

Mode	Constraints			No. of T.F. rep.	Action
Initialization	Time Frame	1	(CLK 0)	1	Ovf = 1
		2	(CLK 1) (LD 1) (IN 0) (RST 1)	1	
		3	(CLK 0)	17	
		4	(CLK 1) (LD 1) (IN 4095) (RST 0) (REG_SLT 1)		

Figure 11: An Adder mode when ADD_IN is 0

Adder (shown in Figure 9) is not possible. The feedback loop in the circuit is exploited to generate the initialization mode for the Adder.

5.1 Using generated test knowledge

The use of test knowledge generated by ATKET can be demonstrated by considering two scenarios. One is when the automatic test generator is generating test patterns for a module say M whose input is connected to the output of a module like the EG_CKT module. In this case, the modes in Figure 10 can be used to avoid backtracking through the EG_CKT module, thereby speeding up the test generation process.

The second scenario is when the automatic test generator is targeting faults inside a module, say the Register module in Figure 7. In this case, the Adder modes (since this is the only module that is connected directly to the Register) shown in Figure 9 can be used to speed up the test generation process.

6 Conclusions

Techniques to automatically generate test knowledge from structural and behavioral information in the VHDL description of a design have been introduced. The results obtained from a research prototype of the test knowledge extraction tool for an example circuit which is difficult to test were provided. The tool took only about 1.0 CPU second for the example circuit which indicates the ease with which test knowledge was generated.

We are in the process of extracting test knowledge from additional VHDL constructs. Also, techniques to generate more complicated test knowledge are being identified.

Acknowledgements

Interesting discussions with Desmond D'Souza are acknowledged. This research was supported in part by the Semiconductor Research Corporation under contract 91-DP-142.

References

[1] M.S. Abadir, "TIGER: Testability Insertion Guidance Expert System," Proc. of the International Conference on CAD, pp. 562-565, 1989.

[2] M.S. Abadir and M.A. Breuer, "A Knowledge Based System for Designing Testable VLSI chips," IEEE Design and Test of Computers, Vol. 2, No. 4, pp. 56-68, August, 1985.

[3] M.J. Bending, "Hitest: A Knowledge-Based Test Generation System," IEEE Design and Test of Computers, Vol. 1, No. 2, pp. 83-92, May 1984.

[4] D. Brahme and J.A. Abraham, "Knowledge Based Test Generation For VLSI Circuits," Proc. of the International Conference on CAD, pp. 292-295, Nov. 1987.

[5] M. Crastes de Paulet et al., "Testability Expertise And Test Planning From High-Level Specifications," Proc. of the International Test Conference, pp. 692-699, 1989.

[6] R.P. Kunda et al., "Speed Up Of Test Generation Using High-Level Primitives," Proc. of the 27th Design Automation Conference, pp. 594-598, June 1990.

[7] A. Miczo and C. Miller, "Test Generator Extracts Design Intent From Behavioral Models," High Performance Systems, pp. 54-66, May 1990.

[8] K. Roy and J.A. Abraham, "High Level Test Generation Using Data Flow Descriptions," Proc. of the 1990 European Design Automation Conference, pp. 313-317, March 1990.

[9] T.M. Sarfert et. al., "Hierarchical Test Pattern Generation Based On High-Level Primitives," Proc. of the International Test Conf., pp. 470-479, 1989.

[10] M. Shirley, "Generating Tests By Exploiting Designed Behavior," PhD thesis, MIT, Jan. 1989.

[11] K.H. Thearling, "Techniques For Automatic Test Knowledge Extraction From Compiled Circuits," PhD thesis, Univ. of Illinois at Urbana-Champaign, Aug. 1990.

[12] "IEEE Standard VHDL Language Reference Manual," IEEE, 1988.

Data Path Allocation using an Extended Binding Model*

Ganesh Krishnamoorthy

Mentor Graphics Corporation
Warren, NJ 07059

John A. Nestor

Illinois Institute of Technology
Chicago, Illinois 60616

Abstract

Existing approaches to data path allocation in high-level synthesis use a binding model in which values are assigned to the same register for their entire lifetimes. This paper describes an extended binding model in which segments of a value's lifetime may reside in different registers if there is a cost advantage in doing so. In addition, the model supports multiple copies of values and the use of functional units to "pass through" unmodified values to reduce interconnect. This model is exploited in an allocation tool that uses iterative improvement to search for low-cost designs. Results show that allocation costs can be substantially reduced using this model.

1. Introduction

Data path allocation [1] is the problem of assigning hardware to a scheduled control/data flow graph (CDFG) to implement a specified behavior while meeting performance and timing constraints and minimizing implementation cost. The CDFG specifies *operators* that manipulate data, data *values* that require storage, and *data transfers* that move information between operators and values.

The traditional approach to allocation of a scheduled CDFG is based on a *binding model* that is described many places in the literature (e.g. [2,3,4]). As shown in Figure 1, the binding model specifies the following assignments between CDFG elements and datapath hardware:

1. *Operators* must be assigned to *functional units* during the control steps in which they are scheduled. Operators which are scheduled in the same control step may not be assigned to the same functional unit.
2. *Values* must be assigned to *registers* that store the values. Each value must be assigned to a register for the duration of its lifetime. Values with overlapping lifetimes may not be assigned to the same register.
3. *Data transfers* must be assigned to some path of connections, buses, and multiplexers so that data may flow between allocated operators in functional units and allocated values in registers as specified by the edges of the CDFG.

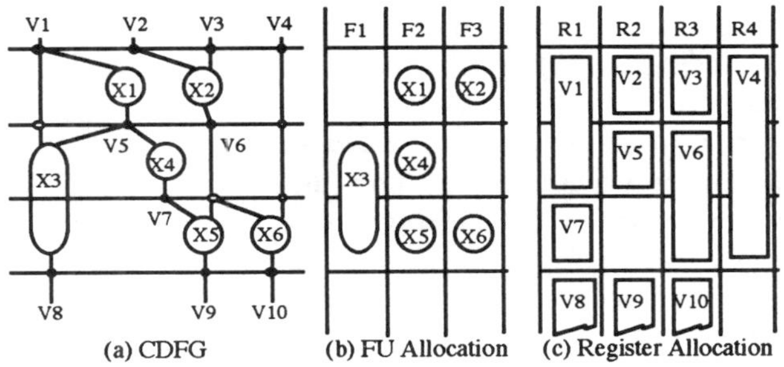

Figure 1 - The Traditional Binding Model

The form of data transfer allocation depends on the connection style used. In a *general* interconnection style [4,5], any interconnection structure of wires, buses, and multiplexers may be used used. In a *bus-oriented* interconnection style [6], a restricted interconnection style is used in which module (register and functional unit) outputs drive one or more buses, which are connected by a single level of multiplexers to module inputs. In a *point-to-point* interconnection style [2], module outputs are connected to module inputs by a single level of multiplexers. The point-to-point style has the advantage that it may be derived directly from functional unit and register allocation. For this reason, it is often used as a starting point for allocation, followed by a postprocessing phase that uses a more general interconnection style (e.g. [2,5]).

The *cost* of a data path allocation is usually taken to be a weighted sum of the number of functional units, registers, and interconnection elements (multiplexers, buses, and wires) used to implement the behavior of the CDFG. The goal of allocation algorithms is to minimize this cost. Since the minimum number of functional units and registers is fixed by scheduling,

* This research was supported by NSF Grant MIP-9010406.

much of the effort in allocation involves minimizing interconnection cost.

Several allocation algorithms and heuristics have been developed that attempt to minimize the cost of an allocation using the traditional binding model. These include constructive, greedy heuristics [7,2,4], iterative approaches such as simulated annealing [8,9,10], simulated evolution [11], and "rip-up-and-reallocate" [12], and exact approaches such as bipartite graph matching [13] and integer programming [14]. These approaches can find optimal or near-optimal allocations for this binding model and cost function.

However, the binding model itself restricts the design decisions that are available in allocation and so prevents the allocation of lower-cost configurations. It does not allow techniques commonly used by human designers, such as the storage of copies of values and the use of ALU's to "pass-through" values during data transfers. The failure to consider these decisions prevents tools from finding the high-quality designs that are demanded by users of high-level synthesis tools.

Some of these problems have been addressed in previous approaches. For example, the HAL allocator [2] uses the concept of "local storage" operators, which allow the use of multiple copies of values to reduce interconnect costs. However, these local copies are restricted to be one control step in length and are placed only at the beginning of a larger value's lifetime. Other approaches (e.g. [5]) allow functional-unit pass-throughs to be considered as part of interconnect allocation. However, these opportunities are only explored after functional units are assigned.

This paper describes an extension of the traditional binding model called the SALSA binding model. The key idea in the SALSA approach is to introduce additional flexibility by breaking the lifetime of each value into *segments* of one control step and to allow different segments of a value to be assigned to different registers if there is a cost advantage in doing so. This approach is similar to the ASYL [15] allocator, which also considers register assignment of value segments independently. However, ASYL stipulates that all segments of a value must be stored contiguously in a single register, so this flexibility is not fully exploited. The SALSA binding model also supports multiple copies of values and allows functional unit pass-throughs to be allocated in a natural way. Results show that this model can find allocations that require fewer interconnections than those supported by the traditional allocation model.

2. The SALSA Binding Model

The SALSA binding model is quite similar to the traditional CDFG binding model described in Section 1 with one key departure: the addition of a new type of node to the CDFG called the *slack node*. Slack nodes explicitly represent slack in the data flow, control flow, and timing constraints implied by CDFG edges. Figure 2(a) shows a CDFG with added slack nodes, each marked with an "S". This explicit representation of slack can be used in scheduling [16] used to rapidly evaluate whether an operator node may be rescheduled without violating data or timing constraints. However, slack nodes on data edges also prove useful in allocation.

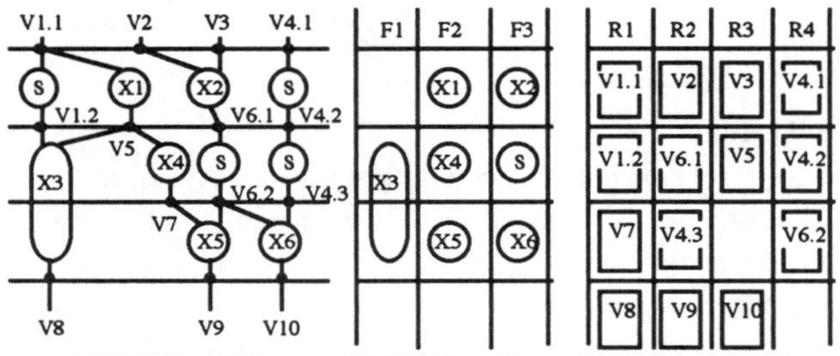

Figure 2 - The SALSA Binding Model

Specifically, slack nodes can be thought of as "No-op" operators that pass an input value unmodified to an output value. This breaks values which may be several control steps in length into *value segments* that are each one control step in length. Thus instead of considering a value as a monolithic element that must be assigned to one register for the duration of its lifetime, a value is divided into multiple segments that may be assigned to separate registers, as shown in Figure 2(c). When segments associated with the same value are assigned to different registers, then a data transfer is created that must be considered during allocation. However, the ability of values to move during their lifetime introduces a new degree of freedom that can reduce the amount of interconnection needed in an allocated datapath. For example, moving a value to a different register during a control step with few data transfers may reduce the demand for new interconnections in another control step with many data transfers.

Slack nodes can also be used to take advantage of "pass-through" operations. In a pass-through operation, an idle functional unit is used to transfer a value from its input to its output without altering the value. This allows existing connections to be combined, reducing the need for additional connections. Exploiting pass-throughs is easily accomplished in the SALSA binding model by interpreting slack nodes as bindable operators that perform a "No-Op" function. Slack nodes which represent data transfers may then be assigned to functional units that can perform this "No-Op" function, as shown in the second step in Figure 2(b). Slack nodes representing contiguous value segments that are stored in the same register do not need to be assigned to functional units and are not considered as part of the allocation.

Pass-through operations can then be used to implement data transfers using existing connections instead of adding a new connection. For example, Figure 3 shows a partially allocated CDFG and datapath in which value segments V1.1 and V1.2 are assigned to different registers - V1.1 to register R2, and V1.2 to register R1. Thus a transfer of data is required. In the normal allocation approach, this assignment requires a multiplexer at the input of R1 to select either the output of functional unit FU1 or register R2, as shown in Figure 3(a). However, since FU1 is idle at the time of the transfer and there are connections from R2 to FU1 and FU1 to R1, the slack node (S) can be assigned to FU1 and FU1 can be used as a "pass-through" device, as shown in Figure 3(b).

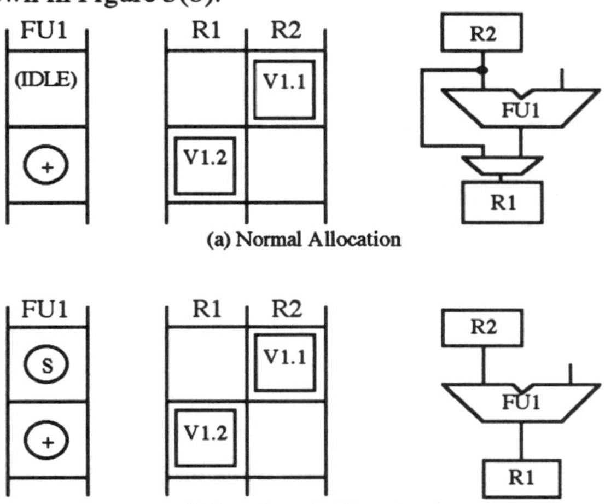

(a) Normal Allocation

(b) Pass-through Allocation

Figure 3 - Pass-through implementation

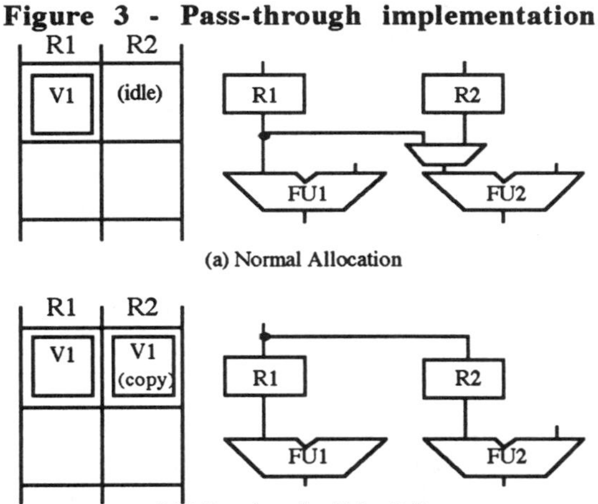

(a) Normal Allocation

(b) Allocation after Value Split

Figure 4 - Value Split

It is sometimes useful to create multiple copies of a value to reduce the global demand for interconnect in an allocated design. For example, Figure 4 shows a value V1 stored in register R1 which is used by operators bound to two functional units, FU1 and FU2. If a copy of V1 is created and stored in register R2, then a connection between R1 and F2 can be eliminated at the expense of an added connection at the input of R2. This can result in a lower cost implementation when the overall number of multiplexers is reduced, as in Figure 4.

The SALSA representation supports value copies by providing a *value splitting* transformation that creates a copy of a value segment. Repeated applications of the value splitting transformation allows a copy to be stored over multiple control steps. In addition, since value splitting may be applied to any value segment, copies of a value may be made at any point in its lifetime. A *value merging* transformation is also available which reverses the effects of value splitting by combining two value segments that represent the same value.

3. Exploring Allocation Alternatives

The new flexibility introduced by the SALSA binding model is exploited using an iterative approach to allocation in which a set of *moves* (transformations) successively modify a complete allocation to produce a new allocation. This approach is similar to earlier simulated annealing approaches (e.g. [8]), but differs in that both functional unit and register allocation are considered simultaneously and additional moves are defined that are unique to the SALSA representation. Two sets of moves were developed: moves that alter functional unit (FU) bindings, and moves that alter register bindings. Table 1 summarizes these moves.

Move	Description
F1	FU Exchange: Exchange binding of 2 FUs
F2	FU Move: Reassign operator to unused FU
F3	Operand Reverse: Switch FU inputs
F4	Bind to Pass-Through: Assign slack operator/data transfer to FU
F5	Unbind Pass-Through: Eliminate pass-through binding.
R1	Segment Exchange: Exchange binding of 2 value segments
R2	Segment Move: Reassign value segment to unused register
R3	Value Exchange: Exchange bindings of two selected values.
R4	Value move: Assign all segments of a value to unused register
R5	Value Split: Copy of a value segment
R6	Value Merge: Eliminate copy of value segment

Table 1 - Allocation Moves

Functional unit moves alter the assignment of operators to functional units. Moves F1, F2, and F3 are similar to moves used previously in simulated annealing [8]. Moves F4 and F5 are new moves that create and remove pass-through bindings. Initially, moves that alter operator scheduling were also included in this set as in [8]. However, in our experience these moves did not lead to better allocations and so were omitted.

Register moves alter the assignment of value segments to registers. While similar moves have been used in previous approaches, the use of value segments means that moves R1 and R2 are local to a single control step. Moves R3 and R4 reassign all segments of a value to the same register. Moves R5 and R6 implement the value split and merge transformations. The combination of segment-oriented and value-oriented moves allows the exploration of tradeoffs between contiguous and non-contiguous register bindings.

4. Allocation Implementation

Allocation is implemented in a two step approach. First, a simple constructive algorithm is used to create an initial allocation. Second, an iterative improvement phase applies the moves described in the previous section to explore the design tradeoffs made possible by the SALSA representation.

The initial allocation is generated using a simple procedure that attempts to find a good starting point for allocation improvement. In this procedure, operators are assigned to functional units in each control step on a first-available basis. Values that represent loop inputs and outputs are then assigned to registers so that consistency is maintained across iterations. Next, values that occur in control steps with maximum register demand are assigned to arbitrary registers. Finally, any remaining values are bound to registers in a way that attempts to avoid adding more interconnections. Segments of each value are initially assigned to the same register unless there is no contiguous space available in a single register. In this case, a value is split into segments that fit in available registers.

Allocation improvement is implemented by applying the move set described in Section 3 and evaluating moves using a cost function that is a weighted sum of functional unit, register, and interconnect costs. Interconnect costs are computing assuming a point-to-point model since this can be quickly computed from functional unit and register allocation. This is important in an iterative approach where costs are recalculated after every move. After allocation improvement, multiplexers are combined using a simple merging procedure.

It was originally thought that allocation improvement would be implemented using simulated annealing. However, attempts to use annealing produced poor results and seldom converged on a good solution. An iterative improvement scheme was developed instead that produced better results for this application. In this approach, several trials are attempted that are analogous to temperature levels in annealing.

Several moves are attempted in each trial. Moves are selected by randomly picking a move type and then randomly picking the CDFG and datapath elements required by the particular move type. The random selection process is weighted to pick complex moves such as value move and value interchange less often to control execution times. After the move is selected and applied, the effect on the cost function is calculated. Moves that decrease the cost function (i.e. "downhill" moves) are always accepted and retained. Moves that increase the cost function (i.e. "uphill" moves) are sometimes accepted (as described below) or else rejected by reversing the move application.

A fixed number of uphill moves are accepted in each trial. After this number is reached, only downhill moves are accepted. The idea behind this strategy is to allow a random search of the configuration space of possible allocations. Uphill moves at the beginning of the trial allow the search to move to a different region of the configuration space. Once the new neighborhood is reached, downhill-only moves search for a local optimum. During the search, the best allocation found so far is recorded. Iterative improvement terminates after no improvement is found in three successive trials or a maximum number of trials is reached. The best allocation found during the search is then returned. While no guarantee can be made that iterative improvement will converge on a high-quality solution, experience has shown that this is generally the case.

After allocation improvement, the number of multiplexers can be reduced by merging together compatible multiplexers. This is done using a simple heuristic in which an arbitrary multiplexer is selected and combined with as many other compatible multiplexers as possible. Then, another multiplexer is selected and merged with other compatible multiplexers. This process continues until merging has been attempted with all multiplexers.

5. Results

The SALSA binding model and allocator have been implemented in C and work with the existing SALSA scheduler [16]. This implementation has been tested with a number of examples.

Table 2 summarizes the results of an experiment allocating the Elliptic Wave Filter (EWF) benchmark [2,17] under a wide variety of conditions. Assumptions about operator delay are equivalent to those in [2]: adders require one control step, multipliers require two control steps, and pipelined multipliers have a latency of one control step. Allocation CPU times for these examples ranged between 8 and 10 CPU minutes on a Sun Sparcstation 1. However, due to the random nature of the iterative improvement scheme, multiple trials are sometimes necessary to find the best result, increasing the actual CPU time required.

In this experiment, schedules were generated for 17 and 19 control steps assuming both non-pipelined and pipelined (marked with "P") multiplier units, and also for 21 control steps assuming only non-pipelined multipliers. In each case, the schedule fixes the minimum number of functional units and registers. Allocations were generated for these schedules using the minimum number of registers, and also with additional registers allowed to trade off storage vs. interconnect. Adder units in these allocations were also used to implement pass-through operations. Table 2 shows these values and the number of equivalent 2-1 multiplexers in the resulting allocations.

Elliptic Wave Filter Benchmark					
Steps	FU *	FU +	Reg	Mux 2-1	Mux Comparison
17	2P	3	10	18	26 [15]
17	2P	3	11	17	16 [11]
17	2P	3	12	16	16 [2,15]
17	3	3	10	21	29 [15]
17	3	3	11	18	18 [15]
17	3	3	12	17	17 [15]
19	1P	2	10	19	20 [15]
19	1P	2	11	16	16 [5]
19	1P	2	12	18	19 [15]
19	2	2	10	20	20 [5]
19	2	2	11	16	16 [5]
19	2	2	12	17	20 [15]
21	1	2	10	15	15 [15]
21	1	2	11	16	16 [11, 15]

Table 2 - Results for EWF Example

For comparison, the best number of 2-1 multiplexers reported by other researchers is presented in the last column. In 5 of the 14 cases, the SALSA allocator finds an allocation with fewer multiplexers than the best result previously reported. In 7 other cases, the result is equal to the best reported results, while in 2 cases the SALSA allocator required 1 more multiplexer than the best reported case. The advantage of the SALSA allocation over other approaches is particularly dramatic for the 17 step schedule with 10 registers for both pipelined and non-pipelined multipliers.

A larger example was also used to demonstrate the effectiveness of the approach with more complex designs. The Discrete Cosine Transform (DCT) [18] is used in image coding and compression applications. The specific implementation of the DCT that we used is drawn from [19] and is shown in CDFG form in Figure 5. This CDFG consists of 25 addition operators, 7 subtraction operators and 16 multiply operators and provides a challenging problem for both scheduling and allocation. This example was scheduled and allocated using hardware assumptions identical to those for the EWF example. As in the EWF example, constants for multiplication were not considered to contribute to the cost of the allocation. Table 3 summarizes schedule and allocation results for four different schedules. Execution times for the DCT example ranged from 10-17 CPU minutes on a Sun Sparcstation 1. Again, multiple trials are sometimes necessary to find the best results.

Discrete Cosine Transform				
Steps	FU *	FU +/–	Reg	Mux 2-1
7	8	6	12	29
10	4	4	15	30
14	3	3	14	29
19	3	2	14	30

Table 3 - Results for DCT Example

7. Conclusions

This paper described an extended allocation model that allows new flexibility through the use of value segments, value copies, and functional unit pass-throughs. An iterative improvement scheme was used to search for allocations that take advantage of this flexibility. Results show that very good allocations can be found using this approach even though a relatively simple point-to-point interconnect model is used. The good solutions found by this approach suggest that the binding model may also be useful in other approaches to allocation.

Future work is possible in a number of areas. First, extensions to interconnection allocation should be investigated to improve on the point-to-point model currently use. Second, the iterative improvement scheme used in allocation improvement could be replaced by a more powerful approach that takes advantage of the added flexibility of the binding model. Finally, further extensions to the binding model should be considered which more accurately model the actual layout.

Acknowledgements

The authors would like to thank R. Cloutier, M. McFarland, and K. Vissers for useful discussions regarding allocation, and J. van Meerbergen for suggesting the DCT example.

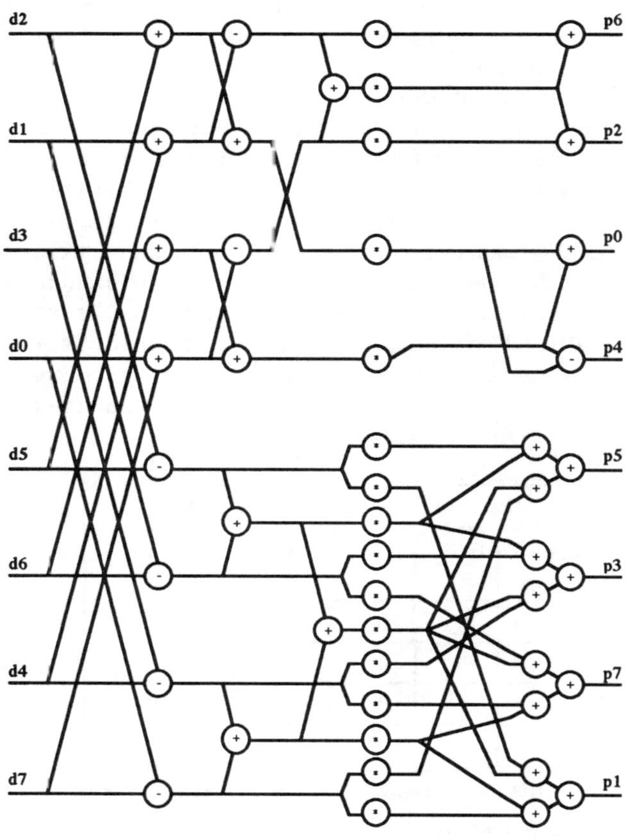

Figure 5 - Discrete Cosine Transform CDFG

References

[1] M. C. McFarland, A. C. Parker, and R. Camposano, "The High-Level Synthesis of Digital Systems", *Proceedings of the IEEE*, Vol. 78, No. 2, Feb. 1990.

[2] P. G. Paulin, "High-Level Synthesis of Digital Circuits using Global Scheduling and Binding Algorithms", Ph.D. Thesis, Carleton University, Jan. 1988.

[3] B. M. Pangrle and D. D. Gajski, "Slicer: A state synthesizer for intelligent silicon compilation", *Proc. ICCAD-87*, Oct. 1987.

[4] D. E. Thomas, et. al., *Algorithmic and Register-Transfer Level Synthesis: The System Architect's Workbench*, Kluwer Academic Publishers, 1990.

[5] T. Ly, W. Elwood, and E. Girczyc, "A Generalized Interconnect Model for Datapath Synthesis", *Proceedings 27th DAC* , ,June 1990.

[6] B. Haroun and M. Elmasry, "Architectural Synthesis for DSP Silicon Compilers", *IEEE Trans. CAD*, Vol. 8, No. 4, pp. 431-447, April, 1989.

[7] A. C. Parker, J. Pizarro, and M. Mlinar, "MAHA: A program for datapath synthesis", in *Proc. 22nd DAC*, July 1986, pp. 461-466.[PAU86] P. G. Paulin, "HAL: A Multi-Paradigm Approach to Automatic Data Path Synthesis", *Proceedings 23rd DAC*, pp. 263-270, June 1986.

[8] S. Devadas and A. R. Newton, "Algorithms for Hardware Allocation in Data Path Synthesis", *IEEE Trans.CAD*, Vol. 8, No. 7, July 1989, pp.768-781.

[9] M. Quayle and L. Grover, "Pipelined and Non-Pipelined Data Path Synthesis using Simulated Annealing", *Progress in Computer Aided VLSI Design*, Vol. 4, Feb. 1990.

[10] L. Stok, "Interconnect Optimisation during Data Path Allocation", *Proc. EDAC-90*, pp. 141-145, March 1990.

[11] T. Ly and J. Mowchenko, "Applying Simulated Evolution to Datapath Allocation in High Level Synthesis", *Proceedings Canadian Conf. VLSI*, pp 6.4.1-6.4.8, Oct. 1990.

[12] F. Tsai and Y. Hsu, "Data Path Construction and Refinement", *Proceedings ICCAD-90*, pp. 308-311, November, 1990.

[13] C. Huang, et. al., "Data Path Allocation Based on Bipartite Weighted Matching", *Proceedings 27th DAC*, pp. 499-504, June 1990.

[14] C. Gebotys and M. Elmasry, "A Global Optimization Approach for Architectural Synthesis", *Proceedings ICCAD-90*, pp. 258-261, Nov. 1990.

[15] A. Mignotte and G. Saucier, "A Generalized Model for Resource Assignment", *Proc. 5th International Workshop on High-Level Synthesis*, pp. 37-43, March 1991.

[16] J. Nestor and G. Krishnamoorthy, "SALSA: A New Approach to Scheduling with Timing Constraints", *Proceedings ICCAD-90*, pp. 262-265, Nov. 1990.

[17] G. Borriello and E. Detjens, "High-Level Synthesis: Current Status and Future Directions", *Proceedings 25th DAC*, pp. 477-482, June 1988.

[18] K. Rao and P. Yap, *Discrete Cosine Transform: Algorithms, Advantages, Applications*, Academic Press, 1990.

[19] R. Woudsma, et. al., "One-Dimensional Linear Picture Transformer", US Patent No. 4,881,192.

ISIS: A System for Performance Driven Resource Sharing

Brent Gregory, Don MacMillen, and Dennis Fogg
Synopsys, Inc.
Mountain View, CA 94043

Abstract

Accurate area and delay estimation is the key to high quality circuits synthesized by resource sharing systems. Bit-level accuracy and integration with a logic optimization system are necessary extensions to any resource sharing algorithm. We describe the implementation of a resource sharing system that uses bit-level modeling, accounts for control delays, and optimizes sharing and resource performance selection simultaneously.

1. Introduction

Register transfer language (RTL) synthesis automatically generates a netlist of components from a description which defines a design's cycle-by-cycle behavior. Resource sharing reduces the cost of the synthesized netlist by reducing the number of expensive components such as adders, subtracters, multipliers, and so on. Operations can be implemented on the same hardware if they never execute in the same cycle. This occurs if they always execute in different cycles or if control signals can activate no more than one operation in a single cycle, such as when the operations lie in different branches of a case statement.

Optimal resource sharing minimizes hardware area while it satisfies the designer's performance requirements. Sharing generally decreases the area consumed by operations, and increases the area devoted to communication. Communication costs include increased multiplexor counts and interconnect area. Sharing generally decreases design performance by introducing delay though multiplexors and by increasing the number of computations that must execute after control values have been computed. This last point is discussed in section 4.2.

2. Contributions

Our resource sharing system, called ISIS (Integrated Sharing Improvement System), differs from previous approaches in three respects:

- Closer integration with logic optimization
- Accurate area and delay modeling
- Control delay modeling

Most other resource sharing systems execute before logic optimization, lump delay into a single value per word, use highly abstract area and timing models, and ignore the delay introduced by the control logic. ISIS, in contrast, is tightly integrated with the logic optimization algorithms, uses more accurate bit-level area and timing models, and includes control delay when evaluating candidate solutions. Integration to the logic optimizer allows ISIS to optimize the performance of sharable operations with a tentative implementation for the remainder of the design. Optimizing in context with more accurate modeling allows ISIS to make better sharing decisions that produce smaller designs that meet the designer's performance requirements. Surprisingly, ISIS's improved results do not come at the expense of unreasonably long execution times.

3. Previous work

It has long been recognized that sharing expensive datapath operations on common resources can result in large circuit area savings. (see [1] and references therein). A wide variety of research has formulated resource sharing into either the Integer Programming (IP) or the Mixed Integer Linear Programming (MILP) frameworks. The work of Hafer and Parker [5] tries to solve the scheduling and allocation tasks simultaneously by solving an MILP problem. Other systems [2][6], have used IP to solve similar combined scheduling and allocation problems. The "integer" part of these formulations typically arise from zero-one decision variables that determine whether a particular operation is bound to a particular resource. Since the size of the problem is determined by the cross product of operations and resources, these methods are not always feasible for large designs.

Another approach has been to cast the problem as a clique partitioning on a graph [4]. Operations in an HDL description are represented by nodes in a graph. An edge connects two operation nodes if those operations are never required to execute simultaneously. Partitioning this graph (sometimes called the compatibility graph) into a minimum number of maximal cliques, where each clique is implemented by a separate resource, generates the minimum area design. Finding a minimum clique partition of the graph is an NP

29th ACM/IEEE Design Automation Conference®

complete problem, so heuristics are used to generate near optimal solutions.

Instead of trying to partition the operations in a top-down manner, as in clique partitioning, BUD [3] constructs clusters from the "bottom-up" by grouping operations based on a distance metric. The distance between operations measures their common functionality, degree of interconnection, and potential parallelism. Sharable clusters are built pair-wise from the bottom up based on minimum distance. MABAL [8] uses a simple heuristic algorithm to investigate the area trade-offs between communication costs and functional unit costs. CHOPIN [2] takes advantage of bit-level timing models to build efficient pipelined data paths, but runs after resource sharing has been completed.

What we call the resource sharing problem in this paper, is often referred to as resource allocation. We use the term "selection of resource performance" instead of the more common module binding, or implementation selection. This is to emphasize that we are able pick from a continuum of resource performance characteristics instead of selecting parts from a module library.

4. Problem Formulation

Accurate delay and area estimation is the key to high quality automatic resource sharing. The following subsections illustrate why resource sharing optimization must work in concert with logic optimization using bit-level area and delay cost estimation.

4.1 Bit-level area and delay estimation

Circuit area and delay estimation must run at the bit-level to accurately evaluate candidate sharing configurations. Here are some examples that illustrate the danger of ignoring bit-level effects.

The following HDL fragment implements twos-complement and unsigned comparison in a microprocessor ALU with an unsigned comparison operator, "<".

```
case unsigned_comparison:
   result = a < b;
case twos_compliment_comparison:
   result = invert_msb(a) < invert_msb(b);
```

The invert_msb function flips the sense of the most significant bit. Traditional, word-level cost estimation either overestimates the cost of sharing the comparisons by assuming that two full multiplexors are required, or underestimates the cost by assuming that no multiplexors are required. Figure 1 shows why only the most significant bits need multiplexing since the other bits are identical.

Another example of the importance of bit-level effects is the strategy of assigning one arrival time to a bus of signals. This has led previous algorithms to generate designs that were optimal only under the simplified model. For example, cascaded ripple adders appear to be much slower than cascaded carry-lookahead adders to a word-level timing model, but when evaluated with a more accurate bit-level timing model, the cascaded ripple adders are much faster.

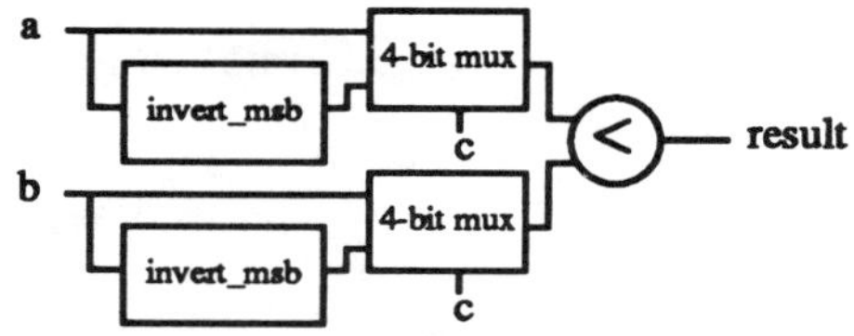

(a) Word-Level Mux Estimation

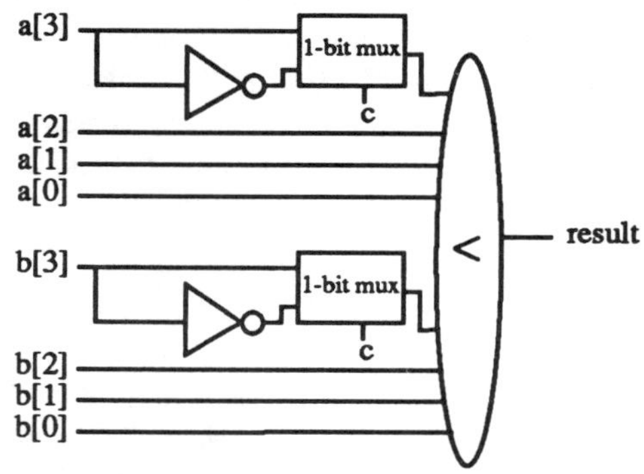

(b) Bit-Level Mux Estimation

Figure 1. Bit-Level vs. Word-Level Multiplexor Estimation

Figure 2 illustrates why logic optimization and resource sharing must be closely integrated at the bit-level. Each needs to account for changes in circuit delay caused by the other. Most designs consist of a mixture of bit-level random logic, which is left untouched by resource sharing, and operations that may be shared. Here, propagation delays must be balanced between operations and the rest of the design. Resource sharing algorithms need surrounding logic circuity to be in near-final form to determine the time remaining for the execution of the sharable operations.

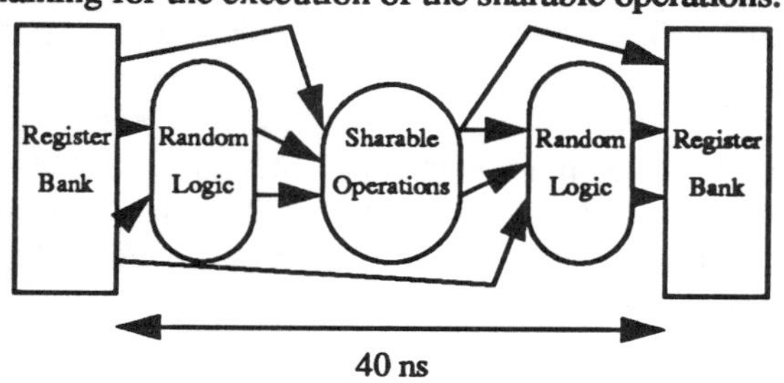

Figure 2. Sharable Operations Imbedded in Random Logic

4.2 Control timing dependencies

Previous research has addressed area optimization of control logic [15]. When performance is an important criteria of the resource sharing algorithm, control timing dependencies become critically important in determining the viability of a particular sharing. When operations are shared on a resource, control values are required before the initiation of the resource's computation. Consider the following example:

```
if r + s < 0 then
   z = a * b;
else
   z = c * d;
end if;
```

Figure 3 shows this circuit before and after the two multiplication operations are shared on the same multiplier.

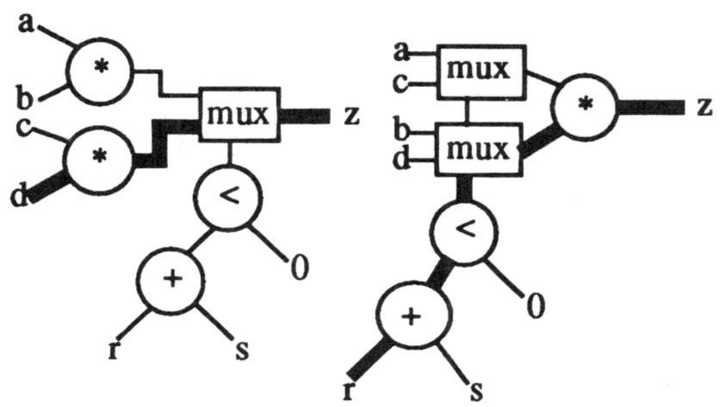

Figure 3. Critical Control Dependency

The highlighted critical path through the unshared circuit passes through a multiplier and a multiplexor. The critical path through the shared circuit includes the additional adder and comparator from the control logic. Depending on the performance constraints, this added delay may not be acceptable. Obtaining accurate delays for the control logic is vital in determining an acceptable sharing.

4.3 Combining sharing and performance selection

The performance of a design is determined by the performance selected for the design's resources and the sharing configuration. For example, when two addition operators share the same ripple carry adder, the circuit may be slowed by the muxing needed to select the data. However, the circuit may be sped back up by replacing the ripple adder by a carry lookahead adder.

When resources are pre-built as megacells, only discrete points on the performance/area trade-off curve are possible. When resources are built with gate-level components, a logic optimization system makes almost any point along the performance/area trade-off curve accessible [9][10]. The ability to individually tailor the performance of resources means that no part of the circuit needs to be over-designed. A resource sharing algorithm must both map operations onto resources and select from the available points on the performance/area trade-off curve for each resource. For resource sharing to achieve the optimal circuit area for a given performance requirement, it must make these decisions together. For example, an algorithm which first shares and then selects resource performance may not produce the ideal design.

Figure 4 illustrates the importance of combining sharing and resource performance decisions. (a) shows a scheduled data-flow graph for a biquadratic filter. (b) shows the smallest design for a selection of performance targets. (c) enumerates the resources used for each solution. Notice that the optimal circuit often includes a mixture of high and low performance resources and is neither maximally nor minimally shared. An algorithm which performs sharing and performance selection separately is highly unlikely to generate such solutions.

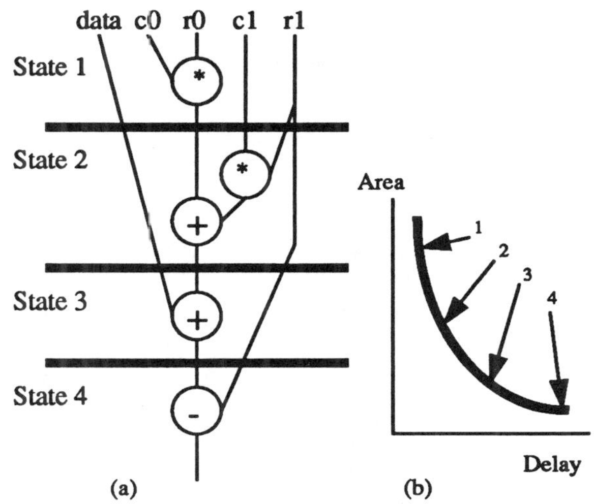

Circuit	Area	Delay	Resources
1	4349	49 ns.	33ns addsub, 11ns add, two 42ns multipliers
2	2764	55 ns.	11ns add, 17ns sub, 42ns mult
3	2335	60 ns.	29ns addsub, 42ns multiplier
4	2267	65 ns.	33ns addsub,42ns multiplier

Note that due to bit level timing effects the total delay is not equal to the sum of component delays.

(c)

Figure 4. Biquadratic Filter Example

5. ISIS Details

This section explains how ISIS achieves high quality results by interacting closely with logic optimization and by using accurate timing and area estimation. Section 5.1 explains the flow of design information between resource sharing, the logic optimizer and the rest of the synthesis system. Section 5.2 explains ISIS's solution management and cost evaluation. Finally, section 5.3 describes the sharing optimization algorithm.

5.1 Design Flow Through ISIS

Figure 5 shows how information flows between the different components of the synthesis system. Fat lines represent circuits, while thin lines represent constraints, models, and estimates. When a circuit first enters the system, it is sent to the logic optimizer. Here, all unmapped random logic is implemented with gates from the target library. Next, the logic optimizer and ISIS take turns optimizing (and re-optimizing) portions of the circuit. The logic optimizer restructures the random logic to achieve the smallest circuit that, when combined with the current sharing configuration, meets the performance constraints. Similarly, ISIS sets the sharing configuration and resource performance to reduce area while meeting the performance constraints. Iteration allows ISIS to run with realistic delays through the random logic, and logic optimization to run with realistic delays through the operations.

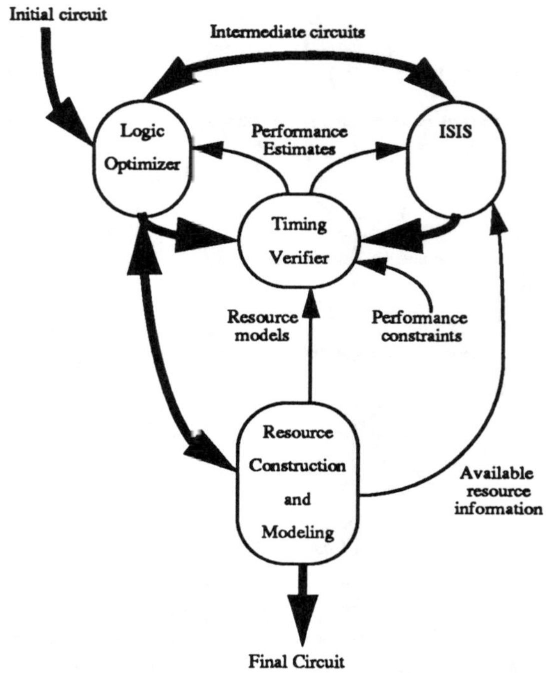

Figure 5. Design Flow

Both the logic optimizer and the resource sharing system use the same (bit-level) timing verifier to evaluate possible solutions against the design's performance goals. The delays through components selected by the logic optimizer are defined in the target technology library. The delays of resources are defined by the resource construction and modeling subsystem. This subsystem reports on the available resources and their ranges of performance and cost. It builds accurate, bit-level timing models for resources in intermediate solutions as well as the actual resource hardware for the final design. The resource construction system may recursively call the optimization subsystem to build the final resources.

Before each call to the timing verifier to measure a candidate solution, ISIS incrementally rebuilds the netlist to reflect the new sharing configuration and resource performance selection. The circuit includes multiplexors with appropriate control signals. This ensures accurate modeling of control delay.

5.2 The Resource Sharing Workbench

The lowest layer of ISIS is a solution management workbench. It provides the following services to ISIS's algorithm layer: Multiple (parallel) solution storage and retrieval, solution transformation (such as changing the performance of a resource or moving an operation from one resource to another), solution evaluation (both area estimation and the interface to the timing verifier) and resource information (an interface to the resource modeling subsystem). With the workbench, new algorithms can be quickly developed and tested.

5.3 Optimization strategy

Since ISIS must process a wide variety of designs, it runs a collection of optimization algorithms on the sharing workbench. For example, a branch and bound search algorithm works well on designs with a small number of operations such as ALUs, many controllers, and simple DSP circuits, but is too slow for designs with a large number of operations. In contrast, incremental improvement and initial placement algorithms handle large circuits in a reasonable time.

The branch and bound algorithm constructs a solution by considering one operation at a time. It bounds whenever the minimum area is larger than the best feasible solution found. When resources can be implemented in a continuous range of performance values, the branch and bound algorithm picks a small number of performance alternatives to reduce the search space. The branch and bound algorithm only calls the timing verifier to analyze solutions that have less area than the best feasible solution known. Because timing is the slowest operation, all optimization algorithms invoke the timing verifier as little as possible.

When designs are too large to search with branch and bound, a greedy initial solution generator rapidly builds feasible solutions. Starting from an unshared configuration, operations are iteratively combined and resource performance is updated until no area improvement is possible within the performance constraints. ISIS also runs several incremental improvement algorithms that use probabilistic techniques similar to those described in [11][12] to escape from local minima. Candidate solutions are rejected if the timing verifier reports that the performance constraints were violated.

Some of ISIS's algorithms mix sharing and the selection of resource performance. For example, during iterative improvement, sharing changes are usually immediately followed by resource performance selection changes. Other ISIS, algorithms concentrate only on sharing. For example, ISIS first generates a minimum area sharing to see if it meets the performance constraints. Other ISIS algorithms concentrate only on performance selection. For example, a clean-up algorithms tries to improve the selection of performance for the final resource configuration.

A top-level executive algorithm controls the optimization of the resource configuration by invoking different strategies based on the problem size, and the performance constraints. The top-level algorithm places execution-time limits on the lower-level algorithms, such as branch and bound, to avoid their potential exponential complexity.

6. Results

The goal of synthesizing high-quality circuits led us to develop a tight interface to logic optimization, and to rely on the most accurate cost measures available. Initially, we were concerned that including a bit-level timing verification and area estimation in the inner loops of our algorithms would result in unacceptably long execution times. The results in Figure 6 show that resource optimization actually consumes a fraction of the amount of time used by logic optimization.

The target library used in these examples implements a 1.5 micron gate array technology. All examples were run on a Solbourne 5E/904. Execution times are in CPU minutes. Area (indicated by G) is measured in two-input nand equivalents. RS stands for time spent resource sharing and includes the time needed for timing verification. Total execution times include logic optimization (and again includes the time needed for timing verification).

The design reg_to_reg is an RTL description of the register to register instructions of the ALU block of the MIPS R3000. At the high performance end, ISIS has created a maximally parallel design with a fast, and large, adder and subtracter. In the intermediate performance region, the comparator is shared, while separate resources of intermediate speed were created for the additions and the subtractions. Finally, at the low performance end, all the operations are maximally shared onto a slow adder-subtracter and a slow comparator.

The second design is the bi-quadratic filter shown in Figure 4. State 2 is responsible for the design's critical path since it must perform both a multiplication and a cascaded addition. In both the 49ns and the 55ns design, the critical operations are performed on the fastest available architecture.

EWF is the famous 5th order elliptic wave filter. The schedule used here was obtained by merging control steps of the 18 cycle schedule presented in [14]. This merging increases the chances for sharing of resources. Notice that the resources in both results have similar performance and area. This is because the critical path through the design passes through three cascaded adders. ISIS consistently selected a ripple architecture because, in this design, it is both the smallest and fastest choice. Optimization of the multiplexing logic caused the speed and area differences.

Figure 6. Execution time and circuit quality examples

Design Name	Delay Area	Final Resources	Run Times
reg_to_reg 2 adds, 2 subtracts, 2 compares	16 ns 1575 G	14 ns, 403 G, 32 bit adder 13ns, 415 G, 32 bit subtracter 13 ns, 159 G, 32 bit comparator 13 ns, 159 G, 32 bit comparator	11 RS 41 total
"	34 ns 1217 G	31 ns, 311 G, 32 bit adder 30 ns,334 G, 32 bit subtracter 19ns, 136 G, 32 bit comparator	10 RS 26 total
"	47 ns 888 G	39 ns, 365 G, 32 bit adder-subt 37 ns, 139 G, 32 bit comparator	4RS 14 total
biquad 2 adds, 1 subtract, 2 multiply	49 ns 4349 G	11 ns, 291 G, 24 bit adder 33 ns, 370 G, 28 bit adder-subt 42 ns, 1541 G, 24 bit multiplier 42 ns, 1541 G, 24 bit multiplier	14 RS 31 total
"	55 ns 2764 G	11 ns, 397 G, 28 bit adder 17 ns, 136 G, 16 bit subtracter 42 ns, 1541 G, 24 bit multiplier	10 RS 44 total
"	65 ns 2267 G	33 ns, 370 G, 28 bit adder-subt 42 ns, 1541, G 24 bit multiplier;	5 RS 35 total
EWF 26 adds, 8 multi-plies,	23 ns 3121 G	15 ns, 320 G, 11 bit multiplier 12 ns, 218 G, 9 bit multiplier 11 ns, 105 G, 11 bit adder 10 ns, 130 G, 11 bit adder 10 ns, 125 G, 11 bit adder 10 ns, 130 G, 11 bit adder	22 RS 211 tot
"	43 ns 2697 G	15 ns, 269 G, 10 bit multiplier 13 ns, 241 G, 10 bit multiplier 13 ns, 86 G, 11 bit adder 12 ns, 87 G, 11 bit adder 12ns, 87 G, 11 bit adder 12 ns, 87 G, 11 bit adder	2 RS 34 total

7. Limitations and Future Work

The circuits generated by ISIS underscore the need for false path elimination in static timing verifiers used for resource sharing and logic optimization. Consider the following example:

```
if c then
   w = 0;
   x = a + 1;
else
   w = b + 1;
   x = 0;
end if;
```

Figure 7 shows the final design when the two additions share a single incrementor. If signal x has a tight required time and signal b arrives late, then most static timing verification algorithms report a critical path from b to the output x. This is a false path. It can never be activated due to the control structure of this circuit.

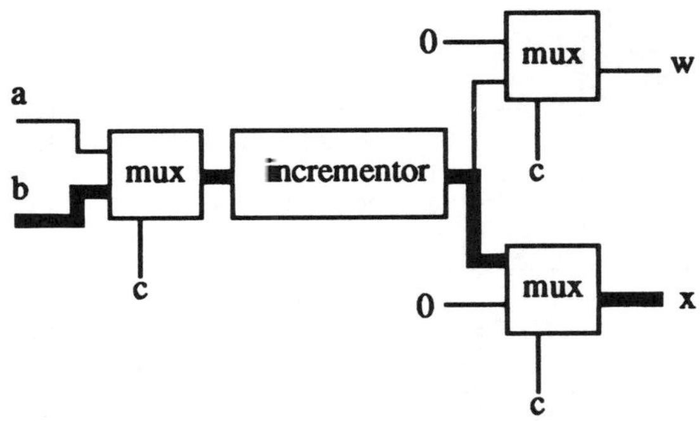

Figure 7. Circuit With False Timing Path

Another area ready for improvement is delay-budgeting. The current implementation of ISIS divides the available time between the delay of the random logic and the delay of the resources by iterating between logic and resource sharing optimization. A more direct delay-budgeting algorithm should allocate the delay more optimally than the current greedy scheme. For example, such an algorithm could direct the logic optimizer to speed up some control logic that was preventing an advantageous sharing.

Today, ISIS only operates on fully scheduled descriptions. We think it is important to consider the simultaneous solution of scheduling and allocation [12][13], and intend to improve ISIS's integration with scheduling optimization as we have already done for logic optimization.

8. Conclusions

We have established the importance of accurate bit-level area and delay modeling to the quality of circuits synthesized by resource sharing systems. We have shown that bit-level accuracy and integration with logic optimization is both desirable and feasible, since the added execution time is a small fraction of the total optimization time. Additionally, we have described the implementation of a resource sharing system that uses bit-level modeling, accounts for control delays, optimizes sharing and resource performance selection together, all to generate high quality circuits from RTL descriptions.

9. References

[1] M. C. McFarland, A. C. Parker and R. Camposano, "The High-Level Synthesis of Digital Systems," *Proc. of the IEEE*, Vol. 78, No. 2, February 1990, pp. 301-317.

[2] S. Note, F. Cathoor, G. Goossens and H. De Man, "Combined Hardware Selection and Pipelining in High Performance Data-Path Design," *Proc. IEEE International Conference on Computer Design*, September 1990, pp. 328-331.

[3] M. C. McFarland and T. J Kowalski, "Incorporating Bottom-up Design into Hardware Synthesis," *IEEE Transactions on Computer-Aided Design*, September 1990, Vol. 9, No. 9, pp. 938-950.

[4] C-J. Tseng and D. P. Siewiorek, "Automated Synthesis of Data Paths in Digital Systems," *IEEE Transactions on Computer-Aided Design*, Vol. CAD-5, NO. 3, July 1986, pp. 379-395.

[5] L. J. Hafer and A. C. Parker, "A Formal Method for Specification, Analysis, and Design of Register-Transfer Level Digital Logic," *IEEE Transactions on Computer-Aided Design*, Vol. CAD-2, No. 1, January 1983, pp. 4-18.

[6] M. Balakrishnan and P. Marwedel, "Integrated Scheduling and Binding: a Synthesis Approach for Design Space Exploration," *Proc. of 26th Design Automation Conf., ACM/IEEE*, June 1989, pp. 68-74.

[7] G. De Micheli and D. C. Ku, "HERCULES - A System for High-Level Synthesis," *Proc. of 25th Design Automation Conf., ACM/IEEE*, June 1988, pp. 483-488.

[8] K. Kucukcakar, A. C. Parker, "Data Path Tradeoffs Using MABAL," *Proc. of 27th Design Automation Conf., ACM/IEEE*, June 1990, pp. 511-516.

[9] K. J. Singh, A. R. Wang, R. K. Brayton and A. Sangiovanni-Vincentelli, "Timing Optimization of Combinational Logic," *Proc. of ICCAD-88, IEEE*, November 1988, pp. 282-285.

[10] J. P. Fishburn, "A Depth-Decreasing Heuristic for Combinational Logic; or How to Convert a Ripple-Carry Adder into a Carry-Lookahead Adder or Anything In-between," *Proc. of 27th Design Automation Conf.*, ACM/IEEE, June 1990, pp. 361-364.

[11] T. A. Ly and J. T. Mowchenko, "Applying Simulated Evolution to Data Path Allocation in High Level Synthesis," *Proceedings of Canadian Conference on VLSI*, October, 1990, pp. 6.4.1-6.4.8.

[12] S. Devadas and R. Newton, "Algorithms for Hardware Allocation in Data Path Synthesis," *IEEE Transactions on Computer-Aided Design*, Vol. 8, No. 7, July 1989, pp. 768-781.

[13] R. J. Cloutier and D. E. Thomas, "The Combination of Scheduling, Allocation, and Mapping in a Single Algorithm," *Proc. of 27th Design Automation Conf.*, ACM/IEEE, June 1990, pp. 71-75.

[14] P.G. Paulin, "High-Level Synthesis of Digital Circuits Using Global Scheduling and Binding Algorithms," Ph.D. thesis, Carleton Univ., February 1988.

[15] D. C. Ku, D. Filo and G. De Micheli, "Control optimization Based on Resynchronization of Operations," *Proc. of 28th Design Automation Conf.*, ACM/IEEE, June 1991, pp. 366-371.

Functional Synthesis Using Area and Delay Optimization

Elke A. Rundensteiner and Daniel D. Gajski
Department of Information and Computer Science
University of California, Irvine, CA 92717

Abstract

It is generally overlooked that designers use functional models more frequently than behavioral or gate-level models. In functional modeling, the functionality of one or more register-transfer level (RTL) components are described as separate concurrent blocks using a hardware description language such as VHDL. In this paper, we present two algorithms that synthesize a netlist of RTL components from a functional description while minimizing hardware costs and delay. Experimental results show that the proposed algorithms produce designs that are comparable to those produced by human designers.

1 Introduction

A design can be modeled at the behavioral or at the register-transfer (RTL) description level. The former corresponds to an algorithmic specification that describes the behavior of a complete system as a set of sequential actions to be executed over time, i.e., a *behavioral* description follows the conventional sequential programming language style. The latter corresponds to a concurrent description which decomposes the design into its major subcomponents and then specifies the functionality of each of these groups using concurrent statements from a hardware description language such as VHDL. A *functional* specification thus describes one or more (generic or technology-specific) components, like arithmetic/logic units, memories, and counters, as separate concurrent blocks.

In Figure 1, we present examples of these two description styles for the Am2901 microprocessor, which is composed of an ALU, several shifters, a RAM, several registers, and selectors for the sources of the ALU (Figure 1(a)). These components operate concurrently; each controlled by a different substring of the 9-bit control word *I*.

The behavioral description of this design (using a sequential process statement in VHDL) is given in Figure 1(b). It consists of one case statement that decodes the control word *I*. Since the different subcomponents of the design operate independently from one another and concurrently, each branch of the case statement must describe the appropriate actions for all parts of the design. Note that the description is not truly sequential; since all statements within a branch executed concurrently. This description is awkward since we must use a case statement with 2^9 cases.

For the functional description style, we group the subcomponents of the Am2901 structure into sets of related functionalities, e.g., we place the RAM_shift and the RAM into one group (Figure 1(a)). The functional description shown in Figure 1(c) describes the functionality of each of these groups by a block of concurrent statements (using block statements in VHDL). In this description: (1) the different parts of the design are explicitly described as operating concurrently, (2) the functionality of each subpart of the design is specified independently from all others, and (3) each block statement corresponds to a different subpart of the design. Clearly, this is a compacter and more natural description of the Am2901.

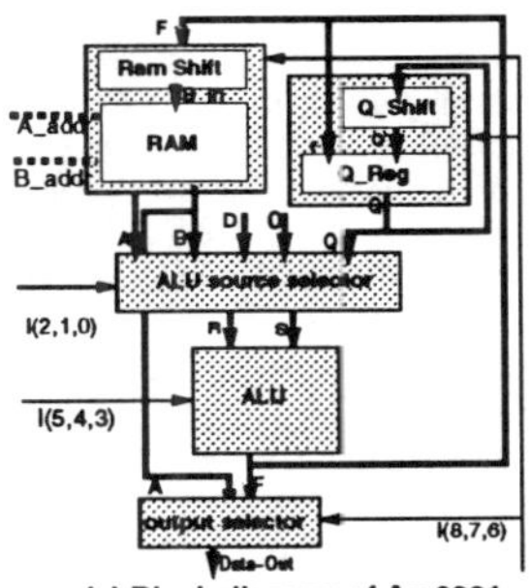

(a) Block diagram of Am2901

```
architecture am2901_behav of am2901 is
begin
process(clk)
begin
case I is
-------------------------------------
  when B"000000000" =>
     ## select source operands for alu
     R_ext <= '0' & A;
     S_ext <= '0' & Q;
     ## select function for alu
     result <= R_ext + S_ext + (B"000" & CO);
     ## output and shift
     if (OE_bar = '0') then
        data_out <= F;
     end if;
     Q <= F ;
-------------------------------------
  when B"000000001" => . . .
-------------------------------------
  when B"000000010" => . . .
-------------------------------------
end case;
end process;
. . .
end am2901_behav;
```

(b) Behavioral Description

```
architecture am2901_func of am2901 is
begin
-------------------------------------
alu_source : block
begin

with I ( 2 downto 0) select
  R <= A          when B"000" | B"001",
       B"0000"    when B"010" | B"011" | B"100",
       data_in    when others;
  S <= A          when B"100" | B"101",
       B          when B"001" | B"011",
       B"0000"    when B"111",
       Q          when others;
end block alu_source;
-------------------------------------
alu: block
signal R_ext, S_ext, result, . . .
begin

R_ext <= '0' & not(R) when I(5 downto 3)=B"001" else
         '0' & R;
S_ext <= '0' & not(S) when I(5 downto 3)=B"010" else
         '0' & S;
with I(5 downto 3) select
  result <= R_ext + S_ext + (B"000" & CO)
                    when B"000" | B"001" | B"010",
            R_ext or S_ext             when B"001",
            R_ext and S_ext            when B"100",
            not(R_ext) and S_ext       when B"101",
            R_ext xor S_ext            when B"110",
            not(R_ext xor S_ext)       when others;
F <= result(3 downto 0);
...
end block alu;
-------------------------------------
Q_reg_shifter : block ((clk = '1') and (not clk'stable))
begin

Q <= guarded F when I ( 8 downto 6) = B"000" else
      Q3_in & Q(3 downto 1)
              when I ( 8 downto 6)=B"100"  else
      Q( 2 downto 0) & Q0_in
              when I ( 8 downto 6)=B"110"  else
      Q;
. . .
end block Q_reg;-----------------------------

. . .
end am2901_func;
```

(c) Functional Description

Figure 1: Description Styles of the Am2901 Design.

As demonstrated above, some designs lend themselves to a functional description style. Furthermore, designers use functional descriptions since they are more accustomed to the RTL level than to the algorithmic level of design. A description at the functional level is a perfect basis for *redesign*, particularly, if some obsolete components of the design are to be replaced with a newer technology while others are to be maintained. We propose to automate this labor intensive and costly process of redesign by automatically mapping functional descriptions into components from a given RTL library.

In contrast, behavioral synthesis maps a *behavioral* description of the desired system to a RTL structure of generic RTL components [4, 8]. Thus, functional synthesis can work in synergism with behavioral synthesis by mapping *functional* descriptions of RTL designs produced by the latter onto actual hardware. Functional synthesis therefore addresses the *technology mapping* problem at the RTL level. Previous work on RTL technology mapping has focussed on translating a generic RTL structure into a technology-specific one [3, 1]. These approaches work on *structural netlists* that capture designs as a set of black boxes and their interconnections rather than *functional descriptions*.

The use of hardware description languages such as VHDL for functional descriptions complicates the problem of functional synthesis, since the languages provide several different ways to describe the same functionality. Also, a single component may be described in a functional description by several, often nested conditional, statements. These statements need to be grouped in order to map them to the same component. What makes mapping even more difficult is that standard operators in the description language may not match the functions supported by the target library components.

To address these problems, functional synthesis is composed of two subtasks: *functionality recognition* and *component mapping*. The first addresses the mismatch between the standard language operators used to specify a functional description and the functions supported by RTL library components, while the second solves the problem that functional descriptions use multiple statements to describe the functionality of a single component. In addition, component mapping exploits the mutual exclusiveness of operators for unit sharing in order to minimize hardware area and delay.

Our Functional Synthesis Algorithms tool (**FSA**) consists of two modules: the *functionality recognizer* and the *component mapper* [7]. In [6], we present an algorithm for functionality recognition which merges expression subtrees into library-specific nodes whenever components are capable of executing those expressions as one function. In this paper, we present two algorithms that solve the second problem by formulating it as a clique partitioning problem - similar to work done for solving the allocation problem in behavioral synthesis [8, 4]. Our algorithms handle nested conditional branches and do not assume a preallocation of functional units and unit binding as done in [4]. Our cost functions incorporate costs for both interconnection and function select logic. Lastly, by combining the branch-and-bound methodology with clique partitioning, we succeed in pruning the search space without losing the guarantee of an optimal solution.

We outline our component mapping approach in Section 2. The cost function and the bounding function (for the branch-&-bound component mapper) are given in Sections 3 and 4. We describe the component mapping algorithms in Sections 5, present results in Section 6, and conclude with Section 7.

2 Our component mapping approach

2.1 Formulation of the problem

First, **FSA** compiles a functional description into an internal representation, a directed flow graph $G=(V,A)$ with V the set of nodes and A the set of data dependency edges. V is composed of three disjoint sets $V = N \cup R \cup D$, with N the set of operators, R the set of storage elements, and D the set of decision nodes that model data selection [5]. Two operator nodes *n1* and *n2* in G are defined to be **mergeable** with respect to a given unit table U (component library) if and only if there is a unit $u \in U$ which implements the functions of both *n1* and *n2*. They are defined to be **mutually exclusive** to one other if the conditions which selects each can never be true at the same time [6]. Finally, they are said to be **compatible** with respect to U, if they are **mutually exclusive** and **mergeable**.

FSA generates a **compatibility graph** $CG = (N,E)$ from G with N the set of operator nodes of G and E a set of undirected edges, called **compatibility edges**. There is an edge $e_i = < n_j, n_k >$ in CG for each pair of **compatible** nodes $n_j, n_k \in N$. See the example in Figure 2.

A collection of operator nodes in G can be mapped to the same functional unit if and only if the nodes are pairwise **compatible**. In other words, a subgraph of CG that is completely connected by compatibility edges, called a clique, can be mapped to one unit. Component mapping corresponds to the problem of finding a minimal cost clique partition of CG given an appropriate cost function.

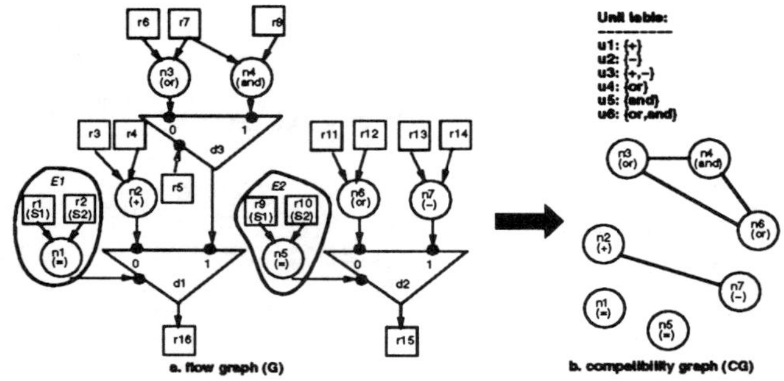

Figure 2: Compatibility Graph Formation Example.

2.2 Compatibility-graph and flow-graph transformations

During the process of incrementally creating clique covers on CG, **FSA** not only adjusts CG but also the flow graph G. This allows for the simultaneous consideration of (1) the connections costs due to the sharing of units and (2) the control logic costs for the selection of the correct function of a multi-functional unit. Our component mapping algorithms explore the two cases of whether or not to map two nodes n_1 and n_2 to the same component expressed by the compatibility edge $e = < n1, n2 >$:

- **case 1:** Map n_1 and n_2 to the same unit, i.e., use edge e in the clique cover:
  - $CG \Rightarrow CG'$ by adjusting the compatibility edges,
  - $G \Rightarrow G'$ by directly reflecting the sharing of units in the flow graph G.
- **case 2:** Do not map n_1 and n_2 to same unit, i.e., do not use edge e in the clique cover:
  - $CG \Rightarrow CG'$ by deleting edge e from CG,
  - $G \Rightarrow G'$ by setting $G' = G$.

The rules for compatibility and flow graph transformations are given in [6], while below we give an example of case 1. Figures 3(a) and 3(b) describes how two operators n_1 and n_2 are mapped to the same component by merging them into one multi-functional operator node n_k in G. In the transformed G, the new node n_k is connected to the original input and output destinations of n_1 and n_2. The *semantic equivalence* of the original and the transformed flow graph is assured by inserting decision nodes that select the correct inputs for the new node. We maintain the *function select logic* of n_k by encoding function select conditions and storing them in the associated decoder node (depicted by a bold outlined box in Figure 3(b)).

Figures 3(c) and 3(d) show how the corresponding CG is transformed. The compatibility edges between the new node n_k and the remaining nodes are updated; e.g., e1 and e5 are combined into the edge e1/5, e3 and e4 are combined into the edge e3/4, and edge e2 is removed. If all decisions for merging operator nodes have been made, i.e., $CG = \emptyset$, then the solution pair (G, CG) is called a *complete* solution. Otherwise, it is a *partial* solution.

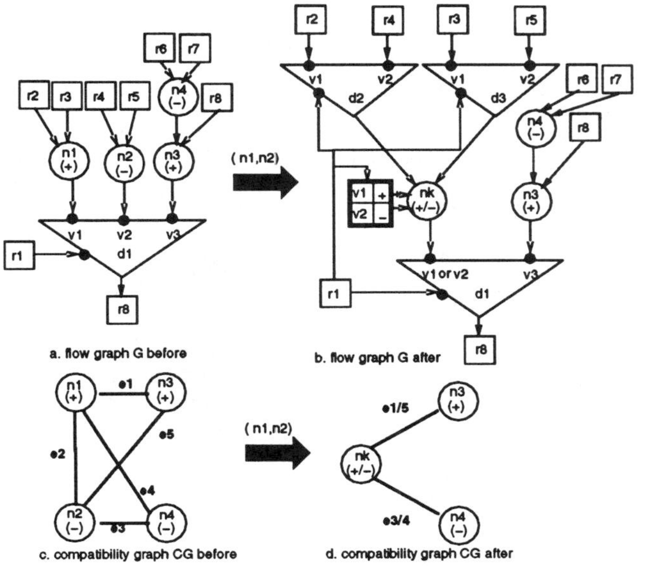

Figure 3: An Example of Merging Operators.

3 The cost function

3.1 The overall cost function

We approximate the quality of design solutions by a weighted sum of design area and delay. The cost of a complete solution (G,CG) for a mapping M_i (with M_i mapping each operator to a different component instance, since unit sharing is taken care of by creating multi-functional operator nodes in G as shown in Section 2.2) is:

$$cost(G, M_i) := \alpha \times area\_cost(G,M_i) + \beta \times delay\_cost(G,M_i)$$

with α and $\beta \in [0{:}1]$ the relative importance of area versus delay optimization entered by the designer and β:=1.0-α. The area and delay for the mapping M_i, the functions *area_cost(G,M)* and *delay_cost(G,M)*, are described in later sections. Then the total cost is:

$$cost(G) := \min_{M_i \in M}(cost(G,M_i)).$$

3.2 The area cost function

We approximate the area cost (*area_cost*) of design solutions by transistor counts. The cost function is parameterized by the chosen component library, i.e., the unit table U. U contains the function **op-cost**: U $\times$ Integer $\rightarrow$ Integer that maps a unit type and its bit width $(u, bw(u))$ to a cost value. For units with operators, such as plus, minus, and comparison, this function is $(bw(u) \times \log(bw(u)) \times$ k) with k some unit constant. For units with multiplication or division operators, the function is $((bw(u))^2 \times$k). Since each operator node n_i is mapped to a functional unit $u \in U$, the cost of an operator node n_i can be expressed by **op-cost**$(M_i(n_i), bw(n_i))$ assuming the mapping M_i. A multi-functional operator node n_i requires a decoder node v_j that captures the function select logic in a truth table as discussed in Section 2.2. The decoder cost, **decoder-cost**(v_j), is approximated by the number of transistors needed to implement the decoding. Interconnection costs are approximated by the cost of the decision nodes d_i in G, since a decision node is mapped to a multiplexor or a bus. The cost of a decision node d_i is calculated by a piece-wise linear function, **conn-cost**$(choices(d_i), bw(d_i))$, which maps the number of choices of d_i and the bit width of its data operands to a cost value. The area cost of G for a mapping M_i is approximated by

$$area\_cost(G, M_i) := \textstyle\sum_{n_j \in N}(\textbf{op-cost}(M_i(n_j), bw(n_j))) + \sum_{(v_j \in V) \wedge (v_j \text{ is a decoder-node})}(\textbf{decoder-cost}(v_j)) + \sum_{d_j \in D}(\textbf{conn-cost}(choices(d_j), bw(d_j))).$$

3.3 The delay cost function

We approximate the performance (*delay_cost*) of design solutions in terms of the maximal delay through the design. The cost function is again parameterized by the chosen component library U. U contains the function **delay**: U $\times$ Integer $\rightarrow$ Integer that maps a unit type and its bit width $(u, bw(u))$ to a delay value. Since each operator node n_i is mapped to a functional unit $u \in U$ by a mapping M_i, the delay of an operator node n_i can be expressed by **delay**$(M_i(n_i), bw(n_i))$. A multi-functional operator node n_i requires a decoder node v_j to select the desired function. Its delay, denoted by **decoder-delay**(v_j), corresponds to the longest delay path through the decode logic. The interconnection delay is approximated by the delay attributed to the decision nodes $d_i \in D$ in G. The delay of a decision node d_i, **conn-delay**$(choices(d_i), bw(d_i))$, is a function of the number of its choices and the bit width of its data operands. Let p_j denote a directed path in G. The delay cost of G for a mapping M_i is approximated by

$$delay\_cost(G, M_i) := \max_{path\ p_j \in G}\Big(\textstyle\sum_{(n_j \text{ on } p_j) \wedge (n_j \in N)}(\textbf{delay}(M_i(n_j), bw(n_j))) + \sum_{(v_j \text{ on } p_j) \wedge (v_j \text{ a decoder-node})}(\textbf{decoder-delay}(v_j)) + \sum_{(d_j \text{ on } p_j) \wedge (d_j \in D)}(\textbf{conn-delay}(choices(d_j), bw(d_j)))\Big).$$

4 The bounding function

4.1 The overall bounding function

The bounding function determines a lower bound on the cost of a partial solution (G, CG), i.e. a bound on the cost of *any* complete solution that can be derived from (G, CG) using further operator merging (Section 2.2). The bounding function used by **FSA** is a weighted sum of area and delay bounds defined by

$$bound(G) := \alpha \times area\_bound(G) + \beta \times delay\_bound(G)$$

with α and β the area and delay parameters and the *area_bound(G)* and *delay_bound(G)* functions defined below.

4.2 The area bounding function

Let (G,CG) represent a partial solution. We denote the minimal cost of implementing an operator node n_j by $bound\_node\_area(n_j) := \min_{M_k \in M}(\textbf{op-cost}(M_k(n_j), bw(n_j)))$. A node n_j in CG is *isolated* if no compatibility edges are connected with this node. We denote an input decision node of an isolated node n_j by $D(n_j)$. A subgraph of operator nodes (not necessarily completely) connected by compatibility edges is called a cluster. Denote the size of a cluster c by $|c|$. *CLUSTER* stands for the set of all maximal clusters c of CG with $|c| \geq 2$. Then define the function *ADDON: CLUSTER* $\rightarrow$ Integer as follows. For $c \in$ *CLUSTER*, if there is one node *n1* $\in c$ with the number of compatibility edges connected to *n1* less than $|c| - 1$, then let *n2* be a node in c with *n1* $\neq$ *n2* that is not connected to *n1* and then set $ADDON(c) := \min(bound\_node\_area(n_1), bound\_node\_area(n_2))$. Otherwise $ADDON(c) := 0$. The area bounding function is defined by

$$area\_bound(G) := \sum_{isolated\ nodes\ n_i \in N}(bound\_node\_area(n_i))$$
$$+ \sum_{isolated\ nodes\ n_i \in N}(\mathbf{conn\text{-}cost}(choices(D(n_i)), bw(D(n_i))))$$
$$+ \sum_{c\ \in\ CLUSTER}(\max_{n \in c}(bound\_node\_area(n)) + ADDON(c)).$$

Example 1. *We assume that all nodes n_i in Figure 4 have a bit width of one and the costs bound_node_area(n_i)=15 if i is even and 10 if i is odd. n1 and n2 are isolated nodes. Therefore, area_bound(CG) is incremented by 10 for n1 and 15 for n2. CG has two maximal clusters: n3 to n5 form the cluster C1 and n6 to n10 form the cluster C2. Thus, area_bound(CG) is incremented by 15 for C1 and by 15 for C2. Cluster C1 is a clique, therefore no ADDON is calculated for C1. C2 is not a clique, and n8 and n10 are not connected by compatibility edges. Hence, area_bound(CG) can be increased by the lesser cost of the two nodes, i.e., ADDON(C2)=15. Thus area_bound(CG)=70.*

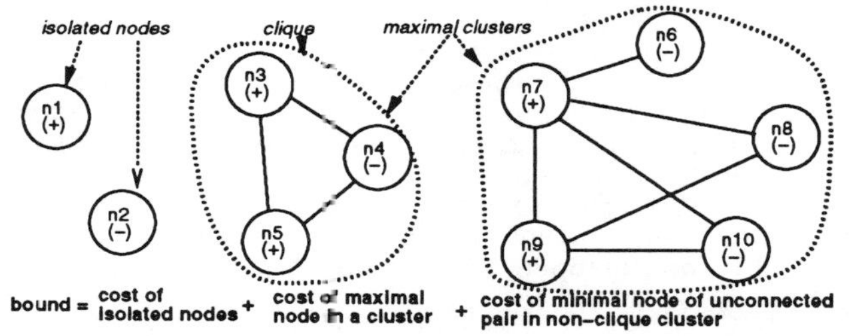

Figure 4: An Area Bounding Example.

4.3 The delay bounding function

We assume the definitions given above. We define a decision node d_j to be *stable* if it has no decision nodes as data inputs and if none of its input operator nodes share compatibility edges among one another (since in both cases the decision node d_j could potentially be optimized away due to future operator merging). We define the function *bound_node_delay: V* $\rightarrow$ Integer with V the set of vertices of *G* as follows:

$$bound\_node\_delay(v_j) := \begin{cases} \min_{M_k \in M}(\mathbf{delay}(M_k(v_j), bw(v_j))) & v_j \in N \\ \mathbf{decoder\text{-}delay}(v_j) & v_j \in V \text{ and } v_j \text{ is decoder-node} \\ \mathbf{conn\text{-}delay}(choices(v_j), bw(v_j)) & v_j \in D \text{ and } v_j \text{ is stable} \\ 0 & otherwise \end{cases}$$

Let p_i denote a path in *G*. Then the delay bounding function is defined by

$$delay\_bound(G) := \max_{p_i \in G}(\sum_{v_j\ on\ p_i}(bound\_node\_delay(v_j))).$$

Example 2. *In Figure 2 we assume that all operator and decision nodes can be implemented with a maximal delay of 2. d_2 is stable because its data inputs n_6 and n_7 are not compatible, i.e. bound_node_delay(d_2)=2. d_1 and d_3 are not stable, i.e., bound_node_delay(d_1)=bound_node_delay(d_3)=0. The delay_bound(G) thus is 4 based on the two longest delay paths $<n_6,d_2>$ and $<n_7,d_2>$.*

4.4 Correctness of the bounding function

Figure 5 illustrates the interplay between the area and delay bounding functions. Let the point *PD*=(Ax, Dx) correspond to a partial design with *area_bound(PD)*=Ax and *delay_bound(PD)*=Dx. The bounding function for *PD* results in the point $PD' = (\alpha \times Ax, \beta \times Dx)$ with *bound(PD)* $= \alpha \times Ax + \beta \times Dx$. The design costs of all solutions derivable by **FSA** from *PD* will be in the solution space in the shaded area above *PD'*. This implies the cutoff function $\alpha \times Ax + \beta \times \text{Dx} = \alpha \times X + \beta \times Y$ depicted by a dark diagonal line. To the lower left part of the line are all design solutions which are at least as good as *PD* or better. If **FSA** finds one complete solution to the left of the cutoff line then the partial solution *PD* and all solutions derivable from *PD* can be discarded. Due to limited space, the correctness proofs of the bounding function can be found in [7].

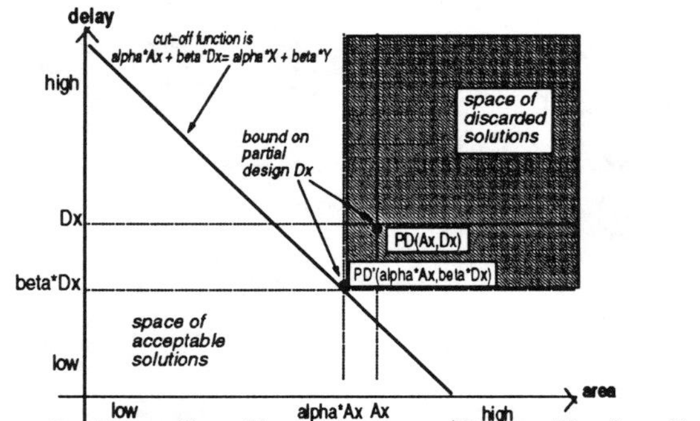

Figure 5: Bounding Fct. for Area/Delay Trade-off.

5 Component mapping algorithms

The **FSA** tool first generates a compatibility graph CG for the given flow graph G as described in Section 2.1. Thereafter, the two component mapping algorithms described below solve the problem of 'maximally' merging mutually exclusive operators of *G* by finding a minimal cost clique partition on *CG*. Both algorithms are based on the heuristic function, *benefit()*, which associates with each compatibility edge *e* a measure of the benefit of using the edge *e* for clique merging (see Section 5.1 below).

5.1 Compatibility edge selection

The heuristic function, *benefit()*, evaluates the gain of mapping two operator nodes to the same component in terms of both design area and delay. The function *benefit()*: N $\times$ N $\Rightarrow$ Real, with N the set of operator nodes in *G*, is a function from a pair of operator nodes $n_1, n_2 \in$ N to a cost value defined by

$$benefit(n_1, n_2) := \alpha \times\ area\_benefit(n_1, n_2) + \beta \times delay\_benefit(n_1, n_2),$$

with α and β the area and delay optimization parameters. Due to space limitations, the reader is referred to [7] for the definition of the *area_benefit()* and *delay_benefit()* functions.

5.2 Heuristic component mapper

Algorithm 5.1. : Heuristic Component Mapper

```
Input: flow graph G, compatibility graph CG, unit table U, α, β.
Output: a flow graph G optimized for area cost and/or for delay.
while (there is an edge e in CG with benefit(e)>0) do
  (1) Select edge e=< n1, n2 > from CG
      with the largest heuristic benefit(e) (Sec.5.1).
  (2) Map n1 and n2 to the same unit
      by transforming G and CG (Sec.2.2).
end while;
```

The component mapper described in Algorithm 5.1 uses a heuristic approach for maximally merging mutually exclusive operators of *G*. It searches for a minimal cost clique partition on *CG* by repeatedly selecting the edge *e* with the largest benefit and utilizing it for operator merging (Section 2.2). It stops when no edge with positive benefit remains.

5.3 Branch&bound component mapper

The component mapper described in Algorithm 5.2 replaces the greedy control strategy by a branch-and-bound one. The algorithm uses the inclusion (exclusion) of a compatibility edge as the branching criterion. At each node in the search tree, the solution space is partitioned into two sets of potential solutions according to whether or not an edge is used in the clique cover as discussed in Section 2.2.

Algorithm 5.2. : Branch&Bound Component Mapper

Input: G, CG, unit table, α, β, iteration count.
Output: a flow graph BEST optimized for area and/or for delay.
active-stack={ (G, CG) }; BEST=empty; BEST-COST=∞;
while ((active-stack $\neq \emptyset$) and (iteration-count>0)) **do**
(1) Pop branching node b (partial design sol) from active-stack
(2) Select edge e with the largest $benefit(e)$ (Sec.5.1).
(3) Generate children(b): child$_1$ $= b + e$, child$_2$ $= b - e$ (Sec.2.2).
(4) **for** i = 2 downto 1 do
  if $bound$(child$_i$) $\geq$ BEST-COST (Sec.4).
  then discard child$_i$;
  else if child$_i$ is a complete solution (Sec. 2.2)
   then
    if $cost$(child$_i$) $<$ BEST-COST
    then child$_i$ replaces the best solution found
     (BEST-COST:=$cost$(child$_i$); BEST:=child$_i$);
    else discard child$_i$;
   else push child$_i$ onto active-stack;
(5) decrement iteration-count.
end while;

The algorithm selects a partial solution $b = (G, CG)$ using a last-in-first-out scheme. Then it applies the heuristic function *benefit()* to determine the next most profitable compatibility edge e (Section 5.1). It expands b using the selected edge e: The first child uses e for operator merging, while the second child eliminates e (Section 2.2). If a child cannot lead to a least-cost solution based on the bounding function (Section 4), it is discarded. If a child represents a complete solution, then it is compared against the current best solution to determine the new best solution. This process is repeated until either no partial solution remains in the active list or the time limit given by the user is exceeded. We utilize several techniques to prune the search space and hence have successfully run the potentially exponential algorithm on typical example descriptions.

6 Experimental results

FSA has been implemented using approximately 14,000 lines of C code; and it currently runs on SUN3/SUN4 workstations under the UNIX operating system. The experiments presented below are based on the following examples. Example 1 is an Adder/Subtractor from [2]; examples 2 and 3 are variations of an ALU from the Mano book; example 4 is a functionally reduced version of the TI 74181 ALU; example 5 is a divide-by-3328 counter from Rockwell; example 6 describes a combination of a counting and a logic unit with partially permuted inputs; example 7 the ALU including input multiplexors of the Am2901; examples 8 and 9 describe several concurrently executing components with input multiplexors and register files.

6.1 Design quality and performance

In this experiment, we ran area optimization ($\alpha = 1.0$) and delay optimization ($\alpha = 0.0$) as indicated in the second column of Table 1. We also ran **FSA** with or without applying the functionality recognition algorithm. Design results achieved by **FSA** using the heuristic and the branch-and-bound component mapper are displayed in the columns **CM1** and **CM2**, respectively. For each, the solution quality of the design is described by the **Area** and **Delay** measures (Section 3). The computation time measured in CPU seconds on a SUN4 is given in the **CPU** column. The last column displays the results achieved by the human designer (meaning the best possible design we could find by hand).

Table 1: Design Quality and Algorithm Performances.

			CM1(heuristic)			CM2(B&B)			CM2	Designer	
Example	α	FR	Area (trans)	Del (ns)	CPU (sec)	Area (trans)	Del (ns)	CPU (sec)	vs CM1	Area (trans)	Del (ns)
AddSub	1.0	on	172	8	1	172	8	1	0%	172	8
AddSub	0.0	on	172	8	1	172	8	1	0%	172	8
AddSub	1.0	off	344	16	2	344	16	7	0%		
AddSub	0.0	off	792	19	2	344	16	2	16%		
AddSub	FR%		50%	58%		50%	58%				
Manof1	1.0	on	578	11	2	408	8	9	29%	408	8
Mano1	0.0	on	408	8	2	408	8	2	0%	408	8
Mano1	1.0	off	726	19	1	716	18	780	1%		
Mano1	0.0	off	828	19	2	716	18	141	5%		
Mano1	FR%		20%	58%		43%	56%				
Mano2	1.0	on	464	15	2	464	15	11	0%	408	8
Mano2	0.0	on	706	17	1	761	14	23	18%	408	8
Mano2	1.0	off	614	24	2	614	24	98	0%		
Mano2	0.0	off	911	20	2	752	19	122	5%		
Mano2	FR%		24%	15%		24%	26%				
TI74181	1.0	on	706	12	2	411	8	127	42%	411	8
TI74181	0.0	on	411	8	2	411	8	2	0%	411	8
TI74181	1.0	off	910	20	2	789	18	672	1%		
TI74181	0.0	off	909	19	2	789	18	532	5%		
TI74181	FR%		22%	58%		48%	56%				
Rockwll	1.0	on	2170	56	3	2170	56	12	0%	2170	56
Rockwll	0.0	on	3223	57	3	2912	33	25	42%	2912	33
Rockwll	1.0	off	2170	56	3	2170	56	12	0%		
Rockwll	0.0	off	3223	57	3	2912	33	25	42%		
Rockwll	FR%		0%	0%		0%	0%				
CntLog	1.0	on	409	11	2	409	11	4	0%	364	11
CntLog	0.0	on	634	15	1	543	10	6	33%	406	8
CntLog	1.0	off	409	11	2	409	11	4	0%		
CntLog	0.0	off	634	15	1	543	10	6	33%		
CntLog	FR%		0%	0%		0%	0%				
Am2901	1.0	on	1038	18	2	964	18	238	7%	964	18
Am2901	0.0	on	1132	16	3	1132	16	255	0%	1132	16
Am2901	1.0	off	1038	18	2	964	18	238	7%		
Am2901	0.0	off	1132	16	3	1132	16	255	0%		
Am2901	FR%		0%	0%		0%	0%				
Concur1	1.0	on	459	13	2	459	13	2	0%	459	13
Concur1	0.0	on	459	13	2	459	13	2	0%	459	13
Concur1	1.0	off	685	23	2	685	23	33	0%		
Concur1	0.0	off	831	24	2	1133	23	14	4%		
Concur1	FR%		33%	46%		33%	43%				
Concur2	1.0	on	379	8	2	379	8	3	0%	379	8
Concur2	0.0	on	452	8	2	379	8	2	0%	379	8
Concur2	1.0	off	579	18	2	579	18	91	0%		
Concur2	0.0	off	790	19	1	579	18	453	5%		
Concur2	FR%		36%	58%		36%	56%				

In the column labeled **"CM2 vs CM1"**, we present the percentage of design quality improvement obtained by **FSA** using the branch-and-bound versus using the heuristic component mapper. The improvement is calculated by $(quality(\mathbf{CM1}) - quality(\mathbf{CM2}))/quality(\mathbf{CM1})$ with $quality$ being **Area** for $\alpha = 1.0$ and **Delay** for $\alpha = 0.0$. We found that **FSA** improves the design quality by about 18% percent in almost half of the thirty six example runs when using the branch-and-bound algorithm (indicated by a **"CM2 vs CM1"** percentage larger than zero).

For each group of example descriptions, the fifth row labeled "FR%" indicates the percentage of improvement obtained by **FSA** when using the **FR** option over when not using the **FR** option. This design quality improvement is calculated by $(quality(not\_FR) - quality(FR))/quality(not\_FR)$. The percentage listed in the **Area** (**Delay**) column is obtained by plugging the **Area** (**Delay**) measures taken from the rows with the parameter setting $\alpha = 1.0$ ($\alpha = 0.0$) into the formula. For six out of nine examples, **FSA** with the **FR** option improves the design quality by an average of 42%.

The hand-produced designs are nearly always equivalent to those of **FSA** (using **FR** and the **B&B** component mapper). The designer produced a better design for two of the examples. The designer improved Mano2 by recognizing its equivalence with the Mano1 description. The designer produced a better design for the CntLog example by reversing the inputs of a commutative operation.

6.2 Technology adaptation

Table 2: Design Quality Using Three Libraries.

Ex.	Using TTL Library: FSA Area	FSA Delay	FSA CR	Designer Area	Designer Delay	Designer CR	Manual vs FSA
AddSub	172	8	yes	172	8	yes	0%
Mano1	408	8	yes	408	8	yes	0%
Mano2	464	15	no	408	8	yes	12%
TI74181	411	8	yes	411	8	yes	0%
Rockwll	2170	56	yes	2170	56	yes	0%
CntLog	409	11	*	364	11	yes	11%
Am2901	964	18	*	964	18	*	0%
Concur1	459	13	yes	459	13	yes	0%
Concur2	379	8	yes	379	8	yes	0%

Ex.	Using Mano Library: FSA Area	FSA Delay	FSA CR	Designer Area	Designer Delay	Designer CR	Manual vs FSA
AddSub	344	20	no	344	20	no	0%
Mano1	408	12	yes	408	12	yes	0%
Mano2	587	17	no	408	12	yes	30%
TI74181	768	15	no	768	15	no	0%
Rockwll	2170	60	yes	2170	60	yes	0%
CntLog	409	11	*	364	11	yes	11%
Am2901	1160	21	*	1160	21	*	0%
Concur1	566	15	*	566	15	*	0%
Concur2	452	10	*	452	10	*	0%

Ex.	Using Genus Library: FSA Area	FSA Delay	FSA CR	Designer Area	Designer Delay	Designer CR	Manual vs FSA
AddSub	344	20	no	344	20	no	0%
Mano1	716	24	no	716	24	no	0%
Mano2	654	16	no	654	16	no	0%
TI74181	826	25	no	826	25	no	0%
Rockwll	2170	60	yes	2170	60	yes	0%
CntLog	409	11	no	409	11	no	0%
Am2901	1160	21	*	1160	21	*	0%
Concur1	566	15	*	566	15	*	0%
Concur2	452	10	*	452	10	*	0%

In this second experiment, we explore **FSA**'s ability to perform technology adaptation by replacing the component-specific information by different libraries: (a) the TTL library, (b) the Mano library, and (c) the Generic Component Library GENUS developed at UC Irvine [2]. The results using **FSA** with the **B&B** Component Mapper and α=1 are given in Table 2. Table 2 shows that **FSA** produces better designs using the TTL library than when using the Mano or the Genus library. This is so because the TTL library has a set of richer components than Mano. For instance, **FSA** using Mano is able to reduce the second description to one unit, whereas **FSA** using Genus is not. There is no unit in the Genus library which directly supports some of the described functions, e.g., "A - B - 1".

The last column of Table 2 indicates the percentage of area improvement of the designer's result over **FSA**'s result calculated by (**Area**(FSA) − **Area**($designer$))/**Area**(FSA). Given a particular library, **FSA** almost always produced the same result as the designer (indicated by 0% improvement).

We also examined whether **FSA** is able to recognize the component(s) being described by the descriptions. The column labeled **CR** (for **C**omponent **R**ecognition) indicates whether **FSA** was able to reduce the functional description to the described component. **CR**='yes' (**CR**='no') means that (no) proper component recognition took place. **CR**='*' indicates that the algorithm succeeded in finding an alternative and even better design implementation. The number of component recognitions decreases for simpler component libraries. **FSA** recognizes (and possibly improves) the design implementation for 88% of the examples for TTL, 66% for Mano, and 44% for Genus.

7 Conclusions

In this paper, we have presented two algorithms to solve the component mapping problem that accomplish both area and delay optimization. Our experiments show that in most cases the algorithms produce a design that is comparable to that of a human designer. We also found that **FSA** improves the design quality (both area and delay) by 18% in about half of the example runs when using the branch-and-bound rather than the heuristic component mapper. Designs synthesized using functionality recognition are smaller in cost than those synthesized without it. **FSA** was able to improve 66% of the example designs when using functionality recognition by a 42% improvement in design quality.

Acknowledgements. We are grateful for support from the NSF grant MIP-8922851, California MICRO grant #89-057, and contributions from TRW Inc. Elke A. Rundensteiner also thanks the IBM Data Systems Division in Poughkeepsie, NY for granting her an IBM Graduate Fellowship. We thank members of the UCI CADLAB for helpful discussions. In particular, we want to thank J. Lis for the development of the VHDL graph compiler and I. Ghosh for the Am2901 description.

References

[1] N. D. Dutt, and J. R. Kipps, Bridging High-Level Synthesis to RTL Technology Libraries, *Proc. 28th Design Automation Conf.*, pp. 526–529, 1991.

[2] N. D. Dutt, GENUS: A Generic Component Library for High Level Synthesis, Tech. Rep. 89-09, Univ. of Cal., Irvine, 1989.

[3] G. W. Leive and D. E. Thomas, A Technology Relative Logic Synthesis and Module Selection System, *Proc. 18th Design Automation Conf.*, pp. 479–485, 1981.

[4] G. P. Paulin and J. P. Knight, Scheduling and Binding Algorithms for High-Level Synthesis, *Proc. 26th Design Automation Conf.*, pp. 1–6, 1989.

[5] E. A. Rundensteiner and D. D. Gajski, A Design Representation Model for High-Level Synthesis, Tech. Rep. 90-27, Univ. of Cal., Irvine, Sep. 1990.

[6] E. A. Rundensteiner, D. D. Gajski, and L. Bic, Component Synthesis: Technology Mapping from Register Transfer Descriptions, *Proc. Int. Conf. on Computer-Aided Design*, pp. 208–211, Nov. 1990.

[7] E. A. Rundensteiner, D. D. Gajski, and L. Bic, Component Synthesis from Functional Descriptions, Tech. Rep. 90-24, Univ. of Cal., Irvine, Dec. 1990, (revised Aug. 1991).

[8] C. Tseng and D. P. Siewiorek, Automated Synthesis of Data Paths in Digital Systems, *IEEE Trans. on Computer-Aided Design*, CAD-5, 3, pp. 379–395, July 1986.

SESSION 19

PANEL: CASE -- WHY IT DOESN'T WORK FOR CAD

Chair: Rick Potter - Mentor Graphics Corp., Wilsonville, OR

As tool makers, CAD developers understand that the use of tools improves quality, increases reliability, reduces maintenance problems and improves time to market for their customers. It seems natural then that CAD developers would be anxious to adopt tools to their own development. However, in the area of CASE tools this does not seem to be the situation.

Although CASE tools are making an impact in some of the data processing industries, they appear to be making little, if any, impact in the CAD industry. There are many theories concerning this lack of adoption of CASE tools. Foremost appears to be the concern about the focus of CASE tools. Traditional CASE is focused on the front end of the design cycle. CAD developers recognize a benefit in doing better requirements capture and analysis for their software development projects, but at this point in time they believe they have more significant problems in the back end of the development cycle. Often the attitude appears to be "You're giving me a solution for the common cold, but I've just learned I have cancer. Can't you help solve my real problem?"

In other words, CAD developers need tools that will help them cut, debug and deliver code in a more productive fashion. Better compilers, linkers, and debuggers along with improved project and configuration management systems are required. This is especially true for those companies working with object oriented languages. Once significant gains have been made in these areas, then the front end of the life cycle becomes more significant.

From the CASE developers viewpoint there is recognition of the importance of the tools requested by CAD developers. However, there is also the belief that front end tools provide significant return. Future CASE systems will attempt to provide both types of tools with the required linkages between them. These systems of the future are not here today, but point tools that are currently available will provide much of the needed functionality.

This panel is made up of representatives of both the CAD and CASE industries. It will examine current CASE tools and discuss why they have or have not been adopted by CAD developers. The panel will end by discussing CAD developers requirements for future CASE tools.

Panel Members:

Scott Baeder - Cadence Design Systems, Inc., Lowell, MA
Eileen Boerger - Mentor Graphics Corp., Wilsonville, OR
Jerry Sullivan - MCC, Austin, TX
George Symons - PROCASE, Santa Clara, CA
Anthony Wasserman - IDE, San Francisco, CA

SESSION 20

TUTORIAL: DIRECTIONS TO WATCH IN DESIGN TECHNOLOGY

Chair: Gerry Langeler - Mentor Graphics Corp., Wilsonville, OR
Organizer: Tom Pennino - AT&T Bell Labs., Holmdel, NJ

While new algorithms and CAD tools continue to have a significant impact on the design process, other factors in the design technology industry now dominate purchase and development decisions, including component libraries, design flows, interfaces, hardware support, and design data management. Factors which will determine future directions in these areas will be presented, along with some predictions for the next 2-5 years. The goal is to help design managers make effective decisions for long-range planning in both product development and acquisition.

Speaker:

A. Richard Newton - Univ. of California, Berkeley, CA

Hcompare : A Hierarchical Netlist Comparison Program

Pradeep Batra
Sun MicroSystems
MS 19-02, 2550 Garcia Ave.
Mountain View, CA 94043
(Currently at Rambus Inc., Mt View, CA)

David Cooke
Sun MicroSystems
MS 19-02, 2550 Garcia Ave.
Mountain View, CA 94043

Abstract

This paper describes a new program for the hierarchical comparison of two netlists. The program differs from other netlist comparison programs in that it does a true hierarchical comparison on the maximal overlap in the two hierarchies. It does not require the hierarchies on the two netlists to be isomorphic. Built in heuristics are used to correct many commonly found errors. The program runs fast (takes about 8 minutes on a Sun4/75 to compare 3 million transistor chip) and has been used to compare layout vs schematic netlists for 2 chips with over 2 million devices each.

Introduction :

One of the crucial tasks in the design of VLSI circuits is that of determining if two views of the design represent the same design, and to identify any differences between the two accurately. Present day designs are done at the behavioral level. This is transformed to the RTL then to the gate level (schematic , netlist) and then to the layout. A successful silicon implementation requires fast and reliable verification between these successive views of the design. For gate level description to schematic and schematic to layout verification, the problem can be reduced to the comparison of two netlists. Most of the comparison tools do a flat device level comparison [2,7,8,9]. This was sufficient in the past as designs were less complex and had only two representations; schematic and the layout. These tools become inefficient and inadequate as the design size grows beyond a few million devices. For most flat approaches verification time grows as O(N**m) where N is the number of devices and m > 1. Memory requirements grow linearly with the size of the design. The problem with the flat approach is that it takes too long (40-50 hours) to verify about 3-4 million devices. The errors or mismatches are reported in terms of devices. Some programs [10, 11] carry the hierarchical path for the device names and net names to aid the user in locating errors in the design. These reports are difficult for the logic designer to understand because they do not directly reference the design in terms of familiar blocks such as registers, ALUs, multipliers, caches etc. As the number of devices in a given design increases the mask designer is increasingly overwhelmed by the size of flattened designs and the error reports generated by such designs. Thus there is a need for a tool that would do a fast reliable and thorough hierarchical comparison of two hierarchical netlists. Historically hierarchical netlist comparison [3,5,6] has been too restrictive. It required the two hierarchies to be exactly the same, or handled only a two level hierarchy [10], or picked hierarchies on its own disregarding the user defined design hierarchy[10].

In this paper, we present "Hcompare", a hierarchical comparison tool. This tool does a true hierarchical comparison with the user defined hierarchies. Errors are reported hierarchically in terms of user defined blocks. Error reports are easy to read as errors are reported in terms of mismatched connections to design blocks. Run time is of the order of minutes even for large blocks with 2-3 million devices. Input to the tool is in the form of two netlists one from the definition, typically schematic, and one from implementation, typically layout.

Requirements for a hierarchical netlist compare :

For a hierarchical comparison program to be useful for actual design verification it should meet the following requirements

i) Compare two hierarchical netlists with no constraint on the hierarchy definition in each netlist. The only constraint allowed is that the two hierarchies represent the same flattened design.

ii) Handle differences in the two hierarchies. In the worst case with no match in the two hierarchies it should do a full flat compare. This is very important as from our experience the hierarchies for RTL design, schematics for the design, and layout for the design have different constraints. Current tools and methodologies make it too restrictive to have the same hierarchical structure for all representations of the design.

iii) Handle different number of ports for the same level of hierarchy in the two representation. For instance, schematics may have extra power and ground ports compared to RTL. While the layout may have feed thru signal ports. A feed thru signal is defined as a signal going through the block without connecting to any device in the block. Physical constraints force feed thru's in the layout for optimal densities.

iv) Handle multiple versions of the same block in one representation. This typically happens in the physical representation. Size and area constraints at times require the same block to be laid out with different aspect ratios.

v) Handle flattened and unflattened instantiations of the same block in a level of hierarchy. This typically happens in the physical representation. It is more common with small blocks like inverters and simple logic gates. In high performance custom designs mask designers tend to explode/(flatten) some instantiations to aid area optimization.

vi) Handle permutablity between schematic and layout for logically equivalent inputs on an instance basis. Logically equivalent inputs have different timing constraints and have different physical location in the layout. This is required as in any design for some instances of a gate logically equivalent inputs can not be permuted. For most instances logically equivalent inputs are permutable between the layout and the schematic and an absolute non-permutability constraint can lead to larger areas in the physical implementation. For example for a two input NAND gate input1 and input2 are logically permutable. The circuit implementation will have different switching times for the two inputs and the physical implementation will have a different phy - ical position for the two inputs. Timing considerations might require some instances of NAND gate to have a direct mapping for inputs between schematic and layout. For other non critical instances the inputs may be reversed in the layout for optimal area.

vii) Handle split and parallel devices and do parameter checking for devices. Typically devices with large sizes are drawn as parallel devices with a smaller size in the layout. For example an n channel transistor with channel width of 50 microns in schematic might be drawn as 10 n channel transistor's connected in parallel in the layout for better aspect ratio and control over the device properties.

Hierarchical Compare Algorithm :

Fig 1.1 shows the flow diagram for the hierarchical compare program. Two hierarchical netlists are read in. Block definitions are ordered such that blocks are defined before they are instantiated. For each block in the list, if the block is defined in both netlists then the two blocks are compared otherwise the block is marked for flattening. "Equalize hierarchies" flattens instances of the blocks marked for flattening. This eliminates extra levels of hierarchies and reduces the two representations for the blocks to have the same hierarchies. The next step compares these two representations. Comparison is done using standard graph isomorphism algorithms [1,4]. Comparison starts with a initial correspondence of nodes which have the same name in the two representations of the block. If no node names match in the two representations then an initial correspondence is generated based on pattern matching of unique nodes.

Every node and every instance is compared and a corresponding match is found for it. If the two representations do not match errors are reported and the comparison stops. If the two representations match then a pin correspondence list is generated. This list is used to compare instances of the block at the next level of the hierarchy. This process is repeated for each level of hierarchy.

The following example further explains the hierarchical compare algorithm. The schematic and the corresponding layout hierarchies for the example design are shown in fig1.2. The hierarchies do not have one to one correspondence. The schematic hierarchy has an extra level at mod1 shown as gate1 which is not present in the layout. This represents the situation of parameterized gates in schematics which are drawn flat in the layout. Layout has cells under submod3 which are not present in

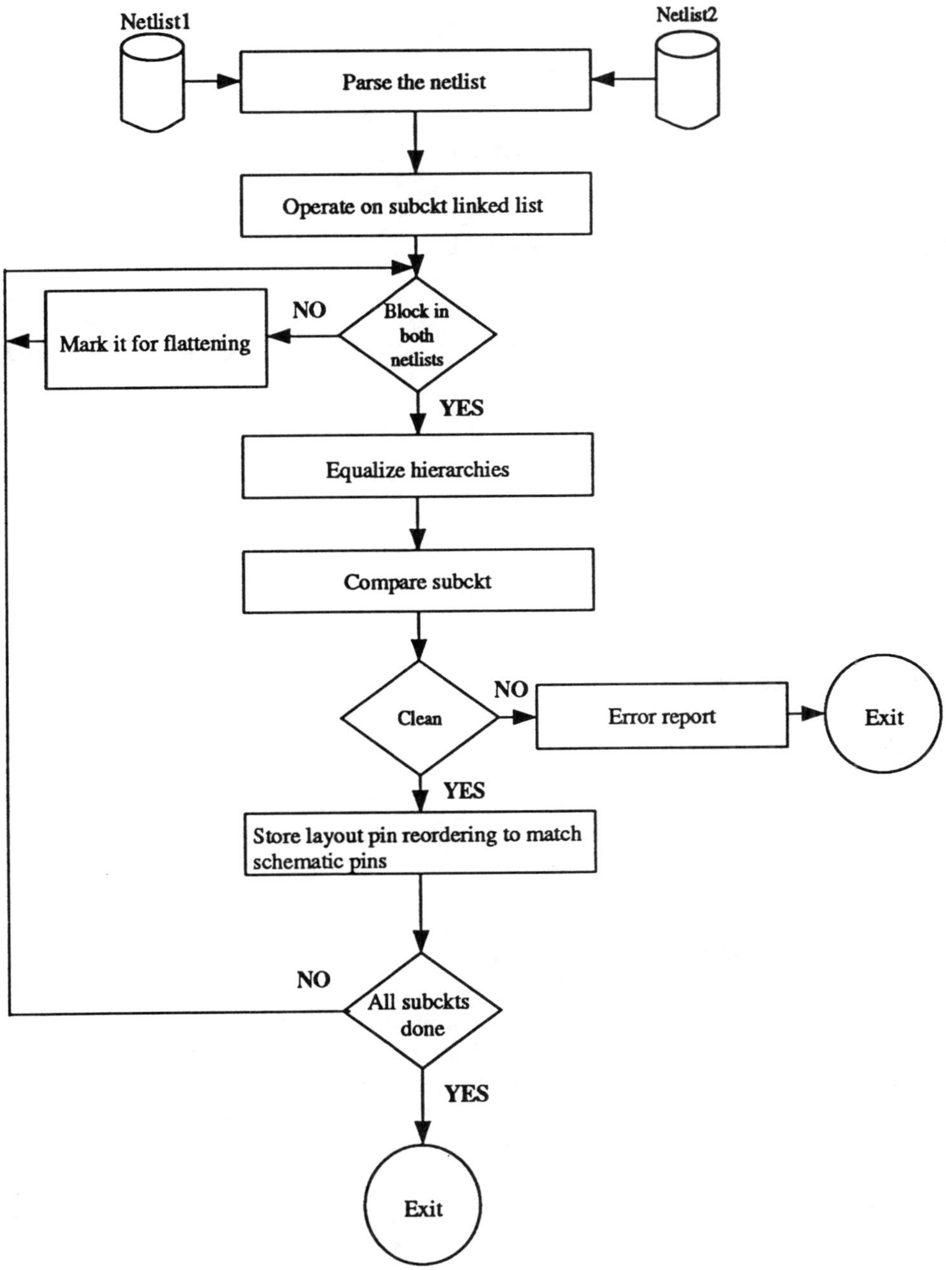

Pseudo Code

```
Parse netlist1
Parse netlist2
Order blocks in netlist1
Order blocks in netlist2
Foreach block in netlist1
{
  If block in netlist2
  {
    Equalize hierarchies for
       block from netlist1
    Equalize hierarchies for
      block from netlist2
    Compare the two blocks
    If errors
    {
      Report errors
      Exit
    }
    else
    Store pin reordering for
     block from netlist2
}
else
  Mark block for flattening
}
Exit
```

Fig1.1 : Flow Diagram For Hierarchical Compare

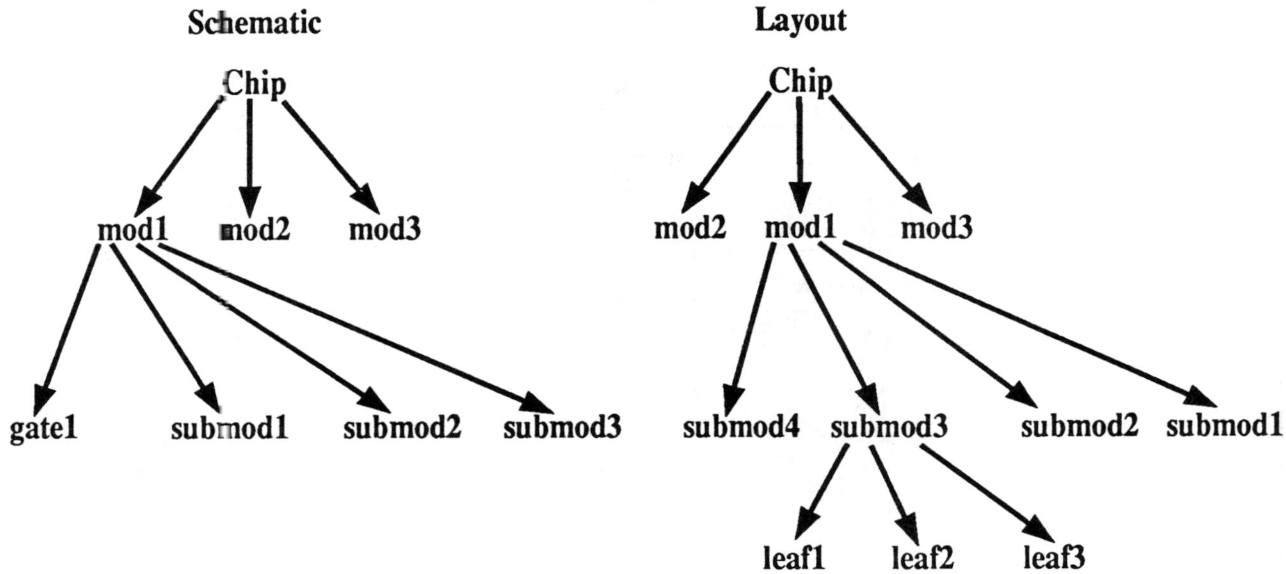

Fig.1.2 : Schematic and layout hierarchy

the schematics. This is representative of the case where layout designer puts the repetitive geometries in a cell. Hierarchical compare will go through this in the following way:

1: Leaf1, leaf2, and leaf3 have no corresponding schematics so mark them for flattening.

2: Compare submod1, submod2, and submod3 and if clean keep track of pin ordering in layout and schematic.

3: Submod4 in layout is not present in schematic so mark it for flattening at the next level.

4: Gate1 in schematic is not present in layout mark it for flattening in next level schematics.

5: Do a compare on mod1 with submod1, submod2, and submod3 as abstract devices. Gate1 is flattened on schematic and submod4 is flattened on the layout.

6: Similarly compare mod2, mod3 and then go a level higher to chip and compare chip with mod1, mod2, and mod3 as devices.

Results & Discussion :

Table 1.1 shows run time statistics for large blocks. Schematic blocks is the number of distinct blocks in the schematic. Layout blocks is the number of distinct blocks in the layout. Flat devices is the number of transistors if the design was flattened all the way down to the transistor level. For all the blocks in the table number of schematic and layout blocks are not identical. This means that the hierarchies in the two representations do not match one to one. Run time is the sum of user time and system time. All the blocks were run on a Sun 4/75 with 64 megabytes of RAM. Run time is not a linear function of the number of flat devices, but depends on the number of distinct levels of hierarchies and number of levels that match in the two representations. In the degenerate case where the two hierarchies are totally different Hcompare would do a full flat compare after flattening the two hierarchies. For example f_unit has the worst run time though it has less flat devices then d-unit, c_unit, and mxcc. This is because f_unit had the least number of levels of hierarchies that matched in the two representations.

Hcompare was used at Sun Microsystems to verify two chips with about 3 million devices each. It was used

Blockname	Schematic Blocks	Layout Blocks	Flat Devices	User Time (secs)	System Time (secs)	Memory (Kbytes)
f_unit	621	481	391853	601	24	108445
d_unit	354	348	1076319	171	6	33732
c_unit	308	335	1224908	65	4	23808
b_unit	197	216	52950	28	6	14322
mxcc	505	431	2263700	434	17	72694

Table 1.1 Hcompare Statistics

by 10-15 layout designers to verify layout with the schematic. The main problems we encountered were related to hierarchical circuit extraction. In our methodology, layout designers were required to put geometries on special layers in the layout that defined the physical ports of a block. At the next level signals could be hooked only to the ports for valid connections. At times this was found to be too restrictive for a custom layout.

The current release of "Hcompare" stops comparison if it finds errors in the current block it is comparing. It does not compare next level blocks even if the block with errors is not instantiated at this level. Ideally "Hcompare" should compare all independent blocks in a hierarchy. When errors are detected in a block any other block which directly or indirectly instantiate the block with errors should not be compared. This will lead to a better error report. This problem is currently being investigated.

We were able to start the top level chip verification even when the lower level blocks were not done. This was possible since we could use Hcompare in a mode where it could just compare top level block interconnection without going inside the blocks.

Conclusion :

Hcompare is a hierarchical comparison tool for comparing two hierarchical netlists. Hcompare determines whether two hierarchical netlists represents the same circuit. Any differences between the two netlists are reported hierarchically. The tool does not require the two hierarchies contained in the netlists being compared to be isomorphic. Heuristics are used to correct many common errors. Hcompare can compare netlists that contain blocks with specific port differences between blocks such as permutations and feed thru ports. User defined hierarchies that exist in both representations are preserved.

Hcompare is a fast comparison tool. This tool has been used successfully to verify schematic vs layout for two large VLSI chips.

Acknowledgments:

Hcompare would not have been possible without the work of Won Kim of Maspar. His flat compare program inspired the development of Hcompare. We would like also to thank Chuck Tucker and Joseph Yang for their support in the development and use of the Hcompare program at Sun Microsystems.

References

[1] N. Kubo, I.Shirokawa, H.Ozaki, "A Fast algorithm for Testing Graph Isomorphism", Proceedings ISCAS Conference,

pp. 641-644 1979.

[2] C.Ebeling and O.Zajicek, "Validating VLSI Circuit Layout by Wirelist Comparison", IEEE Int. Conference on CAD, pp. 172-173, Sept. 1983.

[3] R.L.Spickelmier and A.R.Newton, "Wombat : A New Netlist Comparison Program" IEEE Int. Conference on CAD, pp. 170-171, Sept. 1983.

[4] D.G.Corneil and D.G.Kirkpatrick, "A Theoretical Analysis of Various Heuristics for Graph Isomorphism Problem", SIAM Journal of Computing, vol. 9, no.2, pp. 197-281, May 1980.

[5] S.P.Chao, Y.S.Huang and L.M.Yam, "A hierarchical Apporach for Layout versus Circuit Consistency Check", Proc. 17th Design Automation Conference, pp. 270-276, June 1980.

[6] Todd.J.Wagner, "Hierarchical Layout Verification", Proc. 21st Design Automation Conference, pp. 484-489, June 1984.

[7] P.Losleben and K.Thompson, "Topological Analysis for VLSI Circuits", Proc. 16th Design Automation Conference, pp. 461-473, June 1979.

[8] L.Williams, "Automatic VLSI Layout Verification", Proc. 18th Design Automation Conference, pp. 726-732, June 1981.

[9] A.Kishimoto, H.Kawanishi, H.Yoshizawa, H.Ohno, Y.Fujinami and K.Kani, "An Inter-Connection Check Algorithm for Mask Pattern", Proc, ISCAS Conference, pp. 669-672, 1979.

[10] Dracula III Reference Manual Nov, 1989, pp. 1-39.

[11] Checkmate Netcheck Reference and SVDB utilities reference Release 1.0 Dec 1990.

An Interpreter for General Netlist Design Rule Checking

Georg Pelz

Fraunhofer-Institute of Microelectronic Circuits and Systems
Finkenstr. 61, W-4100 Duisburg 1, Germany

Abstract

A new approach to general netlist design rule checking is presented, which has been successfully applied to design for testability (DFT) rule checking and electrical rule checking. Core of the checker is an interpreter, which performs operations of a set-based calculus. In combination with hierarchy preprocessing by expansion and netlist pattern matching, the interpreter approach turned out to be flexible and efficient.

1 Introduction

Chip level verification is a growing bottleneck for VLSI designs. As long as automated synthesis is restricted to limited applications and manual interference is necessary for most designs, chips have to be completely validated before manufacturing. This has to be done on physical (design rule check) and on electrical level (electrical rule check and simulation).

With the increasing complexity, testing of integrated circuits has become more and more difficult as well. Design for testability (DFT) approaches (e.g. scan path) have been introduced to cope with these problems, by improving controllability, observability and separability of large circuits. DFT techniques require compliance with associated design rules. For instance, a scan path has to be a valid chain of memory elements.

In these two fields, design rules on netlist level have to be checked. The most important properties of tools, that accomplish a netlist audit, are *flexibility* in the checks to be carried out and *performance* of execution. Since netlist design rules strongly depend on factors as the chosen circuit technology, the DFT strategy or the actual implementation of the circuit, they may vary even from design to design. Therefore, a lack of flexibility causes an unacceptable amount of work for implementation and modification of the checks. Since hierarchical processing of circuits is not possible in all cases, efficient execution of a netlist audit is necessary to preserve general applicability. For instance extracted flat circuits (possibly with parasitic elements, e.g. coupling capacitances) may be checked.

Testability rule checking and electrical rule checking have been investigated since years and many algorithms have been proposed. The related work divides to six groups:

1. *hard coded approaches*
   Hard coded tools (e.g. [1, 2]) offer good performance, but suffer from a lack of flexibility. For new checks new programs have to be implemented. Variations of rules require the modification of programs. In almost all cases this programming-task is tedious and error-prone. The amount of work to be done would be intractible, if the applied rules are not restricted to a small subset of general rules.
2. *expert systems and similar approaches*
   Expert systems and similar approaches (e.g. [3, 4, 5, 6, 7]) provide good flexibility, but they have performance problems. Most cited publications restrict their examples (if supplied) to some thousands of devices. An exception is reported by Wenin et al. [7]. Their largest example consists of about 45000 elements. This is achieved by a partitioning of the circuit under check, which gives possibly rise to typical coupling problems for the general netlist rule checking. Additionally, maintainance problems arise, if the implemented checks become complicated [3].
3. *program generators*
   Program generator approaches (e.g. [3]) establish better performance, but the process of generating program text with a given syntax is involved and complicated. Further on, the generated program has to be compiled and linked with a library before it can be applied as a netlist rule checker.
4. *logic simulator approach*
   In [8] a logic simulator is employed to trace paths for DFT rule checking. This requires a considerable amount of modification of its logic models. The achieved performance is good, but the domain of rules to be checked is restricted.
5. *graph-grammar approach*
   In [9] graph-grammars are used to parse a netlist. If this can be accomplished, this netlist obeys the defined syntax. In this way, valid topologies of circuits are specified, rather than defining what might be an error. The problem of this method is that the grammar has to cover the fantasy of the designers in inventing circuits. Especially for complicated circuit technologies it might be difficult if not impossible to define a valid grammar.
6. *design for test calculus*
   A design for test calculus is presented in [10]. For interesting inputs and outputs of components,

0738-100X/92 $3.00 © 1992 IEEE

"signal sets" are calculated. These contain the nets, which determine the value at that component pin, according to a predefined signal flow through components. All checks are carried out by evaluating these "signal sets". This approach does not deal with general sets of components, nets or pins. Since these "signal sets" can easily be extracted by the interpreter's HULL command (see below), the checks of [10] can be emulated.

The aim of our novel approach was to supply a tool, which offers both performance and flexibility without restricting the domain of managable netlist design rules. Almost all publications are related to either electrical rule checking or DFT rule checking. Here, a general netlist checking tool is proposed, which covers both domains. In the following, section 2 will deal with the set-based netlist rule checking by an interpreter and section 3 shows possible applications. In section 4 the hierarchy preprocessing is described and section 5 supplies some information on the implementation and the results achieved.

2 Set-based interpreted rule checking

In our novel approach, sets of items are the basic structure to deal with. Almost all operations use sets as inputs and outputs. According to the wide-spread netlist model (figure 1), three types of items are possible: components, nets and pin-types. Components

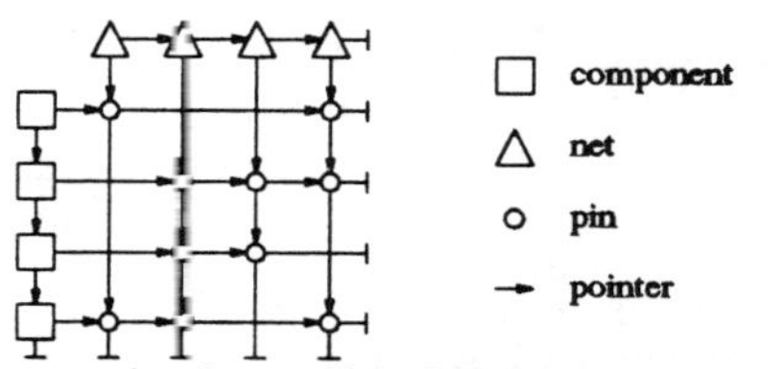

Figure 1: basic netlist data structure.

might be primitives as well as higher-level blocks. Nets are used to interconnect components. Every link between nets and components is implemented by a pin. The three set-types imply three basic sets:

- the set of all instances of components in the cell under check
- the set of all nets in the cell under check
- the set of all distinguishable pin-types in the cell under check

Every set has to be allocated, defining its type. Obsolete sets can be returned, in order to decrease the number of sets used at a time. Initially all sets are empty, except some sets with predefined names (e.g. @transistors, @vdd), which are filled with the respective items. Predefinition of a set is indicated by the first letter, being a '@'.

This problem-modeling leads to a lot of advantages. The sets represent the state of computation, since almost all basic operations have sets as inputs and outputs. Therefore, each operation accomplishes a defined state transition. Complex checking-tasks are broken down into subtasks, which can be directly served. In this way, the opportunity of using an interpreter for netlist rule checking is enabled. Intermediate results are explicitly accessible. Since the subtasks were implemented efficiently, the whole rule checking process is kept efficient. Besides, the set-based approach to netlist rule checking can be easily implemented, since the chosen atomar operations are simple.

The structure of the interpreter allows the execution of scripts as well as interactive working. These modes can be arbitrarily mixed, allowing comfortable debugging. Since the user interface is provided by expressions on the set calculus, it is simple but expressive. Checks are formulated on an abstract level, without keeping in mind the actual netlist data structure. Additionally it is possible to incorporate more set-defining commands (e.g. circle detection etc.), which would be realized by efficient graph algorithms on the netlist data structure.

The interpreter deals with seven types of objects. Three of them are set types: components (COM), nets (NET) and pins (PIN). The other four are types for variables as integers (INT), floating point values (DOUBLE), strings (STRING) and functional expressions (FUNC). Such variables may be attached to components or nets and will then be called properties (PROPERTY). Some properties might be set, when parsing the circuit (e.g. W, L of transistors). Others can be assigned or modified by interpreter commands; for instance net voltages [5] or timing information could be taken into account by properties.

Now, an overview on the basic commands of the set calculus will be given, which distribute to five groups. The first group includes the canonical set operators as INTERSECTION, UNION, DIFFERENCE and COMPLEMENT (to the respective basic set).

- r = INTERSECTION(a,b);
  calculate set-intersection on "a", "b" and write result to set "r"
- r = UNION(a,b);
  calculate set-union on "a", "b" and write result to set "r"
- r = DIFFERENCE(a,b);
  calculate set-minus on "a", "b" and write result to set "r"
- r = COMPLEMENT(a);
  calculate set-complement of set "a" , which refers to the respective basic set

In the second group, connection operators can be found, which detect direct links and their transitive hull.

- r = CONNECT(a,p);
  "r" is the set of items (components, nets) directly connecting to items of "a" (nets, components) over pin-types found in "p"; CONNECT_FORWARD() and CONNECT_BACKWARD() work respectivly, but according or against signal flow given by signal directions of pins
- r = HULL(a,p,<depth>);
  "r" is in result the transitive hull of 1 to <depth> CONNECT operations; members of set "r" are

chosen according to the set-type of "r"; value '*' for <depth> : depth of the hull is not restricted; HULL_FORWARD() and HULL_BACKWARD() work respective to CONNECT_FORWARD() and CONNECT_BACKWARD()

The third group holds operators dealing with properties of sets (e.g. member-count etc.) and items (e.g. W,L of transistors, functional information of nets etc.).

- COUNT(a)
  function, which returns the member-count of set "a"
- NAME(a)
  function which supplies string with name of first member of "a"
- PUT(a,<name>,<value>);
  sets the attributes <name> of all members of "a" to <value>
- GET(a,<name>);
  returns value of attribute <name> of the member of "a" ("a" has to have exactly one element)

The fourth group contains miscellaneous operators which implement for instance the sweep over the members of a set (FORALL). The members are accessible through a one-element set, which is updated in each loop. Other members of that group do the manual filling of sets or the set output. Note that each error output will be described as a set output, containing items, which violate the checked netlist design rule.

- FORALL(x,a) { ... }
  sweep over set "a"; in each loop the one-element set "x" contains the next member of "a"
- ADDMEMBER(a,<name>);
  manual adding of items to set "a" by name
- SHOW(x);
  output of set or variable (named "x") to screen
- OUTPUT(x);
  output of set or variable (named "x") to file

The last group contains typical programming language statements, which allow the convenient formulation of checks and some control statements.

- IF (condition) expression
  WHILE (condition) expression
  DO expression WHILE (condition)
  control statements
- +, -, *, /, ==, >=, <=, !=, ||, &&
  basic literals (C-syntax)
- AND, OR, NOT
  operators on functional expressions
- {, }
  brackets for grouping expressions to one expression
- CALL "x.erc";
  calls another interpreter script ("x.erc") and switches control to it
- INTERACTIVE;
  calls an interactive shell
- RETURN;
  returns to calling shell

The representation of the sets is very important for the resulting consumption of CPU and memory resources. A flag-field (32 bit) is used, indicating set membership. Such a field is attached to each instance, net-record and pin-record representing a pin-type. Note that a strong difference is made between pin-types and instances of pins. Sets with low member-count have an additional explicit representation by a list of pointers to members, attached to the set-record. This economical data-structure turned out to be sufficiently expressive and efficient. The count of (component-, net- or pin-) sets used at a time was under 15 for every check implemented.

This data-structure is motivated by the local and global sight of a netlist scenario. For global operations (e.g. touch all N-transistors, and look whether they are connected to VDD) the sweep over the basic (component-)set does not degrade the performance too much. A "local view"-situation can be observed, when a special item was selected and operations on its neighborhood are performed (e.g. inside a FORALL loop). In almost all cases, the explicit representation is sufficient for representing local sets. In order to indicate the execution-efficiency of the basic operators, the implementation of the INTERSECTION operator and the HULL operator will be sketched.

For the intersection, if both sets have a high member count, a sweep over the basic set has to be performed, since they have no explicit representation. When visiting the members of the basic set, the only thing to be done is a bit-test and an adding of a new member to the resulting set where necessary. If one or both sets have an explicit description, a sweep will be performed over the members of the smaller set, using the list of explicit pointers to set-members. If the output set is one of the input sets, special handling assures, that the minimal necessary amount of work has to be spent.

For the hull-command two temporary sets are defined, one for nets and one for components. These are filled in an alternating manner by running through the lines or columns in the data structure (figure 1), which refer to the members of the previously filled set. Note that each pin has an additional pointer to its adjacent net and component. Where necessary touched elements are added to the result set. The filtering according to the pin types is achieved by a hash table storing unique pin values.

3 Range of Application

The set-approach to universal netlist rule checking offers a wide range of applications. Traditional electrical rule checking can be performed as well as design for testability rule checking or methodology checking. In order to indicate the general applicability of our approach, a suite of examples will be presented. Three interpreter scripts are supplied, dealing with the combinational depth behind scan flip-flops, parameter-driven electrical rules on primitive level and function extraction. Note that keywords are written in upper case and set descriptors in lower case.

The first example-check is to find out, whether the combinational depth behind scan path memory elements is greater than six. To accomplish this check, two sets are manually filled (by name): the set of scan flip-flops (ff) and the set of scan flip-flop input pins (ff_in). The set ff_in is complemented, in order to get a set of pins, over which paths concerning the combinational depth can be traced. The starting point for those directed paths is the net-set startnet, which consists of the nets at the scan flip-flop outputs. The HULL_FORWARD command traces all directed paths of depth 14 or lower and depth 12 or lower. The value of depth is related to the number of component-net and net-component transitions. In theory, the HULL_FORWARD commands could be emulated by 14 or 12 CONNECT_FORWARD commands, but this would degrade the performance. The difference of the resulting net sets (badnet, oknet), contains a net for each path starting at a scan flip-flop output and exceeding the allowed length of 6 (= 12 transitions). If this set (err) is not empty, the set x is added to the the set of all errors (error). This implementation is a little bit involved, because the two hull-commands are used to identify the nets, which have a distance of exactly 7 to the members of startnet. A modified hull-command would accomplish this in one pass.

```
NET startnet, x, oknet, badnet, err, error;
COM latch;
PIN inlatch, allbut;

ADDMEMBER(ff,"SCANFF");
ADDMEMBER(ff_in,"SCANFF:DATA_IN",
                "SCANFF:SCAN_IN");
allbut   = COMPLEMENT(ff_in);
startnet = CONNECT_FORWARD(ff,allbut);
FORALL(x,startnet) {
  badnet = HULL_FORWARD(x,allbut,14);
  oknet  = HULL_FORWARD(x,allbut,12);
  err    = DIFFERENCE(badnet,oknet);
  if (COUNT(err) > 0)
    error = UNION(error,x);
}
```

The second example check deals with the characterization of a buffer fan-out and a judgement on the parameters on the chosen buffer transistors. Note that this script template could as well be used for the dimensioning of buffer transistors according to their load. This check is accomplished at best, if a mixed abstraction level is chosen. This mixed description contains the buffers as transistors, whereas the rest of the circuit consists of gate level objects (figure 2). This leads to an enormous data reduction and the load to be driven could be specified on a more abstract level. Especially the influence of transfer gates must be taken into account. If switched on, they "transfer" the load to be driven. For a worst case checking, all transfer gates are assumed to be switched on.

At first, we have to find out which nets are buffer outputs. This is accomplished by a sweep over all nets and a decision, whether there is a N-transistor connecting it to VSS (over predefined pin-set @nchannel, i.e. N-transistor source and drain), a P-transistor connecting it to VDD and a short at the gates of these transistors. Afterwards, the fanout to be driven is correlated to the buffer transistor parameters, accessible through the two one-element sets an and ap.

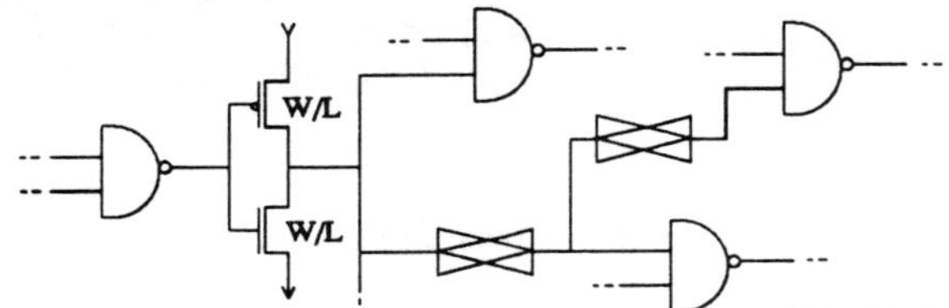

Figure 2: Buffer fanout check on mixed abstraction level.

The error condition could be formulated more sophisticated; for ease of description, here the number of logic gates to be driven (possibly over transfer gates) is taken into account.

```
NET @vdd,@vss,allnets,error,nload,gx,n;
COM load,tn,tp,an,ap,x;
PIN @nchannel,@pchannel,@gate,allpins,tpin;
PROPERTY DOUBLE w,l;

/* empty set -> basic set, by complement */
allnets = COMPLEMENT(allnets);
allpins = COMPLEMENT(allpins);
ADDMEMBER(tpin,"TGATE:PIN1","TGATE:PIN2");
/* N-transistors connected to VSS */
tn = CONNECT(@vss,@nchannel);
/* P-transistors connected to VDD */
tp = CONNECT(@vdd,@pchannel);
/* loop over all nets */
FORALL(n,allnets) {
  an = CONNECT(n,@nchannel);
  an = INTERSECTION(an,tn);
  if (COUNT(an) == 1) {
    /* N-transistor to VSS */
    ap = CONNECT(n,@pchannel);
    ap = INTERSECTION(ap,tp);
    if (COUNT(ap) == 1) {
      /* P-transistor to VDD */
      gx = CONNECT(ap,@gate);
      x = CONNECT(gx,@gate);
      x = INTERSECTION(x,an);
      if (COUNT(x) == 1) { /* gate short */
        nload = HULL_FORWARD(n,tpin,*);
        nload = UNION(nload,n);
        load  = CONNECT(nload,allpins);
        load  = DIFFERENCE(load,an);
        load  = DIFFERENCE(load,ap);
        if ((COUNT(load) > 8) &&
            (GET(an,w) / GET(an,l) < 2 ||
             GET(ap,w) / GET(ap,l) < 4))
          error = UNION(error,n);
      }
    }
  }
}
```

The third example deals with the function extraction of a switch network, which in our case is the P-core of a CMOS circuit. The extracted function expression indicates, whether the switch network allows passing. This depends on the gates of the switch transistors. The analogous extraction of the N-core net functions would enable us to check the duality of a CMOS circuit. The function extraction is performed as follows: starting with the VDD nets, a wavefront is propagated. For each net N, that was visited over a transistor T, which connects N to the previous net P, the new function property (func) of N is calculated as follows (gate(T) is the name of the net connecting to the gate of T):

$$func(N) := func(N)\ OR\ (func(P)\ AND\ gate(T))$$

All net function properties are preset to "0". The result of the operation will be that the function properties of all nets in the P-core are set to the calculated boolean expressions. In the beginning the function properties of the VDD nets are set to "1", since the pass from VDD to VDD is open by definition. A net is not further propagated, if it connects to an N-transistor (i.e. over pin set @nchannel).

```
COM olddev,newdev,devs,cut,z;
NET @vdd,oldspi,newspi,spis,x,g,s;
PIN @pchannel,@nchannel,@gate;
/* function expression as property */
PROPERTY FUNC pfun;

oldspi = @vdd;
/* by definition */
PUT(oldspi,pfun,"1");
DO {
  FORALL (x,oldspi) {
    /* net -> component */
    devs = CONNECT(x,@pchannel);
    devs = DIFFERENCE(devs,olddev);
    newdev = UNION(newdev,devs);
    FORALL(z,devs) {
      /* component -> net */
      spis = CONNECT(z,@pchannel);
      s = DIFFERENCE(spis,x);
      newspi = UNION(s,newspi);
      g = CONNECT(z,@gate);
      PUT(s,pfun,GET(s,pfun) OR
                (NAME(g) AND
                 GET(x,pfun)));
      cut = CONNECT(s,@nchannel);
      if (COUNT(cut) > 0) {
        /* no further propagation */
        newspi = DIFFERENCE(newspi,s);
      }
    }
  }
  olddev = newdev;
  oldspi = newspi;
  CLEAR(newdev);
  CLEAR(newspi);
} WHILE (COUNT(oldspi) > 0);
```

4 Hierarchy Preprocessing

The described approach to netlist design rule checking is non-hierarchical. The way to process different and especially mixed levels of abstraction is to form a flat cell with instances of the required hierarchy level. This might be a netlist of transistors, logic gates or arbitrary blocks, as well as a mixed description. Two methods are available to accomplish an appropriate level of abstraction: expansion and pattern matching.

For checking hierarchical schematics, special expansion routines accomplish the hierarchy preprocessing. The implemented expander uses a list of cell types, which must not be expanded. Operating top down, it skips these cells from further handling.

If flat (e.g. extracted) circuits have to be checked, graph-based netlist pattern matching [11] is employed to achieve the requested hierarchy level in bottom up direction. Arbitrary structures of higher levels are identified in a lower level netlist in a define-by-example manner. The patterns are described by small netlists, usually held within in a library. The general matching routine identifies all occurrences of the pattern in the lower level netlist, and replaces them by the respective subcircuit calls. This is done in an iterative manner. Details can be found in [11]. An example is supplied to illustrate this method. For instance, a scan path has to be verified in a flat transistor netlist. The first level of abstraction would be achieved by matching the basic logic gates (e.g. nands, nors etc.). In the next step all gates, forming flip-flops, are identified and replaced by flip-flop symbols. The final stage extracts the scan path flip-flops, which have additional multiplexers for switching its inputs, according to "scan"- or "normal"-mode. On this level of abstraction, it is easy to follow the sequence of a scan path.

5 Implementation and Results

The interpreter was implemented in C and parses netlists and command/script input in a "yacc-alike" manner. The SPICE syntax was chosen for the netlist format and the system was run under UNIX© on a SUN SPARC2^{TM} station. The set-based netlist rule checking was successfully applied to several real-life examples of different circuit technologies and design styles. Table 1 presents the results for four rule checking examples and two circuits. Further improvements in the implementation could be made by introducing new set defining commands into the interpreter. These would be implemented by efficient graph algorithms on the netlist data structure. Candidates are for instance algorithms for circle detection, longest path, shortest path and flow evaluation.

6 Conclusion

A new approach to general netlist rule checking is proposed, and applications to electrical rule checking and DFT rule checking are presented. Examples indicate the efficiency of execution and the flexible and convenient formulation of checks on an abstract level. Employing an interpreter for the checking management enables comfortable debugging, by allowing arbitrary switches between file-driven and interactive command entry. In a hierarchy preprocessing, a suitable level of abstraction is achieved by expansion (top down) or netlist pattern matching (bottom up). One drawback of the approach is, that it is still non-hierarchical. Further work could be related to develop a mechanism to propagate information between cells in a hierarchy. Other interesting topics would be the mixing of functional and electrical aspects for verification and an advanced device dimensioning.

circuit	#1	#2
# components (transistors, resistors etc.)	18951	203566
# nets	9283	68575
CPU 12 electr. rules (tab. 2)	10 sec	102 sec
virtual memory	8.1 MB	65.3 MB
CPU function extraction (example 3, chapter 3)	20 sec	93 sec
virtual memory	8.7 MB	70.2 MB
additional CPU for parsing the circuit	25 sec	245 sec
# components (logic gates)	2893	54147
# nets	2224	46996
CPU 5 DFT rules (tab. 2) incl. example 1, chapter 3	6 sec	65 sec
virtual memory	1.8 MB	22.1 MB
additional CPU for parsing the circuit	2 sec	120 sec
# components (buffer-transistors and logic gates)	3074	76521
# nets	2225	46994
CPU fanout check (example 2, chapter 3)	3 sec	57 sec
virtual memory	2.1 MB	33.8 MB
additional CPU for parsing the circuit	2 sec	157 sec

Table 1: Performance of netlist rule checking.

Acknowledgements

The author wishes to thank an unknown referee who did a very careful reading and triggered a rewriting of the first example script.

Many thanks also to F.J. Zappe, who did parts of the coding, G. Hess, Dr. U. Roettcher, Dr. R. Lackmann and C. Schmitz for valuable discussions and Prof. Dr. G. Zimmer for the given support.

References

[1] V.D. Agrawal, S.K. Jain and D.M. Singer, "A CAD System for Design for Testability", *VLSI Design*, 46 - 54, Oct. 1984.

[2] K. Karplus, "Exclusion Constraints for Digital MOS Circuits: A New Set of Electrical Design Rules", *Proc. ICCAD*, 244 - 246, 1985.

[3] H.J. De Man, I. Bolsens, E. vanden Meersch and J. van Cleynebreugel, "Dialog: an Expert Debugging System for MOS VLSI Design", *IEEE Transactions on CAD*, Vol. CAD-4, No. 3, 303 - 311, July 1985.

[4] C. Lob, R.S. Spickelmier and A.R. Newton, "Circuit Verification using Rule-based Expert Systems", *Proc. of the ISCAS*, 881 - 884, 1985.

[5] S. Bergquist and R. Sparkes, "QCritic: A Rule-Based Analyzer for Bipolar Circuit Designs", *Proc. of the CICC*, 617 - 620, 1986.

[6] R.L. Spickelmier and A.R. Newton, "Critic: A Knowledge-Based Program for Critiquing Circuit Designs", *Int. Conf. on Computer Design, VLSI in Computers*, 324 - 327, 1988.

[7] J. Wenin, J. Verhasselt, M. Van Camp, J. Leonard and P. Guebels, "Rule-Based VLSI Verification System Constrained by Layout Parasitics", *Proc. 26th Design Automation Conference*, 662 - 667, 1989.

[8] C. Godoy, G.B. Franklin und P.S. Bottorf, "Automatic checking of logic design structures for compliance with testability ground rules", *Proc. 14th Design Automation Conference*, New Orleans, LA, 469 - 478, June 1977.

[9] C. Bamji and J. Allen, "GRASP: A Grammar-based Schematic Parser", *Proc. 26th Design Automation Conference*, 448 - 453, June 1989.

[10] D. Bhavsar, "Design for Test Calculus: An Algorithm for DFT Rules Checking", *Proc. 20th Design Automation Conference*, 300 - 307, 1983.

[11] G. Pelz and U. Roettcher, "Circuit Comparison by Hierarchical Pattern Matching", *Proc. ICCAD*, Santa Clara, 290 - 293, 1991.

Appendix

• all nets have a path to VDD over transistors channels or resistors. • all nets have a path to VSS over transistors channels or resistors. • no transistor gate is connected to SUPPLY. • no transistor's source and drain are shorted. • no N-transistor is connected to VDD. • no P-transistor is connected to VSS. • no N-transistor bulk is connected to VSS. • no P-transistor bulk is connected to VDD. • no diode cathode is connected to VDD. • no diode anode is connected to VSS. • no diode cathode and anode are shorted. • no net is connected to just one device.
• the combinational depth at the output of a scan path memory element is not greater than 6. • the scan memory elements form a valid sequence. • all scan memory elements of a chain share same SCAN-ENABLE signal (possibly over buffers). • all scan memory elements of a chain share same SCAN-CLOCK signal (possibly over buffers). • no reconvergences appear, affecting on the control pins of the scan memory elements.

Table 2: Examples: 12 electr. rules, 5 DFT rules.

Hierarchical Pitchmatching Compaction Using Minimum Design

Cyrus S. Bamji and Ravi Varadarajan
Cadence Design Systems
555 River Oaks Parkway
San Jose, CA 95134

Abstract

A new hierarchical compactor capable of compacting and pitchmatching hierarchically defined layouts is described. The hierarchical compactor can handle most input hierarchies including multi-level hierarchies, over the cell routing and cell rotations and reflections. The compactor simultaneously compacts the contents of all the cells of the layout hierarchy maintaining the hierarchy of the input layout as well as the pitchmatching and abutment constraints between the cells. The hierarchical compactor automatically factors out the regularity in the layout and performs almost all of its operations on a "minimum design". This novel and unique formulation of the hierarchical compaction problem enables compaction time to be a function of the irregularity rather than the size of the layout.

1 Overview

1.1 Background of the Approach

Layout compaction is the process of taking an existing symbolic layout and producing a new layout in which some geometrical aspect (usually size) is minimized. This process preserves the underlying circuit integrity and enforces *design rule* correctness. The compactor frees the layout designer from having to produce an optimized layout thus providing an automated way to generate area efficient layouts. Because the input for compaction does not have to be design rule correct it can also be used to migrate a layout from one set of design rules to another.

Since most IC input layouts are hierarchical, it is essential that the compactor accept hierarchical designs and produce compacted layouts which preserve the input hierarchy. This avoids an explosion in database size, facilitates characterization and allows for easier modification of the compacted layout.

The mathematical formulation of the *true* hierarchical compaction problem is an integer linear programming problem. This class of problems has a high computational complexity and using *brute force* integer linear programming techniques to compact even moderate size layouts is prohibitively expensive.

1.1.1 Hierarchical Compaction using Leaf Cell Compactors

Solving the compaction problem for a layout devoid of hierarchy is called leaf cell compaction. Leaf cells compaction has been extensively studied and reported in the literature [2] [4] [13]. Because leaf cell compaction is comparatively well understood, it is often used to compact layouts with hierarchy. These methods fall into the categories described below.

- Flatten the input hierarchy and compact the resulting layout with a leaf cell compactor. This causes an explosion of the database size for the compacted output and significantly degrades the run time performance of the compaction process. Also since the input hierarchy is lost, characterization and modification of the output layout is more difficult.
- Some researchers have used bottom up approaches to hierarchical compaction [3]. In this approach, leaf cell compaction is invoked bottom up. At each level of the hierarchy all subcells at lower levels of the hierarchy are assumed to be rigid objects. This destroys the connectivity at the next higher level which must then be reestablished by *river routing* before compaction of the next level can proceed. Since cell abutment is not maintained, unacceptable results are produced in designs where cells are assembled predominantly by abutment.
- Other researchers have attempted to de-couple the cell abutment and leaf cell compaction phases. In this approach port positions of all the leaf cells are determined without taking into account the detailed constraints within the leaf cells. A leaf cell compactor is then used to compact the individual leaf cells with user specified constraints to fix the above mentioned port positions. However, there is no guarantee that the leaf cell compactor will find a feasible solution and overconstraints may result due to the fixed constraints on the ports. This methodology is a time consuming trial and error process and does not guarantee a solution due to the XY-*interlock* problem[6]. Moreover, the resulting solution is usually sub-optimal.

1.1.2 True Hierarchical Compaction

- The true hierarchical compaction problem can be posed as an integer linear programming problem and the Simplex algorithm can be used to solve the system of linear inequalities[10]. However, the computational complexity of this approach is extremely high and the size of hierarchical layout that can be handled is severely limited by the capacity of the Simplex algorithm. Moreover, the number of variables and constraints that need to be solved by the Simplex algorithm grows rapidly with the database representation of the hierarchy.
- Due to the computational complexity of linear programming techniques, other researchers [9] have attempted to solve the hierarchical compaction problem in the constraint graph domain using heuristic approaches. This methodology does not guarantee an optimal solution.

0738-100X/92 $3.00

1.2 Definition Of The Hierarchical Compaction Problem

True Hierarchical Compaction is defined as minimizing the area of a hierarchical layout while preserving input hierarchy, design rule correctness and electrical connectivity between components and subcells. Cell instances that are connected by abutment in the input must remain abutted after hierarchical compaction. This feature is often referred to as ***pitchmatching***. The true hierarchical compactor must handle the entire task globally. The constraints within each of the leaf cells must be solved in conjunction with the constraints between various cells across the design hierarchy.

The mathematical formulation of the hierarchical compaction problem can be expressed succinctly by the following integer linear program [10]:

$$\textbf{Minimize} \quad (X_n - X_0)$$

$$\textbf{Subject to:} \quad \begin{array}{rcl} Ax & \geq & b \\ x & \geq & 0 \end{array}$$

x is a vector of positions for all elements in all the cell instances, and X_0 and X_n are the source and sink positions of the root cell. A is the constraint coefficient matrix. Each row A_i of A represents a single constraint. The vector b represents the constraint spacing amounts, where b_i is the value of the constraint A_i.

Leaf cell compaction (or ***flat compaction***) is a special case of the above integer linear program in which the constraint coefficient matrix A is totally unimodular and each constraint A_i has exactly two variables which have unity coefficients. Due to this restricted nature of the constraints the leaf cell compaction problem can be trivially transformed from an integer linear programming problem to finding the shortest paths in a directed graph [8], for which efficient solutions exists.

In hierarchical compaction, the pitchmatching constraints between cells coupled with the constraints that are needed to preserve the input hierarchy results in a very general constraint coefficient matrix. Hence, the hierarchical compaction problem cannot be transformed into an equivalent problem in a constraint graph and integer linear programming techniques are required to solve the problem optimally.

Integer linear programming problems fall into a class of problems that are NP complete [11]. Hence most researchers have used linear programming techniques (LP) to solve the problem and perform integer round off if necessary using heuristics [10]. However existing linear programming methods are also extremely intensive in both space and time, and thus severely limit the complexity and the size of design hierarchy that can be handled. The unique and major thrust of our approach is to drastically reduce the number of constraints that need to be solved by linear programming methods.

1.3 Our Approach

This paper describes a hierarchical compactor, capable of automatically compacting and pitchmatching hierarchically defined layouts. The compactor accepts most input hierarchies and maintains this hierarchy in the output. Also the output hierarchy accurately reflects the input in that cells that are connected by abutment in the input hierarchy remain abutted in the output hierarchy. Connectivity and design rule correctness between cells across levels in the hierarchy, and for over the cell routing at any level in the hierarchy, is maintained. Cell reflections and rotations in the input hierarchy are preserved in the compacted output.

Our approach works on the unique concept of ***minimum design***. From the input layout hierarchy the compactor automatically extracts a reduced representation

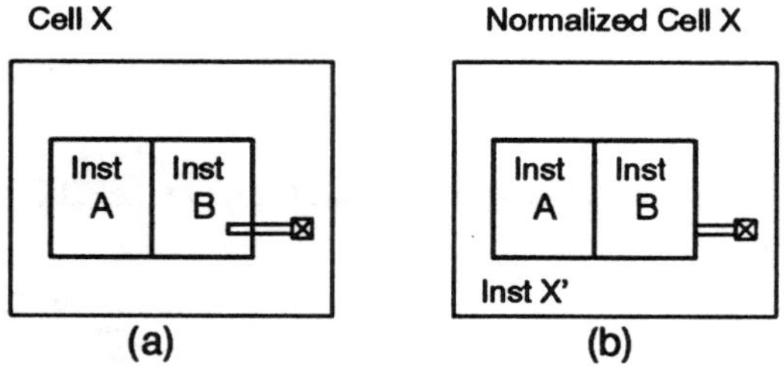

Figure 1: Hierarchy Normalization

of the input called the ***minimum design***. The ***minimum design*** captures all the necessary features of the layout required for performing hierarchical compaction of the entire input hierarchy. From a compacted version of the minimum design a compacted version of the entire hierarchy can be easily constructed. Hence instead of compacting the entire input hierarchy, it suffices to compact its minimum design which, depending on the amount of regularity in the layout, can be orders of magnitude smaller than the input hierarchy. This substantial reduction in the size of the problem that needs to be solved makes compaction of the very large modules that occur in today's layouts feasible.

The ***minimum design*** methodology is based on a strong mathematical foundation. It uses non-heuristic techniques for which efficient solution methods are available. It is guaranteed to find the optimal solution for the compacted hierarchy if it exists. The set of equations extracted via the minimum design methodology that needs to be solved by LP is provably minimal.

2 Hierarchical Pitchmatching Compaction by Minimum Design

The hierarchical compactor accepts most multilevel hierarchies. Each master cell at any level in the hierarchy can contain instances of other cells (which can stretch) as well as primitive geometries such as devices and wires at lower levels of the hierarchy. The primitive geometries are allowed to overlap the instances as occurs in ***over the cell routing***. Electrical connections between cells are established via distinguished objects called ports.

Compaction is performed on a ***normalized 2 level*** version of the input hierarchy. First the hierarchy is modified into its corresponding ***normalized 2 level version***. Compaction constraints are then generated and solved for this version, producing a ***normalized 2 level*** version of the compacted output. The final compacted output hierarchy is then reconstructed from the compacted ***normalized 2 level*** version.

In the ***normalized*** version of an input hierarchy, cells are either pure hierarchy or leaf cells. Pure hierarchy cells contain only instances of other cells. Leaf cells contain only primitive geometrical shapes and devices. Cells in the input hierarchy such as cell X in Figure 1(a) which contains instances as well as primitive shapes and devices are ***normalized*** by creating a new master cell X', shown in figure 1(b), which contains all of the primitive geometrical shapes and devices in Cell X that do not overlap with other cell instances in X. The ***normalized*** hierarchy is shown in figure 1(b).

Cells at different levels in the input hierarchy may be electrically connected together. Figure 2(a) shows a hierarchy tree and figure 2(b) its corresponding hierarchical layout. In this example the instance of cell C in level 1 is electrically connected to the instance of cell E in level 2. For optimum hierarchical compaction, the compaction constraints describing the electrical and geometrical interactions between cell instances across

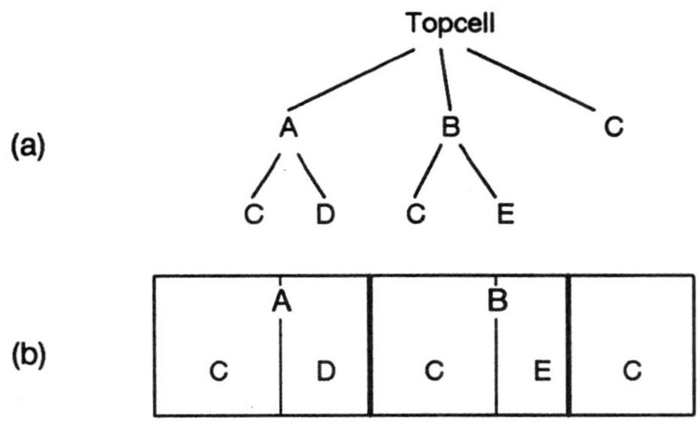

Figure 2: *2 Level* Normalization

all levels of hierarchy should be solved simultaneously. Therefore for the purpose of generating compaction constraints, the design hierarchy is examined as a 2 level hierarchy consisting of instances of the leaf cells. This version of the hierarchy is referred to as the *normalized 2 level* version.

The computation time for hierarchical compaction is heavily dominated by the time required to solve the system of linear inequalities via LP. Because of the large time and space complexity associated with LP, the size of the system of linear inequalities must be kept as small as possible. In our approach only a very small fraction of the total number of linear inequality constraints present in the input layout is actually solved via LP. The vast majority of constraints are either factored out or solved by more efficient graph based methods. Regularity in the input is heavily exploited and is automatically extracted from the layout. Repetitive patterns give rise to a single set of equations regardless of the number of times they appear in the layout. Hence hierarchical compaction time grows proportionally to the *irregularity* in the layout and not the size of the layout.

In our approach the set of linear inequality constraints that describe the hierarchical compaction problem is divided into three classes of equations. Each class is derived from a specific kind of geometrical constraint occurring in hierarchical compaction. Combined, these three classes can yield a set of equations equivalent to the formulation described in section 1.2. With each class is associated a set of efficient reduction techniques that are specifically tailored to exploit the particular characteristics of equations in that class. This partitioning of equations makes regularity in the layout apparent. Regularity appears in the form of linear dependence of the equations within one of the classes and is factored out by selecting a set of linearly independent equations within that class. This process corresponds to extracting the *minimum design.*

The three classes of constraints are:

1. Intracell constraints. These constraints arise from constraints between shapes and devices within each of the leaf cells. These constraints are necessary and sufficient to preserve the integrity of the leaf cells.
2. Intercell constraints. These constraints arise from electrical and geometrical interactions between the leaf cells. These constraints are necessary and sufficient to preserve abutment and maintain connectivity and design rule correctness across the cells in the hierarchy.
3. Loop constraint. These constraints capture the geometric pattern of instances in the design.

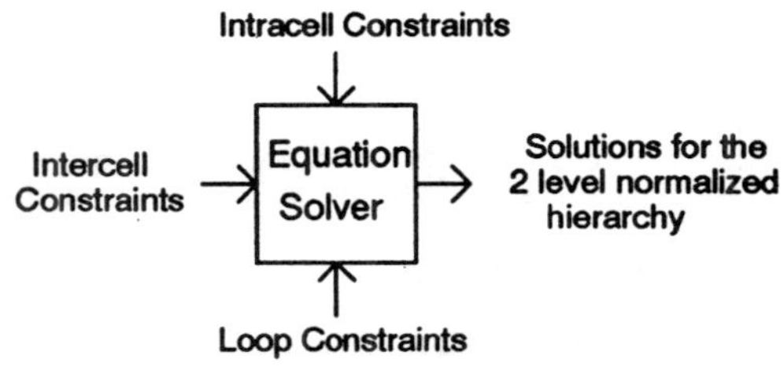

Figure 3: System of Equations

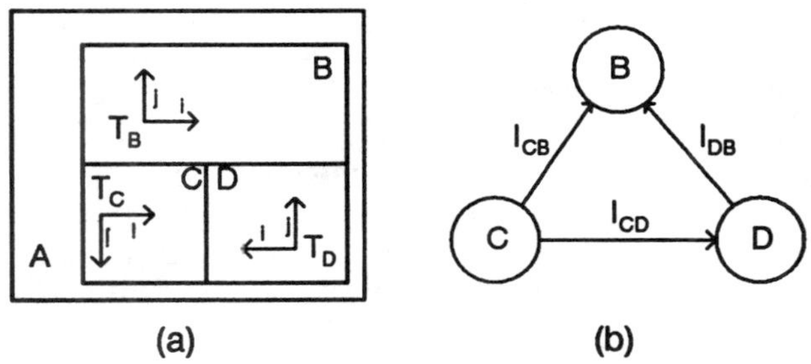

Figure 4: Interface Graph

The three sets of constraints together constitutes the system of linear inequalities that need to be solved by LP as shown in figure 3. The solution to the system yields values to the variables that are required to reconstruct the *normalized 2 level* version of the compacted output.

2.1 Interface Graph

Interface graphs, first introduced in [1], provide a simple yet powerful representation for layouts. The interface graph is especially useful when dealing with the placement of the cells in the layout. The interface graph is a better representation than the layout itself for capturing the relative placement of instances in a master cell. This section quickly reviews the basics of interface graphs.

2.1.1 Interfaces

If instances of cells B and C are instantiated in the same coordinate system (same cell) then B and C have an interface between them. The interface defines a legal relative positioning of cells B and C. The interface is an affine transformation that defines the relationship between the transforms T_B of the instance of B and T_C of the instance of C. The interface defined by the transforms T_B and T_C is $I_{BC} = (T_B)^{-1}T_C$. From of the transform T_B of B the transform T_C of C is obtained by $T_C = T_B I_{BC}$. Cells B and C may have more than one defined interface between them. In this case the interfaces are labeled $I^1_{BC}, I^2_{BC}, I^3_{BC}$ *etc.* Figure 4(a) shows instances of cells B, C and D called within cell A. This layout has the three interface transforms $I_{CB} = (T_C)^{-1}T_B$, $I_{CD} = (T_C)^{-1}T_D$ and $I_{DB} = (T_D)^{-1}T_B$

Interfaces between leaf cells in the *normalized 2 level* version of the hierarchy are called primitive interfaces. The system of equations which describe the *intercell* and *loop* constraints is written in terms of these interfaces and solving the system provides values for them. The other interfaces are non-primitive interfaces and their transforms are computed in terms of the transforms of the primitive interfaces as explained in Section 2.5.

2.1.2 Interface Graphs

A interface graph $G = (V_G, E_G)$ is a directed graph in which the vertices V_G represent instances of cells and the edges E_G represent interfaces between them. Figure 4(b) shows the interface graph for the layout of Figure 4(a). Each vertex corresponds to an instance of a cell and is labeled with the master cell it corresponds to. The edges are labeled with the corresponding interfaces in Figure 4(a).

An interface graph is transformed into a layout equivalent via the interface graph to layout transformation described in [1]. Each non leaf cell in the *normalized hierarchy* is described by an interface graph and the hierarchy is reconstructed from these interface graphs as described in section 2.5. An interface graph for the *normalized 2 level hierarchy* is also constructed. This graph is referred to as the *leaf interface graph.*

2.2 Minimum Intra cell constraints

For every master cell in the *normalized 2 level* hierarchy, a set of constraints called *Intracell* constraints is generated. These constraints are necessary and sufficient to preserve the integrity of the leaf master cells during hierarchical compaction. Since these constraints form a part of the system of linear inequalities that need to be solved by LP, it is essential to keep their number as small as possible.

The electrical and geometrical interactions between the various shapes and devices in each of the master cell is captured by a directed constraint graph $G_c = (V_c, E_c)$ [8] [9]. The vertices in the constraint graph correspond to circuit elements and the directed edges represent the constraints between pairs of elements. These constraints arise from design rule, connectivity and user specified constraints within the leaf cell. An edge e_{ij} of value d from vertex v_i to vertex v_j represents the linear inequality $v_i + d \leq v_j$. In theory the linear inequalities representing all the directed edges of G_c for all the leaf masters should be solved in LP. However, this would be prohibitively expensive and would severely limit the size of the hierarchies that can be efficiently handled. For every leaf master, a series of transformations in the graph domain on the constraint graph G_c is performed. These transformations generate a minimum set of inequalities (*intracell constraints*) which accurately capture the essence of the constraints in G_c. These *intracell constraints* are expressed solely in terms of the *ports* and describe the ways in which the corresponding leaf master can stretch during the hierarchical compaction process.

The electrical and geometrical interactions between various cells in the hierarchy occurs through a small subset of elements *(ports)* within each of the leaf master cell. The vertices in the constraint graph, corresponding to these elements, are called the *port vertices* P_c. For each constraint graph G_c we derive a minimum set of equations between the port vertices P_c, which captures the essence of all the constraints in the graph. This is performed in two steps. In the first step we create a transitive closure graph G_{pc} on the *port vertices* P_c by finding the longest paths between every ordered pair of ports [9] [12]. We then perform a second transitive closure on G_{pc} which further reduces the number of constraints to generate the minimum set of intracell constraints. The number of *intracell constraints* after the two transformations is of the order $O(p)$ where p is the number of ports in P_c, and is in the same range as the number of constraints produced by the greedy method [5].

Figure 5(a) shows the constraint graph for a leaf

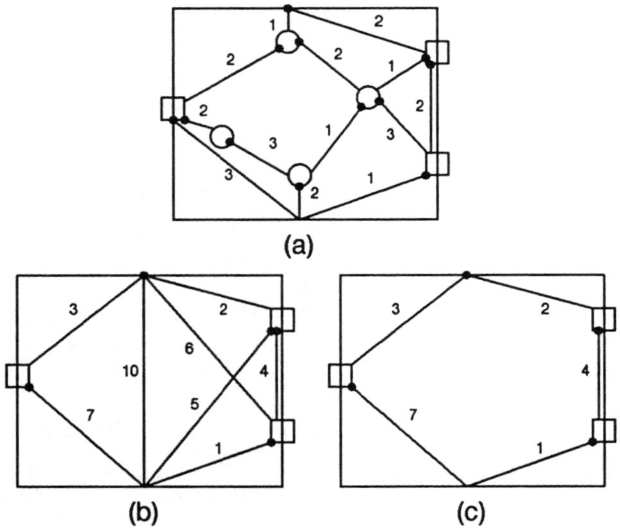

Figure 5: Intracell constraints

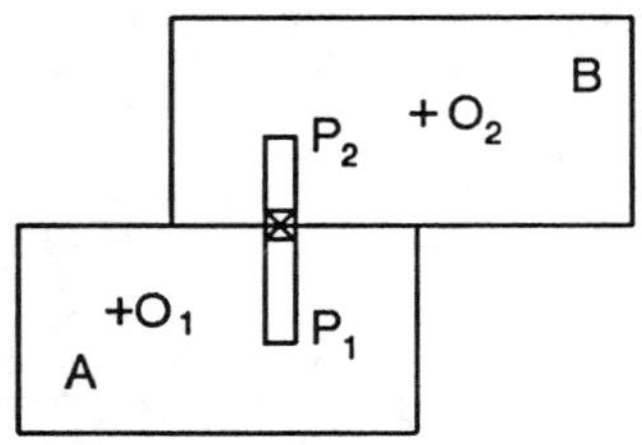

Figure 6: Intercell constraints

cell. The rectangular vertices corresponds to the *ports* of the leaf cell and the circular vertices represents the internals of the cell. The transitive closure graph G_{pc} on the port vertices is shown in figure 5(b). The final graph after performing a second transitive closure on G_{pc} is shown in figure 5(c). The edges in the final graph correspond to the minimum *intracell constraints* for this cell.

The set of all *Intracell constraints* for all the leaf cells along with the constraints described in section 2 are solved in LP. The values of the port variables as found by LP are fixed in the constraint graph for each of the master cells. The graphs are then solved to propagate the port constraints to the interior of the leaf cells. This operation is guaranteed to find a solution [9] and is free of the XY-*interlock* problem [6].

2.3 Minimum Inter cell constraints

For every distinct *primitive interface* in the *normalized 2 level* hierarchy a set of constraints called *intercell constraints* is generated. These constraints are necessary and sufficient to preserve cell abutment and maintain connectivity and design rule correctness across the cells in the design hierarchy. By forcing every occurrence of an interface in the design hierarchy to be identical in the output, only one set of equations per interface need to be generated. Thus the number of *intercell constraints* in the system of linear inequalities that need to be solved by LP is proportional to the number of distinct interfaces and not the number of instances in hierarchy.

Figure 6 shows an interface I_{AB} between leaf cells A and B. P_1 and P_2 are the port variables of A and B respectively that need to be connected. However, P_1 and

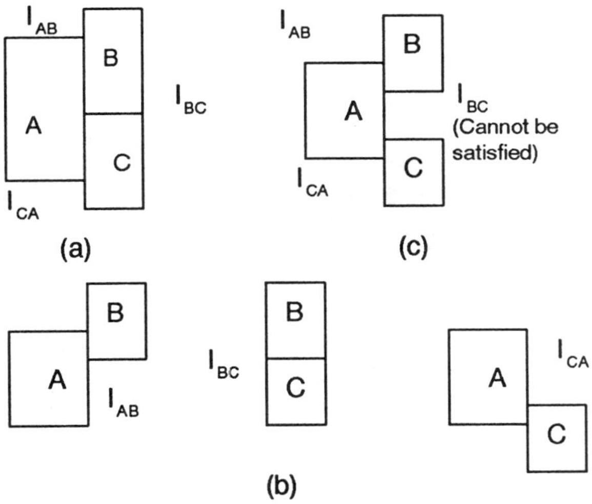

Figure 7: Loop Constraints Problem

P_2 are variables in the coordinate system of the corresponding leaf cells and an ***intercell constraint*** is needed to tie them together. If O_1 denotes the origin of A and O_2 the origin of B, the intercell constraint between P_1 and P_2 can be written as $O_1 + P_1 = O_2 + P_2$. This formulation would cause an explosion in the number of ***intercell constraints*** since a set of equations is generated for every occurrence of all the interfaces. However, by forcing every occurrence of an interface in the design to be identical in the output, the relative position of the two instances in each interface is constrained to be the same. If we denote the offset value of the interface I_{AB} in the direction of compaction to be λ_{AB}, the ***intercell constraint*** between P_1 and P_2 can written in terms of λ_{AB} : $P_1 = \lambda_{AB} + P_2$. Thus, only one set of constraints need to be generated and solved for every distinct ***primitive interface*** in the design hierarchy.

In highly regular layouts such as *datapaths* and PLA's the number of distinct interfaces is very small, regardless of the size. Hence compaction time is small and remains relatively constant with respect to the size of the design.

2.4 Minimum Loop equations

The intercell constraints provide necessary and sufficient constraints for abutment, design rule correctness and connectivity to be maintained between two master cells taken in isolation. However they are not enough to maintain correctness in a real layout where each instance has to simultaneously satisfy constraints for several interfaces.

Figure 7(a) contains the three primitive interfaces I_{AB}, I_{BC} and I_{CA} between leaf cells A, B and C. After solving the ***intracell*** and ***intercell*** constraints a solution is found and the three resulting interfaces are shown in figure 7(b). Finally in figure 7(c) the compacted layout is put back together. However the layout of figure 7(a) cannot be reconstructed because the compacted interfaces I_{AB}, I_{BC} and I_{CA} cannot be satisfied simultaneously. In figure 7(c) I_{AB} and I_{AC} are satisfied but I_{BC} cannot be satisfied.

2.4.1 Definition of Loop Equations

Loop equations are equations between the interface offsets (λ's) that if satisfied guarantee that the compacted layout can be put back together with all interfaces simultaneously satisfied. Figure 8 shows instances A_1,

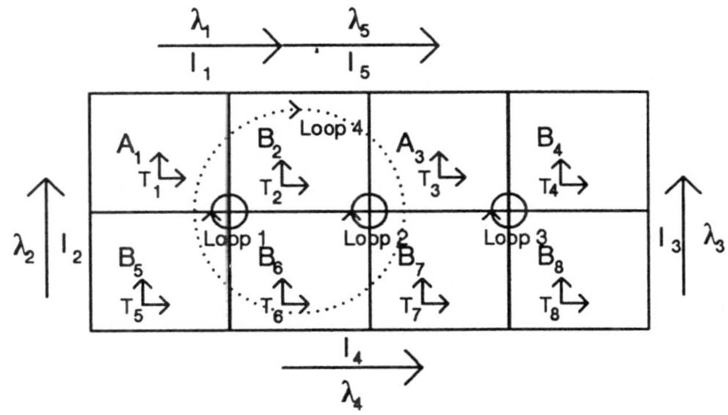

Figure 8: Loop Constraints

B_2 A_3, B_4, B_5, B_6, B_7 and B_8 of master cells A and B with transforms respectively T_1, T_2 T_3, T_4, T_5, T_6, T_7 and T_8. There are five interfaces I_1, I_2, I_3, I_4 and I_5 as indicated on the figure. For compaction in the X direction the corresponding interface offsets are[1] λ_1, λ_2, λ_3, λ_4 and λ_5. Going around loop 1 in figure 8 and expanding transforms using the interface relations from section 2.1.1 we have:

$$\begin{aligned} T_1 &= T_5 I_2 \\ &= (T_6 I_4^{-1}) I_2 \\ &= ((T_2 I_3^{-1}) I_4^{-1}) I_2 \\ &= (((T_1 I_1) I_3^{-1}) I_4^{-1}) I_2 \\ &= T_1 I_1 I_3^{-1} I_4^{-1} I_2 \end{aligned}$$

Hence

$I_1 I_3^{-1} I_4^{-1} I_2 = \textit{Identity}$

The above equation represents the constraint for interface transforms due to loop 1. This form of constraint is true for all loops in the layout. The transform of any instance can be expressed in terms of itself and the product of interface transforms in a loop path starting from and terminating on that same instance. Therefore the product of the interface transforms around all loops in the layout must be ***identity***.

In terms of the interface offsets which are the variables actually solved by LP, the loop constraints for loop 1 give rise to:

$\lambda_1 - \lambda_3 - \lambda_4 + \lambda_2 = 0$

In the presence of instances whose transforms have rotations and reflections, deriving the equations for the interface offsets from the interface transforms is slightly more complex.

2.4.2 Minimum Set of Loops

Loops in the ***normalized 2 level layout*** correspond to cycles in the corresponding leaf ***interface graph***. In theory an equation is needed for every possible loop in the layout hence every possible cycle in the graph. However the set of all possible loops can be derived from a small set of ***primitive loops***. For example loop 4 shown in dotted line in figure 8 can be derived by adding loops 1 and 2. Because the interface graph is planar it can be shown that the minimum set of loops corresponds to the faces of this graph. A modified version of Tarjans graph planarization algorithm [7] is used to find the faces of the planar ***interface graph***. From the faces of the graph the set of constraints on the interface offsets is computed.

---

[1] Note: λ_2 and λ_3 are offsets in the X dimension between the corresponding instance origins

2.4.3 Minimum Set of Loop Equations

Loops 1, 2 and 3 in figure 8 form the minimum set of loops for the layout of figure 8. These loops are independent in the sense that no one loop can be derived by adding the two others. The equations for loops 1 and 3 however both yield the equation $\lambda_1 - \lambda_3 - \lambda_4 + \lambda_2 = 0$. Hence although the set of loops given by the faces of the planar graph are independent the set of equations they correspond to are not.

Finding the minimal set of loop equations corresponds to finding a linearly independent subset of these equations from which all other equations can be obtained by linear combination. This is equivalent to finding the basis in a vector space given a generating set of vectors.

A modified version of the Schmidt orthonormalization algorithm is used for finding the basis equation set. In this procedure the basis set is initially set to null. Each loop equation is then tested to see if it is a linear combination of equations already in the basis set. If it is a linear combination of equations in the basis set, it is thrown away, otherwise it is added to the basis set. At the end of the procedure the basis contains a minimal set of equations from which all others can be generated by linear combination.

An orthonormalized version of the basis set is kept and updated using Schmidt orthonormalization whenever an equation is added to the basis set. The orthonormal basis is used to efficiently determine whether a candidate equation is linear combination of equations in the basis. Instead of testing whether the candidate equation is a linear combination of equations in the basis set the equivalent but much more efficient test of whether it is a linear combination of equations in the orthonormal set is performed.

2.4.4 Extracting the Minimum Design

Computing the minimum set of loops using the procedure described in section 2.4.2 followed by extraction of the minimal set of loop equations described in section 2.4.3 is referred to as ***extracting the minimum design***. Because the procedure of section 2.4.2 produces the smallest set of loop constraints and the procedure of section 2.4.3 produces the smallest set of loop equations the resulting set of loop equations is provably minimal.

2.5 Hierarchical Graph Reversion

Hierarchical graph reversion creates the compacted normalized hierarchy from the compacted ***2 level normalized*** hierarchy. Once values for all the port locations of the leaf cells and all the ***primitive interfaces*** are found via LP the compacted hierarchy must be constructed. Compacted versions of the leaf cells are obtained using the port position via the procedure described in Section 2.2. From the compacted leaf cells and the values of the interfaces between them (the ***primitive interfaces***) the entire normalized hierarchy is reconstructed using the hierarchical graph reversion algorithm described in this section.

2.5.1 Interface Inheritance Relations

Figure 9 shows a layout of cell X which contains instances of cells A and B. Cell A in turn contains an instance of cell C and cell B an instance of cell D. Cell A and B have interface I_{AB} between them. This induces an interface I_{CD} between cells C and D inside A and B. This is the interface C and D will have, if the cells A and B are ***flattened***. It can be shown [1] that

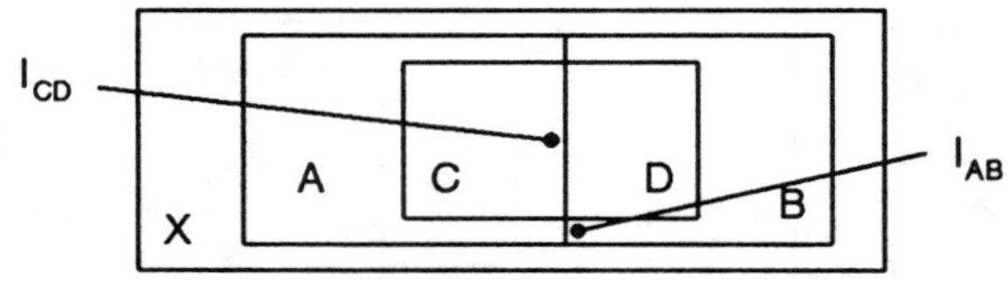

Figure 9: Interface Inheritance

I_{AB} can be expressed in terms of the induced interface I_{CD} by the relation:

$I_{AB} = T_C^{-1} I_{CD} T_B$

where T_C (respectively T_D) is the relative transform of the instance of C (respectively D) in A (respectively B). The above equation provides an ***inheritance relation*** for interface I_{AB}. Each ***non-primitive interface*** has an inheritance relation.

2.5.2 Hierarchical Graph Traversal

Each cell in the hierarchy has an interface graph associated with it as described in section 2.1.2. If compacted versions for all the subcells of a hierarchical cell X exist and values for all the interface transforms in the interface graph of X are known, cell X can be reconstructed using the graph to layout transformation described in [1]. This transformation fixes the transforms of all the instances in X.

Transforms for ***non-primitive interfaces*** in the interface graph of X are not provided by LP. Interface inheritance relations are used to compute these interfaces in terms of the ***primitive interfaces*** whose values are provided by LP.

Hierarchy reconstruction can be viewed as taking place bottom up. The graph to layout transformation of cell X takes place only after the graph to layout transformation for all its subcells has occurred. Using interface transform inheritance relations all ***non-primitive interfaces*** (I_{AB} in Figure 9) in the interface graph of X can be expressed in terms of transforms of instances in subcells of X (cells C and D in Figure 9) and interfaces between those subcells. Since all the subcells of X have been computed (cells A and B in Figure 9) the transforms of all instances in those subcells are known. The interface transforms between those subcells (I_{CD} in Figure 9) may in turn need to be computed in terms of their inheritance relations. The recursion terminates when a primitive interface (whose value is given by LP) is found.

In the actual implementation of the hierarchical reversion algorithm the graph to layout transformation is applied to the interface graph of the root cell of the hierarchical layout. If a vertex whose corresponding master cell has not been created is encountered, the hierarchical graph to layout transformation is first performed on the master cell of that vertex. If an an edge whose interface transform is unknown is encountered, it is evaluated using interface inheritance relations. Using this technique applying the hierarchical graph to layout transformation on the root cellview interface graph recursively triggers this transformation to be applied to all the master cells in the hierarchy. When the process completes on the interface graph of the root cell in the hierarchy, reconstruction of the entire normalized hierarchy is complete.

2.6 Outline of the Algorithm

An outline of the algorithm is shown in figure 10. First the input design is normalized to create a ***normalized 2 level*** version of the hierarchy. Interface graphs

Normalize hierarchy
Create Interface Graph for each master
Generate Loop Constraints
For each direction of compaction
    For each Leaf Master Cell
        Create Leaf Cell Constraint Graph
        Generate Intra cell constraints between ports
    For each distinct primitive interface
        Generate intercell constraints
    Solve the system of equations in LP
    For each Leaf Master Cell
        Propagate port positions to the internals of cell
Perform Hierarchical Graph reversion
De-Normalize Hierarchy to create compacted output

Figure 10: Outline of the Algorithm

are then created for each master cell and the ***normalized 2 level*** hierarchy. The ***leaf interface graph*** described in section 2.1.2 is then examined to extract the ***minimum design*** and generate the minimum set of ***loop equations***. As the hierarchical compactor does not create any new interfaces the set of loop equations for each compaction direction does not change with multiple compaction XY iterations.

The hierarchical compactor is essentially one dimensional, in that it first compacts and pitchmatches globally in one dimension and then performs the same operations in the other dimension. For each leaf master cell the hierarchical compactor first generates the constraint graph describing the constraints between the elements in the cell. The minimum set of ***intracell constraints*** between the port elements of that leaf cell is then generated. Next, for each distinct ***primitive interface*** the set of ***intercell constraints*** are generated. These three sets of equations (***intracell, intercell and loop***) form the system of linear inequalities that is solved by LP. The solution to the system yields values to the port variables of the all the leaf master cells and the offset values (λ's) to the ***primitive interfaces*** in the direction of compaction. The values of the port variables of all the leaf master cells are then propagated to the internals of the leaf cells. This sequence of operations is alternated between X and Y direction to further reduce the size of the layout.

Finally the hierarchical compactor performs hierarchical graph reversion described in section 2.5 to obtain the values of ***non-primitive interfaces***. The hierarchy is then de-normalized to recreate the input hierarchy.

3 Results

The hierarchical pitchmatching compactor has been tried out on several large industrial designs. These designs are multi-level hierarchies containing abutted cells, routing, over the cell routing and rotations and reflections of cell instances. Table 1 summarizes the computation time and the LP problem size for each test case. The compaction was performed on a SPARCstation 1 with 24 Mbytes of main memory. Cell T1 is part of a custom module and cells T1x16 and T1x256 are T1 replicated 16 and 256 times respectively. Cell T2 is a part of a datapath. As expected, replicating portions of a design does not increase the computation time significantly. Also compaction time is largely independent of circuit size.

4 Conclusion

In this paper we have described a new hierarchical pitchmatching compactor that can handle most practically occurring input hierarchies. The hierarchical compactor works on the unique concept of ***minimum design***, whereby it automatically factors out the regularity in the layout and performs almost all of its operation on a minimum representation of the design hierarchy. This novel formulation of the hierarchical compaction problem enables compaction time to be a function of the irregularity rather than the size of the layout.

Name	Xtrs	Insts	Vars	Eqns	CPU (sec)
T1	352	5	30	37	21
T1x16	5632	80	32	42	22
T1x256	90112	1280	32	42	33
T2	384000	20500	78	146	348

Table 1: Compaction Time

References

[1] C. Bamji, C. Hauck, and J. Allen. A design-by-example regular structure generator. In *ACM IEEE 22nd Design Automation Conference*, 1985.

[2] D. Boyer. Symbolic layout compaction review. In *ACM IEEE 25th Design Automation Conference*, 1988.

[3] J. Burns and R. Newton. SPARCS: A new constraint-based IC symbolic layout spacer. In *Proceeding of the IEEE Custom Integrated Circuits Conference*, 1986.

[4] J. Doenhardt and T. Lengauer. Algorithmic aspects of one dimentional layout compaction. *IEEE Transactions on Computer aided Design*, CAD-6:863–878, 1987.

[5] D. Dutt and C. Lo. On minimal closure constraint generation for symbolic cell assembly. In *ACM IEEE 28th Design Automation Conference*, 1991.

[6] P. Eichenberger. *Fast Symbolic Layout Translation for Custom VLSI Integrated Circuits*. PhD thesis, Stanford, 1986.

[7] J. Hopcroft and R. Tarjan. Efficient planarity testing. *Journal of the ACM*, 21-4:549–568, 1974.

[8] Y. Liao and C. Wong. An algorithm to compact VLSI symbolic layouts with mixed constraints. *IEEE Transactions on Computer aided Design*, CAD-2:62–69, 1983.

[9] C. Lo and R. Varadarajan. An $O(n^{1.5}\log n)$ 1-d compaction algorithm. In *ACM IEEE 27th Design Automation Conference*, 1990.

[10] D. Marple. A hierarchy preserving hierarchical compactor. In *ACM IEEE 27th Design Automation Conference*, pages 134–140, 1990.

[11] C. Papadimitriou and K. Steiglitz. *Combinatorial Optimization*. Prentice Hall, Englewood Cliffs, NJ, 1982.

[12] M. Reichelt and W. Wolf. An improved cell model for hierarchical constraint-graph compaction. In *IEEE International Conference on Computer-aided Design*, 1986.

[13] W Wolf. An experimental comparison of 1-d compaction algorithms. In *Chapel Hill Conference on VLSI*, 1985.

Process Independent Constraint Graph Compaction

David G. Boyer
Bellcore
Newman Springs Road
Room 3X327
Red Bank, N.J. 07701
dgb@nyquist.bellcore.com

Abstract

The DASL symbolic layout system is used to create experimental VLSI circuits. It is important that designs created in DASL be created in a process independent fashion, since Bellcore does not have a captive fabrication facility. The DASL constraint graph compactor requires no user intervention to produce a process independent design. This contrasts with other constraint graph compaction systems.

1 Motivation

The goal of the DASL compaction system was to create a constraint graph compactor that produces design rule correct results without user intervention. It also was desired to produce results that were electrically correct, eg. control the placement of substrate contacts. In other constraint graph compaction systems users must control the placement of substrate contacts and sometimes, the placement of the tubs in order to acheive reasonable results. In constraint graph compaction if there is nothing to constrain an element's placement, it will slide as far to the left as possible (in X compaction). This is undesireable for substrate contacts which need to be spaced at reasonable intervals. A technique has been developed that controls the size of the tubs and the placement of the substrate contacts. A given design can be recompacted for a new process, therefore, without any user intervention being necessary.

The DASL system is an outgrowth of our earlier work with the MULGA virtual grid compaction system [1,2]. Initially, tub generation was performed in DASL in the same way that it is done in MULGA. A minimal tub was assumed to surround each diffusion element during compaction. After compaction the tubs are grown until they intersect the "nearest" substrate contact. This approach works well for a virtual grid compactor but does not work for a constraint graph compactor. Substrate contacts tend to slide toward the left (for X-compaction) along a horizontal power rail on which they are placed. This may cause some tubs to be stretched an unnecessarily long distance in order to enclose a substrate contact and can cause design rule violations since the final tubs are different than the tubs that were present during compaction (Figure 1).

The solution developed in DASL was to introduce "symbolic tubs" into the virtual grid design that is the input to the DASL compactor [3]. The symbolic tubs are automatically generated and assure that design rule correct results are acheived because the compactor is dealing with the tubs that will be present in the final layout. The substrate contact sliding problem is constrained by minimizing the area of the tubs in a unique fashion that will be discussed below.

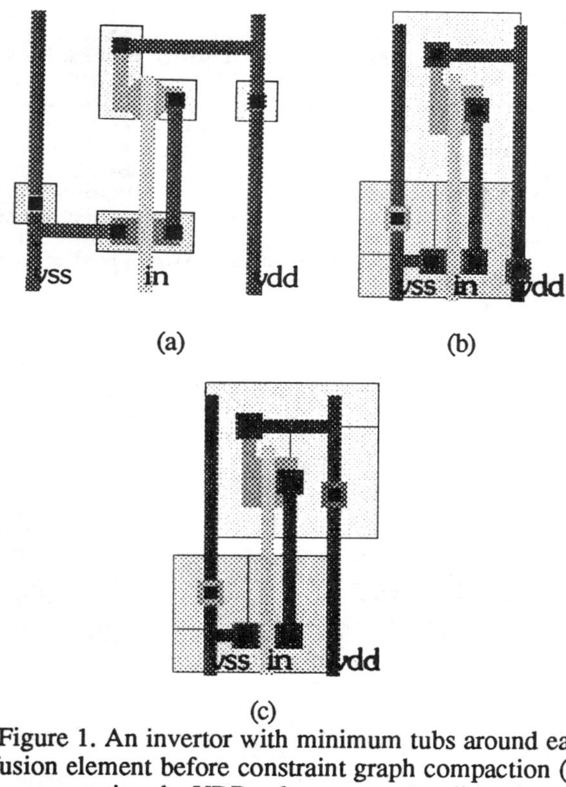

Figure 1. An invertor with minimum tubs around each diffusion element before constraint graph compaction (a). After compaction the VDD substrate contact slides down to the bottom of the VDD rail since there is nothing to stop it (b). When the Ntubs are grown to include a Vdd substrate contact they overlap the Ptubs. In (c) the desired results are shown.

0738-100X/92 $3.00 © 1992 IEEE

3 DASL Symbolic Tub Compaction

DASL compaction begins by splitting a virtual grid symbolic layout into distinct circuit groups [3]. An initial compaction is done during which time a constraint graph is generated and the groups are given their minimal locations. A "newave" constraint graph generation technique is employed [8] that combines the perpendicular plane sweep approach [4] with the Shadowing technique [5] and is an outgrowth of our research with virtual grid compaction [3]. The constraint graph is then used to place groups that are not on the longest path in order to minimize the areas of selected layers (wire minimization). The constraint graph is also used as the input to a pitchmatcher that is used for hierarchical module assembly.

3.1 Group Identification

The first step in the DASL compaction scheme is to identify distinct circuit groups that are on the same grid line [3]. An X compaction group is composed of symbolic circuit elements that are connected along the same vertical grid line and that need to remain together in the mask version of the cell. For example, if a vertical metal wire and a vertical poly wire lie on the same vertical grid line and are connected together with a metal-poly contact, they need to remain together in the mask. The metal wire, the poly wire, and the metal-poly contact are grouped together and placed together in the mask. A metal wire and a poly wire that share the same vertical grid line and are not connected by a metal-poly contact, are not grouped together and receive independent mask locations.

Local connectivity (along the grid line) is used to identify the groups into which the symbolic elements will be placed. As an example, consider a vertical grid line. There are three situations that can result in the elements on this grid line being grouped together. The first situation is if the elements are connected by a vertical wire segment. The second is if the elements are connected to a vertical transistor (the source, gate and drain of the transistor have the same X coordinate). The final grouping case is the grouping of vertical symbolic tub edges. Tub edges that share the same vertical grid line and overlap will be placed in the same group.

In Figure 2b the X groups on vertical grid 4 are identified for the symbolic circuit found in Figure 2a. There are two distinct groups on this grid line. The first group (X group 2) consists of a contact, a vertical wire segment, and a transistor source. These elements will be placed together. The placement of a group is governed by the element in the group which requires the greatest location in order to observe the design rules that affect it. The second group (X group 3) of elements on this line will be placed independently from the first group. Its location will be governed by its member element requiring the greatest location.

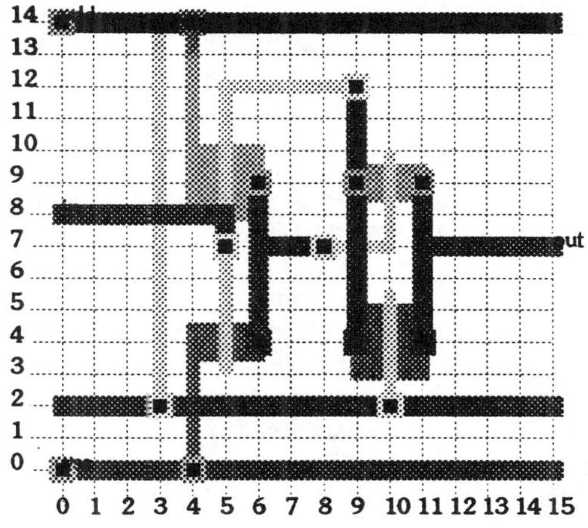

(a)

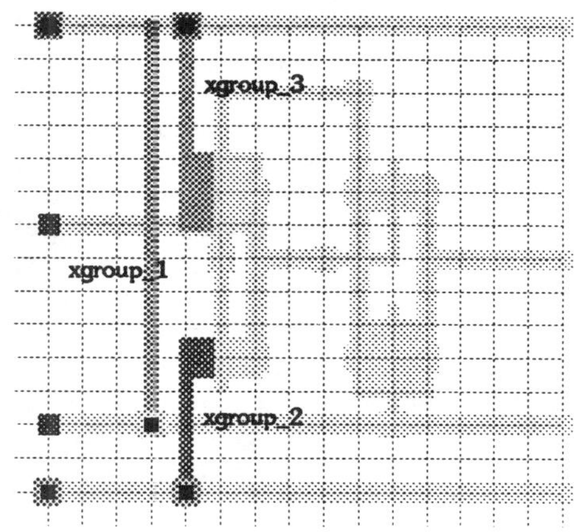

(b)

Figure 2 Virtual grid symbolic layout (a) and some of its X groups (b).

The size of a tub cannot be predetermined because the tub must observe the correct overlap rules with diffusion elements and must also be tied to a substrate contact. Therefore the width and length of a tub is variable and is topology dependent. Since the size of a tub is not predetermined, the edges of a tub are treated as distinct compaction groups and placed separately (Figure 3). The correct overlap of a tub to a diffusion element is accomplished by spacing a diffusion element with respect to the tub edge (tub group).

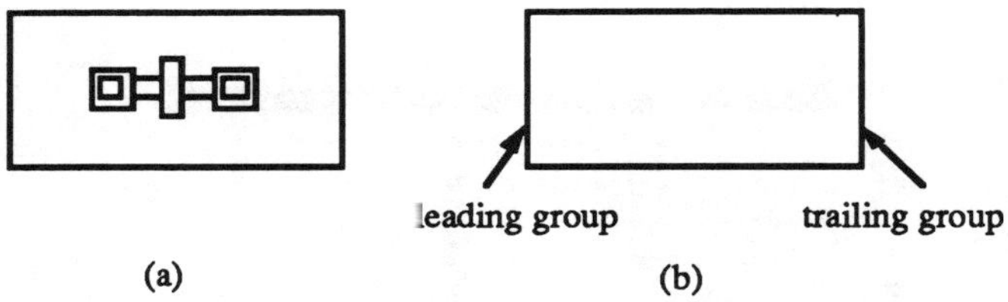

Figure 3. A device surrounded by a symbolic tub is found in (a). The tub groups are found in (b).

3.2 Constraint Graph Generation

The first compaction pass (X in this description) proceeds in a similar way to that of virtual grid compaction The constraint graph edges for a group are generated by determining the spacing necessary for each element in the group with elements that fall on the same horizontal grid line. The leading edge of a rectangle from the group to be placed is typically spaced against the trailing edges of the rectangles in the "wave" (Figure 4a). The "wave" keeps a record of the most recently placed rectangle for each layer of a fabrication technology at each unique location [8]. Edges are created in the constraint graph between the group being placed and the group whose wave rectangle was being spaced against. Tubs are an exception to this since constraint graph edges are generated between diffusion edges and the leading edges of the tubs (Figure 4b). When a diffusion rectangle is being spaced against the wave, it is spaced against the leading edge, not the trailing edge of a tub rectangle of the opposite doping type (Ntub for a P-diffusion rectangle). The diffusion rectangle will be spaced against the trailing edge of the same doping type of tub (Ptub for P-diffusion rectangle), however. The leading and trailing edges of a tub rectangle are given distinct group numbers and treated, essentially, as zero width wires[7]. When a leading tub edge is placed, it is only spaced (receives constraint graph edges) with the trailing edges of diffusion rectangles of the same doping type and tub rectangles of the opposite doping type (an Ntub

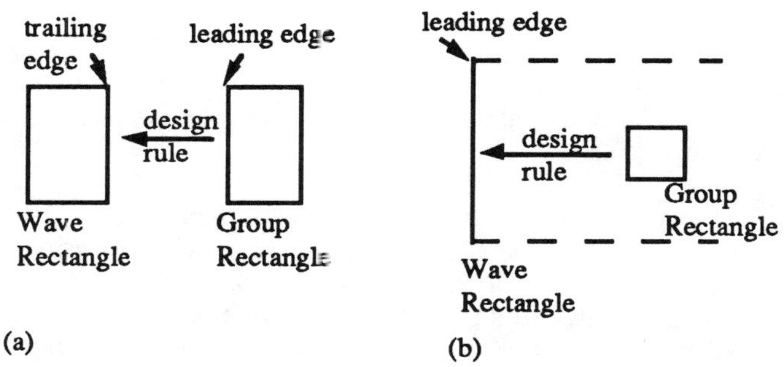

Figure 4. Spacing of a rectangle from the group being placed with a wave rectanfle is found in (a). In (b) the spacing of a diffusion rectangle with a leading edge is shown.

leading edge is spaced against the trailing edge of an N-diffusion rectangle or a Ptub). Trailing tub edges are spaced against trailing edges of diffusion rectangles and leading edges of tubs of the same type (they are not spaced against trailing edges of the same type).

When all the elements of a group have been spaced against (all the constraint graph edges have been generated for a group), the wave is updated. Each element in the placed group is compared with the rectangles in the wave. If the leading edge of the rectangle of the placed group was given a location greater than the leading edge of the rectangle in the wave, it will shadow the old rectangle.

When a tub group is updated, its edge will replace (shadow in the wave) tub edges of the same type (leading or trailing edge) that are already in the wave. A leading edge will shadow only other leading tub edges and will not affect trailing tub edges. The next diffusion rectangle of the appropriate type to be placed receives constraint graph edges with this tub edge. The tub edge that was closest to a diffusion rectangle in the symbolic view will get a constraint graph edge with this tub.

An example of how this works can be found in figure 5. In figure 5a, two symbolic tubs are found. The leading edge of tub A is first placed; then the transistor source and gate are placed (in a virtual grid design a transistor is treated as a three terminal device [1]). A constraint graph edge will be generated between the device source and the leading edge of tub A. Next, the leading edge of tub B is placed. Due to the spacing rules used for a leading tub edge, the leading edge of tub B is not constrained by the device source and is placed coincident with the leading edge of A, shadowing tub A from further interactions. After the leading edge of tub B is placed, the wave is updated and the leading edge of tub A is forgotten. When the MD contact on the drain of the device is placed, a constraint graph edge is generated between it and the leading edge of tub B (the usefulness of this will be seen in the wire minimization section below). A constraint graph edge will also be generated between the substrate contact and the leading edge of tub B. Note that since the leading edge of tub A was shadowed by tub B, the substrate contact and the MD contact do not have constraint graph edges to it. The source of the device does have a constraint graph edge to tub A, however (5c).

Y compaction proceeds in a similar fashion to X compaction. The only difference is that diagonal constraints (corner to corner spacings) are accounted for during Y compaction (see reference [8] for more details on diagonal constraint generation).

4 Wire (Area) Minimization

In a virtual grid symbolic layout transistors span at least three grids. When the groups of a device are placed, the device can be stretched if one of its groups is constrained and the others are not (Figure 6). The minimization of the areas of devices is important for high speed circuits. It is also important to minimize the areas of the tubs so that the placement of the substrate contacts can be controlled.

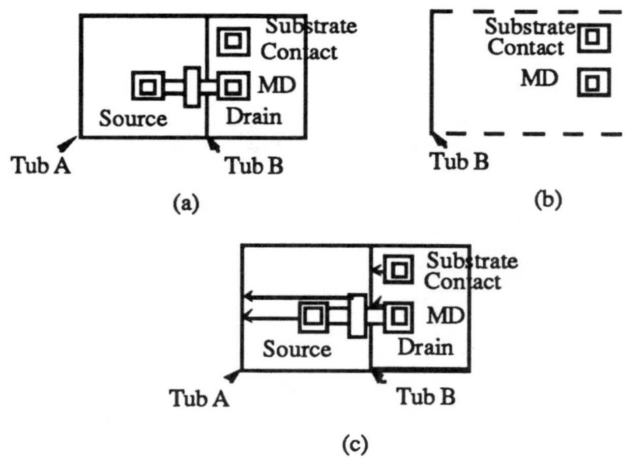

Figure 5. Constraint graph generation during the compaction of symbolic tubs. The symbolic tubs are shown in (a). In (b) the wave is shown with the tub edge before the placement of the contacts. In (c) the MD and the Substrate Contact both get constraint graph edges with Tub B. They do not get edges with Tub A.

The layers are minimized in order of importance for speed considerations. Diffusion area is minimized first, followed by tub areas, then by polysilicon, and finally by the first layer of aluminum. The user selects which layers are to be minimized.

The first step in minimization is to identify groups on the longest path through the cell and remove them as candidates for minimization. Then the following steps are done for each layer to be minimized :

1. Identify the circuit groups that have only the layer under consideration entering them from the left (for example X group 3 of Figure 6a has diffusion that comes from the left and connects to it). These groups are "locked" into place.

2. Identify groups that only have the layer leaving them from the right (X group 1 of Figure 6has diffusion that connects to it and goes to the right). These groups are pulled as far to the right as the constraint graph will allow (Figure 6b).

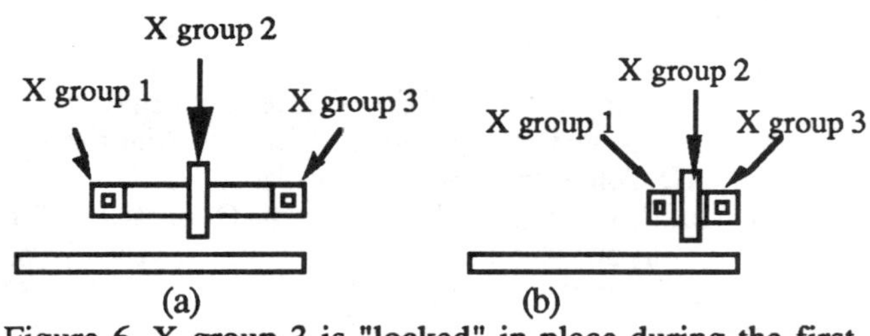

Figure 6. X group 3 is "locked" in place during the first step of diffusion wire minimization (a). In the second step of diffusion minimization, X group 1 is pulled as far to the right as design rules permit, minimizing the diffusino stretch of the device.

This process is repeated for all layers to be minimized. The first layer to be minimized gets preferential treatment over the other layers, the second over the remaining layers, etc. As mentioned above, the minimization of the tub areas controls the placement of the substrate contacts. Unlike other symbolic tub approaches [7] that use bounding polygons or rectangles, the approach presented here allows symbolic tubs to overlap. The advantage to using overlapping tubs is that, when the tub areas are minimized, sliding substrate contacts are pulled back toward the devices that they were near in the symbolic view of the cell.

In figure 7a the results of compacting the example presented in figure 5 can be found before minimization (note that the substrate contact slid to the left). During the second step of the minimization of the tub layers, the leading edges of the tubs are pulled to the right (for X compaction). Since there were no constraint graph edges generated between the leading tub edge of Tub B and the source and gate of the transistor, it is pulled back until it is constrained by the MD contact. The MD contact's location was fixed during diffusion minimization; consequently, it stops the horizontal minimization of Tub B because of the constraint graph edge that it has with it. The tub edge pulls along with it, however, the substrate contact because of the constraint graph edge between them, producing the results found in figure 7b. This technique keeps the substrate contacts close to the diffusion elements that they were near in the symbolic view of the cell and eliminates the need for the designer to manually fix the location of the substrate contacts. In Figure 8 the results of compacting the "afa" benchmark cell with wire minimization are found. The DASL compaction time for the "afa" cell on a VAX 8620 was 3 seconds.

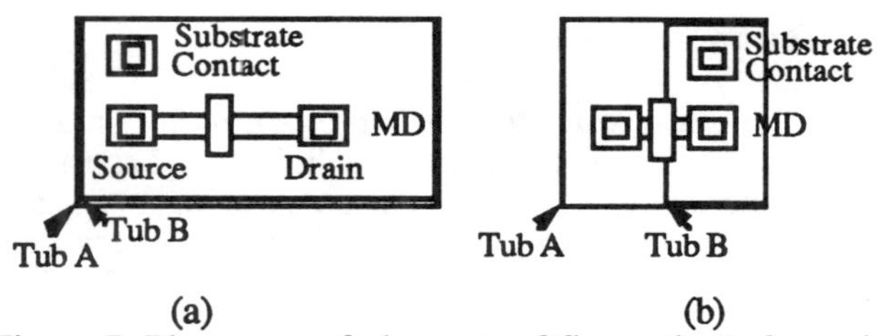

Figure 7. Placement of elements of figure 6. Before wire minimization the substrate contact slides to the left (a). In (b) the substrate contact is pulled back to the right by the leading edge of Tub B. It aligns with the Drain fo the device as it did in the symbolic view of the cell.

5 Conclusion

The DASL compaction system system has been described. It allows a designer to create a process independent design without user intervention. Tub design rule violations are eliminated with the compaction of symbolic tubs and substrate contact placement is controlled by the minimization of the tub areas.

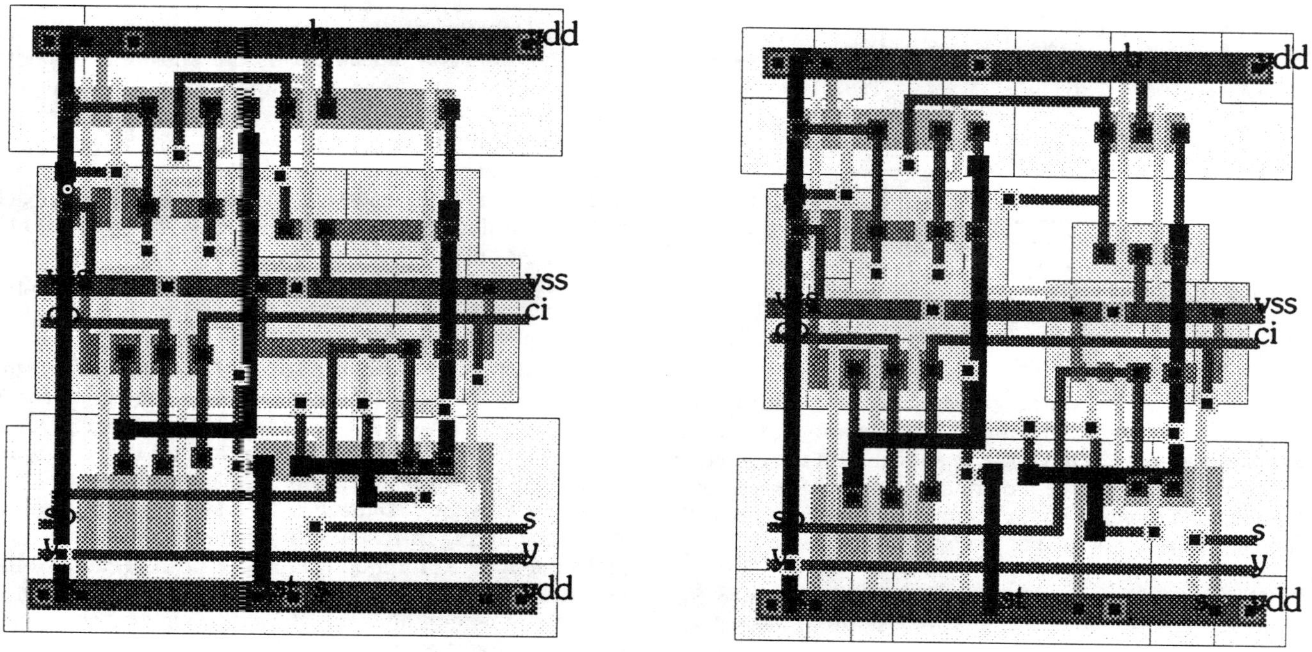

(a) (b)

Figure 8. afa benchmark cell compacted without (a) and with (b) wire minimization. Note that the substrate contacts slid to the left without wire minimization and that the minimization of the tub areas pulled them back.

6 Acknowledgments

Thanks are due to Phil Benjamin for helpful discussions and comments on symbolic tubs and constraint graph generation. Thanks are also due to Bob Cordell, Steve Okun, Anna Reidy, and Deb Rappaport for their contributions to the symbolic tubs research and to DASL.

7 References

1. N.H.E. Weste,"Virtual Grid Symbolic Layout," 18th Design Automation Conference, July 1981.
2. D.G. Boyer and R.R. Cordell, "Symbolic Layout for Rapid Full-Custom Prototyping of High-Speed Telecommunications Chips", IEEE Design & Test, pp 27-39, June 1991.
3. D.G. Boyer, "Split Grid Compaction for a Virtual Grid Symbolic Design System", Proc. of the IEEE International Conference on Computer-Aided Design, pp 134-137, Nov. 1987.
4. Burns, J. and Newton, A.R., "SPARCS: A New Constraint-Based IC Symbolic Layout Spacer", Proc. of the IEEE Custom Integrated Circuits Conference, pp 534-539, May 1986.
5. M.Y. Hsueh and D.O. Pederson, "Computer-Aided Layout of LSI Circuit Building-Blocks", PhD Thesis, UCB/ERL M79/80, University of California at Berkeley, Dec 1979.
6. D.G. Boyer and N.H.E. Weste "Virtual Grid Compaction Using the Most Recent Layers Algorithm", IEEE ICCAD Conference, September 1983.
7. C-Y Lo, "Automatic Tub Region Generation for Symbolic Layout Compaction", Proc. of the 26th ACM/IEEE Design Automation Conference, pp 302-305, June 1989.
8. D.G. Boyer, "DASL Constraint Graph Compaction with Symbolic Tubs", ACM International Workshop on Layout Synthesis, May 1990.

A New Hierarchical Layout Compactor Using Simplified Graph Models

Wonjong Kim*, Joohack Lee**, Hyunchul Shin*

*Dept. of Electronics Engineering
Hanyang University, Korea

**Samsung Electronics Co. Korea

Abstract

A new hierarchy-preserving hierarchical compactor which can be used with either 1D or 2D leafcell compaction techniques has been developed.

The compactor is applicable for hierarchical layouts which consist of a number of arrays of identical cells. The hierarchy is maintained throughout the compaction process so that all the instances of a subarray of identical cells have the same shape after compaction.

Currently a hierarchical compactor based on the suggested simplified graph model has been developed and experimental results on several benchmark examples show that the proposed method is satisfactory.

1. Introduction

As the complexity of integrated circuits increases and the first-time-correct design is required, layout compactors become essential for IC design, especially for ASIC design. In addition to minimizing the chip area at the final stage of physical design, a compactor can generate mask layouts for several different technologies(design rules) from a single symbolic cell/module in a library. It can also remove "minor" design rule violations. Several leafcell compactors have been developed and some of them show excellent results as surveyed in [Cho85, Boy87].

However, less amount of work has been done in the hierarchical compaction area. For example, as far as the authors know, there is no general way reported that can compact arbitrary hierarchical symbolic layouts maintaining the hierarchy and allowing wire-jogging and two-dimensional movements of components.

Marple[Mar90] developed an effective hierarchical compactor based on the Simplex method which can handle difficult constraints like symmetric constraints and wire length minimization simultaneously. However, it does not allow wire-jogging or 2D movements of components.

In this paper, we suggest a new hierarchy -preserving compactor for certain kinds of hierarchical layouts. The layout must be composed of subarrays of identical cells. Then we form a simplified model of the circuit and compact the simplified model with hierarchy-preserving constraints.

In section 2, we summarize several typical hierarchical compaction techniques. In section 3, we describe the simplified model of our approach. In section 4, the overall hierarchical compaction algorithm is described. In section 5, we show experimental results. Finally, we conclude in section 6.

2. Previous Hierarchical Approaches

The hierarchical compactor is important in that

1) "good" leafcell compactors do not guarantee "good" results for hierarchically described circuits and

2) when the hierarchy is destroyed during compaction, several instances(appearances) of an identical cell before compaction may have different geometric shapes after compaction, which bothers efficient management of the design data.

Several hierarchical compaction approaches have been reported. Some methods[Bal82, Bur87] compact leafcells first and then treat them as rigid objects at the higher level. In these methods, routing is necessary to maintain connections among the rigid subcells. Pitchmatching methods[Kin84, Cro87] compact leafcells first using a leafcell compactor and then pitchmatch ports of the cells to preserve the connections by abutment. The pitchmatching methods are efficient but they may not generate an identical mask layout for every instance of a cell. Some other methods[Rei86] partition each cell into clusters of elements and perform hierarchical compaction assuming the clusters as rigid objects. Another method[Shi87] compact the core part of leafcells first, and then the wires connected to the ports of the cells are compacted during the higher level compaction.

In [Mar90], a powerful hierarchy-preserving hierarchical compactor has been reported. It performs hierarchy-preserving compaction, symmetry-preserving compaction, and wire-length minimization using the Simplex method for linear programs. The limitations of this method are in

0738-100X/92 $3.00 © 1992 IEEE

that the Simplex method takes more time and memory than the longest-path method and that automatic wire-jogging and two-dimensional movements of components are not allowed. Note that frequently wire jogging can reduce the layout area significantly, say more than 10%.

When the given layout to be compacted is limited to an array of a single cell, torus compaction[Meh87] can be used.

3. The Simplified Graph Model

In this section, we describe how a simplified graph model is obtained from a certain kind of hierarchical layouts. The compaction problem we are trying to solve is limited to an array of cells whose subarrays are composed of identical cells. This problem is more general than the one used in the "compaction on the torus[Meh87]". The array may contain several types of cells. The restriction is that identical cells should form a row, a column, or an array. This restriction can be relaxed if exact hierarchy-preserving is not required.

For example, Fig.1(a) shows a 4x4 multiplier circuit with 16 instances of 6 types of cells. It is composed of six subarrays and each of the subarrays contains only identical cells as shown in Fig.1(b). In Section 3.1, we describe how to simplify a given layout, and in Section 3.2, we describe the extra hierarchical constraints that must be considered with the simplified circuit.

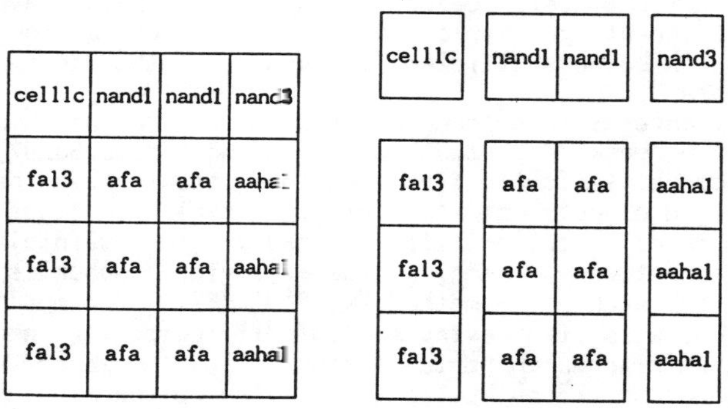

(a) 4x4 multiplier (b) six subarrays

Fig.1. 4x4 multiplier(mul4x4) example

Conventional pitchmatching compactors [Kin84,Cro87] would compact each of six leafcells first and then construct an upper level graph of 16 cells for pitchmatching. In our approach, we model the multiplier as a simplified model shown in Fig.2 and solve the graph of six cells with additional hierarchical constraints at once. The numbers (α,β) under the cell names indicate that the cell is repeated α times in the x-direction and β times in the y-direction.

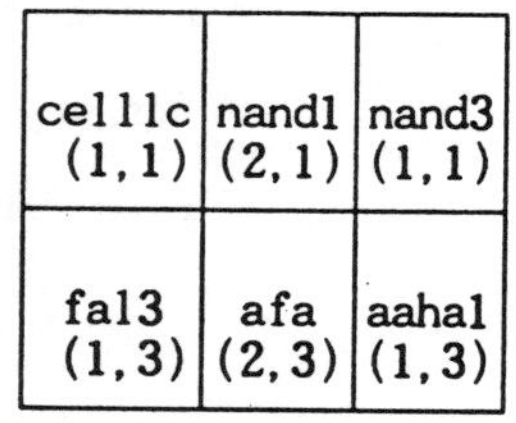

Fig.2 Simplified model

3.1 Formation of the Simplified Model

The simplified model is obtained by collapsing identical rows or columns recursively. For example, the simplified model in Fig.2 is obtained by collapsing rows 2, 3, and 4 of Fig.1 single row and then columns 2 and 3 into a single column.

Note that only one instance for each identical cell group appears in the simplified model. This allows an efficient hierarchical compaction. This modeling method has some limitations since general hierarchical layouts may not be simplified like the above example. Let m be the number of instances of a cell. When n instances ($n>1$) of the cell remain in the final simplified model, then the hierarchy of the circuit may not be preserved, but there may be n different geometric shapes for the cell after compaction. However, this method is still usefull whenever $m>n$.

Now the compaction is performed on the simplified model given in Fig.2. If we compact the circuit in Fig.2 as a flattened layout, we can not reproduce the compacted original circuit by simply reversing the collapsing procedures, since it may produce an illegal(design rule violated) layout. Therefore we need to consider other necessary hierarchical constraints in addition to the "normal" constraints imposed by the simplified model.

3.2 The Hierarchical Constraints

The hierarchical constraints are:

1) pitchmatching constraints on the cell boundaries to align cell ports and

2) spatial constraints near the cell boundaries due to design rules.

To enforce pitchmatching, we simply need to add a fixed constraint (lower bound = upper bound = 0) to each pair of abutted ports on cell boundaries.

The hierarchical constraints due to design rules are more complicated to generate and they are updated iteratively to minimize the area of the circuit. These are explained in detail in the following section.

4. The Hierarchical Compaction Approach

As many other graph-based compactors, our compactor has two major procedures, the constraint graph generation and the event-driven longest path solving. The difference is that the graph generation is based on the simplified hierarchical model and that the hierarchical compaction is performed iteratively.

The following algorithm shows the overview of our one-dimensional hierarchical compactor. Two dimensional compaction is described in section 4.5.

Algorithm: Overall algorithm

ⓐ Read design rules.
ⓑ Input the hierarchical layout.
ⓒ Build simplified model.
ⓓ Extract net numbers.
ⓔ Build permanent graph of the constraint graph. (for electronic connections among wires, jogs, contacts, ports of abutment/repeated cells)
ⓕ (optional)Rotate the layout 90 degrees to change the compaction direction.
ⓖ Generate leafcell level spacing edges.
ⓗ Generate/Update hierarchical constraints.
ⓘ Solve the graph by the longest path method. Insert jogs if necessary.
ⓙ If area is reduced
    Goto step ⓗ.
ⓚ Do wire length minimization.
  If compaction sequence is not completed
    Goto step ⓕ.
ⓛ Build the original circuit from the simplified circuit.
  Output mask layout.

There are two loops in the above algorithm. The main loop ⓕ to ⓚ is performed according to the compaction sequence given by the user. For example, the number of compaction passes and the direction for each pass is defined by the compaction sequence. The inner loop ⓗ to ⓙ is for the hierarchical compaction. During a pass of 1D hierarchical compaction the loop ⓗ to ⓙ may be executed several times until a "satisfactory" distribution of spacing near the cell boundaries is obtained. The more detailed explanations on several important procedures of the overall algorithm are given in the following subsections.

4.1 Hierarchical Constraint Generation/Update

The hierarchical constraints are added to the leafcell-level constraints for the simplified models. For the generation of normal leafcell constraints in the graph-based compaction, refer to [Cho85] or other papers.

The hierarchical constraints due to design rules are generated for all the components which is placed near the cell boundaries, i.e. if the distance from the component edge to the cell boundary may become smaller than the maximum possible design rule for the component.

4.2 Graph Solving with Wire Jogging

The compaction is performed by finding a solution that satisfies all the constraints in the graph using an event-driven longest path algorithm like other 1D compactors[Cho85, Boy87]. During the graph solving, critical paths are found and they are shortened, if possible, by bending wires. (The critical paths are the sequences of instances which define the width or height of the layout.) The automatic jogging of wires are repeated until no further reduction of area is possible and several stacked wires can be jogged into a 'river routed' shape when they are on the critical path.

4.3 Merging & Wire Length Minimization with Resistivity

After each pass of compaction, wires may be unnecessarily stretched. They increase the resistance and bother the next pass of compaction. Hence after each pass of compaction, clusters of components are moved to minimize the total resistance of wires. During the wire-length minimization, mergable components at the ends of the wires are merged, if possible, into a single component and the data structure is updated accordingly.

4.4 Postprocessing after compaction

The graph-solving generates a leagal layout that satisfies all the spacing constraints of the simplified model. Now when there are repeated cells, they are copied to construct the original hierarchical layout. This is the reverse process of the simplification shown in Fig.1 and Fig.2. Since all the pitchmatching constraints are considered during the graph generation, simple tiling is good enough.

4.5 Two - Dimensional Hierarchical compaction

To incorporate 2D compaction, we build the constraint graphs in both (x and y) direction. To reduce the size of the layout in one direction, the critical components are rearranged in the other direction, if possible. The horizontal and/or vertical wires may be jogged whenever necessary during the 2D compaction.

5. Experimental Results

The 1D/2D hierarchical compactor is developed on Sun 4/40 under the UNIX operating system using the 'C' language. We used the design rules given by the MCNC/ICCD benchmarks[Boy87], where 1 λ = 1.5 μm.

Fig.3 shows the hierarchically compacted result of the 4x4 multiplier circuit and Table 1 shows the comparisions of our results with those of several well-known compactors. Our compactor generated the smallest areas by using comparable memory and CPU time when compared with other results. The area reduction is mainly due to the near optimum arrangement of components near the cell boundaries.

Note that the CPU time in our approach does not increase as the example size increases.

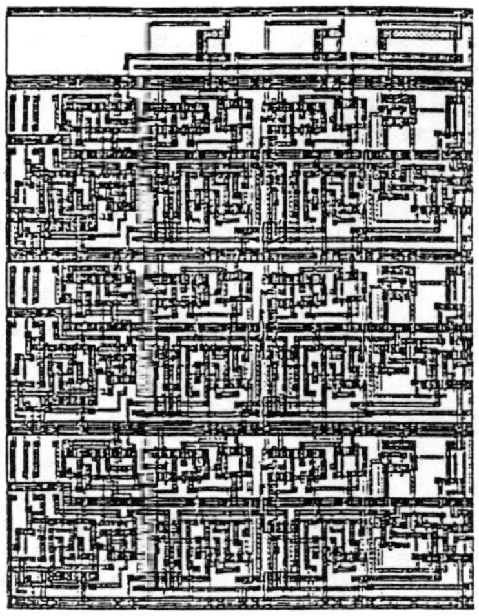

Fig.3 2D hierarchically compacted 4x4 multiplier

Table 1: Comparisons of hierarchical compactors

Cell	Compactor	Width (μm)	Height (μm)	Area (μm^2) (%)	Memory (Kbyte)	CPU (sec)
mul 4x4	Our's (2D)	483	621	299943 100	3064	891
	Our's (1D)	487.5	621	302737.5 100.9	2511	266
	MACS.1D (DAC'90)	570	558	318060 106.0	1712	30
	MACS.2D (DAC'89)	544.5	566.25	308323 102.8	1573	96
	SPARCS(ICCD'87)	649	601	390049 125.2	754	66
	Symbolics(ICCD'87)	654	638	417252 133.9	840	54
	Zorro(ICCD'87)	577	577.5	333217 111.1	741	1904
mul 16x16	Our's (2D)	1905	2745	5229225 100	3064	894
	Our's (1D)	1927.5	2745	5290987.5 101.2	2511	269
	MACS.1D (DAC'90)	2194	2502	5489388 105	6335	229
	MACS.2D (DAC'89)	2128.5	2492.25	5304754 101.4	4915	412
	Symbolics(ICCD'87)	2524	2780	7016720 128.8	14400	1073

(The CPU time is on SUN4/40 for our's and on VAX8650 for others.)

6. Conclusions

We developed a new constraint graph-based hierarchy-preserving hierarchical compactor. Hierarchical layouts which are composed of several subarrays of identical cells are modeled into simpler forms and hierarchical constraints are added to guarantee the correctness after tiling. The complexity of our method depends only on the number of different cell types, i.e., the simplified model size, and not on the total number of instances. Furthermore, this method is efficient since it is based on the longest-path method and is flexible since automatic jogging and 2D motions of components can easily be accomodated.

References

[Bal82] M.W.Bales, "Layout Rule Spacing of Symbolic Integrated Circuit Artwork," *UCB/ERL M82/72*, 1982.

[Boy87] D.Boyer, "Symbolic Layout Compaction Bench-marks," *International Conference on Computer Design*, pp.186-191,209-217, October 1987.

[Bur87] J.Burns and A.Newton, "Efficient Constraint Generation for Hierarchical Compaction," *International Conference on Computer Design*, pp.197-200, October 1987.

[Cho85] Y.E.Cho, "A Subjective Review of Compaction," *Proceedings of 22nd Design Automatin Conference*, pp.396-404, 1985.

[Cro87] W.Crocker, R.Varadarajan, and C.Lo, "MACS: A Module Assembly and Compaction System," *International Conference on Computer Design*, pp.205-208, October 1987.

[Ent85] G.Entenman and S.W.Daniel, "A Fully Automatic Hierarchical Compactor," *Proceedings of 22nd Design Automatin Conference*, pp.69-75, 1985.

[Hse79] M.Y.Hseuh & P.O.Pederson, "Computer-Aided Layout of LSI Circuit Building-Blocks," *U.C.Berkley, UCB/ERL M79/80*, 1979.

[Kin84] C.Kingsley, "A Hierarchical, Error-Tolerant Compactor," *Proceedings of 21st Design Automatin Conference*, pp.126-132, 1984.

[Lan89] A.A.J.de Lange, J.S.J.de Lange, J.F.Vink, "A Hierarchical Constraint Graph Generation and Compaction System for Symbolic Layout," *International Conference on Computer Design*, pp.532-535, 1989.

[Mar90] D.Marple, "A Hierarchy Preserving Hierarchical Compactor," *Proceedings of 27th Design Automation Conference*, pp.375-381, 1990.

[Meh87] K. Mehlhorn and W. Rulling, "Compaction on the Torus," *International Conference on Computer Design*, pp.201-204, 1987.

[Rei86] M.Reichelt and W.Wolf, "An Improved Cell Model for Hierarchical Constraint-Graph Compaction," pp.482-485, *Proceedings of International Conference on CAD*, 1986.

[Shi87] H.Shin, A.Sangiovanni-Vincentelli, and Carlo H.Sequin, " Two-Dimensonal Mudule Compactor Based on 'Zone-Refining'," *International Conference on Computer Design*, pp.201-204, 1987.

On Efficient Concurrent Fault Simulation For Synchronous Sequential Circuits

Dong Ho Lee
Department of Computer Science

Sudhakar M. Reddy
Department of Electrical and Computer Engineering

University of Iowa
Iowa City, IA 52242

Abstract

In this paper, we report on an efficient fault simulation method for synchronous sequential circuits. The method is based on concurrent fault simulation, and it has the simplicity of deductive fault simulation. Several new ideas to reduce computation time and memory requirement are proposed. New fault simulators were developed to simulate transition faults as well as stuck-at faults. The experimental results demonstrate that the proposed method is an effective method to simulate faults in large synchronous sequential circuits in the workstation environment.

1. Introduction

Efficient fault simulators for synchronous sequential circuits are important for the design of large digital circuits/systems. Recently, there has been a surge of publications in this area[2-7]. Most of them utilize the fact that zero delay simulation is adequate for fault simulation in synchronous sequential circuits. [4] and [7] extended critical path tracing to sequential circuits, but they didn't give adequate experimental results to assess the effectiveness of their methods. [5] proposes a fault simulator based on parallel simulation method.

Concurrent fault simulation method has been successfully used in industry for a long time. The success of concurrent fault simulation is due to its flexibility to allow arbitrary delay fault simulation(i.e., the circuit gates may have arbitrary but known propagation delays) and mixed or hierarchical designs. But the concurrent fault simulation method has been criticized to be inefficient as compared to other methods which are specialized to synchronous sequential circuits[5]. In this paper, we introduce efficient fault simulators for synchronous sequential circuits which are based on concurrent fault simulation. The proposed method has the simplicity of deductive fault simulation and flexibility and efficiency of concurrent fault simulation. Several improvements to reduce the computation time and memory requirement are also proposed. The experimental results show that the concurrent fault simulators with the proposed improvements are competitive with other methods both in computation time and memory requirements.

Although test generators and fault simulators for delay faults[10] have been reported for combinational circuits, stuck-at fault model has been the only practical fault model for sequential circuit test generation so far. A simpler delay fault model called transition fault model has been proposed and fault simulators for stuck-at faults have been augmented to accommodate this fault model[11]. But no study of transition faults has been reported for sequential circuits in the literature. In this paper, transition fault model for synchronous sequential circuits is introduced and the concurrent fault simulator developed for stuck-at faults is extended to include transition faults.

The paper is organized as follows. In Section 2, a concurrent fault simulation method is discussed together with several improvements. In Section 3, transition fault model for synchronous sequential circuits is discussed. In Section 4, experimental results are presented followed by a summary in Section 5.

2. Concurrent fault simulation

Concurrent simulation is a simulation paradigm rather than an algorithm. Many algorithmic details should be carefully planned to make this paradigm into an effective fault simulator. In concurrent fault simulation, the circuit state is defined in terms of the collection of gate states. The state of each gate is composed of all input line values and an output line value. The basic idea of concurrent fault simulation is to simulate one good machine and multiple faulty machines at the same time. In Figure 1, good machine elements are shown with the white box and faulty machine elements for a fault f are shown with the gray box. Faulty machines are simulated specifically only where they assume different logic values than the good machine value. Each input line of a gate in a faulty machine gets its value either from the good machine or the corresponding faulty machine gate depending on whether the fault is explicit in its fanin gate. For example, the fault element at gate G4 gets its input from the fault element at gate G1 in Figure 1.1, but it gets its input value from the good machine element in Figure 1.2 because fault f is not explicit at gate G1.

In concurrent fault simulation, an event generated on a machine propagates only within the machine. In other words, when an event is generated for a faulty machine,

This research has been supported by the SDIO/IST Contract No. N00014-90-J-1793 managed by US office of Naval Research

only that faulty machine need be processed to propagate the changed value at its fanout gates. A new fault element is introduced, i.e., *diverged*, when the propagated event makes the faulty machine assume a logic state which is different from the good state at a gate.

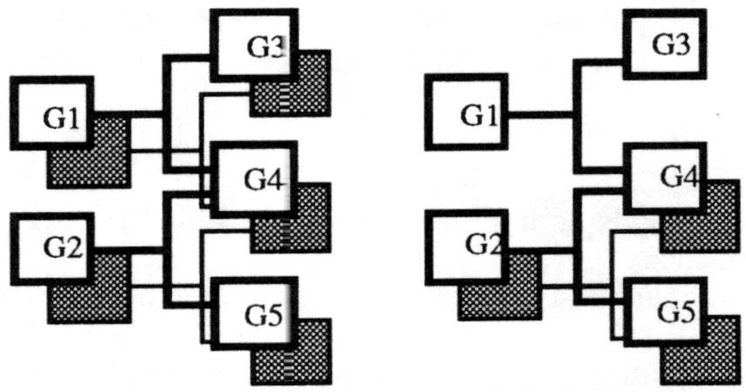

Fig 1.1 fault f explicit in G1 Fig 1.2 fault f implicit in G1

Figure1. Concurrent Fault Simulation

On the other hand, when an existing fault element assumes the same logic state as the good machine state at a gate, the fault element should be removed, i.e., *converged*. In Figure 1, when the faulty machine assumed the same output value as the good machine value at gate G1, this event is propagated to gate G3 and G4. The fault element at gate G3 is removed, but the fault element at gate G4 should remain since the fault effect has also propagated through G2. The convergence and divergence can be determined by simply comparing logic state after gate evaluation. In some case, an event is generated when a faulty machine is converged at the gate. In this case, at the next level of simulation, the fault element of the fanout gate will copy its input value from the good machine element.

In addition, a redundant copy of gate output value is kept at the input pin of its fanouts so that it is only necessary to compare the input fault lists of the gate which have events with its fault list. Furthermore, the state of a gate is packed into a word so that the output can be efficiently evaluated by table look up. Fast evaluation is extremely important in concurrent fault simulation because each faulty gate is explicitly evaluated one by one. Normally this is achieved through table look up.

2.1 Simplification

Unlike in the originally proposed concurrent simulation method[9], the data structure of the proposed fault simulators adopts the simplicity of deductive fault simulation. As shown in Figure 2, each gate has a fault list and each element of the fault list is composed of a fault identifier, a state variable, and a pointer to the next element. The fault identifier is a unique identifier which differentiates a fault from all other faults. Furthermore, a fault descriptor is associated with the fault identifier and the information central to the fault is stored in the fault descriptor and each fault element keeps only the local information, which is specific to the faulty machine state at the location, in the state variable. For example, the fault descriptor holds information about how to evaluate the faulty machine, or whether the fault has already been detected or not. Finally, the fault identifier is a pointer to its fault descriptor and each fault list is terminated with a terminal fault element which has a fault identifier which lies in high end memory location to avoid checking end of list during fault list processing.

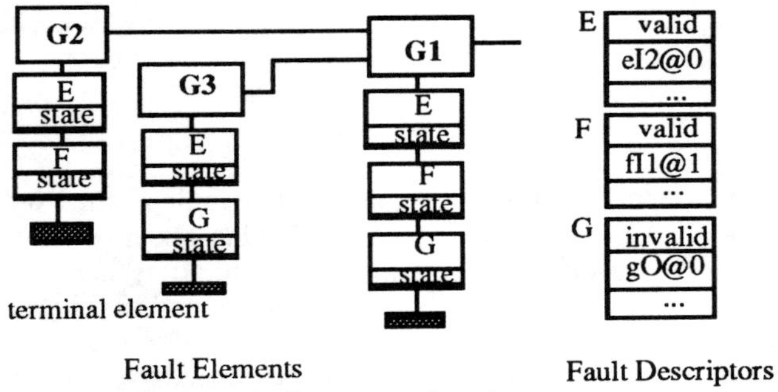

Fault E : input 2 of gate e stuck at 0

Fault G : output of gate g stuck at 0, dropped

Figure 2. Data Structure

In the following, we will briefly discuss arbitrary delay fault simulation using the given data structure. For convenience, we assume two phase fault simulation scheme. Assuming that delays are associated with gates, events are posted for all changing elements after gate evaluation. When delays vary widely among different faulty machines, it is advantageous to queue each faulty machine element independently into the timing queue, but for unit delay simulation, one can use a list event to queue a collection of faulty machine elements whose output values change at the same time.

In the first phase of fault simulation, the matured events are fetched to assign logic values to gate outputs. When list events are used, the actual output values of fault elements are set via list traversal. The fanout gate identifiers are entered into a local queue, *not the timing queue*, for the second phase. When the event, i.e., the gate identifier, is fetched in second phase, the multi-list traversal technique[3] is employed to copy the logic values from the source fault lists to the destination fault list. Then the fault elements are evaluated and when the evaluated output values are different from previous values, events are posted.

Fault simulation procedure for synchronous circuits is much simpler than described above because one can use zero delay simulation. For zero delay fault simulation, only the second phase is necessary since the evaluated value can be assigned directly on the output as long as the gate evaluation is done orderly according to its level, where the level of a gate is assigned so that all its fanins are at the lower levels. Furthermore, the timing queue is no longer necessary and only gate identifiers are "scheduled" into the event queue when there is an event on at least one machine element. Note that, in the general simulation described before, a fault element is treated the same as the good machine gate so that it should be scheduled to be evaluated after a given delay.

2.2 Further improvements

Among the three improvements discussed in this subsection, both fault dropping scheme and invisible fault handling scheme are good for arbitrary delay fault simulation as well as zero delay fault simulation.

Fault dropping is very important in concurrent fault simulation because dropped fault effects should be eliminated as soon as possible for efficient fault simulation. Nevertheless, the fault elements for a detected fault are scattered around the circuit and there is no effective scheme to search them without scanning the whole circuit. We adopted an event driven fault dropping scheme. In this scheme fault elements for detected faults are removed while the fault list to which it belongs is traversed. This scheme can be implemented efficiently in the fault simulation procedure discussed before. The fault identifier of terminal element points to an imaginary fault descriptor which is never dropped. With this provision together with the proposed data structure, only several machine instructions are enough to check if a fault has been detected already.

Most concurrent fault simulators keep invisible faults. Invisible faults are the fault elements which have the same output value as the good machine value. Invisible faults can be avoided if one copies the logic values of all input values when a new fault element is introduced. We use the method which keeps two fault lists, one for visible faults and the other for invisible faults, to avoid examining invisible faults. We found that splitting fault lists help reduce computation time. For further details, we refer to [13].

Concurrent fault simulation uses table look up method to speed up gate evaluation. As a result, for example, evaluating an inverter takes the same amount of time as evaluating a multinput gate. In order to take advantage of table look up mechanism, it is advantageous to partition the circuit into macro modules.

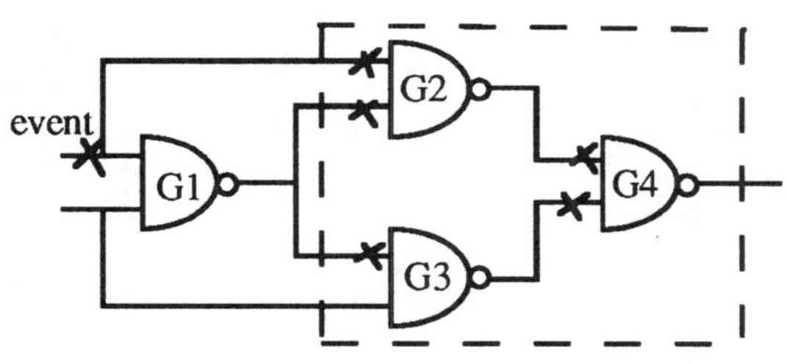

Figure 3. Macro Extraction

In Figure 3, a possible macro is enclosed by a dashed box. Macro extraction collapses many events into an event to save computation time. In Figure 3, the number of gates to evaluate can be reduced from 3 evaluations to 1 evaluation. More importantly, macro extraction reduces the memory requirement because many fault elements are collapsed into one fault element. The number of fault elements is also reduced from 3 elements to 1 element in Figure 3. If the look up table overhead is not too high, macro extraction reduces the total memory requirement. We found that the speed improvement and decrease in memory are both significant for large circuits when macros are used.

Although macros are limited to be fanout free in this paper, they can be any combinational circuits with limited number of inputs. When reconvergent macros are used, stuck at faults may be translated into functional faults which can be represented by look up table entries. The functional faults can be evaluated efficiently because each fault descriptor holds an adequate look up table entry corresponding the fault. Further details can be found in [13].

3. Simulating transition faults

Some physical faults do not show up when test clock cycle is much larger than the operational cycle. The transition fault is a gross delay fault where a gate is experiencing a modestly large delay. In general, any gate delay fault which delays a gate transition slightly longer than its slack time can be considered to be a transition fault. If the delay is extremely large, it acts as if it is a stuck-at fault[12]. Two transition faults are associated with each gate input, i.e., 0 to 1 transition fault and 1 to 0 transition. If a gate input is experiencing a 0(1) to 1(0) transition fault, the 0(1) to 1(0) transition is delayed. This affects both the primary output values and the values sampled into flip-flops. Since the delay is not too large, after the primary outputs and flip-flops are sampled, the combinational part of the circuit is assumed to settle down correctly. Strictly speaking, this implies that the delay defect does not increase the delay at the fault site by more than one "clock cycle", which we believe is reasonable. Table 1 shows the complete logic value relationship of this fault model. In Table 1, PV denotes the value when transition does not occur due to the transition fault and CV denotes the value when the transition is complete.

Table 1. Transition Table

Transition	0 -> 1 Fault		1 -> 0 Fault	
	PV	CV	PV	CV
0 -> 1	0	1	1	1
0 -> 0	0	0	0	0
0 -> X	0	X	X	X
1 -> 1	1	1	1	1
1 -> 0	0	0	1	0
1 -> X	X	X	1	X
X -> 1	X	1	1	1
X -> 0	0	0	X	0
X -> X	X	X	X	X

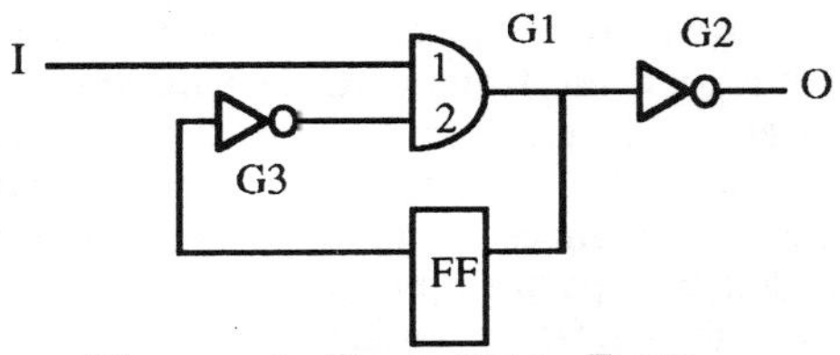

Figure 4. Transition Fault

We illustrate our model through an example given next. In Figure 4, consider a 0 to 1 transition fault at the input 1 of gate G1. Assume that all signal line values are set to X's initially. To detect this fault the 01 input sequence is enough. When the first input 0 is applied, logic value 1 will appear at the output. When the second input 1 is applied, the value of input 2 of gate G1 is changed from X to 1. Since there is no fault there, second input will be immediately set to 1. So there exists a sensitized path from the input I to the output O. The good machine will output 0 at the sampling time, but the faulty machine value remains at logic value 1. Transition fault 1 to 0 at the same gate input is more interesting. When this fault exists at input 1, it is necessary to use the sequence 00 to set the flip-flop. When a logic value 1 is applied as a third input, input 2 will assume logic value 1. The flip-flop input will have a logic value 1. If logic 0 is applied at the input 1, there is a transition, but the fault would not be detected because the latched value 1 will cause input 2 to assume logic value 0 to block the sensitizing path. It is necessary to apply logic 1 at the input again to make input 2 to assume logic value 0. Finally when a logic 0 is applied to input 1, input 2 will be changed to logic 1 before the transition fault affects the gate.

The concurrent fault simulation method as proposed is ideal to simulate the transition faults because all previous input values of all the gates are available. To simulate the transition faults, the combinational part of the synchronous sequential circuit is simulated twice. In the first simulation, it is assumed that all faulty transitions do not fire to determine the propagated faults and, in addition, to latch (master part of) the flip-flops. Then the network is simulated again after firing all the transitions. Note that the fault effects are latched into (master part of) the flip-flops at the first simulation, but (slave part of) the flip-flops should not be used in evaluation to avoid the new flip-flop values affecting the second phase simulation. In actual implementation, it is possible to simulate transition faults in one pass since the second phase simulation can be overlapped with the first phase simulation of the next time step. Further details can be found in [13].

4. Experimental results

To determine the effectiveness of the proposed method, a stuck-at fault simulator based on the procedures discussed in Section 2 was implemented and run on SUN Sparc 2. The final version is called csim-VM, where V means that visible and invisible faults are kept separately at each gate, and M means that macro modules are extracted. csim-M is a version which does not keep separate lists for visible and invisible faults and csim-V is the version that does not use macro extraction. In Table 2, the statistics of some benchmark circuits and the tests applied are given. These tests were used to evaluate the performance of PROOFS[5] and were provided to us by Professor J.H Patel of University of Illinois.

Table 2. Circuit Statistics

ckt	#f/f	#faults	#ptns	cov %
s208	8	215	111	63.7
s344	15	342	91	96.2
s382	21	399	2463	91.0
s400	21	426	1282	82.9
s526	21	555	754	75.3
s641	19	467	133	86.3
s820	5	850	411	81.9
s953	29	1079	16	7.8
s1238	18	1355	349	97.2
s1494	6	1506	469	91.1
s5378	179	4603	408	74.0
s35932	1728	39094	86	88.0

In Table 3, we report the CPU times and memory requirements for stuck-at fault simulation using the proposed simulator and PROOFS, all run on the same Sun Sparc 2 workstation. As one can see from the Table 3, both macro extraction and separation of visible/invisible faults consistently reduce the simulation time.

Table 3. Deterministic Patterns (I)

ckts	csim V		csim -M	csim MV		PROOFS	
	CPU sec	mem meg	CPU sec	CPU sec	mem meg	CPU sec	mem meg
s208	0.88	0.19	0.94	0.8	0.23	0.55	0.21
s344	1.14	0.24	0.82	0.77	0.23	0.76	0.25
s382	14.1	0.27	11.3	10.1	0.29	16.0	0.26
s400	12.4	0.28	9.71	8.39	0.28	11.8	0.27
s526	19.0	0.33	18.9	15.5	0.37	11.4	0.29
s641	1.01	0.25	1.06	1.01	0.30	1.28	0.46
s820	7.15	0.29	7.54	6.63	0.41	4.68	0.35
s953	2.76	0.44	3.26	2.73	0.49	0.42	0.35
s1238	4.27	0.32	4.88	4.22	0.49	4.53	0.43
s1494	20.5	0.51	15.8	14.0	0.61	10.4	0.54
s5378	37.6	1.35	32.6	26.6	1.32	50.9	1.45
s35932	226	16.2	168	137	9.24	172	11.6

Macro extraction increases the memory requirement a little bit for small circuits. But for large circuits, it actually reduces the memory requirement considerably. For example, the memory requirement is reduced from 16.2 M to 9.24 M in s35932. Upto circuit s1494, csim-MV and PROOFS show comparable performance. For circuit s5378, csim-MV is about two times faster than PROOFS.

For s35932, csim-MV performs slightly better than PROOFS. For some of the ISCAS89 benchmark circuits, we were able to obtain higher coverage tests using a sequential test generator developed by us[14]. In Table 4, we report the results of fault simulation using these tests.

Table 4. Deterministic Patterns(II)

			csim-MV		PROOFS	
ckts	#ptns	cvg %	CPU sec	MEM meg	CPU sec	MEM meg
s526	1919	81.3	23.74	0.37	17.95	0.33
s526n	1594	81.4	22.83	0.36	15.95	0.29
s820	930	95.5	8.14	0.43	8.86	0.35
s832	905	94.0	7.65	0.42	8.78	0.37
s1488	1270	97.2	19.75	0.69	23.9	0.53
s1494	1104	96.5	19.87	0.64	20.9	0.54

In Table 5, we report the random pattern simulation results. We chose the larger one of ISCAS-89 benchmark circuits, s35932, for which random patterns gave reasonably high fault coverage. The memory requirement of csim-MV in this experiment is lower than that shown in Table 3 because faults are rather slowly activated.

Table 5. Random Pattern Simulation

		csim-MV		PROOFS	
#ptns	flt cvg %	CPU sec	MEM meg	CPU sec	MEM meg
1024	74	744.6	7.53	1284	11.8
2048	80	1073	7.53	2134	11.8
4096	82	1672	7.53	3706	11.8
8192	84	2813	7.53	6722	11.8

In Table 6 we report the transition fault coverages of the ISCAS89 benchmark circuits. The stuck at tests are not good tests for transition faults. Fault coverages are in general much less than 50%.

Table 6.Transition Fault Simulation

ckts	# flts	MEM meg	CPU sec	flt cvg %
s208	332	0.19	0.63	28.3
s344	448	0.21	1.78	7.2
s382	556	0.21	15.4	4.6
s400	612	0.22	10.0	4.4
s526	890	0.23	8.0	2.7
s641	1080	0.33	2.26	50.8
s820	1514	0.29	14.5	20.6
s953	1486	0.30	1.0	0.0
s1238	2082	0.38	11.9	63.5
s1494	2786	0.47	31.3	48.0
s5378	8424	1.61	37.1	46.7
s35932	57178	8.21	412	6.6

5. Conclusion

In this paper we showed that concurrent fault simulation is a simple and effective method for simulating faults in synchronous sequential circuits. More efficient fault simulation is possible when hierarchical design information is utilized because the concurrent fault simulation method is inherently suited to hierarchical designs[3]. Furthermore, the method can be extended to simulate faults for large circuits which contain both synchronous and asynchronous designs.

Acknowledgment: We would like to thank Professor Janak Patel and his graduate students Vivek Chikermane and Elizabeth Rudnick for answering many questions regarding PROOFS and making PROOFS available to us.

Reference

[1] D. B. Armstrong, "A Deductive Method of Simulating Faults in Logic Circuits," IEEE Trans. on Computers, Vol. C-21, No. 5, pp. 464-471, May, 1972

[2] W. T. Cheng and M. L. Yu, "Differential Fault Simulation - A Fast Method Using Minimal Memory," Proc. 26th Design Automation Conf., pp. 424-428, June, 1989

[3] S. Gai, F. Somenzi and E. Ulrich, "Advanced Techniques for Concurrent Multilevel Simulation," Proc. Intn'l Conf. on Computer-Aided Design, pp. 334-337, November, 1986

[4] P. R. Menon, Y.H. Levendel, and M. Abramovici, "Critical Path Tracing in Sequential Circuits," Proc. Intn'l Conf. on Computer-Aided Design, pp. 162-165, November, 1988

[5] T. M. Niermann, W. T. Cheng, and J. H. Patel, "Proofs: A Fast, Memory Efficient Sequential Circuit Fault Simulator," 27th ACM/IEEE Design Automation Conference, pp. 535-540

[6] K. Son, "Fault Simulation with the Parallel Value List Algorithm," VLSI SYSTEMS DESIGN, pp. 36-43, December 1985

[7] X. Wang, "A Sequential Circuit Fault Simulation by Surrogate Fault Propagation," 1989 International Test Conference, pp. 9-18

[8] S. Seshu, "On an Improved Diagnostic Program," IEEE Trans. on Electronic Computers, Vol. EC-12, no. 12, pp. 76-79, 1965

[9] E. G. Ulrich and T. Baker, "The Concurrent Simulation of Nearly Identical Digital Networks", 10th Design Automation Workshop, pp. 154-155, 1973

[10] V. S. Iyengar, B. K. Rosen, and J. A. Waicukauski, "On Computing the Sizes of Detected Delay Faults, " IEEE Transaction on Computer-Aided Design, March 90, pp. 299-312

[11] G. L. Smith, "Model for Delay Faults Based Upon Paths," Proc. 1985, IEEE International Test Conference, November 1985, pp. 342-349

[12] J. A. Waicukauski and E. Lindbloom, "Transition Fault Simulation by Parallel Pattern Single Fault Propagation, 1986 International Conference, pp. 542-549

[13] D.H. Lee and S.M. Reddy, "On Efficient Concurrent Fault Simulation for Synchronous Sequential Circuits," Technical Memo in preparation, Department of Electrical and Computer Engineering, University of Iowa

[14] D.H. Lee and S.M. Reddy, "A New Test Generation Method for Sequential Circuits," Proc. Intn'l Conf. on Computer-Aided Design, pp. 446-449, November, 1991

Concurrent Fault Simulation of Logic Gates and Memory Blocks on Message Passing Multicomputers

Soumitra Bose, Prathima Agrawal
AT&T Bell Laboratories
Murray Hill, NJ 07974

Abstract

While application of hardware accelerators in VLSI CAD is quite common, fault simulation is a notable exception because of extensive resource requirements, the need for dynamic memory management and the inherently complex nature of its event evaluation mechanism. Other than the Zycad Fault Evaluator, for which there is no technical literature, hardware accelerated fault simulation has hardly met with success. We present a new pipelined algorithm suitable for implementation on message passing multicomputer systems or specialized hardware. It also illustrates, for the first time for pipelined architectures, fault simulation of circuits modeled at mixed functional and gate levels. Our results indicate an order of magnitude speedup compared to a production quality simulator running on a SUN SPARC2.

1. Introduction

This paper presents a concurrent fault simulation algorithm directly applicable to memory limited hardware accelerators and message passing multicomputers with distributed data memory [3]. It shows that a judicious intermixing of processors in parallel and pipelined topologies can not only achieve superior performance compared to previous techniques [1, 5], but can also be used in a scheme for implementing fault propagation through functional blocks. Fault propagation through functionally modeled memories requires minimal enhancements to the gate level fault simulation algorithm. This speeds up the simulation of circuits containing both logic and memory elements. Even though we only consider memory blocks, we believe that many of the ideas are applicable to other functional models also.

This paper is organized as follows: Section 2 contains the architectural overview of our system, Section 3 explains data structures and algorithms for some of the crucial parts of the fault simulation algorithm, Section 4 shows algorithms to implement fault propagation through functional memory blocks, Section 5 presents results of the algorithm on benchmark circuits and Section 6 contains some concluding remarks.

2. Architectural Overview

The underlying hardware organization assumed is that of a special or general purpose multicomputer where the processors can be arranged in a pipeline with implicit communication permitted among them. However, any architecture based on message passing with no shared memory and communicating over a shared network can be used to implement such a scheme. Unlike other message passing systems, the architecture that we consider [2] requires no polling of the input queues to detect new messages due to a merging of network communication and computing functions in each pipeline stage. Results presented in Section 6 are based on software emulation of the second generation MARS [2] currently being developed.

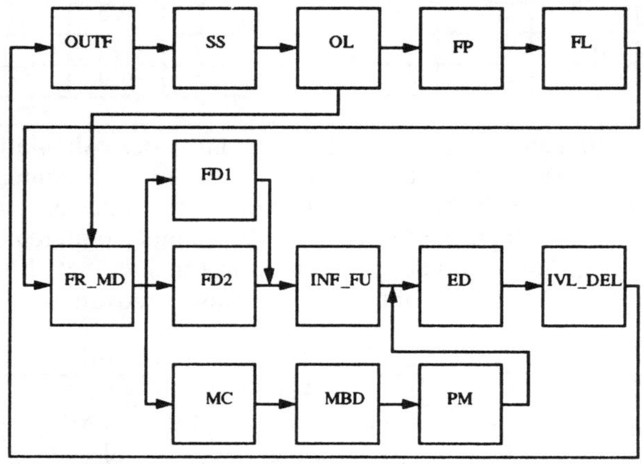

Figure 1: Pipeline Organization

Most of the stages are similar to those reported elsewhere[1], except for the addition of the functional memory pipeline, and the processing of new data structures in some of the stages. Figure 1 shows a fourteen processor pipeline arrangement with two separate parallel paths for gate level modeling (FD1, FD2 and INF_FU) and fault propagation through functional memory blocks (MC, MBD and PM). The following describes all pipeline stages except those concerned with memory modeling, for which the reader is referred to Section 4.

Signal Scheduler (SS): Other than scheduling both good and faulty events, it advances simulation time when appropriate. It controls the scheduling of new vector inputs, by passing appropriate messages to the Input Vector List (IVL) stage, and initiates fault detection by the Output Log (OL) stage.

Output Log (OL): Carries out fault detection by storing good and faulty values for each output gate. At the end of a vector, it informs the Fault Removal (FR_MD) stage of any newly detected faults so that they can be dropped from the fault lists of all active gates and memory blocks.

Fanout Pointer (FP) and Fanout List (FL): Store circuit connectivity and forward any event at the output of a gate to all members of its fanout.

Fault Removal and Memory Detection (FR_MD): This routes event messages to either the logic gate pipeline (FD1

or FD2) or the functional memory pipeline depending on gate type. Also, due to fault detection initiated by the OL PE at the end of each vector, this stage initiates fault dropping at the fault site. Memory faults (on address, data and control lines) are treated as output faults of dummy gates which are added to the circuit connectivity.
Fault Data Structure (FD1 and FD2): These stages maintain the fault list for each gate and perform *convergence-divergence* [6] functions on receiving good and faulty events. These stages are the most crucial in the pipeline and will be explained in greater detail in Section 3. Faults are partitioned among these two processing elements and good events are sent to both. These stages forward all input values for a gate to the next stage.
Input Fault Injection and Function Unit (INF_FU): For input faults, this stage replaces the appropriate input value by the injected value. Depending on the gate type and input values, it evaluates the gate by table lookup and passes on the new output value to the following stage.
Event Detector (ED): This stage detects logic events in the good circuit. For faulty circuits, the algorithm is slightly different and is related to the evaluation of *convergence-divergence* of faulty circuits. This will be explained in greater detail in the following section.
Input Vector List and Delay Unit (IVL_DEL): This stage schedules precompiled new vectors at the end of each vector period as and when indicated by the signal scheduler (SS). For all input messages from the event detector, it adds delay information to the sequence.
Output Fault Injection (OUTF): For gate output faults, this stage replaces the output value evaluated (by INF_FU) by the appropriate stuck-at value before forwarding it to the signal scheduler (SS).

3. Algorithm Partition

In this section, we elaborate on the algorithm and the data structure of some of the more important pipeline stages: Fault Data Structure (FD1 and FD2) and Event Detector (ED). The terminology in this section refers to a source as the gate at whose output an event occurs and a fanout gate is driven by such a source gate. Thus an event transmitted from the source arrives at one of the inputs of the fanout gate. As mentioned previously, and illustrated in Figure 2, FD1 and FD2 store the input values for each faulty circuit visible at a gate. These faults are stored as ordered link lists, sorted by fault identifiers. Sorting enables the processing of good events more efficient. If the lengths of the fault lists at the source and fanout gates are l_s and l_f respectively, processing of a good event for sorted lists has a complexity of $O(l_s + l_f)$. For purposes of efficiency,the fault lists are also hashed. This speeds up the searching process for faulty events. From our experiments, it appears that this speeds up the simulation by a factor of about two.

Each element of the link list stores 4 input pin values (bits 0-7), a fault injection bit (bit 8) and a 16-bit fault identifier (bits 16-31). For the good circuit value, the latter two fields are always zero. The injection bit is set for those members of a fault list of any gate whose fault identifier represents the circuit with either a faulty input or output pin for that gate. When input values of a gate in some faulty circuit are compared to those in the good circuit, the least significant 9 bits are compared. This prevents convergence in faulty circuits at injection sites where the stuck at value is equal to the good value.

The algorithm executed by FD1 and FD2 evaluates the *convergence* or *divergence* [6] of values in faulty circuits. All message sequences involving faulty circuits carry a 2-bit flag, referred to as the cd_flag, signifying the follow-

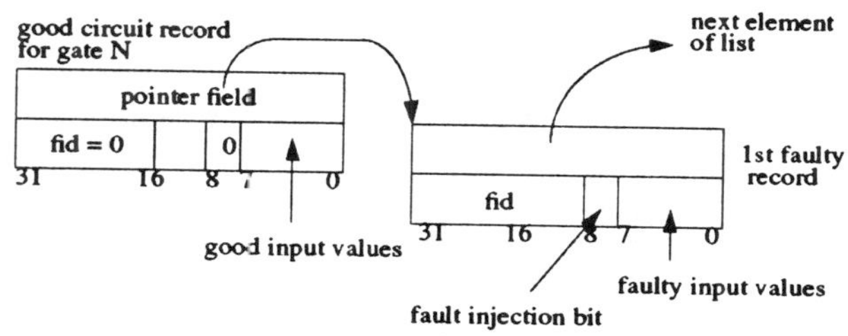

Figure 2: Fault List Data Structure

ing possible situations: convergence (01), divergence (10), detection (11) and normal update of a faulty value (00). Faulty events with the detected flag set arise at the end of a vector period when a fault is detected by the OL stage. Otherwise, the cd_flag is set or changed exclusively by the FD1 and FD2 stages. Figure 3 gives the pseudo code for these stages of the pipeline.

Figure 4 shows the algorithm for the Event Detection stage. As evident from the pseudo-code, this stage detects events for the good circuit and for faulty circuits only if the cd_flag is 00. For all other values of cdflag (DIVERGE, CONVERGE, DETECTED), the ED stage merely forwards its input messages. This permits the updating of fault lists by the FD stages at all fanout gates, recursively.

For existing faulty events, the FD algorithm determines if a fault has to be dropped from the fault list. If a fault is dropped, the cd_flag value of CONVERGE prevents the event detector from suppressing this message so that further checking can be carried out in the recursive fanout of the gate. If a fault does not get dropped, the cd_flag is changed to 00 (normal update) so that normal evaluation and event detection for the faulty circuit can be done.

For new faulty events, the FD algorithm decides whether a new fault should be added to the fault list of the fanout gate. If so, the flag is set to DIVERGE so that this fault can be added to fault lists in the recursive fanout if it indeed becomes visible at these gates. The ED and FD algorithms do not suppress these messages till the faulty value at some gate becomes equal to the good value.

4. Fault Propagation through Functional Memory Blocks

Functional modeling has been shown to be effective for handling memory blocks and register level primitives [6, 7]. However, this is the first reported work on functional memory modeling in a pipelined environment. We will restrict ourselves to memory blocks only rather than general functions.

Concurrent simulation of functional primitives in Multi List Traversal (MLT) [6, 7] involves reasoning of a single Reference (R) machine and a large number of Concurrent (C) machines. Consider a functional memory block which has separate fault lists associated with its address, data and control lines. Figure 5, adapted from [7], shows a memory block with the fault lists for the Write Enable (WE), Read Enable (RE), Address Bus (AB), Input Data Bus (DB), Output Data Bus (OB) and 4 memory locations. Faulty circuit F2 differs from the good circuit in two ways: the address bus and memory location 1 has values 1 and 9, respectively, instead of the good values 2 and 5, respectively. All other line values and memory contents

```
While (TRUE) {
  receive(source,fanout,fid,cd_flag,pin#,value);
  if (good_event){
    update good inputs; /* use pin# */
    send(fanout,old_inps,new_inps,fid=0,X);
    for each member of fault list of fanout gate {
      if (fault absent in source fault list)
        update faulty input values;
      if (faulty inputs == good inputs)
        send(fanout,old_inps,new_inps,fid,CONV);
      else
        send(fanout,old_inps,new_inps,fid,DIV);
    }
  }
  /* if fault exists in fanout fault list */
  else if (existing fault) {
    update faulty record of fanout gate;
    if (cd_flag == DETECTED){
      remove fid from fault list of fanout;
      send(fanout,X,X,fid,DETECTED);
    }
    else if (faulty inputs == good inputs){
      remove fid from fault list at fanout;
      send(fanout,old_inps,new_inps,fid,CONV);
    }
    else
      send(fanout,old_inps,new_inps,fid,00);
  }
  else { /* New fault for fanout gate */
    if (faulty inputs != good inputs) {
      insert new fault record for fanout gate;
      send(fanout,old_inps,new_inps,fid,DIV);
    }
  }
} /* while (TRUE) { */
```

Figure 3: Convergence Divergence Evaluation

```
while (TRUE) {
  receive(gate,old_value,new_value,fid,cd_flag);
  if (good event) {
    if (current_value != previous_value) {
      send(gate,new_value,fid=0,X);
    }
  }
  /* normal update in faulty circuit. */
  else if (cd_flag == 00) {
    if (old_value != new_value)
      send(gate,new_value,fid,00);
  }
  else /* convergence, divergence or detection */
    send(gate,current_value,fid,cd_flag);
}
```

Figure 4: Event Detection Algorithm

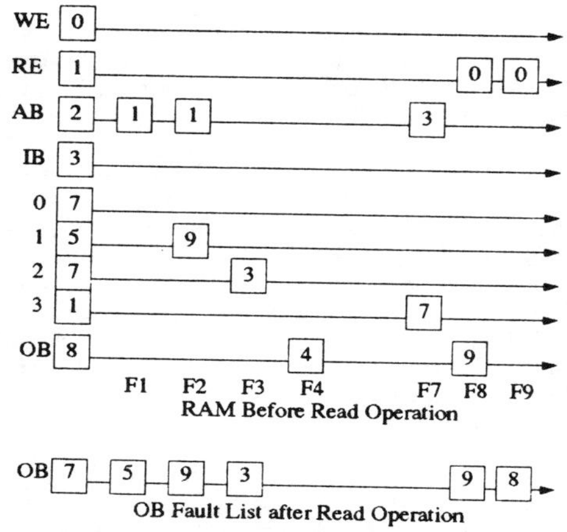

Figure 5: Fault List Data Structure for MLT

for this faulty circuit are identical to the good circuit. Figure 5 also shows the output bus (OB) fault list as a result of a read operation in the good circuit. Processing such a good circuit event requires the coordinated traversal of all relevant fault lists. For the above example, the order of faulty machine evaluation is F1, F2, F3, F4, F7, F8 and F9.

In our scheme, fault propagation through memory blocks is carried out by a three stage pipeline in parallel to the logic evaluation pipeline. The memory pipeline distributes control (WE and RE) and address line values in the Memory Controller (MC) stage, data line values in the Memory Block Descriptor (MBD) and memory contents and output bus values in the Physical Memory (PM) stage. The MC stage controls the memory pipeline by issuing read-write messages, for both good and faulty circuits, to the MBD stage. On receiving write messages, the MBD stage updates the functional memory contents by writing current data line values at the functional address. This is achieved by evaluating the functional address from address line values and sending current data line values to the PM stage. We store a single fault list in all three stages, as shown in Figure 6, and distribute the data among the various stages. Note that the MBD stage also has the same fault list even though none of the data line values are divergent. With this method of storing fault lists, processing of any event becomes essentially the same as functional memory simulation [2], provided the appropriate (faulty or good) values are used. The only additional work is that of setting the cd_flag for faulty read operations. It is set to DIVERGE (10) or normal update (00) if the fault is found to be active in any stage. However, if a fault is found to be inactive in all three stages, the cd_flag is set to CONVERGE and a fault removal process is initiated by PM, as explained below.

Fault removal is initiated by the PM stage whenever it receives a faulty event with the cd_flag set to CONVERGE, indicating convergence in the previous stages. It checks for convergence of memory contents and output line values and accordingly issues fault removal messages to MC. The MC stage verifies that no further change to either the good or faulty memory record has taken place since it last issued the convergence message to MBD.

Fault lists at source gates, necessary for handling good events, are not stored in the MC stage. The cd_flag associated with any faulty event is used to overcome this problem. A "fanin" flag, *one for each line*, indicates mem-

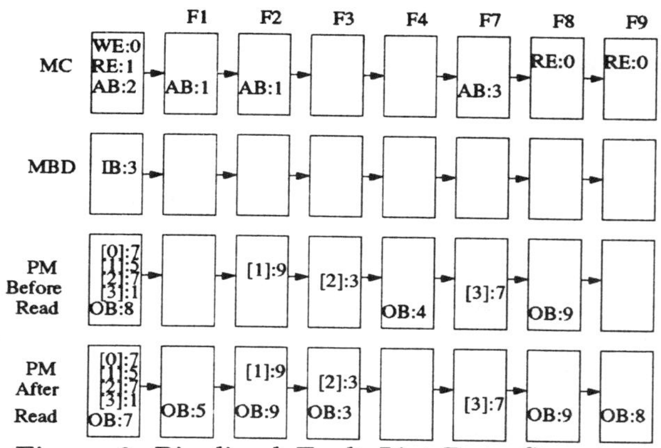

Figure 6: Pipelined Fault List Data Structure

bership in the fault list at the source gate. Whenever a faulty event with the cd_flag set to DIVERGE or normal update (00) is received, the fanin flag in the appropriate faulty record is set. When the cd_flag is CONVERGE, this flag is reset. A good event on any line updates the values in a faulty record only if the appropriate flag is reset. Otherwise, the fault record is left unchanged.

Our scheme only propagates a fault incident at the functional memory block inputs, as in some software fault simulators [4]. We estimate that the time required to process a memory event, whether good or faulty, is typically about 4 to 5 times that of a logic event.

5. Results

Table 1 presents statistics on some ISCAS benchmark circuits, along with performance figures for a software fault simulator[4] running on a SUN SPARC2 station. Table 2 gives results for our fault simulator. Based on the number of instructions for handling events in the FD stages (which is the highest of all stages and is thus the pipeline bottleneck), we estimate the minimum time (column 3 in Table 2) required to carry out the simulation assuming 100% pipeline efficiency and a clock frequency of 20 MHz. The fourth column (max time) is an estimate of the time it would take to carry out the entire simulation on one processor. These figures reveal that algorithmic pipelining gives a speedup factor of about 4 to 5.

Column 5 of Table 1 (MOTIS execution time) and column 4 of Table 2 (max time) reveal another interesting fact. We are able to exploit some amount of parallelism from the architecture [2] due to hardware features like multiple ALUs for arithmetic and bit field operations, pipelining of instruction fetch and execution and pipelined data memory access. Moreover, factors like relative merits of the two algorithms and the inefficiency of compiler generated machine code compared to manually optimized microcode also play an important role in enhancing the performance of our simulator.

At regular instances of time, we measure fault activity by averaging over the entire circuit, the length of the fault list at each gate. The highest value observed over the entire simulation gives us the peak activity. Preliminary results show that this figure is low enough for big circuits (64K gates and above) to allow injection of more than 1000 faults per pass. Our architecture can sustain a peak activity of 28 for circuits with 64K gates.

ckt name	gates	faults	vectors	MOTIS time (s)	cov (%)
c432	244	560	68	10.4	90
c880	630	1000	72	10.3	100
c1355	706	1000	154	25.5	99.3
c1499	463	1000	75	8.8	93.9
c1908	1162	1000	211	56.1	99.7
c3540	2735	1000	218	123	96.1
s344	339	552	85	26.4	80
s641	628	771	124	26.6	73.4
s1196	746	1000	338	57.9	86.5

Table 1: ISCAS Benchmark Circuits

ckt	total events	min (s)	max (s)	actual (s)	peak activity
c432	300900	0.6	2.7	0.9	6.9
c880	236000	0.5	2.1	1.1	4.7
c1355	1599700	3.2	14.4	3.4	13.7
c1499	603000	1.2	4.8	2.3	17.8
c1908	1139600	2.3	10.3	3.8	6.7
c3540	2888500	5.8	26.0	7.4	3.1
s344	1210000	2.4	10.9	5.8	25.7
s641	1452000	2.9	13.6	6.1	15.9
s1196	1474000	2.9	13.6	6.3	3.1

Table 2: Fault Simulation Results

6. Conclusions

We have shown that both unit and multiple delay concurrent fault simulation can be implemented on memory limited hardware accelerators without compromising any desirable feature. Mixed level modeling, which is often thought to be in the domain of software simulators, can also be implemented effectively. We have shown that a primitive memory management scheme, like that implemented with a link list, along with fault list hashing, not only overcomes the deficiencies of fixed space allocation but also has the potential to improve performance substantially.

References

[1] P. Agrawal, V. D. Agrawal, K. T. Cheng, and R. Tutundjian. Fault Simulation in a Pipelined Multiprocessor System. In *Proceedings of the International Test Conference*, pages 727–734, 1989.

[2] P. Agrawal and W. J. Dally. A Hardware Logic Simulation System. In *IEEE Transactions On Computer Aided Design*, pages 19–29, January 1990.

[3] W. C. Athas and C. L. Seitz. Multicomputers: Message passing concurrent computers. In *IEEE Computer*, pages 9–24, August 1988.

[4] C. Y. Lo, H. N. Nham, and A. K. Bose. Algorithms for an Advanced Fault Simulation System in Motis. In *IEEE Transactions on Computer Aided Design*, pages 210–218, March 1987.

[5] S. Patil and P. Banerjee. Fault Partitioning Issues in an Integrated Parallel Test Generation/Fault Simulation Environment. In *Proceedings of the International Test Conference*, pages 718–726, 1989.

[6] E. Ulrich. Concurrent Simulation at the Switch, Gate, and Register Levels. In *Proceedings of the International Test Conference*, pages 703–709, 1985.

[7] E. Ulrich, D. Lacy, N. Philips, J. Tellier, M. Kearney, T. Elkind, and R. Beaven. High Speed Concurrent Fault Simulation with Vectors and Scalars. In *Proceedings of the Design Automation Conference*, pages 374–380, 1980.

HOPE: An Efficient Parallel Fault Simulator for Synchronous Sequential Circuits

Hyung Ki Lee and Dong Sam Ha

The Bradley Department of Electrical Engineering
Virginia Polytechnic Institute and State University
Blacksburg, Virginia 24061

(Phone: 703-231-4942, E-mail: dsha@vtvm1.cc.vt.edu)

Abstract

In this paper, we present an efficient sequential circuit parallel fault simulator, HOPE, which simulates 32 faults at a time. The key idea incorporated in HOPE is to screen out faults with short propagation paths through the single fault propagation. A systematic method of identifying faults with short propagation paths is presented. The proposed method substantially reduces the total number of faults simulated in parallel to achieve high speedup. Our experimental results show that HOPE is about two times faster than PROOFS [5] for most of ISCAS89 sequential benchmark circuits.

I. Introduction

Over a decade intensive research has been conducted for efficient fault simulation [1]-[7]. The parallel pattern single fault propagation technique is known to be the most effective for combinational circuits [1]-[2]. The concurrent fault simulation method has been widely used for sequential circuits [3]. Recently, different types of fault simulators for sequential circuits have been also investigated [4]-[7].

Cheng and Yu introduced a new sequential circuit fault simulator called differential fault simulator (DSIM) based on the single fault propagation [4]. In the single fault propagation, all the faults are simulated, one fault at a time, under the application of a test pattern. After the simulation of a fault, the good values are restored from the good circuit simulation. The key difference between DSIM and the single fault propagation is that, in DSIM, good circuit values are restored based on the previous fault simulation. DSIM substantially reduces simulation time and memory size compared to concurrent fault simulators [4].

Nierman et al. developed a fault simulator, PROOFS, which is a parallel version of the single fault propagation [5]. A group of 32 faults are simulated in PROOFS. Various heuristics are employed in PROOFS to speed up the fault simulation. According to the experiments of [5], PROOFS is 6 to 67 times faster than a concurrent fault simulator for ISCAS89 sequential benchmark circuits [8]. PROOFS also reduces the memory requirement by about five times. An enhancement in speed on the original PROOFS was made through the incorporation of the so called star algorithm [6].

Recently, Gouders and Kaibel introduced a sequential circuit fault simulator, PARIS, which employs the parallel pattern single fault propagation (PPSFP) technique [7]. The experimental results of [7] show that PARIS is faster than PROOFS for most benchmark circuits, but for some large circuits, PARIS is substantially slower than PROOFS. It seems that the high speed of PARIS is attributed to its efficient coding as well as the PPSFP technique. The use of the PPSFP technique for sequential circuits is interesting, and it needs more investigation. All the above three simulators employ the zero gate delay model.

Although PROOFS is efficient in speed, there is still room for further improvement. In this paper, we present a sequential circuit fault simulator, HOPE, which further reduces fault simulation time. Like PROOFS, HOPE is a parallel fault simulator based on the single fault propagation. It adopts the zero gate delay model. The key idea incorporated in HOPE is to screen out faults with short propagation paths, and prevent them from being simulated in parallel. The screening process drastically reduces the number of faults simulated in parallel to achieve substantial speedup. Our experiments demonstrate that HOPE is about two times faster than PROOFS for most of ISCAS89 sequential benchmark circuits.

The organization of the paper is as follows. In Section II, several related terms and a shortcoming of the fault simulator PROOFS is presented. In Section III, our method which addresses the shortcoming is presented. In Section IV, experimental results of HOPE and PROOFS are reported. Finally, Section V summarizes the paper.

II. Background

In this section, we describe definitions, and explain terms. After that, we state a shortcoming of PROOFS. Throughout this paper, three logic values, 0, 1 and x, are considered for signal lines, where x is an unknown state. Single stuck-at faults are assumed for circuits.

A. Terms

The removal of flip-flops (FFs) from a synchronous sequential circuit turns the circuit into a combinational circuit. The outputs of the FFs in the combinational circuit are called pseudo-primary inputs (PPIs) and the inputs of the FFs are called pseudo-primary outputs (PPOs). When fanout stems of a combinational circuit are removed, the circuit is partitioned into fanout free regions (FFRs). The output of each FFR can be a stem, a primary output (PO) or a PPO. For the sake of convenience, POs and PPOs are considered as stems in this paper. Let FFR(s) denote the FFR whose output is s. An example circuit and its partition into FFRs is shown in Figure 1.

Dominators of a signal line are the lines through which all the paths from the signal line to the POs and PPOs of the circuit pass [2]. For all the lines inside a FFR(s), the stem s is a dominator of the lines. However, a stem may or may not have a dominator. For example, consider the circuit given in Figure 1 b). All paths from the stem i pass through lines n and p. Hence, n and p are dominators of the stem i. However, the stem p does not have any dominator.

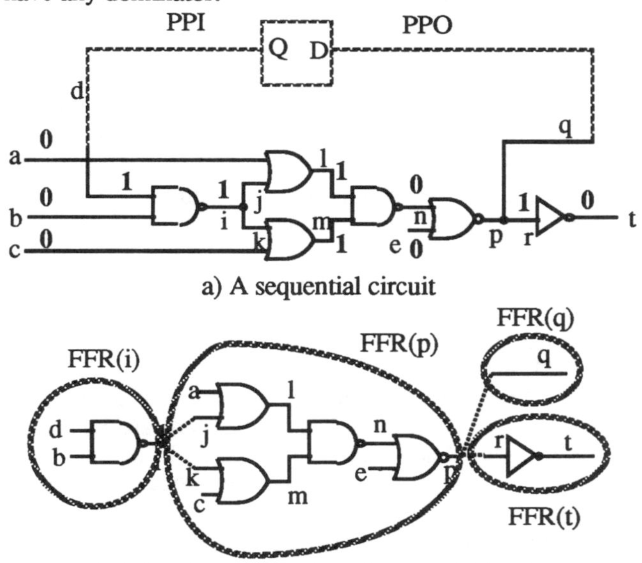

a) A sequential circuit

b) A Partition into Combinational Circuits

Figure 1. An Example Circuit

Under the application of a test pattern, a fault f at a time frame belongs to one of the following two categories depending on the effect of the fault. The first case is that the faulty value originates only at the faulty site as shown in Figure 2 a). We call the fault f **single event fault** in this paper. The second case is that faulty values originate at some PPI(s) as well as the faulty site as shown in Figure 2 b). We call the fault f **multiple event fault**. The effect of a multiple event fault is the same as the occurrence of multiple stuck-at faults at the time frame. A multiple event fault occurs when the effect of the fault has propagated to some PPO(s) under the application of the previous test pattern. It should be noted that a fault f can be a single event fault at one time frame and a multiple event fault at another time frame.

Let x/a* denote a stuck-at-a* fault occurred on a line x, where a*∈ {0,1,x}. Suppose that the fault x/a* propagates to another signal line y whose fault free value is b. Then, let the faulty value of y (after the propagation of the fault x/a*) be b*. Then we can consider a pseudo-fault y/b* which appears on the line y. If the faulty value b* is equal to b, i.e., the fault x/a* does not propagate to y, then the pseudo-fault y/b* is called **insensitive fault** in this paper. Otherwise, the fault y/b* is called **sensitive fault**. The propagation of a fault x/a* at a line x to y can be viewed as a mapping of the fault x/a* to a pseudo-fault y/b*.

Two faults f and g are said to be equivalent at a time frame if they produce the same logic values at all the POs and the PPOs. Then a fault g instead of f can be injected to the circuit at the time frame.

Statement: If the line y is a dominator of the line x, then the two faults x/a* and y/b* are equivalent.

Hence, the fault y/b* instead of the original fault x/a* can be injected to the circuit and simulated.

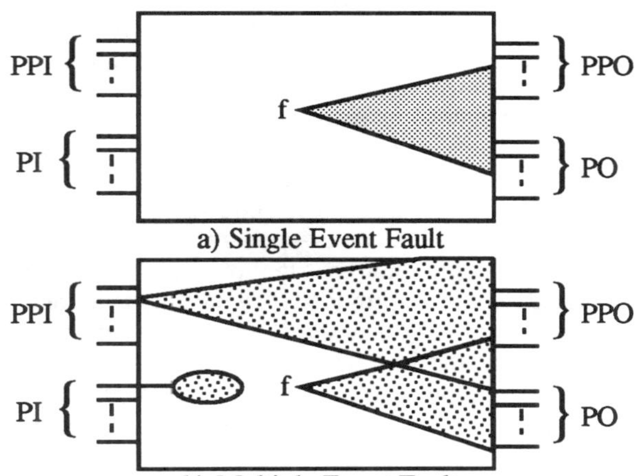

a) Single Event Fault

b) Multiple Event Fault

Figure 2. Single and Multiple Event Faults

B. Problem Statement

Two key issues in fault simulation are 1) reduction of faults to be simulated explicitly, and 2) efficient propagation of faults. In parallel fault simulation, the first issue becomes the reduction of the total number of faults to be simulated in parallel, and is the focus of this paper.

To our knowledge, PROOFS is the only sequential circuit parallel fault simulator reported in the open literature [5]. In the original PROOFS of [5], all the faults with faulty values different from fault-free values are injected and simulated in parallel. Although PROOFS achieves substantial speedup through the parallel simulation, some faults with short propagation zones may not need to be simulated in parallel. Consider the example circuit shown in Figure 1. Suppose that four faults, a/1, b/1, c/1 and d/0, are put into the same packet and simulated in parallel. The effects of the three faults, a/1, c/1 and d/0, stop at the outputs of the immediately following gates. However, the effect of the second fault b/1 propagates all the way to the PO t. This necessitates the packet containing the four faults be simulated to the primary output, even though the three faults, a/1, c/1 and d/0, do not propagate to the outputs of the immediately following gates. Therefore, the bits assigned for the three faults are wasted, as these bits could have been assigned to other faults. If such faults are identified in the early stage and are not put into the packet, the waste can be avoided. An ad hoc method is undertaken to address the problem in the modified PROOFS [6]. In the modified PROOFS, each fault is propagated one or two level(s) further. The fault is put into parallel simulation only if it propagates to the output of any gate at the level(s). In this paper, we fully investigate the issue and present a systematic algorithm which addresses the problem.

III. Proposed fault simulation algorithm

In this section, we present our method which addresses the shortcoming of PROOFS described above. Our strategy is to identify faults with short propagation paths in the early stage using single fault propagation and to prevent them from being simulated in parallel.

As explained in Section II, faults can be categorized into two groups, single event faults and multiple event faults, according to the effect on PPIs. In HOPE, all the single event faults inside FFRs, equivalently to say single event non-stem faults, are mapped into the equivalent stem faults using single fault propagation. Only stem faults are simulated in parallel provided they meet certain conditions described later. All the multiple event faults are simulated in parallel. In the following subsections, we explain the simulation of single event faults. Faults mentioned in the rest of the section, if otherwise specified, are single event faults.

A. Simulation of Non-Stem Faults

Consider a fault f inside a FFR(s) with output as s. The fault f can be mapped into the equivalent fault f_s on the stem line s. In HOPE, the fault f is first simulated to its stem using single fault propagation. If the equivalent stem fault f_s is insensitive, the fault f_s (as well as f) is dropped from further fault simulation. The fault dropping is possible since the effect of a single event fault appears only at the faulty site, and not at any PPIs. If the stem s is a PO or a PPO, the simulation of f_s is trivial. Therefore, only when the fault f_s is sensitive and the line s is a real stem (i.e., neither a PO nor a PPO), the fault is considered for further simulation.

For example, consider the circuit shown in Figure 3. Suppose that there exist seven faults within the FFR(s) as shown in the figure. Each fault can be mapped into one of the three faults, s/1, s/0, or s/x, at the stem s. The s/x fault may occur when the values of some PPIs are at unknown state x. Among the three stem faults, one fault is always insensitive. For example, when the fault free value of the stem s is 1 as shown in the figure, the fault s/1 is insensitive. Therefore, only two faults, s/0 and s/x, are further considered for parallel fault simulation. Stem faults are simulated to POs and PPOs in parallel provided they meet certain conditions. The conditions are described in the following section. In summary, HOPE simulates at most two stem faults in parallel. All the non-stem faults (which are in majority) are not simulated in parallel at all.

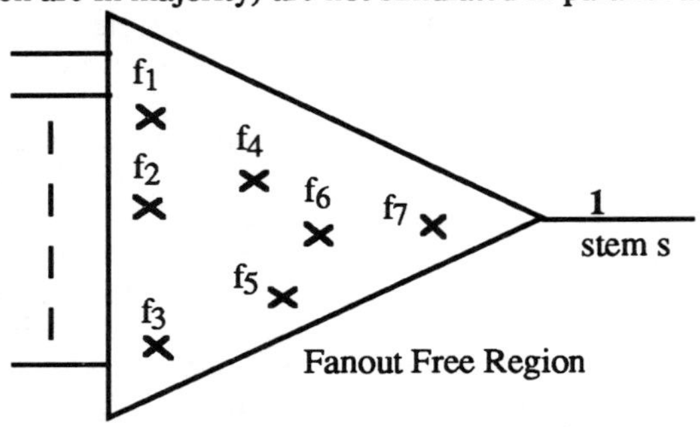

Figure 3. Simulation of non-stem faults

B. Simulation of Stem Faults

A sensitive stem fault f_s is further examined to screen out faults with small propagation paths. The fault f_s is attempted to propagate to a certain node(s) using single fault propagation. If the fault successfully propagates to the node(s), then the fault is simulated in parallel. The examination process is called **candidacy test**. The node(s) to which a stem fault is attempted to propagate, and the fault to be injected into a packet depend on whether the stem has a dominator or not. Details of the candidacy test are explained below.

1) Candidacy Test of Stem Faults with Dominators

When a stem s has a dominator, it is easy to show that all the fanout branches of the stem are inputs of a single fanout free region, say FFR(t). In our candidacy test, the stem fault f_s is attempted to propagate to the line t using single fault propagation. After the fault f_s is mapped into the equivalent fault f_t at the line t, the fault f_t is inserted into a packet and simulated in parallel provided

1) the line t is a real stem (neither a PO nor a PPO),
2) the fault f_t is sensitive, and
3) the fault f_t has not been simulated in a previous pass.

An example circuit is shown in Figure 4. The stem fault, s stuck-at-0 fault, is propagated to the line t using single fault propagation. If the stuck-at-0 fault at the line t has not been simulated in the past, the fault is injected into a packet and simulated in parallel. It is important to note that the equivalent fault f_t, not the original stem fault f_s, is put into a packet for parallel simulation. It avoids the re-simulation of the area of FFR(t).

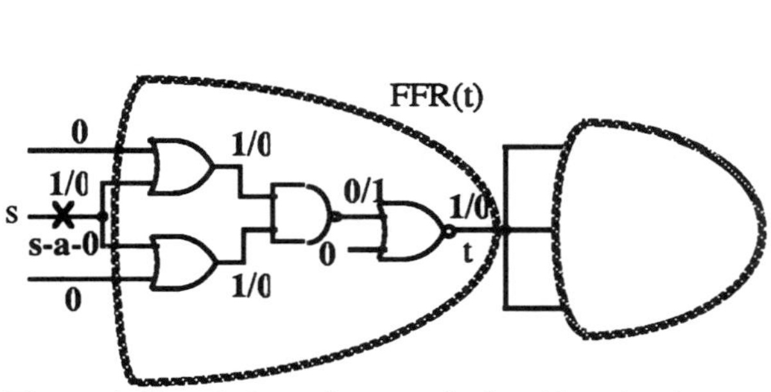

Figure 4. Simulation of a stem fault with a dominators

2) Candidacy Test of Stem Faults without Dominators

In this case, a stem fault can not be mapped into an equivalent single fault on a line. Hence, the candidacy test is performed to the immediately following gates of the stem. If a stem fault f_s propagates to any one output of the gates, the fault f_s is put into a packet for parallel simulation. Otherwise, the fault is dropped from further simulation. An example circuit is shown in Figure 5. Since the fanout branches of the stem s do not reconverge at a gate, a PO or a PPO, the stem s does not have any dominators. As the s-a-0 fault at the stem s propagates to t, the fault s s-a-0 is injected in a packet, and simulated in parallel.

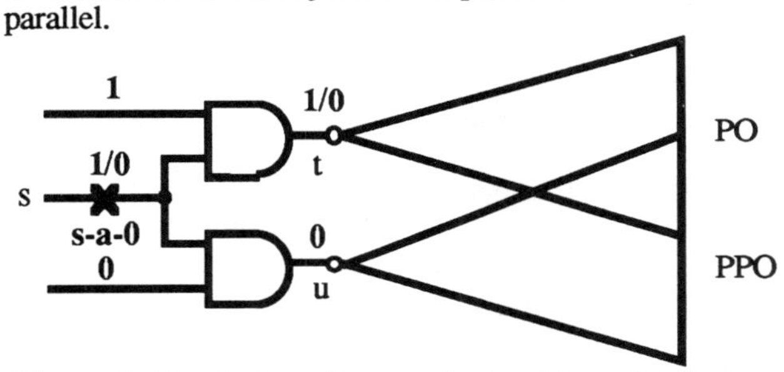

Figure 5. Simulation of a stem fault without dominators

IV. Experimental results

In the previous chapter, we proposed a method of reducing the number of faults simulated in parallel. In order to measure the effectiveness of the method, we conducted experiments using ISCAS89 benchmark circuits. Three major objectives of our experiments were:

1. to measure the portions of single event faults versus multiple event faults,
2. to find the effectiveness of the proposed algorithm in terms of reducing the number of faults simulated in parallel, and
3. to compare the speed of our simulator with a competing fault simulator, PROOFS.

To accomplish the objectives, we have implemented a fault simulator, HOPE, which employs the proposed method. HOPE is written in C language, and runs on Sun Sparc workstations. In the experiments, all the flip-flops of the circuits were initialized to unknown value x. Test patterns generated by a test generator GENTEST of AT&T were applied to each circuit.

As explained in Section III, the method of reducing faults simulated in parallel applies to only single event faults. All the multiple event faults are simulated in parallel in our fault simulator HOPE (as well as in PROOFS). Since the effectiveness of the proposed method depends on the portion of single event faults, we counted the numbers of single and multiple event faults for each time frame under the application of a test pattern. The experimental results are shown in Table 1.

Table 1: Occurrence of Single and Multiple Event Faults

name	no. of gates	no. of FFs	no. of tests	single event faults (%)	multiple event faults (%)
s298	133	14	162	83.53	16.47
s344	175	15	91	79.83	20.17
s382	179	21	2463	64.17	35.83
s444	202	21	1881	60.35	39.65
s526	214	21	754	80.32	19.68
s641	398	19	133	93.01	6.99
s713	412	19	107	91.63	8.37
s820	294	5	411	98.46	1.54
s832	292	5	377	98.49	1.51
s953	424	29	16	80.98	19.02
s1238	526	18	349	97.64	2.36
s1423	731	74	36	67.05	32.95
s1488	659	6	590	98.16	1.84
s1494	653	6	469	97.94	2.06
s5378	2958	179	408	90.54	9.46
s35932	17793	1728	86	85.14	14.86
average	1628	136	521	85.45	14.55

As shown in Table 1, on the average 85% of faults are single event faults under the application of the test patterns. The high percentage of single event faults predicts that the proposed method would be effective. Note that the percentage of single event faults is independent on the circuit size, number of test patterns and fault coverage.

Next, we measured the reduction ratio of faults simulated in parallel for HOPE. The experimental results are given in Table 2. The numbers given under the column headings HOPE and PROOFS are the total numbers of faults put into packets for parallel simulation during the entire fault simulation of each circuit. The reduction ratio in the last column is the number of faults simulated in parallel for HOPE to that of PROOFS.

As shown in Table 2, HOPE achieves on the average 67% reduction of faults simulated in parallel as compared to PROOFS. In other words, 67% of faults are screened out through the less expensive single fault propagation in HOPE. The high reduction ratio indeed speeds up HOPE as can be seen in the next experimental results.

Finally, we compared the performance of HOPE with PROOF in speed. We measured the CPU time of HOPE and PROOFS on a SUN Sparc 2 workstation. The results are shown in Table 3. As shown in the table, the speedup of HOPE over PROOFS is on the average 2.08.

Table 2. Number of Faults Simulated in Parallel

name	no. of tests	no. of faults	PROOFS	HOPE	reduction ratio
s298	162	308	6131	2436	0.40
s344	91	342	2864	1351	0.47
s382	2463	399	83556	59318	0.71
s444	1881	474	106587	76531	0.72
s526	754	555	91694	39617	0.43
s641	133	467	7032	1530	0.22
s713	107	581	7986	1926	0.24
s820	411	850	51512	3404	0.07
s832	377	870	48851	3111	0.06
s953	16	1079	12132	5019	0.41
s1238	349	1355	53633	6713	0.13
s1423	36	1515	30091	18208	0.61
s1488	590	1486	59701	4128	0.07
s1494	469	1506	56227	4208	0.07
s5378	408	4603	307550	86116	0.28
s35932	86	39094	553870	202627	0.37
average	521	3468	92464	32265	0.33

As expected, HOPE performs better for circuits with high percentage of single event faults. For one particular circuit s953, HOPE is slower than PROOFS. This is due to the small number of test patterns. We tried 1000 random patterns (which achieved the fault coverage 8.34%), and the speedup ratio increased to 1.40 from 0.63. The above three tables show that there is a strong correlation between the percentage of single event faults, the reduction ratio of faults simulated in parallel and the speedup ratio. Considering the efficiency of PROOFS, the speedup of HOPE is remarkable.

A method of reducing faults simulated in parallel for PROOFS was presented in [6]. According to the experiment, on the average 17% of CPU time of PROOFS was reduced. In that case, the average speedup ratio of HOPE would be reduced to 1.57.

Table 3. Performance of HOPE and PROOFS

name	no. of tests	no. of faults	fault cover age	CPU time		speed up ratio
				PROOFS	HOPE	
s298	162	308	85.71	1.1	0.4	2.75
s344	91	342	96.20	0.8	0.3	2.67
s382	2463	399	90.98	17.2	9.2	1.87
s444	1881	474	89.45	20.8	10.2	2.04
s526	754	555	75.32	11.0	6.0	1.83
s641	133	467	86.30	1.4	0.6	2.33
s713	107	581	80.90	1.4	0.6	2.33
s820	411	850	81.88	5.1	2.1	2.43
s832	377	870	81.38	4.8	2.1	2.29
s953	16	1079	7.78	0.5	0.8	0.63
s1238	349	1355	94.69	4.9	2.0	2.45
s1423	36	1515	24.42	3.8	2.0	1.90
s1488	590	1486	92.60	12.7	6.3	2.02
s1494	469	1506	91.10	10.7	5.3	2.02
s5378	408	4603	74.02	51.5	26.5	1.94
s35932	86	39094	87.99	173.1	95.3	1.82
average	521	3468	77.55	20.1	10.6	2.08

V. Summary

In this paper, we presented an efficient sequential circuit fault simulator, HOPE. The key idea incorporated in HOPE is to screen out faults with short propagation paths through the single fault propagation. A systematic method of identifying faults with short propagation paths is presented. The method substantially reduces the number of faults simulated in parallel to achieve high speedup. Our experimental results show that HOPE is about two times faster than PROOFS [5] for most of ISCAS89 sequential benchmark circuits.

Acknowledgment and announcement

The authors would like to thank Professor Janak Patel for providing us with PROOFS used in the experiments. The source code for HOPE is released to public. For details, please contact the second author.

References:

[1] J. A. Waicukauski, E. B. Eichelberger, D. O. Forlenza, E. Lindbloom and T. McCarthy, "Fault Simulation for Structured VLSI," VLSI Systems Design, Vol. 6, No. 12, pp. 20-32, Dec. 1985.

[2] H. K. Lee and D. S. Ha, "An Efficient Forward Fault Simulation Algorithm Based on the Parallel Pattern Single Fault Propagation," in Proc. Int. Test Conf., pp. 946-955, Oct. 1991.

[3] E. G. Ulrich and T. Baker, "The Concurrent Simulation of Nearly Identical Digital Networks," 10th Design Automation Workshop, Vol. 6, pp. 145-150, June 1973.

[4] W.-T. Cheng and M.-L. Yu, "Differential Fault Simulation - A Fast Method Using Minimal Memory," 26th Design Automation Conference, pp. 424-428, June 1989.

[5] T. M. Niermann, W.-T. Cheng and J. H. Patel, "PROOFS: A Fast, Memory Efficient Sequential Circuit Fault Simulator," 27th Design Automation Conference, pp. 535-540, June 1990.

[6] E. M. Rudnick, T. M. Niermann and J. H. Patel, "Methods of Reducing Events in Sequential Circuit Fault Simulation," To appear in Proc., in Proc. Int. Conf. on Computer Aided Design, pp. 546-549, Nov., 1991.

[7] N. Gouders and R. Kaibel, "PARIS: a parallel pattern fault simulator for synchronous sequential circuits," in Proc. Int. Conf. Computer-Aided Design, pp. 542-545, Nov. 1991.

[8] F. Brglez, D. Bryan and K. Kozminski, "Combinational profiles of Sequential Benchmark

On the Distribution of Fault Coverage and Test Length in Random Testing of Combinational Circuits

Amitava Majumdar
Dept. of Electrical Engineering
Southern Illinois University
Carbondale, IL 62901-6603

Sarma Sastry
Electrical Engineering-Systems Dept.
University of Southern California
Los Angeles, CA 90089

Abstract

We propose new models for the process of testing faults in combinational circuits under both random and pseudorandom tests. Based on these models, the *probability distribution function* (pdf) of fault coverage is derived. Furthermore, using an important relationship between the pdf of fault coverage and that of test length we derive an expression for the pdf of test length. These analytical results allow us to obtain expressions for expected values of fault coverage and test length as well as other important statistics not obtained by existing techniques. Empirical results validating these models for fault coverage analysis are presented from experiments with several circuits. Techniques for estimation of necessary parameters, based on both statistical as well as probabilistic models, are proposed. These techniques, combined with the theoretical results, define a comprehensive methodology for random testability prediction of combinational circuits.

1 Introduction

Fault coverage (FC) and test length (TL) are the two basic metrics used for evaluating the effectiveness of random or pseudo random testing of digital circuits. In this paper we present a new model for random pattern testing of combinational circuits and derive the *probability distribution* function of FC as a function of TL and the probability distribution function of TL as a function of FC. The analytical results presented in this paper offer a significant improvement over existing results. Whereas prior work in this area only provide formulae for the expected values of FC and TL, we derive their distribution functions under fairly general conditions. This not only allows the computation of their expected values, but also yields estimates of errors in their prediction. Furthermore, knowledge of the distribution functions permits quantitative answers to some practical questions such as when further testing will not produce any significant improvement in FC or how many more test vectors should be applied to obtain a specified improvement in fault coverage.

The main aim of this paper is to provide an analytical solution to the problem of estimating the distribution of FC of a given circuit under random test. We also establish a relationship between the distribution of TL and that of FC which allows us to obtain characteristics of TL as well. Experimental results presented in this paper illustrate the usefulness of the models in predicting FC. Verification of these models in predicting TL is left for future research. For the sake of clarity of the discussion, all our experimental results are presented at the end of the paper.

[0]This work was supported in part by NSF award No. MIP-9111206 and by a grant from DARPA monitored by the FBI under contract No. JFBI 90092.

2 Literature Review

A brief examination of the literature on random testing shows that three different approaches have been taken. We classify each of these three methods as being *statistical*, *empirical* or *probabilistic*.

The statistical method proposed by Agrawal [1] is aimed at evaluating a given test set T^* in terms of its FC. Vectors in T^* are fault simulated with respect to a randomly selected subset F^* of all faults. Using a hypergeometric distribution, confidence intervals on FC with respect to the complete fault set, are obtained.

Goel [3], Williams [13] and later Huang et al [4] use empirical models for estimating FC. Goel constructs a cost model for a two phase testing strategy. In Phase 1, pseudo random patterns are applied and in Phase 2 deterministic test patterns are generated for those faults left untested in Phase 1. Based on empirical evidence, Goel fits a decaying exponential function for the number of untested faults in Phase 1 and a linear function for Phase 2. The parameters of these functions are estimated from simulation results. Williams [13] proposed a similar exponential function as a predictor for FC. Huang et al. [4] propose a three parameter function for the average FC as a function of TL. Iterative methods for estimating the parameters based on experimental values of FC are presented.

A *probabilistic* approach has been adopted by several authors for modelling fault (pattern) coverage and test length [11, 12, 10, 9]. Developing analytically tractable models for the distribution of FC and TL has been considered difficult and most earlier research has focused on predicting their expected values or suitable bounds. The various methods appearing in

0738-100X/92 $3.00 © 1992 IEEE

the literature differ primarily in the amount of circuit specific information that is used.

In [11], Savir and Bardell present a methodology for computing bounds on TL. For simplicity, these bounds are estimated only for faults in a subset A of hard-to-detect faults. Using an upper bound on the probability that at least one fault in A escapes detection, an upper bound on TL for 100% coverage of faults in A is obtained.

Wagner et al. [12] examine both random and pseudo random testing and derive formulae for *expected values* of FC. Faults in a circuit are grouped according to the number of vectors that test each fault thus forming a fault *detectability profile.* A hypergeometric vector sampling model is then used to predict expected FC. This method does not yield accuracy or confidence of predictions. Furthermore, predictions of expected FC are obtained only for the case where each input vector is sampled with equal probability.

3 Model of Testing

For the sake of conciseness and clarity the following notation will be used throughout the remainder of this paper.

$\mathcal{F}$ = set of all faults.

$M = |\mathcal{F}|$.

$\binom{n}{k}$ = binomial coefficient n *choose* k.

$x^{\underline{r}} = x(x-1)\cdots(x-r+1)$.

$\Delta f(x) = f(x+1) - f(x)$.

$\Delta^n f(0) = [\Delta^n f(x)]_{x=0}$

Fault Sampling Model: Associated with each test vector t is a subset $f_t \subset \mathcal{F}$ of faults that are detected by t. Let the number of faults detected by t_i be k_i, i.e., $k_i = |f_{t_i}|$. The application of a randomly selected test vector can be viewed as randomly selecting the corresponding subset of faults. Thus, if we apply a sequence of r test vectors, $t_1, t_2, \ldots, t_r$, with or without replacement, we would observe r random numbers $k_1, k_2, \ldots, k_r$, where $k_i = |f_{t_i}|, 0 \leq k_i \leq M$. Since a test vector t_i detects a subset f_{t_i} of faults, we can represent the application of t_i as sampling the set $\mathcal{F}$ of faults, k_i number of times, without replacement. Now for any two test vectors, t_i and t_{i+1}, $f_{t_i} \cap f_{t_{i+1}}$ is not necessarily empty. Therefore, the faults detected by t_i are *marked* as being detected and replaced back in the population before the application of t_{i+1}.

Let W be a *random variable* (rv) that denotes the number of faults detected on any trial. The application of a sequence of r test vectors $t_1, t_2, \ldots, t_r$ gives rise to r independent realizations $k_1, k_2, \ldots, k_r$ of W.

Definitions:

Definition 1 *The fault coverage X_r is a rv that denotes the number of distinct faults that are detected by the application of r randomly selected test vectors.*

Definition 2 *The test length R_x is a rv that denotes the number of test patterns required to obtain a fault coverage of at least x.*

Definition 3 *For a given sampling distribution of input vectors, let p_i be the probability of detecting fault $f_i \in \mathcal{F}$ and let $P = \{p_j\}$ is the set of all possible detectabilities, i.e. $|P| \leq M$. Also let $h(p_j)$ $(p_j \in P)$ denote the number of faults with a detectability of p_j. Then the detectability profile H is the set of all $h(p_j)$ where $p_j \in P$.*

Having defined the sampling model for random testing we are interested in deriving the distribution of X_r and R_x. In the interest of clarity and conciseness, we omit the proofs of the theorems from the text.

4 Distribution of Fault Coverage

Fault coverage X_r is a measure of the quality of a random sequence of r vectors and is a random variable. We recall that the number of distinct faults tested by any single randomly selected vector is represented by the rv W, i.e. $W = X_1$. The probability mass function (pmf) and the probability generating function (pgf) of X_r conditioned on k_i faults being tested by the ith vector $(1 \leq i \leq r)$ is given below. Here each k_i is an independent realization of W.

Theorem 1 : *The pdf and pgf for X_r, conditioned on $k_i, i = 1, 2, \ldots, r$, are respectively given by*

$$P\{X_r = j \mid k_1, k_2, \ldots, k_r\} = \frac{\binom{M}{j}}{\prod_{i=1}^{r} M^{\underline{k_i}}} \Delta^j \left[\prod_{i=1}^{r} (0)^{\underline{k_i}}\right] \quad (1)$$

$$G_{X_r}(s \mid k_1, k_2, \ldots, k_r) = \frac{1}{\prod_{i=1}^{r} M^{\underline{k_i}}} (1 + s\Delta)^M \left[\prod_{i=1}^{r} (0)^{\underline{k_i}}\right] \quad (2)$$

These equations allow us to examine various distributions of W and obtain much simpler formulae for the special cases of interest. We first provide a formula for the factorial moments of X_r from which the mean and variance are easily obtained.

Lemma 1 : *The jth factorial moment of X_r, conditioned on $k_i, i = 1, 2, \ldots, r$, is given by*

$$E\left(X_r^{\underline{j}} \mid k_1, k_2, \ldots, k_r\right) = \frac{M^{\underline{j}}}{\prod_{i=1}^{r} M^{\underline{k_i}}} (1+\Delta)^{M-j} \Delta^j \left[\prod_{i=1}^{r} (0)^{\underline{k_i}}\right] \quad (3)$$

Evaluating Equation 3 for $j = 1$ and $j = 2$ yields the conditional expectation and variance of FC. Let $\xi_t = \prod_{i=1}^{r} \left(1 - \frac{k_i}{t}\right)$. Then

$$E(X_r \mid k_1, k_2, \ldots, k_r) = M(1 - \xi_M) \quad (4)$$

$$Var(X_r \mid k_1, k_2, \ldots, k_r) = M\xi_M [1 - M\xi_M + (M-1)\xi_{M-1}] \quad (5)$$

Equations 1 through 5 all depend on the statistics of the rv W. To get some insight into the utility of these formulae as well as obtain simplifications, we examine two special cases. These are:

1. Deterministic W : This situation corresponds to assuming that W is a constant k^*, i.e. $k_i = k_*$ $(1 \leq i \leq r)$ implying that every test vector tests exactly k^* faults. Although this assumption seems unrealistic, we show three methods for determining an appropriate value of k^* which yield excellent estimates of FC as a function of test length.

2. W is random variable: Here we assume that W is a

binomial rv with parameters M and p, where p is an unknown constant. This yields a much simpler form for the unconditional distribution of FC and is useful for large circuits where variance in fault detectabilities may be large. In such cases, we partition the fault set into equal (or approximately equal) detectability subsets. Using this as our estimate of the detectability profile we can obtain expected FC values and their confidence levels.

4.1 Deterministic W

Replacing $k_i, 1 \leq i \leq r$ by k^* in Equation 1, we obtain

$$P\{X_r = j \mid k^*\} = \frac{1}{(M^{\underline{k^*}})^r} \begin{pmatrix} M \\ j \end{pmatrix} \Delta^j \left[\left((0)^{\underline{k^*}} \right)^r \right] \quad (6)$$

Similarly we can obtain expressions for the pgf and factorial moments of X_r. Letting $\lambda_t = \left(1 - \frac{k^*}{t}\right)^r$, Equations 4 and 5 yield

$$E(X_r \mid k^*) = M(1 - \lambda_M) \quad (7)$$

$$Var(X_r \mid k^*) = M\lambda_M [1 - M\lambda_M + (M-1)\lambda_{M-1}] \quad (8)$$

The main issue here is selecting a suitable k^* that yields good predictions of FC. We consider three methods of selecting k^* for a given circuit.

Method A: We let $k^* = E[W]$, the expected number of faults tested by a random vector. Given the detectability profile, we can estimate $E[W] = \sum_{p_i \in P} p_i h(p_i)$. An alternative approach is to fault simulate a small number of random vectors and use the average number of faults tested per vector as our estimate of $E[W]$. Using the data provided by Wagner et. al [12] we computed k^* for various circuits. For small circuits (with inputs ranging from 2 to 5), this method yielded almost identical results as in [12]. However, for the largest circuit (the 14 input SN181 ALU) this gave poor estimates of expected FC.The reason for this is that the variance in fault detectabilities is quite large. Method B (described below) works extremely well in such situations.

Method B: This method determines a different value for k^* for each set F_i, where F_i is the set of faults with detectability p_i, i.e. $|F_i| = h(p_i)$.

Let $X_{i,r}$ be the FC in set F_i after r random vectors have been applied. For set F_i, we let $k_i^* = p_i h(p_i)$ and $M_i = h(p_i)$. Using these values we carry out the analysis for the sampling process restricted to the set F_i only. Letting $q_i = (1 - p_i)$, Equation 7 yields

$$E(X_{i,r} \mid k_i^*, M_i) = M_i(1 - q_i^r) \quad (9)$$

$$Var(X_{i,r} \mid k_i^*, M_i) = M_i q_i^r \left[1 - M_i q_i^r + (M_i - 1)\left(\frac{q_i M_i - 1}{M_i - 1}\right)^r\right] \quad (10)$$

For obtaining statistics of total FC, we can use the following formulae.

$$E[X_r] = \sum_{p_i \in P} E[X_{i,r}] \quad (11)$$

$$Var(X_r) = \sum_{p_i \in P} Var(X_{i,r}). \quad (12)$$

Equation 12 assumes that detection of faults from individual sets F_i are independent events, although this is not true in general. We used Methods A and B to estimate expected FC values of SN181ALU given by Wagner et al. [12] and compared them with their predictions. These results are illustrated in Figure 1. It was found that the predictions yielded by Method A were very inaccurate. This is explained by the fact that there are many faults with large differences in their detectabilities. However, using Method B our predictions matched exactly with those given in [12]. The *main advantage* of the analysis presented in this paper lies in the fact that, in addition to the expected values, we can now get estimates of *variance* and *confidence levels*.

For most circuits it is not feasible to find the detectability profile and thus neither Method A nor Method B can be used. For these circuits we propose a statistical method for approximating FC in large circuits using a single value k^*.

Method C: Let us assume that the set $\mathcal{F}$ of faults can be partitioned into two subsets, namely F_{high}, containing M_{high} high detectability faults, and F_{low}, containing M_{low} low detectability faults. Furthermore, of the k^* faults detected by each random vector let k_{high}^* be from F_{high} and k_{low}^* be from F_{low}, i.e. $k^* = k_{high}^* + k_{low}^*$.

Once the test process is started, we expect that most of the faults in F_{high} will be detected within a short time (say r cycles), while few faults from F_{low} would have been detected in that time. As an *idealized situation*, let us assume that at some time r, *all faults* from F_{high} have been detected and *no fault* from F_{low} has been detected. Now suppose we observe that in the $(r+1)$th cycle, j new faults are detected. If the above idealized condition was true, then all j faults would be from F_{low}. Our sampling model dictates that (under these conditions) $k_{low}^* = j$.

In order to put the above reasoning into practice, we fault simulate a small number $(r+1)$ of random patterns. We now associate, with the rth pattern, $\hat{k}_{low}^*(r) = \hat{X}_{r+1} - \hat{X}_r$ and $\hat{M}_{low}(r) = M - \hat{X}_r$, where $\hat{X}_r$ is the observed value of FC X_r. Now we consider the sampling process only on the set F_{low} using the constant $\hat{k}_{low}^*(r)$. In essence we reduce the universe of faults to be considered (for sampling) to the set of $\hat{M}_{low}(r)$ undetected faults.

The following criteria are used for selecting $\hat{k}_{low}^*(r)$ as k_{low}^* and stopping the fault simulation process. $\hat{k}_{low}^*(r)$ and $\hat{M}_{low}(r)$ are used for predicting t more coverage values, i.e. FC values for vectors numbered $r+2$ to $r+t+1$. If the predicted values fall within a tolerance of the observed coverages, then simulations are stopped and $\hat{k}_{low}^*(r)$ and $\hat{M}_{low}(r)$ are accepted as good *estimates* of k_{low}^* and M_{low} respectively. Thus, in order to use $\hat{k}_{low}^*(r)$ and $\hat{M}_{low}(r)$ as our estimates we need to simulate $r+t+1$ vectors.

Depending on circuit size, t can be varied. In our experiments, we fixed t at 10 for all circuits and obtained quite good results even for relatively large circuits. Table 4.1 shows some characteristics of these circuits. Figures 2 through 5 show the expected and observed FC as a function of the TL. It should be noted that the number of vectors simulated for each circuit is a very small fraction of the set of all vectors (2^n for an n input circuit).

Circuit	# of inputs	# of faults
Viterbi Decoder	49	602
C880	60	886
C1908	33	1826
C2670	233	3004

Table 1: Characteristics of Circuits

4.2 Binomially Distributed W

Let F_j be the set of all faults with probability $p_j \in P$ (see Definition 3). Thus $|F_j| = h(p_j)$. We now assume that the number of faults W_j, detected by a random pattern from set F_j is distributed binomially with parameters $M_j = h(p_j)$ and p_j. Let $X_{j,r}$ be the FC in set F_j after r random vectors have been applied. We can then prove the following theorem.

Theorem 2 : *Let W_j be binomially distributed with parameters M_j and p_j. Then, the rv $X_{j,r}$, is also* **binomially distributed** *with parameters M_j and $(1-(1-p_j)^r)$.*

Letting $P_{j,r} = (1-(1-p_j)^r)$ and $Q_{j,r} = 1 - P_{j,r}$, we get

$$E[X_{j,r}] = M_j P_{j,r} \quad (13)$$

$$Var(X_{j,r}) = M_j P_{j,r} Q_{j,r} \quad (14)$$

We can obtain statistics of X_r by using Equations 11 and 12. From Equations 9 and 13, it is clear that the expected value of X_r using the binomial sampling model is the same as that obtained using the constant k sampling model. The variances and other moments are different.

The model described in this section is more useful than the one presented in Section 4.1. This is because more information about fault characteristics (in the form of the detectability profile) is used for the predictions. However, it may not be easy to obtain the exact detectability profile of a circuit.The main issue, therefore, is the estimation of the detectability profile. The method adopted by us involves the following steps.

1. Estimate controllabilities and observabilities of faults in the circuit.
2. Estimate the detectability of a fault as the product of its controllability and observability.
3. Form a frequency histogram of K fault subsets each containing faults of approximately equal detectabilities. Let $\hat{F}_1, \ldots, \hat{F}_K$ denote these subsets and let $\hat{p}_1, \ldots, \hat{p}_K$ denote the **mid-point** detectability of faults in each subset. Thus $\sum_{j=1}^{K} |\hat{F}_j| = M$.
4. Let $M_j = |\hat{F}_j|$ and $p_j = \hat{p}_j$ $(1 \le i \le K)$. Use Equations 13 and 14 to compute expected coverage and its variance.

For step 1 above, we can use one of several available techniques (such as COP [2] or STAFAN [6]) for estimating controllability and observability of faults. These techniques provide simple but fast measures of controllabilities and observabilities although their accuracies of prediction are not known. Another method is to fault simulate a small number of random vectors and determine the proportion of times each fault is detected. The main problem with this method is that faults which are untested by the applied sequence of vectors are considered undetectable. This leads to inaccuracy in coverage predictions for large sequences.

We should note that in step 3 we could have used the lowest (highest) detectability in set j as $\hat{p}_j$ to obtain a pessimistic (optimistic) estimate of expected coverage. We conducted experiments with several of the ISCAS benchmark circuits and the Viterbi decoder circuit. These results are shown in Figures 6 through 9. Note that no fault simulation is done. The only computations that are carried out are in estimating the controllabilities and observabilities. The results show that the binomial assumption for W provides a simple and accurate method for computing the unconditional distribution of FC using Equation 1.

5 Distribution of Test Length

In this section we present a summary of the results leading to the distribution of test length. Experimental results and other details are reported in a subsequent paper.

Our aim is to find the number of random vectors R_x required to obtain a coverage of x *or more* (out of M faults). We use the sampling models presented in Section 3. The assumption is that each fault in set $\mathcal{F}$ is equally likely to be sampled by a random test pattern. In order to satisfy this criterion, we partition the set of faults into equal detectability subsets (detectability profile) and determine the pdf of TL for each subset.

The results of this section hinge on the observation that the *pdf*'s of TL R_x and FC X_r are related in the following manner.

$$P\{R_x \le r\} = P\{X_r \ge x\}. \quad (15)$$

Equation (15) further implies that

$$P\{R_x = r\} = P\{X_r \ge x\} - P\{X_{r-1} \ge x\} \quad (16)$$

This establishes an important relation between the metrics fault coverage and test time. Further, it allows us to use the results of Section 4 to obtain an expression for the pdf of R_x. A special case of interest is the pdf for 100% coverage, i.e. for $x = M$. It can be shown easily that

$$P\{R_M \le r\} = P\{X_r = M\} - P\{X_{r-1} = M\} \quad (17)$$

__Deterministic W:__ Let $\delta_t = \Delta^j \left[\left(0^{\underline{k^*}}\right)^t \right]$. Then Equations 16 and 6 imply that

$$P\{R_x = r \mid k^*\} = \sum_{j=x}^{M} \binom{M}{j} \left[\frac{\delta_r}{(M^{\underline{k^*}})^r} - \frac{\delta_{r-1}}{(M^{\underline{k^*}})^{r-1}} \right]. \quad (18)$$

Given the *pdf* of R_x, we can find all important statistics of this rv. However, due to the nature of the above summation, it is unlikely that a simpler closed form can be found for the *pdf* or its statistics. For this reason we resort to purely numerical methods for computing expected values and other statistics of TL.

Closed form expressions, even for approximate values, are usually difficult to obtain. However, for the special case of $x = M$, Pólya in [7], presented good approximations for the average number of patterns required to achieve 100% coverage. This is given below (see Equation (13) in [7]).

$$E[R_M \mid k^*] = \left(\frac{M + 1/2}{k^*} - \frac{1}{2} \right) (\log M + \gamma) + \frac{1}{2} + \epsilon_M. \quad (19)$$

where $\gamma = 0.5772..$is Euler's constant and $\epsilon_M \to 0$ as $M \to \infty$.

Binomially Distributed W: When W is binomially distributed with parameters M and p, Equation (16) and Theorem 2 can be used to obtain an expression for the pdf of R_x. This can be equivalently represented as

$$P\{R_x = r\} = \\ I_{P_r}(x, M - x + 1) - I_{P+r-1}(x, M - x + 1). \quad (20)$$

where $I_x(a,b)$ is the *incomplete beta function* with parameters a, b and x and $P_t = (1 - (1-p)^t)$. Methods for obtaining excellent approximations to the incomplete beta function are available in [8]. For the special case of 100% coverage, we use Equation (17) to get $P\{R_M = r\} = P_r^M - P_{r-1}^M$. Letting $Q_t = 1 - P_t$ and using the fact that $E[R_x] = \sum_{r=1}^{\infty} P\{R_x \geq r\}$, the expected value of TL can be expressed as

$$E[R_x] = \sum_{i=0}^{x-1} (-1)^i \binom{M}{i} \sum_{\ell=0}^{x-1-i} \binom{M-i}{\ell} \frac{Q_\ell}{Q_\ell - Q_M}. \quad (21)$$

We do not know of any closed form for the above equation. For numerical computation of the above formula, care should be taken to avoid overflow or underflow since it involves an alternating series.

6 Conclusion

In this paper we present several novel results that can be applied to the problems of fault coverage (FC) and test length (TL) estimation in combinational circuits under random or pseudorandom test. A sequential sampling model is used to obtain the pdf of FC and TL. This yields not only the expected values and bounds for these testability metrics (as done in earlier techniques) but also other important statistics such as variance and confidence values.

Two special cases of sequential sampling are explored. In the first case we assume that the number of faults tested by any vector is the same. Based on this model, a statistical method for estimating expected FC is developed and validated. The second case assumes that the fault set is partitioned into equal detectability subsets and the number of faults tested by a random vector in a subset is binomially distributed with appropriate parameters. Based on this method a systematic approach for estimating expected FC, is defined.

Lastly, we establish an important relation between the rv's TL and FC. This allows to use the theoretical results for FC for obtaining the distribution function of TL. Expressions for the expected value of test length for different cases are also presented.

References

[1] Agrawal, V.D., "Sampling Techniques for Determining Fault Coverage in LSI Circuits," *Journal of Digital Systems*, Vol. V, Num. 3, pp. 189-202.

[2] Brglez, F., "On Testability Analysis of Combinational Networks," *Proc. IEEE ISCAS 84*, 1984, pp. 221-225.

[3] Goel, P., "Test Generation Costs Analysis and Projections," *Proc. 17th Design Automation Conf.*, 1980, pp. 77-84.

[4] Huang, W.K., M. Lightner and F. Lombardi, "Predicting Fault Coverage for Random Testing of Combinational Circuits," *Proc. IEEE Int. Test Conf.*, 1987, pp. 843-848.

[5] Graham, R.L., D.E. Knuth and O. Patashnik, Concrete Mathematics, *Addison-Wesley Publishing Co.*, 1989.

[6] Jain, S.K. and V.D. Agrawal, "STAFAN: An Alternative to Fault Simulation," *Proc. 21st Design Automation Conf.*, 1984, pp. 18-23.

[7] Pólya, G., "Eine Wahrscheinlichkeitsaufgabe in der Kundenwerbung," *Z. Angew. Math. Mech.*, 1930, pp. 96-97.

[8] Press, W.H., B.P. Flannery, S.A. Teukolsky and W.T. Vetterling, Numerical Recipes in C, *Cambridge University Press*, 1988.

[9] Sastry, S., "Detectability of CMOS Stuck-Open Faults using Random and Pseudorandom Test Sequences," *IEEE Trans. on Comp. Aided Design*, Vol. 7, Sept. 1989, pp. 933-946.

[10] Sastry, S. and A. Majumdar, "Test Efficiency Analysis of Random Self-Test of Sequential Circuits," *IEEE Trans. on Comp. Aided Design*, Vol. 10, March 1991, pp. 390-398.

[11] Savir, J., G. S. Ditlow and P.H. Bardell, "Random Pattern Testability," *IEEE Trans. on Computers*, Vol C-33, No. 1, pp 79-90. Jan. 1984.

[12] Wagner, K.D., C.K. Chin and E.J. McCluskey, "Pseudorandom Testing," *IEEE Trans. on Computers*, Vol. C-36, Mar. 1987, pp. 332-343.

[13] Williams, T.W., "Test Length in a Self-Testing Environment," *IEEE Design & Test*, April 1985, pp. 59-63.

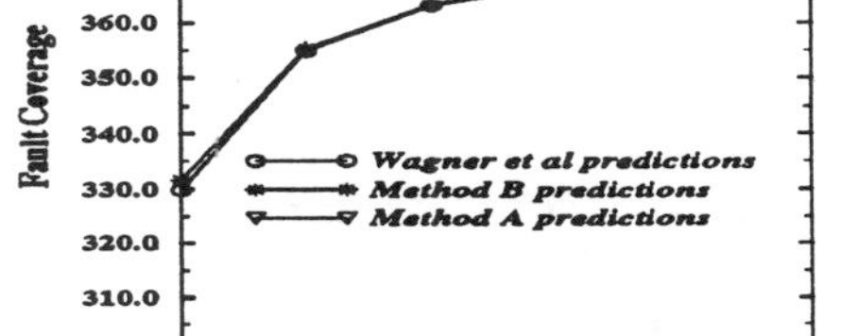

Figure 1: Estimated fault coverage for the SN181 ALU.

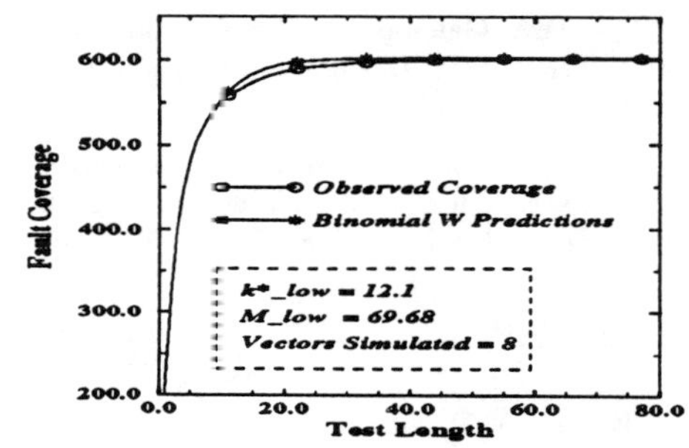

Figure 2: Fault coverage prediction for a Viterbi decoder circuit using Method C.

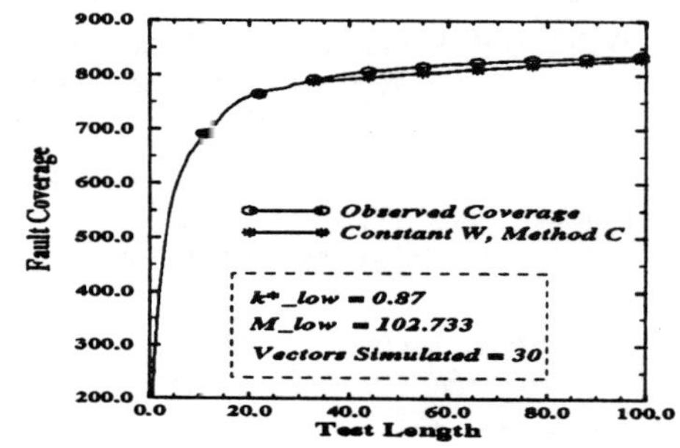

Figure 3: Fault coverage prediction for C880 using Method C.

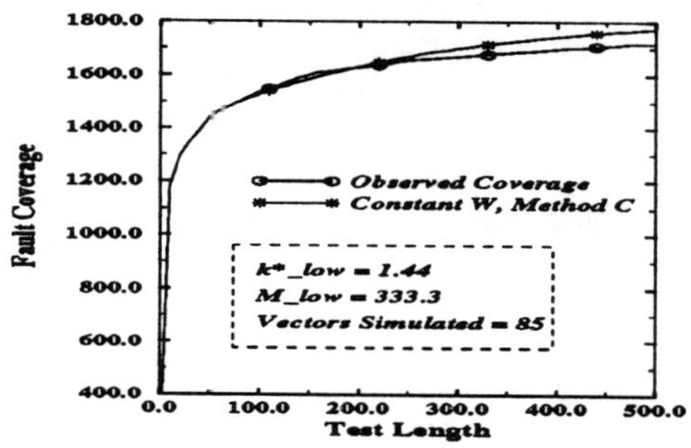

Figure 4: Fault coverage prediction for C1908 using Method C.

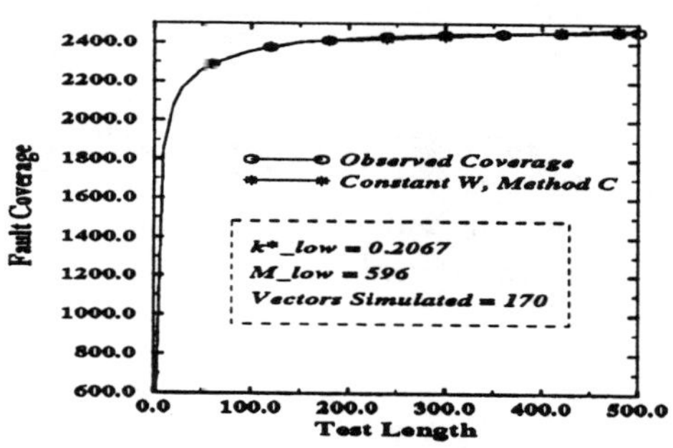

Figure 5: Fault coverage prediction for C2670 using Method C.

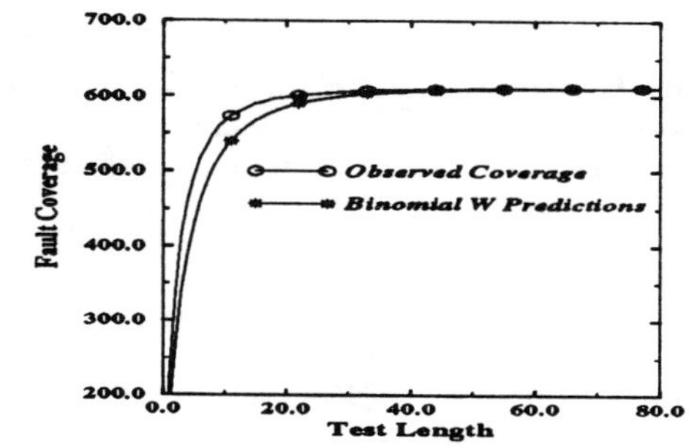

Figure 6: Fault coverage prediction for a Viterbi decoder using Binomial sampling model.

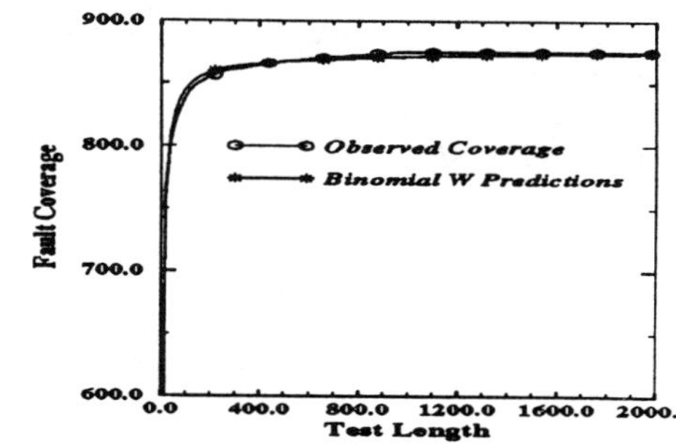

Figure 7: Fault coverage prediction for C880 using Binomial sampling model.

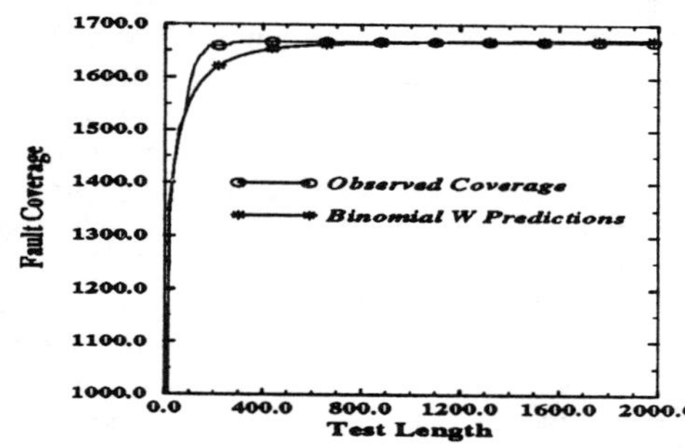

Figure 8: Fault coverage prediction for C1908 using Binomial sampling model.

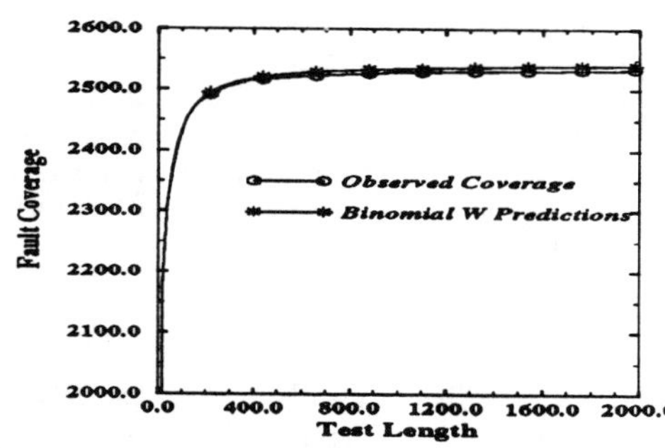

Figure 9: Fault coverage prediction for C2670 using Binomial sampling model.

Exact Evaluation of Diagnostic Test Resolution

Ken Kubiak Steven Parkes W. Kent Fuchs Resve Saleh

Coordinated Science Laboratory
University of Illinois at Urbana/Champaign

Abstract

This paper introduces a new measure of the diagnostic resolution *of a test set: the sizes of all equivalence classes in the circuit under the test set. This measure is a better indicator of the diagnostic capabilities of a test set than single-value metrics based on undistinguished pairs of faults or completely distinguished faults. A symbolic algorithm for computing equivalence class sizes has been used to evaluate the diagnostic resolution of deterministic single-stuck-at fault test sets for ISCAS combinational and sequential benchmark circuits.*

1. Introduction

Diagnosis is the process of determining the cause of a circuit failure. Given a circuit, a set of input vectors, and a set of output vectors, diagnosis attempts to find the faults or combinations of faults which are consistent with the symptom. Fault location is often more difficult than fault detection and usually requires the application of more test vectors.

Recently, diagnostic test pattern generators for combinational circuits have been implemented which extend traditional test generation methods to diagnostic test generation [1, 2]. In order to compare the effectiveness of such techniques, it is necessary to define a measure of diagnostic test resolution and efficiently calculate this value. The metrics which have previously been used for this purpose are based on the *number of undistinguished pairs of faults* and the *number of completely distinguished faults* [1, 2]. Though easy to compute by pairwise diagnostic fault simulation, such metrics may not adequately describe the diagnostic capabilities of a test set.

This research was supported by the Semiconductor Research Corporation under contract SRC 91-DP-109. Ken Kubiak is supported by a Semiconductor Research Corporation Fellowship. Steven Parkes is supported by a Digital Equipment Corporation Fellowship.

As an example, consider a test set which distinguishes all but 36 pairs of faults. For each pair, the two faults belong to the same *equivalence class*; under the given test set, the same output will be observed in the presence of either fault. The 36 pairs may fall into 12 equivalence classes, each containing 3 faults, in which case every failure is resolved to a set of at most 3 faults. However, it may instead be the case that these 36 pairs fall into one equivalence class of size 9 (since $\binom{9}{2} = 36$). If so, some circuits will have 9 possible diagnoses. If logical diagnosis is followed by invasive methods in order to isolate the exact cause of failure, the sizes of the equivalence classes will impact the amount of time spent in physical diagnosis.

In this paper, diagnostic resolution is exactly determined by computing the number and sizes of equivalence classes in the circuit under the given test. This metric more accurately reflects the effort that may be required to physically isolate the fault after it has been logically diagnosed to within an equivalence class. The metric also contains enough information to compute the metrics DR (ratio of number of distinguished pairs of faults to total number of pairs of faults) and DP (ratio of completely distinguished faults to total number of faults) [1].

A symbolic algorithm for determining the sizes of all equivalence classes in a circuit under a test set is presented. A fault variable model is used, and each fault variable is assigned an integral index. Circuit node values are computed as Boolean functions of these indices, encoded as vectors of Boolean variables. The Boolean functions which result at the circuit outputs are analyzed to determine the sizes of the equivalence classes.

This method has been used to compute the diagnostic resolution of deterministic single-stuck-at fault test sets for ISCAS combinational [3] and sequential [4] benchmark circuits.

2. N-ary decision diagrams

A *binary decision diagram* (BDD) [5, 6, 7] is a directed acyclic graph in which each node has exactly 2 out-edges, labeled **0** and **1**. Each node is labeled with an index variable f_i. The terminal nodes are labeled with the Boolean values **0** and **1**. A BDD of k index variables represents a Boolean-valued function of k Boolean arguments, $f(x_0, \ldots, x_{k-1})$.

An *N-ary decision diagram* (NDD) is a directed acyclic graph in which each node has N out-edges, labeled 0 through $N-1$. Each node is labeled with an index variable $\mathcal{X}_i$ which can take on values in the range 0 to $N-1$. Terminal nodes are labeled with the Boolean values **0** and **1**. An NDD of m index variables represents a Boolean-valued function of m integer arguments, $F(\mathcal{X}_0, \ldots, \mathcal{X}_{m-1})$.

In the current implementation, NDDs of m index variables are implemented as binary decision diagrams of $k \cdot m$ (Boolean) index variables, where each NDD index variable $\mathcal{X}_i$ is represented in the BDD by $k = \lceil \log_2 N \rceil$ BDD index variables $\langle x_i^{k-1}, x_i^{k-2}, \ldots, x_i^0 \rangle$ such that

$$\mathcal{X}_i = \sum_{j=0}^{k-1} 2^j x_i^j$$

For the purposes of this paper, the BDD respresentation of $F(\mathcal{X}_0, \ldots, \mathcal{X}_{m-1})$ takes the form

$$f(x_0^0, \ldots, x_0^{k-1}, x_1^0, \ldots, x_1^{k-1}, \ldots, x_{m-1}^0, \ldots, x_{m-1}^{k-1})$$

With this variable ordering, each NDD index corresponds to a layer of nodes in the BDD. For the application at hand, this ordering was found to provide better performance than an interleaved ordering.

The idea of encoding an N-way selector as bits in a BDD was used by Cho and Bryant [8] in generating test sets for sequential MOS circuits. Their application required only one N-ary variable in the decision diagram, along with many Boolean variables. The algorithm in this paper requires that two fault indices be encoded; the NDD representation is a convenient abstraction for such an encoding.

NDDs can be manipulated by Boolean operators; these operations are implemented by Boolean operations on the underlying BDD representation. The NDD cofactor operation, which replaces an NDD index $\mathcal{X}_i$ by an integer constant C, can be implemented in terms of the BDD cofactor operation. If $\langle c^{k-1}, c^{k-2}, \ldots, c^0 \rangle$ are the bits of the binary representation of C, then

$$F(\mathcal{X}_0, \ldots, \mathcal{X}_{i-1}, C, \mathcal{X}_{i+1}, \ldots, \mathcal{X}_{m-1}) = f(x_0^0, \ldots, x_{i-1}^{k-1}, c^0, \ldots, c^{k-1}, x_{i+1}^0, \ldots, x_{m-1}^{k-1}) \quad (1)$$

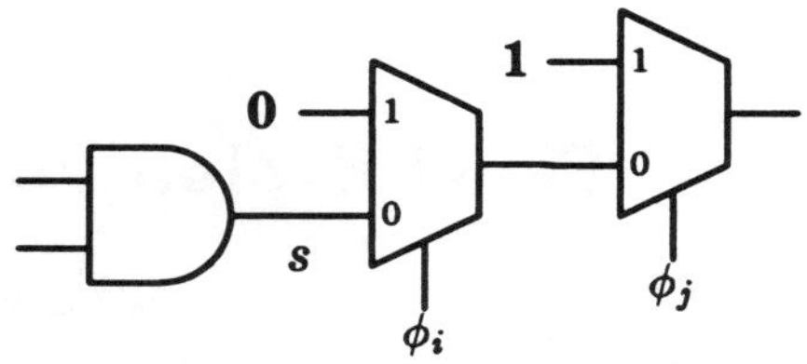

Figure 1: The fault variable model.

The function $[\mathcal{X}_i \equiv C](\mathcal{X}_0, \ldots, \mathcal{X}_{m-1})$ evaluates to **1** iff the index variable $\mathcal{X}_i$ takes on the integral value C. In terms of the underlying BDD representation,

$$[\mathcal{X}_i \equiv C](\mathcal{X}_0, \ldots, \mathcal{X}_{m-1}) = \bigcap_{j=0}^{k-1} \left(x_i^j \equiv c^j \right) \quad (2)$$

where $a \equiv b$ is the Boolean equivalence operation $ab + \bar{a}\bar{b}$. The support of the resulting function is $\{\mathcal{X}_i\}$.

Replacing one NDD index variable, $\mathcal{X}_i$, by another, $\mathcal{X}_j$, is a special case of function composition and can be computed by composing BDDs.

$$F(\mathcal{X}_0, \ldots, \mathcal{X}_{m-1})|_{\mathcal{X}_i \to \mathcal{X}_j} = f(x_0^0, \ldots, x_{i-1}^{k-1}, x_j^0, \ldots, x_j^{k-1}, x_{i+1}^0, \ldots, x_{m-1}^{k-1}) \quad (3)$$

$\mathcal{X}_i$ does not appear in the support of the resulting function. $\mathcal{X}_j$ need not appear in the support of the original function, but will appear in the result if $\mathcal{X}_i$ appears in the original.

3. The Diagnosis function

The algorithm in this paper uses Boolean fault variables to activate faulty behavior in a circuit. Any fault which can be represented by the insertion of additional logic into the circuit can be modeled. A possible fault is injected by inserting a 2-input multiplexer, controlled by a fault variable, which selects either the good circuit behavior or the behavior under the fault. Figure 1 demonstrates stuck-at-**0** and stuck-at-**1** faults modeled on the output of the AND gate s. ϕ_i is a fault variable which activates the stuck-at-**0** fault, while ϕ_j activates the stuck-at-**1** fault. If both fault variables take on the value **0**, the good circuit behavior is manifested.

In this paper, a single-fault model is used; only one fault can be active at a time. All faults are injected simultaneously and symbolic Boolean functions are used to represent sets of faults.

For a circuit C with N fault variables, let $C(\mathcal{F})$, $0 \le \mathcal{F} \le N$, denote all faulty and fault-free circuits on C. In circuit $C(i)$, $\mathcal{F}$ takes on the value i, activating fault i. The fault-free circuit is represented by $C(0)$. All node values in $C(\mathcal{F})$ are Boolean-valued functions of the integer $\mathcal{F}$ and can be represented as NDDs of depth one.

The first step of the diagnostic test evaluation algorithm is to symbolically fault-simulate all possibly faulty circuits $C(\mathcal{F})$ in order to find the value of each output. All single faults are injected simultaneously by setting the value of each fault variable $\phi_i(\mathcal{F}) = [\mathcal{F} \equiv i](\mathcal{F})$ as given by Equation 2. For each test vector k, the output vector of $C(\mathcal{F})$, $\langle Y_0^k(\mathcal{F}), \ldots, Y_{M-1}^k(\mathcal{F})\rangle$, is the response of all possibly faulty circuits $C(\mathcal{F})$ to the test vector k. The simulation resembles parallel fault simulation — if the value of a node is computed to be $X(\mathcal{F})$, the Boolean value of the node in the presence of fault i can be extracted by computing $X(i)$ using Equation 1.

Now consider two possibly faulty circuits $C(\mathcal{A})$ and $C(\mathcal{B})$, which are identical except for the existence of some fault $\mathcal{A}$ in $C(\mathcal{A})$ and some fault $\mathcal{B}$ in $C(\mathcal{B})$. For a test set T, diagnostic test evaluation requires the comparison of the responses of all possible pairs of circuits $\{C(\mathcal{A}), C(\mathcal{B})\}$. The *diagnosis function* for test T, $\delta_T(\mathcal{A},\mathcal{B})$, evaluates to **1** when circuits $C(\mathcal{A})$ and $C(\mathcal{B})$ can be distinguished. Circuits $C(\mathcal{A})$ and $C(\mathcal{B})$ can be distinguished if for some circuit output j and test vector k, $Y_j^k(\mathcal{F})|_{\mathcal{F}\to\mathcal{A}}$ differs from $Y_j^k(\mathcal{F})|_{\mathcal{F}\to\mathcal{B}}$. The test responses of $C(\mathcal{A})$ and $C(\mathcal{B})$ are computed from the response of $C(\mathcal{F})$ using Equation 3. The diagnosis function for test T is computed by

$$\delta_T(\mathcal{A},\mathcal{B}) = \bigcup_{k=0}^{|T|-1} \bigcup_{j=0}^{M-1} Y_j^k(\mathcal{F})|_{\mathcal{F}\to\mathcal{A}} \oplus Y_j^k(\mathcal{F})|_{\mathcal{F}\to\mathcal{B}}$$

$\delta_T(\mathcal{A},\mathcal{B})$ will be represented as an NDD of depth two, having node indices $\mathcal{A}$ and $\mathcal{B}$. Figure 2 depicts one possible $\delta_T(\mathcal{A},\mathcal{B})$ for a circuit with seven fault variables.

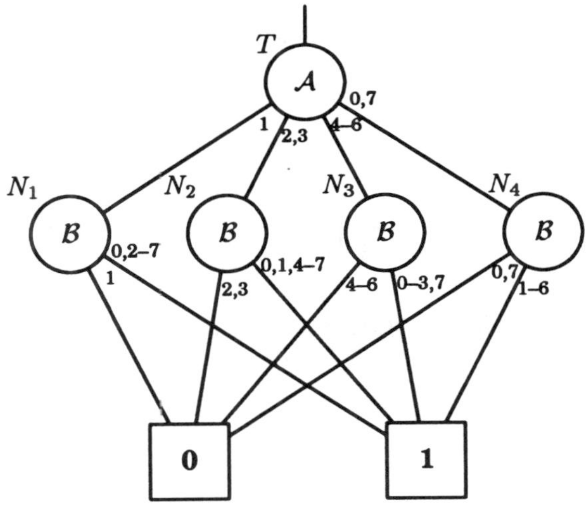

Figure 2: NDD representation of $\delta_T(\mathcal{A},\mathcal{B})$

Property 2 (Symmetric) *If two circuits are not distinguished by some test T, then they remain undistinguished if the faults they contain are exchanged.*

$$\delta_T(i,j) = \delta_T(j,i) \quad \forall i,j$$

Properties 1 and 2 above can be restated in terms of the two-level NDD representation of $\delta_T(\mathcal{A},\mathcal{B})$ as follows:

Property 3 (Reflexive) *If the edge labeled i out of the top node T of the NDD leads to a node N, then the edge labeled i out of N leads to the **0** terminal.*

Property 4 (Symmetric) *If the edge labeled i out of the top node T of the NDD leads to a node N and the edge labeled j out of N leads to the **0** terminal, then the edge labeled j out of T leads to N and the edge labeled i out of N leads to the **0** terminal.*

These properties are demonstrated in Figures 3 and 4. The following counting rules follow from Properties 3 and 4:

Rule 1 *Every node N in the second level of the NDD corresponds to an equivalence class.*

Rule 2 *The number of edges into a node N in the second level is the number of faults in the equivalence class corresponding to N.*

Consider the diagnosis function represented in Figure 2. N_1 is reached from T by an edge labeled 1. Therefore, fault 1 lies in an equivalence class of size

4. Counting equivalence classes

The number of equivalence classes in circuit C under test set T is computed by a traversal of the diagnosis function, $\delta_T(\mathcal{A},\mathcal{B})$. The algorithm is based on two properties of the diagnosis function.

Property 1 (Reflexive) *Two circuits containing the same fault are not distinguished by any test T.*

$$\delta_T(i,i) = \mathbf{0}, \quad \forall i$$

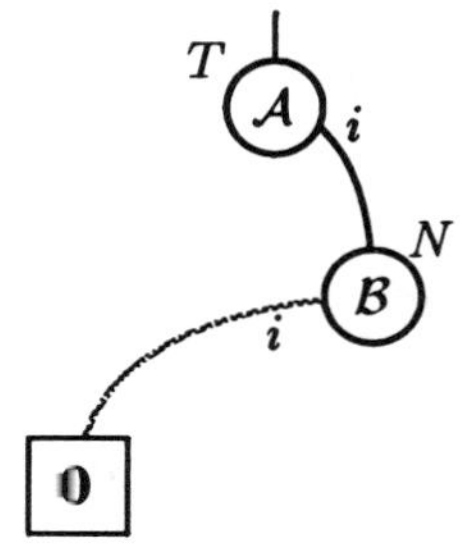

Figure 3: The reflexive property of the diagnosis function.

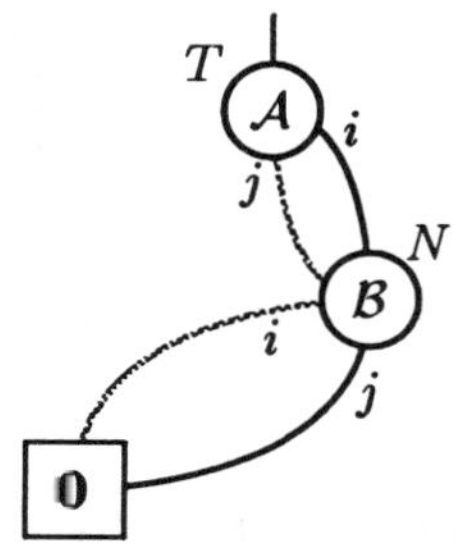

Figure 4: The symmetric property of the diagnosis function.

one; it is completely distinguished from all other faults. N_2, reached by edges labeled 2 and 3, corresponds to the equivalence class consisting of faults 2 and 3. N_3 corresponds the equivalence class consisting of faults 4, 5, and 6. The last node, N_4 is reached by edges labeled 0 and 7. Since $\mathcal{A} \equiv 0$ indicates that no faults are activated in circuit $C(\mathcal{A})$, N_4 corresponds to the equivalence class containing all faults which are not distinguished from the fault-free circuit and are not detected by the test. Therefore, fault 7 lies in its own equivalence class and is not detected by the test.

5. Experimental Results

The algorithm for diagnostic test evaluation presented in this paper has been implemented in C++. The BDD package used is similar to those presented by Bryant *et al.*[9] and Minato *et al.*[10], but uses mark-and-sweep garbage collection in place of reference counting. The application was allowed to allocate up to 2M BDD nodes. Experiments were conducted on a Sun SPARCstation 2 (4/70) with 64M of memory.

Our objective was to evaluate the diagnostic resolution of tests which were intended for fault detection only. The circuits tested included two standard TTL circuits as well as ISCAS combinational [3] and sequential [4] benchmark circuits. Deterministic single-stuck-at fault test sets were generated using the HITEC ATPG program [11], which also generates a collapsed fault list. These test sets were evaluated using the algorithm presented in this paper. The performance of the diagnostic test evaluation program is shown in Table 1. For each circuit, the table gives the number of faults, number of vectors simulated, the execution time, the number of BDD nodes known to be in use after a garbage collection (low), and the number of BDD nodes allocated by the application (high).

Table 1: Diagnostic test evaluation performance statistics on a Sun 4/70 with 64M of memory.

circuit	faults	vectors	execution time	memory usage (in BDD nodes) low	high
7485	123	47	6s	951	131,072
74181	269	44	36s	2531	131,072
c432	560	102	4m03s	6885	393,216
c499	1158	171	26m30s	10,311	786,432
c880	942	131	12m36s	7,656	524,288
c1355	1574	169	37m15s	13,040	917,504
c1908	1879	116	35m59s	20,025	1,310,720
c2670	2595	249	3h18m41s	23,705	2,097,152
c3540	3428	303	2h56m29s	41,342	2,097,152
s298	308	287	3m30s	2948	262,144
s344	342	109	2m00s	4034	262,144
s349	350	77	1m44s	4077	262,144
s526	555	34	17s	3762	131,072
s641	467	219	6m21s	3976	262,144
s713	581	175	7m09s	5444	393,216
s820	850	1135	1h55m23s	9031	2,097,152
s832	870	1027	1h48m90s	9413	2,097,152
s1238	1355	478	46m28s	11,823	1,310,720
s1423	1515	88	2m56s	13,000	262,144

When the initial state of a sequential circuit is unknown, it may be impossible to determine the equivalence classes due to the pessimistic nature of simulation with unknowns. If any output of circuit C_A or C_B is unknown, it is no longer possible for the two circuits to be equivalent. For this reason, the initial state of all latches was taken to be **0** in these experiments.

The diagnostic resolutions provided by the fault-detection test sets are summarized in Table 2. For each equivalence class size, the number of classes

Table 2: Diagnostic test resolution of single stuck-at fault test sets.

circuit	Number of equivalence classes (by size) 1	2	3	4	5	6	7	8	9	> 9
7485	107	8								
74181	225	20		[1]						
c432	431	16	19							[1]×40
c499	710	220							[1]	
c880	823	58	1							
c1355	398	428	104					[1]		
c1908	1091	220	34	16	1		2	3		[1]×139
c2670	1758	205	63	8	9	2		1		1×11 1×14 [1]×130
c3540	2122	254	44	16		22		3		[1]×446
s298	135	21	9	7	1	2	1	1	1	[1]×35
s344	201	18	6	5	4	4				[1]×23
s349	219	18	3	6	4	4				[1]×18
s526	11	9	4	1		1				[1]×498
s641	297	32	1	4	2	3				[1]×59
s713	284	28		12	4	4		6		[1]×101
s820	653	50	10	3	1	1	1			[1]×37
s832	658	50	7	6	1		1			[1]×55
s1238	1106	77	5	2						[1]×72
s1423	103	36	11	6	10	10	2	2	2	1×11 1×13 1×14 1×15 1×17 1×18 1×23 1×29 1×41 [1]×944

found of that size is reported. Boxed entries indicate undetected faults. For most of these circuits, all failures which were detected were diagnosable to within 10 faults, though some circuits contained several large equivalence classes. Whether this is an acceptable resolution is dependent on the cost of physical diagnosis. If this resolution is not acceptable, diagnostic test generation targeted at faults in the same equivalence class could be used to improve the resolution of the test.

For the purpose of comparison, the other measures of diagnostic resolution can be computed from the sizes of the equivalence classes. Three that have appeared in the literature are the number of undistinguished pairs of faults [2], the diagnostic resolution DR (ratio of number of distinguished pairs to total number of pairs of faults), and the diagnostic power DP (ratio of completely distinguished faults to total number of faults) [1]. These metrics are shown in Table 3.

Table 3: Single-value metrics for diagnostic resolution.

circuit	total pairs	undistinguished pairs	DR	DP
7485	1403	8	99.43	87
74181	36,046	26	99.93	84
c432	156,520	853	99.46	77
c499	669,903	256	99.96	61
c880	443,211	61	99.99	87
c1355	1,237,951	768	99.94	25
c1908	1,764,381	10,068	99.42	58
c2670	3,365,715	9121	99.73	68
c3540	5,873,878	100,131	98.30	62
s298	47,278	810	98.29	43
s344	58,311	419	99.29	59
s349	61,075	316	99.48	63
s526	153,735	123,795	19.48	2
s641	108,811	1835	98.31	64
s713	168,490	5418	96.78	49
s820	360,825	810	99.78	77
s832	378,015	1623	99.57	76
s1238	917,335	2660	99.71	82
s1423	1,146,855	447,718	60.96	1

As evidenced by circuits c2670 and s820, it is possible for rather large equivalence classes to exist even when the test attains a very high value of DR. It is also possible for a test with a very low value of DP to provide reasonable distinguishability, as shown by c1355, in which all faults (except 8 undetected ones) fall into classes which contain no more than three faults each.

6. Future Work

The algorithm in this paper is an adaptation of a symbolic multiple-fault simulation algorithm [12]. Traditional fault simulation algorithms can also be adapted to compute the sizes of equivalence classes [13]. Work is currently under way to develop diagnostic test generation algorithms with the goal of minimizing equivalence class sizes.

7. Conclusion

The sizes of the equivalence classes in a circuit under a test set is an accurate indication of the diagnostic resolution of the test. This measure gives a more complete picture of a test's diagnostic capabilities than single-value metrics.

A symbolic algorithm has been presented for computing the sizes of the equivalence classes in a circuit under a test. This algorithm was used to determine the diagnostic resolution of deterministic test sets which are intended for use in fault detection. While these tests are successful in resolving many of the faults, for some circuits, large equivalence classes remain.

References

[1] P. Camurati, D. Medina, P. Prinetto, and M. S. Reorda, "A diagnostic test pattern generation algorithm," *Proceedings of the International Test Conference*, pp. 52–58, 1990.

[2] T. Grüning, H. Koopmeiners, and U. Mahlstedt, "DIATEST: A fast diagnostic test pattern generator for combinational circuits," *Digest of Papers, International Conference on Computer-Aided Design*, pp. 194–197, 1991.

[3] F. Brglez and H. Fujiwara, "A neutral netlist of 10 combinational benchmark circuits and a target translator in FORTRAN," *Special Session on ATPG and Fault Simulation, International Symposium on Circuits and Systems*, 1985.

[4] F. Brglez, D. Bryan, and K. Kozminski, "Combinational profiles of sequential circuit benchmarks," *Proceedings of the International Symposium on Circuits and Systems*, pp. 1929–1934, 1989.

[5] C. J. Lee, "Representation of switching circuits by binary-decision programs," *Bell System Technical Journal*, vol. 38, pp. 985–999, July 1959.

[6] S. B. Akers, "Binary decision diagrams," *IEEE Trans. Computers*, vol. C-27, pp. 509–516, June 1978.

[7] R. E. Bryant, "Graph-based algorithms for Boolean function manipulation," *IEEE Trans. Computers*, vol. C-35, pp. 677–691, Aug. 1986.

[8] K. Cho and R. E. Bryant, "Test pattern generation for sequential MOS circuits by symbolic simulation," *Proceedings of the Design Automation Conference*, pp. 418–423, 1989.

[9] K. S. Brace, R. L. Rudell, and R. E. Bryant, "Efficient implementation of a BDD package," *Proceedings of the Design Automation Conference*, pp. 40–45, 1990.

[10] S. Minato, N. Ishiura, and S. Yajima, "Shared binary decision diagram with attributed edges for efficient Boolean function manipulation," *Proceedings of the Design Automation Conference*, pp. 52–57, 1990.

[11] T. Niermann and J. Patel, "HITEC: A test generation package for sequential circuits," *Proceedings of the European Design Automation Conference*, pp. 214–218, 1991.

[12] K. Kubiak and W. K. Fuchs, "Multiple-fault simulation and coverage of deterministic single-fault test sets," *Proceedings of the International Test Conference*, pp. 956–962, 1991.

[13] E. Rudnick, W. K. Fuchs, and J. Patel, "Diagnostic fault simulation of sequential circuits," manuscript, Jan. 1992.

Algorithms for Current Monitor Based Diagnosis of Bridging and Leakage Faults

Sreejit Chakravarty*
Dept. of Computer Science
State University of New York
Buffalo, NY

Minsheng Liu
Performance Technologies, Inc.
315 Science Parkway
Rochester, NY

Abstract

Current monitor based diagnosis algorithms for bridging and leakage faults are presented. These algorithms do not use fault dictionaries. *Instead, they use SOPS, a novel representation of faultlists. Experimental evaluation of the algorithms are presented.*

1 Introduction

The importance of testing and diagnosing CMOS circuits for bridging and leakage faults is discussed in [9]. We assume that a given bridging fault involves only two lines. On application of any input vector to the good circuit, in the steady state, none of the two FET networks, of a static CMOS gate, conducts. Therefore the steady state supply current (I_{DDQ}) in the good circuit is small enough to be negligible. If there is a bridging or leakage fault it can be excited so that I_{ddq} is large. Thus, *Current Monitors*, built into the circuit, that can detect the large I_{ddq} is used to detect the presence of bridging and leakage faults. This is known as I_{DDQ}-*Testing*[8,10,13]. The tests are known as I_{ddq} tests.

Let $v(L)$ denote the value at L, in the absence of the fault, on application of the input vector v. An input vector v, in a given circuit C does not detect the bridging fault between L_1, L_2 if $v(L_1) = v(L_2)$[11]. This is used to drop faults from the fault list. If $v(L_1) \neq v(L_2)$ it is not necessary that v will detect the bridging fault between L_1, L_2[4]. We assume that v detects the bridging fault between L_1, L_2 iff $v(L_1) \neq v(L_2)$. This forces our algorithm to retain some of the faults that should be droped making our algorithm pessimistic. Tests for Leakage Faults are similarly defined[11].

Diagnosis Problem: Given a circuit C that has been detected to be faulty by an I_{DDQ} test set T identify the fault in C. We assume that: (i) the definition of the input vectors in T; and (ii) for each input vector T_i in T whether C gave a faulty response or not is known.

A number of diagnosis algorithms have been reported in the literature. The algorithms in [1, 2, 6, 7, 12, 14-16] assumes that: the faults are of the stuck-at type; and input/output experiments are used to detect the presence of faults. The algorithm in [17] is for leakage faults only and uses a combination of I_{ddq} Testing and Voltage Monitoring.

In this paper algorithms for the diagnosis problem, for combinational as well as sequential circuits, is described. Experimental evaluation of these algorithms are presented. Besides the work presented here we are unaware of any algorithm for the diagnosis of bridging faults when either current monitors or input/output experiments are used.

Our algorithms, unlike a number of known diagnosis algorithms, do not use fault dictionaries. Instead, a set of ordered-pair of sets (SOPS) is used to represent all two line bridging faults. If m is the size of the circuit then SOPS uses $O(m)$ space to represent these faults. In contrast to this, the space complexity of fault dictionaries equals $O(m^2)$. The resulting algorithm takes time $O(mN)$ where N is the size of the test set, unlike algorithms that use fault dictionary thereby taking time $O(m^2N)$. For lack of space we sketch the algorithm. Details can be found in [5].

2 Combinational Circuit

Algorithm DIAG-BFC

/* Algorithm for diagnosing bridging faults in combinational circuits. */

Notations.
N is the number of input vectors in the test set.

*Research partially supported by NSF Grants No. MIP-8807540 and MIP-9102509.

$T = \{T_1, \ldots, T_N\}$ is the test set.
$R = \{r_1, \ldots, r_N\}$ is the response vector where $\forall i, r_i \in \{0,1\}$ and $r_i = 0(1)$ if there wasn't (there was) a large steady state current on application of T_i.
C is the circuit being diagnosed.
X is the set of all lines in C.
m is the number of gates in C.
$T_i(y)$ is the value at line y on application of T_i.
n is the number of primary inputs of C.
$\{X_1, \ldots, X_n\}$ is the set of primary inputs of C.
$A^j(T_i)$ is the subset of A s.t. $\forall y \in A^j(T_i)$, $T_i(y) = j, j \in \{0,1\}$.
$< A, \Psi >$ is the set of pairs of lines (y_1, y_2) s.t. $y_1, y_2 \in A$.

$< A, B >$, $B \neq \Psi$ is the set of pairs of lines (y_1, y_2) s.t $y_1 \in A$ and $y_2 \in B$.
L_i is the set of pairs of lines, denoting the set of possible faults deduced by the algorithm, after it has processed vectors $T_1, \ldots, T_i$.

Step 1. $L_0 = \{< X, \Psi >\}$.

Step 2. For each $T_i, 1 \leq i \leq N$ do Step 3.

Step 3. Let $L_{i-1} = \{< A_1, B_1 >, \ldots, < A_q, B_q >\}$. For $1 \leq j \leq q$ process $< A_j, B_j >$ to form new ordered pairs as follows. These new ordered pairs are added to L_i and $< A_j, B_j >$ is deleted from L_{i-1}.

Compute $A_j^0(T_i), A_j^1(T_i)$. If $B_j \neq \Psi$, compute $B_j^0(T_i), B_j^1(T_i)$. We have two cases.

Case 3.1. $B_j = \Psi$. If $r_i = 0$ then the new ordered pairs are $< A_j^0(T_i), \Psi >$ and $< A_j^1(T_i), \Psi >$. If $r_i = 1$ then the new ordered pair is $< A_j^0(T_i), A_j^1(T_i) >$.

Case 3.2. $B_j \neq \Psi$. If $r_i = 0$ then the new ordered pair(s) are $< A_j^0(T_i), B_j^0(T_i) >$ and $< A_j^1(T_i), B_j^1(T_i) >$. If $r_i = 1$ then the new ordered pair(s) are $< A_j^0(T_i), B_j^1(T_i) >$ and $< A_j^1(T_i), B_j^0(T_i) >$.

Note that if either A or B is empty then $< A, B >$ denotes an empty set of lines and are therefore not added to the list L_i. Similarly, if A is empty then $< A, \Psi >$ denotes an empty set and is not added to L_i.

Step 4. Let $L_N = \{< C_1, D_1 >, \ldots, < C_p, D_p >\}$. The union of the set of bridging faults represented by these ordered pairs is the possible set of bridging faults.

DIAG-LEAK, our diagnosis algorithm for leakage faults, uses fault dictionaries. It then deletes faults from the fault list after processing each input vector. It is similar to the algorithm in [17].

Example. Refer to Figure 1 and Table 1 which shows respectively the circuit being diagnosed and the test set being used. R is the response vector. To start with all faults are possible so that $L_0 = < \{1, 2 \ldots, 13\}, \Psi >$. $T1$ sets { 1-7, 10 } to 0 and { 8, 9, 11-13 } to 1. There was a good response on application of $T1$. Therefore, $L_1 = \{< \{1-7, 10\}, \Psi >, < \{8, 9, 11-13\}, \Psi >\}$. The reader can verify that: $L_2 = \{< \{8, 9\}, \{11, 12, 13\} >\}$; $L_3 = \{< \{8\}, \{9\} >, < \{11, 13\}, \{9\} >\}$; and $L_4 = L_5 = \{< \{8\}, \{12\} >, < \{13\}, \{9\} >\}$. From this we conclude that the two possible bridging faults are $(8, 12), (13, 9)$. Thus from a total of 78 faults we have narrowed it down to 2.

3 Sequential Circuits

Note that the value at a line in a sequential circuit is dependent on the current values at the primary inputs, the history of the values at the primary inputs and the initial state of the machine. In many systems a master reset, that sets the machine to a known initial state is provided. In this case the initial state of the machine is known. On the other hand the initial state of the machine may not be known and the state variables are assumed to have the value x which could be either a 1 or 0 but not floating. The algorithm for the initial state is similar to Algorithm DIAG-BFC. The following algorithm is used when the initial state is unknown.

DIAG-BFS-UIS

Notations. In addition to the notations used in describing DIAG-BFC we use $A^j(t)$ to denote the subset of nodes in A that has the value j at time t, $j \in \{0, 1, x\}$. The symbol 'x' denotes and unknown value.

Step 1. $L_0 = \{< X, \Psi >\}$. For each time instance $t_i, 1 \leq i \leq N$ do Steps 2, 3.

Step 2. Compute L_i from L_{i-1} as follows. Let $L_{i-1} = \{< A_1, B_1 >, \ldots, < A_q, B_q >\}$. Using three valued fault free simulation and the definition of T_i determine the value, at time t_i, of all lines in X. From this compute $A_j^0(t_i), A_j^1(t_i), A_j^x(t_i), B_j^0(t_i), B_j^1(t_i), B_j^x(t_i)$, $1 \leq j \leq q$. For each $< A_j, B_j >$ construct new ordered pair(s) as follows. Insert the new ordered pairs into L_i and delete $< A_j, B_j >$ from L_{i-1}. There are two cases.

Case 2.1. $B_j = \Psi$. If $r_i = 0$, then the new ordered pairs are: $< A_j^0(t_i), \Psi >$, $< A_j^1(t_i), \Psi >$, $< A_j^x(t_i), \Psi >$, $< A_j^1(t_i) \bigcup A_j^0(t_i), A_j^x(t_i) >$. If $r_i = 1$ then the new ordered pairs are: $< A_j^0(t_i) \bigcup A_j^1(t_i), A_j^x(t_i) >$, $< A_j^0(t_i), A_j^1(t_i) >$, $< A_j^x(t_i), \Psi >$.

Case 2.2. $B_j \neq \Psi$. If $r_i = 0$, then the new ordered pairs are : $< A_j^0(t_i), B_j^0(t_i) >$, $< A_j^1(t_i), B_j^1(t_i) >$,

$< A_j^{\overline{x}}(t_i), \Psi >$, $< B_j^{\overline{x}}(t_i), \Psi >$, $< A_j(t_i), B_j^{\overline{x}}(t_i) >$, $< B_j(t_i), B_j^{\overline{x}}(t_i) >$. If $r_i = 1$ then the new ordered pairs are: $< A_j^0(t_i), B_j^1(t_i) >$, $< A_j^1(t_i), B_j^0(t_i) >$, $< A_j^{\overline{x}}(t_i), \Psi >$, $< B_j^{\overline{x}}(t_i), \Psi >$, $< A_j(t_i), B_j^{\overline{x}}(t_i) >$, $< B_j(t_i), A_j^{\overline{x}}(t_i) >$.

If any of the new ordered pairs represent an empty set then it is not added to L_i.

Step 3. Remove duplicates by comparing all the set of ordered pairs.

Step 4. Let L_N be the list of ordered pairs after processing all the input vectors. Then, the set of possible bridging faults is determined using the semantics of the ordered pairs described in Section 2.

Example. Refer to Figure 2 and Table 2 which shows the circuit being doiagnosed and the test sequence used. The reader can verify that the L_i's computed by the algorithm will be as shown below when we assume an unknown initial state.

$$
\begin{aligned}
L_0 &= \{< \{0-17\}, \Psi >\} \\
L_1 &= \{< \{0-2, 5-7, 9, 10, 17\}, \{3, 5-8, 11-16\} >\} \\
L_2 &= \{< \{0,1,5,9,10,17\}, \{3,5,13,16\} >, < \{2,6,7\}, \\
&\quad \{6,7,8,11,12,14,15\} >\} \\
L_3 &= \{< \{1,9\}, \{13\} >, < \{0,5,10,17\}, \{3,5,16\} >, \\
&\quad < \{2,7\}, \{7,15\} >, < \{6\}, \{6,8,11,12,14\} >\} \\
L_4 &= \{< \{1,9\}, \{13\} >, < \{0\}, \{3,5,16\} >, \\
&\quad < \{2\}, \{7,15\} >, < \{6\}, \{14\}\} \\
L_5 &= \{< \{1\}, \{13\} >, < \{0\}, \{3,16\} >, \\
&\quad < \{2\}, \{7\} >\} \\
L_6 &= \{< \{1\}, \{13\} >, < \{2\}, \{7\} >\}
\end{aligned}
$$

The possible faults are $(1,13), (2,7)$. Thus from a total of 153 faults we have reduced it to 2 faults.

4 Experimental Results

Residual-Faults is the set of faults, determined by the algorithm, to be the possible set of faults. Ideally, the algorithm should terminate when the cardinality of *Residual Faults* is one. However, because of fault equivalence, such a termination condition is often not achievable. In our experiments we terminated the algorithm after the number of residual faults falls below a threshold, denoted by *ResThresh*.

The fault in the faulty circuit may belong to an equivalence class which is very large. In that case even if we process a large number of vectors the cardinality of *Residual Faults* may not fall below *ResThresh*. Two additional termination conditions were used.

(i). *UnChange* is the number of consecutive input vectors for which the cardinality of *Residual Faults* remains unchanged. If *UnChange* exceeds a threshold, denoted by *UnThresh*, then the algorithm is terminated.

(ii). *MaxVec* is an upper bound on the maximum number of input vectors that the algorithm processes. If the number of input vectors processed exceeds *MaxVec* then the algorithm is terminated.

Size is the cardinality of *Residual-Faults*, *Len* is the number of input vectors used and *Time* is the CPU secs on a SUN 4/330 that was required. The values of all the parameters used as well as the statistical distribution obtained is given in [3]. Here we present only the average value of *Size*, *Len*, and *Time*. The ISCAS combinational circuits[5] and the sequential benchmarks from MCNC were used in the experiments. Table 3 is for combinational circuits and Table 4 is for sequential circuits.

Observe that under 1000 randomly generated vectors were enough to effectively diagnose the combinational circuits and in almost all cases used less than 11 CPU secs. Leakaqe faults in combinational circuits and bridging faults in sequential circuits are considerably more difficult to diagnose. Some sequential circuits are almost impossible to diagnose starting with and unknown initial state.

Reference

[1]. M. Abramovici, "A hierarchical, path-oriented approach to fault diagnosis in modular combinational circuits", *IEEE Trans. on Comput.*, Vol. C-31, No. 7, July 1982, pp. 672 - 677.

[2]. M. Abramovici and M. A. Breuer, "Multiple fault diagnosis of MOS Combinational Networks", *IEEE Trans. on Comput.*, Vol. C-29, No. 6, June 1980, pp. 451 - 460.

[3]. F. Brglez and H. Fujiwara, "A Neural Netlist of 10 Combinational Benchmark Circuits and a Target Translator in Fortran", special session on ATPG and Fault simulation, *IEEE Int'l. Symposium on Circuits and Systems*, 1985.

[4]. S. Chakravarty, "On Computing Tests for Bridging and Leakage Faults: Complexity Results and Universal Test Sets", IEEE VLSI DESIGN '92.

[5]. S. Chakravarty and M. Liu, "Algorithms for Current Monitor Based Diagnosis of Bridging and Leakage Faults", Tech. Report No. 92-05, Dept. of Computer Science, State University of New York, Buffalo.

[6]. H. Cox and J. Rajski, "A Method for Fault Analysis for Test Generation and Fault Diagnosis", *IEEE Trans. on Comput.-Aided Design*, Vol. 7, No. 7, July 1988, pp. 813 - 833.

Table 1						
	$X1$	$X2$	$X3$	$X4$	$X5$	R
$T1$	0	0	0	0	0	0
$T2$	1	1	1	1	1	1
$T3$	0	1	0	1	0	1
$T4$	1	0	0	1	1	0
$T5$	1	1	1	0	0	0

Table 2					
	$G0$	$G1$	$G2$	$G3$	R
$V1$	1	1	1	0	1
$V2$	1	1	0	1	0
$V3$	1	0	1	1	0
$V4$	0	1	1	1	1
$V5$	0	0	0	0	0
$V6$	1	1	1	1	1

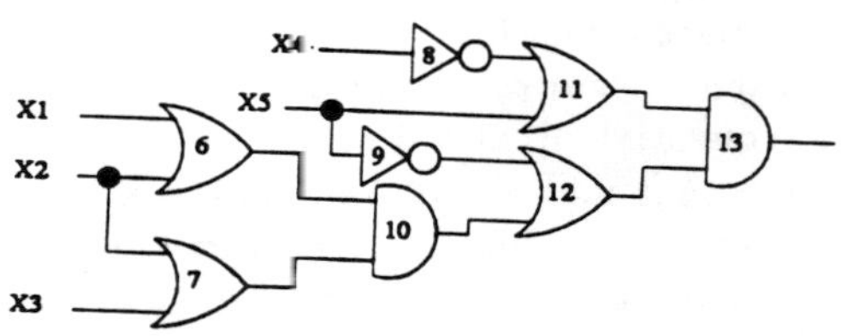

Figure 1:

Table 3						
	Bridging			Leakage		
Name	Size	Len	Time	Size	Len	Time
c880	18	38	0.57	14	243	7.90
c7552	19	443	18.99	148	1575	416.92
c6288	16	19	1.91	74	351	77.19
c5315	35	252	10.65	79	773	139.65
c499	7	80	0.40	28	394	8.83
c432	7	196	0.14	20	117	1.91
c3540	17	718	10.71	75	841	102.12
c2670	10	318	5.46	61	592	55.30
c1908	6	168	1.87	31	336	15.54

[7]. Y. M. El-Zic and S. Y. H. Su, "Fault Diagnosis of MOS Combinational Networks", *IEEE Trans. on Comput.*, Vol. C-31, No. 2, Feb. 1982.

[8]. D. Feltman et. al., "Current Sensing for Built-In Testing of CMOS circuits", *1988 Int'l Conf. on Computer-Aided Design*, pp. 454 - 457.

[9]. F. J. Ferguson and J. P. Shen, "A CMOS Fault Extractor for Inductive Fault Analysis", *IEEE Trans. on Comput.-Aided Design*, Vol. 7, No. 11, pp. 1181 - 1194, 1988.

[10]. W. Maly and P. Nigh, "Built-In Current Testing: Feasibility Study", *1988 Int'l Conf. on Computer-Aided Design*, pp. 340-343.

[11]. P. Nigh and W. Maly, "Test Generation for current testing", *IEEE Design and Test of Computers*, Feb. 1990, pp. 26 - 38.

Table 4						
	Known Init. St.			Unknown Init. St.		
Name	Size	Len	Time	Size	Len	Time
s953	1466	4063	37.07	1845	4270	656.89
s838.1	2293	7519	112.88	40668	13704	1488.74
s832	756	2172	15.47	601	2123	18.99
s820	639	2218	15.44	443	2121	18.86
s713	56	1498	10.53	49	870	12.08
s641	63	1490	9.62	34	928	9.81
s526n	2040	5361	22.55	2166	4964	123.72
S526	2046	5221	22.12	1718	5094	105.20
S510	1816	2102	10.28	2381	2002	91.89
s444	230	3211	11.59	281	3423	39.68
s420.1	105	2252	10.23	4776	5582	149.05
s400	909	3940	12.74	823	3861	42.84
s386	18	618	2.05	17	430	2.19
s382	343	3491	10.91	346	3410	33.32
s349	21	138	0.70	30	355	1.93
s344	25	296	1.13	17	137	1.16
s298	132	1678	4.47	75	1443	8.06
s208.1	26	298	0.83	2206	1895	21.26
s1238	88	631	10.03	104	502	17.25
s1196	75	1227	13.38	89	1092	23.15

[12]. J. Richman, K. Bowden, " The Modern Fault Dictionary", In *Proc. International Test Conference*, pp. 696 - 702, 1985.

[13]. J. Soden and C. Hawkins, "Test Considerations for Gate Oxide Shorts in CMOS ICs", *IEEE Design & Test of Comput.*, Aug. 1986, pp. 56-64.

[14]. J. Waicukauski, J. V. Gupta and S. Patel, "Diagnosis of BIST failures by PPSFP Simulation", In *Proc. International Test Conference*, pp. 480 - 484, 1987.

[15]. J. Waicukauski, E. Lindbloom, "Failure Diagnosis of Structured VLSI", *IEEE Design and Test of Computers*, Vol. 6, No. 4, pp. 49 - 60, August 1989.

[16]. T. Yano, H. Okamoto, "Fast Fault Diagnosis Method using fault dictionary for electron beam tester", In *Proc. Int'l Test Conference*, 1987, pp. 561 - 565.

[17]. R. C. Aitken, "Fault Location with Current Monitoring", In proc. *IEEE International Test Conference*, 1991, pp. 623-632.

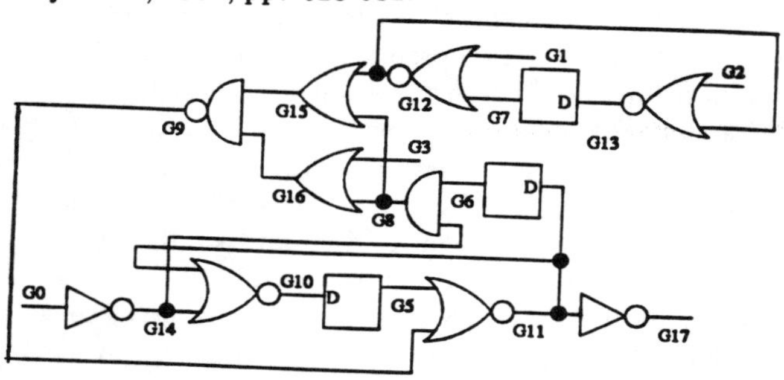

Figure 2:

A Novel Approach to Delay–Fault Diagnosis

P. Girard, C. Landrault, S. Pravossoudovitch

Laboratoire d'Informatique de Robotique et de Microélectronique de Montpellier,
UA CNRS D01480, Université de Montpellier II : Sciences et Techniques du Languedoc
Place Eugène Bataillon, 34095 Montpellier Cedex 05, FRANCE

Abstract

Different approaches to delay fault diagnosis are discussed in this paper. Then a new method, based on critical path tracing from a six–valued simulation, is presented. This method is an alternative to fault simulation based approaches and provides perfectly reliable results. It does not require timing evaluations and can be very accurate.

1: Introduction

In order to ascertain the correct operation of digital logic circuits, the functional behavior, as well as the timing performance, must be verified. Physical defects that may affect the timing behavior without changing the steady–state logical operation of a system are modeled by **delay faults** [8], [15].

The diagnosis of failures affecting the functional behavior of a circuit is generally performed by using a fault dictionary [2], [17]. These failures are modeled by stuck–at faults. As for delay faults, diagnosis is obviously different. This difference arises from the fault model associated with timing failures. A delay fault diagnosis method, that is based on an effect–cause analysis, has been developed by Cox and Rajski [4]. However, this method is unrealistic due to the limitations associated with the transition fault model used by the authors (limitations induced by gross delay faults which are not the most predominant in a circuit [16]). Unlike the stuck–at fault model, a delay fault model has to consider the size of delay defects and, therefore, is often harder to define [3], [9], [14].

In this paper, we shall briefly discuss possibilities of delay fault diagnosis based on fault simulation. Then we will detail the proposed approach based on critical path tracing. Firstly, we will present a path tracing process with information provided by a logic simulation. Subsequently, due to the limitations induced by such a simulation, we will present a reliable approach based on a six–valued logic simulation. It requires no delay size based fault models and considers only the fault–free circuit. Consequently, this approach is faster, reliable and requires less memory than conventional fault simulation based approaches.

2: Main concepts on delay–fault detection

In this study, we use the single fault assumption which is the most widely used, particularly in the context of test generation [11], [13]. Furthermore, the fault type considered is that where the falling and rising transition delay on a faulty element is longer than the modeled value. A delay fault is, therefore, assumed to be local to a gate (**gate delay fault**). However, the method presented can be extended, in a straightforward manner, to the case of **distributed failures** (defects for which the propagation delays along circuit paths exceed the specified values) [15]. In the same way, physical failures affecting only one type of transition (slow–to–rise or slow–to–fall delay faults) [8], are taken into account by considering both falling and rising faulty transitions (which appear, for example, due to a narrow interconnection line). In section 7, we will discuss the degree of diagnosis accuracy according to delay fault models.

Whatever the fault model and the size of a lumped delay fault, two conditions are necessary for the fault to degrade the timing performance of a circuit, and then to be detected during the test.

The ***sensitization condition*** implies that a signal change (transition) must occur on the fault location. This condition will be specified for failures affecting only falling or rising transitions (section 7). On the other hand, we emphasize that it is not necessary for the signal change on the fault site to be a single transition. For example, a delay fault on a line with a transient signal can be detected during the test.

The ***observation condition*** of a delay fault on a primary output is that at least one fault effect propagation path from the fault site to the primary output must exist.

These two conditions which are not sufficient to detect a delay fault (in this case, the size of the fault and the observation time have also to be considered), will be the only ones to be considered during the diagnosis methodology proposed in this paper. In fact, we assume that the only information available for the diagnosis process is the failed test patterns (input values and fault–free and faulty output values). Naturally, the diagnosis accuracy strongly depends on the test sequence itself and on the observation time. This problem which relies on test pattern generation aspects will not be addressed in this paper.

3: Delay–fault diagnosis based on fault simulation

As for stuck–at faults, delay fault diagnosis could be performed by using a delay fault simulation. However, a fault dictionary is harder and longer to establish with a delay size based fault model. Moreover, such an approach requires a very precise timing simulation [5][12]. So, when we are only interested in locating a delay defect, delay fault diagnosis based on delay fault simulation may not be the best solution.

Nevertheless, we can imagine another approach to delay fault diagnosis based on stuck–at fault simulation [10]. This is possible due to the relationship existing between the detectability conditions of a delay fault (section 2) and those of a stuck–at fault.

Firstly, if we remind that a necessary condition for the observation of a stuck–at fault on a primary output is that at least one sensitized path from the fault site to the primary output must exist, we can define the following properties:

Property 1: If the observability condition of a stuck–at fault on a circuit line is verified, then the observability condition of a delay fault on the same site is also verified.

Property 2: If an error, possibly resulting from a stuck–at fault on a circuit line, is observed on a circuit output, then it may also result from a delay fault on the same site, provided that a signal change exists on this site.

If the opposite property were true, i.e. a delay fault observed on an output implies that a stuck–at fault on the same site is observed, the delay fault diagnosis problem would be solved. Indeed, it

0738-100X/92 $3.00 © 1992 IEEE

Paper 22.7

would be sufficient to perform the diagnosis from a stuck–at fault simulation, by considering only the lines with a signal change.

Unfortunately, this opposite property is not valid, particularly in the context of reconvergent fanout circuits. Indeed, some faults in this case can verify the observability condition of a delay fault without affirming the observability condition of a stuck_at fault on the same site. Thus, a delay fault diagnosis based on such an approach can lead to incomplete or erroneous results [6].

4: Delay–fault diagnosis based on path tracing from a logic simulation

In our attempt to develop an alternative method to fault simulation, we take advantage of the critical path tracing algorithm that was originally developed for stuck–at faults [1].

First of all we consider a simple logic simulation that provides the list of transitions on each line in the circuit for the applied test sequence. All this information is used by the diagnostic process, in addition to the logical description of the circuit and the failing patterns and primary outputs (information provided by the tester). Thus, it seems possible to extract some information about fault locations verifying the necessary detectability conditions of a delay fault. This process consists of the following steps:

step 1: After the test, store the generated list of failing patterns and primary outputs.

step 2: Perform a good–machine two–value simulation for one failing pattern and store references about transitions on each line of the combinational circuit.

step 3: From the failing primary outputs associated with the current test pattern, perform a critical path tracing (depth–first–search algorithm) towards primary inputs by marking as critical every sensitized line (line with one or more transitions). In this way, the set of paths having propagated a signal change is determined.

step 4: After this phase, perform a set intersection between the previous failing lines set and the current set constructed with critical lines.

Steps 2, 3 and 4 are repeated for every failing pattern of the test sequence.

The critical lines traced from every failing pattern are possible causes of failure because of the following property:

Property 3: Every delay fault affecting one element belonging to these critical paths verifies the necessary detectability conditions of a delay fault.

Unfortunately, critical path tracing for delay faults based on a two–value logic simulation shows certain deficiencies. These limitations are due to the fact that in real situations output errors may be produced by delay faults affecting internal lines submitted to static hazards which are not handled by a two–valued logic simulation [6]. Thus, the diagnosis process can sometimes provide erroneous results.

However, we can imagine the same approach to diagnosis where the logic simulation is replaced by a six–valued simulation.

5: Delay–fault diagnosis based on path tracing from a six–valued simulation

When we define a set of elements which can be the cause of failure, and if we want to make sure that the fault having induced the error is necessarily on a line belonging to this set, we must have more detailed information about the observability of lines. As we have seen, it is not sufficient to consider transition propagation from the fault–free circuit. All transition propagation possibilities have to be considered, especially when they only exist in the presence of a delay fault.

In order to make transition propagation possibilities appear in fault–free and faulty circuits, we have replaced the classical logic simulation by a six–valued logic simulation based on the H6 algebra [7] with the following symbols: C0 for static 0 signal, C1 for static 1 signal, R1 for 0–to–1 transition, F0 for 1–to–0 transition, P0 for static 0–hazard and P1 for static 1–hazard.

During simulation, the values associated with primary inputs are C0, C1, F0 and R1 for any pattern (for the first vector of the sequence, the values are C0 or C1), provided that the input latches are glitchless.

The first advantage of six–valued simulation is that it takes into account the possibility for a given line to be stable in the fault–free circuit and unstable in the presence of a delay fault. Another advantage is that it is independent of any gate propagation delay or delay fault size. Thus, such a simulation does not require the consideration of timing specifications.

By applying the critical path tracing algorithm (as described in section 4) with 6–valued simulation results, we can assert that every line not in the potential failing lines set is necessarily fault–free. In this process, starting from a failing primary output and trying to extend the critical state of the primary output as far as possible towards the primary inputs, all lines with the values F0, R1, P0, P1 are considered to be on a fault effect propagation path. The lines with the values C0 or C1 stop the depth search process.

Consequently, the proposed delay fault diagnosis method consists of the following steps:

step 1: Preprocessing the combinational circuit model to determine a levelled order.

step 2: 6–valued simulation of the first sequence's pattern with the symbols C0 or C1 on every primary input.

step 3: 6–valued simulation of one failing test pattern and determination of the signal values on each line.

step 4: Critical path tracing for delay faults. This is a backtracking procedure that identifies the critical lines (and hence the potential failing lines) for the test simulated in step 3. For every gate evaluation, the critical path tracing is carried out from inputs having the value F0,R1,P0 or P1. Conversely, the trace is not extended through lines having the value C0 or C1.

step 5: Execute a set intersection between the potential failing lines set and the set constructed with critical lines from the current test (we use the single fault assumption).

Steps 3, 4 and 5 are repeated for every failing test pattern of the test sequence.

Although we can improve diagnosis accuracy (section 6), this 6–valued simulation based approach is reliable. Indeed, the fault location associated with the cause of failure is always contained in the potential failing lines set.

Example: Let us consider the test pair (0111,1010) which, when applied to the circuit of figure 1, gives two failing outputs (O2 and O3).

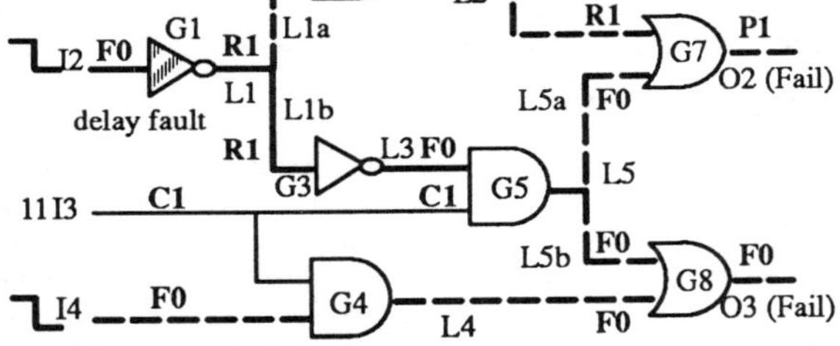

Figure 1: Diagnostic by critical path tracing from a six–valued simulation

The critical path tracing performed from O2 provides the set {I1,I2,L1,L2,L3,L5,O2}, and the one obtained from O3 is {I2,I4,L1,L3,L4,L5,O3} (dotted lines). As a result, the set of potential failing lines generated after the test and resulting from the set intersection is {I2,L1,L3,L5} (bold and heavy lines). The line associated with the cause of failure (L1) is actually contained in the resulting set of potential lines.

6: Improvement of diagnosis accuracy

In order to improve diagnosis accuracy without altering reliability, we have implemented a heuristic whose objective is to reduce the number of critical lines traced during the backward procedure. These improvements are obtained by considering fault propagation possibilities through every gate according to transition polarity on the gate inputs. Let us consider an OR gate with transitions on its inputs (Figure 2).

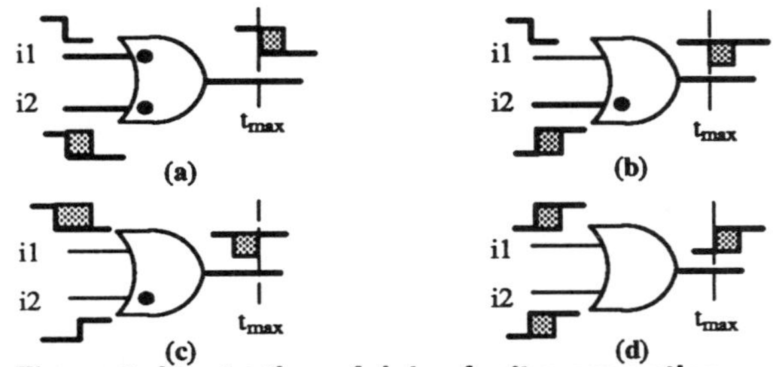

Figure 2: Illustration of delay fault propagation

In the first case (2.a), a slow–to–fall delay fault on one of the two inputs is propagated to the gate output. Due to their fault effect propagation ability, these inputs are called **"sensitive lines"** (they are marked by dots). In the second case (2.b), line i2 is a sensitive input because a delay on this line can be propagated throughout the gate. Inversely, a delay on i1 cannot be observed at the gate output (after the normal stabilization time of the circuit Tmax) due to the dominant final value of input i2 (2.c). During the diagnosis process, the search will be carried out from the sensitive input i2 only, reducing the number of lines to be considered and improving diagnosis accuracy. In the case of gates with no sensitive inputs (2.d), called **"non sensitive gates"**, a delay fault can only be propagated throughout the gate if the two inputs are affected simultaneously by the fault. Due to the single fault assumption, this is only possible if the failure is located on a line belonging to a common part of the paths descending from these inputs. So, in the diagnosis process, the critical path tracing is stopped at this gate, and then, we have to look for a reconvergence of the paths issuing from these inputs (only inputs with signal F0 or P0 /AND,NAND/ and R1 or P1 /OR,NOR/). If at least one such reconvergence exists (associated fanout stem), the search for potential fault locations in the circuit will go on from the fanout stems found. By not marking as critical the lines belonging to the fanout region comprised between the fanout stem and this non sensitive gate, the diagnosis is improved again.

If we generalize the considerations established for an OR gate, we obtain the following rules which allow the evaluation of gate input sensitivity:

Rule 1: If only one input has the dominant final logic value and if the signal on this line is F0, R1, P0, P1 (unstable signal), then the line is sensitive.

Rule 2: If all inputs have the non dominant final logic value, then each input with a signal F0, R1, P0 or P1 is sensitive.

Rule 3: otherwise, no input is sensitive.

We emphasize that in the case of XOR or NXOR gates, there is no dominant logic value. So, all inputs with the value F0, R1, P0 or P1 are sensitive.

From the computed examples, it appears that the fault localization, which closely depends on the applied test sequence and on the number of failing patterns, may be very accurate in many cases.

The search for one or more associated fanout stems (if any), using these concepts of sensitive lines, is performed during the critical path tracing (step 4 of the overall algorithm), detailed as follows.

Detailed description of the critical path tracing procedure

- the failing output is a potential failing line
- determination of the sensitive inputs of the above gate:
  - *if sensitive inputs exist, they are added to the set of potential failing lines, and the research process is carried on from these sensitive lines.*
  - *if there is no sensitive input, the inputs with the value F0,R1,P0 or P1 are labelled if, according to the gate, they have a dominant final value (F0 or P0 for AND and NAND gates, R1 or P1 for OR and NOR gates). The fanout stem(s) associated to the gate is (are) determined by a backward propagation of labels through sensitive paths (width first search algorithm), and by the union of labels propagated on every fanout stem:*
    - *if we can find lines assigned to all the initial labels, these lines are fanout stems associated with the non–sensitive gate. Furthermore, they are added to the potential failing lines set, and the research process is carried on from these lines.*
    - *if the initial labelled lines do not reach a fanout stem, any descendent line of the non–sensitive gate is potentially faulty. Therefore, the critical path tracing stopped at the non–sensitive gate will never start again from one of these descendent lines.*

7: Improvements according to delay–fault models

As mentioned in section 2, the delay fault diagnosis method presented in this paper does not depend on the delayed transition polarity. In other words, a delay fault affecting a falling or a rising transition may be located as well as a delay fault having an effect on both falling and rising transitions. Conversely, we may often distinguish between slow–to–fall (StF) and slow–to–rise (StR) delay faults with respect to delay fault test pattern generation [4], [9]. In this case, we can specify the sensitization condition of a delay fault (section 2) in the following manner: A signal change must occur on the fault location, with the final logic value 0 (1) for an StF (StR) delay fault sensitization.

Due to the symbolic values associated with a 0 (1) final logic value signal, it is only possible to sensitize a StF (StR) delay fault with symbols P0 or F0 (P1 or R1). Thereby, assuming that the delay faults affect only one type of transition, we can improve the diagnosis based on critical path tracing from a symbolic simulation. In this case, the lines that are supposed to have both StF and StR delay faults will be declared fault–free after the set intersection process (due to the single fault assumption). Thus, this reduction process allows us to minimize the final set of potential failing lines. Taking these considerations into account, the intersection process must be modified in order to remove the lines sensitized for both StF and StR delay faults.

Consequently, diagnosis based on critical path tracing from a symbolic simulation can produce different results according to previous considerations. Indeed, diagnosis accuracy is improved by considering only StF or StR delay faults. However, delay faults affecting both falling and rising transitions (Slow–to–Rise&Fall delay faults) are not taken into account in this case. So, if we are interested in considering all types of delay faults (StR, StF and StR&F), we cannot remove any StR&F delay fault during the set intersection process. Here, we can only provide some information about the type of faults associated with lines belonging to the potential failing lines set. This information is given by the following symbols:

R : "slow to rise" delay fault
F : "slow to fall" delay fault
D : "slow to rise and fall" delay fault

and is obtained from the symbolic values through the intersection and association rules given below :

P0 -> F
P1 -> R
F0 -> F
R1 -> R

∩	F	R	D
F	F	D	D
R	D	R	D
D	D	D	D

8: Practical Results

Delay fault diagnosis accuracy is strongly dependent on the test sequence applied to the circuit and on the delay defect size. Moreover, diagnosis accuracy depends on the fault discernability (equivalent delay faults). Nevertheless, we present significant results which show the efficiency of the reduction process based on sensitive lines, compared to a symbolic simulation only process.

The following tables summarize the results obtained with test sequences generated for stuck–at faults by "System–Hilo: Hitest" (GENRAD). Faulty circuits have been simulated using "System–Hilo: Hisim" with random delay faults (location and duration). We have assumed that there is a propagation delay of one unit for every gate in the fault–free circuit, and have explained defect size in terms of unit delay. The CPU time is given in seconds. This logic simulation produces information normally provided by the tester in the presence of real faulty circuits. These results are given for two ISCAS'85 benchmarks in the number of potential failing lines produced by the following diagnostic processes:

D1: Diagnosis based on critical path tracing from a symbolic simulation without considering sensitive lines.

D2: Diagnosis based on critical path tracing from a symbolic simulation with the reduction process based on sensitive lines.

C432 27 test patterns	Patterns "fail"	Outputs "fail"	D1	D2	CPU (D1)	CPU (D2)
Gate: 46 / 1u	2	4	9	2	0.21	0.21
Gate: 29 / 1u	2	4	13	5	0.21	0.27
Gate: 144 / 1u	2	2	17	7	0.20	0.19
Gate: 48 / 2u	6	8	4	2	0.32	0.25
Gate: 27 / 2u	2	3	25	5	0.18	0.24

C7552 137 test patterns	Patterns "fail"	Outputs "fail"	D1	D2	CPU (D1)	CPU (D2)
Gate: 67 / 1u	3	3	90	15	14.26	12.10
Gate: 67 / 2u	8	10	64	6	15.58	13.74
Gate: 141 / 3u	21	23	59	4	26.24	22.61
Gate: 3498 / 3u	8	8	143	42	15.99	21.03
Gate: 3495 / 4u	11	11	302	39	22.02	18.80

9: Conclusion

Delay fault diagnosis is quite different from the stuck–at fault diagnosis. At the beginning of this paper, we showed the deficiencies of delay fault diagnosis based on fault simulation. Then we presented an alternative to the fault simulation based approach. The proposed method for delay fault diagnosis is based on critical path tracing from a symbolic simulation, due to the limitations caused by a logic simulation. The principle of such a simulation and the critical path tracing process have been presented. This diagnosis method is perfectly reliable, and does not need timing evaluations. To improve diagnosis accuracy, we have also developed the concept of sensitive lines. Based on the method presented, a fast algorithm has been implemented in C++ on a SUN–sparc workstation.

References

[1] M. Abramovici, P.R. Menon and D.T. Miller, "Critical Path Tracing – An Alternative to Fault Simulation", *Proc. 20th Design Autom. Conf.*, pp. 214–220, June 1983.

[2] Y. Arzoumanian and J. Waicukauski, "Fault Diagnosis in an LSSD Environment", *Proc. of Int. Test Conf.*, pp. 86–88, Oct. 1981.

[3] J.L. Carter, V.S. Iyengar and B.K. Rosen, "Efficient Test Coverage Determination for Delay Faults", *Proc. of Int. Test Conf.*, pp. 418–427, Sept. 1990.

[4] H. Cox and J. Rajski, "A Method of Fault Analysis for Test Generation and Fault Diagnosis", *IEEE Trans. Computer-Aided Design*, vol. 7, no. 7, pp. 813–833, 1988.

[5] F. Fink, K. Fuchs and M.H. Schulz, "An Efficient Parallel Pattern Gate Delay Fault Simulator with Accelerated Detected Fault Size Determination Capabilities", *Proc. of Euro. Test Conf.*, pp. 171–180, April 1991.

[6] M.L. Flottes, P. Girard, C. Landrault and S. Pravossoudovitch, "A New Reliable Method for Delay–Fault Diagnosis", *Proc. of the 5th International Conference on VLSI Design*, pp. 12–18, Jan. 1992.

[7] J.P. Hayes, "Digital Simulation with Multiple Logic Values", *IEEE Trans. Computer–Aided Design*, vol. 5, no. 2, pp. 274–283, April 1986.

[8] E.P. Hsieh, R.A. Rasmussen, L.J. Vidunas and W.T. Davis, "Delay Test Generation", *Proc. 14th Design Autom. Conf.*, pp. 486–491, June 1977.

[9] V.S. Iyengar, B.K. Rosen and I. Spillinger, "Delay Test Generation 1 – Concepts and Coverage Metrics", *Proc. of Int. Test Conf.*, pp. 857–866, Sept. 1988.

[10] S. Koeppe, "Modeling and Simulation of Delay Faults in CMOS Logic Circuits", *Proc. of Int. Test Conf.*, pp. 530–536, Sept. 1986.

[11] J.D. Lesser and J.J. Shedletsky, "An Experimental Delay Test Generator for LSI Logic", *IEEE Trans. on Computers*, vol. C–29, no. 3, March 1980.

[12] W.W. Mao and M.D. Ciletti, "A Simplified Six–Waveform Type Method for Delay Fault Testing", *Proc. 26th Design Autom. Conf.*, pp. 730–733, June 1989.

[13] E.S. Park and M.R. Mercer, "Robust and Nonrobust Tests for Path Delay Faults in a Combinational Circuit", *Proc. of Int. Test Conf.*, pp. 1027–1034, Sept. 1987.

[14] A.K. Pramanick and S.M. Reddy, "On the Fault Coverage of Delay Fault Detecting Tests", *Proc. of Euro. Design Autom. Conf.*, pp. 334–338, March 1990.

[15] G.L. Smith, "Model for Delay Faults Based upon Paths", *Proc. of Int. Test Conf.*, pp. 342–349, November 1985.

[16] N.N. Tendolkar, "Analysis of Timing Failures due to Random AC Defects in VLSI Modules", *Proc. 22th Design Autom. Conf.*, pp. 709–714, June 1985.

[17] J.A. Waicukausky and E. Lindbloom, "Failure Diagnosis of Structured VLSI", *IEEE Design and Test of Computers*, pp. 49–60, August 1989.

TEMPT: Technology Mapping for the Exploration of FPGA Architectures with Hard-Wired Connections

Kevin Chung & Jonathan Rose
Department of Electrical Engineering
University of Toronto
CANADA M5S 1A4

Abstract

This paper describes TEMPT, a technology mapping algorithm aimed at exploring FPGA architectures with hard-wired connections. Such FPGA architectures may be important because hard-wired connections are much faster and smaller than the programmable connections between basic logic blocks that they replace.

TEMPT maps a network of basic blocks to a netlist of hard-wired logic blocks (HLBs), in which each HLB consists of several basic blocks hard-wire connected in an arbitrary tree topology, and optimizes either speed or area. TEMPT is as effective as the Xilinx 4000 CLB mapper, PPR, when minimizing CLBs to implement a set of MCNC benchmarks. Using TEMPT we demonstrate empirically how many HLBs are significantly faster than FPGAs without hard-wired links. Also, we demonstrate several HLBs that exhibit superior logic density to the Xilinx 4000 CLB[1].

1 INTRODUCTION

Field-Programmable Gate Arrays (FPGAs) are an increasingly popular means of implementing Application Specific Integrated Circuits (ASICs) because designers can manufacture their ASIC cheaply and in minutes. The user configures the FPGA logic and routing connections by a simple programming technology (such as downloading bits into static RAM cells [Hsie90] or blowing anti-fuses [ElGa89]). This user-programmability of routing connections, however, makes FPGAs slower and less dense than Mask-Programmable Gate Arrays because of the significant capacitance, resistance and size of the programmable switches. This routing delay and area can be significantly reduced by replacing some of the programmable connections with hard-wired connections, which are metal wires that have almost-zero delay and consume little area [Chun91] [Sing91]. For example, Figure 1(a) illustrates a network of basic logic blocks with five slow programmable connections in the routing along the critical path[2]. Suppose three basic blocks are hard-wired together, as in Figure 1(b), to make a hard-wired logic block (HLB) with a fast path through 3 basic blocks. Using this HLB to implement the circuit in Figure 1(a) results in the faster circuit shown in Figure 1(c)[3]. This faster circuit has only two

[1]This work was supported by an ITRC Fellowship, NSERC Operating Grant #URF0043298 & a Xilinx Inc. research grant.

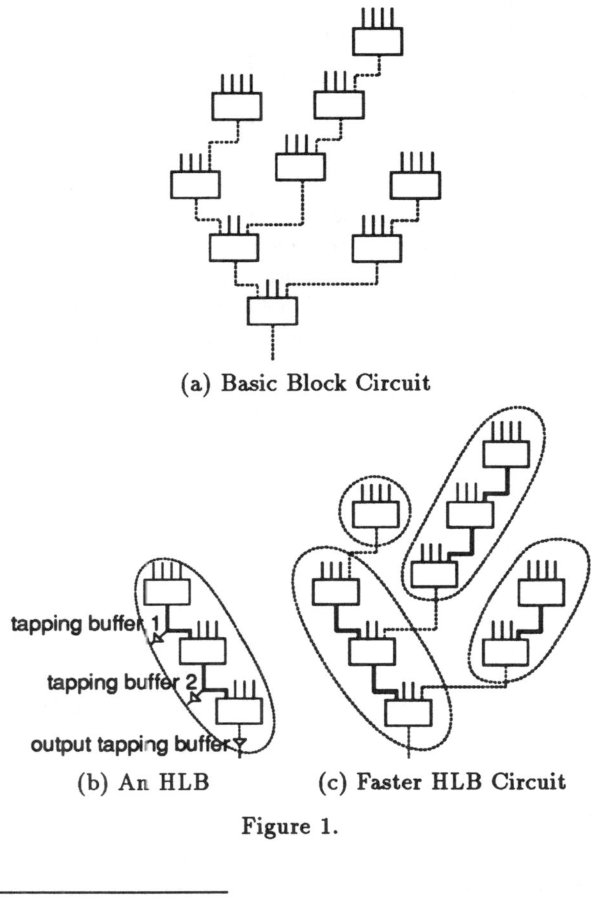

(a) Basic Block Circuit

(b) An HLB

(c) Faster HLB Circuit

Figure 1.

[2]A dotted line represents a programmable connection and only output connections along the path are counted.

[3]The hard-wired links are shown in thick lines and the hard-wire connected sets of basic blocks are circled by dashed ovals.

0738-100X/92 $3.00 © 1992 IEEE

programmable links instead of five along the critical path, giving a significant reduction in routing delay. However, the inflexibility of hard-wired links in the circuit can lead to unusable basic blocks, resulting in lower logic density.

The presence of hard-wired connections provides several reasons for the introduction of a new synthesis tool. First, there already exists a commercial architecture [Hsie90] with hard-wired connected basic logic blocks. Second, hard-wired connected basic blocks present a new synthesis problem because they present a strong link between logic synthesis and routing. The third and most important motivation is to aid in the exploration of FPGA architectures with hard-wire connected basic blocks. In [Chun91] [Sing91] we investigated the speed/area tradeoffs of an FPGA architecture composed of an array of Hard-wired Logic Blocks (HLBs), where an HLB is several basic logic blocks (such as lookup tables or NAND gates) connected by metal wires in a tree topology. This study showed that certain HLB topologies result in a 44% decrease in the average number of programmable connections along the critical path compared to an FPGA without hard-wired connections, yet only reduce the logic density (measured by the total number of basic blocks) by 6% [Chun91]. The results in [Chun91] [Sing91] were obtained empirically using logic synthesis tools to implement benchmark circuits. This paper presents one of the tools, TEMPT, a technology mapper for tree-connected hard-wired logic blocks.

This paper is organized as follows. Section 2 presents some assumptions and terminology. Section 3 describes the TEMPT technology mapping algorithm for both delay and area optimization. Section 4 shows the results of using TEMPT to map to several different HLB architectures.

2 ASSUMPTIONS & DEFINITIONS

An important architectural assumption is that each basic block output in an HLB has a *tapping buffer* that makes the output accessible to the routing. Figure 1(b) shows an HLB with three 4-input basic blocks hard-wired together in a cascade. Tapping buffers can lead to faster HLB circuits since the output of one basic block can be accessed directly instead of propagating it through another basic block. Tapping buffers can also improve logic density because unrelated pieces of logic can be packed together in the same HLB.

The inputs to TEMPT consist of a description of the HLB topology and a network of basic blocks. In the remainder of this paper, we consider only lookup table (LUT) basic blocks since they exhibit reasonable area and delay in FPGAs without hard-wired connections [Rose90] [Sing91]. Also, there is an FPGA [Hsie90] with hard-wire connected LUTs and an associated mapper for testing and comparison. The LUT basic block network is generated

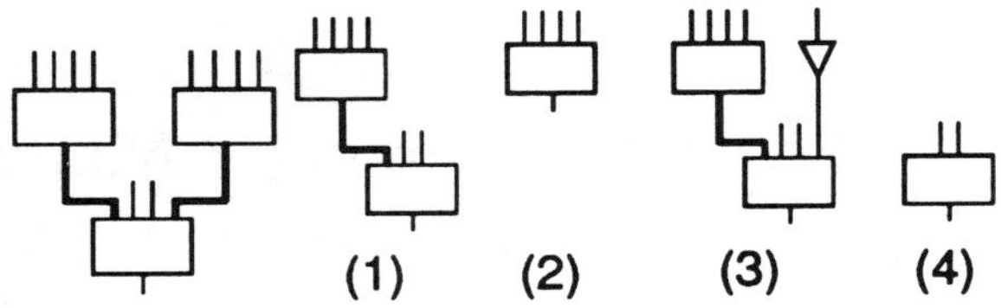

(a) L2-3 HLB (b) L2-3 HLB Fragments

Figure 2: The L2-3 HLB and its Fragments

using the following two steps: (1) logic optimization of the MCNC benchmark circuit using mis2.2 [Bray86], and (2) mapping to LUTs using Chortle [Fran91a] [Fran91b].

Synthesizing HLB circuits is a new problem because HLBs are a combination of logic and routing resources. Thus, the mapping to HLBs can occur during either logic *or* layout synthesis. In TEMPT, we chose to perform the assignment of hard-wired links in the HLBs during logic synthesis because the goal of our previous work [Chun91] [Sing91] was an exploration of FPGA logic block architectures with no assumption of specific routing architectures. The use of a placement and routing algorithm would require more detailed specification of the routing architecture, whereas mapping to HLBs during technology mapping abstracts out physical layout details. This abstraction for technology mapping also allows a tighter focus on optimizing the use of the hard-wired connections for each HLB and may lead to better optimization for HLBs.

2.1 HLB Fragment Patterns

An HLB fragment is a connected subset of the basic blocks in an HLB [4]. Since we assume tapping buffers make every basic block output in an HLB accessible, any HLB fragment that matches a portion of the basic block network can be used to implement that portion. An HLB fragment that matches part of the basic block network rooted at a specific node is said to be *feasible* at that node. Since HLB fragment DAGs are trees, the feasible HLB fragments can be quickly found using tree pattern matching [Keut87].

Given an HLB topology described by a tree, the HLB fragment patterns are generated by one of the following operations on the tree. The first operation deletes a hard-wired link edge between two basic blocks (this assumes that a hard-wired input of the downstream block can be ignored) and the second operation converts a basic block into a buffer. Note that LUTs allow both of these operations, but a basic block such as a 4-input NAND gate does not. Figure 2 shows the L2-3 HLB[5] in Figure 2(a) and its

[4] Note that the complete HLB is also a fragment

[5] The HLB name is the height of the HLB followed by a listing of the sizes of the subtrees from a pre-order traversal. Subtrees of size 0 or 1 are not listed.

generated fragments in Figure 2(b). HLB fragment number 1 is produced by deleting an input edge of the root basic block of the HLB. At the same time the complementary HLB fragment (numbered 2) composed of a single 4-input basic block is also created. Fragment number 3 is made by using one of the basic blocks that feeds the root block to implement a buffer. The deletion of the edge between the two basic blocks of fragment 1 yields a new HLB fragment (number 4), which is a two-input basic block.

3 TEMPT MAPPING ALGORITHM

TEMPT uses two separate algorithms for optimizing the delay or area of the netlist of HLBs that implement the functionality of the input basic logic block network. The goal of the delay algorithm is to find a set of feasible HLB fragments that minimize the delay along the critical path. This delay-optimized set of fragments is then packed together into a minimum number of HLBs to reduce the area overhead. The goal of the area-optimization algorithm is to find the set of feasible HLB fragments that when packed together yields the minimum number of HLBs.

3.1 Delay Optimization

As in [Keut87], the DAG representing the input basic block network is first partitioned into a forest of trees. Each network tree is traversed in postorder fashion and the feasible HLB fragments at each node are found using tree matching. The optimal delay cover of feasible fragments for the tree is determined using dynamic programming [Keut87]. A straightforward merging of the trees to reproduce the final network would give sub-optimal delay performance because optimization across tree roots with fanout > 1 is prevented. This limitation is overcome by selectively replicating portions of the basic block netlist that fanout to two or more nodes. For example, Figure 3(a) shows a circuit that has been implemented in the HLB of Figure 2(a) when no replication across fanout is allowed. The circuit in Figure 3(a) has two HLBs along the critical paths and thus there are two programmable connections in the critical path. Replication of the basic block in the centre of Figure 3(a) into its fanouts leads to the faster implementation in Figure 3(b) with only one programmable connection delay. However, the number of basic blocks in the new network has increased from five to six.

The primary cost function in the dynamic programming and tree matching algorithm is the delay measured as the maximum number of HLBs (and hence the number of programmable connections) between the primary inputs and the node. If two feasible HLB fragment patterns lead to solutions with the same delay, the one with the lower height (or if the heights are also the same, the one with smaller size) is chosen. The height of an HLB fragment is the

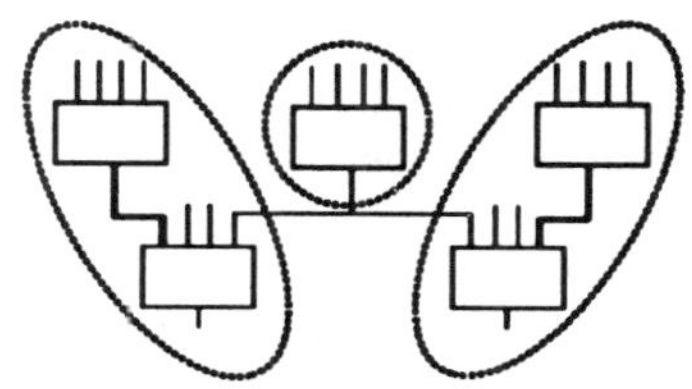

(a) Circuit implementation without Replication

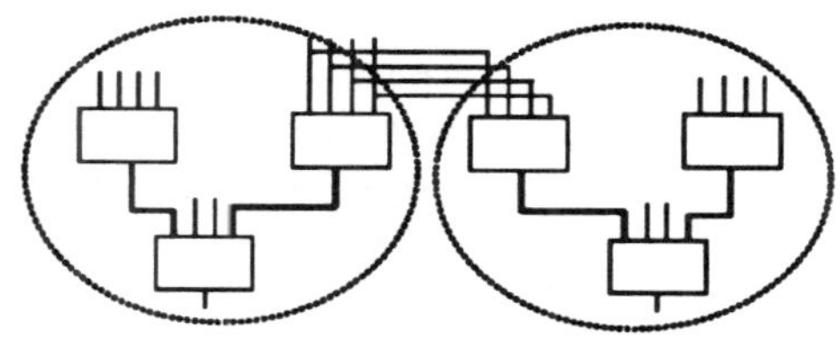

(b) Circuit implementation with Replication
Figure 3.

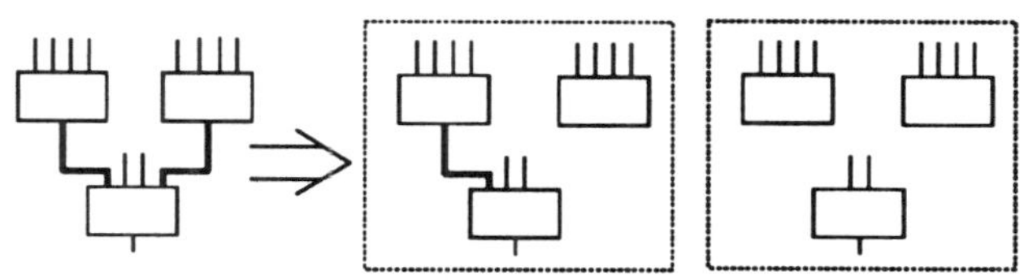

Figure 4: The L2-3 HLB and its Packing Sets

maximum number of basic logic blocks from the fragment inputs to the output. The size is the number of basic blocks in the HLB fragment. These secondary cost functions lead to an HLB fragment at the root of each tree with minimal delay with respect to the tree inputs and minimal height and size. Small root HLB fragments lead to fewer replicated basic blocks when replication at tree roots is needed.

Each HLB fragment produced by the delay optimization algorithm can fit in an HLB. However, allotting one fragment per HLB is excessive, so the delay-optimization algorithm packs the HLB fragments into as few HLBs as possible using the algorithm outlined in Section 3.1.1.

3.1.1 Packing HLB Fragments

The packing algorithm takes a list of HLB fragments and packs them into the smallest number of HLBs. A set of HLB fragments that can be packed together into a single HLB is called a *packing set* of HLB fragments and every packing set is a subset of a maximal packing set.

The maximal packing sets are generated using the same deletion and buffering operations as in Section 2.1. Figure 4 illustrates how the edge deletion operation generates two maximal packing sets for the HLB from Figure 2(a). Each maximal packing set is enclosed in a dotted rectangle. The first edge deletion generates a maximal packing set with two HLB fragments, a two-basic block fragment

```
PackHLBfrags()
    PackedHLB_list ← φ
    UnpackedFrags ← descendingSort(UnpackedFrags)
    while UnpackedFrags not = φ
        packedHLB ← φ
        while UnpackedFrags not = φ
            addedFrag ← next (UnpackedFrags)
            packedHLB ← packedHLB ∪ addedFrag
            if packedHLB is a maximal packing set
                exit inner loop
            else if packedHLB is not a packing set
                return addedFrag to UnpackedFrags
            else if addedFrag is last of UnpackedFrags
                exit inner loop
            end if
        end while
        Packed_list ← Packed_list ∪ packedHLB
    end while
```

Figure 5: HLB Fragment Packing algorithm

plus a single 4-input basic block fragment. Deleting the edge from the two-basic block fragment generates the second maximal packing set, which consists of two 4-input basic blocks and one 2-input basic block. Note that the HLB fragments in a maximal packing set can be packed in the same HLB because every basic block output in an HLB is accessible via tapping buffers.

The pseudo-code for the packing algorithm is illustrated in Figure 5. Since smaller fragments can fit more easily into an HLB with other fragments it is more effective to pack them last; thus, the packing algorithm begins by sorting the unpacked HLB fragments into a descending list according to size. While the unpacked list is not empty, the outer loop of the packing algorithm constructs packed HLBs with the fragments from the list in a greedy manner.

A new packed HLB is initially empty. The inner loop scans the unpacked fragment list from largest to smallest, greedily adding any unpacked fragments that can fit in the unused part of the new packed HLB. The set of fragments can fit in a packed HLB if it is a subset of a maximal packing set. Unpacked fragments are added until one of three conditions holds: (1) The set of fragments form a *maximal packing set* and thus the packed HLB can hold no more fragments; (2) The unused capacity of the packed HLB is too small to accommodate any of the remaining unpacked fragments or (3) There are no remaining unpacked fragments. At this point the "filled" packed HLB is added to the packed HLB list and if there are remaining unpacked fragments, a new packed HLB is started.

```
FindMinimumAreaMapping()
    initially there are only single-block HLB fragments
    pack single-block fragments to find global area cost
    if global area cost satisfies lower bound
        exit mapping algorithm
    end if
    for every node n in the basic block network
    traversed in pre-order
        for every feasible HLB fragment f at n
            pack f with all other HLB fragments
            if f leads to lowest global area cost so far
                record f as best frag. at n and area cost
                if area cost satisfies lower bound
                    exit mapping algorithm
                end if
            end if
        end for
        if the lowest area-cost feasible HLB fragment
        pattern at n is not a single-block fragment
            make the basic blocks in the lowest cost
            HLB fragment into a "fixed HLB"
        end if
    end for
```

Figure 6: Area-optimization algorithm

3.2 Area Optimization

The area-optimization mapping algorithm seeks to minimize the number of HLBs in the final HLB netlist. Each feasible HLB fragment at a node in the basic block network corresponds to a potential assignment of hard-wired links between basic blocks. The goal of the algorithm is to select the set of feasible fragments (and corresponding hard-wired links) that pack together to yield the fewest HLBs. A lower bound on the number of packed HLBs is $\left\lceil \frac{\textit{number of basic blocks in network}}{\textit{number of basic blocks in an HLB}} \right\rceil$ and whenever this lower bound is achieved, the mapping algorithm ends.

In order to do a packing to determine global area cost every node in the basic block network has to be assigned to some HLB fragment. Thus, the algorithm described in Figure 6 starts with each basic block being mapped to a single-block HLB fragment, that is, the initial HLB network has no hard-wired links. We chose this initial assignment because small fragments can fit more easily into unused portions of an HLB during packing. The mapping algorithm then proceeds to add hard-wired links by choosing feasible multi-block HLB fragments at nodes in the basic block network that lead to a better packing. The nodes in the network are accessed in pre-order. Each feasible multi-block HLB fragment at a node is packed with the other previously constructed multi-block fragments and any remaining single-block fragments. Note that since there is no tree partitioning, matches for feasible fragments can occur across fanout. This leads to an increase in the number of feasible fragments at some nodes and can lead to better optimization. After packing, if the best global area cost so far is reduced then this feasible HLB fragment becomes the best fragment at the node. After all feasible HLB fragments for the node have been evaluated, the lowest area cost fragment is fixed permanently so that the nodes in the best fragment cannot be assigned to a different fragment. When all nodes have been visited the algorithm ends.

The primary global area cost function is the number of HLBs after packing. However, if two sets of HLB fragments have the same number of packed HLBs, the set with the largest HLB hole (i.e. number of unused logic blocks in an HLB) is chosen. If the two sets are still tied, then the one with fewer unpacked HLB fragments wins. The largest HLB hole cost function is used because a large hole can be more easily filled, during packing, by an unpacked fragment. Having fewer HLB fragments is also useful because it means a greater total number of hard-wired connections are in the fragments. Since the packing of fragments into the same HLB leads to the wastage of hard-wired connections, more hard-wired connections in the fragments mean that fewer are needed for packing them together.

4 RESULTS

In this section, we first compare TEMPT to the commercial technology mapper for the Xilinx 4000 (X4000) CLB [Hsie90], which is an FPGA with hard-wired connections. To illustrate how TEMPT can be used for FPGA architecture exploration we compare the X4000 CLB to other HLBs for area-efficiency. We also present a comparison of several HLB topologies when optimizing purely for speed. More extensive results comparing HLBs from a speed and area perspective are in [Chun91] [Sing91].

4.1 Comparison with Xilinx mapper

In this experiment, the goal was to minimize the number of X4000 CLBs used to implement a set of 15 MCNC benchmark circuits. The circuits were chosen because they could fit in the largest available X4000 CLB package. Each benchmark was optimized by mis2.2 [Bray86]. For TEMPT, the mis2.2-optimized circuits were first mapped by Chortle-crf [Fran91a] to the minimum number of 4-input LUTs and then TEMPT minimized the number of CLBs to cover the LUT network. The input to the Xilinx mapper, *PPR* was also the mis2.2-optimized circuits. The results quoted for PPR, are the *Packed CLB* utilization number from the PPR report file. The Packed CLB value is the CLB utilization if PPR were to pack the design into as small an area as possible. Table 1 lists the name of each benchmark circuit in Column 1. Columns 2 and 3 list the number of X4000 CLBs for PPR and TEMPT.

Bench Cct	PPR CLBs	TEMPT CLBs
9symml	36	35
alu2	71	69
alu4	122	118
apex7	38	38
b9	20	20
c1355	91	103
c8	17	18
cc	8	10
cm162a	5	6
comp	17	16
count	21	16
decod	10	10
mux	5	7
vda	97	97
z4ml	3	3
Totals	561	566

Table 1: Xilinx mapper vs. TEMPT

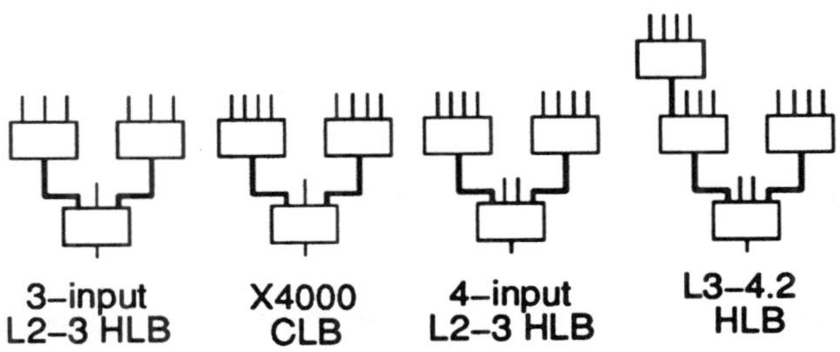

Figure 7: Area Study HLB Topologies

For the set of 15 benchmark circuits, the X4000 mapper required 561 CLBs and TEMPT used 566 CLBs. The two mappers are close in effectiveness even though TEMPT is not well-suited to mapping the X4000 CLB for the following reasons: (1) The X4000 CLB consists of both 3- and 4-input LUTs but since Chortle does the mapping to a homogeneous network of LUTs our methodology is better-suited for homogeneous LUT HLBs, and (2) The X4000 CLB violates the tapping buffer assumption since it allows only 2 tapping buffers for 3 LUTs. TEMPT had to be modified to allow this restriction.

4.2 Area-efficiency Results

In this section, we demonstrate how TEMPT may be used to evaluate the area-efficiency of the different HLB architectural alternatives illustrated in Figure 7. The X4000 CLB is an example of an L2-3 HLB topology with two 4-input LUTs and one 3-input LUT. The other L2-3 HLBs investigated are composed of three 3-input LUTs and three 4-input LUTs respectively. Comparing these three L2-3 HLBs is interesting because the functionality of the L2-3 HLB with 3-input LUTs and the functionality of the L2-3 HLB with 4-input LUTs bracket the functionality of

Bench Cct	3-inp L2-3	X4000 CLB	4-inp L2-3	4-inp L3-4.2
9symml	40	36	26	19
alu2	77	71	48	37
alu4	126	122	82	61
apex7	44	38	27	20
b9	22	20	16	12
c1355	131	91	80	59
c8	21	17	15	11
cc	17	8	9	7
cm162a	8	5	5	4
comp	27	17	14	13
count	21	21	14	10
decod	10	10	8	8
mux	10	5	5	5
vda	96	97	70	52
z4ml	3	3	3	2
Tot HLBs	653	561	421	320
LUT Bits	15672	22440	20208	20480
Ratios	0.70	1	0.90	0.91
HLB pins	6530	6171	5473	5440
Ratios	1.06	1	0.89	0.88

Table 2: Area Measures of Different HLBs

the X4000 CLB. The L3-4.2 HLB topology was also evaluated. This topology is interesting because the three 4-input LUTs nearest the root LUT form an L2-3 HLB and the fourth LUT makes a fast three basic-block level path.

Table 2 compares the HLBs in Figure 7 when Chortle and TEMPT were used in area-optimizing mode to map the benchmark circuits. For the X4000 CLB, the PPR numbers in Table 1 were used. Each row in the table corresponds to individual circuit data and the last 4 columns gives the number of HLBs needed by each HLB topology. The last five rows of Table 2 summarize the individual data. The first summary row of Table 2 shows the total number of each HLB required to implement the set of benchmark circuits. The second and fourth row are interesting architectural measures derived from the number of HLBs. The second row compares the total number of bits in the LUTs of the HLBs needed to implement the benchmark set. This is a measure of the active silicon area required to implement the logic. The fourth row is the total number of pins (one pin for each LUT input and tapping buffer output in the HLB) for the required HLBs. A previous architectural study [Rose90] demonstrated a correlation between the total number of pins and the routing area. The third row shows the ratio of the number of bits relative to the X4000 CLB. The last row shows the ratio of the number of pins relative to the X4000 CLB.

These results show that the L2-3 HLB composed of only 4-input LUTs (Column 4) uses fewer LUT bits and fewer pins than the X4000 CLB (Column 3). These architectural measures indicate that this HLB will have superior logic density. Also, the L2-3 HLB composed of 3-input LUTs

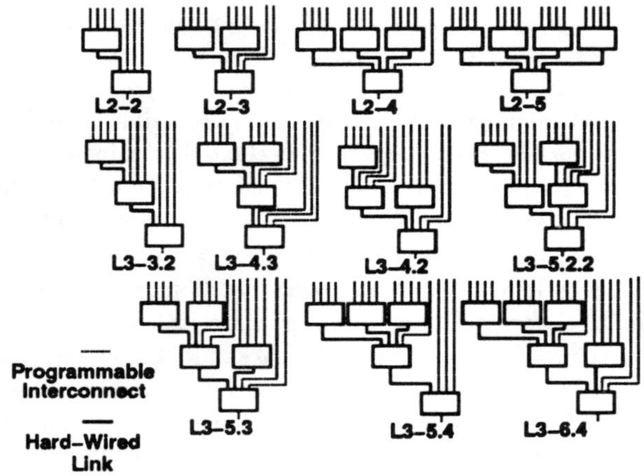

Figure 8: Delay Study HLB Topologies

(Column 2) uses much fewer bits and slightly more pins than the X4000 CLB. The architectural measures for the L3-4.2 HLB (Column 5) show it is about as area-efficient as the L2-3 HLB composed of 4-input LUTs. However, the delay results given in Section 4.3 and [Chun91] show that, when optimizing for delay, the L3-4.2 architecture has on average 12% fewer programmable connnections along the critical path than the L2-3 HLB, and so the L3-4.2 architecture is the better overall architecture.

The absolute effectiveness of the area-optimization algorithm can be evaluated by comparing the results in Table 2 to the lower bound in Section 3.2. The number of HLBs using TEMPT is within 3% and 4% of this lower bound for the L2-3 and L3-4.2 HLBs consisting of 4-input LUTs. For the L2-3 HLB consisting of 3-input LUTs and the X4000 CLB, the differences are 13% and 33% respectively. However, it should be noted that the lower bound is optimistic for the X4000 CLB because it consists of two 4-input and one 3-input LUT and the lower bound assumes all HLB LUTs have the same functionality.

4.3 Pure Delay Optimization Results

In this section, we demonstrate how TEMPT may be used to evaluate the speed improvement for different HLB topologies. A sampling of the many HLB topologies considered are illustrated in Figure 8. Each HLB consists of several 4-input lookup tables hard-wired together. The depth-optimized LUT networks were generated by Chortle-d [Fran91b] and the HLB netlists were generated using the delay-optimization algorithm of TEMPT.

The total critical path delay, D_{tot}, is given as $D_{tot} = N_{LB} \times D_{LB} + N_R \times D_R$. The first product in this expression is the delay due to the basic blocks along the critical path. Here N_{LB} is the number of basic logic blocks and D_{LB} is the delay of each basic block. For our $1.2\mu m$

Logic Block	$\overline{N_R}$	% decr in $\overline{N_R}$	$\overline{D_{tot}}$ (ns)	% decr in $\overline{D_{tot}}$
K4	5.4	0	30	0
L2-2	4.2	22	26	13
L2-3	3.4	37	22	27
L2-4	3.1	43	21	30
L2-5	3.0	44	21	30
L3-3.2	4.0	26	25	17
L3-4.2	3.0	44	21	30
L3-4.3	3.1	43	21	30
L3-5.2.2	3.1	43	21	30
L3-5.3	3.0	44	21	30
L3-5.4	2.9	46	20	33
L3-6.4	2.8	48	20	33

Table 3: Delay Performance of Different HLBs

CMOS technology, the delay for a 4-input lookup table, $D_{LB} = 1.7ns$. The second product is the routing delay. Here N_R is the number of programmable connections in the critical path and D_R is the average delay per routing connection. D_R varies widely according to interconnection architecture and in [Chun91] and [Sing91] we explored the effects of various values of D_R. For Table 3 an intermediate value of $D_R = 4\ ns$ was chosen. In Table 3 the % decrease in $\overline{N_R}$ and $\overline{D_{tot}}$ are relative to K4, an FPGA with 4-input LUT basic blocks with no hard-wired connections.

From Table 3, the L2-3 HLB reduces $\overline{N_R}$ to 3.4, while L3-4.2 HLB reduces $\overline{N_R}$ to 3.0, a difference of 12% in the average number of programmable connections in the critical path. Compared to an FPGA without hard-wired links, the L2-3 HLB and L3-4.2 HLB reduce $\overline{N_R}$ by 37% and 44% respectively and reduce average total critical path delay by 27% and 30%. The greatest speed gains per added basic block are realized by the L2-2 and L2-3 HLB and the speed gains are very small, per added basic block, for HLBs with more than four basic blocks. In general, for HLBs with a specific number of basic blocks, symmetric topologies attained superior speed than asymmetric topologies. This is likely because speed-optimized basic block networks tend to exhibit more balanced tree-structures.

5 CONCLUSION

A technology mapping algorithm that can be used for exploring FPGA architectures with hard-wired connections has been presented. It is capable of optimizing for speed or area and can map to any tree-topology hard-wired logic block FPGA. The mapper is comparable to a commercial CAD tool when mapping to the Xilinx 4000 CLB, a commercial FPGA with hard-wired connections.

The utility of this mapper for FPGA architectural exploration was shown using two studies, one that compared the area-efficiency of several HLB topologies and another that compared the speed performance of several HLB topologies. In the area-efficiency study, it was demonstrated that a hard-wired logic block consisting of three 4-input LUTs is superior to the Xilinx 4000 CLB in terms of both logic and routing area. The mapper was also used to show that another topology, the L3-4.2 HLB topology consisting of four 4-input LUTs, was similar in area-efficiency to the best of the 3-LUT HLB topologies. Also, the L3-4.2 HLB topology was shown to be faster than the 3-LUT HLBs in the speed performance study.

A future improvement to TEMPT is to integrate the delay and area algorithms to allow trade-offs between delay and area. Placement costs will also be considered in the packing step. More architectural studies on HLBs with basic blocks other than 4-input LUTs will be performed.

REFERENCES

[Bray86] R. Brayton, et al., "Multiple-Level Logic Optimization System," *Proc. of Int. Conf. on CAD 1986 (ICCAD-86)*, Nov. 1986, pp. 356-359.

[Chun91] K. Chung, S. Singh, J. Rose, P. Chow "Using Hierarchical Logic Blocks to Improve the Speed of Field-Programmable Gate Arrays." in *FPGAs - Proc. of 1st Int. Workshop on Field Programmable Logic and Applications*, Sept. 1991, pp. 103-113.

[ElGa89] A. El Gamal, et. al, "An Architecture for Electrically Configurable Gate Arrays," *IEEE Jour. of Solid State Ccts. (JSSC)*, Vol. 24, No. 2, Apr. 1989, pp. 394-398.

[Erco91] S. Ercolani, G. De Micheli, "Technology Mapping for Electrically Programmable Gate Arrays," *Proc. of 28th Design Automation Conf. (DAC-28)*, June 1991, pp. 234-239.

[Fran91a] R. Francis, J. Rose, Z. Vranesic "Chortle-crf: Fast Technology Mapping for Lookup Table-Based FPGAs," *Proc. of DAC-28*, June 1991, pp. 227-233.

[Fran91b] R. Francis, J. Rose, Z. Vranesic "Technology Mapping of Lookup Table-Based FPGAs for Performance," *Proc. of ICCAD-91*, Nov. 1991, pp. 568-571.

[Hsie90] H. Hsieh, et al., "Third-Generation Architectures Boosts Speed and Density of Field-Programmable Gate Arrays," *Proc. of Custom IC Conf. 1990 (CICC-90)*, May 1990, pp. 31.2.1-31.2.7.

[Keut87] K. Keutzer, "DAGON: Technology Binding and Local Optimization by DAG Matching," *Proc. of DAC-24*, June 1987, pp. 341-347.

[Rose90] J. Rose, R. J. Francis, D. Lewis, P. Chow, "Architectures of Field-Programmable Gate Arrays: The Effect of Logic Block Functionality on Area Efficiency," *IEEE JSSC*, Vol. 25, No. 5, Oct. 1990, pp. 1217-1225.

[Sing91] S. Singh, J. Rose, D. Lewis, K. Chung, P. Chow "Optimization of Field Programmable Gate Array Logic Block Architecture for Speed," *Proc. of CICC-1991*, May 1991, pp. 6.1.1-6.1.6.

Area and Delay Mapping for Table-Look-Up Based Field Programmable Gate Arrays

Prashant Sawkar and Donald Thomas

Electrical and Computer Engineering Dept.
Carnegie-Mellon University
Pittsburgh, PA 15213

Abstract

In this paper we present a new approach to technology mapping for area and delay for truth table based Field Programmable Gate Arrays. We view the area and delay optimizations during technology mapping as a case of clique partitioning for which we have developed an efficient heuristic. Additionally, we also explore alternate decompositions using Shannon Expansion. Experimental results indicate that on the average our algorithm for area optimization performs 10% - 19% better compared to previous approaches [3, 6]. Our approach to delay optimization on the average produces 3% fewer number of levels with 44% less area compared to other approaches [4].

1 Introduction

Field Programmable Gate Array (FPGA) architectures pose an interesting technology mapping problem due to the scarce but more powerful building blocks and limited wiring resources. A basic building block in truth table based FPGA (also referred to as a Configurable Logic Block – CLB [9]) can realize any function of upto k inputs (where k is typically 5). In the Xilinx 3000 series architecture [9] a CLB can realize two functions of 4 inputs, when the number of common inputs is 3 and the total number of inputs to the CLB is 5.

Recently several approaches have been reported for table lookup based FPGA mapping [6, 2, 3, 10, 5]. The *mispga* mapper [6] produces a circuit of lookup tables as an intermediate result and uses a covering approximation to reduce the number of CLBs. The *chortle* [2] mapper divides the input network into a forest of trees and determines an optimal mapping for each tree. Due to the tree boundary limitation, the *chortle* mapper does not exploit reconvergences effectively. This fact is reflected in the better results obtained by *mispga*. The *chortle-crf* mapper [3] uses a bin packing approach to select gate level decompositions. It exploits reconvergent paths and duplicates multi-fanout nodes as necessary, and produces results better than *chortle* and *mispga*. The *XMAP* [10] mapper constructs an **if-then-else dag** of the function and uses a covering procedure to map to CLBs. The delay optimization technique presented in *chortle-d* [4] uses a bin packing approach to realize each node in the network with as few levels as possible.

In this paper we present new approaches to technology mapping for area and delay optimization for truth table based FPGAs using clique partitioning. In section 2 we present the basic terminology used in the paper. In section 3 we present our approach to FPGA mapping for area optimization. In section 4 we present our approach to FPGA mapping for delay optimization. In section 5 we present the results obtained by our approach for area and delay optimization on a number of benchmark examples. In section 6 we present the conclusions.

2 Terminology

In this section we will define some basic terms which are used later in describing the mapping approach. The input network for technology mapping consisting of a set of boolean functions and sub-expressions may be looked upon as a directed-acyclic-graph (DAG). The vertices (or nodes) are the primitive boolean operators (such as AND, OR etc.) and the directed edges (from output to input) are the connections between operators. Edges also carry phase information indicating whether an operator's output must be complemented. The mapping process assigns one or more

0738-100X/92 $3.00 © 1992 IEEE

CLBs to each node visited to realize the node's function.

The Frontier Signals (FS) at a CLB assigned to an operator (or node) of the input network is defined as the set of signals that form the inputs to the CLB. The Extension Potential (EP) of a CLB assigned to an operator of the input network is defined as the number of unused inputs of the CLB. The Total Cost (TC) of realizing an operator of an input network is defined as the number of CLBs required to support the boolean function of the node in terms of its primary inputs (or common sub-expressions). The *level* of a CLB is recursively defined as the largest level among its input CLBs plus 1 (the primary inputs are considered to be at level 0). The *slack* at each CLB in the input network is defined as the difference between the required number of levels at that CLB and the actual level of the CLB. The required number of levels are specified by the user for primary outputs (or the CLBs producing them) and these are in turn used to derive the required number of levels for the internal CLBs.

2.1 Compatibility Types

The mapping process visits each node (or operator) in the input network in post order. The inputs to the node being mapped (visited) may consist of other nodes that have CLBs assigned to them (also referred to as input CLBs) or primary inputs (which do not require any CLBs). During the mapping process of a node, it may be possible to merge the node's input CLBs depending upon their reconvergence characteristics and extension potentials. In this section we will examine the various types of compatibilities that can exist between any pair of CLBs (*say i and j*) that form inputs to an operator N of the network being mapped.

Compatibility Type 1: A compatibility of type 1 (CT1) exists between CLBs i and j at operator N when the FS at CLB_i is a subset of FS at CLB_j. This may be graphically depicted by two nodes for CLBs i and j, and an edge (i,j) between these nodes.

When edge (i,j) is selected by the mapping process, the result of merging CLBs i and j is to reduce the number of CLBs necessary to realize the function of N by 1, and also to reduce the fanin at N by 1. The level of the merged CLB produced by CT1 is the larger of the two levels of CLBs i and j.

Compatibility Type 2: A compatibility of type 2 (CT2) exists between CLBs i and j at N when the FS at CLB_i is not a subset of FS at CLB_j, but, the total number of distinct inputs to CLBs i and j is less than or equal to the maximum CLB fanin (which is assumed to be 5 here). This may be graphically represented by two nodes for CLBs i and j and an edge (i,j) between these nodes.

When edge (i,j) is selected by the mapping process, the result of merging CLBs i and j is to reduce the number of CLBs necessary to realize the function of N by 1, and also to reduce the fanin at N by 1. The level of the merged CLB produced by CT2 is the larger of the two levels of CLBs i and j.

Compatibility Type 3: A compatibility of type 3 (CT3) exists between CLBs i and j at N when compatibility types "CT1" and "CT2" do not apply, and the extension potential of CLB_i or CLB_j or both is non-zero.

In this case no CLB reduction is possible in the realization of N. However, it is possible to reduce fanin at N by subsuming the output of one CLB (say CLB_j) into another CLB (say CLB_i) having a non-zero EP. The CLBs i and j participating in the CT3 merge are referred to as the *successor* and *predecessor* CLBs.

The level of the merged CLB produced by CT3 depends on the levels of the 2 CLBs participating in the merge. The CT3 merge is a level preserving transformation when the successor CLB's level is larger than the predecessor CLB's level. A level increase by one results when the successor CLB's level is not larger than the level of the predecessor.

3 Our Approach to Area Optimization

The input network (DAG) used for technology mapping is optimized to identify common subexpressions and common factors using MIS [1]. Our approach to area optimization is to minimize the number of CLBs needed to realize the function of each node (or operator) in the DAG. We visit the nodes in the DAG in post order, and use two techniques: a clique partitioning based approach, and a cofactoring approach using repeated Shannon Expansion to minimize the number of CLBs required to realize the function of each node in terms of its primary inputs (or common subexpressions). These techniques are described next.

Clique Partitioning based Approach: Our objective at each node of the input network is to minimize the number of CLBs required to realize the node. In this process we may be able to merge the CLBs assigned to the node's inputs. We use a three step approach to minimize the number of CLBs.

In the first step we apply *clique partitioning* to obtain a minimal clique partition of the CLBs that form inputs to the node being mapped. The input CLBs that can be merged can be characterized by the merges of type CT1 and CT2 described earlier. The clique partitioning algorithm used to accomplish this is described later. The number of cliques in the clique partition represents the lower bound on the number of CLBs required to realize the node. It should be noted here that our clique partitioning formulation is a special case of the general clique partitioning problem since we restrict each of the cliques contained in the partition to be realizable with a single CLB. The cliques (or CLBs) contained in the clique partition now form the new inputs to the node being mapped.

The second step takes the output of the clique partitioning step as input that may consist of one or more CLB(s). When the clique partition contains only one CLB, the mapping of the node is complete since the original CLBs that formed inputs to the node were all merged into a single CLB. When the clique partition contains multiple CLBs it is necessary to combine these CLBs in order to realize the node being mapped. However, since we have already performed clique partitioning, it is not possible to merge these CLBs any further using compatibility types CT1 and CT2. Our objective in this step is to combine these CLBs using CT3 merges and avoid adding further CLBs whenever possible. In addition, we attempt to maximize the extension potential of the lead CLB that realizes the function of the node being mapped so that the lead CLB can be extended further extension at the node's successors (nodes on the fanout of the node being mapped). This can be achieved by assigning the least priority to CT3 merges involving a CLB with the largest extension potential.

The third step takes the output of the second step as input and may have to combine the CLBs that could not be combined in the second step. Since we cannot perform any further merges of type CT1, CT2 and CT3, we have to add new CLB(s) in order to achieve this. The minimum number of additional CLBs as determined by the number of inputs to the node being mapped are added to accomplish this step.

The clique partitioning (used in step 1 above) is performed as follows:

1. Determine the types of merges (CT1, CT2, CT3) that are possible with the CLBs that form inputs to the gate being mapped.

2. Construct a graph consisting of one vertex for each CLB that forms an input to the gate, and edges between CLBs that can participate in merges of type CT1 and CT2.

3. For each edge (i,j) in the graph compute the number of edges removed (icount) due to the selection of edge (i,j).

4. while there are edges in the graph do the following:

5. Select the best edge (i,j).

6. Merge CLBs at vertices i and j and form a new vertex k.

7. Update the graph removing edges from vertices i and j, and adding edges from vertex k to other vertices wherever possible.

In the computation of icount in step 3 an edge (i,k) is considered to be incompatible with edge (i,j) if the combined realization of the functions of CLBs at vertices i,j, and k exceeds the maximum fanin of a CLB.

The clique partitioning heuristic proceeds by repeated selection of the best edge. The criterion we use in deciding on the best edge includes the incompatibility count (icount), the type of merge (CT1, CT2), and the number of new signals added when the edge is of type CT2. The CT1 merges are accorded higher precedence than CT2 merges as they do not consume any additional fanin signals in the resultant CLB. The icount information is used to guide the selection of edges that cause the least disruption from the point of view of formation of maximal cliques, and also from the point of view of promoting the maximum number of merges. We have observed the heuristic perform optimally on several small to moderate sized examples.

Cofactoring: The clique partitioning approach described above works quite well in general, but for some input networks, the large number of common sub-expressions that may be present in the network may preclude effective mapping. Hence, we explore alternative decompositions at each node (applying repeated Shannon expansion) and accept the decomposition when it is beneficial. Due to the exhaustive nature of this technique we limit its application to nodes with fewer than 10 primary inputs (or common sub-expressions).

The mapping process described above produces a network of single output CLBs. It is possible to further reduce the number of CLBs by merging underutilized CLBs into 2 output CLBs as permitted in the Xilinx 3000 series FPGAs. We use a greedy approach for merging CLBs by repeatedly selecting a

CLB with the least (but nonzero) merger potential (number of other CLBs with which it can merge). Alternately, this problem may be formulated as one of maximum cardinality matching [6] and solved using standard techniques.

4 Our Approach to Delay Optimization

In order to achieve the required timing performance in FPGAs with fixed block delays it is important to reduce the number of levels of CLBs producing the output(s). The dominancy of the wiring delays in FPGAs due to the programmable interconnect make it further necessary to achieve level reductions with minimal area increase. In this section we will discuss an approach to level reduction using circuit restructuring confined to the critical portions of the circuit being mapped.

The input network and the required number of levels form inputs to the technology mapper for delay optimization. As in area optimization the input network is first optimized to identify common sub-expressions and common factors using MIS [1]. The level reduction in the mapped network is achieved using the following approach:

1. Map the network using the clique partitioning approach for area optimization (described in section 3).
2. Compute slacks at primary outputs and the intermediate nodes of the area optimized mapped network produced in step 1, and identify the critical sections (portion of the newtork with negative slack).
3. Remap the network using clique partitioning, reducing the number of levels in the critical sections of the area optimized mapped network of step 1.
4. Recompute Slacks at primary outputs and the intermediate nodes of the mapped network produced in step 3, and identify the critical sections.
5. Perform level reduction using cofactoring on the critical portion of the network produced in step 4.

The slack computation at the primary outputs and intermediate nodes is performed by determining the difference between the actual and the required levels. The actual levels of each node are available after completion of mapping (in steps 1 and 3). The required number of levels for primary outputs are given at the time of mapping, and can be derived for intermediate nodes once the mapping (in steps 1 and 3) is complete. In the following sections we will examine the details of the clique partitioning approximation and cofactoring for level reduction.

Clique Partitioning based Approach for Level Reduction: As in the case of area optimization, we visit the nodes in the input network in post order during clique partitioning. A node is termed to be *critical* if its slack is negative. Our objective at each critical node of the input network is to minimize the number of levels and also the number of CLBs required to realize the node. In this process we may be able to merge the CLBs assigned to the node's inputs. The noncritical nodes are handled as explained earlier in the clique partitioning approach for area optimization.

The clique partitioning approach for level reduction is essentially the same as for area optimization (described in section 3). It should be noted that the clique partitioning process involving CT1 and CT2 merges preserves the level of the merged CLBs. The difference is in the assignment of priorities for CT3 merges, addition of new CLBs, and in the replication of multi-fanout nodes.

Since our objective in delay optimization is to realize each critical node with as few levels as possible, we assign higher priority to CT3 merges which do not increase the levels over other CT3 merges. Also, when adding new CLB(s), critical inputs are connected closer to the output of the node being mapped. When the node being mapped is a multi-fanout node, we replicate the lead CLB realizing the node along the critical paths so that may participate in merges without adding more levels.

Cofactoring based Level Reduction: The level reduction technique using clique partitioning described earlier works by locally decomposing and minimizing the level of each critical node, and replicating critical nodes at multi-fanout points as necessary. While these techniques are effective, some critical sections may not be correctable using this technique alone due to the common sub-expressions present. Also the cost of suppressing these common sub-expressions and resynthesizing entire critical sections with desired number of levels may be very high in terms of extra CLBs added and the resultant additional wiring. Hence, we explore alternate decompositions with fewer number of levels for critical outputs (applying Shannon Expansion repeatedly).

example	TechMap	Chortle-crf	mispga (old)	mispga+ (new)
5xp1	16	20	23	18
9symml	9	41	43	7
apex6	157	161	191	182
apex7	41	42	50	60
duke2	88	89	105	110
rot	126	131	153	181
des	694	743	-	904
C499	51	50	50	68
z4ml	4	3	7	5
misex1	12	14	10	11
vg2	19	18	21	20
9sym	9	42	59	7
rd84	33	53	32	10
apex2	82	93	70	67

+ single output CLBs

Figure 1: Result from area optimization

Due to the exponential number of partitions that are possible of the input variable set, we limit the application of this technique to functions with 10 or fewer variables. When a function with more than 10 variables is encountered, we search for intermediate functions that can be substituted for the original input variables so that the total number of inputs of function can be reduced to 10.

5 Results

The techniques described above for area optimization are implemented in a program called TechMap. The techniques for delay optimization are implemented in a program called TechMap-L. We have conducted experiments on several circuits from the MCNC logic synthesis benchmark set. The results presented are for a CLB that can implement any 5 input function or any two functions of 4 inputs when the total number of inputs to a CLB is 5. Our experiments were run on a DECstation 3100 The runtimes for our experiments ranged from 10-300 seconds over the benchmark set for both TechMap and TechMap-L.

The results obtained for area optimization are summarized in figure 1 alongside earlier results [3, 6].

We observe that results from TechMap were on the average 10% better than *chortle-crf* [3] and 19% better than *mispga* [6] (Recent results [8] indicate better performance by *mispga(new)*) respectively.

We attribute the improvement in results produced by TechMap over *chortle-crf* to the clique partitioning approach used for minimization of number of CLBs at each node. The clique partition produced contains a smaller number of CLBs in contrast to the bin-packing and the greedy reconvergence optimizations used by *chortle-crf*. Additionally, our approach to the clique partitioning which prioritizes CT1 edges over CT2, and CT2 edges with the least number of new literals added, attempts to maximize the extension potential of the clique partition. The smaller clique partition coupled with the maximization of CT3 type of merges produces more efficient realization of each node, with maximum extension potential at the leading CLB. This inturn leads to better overall mapping of the circuit as a whole as shown in figure 1 for *apex6, apex7, duke2, rot, des, misex1, rd84 and apex2.*

Due to the structure of the input network it may not always be possible to fully utilize some of the CLBs. The clique partitioning approach described earlier tends to produce structures with the lead CLBs (at common subexpressions and primary outputs) having the maximum extension potential which inturn provides more opportunities for merging CLBs.

The cofactoring technique described earlier in section 3 is useful in considering alternate decompositions at each node. We found cofactoring to produce a significant impact on two testcases *9symml* and *9sym* which happen to be single output functions and where the existing common subexpressions could be discarded without additional costs.

The results obtained for level reduction optimization are summarized in figure 2 alongside earlier results [4]. In figure 2 we have also included the single output CLB results from *mis-pga* [7]. The results from TechMap-L and *mis-pga* are not directly comparable except in the case of *9symml, z4ml, 5xp1 and 9sym* where the TechMap-L did not use any 2-output CLBs.

We observe that results from TechMap-L were on the average 3% fewer number of levels and 44% fewer number of CLBs compared to *chortle-d* [4].

We attribute the improvement in results produced by TechMap-L over *chortle-d* to three interrelated factors:

1. Selective application of level reduction.
2. Clique partitioning based level reduction.
3. Cofactoring based level reduction.

The selective application of level reduction to critical nodes during clique partitioning and the area optimization of non-critical nodes in TechMap-L results in level reduction with minimal area increase.

example	TechMap-L		Chortle-d		mis-pga *	
	levels	clbs	levels	clbs	levels	clbs
z4ml	2	6	3	19	2	10
5xp1	2	21	3	23	2	21
apex6	6	170	4	275	5	274
apex7	5	73	4	84	4	95
vg2	5	27	4	41	4	39
9symml	3	9	5	54	3	7
rd84	3	14	4	64	3	13
9sym	3	9	5	52	3	7
misex1	2	16	2	14	2	17

* - single output clbs

Figure 2: Result from Level Reduction

The clique partitioning approach for level reduction in TechMap-L attempts to minimize the size of the clique partition and also the level of each clique in the partition simultaneously. The clique partition produced by TechMap-L contains a smaller number of CLBs in contrast to the bin-packing based level reduction optimizations used by *chortle-d*.

The cofactoring technique for level reduction described in section 4 is useful in exploring alternative decompositions of critical nodes with fewer number of levels and minimal area increase. The restricted application of this technique to critical nodes and nodes which are functions of 10 or less inputs makes it an attractive alternative despite its exponential complexity.

The effect of the selective application of clique partitioning and cofactoring for level reduction is evident for several test cases (*z4ml, 5xp1, 9symml, rd84, 9sym*) shown in figure 2. In the case of *apex6, apex7* and *vg2* we could not achieve the desired level reduction but, our mappings were significantly smaller (38%, 13% and 31% respectively) as compared to *chortle-d*. We believe that the significantly smaller mappings produced by TechMap-L would be relatively easier to place and route compared to the ones produced by *chortle-d*, incurring fewer extra wiring delays and thereby compensating at least in part for the extra levels that may be present.

6 Conclusions

In this paper we have presented new approaches to technology mapping for area and delay for truth table based FPGAs based on clique partitioning and cofactoring. The clique partitioning based approach effectively exploits reconvergences and produces area efficient mappings. Our approach to selective application of clique partitioning and cofactoring for level reduction at critical nodes reduces levels with minimal area increase. Experimental results demonstrate the significant promise of the presented approaches.

Acknowledgements

The authors gratefully acknowledge the Resident Study Grant received from IBM Corporation and other support received from Xilinx and Motorola Corporations.

References

[1] R. Brayton, et. al, "Multiple-level logic optimization system," ICCAD, 1986.

[2] R.J. Francis, J. Rose, K. Chung, "Chortle: A Technology Mapping Program for Lookup Table-Based Field Programmable Gate Arrays," 27th DAC, 1990.

[3] R.J. Francis, J. Rose, Z. Vranesic, "Chortle-crf: Fast Technology Mapping for Lookup Table-Based FPGAs," 28th DAC, 1991.

[4] R.J. Francis, J. Rose, Z. Vranesic, "Technology Mapping for Delay Optimization of Lookup Table-Based FPGAs," International Workshop on Logic Synthesis, 1991.

[5] D. Filo, J. C. Yang, F. Mailhot, G. D.Micheli, "Technology Mapping for a Two-Output RAM-based FPGAs" *EDAC*, February 1991.

[6] R. Murgai, Y. Nishizaki, N. Shenoy, R. K. Brayton, A. Sangiovanni-Vincentelli, "Logic Synthesis for Programmable Gate Arrays," 27th DAC, 1990.

[7] R. Murgai, N. Shenoy, R. K. Brayton, A. Sangiovanni-Vincentelli, "Performance Directed Synthesis for Table Look Up Programmable Gate Arrays," ICCAD, 1991.

[8] R. Murgai, N. Shenoy, R. K. Brayton, A. Sangiovanni-Vincentelli, "Improved Logic Synthesis Algorithms for Table Look Up Architectures," ICCAD, 1991.

[9] Xilinx Programmable Gate Array Data Book, Xilinx Corporation, 1989.

[10] K. Karplus, "Xmap: A Technology Mapper for Table-Lookup FPGAs," *28th DAC*, June 1991.

Characterization of Boolean Functions for Rapid Matching in EPGA Technology Mapping

Ulf Schlichtmann[1], Franc Brglez[2], Michael Hermann[1]

[1] Institute of Electronic Design Automation
Technical University of Munich
Munich, Germany

[2] MCNC
Center for Microelectronics
Research Triangle Park, N.C., U.S.A.

Abstract

This paper introduces characteristic signatures for Boolean functions. The signatures do not exhibit sensitivity to permutations of input variables. We use these signatures to develop a method of rapidly matching subcircuits with cells in a (large) library. The procedure is analogous to hashing. This approach promises significant improvements for library matching with large libraries, for example in EPGA technology mapping.

1 Introduction

An important step in the process of converting an abstract specification of a digital system into hardware is *technology mapping*. Technology mapping converts a technology-independent representation of a circuit into a representation that can directly be implemented with elements of the target technology.

A core problem of technology mapping is to determine whether the given library contains a cell that can implement a particular subcircuit. Traditionally this problem has been solved using tree- or graph-based pattern matching techniques [1], [2]. However, not all cells can be represented as trees and graph-matching exhibits a complexity of $O(n!)$ to determine for a cell of n inputs whether it can implement a subcircuit of n inputs, as all permutations of inputs need to be considered.

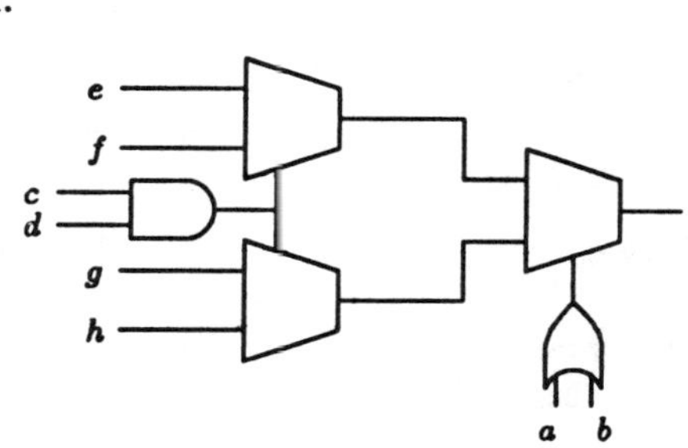

Figure 1: Unpersonalized logic module ACT2

As *Field Programmable Gate Arrays* (FPGAs) are becoming commonplace in digital system design, the complexity of the libraries that technology mapping has to deal with is increasing dramatically. FPGAs in which the logic module implements a *specific* Boolean function of n inputs are called *Electrically Programmable Gate Arrays* (EPGAs). Figure 1 depicts an example of the unpersonalized logic module (ACT2) developed by Actel Inc [3]. By forcing any input of the module to low or high or by bridging inputs together, the module can be *personalized* to implement other functions of less than n variables. Thus a library can be created that characterizes all cells that the module can represent. For the ACT2 module, this library comprises 780 cells, an order of magnitude higher than with standard cell libraries used traditionally. In addition, the average number of cell inputs is 4.75 for the ACT2 library, also significantly higher than with traditional libraries.

For these reasons, research has focussed on other approaches. One method [4] is a specialized technology mapper for EPGAs with logic modules that are based on a multiplexor structure, such as Actel's architecture. Both the network and the library are transformed into a representation that employs multiplexors as basic primitives. By thus exploiting the Actel-architecture, the library can be modeled with a few basic primitives. While the implementation of this approach is limited to multiplexor-based EPGAs, it achieves good results in acceptable CPU time.

An approach to technology mapping for EPGAs that can deal with any single-output logic module is described in [5]. It uses BDDs to perform the matching. BDDs need to be constructed for all possible permutations of the module's inputs. A novel data structure called Global BDD is used to efficiently combine all these BDDs. Still, this approach is of exponential complexity in the number of different functions that can be realized by the logic module.

The problem of determining the correspondence of the inputs of two circuits was first addressed in [6]. The goal was to verify the equivalence of two circuits without knowing how the circuits' inputs corresponded with each other. A simple *signature* was introduced that yields a characteristic value for each input of a circuit. Only inputs with the same characteristic value could be matched. This method turned out to be very powerful, despite it's simplicity. A refinement could resolve some cases where more than one input turned out to have the same characteristic value. However, a conclusive solution could not be guaranteed for every case. Benchmark results showed the method to be very efficient for practical problems.

We propose to solve the task of matching a subcircuit with a cell in the library by use of *characteristic signatures.*

We develop our concept of matching in analogy with hashing in Section 2 and introduce the characteristic signatures that eliminate or minimize sensitivity to permutation of input variables in Section 3. Section 4 discusses filters we found useful in improving the matching of variables before applying the equivalence verification. Section 5 summarizes experimental results and focuses on the figure of merit of various signatures.

2 New Approach to Matching

Our approach to matching subcircuits with cells in the library is analogous to the well-known and efficient data storage and retrieval method known as *hashing.* This method directly determines the location of desired data by applying a *hash function* h to the retrieval key. This results in an index $i = h(\text{key})$ into the storage table. However, it cannot be guaranteed that the hash function will return distinct indices for two different keys. Therefore, the retrieval key needs to be compared with the key in the retrieved record. If they are not equal, a *collision* has occurred. Various strategies have been researched to efficiently deal with collisions [7].

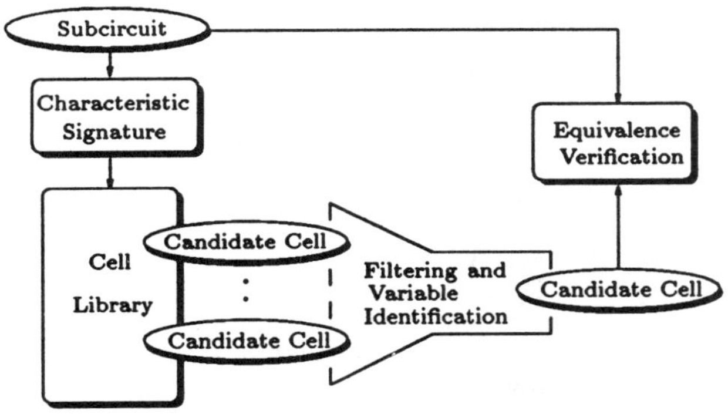

Figure 2: Library matching with signatures

Our approach to matching is shown in Figure 2. As with hashing, it relies on the quick computation of a *characteristic signature* for the subcircuit's function. Using a hash function, we then derive from the characteristic signature an index into our library. This index points to the *candidate cells* that have the same characteristic signature. As the characteristic signature is not unique, there may be more than one cell with the same characteristic signature. We will propose and analyze different signatures.

To verify that the candidate cell is indeed capable of implementing the subcircuit, we need to perform an equivalence verification. For this task we need to know the correspondence of the subcircuit's inputs with the cell's inputs. We will propose ways to determine this correspondence while at the same time filtering out candidate cells that do not match the circuit. If the library contains a matching cell, this procedure will always find it. We will give experimental data demonstrating that, in most cases, very few steps are required to accomplish this task.

3 Definitions and Signatures

We are dealing with Boolean functions of n variables. Let $\underline{x}$ denote a vector of Boolean variables $(\underline{x} \equiv (x_1, x_2, \ldots x_n))$.

- Two functions $f_1(\underline{x})$, $f_2(\underline{x})$ are considered to be *equal* $(f_1(\underline{x}) = f_2(\underline{x}))$, if they are identical in any canonical representation.
- Two functions $f_1(\underline{x})$, $f_2(\underline{x})$ are considered to be *equivalent*, if there exists a permutation $\Omega(\underline{x})$ of their variables $\underline{x}$, such that $f_1(\underline{x}) = f_2(\Omega(\underline{x}))$.
- Given the universal set of n-input functions, it is possible to select a minimal subset of functions such that by permuting their variables, all n-input functions can be derived. We will call the functions contained in such a subset *basis functions.* Note that if $f_1(\underline{x})$ and $f_2(\underline{x})$ belong to the basis set, they can never be equivalent.
- A *signature* is a function that is applied to a Boolean function and yields as a result a *signature value* that characterizes some properties of the function. For convenience, we will often refer to *signature* both to denote the function as well as it's value.
- A *characteristic signature* of a Boolean function is obtained by applying one or more signatures to the same function.
- *Aliasing* occurs when two or more different basis functions share the same characteristic signature.

The presented signatures will have the following property:

If two functions f_1, f_2 are equivalent, then for any signature $S(f)$, $S(f_1)$ must be equal to $S(f_2)$. All signatures that will be presented here determine a set of values for a function. Two signatures are equal, if they contain the same elements.

Most of the signatures are based on the notion of a weight. $S_{FW}(f)$ for example is derived from the *function weight*, the number of minterms in a canonical sum-of-products (CSOP) representation of a function.

Generally a weight compresses information about some characteristics of a function into a single value.

This operation usually leads to some loss of information. E.g. $f_1 \neq f_2 \Rightarrow S_{FW}(f_1) \neq S_{FW}(f_2)$ does not hold for all f_1, f_2. Therefore, we will combine various signatures and filters to uniquely characterize functions.

We will now use a CSOP representation of an example function to illustrate various signatures that can be formed readily from this representation. Afterwards we will formally define some signatures. Other signatures have been considered also, see [8].

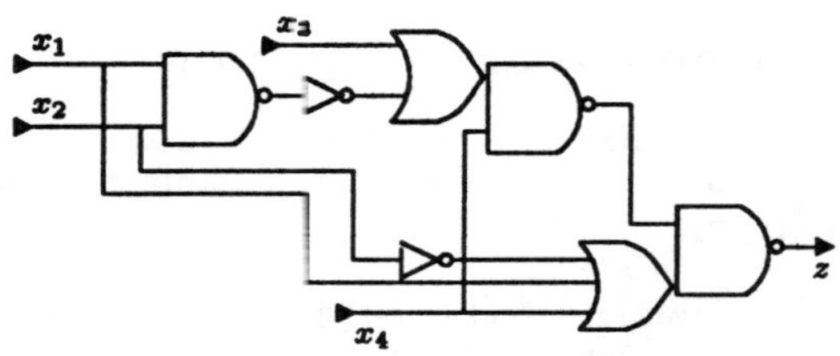

Figure 3: Circuit that maps into single ACT2 module

Consider the circuit in Figure 3 that has inputs x_1, x_2, x_3, x_4 and output z. The minterms of the corresponding CSOP representation are shown in Figure 4. The values for three respective signatures are shown in the border columns and rows: minterm weight S_{MW}, column weight S_{CW} and single fault propagation weight S_{SFP}. Both S_{MW} and S_{CW} can be formed by inspection. For example, $w(m_3) = 2$, $w(m_4) = 1$, etc. represent the weight of each minterm.

Similarly, we can determine for each column the number of 1's, the 1-signature $s_{x_i=1}$: $s_{x_1=1} = 3$, $s_{x_2=1} = 5$, ..., as suggested in [6]. The variables are grouped into variable equivalence classes EC_{CW_j} according to the results of these computations. A variable equivalence class EC_{CW_j} contains all variables, that yield a 1-signature of j. If all non-empty variable equivalence classes contain just one variable, the variables are uniquely distinguished.

The signature S_{CW} consists of the value and size of all non-empty equivalence classes. The signature must be invariant to variable permutations, therefore it cannot contain any variables directly. For our example we get $EC_{CW_3} = \{1\}$, $EC_{CW_5} = \{2,3,4\}$, which yields $S_{CW} = \{(3,1),(5,3)\}$.

Less obvious are the values shown for S_{SFP}. Here we associate with each variable a *triple*, e.g. $(4,2,4)$ with x_1. This signature is directly related to the number of patterns that would sensitize a fault on each of the inputs or inhibit this sensitization, respectively. For example, in Figure 3, a stuck-at fault on input x_1 can be propagated to the output on paths with even parity by the patterns (1110), (1100), (0110) or (0100) and on paths with odd parity by the patterns (1101), (0101). Out of the minterms of f, patterns (0011), (0111), (1011) and (1111) inhibit propagation. The values associated with each input are the total number of these patterns.

MW:	x_1	x_2	x_3	x_4
$w(m_3) = 2$	0	0	1	1
$w(m_4) = 1$	0	1	0	0
$w(m_6) = 2$	0	1	1	0
$w(m_7) = 3$	0	1	1	1
$w(m_{11}) = 3$	1	0	1	1
$w(m_{13}) = 3$	1	1	0	1
$w(m_{15}) = 4$	1	1	1	1
CW:	(3)	(5)	(5)	(5)
SFP:	$(4,2,4)$	$(0,6,4)$	$(0,6,4)$	$(2,8,2)$

Figure 4: Signatures for the circuit of Figure 3

As can be seen clearly from the example, the signatures are not perfect both with respect to function and input variable resolution. But they can be refined so that different basis functions can be distinguished. This is the purpose of filters which will be introduced in section 4.

Minterm Weight Signature:
Given a minterm m_i, the minterm weight $w(m_i)$ is equal to the number of uncomplemented variables in the minterm. The minterms are grouped into equivalence classes EC_{MW_j} according to the results of this computation. The signature S_{MW} as such comprises the size and value of all non-empty equivalence classes.

$$EC_{MW_j} = \{m_i \mid w(m_i) = j, \qquad j = 0, \ldots, n\}$$

$$S_{MW} = \{(j, |EC_{MW_j}|) \mid EC_{MW_j} \neq \emptyset \quad j = 0, \ldots, n\}$$

Figure 4 shows the weights for all minterms. Thus the following equivalence classes would be created: $EC_{MW_1} = \{m_4\}$, $EC_{MW_2} = \{m_3, m_6\}$, etc. This would result in $S_{MW} = \{(1,1),(2,2),(3,3),(4,1)\}$.

The minterm weight signature does not distinguish between variables in a function. However, it can distinguish between functions where variables may be indistinguishable due to symmetry or otherwise. For example, minterm weight signature is distinct for XOR and XNOR functions. Experimental results also show that this signature complements well the variable-distinguishing signature introduced next.

Single Fault Propagation Weight Signature:
We expand a function into three terms with respect to variable x_i in the following way:

$$f = \overline{x_i} \cdot w_0 + x_i \cdot w_1 + w_2$$

where

$$w_{0_i} = f_{\overline{x_i}} \cdot \overline{f_{x_i}}, \; w_{1_i} = \overline{f_{\overline{x_i}}} \cdot f_{x_i}, \; w_{2_i} = f_{\overline{x_i}} \cdot f_{x_i}$$

The signature S_{SFP} associates with each variable x_i the number of minterms in the disjoint components of the respective observability function $O_{x_i}^{f(\underline{x})} = w_{0_i} + w_{1_i}$ as well as in the term w_{2_i}, the number of patterns that inhibit fault propagation.

The variables are grouped into equivalence classes $EC_{SFP_{j,k,l}}$ according to the number of minterms in w_0, w_1 and w_2. The signature S_{SFP} consists of the value and size of all non-empty equivalence classes.

$$\begin{aligned} EC_{SFP_{j,k,l}} &= \{i \mid \ |w_{0_i}| = j \wedge |w_{1_i}| = k \wedge |w_{2_i}| = l, \\ &\qquad \textstyle\sum j + k + l = 0, \ldots, 2^n\} \\ S_{SFP} &= \{(j, k, l, \ |EC_{SFP_{j,k,l}}|) \mid EC_{SFP_{j,k,l}} \neq \emptyset, \\ &\qquad \textstyle\sum j + k + l = 0, \ldots, 2^n\} \end{aligned}$$

In the example from Figure 4, we get $EC_{SFP_{0,6,4}} = \{2, 3\}$, $EC_{SFP_{4,2,4}} = \{1\}$, $EC_{SFP_{2,8,2}} = \{4\}$. The signature then is $S_{SFP} = \{\ (0, 6, 4,\ 2),\ (4, 2, 4,\ 1),\ (2, 8, 2,\ 1)\ \}$. In this example S_{SFP} distinguishes the variables better than S_{CW}. As can be seen in Section 5, this is true in general also.

4 Signature Refinement with Filters

Filters are invoked if the use of signatures returns at least one candidate cell. Their purpose is twofold:

- to filter out candidate cells that have been selected due to aliasing of signatures. If $k > 1$ candidate cells have been selected, obviously at least $k - 1$ functions need to be filtered out. Even if only $k = 1$ candidate cell has been discovered, it may have been selected incorrectly due to aliasing.

- to furnish information that distinguishes the variables, so that a equivalence verification can be performed to ascertain the equivalence of the remaining candidate cell and the function that is to be matched.

The filtering process is shown in Figure 5. Filtering is always performed on a pair of functions $f_1(\underline{x})$, $f_2(\underline{x})$, one of them being the function of the subcircuit and the other one being the function of a candidate cell from the library. The signature S_k refers to one of S_{CW}, S_{SFP} introduced earlier or to one of the filters presented in this section.

A filter either determines a characteristic value for each variable of a function (type I) or explores the relationship between variables in another way (type II). The goal is always to uniquely characterize each variable. The result of filtering can be measured by the number e of non-empty variable equivalence classes. The smaller the difference $n - e$, the better the filtering.

The application of a filter to two functions can have three different results:

- If $e = n$ for both functions, every variable is uniquely characterized for both functions. The correspondence between the variables of the two functions is known. An equivalence check can be performed to determine whether the functions are equivalent.

- If the equivalence classes that have been determined for the two functions and their respective sizes do not match, the two functions cannot be equivalent.

- If the equivalence classes match, but $e < n$, no conclusive result has been achieved. The next filter will be applied.

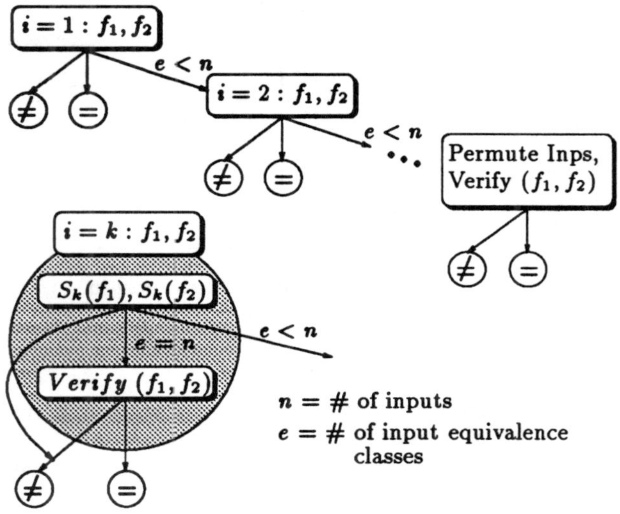

Figure 5: Filtering two functions

The number of filters as well as the order in which filters are applied is a heuristic which is still being evaluated. It is possible that all filters have been exhausted and no conclusive result has been obtained. In this case we need to perform equivalence verification between the two functions with an arbitrary ordering of the variables of one function and all possible permutations of the variables of the other function. In the worst case this requires an effort of order $O(n!)$. However, our results clearly indicate, that we need not worry about this since exhaustive permutation is performed extremely rarely. If invoked, it usually involves only the permutations of two or three variables, since the remaining variables are uniquely characterized.

Filters of type I:
These filters generally refine a signature by (recursively) applying it to a subfunction.

For example the original function f can be split into subfunctions f^i, where each subfunction contains only minterms that have the same weight. Then all variable-distinguishing signatures can be applied separately to the subfunctions as well [8].

Another approach is to apply S_{SFP} to a circuit that has been modified by shorting two or more inputs to-

Characteriz.	Figure of merit		size (bit)
	ACT2	All 4-inp.	
S_{FW}	0.1026	0.0035	n
S_{MW}	0.5218	0.1743	$(n+1)\cdot n$
S_{CW}	0.6474	0.0492	n^2
$\{S_{CW}, S_{FW}\}$	0.8026	0.1607	$2n^2$
$\{S_{MW}, S_{CW}\}$	0.8641	0.5475	$2n^2+n$
S_{SFP}	0.8885	0.5337	$2n^2$
$\{S_{MW}, S_{SFP}\}$	0.9346	0.9216	$4n^2+n$

Table 1: Figure of merit for various signatures

gether.

Filters of type II:
These filters explore symmetry relationships between variables of a function. If initial application of signatures has not been able to distinguish some variables of a function, the reason might well be that these variables are symmetric and thus cannot be distinguished.

The filters S_{TS} and S_{PS} check a function for total or partial symmetry, respectively [8].

5 Experimental Results

As shown in Figure 2, our approach to matching subcircuits with cells in the library is analogous to hashing. It relies on the quick computation of the characteristic signature of the function represented by the subcircuit. As with hashing, we have to devise an efficient collision strategy when we encounter a signature that is aliased to two or more cells in the library. This section provides experimental results that point out that it is possible to devise a characteristic signature of a Boolean function that will minimize aliasing to an acceptable level and that successive refinements of this characterization also offer an effective collision strategy. The functions that are stored in the library are all 780 basis functions that we derived from the ACT2 module. We first examine effects of generating this library with various signatures. We then analyze the matching of the universal set of all 4-input functions to the same library of 780 basis functions using a specific characteristic signature.

Ideally, all basis functions in the library should have a unique characteristic signature. We can closely approach but not quite meet this objective for the 780 basis functions derived from the ACT2 module. We introduce a generic *figure of merit* u that measures the distinguishability of the signature function as follows:

$$u = \frac{\#\text{ of different signatures in the library}}{\#\text{ of basis functions in the library}}$$

The value of $u = 1$ implies that every basis function in the library is characterized with a unique signature. Table 1 lists values of u for all signature functions discussed in this paper and for a variety of functions: all 780 basis functions associated with the ACT2 module and all basis functions representing the universal set of 4 input functions. In addition, we list the storage requirements for each of the signatures, given the maximum number of inputs (n) used in the cell library. For some signatures this value is exact, for others it represents the worst case bound.

The signature $\{S_{MW}, S_{SFP}\}$ is an example of a composite signature that enhances the distinguishability and has only modest storage requirements for the problem at hand.

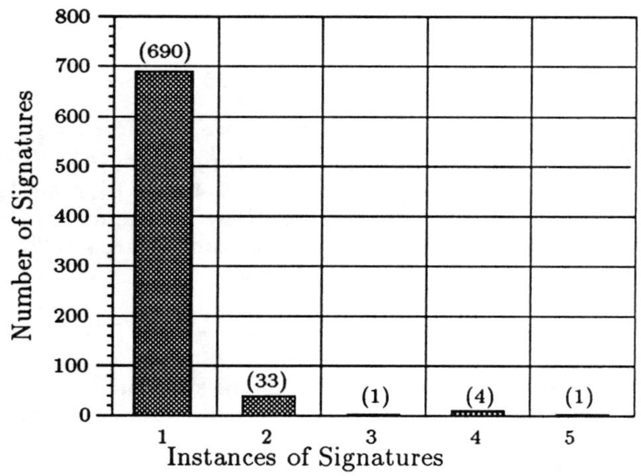

Figure 6: Distribution of signatures for ACT2 library

Figure 6 shows the distribution of characteristic signatures for the set of all basis functions derived from the ACT2 module. The library was created with the composite signature $\{S_{MW}, S_{SFP}\}$. The distribution visually emphasizes the merits of the composite signature even better than the results shown in Table 1. Over 88% of the cells have a unique signature. The worst case is one signature that aliases five different cells.

Whereas Figure 6 demonstrates that some of our signatures offer very powerful function distinguishability, Figure 7 shows the filtering and variable distinguishing capability. In this experiment, we have matched the universal set of all 4-input functions onto the ACT2 library. In the figure, shaded or dark areas represent those functions that have been definitely rejected or conclusively matched with a cell, respectively. Only those functions, that are represented by the white parts of the bar, need to be evaluated further. The left bar shows the result of matching the set of all 4-input functions with the ACT2-library. On the right the effects of various filtering steps are shown for those functions that are not immediately rejected by the library check.

The figure shows that

- the composite signature $\{S_{MW}, S_{SFP}\}$ distinguishes functions well enough to immediately reject more than 92% of the functions. These are 99.5% of all functions that will eventually be rejected. This is an important result: it shows that

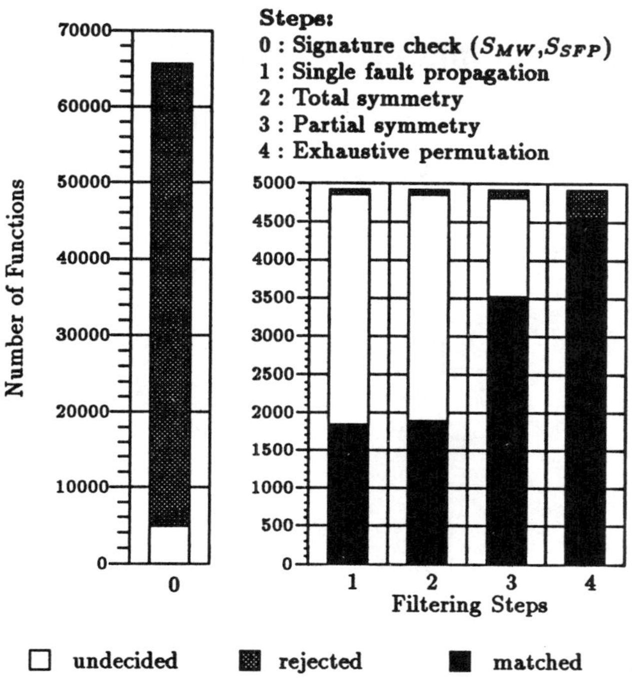

Figure 7: Filtering the universal set of 4-input functions

filtering effort will seldom be targeted at a function that does not have a matching cell in the library.

- when filtering is invoked on two functions $f_1(\underline{x})$, $f_2(\underline{x})$, it rapidly determines the correspondence of the variables of f_1 and f_2, so that an equivalence verification can be performed. In 93% of those cases the equivalence verification succeeds. This again shows, that little filtering effort is "wasted" on pairs of functions that eventually turn out not to be equivalent.

S_{SFP} is used as the first filter. As this was used to determine the signature in the library, we use it's filtering capabilities to minimize the required additional effort. It succeeds in uniquely characterizing the variables of 38% of the functions that enter the filtering process. We then use other filters. It can be seen that they quickly determine a unique characterization for the inputs of most functions. In 1062 cases an exhaustive permutation of the variables needs to be performed. This permutation, however, mostly involves two variables only, as the other variables have already been characterized. Filters of type I could be used to drastically reduce the number of functions that need to be evaluated by exhaustive permutation. Our experience indicates that this is not worth the computational effort, however.

The above results have been obtained with the first implementation of the techniques presented in this paper that was done in LISP. An implementation in C has recently been completed and is currently being integrated with a new technology mapper for EPGAs. First results indicate that we can decide for up to 1000 subcircuits per second whether they have a matching cell in the library or not. These results, obtained on a DEC 5000/200, clearly indicate that the proposed approach is fast enough to be used as the core of a technology mapper.

6 Conclusion

We have introduced an efficient and effective technique to compress properties of basis functions into a characteristic signature that will be unique for most of the functions regardless of the permutation of inputs. We have demonstrated the use of the characteristic signature as a key into the library for rapid matching of subcircuits to library cells. An efficient aliasing strategy has been devised with various filters based on extensions of basic signatures, significantly reducing the effort to resolve the aliased signatures. The approach is generic and will enhance technology mappers both for EPGAs and standard cells.

This matching procedure is currently being integrated with a new technology mapping tool that will be able to target all EPGAs with a single output basic logic module. Other work in progress includes further research about characteristics of signatures and favorable representations of Boolean functions for our purposes.

References

[1] K. Keutzer, "DAGON: Technology binding and local optimization by DAG matching," *24th Design Automation Conference DAC*, pp. 341 – 347, 1987.

[2] R. Rudell, "Logic synthesis for VLSI design," *Ph. D. thesis, U.C. Berkeley Memorandum UCB/ERL M89/49*, Apr. 1989.

[3] A. E. Gamal, J. Greene, J. Reynery, E. Rogoyski, K. A. E. Ayat, and A. Mohsen, "Architecture for electrically configurable gate arrays," *IEEE Journal of Solid-State Circuits*, vol. 24, pp. 394 – 398, Apr. 1989.

[4] R. Murgai, Y. Nishizaki, N. Shenoy, R. K. Brayton, and A. Saniovanni-Vincentelli, "Logic synthesis for programmable gate arrays," *27th Design Automation Conference DAC*, pp. 620 – 625, 1990.

[5] S. Ercolani and G. D. Micheli, "Technology mapping for electrically programmable gate arrays," *28th Design Automation Conference DAC*, pp. 234 – 239, June 1991.

[6] D. S. Reeves and G. Kedem, "Variable matching for verification," *MCNC Technical Report TR90-15*, 1990.

[7] D. E. Knuth. The Art of Computer Programming, Reading: Addison-Wesley Publishing Company, 1973.

[8] U. Schlichtmann, F. Brglez, and M. Hermann, "Rapid matching of large libraries by characterization of Boolean functions," *MCNC Technical Report (in preparation)*, 1992.

An Improved Synthesis Algorithm for Multiplexor-based PGA's*

Rajeev Murgai Robert K. Brayton Alberto Sangiovanni-Vincentelli

Department of EECS, University of California, Berkeley, CA-94720

Abstract

We address the problem of synthesis for a popular class of Programmable Gate Array (PGA) architectures - the multiplexor-based (MB) architectures. We present improved techniques for minimizing the number of basic blocks used to implement a combinational circuit. One source of improvement is the use of if-then-else dags as subject graphs along with BDD's. An important contribution is a very fast algorithm which always gives a match for a function onto the basic block of the architecture, when one exists. On average, the results obtained on a set of benchmarks are 22% better than those of a library-based technology mapping [10] and *Amap* [4], 31% better than those of *Proserpine* [5] and 17% better than those of *mis-pga* [2].

1 Introduction

The basic PGA architectures share a common feature: they consist of repeated arrays of identical logic blocks. A **logic block** (also called a **basic block**) is a versatile configuration of logic elements which can be programmed by the user. There are two dominant categories of block structures, namely **Table-Look-Up (TLU)** and **Multiplexor-Based (MB)**; the architectures resulting from them are called the TLU architectures and the MB architectures respectively. In a TLU architecture each block of the architecture can implement any Boolean function with m inputs ($m \geq 2$). An example is the Xilinx(3000) architecture, in which $m = 5$. In the MB architecture, the basic block is a configuration of multiplexors. Actel's architectures are two such examples [1]. Their basic blocks - *act1* and *act2* - are shown in Figure 1.

Previously we gave algorithms for synthesis for both TLU and MB architectures [2, 3]. We present an improved technique for minimizing the number of blocks for MB architectures. For clarity, we concentrate on the Actel architectures. The problem is stated as: *given a combinational circuit described in terms of Boolean equations, realize it using the minimum number of basic blocks of the target MB architecture.*

The outline of the paper is as follows. Previous approaches are briefly discussed in Section 2. Our new approach is presented in Section 3. Finally, we give results on a number of benchmark examples in Section 4.

2 Previous Work

We are aware of four approaches to the synthesis for such architectures:

DAGON [9], misII [10] (library-based approach): A library is created which has gates that represent functions obtained from the basic block by either tying some inputs to constants (0 or 1), or to each other. The network is represented as a subject graph in terms of a base function, typically a 2-input NAND gate. All the functions in the library are also represented in terms of NAND gates. These are the pattern-graphs. The problem then is to cover the subject-graph using the minimum number of pattern graphs. This problem is NP-complete, though efficient heuristics exist. A commonly used heuristic divides the subject-graph into trees and covers the trees optimally by tree patterns in polynomial time using a dynamic programming approach.

* This work is supported in part by NSF under contract number EMC-84-19744, DARPA under contract number 44-25611, AT&T, IBM, Intel and Micro.

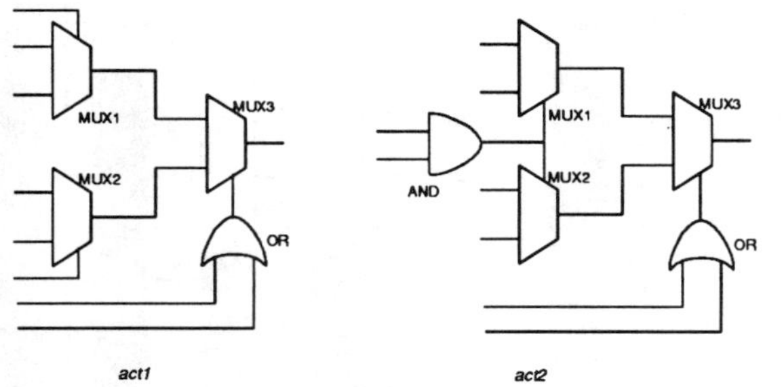

Figure 1: Actel architectures: *act1* and *act2*

mis-pga [2] (using BDD's): Here the base function in the subject graph is a 2-1 multiplexor instead of a NAND gate. Since a BDD vertex corresponds to a 2-1 multiplexor, the function at each node of the network is represented as a BDD in terms of its immediate fanins. A small set of 8 pattern graphs is used. Covering of the subject-graph by patterns is done using the dynamic programming algorithm. After an initial mapping, an iterative improvement phase is used. This phase tries to exploit the relationship among nodes of the network. It consists of three main operations: partial collapsing, decomposition and quick-phase.

Amap [4] (using if-then-else (ITE) dags): If-then-else dags are a generalization of BDD's. Here, the selector function at each vertex can be a function of inputs, rather than being an input. There are three children of an ITE vertex: **if, then** and **else**. The **if** child corresponds to the select line of the multiplexor. The **then** and **else** children correspond to the branches taken when the **if** child evaluates to 1 and 0 respectively. The main advantage of ITE's over BDD's is that duplication of cubes can be avoided [4]. This is explained in Section 3. *Amap* constructs an ITE for each function and does the mapping in a greedy fashion.

Proserpine [5] (boolean matching): This approach is based on answering a fundamental problem: determine if a given function f can be implemented by a gate-function GF, where some of the inputs of the gate may be tied to constants (0 or 1) or to other inputs. It constructs reduced ordered BDD's for f and for the gate-function GF and checks if the reduced ordered BDD for f is a subgraph of the BDD for GF. In the worst case, all possible variable orderings for GF have to be considered. The problem is further complicated by the fact that a few of the inputs of GF may have to be tied to constants (0 or 1) or to other inputs to get the desired f. We conjecture that the matching check is expensive in terms of CPU time. This matching is used as a subroutine in the mapping algorithm.

The last three approaches have the benefit of not reading in a library.

3 Proposed Mapping Algorithm

Following [4], we represent functions with ITE's. Reduced ordered BDD's are used optionally at the end to detect a better solution. We use an iterative improvement phase after the initial mapping phase exactly as in [2] (and hence do not describe it here). In this section, we describe the following: outline of the overall algorithm, the matching problem (given a function f, is it realizable

0738-100X/92 $3.00 © 1992 IEEE

by one basic block (*act1* or *act2*)?), the construction of an ITE of a function, and the pattern graphs.

3.1 Overall Algorithm

The overall algorithm works as follows.

1. Each node of the network is first mapped. If the function f at the node can be implemented by one block (using the matching algorithm to be described in Section 3.2), the corresponding match is saved. Else, an ITE of f is constructed. The ITE is covered by the pattern graphs of Figure 6 using a dynamic programming algorithm [9].

2. After obtaining this initial mapping, an iterative improvement phase is entered. Two operations, namely partial collapse and decomposition, are tried. Partial collapse tries to collapse a node into all its immediate fanouts. If the sum of the new costs of the fanouts is less than the sum of the old cost of the node and the fanouts, the collapse is accepted. The new cost of a fanout is determined by remapping it using step 1. This process is repeated for all nodes of the network. Decomposition decomposes a node using `decomp -g` of misII [8] and the decomposed nodes are mapped. If the cost improves, the original node is replaced by its decomposition. Partial collapse and decomposition are repeated for some number of iterations.

3. Each node of the network is replaced by a set of nodes, each of which can be implemented using one basic block of the architecture. This is done by using the cover of ITE at each node.

4. Next, we perform one iteration of partial collapse. This sometimes gives additional improvement because partial collapse is now being performed on smaller nodes, each having a cost of 1 block. Also, inside an ITE, there might have been a multiple-fanout vertex v, such that v and all its fanin and fanout vertices were roots of some pattern graphs in the cover of the ITE. This may happen, for instance, when the fanin and the fanout vertices of v have multiple-fanouts. Since the covering algorithm maps an ITE by mapping each tree in it separately, v formed a single-vertex tree. So the node corresponding to v in the network can be collapsed into all its fanouts, thereby decreasing the cost of the network by 1. This step was inspired by [6] and [3].

5. Finally, we perform a check: if the number of primary inputs of the network is small (say less than 10), we collapse the network and construct a reduced ordered BDD for this. We map this BDD using the method described in [2]. If this mapping result is better, we accept it. Construction of ordered BDD's helps when the circuit is symmetric.

The algorithm is applicable for both *act1* and *act2* architectures. However, some architecture-specific changes have to be made. The main differences are in the matching algorithm and the pattern graph construction. The ITE-construction procedure is basically the same with a few minor changes. Since the matching algorithms for *act1* and *act2* are quite different, we describe both of them. For pattern graphs, we focus on *act1*. For *act2*, they can be similarly derived.

3.2 Matching Problem

In this subsection, we address a fundamental problem: *given a completely specified function f and the basic block of the architecture, is f realizable by a single block?* A more general version of this problem is addressed in [5]: given a function f and a gate-function GF, is f implementable by GF? We restrict ourselves to a special GF - the gate function corresponding to the *act1* or *act2* block. By this restriction, we hope to have a much faster algorithm.

3.2.1 Matching problem for *act1*

Let us assume that a function f is implementable by an *act1* block. Let us also assume that f is expressed in minimum support[1]. Let $S(f)$ be the set of inputs of f. Since the basic block *act1* has 8 input pins, assume that $1 \leq |S(f)| \leq 8$. If $|S(f)| = 8$, both the OR gate inputs have to be tied to input variables. Next we observe that, in general, we can tie one of the input-pins of the OR gate to an input of f. This is because otherwise, the select input of MUX3 is either constant 0 or 1, and in either case, if f is implementable by *act1*, it is implementable by a 2-1 mux and hence by MUX3 with a non-constant input at its select line. Refer to Figure 2 (A) for the rest of the discussion. There are then two types of matches:

1. One input of the OR gate is a, $a \in S(f)$, and the other input is 0. Function f then has a decomposition of the form

$$f = ag + \overline{a}h \tag{1}$$

for some input a of f, where g, h are each implementable by a 2-1 mux. The problem is to find g and h. We may assume without loss of generality that g and h are independent of a. This follows from the following proposition:

Proposition 3.1 *If f has a decomposition as in (1), where g, h are each implementable by a 2-1 mux, then it also has a decomposition of the form*

$$f = ag_1 + \overline{a}h_1 \tag{2}$$

where g_1, h_1 are each implementable by a 2-1 mux, and are independent of a.

Then, it easily follows that

$$g = f_a, h = f_{\overline{a}} \tag{3}$$

2. The input pins of OR gate are tied to a and b, for some inputs a, b of f. Then f has decomposition of the form

$$f = (a + b)g + \overline{a}\overline{b}h \tag{4}$$

where g, h are implementable by a 2-1 mux.

The problem is to find g and h. The following proposition gives a way to find h.

Proposition 3.2 *If there exists a decomposition of f as in (4), g, h each implementable by a 2-1 mux, then there exists another decomposition of f of type (4) with h replaced by h_1, h_1 independent of a and b, and implementable by one 2-1 mux.*

Hence h may be assumed independent of a, b. Then $h = f_{\overline{a}\overline{b}}$. The problem reduces to finding g which is 2-1 mux implementable. We divide it into two cases.

<u>Case A:</u> g is independent of a, b. Then the following proposition gives the value of g.

Proposition 3.3 *A decomposition of f as in (4), with g independent of a and b, exists if and only if $f_a = f_b$. Then $g = f_a$.*

<u>Case B:</u> g depends on a or b or both. Cofactoring (4) with respect to ab, $\overline{a}b$ and $a\overline{b}$, we get $g_{ab} = f_{ab}$, $g_{\overline{a}b} = f_{\overline{a}b}$, $g_{a\overline{b}} = f_{a\overline{b}}$ Then, from Shannon expansion of g,

$$g = f_{ab}ab + f_{\overline{a}b}\overline{a}b + f_{a\overline{b}}a\overline{b} + g_{\overline{a}\overline{b}}\overline{a}\overline{b} \tag{5}$$

$$g = G + H\overline{a}\overline{b} \tag{6}$$

Here $G = f_{ab}ab + f_{\overline{a}b}\overline{a}b + f_{a\overline{b}}a\overline{b}$ and $H = g_{\overline{a}\overline{b}}$. Note that

$$G_{\overline{a}\overline{b}} = 0 \tag{7}$$

$$H_{\overline{a}\overline{b}} = H \tag{8}$$

[1] A prime cover of f guarantees this.

Given f, and a choice of a and b, we are interested in finding g which is mux-implementable. We know the component G of g, but not H. Note that any choice of H does not change f, since H is a don't care for f. However, we are interested in finding only those choices of H which make g implementable by a 2-1 mux. Writing g as a mux implementation, we get $g = CA + \overline{C}B$, where $A, B, C \in \{0, 1\} \cup S(f)$. Using (6), (7) and (8), we get,

$$H = (CA + \overline{C}B)_{\overline{a}\overline{b}} = C_{\overline{a}\overline{b}}A_{\overline{a}\overline{b}} + \overline{C}_{\overline{a}\overline{b}}B_{\overline{a}\overline{b}} \quad (9)$$

If we use this formula as such to compute H, there may be 224 possibilities (when $|S(f)| = 8$). However, we observe that at least one of $A, B, C \in V = \{a, b\}$. This is because by assumption g depends on at least a or b. If $C \in V$, then $H = B_{\overline{a}\overline{b}}$. If $A \in V$, $H = \overline{C}_{\overline{a}\overline{b}}B_{\overline{a}\overline{b}}$. If $B \in V$, $H = C_{\overline{a}\overline{b}}A_{\overline{a}\overline{b}}$. Thus

$$H \in \{0, 1, c, \overline{c}, cd, \overline{c}d\},\ c, d \in S(f) - \{a, b\},\ c \neq d. \quad (10)$$

This reduces the possible choices of H to at most 59 (when $|S(f)| = 8$). These can be reduced further by examining the cases when $H = cd$ and when $H = \overline{c}d$. When $H = cd$, $g = G + cd\overline{a}\overline{b}$. Since g is mux-implementable, it has at most 3 inputs. Hence, g does not depent on at least one of a, b, c, d. We can then show that either $f_{\overline{a}b} = cd$ or $f_{a\overline{b}} = cd$. Similarly, if $H = \overline{c}d$, either $f_{a\overline{b}} = \overline{c}d$ or $f_{\overline{a}b} = \overline{c}d$. Hence, for the case of a two-input H, we just look at $f_{a\overline{b}}$ and $f_{\overline{a}b}$. If any of these are of the form cd or $\overline{c}d$, we consider that form for H. This reduces the number of possibilities from 59 to at most 16.

Matching algorithm: Based on this theory, we can describe the matching algorithm. We first give a subroutine that checks if a function f, with a given prime and irredundant cover $\mathcal{C}$, is implementable by a 2-1 multiplexor. This is an easy and a fast check. Such an f should have at most 3 inputs. Also $|\mathcal{C}| \leq 2$. We perform this check based on the number of inputs $|S(f)|$ of f, as follows.
1. If $|S(f)| \leq 1$, the function is implementable.
2. If $|S(f)| = 2$, say x and y are the inputs, then the possible functions are $x + y, x + \overline{y}, \overline{x} + y, xy, x\overline{y}, \overline{x}y$ and these are easy to check.
3. If $|S(f)| = 3$, then f should be of the form $xy + \overline{x}z$, i.e., there should be exactly two cubes in $\mathcal{C}$, each with two literals, with one input x occurring in both positive and negative phases, and the other two inputs y and z occurring exactly once in positive phase.

We are now ready to describe the algorithm that checks whether f is implementable by one *act1* block.

(0) If $|S(f)| = 0$, return match for the constant function f. If $|S(f)| > 8$, no match exists, quit.

(i) Compute $T_3(f) = \{a | a \in S(f), |S(f_a)| \leq 3\}$. We can restrict the potential OR gate inputs to $T_3(f)$, i.e., if a is an input to the OR gate ($a \in S(f)$) for an *act1*-implementable function f, then $a \in T_3(f)$. This follows from (1) and (4), by noting that $f_a = g_a$ and g, being implementable by 2-1 mux, has at most 3 inputs.

(ii) (Single input at the OR gate) Check if an input $a \in T_3(f)$ may be put along with a 0 at the OR gate. The check is performed by seeing if f_a and $f_{\overline{a}}$ are each implementable by a 2-1 mux (using (3)). If so, report a match and quit. Else, pick another $a \in T_3(f)$ and check likewise.

(iii) (Two inputs at the OR gate and g independent of these two inputs) Select a pair of inputs a and b, $a, b \in T_3(f)$, at the OR gate. Then f_a should be equal to f_b. (Note that this check is fast, since f_a and f_b are functions of at most 3 inputs.) If they are equal, and f_a and $f_{\overline{a}\overline{b}}$ are each implementable by a 2-1 mux, we know f is implementable by the *act1* block. Else, pick another pair a, b and perform the check.

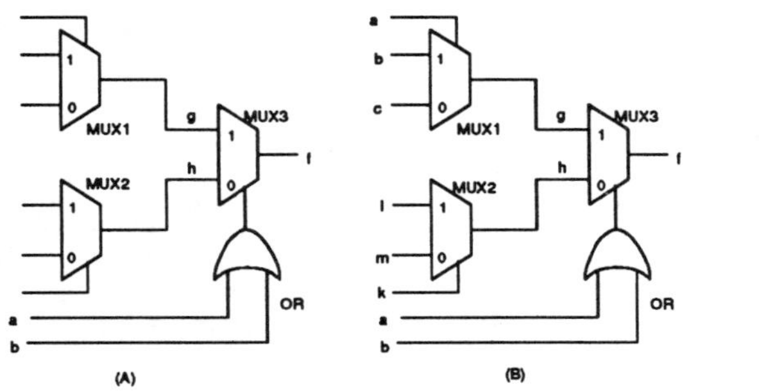

Figure 2: (A) An *act1*-implementable function, (B) An *act1*-implementable function that escapes steps (ii) and (iii) of the matching-algorithm

(iv) (Two inputs at the OR gate and g depends on at least one of these inputs) For each pair $a, b \in T_3(f)$, check for mux-implementation of $f_{\overline{a}\overline{b}}$ and then the possibility of existence of the function g which is implementable by a 2-1 mux and depends on either a or b. For that go through all the possibilities (at most 16) for H one by one and see if $g = G + H\overline{a}\overline{b}$ is mux-implementable.

Note that in steps (ii), (iii) and (iv), we obtain a prime and irredundant cover for possible candidates for g and h before performing any checks. This is needed because this is a precondition for our 2-1 mux implementability subroutine.

Proposition 3.4 *If f is expressed in minimum support, then the above procedure is complete for act1, i.e., it generates a match if and only if f is implementable by one act1 block.*

Note that if f is not expressed in minimum support, the algorithm may not find a match, when it existed. A simple example is the tautology function $f = 1$ expressed as a function of greater than 8 variables.

Using just the steps (i), (ii) and (iii), we may miss a match. For example, $f = (a + b)(\overline{a}c + ab) + \overline{a}\overline{b}(kl + \overline{k}m)$ is implementable by *act1*, as shown in Figure 2 (B). However, (ii) and (iii) fail to find a match, because f can be matched to *act1* only when g is dependent on a, b. This example also demonstrates that a method based on cofactoring cannot guarantee optimality for mapping a function onto minimum number of *act1* blocks. This is because a cofactoring method will never consider step (iv).

3.2.2 Matching problem for *act2*

The *act2* architecture (Figure 1) is slightly more difficult to handle than *act1*, because MUX1 and MUX2 cannot be treated separately, unlike *act1*, and the extra AND gate complicates the matching check. This is similar to the complications arising from the presence of the OR gate in the *act1* block.

Definition 1 *Two functions g and h are* **common-select mux implementable** *if each can be realized with a 2-1 multiplexor, with the same select line (it may be a constant).*

Such functions have a realization as shown in Figure 3 (A). For example, if g and h have at most one input each, they are common-select mux implementable.

Definition 2 *Two functions g and h are* **common_AND-select mux implementable** *if each can be realized with a 2-1 multiplexor, such that the select line for both the multiplexors is the same and is equal to AND of some inputs c, d of g or h, $c \neq d$ (so constants 0, 1 are excluded[2]).*

[2]The cases when constants or $c = d$ are allowed are handled by common-select mux implementable functions.

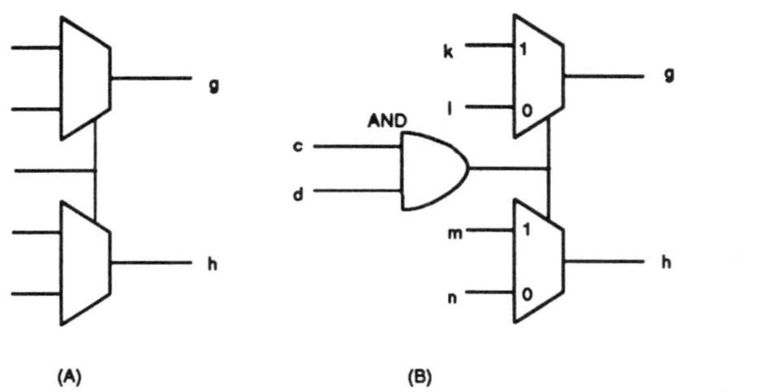

Figure 3: common-select and common_AND-select mux implementable functions

Such functions have a realization as shown in Figure 3 (B).

Problem 1: *Given two functions g and h, are they common-select mux implementable?*

Solution: First find out all possible select line inputs $\mathcal{C}_g$ under which g can be realized by a multiplexor. To find out all possible mux implementations for g is easy. If a function g is mux-implementable, either 1) it imposes no restriction on the kind of selector function (g is 0, 1 or some input; then tie g to 0 and 1 pins of the mux), or 2) there is exactly one candidate, x, for the select line (i.e. g is of the form $\overline{x}, \overline{x}y, \overline{x}+y, xy+\overline{x}z$), or 3) there are two choices for select line (g is of the form $x+y$ or xy). Similarly find $\mathcal{C}_h$ for h. $\mathcal{C}_g \cap \mathcal{C}_h \neq \phi$ if and only if g and h are common-select mux implementable.

Problem 2: *Given two functions g and h, are they common_AND-select mux implementable?*

Solution: Let $S(g)$ and $S(h)$ be the set of inputs of g and h respectively. Let $U = S(g) \cup S(h)$. Assume g and h are common-AND_select mux implementable (Figure 3 (B)). Since $c, d \notin \{0,1\}, |U| \geq 2$. Then,

$$g = cdk + (\overline{c} + \overline{d})l, \quad (11)$$

$$h = cdm + (\overline{c} + \overline{d})n, \quad (12)$$

$$c, \quad d \in U, c \neq d, k, l, m, n \in U \cup \{0,1\} \quad (13)$$

We let c, d vary over U. Given g, h, c and d, we may assume k, m to be independent of c and d (i.e. $k, m \notin \{c, d\}$). If they are not, we can replace them by 1 without changing g, h. We now present necessary and sufficient conditions for the functions g and h to be common_AND-select mux implementable for a given c, d pair. The corresponding values of k, l, m and n are also given.

1. $g_{cd} \in \{0,1\} \cup S(g)$. (Set $k = g_{cd}$).
2. $h_{cd} \in \{0,1\} \cup S(h)$. (Set $m = h_{cd}$).
3. Exactly one of the following should hold:
   (i) $g_{\overline{c}} = 0$ and $g_{\overline{d}} = c$. (Set $l = c$).
   (ii) $g_{\overline{c}} = 0$ and $g_{\overline{d}} = 0$. (Set $l = 0$).
   (iii) $g_{\overline{c}} = 1$ and $g_{\overline{d}} = 1$. (Set $l = 1$).
   (iv) $g_{\overline{c}} = d$ and $g_{\overline{d}} = 0$. (Set $l = d$).
   (v) $g_{\overline{c}} = z$ and $g_{\overline{d}} = z$. Here $z \in S(g) - \{c, d\}$. (Set $l = z$).
4. Exactly one of the following should hold:
   (i) $h_{\overline{c}} = 0$ and $h_{\overline{d}} = c$. (Set $n = c$).
   (ii) $h_{\overline{c}} = 0$ and $h_{\overline{d}} = 0$. (Set $n = 0$).
   (iii) $h_{\overline{c}} = 1$ and $h_{\overline{d}} = 1$. (Set $n = 1$).
   (iv) $h_{\overline{c}} = d$ and $h_{\overline{d}} = 0$. (Set $n = d$).
   (v) $h_{\overline{c}} = z$ and $h_{\overline{d}} = z$. Here $z \in S(h) - \{c, d\}$. (Set $n = z$).

These conditions are easy to check. Note that in the worst case, 15 possible c, d pairs have to be checked.

We are now ready to answer the original problem: given f, is f realizable by one *act2* block? The strategy is to assume that f is realizable by one *act2* block and derive all possible pairs of functions g and h (Figure 4). Then depending on the number of proper (i.e. non-constant) inputs at the AND gate, check if g and h are common-select mux implementable or are common_AND-select mux implementable. Refer to Figure 4 for the rest of this subsection. It suffices to consider the following five cases:

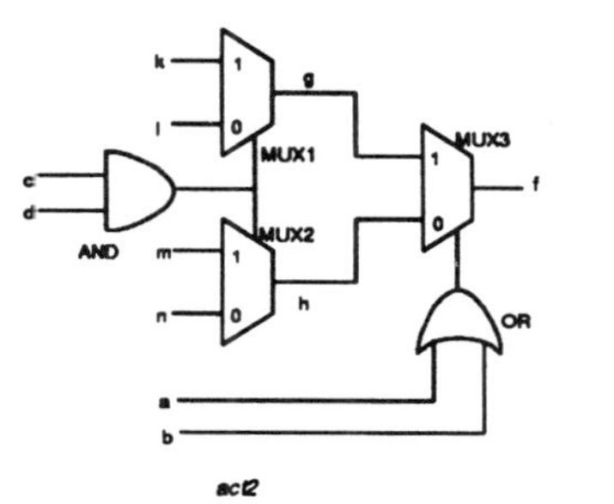

Figure 4: An act2-implementable function f

1. One proper input at the OR gate, a, and at most one proper input at the AND gate, c. If f is realizable by a single *act2* block, then $f = ag + \overline{a}h$, $g = ck + \overline{c}l$, $h = cm + \overline{c}n$. Note that g and h are common-select mux implementable. It is sufficient to check for the case when g and h are independent of a. For if any of $k, l = a$, replace them by 1. If any of $m, n = a$, replace them by 0. Then, if $c = a$,

   $$f = a(ak + \overline{a}l) + \overline{a}(am + \overline{a}n) = ak + \overline{a}n$$

   If $g_1 = k$ and $h_1 = n$, then g_1, h_1 are common-select mux implementable, are independent of a, and can replace g and h without changing f. Then, from (3), $g = f_a, h = f_{\overline{a}}$. Finally check if g and h are common-select mux implementable.

2. One proper input at the OR gate, a, and two proper inputs at the AND gate, c and $d, c \neq d$. If f is realizable by a single *act2* block, $f = ag + \overline{a}h$, $g = cdk + (\overline{c} + \overline{d})l$, $h = cdm + (\overline{c} + \overline{d})n$. Again, it is sufficient to assume that g, h are independent of a. Hence, $g = f_a, h = f_{\overline{a}}$. Finally check if g and h are common_AND-select mux implementable.

3. Two proper inputs at the OR gate, $a, b, a \neq b$, and at most one proper input at the AND gate, c. If f is realizable by a single *act2* block, $f = (a+b)g + \overline{a}\overline{b}h$, $g = ck + \overline{c}l$, $h = cm + \overline{c}n$. h may be assumed independent of a, b. For if either m or $n \in \{a, b\}$, it may be replaced by 0. Then if $c \in \{a, b\}$, h may be replaced by $h_1 = n$, which is independent of a, b and is common-select mux implementable with any other mux implementable function. This implies $h = f_{\overline{a}\overline{b}}$. For g, we need to consider two cases:

   - g is independent of a, b. Then f_a should be equal to f_b, and $g = f_a$ (Proposition 3.3). We then check if g and h are common-select mux implementable.
   - g depends on either a or b. As shown earlier in (6), $g = G + H\overline{a}\overline{b}$. Then, as before, for g to be mux-implementable, we have at most 16 choices for H. For each such choice, we compute the corresponding g and check if g and h are common-select mux implementable. We quit if we get a YES, else we consider the next choice of H and repeat this step.

4. Two proper inputs at the OR gate, $a, b, a \neq b$ and two proper inputs at the AND gate, $c, d, c \neq d$. If f is realizable by a single *act2* block,

   $$f = (a+b)g + \overline{a}\overline{b}h \quad (14)$$

$$g = cdk + (\overline{c} + \overline{d})l \quad (15)$$
$$h = cdm + (\overline{c} + \overline{d})n \quad (16)$$

h may be assumed independent of a, b. This implies $h = f_{\overline{a}\overline{b}}$. For g, we need to consider two cases:

- g is independent of a, b. Then f_a should be equal to f_b and $g = f_a$ (Proposition 3.3). We check if g and h are common_AND-select mux implementable.
- g depends on either a or b. As done earlier, we see that $g = G + H\overline{a}\overline{b}$. Also, from (15), $g = cdk + (\overline{c} + \overline{d})l$. Here, $c, d, k, l \in \{0, 1\} \cup S(f)$. Since $G_{\overline{a}\overline{b}} = 0$, and H is independent of a, b, we get

$$H = c_{\overline{a}\overline{b}} d_{\overline{a}\overline{b}} k_{\overline{a}\overline{b}} + (\overline{c}_{\overline{a}\overline{b}} + \overline{d}_{\overline{a}\overline{b}}) l_{\overline{a}\overline{b}}$$

Since g depends on $\{a, b\} = V$, we consider the following cases:

(a) If $c \in V, H = l_{\overline{a}\overline{b}}$. Similarly handle the case $d \in V$.

(b) Else,
   - if $k \in V, H = (\overline{c}_{\overline{a}\overline{b}} + \overline{d}_{\overline{a}\overline{b}}) l_{\overline{a}\overline{b}}$.
   - if $l \in V, H = c_{\overline{a}\overline{b}} d_{\overline{a}\overline{b}} k_{\overline{a}\overline{b}}$.

We then get

$$H \in \{0, 1, x, \overline{x}, \overline{x}y, \overline{x} + \overline{y}, (\overline{x} + \overline{y})z, xy, xyz\} \quad (17)$$
$$x, y, z \in S(f) - V, \quad (18)$$
$$x \neq y, y \neq z, x \neq z \quad (19)$$

As in the algorithm for *act1*, we investigate the cases $H = xyz$ and $H = (\overline{x} + \overline{y})z$. Let $H = xyz$. Then $g = G + xyz\overline{a}\overline{b}$. But g has at most 4 inputs. So, g must be independent of at least one of x, y, z, a, b. We can then show that either $f_{\overline{a}\overline{b}} = xyz$ or $f_{\overline{a}b} = xyz$. Similarly, if $H = (\overline{x} + \overline{y})z$, either $f_{\overline{a}\overline{b}} = (\overline{x} + \overline{y})z$ or $f_{\overline{a}b} = (\overline{x} + \overline{y})z$. Hence, for a three-input H, we just need to look at $f_{\overline{a}\overline{b}}$ and $f_{\overline{a}b}$. Only if any of these are of the form xyz or $(\overline{x} + \overline{y})z$, we consider that form for H. This reduces the possible choices of H to at most $1 + 1 + 6 + 6 + 30 + 15 + 1 + 15 + 1 = 76$. For each such H, we get a g, and we check if g and h are common_AND-select mux implementable.

5. Output of OR gate is constant (0 or 1): So $f = g$ or $f = h$. Then if f is realizable by a single *act2* block, either $f = x$ ($x \in S(f)$), or $f = ck + \overline{c}l$ or $f = cdk + (\overline{c} + \overline{d})l = c(dk + \overline{d}l) + \overline{c}l$. All these subcases are covered by case 1.

Note that a, b vary over $T_4(f) = \{x | x \in S(f), |S(f_x)| \leq 4\}$, and c, d vary over $S(f)$. Also note that as for *act1*, each time we obtain a subfunction of f, we first obtain a prime and irredundant cover for it before performing any checks on it. From above discussion it follows then that

Proposition 3.5 *If f is expressed in minimum support, then the above procedure is complete for act2.*

3.3 Construction of ITE's

We first state two propositions from [2]. These give the best possible variable ordering for reduced ordered BDD's for certain kinds of functions, i.e., this ordering generates an ROBDD which after mapping on *act1* blocks results in the least number of blocks as compared to all other variable orderings. We shall make use of these propositions while constructing the ITE.

Proposition 3.6 *If a function $f = f(f_1, f_2, \ldots, f_k, g_1, g_2, \ldots, g_l)$ is a single cube with input variables $f_1, f_2, \ldots, f_k$ occurring in the positive phase and $g_1, g_2, g_3, \ldots g_l$ in the negative phase, then the reduced ordered BDD corresponding to the ordering $f_1, f_2, g_1, g_2, f_3, g_3, g_4, f_4, g_5, g_6, \ldots$ of input variables results in the minimum number of act1 blocks after covering as compared to other orderings.*

Proposition 3.7 *If a function $f = f(f_1, f_2, \ldots, f_k, g_1, g_2, \ldots, g_l)$ consists of only single literal cubes with input variables $f_1, f_2, \ldots, f_k$ occurring in the positive phase and $g_1, g_2, \ldots, g_l$ in the negative phase, then the ROBDD corresponding to the ordering $f_1, g_1, f_2, f_3, g_2, f_4, f_5, g_3, f_6, f_7, \ldots$ of input variables results in the minimum number of act1 blocks after covering as compared to other orderings.*

Comments: 1) When inputs of a phase finish, the remaining inputs are concatenated to complete the ordering. 2) Both orderings are listed so that they start from the terminal vertices and end at the root of the reduced ordered BDD. 3) The covering algorithm used is the dynamic programming algorithm.

We construct an unordered ITE for a function. The way ITE is constructed is very similar to the heuristic we proposed for unordered BDD's in [2]. We describe the full procedure here.

Suppose we are given the cover of a function f (i.e., a sum-of-products representation). Also suppose the cover is not unate. We cofactor f with respect to the input variables until we reach a unate cover. The selection of the cofactoring variable at each stage is done as follows. We first select those variables that occur in all the cubes in the same phase (in any order). Next, at each step the most binate variable x is selected. This is the variable that appears in the maximum number of cubes, either in positive or negative phase. Ties are broken by selecting those variables first that occur in the two phases nearly the same number of times. We factor the function f as follows:

$$f = C_1 x + C_0 \overline{x} + C_2 \quad (20)$$

C_2 is the set of cubes independent of x. If we were to construct a BDD for f, then the set of cubes C_2 is replicated in f_x and $f_{\overline{x}}$. However, ITE's avoid the duplication, as pointed out in [4]. We can represent f as **if** C_2 **then** TRUE **else** (**if** x **then** C_1 **else** C_0). This way we do not have to duplicate any literal of C_2. We recursively construct ITE's for C_0, C_1 and C_2 until we reach a unate cover. For two special cases, we use a slightly different realization. When $C_2 = \overline{a}$ or $C_2 = \overline{a}\overline{b}$, where a, b are inputs of f, we realize f as **if** $\overline{C_2}$ **then** (**if** x **then** C_1 **else** C_0) **else** TRUE. This allows us to save an inverter (when $C_2 = \overline{a}$) or use the OR gate of the basic block (when $C_2 = \overline{a}\overline{b}$).

At a unate leaf with unate cover UC, we do the following: If the leaf is constant function, 1 or 0, we return the corresponding ITE constant 1 or 0. If there is just one cube, we construct ITE for a single cube using the variable ordering of Proposition 3.6. If the leaf is composed of single literal cubes, we use the variable ordering of Proposition 3.7 to construct an ITE for it. Else, we construct a *0-1* matrix $B = (b_{ij})$ for the unate cover UC, where

$$b_{ij} = \begin{cases} 1 & \text{if } i^{th} \text{ cube of } UC \text{ contains variable } x_j, \\ 0 & \text{otherwise.} \end{cases} \quad (21)$$

A minimal weight column cover CC for B is then obtained [7]. CC contains some variables of the unate leaf. For each variable x_j in CC, a sub-cover SC_j is constructed using cubes of UC that depend on x_j. Each cube is put in exactly one such sub-cover (ties are broken arbitrarily). From the sub-cover SC_j, x_j is extracted out to yield a modified sub-cover MSC_j. In MSC_j, x_j is a don't care. This corresponds to creating an ITE vertex v_j for SC_j with the **if** child corresponding to x_j, the **then** child corresponding to MSC_j and the **else** child corresponding to 0 (this is an

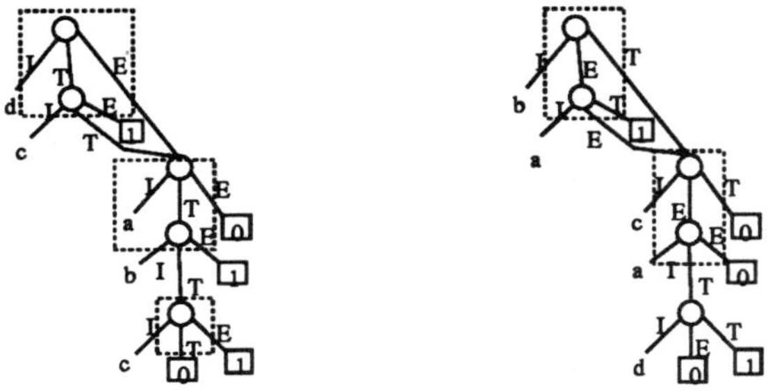

(A) (B)

Figure 5: Positive unate variables should have higher weights

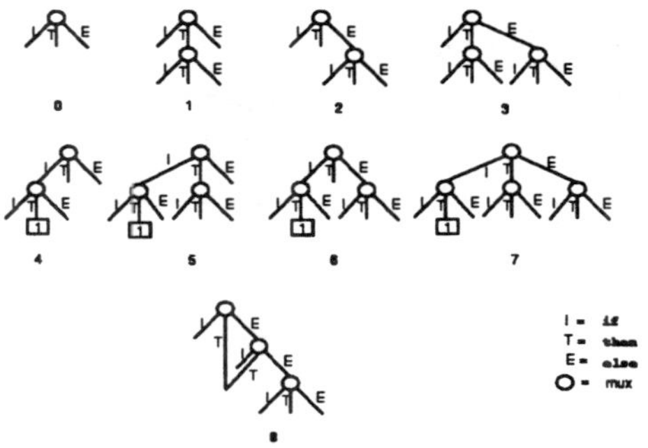

Figure 6: Pattern graphs for *act1*

AND operation). This is the case if x_j appears in positive phase. If it appears in negative phase, the **then** and **else** children of v_j are swapped. The column covers for MSC_j are recursively obtained. The ITE of UC is then obtained by ORing repeatedly pairs of ITE's corresponding to sub-covers SC_j. ORing of two ITE's (corresponding to two subcovers), say A and B, is performed by changing all the pointers in ITE A that pointed to constant 0 to point to root B[3].

The weight of a positive unate variable is chosen slightly more than the weight of a negative unate variable. To see why, suppose at a unate leaf, we have function $f = a\overline{b} + a\overline{c} + \overline{c}d$. If all the variables were given same weight, we may have ended up with column cover $CC_1 = \{a, d\}$. Then we factor f as $f = a(\overline{b} + \overline{c}) + d(\overline{c})$. The corresponding ITE is shown in Figure 5 (A). After mapping using the patterns to be described in Section 3.4, we get the cost as 3 blocks (the dotted rectangle stands for an *act1* block). The reason is that we end up with $\overline{c}$, an inverted input, at the leaf of the ITE. However, if the weight given to a positive unate variable is slightly more than that given to a negative unate variable, we obtain column cover $CC_2 = \{\overline{b}, \overline{c}\}$. These variables then appear as **if** children of some ITE vertex and hence their complements are automatically realized. The function f is then factored as $f = \overline{b}(a) + \overline{c}(a + d)$. The corresponding ITE is shown in Figure 5 (B). A cost of 2 is incurred after mapping[4]. By giving lower weights to the negative unate variables, negative variables tend not to be at the leaf of the ITE for a subcover. We choose the weights such that if the cardinality of a column cover is t, then each cover of size greater than t should have weight greater than t. It is easy to see that weight(negative unate variable) = 1, weight(positive unate variable) = $1 + 1/n$ where f is a function of n variables, satisfy this property.

Note that there are two main differences between our method of construction of ITE's and that of Karplus [4]: 1) we handle unate functions in a special manner. Karplus instead uses (20) for all functions, and 2) we employ Propositions 3.6 and 3.7 to construct ITE's for two special cases.

3.4 Pattern Graphs

The pattern graphs for the "library" are also ITE's. We use 9 pattern graphs for matching onto *act1* (Figure 6). The first four pattern graphs (from 0 to 3) correspond to the *act1* without OR gate using one, two or all three multiplexors. The next four patterns (from 4 to 7) are the same as the first four except for an OR function at the **if** child of the root vertex. The OR function is detected by the presence of constant 1 at the **then** child of the **if** child of the root vertex. Pattern-graph 8 represents a special case derivable from the basic block and uses the OR gate at the select line of MUX3.

4 Results and Conclusions

The techniques presented above have been incorporated into **mis-pga (new)**, a version of misII[5]. The input to the program is a network optimized by standard methods. In Table 1, we compare our results with *misII* (a library-based approach), *Amap* [4] and *mis-pga (old)* [2] using the same starting networks for all the systems. We also compare with *Proserpine* [5]. However, in this case, we do not know if the starting points are the same[6]. *mis-pga (new)*, *misII* and *mis-pga (old)* were run on DEC5500 (a 28 mips machine), whereas *Amap* was run on SUN4/370 (a 12.5 mips machine). The option used for *Amap* was -an3. The command used for *mis-pga (old)* was act_map -h 3 -n 1 -q -d 4 -f 3 -M 4 -l -g 0.001. This means that both reduced ordered BDD (ROBDD) and unordered BDD are constructed for each function. An optimum ROBDD is constructed for any function with at most 4 inputs. One iteration of the iterative improvement phase is executed. In the partial collapse routine, the only nodes with fanin at most 3 are considered for collapsing. In the decomposition routine, all nodes with fanin at least 4 are considered. The quick phase is also executed. Finally, a last-gasp routine is entered at the end [2]. For *mis-pga (new)*, we report results for two sets of experiments. In both, we used two iterations:
NLG: did not perform last-gasp.
LG: performed last-gasp, in which an ROBDD is constructed and checked for better results.

Table 1 shows the number of *act1* basic blocks needed to implement the benchmark, and the time taken (in sec.) in columns *b.c.* (block count) and *sec.* respectively. The minimum block count for each example is highlighted. On a set of 28 benchmarks, *mis-pga (new):NLG* outdoes *misII* technology mapper (using library-based approach) by 22.7%, *Amap* by 22.3%, *mis-pga (old)* by 17.3% and *Proserpine* by 31.0% (on a set of 14 benchmarks). Here, the average is calculated as the arithmetic mean of the percentage improvements for each example. Alternatively, using total number of blocks for all the examples as the basis, *mis-pga (new):NLG* outdoes *misII* technology mapper by 19.2%, *Amap* by 19.9%, *mis-pga (old)* by 18.6% and *Proserpine* by 36.0%. *mis-pga(new): LG* gives slightly better quality results than *mis-pga(new): NLG* but sometimes at the expense of longer runtimes (2-3 times).

The reason why *mis-pga (new)* outperforms *misII* technology mapper is because the multiplexor representation of a function fits nicely into the architecture. Though *Amap* also uses ITE's, our way of construction of ITE is different. Also, we use the matching algorithm and do an iterative improvement, thereby sacrificing speed for quality. Finally, *mis-pga (old)* constructs BDD's and hence replicates parts of cubes in 0 and 1 branches.

[3] It should be noted that for general ITE's, changing pointers may not always work. However, we have built ITE's for the unate subcovers in a special way. The **if** child of each vertex is a variable in positive phase. We can then prove, by induction on the number of vertices of the ITE A, that changing pointers correctly implements the OR.

[4] Incidentally, this function can be realized by one block. The match is discovered by the matching algorithm described in Section 3.2.1.

[5] now sis

[6] *Proserpine* is not being distributed yet.

example	mis-pga (new): NLG		mis-pga (new): LG		misII		Amap		mis-pga (old)		Proserpine
	b.c.	*sec.*	*b.c.*	*sec.*	*b.c.*	*sec.*	*b.c.*	*sec.*	*b.c.*	*sec.*	*b.c.*
z4ml	18	19	**14**	24	20	2	20	1	16	20	-
misex1	17	13	**16**	20	22	2	25	1	20	11	25
vg2	31	5	**30**	24	47	4	44	1	36	18	46
5xp1	39	28	**35**	48	51	4	42	2	45	60	53
count	**39**	9	**39**	14	63	7	41	1	46	13	-
9symml	**17**	55	**17**	1491	73	10	74	3	80	3124	-
9sym	**17**	78	**17**	4109	99	14	106	4	119	17583	-
apex7	92	53	**91**	60	113	11	104	4	96	42	121
C1908	159	208	159	213	188	19	**158**	8	175	646	-
rd84	**36**	67	**36**	95	62	6	62	3	61	151	70
e64	**94**	12	**94**	12	95	8	105	3	**94**	4	-
C880	**159**	75	**159**	127	175	18	190	7	171	77	-
apex2	100	70	**99**	86	106	12	122	5	124[1]	8	-
alu2	**175**	103	**175**	150	193	24	188	8	208	824	-
duke2	**158**	62	**158**	131	176	18	175	7	166	403	177
C499	166	116	166	119	174	18	**136**	7	166	35	170
rot	259	129	**255**	273	313	29	335	12	288	1072	465
apex6	275	115	**272**	265	360	34	392	15	289	255	396
alu4	**110**	34	**110**	71	149	32	160	6	132	146	350
des	1318	882	**1315**	2352	1571	757	1634	67	1749[1]	763	-
sao2	**49**	13	**49**	53	52	8	56	2	62	54	-
rd73	**25**	27	**25**	32	32	3	32	2	31	37	-
misex2	**38**	7	**38**	10	46	4	47	1	41	6	45
f51m	**39**	30	**39**	62	52	5	56	2	48	37	63
clip	**43**	63	**43**	88	57	5	60	2	51	92	73
bw	**54**	32	**54**	48	81	7	83	4	65	20	67
b9	63	20	**60**	43	64	6	81	3	65	32	-
C5315	560	505	**558**	611	704	88	653	27	656	673	-
total	4150		4123		5138		5181		5100		
subtotal	1357										2121

[1]timeout: so we used the script act_map -n 2 -h 2 -d 4 -f 3 -g 0.001, act_map -h 3 -M 4

Table 1: Number of *act1* blocks

Though not shown in the table, we saw that without much difference in run time, the matching procedure of Section 3.2.1 gives 3-4% improvement on average. To give an idea about the speed of the matching algorithm of Section 3.2.1, we found that for each function, the algorithm takes less than one second to complete. This includes cases when a match existed and when it did not. The experiment was performed on around fifty 5-8 input functions.

Acknowledgement

The authors wish to thank Prof. Kevin Karplus (for pointing out that the original version of matching algorithm (described in Section 3.2.1) could miss a match) and William Lam and Hamid Savoj (for helpful discussions).

References

[1] A. Gamal, J. Greene, J. Reyneri, E. Rogoyski, K. A. El-Ayat, and A. Mohsen, "An Architecture for Electrically Configurable Gate Arrays", *IEEE Journal of Solid State Circuits*, Vol. 24, No. 2, April 1989.

[2] R. Murgai, Y. Nishizaki, N. Shenoy, R. K. Brayton and A. Sangiovanni-Vincentelli, "Logic Synthesis for Programmable Gate Arrays", *Proc. 27th Design Automation Conference*, June 1990, pp. 620-625.

[3] R. Murgai, N. Shenoy, R. K. Brayton and A. Sangiovanni-Vincentelli, "Improved Logic Synthesis Algorithms for Table Look Up Architectures, *Proc. ICCAD*, Nov. 1991.

[4] K. Karplus, "Amap: A Technology Mapper for Selector-based Field-Programmable Gate Arrays", *Proc. 28th Design Automation Conference*, June 1991.

[5] S. Ercolani and G. D. Micheli, "Technology Mapping for Electrically Programmable Gate Arrays", *Proc. 28th Design Automation Conference*, June 1991.

[6] T. Besson, H. Bouzouzou, M. Crastes and G. Saucier, "Synthesis on Multiplexor-based Programmable Devices using (Ordered) Binary Decision Diagrams", *Unpublished manuscript*, 1992.

[7] R. K. Brayton, C. McMullen, G. D. Hachtel and A. Sangiovanni-Vincentelli, *Logic Minimization Algorithms for VLSI Synthesis*, Kluwer Academic Publishers, 1984.

[8] R. K. Brayton, R. Rudell, A. Sangiovanni-Vincentelli, and A. R. Wang, "MIS: A Multiple-Level Logic Optimization System", *IEEE Transactions on CAD*, November 1987.

[9] K. Keutzer, "Dagon: Technology Binding and Local Optimization by DAG Matching", *Proc. 24th Design Automation Conference*, June 1987.

[10] E. Detjens, G. Gannot, R. Rudell, A. Sangiovanni-Vincentelli, and A. R. Wang, "Technology Mapping in MIS", *Proc. of ICCAD*, 1987.

Acquiring and Maintaining State-of-the-Art DA Systems

Patrick Hefferan
Consultant
1681 Princeton Ave.
St. Paul, MN 55105

Steve Sapiro
Consultant
97 Mountainview Rd.
Santa Cruz, CA 95065

ABSTRACT

This paper describes methods and approaches to successfully acquire and maintain a state-of-the-art Design Automation (DA) environment. Information is presented on strategies for evaluating DA technologies, assessing and justifying their acquisition, negotiating their purchase, and resolving post-sales issues with vendors. Also, this paper outlines a practical strategy for managing and operating a working DA environment. The strategy is based on the use of Standards and Frameworks to create a 'plug and play' environment. A structured plan for managing and expanding the design environment is presented, with emphasis on developing and implementing new capabilities.

1. Successfully Acquiring DA Technology

Acquiring DA technology is both an art and science. Objective measures and criteria must be used in evaluating the technology's capabilities, but a high degree of art is required to justify its cost and negotiate tool purchases from a vendor. This makes the job of the CAE/CAD manager especially demanding.

1.1 Make-or-Buy Decisions

The most fundamental decision to make about acquiring DA technology is whether to buy tools or to develop them. In earlier times, developing the tools and technology in-house was the only alternative. But in the early 1970's, a vigorous and competitive CAE/CAD industry emerged which drove the cost of tools down as they recaptured their development investments over a wide customer base. So today, the cost of internally developing DA technology is prohibitive when compared with purchasing commercial alternatives.

An example helps illustrate this. A typical schematic entry tool costs from $500 to $10,000 per copy depending on the vendor. Developing even a simple capture facility would require at least 3 man-months, and at $8,000/man-month, cost $24,000, plus continual maintenance.

But seldom are make-or-buy decisions so simple. Often the development costs have already been sunk, and the decision is whether to incrementally enhance the in-house tool or pay more for a commercial one. The economics still favor the commercial option because the in-house tool's support costs can only be amortized over a single user group. Thus, over a longer timeframe, the commercial tool will accrue less support cost.

An instance where developing a tool is likely to be justified is where the tool provides a unique capability which gives an organization a competitive advantage. For example, a control systems company might develop a special fluid simulator, integrated with its electrical HDL simulator to accelerate its development effort.

1.2 Evaluating Commercial DA Technology

Evaluation of CAE/CAD tools needs to be a structured process driven by the needs and requirements of the development groups. It is essential to get input from working designers, product management, and the CAD group, and then capture it as a formal requirements document. The CAD group contributes input on systems and compatibility. The requirements document should drive the evaluation criteria for each technology evaluated. A formal benchmark or in-house evaluation should be the final step in the process to ensure that there is a good *fit* with the technology. This process is illustrated in Figure 1:

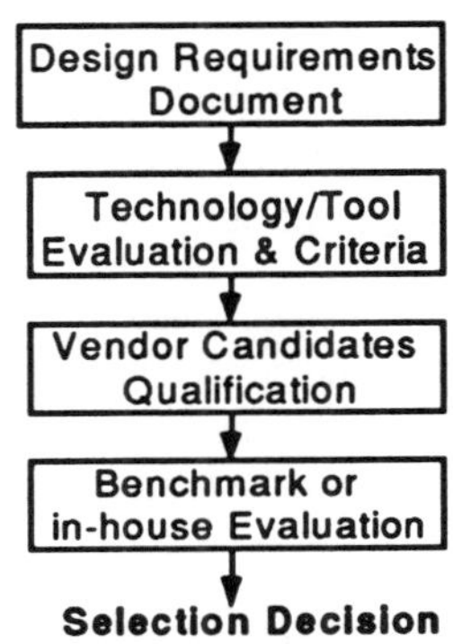

Figure 1 - Technology Evaluation Process

Written evaluation criteria along with a formal evaluation plan helps ensure an unambiguous end result, and gets all parties' buy in.

1.3 Evaluating the Vendor

As important as the product features are, the vendor is equally as important. The vendor's support, responsiveness, and even viability are critical to a product's success. Joe Costello, the CEO of Cadence Design Corporation, made the observation that no CAD/CAE company has survived for more than a decade[1]. There have been mergers, acquisitions, and outright disappearances of major CAE/CAD companies. These facts have significant bearing on the usability of an purchased DA technology. Key issues that should be evaluated prior to purchase include:

- Financial viability of vendor (annual report)
- Location of support (local or distant?)
- Ratio of technical/non-technical staff
- Responsiveness to requests/bugs
- Vendor's technical commitment
- Vendor's development system

The last item is especially important if the designers use platforms different from the vendor's primary development platform. Then bug fixes and feature releases may be delayed beyond normal release dates.

1.4 Negotiating with Vendors

Negotiating with the vendor is perhaps the least scientific and most intuitive part of the acquisition process. Most often the vendor's sales people are well schooled in the art of negotiating. Many vendors provide their sales staff with negotiating seminars and training several times per year. Yet CAE managers seldom have any formal negotiations training. However, the CAE manager has a place to turn. The corporate purchasing department often contains individuals trained in the art of negotiating.

Generally, the CAD Manager is the first point of contact with the vendor and will perform most of the important negotiations for the purchase. Some factors to consider in negotiating are:

- Use all variables in the negotiation
- Ask for discounts, price protection, etc.
- Buy near end of vendor's quarter/year
- Setup a long term relationship/partnering
- Don't negotiate with yourself (propose/counter)

The CAD manager should get the process started early enough to provide sufficient time for effective negotiations. If the negotiations seemed stalled, ask to have the vendor's sales manager or VP of sales involved. These individuals are often empowered to make concessions beyond those authorized by field sales, and they will want to have results generated when they commit their time.

1.5 Post-Sales Vendor Relationships

The need to continue a relationship with the vendor after the sale is essential if future needs are to be met. The user will want to feedback to the vendor requests for bug fixes and feature enhancements. Also, the user may want to influence the direction of vendor's development effort to mesh with his own technology goals. The vendor typically wants to remain in contact with the DA user to sell more products and services. Thus, both parties need each other and need to commit time toward building a relationship. Recently, the term *partnering* has come into vogue as the direction of vendor/user relationships.

But what happens when something goes seriously wrong, like when a DA tool has a defect that halts the progress of a design. It is often difficult to make a distant vendor appreciate the importance of the problem. An effective way of communicating this is to show up on the vendor's door step with the problem in hand. You won't be ignored, and your level of need will be obvious.

2.0 Justifying DA Purchases

The value of DA technology is often unquestioned by the design community and CAD developers, but is looked at quite critically by business unit managers and others responsible for the overall success of a business enterprise. The benefit of a CAD tool might be compared with the net economic benefit of purchasing a new delivery truck. Thus, a purchase tends to be evaluated on the economic return it will generate for the company.

Generally, there are two major methods[2] for justifying investments. The first is a traditional method based on the technology providing definable cost reductions in the development process. The second assesses the revenue impact of the technology investment. This second method is gaining acceptance by many organizations, especially where time-to-market considerations and market share issues are paramount.

2.1 Cost-Based Justification

In cost-based justification, the DA technology's benefits are translated to enhanced design efficiencies which can be reduced to cost savings in the development process. Methods for quantifying these cost savings include:

- Cost Reduction Assessment
- Net Payback Analysis
- Internal Rate of Return (IRR)

A new technology's benefits, offset by implementation costs are weighted against the cost of continuing the old methods. Cost savings then can be expressed either as improved efficiencies, pay-back of the initial investment in months, or a percent return over the investment's lifetime. Often such comparisons are adjusted for the cost of money (at the prevailing interest rate).

For example, the costs savings for a schematic entry tool, costing $5000, could be calculated based on the following facts. It is a multi-user tool shared by 6 users who each use it approximately 20% of their time. It is projected to increase productivity by 10% after training, which requires 10 hours of designer time upon first use, and yearly update training of 10 hours/year for the team. The following table details the costs and net savings over the life of the investment:.

	Year1	Year2	Year3
Capture hrs.	2160	2484	2856
Enhancement	8%	10%	10%
Savings hrs.	162	248	285
Training hrs.	60	10	10
Net hrs. Saved	102	238	276
Wage Rate $/hr	30	32	33
Yearly Benefits	3060	7510	9117

TOTAL SAVINGS ($19,687 - $5000) = $14,687

The calculation could be made more exact by adjusting the $5000 investment for the net cost of money over the investment period. The tool has a payback period of about 15 months, again unadjusted for the cost of money.

In the late 1970's and early 1980's, this method gave many companies insight into the economic benefits of autolayout technology. Significant gains were realized as automated layout methods replaced labor intensive and error-prone manual methods. Logic synthesis and layout synthesis technologies may be candidates for similar cost savings in the 1990s.

2.2 Revenue-Enhancing Justification

As product lifetimes decrease and premiums for the latest technology escalate, a higher value is placed on quickly moving products from concept to market. Thus reducing design cycle time and reducing the number of design iterations required to get a product to market become prime goals of the design environment.

One significant economic advantage to improving time-to-market is the opportunity to enter a market earlier and benefit from the early (higher) pricing. This effect is called *learning curve pricing*, where the marketplace pays a premium for leading-edge technology before it begins its decent toward commodity pricing levels. Figure 2 shows this effect.

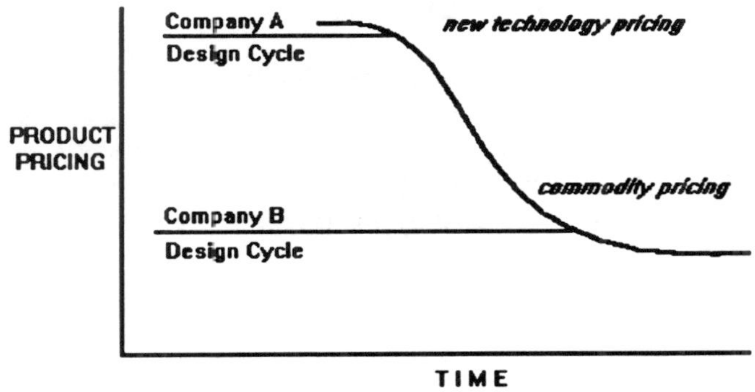

Figure 2. Time-To-Market Advantages

Another benefit that can result from adoption of DA technology is increased market share based on a product's richness of features, enhanced performance, or improved quality/reliability. Simulation technology offers the prospect of developing and verifying more features in the design cycle, and offers the additional capability of optimizing performance. Fault simulation and ATPG technologies offer the prospect of improving quality and reliability. All of these can contribute to a larger market share.

Capturing even a fraction of a market share point can result in significant accrual of revenue. For example, moving from a 20% to 20.5% market share of a $40M market results in $200,000 of additional revenue.

2.3 Justification & Expectations - Final Notes

All justification methods rely on a significant amount of judgement as well as assumptions about how it will be used. Often the technology changes how a product is developed or how the design group works. Then adoption of the new technology does not provide the expected results. More design alternatives may be explored, more features added to the product, and higher performance included in the final design, rather than faster time to market. Or increasing complexity of the designs produced might mask the improved efficiency provided by the technology. The value of the technology is still likely to be very high, and is in fact likely to be enhanced when modified through interaction with the working design environment.

There is a synergy that goes beyond the value of any individual technology or tool. Often, design changes and enhancements are severely constrained by the amount of design overhead required to implement them. Lowering the threshold of design effort can have the effect of encouraging exploration of design alternatives. If such efforts are channelled to innovation and improved product capabilities, this can yield great benefit.

3. Managing the DA Environment

Managing a Design Automation Environment is one of the biggest challenges that a company faces as it uses more technology to deliver its product to market on time. Many times the DA Environment, if not managed properly, actually causes more delay than necessary and may seriously impede time to market product delivery.

Many issues are critical to the effective operation of a DA Environment. Two key items - Standards and detailed methodology will be discussed here.

3.1 Standards: What are they & what do they do for you?

A few quotes may help define the role and value of standards in the design environment

STANDARD

"substantially uniform and well established" *(Webster's Dictionary)*

"*accepted, established, proven, sound, traditional*" (*WORDFINDER™*)

"*Standards acceptance will be driven by availability of tool technology rather than committees. Effective solutions will become de facto standards, in spite of IEEE standards.*" [3]

At a recent gathering of test experts, the subject of standards came up. Some of their comments were:

- *everyone has a different scan scheme, making it difficult to automate*
- *there is no standard test language*
- *must we depend on the agonizingly slow pace set by committees?*
- *effective solutions will become de facto standards, in spite of IEEE standards*

The biggest barrier to the use of new technology is NOT its availability but its rate of absorption. Standards allow you to absorb new technology in a transparent way. This can lead to productivity gains of major proportions.

After several years of gestation, standards have finally emerged in the EDA world. The test world should not be far behind. As seen in the past, it takes the CAE/CAD USER to get the vendors to discuss standards and to eventually adopt them. Some of the important standards that directly affect CAE/CAD users are detailed in the following sections.

3.2 Frameworks & CAD Framework Initiative (CFI)

WHAT IS A CAD FRAMEWORK?

"*... a collection of mechanisms ... (programs, libraries, languages, data management, user interface, etc.) at many levels of abstraction ...*" [4]

HOW DO YOU TELL A GOOD ONE?

"*... it reduces the time and cost needed to develop or modify a CAD system so that it meets the needs of a user.*" [4]

A framework is a new solution for an old problem. The issue - how do you have different tools access the same product information. An important result of tool access is the ability to "plug and play" - to transparently replace one tool with any other tool. This means you have the ability to pick the tool that solves your particular problem (not the one that the vendor of the total solution supplies) and plug it into your DA environment. Any DA environment can then be customized to support the experimentation and evolution required to specialize it. It can be made to meet the exact needs of the product group using it. In addition, as tools become more granular to deal with more discrete steps of the product process, the framework becomes the glue to bind everything together. As time-to-market becomes more critical in the 1990s, this will be a requirement for success.

Dealing with product information implies a three dimensional view of the world - of data, tools and process. Product information is usually a combination of several of the following:

- design data
- design methodology(s)
- field service information
- design support data
- specials, folklore
- tools
- diagnostics
- training material
- userware, macros

Obviously, each major issue has many sub issues. All may not be critical to a DA user now. A great deal depends on the maturity of the design process and the types of designs that you do. For example, if a design group is doing small to medium sized designs that are one or two person projects, many of the design methodology and support task requirements may not have reared their ugly heads yet. But as a design group moves toward larger, more complex design, it will see the light.

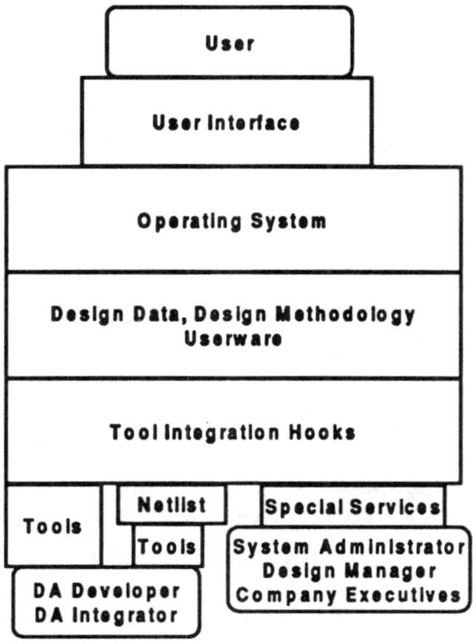

Figure 3: DA Framework

Figure 3 shows that a framework is not a tool or a database, but a process that includes users, tools, design information, methodologies, knowledge and folklore. No one can pick the one that is RIGHT for you. What is good for someone else, even another group of your company, may NOT be good for you. Because this is a subjective issue, there will never be a right answer - just one that works now for your given scenario.

CFI Mission statement: "*To develop industry acceptable guidelines for design automation frameworks which will enable the coexistence and cooperation of a variety of tools.*"

CFI is a volunteer organization (supported by vendors, universities and users) that is trying to provide as much information and standardization as possible in the design process. In that way, you can pick and choose the things, whether they are methods, tools, databases or knowledge, that make sense for you now knowing that as your needs evolve you can add on to what you have. This is a key point - because of their (your) standardization you should always be able to update your design process with a minimum of pain, loss of time and loss of data. In addition, you can experiment with new (to you) concepts by bolting them on to your existing design process, again with a minimum of pain.

3.3 Electronic Design Interchange Format (EDIF)

EDIF is a standard format that makes it easy to exchange design information. Translators can convert data between EDIF and compatible tools. Its major benefits are that you can move design data back and forth between different vendors' tools and between different vendors' CAE systems. This allows you to choose tools that you want rather than those that talk to each other. In addition, you can have a heterogeneous computing environment - for example, schematic capture on PCs that send data to simulators on workstations.

While VHDL describes a design at a high level, EDIF is low level, detail oriented. It can be used for exchanging:

- netlists (both flat and hierarchical)
- schematic drawings and symbols
- mask layouts
- symbolic layouts
- logic models

Be aware, however, that EDIF does have some problems translating design data with certain properties.

EDIF is here now and is here to stay. Many commercial tool vendors support EDIF as well as numerous universities (for example, UC Berkeley).

3.4 IEEE 1149.1 (JTAG)

Component level test of assembled boards is critical to getting a product to market quickly and correctly. Yet until recently there was very little effort devoted to this issue. That has changed with IEEE 1149.1. This standard defines test structures that are embedded in your ICs to provide information about their operation in both stand alone mode and when connected on the board. When used on a board, they obviate the need for in-circuit testers or bed-of-nails testers.

The 1149.1 architecture includes a Test Access Port (TAP), a TAP controller, a scannable instruction register and some scannable test data registers. The TAP is a port that provides access to the test structures and is required to be a dedicated connection (that is not used by the normal logic of your circuit). This allows you to examine the test information at any time, without resetting or presetting your internal logic.

3.5 Other Standards

Other standards that can help during the product development are Verilog HDL and WAVES.

4. Managing The Design Environment

Installing or updating capability into a DA environment requires a detailed plan and strategy to be successful. An outline for a sample plan might include the phases detailed in the following sections:

4.1 Phase 1: *Determine What Currently Exists*

Find out about the existing tools and methods. Without making any judgments, subjectively document the good and bad points from the users', managements', support groups' and customers' perspective. Put together the mechanics of the plan creation - who to work with, when it must be done, and what the deliverables must be, including reports, presentations, black box transfer functions, etc.

4.2 Phase 2: *Plan Preparation*

Determine if there any current or planned strategic alliances; how can they be used to expedite, empower or simplify the process. Review market trends and company business goals to put stakes in the ground. Based on that review, determine what is the problem space for this product group going to be over the next one to five years. Then predict how circuits and methods might change.

Define in as much detail as possible the designs, methodologies, environments that will be required to support the business plan, including:

- distributed verses central computing, databases
- design centers necessary
- internal user profile & how it might change over time
- outside user profile of design centers
- data management issues
- design concurrency: interaction with test, manufacturing...
- design for testability
- others???

Talk with all groups involved. At this point, look not for a consensus, but for all points of view.

4.2.1 Develop an IDEAL Plan

Develop an IDEAL plan including product information flow, design data, user and tool capabilities. Do this without respect to what is possible to deliver in your required time frame. Take a first cut at benefits analysis and how things might evolve. This is the golden model to which everything is compared.

4.2.2 Determine the REAL Tools & Implementation

Enhance the IDEAL plan/diagram to include the realities of the requirements from the business plan. Put together detailed timing charts to show all dependencies and critical paths. Play out several different scenarios of methods, tools, times to see which ones work the best. Include information about what is commercially available and what is coming soon. Develop a generic tools list, including ballpark costs, and a list of suggested vendors. Then pare it down based on what fits with the existing environment and tools.

Look at what risks might be associated with the tools and methods chosen. For high risk areas, look at safer alternatives with higher risk/reward ratios. Define what is not available and will have to be developed. Remember to set a reasonable range of expectations for the payback from the changes.

4.3 Phase 3: *Plan Realization - Select Tools to Buy*

Develop a product requirements spec including issues such as environment, users, cost, timeframe, selection criterion, benchmark and evaluation issues. Do a vendor search including strategic alliance choices. Select two to three choices that meet requirements. Have meetings, benchmarks, and technical discussions to narrow the choice to one vendor for each tool.

4.4 Phase 4: *Implementation*

Finalize the methods and environments developed. Set the project up to succeed by leaving nothing to chance, including the incoming environment, the first project, a champion and wizard to make the project work and help from upper management.

4.5 Phase 5: *Review Progress and Correct*

Review progress each quarter (sooner if the project is just getting started). Look for any changes that need to be made. Talk with vendors about new products.

References

[1] Costello, Joe, CEO Cadence Design Systems, Keynote Address, *1991 Design Automation Conference.*
[2] Powell, R. E., Justification & Financial Analysis for CAD, *Design Automation Conference Proceedings,* 1980.
[3] Probu Goel, President, Systems Division, *Cadence Design Systems, Inc.*, 1/90.
[4] Richard Newton and Alberto Sangiovanni-Vincentelli, *ICCAD 90 Tutorial.*

Analyzing Cycle Stealing on Synchronous Circuits with Level-Sensitive Latches

Ichiang Lin   John A. Ludwig   Kwok Eng
IBM Corporation
Enterprise Systems, Kingston, New York 12401

Abstract

A new analytic method for timing analysis on level-sensitive synchronous circuits is presented. Because a level-sensitive latch acts as combinational logic when the clock signal is active, a data signal arriving at the latch during the clock active interval may propagate through the latch to the next latch stage. In a single phase clocking scheme, a signal propagation that starts at the leading edge of the clock may end before the trailing clock edge but after the leading clock edge of the next cycle. Therefore, a signal propagation delay through combinational logic can be longer than a clock cycle (called cycle stealing) if the next cycle's data launch time can be delayed. Cycle stealing may propagate through several latches without any timing problem if a delay of the data launch time in each stage is allowed.

The proposed method can fully discover the cycle stealing effect on level-sensitive synchronous designs. The algorithm first constructs a latch graph from a timing analysis on the combinational logic, then it analyzes cycle stealing based on overlay timing relationships among latch nodes. A breadth-first search examines all possible cycle stealing among latches. The algorithm also considers the fact that cycle stealing is topology dependent.

The timing analysis program also takes into account the variation of clock width and leading and trailing clock edges in each latch, so the data can be used to assist physical design. The accurate timing information benefits both clock tree and combinational circuit design using level-sensitive clock schemes. The program has been implemented on an IBM RISC System/6000 [1] coupled with an IBM integrated circuit (IC) design system. The results show the benefit of using cycle steal opportunities in the design.

1 Introduction

An accurate timing analysis tool allows the IC design process to detect timing errors and to make proper design decisions. It is vital for a timing-driven design environment to have such tools to capture timing problems in the early design stage. One main concern of current timing analysis tools on level-sensitive synchronous circuits is the inaccurate timing analysis on transparent latches. As a signal arrives at a latch during its active interval, the signal may pass through that latch as combinational logic. Because its arrival time is after the next cycle leading clock edge, the signal propagation delay can be longer than the clock cycle time. A delay of the next cycle signal launch time is allowed if the signal propagation delay time in the next cycle is not critical. Various terms can be found in literature to describe the extension of arrival times, such as retardation, slack borrowing, slack steal, or transparent latch. We use the term "cycle stealing" to emphasize the signal propagation which steals a portion of the next clock cycle time. With the cycle stealing technique employed, a signal propagation which needs more than one clock cycle time is not a late path if its latch time can be extended to the next cycle. Note that other design issues, such as fast paths or early mode delays, need be considered when level-sensitive latches are used in the design.

A cycle stealing may propagate through several latch stages, as shown in a two-stage cycle stealing in Figure 1. Because there is a delay on the latch time in the first clock cycle, the trigger time in the second clock cycle is postponed and so is the latch time in the second clock cycle. Hence, the second clock cycle may steal some clock cycle time from the following clock cycle to allow the signal propagation.

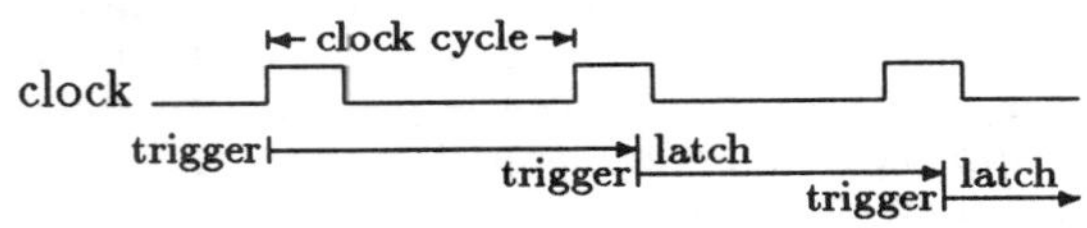

Figure 1: A two-stage cycle stealing.

In the existing timing analysis tools, which tend to look at only a single cycle, all signal propagation delay times between any two latches are bounded in some fixed time. The analysis results tend to be conservative because the cycle steal opportunities are not explored at all. For example, a timing verification, called "trigger-to-trigger test", computes path slacks as the differences between path delays and the cycle time, which is the time between the clock leading edge and its next cycle clock leading edge. The result of this analysis is too pessimistic because some failed paths under this test do not actually violate timing constraints because of the cycle steal effect. Another test, called "trigger-to-latch test", computes path slacks as the difference between the path delays

[1] RISC System/6000 is a trademark of International Business Machines Corporation.

29th ACM/IEEE Design Automation Conference®

0738-100X/92 $3.00 © 1992 IEEE

and the longest possible capture time, which is the time between the clock leading edge and the capturing clock trailing edge. The results tend to be too optimistic because some safe paths under this test actually violate timing constraints. Using both tests, two slack reports may expose these safe paths which pass the trigger-to-trigger test and those failed paths which fail the trigger-to-latch test. However, the results cannot discriminate timing violations from non-violations on these paths which fail the trigger-to-trigger test but pass the trigger-to-latch test.

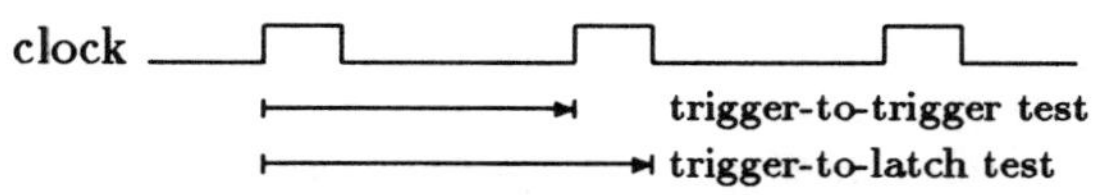

Figure 2: Trigger-to-trigger and trigger-to-latch tests

The difficulty of timing analysis on level-sensitive synchronous circuits is a challenge to both the industry and academia. Earlier study on cycle stealing can be found in [1][2]. The timing rules and properties are investigated in [3][4]. An event propagation method is one which treats a signal propagation in the circuit as an event. Each event propagates through logic gates and then compares its arrival time against the clock constraints. The variations of using this approach can be found in [5]-[8].

An arrival time assertion method is used in several timing analysis tools [9]-[11]. In this method, the arrival time on each latch is initially asserted. The asserted arrival time can be set, for example, at the middle of the active clock interval. Then the asserted arrival times are corrected by an iterative procedure. The change in each latch arrival time assertion is determined by the slack associated with the latch. The slack information is then decided by the asserted arrival times of the latch and its neighboring latches, and the arrival time assertions are adjusted by the slack information on the paths connected to the latch. The iteration may terminate if there is no further change in the assertions or other threshold conditions that are met. So far, no theoretical proof is provided to show that this approach does not sometimes treat a safe path as a timing violation.

The features of this paper are the following:

1. The block-oriented timing analysis method derives a slack report based on the given timing assertions on sources (arrival times) and sinks (required times) of the combinational logic network [12]. Because the accuracy of assertions needs to be ensured, the clock skews among all clock pins have to be minimized. Due to the cycle steal effect, the assertions at the beginning of timing analysis may no longer be valid. Arrival times through combinational logic need to be evaluated for every assertion adjustment. To avoid this problem, we use a latch graph to depict delays among latches. The latch graph contains combinational logic path delays in addition to the slack information. Any change in the arrival time at one latch is directly reflected on its neighboring latch nodes by the latch graph without re-computing combinational logic delays.

2. Most approaches mix the combinational logic path delay calculation with the cycle steal analysis at the same time [5]-[8]. We distinguish the data paths from the clock distribution tree. The path delays from latch to latch through combinational logic are first calculated to construct the latch graph. The cycle steal analysis which follows is based on the given latch graph. This provides a convenient method of synchronous circuit design with cycle steal analysis because the clock waveforms on latches may be adjusted by various design techniques on the clock distribution tree, and the latch graph can serve this need on a higher level basis.

3. We view the cycle steal problem as active interval propagation in contrast to the event-driven or arrival time assertion methods mentioned above. Overlay timing relationships which are used to depict the cycle steal behaviors among latches are derived for latch nodes of the latch graph. The cycle steal behavior on each latch can be clearly observed by its overlay timing diagram. A multiple stage cycle steal is also easily analyzed by this approach. The breadth-first tracing technique is applied for this purpose.

4. This work considers the nature of cyclic graphs which exist in synchronous circuits. In fact, the cycle steal timing analysis is topology dependent [6][13]. The feedback loop structure in synchronous circuits has a timing property which can be reflected in the overlay timing relationship. The tracing procedure analyzes feedback loops in circuits according to this property.

In this paper, the concept of an overlay timing diagram and latch graph is introduced in Sections 2 and 3. The property of feedback loops and their relation to overlay timing diagrams are discussed in Section 4. A breadth-first cycle steal algorithm and its experimental results are also presented in Sections 5 and 6.

2 Overlay Timing Diagram

Consider the latch $L(x)$ which receives the signal from the incoming latch $L(i)$, and sends the signal to the outgoing latch $L(j)$ as shown in Figure 3. Each clock waveform may be different. We assume that the setup time and the hold time hereinafter have been considered in the clock waveforms. A signal which starts at time A in latch $L(i)$ and passes through the combinational logic with delay $d(i,x)$ arrives at time B in latch $L(x)$. The signal which is triggered at time B in latch $L(x)$ passes through the combinational logic with delay $d(x,j)$ and arrives at time C in latch $L(j)$.

In order to illustrate the timing relationships between $L(x)$ and $L(i)$, the incoming latch of $L(x)$, the $L(i)$ clock waveform is shifted as

$t'(i) = t(i) + d(i,x)$,

where $t(i)$ is the time the signal is launched at latch

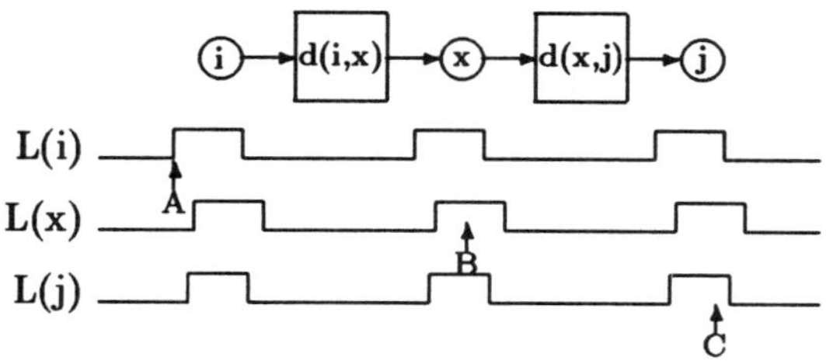

Figure 3: Three latches and their clock waveforms.

$L(i)$, $t'(i)$ is the time the signal arrives at latch $L(x)$, and $d(i,x)$ is the worst case signal propagation delay time from $L(i)$ to $L(x)$ through the combinational logic. In Figure 3, $t'(i)$ is the time B if $t(i)$ is the time A. Figure 4 shows the clock waveform of $L(i)$ overlaid with the clock waveform of $L(x)$ with this shift. Because of the shift of the clock waveform of $L(i)$, the time A in $L(i)$ is aligned with the time B in $L(x)$. We may observe that the amount of time $CS1$ is stolen by the signal launched at time A. In the overlay timing diagram, we define a window which is the clock active-interval of $L(x)$. The launch time D which is located outside of the window fails to arrive before $L(x)$ closes.

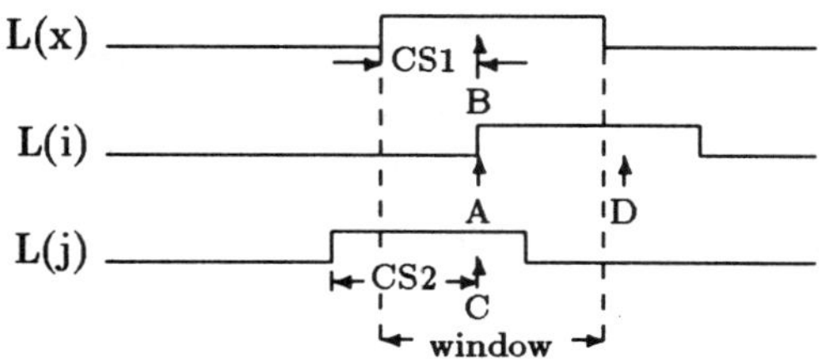

Figure 4: Overlay clock waveforms.

We may also construct the overlay timing relationships between the clock waveforms of $L(x)$ and its outgoing latch $L(j)$ with
$t'(j) = t(j) - d(x,j)$,
where the definitions of $t'(j)$, $t(j)$ and $d(x,j)$ are similar to the above. In Figure 3, $t'(j)$ is the time B if $t(j)$ is the time C. The time C, which is the arrival time of the signal propagation to $L(j)$, is aligned with the times A and B. The signal which launches at the time A in $L(i)$ must steal $CS2$ time from the clock cycle in $L(j)$. Note that the information of delay times and clock waveforms has been embedded in the overlay timing diagram.

For any incoming latch $L(i)$, the overlay clock waveform of $L(i)$ moves to the right with respect to the clock window of $L(x)$ if the delay $d(i,x)$ increases. On the other hand, the overlay clock waveform of the outgoing latch $L(j)$ moves to the left if the delay $d(x,j)$ increases. We can identify the case in which the clock active-interval of $L(i)$ appears to the left of the window of $L(x)$. Figure 5(a) shows no timing violation, labeled *Safe*, between $L(i)$ and $L(x)$, and no cycle steal is necessary. We may also identify the case of a timing violation if the clock active-interval of $L(i)$ appears to the right of the window of $L(x)$. Figure 5(b) shows a late-mode timing violation, labeled *Failed*, between $L(i)$ and $L(x)$, and the cycle stealing doesn't help. A similar way of identifying the timing status between $L(x)$ and $L(j)$ is also shown in Figure 5.

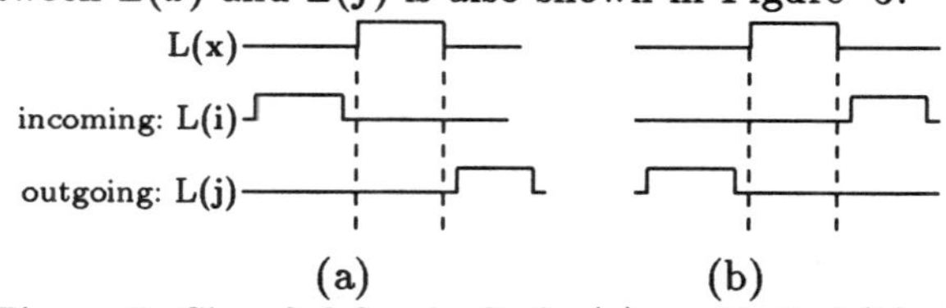

Figure 5: Signal delay is *Safe* (a) or *Failed* (b).

If the $L(i)$ clock active-interval overlaps with the window of $L(x)$, it may or may not violate the timing constraints depending on the relationship of $L(i)$ with the $L(j)$ clock waveform. Figure 6(a) shows that $L(i)$ can do cycle stealing from $L(x)$ and does not need to steal any cycle time from $L(j)$. Figure 6(b) shows that $L(i)$ can do cycle stealing from both $L(x)$ and $L(j)$. Figure 6(c) shows that even though $L(i)$ can steal cycle time from $L(x)$, the signal fails to steal cycle time from $L(j)$ because the signal arrival time on $L(j)$ is after the trailing clock edge of $L(j)$.

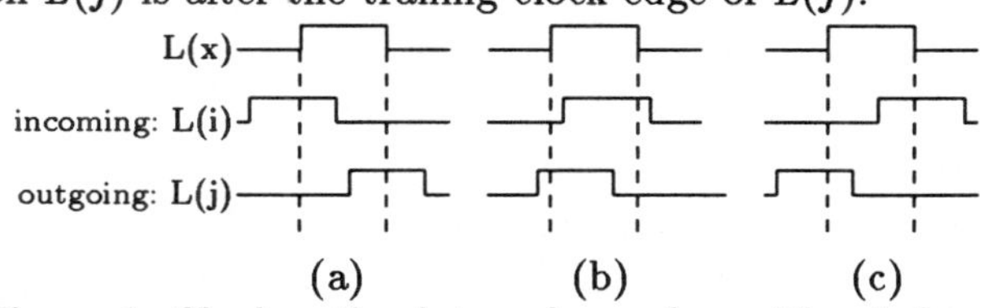

Figure 6: Clock active-intervals overlap with window.

3 Latch Graph Construction

A latch graph is first constructed before applying the cycle stealing timing analysis. The latch graph is a directed graph in which the vertices are the latches, primary inputs and primary outputs, and the edges are the delay times among them. Each edge represents a path from a latch (or primary input), to a latch (or primary output) through the combinational logic network. For each edge, we compute the late-mode worst case delay time from the output pin of the latch (or primary input) to the input pin of the latch (or primary output) through the combinational logic network. For each vertex, we compute the times of the leading and trailing clock edges if the information is available.

From a latch graph, we can construct the overlay timing relationships with incoming latches $L(i)$'s and outgoing latches $L(j)$'s in each latch $L(x)$. Consider the latch graph in Figure 7(a), in which latch $L(b)$ has three incoming latches, $L(a)$, $L(d)$, $L(e)$, and two outgoing latches, $L(c)$ and $L(e)$. Let the clock cycle time be 10 ns. The overlay timing relationships in $L(b)$ are constructed as shown in Figure 7(b). We may label the edge of $L(a)$ to $L(b)$ to be *Safe*, as we mentioned previously. Since the trailing clock edge of $L(c)$ is the earliest time among all trailing clock edges of outgoing latch waveforms, we define a new restricted window as shown in Figure 7(b). We can not determine the status of the edges, $L(d)$ to $L(b)$

and $L(e)$ to $L(b)$, from the overlay timing relationships because their clock active-intervals overlap with the restricted window. It is possible that the trigger time (or latch time) may be set at any point of the overlap.

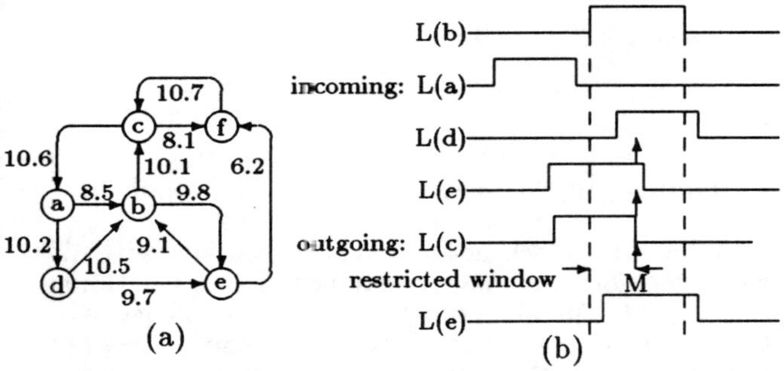

Figure 7: Latch graph and overlay timing diagram.

A backtracking method using the breadth-first search technique is applied to verify the overlay relationships through multiple latch stages. In the backtracking process, the restricted window information, i.e. the time M in Figure 7(b), propagates backward in the graph in order to verify the feasibility of cycle stealing. The backtracking process will verify one latch at a time. The latch under investigation, latch $L(b)$ in Figure 7, is called the root latch.

Consider the example in Figure 7 where a signal launched from $L(d)$ through $L(b)$ to $L(c)$ may be captured by $L(c)$, depending on the trigger time on $L(d)$. Figure 8 shows that the signal launched from the leading clock edge on $L(a)$ arrives at $L(d)$ during the clock active interval, and hence it arrives at $L(c)$ after the trailing clock edge on $L(c)$. The multiple stage cycle steal will be verified by the backtracking process.

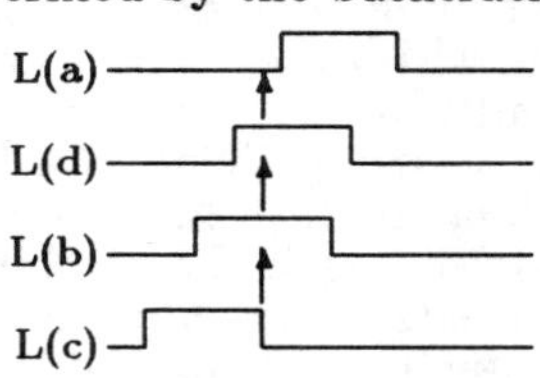

Figure 8: Multiple-stage cycle steal and backtracking process.

4 Feedback Loop Detection in Latch Graph

In the backtracking process, a feedback loop starts at a root latch $L(x)$, traverses the latch graph, and ends at $L(x)$. Every latch in the feedback loop is visited once but the root latch $L(x)$ is visited twice. The delays in some feedback loops cannot satisfy the timing requirements, and the backward propagation of the overlay timing information can detect these timing violations. Let us first examine the case of cycle steal in a single-stage feedback loop. Consider the latch graph in Figure 9(a). Assume the clock cycle time, clk, is known. If the delay time $d(x,x)$ is less than clk, the signal propagation through the feedback loop can be captured by $L(x)$ itself, as shown in Figure 9(b). Figure 9(c) shows the overlay timing relationship if $d(x,x)$ equals clk. Figure 9(d) illustrates the timing violation in which $d(x,x)$ is greater than clk. In Figure 9(d), the restricted window is initially set at

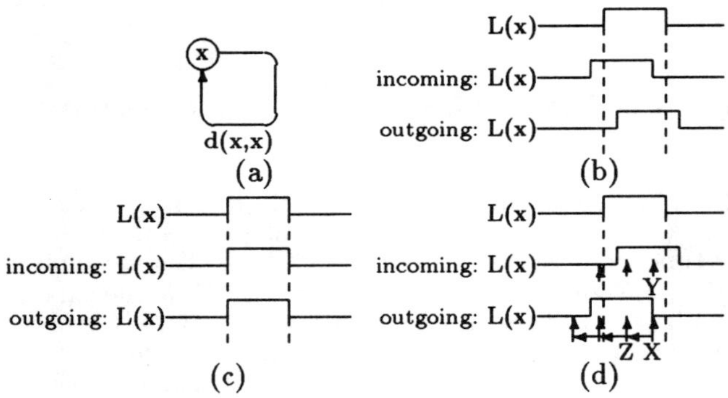

Figure 9: Overlay timing diagrams with a feedback loop.

the time X. The trigger time for the incoming latch $L(x)$ has to be set before the time Y. Because the incoming and outgoing latches actually are the same latch $L(x)$, the latch time for the outgoing latch $L(x)$ follows the time Y to be set on the time Z. We call the new setting on the restricted window 'feedback change'. Hence the restricted window becomes narrow, and it forces the trigger time of the incoming latch to occur earlier. After several iterations, the signal from the incoming latch will not arrive before the restricted window closes.

We may classify the timing violations into two types. The first type is the violation labeled *Failed*, as shown in Figure 5(b). The second type of timing violation is caused by signal propagation in a feedback loop. If the total signal propagation delay time in an N-stage feedback loop exceeds $N \times clk$, where clk stands for clock cycle time, it cannot satisfy the timing constraints.

Lemma 1 When the backtracking process detects a feedback loop, the total delay time in the feedback loop exceeds $N \times clk$ if a feedback change occurs in the overlay timing diagram of the root latch.

Proof: In the backtracking process, the restricted window limit, the time M in Figure 7(b) or the time X in Figure 9(d), is propagated backward through the feedback loop with one clock cycle time ahead through each latch stage. A feedback change implies that the $L(x)$ trigger time at N clock cycle time ahead must launch the signal earlier than the time of the window limit. For instance, the latch $L(x)$ in Figure 9(d) in the previous cycle must launch the signal at the time Z in order to let $L(x)$ capture the signal in the next cycle time. Hence the total delay time through a feedback loop is greater than $N \times clk$.

Lemma 2 For a feedback loop found in the backtracking process, a feedback change occurs in the root latch if the total delay time in the feedback loop exceeds $N \times clk$.

Proof: If the total delay time in a feedback loop is greater than $N \times clk$, the root latch $L(x)$ at N clock cycle ahead must set the trigger time earlier than the time of the window limit, such as the time Z in Figure 9(d). The backtracking calculates the trigger time after the N stage backward search and reports the feedback change.

Theorem 1 For a feedback loop detected by the backtracking process, the total delay time in the feedback loop exceeds $N \times clk$ if and only if a feedback change occurs.

It follows from the Lemmas 1 and 2.

The significance of the above theorem is as follows. The decrement of the feedback change, X to Z in Figure 9, may be very small so that the backtracking may theoretically explore the same feedback loop infinite times before the timing violation occurs. It is unnecessary to actually find the timing violation in which the window limit is set before the leading edge of the incoming clock waveform. In our algorithm, the backtracking process stops when a feedback change in a feedback loop is detected.

5 Cycle Stealing Based Timing Analysis Algorithm

The breadth first backtracking technique is applied to the algorithm. We defined *Safe* and *Failed* for the timing status of edges as mentioned. Here we define three other labels for them. If the signal propagation delay on an edge satisfies the timing requirement by stealing cycle time from the succeeding latch stages, we label it *CS*. If the signal delay can steal cycle time from one latch stage but not multiple latch stages, we label it *CSfailed*. If the timing violation is because of its feedback loop delay exceeding $N \times clk$, we label it *CSloop*. *CSloop*, *CSfailed* and *Failed* violate timing constraints.

Theorem 2 If neither the first type of the timing violations, *Failed* or *CSfailed*, nor the second type of the timing violations, *CSloop*, is found by the algorithm, no timing violation occurs.

Proof: Since both latches and primary outputs will be investigated by the algorithm once, all the timing violations will be explored and labeled. If there is a timing violation at latch $L(x)$, the timing violation will be detected when $L(x)$ is under investigation by the algorithm.

The backtracking process should explore all the feedback loops which are associated with the root latch, and discard the feedback loops which do not belong to the root latch. But in the backtracking process, it may be trapped in the feedback loops which do not go through the root latch, and result in tracing the same feedback loop infinite times. To avoid such a situation, each latch vertex x has a descendant list which records all latches on the paths from latch x to the root latch. If a new explored latch vertex i is in the descendant list of x, a feedback loop which does not belong to the root latch is found, and it is excluded from the search process.

```
Read in data file, initialize data array.
Construct overlay timing information.
Check each edge, label Safe and Failed if found.
Label other edges CS.
Initialize descendant list in each vertex latch.
For each latch ID, called root, do the following:
  Put root into FIFO queue.
  While the queue is not null, do the following:
      Get latch ID, called x, from the queue.
      For each incoming i of latch x, do the following:
         If latch i is not in the descendant list of x,
         do the following:
           Check if (i, x) is CSfailed,
                      update negative slack if necessary.
           If the window limit of i is changed,
             update the window limit and tracing info.
             If latch i is root, found feedback loop.
             Else put latch i in the queue.
                Maintain the descendant list of i.
      If found feedback loop
        report feedback loop information,
                update CSloop and slack information.
  Reset descendant lists.
Generate slack report.
```

The number of feedback loops is theoretically exponential to the size of the latch graph. In our current implementation, the program does not report all feedback loops in the design. Instead, it averages the negative slack at each latch stage in the feedback loop. For each edge, only the maximum negative slack is recorded.

6 Implementation and Results

The algorithm has been implemented on an IBM RISC System/6000 in C code. The timing analysis program consists of two steps. In the first step, timing information is gathered from the circuit design, and a latch graph is constructed. All circuit design data are retrieved from a design database, and timing information is computed by an IBM timing tool. The second step takes the data, performs the cycle stealing timing analysis and generates a slack report.

For the available test circuits, the experimental results are based on various clock cycles and widths of the clock active-interval. Table 1 shows that the number of failed paths decreases when the width of the clock active-interval increases. As the width of the clock active-interval increases, the number of paths which successfully steal cycle time increases. Although the experiments do not include the early mode timing violations, we can expect that the number of early mode violations increases as the width of the clock active-interval increases. The decision on the width of the clock active-interval needs to take into account

ckt name	# of latches	# of PI's/PO's	clock cycle (ns)	active interval (ns)	# of Safe	# of CS	# of CS loop	# of CS failed	# of Failed	1st step runtime (sec)	2nd step runtime (sec)
decred	10	46/6	20.0	1.0	262	101	0	42	113	6.0	0.38
				2.0	200	243	1	50	24	6.0	0.38
				4.0	178	275	1	50	14	6.0	0.37
bbbtest	53	11/37	25.0	1.0	1778	64	0	5	15	10.0	1.33
				2.0	1717	134	0	1	10	10.0	1.31
				4.0	1514	345	1	2	0	10.0	1.31
aa2gor	237	123/107	80.0	4.0	5669	81	0	0	64	59.0	4.75
				8.0	5641	145	0	0	28	59.0	4.21
				16.0	5580	234	0	0	0	60.0	4.22
sma	251	100/121	50.0	4.0	11643	488	0	3	677	114.0	9.23
				8.0	11022	1250	0	2	537	114.0	9.25
				16.0	9071	3309	0	2	429	113.0	9.42

Table 1: Experimental Results.

both early- and late-mode timing violations.

7 Conclusions

We have presented a new method to fully explore cycle steal opportunities in the timing analysis for level-sensitive synchronous circuit designs. In this paper, we use a single-phase clock discipline to explain the proposed analytic method to the cycle steal problem. It is true that two-phase overlapping clock designs can be analyzed in a similar way. For the application of multiple-phase clocking designs, it is understandable that the information of the combinational logic path delay between any two latches and the interval between the two phases can be reflected on the overlay timing relationships as we mentioned. The parameters, such as clock skew, hold time and setup time, can be represented in the clock waveform, and other parameters, such as clock delay and latch delay can be lumped with combinational delay in the latch graph. The timing analysis also considers the variation of clock width and leading/trailing edges in each latch, so the program is useful in the physical design environment. The scope of this paper does not cover the false path problem in which some critical paths in design are never sensitized by input patterns. The paper does not distinguish the types of signal transitions, such as rising, falling, stable and change. The accurate timing analysis comes from a latch graph construction which may take a higher order computation time than the one of slack computation in the block-oriented approach. This approach, however, is efficient for the dynamic change in circuit designs.

Acknowledgements

The authors would like to thank the referees for their valuable suggestions. The support from Donald R. Thompson, Robert F. Hatch, Cyril A. Price, and our colleagues is gratefully acknowledged.

References

[1] V. D. Agrawal, "Synchronous Path Analysis in MOS Circuit Simulator", Proc. 19th Design Automation Conf., pp. 629-635, 1982.

[2] J. K. Ousterhout, "A Switch-Level Timing Verifier for Digital MOS VLSI", IEEE Trans. Computer-Aided Design, vol. CAD-4, no. 3, pp. 336-349, July 1985.

[3] S. H. Unger and C. Tan, "Clocking Schemes for High-Speed Digital Systems", IEEE Trans. on Computers, vol. C-35, no. 10, pp. 880-895, October 1986.

[4] M. R. Dagenais and N. C. Rumin, " On the Calculation of Optimal Clocking Parameters in Synchronous Circuits with Level-Sensitive Latches", IEEE Trans. Computer-Aided Design, vol. 8, no. 3, pp. 268-278, March 1989.

[5] T. G. Szymanski, "LEADOUT: A Static Timing Analyzer for MOS Circuits", Proc. 1986 IEEE Int. Conf. on CAD, pp. 130-133, Nov. 1986.

[6] M. Glesner, J. Schuck, R. B. Steck, "SCAT - A New Statistical Timing Verifier In a Silicon Compiler System", Proc. 23rd Design Automation Conf., pp. 220-226, 1986.

[7] D. E. Wallace and C. H. Sequin, "ATV: An Abstract Timing Verifier", Proc 25th Design Automation Conf., pp. 154-159, 1988.

[8] J. J. Cherry, "Pearl: A CMOS Timing Analyzer", Proc. 25th Design Automation Conf., pp. 148-153, 1988.

[9] N. Weiner and A. Sangiovanni-Vincentelli, "Timing Analysis In A Logic Synthesis Environment", Proc. 26th Design Automation Conf., pp. 655-661, 1989.

[10] K. A. Sakallah, T. N. Mudge and O. A. Olukotun, "Analysis and Design of Latch-Controlled Synchronous Digital Circuits", Proc. 27th Design Automation Conf., pp. 111-117, 1990.

[11] D. Hathaway and J. Goetz, "Timing Analysis Using Slack Stealing", IBM Invention Disclosure, 1990.

[12] R. B. Hitchcock, "Timing Verification and the Timing Analysis Program", Proc. 19th Design Automation Conf., pp. 605-615, 1982.

[13] J. Beausang and A. Albicki, "A Method to Obtain An Optimal Clocking Scheme For A Digital System", Proc. Int. Conf. Computer Design: VLSI Comput., pp. 68-72, Oct. 1985.

Computing Optimal Clock Schedules

Thomas G. Szymanski
AT&T Bell Laboratories, Murray Hill, NJ 07974
Email: tgs@research.att.com

Abstract

Consider the problem of optimizing the parameters of a multiphase clock for a circuit containing both edge-triggered flipflops and level-sensitive latches. We demonstrate that recently proposed linear programming approaches to this problem require excessive computation time. We give an alternative method in which LP constraints are generated selectively, thus allowing fast solution. Various formulations of short path constraints are discussed, as are experimental results for large circuits.

1 Introduction

Consider a circuit containing edge-triggered flip-flops and level-sensitive latches, collectively referred to as *memory elements*, interconnected with combinational logic. The circuit is driven by a multiphase clock. We want to determine those values of the various clock parameters that minimize the overall clock period.

Many previous papers, including [1, 2, 3, 4, 5, 6], have demonstrated fast methods for verifying the correctness of clock schedules. Armed with such a verification algorithm, a user might then hunt for an optimum schedule by trial-and-error or multidimensional binary search. Trial-and-error has obvious disadvantages, and binary search is prohibitively expensive. If there are k clock parameters and we need to perform σ probes along each dimension, then σ^k verification runs would be needed.

A more elegant and systematic approach treats the problem as a mathematical optimization problem. Several papers, including [7, 1], have mentioned this possibility without pursuing it. More recently, Sakallah and his colleagues have written a series of papers [8, 9, 6] exploring and refining the approach.

In this paper, we describe our experiences using linear programming methods to optimize clock schedules. Initially, we were disappointed with the resulting computation times. Too many constraints were needed to model a circuit and the bulk of these constraints were superfluous. Eventually we discovered ways to determine just which constraints are actually necessary. A minimal such set of constraints can be generated quite rapidly; our implementation spends more time reading the circuit description than generating constraints. The generated constraints may then be solved directly, or combined with additional constraints arising at the system level and then used for timing analyses at higher levels in the design hierarchy. Our methods are built on those of [1, 9, 3].

2 The Circuit Model

Our model borrows heavily from [8]. We assume that a circuit is clocked by a linearly-ordered sequence of k clock events, $E_1, \ldots, E_k$. A *clock schedule* with *period* π consists of a vector of k real numbers $0 \leq t_1 \leq t_2 \ldots \leq t_k = \pi$. When the circuit is run according to the schedule, clock event E_i occurs at time $t_i + j\pi$ for every integer j. A clock schedule is said to be *feasible* if the circuit operates correctly under these conditions. Given a schedule $\langle t_1, \ldots, t_k \rangle$, denote the *separation* between clock events within the same cycle by $\phi_{i,j} = t_j - t_i$. Note that $\phi_{i,j}$ is nonpositive when $i \geq j$ and that $\phi_{i,j} = -\phi_{j,i}$.

Each memory element M has setup and hold delays which are denoted by $setup(M)$ and $hold(M)$ respectively. Each flip-flop F is triggered $skew(F)$ time units after its associated clock edge, denoted $\updownarrow F$. Each latch L is associated with two clock edges, an opening edge and a closing edge, denoted $\uparrow L$ and $\downarrow L$ respectively, which specify that portion of the cycle when the latch is transparent. The corresponding skews are denoted by $oskew(L)$ and $cskew(L)$. The combinational logic interconnecting the memory elements is assumed to be acyclic with $\Delta(M, M')$ and $\delta(M, M')$ denoting the maximum and minimum combinational delay between memory elements M and M'.

Next we must define the intended behavior of the circuit. Accordingly, we stipulate that each memory element M *assumes* exactly one value per clock cycle. This value is assumed at $\downarrow M$ for latches and $\updownarrow M$ for flipflops and is a combinational function of the values last assumed by those memory elements M' for which $\Delta(M', M)$ is defined. This provides us with a delay-independent simulation model; it is now the job of timing analysis to find those conditions under which the real circuit will have equivalent behavior.

Our clocking model differs somewhat from that of [8] in that we bind opening edges of latches to specific events in the linear clock order, whereas they allow opening edges to float free and solve for their position. In our environment, multi-phase clocks are usually implemented by distributing a few master clocks in quadrature from which specific local clocks are decoded. As such, one latch's closing edge quite naturally becomes another latch's opening edge. This practice is prevalent in precisely those leading edge CMOS designs which make most heavy use of level-sensitive latches. In such an environment, *all* clock events must be linearly ordered. In any case, it is easy to extend either model to accommodate the other's perspective.

Many circuits will require preprocessing to fit either

0738-100X/92 $3.00 © 1992 IEEE

model, especially circuits in which the clock signals themselves are generated by combinational logic. Latches that can open more than once per cycle, or that can produce multiple values during a single open period, must be detected and modeled as distinct events using symbolic simulation techniques, e.g., [4]. Qualified clocks must be detected, and appropriate setup and hold requirements fabricated to check stability of qualifying logic signals while the clock is active.

For the sake of clarity, we have omitted such complicating details as distinct rising and falling delays or different "flavors" of internal latch delay (e.g., different delays from clock to output, or data to output). Treatment of these details is straightforward. Circuits employing wave pipelining fall beyond the scope of our work.

3 Short Path Constraints

Two different formulations of short path or hold constraints were presented in [8]. Suppose that a_L and d_L respectively represent the earliest time when data arrives at, and departs from latch L. Omitting details of propagation times through latches and time zone translation, the more aggressive approach is given by

$$\begin{aligned} d_L &= \max(a_L, t_{\uparrow L}) \\ a_L &\geq hold(L) \\ a_L &= \min_{M \to L}(d_M + \delta(M, L)) \end{aligned}$$

and the more conservative approach by

$$\begin{aligned} d_L &= t_{\uparrow L} \\ a_L &\geq hold(L) \\ a_L &= \min_{M \to L}(d_M + \delta(M, L)) \end{aligned}$$

where $\min_{M \to L}$ ranges over all M for which $\delta(M, L)$ is defined. The aggressive formulation yields solutions in which the a_i's and d_i's can be larger than in the conservative formulation, and therefore less likely to violate hold constraints. In principle, the aggressive formulation might yield a solution with a shorter period, although this did not happen in any of the circuits we studied.

Unfortunately, there are common situations, such as a latch driven by a qualified clock signal, in which the aggressive formulation is incorrect as stated. A similar problem arises in circuits which permit the clock to be stopped between adjacent E_i's to save power. A "fix" is possible, but it involves modifying all constraints for signals that might be "in flight" during the pause. Another difficulty with the aggressive approach is that the set of feasible solutions for the equations might not be convex or even connected. Indeed, we can show that clock schedule optimization under aggressive hold constraints is NP-complete.

We therefore chose the conservative approach. Within this model, each pair of memory elements gives rise to a short path constraint of either the form $\phi_{i,j} \geq c$ with $i < j$ or $\phi_{j,i} \leq c$ otherwise. The constant c depends only on the minimum delay between the elements and the hold time requirement of the destination. Thus all the short path constraints may be reduced to roughly k^2 inequalities.

4 The Lower Bounds Graph

We can express lower bounds on the required separation between circuit events using a graph we call the *lower bound graph*, or *LBG*. The primary use of the graph is to encode long path or setup constraints, but we will also use it to express system level requirements imposed by the user, some of the short path constraints of §3, minimum pulse width requirements on clock lines, etc. The LBG contains one vertex e_i for each clock event E_i, and one vertex l_i for each level sensitive latch L_i. Vertices thus correspond to circuit events, specifically, the occurrence of a clock edge or the assumption of a latch value.

Each arc in the graph has a weight and a color. An arc weight has units of time and corresponds to a requirement that two events must be separated by at least that much time for the circuit to operate correctly. The color of an arc is either black or red, with black meaning that the connected events occur in the same (abstract) clock cycle, and red meaning they occur in adjacent cycles.

Arcs are placed in the LBG as described in Figure 1. The graph may contain several arcs between the same pair of vertices, but for computational purposes we can discard all but the heaviest such arc. Note that an LBG is strongly connected, and the subgraph consisting of just the black arcs is acyclic. Note further that although flipflops are not represented explicitly as vertices in the LPG, they are represented in the arc set.

Figure 2 shows a simple circuit and its corresponding LBG. Delays and skews are shown on the circuit's wires and we assume that all latches have a setup time of 1.5 time units. There are two events per clock cycle, E_1 when signal CK rises, and E_2 when it falls. Event E_1 causes latches A and C to open, and B to close, whereas E_2 does the opposite. Latch A assumes a new value at E_2 of each clock cycle, and latch B then uses this value to compute its next value at E_1 of the subsequent cycle. Accordingly, there is a red arc of weight 10 from a to b in the LBG. Because of clock skew and setup requirements, the data assumed by B at E_1 must arrive at B no more than 0.5 time units after E_1 occurs. We therefore place a black arc of weight -0.5 between b and e_1 in the LBG.

Now we need an additional definition and some notation. A *path* P is a sequence of alternating vertices and arcs, $v_0, a_0, v_1, a_1, \ldots, v_n$, $n \geq 0$, such that each a_i leads from vertex v_i to v_{i+1}. This is denoted by $P\colon v_0 \rightsquigarrow v_n$. Paths are allowed to pass through the same vertex more than once. The sum of the arc weights in P is denoted $w(P)$, and $r(P)$ is the number of red arcs in P.

5 Relevant Constraints

In this section we shall show that every path in a LBG gives rise to a constraint, more specifically a linear inequality, involving the variables of a clock schedule for the circuit. In order for the circuit to operate correctly, every constraint implied by every path must be satisfied by the clock schedule. Fortunately, only those paths that we call *relevant* matter for purposes of testing whether a schedule is feasible. The set of relevant paths turns out to be

Arc	Condition	Black?	Weight
$l_i \rightarrow l_j$	true	$\downarrow L_i < \downarrow L_j$	$\Delta(L_i, L_j)$
$e_i \rightarrow l_j$	$E_i = \updownarrow F$	$i < \downarrow L_j$	$\Delta(F, L_j) + skew(F)$
$e_i \rightarrow l_j$	$E_i = \uparrow L_j$	$i < \downarrow L_j$	$oskew(L_j)$
$l_i \rightarrow e_j$	$E_j = \updownarrow F$	$\downarrow L_i < j$	$\Delta(L_i, F) + setup(F) - skew(F)$
$l_i \rightarrow e_j$	$E_j = \downarrow L_i$	true	$setup(L_i) - cskew(L_i)$
$e_i \rightarrow e_j$	$E_i = \updownarrow F$ & $E_j = \updownarrow F'$	$i < j$	$\Delta(F, F') + skew(F) - skew(F') + setup(F)$
$e_i \rightarrow e_j$	true	$i < j$	any lower bound on time between E_i and next E_j

Figure 1: Lower Bounds Graph Construction Rules. The indicated arc is added to the graph whenever flipflops F and F' can be found that satisfy the specified condition. The arc's color and weight are then given by the two remaining columns.

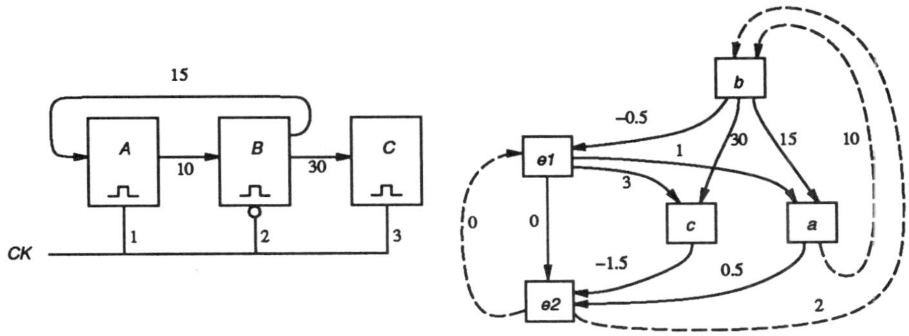

Figure 2: A circuit and its LBG. All setup times are 1.5. Red arcs shown dashed.

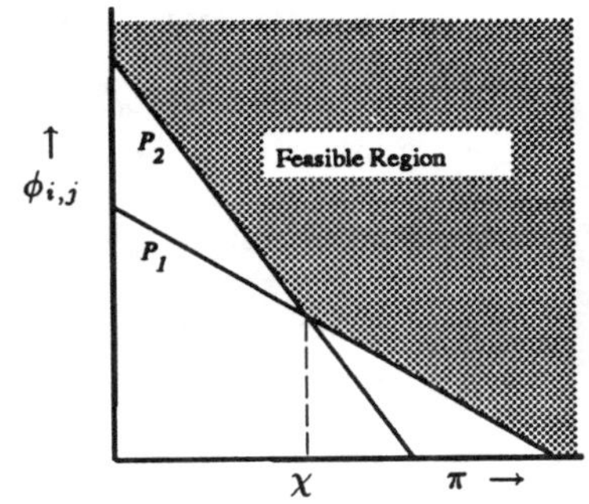

Figure 3: Two Constraints.

a function of the clock period, with more and more paths becoming relevant as the period decreases. However, the period can never decrease below the critical value ψ defined below. The rest of the section develops those properties of relevant paths that make them computationally useful.

Theorem 5.1 *Suppose that P: $e_i \leadsto e_j$ is a path in an LBG. Let S be a feasible clock schedule with period π. Then $w(P) \leq r(P)\pi + \phi_{i,j}$.*

We henceforth refer to the inequality $w(P) \leq r(P)\pi + \phi_{i,j}$ as the *constraint implied by path P*. For example, in Figure 2, the path from e_1 to a to b to c to e_2 has weight 39.5 and 1 red arc, so the corresponding constraint is $39.5 \leq \pi + \phi_{1,2}$. In linear programming terminology, such a constraint constitutes a bounding hyperplane on the set of feasible clock schedules. Figure 3 relates $\phi_{i,j}$ to π for the constraints implied by two paths, P_1 and P_2, with the same endpoints, e_i and e_j, but differing numbers of red arcs. For $\pi < \chi$, P_1 is less constraining than P_2, and the reverse is true for $\pi > \chi$. This notion that certain paths are most constraining for certain ranges of the clock period motivates the next definitions.

A path P is said to be *relevant for period λ* if for any other path P' with the same endpoints as P, $w(P') - r(P')\lambda < w(P) - r(P)\lambda$ if $r(P') < r(P)$ and $w(P') \leq w(P)$ if $r(P') = r(P)$. The set of all paths relevant for period λ will be denoted by $\mathcal{R}(\lambda)$ and the set of all paths in an LBG will be denoted by $\mathcal{P}$. Finally, define ψ to be the maximum of $w(C)/r(C)$ taken over all cycles C in the LBG.

Lemma 5.2 *If path P is relevant for period λ, so is every prefix of P.*

Theorem 5.3 *Let S be a clock schedule with period π. Then S satisfies all constraints implied by $\mathcal{P}$ if and only if S satisfies all constraints implied by $\mathcal{R}(\pi)$.*

Lemma 5.4 *$\mathcal{R}(\rho) \supseteq \mathcal{R}(\lambda)$ if $\rho \leq \lambda$.*

Theorem 5.5 *A clock schedule with period π is not feasible unless $\pi \geq \psi$.*

Corollary 5.6 *No path in $\mathcal{R}(\psi)$ contains a cycle.*

Theorem 5.7 *Given a schedule S with period π and any λ with $\lambda \leq \pi$. Then the following are equivalent:*
(1) S satisfies all constraints implied by $\mathcal{P}$,
(2) S satisfies all constraints implied by $\mathcal{R}(\pi)$,
(3) S satisfies all constraints implied by $\mathcal{R}(\lambda)$,
(4) S satisfies all constraints implied by $\mathcal{R}(\psi)$.

6 Algorithms

To test whether a proposed clock schedule S with period π is feasible, it suffices, according to Theorem 5.7, to check whether S satisfies all constraints implied by $\mathcal{R}(\lambda)$ for any $\lambda \leq \pi$. By Lemma 5.4, picking λ as large as possible will yield the smallest constraint set, so we should generate the constraints implied by $\mathcal{R}(\pi)$.

On the other hand, suppose that we are trying to find an optimum clock schedule for the circuit, possibly constrained by additional requirements arising from system level considerations. The set of constraints U implied by $\mathcal{R}(\psi)$ is ideal for this purpose because, again referring to Theorem 5.7, a clock schedule is feasible if and only if it satisfies U. Thus U represents *all* the timing constraints

```
subroutine update (m, w, r)
   if w − rλ > w_m − r_m λ then
      w_m ← w;    r_m ← r;
end subroutine;

for each i with 1 ≤ i ≤ k
   for each vertex n
      w_n ← −∞;    r_n ← 0;
   w_{e_i} ← 0;
   r ← 0;
   while r_n = r for at least one vertex n
      for each vertex n with r_n = r in topological
               order according to the black arcs
         prevw_n ← w_n;
         for each black arc a from vertex n to any m
            update (m, w_n + w(a), r);
      for each j with 1 ≤ j ≤ k and r_{e_j} = r
         emit constraint w_{e_j} ≤ rπ + φ_{i,j} using the
               current values of w_{e_j} and r;
      for each vertex n with r_n = r
         for each red arc a from vertex n to any m
            update (m, prevw_n + w(a), r+1);
      r ← r + 1;
```

Figure 4: Generate all constraints implied by $\mathcal{R}(\lambda)$.

implicit in the LBG. Typically U is a small set, as we shall see in §7.

6.1 Generating Relevant Constraints

As we have seen, constraints are implied by paths between the e_i's in the LBG. For each e_i, we use a single source, longest path algorithm to successively consider all relevant paths from e_i that contain exactly 0 red arcs, then 1 red arc, then 2 red arcs, etc. The correctness of this algorithm is based on Lemma 5.2 and the following assertion, which holds at the start of the for loop on the variable j, and which states the relationship amongst the program variables r, i, w_n and r_n:

> w_n and r_n give the weight and number of red arcs in that path for which $w(P) - r(P)\lambda$ is a maximum, taken over all paths P: $e_i \leadsto n$ with P relevant for λ, and $r(P) \leq r$.

Suppose that v and a are the number of vertices and arcs in the LBG. By Theorem 5.5, we can assume $\lambda \geq \psi$. Corollary 5.6 tells us the algorithm will terminate with $r \leq v$, and so the worst case running time is $O(kva)$. At most vk^2 constraints are generated.

6.2 Computing ψ

The problem of computing ψ is a variant of the minimum mean-weight cycle problem studied by Lawler [10] and Karp [11]. Karp's algorithm produces an exact answer in $\Theta(va)$ time, whereas Lawler's algorithm produces an answer with b bits of precision in $O(bva)$ time. Lawler's algorithm was better suited to our purposes because it could be augmented with heuristics to accelerate convergence.

```
subroutine checkforcycles (λ)
   for each vertex n
      d_n ← −∞;
      b_n ← a dummy arc from e_1 to n with weight −∞;
   d_{e_1} ← 0;
   for pass ← 0 to v
      if d_{e_1} > 0 then
         return "positive cycle through e_1";
      if no d_n changed during the last pass then
         return "no positive cycles";
      for each vertex n for which d_n did change
         for each arc a from vertex n to any m
            if a is black then
               t ← w(a) else t ← w(a) − λ;
            if d_n + t > d_m then
               d_m ← d_n + t;    b_m ← a;
      every so often
         examine the graph B whose arcs consist
               of {b_n | n is a vertex};
         if B contains a positive weight cycle then
            return "positive cycle";
   return "positive cycle exists";
end subroutine;

u ← large number;
l ← 0;
while u − l > ε
   ψ ← (u − l)/2;
   checkforcycles (ψ);
   if a positive cycle was found then
      l ← ψ (or something larger: see text);
   else
      u ← ψ (or something smaller: see text);
```

Figure 5: Algorithm to determine ψ.

Lemma 6.1 (Lawler) *Let G be a LBG and let $G(\lambda)$ be a directed graph constructed from G by subtracting λ from the weight of every red arc and leaving the black arcs unmodified. Then G contains a cycle C with $w(C)/r(C) > \lambda$ if and only if $G(\lambda)$ contains a positive weight cycle.*

It is easy to test whether a directed graph contains a cycle of positive weight using the Bellman-Ford algorithm. Calculating ψ then becomes a simple binary search for the maximum value of λ for which $G(\lambda)$ contains a positive weight cycle, as shown in Figure 5.

The subroutine checkforcycles computes the length of the longest path to each vertex in $G(\lambda)$ starting from a single source vertex. The code uses "selective tracing," that is, it relaxes just those arcs that leave a vertex whose distance increased during the previous pass. This helps because most of the d_n assume their ultimate values during the first few passes. More importantly, keeping track of the active vertices allows early detection of convergence.

We use e_1 as the source. Once per relaxation pass we check whether a positive weight path to e_1 has been discovered. This test is cheap, and often effective because the

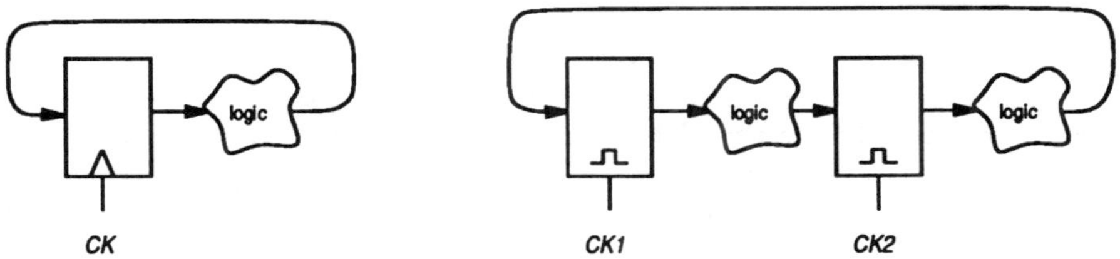

Figure 6: Bipartite transformation to convert edge-triggered circuit into two phase, level-sensitive circuit.

e_i's are so tightly coupled to the rest of the graph.

Every time a new longest path to a vertex n is discovered, we store in b_n the last arc on that path. Thus, the b_n together define a subgraph in which each vertex has indegree exactly one. It is easy to find and examine all cycles in this graph in $O(v)$ time using the following method. Associate the null tag with each vertex. Pick an untagged vertex m and trace back through the b_n's tagging vertices with m as you go. If you encounter a vertex already tagged with m, you have found a cycle and can compute its weight. Otherwise, pick some other untagged vertex and continue. If this process uncovers a positive weight cycle, then the subroutine can terminate immediately.

Whenever a positive weight cycle is detected by `checkforcycles`, we can use that cycle to accelerate the convergence of the surrounding binary search for ψ. In a normal binary search, detecting a cycle in $G(\lambda)$ would cause λ to become a new lower bound on ψ. However, if the cycle detected in $G(\lambda)$ had weight w with r red arcs, then a better lower bound on ψ is provided by $(w + r\lambda)/r$.

The movement of the upper bound on ψ can also be accelerated by noting the minimum decrease in λ that would be needed for the longest path computation to have a different outcome. More specifically, in the case where `checkforcycles` terminates without finding any positive cycles, the B graph is a spanning tree rooted at e_1 with one non-tree arc added, namely the dummy arc from e_1 to e_1. In this case we calculate the maximum value of λ for which the routine would have produced a different tree.

7 Experimental Results

We wrote three programs. The program `lps` optimizes clock schedules using the linear programming framework of [8]. The program `solve` uses the reduced constraints set approached introduced in this paper. The program `check` is a clock schedule verifier using relaxation, as described in [6]. This program was used to double check the other two. An option on `check` determines whether aggressive or conservative short path constraints are used.

7.1 The Data

Since flipflops essentially disappear during the construction of the LBG, we needed circuits with lots of level-sensitive latches to stress our techniques. The natural place to look was in full-custom designs. Although we had access to several such chips, they had all been captured in ways which were unsuitable for our needs.

transformed circuit	D	B	s38417	s38584
# of memory elements	512	1890	3272	2904
# of connections	20496	30122	67704	35956
cycletime using flipflops	39	26	94	112
cycletime using latches	23	17	63	70
time to read circuit	4.9	7.2	15.5	8.6
aggressive `lps` time	235	1822	7718	4880
conservative `lps` time	48	284	1432	761
`solve` time	1.4	2.1	6.0	4.1
# generated constraints	19	13	21	21

Figure 7: Experimental results.

We then turned to standard cell designs. Since we did not want to create software to perform the extensive preprocessing mentioned in §2 until we were sure that our algorithms were practical, we obtained some netlists for circuits built entirely using edge-triggered flipflops with a single clock. Two such circuits had been designed in house, and two were taken from the ISCAS'89 benchmark set. We used a unit delay model for the ISCAS circuits. We converted these designs to level-sensitive circuits using the transformation shown in Figure 6. The transformation consists of duplicating the circuit and cross-connecting the data inputs of just the memory elements. Thus if the output of circuit element x is input to element y, then x_i is an input to y_i unless y is a memory element, in which case x_i is an input to y_{1-i}. Each copy of the circuit is given its own clock and all flipflops are changed to latches. The characteristics of the test circuits are shown in Figure 7. Note that these circuits all run substantially faster after transformation, which indicates that the optimal clocking schedule is making good use of the transparent phase of the latches.

7.2 The Programs

The program `lps` takes a circuit description and a linear order for clock events and generates a linear program for the optimal clock schedule. An option specifies whether aggressive or conservative constraints should be generated. The max and min operators in the constraints are, of course, *relaxed* to linear inequalities in the process. We tried three different LP packages (two using the simplex algorithm and one using an interior point method) on the resulting linear programs. We soon learned that solving the dual rather than primal problem was orders of magnitude faster regardless of which solver was used. Although one solver was generally the fastest of the three, it was not uniformly so; on occasional problem instances it was as much as five times slower than one of the other solvers.

The program `solve` reads in a circuit description, a linear order for the clock events, and a set of user provided constraints. User constraints can be any linear inequalities involving the variables in the clock schedule, for example, "no interval between events is more than 25% of the period," or "certain intervals must be equal." `Solve` then constructs a set of short path constraints using the conservative criteria in §3. The LBG is then built incorporating any short path or user constraints that give lower bounds on the separation between clock events. Then the program calculates ψ and generates long path constraints as described in §6. The resulting constraints are combined with any user or short path constraints that give upper bounds on event separation, as well as a few constraints that are implicit in our formulation, such as $\pi \geq \psi$. This final constraint set is then passed on to an LP solver. The whole program is about 1300 lines of C.

7.3 Results

The performance of the algorithms is shown in Figure 7 using one processor on a Silicon Graphics 4D/440 processor. Both programs `lps` and `solve` use a common routine to read a circuit description into memory and build a minimal data structure, so we list the time for that task separately. We then show the time used by `lps` to solve the full set of constraints prescribed in [8] under both the aggressive and conservative formulations of hold constraints. The time shown is actually the best time over the three LP solvers we tried. Finally, we list the time used by `solve` to compute ψ, generate $\mathcal{R}(\psi)$, and solve the resulting LP, along with the size of $\mathcal{R}(\psi)$. Quite clearly, `solve` is orders of magnitude faster than `lps` and spends more of its time absorbing input than producing answers.

Another experiment was to determine how much the optimal clock period of a circuit was affected by the use of conservative rather than aggressive constraints. It has been shown[9] that relaxing the conservative constraints will always yield the optimum answer, but the relaxed aggressive constraints might produce an overly optimistic (i.e., wrong) answer. We used the `lps` program to generate and solve both sets of constraints for each of our test circuits, as well as bipartite-transformed versions of every circuit in the ISCAS'89 suite. In all but one case the answers were the same. The exception was s15850, for which the relaxed aggressive constraints yielded a period that was 0.3% faster than the optimal conservative schedule. The aggressive schedule, when verified with `check`, turned out to have several hold violations that exceeded this "improvement." So for our test data, the use of aggressive constraints did not yield any faster schedules than the conservative ones. They did, however, slow the LP solver down dramatically. For this and the other reasons listed in §3, we suggest that the conservative formulation is more useful in practice than the aggressive formulation.

Acknowledgments

David Gay was an invaluable source of help regarding both linear and non-linear programming. I used his AMPL mathematical programming language as a front end to the various solvers used in the experimental phase of this research. Rae McLellan taught me a lot about sophisticated clocking schemes.

References

[1] D. E. Wallace and C. H. Sequin, "ATV: An abstract timing verifier," in *Proceedings of the IEEE/ACM Design Automation Conference*, vol. 25, pp. 154–159, June 1988.

[2] N. Weiner and A. Sangiovanni-Vincentelli, "Timing analysis in a logic synthesis environment," in *Proceedings of the IEEE/ACM Design Automation Conference*, vol. 26, pp. 655–661, June 1989.

[3] A. T. Ishii and C. E. Leiserson, "A timing analysis of level-clocked circuitry," in *Proceedings of the Sixth MIT Conference* (W. J. Dally, ed.), Advanced Research in VLSI, pp. 57–69, MIT Press, 1990.

[4] T. G. Szymanski, "LEADOUT: A static timing analyzer for MOS circuits," in *Digest of Technical Papers of the IEEE International Conference on Computer-Aided Design*, pp. 130–133, Nov. 1986.

[5] J. J. Cherry, "Pearl: A CMOS timing analyzer," in *Proceedings of the IEEE/ACM Design Automation Conference*, vol. 25, pp. 148–153, June 1988.

[6] K. A. Sakallah, T. N. Mudge, and O. A. Olukotun, "$checkT_c$ and $mint_c$: Timing verification and optimal clocking of synchronous digital circuits," in *Digest of Technical Papers of the IEEE International Conference on Computer-Aided Design*, pp. 552–555, Nov. 1990.

[7] M. R. Dagenais and N. C. Rumin, "On the calculation of optimal clocking parameters in synchronous circuits with level-sensitive latches," *IEEE Trans. Computer-Aided Design*, vol. 8, pp. 268–278, Mar. 1989.

[8] K. A. Sakallah, T. N. Mudge, and O. A. Olukotun, "Analysis and design of latch-controlled synchronous digital circuits," in *Proceedings of the IEEE/ACM Design Automation Conference*, vol. 27, pp. 111–117, June 1990.

[9] K. A. Sakallah, T. N. Mudge, and O. A. Olukotun, "Algorithms for timing verification and optimal clocking of synchronous digital circuits," Tech. Rep. CSE-TR-71-90, University of Michigan, Ann Arbor, Sept. 1990.

[10] E. L. Lawler, *Combinatorial Optimization: Networks and Matroids*. Holt, Rinehart and Winston, 1976.

[11] R. M. Karp, "A characterization of the minimum cycle mean in a digraph," *Discrete Mathematics*, vol. 23, pp. 309–311, 1978.

On the Temporal Equivalence of Sequential Circuits

Narendra V. Shenoy Kanwar Jit Singh Robert K. Brayton Alberto L. Sangiovanni-Vincentelli
Department of EECS
University of California, Berkeley, CA-94720

Abstract

Performance driven synthesis of sequential circuits relies on techniques such as retiming and resynthesis [5] [6] [10]. These operations change the topological structure of the circuit, but should guarantee that the temporal behavior between the inputs and outputs of the circuit is preserved. In this paper, we consider properties that remain invariant under certain transformations. We derive constraints that permit a legal retiming in the case of multi-phase sequential circuits with edge triggered and/or transparent latches. For a particular design style we give an efficient procedure to check for temporal equivalence of sequential circuits.

1 Introduction

Circuit verification has received great attention because of its role in detecting design errors at an early stage. Functional (logic) verification of sequential circuits has been well investigated ([4] [2] [11]). However given two functionally equivalent circuits, they may not have identical temporal properties *e.g.*, the clock cycle often depends on the circuit implementation.

The verification of temporal properties of a sequential circuit is still in its infancy. Leiserson *et al.* ([5]) defined the behavior of a single phase edge-triggered circuit. They related the behavior to the number of memory elements in each cycle in the circuit. They also proved that it was necessary to preserve this number in order to preserve the behavior. Shenoy *et al.* ([10]) extended this concept to include single phase level-sensitive circuits. Sakallah *et al.* used the notion of correct behavior of a circuit for a given clocking scheme implicitly ([8]), as the existence of early (and late) arrival and departure times at each latch and showed that these times must satisfy clocking constraints for correct operation. In this paper we extend the abstract notion of temporal behavior to compare arbitrary circuits with arbitrary multi-phase clocking schemes.

We are concerned about the input/output behavior of circuits with respect to time. We assume the logical equivalence of two circuits (checked by existing methods) and focus primarily in the temporal behavior of the circuit. A multi-phase clocking scheme is [10] assumed. We define **temporality** to be the temporal behavior of the circuit. In the next section we provide a formal definition of this term. We prove conditions under which this property is preserved. In Section 2, we introduce the basic terminology. Section 3 deals with multi-phase edge-triggered systems. This provides a means of extending the conditions to multi-phase mixed systems (Section 4). Results of the implementation on a few circuits are given in Section 5 and Section 6 provides the conclusions.

2 Definitions

A single clock multi-phase circuit can be considered to consist of two kinds of elements, *functional* elements (gates) and globally clocked *memory* elements. It is modeled as a finite ,vertex-weighted, edge-weighted, directed multi- graph $G = (V, E, D, d, w)$. Every vertex represents a functional element(gate). The vertex-weights D_i, d_i are the maximum and minimum delays through the functional element at vertex i[1]. There is a directed edge e_{ij} from vertex v_i to vertex v_j if the output of the gate at v_i is an input the gate at v_j. Vertex v_i (v_j) is called the in-node (out-node) of the edge, denoted by IN(e_{ij}) (OUT(e_{ij})). The edge weight on edge e_i is referred to as w_i. A path $u \xrightarrow{p} v$ is a set of alternating vertices and edges $\{u, e_{u1}, v_1, e_{12}, \cdots e_{(n-1)v} v\}$, such that every pair of successive edges form an input-output pair to the vertex between. A path may have repeated edges.

Memory elements can be either edge-triggered or level-sensitive. The *retardation*([3]) at a level-sensitive latch is the amount of time by which the data input is delayed since the beginning of the active period. For correct operation of both memory elements, the data signal should be stable at the input before the latching edge occurs by an amount called the *set-up* time and should remain stable after the latching edge has occurred by an amount called the *hold* time. We assume, without loss of generality, that edge-triggered elements sample input data on the falling transition and the level-sensitive latches use the time when the phase is high as the active period. Thus the falling edge of each phase is the latching edge with respect to which set-up and hold constraints must be satisfied. Henceforth we shall assume that all circuits satisfy the set-up and hold requirements and that the circuit operates in the *fundamental mode*. This ensures that there are no "wave-pipelining" like effects([12]) in the circuit.

A special vertex in G, called the host vertex v_h, represents the environment. The host vertex has edges into all primary inputs and has edges from all primary outputs. The graph G is cyclic, not only with cycles through the host v_h, but also possibly internal cycles. The set of cycles internal to G, due to the circuit structure are called *internal cycles*. The set of cycles that contain the host vertex are called *external cycles*[see Figure 1]. We use the terms circuit and graph interchangeably.

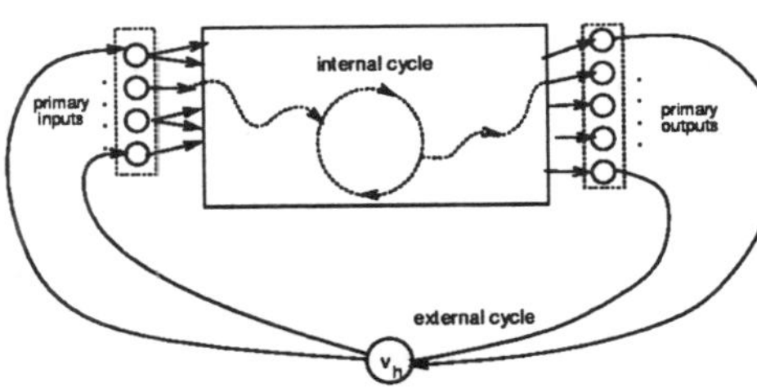

Figure 1: External and Internal Cycles

The clock $\hat{\phi}$ has k phases *i.e.* $\hat{\phi} = (\phi_1, \phi_2, \cdots, \phi_k)(|\hat{\phi}| = k)$. The edge weight w_{ij} is an integer, $w : E \rightarrow \{0, 1, \cdots, k\}$, which reflects the phase of the memory element (if any) on edge e_{ij}

$$w_{ij} = \begin{cases} 0 \text{ if there is no memory element on } e_{ij} \\ \text{phase of memory element on edge } e_{ij} \end{cases} \quad (1)$$

[1] We assume that wiring delay (delay through the edges) is insignificant or can be approximated as a function of the fanout of the vertex v_i and included in the vertex delay d_i.

If the output of a memory element is an input to another memory element, we introduce a dummy vertex with zero delay between the two. This forces every edge to have only one memory element. Let c denote the clock cycle. Associated with each phase ϕ_i are s_i, π_i: $0 \leq s_i, \pi_i \leq c$, the times at which the phase rises and falls respectively. We order the phases so that $\pi_1 \leq \pi_2 \cdots \leq \pi_k = c$. We say $\phi_i \leq \phi_j$ if $\pi_i \leq \pi_j$.

All memory elements (henceforth referred to as a latch) are assumed to be edge-triggered to begin with. We later relax this restriction to include mixed circuits.

Define the *temporality* of a vertex v_j along a path p from the host vertex, to be as follows:

$$\varphi(v_j, p) = \left\{ \begin{array}{l} \text{least number of clock phases after which signal} \\ \text{at output of } v_j \text{ is dependent on an input along} \\ p \text{ after the circuit has started computation} \end{array} \right. \quad (2)$$

We restrict the path to pass through the host vertex only once. Similarly define the *temporality* of an edge e_{ij} along a path p to be the least number of clock phases after which the signal on the edge (and at the output of the latch, if e_{ij} has a latch) is dependent on an input along the path p. We can now recursively define $\varphi(x, p), x \in V \bigcup E$ as follows.

$$\varphi(v_j, p) = \left\{ \begin{array}{ll} 0 & \text{if } v_j \text{ is a primary input} \\ \varphi(e_{ij}, p) & \text{otherwise, } v_i = \text{IN}(e_{ij}),\ v_i \in p. \end{array} \right. \quad (3)$$

Let $\rho(i, p)$ denote the phase of the last latch encountered along path p, when we reach vertex v_i.

$$\varphi(e_{ij}, p) = \left\{ \begin{array}{ll} \varphi(v_i, p) & \text{if } w_{ij} = 0 \\ \varphi(v_i, p) + w_{ij} - \rho(i, p) & \text{if } \rho(i, p) < w_{ij} \\ \varphi(v_i, p) + |\hat{\phi}| + w_{ij} - \rho(i, p) & \text{if } \rho(i, p) \geq w_{ij} \end{array} \right. \quad (4)$$

Intuitively, the *temporality* represents the latency of the latch or gate along the path p; assuming that all latches have been initialized correctly, $\varphi(x), x \in \{V \bigcup E\}$ is the number of phases after which x has a valid signal (dependent on the primary input in p). Consider the circuit in Figure 2. Let p be the path $\{c, e_{ce}, e, e_{ef}, f, e_{fg}, g, e_{go_1}, o_1\}$. Then

$$\begin{array}{rclrcl} \varphi(c, p) & = & 0 & \rho(f, p) & = & 2 \\ \rho(c, p) & = & 0 & \varphi(e_{fg}, p) & = & 3 \\ \varphi(e_{ce}, p) & = & 2 & \varphi(g, p) & = & 3 \\ \varphi(e, p) & = & 2 & \rho(g, p) & = & 3 \\ \rho(e, p) & = & 2 & \varphi(e_{go_1}) & = & 4 \\ \varphi(e_{ef}, p) & = & 2 & \varphi(o_1, p) & = & 4 \\ \varphi(f, p) & = & 2 & \rho(o_1, p) & = & 4 \end{array}$$

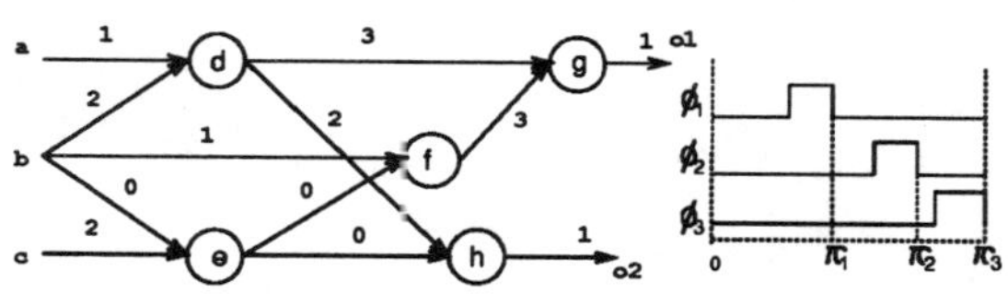

Figure 2: Temporality along a path

In short a signal from c takes 4 phases to reach the output. The *temporality* of a circuit $\varphi(G)$ is said to be preserved if the *temporality* of all outputs is preserved for all paths from primary inputs to primary outputs. Mathematically $\varphi(G)$ is a map from the set consisting of the product of primary outputs (PO) and valid paths to them ($\mathcal{P}$) to integers. A path is valid to a primary output if the primary output lies on the path.

$$\varphi(G) : PO \times \mathcal{P} \rightarrow \mathbf{Z} \quad (5)$$

The output of G, at any time can be interpreted as the result of the causal interaction of several paths in $\mathcal{P}$. The set $\mathcal{P}$ can be infinite (it is finite when we have a acyclic pipelined circuit).

3 Multi-Phase Edge Triggered Systems

For the case of sequential circuits with edge-triggered and level-sensitive latches we assume that gates are characterized by the maximum and minimum delays through them. Correct operation requires the phases to separated appropriately.

For any path p in G define the number of tokens $\tau(p)$ to be the number of clock cycles needed for computation along p. $\tau(p)$ can be computed by scanning the path p for successive latches, l_m followed by l_n with $\phi_{l_n} \leq \phi_{l_m}$. In other words, during the computation of $\varphi(v_i, p)$, $\tau(p)$ is the number of times we encounter the last case in the definition of $\varphi(v_i, p)$. The theorem given below provides the invariant that preserves temporality in the case of retiming. We use the subscript r to denote the variables after retiming.

Theorem: Let $r : G \longrightarrow G_r$ represent an operation of retiming. Assume all primary outputs are latched. The temporality of G and G_r is the same *if and only if* the number of tokens in each cycle is preserved and $r : G \longrightarrow G_r$.

Assumptions: The clocking scheme does not change under this operation and the primary output latches do not move.

Proof: Consider any primary output in G and the corresponding output in G_r, say P_o. Let P_o be latched on phase ϕ_j. Then we can write for any path p from v_h to P_o

$$\varphi(P_o, p) = \tau(p)|\hat{\phi}| + j \quad (6)$$

where $|\hat{\phi}|$ is the total number of phases.

$\Rightarrow$) (If part): Let the number of tokens be preserved along all cycles. Now any path from v_h to any primary output P_o can be seen as a part of a cycle in the graph. To obtain a cycle from such a path, we only need to add the edge from the primary output to v_h. Since the tokens along all cycles are conserved then clearly the value of $\varphi(P_o, p)$ remains the same for all primary outputs P_o along all paths p.

$\Leftarrow$) (Only if part): Assume there is a cycle C in which the number of tokens is not preserved then the temporality is not preserved. There arise two cases.

case 1: C is an *external cycle*. Without loss of generality assume that the number of tokens after retiming in a cycle C, denoted by $\tau_r(C)$ increases. Let P_o be the primary output on C. Let p_o be the path from v_h to P_o. Since

$$\tau(C) < \tau_r(C). \quad (7)$$

and the output of the latch has not changed during retiming, then

$$\varphi(P_o, p_o) < \varphi_r(P_o, p_o). \quad (8)$$

Hence temporality is not preserved.

case 2: C is an internal cycle. Since the presence of a clock implies periodicity, an internal cycle represents a set of infinite paths from at least one primary input to at least one primary output. Consider an internal cycle shown in Figure 3.

Let a be the vertex where a path $(p_{v_h,a})$ from v_h meets C. Let b be the vertex in C where a path (p_{b,v_h}) exists from b to a primary output. In case 1 we showed that tokens in all external cycles must be preserved. Consider the set of extended cycles C'_m obtained by traversing $p_{v_h,a}, p_{b,v_h}$ and m times through the cycle C.

$$C'_m = p_{v_h,a} \bigcup_{i=0}^{m} C \bigcup p_{a,b} \bigcup p_{b,v_h}. \quad (9)$$

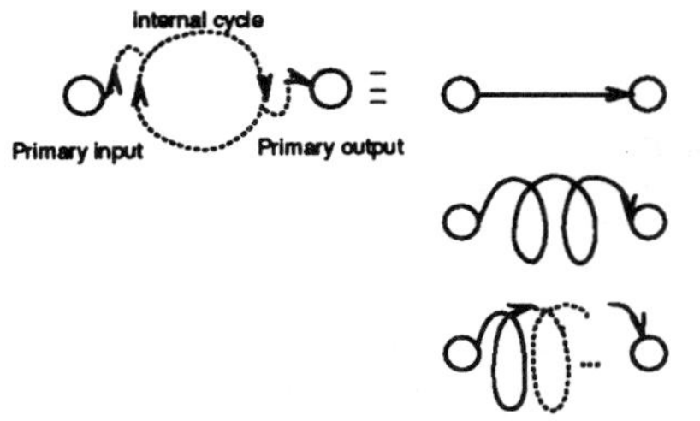

Figure 3: Unrolling an Internal Cycle

$C_m^{'}$ is an external cycle. From case 1 we get one constraint for every natural number m.

$$\tau(p_{v_h,a}) + \tau(p_{a,b}) + m\tau(C) + \tau(p_{b,v_h}) \quad = \quad \tau_r(p_{v_h,a}) \; \tau_r(p_{a,b}) + m\tau_r(C) + \tau_r(p_{b,v_h})$$

This is impossible since

$$\tau(C) < \tau_r(C). \square \tag{10}$$

Observation : Let $r : G \longrightarrow G_r$ be a retiming of G, where G is a single phase edge-triggered circuit. Leiserson's formulation of retiming is an integer-valued vertex labeling. A retiming is legal if after retiming no edge has a negative memory element count. Leiserson's [5] formulation is equivalent to preserving the number of latches (the number of tokens is the same as the number of latches for single phase edge-triggered circuits) in each cycle.

The temporality defined in the previous section depends on the number of phases in the clock set $\hat{\phi}$. Let us define the τ-temporality to be the temporality using the clock cycle c as a parameter *i.e.*

$$\varphi_\tau(x,p) = c\lfloor \frac{\varphi(x,p)}{|\hat{\phi}|} \rfloor + \pi_{(\varphi(x,p) \bmod |\hat{\phi}|)}, \qquad x \in V \bigcup E. \tag{11}$$

The tokens along a path to x from y can be related as

$$\tau(p) = \lfloor \frac{\varphi(x,p)}{|\hat{\phi}|} \rfloor, \text{where } y \xrightarrow{p} x. \tag{12}$$

Let $\pi_{(\varphi(x,p) \bmod |\hat{\phi}|)}$ be called the remainder $r(p)$. This is the fall time of the phase for the last latch encountered. The τ-temporality just gives the temporality in terms of time rather than the phase, *i.e.* a sum of a multiple of the clock cycle c and the phase of the last latch along the path.

We now introduce the concept of generalized retiming. In generalized retiming, we permit the clock set $\hat{\phi}$ to change. Look upon the τ-temporality as a composition of the maps from (x,p) to $\varphi(x,p)$, and from $\varphi(x,p)$ to the pair $(\tau(p), r(p))$. Let

$$\varphi_\tau(G) : PO \times \mathcal{P} \rightarrow (\tau(p), r(p)), \qquad p \in \mathcal{P}. \tag{13}$$

Define $R : (G, \hat{\phi}) \longrightarrow (G_R, \hat{\phi}_R)$ to be an generalized retiming if,

- $\forall P_o \in PO$, $w_{P_o v_h} \in \hat{\phi} \bigcap \hat{\phi}_R$,
- the corresponding outputs in G and G_R are latched by elements in the same phase,
- τ-temporality is preserved along all paths from the host vertex to all primary-outputs.

Note, in an generalized retiming we cannot relate to the temporality defined earlier in this section since the number of phases of the clock set has changed. However we can still make use the concept of τ-temporality. If the output latches remain fixed this implies that $\lfloor \frac{\phi(P_o,p)}{|\hat{\phi}|} \rfloor$ is the same. Note that this is nothing but $\tau(p)$. In theorem 1 above there was no reference to the clocking scheme, except in relating the temporality of the circuit to the tokens (equation 6). If we replace $\varphi(P_o, p)$ by $\varphi_\tau(P_o, p)$ and carry out the argument, we see that tokens along all cycles need to be preserved for generalized retiming. We get the following lemma:

Lemma: Let $R : G \longrightarrow G_R$ represent an generalized retiming. Assume all primary outputs are latched. The τ-temporality of G and G_R is the same *if and only if* the number of tokens in each cycle is preserved. □

Hence generalized retiming in the multi-phase scenario, implies not only shifting latches but also assigning phases to latches, possibly extending the clock set if necessary. However almost all designers fix the clocking scheme *a priori* to the design and rarely tolerate any deviations. The main reason for introducing the concept of generalized retiming is that it greatly facilitates the extension to mixed circuits.

4 Multi-Phase Mixed Systems

We now extend the algorithm to circuits that have both edge-triggered and level-sensitive memory elements. For every circuit G with mixed memory elements and a clock set $\hat{\phi}$, we shall show that there exists a temporally equivalent circuit G_{eq} with the same topology, and an extended clock set $\hat{\phi}_{eq}$. G_{eq} consists of edge-triggered elements only.

Let $retard(l_i)$ denote the retardation at the latch on edge e_i [3]. If the signal arrives before the active phase, $retard(l_i)$ is equal to zero. For all edge-triggered elements the retardation is zero. There are well known algorithms to compute the retardation at a latch ([9] [8]).

We now describe the transformation (γ) to an edge-triggered system without changing the temporality. Replace every level-sensitive element in the circuit by edge-triggered ones. For each latch l_i with retardation greater than zero, add a phase $\phi^{'}_{l_i}$ to the clock set and use it to latch element l_i.

$$\pi_{\phi^{'}_{l_i}} = (\pi_{\phi_{l_i}} + retard(l_i)) \bmod c. \tag{14}$$

A phase which is not present in the original clock set but has been added to it due to the procedure above is called a pseudo-phase. We could potentially end up with a clock set with as many phases as the number of latches. In the above process the circuit behavior has not changed. This implies that the τ-temporality is preserved.

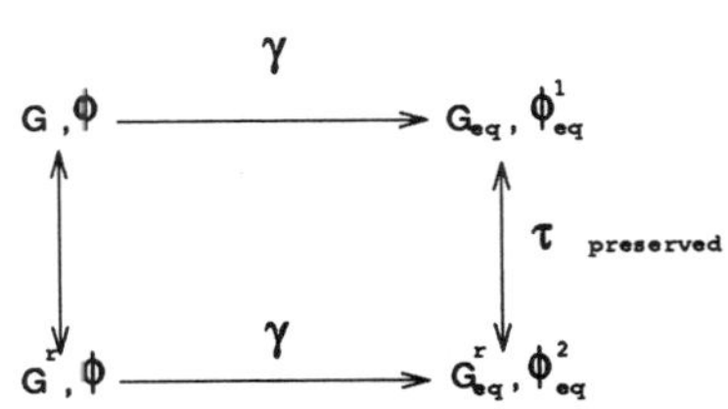

Figure 4: Equivalent Behavior

We have the commutative diagram shown in figure 4. Note that the transformation γ may result in different clock sets, $\hat{\phi}^1_{eq}$

and $\hat{\phi}^2_{eq}$, but the counting of tokens is independent of the sets as long as $w_{P_o v_h} \in \hat{\phi}^1_{eq} \bigcap \hat{\phi}^2_{eq}, \forall P_o \in PO$. We know G^r_{eq} is an extended retiming of G_{eq}. As a result we need to conserve the tokens in all cycles of the circuit with mixed memory elements.

4.1 Temporal Verification

The problem posed is, given two circuits, $G_1(V_1, E_1, d_1, w_1)$ and $G_2(V_2, E_2, d_2, w_2)$ with the same clock set $\hat{\phi}$ verify if one is a legal retiming of the other. First we need to verify that the two circuits have the same graph structure (modulo the dummy vertices and edges introduced). We assume that the primary inputs and primary outputs have an existing correspondence, *i.e.* the same names. This is an example of directed graph isomorphism. We restrict the retiming algorithm to preserve the gate names. Once an isomorphism has been established, we proceed to check the τ-temporality. Since the graphs are actually multi-graphs we need to take into account the isomorphism of edges when two vertices have more than one edge between them. In such a case the order of latches on the two edges will be different. After a retiming this difference must still be present [2]. So when there is more than one edge connecting two vertices, we use the order of latches on the edges to distinguish them.

The second problem is to verify that τ-temporality is maintained. The number of cycles in a circuit could be exponential in the size of the circuit. Hence we face the task of examining all cycles in a graph. However under a restriction on latch placement, we can reduce the problem to a linearly independent cycle set. Assume all primary outputs are latched on the last phase ϕ_k and all primary inputs are latched in the first phase ϕ_1. Let p, $u \xrightarrow{p} v : u, v \in V_1(\text{and } V_2)$ be any path in the graph G_1 (isomorphic to G_2). The restriction requires that the phases of the latches (in the order of traversal along p) appear in the same order as a cyclic ordering of the clock phases. Under this assumption, if there are τ tokens in a cycle, then there are τk latches in that cycle. Some of the popular design styles fall in this category.

Proposition 1: Under the restriction described above, the number of tokens in any cycle is preserved *if and only if* the number of latches in all cycles is preserved.

The number of latches in a cycle is a property dependent on the cycle edges. Let C_{ce} represent the cycle edge incidence matrix, *i.e.* $C_{c_i e_j} = 1$ if edge e_j is in cycle c_i. Let C^*_{ce} represent a linearly independent set of cycles. Any other cycle say C_0 with w_0 latches can be written as $C_0 = \alpha^T C^*_{ce}$ $\alpha \in R^p$, where p is the rank of C_{ce} (= rank of C^*_{ce}). It is easy to see $w_0 = \alpha^T w^*$, where w^* is a column vector whose i^{th} entry is the number of latches in the cycle $C^*_{c_i}$. Hence if w^* is preserved across retiming, the value of w_0 will also be preserved. Intuitively we are finding the largest submatrix of non-zero rank in the matrix $[C_{ce}|w]$, where w_i is the number of latches in C_{c_i}.

5 Implementation

The algorithm has been implemented in SIS, the sequential synthesis system at Berkeley. This provides a means of formally verifying the results of optimization algorithms (like retiming) once they are designed and implemented. The pseudo-code is provided below.

```
G1, G2 = circuits to be verified.
C = cycle set
timing_analyze(G1)
timing_analyze(G2)
check_graph_isomorphism(G1, G2)
C = enumerate_cycles(G1, G2)
foreach cycle in C {
  check tokens
}
```

[2] If this difference is not maintained across retiming then it is easy to show that the τ-temporality of at least one path is not preserved

Note that if we are enumerating all cycles, we check the token count as each cycle is found. So we never have to store all the cycles at any time. Often under restrictive retiming, the "all cycles" problem can be reduced to a set of linearly independent cycles called the "fundamental cycle" set. The latter can be enumerated in polynomial time (complexity of $O(m^2 \log(n))$). The algorithm is described (without proof) in the appendix. In this case, the temporal verification problem of a retimed circuit with the original circuit has the complexity of the timing verification problem. We present some results on checking if two circuits are equivalent in the table(1). Examples 1 and 2 are taken from [7], 3 and 4 are modified ISCAS benchmarks and 5 is a simple biquadratic filter. The third column specifies the nature of the clocking scheme. The L or E in brackets gives the type of latches (level-sensitive or edge-triggered). The time reported in column four is the total time taken by the algorithm on a DEC 5000, *i.e.* the time for the graph isomorphism check, the time for doing timing analysis on both circuits, and the time taken for the cycle enumeration. The last column gives the number of cycles in each circuit graph. The numbers in brackets in the fourth column are obtained by using the algorithm described in the appendix to generate a fundamental cycle set. The size of the fundamental cycle set is shown in brackets in the fifth column. The main advantage of the algorithm given in the appendix is only when we need to have the fundamental cycle set explicitly. For the verification problem there is no need to carry the set of cycles throughout the algorithm. Consequently the brute force enumeration is comparable with the algorithm in the appendix for moderately sized circuits.

no.	*name*	*Clocking scheme*	*time (s)*	# cycles
1	K1	2 phase (L)	0.2	2
2	K2	2 phase (L)	0.1	1
3	s27.blif	1 phase (L)	0.1 (0.1)	28 (10)
4	s208.blif	1 phase (E)	1.7 (1.3)	10676 (73)
5	filter	2 phase (L)	0.1	5

Table 1: Table of Results

6 Conclusion

To summarize, we have provided a model and a formal definition for the temporal behavior of a arbitrary multi-phase circuit. We have also described an algorithm to do formal verification of the temporal behavior of circuits.

The authors thank A. Wang for several useful discussions and an anonymous referee for several constructive comments. This research was supported by NSF under contract number EMC-8419744 and DARPA under contract number JFB-190-073. Their support is gratefully acknowledged.

References

[1] N. Balabanian and T. A. Bickart. *Electrical network theory*. John-Wiley and Sons, Inc., 1969.

[2] O. Coudert, J. C. Madre, and J. P. Billon. Automating Diagnosis and the Rectification of Design Errors with PRIAM. In *Proceedings of the International Conference on Computer-Aided Design*, pages 30–33. IEEE, 1989.

[3] M. R. Dagenais and N. C. Rumin. Automatic Determination of Optimal Clocking Parameters in MOS VLSI Circuits. In *Advanced Research in VLSI:Proc. of the 5th MIT Conference*, pages 19–33, 1988.

[4] A. Ghosh, S. Devadas, and A. R. Newton. Verification of Interacting Sequential Circuits. In *Proceedings of the Design Automation Conference*, pages 213–219. IEEE/ACM, June 1990.

[5] C. E. Leiserson and J. B. Saxe. Optimizing Synchronous Systems. In *Journal of VLSI and Computer Systems*, pages 41–67, 1983.

[6] S. Malik, K. J. Singh, R. K. Brayton, and A. Sangiovanni-Vincentelli. Performance optimization of pipelined circuits. In *Proceedings of the International Conference on Computer-Aided Design*, pages 410–413. IEEE, 1990.

[7] K. Sakallah, T. Mudge, and O Olukotun. Analysis and Design of Latch-Controlled Synchronous Circuits. In *Proceedings of the Design Automation Conference*, pages 111–117. IEEE/ACM, 1990.

[8] K. Sakallah, T. N. Mudge, and O. A. Olukotun. Timing Verification and Optimal Clocking of Synchronous Digital Circuits. In *Proceedings of the International Conference on Computer-Aided Design*, pages 552–555. IEEE, 1990.

[9] N. Shenoy, R. K. Brayton, and A. Sangiovanni-Vincentelli. A Pseudo-Polynomial Algorithm for Verification of Clocking Schemes. Submitted for review.

[10] N. Shenoy, R. K. Brayton, and A. Sangiovanni-Vincentelli. Retiming of Circuits with Single Phase Transparent Latches. In *Proceedings of the International Conference on Computer Design*, pages 86–89, 1991.

[11] H. Touati, H. Savoj, B. Lin, R. K. Brayton, and A. Sangiovanni-Vincentelli. Implicit State Enumeration of Finite State Machines using BDD's. In *Proceedings of the International Conference on Computer-Aided Design*, pages 130–133. IEEE, 1990.

[12] D. Wong, G. De Micheli, and M. Flynn. Inserting Active Delay Elements to Achieve Wave Pipelining. In *Proceedings of the International Conference on Computer-Aided Design*, pages 270–273. IEEE, 1989.

Appendix: Computing the Fundamental Cycle Set

We present here, an algorithm to compute the fundamental cycle set in a strongly connected directed multi-graph. The circuit graph described in the paper often satisfies this requirement. If it is not we can break it into strongly connected components and verify the behavior of each component. Let the graph be represented by $G(V, E)$, where V is the set of vertices and E the set of edges. Let the number of vertices be n $(= |V|)$, and the number of edges be m $(= |E|)$. We use the terms node and vertex interchangeably. The terms fanin and fanout are used as defined in Section 2.

We proceed as follows. First we shall derive an upper bound on the cardinality of the fundamental cycle set. Then the algorithm will be outlined and we shall show that the upper bound is actually achieved. Finally we shall analyze the complexity of the algorithm

Let C represent the cycle-edge incidence matrix.

$$c_{ij} = \begin{cases} 1 & \text{if edge } e_j \text{ is in cycle } C_i \\ 0 & \text{otherwise.} \end{cases} \qquad (15)$$

We shall show that the rank of C is $m - n + 1$. Readers familiar with network theory will recognize the correspondence of this proof with the technique for generating a linearly independent set of Kirchoff Voltage Law equations in a circuit [1]. Let the node incidence matrix A, be defined as

$$a_{ij} = \begin{cases} +1 & \text{vertex } v_i \text{ is the source of edge } e_j \\ -1 & \text{vertex } v_i \text{ is the sink of edge } e_j \\ 0 & \text{otherwise.} \end{cases} \qquad (16)$$

Each column of A has only exactly one +1 entry and one -1 entry. The rank of A, denoted by $R(A)$ is $n - 1$ (adding all the rows to the last row we get a zero row vector).

Consider the matrix product AC^T. It is easy to show that this product is identically zero.

By Sylvester's Law we have $R(A) + R(C) - m <= 0$. But $R(A) = n - 1$. Hence we conclude $R(C) \leq m - n + 1$. The rank of C is the number of linearly independent cycles.

Algorithm

The algorithm uses a depth first search (DFS) on a directed graph. The host vertex as the starting vertex. The edges are classified as

1. Tree edge: Edge that is a part of the DFS tree.
2. Back edge: Edge that is directed from the current node during DFS to a node which has been already visited and this node is still on the DFS stack.
3. Cross edge: Edge that is directed from the current node during DFS to a node which has been already visited and this node is no longer on the DFS stack.

```
n = number of nodes
m = number of edges
sink(e) = sink vertex of edge e
source(e) = source vertex of edge e
S = cycle set
T, L, B, X = sets of edges
i)    depth_first_search(G)
      T = tree edge set
      B = back edge set
      X = cross edge list (in order in
      which they are obtained during DFS)

      L = tree edge set
ii)   foreach back edge e in B do{
        find path (p) from source(e) to
          sink(e) using edges in L
        cycle = {b} U {p}
        add cycle to S
      }

iii)  L = L + B
      for(i=0; i<num_cross_edges; i++){
        find path (p1) from host vertex
          to source(X(i)) using edges in L
        find path (p2) from sink(X(i))
          to host vertex using edges in L
        cycle = {p1} U {X(i)} U {p2}
        add cycle to S
        L = L U {X(i)}
      }
```

It is easy to show that $|S|$ is $m-n+1$ and S is linearly independent.

Part (i) of the algorithm takes $O(m)$ time. Each path is found by doing a shortest path algorithm on the graph consisting of edges in L. This has a complexity of $O(m \log(n))$ using a priority queue. Hence parts (ii) and (iii) have a total complexity of $O(m^2 \log(n))$.

An Approach to Symbolic Timing Verification

Tod Amon and Gaetano Borriello

Department of Computer Science and Engineering
University of Washington
Seattle, WA 98195

Abstract

Symbolic timing verification is a powerful extension to traditional constraint checking that allows delays and constraints to be expressed as symbolic variables. In this paper, we present an approach to symbolic timing verification using constraint logic programming techniques. The techniques are quite powerful in that they yield not only simple bounds on delays but also relate the delays in linear inequalities so that tradeoffs are apparent. We model circuits as communicating processes and our current implementation can verify a large class of mixed synchronous and asynchronous specifications. The utility of the approach is illustrated with some examples.

I. Introduction

Symbolic timing verification can be used to answer many questions about a design specification and implementation. When circuit delays are known, a symbolic verifier can provide a "yes/no" answer as to whether or not a particular implementation meets all the timing constraints in the specification (the fundamental question for non-symbolic verifiers).

More generally, however, using variables for delays can provide an answer about the range of values assignable to that variable that will still meet the constraints. This can be valuable information for a synthesis tool that must decide how to implement that particular function. When many delays are represented by variables, the answer may be a set of linear inequalities constraining the variables. These relations provide synthesis tools with information about tradeoffs between circuit delays and how implementation choices for circuit functions may affect each other.

The utility of symbolic timing verification can be further extended if we consider symbolic timing constraints. In this case, the verification tool serves as an analysis tool that can determine how circuit delays relate to the symbolic constraints. For example, if we wish to determine the maximum throughput of a circuit, a variable can be used to express a throughput constraint and the verifier can determine its range of values given circuit delays and other constraints.

It is these uses of symbolic timing verification, determining delay flexibility in synthesis and how it affects the resulting circuit architecture, that motivates this work.

II. Verification model

We model our behavioral specifications using a bipartite directed graph whose two node types are: events, which serve as reference points for timing relationships; and operations, which encapsulate different ways to interconnect events. Events can represent changes in logic level on circuit wires (from a structural perspective) or data and control flow in a hardware description language (from a behavioral perspective). Operations represent circuit functionality by taking events as inputs and determining which event outputs to generate. Events and operations are connected in a bipartite graph to form a single process loop with one thread of control. The representation is formally based on a restricted form of *operation/event graphs* (OEgraphs) [1, 2].

Timing constraints are expressed using the notation: "constraint(*event1*, *event2*, {≤, ≥, =}, {*integer*, *integer variable*})" and specify minimum or maximum separation times between two events. Every timing constraint can alternatively be expressed as an inequality (or a conjunction/disjunction of inequalities) involving operations' propagation delays (expressed as values or variables). We present an example of some timing constraints and their equivalent inequalities in Figure 2.

III. Verification methodology

Verification could be thought of as a two-step process: translate each timing constraint into a non-linear inequality and then simultaneously solve the set of inequalities to find a solution. Our methodology combines these two steps into a single integrated approach.

We use a constraint transformation technique which reduces timing constraints to other "simpler" constraints and zero or more additional linear inequalities. These inequalities are immediately added to the solution using an incremental linear programming algorithm. If the set of inequalities is not solvable (e.g., $X > 10$, $Y > X$, $Y < 5$) this is detected and verification fails. The transformation strategy essentially consists of a list of rules which specify how to decompose constraints.

To handle delay operations, only two rules are needed. The first rule for "constraint(e1, e2, ≤, X)" applies when e1 is directly connected to e2 via a delay operation, and requires that δ_{max}e1e2 be less than or equal to X. The second rule applies when e1 is not directly connected to e2

0738-100X/92 $3.00 © 1992 IEEE

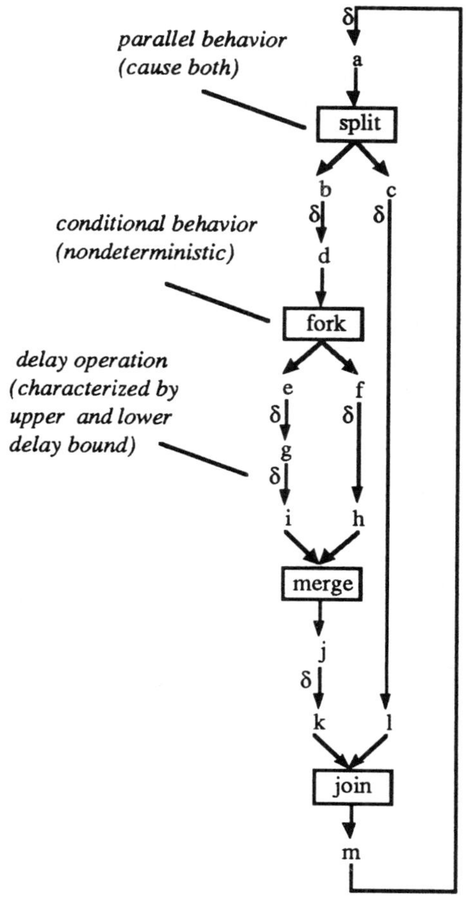

Simple constraints across delay operations:
constraint(e, i, ≤, 20) ≡ $\delta_{max}eg + \delta_{max}gi \leq 20$
constraint(e, i, ≥, X) ≡ $\delta_{min}eg + \delta_{min}gi \geq X$

A constraint across a fork operation:
constraint(b, k, ≥, 15) ≡ $\delta_{min}bd + \min(\delta_{min}eg + \delta_{min}gi, \delta_{min}fh) + \delta_{min}jk \geq 15$
In this case, a minimum is necessary since the constraint applies to both ways that j could occur relative to d.

A constraint across a split operation:
constraint(a, m, ≥, 10) ≡ $\max(\delta_{min}cl, (\delta_{min}bd + \min(\delta_{min}eg + \delta_{min}gi, \delta_{min}fh) + \delta_{min}jk)) \geq 10$
In this case a maximum is used since m will not occur until both k and l occur and thus only one path must be at least 10.

Figure 1. Several constraints and their equivalent inequalities for the process shown above.

but some other event e3 is. The inequality "$\delta_{max}e3e2 + T \leq X$" is added to the solution and "constraint(e1, e3, ≤, T)" must be proven.

Now consider "constraint(e, i, ≤, 20)" with respect to the process shown in Figure 2, i.e., e→g→i. We see that it is trivially solved using these two rules. The second rule is applied first and results in the inequality $T + \delta_{max}gi \leq 20$ and the new constraint "constraint(e, g, ≤, T)". This new constraint is handled by the first rule and results in $\delta_{max}eg \leq T$. If the actual values of $\delta_{max}eg$ and $\delta_{max}gi$ are known, we can report "yes" or "no", otherwise we report our answer symbolically by removing internal variable T (i.e., $\delta_{max}eg \leq T \leq 20 - \delta_{max}gi$) yielding: $\delta_{max}eg + \delta_{max}gi \leq 20$.

The rules and verification methodology have been implemented in a CLP(R) program. CLP(R) is a constraint logic programming language which provides a framework to experiment with this approach to symbolic timing verification ([5] contains an excellent summary of these languages). CLP(R) incorporates an incremental version of the standard *simplex* linear programming algorithm and a goal based programming language akin to Prolog. Its use has enabled us to focus our attention on the verification rules instead of the symbolic linear programming, unification, and search/backtracking aspects of the methodology. It is possible that a more efficient version of the verifier could be constructed but the version of CLP(R) we are using [6] has proven quite adequate.

The verification tool takes as input a description of the process and a list of constraints to be verified. Additional inequalities can also be supplied to further constrain the variables. The verifier currently consists of over 2500 lines of CLP(R) code implementing well over one hundred transformation rules. A large number of rules are needed to account for: the quality of the verification tool (i.e., its ability to detect and correctly process complex behaviors), the different types of operations (and the resultant behaviors which they allow), and the different types of constraints and their semantics (we have simplified the constraint representation for this paper).

Most of the rules are much more complicated than the simple rules used for delay operations and there are a number of subtleties that space constraints prohibit us from describing in detail. Some of the complexities arise due to the formal semantics of our constraints (e.g., what if e1 and e2 occur at the same time?), others arise due to the inherent complexity of the constraint with respect to the functional specification (e.g., a constraint from an event in one side of a split to an event on the other side, or constraints into and out of loops).

In some cases, multiple rules apply for a given situation. This is the case for "constraint(e1, e2, ≥, X)" where e2 is the output of a join. Every join serves to bring together two parallel paths divided by a matching split operation. The choice of transformation rules is very dependent upon whether or not e1 is an event inside the split (present in either the left or right parallel branch) or outside the split. For "constraint(e1, e2, ≥, X)" if e2 joins c1 and c2 and a matching split on e3 causes p1 and p2, and if e1 is outside the split/join then:

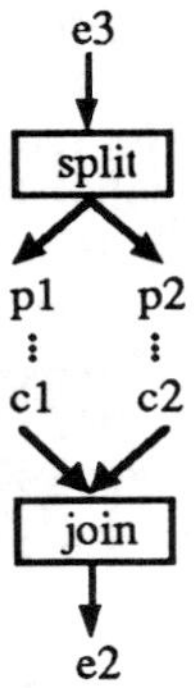

prove: constraint(e1, e3, ≥, T1),
and either:
  add the new inequality T1 + T2 ≥ X,
  prove: constraint(p1, c1, ≥, T2).
or, alternatively:
  add the new inequality T1 + T3 ≥ X,
  prove: constraint(p2, c2, ≥, T3).

Since e2 will not occur until both c1 and c2 occur, a minimum separation time may be ensured even if only one path actually satisfies the constraint. This represents an implicit choice for the verifier: it must decide which path to try and verify. If it succeeds, it does not need to look at the other path. If it fails, it must backtrack to this point and try the other.

The constraint could be expressed as an inequality involving a maximum function (e.g., see Figure 1). The verification problem is thus essentially nonlinear, although it can be reduced to a potentially exponential number of linear ones. Our methodology relies on a search paradigm — the search space being the many linear solutions to the non-linear formulation. Our approach is more efficient than the alternative generate and test methodology (i.e., convert constraints into nonlinear inequalities and then solve each expansion of the nonlinear inequalities until a solution is found or all expansions fail). In our methodology, an incremental linear programming algorithm is used. This means that solutions which fail can be detected without requiring a full expansion of the nonlinear inequalities (i.e., failure can occur even if many of the constraints have not yet been transformed). This early termination can prune large segments of the search space for which there is no solution. The incremental transformation process is also easy to understand (and extend) since it consists of a number of small rules used for different constraint semantics and different operations.

We now present results of symbolic timing verification for a single constraint: "constraint(g, a, ≤, 30)" for the process shown in Figure 1. Two different sets of propagation delays were used:

Case 1:	Case 2:
$\delta_{min}eg = 70$,	$\delta_{max}bd = 30$,
$\delta_{min}cl = \delta_{max}cl = 50$,	$\delta_{min}eg = \delta_{max}eg = 5$,
(all other δs symbolic)	$\delta_{min}gi = \delta_{max}gi = 5$,
	$\delta_{min}jk = \delta_{max}jk = 5$,
	$\delta_{min}cl = \delta_{max}cl = 50$,
	$\delta_{min}ma = \delta_{max}ma = 5$,
	(all other δs symbolic)

The solution for the first case is a linear inequality which closely resembles the constraint being proven in that the delay operations along the path from 'g' to 'a' are constrained: $\delta_{max}gi + \delta_{max}jk + \delta_{max}ma \leq 30$. This solution was obtained by noticing that 'k' will always occur after 'l' when occurring via 'g'. The second solution: $\delta_{min}bd \geq 20$ is less obvious. In this case, it is not possible to show that 'k' will always occur after 'l' (if it was, the constraint from 'g' to 'a' would be satisfied since the delay along that path adds up to 15 which is ≤ 30). The solution requires that 'g' does not occur too early with respect to 'l' since 'k' will end up waiting. This is accomplished by delaying event 'd' (and thus delaying 'g'). Note that many of the propagation delays specified as variables do not appear in the two solutions (e.g., the delay from 'f' to 'h'). This indicates that these delays are irrelevant with respect to the constraint being verified.

From a behavioral perspective, most circuits are expressed not as a single process but rather as a set of communicating processes. Extensions for dealing with some types of inter-process communication and also for modeling iterative behavior (e.g., loops) have been defined but are outside the scope of this paper (see [4]).

We have successfully applied our methodology to a small number of examples that involve cross-process constraints. As an example of the type of symbolic timing verification that we can currently perform, we present a moderately sized verification of a partial specification for the Intel Multibus. The specification consists of five communicating processing which contain a total of approximately 75 events, 65 operations, and 35 timing constraints to be verified.

There is a constraint that states that the arbitration process should not release the bus until the read/write process has finished (technically, this is expressed using two constraints and an inequality). This constraint could be eliminated if the read/write process sent a message to the arbitration process indicating completion, but this specification is less rigid; if an upper bound on the time taken by the read/write process to complete a transaction can be determined, the arbitration process need only wait a fixed amount of time before releasing the bus (leading to less circuitry). We have verified the Multibus specification in a variety of ways that help demonstrate the flexibility of our methodology.

First, we assigned delay values (intervals characterized by a minimum and maximum, not simply fixed values) to all of the operations in the specification. Before performing verification, we simulated the specification using a behavioral simulator [2] to validate our specification and ensure that what we specified was what we wanted. We then ran the verifier which reported "yes" — it was able to successfully prove that all of the timing constraints were satisfied with the given delay ranges. We then experimented with changing the delay values. We introduced a setup violation and the verifier failed (i.e., "no") when attempting to prove the setup constraint. We also experimented with setting the amount of time before the bus was released to too small a number, this also resulted in failure.

We then expressed the delay through the operation that holds up the release of the bus symbolically, using variables to represent the upper and lower bounds of the operation. The verifier reported "yes" along with constraints on the values of the upper and lower bounds. The minimum delay was constrained ("time to release bus > 300") via successfully determining the upper bound on the time taken in the read/write transaction with the memory and the lower bounds on the other delays incurred in the arbitration process (in other operations) before the bus was released. An upper bound was also given ("time to release bus ≤ 11870") due to a constraint which requires that the bus not be tied up for more than a few milliseconds. We then performed a similar experiment in which we fixed the release delay and represented the delay in the memory symbolically. In this case, the verifier reported a number of constraints on the delays which essentially answered the question "how slow can the memory be?" if we release the bus after some number of cycles.

As a final experiment, we expressed both delays symbolically and ran the verifier. It reported the relationship between the two; namely, that the time to release the bus is: ≤ 11870, > 170 + memory read access time, and > 150 + memory write access time. It also reported additional constraints with respect to both memory delays which cannot be too fast otherwise constraints in the read/write process are violated.

V. Conclusion

We believe that symbolic timing verification is a critical tool in the development of higher-level synthesis tools. The information obtainable from a symbolic verification process has three principal uses: (1) implementation verification, (2) obtaining constraints for synthesis, and (3) design evaluation. Implementation verification confirms that an implementation of a design and its associated delays will meet the constraints in the original specification. Synthesis tools can use symbolic delay values to determine the degree of flexibility that is available while still satisfying the timing constraints. This can lead to much more efficient use of resources in the final implementation. Design evaluation can be performed by using symbolic values in the constraints and determining bounds on the values of these variables given circuit delays. This can provide information about how well a design will perform and also relate the constraint variable to circuit delays.

In this paper, we have described an approach to symbolic timing verification using constraint logic programming as a framework. CLP(R) is an excellent test-bed for many of the algorithms. Its unification and backtracking features provide us with a mechanism for applying transformation rules and searching the space of possible solutions. We have made no attempt to optimize execution speed but have found it to be quite adequate using compiled CLP(R). The examples presented for Figure 1 take approximately five seconds of CPU time on a standard workstation (this includes compile time) and each Multibus example uses approximately two minutes of CPU time. Execution speed could be dramatically improved through the use of caching since the results of graph analysis are not currently reused. We believe that execution speed will scale well to larger specifications due to the fact that most constraints describe relationships that are local and interrelate a small number of symbolic delays. Some constraints can be quite complicated but it is unlikely that larger specifications will contain many such complex interrelated constraints. In any case, the complexity of our approach is dominated not by graph size, but by the complexity of the interrelationships and how they affect the solution space.

Our current directions are to generalize the operations and rules we currently employ in our system, especially with regard to inter-process communication and cross-process constraints. The use of symbolic timing verification in higher-level synthesis is also being explored with the synthesis of real-time controllers as a primary focus. We have already looked into queue sizing as a problem in synthesis [3] and the approach outlined here will be instrumental in obtaining the rate of send/receive events required as input to that algorithm.

Acknowledgments

The authors gratefully acknowledge numerous discussions with A. Borning and R. Yap regarding constraint logic programming and thank IBM for providing the new compiled version of CLP(R) we have been fortunate enough to use.

This work was supported by an NSF PYI Award (MIP-8858782) and by the DARPA/ISTO Microsystems Program under an ONR monitored contract (N00014-91-J-4041).

References

[1] T. Amon, G. Borriello, "Operation/Event Graphs: A Design Representation for Timing Behavior", 10th Int'l Conference on Computer Hardware Description Languages, April 1991.

[2] T. Amon, G. Borriello, "OEsim: A Simulator for Timing Behavior", Proc. of the 28th DAC, June 1991.

[3] T. Amon, G. Borriello, "Sizing Synchronization Queues", Proc. of the 28th DAC, June 1991.

[4] T. Amon, G. Borriello, "An Approach to Symbolic Timing Verification", 2nd Workshop on Timing Issues in the Specification and Synthesis of Digital Systems, March 1992.

[5] J. Cohen, "Constraint Logic Programming Languages", Communications of the ACM Vol. 33, No. 7, July 1990.

[6] J. Jaffar, S. Michaylov, P. Stuckey, R. Yap, "The CLP(R) Language and System", ACM Transactions on Programming Languages and Systems, to appear.

Validating Discrete Event Simulations Using Event Pattern Mappings *

Benoit A. Gennart
NTT LSI Laboratories
3-1 Morinosato-Wakamiya
Atsugi-shi, Kanagawa-ken
243-01 JAPAN

David C. Luckham
Computer System Laboratory
Stanford University
Stanford, CA 94305

Abstract

This paper introduces a new concept for the validation of discrete event simulations, based on recursively detecting and naming patterns of events. This paper introduces language constructs for defining event patterns (VAL+ mappings), explains the methodology for validating a design using event patterns, describes a software tool based on mappings (VAL+ debugger) and lists results of using the debugger on three larger examples (a 16-bit CPU, a systolic array, and a computer).

1 Introduction

Asserting the correctness of (*validating*) a hardware model is a notoriously difficult problem. The increasing complexity of hardware designs (pipelined, superscalar and parallel computer architectures), requires ever larger and more detailed models to ensure that a design will work as expected, making the validation problem ever more difficult.

Formal methods may show the correctness of a design, but in the case of incorrect designs (the common case), they do not help the designer to pinpoint errors. Moreover, formal methods by themselves don't ensure that a specification conforms to the designer's intention. Because of the current shortcomings of formal methods, the hardware industry relies on simulation to ensure the correctness of its designs.

The process of validating a design by simulation is very time consuming. Hardware accelerators speed up simulations. But browsing through simulation results is often tedious and always beyond the attention span of the designer. Inspecting simulations is also the most time consuming part of the validation process, and is prone to errors. However, little or nothing is done to help the designer deal with large, low level simulation results.

This paper presents a new methodology to help the designer handle the complexity of simulation results, speed up the validation of simulations and speed up the detection of errors. The methodology is based on recursively recognizing and naming patterns of events in a simulation. These patterns of events are called ***event pattern mappings***, (mapping, for short) and are specified by the designer. Each of these patterns maps up to a new event, describing higher level behavior. Event pattern mappings :

- are a new design language construct that allow the designer to define detailed relationships between various levels of a hierarchical design ;
- allow the interpretation of simulation results at any desired level of detail or abstraction ;
- are composable, i. e. the result of applying event pattern mappings to a simulation is itself a simulation, to which new event patterns can be applied ;
- handle synchronous as well as asynchronous behavior ;
- are applicable not only to simulation based validation methodologies, but also to more formal validation methodologies.

In this methodology, simulation results are presented as a small set of easy to understand high level events. Each of these high level events consists of a (small) number of lower level events. Each of these lower level events consists itself of even lower level events. This hierarchical presentation of simulation results (generated automatically using mappings) greatly reduces the designer's work in browsing through simulation results and detecting errors. This paper refers interchangeably to this simulation result presentation under two names, ***hierarchy of simulations***, and ***hierarchies of events***.

To test this methodology, the authors have implemented a simulation debugger based on event pattern mappings, and tested it on three significant designs, a 16-bit CPU with four levels of abstraction, a string pattern matching chip with three levels of abstraction, and a computer with two levels of abstraction.

This paper uses VHDL as a simulation language. The language constructs supporting the specification of mappings are introduced as an annotation language for VHDL called VAL+. The tool that extracts hierarchies of simulations is the VAL+ debugger. This paper limits itself to the standard abstraction levels

[0] **Research supported by the Air Force Office of Scientific Research under Grant AFOSR91-0354, and by the Defense Advanced Research Projects Agency/Information Systems Technology Office under the Office of Naval Research contract N00014-90-J1232.***

of usual hardware designs (gate level, register transfer level, and instruction level). However, the VAL+ debugger is by no means limited to these abstraction levels.

Structure of the paper Section 2 discusses prior work in the testing and verification area. Section 3 presents a debugging methodology based on event pattern mappings. It introduces VAL+, a language for specifying event pattern mappings, explains how to build simulation hierarchies, and how to debug simulations using simulation hierarchies. Section 4 shows the environment of the VAL+ debugger, a software tool based on event patterns. Section 5 describes the results of applying the debugger to three larger examples.

2 Background

Many previous papers have been devoted to the validation of hardware designs. The validation strategies include formal verification methods (theorem proving [9], state machine equivalence [8], symbolic simulation [4, 6, 5]), and comparative validation [1, 2, 3]. Formal methods aim at proving the correctness of a design. There are a few reported cases of circuits validated by theorem proving. However, theorem proving requires a large amount of effort on the part of the designer in guiding the theorem prover [4]. State machine equivalence is limited by the number of states such a methodology can handle. Early versions of state machine equivalence programs required space proportional to the number of states, and were limited to 10^6 states (a mere 20 bits). Encoding states symbolically would allow such methods to handle machines with up to 10^{20} states. The basic principles of symbolic simulation have been used by CAD programs for many years. However, symbolic simulation is complicated by the presence of conditional branching and loops in a design. When a conditional branch is encountered, all the alternative execution traces need to be considered, which causes case splitting. Loops that are conditioned by symbolic values lead to infinite execution traces. Techniques introduced to deal with the complexity caused by case splitting (case merging [7]) increase the cost of expression evaluation.

All the methods described above have in common an important problem. They help the designer to determine the correctness of a design. But they are of little help to the designer when designs are not correct. Nothing is done to help the designer pinpoint errors, once a circuit is shown incorrect.

VHDL, in order to help the designer check the correctness of a simulation, provides the **assert** construct. Assertions help the designer check the value of signals at specific points in time. The mechanism is however insufficient when it comes to checking the behavior of a signal (or a set of signals) over time.

Comparative simulation helps the designer automate the process of checking the correctness of large simulations. VAL [1, 2, 3], an annotation language for VHDL, formalizes the comparative simulation process. VAL allows abstract specifications to be associated with VHDL architectures, so that a tightly coupled comparison of the results of the two simulations (VAL abstract specification and VHDL architecture) can be made automatically at each output event and even at certain internal events. The ability to associate internal events at a high level of abstraction with sets of events at lower levels of abstraction is greatly enhanced by event pattern mappings, thus increasing the comparative simulation capabilities of VAL.

The major benefits of the event pattern based methodology over current formal methods are that :

- it produces structured, high level, easy to understand simulation results ;
- it helps the designer in locating errors ;
- it does not require a complete change in design methodology. It automates a part of the current simulation-based validation process ;
- it is applicable to any level of abstraction from the gate level up ;
- because mappings are composable, it 'scales up'. The larger the design, the more the designer benefits from the use of event pattern mappings.

3 Event pattern mapping validation methodology

This section presents the validation methodology based on event patterns. The methodology has three steps :

- The designer specifies explicitly event patterns in VAL+, a language we have developed to describe event patterns. Both event pattern mappings and the VAL+ language are introduced in section 3.2.
- A software tool, the VAL+ debugger, recognizes in simulations sets of events that match the event patterns specified in the mappings, and produces corresponding higher level events, which form higher level simulations (section 3.3).
- The designer then inspects the hierarchy of simulations to check the simulation results (section 3.4).

3.1 Definitions

A *design* consists of several *specifications* of a *system*, specifications made at various levels of abstraction. The specification of a system consists of its *interface* (how it communicates with other systems), its *state*, and its behavior, consisting of a set of *actions*, i. e. the acceptable transitions between states. Large specifications are hierarchical, and consist of a network of lower level specifications. A *simulation* is an ordering of *actual events*, i. e. instances of actions, resulting from 1) applying a *stimulus* to the interface of a specification, and 2) executing the specification according to some algorithm. This paper assumes that simulations are total orderings of events, and that the simulation algorithm is the discrete event simulation algorithm. The ordering of events in a discrete event simulation can be fully described by associating to each event one integer, called *time stamp*. The values of time stamps define the ordering relationship between two events.

```
--| event ActionName
--|    ( FormalParameters ) is
--|      PatternOfActions
--| end ActionName ;
```

FIGURE 1 : Mapping interface and body

3.2 The VAL+ event pattern language

We have developed VAL+, a language suitable for the definition of event pattern mappings. VAL+ is an annotation language for VHDL. VAL+ annotations are introduced into the VHDL specification as formal comments, distinguished by the symbol --|. The main construct in VAL+ is the event pattern mapping. An event pattern mapping consists of an action name and its parameters (the *interface* of the mapping), followed by a pattern of actions (the *body* of the mapping) :

```
--| event Load ( Data : Bit ) ;

--| event Fetch
--| ( Data : Bit_Vector (1 to 32) ;
--|   Reg  : Bit_Vector (1 to 5) ;
--|   Addr : Bit_Vector (1 to 32) ) ;
```

FIGURE 2 : Example of mapping interfaces

The interface of a mapping declares a *VAL+ action*, i.e. it gives a name to a specific aspect of the behavior of a specification. The interface describes *what* a specification can do, rather than how it does it. Examples of mapping interfaces are given in figure 2.

In figure 2, the first mapping interface declares the behavior of a one–bit register. This one–bit register can **Load** a one–bit value, **Data**. The second mapping interface declares one of the instructions of a microprocessor. This microprocessor can **Fetch** some **Data** from address **Addr** in memory, and store it internally in register **Reg**.

The body of a mapping defines the pattern of low level actions corresponding to the high level action declared in the mapping interface. To achieve this goal, the mapping body describes *pattern variables*, *event templates* and *ordering relationships* between event templates.

Pattern variables are parameters of mapping descriptions. They are initially *unbound*. They become *bound* when event templates in which they appear are *matched* to actual events (see next paragraph).

Event templates consist of a label, and the name and parameters of an action. There are two kinds of event templates. The first kind of event template consists of the name of a VHDL signal, followed optionally by an expression for the value of the VHDL signal. The second kind of event template consists of the name of a VAL+ action and optionally, expressions for values of the action parameters. Event templates are matched to actual events in a simulation. In the matching process, each pattern variable in the parameter expressions of an event template becomes bound to a value.

```
--| Cl : Clock ('1') ;
--| Ld : Load (Data) ;
```

FIGURE 3 : Event templates

In figure 3, the first template is labeled **Cl**, refers to the VHDL signal **Clock**, taking the value **('1')** ; the second template is labeled **Ld**, refers to the VAL+ action **Load**. The parameter expression lists a pattern variable named **Data**. When this event template is matched to an actual event (e.g, **Load ('1')**), the pattern variable becomes bound to the value **'1'**.

```
entity Register is
    port ( Clock : in Bit ; ... ) ;
--| event Load ( Data : Bit ) ;
end Register ;

architecture Behavior of Register is
    signal State : Bit ;
    ... -- declarations
begin
    ... -- behavior
--| event Load ( Data : Bit ) is
--| nodes
--|     Cl : Clock ('1') ;
--|     St : State (Data) ;
--| pattern
--|     Cl -> St ;
--| end Load ;
end Behavior ;
```

FIGURE 4 : VHDL/VAL+ spec. of the register

Mappings allow the definition of any partial ordering relationship among event templates. Ordering relationships among event templates must match ordering relationships between actual events in a simulation. Figure 4 shows an example of a mapping.

Let's focus first on the mapping definition of **Load**, **event Load ...** till **end Load**. This mapping contains one pattern variable, **Data**, the parameter. It lists two event templates, **Clock ('1')**, labeled **Cl**, and **State (Data)**, labeled **St**. The **Load** action, for any value **V** of the parameter **Data**, corresponds to a sequence of these two event templates, with value **V** replacing **Data**. Much more sophisticated patterns of events are of course possible.

Figure 4 also shows how mappings are integrated in

the VHDL language. The mapping interface is present in both the VHDL entity interface and the VHDL architecture. This allows the mapping to be made visible both in the component architecture and in higher level components in which the **Register** entity is instantiated. The mapping body, in the architecture, has visibility only on the entity ports and the architecture signals.

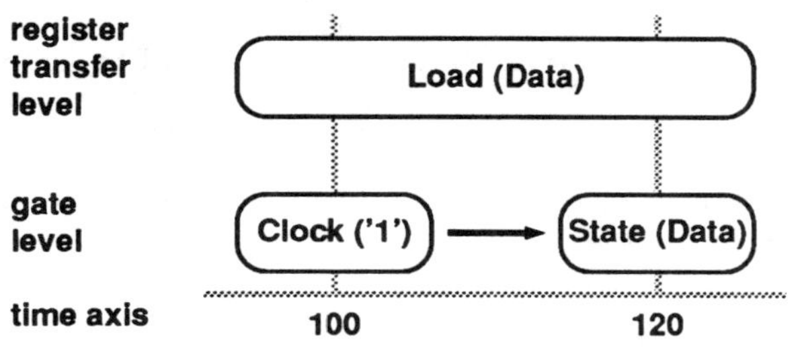

FIGURE 5 : Register mapping

The graphic representation of the mapping (Figure 5) illustrates an important aspect of event pattern mappings. The high level action in a mapping exhibits *duration*. It starts when the first event in the pattern occurs (time 100), and finishes when the last event in the pattern occurs (time 120).

3.3 Building hierarchies of events

We have developed a new algorithm, to detect all instances of event patterns in a simulation. Consider first the case of a non hierarchical VHDL specification (i.e., an entity without subcomponents). The debugger applies the procedure specified in figure 6. The procedure has two parameters : the mapping to be applied, and the simulation to which the mapping is applied. At first, the **Apply** procedure selects in the simulation (**Select** function) events on only the signals which are referred to in the mapping (**PartOfSim**). It then detects all instances of the pattern described in the body of the mapping, and for each such instance, generates a high level event (**Match** function). Each high level event name is the name of the mapping, whose parameters are bound during the pattern matching operation. A set of actual events matching a mapping body (resulting in the generation of a high level event) is called the *shadow* or the *domain* of that high level event. The set of high level events generated by the **Match** routine constitutes a high level simulation (**HighLevelSim**). The **Apply** procedure then appends all high level events to the **WholeSim** simulation. None of the original events of the **WholeSim** simulation is modified in the process.

Figure 7 shows graphically the effect of applying a linear and a diamond-shaped pattern to a low level simulation. In the low level simulation, events are represented as ovals. Significant ordering relationships are represented as thick arrows. The two instances of the diamond pattern and the instance of the linear pattern of events are indicated by the shaded areas. These three patterns are mapped up (thin lines) to three high level events. The ordering relationships between these three events are derived from the ordering relationships between lower level events. The three high level events constitute the high level simulation corresponding to the low level simulation.

Details of the pattern matching semantics and the timing model (how to deal with common lower events in pattern matches, with overlapping higher level events, how to determine timing relationships between high level events) are described in a forthcoming paper by Gennart & Luckham.

```
procedure Apply (
  Map : Mapping ;
  WholeSim : Simulation )
is
  PartOfSim : Simulation ;
  HighLevelSim : Simulation ;
begin
  PartOfSim :=
    Select (WholeSim, Map) ;
  HighLevelSim :=
    Match (Map, PartOfSim) ;
  Append (HighLevelSim, WholeSim) ;
end ;
```

FIGURE 6 : Mapping algorithm

Consider now what happens if the VHDL specification is hierarchical (figure 8). The debugger recursively applies, from the bottom up, mappings to a simulation. By applying mappings in a bottom-up fashion, the debugger can take advantage of the events generated by lower level mappings. In this way, the VAL+ debugger builds hierarchies of events. Top level events consists of a small number of lower level events. Each of these lower level events consists of a small number of even lower level events. The higher the level of abstraction, the smaller the number of events, and the easier it is to validate the simulation.

Experience shows that by going up one level of abstraction, the number of events is reduced by a factor of about 10 (see section 5). For the CPU example discussed in section 5, in just three levels of mapping, a 8,073-event gate level simulation has been reduced to a 5-event instruction level simulation.

3.4 Validating and debugging

The designer traditionally validates his designs by manually comparing simulation results against an informal specification. Occasionally, a designer also has a high level specification whose output can be compared to the output of the simulator. As designs become more complex, this verification process becomes less satisfactory. It is simply too cumbersome to browse through simulation results after each simulation run.

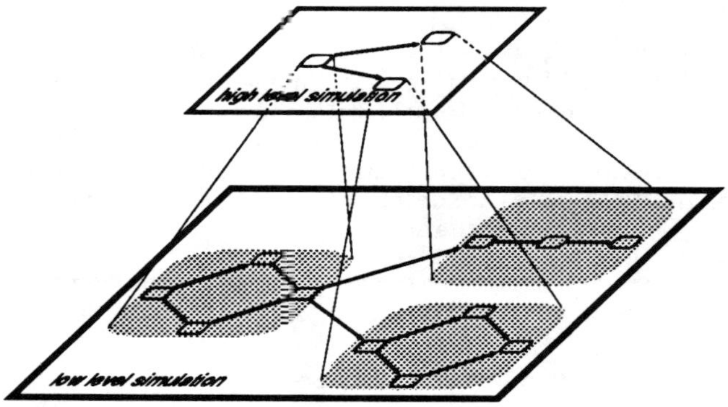

FIGURE 7 : Building simulation hierarchies

This is where event pattern mappings, and the hierarchy of simulations generated by them, come into the picture. By inspecting the simulation at the highest level of abstraction (the events in the highest level simulation have been generated by recursively extracting patterns of events), the designer can rapidly, with a high level of confidence, determine whether or not the simulation at the lowest level of abstraction is correct. In effect, the VAL+ debugger has looked at the entire simulation results, and summarized it in a compact, high level simulation. No information is lost in the process. Information has been structured for the convenience of the designer.

```
procedure BuildSimulationHierarchy
   ( Spec : Specification ;
     Sim : Simulation ) is
begin
   for each Component of Spec do
     BuildSimulationHierarchy
         (Component, Sim) ;
   end do ;
   for each Mapping of Spec do
     Apply (Mapping, Sim) ;
   end do ;
end ;
```

FIGURE 8 : mapping algorithm (2)

Debugging a large simulation using simulation hierarchies works as follows. An *erroneous event* in a top level simulation is either a missing event, or an event with incorrect parameter values. The erroneous event in the top level simulation indicates either a lower level error in its domain, or an error in the mapping description. The pinpointing method is to search the domain of the lower level events that matched (or failed to match) a particular instance of a mapping. An error in a lower level simulation can itself be traced down to lower levels of abstraction, all the way to the lowest level.

Such a debugging mechanism is of course most useful when large number of events are involved, but what this section has described is a structured way of debugging simulations. The designer need only look at the top level simulation (with comparatively few, easily understandable, high level events) to acquire confidence in the correctness of his low level simulation (with a comparatively large number of hard to grasp, low level events). Using the mapping definitions, the designer can easily trace errors down to lower levels of abstraction. The debugging can be conducted without ever having to look at more than a limited number of events at a time.

4 The VAL+ debugger

This section presents the environment of the VAL+ debugger. Traditional validation environments consists only of a simulator to which both a specification and a stimulus are fed. This environment augments the traditional validation environment with an event pattern based debugger.

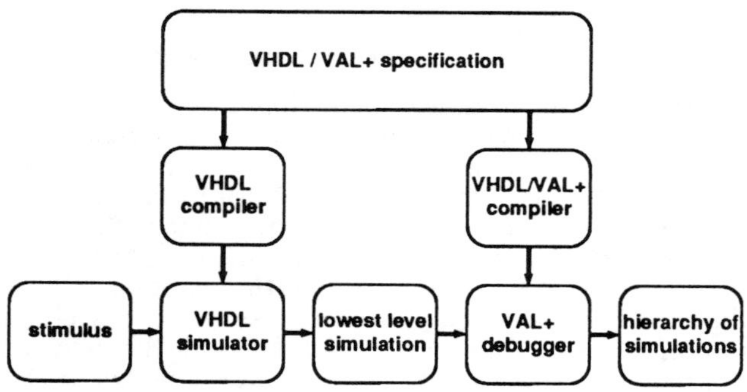

FIGURE 9 : The VAL+ debugger

The VHDL / VAL+ specification consists of 1) a (possibly hierarchical) VHDL specification of a system, made at a given (low) level of abstraction ; and 2) the VAL+ mapping specifications. The mappings define actions at higher levels of abstraction as well as the relationships between actions at adjacent levels of abstraction.

The VHDL / VAL+ specification is compiled in two programs, a simulator and a debugger. The simulator, compiled from the VHDL low level specification, takes as input a stimulus and produces a simulation at the lowest level of abstractions. The debugger, compiled from the VHDL / VAL+ specification, takes as input the low level simulation, recursively applies mappings to the simulation, and generates simulations at higher levels of abstraction.

5 Results

The VAL+ debugger has been implemented. The debugger was tested on three examples, a 16-bit CPU, a string pattern matching chip, and a computer.

The simulated CPU example has four levels of abstraction, gate level, elementary register transfer level (one–bit buffer, one–bit register, ...), register transfer level (multiple bit buffer, multiple bit register), and instruction level (the CPU itself). We defined event pattern mappings to generate simulations at each level of abstraction from the preceding lower level, starting with the gate level behavioral simulation. For a five instructions stimulus, the gate level simulation consists of 8073 events, the elementary register transfer simulation consists of 334 events, the register transfer level simulation consists of 73 events, and the instruction level simulation consists of 5 events. These numbers show how well the event pattern mappings help structure the gate level simulation. To understand the problem of browsing through the gate level simulation, printing the gate level simulation, with as horizontal coordinate the VHDL signals and vertical coordinate the time points, would require a rather unmanageable page 4880 characters wide and 680 lines long.

The debugger helped find five bugs in the CPU gate level specification (including errors in the logic of the clock enable circuitry of the registers, in the logic unit, and in pin connections between components). Four out of five bugs were found inspecting the top level (instruction level) simulation. The fifth bug was found inspecting the register transfer simulation, as a consequence of another bug at the instruction level. In the CPU instruction level simulation generated by the VAL+ debugger shows obvious errors – failure of an instruction to execute properly or incorrect high level event parameters for the others. On the other hand, discovering these errors by inspecting the VHDL gate level simulation would have been extremely tedious. The gate level bugs were found in that part of the gate level design that generated the events in the domain of the erroneous high level events.

The debugger was also used in debugging two other designs : (1) a string matching chip, with specifications at the register transfer and chip levels ; and (2) a computer, consisting of two memory boards, a central processing unit, and a data transfer bus based on the VME protocol. The register transfer level specification of the string matching chip is a systolic array that demonstrates the ability of the pattern language to handle pipelined behavior. The computer was specified at the board and system levels, and shows the benefit of using event pattern mappings in an asynchronous design (the VME bus protocol is asynchronous).

6 Conclusion

This paper has presented a new methodology for validating simulation results. The VAL+ debugger was implemented according to this methodology and has demonstrated the benefits of the methodology on three examples. Further work includes 1) performing comparative simulation using event pattern mappings (which requires a simulation format different from the discrete simulation), 2) studying the benefits of applying patterns to more formal validation methods, and 3) studying the benefits of alternate simulation formats (partial orders and duration events).

References

[1] L. M. Augustin, B. A. Gennart, Y. Huh, D. C. Luckham, and A. G. Stanculescu. VAL: An annotation language for VHDL. In *International Conference on Computer-Aided Design (ICCAD '87) Digest of Technical Papers*, pp. 418–421, Santa Clara, CA, November 1987.

[2] L. M. Augustin, B. A. Gennart, Y. Huh, D. C. Luckham, and A. G. Stanculescu. Verification of VHDL designs using VAL. In *Proceedings of the 25th Design Automation Conference (DAC)*, pp. 48–53, Anaheim, CA, June 1988.

[3] L. M. Augustin, D. C. Luckham, B. A. Gennart, Y. Huh, and A. G. Stanculescu. *Hardware Design and Simulation in VAL/VHDL*. Kluwer Press, October 1990. 322 pages.

[4] Randal E. Bryant. Formal hardware verification by symbolic simulation. In R. W. Dutton, editor, *VLSI Logic Synthesis and Design*, pp. 125–132. IOS Press, 1991. Proceedings of the 1990 Synthesis and Simulation Meeting and Interchange.

[5] J. R. Burch, E. M. Clarke, and D. E. Long. Representing circuits more efficiently in symbolic model checking. In *Proceedings of the 28th Design Automation Conference (DAC)*, pp. 403–407, July 1991.

[6] J. R. Burch, E. M. Clarke, K. L. McMillan, and David L. Dill. Sequential circuit verification using symbolic model checking. In *Proceedings of the 27th Design Automation Conference (DAC)*, pp. 46–51, Orlando, Florida, June 1990.

[7] W. E. Cory. *Verification of Hardware Design Correctness : Symbolic Execution Techniques and Criteria for Consistency*. PhD thesis, Electrical Engineering Department, Stanford University, 1983.

[8] O. Coudert, J.-C. Madre, and C. Berthet. Verifying temporal properties of finite finite state machines without building their state diagram. In *Workshop on Computer-Aided Verification*, Rutgers, NJ, June 1990.

[9] M. Gordon. Why higher order logic is a good formalism for specifying and verifying hardware. In G. Milne and P. A. Subrahmanyam, editors, *Formal Aspects of VLSI Design*, pp. 153–177. North-Holland, 1986.

Two New Techniques for Compiled Multi-Delay Logic Simulation*

Yun Sik Lee
Peter M. Maurer
Department of Computer Science and Engineering
University of South Florida
Tampa, FL 33620

Abstract

This paper describes two techniques for compiled event driven multi-delay logic simulation that provide significant performance improvements over interpreted multi-delay logic simulation. These two techniques are based on the concept of retargetable branch instructions that can be used to switch segments of code into and out of the instruction stream. Our second algorithm, called the shadow technique, has been designed especially for systems with instruction caches. Benchmark experiments shows these two techniques are up to 15 times faster than our interpreted multi-delay simulator, with an average improvement of about 5 times for our fastest method.

1. Introduction.

As the design of a circuit proceeds, more accurate simulation is required. In particular, more accurate models than zero or unit-delay timing are usually needed during the final phases of the design. The many well-known techniques for compiled simulation [1-6], are based on zero and unit-delay which do not provide an accurate picture of a circuit's timing behavior. For some circuit elements, delay is the essential nature of the function, and a accurate timing model is necessary to model them.

This paper focuses on multi-delay timing [7], in which the delay of each gate is an integral multiple of some basic unit. Delays may be the same for each instance of a gate type or delays can be assigned based on layout parameters. There are several types of delay which could be modeled, such as transport delay, which is the amount of time taken for changes in a gate's input to reach the output, ambiguous delay, which is a short interval in which a net is undefined, and rise-fall delay, which is the amount of time a signal takes to change from low to high and vice-versa[1]. For simplicity, our simulators deal only with transport delay, but could easily be extended to other delays.

The problem we encountered with compiled multi-delay simulation is that simple adaptations of unit-delay techniques have usually been unsuccessful in providing significant increases in performance. In Section 2 we describe work that we have done in oblivious multi-delay, simulation and show why it did not provide the expected performance improvements. In Section 3 we describe how we applied the concepts of threaded code[2] to multi-delay simulation, and why these were also unsuccessful. Sections 4 and 5 describe two new techniques that *do* provide significant performance improvements. The results for these two algorithms are presented in Section 6. During this research, we replaced our existing benchmark machine with a new, faster machine that also possesses an instruction cache. We discovered that locality of reference has a profound effect on simulators run on such machines, much more than we had anticipated. The algorithm presented in Section 5 was designed to provide good locality of reference at the expense of using somewhat slower operations.

2. Oblivious compiled simulation.

In both zero-delay and unit-delay simulation, compiled oblivious techniques have demonstrated performance improvements over interpreted event driven simulation, as long as the activity rate is above a certain threshold. The reason for this performance improvement is that the number of instructions executed per gate simulation is reduced to a minimum. Although oblivious simulatators usually simulate many more gates than event-driven simulators, the gate simulations are so efficient that a performance improvement can be realized.

However, the problem with oblivious techniques is that they must provide for every eventuality. In zero-delay simulation, this means simulating every gate for every input vector. However in unit-delay and multi-delay simulation this means the simulation of a gate at every time that an event-driven simulator could schedule the simulation. Thus if there is a path of delay d between the primary inputs and gate G, then G must be simulated at time d. The number of required simulations depends on the circuit structure, the the number of different delays, and the circuit size.

Two oblivious methods of unit-delay simulation that can be adapted to multi-delay are the PC-Set Method and the Parallel Technique[4,5]. We concentrated on the PC-Set Method, because it allowed us to pack several vectors

* This work was supported in part by the National Science Foundation under grant number MIP-906444 and the USF Center for Microelectronics Research (CMR).

into a single word, thus performing many simulations for the price of one. Our preliminary studies of the PC-Set method have suggested that oblivious simulation will not provide significant speedups over interpretive techniques for realistic circuits. In our studies, we used the ISCAS85 benchmarks, and assigned a random delay of 1-8 to each gate. Without vector packing, there was no significant difference between the performance of the PC-Set method and our interpreted simulator. Although vector packing gave a performance improvement of over 90%, similar improvements could probably be realized with the interpreted simulator. Furthermore, the size of the generated code was unacceptably large for many circuits, almost 50 megabytes for c6288. Experiments with random delays over a larger range of values have shown that the size of the generated code expands significantly with an increase in the number of delays.

Although the Parallel Technique has been shown to be substantially more efficient than the PC-Set Method, the it does not permit vector packing, and can probably not compete with an interpreted simulator using packed vectors. It seems that the number of gate simulations that must be performed by an oblivious multi-delay simulator is so large that little or no performance gains can be realized, even though the gate simulations can be done more efficiently than those of an interpreted simulator. Therefore, we have abandoned the oblivious approach and concentrated on event-driven techniques.

3. Threaded code techniques.

Lewis has demonstrated that using threaded code in a compiled unit-delay simulation can provide performance improvements over interpreted unit-delay simulation[2]. Apart from the distributed scheduler, the technique is much the same as that used by an interpreted simulator. Simulation is done in two phases, the event phase which processes events and produces a queue of gates, and the simulation phase which simulates gates and produces a queue of events. Because neither events nor gates are queued for more than one phase, event generation and storage is relatively simple, and it is possible to suppress event generation if a new value produced by a gate simulation identical to the existing value of the net.

Event-driven multi-delay simulation is also done in two phases, but because of different gates may have different delays, events must often remain queued for several iterations of the event phase. The event phase processes only those events for the current time. It is not possible to suppress event generation. If there is already an event queued for the net, it may change value before the value produced by the current simulation becomes effective.

Since only the events for the current time are processed during the event-phase, it is necessary to sort events by time. The simplest mechanism for doing this is the timing wheel, which is an array of event queues, each of which corresponds to a single time. If the number of queues is less than the total number of times at which events are processed, the queue may be reused by indexing the array modulo its size.

We have implemented a multi-delay simulator using threaded-code techniques similar to those used by Lewis[2]. This simulator provided an improvement of about 45% over our interpretive simulator. Although this is a significant improvement, it is less than we had expected. Subsequent investigations of the interpreted and compiled algorithms showed that the threaded-code technique was able to eliminate only a small part of the code used by the interpreted algorithm, hence the unexpectedly low performance.

Although multi-delay simulation demands the use of an event-driven algorithm, our experiments with threaded code demonstrate that a simple adaptation of an interpretive algorithm will not provide the significant performance improvements that we are seeking.

4. Gateways for multi-delay simulation.

We have recently investigated a concept called "gateways" that can be used to introduce event-driven behavior into oblivious simulations [6]. A gateway is a retargetable branch instruction that can be used to switch portions of the generated code into and out of the instruction stream. Although this concept is not directly applicable to multi-delay simulation because of the code size, we have discovered an adaptation of the gateway technique that can be used to substantially improve the performance of an event-driven multi-delay simulation. In the gateway technique, the branch target for a particular gateway is stored in a fixed location which are dynamically altered to construct an instruction stream. The generated code contains a simulation routine for each gate and a net handler for each net. Each routine is terminated by a gateway.

The primary difficulty in adapting the gateway technique to multi-delay simulation is that several events may be queued for a net at one time, making fixed branch targets infeasible. To get around this problem, our technique stores all branch targets in a collection of stacks. The gateway instructions pop the stack to obtain the address of the next block of code to be executed. (Lewis used a similar technique for unit-delay simulation in [2].) Our current implementation uses $n+1$ stacks where n is the number of slots in the timing wheel. The first n stacks implement the timing wheel, while the $n+1$st stack is used to schedule gate simulations.

The timing wheel structure is illustrated in Fig. 1, along with the structure for the gate-simulation routines. Pseudo-Code for the net and gate trailer routines is illustrated in Fig. 2. Gateway operations are represented by "jump *label" statement. Because this is not legal C, assembly language must be used to generate this operation, otherwise, C is generated.

Initially, the primary input tests, illustrated in Fig. 3, are used to detect changes in the input vector. In Fig. 3, "index" is the ID of a gate, which is assigned at compile time. The flag is used to avoid switching the simulation routine into the instruction stream twice.

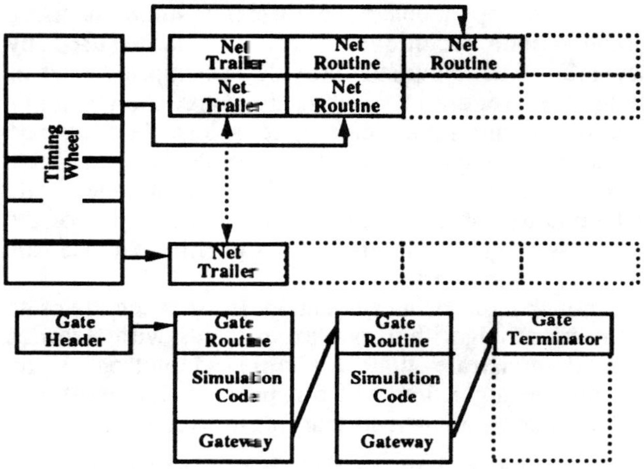

Fig. 1. The gateway stack structures.

```
Net_Trailer:
   jump *Gate_Header;  /* Gateway Operation */

Gate_Trailer:
   Restore Gate_Header to ADDRESS_of_Gate_Trailer;
   Clear TIMING_WHEEL[current_time];
   if event_count = 0 then
      return;
   Increment current time by 1;
   jump *TIMING_WHEEL[current_time].addr;
```

Fig. 2. Net and gate trailer routines.

Initially, the top of the simulation stack will cause an immediate exit. The branch address of the gateway instruction us used to form a singly linked list of simulation routines. The link address is conveniently located in the address field of a branch instruction..

After all input tests have been performed, the Net_Trailer branches to the first simulation routine in the stack. The last routine branchs to the gate trailer, which advances the current time, and jumps indirectly to the top of the new event stack.

```
/* index = fanout gate's ID at compiled time; */
   if (new_net_k != old_net_k) then
      if (flag[index] = 0) then
         old_net_k = new_net_k;
         Gateway[index] = Gate_Header;
         Gate_Header = ADDRESS_of_Net_k;
      endif
   endif
```

Fig. 3. An input test.

Four basic routines are used to simulate a single state at certain time *t*, the trailer routines illustrated in Fig. 2, and the gate-simulation and net-handling routines illustrated in Fig. 4. The trailer routines do the bridge operations between the net handlers and the gate simulations, simulation is performed by the gate-simulation routines, and scheduling is done by the net handlers.

```
/* Net handling routine */
Net_k_Handler:
   if (new_net_k != old_net_k) then
      old_net_k = new_net_k;
      if (flag[index] = 0) then
         Gateway[index] = Gate_Header;
         Gate_Header = Address_of_Gate_Routine;
      endif
   endif
   Decrement event count;
   Timing_Wheel[current_time] =
      Timing_Wheel[current_time] - Address_Size;
   goto *(Timing_Wheel[current_time]);
/* Gate simulation routine */
Gate_index_Simulator:
   Reset flag[index];
   new_net_k = old_net_x & old_net_y;
   Add_Time =( Current_Time + Gate_Delay) MOD
      Timing_Wheel_Size;
   Timing_Wheel[Add_Time] = Addr. of Net_k_Handler;
   Timing_Wheel[Add_Time] =
      Timing_Wheel[Add_Time] + Address_Size;
   Increment the event-count;
   goto *Gateway[index];          /* the next stream */
```

Fig. 4. Generated code structure.

5. The shadow technique.

The shadow technique improves performance on machines with instruction caches, because it generates a smaller amount of code than the gateway method. All information for a gate or a net is placed a descriptor called a shadow . All values are accessed indirectly using the shadow but, the reduced amount of code increases the performance of the instruction cache.

Only a minimal number of generic gate and net processing routines are generated, but a descriptor is generated for each gate and each net. A gate descriptor has the address of a simulation routine, a flag to prevent duplicate scheduling of gates, the gate delay, and pointers to input and output values. Each net descriptor has the address of a net handling routine, pointers to old and new values, and the addresses of fanout gate descriptors. Fig. 5 illustrates these descriptors.

Gate simulators are generated for each combination of gate-type and input-count in the circuit. Thus if the circuit contains only two and three-input NANDs, then two routines will be generated, one for two-input NANDs and one for three-input NANDs. Each of these routines assumes that a register variable called the "Shadow Pointer" contains the address of the current descriptor. One net-handler is generated for each different fan-out in the circuit.

Each of the routines terminates with load of the Shadow Pointer, and a branch to the routine of the new descriptor. The net handling routines use a timing-wheel that is identical to that used by the gateway algorithm, except the stacks contain descriptor addresses rather than routine addresses. The gate-simulation stack is also similar to that used by the gateway algorithm, except that a linked list of descriptors is used instead of a list of routines.

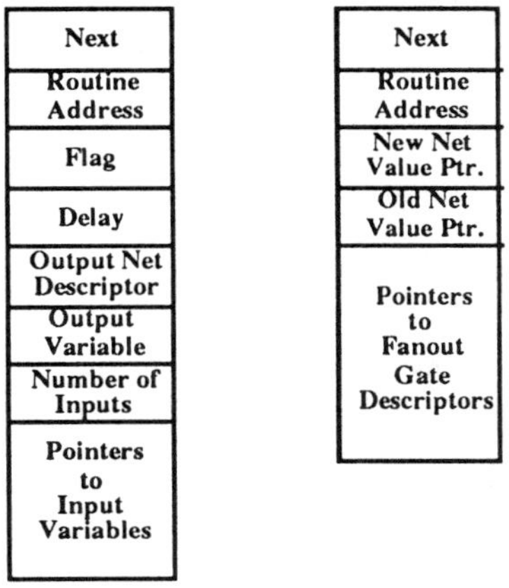

Fig. 5. Net and gate descriptors.

As usual, simulation is a two-phase process that alternates between gate-simulation and net-handling. At the end of each phase either a termination routine is executed, which is accessed through its own descriptor.

6. Experimental results.

A random delay from 1 to 8 was added to each gate of the ISCAS-85 benchmarks without regard to type. Because the same gates were simulated by all simulators, the range of delays has no effect on the conclusions drawn in this paper. Tests were run on a SUN 4-IPC with 12 megabytes of memory and a dedicated disk. The values of Fig. 6 are in seconds of execution time for 5,000 random vectors. The time required for reading inputs and printing outputs is excluded. Two interpreted simulators were designed for the purpose of comparing benchmark performance, one based on the enhanced two-valued logic model described below, and one based on a more conventional 3-valued logic model. The results for these simulators are reported in Fig. 6 under "2-Interp" and "3-Interp." The performance improvement percentages are computed using the results from the 2-valued simulator in the formula (interp_time - comp_time)/interp_time.

The logic model of our compiled simulators uses an additional bit to force every gate to be simulated at least once on the first input vector. The value of each net is initialized to 2 or 3, which represent values 0 and 1. The simulation routines set this second bit to zero, producing a value of either 0 or 1, allowing us to use a simpler 2-valued logic model, and still guarantee that nets are set to consistent values. It would be a simple matter to extend our simulators to 3-valued logic.

The Gateway Algorithm gives an average performance improvement of 67% running in about 1/3 the time of an interpreted simulation, while the Shadow Algorithm gives a performance improvement of 81%, running in about 1/5 the time of an interpreted simulation.

Multi-Delay				
Ckt	2-Interp	Gateways	Shadows	3-Interp
c432	35.9	6.8	6.4	40.9
c499	72.1	12.4	10.3	83.9
c880	55.9	23.9	13.4	61.8
c1355	196.0	46.5	26.6	273.4
c1908	219.5	85.8	43.9	248.0
c2670	253.5	114.3	59.9	280.7
c3540	467.6	161.6	83.2	523.2
c5315	644.6	288.0	151.8	716.2
c6288	12330.1	1677.2	788.7	13850.5
c7552	1019.0	484.9	256.3	1122.4
MaxImprov.(%)		86	94	
MinImprov.(%)		52	75	
AvgImprov.(%)		67	81	

Fig. 6. Performance results.

The shadow algorithm executes essentially the same instructions as the gateway algorithm. On a machine with no cache, we would expect the gateway algorithm to perform slightly better than the shadow algorithm. The performance improvement in Fig. 6 is due entirely to the instruction cache. Because of the steady advance in the today's technology, we expect caches to become more common in the future. Many of today's compiled simulators do not perform well with caches, because the generated code is one long sequence of instructions without loops or branches.

7. Conclusion.

Our experimental results show that the average performance improvement over interpreted simulation is 67% for the gateway algorithm and 81% for the shadow algorithm. This research demonstrates the importance of designing compiled simulators that work well with instruction caches. The differences between the gateway and shadow algorithms demonstrate the effect that locality of reference can have on simulation performance.

References

1. M. Breuer and A. Friedman, *Diagnosis & Reliable Design of Digital Systems*, Woodland Hills, CA: Computer Science Press, 1976.
2. D. Lewis, "A Hierarchical Compiled Code Event-Driven Logic Simulator,", *TCAD*, vol 10, 1991, pp. 726-737.
3. R. Bryant, D. Beatty, K. Brace, K. Cho and T. Sheffler, "COSMOS: A Compiled Simulator for MOS Circuits", *DAC- 24* 1987, pp. 9-16.
4. P. Maurer and Z. Wang, "Techniques for unit-delay compiled simulation", *DAC-27*, 1990, pp. 480-484.
5. P. Maurer, "Optimization of the Parallel Technique for Compiled Unit-Delay Simulation," *ICCAD-90*, 1990, pp. 70-73.
6. P. Maurer, "Gateways: A Technique for Adding Event-Driven Behavior To Compiled Unit-Delay Simulations," Submitted for Publication.
7. S. Szygenda, et. al, "A Model and Implementation of a Universal Time Delay Simulator for Large Digital Nets," SJCC, 1970, pp. 207-216.

ZERO DELAY VERSUS POSITIVE DELAY IN AN INCREMENTAL SWITCH–LEVEL SIMULATOR

Larry G. Jones[1]

Department of Computer Science

University of Illinois at Urbana–Champaign

ABSTRACT

We present the methods used in the implementation of an incremental zero/integer–delay switch–level logic simulator for MOS circuits based on the MOSSIM II switch–level model. Zero–delay timing reduces spurious reevaluations caused by minor changes to signal timing that do not affect logic, while integer–delay timing provides an ability to model race conditions that do affect the logic. In experiments run on switch–level versions of the ISCAS combinational and sequential benchmarks, incremental switch–level simulation with mixed zero/integer delay was 4 times faster (on average) than incremental switch–level simulation with only postive integer delays.

1. INTRODUCTION

As MOS feature sizes continue to shrink and the complexity of circuits integrated on a single silicon chip grows, simulation becomes an ever more important step in the design process. After the initial design phase, designs are often subjected to small changes in an effort to fix design flaws or tune performance. The design then must be resimulated to check the success of the changes and to ensure that no new problems have been introduced. Many of the results of the computations performed in resimulating a modified design are identical to the results obtained during previous runs of the simulator. By reusing previous results *incremental simulators* often complete their task in a fraction of the time required by conventional simulation techniques [9] [7] [5] [11].

Previous incremental simulators have modeled signal delay as strictly positive. This limited the usefulness of the simulators since even slight modifications to the circuit (such as transistor sizing for performance) may cause all directly and indirectly dependent signals to experience a *time–shift* requiring massive resimulation, even when neither the signal values nor their relative arrival order has changed [11]. The incremental switch–level simulator we present (ICESIM) uses a delay model which allows the mixing of zero–delay with positive integer–delay elements. Zero–delay timing provides an abstraction that can be used for modeling digital circuits where signal delay does not affect logic. Positive integer–delay provides a capability for modeling race conditions that affect the logic of the circuit. The logical behavior of most MOS digital circuits can be modeled using a combination of both delay techniques.

It is well known that zero–delay simulation often reduces simulation overhead by reducing the number of events requiring scheduling and by reducing scheduling costs [13]. With respect to incremental simulation, zero–delay also reduces the simulators sensitivity to time–shifts. The result is that incremental simulation with zero–delay is often much faster than incremental simulation with only positive delays.

ICESIM operates on a flattened representation of the circuit graph but provides a high–level hierarchical interface by being tightly coupled to a schematic capture system (ICE) using an incremental netlist compiler[10]. Manipulations of the design at any level in the structural design hierarchy are quickly and automatically mapped into manipulations of the flat netlist by the compiler. Other differences which distinguish this system from previous incremental simulators are its simulation model (based on the MOSSIM II switch–level model [3]), and the mixing of static and dynamic subnetwork partitioning techniques to reduce overhead.

2. EVENT HISTORY

Before the incremental simulator can be used, the design must have been simulated at least once using the nonincremental simulator. During this initial simulation the nonincremental simulator produces a *history* of simulation events which, for each node, records each change in state and the time the event occurred. Subsequent incremental simulations update the event history to be consistent with any changes that have occurred in the logic since the previous (nonincremental or incremental) simulation. The incremental simulator does not make use of the state of every node for every moment in time. Instead, the states of nodes are *extracted* from the event history *on demand* as they are required to determine and evaluate nodes affected by the modification.

3. TIMING

The simulator models time using a zero/integer–delay timing model that assumes each node x has an integer transport delay $D(x) \geq 0$ associated with it. Node x reaches its steady–state $D(x)$ units of time after all inputs to the

[1] This work supported in part by the National Science Foundation (grant MIP–91–01051) and by the Texas National Research Laboratory Commission (project #047).

channel connected **subnetwork** containing x become stable. The inclusion of zero-delay reduces the simulators sensitivity to shifts in signal arrival time, but requires special handling during simulation to prevent spurious applications of the evaluation procedure. Further, when the circuit contains feedback paths it is necessary to introduce positive delays on specific nodes to remove zero-delay cycles (which have ill-defined semantics [6]).

4. INCREMENTAL STATE EQUATION

Let $S_i(x,t)$ represent the state of node x at time t during simulation i. Let $D_i(x)$ represent the delay assigned to node x during simulation i. For a set of nodes, X, let $S_i(X,t)$ represent the states of the nodes in X at time t during simulation i. In nonincremental switch-level simulation the state of x at time $t+D_i(x)$ during simulation i is given by

$$S_i(x,t+D_i(x))=F_x(S_i(A,t)) \quad (1),$$

where A is the set of nodes containing all nodes in the subnetwork containing x, as well as all nodes driving the gates of transistors in the subnetwork containing x. The function, F_x, indicates the result, with respect to x, of applying the evaluation procedure to the subnetwork containing x. We refer to the nodes in A as the ***arguments*** of x and the subnetwork containing x. We refer to nodes that drive gates of transistors whose source or drain is connected to x as the *fan-in* of x. Similarly, we refer to nodes that serve as the source or drain of transistors whose gate is driven by x as the *fan-out* of x.

We say that the state of node x at time t ***deviates*** during simulation i if and only if the state of x at time t during simulation i differs from its state at time t in simulation $i-1$, i.e., $S_i(x,t)\neq S_{i-1}(x,t)$. The state of node x at time t can deviate during simulation i only if one or more of the following three conditions are true:

(a) the subnetwork containing x has been modified, i.e., F_x has changed,

(b) the delay assigned to a node in the subnetwork containing x has been modified, i.e., $D_i(y)\neq D_{i-1}(y)$ for $y \in subnet(x)$, or

(c) an argument of x deviates at time t.

Conditions (a) and (b) represent explicit changes to a subnetwork by the user that occur between simulation runs. Condition (c) is a result of state changes propagating through the circuit from ***touched*** nodes (nodes involved in explicit changes by the user). We say node x is ***active*** at time t for simulation i if it is touched, it deviates at time t, or the state of a fan-in of x deviates at time t. Otherwise the node is considered to be ***inactive***. We say a subnetwork (dynamic or static) is touched if it contains a touched node, active at time t if it contains an active node at time t, and inactive at time t if it does not contain any active nodes at time t.

The state of a node whose subnetwork is inactive at time t for simulation i is identical to its state at time t for simulation $i-1$. We can avoid reevaluating inactive subnetworks during resimulation by accepting the state of their nodes from the previous simulation. The value of $S_i(x,t+D_i(x))$ can therefore be determined from:

$$\begin{array}{ll} F_x(S_i(A,t)) & \text{if } subnet(x) \text{ active} \\ S_{i-1}(x,t+D_i(x)) & \text{if } subnet(x) \text{ inactive} \end{array} \quad (2),$$

that is, whenever a node is in an active subnetwork its state is derived from the state equation applied to the current state of the arguments of the node. Whenever a node is in an inactive subnetwork, its state is derived from the results of the previous simulation.

For any node x during simulation i the state of x at time $t + D_i(x)$ can differ from its state at time $t + D_i(x) - 1$ only if the state of one of its arguments changes at time t. We say a subnetwork is ***stimulated*** at time t if one of its arguments change state at time t, i.e., for some y in A, $S_i(y,t) \neq S_i(y,t-1)$. The value of $S_i(x,t+D_i(x))$ is now given by:

$$\begin{array}{ll} F_x(S_i(A,t)) & \text{if } subnet(x) \text{ active and stimulated} \\ S_i(x,t+D_i(x)-1) & \text{if } subnet(x) \text{ active but not stimulated} \\ S_{i-1}(x,t+D_i(x)) & \text{if } subnet(x) \text{ inactive} \end{array} \quad (3).$$

Whenever the subnetwork of a node is active and stimulated its state is derived from the evaluation procedure applied to the current state of the arguments of the node. Whenever the subnetwork of a node is active but not stimulated, its state can be derived from the previous time. Whenever the subnetwork of a node is inactive, its state can be derived from the history of the previous simulation.

5. INCREMENTAL SIMULATION

Equation (3) shows that during incremental simulation state values can be obtained from either the current simulation state history or the previous simulation state history, depending on the active status of the subnetwork. To ensure correct results and to minimize the number of subnetwork evaluations that occur during incremental simulation it is important to recognize the active status of subnetworks as soon as a change from active to inactive or inactive to active occurs.

There are two ways that a subnetwork can change its active status: either some argument changes state during the current simulation, or some argument changes state during the previous simulation. In the former case, whenever a node is assigned a new state, it is necessary to redetermine the active status of the fan-outs of that node. In the latter case, an active subnetwork must have its active status rechecked for each time in the history trace of the previous simulation that an argument changes state (but only as long as the subnetwork remains active).

By equation (3) the only time a subnetwork requires reevaluation during incremental simulation is when it is both active (it contains an active node) and stimulated (some argument changes state). During the period in which a subnetwork is active it is necessary to reevaluate the subnetwork whenever there is a change in state on any argument. An argument which stimulates the active subnetwork can itself be either active or inactive. If it is inactive, the same state change also occurs at time t in the previous simulation, and an evaluation of the argument is not required even though an evaluation of the stimulated subnetwork is. If the argument is active, the state change has occurred due to an evaluation of the argument during the current simulation. During the span of time that a subnetwork is active, the incremental simulator must reevaluate the subnetwork whenever there is a change in state with either an inactive argument during the previous simulation or an active argument during the current simulation.

Since we cannot predict the future active status of a subnetwork argument, during incremental simulation as soon as a subnet becomes active it is scheduled for an *evaluation check* for the earliest future event (EFE) occurring on any argument in the history trace of the previous simulation. At each evaluation check, the subnetwork is examined and if it is found to be both active and stimulated it is evaluated. At the end of the evaluation check, if the subnetwork remains active it is scheduled for another evaluation check at the next EFE occurring in the history trace of the previous simulation. This strategy ensures that an active subnetwork will be reevaluated whenever it is stimulated by change in state on an inactive argument. The case where an active subnet is stimulated by a change in state on an active argument is handled by scheduling an evaluation check of the subnetwork whenever one of its arguments is evaluated in the current simulation and found to change state, in a manner similar to nonincremental event-driven simulators.

Shown in Figure 1 is a high level view of the incremental switch-level simulation algorithm. The algorithm is event-driven [4] and uses a levelized two-pass approach [6] to suppress delta-spikes (multiple state transitions occurring on a zero-delay node during a single time-slice). Two event queue mechanisms are employed. A time-wheel [12] is used to synchronize future events A level-queue implemented as an array of list [6] is used to synchronize simultaneous events and suppress delta-spikes. The simulation progresses monotonically through time beginning with the initial time-slice and proceeding until either there is no more activity within the circuit or a user controllable time limit is exceeded. During each time-slice, the simulator processes two types of pending events: settle events, and check events. Settle events are similar to events scheduled in conventional nonincremental event-driven simulators. The incremental simulator only generates settle events for active positive-delay nodes, and then only when their subnetworks are evaluated. Check events are used to ensure subnetworks are considered for evaluation whenever active or stimulate conditions are in flux. A more extensive description of the incremental simulation algorithm may be found in [8].

```
/* touched nodes are always considered active */
/* all nodes are considered stimulated at time zero */

schedule check event at time zero for each touched node;
while there is a scheduled event do
  let t be the time of the earliest scheduled event;
  for each settle event scheduled for time t do
    schedule check event for this node at time t;
    extract state of node;
    if node changed state then
      update history;
      schedule check event for fan-outs at time t;
    endif
  endfor

  for all time t check events (in topological order) do
    expand the subnet;
    cancel all pending check events for subnet;
    extract states of subnet arguments;
    if subnet is active and stimulated then
      evaluate subnet;
      schedule settle events for positive delay subnet nodes;
      for each zero delay node in subnet do
        if node changed state then
          update history;
          schedule check event at time t for fan-outs;
        endif
      endfor
    endif
    schedule check event at time t for fan-outs of all subnet
    nodes that started/stopped deviating;
    let r be the time of the EFE of the subnet;
    schedule check events at time r for active nodes in sub-
    net;
  endfor
endwhile
```

Fig. 1. The incremental simulation algorithm.

6. PERFORMANCE

ICESIM is implemented in C on a SUN SPARCstation 2. The system supports both nonincremental and incremental simulation, and applying either simulation method yields identical results. Experiments were carried out on the ISCAS combinational [2] and sequential [1] benchmark test circuits. Although these are gate logic designs, for our purposes, the gates were defined using standard CMOS design techniques. The size of the resulting circuits ranged from 414 transistors to 89,046 transistors.

One hundred randomly generated test vectors were applied to the primary inputs of each circuit (the sequential circuits also required a clock signal). Each design was then subjected to a series of ten randomly generated design flaws. Each flaw was generated as follows: A single subcomponent pin was selected at random. The pin was then (randomly) either disconnected from whatever wire it was connected to leaving it unconnected, or disconnected and reconnected to some other (randomly chosen) wire in the same schematic. After each design flaw was introduced, the design was simulated for all one hundred test vectors, the flaw was corrected, the design was incrementally resimulated, and then (for comparison) nonincrementally simulated. The statistics presented in this section were gathered from the latter two simulations for each corrected flaw. The experiments were carried out twice for each design. The first using zero-delay wherever possible, the second using unit-delay on every node. Both sets of experiments used the same test vectors, and the same design flaws. For the zero-delay experiments, zero-delay cycles were broken by introducing unit-delays on feedback nodes (necessary for the sequential circuits).

Shown in Figure 2 is a summary of the relative speedup encountered using the incremental simulator on the ISCAS test circuits. Due to lack of space, we give only the results for the larger circuits (those having more than 10,000 transistors). The reader is refered to the complete paper[8] for more detailed results of the experiments. The relative speedup for a particular application of the incremental simulator is determined by dividing the CPU time used by the nonincremental simulator by the CPU time used by the incremental simulator on the same circuit. For each design and simulation delay, the table gives the ***median*** relative speedup which occurred during the ten resimulations. The median relative speedup is indicative of the typical speedups achieved using incremental simulation.

The speed advantage zero-delay has over unit-delay for nonincremental simulation is well known. The performance difference comes from two factors. First, zero-delay event processing is simpler than with unit-delay. Second, zero-delay tends to generate fewer events due to the rank-ordered synchronization of evaluations. In our experiments the combination of these factors meant that zero-delay nonincremental simulation was about 1.6 times faster (on average) than unit-delay for the same circuit. These differences are more pronounced when considering incremental simulation due to the higher complexity of processing positive-delay nodes in the incremental evaluator and the increase in events caused by time-shifts. Overall, zero-delay incremental simulation was about 4 times faster (on average) than unit-delay incremental simulation for the same circuit modification.

test	fets	zero delay	unit delay
c6288	10112	15.66x	3.18x
c5315	11270	201.05x	132.75x
c7552	15396	118.74x	58.74x
s9234	19854	714.33x	49.06x
s13207	28927	3427.20x	68.82x
s15850	34933	1860.00x	14.98x
s35932	75546	267.72x	98.14x
s38417	80996	4620.32x	341.42x
s38584	89046	503.32x	27.79x

Fig. 2. Relative speedup using incremental simulation.

7. REFERENCES

[1] F. Brglez, D. Bryan and K. Kozminski. "Combinational Profiles of Sequential Benchmark Circuits," *Proceedings of the 1989 International Symposium on Circuits and Systems* (May 1989), pp. 1929–1934.

[2] F. Brglez and H. Fujiwara. "A Neutral Netlist of 10 Combinational Bench Mark Circuits and a Target Translator in FORTRAN," *Proceedings of the 1985 International Symposium on Circuits and Systems* (June 1985), pp. 151–158.

[3] R. E. Bryant. "A Switch-Level Model and Simulator for MOS Digital Systems," *IEEE Transactions on Computers* (February 1984), Vol. C-33, No. 2, pp. 160–177.

[4] P. W. Case, H. H. Graff, L. E. Griffith, A. R. LeClercq, W. B. Murley and T. M. Spence. "Solid Logic Design Automation," *IBM Journal* (April 1964), pp. 127–140.

[5] K. Choi, S. Y. Hwang and T. Blank. "Incremental-in-Time Algorithm for Digital Simulation," *Proceedings of the 25th ACM/IEEE Design Automation Conference* (June 1988), pp. 501–505.

[6] S. Gai, F. Somenzi and M. Spalla. "Fast and Coherent Simulation with Zero Delay Elements," *IEEE Transactions on Computer-Aided Design* (January 1987), Vol. CAD-6, No. 1, pp. 85–92.

[7] S. Y. Hwang, T. Blank and K. Choi. "Fast Functional Simulation: An Incremental Approach," *IEEE Transactions on Computer-Aided Design of Integrated Circuits and Systems* (July 1988), Vol. 7, No. 7, pp. 765–774.

[8] L. G. Jones. "An Incremental Zero/Integer-Delay Switch-Level Simulation Environment," *IEEE Transactions on Computer-Aided Design* (to appear).

[9] L. G. Jones. "Incremental VLSI Design Systems Based on Circular Attribute Grammars", Ph.D. thesis, Department of Computer Science, Pennsylvania State University, December 1986.

[10] L. G. Jones. "Fast Batch and Incremental Netlist Compilation of Hierarchical Schematics," *IEEE Transactions on Computer-Aided Design* (July 1991), Vol. 10, No. 7, pp. 922–931.

[11] A. Salz and M. Horowitz. "IRSIM: An Incremental MOS Switch-Level Simulator," *Proceedings of the 26th ACM/IEEE Design Automation Conference* (June 1989), pp. 173–178.

[12] E. G. Ulrich. "Time-Sequenced Logical Simulation Based on Circuit Delay and Selective Tracing of Active Network Paths," *Proceedings of the ACM 20th National Conference* (1965), pp. 437–447.

[13] E. G. Ulrich. "Event Manipulation for Discrete Simulations Requiring Large Numbers of Events," *Communications of the ACM* (September 1978), Vol. 21, No. 9, pp. 777–785.

Performance Evaluation of an Event-Driven Logic Simulation Machine

Fumiyasu Hirose

Fujitsu Laboratories LTD.

1015 Kamikodanaka, Nakahara-ku, Kawasaki 211, Japan

ABSTRACT

This paper evaluates the performance of an event-driven logic simulation machine, the SP. Since an event-driven machine only schedules gates that have signal-changes on their inputs, it processes fewer gates than the level-sort machine does. However, if the event-driven machine spends too many clocks on dynamic scheduling, the simulation time can not be reduced. We measured the overhead for dynamic scheduling, and found that it only averages 2% over the total process. The evaluation was done using the ISCAS'89 benchmark circuits, and the results are shown on a machine cycle basis We individually evaluated some special functions of the SP for acceleration. Furthermore, we compared the simulation speed with that of a software simulator that uses the same data structure and algorithm as the SP.

1 Introduction

Logic simulation is essential for the quick development of design-faultless digital systems. Logic simulation machines[1] have made great progress in accelerating this process. The origin of the machines is said to be the Boeing Computer Simulator in 1970[2]. However, practical developing machines have became popular since the 1982 YSE[3]. The logic simulation machines are one of the most successful CAD engines in both technical and economic effects.

The logic simulator evaluates the output values of gates in a circuit. In the YSE, the gates are evaluated according to a **static** pre-defined order, where each gate is evaluated exactly once during each simulated clock cycle. This makes the hardware simple, but involves useless evaluation. An event driven machine employs **dynamic** scheduling. It only schedules gates that have signal-changes (events) on their inputs. Far fewer gates need to be evaluated if the circuit activity is low, and the average event rate is estimated to be 15%[1]. This means that event-driven machines could reduce the number of evaluated gates by five-sixths. However, if the machine spends a lot of clocks on dynamic scheduling, this advantage can not reduce the simulation time.

In the event-driven machine architecture of the SP[4], the gates that should be evaluated during the next simulation cycle can be scheduled in parallel to the main process of the evaluation and the update of gates for the present cycle. We have measured the overhead of dynamic scheduling, and found that it only averages 2% over the total process.

A performance evaluation of the level-sort machine has been reported in detail[5]. However, there is little information on event-driven machines. Furthermore, conventional reports have not used circuits publicly available, and the simulation speed is often compared with a software simulator that does not necessarily use the same data structures, models, or algorithms. We do not consider these evaluations sufficient to evaluate a machine architecture.

This paper evaluates the performance of the SP using the ISCAS'89[6] benchmark circuits, and reports it on a machine cycle basis. Special functions of the SP are evaluated one by one. Furthermore, the SP's speed is compared to a software simulator that uses identical data structures and algorithms.

2 Characteristics of benchmark circuits

2.1 Static characteristics

The five largest ISCAS'89 benchmark circuits were selected. These were originally synchronous circuits. However, we removed registers and made them combinational so that we could easily control the circuit event rate. The static characteristics of the circuits we obtained are shown in Table 1.

"#PO" is the number of primary output buffers, "#PI" is the number of primary input buffers, and "#Gates" is the number of internal gates. The internal gates do not contain the POs and the PIs. "Ave. #FO" is the average number of fanout gates of the PIs and internal gates. "Ave. #FI" is the average number of fanin gates of the POs and internal gates.

2.2 Simulation conditions

We generated input patterns for the circuits with a random pattern generator that generates input sequences at a given rate. Each input is independent of the other inputs. A single processor of the SP is used for the simulation. The unit delay model is used. When the circuit becomes stable for the current input pattern, the next

Table 1 Static characteristics

Name	#Gates	#PIs	#POs	#FOs	Ave.#FO	Ave.#FI
s38584	22069	1464	1730	37302	1.59	1.57
s38417	24101	1664	1742	35692	1.39	1.38
s35932	17793	1763	2048	32045	1.64	1.62
s15850	10649	611	684	15206	1.35	1.34
s13207	9116	700	790	13120	1.34	1.32
Average	16746	1240	1399	26673	1.48	1.47

one is applied. The period of the simulation for one input pattern is called a **cycle**, while the word **clock** refers to the machine clock cycles of the SP.

2.3 Dynamic characteristics

Gates that had events in their inputs are called **active gates**. The event-driven machine only evaluates the active gates. The ratio of active gates to all gates is the **gate event rate**. The ratio active gates whose output will change to all active gates is the **event through rate**.

Figure 1 shows the relation between the gate event rate and the event rate at the primary inputs(PIs). The relation is almost linear for all circuits. The gate event rate can exceed 100% because some gates are evaluated more than once based on the existence of circuit hazards. Figure 2 shows the relation between the event through rate and the event rate at the PIs. The event

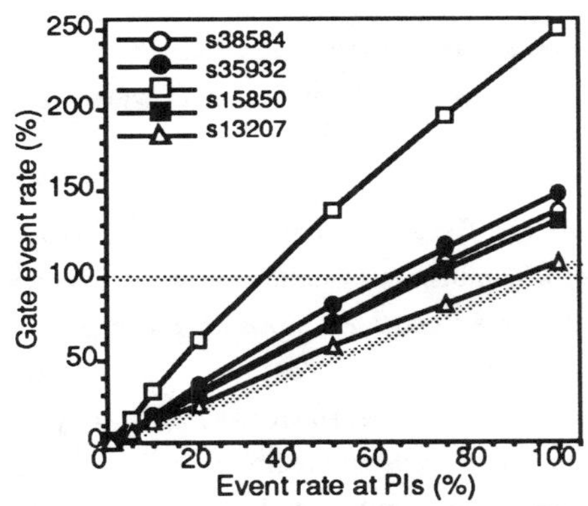

Fig. 1 Gate event rate vs. event rate at PIs

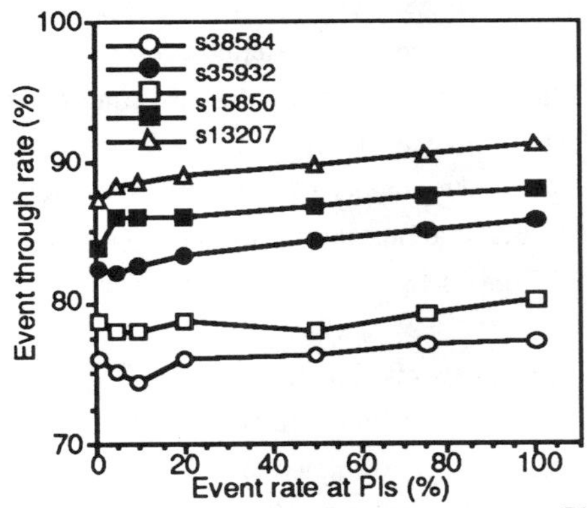

Fig. 2 Event through rate vs. event rate at PIs

through rate is almost constant. On average, 82% of the active gates generate events on their output.

3 Basic performance evaluation

3.1 Ideal clocks

All logic simulator have data structures that hold the status of the circuit signals. This is called net status memory (NSM). The simulator computes the NSM from unit time to unit time. The NSM is read for gate evaluations, and written for status updates.

In the level-sort simulation, every gate is evaluated and updated exactly once during each simulation cycle. Let the number of gates G. Then, the NSM is accessed $2G$ times. On the other hand, the event-driven simulator only evaluates active gates, updates gates that will have events on their outputs, and schedules active gates for the next unit time. This scheduling is called fanout-fetch. Let the gate event rate be E, and the event through rate be R. Then, the NSM is read EG times for evaluation, and is written REG times for update. Therefore, the NSM is accessed $E*(1+R)*G$ times during each simulation cycle. Remember that the average event through rate is 0.82. When we use a gate event rate of 15%[1], the event-driven NSM is accessed $0.15*1.82*G=0.273G$ times, which is about one-seventh as often as the level-sort simulator.

In the SP architecture, we partitioned the simulation process into three functions, Evaluate, Update, and Fanout-fetch, as shown in Figure 3. Each function is processed by a pipeline. Furthermore, we intended to mask the Fanout-fetch process by running it in parallel to Evaluate and Update. Evaluate and Update cannot run in parallel because they both access the NSM. Each stage of the pipeline can access its local memory once per clock in the SP. Therefore, the pipelines ideally process one gate per clock. If the Fanout-fetch is also completely masked, the machine can process a simulation cycle in $E*(1+R)*G$ clocks, which is the number of accesses to the NSM. We define this as the number of **ideal clocks** for a simulation cycle. Also, $(1+R)$ clocks is called the number of ideal clocks for an active gate.

3.2 Relation between ideal and measured clocks

We calculated the number of ideal clocks and measured the actual machine clocks of the SP for a simulation cycle. Figure 4 shows the relation. This rough observation shows that the SP operates in

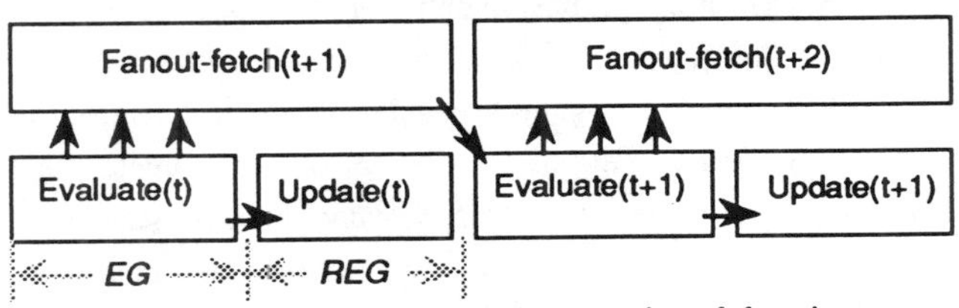

Fig. 3 Pipeline and parallel processing of functions

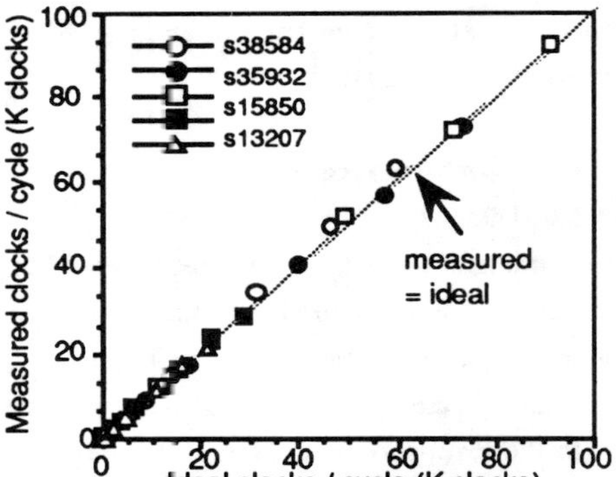

Fig. 4 Ideal clocks vs. measured clocks

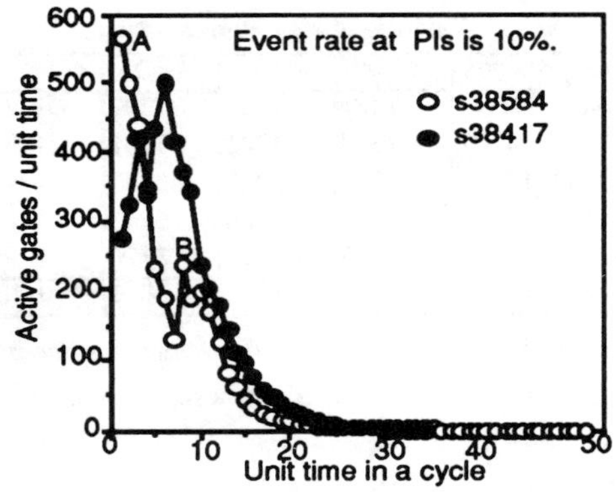

Fig. 6 Circuit activity

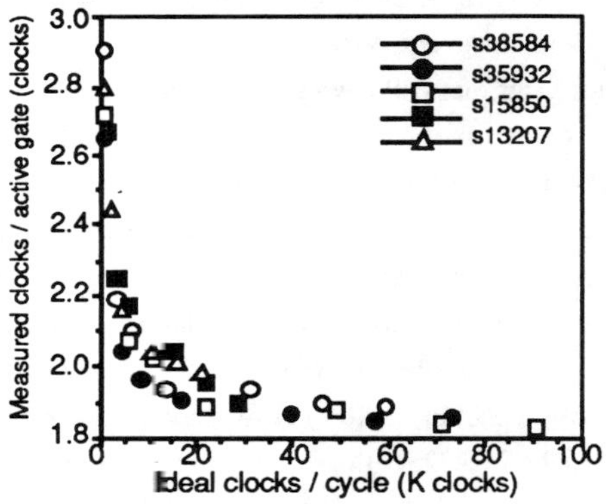

Fig. 5 Measured SP clocks to process an active gate

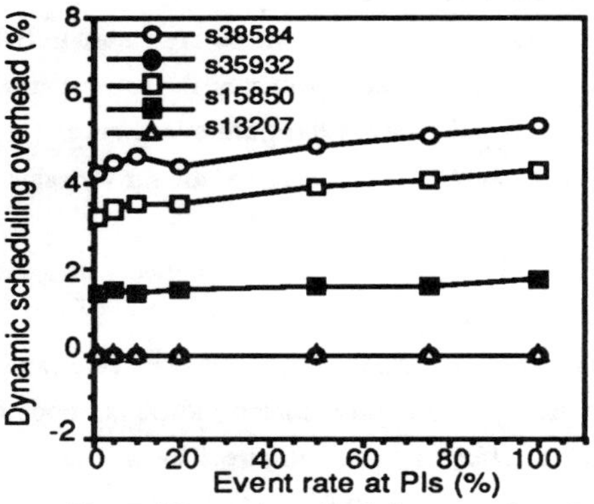

Fig. 7 Dynamic scheduling overhead

close to the number of ideal clocks.

We measured the number of clocks, "*Y*", spent to process an active gate. Figure 5 shows the relation between *Y* and the number of ideal clocks for a simulation cycle, "$X=E*(1+R)*G$". Ideally, the relation should be $Y=(1+R)$, that is almost constant. However, it is observed that if *X* is small, in other words, if the number of gates to be processed by the pipelines is small, the SP needs more machine clocks.

The SP has overhead for pipelining and dynamic scheduling. The pipeline overhead is in inverse proportion to *X*. Figure 5 shows this relation. The pipeline overhead also exists in a level-sort machine.

3.3 Dynamic scheduling overhead

Figure 6 shows the number of gates evaluated per unit time in a simulation cycle for the two largest circuits, s38584 and s38417. The event rate at the primary inputs is 10%. About 80% of the active gates for a cycle are evaluated by unit time 10.

If the number of gates scheduled in Fanout-fetch(t+1) in Figure 3 is greater than the total number of gates in Evaluate(t) and Update(t) at unit time t, then there is some dynamic scheduling overhead. We found two points where overhead exists in the case of s38584. These are denoted in Figure 6 as A at unit time 1 and B at unit time 8. No overhead was observed in s38417. Receiving external events as PI events at unit time 1 is considered to be a part of the Evaluate pipeline process.

We added the number of overhead clocks for each time unit for a cycle. The total number is divided by the number of ideal clocks of a simulation cycle to obtain the dynamic scheduling overhead. These figures are shown in Figure 7. The overhead is almost independent of the event rate at the primary inputs. The maximum overhead is 5%, and the average is only 2%. Except for s38584, the only overhead observed was at unit time 1. s38584 only had overhead at unit times 1 and 8, and the overhead at unit time 1 was much heavier. This means that almost all of the overhead is caused when the external events are received at unit time 1. There is little overhead for the simulation of internal circuits. The overhead of dynamic scheduling in the SP is thus very small, so the number of machine clocks spent to process individual gates is almost the same as that of a level-sort machine, while the SP only selects the active gates.

4 Effect of special functions

We evaluated the effects of two SP functions. One is a **registration filter** for avoiding double registration of active gates. The other is **input invertor** that inverts input values without using

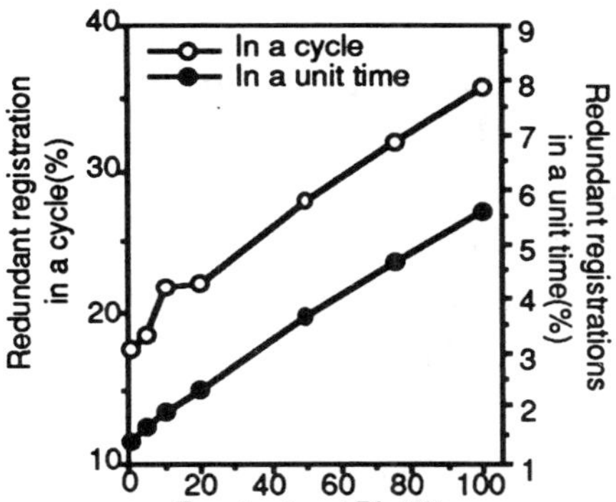

Fig.8 Effect of registration filter

inverting gates. The registration filter is a built-in SP function. Its effect is included in the basic performance evaluation in Chapter 3. On the other hand, the input invertor is used by the circuit compiler. Its effect is not included in the basic performance.

We measured the number of gates that are scheduled more than once per unit time. The ratio of this number to the total number of active gates is shown in Figure 8 using black circles. This is the average ratio of the five circuits. When the event rate at the primary inputs increases, the double registration ratio per unit time also increases. The increase in number is trivial, but the ratio also increases. When the event rate at the primary inputs is 10%, about 2 % of the active gates are scheduled more than once, if the SP does not have the registration filter.

If the registration filter is not used, extra gates are caused and processed in the following ways. First, if a gate is scheduled more than once, it must be evaluated more than once with identical input vectors. Second, if an event is detected at the output of the gate, it must be updated more than once by the same value. Third, the same fanout-fetch process must be done more than once. Then, the rescheduled gates will generate active gates that are registered more than once for the next cycle. The ratio of the extra gates that would be registered to that of the active gates actually registered in the SP is shown in Figure 8 using white circles. This is also the average ratio of the five circuits. When the primary input event rate is 10%, about 22% of the active gates are scheduled more than once per simulation cycle.

Some machines can invert input vectors without using invertors. It is especially important when ECL circuits are simulated. By using the input invertor, we can eliminate invertors from a circuit. The simulation may be accelerated, if few hazards are generated. Table 2 shows this effect. The size effect column shows the ratio of the original circuits to the new circuits. On average, the circuits are reduced by half. The speed effect column shows the acceleration of the input invertor. The event ratio at primary inputs is 10%. On average, the speed is accelerated 1.7 times.

Table 2 Effects of input inversion

Name	#Gates		Effect	
	Old	New	Size	Speed
s38584	22069	12130	1.8	1.5
s38417	24101	8852	2.7	2.0
s35932	17793	12204	1.5	1.4
s15850	10649	3588	3.0	2.2
s13207	9116	2821	3.2	2.3
Average	16746	7919	2.1	1.7

5 Comparison to software simulator

We compared the SP's simulation speed with a software simulator that uses the same data structures and algorithms. When the event rate at the primary inputs is 10%, a single SP processor simulates 170 times faster than the software simulator does on average. The maximum speed-up is 181 times for s38417, and the minimum is 149 times for s13207. The software simulator runs on a 25 MHz SUN4/370 with 32 megabytes of memory.

6 Conclusion

We evaluated the performance of the SP event-driven simulation machine. The dynamic scheduling overhead averaged 2%. We used the ISCAS'89 benchmark circuits, and reported the performance on a machine cycle basis for objectivity. We also evaluated special functions of the SP, and compared the speed to a software simulator that uses identical data structures and algorithms.

SPs are being used as a main tool of logic verification for large digital system designs. We would like to extend this research to establish a basis of comparing the performance of other machine architectures, and discuss the tradeoffs of simulation algorithms.

References

[1] T. Blank, "A Survey of Hardware Accelerators Used in Computer-Aided Design," IEEE Design and Test of Computers, vol. 1, no. 4, pp. 21-39 (1984).

[2] J. K. Howard, R.L. Malm, and L. M. Warren, "Introduction to the IBM Los Gatos Logic Simulation Machine," Proc. of IEEE Int'l Conf. on Computer Design, pp. 580-583 (1983).

[3] M. M. Denneau, "The Yorktown Simulation Engine," Proc. of 19th Design Automation Conf., pp. 55-59 (1982).

[4] Hirose, F. et al., "Simulation Processor 'SP'," Proc. of IEEE Int'l Conf. on Computer Aided Design, pp. 484-487 (1987).

[5] G. F. Pfister, 'The IBM Yorktown Simulation Engine," Proc. of IEEE, vol. 74, no. 6, pp. 850-860 (1986).

[6] E. Brglez, D. Bryan, and K. Kozminski, "Combinational Profiles of Sequential Benchmark Circuits, " Proc. Int'l Symp. Circuits and Systems, pp. 1929-1934 (1989).

HLSIM - A New Hierarchical Logic Simulator and Netlist Converter

D. A. Zein, O. P. Engel and Gary Ditlow*
IBM, East Fishkill, Hopewell Junction, N.Y. 12533
*** IBM, T. J. Watson, Research Center, Yorktown Height, N.Y.**
Tel:(914)892-2030, Fax:(914)892-4316

Abstract

HLSIM is a Hierarchical Logic SIMulator that can deal with nested models. The program has facilities to handle large VLSI circuits with complicated sequential logic which can be used to simulate the logical behavior of complete chips. The paper discusses two aspects of the simulator, as an analog to digital netlist converter and as a new digital simulator. As a converter, it takes a hierarchical analog ASTAP netlist and supplementary files for the leaf models and creates a digital model of the VLSI circuit. The simulator is discussed in terms of use and features, including how special cases such as emitter and collector dots and differential pairs are handled. Various examples of simulations are given.

1.0 HLSIM in an Automated Design Environment

HLSIM is integrated in an analog and digital simulation and design verification environment demonstrated in Figure 5. First, the function and characteristics of the VLSI circuit are defined. The circuit block diagram is laid out using SLED (an IBM schematic editor/capture program). It is then subdivided hierarchically into smaller functional blocks that form a tree with the basic circuits (books) or macro models at the leafs of the tree. A simulation library SIMLIB is created. Each member of which belongs to a leaf model. The program uses the SIMLIB and the analog ASTAP netlist library created by the SLED's netlist capture facilities to create automatically a digital model of the VLSI circuit. The model is then simulated to verify the design in a bottom-up fashion, starting with the leaf models and ending with the whole VLSI circuit. In this environment, the user only specifies the names of two libraries, the SIMLIB and the ASTAP analog library. A third library is used to store the output results. In this architecture, the logic functions of the leaf models are verified by the analog simulator [1]. The simulator uses a simulation order [2] and conducts 2^N dc reruns on each book (N is the number of fan-ins) creating a voltage matrix that is converted to a truth table using thresholding values. The truth table thus created is compared to the user's tables obtained from the logic functions in the SIM files.

1.1 Program Features

HLSIM is intended to simulate ultra VLSI circuits including chips. It has a netlist convertersion facility that converts analog nested models to digital models as is described above. Its language syntax conforms to that of ICD or ASTAP [1,5]. Wide buses are represented compactly using strand notation, and can be outputted in either binary or hexadecimal in a zero delay mode. The program accepts primitive logic gates such as NOT, OR, AND, NOR, NAND, XNOR as well as truth tables and arbitrary user functions. These are useful to model complicated logic blocks including arrays and sequential circuits. HLSIM has two modes, zero and variable, including unit delay. It uses 2-state for zero and 3-state logic for variable delay. Emitter and collector dots in bipolar circuits are handled easily, efficiently, and uniquely. Here, the program automatically creates OR, AND gates when the dotted nodes are flagged in SIM or supplementary files, (see Figure 1). The program has been implemented in APL (A Programming Language) [6] because of its powerful nested arrays that are ideal for gates representation. Furthermore, since APL is interpretive, it is conducive to the use of arbitrary user models since these models are evaluated on the fly without having to augment them to a program's load module as is the case in compiler languages [5].

1.2 Simulation Files

To automate logic simulation and design verification, a new libray of simulation, SIM, files is created. Each file belongs to a leaf model or book. An example of a SIM file can be seen in Figure 1 for an L1 latch. in the COUNTER circuit of Figure 2 shown below. The file contains information on the type of circuit, whether it is sequential or combinational, the name of the feedback variable, the input, output nodes. These are flagged by the word EDOT, CDOT if the node is emitter or collector dotted. Differential pairs are represented syntactically by two node names separated by a colon. SETUP statements are used for latches to represent the decision tree of the feedback variable, (see Figure 2). Finally, Boolean expressions are given for each output variable.

```
MODEL ADR4     = TYPE (SEQUENTIAL,SRL_L1,SL1,AK);
* PRODUCED BY TXT2SVM ON 02/19/91 AT 11:47:28 BY STOM
* EDITED BY HUOTT ON 2/25/91
* SLED LASTMOD 11:17:0301/25/91WVH52C
SO = FEEDBACK(VOLTAGE);
SETUP0 (AND (NOT SL1) (NOT AK) (NOT SO));
SETUP1 (AND (NOT SL1) (NOT AK) SO);
SETUP0 (AND (NOT SL1) (AK) (NOT SI));
SETUP1 (AND (NOT SL1) (AK) (SI));
SETUP0 (AND (SL1) (NOT AK) (NOT IN));
SETUP1 (AND (SL1) (NOT AK) (IN));
AK        = INPUT  (PIN_X);
IN        = INPUT  (PIN_X);
SI        = INPUT  (PIN_X);
SL1       = INPUT  (PIN_X);
SO        = INPUT (PIN_X);
IP1:OP1   = OUTPUT (PIN_X,EDOT);
IP2:OP2   = OUTPUT (PIN_X,EDOT);
IP3:OP3   = OUTPUT (PIN_X,EDOT);
IP4:OP4   = OUTPUT (PIN_X,EDOT);
GND       = VOLTAGE(0.0);
VCC       = VOLTAGE(1.4);
VEE       = VOLTAGE(-2.2);
VTR       = VOLTAGE(-0.75);
IP1:OP1 =OR (AND AK SI) (AND SL1 IN) (AND (NOT AK) (NOT SL1) SO);
IP2:OP2 =OR (AND AK SI) (AND SL1 IN) (AND (NOT AK) (NOT SL1) SO);
IP3:OP3 =OR (AND AK SI) (AND SL1 IN) (AND (NOT AK) (NOT SL1) SO);
IP4:OP4 =OR (AND AK SI) (AND SL1 IN) (AND (NOT AK) (NOT SL1) SO);
```

Figure 1. A simulation (SIM) file for an L1 latch

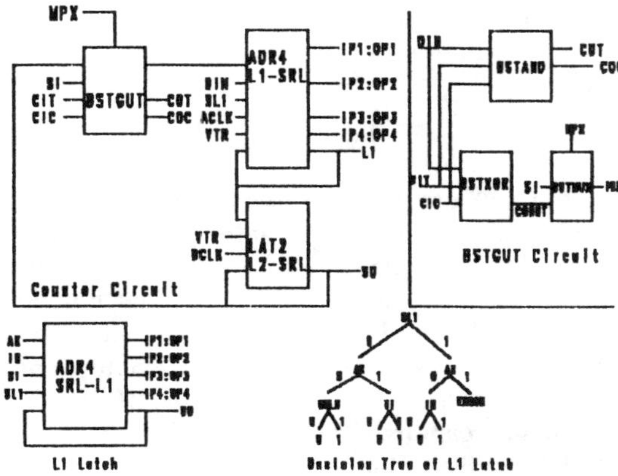

Figure 2. Bock diagram of counter circuit of a BICMOS decoder

1.3 The Netlist Converter

The input to the program comes basically from two libraries: The first one is the SIMLIB described above. The second is the analog netlist from a schemetic editor.

The netlist converter takes the analog hierarchical netlist and keeps only two types of statements, the model reference (MRS) and model definition (MDS) statement. For leaf models, it uses the corresponding SIM file to augment the MDS with a digital model. For example, for the L1 latch mentioned above, the netlist converter created three types of gates: 1. Looking at the SIM file of Figure 1, the SETUP statements in the SIM file are represented by the decision tree of Figure 2. This is implemented by the user function or behavioral model:

```
    Z←ADR4
[1] Z←(Y1[5 3 2],¯1)[1+2⊥Y1[4 1]]
    ∇
```

In the above function, Vector Y1 is the state of the input nodes, (AK,IN,SI,SL1,SO), (see statement # 4 in Figure 3). Vector Y1 is made available to the user by the program whenever user functions or truth tables are used. In case of truth tables it is decoded into a decimal integer to pick the proper row of the truth table that contains the states of the output nodes. 2. The TRUES of the four differential pairs are represented by a truth table, 3. while their complements are implemented by a NOT or inverter gates. The three statements are numbered 4, 5, (6,7,8,9), respectively, in the digital model of Figure 3 created by the converter. The truth table is augmented to a TRUTHTABL statement in the netlist. The program automatically creates those truth tables from the Boolean expressions in the SIM file. The name of Function ADR4 is added to an EXPRESSION in the digital netlist, (see Figure 7).

1. ADR4 (DIN,MUX,SL1,ACLK,IP1,IP2,IP3,IP4,OP1, OP2,OP3,OP4,L1) MODEL ADR4
2. MODEL ADR4 (DIN,MUX,SL1,ACLK,IP1,IP2,IP3, IP4,OP1,OP2,OP3,OP4,L1)
3. EDOT:IP1,OP1,IP2,OP2,IP3,OP3,IP4,OP4;
4. Q2 (AK,IN,SI,SL1,SO;SO) X1
5. Q1 (AK,IN,SI,SL1,SO;IP1,IP2,IP3,IP4) TT1
6. D1 (IP1;OP1) NOT
7. D2 (IP2;OP2) NOT
8. D3 (IP3;OP3) NOT
9. D4 (IP4;OP4) NOT

Figure 3. Digital Model for an L1 Latch

1.3.1 Emitter and Collector Dotting

Emitter and collector dots are unique with bipolar circuits. An emitter, collector dot is, in general, an OR, AND gate. If an output node is emitter or collector dotted, it is flagged in the SIM file. The four differential pairs of the L1 latch are emitter dotted and are flagged as such in the SIM file of Figure 1. The dotting can be seen in the output of the first stage decoder of a 1-megabit BICMOS chip of Figure 6. In that figure, there are 24 emitter dots to account for three 8-bit buses that are the output of the first stage decoder. The program created the 24 OR gates as follows: It first detects the EDOT flags in the SIM file of Figure 1. It then augmented the EDOT statement (#3) in the digital model, and the program automatically creates the OR gates while dissolving the models of the digital netlist. For example, the logic block of Figure 4 has four emitter dots. After dissolving the models, the gates are those of Figure 4, (b) and (c) shown as tuplets. They are the input, output node numbers of each gate occurring in APL as nested arrays. The last four gates are the four emitter dots created automatically by the program. (This logic block is used to choose one of 4 sectors in each half of the chip of Figure 11).

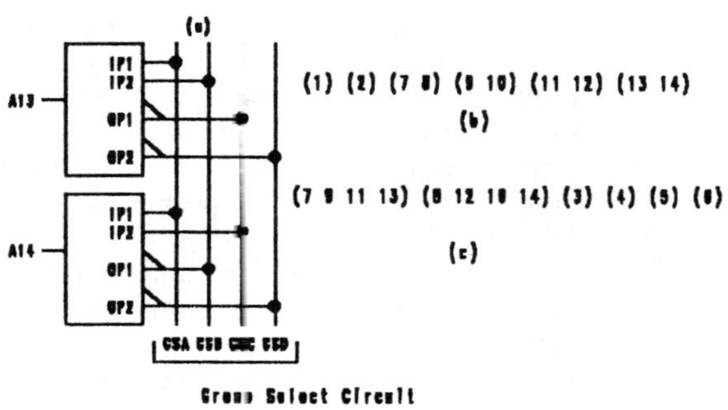

Figure 4. A Group Select Circuit of a 48k-bit Bipolar Array Chip

(a) Block diagram of GRPSEL circuit
(b) Input nodes to gates
(c) Output nodes of gates

1.3.2 Differential Pairs

The SIM file of Figure 1 shows how differential pairs are used. A differential pair is two nodes that have complementary logic states. A zero, one of a differential pair is the ordered pair (0,1), (1,0), respectively. When the netlist converter sees differential pairs, the truth tables created from the Boolean expressions in the SIM file has row dimension equal to 2^N, where N is the number of differential pairs - This reduces the size of the truth table exponentially. Furthermore, in verifying a book's logic using analog methods, 2^N dc simulations or reruns are used. Treating a differential pair as one variable reduces the number of reruns exponentially [2].

1.4 Algorithms

The program has three basic algorithms: First, since the digital netlist has the same structure as the analog netlists of [1,5], the same recursive algorithm is used in HLSIM to dissolve the models as that used in ICD [1]. However, the gates in the expanded circuit look like two huge APL nested arrays where the input nodes and output nodes are in vectors of vectors. Each element of a vector is a set of indices into the solution vector. The solution vector contains the states of the unique nodes of the VLSI circuit. The second algorithm deals with the way the zero delay problem is solved in the execution part of the program. The input part essentially creates a scheduler that splits the VLSI circuit into a combinational and sequential parts. For each input pattern, the combinational part is evaluated once while a Gauss-Jacobi type of relaxation (see Eq, (1) below) is used to solve the sequential part.

$$x_{i+1} = f(x_i) \qquad (1)$$

Where in Eq. (1), x is the solution vector containing the logic states of the sequential part of the circuit, f is a vector of logic functions, and i is the iteration number. Equation (1) is applied iteratively until $x_{i+1} = x_i$. The third algorithm deals with the way the variable delay problem is solved. The algorithm does not differ from the basic algorithm that was first introduced in [4] and is essentially used in most simulators such as Verilog [7] with slight variations. However, we use residue theory to choose the minimum step size for the time queue. [4]. This is described in [3].

1.5 Examples

We will use three examples to demonstrate how the program works in an automated design environment. The first one is a word decoder of a 1-megabit BICMOS chip, the second is a 48k-bit bipolar array chip, and the third is a ring oscillator. In this environment, the IBM schematic editor/capture program SLED creates the schematic of the VLSI circuit. Its schematic capture facility is used to produced the ASTAP netlist. This is FED to HLSIM which creates the digital netlist and performs the simulation. The details of the simulation, including the syntax of the input language can be seen in [8].

1.5.1 Example 1 - A word decoder of a 1-megabit BCMOS chip

The block diagram of this circuit is shown in Figure 6. The most important aspect of the decoder is that each of its nine address lines feeds into circuits each one of which contains the counter circuit of Figure 2. We have discussed how the netlist converter creates the model for the L1 latch using the SIM file of Figure 1. The complicated sequential logic is evident from the fact that the feedback variable is a function of itself, the scan line the input, and the clocks. The four differential pairs are themselves very complicated Boolean functions of the feedback variable, the input, the clocks and the scan variables, (see Figure 1 and Figure 2). The sequential logic and the feedback loops in the COUNTER circuits were handled efficiently by the algorithm described above. The program verified the three modes of operation of the decoder: system, self and test modes using 70 input patterns. Sample outputs contain two sets of input/output patterns. The output of the first stage is in binary consisting of three 8-bit buses, while the 512-bit output of the decoder is in hex. Note that the decoder has the 24 emitter dots as outputs of the first stage decode (see Figure 6). The program automatically created the 24 OR gates from just using the flags in the SIM file of Figure 1.

1.5.2 Example 2 - A 48k-bit Bipolar Array Cache Chip

The second example is the 48k-bit bipolar array cache chip represented in the block diagram of Figure 11. The most important characteristics of this simulation is that the chip has the same type of emitter dotting for its bit (column) and word (row) decoders as the decoder of the first example. However, there are 192 collector dots that are coming out of the sense amplifiers. But the most important aspect of the simulation is the way the 48k-bit of array cells is represented and modeled. Here, the array itself is represented as a 4-dimensional APL array whose rank is (2x4x48x144) standing for 2 halves of the array, odd and even, (see Figure 11); 4 sectors in each half, and 48 bit lines in each sector and 144 word lines. The array itself is modeled as a user function or behavioral model that identifies the correct cells selected and the mode of operation, namely, deselect the array, read/write 1 or zero. The program was also used to verify redundancy switching, among other things. A sample I/O set is given below.

1.5.3 Example 3 - A 5_Stage Ring Oscillator

The last example shows a simple 5-stage ring oscillator to demonstrate unit delay and 3-state logic. Note that at the very beginning, the output is in an unknown state. Oscillations are triggered when the input goes down and are halted when it goes up again. The circuit schematic is shown in Figure 9 and the output from HLSIM is shown in Figure 10.

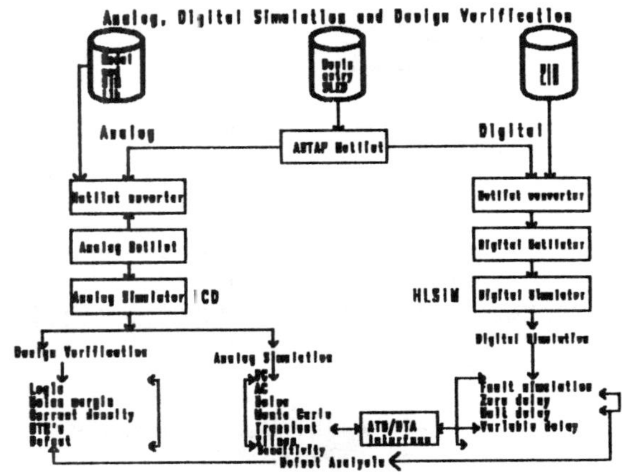

Figure 5. Analog & digital design environment

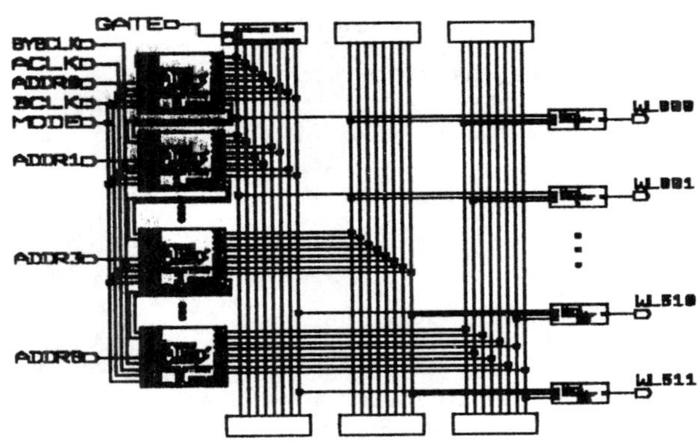

Figure 6. Block diagram of 1-megabit BICMOS DECODER

```
DECLARE BIN O1=W<8:0>,CK0,INH,AMPX,SLOW,AK0,BK,SI;
DECLARE BIN O2=WAD<0:23>;
DECLARE BIN O3=L1.CNTW2.ADR4X32.WADD9.LOG1A,
               L1.CNTW1.ADR4X32.WADD9.LOG1A,
               L1.CNTW0.ADR4X32.WADD9.LOG1A,
               L1.CNTW2.ADR4X31.WADD9.LOG1A,  <== fully qualified name
               L1.CNTW1.ADR4X31.WADD9.LOG1A,      of the feedback variable
               L1.CNTW0.ADR4X31.WADD9.LOG1A,      of the 9 L1 latches.
               L1.CNTW2.ADR4X30.WADD9.LOG1A,
               L1.CNTW1.ADR4X30.WADD9.LOG1A,
               L1.CNTW0.ADR4X30.WADD9.LOG1A;
DECLARE HEX O4=WL<0:511>;
SIGNALS:
W8=0,W7=0,W6=0,W5=0,W4=0,W3=0,W2=0,W1=0,W0=0,CK0=0,INH=0,AMPX=0,
SLOW=0,AK0=0,BK=0,SI=0,VE''=0,VUP=1;
LOG1A (W<8:0>,CK0,INH,AMPX,SLOW,AK0,BK,SI,+
WAD<0:23>,WL<0:511>,VE'',VUP) MODEL LOG1A
MODEL LOG1A (W<8:0>,CK0,INH,AMPX,SLOW,AK0,BK,SI,+
WAD<0:23>,WL<0:511>,VE'',VUP)
      .    ==> rest of netlist <==
      .
EXPRESSIONS:X1=ADR4,X2=LAT2;
TRUTHTBLS:TT1=00001111000000000000001110000000000000011111111111110000111
1111111000000000000000001111111111111111100000000111111111111111111111111
11,TT2=11111111111111111000000001111111,TT3=0001,TT4=00011101,
TT5=0110,
TT6=01111111111111111111111111111111111111111111111111111111111111111111
111111111111111111111111111111111111111111111111111111111111111111111111
111111111111111111111111111111111111111111111111111111111111111111111111
111111111111111111111111111111111111111111111111111,TT7=01,TT8=01,
TT9=1100000000000000,TT10=1100000000000000;
OUTPUTS:
```

Figure 7. A Segment of the 1M BICMOS Netlist (Created by Netlist Converter)

O1 = 1111111001001000 < = = input pattern in binary
O2 = 111101111111111101 111110 < = = 1st. stage (3 8-bit buses)
O3 = 111111100 outputs from the 9 L1 latches
decoder output in hex - 512-bit line bus
O4 = 00000000000000000000000000000000000000000000000
000000000000000000000000000000000000000000000000
000000000000000000000000000000000000000008

O1 = 0000000001000100
O2 = 111111101111111101111111 0
O3 = 111111111
O4 = 000000000000000000000000000000000000000000000
000000000000000000000000000000000000000000000000
000000000000000000000000000000000000000000001

Two Sets of I.O Patterns for 1M BICMOS Decoder

```
DECLARE BIN INPUT=A;
DECLARE BIN OUTPUT=FB;
SIGNALS:
A=T1(TIME);
OSCIL(A,FB)MODEL OSCIL
MODEL OSCIL (A,FB)
STG1(FB,A,B)MODEL HP2CSO (TR=1,TF=1)
STG2(B,B,C)MODEL HP2CSO (TR=1,TF=1)
STG3(C,C,D)MODEL HP2CSO (TR=1,TF=1)
STG4(D,D,E)MODEL HP2CSO (TR=1,TF=1)
STG5(E,E,FB)MODEL HP2CSO (TR=1,TF=1)
MODEL HP2CSO(IN1,IN2,OUT)
Q1(IN1,IN2;OUT)NOR
RERUNS:
PARMS:
A,FB.
VALUES:0,1;
TABLES:
T1=0,1,5,1,6,0,100,0,101,1,125,1;
RUN CONTROLS:
START TIME=0,
STOP TIME=120,
PRINT INTERVAL=1,
STEP SIZE=1,
ANALYSIS=TIME;
OUTPUTS:
PLOT A,FB;
```

Figure 8. Digital Netlist Input to HLSIM for Oscillator

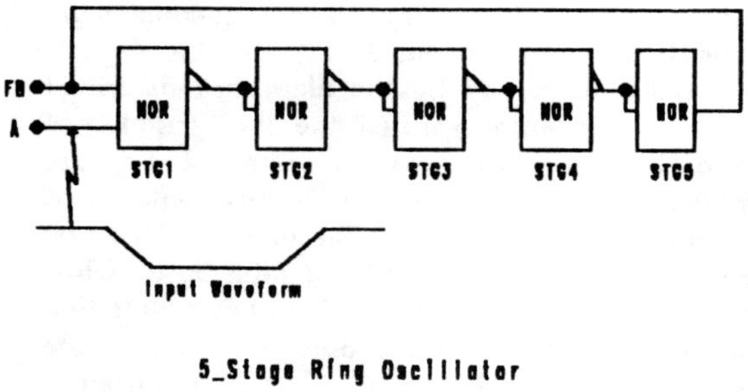

Figure 9. Block Diagram of 5_Stage Ring Oscillator

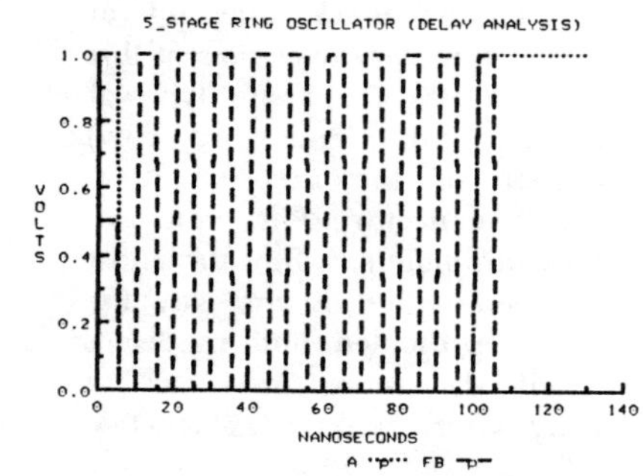

Figure 10. Ring Oscillator - Output from HLSIM

```
* RRRR CCPD CC A RR DD AAAAAAA AAAAAA AA
* 012E ISDC CI C WW II 0123456 789111 11
*          EN L EO EO           012 34
*           H
 0001 1100 00 1 00 11 0000000 000000 00 < = = Input pattern
```

ACL_CICSOC = 1111,DHS = 1,DHU = 0,DHO = 0,GW = 1,WC = 1,
RW = 0,DI = 1,
WRTOO = 0,WRT1O = 1,PDLO = 1111,PDRO = 1111,
CMP89 = 1,DO = 0,GS = 7,
WL = FFFFFFFFFFFFFFFFFFFFFFFFFFFFFFFFFFFFF0000,
BD = 000000000000,
BLO = 000000000000000000000000000000000000000000000000,
BRO = 000000000000000000000000000000000000000000000000
ACL_CICSOC = 0111,
HS = 1,DHU = 0,DHO = 0,GW = 0,WC = 0,RW = 0,DI = 1,
WRTOO = 0,WRT1O = 1,
PDLO = 0000,PDRO = 1000,CMP89 = 1,DO = 0,GS = 7,
WL = 7FFFFFFFFFFFFFFFFFFFFFFFFFFFFFFFFFFFF0000,
BD = 000000000000,
BLO = 000000000000000000000000000000000000000000000000,
BRO = 000000000000000000000000000000000000000000000000

I/O Patterns from the 48k-bit Bipolar Array Chip

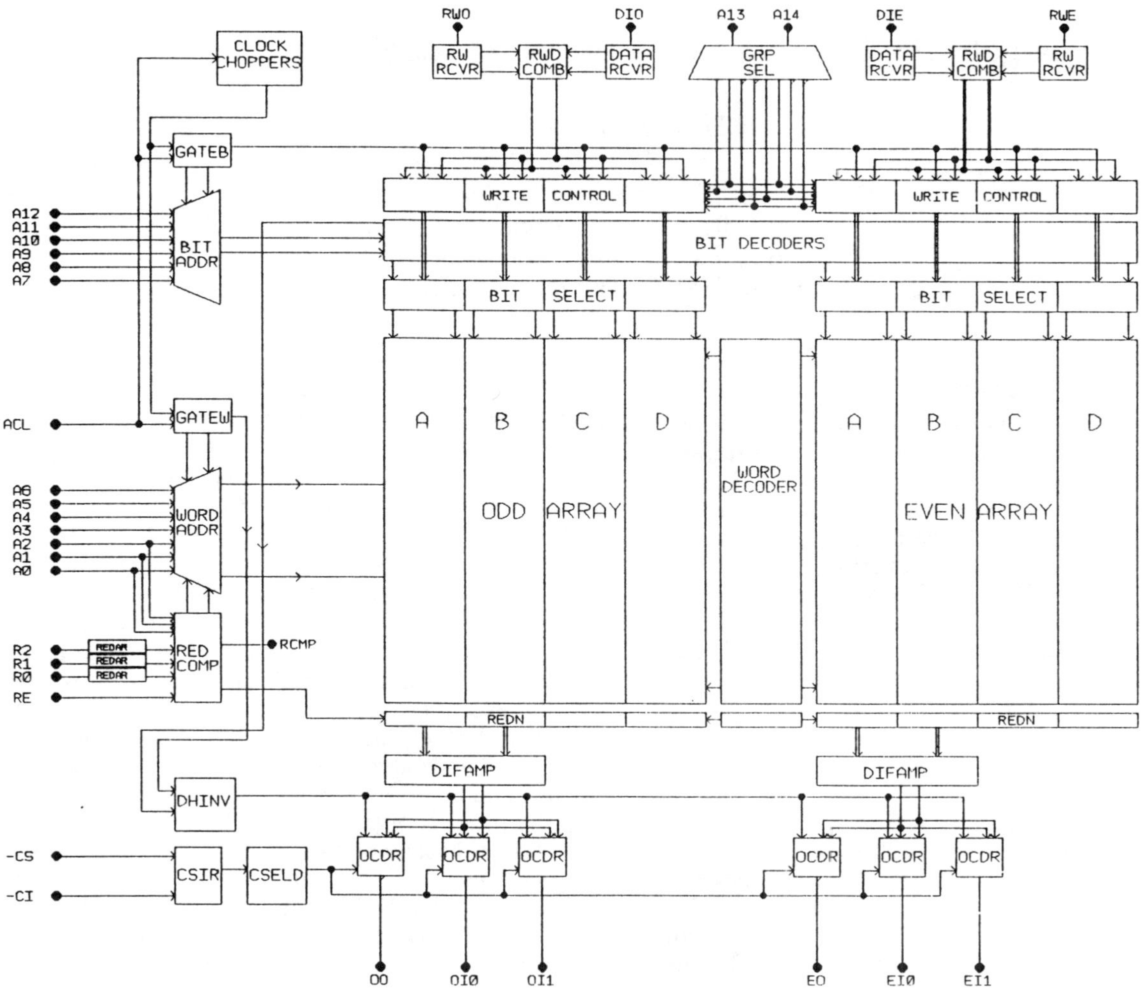

Figure 11. A Block Diagram of a 48k-bit Bipolar Array Chip

Acknowledgement

The authors would like to thank John Colclough and Bill Stinson for their constant support and encouragement. Thanks are also due to Bill Huott for giving the decoder example of the 1M BICMOS chip.

References

1. D. A. Zein, and C. W. Ho, "APL Simplifies Interactive Circuit Design," Electronic Design Magazine, March 31, 1982, pp.16-31.
2. D. A. Zein O. P. Engel and G. Ditlow, "Algorithms for the Design Verification of Bipolar Array Chips," (to appear in IEEE 10th VLSI Test Symposium, Atlantic City NJ April, 1992)
3. D. A. Zein, O. P. Engel and G. Ditlow, "HLSIM - A New Hierarchical Logic Simulator in APL," to appear in IEEE 10th VLSI Test Symposium, Atlantic City, April, 1992.
4. S. A. Szygenda, "TEGAS2 - Anatomy of a General Purpose Test Generation and Simulation System for Digital Logic," Proc. 9th Design Automation Workshop.
5. W. T. Weeks et al., "Algorithm for ASTAP, A Network Analysis Program," IEEE Trans. Circuit Theory, Vol.CT-20, pp. 628-634, Nov. 1973.
6. APL2 Programming Language Reference, Report #SH20-9227-2, IBM Corp., pp.188-193. Available through IBM Branch Offices.
7. Verilog - XL Interface Reference Manual, Cadence Design Systems Inc., Version 4, Sep., 1990.
8. D. A. Zein et. al., "HLSIM - Hierarchical Logic Simulator, User's Guide," (Unpublished IBM manuscript).

Coalgebraic Division for Multilevel Logic Synthesis

Wen-Jun Hsu and Wen-Zen Shen

Department of Electronics Engineering
& Institute of Electronics
National Chiao Tung University
Hsinchu, 30050, Taiwan, R.O.C.

Abstract

By introducing two Boolean properties ($a \circ a = a$, $a \circ a' = \emptyset$) into algebraic division operation, we can perform a subset of Boolean division with approximately the same complexity as the algebraic division implemented in misII. With the extended algebraic division algorithm called coalgebraic division, we can perform synthesizing procedures more generally within acceptable time. With a simple restriction during division, coalgebraic division can also preserve the network testability as well as the test patterns.

I. Introduction

When designing a logic function, it is more economic to implement the function with multilevel circuitry. In order to increase the efficiency of system design, various multilevel synthesizing tools have been implemented. One category of synthesizing system refines the circuit structure with rule-based local transformations[1,2]. The rule-based local transformation methods use a set of ad-hoc rules which are activated when certain patterns are found in the network of logic gates. Each rule transforms a pattern of a local set of gates and interconnections into another equivalent one. In general, the transformations have limited optimization capability since they are local in nature and do not have a global perspective on the design. The other category of synthesizing system optimizes network with algorithmic transformations[3-5]. The algorithmic transformation can be divided into two phases: a technology-independent approach based on algorithms for manipulating general Boolean functions and a technology-mapping step where the design described in terms of generic Boolean functions is mapped into a set of gates. In technology independent synthesizing phase, the most important operation is to identify common sublogic and then the overall architecture of the given logic is modified to produce near optimal structure. In algorithmic approach, identifying and substituting common sublogic is performed via 'division' operation. According to the property of Boolean algebra used, the division operation is divided into algebraic and Boolean approaches. Algebraic division has lower complexity and can be performed quickly. However, this may lose circuit optimality. On the other hand, Boolean division can produce smaller circuits at the expense of larger computation time. In order to further improve the quality of division operation while keeping lower complexity, we propose the coalgebraic division. By introducing two Boolean properties ($a \circ a = a$, $a \circ a' = \emptyset$) into division operation, we can further improve the quality of division. We have fitted the coalgebraic division into resubstitution operation in misII environment. From the experimental results, it is shown that the execution time of coalgebraic division is very close to that of algebraic division. And the literal count of synthesized network can be reduced by using coalgebraic resubstitution.

The paper is organized as follows: In section II, we will define basic notation and the representation of combinational logic. Then the proposed coalgebraic division algorithm is described in section III. In section IV, some experimental results are given to verify the efficiency of coalgebraic division. And at the last section, we will give a conclusion.

II. Basic definitions

We begin with some formal definitions. A literal is a Boolean variable or its complement. We formally define a cube as a set of literals and we interpret the cube as a product of literals. A cover is a set of cubes and we interpret the cover as a sum-of-products expression. A Boolean Network η is a directed-acyclic graph(DAG) such that for each node n_i in η there is an associated cover F_i , and a Boolean variable y_i representing the output of F_i . As there is a one-to-one correspondence between a node and its cover, we will use the terms interchangeable. There is a directed edge from n_i to n_j iff F_j explicitly depends on y_i or y_i'. Furthermore, some variables in η may be classified as primary inputs or primary outputs. In the sequel, literals are denoted with lowercase characters, cubes are denoted with boldfaced characters and covers are denoted with uppercase characters.

0738-100X/92 $3.00

Definition 2-1 : F is an **algebraic expression** if F is a set of cubes such that no one cube contains another.

Definition 2-2 : The **product** of two cubes **c** and **d**, denoted as **c∘d,** is the product of literals. The product of literals is defined as : $a \circ a = a$, $a \circ a' = \emptyset$.

Definition 2-3 : The **product** of two expression F and G is a set defined by
$F \circ G = \{ c \circ d \mid c \in F \text{ and } d \in G \text{ and } c \circ d \neq \emptyset \}$

Definition 2-4 : F∘G is an **algebraic product** if F and G are algebraic expressions and have disjoint support. Otherwise, F∘G is a **Boolean product.**

In multilevel synthesizing system[4,5], some basic operations including decomposition, resubstitution and common term extraction are used. The decomposition procedure is to find and extract out common terms in an expression. The extraction procedure is to find and extract out common terms between different expressions. The resubstitution procedure is to substitute an existing node into other nodes. Among these procedures, "division" plays a key role.

Definition 2-5 : **Algebraic division** is an operation, given F,P returns Q,R such that $Q \neq \emptyset$, P∘Q is an algebraic product and P∘Q+R and F are the same expression(having the same set of cubes). Expression P, F, Q, R are called the divisor, the dividend, the quotient and the remainder respectively.

Example 2-1 : Given the dividend expression F = abd + abe + acd + ce and divisor P = ab + c. The quotient Q of algebraic division is Q = e , and the remainder R is R = abd + acd .

Definition 2-6: The **divisor set** U is the set of cubes in divisor P. Each element $\mathbf{u_i}$ in U is a cube in P.

Definition 2-7 : A dividend cube **c** in dividend F is said to be divisible with respect to divisor cube $\mathbf{u_i}$ if all literals in cube $\mathbf{u_i}$ are contained in cube **c**.

Definition 2-8 : The **cofactor cube** of a cube **c** with respect to cube $\mathbf{u_i}$, which is denoted as $(\mathbf{c})_{\mathbf{ui}}$, is the remained parts of **c** by removing literals in $\mathbf{u_i}$ from **c**.

In Example 2-1, the divisor set U is U= { ab, c}. Cube $\mathbf{c_1}$ = abd is divisible with respect to divisor cube $\mathbf{u_1}$ = ab and the cofactor cube $(\mathbf{c})_{\mathbf{u1}}$ is $(\mathbf{c})_{\mathbf{u1}}$ = d.

Definition 2-9 : The **divided set** V_i of divisor cube $\mathbf{u_i}$ is a set of cofactor cubes of those dividend cubes which are divisible with respect to divisor cube $\mathbf{u_i}$.

In Example 2-1, the divided sets are V_1 = { d, e}and V_2 = {ad, e}.

With these definitions, a typical algebraic division algorithm is shown as follows :

Algebraic_division(F,P)
  U = set $\{\mathbf{u_i}\}$ consists of cubes in divisor P.
  $V_i = \{ (\mathbf{c})_{\mathbf{ui}} \mid \mathbf{c} \in F, \mathbf{c}$ is divisible with respect to $\mathbf{u_i} \}$
  $Q = \cap V_i$.
  R = F - P∘Q
return (Q,R).

From the algebraic division algorithm, the quotient Q of Example 2-1 is $Q = V_1 \cap V_2$ = e , and the remainder R is R = F - P∘Q = abd + acd.

III. Coalgebraic division

Considering Example 2-1, it is apparently that F can be expressed as F= (ab + c)(ad + e). With algebraic division, the division quality is scarified. Let us consider another example of division.

Example 3-1 : Given F_2 = a'be + acd + ce and G = ad + e . The algebraic division returns the result that Q = c, and R = a'be . However expression F can be expressed as F_2 = (ad+ e)(a'b + c).

In Example 2-1 and Example 3-1, the products (ab + c)(ad+e) and (ad+ e)(a'b + c) have the same set of cubes as original expressions F and F_2 . However, they are not algebraic products.

Definition 3-1: **Coalgebraic division** is an operation, given F,P returns Q,R such that $Q \neq \emptyset$, , P∘Q+R and F are the same expression(having the same set of cubes).

For division, each cube **c** in F can be either in remainder R or a product of divisor cube $\mathbf{u_i}$ and quotient cube **q**. If **c** is a product of divisor cube $\mathbf{u_i}$ and quotient cube **q**, the only requirement is $\mathbf{c} = \mathbf{u_i} \circ \mathbf{q}$. Quotient cube **q** can be the divided cube $(\mathbf{c})_{\mathbf{ui}}$ or dividend cube **c.** With this fact, we say that $\mathbf{u_i}$ is the **tolerable part** of $(\mathbf{c})_{\mathbf{ui}}$. Considering Example 2-1, for dividend cube $\mathbf{f_1}$=abd and divisor cube $\mathbf{u_1}$=ab, the tolerable part **t** of divided cube $\mathbf{v_{11}}$ = d is **t** = ab. That is, the quotient cube can be any one of the cubes{d, ad, bd, abd}.

Definition 3-2 : A cube **c** with tolerable part **t** is a **quotient candidate** of divisor cube $\mathbf{u_i}$ if either $c \circ \mathbf{u_i} = \emptyset$ or there exists a divided cube $\mathbf{v_{ik}}$ in V_i such that $(c)_{ui} = (v_{ik})_t$.

Considering Example 2-1, cube **c** = ad is a quotient candidate of divisor cube $\mathbf{u_1}$ = ab, because cofactor cube $(c)_{u1}$ = d is equal to one of the divided cube in V_1. For the Example 3-1, cube **c** = a'b is also a quotient candidate of divisor cube $\mathbf{u_1}$ = ab, because $\mathbf{c} \circ \mathbf{u_1} = \emptyset$.

Definition 3-3 : The intersection of two cubes, denoted as $\mathbf{c} \wedge \mathbf{d}$, is defined as the intersection of the two literal sets.

Let cube **c** be **c** = abc' and cube **d** be **d** = a'bd, then the intersection of **c** and **d** is $\mathbf{c} \wedge \mathbf{d}$ = b.

Lemma 3-1 : If a cube $\mathbf{v_{ik}}$ in divided set V_i is a quotient candidate of $\mathbf{u_j}$, a cube $\mathbf{v_{jl}}$ in divided set V_j is a quotient candidate of $\mathbf{u_i}$, and $(v_{ik})_{uj} = (v_{jl})_{ui}$ then the product of $\mathbf{v_{ik}} \circ \mathbf{v_{jl}}$ is a quotient candidate of $\mathbf{u_i}$ and $\mathbf{u_j}$. The tolerable part **t** of $\mathbf{v_{ik}} \circ \mathbf{v_{jl}}$ is $\mathbf{t} = \mathbf{u_i} \wedge \mathbf{u_j}$.

Definition 3-4 : The **quotient candidate** of two divided set V_i and V_j denoted as $V_i \sqcap V_j$ is defined as :

$$V_i \sqcap V_j = \{v_{ik} \mid v_{ik} \in V_i, v_{ik} \circ t_j = \emptyset\} \cup \{v_{jl} \mid v_{jl} \in V_j, v_{jl} \circ t_i = \emptyset\} \cup \{v_{ik} \circ v_{jl} \mid \text{for any } v_{ik} \in V_i \text{ and } v_{jl} \in V_j, (v_{ik})_{tj} = (v_{jl})_{ti}\}$$

Tolerable part of elements in the first set is $\mathbf{t_i}$, tolerable part of elements in the second set is $\mathbf{t_j}$, tolerable part of elements in the last set is $\mathbf{t_i} \wedge \mathbf{t_j}$.

Definition 3-5 : A cube **c** is a **quotient cube** of divisor G if **c** is a **quotient candidate** for each $\mathbf{u_i}$ in divisor set U. That is, $\mathbf{c} \in \sqcap V_i$.

With these Definitions, the algorithm for coalgebraic division is shown as follows :

```
Coalgebraic division(F,P)
    U = setup_divisor_set(P) ;
    (V, R) = setup_divided_set(F,P) ;
    Q = ⊓ Vi .
    R' = original cube of (V - Q)
    R = R + R'
return ( Q,R).
```

In the coalgebraic division algorithm, procedure setup_divided_set will form the divided set Vi for each divisor cube $\mathbf{u_i}$. When setting up divided set, if a dividend cube is not divisible with respect to any of the divisor cube, then the cube must be in the remainder set. The cube can be put into remainder R. After finding the quotient cube, dividend cube with its divided cube not in quotient set is put into remainder set.

Example 3-2 : Considering the division problem in Example 2-1, F = abd + abe + acd + ce and P = ab + c, the divisor set U is U = {ab, c}, the divided set of cube ab is V_1= { d, e } , the divided set of c is V_2 = {ad, e }. With set U, V_1 and V_2, we know that

$(v_{11})_{u2} = (v_{21})_{u1}$ = d and $(v_{12})_{u2} = (v_{22})_{u1}$ = e.

$V_1 \sqcap V_2$ = { ad , e }

Q = ad + e, R = ∅ .

Example 3-3 : Given dividend F = abc'd' + abe'f' + a'b'cd + a'b'ef + c'd'ef + cde'f' and divisor P=ab + cd+ ef. With algebraic division, the quotient Q = ∅, and remainder R = F . With coalgebraic division, we can obtain quotient Q = a'b' + c'd' + e'f' and R = ∅. The division process is shown as follows :

U = { ab, cd, ef }, V_1 = {c'd', e'f'}, V_2 = {a'b', e'f'}, V_3 = {a'b', c'd'}

$V_1 \sqcap V_2$ = {c'd', e'f', a'b' }

$(V_1 \sqcap V_2) \sqcap V_3$ = { c'd', e'f', a'b' }

After coalgebraic division, the quotient Q = a'b' + c'd' + e'f', and the remainder R = ∅.

According to recent researches[6-8], it is shown that algebraic division can preserve network testability as well as test patterns during synthesis. This property may come from the fact that all literal lines will not be blocked due to the restriction of algebraic product of nodes. However, from the practical view of testing, it is sufficient to keep one unblocked path for each line. Thus, with some restriction on coalgebraic division, we can preserve the network testability with coalgebraic division.

Definition 3-6 : The product of two algebraic expressions F ∘ G is a **restricted coalgebraic product** if for every cube **c** in F, there exists at least one cube **d** in G such that **c** and **d** have no input variables in common, and for every cube **a** in G, there also exists at least one cube **b** in F such that **a** and **b** have no input variable in common.

Lemma 3-2 : For a coalgebraic division, F = P ∘ Q + R, if P ∘ Q is a restricted coalgebraic product then the single stuck_at fault testability of synthesized network after applying the division is kept unchanged. That is, test patterns for original network will also be tests for the synthesized network.

IV. Experimental results

The proposed coalgebraic division algorithm have been implemented and fitted into our misII environment. The coalgebraic division can be invoked with **resub** operation using the option **-c**. This command stands for performing resubstitution operation with coalgebraic division. Some experimental results using coalgebraic division will be presented in this section.

In order to compare the complexity and quality between algebraic and coalgebraic division, a simple script shown in Appendix is applied. The synthesizing results are shown in Table 1. The second column is the literal count after initial algebraic resubstitution. The third column is the literal count after second algebraic resubstitution. The fourth and fifth column are the literal count after coalgebraic resubstitution. The last four columns show execution time of the four resubstitution operations respectively. Because the execution time of division algorithm depends on the cover size of divisor and dividend node, resubstitution from initial network can provide more information on complexity and effectiveness of algorithm. From the sixth column and the eighth column, it shows that the execution time of algebraic and coalgebraic division are very close. From the second column and the fourth column, it shows that coalgebraic division can obtain better results than algebraic division. The third column and the fifth column show the results after decomposition and second resubstitution. After decomposition, the number of nodes in network is increased, and the size of each node is reduced. From the results, we can see that after decomposition, the execution time of resubstitution with coalgebraic division is slightly less than that of resubstitution with algebraic division. The total number of literals of network after resubstitution with coalgebraic division is still less than that of resubstitution with algebraic division. In order to show the overall effect of coalgebraic division, a more complicated script file shown in Appendix is applied. The results are shown in Table 2. The script file is similar to the typical script file described in [3]. Restricted coalgebraic division defined in section III is used. The literal count of resubstitution with restricted coalgebraic division is less than that of resubstitution with algebraic division. Execution time for the two method are very close. From the results, we can conclude that the time complexity of coalgebraic division is very close to the complexity of algebraic division implemented in misII. Although there exists a linear time algorithm for algebraic division[9], the complexity is not so significant when the size of node is small. After several iterations of synthesizing procedure, resubstitution with algebraic division can obtain good results. However, the total number of literal count is still larger than that of resubstitution with coalgebraic division.

Table 1: Comparisons between algebraic division and coalgebraic division.

	Literal count				Exec. Time(sec)			
PLA	R_a 1	R_a 2	R_c 1	R_c 2	R_a 1	R_a 2	R_c 1	R_c 2
add6	2196	104	1564	117	0.2	0.7	0.4	0.5
adr4	340	59	236	61	0	0.2	0.1	0.1
alu1	41	41	41	41	0	0	0	0
alu2	283	107	267	106	0.1	0.4	0.2	0.4
alu3	278	107	278	107	0.1	0.4	0.1	0.4
apla	263	185	263	185	0.1	0.5	0.1	0.5
chkn	1598	509	1598	509	0.1	2.2	0.1	2.2
dc1	66	52	66	52	0.1	0.1	0.1	0.1
dc2	253	144	252	144	0.1	0.4	0.1	0.4
dist	831	486	777	462	0.1	3.3	0.2	3
dk17	125	98	125	98	0.1	0.2	0.1	0.2
dk27	38	30	37	30	0.1	0.1	0	0.1
dk48	113	111	113	111	0.2	0.3	0.2	0.3
f51m	315	161	315	161	0.2	0.5	0.2	0.5
gary	1488	627	1488	625	0.3	6.7	0.3	6.7
in0	1488	627	1488	625	0.3	6.8	0.4	6.7
in1	5367	1903	5367	1903	2.1	277.5	2.9	276.1
in2	1987	646	1987	646	0.3	7	0.3	6.9
in3	1190	545	1190	545	1.1	5.2	1.2	5.2
in4	3067	513	2599	540	1.3	7.1	1.3	6.2
in5	1441	479	1410	475	0.4	3.1	0.5	3.1
in6	675	416	675	416	0.5	1.1	0.5	1.1
in7	473	167	473	167	0.1	0.5	0.1	0.5
misg	180	102	180	102	0.1	0.1	0	0.1
mish	137	126	137	126	0.1	0.2	0.1	0.2
mlp4	736	393	703	380	0.3	2.9	0.4	2.8
radd	340	58	236	60	0.1	0.1	0.1	0.2
rckl	1143	237	1143	237	0.2	0.3	0.2	0.3
risc	129	118	129	118	0.7	0.5	0.7	0.5
root	347	176	329	173	0.1	1.1	0.1	0.9
sqn	201	130	201	130	0	0.4	0	0.3
sqr6	213	157	208	156	0.2	0.5	0.2	0.5
wim	31	29	31	29	0.1	0.1	0.1	0.1
x1dn	536	106	536	106	0.2	0.6	0.2	0.5
x9dn	518	106	518	106	0.1	0.5	0.1	0.5
z4	252	47	188	48	0	0.1	0.1	0.1
Z5xp1	289	154	265	145	0.2	0.5	0.3	0.4
Z9sym	510	273	510	273	0	0.4	0	0.4
Total(38)	29478	10329	27923	10315	10.3	332.6	12	329

V. Conclusion

In this paper, we propose a more general division algorithm for multilevel logic synthesizing system. By introducing two Boolean properties ($a \circ a = a$, $a \circ a' = \emptyset$) into algebraic division algorithm, we can perform division operation more flexibly. With the extended algebraic division called coalgebraic division, we have more freedom to synthesize network within acceptable time. From the experimental results, it is shown that algorithm complexity of coalgebraic division is very close to complexity of algebraic division implemented in misII. The literal count of resubstitution with coalgebraic division is smaller than that of resubstitution with algebraic division. By applying some restrictions on division, the coalgebraic division can also preserve the network testability as well as the test patterns of network.

Table 2: Synthesizing results of multilevel logic synthesis with typical script.

	literal count		Exec. Time(sec)	
PLA name	Resub -a	Resub -c	Resub -a	Resub -c
add6	84	80	13.6	12.9
adr4	55	48	1.4	1.4
alu1	41	41	0.4	0.4
alu2	106	106	2.3	2.3
alu3	105	105	2.2	2.3
apla	155	156	2.8	2.8
chkn	462	457	47.7	51.8
dc1	49	49	0.8	0.8
dc2	135	134	3.7	3.6
dist	410	409	25.8	26.1
dk17	93	93	1.8	1.8
dk27	30	30	0.5	0.5
dk48	76	79	1.3	1.4
f51m	160	160	4.3	4.3
gary	554	548	28.6	29.8
in0	554	548	28.9	29.7
in1	1666	1657	325.2	321.6
in2	493	490	28.4	27.8
in3	446	446	17.7	17.8
in4	399	395	22.4	22.4
in5	344	344	12.8	12.7
in6	315	315	8.2	8.3
in7	149	149	3.6	3.6
misg	101	101	3.7	3.7
mish	126	126	2.1	2.2
mlp4	370	363	19	19
radd	52	48	1.4	1.3
rckl	193	193	10.8	10.7
risc	102	102	2.6	2.7
root	165	167	6.5	6.3
sqn	122	122	3.2	3.2
sqr6	147	151	4	3.9
wim	27	27	0.5	0.5
x1dn	101	101	2.8	2.8
x9dn	101	101	2.7	2.7
z4	50	43	1.1	1
Z5xp1	152	151	4.8	4.8
Z9sym	234	234	21.7	21.6
Total(38)	8930	8869	671.3	672.5

Reference

[1] J. Darringer, D. Brand, J. Gerbi, W. Joyner, and L. Trevillyan, "Logic synthesis through local transformations," IBM J. Res. Develop. Vol-25, pp.272-280, July 1981.

[2] D. Gregory, K. Bartlett, A. de Geus and G. Hachtel, "Socrates: A system for automatically synthesizing and optimizing combinational logic", In Proc. 23th Design Automation Conference, PP. 79-85, June 1986.

[3] R. Brayton, R. Rudell, A. L. Sangiovanni-Vincentelli and A. Wang, "MIS: A Multiple-Level Logic Optimization System," IEEE Trans. Comput. Aided Design Integr. Circuits Syst. vol. CAD-6, No.6, pp.1062-1081, Nov. 1987.

[4] D. Bostick, G. Hachtel, R. Jacoby, M. Lightner, P. Moceyunas, C. Morrison and D. Ravenscroft, "The boulder optimal logic design system", in Int. Conf. on Computer Aided Design Nov. 1987.

[5] R. K. Brayton, G.D. Hachtel and A. L. Sangiovanni-Vincentelli, "Multilevel Logic Synthesis," Proceedings of the IEEE, Vol.78 No.2, pp.264-300, Feb. 1990.

[6] Gary Hachtel, Reily Jacoby, Kurt Keutzer and Chris Morrison, "On the Relationship Between Area Optimization and Multifault Testability of Multilevel Logic," Proceedings of International wrokshop on Logic Synthesis, 1989.

[7] Janusz Rajski and Jagadeesh Vasudevamurthy, "Testability Preserving Transformations in Multi-level Logic Synthesis," Proceedings of International Test Conference , pp.265-273, 1990.

[8] Michael J. Bryan, Srinivas Devadas and Kurt Keutzer, "Testability-Perserving Circuit Transformations", Proceedings of ICCAD, pp.456-459, Nov. 1990 .

[9] P. McGeer and R. Brayton, "Efficient, stable algebraic operations on logic expressions", In Proceedings of International Conference on VLSI, Aug. 1987.

Appendix

Script file used for Table 1 :

```
collapse
simplify -m espresso
resub -a (-c)      /* Perform esubstitution operation */
sweep              /* remove inverters and buffers  */
decomp -g
resub -a(-c)
sweep
decomp -g
```

Script file used for Table 2 :

```
collapse
simplify -m espresso
resub -a (-c)
gkx -bc  /* Extract common multiple-cube divisors*/
gcx -bc   /* Common cube extraction  */
resub -a (-c)
decomp -g
sweep
eliminate 0  /* eliminate nodes with threshold 0 */
gkx -bc
gcx -bc
resub -a (-c)
eliminate 0
decomp -g
resub -a (-c)
liminate 0
simplify -m espresso
gkx -bc
gcx -bc
resub -a (-c)
decomp -g
```

Efficient Sum-To-One Subsets Algorithm for Logic Optimization

Kuang-Chien Chen
Fujitsu America Inc., San Jose

Masahiro Fujita
Fujitsu Labs. Ltd., Japan

Abstract

Given a set of functions $G = \{g_1, g_2, ..., g_k\}$, where $\sum_{i=1}^{k} g_i = 1$ and each g_i, $i = 1, ..., k$, is a function of $X = \{x_1, x_2, ..., x_n\}$. The **sum-to-one subsets** problem is to find all minimal subsets of the functions in G which add up to 1. A related problem, i.e., finding a **minimum-cost sum-to-one subset** based on a given cost function, is an important problem that often occurs in logic optimization algorithms. In this paper, we presents efficient procedures for solving both sum-to-one subsets and minimum-cost sum-to-one subset problems. We shall apply these techniques to multi-level network optimization algorithms, and will demonstrate that both the efficiency and quality of these algorithms are greatly improved. The application of these techniques to muti-node minimization using Boolean relations will also be discussed.

1. Introduction

Algorithms for optimizing multi-level combinational networks have been greatly improved in recent years [1][3][7][8][12][13]. In [7], an optimization algorithm RENO (REsynthesis for Network Optimization) was proposed in which a given network is minimized by optimally resynthesizing each gate in the network.

As will be discussed, the resynthesis problem in RENO can be transformed into a **minimum-cost sum-to-one subset** problem. Therefore, the efficiency and quality of the RENO algorithm depend largely on how well this problem can be solved. In the next section we will formally define the minimum-cost sum-to-one subset problem and review the existing approaches for solving it. Then, we discuss an innovative technique for solving the sum-to-one subsets problem by using Ordered Binary Decision Diagrams [6]. An algorithm for solving the minimum-cost sum-to-one subset problem is then developed by incorporating this technique. We have applied these techniques to RENO, and will demonstrate that they can greatly enhance RENO's capabilities in both area and delay minimization. Comparisons with algorithms in SIS-1.0 will be presented. We shall also discuss the application of these techniques to solve the problem of multi-node minimization using Boolean relations [4].

2. The Minimum-Cost Sum-To-One Subset Problem

We give basic definitions and explain the problem to be solved more precisely.

Definition 1: Given a set of functions $G = \{g_1, g_2, ..., g_k\}$, where $\sum_{i=1}^{k} g_i = 1$ and each g_i, $i = 1, ..., k$, is a function of $X = \{x_1, x_2, ..., x_n\}$. The **sum-to-one subsets** problem is to find *all* minimal (i.e., irredundant) subsets of the functions $g_1, g_2, ..., g_k$ which add up to 1 (or tautology). Suppose each g_i, $i = 1, ..., k$, has a given cost. The **minimum-cost sum-to-one subset** problem is to find a sum-to-one subset from G which has a minimum cost. Henceforth, we shall refer to the sum-to-one subsets problem as **STSOs**, and refer to the minimum-cost sum-to-one subset problem as **MC-STOS**.

STOSs and MC-STOS problems can arise in different settings. But here we shall mainly consider solving the STOSs and MC-STOS problems that arise in optimizing a multi-level combinational network N. N has m primary outputs, n primary inputs $X = \{x_1, x_2, ..., x_n\}$, and the function realized by a gate v_i in N *with respect to the primary inputs* is denoted by $f(v_i)$.

An ordered binary decision diagram (OBDD) is a rooted directed graph representation of a Boolean function $f(x_1, x_2, . . . , x_n)$ with respect to a given ordering of the variables. Without loss of generality, suppose the ordering of the variables is $x_1, x_2, . . . , x_n$, then each variable x_i is associated with a unique index i, $1 \leq i \leq n$. Each non-terminal vertex w in an OBDD is associated with a variable $x_i \in \{x_1, x_2, . . . , x_n\}$ and is labeled with the index of that variable. A non-terminal vertex w has two children vertices $low(w)$ and $high(w)$. The edge between w and $low(w)$ is labeled with the negative literal $\bar{x}_i$ and the edge between w and $high(w)$ is labeled with the positive literal x_i. Each OBDD also has two terminal vertices associated with constant 0 and 1 respectively. A **1-path** in an OBDD is a sequence of connected vertices, $\{w_1, w_2, ..., w_t\}$, where w_1 is the root of the OBDD, $w_2, ..., w_{t-1}$ are non-terminal vertices and w_t is the constant-1 terminal vertex. A **positive term** associated with a 1-path consists of all the *positive literals* appearing on the edges of the path. The **depth** of a 1-path is the largest index of the non-terminal vertices in the path.

When resynthesizing a gate v_i in the RENO algorithm, we first choose a set of candidate cubes $c_1, c_2, ..., c_k$ whose functions with respect to the primary inputs of the network are $f(c_1), f(c_2), ..., f(c_k)$, respectively, and a function h to be resynthesized using those candidate cubes (e.g., h can simply be $f(v_i)$). Each function $f(c_i)$ has an associated cost (e.g., for area minimization, the cost can be the number of literals in the corresponding candidate cube c_i), and the goal is to choose a minimum-cost subset of functions from $f(c_1), f(c_2), ..., f(c_k)$ which can realize h (a basic assumption here is $h \subseteq \sum_{i=1}^{k} f(c_i)$, which is always satisfied by the candidate cubes chosen in RENO [7]).

Theorem 1: Let $G = \{t_1, t_2, ..., t_k\}$ and h be Boolean functions with respect to variables $X = \{x_1, x_2, ..., x_n\}$. Then

$$h \subseteq \sum_{i=1}^{k} t_i \quad \text{if and } \textit{only} \text{ if} \quad \sum_{i=1}^{k} g_i = 1,$$

$g_i = t_{i_h}$ is the **generalized cofactor** of t_i w.r.t. h [13].

Base Theorem 1, the gate resynthesis problem in RENO can be formulated as a MC-STOS problem:

1. Since $h \subseteq \sum_{i=1}^{k} f(c_i)$, we have $\sum_{i=1}^{k} f(c_i)_h = 1$.

0738-100X/92 $3.00 © 1992 IEEE

2. Since $\sum_{i=1}^{k} f(c_i)_h = 1$, we can find all the STOSs of the cofactors $f(c_1)_h$, $f(c_2)_h$, ..., $f(c_k)_h$. Each STOS represents a subset of the $f(c_i)$'s which can realize h, and thus represents a subset of the c_i's which can be used for resynthesizing the gate v_i. Therefore, a solution of the optimal gate resynthesis problem corresponds to a STOS with the minimum cost.

3. Solving the MC-STOS Problem

A naive way to solve the MC-STOS problem is to generate all the STOSs, and then choose a subset with a minimum cost. But in practice, all the STOSs are seldom generated. Instead, since $\sum_{i=1}^{k} f(c_i)_h = 1$, a **tautology-checking** algorithm [7][9] has been used to derive a covering table in which a set of columns that cover all the rows corresponds to a STOS of $f(c_1)_h$, $f(c_2)_h$, ..., $f(c_k)_h$. A MC-STOS can thus be found by finding a minimum-cost column cover of the covering table. In RENO, in order to handle large networks, functions are represented in OBDDs and an OBDD-based tautology-checking algorithm has been developed [7] which is briefly discussed in the next section.

3.1 OBDD-based Tautology-Checking

Let the set of functions in a given MC-STOS problem be $G = \{g_1, g_2, ..., g_k\}$, where $\sum_{i=1}^{k} g_i = 1$ and g_i's are represented in OBDDs. The solution can be obtained by the OBDD-based tautology-checking procedure Shown in Figure 1 [7].

Example 1: Suppose given 6 functions:

$$g_1 = x_1x_2 + \bar{x}_1\bar{x}_2 \qquad g_2 = \bar{x}_1x_2$$
$$g_3 = \bar{x}_2x_3 + x_1x_2 \qquad g_4 = x_1\bar{x}_3$$
$$g_5 = \bar{x}_1\bar{x}_2 \qquad g_6 = x_1x_2 + \bar{x}_1\bar{x}_2\bar{x}_3 + x_3$$

By applying the tautology-checking procedure, we get an expression which represents a covering table:

$$(g_1+g_3+g_6)(g_3+g_6)(g_1+g_5+g_6)(g_2)(g_4)$$

which is simplified to $(g_3+g_6)(g_1+g_5+g_6)(g_2)(g_4)$. Suppose each g_i has a given cost, then a MC-STOS can be obtained by finding a minimum-cost column cover of this covering table. □

```
OBDD_tautology_checking(G, table) {
/*
G = {g1, g2, ..., gk}: functions in OBDDs.
table: the covering table to be built.
*/
     start:
1.   if (all OBDDs in G are identities) {
2.        let g1, g2, ..., gr be the OBDDs in G;
3.        add a row covered by columns g1.idx,
                    g2.idx, ...,gr.idx to table;
          return;
     }
4.   if (OBDD_unate_reduction(G) succeed) {
          goto start;
     }
5.   choose a binate variable x in G;

     /* recur on the left branch */
6.   Gx = restriction(G, x);
7.   OBDD_tautology_checking(Gx, table);

     /* recur on the right branch */
8.   Gx̄ = restriction(G, x̄);
9.   OBDD_tautology_checking(Gx̄, table);
     return;
}
```

Figure 1. Tautology-checking based on OBDDs.

Experiments show that the efficiency of Procedure *OBDD_tautology_checking* is mainly determined by the number of variables that the functions g_i's depends on (i.e., $x_1, x_2, ..., x_n$). Since tautology-checking applies Shannon expansions repeatedly, we need to partition using all the n variables in the worst case. In the next section, we shall discuss how to improve the efficiency of this process.

3.2 An OBDD-based STOSs Algorithm

We discuss a technique for solving the STOSs problem which will lead to a significant speedup of the OBDD-based tautology-checking process. This following lemma is a generalization of Lemma 5.10.1 in [5].

Lemma 1: Let $G = \{g_1, g_2, ..., g_k\}$ be a set of functions of variables $X = \{x_1, \ldots, x_n\}$, and let $A_1, A_2, \ldots, A_k$ be Boolean variables (they will be referred to as **A-variables**). Let $S = \{g_1, g_2, ..., g_s\}$ $(1 \leq s \leq k)$ be a subset of G. Then the following conditions are equivalent:

(i) $g_1 + g_2 + \ldots + g_s = 1$

(ii) $A_1A_2 \cdots A_s$ implies $A_1g_1 + A_2g_2 + \cdots + A_kg_k$.

So the problem of finding STOSs (not necessarily minimal) of a set of functions G is reduced to that of finding the terms which consist of only positive literals of the A-variables and imply the summation $A_1g_1 + A_2g_2 + \cdots + A_kg_k$. In the following, we shall refer to such terms as **A-terms**. Lemma 1 leads to a method for deriving all the minimal STOSs.

Theorem 3: Let $G = \{g_1, g_2, ..., g_k\}$ be a set of functions of variables $X = \{x_1, \ldots, x_n\}$, and let $A_1, A_2, \ldots, A_k$ be Boolean variables. Let $S = \{g_1, g_2, ..., g_s\}$ $(1 \leq s \leq k)$ be a subset of G. Then S is a minimal sum-to-one subset of G if and only if the product term $A_1A_2 \cdots A_s$ is a prime implicant of $A_1g_1 + A_2g_2 + \cdots + A_kg_k$ (such terms shall be referred to as **prime A-terms**).

Example 2.: Consider the 6 functions used in Example 1. The following summation expression is obtained:

$$A_1g_1 + A_2g_2 + A_3g_3 + A_4g_4 + A_5g_5 + A_6g_6 =$$
$$A_1x_1x_2 + A_1\bar{x}_1\bar{x}_2 + A_2\bar{x}_1x_2 + A_3\bar{x}_2x_3 + A_3x_1x_2 +$$
$$A_4x_1\bar{x}_3 + A_5\bar{x}_1\bar{x}_2 + A_6x_1x_2 + A_6\bar{x}_1\bar{x}_2\bar{x}_3 + A_6x_3. \quad (*)$$

There are totally 31 prime implicants for the expression above, but only 3 of them consists of only the A-variables: $A_2A_4A_6$, $A_1A_2A_3A_4$ and $A_2A_3A_4A_5$. Therefore, the STOSs corresponding to the prime A-terms are $\{g_2, g_4, g_6\}$, $\{g_1, g_2, g_3, g_4\}$, and $\{g_2, g_3, g_4, g_5\}$ □

In the process of resynthesizing a gate in RENO, usually we can find many candidate cubes. Therefore, if we apply Theorem 3, there are many functions in G and we need to use many A-variables which means it will be time-consuming to derive all the prime implicants for the summation expression. Also, how to order all the X- and A- variables is a problem. These problems, however, can be alleviated by observing that we don't need to derive all the prime implicants for the expression. What we need are only those prime A-terms. Also, all the A-variables are *positive unate* in the expression (i.e., only A_i's, but not $\overline{A_i}$'s appear). Therefore, all the prime A-terms consist of only positive literals. Base on those observations, the procedure shown in Figure 2 enables us to efficiently derive all the prime A-terms without generating all prime implicants for the summation expression.

Theorem 4: Procedure OBDD_STOSs derives all the prime A-terms corresponding to the set of all minimal sum-to-one subsets.

Example 3.: Consider the functions in Example 2 again. A partial OBDD for the expression (*) is shown in Figure 3 where the indices for the A-variables are from 1 to 6 (a circle in Figure 3 represents a non-terminal vertex, and sharing among vertices are not shown). On the 1-paths, edges labeled with positive literals are shown in solid bold lines and edges labeled with negative literals are shown in dashed bold lines (labels are not shown). If we write down the positive terms for the 1-paths, we have:

$\{A_1A_2A_3A_4,\ A_1A_2A_4A_6,\ A_2A_3A_4A_5,\ A_2A_3A_4A_6,\ A_2A_4A_5A_6,\ A_2A_4A_6\}.$

After removing redundant A-terms, the set of prime A-terms is:

$$\{A_1A_2A_3A_4,\ A_2A_3A_4A_5,\ A_2A_4A_6\}. \qquad \square$$

Experiments show that Procedure *OBDD_STOSs* is very efficient even when those g_i's are functions of a large number of variables. However, since the A-variables are ordered before the X-variables, the first k levels in the OBDD for the summation are close to a complete binary expansion tree as we can see in Figure 3. If there are many functions in G, the OBDD will be large and it may become infeasible to enumerate all the A-terms from the 1-paths.

3.3 A New Algorithm for Solving the Minimum-Cost STOS Problem

So far, we have discussed two methods for solving the MC-STOS problem:

(1) Use OBDD-based tautology-checking procedure.

(2) Use OBDD-based sum-to-one subsets procedure.

Method (1) is fast even when there are many candidate functions and method (2) is fast even for a large number of variables. To utilize the advantages of both methods,

OBDD_STOSs(G) {

/*

$G = \{g_1, g_2, \ldots, g_k\}$: functions of $X = (x_1, x_2, \ldots, x_n)$.

This procedure derives all minimal STOSs of G.

*/

1. create k new variables $A_1, A_2, \ldots, A_k$;
2. order A-variables before X-variables;

/* build the OBDD for summation */

3. $B = A_1g_1 + A_2g_2 + \ldots, + A_kg_k$;

/* enumerate positive terms of depths $\leq$ k */

4. *STOSs* = positive_terms_of_1_paths(B, k);

/* irredundant *STOSs* contains all prime A-terms */

5. *STOSs* = irredundant(*STOSs*);

return *STOSs*;

}

Figure 2. A STOSs algorithm based on OBDDs.

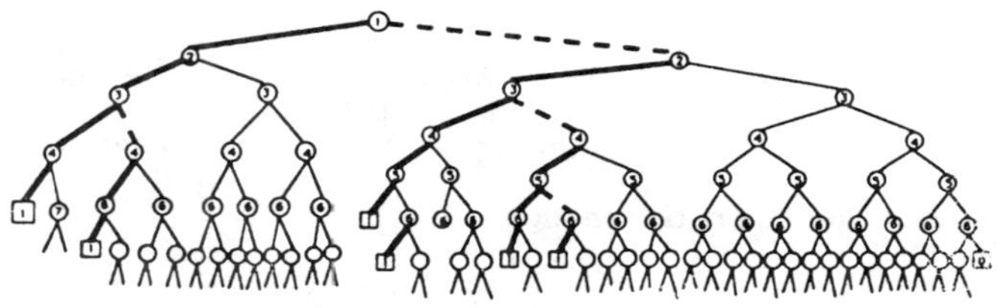

Figure 3: A partial OBDD for the summation expression in Example 2.

a new algorithm is shown in Figure 4 which combines both methods in a natural way.

Procedure *TC_with_STOSs* (tautology-checking with sum-to-one subsets) is still based on Shannon expansion. But when the number of functions at a node is not too large (the threshold is 12 in Figure 4), the OBDD-based STOSs algorithm is used to derive all the minimal STOSs efficiently (line 5). Then, a sub-table is derived from the set of minimal STOSs (line 6) and it is combined with an overall covering table (line 7).

4. Generation of Candidate Cubes by STOSs Algorithm

In [7], a Procedure *generate_candidate_cubes* was proposed as a way to generate many candidate cubes for resynthesizing a gate. However, That idea was regarded as not practical because it involved tautology-checking and also branch-and-bound processes which have to enumerate many irredundant column covers. Therefore, only existing cubes in the network were used for resynthesis in the original RENO algorithm,

We have found that Procedure *OBDD_STOSs* is not only useful for speeding up the tautology-checking process, it is also very useful for generating many candidate cubes (the details will be shown in a full-length paper). Resynthesis using these new candidate cubes has greatly improve the ability of RENO for area minimization. Also, by generating many candidate cubes which are good for reducing the levels (counted from the primary inputs) of those gates on the longest paths of a technology-mapped network [10], the delay minimization ability of RENO is also grealy enhanced.

5. Multi-Node Minimization Using Boolean Relations

Boolean relation, as discussed in [4], represents a relation that exists among interacting circuit modules. When the concept of Boolean relation is applied to multiple gates in a multi-level network, it represents a larger degree of freedom than don't-cares. Synthesis methods for Boolean relations based on multi-level representations have been proposed [11], but so far there is no effective way to directly minimize multiple gates using Boolean relations. In this section, we shall address this problem and propose an approach base on the techniques discussed earlier.

Without loss of generality, let $V = \{v_1, v_2, \ldots, v_t\}$ be a set of gates in N. We consider the minimization of gates in V using their Boolean relations.

Definition 2: The **smoothing** operator, $S_{x_i}f$ is defined as:

$$S_{x_i}f = f_{x_i} + f_{\overline{x}_i}$$

where f_{x_i} and $f_{\overline{x}_i}$ are the cofactors of f with respect to variables x_i and $\overline{x}_i$ respectively. Let $x = \{x_{i_1}, \ldots, x_{i_k}\}$, then

$$S_{x_J} f = S_{x_{i_1}} \cdots S_{x_{i_k}} f$$

Definition 3: Let $z_i = h_i(X)$ $(1 \le i \le m)$ be the function realized at the primary output z_i of network N with respect to the primary inputs X. The specification of N, denoted by $S(Z, X)$, is represented by the following characteristic function:

$$S(Z, X) = \prod_{i=1}^{m} (z_i \equiv h_i(x))$$

The function realized at z_i can also be expressed as a function of the primary inputs X and the intermediate variables corresponding to gates in V, i.e., $z_i = g_i(X, V)$. Then, the specification of N can be written as:

$$S'(Z, X, V) = \prod_{j=1}^{m} (z_j \equiv g_j(X, V)).$$

Therefore, the Boolean relation for the gates in V is:

$$R(X, V) = S_Z(S(Z, X)S'(Z, X, V)).$$

A realization of the gates in V is *legal* if the functions of those gates satisfy the condition $R(X, V) = 1$, and our goal is to find a legal realization with a minimum cost.

In [5], specifications in the form of $R(X, V) = 1$ were discussed extensively. It has been shown that if for each $X \in \{0,1\}^n$, $R(X, V)$ reduces to zero or a *single term* of the V-variables, then we will be able to derive the lower and upper bounds for the functions realized by gates in V *independently* (such specifications are called **tabular specifications**, and they represents specifications which can be specified using don't-cares). In such cases, we can synthesize all the gates in V base on their bounds simultaneously (for example, using the concurrent resynthesis algorithm discussed in [8]). However, if for some $X \in \{0,1\}^n$, $R(X, V)$ reduces to several terms of the V-variables, then the specification is a Boolean relation, and more powerful methods should be used.

```
TC_with_STOSs(G, table) {
/*
G: a set of functions represented in OBDDs.
table: the covering table to be built.
*/
     start:
1.   if (all OBDDs in G are identities) {
2.        let g1, g2, ..., gr be the OBDDs in G;
3.        add a row covered by columns g1.idx,
               g2.idx, ...,gr.idx to table;
          return;
     }
4.   if(number of OBDDs in G ≤ 12) {
5.        STOSs = OBDD_STOSs(G);
6.        sub_table = table_from_prime_A_terms(STOSs);
7.        table = table_merge(table, sub_table);
          return;
     }
8.   if (OBDD_unate_reduction(G) succeed) {
               goto start;
     }
9.   choose a binate variable x in G;

     /* recur on the left branch */
10.  Gx = restriction(G, x);
11.  TC_with_STOS(Gx, table);

     /* recur on the right branch */
12.  Gx̄ = restriction(G, x̄);
13.  TC_with_STOS(Gx̄, table);
     return;
}
```

Figure 4. Combining tautology-checking and STOSs.

Let $C = \{c_1, c_2, \ldots, c_k\}$ be a set of candidate cubes (they could be existing cubes in the network N or cubes generated by procedures such as OBDD-based STOSs algorithm). Then, we can express the realization of the gates in V using these candidate cubes by a general form:

$$v_1 = A_{1,1}c_1 + A_{1,2}c_2 + \cdots + A_{1,k}c_k$$
$$v_2 = A_{2,1}c_1 + A_{2,2}c_2 + \cdots + A_{2,k}c_k$$
$$\cdots$$
$$v_t = A_{t,1}c_1 + A_{t,2}c_2 + \cdots + A_{t,k}c_k$$

where $A_{i,j}$, $1 \le i \le t$ and $1 \le j \le k$ are all Boolean variables. By representing these functions of the gates and cubes in terms of primary inputs, we have the following expressions:

$$f(v_1) = A_{1,1}f(c_1) + A_{1,2}f(c_2) + \cdots + A_{1,k}f(c_k)$$
$$f(v_2) = A_{2,1}f(c_1) + A_{2,2}f(c_2) + \cdots + A_{2,k}f(c_k)$$
$$\cdots$$
$$f(v_t) = A_{t,1}f(c_1) + A_{t,2}f(c_2) + \cdots + A_{t,k}f(c_k)$$

Substituting these expressions into $R(X, V)$, we obtain an expression in terms of the $A_{i,j}$'s and X.

$$R(X, A_{i,j}) = R(x_1, \ldots, x_n, A_{1,1}, \ldots, A_{1,k}, A_{2,1}, \ldots, A_{2,k}, A_{t,1}, \ldots, A_{t,k}).$$

Theorem 5: A prime implicant of the expression $R(X, A_{i,j})$ which only consists of literals of the $A_{i,j}$'s corresponds to *a set of legal realizations* of the gates in V that satisfies the relation $R(X, V)$. We shall also call these prime implicants **prime A-terms**, but notice that they may contain both complemented and uncomplemented literals of the $A_{i,j}$'s.

Therefore, the problem of minimizing a set of gates in V using their relation $R(X, V)$ is reduced to the problem of finding a prime A-term of the expression $R(X, A_{i,j})$ whose corresponding set of legal realizations contain a legal realization that has a minimum cost.

Example 4: For the network in Figure 5(a), if we only use network don't-cares to minimize it by RENO or the *full_simplify* in SIS-1.0, we get a network of 3 gates and 6 literals. However, if we choose $V = \{v_1, v_2\}$ and minimize the network using Boolean relations, then we have:

$$S(Z, X) = z\bar{x}_1x_2 + zx_1\bar{x}_2 + \bar{z}x_1x_2 + \bar{z}\bar{x}_1\bar{x}_2$$

$$S'(Z, X, V) = z\bar{v}_1v_2 + zv_1\bar{v}_2 + \bar{z}v_1v_2 + \bar{z}\bar{v}_1\bar{v}_2$$

$$R(X, V) = S_Z(S(Z, X)S'(Z, X, V)) = x_1\bar{x}_2v_1\bar{v}_2 + x_1\bar{x}_2\bar{v}_1v_2 + \bar{x}_1x_2v_1\bar{v}_2 + \bar{x}_1x_2\bar{v}_1v_2 +$$

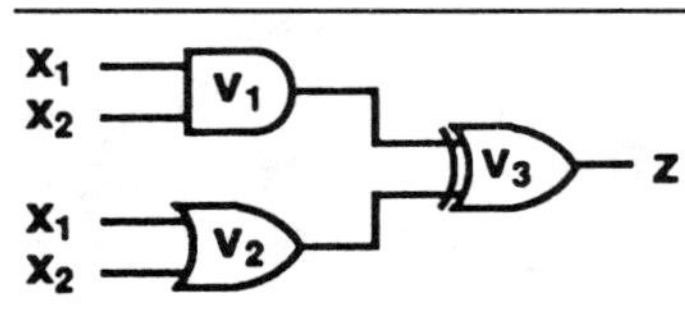

Figure 5(a): The original network.

Figure 5(b): Minimized network.

$$\bar{x}_1\bar{x}_2\bar{v}_1\bar{v}_2+\bar{x}_1\bar{x}_2v_1v_2+x_1x_2\bar{v}_1\bar{v}_2+x_1x_2v_1v_2$$

Note that $R(X,V)$ is not a tabular specification, and it shows the condition that must be satisfied by any legal realization of v_1 and v_2. Suppose the candidate cubes $c_1=x_1$, $c_2=x_2$, $c_3=x_1+x_2$, and $c_4=x_1x_2$ are derived, then we can express the functions of v_1 and v_2 as:

$$f(v_1)=A_{1,1}x_1+A_{1,2}x_2+A_{1,3}(x_1+x_2)+A_{1,4}(x_1x_2)$$
$$f(v_2)=A_{2,1}x_1+A_{2,2}x_2+A_{2,3}(x_1+x_2)+A_{2,4}(x_1x_2)$$

Substituting these two expressions into the relation $R(X,V)$, a relation $R(x_1,x_2,A_{i,j})$, $1\le i\le 2$ and $1\le j\le 4$, is obtained. The prime A-terms of $R(x_1,x_2,A_{i,j})$ are: $\bar{A}_{1,1}\bar{A}_{1,2}\bar{A}_{1,3}A_{1,4}A_{2,3}$, $A_{1,3}\bar{A}_{2,1}\bar{A}_{2,2}\bar{A}_{2,3}A_{2,4}$, $\bar{A}_{1,1}A_{1,2}\bar{A}_{1,3}\bar{A}_{1,4}A_{2,1}A_{2,2}$, $A_{1,1}A_{1,2}\bar{A}_{1,3}\bar{A}_{2,1}\bar{A}_{2,2}\bar{A}_{2,3}$, $\bar{A}_{1,1}A_{1,2}\bar{A}_{1,3}A_{2,1}\bar{A}_{2,2}\bar{A}_{2,3}$, and $A_{1,1}A_{1,2}\bar{A}_{2,1}\bar{A}_{2,2}A_{2,3}A_{2,4}$. Each of the prime A-term corresponds to a set of legal realizations of v_1 and v_2. For example, the prime A-term $\bar{A}_{1,1}\bar{A}_{1,2}\bar{A}_{1,3}A_{1,4}A_{2,3}$ specifies that if v_1 only contains c_4 (since $A_{1,1}$, $A_{1,2}$, and $A_{1,3}$ should be 0 and $A_{1,4}$ should be 1) and as long as v_2 contains c_3 (since $A_{2,3}$ should be 1), then such realizations will be legal, no matter whether v_2 contains c_1, c_2, and c_4 or not. □

In the case of area minimization where usually the number of literals is minimized, the cost of a legal realization is the total costs of the cubes being used. Therefore, although each prime A-term corresponds to a set of legal realizations, each variable $A_{i,j}$ can be assigned the cost of its associated cube (i.e., c_j), and the cost of each prime A-term can be defined as the sum of the costs of the $A_{i,j}$-variables which are 1's. Then, our goal is to find a prime A-term which has the minimum cost.

Example 5: Consider the network shown in Figure 5(a), where the relation $R(x_1,x_2,A_{i,j})$, $1\le i\le 2$ and $1\le j\le 4$, has been derived in Example 4. Because $R(x_1,x_2,A_{i,j})=1$ if and only if $R(0,0,A_{i,j})=1$, $R(0,1,A_{i,j})=1$, $R(1,0,A_{i,j})=1$, and $R(1,1,A_{i,j})=1$ are all true, the following set of clauses can be obtained by tautology checking on x_1 and x_2:

$$(\bar{A}_{1,3}+\bar{A}_{2,2})(\bar{A}_{1,3}+\bar{A}_{2,3})(\bar{A}_{1,2}+\bar{A}_{2,2})(\bar{A}_{1,2}+\bar{A}_{2,3})$$
$$(\bar{A}_{1,2}+\bar{A}_{1,3}+\bar{A}_{2,2}+\bar{A}_{2,3})(A_{1,3}+A_{2,1})(A_{1,3}+A_{2,3})$$
$$(\bar{A}_{1,1}+\bar{A}_{2,1})(\bar{A}_{1,1}+\bar{A}_{2,3})(A_{1,1}+A_{1,3}+A_{2,1}+A_{2,3})$$
$$(A_{1,1}+A_{1,2}+A_{1,3}+A_{1,4}+\bar{A}_{2,1})(A_{1,1}+A_{1,2}+A_{1,3}+A_{1,4}+\bar{A}_{2,2})$$
$$(A_{1,1}+A_{1,2}+A_{1,3}+A_{1,4}+\bar{A}_{2,3})(A_{1,1}+A_{1,2}+A_{1,3}+A_{1,4}+\bar{A}_{2,4})$$
$$(A_{2,1}+A_{2,2}+A_{2,3}+A_{2,4}+\bar{A}_{1,1})(A_{2,1}+A_{2,2}+A_{2,3}+A_{2,4}+\bar{A}_{1,2})$$
$$(A_{2,1}+A_{2,2}+A_{2,3}+A_{2,4}+\bar{A}_{1,3})(A_{2,1}+A_{2,2}+A_{2,3}+A_{2,4}+\bar{A}_{1,4})$$

These clauses form a **binate-covering** problem and they can be solved by existing procedures. A satisfying assignment which has a minimum cost is $A_{1,2}=A_{2,1}=1$ and all other $A_{i,j}$'s are 0's. It corresponds to the legal realization $v_1=x_2$ and $v_2=x_1$ shown in Figure 5(b). If we check the results of Example 4, this minimum-cost legal realization is in the set of legal realizations associated with the prime A-term $\bar{A}_{1,1}A_{1,2}\bar{A}_{1,3}A_{2,1}\bar{A}_{2,2}\bar{A}_{2,3}$. Another minimum-cost legal realization can be obtained by assigning 1 to $A_{1,1}$ and $A_{2,2}$ and 0 to all other $A_{i,j}$'s. This solution is in the set of legal realizations associated with the prime A-term $A_{1,1}\bar{A}_{1,2}\bar{A}_{1,3}\bar{A}_{2,1}A_{2,2}\bar{A}_{2,3}$. □

6. Experimental Results

The techniques discussed in Sections 2-4 have been implemented in the RENO algorithm and the following experiments were conducted:

(1) Comparison with the original RENO algorithm: some results are shown in Table 1. The initial networks are synthesized by SIS-1.0 using *script.rugged* and then mapped to MCNC standard-cell library using area-mode mapping command *map −m* 0. Then, the original RENO and improved RENO algorithms are applied to reduce the number of literals. When resynthesizing a gate, its *maximum set of permissible functions* [12] is used if it can be derived within a time limit. Otherwise, the original function realized by the gate is used. The area-minimized networks are then mapped to library again using *map −m* 0.
As can be seen, due to the use of OBDD-based STOSs algorithm, the processing speed of RENO has been greatly improved while achieving comparable or better minimization quality. In cases such as C880, for which the old RENO can finish without being stopped at time limit, about 20 times speedup has been obtained. The initial networks generated by SIS-1.0 using *script.rugged* and area-mode mapping command are usually high-quality networks of minimal area. Yet on average RENO can still obtain 8.1% area reduction. In cases like *t*481, significant area reduction of 85% has been obtained. This shows that in order to obtain the best results, the RENO algorithm should be used even when technology-independent minimizations have been done using algorithms such as *full_simplify*.

(2) Application of RENO for delay minimization: the results are shown in Table 2. The initial networks are synthesized by SIS-1.0 using *script.rugged* and then mapped to the MCNC standard-cell library using the speed-mode mapping command *map −m* 1. Then, RENO is applied to reduce the maximum depth of the mapped networks, and those level-reduced networks are mapped to library again using *map −m* 1. On the average, delay of the networks is reduced by 18% and area is also *reduced* by 2%. In the best case (e.g., *sct*), 63% delay reduction has been obtained.
The results of applying *speed_up* in SIS-1.0 are also reported in Table 2. In this experiment, the initial networks are processed by *speed_up −m mapped*, and then the mapping command *map −m* 1 is applied. On the average, *speed_up* reduces the delay of the networks by 10.4% but area is increased by 13.4% (however, *speed_up* is in general more efficient).

7. Conclusions

We discussed techniques for solving the sum-to-one subsets problem using properties of OBDDs and unate functions. By combining these techniques with tautology-checking, an efficient algorithm for solving the MC-STOS problem has been developed, and it greatly improved the processing speed of the RENO algorithm. The ability to generate many candidate cubes efficiently has also enhanced RENO's ability for area and delay minimization. An algorithm for multi-node minimization using Boolean relations is also proposed. We expect that the techniques discussed in this paper will be useful in solving other minimization problems.

References

[1] D. Bostick, G. D. Hachtel, R. Jacoby, M. R. Lightner, P. Moceyunas, C. R. Morrison, and Ravenscroft, "The Boulder Optimal Logic Design System," Proc. ICCAD, 1987, pp. 62-65.

[2] R. K. Brayton, G. D. Hachtel, C. T. McMullen, and A. L. Sangiovanni-Vincentelli, *Logic Minimization Algorithms* for *VLSI Synthesis*, Kluwer Academic Publishers, 1984.

[3] R. K. Brayton, R. Rudell, A. L. Sangiovanni-Vincentelli, and A. R. Wang, "MIS: A Multiple-Level Logic Optimization," IEEE TCAD, Nov. 1987, pp. 1062-1081.

[4] R. K. Brayton and F. Somenzi, "Boolean Relations

and the Incomplete Specification of Logic Networks," Proc. VLSI 1989.
[5] F. M. Brown, *Boolean Reasoning: The Logic of Boolean Equations*, Kluwer Academic Publishers, 1990.
[6] R. E. Bryant, "Graph-Base Algorithms for Boolean Function Manipulation," IEEE TCAD, Vol. C-35, No. 8, Aug. 1986.
[7] K. C. Chen, Y. Matsunaga, S. Muroga and M. Fujita, "A Resynthesis Approach for Network Optimization," Proc. DAC, 1991, pp. 458-463.
[8] K. C. Chen and M. Fujita, "Concurrent Resynthesis for Network Optimization," Proc. ICCD 1991, pp. 44-48.
[9] K. C. Chen and S. Muroga, "Sequential and parallel multiple-output ratio-set algorithms for the speedup of PLA minimization," Proc. CAD & CG, 1989, China, pp. 656-662.
[10] K. C. Chen and S. Muroga, "Timing Optimization for Multi-Level Combinational Networks," Proc. DAC, 1990, pp. 339-344.
[11] Y. Kukimoto, M. Fujita, and K. C. Chen, "Application of Boolean Unification to Combinational Logic Synthesis," Proc. ICCAD, 1991.
[12] S. Muroga, Y. Kambayashi, H. C. Lai, and J. N. Culliney, "The Transduction Method – Design of Logic Networks based on Permissible Functions," IEEE TC, Oct., 1989.
[13] H. Savoj and R. K. Brayton, "The Use of Image Computation Techniques in Extracting Local Don't Cares and Network Optimization," Proc. International Workshop on Logic Synthesis, 1991, 6.2a.

Table 1: Comparison of area minimization by old and new RENO algorithms

Function	SIS-1.0	Old RENO			New RENO		
	Area	Area	Time	ΔArea	Area	Time	ΔArea
9symml	2952	2976	74.5	+1.0	2976	35.2	+1.0
C1355	6960	timeout	-	-	7104	1237.9	+2.0
C1908	7632	7392	7162.3	-3.1	7424	1174.5	-2.7
C432	3064	3024	2020.6	-1.3	2968	1185.4	-3.0
C5315	24616	timeout	-	-	24522	4512.3	-1.0
C880	6256	6184	15954.5	-1.2	6184	858.8	-1.2
alu4	5928	5520	175.6	-6.9	5664	74.5	-4.5
apex1	18272	memory	-	-	17904	560.5	-2.0
apex6	11816	11600	1814.4	-1.8	11568	662.8	-2.0
apex7	3720	3664	115.4	-1.5	3608	47.4	-3.0
f51m	1448	1042	3.3	-28.0	1040	4.0	-28.0
frg1	2328	2144	100.1	-7.9	2240	62.4	-3.7
frg2	13112	12360	2601.8	-5.7	11848	1077.9	-9.6
ldd	1416	1376	4.1	-2.8	1352	4.2	-4.5
pcle	1152	1136	6.8	-1.4	1136	9.8	-1.4
rot	10872	timeout	-	-	10376	975.1	-4.5
sct	1296	1272	4.5	-1.8	1264	3.8	-2.5
t481	11520	4184	212.9	-63.7	1640	85.6	-85.8
term1	2656	2328	62.1	-12.3	2328	22.0	-12.3
z4ml	680	520	1.1	-23.5	520	1.2	-23.5

Table 2: Comparison of delay minimization by speed_up and RENO

Functions	Initial		*speed_up*				New RENO			
	Area	Delay	Area	Delay	Time	ΔDelay	Area	Delay	Time	ΔDelay
9symml	3048	18.4	3824	16.6	79.7	-9.8	3160	15.0	35.3	-18.5
C1355	7600	20.0	7600	20.2	490.4	+1.0	7600	18.8	3866.6	-6.0
C1908	7984	35.0	10272	30.2	233.4	-13.7	8272	31.8	1566.4	-9.1
C432	3224	34.4	3632	34.0	401.1	-1.2	4728	30.8	1469.9	-10.5
C5315	24912	34.0	29504	34.4	520.7	+1.2	24720	33.4	3673.4	-1.8
C880	6568	41.8	10576	31.2	202.2	-25.4	6888	32.4	657.7	-22.5
alu4	6256	34.2	7488	33.6	128.8	-1.8	6152	25.4	74.1	-25.7
apex1	18744	29.8	19928	23.2	356.2	-22.0	18904	28.0	654.2	-6.0
apex6	12128	24.6	12424	24.4	208.8	-1.0	12072	20.0	715.9	-18.7
apex7	3864	19.0	4240	15.6	83.2	-17.9	3896	15.6	45.6	-17.9
f51m	1512	29.4	2760	26.6	71.3	-9.5	1376	16.0	5.7	-45.6
frg1	2408	11.8	2768	11.2	62.1	-5.0	2552	11.0	49.8	-6.7
frg2	13736	31.2	14624	27.0	399.6	-13.5	13904	27.8	940.8	-10.9
ldd	1472	21.0	2392	13.6	53.9	-35.2	1904	12.0	5.4	-43.0
pcle	1224	11.6	1488	8.6	51.2	-25.9	1544	8.2	6.9	-29.3
rot	11416	26.6	12264	23.2	201.9	-12.8	11096	27.0	944.6	+1.5
sct	1432	26.4	2784	21.8	58.0	-17.4	1672	9.6	6.5	-63.6
t481	12192	21.6	11736	20.0	187.5	-7.4	3112	14.2	193.9	-34.3
term1	2792	13.4	3632	13.8	90.5	+2.9	2840	12.8	23.8	-4.5
z4ml	736	13.0	1016	11.4	21.6	-12.3	816	8.0	1.3	-38.5

Optimization of Primitive Gate Networks Using Multiple Output Two-Level Minimization

Abdul A. Malik
IBM Corporation, T. J. Watson Research Center
P. O. Box 218, Yorktown Heights, N.Y 10598

Abstract

Two-level minimization has been very effective for the optimization of multi-level logic in MIS. The use of two-level minimization in MIS depends primarily on the representation used in MIS for a multi-level network. However, most other logic synthesis systems use a different representation consisting of primitive gates. We explain why a primitive gate representation may be necessary in certain situations and describe the problems with using two-level minimization in that case. We then describe a novel method for applying the two-level minimization for the optimization of primitive gate network. Our approach is based on multiple output two-level minimization as opposed to single output two-level minimization as in MIS. It has been effective at reducing the number of gates and connections in the network while maintaining the primitive gate representation.

1 Introduction

The application of two-level minimization has proved very useful for the optimization of multi-level Boolean networks in MIS [8]. MIS represents the network as a Directed Acyclic Graph (DAG). Each node of the graph represents a two-level logic expression. The number of literals in each expression summed over all the nodes is taken as a measure of the size of the network. The smaller the number of literals, the smaller the network. The idea being that a network with smaller number of literals is likely to have a smaller area layout.

An effective operation for reducing the number of literals in the network is two-level minimization. The logic expression at each node is minimized using the implicit don't cares which are derived from the structure of the network [7]. If the two-level minimization results in an expression with a smaller number of literals, the original expression is replaced by the minimized expression. The result is a reduction in the number of literals in the network.

There are many logic synthesis systems that use primitive gates: OR, AND, NOR, NAND, and INV rather then a MIS like DAG representation. Examples of such systems are LSS [2], BOLD [3] and SYLON [9]. Our system called BooleDozer also uses a primitive gate representation. The objective of optimization of the primitive gate network is usually to reduce the number of gates and connections. The application of two-level minimization for the optimization of primitive gate network is not obvious.

Our first attempt at applying two-level minimization to a primitive gate network was to treat each gate in the network as a node in the DAG representation of MIS and apply two-level minimization as done in MIS. However, no noticeable optimization resulted. We attribute the failure to the fact that two-level minimizer had very small expressions to work with. For example, the two-level expression for a NOR gate is a single product term. In a DAG representation, the two-level expression usually consists of several product terms which gives the two-level minimizer more flexibility.

Another approach to consider was to selectively collapse parts of the network to two levels and apply the two-level minimization to each collapsed part. There are several problems with that approach. First, the collapsing of a part of the network usually increases its size. The two-level minimization is often not able to reduce the size to less than the pre-collapsed size. Secondly, the quality of the optimization will depend heavily on how the parts to be collapsed are selected. Finally, it may become difficult to collapse parts of the network if the network contains gates that the user may not like to be altered.

We have developed an effective method for applying the two-level minimization to primitive gate network which we describe in this paper. With our method, we obtained significant reduction in the sizes of industrial examples. Our approach does not collapse the network and allows for the presence of gates that should not be altered. For each gate in the network, a two-level expression is set up. In our method, a reduction in the size of the two-level expression guarantees a reduction in the number of connections and possibly in the number of gates in the network. Unlike MIS, our approach uses *multiple output* two-level minimization. However, we encountered some of the same problems that were faced when *single output* two-level minimization was introduced in MIS.

The setup for our method is described in Section 2. Section 3 provides some definitions. In Section 4, a simplified optimization technique is described that relies on gates with single fanouts. The technique is extended in Section 5 to take advantage of multiple fanout gates. Section 9 provides some experimental results and Section 10 concludes.

2 The setup

The object of the optimization is a primitive gate network consisting solely of NOR gates or NAND gates. We will first describe our scheme for the NOR-network. Later, we will describe how our methodology can be adopted for a NAND-network.

Each net in the network is driven by a gate or is a primary input. If a net is driven by an inverter, it does not drive an inverter. This condition does not result in a loss of generality: if a net n is driven by inverter g_1 and drives inverter g_2 then g_2 can be removed and all of its fanouts can be connected to n.

A variable is associated with every net not driven by an inverter. These variables are called **internal variables**. In addition, a literal is associated with every net. If a net is not driven by an inverter, the literal associated with it is the asserted literal of the variable. If an inverter drives a net then the literal associated with the net is the complement of that associated with the input of the inverter.

We allow for the presence of gates that should not be altered. They are called **non-minimizable** gates. A two-level function is associated with each NOR gate. The NOR gate is called the **reference gate** for the function. The optimization is carried out by applying the two-level minimization to the function for each minimizable NOR gate and modifying the network accordingly.

3 Definitions

The definitions in this section can be found elsewhere e.g [1, 7]. New definitions will be presented later as they are needed.

An **incompletely specified multiple output function** $\mathcal{F}$ with n inputs and m outputs is represented by the triplet (f, d, r) where f is the **onset**, d is the **don't care set** and r is the **offset** of the function. Each one of them is a mapping from B^n to B^m where $B = \{0, 1\}$. f, d and r partition B^n. A point in B^n is called a **minterm**. Each output $\mathcal{F}_i$ of $\mathcal{F}$ represents a single output incompletely specified function (f_i, d_i, r_i) where f_i, d_i and r_i are the onset, don't care set and the offset for output i. Each of them is a mappings from B^n to B.

The set of variables present in a sum-of-products g is called the **variable support** of g. Similarly, the set of literals present in g is called the **literal support** of g.

A multiple output sum-of-products h with n inputs and m outputs is called a **cover** for $\mathcal{F}$ if $f \subseteq g \subseteq f \cup d$ i.e $f_i \subseteq g_i \subseteq f_i \cup d_i$ for each output i. $f_i \subseteq g_i$ means that if a minterm m is mapped to 1 by f_i, it is mapped to 1 by g_i as well.

Some don't cares for the two-level minimization are derived from the structure of the network [7]. They are called implicit don't cares. There are two types of implicit don't cares: the **satisfiability don't cares** DC_{sat} and the **observability don't cares**. We only use DC_{sat} at present. DC_{sat} are conditions in terms of internal variables that can not occur. Such conditions exist because the internel variables are not independent.

4 Minimization for single fanout gates

We now provide an application of the two-level minimization that relies on the presence of gates with one fanout each. This restriction will be removed later.

Definition 4.1 *A product term p corresponding to a NOR gate g consists of an input part $In(p)$ and an output part $Out(p)$. $In(p)$ consists of the product of the complements of literals associated with the fanins of g. $Out(p)$ part consists of the set of fanout NOR gates of g.*

$In(p)$ is simply the function at the output of the NOR gate in terms of its fanins. We will often use g to mean p and vice versa.

Definition 4.2 *A product term p corresponds to every net n in the network such that $In(p)$ consists of the literal associated with the net and $Out(p)$ consists of the fanout gates of n.*

Theorem 4.1 *Let g be the reference gate. Let $p_1, p_2, \cdots p_i$ be the input parts of the product terms for fanin gates of g that only fanout to g. Let $q_1, q_2 \cdots q_j$ be those for the rest of the fanin NOR gates of g. Let $f = p_1 + p_2 + \cdots + p_i$, $d = DC_{sat} + q_1 + q_2 + \cdots + q_j$, and $h = \tilde{p_1} + \tilde{p_2} + \cdots + \tilde{p_k}$ such that h is a cover for $\mathcal{F} = (f, d, r)$. Then the gates for $p_1, p_2, \cdots p_i$ can be replaced by those for $\tilde{p_1}, \tilde{p_2}, \cdots \tilde{p_k}$ without altering the input/output behavior of the network.*

The minimization is carried out by first obtaining f and d and then finding the smallest cover h for $\mathcal{F} = (f, d, r)$ provided by ESPRESSO. Obtaining f and t is straight forward; however, obtaining d requires finding DC_{sat} which will be described later.

The minimization based on Theorem 4.1 allows for the reduction in the number of the NOR gates that fanout only to the reference gate g and in the number of their fanins. However, if g does not have any fanin gates with single fanouts, no minimization can occur. This is a sever restriction since most gates in a primitive gate network have multiple fanouts.

5 Minimization for multiple fanout gates

We now see how two-level minimization can be applied to reduce the fanin and fanouts of the gates with multiple fanouts and even eliminate them.

Theorem 5.1 *Let g be the reference gate. Let $p_1, p_2, \cdots, p_j$ be the product terms corresponding to the fanin gates of g such that the fanin gates are* *not inverters and they don't fan out to primary outputs or to other fanins of g. Let* $S = \bigcup_{i=1}^{j} Out(p_i)$. *Let* $q_1, q_2, \cdots q_k$ *be the product terms for the fanin gates of those in S other than* $p_1, p_2, \cdots p_j$. *Let* $\tilde{q_i}$ *for* $1 \leq i \leq k$ *be such that* $In(\tilde{q_i}) = In(q_i)$ *and* $Out(\tilde{q_i}) = Out(q_i) \cap S$. *Let* $f = p_1 + p_2 + \cdots + p_j$, $d = DC_{sat} + \tilde{q_1} + \tilde{q_2} + \cdots + \tilde{q_k}$, *and* $h = \tilde{p_1} + \tilde{p_2} + \cdots + \tilde{p_l}$ *such that h is a cover for* $\mathcal{F} = (f, d, r)$ *and contains only literals whose complements are present in the network. Then the gates corresponding to* $p_1, p_2, \cdots p_j$ *can be replaced by those corresponding to* $\tilde{p_1}, \tilde{p_2}, \cdots, \tilde{p_l}$ *without altering the input/output behavior of the network.*

Theorem 5.1 allows the replacement of gates corresponding to the product terms in f by those in h. The Theorem requires that h only contains literals whose complements are present in the network. If this condition was relaxed, additional logic could be needed to provide the fanins for the gates corresponding to the product terms in h. As a result, the size of the network could increase even if h had fewer literals than f.

To avoid any additional logic, Theorem 5.1 also imposes some restrictions on f. The restrictions ensure that f does not contain a gate whose output literal may be present in h so that all literals needed for the gates in h are available even after the removal of all gates in f. The requirements are that the gate must not be an inverter and must not fanout to another gate whose product term is in f.

Theorem 5.1 also requires that any gate whose associated product term is present in f does not fanout to a primary output. This is becasue the removal of such a gate will leave the primary output unconnected.

Further restrictions are necessary if the network contains non-minimizable gates. The application of Theorem 5.1 alters the gates corresponding to the product terms in f. Hence, the product terms associated with non-minimizable gates should not be allowed in f. The Theorem may also alter the fanin of any gate is S. If a gate in S is non-minimizable then none of its fanin gates should be in f.

Subroutine *apply_TLM*() shown in Algorithm 1 is used to minimize the fanins of the reference gate g. In order to minimize the entire network, the subroutine is called repeatedly with each minimizable gate g in the network as its argument.

The network is modified as a result of the two-level minimization if the quality of the cover h is better than that of the onset f. The following definition is used for measuring the quality.

Definition 5.1
For a multiple output sum-of-products g, $|g|$ *is the number of product terms in g and* $||g||$ *is the sum over all product terms of the number of literals in their input parts and the number of gates in their output parts.*

EXAMPLE 1 :
Figure 5 shows a part of a network consisting of NOR gates and inverters, all of which are minimizable. The literals associated with the nets are shown for selected nets only. The gates are labeled 1 through 9. Symbol g_i will be used to refer to the gate with label i. The rectangles with crosses represent primary outputs. We now illustrate Algorithm 1 taking g_1 as the reference gate.

From the figure, $P = \{g_6, g_8\}$, $S = \{g_1, g_2\}$, and $Q = \{g_4, g_5, g_7, g_9\}$.

Normally, the output parts of product terms are represented as bit vectors with a bit corresponding to each gate in S. The bit value is 1 if the gate is present in the output part and 0 otherwise. Here, we represent the output part using two bits; the first corresponding to g_1 and the second to g_2. Hence, $f = p_6 + p_8 = \overline{D}\,\overline{E}|11 + A\overline{C}|10$. Similarly, $t = \tilde{p_4} + \tilde{p_5} + \tilde{p_7} + \tilde{p_9} = \overline{C}E|01 + \overline{D}|10 + B\overline{C}|10 + \overline{A}\,\overline{B}|10$.

For simplicity, we take $DC_{sat} = \emptyset$; hence, $d = t$. We also assume that nets corresponding to the complements of all variables are present in the network. ESPRESSO provides $h = \overline{D}\,\overline{E}|01 + \overline{C}|10$. Since a net corresponding to C is already present, no new gate need to be created for the second product term in h. The replacement of gates in S by those in h results in the reduction of the number of gates by 1 and of connections by 3.

Algorithm 1
****

Input: A Boolean NOR network η *and a gate g in* η.
Effect: Reduction in the size of the network.

apply_TLM(η, g) {
  $P = \{g_i | g_i$ *is a minimizable fanin gate of g,*
    g_i *does not fan out to primary outputs,*
    g_i *is not an inverter,*
    g_i *does not fan out to fanins of g, and*
    all fanouts of g_i *are minimizable gates.* $\}$.
  $S = \{j |$ *gate j is a fanout of a* $g_i \in P\}$.
  $Q = \{g_i | g_i$ *is a fanin of a* $g \in S$ *and* $g_i \notin P\}$.
  Set $f = p_1 + p_2 + \cdots + p_j$ *where* $p_i \in P$.
  For each $g_i \in Q$, *obtain a product term* q_i.
  For each q_i, *obtain a product term* $\tilde{q_i}$ *such that* $In(\tilde{q_i}) = In(q_i)$ *and* $Out(\tilde{q_i}) = Out(q_i) \cap S$.
  Set $t = \tilde{q_1} + \tilde{q_2} + \cdots + \tilde{q_k}$.
  Obtain the satisfiability don't care set DC_{sat}.
  Form $d = DC_{sat} + t$.
  Obtain a list L of the complements of literals whose corresponding nets are present in η.
  Get the smallest cover h of $\mathcal{F} = (f, d, r)$ *from ESPRESSO with only those literals in L.*
  If $(||h|| < ||f||)$ {
    Replace gates in P by those corresponding to the product terms in h.
  }
}

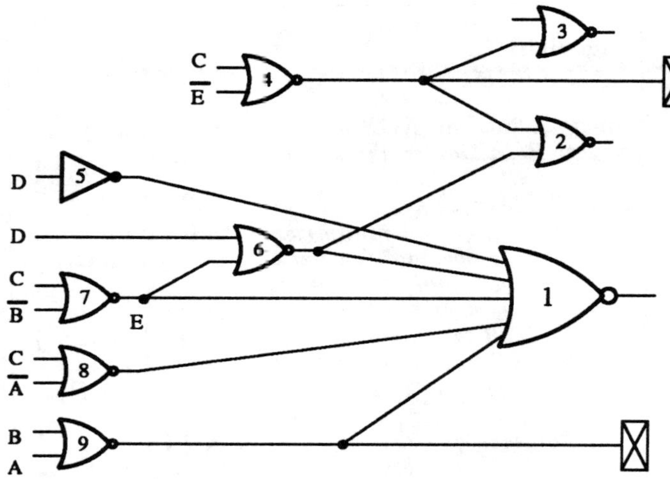

Figure 1: A partial primitive gate network.

We now show that the quality of the two-level minimization in Algorithm 1 directly affects the reduction in the size of the network.

Theorem 5.2 *The replacement of gates in the network in accordance with Theorem 5.1 results in the reduction of the number of gates by at least* $|f| - |h|$ *and the reduction of the number of connections by at least* $||f|| - ||h||$ *in the network.*

6 Obtaining the satisfiability don't cares

The internal variables of the network are not independent. The gates in the network ensure that these variables don't take on certain logic values. For instance, the presence of a NOR gate guarantees that when any of its inputs is 1, the output is not 1. In particular, for a NOR gate $y = \overline{x_1 + x_2 + \cdots + x_l} = \overline{x}_1\overline{x}_2\cdots\overline{x}_l$, the sum-of-products $y \oplus \overline{x}_1\overline{x}_2\cdots\overline{x}_l = \overline{y}\,\overline{x}_1\,\overline{x}_2\cdots\overline{x}_l + yx_1 + yx_2 + \cdots yx_l$ is a contribution to DC_{sat}. Similarly, the contribution to DC_{sat} by a NAND gate $y = \overline{x_1x_2\cdots x_l}$ is $yx_1x_2\cdots x_l + \overline{y}\,\overline{x}_1 + \overline{y}\,\overline{x}_2 + \cdots \overline{y}\,\overline{x}_l$.

Theorem 6.1 *Let* g_i *be a gate such that the product term* p_i *corresponding to* g_i *is in* f *in Theorem 5.1. Then the contribution of* g_i *to* DC_{sat} *can not be relied upon.*

The DC_{sat} can be obtained by adding the contributions of all the gates in the network except those whose product terms are present in the onset f. This usually results in a large DC_{sat}. The computation time for the two-level minimizer is adversely affected if a large number of variables are present in the don't care set d. On the other hand if a very small part of the don't care set is used, the quality of the minimization may suffer. We discovered that a good compromise between the speed and quality of the minimization is reached when all those product terms are deleted from d whose support is not a subset of the support of f. The same conclusion was reached for limiting the size of DC_{sat} for two-level minimization in MIS [10]. However, we found that this holds for the entire don't care d; not just for DC_{sat}.

7 Adaptation for NAND network

Definition 7.1 *A product term* p *associated with a NAND gate* g *has an input part* $In(p)$ *and an output part* $Out(p)$. $In(p)$ *is the product of the literals corresponding to its fanins.* $Out(p)$ *consists of the set of fanout gates of* g.

$In(p)$ is now the complement of function at the output of the NAND gate in terms of its fanins. The methodology described so far is also applicable to a NAND-network. Theorem 5.1 can be used for a NAND-network with Definition 7.1 with a minor modification: h is now required to have only those literals which are present in the network. The relationship between the quality of the two-level minimization and the reduction in the size of the network as stated in Theorem 5.2 also holds for NAND-network.

8 Problems with large offsets

Two-level minimizers like ESPRESSO need the offset r of the function to be minimized. Often r is computed by the minimizer as $\overline{f \cup d}$ because it is not provided by the user. It has been known that there are several functions that have reasonable size representations for f and d but unreasonably large representation for r [1]. When two-level minimization was incorporated in MIS, it was found that functions with large offsets occured frequently. The minimization in those cases either took very long time or ESPRESSO ran out of memory. This was a major obstacle in using two-level minimization in MIS. The problem was solved by the invention of the reduced offsets that obviated the need for using the offsets without compromising the quality of the minimization [4, 12, 6].

We faced the same problem with our scheme. However, the reduced offsets for *single output* minimization used in MIS were of little use to us since our two-level functions have *multiple outputs*. ESPRESSO treats all outputs together as a single multi-valued variable and apply multi-valued minimization to the function [11]. This enabled us to use multi-valued reduced offsets [5]. This solved the problem for us without any loss of quality in the minimization.

9 Experimental results

Before TLM		After TLM		
G	*C*	*G*	*C*	*Time*
951	3397	924	2677	141
1439	4689	1399	4378	4957
542	1248	529	1157	105
876	2294	868	2224	120
1424	4585	1385	4289	1722

Table 1: Expermimental Results for some industrial networks.

The method described here has been implemented our logic synthesis system: BooleDozer. Table 1 presents results for some representative networks taken from an industrial environment. The optimization was done using the two-level minimization along with other operations present in BooleDozer. The first two columns in the table show the number of gates G and the number of connections C in the network just before the two-level minimization (TLM) is applied. The last two columns show the same just after the two-level minimization is applied. The last column shows the CPU time in seconds taken by the two-level minimization on an IBM RS/6000. The results show that the two-level minimization is able to reduce the size of the network significantly (upto 21% connections) in a reasonable time. The quality of the optimization is expected to further improve if the observability don't cares are used along with the satisfiability don't cares.

10 Conclusion

An novel method for the optimization of primitive gate network has been presented. A multiple output two-level function is used to abstract part of the network. The reduction in the number of literals in the function by a two-level minimizer directly translates into a reduction in the number of connections in the network. An advantage of this scheme is that the network does not have to be even partially collapsed. The experimental results with some industrial examples are encouraging. The quality of the results is likely to improve when observability don't cares are incorporated in the system.

Acknowledgements

The autor would like to thank his colleagues Dr. Daniel Brand, Mr. Anthony Drumm and Dr. Daniel Ostapko for their suggestions.

References

[1] R. Brayton, G. Hachtel, C. McMullen, and A. Sangiovanni-Vincentelli. *Logic Minimization Algorithms for VLSI Synthesis*. Kluwer Academic Publishers, 1984.

[2] J.A Darringer, D. Brand, J.V. Gerbi, Jr. W.H. Joyner, and L. Trevillyan. LSS: A system for Production Logic Synthesis. *IBM Journal of Research and Development*, 28(5), September 1984.

[3] G. Hachtel, R. Jacoby, P. Moceyunas, and C. Morrison. Performance Enhancements in BOLD using Implicants. In *Proceedings of International Conference on Computer-Aided Design*, Santa Clara, CA, Nov. 1988.

[4] A. Malik, R. Brayton, A. Newton, and A. Sangiovanni-Vincentelli. A Modified Approach to Two-Level Minimization. In *Proceedings of International Conference on Computer-Aided Design*, Santa Clara, CA, Nov. 1988.

[5] A. Malik, R. Brayton, A. Newton, and A. Sangiovanni-Vincentelli. Reduced Offsets for Two-Level Multi-Valued Logic Minimization. In *Proceedings of Design Automation Conference*, Orlando, FL, June 1990.

[6] A. Malik, R. Brayton, A. Newton, and A. Sangivanni-Vincentelli. Reduced Offsets for Minimization of Binary Valued Functions. *IEEE Transactions on Computer-Aided Design*, 10(4):413–426, April 1991.

[7] K. Bartlet and R. Brayton and G. Hachtel, R. Jacoby, C. Morrison, R. Rudell, A. Sangiovanni-Vincentelli, and A. Wang. Multilevel Logic Minimization Using Implicit Don't Cares. *IEEE Transactions on Computer-Aided Design*, 7(6):723–740, June 1988.

[8] R. Brayton and R. Rudell and A. Sangiovanni-Vincentelli and A. Wang. MIS: Multiple-Level Logic Optimization System. *IEEE Transactions on Computer-Aided Design of Integrated Circuits and Systems*, CAD-6(6):1062–1081, November 1987.

[9] S. Muroga, X. Xiang, J. Limqueco, L. Lin, and K. Chen. A Logic Network Synthesis System, SYLON. In *Proceedings of International Conference on Computer Design*, Cambridge, MA, Oct. 1989.

[10] A. Saldanha, A. Wang, and R. Brayton. Multi-Level Logic Simplification using Don't Cares and Filters. In *Proceedings of 26th Design Automation Conference*, Las Vegas, June 1989.

[11] T. Sasao. An Application of Multiple-Valued Logic to a Design of Programmable Logic Arrays. In *Proceedings of 8th International Symposium on Multiple Valued Logic*, 1978.

[12] H. Savoj, A. Malik, and R. Brayton. Fast Two-Level Logic Minimizers for Multi-Level Logic Synthesis. In *Proceedings of International Conference on Computer-Aided Design*, Santa Clara, CA, Nov. 1989.

Test-Set Preserving Logic Transformations

Michael J. Batek John P. Hayes

Advanced Computer Architecture Laboratory
Department of Electrical Engineering and Computer Science
University of Michigan, Ann Arbor, MI 48109

Abstract

Logic transformations that preserve a combinational circuit's minimal or complete test sets are examined. Some basic transformation types are rigorously defined and characterized with respect to test-set preservation. We apply our transformations to adder design and show that any complete test set for a two-level adder is preserved on transformation to ripple-carry and carry-lookahead designs, thus verifying some recent simulation results.

1. Introduction

As a circuit is transformed during logic synthesis to improve area or performance, its testing properties change, often in unpredictable ways. Unlike the estimation of area and delay, performing test generation at each design iteration is prohibitively expensive. We are investigating ways in which changes to a circuit's test sets can be predicted and controlled during the design process without explicit test generation.

If a transformed logic circuit can be tested by a set of tests designed for the original circuit, then we say that the transformation is "test-set preserving" or TSP. Test-set preservation ensures that each iteration of the design is testable and contains no undetectable faults. Moreover, TSP sequences guarantee that any transformed design can be fully tested using the test sets for the initial circuit. Hence, test-generation effort is reduced since it must only be expended for the initial circuit, not for derived designs. Our goal is to develop rigorous and practical methods of incorporating TSP into logic synthesis.

The first mention of test-set preservation appears to be by Kohavi and Kohavi [3]. More recently, transformations used in logic synthesis programs have been analyzed with respect to testability preservation. In most cases, "testability" refers only to the preservation of the ability to detect faults (irredundancy), and the faults of interest are usually single stuck-line (SSL) faults. However, testability preservation has also been used to denote the preservation of complete test sets. It has been shown that a restricted class of the transformations used in MIS [1] preserve complete test sets for multiple stuck-line (MSL) faults when the initial circuit is a minimal two-level design [4]. The transformations used by Pendulum [2] preserve complete SSL fault test sets, provided the initial circuit is irredundant. Like MIS, Pendulum is limited to complete test sets.

There is considerable experimental evidence demonstrating a strong correlation among the test sets for different implementations of the same circuit. For instance, Davé and Patel have recently carried out simulation experiments that show that a complete SSL fault test set for a two-level implementation of a circuit very often covers all SSL faults in a multi-level implementation [5]. The authors do not suggest any analytical reasons for this phenomenon.

We present an approach to TSP that analytically explains empirical results of this kind. It is based on a precise definition of the transformations of interest, and considers the preservation of minimal and complete test sets. We also introduce some useful new types of TSP transformations. We then apply our approach to adder design and obtain analytical verification for the experimental results of [5].

This research was supported by the National Science Foundation under Grant No. MIP-8805517.

2. Definitions and Notation

For a particular SSL fault $f = h/s$ denoting line h stuck-at-s, let $z_f(x)$ specify the function of a faulty circuit whose fault-free output is $z(x)$. In the case of the fault $f = a/0$ in the circuit C_1 of Fig. 1a, $z_{a/0}(x) = x_1x_3$. The fault f is *detectable* if and only if there exists some input combination t for which $z(t) \neq z_f(t)$. Then t *detects* f and t is a *test* for the fault f; $t = 110$ is a test for $a/0$ in C_1.

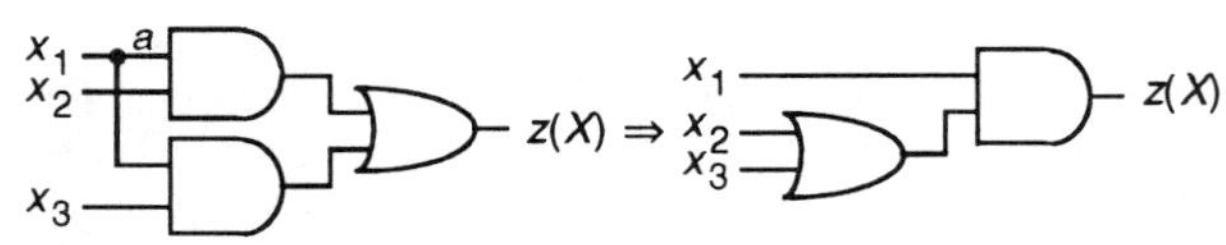

(a) $z(x) = ((x_1x_2) + (x_1x_3))$ (b) $z(x) = (x_1(x_2 + x_3))$

Figure 1. (a) Circuit C_1; (b) Transformed circuit C_2.

Test set T is *complete* if and only if for every detectable fault f in C, there exists some $t \in T$ such that $z(t) \neq z_f(t)$;

for the circuit C_1, a complete test set is {001, 011, 100 101, 110}. A test set T is *minimal* if and only if it is complete and there is no complete test set T' such that $|T'| < |T|$; a minimal test set for C_1 is {011, 100, 101, 110}. Note that if T is minimal, it is complete. Let ***CTS*** and ***MTS*** be the set of all complete and minimal test sets. Clearly, $\boldsymbol{MTS} \subseteq \boldsymbol{CTS}$. For all SSL faults in circuit C_1 of Fig. 1a, $\boldsymbol{MTS}$ = {{011, 100, 101, 110}}. ***CTS*** for the same circuit contains 20 test sets, of which {001, 010, 100, 101, 110} and {000, 011, 100, 101, 110} are examples.

In the usual sense, $\boldsymbol{T_1}$ is *included in* $\boldsymbol{T_2}$, or $\boldsymbol{T_1} \subseteq \boldsymbol{T_2}$, if and only if for any test set $T_1 \in \boldsymbol{T_1}$, there exists some test set $T_2 \in \boldsymbol{T_2}$ such that $T_2 = T_1$. $\boldsymbol{T_1}$ is *equal to* $\boldsymbol{T_2}$, or $\boldsymbol{T_1} = \boldsymbol{T_2}$ if and only if $\boldsymbol{T_1} \subseteq \boldsymbol{T_2}$ and $\boldsymbol{T_2} \subseteq \boldsymbol{T_1}$. We also introduce a special inclusion relation called covering. $\boldsymbol{T_1}$ *covers* $\boldsymbol{T_2}$, or $\boldsymbol{T_1} \subseteq_c \boldsymbol{T_2}$, if and only if for any test set $T_1 \in \boldsymbol{T_1}$, there exists some test set $T_2 \in \boldsymbol{T_2}$ such that $T_2 \subseteq T_1$. In Fig. 2a, $\boldsymbol{T_1}$ is included in $\boldsymbol{T_2}$. $\boldsymbol{T_1}$ also covers $\boldsymbol{T_2}$ by implication. Figure 2b illustrates the covering relationship between $\boldsymbol{T_1}$ and $\boldsymbol{T_2}$ that allows us to relate the smaller test sets for $\boldsymbol{T_2}$ to $\boldsymbol{T_1}$. Each of these relationships is transitive, hence a sequence of TSP relationships is also TSP.

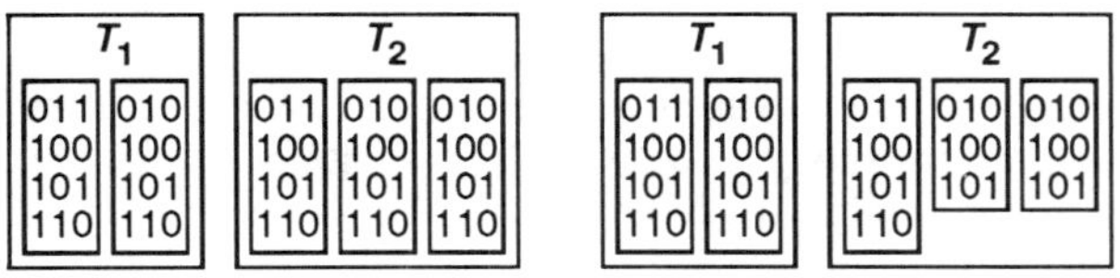

(a) $\boldsymbol{T_1} \subseteq \boldsymbol{T_2}$, $\boldsymbol{T_1} \subseteq_c \boldsymbol{T_2}$ (b) $\boldsymbol{T_1} \not\subseteq \boldsymbol{T_2}$, $\boldsymbol{T_1} \subseteq_c \boldsymbol{T_2}$

Figure 2. Examples of inclusion relationships among sets of test sets.

We define a general *logic transformation* to be a function $\mathcal{T}$ that takes a circuit (or corresponding expression) C_1 as input, and returns the circuit $\mathcal{T}(C_1) = C_2$. Such a transformation is *minimum test-set preserving* if $\boldsymbol{MTS}(C_1) = \boldsymbol{MTS}(C_2)$, $\boldsymbol{MTS}(C_1) \subseteq \boldsymbol{MTS}(C_2)$, or $\boldsymbol{MTS}(C_1) \subseteq_c \boldsymbol{MTS}(C_2)$. Complete TSP transformations are defined similarly. For example, we may define the transformation $\mathcal{E}_1$ that transforms the circuit C_1 of Fig. 1a to C_2 of Fig. 1b as $\mathcal{E}_1(((x_1x_2) + (x_1x_3))) = (x_1(x_2 + x_3))$. This transformation is an example of single-cube extraction [2] and algebraic division [1]. Since $\boldsymbol{MTS}(C_2)$ = {{001, 100, 101, 110}, {010, 100, 101, 110}, {011, 100, 101, 110}}, $\boldsymbol{MTS}(C_1) \subseteq \boldsymbol{MTS}(C_2)$. Hence $\mathcal{E}_1$ is minimum TSP. We will also distinguish the foregoing *single-output* transformations from *multiple-output* transformations of the form $\mathcal{T}(C_1, C_2, \ldots, C_k)$. In this paper, we will only deal with circuits that are irredundant in the usual sense.

3. Logic Circuit Transformations

Our set of basic logic transformations contains three single-output transformation types, which we call fanout-free, extraction and De Morgan transformations, and one multiple-output case, resubstitution. We deal only with circuits of AND, OR, NAND, NOR and NOT gates.

Fanout-Free Transformations A fanout-free or tree circuit has a unique path from every primary input to the primary output. A *fanout-free transformation* $\mathcal{F}(F) = G$ takes a fanout-free circuit to another fanout-free circuit G that is functionally equivalent to F. An example appears in Fig. 3. These transformations are primarily useful for grouping related items for later transformations.

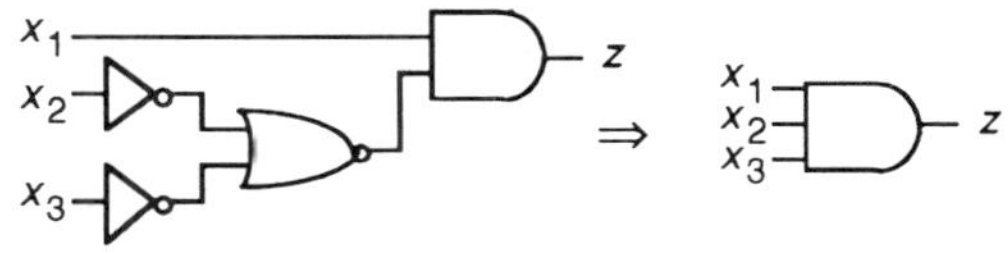

Figure 3. The fanout-free transformation $\mathcal{F}((x_1((x_2)'+(x_3)')')) = (x_1x_2x_3)$.

Theorem 1: The fanout-free transformation $\mathcal{F}(F) = G$ has the TSP properties $\boldsymbol{MTS}(F) = \boldsymbol{MTS}(G)$ and $\boldsymbol{CTS}(F) = \boldsymbol{CTS}(G)$.

This theorem may be proven by showing that a minimal or complete test set for input faults is a minimal or complete test set for all faults and that a test set for an input fault depends only on the function of the circuit.

Extraction Transformation We saw a simple version of the extraction transformation in Fig. 1, namely $\mathcal{E}_1$. In general, extraction transforms an AND/OR circuit to a simpler OR/AND circuit. The *extraction transformation* has the form $\mathcal{E}(P) = S$, where $P = ((p_1{}^*) + (p_2{}^*) + \ldots + (p_q{}^*))$ and $S = ((s_1{}^+)\,(s_2{}^+) \ldots (s_m{}^+))$. The expressions (E^*) and (E^+) denote an arbitrary number of product or sum inputs, respectively, and are illustrated in Fig. 4. Each product term $(p_i{}^*)$ consists of m variables or "single-output" expressions, and each sum term $(s_j{}^+)$ consists of n_j variables or single-output expressions. Distinct sum terms depend on disjoint sets of primary input variables, and no input expression to a sum term implies any other input expression in that sum term. Extraction combined with fanout-free transformations generalizes single- and double-cube extraction [2], and algebraic division [1].

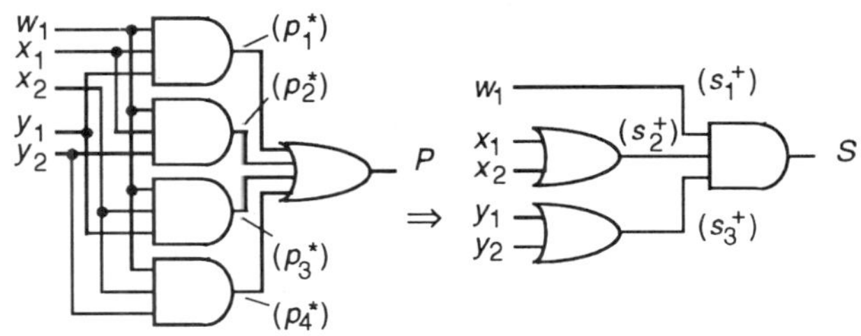

Figure 4. The extraction transformation $\mathcal{E}(P) = S = \mathcal{E}(((w_1x_1y_1) + (w_1x_1y_2) + (w_1x_2y_1) + (w_1x_2y_2)) = (w_1(x_1 + x_2)(y_1 + y_2))$.

Theorem 2: The extraction transformation $\mathcal{E}(P) = S$, where $P = ((p_1{}^*) + (p_2{}^*) + \ldots + (p_q{}^*))$ and $S = ((s_1{}^+)\,(s_2{}^+) \ldots (s_m{}^+))$, has the TSP properties $\boldsymbol{MTS}(P) \subseteq_c \boldsymbol{MTS}(S)$ and $\boldsymbol{CTS}(P) \subseteq_c \boldsymbol{CTS}(S)$.

This theorem may be proven by induction on m, the

number of sum terms in S, and on n_j, the number of inputs per sum term in S. We assume that the covering relationship $T(P) \subseteq_c T(S)$ holds for some P and S with test sets $T(P)$ and $T(S)$. Let P' and S' denote the circuits resulting from either inductive step. The test sets for P' and S', denoted $T(P')$ and $T(S')$, are obtained from $T(P)$ and $T(S)$ by expanding and adding the same tests to every test set in $T(P)$ and $T(S)$. These modifications result in the relationship $T(P') \subseteq_c T(S')$, hence the covering relationship holds for the inductive step.

De Morgan Transformation This transforms an AND/OR circuit into a dual OR/AND circuit, modifying the number and location of input/output inverters, as in Fig. 5. $\mathcal{M}(P) = S$ is a *De Morgan transformation* if $P = ((p_1^*) + (p_2^*) + ... + (p_n^*))$ and $S = (((s_1^+)\ (s_2^+)\ ...\ (s_n^+)))'$, where each input in P is the complement of the corresponding input in S. Inputs may be variables or single-output expressions as long as P is irredundant.

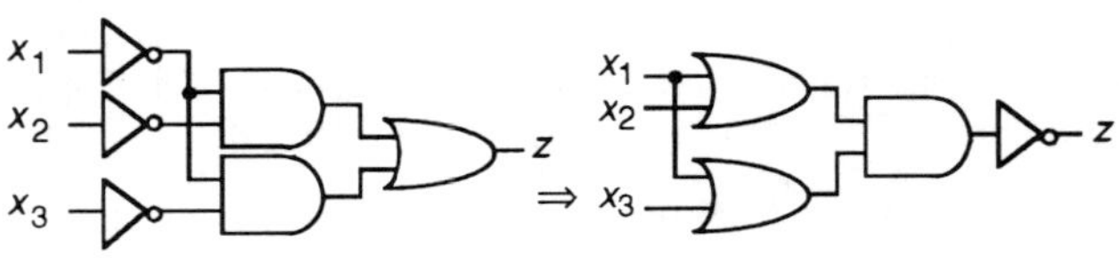

Figure 5. The De Morgan transformation $\mathcal{M}((((x_1)'(x_2)') + ((x_1)'(x_3)'))) = (((x_1 + x_2)(x_1 + x_3)))'$.

Theorem 3: The De Morgan transformation $\mathcal{M}(P) = S$, where $P = ((p_1^*) + (p_2^*) + ... + (p_n^*))$, $S = (((s_1^+)\ (s_2^+)\ ...\ (s_n^+)))'$, and input x_i in $P = x_i'$ in S, has the TSP properties $\boldsymbol{MTS}(P) = \boldsymbol{MTS}(S)$ and $\boldsymbol{CTS}(P) = \boldsymbol{CTS}(S)$.

This theorem may be easily proven by showing that P and S have the same fault equivalence classes and hence the same test sets.

To transform more general circuits or expressions, the above transformations may be applied iteratively to parts of a circuit. A key TSP property of transformations of single-output subcircuits has been determined by Rajski and Vasudevamurthy. We restate their result using our terminology as follows.

Theorem 4 [2]: Let F be a circuit with a single-output subcircuit F'. Let $\mathcal{T}(F') = G'$ for some transformation $\mathcal{T}$, and let G be the circuit obtained by replacing F' by G'. If $\boldsymbol{CTS}(F') \subseteq_c \boldsymbol{CTS}(G')$, then $\boldsymbol{CTS}(F) \subseteq \boldsymbol{CTS}(G)$.

4. Multiple-Output Transformations

Unlike the previous transformations, the next type is defined for multiple-output (sub)circuits. It involves substituting a single copy of a logic element E whose output fans out to k places (denoted $E, E, ..., E$) for k separate copies of E (denoted by $[E]_1, [E]_2, ..., [E]_k$) that do not fan out. This is referred to as "resubstitution" in the literature. Formally, the *resubstitution* transformation is defined as $\mathcal{R}([E]_1, [E]_2, ..., [E]_k) = E, E, ..., E$, and is illustrated in Fig. 6.

Theorem 5: The resubstitution transformation $\mathcal{R}(F) = G$ where $F = [E]_1, [E]_2, ..., [E]_k$ and $G = E, E, ..., E$, has TSP properties $\boldsymbol{MTS}(F) = \boldsymbol{MTS}(G)$ and $\boldsymbol{CTS}(F) = \boldsymbol{CTS}(G)$.

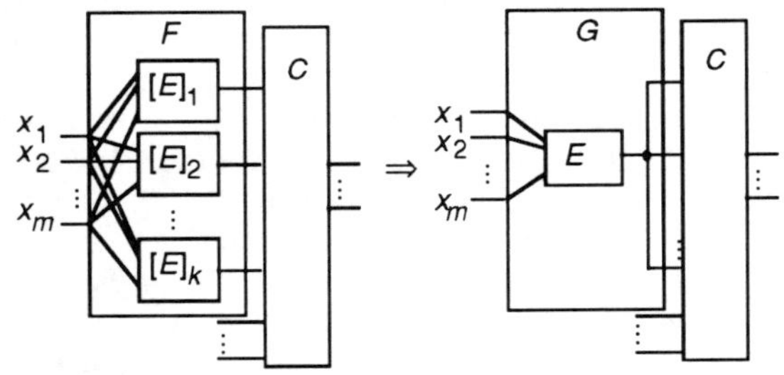

Figure 6. The resubstitution transformation.

Theorem 5 may be proven by showing that any test that detects a fault f in F (i.e. propagates an error to the output of F) also detects the corresponding fault g in G and vice versa.

Theorem 4 above does not extend, in general, to transformations of multiple-output subcircuits [4]. Suppose that the propagation of a fault on some line $[E]_i$ in F depends on the states of the other $[E]_j$ output lines for all tests for that fault. The corresponding fault in G produces different values on these branches that can interact in C to prevent error propagation to a primary output. Hence the fault may be undetectable in G.

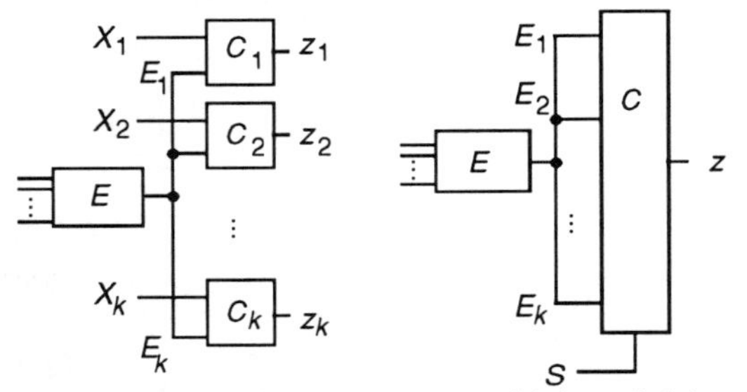

(a) space-disjoint: $z_i = E_iX_i$ (b) time-disjoint: $z = E_1V_1 + E_2V_2 + .. + E_kV_k$

Figure 7. Conditions for applying resubstitution.

Rajski and Vasudevamurthy's solution to the foregoing problem is to allow resubstitution only if each $[E]_i$ branches out to different primary outputs. We refer to this as *space-disjoint* resubstitution; see Fig. 7a. The problematic interference may also be avoided by *time-disjoint* resubstitution, where the signals on the fanout branches from E are separated by time-multiplexing them at the output of C by an auxiliary set of variables S. For each fanout signal E_i, there is a specific set of values V_i of S such that $z = E_i$ whenever $S = v \in V_i$. Clearly these conditions are satisfied when C is a multiplexer with select inputs S. However, more general multiplexer-like structures that support time-multiplexing of resubstituted signals exist in common circuits like adders. These concepts lead to the next theorem, which may be proven by showing that any test for a fault f in F must propagate the corresponding fault g in G to the output of C (or C_i).

Theorem 6: The space- or time-disjoint resubstitution transformation $\mathcal{R}(F) = G = \mathcal{R}([E]_1, [E]_2, ..., [E]_k) = E, E, ..., E$ has the TSP property $\boldsymbol{CTS}(F) \subseteq \boldsymbol{CTS}(G)$.

5. Application to Adder Design

We now use the set of TSP transformations introduced above to analyze the experimental data presented by Davé and Patel for adders. They began with an optimal 4-bit two-level adder and derived a complete test set T consisting of 189 tests for over 1600 SSL faults. They showed by simulation that T is complete for ripple-carry and carry-lookahead implementations, which contain an average of 12% of the number of faults in the two-level circuit [5].

Ripple-Carry Adder The minimal form for the n-bit two-level adder can be determined from the following recursive expressions for the carry and sum functions for a two-level full adder.

$$c_i = ((a_ib_i) + (a_ic_{i-1}) + (b_ic_{i-1}))$$
$$s_i = ((a_ib_ic_{i-1}) + (a_i'b_i'c_{i-1}) + (a_ib_i'c_{i-1}') + (a_i'b_ic_{i-1}'))$$

These expressions define minimal sum-of-products (SOP) forms. By recursively expanding them for $i = 1, 2, ...$ we can directly obtain minimal SOP forms for c_n and $s_1, s_2, ..., s_n$ for any n, in terms of the adder's primary inputs. These $n + 1$ expressions define a two-level adder which is the starting point in our TSP transformation process for adder design. This two-level adder is also the source of the basic test set T used in [5].

The process of transforming the two-level adder to a ripple-carry adder is summarized in Table 1, which lists the sequence of transformations applied to the expanded SOP expressions for c_n and s_n. Let $((C_n)^+)$ and $((S_n)^+)$ denote these SOP expressions for c_n and s_n, and let $((D_n)^+)$ be the SOP expression for c_n' obtained by complementing each primary input a_i, b_i of c_n. In step 1, the fanout-free transformation $\mathcal{F}_1$ is used to group the terms of $((C_n)^+)$ in which only a_n appears. A series of fanout-free transformations $\mathcal{F}_2$ is then used to set up the carry subexpression $((a_nC_{n-1})^+)$ for the extraction transformation $\mathcal{E}_2$. Figure 8a illustrates the resulting circuit. The extraction transformation $\mathcal{E}_2$ produces the SOP expression $((C_{n-1})^+)$; see Fig. 8b. Returning to Table 1, step 2 repeats the composite transformation of step 1 for terms of $((C_n)^+)$ in which only b_n appears, producing another copy of the SOP subexpression $((C_{n-1})^+)$ and giving the circuit of Fig. 8c. Time-disjoint resubstitution $\mathcal{R}_{T,1}$ is then applied to the repeated subexpressions $[((C_{n-1})^+)]_i$ in step 3, resulting in the usual carry expression for the n-th (most significant) ripple-carry cell. From Fig. 8d, it can be seen that the time-disjoint criterion is satisfied given select input $S = a_nb_n$ and select values $V_1 = 10$ and $V_2 = 01$.

The SOP expression $((S_n)^+)$ is transformed in a similar fashion, as shown by steps 4-10 in Table 1. Step 4 involves a composite transformation in which the terms containing a_nb_n are grouped using $\mathcal{F}_3$, and isolated using $\mathcal{F}_4$, $\mathcal{F}_5$, and the time-disjoint resubstitution transformation $\mathcal{R}_{T,2}$. The extraction transformation $\mathcal{E}_3$ produces the subexpression $((C_{n-1})^+)$. Step 5 is similar. In steps 6 and 7, $((D_{n-1})^+)$ is obtained from the terms containing a_nb_n' and $a_n'b_n$ respectively; see Fig. 8e. A De Morgan transformation $\mathcal{M}_1$ eliminates the input inverters to $((D_{n-1})^+)$ in step 8, resulting in an OR/AND subexpression; see Fig. 8f. $((C_{n-1})^+)$ is the dual expression of $((D_{n-1}')^*)$, and is obtained by applying the extraction transformation $\mathcal{E}_4$ in step 9. Finally in step 10, time-disjoint resubstitution $\mathcal{R}_{T,3}$ is applied to s_n to eliminate repeated subexpressions, and then space-disjoint resubstitution $\mathcal{R}_{S,1}$ is applied to both c_n and s_n. The resulting expressions define the structure of the n-th stage of the ripple-carry adder. The other n - 1 stages may be obtained by repeating the foregoing procedure. All the transformations used are TSP, therefore the final design, which is a conventional n-bit ripple-carry adder, is completely tested by any complete test set for the initial two-level circuit.

Step	Transformed Expression	Transformation
	$c_n = ((a_nb_n) + (a_nC_{n-1})^+ + (b_nC_{n-1})^+) = ((C_n)^+)$	Initial Subcircuit
1.	$c_n = ((a_nb_n) + (a_n((C_{n-1})^+)) + (b_nC_{n-1})^+)$	$\mathcal{F}_1(...)$, $\mathcal{E}_2(\mathcal{F}_2(...\ \mathcal{F}_2(...)\ ...))$
2.	$c_n = ((a_nb_n) + (a_n[((C_{n-1})^+)]_1) + (b_n[((C_{n-1})^+)]_2))$	$\mathcal{E}_2(\mathcal{F}_2(...\ \mathcal{F}_2(...)\ ...))$
3.	$c_n = ((a_nb_n) + (a_n((C_{n-1})^+)) + (b_n((C_{n-1})^+)))$	$\mathcal{R}_{T,1}(...)$
	$s_n = ((a_nb_nC_{n-1})^+ + (a_n'b_n'C_{n-1})^+ + (a_nb_n'D_{n-1})^+ + (a_n'b_nD_{n-1})^+) = ((S_n)^+)$	Initial Subcircuit
4.	$s_n = ((a_nb_n[((C_{n-1})^+)]_2) + (a_n'b_n'C_{n-1})^+ + (a_nb_n'D_{n-1})^+ + (a_n'b_nD_{n-1})^+)$	$\mathcal{F}_3(...)$, $\mathcal{F}_6(\mathcal{E}_3(\mathcal{R}_{T,2}(\mathcal{F}_5(...\mathcal{F}_4(...)...))))$
5.	$s_n = ((a_nb_n[((C_{n-1})^+)]_2) + (a_n'b_n'[((C_{n-1})^+)]_3) + (a_nb_n'D_{n-1})^+ + (a_n'b_nD_{n-1})^+)$	$\mathcal{F}_6(\mathcal{E}_3(\mathcal{R}_{T,2}(\mathcal{F}_5(...\mathcal{F}_4(...)...))))$
6.	$s_n = ((a_nb_n[((C_{n-1})^+)]_2) + (a_n'b_n'[((C_{n-1})^+)]_3) + (a_nb_n'((D_{n-1})^+)) + (a_n'b_nD_{n-1})^+)$	$\mathcal{F}_6(\mathcal{E}_3(\mathcal{R}_{T,2}(\mathcal{F}_5(...\mathcal{F}_4(...)...))))$
7.	$s_n = ((a_nb_n[((C_{n-1})^+)]_2) + (a_n'b_n'[((C_{n-1})^+)]_3) + (a_nb_n'[((D_{n-1})^+)]_1) + (a_n'b_n[((D_{n-1})^+)]_2))$	$\mathcal{F}_6(\mathcal{E}_3(\mathcal{R}_{T,2}(\mathcal{F}_5(...\mathcal{F}_4(...)...))))$
8.	$s_n = ((a_nb_n[((C_{n-1})^+)]_2) + (a_n'b_n'[((C_{n-1})^+)]_3) + (a_nb_n'[(((D_{n-1}')^*))']_1) + (a_n'b_n[(((D_{n-1}')^*))']_2))$	$\mathcal{M}_1(...)$
9.	$s_n = ((a_nb_n[((C_{n-1})^+)]_2) + (a_n'b_n'[((C_{n-1})^+)]_3) + (a_nb_n'[(((C_{n-1})^+))']_1) + (a_n'b_n[(((C_{n-1})^+))']_2))$	$\mathcal{E}_4(...)$
10.	$s_n = ((a_nb_n((C_{n-1})^+)) + (a_n'b_n'((C_{n-1})^+)) + (a_nb_n'(((C_{n-1})^+))') + (a_n'b_n(((C_{n-1})^+))'))$	$\mathcal{R}_{T,3}(...)$, $\mathcal{R}_{S,1}(...)$

Table 1. Transformation sequence for the n-th stage of a ripple-carry adder.

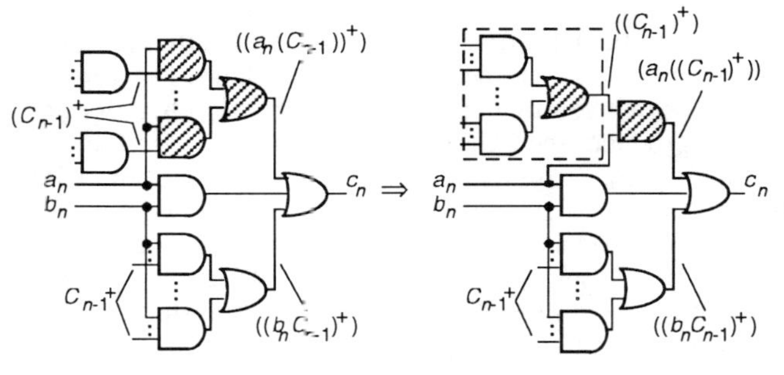

(a) (b)

The extraction transformation $\mathcal{E}_2$.

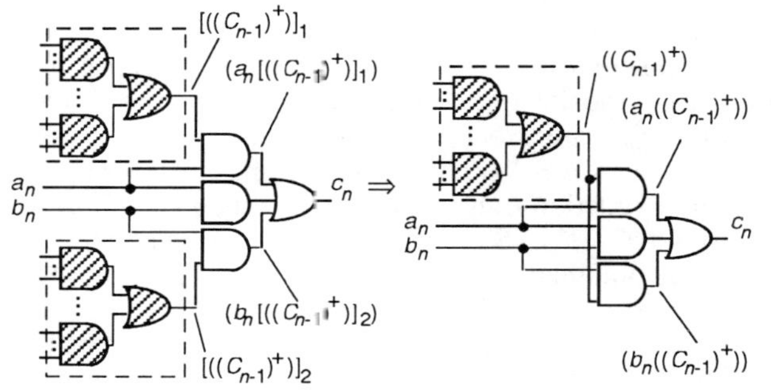

(c) (d)

The time-disjoint resubstitution transformation $\mathcal{R}_{T,1}$.

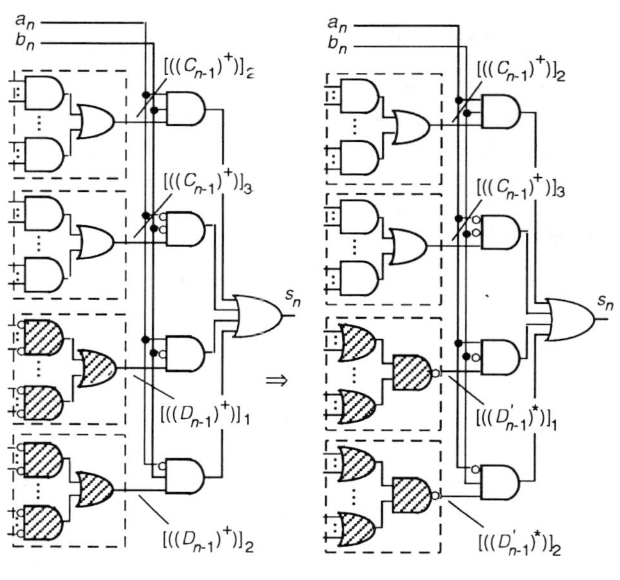

(e) (f)

The De Morgan transformation $\mathcal{M}_1$.

Figure 8. Transformations used in ripple-carry design.

Carry-lookahead Adder The process for transforming the n-bit, two-level adder to the carry-lookahead form begins by applying steps 4-10 of Table 1 to $((S_n)^+)$, $((S_{n-1})^+)$, ... $((S_2)^+)$, producing the SOP expressions $((C_n)^+)$, $((C_{n-1})^+)$, ..., $((C_2)^+)$. The remaining steps involve grouping the appropriate terms in $((C_i)^+)$ and applying an extraction transformation to each. A group corresponds to a carry generate $g_i = (a_ib_i)$ and propagate $p_i = (a_i + b_i)$ subexpression of the form $(p_np_{n-1}\ldots p_{n-k+1}g_{n-k})$. Time-disjoint resubstitution is applied in the final step to eliminate repeated subexpressions, and produce the desired carry-lookahead form. We have only used complete TSP transformations at each step, so again complete test sets are preserved.

6. Conclusions

We have presented a rigorous approach to the transformation of combinational circuits with respect to test-set preservation. We have defined a powerful set of logic transformations and shown that each preserves a precise inclusion relationship between the minimal and complete test sets of the initial and transformed circuits. Our transformations extend earlier work and introduce some new ones, notably time-disjoint resubstitution.

TSP transformations allow us to control the unpredictable effects on test sets made by area- and performance-driven transformations and avoid designs that are subsequently found to be difficult to test. Test-generation effort is reduced and designs are produced that can be fully tested using test sets for the initial circuit. We have illustrated this by demonstrating a sequence of TSP transformations that produce ripple-carry and carry-lookahead designs. Our results substantiate and generalize an earlier experimental demonstration of close relationships among the test sets of different adder designs. We are currently exploring the systematic identification of useful sequences of transformations and the application of our transformations to other common circuit types.

References

[1] R. Brayton et al., "MIS: a multiple-level logic optimization system", *IEEE Transactions on CAD*, Vol. CAD-6, November 1987, pp. 1062-1081.

[2] J. Rajski and J. Vasudevamurthy, "Testability preserving transformations in multi-level logic synthesis", in *Proceedings of the International Test Conference*, 1990, pp. 265-273.

[3] Z. Kohavi and I. Kohavi, "Detection of multiple faults in combinational logic networks", *IEEE Transactions on Computers*, Vol. C-21, June 1972, pp. 556-568.

[4] G. Hachtel et al., "On properties of algebraic transformations and the multifault testability of multilevel logic", in *Proceedings of the International Conference on CAD*, 1989, pp. 422-425.

[5] U. Davé and J. Patel, "A functional-level test generation methodology using two-level representations", in *Proceedings of the Design Automation Conference*, 1989, pp. 722-725.

SESSION 28

PANEL: DESIGN AND INTEGRATION SERVICES

Chair: Ron Collett - Dataquest, Inc., San Jose, CA
Organizer: Ken Coley - Digital Equipment Corp., Marlboro, MA

Electronic product design is becoming increasingly complex, yet market windows are narrowing and product life cycles are shrinking. The dynamics are forcing manufacturers to compress design cycles further and further. How will manufacturers reconcile these competing factors? It is becoming clear that an optimized product development process will become essential to establishing a competitive advantage in the electronics industry. Such an optimized product development process has many facets, including not only the creation of an efficient design flow, but also development of the product data management process at the work group, project and enterprise levels. Project planning, which includes integration of the design process solution with the entire enterprise, will also become critical to a manufacturers/'s product development strategy.

But what is the cost of establishing this kind of optimized product development process? Moreover, what are its tangible benefits, and can they be quantified? What kind of support and consulting services do electronics manufacturers envision will be needed to implement future product development environments? Can design automation vendors supply the requisite services to manufacturers? What kind of services are these vendors currently providing and what will they offer in the future? How much are electronics manufacturers spending on integration services in the development of their design processes? What role do the present framework technologies play in current and future product development environments? Do the frameworks address the needs of today's electronics manufacturers? Indeed, much has been discussed about framework technology, but the benefits remain amorphous.

Panelists will debate these and other issues surrounding design and integration services, framework technology and their overall impact on electronic product development.

Panel Members:

Deme Clainos - Mentor Graphics Corp., Wilsonville, OR
Don Jenkins - Digital Equipment Corp., Marlboro, MA
Kevin D. Kolwicz - AT&T Bell Labs., Allentown, PA
John Payne - Cadence Design Systems, Inc., San Jose, CA
John Sweet, Jr. - Rockwell, Anaheim, CA
H. Yoshi Yoshizawa - NEC Corp., Kawasaki, Japan

Challenges and Advances in Electrical Interconnect Analysis

Albert E. Ruehli and Hansruedi Heeb
IBM T. J. Watson Research Center
Yorktown Heights, NY 10598

Abstract

A growing need exists for electrical interconnect analysis (EIA) for chips, packages and printed circuit boards. In this short tutorial we review key issues regarding EIA for VLSI parasitic circuits. We give a general introduction of important aspects for technologies with different performances and then we review some issues of concern for state of the art high performance chips and packages.

1 Introduction

1.1 General aspects

Electronic Design Automation (EDA) is an essential part of VLSI design. Advances in EDA are driven by the requirements of the next generation technologies which demand a continuous evolution of new tools and techniques. These tools can be divided into two different classes. The primary set of EDA tools are physical design and logic synthesis tools. A secondary, less obviously needed set is design and verification tools in areas like testing, timing verification and electrical design. Electrical interconnect analysis (EIA) belongs to this second class of tools. Generally, VLSI designers focus on these tools only if there exists an obvious problem or if there is much evidence from previous designs that such tools must be used. This has resulted in a slower development of the EIA-CAD area than tools in the first class.

Perhaps the most well understood techniques in the electrical aspects of system design are general purpose circuit simulators like SPICE[7] or ASTAP[21]. Today, these tools are the cornerstones of electrical design since they are tied in with measurable quantities which relate to all aspects of circuit design. All other generally available and company proprietary programs for EIA are much more specialized in their application. As a consequence, no other EIA tool has reached even a small fraction of the utility of a circuit simulator. In fact, the capabilities exist in good circuit solvers for the modeling of interconnects in terms of lumped models and transmission lines. However, this still leaves the calculation of the circuit parameters (R,L,C) for the physical geometries to other EIA programs. Also, transmission lines are mostly needed for the design of PC boards only. Moderate size circuits with a few thousand transistors and only few parasitics may take hours of simulation time for transient analysis. Hence, these tools are lagging badly behind in capacity especially for high performance circuits. Circuits must be broken into very small pieces for the analysis which is a very time consuming and error prone process. The main problem is that circuit simulation is one of the most time consuming processes in EDA in terms of compute time and storage needed per transistor. It is clear that this will be worse once more complicated parasitic models are included.

Historically, the first work in EIA was done for high performance bipolar mainframe computers. Techniques developed for the electrical design of such systems are given in [14] and in the many references cited there. The performance of MOSFET transistor logic has only recently become competitive with bipolar and GaAs logic so that extensive EIA is necessary[1]. The first key problem with this was discovered by some of the early RAM designers in the early eighties. Several designs did not work since they did not take the increase in capacitance due to fringing into account. Since then, much work was done in the area of VLSI capacitance calculators and other CAD tools for EIA.

1.2 High Performance and High Frequency Systems

In the last two years, EIA and interconnect design has become one of the focus areas in CAD for high performance systems. The main reason for the increased activities in this area is the desire to decrease the cycle time in order to improve the performance of the next generation high performance systems. Today, many companies have tools for delay characterization of their circuits which are an efficient approach for most technologies with the exception of the highest performance designs. However, in future high end technologies, the interconnects and transistor parasitic elements contribute more to the circuit delay than the intrinsic transistors. Hence, the transistor based delay which is easily predictable, is decreasing while the part which depends on the external connections is increasing. Then the design effort needed to utilize these technologies will increase drastically especially as we reach the limit of a particular technology. We conclude from this observation that the highest performance chips and systems will be designed at the circuit level with custom design tools, rather than today's more simple delay equation based structured tools.

Two dimensional transmission line models are of importance for printed circuit board (PCB) modeling where many of the wires are still long compared to the cross-sectional dimensions and where in most cases ground planes provide a reasonably good, continuous ground return path. Exceptions include cuts in ground planes, boards without ground or voltage supply planes and very lossy planes which do not support simple transverse mode (TEM) models. Recently, quite a number of researchers have made progress in the implementation of models for transmission lines e.g. [17]. A discussion of these techniques is beyond the scope of this paper.

An excellent summary of commercially available electromagnetic field solvers and EIA tools has been published in the IEEE Spectrum[18]. A wide range of applications is addressed by the tool vendors. Some of the applications are circuit parameter calculations for coupled transmission lines (L and C matrices), microwave printed circuit analysis, simulators for electrical machines, magnetic shielding, crosstalk analysis and linear time and frequency domain circuit simulators, as well as electromagnetic field or Maxwell's equation solvers. The solution methods are based on differential equations, finite element formulations and integral equation techniques. All but the most simple computations are extremely time consuming and the more complex computations have been made widely possible only through the availability of new more powerful workstations.

The state of the art in the solver tools and fundamental analysis formulations are evolving at a slow pace. To solve the various EIA problems listed below, advances are necessary. The formulation techniques today are still evolving, but most of the fundamental aspects are well understood. The main advances are in details of the solution methodologies and the clever exploitation of different properties of the special problem at hand. It is well known that special purpose solvers, which exploit specific properties of a problem, can be much faster than general purpose tools. The authors have pursued a different approach for model simplification. Circuit theorists and practitioners have for years been able to find simplified "engineering" solutions to problems. This has motivated us to formulate the solution of Maxwell's equations in terms of circuits so that similar simplification techniques can be applied for both the time and frequency domain[3][16].

Here, we give a list some of the EIA problems which may occur for high speed systems:

- Signal waveform distortion, delay prediction
- Ground and voltage supply bounce due to L and R voltage drops
- Coupled interconnect noise signal waveforms
- Substrate noise coupling for analog chips, or low signal grounds
- Electromagnetic Interference EMI radiation (EMI)

2 Modeling of High Performance Systems

In the second part of this short survey we explore techniques for the modeling of the three dimensional aspects of high performance chip and package interconnects. It is clear that smaller geometries imply that three dimensional aspects are going to dominate in several ways. Today, research EIA efforts are under way to find new ways for the modeling of such structures.

The use of differential and integral equation techniques for the solution of EM problems has evolved over the last twenty years to the point where moderate size problems can be solved rather routinely using *full wave* formulations. Differential equation (and finite element) techniques have an advantage for interior problems (inside a metal box) and for non-homogeneous regions as they occur in device modeling and perhaps the computation of interconnect resistances. e.g. [10, 5]. Another example is transient analysis using the finite difference time-domain (FDTD) method[19]. Integral equation methods have an advantage for problems with open regions. The most popular technique in this area is the method of moments (MoM) e.g. [2, 12]. So far MoM solvers have only addressed the frequency-domain even though time-domain MoM is possible in principle[11]. One of the main reasons for the popularity of integral equations for EIA is the fact that most of the important problems involve open regions. The fundamental solution utilized is the free space Green's function, which is a solution to the open problem in the infinite domain.

Another integral technique is based on circuit models. Circuit interconnect models have the advantage over MoM that they can easily be combined with traditional circuit models (like transistors) and that the same model can be used for both frequency- and time-domain. In the following section we will therefore take a closer look at an interconnect analysis system based on the circuit model. A direct connection exists between the mathematical formulation techniques like MoM and the circuit based approaches. Certainly, the techniques for the discretization into finite cells are the same for both approaches and the numerical solution of the differential MoM equations can be viewed as a lumped model solution, where the lump size is small. One of the interesting aspects of the circuit model solution is that circuit simplification techniques may be applied.

3 Circuit Models

Circuit based EIA modeling is a emerging area. As an sample technique in this area we describe the rPEEC (retarded partial element equivalent circuit) approach in some detail.

3.1 Lumped element model

The connection between the electrical model and the physical geometry is of key importance for the accurate modeling of the behavior of the interconnects. In the PEEC (partial element equivalent circuit) approach [13, 15] the conductors and finite dielectrics of the interconnects are subdivided into small surface and volume elements similar to the ones used for method of moments. Partial capacitances and inductances are calculated from these elements[6][22]. Different elements or cells which are offset by half the grid-size are used for the calculation of partial inductances and capacitances. The resulting circuit elements are combined with each other into a complete PEEC circuit. This interconnect circuit can now be combined with other circuits like transistors. While the original PEEC circuit theory is a good model of the interconnect at lower frequencies, it lacks the inclusion of *retardation* (the effect of the finite speed of light) which is necessary for problems like EMI. Also, regions of different dielectrics were taken into account in the capacitances only. This resulted in inaccuracies for high frequencies.

In our new circuit models retardation is included [3]. These new rPEEC circuits were shown to be effective for EMI problems. In the frequency domain the retardation is included explicitly as a phase factor of the form e^{-ikR} in the capacitive and inductive cross-couplings. Hence, the PEEC approach has been extended to geometries where the size of the critical coupling distances is no longer short compared to the wavelength. Hence, hopefully the most simple PEEC model can be utilized, which within engineering accuracy will give the correct answer. This is one of the advantages of the circuit domain formulation. It should be noted that this approach represents what is known in electromagnetics as a *full wave solution* in that all the possible modes are included. It has the advantage over the conventional frequency domain implementation of the method of moments or finite element approach that nonlinear circuit models of transistors can be included with a special circuit simulation tool like ASTAP. This also has the advantage that results can be compared for both the frequency as well as the time domain. Another advantage is that the approach can be used for just a part of the interconnects while the remaining interconnects are represented by traditional transmission line models.

Recently, a second advance has been made in the PEEC approach. As indicated above, the dielectrics are included in the original formulation using capacitances only. However, it is well known that at sufficiently high frequencies, the E- and the H-fields are of importance for the dielectrics. A model has been developed[15] which correctly takes the finite dielectrics into account by a mesh of partial inductance and capacitances. Again, this type of model represents an increase in complexity which should only be taken into account if the field solution requires the exact modeling of dielectrics. With these advances, the circuit modeling approach is equivalent to the full-wave solution for the general finite dielectrics case.

Our present work is in the area of retardation in the time domain. In the time-domain a *history mechanism* must be implemented which stores the delayed voltages which then must be accessed efficiently at the current time-step[3]. Not only does this approach represent a full-wave analysis capability for the time-domain but it also may result in *reduced simulation time*. Also, with the retarded PEEC models, a new class of problems can be solved. The prediction of the radiation from VLSI systems is one of the most important EIA problems in this class.

3.2 Circuit Analysis

PEEC and rPEEC circuits are simulated with a general purpose circuit simulator like SPICE or ASTAP, which has to be extended for retarded time-domain simulation. This has been shown to work well for models with several hundred partial inductances. However, it may be somewhat wasteful because traditional circuit simulators were written for sparse circuits while the PEEC circuits part of the matrix is non-sparse. Another approach which may be promising is the AWE approach for the reduction of model complexity[9, 23, 20].

4 Electromagnetic Interference Analysis

A relatively new field for EIA is the computation of electromagnetic radiation[4]. The purpose of this is to insure that a product does not exceed the amount of radiation allowed by government regulation over the frequency range of 30 MHz to 1 GHz at a distance of 3 meters. Most of the early work in this area has been limited to rather simple structures[8]. However, the main problem with CAD tools for the prediction of radiation is the complexity of the structures at hand. So far, little progress has been made in prediction of the radiation for such complex structures. Our attempt at predicting radiation from PC boards[4] can be summarized as follows:

1. Breakup of geometry (including ground planes) into cells for rPEEC models.

2. Computation of partial capacitance (potential) and inductance matrix coefficients.

3. Solution of nonlinear transistor circuit models with linear rPEEC models using ASTAP.

4. Fast Fourier Transform of waveforms to get spectral content and field calculation.

This method is quite general. Besides its usefulness for calculating EMI it can be used for other difficult problems like time-domain antenna simulation or prediction of the ground-bounce of printed circuit boards.

5 Example

The example we are presenting consists of a trace on a printed circuit board driven by a standard TTL 74F04 inverter chip. The load is another 74F04 inverter. The model contains the trace, the board including ground- and power-planes, the two inverters and the IC sockets. Figure 1 shows the driving inverter chip. Figure 2 shows the rPEEC equivalent circuit for a part of the setup shown in Figure 1. The distortion of a 50 MHz signal can be seen in Figure 3 which shows the waveform before and after the trace. Figure 4 finally shows the three meter radiation spectrum of the example.

6 Conclusions

Today's techniques for EIA are fully adequate for medium performance technologies. On the other hand, for high performance technologies, the state of the art tools leave a lot to be desired. The methods of choice for EIA are integral formulations like the method of moments or circuit based techniques because the regions to be analyzed are usually open. Circuit based formulations like rPEEC are advantageous because

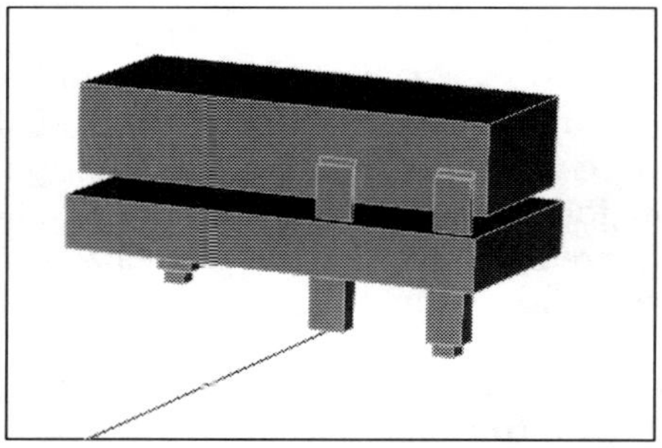

Figure 1: *The driving chip of the example. The power- and ground-planes and the board epoxy have been deleted from the picture for better visibility. The figure shows the 74F04 (with pins 1, 12 and 14) and its socket, part of the trace on the PCB and the two vias to the ground- and power planes.*

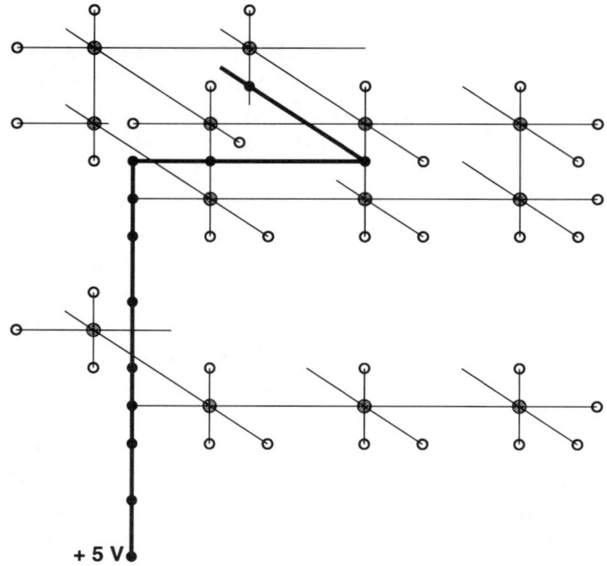

Figure 2: *Part of the rPEEC model of Figure 1. Fat lines are inductances (conductors), thin lines are inductances with a series capacitance (dielectrics). Not shown are the cross-couplings. Solid nodes are metal nodes, grey nodes are internal dielectric nodes and white nodes are dielectric surface nodes. The figure shows the power pin and some internal wiring, the corner of the chip case and the corner of the socket.*

they can be combined with other circuit models like transistors and because they can be used in time- and frequency-domain.

References

[1] H. B. Bakoglu *Circuits, Interconnections and Packaging for VLSI.* Addison Wesley Publishing Company, 1990.

[2] R. F. Harrington. *Field Computation by Moment Methods.* Macmillan, 1968.

[3] H. Heeb, A. Ruehli. Retarded models for PC board interconnects – or how the speed of light affects your SPICE circuit simulation. In *Proc. of the IEEE Int. Conf. on Computer Aided Design,* pages 70–73, November 1991.

[4] H. Heeb, A. Ruehli, J. Janak and S. Daijavad. Simulating electromagnetic radiation of printed circuit boards. In *Proc. of the IEEE Int. Conf. on Computer Aided Design,* pages 392–395, 1990.

[5] G. Heiser, C. Pommerell, J. Weis and W. Fichtner. Three-dimensional numerical semiconductor device simulation: Algorithms, architectures, results. *IEEE Transactions on Computer-aided Design,* 10(10):1218–1230, 1991.

[6] J. F. Janak, D. D. Ling and H.-M. Huang. C3DSTAR: A 3D wiring capacitance calculator. In *Proc. of the IEEE Int. Conf. on Computer Aided Design,* pages 530–533, 1989.

[7] L. W. Nagel. SPICE2: A computer program to simulate semiconductor circuits. Electr. Res. Lab. Report ERL M520, University of California, Berkeley, May 1975.

[8] C. R. Paul and D. R. Bush. Radiated emissions from common-mode currents. In *Proc. of the IEEE Int. Symp. on Electromagnetic Compatibility,* pages 197–203, 1987.

[9] L. T. Pillage and R. A. Rohrer. Asymptotic waveform evaluation for timing analysis. *IEEE Transactions on Computer-aided Design,* 9(4):352–366, 1990.

[10] M. R. Pinto, C. S. Rafferty and R. W. Dutton. PISCES II: Poisson and continuity equation solver. Technical report, Stanford University, 1984.

[11] S. M. Rao and D. R. Wilton. Transient scattering by conducting surfaces of arbitrary shape. *IEEE Transactions on Antennas and Propagation,* 39(1):56–61, 1991.

[12] B. J. Rubin and S. Daijavad. Radiation and scattering from structures involving finite-size dielectric regions. *IEEE Transactions on Antennas and Propagation,* 38(11):1863–1873, 1990.

[13] A. E. Ruehli. Equivalent circuit models for three dimensional multiconductor systems. *IEEE*

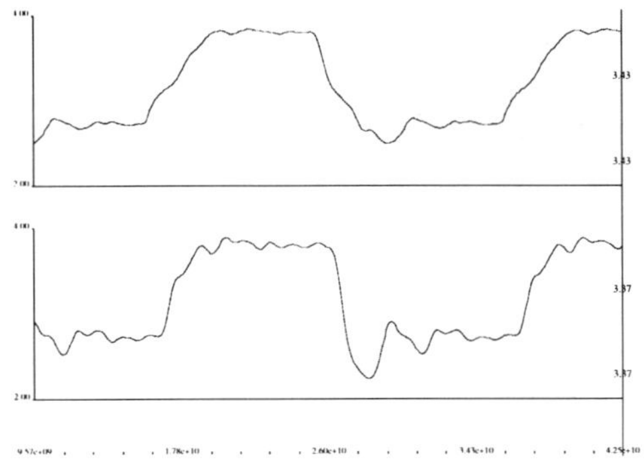

Figure 3: *The waveforms at the source and load end of the trace.*

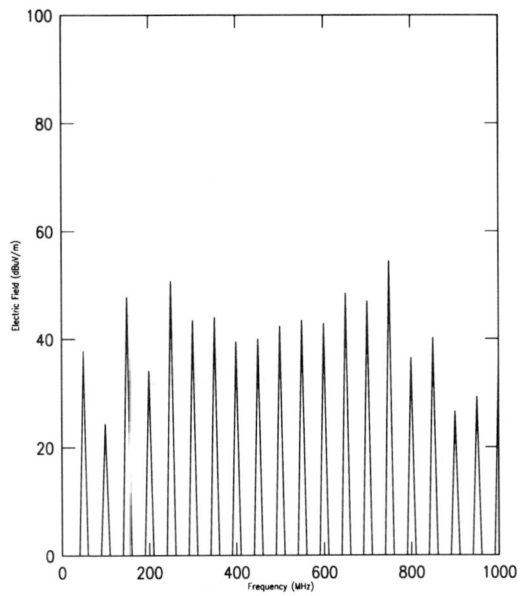

Figure 4: *The simulated radiation of the example in dB/μV three meters from the PCB.*

Transactions on Microwave Theory and Techniques, MTT-22(3):216–221, March 1974.

[14] A. E. Ruehli. Survey of computer-aided electrical analysis of integrated circuit interconnections. *IBM Journal of Research and Development*, 23(6):626–639, November 1979.

[15] A. E. Ruehli and H. Heeb. Circuit models for three-dimensional geometries including dielectrics. Accepted for publication in IEEE Transactions on Microwave Theory and Techniques, 1992.

[16] A. E. Ruehli, H. Heeb, E. Bracken and R. Rohrer. Three dimensional circuit oriented electromagnetic modeling for VLSI interconnects. In *Proceedings of the Packaging, Interconnects, Optoelectronics for the Design of Parallel Computers Workshop*, March 1992.

[17] J. E. Schutt-Aine and R. Mittra. Nonlinear transient analysis of coupled transmission lines. *IEEE Transactions on Circuits and Systems*, 36(7):959–967, July 1989.

[18] D. G. Swanson Jr. Simulating EM fields. *IEEE Spectrum*, 28(11):34–37, November 1991.

[19] A. Taflove and M. E. Brodwin. Numerical solution of steady-state electromagnetic scattering problems using the time-dependant Maxwell's equation. *IEEE Transactions on Microwave Theory and Techniques*, 23(8):623–630, August 1975.

[20] T. K. Tang and M. S. Nakhla. Analysis of high-speed VLSI interconnects using the asymptotic waveform evaluation technique. *IEEE Transactions on Computer-aided Design*, 11(3):341–352, 1992.

[21] W. T. Weeks, A. J. Jimenez, G. W. Mahoney, D. Mehta, H. Quasemzadeh and T. R. Scott. Algorithms for ASTAP – a network analysis program. *IEEE Transactions on Circuits Theory*, CT-20(11):628–634, November 1973.

[22] P. K. Wolff and A. E. Ruehli. Inductance computations for complex three dimensional geometries. In *Proc. of the IEEE Int. Symp. on Circuits and Systems*, pages 16–19, 1981.

[23] D. H. Xie and M. Nakhla. Delay and crosstalk simulation of high-speed VLSI interconnects with nonlinear terminations. In *Proc. of the IEEE Int. Conf. on Computer Aided Design*, pages 66–69, November 1991.

Tools to Aid in Wiring Rule Generation for High Speed Interconnects

Paul Franzon Slobodan Simovich Michael Steer
Mark Basel Sharad Mehrotra Tom Mills

Picosecond Digital Electronics Laboratory
Department of Electrical and Computer Engineering
North Carolina State University
Box 7911
Raleigh, NC 27695

Abstract

This paper discusses the construction and application of two tools. The first tool, MetaSim, allows multiple simulation studies to be automatically specified, conducted, and analyzed in such a fashion as to produce wiring rules for layout tools. One of MetaSim's current engines, Transim is a convolution simulator that incorporates a robust, rapid, and accurate method for the simulation of coupled, lossy interconnect structures specified with frequency-dependent parameters.

1 Introduction

The high speed performance of most digital systems is now limited by distributed discontinuities and transmission line effects of packaging and interconnects, and not by the switching speed of semiconductor devices. Among the effects of these distributed circuit elements are: Delay and rise time degradation, ringing, crosstalk noise, metastates, and simultaneous switching noise.[1]

In many cases, the quest for higher performance is centering on choosing the most appropriate packaging technology, as well as layout considerations, not just increasing device clocking rates.

However, a number of difficulties are encountered as the circuit designer, layout engineer, and signal integrity engineer seek to design high speed systems.

One primary difficulty is the need for the signal integrity engineer to develop wiring rules to be used in the router [2]. These wiring rules constrain the lengths and spacings (and sometimes widths) on the board or multichip module so that delay and noise requirements are met. Care has to be taken in doing this. If the wiring rules are too lenient, many signal integrity problems will be detected during verification and the board will need extensive redesign. If the rules are too conservative, the router will be overconstrained and available routing area will be used inefficiently. Until recently rules of thumb often sufficed. However, as the rise times in complex CMOS boards and MCMs approach sub-1 ns ranges, the need for more precise wiring rules will dramatically increase.

Precise wiring rules can not be generated from simple analytic expressions for propagation delay as no expression available today can accurately predict the effects of reflection and simultaneous switching noise on delay. Instead, the generation of wiring rules requires multiple simulation studies to be performed across an appropriate range of design and parametric variations. The simulation studies are analyzed in order to characterize the transformation between the physical design space (P – lengths, spacings, etc.) and the electrical design space (E – delay, noise). This transformation is then inverted and applied to the placement in order to check it and to formulate wiring rules for the router. We will refer to this set of transformations as the "P↔E transformation".

Carrying out such studies traditionally requires a large amount of engineer time and can be error prone. MetaSim (Meta Simulator) automates the generation, conduct, and analysis of such simulation studies. In this paper, we describe the operation of MetaSim and give an example of its use.

MetaSim interfaces with three simulators, Spice, Cazm, and Transim. Transim, solves the problem of

[1] This work was supported, in part, by the National Science Foundation under contract MIP-9017054

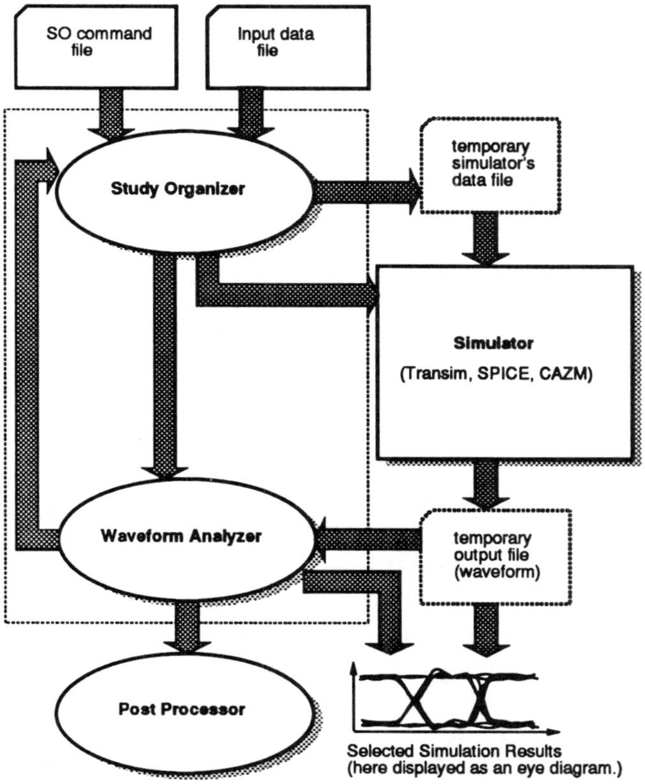

Figure 1: Organization of MetaSim.

incorporating transient circuit simulation with nonlinear devices, with transmission line simulation in the frequency domain. For example, this capability is a requirement for thin film MCM interconnect once the skin effect starts becoming important (ie. when rise times start approaching around 200 ps [4].)

2 MetaSim: A Simulation Generator and Analyzer

MetaSim (figure 1) automates the <simulation set up → simulation execution → simulation results analysis> loop. It thus allows complex relationships between delay and noise requirements on one hand, and technology selection and physical design issues, on the other hand, to be explored with a minimum of human time. MetaSim consists of the following elements (Figure 1):

- A study organizer, which allows variational studies requiring multiple simulation runs to be concisely specified and generated. It also includes specialized simulation file generators for Cazm, SPICE and Transim.
- A waveform analyzer that carries out signal integrity analysis on each set of simulation results and extracts delay and noise data as required.
- Post-processing software that enables relationships to be readily extracted from the analyzed results, and selected waveforms to be retained and viewed.

The nature of these tools will be described further, in section 3, primarily by the way of an example.

In the following sections, we will first describe related work conducted elsewhere, and present an example of MetaSim's use.

2.1 Related Work

The existing approach to wiring rule generation is to make initial estimates and refine on them through multiple iterations of the <specify rules → carry out route → verify → analyze simulations> loop. Although some commercial tools automate part of the analysis they are not able to suggest how to improve the rules. Attempts at this, such as TLTS [7] make very simple suggestions such as 'shorten net' and 'remove receivers'. As better suggestions require that a characterization of the P↔E transformation exist anyway, it makes sense to use this transformation to form good rules initially for as many nets as is reasonable. This will greatly reduce the number of design iterations required, hopefully to one.

The closest existing tool to MetaSim is Adline, which was developed at NTT [5]. It is only available in Japan. Adline includes in its simulator, the ability to specify multiple simulations for some structures. With Adline, as with MetaSim, it is also possible to study the effect of process variations by automatically carrying out multiple simulations. The main distinction between Adline and MetaSim is that MetaSim offers far greater flexibility in the range of studies that can be undertaken. MetaSim can also be easily integrated with any simulator while Adline is integral with one simulator.

3 Example of the Application of MetaSim

MetaSim is applied before layout by taking a net, or set of nets, that are likely to require layout constraints and conducting studies on them. For example, here, we will consider the set of nets that have one output and four inputs, and the output may, or may not be,

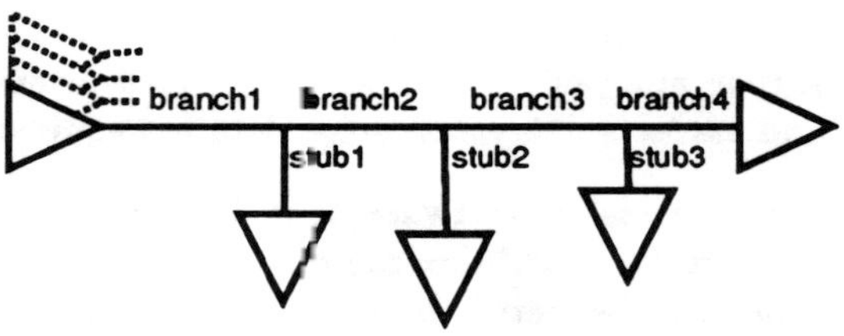

Figure 2: Network topology to be analyzed. The output driving the net might have up to three other outputs switching with it. The user can specify any circuit model for the ground and power connections and can specify driver skew.

part of four simultaneous switching outputs. The interconnection network topology is described as a set of branches and stubs, in a daisy chain configuration, as shown in figure 2. The circuit technology is 0.8μm CMOS, and the packaging technology, a polyimide-dielectric thin film MCM, with copper interconnect, and non-fan-out-TAB chip attach.

For the purposes of this paper all of the branches shown in figure 2 will be set at the same length as will the stubs. This is only done to make the results presentable. Part of the study command file that specified the studies that produced the results presented here is shown in figure 3. Part of the 'data file' is also presented. In this study, three network related components are varied independently, the driver circuit model, the branch length and stub length, in order to specify a total of 32 simulation runs. The command language is actually a formally defined programming language with many constructs other than the ones shown including random generation constructs.

The 'data file' shown in figure 1 is a generic simulation specification file that lists all of the elements to be simulated as a network of objects (eg. a branch in a transmission line is an object) with certain properties associated with them. The "$" variables that appear in the command file also appear in the data file. The study organizer then produces appropriate substitutions for each simulation run and passes the modified data file to the program that generates the actual simulation file (or directly to Transim, as a Transim file is already described in object-orientated terms). Discontinuities, such as vias, bends, and crossovers can also be specified in the data file, as can coupled lines and connectors.

The 'do' statement specifies what analysis needs to be carried out on the simulation results. In this case we wish to know the peak undershoot at node "5" (the last receiver in the chain) and the delay from the driver output, node "1" to this receiver. We are

Extract from command file:

```
contr
  for ($branch = 1e-2 4e-2 1e-2 incr)
  for ($stub= 1e-3 2.9e-2 4e-3 incr)
  do{
    undsht = "5"
    delay  = "1" "5" 0.23
    delay  = "1" "5" 0.734
    }
```

Extract from 'data' (generic simulation specification) file:

```
Branch 1:

Brch_type_1 bstrip Brch_length_1 $branch
Height1A_1 10e-6 Height1B_1 20e-6 Thick1_1 5e-6
Width1_1 17e-6 Er1_1 3.5 Er2_1 3.5
Brch_load_1 non-loaded
```

Figure 3: Part of the command and data files used to specify the study described here.

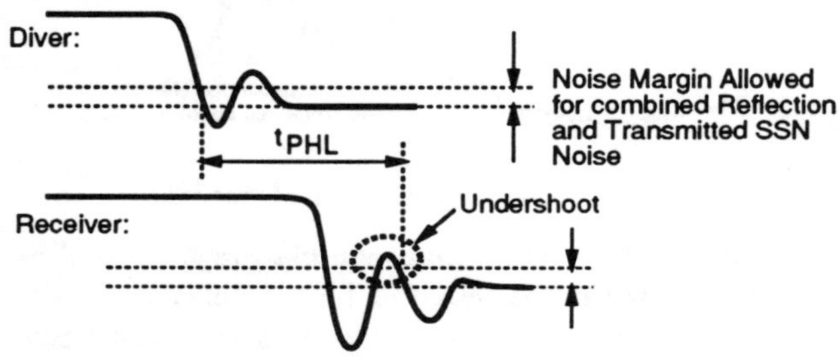

Figure 4: Definition of delay to include settling time for noise.

not interested in the 50% delay, however. As this is a data line, we wish to know when it is safe to latch the data, that is when the noise on the signal line has settled to the point that the combined reflection and transmitted simultaneous switching noise has settled to less than that fraction of the noise margin that the noise budget allows for it, as shown in figure 4. In the command file shown in figure 3, the maximum allowed combined reflection and transmitted simultaneous switching noise for a high-to-low transition is 23% of the voltage swing. Many other analyses can be done, including overshoot and metastate analysis. It is also possible to specify other noise budget and delay specifications.

Figure 5 shows some of the results obtained from running MetaSim with Cazm as the simulation engine. An important result demonstrated through these examples is that despite the non-regular shapes generally

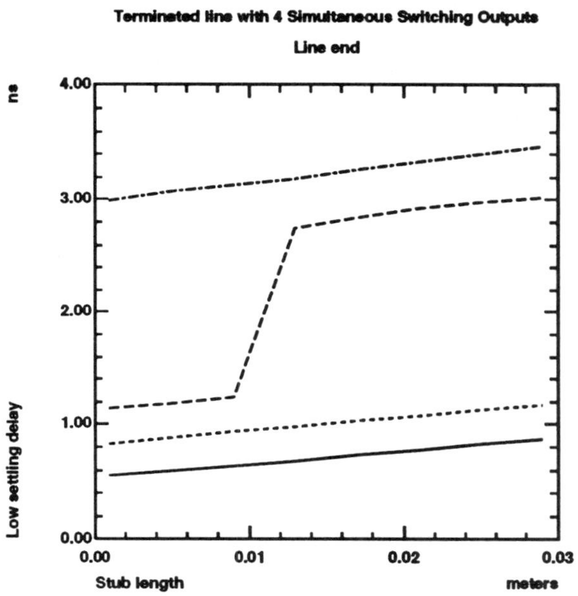

Figure 5: Total high-to-low delay between the ends of the line for different branch and stub lengths, for series terminated lines. (Solid line = 1cm branch lengths, short dashes = 2cm, long dashes = 3 cm, dash-dot = 4 cm.)

found in simulation waveforms, the trends in the settling time plots are generally regular. The steps in the plots are the only irregularities. The steps arise when a change in length suddenly makes the peak undershoot voltage edge past the noise threshold.

This result is important because it means that, as long as these steps are located, it is not necessary to carry out a large number of simulations in order to characterize the physical design space being considered. It also means that piecewise linear equations could be fitted. All of the studies we have carried out so far verify this result (sometimes piecewise polynomial equations would be required, however) though we can not claim to have analyzed every possible situation. Sensitivities would be expected to increase with pulse edge rate, for example.

The physical design tools would then use these results to adjust the placement and specify wiring rules. For example, if the total high-to-low delay between two chips at opposite ends of a net had to be less than 1.0 ns, then these results can be used by the placement tool to indicate a requirement for a parallel termination in order to control the excessive noise. If the two pins in question were just under 8 cm apart then a maximum stub length of 1 cm is permitted as long as the first to last pin length is 10 cm. The placement tool can then determine if it is likely that the router can meet these constraints. If there is a high likelihood that it can, then these constraints are passed to the router as a wiring rule. If it is likely that it can not then the placement tool suggests an improved placement. Automating this procedure should be straightforward.

4 Transim

Transim uses a convolution technique to allow mixed simulation of linear elements specified in the frequency domain and non-linear elements that need to be simulated in the time domain. In this section of the paper, we will first briefly discuss the operation of the convolution simulation core before presenting its operation as a complete simulator.

4.1 Related Work

Several approaches to this problem have been tried with varying degrees of success. The Laplace Transform technique is good at handling multiconductor transmission line systems and while it is more efficient than the convolution technique, its limited ability to handle frequency dependent effects and problems dealing with nonlinear terminations restrict its' usefulness. The Method of Characteristics [1] is relatively easy to use with conventional circuit simulators, but it is restricted to lines described by RLGC parameters. Other techniques for simulating distributed element systems have recently been reviewed by Djordjevic et al [3] and Schutt-Aine and Mittra [6].

In one of the convolution techniques the frequency domain network description is converted to a time-domain description using a Fourier transform. While this technique can handle arbitrary lossy coupled networks, aliasing in the frequency-domain to time-domain transformation can cause appreciable errors in the simulated transient response.

4.2 Convolution Simulation Core

Our approach is similar in some respects to that of Djordjevic et al, expect that where Djordjevic et al use admittance (Y) parameters to characterize the frequency dependent network, we use the scattering (S) parameters. As the S parameters vary only between -1 and +1, we overcome the dynamic range problems with the other convolution method's approach. S parameters of a transmission line system describe the relative amplitude and phase of the forward and backward traveling waves at each port and at each frequency on a transmission line of characteristic impedance Z_m.

Integration into a transient circuit simulator requires that the description of the transmission line system be in terms of total voltages rather than the traveling wave components implied by use of S parameters. This can be achieved by terminating each port in its reference impedance so ensuring that there are no reflections at the ports of the distributed network. Now the total transient response of any linear network to voltage sources at the ports of the network can be determined by convolving in time these voltages with the Dirac impulse voltage responses at each external port of the network. More explicitly, the total transient response at port i of a transmission line system to a total voltage $E_j(t)$ with output impedance Z_m at each port is [8]

$$\begin{aligned} V_i(t) &= \sum_{j=1}^{N} \int_{-\infty}^{t} g_{ij}(t-\tau)E_j(\tau)d\tau \\ &= \sum_{j=1}^{N} g_{ij}(t) \star E_j \end{aligned} \qquad (1)$$

Here the symbol $\star$ denotes convolution, N is the number of external ports in the transmission line system and

$$g_{ij}(t) = \frac{1}{\Delta t}\mathbf{F}^{-1}\left[G_{ij}(\omega)\right] = V_i^-(t)/E_j(0) \qquad (2)$$

is the impulse response at port i and at time t to a Dirac delta source at port j and time zero, In (2)

$$G_{ij}(\omega) = \begin{cases} (1+S_{ij}(\omega))/2 & i=j \\ S_{ij}(\omega)/2 & i \neq j \end{cases} \qquad (3)$$

is the frequency-domain Green's function from port j to port i at radian frequency ω and $\mathbf{F}^{-1}[\ ]$ is the inverse Fourier transform. Finally, $\Delta t = 0.5/f_{max}$ is the time increment of $g_{ij}(t)$ where f_{max} is the highest frequency component of $G_{ij}(\omega)$.

Based on these techniques, Transim has been implemented [8]. Essentially, the interconnect structures to be simulated, are partitioned into a collection of linear, frequency dependent parameter structures, and non-linear circuit structures (including time dependent voltage and driver sources). Simulation proceeds by minimizing the difference between the estimated voltages at the terminals of the non-linear elements and the voltages at the terminals of the linear elements, as determined via convolution.

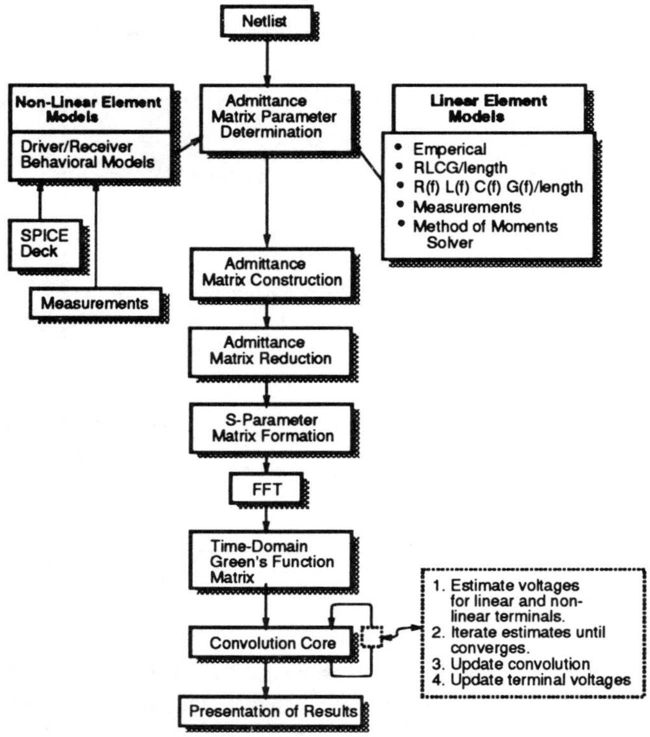

Figure 6: Complete Structure of Transim Simulator.

4.3 Operation as a Packaging Simulator

The complete structure of Transim is shown in Figure 6. Transim is a model-independent simulator. Instead modeled elements are represented internally as admittance matrices. Thus new models for both interconnect structures and drivers and receivers can be added quickly. From the Y-matrices corresponding to all of the modeled elements a complete Y-parameter matrix is constructed. An important step is the reduction of this matrix to a $M \times M$ admittance matrix, where M is the number of nonlinear nodes. After reduction, the Y-matrices are converted to S-matrices and then an FFT is carried out in order to obtain the Green's functions. These are then used in the convolution.

5 Conclusions

MetaSim can be used to characterize the transformation between the physical design space (length, spacing, etc.) and electrical design space (delay, noise) before any layout is done. The layout tools can then apply this transformation to the placement in order to suggest placement improvements and also to formulate wiring rules (routing constraints) for use by the router. MetaSim could also be used for other physical design scenarios where analytic delay and noise models are inadequate, such as high speed DRAM design, and packaging technology selection. It is very flexible and could be integrated with any simulator.

As it is used for a-priori analysis, MetaSim would conduct more simulations than a post-layout simulator would be required to. This suggests that increased run time might be an issue of concern. There are three counterpoints, however. First, in high speed designs, avoiding the need for multiple iterations of the design will save more user time and computer time than would be spent in using MetaSim up front, and will also lead to more efficient use of board space. Second, MetaSim is very efficient with the human designer's time. Third, initial results indicate that the P↔E transformation is generally regular and thus a good characterization can be obtained with a small number of randomly generated results, as long as the irregular points are found.

Transim is one of the three simulation engines currently in MetaSim. It is a convolution-based simulation that has solved a number of the problems usually found in such simulators. It has been developed as a complete packaging simulator.

Acknowledgements

The authors would like to thank Jack Gregenkemper of Tandem, George Katopis and Jay Diepenbrock, both of IBM, and Steve Lipa of NCSU for discussions about circuit signal integrity requirements. Thanks also to Zaki Rakib, of Cadence, and Wayne Dai, of the University of California (Santa Cruz), for discussions about CAE integration and physical design issues. We also gratefully acknowledge MCNC, particularly Wayne Detloff, for providing us with their 0.8μm CMOS driver and receiver simulation files.

References

[1] F.Y. Chang. The generalized method of characteristics for waveform relaxation analysis of lossy coupled transmission lines. In *IEEE MTT-S Digest*, pages pp.821–826, 1989.

[2] E.E. Davidson and G.A. Katopis. Package electrical design. In R.R. Tummala and E.J. Rymaszewski, editors, *Microlectronics Semiconductor Handbook*, chapter 3. Van Nostrand Reinhold, 1989.

[3] A.R. Dhordjevic, T.K. Sarkar, and R.F. Harrington. Time-domain response of multiconductor transmission lines. *Proc. IEEE*, 75:pp.743–764, June 1987.

[4] B.K. Gilbert and W.L. Walters. Design options for digital multichip modules operating at high system clock rates. In *Proc. 1st ISHM MCM Conference*, 1992.

[5] T. Mikazuki and N. Matsui. Statistical design technqiues for high-speed circuit boards. In *IEEE/CHMT'90 IEMT Symposium*, number IEEE no. CH2864-7/90/-0000-0185, pages 185–191, 1990.

[6] J.E. Schutt-Aine and R. Mittra. Nonlinear transient analysis of coupled transmission lines. *IEEE Trans. Circuits and Systems*, Vol. CAS-36:pp. 959–967, July 1989.

[7] E. Simoudis. A knowledge-based system for the evaluation and redesign of digital circuit networks. *IEEE Trans on CAD*, 8(3):pp.302–315, March 1989.

[8] D. Winkelstein, M.B. Steer, and R. Pomerleau. Simulation of the transient response of arbitrary transmission line networks with nonlinear terminations. *IEEE Trans. Circuits and Systems*, pages pp.418–422, April 1991.

IPDA: Interconnect Performance Design Assistant

Norman H. Chang, Keh-Jeng Chang, John Leo, Ken Lee, and Soo-Young Oh

Hewlett-Packard Company, Palo Alto, CA 94304

Abstract

IPDA is a generic interconnect performance design assistant that integrates a finite-difference numerical simulation method, linear interpolation algorithm, interactive performance synthesis methodology, and lossless/lossy transmission-line SPICE modeling capability into a spreadsheet-style graphical user interface. It assists users in selecting interconnect technologies for design-for-performance goals and also in optimizing interconnect performance designs for the full hierarchy of packaging including IC/MCM/PCB/lead-bonding/via interconnect designs. Although this paper describes electrical performance measures for given geometry and material parameters of conductors and dielectrics, IPDA can also be customized for other packaging measures such as reliability, thermal and cost effects through spreadsheet interface and calculations.

1 Introduction

Multi-level interconnect technologies (MLITs) have been widely used for on-chip and off-chip interconnections in VLSI/ULSI systems. Interconnect delays have constituted a large percentage of the total system delays. In high-frequency cases such as GaAs digital design, the interconnect delay is above 70% of the total delay [1]. However, we observe that MLIT modeling is still an ad-hoc process in the IC/packaging industry with very few CAD tools available to increase the productivity.

There are three major obstacles for efficient MLIT designs. First, the scaling of recent MLITs has made analytical and empirical formula obsolete [2,3]. 3-D structures such as dual signal stripline, via, lead attachment, and trace bends require 3-D numerical simulations for accurate impedance control, noise margin and delay simulation, and so on. However, the drawbacks of ad-hoc 2-D/3-D numerical simulations are time-consuming and tedious [4]. A solution is needed to provide MLIT designers accurate and immediate performance information.

Second, more and more high speed system designs demand that engineers consider the full hierarchy of packaging designs. In particular, chip to chip communication through the hierarchy of packaging structures is crucial to system performance and cost. Figure 1 shows a typical view of such an interconnection [5]. A signal travels from an output driver pad to a receiver pad through a solder bump, short transmission line on the carrier, via through the carrier, solder bump to the substrate, transmission line on the substrate, and the same connection through the receiver carrier in the reverse order. An immediate question posed to the designer is what combination of technologies from the full hierarchy of packaging categories (i.e. IC processes, MCM/PCB/via/lead attachment technologies) will satisfy the system performance specification such as delay, and noise margin requirements. A common interconnect modeling methodology is therefore needed to integrate and optimize the design of disparate interconnect domains such as IC, MCM, PCB, via, and lead attachment. For long interconnects, lossless and/or lossy transmission line effects have to be considered for dramatic changes of interconnect pitches which greatly affect the trace resistance. Although MLIT performance simulation with non-linear driver/receiver characteristics is time-consuming, it has been indispensable for designers of high speed systems [6]. CAD tools have to be integrated to handle this easily and efficiently.

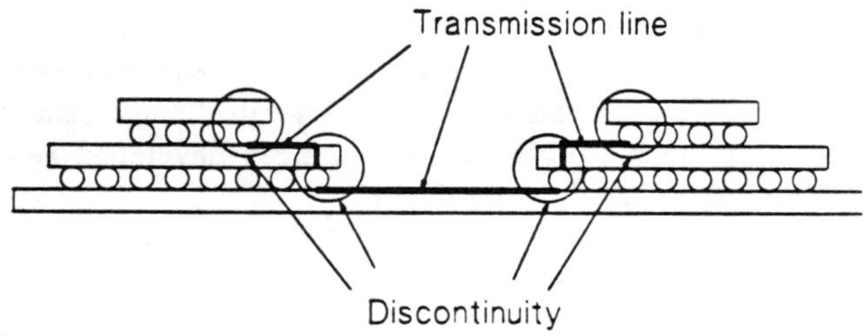

Figure 1. Chip to chip signal communication through the full hierarchy of packaging elements.

Third, MLIT designs require multi-disciplinary interactions such as material science, electrical engineering, and mechanical engineering [7], etc. System designers therefore need to search for an optimum design given numerous constraints from different engineering requirements [8]. A unified environment is therefore needed to help designers do system-level cost-performance tradeoff.

To address the problems above, we have devised and implemented a generic interconnect performance design assistant (IPDA) for MLITs designs. The algorithm, implementation and methodology of IPDA are described in Section 2. An application example is demonstrated and discussed in Section 3. Conclusions and future directions are given in Section 4.

2 Overview of IPDA

IPDA integrates four subsystems: (1) a batch-mode computation that combines 2-D/3-D finite difference numerical simulations and a fast interpolation algorithm; (2) an interactive design of MLITs through performance browser, synthesis, and on-line evaluation panel; (3) an interactive SPICE subcircuit/circuit generations and simulation; and (4) a spreadsheet-style graphical user interface environment for customization. The functionalities of the four sub-

0738-100X/92 $3.00 © 1992 IEEE

systems of IPDA are described below.

2.1 Batch Mode Computation

The batch mode computation algorithm is derived from the HIVE algorithm [9], which computes selective lateral and vertical capacitances of sub-micron IC interconnect using an internally developed tool based on finite-difference methods [10]. The non-calculated capacitances can then be obtained through fast interpolation.

We decided to take advantage of HIVE after we found that the assumptions of extracting submicron on-chip interconnect capacitances reported in [9] are also valid for MCM and PCB. To be specific, first, interline ("Cil") and vertical ("Cg") capacitances linearly vary with trace width when the interline spacing is fixed. Second, with enough points of numerical simulations on capacitances when interline spacing is varying and width fixed, other capacitances for any spacings bounded by the simulated spacings can be interpolated.

For example, Figure 2 illustrate that both the changes of "Cil" and "Cg" of a stripline for a thin-film MCM behave linearly when the width is changed but the spacing is fixed. Therefore, for a fixed spacing, we only need to calculate two capacitances using numerical simulation and are able to interpolate the capacitance values for different widths. However, due to slightly non-linearity of capacitance curves vs. spacing when the width is fixed as shown in Figure 3, we need to take more points using numerical simulations and do a linear, polynomial, or cubic spline interpolations for different spacings depending on the accuracy wanted. In either way, the interpolated capacitances are carefully checked to have the maximal induced error to be around 3% of the capacitance.

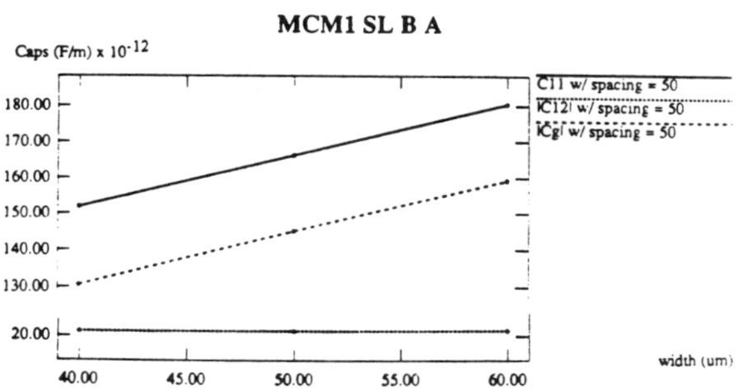

Figure 2. Illustrates the linearity of capacitances vs. width

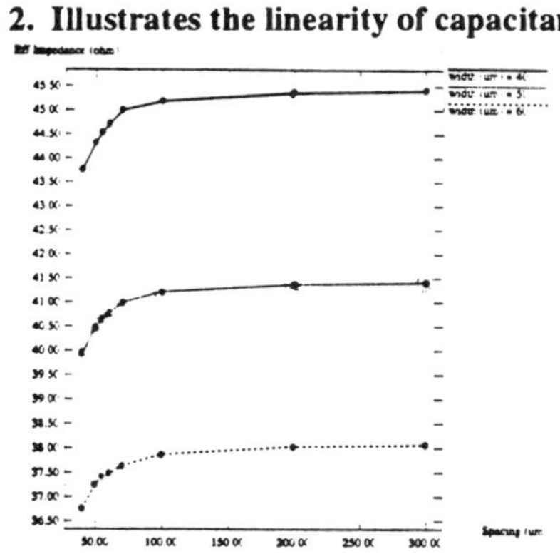

Figure 3. Solid points are values from numerical simulations, any other points within the bounded region formed by width/spacing ranges can be interpolated.

IPDA enhances the HIVE's batch-mode by automating the setup of batch-mode computation for new constructions and extending the application to packaging domains. Each new MLIT construction is specified with a common set of characteristic parameters, such as number of layers, dielectric constants, dielectric thicknesses, conductor type, soldermask or overcoat dielectric constant and thickness, outer layer and inner layer conductor thicknesses, and number of coupling traces. Since different interconnect domains have different set of characteristic parameters, the current IPDA has three construction worksheets. They are IC, MCM thin film, and PCB/MCM thick film worksheets, respectively. The Construction worksheet for MCM thin film shown in Figure 4 contains three construction examples: MCM_1, MCM_2, and MCM_3.

				User Settable Parameters					
width range	um	10	20	30	40	50	60		
spacing range	um	40	50	55	60	70	100	200	300
Models Available		SM	SM_M2	BMO	BMO_M2	SL	USLP	USLC	BMI_MASK
Models Selection	Yes(1)/No(0)	1	0	1	0	1	0	0	0

			Assumptions for the Performance Database
Unit		um	
Rise Time	ns	2.00	
Temp	centigrade	25.00	
Coupled length	inch	1.00	

				MCM Construction Database					
Type	Layer	Diel	Core	Prepreg	O/L_Metal	I/L_Metal	Metal_R	Mask_Diel	Mask_Thick
			um	um	um	um	ohm*meter		um
MCM1	4	4.00	50.00	50.00	18.00	18.00	3.76E-08	4.00	50.00
MCM2	4	4.00	55.00	55.00	18.00	18.00	3.76E-08	4.00	50.00
MCM3	4	4.00	60.00	60.00	18.00	18.00	3.76E-08	4.00	50.00

Figure 4. A construction worksheet for MCM thin film.

Representative data points for interconnect capacitances are extracted numerically for all the conceivable models of a new construction in the batch mode. For example, a 4-layer MCM construction can have models including stripline (SL), dual stripline (DSS), surface microstrip (SM), and buried microstrip outer (BMO), etc. A linear interpolation algorithm can then be applied to generate any other points bounded by the width/spacing ranges of the simulated points. The simulated capacitances are subsequently used to derive inductances L and transmission line properties, including Z_o, Z_e, Z_{eff} (odd, even, effective mode impedances), Pd_o, Pd_e, Pd_{eff} (odd, even, effective propagation delays) [11], and V_B, V_F (backward, forward crosstalks) [12,13] for all the interconnect models. Below we show the key equations for obtaining the performance parameters. First, C and C^U (capacitance obtained by setting all $E_r = 1$) matrices are obtained from simulation or interpolation.

$$L = \left(\frac{c}{\sqrt{E_r}}\right)^{-2} (C^U)^{-1} \quad (1)$$

where c is the speed of light, E_r is the relative permeability of the dielectric, assuming TEM or quasi-TEM mode.

$$Z_e = \sqrt{\frac{L_{11}+L_{12}}{C_{11}+C_{12}}} \quad (2)$$

$$Z_o = \sqrt{\frac{L_{11}-L_{12}}{C_{11}-C_{12}}} \quad (3)$$

$$Z_{eff} = \sqrt{Z_e Z_o} \quad (4)$$

$$Pd_e = \sqrt{(L_{11}+L_{12})\,(C_{11}+C_{12})} \quad (5)$$

$$Pd_o = \sqrt{(L_{11}-L_{12})\,(C_{11}-C_{12})} \quad (6)$$

$$Pd_{eff} = \sqrt{Pd_e Pd_o} \quad (7)$$

$$V_B = \left(\frac{\frac{C_{12}}{C_{11}}+\frac{L_{12}}{L_{11}}}{4}\right)\left(\frac{2L_{cp}Pd_{eff}}{T_r}\right)\Delta V_S \quad (8)$$

for coupled line lengths from $L_{cp} = 0$ to $L_{cp} = t_r / 2Pd_{eff}$, and

$$V_B = \left(\frac{\frac{C_{12}}{C_{11}}+\frac{L_{12}}{L_{11}}}{4}\right)\Delta V_S \quad (9)$$

for coupled line lengths of $L_{cp} = t_r / 2Pd_{eff}$ or greater.

$$V_F = \left(\frac{\frac{C_{12}}{C_{11}}-\frac{L_{12}}{L_{11}}}{2}\right)\left(\frac{L_{cp}Pd_{eff}}{T_r}\right)\Delta V_S \quad (10)$$

where ΔV_s is the driving signal transition amplitude.

2.2 Interactive Mode

The cost-performance parameters to be considered in optimizing a high speed, fast delivery, and low cost RISC system include not only "conventional" design parameters (e.g. loaded gate delays, interconnect delays, EM interference, power dissipation, and clock skew) but also cost and material concerns (e.g. cost, yield, device technologies selection, MLIT technologies selection, MLIT pitch optimization, lead attachments selection, chip footprints, and packaging sizes). The interactive mode of IPDA represents our first step to assist designers making trade-off. It provides designers an on-line tool to help optimize multi-level interconnect design and/or to obtain appropriate geometries which can satisfy the desired performance goals. The following interactive functions are provided by IPDA: goal-directed interconnect synthesis, design optimization through X-Y graphing, and on-line evaluation. IPDA provides *immediate* results to the users and the average execution times of the three functions in an HP/Apollo 720 workstation are within 1 second while 2-D/3-D add-hoc simulations take about 20 to 1000 seconds. Therefore, the time for IPDA to calculate tens to hundreds of data points of all the user-specified models for a construction is much faster than 2-D/3-D ad-hoc simulations which provide single data point for a specified geometry.

2.2.1 Goal-directed Interconnect Synthesis

Interconnect synthesis helps users to narrow down possible interconnect designs quickly. The domain of search is all processes/technologies that have been simulated in the batch-mode. The IPDA interconnect synthesis subsystem extracts a subset of designs that satisfy multiple performance goals such as impedance and noise margin specified by the user.

2.2.2 Viewing through X-Y Graph

The user can also look at the effect of changing trace width/spacing on any performance variable. This is done by showing any performance variable versus trace width and/or trace spacing. An arbitrary number of graphs can be generated so a user can simultaneously visualize trade-off for different trace width/spacing designs as well as their impact on several performance variables.

2.2.3 On-line Evaluation

An evaluation panel is provided for users to do stackup design. Stackup design involves assigning signal or reference planes to each layer of an interconnect construction. Once a user specifies layer assignment for a selected construction, IPDA can determine layer models for each layer. Furthermore, on-line performance evaluation can be done for each signal layer given the arbitrary trace width and spacing within the interpolation range.

To summarize for the use of interactive mode, users can interactively use interconnect synthesis, flexible X-Y graphing, and on-line evaluation simultaneously and iteratively to identify optimized interconnect designs in a timely fashion.

2.3 SPICE Generation and Simulation

With the scaling of MLITs, the multi-line busses carrying high frequency signals can no longer be accurately analyzed through the lossless transmission line performance parameters since the line resistance is comparable to the line impedance [11]. Using the SPICE ladder circuit model, the capacitive and inductive coupling in the multiple transmission lines can be included in the simulation and the resistive loss can be included with serial resistance in the ladder circuit. RLC SPICE subcircuits/circuits are therefore automatically generated so that distributed RLC interconnect simulation can accurately predict lossy-transmission line effects. An option is available on the Evaluation worksheet so that a user can choose any signal layer and the corresponding RLC netlists and circuit will be provided. Shown below is a netlist for MCM1 stripline. The corresponding subcircuit schematic is shown in Figure 5.

```
# For MCM1, Stripline model, 2 traces,  T/W = 50 µms,
T/S = 50 µms
.subckt sub 1 2 3 4 99 (length)
r1 1 101 value = 1.64604 * length
r2 3 103 value = 1.64604 * length
l1 101 2 value = 2.72263e-07 * length
l2 103 4 value = 2.72263e-07 * length
k12 l1 l2 value = 0.132351
c1g1 101 99 value = 7.244e-11 * length
c1g2 101 99 value = 7.23925e-11 * length
c2g1 103 99 value = 7.244e-11 * length
c2g2 103 99 value = 7.23925e-11 * length
c12 101 103 value = 2.14015e-11 * length
.ends
```

The above format is used for both stripline and microstrip lines. To correctly represent distributed RLC networks for accurate SPICE simulation for lossy transmission lines, the length of each segment should be less than 10% of the wave length of the highest frequency component of significance. In terms of equation, the number of segments "N" can be obtained from

$$N \geq 10L_{cp}\left(\frac{0.35}{T_r}\right)\left(\frac{\sqrt{E_r}}{c}\right) \quad (11)$$

where L_{cp} is the coupled line lengths and c is the speed of light. The SPICE circuit representation for the same geometry is shown below. Here nine segments are required for simulating this lossy transmission line from Eq. 11 with 0.2 ns risetime, and 3 in coupling lengths.

```
#MCM1, Stripline model, 2 traces, T/W= 50 µms,
T/S=  50 µms, T= 20 C
.param r open=100meg term=40.48
X001 101 201 102 202 %plane sub (0.0085)
X002 201 301 202 302 %plane sub (0.0085)
X003 301 401 302 402 %plane sub (0.0085)
X004 401 501 402 502 %plane sub (0.0085)
X005 501 601 502 602 %plane sub (0.0085)
X006 601 701 602 702 %plane sub (0.0085)
X007 701 801 702 802 %plane sub (0.0085)
X008 801 901 802 902 %plane sub (0.0085)
X009 901 1001 902 1002 %plane sub (0.0085)
r1in 101 %invalue = 40.48
r1out 1001 %plane value = 40.48
r2in 102 %planevalue = 40.48
r2out 1002 %plane value = 40.48
vconl %plane  0  0
vin %in 0 pulse(0 1 0 0.2N 0.2N 10N 20N)
.include MCM1.SL.2trace.static.subckt
.tran 0.1N 6N
.post tran v(101) v(1001) v(102) v(1002)
.end
```

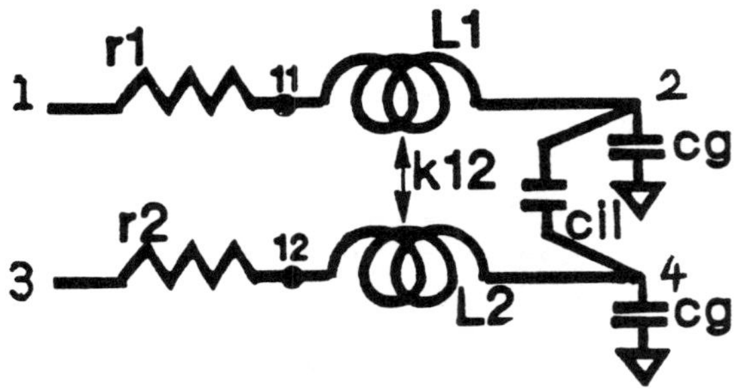

Figure 5. Schematic of RLC Netlist for SPICE Simulation

The users can set up different signal inputs, termination, and node running conditions through the interactive SPICE setup worksheet. HPSPICE is therefore tightly integrated into IPDA. Here IPDA treats the length of multi-line buses as a parameter so that the user can reuse the netlist for different frequencies/rise-times. We have also parameterized other geometry variables such as trace width/spacing, etc. as parameters for the SPICE subcircuit generations. The parameterized subcircuit generation for IC is reported elsewhere [14].

SPICE format is preferred by the users because all the non-linear driver/receiver models of our IC's are available from our HPSPICE and it is likely to be readable by newly developed circuit simulators [15]. Some circuit simulators that are faster than SPICE have been developed and are currently being evaluated since distributed RLC SPICE simulation is very time-consuming.

2.4 Application Example

A simple chip to chip signal communication optimization problem is chosen to demonstrate how IPDA helps designers analyze and optimize the interconnect design. In this example, a signal transmits from an output driver pad to a receiver pad through a multi-line transmission lines on an MCM, ignoring solder bumps required between chips and MCM. To further simplify the example, output driver signal is represented by a single low-to-high pulse with 0.2 ns risetime and 40 Ω source resistance. Receiver is assumed to have a high input impedance. The distance between driver and receiver is required to be 8 cm on a MCM. Interconnect choices include stripline or microstrip coupler configuration on several MCMs with different dielectric thicknesses and with/without coatings. The goal is to achieve the maximum of 1 ns delay and 100 mV/V crosstalk possible. Figure 6 shows the schematic for the example with open-ended termination.

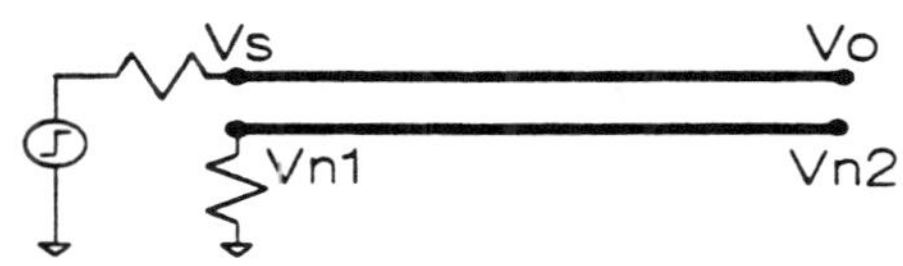

Figure 6. The schematic for the application example

To use IPDA to optimize the design in this case, first one will set up batch mode simulations for possible interconnect design configurations. Figure 4 shows several construction choices for the MCMs. There are polyimide and ceramic materials with dielectric constants of 4.0 and 9.0, respectively. Coatings of various thicknesses are available for microstrips. Also trace width ranges from 10 to 60 µms, and trace spacing ranges from 40 to 500 µms. The result of the batch mode simulation is the interconnect R, L, C, impedance, propagation delay, and crosstalk information in the database for surface microstrip, buried microstrip outer (with coating), and stripline configurations for various technologies.

Once the database is built, the user can use IPDA's performance browser and graphics capability to look at different curves of interest. Figure 7 shows the impedance vs. dielectric thickness. Figure 8 shows the crosstalk vs. spacing. With coating thickness ranging from 10 to 50 ums, microstrip impedance and crosstalk changes about 5% and 10%, respectively. Now we could enter impedance and noise margin goals in the synthesis worksheet as shown in Figure 9, and IPDA will return all the choices that satisfy the requirements. The user could then use the evaluation worksheet to do more detailed design by specifying layer assignments for the chosen MCM technology. Arbitrary choices of trace width/spacing, coupling length, and risetime can be specified, and performance information can subsequently be derived. Figure 10 shows this example on the evaluation worksheet.

By choosing a specific signal layer from the evaluation worksheet, the user can generate and simulate SPICE subcircuits/circuits through the interactive SPICE setup worksheet and HPSPICE window. Figure 11 shows the waveform of near end and far end crosstalk for an open-ended terminated coupler.

IPDA helps designers to analyze and optimize interconnect designs in an interactive way, but this procedure certainly could be done automatically as well. However, most designers would like to understand the results and options. There are other design constraints such as cost, thermal management, power consumption, manufacturability, etc that determine the final choice of a certain technology.

For this application example, an open ended termination coupler has less power consumption than well-terminated coupler. The cost of 9.0 dielectric material is cheaper than 4.0 dielectric material and may as well satisfy performance requirements. Microstrip with coating is more manufacturable than striplines. It would be difficult to know all the constraints up front before you explore possible alternatives. Since we use a driver with 0.2 ns risetime and 40 Ω source resistance, an open-ended termination microstrip (with coating) coupler with length equals to 8 cms is used on a 4.0 dielectric material with T/W = 40 µm, T/S = 60 µm, and dielectric thickness equals to 20 µm. For this choice, the final performance result are 0.54 ns delay, and 31 and 61 mV/V reduced noise margin due to crosstalk and reflection for near and far ends, respectively.

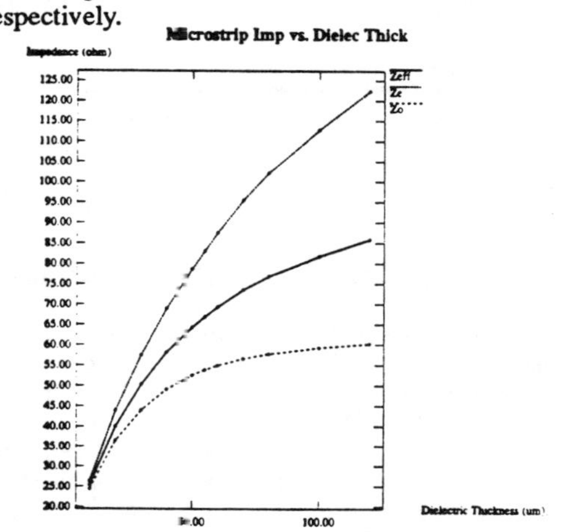

Figure 7. Microstrip impedances vs. dielectric thickness for Er = 4, T/W = 50 µms, and T/S = 50 µms.

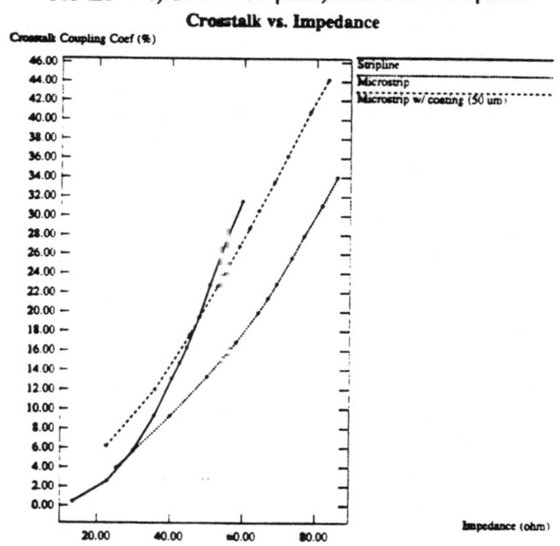

Figure 8. Crosstalk vs. impedance for Er = 4.

PARAMETER	UNIT	RELATIONSHIP	MINIMUM	MAXIMUM	INPUT
MODEL	input a model				
LAYERS					
TRACE WIDTH	um				
TRACE SPACING	um				
IMPEDANCE	ohm		38.00	41.00	
PROP DELAY	fs/um			2.00	
NOISE MARGIN	mV/V		0.00	100.00	
LATB-XTALK	mV/V	38% NOISE MARGIN	0.00	38.00	
VERT-XTALK	mV/V	20% NOISE MARGIN	0.00	20.00	
COUPLING COEF					

Figure 9. Arbitrary performance goals can be input on the synthesis worksheet.

Construction Type	Rise Time	Max Temp								
	ns	centigrade								
MCM10	0.20	20.00								
Layer Type	Layer Assign	Layer Model	Ntrace	T/W	T/S	Coupled L	Eff_Imp	Eff_PD	latB-Xtalk	latF-Xtalk
Upper Soldermask	Mask			um	um	in	ohm	fs/um	mV/V	mV/V
Layer 1	Signal	BMO	2.00	40	60	3.15	40.1	6.88	51.4	70.2
Layer 2	Reference	PP								
Layer 3	Reference	PP								
Layer 4	Reference	PP								
Lower Soldermask	Mask									

Figure 10. MCM1 on the evaluation worksheet.

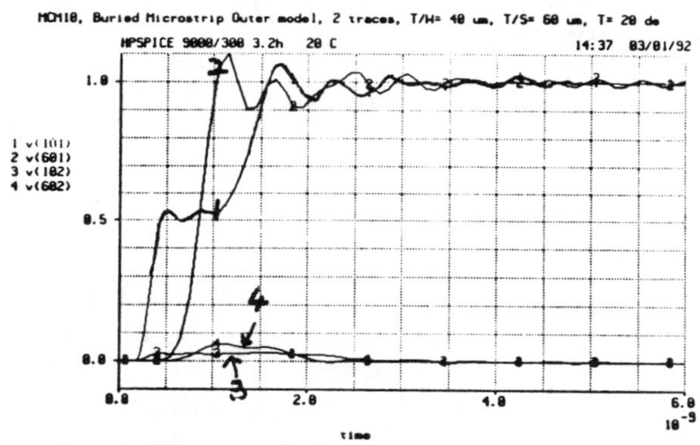

Figure 11. The near and far end signal and crosstalk waveforms for an open-ended coupler

3 IPDA Customization

We have observed major disparities between the interconnect modeling methods of IC, MCM, and PCB designers. Even for the same MLIT design such as PCB, microwave engineers may use modeling techniques very different from digital system designers. Although this paper primarily describes the electrical performance measures of IPDA, other MLIT measures such as electromigration design rules, thermal dissipation, manufacturability, and cost analysis are also very important. We therefore implement IPDA using a spreadsheet interface so that users can do customization by entering their analytic equations or empirical rules regarding reliability design, thermal management, cost, or other manufacturability measures. Although progress has been made [16], software that can provide a total solution for system-level cost-performance trade-off is still missing. We hope the continuous improvement of

IPDA can achieve this goal.

4 Conclusions and Future Directions

In this paper, we present an interconnect performance design assistant (IPDA) that provides a unified way for hierarchically analyzing and designing multi-level interconnect technologies (MLITs) including IC, MCM, PCB, via and lead attachment. IPDA provides three major functionalities to achieve this.

First, for completeness and speed, combinations of the whole range of parameters (width, spacing) for possible interconnect configurations are numerically simulated or interpolated in batch mode and are saved into a data file for each MLIT of interest. Furthermore, batch mode simulation for a new interconnect configuration can be easily setup and run. Results are then automatically added to the existing database.

Second, optimization of an interconnect design can be achieved interactively through the goal-directed interconnect synthesis given performance goals, flexible X-Y graphing, and fast on-line performance evaluation for arbitrary layer assignment of an interconnect structure with the user-specified risetime, temperature, width, spacing, and coupling length.

Third, complete SPICE subcircuit/circuit generation and simulation for an interconnect structure design including the layer assignment, risetime, width/spacing, and coupling length can be generated and simulated automatically for signal integrity analysis. Study of the chip to chip communication for optimizing driver, interconnect, and receiver design in terms of total delay, noise margin, etc. can be done much easier and quicker.

Besides the functionality above, IPDA's ease of use is also very important for its acceptance to engineers. We have implemented the user interface of IPDA using a spreadsheet running on UNIX [17], with the bulk work done in C.

Therefore, IPDA provides engineers with common access to company-standard interconnect models for the full hierarchy of MLITs. This makes the different levels of packaging design more compatible and eases the task of system interconnect simulation even within the same company.The current IPDA is our first step toward unifying the performance simulations of MLITs at our company. Following are some of the future work.

With the performance database, the task of system-level cost-performance trade-off dealing with the full hierarchy of packaging and various bipolar or CMOS off-chip driver/receiver models becomes feasible.

IPDA database can be used not only for pre-layout system-level cost-performance analysis but also for schematic capturing and post-layout parasitic circuit extraction. However, both the database format and the layout extraction methods for the IC/MCM/PCB tools have to be modified before the interconnect performance extraction routines and database can be integrated and used.

Acknowledgments

We would like to thank M. Peterson and S. Bird for many of their suggestions and stimulating discussions. Thanks also go to W. Whiteley, R. Liu, D. Kuokka, S. Jefferson, F. Frayman, J. Bradford, R. Coverstone, and B. Shreve for their comments and suggestions.

References

[1] IPC, *Design Guidelines for Electronic Packaging Utilizing High-speed Techniques, IPC-D-317 Report,* American National Standards Institute, 1991.

[2] T. Sakurai et al, ''Simple Formulas for Two- and Three-Dimensional Capacitances,'' *IEEE Trans. Electron Devices,* pp. 183-185, Feb 1983.

[3] Kit Cham, et al, *Computer-Aided Design and VLSI Device Development,* 2nd Edition, Kluwer Academic Publishers, Norwell, MA, pp. 348-351, 1988.

[4] N. P. van der Meijs, et al, ''An Efficient Finite Element Method for Submicron IC Capacitance Extraction,'' *26th ACM/IEEE Design Automation Conference,* pp. 678-681, June 1989.

[5] Al Barber, and Ken Lee, ''Modeling and Analysis of High Speed VLSI Interconnects for Digital Applications,'' *HP Labs Report HPL-92-03,* January, 1992.

[6] Vivek Raghavan and Ron Rohrer, ''A New Nonlinear Driver Model for Interconnect Analysis,'' *28th ACM/IEEE Design Automation Conference,* pp. 561-566, June 1991.

[7] Olgierd A. Palusinski et al, ''Modeling of Performance-Related Design Trade-offs in Multi-Chip Assemblies,'' *Proceedings of IEEE/CHMT IEMT Symposium,* pp. 333-336, March 1991.

[8] Larry L. Moresco, ''Electronic System Packaging: The Search for Manufacturing the Optimum in Sea of Constrains,'' *IEEE Trans. Comp., Hybrids, Manuf. Technol.,* vol. 11, pp. 494–508, September 1990.

[9] Keh-Jeng Chang, Soo-Young Oh, and Ken Lee, ''HIVE: An Efficient Interconnect Capacitance Extractor to Support Submicron Multilevel Interconnect Designs,'' *Proceedings of ICCAD,* Nov. 1991.

[10] TMA, Inc., TMA Raphael User's Manual, Oct. 1991.

[11] H. B. Bakoglu, *Circuits, Interconnections, and Packaging for VLSI,* Addison-Wesley, Reading, MA, 1990.

[12] A. Feller, et al, ''Crosstalk and Reflections in High-speed Digital Systems,'' *Proceedings of Fall Joint Computer Conference,* 1965.

[13] J. Buchanan, *BiCMOS/CMOS Systems Design,* McGraw-Hill, Inc., 1991.

[14] Keh-Jeng Chang, Soo-Young Oh, Norman Chang, and Ken Lee, ''Parameterized SPICE Subcircuits for Submicron Multilevel Interconnect Modeling,'' *Proceedings of IEEE Symposium on VLSI Technology,* June, 1992.

[15] Curtis L. Ratzlaff, et al, ''RICE: Rapid Interconnect Circuit Evaluator,'' *28th ACM/IEEE Design Automation Conference,* pp. 555-560, June 1991.

[16] Peter A. Sandborn, ''A Software Tool for Technology Trade-off Evaluation in Multichip Packaging,'' *IEEE/CHMT IEMT Symposium,* pp. 337-340, 1991.

[17] Informix Software, Inc., Wingz Reference Manual, 1991.

On The Circuit Implementation Problem *

WingNing Li† Andrew Lim¶‡ Prathima Agrawal§ Sartaj Sahni‡

† Dept. of Computer Science, University of Arkansas
¶ Information Technology Institute, Republic of Singapore
‡ Dept. of Computer and Info. Sciences, University of Florida
§ AT&T Bell Lab., Murray Hill, New Jersey

Abstract

We consider the problem of selecting an implementation of each circuit module from a cell library so as to satisfy overall delay and area (or delay and power) requirements. Two versions of the circuit implementation problem, *basic circuit implementation problem* and the *general circuit implementation problem* are shown to be *NP*-Hard. A pseudo-polynomial time algorithm for the basic circuit implementation problem on series-parallel circuits is developed, and heuristics for the *basic circuit implementation problem* on general circuits are formulated and experimented with.

1 Introduction

In this paper we examine the problem of selecting, from a precharacterized library, an implementation of each module in a combinatorial logic circuit such that the resulting circuit implementation satisfies the timing and area (power) requirements. While our development applies equally to the cases of satisfying timing and area requirements and satisfying timing and power requirements, in what follows, we explicitly refer only to the case of satisfying timing and area requirements.

We define the *circuit implementation problem* as that of selecting an implementation for each of the circuit modules so that the area and delay of the overall circuit do not exceed some specified limits α and δ respectively. In the *basic circuit implementation problem* (BCI), the delay between every pair of input-output terminals of a module is the same. In the *general circuit implementation problem* (GCI), the delay between different pairs of input-output terminals of a module maybe different. The former model is more applicable for simple gates and the latter for compound gates.

In this paper, we show, in Section 3, that the basic circuit implementation problem is NP-hard even when the circuits are chains of components. In the same section we show that the general circuit implementation problem is NP-hard even if all implementations of all modules have the same area. By using suitable decomposition techniques and dynamic programming, we are able to develop (Section 4) a pseudo-polynomial time algorithm that obtains optimal solutions for basic series-parallel circuits. This result contrasts with that of McNall and Casvant[1] who have developed a similar dynamic programming algorithm for the synthesis of pipelined architectures.

In Section 5, we propose several heuristics for basic circuits. These heuristics of Section 5 are evaluated experimentally using ISCAS benchmark circuits. The results of this evaluation are presented in Section 6.

Chan[2] has studied a circuit implementation problem that is closely related to ours. In his version of the problem, there is a minimum and maximum delay requirement for the realized circuit (rather than just a maximum delay). He has obtained a pseudo-polynomial time algorithm to obtain a minimum area realization of a tree. The obtained realization satisfies the delay constraints. Our pseudo-polynomial time algorithm is easily generalized to handle the additional delay constraint and applies not only to trees but also to series-parallel graphs. Chan[2] also proposes a backtracking algorithm for circuit modeled by dags. While this algorithm obtains circuit realizations that satisfy the delay constraint (when such a realization exist), it does not guarantee optimal area realizations. In addition to generating suboptimal circuit implementations, this backtracking algorithm is of exponential complexity and so is applicable only to relative small circuits. Chan[2] also shown that the circuit realization problem with minimum and maximum delay constraints is NP-hard for basic circuits with tree topologies. Our NP-hard result for chains is a stronger result as a chain is a special case of a tree.

The circuit implementation problem studied here is also related to the technology mapping problem studied earlier by Darringer et al.[3], Gregory et al.[4], Keutzer [5], Detjens et al.[6], Touati[7], and others. Because of space limitations, we shall not elaborate on this relationship.

*This research was supported in part by the National Science Foundation under grants DCR 84-20935 and MIP86-17374.

2 Terminology and Notation

A *circuit* is an interconnected set of modules. A p input q output module has p input pins and q output pins. Such a module is called a (p,q)-module. The interconnects of the circuit connect an output pin of one module to an input pin of another module. Some input pins are labeled as ***primary inputs*** and some output pins are labeled as ***primary outputs***. We shall limit ourselves to circuits in which the modules can be ordered such that every interconnect connects an output pin of a module to an input pin of another module to its right. So, all interconnects are directed left to right. The graph for such a circuit is directed and acyclic, i.e., it is a dag.

Suppose there are n_i possible implementations of module i and that module i is a (p_i, q_i)-module. Let $A(i,j)$ denote the area of the jth implementation of this module, $1 \le j \le n_i$ and let $D(i,j,u,v)$ be the delay from input u to output v of the jth implementation of module i, $1 \le j \le n_i$, $1 \le u \le p_i$, $1 \le v \le q_i$. For basic circuits, $D(i,j,u,v) = d_{ij}, 1 \le u \le p_i, 1 \le v \le q_i$. The *area* of the implementation $S = (s_1, s_2, \ldots, s_n)$ is $\sum_{i=1}^{n} A(i, s_i)$

3 Complexity Results

We first define the following known *NP*-complete problem [8]:

PARTITION
Input : A finite set B and a size $s(b) \in Z^+$, for each $b \in B$.
Output: "Yes" iff there is a subset $B' \subset B$ such that $\sum_{b \in B'} s(b) = \sum_{b \in B-B'} s(b)$.

Theorem 1 *The BCI problem is NP-hard even when all modules are $(1,1)$-modules interconnected to form a chain and each module has only two possible implementations.*

Proof We shall show that any instance of the *partition* problem can be transformed, in polynomial time, into an instance of the BCI problem for which there is an (α, δ) implementation iff the output to the partition instance is "yes". The constructed BCI instance satisfies the remaining requirements of the theorem. The BCI instance is obtained from the *partition* instance as follows. For each $b \in B$ we have a distinct $(1,1)$-module, i, with two possible implementations.

$$A(i,1) = s(b), D(i,1,1,1) = 0$$

$$A(i,2) = 0, D(i,2,1,1) = s(b)$$

The modules are connected in any order to form a chain and the values of α and δ are $\alpha = \delta = \sum_{b \in B} s(b)/2$. One may easily verify that the constructed BCI instance has an (α, δ) implementation iff the corresponding *partition* instance has output "yes". □

Theorem 2 *The GCI problem is NP-hard even if all implementations of all modules have the same area.*

Proof See [9]. □

4 Series-Parallel Circuits

A ***simple parallel circuit*** is a basic circuit that is comprised of several chains that have the same first and last module. A ***series-parallel circuit*** is a basic circuit recursively defined as:

- A chain of basic modules is a series-parallel circuit
- A simple parallel circuit is a series-parallel circuit
- A circuit obtained from a series-parallel circuit C by replacing any interconnect of C by another series-parallel circuit is also a series-parallel circuit.

We shall assume that the area and delay values associated with each implementation of each module are integers. An algorithm for the basic circuit realization problem is a *pseudo-polynomial* time algorithm if its complexity is some polynomial in n (the number of modules), m (maximum possible implementations of any module), *area* (the maximum allowable area), and *delay* (the maximum allowable delay).

An implementation that has area a and delay d is an *optimal implementation* iff there is no implementation with area $\le a$ and delay $< d$ or with area $< a$ and delay $\le d$.

4.1 Chains

The optimal implementations for a chain may be obtained by enumerating the different possibilities using a dynamic programming type approach (Chapter 5, [10]). Suppose the chain has c modules $M_1, M_2, \ldots, M_c$ where M_i is to the left of M_{i+1}, $1 \le i < c$. Let L_r be an ordered list of pairs (a_i, d_i) where a_i and d_i are, respectively, the area and delay associated with an optimal implementation of the chain $M_1, M_2, \ldots, M_r$. L_r has the form:

$$L_r = \{(a_1, d_1), \ldots, (a_k, d_k)\} \quad \text{where}$$

$$a_i < a_{i+1} \text{ and } d_i > d_{i+1}, 1 \le i < k \qquad (1)$$

L_r cannot have two pairs (a_i, d_i) and (a_j, d_j) such that

$$(a_i \le a_j \text{ and } d_i < d_j) \text{ or } (a_i < a_j \text{ and } d_i \le d_j) \qquad (2)$$

The set L_1 is simply those implementations of M_1 that are optimal. Hence if the set of implementations of M_1 includes pairs (a_i, d_i) and (a_j, d_j) that satisfy (2) then the pair (a_j, d_j) is eliminated. Note that a suboptimal implementation of M_1 cannot be used in any optimal implementation of any of the chains M_1, M_2, ..., M_s, $1 \le s \le c$. Similarly no suboptimal implementation of the chain $M_1, M_2, \ldots, M_s$ can be used to obtain an optimal implementation of $M_1, M_2, \ldots, M_{s+1}$, $1 \le s < c$.

From L_r we may obtain L_{r+1} by considering each of the possible optimal implementations of M_{r+1}. For each optimal implementation (a_e, d_e) of M_{r+1} we create an ordered list L_{r+1} as below:

```
L^e_{r+1} = ∅
for i = 1 to |L_r| do
        append (a_i + a_e, d_i + d_e) to L^e_{r+1};
        {(a_i, d_i) is the ith pair in L_r}
end;
```

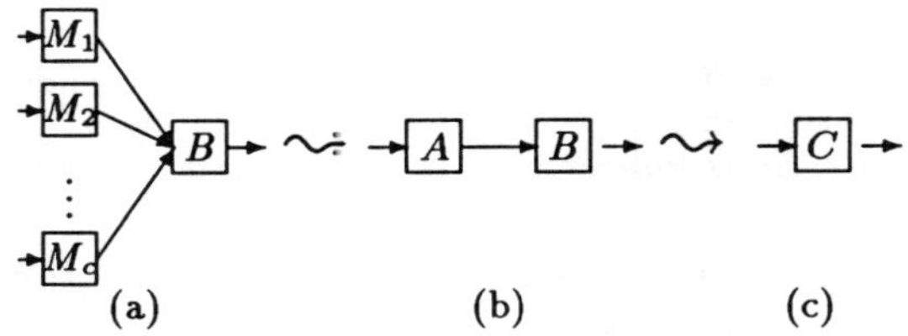

Figure 1: A basic tree to a module transformation

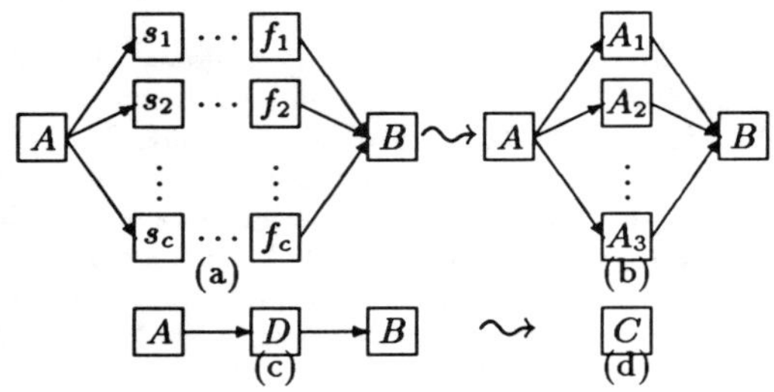

Figure 2: A simple parallel graph to a module transformation

Next, the lists L_{r+1}^e, $1 \le e \le$ (number of optimal implementations of M_{r+1}) are merged to obtain L_{r+1}. The merging is done so that L_{r+1} satisfies (1). This requires the elimination of pairs (a_j, d_j) that satisfy (2).

Using this strategy we begin with L_1, then obtain L_2, then $L_3, \ldots$, and finally L_c. For each pair (a_i, d_i) in L_c the actual implementation can be obtained using a traceback as in Section 5.5 of [10].

The above algorithm for a chain also allows us to obtain a chain to module transformation in which any chain of modules can be replaced by a single module, the delay and area of whose possible implementations are given by the pairs in L_c.

4.2 Simple Parallel Circuits

A *basic tree circuit* has two columns of basic modules as in Figure 1(a). The second column has exactly one module. The area and delay pairs of the optimal implementations of a basic tree circuit can be obtained by first obtaining those for the column one modules. This allows us to replace the circuit of Figure 1(a) with the two module chain of Figure 1(b) where the possible implementations of module A have area and delay corresponding to the optimal implementations of the column 1 modules. The optimal implementation of the two module chain of Figure 1(b) may be obtained using the algorithm for a chain. In this way, the basic tree of Figure 1(a) is transformed into a single module as in Figure 1(c).

By beginning at the leaves and using this transformation repeatedly, we can obtain the optimal implementations of any tree.

The c modules of column 1 of the basic tree of Figure 1(a) may be transformed into the equivalent single module A of Figure 1(b) by constructing the ordered lists L_1, $L_2, \ldots$, L_c as was done for a chain. In the code to construct L_{r+1}^e we replace $(a_i + a_e, d_i + d_e)$ by $(a_i + a_e, max\{d_i, d_e\})$.

The optimal implementations of a simple parallel graph can be obtained by using the transformations shown in Figure 2. The steps are:

Step 1 :
: The modules from s_i to f_i of Figure 2(a) form a chain. Using the chain transformation of Section 4.1 these are replaced by an equivalent single module A_i

Step 2 :
: The modules $A_1, A_2, \ldots, A_c$ of Figure 2(b) are replaced by a single module D using the transformation used to go from Figure 1(a) to Figure 1(b).

Step3 :
: Use the chain transformation of Section 4.1 to go from Figure 2(c) to Figure 2(d).

4.3 Series-Parallel Circuits

The optimal implementation of any series-parallel circuit can be obtained by repeatedly using the chain and simple parallel circuit transformations on sub-circuits of the given series-parallel circuit. The time complexity of the algorithm remains pseudo-polynomial.

In case the series-parallel decomposition of the circuit isn't already known, it can be obtained in linear time using the algorithm of Valdes[11].

5 Heuristics For Basic Circuits modeled by DAGs

In this section, we develop six heuristics to obtain minimal area circuit implementations given a delay constraint δ. All are for the *basic circuit implementation problem.*

Common to all of the heuristics is a preprocessing step in which the following transformations are made:

T1: [Elimination of module chains]
: Each chain of modules is replaced by a single module whose possible implementations are the optimal implementations of the chain. This is done using the algorithm in section 4.1.

T2: [Elimination of redundant edges]
: An edge $< u, v >$ of the circuit graph is *redundant* if the graph contains a u to v path that does not use this edge. Since the delay on such a path is no less than the delay on the edge $< u, v >$, the edge $< u, v >$ may be removed from the circuit. All redundant edges are removed in this transformation.

T3: [Elimination of suboptimal implementations]
: Let (a_i, d_i) and (a_j, d_j) be the areas and delays of two different implementations of the same module. If $a_i \le a_j$ and $d_i \le d_j$ then the implementation (a_j, d_j) is suboptimal and is eliminated.

Step 1: Select an initial implementation for each module.

Step 2: Consider the modules, one at a time, in some order and revise the selected implementation of each.

Step 3: This is a postprocessing step used by only one of our heuristics to further revise the selected implementation of each module.

Figure 3: Steps in our heuristics

The general structure of our heuristics is described in Figure 3. The details of each are provided in the next six subsections.

5.1 Random Heuristic

In this, the initial module implementations of step 1 are selected at random and the implementations revised in step 2 by considering the modules in some random order. When a module is considered for revision in step 2, an implementation with the least area that causes this module to be on no path with length $> \delta$ is chosen, If no such implementation exists, then the least delay (i.e. fastest) implementation for this module is selected. For this heuristic, step 3 is null. This heuristic is easily implemented to run in $O(|V|(|V| + |E|))$ time where $|V|$ is the number of vertices in the circuit graph and $|E|$ is the number of edges in this graph. When a module is being considered in step 2, we need to know the length l of the longest delay path it is on. Let the delay of this module's current implementation be d. Then for this module to be on no path with delay $> \delta$, its implementation should have delay $\leq \delta - l + d$.

5.2 Topological Heuristic

In step 1, the fastest implementation for each heuristic is selected. In step 2, the modules are considered in two different orders: topological and reverse topological. The better of the two resultant circuit implementations is used. When a module is considered in step 2, the implementation with the least area that results in no path of length $> \delta$ is selected. Step 3 is null. The asymptotic complexity of this heuristic is the same as that of the random heuristic.

5.3 Static Delay Heuristic

The initial step 1 implementations are the slowest ones for each module. To obtain the order in which modules are considered in step 2, we compute the number of input to output paths that pass through each module. This can be done in $O(|V| + |E|)$ time by considering the modules in topological and reverse topological order. In step 2, the modules are considered in nonincreasing order of the number of input to output paths through them. The implementation revision scheme is the same as that for the random heuristic.

5.4 Greedy Heuristic

For the initial implementation of step 1, this heuristic uses the fastest implementation of each module. In step 2, the order in which the modules are considered is obtained dynamically. The next module to consider is one which results in the maximum reduction in the circuit area. For this, each of the modules not yet considered is evaluated to determine the area reduction possible by changing its selected implementation. If the delay of the longest path through the module is d, then we can choose an implementation with delay upto $\delta - d$ larger than the delay of its currently selected implementation. If the possible implementations are ordered by delay (and hence by area), a binary search can be performed to obtained the implementation that results in the maximum area reduction.

As for the other heuristics described so far, step 3 of this heuristic is also null. The complexity of the greedy heuristic is $O(|V|(|V| \log m + |E|))$ where m is the maximum number of possible implementations for any module.

5.5 Static Area Heuristic

The fastest implementation of each module is used, in step 1, as the initial implementation. In step 2, the modules are considered in nondecreasing order of the number of input to output paths through them. When module i is being considered in step 2, each of its possible implementations is given a score and the implementation with highest score is selected. The score of implementation j of module i is zero if using implementation j results in a circuit delay $> \delta$. Otherwise, its score is obtained using the formula:

$$score(i,j) = \Delta a_j(\alpha + \beta/(f(i,j) + \gamma))$$

where Δa_j is the area reduction that results from using implementation j over the currently selected implementation, α, β, and γ are constants that were set to 4, 1, and 1 respectively. These values for α, β, and γ were determined experimentally using half of our test circuits. While the performance of the heuristic is sensitive to the choice of these constants, the chosen values tended to produce uniformly good solutions over the range of circuits tested. The function f measures the undesirability of implementation j of module i, f is given by:

$$\begin{aligned} f(i,j) &= \sum_{<k,i>\in E} max\{0, right(i) + d_{ij} - right(k)\} \times LP(k) \\ &+ \sum_{<i,k>\in E} max\{0, left(i) + d_{ij} - left(k)\} \times RP(k) \end{aligned}$$

where

$right(i)$ is the length of the longest delay path from module i to a primary output, this length does not include the delay of module i;

$left(i)$ is the length of the longest delay path from a primary input to module i;

d_{ij} is the delay of the jth implementation of module i;

$LP(k)$ is the number of paths from the primary inputs to module k;

$RP(k)$ is the number of paths from the module k to primary outputs;

The asymptotic complexity of this heuristic is $O(|V|(|V|m + |E|))$.

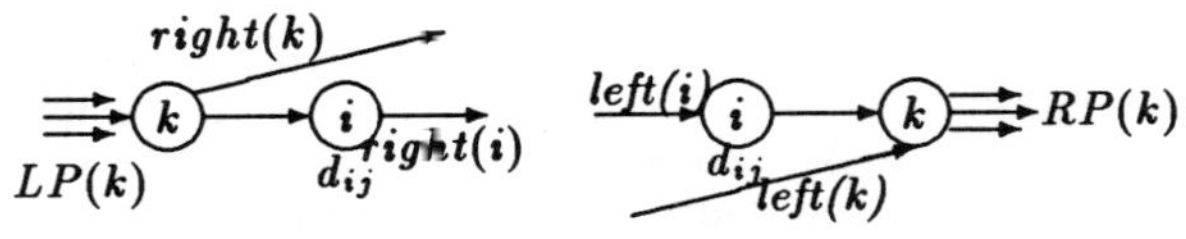

Figure 4: Terms in the definition of f

E is the set of edges in the circuit graph. Note that $\Delta a_j > 0$ as the initial implementation is the fastest one.

Let us consider the rationale for the function f. Figure 4 illustrates the terms used in the definition of f. If $right(k) \geq right(i) + d_{ij}$, then the longest delay path through module k is unaffected. On the other hand, if $right(k) < right(i) + d_{ij}$ then the $LP(k)$ paths from primary input to module k now prefix longer paths than before. The increase in path length is $right(i) + d_{ij} - right(k)$. So, the product $max\{0, right(i) + d_{ij} - right(k)\} \times LP(k)$ measures the undesirability of the implementation j. The terms in the second summation also measure the undesirability of the implementation j. Hence, the larger the f value, the less desirable the implementation j.

Let m be the maximum number of possible implementations for any of the modules. The complexity of the static area heuristic is $O(|V|(m + |E|))$.

5.6 Dynamic Area Heuristic

The initial implementation for step 1 is the same as for the static area heuristic. I.e., the fastest implementation of each module is selected. In step 2, the module order is determined in a dynamic fashion. To determine the next module to consider in step 2, the modules not yet considered are *scored*. For the jth possible implementation of module i, the score is 0 if using implementation j results in a circuit delay that exceeds δ. Otherwise, the score is :

$$score(i,j) = \Delta a_j(\alpha z + 1/(f(i,j) + \beta z + \gamma))$$

where Δa_j and $f(i,j)$ are as defined for the static area heuristic; α, β, and γ are constants which were experimentally tuned to be 2, 1, and 0.1 respectively; and $z = (s - \delta)/(s - q)$ where s and q are, respectively, the delays of the slowest and fastest circuits that can be realized (i.e., q is the delay of the circuit that results from selecting the fastest implementation of each module and s the delay when the slowest implementation of each module is selected).

The parameter z measures the "tightness" of the delay constraint δ. A small z (close to 0) means that delay increases need to be done sparingly while a z close to 1 means greater room for using larger delay implementations.

In step 3 of the dynamic area heuristic, we run step 2 of the static area heuristic using the circuit realization just obtained as the initial realization. For this, the parameter β in the score function is set to 0 and α is set to 1.

6 Experimental Results

Our algorithms were programmed in *C* and the programs run on a SUN SPARCstation *1*. For test data, we used the ten ISCAS circuits described in Brglez [12] and another circuit (C17) described in Chan [2] together with our precharacterized cell library.

The run times of the six heuristics are tabulated in the last six columns of Table 1. The times are in seconds and represent the average time needed to find a solution for each of the δ values within the range in column 5. Table 2 provides the average area obtained for each δ value and the percentage of improvement when compared to the random heuristic. The dynamic area heuristic was able to reduce area by as much as 60% on some of the test circuits and for some δ values. The average area reduction was as much as 35% on some circuits. The dynamic area heuristic obtained optimal solutions for all but one of the δ values for that circuit.

From the experiments, we made several observations:

- All the heuristics are fast and can easily handle large circuits.
- The run time of the heuristics is highly dependent on the depth of the circuit. This is due to the amount of time required to update the circuit once a module changes its delay and area.
- Simple methods like the first four heuristics are not as effective as more sophisticated heuristics like the *static area* and *dynamic area* heuristics.
- Starting with the minimum area circuit with a large delay and then trying to reduce the delay of paths that violate timing requirement requires more sophisticated techniques as the delay of a circuit is the maximum delay of all paths in the circuit whereas the area of the circuit is the summation of all the modules in the circuit.

7 Conclusion

We have shown that the *basic circuit implementation* problem and the *general circuit implementation* problem are *NP*-hard. We developed a pseudo-polynomial time algorithm for the *basic circuit implementation* problem on series-parallel circuits. Our algorithm can easily be generalized to circuits in which a minimum and maximum source to module delay is associated with each module. This includes the case when a minimum and maximum delay for the entire circuit is provided. Thus the results of Chan [2] for trees are subsumed by our algorithm. Furthermore, we develop 2 fast and effective heuristics (*static area* and *dynamic area*) for the *basic circuit implementation* problem and compare them exhaustively with 4 other reasonable approaches. Experimental results confirmed the superiority of our *static area* and *dynamic area* heuristics. While our discussion considered the circuit implementation problem in terms of satisfying timing and

Circuit Name	No. of Nodes	No. of Edges	Depth of Circuit	Delays Tested	Run times for heuristics					
					Random	Topo.	Static D	Greedy	Static A	Dyn. A
C17	13	14	5	19-29	.003	.001	.002	.003	.001	.003
C432	250	426	22	22-44	.08	.07	.07	.37	.07	.36
C499	555	928	24	24-48	.27	.20	.22	.53	.18	1.33
C880	443	729	25	25-50	.13	.15	.11	.33	.13	.75
C1355	587	1064	25	25-50	.24	.25	.22	.59	.19	1.90
C1908	913	1498	41	41-82	.43	.59	.42	1.38	.38	3.33
C2670	1426	2076	33	33-66	.52	.69	.42	2.84	.47	5.40
C3540	1719	2939	48	48-96	.93	1.53	.81	4.45	.81	9.39
C5315	2485	4386	50	50-100	1.09	1.55	.90	8.54	1.04	15.8
C6288	2448	4800	125	125-250	4.03	7.11	3.75	15.5	3.20	31.0
C7552	3719	6144	44	44-88	1.73	2.59	1.25	20.4	1.67	36.5

Table 1: Characteristics of the test data and run times

Cir. ISCAS	Random	Topo Order		Static D		Greedy		Static A		Dyn. A	
	ave	ave	%	ave	%	ave	%	ave	%	ave	%
C17	25.57	24.65	3.6	25.68	-0.4	25.24	1.3	24.73	3.3	24.39	4.6
C432	810	596	26.4	659	18.6	591	27.0	568	29.9	523	35.4
C499	1905	1488	21.9	1655	13.1	1481	22.2	1416	25.7	1351	29.1
C880	1008	901	10.6	905	10.2	917	9.03	797	20.9	753	29.1
C1355	2078	1732	16.6	1803	13.2	1734	16.6	1622	20.0	1589	23.5
C1908	2298	1871	18.6	2038	11.3	1871	18.6	1722	25.1	1579	31.3
C2670	3026	2748	9.19	2511	17.1	2733	9.68	2188	27.7	2144	29.2
C3540	4255	3264	23.3	3608	15.2	3261	23.3	2988	29.8	2768	35.0
C5315	3893	3628	6.8	3543	8.99	3613	7.19	3324	14.62	3147	19.2
C6288	7672	6941	9.53	8396	-9.4	6933	9.60	6617	13.8	5489	28.5
C7552	6548	6405	2.18	5947	9.18	6377	2.6	5377	17.9	5148	21.38

Table 2: Results for *Basic Circuit Implementation* problem

delay requirements, the discussion applies equally to the case of satisfying timing and power requirements.

References

[1] K. McNall and A. Casavant, "Automatic Operator Configuration in the Synthesis of Pipelined Architectures," in *27th Design Automation Conference*, pp. 174–179, 1990.

[2] P. Chan, "Algorithms for library-specific sizing of combinational logic," in *ACM/IEEE Design Automation Conference*, pp. 353–356, 1990.

[3] J. Darringer and W. Joyner and L. Bermen and L. Trevillyan, "Logic Synthesis Through Local Transformations," *IBM Journal of Research and Development*, vol. 25, pp. 272–280, July 1981.

[4] D. Gregory and K. Bartlett and A. Gues and G. Hachtel, "SOCRATES: A System for automatically synthesizing and optimizingcombinational logic," in *23rd Design Automation Conference*, pp. 79–85, 1986.

[5] K. Keutzer, "DAGON : technology binding and local optimization by DAG matching," in *24th Design Automation Conference*, pp. 341–347, 1987.

[6] E. Detjens and G. Gannot and R. Rudell and A. Sangiovanni-Vincentelli and A. Wang, "Technology Mapping in MIS," in *International Conference in Computer-Aided Design*, pp. 116–119, 1987.

[7] H. Touati, C. Moon, R. Brayton, and A. Wang, "Performance-Oriented Technology Mapping," in *Advanced Research in VLSI*, pp. 79–97, 1990.

[8] M. Garey and D. Johnson, *COMPUTERS AND INTRACTABILITY A Guide to the Theory of NP-Completeness*. W.H. Freeman and Company, 1979.

[9] W. Li, A. Lim, P. Agrawal, and S. Sahni, "On the Circuit Implementation Problem," Tech. Rep. TR90-33, Dept. of Computer and Info. Sciences, University of Florida, 1990.

[10] E. Horowitz and S. Sahni, *Fundamentals of Computer Algorithms*. Computer Science Press, 1978.

[11] J. Valdes, R. Tarjan, and E. Lawler, "The Recognition of Series and Parallel Digraphs," *SIAM Journal on Computing*, vol. 11, no. 2, pp. 298–313, May 1982.

[12] F. Brglez and H. Fujiwara, "Neutral netlist of ten combinational benchmark circuits and a target translator in FORTRAN," in *Proc. IEEE Info. Symp. Circuits and Systems*, 1985.

BDDMAP : a technology mapper based on a new covering algorithm

D S. Kung, R. F. Damiano, T. A. Nix
IBM T. J. Watson Research Center, Yorktown Heights, NY 10598

D. J. Geiger
IBM Corporation, P.O. Box 950, Poughkeepsie, NY 12602

Abstract

We present a technology mapper which combines the strengths of rule-based heuristics and algorithmic techniques. Matching is mainly performed by rule-based heuristics which is complemented by functional matching based on OBDD's. The novel aspects of the covering algorithm includes using an anticipative cost function, global cost propagation and handling of multiple output technology gates. We obtain favorable results when BDDMAP is benchmarked against the mapper of MisII, release 3.5, using a complete set of recommended MCNC examples. Topic number 4.1

1 Introduction

Technology mapping is the implementation of a Boolean network, henceforth referred to as the target network, using technology dependent gates from a prescribed library of primitives. The various approaches to this difficult mapping problem can be broadly divided into 4 categories: rule-based mapping [1], graph matching [2], direct mapping [3] and functional matching [4]. Technology mappers that use rule-based techniques [1] are usually fast but the rules tend to be ad hoc and heavily dependent on the library. In graph matching [2] the target network and the library cells are usually expanded into a two-input NAND/NOT representation. Since a two-input representation of a logic function is not canonical, library elements could have a large number of topologically inequivalent but functionally equivalent representations even if the target design is assumed to be free of redundancies. The target network is frequently partitioned into trees and then tree matching and dynamic programming algorithms can be applied to each tree partition to obtain a local optimal solution. As a post-process, some further local transformations such as inverter optimization can be applied across the partitions to improve the mapping results. While this algorithm is efficient on designs in which the tree partitions are relatively large and deep, the library is not fully utilized if the tree partitions contain very few gates. Moreover, more complex library gates (eg. XOR) do not have a tree representations of NAND/NOT gates so a separate process must be used. In the functional matching approach [4] a library gate is a candidate for implementing a pattern of logic in the target network if both of them compute the same logic function. This boolean equivalence checking problem can be expedited by employing a canonical form to represent the logic functions. One disadvantage is that the canonical form is often storage expensive.

In this paper, we present a technology mapper, BDDMAP, which uses rule-based and functional matching. The covering algorithm is an improvement over the dynamical programming - tree partitioning method used in DAGON. Rule-based heuristics or functional matching is invoked to match the type of technology gates for which it is most efficient. We find this judicious division of labor provides the most efficient and effective matching process. The algorithmic part of BDDMAP lies in the covering process. The novel aspects of the covering algorithm are using an anticipative cost function, global cost propagation and handling of multiple output gates. In section 2 we discuss the mapping problem in the context of matching of patterns and covering of the target network. An overview of BDDMAP is given in section 3. The matching process, namely, the functional matching technique and the rule-based heuristics are discussed in section 4. This is followed by benchmark results and concluding remarks.

2 The Mapping Problem

The target logic network is assumed to consist of single output gates with standard Boolean functions. The basic entity for matching is a pattern of logic which is defined as follows. A pattern of logic, C, is a subgraph of the target network. The nets of the pattern with no source from within the pattern are the input nets of the pattern. The nets of the pattern with no sinks within the pattern are outputs of the pattern although other nets with fanouts outside the pattern could also be designated as outputs. Nets of the pattern with fanouts outside the pattern but are not themselves designated outputs of the pattern are copy nets. A pattern with n inputs and m outputs computes a logic function with the same inputs and outputs. If such a logic function is equivalent to a logic function available in the library, then a match is found in the pattern. A cover of the target logic network is a collection of matched patterns such that the union of these matches includes the target network with the following constraint. Every input of a match is either a primary input of the target network or connected to the output of some match and every output of a match is either a primary output of the target network or connected to the input of some match. Overlapping among the matched patterns could occur at the copy nets and are resolved by copying the logic gates in the overlap. The technology mapping problem thus consists of the matching problem and the covering problem. The matching problem is to select a set of patterns and identify the matches among these patterns. The covering problem is to select a cover which minimizes the area or power or delay or a combination of these subject to constraints like testability and fanin/fanout restriction.

3 BDDMAP

BDDMAP is a general purpose technology mapper with good performance and high efficiency. BDDMAP is outlined as follows:

A. Matching	B. Covering
i. pre-processes	i. cost propagation
ii. functional matching	ii. technology binding
iii. post-processes	iii. post-processes.

It combines both rule-based and algorithmic techniques to handle the mapping problems in which libraries exhibit a wide varieties of characteristics. The philosophy we adopt is to separate the algorithmic operations, namely, functional matching and covering and the rule-based operations, namely, the pre- and post-processes. Matching is done first and then the target network is covered after all the matches have been gathered. The separation of the mapping into these two stages allows a more global covering algorithm to be used.

The covering algorithm incorporates heuristics to attain a more global optimization. The cost of a match depends on the implementations adjacent to the match. The cost propagation from inputs to outputs are not terminated at fanout points so that the covering algorithm is more global in nature. Another new aspect of the covering algorithm is that multiple output matches are handled as an integral part of the algorithm. A detailed description of the covering appears in section 5.

4 The Matching Algorithm

In BDDMAP, we optimize the division of labor between functional matching and rule-based heuristics. The complex unsymmetric technology gates, the basic gates (NAND, NOR, AND, OR), the multiple output gates and the non-standard gates such as those in bipolar technologies are handled by the pre- and post-processes. Functional matching complements the rule-based techniques by matching some of the basic gates missed by the rule-based approach, such as XOR's, XNOR's and small unsymmetric gates like the 2-1 selector.

For each library element handled by functional matching, OBDD's are created to represent the logic function of that element. For library functions that are not totally symmetric, we create all the OBDD's which represent the inequivalent input orderings. Functional matching is only applied to technology gates with large amount of symmetry or small number of inputs to avoid the possible exponential blowup of OBDD's. As for generating OBDD's for the target network, we limit ourselves to subset of patterns of logic generated by a breadth first traversal of the network. This subset of patterns is chosen because functions that are symmetrically decomposed can be matched easily. If the OBDD for a pattern is isomorphic to an OBDD in the library, a match is added to the list of matches associated with the gate at the root of the pattern.

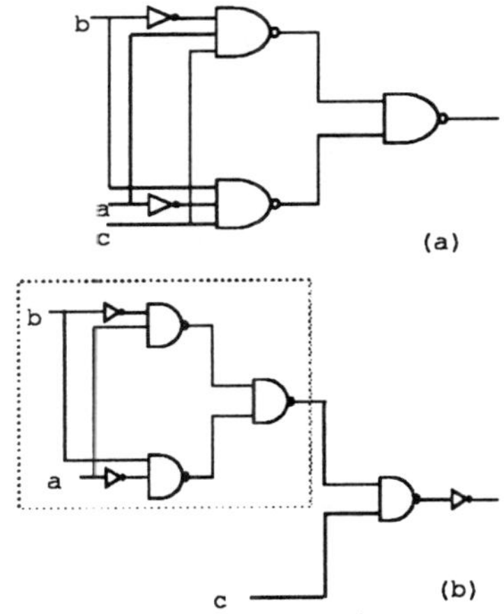

Figure 1: Two network decompositions.

The main functions of the pre- and post-processes for matching are rule-based matching and network priming. Complicated library elements such as adders, large multiplexors and decoders are matched efficiently using rule-based techniques such as those used in LSS [5]. For multiple output matches, the match is added to the lists of matches associated with the source box of every output net of the match. Matches for basic gates are most easily obtained while decomposing large basic gates into smaller ones. The matching of AO's, OA's, AOI's and OAI's are best done as a post-process based on the NAND, NOR, AND and OR matches obtained thus far.

Network priming refers to selecting a decomposition of the target network so that better matches are exposed. This is particularly important for the matching of XOR's, AO's, OA's and their complements. No patterns in the target network in Fig. 1(a) can be matched to a two-input XOR. However, if the target network is decomposed as in Fig. 1(b) the XOR match (inside dash box) is exposed.

Some of the most important matches are obtained by adjusting the phase of the inputs and outputs of a pattern. The standard approach is to insert double inverters in the network, which increases the number of target gates substantially. In BDDMAP we circumvent this increase in complexity by the use of decomposition matches. A decomposition match is a network whose internal nodes consist of library elements. In this application, the decomposition match consists of a main match and inverters to the some of the inputs and outputs of the main match. As an example, the function $d = AO21(a, b, c)$ can be matched by the decomposition match $\{p = AOI21(a, b, c),\ d = INV(p)\}$ where p is an internal net which will be realized if the match is used. In the cost calculation during covering, the cost of a decomposition match is assigned in such a way as to anticipate the cost upon implementation of the technology gates. A detailed description of the anticipated cost appears in section VII. These decomposition matches are generated after the the preprocesses and functional matching and are often derived from matches already found.

5 The Covering Algorithm

The covering algorithm is an extension of the dynamic programming technique for tree covering [2] and is optimal if the target network is a forest of trees. We emphasize that the network is not partitioned into trees so the matches are separated into two different classes. Matches with copies are the ones in which some nets internal to the match have fanouts to gates outside of the match and these nets are referred to as copy nets. Otherwise, the matches are said to have no copies.

In this paper we will concentrate on area optimization so the cost function is related to the area of the matches. We proceed to discuss the cost calculation algorithm for covering. If the target network is a tree, the principle of optimality implies that for every match M at the root of the tree, the cumulative cost, C_M, of an optimal cover containing M is the sum of the cost of M and the cost of an optimal cover of each subtree rooted at the inputs of M. The best match at the root is the match M^b such that the cumulative cost C_{M^b} at the root is minimum and C_{M^b} is the cost of an optimal cover for the tree. Therefore, computing the best match and minimum cumulative cost at each net gives the cost of an optimal cover for a tree. The following is the extension of the cost calculation to a general DAG. The cumulative cost at a net j, C_j, of a match with j as an output and with no copies is simply

$$C_j = \frac{W}{m} + \sum_{i \in I} \frac{B_i}{f_i}, \qquad (1)$$

where I is the set of input nets of the match, W is the anticipated cost of the match, m is the number of outputs of the match, B_i is the best cumulative cost at net i and f_i is the fanout of net i. If a match has copies, we compute for each copy net, l, the set, S_l, of input nets of the match that are in the cone of influence of l. Let n_i be the number of times i appears in the sets S_l for each copy net. Then the cumulative cost of the match is

$$C_j = \frac{W}{m} + \sum_{i \in I} \frac{B_i}{f_i + n_i}, \qquad (2)$$

The best match is the match with the minimum cumulative cost and this minimum cost is associated with net j as the best cumulative cost. The initial condition is that the best cumulative cost at the primary input nets and register output nets is zero. The anticipated cost of a match is the cumulative cost at the output of the match if the best matches at the input nets of the match are realized and double inverters are removed. The use of decomposition matches and anitcipated costs obviates the use of double inverter insertion at every net.

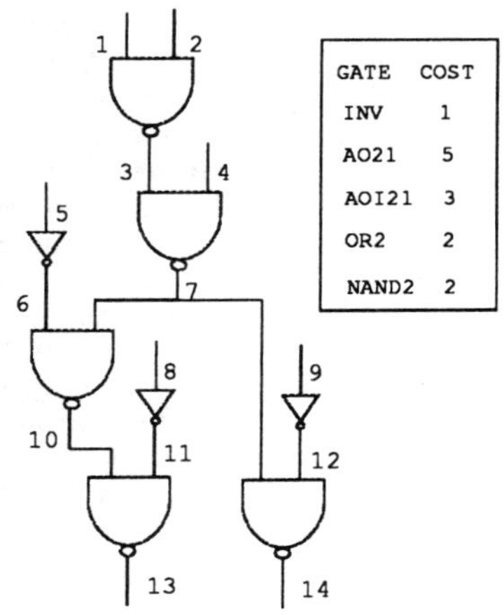

Figure 2: Model network and library.

The above cost function has the following nice property:

Let the target network be transformed into a DAG by removing the registers. If all the matches for the target network have no copies, the cost of an optimal cover for the DAG is the sum of the best cumulative cost of all the primary outputs, using the cost calculation of eqn. 1.

For our case, matches are allowed to have copy nets and the fanout of the nets changes as gates are being copied. The optimal cover can no longer be easily derived from the best cover of subnetworks. Eqn. 2 attempts to attain a reasonable approximation to the optimal case by estimating the effect of using a match with copies. Note that eqn. 2 disfavors matches which contain copy nets with a high external fanout, which is in agreement with our intuition.

The following example serves to illustrate the above concepts. Let a model network and a toy library be displayed in Fig. 2. The best match and the best cumulative cost at some nets are as follows:

B_3 - NAND2(1,2), $C_3 = 2$;
B_7 - NAND2(3,2), $C_7 = 4$;
B_{10} - $\{p = AOI21(3,4,5), 10 = INV(p)\}$, $C_{10} = 5$;
B_{13} - $\{13 = OR2(8,p), p = INV(10)\}$, $C_{13} = 6$.

In the calculation of C_{10}, B_{10} is a decomposition match with copy net 7. so $W = 4$, $n_3 = 1$ and $f_3 = 1$. In the calculation of C_{13}, the anticipated cost $W = 3 - 2 = 1$ due to the anticipated double inverter removal.

The gates in the combinational network are topologically sorted from inputs to outputs. The best cumulative cost for each net is computed in this topological order. With all the best cumulative cost and best matches in place, the target network is bound to the technology gates chosen as best matches. The order of binding is according to the following queue. All the primary output nets and input nets to registers are put on the queue. The rest of the nets are put whenever all the gates to which they fanout are bound to technology. In the course of technology binding, matches that can no longer be realized are invalidated. In addition, the technology binding affects the fanout of the nets which in turn changes the cumulative cost of the matches. Therefore, we constantly update the cumulative costs and the best matches are replaced as more favorable ones take over.

6 Results

We present the results of BDDMAP on some benchmark examples [6] for the 1991 International Workshop on Logic Synthesis. The library we used is an IBM CMOS library which is similar to but more extensive than the MOSIS 2u standard cell library in the MCNC distribution [6]. We compare the area results (in terms of internal cells) of BDDMAP with the technology mapper of MisII [7]. The technology independent optimizations are done using BooleDozer so the starting point for the MIS technology mapper and BDDMAP is exactly the same. Therefore only the performance of the technology mapping stage is being compared. The parameters for the MisII mapper is set such that logic duplication is allowed and multiple fanouts within the match is allowed. This setting should bring out the full strength of the MisII mapper for area optimization. All the BDDMAP runs are performed using the same scenario (or script) so that no fine tuning is is being performed. In addition, instead of limiting ourselves to examples which are favorable to BDDMAP, we benchmark all the recommended multi-level examples.

The recommended benchmarks cover a variety of model size and model types. There are ALU's, control logic, error correction logic, data encryption logic, multiplier and priority encoder. The mapping results are displayed in Table 1. In the thirteen examples, BDDMAP clearly performs better in 10 examples, 2 examples have a difference of 2 cells or less and MisII clearly performs better on 1 example, rot. On the average, BDDMAP performs better in area by 3.55 percent. The runtime of BDDMAP takes approximately twice as long as the MisII mapper but the largest model, des, takes only 10 minutes or so to run on a RISC6000 530 for BDDMAP.

7 Summary and Conclusions

It is common for technology libraries to contain gates which are as simple as a two input NAND or as complex as a 32 bit adder. Pure algorithmic approaches, though aesthetically pleasing and conceptually unifying, are impractical in handling complex gates. Rule-based techniques serves well to complement the weaknesses of algorithmic techniques. BDDMAP aims to strike a balanced compromise between these two approaches. Multiple output gates and connectivity constraints for bipolar technologies are handled efficiently by the mapper. The competitiveness of BDDMAP is shown by the benchmark comparison with the mapper of MisII. The current mapper is primarily optimized for area and delay optimization will be incorporated in the future.

Acknowledgements

We would like to thank Louise Trevillyan and Len Berman for useful discussions and Lukas van Ginneken for a critical reading of the manuscript.

	BDDMAP	MisII
C432	412	416
C1355	1150	1184
C1908	1070	1115
C2670	1447	1554
C3540	2312	2311
C6288	6107	6129
C7552	4808	4897
t481	1781	1844
rot	1435	1412
b9	241	239
dalu	2887	3037
des	6676	6903
k2	2065	2190

Table 1: Results from ISCAS Multi-level Examples

References

[1] D. Gregory, K. Bartlett, A. de Geus, and G. Hachtel, "Socrates: a system for automatically synthesizing and optimizing combinational logic," in *Proceedings of the 23rd ACM/IEEE Design Automation Conference*, pp. 79–85, ACM/IEEE, June 1986.

[2] K. Keutzer, "Dagon: technology binding and local optimization by dag matching," in *Proceedings of the 24th ACM/IEEE Design Automation Conference*, pp. 341–347, ACM/IEEE, June 1987.

[3] M. Lega, "Mapping properties of multi-level logic synthesis operations," in *Proceedings of the ICCD'88*, pp. 257–260, ACM/IEEE, 1988.

[4] F. Mailhot and G. De Micheli, "Technology mapping using boolean matching and don't care sets," in *Proceedings of the EDAC'90*, pp. 212–216, 1990.

[5] W. Joyner, L. Trevillyan, D. Brand, T. Nix, and S. Gundersen, "Technology adaptation in logic synthesis," in *Proceedings of the 23rd ACM/IEEE Design Automation Conference*, pp. 94–100, ACM/IEEE, June 1986.

[6] S. Yang, "Logic synthesis and optimization benchmarks user guide," 1991. Version 3.1.

[7] E. Detjens, G. Gannot, R. Rudell, A. Sangiovanni-Vincentelli, and A. Wang, "Technology mapping in mis," in *Proceedings of the ICCAD*, pp. 116–119, ACM/IEEE, November 1987.

LATTIS: An Iterative Speedup Heuristic for Mapped Logic

John P. Fishburn
AT&T Bell Laboratories
Murray Hill, New Jersey 07974
jack@allegra.att.com

Abstract

This paper describes heuristic procedures for performance optimization of mapped combinational logic implemented in the system LATTIS (Logic Area-Time Tradeoff for Integrated Systems). LATTIS currently has six transform types: 1) Gate repowering, 2) Buffer insertion, 3) Downpowering of non-critical fanouts of the critical path, 4) Gate duplication 5) DeMorgan's laws, and 6) Timing-directed factorization and remapping of subcircuits. From among the transforms applicable on the critical path, LATTIS chooses the one with maximum benefit/cost. Cost is increase in area, and benefit is improvement in local slack, weighted by the number of primary I/O's affected. The delay-area curves produced by LATTIS for the 70 largest circuits of the 1991 MCNC multi-level combinational logic benchmark set are given.

1 Introduction

In recent years, significant progress has been made in automatic procedures for the joint minimization of area and delay of combinational logic. Several kinds of techniques have been explored:

1. Repowering, or changing the size of a logic gate without changing its function [14] [7].
2. Algorithms for the amplification and distribution of a signal to its fanouts [8] [3] [14]. This is accomplished by the insertion of buffer trees, which may be asymmetric to accommodate differing required times.
3. Dynamic programming techniques for mapping a technology-independent representation of a logic network into a circuit made up of library gates. The original area-minimization formulation [9] has been extended to minimize area while simultaneously meeting delay constraints [13] [16].
4. Restructuring of logic to reduce the number of gates along a path [6] [15] [11] [2].

LATTIS uses all of the above techniques for speeding up combinational logic to meet delay constraints, while minimizing area. Unlike MIS's `speedup` command [15], which outputs in a technology-independent format, the output of LATTIS is mapped. In contrast to SOCRATES [1] and a previous version of LATTIS [6], transforms are algorithmically derived on the fly, rather than taken from a library. Unlike `speedup` and the system described in [17], the delay and area effects of each candidate transform are calculated exactly. Consisting of 5,400 lines of C code, LATTIS has been added as a command to the MIS system [4], and is in production use as part of the AT&T High Level Design System [5].

2 Area-to-Speed Tradeoff Heuristic

An area-minimized mapped circuit is the appropriate starting point for LATTIS. A static timing analysis using a linear delay model [14] (within MIS, called "DELAY_MODEL_LIBRARY") is performed. Every subsequent change maintains the circuit in the mapped state, and updates the timing analysis. Three passes are performed: 1) power-down, 2) speedup, and 3) area reclamation.

2.1 Power-Down Pass

It is sometimes the case (as in MCNC's `lib2.mis2lib`, or NEC's gate-array library) that two gates within a class both have minimum area, but differ in capacitive load. In the initial power-down pass, each higher-power gate is replaced by the lower-power version.

2.2 Speedup Pass

This pass consumes almost all the cpu time. For each node along the critical path, and for each transform, LATTIS investigates the consequences of applying that transform to that node. The (transform, node) pair having maximum BENEFIT/COST is selected. COST is the increase in total cell area. BENEFIT is the increase in the worst slack among the I/O of the transformed subcircuit, multiplied by INFAIL and OUTFAIL. INFAIL (OUTFAIL) is the number of failing primary inputs (outputs) of the global circuit whose required (arrival) time would be increased (decreased) by the application of the candidate transform. This weighting of BENEFIT by the number of affected inputs and outputs causes the heuristic to try harder at 'choke points' where leverage is greatest.

```
SPEEDUP_PASS( ):
  while(some nodes have negative slack){
    do critical path transform with greatest BENEFIT/COST;
    update timing analysis;
  }
```

The selected transform is applied, and the static timing analysis is updated with an incremental delay trace, which performs work proportional only to the number of nodes whose arrival or required times change. The process is repeated until all nodes have non-negative slack, or until no further improvement is possible. Due to the application of the transform, the critical path might change, and LATTIS will next operate on the new critical path. LATTIS is in this respect different from MIS's `speedup` command and the system described in [17], which may perform many transforms between successive delay traces.

2.3 Area Reclamation Pass

After the user specification is achieved, a sweep is made of all the gates in the circuit, to reclaim area without producing negative slack. This too is done on the basis of maximum BENEFIT/COST, except that now BENEFIT is area improvement, and COST is decrease in slack. Currently, area reclamation involves only a single transform, REPOWER.

3 Transform Set

In this section, when we say that a transform is "considered" we mean that BENEFIT and COST, as defined in the previous section, are calculated. After all transforms are considered, the single one with maximum BENEFIT/COST is performed. The transform set has evolved over time in response to monitoring of the optimization procedure on example circuits. When a need is perceived, a new transform can be added or an existing one made more general. At the current time, LATTIS has six transform types available to it:

1. REPOWER considers replacement of a gate by each of the other gates in its logic class.

0738-100X/92 $3.00

2. INSERTBUF divides the fanouts of a gate into critical and non-critical parts, and drives the non-critical fanouts with a double inverter. The division is made by sorting the fanouts according to worst slack of the two polarities, and considering, for $0 \leq m < number\_of\_fanouts$, driving the m most critical fanouts directly, and the remainder through the double buffer. All possible combinations of sizes for the two inverters are considered.
3. Like INSERTBUF, SPLIT considers all possible divisions of the fanouts of a gate into critical and non-critical parts. Each part is driven with a copy of the original gate. All combinations of sizes of the two gates are considered. SPLIT & INSERTBUF can interact to generate buffer trees for distributing a signal to its fanouts. Three observed extremes for a high-fanout node are the following: If the overall delay specification is lax enough, no transforms are applied. If the overall delay specification is tight, and the required times of the fanouts are similar, a symmetric fanout tree results. If the overall delay specification is tight, and the required times of the fanouts are dissimilar, an asymmetric fanout tree results.
4. DEMORGAN replaces a gate with its dual, and reverses the polarity of inputs and output. For each input, a gate generating the complement of the signal is used if it already exists, otherwise a minimum-area inverter is added. DEMORGAN considers adding a new output inverter, and if an output inverter already exists, considers eliminating it. All possible sizes for the dual gate, and output inverter (if added) are considered.
5. DOWNPOWER considers reducing gate size of a non-critical fanout of the critical path.
6. PEEPHOLE collapses a section of the critical path and resynthesizes it for speed, using timing-directed weak factorization [15] and mapping [13] routines already present in MIS. After the subcircuit is mapped, PEEPHOLE, like the other transforms, calculates BENEFIT and COST exactly. Starting with a single gate G, PEEPHOLE considers the collapsing and mapping of the n-gate segment of the critical path leading back from G, for n=1, 2, ..., until the number of literals exceeds a default threshold.

4 Timing Analysis Diagraming Method

An AR-diagram of a circuit is a two-dimensional display in which each signal S is represented by the point (S.arrival, S.required), and each point is connected with a line to each fanin and fanout. AR-diagrams have several desirable properties that quickly communicate circuit timing characteristics. The most important of these is that the locus of points having a given slack value, or *isoslack contour*, is a 45-degree line going from lower-left to upper-right. The critical path stands out from the rest of the circuit as the lowest isoslack contour. The object of the optimization is to push the circuit above the slack=0 contour, with minimum area. Non-critical outputs of the critical path, which indicate potential for speed optimization through insertion of buffers and parallel gates, stand out as near-vertical lines. Non-critical inputs of the critical path, which indicate potential for speed optimization through re-factorization and re-mapping, stand out as near-horizontal lines. AR-Diagrams have been helpful as a window to monitor the actions of LATTIS. For example, Figure 1 shows the AR-diagram of a critical path. A dotted line represents the zero-slack contour. Each signal on the critical path is labeled with its name, polarity (v or ^), gate name, and number of fanouts. It is apparent from the diagram that A and B are in need of buffering or higher power. G, which currently is a 4-input NOR gate, can be broken into a 3-input OR combining the 3 early signals, and a 2-input NOR.

5 Experimental Results

This section gives conclusions made from empirical comparisons between various versions of LATTIS. A benchmark set, made up of 23 circuits ranging in size from 100 to 5000 grids, was used for these experiments. (A minimum-size gate with N inputs generally has N+1 grids.) For each item in the following list, each of two versions of the software produced a delay-area curve for each circuit. The 23 pairs of curves provided a sensitive measure of the difference in capability between the two versions of the software.

1. Timing-directed weak factorization [15] is generally superior to Shannon factorization [6] as the restructuring method for PEEPHOLE, although neither method always dominates the other.

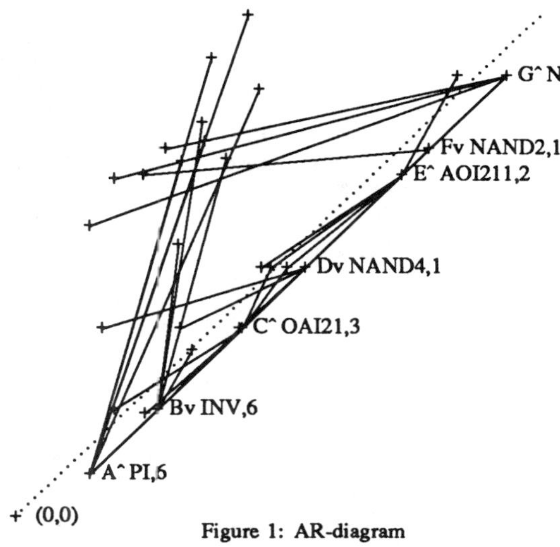

Figure 1: AR-diagram

2. Maximizing slack is superior to maximizing required time or minimizing arrival time. The first is different from the other two because each node has both a rise and a fall polarity.
3. If two logically equivalent gates have different powers but the same area, it is beneficial for the final delay-area curve if an initial pass replaces each higher-power gate by the lower-power version. If necessary, the REPOWER transform will then change back. Some gates not changed back might be non-critical fanouts of a critical path, which would be sped up by the smaller loads.
4. It is essential, in PEEPHOLE, to follow the restructuring and remapping steps with a recursive LATTIS call, with REPOWER and DOWNPOWER turned on. This power tuning adds little to the CPU time and greatly reduces the area of the subcircuit without increasing its delay.
5. With PEEPHOLE turned on, LATTIS is able to achieve $N \rightarrow \log N$ depth reduction starting with AND, OR, XOR and XNOR ripple-prefix circuits, and ripple-carry adders. The result is competitive with classic log-depth structures [10] in all but the XOR and XNOR cases, where problems with the weak factorization step cause significant area bloat.
6. Weighting BENEFIT by the number of failing inputs and outputs is beneficial. BENEFIT = $slack\_improvement \times INFAIL \times OUTFAIL$ is better than BENEFIT = $slack\_improvement \times (INFAIL + OUTFAIL)$.
7. BENEFIT/COST is generally superior to BENEFIT as a criterion for transform selection. Figure 2.1 shows a typical pair of curves for an AT&T circuit, comparing the standard LATTIS algorithm with the same algorithm modified to be similar to the SOCRATES [1] approach: repeatedly picking the transform with greatest BENEFIT until the delay specifications are met, then repeatedly picking the transform with greatest area improvement that does not violate the delay specification. In all 23 circuits, the BENEFIT/COST curve was lower than the BENEFIT curve. In 7 out of 23 cases, the BENEFIT speed champion was faster and larger than the BENEFIT/COST speed champion. In the other 16 cases, the BENEFIT/COST speed champion was both faster and smaller than the BENEFIT speed champion.

6 Benchmark results

Table 1 shows the results of applying LATTIS to the 70 largest circuits of the MCNC-91 multi-level benchmark set, minus i3, which LATTIS could not speed up. CPU times are in seconds on a MIPS RC6280 (about 50x faster than a VAX-780). All circuits were mapped to the MCNC-91 library `lib2.mis2lib`. Table 1 gives delay and (sum of cell areas)/464 for the seven points at the high-speed end of each curve. The timing goals that produced the table values were $initial\_delay * 0.92^{n+A}$, for n=1,2, ... 7. The table gives the value of A for each circuit. "-" indicates that

Circuit Name	A	Initial Delay/Area	Greatest Speedup	Total CPU	Delay/Area After LATTIS with Goal = (Initial DELAY) * 0.92^{n+A}						
					n=1	n=2	n=3	n=4	n=5	n=6	n=7
9symml	1	28.44/403	1.945	146	23.7/415	21.8/423	20.3/427	18.7/435	17.2/463	15.8/506	14.6/607
C1355	0	44.71/860	1.434	1200	40.9/880	37.8/998	34.8/1154	32/1351	31.2/2030		
C1908	1	68.98/1279	1.878	2004	57.8/1295	53.7/1320	48.9/1337	45.4/1378	41.8/1510	38.5/1873	36.7/2409
C2670	3	72.06/2072	2.221	802	51/2082	47.5/2087	43.6/2101	40.2/2128	36.9/2165	34/2283	32.4/2394
C3540	4	115.3/2941	2.396	3052	75.8/2970	69.8/2991	64.1/3016	59.2/3056	54.4/3205	50.1/3554	48.1/4003
C432	5	97.25/626	2.604	1309	58.5/641	54.2/652	49.4/663	45.8/709	42.2/793	38.8/967	37.3/1309
C499	0	44.99/860	1.428	659	40.9/875	38/939	35/1069	32.2/1260	31.5/1526		
C5315	0	72.71/4514	1.654	1393	66.8/4524	61.4/4548	56.6/4574	52/4629	47.9/4721	44.1/4966	44/4979
C6288	1	221.5/8178	1.935	45676	187/8214	172/8318	159/8496	146/8847	134/9336	124/10800	114/15444
C7552	10	165.1/4497	4.127	2972	65.9/4557	60.7/4580	55.8/4616	51.3/4681	47.2/4771	43.5/4930	40/5496
C880	0	63.15/823	1.727	402	57.7/826	52.2/834	47.9/838	44.7/849	41.2/873	38.3/938	36.6/1178
alu2	7	103.2/903	3.076	861	52.9/960	48.5/971	44.6/997	41.2/1042	37.9/1129	34.9/1278	33.6/1368
alu4	6	106.4/1736	2.725	2313	59.3/1834	54.5/1863	50.2/1920	46.1/2023	42.5/2228	39.1/2779	39/2813
apex6	3	32.46/1583	2.221	163	23/1605	21.3/1627	19.5/1642	18.1/1665	16.6/1710	15.3/1774	14.6/1822
apex7	0	23.98/539	1.595	122	21.9/544	20.2/551	18.6/563	17.2/600	15.8/664	15/743	
b9	0	15.77/279	1.657	46	14.4/280	13/287	12.1/291	11.1/299	10.4/314	9.56/392	9.52/412
c8	0	13.31/264	1.292	21	12.2/270	10.9/282	10.3/295	10.3/299			
cc	2	16.91/150	2.033	17	12/154	-	10/162	-	9.23/171	8.6/175	8.32/265
cht	3	17.22/353	2.221	21	12.1/363	10.5/369	10.4/369	9.39/374	8.81/383	8.06/387	7.75/478
cm138a	0	10.25/64	1.588	8.6	8.89/60	7.9/63	-	7.23/70	6.68/79	6.46/81	
cm150a	0	14.02/86	1.317	8.1	12.6/87	11.7/91	10.9/93	10.6/104			
cm162a	1	17.03/101	1.846	20	13.8/105	12/109	-	11.2/109	10.3/114	9.47/181	9.23/203
cm163a	0	12.65/82	1.525	6.8	10.2/85	-	9.52/93	9.03/103	8.3/124		
cm82a	0	12.33/46	1.451	11	11.1/53	9.92/53	9.59/58	8.81/66	8.5/78		
cm85a	0	11.78/102	1.177	11	10.8/110	10/125					
cmb	0	13.01/127	1.339	10	11.8/127	10.9/127	10.1/137	9.71/149			
comp	0	20.24/312	1.198	35	18.6/320	17.1/333	16.9/344				
cordic	0	13.99/224	1.240	15	12.8/240	11.7/271	11.3/276				
count	5	35.31/278	2.722	118	21.2/312	19.6/319	17.7/328	16.7/342	15.2/369	14.1/411	13/516
cu	0	11.82/109	1.426	13	10.6/116	9.52/126	9.18/136	8.41/156	8.29/165		
dalu	7	161.4/3413	3.153	1618	81.8/3458	76.2/3471	70/3478	64.4/3512	59.3/3559	54.6/3685	51.2/3959
decod	0	10.77/94	1.455	6.7	9.21/100	9.03/101	8.36/103	7.63/122	7.4/147		
des	19	213.8/8989	8.584	1228	40.3/9157	36.7/9207	34.1/9263	31.4/9326	28.9/9577	26.6/9899	25/10524
example2	4	37.4/768	2.507	114	24.3/790	21.7/791	20.8/795	19/810	17.6/833	16.2/868	14.9/906
f51m	0	19.95/261	1.735	38	18/263	16.6/269	15.3/273	14.2/282	13.1/291	12.1/303	11.5/359
frg1	0	18.89/249	1.514	36	17.3/250	15.9/255	14.6/275	13.5/299	12.5/343		
frg2	16	109.7/2145	6.700	354	26.6/2258	24.1/2291	22.5/2310	20.7/2336	19/2407	17.5/2649	16.4/2744
i1	0	11.21/117	1.684	9.9	9.89/120	8.88/122	8.62/127	8.03/133	6.91/142	6.66/144	
i10	8	163.9/5514	3.357	3678	77.2/5624	71.1/5656	65.4/5711	60.2/5788	55.4/6018	51/6263	48.8/6660
i2	0	12.18/370	1.107	3.7	11.2/380	11/386					
i4	0	9.959/390	1.125	22	9.05/538	8.85/614					
i5	1	22.88/454	1.825	159	19/467	17.8/474	16.4/486	15.1/519	13.8/615	12.7/783	12.5/837
i6	13	62.31/966	5.306	30	19.3/986	16.9/996	16.2/1000	15.1/1007	13.5/1023	12.7/1026	11.7/1062
i7	20	106.9/1353	9.509	121	17.8/1391	17/1392	15.7/1413	14.3/1416	13/1435	12.2/1458	11.2/1607
i8	19	168.2/2479	8.744	508	31.5/2575	29.2/2584	26.6/2613	24.7/2636	22.7/2677	20.9/2773	19.2/2984
i9	21	183.1/1555	9.614	480	29.1/1607	26.9/1618	24.5/1643	22.7/1662	20.9/1713	19.2/1886	19/2501
k2	12	94.21/4614	4.877	1818	31.9/4721	29.3/4755	27/4821	24.8/4909	22.8/5141	21/5446	19.3/5998
lal	0	13.5/250	1.337	48	12.3/273	11.3/289	10.5/346	10.1/426			
mux	0	15.02/88	1.528	10	13.5/87	12.1/88	11.4/91	10.4/93	9.83/112		
my_adder	0	80.67/352	1.667	967	74/388	68.3/446	62.8/538	57.8/599	53.1/685	48.9/927	48.4/966
pair	5	54.34/3372	2.576	1058	32.4/3415	30.2/3443	27.8/3477	25.6/3566	23.6/3722	21.7/4104	21.1/4452
parity	0	16.13/93	1.289	40	14.8/111	13.6/157	12.5/219				
pcle	0	16.18/154	1.680	39	14.8/156	13.5/165	12.4/171	11.6/183	10.7/199	9.72/209	9.63/240
pcler8	2	22.74/204	2.122	22	16.7/206	15.7/208	14.9/210	13.6/208	12.5/222	11.6/237	10.7/265
pm1	0	9.942/113	1.489	11	8.73/117	8.32/121	7.61/126	7.09/144	6.68/172		
rot	3	52.93/1552	2.177	377	37.9/1585	34.8/1597	32/1612	29.4/1644	27.1/1746	25/1876	24.3/1913
sct	0	12.13/225	1.462	32	11.1/228	10/240	9.41/251	8.63/275	8.29/326		
t481	0	15.44/85	1.324	23	13.7/90	13.1/96	11.9/121	11.7/129			
tcon	0	9.201/90	1.888	1.8	8.23/90	6.93/91	-	6.04/95	-	5.45/95	4.87/100
term1	0	22.64/503	1.662	55	20.7/503	18.8/510	17.3/519	16.1/531	14.9/550	13.7/602	13.6/622
too_large	3	39.09/1474	2.176	324	27.9/1509	25.4/1529	23.7/1552	21.8/1576	20/1626	18.4/1693	18/1767
ttt2	1	19.51/454	1.836	104	16.4/464	14.8/472	14/485	12.8/517	11.8/559	10.9/638	10.6/694
unreg	2	16.8/210	2.120	3.5	12.3/215	11.9/218	11/219	10/220	9.2/221	8.42/230	7.93/233
vda	8	57.85/2223	3.341	1806	27.3/2295	25.1/2325	23.1/2406	21.3/2497	19.6/2582	18/2813	17.3/3691
x1	0	18.84/720	1.721	77	17.1/721	15.3/730	14.6/746	13.5/762	12.4/786	11.4/840	10.9/917
x2	0	10.67/99	1.401	17	9.72/101	8.99/110	8.26/129	7.63/164	7.61/171		
x3	6	43.59/2195	2.802	69	23.6/2230	21.6/2231	20.4/2236	18.9/2250	17.2/2267	16/2304	15.6/2357
x4	9	41.89/939	3.643	128	18.2/1017	16.5/1035	15.4/1063	14.1/1081	13/1096	12/1225	11.5/1294
z4ml	0	14.68/92	1.606	18	12.4/94	-	11.3/109	10.4/115	9.56/127	9.14/164	

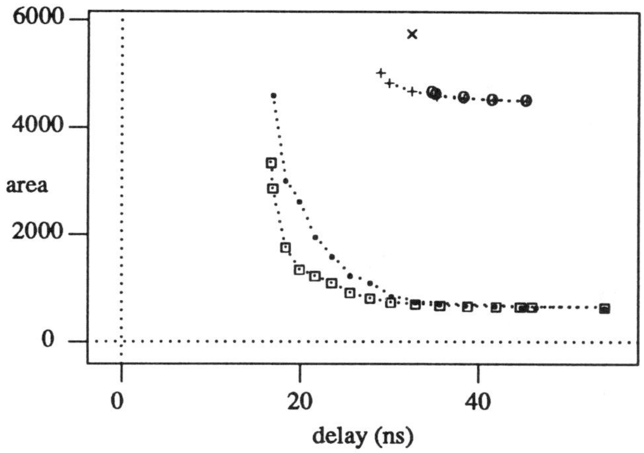

Figure 2: 1) Delay-area tradeoff curves for an AT&T circuit when transform is selected on basis of BENEFIT/COST (□) and on basis of BENEFIT (•). 2) LATTIS applied to C5315 with (+) and without (o) PEEPHOLE. "`map -n 1 -AF`" (x) applied to C5315.

this goal was met by the result for a smaller n. Primary output load and primary input drive were set to the load and drive of `lib2.mis2lib`'s `inv2x`, and wire capacitance was 0.1pf (`set_delay -d 1.98 i() -l 0.1009 o() -S 0.1`). The area-minimized-and-mapped starting point for each circuit was produced by MIS Release #2.2 with the script "`sw; el 5; sim0 *; asb; gkx -t 100; asb; sw; gcx -t 100; asb; sw; el 0; gd *; map; map_wires`". The only non-MIS command in this script is `map_wires`, which maps the few noninverting buffers left unmapped by `map` into double inverters. For a given goal, the starting circuit was taken to be the result from the next-larger goal. Thus 'Total CPU' is approximately the time it would take to produce the fastest circuit in the row. 'Greatest Speedup' is the delay of the area-minimized circuit, divided by the smallest delay achieved by LATTIS for that circuit.

Figure 2.2 shows delay-area curves for C5315 with all the transforms turned on, and with all transforms minus PEEPHOLE. Also shown is C5315 mapped with sis's "`map -n 1 -AF`" (technology mapping, buffer insertion, and area recovery to minimize delay). The experimental conditions were the same as for Table 1, except that the wire capacitance per fanout was set to zero.

7 Conclusions and Future Work

The LATTIS transforms are not new; Logic designers were using most of them before VLSI CAD existed. The contribution of the present work has been to identify, through extensive experimentation, an eclectic heuristic and set of transforms that can generate a high-quality delay-area tradeoff curve, with CPU time proportional to circuit size. To enable comparisons between other performance optimizers and LATTIS, benchmark results and their area-optimized starting points have been given. As the state of the art progresses, we can expect these delay-area curves to drop and move further to the left.

The set of transforms is sufficiently rich that many known high-speed circuit structures can be achieved: As needed, asymmetric or symmetric fanout trees can be grown, a linear chain of any associative logic gate can be turned into a log-depth circuit, and a ripple-carry adder can be compressed to a carry-lookahead adder. At the same time, the transforms are sufficiently fine-grained that wasteful application can be avoided. For example, if a path has several sites at which delay reduction has superlinear cost, LATTIS partitions the required delay reduction equally among the sites. By contrast, a coarser-grained approach might perform extreme optimization at one site after another until the path requirement is met, thus incurring higher cost.

The BENEFIT/COST figure-of-merit combines greed and stinginess to provide a level playing field on which diverse candidate transforms compete for application. Any proposed transform scheme, for which timing and area changes are calculable, could be added to the current set.

The REPLUGGING problem is to determine the assignment of input signals to gate inputs that achieves minimum arrival time at the gate output. Matching earlier arrival times with larger gate delays is not always the best solution, because it does not account for non-uniform signal drives and input capacitances. A better approach would be to construct an N-by-N table in which entry (i,j) is the gate output arrival time due to gate input j, if input signal i is attached to gate input j. The problem then reduces to picking a set of N elements from the table, such that no two elements in the set agree in either coordinate, and such that the maximum value in the set is as small as possible. This is the *Bottleneck Matching Problem* [12]. We are currently working on installing the REPLUGGING transform.

PEEPHOLE is the only one of the six transforms capable of logic restructuring, but its CPU usage is more than an order of magnitude greater than the other five put together. PEEPHOLE needs to be sped up and improved. Many other improvements need to be implemented, such as using non-inverting buffers in INSERTBUF, code speedup, and rebalancing operations that reconsider the partitions made in INSERTBUF and SPLIT.

8 Acknowledgments

I would like to thank J. Dussault, M. Lega, H. Nham, A. Prabhu, and A. Such for daily advice and help. I am also grateful to the authors of MIS, who created a very useful platform for logic optimization experimentation.

References

[1] K. Bartlett, W. Cohen, A. DeGeus, and G. Hachtel. Synthesis and optimization of multilevel logic under timing constraints. *IEEE Trans. CAD*, 5:582–596, 1986.

[2] C. Berman, D. Hathaway, A. LaPaugh, and L. Trevillyan. Efficient techniques for timing correction. In *ISCAS*, pages 415–419, 1990.

[3] C. L. Berman, J. L. Carter, and K. F. Day. The fanout problem: from theory to practice. In *Conf. on Advanced Research in VLSI*, pages 69–99, 1989.

[4] R. K. Brayton, R. Rudell, A. L. Sangiovanni-Vincentelli, and A. R. Wang. Mis: A multiple-level logic optimization system. *IEEE Trans. CAD*, 6:1062–1081, 1987.

[5] K. T. Cheng, J. Dussault, J. P. Fishburn, J. Y. Jou, M. C. Lega, and M. M. Tong. Behavioral and logic synthesis for performance and testability. In *4th Int. Symp. on IC Design, Manufacture and Applications ISIC-91*, pages 23–28, Sept. 1991.

[6] J. P. Fishburn. A depth-decreasing heuristic for combinational logic; or how to convert a ripple-carry adder into a carry-lookahead adder or anything in-between. In *DAC*, pages 361–364, 1990.

[7] D. Hill, D. Shugard, J. Fishburn, and K. Keutzer. *Algorithms and Techniques for VLSI Layout Synthesis*. Kluwer, 1989.

[8] H. J. Hoover, M. M. Klawe, and N. J. Pippenger. Bounding fan-out in logical networks. *JACM*, 31:13–18, 1984.

[9] K. Keutzer. Dagon: technology binding and local optimization. In *DAC*, pages 341–347, 1987.

[10] R. E. Ladner and M. J. Fischer. Parallel prefix computation. *JACM*, 27:831–838, 1980.

[11] P. C. McGeer, R. K. Brayton, A. L. Sangiovanni-Vincentelli, and S. J. Sahni. Performance enhancement through the generalized bypass transform. In *Int. Workshop on Logic Synthesis*, 1991.

[12] C. H. Papadimitriou and K. Steiglitz. *Combinatorial Optimization*. Prentice-Hall, 1982.

[13] R. L. Rudell. *Logic synthesis for logic design*. PhD thesis, University of California, Berkeley, 1989.

[14] K. J. Singh and A. Sangiovanni-Vincentelli. A heuristic algorithm for the fanout problem. In *DAC*, pages 357–360, 1990.

[15] K. J. Singh, A. R. Wang, R. K. Brayton, and A. Sangiovanni-Vincentelli. Timing optimization of combinational logic. In *ICCAD*, pages 282–285, 1988.

[16] H. J. Touati, C. W. Moon, R. K. Brayton, and A. Wang. Performance-oriented technology mapping. In *Conf. on Advanced Research in VLSI*, pages 79–97, 1990.

[17] K. Yoshikawa, H. Ichiryu, H. Tanishita, S. Suzuki, N. Nomizu, and A. Kondoh. Timing optimization on mapped circuits. In *DAC*, pages 112–117, 1991.

A Near Optimal Algorithm for Technology Mapping Minimizing Area under Delay Constraints

Kamal Chaudhary
Department of Electrical Engineering and Computer Sciences
University of California, Berkeley, CA 94720

Massoud Pedram
Department of Electrical Engineering – Systems
University of Southern California, Los Angeles, CA 90089

Abstract

We examine the problem of mapping a Boolean network using gates from a finite size cell library. The objective is to minimize the total gate area subject to constraints on signal arrival time at the primary outputs. Our approach consists of two steps: In the first step, we compute delay functions (which capture arrival time – gate area tradeoffs) at all nodes in the network, and in the second step we generate the mapping solution based on the computed delay functions and the required times at the primary outputs. For a NAND-decomposed tree, subject to load calculation errors, this two step approach finds the minimum area mapping satisfying all delay constraints if such solution exists. The algorithm has polynomial run time on a node-balanced tree and is easily extended to mapping a directed acyclic graph (DAG). Our results compare favorably with those of MIS2.2 mapper.

1 Introduction

The goal of logic synthesis is to produce a circuit which satisfies a set of logic equations, occupies minimal area and meets the timing constraints. Most logic synthesis systems currently available split this task into two phases – a technology independent phase and a technology dependent phase. In the first phase, transformations are applied on a Boolean network to find a representation with the least number of literals in the factored form. Additional timing optimization transformations are applied on this minimal area network to improve circuit performance. The role of the technology-dependent phase is to finish the synthesis of the circuit by performing the final gate selection from a target library.

1.1 Problem Definition

The technology mapping problem can be stated as follows: Given a Boolean network representing a combinational logic circuit optimized by technology independent synthesis procedures and a target library, we bind nodes in the network to gates in the library such that area of the final implementation is minimized and timing constraints are satisfied. A successful and efficient solution to the minimum area mapping problem was suggested in [3] and implemented in programs such as DAGON and MIS. The idea is to reduce technology mapping to DAG covering and to approximate DAG covering by a sequence of tree coverings which can be performed optimally using dynamic programming. [9] extended this approach to solve the technology mapping problem minimizing delay (subject to error due to unknown load values during mapping) and the technology mapping problem minimizing area under delay constraints. The solution to the latter problem, however, is not satisfactory.

The procedure for converting an optimized Boolean network into a two-input NAND and inverter decomposed (for short, NAND-decomposed) DAG is not unique and it is an open problem to determine which of the possible subject DAGs yields an optimum solution when an optimum covering algorithm is applied [2]. Different decomposition schemes for minimizing area [6], minimizing delay [11], or reducing routing complexity [5] have been introduced by various authors. In this paper, we assume that the DAG has already been decomposed into two-input NAND and inverter gates.

1.2 Prior Work

In [1], the author gives a pseudo-polynomial time algorithm for computing the min-max slacks (given the minimum and maximum required times at the primary outouts) at all internal nodes of a tree. The technique, however, can be used only for gate sizing (i.e., the tree has already been mapped, the problem is to assign the optimal size to each gate from a discrete number of sizes in the target library while satisfying the timing constraints).

In [9], the authors first compute a range of "interesting" values for the required times at each node (by finding the minimum area and the minimum delay mapping solutions) and then divide this range into equal intervals. The best mapping solution for each of the required times are generated and stored at the node during a postorder traversal (from primary inputs to primary outputs) of the tree. The final mapping solution is generated during a preorder traversal (from primary outputs toward primary inputs) of the tree. In order to obtain high-quality mapping solutions, this method requires a small time step resulting in large number of delay-area points. In contrast, our method works with the arrival times (as opposed to the required times), keeps all (and only) non-inferior delay-area points, and does not need an a priori range of interest for arrival times.

1.3 Overview and Organization of the Paper

In this paper, we present an efficient algorithm for generating a technology mapping solution with minimum gate area subject to given delay constraints. Our approach consists of two steps: In the first step, we compute delay functions (which capture arrival time – gate area tradeoffs) at all nodes of a NAND-decomposed network, and in the second step we generate the mapping solution based on the computed delay functions and the required times at the primary outputs.

The paper is organized as follows. In Section 2, we introduce some terminology and describe the timing model. Section 3 presents details of our algorithm. Sections 4 and 5 are devoted to the extension from trees to general DAGs and the complexity analysis. We present our results and concluding remarks in Sections 6 and 7.

2 Terminology and Timing Model

Consider a match g at node n of a NAND-decomposed tree. The inputs to node n consist of nodes n_i which fanout to node n (that is, $n = n'_1 + n'_2$ if n has two inputs or $n = n'_1$ if n has a single input). The nodes which are covered by match g are denoted by $merged(n, g)$. The nodes which are not in $merged(n, g)$ but fanin to $merged(n, g)$ are denoted by $inputs(n, g)$. The *mapped-parent*(n_i) is some node n for which there exists a matching gate g such that $n_i \in inputs(n, g)$. Note that given node n and gate g matching at n, $inputs(n, g)$ are uniquely determined. However, n_i may have many distinct mapped parents (Figure 2.1).

With each node in the network we store a delay curve. A point on the curve represents the arrival time at the *output* of the node and the total gate area which is required to map its transitive fanin cone up to (and including) the node. In addition to the area and delay value, the matching gate and input bindings for the match are also stored with each point on the

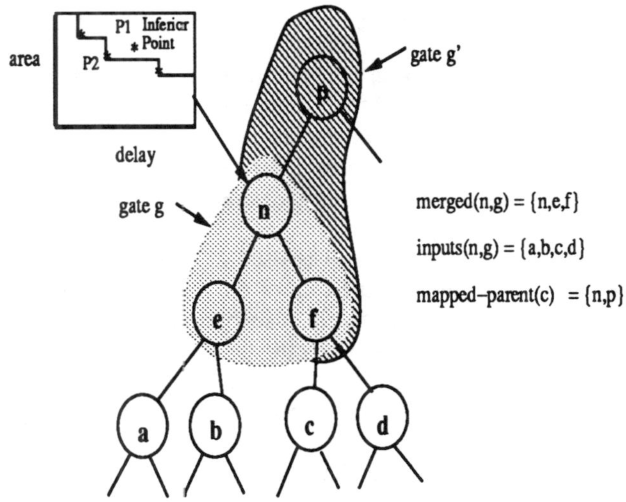

Figure 2.1: Terminology

curve. Points on the curve represent various mapping solutions with different tradeoffs between area and speed. We are interested in a mapping with minimum area satisfying delay requirements. Consequently, we can drop point P_1 on the curve if there exists another point P_2 on the curve with lower area but equal or lower delay. This is possible because the solution associated with P_2 is superior to the solution associated with P_1 in terms of area, delay or both. By dropping points, the delay curve can always be made monotonically non-increasing without loss of optimality. We would refer to P_1 as an *inferior point*. Point $P^* = (t^*, a^*)$ is a *non-inferior point* if and only if there does not exist a point $P = (t, a)$ such that either $t \leq t^*, a < a^*$ or $t < t^*, a \leq a^*$.

Lemma 2.1 *The delay function for a node contains the set of all non-inferior points and is monotonically non-increasing.*

In addition, if the difference in delay among two points is small (according to some user-specified parameter ϵ), we drop the point with higher area without any noticeable impact on the quality of the result. Similarly, points which are close in terms of their areas are merged together.

We have adopted the pin-dependent MIS library delay model as follows. Suppose that gate g has matched at node n, then the output arrival time at n is given by:

$$arrival(n, g, C_n) = max_{n_i \in inputs(n,g)}(\tau_{i,g} + R_{i,g}C_n + arrival(n_i, g_i, C_i))$$

where $\tau_{i,g}$ is the *intrinsic* gate delay from input i to output of g, $R_{i,g}$ is the *drive* resistance of g corresponding to a signal transition at input i, C_n is the load capacitance seen at n, $arrival(n_i, g_i, C_i)$ is the arrival time at input i corresponding to load C_i seen at that input, and g_i is the best match found at input i.

3 Tree Mapping

In this section, we focus on tree mapping. Later, we shall describe extensions to DAG mapping. In particular, we describe two tree-traversal operations which are applied to a NAND-decomposed tree in order to obtain a technology mapping solution which minimizes the total gate area while satisfying the timing constraints.

First, a postorder traversal is used to determine a set of possible arrival times at the root of the tree. Once the user specifies a single required time, a second, preorder traversal is performed to determine a specific technology mapping solution. This scheme is similar to that proposed in [8] in order to solve the optimal orientation problem for a slicing tree of macro-cells.

We begin by stating that the possible accumulated gate areas at each node can be described as a function of the arrival times at the node. The accumulated gate area is the total area used by the gates which have matched nodes in the transitive fanin cone of the node. The arrival time is the earliest time at which the signal at the output of the node settles within 50% of its final value (due to a signal transition at some primary input). The delay function is therefore represented by a set of ordered pairs of real positive numbers (t, a), where a piecewise linear function $a = f(t)$ can be constructed which describes the graph of all possible accumulated gate areas. This function describes all possible arrival time-area tradeoffs at a given node. The delay function at an input node of the NAND-decomposed tree consists of ordered pairs (t, a) where t and a have been specified by the user (in case of primary inputs of the network) or have been previously computed (in case that inputs to this tree are outputs of other trees).

3.1 Postorder Traversal

On the first traversal, we begin at the leaf nodes of the NAND-decomposed tree. Since each leaf node possesses a set of possible arrival time-area points which are reflected in its delay function, the delay function at any *mapped-parent*(n) must also reflect these possible arrival time-area tradeoffs. A postorder traversal of the NAND-decomposed tree is performed, where for each node n and for each gate g matching at n, a new delay function is produced by appropriately merging the delay functions at the $inputs(n, g)$. Merging must occur in the common region in order to ensure that the resulting function reflects feasible matches at the $inputs(n, g)$. The delay functions for successive gates g matching at n are then merged by applying a *lower-bound merge* operation on the corresponding delay functions. At a given node n, the resulting delay function will describe the arrival time-area tradeoffs in propagating a signal from the tree inputs to the output of n.

To illustrate the delay function computation procedure, consider example in Figure 3.1. It shows the computation

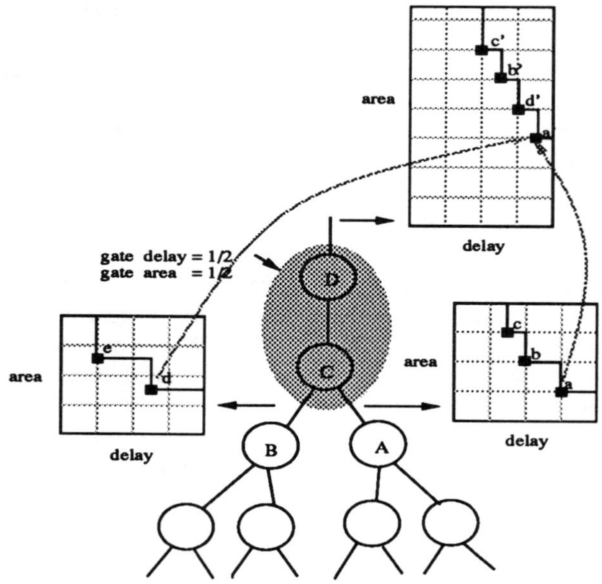

Figure 3.1: Generating the delay curve for a given match

of delay function for match g at node D. The inputs to the match are nodes A and B. The delay functions for A and B are known at this time. To compute a point on the delay function for node D, we select a point from delay function of inputs, say point a on delay curve of node A. The delay of point a is 3 units. So, we look for a point on the delay function of node B with delay less than 3 which has the minimum area. In this example, d is the desired point. We therefore combine points a and d to generate point a' on delay-curve(D), with

$$arrival(a') = arrival(a) + delay(g)$$

$$area(a') = area(a) + area(d) + area(gate).$$

Similarly, we generate all other points on the curve. Note, that there is no point on delay-curve(D) corresponding to the point e on delay-curve(B), as there is no point on delay-curve(A) which has delay less than or equal to delay(e).

To illustrate the lower bound merging procedure, consider example in Figure 3.2. Here, we have already generated the delay-curves for the matching gates g_1 and g_2 at node n. In order to obtain the composite delay curve at n, we must merge the two delay curves into one. This operation is simple since we only need to keep the non-inferior points on either curve. The minimum of the two delay-curves is computed, and information is attached to each point on the resulting delay curve indicating which gate alternative yields that point.

The delay function computation and merging are performed recursively until the root of the tree is reached. The resulting function is saved in the tree at its corresponding node. Thus, each node of the tree will have an associated delay function. The set of (t, a) pairs corresponding to the composite delay function at the root node will define a set of arrival time-area tradeoffs for the user to choose from.

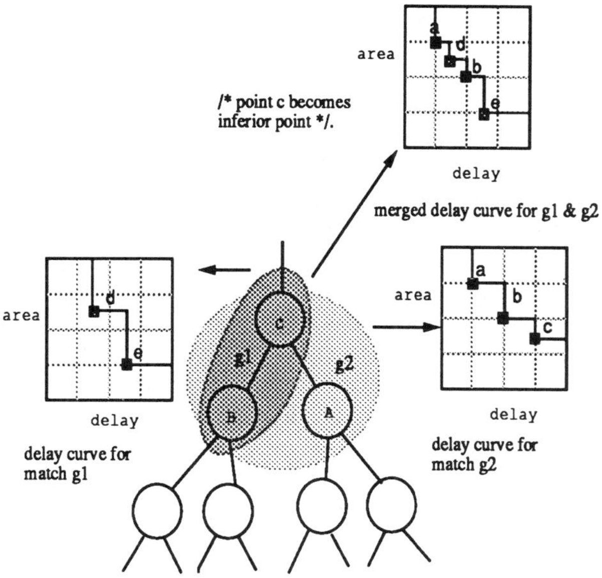

Figure 3.2: Lower bound merging of delay curves

3.2 Preorder Traversal

The user is allowed to select the arrival time-area tradeoff which is most suitable for his application. Given the required time t at the root of the tree, a suitable (t, a) point on the delay function for the root node is chosen. The gate g matching at the root which corresponds to this point and $inputs(root, g)$ are, thus, identified. The required times t_i at $inputs(root, g)$ are computed from t, g, and the observation that $inputs(root, g)$ must now drive gate g. The preorder traversal resumes at $inputs(root, g)$ where t_i is the constraining factor and a matching gate g_i with minimum a_i satisfying t_i is sought.

3.3 Timing Recalculation

The gate delay is a function of the load it is driving. During the postorder tree traversal, the output of current node n_i, is not mapped hence the load capacitance is unknown (unless, all the gates in the library have identical pin capacitances). At this time the load value is assumed to be that offered by the smallest two-input NAND gate in the library. When we come to a node n =*mapped-parent*(n_i) with matching gate g, we know the exact load seen by n_i. This load is equal to the input capacitance of g and is, in general, different from the default load. Therefore, in order to calculate the arrival time at node n, the output arrival times for all nodes in $inputs(n, g)$ must be adjusted to account for the change in the load capacitance [4]. Similarly, during the preorder tree traversal, when a gate g is selected to match at n, the load seen by $inputs(n, g)$ must be recalculated.

In order to account for this load change ($_i$), the delay curves at the inputs have to be appropriately shifted. In particular, since the drive resistance of gate matching at n_i and giving rise to a point p_j on delay-curve of n_i is stored with that point, the delay shift is computed as $_i \times p_j.gate.drive$. (See pseudo-code for compute_delay_curve and assign_best_gate algorithms. Details of timing recalculation are given for compute_delay_curve.)

function compute_delay_curve(n)
begin
  for each candidate match g at the node n **do**
    for each input i of match g **do**
      $\Gamma_i = g.intrinsic_i + g.drive_i \times n.load$
      Δ_i = calculate change in load
    end
    for each input i of match g **do**
      for each point p_j on the delay curve of input i **do**
        $np.delay = p_j.delay + \Delta_i \times p_j.gate.drive + \Gamma_i$
        feasible-flag = true
        for each input k of match g **do**
          if $k = i$, continue
          find $p_k^\star$ where $p_k^\star.area$ is minimum and
            $p_k^\star.delay \leq np.delay - _k$
          if no such $p_k^\star$ found
            feasible-flag = false
            break;
          end
        end
        if feasible-flag = true
          $np.gate = g$
          $np.binding = inputs(n, g)$
          $np.area = g.area + \sum_k p_k^\star.area$
          insert np to the delay-curve(n)
        end
      end
    end
  end
  sort delay curve of n based on the delay
  delete the inferior points on the curve
  reduce the non-inferior points by merging
end

function assign_best_gate(n, t)
begin
  calculate the current load based on the partial mapping
  shift delay-curve(n) to reflect the change in load
  find a point $p^\star$ on delay-curve(n) which satisfies the
    required time t and has minimum area
  $n.best\_gate = p^\star.gate$
  $n.best\_binding = p^\star.binding$
  for each input $i \in inputs(p^\star.gate, n)$ **do**
    $_i = g.intrinsic_i + g.drive_i \times n.load$
    assign_best_gate($n_i, t - _i$)
  end
end

3.4 Taking the Load Values into Account

The shift in delay for a point is a function of change in the load and the matching gate's driving resistance at the point. Different points on a curve may shift by different amounts depending on the matching gate. Differential shift may make the curve non-monotonic. In the worst case, a previously inferior point on the curve might have become non-inferior, had it not been dropped earlier. This may cause an optimal mapping being rejected. A possible solution is not to drop inferior points from the delay curves till we reach the output node. This will require a large number of points being stored for each curve without much gain.

Theorem 3.1 *Let R_{max} and R_{min} be the maximum and minimum driving resistances, C_{max} and C_{min} the maximum and minimum loads among gates in the library, and ϵ the error tolerance. If $(R_{max} - R_{min}) \times (C_{max} - C_{min}) \leq \epsilon$, then no optimal solution is dropped.*

Proof Maximum delay shift among the points on the curve is given by $(R_{max} - R_{min}) \times (C_{max} - C_{min})$. If it is $\leq \epsilon$, then no significant loss of optimality. ∎

Corollary 3.2 *If all the gates have the same pin capacitances then the tree traversals will produce the optimum solution.*

Corollary 3.3 *If all the gates have the same drive resistances then the tree traversals will produce the optimum solution.*

The other possible solution is to use a load bin method similar to that of MIS2.2. For each load bin, we store a delay curve. If the load bins are separated by less than $\epsilon/(R_{max} - R_{min})$, then the timing error will be less than ϵ. In practice, most of the libraries have a small number of gate series (e.g., performance- versus area-optimized, low-power versus high-power series). Within each series, the gates tend to have almost the same pin capacitances. Therefore, use of one load bin per gate series should be sufficient.

Note that during delay estimation we ignore the wire load (or alternatively, approximate it based on the fanout count, the expected average interconnect length and capacitance per unit length of interconnect). In fact, wire load can vary by a large amount (compared to the variance in pin capacitance) depending on the placement and routing. Therefore, it does not pay much to improve the accuracy of computing gate loads while ignoring (or only roughly capturing) the wire loads.

4 DAG Mapping

Most of the real circuits are not trees, but general DAGs. The problem of mapping a DAG even for the constant load model is NP-hard [2]. Therefore, we resort to heuristics. One heuristic is to decompose the DAG into a number of trees such that the inputs for each tree come from other tree outputs or the primary inputs. During the delay curve computation step, entire trees are processed in postorder and delay curves are computed for each primary output of the DAG. During the gate assignment step, entire trees are mapped in preorder. This heuristic which does not allow mapping across tree boundaries is similar to that used by DAGON.

Alternatively, we could avoid decomposing the DAG into trees as follows. During the delay curve computation step, nodes are visited in postorder. For each node, we compute the delay curve as in case of trees. However, if the input for a candidate match at the node is coming from a multiple fanout node we divide the area contribution of that input by the fanout count of the input node. By reducing the area contribution we tend to favor a solution in which multiple fanout nodes are preserved after mapping, which reduces logic duplication and improve the final mapped area. This heuristic which permits tree boundary crossing only for nodes with relatively few fanouts was also adopted by the MIS mapper. During the gate selection step, if we come to a node which has already been mapped, we check if the mapped solution at the node satisfies the timing requirement. If so, we keep the mapping; otherwise, we replace it with another solution from the delay curve which satisfies the current timing requirement and has minimum area. The new solution may have higher area compared to the previous solution. Note that satisfying the current timing can only decrease the delay for the previous cones, although it may increase the total gate area.

The solution for circuits with multiple outputs also depends on the order in which the output cones are processed. During the delay curve generation step, when we are computing the signal arrival time for a match g at node n, we need to recalculate the load seen at $inputs(n, g)$. For $n_i \in inputs(n, g)$, some of the fanouts of node n_i (other than g) may have already been mapped (because they are part of a logic cone which has been processed), and hence, the contribution of these fanouts to the load can be calculated exactly. This incremental load recalculation will result in more accurate arrival time calculation at the output of n. Similar incremental load recalculation is applied during the gate assignment step.

5 Complexity Analysis

Consider a gate g (with k inputs) matching at node n where input i has N_i points on its delay curve. The delay curve corresponding to match g at node n has $N = \sum_{i=0}^{k} N_i$ points in the worst case. The time required to generate each point, assuming that delay curve for each input is sorted, is $O(klog(N_{max}))$ (time for binary search) where N_{max} is the maximum N_i. It will require another $Nlog(N)$ to sort the

N points. Thus, the total time for generating delay curve per candidate match is $O(N^2 log(N) k log(N_m))$. A more efficient implementation based on merging of sorted lists will have $O(N)$ run time.

For a finite size library, the maximum number of gates that can match at a node n is bounded which means that the number of points on the delay-curve(n) will remain linear in the total number of points on the delay-curve of $inputs(n, g)$ for various matching gates. Therefore, the number of points increases linearly from one level to another. Despite this, the number of points could still grow exponentially in terms of the number of levels in the tree. However, if the tree is node-balanced (its height is logarithmic in the number of its leaf nodes), then the number of points will remain polynomial. In practice, the increase in number of points is even lower due to the fact that a large number of points generated are inferior points which are dropped and not propagated to higher levels.

It is observed that the range of areas generated for various solutions varies only by a factor of two, which means that if we use only 50 points at each node, the solutions produced will be at most 2% poorer in the area compared to the case where unbounded number of points are allowed. With a fixed upper bound P on the number of points the time to generate delay curve becomes $O(k^2 P^2 log(kP) log(P))$ or $O(k^2 log(k))$, which is a constant since the number of inputs for any gate in the library is bounded.

6 Experimental Results

The procedure has been implemented in a computer program named ADIEU. We have run the MCNC benchmarks using ADIEU and compared the results to the MIS2.2 technology mapper [9]. The same technology independent optimized *blif* files were used as input in both cases. The circuits were first optimized using *script.rugged* [7] and then delay optimized using the MIS2.2 new delay script [10]. Finally, they were decomposed into NAND gates. We mapped the circuits using the MIS2.2 and ADIEU mappers (without fanout optimization). We used the *lib2* library of the MIS2.2 package.

Table 1 presents the total gate area and the longest path delay through the circuit in the area mode of ADIEU. In Tables 2 and 3, all numbers have been normalized to the area mode of ADIEU. On average, ADIEU's area mode produces circuits which are faster by 6%, but larger by 3% (compared to MIS2.2's area mode); ADIEU's timing mode produces circuits as fast as MIS2.2's timing mode, but with 17% less area.

The graph in Figure 7.1 shows a range of mapped solutions produced for *C432* benchmark when using with different cycle times. These plots serve to illustrate the universality of our method in obtaining a range of solutions with different tradeoffs under user control. Our algorithm subsumes technology mapping techniques for minimum area or maximum performance.

Example	ADIEU	
	Area mode	
	area $*10^4\ \mu^2$	delay (ns)
9symml	15.5	18.8
apex6	76.4	37.0
apex7	25.5	16.2
b9	12.8	8.7
des	329.7	66.5
rot	68.9	23.4
z4ml	4.1	11.0
C1908	61.8	36.2
C1355	55.4	23.8
C432	25.9	33.7
C880	43.4	36.4
C3540	118.5	49.2
C5315	168.1	39.6
C7552	247.8	86.9

Table 1: Technology mapping results

7 Concluding Remarks

We have presented a powerful technique for technology mapping which generates solutions with different area/delay tradeoffs. Our technique unifies techniques for technology mapping with different objectives (minimum area, maximum performance, and minimum area under delay constraints) and is based on principles of dynamic programming and computation of delay curves. For a node-balanced NAND-decomposed tree, our algorithm finds the optimum area solution under delay constraints (subject to error due to unknown loads during delay computation step) in polynomial time and space. For the general problem of mapping DAGs, the algorithm retains its efficiency and produces results which are superior to those produced by other mappers.

We plan to combine this technique with the layout driven mapping technique of [4] in order to include and improve the wire load and routing area estimation during generation of the delay curve and gate selection.

Acknowledgements

The authors would like to thank Professor Ernest S. Kuh for his support and encouragement and Narasimha Bhat for valuable discussions. This research was supported in part by the National Science Foundation under grant number MIP 88-03711 and by the Defense Advanced Research Projects Agency under contract number JFBI90092.

References

[1] P. K. Chan, "Algorithms for library-specific sizing of combinational logic," *Proc. 27th ACM/IEEE Design Automation Conference*, pp. 352-356, 1990.

Example	MIS2.2			
	Area mode		Timing mode	
	area	delay	area	delay
9symml	1.00	0.99	1.35	0.91
apex6	0.94	0.92	1.35	0.90
apex7	0.97	0.97	1.26	0.88
b9	0.96	0.93	1.06	0.78
des	0.98	2.19	1.43	1.24
rot	0.99	0.98	1.18	1.02
z4ml	0.93	1.01	1.24	0.87
C1908	0.93	1.16	1.27	1.13
C1355	0.97	0.92	1.18	0.86
C432	1.00	1.00	1.26	0.81
C880	0.98	0.99	1.17	0.83
C3540	0.99	1.00	1.27	0.96
C5315	0.98	0.89	1.28	0.84
C7552	0.97	1.03	1.31	0.74
average	0.97	1.06	1.26	0.91

Table 2: Normalized mapping results for MIS2.2

Example	ADIEU			
	Area mode		Timing mode	
	area	delay	area	delay
9symml	1.00	1.00	1.09	0.90
apex6	1.00	1.00	1.03	0.75
apex7	1.00	1.00	1.01	0.93
des	1.00	1.00	1.00	0.89
b9	1.00	1.00	1.06	0.92
rot	1.00	1.00	1.01	0.95
z4ml	1.00	1.00	1.10	0.87
C1908	1.00	1.00	1.04	1.01
C1355	1.00	1.00	1.10	0.93
C432	1.00	1.00	1.07	0.83
C880	1.00	1.00	1.07	0.90
C3540	1.00	1.00	1.04	0.91
C5315	1.00	1.00	1.05	0.85
C7552	1.00	1.00	1.03	0.90
average	1.00	1.00	1.05	0.91

Table 3: Normalized mapping results for ADIEU

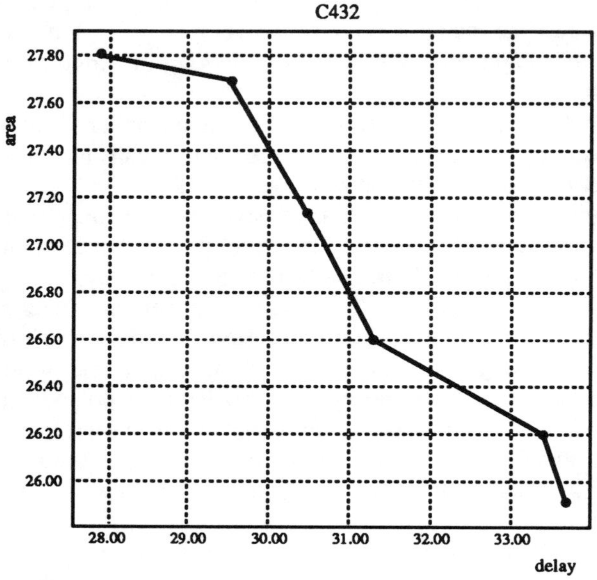

Figure 7.1: Delay-curves for *C432* benchmark

[2] R. K. Brayton, G. D. Hachtel and A. L. Sangiovanni-Vincenntelli, "Multilevel logic synthesis," *Proc. of the IEEE*, vol 78, no. 2, pp. 264-300, February 1990.

[3] K. Keutzer, "DAGON: technology binding and local optimization by DAG matching," *Proc. 24th ACM/IEEE Design Automation Conference*, pp. 341-347, 1987.

[4] M. Pedram and N. Bhat, "Layout driven technology mapping," *Proc. 28th ACM/IEEE Design Automation Conf.*, pages 99-105, 1991.

[5] M. Pedram and N. Bhat, "Layout driven logic restructuring / decomposition," *Proc. IEEE Int. Conf. Computer-Aided Design*, pages 134-137, 1991.

[6] R. Rudell, "Logic synthesis for VLSI design," Ph.D. dissertation, University of California, Berkeley, April 1989.

[7] H. Savoj, H. -Y. Wang, "Improved scripts in MIS-II for logic minimization of combinational circuits," *Proc. Int. Workshop on Logic Synthesis*, Vol. 3, 1991.

[8] L. Stockmeyer, "Optimal orientation of cells in slicing floorplan designs," *Information and Control*, Vol. 57, pages 91-101, 1983.

[9] H. J. Touati, C. W. Moon, R. K. Brayton and A. Wang, "Performance-oriented technology mapping," *Proc. 6th MIT Conf, Advanced Research in VLSI*, W. J. Dally ed., pp. 79-97, 1990.

[10] H. J. Touati, H. Savoj, and R. K. Brayton, "Delay optimization of combinational logic circuits by clustering and partial collapsing," *Proc. IEEE Int. Conf. Computer Design*, pages 188-191, 1991.

[11] A. Wang, "Algorithms for multi-level logic optimization," Ph.D. dissertation, University of California, Berkeley, April 1989.

SESSION 31

PANEL: WHY IS TODAY'S CAD INADEQUATE FOR DESIGNING TOMORROW'S COMPUTERS

Chair: Arnie Goldfein - Cadence Design Systems, Inc., San Jose, CA

Moore's law is not dead. This means, if you are designing computers, that the complexity of your competitor's constant-price machine will go up by a factor of four every two and a half years, and its speed will double. Whether the machine design consists of one chip, some custom chips, or a collection of merchant semiconductor-supplied devices, to stay in the game, you have to shoot for competitive speed and complexity targets. Because of the growth in the complexity of what you need to specify, when building a competitive computer, the sizes of design teams have tended to double every two and a half years, as well. In addition, as the industry matures, the need to design for testability, reliability, backward compatibility, and standards are adding to the complexity of our designs.

Todays computers are being developed by teams of, sometimes hundreds of, engineers collaborating over networks of workstations, using many different tools. The tools are developed by CAD vendors, the company's in-house CAD group, and the computer's designers themselves. A modern design methodology requires that engineers be able to operate these tools together as if they came from a single supplier, yet today's tools leave a lot to be desired in terms of interoperability.

Developing the behavioral description of new machine can approach in complexity the development of an operating system. Behavioral modeling systems are too slow, and the ability to track and verify the impact of changes at one level of abstraction, on representations of the same design at lower-levels of abstraction, is non-existent. Keeping engineers files of design data in sync as they jointly evolve the design is very difficult, and Configuration management tools are lacking or inadequate for the size of teams involved.

If the teams are not to grow impossibly large, we need more effective synthesis at higher levels of abstraction, and we to allow users to express themselves in a variety of representations that more closely reflect the design issues than what one panelist has described as "some new write-only language". We also need the ability to validate the high-level descriptions and extrapolate the impact of high level design decisions, before the team chases itself down blind design allies. We currently have no practical way to explore the feasibility of using a variety of technical implementation strategies before committing to one.

Panel Members:

Mike Baird - Pyramid Computers, Beaverton, OR
Bob Colwell - Intel Corp., Hillsboro, OR
Dave Ditzel - Sun Microsystems, Inc., Mountain View, CA
Steve Murphy - Motorola, Inc., Austin, TX
Charlie Evans - IBM Corp., Austin, TX
Hiroshi Ichiryu - NEC Corp., Tokyo, Japan

Design Process Management for CAD Frameworks*

Margarida F. Jacome and Stephen W. Director
Electrical and Computer Engineering Department
Carnegie Mellon University
Pittsburgh, PA 15217

Abstract

We introduce a new mechanism for planning and managing the VLSI design process. This mechanism significantly enhances the capabilities of CAD frameworks, relieving designers from dealing with low level details, thereby allowing them to concentrate on the more innovative aspects of design. A model for representing design processes is described. A prototype design process manager, called Minerva, that uses this model is presented.

1 Introduction

VLSI design has become a complex undertaking that involves a variety of activities and the use of a large, diverse set of CAD tools. During the course of design many data files are generated and the designer is faced with the immense task of keeping track of myriad details. In order to make the design process itself more tractable, CAD frameworks have been developed. Among other things, the purpose of these frameworks, from the user's point of view, has been to provide basic mechanisms for tool and data management. A few frameworks also provide some higher level facilities to help designers manage the design processes. Typical of such facilities are means for executing static design flows and capture of design history. From the point of view of tool developers and framework maintainers, frameworks should provide convenient mechanisms for tool and data encapsulation in order to facilitate the integration of new CAD resources. These issues have also been addressed in several framework implementations. Unfortunately, CAD frameworks still fall short in the area of true design process management. Consequently, designers still expend a significant amount of time dealing with cumbersome and error prone management details. In this paper we describe an approach to enhance CAD frameworks that can significantly aid the designer in managing the entire VLSI design process.

As a means for facilitating the discussion that follows, it is convenient to view the services that can be provided by a framework in terms of four basic levels of abstraction [1]. In this view, the *component level* comprises the CAD tools and databases available to the framework, exactly as they come from their developers or distributors. Designers "working" at the component level must therefore be familiar with all the details associated with specific idiosyncrasies of each particular tool.

At the next higher level of abstraction, called the *resource level*, all component level tools and data are encapsulated, thereby providing them will well defined and consistent interfaces. Such interfaces allow resources to interact with each other through a framework. While designers working at this level are still basically manipulating individual CAD tools and data objects, the specifics of each resource are hidden by a stable, and consistent "front-end", provided by the encapsulation.

In general, CAD tools are developed as a means for performing specific *design functions* or *CAD tasks* (e.g. synthesis, verification and optimization). CAD tasks constitute therefore the first real "design management abstraction" to be implemented in a framework. Accordingly, the next abstraction level is the *CAD task level*. At this level, the semantics of specific design functions are actually modelled and represented, independent of specific CAD resources, and are further dynamically mapped, by the *CAD Task Manager*[2], to these resources. Thus, the designer interacting with the framework at the CAD task level is allowed to request execution of abstract tasks such as "verify circuit performance" rather than directly requesting tool invocations such as "execute SPICE".

Although working at the CAD task level relieves designers from low level details concerned with CAD tool invocation, the *design process itself* is still not represented. Thus, while a CAD Task Manager provides a CAD framework with the capability to perform design functions, the actual reason for the need to perform these functions is not explicitly known by the framework. In other words, a framework whose highest level of abstraction is the CAD task level "knows how to do things" but does not know "why these things are being done". So, designers interacting with the framework directly at the

* This work is supported in part by the Semiconductor Research Corporation under contract no. 91-DC-068 and the Engineering Design Research Center, Carnegie Mellon University, under contract no. EDC-8943164.

CAD task level are still required to map the "design problem" being solved onto the specific collection of CAD tasks available in the framework. Further, the designer must still manually guarantee consistency among decisions made during the course of design; has to sequence the execution of CAD tasks; must manage all generated data objects; and has to coordinate activities with the group of designers working simultaneously on the same project.

The highest level of abstraction is the ***design process level*** (or ***problem level***). At this level the designer carries out design directly in terms of *design problems*, such as "design an operational amplifier to meet a set of specifications", or "verify the performance of this ALU". The *Design Process Manager* is responsible for automatically and dynamically mapping this "natural" (designer oriented) representation of problems to executable CAD tasks. In other words, it is responsible for planning the design process in the sense that it will derive the set of CAD tasks (seen as atomic operators) that have to be executed, together with the data objects involved on such executions, in order to solve a given design problem. Since the Design Process Manager relieves designers from all design process *planning* and *data management* activities, the obvious advantage of interacting with a framework at the proposed abstraction level is precisely to allow designers to concentrate their effort on the creative and exploratory aspects of design. Exploring alternative implementations, choosing appropriate specifications, and investigating trade-offs among performances while deriving those components, are examples of such creative/exploratory activities to which designers working at the design process level focus their attention.

In our view, the planning and managing capabilities provided by a Design Process Manager are useful to both the novice and experienced designers, so it should not in anyway limit the designer's flexibility or autonomy in implementing a design. Thus, the Design Process Manager must be able to cope with non-routine design problems, which are typically loosely defined at the beginning of the design process, as well as support any design methodology (e.g., bottom-up, top-down).

In order to demonstrate the suitability of our design process model, a prototype design process manager, called ***Minerva***, has been developed. While Minerva can be used as a stand alone meta-tool, its true benefit is only fully realized when it is incorporated into a CAD framework, such as the ***Odyssey*** framework [1], that supports resource and CAD task management.

Before proceeding it is worthwhile to place our work within the context of other related work. Knapp's Design Planning Engine (DPE) [3] generates plans for invoking CAD tools to realize design functions. Thus, DPE employs planning at the *design task level*, rather than at the *design problem level*. Harjani, et. al., [4] employs hardcoded plans for encoding multi-variable numerical optimization techniques in an analog circuit synthesis framework. These hardcoded plans are conceptually similar to complex CAD tasks (they statically encode a given design function), and therefore [4] also focuses on the design task level. Dewey's *conceptual design* [5] work, developed to aid the designer in the decision making process that occurs during the early stages of the design process, is most closely related to the work described here. However, Dewey did not attempt to explicitly represent design problems or to map the set of design decisions developed during conceptual design into an actual executable design plan.

The remainder of this paper is organized as follows. We first describe the Minerva Design Process Representation, or MDPR, which is our formal model for representing VLSI design processes. Then, in Section 3, we describe the dynamic modeling of the design process implemented by the Design Process Manager. Minerva, the prototype design process manager, is described in Section 4. Some conclusions are given in Section 5.

2 Minerva Design Process Representation (MDPR)

We begin this section by briefly reviewing the main phases of a typical VLSI design process. Of special interest are the strategies employed by designers in order to control complexity, since it is these strategies, and how they are used, that actually define a "design methodology".

The initial phase of design is *problem definition.* During this phase, the designer develops detailed specifications of the system or circuit to be designed. Such specifications may include the desired behavior and the functionality of the circuit, the general timing constraints, as well as environmental conditions such as the temperature range over which the circuit is expected to operate. As the specifications stabilizes, the next phase, conceptual design, begins.

Conceptual design involves the investigation of alternative, often high level, design options in terms of their consequences on performance, without actually undertaking design and fabrication steps [5]. During this phase, designers may need assistance in the form of qualitative advice and/or quantitative information that can be used to discriminate among the various options.

Upon completion of conceptual design, designers are typically faced with two options. If the size of the design problem is such that it can be solved using an available CAD tool suite, the designer may choose to attack the design problem as formulated. Alternatively the designer may employ some type of decomposition strat-

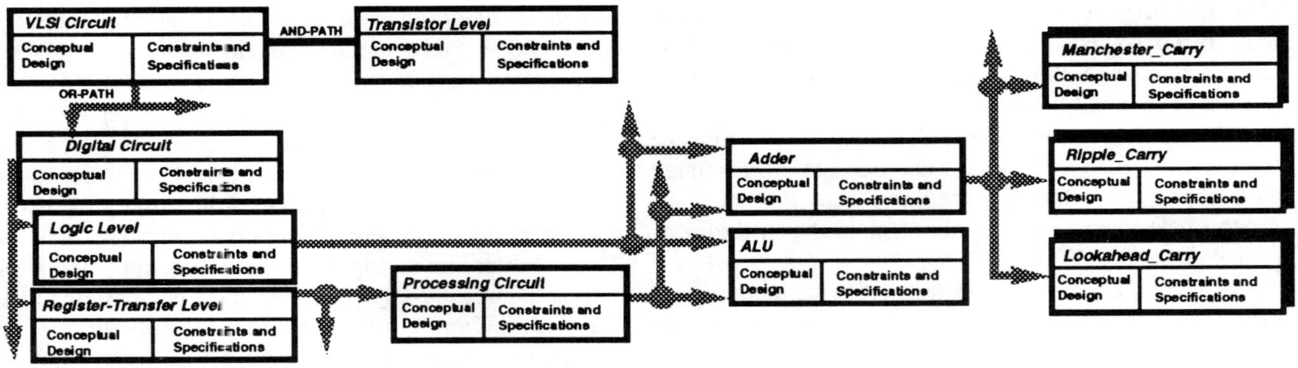

Figure 1 Example of hierarchical definition of design domains

egy (either structurally, or functionally), in order to create a number of smaller design problems. Specifications for each of these design sub-problems must then be developed, reflecting the restrictions imposed on the original circuit by the original specifications. Conceptual design is then performed for each of these sub-problems taking into account decisions made for the original circuit. Note that, because of the nearly decomposable nature of VLSI design problems, different designers may be simultaneously working on each of the sub-problems. Depending on the complexity of the resultant sub-problems, the designer may choose to undertake further decomposition. An important observation is that different design domains may lend themselves to different design methodologies or patterns of design activity. For instance, within the domain of analog circuits, parametric yield optimization procedures may be employed, while in the domain of digital circuits such techniques may not be necessary.

Let us now introduce a formal means for representing the VLSI design process.[1] We begin by defining the VLSI design problem in terms of three components: *design domains*, *design objectives* and *design specifications*. In what follows, we discuss each of these components and discuss the role that decomposition plays in controlling complexity.

2.1 Design Domains

A *design domain* is a self contained design context that may be broad, such as digital signal processing systems, or more limited, such as operational amplifiers, multipliers, adders, etc. Design domains can be defined in terms of *a static hierarchy* of *design domain knowledge templates* where each template contains two basic categories of knowledge: conceptual design knowledge and domain constraint knowledge.

1. While we are specifically focusing on the domain of VLSI design processes in this paper, our means of representing the design process may be generalized to many other domains as well.

Conceptual design knowledge includes: the set of specifications relevant for the particular domain; the set of *design issues* and related *design options* relevant to the particular design domain; *ordering* and *consistency constraints*[2] between design issues and related design options, respectively; and the set of relevant *discriminating factors* that are used to differentiate among the design options. Domain constraint knowledge includes: definition of formal languages for describing behavioral and structural constraints on the domain object; and actual constraints on behavioral, structural and physical characteristics for the domain of interest.

Figure 1 illustrates a hierarchy of design domain templates. In this hierarchy, descendent templates *inherit* knowledge (e.g. specifications, design issues, etc.) from their parent templates. Leaf templates that have predefined constraints describing their behavior are called *closed design domains* while other templates are called *open design domains*. Note that the knowledge contained in this representation of design domains[3] allows designers to specify, during run time, the behavior of circuits, or to specify functional or structural decompositions of circuits. This feature supports the development of new (non-routine) systems and circuits, treated as open domains.

In order to facilitate dynamic decomposition of design domains during the design process, as is often done to control the complexity of the circuit being designed, design domain knowledge templates can contain a *decomposition design issue* whose design options correspond to the possible decompositions, expressed in terms of subdomains available to the designer. Thus, for example, a digital filter may be designed by designing multipliers, adders, registers, etc. When a design domain is an option associated with the decomposition design issue of an

2. Ordering constraints guide the designer towards addressing the most pervasive design issues first. Consistency constraints prevents the designer from choosing a design option that is inconsistent with previous design decsions.
3. Namely, the description languages defined in the design domain knowledge templates.

higher level design domain, ordering and consistency constraints can be stated between design issues and related design options of both domains. These constraints are included in the knowledge template of the parent design domain.

2.2 Design Objectives

Design objectives refer to, and characterize, the general types of problems that are typically addressed during the design process, e.g. "design", "verify", "synthesize", "conceptual design" and "optimize." Design objectives have varying degrees of complexity. In order to control the complexity of the design problem itself, designers often decompose design objectives into a set of sub-objectives. For example, the objective "design" may be decomposed into the sub-objectives "synthesize&optimize" and "verify". Similarly, the objective "synthesize&-optimize" may be decomposed into the sub-objectives "conceptual design", "synthesize" and "optimize".

Ignoring for the moment design objective decomposition, realization of a design objective involves typically four steps: *design plan generation, design plan validation, design plan execution* and, finally, *execution validation.*

Design plan generation transforms a design problem into a partially ordered sequence of CAD tasks that reflects the design decisions made during conceptual design, the specifications imposed by the designer, and the design objective itself. This step is carried out in the *planning space,* viz. the space of all possible task sequences that can be generated to solve the same basic VLSI design problem.

The *design plan validation* step determines the existence of resources required by the individual tasks in the plan. If the plan cannot be validated there are two alternatives: backtracking, by reactivating the conceptual design objective in order to re-evaluate basic design decisions, or design objective decomposition, as discussed below.

Design plan execution is performed in the *design space,* viz. the space of all possible designs that totally or partially satisfy a set of design constraints with each point in the space representing a different trade-off between performance characteristics. Successful execution of a design plan results in an actual candidate solution to the design problem. Validation of the solution must then occur. If performance is unacceptable, again several different strategies may be pursued. Backtracking can be performed or, alternatively, the problem currently being addressed (or others) can be reactivated. Finally, a design plan may also fail because the selected tasks (or the related tools) failed during execution. In this case backtracking can be performed, or alternatively, the design objective can be decomposed.

As indicated above, design objective decomposition, may occur for several reasons. In order to achieve decomposition, the objective has to generate an appropriate set of sub-objectives, as well as a possible set of ordering restrictions for addressing these sub-objectives. It is important to note that there is no unique decomposition associated with a VLSI design problem. Therefore our model of the design process has to capture and represent expressively the nature of such diversity. Thus, when objectives are defined and represented, their decomposition can be made design domain dependent and/or dependent on *specifications defined for the objective itself,* as would be the case for enforcing a particular *design methodology.*

As was the case for design domains, each design objective can be defined in terms of a knowledge template that contain information such as: the set of specifications that can be defined for the objective; the steps during which the design objective is to be accomplished, and what to do in case of success and/or failure; possible decompositions for the objective, conditions required to activate and sequence the sub-objective, and what to do in case of sub-objective failure;

2.3 Design Specifications

After a given design objective is selected, within a particular design domain, the problem description is completed by a statement of *design specifications* (or a priori constraints) to be satisfied by the final solution. Design specifications can either be design objective dependent or design domain dependent. Examples of design objective dependent specifications are minimum area and minimum dissipated power, which may be associated with an optimization objective, and design methodology, that may be associated with a "design" objective. Examples of design domain dependent specifications are gain, bandwidth and slew rate, which may be associated with the domain of operational amplifiers.

Specifications can be defined in terms of *specification knowledge templates*, that contain the following information: indication of whether the specification is design domain dependent or design objective dependent; references to templates of corresponding design domains or design objectives; and restrictions or bounds on design specification assignments.[1]

Whenever problem decomposition occurs, specifications related to the original problem have to be propagated to the newly generated sub-problems. In order to accommodate this, our problem model contains *propagation constraint knowledge templates* that properly transform high level constraints into constraints associated with each sub-problem.

1. Assignments may be values, behavioral descriptions, etc.

3 The Design Process Manager

We now describe how a design process can be planned and managed given the representation of design problems described above. The design process begins with the definition of a specific VLSI design problem, called the *target problem*. Towards this end, the designer selects an appropriate design domain from among the entire set of design domains represented in the knowledge base. This choice causes the creation of an *instance* of the selected design domain knowledge template. The designer then selects a primary design objective and an *instance* of the associated design objective knowledge template is also created. Finally the designer chooses an appropriate set of specifications and assigns values to each specification in the set.

After the *target problem* has been defined, the design process can commence. The design process is basically a repetitive (quasi-cyclic) process involving two basic steps: *sub-problem selection* and *sub-problem solution*. We use a *decision tree* to represent the state of the design process. Decomposition relations among instances of design domain templates are implicitly represented by the branches of the tree. Thus, the decision tree represents the dynamic hierarchy of instances of design domain knowledge templates and associated design objective knowledge templates generated during the design process.

Sub-problem selection involves the choice of a design domain instance, i.e., a node in the decision tree, the choice of a design objective associated with that instance and, possibly, the choice of a set of design specifications to be satisfied by the solution. The sub-problem so selected by the designer is solved during the next *sub-problem solution step*. Depending on the design objective and the current step of this objective, the problem solving process involves one of five basic activities: conceptual design; building sub-plans; validating generated sub-plans; requesting execution of sub-plans or validating the prior execution of sub-plans.

Upon conclusion of a sub-problem solving step, a new sub-problem solving cycle is started. Notice that this basic "sub-problem selection - sub-problem solution" cycle is performed in a consistent and homogeneous fashion throughout the entire design hierarchy, independent of the abstraction level at which the design domain instances are defined (i.e. system, algorithmic, register-transfer level, logic, etc.). Also notice that our model of the design process does not enforce any specific patterns for decision tree generation and traversal.[1] Thus designers have the freedom to determine the critical components or paths of the design, and can focus on those elements first.

1. The decision tree represents at each moment the sub-problems that can be addressed concurrently.

Whenever design domain decomposition occurs, the top-level objective associated with the parent domain must, in general, be propagated to the sub-domains. This is the way domain decomposition impacts the design objective hierarchy that is simultaneously being developed. Also, a design objective will typically be generated (at the level of the parent domain) whose goal is to rebuild the original element from the cells or components into which it was decomposed.

4 Minerva - A Prototype Design Process Manager

A prototype design process manager has been implemented, called Minerva. [2]In order to illustrate the Design Process Manager capabilities we briefly, by example, illustrate how a designer interacts with Minerva during a "problem solving cycle". When a designer selects a sub-problem, a window showing the hierarchy of design domain instances for the current decision tree is created, as shown in Figure 2. The designer is allowed to browse the decision tree, inspecting the *status* and *current design decisions* for the existing design domain instances.

Each design domain instance will have an identifiable status such as: "*non-available selectable objectives*", meaning that there are no sub-problems currently ready to be solved within this design domain instance; "*busy*", meaning that another designer is concurrently working within this design domain instance (this allows multiple designers to be working on different aspects of the same design problem simultaneously); and "*ready for selection*", meaning that there is at least one sub-problem within the design domain instance that is ready to be addressed, and that no other designer is addressing it simultaneously.

In Figure 2, design domain instances that are available for selection are indicated by having the button "Select Domain" active. A designer who wants to know why a particular domain is not available for selection can click on the button "Show Status". Also, in order to have a better insight on the design domain instance, the design can click on the button "Inspect Decisions", and the design decisions that had been made for the instance will be displayed, although they cannot be modified at this point.

After selecting a given design domain instance, a window showing the associated hierarchy of objective instances is created, as shown in Figure 2. The designer is also allowed to browse the objectives hierarchy, inspecting the *status* of the existing objective instances.

The status of a design objective instance may be: "*accomplished*", meaning that the associated sub-problem

2. Minerva has been implemented in C++.

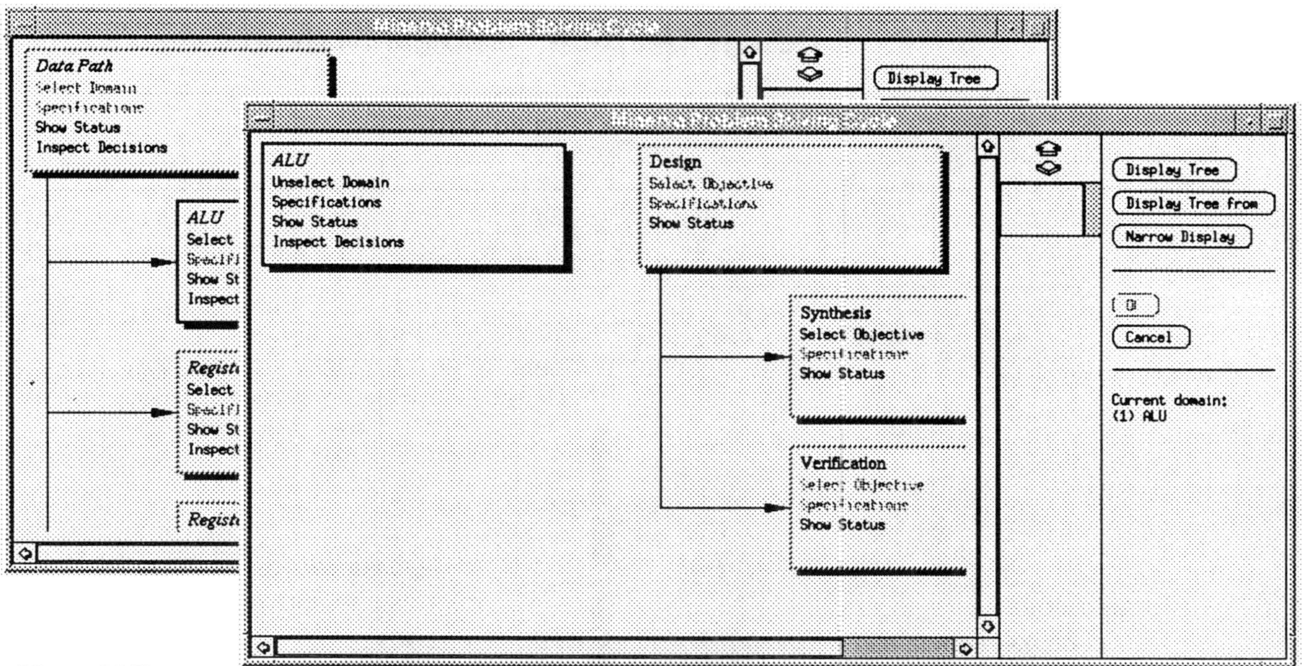

Figure 2 Minerva sub-problem selection step: selecting a design domain instance and an objective instance

was successfully solved; *"waiting for sub-objective accomplishment"*, meaning that the associated sub-problem depends on the results of other sub-problems that have not yet been addressed; *"waiting for design space event"*, meaning that the results of executin a plan are still not available; and *"ready for selection"*, meaning that the sub-problem associated with the sub-objective has not yet been addressed, and that there is no active sequencing constraints for it.

In Figure 2, design objective instances that are available for selection are indicated by having the button "Select Objective" active. The designer can determine why a particular objective instance is not available for selection by clicking on the button "Show Status".

Upon completion of "sub-problem selection", the designer enters one the possible "problem solving" steps, conceptual design, building sub-plans, validating generated sub-plans, requesting execution of sub-plans or validating the prior execution of sub-plans. For each step an appropriate window opens directing the designer towards solution of the selected problem (space limitations preclude us from showing these windows here). Once this sub-problem has been solved, the sub-problem selection process resumes until all sub-problems have been addressed and the target problem is solved.

5 Conclusions

In this paper we have extended the CAD framework paradigm to support Design Process Management allowing designers to concentrate exclusively on those issues that concern the creative and exploratory phases of design. A prototype Design Process Manager, Minerva, has been realized. Currently the Minerva knowledge base supports the design of datapaths, and to perform conceptual design on DSP Filters. The framework executive component interacting with Minerva currently has available the Berkeley Synthesis Tools for logic synthesis [6]. Expansion of the knowledge base, in order to cover the analog circuit domain is underway.

6 Bibliography

[1] J. Brockman, T. Cobourn, M. Jacome and S. Director. "Odyssey CAD Framework". To appear in *IEEE DACT Newsletter on Design Automation.*

[2] J. B. Brockman and S. W. Director, "The Hercules CAD Task Management System". In *Proceedings of the IEEE International Conference on Computer-Aided Design*, IEEE, 1991.

[3] D.W. Knapp and A.C. Parker. A Design Utility Manager: The ADAM Planning Engine. In *Proceedings of the 23th ACM/IEEE Design Automation Conference*, ACM Press, 1989, pages 48-54.

[4] Ramesh Harjani. "OASYS-A Framework for Analog Circuit Synthesis". *Ph.D. thesis, Dept. of ECE, Carnegie Mellon University, CMUCAD-89-24*. March 1989.

[5] A. M. Dewey and S. W. Director. "YODA-A Framework for the Conceptual Design of VLSI Design Systems." In *Proceedings of the 26th ACM/IEEE Design Automation Conference*, ACM Press, 1989, pages 380-383.

[6] G. DeMicheli. "Computer-Aided Synthesis of PLA-Based Systems". *UCB/ERL M84/31*. 1984.

Automated Design Decision Support System

Robert Beggs
Human Factors Engineering
Boeing Helicopters
Philadelphia, PA 19142-0858

John Sawaya, Catharine Ciric, Julius Etzl
Advanced. Computing Technologies Center
Boeing Helicopters Computing
Philadelphia, PA 19142

Abstract

The Automated Design Decision Support System is an ongoing Boeing Helicopters IR&D program to provide a concurrent engineering (CE) design environment for preliminary design. ADDSS research has focused on three components necessary for effective CE: a design representation for structuring and sharing design knowledge; design assessment using attributes as a design language; and an advanced computing architecture for the integration of design tools.

Concurrent Engineering: Design For the 'Ilities

Concurrent Engineering is a design philosophy that will improve design quality by integrating all the required ***disciplines*** into the design process, e.g. design for producibility or design for supportability. At Boeing Helicopters, design for supportability is a formidable challenge to today's rotorcraft design engineer. Both commercial and military customers have come to realize that the supportability of a system plays a role equal to performance, not only because of its effect on life cycle cost, but also because of its profound implications for fleet availability. All large airframe design organizations maintain a staff of advisors and/or evaluators (domain-specific engineers and experts) to address Reliability, Maintainability, Human Factors, Safety, Training, Design to Cost, and other supportability issues. However, the design engineer, being responsible for the final product, is ultimately in the best position to influence the supportability aspects of their systems.

Addressing supportability requirements early in the design process is recognized as an effective method of achieving a supportable end product. Supportability specialists have devised a number of methods to influence the preliminary design process, each with its own level of acceptance and success. Some of the tried methods include "board leaning" i.e. working alongside the designer, circulation of design guidelines and handbooks, and distribution of design for supportability checklists. Unfortunately, there are far too few specialists to address all the intricacies of a typical rotorcraft design, and this situation is further complicated by an overwhelming amount of supportability documentation, criteria, and guidelines.

CE Approaches

Design-Build Teams (DBT) and Product Development Teams represent one successful approach to practicing Concurrent Engineering. Design-Build-Teams are characterized as co-located, multi-discipline teams typically consisting of Engineering, Product Support and Manufacturing personnel, focused on addressing all aspects of a product early in the design process. This approach promotes open communication between organizations during preliminary design. DBT participants must be able to quickly evaluate technology alternatives in light of their disciplinary requirements and to draw accurately and rapidly on lessons learned and military specifications and guidelines. The low ratio between the number of engineering support specialists as compared to the size of the typical design engineering staff creates a fundamental problem. It is virtually impossible to adequately address all aspects of a complex system design in a timely manner with limited staff. As a result, DBTs tend to focus on top-level design concerns, leaving a multitude of design decisions solely to the responsible design engineer.

Recent studies of the design process have revealed other problems in the current approach to design team interaction. Participants have diverging ideas of the design process, a very large volume of data must be managed and shared among participants, and design guidelines and other important design-related data are difficult to retrieve. The process of engineering design has become so complex that the current design state is not easily characterized. An initial goal of Boeing Helicopters' Concurrent Engineering research and development effort was to develop the framework for a preliminary design environment that provided the design team with the following facilities:

- Fast and easy access to alternative components that satisfy certain functional requirements;
- Easy access to design guidelines from corporate, industry, or government sources;

- Rapid evaluation of the merits of design alternatives on performance and supportability scales;
- Verification of compliance of both function and component attributes to pre-assigned design requirements or standards;
- Identification of interactions between design functions and components;
- A record of design progress and design evolution;
- Enhancement of communications between and extending beyond the design team.

Designers explore, innovate, define and refine new design concepts, and search for solutions that are affected by numerous technical and non-technical constraints. During preliminary design, constraints change frequently. It is not possible to work out the details of each alternative solution. Instead, the design team must select alternatives by considering the information available during preliminary design. This leads to the designer's need for approximate methods to assess design solutions. Provisional design decisions evolve into guidelines, goals and commitments. Uncertain assumptions and the lack of complete information makes the designer's task of evaluating and coordinating preliminary design decisions very difficult. All the factors that influence and shape the design form the context in which decisions are made. Identifying and capturing interdependencies among design decisions are essential for concurrent engineering during preliminary design.

The goal of computer-based concurrent engineering is to ensure that all design decisions are based on the best collective knowledge available from all pertinent information sources, while recording and evaluating each significant design decision. This goal requires that all design information be extremely accessible, hence the special emphasis on design representation. The current lack of any standards for preliminary design information means that it is impossible to capture, share, review and build on design information. Concurrent engineering involves diverse groups of people. During the preliminary design process, these people need efficient methods for recording, communicating and sharing relevant information on complex and abstract design issues. By capturing design decision information, designers can take advantage of previous experience and make better design choices.

Automated Design Decision Support System (ADDSS)

Boeing Helicopters has established a Concurrent Engineering Lab which supports the diversities of a large engineering organization. The lab is structured around the concept of providing an integrated engineering environment to support multiple engineering domains as required for concurrent engineering. In order to support a computer based concurrent engineering environment, the research and development of computer systems must support the existing computing environment for all the design/build/support team members, as well as, the future facilities for these disciplines. The BH CE lab was designed with this philosophy in mind.

The CE Lab and the Advanced Computing Technologies group at Boeing have undertaken a multi-year Independent Research and Development (IR&D) program to research and identify methods to improve the Supportability/Operability of systems, equipment, or components through the timely influence of the preliminary design process. The Automated Design Decision Support System (ADDSS) and the Electronic Document Retrieval and Annotation System (EDRAS) are the products of multi-year research and development projects aimed at investigating computer-based Concurrent Engineering (CE). The objective of ADDSS is to create a computer-based environment for preliminary design that encourages and assists the design engineer to account for supportability, operability and producibility requirements early in the design process. ADDSS takes advantage of the flexible nature of preliminary design to add a structure that encourages the incorporation of Reliability, Maintainability, Safety, Human Factors, and Producibility features without sacrificing traditional system performance goals. In ADDSS, the representation and modeling environment will permit designers to explore alternative formulations of the design task and to compare alternative solutions generated in terms of a given formulation. The interface between high-level concept models and their underlying computational tools(agents) enables a systematic analysis of evolving solutions. Feedback from these agents, such as design assessments, will lead designers to reformulate and generate alternative solutions until all design objectives are satisfied. ADDSS research has pursued the development of three components necessary in creating a powerful preliminary design environment: design representation for structuring and sharing design knowledge; design assessment using attributes, which form a common language for design; and an advanced computing architecture for integration of existing and future design tools. The ADDSS project is working on three facets of this problem, (1) the conceptual model of design information, (2) system integration issues and (3) the user interface, which are all critical for a successful implementation.

Design Representation

Extended Function Logic Model

In 1989, Carnegie-Mellon University-Engineering Design Research Center (CMU-EDRC), Pittsburgh, PA was contracted to research and identify a common language for preliminary design that is responsive to multiple engineering disciplines. CMU proposed the Extended Function Logic Model (EFLM) as the design representation language for ADDSS. Based on a Value Analysis and Engineering approach, this strategy focuses on identifying a product's functions rather than conventional hardware/software descriptions, implying what each element does or performs rather than what it is. This approach provides a common language for all specialists and allows the discovery of redundant activities at the early stages of the design. Further, the functional language motivates the generation of alternative design solutions that might accomplish the same thing with better quality or reduced costs. It is this design approach which most markedly affects the preliminary design process and separates it from much of the current research work being done across industry and academia in design synthesis [2].

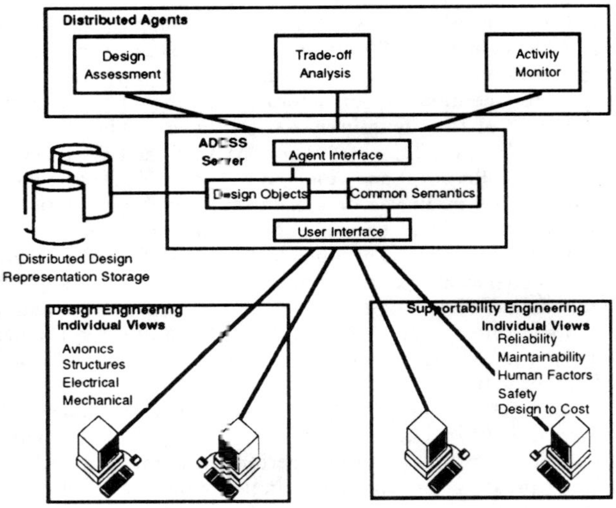

Figure 1 ADDSS Architecture

Boeing has translated EFLM theory into practice and created a design representation approach for ADDSS that is powerful yet flexible. Early user evaluations had indicated that not all designers were willing to work within the structure of the EFLM. Traditional approaches of creating system block diagrams without regard to functionality or with only mental images of the required functionality remain prevalent, hence the ADDSS design representation module has been designed to accept and operate effectively with a wide range of system representation styles. In addition, the ADDSS implementation takes into account that strict hierarchical representations of a system prove awkward at best and can often be incomplete or misleading. ADDSS offers sophisticated block linking mechanisms that capture the temporal or spatial relationships, including side effects, that may exist in a complex system design. The form of some of these linking mechanisms are the subject of continuing ADDSS research and development. The design representation allows a designer to build a system description that can be automatically or manually acted upon by a set of design decision support tools.

Object Representation

The ADDSS Design Representation will exploit object-oriented concepts for development and implementation. The object approach will enable designers to intuitively structure knowledge as it is developed, while retaining the properties of that knowledge, and to model non-hierarchical, complex relationships between "atoms of knowledge" . Object-oriented technology supports complex data types, relations, modularity and abstraction. Classes define the internal structural properties, model the behavior and declare the external interface protocol (accessor functions) for communication between objects. Classes may be derived from parent classes, i.e. forming a more specific type from a more general type. This expresses levels of abstraction and provides the means for inheritance of class attributes and

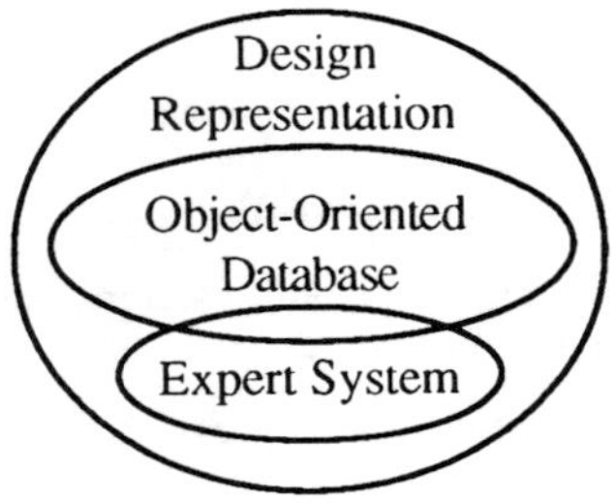

Figure 2 Design Representation Technologies

behavior. Individual objects are instances of a class and they encapsulate their own internal state information. An object-oriented concept, known as polymorphism, describes how different types of objects may respond similarly to the same message. Object-oriented technology is currently the best way to model the preliminary design information, given the need to assimilate large amounts of related textual, graphical, numerical, logical, structural and procedural information during preliminary design. Traditional methods, such as relational databases, just aren't adequate for this purpose. The preliminary design information evolves from abstract concepts into very complex and interrelated design information.

The Workspace

The conceptual model of design information is based on an object-oriented schema. At the highest level of abstraction, we define a "workspace" that can represent literally anything. A workspace is a kind of container. Each workspace contains those objects needed to describe a particular concept. A workspace may contain other workspaces. This provides a way to describe and reference concepts at various levels of detail. The other types of ADDSS objects include elemental types and other container types. Detailed information exists in ADDSS elemental objects. The elemental types are used as building blocks to create more complex forms of design concepts. Elemental types include: text, images, numbers, rules, procedures and links. Container objects represent structural or organizational information. They aggregate or associate groups of objects, such as: sets, lists, arrays, networks and dictionaries. By combining ADDSS objects, the user can dynamically create composite objects to express virtually any kind of preliminary design information. For example, links can express point to point relationships (how/why links, revision links, hypertext links). Sets can model groups of people in an organization, or attribute assignments. Lists show sequences of tasks, or prioritize relevant design information. Further research is needed to identify and define what extended types are needed to fully support concurrent engineering.

Through the ADDSS user interface, an empty workspace looks like a large drawing area (20K by 20K pixels). The user can draw blocks on the screen using a mouse and scroll a window over this drawing area. These blocks are a visual representation of objects that the workspace contains. The user may create a new block in the workspace, indicate what type it represents (any elemental or container type), then proceed to specify more information about that block. Existing objects can simply be referenced using these blocks. For example, an engineer may create a functional block diagram of the major functions that a particular design must perform. The workspace that represents this design could include blocks for all the required functions and blocks that define how each of the functions are associated. There may be alternative ways to implement the functions. The user creates a workspace to represent each alternative. Each alternative carries with it certain attributes (see below), such as features, cost, material properties, fabrication (or procurement) information, and historical records. A workspace for each alternative could contain blocks representing design sketches, references to similar designs, performance data, industry standards and specifications, supportability assessments, and other analyses.

Design Assessment and Attributes

Attributes

ADDSS research has developed a concept simply called attributes. Attributes express relevant characteristics of design objects. Here, the term attribute does not equate to a "field", frame slot, nor a column in a relational table. Attributes are not simply properties of a design object. From an object-oriented perspective, attributes are partial types (abstract classes) that encapsulate operations and state information to describe certain essential design characteristics. Attributes do not use multiple inheritance to bring characteristics into a class definition. Instead, attributes use inheritance only to build other attributes. Attributes are not inherited by non-attribute objects. ADDSS types, such as links and sets, describe the relationship between attributes and other design objects. An attribute is a kind of design object that is referenced by other design objects. This is a better way to specify a "has-attribute" relationship than using inheritance because attributes can describe characteristics which may be only temporarily associated with particular design objects. Therefore, this association may also include certainty factors (fuzzy sets). Attributes form the basis for common language in the design representation. Attributes capture the semantics of particular design concepts and discipline specific terminology. Attributes allow for indexing design objects. This attribute indexing mechanism allows interactive access, as well as procedural access, to retrieve relevant design objects. Furthermore, attributes and logical operators may be combined to form complex queries. Attributes also serve as indexes for other external systems that designers may access indirectly.

Design Assessment

The goal of design assessment is to provide an automated means to flag potential problem areas early in the preliminary design process. Supportability assessment evaluates design alternatives with respect to supportability criteria. Using ADDSS, designers associate attributes with alternative solutions. The risks and benefits associated with each attribute are, in turn, associated with the alternative solutions. Supportability analysts and designers have defined attributes and relationships among attributes. They will be able to add more attributes as the design process progresses. Attributes associated with actual designs form the basis for assessments. ADDSS is able to assess designs based on metrics defined for specific attributes and certain combinations of attributes. This makes the rationale for design decisions more visible and trackable. If assumptions for a given design approach are either relaxed or further constrained, the ADDSS user can recheck the design using new attributes.

Design assessment is a form of decision making under uncertainty. Early design decisions may be based on uncertain information. Attribute associations between the rationale and the design option will model this relationship. ADDSS will apply appropriate mathematical techniques in the methods of attribute assignment under uncertainty. Rule-based knowledge stored in the design representation may also be applied to cross-check the preliminary design for compliance with specifications, interface requirements and ever-changing design goals. Again, attributes will be used to associate rule objects with design options so that inferencing techniques may be applied to the design assessment. Attributes allow designers to quickly recreate the design assessment for later review or correction. Furthermore, attributes are an integral part of the design representation.

ADDSS Class Hierarchy

ADDSS uses the ONTOS object-oriented database, from Ontos Inc. as the data management system for the capture, storage and retrieval of design information. The root of all ONTOS persistent objects is the "Object" class which has member functions for constructing, naming, storing, deleting and other database functions.

The root ADDSS class "A_Concept" inherits from the ONTOS "Object". Note, all ADDSS class names begin with "A_" to eliminate naming conflicts. The constructor for A_Concept makes three private objects: a reference to alias names, a timestamp, and a reference to the user. This class defines common functions that all ADDSS objects share. The accessor functions alias(), time(), and user() are available to any derived class. A_Concept also defines virtual functions (suffered operations) that each leaf class must implement, such as: show() and edit(). The show() function generally shows the user something; display a dialog box, open windows, or perform a procedure. The edit() function allows a user to modify the object's internal state. Other functions are available for ADDSS messages.

The abstract class A_Container is derived from A_Concept. It is a base class for all the ADDSS container classes: A_Workspace (ADDSS composite object), A_Collection (unordered set), A_Sequence (list/queue), A_Matrix (vector/array), and A_Network (association table).

A_Container includes the following public virtual functions:

```
virtual void member_names(SLIST& the_slist);
virtual Aggregate& container();
virtual Boolean is_member(A_Concept* the_concept);
```

The member_names() function returns the name of each object in the container, where SLIST is a convenient type for interfacing with XVT dialog listboxes (a user interface development toolkit from XVT Software Inc.). The container() function returns a reference to the internal ONTOS aggregate object. This function is particularly effective when used in conjunction with the ONTOS SQL aggregate query capability. The is_member() function searches for a particular concept either directly or indirectly referenced in a container. ADDSS SQL queries result in the creation of container objects.

ADDSS supports user definable messages that can be sent to a container. The container, in turn, sends the message on to each member. Nested containers propagate messages down to each leaf object. Each container must track the messages it sends so that circular references don't cause infinite loops. This is one reason why ADDSS messages are persistent objects.

The abstract and (ungrammatical) class A_Element is also derived from A_Concept. It is the base class for all the ADDSS elemental classes: A_File (text), A_Block (on a workspace), A_Image (graphics), A_Numeric (integer, real...), A_Rule (knowledge-base), and A_Procedure (ADDSS macro command language).

A_File class holds Unix files. Typically, these are text files that represent documents. Functions in this class can read, write, show and edit files.

A_Block is a class of visual objects associated a workspace or a file. A block shows itself either inside a workspace or text file. In a workspace, a block is a rectangular box that can be moved around and linked with other blocks (shown as an arrow). In a text file, a block can highlight and scroll to a range of text given offsets from the beginning of the file. Attributes can be associated with particular blocks.

The A_Image class holds graphic images in GIF format files, with functions for converting from TIFF or PICT Macintosh file format. The show() function opens an X-Window and displays the image.

The A_Procedure class provides a macro command language capability with persistent storage. Procedural knowledge is stored in the ADDSS design representation as REXX language programs. REXX is a high-level interpreted language and SAA standard. REXX has two components: (1) a powerful interpreted macro command processor and (2) a set of C language interface routines. These tools permit ADDSS users to write application programs that dynamically link with ADDSS compiled (C++) routines. The REXX interface also allows external systems to interact closely with ADDSS object-oriented database.

A_Rule class encapsulates "knowledge-based" information. In ADDSS, the expert system provides inference techniques to process rule-based knowledge contained in the design representation. Design objects can contain logical information to be shared. Rules would be stored in the object-oriented database as elemental type objects. Each rule object will have an inference operation defined that tests its hypothesis by invoking backward chaining on an integrated expert system. The methods for implementing this are contained within the the rule class. Since rule objects encapsulate the methods for communicating with the expert system, the object-oriented database will automatically handle locking and serialization of the inference requests for the expert system. In this manner, the expert system may have multiple users, even though typical expert system shells are memory based (disk storage is only used to load the knowledge-base) and serve only one user at a time. It is also possible for multiple expert systems to service the ADDSS object-oriented database. ADDSS uses the Nexpert/Object expert system shell from Neuron Data Inc. This product has a C language callable interface that permits an embedded inference capability for the design representation. Rule-based knowledge stored in the design representation will be applied to cross-check the preliminary design for compliance with specifications, interface requirements and ever-changing design goals. Again, attributes will be used to associate rule objects with design options.

ADDSS Architecture

System Agency

Another facet of ADDSS research focuses on system integration issues. As a result, ADDSS employs an advanced computing architecture founded upon a System Agency. The System Agency is a new model for system integration built on the following concepts:

- Client/Server architecture - basic process control methodology
- Network Computing - parallel/distributed processing technology

ADDSS uses a multiple-client/multiple-server architecture to provide access to distributed heterogeneous data sources and applications. Client/server architecture here refers to a system in which the client performs local applications processing and user access, and the server retrieves data from the appropriate location, accesses the appropriate analysis tools as required, and returns the results to the client. Client programs request services of server programs. Server programs are started tasks waiting for requests from their clients. Multiple clients may request services from a single server. Or conversely, multiple servers may be required to support a single client program. A client/server architecture provides the basic control for interactions among processes.

The System Agency extends the client/server model and implements network computing techniques. The function of the System Agency is somewhat analogous to a real-world agency (e.g. employment agency). An agency is a set of cooperative independent agents that serve as surrogates for clients. Instead of integrating each pair of N processes (or systems) with N*N-1 separate lines of communication, with only N lines the System Agency is responsible for making connections, synchronizing client/server processes, broadcasting general messages and buffering real-time information. Parallel routines communicate with client and server processes. The System Agency is based on Network/Linda, TCP/IP, X/Open Transport Interface library and NFS shared files. ADDSS handles the coordination and communication between the software components and external systems through the System Agency.

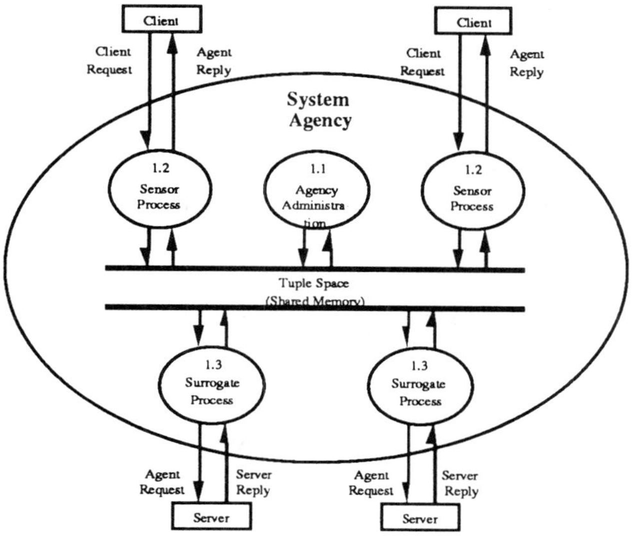

Figure 3 Agency View

User Interface

The ADDSS user interface is accessible through Unix workstations and X-window terminals. This will be a prerequisite of the ADDSS operating environment. The user interface look and feel is consistent with OSF/Motif standards. The XVT library of C routines has helped build the user interface and supplements well the ONTOS library. XVT tools support user dialogs: scrolling lists, requesters, text display, windows, menus, fonts, mouse actions, keyboard handling, drawing, and printing. Lower level X-Lib routines are used to display bit-mapped images. The user interface is also linked with REXX and the X/Open Transport Interface libraries. The user interface interacts with the System Agency for asynchronous alerts and real-time communication. The user interface also includes a sensor process which informs and alerts the user as necessary. This server process responds well to asynchronous events in a concurrent engineering environment.

A non-interactive user interface through E-mail allows users forward copies of their relevant engineering design correspondence to the ADDSS system -- an important aspect of tracking the decision making process. Once the correspondence is in the system, it is then available to users through attribute assignments -- just like any other design information in the system.

External Processes

The external system interface permits other systems to work concurrently and interactively with the ADDSS system. ADDSS defines a **Virtual User** as an agent process that supports external client requests, as well as internal background processes. The Virtual User concept provides external systems with much of the functionality of the interactive user interface. External processes may also implement REXX procedures to "run" ADDSS procedures remotely. In addition, the ONTOS SQL interface allows ADDSS to act as a passive server for external clients. Using the POSIX compliant X/Open Transport Interface, Internet TCP/IP provides the network communications between ADDSS and external systems.

Future Direction

At Boeing Helicopters, ADDSS research will continue to investigate and demonstrate the mechanisms necessary for providing a foundation for a computer-based concurrent engineering environment. As these foundations are laid, the techniques, technologies and systems will be moved into production through incremental development of the ICE (Integrated Concurrent Engineering) system. The overall approach to implementing computer-based concurrent engineering at Boeing Helicopters will be a multi-phase effort to allow a growth process that will utilize existing tools, techniques, and methodologies and expand with the results of new techniques and technologies from research and standards. As the ADDSS research and other Boeing, university, and industry efforts produce useful CE techniques, they will be incorporated into the ICE project. A multi-phase approach will minimize the risk and insure that the proven methodologies and technologies are incorporated into the ICE system as required.

Application-driven Design Automation for Microprocessor Design*

Iksoo Pyo, Ching-Long Su, Ing-Jer Huang, Kuo-Rueih Pan, Yong-seon Koh,
Chi-Ying Tsui, Hsu-Tsun Chen,† Gino Cheng,‡ Shihming Liu, Shiqun Wu, Alvin M. Despain
Department of Electrical Engineering - Systems
University of Southern California
aspian@zelea.usc.edu

Abstract

ADAS is an Application-driven Design Automation System for microprocessor design. The goal of ADAS is to automatically explore the design space and synthesize a single chip VLSI processor from a high-level specification of the Instruction Set Architecture (ISA) written in a subset of standard Prolog. Our idea is to develop a design automation system which considers both microprocessor hardware design and design of the corresponding language compiler concurrently. Benchmark programs are used to motivate design decisions and optimize performance. Compiler optimizations are considered during the design of hardware. Our system spans language design, compiler design, instruction set design, microarchitecture, and VLSI implementation. Another goal of our project is to determine the feasibility of applying formal methodology to design automation and the usefulness of formal syntax and semantics to define the meaning of specifications. We have exercised our system on a real industrial example, the TDY-43 processor.

1 Introduction

In this paper, we present an overview of the ADAS system. ADAS is an extension of the ASP system developed at Berkeley [1]. Many changes were necessary in order to adapt ASP to the needs of our industrial partners. In addition, many other design tools were integrated to enhance the utility and to improve the efficiency and performance of ASP. Throughout ADAS, we have employed formal notations and methods to aid us in controlling the system development and to aid us in formal validation of designs.

ADAS accepts a specification of the Instruction Set Architecture (ISA) as input, and produces both layout specified in Caltech Intermediate Form (CIF), and a re-order table for the language compiler as output. In ADAS, the design process has been partitioned into 45 design stages. These design stages are categorized into three major domains - the behavioral domain, the logic and circuit domain, and the geometric domain. The ADAS system is shown in Figure 1.

From an abstract ISA specification, Fiper and Piper determines the hardware resources and the control necessary to implement the instructions. Important features of our behavioral synthesis system will be discussed in sections 3, 6 and 11.

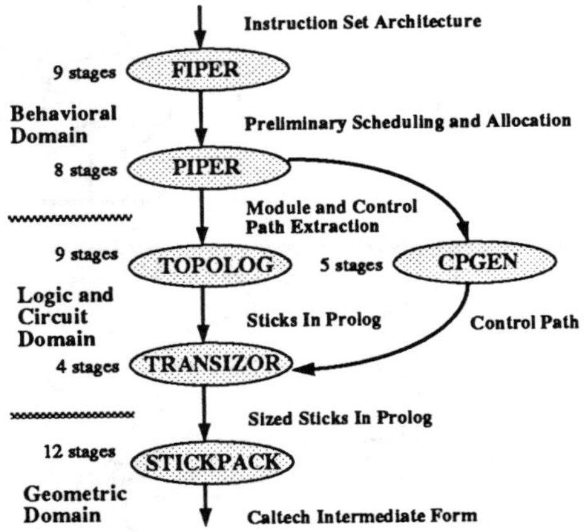

Figure 1: The Major Domains of ADAS

CPGen performs logic minimization on the controller specification and creates a symbolic layout implementation for the controller. Details of CPGen will be discussed in section 7.

In Topolog the hardware functions are implemented in a datapath which is translated into a symbolic layout. The major tasks in Topolog are cell allocation, cell placement, and routing. In section 8, we describe this process.

Circuit performance is enhanced by TranSizor, an automatic transistor sizing program. Transistor sizes are adjusted such that the delay constraints can be satisfied. In Section 9, a detailed explanation of TranSizor is presented.

Symbolic layout is translated into mask geometries, placed on a die, and routed by StickPack. A pad frame for the die is also created automatically. These procedures will be discussed in section 10.

2 Design Philosophy and Framework

2.1 ADAS Structure

The design issues of ADAS are:

1. How do high level implementation methods influence low level design decisions?
2. How is new design knowledge incorporated into an automated design environment?

*This research was partially supported by the California Competitive Technologies Program Grant No. C88-150, the Defense Advanced Research Projects Agency under Contract No. N00014-88-K-0579, TRW/ESL, and Teledyne Systems Company.

†FM3-95, Intel, 1900 Prairie City Road, Folsom, CA 95630
‡visitor from University of California, Berkeley

0738-100X/92 $3.00 © 1992 IEEE

3. How can information derived at high levels of design be used at lower levels to improve design quality?

4. How can an automated design environment efficiently redesign an implementation which does not meet specification?

In order to better understand these problems, we have developed a hierarchy of formal representation languages which span the entire gamut of processor design. Traditionally, the design problem has been decoupled into relatively independent design tasks, each of which is automated separately. Tools focus upon very specific design problems such as detailed routing, compaction, logic synthesis, and resource allocation without regard for the overall design. In ADAS, the design task is viewed as a sequential series of design transformations and is optimized as such. Design techniques at one stage of design may be used to simplify tasks at other stages. For example, higher level design problems such as resolving scheduling conflicts can be solved by providing new lower level components such as a priority generation module.

At each stage of the design process, the design specification is successively transformed into equivalent, more detailed representations, and ultimately to one which can be manufactured. Breaking the design process into several stages simplifies the overall design problem and makes it more manageable. Conceptually, each stage consists of a Semantic Design Generator (SDG), a simulator, an evaluator, a translator, a design library, and a design rule module as shown in Figure 2.

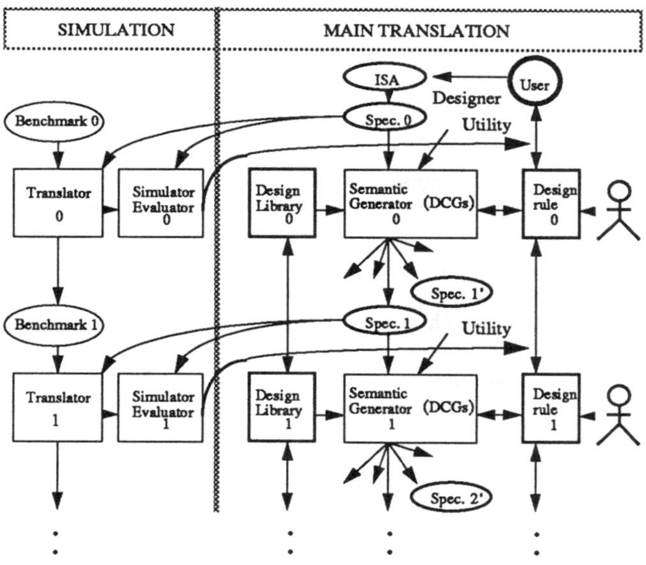

Figure 2: The Conceptual Structure of ADAS

The semantic design generator decomposes a design specification into more primitive components by applying transformations specified in the design rule module. For a given design, many alternatives may be possible. The most favorable transformation is selected by evaluating each alternative with the metrics important to that design stage.

2.2 Benchmark Driven

ADAS is a benchmark driven synthesis system. A collection of benchmark programs [2] are provided with the initial design specification. These programs are compiled into machine code and ultimately to binary test sequences. Benchmark program representations at the behavioral, RTL, and switch level are applied to the design specification at the appropriate level. By comparing simulation results, design consistency can be verified. Information extracted from benchmark programs is also used to motivate design decisions. Performance evaluation from benchmarks is more realistic than the best and worst case analytic estimation models.

3 Formal Methods

To ensure that a design specification implements the behaviors intended by the designer, it is important that the design specification languages throughout the design process be formal. In order to formalize a language, the syntax and semantics of the language must be formally defined.

In ADAS we used the Definite Clause Grammar (DCG) - a form of attribute grammar - to formally specify both the syntax and semantics of a design language. The syntax of these languages is specified by a BNF (Backus Normal Form) grammar. The semantic specification of a design is defined by attaching well defined "semantic attributes" to each production of the BNF grammar.

In a DCG system the correctness of a design specification may be verified by parsing. During parsing, a design which is represented in terms of "non-terminal" structures, is decomposed to a collection of "terminals". Terminals are objects for which implementation methods exist at the next stage of the design process. The decomposition process is guided by well defined translation rules.

As an example, valid behavioral specifications are verified by a parser which decomposes the specification into a list data structure, checks the syntax, verifies linear recurrence, and applies rules which translate the specification into one which is read at the next design stage. A fragmentation of the ISA parsing code is described in [3].

4 The Specification of Behavior

The design description for the ADAS system is specified in subset of Prolog. In this manner, instruction set architecture specifications can be both *simulated* and *synthesized*. This general approach is similar to that taken with the MacPitts system for LISP [4] and with the Flamel system for Pascal [5]. The descriptive power of ADAS corresponds to that of ISPS [6].

The specification of a design, the SM1 (a simple eight instruction processor) is illustrated in Figure 3. During each instruction cycle, an opcode and address are fetched, the PC is incremented, and the instruction specified by the opcode is decoded (implicitly) and executed. The behavior of each instruction is specified in an *execute* clause. The specification is directly *executable* in Prolog, thus special simulators are not necessary.

```
% Symbolic representation of Instruction Set Architecture

sm1( AC, PC , Memory ):-
      fetch( PC, Op, Adr, Memory ),
      P1 is PC + 1,
      execute( [Op], Adr, AC, P1, Memory, AC_next, PC_next,
                  Memory_next ),
      sm1( AC_next, PC_next, Memory_next ).

fetch( PC, Op, Adr, Memory ):-
      memory_read( PC, Memory, [Op,Adr] ).

execute( [add], X, AC, PC, Memory, AC_next, PC, Memory ) :-
      MemAR <- [constant(0)/15-13,X/12-0],
      memory_read( MemAR, Memory, MemDR ),
      AC_next is AC + MemDR.

execute( [and], X, AC, PC, Memory, AC_next, PC, Memory ) :-
      MemAR <- [constant(0)/15-13,X/12-0],
      memory_read( MemAR, Memory, MemDR ),
      AC_next is AC ∧ MemDR.

execute( [shr1], X, AC, PC, Memory, AC_next, PC, Memory ) :-
      AC_next is AC >> 1.

execute( [store], X, AC, PC Memory, AC, PC, Memory_next ):-
      MemAR <- [constant(0)/15-13,X/12-0],
      memory_write( MemAR, AC, Memory, Memory_next ).

execute( [load], X, _, PC, Memory, AC_next, PC, Memory ):-
      MemAR <- [constant(0)/15-13,X/12-0],
      memory_read( MemAR, Memory, MemDR ),
      AC_next = MemDR.

execute( [jump], X, AC, _, Memory, AC, PC_next, Memory ):-
      PC_next <- [constant(0)/15-13,X/12-0].

execute( [brn], X, AC, PC, Memory, AC, PC_next, Memory ) :-
      (AC < 0 -> PC_next <- [constant(0)/15-13,X/12-0];PC_next = PC).

class([reg,master_slave,r1,15]).
class([memory,ram,m1,[16,64]]). % 16bit x 64K

attributed([r1,ac,r,sm1,1]).
attributed([r1,pc,r,sm1,2]).
attributed([m1,m, m,sm1,3]).
```

Figure 3: Instruction Set Architecture of SM1

Instructions must be deterministic (with only shallow backtracking), and linear tail-recursive. True recursion is not allowed. Furthermore, only a limited set of built-ins and user-defined predicates are currently supported. However, special predicates are provided for creating and manipulating registers, bit fields, and for managing the state those registers contain.

5 Simulation, Analysis and Verification

Benchmark programs [2] specified in the ISA of a processor are compiled into target machine code which is applied to simulate the design specification through different abstract design levels.

Simulation of designs at the functional and behavioral level are based upon an event-driven finite-state machine model. Benchmark programs are simulated cycle by cycle. Clock ticks and interrupts represent events.

An analyzer calculates the frequency of both individual instructions and sequences of contiguous instructions. Results are used to optimize scheduling and resource allocation in behavioral synthesis. Frequently used instructions are given priority during scheduling while infrequent instructions are scheduled to maximize resource sharing. Data dependencies determined by instruction pattern analysis are used to evaluate latency in pipelined designs.

Test vectors automatically produced from the benchmarks are applied to networks extracted from the layout. Machine state traces are recorded. Verification programs compare these traces with the results predicted from the high level specification using the circuit-level simulation tools - *ESIM*, *IRSIM* and *SPICE*.

6 Behavioral Level Synthesis

6.1 Behavioral Level System Overview

After parsing an ISA in section 3, Fiper creates a global control/data graph describing the execution flow of each instruction by dependency analysis. Based upon this graph, instructions are scheduled by applying the scheduling/allocation algorithm illustrated in Figure 4. This algorithm attempts to minimize the number of states necessary to traverse the most frequently used RTL paths based upon the given *latency*, *stage-time limit* and *the available resources*. Current scheduling/allocation algorithms focus on the reduction of hardware resources and the cycle time independent of program application. By using instruction frequency information, resources are allocated and scheduled to reduce time for critical instructions. Fiper finally produces an optimized state transition graph which describes the sequential behavior of the machine under design.

```
Step 1: Evaluate frequencies of RTL patterns of fixed latency
        and Create an allocation table.

N = 1.
Repeat until all RTL patterns are scheduled
Step 2: Find path(s) which has the most frequently used RTL pattern
Step 3: Try to schedule the pattern within N stage(s).
        If fail, N is N + 1 and go to Step 3.
Step 4: Update the allocation table.
        Delete the scheduled path (RTL sequence).
        N = 1.
End Repeat
```

Figure 4: ADAS Scheduling/Allocation Algorithm

From the state transition graph produced by Fiper, Piper performs pipeline dependency analysis and decouples the state graph into stages, each stage implementing concurrent operations (pipeline stage assignment). Dependencies between pipeline stages are detected and resolved. Operations at each stage are then bound to functional units for a data path allocated. Additional hardware resources are allocated to resolve scheduling conflicts and to control buses.

Functional units in the data path are described in terms of a netlist of modules. The control is represented by symbolic boolean equations and special hardware modules which resolve scheduling conflicts. Additional hardware is also needed to interface the data path with the control path.

6.2 Important Features

Different pipeline latencies will result in designs which vary in size and performance. Piper applies the benchmark set to evaluate both the peak performance and the average performance of a pipelined processor for several latencies and selects the latency that is most cost and performance effective. Examples are described in section 11.

In a pipelined design, the dependency between consecutive instructions may prevent the processor from operating at its maximal instruction firing rate [7, 8]. These dependencies are also dependent upon the latency of the pipeline. The performance of a pipelined design may be improved by properly inserting NOPs between instructions or, preferably, reordering instructions to eliminate dependencies. Piper solves the dependency problem by creating priority modules to handle the physical signal conflicts and generating reorder information for the language compiler.

More details of benchmark evaluation and the derivation of micro-architecture dependent back-end compilers can be found in [9].

7 Control Path Synthesis

CPGen is the control path generator for both pipeline and non-pipeline designs. A data stationary control model [10] is used for pipeline designs. In this model the instruction opcode is transmitted through each stage of the pipeline in synchronization with the data. Control signals for each stage are generated by decoding the instruction which corresponds to that stage.

Rather than encoding the next state address of a state transition graph in the combinational logic, the address may be encoded with a counter triggered by the clock. In this manner, state changes as the counter is incremented. The number of counter cycles is determined by maximum number of clock cycles necessary to implement an instruction. This scheme is particularly useful for pipeline designs which use separate controllers for each pipeline stage, they may be synchronized by the sequence counter.

Pipelined designs provide a natural partitioning for large controllers. Unfortunately, a controller may still be too large to be implemented in a single PLA. Another level of logic partitioning is done to further subdivide the control logic such that a multiple PLA implementation is possible. Also, by partitioning a PLA into smaller components, a better chip floorplan is possible.

8 Data Path Synthesis

Given a register-level specification of functional units, Topolog produces a symbolic layout in Sticks in Prolog (SIP). This is achieved by reducing the functional units to gates which are defined by transistor netlists, placing the transistors onto a bit slice, and routing together signals which are electrically equivalent [11]. Figure 5 shows an overview of Topolog.

For each function supported by Piper, a module specification exists in Topolog which translates a RTL representation of the function into a gate-level description. In addition to defining the gate structure, the module specification also distributes gates into the appropriate bit-slices. The modules supported by ADAS include logic gates, tristates, ALUs, adder/subtractors, multiplier, divider, encoder/decoder, mux/demux, priority encoder, shifter, barrel shifter, comparators, registers, shift registers, clock divider, etc.

By representing gates in terms of transistor netlists rather than layout cells, greater layout density is possible by merging connected transistor chains between gates.

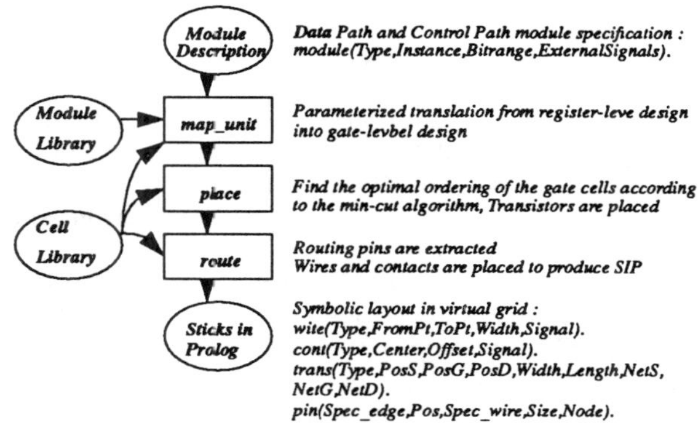

Figure 5: Topolog System

Transistor islands are placed in a bit slice to minimize the routing density. A placement tool based on the Min-Cut algorithm [12] has been implemented. The global-max gain pair is obtained by searching all possible pairs from partitioned groups. The local-max gain pair is obtained by searching the max-gain cell from each partitioned group. The performance of the algorithm has been improved by dynamically determining the pair of cells to interchange [13]. Consequently, the complexity of the placement algorithm improves from $O(N^2)$ to O(N) while still producing acceptable solutions. Layout quality is further refined by applying post-placement which is non-pair-wise-exchange after transistor islands have been placed by Min-Cut. Results are shown in Table 1.

In addition to the number of routing tracks, timing is also considered during placement. The critical paths through the datapath are identified from the benchmark programs and placed together to reduce delay.

IMPROVED MIN-CUT	MC	TC	CPU Time
Global Max.	15	1294	756s
Local Max.	24	1844	11s
Dynamic GL Max.	21	1670	96s

a) Modified Min-Cut on SM1 bit-5 (134 cells)

PLACEMENT	MC	TC
Min-cut Placement	21	1670
Post-Placement	17	1306

b) Post-Placement on SM1 bit-5 (134 cells)

GL: Global/Local, MC: Max Cut, TC: Total Cut
MC and TC are defined in [14].

Table 1: Improvement result

After placement, gate netlists are expanded into symbolic layout and routed. The signals within each bit slice are routed by a modified left edge first channel router [15] while control signals between adjacent bit slices are connected together by river routing [16]. During placement, location of feed-through blocks is placed to reduce the number of routing tracks necessary for river routing.

9 A Transistor Sizing

ADAS includes a transistor sizing program named TranSizor for optimizing the delay of symbolic layout components within an area constraint. Given a layout specification in a sticks language, TranSizor produces a symbolic layout specification with sized devices.

TranSizor uses high level information provided by Piper in order to determine the critical paths of a circuit. This is computationally less expensive and more accurate than deriving all possible circuit paths as many false paths, critical paths which do not occur, are not considered.

TranSizor uses PTA (Prolog Timing Analyzer) to calculate the time delay of each path. Using a lumped RC delay model [17], PTA produces symbolic equations for each delay path. Equations are differentiated symbolically to avoid expensive numerical computation. The resulting equations are solved to find the transistor sizes which yield the minimum delay. Two techniques, simulated annealing [18] and critical-path heuristic [19] are used to resize the transistors along each critical path.

10 Geometrical Level Synthesis

Given a collection of symbolic layout modules [20] created by Topolog and CPGen and modified by TranSizor, StickPack produces, a technology dependent mask layout description of the chip in CIF. StickPack contains a number of design tools: a technology-independent compactor, a power/ground bus size estimator, a joiner that joins together cells by pitchmatching and river routing, a global placer, and a global router and a channel router.

A technology independent split-grid compactor [21] is used to compact symbolic layout cells before they are placed and routed. Novel features such as automatic wire jog insertion, contact offsetting, and wire minimization are incorporated in the compactor. Hierarchical compaction is supported. The chip design is described in a hierarchical manner. Connecting pins between cells are identified. Constraints are then added during the compaction process to make sure that the abutting pins are in the same physical location after compaction. A hierarchical symbolic editor is developed to help the user the describe the design hierarchically.

Once the symbolic layout modules have been compacted, they are placed and routed to form the final chip, a constructive initial placement algorithm is used for the macromodule placer. Slicing structure [22] is maintained during the placement phase to facilitate the later channel definition and channel ordering in the routing stage. After all the blocks are placed, a global router which uses a depth first search algorithm to find the shortest path of connecting each net is called to assign nets to the channels. A detail router based on left edge greedy algorithm is called to route each channel in the order pre-determined by the slicing structure.

11 Synthesis Example

Figure 6, Figure 7, and Figure 8 show examples of two-stage pipelined SM1, six-stage pipelined SM1, and two-stage pipelined SM3 which are synthesized by ADAS, where SM stands for simple machine. SM1 has eight instructions with about 6K transistors. SM3 has forty instructions with about 10K.

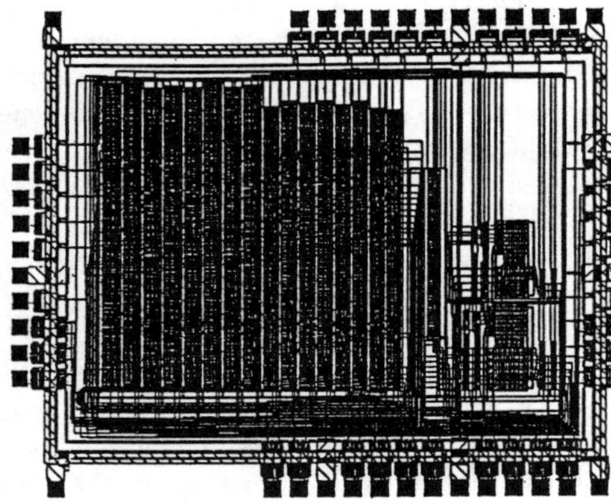

Figure 6: SM1 2 Pipeline-Stage Plot

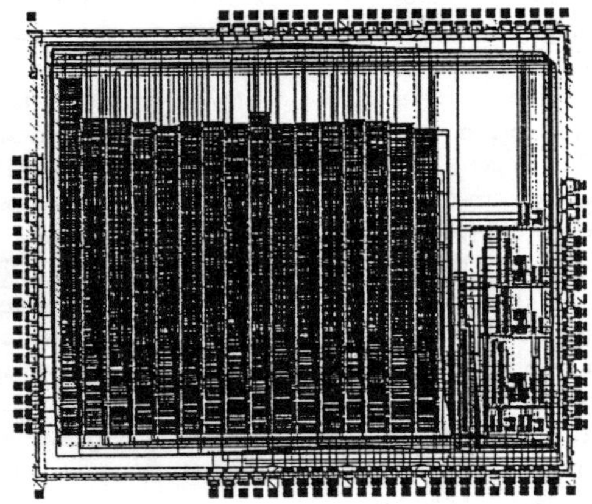

Figure 7: SM1 6 Pipeline-Stage Plot

Figure 9 shows the estimated hardware cost (transistor count), performances (peak and average), and performance/cost ratios for six different designs of SM1 produced by Piper. The pipeline configuration of each design is denoted by C/L, where C and L stand for the elapsed time of instruction and the latency respectively. We chose speed up as the performance metric, assuming the same system clock rate. A straight-line instruction benchmark with conditional branches is used to estimate average-case performance. With proper scheduling and allocation, it is possible to increase the throughput (instruction firing rate) without a significant increase of cost, and even at no extra cost. The design 6/1 has the highest peak-performance per unit cost, while designs 6/1 and 6/2 (which is a 3 stage pipeline with instruction latency of two clocks) have the highest average-performance per unit cost.

An implementation of the SM1 processor has been submitted to MOSIS for fabrication. ADAS has also successfully

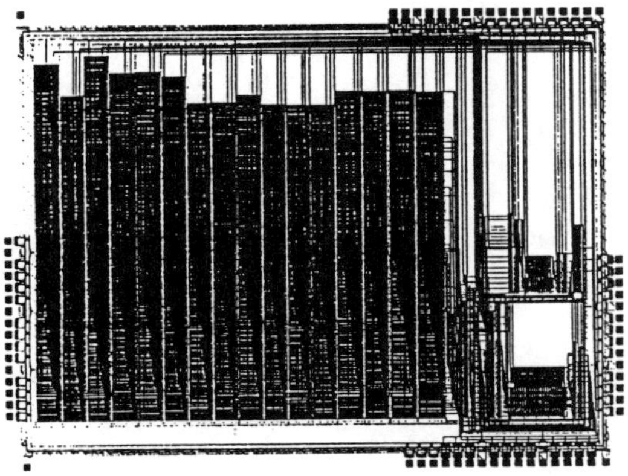

Figure 8: SM3 2 Pipeline-Stage Plot

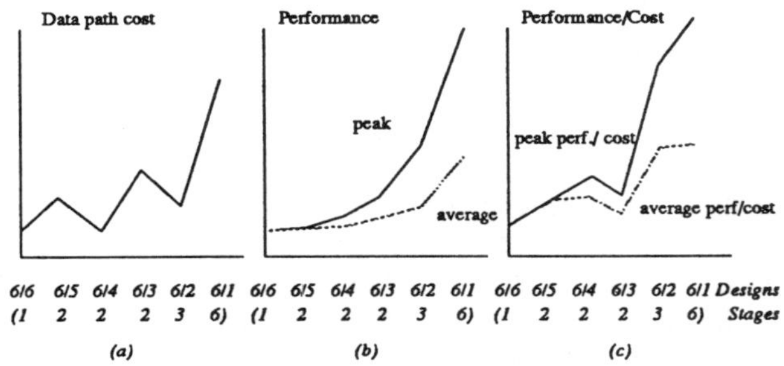

Figure 9: Cost/Performance

implemented the TDY-43, a 250 instruction processor which supports eight addressing modes. From an ISA specification, ADAS produced a gate-level VHDL netlist for implementation by gate array. The design also implements a JTAG standard scan testing protocol. The implementation produced by ADAS requires 34K gates or 380K transistors and has been verified at the gate-level using benchmark-driven test vectors.

12 Conclusion

In this paper, we have presented a full range microprocessor design automation system. It is demonstrated that by giving an instruction set, ADAS can automatically generate layout of a single chip microprocessor. By utilizing the benchmark simulation data, high level synthesis tools provide valuable information to guide the low level synthesis. Future research on ADAS includes further exploration of the interaction between different levels of design tools, capturing design knowledge for the redesign to meet constraints and extending the scope to multi-chip module (MCM) design.

Acknowledgements

We would like to thank Dave Boyer, Bill Bush, Angela Cheng, Rick McGeer, Dylan McNamee, Jonathan Pincus, Wes Reed, and Steve Schoetler for their contributions to this research and Paul A. Suhler for his contributions to this paper.

References

[1] Alvin M. Despain, "The Design System (ASP) of the Aquarius Project," *Proceedings of the Second International Workshop on VLSI Design*, pp. 149-166, Dec. 1988.

[2] Ralph Haygood, "A Prolog Benchmark Suite for Aquarius," *Report No. UCB/CSD 509*, 1989.

[3] Gino Cheng, et al., "A Full-Range Design Automation for Instruction Set Processors," *First International Conference on the Practical Application of PROLOG*, 1992.

[4] Jay R. Southard, "MacPitts: An Approach to Silicon Compilation," *IEEE Computer*, pp. 74-82, Dec. 1983.

[5] Howard Trickey, "Flamel: A High-Level Hardware Compiler," *IEEE Trans. CAD*, pp. 259-269, March 1987.

[6] Mario R. Barbacci, "Instruction Set Processor Specifiations (ISPS): The Notation and Its Applications," *IEEE Trans. Computer*, pp. 24-40, Jan. 1981.

[7] Nohbyung Park, et al., "Sehwa: A Program for Synthesis of Pipelines," *23rd IEEE Design Automation Conference*, pp. 454-460, March 1986.

[8] Michael C. McFarland, et al., "Tutorial on High-Level Synthesis," *25th IEEE Design Automation Conference*, pp. 330-336, Feb. 1988.

[9] Ing-Jer Huang and Alvin Despain, "High Level Synthesis of Pipelined Instruction Set Processors and Back-End Compilers," *29th IEEE Design Automation Conference*, June 1992.

[10] P.M. Kogge, "The Architecture of Pipelined Computers," *McGraw-Hill Book Company*, 1981.

[11] P. Kollaritsch and N. Weste, "A Rule-Based Symbolic Layout Expert," *VLSI Design*, pp. 62-66, Aug. 1984.

[12] B. W. Kernighan, et al., "An efficient heuristic procedure for partitioning graphs," *Bell Sys. Tech. J.*, pp. 291-308, Feb. 1970.

[13] C. M. Fiduccia and R. M. Mattheyses, "A Linear-Time Heuristic for Improving Network Partitions," *19th IEEE Design Automation Conference*, pp. 175-181, 1982

[14] Melvin A. Breuer, "Min-cut Placement, VLSI circuit layout theory and design," *IEEE press*.

[15] Takeshi Yoshimura, et al., "Efficient Algorithms for Channel Routing," *IEEE Trans. on CAD of Integrated Circuit and Systems*, Vol. CAD-1, No. 1, Jan. 1982.

[16] C. P. Hsu, "General River Routing Algorithm," *20th IEEE Design Automation Conference*, pp. 578-583, 1983.

[17] J. Ousterhout, "A Switch-Level Timing Verifier for Digital MOS VLSI," *IEEE Trans. CAD*, Vol. CAD-4, No. 3, July 1985.

[18] Jonathan D. Pincus and Alvin M. Despain, "Delay Reduction Using Simulated Annealing," *23rd IEEE Design Automation Conference*, pp. 1986.

[19] W. Kao, "Algorithms for Automatic Transistor Sizing in CMOS Digital Circuits," *22nd IEEE Design Automation Conference*, pp. 781-784, 1985.

[20] N. Weste, "Virtual Grid Symbolic Layout," *18th IEEE Design Automation Conference*, pp. 225-233, 1981.

[21] David G. Boyer, "Split Grid Compaction for a Virtual Grid Symbolic Design System," *ICCAD*, pp. 134-137, 1987.

[22] R. Otten, "Automatic Floorplan Design," *19th IEEE Design Automation Conference*, pp. 261-267, 1982.

Zero Skew Clock Net Routing

Ting-Hai Chao

Computer & Communication Research Laboratories
ITRI
Hsin-Chu, Taiwan 31015
R.O.C.

Yu-Chin Hsu

Department of Computer Science
University of California
Riverside, CA 92521
U.S.A.

Jan-Ming Ho

Institute of Information Science
Academia Sinica
Taipei, Taiwan 11529
R.O.C.

Abstract

In this paper, we present an algorithm, called Zero Skew Segment Tree Method(ZSTM), for the clock net routing problem. To eliminate the clock skew and minimize the total wire length, ZSTM recursively partitions the sink nodes into two subsets which have equal loadings and minimum sum of diameters, and then constructs a zero skew segment tree according to the partitioning result. The final layout of the clock net can be decided by the channel information of the routing region. Experiments show that ZSTM improves 15% in wire length and 3% in maximum delay over the published results. It also completely eliminates the clock skew.

1. Introduction

Circuit speed is a major consideration in high performance VLSI designs. In a synchronous VLSI design, data transfer between functional units is synchronized by the clock signal. The clock skew is the maximum difference among the delays from the clock source to sink nodes. The clock period C_p of a clock signal must satisfy the following inequality: $C_p \geq t_d + t_{skew} + t_{su} + t_{ds}$, where t_d is the delay on the longest path through combinational logic, t_{skew} is the clock skew, t_{su} is the set up time of the synchronizing elements, and t_{ds} is the propagation delay within the synchronizing elements. As VLSI feature sizes become smaller, t_d, t_{su}, and t_{ds} are decreased significantly. Therefore clock skew becomes a dominant factor in determining circuit performance.

Several clock distribution schemes have been proposed to avoid the clock skew, including buffer optimization[1,2], clock assignment[3,4] and clock net routing [5,6,7,8]. Among the proposed clock net routing schemes, the H-tree[5] is effective if the clock pins are distributed on a $2^k \times 2^k$ array. The Method of Means and Medians(MMM)[6] algorithm is a generalization of H-tree. It splits a set of sink nodes into two subsets with equal cardinality, routes from the center of the set into the centers of the two subsets, and ensures an exact balance and no length skew at the current level of the clock tree. The recursive matching-based clock tree algorithm[7] eliminates length skew while using 5%-7% less wire length than those by MMM. These algorithms, however, focus only on local wire length balancing rather than real clock skew which includes the effects of the connection lines and loadings. Tsay proposed an algorithm[8] to pursue an exact zero skew clock routing in which RC tree is used as the delay model. However, the optimization of total wire length was not considered in that paper.

In this paper, we propose a zero skew clock net routing algorithm. We adopt the same zero skew merging scheme in [8]. However, to minimize the total wire length, a balanced partition and a segment tree method are introduced. The remainder of this paper is organized as follows. We define the clock net routing problem in Section 2. Section 3 contains the analysis of the problem and the theoretical foundation of our work. The proposed algorithm is described in Section 4. The experimental results are shown in Section 5. Finally, concluding remarks are made in Section 6.

2. Problem Definition

Physical design of VLSI consists of three steps: placement, global routing and detail routing. After placement, the positions of clock pins, which include a source node and a number of sink nodes, are determined. During the global routing stage, the clock nets and power/ground nets are considered first before the other signals. Usually, there are two signal routing layers: one for horizontal segments and the other for vertical segments. With multi-metal processing technologies, we assume that the parameters of interconnection lines are the same for both routing layers and the resistances of vias are neglected. So the wire length is used to measure the RC parameters of connections.

Let $N = \{s_0\} \cup S$ denote a clock net, where s_0 is the source node and $S = \{s_1, s_2, \ldots, s_n\}$ is the set of sink nodes. A rooted rectilinear Steiner tree $T(N)$ of N is called a *clock tree* of N if s_0 is the root and S is the set of leaf nodes. The wire length of *edge* $e(v_i, v_j)$ is denoted by $|e(v_i, v_j)|$. The total length of the edges of $T(N)$ is the *cost* of the tree. For each sink node s_i, let $t_d(s_0, s_i)$ be the delay time from the root s_0 to s_i. The *skew* of $T(N)$ is the maximum of $|t_d(s_0, s_i) - t_d(s_0, s_j)|$ among all s_i and s_j in S. A clock tree is called a *zero skew clock tree*(ZST) if its skew is zero. Now, we define the clock net routing problem as follows:

Given a clock net N, construct a zero skew clock tree T(N) with minimal cost.

3. Preliminaries

3.1. Models

3.1.1. Circuit Model

Let $e(v_i, v_j)$ be an edge of clock tree $T(N)$. If there is a path from the root to v_i and then v_j, we say that v_i is the *father* of v_j and v_j is the *son* of v_i. Since the father of a vertex is unique in a clock tree, we use e_{v_j} to represent the edge connecting v_j to its father. Let α and β be the resistance and the capacitance per unit length, respectively. The resistance $r_{e_{v_j}}$ and capacitance $c_{e_{v_j}}$ of e_{v_j} are calculated by $\alpha \mid e_{v_j} \mid$ and $\beta \mid e_{v_j} \mid$, respectively. Here, we use a π-model [9] to approximate the equivalent circuit of the distributed RC line. For each sink node s_i of $T(N)$, there is a loading capacitance, c_{L_i}, which is the input capacitance of the function unit driven by the s_i. For example, Fig. 1(a) is a clock net tree, Fig. 1(b) shows its equivalent RC tree, and Fig. 1(c) shows the equivalent circuit.

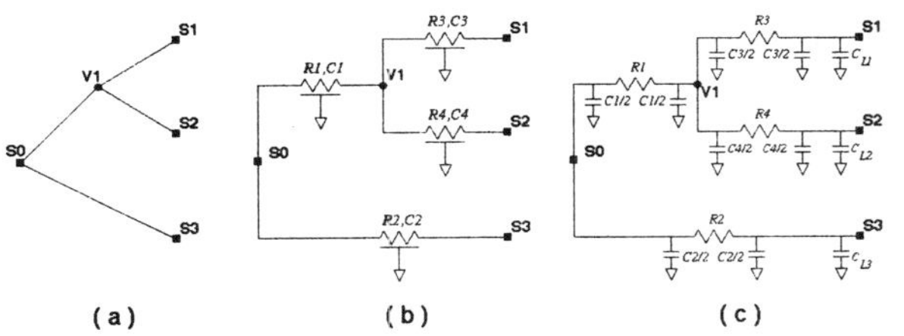

Fig. 1 (a) A clock tree (b) The RC tree model (c) The equivalent circuit model

Let c_{v_i} denote the node capacitance of node v_i and $Son(v_i)$ denote the set of sons of v_i. Then we have

$$c_{v_i} = \begin{cases} \frac{1}{2}c_{e_{v_i}} + \frac{1}{2} \sum_{v_j \in Son(v_i)} c_{e_{v_j}} & \text{if } v_i \text{ is an internal node} \\ \frac{1}{2}c_{e_{v_i}} + c_{L_i} & \text{if } v_i \text{ is a sink node.} \end{cases}$$

Let T_{v_i} denote the subtree of $T(N)$ using v_i as its root node and C_{v_i} denote the *tree capacitance* of T_{v_i} which is equal to the total node capacitance of T_{v_i}. Since C_{v_i} can be calculated from its own node capacitance and the summation of the capacitance of its sons, so we have the following recursive formula:

$$C_{v_i} = c_{v_i} + \sum_{v_j \in Son(v_i)} C_{v_j}.$$

3.1.2. Delay Model

Let $path(s_0, s_i)$ denote the set of edges in the path from source s_0 to a sink node s_i. According to [9,10,11], the delay time $t_d(s_0, s_i)$ can be calculated by the following formula:

$$t_d(s_0, s_i) = \sum_{e_{v_i} \in path(s_0, s_i)} r_{e_{v_i}} C_{v_i}.$$

So the delay time $t_d(v_i, v_j)$ between any two vertices v_i and v_j in a path from s_0 to s_i can be defined as:

$$t_d(v_i, v_j) = \sum_{e_{v_k} \in path(v_i, v_j)} r_{e_{v_k}} C_{v_k}.$$

If v_k is a vertex in the path from v_i to v_j, then $t_d(v_i, v_j) = t_d(v_i, v_k) + t_d(v_k, v_j)$. In a special case, if v_k is a son of v_i and v_j is a sink node s_i, then

$$t_d(v_i, s_i) = r_{e_{v_k}} C_{v_k} + t_d(v_k, s_i). \qquad (1)$$

In formula (1), C_{v_k} is the tree capacitance of T_{v_k} and $t_d(v_k, s_i)$ is the delay time from v_k to s_i. If T_{v_k} is a ZST, then for all the sink nodes s_i of T_{v_k}, $t_d(v_k, s_i)$ is a constant. So T_{v_k} can be modeled as a two port element. One port is the root node v_k with a node capacitance equal to C_{v_k} and the other is a leaf node s_i. The only edge between two ports is a delay element with delay $t_d(v_k, s_i)$. Consequently, a sink node, s_i, can be treated as a ZST with capacitance c_{L_i} and delay zero.

3.2. Bottom-Up Merging Scheme

3.2.1. Merging Two Zero Skew Trees

Let T_{v_1} and T_{v_2} be two ZST's with capacitance C_1 and C_2, respectively, and delay t_1 and t_2, respectively. If there is a point v_3, and two edges $e(v_3, v_1)$ and $e(v_3, v_2)$ such that $t_d(v_3, v_1) + t_1 = t_d(v_3, v_2) + t_2$, then v_3 is called a *zero skew merging point* of T_{v_1} and T_{v_2}, and e_{v_1} and e_{v_2} are called the *merging edges*. The tree T_{v_3}, rooted at v_3, connecting T_{v_1} and T_{v_2} by the two edges $e(v_3, v_1)$ and $e(v_3, v_2)$ to v_3 is obviously a ZST. Since $cost(T_{v_3})$ equals $cost(T_{v_1}) + cost(T_{v_2}) + \mid e_{v_1} \mid + \mid e_{v_2} \mid$, $\mid e_{v_1} \mid + \mid e_{v_2} \mid$ is called the *merging cost* of T_{v_1} and T_{v_2}.

From $t_d(v_3, v_1) + t_1 = t_d(v_3, v_2) + t_2$ and formula (1), a zero skew merging point v_3 satisfies the following equation:

$$r_{e_{v_1}}(\tfrac{1}{2}c_{e_{v_1}} + C_1) + t_1 = r_{e_{v_2}}(\tfrac{1}{2}c_{e_{v_2}} + C_2) + t_2. \qquad (2)$$

Let $d_r(v_1, v_2)$ denote the *rectilinear distance* between v_1 and v_2 and $d_r(v_1, v_2) = \kappa$. The *boundaries* of v_1 and v_2 is the minimum enclosing box using v_1 and v_2 as its opposite corners. The $d_r(v_3, v_1)$ is called *the merging distance*. Suppose there is a merging point v_3 located within the boundaries of v_1 and v_2 with the merging distance x, and $\mid e_{v_1} \mid = x$ and $\mid e_{v_2} \mid = \kappa - x$. Then the resistances for $r_{e_{v_1}}$ and $r_{e_{v_2}}$ are αx and $\alpha(\kappa - x)$, respectively. Also, the capacitances for $c_{e_{v_1}}$ and $c_{e_{v_2}}$ are βx and

$\beta(\kappa-x)$, respectively. By substituting these equalities into Eq. 2 and solving the equation, we have

$$x=\frac{t_2-t_1+\alpha\kappa(C_2+\frac{1}{2}\beta\kappa)}{\alpha(C_1+C_2+\beta\kappa)} \qquad (3)$$

case 1: If $0\leq x\leq\kappa$, then there exists a zero skew merging point of T_{v_1} and T_{v_2} located within the boundaries of v_1 and v_2. The merging distance is solved by formula (3). The merging cost is κ.

case 2: If $x<0$ or $x>\kappa$, then the solution contradicts the assumption that v_3 is within the boundaries of v_1 and v_2. In this case, an extended distance κ' is required to balance the difference of the delay of the two trees. If $x<0$, which means $t_1>t_2$, we can choose v_1 as the zero skew merging point and assume $|e_{v_1}|=0$ and $|e_{v_2}|=\kappa'$. By Eq. 2, we have $t_1=\alpha\kappa'(\frac{1}{2}\beta\kappa'+C_2)+t_2$, hence

$$\kappa'=\frac{((\alpha C_2)^2+2\alpha\beta(t_1-t_2))^{1/2}-\alpha C_2}{\alpha\beta}.$$

Similarly, if $x>\kappa$, we choose v_2 as the zero skew merging point, and we have

$$\kappa'=\frac{((\alpha C_1)^2+2\alpha\beta(t_2-t_1))^{1/2}-\alpha C_1}{\alpha\beta}.$$

Clearly, κ' is greater than κ. To route a wire of specified length κ' between two points at a distance κ, the method of specified-length routing can be used [12]. In this case, the merging distance is 0 (κ), the merging point is v_1 (v_2) if $x<0$ (if $x>\kappa$) and the merging cost is κ'.

From the above analysis, a zero skew merging point is always solvable with merging distance between 0 and κ. This method is called *the zero skew merging scheme*. The merging cost depends on both the distance of two roots of ZSTs and the total node capacitance of two ZSTs. To minimize the merging cost, we should choose the ZSTs which have the minimum distance between roots and have equal tree capacitance to avoid an extra cost $\kappa'-\kappa$.

3.2.2. Tree of Segments

In Manhattan space, the collection of all points with fixed distance from a point is called a *Manhattan circle*, the shape of Manhattan circle is a square rotated by 45 degree. For example, Fig. 2(a) shows a Manhattan circle with radius 3. A line segment of a Manhattan circle is called a *Manhattan arc* which is a segment with a slope of +1 or -1.

Let a *merging segment* be a collection of all the zero skew merging points of two sink nodes, then the segment is the intersection of two Manhattan circles and is a Manhattan arc. For example, in Fig. 2(b), $\kappa=d_r(p_1,p_2)=8$ and $x=4$. The segment represented by the real line in the figure is the merging segment. Fig. 2(c) shows a merging segment with the slope of -1. Since each point on the merging segment can be used as the root of the merged ZST, the selection of the exact position of the merging point of the merged tree affects the cost charged to future mergings steps during the bottom-up merging process. In order to minimize the cost of a clock tree, an appropriate selection strategy is desired.

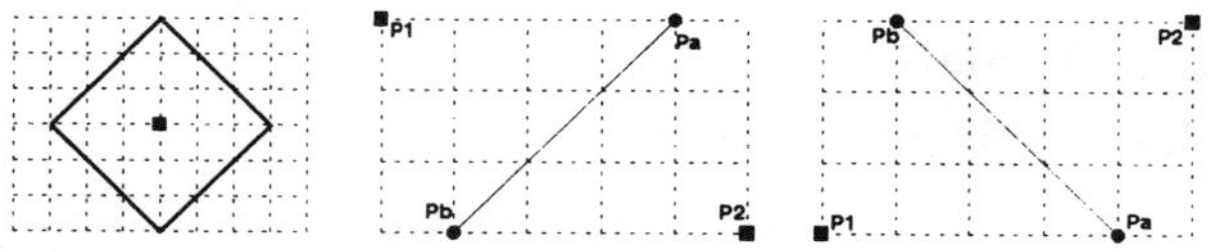

(a) (b) (c)

Fig. 2 (a) A Manhanttan circle (b) A merging segment with slope=1 (c) A merging segment with slope=−1

Let S_{p_1,p_2} denote a merging segment of two ZST's with p_1 and p_2 as its two end points. The *segment distance*, $d(S_{p_1,p_2},S_{p_3,p_4})$, between S_{p_1,p_2} and S_{p_3,p_4} is defined as $\min\{d_r(v_a,v_b) \mid v_a\in S_{p_1,p_2}, v_b\in S_{p_3,p_4}\}$. If $d_r(v_a,v_b)=d(S_{p_1,p_2},S_{p_3,p_4})$, then v_a and v_b are called a *nearest pair* of S_{p_1,p_2} and S_{p_3,p_4}. To merge two ZST's which are each rooted at a merging point on a merging segment, the selection of the nearest pair as the roots of two respective subtrees will result in a tree with the minimal cost.

Lemma 1: If two merging segments S_{p_1,p_2} and S_{p_3,p_4} do not intersect, then there exist a nearest pair $p_a\in S_{p_1,p_2}$ and $p_b\in S_{p_3,p_4}$, such that at least one of the points p_a or p_b is an end point of S_{p_1,p_2} or S_{p_3,p_4}.

According to Lemma 1, $d(S_{p_1,p_2},S_{p_3,p_4})$, equals to the minimum of $d(p_1,S_{p_3,p_4})$, $d(p_2,S_{p_3,p_4})$, $d(p_3,S_{p_1,p_2})$, and $d(p_4,S_{p_1,p_2})$, where $d(p_i,S_{p_j,p_k})$ is the minimum distance between point p_i and the segment S_{p_j,p_k}, which can be easily computed.

Lemma 2: The collection of all merging segments of nearest pairs of two merging segments is also a Manhattan arc.

According to Lemma 2, the merging segment of two merging segments is also a Manhattan arc. The two end points of the Manhattan arc can be determined as follows. First, we evaluate the four combinations of $d(p_i,S_{p_j,p_k})$, where p_i is an end point of one segment and S_{p_j,p_k} is the other segment. We then record the minimal distance as the segment distance and all the pairs of points achieving the minimum distance. For each pair of points, we compute the merging segment of the pair and put its two end points into a set P. The process is repeated until all pairs are processed. The points in P are then sorted according to their x coordinates. The points with the minimum x-coordinate and maximum x-coordinate are the two end points. For example, in Fig. 3(a), $d(S_{p_1,p_2},S_{p_3,p_4})=10=d_r(p_1,p_3)=d_r(p_1,p_4)=d_r(p_2,p_3)=d_r(p_2,p_4)$. Suppose $x=5$, from the pairs (p_1,p_3), (p_1,p_4), (p_2,p_3) and (p_2,p_4), six points p_c, p_d, p_a, p_b, p_e and p_f are computed and stored in P. Points p_a and p_b are the two end points of merging segment of S_{p_1,p_2} and S_{p_3,p_4}. Fig. 3(b) and Fig. 3(c) show other cases of the merging segments.

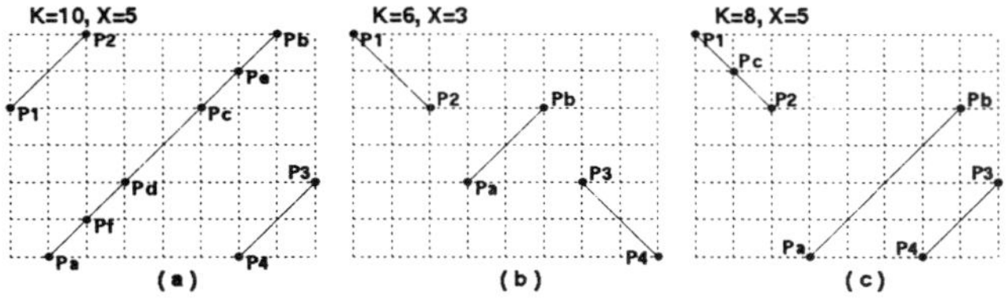

Fig. 3 The merging segment of S_{p_1,p_2} and S_{p_3,p_4}

When extended distance is required, the merging segment can be similarly determined as above with the following adjustment. For example, in Fig. 3(c), if κ=8 and $|e_{v_2}|$=κ′=10, after processing all the nearest pairs, the merging segment is a point p_2. Since the extended distance is 10, all points in S_{p_1,p_2} with segment distance less than 10 can be a merging point, so the merging segment is extended to p_c where $d(p_c, S_{p_3,p_4})$=10, so the final merging segment is $\overline{p_c p_2}$ as shown in Fig. 3(c).

A binary tree is called a *tree of segments* if it consists of merging segments; i.e. each vertex is the merging segment of its two sons. Given a set of points S, denote $T(S)$ as a tree of segments with S as its leaf nodes. Let p_0 be a point on the root of $T(S)$ and segments S_1 and S_2 be the two sons of the root. According to the definition, p_0 is a zero skew merging point of S_1 and S_2. So there is a corresponding pair of points p_1 and p_2 on S_1 and S_2, respectively, such that p_0 is a zero skew merging point of T_{p_1} and T_{p_2}. By connecting p_1 and p_2 as the sons of p_0, and recursively identifying the corresponding pair of points for the two subtrees until the leaf nodes are reached, a *routing tree* $RT(S)$ of S is defined. Clearly a routing tree $RT(S)$ is a ZST. Since there may exist several corresponding pairs for a root, the routing tree is not unique. For example, Fig. 4(a) shows a tree of segments. Fig. 4(b) and Fig. 4(c) are two different routing trees with the same cost of 438.

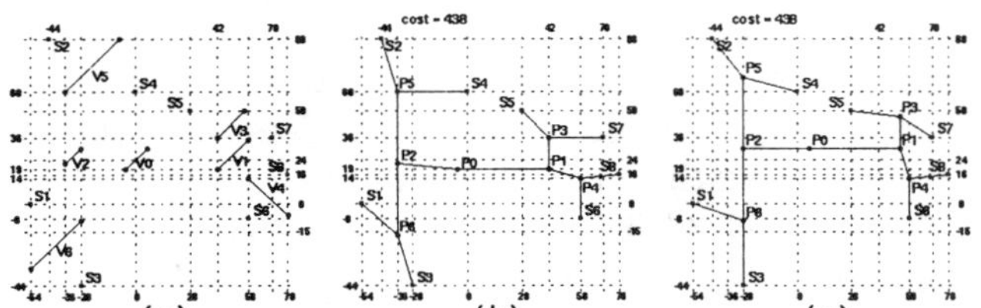

Fig. 4 (a) A tree of segments (b) A routing tree of (a)
(c) Another routing tree of (a)

In a tree of segments, the merging cost is minimized by using the nearest pair as the roots of two sub-ZST's. Since the merging segment consists of all the zero skew merging points, it implies that the total cost of the tree is minimal among all its topologically equivalent trees. Since the selection of the roots of the merged tree is made later during the routing tree construction stage, this method is called *delayed merging method*.

3.3. Top-Down Partition

Given a set S of sink nodes and their node capacitance c_{L_i}, a bi-partition is called a *balanced partition* if the difference of the total node capacitance of the two subgroups is at most $\max\{c_{L_i}\}$. Suppose S is recursively balanced-partitioned until each subgroup contains only one node, then a tree of segments may be generated according the result of the partitioning. As one can observe in subsection 3.2.2, extra length of routing is unavoidable when two component subtrees are unbalanced. By imposing the balanced condition in the partition phase, the necessity of extra length is minimized. In order to minimize the cost of a routing tree, the balanced partition must be a *clear cut* and have the least *partition cost*. A clear cut refers to the existence of a simple curve such that the two subgroups of points lies on either side of the curve.

Define the *diameter*, $dia(S)$, of S as $\max\{d_r(p_i, p_j) \mid p_i, p_j \in S\}$. Since the cost of any routing tree of S is greater than the $dia(S)$ and less than $|S| \times dia(S)$, so the cost of a routing tree is bounded by a function of the diameter. This suggests that the diameter can be used to estimate the cost of a routing tree.

Suppose a balanced partition P divides a set G into G_1 and G_2. We define the *partition cost* of P as $dia(G_1)+dia(G_2)$. A balanced partition is optimal if it has the smallest partition cost.

Let G be a set of points, the *octagon* of G is the region enclosed by the following eight lines x=$\min\{p_i.x \mid p_i \in G\}$, y−x=$\min\{p_i.y-p_i.x \mid p_i \in G\}$, y=$\max\{p_i.y \mid p_i \in G\}$, y+x=$\max\{p_i.y+p_i.x \mid p_i \in G\}$, x=$\max\{p_i.x \mid p_i \in G\}$, y−x=$\max\{p_i.y-p_i.x \mid p_i \in G\}$, y=$\min\{p_i.y \mid p_i \in G\}$, y+x=$\min\{p_i.y+p_i.x \mid p_i \in G\}$. Also, let the *octagon set*, $Oct(G)$, of G be the set of points which belong to G and lie on the octagon.

Fig. 5(a) shows the octagon of 16 points and $\{s_1, s_3, s_5, s_{10}, s_{14}, s_{16}\}$ as the octagon set. The points in the octagon set have a natural circular ordering along the octagon lines. For example, s_1-s_5-s_{10}-s_{16}-s_{14}-s_3-s_1 is the circular order of the octagon set of Fig. 5(a). By pigeon-hole principle, one of the subgroups of an optimal partition must contain $\frac{1}{2}|Oct(G)|$ consecutive points of $Oct(G)$. Based on this property, a heuristic balanced partition is design as follows: First, compute $Oct(G)$ and sort $Oct(G)$ in circular order. Select $\frac{1}{2}|Oct(G)|$ consecutive points as the reference set, denoted by REF. For each point p_i in G, define the weight of p_i to be $\min\{d_r(p_i, p_{oi}) \mid p_{oi} \in REF\}$ +

$\max\{d_r(p_i, p_{oi}) \mid p_{oi} \in REF\}$. Sort the points in ascending order by their weights, and move points according the sorted order into a group G_1 until the difference between the sum of capacitances in G_1 and one half the total capacitance is minimized. The remaining points are collected into G_2. The partition cost dia(G_1)+dia(G_2) is then obtained. After all consecutive combinations of reference set are calculated, the partition with the least cost is selected. For example, there are 6 octagon points in Fig. 5(a). Three consecutive octagon points are picked as reference set. Let $REF_1=\{s_5, s_{10}, s_{16}\}$. The partition cost of REF_1 is 280 as shown in Fig. 5(b). Let $REF_2=\{s_{10}, s_{16}, s_{14}\}$. The partition cost of REF_2 is 270 as shown in Fig. 5(c). After 6 different reference sets are evaluated, the optimal one is REF_2 which has a cost 270.

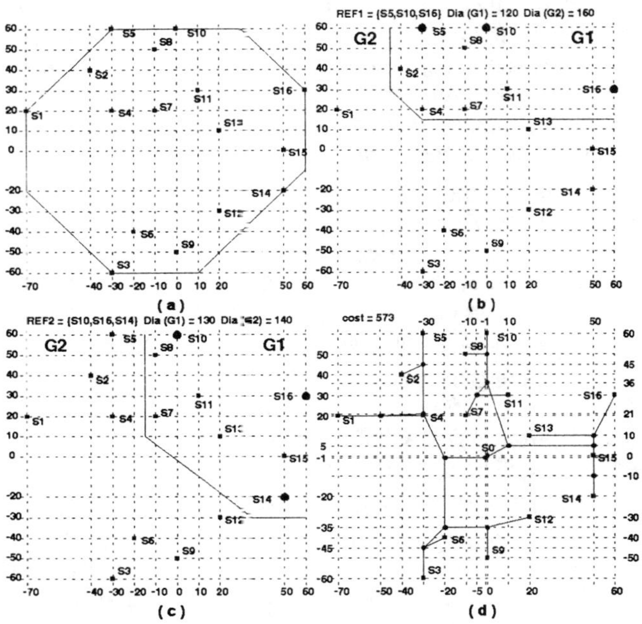

Fig. 5(a)Octagon lines of 16 points (b)Partition result of REF_1 (c)Partition result of REF_2 (d)The routing tree of ZSTM

4. Algorithm

Our algorithm is called the Zero Skew Segment Tree Method(ZSTM). The following is a pseudo-code description of the algorithm.

Clock_Net_Routing_Algorithm
begin
For i=1 to n
  assign s_i to a ZST, T_i;
G = { $T_1, T_2, \ldots, T_r$ }
SinkTree = ZSTM(G);
ClockTree = Connect(s_0, SinkTree);
Select_Layout(ClockTree);
end.

Procedure ZSTM(G);
begin
  if |G| = 1 **return**(T_1);
  else { **if** |G| = 2 **return**(Merge_ZST(T_1, T_2));
      else { partition G into G_1 and G_2;
          R_ZST = ZSTM(G_1);
          L_ZST = ZSTM(G_2);
          return(Merge_ZST(R_ZST,L_ZST));
        }
    }
end.

ZSTM uses the balanced partition algorithm described in the Section 3.3 to separate G into two subgroups G_1 and G_2, and recursively calls ZSTM to generate a tree of segments. The procedure Merge_ZST(T_i, T_j) returns a ZST with a merging segment computed from the merging segments of two roots of T_i and T_j as the root and T_i and T_j as the sons of the root. Procedure Connect(s_0, SinkTree) calculates the segment distance $d(s_0, S_{p_a,p_b})$, where S_{p_a,p_b} is the root of the SinkTree, and determines a subsegment $S_{p_{a'},p_{b'}}$ of S_{p_a,p_b} such that each point p_0 in $S_{p_{a'},p_{b'}}$ has $d_r(s_0, p_0) = d(s_0, S_{p_a,p_b})$. As described in the last paragraph of Section 3.1, for each point p_0 in $S_{p_{a'},p_{b'}}$, there are many routing trees with the same cost. Procedure Select_Layout(ClockTree) selects one of the routing trees and decides its final layout according the channel information of the routing region. The tree of segments is very useful in the detailed routing stage. Based on its descriptions and the channel information, we can reroute a clock net without having to change its wire length and clock skew. Fig. 5(d) shows one of the routing tree of Fig 5(a).

5. Experimental Results

We have implemented the clock net routing algorithm in C on a SUN SPARC workstation. The ZSTM algorithm has been tested extensively on a number of random point sets in (−2500,2500)×(−2500,2500) and source node at (0,0). The cardinality of sink nodes ranges from 8 to 64. We generate 20 examples for each cardinality. Clock net trees are constructed for each example. For comparison purposes, we have implemented MMM algorithm as is described in [6]. We also compare our result with approximated rectilinear Steiner trees(RST) which are constructed by the TR algorithm[13]. For all the cases, the cost (length) of a tree is calculated. Table 1 shows the average cost of three approaches. As shown in the table, RST has the least cost, but it is not suitable for clock net routing because it does not consider clock skew problem. The average cost of ZSTM is 85% of that by MMM.

We also use circuit simulator SPICE[14] to evaluate clock skew. The SPICE files of the clock net trees are generated for both MMM and ZSTM result. The routing area is assumed to be 0.5cm×0.5cm, and all the parameters are based on a 1.2μm CMOS technology. The input clock frequency is 100MHz and a superbuffer is driven by the source node. The delays between the source and the sink nodes are measured at the output node of the inverter which drives the sink nodes. Table 2 shows the average maximal delays, minimal delays and clock skews. The average maximum delay of ZSTM is 97% of that of MMM. The average skew of MMM is 9.2 pico-second while ZSTM is 0.5 pico-second. We have also tested our algorithm on the benchmark introduced in [8]. The experimental result is illustrated in table 3.

6. Conclusion

In this paper, we present ZSTM, a new algorithm for clock net routing. It is based on balanced partition and a zero skew merging scheme. The minimum sum of diameter is used as the cost function in the balanced partition to reduce the total wire length. The delayed merging method constructs a tree of segments which keeps all possible layouts of the clock tree with the same cost. The final layout of the clock tree can be determined in later stages of routing to reduce the routing area. Experiments show that ZSTM has 15% and 3% reductions of the cost and the maximum delay respectively over the MMM approach and it almost eliminates the clock skew completely. As compared to the result of Tsay, a reduction of 10% in cost is observed.

Acknowledgement

The authors would like to thank Dr. Ren-Song Tsay for providing the valuable benchmark data.

REFERENCES

[1] P. Ramanathan and Kang G. Shin, "A Clock Distribution Scheme for Non-Symmetric VLSI Circuits," *IEEE Int. Conference on Computer-Aided Design (ICCAD-89)*, pp. 398-401, Nov. 1989.

[2] S. Boon, S. Butler, R. Byrne, B. Setering, M. Casalanda and Al Scherf, "High Performance Clock Distribution For CMOS ASICS," *IEEE Custom Integrated Circuits Conference*, pp. 15.4.1-15.4.4, 1989.

[3] John P. Fishburn, "Clock Skew Optimization," *IEEE Transactions on Computers, vol. 39. No 7, pp. 945-951, July 1990.*

[4] Masato Edahiro, "A Clock Net Reassignment Algorithm Using Voronoi Diagram," *IEEE Int. Conference on Computer-Aided Design (ICCAD-90)*, pp. 420-423, Nov. 1990.

[5] D. F. Wann and M. A. Franklin, "Asynchronous and Clocked Control Structures for VLSI Based Interconnection Networks," *IEEE Transactions on Computers*, vol. c-32, No. 3, pp 284-293, March 1983.

[6] Michael A. B. Jackson, Arvind Srinivasan and E. S. Kuh, "Clock Routing for High Performance ICs," *27th ACM/IEEE Design Automation Conference*, pp. 573-579, 1990.

[7] Andrew Hahng, Jason Cong, and Gabriel Robins, "High-Performance Clock Routing Based on Recursive Geometric Matching," *28th ACM/IEEE Design Automation Conference*, pp. 322-327, 1991.

[8] Ren-Song Tsay, "Exact Zero Skew," *IEEE Int. Conference on Computer-Aided Design (ICCAD-91)*, pp. 336-339, 1991. Nov. 1991.

[9] Takayasu Sakurai, "Approximation of Wiring Delay in MOSFET LSI," *IEEE Journal of Solid-State Circuits*, vol sc-18, No. 4 pp. 418-426, August 1983

[10] J. Rubinstein, P. Penfield, and M. A. Horowitz, "Signal Delay in RC Tree Networks," *IEEE Transactions on Computer-Aided Design*, vol . CAD-2, No. 3, pp. 202-211, July 1983.

[11] W. C. Elmore, "The transient response of damped linear networks with particular regard to wide-band amplifiers," *Journal of Applied Physics*, vol.19, no.1, pp. 55-63, Jan. 1948.

[12] Akihiko Hanafusa, Yasuhiro Yamashita and Mitsuru Yasuda, "Three-Dimensional Routing for Multilayer Ceramic Printed Circuit Boards," *27th ACM/IEEE Design Automation Conference*, pp. 386-389, 1990.

[13] Ting-Hai Chao and Yu-Chin Hsu, "Rectilinear Steiner Tree Construction by Local and Global Refinement," *IEEE Int. Conference on Computer-Aided Design (ICCAD-90)*, pp. 432-435, Nov. 1990.

[14] L. Nagel, "SPICE2: A Computer Program to Simulate Semiconductor Circuits," *ERL Memo. No. UCB/ERL M75/520*, May 1975.

#Pts	RST	MMM	ZSTM	Time(sec)	ZSTM / MMM
8	11043	18286	15087	0.0054	0.83
16	15189	26513	23381	0.0400	0.88
24	18355	34821	29342	0.0792	0.84
32	21972	42573	36873	0.1392	0.87
40	24112	48940	40719	0.2050	0.83
48	26569	54862	45765	0.2825	0.83
56	28158	57964	48178	0.4083	0.83
64	30269	62964	54041	0.5083	0.86
average					0.85

TABLE 1. Mean Ratios of Cost.

#Pts	MMM			ZSTM			ZSTM / MMM	
	delay		clock	delay		clock	delay	
	max	min	skew	max	min	skew	max	min
8	769.3	763.2	6.1	746.6	746.2	0.4	0.970	0.978
16	801.8	797.0	4.8	783.2	782.5	0.7	0.977	0.982
24	836.6	826.2	10.4	808.7	808.3	0.4	0.967	0.978
32	863.5	855.6	7.9	837.3	836.5	0.8	0.970	0.978
40	885.6	876.3	9.3	857.0	856.5	0.5	0.968	0.977
48	908.9	896.4	12.5	876.8	876.3	0.5	0.965	0.978
56	926.2	914.4	11.8	890.2	889.7	0.5	0.961	0.973
64	940.6	930.1	10.5	910.7	910.2	0.5	0.968	0.979
ave	866.6	857.4	9.2	838.8	838.3	0.5	0.968	0.978

time unit : pico-second

TABLE 2. Mean of Delay Time and Clock Skew

case	#Pins	MMM	Tsay	ZSTM	ZSTM / MMM	ZSTM / Tsay
r1	267	1815022	1697021	1534786	84.6%	90.4%
r2	598	3624740	3431844	3065288	84.6%	89.3%
r3	862	4642628	4406914	3961817	85.3%	89.9%
r4	1903	9375972	8865482	8054162	85.9%	90.8%
r5	3101	13805340	13197672	11837293	85.7%	89.6%

TABLE 3. The cost of benchmark.

Power and Ground Network Topology Optimization for Cell Based VLSIs

Takashi Mitsuhashi* and Ernest S. Kuh
Electronics Research Laboratory
University of California, Berkeley, CA 94720

Abstract

A new power and ground network design problem for cell-based VLSIs is discussed. In contrast to the conventional method, the network topology is optimized, or wiring resource consumption subject to electromigraton and voltage drop constraints is minimized. The results obtained show that this problem is a critical issue in the coming VLSIs and the proposed method will be a good solution.

1 Introduction

A new power and ground routing method for cell-based VLSIs is presented that minimizes the wiring resource consumption under the constraints of current density and voltage drop. In contrast to conventional methods, the network topology, not the wire width of given networks, is optimized because it has dominant effect on the resource consumption[8].

Undesirable wear-out of metal wiring caused by electromigration[1], and narrowing margins caused by voltage drops are two major problems in power and ground network design. To reduce the possibility of wear-out, maximum current density constraints are enforced on power and ground distribution networks. Although some of the above mentioned problems can be solved by using sufficient wiring width, the wiring resources are still too expensive to use freely. The minimization of wiring resource consumption, before the signal routing is very important for complex VLSI designs.

Several studies on power and ground routing have been published [2-8]. Many of those papers assume single metal layer technologies and building block layout styles that consist of well characterized large functional blocks and inter-block wiring. Conventionally the problem has been treated as a two step decision process, namely finding non-crossing power and ground topology on a single metal layer[2-5], and the wiring area minimization of a given topology under voltage drop and current density constraints[6-7]. Because of recent progress, the above mentioned assumptions are becoming unsuitable for many current VLSIs. Although the introduction of double and triple metal layer wiring technologies has relatively weakened the demand for single metal layer routing method, the wiring resource is still expensive. The conventional methods are difficult to apply to modern cell-based designs like Sea of Gates and Standard Cells. This inconvenience comes from the rigid building block concepts used to characterize power consumption and terminal information. For cell-based VLSI the power consumption on a chip should be characterized in a distributed manner.

In the proposed approach, the VLSI chip area is divided into a set of non-overlapping rectangular elements. Given the power consumption of each rectangular, the buses are initially modeled by a mesh net topology that supplies current to each rectangular element. Power consumption of the circuit can be estimated by some tools[9], the only thing we need to do for the model is mapping each gate power consumption to the rectangular element covering the chip. This model accurately represents the structure of power and ground buses for cell-based VLSIs. The current flow is controlled by conductance of the mesh net edges. In the proposed method, the routing is performed in two steps. In the first step, the global current flow on a chip is determined to minimize the objective function or, minimize the wiring resource consumption under design constraints. In the second step, the global current flow is mapped to new bus topology that supports the current flow. The detail of the global current flow optimization method is described in this paper.

2 The problem

As shown in Figure 1, power and ground wiring for cell-based VLSI consists of pre-wired or mandatory wiring such

*on leave from Toshiba ULSI Research Center, Kawasaki 210, Japan

0738-100X/92 $3.00

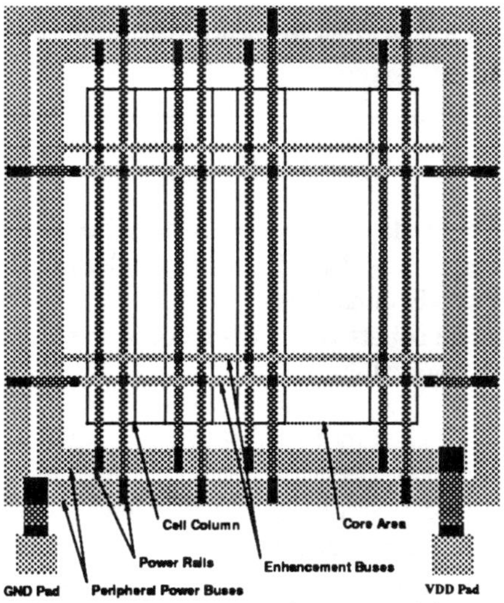

Figure 1: Power and Ground Wiring

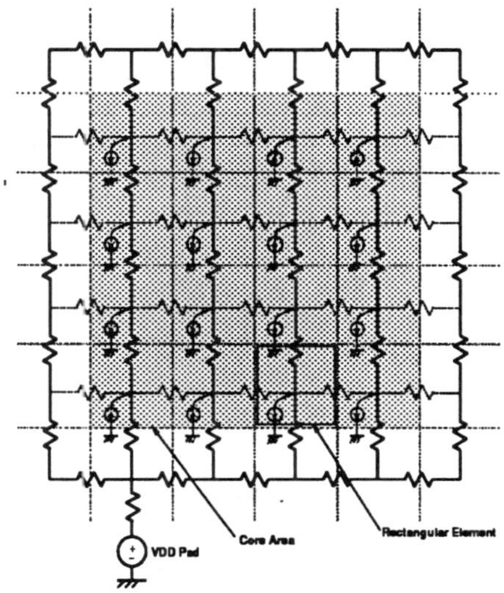

Figure 2: Equivalent Circuit Model

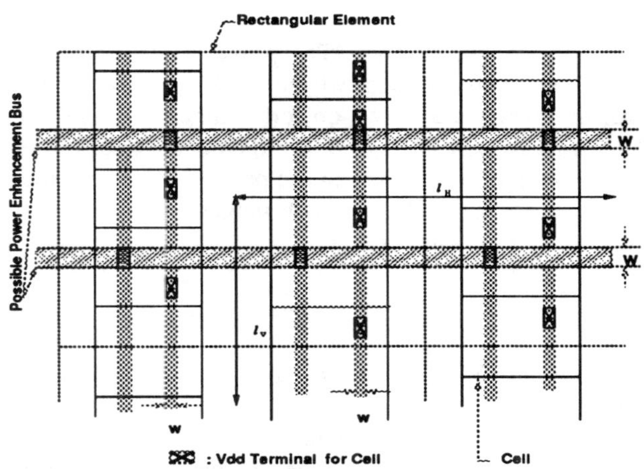

Figure 3: Rectangular Element

as peripheral power buses, pads, and power rails, and additional or optional power enhancement buses. Cells are placed on the rail so that power is fed from the rail. In Sea of Gates, mandatory wiring is predetermined by its die size and chip architecture. Because of consistency between the cell library and power rail, and wiring resource economy, the line width of power rail may not be increased with an increase in the chip size. Power supply current for more than a hundred cells is to be fed through this thin rail. The power enhancement buses are used to improve the reliability and quality of power buses as shown in Figure 1. From wiring resource points of view, the power enhancements mean consumption of wiring resources. Based on estimated power dissipation distribution, an optimal design that minimizes the use of enhancement buses or wiring resource consumption becomes possible.

2.1 A power and ground bus model

The power and ground bus model is represented by an equivalent circuit as shown in Figure 2. The core area of the VLSI chip where cells are placed, is divided by a set of rectangular elements that covers the area. Using a rectangle element model, the power dissipation distribution can be represented with appropriate accuracy. The value of the current source of a rectangular element is the sum of average current that flows through the Vdd terminals in the rectangular element. The current dependence on the voltage is ignored because the variation of voltage is small enough. A power pad is represented by a voltage source and connected with peripheral power buses by a pad conductor. In Figure 2, pre-wired power buses are depicted by solid line resistors whose conductance values are determined by actual bus geometries. Each resistor depicted by a dashed line indicates decision variable resistor or power enhancement bus whose conductance value takes either real enhancement bus conductance or zero conductance that cut the current.

Figure 3 depicts the relation between the equivalent circuit and the rectangular element. The value of the current source of a rectangular element is the sum of average current that flows through the Vdd terminals in the rectangular element. The conductance value of a pre-wired power bus is determined based on the distance l_V between centers of the adjacent elements, line width w, and number of power rails crossing the boundary of the element. The decision variable resistor corresponds to a possible enhancement bus. The optimization program selects the value of decision variables from either an arbitrary small conductance value that has small impact on the current distribution, or the actual conductance value determined based on the distance l_H between center of the adjacent elements and line width W.

The following notations will be used in a formulation of the problem and a description of the solution method:

N_n : Set of indices of all nodes in the equivalent circuit $\mathcal{N}$.
N_b : Set of indices of all branches in $\mathcal{N}$.
N_x : Set of indices of decision variable resistors in $\mathcal{N}$.
N_c : Set of indices of pre-wired bus resistors.
v_i : Voltage across branch $i \in N_b$.
j_i : Current through branch $i \in N_b$.
e_i : Voltage of node $i \in N_b$.
g_i^x : Conductance of decision variable resistor, $i \in N_x$.
g_i^c : Conductance of pre-wired bus resistor $i \in N_c$.
j_i^{max} : Maximum allowed current through branch $i \in N_x \cup N_c$.
w_i : Weight of branch $i \in N_x$.

2.2 Formulation

Objective Function: The primary intention of this problem is to minimize the wiring resources used for power enhancement buses. The objective function z is the total wiring resources required by the enhancement buses, that is proportional to sum of the product of enhancement bus conductances of branch with weighting. The objective function is expressed as,

$$z = f(g_1^x, g_2^x, g_3^x, ..., g_{n_x}^x) = \sum_{i \in N_x} w_i g_i^x, \qquad (1)$$

where $n_x = |N_x|$. In this formulation, g_i^x takes the actual conductance value(≥ 0) if the enhancement bus is realized, or zero if the bus is not implemented. This means the z is directly proportional to the number of enhancement buses. The weight of branch w_i represents the relative cost of wiring resources at the enhancement bus $i \in N_x$. Given a cell placement, w_i is easily determined using the estimated data of wiring congestion.

Voltage Drop Constraints: For some node pairs (p, q), $p, q \in N_n$, the voltage difference is bounded by a designer's specifications, ϵ_{pq}. Let Q be a set of (p, q) pairs, for each of which, voltage constraints,

$$|e_p - e_q| \leq \epsilon_{pq}, \qquad (2)$$

are enforced for every $(p, q) \in Q$. This inequality is rewritten as follows for convenience,

$$e_p - e_q \leq \epsilon_{pq}, \ e_q - e_p \leq \epsilon_{pq}, \ (p, q) \in Q. \qquad (3)$$

Electromigration Constraints: For each branch $i \in N_x \cup N_c$, the maximum current constraints due to electromigration are determined as follows,

$$|j_i| \leq j_i^{max}, i \in N_x \cup N_c. \qquad (4)$$

Using Ohm's law, $j_i = g_i v_i$, the constraints are rewritten as follows,

$$v_i \leq \frac{j_i^{max}}{g_i}, \quad -v_i \leq \frac{j_i^{max}}{g_i}, \qquad (5)$$

where j_i^{max} is determined by considering electromigrations. For g_i^x, $i \in N_x$, (5) is meaningless when $g_i^x = 0$, we must directly consider (4).

Conductance constraints: Each decision variable g_i^x may take one of two values,

$$g_i^x \in \{0, g_i^X\}, \ i \in N_x, \qquad (6)$$

where g_i^X is the conductance value that is determined if the enhancement bus is implemented. The conductance g_i^x takes 0, if no enhancement bus is implemented at i.

Circuit Constraints: Let $\mathbf{v} = (v_1, v_2, ..., v_i, ..., v_{n_b})$, $\mathbf{j} = (j_1, j_2, ..., j_i, ..., j_{n_b})$, $i \in N_b$, and $n_b = |N_b|$, be edge voltage vector and edge current vector of circuit $\mathcal{N}$ respectively. As an electrical circuit, Kirchhoff's current law(KCL), $\mathbf{Cj} = 0$, and voltage law(KVL), $\mathbf{Bv} = 0$, are required, where $\mathbf{B}$ and $\mathbf{C}$ are fundamental loop matrix and fundamental cut set matrix with respect to a tree of circuit $\mathcal{N}$. Let $\mathbf{G}$ be a conductance matrix those diagonal components are g_i^x and g_i^c. The Ohm's law is expressed by $\mathbf{j} = \mathbf{Gv}$.

2.3 Problem Formulation

Given a cell placement, the problem may be formulated as follows:

$$\begin{array}{lll} minimize & z = f(\mathbf{g}^x), & (7) \\ subject\ to & g_i^x \in \{0, g_i^X\}, \ i \in N_x, & (8) \\ & h_j(\mathbf{g}^x) \leq o, \ j \in I^h, & (9) \\ and & circuit\ constraints & \end{array}$$

where z is defined by (1), domain of g_i^x is determined by (6), inequalities $h_j(\mathbf{g}^x) \leq o$ come from (3) and (5), I^h is a index set for constraint inequalities. Circuit constraints consist of KVL, KCL and Ohm's law. $\mathbf{v}$ and $\mathbf{j}$ are function of conductance or decision variable vector $\mathbf{g}^x = (g_1^x, g_2^x, ..., g_{n_x}^x)^T$. This problem is a combinatorial optimization problem, because the decision variables take discrete values.

3 Solution method

The problem is a (nonlinear) combinatorial optimization problem. It is very difficult to solve the problem directly.

We solve an approximate problem that relaxes the discrete variable constraints(9) as follows,

$$g_i^0 \le g_i^x \le g_i^X, \quad i \in N_x, \tag{10}$$

where g_i^0 is small enough that it has negligible impact on the current distribution of $\mathcal{N}$. By this relaxation, the variables are allowed to take arbitrary values in the relaxed decision variable domains, and the problem is to be considered on a single network topology containing every possible network topology. In addition to the topology issue, the constraints (10) can be treated in the framework of (9). The extended index set for constraint inequalities including (10) is referred as $I^{h'}$ from now on. After obtaining an approximate solution, an exact or integer solution is locally searched in a neighborhood of the approximate solution.

In the first step, an initial feasible solution that satisfies every constraint is obtained. In the following steps, the solution is improved by moving to the direction $\mathbf{d}$ that decreases the objective function without causing violation of any constraint. This improvement step is iterated until no improvement is obtained. The general procedure will be described after the description of feasible direction method and gradient calculation.

3.1 Feasible direction method

The feasible direction method is an optimization algorithm that improves the objective function without violating constraints [10,11]. The algorithm is used both in the initial solution search phase and in the improvement phase as their algorithmic bases.

Let $\mathbf{g}_k^x$ be a decision variable vector at the k_{th} step of iteration, the next decision variable vector $\mathbf{g}_{k+1}^x$ is determined by

$$\mathbf{g}_{k+1}^x = \mu_k \mathbf{d}_k + \mathbf{g}_k^x, \tag{11}$$

where $\mathbf{d}_k$ is a direction vector determined by solving the following linear programming problem:

$$\begin{aligned} &minimize && \varepsilon && (12)\\ &subject\ to && < \nabla f(\mathbf{g}_k^x), \mathbf{d}_k > \le \varepsilon, \\ & && < \nabla h_i(\mathbf{g}_k^x), \mathbf{d}_k > \le \varepsilon, \ i \in I_k^a, \\ &and && normalization\ \ requirement, \end{aligned}$$

where I_k^a is a set of indices that satisfies following near active constraint conditions to avoid the zigzagging[10,11],

$$I_k^a = \{\ i \ | h_i(\mathbf{g}_k^x) \ge -\alpha\} \tag{13}$$

where α is the near active constant. As a normalization requirement, several equations are proposed[11]. Followings are examples:

$$\begin{aligned} N1: &\quad \mathbf{d}^T\mathbf{d} \le 1 \\ N2: &\quad -1 \le d_i \le 1 \ \ for \ \ i \in N_x \\ &\quad \dots \end{aligned}$$

where T denotes a transpose of the vector.

According to our experiences, the convergence characteristics of the feasible solution method is influenced by the selection of normalization requirement. Figure 4 shows typical convergence characteristics of N1.

3.2 Gradient calculation

To implement the feasible direction method, gradient calculation of $f(\mathbf{g}^x)$ and $h_i(\mathbf{g}^x)$ are required. $\nabla f(\mathbf{g}^x)$ is easily calculated because $f(\mathbf{g}^x)$ is linear function in our formulation. But $\nabla h_i(\mathbf{g}^x)$ is difficult to calculate, because $h_i(\mathbf{g}^x)$ is a function of node voltage $\mathbf{e}$ or edge voltage $\mathbf{v}$ of circuit $\mathcal{N}$, like (3) and (5), and not expressed in explicit form in many cases. We need $\nabla v_i(\mathbf{g}^x)$ or $\nabla e_i(\mathbf{g}^x)$, for $\nabla h_i(\mathbf{g}^x)$. One way to obtain $\nabla v_i(\mathbf{g}^x)$ and $\nabla e_i(\mathbf{g}^x)$ is numerical differentiation[7]. But in this case, the wide value range of g_i^x requires accurate calculation of $\nabla v_i(\mathbf{g}^x)$ and $\nabla e_i(\mathbf{g}^x)$ to avoid the accumulation of errors by iteration.

The method of adjoint network[12] is adopted to calculate $\nabla v_i(\mathbf{g}^x)$. Let $\mathcal{N}'$ be the adjoint network related to $\mathcal{N}$. $\mathcal{N}$ and $\mathcal{N}'$ must have the same topology, but need not have the same element type or values in corresponding branches. Let ψ_i and ϕ_i be the branch voltage and branch current of $\mathcal{N}'$ and $\psi_i g_i = \phi_i, \quad i \in N_x \cup N_c$. Based on Tellegen's theorem[12] and assuming appropriate terminal conditions, the following equation is obtained for a specific edge voltage v_l,

$$\delta v_l = \sum_{i \in N_x} \psi_i v_i \delta g_i^x \tag{14}$$

Thus the gradient or sensitivity of v_l with respect to g_i^x is as follow,

$$\nabla v_l(\mathbf{g}^x) = (\psi_{l_1} v_{l_1}, \psi_{l_2} v_{l_2}, \dots, \psi_{l_i} v_{l_i}, \dots, \psi_{l_{n_x}} v_{l_{n_x}}), \ \ l_i \in N_x. \tag{15}$$

3.3 Initial Feasible Solution

A robust initial solution finding algorithm is indispensable because the problem is very complex. For example, we can generate a feasible solution by removing some enhancement power buses, from a circuit that violates constraints with fully enhanced by extra power enhancement buses. The procedure is as follows:

1) Generate Initial Solution.

2) Check constraint violation
a) Check all constraints $h_i(\mathbf{g}^x)$, $i \in I^h$.
b)Update index set I^v, if violated constraint is found.
c)Update near active index set I^a, if near active constraint is found.

3) Find optimal direction $\mathbf{d}$ by solving a feasible direction problem described in section 3.1.

4) Minimize $h_i(\mathbf{g}^x)$, $i \in I^v$ under constraints so that no increase of constraint violation occurs.

5) If there remains violated constraint, go to step (2), otherwise stop.

3.4 Improvement

Using an initial feasible solution obtained by the procedure described in Section 3.3, the improvement program minimizes the object function $z = f(\mathbf{g}^x)$. The linear characteristics of $z = f(\mathbf{g}^x)$ simplify the problem. The procedure is as follows:

1) Analyze the circuit $\mathcal{N}$.

2) Check constraints including voltage drop and electromigration constraints, and then generate a near active constraint index set, $I_k^{na} = \{i \mid h_i(\mathbf{g}_k) \geq -\epsilon\ ,\ i \in I^h\}$.

3) Calculate gradient of near active constraints, $\nabla h_i(\mathbf{g}_k^x)$, $i \in I^{na}$, by the method of adjoint networks described in section 3.2.

4) Determine the best direction $\mathbf{d}_k$ that will not violate the constraints and minimize object function. The method is described in section 3.1.

5) If $\epsilon < 0$ determine the next point, $\mathbf{g}_{k+1}^x = \mu_k \mathbf{d}_k + \mathbf{g}_k^x$, where μ_k minimizes $f_i(\mathbf{g}_k^x + \mu_k \mathbf{d}_k)$ and satisfies following conditions, $h_i(\mathbf{g}_k^x + \mu_k \mathbf{d}_k) \leq o$, $i \in I^h$.

6) If the termination condition, $\epsilon \geq 0$ is satisfied, the improvement is stopped. If not satisfied $k = k + 1$ and go to step 1).

3.5 Integer Point Search

The proposed method approximates the combinatorial optimization by continuous optimization method. So after obtaining a solution in the continuous decision variable space, we must search an integer point that satisfies all constraints of the original problem in the neighborhood of the approximated solution. A greedy algorithm exhaustively searches the neighborhood is adopted.

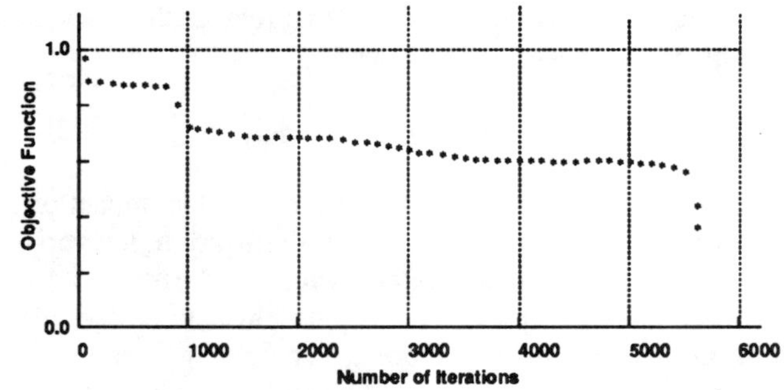

Figure 4: Convergence Characteristics of Improvement

Table 1: Convergence of Improvement Phase

Circuit	Branches	Variables	Rectangles	Iteration
A	35	2	-	8
B	78	12	9	5601
C	179	42	36	203

4 Experimental results

The proposed method has been implemented. Using several examples, we have confirmed the validity of the problem formulation and the solution method.

The size of circuit examples and needed iteration numbers are shown in Table-1. A convergence characteristic example is shown in Figure 4. These data show that the required cpu times for solution improvement vary widely. A conjectural reason for this variety is strong dependence on the quality of initial solutions.

From some experimental results, we can learn some lessons for power and ground wiring design. For an example, indiscriminate power bus enhancement sometimes causes unexpected current crowding that becomes a cause of electromigration failure. Figure 5 shows current and voltage distribution in a small but interesting example. As an initial solution, shown in (A) of Figure 5, all power enhancement buses are used, but an electromigration constraint is violated at the enhancement bus near pad. The current through this enhancement bus is 2.948mA, but the maximum current of this bus is 2.78mA. The described program solved this violation by means of blocking the current by removing the adjacent enhancement buses. The current is decreased to 2.597mA after the improvement as shown in (B) of Figure 5. Ironically, the electromigration violation is resolved by removing the enhancement buses. Although this is quite a simple example, it teaches us that power and ground routing is not so simple. It needs global consideration.

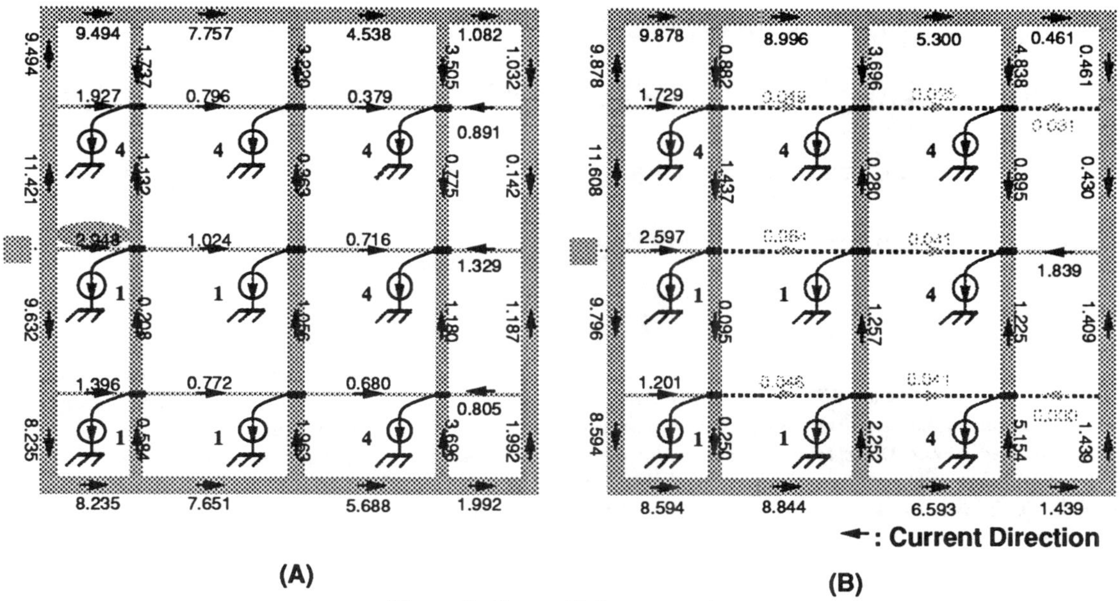

Figure 5: Power Enhancements

5 Conclusions

A new power and ground network topology optimization problem for cell-based VLSIs has been described. The problem has been formulated as a combinatorial optimization problem and approximated by constraint relaxation. A prototype program based on the feasible direction method and local search has been implemented.

Some experimental results show that brute power bus enhancement is meaningless and smart power and ground topologies significantly reduce the consumption of wiring resources. Our study shows that the topological optimization of power and ground routing will be an important design issue in the next generation VLSI chip design. An example assuming practical 5mm by 5mm chip size was successfully handled by the prototype program.

With a little modification, this formulation and algorithm of the proposed problem is applicable to general power and ground routing problems in the sense of global routing.

Reference

[1] J. R. Black, "Electromigration failure modes in aluminum metalization for semiconductor devices," Proceedings of the IEEE, vol.57,No.9, pp.1587-1594,Sept., 1969.

[2] H-J. Rothermel and D. A. Mlynski, "Computation of power and ground supply nets in VLSI layout," Proc. of 18th DA Conf. pp. 37-42, June,1981.

[3] Z. A. Syed and A. El Gamal, "Single layer routing of power and ground networks in integrated circuits," Journal of Digital Systems, Vol.6, No.1, pp.53-63, 1982.

[4] Xiao-Ming Xiong and E. S. Kuh, "The Scan Line Approach to Power and Ground Routing", Proc. of ICCAD, pp.6-9,1986.

[5] S. Haruyama and D. Fussell, "A New Area-Efficient Power Routing Algorithm for VLSI Layout," Proc. of ICCAD, pp.38-41, 1987.

[6] S. Chowdhury and M. A. Breuer, "Minimal area design of power/ground nets having graph topologies," IEEE Trans. on Circuits and Systems, pp.1441-1451, December 1987.

[7] R. Dutta and M. Marek-Sadowska, "Automatic Sizing of Power/Ground (P/G) Networks in VLSI," Proc. of 26th Design Automation Conference, pp. 783-786,June 1989.

[8] K-H. Erhard and F. Johannes, "Area Minimization of IC Power/Ground Nets by Topology Optimization," Proc. of the VLSI 91, pp. 3b.4.1-3b.4.8, August 1991.

[9] S. Chowdhury and J. S. Barkatullah, "Current estimation in MOS IC logic circuits," Proc. ICCAD, pp.212-215, November 1988.

[10] R. K. Brayton, H.G. Hachtel and A. Sangiovani-Vincentelli, "A Survey of Optimization Techniques for Integrated Circuit Design," Proc. of IEEE, pp. 1334-1363,Oct. 1981.

[11] G. Zoutendijk "Methods of Feasible Directions", Amsterdam, Elsevier Publishing 1960.

[12] S. W. Director and R.A. Roher, "A Generalized Adjoint Network and Network Sensitivities," IEEE Trans. on Circuits and Systems, Vol. CT-16, pp. 318-323, August 1969.

FARM: An Efficient Feed-Through Pin Assignment Algorithm

Xian-Long Hong, Jin Huang, Chung-Kuan Cheng† and E.S.Kuh
Dept. of EECS, University of California at Berkeley, CA 94720
†Dept. of Computer Science and Engineering, UC San Diego, La Jolla, CA 92093

Abstract

In this paper, we propose an efficient feedthrough pin assignment algorithm, FARM, to minimize the maximum channel density, and at the same time to reduce the wire length and via number. A novel vertical channel routing formulation is devised to model the pin assignment problem. A multi-row density minimization followed with a single row pin assignment is proposed to complete the assignment process.

We have connected the FARM with a multi-terminal multi-commodity flow based global router proposed by Carden et al[IC91]. The experiments on several benchmarks demonstrate that [IC91] and FARM achieve comparable or better results comparing with TimberWolf 5.6.

1 Introduction

Feedthrough pin assignment is a process connecting global router with detailed router. The global router determines paths for each net on a global routing graph (GRG) which usually has the vertical edge representing a set of feedthroughs of macro cells and feedthrough cells, and the horizontal edge representing the channel between the adjacent grids [IC91][CCW90][CH90]. In order to interface detailed routers, we need to assign vertical paths of global routing tree into feedthroughs. Because in general a good channel router can achieve a layout with a channel height equal to the routing density, the pin assignment which determines the final routing density becomes a critical step to minimize the layout area.

For gate array design, we minimize the maximum channel density which determines the routability. For channel-type sea-of-gates or standard cell designs, we need to minimize the total tracks of all the channels. However the minimization of the channel density may not coincide with the goal of the wire length minimization.

In some previous global routing algorithms, creating global routing trees and assigning feedthrough pins were implemented simultaneously [PT89][LS91][ML90][Con89]. These feedthrough pin assignment algorithms utilize the approach of minimizing wire length and via and achieve very good results.

We propose an efficient feedthrough pin assignment algorithm, FARM, with an objective to minimize the maximum channel density or the total number of tracks, and at the same time to try to reduce wire length and via number. FARM consists of three parts, global routing tree mapping, multi-row feedthrough pin assignment and single-row feedthrough pin assignment.

We devise a vertical channel routing formulation to model the feedthrough assignment problem. The chip is rotated by 90 degrees. Thus, one GRG grid column is viewed as a horizontal channel, where the feedthroughs are functioning as the horizontal tracks of a classical channel routing problem. We assign signals to the tracks so that a proposed vertical channel density is minimized.

Deriving from the formulation of the vertical channel routing, we realize that a good feedthrough pin assingment needs to involve the assignment of multiple rows simultaneously. We propose a multi-row feedthrough pin assignment approach to seek the common x-position feedthroughs in multiple rows for the continuous vertical paths, and thus reduce the horizontal paths in the congested area.

After multi-row feedthrough pin assignment a single-row feedthrough pin assignment approach is applied to determine the sequence and positions for all vertical paths in a single-row. Some heuristic techniques, such as exchanging cross and path shifting, are used to further reduce the maximum channel density.

FARM can be used for gate array, standard cell and sea-of-gates layout environment if only the global routing graph has a regular rectangular grid.

The remainder of this paper is organized as follows. In section 2, some definitions and problem formulation are given. Two main techniques of our algorithm, multi-row feedthrough pin assignment and single-row feedthrough pin assignment are described in section 3 and 4. Finally, some experimental results and comparison with previous work are given.

2 Problem Formulation

2.1 The Feedthrough Assignment Formulated as a Vertical Channel Routing Problem

For each grid column of GRG, we formulate the feedthrough pin assignment as a vertical channel routing problem. In a classical channel routing problem, terminals are on the top and bottom boundary of the channel. The horizontal routing segments are ordered into tracks. The routing problem is to assign the signals to the tracks with a certain object function.

Suppose we rotate one single GRG grid column (1 column by n rows) by 90 degrees, and interface the signals connecting from its two adjacent columns with terminals on the edges, the whole column has a shape of a channel with terminals on the top and bottom of the channel boundary(Fig.1). The channel is striped with alternating vertical bands of cell area (cell-band) and free routing space (free-band). The terminals are on the top and bottom edges of the free-bands. The order of the terminals in each free-band are not specified. The feedthroughs in cell-band become the horizontal tracks. A fixed pin 'A' of a macro-cell denotes that the corresponding track in cell-band is assigned to signal 'A'. Let us define the vertical density of a free-band be the maximum number of vertical segments in the free-band crossing the same Y coordinate. The vertical channel routing problem is to assign the signals to the tracks in cell-bands so that the maximum vertical density in free-bands is minimized (Fig.1(a) and (b)).

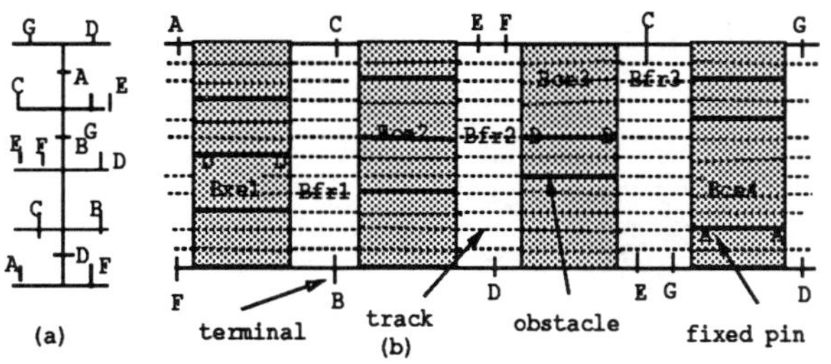

(a) The Original Single Column Feedthrough Pin Assignment on GRG

(b) The Vertical Channel Routing corresponding to (a)

Fig. 1 The Vertical Channel Routing Problem

For each GRG grid column, we can construct a vertical channel Vch. Let us denote D(Vch) be the maximum vertical density for the vertical channel Vch; D(chip) be the global maximum density in the whole chip. Let us denote DD(chip) be the sum of total tracks in the chip. Thus, if we denote a cell-band as Bce, a free-band as Bfr, D(Bfr) as the local vertical density of Bfr in a vertical channel, and DD(Bfr) as the maximum vertical density of Bfr in all vertical channels, we can express the above definitions as follows:

$$D(Vch) = \max\{D(Bfr) \mid \forall Bfr \in Vch\} \quad (1)$$

$$D(chip) = \max\{D(Vch) \mid \forall Vch \in chip\} \quad (2)$$

$$DD(Bfr) = \max\{D(Bfr) \mid \forall Vch \in chip\} \quad (3)$$

$$DD(chip) = \sum\{DD(Bfr) \mid \forall Bfr \in chip\} \quad (4)$$

2.2 The Properties of The Vertical Channel Routing Problem

We first present a formal statement of the vertical channel routing problem using an objective of minimizing the maximum density in free-bands, D(Vch). The horizontal and vertical constraints, which correspond to the constraints in the classical channel routing problem, are then derived based on the formulation. Using the properties of these constraints, we show that the classical channel routing can be transformed into a special case of the vertical channel routing problem. Therefore, we deduce that the vertical channel routing problem is NP complete.

2.2.1 Statement of the Vertical Channel Routing Problem

Given a rectangular channel, the area is covered with vertical strips of cell-bands and free-bands alternating in the x direction. The terminals locate on the top and bottom of the channel, and in the interval of free-bands. For each cell-band, there are a set of horizontal tracks connecting between its two adjacent free-bands. Some of the tracks may be designated for certain signals. The routing problem is to assign signals into the tracks such that the maximum vertical density of free-bands is minimized.

2.2.2 Horizontal Constraint

The horizontal interval of a signal is determined by its leftmost and rightmost terminals or its designated tracks. Let S(Bce) be the set of signals whose horizontal intervals intersect cell-band Bce. Since horizontal segments of distinct signals must not overlap, any two signals in S(Bce) must not be assigned to the same track in cell-band Bce;

The above constraint is identical to the horizontal constraint in a classical channel routing problem. Therefore, the zone representation technique [YK82] proposed in channel routing can also be applied to the vertical channel routing problem.

2.2.3 Vertical Constraint

In classical channel routing problem, the vertical segment of any two signals must not overlap, which constitutes the vertical constraints for the track assignment. However, in the vertical channel routing problem, the terminals locate in the free-bands without a specific order. Therefore, the vertical routing segments are not required to follow the vertical constraints as the vertical segments in a classical channel routing problem are.

However, in the vertical channel routing, if the intervals of vertical segments in a free-band overlap, the overlapping contributes extra vertical density to the interval in the free-band. Therefore, the violation of the vertical constraint is translated into the local vertical density.

2.2.4 The Classical Channel Routing As A Special Case Of The Vertical Channel Routing Problem

We first state the classical channel routing problem and then construct the corresponding problem in vertical channel routing.

A classical channel routing problem: Given a rectangular channel, the terminals are on the top and bottom of the channel. Two layers are available for routing. We assume horizontal tracks on one layer and vertical tracks on the other. Given a constant k, the problem is to route the channel with k horizontal tracks. Two theorems are given and their proofs can be found in [XJ92].

Theorem 1

A classical channel routing problem can be transformed into a vertical channel routing problem with a polynomial number of operations.

Theorem 2

The vertical channel routing problem is NP complete.

2.3 Single GRG Grid Column Feed-Through Assignment

Given a single column of GRG and the tree paths related to this column we treat the column as a vertical channel. The single column feedthrough pin assignment problem is formulated as a vertical channel routing problem.

To simplify the explanation, we rotate the vertical channel by -90 degrees. Therefore, for the rest of the paper, the vertical channel has terminals on left and right and tracks running vertically. The cell-bands and free-bands alternate in the vertical direction. The following terms: the tracks of the cell-bands, the feedthrough pins of the cells, and the feedthroughs are used interchangeably.

2.4 Global Vertical Channel Routing Problem

Given a chip of n columns of GRG, we treat the n columns as n vertical channels. For gate array, the objective for the n vertical channel routing is to minimize **D(chip)**. For sea-of-gates and standard cell, the objective is to minimize **DD(chip)**. A secondary goal of the process is to reduce wire length and via number, if possible.

3 Multiple Cell-Band Feed-Through Assignment

Since the density minimization involves the pin assignment of more than one single cell-band, we start with a multiple cell-band track assignment operation. The algorithm is based on a horizontal segment reduction between two adjacent cell-bands. For each two adjacent cell-bands, we search all possible pin alignment configurations. The relation of different assignment configurations among multiple cell-bands is represented by a graph structure. The longest path of the graph determines the feedthrough assignment.

In the following, we discuss the reduction of the horizontal segment according to its classification. The operation is then extended to the case of multiple cell-bands. The algorithm utilizing these operations as basic routines is presented.

3.1 The Classification of Horizontal Segments

The misalignment of the feedthroughs in adjacent cell-bands causes extra horizontal segment which contributes to the local density of the free-band. Except those segments predefined by the users, all the horizontal segments can be conceived as removable by aligning the feedthroughs in the adjacent cell-band. The removable horizontal segment (RHS) can be classified into three categories: Free Removable Horizontal Segment (Free RHS), Semi-Free Horizontal Segment (Semi-Free RHS), and Fixed Removable Horizontal Segment (Fixed RHS) according to the connected vertical segments' features. If neither end of the horizontal segment in the free-band is connected through a vertical segment to a fixed pin on the adjacent cell-band, the horizontal segment is a Free RHS (Fig.2a).

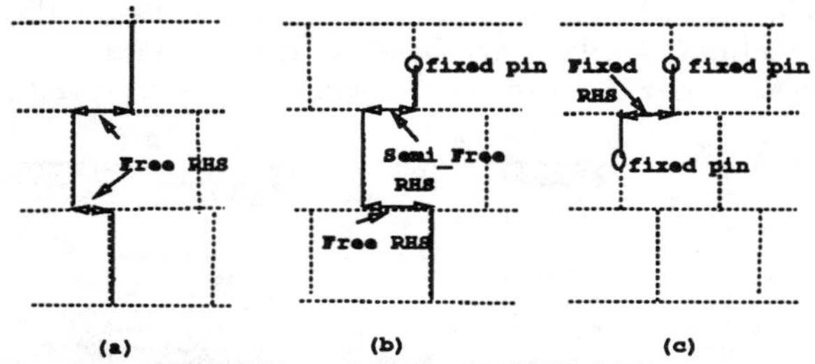

Fig. 2 Free, Semi_Free and Fixed RHS

If one end is connected to a fixed pin, the horizontal segment is called a Semi-Free RHS (Fig.2b). If both of ends are connected through vertical segments to the fixed pins, the horizontal segment is referred to as a Fixed RHS (Fig.2c).

3.1.1 The Reduction of Free RHS

For a Free RHS, because its connected two vertical segments in the adjacent cell-bands are movable, we seek a common feedthrough X position in both cell-bands and then shift the vertical segments to this aligned position (Fig.3(a)). The searching of the common X position feedthroughs are first performed inside the cells to which the original feedthroughs belong. If no position is found, we extend the searching range to their neighboring cells on left and right hand sides. Extending the searching range may cause a new RHS in the extended direction (Fig.3(c)).

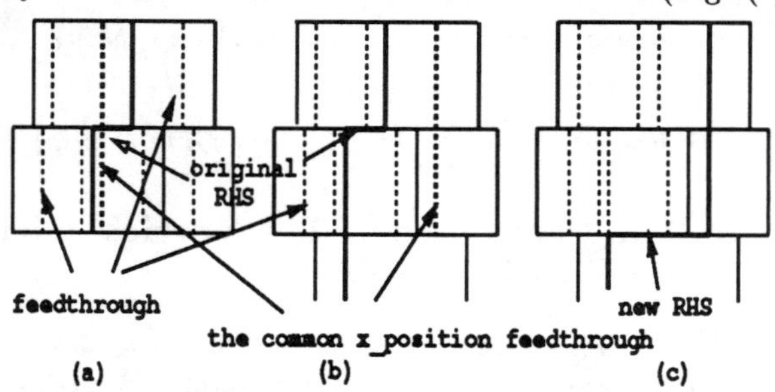

Fig. 3 Reduction of Free RHS

If the new RHS does not affect the maximum density,

we accept the move, otherwise the original RHS remains. Note that once the assignment is performed, the feedthrough becomes occupied.

3.1.2 The Reduction of Semi-Free RHS

For the Semi-Free RHS, only one of its two connected vertical segments in cell-bands is not fixed. If the cell-band which contains the movable vertical segment, has a feedthrough at the same X position as the fixed pin, we assign the segment to this feedthrough (Fig.4(a)), otherwise we search a common X position feedthrough in both cell-bands. The strategy of the searching range is the same as strategy for the Free RHS. If the aligned feedthroughs are found, we assign both vertical segments to this position and connect the fixed pin from the other side of the cell-band (Fig.4(b)(c)). However, this arrangement creates a new RHS in another free-band.

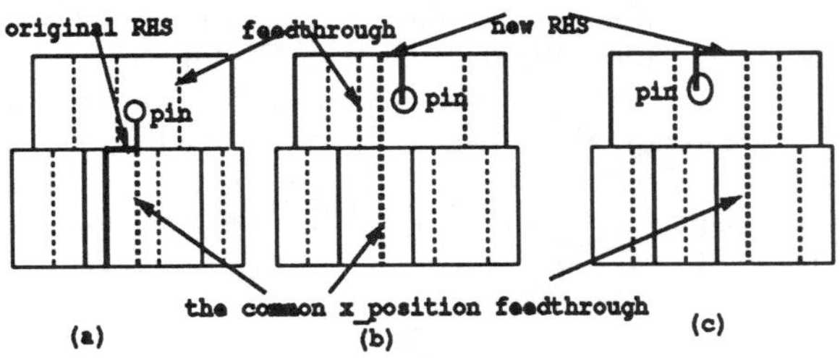

Fig. 4 Reduction of Semi_Free RHS

3.1.3 The Reduction of Fixed RHS

For a Fixed RHS, both of its connect vertical segments are fixed on the two cell-bands. We remove RHS by using extra RHSs and making connection to the equivalent pins on the other side of the cell-bands. The RHS transfer technique benefits to minimize the maximum channel density. Fig.5 shows 3 kinds of Fixed RHS transfer technique. The original problem is shown in Fig.5(a). Fig.5(b) shows the case that there exist aligned feedthroughs between the fixed pins. We remove the Fixed RHS by connecting the equivalent pins on the other side of the cell-bands. However, the connection causes two new RHS on the other sides of the cell-bands.

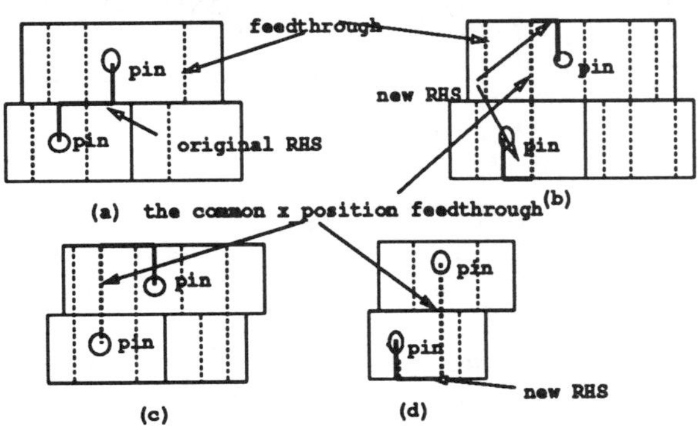

Fig. 5 Reduction of Fixed RHS

Fig.5(c) and 5(d) shows that case there exists a feedthrough aligned with the fixed pin on the other side. The transfer operation is executed only on the condition that the maximum density is reduced.

3.2 The Extension of RHS Reduction to Multiple Free-Bands

If there is a signal spanning over several cell-bands, the misalignment of the feedthroughs in the cell-bands would cause multiple RHSes in the free-bands. On each free-band, we perform an exhaustive RHS reduction process. We then construct a directed graph to present the relation of all the possible feedthrough pin assignments of the different free-bands. For each assignment in a free-band, we create a node. The node is assigned a weight according to the density reduction plus a small fraction of the wire length reduction. We construct an edge to connect two nodes on the adjacent free-band, if the two nodes have the same X position feedthrough in the common middle cell-band being identical. The direction of the edge is based on the index of the free-bands. It directs from the node in high index free-band to the node in low index free-band. We then find the longest path on the constructed graph from A to A'. The nodes on the longest path determine the pin assignment of the signal. Fig.6a shows the initial feedthrough assignment and Fig.6b shows final result for multiple free-band RHS reduction.

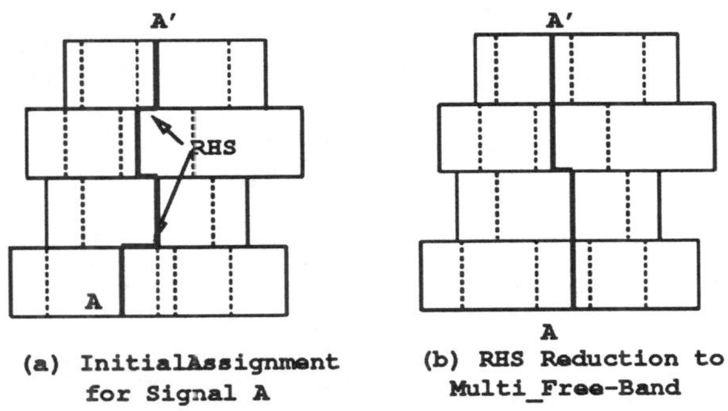

Fig. 6 Multiple Free-Band RHS Reduction

3.3 Multiple Cell-Band Feed-Through Assignment Algorithm

The algorithm assigns vertical paths of trees into proper feedthroughs to minimize D(Vch) or DD(Vch). First, we find the most congested free-band area Hfr in vertical channel Vch, then search the RHS which contributes to the density most in Hfr and perform the RHS reduction in vertical channel Vch. If any new horizontal path is generated, it is added to the related free-band and the density is updated. The procedure is repeated until the current D(Vch) or DD(Vch) can not be reduced further.

3.3.1 The Algorithm That Minimizes The Total Tracks

For channel-type sea-of-gates or standard cell designs, the objective of the algorithm is to minimize the total number of tracks. Thus, the congested region becomes the free-band Bfr that the density reduction of D(Bfr) will bring the total density down. For the RHS reduction, the process is accepted only if the total density is decreased. Otherwise the procedure of the algorithm remains the same.

4 Single Cell-Band Feed-Through Assignment

The multiple cell-band feedthrough alignment assigns feedthroughs on the paths involving multiple cell-band in the congested region. The single cell-band feedthrough assignment process the rest of vertical paths. The objective is to further reduce the maximum channel density. However the operation at this stage also improves the wire length and via number reduction.

4.1 Classification of Paths in Cell-Band

With respect to each cell-band, we classify the paths into 64 types according to the local pattern of their tree structure. We use two digits to represent the path type, 00. 01. ..07, 10, ..17,..77. Fig 8 shows a part of path types. In the algorithm, the process on each signal is performed according to the class of the path. The path with type (x0) or (0x) has a fixed pin in the cell-band. Thus it is not included in the process. For paths with type (2x) and (x2) their top and/or bottom terminal is connected with a path which has a type (x2) or (2x) in the neighboring cell-band. To avoid generating a new extra horizontal path the best way is to keep them in the same X position. Thus, they are processed prior other types of paths.

4.2 Sequence Constraint and Zone

We create the sequence constraint of path passing through the macro cell or feedthrough cell. For example, the type (11) should lie on the left of type (44). In this case both horizontal segments of type (11) and (44) can occupy the same track in both free-bands. Type (14), (41), (16) and type (61) can be put in any sequence and the number of occupied tracks are the same. We define the sequence constraint of net paths as follows:

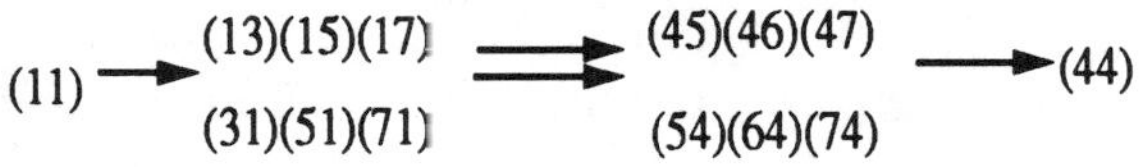

Fig. 7 Sequence Constraint of Path

In the Fig.7, (xx) represents the node, the signs → and ⇒ represent the sequence constraint of the relative nodes, The sign "→" represents the weak constraint and "⇒" strong constraint. The (11) →(13) means (11) should be put on the left of (13). Otherwise, the via number will be increased. The (11) ⇒ (44) means (11) should be put on the left of (44). Otherwise, The density, wire length and via number will be increased. For the path types which are not included in the sequence constraint(Fig.7), the assignment of these paths does not change the density. Therefore, these paths are assigned at the end of the process.

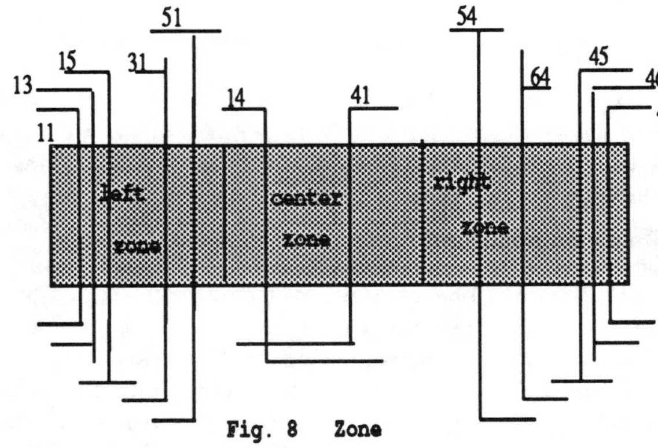

Fig. 8 Zone

According to the sequence constraint we divide the routable area of macro cell and feedthrough cell into three parts: Left Zone, Right Zone and Center Zone. The zone may not be the continuous region but contains several feedthroughs which can be passed through by vertical paths. In order to guarantee the sequence constraint, in general, type (11), (13)), (15), (17), (31), (51) and (71) are assigned into the Left Zone, type (44), (45), (46), (47), (54), (64) and (74) are assigned into the Right Zone and type (14), (34), (43), (55), (77), (57), (75) and (41) are assigned into the Center Zone as shown in Fig.8.

4.3 Exchanging Path Cross

In Center Zone, there is a cross in the free-bands for type (14) and type (41) paths as shown in Fig.9(a) and Fig.9(b) if both exist within the same cell-band of a vertical channel. We can assign the cross into either of the two adjacent free-bands.

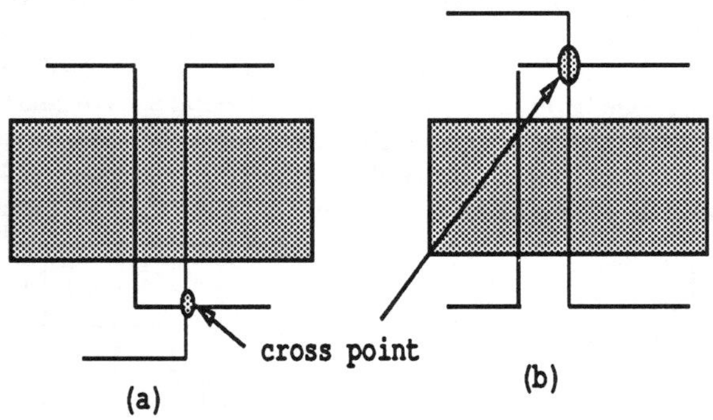

Fig. 9 Exchanging Path Cross

To control maximum density of the free-bands, we try the two possible choices and select the better arrangement.

4.4 Determine Sequence of Pins in Left and Right Zone

We have determined the sequence of pins for most types of vertical paths. However, the type (11), (13), (15), (17),

(31), (51) and (71) in Left Zone and type (44), (45), (46), (47), (54), (64) and (74) in Right Zone have only been assigned into the zone as a group. In this subsection, we determine the sequence of paths in the left zone and right zone with an objective to minimize the wire length and number of vias. Let C be a circle, Cupper be the upper half circle, Cunder be the under half circle and Dim be the horizontal diameter as shown in Fig.10. The available feedthroughs in the cell-band is mapped into Dim and upper cell-band and under cell-band are mapped into Cupper and Cunder respectively. Pins on the upper cell-band

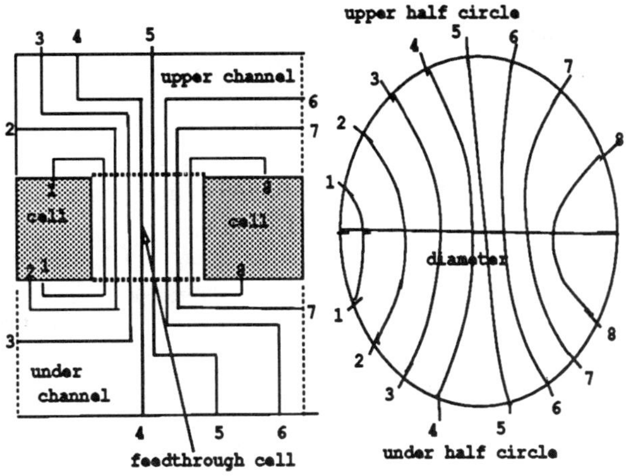

Fig. 10 Sequence Determination of Feedthrough Pins in Left and Right Zone

boundary are arranged on Cupper clockwise and vertical paths are arranged on Dim from left to right corresponding to pin's sequence. Similarly, pins on the under cell-band boundary are put on Cunder counterclockwise. A simple matching approach is used to connect the pins which belong to the same net. The sequence of the crossing point on the diameter is the vertical path sequence. According to this sequence we have got the reasonable assignment of vertical paths in minimizing wire length and via number.

So far, we have determined the sequence of all vertical paths. If there is no spare room for feedthrough in the macro cell or feedthrough cell, the position of each vertical path is fixed. Otherwise, the assignment will be processed from the two ends toward the center according to the sequence.

5 Experimental Results

FARM is implemented in C language running on DECstation 3100. we have connected FARM with the global router developed by Carden et al[IC91] and tested several benchmarks. Table 1 gives the comparison with TimberWolf Version 5.6. Our results are comparable to or better than TimberWolf Version 5.6.

Table 1 Benchmark Examples

Test Cases	cells	nets	TimberWolf Max.Den.	FARM Max.Den.
Primary1	750	915	7	7
Primary2	2907	3050	13	12
C2	590	963	12	12
C7	2150	2465	14	13

Acknowledgements

The authors would like to acknowledge the support received for this research from NSF grant MIP88-03711.

References

[CCW90] Majid Sarrafzadeh Charles Chang and C.K. Wong. Global routing based on steiner min-max trees. *IEEE Trans. on CAD, vol.9, No.12*, pages 1318–1325, 1990.

[CH90] Ting-Hai Chao and Yu-Chin Hsu. Rectilinear steiner tree construction by local and global refinement. *ICCAD*, 1990.

[Con89] Jingsheng Cong. Pin assignment with global routing. *ICCAD*, 1989.

[IC91] Robert C. Carden IV and C.K. Cheng. A global router using an efficient approximate multicommodity multiterminal flow algorithm. *28th ACM/IEEE Design Automation Conference*, pages 316–319, 1991.

[LS91] Kai-Win Lee and Carl Sechen. A global router for sea-of-gates circuits. *EDAC*, 1991.

[ML90] G. Meixner and U. Lauther. A new global router based on a flow model and linear assignment. *ICCAD*, 1990.

[PT89] Tai-Ming Parng and Ren-Song Tsay. A new approach to sea-of-gates global routing. *ICCAD*, 1989.

[XJ92] X.Hong and J.Huang. An efficient feed-through pin assignment algorithm for gate array, standard cell and sea-of-gates. *Research Report, EECS/ERL, UC Berkeley*, 1992.

[YK82] T. Yoshimura and E.S. Kuh. Efficient algorithms for channel routing. *IEEE Trans. on CAD, Vol.1, No.1*, pages 25–35, 1982.

Iterative and Adaptive Slack Allocation for Performance-driven Layout and FPGA Routing

Jon Frankle
Xilinx, Inc.
2100 Logic Drive
San Jose, CA 95124

Abstract

We give a generalization, called the limit-bumping algorithm (LBA), of a procedure of Youssef et. al. [1] that transforms initial connection delays into upper limits on delay suitable for performance-driven layout. LBA is a simple way to distribute slacks using arbitrary allocation functions. We then show how lower and upper bounds on connection delays can be used in the computation of upper limits for initial layout and for layout improvement.

The methods have been integrated into a delay-sensitive router for FPGAs. In 22 standard benchmark designs (with placements fixed), feasible system clock periods were reduced in every case, by an average of 14% and as much as 32%.

1. Introduction

In recent years, increasing attention has turned to the problem of performance-driven layout. This is largely because with increased chip complexity and switching speeds, wiring delay accounts for an increasing proportion of overall system delay.

The challenge for performance-driven layout tools is to implement designs so that total path delays from start pins (primary inputs or register outputs) to end pins (primary outputs or register inputs) satisfy specified limits. This presentation will focus on register transfers, for which the performance measure is the clock period T. We treat the long-path problem and ignore the related short-path problem. For a completed layout, the minimum feasible period $T_{achieved}$ is given by

$$T_{achieved} = \max_{\pi} \left[clock\_to\_out(start) + \sum_{block\ b \in \pi} Delay(b) + \sum_{connection\ c \in \pi} Delay(c) + setup(end) \right]$$

over data transfer paths π from start to end registers synchronized by the clock. We assume that block delays are fixed, and that the design is free of purely combinational cycles, *i.e.*, has no unclocked feedback.

The difficulty is that layout tools ordinarily work on individual nets, without attending to path requirements. Automated performance optimization demands interaction between the layout process and timing analysis of paths. For example, the results of timing analysis can be used to produce suggested upper limits on connection delays for the layout process, such that any layout that meets the limits would satisfy the performance requirements.

This work makes three contributions: (1) We give a generalization, called the limit-bumping algorithm (LBA), of a procedure of Youssef *et. al.* [1] that transforms initial connection delays into upper limits suitable for performance-driven layout. LBA is a simple way to distribute slacks using arbitrary allocation functions. (2) We show how lower and upper bounds on connection delays can be used in the computation of upper limits for initial layout and for layout improvement. (3) We demonstrate that major improvements in system performance can be achieved by integrating these techniques in an FPGA router.

1.1. General background

Early work [2,3] performed timing analysis using delay estimates, and gave critical signals higher weight during partitioning or higher priority during routing. The importance of interactions between timing analysis and place-and-route steps was highlighted in [4], but the weighting function used to mediate the interactions was not described. At each stage of recursive mincut in [5], non-critical connections got weights in inverse proportion to their slacks, and critical connections got slightly higher weights. In [6], improved performance was obtained by recursive partitioning and global routing with a re-weighting scheme in which a net's weight is simply incremented by one at any stage it is found to be critical; the evolving weights influence both partition costs and routing order.

A common problem with weight adjustment is that at the same time critical connections are improved, other connections can become critical. Another approach has been to perform layout with continuous guidance from path constraints. In [7], incremental timing analysis is used to evaluate individual moves during placement. Linear programming has been used at each stage of recursive partitioning to track path constraints dynamically during placement [8]. Related work in [9, 10] used quadratic programming and reduced sets of active constraints to produce high-performance placements more efficiently.

The task of transforming the results of timing analysis into guidance (typically weights) for placement is nontrivial, as evidenced by the many heuristics that have been proposed. An interesting formulation in [11] allows the derivation of an *optimum* formula for connection re-weighting; it requires as inputs from timing analysis a set of precise limits on individual connection delays. We will see later that the routing process in some technologies can utilize connection delay upper limits more directly, to good effect. This increases the importance of developing effective techniques to compute such limits. Work on computing limits is reviewed below.

1.2. Integration of path analysis with layout

The slack of a directed path π is defined as $R(\pi)-A(\pi)$, where $R(\pi)$ and $A(\pi)$ are the required and actual total propagation times along π, respectively. A system is said to have a long-path timing problem if for some path, signals fail to propagate through the constituent logic blocks and interconnect within the required time. (This corresponds to a negative slack.) We also define a slack for each individual source-to-load connection c:

$$slack(c) = \min_{paths\ \pi : c \in \pi} [slack(\pi)] \qquad (1)$$

Let R(c) and A(c) be the earliest required and latest actual arrival times at the load pin of c, respectively. An equivalent definition of the slack on connection c is then

$$slack(c) = R(c) - A(c)$$

Computing slacks is straightforward [12]. Two linear-time computations are performed, one in which A(c)'s propagate forward, and one in which R(c)'s propagate backward.

When layout is complete, slacks are readily computed; but during the layout process, connection delays are not yet known. Still, a performance-driven layout system can make use of path analysis by iterating the following steps: estimate delays --> compute resulting slacks --> suggest delay upper limits --> aim to meet upper limits. Our focus is the third step of this loop, in which upper limits are derived from slacks.

1.3. The zero-slack algorithm and variations

Hauge *et. al.* introduced the zero-slack algorithm (ZSA) [13]. This algorithm begins by computing slacks based on a tentative set of connection delays chosen so that they meet the timing requirements. The algorithm increases the delays in this set until they are maximal in the sense that they still meet the requirements, but a further delay increase on any connection would produce a violation. These delays are provided to the layout tool as upper limits.

The ZSA identifies a continuous path segment with minimum non-zero slack. Excess delay is distributed uniformly among connections on that path segment, slacks are updated on other connections that are affected, and the process is repeated until every connection has zero slack.

Excess delays can also be distributed in proportion to physical measures, e.g., capacitance per fanout or capacitance change per fanout [14]. In [1], the slack of a path is budgeted to its connections in proportion to the function

$$weight(c) = LF(c) * AcL(c) \qquad (2)$$

where $LF(c)$ is the delay per unit load on c's source pin, and $AcL(c)$ is the external capacitance of c's load pin; other weighting criteria appear in [15].

Luk [14] sped up the ZSA by omitting the recomputation of slacks on connections whose slacks are altered by delay increases on the minimum-slack segments. This can create intermediate slacks that are negative, unless the increases are appropriately bounded. In practice, all slacks converge to near zero in a few iterations.

Another iterative procedure to budget slacks, called *Iterative-Minimax-PERT*, is described in [1,15]. The weight function in (2) is used to define multipliers f(c):

$$f(c) = \frac{weight(c)}{\max_{\pi : c \in \pi} [weight(\pi)]}, \text{ where } weight(\pi) = \sum_{c : c \in \pi} weight(c). \qquad (3)$$

In each iteration, the delay of every connection c is incremented by $f(c) * slack(c)$. It is proven that all slacks decline monotonically and converge toward zero.

In the next section, we give a generalization of *Iterative-Minimax-PERT*, called the limit-bumping algorithm, that accepts arbitrary multipliers $f(c)$ as inputs. We identify simple conditions on the $f(c)$'s that are necessary and sufficient to guarantee that slacks go monotonically to zero. Finally, we introduce new methods for setting both initial delays and f(c)'s. Section 3 describes the application of these procedures to an FPGA routing program. Section 4 presents experimental results.

2. The limit-bumping algorithm

The inputs to the limit-bumping algorithm ("LBA") are a netlist, timing constraints, block delays, and for each connection c, an initial delay I(c) and a multiplier f(c). The outputs of LBA are delays U(c), such that any layout whose connection delays are less than or equal to U(c) will satisfy the timing constraints.

Following [1], at each connection c, the product of f(c) and c's current slack is added to the current delay in each iteration.

Limit-bumping algorithm

```
/* Step 1: initialize. */
U(c) = I(c);

/* Let slack(c, U) denote the slack on connection c, given the
   set of delays U. Increase U until slacks are near 0. */
do {
   /* Step 2: */
   compute slack(c, U) for all connections c;

   close_enough = near_zero (slacks); /* (stop criterion). */
   if (!close_enough)
       /* Step 3: distribute slacks. */
       for every connection c
           U(c) = U(c) + (f(c) * slack(c, U));
   }
while (!close_enough);
```

2.1. Convergence conditions

THEOREM 1: Every term slack(c, U) generated by LBA decreases monotonically to zero for arbitrary inputs with non-negative slack if and only if the multipliers f(c) satisfy

(a): f(c) > 0 for every connection c.

(b): $\sum_{c:c \in \pi} f(c) \leq 1$ for every path π.

Proof.

(NECESSITY): If either condition is violated, there will be inputs for which some slack does not decrease monotonically to zero. If (a) is violated, slack(c) may not decrease; if (b) is violated, slack(π) can become negative.

(SUFFICIENCY): If (a) is satisfied, then every positive slack decreases by at least a fixed fraction in each iteration, and hence converges to zero. We need only show that no slack can become negative, *i.e.*, that slack *decrease* ≤ slack.

Decrease in slack (path π)

$$= \sum_{c:c \in \pi} f(c) * slack(c)$$

$$\leq \sum_{c:c \in \pi} f(c) * slack(\pi) \quad \text{(From (1): slack(c) = min slack of all paths that include c)}$$

$$\leq slack(\pi) \quad \text{(by condition (b))}$$

Q.E.D.

From now on, we restrict ourselves to f(c) values that meet the conditions of theorem 1. We can thus safely refer to the multipliers f(c) as fractions.

2.2. Adapting known slack budgeting heuristics

Each heuristic in past implementations of ZSA has distributed path slack among connections in proportion to one or more functions weight(c). Alternatives already mentioned for weight(c) include:

$w_1(c) = 1$;	[13]
$w_2(c)$ = source fanout (c);	[13]
$w_3(c)$ = LF(c) * AcL(c);	[1]
$w_4(c)$ = fanout(c) * k(c);	[14]
$w_5(c)$ = fanout(c) * k(c) * capacitance(c);	[14]

where k(c) is a delay sensitivity (dt/dC for nets or dt/dR for source-sink pairs).

Any positive function *weight*(*c*) determines a corresponding set of fractions f(c) according to equation (3), which we restate here:

$$f(c) = \frac{weight(c)}{\max_{\pi:c \in \pi} [weight(\pi)]}, \text{ where } weight(\pi) = \sum_{c:c \in \pi} weight(c). \quad (3)$$

Furthermore, fractions defined this way satisfy the conditions of theorem 1 *by construction*.†

LBA is particularly efficient when using fractions $f(c)$ given by equation (3) and values *weight*(*c*) that do not change during the layout process. The fractions then need be computed only once. This can be done in linear time, *e.g.*, using depth-first searches. From then on, it is trivial to distribute slacks as soon as they are computed, using a single multiply-and-add per connection. No sorting or recomputing of slacks is necessary during delay distribution, as is required in most previous methods.

When we determine an $f(c)$ *a priori* by equation (3), we divide by the total weight on the static maximum-weight path through *c* rather than by the total weight on a path segment of minimum slack. Static fractions are thus "conservative", *i.e.*, sometimes smaller than what could be used. The only apparent downside of pre-computing the fractions f(c) is a possible increase in iterations for convergence - but we do not expect convergence to be much slower. Different paths through the same connection typically have similar weight. And when some path *does* have less weight (where pre-computation would be conservative) it is less likely to be critical.

† Consider a path π. If it is the max weight path through every one of its connections, then $\sum_{c:c \in \pi} f(c) = 1$ exactly; if it is not, then any connection *c* that is on a path with greater weight would contribute correspondingly *less* to the sum.

2.3. Uses of realistic lower bounds

The best available lower bounds on connection delays, which we denote L(c), are a reasonable choice of settings for inputs I(c) to LBA. One benefit of using L(c) can be seen by considering what can happen when the input I(c)'s are set to zero, as in [1]. An unsatisfiable upper limit U(c) (< L(c)) could result, even in designs for which layouts exist that meet the timing constraints. To avoid this possibility, we should at least ensure that I(c) ≥ L(c) for every connection.

It is worthwhile to compute slacks based on the delays L(c) in any case. The computation yields a "best conceivable" clock period T_{low}, *i.e.*, what would result if every connection achieved its lower bound simultaneously. This is useful, independent of the time T_{wish} that the designer wants. Even if $T_{wish} < T_{low}$ (what the designer wants is impossible) we probably still want to push the design's performance. The request often reduces to: "make it run as fast as it can".

When timing constraint values have not been specified, we need a way to set them automatically. We set the system goal

$$T_{goal} = G * T_{low} \qquad (G \geq 1) \qquad (4)$$

where the constant G is based on experience with the layout tool. Starting with a conservative T_{goal} can have practical advantages - even if $T_{wish} < T_{goal}$. First, the performance achieved often beats the goal. Second, a solution in hand whose performance falls short may be valuable, especially when improvement methods are available.

2.4. Use of realizable delays

Suppose our tool has produced a layout. We will use upper bounds $D(c)$, namely

$D(c)$ = connection delays achieved in existing layout,

to derive new upper limits U(c) that (1) are consistent with a specified improvement in system performance, and (2) optimize the chances that our layout tool will meet them.

Our approach to performance improvement uses the following new weight function:

$$w_6(c) = D(c) - L(c) \qquad (5)$$

Unlike the other weight functions we have considered, w_6 depends on the current state of the layout, and thus cannot be pre-computed. $w_6(c)$ is the gap between the delay achieved on a connection and its lower bound. It thus measures c's maximum "potential for improvement". Our idea is to set the upper limits U(c) so that the amount by which each delay is asked to improve is proportional to this room to improve. We first show how to use w_6 in the LBA, and then examine a tiny example in detail.

First, set the desired new period T_{goal}, and compute every slack(c, D). Then, compute fractions f(c) according to equation (3), using function $w_6(c)$ for weight(c). Using the resulting f(c) values, apply the variant of LBA described below.

Since we start with the old delays, a requirement of improved performance means that one or more slacks will initially be negative. This destroys the monotonic change property from theorem 1. Although in practice convergence is fast even when negative slacks are allowed [14], we prefer to preserve monotonic change by running LBA twice (in two stages). In stage 1, the delay upper limits U(c) are bumped only for those connections with negative slack. The same multiply-and-add operation is used, with one precaution. Since in this stage the limits are *decreasing*, we must check not to allow any connection's upper limit to be bumped below its lower bound. Step 3 of LBA as revised for stage 1 is thus:

```
/* Step 3: distribute NEGATIVE slacks,
   don't violate lower bounds. */
for every connection c
    if (slack(c, U) < 0)
        U(c) = max (U(c) + f(c) * slack(c, U), L(c));
```

Stage 2 is the ordinary LBA, with input delays taken from the U(c)'s produced by stage 1.

Example: Suppose $D(c_1)$=20, $L(c_1)$=10; $D(c_2)$=20, $L(c_2)$=15; and we have slack -6 to distribute between c_1 and c_2. Since $w_6(c_1)$=10 and $w_6(c_2)$=5, c_1 gets twice as much delay decrement as c_2. The resulting values are $U(c_1)$=16, $U(c_2)$=18.

The delay decrement $\Delta(c)$ asked of connection c is $D(c) - U(c)$. Equation (3) makes these decrements proportional to weight. With weight w_6, the effect is to equate the "approach fractions" of different connections' U(c) values from their achieved delays towards their lower bounds, where

$$approach\_fraction(c) \equiv \frac{(D(C) - U(c))}{(D(C) - L(C))} = \frac{\Delta(c)}{w_6(c)} \qquad (6)$$

(The approach fraction in the example is 0.4 for both c_1 and c_2.)

The approach fraction seems a reasonable measure of how difficult it will be for the layout process to meet a new limit U(c). Weight function w_6 attempts to balance this difficulty among connections as much as possible. It is in this sense that the layout tool's chances for success are optimized.

3. Application to FPGA routing

3.1. Performance-driven FPGA routing

As noted in section 1.1, past work on performance-driven place-and-route has focused more on placement than on routing, because for most technologies, improvements to the placement algorithms have much more impact on performance. In field-programmable gate arrays (FPGAs), however, routing is unusually important. The fraction of delay due to routing in FPGAs is typically from 40% to 60%, which is much greater than that for mask-programmed gate arrays.

The Xilinx XC4000 architecture [16] offers a hierarchy of routing resources distinguished by their path lengths and delay characteristics. General-purpose interconnects (figure 1) pass through switching matrices between the blocks and are best for medium-length, local interconnects.

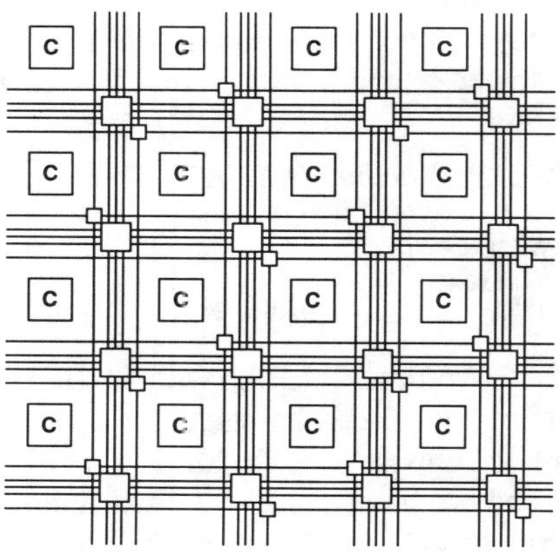

Figure 1. General-purpose wiring.

Other segments bypass alternate switchboxes, allowing longer paths with fewer switches (also shown in figure 1). Long lines (figure 2), which span the entire height or width of the array, are ideal for high-fanout nets and

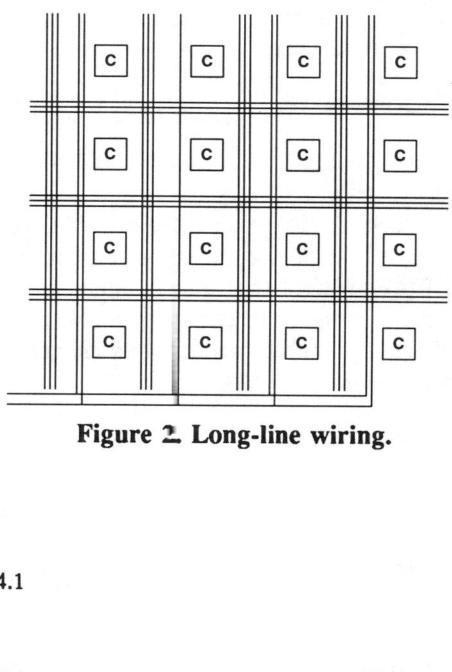

Figure 2. Long-line wiring.

time-critical signals. Special buffers can also be inserted in routing paths to speed up connections. The freedom to select the most appropriate resources for each connection means the router has more influence on delays than in traditional technologies. Improved routing algorithms offer great potential for enhanced performance in FPGAs.

To realize this potential for improvement, a router should be able to:

(1) Route connections within a specified delay limit when possible.

(2) Use different cost functions in different situations. For example, most routing choices must be made not according to delay alone, but according to a combination of delay and resource costs. Special nets such as power/ground signals should be oblivious to delay.

(3) support a retry mode to complete unrouted connections or improve existing routes. In this mode, previously completed connections can become temporarily unrouted.

Such capabilities are supported in existing routers, including a router for the Xilinx XC4000 family. Although this router is sensitive to delays, it has lacked the capability to decide what connection delay limits to aim for. Until now, a single upper limit has been used for all connections. What is needed is an optimized *set* of upper limits on connection delays. Guidance from path analysis lets the router realize its full potential to enhance system performance.

3.2. Initial routing solution

Following section 2.3, slacks are analyzed assuming that connection delays equal their lower bounds L(c), to find the best conceivable clock period T_{low}. Using (4) with G=1.5, our initial goal clock period is

$$T_{goal} = 1.5 * T_{low}.$$

Fractions f(c) are determined from equation (3) using weight function $w_1(c)=1$ (uniform slack distribution). We run LBA and use the resulting upper limits U(c) both to constrain the connections and to determine the order in which they are routed. As in the current delay-sensitive router, connections that require special resources are routed first. The others are routed in decreasing order of L(c)/U(c), so connections with tighter limits go first.

Connections that fail to meet their prescribed limits U(c) are left unrouted. If any connections fail, the limits are relaxed by 20% and the retry procedure is called. If failures still remain, the two steps of relaxing limits and calling retry are repeated until every connection is routed.

3.3. Improvement of an existing routing

We aim to reduce the clock period by one third of the maximum possible decrement:

$$T_{goal} = T_{achieved} - \frac{(T_{achieved} - T_{low})}{3}$$

Fractions f(c) are determined from equation (3) using the new weight function $w_6(c) = D(c) - L(c)$. We run the two-stage LBA from section 2.4 to generate new upper limits U(c). Every connection for which the already achieved delay D(c) exceeds the new U(c) is ripped up.

The connections are sorted in decreasing order of $D(c)/U(c)$, so that those whose delays most exceed their new limits go first. The retry procedure is then called. If unrouted connections remain, a loop is repeated as above in which their limits are relaxed by 20% and retry is called, until all are routed.

4. Experimental results

We experimented with 22 designs: *primary*1, from the MCNC layout benchmarks, and 21 MCNC finite state machine (fsm) examples. No logic optimization was attempted, so comparisons of absolute sizes with published optimization results are not meaningful.

Statistics on the designs are presented in Table 1.

design	elements	connections	ffs	ff transfer pairs	max levels	array size
bbara	137	392	12	45	5	8-by-8
bbsse	132	356	13	39	6	8-by-8
beecount	92	270	9	31	4	8-by-8
cse	240	679	16	55	7	10-by-10
dk14	164	523	7	27	5	8-by-8
dk15	104	323	4	12	5	8-by-8
dk16	459	1332	53	207	5	14-by-14
dk17	256	730	28	89	5	10-by-10
ex2	68	188	18	40	4	8-by-8
ex4	71	157	14	18	4	8-by-8
ex6	181	517	12	48	5	8-by-8
keyb	416	1217	26	66	7	12-by-12
mark1	95	201	16	32	4	8-by-8
opus	91	230	11	32	4	8-by-8
planet	388	1035	50	75	4	12-by-12
primary1	988	2426	269	2318	12	18-by-18
s1	389	1138	20	80	6	12-by-12
sand	630	1762	32	90	8	14-by-14
scf	579	1334	123	171	4	14-by-14
sse	132	356	13	39	6	8-by-8
styr	519	1494	31	95	7	14-by-14
tav	588	1769	18	32	7	16-by-16

Table 1. Statistics on the test designs after conversion to the Xilinx XC4000 family

For each design, we give counts of total elements (pads, function generators, and flip-flops), connections, flip-flops, and pairs of flip-flops involved in data transfer. We also indicate the maximum levels of connections on any one transfer path, and the size of the array of configurable logic blocks (CLBs)† used for layout.

Table 2 shows the minimum feasible clock period achieved for each design obtained by running the router three different ways. $T_{resource}$ is the result when delays are ignored, so that resource utilization alone is optimized. T_{delay} is the result with delay-sensitive routing, when every connection is given an identical upper limit on delay. T_{LBA} is the result using LBA to compute upper limits U(c) for initial routing and for one pass of routing improvement as described in sections 3.2 and 3.3. For comparison, the last column gives T_{low}, the feasible period if every connection could be routed at its lower bound delay. Table 2 also shows percentage improvements: from $T_{resource}$ to T_{delay}, from T_{delay} to T_{LBA}, and from T_{LBA} to T_{low}.

design	T_resource	T_delay (%improve) vs. T_resource	T_LBA (%improve) vs. T_delay	T_low (%improve) vs. T_LBA
sand	94.4 ns	101.0 (-7%)	68.7 (32%)	56.6 (18%)
s1	90.1 ns	66.4 (26%)	50.9 (23%)	44.6 (12%)
styr	83.3 ns	77.8 (7%)	59.8 (23%)	48.2 (19%)
scf	63.4 ns	47.7 (25%)	36.7 (23%)	30.0 (18%)
tav	122.7 ns	96.4 (21%)	73.8 (23%)	57.1 (23%)
dk17	63.2 ns	58.5 (7%)	45.9 (22%)	39.0 (15%)
planet	51.8 ns	48.3 (7%)	39.8 (18%)	30.3 (24%)
dk15	67.3 ns	50.7 (25%)	41.9 (17%)	36.0 (14%)
keyb	96.3 ns	73.5 (24%)	61.2 (17%)	51.0 (17%)
primary1	165.0 ns	131.1 (21%)	110.6 (16%)	82.4 (25%)
dk14	63.5 ns	51.2 (19%)	43.5 (15%)	36.6 (16%)
ex6	60.0 ns	47.7 (20%)	41.7 (13%)	36.6 (12%)
opus	36.7 ns	39.0 (-6%)	34.6 (11%)	29.3 (15%)
dk16	69.9 ns	49.6 (29%)	44.8 (10%)	38.3 (15%)
ex2	41.0 ns	30.4 (26%)	27.8 (9%)	23.9 (14%)
bbara	53.8 ns	42.4 (21%)	39.0 (8%)	35.2 (10%)
bbsse	45.5 ns	43.2 (5%)	40.3 (7%)	36.5 (9%)
beecount	43.8 ns	36.7 (16%)	34.3 (7%)	26.5 (23%)
sse	42.0 ns	42.6 (-1%)	40.1 (6%)	35.6 (11%)
cse	74.9 ns	63.0 (16%)	59.0 (6%)	47.9 (19%)
mark1	41.5 ns	32.8 (21%)	31.2 (5%)	27.8 (11%)
ex4	32.3 ns	29.8 (8%)	28.9 (3%)	24.3 (16%)
Average		15.0%	14.2%	16.2%

Table 2. Effect of router improvements on clock period

Not surprisingly, table 2 shows that delay-sensitive routing with a single upper limit for all connections gives substantial improvement in clock period (15% shorter, on average) vs. routing to minimize resource usage alone. There are cases in which delay-sensitive routing produces a longer clock period. This is understandable, because with a single limit, non-critical connections will compete for and sometimes win fast rout-

† Each CLB contains two flip-flops, two function generators capable of generating arbitrary functions of four Boolean variables, and other special-purpose logic.

ing resources, which can hurt the performance on critical paths.

Table 2 also shows that path-sensitive routing, guided by limits U(c) determined by LBA, yields *further* improvements to clock period for every design: 14% on average, and more than 20% in six cases. These results demonstrate that path analysis is especially valuable in improving system performance in FPGAs.

The percentages associated with the T_{low} column of table 2 tell us that not much more can be achieved solely through improvements to the router. We cannot expect every connection to achieve its minimum possible delay in the same layout. An average of at most 16% further performance improvement is possible, and in most cases even less. To improve performance further, we can apply LBA to earlier stages of layout.

5. Conclusions

(1) We have introduced LBA, a generalization of a known procedure to produce upper limits on connection delay, and demonstrated its utility to implement new approaches to slack allocation for performance-driven layout. (2) We showed how the process of generating upper limits can make use of realistic lower bounds and realizable upper bounds on connection delays. (3) We showed that integrating these techniques in an FPGA router yields major performance improvements.

Future work will study the effectiveness of LBA for other path types and for systems with multiple clocks. It will also investigate the use of LBA for performance improvement during FPGA technology mapping and placement.

References

[1] H. Youssef and E. Shragowitz, "Timing constraints for correct performance," *Proc. of ICCAD '90*, pp. 24-27, 1990.

[2] A.E. Dunlop, V.D. Agrawal, D.N. Deutsch, "Chip layout optimization using critical path weighting," *Proc. of the 21st Design Automation Conference*, pp. 278-281, 1984.

[3] M. Burstein and M.N. Youssef, "Timing influenced layout design," *Proc. of the 22nd Design Automation Conference*, pp. 124-130, 1985.

[4] S. Teig, R.L. Smith, and J. Seaton, "Timing-driven layout of cell-based ICs," *VLSI Systems Design*, pp. 63-73, May 1986.

[5] M. Marek-Sadowska and S.P. Lin, "Timing driven placement," *Proc. of ICCAD '89*, pp. 94-97, 1989.

[6] J. Garbers, B. Korte, H.J. Promel, E. Scheietzke, A. Steger, "VLSI-placement based on routing and timing information," *Proc. of the European Design Automation Conference*, pp. 317-321, 1990.

[7] W.E. Donath, R.J. Norman, B.K. Agrawal, S.E. Bello, S.Y. Han, J.M. Kurtzberg, P. Lowy, and R.I.McMillan, "Timing driven placement using complete path delays," *Proc. of the 27th Design Automation Conference*, pp. 84-89, 1990.

[8] M.A.B. Jackson and E.S. Kuh, "Performance-driven placement of cell based IC's," *Proc. of the 26nd Design Automation Conference*, pp. 370-375, 1989.

[9] M.A.B. Jackson, A. Srinivasan, and E.S. Kuh, "A fast algorithm for performance-driven placement," *Proc. of ICCAD '90*, pp. 328-331, 1990.

[10] A. Srinivasan "An algorithm for performance-driven initial placement of small-cell ICs," *Proc. of the 28th Design Automation Conference*, pp. 636-639, 1991.

[11] R.-S. Tsay and J. Koehl, "An analytic net weighting approach for performance optimization in circuit placement," *Proc. of the 28th Design Automation Conference*, pp. 620-625, 1991.

[12] R.B. Hitchcock, Sr., G.L. Smith, and D.D. Cheng, "Timing analysis of computer hardware," *IBM J. Res. Devlop.*, Vol. 26, No. 1, pp. 100-108, 1982.

[13] P.S. Hauge, R. Nair, E.J. Yoffa, "Circuit placement for predictable performance," *Proc. of ICCAD '87*, pp. 88-91, 1987.

[14] W.K. Luk, "A fast physical constraint generator for timing driven layout," *Proc. of the 28th Design Automation Conference*, pp. 626-631, 1991.

[15] H. Youssef, "Timing analysis of cell based VLSI designs," Computer and Information Sciences, University of Minnesota, Ph.D. Thesis, January 1990.

[16] Xilinx, *The XC4000 data book*, Xilinx, Inc., 1991.

The Role of Long and Short Paths in Circuit Performance Optimization

Siu Wing Cheng* Hsi-chuan Chen†‡ David H.C. Du†
Department of Computer Science
University of Minnesota

Andrew Lim§¶
Dept of Computer and Information Sciences
University of Florida

Abstract

We consider the problem of determining the smallest clock period for a combinational circuit by considering both the long and short paths. To bring in the timing of the circuit, we make use of a new class of paths called the shortest destabilizing paths as well as the longest sensitizable paths. The bounds on the clock period can alternatively be viewed as optimization objectives. At the physical level, the problem becomes applying transistor sizing and delay buffer insertion to achieve specified upper bounds on clock period and latency. We present experimental results that reflect the complexity of the optimization problem. The clock period determination can also be extended to circuits with feedbacks.

1 Introduction

The performance of a combinational circuit is usually measured by the delay between the availability of successive outputs. Under the condition that the outputs are latched together at the rise of the clock, the performance can then be measured by the clock period. In order to latch the outputs properly, the outputs must have been stable for at least t_s time, which is called the *set-up time*, when the clock signal arrives. Moreover, the outputs are also required to remain stable for at least t_h time, which is called the *hold time*, after the clock signal arrives. In a conservative design, the latches at the primary inputs and the primary outputs may be triggered by the same clock signal and the delay of the longest path from the primary inputs to the primary outputs is used to determine the clock period. Referring to Figure 1, the clock rises at time 0 to feed an input vector to the circuit and the outputs become stable at time T, where T is the delay of the longest path. Then the next clock signal is set to rise at time x, where $x \geq T + t_s$, to latch the first set of outputs and to apply the next input vector. The outputs are also required to remain stable until time $x+t_h$. In this clocking scheme, if it is desired to reduce the clock period to improve the circuit performance, then the delay of the longest paths must be reduced. However, since there are certainly many gates and leads on the longest paths, there can be a fairly large inherent lower bound on the clock period.

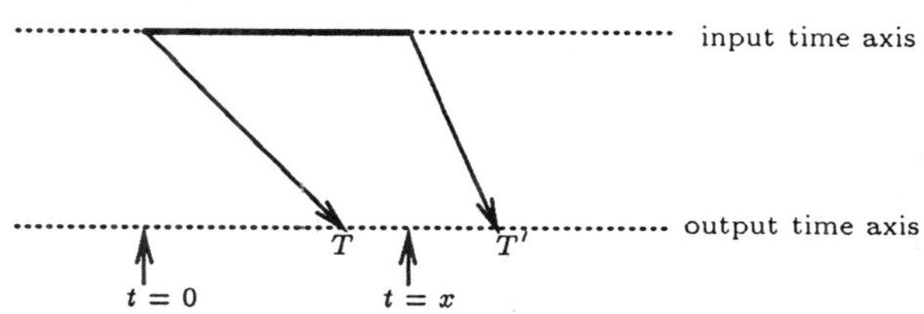

Figure 1. T is the time at which the output signal becomes stable and T' is the time at which the output signal becomes unstable. The clock signal should arrive at time x such that $T + t_s \leq x \leq T' - t_h$.

The major deficiency of the above clocking scheme is that both the *latency* and the clock period are tied up with the same characteristic of the circuit, namely,

*Current address: Dept of Computer Science, HKUST, HK.
†Research supported in part by NSF grant MIP-9007168.
‡Current address: AT&T Bell Labs., Murray Hill, NJ 7974.
§Research supported in part by NSF grant MIP86-17374.
¶Current address: ITI, 71 Science Park Dr., Singapore 0511.

the delay of the longest path. The latency of a combinational circuit is defined to be the time one needs to wait before the first set of outputs is latched. Clearly, the latency of a circuit is tightly related to the delay of the longest path, but it also partly depends on the clock period because the clock period determines when the outputs will be latched. If input vectors are being applied to the circuit in a fairly uninterrupted manner, then we may have a lower requirement on the latency than on the clock period. That is, it may be acceptable to wait for a while for the first set of outputs as long as the future outputs come in rapid succession. Moreover, it is unnecessary to specify exactly when the clock signal will arrive at the primary outputs. As long as there is a time interval of width at least $t_s + t_h$ during which the outputs will remain stable, the clock skew can be adjusted so that the clock signal will arrive at the right time to latch the outputs.

In this paper, we consider the problem of determining the smallest clock period for a combinational circuit by considering both the long and short paths. To bring in the timing of the circuit, we make use of a new class of paths called the *shortest destabilizing paths* as well as the longest sensitizable paths. We show that considering both the longest sensitizable and shortest destabilizing paths together does not always give you a valid bound. The bounds on the clock period can alternatively be viewed as optimization objectives. At the physical level, the problem becomes applying transistor sizing and delay buffer insertion to achieve specified upper bounds on clock period and latency. We present experimental results that reflect the complexity of the optimization problem. The clock period determination can also be extended to circuits with feedbacks.

2 Destabilizing paths

We assume that the reader is familiar with the concepts like *stable value* and *stable time* of a gate, *controlling* and *non-controlling* inputs to a gate, and *sensitizable paths.* The notion of sensitizability is very useful in determining the stable time of the primary outputs. However, given a circuit that is stabilized under an input vector, it is also important to study how soon the primary outputs will be affected once that input vector changes because the output must be latched before it disappears. The time needed for the signals to change from stable to unstable at a lead and the output of a gate are called the **destabilizing time** of the lead and the gate, respectively.

Given both the stable value and the destabilizing time of each input to a gate G, the destabilizing time of G can be computed as follows. If there are one or more controlling inputs to G under v, then the controlling input with the latest destabilizing time is considered to destabilize G. The destabilizing time of G is equal to the destabilizing time of this controlling input plus $d(G)$. On the other hand, if all the inputs to G are non-controlling inputs under v, then the input with the earliest destabilizing time is considered to destabilize G. The destabilizing time of G is then equal to the destabilizing time of this input plus $d(G)$. Hence, an input f to a gate G **destabilize** G if and only if f is the controlling input with the latest destabilizing time, or all the inputs to G are non-controlling and f is the one with the earliest destabilizing time.

For a path $P = (G_0, f_0, G_1, f_1, \cdots, f_{m-1}, G_m)$, we say that P is **destabilizing** under an input vector v if each on-input f_i of P destabilizes its succeeding gate G_{i+1}. A path is defined to be a **destabilizing path** if it is destabilizing under some input vector. The delay of the **shortest destabilizing path**[1] is the shortest time that a circuit takes for its primary outputs to change from stable to unstable. Due to uncertainties, it is usually not possible to determine the exact delay of a gate. Therefore, the *minimum* possible delay of each gate is used instead (in contrast to using the maximum possible delay in determining sensitizability).

The delay of the shortest destabilizing path is a lower bound of how soon the primary outputs will change when the input vector changes. The similarity between the definitions of the destabilizing path and the sensitizable path also allows the different criteria proposed (like those in [1, 2, 6]) to compute the delay of the longest sensitizable path to be used, with no or minor modification, to compute the delay of the shortest destabilizing path. An algorithm to find the delay of the shortest destabilizing path is described in [5].

3 Feasible settings of the clock period

Suppose an input vector v_1 is applied to a combinational circuit at time 0. The signal for v_1 takes T_1 time to propagate to and stabilize at the output. The second input vector v_2 is applied to the circuit one clock period, C_p, after v_1. Suppose that the switch from v_1 to v_2 will make the output first become unstable at time T_2. Then there is a sufficiently large stable range of the output for v_1 if and only if $T_2 \geq T_1 + t_c$, where $t_c = t_s + t_h$.

[1] The same concept is also defined in [5], but the term *shortest sensitizable path* is used instead. We find it confusing to overload the meaning of the adjective *sensitizable*, and therefore we change it.

By definition, $T_1 \leq \tau_{\max}$ and $T_2 \geq C_p + \tau_{\min}$, where $\tau_{\max}$ and $\tau_{\min}$ are the delay of the longest path and the shortest path, respectively. Hence, if C_p is selected such that $C_p + \tau_{\min} \geq \tau_{\max} + t_c \Leftrightarrow C_p \geq \tau_{\max} - \tau_{\min} + t_c$, then we are guaranteed to obtain a sufficiently large stable range at the output (i.e., $T_2 \geq T_1 + t_c$). This clocking scheme has been studied before [3, 4].

However, $\tau_{\max}$ may be a poor upper bound for T_1 because the longest paths may not be sensitizable. Thus, it may be possible to improve the bound for C_p by replacing $\tau_{\max}$ by the delay of the longest sensitizable path $\tau^*_{\max}$. Similarly, it may be possible to improve the bound for C_p by replacing $\tau_{\min}$ by the delay of the shortest destabilizing path $\tau^*_{\min}$. Therefore, there seems to be three other potentially better estimates of C_p, namely, $\tau^*_{\max} - \tau_{\min} + t_c$, $\tau_{\max} - \tau^*_{\min} + t_c$, and $\tau^*_{\max} - \tau^*_{\min} + t_c$. We prove the correctness of the bounds $\tau_{\max} - \tau_{\min} + t_c$, $\tau^*_{\max} - \tau_{\min} + t_c$, and $\tau_{\max} - \tau^*_{\min} + t_c$ in the following. We shall point out the fallacy of the bound $\tau^*_{\max} - \tau^*_{\min} + t_c$ in Section 4.

Theorem 1 *Any combinational circuit can function correctly if the clock period is set to be larger than or equal to one of the following:* $\tau_{\max} - \tau_{\min} + t_c$, *or* $\tau^*_{\max} - \tau_{\min} + t_c$, *or* $\tau_{\max} - \tau^*_{\min} + t_c$. *The latency of the circuit will be* $\tau_{\max} + t_s$, $\tau^*_{\max} + t_s$, *and* $\tau_{\max} + t_s$, *respectively.*

Proof Given a combinational circuit, we first apply an arbitrary input vector v to the circuit at time 0 and then apply another arbitrary input vector C_p time later, where C_p is set to be one of the three bounds. Consider any gate G with delay $d(G)$. Suppose *long*(G) and *short*(G) are the longest and shortest partial path from the primary inputs to G (excluding G itself). Let $\tau_{\max}(G)$ and $\tau_{\min}(G)$ be the delay of *long*(G) and *short*(G), respectively. Similarly, we define $long_v(G)$ and $short_v(G)$ to be the longest sensitizable and the shortest destabilizing partial paths, under v, from the primary inputs to G (excluding G itself). Let $\tau^v_{\max}(G)$ and $\tau^v_{\min}(G)$ be the delay of $long_v(G)$ and $short_v(G)$, respectively. Consider the input leads of the gate G. If some input leads hold stable values that are controlling, then we ignore all the input leads that hold the non-controlling value. As a result, we are left with a set, *active_leads*(G), of input leads all of which hold the controlling value or the non-controlling value. We prove the validity of each bound separately in the following.

Consider $C_p \geq \tau_{\max} - \tau_{\min} + t_c$. We claim that there exists a time interval $[\tau_{\max}(G) + d(G), \tau_{\min}(G) + d(G) + C_p]$ during which the output of G is stable. Note that this time interval has a non-zero width of at least t_c because $\tau_{\min}(G) + C_p - \tau_{\max}(G) \geq (\tau_{\max} - \tau_{\min}) - (\tau_{\max}(G) - \tau_{\min}(G)) + t_c \geq t_c$. Among all the input leads in *active_leads*(G), let the latest time at which an input signal becomes stable be T_1 and let the earliest time at which an input signal becomes unstable be T_2. Clearly, $T_2 \geq \tau_{\min}(G) + C_p$ and $T_1 \leq \tau_{\max}(G)$. Also, $\tau_{\min}(G) + C_p - \tau_{\max}(G) \geq t_c > 0$. Thus, the intersection $[T_1, T_2]$ of the stable ranges of all the input leads in *active_leads*(G) is non-empty and it contains $[\tau_{\max}(G), \tau_{\min}(G) + C_p]$. Hence, $[\tau_{\max}(G) + d(G), \tau_{\min}(G) + d(G) + C_p]$ must be a time interval during which the output of G is stable. This proves the claim. Now, consider all the gates that are connected to the primary outputs. By the claim, the intersection of the stable ranges at the outputs of all these gates must contain the interval $[\tau_{\max}, \tau_{\min} + C_p]$. The width of $[\tau_{\max}, \tau_{\min} + C_p]$ must be at least t_c as $C_p \geq \tau_{\max} - \tau_{\min} + t_c$. Therefore, it suffices to adjust the clock skew so that the clock signal arrives at time $\tau_{\max} + t_s$ and the output signal for the input vector v can then be latched properly. The latency is clearly equal to $\tau_{\max} + t_s$.

Consider $C_p \geq \tau^*_{\max} - \tau_{\min} + t_c$. We claim that if G lies on a sensitizable path under v, then there exists a time interval $[\tau^v_{\max}(G) + d(G), \tau_{\min}(G) + d(G) + C_p]$ during which the output of G is stable. Note that $\tau_{\min}(G) + C_p - \tau^v_{\max}(G) \geq (\tau^*_{\max} - \tau^v_{\max}(G) + \tau_{\min}(G)) - \tau_{\min} + t_c$. Since $long_v(G)$ can be extended to become a sensitizable path, $\tau^*_{\max} \geq \tau^v_{\max}(G)$. Moreover, the length of the concatenation of *short*(G) and this extension of $long_v(G)$ must be at least $\tau_{\min}(G)$. Therefore, $(\tau^*_{\max} - \tau^v_{\max}(G) + \tau_{\min}(G)) - \tau_{\min} \geq 0$ and thus the width of $[\tau^v_{\max}(G) + d(G), \tau_{\min}(G) + d(G) + C_p]$ is at least t_c. We prove the claim by induction and the basis step is trivial. Suppose that all the input leads in *active_leads*(G) hold the non-controlling value. Let T_1 be the latest time at which an input signal becomes stable and let T_2 be the earliest time at which an input signal becomes unstable. By definition, $T_1 = \tau^v_{\max}(G)$ and $T_2 = \tau^v_{\min}(G) + C_p \geq \tau_{\min}(G) + C_p$. Therefore, the intersection $[T_1, T_2]$ of the stable ranges of all input leads in *active_leads*(G) is non-empty and contains $[\tau^v_{\max}(G), \tau_{\min}(G) + C_p]$. Hence, $[\tau^v_{\max}(G) + d(G), \tau_{\min}(G) + d(G) + C_p]$ must be an interval during which the output of G is stable. Suppose that all the input leads in *active_leads*(G) hold the controlling value. Consider the union of the stable ranges of these input leads and let $[T_1, T_2]$ be the first interval in the union. By definition, $T_1 = \tau^v_{\max}(G)$ and $T_2 \geq \tau_{\min}(G) + C_p$. Therefore, all the input signals are stable within $[\tau^v_{\max}(G), \tau_{\min}(G) + C_p]$. Thus, the output of G must also be stable within $[\tau^v_{\max}(G) +$

$d(G), \tau_{\min}(G) + d(G) + C_p]$. This completes the induction step and the claim is proved. Now, consider all the gates that are connected to the primary outputs. By the claim, the intersection of the stable ranges at the outputs of all these gates must contain the interval $[\tau^*_{\max}, \tau_{\min} + C_p]$. The width of $[\tau^*_{\max}, \tau_{\min} + C_p]$ must be at least t_c as $C_p \geq \tau^*_{\max} - \tau_{\min} + t_c$. Therefore, the output signals for the input vector v can be latched properly. (An appropriate amount of clock skew may need to be introduced as discussed before.) The latency is clearly equal to $\tau^*_{\max} + t_s$.

Consider $C_p \geq \tau_{\max} - \tau^*_{\min} + t_c$. We claim that if G lies on a destabilizing path, then there exists a time interval $[\tau_{\max}(G) + d(G), \tau^v_{\min}(G) + d(G) + C_p]$ during which the output of G is stable. Note that $\tau^v_{\min}(G) + C_p - \tau_{\max}(G) \geq \tau_{\max} - (\tau_{\max}(G) + \tau^*_{\min} - \tau^v_{\min}(G)) + t_c$. Since $short_v(G)$ can be extended to become a destabilizing path, $\tau^v_{\min}(G) + x \geq \tau^*_{\min} \Rightarrow x \geq \tau^*_{\min} - \tau^v_{\min}(G)$, where x is the length of such an extension. Moreover, $\tau_{\max} \geq \tau_{\max}(G) + x \geq \tau_{\max}(G) + \tau^*_{\min} - \tau^v_{\min}(G)$ and thus the width of $[\tau_{\max}(G)+d(G), \tau^v_{\min}(G)+d(G)+C_p]$ is at least t_c. We prove the claim by induction and the basis step is trivial. Suppose that all the input leads in *active_leads*(G) hold the non-controlling value. Let T_1 be the latest time at which an input signal becomes stable and let T_2 be the earliest time at which an input signal becomes unstable. By definition, $T_1 = \tau^v_{\max}(G) \leq \tau_{\max}(G)$ and $T_2 = \tau^v_{\min}(G)+C_p$. Therefore, the intersection $[T_1, T_2]$ of the stable ranges of all input leads in *active_leads*(G) is non-empty and contains $[\tau_{\max}(G), \tau^v_{\min}(G) + C_p]$. Hence, the output of G must be stable within $[\tau_{\max}(G)+d(G), \tau^v_{\min}(G)+d(G)+C_p]$. Suppose that all the input leads in *active_leads*(G) hold the controlling value. Consider the union of the stable ranges of these input leads and let $[T_1, T_2]$ be the last interval in the union. By definition, $T_1 \leq \tau_{\max}(G)$ and $T_2 = \tau^v_{\min}(G)+C_p$. Therefore, $[\tau_{\max}(G), \tau^v_{\min}(G)+C_p]$ is an interval in which all the input signals are stable. Thus, the output of G must be stable within $[\tau_{\max}(G)+d(G), \tau^v_{\min}(G)+d(G)+C_p]$. This completes the induction step and the claim is proved. Now, consider all the gates that are connected to the primary output. By the claim, the intersection of the stable ranges at the outputs of all these gates must contain $[\tau_{\max}, \tau^*_{\min} + C_p]$. The width of $[\tau_{\max}, \tau^*_{\min} + C_p]$ must be at least t_c as $C_p \geq \tau_{\max} - \tau^*_{\min} + t_c$. Therefore, the output signals for the input vector v can be latched properly. (An appropriate amount of clock skew may need to be introduced as discussed before.) The latency is clearly equal to $\tau_{\max} + t_s$. □

Remark 1 : The bound $\tau_{\max} - \tau_{\min} + t_c$ is the most conservative because the C_p will then be large enough so that the stable ranges of the inputs to each gate must overlap with each other. Thus, the stable range of the output of each gate consists of a **single** time interval.

Remark 2 : The bounds $\tau^*_{\max} - \tau_{\min} + t_c$ and $\tau_{\max} - \tau^*_{\min} + t_c$ are more aggressive. If some input leads hold the controlling value, C_p is no longer large enough to guarantee overlapping of the stable ranges of any two input leads. Therefore, it is possible that the stable range of the output of each gate consists of several disjoint time intervals. Using the bound $\tau^*_{\max} - \tau_{\min} + t_c$, the outputs will be latched within the subinterval $[\tau^*_{\max}, \tau_{\min}+C_p]$ of the **first** time interval in the stable range of each output gate. Using the bound $[\tau_{\max}, \tau^*_{\min}+C_p]$, the outputs will be latched within the subinterval $[\tau_{\max}, \tau^*_{\min} + C_p]$ of the **last** time interval in the stable range of each output gate.

We believe that the bound $\tau^*_{\max} - \tau_{\min} + t_c$ will be the best of the three in practice because:

1. it produces the smallest possible latency, and
2. $\tau_{\min}$ should be very close to $\tau^*_{\min}$ in most cases and therefore $\tau^*_{\max} - \tau_{\min} + t_c$ is usually a good approximation of the bound $\tau^*_{\max} - \tau^*_{\min} + t_c$.

By our experimental result in Section 5, the prediction $\tau_{\min} \approx \tau^*_{\min}$ is at least true for several typical circuits in ISCAS85.

4 On the bound $\tau^*_{\max} - \tau^*_{\min} + t_c$

In this section, we describe two examples in which the bound $\tau^*_{\max} - \tau^*_{\min} + t_c$ is invalid.

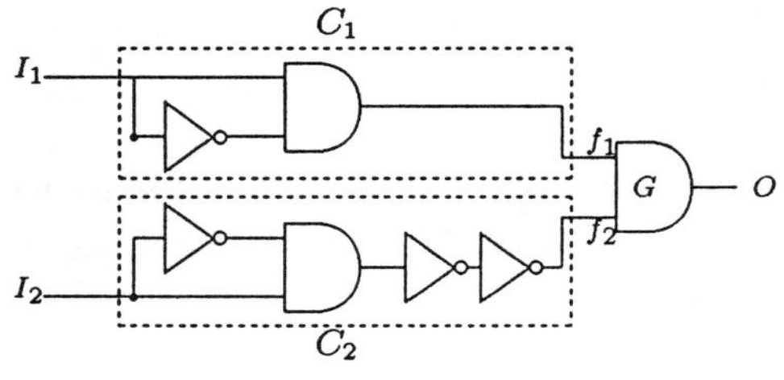

Figure 2.

Consider the circuit in Figure 2. The stable range of the output of C_1 is a single interval and its width is equal to the clock period. This is also true for C_2. The subcircuits C_1 and C_2 generate a logic 0 independent of the input vector. It is important to realize that this does not imply that the signals propagated along the leads f_1 and f_2 are always equal to logic 0. When the input vector changes, hazard may occur at f_1 and f_2 (after some propagation delay) that will make the signals at f_1 and f_2 unstable. Furthermore, by using inverters of sufficiently large delay in C_2, it can be guaranteed that the delay of every path in C_2 is larger than the delay of any path in C_1 by at least t_c. Since logic 0 is the controlling input to the AND gate G, a longest sensitizable path must pass through C_1 and then through G along f_1. On the other hand, the output of G will remain stable as long as the signal at f_2 is stable. Therefore, a shortest destabilizing path must pass through C_2 and then through G along f_2. Hence, we conclude that $\tau^*_{\text{min}} > \tau^*_{\text{max}} + t_c$ for this particular circuit. Thus by the inequality $C_p \geq \tau^*_{\text{max}} - \tau^*_{\text{min}} + t_c$, it would then be possible to set C_p to be smaller than t_c. In this case, the stable ranges of the two inputs of G will not overlap at all. Therefore, the output stable range of G will consist of two disjoint intervals and each of them has a width less than t_c. Hence, it is impossible to latch the output properly.

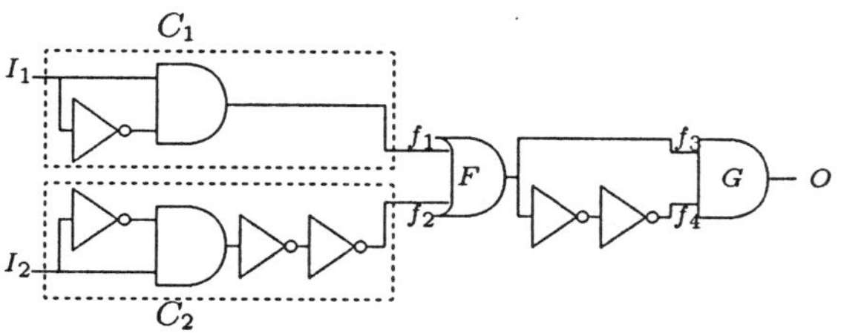

Figure 3.

It can be observed that the bound $\tau^*_{\text{max}} - \tau^*_{\text{min}} + t_c$ does not hold for the circuit in Figure 2 because $\tau^*_{\text{min}} > \tau^*_{\text{max}}$. Thus, it is natural to ask whether C_p can be set to be $\tau^*_{\text{max}} - \tau^*_{\text{min}} + t_c$ for a circuit in which $\tau^*_{\text{max}} \geq \tau^*_{\text{min}}$. It turns out that the bound can still be invalid. Consider the circuit in Figure 3. The sub-circuits C_1 and C_2 are as in Figure 2. The outputs of C_1 and C_2 are connected to an OR gate F. F has two fan-outs which are connected to an AND gate G. Since logic 0 is a non-controlling value to F and the signal at f_1 must stabilize before the signal at f_2, a longest sensitizable path must pass through C_2. Similarly, a shortest destabilizing path must pass through C_1. Since logic 0 is a controlling value to G, the inverters at the lead f_4 serve to lengthen the shortest destabilizing paths. By adjusting the delay of the inverters at the lead f_4, we can lengthen the shortest destabilizing paths until $\tau^*_{\text{max}} = \tau^*_{\text{min}}$. According to the inequality $C_p \geq \tau^*_{\text{max}} - \tau^*_{\text{min}} + t_c$, we can then set the clock period to $\tau^*_{\text{max}} - \tau^*_{\text{min}} + t_c = t_c$ to obtain the best performance. However, if the clock period is equal to t_c, then the stable ranges at the input leads f_1 and f_2 do not overlap at all. (Recall that the delay of every path in C_2 is larger than the delay of any path in C_1 by at least t_c.) This implies that there is no stable range at the output of F, and thus there must be no stable range at the output of G.

5 The optimization problem and experimental results

The optimization problem is to modify the circuit so as to achieve a desired clock period C_p. Very often, there will also be an additional constraint on the latency LT. We focus on optimization at the physical level, where transistor sizing and delay buffer insertion techniques are employed. Our goal is to meet the performance criteria specified by C_p and LT, while minimizing the circuit area. We have experimented with several typical circuits in ISCAS85 to reflect the complexity of the optimization problem. We are mainly interested in the numbers of long (sensitizable) and short (destabilizing) paths and the numbers of gates covered by them because the optimization process is tightly related to them. The results are tabulated in Table 1, Table 2, and Table 3. For simplicity, we assume that $t_c = t_s + t_h$ is zero. We used a LT equal to 95 percent of the delay of the longest path of each circuit, and for each circuit we tried C_p equal to 80 percent of the longest path delay. In Table 1, we list out the number of long paths, the number of gates covered by the long paths, the number of long sensitizable paths, and the number of gates covered by the long sensitizable paths. As expected, there are significantly more long paths (gates covered by long paths) than long sensitizable paths (gates covered by long sensitizable paths). This indicates that the optimization problem is probably harder for the bounds $\tau_{\text{max}} - \tau_{\text{min}}$ and $\tau_{\text{max}} - \tau^*_{\text{min}}$ when compared with $\tau^*_{\text{max}} - \tau_{\text{min}}$. In Table 2, we set a bound of $LT - C_p$ on the length of a short (destabilizing) path. It is interesting to observe that the number of short paths is the same as the number of short destabilizing paths in all circuits except C5315. This agrees with the prediction that $\tau_{\text{min}} \simeq \tau^*_{\text{min}}$. This indicates that the bound $\tau^*_{\text{max}} - \tau_{\text{min}}$ can be expected to be as tight as $\tau^*_{\text{max}} - \tau^*_{\text{min}}$ in most cases. Moreover, Table 2 shows that there are fairly few short (destabilizing) paths that need to be lengthened. The potential of the optimization under our formulation is also highly dependent on the degree of sharing of the long (sensitizable) and short (destabilizing) paths. In general, we would like them to share as few gates as possible. In Table 3, this can be seen to be true for our test cases.

Circuit	LT(ns.)	$\#(L)$	$\#(G_L)$	$\#(L^*)$	$\#(G_{L^*})$
C432	54.50	20,412	154	6,149	144
C1908	72.70	2,128	210	0	0
C2670	82.50	2,848	191	320	88
C3540	93.70	16,996	427	267	146
C5315	94.30	2,280	210	26	154
C7552	81.00	423	301	5	74

Table 1. $\#(L)$ (resp. $\#(L^*)$) is the number of long (resp. long sensitizable) paths, and $\#(G_L)$ (resp. $\#(G_{L^*})$) is the number of gates covered by long (resp. long sensitizable) paths.

Circuit	$LT - C_p$(ns.)	$\#(S)$	$\#(G_S)$	$\#(S^*)$	$\#(G_{S^*})$
C432	8.60	60	38	60	38
C1908	11.50	75	146	75	146
C2670	13.00	274	216	274	216
C3540	14.80	68	143	68	143
C5315	14.90	709	766	708	766
C7552	12.80	115	140	115	140

Table 2. $\#(S)$ (resp. $\#(S^*)$) is the number of short (resp. short destabilizing) paths, and $\#(G_S)$ (resp. $\#(G_{S^*})$) is the number of gates covered by short (resp. short destabilizing) paths.

6 Extension to circuits with feedbacks

Our results can also be extended to circuits with feedbacks. In general, a feedback will appear as in Figure 4. The latch on the feedback connection serves two purposes: (1) to store the current output, and (2) feed it back into the circuit at the next clock signal. As indicated before, we may need to introduce skew between the clock signal that triggers the input latches and the clock signal that triggers the output latches. Therefore, we first duplicate the latch on each feedback connection, so that one latch serves as the output latch and the other serves as the input latch. These two latches can now be triggered by two different clock signals. Consider gate G. In order to ensure the correctness of the output of G, the signals at input leads f_2 and f_3 under the current input vector must remain stable until the output signal for the previous input vector propagates back to G at f_1. In particular, if G is connected to the primary inputs, then the next input vector must not be fed into the circuit until certain outputs for the previous input vector are available. This motivates the following theorem.

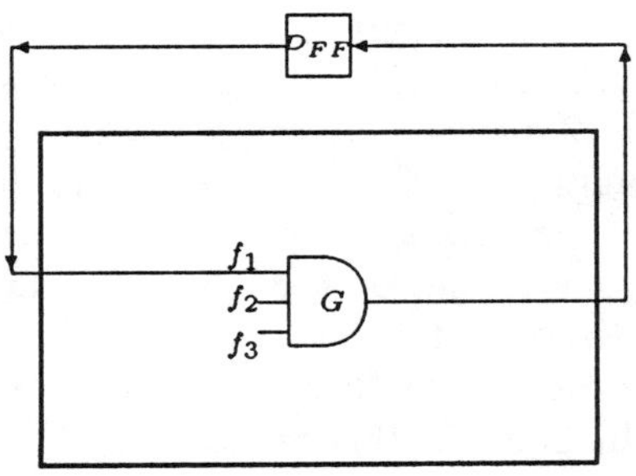

Figure 4.

Circuit	$\#(G_{LS})$	$\#(G_{LS^*})$	$\#(G_{L^*S})$
C432	16	16	16
C1908	1	1	0
C2670	15	15	14
C3540	26	26	8
C5315	33	33	23
C7552	5	5	5

Table 3. $\#(G_{LS})$ is the number of gates covered by long and short paths. $\#(G_{LS^*})$ is the number of gates covered by long and short destabilizing paths. $\#(G_{L^*S})$ is the number of gates covered by long sensitizable and short paths.

Theorem 2 *After duplicating the latch on each feedback connection, a circuit can function correctly if the clock period is set to be at least* $\max(\Delta\tau, \tau_b)$. *The parameter* $\Delta\tau$ *can be* $\tau_{\max} - \tau_{\min}$, *or* $\tau^*_{\max} - \tau_{\min}$, *or* $\tau_{\max} - \tau^*_{\min}$. *The parameter* τ_b *is the delay of the longest sensitizable path whose beginning and end are connected by a feedback.* □

References

[1] D. Brand and V. Iyengar. Timing Analysis using Functional Analysis. Technical report, IBM Thomas J. Watson Research Center, 1986.

[2] H. Chen and David Du. Path Sensitization in Critical Path Problem. In *IEEE International Conference on Computer-Aided Design*, 1991.

[3] L. Cotten. Maximum-rate pipeline systems. In *AFIPS Proc. 1969 Spring Joint Computer Conference*, volume 34, pages 581–586, 1969.

[4] B. Ekroot. *Optimization of pipelined processors by insertion of combinational logic delay.* PhD thesis, Department of Electrical Enginerring, Stanford University, September 1987.

[5] L. Liu and H. Chen D. Du. The calculation of signal stable ranges in combinational circuits. In *IEEE International Conference on Computer-Aided Design*, 1991.

[6] P. McGeer and R. Brayton. Efficient Algorithms for Computing the Longest Viable Path in a Combinational Network. In *26th Design Automation Conference*, pages 561–567, 1989.

Certified Timing Verification and the Transition Delay of a Logic Circuit

Srinivas Devadas
Department of EECS
MIT, Cambridge

Kurt Keutzer
Synopsys, Inc.
Mountain View

Sharad Malik
Department of EE
Princeton University

Albert Wang
Synopsys, Inc.
Mountain View

Abstract

Most research in timing verification has implicitly assumed a single vector *floating* mode computation of delay which is an approximation of the multi-vector transition delay. In this paper we examine the transition delay of a circuit and demonstrate that the transition delay of a circuit can differ from the floating delay of a circuit. We then provide a procedure for directly calculating the transition delay of a circuit. The most practical benefit of this procedure is the fact that it not only results in a delay calculation but outputs a vector sequence that may be timing simulated to *certify* static timing verification.

1 Introduction

The longest-path delay of a circuit is simply the sum of the cumulative delays of a circuit along the longest graphical path. This measure of delay is still used in most static timing verifiers but has the deficiency that it does not take into account *false paths*. To remedy this deficiency the *floating delay* of a circuit may be analyzed. The floating delay of a circuit is the delay under a single-vector static analysis condition that considers the Boolean behavior of the circuit but makes conservative assumptions about the state of the circuit before the single vector is applied. A number of techniques have been proposed for computing the true floating delay of circuit, but a significant step was taken in [10] where a technique that provided correctness in the light of *monotone speed-ups* was demonstrated. Further improvements were made in [2, 5] where techniques that more precisely identified the critical path were presented. The transition delay of a circuit is the delay under a multi-vector dynamic-analysis condition that makes no assumptions about the state of the circuit before the vector sequence is applied. A circuit is presented in this paper whose true floating delay is greater than its transition delay; thus, the floating analysis condition itself has some deficiencies.

A full consideration of transition delay requires examination of fundamental assumptions about delay modeling (e.g. bounded or fixed) and the relationship between the computed delay of a circuit and the minimum clock period. Due to space limitations, in this paper we limit ourselves to a discussion of the differences between floating and transition delay in Section 4.2, a technique for computing the transition delay in Section 5, and results on these techniques in Section 6. A more comprehensive consideration of transition delay is given in [6].

2 Definitions and Notation

In this section we introduce terminology that will allow us to discuss timing issues as well for temporal behavior.

A **path** in a combinational circuit is an alternating sequence of vertices and edges, $\{v_0, e_0, ..., v_n, e_n, v_{n+1}\}$, where edge e_i, $0 \leq i \leq n$, connects the output of vertex v_i to an input of vertex v_{i+1}. For $1 \leq i \leq n$, v_i is a gate; v_0 is a primary input and v_{n+1} is a primary output. Each e_i is a net. With each vertex v (edge e) we associate a delay $d(v)$ ($d(e)$).

The **length** of a path $P = \{e_0, v_0, e_1, ..., e_n, v_n, e_{n+1}\}$ is defined as $D(P) = \sum_{i=0}^{n} d(v_i) + \sum_{i=0}^{n+1} d(e_i)$. Ignoring sensitization conditions, the delay of a circuit as given strictly by the length of the longest path is called the **topological** or **graphical** delay.

An **event** is a transition $0 \rightarrow 1$ or $1 \rightarrow 0$ at a gate. Consider a sequence of events, $\{r_0, r_1, ..., r_n\}$ occurring at gates $\{g_0, g_1, ..., g_n\}$ along a path, such that r_i occurs as a result of event r_{i-1}. The event r_0 is said to propagate along the path.

A **controlling value** at a gate input is the value that determines the value at the output of the gate independent of the other inputs. For example, 0 is a controlling value for an AND gate. A **non-controlling value** at a gate input is the value which is not a controlling value for the gate. For example, 1 is a non-controlling value for an AND gate. We say a gate g has a **controlled value** if one of its inputs has a controlling value; otherwise, we say g has a **non-controlled value**.

Let $\pi = \{v_0, e_0, ..., v_n, e_n, v_{n+1}\}$ be a path. The inputs of v_i other than e_{i-1} are referred to as the **side-inputs** to π. If there exists an input vector v such that all the side-inputs along π settle to non-controlling values on v then π is **statically sensitizable**.

The **critical path** is the longest sensitizable path in the circuit under the stated delay model. If a path is not sensitizable under the stated delay model then it is a **false path**.

3 Circuit History: How Much is Enough?

We are interested in determining the delay of a circuit for a given delay model, but the real motivation is to determine what is the frequency at which a circuit can be clocked. There are a number of possible definitions of transition delay and each definition has implications on the issue of clocking frequency. A full consideration of these problems is beyond the scope of this paper, but in this section we introduce our notion of transition delay and show under what conditions it results in a valid clock clocking frequency.

Consider the operation of a synchronous digital circuit being clocked at period τ. At every clock period, the outputs are latched and a new set of inputs presented to the circuit. Let us examine the operation of a circuit over the period of application of a sequence of input vectors. Let v_0 be the vector applied at the present clock cycle, v_{-1} be the vector applied at the previous clock cycle and so on. In the **floating** mode of operation, the nodes are not assumed to be ideal capacitors and hence their state is unknown till it is set by the current vector. Thus, the timing behavior for v_0 is independent of all previous vectors. In the **transition** mode of operation, the circuit nodes are assumed to be ideal capacitors and retain their value set by the previous

0738-100X/92 $3.00 © 1992 IEEE

vectors till the current vector forces the voltage to change. Thus the timing response for v_0 is also a function of v_{-1} and possibly other previously applied vectors. In analyzing the timing response of the circuit, we would like to deal with as little history as possible while making no compromises on the accuracy. In this direction we first propose the following model of measuring the delay in the transition mode and subsequently justify it.

Let us assume that when vector v_0 is applied, all circuit nodes have stabilized to their values under v_{-1}. In this case the effect is the same as if v_{-1} is given an arbitrary amount of time to settle. This mode of operation will be referred to as the *single stepping* transition mode and for the remainder of this paper and whenever we refer to transition delay it will be relative to this mode of operation. The input transition from v_{-1} to v_0 will result in some transitions at the circuit nodes. Let δ be the time taken for the last transition at any of the output nodes for all possible vector changes v_{-1} to v_0. Thus, for all $\tau > \delta$ no transition will ever be observed at any of the outputs in the single stepping mode.

Let us now use this value of τ to clock the synchronous circuit. At time 0, when v_0 is applied it is possible that the circuit nodes may not have stabilized to their values under v_{-1}. (Note that the fact that the outputs have stabilized does not imply that all the circuit nodes have stabilized.) A simple sufficiency condition for a transition delay τ to be a valid clock period is for the state of the circuit to be the same whether 1) v_0 is applied only an interval of τ after v_{-1} or 2) v_0 is applied at an arbitary interval after v_{-1}. This is expressed in the following theorem.

Theorem 3.1 *Let C be a combinational sub-circuit of a synchronous digital circuit. Let τ be the transition delay of the circuit derived using the single stepping mode of operation. Let ω be the length of the longest graphical path in C. If $\tau > \omega/2$ then τ is a valid clocking period for C.*

Proof: Let v_{-1} be the vector applied at time $-\tau$ and let v_0 be the vector applied at time 0. Let e_{-1} be an event in the circuit that occurs after time 0 and is caused by v_{-1}, i.e. an event that is caused by the v_{-1} but is still propagating at time 0 after v_0 is applied. Event e_{-1} has traversed a distance corresponding to delay τ from the primary inputs and has at most a distance corresponding to $\omega-\tau$ to traverse to reach the primary outputs. Let e_0 be an event caused by v_0 that propagates along a path π of delay at least τ. Single-stepping transition analysis has analyzed the delay assuming that all events at side inputs of a gate g along π have settled when e_0 arrives at g. This analysis will be valid if each e_{-1} has settled to its final value under v_{-1} when e_0 reaches that gate. This is equivalent to the condition that e_{-1} not present itself at the input of a gate g along π after e_0. This will be true as long as $\omega - \tau < \tau$ or equivalently $\tau > \omega/2$. ∎

When this condition is true, then for events propagating along all paths of length at least τ, each gate will have settled to its value under v_{-1} by the time the event gets to that gate, which is the same as it would be in the single stepping mode. The restriction $\tau > \omega/2$ is not very stringent. Based on practical experience this property holds for most circuits. This is significant since it enables us to consider only the two vectors involved in the change at the inputs in the analysis of the timing response of this transition.

4 Component Delay Models

All simulation techniques as well as timing analyzers make some assumptions about the possible variations in the delay of various circuit components. To avoid further confusion with other uses of the term "delay model" we call these component delay models. We now examine some of the models used and then specify the domain of this paper in terms of these.

The most common component delay model for a circuit component is one in which the delay is assumed to be a fixed number d. This is referred to as the **fixed delay** model. In this model a delay of 2 units on a gate indicates that the gate switches instantaneously to a logical 0 or 1 value but that the communication of this event to the output of the gate is delayed by 2 units. In reality this number is typically an upper bound on the expected delay, so in fact the actual delay may be any number bounded above by d. This potential speedup is incorporated in the **monotone speedup** model [10], which assumes that the delay for each component lies in the range $[0, d]$.

The **bounded delay** model is more realistic about how much each gate can in fact be sped up. It specifies the delay as a range, $[d^l, d^u]$, given by the lower and upper bounds on the actual delays. There is some ambiguity in the literature as to what the bounded delay model means. It has been interpreted as either the switching delay or as the propagation delay. With the former interpretation, a gate delay of $[2, 4]$ would imply that the gate would take somewhere between 2 and 4 units to make the transition between the two logical values. Ternary algebras [11] have been used to accommodate the unknown value of the gate output (which is neither a 0 or a 1) in the interval of uncertainty. If the bounded delay relates to the propagation delay then this implies that the gate switches instantly, but it is uncertain as to the time in the $[2, 4]$ interval at which it will switch. Note that unlike the previous case, the gate output is always a 0 or a 1 and ternary algebras are not needed. In this paper, we use the propagation delay interpretation since we feel it is more realistic with respect to current technologies.

To simplify the presentation of this paper we consider only the fixed delay model and we restrict the delays in circuits to gates. This is still general enough to accommodate other delay quantities such as wire delays and pin-to-pin delays by introducing buffers with appropriate delays in the circuit. While more sophisticated delay parameters such as slope delays and separate rise and fall delays are not directly accommodated into the "delay lumped at a gate" paradigm, it can be shown with little effort that the results in this paper hold for even those models. Bounded delays are treated extensively in [6].

4.1 Instantaneous Glitches

Suppose that in a circuit simulation a rising event and a falling event arrive at at precisely the same time at an AND gate with delay "2". In static timing verification, which aims at a robust though possibly pessimistic model of delay, it is commonly assumed that an "instantaneous glitch" event immediately occurs which is then communicated at the output of the AND gate after a delay of 2. In this paper our general assumption will be that such instantaneous glitches will not be able to overcome the inertial delay of a gate. This is more realistic given existing technologies.

4.2 Differences in the Floating Delay and the Transition Delay

It is easy to show that the floating and transition delay modes can give different results. In Figure 1 we show a two-level circuit, resulting from a prime and irredundant cover, for which the transition delay (under fixed gate delays) is different from the floating delay. The number inside a gate

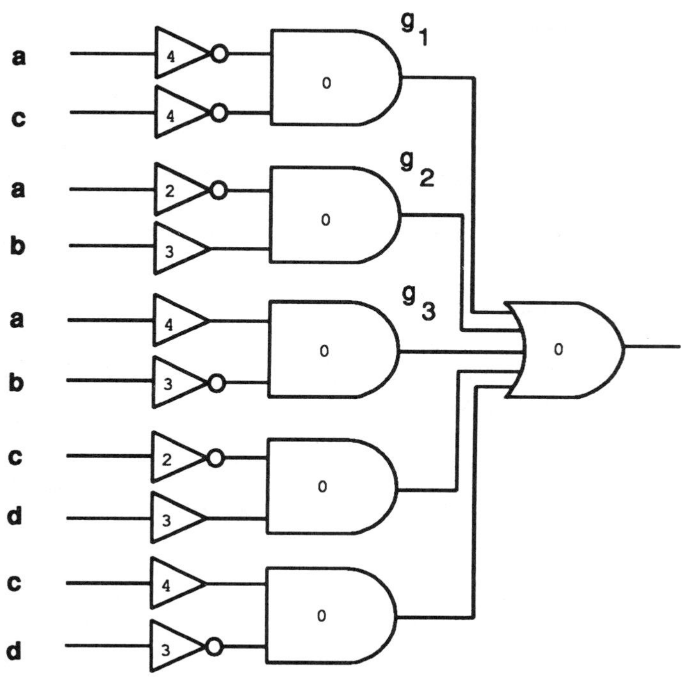

Figure 1: Floating and Transition Delays Differ

corresponds to the delay of the gate.

Figure 1 shows one of the ways that the transition delay of a circuit, assuming no monotonic speed-ups and no instantaneous glitches, can differ from the floating delay of a circuit. The floating delay of this circuit for a rising transition at the output is 4. In this example, on each transition on g_1 for $0 \rightarrow 1$ at least one of the other gates makes a $0 \rightarrow 1$ transition sooner. For example, let us look at the vector pair $< 1100,\ 0000 >$. The gate g_2 glitches to a 1 during the time interval [2,3] then the gate g_3 glitches to to a 1 during the time interval [3,4] and as a result the by the time the gate g_1 makes a $0 \rightarrow 1$ transition (at time 4), the output OR gate is already a 1.

However, since the work of [10] it has been considered to be necessary that speed-ups in the circuit should not result in increasing the delay of the circuit. In Figure 1, if the input buffers/inverters to gates g_2 and g_3 speed up then the $0 \rightarrow 1 \rightarrow 0$ glitches of g_2 and g_3 on $< 1100,\ 0000 >$ settle to 0 before g_1 makes its $0 \rightarrow 1$ transition. As a result the transition delay of the circuit becomes equal to the floating delay.

Because of examples like this it has been conjectured that for any combinational circuit there exists a monotone speed-up such that the transition delay equals the floating delay; however, in Figure 2 we give a circuit in which the transition delay, with or without monotone speedup, is less than the floating delay.

The floating delay of the circuit in Figure 2 is 5, and the associated floating delay vector is $a =< 1 >$. We could give a full analysis of the transition delay for this circuit, but for our purpose here it suffices to give an intuitive argument why the transition delay is strictly less than the floating delay even with monotone speedup.

In single stepping mode, we apply an input vector when all circuit nodes are stabilized to their final values, in this case signal d and e to logic 1. Consider the case where next input vector generates a rising transition at the input: It immediately forces signal d, which in turn forces e, to remain at logic 1. If the next input vector generates a falling transition, it immediately causes a rising transition at signal c which forces e to remain at 1 also. Therefore, no transition can be observed at the output in the single stepping mode of operation. The above argument remains valid under arbitrary monotone speedup of the circuit as will be demonstrated in the discussion in Section 4.3.

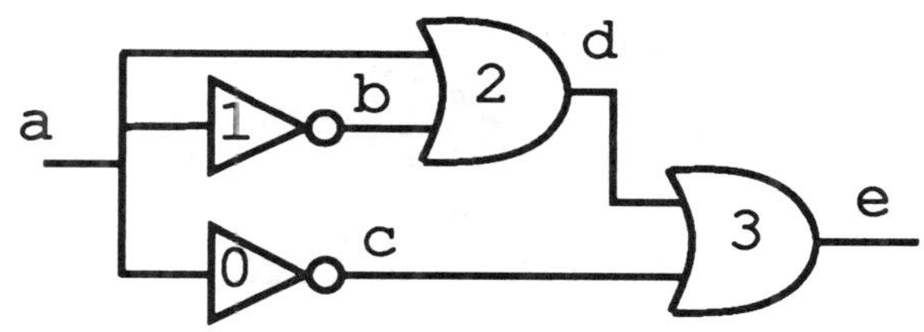

Figure 2: Floating and transition delays differ - even with monotone speedup

Thus, transition delay analysis using the single stepping mode returns a delay of 0. However, Theorem 3.1 only guarantees the validity of a clock period greater than $\omega/2 = 3$. For example with a clock period of 4, less than the floating delay of 5, the output of the circuit stays a stable 1.

It is interesting to note that the path {a,d,e} of length 5 in Figure 2 is statically sensitizable. While it has been known for some time that static sensitization can be too optimistic a condition to determine the delay of a circuit this example demonstrates that it can also be too pessimistic.

Circuits such as those in Figures 1 and 2 motivate a further enquiry into the relationship between transition delay and floating delay.

4.3 Sources of Difference Between Floating and Transition Delay

In the floating delay mode the delay of a path π is computed on a single vector v_0. At a gate g, along π, that is controlled by v_0 it is implicitly assumed that if v_0 results in a a static non-controlling value at a side input e along π, then there always exists a vector sequence ending in v_0 that will result in that non-controlling value on e when an event propagates along π due to v_0. In other words the vector sequence causes the non-controlling value to be available when an event occurs along π, even if a speed-up is required to deliver that non-controlling value. It will be useful to understand precisely why this assumption is not always valid, and why a violation of this assumption can result in the transition delay of a circuit being less than the floating delay.

It will be interesting to understand precisely why the transition delay of a circuit can differ from the floating delay. For an event traveling down a path to be affected by the delay of a gate the event must propagate through the gate. The floating delay mode makes conservative assumptions about the initial state of a circuit when a vector is applied. Let us assume that a path π is determined to be responsible for the delay of the circuit in the floating delay mode. Let v_0 be the vector that sensitizes π to result in this delay in the floating delay mode. Let g be a gate along π whose output is a controlled value on v_0 and let e be a side-input of π at g that has a statically non-controlling value on v_0. In floating delay analysis it is implicitly assumed that there exists a vector v_{-1} that will result in a

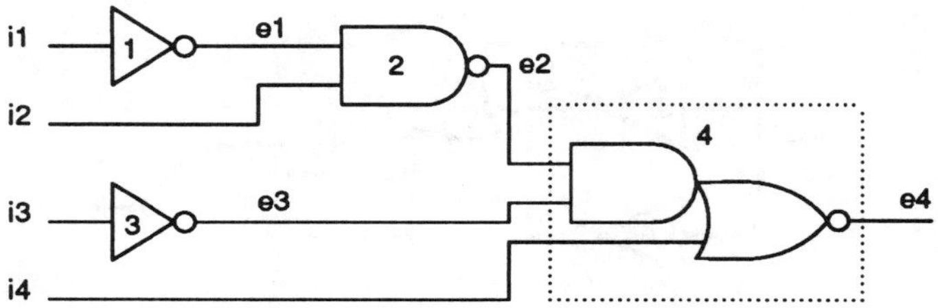

Figure 3: A multilevel combinational circuit

non-controlling value on e when an event propagates along π due to v_0; however, it may be the case that for any vector v_{-1}, e may be slow to transition to a non-controlling value and thus the event along π is blocked at g.

Figure 2 demonstrates this point. The vector $v_0 = a = < 1 >$ statically sensitizes the path $\{a, d, e\}$ and thus results in a delay of 5 in the floating delay mode. Consider the activity at gate d on the input pair $v_{-1} = \overline{a} = < 0 >$ and $v_0 = a = < 1 >$. Gate b settles to a 0 on v_0 but only after the rising event associated with input a has reached the input of gate d. Thus the event on the path $\{a, d, e\}$ gets stopped at d. If the delay of gate b is reduced to 0, through a monotone speed-up, then an *instantaneous glitch* (see Section 4.1) occurs at the inputs to d, but such instantaneous glitches are assumed not to change the output of a gate. (We have a more complex example for which it is not possible to transmit even instantaneous glitches.) Thus, the floating delay assumptions are too conservative for this circuit.

5 Symbolic Simulation Under Fixed Gate Delays

5.1 Introduction

Having motivated transition delay let us now consider its computation. In order to compute the delay of a circuit under the transition mode, a strategy of symbolically simulating all possible input vector *pairs*, that can be applied to the circuit inputs, can be adopted. The possible resulting waveforms at each gate are encoded by a set of Boolean functions, one for each discrete time point. Each Boolean function is defined over the Boolean variables corresponding to the circuit inputs for the first and second vectors. We will consider the details of symbolic simulation for fixed gate delays in the following sections. For a symbolic simulation method for bounded gate delays, the reader is referred to [4].

In the fixed delay case, a single set of Boolean functions at each gate in the circuit suffices to capture all the information regarding the transitions occurring at the gate. We will begin with an illustrative example.

5.2 An Example

Consider the multilevel combinational circuit shown in Figure 3.

It has four primary inputs and one output and consists of four CMOS gates. (The gate surrounded by the dotted line is a complex gate having series-parallel connections only.) The circuit has a total of four nodes: e_1, e_2, e_3, and e_4, in addition to the four primary input nodes. Figure 4 shows the signal transitions at the primary input nodes as well as the possible transitions at the internal nodes of the network. The time points are in normalized units. The first three inputs, $i1$, $i2$, and $i3$, switch simultaneously between time periods 0 and 1. The fourth input, $i4$, is a late arriving

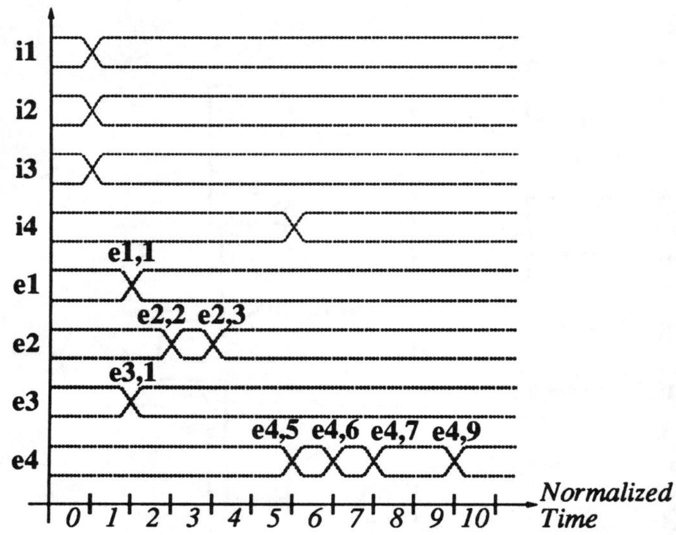

Figure 4: Signal waveforms for the primary inputs and the outputs of the logic gates

signal that switches between the time points 5 and 6. In this example, the delays of both Gate 1 and Gate 3 are one time unit. Gate 2 has a delay of two units while Gate 4 has a delay of four units. $i4$ arrives five time units after the other three inputs.

Also shown in Figure 4 are the waveforms, e_i's, representing the signals at the outputs of ith logic gates. Each of the possible transitions, $e_{i,j}$, represents either a low-to-high or high-to-low signal transition between $[j]^{th}$ and $[j+1]^{th}$ time points. The number of all possible transitions at a gate output may equal the sum of all possible transitions at the gate inputs. These transitions are delayed by the gate's propagation delay. If a gate is driven by the primary input signals, then the transitions at the gate output will be determined by the transitions of the primary input signals. Due to the combinational nature of the network, the number of all possible transitions at the output node of a given gate for all time points is

$$N_t = \sum_{i \in inputs\ of\ gate} e_{i,j} \quad (1)$$

for all $[j]$ distinct time points. Referring to the example, e_1 has only one possible transition, $e_{1,1}$, between the time points 1 and 2 because the total number of transitions at the input of Gate 1 at different time points is one. The earliest signal event will arrive at the gate output one time unit after $i1$ switches because the delay of Gate 1 is one time unit. Similarly, e_4 has a total of four possible transitions between the time points 5 and 10 because the number of transitions at the inputs of Gate 4 at distinct time points is four. The earliest transition, originated from node e_3, arrives at the gate output between the time points 5 and 6, since Gate 4 has a delay of four time units.

5.3 Unit-Delay Model

In this section, we describe how symbolic simulation can be used to compute signal transitions in a circuit under the unit-delay model. The unit-delay restriction will be relaxed in Section 5.5.

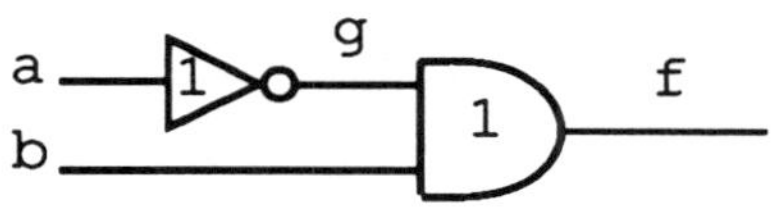

Figure 5: An Inverter-And Circuit

Under the unit-delay model, the circuit activity caused by applying an input vector pair can be seen as occurring at discrete points in time implied by the unit-delay model, and all signals are stable between the time points. The central idea in our formulation is to describe the stable values of a signal in a particular time interval symbolically as a Boolean function of input variables over two input vectors (i.e., each implicant of f_i corresponds to an input vector pair which generates a 1 at signal f in time interval i). Determining whether a particular signal changes from interval i to $i+1$ now reduces to checking for logical equivalence of the functions in the two intervals. As a by-product, the input assignment that causes the two functions to differ gives us the input vector pair that generates a transition at time point between intervals i and $i+1$.

As a notational convention, we use subscripts to denote the time interval a variable is associated with. A special subscript "$-$" is used to denote input variables associated with the first vector [1]. The time at which the second vector is applied is a common reference point 0. So, for an input variable a, all a_i with $i < 0$ map to a_- and all a_i with $i >= 0$ map to a_0. Furthermore, for every function $f(g,\ h,\ \cdots)$ we construct a new function $f_i(g_{i-1},\ h_{i-1},\ \cdots)$ for a chosen i.

For example, consider the network in Figure 5. For this network in the first time interval,

$$g_0 = \overline{a_-}$$

$$f_0 = g_{-1}b_{-1} = \overline{a_{-2}}b_{-1} = \overline{a_-}b_-$$

In the next time interval,

$$g_1 = \overline{a_0}$$

$$f_1 = g_0b_0 = \overline{a_-}b_0 = \overline{a_-}b_0$$

Finally for the last interval,

$$f_2 = g_1b_1 = \overline{a_0}b_1$$

For this example there are three possible transitions:

- g changes state from time interval 0 to 1,
- f changes state from time interval 0 to 1,
- f changes state from time interval 1 to 2.

The symbolic formulae for these transitions are respectively:

$$e_1 = g_0 \oplus g_1 = \overline{a_-}a_0 + a_-\overline{a_0}$$

$$e_2 = f_0 \oplus f_1 = \overline{a_-}b_-\overline{b_0} + \overline{a_-}\overline{b_-}b_0$$

$$e_3 = f_1 \oplus f_2 = \overline{a_-}a_0b_0 + a_-\overline{a_0}b_0$$

Each implicant in e_3 gives an input vector pair which generates a transition of signal f at time 2 (between time interval 1 and 2). For example, implicant $\overline{a_-}a_0b_0$ corresponds to vector pair $v_1(a,b) = (0,X)$ and $v_2(a,b) = (1,1)$.

[1] Keep in mind that we assume the single stepping mode of operation

It should be pointed out that our formulation is more powerful than a procedure that just determines transitions of a signal at a particular time point. For example, finding an input vector pair that generates transition of f at both time 1 and 2 amounts to finding an implicant of e_2e_3 (e.g., $\overline{a_-}a_0\overline{b_-}b_0$). Also, the inputs need not be clocked at the same time. For example, if the second value of a is clocked at time t, all a_i with $i < t$ map to a_- and all a_i with $i >= t$ map to a_t. This can be done on a per input basis.

As an efficiency concern, it is not necessary to generate function f_i for all time points. The following lemma gives a simple bound on the time points needed for a signal in the circuit.

Lemma 5.1 *Let Δ and δ be the longest and shortest graphical delay to a signal f. The set $f_0, f_\delta, f_{\delta+1}, \ldots f_{\Delta-1}, f_\Delta$ is sufficient to determine all possible transitions in the circuit.*

The above lemma follows directly from the unit delay model. Signal f cannot change until the change in the nearest input propagates through, and will stop changing when the input furthest away finally arrives. All references to f_i map to f_Δ if $i > \Delta$ and to f_0 if $i < \delta$.

5.4 Symbolic Event Suppression

To compute the transition delay of a circuit, it may not be necessary during symbolic simulation to store, or even generate, all the Boolean functions corresponding to each gate and each time point. Typically, one is interested in answering the question: "Is the delay of the circuit $\geq \delta$?" In this case, we only require the f_t's where $\delta_t \geq \delta - 1$ and f is the circuit output. We XOR $f_{\delta-1}$ with each such f_t and check the XOR'ed function for satisfiability to see if there is indeed a transition at time δ.

Techniques similar to the event suppression techniques described in [8] for the efficient simulation of a vector pair on a circuit can be used in the symbolic simulation procedure as well. For example, given a gate g in the circuit, let w_g be the length of the longest path from the gate output to the circuit output. We only need to compute the g_t's such that $t + w_g \geq \delta - 1$.

5.5 General Delay Model

A gate with a large fanin may have several times the delay of an inverter. If one uses normalized time units, one can always introduce unit-delay buffers at the output of gates in a circuit, which have a delay greater than unity, in order to model differing delays among logic gates.

5.6 Checking Satisfiability

We maintain the various Boolean functions as multilevel logic networks during the course of symbolic simulation. The size of these networks is not much larger than the circuit itself. Alternatively, we could have used reduced, ordered Binary Decision Diagram (ROBDD) [1] representations for these functions, and propagated ROBDDs through the circuit.

Once we have the $f_{\delta-1} \oplus f_\delta$ function for the circuit output, we can determine if the function is satisfiable by constructing a reduced, ordered Binary Decision Diagram for the function, or using the satisfiability checking procedure of Larrabee [9]. In the case of circuits like multipliers, constructing ROBDDs for the Boolean functions is infeasible, but our method of maintaining the Boolean functions as multilevel networks, and the use of Larrabee's satisfiability checking algorithm succeeds in computing the transition delay of the circuit (*cf.* Section 6).

EX	inputs	outputs	literals	longest
cbp.12.2	25	13	331	40.00
cbp.16.4	33	17	413	44.00
C17	5	2	19	5.00
C432	36	7	405	19.00
C499	41	32	977	25.00
C880	60	26	718	20.00
C1355	41	32	1121	27.00
C1908	33	25	1225	34.00
C2670	233	140	1764	25.00
C3540	50	22	2332	41.00
C5315	178	123	3923	46.00
C6288	32	32	4752	123.00
C7552	207	108	5488	38.00

Table 1: Statistics of Benchmark Examples

EX	val	l. d.	f. d.	#check	CPU secs	t. d.
C1355	1	27	27	1	10	27
C1908	1	34	31	21	3675	31
C2670	0	25	24	2	232	25
C3540	0	41	39	10	182	39
C5315	1	46	45	9	11	45
C6288	1	123	122	2	812	122
C7552	1	38	37	9	16	37

Table 2: Transition Delay Computation on Benchmark Examples for Fixed Gate Delays

6 Experimental Results

In this section, we present preliminary experimental results in determining the transition delay of a circuit. Transition delay computation results in the procurement of vector pairs that each propagate a transition along a longest true path to the output. The set of vector pairs can be used to perform timing analysis under more sophisticated delay models, (*e.g.* using **SPICE**).

The techniques described in the previous sections have been implemented in the program **TrueD-T**. The statistics of the chosen benchmark circuits are shown in Table 1.

Results on applying the fixed delay simulation calculus to compute the transition delay of a circuit are given in Table 2. We were able to exactly compute the transition delay under the fixed unit gate delay model for all the benchmark circuits, using the symbolic simulation algorithm described in Section 5. In the table, val corresponds to the logical value the path was sensitized to, l. d. is the longest path delay, f. d. is the floating mode delay, and t. d. is the transition delay.

The fixed delay model may not be realistic for use in certified timing verification since there will always be some statistical variation in the predicted gate delays. We next experimented with the bounded gate delay model, where all gates have upper bounds equal to unity, and lower bounds equaling zero.[2] We have been able to obtain vector pairs that validate the floating delay for all the ISCAS-85 benchmark circuits under the bounded gate delay model. The results are shown in Table 3.

EX	val	l. d.	f. d.	#check	CPU secs	t. d.
C1355	1	27	27	1	15.76	27
C1908	1	34	31	21	3675	31
C2670	0	25	24	2	210	24
C3540	0	41	39	10	182	39
C5315	1	46	45	9	9	45
C6288	1	123	122	2	812	122
C7552	1	38	37	9	20	37

Table 3: Transition Delay Computation on Benchmark Examples for Bounded Gate Delays

The CPU times in Tables 2 and 3 are on a SUN-4 workstation and correspond to the time required for floating delay computation using the method of [7] plus transition delay computation. The time required for transition delay calculation is typically a small fraction of the time required for floating delay computation.

We are currently experimenting with random-logic circuits to see if logic optimization affects the transition delay of a circuit.

7 Certified Timing Verification

In Section 4.2 we demonstrated that there can be a difference in the floating delay and transition delay of a circuit. In Section 5 we gave a procedure that actually computed the transition delay of a circuit assuming fixed delays and no monotonic speed-up. This procedure produces a vector pair that sensitizes the critical path in the transition delay mode. Finally, in Section 6 we presented results on applying the transition delay computation procedure to a number of benchmark examples. As Table 2 indicates, on benchmark circuits the transition delay typically does not differ from the floating delay, but we claim that it is still useful to compute the transition delay because of the utility of the resulting vector set for certifying the delay of the circuit. In this section we briefly outline a procedure for certifying the results of static timing verification.

The transition delay procedure works on the question "Is there a path with transition delay greater than δ?" the first step is to identify the upper bound on the delay of the circuit. As the transition delay of a circuit is bounded above by the true floating delay [2] the natural value for δ is the true floating delay. This can efficiently be computed using techniques described in [5, 7]. The derived value of δ is then passed to the symbolic simulation procedure described in Section 5, or for computation using bounded delays the procedure described in [6] may be employed. The nominal use of these transition delay computation procedures is to return a single vector sequence which sensitizes an event along some path, perhaps only one, of length at least δ. Whichever approach is applied let us call the resulting vector sequence V.

The circuit model, perhaps with speed-ups required by the transition delay computation procedure, is then given to the timing simulator of choice. The vector sequence V is then applied. In general the results of the timing simulation should not give delay values that are worse than the results of the transition delay calculation. If this happens it means that the delays used in the transition delay calculation were not pessimistic enough; these should be modified and the delay calculation re-run.

[2] The monotone speed-up condition.

The most complex case is when the timing simulation of V reports a delay γ that is less than δ. If there is sufficient confidence in the coverage of the vector set then an aggressive designer may opt to clock the circuit at γ. Another approach is to further investigate the range of possible clocking speeds using statistical methods [4]. The hope here is that the statistical analysis procedures will give a quantitative notion of what percentage of parts are likely to run at each speed in the range between γ and δ.

Using a combination of transition delay computation and timing simulation in this way gives a greater predictability to the post-manufacture delay of the circuit.

8 Conclusions

In this paper we demonstrated for the first time that the transition delay of a circuit *can differ* from the floating delay even in the presence of arbitrary monotonic speed-ups in the circuit. This result is used to motivate the derivation of a procedure which directly computes the transition delay of a circuit. A fuller development of this procedure required the development of a number of different simulation calculi, each suited to assumptions regarding particular delay modeling such as bounded delay or monotonic speed-ups. Space limitations have reduced this paper to an exposition of a symbolic simulation procedure for computing the transition delay assuming fixed gate-delays and no speed-ups. The output of the transition-delay computation procedure is a vector sequence which excites an event along the longest sensitizable path of the circuit under consideration.

While this theoretical framework for the analysis of transition delay is in itself useful for understanding the relationship between static and dynamic delay analysis, we envision the most practical application of these results in *certified timing verification.* In such a scenario the upper bound on circuit delay is first derived by means of a floating delay calculation. The transition delay of the circuit is then derived using the transition delay calculator, and a vector sequence for sensitizing the critical paths of the circuit is produced as a by-product of this delay calculation. This vector sequence can then be applied using a timing simulator equipped with more accurate timing models. Such a procedure promises to give the high accuracy of timing simulation with the computational efficiency and the comprehensive path coverage of static timing verification.

8.1 Work in Progress

A great deal of work remains to be done in order to understand completely the relationship between transition and floating delay. Even the definition of transition delay requires further examination. The common *single-stepping* definition of transition delay together with a simple sufficiency condition for a valid clock frequency was presented in Section 3, but a full consideration of the relationship between transition delay and clocking frequency remains to be done. Encouraging progress toward resolving this question with regard to floating delay will be presented at this conference [3]. Correct computation of transition delay seems to become enmeshed in many technology specific issues, such as the instantaneous glitches discussed in Section 4.1, and these issues also require further resolution. Furthermore, while a distinction between floating delay and transition delay has been drawn in this paper we presently have no clear idea of how fundamental this difference is. We have presented circuits in which a difference occurs and we have derived a number of circuit properties that give sufficiency conditions under which the two delay modes give the same value [6], but we have not closed the gap with precise necessary and sufficient conditions under which those two delay modes give the same value. Finally, while we see the immediate practical applications of this work in certified timing verification and delay fault testing, we hope that resolution of the issues discussed in this section will ultimately eliminate the need of timing simulation for synchronous digital circuits altogether.

References

[1] R. Bryant. Graph-Based Algorithms for Boolean Function Manipulation. In *IEEE Transactions on Computers*, volume C-35, pages 677–691, August 1986.

[2] H-C. Chen and D. Du. Path Sensitization in Critical Path Problem. In *Proceedings of Int'l Conference on Computer-Aided Design*, November 1991.

[3] S-W Cheng, H-C Chen, David Du, and A. Lim. The role of long and short paths in circuit performance optimization. In *The Proceedings of the Design Automation Conference*, 1992.

[4] S. Devadas, H-F. Jyu, K. Keutzer, and S. Malik. Statistical timing analysis of combinational logic circuits. In *Proceedings, Tau 92: 1992 ACM Workshop on Timing Issues in the Specification and Synthesis of Digital Systems*, 1992.

[5] S. Devadas, K. Keutzer, and S. Malik. Delay Computation in Combinational Circuits: Theory and Algorithms. In *Proceedings of the International Conference on Computer-Aided Design*, November 1991.

[6] S. Devadas, K. Keutzer, S. Malik, and A. Wang. Certified Timing Verification and the Transition Delay of a Combinational Logic Circuit. In *MIT Technical Report, Massachusetts Institute of Technology*, October 1991.

[7] S. Devadas, K. Keutzer, S. Malik, and A. Wang. Computation of Floating Mode Delay in Logic Circuits: Practice and Implementation. In *MIT Technical Report, Massachusetts Institute of Technology*, October 1991.

[8] S. Devadas, K. Keutzer, S. Malik, and A. Wang. Efficient Timing Simulation for Delay Computation. In *Proc. of Brown/MIT Conference on Advanced Research in VLSI and Parallel Systems*, March 1992.

[9] T. Larrabee. Test Pattern Generation Using Boolean Satisfiability. In *IEEE Transactions on Computer-Aided Design*, volume 11, pages 4–15, January 1992.

[10] P. C. McGeer and R. K. Brayton. *Integrating functional and temporal domains in logic design.* Kluwer Academic Publishers, 1991.

[11] C-J. Seger and R. E. Bryant. Modelling of Circuit Delays in Symbolic Simulation. In *IFIP International Workshop on Applied Formal Methods for Correct VLSI Design*, pages 625–639, November 1989.

Recurrence Equations and the Optimization of Synchronous Logic Circuits.

Maurizio Damiani Giovanni De Micheli

Center for Integrated Systems
Stanford University

Abstract

In this paper we present a formulation for the problem of optimizing synchronous logic across register boundaries. We describe the degrees of freedom (i.e. the don't care conditions) of an embedded subnetwork by means of **sets of execution traces,** *described implicitly by Synchronous Recurrence Equations. The optimization problem reduces to that of finding minimum-cost solutions to such equations. An exact solution algorithm for this problem is presented, along with approximations that improve its computational efficiency. Eventually, we demonstrate the feasibility and effectiveness of the approach on synchronous benchmark circuits.*

1 Introduction.

Synthesis and optimization problems for combinational and synchronous logic circuits are currently the object of intense investigation. Boolean methods for combinational networks [1, 2, 3, 4, 5, 6, 7, 8] have matured both from a theoretical and application viewpoint. Such methods allow incomplete specification of the global behavior of a network, and work directly on its structural, hierarchical representation. They improve iteratively on the initial network by extracting subnetworks to be optimized and identifying precisely the degrees of freedom in their desired terminal behavior. Two-level optimization algorithms are then used for their optimization. In particular, in the case of single-output subnetworks, it has been proved that such degrees of freedom are fully represented by a *don't care set* [2].

By contrast, synthesis and optimization of synchronous circuits have so far relied on procedures based on (possibly iterative) manipulations of their *behavioral models*, typically in terms of *state transition graphs* [12, 13, 9, 10, 14, 11].

There are several disadvantages associated with this approach. First, a state diagram description does not allow us to represent any degree of freedom in the global behavior of a synchronous circuit. It is indeed often the case (see, for example, the control synthesis problem from high-level specifications [19, 20]) where the cycle-by-cycle behavior of a synchronous machine is only incompletely specified, and several *not equivalent* finite-state machines may meet the specifications. The second drawback of this approach is the remoteness of the state diagram model from the final implementation, that makes it difficult to evaluate the key figures of merit, such as area, testability and performance, during the optimization process.

It is then desirable to have available methods that, similarly to the combinational case, can optimize synchronous circuits directly from *structural* models, *i.e.* netlists, (according to some given area/performance metrics) and can take into account incomplete specifications [15, 24].

We show in this paper that, unlike the combinational case, capturing the degrees of freedom (the "*don't care conditions*") associated to a synchronous subnetwork entails being able to reason about *sets of sequences* of values in the synchronous network. This capability would allow the automation of logic transformations otherwise impossible. The following example is taken from [25]:

Example 1. The circuit shown in Fig. (1) computes a second order-moment M of the luminance of an image:

$$M = \sum_{i,j} i\ j\ a_{ij}$$

where a_{ij} is the luminance of the pixel of coordinates i, j. The binary counters C_i, C_j scan by rows the space of coordinates (i, j). C_j is clocked faster than C_i, and the product $p = i\ j$ is first computed by the multiplier M_1. Its output feeds a second multiplier M_2 along with the input a_{ij} to compute the term $i\ j\ a_{ij}$. These terms are eventually added up in the accumulator register *acc*. M_1 cannot be optimized by any combinational method. All input combinations are, in fact, asserted at its inputs by the two counters, so that ther are no external *controllability don't cares*. Moreover, any change at its outputs would be revealed by an error in the final result, and therefore we have no external *observability don't cares*.
It is possible, however, to optimize M_1 by observing that its input signals are fed in a precise order by the counters. Since during the scan of a row only C_j is incremented, the outputs of M_1 at two consecutive time-points, n and $n+1$, are *related* by the recurrence:

$$p_{n+1} = i\ (j+1) = i\ j + i = p_n + i$$

suggesting the simpler realization shown in Fig. (2). This optimization would have been impossible without knowledge of the recurrence. □

In this paper we consider first a behavioral representation of a synchronous circuit in terms of *execution traces*, as these allow us to capture degrees of freedom at the sequential level, and then consider the ensuing optimization problem.

A first attempt to the structural optimization of general synchronous circuits is the retiming/resynthesis strategy (considered, for example, in [18, 16, 17]). It essentially consists of identifying pipeline-like subcircuits, pushing the registers to their periphery, and then optimizing the resulting combinational subcircuit. The following very simple example shows the limitations of the method.

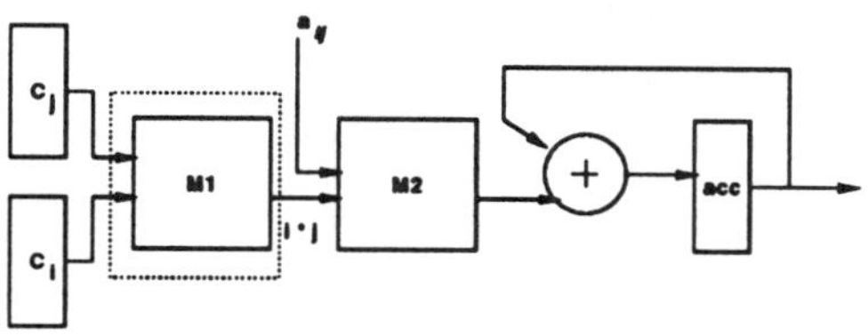

Figure 1: A circuit for a second-order moment computation.

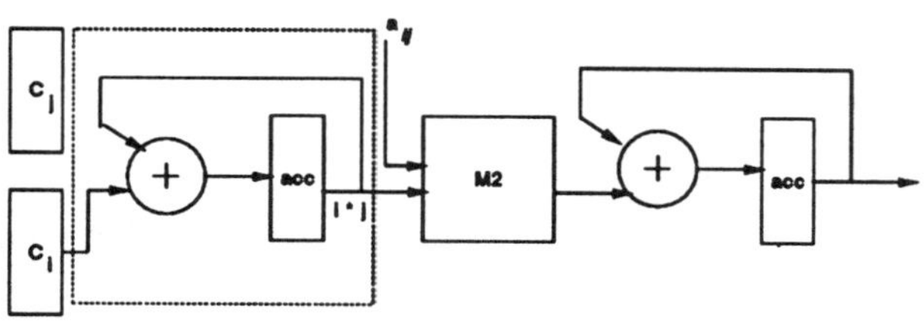

Figure 2: Optimization involving sequential information.

Example 2. Consider the circuit shown in Fig. (3). It can easily be verified that the inverter driving the variable y can be replaced by a simple interconnection, *i.e.* that the function $f(x) = x'$ can be replaced by $g(x) = x$. Since there are no pipeline-like subcircuits, no retiming operation is possible on the circuit, and consequently retiming would not remove the inverter. The inverter can be removed even though there are no *don't care* conditions associated to it in the combinational sense. To check this, it suffices to observe that any *don't care* condition on $f(x) = x'$ would result in the possibility of replacing the inverter with a constant 1 or 0, which is clearly incorrect. □

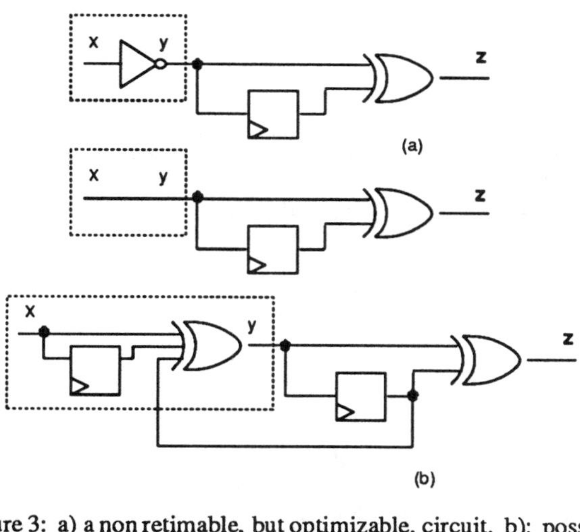

Figure 3: a) a non retimable, but optimizable, circuit. b): possible circuits replacing the inverter in part a).

In this paper we tackle the problem of optimizing an arbitrary subnetwork in a synchronous system. We first determine a description of the acceptable terminal behaviors for the subnetwork in terms of a **recurrence equation.**

Logic optimization is then reduced to finding the minimum cost circuit whose terminal behavior satisfies the recurrence equation. We review an exact solution algorithm, presented in [22], and show that the general problem involves in particular a difficult binate covering step [8, 29].

We then present a strategy for improving the efficiency of the optimization strategy by avoiding the binate covering step, so that conventional covering approaches can be considered. We conclude by presenting experimental results on synchronous logic benchmarks.

2 Synchronous Recurrence Equations.

We first introduce the conventional terminology associated to the manipulation of finite sequences (or **strings**) of Boolean values.

Let $\mathcal{B}$ denote the Boolean set $\{0,1\}$. A k-dimensional Boolean vector $\mathbf{x} = [x_1, \cdots, x_k]^T$ is an element of the set $\mathcal{B}^k$.

The set of all finite sequences over a finite set S (the *Kleene closure* of S) is conventionally denoted by the symbol S^* [21]. We thus denote by $(\mathcal{B}^k)^*$ the set of all finite sequences of k-dimensional Boolean vectors. An element of $(\mathcal{B}^k)^*$ is termed a **synchronous sequence** and denoted by $\mathbf{x}(\cdot)$. The nth element of the sequence is denoted by $\mathbf{x}_n$.

The terminal behavior of a n_i-input, n_o-output synchronous circuit is described by the correspondence it establishes between input and output sequences, each pair representing a possible *execution trace* [27, 28] for the circuit. An execution trace is thus identified as an element of $(\mathcal{B}^{n_i+n_o})^*$.

In general, external specifications do not impose a unique correspondence between input and output sequences, but rather a *relation* between them, *i.e.* an arbitrary set of traces. Intuitively, this is due to: a) not all sequences are usually possible at the inputs of a synchronous circuit, and b) for a given input sequence, usually more output sequences are permitted.

Definition 2.1 *A* **trace set specification** *of an n_i-input, n_o-output synchronous circuit is a subset $\mathcal{T} \subseteq (\mathcal{B}^{n_i+n_o})^*$.*

A sequential synthesis problem could be formulated as follows:

Synchronous synthesis problem.

given	a trace set $\mathcal{T}$,
determine	an optimum circuit whose terminal behavior is contained in $\mathcal{T}$.

Given an arbitrary network, the extraction of the individual trace sets associated to its subnetworks is in general a complex task. For this reason, we first focus on **definite** (or feedback-free) networks. Any network can be decomposed into a definite portion, containing all the logic and delay elements, and a set of feedback interconnections. We first consider the problem of optimizing the definite portion, by assuming the feedback inputs and output to be ordinary primary inputs and outputs, respectively. In Sect. (3.3) we take feedback into account.

The terminal behavior of a definite network at time n is **specified** by a function $S(\mathbf{x}_n, \mathbf{x}_{n-1}, \ldots, \mathbf{x}_{n-d})$. [1] Usually, this specification is implicitly provided by the initial network.

We also assume that the external *don't care* conditions associated to the network can be described by an expression $DC(\mathbf{x}_n, \mathbf{x}_{n-1}, \ldots, \mathbf{x}_{n-d})$. Such *don't cares* represent the input sequences (of length up to d) that either do not occur or such that the network output is not observed [23].

The knowledge of S and DC identifies precisely the set of possible terminal behaviors for the network. A network realizing a function $F(\mathbf{x}_n, \ldots, \mathbf{x}_{n-d})$ meets such specifications if and only if $F = S$ for every sequence not in DC.

Another, equivalent, description of the set of terminal behaviors is in terms of the functions $F_{min} = S \cap DC'$ and $F_{max} = S \cup DC$. Specifications are met by F if

$$F_{min} \subseteq F \subseteq F_{max}$$

[1] Boolean functions can be seen as describing sets. Therefore they are referenced also as sets in this paper. Union and intersection denote sum and product, respectively. $\supseteq$ and $\subseteq$ denote containment.

The following examples show some contexts in which trace sets arise naturally in the optimization of synchronous circuits.

Example 3. In the circuit in Fig. (3a), we attempt to replace the input inverter by a simpler logic gate, producing the internal signal y. The replacement is possible as long as the global terminal behavior is unaffected. The desired terminal behavior for the network is described by $z_n = S(x_n, x_{n-1}) = x'_n \oplus x'_{n-1}$. The primary output z_n can also be expressed in terms of the internal signal y (to be re-synthesized) as $z_n = y_n \oplus y_{n-1}$. The signal y must therefore satisfy the constraint:

$$x'_n \oplus x'_{n-1} \subseteq y_n \oplus y_{n-1} \subseteq x'_n \oplus x'_{n-1}, \quad \forall n \geq 0$$

In this case, $n_i = n_o = 1$, and the above equation represents the constraint on the sequences of (x, y) pairs in the circuit.
For any given input sequence $x(\cdot)$, there exist more than one output sequence $y(\cdot)$ that satisfies the equation. Two possible solutions are

$$y_{-1} = x_{-1}; \quad y_n = x_n \quad \forall n \geq 0; \qquad y_{-1} = 0 \quad y_n = x_n \oplus x_{n-1} \oplus y_{n-1} \quad \forall n \geq 0.$$

(The second solution is obtained simply by adding y_{n-1} to both terms of the equation).
The solutions correspond non equivalent circuits replacing the original inverter, shown in Fig. (3b). The assignments of y_{-1} correspond to the assignment of the *initial conditions* for the subcircuit. Although in this case the original circuit is combinational, the second solution is not, and the new network contains a feedback interconnection. □

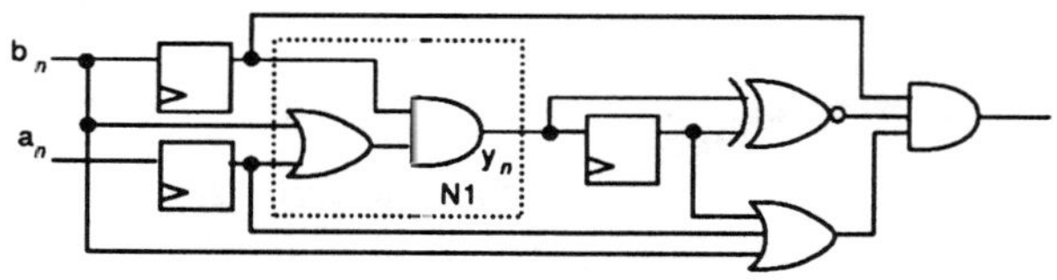

Figure 4: Circuit for the optimization problem of Example (4)

Example 4. As a more complex example, consider the optimization of the subnetwork N1 in Fig. (4). The desired terminal behavior of the entire network can be described by

$$S = b_{n-2}b_{n-1}(a_{n-1} + b_n)$$

Its output is expressed in terms of the internal signal y by:

$$b_{n-1}[y_n \overline{\oplus} y_{n-1}](b_n + a_{n-1} + y_{n-1})$$

Therefore, for every input sequence, y must satisfy

$$b_{n-1}[y_n \overline{\oplus} y_{n-1}](b_n + a_{n-1} + y_{n-1}) = S$$

which represents the recurrence equation to be satisfied by any subnetwork generating the signal y.
Suppose that a *don't care* information (say, $DC = a_n(a_{n-1} + b_n)$) is added. The above equality then would have to be satisfied only for those sequences not in DC. Consequently, y must satisfy only the recurrence:

$$S\ DC' \subseteq b_{n-1}[y_n \overline{\oplus} y_{n-1}](b_n + a_{n-1} + y_{n-1}) \subseteq S + DC$$

□

In all the above examples trace sets are most naturally represented implicitly as solutions to a *recurrence equation*, involving the elements $\mathbf{y}_n, \ldots, \mathbf{y}_{n-d}$ of the output sequences of the circuit to be synthesized. We thus introduce the following definition:

Definition 2.2 *A* **Synchronous Recurrence Equation** *(SRE) is a Boolean equation of type*

$$F_{min} \subseteq F(\mathbf{x}_n, \cdots, \mathbf{x}_{n-2d}, \mathbf{y}_n, \cdots, \mathbf{y}_{n-d}) \subseteq F_{max}; \quad \forall n \geq 0.$$

A **feasible solution** *of the SRE is a function*

$$\mathbf{f}(\mathbf{x}_n, \cdots, \mathbf{x}_{n-d}, \mathbf{y}_{n-1}, \cdots, \mathbf{y}_{n-d})$$

and an initial value specification

$$\mathbf{y}_{-d} = \mathbf{g}_{-d}(\mathbf{x}_{-d}, \cdots, \mathbf{x}_{-2d})$$

$$\vdots$$

$$\mathbf{y}_{-1} = \mathbf{g}_{-1}(\mathbf{x}_{-1}, \cdots, \mathbf{x}_{-d})$$

such that if

$$\mathbf{y}_n = \mathbf{f}(\mathbf{x}_n, \cdots, \mathbf{x}_{n-d}, \mathbf{y}_{n-1}, \cdots, \mathbf{y}_{n-d}) \quad \forall n \geq 0$$

then Eq. 2.2 holds true.

3 Solving the SRE.

We consider in this paper terminal specifications for a synchronous circuit provided in form of a SRE. Each solution **f** to the SRE corresponds to a possible realization of such specifications, with an associated cost. The task of logic synthesis is in this case to determine the minimum cost (typically, minimum-hardware) synchronous circuit whose terminal behavior satisfies the SRE:

Synthesis from SRE.

given	an SRE,
determine	its minimum cost feasible solution (if one exist).

A synchronous network realizing a function as in Eq. 2.2 may in general contain feedback interconnections, as $\mathbf{y}_n$ is expressed in terms of the past values $\mathbf{y}_{n-1}, \cdots, \mathbf{y}_{n-d}$ (one such example is shown in Fig. (3b)). We are not interested in this type of solutions, as they alter the network topology during optimization.
We therefore focus our attention on simpler solutions, in the form $\mathbf{f}(\mathbf{x}_n, \cdots, \mathbf{x}_{n-d})$ only. These solutions, yielding feedback-free (or **definite**) realizations, are hereafter termed **definite**.

3.1 Definite solutions.

We focus now on the optimization of a single-output function f, the extension to the multiple-output case being straightforward. An exact solution procedure for two-level expressions of f was outlined in [22]. We define a **cube** $c(\mathbf{x}_n, \cdots, \mathbf{x}_{n-d})$ on the variables of $\mathbf{x}_0, \cdots, \mathbf{x}_{-d}$ as the product of some of such variables, in either true or complemented form (in practice, we allow also outputs of other internal gates to appear as factors of a cube).
We seek expressions of f as a sum of cubes:

$$f = \sum_{k=1}^{N} c_k.$$

A cube c is an **implicant** if there exists a feasible solution f containing c. An implicant c is a **prime** if there exists a feasible solution f for which c is prime, *i.e.* for which no implicant c' of f strictly contains c. A procedure is outlined in [22] to determine the prime implicants of a SRE.

Example 5. The primes for Example (4) are shown in Table (1). □

Once the set S of primes has been built, Petrick's method is used to construct the subsequent covering problem [26, 7, 8] as follows. The general solution is written as

$$f = \sum_{r=1}^{|S|} \alpha_r c_r$$

Primes
$c_1 = b'_{n-1}$
$c_2 = a_{n-1}b_{n-1}$
$c_3 = b'_n a_{n-1}$
$c_4 = b_n b_{n-1}$

Table 1: List of primes for the problem of Example (4).

where the parameter variable α_r is 1 if c_r is present in the solution, and $\alpha_r = 0$ otherwise.

This expression of f replaces y in the SRE, to obtain a new equation in terms of the input variables $\mathbf{x}_i$ and the parameters α.

Example 6. Table (1) contains the primes for Example (4). Corresponding to the assignment $b_n = b_{n-1} = 1; a_{n-1} = a_{n-2} = b_{n-2} = 0$ the SRE reduces to

$$0 \subseteq y_n \overline{\oplus} y_{n-1} \subseteq 0$$

On the other hand, corresponding to that assignment, $y_n = \alpha_4$ and $y_{n-1} = \alpha_1$. It must therefore be $\alpha_1 \oplus \alpha_4 = 1$.
By repeating the same process over all assignments, and by forming the product of all the resulting constraints on the α_i, the global constraint equation (already in conjunctive normal form) follows:

$$SRE_\alpha = \begin{array}{l}(\alpha_1 + \alpha_2 + \alpha_3)(\alpha'_1 + \alpha'_3)(\alpha'_1 + \alpha'_2) \\ (\alpha_1 + \alpha_4)(\alpha'_1 + \alpha'_4) = 1.\end{array}$$

In particular, the last two products correspond to the factors of $\alpha_1 \oplus \alpha_4$.
The minimum-cost solution to $SRE_\alpha = 1$ is represented by $\alpha_1 = 1, \alpha_2 = \alpha_3 = \alpha_4 = 0$, corresponding to $f = b'_{n-1}$. □

The synthesis problem is thus eventually reduced to that of finding the minimum cost assignment to the parameters α_r such that the new equation holds for all assignments of the variables x_i. This problem is known in the literature as **Minimum Cost Satisfiability** or **Binate Covering** problem.

Its binate nature comes from the possibility for some of the parameter variables α_i to appear in both forms, true and complemented, in the conjunctive form of SRE_α. For instance, the variable α_1 appears in both forms in the SRE_α of Example (6).

3.2 Unate Covering Problems

The potentially exponential number of primes and the binate nature of the covering step represent the difficulties associated with the optimization problem.

Binate covering may arise simply because of the nature of the function F in the SRE. In particular, if F contains $y_n, \cdots, y_{n-d}$ in both true and complemented forms, then the coefficients α_i will generally appear in SRE_α with mixed polarity.

The binate nature of the covering problem reflects an intrinsic difficulty in the covering step. In the unate case, the effect of adding / removing a prime to a partial solution is always predictable: we increase / decrease the cover given by that partial solution. This is not necessarily true in the present case. Since F is of mixed polarity w.r.t. $y_n, \cdots, y_{n-d}$, the effect on F of adding a cube to y becomes unpredictable.

It is thus desirable to remove the mixed-polarity dependency of F on y. Such a dependency occurs only if there are paths from the gate under optimization to the primary outputs with different parity of inversions.

Definition 3.1 *A function* $F(\mathbf{y}_n, \ldots, \mathbf{y}_{n-d})$ *is termed* **positive (negative) unate** *in* $\mathbf{y}$ *if it is positive (negative) unate in each of* $\mathbf{y}_n, \ldots, \mathbf{y}_{n-d}$.

Definition 3.2 *A network is termed* **unate** *w.r.t. a gate* **g** *if all reconvergent paths from* **g** *have the same parity of inversions. A network is* **internally unate** *if it is unate w.r.t. each of its gates.*

The output of a network unate w.r.t. a gate **g** is a unate function of the gate output. Any network can be made internally unate by duplicating the gates with different parity of inversions in their fanout paths. The new network is at most twice the size the original one. In practice, the increase is generally much smaller.

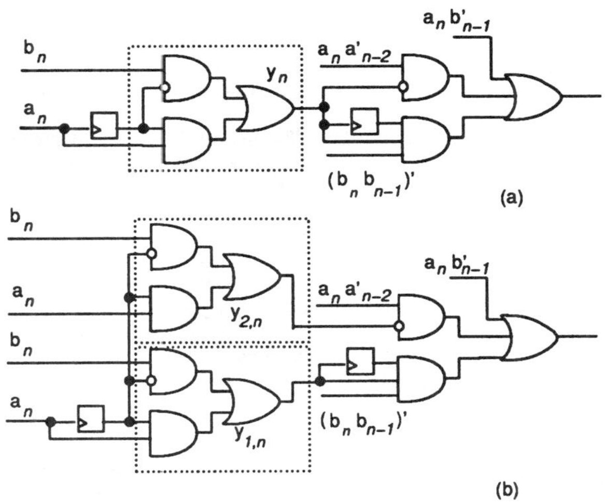

Figure 5: Unate transformation of a network.

Example 7. In the circuit of Fig. (5a), we seek the optimization of the gate with output y. The network realizes the function

$$S = a_n b'_{n-1} + a_n b'_n (a_{n-1}(a_{n-2} + b_{n-1}) + a'_{n-1} a'_{n-2})$$

We assume no external *don't care* conditions. Consequently, $F_{min} = F_{max} = S$. The output is expressed in terms of the y by

$$F = a_n b'_{n-1} + a_n a'_{n-2} y'_n + (b_n b_{n-1})' y_n y_{n-1}$$

Fig. (5b) shows the unate version of the example in Fig. (5a). The gate to be optimized has been duplicated, the two new outputs being y_1, y_2. The output is now a unate function of each of y_1, y_2:

$$F = a_n a'_{n-2} y'_{2,n} + (b_n b_{n-1})' y_{1,n} y_{1,n-1}$$

□

Primes
$c_1 = a_n$
$c_2 = b_n$
$c_3 = a_n a_{n-1}$
$c_4 = b_n b_{n-1}$
$c_5 = a_n b'_{n-1}$
$c_6 = b_n a'_{n-1}$

Table 2: List of primes for the problem of Example (7).

The following example shows that the unateness of F alone, however, is not a sufficient condition to insure unateness of covering problems.

Example 8. Table (2) contains a list of primes of y_1. The prime b_n, however, can never appear in the same cover with, for example, $a_n a_{n-1}$. This can be verified as follows. Consider, for instance, a cover of y containing $a_n a_{n-1} + b_n$. The product $y_n y_{n-1}$ would then contain the cross-product $b_n a_{n-1} a_{n-2}$. Corresponding to the assignment $b_n = a_{n-1} = a_{n-2} = 1; a_n = b_{n-1} = 0$ we have $F = 1$ and $F_{max} = 0$, thereby violating the constraint $F \subseteq F_{max}$. □

In the case of unate covering problems, two partial covers can always be merged to obtain a larger valid cover. In Example (8) instead, the appearance of the cross-product $b_n a_{n-1} a_{n-2}$ invalidates the union of any two covers containing $a_n a_{n-1}$ and b_n, respectively, by violating $F \subseteq F_{max}$.

In order to construct unate covering subproblems, a more accurate understanding of the effect of adding / expanding an implicant to a partial cover is therefore necessary.

Assume, for simplicity, that F is positive unate in **y**, and consider the effect on F of elementary operations, such as the addition / expansion of an implicant to **y**. Due to the positive unateness of F in **y**, corresponding to each assignment of $\mathbf{x}_n, \ldots, \mathbf{x}_{n-d}$, an elementary operation can at most change $F = 0$ to $F = 1$, but not viceversa. Each elementary operation therefore preserves the inequality $F_{min} \subseteq F$, and only the second inequality $F \subseteq F_{max}$ needs be checked. In Example (8), precisely this constraint is violated by the cover $a_n a_{n-1} + b_n$. This leads to the following definition:

Definition 3.3 *(***Maximal set of primes.***) A set of implicants is said to be* **maximal** *if no implicant can be expanded / added to it without violating* $F \subseteq F_{max}$.

As Example (9) implicitly points out, there may be more than one maximal set of primes.

Example 9.
For the circuit shown in Fig. (5), the following are maximal sets of primes for the variable y_1:

$$S_1 = \{a_n b_n, a_n c_{n-1}, b_n b_{n-1}, a_n b'_{n-1}, b_n a'_{n-1}\};$$

$$S_2 = \{b_n\}.$$

$$S_3 = \{a_n, b_n a'_{n-1}\};$$

□

Once a maximal set of primes has been built, the removal of any primes can only change $F = 1$ to $F = 0$ for some assignments of $\mathbf{x}_n, \ldots, \mathbf{x}_{n-d}$. Therefore, removal of any primes can never result in a violation of $F \subseteq F_{max}$. The covering step can thus be formulated as

minimize the number of implicants/literals of y;
constraint: $F_{min} \subseteq F$.

To achieve exact minimization, all possible maximal sets of primes should be generated and the subsequent covering step solved. This is obviously inefficient and costly in terms of CPU time. A useful approximation is instead that of building only one maximal set of primes upon the original set. For example, for the network of Fig. (5) only S_1 or S_3 would be derived.

Once the internally unate network is optimized, it can be folded back into an arbitrary, internally binate, network.

Example 10. Beginning from the set S_1 of Example (9), the optimum two-level realization of y_1 is $a_n a_{n-1} + b_n b_{n-1}$. It could then be shown that an optimal realization of y_2 is b'_n. The network after the above transformations is shown in Fig. (6). □

3.3 Adding Feedback.

In the optimization process followed so far, we treated the feedback inputs and outputs as ordinary primary inputs and outputs,

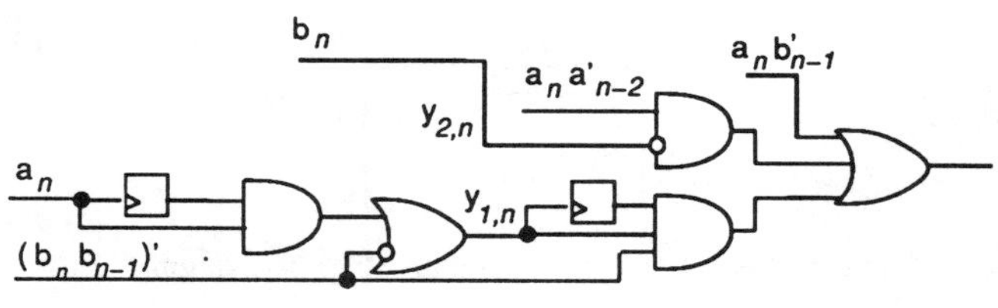

Figure 6: The optimized version of the network.

respectively. In practice, at any point in time the value at a feedback input is fixed by the network, and we expect that not all feedback sequences will be possible. Moreover, the feedback outputs are not perfectly observable. We can model both effects by means of suitable *don't care* conditions, as shown by the following example.

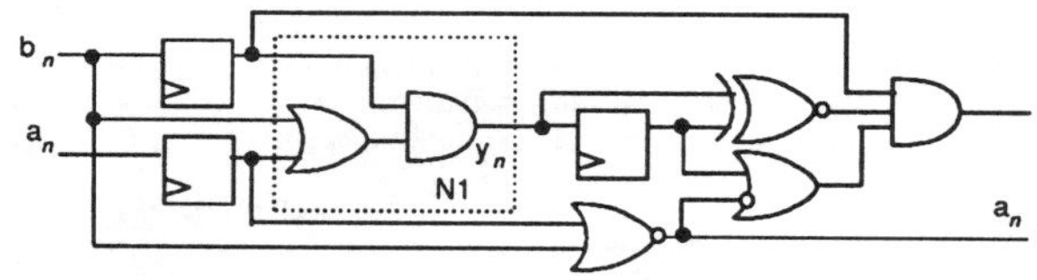

Figure 7: *Don't care* conditions for a circuit with feedback.

Example 11. Consider the network of Fig. (7). It represents essentially the same network of Fig. (4). In this case, however, a_n is not an external signal, but rather a feedback input. In particular, $a_n = (a_{n-1} + b_n)'$. We can therefore add an external controllability *don't care* set $DC = a_n \oplus (a_{n-1} + b_n)'$ to the specifications.
The input a_n is also not observed at any time in the future if $b_{n+1} = 1$. We can therefore associate an observability *don't care* set b_{n+1} to the feedback output. □

Unlike the combinational case, however, the controllability and observability *don't care* conditions derived from the feedback interconnection cannot generally be used simultaneously in the optimization of the network. Indeed, using the observability *don't care* set implies changing the feedback function, and possibly invalidating the corresponding controllability *don't care* set.

3.4 Experimental Results

The algorithms outlined in this paper have been written in C and tested on standard synchronous logic benchmarks. The results of applying the exact method of Sect. (3.1) are shown in Table 3.4. They were obtained on a DEC 5000 workstation. In particular, the first four columns refer to the initial benchmark statistics, in terms of inputs, outputs, literal, and register counts, respectively. Column *optl* reports the final number of literals and registers obtained, while *cpu* shows the CPU time in seconds.

4 Summary and Conclusions.

In this paper we outlined a model for the optimization of synchronous circuits across register boundaries. We use the concept of **sets of execution traces**, rather than state diagrams, to specify the desired terminal behavior (and its associated degrees of freedom) of a synchronous circuit.

For synchronous logic optimization problems, **Synchronous Recurrence Equations** represent an efficient way of representing trace sets. We can thus cast the synthesis problem of a syn-

Circuit	inputs	outputs	lits	regs	optl	cpu
s208	11	2	166	8	108	3
s298	3	6	244	14	155	14
s344	9	11	269	15	186	25
s420	19	2	336	16	251	258
s444	3	6	352	21	202	142
s641	35	24	539	19	241	302

Table 3: Experimental results for some logic optimization benchmarks.

chronous circuit as that of finding the minimum-cost solution to such equations.

An exact two-step solution algorithm has been proposed. The first step transforms the synchronous problem into a combinational one, which we have shown to differ from those previously considered in the literature. An exact algorithm for the latter problem is then presented.

We have shown that the solution of this problem involves a difficult **binate covering** step, for which efficient algorithms are under investigation [29]. To overcome this difficulty, we have shown that by making the network **internally unate** it is possible to reduce the covering step to a much simpler unate one.

Experimental results show that this approach is viable for the hierarchical optimization of synchronous circuits, with a solution space not otherwise reachable.

5 Acknowledgements.

This research was sponsored by DEC, jointly with NSF, under a PYI Award and by NSF/DARPA, under contract # MIP 8719546 and its renewal.

References

[1] R. Brayton, R. Rudell, A. Sangiovanni-Vincentelli, A. Wang , "MIS: A Multiple-Level Logic Optimization System", *IEEE Transactions on CAD/ICAS*, Vol. CAD-6, No. 6, pp. 1062-1081, November 1987.

[2] K. A. Bartlett, R. K. Brayton, G. D. Hachtel, R. M. Jacoby, R. Rudell, A. Sangiovanni-Vincentelli, and A. Wang, "Multilevel Logic Minimization Using Implicit Don't Cares", *IEEE Transactions on CAD/ICAS*, vol. CAD-7, No. 6, pp. 723-739, June 1988.

[3] S. Muroga, Y.Kambayashi, H.Lai and J.Culliney, "The Transduction Method - Design of Logic Networks Based on Permissible Functions", *IEEE Trans. Comp.*, vol. 38, No. 10, pp. 1404-1424, 1989.

[4] E. J. McCluskey, " Minimization of Boolean Functions," *Bell Syst. Tech. Jour.*, vol. 35, pp. 1417-1444, 1956.

[5] W. V. Quine, " The Problem of Simplifying Truth Functions ", *Am. Math. Monthly*, vol. 59, pp. 521-531, 1952.

[6] R. K. Brayton, G. D. Hachtel, C. T. McMullen, and A. L. Sangiovanni-Vincentelli, *Logic Minimization Algorithms for VLSI Synthesis*, Boston, Kluwer Academic Publishers, 1984.

[7] R. K. Brayton and F. Somenzi, " Boolean Relations and the Incomplete Specification of Logic Networks", *IFIP VLSI 89 Int. Conference*, pp. 231-140, Munich, 1989.

[8] R. K. Brayton and F. Somenzi, " An Exact Minimizer for Boolean Relations", *Proc. ICCAD 1989*, pp. 316-319, S. Clara, Nov. 1989.

[9] S.Devadas, T.Ma, A. Newton, and A.Sangiovanni-Vincentelli, "A Synthesis and Optimization Procedure for Fully and Easily Testable Sequential Machines", *IEEE Transactions on CAD/ICAS*, Vol. CAD-8 No. 10, pp. 1100-1109 October 1989.

[10] S. Devadas and A. R. Newton, " Decomposition and Factorization of Sequential Finite State Machines", *IEEE Trans. on CAD*, vol. 8, pp. 1206-1217, 1989.

[11] G. Saucier , M. Crastes de Paulet and P. Sicard, "ASYL: A Rule-Based System for Controller Synthesis", *IEEE Transactions on CAD/ICAS*, Vol. CAD-6, pp. 1088-1097 November 1987.

[12] J. Hartmanis and H. Stearns, *Algebraic Structure Theory of Sequential Machines*, Englewood Cliffs, N.J., Prentice-Hall, 1966.

[13] Z. Kohavi, *Switching and Finite Automata Theory*, 2nd ed., New York, Englewood Cliffs, N.J., Prentice-Hall [1966]) McGraw-Hill, 1978.

[14] K. T. Cheng and V. D. Agrawal, "State Assignment for Initializable Synthesis", *Proc. ICCAD 1989*, S. Clara, Nov. 1989.

[15] G. De Micheli, " Synchronous Logic Synthesis: Algorithms for Cycle-Time Optimization", *IEEE Trans. on CAD*, vol. 10, pp. 63-73, 1991.

[16] S. Malik, E. M. Sentovich, R. K. Brayton, and A. Sangiovanni-Vincentelli, " Retiming and Resynthesis: Optimizing Sequential Networks with Combinational Techniques", *IEEE Trans. on CAD*, vol. 10, pp. 74-84, 1991.

[17] S. Dey, F. Brglez, and G. Kedem, " Partitioning Sequential Circuits for Logic Optimization", *Proc.* 3rd *Int'l Workshop on Logic Synthesis* , Research Triangle Park, 1991.

[18] C. E. Leiserson, F. M. Rose, and J. B. Saxe, " Optimizing Synchronous Circuitry by Retiming", in R. Bryant, ed., *Proc.* 3rd *CALTECH Conference on Large Scale Integration*, Computer Science Press, Rockville, 1983.

[19] W. Wolf, A. Takach, and T. C. Lee, " Architectural Optimization Methods for Control-Dominated Machines", in Camposano and Wolf (editors): *High-Level VLSI Synthesis*, pp. 231-254, Kluwer, 1991.

[20] D. Ku, D. Filo and G. De Micheli, " Control Optimization Based on Resynchronization of Operations", *Proc. 1991 DAC Conf.*, S. Francisco, June 1991.

[21] A. Aho, R. Sethi, and J. Ullman, *Compilers: Principles, Techniques, and Tools*, Addison-Wesley, 1986.

[22] M. Damiani and G. De Micheli, " Synthesis and Optimization of Synchronous Logic Circuits from Recurrence Equations", *Proc. EDAC 1992*, Bruxelles, March 1992.

[23] M. Damiani and G. De Micheli, " *Don't care* conditions in Combinational and Synchronous Logic Networks", To appear on *IEEE Trans. on CAD*.

[24] M. Damiani and G. De Micheli, "The Role of *don't care* conditions in Synchronous Logic Optimization", in *Proceedings of Synthesis and Simulation Meeting and International Interchange (SASIMI)*, pp. 55–62, Tokio, 1990.

[25] N. Weste and K. Eshraghian, *Principles of CMOS VLSI Design*, Addison-Wesley, 1988.

[26] S. R. Petrick, " A Direct Determination of the Irredundant Forms of a Boolean Function from a set of Prime Implicants, " A. F. Cambridge Res. Center Report AFCRC-TR-56-110, Bedford, Mass., 1956.

[27] D. L. Dill, *Trace Theory for Automatic Verification of Speed Independent Circuits*, MIT Press, Cambridge, 1988.

[28] M. Rem, J. L. A. VanDeSnepscheut and J.T. Udding, "Trace Theory and the Definition of Hierarchical Components", in R. Bryant, ed., *Proc.* 3rd *CALTECH Conference on Large Scale Integration*, Computer Science Press, Rockville, 1983.

[29] M. Pipponzi and F. Somenzi, " An Iterative Algorithm for the Binate Covering Problem", *Proc. EDAC 1989*, pp. 208-211, Glasgow, March 1989.

Finite State Machine Synthesis with Fault Tolerant Test Function

Srimat T. Chakradhar
C&C Research Laboratories, NEC, 4 Independence Way, Princeton, NJ 08540
Suman Kanjilal*
Rutgers University, New Brunswick, NJ 08903
Vishwani D. Agrawal
AT&T Bell Laboratories, 600 Mountain Avenue, Murray Hill, NJ 07974

ABSTRACT: In a recent proposal a test function, which is a finite state machine with the same number of state variables as the given object machine, is incorporated into the state diagram of the object machine. Since a complete verification of the test machine is impractical, an often used heuristic sets and observes each state variable. As the two machines share logic, a fault can result in partial or total loss of the test function. Tests based on the assumption that the entire test function is intact can become invalid. We propose a new method of synthesizing PLA-based finite state machines with fault tolerant test machines. This approach eliminates testing of the test function. We constrain the logic minimization phase so that a fault has predictable effect on the state diagram of the composite machine (object machine embedded with the test function). This allows effective use of the test function during test generation, even in the presence of faults that affect both object and test machines. Only a combinational test generator is required for test generation. Each combinational vector is augmented by appropriate initialization and propagation sequences. Unlike prior approaches, our $O(\log_2 n)$ length test sequence is *guaranteed* to detect any targeted crosspoint fault. Experimental results on the MCNC Logic Synthesis Workshop finite state machine benchmark set show the practicality of the proposed approach.

1. INTRODUCTION

Testing sequential circuits is more difficult than testing combinational circuits due to the presence of memory elements. These elements are usually embedded in the sequential circuit and, unlike the primary inputs of the circuit, may not be easily set or *controlled* to arbitrary logic values. Furthermore, unlike primary outputs, values stored in the memory elements may not be directly *observable*. In combinational circuits, a test for a fault consists of a single primary input vector. In a sequential circuit, a sequence of primary input vectors may be required to detect a fault. Solutions to the sequential circuit testing problem range from making all memory elements observable and controllable by post-synthesis modifications to practically no hardware modifications. Often encountered difficulties with these approaches are high overhead or high test generation complexity.

The checking experiment [7] approach, though complete, is impractical due to the long test sequence required. Such an experiment verifies that a given finite state machine is different from any other finite state machine with same or fewer states. In comparison, sequential test generation is a simpler problem. It involves generation of test sequences that differentiate the object machine from a *finite* set of faulty machines. The number of faulty machines is at most equal to the number of modeled faults. Therefore, for a faulty finite state machine that passes these tests there may still exist input sequences for which the terminal behavior of the machine is different from the specified behavior. The fault in such a case does not map onto the set of modeled faults.

A viable approach is to consider testability during synthesis. Typically, several logic circuit implementations are possible for any given state diagram and synthesis selects an implementation that is optimized with respect to performance metrics like area, speed and testability. From a practical standpoint, an easily testable implementation can be viewed as one in which all modeled faults are either detectable by short test sequences or are provably redundant. Furthermore, the computational resources required to generate the test sequences should not be prohibitive.

One possible way to ensure testability is to synthesize a composite machine in which a test function is embedded in the object function [1]. Since the object and test functions can share the same logic elements, a fault in the implementation can result in partial or

*Supported by NEC during the summer of 1991.

total loss of the test function. Therefore, tests that are generated based on the assumption that the entire test function is intact may be invalid. Ideally, a complete verification (e.g., through checking experiment) of the test machine should precede the testing of the object machine. Such verification test can be quite long and is usually not used.

Several VLSI circuits have combinational portions whose sum-of-product descriptions require exponential amounts of storage. Since a state diagram is a flattened sum-of-products representation, it may not be practical to represent such circuits as state diagrams. In a companion paper [3], we proposed a new design for testability method for such circuits that are described as a netlist of logic gates and flip-flops. Again, for testing we rely only on combinational test generation techniques. We guarantee short test sequences for any crosspoint fault in the implementation. No explicit verification of the test machine is required.

A procedure in which the output and state equations of the finite state machine are modified to incorporate the scan function in a PLA was described by Reddy and Dandapani [8]. Simple read and write tests perform approximate functional verification of the scan function. Therefore, the validity of test sequences that are derived assuming that the scan function is not corrupted by the fault is not guaranteed. Alternatively, a checking experiment as a guaranteed test will require a very long test sequence.

Some approaches [5, 6] constrain state assignment and logic optimization to guarantee testability of all combinationally irredundant crosspoint faults in a PLA-based finite state machine. Finite state machines are assumed to have a reset state. Test generation requires the traversal of the state transition diagram. Furthermore, the test sequence for a crosspoint fault may be long. In contrast to this work, our procedure allows arbitrary state encoding. This can be exploited to achieve other objectives like minimizing area of the PLA. Also, we do not assume an explicit reset state and test generation does not require traversal of state transition diagrams. Furthermore, our procedure guarantees bounded-length test sequences for combinationally irredundant crosspoint faults.

2. FAULT MODEL

Due to the special layout of PLAs, it is necessary to consider faults other than the conventional stuck-at faults. The *crosspoint* fault model considers the following faults of the PLA:

1. Extra or missing contact faults in the input (AND) plane.

2. Extra or missing contact faults in the output (OR) plane.

A missing contact in the AND plane of the PLA results in a missing literal in exactly one product term. Similarly, an extra contact in the AND plane results in an extra literal in exactly one product term. A missing contact in the OR plane results in a missing product term from exactly one of the PLA outputs. Similarly, an extra contact in the OR plane results in an extra product term in exactly one output.

These faults can be easily modeled as stuck-at faults in the AND-OR logic model of the PLA [2].

3. EFFECT OF FAULT ON TEST MACHINE

Consider a test function with two input symbols p_1 and p_2. A four state test machine with two input symbols is shown in Figure 1. To study the effect of a fault,

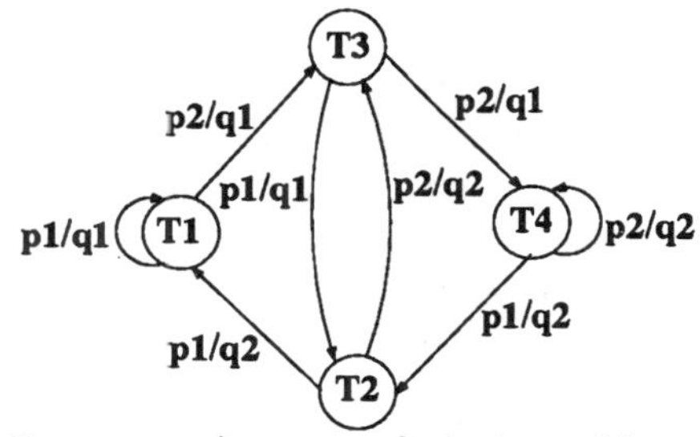

FIGURE 1: An example test machine.

it is convenient to classify the transitions in the composite machine into three groups: E_{p_1}, E_{p_2} and E_{others}. Transitions labeled by the input symbol p_1 belong to group E_{p_1}. Similarly, transitions labeled by the input symbol p_2 belong to group E_{p_2}. All the remaining transitions are grouped into E_{others}. A single crosspoint fault may corrupt an arbitrary subset of transitions in the composite machine.

If the corrupted transitions belong only to group E_{others}, then the test machine is not affected by the fault. Hence, test generation can be performed assuming that the entire test function is intact and tests generated using Agrawal and Cheng's heuristic [1] are guaranteed to detect the fault.

If the corrupted transitions belong to both E_{p_1} and E_{p_2}, then the test machine is faulty. In particular, we may lose the ability to initialize the faulty machine using only the sequences that initialize the test machine. We may no longer be able to set the faulty machine to all its states or observe its states at the output by applying short predetermined sequences. As an example, consider the case when the fault corrupts the transition labeled p_1/q_2 out of state T_4 and the transition labeled p_2/q_1 out of state T_1 (see Figure 1). We show that the faulty machine may not be initializable using only those transitions in the test machine that are unaffected by the fault. Any synchronizing sequence starting with input symbol p_1 (p_2) may not initialize the machine to a known state if the initial state of the faulty machine is $T4$ ($T1$). This is because the transition labeled p_1 (p_2) out of state $T4$ ($T1$) may be corrupted. Therefore, the faulty machine cannot be initialized using only uncorrupted transitions in the test machine. Since the fault affects the test function, use of test function during test generation is suspect. As illustrated by the example in a companion paper [3], tests generated assuming the sanity of the test machine may not detect the fault.

If the corrupted transitions belong to either E_{p_1} or E_{p_2} but not both (some of the corrupted transitions may belong to E_{others}), then also the test machine is affected. However, as we show in Section 4, it is possible to effectively use uncorrupted transitions in the test machine during test generation. The main idea of the present work is to be able to meaningfully use the test machine even when some of its transitions are corrupted by the fault. Among the several implementation alternatives available for the composite machine, we chose the one in which any crosspoint fault will corrupt only the transitions belonging to either E_{p_1} or E_{p_2} but not both. A synthesis method to produce such an implementation is presented in Sections 5 and 6.

4. TEST GENERATION

We consider faults that corrupt transitions belonging only to E_{p_1} or E_{p_2} but not both. However, some of the corrupted transitions may belong to group E_{others}. Only combinational test generation techniques are required to construct input sequences that detect such faults in the composite machine. Let n be the number of states in the test machine and we assume that the fault corrupts a subset of transitions belonging to group E_{p_1}. We also assume that a combinational test for the fault has been obtained using a combinational test generation technique. Suppose, this test requires the present state of the good machine to be S_c. Then the following test sequence is guaranteed to detect the fault:

1. Apply the synchronizing sequence consisting of only input symbol p_2. Such a sequence exists and its length is $\lceil \log_2 n \rceil$ [3]. Furthermore, this sequence puts the good and faulty machines in the same final state, say S_k.

2. Apply a transfer sequence that puts the good machine in state S_c, given that the initial state of the good machine is S_k. However, in the faulty machine, this transfer sequence may not necessarily put the machine in state S_c. This is because the transfer sequence may use transitions in the test machine that are corrupted by the fault.

3. Apply the distinguishing sequence consisting of only input symbol p_2. If the transfer sequence did put both good and faulty machines in the state S_c, then the two machines will yield identical responses to the distinguishing sequence. However, if the transfer sequence puts the good and faulty machines in different states, then the two machines will yield different responses and the fault is detected [3]. Therefore, the remaining steps are not required. If, however, the fault is not detected then both machines are again in state S_k.

4. Apply the transfer sequence of Step 2. Application of this transfer sequence guarantees to put both good and faulty machines in state S_c.

5. Apply the combinational test vector. This will either detect the fault at a primary output or put the good and faulty machines in different states.

6. Apply the distinguishing sequence of Step 3. Since the good and faulty machines are in different states, they will yield different responses and the fault is detected.

All steps except Step 5 require application of a sequence of length $\lceil \log_2 n \rceil$ vectors. Step 5 requires only one vector. Hence, the test sequence length is at most $5\lceil \log_2 n \rceil + 1$ vectors.

5. CROSSPOINT FAULT TESTABILITY

We make the following assumptions about the implementation of the composite machine:

1. Assumption 1: The ON-set of any product term consists of only transitions that belong to either E_{p_1} or E_{p_2} but not both. However, the ON-set may contain transitions from the group E_{others}.

   The ON-set (OFF-set) of a product term consists of all transitions that make the product term evaluate to 1 (0). As an example, consider the product term NS_1 in Figure 2. We assume that input symbols p_1 and p_2 are encoded as ($i_1 = 1, i_2 = 0$) and ($i_1 = i_2 = 1$), respectively. The ON-set consists of all transitions labeled with input symbol p_2 since this symbol is coded as 11 and this is the only combination of input values that forces the AND gate NS_1 to a 1.

2. Assumption 2: If a pair of transitions, one from E_{p_1} and the other from E_{p_2}, belong to the OFF-set of a product term and exactly one primary input literal of the product term is set to 0 in either transition, then the two transitions must set different literals to 0.

   As an example, consider a product term $i_1 i_2 PS_1 PS_2$ where i_1 and i_2 are the primary inputs and PS_1 and PS_2 are the present state variables. Also, let the input symbol p_1 be coded as

$(i_1 = i_2 = 0)$ and the input symbol p_2 be coded as $(i_1 = 0,\ i_2 = 1)$. Here, the OFF-set of the product term contains transitions from E_{p_1} and E_{p_2} since either of the input symbols can force the product term to 0. Since any transition from E_{p_1} makes more than one primary input literal to assume the value 0, such a product term does not violate the assumption. However, the product term $i_1 PS_1 PS_2$ violates the assumption. This is because transitions from E_{p_1} or E_{p_2} force exactly one primary input literal to 0 and they force the same literal $i1$ to 0.

We show that any single crosspoint fault in an implementation that satisfies Assumptions 1 and 2 corrupts a subset of transitions in either E_{p_1} or E_{p_2} but not both. Therefore, test sequences for these faults can be obtained by using the test generation procedure discussed in Section 4.

Let P be a product term that is a logical AND of primary input literals $i_1, i_2, \ldots, i_k$ and present state literals $S_1, S_2, \ldots S_l$. Furthermore, let O be a PLA output that is a logical OR of product terms $P_1, P_2, \ldots, P_m$.

5.1 Missing Contact Fault in AND Plane

Without loss of generality, assume that primary input literal i_1 is missing from the product term P because of a missing contact fault. Let P' be the new product term. It is a logical AND of primary input literals $i_2, \ldots, i_k$ and present state literals $S_1, S_2, \ldots S_l$.

To activate the fault, we require a set values for the primary input literals and the present state literals such that P and P' assume different values. Since the OFF-set of P' is also included in P, none of the transitions in the OFF-set of P' will activate the fault. Of all the transitions in the ON-set of P', only transitions with $i_1 = 0$ can activate the fault since all other transitions will force P and P' to the same value 1. However, from Assumption 2, if a transition belonging to E_{p_1} (E_{p_2}) satisfies the above condition, all transitions belonging to E_{p_2} (E_{p_1}) that are in the OFF-set of P must set at least one literal in P to 0. Therefore, these transitions belong to the OFF-set of P' also and they cannot activate the fault. Since only transitions that activate the fault may be corrupted, only a subset of transitions in either E_{p_1} or E_{p_2} but not both may be corrupted by the missing contact fault in the AND plane.

5.2 Missing Contact Fault in OR Plane

Without loss of generality, assume that product term P_1 is missing from O. Let O' be the logical OR of product terms $P_2, P_3, \ldots, P_m$.

For any transition that activates the fault, O and O' assume different values. Since the ON-set of O' is also included in O, none of the transitions in the ON-set of O' will activate the fault. Of all transitions in the OFF-set of O', only transitions with $P_1 = 1$ can activate the fault since all other transitions will cause O and O' to assume the same value 0. However, from Assumption 1, all transitions in the ON-set of P_1 belong to either E_{p_1} or E_{p_2} but not both. Therefore, only a subset of transitions in either E_{p_1} or E_{p_2} but not both may be corrupted by the missing contact fault in the OR plane.

5.3 Extra Contact Fault in AND Plane

Let x be the extra literal included in the product term P and let P' be the product term that is a logical AND of primary input literals $i_1, i_2, \ldots, i_k$, present state literals $S_1, S_2, \ldots, S_l$, and x. The extra literal may be a primary input literal or a present state literal.

To activate the fault, we require a set of values for the primary input literals and the present state literals such that product terms P and P' assume different values. Since the ON-set of P' is also included in P, none of the transitions in the ON-set of P' will activate the fault. This is because both P and P' will assume the same value 1. Of all the transitions in the OFF-set of P', only the transitions with $x = 0$ and all other literals in P' are 1 can activate the fault. This is because any other transition in the OFF-set of P' will force P and P' to the same value 0. However, all transitions in which $i_1 = 0$ and the rest of the literals in P assume the value 1 belong to either E_{p_1} or E_{p_2} but not both (see Assumption 1). Since these are the only transitions that may be corrupted by the fault, only a subset of transitions in either E_{p_1} or E_{p_2} but not both are corrupted by the extra contact fault in the AND plane.

5.4 Extra Contact Fault in OR Plane

Without loss of generality, assume that the fault causes O to depend on some additional product term x. Let O' be the logical OR of product terms $x, P_1, P_2, \ldots, P_m$.

Again, for any transition that activates the fault, O and O' must assume different values. Since the OFF-set of O' is also included in O, none of the transitions in the OFF-set of O' will activate the fault. Of all transitions in the ON-set of O', only transitions with $x = 1$ and all other product terms equal to 0 can activate the fault since for any other transition, O and O' assume the same value 1. However, from Assumption 1, all transitions that cause the product term x to assume the value 1 belong to either E_{p_1} or E_{p_2} but not both. Therefore, only a subset of transitions in either E_{p_1} or E_{p_2} but not both may be corrupted by the extra contact fault in the OR plane.

6. SYNTHESIS PROCEDURE

In a recent paper [3], we gave two synthesis procedures to realize finite state machines with an embedded fault tolerant test function. In both cases, the two-level combinational logic for the composite finite state machine satisfies the assumptions outlined in Section 5. We illustrate a synthesis method that embeds fault tolerant test function into the state transition diagram of an object machine.

Figure 2 shows a PLA modified to achieve the fault-tolerant test function. This circuit contains three extra AND gates added to the original circuit synthesized after embedding the test function of Figure 1 into an object machine [3]. These extra gates are the one with output labeled g_1' and the last two AND gates feeding into the OR gate producing the signal *Out*. The original circuit [3] did not contain the signal g_1' and *Out* was produced by an OR of g_2 and $\overline{PS_2}$. In that implementation, a crosspoint fault on input $\overline{PS_1}$ of gate g_1 would have corrupted both E_{p_1} and E_{p_2} transitions of the test machine. This is because both p_1 and p_2 symbols are encoded with $i_1 = 1$ and the product term g_1 assumes a value based only on present state of the machine, regardless of what input vector is applied. Similarly, a crosspoint fault on input PS_2 of gate *Out* may corrupt output labels of both E_{p_1} and E_{p_2} transitions. Since product terms g_1 and $\overline{PS_2}$ violate Assumption 1, they are suitably modified as shown in Figure 2. We picked primary input i_2 because the codes for p_1 and p_2 differ in this bit.

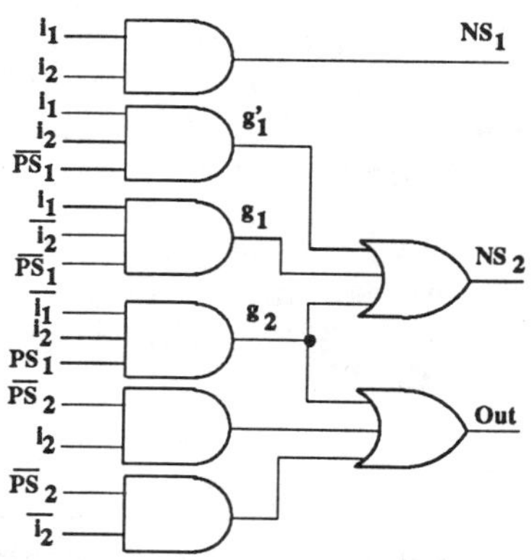

FIGURE 2: Modified PLA of composite machine.

7. UPPER BOUND ON TEST LENGTH

We derive an upper bound on the number of vectors required to detect a set of crosspoint faults. Here, *test* refers to the sequence of vectors required to detect all faults and *test sequence* refers to the sequence of vectors obtained by augmenting a combinational test vector.

Lemma 1: *If f is the number of combinational test vectors required to detect all combinationally irredundant crosspoint faults, n is the number of states in the composite machine and $k = \min(f, n-1)$, then a test of at most $2f + (2 + 4f + 4k)\log_2 n$ vectors can detect all crosspoint faults in the composite machine.*

Proof: Any combinational vector that is augmented to a test sequence consists of the following vectors:

1. A synchronizing sequence of $\lceil \log_2 n \rceil$ vectors.

2. A transfer - distinguishing - transfer sequence of $3\lceil \log_2 n \rceil$ vectors. This sequence of vectors transfers the machine to a particular state S, verifies that the initial state is S and transfers the machine back to state S.

3. A combinational test vector.

4. A distinguishing sequence of $\lceil \log_2 n \rceil$ vectors.

A combinational test typically detects several faults. Some of these faults may corrupt transitions in E_{p_1} and some may corrupt transitions in E_{p_2}. Therefore, every combinational test may have to be augmented into at most two test sequences. One test sequence detects faults that corrupt E_{p_1} transitions and the other sequence detects faults corrupting E_{p_2} transitions. Therefore, f combinational test vectors can result in at most $2f$ test sequences.

Any test sequence consists of exactly one combinational test vector. Therefore, the test consists of at most $2f$ combinational test vectors. Also, every test sequence terminates with a distinguishing sequence of at most $\lceil \log_2 n \rceil$ vectors. Therefore, the test consists of at most $2f\lceil \log_2 n \rceil$ vectors corresponding to distinguishing sequences.

All test sequences are terminated by distinguishing sequences. If a test sequence detects faults that affect only E_{p_1} (E_{p_2}) transitions, the terminating distinguishing sequence can be used as a synchronizing sequence for another test sequence that also detects faults corrupting only E_{p_1} (E_{p_2}) transitions. Therefore, we require only two synchronizing sequences, each consisting of at most $\lceil \log_2 n \rceil$ symbols.

Since each test sequence consists of a transfer - distinguishing - transfer sequence of $3\lceil \log_2 n \rceil$ vectors, and there are at most $2f$ test sequences, the test consists of at most $6f\lceil \log_2 n \rceil$ such vectors. However, if $f > n-1$, this upper bound can be refined to $(2f+4k)\lceil \log_2 n \rceil$ vectors, as follows. Since every test sequence will require a transfer sequence, we require $2f\lceil \log_2 n \rceil$ vectors. After a synchronizing sequence, we may have to transfer the machine to at most $n-1$ states. Therefore, irrespective of how many faults are being considered, we only have to verify transfer to $n-1$ states. Since we use two synchronizing sequences, we have to verify transfer to $2(n-1)$ states. Verification of any initial state and subsequent transfer of the machine to that initial state requires at most $4(n-1)\lceil \log_2 n \rceil$ or $4k\lceil \log_2 n \rceil$ vectors. Hence, the test for all faults may contain at most $2f + (2 + 4f + 4k)\log_2 n$ vectors.

∎

8. RESULTS

Logic implementations of several benchmark examples were obtained using the program NOVA [9] which yields a state encoded and optimized two level PLA implementation of the object machine. The object machine was augmented by a fault tolerant test machine as discussed in Sections 5 and 6.

Test generation was performed for crosspoint faults in the object machine without the test machine and for the composite machine. For crosspoint faults in the object machine, we used a sequential test generator called Gentest [4]. Only missing contact faults in the input and OR plane were modeled. A missing contact fault in the AND plane, also referred to as *growth* fault, is equivalent to an input stuck-at-1 fault on the AND gate corresponding to the affected product term. Similarly, a missing contact fault in the OR plane (*disappearance fault*) is equivalent to an input stuck-at-0 fault on the OR gate corresponding to the affected PLA output. Accordingly, the targeted crosspoint faults were modeled as stuck faults in a two-level gate implementation of the circuit for Gentest which is a gate-level test generator. Test generation required 30 to 35000 seconds on a SUN Sparc2 workstation for a selected set of MCNC benchmarks [3] and the fault coverage varied from 55.8% to 100%.

For the composite machine, we used Gentest only to generate combinational tests for the growth and disappearance faults, again modeled at the gate level. The vectors were suitably augmented, as described in Section 4, to yield test sequences. In Table 1, *Faults-GD* are the modeled growth and disappearance faults and *Faults-SA* are the stuck faults in the two level AND-OR logic implementation of the composite machine. Here, *% A* is the percentage area increase in the composite testable machine compared to the object machine synthesized alone. The testable designs contain one extra test-mode input signal. *Vectors-C*, is the number of combinational test vectors required to detect all modeled crosspoint faults. We augmented each combinational vector to form two test sequences and *Vectors-S* gives the combined length of these test sequences obtained from all combinational vectors. *R* is the number of redundant faults in the fault-tolerant design and *CPU* is the test generation time in seconds on SUN Sparc2. Our sequences provided 100% coverage for the modeled crosspoint faults and stuck-at faults. As an example, in the finite state machine *sand*, we modeled 1225 growth and disappearance faults and 1505 stuck-at faults. Using Gentest, we obtained 334 combinational vectors for the growth and disappearance faults and these vectors were augmented to 7978 sequential test vectors that provided 100% fault coverage.

TABLE 1: Testing of composite machine.

FSM	% A	Faults		Vectors		R	CPU
		GD	SA	C	S		sec.
dk16	31	657	799	130	3490	27	5.3
kirk.	14	581	679	168	3906	8	5.9
s820	25	821	1031	209	5228	29	15
s832	11	853	1065	188	4766	35	15
sand	18	1225	1505	334	7978	19	42
scf	13	2029	2523	502	18630	45	85
tbk	12	2008	2246	677	15524	10	80

9. CONCLUSION

In this paper, we have introduced the concept of fault-tolerant test function in synthesis for testability. Compared to the scan design, our designs do not require specific testing of the test logic. The tests are still generated by combinational test generation method. With improved synthesis methods, that we are currently investigating, we believe the overhead of our method will be close to, or lower than, that of scan. It may also be possible to extend the procedure to multi-level logic synthesis.

REFERENCES

[1] V. D. Agrawal and K. T. Cheng. Finite State Machine Synthesis with Embedded Test Function. *Journal of Electronic Testing: Theory and Applications*, 1:221–228, 1990.

[2] V. D. Agrawal and D. D. Johnson. Logic Modeling of PLA Faults. In *Proceedings of IEEE International Conference on Computer Design*, pages 86–88, October 1986.

[3] S. T. Chakradhar, S. Kanjilal, and V. D. Agrawal. A Synthesis for Testability Technique for PLA-Based Finite State Machines. In *European Conference on Design Automation*, March 1992.

[4] W. T. Cheng and T. J. Chakraborty. GENTEST: An Automatic Test Generation System for Sequential Circuits. *IEEE Computer*, 22:43–49, April 1989.

[5] S. Devadas and K. Keutzer. A Unified Approach to the Synthesis of Fully Testable Sequential Machines. *IEEE Transactions on Computer-Aided Design*, 10(1):39–50, January 1991.

[6] S. Devadas and H. T. Ma. Easily Testable PLA-based Finite State Machines. *IEEE Transactions on Computer-Aided Design*, 9(6):604–611, June 1990.

[7] F. C. Hennie. Fault Detecting Experiments for Sequential Circuits. In *Proceedings 5th Annual Symposium on Switching Theory and Logical Design*, pages 95–110, November 1963.

[8] S. M. Reddy and R. Dandapani. Scan Design using Standard Flip-Flops. *IEEE Design and Test of Computers*, 4:52–54, February 1987.

[9] T. Villa and A. Sangiovanni-Vincentelli. NOVA: State Assignment of Finite State Machines for Optimal Two-Level Logic Implementations. In *ACM/IEEE Proceedings of the 26th Design Automation Conference*, pages 327–332, June 1989.

Solving the State Assignment Problem for Signal Transition Graphs*

Luciano Lavagno, Cho W. Moon, Robert K. Brayton, Alberto Sangiovanni-Vincentelli

Department of Electrical Engineering and Computer Sciences
University of California, Berkeley, CA 94720, USA

Abstract

*We propose a novel framework to solve the state assignment problem arising from the signal transition graph (*STG*) representation of an asynchronous circuit. We first establish a relation between* STG*s and finite state machines (*FSM*s). Then we solve the* STG *state assignment problem by minimizing the number of states in the corresponding* FSM *and by using a critical race-free state assignment technique. State signal transitions may be added to the original* STG*. A lower bound on the number of signals necessary to implement the* STG *is given. Our technique significantly increases the applicability of* STG*s for specifying asynchronous circuits.*

1 Introduction

Asynchronous circuits are playing an increasingly important role in digital designs for two main reasons. First, interface circuits, which are inevitably asynchronous, are becoming a bottleneck in the design process. Second, the clock skew problem, which is increasingly difficult to handle in today's high-speed synchronous designs, can be completely eliminated with asynchronous designs, resulting in more modular, faster, and possibly less power-consuming circuits.

Despite this, CAD support for asynchronous design remains weak. The main reason is the difficulty in dealing with the *hazards* which cause a system to deviate from its specified behavior. In [3] and [6] techniques were proposed to produce a hazard-free implementation from a graphical specification called the Signal Transition Graph (STG, [1]) under either bounded wire delay or unbounded gate delay models. The STG is based on a type of Petri net called the free-choice net, which is expressive enough for specifying concurrency, sequencing, and conflict, yet simple enough for analysis. The specification is complete in that both the system and the environment behaviors are specified.

Sufficient conditions for an STG to be implementable have been given by [1]. The most restrictive condition is the Complete State Coding (CSC) property ([6]), which requires that the signals specified by the STG completely define the circuit state. Until now the burden of satisfying the CSC property has been placed mostly on the designer. [13] and [10] address this problem but only in the context of marked graphs, which are a subset of STGs.

In this paper we propose a new framework to satisfy the CSC property on any STG by formulating the problem as the conjunction of a constrained state minimization problem and a critical race-free state assignment problem. We first give a procedure to derive from an STG an equivalent FSM representation that completely captures the state information implicit in the STG. Minimization of this FSM allows us to prove *necessary conditions* on the number of state signals required to implement it. This proposed framework is general enough to embed also previous methods to solve the CSC problem for some specific sub-classes of STGs, such as [11].

We also address how to extract from the minimized FSM *sufficient conditions* for the STG to have the CSC property. We propose to apply a critical race-free state assignment algorithm (such as the one proposed by Tracey [8]). Using the new states codes, we insert appropriate signal transitions in the STG. At this point, techniques such as [3] [6] can be used to obtain a hazard-free implementation.

This paper is organized as follows. Section 2 defines the terms used in this paper and discusses some previous work. Section 3 gives an informal overview of the whole approach. Section 4 describes how STGs are translated into FSMs. Section 5 describes our state minimization process. Section 6 describes our critical race-free state assignment technique. Section 7 describes how to insert state signals in the STG so that it can be implemented. Section 8 gives some experimental results. Section 9 concludes this paper.

*This work is supported by National Science Foundation under grant number UCB-BS16421 and by Defense Advanced Research Projects Agency under contract number N00039-87-C-0182 and by various grants from DEC, IBM, Intel, Motorola, AT&T, BNR and Cirrus Logic.

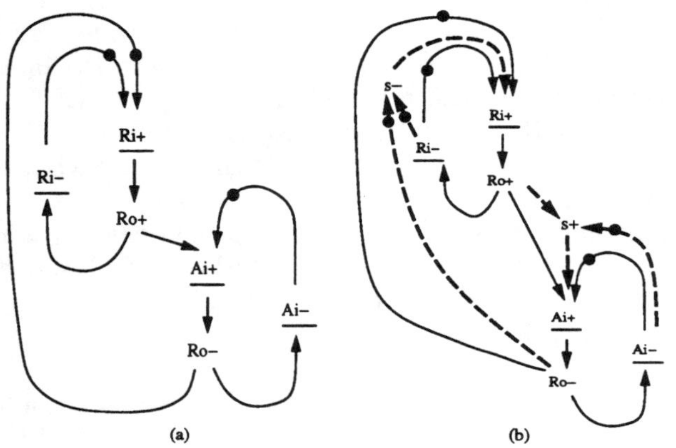

Figure 1: STG Specification before and after state assignment

2 Preliminaries

2.1 Signal Transition Graph

An STG is an interpreted free-choice Petri net introduced by Chu [1] for specifying asynchronous control circuits. It is a 4-tuple, $\langle T, P, F, m_0 \rangle$. T is a set of transitions described by $s \times \{+, -\}$, where $s+$ represents a 0 to 1 transition on signal s and $s-$ a 1 to 0 transition. P is a set of places which can be used to specify conflict or choice. F represents the flow relation between transitions and places: namely, $F \subseteq (T \times P) \cup (P \times T)$. A marking is an assignment of nonnegative integers (denoting the number of tokens) to each place. m_0 defines the initial marking or the initial state of the STG. A transition is *enabled* if all its fanin places are marked. When it *fires*, a token is removed from each fanin place and added to each fanout place. A *Marked Graph* (MG) is a Petri net where each place p has exactly one fanin and one fanout transition. A Petri net is *free-choice*, if for each place p with multiple fanout transitions, p is the unique fanin place for all its fanout transitions. An STG is an *interpreted* free-choice Petri net because each transition $t \in T$ is interpreted as a physical signal change on some input or output signal. An example of an STG is shown in Figure 1(a). Places with only one fanin and one fanout transition are omitted, and input signal transitions are underlined.

An STG is defined as *live* if the underlying net is deadlock-free, no place in it can ever be assigned more than one token and for each signal t, for all firing sequences, rising and falling transitions of t alternate.

2.2 State Graph

The state graph (SG) is a directed graph obtained by performing a reachability analysis on the STG starting from the initial marking m_0, where markings

0738-100X/92 $3.00 © 1992 IEEE

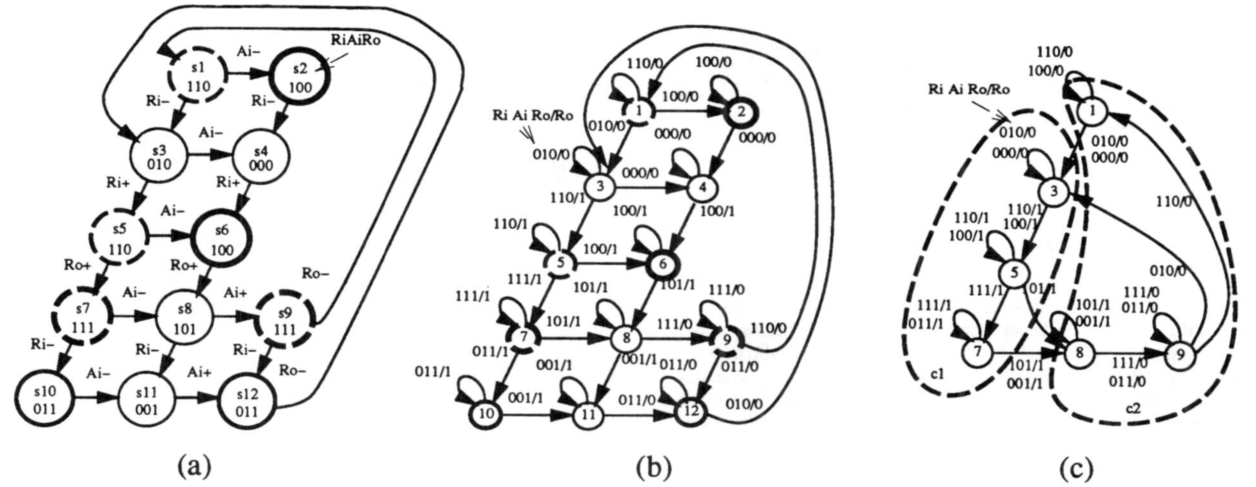

Figure 2: State Graph, Finite State Machine and minimized Finite State Machine

correspond to states (we will often identify a state with the corresponding marking). A state s_1 is connected to a state s_2 by an edge e, if there exists a marking m_2 (corresponding to state s_2) which is reached from marking m_1 (corresponding to state s_1) by firing a single transition t. The edge e is labeled with t.

2.3 Complete State Coding

The procedures given by [1, 3, 6] to produce a circuit implementing an STG specification require that each SG state is assigned a binary label, which is consistent with the firing transitions on the SG edges. For each edge $s_1 \rightarrow s_2$ labeled with transition t^+, the value of signal t is 0 in the label of s_1 and 1 in the label of s_2 (and vice-versa for t^-). All other signals must have the same value in both labels. Such a labeling always exists if the STG is *live*.

The boolean function for each output signal t of the STG maps the binary label of each SG state s into the *implied value* of t in s, that is the complement of the value of t in s if a transition of t is enabled in s, and the value of t in s otherwise. A necessary and sufficient condition for the existence of an implementation, with the structure of Figure 3(a), of a live STG is called the Complete State Coding (CSC) property ([6]). The CSC property is satisfied if *when the same binary label is assigned to two different states, the transitions of the output signals enabled in the two states are identical.* Figure 2(a) gives an example of CSC violation. States $s1$ and $s5$ are assigned the same binary label 110 but $s5$ has an enabled output transition $Ro+$ which $s1$ does not have. The CSC violation also occurs due to states $(s2, s6)$, $(s7, s9)$ and $(s10, s12)$.

2.4 Finite State Machines

A Finite State Machine (FSM) model of a sequential circuit is a six-tuple (S, X, Z, N, O, R), where: S is a set of *internal states*, X is a set of *input states*, Z is a set of *output states*, $N(i, s)$ is the *next-state function*, $O(i, s)$ is the *output function* and R is the initial state.

A state s is called *stable* if there exists an input i such that $N(i, s) = s$, i.e. a self-loop in the graph representation. A set c of states of an FSM is *output compatible* if $s_1, s_2 \in c, O(i, s_1) \neq \phi, O(i, s_2) \neq \phi \Rightarrow O(i, s_1) = O(i, s_2)$. A set c of states *implies* a set c' of states if $c' = \bigcup_{s \in c}\{N(i, s)\}$. A set c of states is *compatible* if it is output compatible and it does not imply an output incompatible set through any sequence of implications. A set C of *compatibles* of an FSM is a set of *maximal compatible sets* of states such that every state belongs to at least one compatible. A set C of compatibles is *closed* if for each input state i, for each compatible $c \in C$, for each pair of states $s_1, s_2 \in c$ there exists a $c' \in C$ such that $N(i, s_1) \subseteq c'$ and $N(i, s_2) \subseteq c'$. A closed set π of compatibles is a *closed partition* of the states of the FSM if every state belongs to one and only one compatible (called block). A closed partition π is *derived* from a closed set of compatibles C if every block of π is a subset of a compatible of C. An FSM is *minimized* if there exists no FSM with the same behavior but fewer states. FSM minimization amounts to finding a closed set of compatibles of minimum cardinality, and then creating one state of the minimized FSM for each compatible ([9]).

A *race* is a transition of an encoded FSM where more than one state signal changes. The race is *critical* if the behavior of the FSM depends on the outcome of the race, i.e. on which signal changes first.

2.5 Previous Work

Early work in the area of CSC enforcement (for example [13] and [10]) concentrated on the introduction of constraints within an MG, using a sufficient condition as a guidance. Namely both [13] and [10] recognized that if all pairs of signals in the STG are *locked* using a chain of handshaking pairs, then the MG satisfies the CSC condition. We will briefly show in Section 5 how to interpret this sufficient condition within our framework, and how to use it to *reduce* the number of state signals.

Also the synthesis methodology developed by Martin et al. ([5]), even though it starts from a different specification formalism, basically guarantees CSC (a sub-case of what they call *non-interference*) by handshaking, if possible, otherwise by heuristic state variable insertion.

More recently Vanbekbergen et al. ([11]) proposed a technique to transform an MG so that it satisfies CSC. The technique is based on the identification of pairs of SG states that cause a CSC violation. Each pair of such states is connected by an edge in an undirected graph called the *constraint graph* (Vanbekbergen also adds constraints to allow optimization of the combinational logic implementing some particular signal, but this has no direct relevance to the CSC problem per se). Then coloring this graph provides, according to Vanbekbergen, a lower bound on the number of state signals required for CSC. This approach, even though it was originally proposed only for MGs, bears some resemblance with the framework that we propose for general STGs. In fact graph coloring has been used since long ago as a heuristic for FSM minimization ([12]). Furthermore Vanbekbergen did not recognize the need to use critical race-free encoding. Critical races would show up as further violations of CSC in the encoded STG. Thus the main advantage of our framework over [11] is the recognition that CSC falls within a much more general problem, previously known as *state minimization/critical race-free encoding* ([9]), and that constraint graph coloring can be considered only as a *heuristic technique* to solve the general problem.

Kishinevsky et al. ([2]) also presented a complete synthesis methodology for asynchronous circuits starting from an event-based specification, called Transition Diagrams, that is very similar to MGs. They also used graph coloring to identify regions of the state graph that must be distinguished using state signals. They recognized the problem of critical races, but solved it through *iteration* of the encoding procedure.

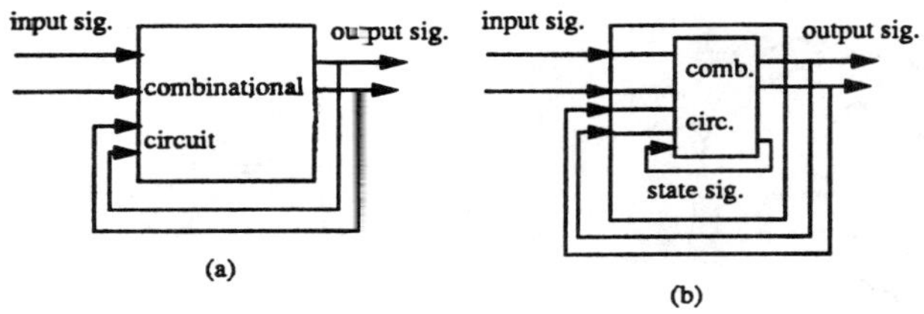

Figure 3: Circuit models

Another trade-off between MGs and general STGs based on free-choice nets is due to the fact that MGs do not allow the behavior of the specified circuit to depend on the value of external signals (e.g. a bus interface with different read and write protocols cannot be specified with an MG). The analysis and synthesis methods for MGs, on the other hand, have often polynomial worst-case behavior, while analogous algorithms for free-choice nets have in general exponential worst-case behavior. For example, the CSC analysis in [13] and [10] is $O(n^3)$ in the worst case for n signals, while the algorithms presented in this paper work on the SG, which is exponential in the number of signals in the worst case.

3 Overview of the Proposed Approach

We transform a given STG into an STG that satisfies CSC as follows. First we interpret the SG derived from the STG as an FSM, and apply to it state minimization and state encoding techniques. From this minimized encoded FSM we extract information sufficient to define how many state signals are needed and where their transitions must be inserted.

The model behind STG synthesis techniques ([1]) is depicted in Figure 3(a): a combinational circuit with inputs *all the signals* specified by the STG and outputs *the output signals* specified by the STG. If the STG does not have the CSC property, then the model fails, because the boolean function that must be implemented by the combinational circuit is not well defined (see Section 2.3). We then propose to conceptually replace the combinational circuit with a *sequential circuit*, that will be implemented in turn with a combinational circuit and some *state signals* (Figure 3(b)).

The key point in the approach is the correspondence between classical *state compatibility* and CSC. Distinguishing between incompatible markings amounts to forcing a state signal change between states in the minimized FSM. There is an important difference between the approach presented here and classical FSM minimization techniques. We consider also the *output* signals as feedback signals, rather than only state signals. This allows drastic reduction in the number of states in the minimized FSM.

4 Translating an STG into an FSM

We can obtain an FSM machine representation of the labeled SG by creating one FSM state for each SG state. The binary state labels and implied value labels in the SG become the input and output states of the FSM, respectively. For each internal state s of the FSM, let i be the binary label and o be the implied value label (i.e. the "next value" of the output signals) of the corresponding SG state. Then $N(i, s) = s$ and $O(i, s) = o$, i.e. every state has itself as next state as long as no external signal changes its value. For each edge among two SG states s' and s'', let i'' be the binary label of state s''. Then $N(i'', s') \doteq s''$. Also let o'' be the implied value label of s''. Then $O(i'', s') = o''$.

An FSM for the state graph of Figure 2(a) is given in Figure 2(b). For each state s_i in Figure 2(a) there exists a corresponding state i in Figure 2(b) with appropriate input and output states. For example, state $s1$ of Figure 2(a), has binary label 110 (Ri = 1, Ai = 1, Ro = 0) and enables transitions Ai^- and Ri^-. It becomes state 1 in Figure 2(b), with a self-loop edge labeled 110/0, an edge to state 2 labeled 100/0 (corresponding to the firing of Ai^-), and an edge labeled 010/0 (corresponding to the firing of Ri^-). State 1 must produce output 0 on inputs 110, 100 and 010 because the output value for signal Ro is 0 in all the binary labels of states $s1$, $s2$ and $s3$, none of which have transition Ro^+ enabled. Note that CSC violations translate into output-incompatibilities among states, namely (1,5), (2,6), (7,9), (10,12). For example, state 1 must produce output 0 on input 110, while state 5 must produce output 1 on the same input.

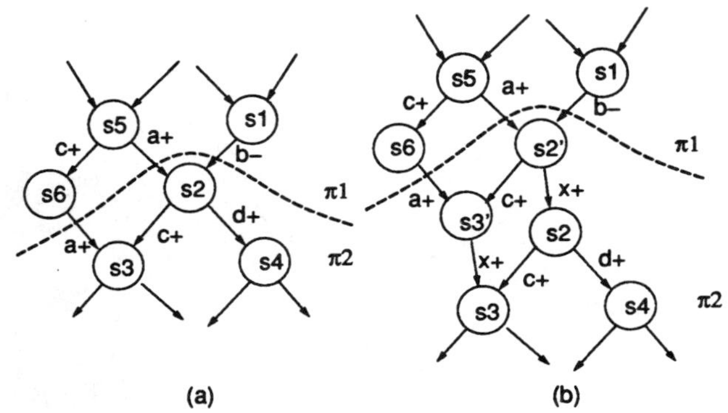

Figure 4: Locally distinguishable states

The relationship between this FSM and the STG can be seen in the following theorem:

Theorem 1 *An* STG *satisfies the* CSC *property if and only if the* FSM *derived from the* STG *by the above procedure can be minimized to a single state.*

Proof: see [4].

5 FSM Minimization

We now describe how the FSM obtained by direct translation of the SG, as shown in Section 4, can be minimized taking into account the need to derive an STG with CSC. The following result (Theorem 5.2 of [9]) shows that using this framework we obtain *a lower bound on the number of state signals*:

Theorem 2 *Given any minimized* FSM *with s_i stable states under input label i, any sequential circuit realizing this* FSM, *in which stable* FSM *states are represented by stable circuit states, must have* at least $\lceil \log_2(\max_i(s_i)) \rceil$ *feedback signals.*

Proof: see [9] and [4].

Note that we can also reduce the number of state signals below the given bound if we are allowed to *declare some states invalid*, because then we are modifying the FSM before realization. For example, the method described in [11] adds constraints to the STG to remove from the FSM states that can cause incompatibilities, such that the resulting FSM has only one compatible and, by Theorem 1, the STG has CSC. However, this is not general, it cannot always be done without adding state signals, and it may overly reduce concurrency.

We now formulate the constrained FSM minimization problem as follows. Let G be a live STG and let F be the corresponding FSM, derived as shown in Section 4. Let D be a subset of the signals appearing in G. Signals in D will be constrained, as shown in Section 7, to follow a state signal transition. Let π be a closed partition of the states of F. Let π_i denote the block of π to which state s_i belongs.

Definition 1 *A pair of* adjacent *states s_1, $s_2 \in F$ belonging to* distinct blocks *$\pi_1 \neq \pi_2$ is* **locally distinguishable using** *D if for every transition t_i^*, such that t_i^* is enabled in s_2, and t_i^* is not enabled in any predecessor of s_2, then the corresponding signal t_i is in D.*

Definition 2 *A pair of states s_1, $s_2 \in F$ belonging to* distinct blocks *$\pi_1 \neq \pi_2$ is* **distinguishable using** *D if for every path from s_1 to s_2 and for every path from s_2 to s_1 there exists a pair of states locally distinguishable using D.*

In Figure 4(a) we require only signal d to be in D, for s_5 and s_2 to be locally distinguishable, because c^+ was enabled in a predecessor of s_2, namely s_5. So in order for the binary labels of s_5 and s_2 to have a different value of the state signals, d^+ must be enabled only after some state signal transition has fired (x^+ in Figure 4(b)).

Definition 3 *A set of* STG *signals D is a* **partitioning set** *of signals with respect to π if every pair of states s_1, s_2 of F belonging to* distinct blocks *in π is* distinguishable *using D.*

Definition 4 *A partitioning set D is* **minimal** *if no signal can be removed from it, while remaining a partitioning set.*

Definition 5 *A partitioning set D is* **optimal** *with respect to a closed set of compatibles C if it is minimal and it has:*

1. *the minimum number of input signals n_i among all partitioning sets with respect to some closed partition π derived from C, and*
2. *the minimum number of output signals n_o among all such partitioning sets with n_i input signals.*

This definition of optimality obviously takes into account the need not to constrain input signals (because this amounts to changing the environment specification, which may not be always legal) and the need to constrain the least number of output signals (because this reduces the overall concurrency).

To obtain the initial closed set of compatibles we are currently using a heuristic classical FSM minimization procedure ([12]). We are interested in minimizing the cardinality of this set, because the minimized FSM will have one state for each block in π. So minimizing the cardinality of C minimizes the cardinality of π, and hence the number of state signals.

The main caution in using a classical FSM minimization procedure is to make sure that *pairs of adjacent states s_1, s_2, where every transition enabled in s_2 is also enabled in s_1*, belong to the *same compatibles*. Otherwise the partitioning set algorithm will not be able to find a solution, because s_2 has no transitions not enabled in s_1 and thus we cannot insert a state signal transition to enforce a state change from s_1 to s_2. This can be accomplished by a pre-minimization step that merges all such state pairs. The pre-minimization merges only compatible state pairs, because s_1 and s_2 are obviously output compatible, and the set of possible implications from s_2 is a subset of those from s_1.

For example, in the FSM of Figure 2(b), in state 1 both Ri and Ai can have a falling transition. In successor state 2 only Ri can have a falling transition, so the two can never be distinguished and they *must* belong to the same π_i in the final partition. Then we merge them before performing the actual FSM minimization. On the other hand, in state 3 Ai can have a falling transition, and Ri can have a *rising* transition, different from the falling transition of Ri enabled in state 1. So states 1 and 3 cannot be merged. The result of merging $(1,2) \rightarrow 1, (3,4) \rightarrow 3, (5,6) \rightarrow 5, (7,10) \rightarrow 7, (8,11) \rightarrow 8, (9,12) \rightarrow 9$, is shown in Figure 2(c).

We will now give a heuristic algorithm to find a *minimal* set of partitioning signals D and an associated closed partition π starting from an initial closed set of compatibles C. See [4] for an exact algorithm to find an *optimal* such set.

5.1 Minimal Partitioning Set Derivation

A minimal partitioning set of signals is found by the following greedy algorithm (let D be the set of all signals in G initially):

Procedure 1

1. **for each** *signal t, beginning from input signals*
   (a) **if** $D - \{t\}$ *is a partitioning set (checked using Procedure 2)*
   then let $D = D - \{t\}$

We take advantage of the monotonicity of the problem, to state the following:

Theorem 3 *Let F be an* FSM *derived from a live* STG *as described in Section 4, let C be a closed set of compatibles of F.*

If there exists an optimal partitioning set D' of F (with respect to any closed partition π derived from C) such that D' does not contain any input signal, then Procedure 1 finds a partitioning set D that does not contain any input signal.

Proof: see [4].

So, if it is possible at all given the initial closed set of compatibles, the heuristic algorithm does not change the environment specification.

Step 1a, checking if a set of signals is a partitioning set, is performed by repeating the following procedure as long as some compatible is removed by some p_s. Let $p_s = \{c \in C \mid s \in c\}$ be the set of initially allowed compatibles for state s according to the closed set of compatibles C.

Procedure 2

1. **for each** *state s*
   for each *compatible $c \in p_s$*
   if *s cannot be distinguished with D from all states s'* **s.t.** $c \notin p_{s'}$
   then *remove c from p_s*
2. **for each** *input state i*
   for each *compatible $c \in C$*
   (a) **let** $p' = C$
   (b) **for each** *state s* **s.t.** $c \in p_s$ **and** $\exists$ *an edge $s \rightarrow s'$ labeled i*
   let $p' = p' \cap p_{s'}$
   (c) **for each** *state s* **s.t.** $c \in p_s$ **and** $\exists$ *an edge $s \rightarrow s'$ labeled i*
   let $p_{s'} = p'$

Procedure 2 terminates with a non-empty p_s for all s if and only if D is a partitioning set for some π that can be derived from C (see [4]).

The output of this procedure is a closed set of compatibles, where every pair of states that *cannot be distinguished* with D are forced to belong to the same subset of compatibles. This is because step 1 guarantees that every state can be distinguished from any other state that does not belong to the same set of compatibles. Furthermore step 2 ensures that the compatibles satisfy the closure condition.

Figure 2(c) shows the result of applying Procedure 1 to the FSM of Figure 2(b). The minimal partitioning set is $D = \{Ri, Ai\}$. In this case each state belongs only to one compatible, so there is no choice. The closed partition has two blocks, π_1 and π_2, corresponding to compatibles c_1 and c_2 respectively.

A worst case running time bound for Procedure 1 is $O(mc^2n^4)$ (m is the number of signals, n is the number of states, c is the number of compatibles).

6 Critical Race-free State Assignment

After the FSM is minimized, a state assignment is performed. This step must not introduce any critical races and should allow state signals to change as soon as possible. We employ Tracey's single transition time (STT) state assignment technique [8] because it satisfies both criteria.

Tracey's technique is based on a set of dichotomy constraints extracted from the FSM. Each dichotomy constraint represents a two-block partition in which the states in one block need to be distinguished from the states in the other block by at least one state signal. After the set of dichotomy constraints is extracted from the FSM, we try to find a minimum length encoding which satisfies all the constraints. We use an exact constraint solver ([7]) which generates all prime dichotomies and then finds a minimum cover of the initial dichotomies by the prime dichotomies.

The complexity of satisfying the dichotomy constraints by the above technique is $0(2^n)$, where n is the number of states. Because the number of the states in the minimized FSM is usually quite small, we were able to find an exact solution to the encoding problem for all of our examples. For large FSMs we need to develop a technique which does not generate all prime dichotomies.

7 State Signal Insertion

In this Section we will show how an appropriate insertion of state signal transitions into the original STG can guarantee that it still satisfies the original specification, has the CSC property, and that logic can be synthesized from it.

In brief for each pair of adjacent un-minimized FSM states s_1, s_2 that were assigned to different states in the minimized FSM (say s_1' and s_2' respectively), we condition the external signal transitions, which are enabled in s_2 but not in any of its predecessors, to be enabled only after the state signals that change from s_1' to s_2' have changed.

Here we will use an algorithm that gives only *sufficient* conditions for the STG to have the CSC property. The algorithm takes as input the original STG G, with initial marking m_0, the pre-minimized FSM F derived from it, the partitioning set D, the associated closed partition of the states of F, and the encoding of each state of the minimized FSM F'. Let π_i denote the block to which state s_i was assigned, and let s_i' denote the corresponding minimized FSM state.

Procedure 3

1. **for each** *pair of maximal connected sets* (S_1, S_2) *of states of* F **s.t.** *all their elements belong to the same states* $s'_1 \neq s'_2$ *of* F' **and** *each state in* S_1 *is a predecessor of at least one state in* S_2 **and** *each state in* S_2 *is a successor of at least one state in* S_1
   (a) **let** T_2 *be the set of external signal transitions enabled in some state* $s_2 \in S_2$ *but not in any predecessor of* s_2
   (b) **let** T_1 *be the set of external signal transitions that are predecessors to all the transitions in* T_2
   (c) **for each** *state signal* x *that changes value between* s'_1 *and* s'_2
      i. **let** x^* *be a transition of* x *in the appropriate direction*
      ii. **for each** *pair* $t_2^* \in T_2$, $t_1^* \in T_1$, *connected by place* q *add the fanin and fanout transitions of* q *to the fanin and fanout place of* x^* *respectively*
2. **for each** *state transition* x^* **s.t.** q *in step 1(c)ii is marked in* m_0
   if x^* *has more than one predecessor*
   then *mark the new fanin place corresponding to* q
   else *mark the new fanout place corresponding to* q

Step 1b assumes that, given T_2, we can find at least one common predecessor to *all* the transitions in it. In [4] we show that this can always be achieved, possibly at the expense of a reduction in the STG concurrency. We are investigating ways to couple the closed compatibles generation with the partitioning set generation to avoid this loss in concurrency.

Figure 1(b) describes the result of Procedure 3 applied to the STG of Figure 1(a).

We can now state that:

Theorem 4 *Let* G *be a live* STG, *with initial marking* m_0, *let* G' *be an encoded* STG *derived from* G *as shown in Sections 4, 5, 6 and 7, with initial marking* m'_0.

Then G' *is live, has the* CSC *property, and specifies a subset of the possible firing sequences of by* G *with respect to the external signals (i.e. it satisfies the original specification).*

Proof: see [4].

8 Experimental Results

The algorithms described above have been implemented within the SIS sequential synthesis system developed at U.C. Berkeley. Table 1 contains the result of our procedure on a set of industrial and literature examples, some of which are free-choice STGs. "Lit." is the number of factored form literals in a hazard-free implementation of the encoded STG. The CPU time was measured on a DEC5000/125. Example "master-read" took a relatively long time to complete due to the fact that most states were assigned to all the compatibles by the initial FSM minimization procedure, and then Procedure 2 was slow to converge.

For the sake of comparison, [11] obtained a result with 6 signals, 12 transitions and 5 literals for vbe4a. Vanbekbergen removed those states that can cause incompatibilities, without requiring the addition of state signals, but reducing the overall concurrency. Our procedure obtained a larger circuit because we use state signals to remove CSC violation but preserve the concurrency of the specification as much as possible. Applying the *state removal technique* of [11] *within our framework* by hand (without too much attention to optimality) led to a result that *did not require any state signal and had 13 literals* but more concurrency than the result of [11].

name	initial / final			lit.	CPU
	signals	transitions	states		sec
alloc-outbound	7 / 9	18 / 22	17 / 3	19	6.0
nak-pa	9 / 10	18 / 22	18 / 2	30	9.9
pe-rcv-ifc	8 / 9	38 / 45	27 / 2	50	12.1
pe-send-ifc	8 / 11	41 / 54	54 / 4	33	82.2
ram-read-sbuf	10 / 11	20 / 22	16 / 2	20	8.0
sbuf-ram-write	10 / 12	20 / 24	24 / 3	30	11.5
sbuf-read-ctl	6 / 7	12 / 14	10 / 2	13	5.2
sbuf-send-ctl	6 / 8	18 / 24	14 / 3	34	8.4
sbuf-send-pkt2	6 / 7	20 / 24	15 / 2	11	5.7
sendr-done	3 / 4	6 / 8	5 / 2	5	3.5
atod	6 / 7	12 / 14	11 / 2	14	5.2
nousc	3 / 4	6 / 8	6 / 2	9	4.1
nousc.ser	3 / 4	6 / 8	4 / 2	8	3.8
master-read	13 / 16	26 / 36	132 / 5	77	1635.1
vbe4a	6 / 8	12 / 16	34 / 4	22	10.1
vbe6a	8 / 10	16 / 20	16 / 4	30	18.4

Table 1: Results of the proposed approach

9 Conclusions

We have introduced a novel and general technique to satisfy the CSC property on STGs. The technique, unlike previous results, *can handle any* STG, *without restriction to marked graphs or limits on the number of transitions for each signal.* Furthermore we believe that the proposed framework can be considered general enough to *analyze the complete problem of state assignment for event-based specifications*, unlike the various "special cases" described and solved in the literature.

We have established a relation between an STG and an FSM and formulated the state assignment problem as the conjunction of the constrained FSM minimization and critical race-free state assignment problems. This allowed us to prove a lower bound on the number of state signals to be added in order to implement the STG specification without substantially reducing its concurrency.

We are investigating ways to use the state-removal technique in [11] in our algorithms, to fully exploit the trade-off between reduction in concurrency and insertion of new signals.

The authors thank Tiziano Villa, Alex Saldanha, Ken Stevens and Kurt Keutzer for many helpful discussions.

References

[1] T. A. Chu. *Synthesis of Self-timed VLSI Circuits from Graph-theoretic Specifications.* PhD thesis, MIT, June 1987.

[2] M. A. Kishinevsky, A. Y. Kondratyev, and A. R. Taubin. Formal method for self-timed design. In *Proceedings of the European Design Automation Conference*, 1991.

[3] L. Lavagno, K. Keutzer, and A. Sangiovanni-Vincentelli. Algorithms for synthesis of hazard-free asynchronous circuits. In *Proceedings of the Design Automation Conference*, June 1991.

[4] L. Lavagno, C. W. Moon, R. K. Brayton, and A. Sangiovanni-Vincentelli. A novel framework for solving the state assignment problem for event-based specifications. Technical Report UCB/ERL M92/19, U.C. Berkeley, 1992.

[5] A. Martin. Formal program transformations for VLSI synthesis. In E. W. Dijkstra, editor, *Formal Development of Programs and Proofs*, The UT Year of Programming Series. Addison-Wesley, 1990.

[6] C. W. Moon, P. R. Stephan, and R. K. Brayton. Synthesis of hazard-free asynchronous circuits from graphical specifications. In *Proceedings of the International Conference on Computer-Aided Design*, November 1991.

[7] A. Saldanha, T. Villa, R. K. Brayton, and A. L. Sangiovanni-Vincentelli. A framework for satisfying input and output encoding constraints. In *Proceedings of the* 28th *Design Automation Conference*, pages 170–175, June 1991.

[8] J. H. Tracey. Internal state assignments for asynchronous sequential machines. *IEEE Transactions on Electronic Computers*, EC-15(4):551–560, August 1966.

[9] S. H. Unger. *Asynchronous Sequential Switching Circuits.* Wiley Interscience, 1969.

[10] P. Vanbekbergen. Optimized synthesis of asynchronous control circuits from graph-theoretic specifications. In *Proceedings of the International Conference on Computer-Aided Design*, pages 184–187, November 1990.

[11] P. Vanbekbergen, G. Goossens, and H. De Man. A local optimization technique for asynchronous control circuits. In *Proceedings of the International Workshop on Logic Synthesis*, May 1991.

[12] T. Villa. A heuristic incompletely specified finite state machine minimizer. Personal communication, 1985.

[13] A. V. Yakovlev and A. Petrov. Petri nets and parallel bus controller design. In *International Conference on Application and Theory of Petri Nets, Paris, France*, June 1990.

State Assignment Using Input/Output Functions

Irith Pomeranz+
Electrical & Computer Eng. Dept.
University of Iowa
Iowa City, IA 52242

Kwang-Ting Cheng
AT&T Bell Laboratories
Murray Hill, NJ 07974

Abstract
A finite state machine (FSM) synthesis procedure, specifically aimed at using primary inputs and primary output functions as state variables, is proposed. The number of next state functions that have to be implemented is thus reduced, potentially reducing the area of the synthesized circuit. Also, as more of the state variables are directly observable and controllable, the testability of the implementation is increased. Experimental results are given to demonstrate the effectiveness of the procedure proposed in reducing area.

1. Introduction
The problem of synthesizing Finite State Machines (FSMs) has been widely studied, and numerous solutions have been proposed [1-8]. The specific problem formulation under consideration is the following: Given a state table description of a sequential machine, find a state encoding (or state assignment) such that after (multi-level) logic minimization is performed, the resulting circuit implementing the machine would have minimum area.

In practice, due to the complexity of the state assignment problem and the difficulty in estimating the effect of state assignment on the area obtained, the minimum area requirement is replaced by weaker conditions. In [2,3,8], a two-level implementation of the combinational circuit is aimed at. Some approaches [4,6,7] estimate the literal savings in a multi-level structure based on common cube extraction. Another synthesis consideration is the testability of the resulting circuit. Testability as a synthesis criterion has been addressed in [5,10].

All existing solutions to the state assignment problem result in a circuit having the structure given in Figure 1, known as the Huffman model. In Figure 1, I is the set of primary inputs, O is the set of primary outputs, y is the set of present state variables and Y is the set of next state variables. For every primary output in the state table of the machine there is a primary output in the synthesized circuit, and for a machine having n states, there are at least $\lceil \log_2 n \rceil$ state variables.

In [1], it was suggested to use k delayed versions of the primary inputs or k delayed versions of the primary outputs as state variables to minimize the number of state variables which are neither primary inputs nor primary outputs (referred to here as *additional state variables*). The structure for $k = 1$ is shown in Figure 2. The motivation is that primary outputs must always be implemented, while primary inputs require no extra logic. The additional area required for such state variables is therefore zero. Although the Huffman model does not preclude the possibility of having a single function implementing both a primary output and a next state variable, or the use of primary inputs as state variables, due to the complexity of the state assignment problem, the possibility of such sharing is rarely identified in practice, even if it may result in reduced area. No investigation of the effect of targeting the structure of Figure 2 on the resulting area was carried out in [1], and no minimization of the number of state variables was carried out for the case of multi-input or multi-output machines. In this work, we propose a state assignment procedure based on the concept presented in [1], which can compete with existing synthesis procedures in the area of the resulting circuit. The case of single delayed versions of primary inputs and primary outputs is investigated (cf. Figure 2). We stress that the structure given in Figure 2 is included in the general structure of Figure 1, however, it is not targeted by existing procedures, that in most cases miss the possibility of sharing primary outputs and primary inputs with state variables.

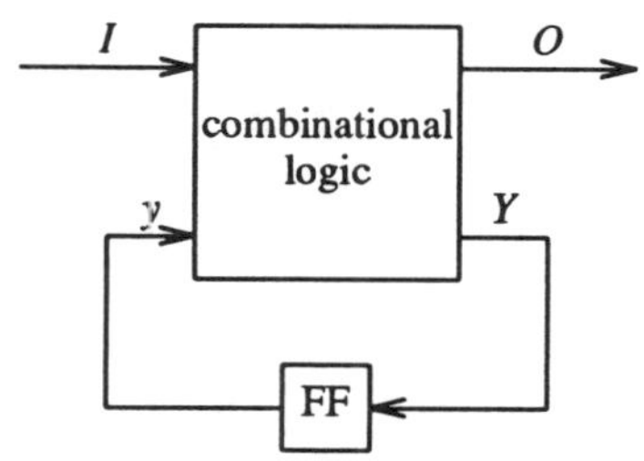

Figure 1: The Huffman model of a sequential circuit

The paper is organized as follows. In Section 2, we formulate the state assignment problem in accordance with our target structure, Figure 2. A solution is presented in Section 3. Experimental results are presented in Section 4. Section 5 gives concluding remarks.

2. Preliminaries
The aim of the synthesis procedure proposed here is to combine some state variable functions with primary inputs and primary output functions, the former requiring zero area and the latter always having to be implemented. The following examples illustrate how a significant reduction in area may result from this approach. For simplicity, the examples treat inputs and outputs separately. A general procedure is presented in Section 3.

Example: Consider machine M_1 given in Table 1. The values of the primary outputs, z_1 and z_2, constitutes a valid state assign-

+ Research performed while the author was a consultant to AT&T Bell Laboratories, Murray Hill, in summer 1991 and was supported in part by NSF Grant No. MIP-9109568.

0738-100X/92 $3.00 © 1992 IEEE

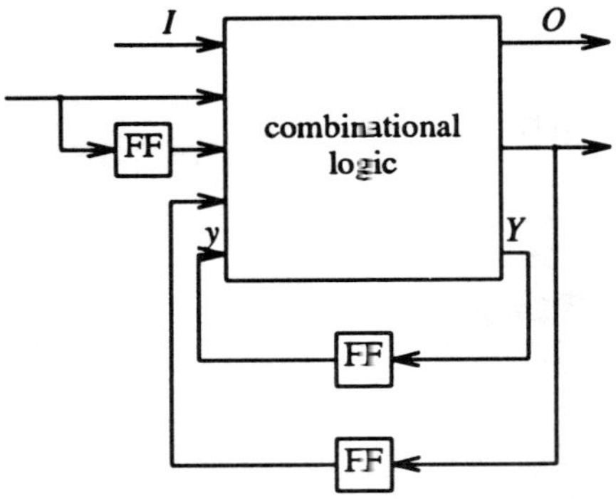

Figure 2: The target structure

ment for M_1, with the following encoding. A - 00, B - 01, C - 10 and D - 11. Instead of two state variables to encode the four states in addition to two output functions, the resulting circuit would have only two functions to implement both states and outputs. □

The previous example illustrates an extreme case, where all primary outputs are used to replace all state variables. In general, primary inputs and primary outputs are used to replace a subset of the state variables, as illustrated by the following example for primary inputs.

Example: Consider machine M_2 given in Table 2. The outputs are omitted. Let us assign the following codes. A - 00, B - 01, C - 10 and D - 11. Under this state assignment, all the next states in the column $x = 0$, $\{A,B\}$, have a code that starts with 0, while all the next states in the column $x = 1$, $\{C,D\}$, have a code that starts with 1. One of the next state variables, say Y_1, can therefore be set equal to x. Instead of using two state variables to encode the four states, only one function has to be implemented, while the other state variable is a primary input. □

Table 1: Machine M_1

PS	NS x=0	NS x=1	$z_1 z_2$
A	A	B	00
B	B	C	01
C	D	B	10
D	C	A	11

Table 2: Machine M_2

PS	NS x=0	NS x=1
A	A	C
B	B	C
C	B	D
D	A	C

In spite of the reduction in the number of functions to be implemented, it should be noted that the resulting area may not be reduced. The reason for this is that primary inputs and primary outputs have fixed, predetermined values in the state table, while symbolic states can be encoded arbitrarily. The use of inputs and outputs to encode states therefore implies loss of flexibility, that may have an undesirable effect on area. However, in many cases, the reduction in the number of functions is the major factor contributing to a significant reduction in area.

To take full advantage of the reduction in the number of functions to be implemented, the criteria for synthesis were selected as follows (in order of importance). (1) A minimum number of functions to implement, implying a minimum number of state variables which are neither inputs nor outputs (*additional state variables*). (2) A minimum number of state variables, to reduce the number of flip-flops required and, consequently, reduce the number of inputs to the combinational logic. (3) For the additional state variables, conventional state assignment is performed to achieve minimal area.

In addition to reduction in area, an increase in testability may also result from the target structure, as more of the state variables are directly observable (the primary outputs used as state variables) and directly controllable (the primary inputs used as state variables).

3. The State Assignment Procedure

In this section, the state assignment procedure used to obtain circuits that have the structure of Figure 2 is described, first for completely specified machines and then for incompletely specified machines.

3.1 Completely Specified Machines

The synthesis procedure for completely specified machines has three steps, as follows. In the first step, the maximum state encoding information is extracted from the primary input and primary output values associated with each next state entry, and the minimum number of additional state variables is computed. In the second step, a minimal number of primary inputs and primary outputs that can be used to yield the same amount of state encoding information as the complete set of inputs and outputs is computed. In the third step, state assignment for the additional state variables is performed. A detailed description follows, illustrated by machine M_3 given in Table 3.

Table 3: Machine M_3

PS	NS, z_1, z_2, z_3 x=0	NS, z_1, z_2, z_3 x=1
A	A, 000	B, 101
B	C, 111	C, 101
C	B, 000	D, 010
D	D, 101	A, 011

Step 1: To make full use of the primary inputs and primary outputs for state encoding, we first assume that all primary inputs and primary outputs are used as state variables. This induces a partition on the states of the machine, in which every next state is associated with the primary input and primary output values with which it is reached (similar to the first step of the Mealy to Moore transformation [9] and to the first step of the procedure described in [1]).

Example: For M_3, the following input-output combinations exist. State A is reached (from state A) with input 0 and output 000. We therefore define a set corresponding to the input-output combination (0000), that contains state A. Also included in this partition is state B, due to the transition from C. The complete partition is given in Table 4(a). Using all inputs and outputs as state variables gives a unique code to all but four next states. The four states are $\{A,B\}$ under input-output combination (0000) and $\{B,C\}$ under input-output combination (1101). For the uniquely encoded next states, the primary input and primary output values are sufficient to determine the next state, e.g., (0111) uniquely identifies state C. For the remaining four states, a single state variable is required to distinguish the states contained in each of the two pairs of states. □

The input-output partition provides a partial state encoding, in which next states that differ in the primary input or primary output values associated with the transition into them are assigned different codes. Additional state variables are required to distinguish among the states in the same set of the input-output partition. Their number is determined by the maximum size of a set in the input-output partition.

Table 4: Input-output partition for M_3

(a) Step 1

x	z_1	z_2	z_3	states
0	0	0	0	A,B
0	1	1	1	C
0	1	0	1	D
1	1	0	1	B,C
1	0	1	0	D
1	0	1	1	A

(b) Step 2

x	z_1	states
0	0	A,B
0	1	C,D
1	1	B,C
1	0	A,D

Step 2: Once the number of additional state variables is determined, some of the inputs and outputs may be unnecessary for distinguishing among the states of the machine. Specifically, sets of states can be combined by removing some inputs or outputs used as state variables, as long as the number of additional state variables required to encode the combined set does not exceed the number of additional state variables determined in Step 1. Reducing the number of inputs and outputs used as state variables also reduces the number of memory elements required for the circuit, without increasing the number of additional state variables required.

Example: Consider M_3 and the partition of Table 4(a). Let us remove x from the set of state variables used. This causes the sets (0101) and (1101) (that differ only in the value of x) to be combined into a single set, with states $\{B,C,D\}$. This requires two additional state variables, instead of one determined in Step 1. x is therefore retained. Removing z_3 causes (1010) and (1011) to be combined, but does not increase the number of state variables. Therefore, z_3 need not be used as a state variable. Removing z_2 as well results in the partition of Table 4(b). No other outputs can be removed without increasing the number of additional state variables. We have thus found a minimum set of inputs and outputs for state encoding, comprised of x and z_1. The synthesized circuit has three state variables, x, z_1 and a single additional state variable. The values of x and z_1 are determined by the partition of Table 4(b). Since an additional state variable is required, the state encoding at this stage is partial. □

The following heuristic is used for selecting a minimal number of inputs and outputs for encoding, given the number of additional state variables to be used. Each primary input and primary output is considered separately. If the input (output) can be eliminated without increasing the number of additional state variables, it is eliminated, and the partition is updated accordingly. Different orders of the primary inputs and primary outputs were tried experimentally, however, similar results were obtained in all cases. The order was therefore fixed to be the order of increasing index.

Step 3: For the additional state variables, a constrained state assignment has to be performed, in which the states in a single set, which are not differentiated by an input-output combination, are differentiated by the additional state variables. To solve the constrained state assignment problem, we use the following approach. We derive a state table, in which every state is assigned a different symbolic name, and the input-output encoding is made a part of the primary input and primary output list of the machine. We apply NOVA [3] to obtain a set of face-embedding and dominance constraints [11] for a minimal two-level implementation. We then attempt to satisfy a maximal set of constraints by assigning the same code to states which are already differentiated by the input-output combination. An assignment of codes that satisfies these constraints follows. This approach is illustrated by the following example.

Example: We consider machine M_3. For M_3, we obtained the input-output partition given in Table 4(b). Step (3) starts by assigning new symbols to the states, as follows. If state S of M_3 appears as the j-th state in input-output set i, S is assigned the new symbol S_{ij}. The new state names for M_3 are the following.

x	z_1	S	new symbol	S	new symbol
0	0	A	S_{11}	B	S_{12}
0	1	C	S_{21}	D	S_{22}
1	1	B	S_{31}	C	S_{32}
1	0	A	S_{41}	D	S_{42}

The state table is then written in terms of the new state symbols, as follows. The present state field is replaced by the new symbol given to the state. If a state is assigned more than one new symbol (e.g., A is assigned the new symbols S_{11} and S_{41}), the transitions originating from state A are duplicated, and in each copy, the present state field is replaced by a different new symbol of A. The partially determined state encoding corresponding to the input-output set of the present state is added to the input part of the state table. The next state field is also replaced by its new symbol. The appropriate symbol is determined by the input-output combination under which a next state is obtained. This should match the input-output set of the new symbol, e.g., from Table 3, under input 1, M_3 goes from state A to state B with an output 101. This next state B belongs to the input-output set with $(xz_1) = (11)$. The new symbol for B under this input-output set is S_{31}. We thus obtain the new symbol S_{31} for B in this case. The output field is then augmented by adding the partially determined code of the next state (11 in the case of S_{31}). The state table obtained for M_3 is the following. Note that we write the input combinations as well as the present states in the rows of the state table (instead of writing the input combinations in the columns).

I		PS	NS	O	
0	00	S_{11}	S_{11}	00	000
1	00	S_{11}	S_{31}	11	101
0	10	S_{41}	S_{11}	00	000
1	10	S_{41}	S_{31}	11	101
0	00	S_{12}	S_{21}	01	111
1	00	S_{12}	S_{32}	11	101
0	11	S_{31}	S_{21}	01	111
1	11	S_{31}	S_{32}	11	101
0	01	S_{21}	S_{12}	00	000
1	01	S_{21}	S_{42}	10	010
0	11	S_{32}	S_{12}	00	000
1	11	S_{32}	S_{42}	10	010
0	01	S_{22}	S_{22}	01	101
1	01	S_{22}	S_{41}	10	011
0	10	S_{42}	S_{22}	01	101
1	10	S_{42}	S_{41}	10	011

Notice that there are eight symbolic states in the modified state table while only one additional state variable is required. The use of a single additional state variable is ensured by letting some of the new states share the same code. For example, S_{11} and S_{12} must have different codes, since they are contained in the same input-output set (and therefore they are not distinguished by their partial state assignment), however, S_{11} is allowed to have the same code as any other state. NOVA is called with the modified state table to determine which states should be assigned the same value of the additional state variables. A set of face embedding and dominance constraints is generated for area minimization. A face embedding constraint specifies a set of symbolic states to be assigned to one face of a binary n-dimensional cube (where n is the number of states in the modified state table), without any other symbolic state sharing the same face. A dominance constraint requires that the code of a symbolic state would bitwise cover the code of another symbolic state. If two symbolic states are specified in a constraint and they are allowed to have the same code, we combine these two states into one. Combining two states into one implies that these two states will be assigned the same code. Assigning the same code to two states automatically satisfies both face embedding and dominance constraints. By combining states, we reduce the number of states to 2^k where k is the number of additional state variables. Any existing state assignment program, that assigns distinct codes to the resulting sets of states, can then be called for the final state table. In our experiments, state assignment at this stage is performed using NOVA. □

3.2 Machines with Incompletely Specified Outputs

When the machine has unspecified outputs, the outputs can be specified to increase the number of input-output sets and reduce the size of each set, thus reducing the number of additional state variables required, as illustrated by the following example.

Example: Consider machine M_4 given in Table 5. The symbol – stands for an unspecified value. It can be seen from Table 5 that the set of states for input-output combination (000) must contain states A and B. If the unspecified outputs for either C or D in column $x = 0$ is specified to be 0, the size of the input-output set corresponding to (000) increases to three or four, and an additional state variable is required. To keep the number of state variables to two, both unspecified values in the column of $x = 0$ should be specified to 1. In the column of $x = 1$, we prefer to set the unspecified value for present state C and input 1 to 0, since then, both B entries in the column are contained in the same input-output set (100), resulting in three input-output sets. □

The specification of the unspecified values is performed while the sets for each input-output combination are being formed, as follows. Suppose that a subset of the transitions in the given state table has already been considered, and that the next states have been partitioned according to the input-output combinations associated with them. Suppose also that the current number of additional state variables is k. Consider the next transition, t. (1) If the primary output values for t and the output values of an existing input-output set can be specified such that the next state under t is added to the existing set without increasing its size beyond 2^k, the specification is made and the state is added to that set. (2) Otherwise, if the outputs for t can be specified to be different from any existing set, the outputs are specified in that way, and a new set is formed, that contains the next state under t. (3) Otherwise, the transition is added to a compatible set. The transitions with the minimum number of unspecified outputs are considered first, to allow maximum flexibility as the sets increase in size.

3.3 Machines with Incompletely Specified Inputs

In contrast to unspecified outputs, which can be treated as *don't–care* combinations, unspecified inputs cannot be arbitrarily specified. To see how unspecified inputs are created, let us consider machine M_5 given in Table 6. M_5 is given in a compact form, where two transitions that differ only in the value of a single input are combined, and are given by a single entry, e.g., the transitions from A to B under inputs 00 and 10 are combined to a single entry under input –0. Only the relevant parts of the state table are shown. The input-output sets corresponding to the part of M_5 shown in Table 6 are $\{B,D\}$ for (000), $\{B\}$ for (100) and $\{D\}$ for (010). To derive these sets, the input combinations must be sufficiently expanded, to yield the three disjoint sets 00, 10 and 01. To avoid having to expand input combinations, we use the following approach. Whenever two transitions have compatible inputs and the same outputs, they are put in the same set, and the input is specified such that it would cover all input combinations forming it, e.g., for the two transitions of M_5, the input combinations –0 and 0– are combined into ––, resulting in the input-output combination (––0), containing states $\{B,D\}$. Although the set is made larger than necessary, possibly requiring more additional state variables to differentiate states which can be differentiated by the inputs, it is justified by the reduced complexity of handling the original inputs.

Table 5: Machine M_4

	NS,z	
PS	x=0	x=1
A	A, 00	B, 00
B	B, 00	C, 11
C	C, –0	B, 0–
D	D, 0–	A, 10

Table 6: Machine M_5

PS	X	NS	Z
A	–0	B	0
C	0–	D	0
.	.	.	.

4. Experimental Results

The state assignment procedure as described above was implemented and applied to MCNC Finite-State Machine Benchmarks. The state assignment program NOVA was used to derive the state assignment for the additional state variables. For purposes of area comparison, NOVA was also applied to the original circuits. After state assignment, the AT&T Logic Synthesis System was used to produce standard cell implementations.

Several versions of the state assignment program were run, as follows. Version *O* uses primary outputs only for state encoding. Version *I* uses primary inputs only for state encoding. Version *A* uses both inputs and outputs, but gives preference to the use of outputs. Version *B* uses both inputs and outputs, and gives preference to the use of inputs. Version *PiA* (*PiB*), where $i \in \{1,2\}$, is based on version *A* (version *B*), except that the number of additional state variables is allowed to be i more than the minimum number of state variables used in version *A* (version *B*). These versions allow us to walk through the possible solutions, from full utilization of primary inputs and primary outputs for state encoding, to encodings that use a minimum number of state variables. The results are given in Table 7. Following circuit name, the numbers of primary inputs, primary outputs and states are given. The version that uses input-output encoding and gives the best result in terms of area is given next, including number of inputs, outputs and additional state variables used for state encoding. The area, measured in grids, follows, followed by grid count obtained by applying NOVA to the original machine. The area of the flip-flops is not included in the grid count (A D-flip-flop can be implemented using 13 grids). It can be seen that in most cases where the synthesis procedure proposed here yields the minimum number of state variables, the area is lower than the one achieved by NOVA. Note in particular the case of *shiftreg*, which is a 3-bit shift register, that requires no combinational logic. While our procedure identifies this implementation, NOVA requires 22 grids for the combinational logic. In addition, in many cases where the number of state variables is higher than the minimum, the results are better than NOVA. The low run time (the largest example is completed in 9 CPU seconds on a SUN Sparc 2 workstation), allows this procedure to be applied in all cases, and the best solution obtained to be selected.

Table 7: Experimental results

(a)

			st.encd.			grids	
circ.	i o st	vers	i	o	st.b	IO	NOVA
ex2	2 2 19	O	0	1	4	258	251
ex3	2 2 10	I	1	0	3	127	129
ex5	2 2 9	O	0	1	3	104	129
ex6	5 8 8	P1A	0	1	2	183	187
ex7	2 2 10	I	1	0	3	127	140
keyb	7 2 19	O	0	2	3	310	413
lion9	2 1 9	A	2	1	1	27	47
shiftreg	1 1 8	A	1	0	2	0	22
train11	2 1 11	I	2	0	2	39	81

(b)

			st.encd.			grids	
circ.	i o st	vers	i	o	st.b	IO	NOVA
bbsse	7 7 16	P1B	2	1	3	224	271
dk27	1 2 7	A	1	1	2	47	48
ex1	9 19 20	O	0	5	2	350	414
ex4	6 9 14	P1A	0	3	3	123	158
opus	5 6 10	A	0	6	1	119	155
pma	8 8 24	P1A	0	5	3	356	414
s1	8 6 20	A	0	6	0	223	522
sse	7 7 16	P1B	2	1	3	200	232
tma	7 6 20	A	4	6	0	226	315

(c)

			st.encd.			grids	
circ.	i o st	vers	i	o	st.b	IO	NOVA
bbara	4 2 10	A	3	2	3	216	120
cse	7 7 16	O	0	6	3	517	402
dk14	3 5 7	P2A	2	2	2	172	170
dk15	3 5 4	I	3	0	1	198	118
dk16	2 3 27	P1B	2	0	4	677	483
dk17	2 3 8	A	1	3	2	149	115
dk512	1 3 15	A	1	2	3	134	113
lion	2 1 4	A	1	1	1	34	28
mc	3 5 4	A	0	4	0	71	57
planet	7 19 48	P1B	0	3	5	958	949
train4	2 1 4	A	2	0	1	34	23

(a) Circuits for which the proposed synthesis procedure requires the minimum number of state variables. (b) Circuits for which the proposed synthesis procedure uses more than the minimum number of state variables, but yields lower area than NOVA. (c) The remaining circuits.

5. Conclusion

In this paper, a state assignment method was proposed, that aims at maximizing the use of primary inputs and primary output functions as next state functions. The main benefits of the proposed method are the following. (1) Primary outputs must always be implemented, while primary inputs require no extra logic. Therefore, if some state variable functions can be combined with primary inputs and primary output functions, the number of next state functions that have to be implemented is reduced, potentially reducing the area of the synthesized circuits. (2) As more of the state variables are directly observable (the primary outputs used as state variables) and directly controllable (the primary inputs used as state variables), the testability of the synthesized circuit is increased. The proposed method can handle both completely specified and incompletely specified machines. Experimental results showed that the proposed method produces more area-efficient implementations than NOVA for 70% of the MCNC benchmarks FSMs. We are now conducting experiments to study the impact of the proposed method on testability.

Acknowledgement

The authors would like to acknowledge the fruitful discussions with Professor S. M. Reddy of the University of Iowa.

References

[1] E.J. McCluskey, "Reduction of Feedback Loops in Sequential Circuits and Carry Leads in Iterative Networks", Information and Control, June 1963, pp. 99-118

[2] G. De Micheli et al, "Optimal State Assignment for Finite State Machines", IEEE Trans. CAD, pp. 269-285, July 1985.

[3] G. De Micheli, "Symbolic Design of Combinational and Sequential Logic Circuits Implemented by Two-Level Macros", IEEE Trans. CAD, pp. 597-616, Sept. 1986.

[4] S. Devadas et al, "MUSTANG: State Assignment of Finite State Machines Targeting Multi-Level Logic Implementations" IEEE Trans. CAD, pp. 1290-1300, Dec. 1988

[5] K.-T. Cheng and V.D. Agrawal, "Design of Sequential Machines for Efficient Test Generation", Proc. Int'l Conf. Computer-Aided Design, Nov. 1989, pp. 358-361.

[6] X. Du et al, "MUSE: A MUltilevel Symbolic Encoding Algorithm for State Assignment", IEEE Trans. CAD, Jan. 1991, pp. 28-38.

[7] B. Lin and A.R. Newton, "Synthesis of Multiple Level Logic From Symbolic High-Level Description Languages", Proc. IFIP Int'l Conf. on VLSI. Aug. 1989, pp. 187-196.

[8] T. Villa and A. Sangiovanni-Vincentelli, "NOVA: State Assignment of Finite State Machines for Optimal Two-Level Logic Implementations", Proc. 26th Design Automation Conf., June 1989, pp. 327-332.

[9] Z. Kohavi, *Switching and Finite Automata Theory* McGraw-Hill, New York, Second Edition, 1978.

[10] S. Devadas, H.-K.T. Ma and A.R. Newton, "Redundancies and Don't Cares in Sequential Logic Synthesis", J. Electronic Testing: Theory and Application (JETTA), Feb. 1990, pp. 15-30.

[11] A. Saldanha et al, "A Framework for Satisfying Input and Output Encoding Constraints", Proc. 28th Design Automation Conf., June 1991, pp. 170-175

SESSION 36

PANEL: FRAMEWORKS - USER'S PERSPECTIVE

Chair: Lucio Lanzo - U.S. Venture Partners, Menlo Park, CA
Organizer: Jayaram Bhat - Cadence Design Systems, Inc., San Jose, CA

With the availability of frameworks from several vendors and the success of standards like the CAD Framework Initiative and EDIF, the interest in CAD frameworks has increased dramatically in the user community. This panel brings together a cross section of users (from various industry segments) of various frameworks. They will discuss and debate the users' requirements, experiences, and their future needs and expectations in terms of products and technologies as well as standards like EDIF and CFI.

Panel Members:

Tamotsu Hiwatashi - Toshiba Corp., Kawasaki, Japan
Ragnar Kling - Ericsson Telecom AB, Stockholm, Sweden
Nancy Nettleton - Intel Corp., Santa Clara, CA
Gary Porter - Harris Corp., Melbourne, FL
Brian Taylor - IBM Corp., Poughkeepsie, NY
Jim Toler - Motorola Inc., Tempe, AZ
Yoshihide Sugiura - Fujitsu, Ltd., Kanagawa, Japan
James Whitmore - Hewlett-Packard Co., Santa Clara, CA

A New Efficient Approach to Multilayer Channel Routing Problem

Sung-Chuan Fang, Wu-Shiung Feng and Shian-Lang Lee
Department of Electrical Engineering
National Taiwan University
Taipei, Taiwan, R.O.C.

Abstract

Among many successfully published papers for solving multilayer channel routing problem, they usually allowed doglegs in their solutions. As a concequence, they correspondingly will introduce more vias (or contacts) and longer wire length in their results than the ones using dogleg-free routing model. From the view of circuit performance, larger via count and longer wire length will result in longer timing delay and lower yield of the design. In this paper, a very efficient multilayer channel router, the M^3CR, will be well-described. The goal of M^3CR is to use optimum number of tracks for any multilayer routing instance in dogleg-free routing model. As well M^3CR has simple time complexity because it is not a maze-running-based router. From the experiments, M^3CR has produced optimum numbers of tracks for all the test examples from 4- through 8- layer in very short running time. Moreover, the variable-wire-width consideration according to the design rules is also discussed and resolved in this paper.

1 Introduction

The channel routing problem is one of the key phases in VLSI automatic layout design and implementation. The traditional channel routing problem assumes that there are two or three layers of materials available for routing. With the advance of manufacturing technology, more layers for routing can be expected in the near future. Thus it is getting more and more important to extend channel routing to handle multilayer regions.

A few multilayer channel-routing algorithms have been developed. *Enbody and Du* [1] proposed a multilayer router using leading column heuristics and limited backtracking. They determine the routing by using a *Five-Column-Window* that moves from left to right through the channel. *Hambrusch* [2] developed an n-layer channel router which puts severe constraints on the arrangement of the wiring and deals with only two terminal nets. Such constraints make the algorithm inapplicable. *Cong, Wong and Liu* [3] proposed a general technique which transforms a two-layer routing solution systematically into a three-layer or four-layer routing solution. But the transformation considered for only three and four layers. *Braun et al.* [4] implemented a multilayer channel router which divides the problem into two-layer and three-layer subproblems and completes each subproblem by the extended two-layer router *YACR2* [5] and a three-dimensional *maze router*.

All previous works introduced many doglegs in their routing solutions. Based on electrical standpoint, each dogleg adds one or two additional vias, increases the wire length, and decreases the performance of the circuit. The idea of introducing dogleg is to remove the heavy constraint of longest vertical constrained paths in vertical constraints graph [6]. However, if more than two layers are available for vertical wire segments, the longest vertical constrained paths are no longer related to a lower bound of channel width, which is mainly decided by channel density now. Thus a *dogleg-free* (or *restrictive*) multilayer channel router is valuable to be developed.

This paper presents a new approach to *multilayer channel routing problem*. Only the dogleg-free solutions are allowed in our router. From some theorems and the experiments, we can show that the dogleg is no longer need to improve the minimum channel width in multilayer routing enviroments. The router divides layers into some partitions. Each partition contains either two or three layers. Layers dedicated to horizontally oriented wire segments are called horizontal layers (H-layers); layers dedicated to vertically oriented wire segments are called vertical layers (V-layers). The VH (or HV) pattern is used by two-layer partition. The HVH is used by three-layer partition.

2 Preliminaries

2.1 Channel Description

A rectangle channel is described by two ordered list of terminals along its top and bottom boundaries. A nonnegative integer is assigned to each terminal. terminals are ordered in top and bottom lists: $TOP = (T_1, T_2, ..., T_L)$ and $BOT = (B_1, B_2, ..., B_L)$, where L is the length of the channel. T_i (or B_i) is a terminal at the top (or bottom) boundary in the i-th column. Terminals with the same positive label have to be connected. Zeros in two lists mean that there is no connection should be made in those points. A simple example is shown in Figure 1. The length of the channel is determined by the positions of the leftmost and rightmost terminals on the top and bottom boundaries. The channel width, the number of horizontal tracks used, is determined by the router. If a router wants to minimize the channel area, it must minimize the number of horizontal tracks.

0738-100X/92 $3.00 © 1992 IEEE

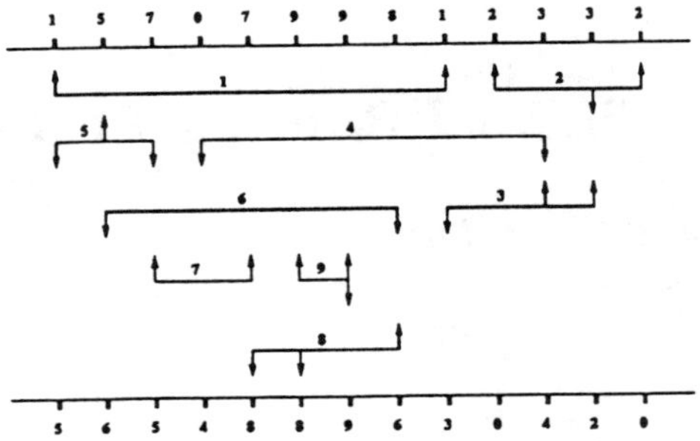

Figure 1: A simple channel routing problem.

2.2 Some Basic Definitions

To describe our approach more conveniently, we need the following definitions.

Definition 2.1 : Let L_i and R_i denote a column number of the leftmost and rightmost terminals for a net i, respectively. The *x_region* of a net i is defined by an interval (L_i, R_i), and the *x_length* of a net is the number of columns in its x_region.

Definition 2.2 : Let $S(j)$ be the set of unassigned nets in which their horizontal segments intersect column j. The unassigned nets are those nets which haven't been physically located in the channel.

Definition 2.3 : The *local density* of a channel at a given column j, d_j, is the number of different nets in $S(j)$. The maximum local density, denoted by $dmax$, is the well-known *channel density* classically . And let D_{dmax} is dynamic channel density, it is updated during the routing process.

Definition 2.4 : The *d_region* is an interval (x_1, x_2) in the channel, where

$$d_j = D_{dmax}, x_1 \le j \le x_2;$$
$$x_1 = L_i \text{ of some net } i;$$
$$x_2 = \min_{net\ k \in S(x_1)} R_k.$$

That is, the dynamically local density of each column in a d_region is D_{dmax}. For example, in Figure 1, there are two d_regions, (5, 5) and (6, 7), and $D_{dmax} = 5$.

Definition 2.5 : Let N_H (N_V) be the number of layers for horizontal (vertical) wires.

2.3 Graph Models of Channel Routing Problem

Two wiring conflicts, *vertical constraint* and *horizontal constraint*, are neceaary to be avoided in the routing process. As the vertical wire of a net must be placed below the vertical wire of another net, there is a vertical constraint between these two nets. If the x_region of a net overlaps the x_region of another net, then there is a horizontal constraint between these two nets. Both graphs which are *vertical constraints graph* (VCG) and *horizontal constraints graph* (HCG) are used to describe the vertical and horizontal constraints among nets, respectively. These two graphs are described as follows.

(1) VCG :
Each node in VCG corresponds to a net. The directed edge from node A to node B means that net A must be placed below net B.

(2) HCG :
HCG and VCG have the same set of nodes. The edge with end nodes A and B means that two x_regions of net A and net B are not overlapping each other. Each clique in HCG forms a group of nets which can be routed in the same track.

Definition 2.6 : Let $vmax$ be the number of nodes on the longest directed path in VCG. And let D_{vmax} is the dynamic $vmax$, it is updated during the routing process.

Definition 2.7 : Let $depth_i$ be the longest directed path length from node i to some vertices which has no descendant in VCG.

Definition 2.8 : In a channel routing problem, the routing is said to be dogleg-free if the number of tracks on which any net's horizontal segments can be positioned is limited to one.

2.4 Lower Bound of Channel Width

The lower bound of channel width in HV alternate model can be expressed in terms of two parameters "*dmax*" and "*vmax*". The lower bounds of the routing instances can be classified into the following two types.

(1) The number of vertical layers is only one:
Let the longest directed path in VCG is $n_1 \rightarrow n_2 \rightarrow \ldots \rightarrow n_k$ ($k = vmax$), no two nets among $n_1, n_2, \ldots, n_k$ can be assigned on the same track. Therefore, the channel routing needs at least $vamx$ horizontal tracks. In addition, as the horizontal segments are placed on N_H layers, in an ideal case the maximum density at each horizontal layer can be reduced to $\lceil dmax/N_H \rceil$, so the number of tracks must also be equal to or larger than $\lceil dmax/N_H \rceil$. We can conclude that the lower bound of channel width is equal to $max(\lceil dmax/N_H \rceil, vmax)$.

(2) The number of vertical layers is more than one:
In this type there are two or more vertical layers, we can eliminate the major part of vertical constraints. The vertical constraints can be ignored theoretically, therefore, the channel width will have the lower bound equal to $\lceil dmax/N_H \rceil$.

3 Mixed Mode Multilayer Channel Routing

As in Section 1 stated, the available layers will be divided into some partitions. In each partition, there are two or three layers (VH or HVH pattern) for routing the nets. In the simplest case, there is only one partition divided for routing. Of course, our router M^3CR can easily handle this basic case: the two- or three-layer channel routing.

3.1 Routing Algorithm for Two- and Three-Layer Channel Routing

As any vertical constraint exists, the ordering of nets to be routed can be derived from VCG if we route every net from the bottom track to the top track in the channel. The net A will be routed preceding the net B if there is a directed path from node A to node

B in VCG. Hence the nodes which have no ancestors in VCG (or all its ancestors have been routed) can be routed close to the bottom track (i.e., above the horizontal wires of assigned nets). And these nodes will construct a subgraph of HCG, denoted as H_{sub}. To avoid the horizontal conflicting in the routing, only nets with corresponding nodes that form a clique in the H_{sub} can be routed in the same track. The nodes in H_{sub} can form some cliques. In two-layer routing N_H is equal to one, therefore, the router selects a maximal weighted cost clique in H_{sub} to assign appropriate track for the nets corresponding to the nodes of the maximal weighted cost clique. *The cost function will be defined and described in Subsection 3.3.* In three-layer routing N_H is equal to two, proportionally, the router selects two maximal weighted cost cliques in H_{sub}. The nets corresponding to the nodes in first maximal weighted cost clique in H_{sub} will be assigned to the first horizontal layer and the others for second horizontal layer on the same track.

3.2 Routing Algorithm for Multilayer Channel Routing

In Chameleon [4], it divides the layers into some sets, where each set contains either two or three layers, and also partitions the nets into these sets. Each set of layers and nets are then solved by a detailed router YACR2. Our routing algorithm doesn't need a preprocessor to partition the nets into sets. And the dogleg-free routing model is employed by our router, the problem of channel routing is to determine a horizontal track number for each net.

3.2.1 Layer Partitioning

Our router divides layers by three cases. Let P be the number of partitions.

- Case 1 (N modulus $3 = 0$ and $N \geq 4$):
  The case has $N/3$ HVH partitions as below.
  HVH,HVH, ...,HVH ($P = N/3$)
- Case 2 (N modulus $3 = 1$ and $N \geq 4$):
  The case has $\lfloor N/3 \rfloor - 1$ HVH partitions and two VH partitions as below.
  HVH,HVH, ...,HVH,VH,VH ($P = \lfloor N/3 \rfloor + 1$)
- Case 3 (N modulus $3 = 2$ and $N \geq 4$):
  The case has $\lfloor N/3 \rfloor$ HVH partitions and one VH partition as below.
  HVH,HVH, ...,HVH,VH ($P = \lfloor N/3 \rfloor + 1$)

3.2.2 Free Horizontal Layer

In Chameleon [4], the routable layers are divided into some partitions, where each partition is independent, i.e., if net A is assigned in partition B, all vertical and horizontal segments of net A are entirely routed in partition B. In fact, vias are allowed between adjacent layers in the different partitions. Therefore, the vertical segments of net A can be routed in different vertical layers so long as these vertical layers are adjacent to the horizontal layer, where the horizontal segment of net A are routed. For example, the four-layer channel router arranges the four layers as $V_1H_1V_2H_2$. Obviously, if net A is assigned in partition 1, vertical segments of net A can be routed in V_1 (belongs to partition 1) and V_2 (belongs to partition 2).

A multilayer channel router which allows nets' vertical segments to be routed in different partitions is called a ***mixed mode multilayer channel router***. The mixed mode multilayer channel router can utilizexs the vertical layers more effectively.

Definition 3.1 : A horizontal layer which is adjacent to two vertical layers is called a ***free horizontal layer***.

Doglegs are allowed in Chameleon. Although doglegs can reduced the vertical constraints, but doglegs have the penalty of increasing the number of vias and the wire length. If VCG has directed cycles, the usage of doglegs may also break the directed cycles. But no published algorithm can guarantee that their introduced doglegs can break all directed cycles. If there are two or more vertical layers available, we can eliminate the major part of vertical constraints by using these vertical layers. And any directed cycle can be broken if any two nets in the directed cycle are routed in the different partitions. We can have some theorems described as follows.

Theorem 3.1 : If the horizontal segments of the nets are routed in the free horizontal layer, they have no ***vertical constraints*** among them.

Theorem 3.2 : If there is any free horizontal layer, all the directed cycles in VCG can be broken successfully, i.e., no cyclic constraint exists.

The proofs are omitted here. There are free horizontal layers in Case 2 and Case 3 (see Subsection 3.2.1). Therefore, doglegs are no longer used in our routing algorithm when two or more vertical layers are available. The lower bound of channel width will only depend on the channel density (*dmax*) when there are two or more vertical layers.

3.2.3 The Algorithm

Let MC be the maximal weighted cost clique in HCG. Every time our router selects one MC to be assigned to an appropriate track, we hope that maximum density of the channel can be reduced by one after removing the corresponding nets from both the VCG and HCG. It is very clear that if our router can satisfy such condition in each processing, we can get the optimum solution consequently.

Our router divides layers into some partitions. Each partition contains either two or three layers. The VH pattern is used by two-layer partition. The HVH pattern is used by three-layer partition. Then our router treats all nodes in VCG as candidates to find the corresponding MC for each partition. In two-layer partition it selects one MC. In three-layer partition it selects two MCs. Two types of constraints are introduced after router assigns appropriate track for the nets corresponding to the nodes of the MC. The first constraint is called ***father_constraint***.

Definition 3.2 : If MC_i is assigned in the j-th partition, the nodes which are the fathers of the nodes of MC_i in VCG cannot be routed in the j-th partition. We define that such nodes have father_constraints.

During the routing process, the node with father_constraint in j-th partition are therefore assigned an *j-color* for avoiding the conflicts of such constraints. Next time when router finds MC(s) for the j-th partition, it treats all nodes as candidates in VCG except those nodes with *j-color*. Note that a node in VCG may have more than one father_constraints if there are two or more partitions divided.

The second constraint is called *partition_constraint*.

Definition 3.3 : If an unassigned net has no partition for track assignment after we assign an *j-color* to itself, we define that this node has the partition_constraint.

If any node has partition_constraint in j-th partition, router must remove such node's children in VCG from candidates and find new MC(s) for enforcing no partition_constraint occurred. We summarize the algorithm as follows.

```
ALGORITHM #3.1
Step 1: input the TOP and BOT lists and construct
        the corresponding HCG and VCG.
Step 2: divide layers into some partitions.
Step 3: ROUTE_NORMAL_H(LB_N).
Step 4: ROUTE_FREE_H(LB_N).
Step 5: while HCG and VCG are not null do
Step 6:         ROUTE_NORMAL_H(1),
Step 7:         ROUTE_FREE_H(1).
Step 8: end
END ALGORITHM #3.1

PROCEDURE ROUTE_NORMAL_H(num_of_tracks)
Step 1: for j=1 to num_of_tracks do
Step 2:     for i=1 to P do
Step 3:         treat all nodes as candidates in VCG except
                those nodes with i-color.
Step 4:         if any unassigned net has partition_constraint
                in the current partition
                then remove the corresponding node's
                    children from candidates.
Step 5:         select the corresponding MC(s) for horizontal
                layer(s) except free horizontal layer(s).
Step 6:         assign the nodes which have father_constraints
                an i-color.
Step 7:         assign appropriate track for the nets
                corresponding to the nodes of the MC(s).
Step 8:         remove the nodes of the MC(s) from HCG
                and VCG
Step 9:         update cost function.
Step 10:        if HCG and VCG are both null then exit.
Step 11:    end
Step 12: end
Step 13: return.
END PROCEDURE ROUTE_NORMAL_H

PROCEDURE ROUTE_FREE_H(num_of_tracks)
Step 1: for j=1 to num_of_tracks do
Step 2:     for all free horizontal layers do
Step 3:         treat all nodes in VCG as candidates.
Step 4:         select the corresponding MC for the free
                horizontal layer
Step 5:         assign appropriate track for the nets
                corresponding to the nodes of the MC.
Step 6:         remove the nodes of the MC from HCG and
                VCG.
Step 7:         update cost functions.
Step 8:         if HCG and VCG are both null then exit.
Step 9:     end
Step 10: end
Step 11: return.
END PROCEDURE ROUTE_FREE_H
```

3.3 Cost Function

The cost function of a net is discussed in two conditions.

(1) Condition 1 (N_V=1):
The cost estimated on each nodes in VCG does not only involve the x_length of the corresponding net but also consider its longest path length in VCG. The component, x_length, of cost function aims for maximizing the occupied space of horizontal track. The *vamx* is a lower bound of channel width in this case. Therefore, for any net i, the $depth_i$ is an important component of cost function. The d_region is an crucial interval. If all x_regions of nets of an MC can cover all d_regions, the D_{dmax} can be reduced by one. Let nd_i be the number of d_region covered by x_region of a net i. And let $cost\_d_i$ be equal to nd_i multiplied by 500. The cost function for a net i can be summarized as follows.

(a) if ($D_{vmax} * N_H \geq D_{dmax}$ and $depth_i = D_{vamx}$) then
$cost(i) = 15000 + w_1 * depth_i + w_2 * (R_i - L_i)$;

(b) if ($D_{vamx} * N_H \geq D_{dmax}$ and $depth_i < D_{vmax}$) then
$cost(i) = w_1 * depth_i + w_2 * (R_i - L_i)$;

(c) if ($D_{vamx} * N_H < D_{dmax}$) then
$cost(i) = w_1 * depth_i + w_2 * (R_i - L_i) + w_3 * cost\_d_i$;
where w_1, w_2 and w_3 are weighted parameters which can be adjusted for each experiment. By default, $w_1 = 20$, $w_2 = 1$ and $w_3 = 1$. And the numbers, 500 and 15000, are experienced constants to enforce the influences of the coresponding situations.

(2) Condition 2 ($N_V \geq 2$):
In this case *vmax* is no longer a lower bound of channel width. The $depth_i$ of any net i can be removed from cost function in Condition 1. The lower bound of channel width only depends on *dmax* now. Every time the router selects an MC, it is important to reduced D_{dmax} by one. Therefore, the component, $cost\_d_i$ of any net i, of cost function is very essential. In routing process we must keep the number of nodes which have father_constraints as small as possible. Hence, we give a penalty in cost function to reduce father_constraints. Let $indegree_i$ be the number of fathers of a net i in VCG. The cost function for a net i can be described as follows.
$cost(i) = a_1 * (R_i - L_i) + cost\_d_i - a_2 * indegree_i$;
where a_1 and a_2 are weighted parameters which can be adjusted for each experiment. By default, $a_1 = 1$ and $a_2 = 50$.

3.4 Maximal Weighted Cost Clique Algorithm

The maximal weighted cost clique problem in the general graph is NP-complete. However, the problem becomes tractable when the graph is a comparability graph [7]. We define the comparability graph and prove that HCG is a comparability graph as follows.

Definition 3.4 : A directed edge $< i, j >$ in directed graph is an edge connecting from node i to node j.

Definition 3.5 : In undirected graph an edge connectinged node i and node j is denoted by (i, j) or (j, i).

Definition 3.6 : An undirected graph $G = (V, E)$ is a comparability graph if there exists an orientation (V, F) of G satisfying the following transitive condition :

$$< a, b > \in F \textit{ and } < b, c > \in F \textit{ imply } < a, c > \in F, \quad \textit{for every } (a, b), (b, c), (a, c) \in E.$$

Theorem 3.3 : The horizontal constraints graph (HCG) is a comparability graph. (The proof is omitted.)

Theorem 3.4 : The time complexity of the algorithm for finding a maximal weighted cost clique in HCG=(V, E) is linear in the size of HCG. That is, the time complexity is $O(|V|)$.

Proof : The time complexity of finding a maximal weighted cost clique in a comparability graph is linear in size of itself [7]. According to *Theorem 3.3*, the proof is completed, therefore.#

Theorem 3.5 : The time complexity of ALGORITHM #3.1 is $O(L * dmax)$, where L is the channel's length.

Proof : The crucial factor in the analysis of our algorithm is the time required to search MCs and update cost functions. The time for searching an MC is $O(n)$ known from *Theorem 3.4*, where n is the number of nodes in VCG and also is the number of nets. The time for updating cost functions include the reconstructions of d_region and VCG. The reconstruction of d_region takes $O(L)$ time. The reconstruction of VCG= (V, E) takes $O(|E|)$ time. There are L vertical constraints in the worst case. One edge in VCG represents one vertical constraint so that $O(|E|) \le O(L)$. We conclude that $O(L)$ time is required to update cost functions. The algorithm may perform searching MCs and updating cost functions for $O(dmax + c)$ times, where c is a small constant. If the router produces any optimal solution, c is equal to zero. Therefore, the time complexity of ALGORITHM #3.1 is $O((n + L) * (dmax + c)) \doteq O((n + L) * dmax) \doteq O(L * dmax)$.#

3.5 Variable-Wire-Width Consideration According to Design Rules

As more available routing layers are used, the minimum wire width and the size of the vias are increased as in top layers due to the design rules; thus, even if a channel can be routed with fewest tracks, the actual routing area is still large providing that the worst case design rules in toppest layer are employed. We suggest that the wire width and the size of the vias are invariant for the layers in a certain partition, but they are variant in the different partitions. The variable-wire-width approach can take the advantages of saving routing area and considering the performance of the circuits, e.g., critical nets were routed on the top layers, where the wire width is wider so that the wires have lower resistivity.

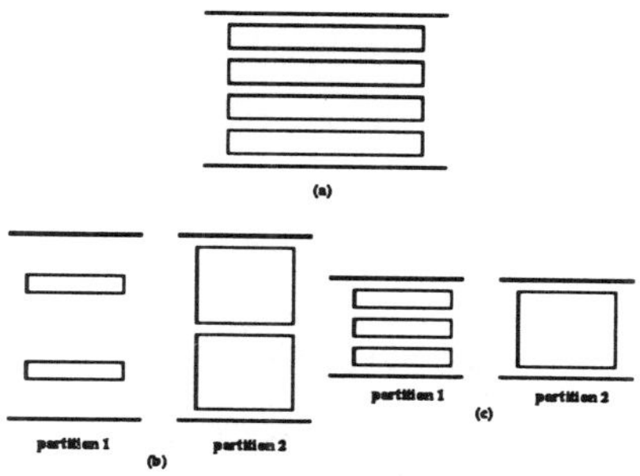

Figure 2: An example to illustrate the advantage of variable-wire-width.

Figure 2 is used to illustrate how the approach saves the routing area. We suppose that a simple channel requires 4 tracks to complete the routing problem in two-layer channel routing as shown in Figure 2(a). In four-layer channel routing, there are two partitions, where partition 2 has the minimum track width which is equal to 3 times of the track width of partition 1. If the router distributes 4 tracks equally to two partitions, i.e., each partition contains 2 tracks as shown in Figure 2(b), the channel width is larger than the one as shown in Figure 2(c). From the discussions above, our router can be easily modified to handle such consideration. We define that the minimum *track-to-track* spacing of j-th partition is $unit_j$ and the $height_j$ of jth partition is equal to $unit_j$ multiplied by the number of tracks used in j-th partition. Our router tries to uniformly distribute the nets to all partitions, and makes each partition with the same $height_j$ as possible.

4 Experimental Results

The Mixed Mode Multilayer Channel Router (M^3CR) was programmed in C language and implemented on the SUN4 workstation. We use the same examples as in [8]. The router generates optimum results in all test examples from 4- through 8-layer. Also the running time for each test example is very short (from 0.03 to 1.27 seconds) because no maze-running algorithm is invoked. Table 1 – Table 5 show the routing results from 4- through 8-layer, respectively. The well-known *Deutsch's difficult example* is denoted as *diff.* Table 6 and Table 7 show some results produced by the variable-wire-width approach.

Table 8 shows the comparison with Chameleon [4]. For all the examples in Table 8, our router shows equal results of Chameleon. The authors of Chameleon declared that they can use minimum number of tracks. However, the lower bound of channel width derived by them is not proper since wire overlaps are allowed in their solutions. From the theoretical point of view, the results of Chameleon are not very surprising. Moreover, the doglegs are also introduced in their solutions, and more vias and longer wire length are used, therefore. From the view of circuit performance, our router is expectably superior to Chameleon. Further M^3CR with time complexity $O(L*dmax)$ will be much faster than other maze-running-based routers.

5 Conclusions

A new efficient approaches to multilayer channel routing problem was developed in this paper. Due to the dogleg-free wiring model employed by M^3CR, the router not only minimizes the number of tracks but also uses less number of vias as well as shorter wire length. The router doesn't need a preprocessor to partition the nets and the maze router as done in Chameleon [4]. In short words, if router invokes a maze-running algorithm in the routing algorithm, the routing algorithm is becoming more complex and requires more running time. The time complexity of M^3CR is $O(L*dmax)$, where L is the channel's length and $dmax$ is the channel density. The router generated optimum results in all test examples from 4- through 8-layer. Additionally the variable-wire-width consideration according to the design rules was discussed and resolved for more compacted channels under the practical multilayer design environments.

Acknowledgements

Here we want to express our gratitude to National Science Council, Taipei, Taiwan, R.O.C., for the supports of this work.

References

[1] R. Enbody and H. Du, "Near-optimal n-layer channel routing," *Proc. 23rd Design Automation Conference*, Las Vegas, NV, pp. 708-714, June 1986.

[2] S. Hambrush, "Channel routing algorithms for overlap models," *IEEE Trans. on Computer-Aided Design*, vol. CAD-4, no.1, pp. 23-30, January 1985.

[3] J. S. Cong, D. F. Wong and C. L. Liu, "A new approach to three- or four-Layer channel routing," *IEEE Trans. on Computer-Aided Design*, vol. CAD-7, no.10, pp. 1094-1103, October 1988.

[4] D. Braun, J. L. Burns, S. Devadas, H. K. Ma, K. Mayaram, F. Romeo, and A. Sangiovanni-Vincentelli, "Chameleon: A new multi-layer channel router," *Proc. 23rd Design Automation Conference*, Las Vegas, NV, pp. 495-502, June 1986.

[5] J. Reed, A. Sangiovanni-Vincentelli, and M. Santomauro, "A new symbolic channel router: YACR2," *IEEE Trans. on Computer-Aided Design*, vol. CAD-4, no.3, pp. 208-219, July 1985.

[6] D. N. Deutsch, "A dogleg channel router," *Proc. 13rd Design Automation Conference*, pp. 425-433, 1976.

[7] M. C. Golumbic, *Algorithmic Graph Theory and Perfect Graphs*, Academic Press, New York, U.S.A., 1980, pp. 133-134.

[8] Y. K. Chen and M. L. Liu, "Three-Layer Channel Routing," *IEEE Trans. on Computer-Aided Design*, vol. CAD-3, no.4, pp. 156-163, April 1984.

Table 1: Four-layer channel routing results in mixed mode

Four-Layer Channel Routing Results (VH VH)						
Example	$dmax$	LB_4	#Track	#Via	Wire length	Time(sec)
ex1	12	6	6	57	513	0.07
ex3a	15	8	8	68	868	0.15
ex3b	17	9	9	107	1363	0.35
ex3c	18	9	9	125	1727	0.48
ex4b	17	9	9	179	2205	0.47
ex5	20	10	10	150	2200	0.45
diff	19	10	10	290	4100	0.78

Table 2: Five-layer channel routing results in mixed mode

Five-Layer Channel Routing Results (HVH VH)						
Example	$dmax$	LB_5	#Track	#Via	Wire length	Time(sec)
ex1	12	4	4	57	459	0.07
ex3a	15	5	5	68	752	0.18
ex3b	17	6	6	107	1192	0.37
ex3c	18	6	6	125	1568	0.45
ex4b	17	6	6	179	1902	0.43
ex5	20	7	7	150	2013	0.47
diff	19	7	7	290	3697	1.00

Table 3: Six-layer channel routing results in mixed mode

Six-Layer Channel Routing Results (HVH HVH)						
Example	$dmax$	LB_6	#Track	#Via	Wire length	Time(sec)
ex1	12	3	3	57	407	0.07
ex3a	15	4	4	68	707	0.27
ex3b	17	5	5	107	1127	0.52
ex3c	18	5	5	125	1494	0.57
ex4b	17	5	5	179	1839	0.58
ex5	20	5	5	150	1870	0.67
diff	19	5	5	290	3360	1.27

Table 4: Seven-layer channel routing results in mixed mode

Seven-Layer Channel Routing Results (HVH VH VH)						
Example	$dmax$	LB_7	#Track	#Via	Wire length	Time(sec)
ex1	12	3	3	57	429	0.03
ex3a	15	4	4	68	723	0.20
ex3b	17	5	5	107	1127	0.37
ex3c	18	5	5	125	1474	0.45
ex4b	17	5	5	179	1862	0.42
ex5	20	5	5	150	1866	0.45
diff	19	5	5	290	3451	0.90

Table 5: Eight-layer channel routing results in mixed mode

Eight-Layer Channel Routing Results (HVH HVH VH)						
Example	$dmax$	LB_8	#Track	#Via	Wire length	Time(sec)
ex1	12	3	3	57	415	0.07
ex3a	15	3	3	68	679	0.20
ex3b	17	4	4	107	1059	0.40
ex3c	18	4	4	125	1418	0.53
ex4b	17	4	4	179	1732	0.50
ex5	20	4	4	150	1774	0.53
diff	19	4	4	290	3255	1.03

Table 6: Four-layer channel routing results by variable-wire-width approach

Four-Layer (VH VH) : $unit_1/unit_2 = 1u/2u$				
Example	$dmax$	$height_1/height_2$	$height_I^{(1)}$	$R^{(2)}$
ex3a	15	10u/10u	16u	38%
ex3b	17	12u/12u	18u	33%
diff	19	14u/14u	20u	30%

(1) The $height_I$ is the minimum channel width used by uniform approach.
(2) The R is the reduced area ratio equal to

$$(height_I - \max_{i=1}^{p} height_i)/height_I.$$

Table 7: Seven-layer channel routing results by variable-wire-width approach

Seven-Layer (HVH VH VH) : $unit_1/unit_2/unit_3 = 1u/2u/4u$				
Example	$dmax$	$height_1/height_2/height_3$	$height_I$	R
ex3a	15	6u/4u/4u	16u	63%
ex3b	17	7u/6u/4u	20u	65%
diff	19	10u/10u/8u	20u	50%

Table 8: Comparison with Chameleon

Comparison with Chameleon for some benchmarks							
Example	$dmax$	Four-Layer #Track		Five-Layer #Track		Six-Layer #Track	
		Cham.	Ours	Cham.	Ours	Cham.	Ours
ex3a	15	8	8	5	5	4	4
ex3b	17	9	9	6	6	5	5
ex3c	18	9	9	6	6	5	5
diff	19	10	10	7	7	5	5

A Multi-Layer Channel Router with New Style of Over-the-Cell Routing

Takashi FUJII, Yoko MIMA, Tsuneo MATSUDA and Takeshi YOSHIMURA

NEC Corporation, Kawasaki, Kanagawa, Japan

Abstract

This paper discusses a new style of over-the-cell routing, where a new cell structure is introduced. Terminals are located around the horizontal center line in the cell. We regard the region between the central terminals in the upper cell row and ones in the lower cell row as an expanded channel. A new channel router handles that region in a lump where connections to terminals are perfectly achieved by avoiding obstacles in cells. Furthermore, the router shortens the length of polysilicon wires so as to achieve the high-performance circuits. Experimental results show that the router achieves a goal of the size reduction which is predicted according to the theoretical estimation, and that the length of polysilicon wires is dramatically shortened.

1 Introduction

With recent advances in manufacturing technology, it has been made possible to use three or even four layers for interconnections. In the standard cell approach, it is indispensable to extend channel to handle multi-layered regions. Several multi-layer channel routers [1]-[6] have been developed because of effectiveness and high-speed processing of channel routing. In order to reduce channel height still more, the over-the-cell channel routers have been discussed [7]-[14]. In their routing models (except for [14]), terminals have been limited to being placed on outlines of cells, and hence routing region was divided into two regions: channels and regions over the cells. Since the routers should be applied to each of regions individually, the results deeply depend on the selection of nets to be routed over the cells [12]. This selection is a key problem, however, very difficult.

This paper proposes a new style of over-the-cell routing, where a new cell structure is introduced to make good use of cell regions. Terminals exist not only on the top and bottom edges of cells but also around the horizontal center line. We regard the region between the central terminals of the upper and lower cell rows as an expanded channel. The router handles that region in a lump, where connections to central terminals are perfectly achieved by avoiding obstacles in cells. Since it is not necessary to beforehand select a set of nets which will run over the cells, our router can globally perform the routing with minimal channel height. Moreover, from the industrial point of view, the length of polysilicon wires should be shortened in order to achieve the high-performance circuits. Experimental results show that our router achieves a goal of the size reduction which is predicted according to the theoretical estimation, and that the length of polysilicon wires is dramatically shortened as compared with that without the consideration about polysilicon terminals.

2 New cell model

It is assumed that four layers (polysilicon layer: P and three metal layers: M1, M2 and M3) are available to routing. P and M2 (M1 and M3) are used for vertical (horizontal) wiring. For each layer, there is a grid superimposed over the routing region. Wires will run along grid lines in rectilinear fashion. An electrical connection between two wires on adjacent layers is performed by a via. The vias are placed on the cross points of grid lines for layers to be connected.

We introduce a new cell structure shown in Fig. 1. Obstacles may exist in upper and lower outsides of the cell. Its shape is regarded as a rectilinear polygon. Terminals are located not only on the outline of the cell but also around the horizontal center line. Equi-potential terminals are only allowed to be on the same vertical column. In fact, it is rather easy to modify the existing cells into cells of the new structure.

Terminals around the center line are referred to as *the central terminals*. The terminal of layer L is denoted by *the L-terminal*. There exist P-terminals only on the outline of the cell. All of central terminals are of M2. An M3 wire can run along any M3 grid lines, since M3 is not used in the cell layout. On the other hand, it is necessary that an M2 wire is routed with avoiding obstacles of M2.

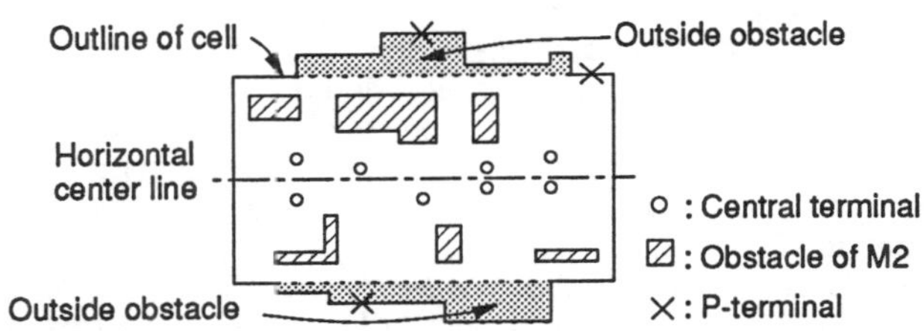

Fig. 1 New cell structure.

3 Expanded channel

On the basis of the cell model, we consider a new routing region as shown in Fig. 2. This region is called *an expanded channel*. It is defined to be the maximal rectilinear polygon surrounded by the central terminals and horizontal center lines in the upper and lower cell rows. The lower (upper) half part of the upper (lower) cell row is referred to as *the over-the-cell region*.

It is assumed that, in the over-the-cell region, only wires of M2 and M3 can be used. The grid lines of M1 and M3 (P and M2) are referred to as *tracks* (*columns*). Especially, the partial grid lines of M2 (M3) included in the cell regions are referred to as *the cell columns (tracks)*.

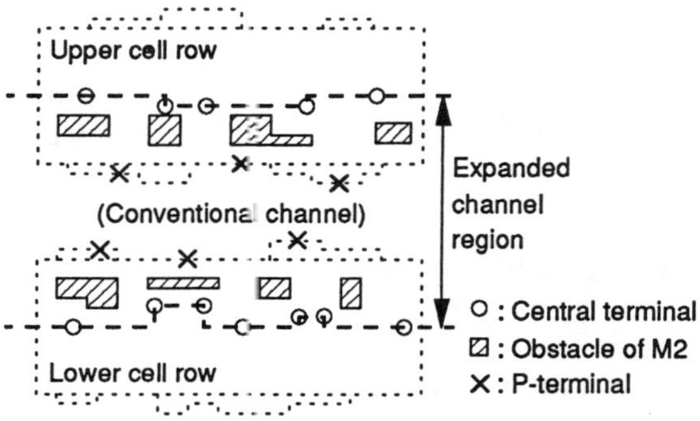

Fig. 2 Expanded channel.

The routing problem for the expanded channel is different from one for the conventional channel in the following points:

(1) There may exist obstacles in the routing region.
(2) The shape of the expanded channel is not rectangular, since the central terminals are not aligned horizontally.
(3) M1 and M3 are permitted to be used by horizontal wires in the conventional channel, however only M3 can be used in the over-the-cell region.
(4) P-terminals are located inside the expanded channel.
(5) The grid interval of M3 in the over-the-cell region need not be equal to one in the conventional channel.

The difference (1) makes the connection to central terminals difficult. For this routing problem, our router finds the points where wires from central terminals can vertically pass to the outlines of cells without hitting any obstacles.

4 Over-the-cell multi-layer routing

4.1 Routing for the expanded channel

Our router acts on the following 4 steps.

Step 1: Specification of channel
Step 2: Routing from central terminals
Step 3: Changing P-terminals to M2
Step 4: Track assignment

In ***Step 1***, the expanded channel is defined based on the cell data. ***Step 2*** performs the connections to central terminals with avoiding obstacles in cells. (We will explain the algorithm of ***Step 2*** in 4.2.) In ***Step 3***, the P wires connecting to P-terminals are routed to M2 wires, for as many P-terminals as possible. This layer changing is achieved by using a pattern which is fittingly chosen out of three patterns shown in Fig. 3. The changing is required to make good use of M3 tracks, since an M2 wire can immediately connect to an M3 wire by setting only one via. Moreover, as a result of the changing, the total length of P wires will be shortened. Finally, according to vertical constraints, horizontal wires of subnets are assigned to tracks of M1 and M3 in ***Step 4***. The algorithm for the track assignment based on [15] is modified to take account of 3.5-layer routing.

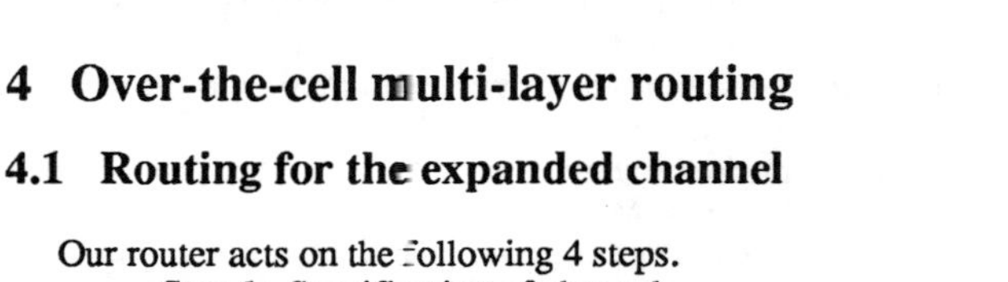

Fig. 3 Patterns of changing P-terminal to M2.

4.2 Routing from central terminals

In order to completely perform the routing from each central terminal, the following two processes are required. The cell column on which a central terminal *t* is located is denoted by col(*t*).

<u>Selection of drawing direction</u>: If col(*t*) intersects an obstacle *R* for a central terminal *t*, then we must determine the direction (left or right) to which the routing from *t* detours and avoids the obstacle *R*.

<u>Selection of cell track</u>: We must determine a cell track on which the detour runs in order to avoid an obstacle with taking account of the shape of the obstacle.

For example, consider central terminals ***a***, ***b*** and ***c*** in Fig. 4. The detours for terminals *a*, *b* and *c* will run along cell tracks T_2, T_1 and T_1, respectively. The direction of routing from *a* and *b* is to be the left. On the other hand, the direction of routing from *c* is to be the right.

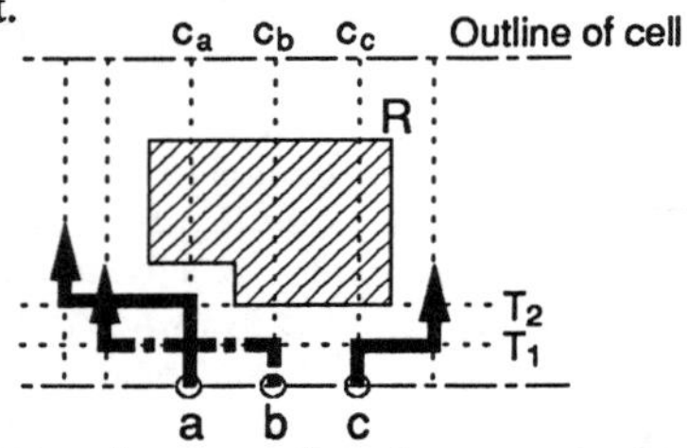

Fig. 4 An example of routing from central terminals.

(A) Routing method: Let $c_L(R)$ ($c_R(R)$) be the left-most (right-most) cell column which intersects an obstacle *R*. Then, *R* is said to *obstruct an interval* $[c_L(R), c_R(R)]$. Let's assume that an obstacle R_1 is adjacent to an obstacle R_2. R_1 and R_2 are said to be *continuous* if at least one condition out of the following conditions 1) and 2) is satisfied: 1) Intervals of R_1 and R_2 are overlapped, and 2) no wire of M2 can pass through the space between obstacles R_1 and R_2.

A maximal set of continuous obstacles is called *an obstacle region*. A set *U* of central terminals obstructed by any obstacle is partitioned into disjoint subsets $F_i \subseteq U$ such that terminals in each F_i obstructed by a same obstacle region.

The routing from central terminals is processed for each F_i by the following steps ***S.1-S.4***. Let *F* be a set of central terminals which are obstructed by an obstacle region *R*.

S.1: Let F_L (F_R) be a set of central terminals which are located on the left (right) side of the center of the obstacle region *R*. Insert central terminals in F_L into queue *QL* according to the following manner: If $|c_L(R) - col(t)| < |c_L(R) - col(s)|$ $(s, t \in F_L)$ then $QL(i)=t$ and $QL(j)=s$ and $i < j$. In the same way, insert central terminals in F_R into queue *QR*.

S.2: Get the first element *t* from *QL*. For each cell track T_i between central terminal *t* and the obstacle region *R*, check upward whether T_i intersects a cell column *c* such that $c < c_L(R)$, and then list up the available cell tracks.

S.3: For the first cell track T_i listed up in ***S.2***, search a cell column *c* such that 1) *c* crosses T_i, 2) $c < c_L(R)$, 3) *c* does not intersect any other obstacle regions, 4) a vertical M2 wire from an intersection between *c* and T_i to the outside of the cell can run along *c*, and 5) no other terminal is located on *c*. If no cell column for T_i is found, then the same process is applied to the next cell track listed up in ***S.2***. If no cell column for all cell tracks listed up in ***S.2*** is found, then insert *t* into *QR* (*i.e.*, *t* is the last element of *QR*).

S.4: If queue *QL* is not empty, then go to ***S.2***, else the similar steps as ***S.2*** and ***S.3*** are applied to queue *QR*.

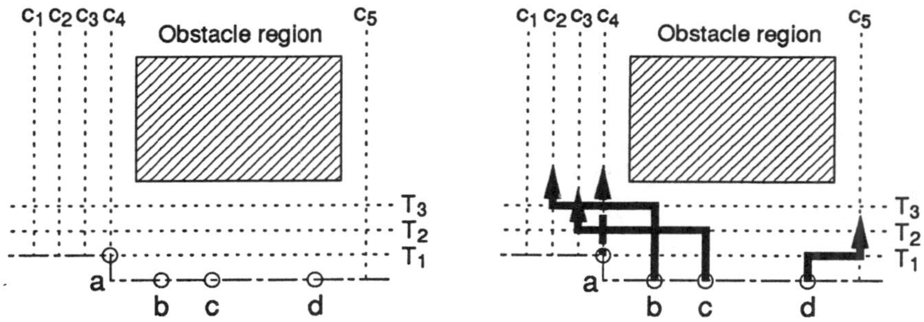

(a) An example **(b) A routing result**

Fig. 5 Process of routing from central terminals.

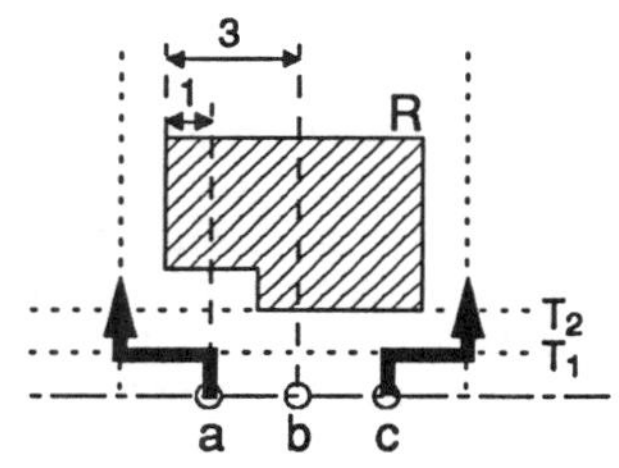

Fig. 6 A selection of cell track.

Consider the routing from central terminals a, b, c and d in Fig. 5(a). In ***S.1***, $QL=(b,c)$, $QR=(d)$. It is noted that cell track T_1 is broken by central terminal a. Thus, the detours from central terminals b and c to the left side of R can not be achieved by using T_1. In ***S.2***, for b, two cell tracks T_2 and T_3 are listed up. Four cell columns c_1-c_4 intersect T_2 in the left side of R. Since col(a)=c_4, three cell columns c_1-c_3 are considered as candidates for the cell column used to detouring from b. In ***S.3***, for b, c_3 is chosen, since c_3 is the nearest to R. In the similar way, for c, T_3 and c_2 are chosen. As shown in Fig. 5(b), in ***S.4***, the routing from d is achieved by using T_1 and c_5 in the right side of R.

(B) Processing priority of central terminals: In the process for the routing from central terminals, we introduce a priority. Then, the routing from central terminals will be performed completely.

Consider an example shown in Fig. 6. Let $F_L=\{a,b\}$ and $F_R=\{c\}$. If the routing from central terminals a and b is achieved according to the order in the queue QL, then the detour from b can not be obtained as shown in Fig. 6.

When central terminals are processed in ***S.2***, it is required to consider not only the distance from the left or right edge of an obstacle region R to each central terminal but also the number of available cell tracks between each central terminal and R.

Let t be a central terminal and R be an obstacle region obstructing t. The estimation function P(t) for central terminal t is defined as follows:

$$P(t) = k_1 \cdot N_T(t) + k_2 \cdot L_X(t),$$

where $N_T(t)$ is the number of available cell tracks between central terminal t and the minimum convex rectilinear polygon surrounding obstacle region R, $L_X(t)$ is the horizontal distance from the left (or right) edge of obstacle region R to central terminal t, and k_1 and k_2 are coefficients ($k_1>k_2>0$). (We say a rectilinear polygon D is convex if any two points in D are joined by a rectilinear path in D with at most one bend.) The central terminals are processed according to the increasing order of values of P(•).

Let's consider the example in Fig. 6 described before, and assume that $N_T(a)=2$, $N_T(b)=1$, $L_X(a)=1$, $L_X(b)=3$, $k_1=10$ and $k_2=1$. Then, P(a)=21 and P(b)=13. Thus, central terminal b is prior to a. As a result, we completely accomplish the routing from both a and b with avoiding R as shown in Fig. 4.

5 Experimental results

The proposed channel router runs on NEC EWS4800/220 workstation (30MIPS). The router has been applied to Deutsch's difficult example (DIFF). Positions of terminals are vertically moved to the inside of cells. The grid intervals of M1, M3 in the conventional channel and M3 in the over-the-cell region are 60, 90 and 120 [unit], respectively. The numbers of the upper and lower cell tracks are 5 and 6, respectively. We generate four patterns (P1-P4). Experimental results are summarized in Table 1. In this table, "no_change" represents no "changing P-terminals to M2" process. The router reduces the channel height by 18.8%-50.0% as compared with no over-the-cell routing. Furthermore, by changing P-terminals to M2, the channel height is reduced by 7.1%-15.4% and the total length of P wires is dramatically shortened by 86.8%-88.3% compared with "no_change" routing. The result for pattern P2 is shown in Fig. 7, where obstacles are denoted by shaded rectilinear polygons.

The channel router has also been applied to several practical circuits. The grid interval of M3 in the conventional channel (the over-the-cell region) is 1.5 times (2 times) wide as one of M1. Table 2 summarizes the experimental results. Note that, for each circuit, the results of the placement and global routing [16] are common to every routing mode. The over-the-cell routing reduces the die size by 22.6%-33.3% and 11.2%-12.9% compared with 2.5-layer routing and 3.5-layer routing without over-the-cell routing, respectively. This reduction is approximately equal to the predicted value. The prediction is made according to the estimation based on the number of necessary tracks in 2.5- and 3.5-layer routing, the number of cell tracks, the rate of obstructed cell columns and the rate of P-terminals. The comparison of the routing modes for DATA3 is shown in Fig. 8.

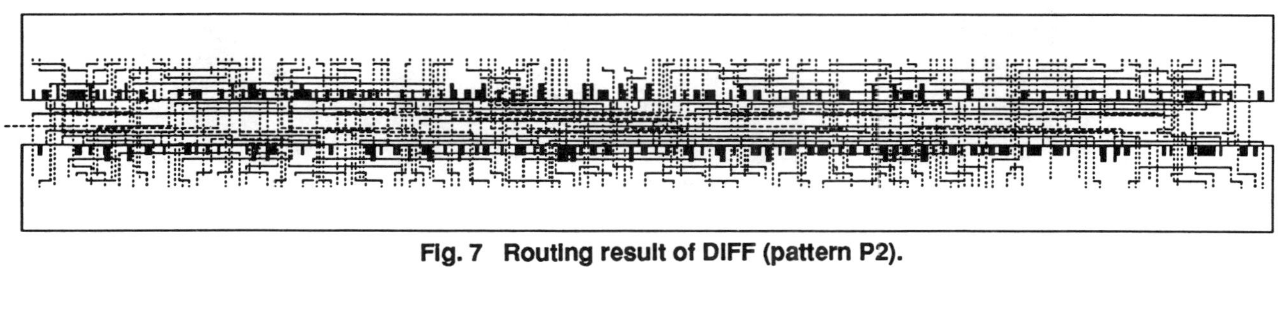

Fig. 7 Routing result of DIFF (pattern P2).

6 Conclusion

In order to make good use of the over-the-cell regions, we introduced the new cell structure in which terminals are located around the horizontal center line. The proposed router processes the expanded channel between the central terminals in the upper cell row and ones in the lower cell row. The experimental results have proven that the proposed router is sufficiently effective and useful to reduce the die size and to shorten the length of polysilicon wires in a practical point of view.

References

[1] Y.K.Chen and M.L.Liu, "Three-layer channel routing," *IEEE Trans. CAD*, CAD-3, 2, pp.156-163 (1984).

[2] P.Bruell and P.Sun, "A 'greedy' three layer channel router," *Proc. ICCAD-85*, pp.298-300 (1985).

[3] D.Braun *et al.*, "Chameleon: A new multi-layer channel router," *Proc. 23rd DA Conf.*, pp.495-502 (1986).

[4] R.J.Enbody and H.C.Du, "Near-optimal n-layer channel routing," *Proc. 23rd DA Conf.*, pp.708-714 (1986).

[5] J.Cong *et al.*, "A new approach to three- or four-layer channel routing," *IEEE Trans CAD.*, 7, 10, pp.1094-1104 (1988).

[6] R.Gidwani and N.A.Sherwani, "MISER: An integrated three layer gridless channel router and compacter," *Proc. 27th DA Conf.*, pp. 698-703 (1990).

[7] D.N.Deutsch and P.Glick, "An over-the-cell router," *Proc. 17th DA Conf.*, pp.32-39 (1980).

[8] H.E.Krohn, "An over-cell gate array channel router," *Proc. 20th DA Conf.*, pp.665-670 (1983).

[9] Y.Shiraishi and J.Sakemi, "A permeation router," *IEEE Trans. CAD*, CAD-6, 3, pp.462-471 (1987).

[10] J.Cong and C.L.Liu, "Over-the-cell channel routing," *Proc. ICCAD-88*, pp.80-83 (1988).

[11] J.Cong *et al.*, "General models and algorithms for over-the-cell routing in standard cell design," *Proc. 27th DA Conf.*, pp.709-715 (1990).

[12] M.-S.Lin *et al.*, "Channel density reduction by routing over the cells," *Proc. 28th DA Conf.*, pp.120-125 (1991).

[13] N.D.Holmes *et al.*, "New algorithm for over-the-cell channel routing using vacant terminals," *Proc. 28th DA Conf.*, pp.126-131 (1991).

[14] M.Terai *et al.*, "A new model for over-the-cell channel routing with three layers," *Proc. ICCAD-91*, pp.432-435 (1991).

[15] T.Yoshimura, "An efficient channel router," *Proc. 21st DA Conf.*, pp.38-44 (1984).

[16] M.Edahiro and T.Yoshimura, "New placement and global routing algorithms for standard cell layouts," *Proc. 27th DA Conf.*, pp. 642-645 (1990).

Table 1 Experimental results for DIFF

	Obstructed cell columns	Polysilicon terminals	Over-the-cell			3.5 layer			Over-the-cell (no_change)			Height reduction	Height reduction	Length reduction
			Channel height HO [unit]	Length of P wire LO [K unit]	CPU time [sec]	Channel height H3 [unit]	Length of P wire L3 [K unit]	CPU time [sec]	Channel height HN [unit]	Length of P wire LN [K unit]	CPU time [sec]	HO v.s. H3	HO v.s. HN	LO v.s. LN
P1	20%	20%	660	1.8	5.5	960	1.8	4.2	780	15.4	9.7	31.3 %	15.4 %	88.3 %
P2	40%	20%	780	2.0	6.2	960	2.0	4.3	840	15.2	10.4	18.8 %	7.1 %	86.8 %
P3	20%	0%	420	0	5.1	840	0	3.8	420	0	5.1	50.0 %	—	—
P4	40%	0%	540	0	8.4	840	0	4.0	540	0	8.4	35.7 %	—	—

Table 2 Experimental results for practical circuits

Data	#cell	#net	2.5 layer routing		3.5 layer routing		3.5 layer over-the-cell routing		Size reduction	Size reduction
			Size S2 [mm x mm]	CPU time [sec]	Size S3 [mm x mm]	CPU time [sec]	Size SO [mm x mm]	CPU time [sec]	SO v.s. S2	SO v.s. S3
DATA1	130	185	0.5576 x 0.4770	11.4	0.5576 x 0.4210	11.1	0.5576 x 0.3690	13.8	22.6 %	12.4 %
DATA2	1000	1111	1.3208 x 1.1710	68.9	1.3208 x 0.9290	70.9	1.3208 x 0.8090	84.1	30.9 %	12.9 %
DATA3	2934	3392	3.4784 x 3.3330	396.6	3.4784 x 2.5030	397.0	3.4784 x 2.2230	492.9	33.3 %	11.2 %

2.5-layer routing

3.5-layer routing without over-the-cell routing

Over-the-cell 3.5-layer routing

Fig. 8 Comparison of routing modes for DATA3.

New Models for Four- and Five-Layer Channel Routing

Tai-Tsung Ho

Department of Mathematics, Computer Science, and Statistics
McNeese State University
Lake Charles, LA 70609

Abstract

In this paper, we propose two new routing models for the four- and five-layer routing problems based on the HHVH (or HVHH) and HHVHH models respectively instead of the HVHV (or VHVH) and HVHVH models, which are traditionally used by most multi-layer channel routers [1-5]. Since more layers are available for horizontal routing in the proposed models and the introduced via-violations in the new models are appropriately handled, the routing performance of the proposed multi-layer router has outperformed all existing multi-layer routers in the four- and five-layer routing environments from our understanding. The experimental results are shown and one detailed five-layer routing solution is demonstrated. The extension to a general multi-layer router is discussed.

I. Introduction

The channel routing problem in layout is to realize a specified set of connections between two modules on opposite sides in as small an area as possible. Traditionally, two layers are assumed for routing, one for horizontal and the other for vertical routing, and the connections between the wire segments in these two layers are through some electrical throughs, which are referred to as *vias*. However, advances in manufacturing technology have made it possible to use three or more layers for interconnections. Many multi-layer routing algorithms with different restrictions have been proposed [1-5,6-8,11-16,18-19].

For most of the proposed multi-layer routers, the following convention is commonly adopted. First, all the layers are divided into two types: one for horizontal routing and the other for vertical routing, and are referred to as H and V layers respectively. Second, the layers are ordered in a way such that all the H and V layers are adjacent to V and H layers respectively. Under the above convention, to maximize the horizontal routing space, which is strictly related to the lower bound of the routing solutions, obviously, the HVHV (or VHVH) and HVHVH models will be selected for the four- and five-layer routing environments respectively. Under the traditional models, no vias will intersect with the horizontal wire segment since there exists at least one V layer adjacent to each H layer. In the multi-layer routing environment, if the above convention is not adopted, then possibly a via may be blocked by an existing wire segment or may block a wire segment to be routed in an inner H layer. Since the simplest two-layer routing problem without considering the blocking of vias has been proved as NP-complete, the introduced via-blocking problem will make the routing problem more complicated if the traditional models are not adopted.

In this paper, we propose two routing models for the four- and five-layer routing environments based on the HHVH and HHVHH models respectively. Let the routing solutions in the two-layer restricted-Manhattan (2-RM) and the routing solutions in the HHVH or HHVHH models be denoted as R and R′ respectively. The strategy to construct the R′ is briefly described as follows. In the first step, we find an R such that for all the wire segments in R, the vias for different nets are not allowed to occur in the same column if the wire segments containing these vias will be transferred to the same track in R′ later. The creation of R with the above restriction can be easily achieved by our two-layer router [19] if the mapping of tracks and layers in R and R′ is specified. After the R is constructed, in the second step the H layers of R are transferred to R′ according to a specified mapping, and the V layer of R′ is constructed accordingly. Let the overlaps of vertical and/or horizontal wire segments for different nets be referred to as *wire-violations* and the overlaps of vias and wire segments for different nets be referred to as *via-violations*. Since R is a valid solution, which contains no wire- and via-violations, it is guaranteed that R′ constructed in the second step contains no wire-violations, but may contain via-violations, which is the one of the major problems that need to be overcome in the proposed models. To alleviate this problem, in the third step we reassign the wire segments for H layers in each track of R′ according to some density-related criterion so that the effect of via-violations on the resulting channel width of R′ is minimized. Finally, we reroute the wires in the H layers to resolve the via-violations.

The experimental results show that the proposed models perform better than the traditional ones in terms of number of tracks used for routing. At the end of this paper, a detailed routing solution in the five-layer routing environment is demonstrated and the extension to a general multi-layer routing algorithm is also discussed.

II. Preliminaries

A *channel* is a layered rectangular routing area with terminals, coded as non-zero integers, placed at the top and bottom edges. Vertical and horizontal routing areas of unit width are referred to as *columns* and *tracks* respectively. A *net* i, denoted as N_i, is a set of terminals with assigned number i. Let

0738-100X/92 $3.00 © 1992 IEEE

t_k and b_k represent the top-side and bottom-side terminals in column k respectively, and c denote the column number. The *left bound* and *right bound* of net i, denoted as $l(N_i)$ and $r(N_i)$, are defined as the minimum k (or 1 if net i will be connected to the left) and maximum k (or c if net i will be connected to the right) with t_k=i or b_k=i respectively. The *span* of a net i is defined as the set of columns k with $l(N_i) \le k \le r(N_i)$ if $l(N_i) < r(N_i)$; otherwise it is defined as an empty set. The *local density* of a column k, denoted as d_k, is defined as the number of nets whose span contains column k. The *channel density*, denoted as d_{max}, is simply defined as the maximum local density. The *channel routing problem* (CRP) is to connect terminals in each net by using a minimum number of tracks without containing any wire- or via-violation.

A CRP is in the *two-layer restricted-Manhattan* (2-RM) model if only two layers are available, and one layer is restricted for vertical routing and the other is restricted for horizontal routing. A CRP in the 2-RM model is also referred to as a 2-RM-CRP. For a 2-RM-CRP, it is easy to see that the endpoints of the topmost horizontal wire segments for net i should be in column k such that t_k=0 or t_k=i. Such an endpoint is referred to as a *feasible endpoint* and a wire segment for net i with two feasible endpoints is referred to as a *feasible wire* for net i and is denoted as f_i. For an f_i, if there exists a column k which is crossed by the f_i with b_k=i and $t_k \ne 0$ and $t_k \ne i$, then the b_k is referred to as an *unconnectable terminal* for f_i. A set of non-overlapped feasible wires is denoted as an F. Let d_{max}' denote the channel density of the new 2-RM-CRP instance which is defined after the F is routed. According to d_{max}', the F can be divided into type 1, type 2, or type 3 if $d_{max}'=d_{max}-1$, $d_{max}'=d_{max}$, or $d_{max}'=d_{max}+1$ respectively.

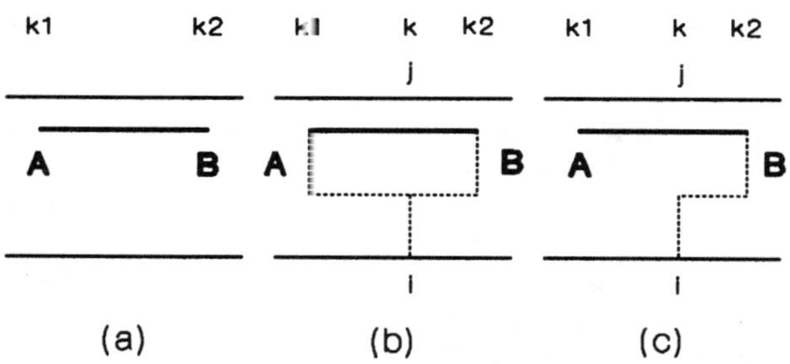

Figure 2.1 unsafe wires

A feasible wire is referred to as an *unsafe wire* if whenever the feasible wire is included in an F, either the F will never be a type 1 F or there exists no type 1 F in the new 2-RM-CRP instance; otherwise, it is referred to as a *safe wire*. A feasible endpoint is referred to as an *unsafe endpoint* if whenever a feasible wire has this kind of endpoint, the feasible wire will always be an unsafe wire; otherwise, it is referred to as a *safe endpoint*. Three brief examples of the unsafe wire f_i are shown in figure 2.1. In (a), the f_i is unsafe if $d_{k1}=d_{max}$ and $k1 \ne l(N_i)$, or $d_{k2}=d_{max}$ and $k2 \ne r(N_i)$. In (b), the connection of the unconnectable terminal b_k=i with A or B is forced to pass the column k3 or k4 if there exist column k3 and k4 such that $d_{k3}=d_{k4}=d_{max}$ and $k1 \le k3 \le k$ and $k \le k4 \le k2$. In (c), suppose $k=r(N_i)$. The unconnectable terminal b_k=i should be connected with B. If there exists a column k3 such that $k<k3 \le k2$ and $d_{k3} \ge d_{max}-1$, then this connection will increase d_{k3} up to d_{max} or $d_{max}+1$.

The strategy used to find an R in the developed two-layer router is to identify all the unsafe endpoints in the set of feasible endpoints. Then, through the connection of remaining safe endpoints, all safe wires can be identified. Finally a type 1 F for each track will be found through the developed dynamic programming technique, if it exists. For details of this section, please refer to [19].

III. Routing Strategy

In this section, we will discuss the strategy used to construct the four- and five-layer routing solutions based on the HHVH and HHVHH models respectively. In both models, V is located in the third layer. Let w and w′ denote the channel width of R and R′ respectively, and m denote the number of layers available for routing, which is restricted to no more than 5 in this paper. The procedure to construct the R′ is simply divided into the following steps:

(A) Construction of a specified R

Under our routing strategy, the horizontal wire segments of each track of R will be transferred to the H layers of R′ from the top to the bottom. To make sure no wire-violation (especially referencing the overlaps of vertical wire segments) will occur in R′, two vias for different nets in the same column are not permitted in each set of (m-1) tracks in R from the top to the bottom. Since the H layers in track 1 and track w′ of R′ can be used for vertical routing to connect the horizontal wire segments with the top-side and bottom-side terminals respectively without blocking other horizontal wire segments, the above restriction in some column in the first and last set of (m-1) tracks in R can be released. As we have mentioned, the via-violation is the major problem introduced in the proposed models. To minimize the number of vias in R′, we simply enforce another restriction that allows no wire segment in R to contain an unconnectable terminal. To obtain the specified R with the above two restrictions, we can simply redefine the endpoint for net i in column k as not a feasible endpoint if a via for other than net i has been created in column k in a track of the same set, and define a wire segment as not a feasible wire if it contains an unconnectable terminal. For better understanding of this portion, please read [19].

(B) Transfer of the H layers from R to R′

The procedure to transfer the H layer of R to R′ is trivial. The wire segments of each track in R is transferred to the H layers in R′ from the top to the bottom according to the specified track mapping. it is easy to see that the resulting $w'=\lceil w/(m-1) \rceil$

(C) Via reduction

In R′, it possible that there may exist more than one wire segment for a certain net in a track. Since the exchange of wire segments in different layers and the same track will not cause extra wire-violations, it is guaranteed that all the wire segments for a net in the same track in R′ can be combined into a single wire segment. Furthermore, two wire segments for a net in different tracks can also be combined together if it will not cause extra wire-violations. In figure 3.1, the wire segments split in different layers of the same track in the HHVHH model as shown in (a) can be combined together without causing any wire-violation, as shown in (b).

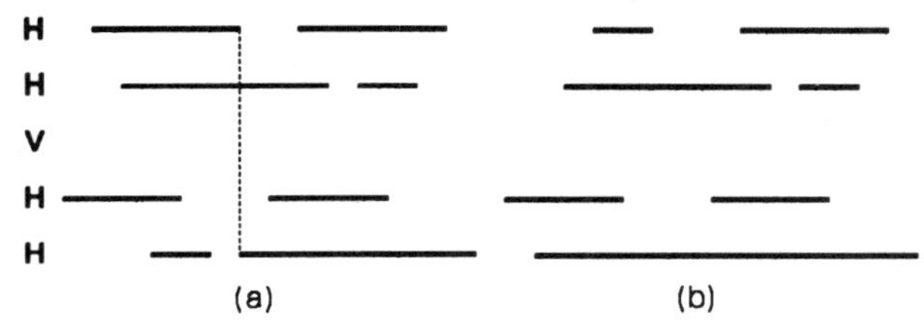

Figure 3.1 via reduction

(D) Layer reassignment

In this step, we try to exchange the wires in H layers in each track so that the effect of generated via-violations on w' will be minimized.

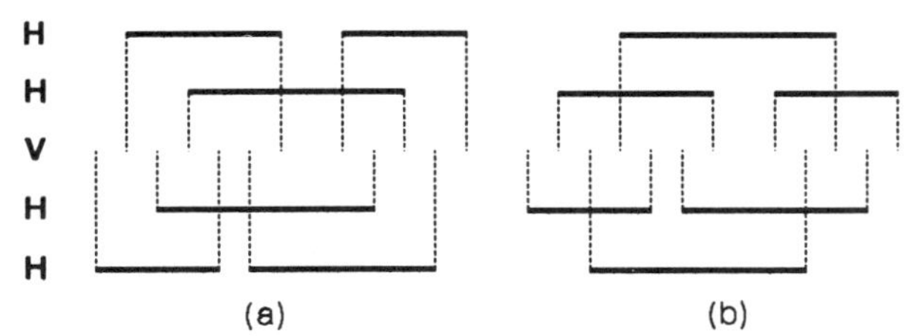

Figure 3.2 layer assignments

Let a layer assignment of a certain track in the HHVHH model be shown in figure 3.2 (a), where the intersections of the dotted line and solid line represent the via-violations. By exchanging the wire segments in the first layer with the second layer and the fourth layer with the fifth layer, a new layer assignment can be obtained as shown in figure 3.2 (b). which one is better depends on where the via-violations are located. We prefer the via-violations to occur in the low density columns. Since there exists (m-1)! permutations for the (m-1) H layers in each track, with the exhaustive search, $(m-1)!^{w'}$ cases need to be evaluated to find the optimal solution. If m=5 and w=20, the possible cases will be 7,962,624. It is quite expensive. Here we simply use a strategy which finds a layer assignment for each track from top to the bottom so that the via-violations generated from these layer assignments will likely cause the least effect on w' based on some density-related criterion. Let c denote the column number. The major ideas and data structures used to construct the procedure for layer reassignment is shown as follows.

In each track t of R', the local density of each column k, denoted as d_k, is defined as the number of the wire segments in H layers between track t+1 and track w' which overlap with column k. Similarly, d_{max} is defined as the maximum d_k

For a given layer assignment in track t, which is one of the (m-1)! permutations and is denoted as LA_t, a two-dimensional array M_t (m by c) is defined as follows. $M_t[i][j]=1$ if a via in column j intersects with a horizontal wire segment in layer i in LA_t; $M_t[i][j]=-1$ if the routing space in layer i column j is not used in LA_t; otherwise $M_t[i][j]=0$. Here $M_t[i][j]$ stands for the number of via-violations in layer i and column j if it is positive, or stands for the potential to solve the via-violations if it is negative. Given an M_t, the weight of M_t is defined as a vector

$W(M_t)=(w(M_t,d_{max}+1),w(M_t,d_{max}),....,w(M_t,1),w(M_t,0))$, where

$w(M_t,x)=$ the number of 1's in $M_t[i][k]$ -the number of -1's in $M_t[i][k]$ with $d_k=x$ if $x\leq dmax$, or

$w(M_t,x)=$the number of $M_t[i][k]\geq 2$ if $x=d_{max}+1$

In the above equation, i=2 if m=4, and i=2 or 4 if m=5. The inequality of the weights for different M_t's is defined as follows: We say $W(M'_t)>W(M_t)$ if there exists an x such that

$w(M'_t,x)>w(M_t,x)$, where $0\leq x\leq d_{max}+1$ and

$w(M'_t,y)=w(M_t,y)$, where $x+1\leq y\leq d_{max}+1$

The basic procedure for layer reassignment is shown below.

procedure *LAYER_REASSIGNMENT*
  M_0=**0**
  for t=1 **to** w' **do**
    construct_density()
    Find an LA_t which minimizes $W(M_t+M_{t-1})$
    reassign the wire segments in track t according to LA_t
    $M_t=M_t+M_{t-1}$
  end_for
end_*LAYER_REASSIGNMENT*

In the above procedure, we simply use the defined weight to compare the effect of introduced via-violations on w' for different LA_t's, and apply a dynamic programming technique from top to the bottom to minimize the cumulative effect of via-violations caused by the LA_t's that have been assigned. Let x be the maximum value in $M_{w'}$. The resulting channel width of R' we can achieve from the above procedure is $w'= w'+ max(0,x)$.

(E) Rerouting of H layers and Construction of the V layer in R'.

In this step, we connect all the necessary vias in R' from the top to the bottom. If a horizontal wire segment is encountered, it will be pushed up if there exists a "hole" on top; otherwise it will be pushed down, which may push another horizontal wire segment down, and etc. After all the via violations have been resolved, we can adjust the wire segments in H layers to minimize the "zigzag" and the wire length. The vertical wire segments in the V layer can be constructed easily according to the relative positions of the horizontal wire segments which need to be connected.

IV. Time Complexity Analysis

According to [19], the time complexity to find the specified R is O(ncw), where n represents the net number, c represents the column number, and w represents the channel width. After the R is constructed, first we transfer the H layers of R to R' from the top to the bottom, which takes O(cw) time. Second, we reduce the number of vias by combining the separate wire segments for the same net in R' if possible, which takes O(cw) time. Third, to minimize the effect of via-violations in R', we reassign the H layers in each track t of R' from top to bottom. The process in each track is shown as follows: first we calculate the density of each column, which takes O(cw) time; then all (m-1)! possible LA_t's are evaluated to find an optimal one, where it takes O(mc) time to construct M_t and calculate its weight for each LA_t. Therefore the process in each track takes O(cw+cm!) time. Since we restrict $m\leq 5$ in this paper, m! can be treated as a constant. So the time complexity in each track is simplified as O(cw) and the resulting time complexity to reassign the H layers in R′ is $O(cw^2)$. Fourth, after the H layers have been reassigned, we connect the necessary vias and vertical wire segments. During this process, if a horizontal wire segment blocks a via, it is pushed up or down. This procedure takes O(cw) time. Finally, in the post processing, we readjust the wire segments in the H and V layers to minimize the "zigzag" of wire segments and further reduce the via number and wire length if possible, which takes O(cw) time. So the overall time complexity of the proposed routing algorithm is still O(ncw).

V. Experimental Results

We implemented the proposed routing algorithm in C on the Vax 3800. The experimental results for the four- and five-layer models are shown in tables 5.1 and 5.2 respectively, where 3a, 3b, and 3c are from [9] and diff represents Deutsch's difficult example; each number represents the number of tracks used; the number with * represents the optimal routing solution in the proposed models. To familiarize the readers with the detailed routing generated by the proposed router, the routing solution of example 3b based on the HHVHH model is demonstrated in figure 1, where the layers are separated by dotted lines and all the diagonal wires are printed as right angles if possible for printing convenience.

Example	Density	Our router	[5]1988
3a	15	6	8
3b	17	6*	9
3c	18	6*	9
diff.	19	7*	10

Table 5.1 Performance comparison in the four-layer model

Ex.	Density	Our router	[2] 1986	[5] 1988
3a	15	4*	5	5
3b	17	5*	6	6
3c	18	5*	6	6
diff.	19	6	7	7

Table 5.2 Performance Comparison in the five-layer model

In tables 5.1 and 5.2, we can see that the performance of our router has outperformed all the existing multi-layer routers in the four- and five-layer routing environments from our understanding. For most examples, the optimal solutions in our models are also achieved.

The four- and five-layer router presented in this paper is originally intended for a general multi-layer router which is being developed. Since its current routing performance in the four- and five-layer routing environments has outperformed other routers, we simply report the current results in their early stage. We believe that if more routing rules which are implied in the m-layer routing solution R′ are considered in constructing the R, possibly not only the diagonal wire segments generated by pushing wires up or down can be removed without rerouting, but also the routing performance can be further improved. This will be discussed in the next section.

VI. Extension to A General Multi-Layer Router

Intuitively, we may consider extending the proposed models to a general multi-layer router, where m is not restricted, by adding more H layers. According to the strategy shown above to construct R and R′, two problems can be expected in this extension. First, as m increases, the restriction of the wire segments of different tracks by each other in each set of (m-1) tracks will also increase. Therefore, the chance to find a type 1 F will be decreased and the via-violation problem will get severer for each track, especially for the tracks in the inner H layers. Possible ways to tackle these problems are suggested as follows. First we simply divide the m-layer routing problem into four- and five-layer subproblems; then use the proposed strategy to solve each sub-problem as done in [5]. Second, We can restrict the via length, which will alleviate the introduced via-violation problem. Third, do not specify the H or V layer at the beginning, and use the available space for vertical or horizontal routing if possible. Of course, the incorporation of this thought into the developed two-layer router will be more complicated. Currently, all alternatives are being investigated. It is anticipated that the final results will be reported soon.

VII Concluding Remarks

In this paper, we propose two new models for the four- and five-layer routing problems respectively. Although the performance of the proposed models are excellent, further improvements are required for applications and to extend to a general m-layer router. First, the diagonal wire segments in the generated routing solutions may present some problems in the practical wiring implementation, which may be solved by rerouting some wire segments through a maze routing strategy. In fact, the diagonal wire in figure 1 can be removed by exchanging some wire segments. Second, direct extension of the proposed models to a general m-layer router is prohibited not only by the problems shown in section VI, but also by its exponential time complexity in the layer reassignment. Possible ways to tackle above problems for better performance are shown as follows. First, if all the via-violations can be considered in constructing the R, the above two problems are automatically solved. Second, since the set of wires in each track has different restrictions, the criteria to evaluate the wire segments in different tracks should be different for better routing solutions. Although the idea of the proposed routers is straight-forward and is simply implemented, the current experimental results are encouraging for further investigation. We are optimistic about developing a general multi-layer router based on the frame work presented in this paper.

Acknowledgement

We sincerely express our thanks to Dr. Yue-Sun Kuo, the acting director of the Institute of Information Science, Academia

Sinica, for giving the author the opportunity and providing necessary supports to start the implementation of the proposed router on the SUN work station when he was on leave in Taiwan.

References

[1] J.F. Naveda, K.C. Chang and H.C. Du, "A New Approach to Multi-Layer PCB Routing with Short Vias", *23rd Design Automation Conference*, pp. 696-701, 1986

[2] R. J. Enbody and H.C. Du, "Near-Optimal n-Layer Channel Routing", *23rd Design Automation Conference*, pp. 708-714, 1986

[3] R. Eric Lunow, "A Channelless, Multilayer Router", 25th ACM/IEEE Design Atomation Conference, 1988

[4] Patrick Groeneveld, "A Multiple Layer Contour-Based Gridless Channel Router" *IEEE Transactions on Computer-Aided Design*, Vol. 9, No. 12, December 1990

[5] Douglas Braun, Jeffrey Burns, Fabio Romeo, Alberto Sangiovanni-Vincentelli, Kartikeya Mayaram, Srinivas Devadas, Hi Keung Ma, "Techniques for Multilayer Channel Routing", *IEEE Transactions on Computer-Aided Design*, Vol. 7, No. 6, pp. 698-712, June 1988

[6] P. Bruell and P. Sun, "A Greedy Three Layer Channel Router," *Proc. ICCAD-85*, pp. 298-300, 1984

[7] Y. K. Chen and M. L. Liu, "Three-Layer Channel Routing," *IEEE Trans. CAD of ICs and Systems*, Vol. CAD-3, No 2, pp. 156-163, 1984

[8] S. E. Hambrusch, "Channel Routing Algorithm for Overlap Models," *IEEE Trans. on Computer Aided Design of ICAS*, Vol. CAD-4, No 1, pp. 23-30, 1985

[9] Takeshi, Yoshimura, and Ernest S. Kuh, "Efficient Algorithm for Channel Routing," *IEEE Transactions on Computer Aided Design of Integrated Circuits and Systems*, Vol. CAD-1, No. 1, pp. 298-307, January 1982

[10] David N. Deutsch, "A Dogleg Channel Router," *Proc. 13th Design Automation Conference*, pp. 425-433, 1976

[11] J. Cong, D.F. Wong, and C.L. Liu, "A New Approach to Three- or Four-Layer Channel Routing", *IEEE Transactions on Computer-Aided Design*, Vol. 7, No. 10, pp. 1094-1104, Oct. 1988

[12] J.M. Jou, J.Y. Lee, and J.F. Wang, "A New Three-Layer Detailed Router for VLSI Layout," *Proc. Int. Conf. on CAD*, pp.382-385, 1987

[13] Uzi Yoeli, "A Robust Channel Router," *IEEE Trans. on CAD of ICAS*, Vol. 10, No. 2, pp.212-219, Feb. 1991

[14] M.L. Brady and D.J. Brown, "VLSI Routing: Four Layers Suffice", *Advances in Computing Research*, Vol.2, 1984

[15] W.Jr. Lipski and F.P. Preparata, "An Unified Approach to Layout Wirability", *Mathematical Systems Theory*, Vol. 19, 1987

[16] F.P. Preparata and W.Jr. Lipski, "Optimal Three-Layer Channel Routing", *IEEE Trans. Comput.*, 33, 5, 1984

[17] T.T. Ho, S.S. Iyengar and S.Q. Zheng, "The General Greedy Channel Routing Algorithm", *IEEE Transactions on Computer-Aided Design*, Vol. 10, No. 2, pp. 204-211, February 1991

[18] T.T. Ho and S.S. Iyengar, "New results on Channel Routing", *4th CSI/IEEE International Symposium on VLSI Design*, pp. 174-179, January 1991

[19] T.T. Ho, "A Density-Based Greedy Router", Technical Report #91-001, DMCS, McNeese State University, 1991

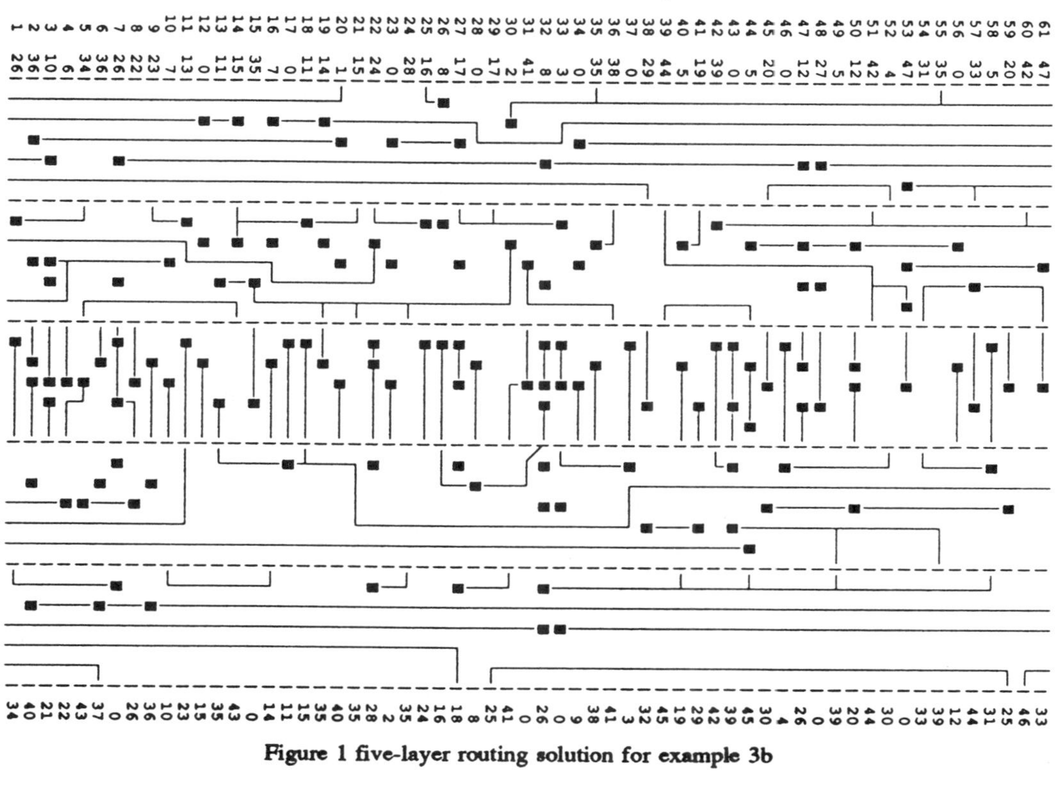

Figure 1 five-layer routing solution for example 3b

A Pin Permutation Algorithm for Improving Over-the-Cell Channel Routing

Cliff Yungchin Hou and C.Y. Roger Chen
Department of Electrical and Computer Engineering
Syracuse University, Syracuse NY 13244-1240

Abstract

In standard cell design, some of the cell terminals and gates are permutable. Therefore, it is important for an over-the-cell channel router to take advantage of this so as to obtain better results. A dynamic programming based algorithm is presented to determine proper gate and terminal positions such that, when over-the-cell routers are used, the area above and below the channel can be utilized more effectively and the channel density can be greatly reduced. Experimental results show that our proposed algorithm indeed considerably reduces the channel density.

1 Introduction

In standard cell design, cells are placed in rows, necessary feedthroughs are inserted, and channel routers are used to complete interconnections between cells. Two types of channel routers are used to complete interconnections, the *conventional channel router* [1, 2, 3, 4, 5] and the *over-the-cell channel router* [6, 7, 8, 9, 10]. The conventional channel router only uses the area in the channel to complete interconnections and has been studied extensively in recent years. To further reduce channel routing area, the over-the-cell channel router is proposed. The area above and below the channel is also used for interconnections with single layer routing. Actual cell layouts to allow such over-the-cell channel routing have been demonstrated in [10]. In most cases, over-the-cell routers generate better results than those generated by the conventional routers.

In most VLSI chips, a large portion of area is used for channel routing; thus, the saving of area by using over-the-cell routing is significant. For the example shown in Figure 1, a conventional channel router would require 6 tracks in the channel to complete the interconnections. By routing over the cell, the number of tracks is reduced to 3 as shown in Figure 2.

In standard cell design, many gates and terminals can be interchanged without changing the functions of the cells. The interchange of gates and terminals may increase the number of nets which can be routed over the cells such that the connections in the channel are reduced. Thus, the density in the channel is also reduced. For the example shown in Figure 2, by interchanging gates and terminals as shown in Figure 3, the number of tracks in the channel is further reduced from 3 to [illegible]. None of the pervious work in over-the-cell channel routing takes this flexibility (i.e., pin permutation) into consideration. For convenience, the interchange of gates and terminals is also referred to as pin permutation in the following discussions.

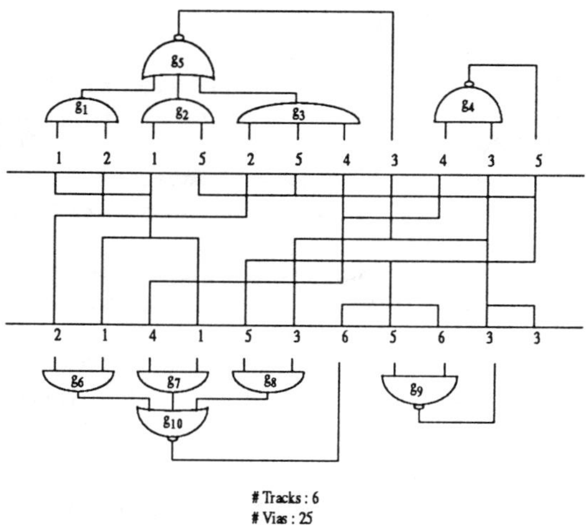

Figure 1: An example of the conventional channel routing.

The goal of this paper is to determine proper positions of gates and terminals such that, when over-the-cell routing is performed, channel density is minimized. This paper is organized as follows. In Section II, the problem formulation and previous work are discussed. In Section III, the terminology and notation are introduced. In Section IV, a polynomial time algorithm is presented. Experimental results and comparisons are given in Section V. Finally, conclusions are given in Section VI.

2 Problem Formulation and Previous Work

A common approach to the over-the-cell channel routing problem was first proposed by Shiraishi and Sakemi [8]. The three steps are :

1. routing over the cell;
2. choosing net segments within the channel;

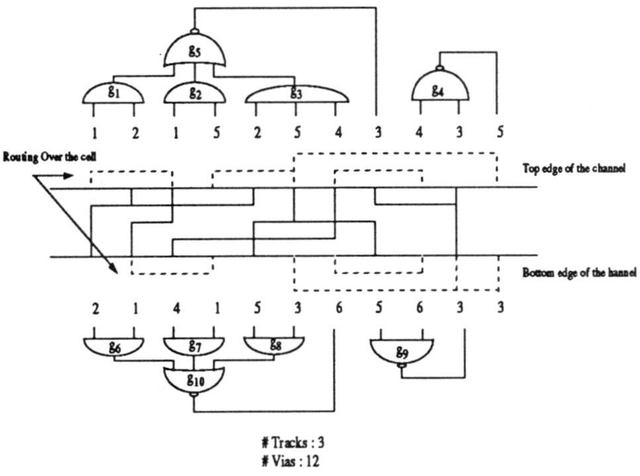

Figure 2: An example of the over-the-cell channel routing.

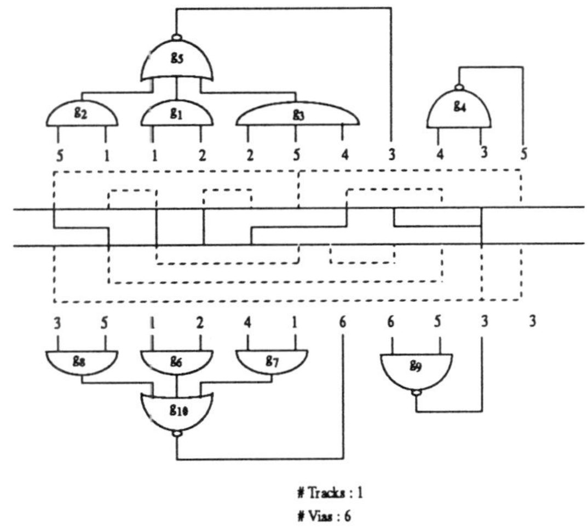

Figure 3: An example of the over-the-cell channel routing with pin permutations.

3. routing within the channel.

In step 1, some nets are routed over the cell in a single layer such that the number of connections in the channel is minimized. In step 2, proper connections between terminals are selected such that the connectivity of nets is preserved and the channel density is minimized. In step 3, a conventional channel router is used to complete the routing.

Cong and Liu [9] have proposed a polynomial time algorithm for step 1 to route a maximal number of nets over the cell. They have also proved that the problem of step 2 is $NP-Hard$ and proposed a heuristic for this step. In [10], Cong, Preas, and Liu mention the use of variable chord weights in performing over-the-cell routing, which allows critical net segments to be routed over the cell with a higher priority. Lin *et al.* [12] divide a channel into zones and use a linear programming approach to reduce the channel density; better results have been reported at the expense of long CPU time.

As mentioned earlier, in standard cell design, many cells have permutable structure, i.e., both gates and terminals are interchangeable as long as the logic function is not changed. For example, complex cells such as those shown in Figure 4 can easily be found in many standard cell libraries [13]. Some examples of logically equivalent permutations of the cell shown in Figure 4(a) are given in Figure 5. Note that Figures 5(b), 5(c), 5(d), and 5(e) are logically equivalent permutations of Figure 5(a). By taking advantage of the permutable structure, the number of nets which can be routed over the cell can be greatly increased. Thus, the channel density can be reduced considerably. Therefore, in this paper our goal is to determine a proper pin permutation, by suitably interchanging gates and terminals, such that when existing over-the-cell routers are used, the channel density can be greatly reduced.

3 Terminology and Notation

In this paper we assume that there are two layers for channel routing and a single layer for over-the-cell routing. Terminals of a net are connected by horizontal and vertical wires. Terminals at the top edge of a channel are labeled as t_1, t_2, ..., t_n and terminals at the bottom edge of a channel are labeled as b_1, b_2, ..., b_n from left to right. The *position set* of g (where g can be either a gate or a terminal), $P(g)$, is recursively defined as follows: If g is a terminal, $P(g) =$ {the position of g}. If g is a gate, $P(g)$ is the union of the position sets of all the inputs of g. For a gate g with $P(g) = \{ p_1, p_2, ..., p_l\}$, the *net set* of g, $N(g)$, is defined to be $\{ n_1, n_2, ..., n_l\}$, where n_i $(1 \leq i \leq l)$ is the net number of the terminal at position p_i. Two gates (terminals) are permutable, if the interchange of their net sets will not change the logic function of the cell. A gate (terminal) is *non−permutable* if it cannot be permuted with any other gate (terminal) without changing the logic function of the cell.

For a gate g with l inputs, $g_1, g_2, ..., g_l$, where each g_i $(1 \leq i \leq l)$ can be a terminal or a gate, the *level* of g, $L(g)$, is defined to be the highest level of its inputs plus 1, i.e., $L(g) = MAX\{L(g_1), L(g_2), ..., L(g_l)\} + 1$. All terminals are of level 0.

A *neighbor terminal pair* of a net x is a pair of terminals, t_i and t_j, of net x such that t_i and t_j are on the same side of the channel and there does not exist any terminal of net x (which is also on the same side of the channel) between t_i and t_j, i.e., there does not exist any terminal t_k of net x on the same side of the channel such that $i < k < j$ or $j < k < i$. A single layer wire s which is used to connect the terminals in a neighbor terminal pair $\{t_i, t_j\}$ of net x is called a *candidate seg-*

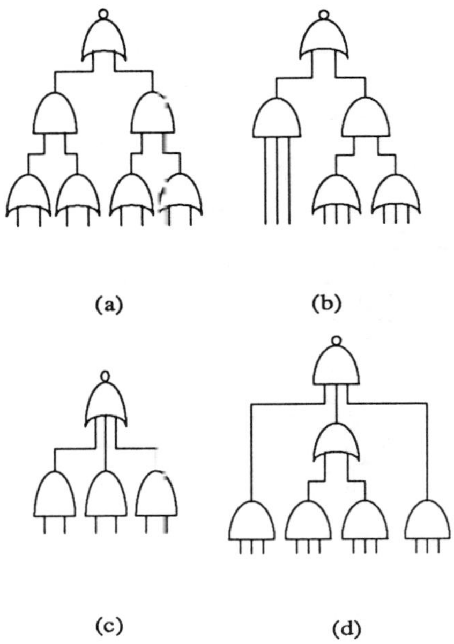

Figure 4: Examples of complex library cells.

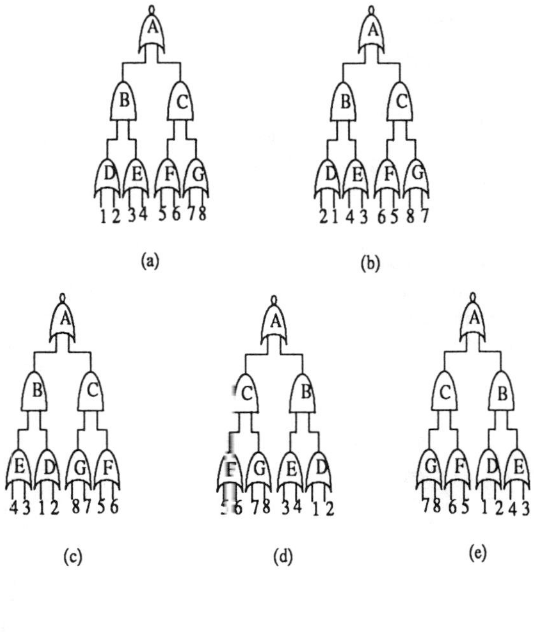

Figure 5: Some examples of logically equivalent permutations.

ment of net x, where t_i and t_j are two endpoints of s. It is clear that a terminal has at most two candidate segments incident to it. A multi-terminal net with m top terminals and l bottom terminals has $m-1$ top neighbor terminal pairs and $l-1$ bottom neighbor terminal pairs. For example, there are two top neighbor terminal pairs and one bottom neighbor terminal pair for net 5 in Figure 2, i.e., $\{t_4,t_6\}$, $\{t_6,t_{11}\}$, and $\{b_5,b_8\}$. Two candidate segments s_m and s_n with endpoints (t_i,t_j) and (t_k,t_l), respectively, are *conflicting* if they are of different nets and $t_i < t_k < t_j < t_l$ or $t_k < t_i < t_l < t_j$. Therefore, the over-the-cell channel routing problem is equivalent to the problem of finding a set of non-conflicting candidate segments in the channel. Candidate segments are routed over the cell if they are in the set of non-conflicting candidate segments.

In order to more effectively reduce the channel density, a candidate segment that spans a denser columns should be routed over the cell with a higher priority. Thus, different weights are assigned to candidate segments. To ensure that a candidate segment which spans denser columns will have a higher priority to be routed over the cell, the weight of a candidate segment s, $W(s)$, is defined to be the sum of squares of the column densities between its two endpoints, i.e.,

$$W(s) = \sum_{i=t_l}^{t_m} D(i)^2,$$

where t_l and t_m are endpoints of s and $D(i)$ is the density of column i.

The way the weight function is selected is to distinguish those candidate segments spanning denser columns from those spanning less dense columns, and is actually determined based on extensive experimental data. By bending the top (bottom) edge of a channel and joining the two ends of the edge, the problem of finding a solution over the cell is then equivalent to the problem of finding the maximal weight of an independent set of chords in a circle graph. For example, the circle graph of the top edge and bottom edge in Figure 2 are shown in Figures 6(a) and 6(b), respectively. The chords of the circle graphs are the candidate segments in the channels. It is known that the problem of finding a maximal weight of an independent set of chords in a circle graph can be solved in polynomial time by using the dynamic programming approach [11]. (Note that although the formulation in [11] assumes uniform chord weight, it can be easily generalized to the case where variable chord weights are assumed.) In this paper, without lose of generality, we assume that the dynamic programming algorithm is performed clockwise starting from terminal 1 to terminal n, where n is the number of terminals in the circle graph. In a circle graph G, the *arc* $A(t_i,t_j)$ of G is defined to be the arc formed by traversing G from t_i to t_j clockwise. The weight of an independent set of $A(t_i,t_j)$ is defined to be the sum of the weights of all the edges in the independent set.

We say that $A(t_i,t_j)$ contains a gate g (or equivalently g is in $A(t_i,t_j)$) if all the terminals which are

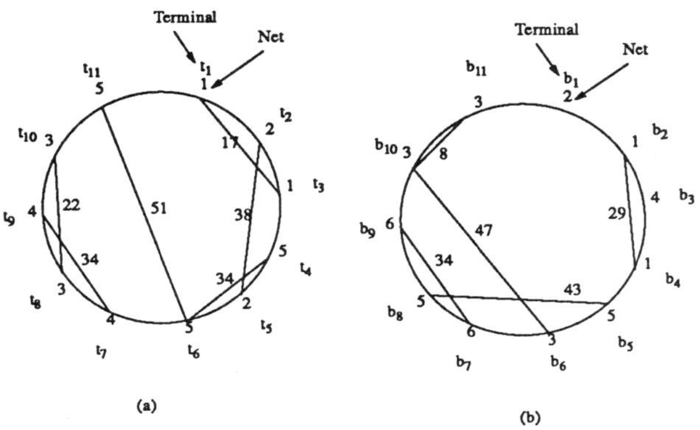

Figure 6: The circle graph of the channel in Figure 2 and the weights of candidate segments.

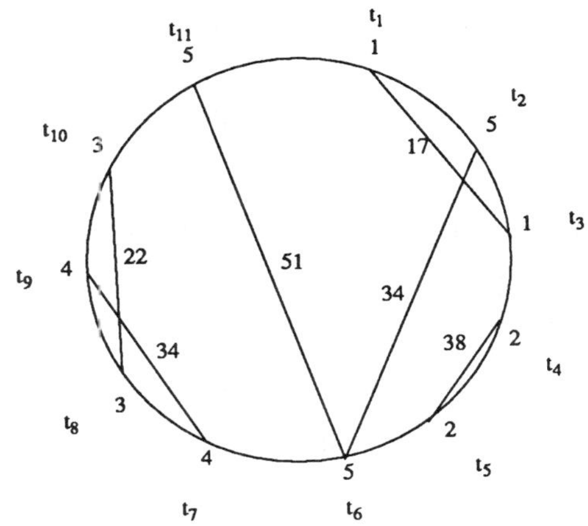

Figure 7: The new circle graph of the top channel in Figure 2 with g_1 and g_2 interchanged.

in the position set of g are in $A(t_i, t_j)$. Similarly, the arc of a gate g is the minimal arc which contains g. The arc of a set of gates (terminals) is the minimal arc which contains all the gates (terminals). For the example shown in Figure 2, gates g_1, g_2, g_3, and g_5 are all in $A(1,7)$ (see the circle graph shown in Figure 6(a)); the arcs of g_1, g_2, g_3, and g_5 are $A(1,2)$, $A(3,4)$, $A(5,7)$, and $A(1,7)$, respectively.

The interchange of gates (terminals) will change the net sequence in G and form a new circle graph G'. For example, in Figure 2, if g_1 and g_2 are interchanged, the new circle graph is shown in Figure 7. The net sequence on $A(t_1, t_4)$ is changed. Thus, the maximal weight of the circle graph is changed. Since it is desirable to increase the maximal weight of the circle graph by pin permutation, the gain of the circle graph by interchanging gates g_x and g_y, $C\_GAIN(g_x, g_y)$, is defined to be the difference of the maximal weight of the circle graph after and before the interchange of g_x and g_y, i.e.,

$$C\_GAIN(g_x, g_y) = MW(G') - MW(G),$$

where $MW(G)$ is the maximal weight of the circle graph G and $MW(G')$ is the maximal weight of the new circle graph G' if g_x and g_y are interchanged.

For a cell C, the interchange of its gates (terminals) will change the local density of C. A decrease (increase) of the column density can be viewed as a removal (addition) of some connections in the channel. Since the weight of a candidate segment is defined to be the sum of squares of the column densities between its two endpoints, the *weighted density gain* of the interchange of two gates g_x and g_y, $D\_GAIN(g_x, g_y)$, is defined as follows:

$$D\_GAIN(g_x, g_y) = \sum_{i=j}^{k}(BD(i) - AD(i)) * BD(i)^2,$$

where j and k are the positions of the two endpoints of the arc of g_x and g_y. $BD(i)$ ($AD(i)$) is the density of column i before (after) g_x and g_y are interchanged. Similarly, the density gain of the interchange of two terminals is defined. Note that the weighing factor $BD(i)^2$ is introduced here in order to make the definition consistent to that of the weight of a candidate segment (which is defined earlier). For the example in Figure 2, the column densities between columns 1 and 4 before g_1 and g_2 are interchanged are 2, 2, 3, and 4, respectively. The column densities between columns 1 and 4 after g_1 and g_2 are interchanged are 2, 3, 4, and 4, respectively. Thus, $D\_GAIN(g_1, g_2) = -4 - 9 = -13$.

The final gain of the interchange of two gates (terminals) is defined to be the sum of the circle graph gain and the weighted density gain.

4 Algorithm

The pin permutation algorithm is performed as follows: Start from the highest level (H) gates. For a permutable gate (terminal) g_i (t_i) which is an input of a level-H gate and each gate (terminal) which is permutable with g_i (t_i), calculate the weighted density gain and the circle graph gain if they are interchanged. Find the gate (terminal) g_j (t_j) with the maximal gain. Interchange the net sets of g_i and g_j (t_i and t_j). Perform the same operation until all the permutable gates and terminals are processed. Then go to the next level and perform the same operation until all the permutable gates (terminals) are processed.

Since the calculation of the density gain can be performed by checking the column density between the two endpoints of the arc of g_i and g_j, the calculation of the new maximal weight of a circle graph with some of its terminals being interchanged is a critical part of this algorithm.

Since an interchange of gates or terminals changes the net sequence of the circle graph, the maximal weight of the circle graph is also changed. Thus, in order to calculate the maximal weight of the new circle graph, the dynamic programming method has to be applied every time two gates or terminals are permuted. Since the time complexity of finding the maximal weight of a circle graph is $O(N^2)$ (using the technique presented in [11]), where N is the number of terminals in the circle graph, the time complexity of the algorithm seems to be at least $O(N^3)$. However, through careful investigation, we have found that an interchange of gates or terminals changes only some specific part of the net sequence in a circle graph G. For example, if g_p and g_q are interchanged and the arc of g_p and g_q is $A(t_i, t_j)$, then only the net sequence in $A(t_i, t_j)$ will be changed while the net sequence in $A(t_{j+1}, t_{i-1})$ will remain the same. If the maximal weight of $A(t_{j+1}, t_{i-1})$ is known, it does not need to be recalculated and can be directly used to calculate the maximal weight of the new circle graph. Based on this concept, an algorithm of $O(N^2)$ time complexity is then proposed, which will be described in this section.

In a circle graph G the maximal weight of G will not be changed if different starting terminals are selected. Thus, if the maximal weight of $A(t_{j+1},\ t_{i-1})$ is known, the maximal weight of the circle graph G can be obtained by performing the dynamic programming method for $j - i + 1$ steps on arcs $A(t_{j+1}, t_i)$, $A(t_{j+1}, t_{i+1})$,..., $A(t_{j+1}, t_{j-1})$, and $A(t_{j+1}, t_j)$. In standard cells, $j - i < m$, where m is the maximal number inputs of a cell. However, if only the maximal weight of $A(t_{j+1}, t_k)$ is known (where t_k is an arbitrary terminal), the maximal weight of $A(t_{j+1}, t_{i-1})$ must be calculated first before we can apply the dynamic programming method in $A(t_{j+1}, t_i)$, $A(t_{j+1}, t_{i+1})$,..., $A(t_{j+1}, t_{j-1})$, and $A(t_{j+1}, t_j)$. The number of extra steps required before we can perform the dynamic programming method on $A(t_{j+1}, t_i)$, $A(t_{j+1}, t_{i+1})$, ..., $A(t_{j+1}, t_{j-1})$, $A(t_{j+1}, t_j)$ is equal to $| A(t_{k+1}, t_{i-1}) |$ if t_k is in $A(t_{j+1}, t_{i-2})$, or 0 , otherwise. $| A(t_{k+1}, t_{i-1}) |$ is the number of terminals in $A(t_{k+1}, t_{i-1})$. This is the term which we want to minimize. If the permutation sequence of gates and terminals is not properly selected, in the worst case, this term can be close to N, where N is the number of terminals in the channel. Therefore, in order to minimize the complexity of the algorithm, the permutation of gates and terminals should be performed in a way that the number of extra steps is minimized. By always selecting t_{j+1} as the starting terminals when the the dynamic programming method is applied and performing the permutation of gates and terminals in the channel from left to right sequentially, the maximal weight of $A(t_{j+1}, t_{i-1})$ is always calculated in the previous iteration. Thus, no extra step is required for each permutation.

Example 1 Consider the top edge of the channel in Figure 2. The corresponding circle graph G is shown in Figure 6(a). The maximal weight of G without interchanging any gate and terminal is $17 + 34 + 51 + 34 = 136$. After the interchange of gate g_1 and g_2, the new circle graph G' is shown in Figure 7. The net sequence in $A(5, 11)$ is not changed. The maximal weight of $A(5, 11)$ was calculated in the previous iteration and is equal to 85. Then, the maximal weight of G' can be calculated by repeatedly applying the dynamic programming method in [11] for $A(5, 1)$, $A(5, 2)$, $A(5, 3)$, and $A(5, 4)$. The maximal weight of G' is equal to 38+34+51+34=157. Thus, $C\_GAIN(g_1, g_2)$ = 157 - 136 = 21. $D\_GAIN(g_1, g_2)$= -13. Since $C\_GAIN(g_1, g_2) + D\_GAIN(g_1, g_2) > 0$, g_1 and g_2 are interchanged. Continue the same process for the permutable gates of level 1 in the bottom region and the terminals in both regions, the final pin positions determined by our algorithm are shown in Figure 3. If the over-the-cell router in [9] is used, the final layout is shown in Figure 3. The final channel density has been reduced from 3 to 1.

□

5 Experimental Results

The algorithm has been implemented in the programming language C on a Sun SPARCstation 1. In order to evaluate the performance of our algorithm, the over-the-cell algorithm in [9] and the greedy algorithm in [2] are implemented. The upper cells and lower cells are randomly selected from the cells in [13]. First, our algorithm is applied to determine all pin positions. Then, the over-the-cell router [9] and the channel router [2] are used to complete the routing. The experimental results of the benchmark circuits in [3] and the Deutsch's difficult problem (De) are shown in Table 1. The comparisons of the results of our pin permutation algorithm+[9]+[2] with that of [9]+[2] (i.e., the comparisons of results with and without using our pin permutation algorithm) are shown in Table 2. From Table 2, it is clear that by including our pin permutation algorithm, the channel densities of all the examples are considerably reduced. In a similar way, our algorithm can also be applied to improve the results of other over-the-cell routers (such as [10] and [12]). Note that our pin permutation is performed to allow as many critical net segments as possible to be routed over the cell. The routing density over the cell may not be a good indication of the actual channel density (which is the actual one to be minimized). This is especially true when an over-the-cell router which assumes uniform chord weight (such as [9]) is used, because such over-the-cell routers cannot fully take advantage of the results of our pin permutation algorithm. For example, the result of De[5] using [9]+[2] has a higher over-the-cell routing density than that using the proposed pin permutation algorithm+[9]+[2]. However, the overall channel densities using our pin permutation algorithm+[9]+[2] are always smaller than that without using our pin permutation algorithm. This indicates that if other over-the-cell routers which assume variable chord weights (such as [10] and [12]) are used, the improvement of channel density by including our pin permutation algorithm can be even more significant.

Table 1. The experimental results of our algorithm+[9]+[2].

Test circuit	Channel density	# of tracks	CPU time (Sec.)
Ex1[9]	1	1	0.1
De[5]	14	15	4.8
Ex1[3]	8	8	0.2
Ex3a[3]	11	11	0.5
Ex3b[3]	11	11	0.7
Ex3c[3]	13	13	2.4
Ex4b[3]	12	12	2.8
Ex5[3]	12	12	2.3

Table 2. Comparison of the results.

Test Circuit	Density over the upper cells		Density over the lower cells		Density within the channel	
	[9]+[2]	ours+ [9]+[2]	[9]+[2]	ours+ [9]+[2]	[9]+[2]	ours+ [9]+[2]
Ex1[9]	2	3	2	2	3	1
De[5]	7	6	8	4	16	14
Ex1[3]	3	2	4	4	9	8
Ex3a[3]	3	5	6	5	12	11
Ex3b[3]	2	6	5	4	13	11
Ex3c[3]	3	3	4	6	14	13
Ex4b[3]	5	3	4	4	16	12
Ex5[3]	4	6	3	3	14	12

6 Conclusions and Future Work

A polynominal time pin permutation algorithm has been presented for the over-the-cell channel routing such that, when included in the over-the-cell routing, the area above and below the channel can be utilized more effectively. The technique of dynamic programming has been applied to calculate the gains of gates and terminals permutation. Experimental results have shown that considerable improvements on the channel density have been achieved by including our pin permutation algorithm.

In this paper, in order to make the basic concept clear and understandable, we only show how to perform pin permutation according to the final gain on one side of a cell row. In standard cell design, the change of pin positions on one side of a cell will change the pin positions on the other side. Thus, the effects on both sides of the cell row should be taken into account when pin permutation is performed. This can be easily performed by using two circle graphs, one for each side of a cell row. Then, the final pin positions are determined according to the gains of the two circle graphs and the density gains of the two channels which are adjacent to this cell row.

The pin permutation algorithm proposed in this paper can be applied to cell rows in standard cell design as follows. First, select a cell row which is either a boundary cell row or a cell row adjacent to the most congested channel. Perform pin permutation on this cell row. Then, select a cell row adjacent to the current cell row. Continue the same operation until all the cell rows are processed.

Note that, in this paper, delay minimization has not been taken into account during pin permutation. For gates or terminals in the critical path, one can easily modify the proposed algorithm and treat those gates or terminals as nonpermutable. Thus, pin permutation is applied only to the noncritical part.

Although conceptually the proposed pin permutation algorithm can also be used for multiple layer over-the-cell routing, currently, it is more effective for single layer routing. A possible future work is to generalize the algorithm such that it can also be effective for multiple layer over-the-cell routing. The proposed algorithm only allows pins within a cell to be permuted. Since in many cases, especially when a lot of primitive cells are used, it is important that pins in one cell can permute with pins in another cell, another possible future work is to generalize our current technique to allow such permutations for over-the-cell routing.

References

[1] A. Hashimoto and J. Stevens, "Wire Routing by Optimizing Channel Assignment within Large Apertures", *Proc. 8th Design Automation Workship*, pp. 214-224, 1971.

[2] R. L. Rivest and C. M. Fiduccia, "A 'Greedy' Channel Router", *Proc 19th Design Automation Conf.*, pp. 418-424, 1982.

[3] T. Yoshimura and E. S. Kuh, "Efficient Algorithms for Channel Routing", *IEEE Transactions on Computer-Aided Design*, vol. CAD-1, pp. 25-35, 1982.

[4] M. Burstein and R. Pelavin, "Hierarchical Channel Router", *Proc. 19th Design Automation Conf.*, pp. 591-597, 1983.

[5] D. N. Deutsch, "A 'Dogleg' Channel Router", *Proc. 13th Design Automation Conf.*, pp. 425-433, 1976.

[6] D. N. Deutsch and P. Glick, "An Over-the-Cell Router", *Proc 17th Design Automation Conf.*, pp. 32-39, 1980.

[7] H. E. Krohn, "An Over-the -Cell Gate Array Channel Router", *Proc. 20th Design Automation Conf.*, pp. 665-670, 1983.

[8] Y. Shiraishi and Y. Sakemi, "A Permeation Router", *IEEE Transactions on Computer-Aided Design*, Vol. CAD-6, pp. 462-471, May 1987.

[9] J. Cong and C. L. Liu, "Over-the-Cell Channel Routing", *IEEE Transactions on Computer-Aided Design*, Vol. 9, No. 4, April 1990.

[10] J. Cong, B. Preas, and C. L. Liu, "General Models and Algorithms for Over-the-Cell Routing in Standard cell design", *Proc. 27th Design Automation Conf.* , pp. 709-715, 1990.

[11] K. J. Supowit, "Finding a Maximum Planar Subset of a Set of Nets in a Channel", *IEEE Transactions on Computer-Aided Design*, Vol. CAD-6, pp. 93-94, Jan. 1987.

[12] M. S. Lin, H. W. Perng, C. Y. Hwang, and Y. L. Lin, "Channel Density reduction by Routing Over the Cells", *IEEE Transactions on Computer-Aided Design*, Vol. 10, No. 8, pp. 1067-1071, August 1991.

[13] Texas Instruments; "2-μM CMOS Standard Cell Data Book: SystemCell Series", 1986.

Over-the-Cell Channel Routing For High Performance Circuits *

Sivakumar Natarajan, Naveed Sherwani, Nancy D. Holmes†, and Majid Sarrafzadeh‡

Dept. of CS, Western Michigan University, Kalamazoo, MI 49008
†Dept. of Info. Sciences and CS, University of California, Irvine, CA 92717
‡Dept. of EE and CS, Northwestern University, Evanston, IL 60108

Abstract

In this paper, we present a new three layer OTC channel routing algorithm for high performance circuits. This router not only minimizes the channel height by using OTC areas but also minimizes the net lengths. Our algorithm is novel in two respects, one, we compute the track bound for each net and the other is that we use 45^0 segments for routing. We have implemented our router in C on SUN Sparc 1+ workstation and tested it on MCNC benchmarks Primary I and II. Experimental results show that WILMA3 can achieve results which are 75% better (on the average) than the conventional two layer channel router, 66% better than two layer OTC router and 56% better than three layer greedy channel router. We have also developed a two-layer version of our routing algorithm called WILMA2. Experimental results with WILMA2 indicate that, WILMA2 can produce routings, which are comparable to existing 2-layer over-the-cell routers. Both WILMA2 and WILMA3 achieve these results, while maintaining the net length bound.

1 Introduction

Several algorithms for over-the-cell channel routing have been presented, and the technique has been proven to be very effective [2, 4, 6]. In [4], net classification schemes are used to indicate the relative ease with which nets may be routed over the cells, and it is shown that these schemes lead to an average reduction in channel height of 35% as compared to a traditional two-layer channel router [7]. Despite the dramatic performance of OTC routers, a major shortcoming of the existing routers is the increase in the total wire length and the length of the longest net. Careful analysis of existing results shows that the total wire length may be increased by as much as 35% in [4]. Although no results on wire length are reported, it is very likely that the net length also increases in case of [6]. However, it is possible that the net length in [6] is less than the corresponding net lengths reported in [4]. This may be due to the fact that the main objective of their router is to minimize the number of routing tracks used in the over-the-cell area, as well as in the channel. However, no direct attempt at wire length minimization is presented, and it is possible that critical wires may be rather long in completed over-the-cell routings.

In this paper, we present a new three layer over-the-cell channel routing algorithm, WILMA3 (WIre Length Minimization Algorithm), for high-performance circuits. This router not only minimizes the channel height by OTC routing but also minimizes the net lengths. Our algorithm is novel in two aspects. First, we optimize the track assignment of each net with respect to delay. We compute the track bound for each net which ensures that the wire length is no greater than the length of the net if routed in the channel. Using this track bound, nets are selected for OTC routing. For this purpose, we develop an $O(dn)$ time algorithm for finding the optimal subset of nets, where d is the density and n is the number of nets. Second, we use 45° segments to route the nets over-the-cells to further reduce the net length.

We have also developed a two-layer router, WILMA2, which produces routings which are comparable to existing 2-layer OTC routers. Both WILMA2 and WILMA3 achieve reduction in channel height while maintaining the net length bound. That is, no net is routed with length greater than its length in the 2 layer channel routing. This puts a bound on net lengths which is an important factor in high performance circuits.

2 Algorithm WILMA3

Algorithm WILMA3 consists of eight basic steps which may be described, briefly, as follows.

1. **Net Decomposition & Classification:** We decompose all m-terminal nets into exactly $m-1$ two-terminal nets at adjacent terminal locations. Clearly, $m-1$ two-terminal nets are sufficient to preserve the connectivity of the original m-terminal net. After decomposition, we classify all nets as one of the three basic types as shown in Figure 1. The net decomposition and classification are done based on the method described in[4].

2. **Estimation:** In this phase of the algorithm, a channel router is used to obtain an initial density

*Research supported in part by ACM/SIGDA scholarship 1991, National Science Foundation Grant USE-9052346, and National Science Foundation Grant MIP-89-21540

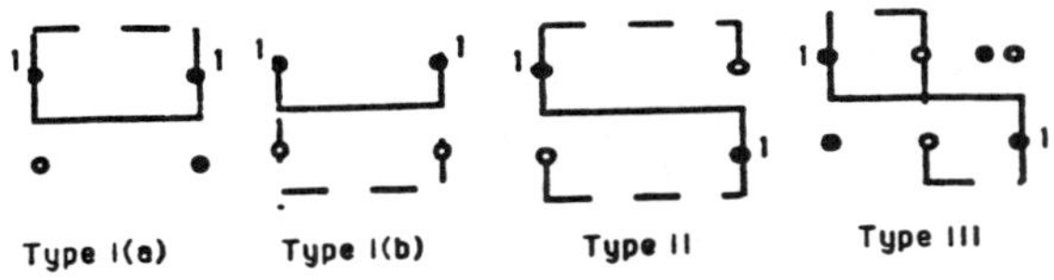

Figure 1: Three basic net types

for the input channel. For each net n_i, we record the track in which n_i is routed. This information is used to compute track bounds.

3. **Track Bound Computation:** For each net n_i, we compute the track bound k_i, which ensures that if the net is routed over-the-cell at a track less than or equal to k_i, it will have a wire length less than or equal to the net length when routed in the channel.

4. **Net Weight Assignment:** The weight of a net intuitively indicates the improvement in channel congestion possible if this net can be routed over the cells. Selecting the nets with maximum weight is necessary to achieve maximal reduction in channel height. Each two-terminal net $n_j = (l_j, r_j)$, where l_j and r_j denote the column locations of the left and right terminals of n_j respectively, is assigned a *Net Weight*. The Weight $W(n_j)$ of net n_j indicates the reduction in channel height possible if n_j is routed over the cell and can be computed based on the relative density of the channel in the interval $[l_j, r_j)$ and the relative path position of n_j in the vertical constraint graph (VCG).

5. **Net Selection with Track Bound Constraint:** Among all the nets which are suitable for routing over the cells, we select four(two) maximum-weighted, planar subsets, subject to the track bound constraint. Once the nets are selected, a set of vacant terminals (or vacant abutments in the case of Type III nets) are assigned to each net n_i depending on its type and weight. These vacant terminal/abutment locations will later be used to determine an over-the-cell routing for n_i.

6. **Over-The-Cell Routing with 45° Segments:** Perform hybrid over-the-cell routing with 45° segments and rectilinear segments. In order to avoid design rule violations, any net n_i routed over-the-cell on track t_i must contain a vertical segment of length ρ_i before 45° segments can be used.

7. **Channel Routing:** As it is possible that some net segments have not been routed in the area over the cells, channel routing must be performed to complete the routing. In algorithm WILMA2 and WILMA3, we use classical channel routers.

   After the channel routing is done the channel density(d_θ) due to hybrid over-the-cell routing of nets is obtained. If $d_\theta > d_o$, we set $d_o = d_\theta$ and goto Step 3.

The iterative process mentioned above takes place very rarely as for most examples our algorithm can complete the routing using no more tracks than d_o. However, in real cases when d_θ is indeed greater than d_o we have noticed that it is usually one or atmost two tracks.

Steps 3,5 and 6 are our contribution. We will discuss these steps in detail in the following sections.

3 Track Bound Computation

In WILMA3, the channel density for over-the-cell routing and channel routing, d_c and d_o respectively, are estimated. Let δ be the track separation and θ the orientation of the angular segments used in routing with respect to the horizontal. For a net n_i, there are three possibilities with respect to track assignment: in the channel(t_i^c), in the upper over-the-cell area(t_i^u), or in the lower over-the-cell area(t_i^l). It is possible that the horizontal segment h_i of the net routed over the cell could be split into two segments, one routed on the upper cell at track k_i^u and the other routed in the lower cell at track k_i^l as in Type III nets. For the purpose of finding the track bound, we are interested in the total length of the net when routed in the channel denoted by L_i^c, the total length when routed over-the-cell denoted by L_i^o and the total length when routed over-the-cell using angular segments denoted by L_i^θ.

The length L_i^c of a net $n_i = (l_i, r_i)$ is given by the sum of the length of horizontal segment h_i and length of vertical segment v_i. The length of the net when routed in the channel is given by $L_i^c = h_i + v_i$. The length of the horizontal segment h_i, is equal to $(r_i - l_i)$. The length of the vertical segment depends on whether the terminals (l_i, r_i) lie on the top or bottom of the channel. Thus v_i is given by:

$$v_i = \begin{cases} 2t_i^c & if\ (l_i, r_i)\ on\ R_t \\ 2(d_c - t_i^c) & if\ (l_i, r_i)\ on\ R_b \end{cases}$$

Without loss of generality, with respect to v_i, we will assume (l_i, r_i) to be on R_t. Hence, the length of a net n_i routed in the channel is given by $L_i^c = h_i + 2t_i^c$. This would be the same irrespective of the net types. Whereas, the L_i^o and L_I^θ would vary depending on the type of net.

Since the length increase due to OTC routing varies with the type of net, we identify for each type of net the track bound in terms of the above mentioned parameters.

Lemma 1 *Given a net n_i of Type I(a), $L_i^\theta \le L_i^c$ if*

$$k_i \le \frac{t_i^c + 3\lambda\sin\theta + 3\lambda\cos\theta - 3\lambda}{3\lambda\sin\theta + 3\lambda\cos\theta + \frac{\delta}{\cos\theta} - \delta\tan\theta - 3\lambda}$$

Lemma 2 *Given a net n_i of Type I(b), $L_i^\theta \le L_i^c$ if*

$$k_i \le \frac{t_i^c + 3\lambda\sin\theta + 3\lambda\cos\theta - 3\lambda - 2d_o}{3\lambda\sin\theta + 3\lambda\cos\theta + \frac{\delta}{\cos\theta} - \delta\tan\theta - 3\lambda}$$

Lemma 3 *Given a net n_i of Type II, $L_i^\theta \le L_i^c$ if*

$$k_i \le \frac{3\lambda sin\theta + 3\lambda cos\theta - 3\lambda - \frac{d_o}{2} + \frac{d_c}{2}}{3\lambda sin\theta + 3\lambda cos\theta + \frac{\delta}{cos\theta} - \delta tan\theta - 3\lambda}$$

Lemma 4 *Given a net n_i of Type III, $L_i^\theta \le L_i^c$ if*

$$k_i^u + k_i^l \le \frac{6\lambda sin\theta + 6\lambda cos\theta - 6\lambda - \frac{d_o}{2} + \frac{d_c}{2}}{3\lambda sin\theta + 3\lambda cos\theta + \frac{\delta}{cos\theta} - \delta tan\theta - 3\lambda}.$$

4 Net Selection with Track Bound Constraint

We present a new algorithm for finding the independent set in a unweighted graph with track bound constraint. This algorithm runs in $O(dn)$ time, where d is the density of nets and n is the number of nets. For finding the weighted independent set we can use the $O(kn^2)$ time algorithm presented in [3].

Consider a set $\mathcal{I} = \{I_1, \ldots, I_n\}$ of intervals, where $I_i = (l_i, r_i)$ is specified by its two end-points. The *left point* l_i and the *right point* $r_i, l_i < r_i$. We assume all end-points are distinct real numbers (the assumption simplifies our analysis and can be trivially removed). Two intervals $I_i = (l_i, r_i)$ and $I_j = (l_j, r_j)$ are *independent* if $r_i < l_j$ or $r_j < l_i$; otherwise, they are *dependent*. Two dependent intervals are *crossing* if $l_i < l_j < r_i < r_j$ or $l_j < l_i < r_j < r_i$. If $l_i < l_j < r_j < r_i$ then we say I_i *contains* I_j.

Assume track $i+1$ is above track i, $1 \le i \le t-1$. Given $\mathcal{I}$ and $\mathcal{C}$, the restricted track assignment problem (RTAP) is the problem of assigning intervals into tracks 1 to t such that:

p1) In each track, intervals are pairwise independent.

p2) If an interval I_i contains another interval I_j then $\tau_i > \tau_j$, where τ_a is the track to which I_a is assigned.

RTAP has been studied by Sarrafzadeh-Lee [8]. They showed it is NP-hard to minimize the number of tracks and proposed a 2-approximation algorithm running in $O(nlogn)$ time. That is a restricted tracks assignment in less than $2d$ tracks, where d is the problem density.

Lemma 5 *An unweighted instance $\mathcal{I}$ of BMIS-K for overlap graphs can be solved in $O(dn + n\log n)$ time, where d is the density of $\mathcal{I}$ and $n = |\mathcal{I}|$.*

Note that the proposed technique cannot be used for the weighted version of the problem, for even one interval in higher tracks may have a large weight and thus replaces (some or all) the previous intervals (the argument in Case 1 does not hold any more). The weighted version can be readily solved employing a straight-forward modification of the algorithm proposed by [2].

Lemma 6 *An arbitrary instance $\mathcal{I}$ of BMIS-K for overlap graphs can be solved in $O(Kn^2)$ time, where d is the density of $\mathcal{I}$ and $n = |\mathcal{I}|$.*

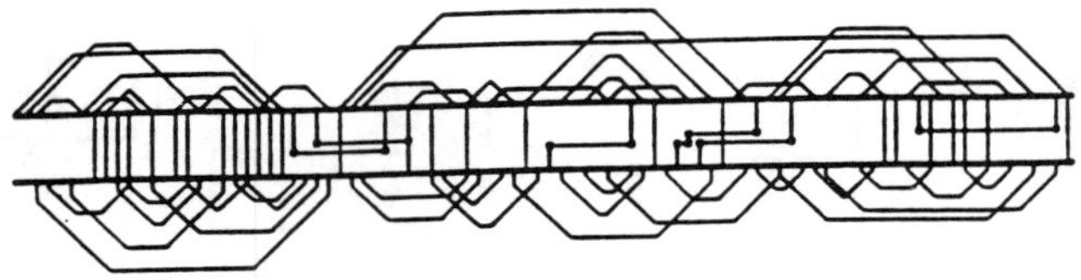

Figure 2: Portion of a channel of Primary I routed by WILMA3

5 Over-The-Cell Routing with 45° Segments

To compare the routing due to 45° segments with those at various angles to the horizontal, we analyze the corresponding lengths of a two terminal net: The rectilinear routing is the one with $\theta = 90^o$. If v_i is the height of the vertical segment and h_i is the length of the horizontal segment, the length due to rectilinear routing is $2v_i + h_i$ whereas, the length due to routing for $\theta = 45^o$ is $0.12v_i + h_i$. Although, smaller angles yield a smaller length, the routing becomes difficult due to the proximity of the terminals. Therefore, we use 45° routing.

To perform hybrid routing for a net n_i with terminals l_i and r_i at track k_i, we first compute the stepper ρ_k^i for that track by calculating $(k_i - 1) \times 3\lambda \cos\theta$. From each of these terminals, a vertical segment of length ρ_k^i is routed over the cell followed by a angular segment at angle 45° to the horizontal with inclination towards the terminal pair, until it reaches the track k_i after which these two angular segments are connected by a horizontal segment on the track.

6 Experimental Results

Algorithm WILMA3 has been implemented on a SPARC 1+ workstation in C, and has been tested on several benchmarks including Primary I and Primary II. For an example, we show the performance of WILMA3 on the Primary I benchmark. The placement of Primary I was obtained from TimberWolfSC Version 5.1 [5], and the global routing is from [1].

Results for the Primary I, indicating the number of horizontal tracks used in the channel are given in Table I in comparison with results obtained by a greedy channel router and a conventional channel router. As can be seen in Table I, WILMA3 reduces the channel height by an average of 52% as compared to best available three-layer greedy channel router. On an average, WILMA3 performs 75% better than a conventional two-layer channel router, 66% better than two-layer OTC router, and 56% better than three layer greedy channel router.

In Figure 2, we show a portion of channel 14 of Primary I, routed by WILMA3. As can be seen, that while preserving the channel reduction due to OTC routing, it reduces the net lengths considerably by utilizing 45° segments for routing over the cell.

7 Algorithm WILMA2

The algorithm WILMA2 differs from WILMA3 in two aspects: net selection and over the cell routing. Therefore all phases of WILMA2 are similar to WILMA3 except Phase 5 and 6. In Phase 5, instead of finding four independent sets, we find only two independent sets, one for the top and other for the bottom over-the-cell area of the channel. In Phase 6; the routing is done for just one layer, that is, M2. Because of the reduction in the complexity of algorithm, WILMA2, has better running times than WILMA3. (We omit details for sake of brevity). The experimental results for WILMA2 are shown in Table II.

Acknowledgements

We would like to acknowledge Ken Wan and Roshan Gidwani of the CAD Technology & Systems group at AMD for their feedback and support.

References

[1] C. Chiang, M. Sarrafzadeh, and C. K. Wong, "A Global Router Based on Steiner Min-Max Trees," ICCAD-89, pp. 2-5.

[2] J. Cong, C.L.Liu, "On the Over-the-cell Channel Routing Problem," *IEEE Trans. on Computer Aided Design,* April 1990, pp. 408-418.

[3] J. Cong, B. Preas, and C. L. Liu, "General Models and Algorithms for Over-the-Cell Routing in Standard Cell Design," *Proc. 27th ACM/IEEE Design Automation Conference,* June 1990, pp. 709-715.

[4] N. Holmes, N. Sherwani, and M. Sarrafzadeh, "New Algorithm for Over-the-Cell Channel Routing Using Vacant Terminals," *Proc. 28th ACM/IEEE Design Automation Conference,* June 1991, pp. 126-131.

[5] K.-W. Lee and C. Sechen, "A New Global Router for Row-Based Layout," *Proceeding of IEEE International Conference on Computer-Aided Design,* November 7-10, 1988, pp. 180-183.

[6] M.-S. Lin, H.-W. Perng, C.-Y. Hwang, and Y.-L. Lin, "Channel Density Reduction by Routing Over the Cells," *Proc. 28th ACM/IEEE Design Automation Conference,* IEEE-91, pp. 120-125.

[7] R. L. Rivest and C. M. Fiduccia, "A 'Greedy' Channel Router," *Proc. 19th ACM/IEEE Design Automation Conf.,* IEEE-82, pp. 418-424.

[8] M. Sarrafzadeh and D.T Lee, "Restricted Track Assignment with Application," *Technical Report # AC-91-03,* Northwestern University, October 1991.

TABLE I: WILMA3
EXPERIMENTAL RESULTS FOR PRIMARY I
CHANNEL HEIGHT

Chan.	No. of Tracks Produced				% Impro. by WILMA3		
	2CR	2OT	3CR	WILMA3	2CR	2OT	3CR
1	11	5	6	2	82	60	66
2	16	12	9	5	69	58	44
3	21	16	11	6	71	62	45
4	24	21	15	9	62	57	40
5	21	17	11	7	67	58	36
6	23	18	12	7	69	61	41
7	22	14	14	5	77	64	64
8	24	18	16	7	71	61	56
9	21	16	13	6	71	62	53
10	15	11	8	5	67	54	37
11	17	12	12	5	70	58	58
12	15	11	8	4	73	64	50
13	13	9	8	3	77	67	62
14	13	9	7	4	69	55	42
15	11	7	6	2	81	71	66
16	11	7	6	2	81	71	66
17	14	8	7	3	78	62	57
18	6	3	3	1	83	66	66
Total	298	214	172	83	72	61	52

TABLE II : WILMA2
EXPERIMENTAL RESULTS FOR PRIMARY I
CHANNEL HEIGHT

Chan.	No. of Tracks Produced			% Impro. by WILMA2	
	2CR	2OT	WILMA2	2CR	2OT
1	11	5	2	82	60
2	16	12	11	31	8
3	21	16	16	23	0
4	24	21	20	16	5
5	21	17	17	19	0
6	23	18	17	26	5
7	22	14	13	41	7
8	24	18	17	29	5
9	21	16	16	24	0
10	15	11	10	33	10
11	17	12	11	35	9
12	15	11	10	26	9
13	13	9	9	31	0
14	13	9	9	31	0
15	11	7	6	45	14
16	11	7	6	45	14
17	14	8	8	43	0
18	6	3	3	50	0
Total	298	214	201	33	6

Over-the-Cell Routers for New Cell Model*

Bo Wu, Naveed A. Sherwani
Dept. of CS
Western Michigan Univ.
Kalamazoo, MI 49008

Nancy D. Holmes
Dept. of Info. & CS
Univ. of California
Irvine, CA 92717

Majid Sarrafzadeh
Dept. EE & CS
Northwestern Univ.
Evanston, IL 60108

Abstract

In this paper, we present new over-the-cell routing techniques for standard cell design style. We have developed both a two-layer and a three-layer router. The key feature of our routers is use of new cell model, in which the terminals are located in the middle of the cells in the second metal layer. This model is similar to one currently being developed for three-layer cell libraries in industry.

We have implemented both of our routers in C on a SUN SPARCstation 1+ and tested them with several benchmarks including PRIMARY I and PRIMARY II from MCNC. Both of our routers out-perform all existing routers. In addition, both ICR-2 and ICR-3 are very fast, for example, ICR-3 completes the routing of entire PRIMARY I benchmark in 5.8 secs.

1 Introduction

Recently, the standard cell design style has increased in popularity as it offers a good compromise between time-to-market and area. Several researchers have investigated the use of area over the standard cells to obtain further reduction in channel height [2, 4, 5, 6, 8, 11], and obtained up to a 35% reduction in channel height as compared to conventional channel routers [5]. This routing technique is referred to as *over-the-cell* routing.

Despite these excellent results, all existing over-the-cell routers suffer from one major drawback. They use a cell layout style (cell model) which assumes that terminals are available in all layers on the cell boundary. This cell model does not correspond to the new three-layer cell model currently under development in industry, as most industrial models provide terminals in the second metal layer (M2) in the middle of the cell. This difference between existing and new cell models makes adaptation of existing over-the-cell routers to suit the new cell model difficult.

In this paper, we develop an new cell model and two corresponding over-the-cell routing algorithms (ICR-2 and ICR-3). Our three-layer cell model is representative of the three-layer cell libraries currently under development in industry, and hence the ICR-3 router is suitable for industrial designs.

*Research supported in part by ACM/SIGDA scholarship 1991, National Science Foundation Grant USE-9052346, and National Science Foundation Grant MIP-89-21540

2 The Existing and New Cell Models

In this section, we describe the differences between the existing and the new cell model and their effects on routing techniques.

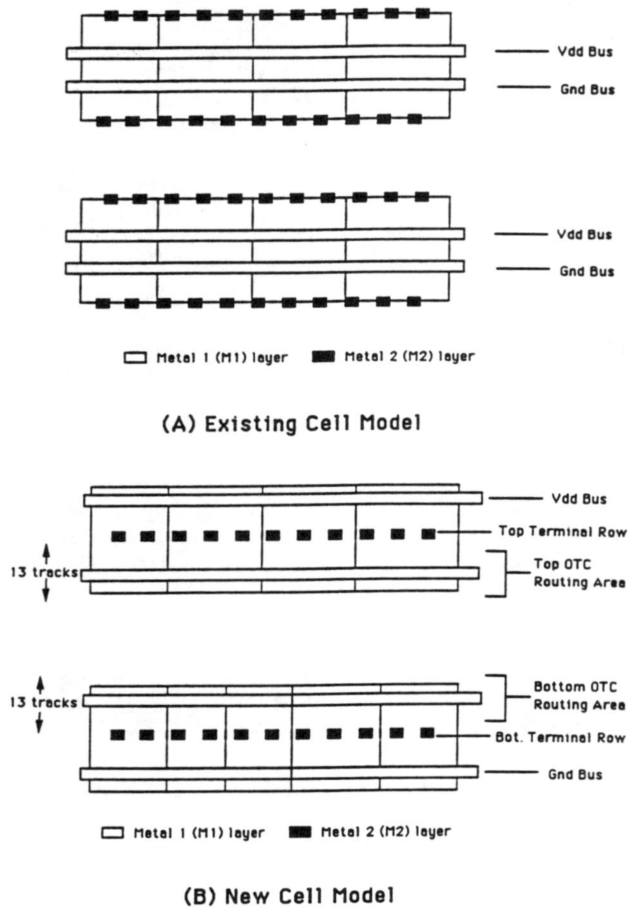

Figure 1: Cell Models.

2.1 Existing Model

In the traditional cell layout style, there are two parallel horizontal diffusion rows, one for the P-type transistors and the other for N-type transistors. The first metal layer (M1) is used to complete connections which are internal to the cells. Feedthrough routing is also done using the M1 layer. Power and ground lines are routed in the second metal layer (M2) in the center of the area over-the-cell rows. Terminal rows are available in all layers and are located on the boundaries of the cells [5]. This model is used by most existing over-the-cell routers [2, 5, 8].

2.2 New Cell Model

The cell model currently being adopted by the industry for the three metal layer process is quite different in terms of terminal locations than the existing cell model described above. In the new model, the terminals are located in M2 in the middle of the cells. The power and ground rails are in M1 near the top and bottom cell boundaries respectively. The connections within the cell are completed in M1. Thus, M2 is only blocked by terminals, and M3 is completely unblocked. Over-the-cell routers may use two rectangular regions (about thirteen tracks wide) in M2 and M3. Vias may or may not be allowed in over-the-cells areas depending upon the process.

The key differences between this model and the existing model is the location of terminals and the assumption of availability of terminals in all metal layers. We observe that these differences make it very difficult to adapt existing over-the-cell routers to new cell models. Examples of each of these models are shown in Fig.1.

3 Algorithm ICR-3

In this section, we present our three-layer over-the-cell router ICR-3. The basic procedure of our two-layer router ICR-2 is similar to ICR-3.

In three metal layer environment, M2 and M3 layers are available for routing over the cells, as it is assumed that M1 is used for routing within each cell. Although two layers are available for routing, their use is somewhat restricted due to the restriction on the use of vias. We do not use vias in the *Over-The-Cell (OTC)* area, as many technologies forbid use of vias over cells and active elements due to fabrication problems. As a result, the routing in OTC areas on both layers must be planar. The terminals cannot be 'brought-up' to M3, as that would require a via, and wires from all the terminals must be routed on M2. However, vias may be used to complete the connections in the channel. This planarity condition is the most important consideration in development of OTC routing algorithms for the new cell model. The basic steps of the ICR-3 algorithm are as follows and a routing example in given in Fig.2.

1. **Net Classification:** In this step, all the nets are classified into three types. *TYPE I nets* are the nets whose terminals are on a same terminal row. *TYPE II nets* are two terminal nets whose terminals are on different terminal rows (One on the top terminal row and one on the bottom terminal row). *TYPE III nets* are multi-terminal nets whose terminals are on both top and bottom terminal rows.

2. **Net Weighting and Net Selection for M2 Layer:** In this step, all TYPE II nets are assigned weights. The weight of a net is based on several factors, such as, criticality of the net, its contribution to channel congestion, among others. Based on the weights assigned to the nets, we find a maximum independent set of the nets. We call them *critical nets*. All critical nets are routed in M2 and pass through the channel as straight wires. These nets partition M2 into several *regions*, and do not contribute to channel density.

3. **Decomposition of Multi-terminal Nets:** After obtaining the initial regions, we partition the multi-terminal nets which have terminals on both top and bottom terminal rows (TYPE III nets). The objective of this operation is to increase the number of wires which can be routed straight in the channel on M2.

4. **Terminal Assignments of Critical Nets:** After the routing regions in M2 is topologically fixed, we assign the geometric terminal positions to the critical nets, to obtain the geometric definitions of the regions. This is done to maximize the routability of the nets in the M2, as well as to satisfy the congestion requirement of each region. We develop an optimal algorithm for this problem based on dynamic programming [12].

5. **Terminal Assignments of Non-Critical Nets:** Within each region, we assign the terminal positions to the remaining nets, to achieve three objectives: First, to eliminate vertical constraints, second, to minimize horizontal constraints and third to maximize the maximum bipartite subgraph in a special graph called overlap graph defined by the intervals of the nets.

6. **M2 River Routing:** After all the terminal positions are fixed, we use a river router to route the nets in a planar fashion in M2 OTC area.

7. **Net Selection for M3 Layer:** Among the remaining nets, we select a set of nets which forms a maximum bipartite subgraph in a circle graph. Since this problem is known to be NP-Hard, we use an approximate algorithm.

8. **M3 OTC Routing:** In this step, the nets selected in the previous step are routed in a planar fashion in M3 OTC areas on both sides of the channel. This step is similar to existing OTC routers.

9. **Channel Routing:** We use a HVH routing model to route the remaining nets in the channel.

It must be noted that since all vertical constraints are eliminated, we use a router based on Left-Edge Algorithm, adapted for three layer routing environment.

10. **Clean up Routing:** Finally, we consider all multi-terminal nets and re-compose the net as much all possible. This step removes the redundant wiring and improves the layout.

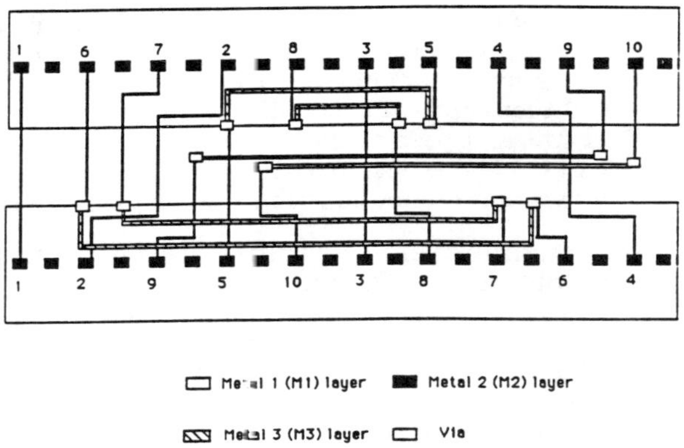

Figure 2: A Routing Example Created by ICR-3.

4 The Performance Bounds of ICR-3

In this section, we present the performance bounds of ICR-3 in terms of channel height and number of vias.

Theorem 1 *In the layout created by ICR-3, the number of vias is $T + B - 2K$, where T is the total number of top terminals, while B is the total number of bottom terminals, and K is the size of the Maximum Independent Set of critical nets for M2.*

Proof: Omitted for sake of brevity. □

Theorem 2 *In the layout created by ICR-3, the number of tracks needed in the channel is $H_{max}/2$, where H_{max} is the size of the maximum clique in the Horizontal Constraint graph defined by the nets in set $L - MIS_2 - MIS_3$. Here, L is a given set of nets, MIS_2 is the Maximum Independent Set of critical nets for M2, and MIS_3 is the two Maximum Independent Set of nets for M3.*

Proof: Omitted for sake of brevity. □

5 Experimental Results

Both of ICR-3 and ICR-2 have been implemented on a SUN SPARCstation 1+ in the C programming language. They were tested extensively on benchmark examples including PRIMARY I and PRIMARY II. For an example, we show the performance of ICR-3 on the PRIMARY I benchmark in comparison with a conventional two-layer router [9], a two-layer over-the-cell router [2], and a conventional three-layer channel router [3]. The placement of the PRIMARY I example was obtained from TimberWolfSC [7], and the global routing is from [1].

Algorithm ICR-3 on the average, produces a 73% reduction in channel height as compared to a conventional two-layer channel (2CRP) router, a 58% reduction as compared to a two-layer over-the-cell (2OTC) router, and a 47% reduction as compared to a conventional three-layer channel (3CRP) router. For PRIMARY I, it produces a routing which are 65%, 51% and 40% better than 2CRP, 2OTC and 3CRP routers, respectively. The channel heights obtained by 2CRP, 2OTC, 3CRP and ICR-3 for the channels of PRIMARY I is listed in TABLE 1. An example routing of channel 14 of PRIMARY I is shown in Fig.3.

TABLE 1
Experimental Results for PRIMARY 1

Chan.	No. of Tracks Produced			
No.	2CRP	2OTC	3CRP	ICR-3
1	11	5	6	3
2	16	12	9	6
3	21	16	11	8
4	24	21	15	10
5	21	17	11	8
6	23	18	12	10
7	22	14	14	7
8	24	18	16	9
9	21	16	13	7
10	15	11	8	7
11	17	12	12	6
12	15	11	8	5
13	13	9	8	4
14	13	9	7	4
15	11	7	6	3
16	11	7	6	3
17	14	8	7	3
18	6	3	3	1
Total	298	214	172	104

6 Conclusions

In this paper, we have developed a two layer and a three layer over-the-cell router. The key feature of these routers is use of new cell model. Both of our routers out-perform all existing routers not only in terms of channel height but also in terms of vias. In addition, our routers are very efficient and can route real world channels in few seconds.

7 Acknowledgements

We would like to acknowledge Ken Wan and Roshan Gidwani of the CAD Technology & Systems group at Advanced Micro Devices for their feedback and support.

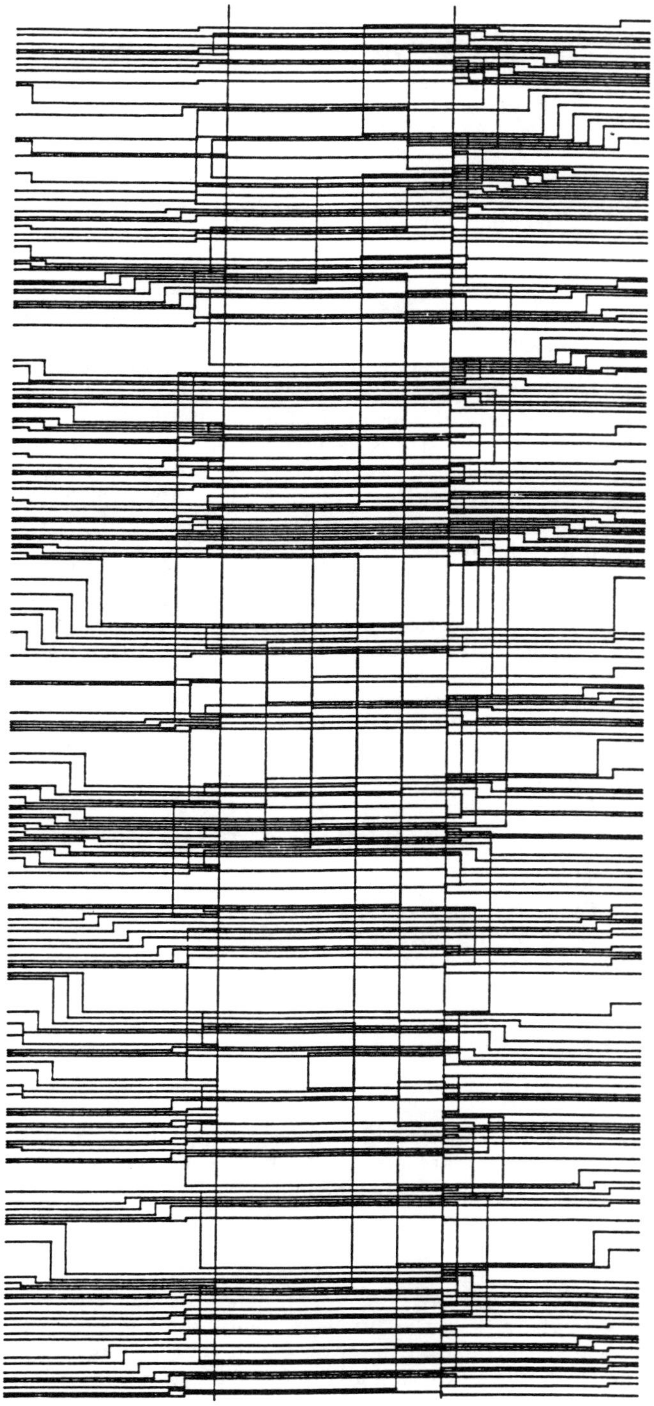

Figure 3: The Complete Routing of Channel 14 in PRIMARY I Created by ICR-3.

References

[1] C. Chiang, M. Sarrafzadeh, and C. K. Wong, "A Global Router Based on Steiner Min-Max Trees," *Proc. Intl. Conference on Computer-Aided Design,* pp. 2-5, 1989.

[2] J. Cong, B. Preas, and C. L. Liu, "General Models and Algorithms for Over-the-Cell Routing in Standard Cell Design," *Proc. 27th ACM/IEEE Design Automation Conference,* pp. 709-715, 1990.

[3] J. Cong, D. F. Wong, and C. L. Liu, "A New Approach to the Three Layer Channel Routing Problem," *Proc. of Intl. Conference on Computer-Aided Design,* pp. 378-381, 1987.

[4] G. Gudmundsson and S. Ntafos, "Channel Routing with Superterminals," *Proc. 25th Allerton Conference on Computing, Control and Communication,* pp. 375-376, 1987.

[5] N. Holmes, N. Sherwani, and M. Sarrafzadeh, "New Algorithm for Over-the-Cell Channel Routing Using Vacant Terminals," *Proc. 28th ACM/IEEE Design Automation Conference,* pp. 126-131, 1991.

[6] H. E. Krohn, "An Over-the-Cell Gate Array Channel Router," *Proc. 20th ACM/IEEE Design Automation Conference,* pp. 665-670, 1983.

[7] K.-W. Lee and C. Sechen, "A New Global Router for Row-Based Layout," *Proc. of IEEE Intl. Conference on Computer-Aided Design,* pp. 180-183, 1988.

[8] M.-S. Lin, H.-W. Perng, C.-Y. Hwang, and Y.-L. Lin, "Channel Density Reduction by Routing Over the Cells," *Proc. 28th ACM/IEEE Design Automation Conference,* pp. 120-125, 1991.

[9] R. L. Rivest and C. M. Fiduccia, "A 'Greedy' Channel Router," *Proc. 19th ACM/IEEE Design Automation Conference,* pp. 418-424, 1982.

[10] M. Sarrafzadeh and R. D. Lou, "Maximum Weighted K-Coverings in Transitive Graphs," *Proc. 1990 IEEE Intl. Symp. on Circuits and Syst.*; also to appear in *Algorithmica.*

[11] Y. Shiraishi and Y. Sakemi, "A Permeation Router," *IEEE Trans. on Computer Aided Design,* Vol. CAD-6, No. 3, pp. 462-471, May 1987.

[12] B. Wu, N. A. Sherwani, N. D. Holmes and M. Sarrafzadeh, "A Three Layer Over-the-Cell Router for New Cell Model," *Technical Report, Dept. of Computer Science, Western Michigan Univ.,* TR91-35.

Edge-Valued Binary Decision Diagrams for Multi-Level Hierarchical Verification

Yung-Te Lai and Sarma Sastry

Electrical Engineering - Systems Department
University of Southern California, Los Angeles, CA 90089-2562

Abstract

In this paper we present a new data structure called *edge-valued binary decision diagrams* (EV) as a representation of functions. An EV is an extension of ordered binary decision diagrams that allows for multi-level and hierarchical verification. We show that an EV is a compact and canonical representation for arbitrary integer functions. Hence the specification can be at a higher level than the implementation. Furthermore, the variable ordering strategy for an EV can be derived from a higher level functional specification instead of the gate level specification. Examples shown in this paper includes SN74L85[7], SN74181[7], a 64-bit comparator built from SN74L85 and a 64-bit adder built from SN74181.

1 Introduction

Ordered Binary Decision diagrams (OBDDs), proposed by Bryant [2], are a canonical and compact representation of boolean functions. Since their introduction, many improvements and extensions have been reported. Examples include sharing and attributed edge [8], complemented edges [8, 1], breadth-first manipulation [9] and heuristic variable ordering strategies [4, 6, 8]. Besides being able to show equivalence of boolean functions, OBDDs have also been used in multi-level logic synthesis [10], resynthesis for network optimization [5] and verification of finite state machines [3, 12].

MDDs [11] are an extension of OBDDs which allow non-terminal nodes to have more than two children and terminal nodes to assume integer values. Although the methods using MDDs [1, 11] work for arbitrary discrete functions, MDDs are too general to capture certain properties of functions in a specific domain, such as arithmetic functions.

In this paper, we present a new extension to OBDDs called *edge-valued binary decision diagrams* (EV) which allow for efficient manipulation of arithmetic functions. An EV provides some additional advantages over OBDDs. First, an EV can be derived from a higher level specification (e.g., '$x + y$' as a specification of an adder) instead of requiring a vector of boolean expressions. Second, EVs allow for hierarchical verification (e.g., designs need not be flattened).

EVs preserve both the canonical and compactness properties from OBDDs. An EV usually is much more compact than a OBDD when it is used to describe an arithmetic function. For example, an EV requires only 129 nodes to represent the specification of a 64-bit adder. An EV shares some of the disadvantages as an OBDD, namely, the number of nodes may grow exponentially and the size is very sensitive to the ordering of input variables. However, since we can specify a design at a higher level, we can order the input variables based on the higher level specification instead of the structure of a design.

In Section 2 we introduce EVs. In Section 3 we show how to use EVs to carry out verification and synthesis. In Section 4 we describe an extension of EVs, called *structural EV* (SEV) which allows the specification of conditional expressions (control signals) directly. Hierarchical verification using SEVs is described in Section 5. In Section 6 we describe an ordering strategy for SEVs based on the function of the circuit. Note: Proofs are omitted due to limitations on space.

2 Edge-Valued Binary Decision Diagrams

Definition 2.1 Given a set of binary variables $\mathbf{V}$ and a total ordering on these variables, an EV is a tuple $\langle c, \mathbf{f} \rangle$ where c is a constant integer and $\mathbf{f}$ is an element of the set S which is defined recursively as:

1. $\mathbf{0}$ is in S.
2. $\langle x, \mathbf{l}, \mathbf{r}, v \rangle$ is in S if the following conditions hold:
   (a) All the variables, if they exist, in $\mathbf{l} \in S$ and $\mathbf{r} \in S$ are greater than $x \in \mathbf{V}$ with respect to the given ordering.
   (b) It is not the case that $\mathbf{l} \equiv \mathbf{r}$ and $v = 0$ where v is an integer.

Definition 2.2 An EV $\langle c, \mathbf{f} \rangle$ denotes the function $c + f$ where f is the function denoted by $\mathbf{f} \in S$. $\mathbf{0}$ denotes the constant function 0 and $\langle x, \mathbf{l}, \mathbf{r}, v \rangle$ denotes the arithmetic function $x(v + l) + (1 - x)r$.

In the graphical representation of an EV $\langle c, \mathbf{f} \rangle$, $\mathbf{f}$ is represented by a rooted, directed, acyclic graph and c by a *dangling* incoming edge to the root node of $\mathbf{f}$. $\mathbf{0}$ is depicted by a square node labelled 0 and this is the only terminal node. A non-terminal node is a quadruple $\langle x, \mathbf{l}, \mathbf{r}, v \rangle$, where x is the node label, $\mathbf{l}$ and $\mathbf{r}$ are the two subgraphs (called *left* and *right* children) rooted at x, and v is the label assigned to the left edge of x. Figure 1 shows the EVs for the *sum* and *carry* function which implement a full adder. If x, y and z are boolean variables, the carry function is $(x \wedge y) \vee (y \wedge z) \vee (z \wedge y)$. If x, y and z are viewed as integers then the carry function is $xy + yz + zx - 2xyz$, whose EV is shown in Figure 1a. The function *carry* can be derived from the following equations:

$$
\begin{array}{|lcl|}
\hline
carry & = & 0 + f_x \\
f_x & = & x(0 + f_{y_l}) + (1-x) f_{y_r} \\
f_{y_l} & = & y(1+0) + (1-y) f_z \\
f_{y_r} & = & y(0 + f_z) + (1-y)0 \\
f_z & = & z(1+0) + (1-z)0 \\
\hline
\end{array}
$$

An EV can represent both Boolean and arithmetic functions. Two examples of arithmetic functions represented by an EV are shown in Figure 2. From Figure 2, it is easy to see that it requires only n non-terminal nodes to represent an n-bit integer and $2n$ non-terminal nodes to represent the addition of two n-bit integers.

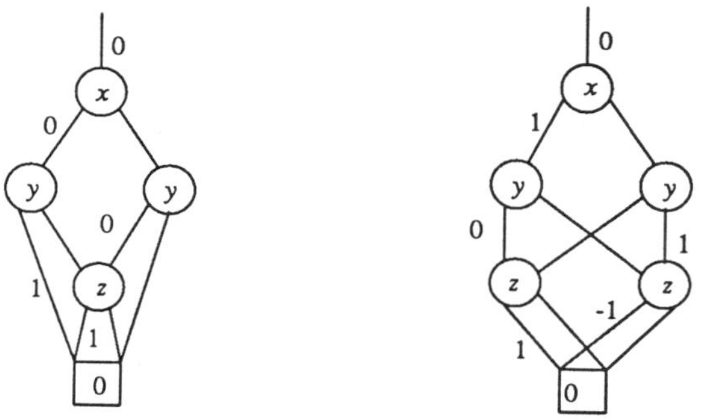

(a) carry (b) sum

Figure 1: EVs for full-adder

The EV is a canonical representation of the functions from $\{0, 1\}^n$ to the set of integers. This is stated in the following lemma.

Lemma 2.1 Two EVs $\langle c_f, \mathbf{f} \rangle$ and $\langle c_g, \mathbf{g} \rangle$ denote the same functions, $c_f + f \equiv c_g + g$, if and only if $c_f = c_g$ and $\mathbf{f}$ and $\mathbf{g}$ are isomorphic.

The following algorithm describes the function *apply* which takes $\langle c_f, \mathbf{f} \rangle, \langle c_g, \mathbf{g} \rangle$ and *op* as arguments and returns $\langle c_h, \mathbf{h} \rangle$ such that $c_h + h \equiv (c_f + f)\ op\ (c_g + g)$ where *op* can be any operator which is closed over the integers.

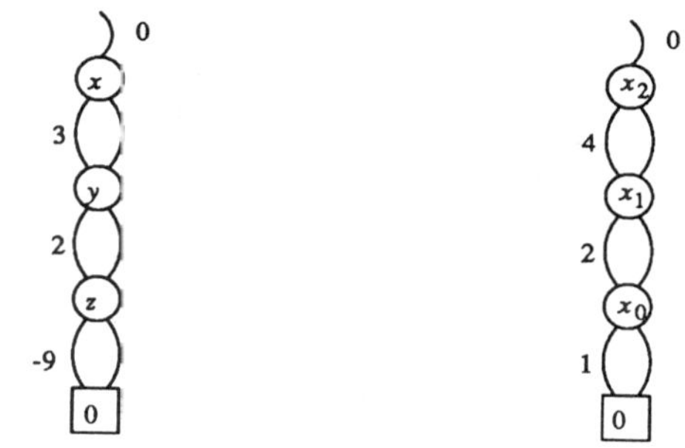

(a) (b)

Figure 2: (a) $f(x, y, z) = 3x + 2y - 9z$. (b) $g(x_0, x_1, x_2) = x_0 + 2x_1 + 4x_2$

Algorithm A:

1 $apply(\langle c_f, \mathbf{f} \rangle, \langle c_g, \mathbf{g} \rangle, op)$
2 {
3 if (*terminal case*) return($(\langle c_f, \mathbf{f} \rangle\ op\ \langle c_g, \mathbf{g} \rangle)$);
4 else if ($computed(\langle c_f, \mathbf{f} \rangle, \langle c_g, \mathbf{g} \rangle, op)$)
   return($comp\text{-}table(\langle c_f, \mathbf{f} \rangle, \langle c_g, \mathbf{g} \rangle, op)$);
5 else {
6   $var = min\_var(root(\mathbf{f}), root(\mathbf{g}), ordering)$;
7   $\langle c_l, \mathbf{h}_l \rangle = apply(L(\langle c_f, \mathbf{f} \rangle, var), L(\langle c_g, \mathbf{g} \rangle, var), op)$;
8   $\langle c_r, \mathbf{h}_r \rangle = apply(R(\langle c_f, \mathbf{f} \rangle, var), R(\langle c_g, \mathbf{g} \rangle, var), op)$;
9   if ($equal(\langle c_l, \mathbf{h}_l \rangle, \langle c_r, \mathbf{h}_r \rangle)$) $ans = \langle c_l, \mathbf{h}_l \rangle$;
10   else if ($exist(var, \mathbf{h}_l, \mathbf{h}_r, c_l - c_r)$)
     $ans = \langle c_r,\ node\text{-}table(var, \mathbf{h}_l, \mathbf{h}_r, c_l - c_r) \rangle$;
11   else {
12     $insert(node\text{-}table, create\text{-}node(var, \mathbf{h}_l, \mathbf{h}_r, c_l - c_r)$);
13     $ans = \langle c_r,\ node\text{-}table(var, \mathbf{h}_l, \mathbf{h}_r, c_l - c_r) \rangle$;
14   }
15   $insert(comp\text{-}table, \langle c_f, \mathbf{f} \rangle, \langle c_g, \mathbf{g} \rangle, op, ans)$;
16   return(ans);
17 }
18 }

In the above algorithm, terminal cases (line 3) occur when both f and g are $\mathbf{0}$, or when for example, *op* is $\times$, c_f is 1, and f is $\mathbf{0}$. The latter is a terminal case because $\langle 1, \mathbf{0} \rangle = 1 + 0 = 1$, $\langle c_g, \mathbf{g} \rangle = c_g + g$, and $1 \times (c_g + g) = (c_g + g) = \langle c_g, \mathbf{g} \rangle$, thus the result can be returned immediately without traversing the graph. Two tables are used in Algorithm A: a *comp-table* to achieve computation efficiency (entries are stored in line 15 and used in line 4), and a *node-table* to ensure there are no duplication of nodes (line 10). If it is neither one of the terminal cases nor been computed before, then the operation requires a traversal of the graph based on the ordering of the root variables associated with $\mathbf{f}$ and $\mathbf{g}$. The function $L(\langle c_f, \mathbf{f} \rangle, var)$ and $R(\langle c_f, \mathbf{f} \rangle, var)$ both return $\langle c_f, \mathbf{f} \rangle$ if $root(\mathbf{f}) > var$, otherwise they return $\langle c_f + v_f, \mathbf{f}_l \rangle$ and $\langle c_f, \mathbf{f}_r \rangle$ respectively.

After the left and right children have been computed resulting in $\langle c_l, \mathbf{h}_l \rangle$ and $\langle c_r, \mathbf{h}_r \rangle$ (lines 7 and 8), the algorithm returns the pair $\langle c_h, \mathbf{h} \rangle$ determined in lines 9, 10 or 13. Line 9 is to guarantee that the case of $\langle x_i, \mathbf{k}, \mathbf{k}, 0 \rangle$ will not occur. In lines 10 and 13, the values c_h and v_h are set to c_r and $c_l - c_r$. This is to ensure that the value of right edge remains 0.

Example 2.1 An example of $apply(\langle 0, \mathbf{f}\rangle, \langle 0, g\rangle, +)$ is shown in Figure 3. Let the variable ordering be $x_1 < x_0$. Figure 3 (a) shows the initial arguments of *apply*; (b) is the recursive call of *apply* on line 7 whose result is (c). Similarly, another call to *apply* on line 8 and its results are shown in (d) and (e) respectively. The final result is shown in (f) which is generated from either line 10 or 13. □

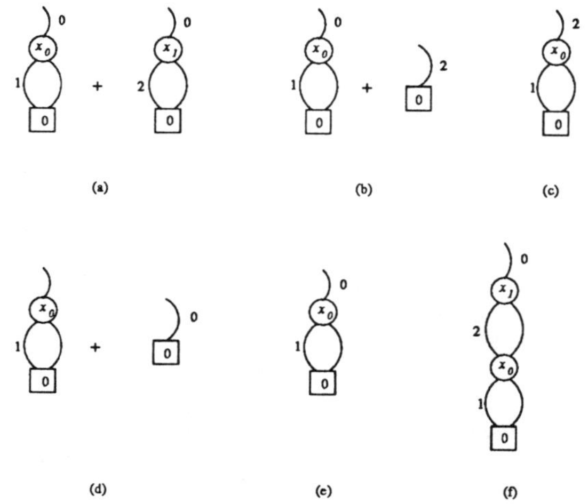

Figure 3: Snapshots of $apply(\langle 0, \mathbf{f}\rangle, \langle 0, \mathbf{g}\rangle, +)$

Since an EV is an acyclic directed graph and there is no backtracking in the above algorithm, the worst case complexity is the product of the numbers of nodes of **f** and **g**, and the size of the ranges of $c_f + f$ and $c_g + g$. In many practical applications, the number of nodes of an EV is small, but the size of the range of $c + f$ is very large. For example, to represent an n-bit integer in $\langle c, \mathbf{f}\rangle$, the number of nodes of **f** is n while $c + f$ can be any one of 2^n different values. In the following subsections, we present several properties of EVs which can be used to reduce the computation complexity of *apply* in many situations.

2.1 Additive Property

An operator *op* is said to satisfy the *additive property* if

$$(c_f + f)\ op\ (c_g + g) = (c_f\ op\ c_g) + (f\ op\ g)$$

Two such examples are + and −:

$$(c_f + f) \pm (c_g + g) = (c_f \pm c_g) + (f \pm g)$$

Because the values c_f and c_g can be separated from the functions f and g, the entries of *comp-table* can be $\langle\langle 0, \mathbf{f}\rangle, \langle 0, \mathbf{g}\rangle, op\rangle$ for this class of operators. After the computation of $\langle\langle 0, \mathbf{f}\rangle, \langle 0, \mathbf{g}\rangle, op\rangle$ resulting $\langle c_h, \mathbf{h}\rangle$, we then add $c_f\ op\ c_g$ to c_h to have the complete result of $\langle\langle c_f, \mathbf{f}\rangle, \langle c_g, \mathbf{g}\rangle, op\rangle$. Hence, every operation $\langle\langle c'_f, \mathbf{f}\rangle, \langle c'_g, \mathbf{g}\rangle, op\rangle$ can share the computation result of $\langle\langle 0, \mathbf{f}\rangle, \langle 0, \mathbf{g}\rangle, op\rangle$.

2.2 Subdomain Property

An operator *op* is said to satisfy the *subdomain property* if

$$(c_f + f)\ op\ (c_g + g) = (c'_f + f)\ op\ (c_g + g)$$

where the domain of c'_f is a subset of c_f's. An example of this is when an expression is modulo a constant:

$$(c_f + f)\ mod\ c = ((c_f\ mod\ c) + f)\ mod\ c$$

In this case, $\langle c_f + kc, f\rangle$ can share the computation result of $\langle c_f, f\rangle$ for any integer k. Thus, we reduce the size of the range of c_f to the value of c. When c is small, computation sharing is large; when c is large, then *boundary property* (next section) can be applied.

2.3 Boundary Property

The additive and subdomain properties often result in a significant savings in computation time. The *boundary property* also results in a reduction of computation time by increasing terminal cases (line 3 of procedure *apply*). As an example, consider the case of $op = mod$ and $c_g + g = c$, for some constant c. If we know the maximum value of f, then for all c_f such that $(c_f + f) < c$, we can have $(c_f + f)\ mod\ c = (c_f + f)$ without further computation.

Another example is from relational operators. Suppose the following result has been computed

$$(c_f + f) > (c_g + g) = 1$$

Then for all c'_f and c'_g such that $(c'_f - c'_g) > (c_f - c_g)$ we have $(c'_f + f) > (c'_g + g) = 1$. To handle this class of behavior, we can use $\langle\langle 0, \mathbf{f}\rangle, \langle 0, \mathbf{g}\rangle, >, c_f - c_g, 1\rangle$ as an entry of *comp-table*. Alternatively, we can keep track of the maximum and minimum value of each non-terminal node and then include the following tests in the terminal conditions for operator >.

$$\begin{array}{lcll} (c_f + min(f)) & > & (c_g + max(g)) & \Rightarrow\ 1 \\ (c_f + max(f)) & < & (c_g + min(g)) & \Rightarrow\ 0 \end{array}$$

3 Verification and Synthesis Paradigm

In this section, we show how EVs can be used to perform functional verification and synthesis. First, we show as an example $sum(x, y, z)$ and $carry(x, y, z)$ implement the full adder $x + y + z$. Given

$$\begin{array}{l} imp(x, y, z) = \langle sum(x, y, z), carry(x, y, z)\rangle \\ spec(x, y, z) = w_1 + 2w_2 \end{array}$$

we want to show $x+y+z = w_1+2w_2 = sum(x, y, z)+ 2carry(x, y, z)$. The left hand side of the above expression implemented in EV is shown in Figure 4, and the EV for the right hand side is shown in Figure 5. By comparing the two rightmost EVs in Figure 5 and Figure 4, we verify that $sum(x, y, z)$ and $carry(x, y, z)$ implement $x + y + z$.

The above procdure can be reversed to become a procedure for functional synthesis. Again, we use the full adder as an example except now the goal $imp(x, y, z)$ is not given. From the description of *spec*, we have

$sum(x, y, z)$	=	w_1	=	$fa\ mod\ 2$
$carry(x, y, z)$	=	w_2	=	$(fa - (fa\ mod\ 2))/2$

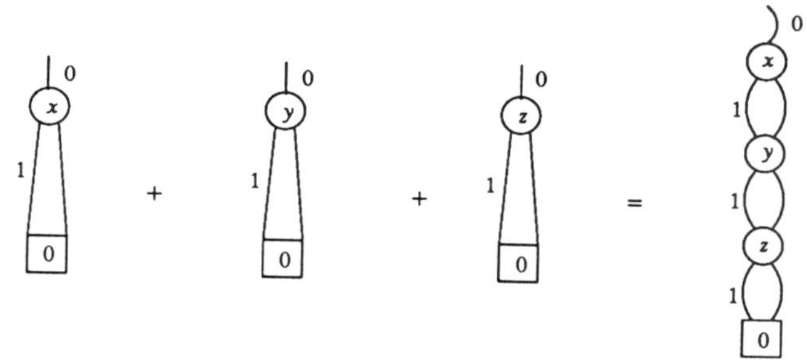

Figure 4: The EV derived from the specification of a full-adder

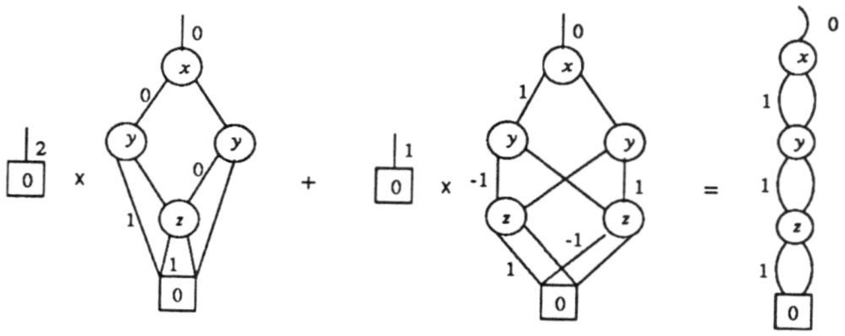

Figure 5: The EV derived from the implementation of a full-adder

where $fa = x+y+z$. The following sequence of *apply* operations on EVs then can produce *sum* and *carry* automatically.

$$\begin{aligned}
\langle 0, \mathbf{xy}\rangle &= apply(\langle 0, \mathbf{x}\rangle, \langle 0, \mathbf{y}\rangle, +);\\
\langle 0, \mathbf{fa}\rangle &= apply(\langle 0, \mathbf{z}\rangle, \langle 0, \mathbf{xy}\rangle, +);\\
\langle 0, \mathbf{sum}\rangle &= apply(\langle 0, \mathbf{fa}\rangle, \langle 2, 0\rangle, mod);\\
\langle 0, \mathbf{temp}\rangle &= apply(\langle 0, \mathbf{fa}\rangle, \langle 0, \mathbf{sum}\rangle, -);\\
\langle 0, \mathbf{carry}\rangle &= apply(\langle 0, \mathbf{temp}\rangle, \langle 2, 0\rangle, /);
\end{aligned}$$

Example 3.1 The design (IMP) is a 64-bit 3-level carry lookahead adder which has 129 inputs, 65 outputs, and 420 logic gates. The intended behavior (SPEC) is specified as:

```
unsigned(65) add64(x, y, c)
        unsigned(64) x, y;
        unsigned c;
{
        return(x + y + c);
}
```

where (64) and (65) declare the number of bits. In our experimental implementation on SUN 4, the generation of 65 EVs of IMP (575 nodes in total) takes 1.47 seconds and the generation of one EV of SPEC (129 nodes) takes 0.17 seconds. The verification process which converts 65 EVs into one, performing $2^0 \times b_0 + \ldots + 2^{64} \times b_{64}$, and then compares with the one of SPEC takes 4.48 seconds. Thus, in our implementation, it takes less than 5 seconds to show 65 Boolean expressions are really carrying out an addition. □

4 Structure Edge-Valued Binary Decision Diagrams

As shown in previous section, EVs are useful for verifying functions performed by a data path. These functions include both arithmetic functions and relational operators. To include conditionals, for verifying control paths, we define a new data structure called **SEV**, as follows:

Definition 4.1 The set SEV is defined recursively as:

1. An EV is an SEV.
2. $(p \rightarrow t; e)$ is a (*conditional*) SEV if p is an SEV with the range {0, 1}, and t and e are SEVs. The function denoted by $(p \rightarrow t; e)$ is that for any input assignment a, if $p(a) = 1$ then the function returns the value $t(a)$, otherwise it returns $e(a)$.
3. $[f_1, \ldots, f_m]$ is an SEV if $f_1, \ldots, f_m$ are SEVs. For any input assignment a, $[f_1, \ldots, f_m]$ returns the vector $\langle f_1(a), \ldots, f_m(a)\rangle$.

To develop the notion of a canonical SEV, we need to define reduction rules. To this end, we use two operators, *ite* and *cofact* which can transform the forms of SEVs without changing the functions they denote. *ite* takes a conditional SEV such as $(p \rightarrow t; e)$ as argument and returns an SEV f which is similar to t and/or e in form. Furthermore, $(p \rightarrow t; e)$ and f will denote the same function. Operator *ite* is essentially the one described in [1]. The other operator, *cofact*, does the reverse operation of *ite*. *cofact* takes an SEV p in the range {0, 1} and an arbitrary SEV f as arguments and returns two SEVs f_1 and f_0 so that $(p \rightarrow f_1; f_0)$ denotes the same function as f. In the sequel, we'll use $cofact_1$ and $cofact_0$ to replace $cofact$ so that $cofact_1(p, f) = f_1$ and $cofact_0(p, f) = f_0$. A description of the algorithm $cofact_1$ can be found in [12]. With these two operators, we can easily change SEVs from a conditional form to a non-conditional form and vice versa.

Now, we define the following restricted form and relation on SEVs so that SEV can become canonical.

Definition 4.2 An SEV is *reduced* if

1. it is an EV, or
2. it is a conditional form of $(p \rightarrow t; e)$ where p is not $\langle c, 0\rangle$, $cofact_1(p, t) = t$, $cofact_0(p, e) = e$, and t and e are reduced, or
3. it is $[f_1, \ldots, f_m]$ and every f_i is reduced.

Definition 4.3 Two SEVs sev_f and sev_g are in the same form if

1. both sev_f and sev_g are EVs, or
2. $sev_f = (p \rightarrow t_f; e_f)$, $sev_g = (p \rightarrow t_g; e_g)$, t_f and t_g are in the same form, and e_f and e_g are in the same form, or
3. $sev_f = [f_1, \ldots, f_m]$, $sev_g = [g_1, \ldots, g_m]$ and every pair of f_i and g_i is in the same form.

Lemma 4.1 Given two SEVs sev_f and sev_g which are reduced and in the same form, then sev_f and sev_g are isomorphic if and only if they denote the same function.

The following two algorithms can be used to convert an SEV derived from an implementation to the form derived from the specification. Note, both algorithms employ operators *ite* and *cofact* which can be implemented in polynomial time in the number of nodes of SEVs.

Algorithm B: Converting f to a reduced form.

$redu(f)$ {
 if (f is an EV) return(f);
 else if ($f = (\langle 1, 0\rangle \rightarrow t; e)$) return($redu(t)$);
 else if ($f = (\langle 0, 0\rangle \rightarrow t; e)$) return($redu(e)$);
 else if ($f = (p \rightarrow t; e) \wedge p = (p' \rightarrow t'; e')$)
  return($redu(ite(p) \rightarrow t; e)$);
 else if ($f = (p \rightarrow t; e)$)
  return($p \rightarrow redu(cofact_1(p, t)); redu(cofact_0(p, e))$);
 else return($[redu(f_1), \ldots, redu(f_m)]$) }

Algorithm C: Converting sev_g to the same form as sev_f assuming sev_f and sev_g have the same number of outputs.

$convert(sev_f, sev_g)$
{
 if (sev_f is an EV)
  if (sev_g is an EV) return(sev_g);
  else return($convert(ite(sev_g))$);
 else if ($sev_f = (p \rightarrow t; e)$)
  return($p \rightarrow convert(t, cofact_1(p, sev_g))$;
    $convert(e, cofact_0(p, sev_g))$);
 else /* $sev_f = [f_1, \ldots, f_m]$ */
  if ($sev_g = (p \rightarrow t; e)$)
   return($convert(sev_f, ite(sev_g))$);
  else return($[convert(f_1, g_1), \ldots, convert(f_m, g_m)]$)
}

Example 4.1 The implemented design (IMP) is the SN74L85 chip [7] which is a 4-bit comparator. This chip has 11 inputs, 3 outputs and 33 gates. The specification (SPEC) of the design may be described as:

```
unsigned(3) comp4(x, y, gt, lt, eq)
    unsigned(4) x, y;
    unsigned gt, lt, eq;
{
    if (x > y) return(⟨1, 0, 0⟩);
    else if (x < y) return(⟨0, 1, 0⟩);
    else return(⟨gt, lt, eq⟩)
}
```

It takes 0.05 seconds to generate the SEV of IMP which has 39 nodes and it takes 0.02 seconds to construct the conditional SEV of SPEC which has 25 nodes. The conversion from the SEV of IMP to the one of SPEC and then the comparison take 0.02 seconds. □

Example 4.2 The implementation is the SN74181 chip which is a 4-bit ALU [7]. A partial specification is given below. Note *un_comp*, *two* and *unsigned* perform type coercion. *un_comp* results in an unsigned integer, with the most significant bit being complemented. *two* means that the result is to be a two's complement integer.

```
SN74181(M, S, A, B, Cin)
 unsigned M, Cin;
 unsigned(4) S, A, B;
{
 if (M = 0)
   if (S = 0) return((un_comp (5)) A + (− Cin));
        ⋮
   else if (S = 3) return((two(5)) − Cin);
        ⋮
   else if (S = 6) return((un_comp(5))A−B−Cin);
        ⋮
 else
   if (S = 0) return((unsigned (4)) not(A));
   else if (S=1) return((unsigned(4))not(A or B));
        ⋮
}
```

Note that different output decodings including the difference of sizes are allowed for the same outputs.

The implementation SEV has 765 nodes and can be generated in 0.31 seconds. The specification SEV has 187 nodes and can be constructed in 0.13 seconds. And the verification process takes 0.35 seconds to complete. □

5 Hierarchical Verification

In addition to providing the capability for multi-level verification, SEV is suitable for hierarchical verification, i.e., without having to completely flatten a component that has already been verified. Briefly, hierarchical verification is done by first deriving SEVs from the specification of each component. Now the inputs to each component are SEVs themselves. Therefore, the effect of interconnecting a set of components is realized by deriving an SEV given the component SEVs and the SEVs corresponding to the input arguments. The net result of this process is an SEV which is compared with the SEV derived from the specification of the entire system. Below we present two examples.

Example 5.1 The design is a 64-bit comparator implemented through serial connection of 16 SN74L85s. A net list description of this design we use is:

```
out1 SN74L85 a0 a1 a2 a3 b0 b1 b2 b3 gt lt eq
out2 SN74L85 a4 a5 a6 a7 b4 b5 b6 b7 out1
  ⋮
out16 SN74L85 a60 a61 ... b62 b63 out15
Output : out16
```

where a net list has the form of : *output_name module_name input_name_list*. The specification of this design is the same as the one in Example 4.1 except that the size declaration is changed from 4 to 64. The generations of implementation and specification SEVs take 0.26 and 0.39 seconds respectively, and the proof takes 3.35 seconds to finish. □

Example 5.2 The design is a 64-bit ripple-carry adder implemented through serial connection of 16 SN74181s. The specification of this design is exactly the same as the one used in Example 3.1.

The times to generate the SEVs for implementation and specification are 2.09 and 0.16 seconds respectively and the time to verify they are equivalent takes 0.98 second. Note that the generation of implementation SEV takes longer time while the verification takes less time than the case in Example 3.1. This is because 16 SEVs each with the sum of 4 bits instead of 64 SEVs each with the sum of 1 bit are generated in this example. □

6 Ordering Strategy

In our implementation, variable ordering is based on the following rules:

1. Use the specification rather than implementation to determine the ordering.

2. Variables occurring in p have lower indices than the ones in t or e for a $(p \rightarrow t; e)$ SEV.

3. Variables having larger coefficient have lower indices than the ones having smaller coefficients. For example, in $x_0 + 2x_1 + 4x_2$, $x_0 > x_1 > x_2$.

The advantage of rule 1 is that it is much easier to determine a good ordering at the functional level. Rule 2 matches the experimental results reported in [2]. The main reason of rule 3 is to be able to utilize the boundary property described in Section 2.3.

7 Concluding Remarks

As shown in the examples, EVs can be used to perform equivalence checking between boolean expressions and arithmetic expressions with relational operators. SEVs allow for the verification of both data path and control path. Furthermore, the time required to carry out the equivalence checking between different levels is comparable to the time required for tautology checking in OBDDs. As with OBDDs, the size of SEVs may grow exponentially. From an implementation point of view, OBDDs require only 22 bytes [1] to represent a node, whereas EVs require a bit more space. Memory management problems can usually be avoided by using arithmetic expressions as the specification and hierarchical verification whenever it is possible.

Acknowledgment

This research was supported in part by a grant from the National Science Foundation, award No. MIP-9111206.

References

[1] K. Brace, R. Rudell and R.Bryant, "Efficient Implementation of a BDD Package," *Proc. of 27th Design Automation Conf.*, pp. 40-45, 1990

[2] R. Bryant, "Graph-Based Algorithms for Boolean Function Manipulation," *IEEE Transactions on Computers*, C-35(8): 677-691, August 1986

[3] J.R. Burch, E.M. Clarke and D.E. Long, "Representing Circuits More Efficiently in Symbolic Model Checking," *Proc. of 28th Design Automation Conf.*, pp. 403-407, 1991

[4] K. Butler, D. Ross, R. Kapur and M. Mercer, "Heuristics to Compute Variable Orderings for Efficient Manipulation of Ordered Binary Decision Diagrams," *Proc. of 28th Design Automation Conf.*, pp. 417-420, 1991

[5] K-C. Chen, Y. Matsunaga and S. Muroga, "A Resynthesis Approach for Network Optimization," *Proc. of 28th Design Automation Conf.*, pp. 458-463, 1991

[6] M. Fujita, H. Fujisawa and N. Kawato, "Evaluation and Improvements of Boolean Comparison Method Based on Binary Decision Diagrams," *Proc. Int. Conf. CAD*, pp. 2-5, 1988

[7] Texas Instruments, "The TTL Data Book for Design Engineers," *Texas Instruments*, 1984

[8] S. Minato, N. Ishiura and S. Yajima, "Shared Binary Decision Diagram with Attributed Edges for Efficient Boolean Function Manipulation," *Proc. of 27th Design Automation Conf.*, pp. 52-57, 1990

[9] H. Ochi, N. Ishiura and S. Yajima, "Breadth-First Manipulation of SBDD of Boolean Function for Vector Processing," *Proc. of 28th Design Automation Conf.*, pp. 413-416, 1991

[10] H. Sato, Y. Yasue, Y.Matsunaga and M. Fujita, "Boolean Resubstitution With Permissible Functions and Binary Decision Diagrams," *Proc. of 27th Design Automation Conf.*, pp. 284-289, 1990

[11] A. Srinivasan, T. Kam, S. Malik and R. Brayton, "Algorithms for Discrete Function Manipulation," *Proc. Int. Conf. CAD*, pp. 92-95, 1990

[12] H. Touati, H. Savoj, B. Lin, R. Brayton and A. Sangiovanni-Vincentelli, "Implicit State Enumeration of Finite State Machines using BDD's," *Proc. Int. Conf. CAD*, pp. 130-133, 1990

A New Model for Improving Symbolic Product Machine Traversal

G. Cabodi P. Camurati F. Corno S. Gai P. Prinetto M. Sonza Reorda

Politecnico di Torino
Dipartimento di Automatica e Informatica
Turin Italy

Abstract

This paper presents algorithms for traversing product machines which improve on [6] of a speedup ranging from 3 up to 6. New features include a model that generalizes the product machine, resulting in simpler and more efficient representations and computations, as well as optimizations in symbolic image computation. In the latter case, the speed-up ranges from 1.5 to 4.

1 Introduction

A major class of hardware devices is represented by *Finite State Machines* (FSMs) that are in extensive use as building blocks in a variety of applications, ranging from VLSI circuits to network controllers. Many ASICs are "*control-dominated*", i.e., they are best modeled in terms of FSMs, rather than of data path and control part. As high-level synthesis becomes increasingly popular, but nevertheless does not guarantee a priori functional correctness, formal proof of equivalence between *implementations* of the same *specification* is an important means for the designer to increase confidence in the toolset. Moving beyond the design phase, sequential Automatic Test Pattern Generation (ATPG) remains a challenging issue, as few companies have the convenience to impose rigid scan design rules. An important application of equivalence proof is the exact identification of untestable faults [7].

The theoretical framework for proving the equivalence of two FSM's consists in building and traversing their *product machine* [14]. Storing all or part of the product machine for later traversal [11] requires a huge amount of memory and is thus applicable only to small circuits. Explicit enumeration techniques to build it are computationally expensive [4, 13]. Two aspects must be considered if real circuits are to be handled: efficient ways to represent boolean functions and symbolic manipulations that allow to tame state explosion. *Binary Decision Diagrams* (BDD's) [3, 2] are currently the most popular answer to the first question, whereas *symbolic state exploration techniques* seem to be the solution to the second problem [9].

The use of symbolic techniques for traversing FSM's was first reported in [8] as a breadth-first implicit enumeration of the states that are reachable by a machine starting from a set of reset states. The same research group at Bull developed a technique for symbolic image computation resorting to *characteristic functions* to represent sets of states. Working on characteristic functions, they proposed an algorithm to find the image of a set of states, i.e., the set of future states. They introduced the concept of function simplification under a constraint and algorithms that convert the representation of a set of states between its characteristic function and its implicit definition as image of a boolean functional vector. By computing the transitive closure of the *image* relation, they find the set of reachable states, in which to look for a state where a different behaviour can be observed on the machines' primary outputs. Recent advances propose an efficient way of simplifying functions and computing images, with some heuristics based on domain or co-domain partitioning [10].

Extending this method, [6] propose some heuristics to reduce the complexity. In particular, they define a hash table to avoid redundant image computations and they devise a simplification strategy based on PODEM-like forward implications, together with exhaustive checks for special cases. Experimental results on ISCAS'89 circuits [1] are given.

All these methods are based on the exploration of the product machine by means of symbolic computations. None of them, to the best of our knowledge, addresses the issue of optimizing the traversal of the product machine, trying to keep its complexity as near as possible to the traversal of each single machine. Empirical evidence shows that, in most cases, the two machines that compose the product one do not differ significantly, thus this similarity can be exploited to reduce the computational burden. In this paper we present a new model, the *General Product Machine* (GPM), that simplifies the search for all reachable states. This model benefits from some known relation between a state of the first machine and a corresponding state of the second one. In the case of ATPG this relation is a simple structural correspondence between the good and faulty machines' state elements. In the case of equivalence proof of FSM's, the assumption of known state assignments is justified, being a common subproduct of the synthesis and optimization processes.

Section 2 presents the model of the *General Product Machine*, Section 3 describes the algorithms used in the symbolic traversal of the state space and the advantages that are obtained by using the aforementioned model; Section 4 shows experimental data on the ISCAS'89 benchmark set that allow a comparison

0738-100X/92 $3.00 © 1992 IEEE

with similar approaches and justify this work. Section 5 eventually draws some conclusions.

2 The General Product Machine

Suppose that for two completely specified FSM's $M = (\mathcal{I}, \mathcal{O}, \mathcal{S}, \delta, \lambda, \mathcal{S}_0)$ and $M' = (\mathcal{I}, \mathcal{O}, \mathcal{S}', \delta', \lambda', \mathcal{S}'_0)$ a correspondence between the state spaces $\mathcal{S}$ and $\mathcal{S}'$ exists. In order to compare elements $s \in \mathcal{S}$ and $s' \in \mathcal{S}'$, we define two functions, ψ and φ, that map s and s' in a common image space.

A subset $\mathcal{E} \subseteq \mathcal{S} \times \mathcal{S}'$ is defined as:
$\mathcal{E} = \{(s, s') \in \mathcal{S} \times \mathcal{S}' : \psi(s) = \varphi(s')\}$.
If most of the states reachable by the product machine $M \times M'$ belong to $\mathcal{E}$, then it is possible to find for them a simple and efficient representation.

We propose to express the product machine $M \times M'$ in terms of a new state space, in which the condition $(s, s') \in \mathcal{E}$ can easily be stated. Such a new space is defined through a change in the coordinate system. A partial coordinate system transformation from space $\mathcal{S} \times \mathcal{S}'$, described by variables s and s', to space $\mathcal{S} \times \mathcal{D}$, with variables s and d, is defined by a transformation function $T : \mathcal{S} \times \mathcal{S}' \rightarrow \mathcal{D}$, with the property that, given any couple (s, d), there exists a unique s' satisfying $d = T(s, s')$:
$\forall s \in \mathcal{S},\ \forall d \in \mathcal{D},\ \exists s' \in \mathcal{S}' : d = T(s, s')$
$\forall s \in \mathcal{S},\ \forall s'_1, s'_2 \in \mathcal{S}' : T(s, s'_1) = T(s, s'_2) \Rightarrow s'_1 = s'_2$
The existence and uniqueness of s' define a new function $\tilde{T} : \mathcal{S} \times \mathcal{D} \rightarrow \mathcal{S}'$, i.e., the partial inverse of T with respect to s'.

The functions ψ and φ can be interpreted as suitable mappings that convert the values of $s \in \mathcal{S}$ and $s' \in \mathcal{S}'$ to a common space in which they can be compared. A good choice for T is the difference between the remapped values of s and s': $T(s, s') = \psi(s) \oplus \varphi(s')$.
The existence of the partial inverse $\tilde{T}$ imposes a necessary condition on function φ. In fact:
$d = T(s, s') = \psi(s) \oplus \varphi(s')$
$s' = \tilde{T}(s, d) = \varphi^{-1}(\psi(s) \oplus d)$,
i.e., in order to have a suitable T, φ must be invertible, while no restriction is set on ψ. As a corollary, we derive that $\varphi(s')$ must be a one-to-one mapping, thus spaces $\mathcal{D}$ and $\mathcal{S}'$ must have the same dimension. On the other hand, $\psi(s)$ is an arbitrary function, therefore $\mathcal{S}$ may have a cardinality other than $\mathcal{D}$. We may conclude that, as this model supports the comparison of FSM's with different state space cardinalities, takes into account many-to-one relations between $\mathcal{S}$ and $\mathcal{S}'$.

In ATPG applications, we choose $\psi(s) = s$ and $\varphi(s') = s'$, so d records the information about which memory elements of the fault-free and faulty machines differ. In the verification of a resynthesis or optimization process, ψ can take into account redundant state suppression and it will not be an injective fuction. φ reflects different state assignments and will be a transcoding function, thus invertible. Our experiments show that the ISCAS'89 pairs of equivalent circuits have state assignments differing only for a permutation of their state variables.

We can now define the *General Product Machine* (GPM) M_Δ of two FSM's M and M' in connection

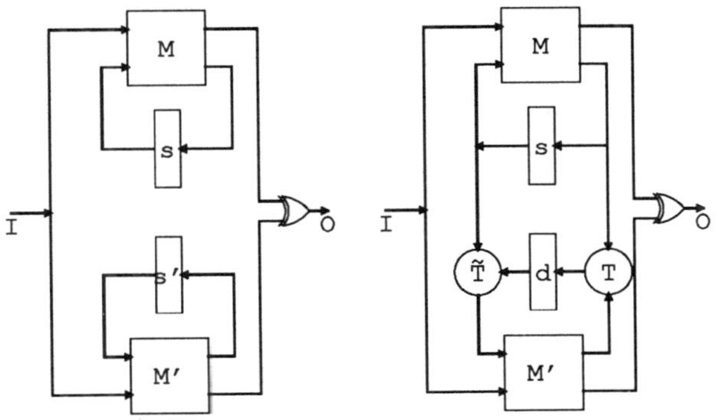

Figure 1 – The models for the PM and the GPM

with the transformation T (Fig. 1):
$M_\Delta = (\mathcal{I}, \mathcal{B}, \mathcal{S} \times \mathcal{D}, \Delta, \Lambda, \mathcal{S}_{0\Delta})$
where:
$\mathcal{B} = \{0, 1\}$
$\delta_d : \mathcal{S} \times \mathcal{D} \times \mathcal{I} \rightarrow \mathcal{D}$
$\delta_d((s, d), x) = T\left(\delta(s, x), \delta'(\tilde{T}(s, d), x)\right)$
$\Delta : \mathcal{S} \times \mathcal{D} \times \mathcal{I} \rightarrow \mathcal{S} \times \mathcal{D}$
$\Delta((s, d), x) = (\delta(s, x), \delta_d((s, d), x))$
$\Lambda : \mathcal{S} \times \mathcal{D} \times \mathcal{I} \rightarrow \mathcal{B}$
$\Lambda((s, d), x) = \lambda(s, x) \not\equiv \lambda'(\tilde{T}(s, d), x)$
$\mathcal{S}_{0\Delta} = \{(s_0, d_0) \in \mathcal{S} \times \mathcal{D} : s_0 \in \mathcal{S}_0 \wedge \tilde{T}(s_0, d_0) \in \mathcal{S}'_0\}$
The state transition function $\Delta = (\delta, \delta_d)$ of the newly defined GPM thus consists of a part that mimics the behaviour of the first machine M (δ), and of a part recording the *difference* between M and M' (δ_d).

Let us define $\mathcal{R}_\Delta \subseteq \mathcal{S} \times \mathcal{D}$ as the set of states (s, d) reachable by the GPM starting from its set of initial states $\mathcal{S}_{0\Delta}$. The equivalence $M \cong M'$ is expressed as:

$$\forall (s, d) \in \mathcal{R}_\Delta, \forall x \in \mathcal{I},\ \Lambda((s, d), x) = 0 \qquad (1)$$

The GPM model is more general than the product machine, in fact we can express the latter as a particular case of the former, by choosing $\psi(s) = 0$ and $\varphi(s') = s'$. The newly introduced T acts as a further degree of freedom that can be exploited to reduce the complexity of the representation. In particular, we can state that: " *if some one-to-one relation $s \leftrightarrow s'$ holds on the set of reachable states of the product machine $M \times M'$, then there exists some choice of T, such that over the entire set $\mathcal{R}_\Delta$ of states reachable by the GPM, we have $d = 0$.*" The aforementioned result indicates that we can really simplify the representation of $\mathcal{R}_\Delta$, whenever the previous assumption holds. The model also deals with reachable states not included in $\mathcal{E}$ without introducing additional complexity.

3 Algorithms

Following the symbolic state exploration technique of [8] and using the symbols proposed in [5], we define an operator, called *generalized cofactor and image restrictor*, that will be extensively used to simplify functions and to compute images. Given a vectorial boolean function $F : \mathcal{X} \rightarrow \mathcal{Y}$, a characteristic function $c : \mathcal{X} \rightarrow \mathcal{B}$, and domain values $x \in \mathcal{X}$, we define the operation $F \updownarrow c$ as a new vectorial boolean function

```
var Reached, From, New, Image: set of S × D;
begin
   From := Reached := S_0Δ ;
   repeat
      if Λ ↓ From ≠ 0 then return Different ;
      Image := IMG(Δ, From) ;
      New := Image ∩ ¬(Reached) ;
      Reached := Reached ∪ Image ;
      From := BETWEEN(New, Reached) ;
   until New = ∅ ;
   return Equivalent ;
end
```

Figure 2 – Symbolic Breadth-First Execution

satisfying the following identities:

$F = c \cdot F \downarrow c + \bar{c} \cdot F \downarrow \bar{c}$

$$\text{IMG}(F, c) = \text{IMG}(F \downarrow c, 1) \qquad (2)$$

The term "1" in (2) is the characteristic function of the entire domain of F. We can deduce that $F \downarrow c$ must have the form:

$$(F \downarrow c)(x) = \begin{cases} F(x) & \text{if } x \text{ is such that } c(x) = 1 \\ y & \text{if } x \text{ is such that } c(x) = 0 \end{cases}$$

where y is some arbitrarily chosen element of the image of F over the set described by c. [10] proposed a way of choosing $y = F(x')$ where x' is the element of the on-set $\{x' : c(x') = 1\}$ closest to x.

Using this new operator, we can check for simple sufficient conditions of equivalence, avoiding the traversal of the GPM whenever there exists a combinational equivalence, as for many optimization processes and for many undetectable faults in ATPG applications.

Suppose that, starting from the subset $\mathcal{E}$, here defined by $d = 0$, the GPM always evolves within $\mathcal{E}$: then the machines are equivalent if the outputs don't differ in $\mathcal{E}$. This condition is expressed as:

$(\delta_d \downarrow \mathcal{E} = 0) \wedge (\Lambda \downarrow \mathcal{E} = 0) \Rightarrow M \cong M'$

This condition can be checked before attempting any state traversal, as it bounds the set reachable by $\mathcal{E}$. When this fails, we can check the same condition over the precomputed set R of the states reachable by machine M:

$((\delta_d \downarrow \mathcal{E}) \downarrow R = 0) \wedge ((\Lambda \downarrow \mathcal{E}) \downarrow R = 0) \Rightarrow M \cong M'$

When this condition holds, it is usually faster to build R and do this check than attempting the traversal of the product machine. This is particularly effective in ATPG, where the same fault-free machine must be checked against many faulty ones: most of the undetectable faults are found by checking the above condition [7].

When the previous tests fail, we have to build the reachable set of M_Δ in order to verify whether (1) holds. We explore the GPM in breadth-first with an image computation technique based on implicit enumeration of sets of states described by characteristic functions. The pseudo-code, similar to that of [10], is shown in Fig. 2.

All the set operations that appear in Fig. 2 are implemented as boolean operations on the BDD's representing the characteristic functions of those states. The operator BETWEEN returns the simplest BDD such that:

New $\subseteq$ BETWEEN(New,Reached) $\subseteq$ Reached.

State-of-the-art image computation In the algorithm of Fig. 2 the heaviest operation is the computation of the image of the vectorial function Δ on the set From.

The algorithm we adopt [9] is based on the recursive Shannon expansion of the image characteristic function. If $F = [f_1, \ldots, f_n]$, we define $F_k = [f_k, f_{k+1}, \ldots, f_n]$. We can write:

$$\begin{aligned} \text{IMG}(F_k, 1) &= \{0\} \times \text{IMG}(F_{k+1} \downarrow \bar{f}_k, 1) \\ &\cup \{1\} \times \text{IMG}(F_{k+1} \downarrow f_k, 1) \end{aligned} \qquad (3)$$

In image computation, according to (2), two steps can be identified:

1. operation $\Delta' = \Delta \downarrow$ From: the vectorial boolean function Δ is simplified by a generalized cofactor operation which has also the effect of restricting the full image of Δ to its image over the set From;
2. Image = IMG$(\Delta', 1)$ is the recursive expansion algorithm of (3).

The following techniques have been proposed in order to speed up image computation:

domain partitioning: [10] use a divide and conquer strategy in which the domain From is decomposed in smaller parts, each contributing to a subset of the entire image. An alternative partitioning method is based on the decomposition of the co-domain;

image cache table: a hash table [5] is used to store the results of image subproblems, avoiding redundant computations; it can be extended to recognize a problem despite the presence of negations and permutations;

implications: the algorithms in [5] improve the termination conditions by taking into account implications among inputs and outputs, according to a PODEM-like technique [12];

full partitioning: at every recursion step, vector F_k is possibly partitioned in two or more subsets consisting of functions with disjoint support variables: an effective optimization consists in calculating the subimages, and then taking their cartesian product;

reached set pruning: whenever the image computation procedure enters a subspace entirely contained in the set of already reached states, recursion can stop; this check is done by traversing the Reached set in parallel to the recursive construction of Image.

Our approach can be characterized by several optimizations, some having already been introduced by the Bull and/or Boulder groups, some being novel. The following ones belong to the former class:

- recursion is based on co-domain partitioning;
- the image is pruned over the set of reachable states;
- generalized cofactor operations take into account implications among variables;
- the F_k vector is fully partitioned at each step according to its support variables;
- no image table is used.

Novel features are the presence of a *cofactor hash table* which records the results of past cofactorizations and the *subtree recombination check*: whenever $F_{k+1} \downarrow f_k$ equals $F_{k+1} \downarrow \bar{f}_k$ on the current subspace, we recognize subtree recombination, thus saving a recursion.

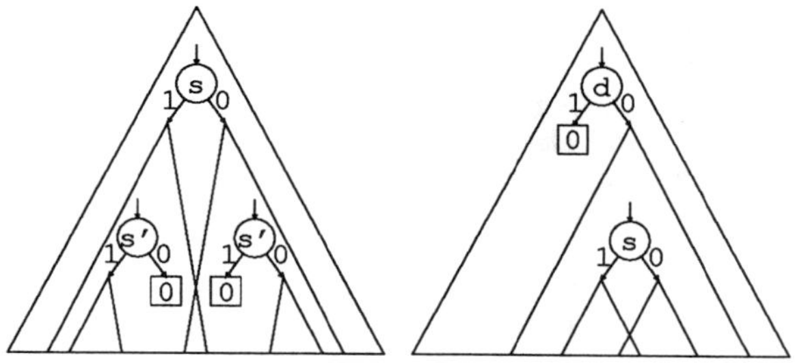

Figure 3 – BDD's in the PM and in the GPM

The contribution of the GPM model to simplified image computation Representation and computation for the product machine is inherently much more complex than for the single machine M, although empirical observations show that in most product machines the reachable states satisfy, at least approximately, the relation $\psi(s) = \varphi(s')$. This has major consequences on the BDD's representing the sets From, Image, Reached and New in the $\mathcal{S} \times \mathcal{S}'$ space and on the complexity of the image computation algorithm. Let us assume a simple relation: $s = s'$ to make the comprehension of the results easier and to give pictorial examples. Note, anyway, that the results drawn from the following considerations are general.

Whenever $s = s'$ holds, a BDD representation that doesn't take this explicitly into account will be necessarily redundant. In fact, for every couple of components s_i and s'_i of the state vectors s and s', the BDD reflects that $s_i = s'_i$. In particular, the subtree rooted at $s_i = 0$ must contain a node assessing $s'_i = 0$, and the subtree rooted at $s_i = 1$ will contain $s'_i = 1$ (Fig. 3). We notice that both s_i and s'_i must explicitly appear on em every path from the BDD root to any "1" terminal node. Moreover, as the two subtrees of a s_i node are distinct at least in their s'_i children, they will never be shared. The problems outlined above may be reduced by choosing a proper variable ordering, for example $s_1, s'_1, s_2, s'_2, \ldots$, but we can't find a suitable order in the general $\psi(s) = \varphi(s')$ case: we really need a model exploiting the known relation instead of being conditioned by it.

The above paragraph showed the inherent complexity of sets From, Reached, Image, and New of the product machine $M \times M'$ compared with the one of the corresponding sets From*, Reached*, Image* and New* of the single machine M. In particular, the representation of From will influence the run-time performance in two aspects:

1. the cofactoring $(\delta, \delta') \Updownarrow$ From is much harder to compute than $\delta \Updownarrow$ From* due to the doubled number of δ components and to the complexity of From;
2. if we compare $\textsc{Img}((\delta, \delta') \Updownarrow \text{From}, 1)$ with $\textsc{Img}(\delta \Updownarrow \text{From}^*, 1)$, we see that, in the many cases in which δ and δ' don't differ significantly, we are performing several redundant computations:

- $\delta \Updownarrow$ From and $\delta' \Updownarrow$ From share many identical subproblems: without the help of a cofactor table, we would double the cofactor operations;
- while applying full partitioning, we must discover that δ and δ' have disjoint state variable sets and that most of the support variable dependencies present in δ are replicated in δ': the partitioning will be significantly slowed;
- the same arguments also apply to the propagation of implications: whenever an implication concerning δ is present also in δ', we should apply it twice.

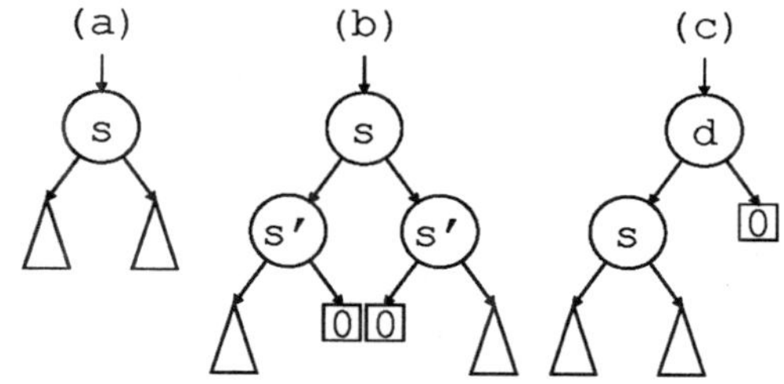

Figure 4 – Local BDD equivalences

We conclude that the product machine representation leads to an increase in CPU time of at least two with respect to single machine traversal, as confirmed by experimental results. We may overcome this limit only by explicitly taking into account the similarities between M and M'. The GPM model helps in simplifying representation and computation as its δ_d function records the behavioural difference of M' with respect to M: we are explicitly factoring out the similarities, i.e., reducing redundancy.

Let us suppose to build the GPM M_Δ from a coordinate system transformation $T(s, s') = \psi(s) \oplus \varphi(s')$. We expect a gain in efficiency directly related to the fraction of reachable states satisfying the assumed many-to-one mapping. In the following sections, we show that the representation of the state sets, in particular From, is simplified, the overhead introduced in the computation of Δ and Λ is negligible, and that, introducing a simple check, the computation of $\Delta \Updownarrow$ From and $\textsc{Img}(\Delta \Updownarrow \text{From})$ boils down to a problem comparable to single machine traversal.

3.1 Representation of From

Over the entire set of states for which $\psi(s) = \varphi(s')$ we have $d = 0$: the condition $s_i = s'_i$ becomes $d_i = 0$ (Fig. 3). In Fig. 4 we see the mapping between each node (a) of the single machine's characteristic functions and the corresponding nodes in the product machine (b) and in the General Product Machine (c), when we choose an interleaved ordering of variables and $s_i = s'_i$ holds. The presence of d variables is structurally simple, showing that every subtree sharing in (a) will be present in (c). Exceptions to this simple BDD structure are due to the states $(s, d) \in$ From for which $d \neq 0$, therefore the gain in representation is directly related to the degree of exactness of the assumed mapping.

Recalling that $\mathcal{E}$ contains all the (s, d) states satisfying the mapping, we may write:

$$\text{From} = ((\text{From} \Updownarrow \mathcal{E}) \cap \mathcal{E}) \cup ((\text{From} \Updownarrow \bar{\mathcal{E}}) \cap \bar{\mathcal{E}}) \quad (4)$$

where From $\Updownarrow \mathcal{E}$ is described only in terms of s variables. As shown in (4), From consists of two subsets, one included in $\mathcal{E}$ and the other outside $\mathcal{E}$. In most cases the second term is the empty set, thus we will benefit of domain partitioning techniques on the set $\mathcal{E}$.

3.2 Computation of Δ and Λ

According to the definition of the GPM, given a transformation function T the state transition and output functions Δ and Λ are expressed by means of simple function composition of corresponding M and M' functions. The overhead imposed by these compositions is negligible if compared with the computational simplifications that the GPM model supports, provided that a suitable ordering for the variables is used, e.g., $d_1, s_1, d_2, s_2, \ldots$.

3.3 Computation of $\Delta \updownarrow \mathsf{From}$

The $\Delta = (\delta, \delta_d)$ function of the GPM M_Δ is to a certain degree more complex than the product machine state transition function (δ, δ'): as a result of the compositions, the δ_d components of Δ depend on all the state variables s and d, while δ' clearly contains only s', not s. We might therefore expect that $\delta_d \updownarrow \mathsf{From}$ will be harder to compute than $\delta' \updownarrow \mathsf{From}$. In order to overcome this limit, as we often explore the $\mathcal{E}$ subset of $\mathcal{S} \times \mathcal{S}'$, we are facing a From set such that $\mathsf{From} \subseteq \mathcal{E}$. Using a property of the generalized cofactor operator we may write:

$$\mathsf{From} \subseteq \mathcal{E} \Rightarrow \begin{cases} \mathsf{From} = (\mathsf{From} \updownarrow \mathcal{E}) \cap \mathcal{E} \\ \Delta \updownarrow \mathsf{From} = (\Delta \updownarrow \mathcal{E}) \updownarrow (\mathsf{From} \updownarrow \mathcal{E}) \end{cases}$$

The term $\Delta \updownarrow \mathcal{E}$ can be computed in advance and its complexity is comparable to that of the original δ and δ' functions, as it doesn't contain d variables. This simple check can reduce the computation of $\Delta \updownarrow \mathsf{From}$ to a problem comparable to single machine traversal.

3.4 Computation of $\textsc{Img}(\Delta \updownarrow \mathsf{From}, 1)$

While discussing the representational power of the GPM model, we focused on the fact that the characteristic functions of From, Image, New and Reached have a high degree of subtree sharing compared to the product machine. We devised a first heuristic, the *subtree recombination check*, to avoid computing two sub-images when we already know they are equal: this technique constantly reduced the number of recursive calls, as shown in the following section.

A greater simplification is due to the easy way of checking if $\mathsf{Image} \subseteq \mathcal{E}$ before computing it. In particular:

$\delta_d \updownarrow \mathsf{From} = 0 \Rightarrow \textsc{Img}(\Delta \updownarrow \mathsf{From}, 1) \subseteq \mathcal{E}$

When the previous relation is satisfied, we may compute:

$\mathsf{Image} = (\textsc{Img}(\delta \updownarrow (\mathsf{From} \updownarrow \mathcal{E})) \times \mathcal{D}) \cap \mathcal{E}$

meaning that we simply compute the image of the original δ function over a set obtained from From by suppressing the d variables, and then append to the resulting BDD the nodes corresponding to $d = 0$.

The previous techniques work whenever the From set is completely contained in $\mathcal{E}$. We can extend their effectiveness by first decomposing From in the part contained in $\mathcal{E}$, upon which we apply the heuristics, and the (usually smaller) remainder. This is a good application of the *domain partitioning* strategy, and can be described by the following identity:

$$\begin{aligned} \mathsf{Image} &= \textsc{Img}((\Delta \updownarrow \mathcal{E}) \updownarrow (\mathsf{From} \updownarrow \mathcal{E}), 1) \\ &\cup \textsc{Img}((\Delta \updownarrow \bar{\mathcal{E}}) \updownarrow (\mathsf{From} \updownarrow \bar{\mathcal{E}}), 1) \end{aligned} \quad (5)$$

```
function IMAGE(F, k, R): set of S × D ;
var found: boolean; Img: set of S × D ;
begin
   (found, Img) := SPECIALCASES(F, k, R) ;
   if found then return Img ;
   (F0, k0, R0) := IMPLICATIONS(F, k, R, 0) ;
   (F1, k1, R1) := IMPLICATIONS(F, k, R, 1) ;
   if F1 = F0 then
       return IMAGE(F1, k1, R1 ∩ R0)
   else return yk · IMAGE(F1, k1, R1) +
               ȳk · IMAGE(F0, k0, R0) ;
end ;
```

Figure 5 – Image computation function

Being $\Delta \updownarrow \mathcal{E}$ computed in advance and as $\mathsf{From} \updownarrow \mathcal{E}$ simply means the suppression of d variables, the first part of the image is easily computed, whereas we use the full Δ representation only over a smaller set.

A pseudo-code description of the complete image computation algorithm can be found in Fig. 5, where we use the symbology presented in [5]. The function SPECIALCASES checks for terminal cases, prunes over the reached set R and implements full partitioning. The main task of IMPLICATIONS is to simplify F under constraints $f_k = 0$ and $f_k = 1$ by means of cofactoring. The subtree recombination check is represented by the test $F_1 = F_0$. The IMAGE function is called as:

$\mathsf{Image} = \textsc{Image}(\Delta \updownarrow \mathsf{From}, 1, \mathsf{Reached})$

4 Experimental Results

The concepts presented in the above sections were implemented in a in-house software package written in C (about 4000 lines of source code), including:

- elementary BDD operations, based on the canonical form and ITE operators presented in [2];
- generation of δ and λ functions from circuit netlist to BDD format; some variable ordering heuristics are included, available in the literature;
- reachable state computation, through symbolic breadth-first execution;
- product machine and GPM verification, as described in the previous section.

Experiments were performed on a subset of the ISCAS'89 circuits, with two objectives: to measure the performance of the state generation module in single machine processing, and to compare the verification of couples of equivalent machines to the above case.

The results obtained on a 3-Mips VAXstation 3100 are summarized in two tables. Tab. 1 shows the state generation module: numbers of image computation recursions (IMG) and CPU times (in seconds) are reported without and with the cofactor-table and subtree recombination optimizations. The speed-up with respect to [6] ranges from 1.5 to 4. Tab. 2 shows verification data for the seven couples of equivalent ISCAS'89 circuits: product machine (PM) traversal without and with cofactor table, and GPM traversal are described. The first case is two to three times slower than state generation for the single machine (as in [6]), the second one has a better performance due to the cofactor table and the last one has very little overhead, compared to single machine processing. The speed-up of the optimized case with respect to [6] ranges from 3 to 6.

Circuit	States	Levels	without optimiz.		with optimiz.	
			IMG	time	IMG	time
s208	17	17	33	0.2	33	0.2
s298	218	19	184	0.3	174	1.0
s344	2625	7	1835	6.1	1767	7.7
s349	2625	7	1835	6.3	1767	7.5
s382	8865	151	11447	46.1	5485	32.4
s386	13	8	31	0.2	30	0.1
s400	8865	151	11447	46.7	5485	32.5
s420	17	17	33	0.4	33	0.4
s444	8865	151	8268	44.9	5994	17.8
s510	47	47	72	0.5	72	0.6
s526	8868	151	7925	33.5	6389	18.7
s526n	8868	151	7925	33.3	6389	19.0
s641	1544	7	2678	54.3	2521	23.1
s713	1544	7	2853	56.8	2521	20.7
s820	25	11	66	0.2	66	0.3
s832	25	11	66	0.2	66	0.3
s838	17	17	33	1.1	33	1.0
s953	504	11	1038	8.9	965	9.1
s1196	2616	3	5835	231.8	4937	67.6
s1238	2616	3	5827	225.5	4932	68.8
s1488	48	22	103	0.4	98	0.5
s1494	48	22	104	0.4	98	0.7

Table 1 – Reachable state set computation

Circuit pairs	PM not opt		PM opt		GPM opt	
	IMG	time	IMG	time	IMG	time
s344-s349	3288	25.1	3288	20.9	1767	8.1
s382-s400	15681	267.3	15681	134.3	8491	46.3
s526-s526n	9896	154.1	9896	54.0	6389	23.0
s641-s713	4566	145.0	4566	36.5	2521	23.2
s820-s832	89	1.0	89	1.1	66	0.4
s1196-s1238	8482	481.7	8482	91.2	4983	68.5
s1488-s1494	111	0.9	111	1.3	98	1.2

Table 2 – Equivalence verification

5 Conclusions

The main theoretical contribution of this paper is a generalization of the product machine model. Defining a suitable reference frame transformation, a *General Product Machine* (GPM) can be obtained, that entails considerable simplifications in representation and computation. This model covers the traditional product machines as a particular case.

Optimized forward-time processing symbolic image computation procedures allow us to obtain a speed-up ranging from 1.5 to 4 with respect to the experimental results of [6] on the same set of ISCAS'89 circuits [1]. The combined use of such techniques and of the GPM model yields a speed-up ranging from 3 to 6 for FSM verification.

Symbolic FSM and product machine traversal techniques are thus a suitable answer to the need for verifying synthesized synchronous circuits and may be of considerable help in ATPG, e.g., for redundant fault identification. Their limits are due to the complexity of the circuits and to the number of states, but, as the same applies for synthesis, this is yet one more proof of the parallel and interleaved evolutions of verification and synthesis.

References

[1] F.Brglez, D.Bryan, K.Koźmiński: "Combinatorial Profiles of Sequential Benchmark Circuits," ISCAS'89: IEEE Intl Symposium on Circuits and Systems, Portland, OR (USA), May 1989, pp. 1929-1934

[2] K.S.Brace, R.L.Rudell, R.E.Bryant: "Efficient Implementation of a BDD Package," 27th ACM/IEEE Design Automation Conference, pp. 40-45

[3] R.E.Bryant: "Graph-based Algorithms for Boolean Function Manipulation," IEEE Transactions on Computer, Vol. C-35, No. 8, August 1986, pp. 677-691

[4] P.Camurati, M.Gilli, P.Prinetto, M.Sonza Reorda: "Proving FSMs correct with an automaton-based method," IEEE 1st Great Lakes Symposium on VLSI, Kalamazoo, MI (USA), March 1991, pp.255-258

[5] H.Cho, G.Hachtel, S.W.Jeong, B.Plessier, E.Schwarz, F.Somenzi: "ATPG Aspects of FSM Verification," ICCAD-90: IEEE Intl Conference on Computer Aided Design, Santa Clara, CA (USA), November 1990, pp. 134-137

[6] H.Cho, G.Hachtel, S.W.Jeong, B.Plessier, E.Schwarz, F.Somenzi: "Results on the Interface Between Formal Verification and ATPG," DIMACS Series in Discrete Mathematics Science, Volume 3, 1991, Clarke, Kurshan Editors, pp. 615-628

[7] H.Cho, G.D.Hachtel, F.Somenzi: "Redundancy identification and removal based on implicit state enumeration," ICCD-91: IEEE International Conference on Computer Design: VLSI Computers & Processors, Cambridge, MA (USA), October 1991, pp. 77-80

[8] O.Coudert, C.Berthet, J.C.Madre: "Verification of Sequential Machines Using Boolean Function Vectors," IFIP Intl Workshop on "Applied Formal Methods for Correct VLSI Design", Leuven (Belgium), November 1989, Vol. 1, pp. 111-128

[9] O.Coudert, J.C.Madre: "A Unified Framework for the Formal Verification of Sequential Circuits," ICCAD-90: IEEE Intl Conf. on Computer Aided Design, Santa Clara, CA (USA), November 1990, pp. 134-137

[10] O.Coudert, J.C.Madre: "Symbolic Computation of the Valid States of the Sequential Machine: Algorithms and Discussion," 1991 Intl Workshop on Formal Methods in VLSI Design, Miami, FL (USA), January 1991

[11] S.Devadas, H.K.T.Ma, A.R.Newton: "On the Verification of Sequential Machines at Differing Levels of Abstraction," IEEE Trans. CAD, Vol. 7, No. 6, June 1988, pp. 713-722

[12] P.Goel: "An Implicit Enumeration Algorithm to Generate Tests for Combinatorial Logic Circuits," IEEE Trans. Computer, Vol. C-30(3), 1981, pp. 215-222

[13] S.H.Hwang, A.R.Newton: "An Efficient Verifier for Finite State Machines," IEEE Trans. CAD, Vol. 10, n. 3, March 1991, pp. 326-334

[14] Z.Kohavi: "Switching and Finite Automata Theory," Computer Science Series, Mc Graw Hill, New York, NY (USA), 1970

Exact Calculation of Synchronization Sequences Based on Binary Decision Diagrams

Carl Pixley [1]
Mitsubish Electric Research Lab. Inc.
Cambridge, MA.

Seh-Woong Jeong Gary D. Hachtel
Dept. of Electrical & Computer Engineering
University of Colorado at Boulder

Abstract

A synchronization sequence for a synchronous design $\mathcal{D}$ is a sequence of primary input vectors which when applied to any initial state of $\mathcal{D}$ will drive $\mathcal{D}$ to a single state, called a reset state [10]. We present efficient methods based upon Universal Alignment Theorem [4] and Binary Decision Diagrams (BDD's) [6, 7] to (1) decide whether a design has an exact synchronization sequence (i.e., a design is resettable), (2) calculate a synchronization sequence for a design, (3) calculate the set of all reset states, and (4) decide whether a specific state is a reset states. New results on the resettability of some benchmark circuits [11, 2] are presented.

1 Introduction

Synchronous circuits $\mathcal{D}$ considered here are interconnections of D flip-flops and combinational logic elements. In order to reliably predict the behavior of $\mathcal{D}$, it is necessary to understand a way to drive $\mathcal{D}$ into a predictable state. This can be accomplished without the expense of special reset circuitry by designing $\mathcal{D}$ so that some sequence of inputs (called a **synchronization** or **reset** sequence) will drive any initial state of $\mathcal{D}$ into a single state, called a **Reset State**.

Each of three previously known methods for checking resettability suffers from some drawback. The synchronization tree method [10] cannot handle large designs because each state and input pattern is handled explicitly. The method of simulation with x-values is not exact due to loss of information. The methods of [5], to which the present methods are closely related, involve an expensive Equivalent State Pair (ESP) calculation. The techniques presented in this paper are superior to the methods in [5] in the following ways. The new methods (1) avoid the expensive Equivalent State Pair calculation [4], (2) optimize the reset sequence computation, and (3) replace computations based upon the state transition relation with more efficient function vector calculations. An improved pre-image calculation based upon functional vectors is presented. Experimental results illustrate the efficiency of this calculation.

[1] This work was performed when the first author was at Microelectronics and Computer Technology Corporation (MCC) in Austin, Tx.

2 Definitions and Theory

Let $\mathcal{D}$ be a synchronous design (a Mealy machine in general). A **state** of $\mathcal{D}$ is an association of Boolean values to the outputs of the flip-flops of $\mathcal{D}$, i.e., a vector $\vec{ps}$ of n many Boolean values. Suppose that $\mathcal{D}$ has p many inputs. The **state transition function** of $\mathcal{D}$ is a mapping $\delta : B^{n+p} \rightarrow B^n$ where $\delta = [\delta_1, \delta_2, \cdots, \delta_n]$ and $B = \{0, 1\}$. Then $\delta_i(\vec{ps}, \vec{in})$ represents the value of the ith flip-flop after one clock event starting from state $\vec{ps}$ with input $\vec{pi}$. Binary Decision Diagrams [6] [7] are used to represent the coordinate functions of δ The vectors of BDD's like δ are called **Boolean Function Vectors (BFVs)** [2].

The domain B^{n+p} of δ is an $(n + p)$-dimensional Boolean space spanned by the present state variables $ps_1, ps_2, \ldots, ps_n$ (abbreviated, PS) and the primary input variables $pi_1, pi_2, \ldots, pi_p$ (abbreviated, PI). The co-domain B^n of δ is an n-dimensional Boolean space defined by the next state variables $ns_1, ns_2, \ldots, ns_n$. The characteristic function of a subset C of Boolean n-space is a Boolean function whose onset is the set elements in C. Therefore a BDD in present state variables PS represents a subset (sometimes called a constraint) of the domain of δ. Sets and characteristic functions are used interchangeably when the context is clear.

The **image** of δ over a subset C of the domain B^{n+p} is defined as $Image(\delta, C) = \{v \mid \exists x \in C \text{ such that } \delta(x) = v\}$. The **pre-image** of δ over a subset C of the co-domain B^n is defined as $Preimage(\delta, C) = \{w \in B^{n+p} \mid \delta(w) \in C\}$, or $\delta^{-1}(C)$ in short. Some details are presented in [8, 9, 2].

If f is a function and x is a variable, **existential abstraction** $(\exists x)f$ is defined to be $f_x + f_{\bar{x}}$. Furthermore, $(\exists x_1, x_2, \cdots)f$ is defined recursively as $(\exists x_1)((\exists x_2, \ldots)(f)\ldots)$. Let D be the set of states of design $\mathcal{D}$. Let s_0 be a state and SEQ be any finite sequence of primary input vectors. Then $SEQ(s_0)$ is defined be the state to which $\mathcal{D}$ is driven from state s_0 by applying SEQ.

Definition 1 (Closed Partition)
A **closed partition** *over D is a partition $\pi = \{b_1, b_2, \cdots, b_m\}$ (i.e., $b_i \cap b_j = \phi$ for $i \neq j$, $\bigcup_{i=1}^m b_i = D$) such that for any two states s, r in the same block b_i and any SEQ, $SEQ(s)$ and $SEQ(r)$ are also in the same block, say b_j.*

Definition 2 (Universal Alignment Sequence) *SEQ is a* **universal alignment sequence** *of $\mathcal{D}$ with respect to a closed partition π, if and only if there is one block b_i of π such that for any state $s \in D$, $SEQ(s) \in b_i$.*

Theorem 1 (Universal Alignment) *There is a universal alignment sequence for a closed partition π of $\mathcal{D}$* if and only if *for any pair of states $p = (s, r)$ there is an input sequence SEQ_p such that $SEQ_p(s)$ and $SEQ_p(r)$ are in the same block of π.*

Suppose π is defined such that two states s and r are in the same block if and only if for each SEQ, corresponding outputs of $\mathcal{D}$ in states $SEQ(r)$ and $SEQ(s)$ are the same. Then a universal alignment sequence with respect to π is an *essential reset sequence*[5]. However, if π is trivial in a sense that each block contains exactly one state, then a universal alignment sequence is a *synchronization sequence*, since every block has only one distinct state in it.

A **product machine** ($\mathcal{D} \times \mathcal{D}$) of $\mathcal{D}$ is a disjoint union of two copies of $\mathcal{D}$ sharing their primary inputs. The state space of $\mathcal{D} \times \mathcal{D}$ is the Cartesian product of the state spaces of each machine, i.e., $D \times D$. Furthermore, the next state BFV for a product machine is concatenation of the next state BFVs of its components. The **diagonal**, $diag(\mathcal{D} \times \mathcal{D})$ is defined to be $\{(s, s) \mid s \in D\}$. Let A be a subset of states of $\mathcal{D} \times \mathcal{D}$ and δ be the next state BFV. Let the **state-preimage of A, State_Preimage(δ, A)**, be $\{s \mid \vec{in}(s) \in A \ for \ some \ \vec{in}\}$. Let the **transitive state-preimage, State_Preimage***(δ, **A**), be the set of states such that $s_0 \in A$ or there exists some input sequence SEQ, such that $SEQ(s_0) \in A$.

Corollary 1 *(to Theorem 1)* *Given a design $\mathcal{D}$ and a closed partition $\pi = \{b_1, b_2, \cdots, b_m\}$, let δ be a next state BFV of $\mathcal{D} \times \mathcal{D}$ and $A = \{(s, r) \mid s, r \in b_i \ for \ some \ i\}$. Then there exists a universal alignment sequence if and only if*

$$State\_Preimage^*(\delta, A) = D \times D \tag{1}$$

As a special case, there exists a synchronization sequence if and only if

$$State\_Preimage^*(\delta, diag(\mathcal{D} \times \mathcal{D})) = D \times D.$$

Given a synchronization sequence SEQ, let **R-state(SEQ)** be the reset state to which SEQ drives every state. When SEQ is clear from the context, we will just use the term **R-state**. It is easy to see that the set of all reset states is just the set of states reachable from R-state. Therefore, we can calculate the set of all reset states by the forward traversal (from R-state) technique in [8, 9].

3 Algorithms

3.1 Synchronization Sequence

The input parameter δ of the algorithm in Fig 1 is a BFV of the next state functions of $\mathcal{D} \times \mathcal{D}$. The first **while** loop (lines 7-13) computes $State\_Preimage^*(\delta, diag)$. When the loop is exited, sum_i is $State\_Preimage^*(\delta, diag)$. Note that the following relationship is satisfied:

$$(\exists PI) ring_{i+1} \quad = \quad sum_{i+1} - sum_i \tag{2}$$

Any onset minterm represented by $\vec{pi}$ and $\vec{ps}$ of $ring_{i+1}$ tells that $\vec{ps}$ is a state of $\mathcal{D} \times \mathcal{D}$ such that $\vec{ps} \in sum_{i+1} - sum_i$ and if we apply $\vec{pi}$ to $\vec{ps}$, the next state will be in $sum_i - sum_{i-1}$. Note that $\vec{ps}$ can be viewed as a state pair of a single design $\mathcal{D}$. If sum_i is **1** when the loop is exited, we know two facts: (1) the design is resettable and (2) any synchronization sequence has length at least $i - 1$ hence $i - 1$ is a lower bound for the synchronization sequences. Assuming that every state pair is alignable to the diagonal, the remainder of the program beginning at line 19 computes an exact synchronization sequence. The $leftover$ set represents the state pairs that have not already been aligned to the diagonal by applying $sync\_seq$ to $\mathcal{D} \times \mathcal{D}$. Because the diagonal is closed under the application of any possible input, $leftover$ will eventually become empty. Due to Theorem 1,

```
Procedure synchronization_seq(δ)
begin
1  diag = characteristic function for the set {(s, s) | s ∈ D}
2  ring_-1 = 0
3  ring_0 = diag
4  sum_-1 = 0
5  sum_0 = diag
6  i = 0
7  while(ring_i != 0)
8  begin
9      c = sum_i − sum_{i−1}
10     ring_{i+1} = δ^{−1}(c) − sum_i
11     sum_{i+1} = sum_i + (∃PI)ring_{i+1}
12     i = i + 1
13 end
14 if(sum_i != 1)
15 begin
16     print("Design is not resettable")
17     return NIL
18 end
19 leftover = ¬diag
20 sync_seq = NIL
21 while(leftover != 0)
22 begin
23     Let k be minimum such that leftover · ring_k != 0
24     Choose a primary input assignment pi⃗ such that
       pi⃗ · leftover · ring_k != 0
25     leftover = Image(δ, leftover · pi⃗) − diag
26     sync_seq = concatenate(sync_seq, pi⃗)
27 end
28 return sync_seq
end
```

Figure 1: Synchronization Sequence Calculation

Procedure Preimage$(\delta, \mathbf{c})$
begin
1 Canonicalize δ and c
2 **if**(c is **0** or **1**) return c
3 $preimg = lookup\_cache\_table(\delta, c)$
4 **if**($preimg$! = **NIL**) return $preimg$
5 Choose a next state variable ns_j
6 $root = \delta_j$
7 $preimg_0 = Preimage(\delta', c_{ns_j})$
8 $preimg_1 = Preimage(\delta', c_{\overline{ns_j}})$
9 $preimg = root \cdot preimg_0 + \overline{root} \cdot preimg_1$
10 $store\_cache\_table(\delta, c, preimg)$
11 return $preimg$
end

Figure 2: Pre-image Calculation

sync_seq returned is a synchronization sequence since any state of $\mathcal{D} \times \mathcal{D}$ (or equivalently any state state pair of $\mathcal{D}$) can be aligned to a state in the diagonal(or equivalently the state pair is aligned to a state). More details can be obtained in [5] where the closed partition in use is formed by equivalent state pairs instead of the diagonal. In line 24, a primary input pattern $\vec{pi}$ is selected in such a way that we can drive into sum_{k-1} as many state pairs of $leftover$ as possible by finding a *"shortest" path* from the root node to the **1** node of the BDD for $leftover \cdot ring_k$. Lengths of an edges coming out of a BDD nodes corresponding to present state variables are assigned length 1; all others (i.e., out of primary input variables), are assigned 0.

3.2 Efficient Pre-Image Calculation

In [3], a technique is proposed to compute a pre-image followed by an abstraction. This section presents a method for pre-image calculation without building a transition relation. The following are some facts about pre-images.

Theorem 2

$$\delta^{-1}(c_1 \cdot c_2) = (\delta^{-1}(c_1)) \cdot (\delta^{-1}(c_2)) \quad (3)$$
$$\delta^{-1}(c_1 + c_2) = \delta^{-1}(c_1) + \delta^{-1}(c_2) \quad (4)$$
$$\delta^{-1}(\bar{c}) = \overline{\delta^{-1}(c)} \quad (5)$$

Corollary 2

$$\delta^{-1}(c) = \delta_j \cdot (\delta')^{-1}(c_{ns_j}) + \overline{\delta_j} \cdot (\delta')^{-1}(c_{\overline{ns_j}}) \quad (6)$$

for any next state variable ns_j where δ_j is the corresponding next state function and δ' is the new BFV after deleting δ_j from δ.

Fig 2 is a procedural description of our pre-image calculation based on Eq 6. It uses a cache table to avoid identical subproblems being solved again as in [8, 12]. In line 1, δ and c are modified in such a way that when δ contains constant components, we simplify c using the constant components and when c doesn't depend on some components of δ, we delete those components out of δ.

Definition 3 *Let f be a function; let c represent a set; and let vars be the set of variables in c but not in f. We define a specific* **generalized cofactor,** $f \downarrow c$, *as follows.*

$$f \downarrow c = f \Updownarrow ((\exists\, vars)c)$$

The operator $\Updownarrow$ is called *image restricting generalized cofactor* and precise details can be found in [2, 8, 9]. Existentially abstracting out the variables $vars$ from c guarantees that $f \downarrow c$ will not introduce any new variable which doesn't appear in f. In [3], a technique which we call **Constrained Pre-image calculation** is proposed to reduce the BDD size for the components of δ. It is restated in the following equation:

$$A \cdot \delta^{-1}(c) = A \cdot (\delta \downarrow A)^{-1}(c) \quad (7)$$

where A is a characteristic function representing a subset of the domain we are concerned about and $\delta \downarrow A = [\delta_1 \downarrow A, \delta_2 \downarrow A, \cdots, \delta_n \downarrow A]$. As an example, line 10 of Figure 1 can be rewritten as below:

$$ring_{i+1} = (\delta \downarrow \overline{sum_i})^{-1}(c) \cdot \overline{sum_i} \quad (8)$$

Similarly, we can apply Eq. 7 to lines 7 to 9 of Fig 2 as follows:

$$preimg = \delta_j \cdot (\delta' \downarrow \delta_j)^{-1}(c_{\mathrm{ns}_j}) + \overline{\delta_j} \cdot (\delta' \downarrow \overline{\delta_j})^{-1}(c_{\overline{\mathrm{ns}_j}})$$

The problem with this approach is that even though we may have a simplified functional vector, we degrade the cache hit ratio. However, for some circuits, this technique saves much time and we will discuss it in the experimental result section.

4 Experimental Results

All the circuits we experimented with come with a specified initial state (or initial states) in the circuit description. Basically the experiments consist of three phases. First we check whether a given design is resettable or not. If it is resettable, we compute a synchronization sequence and the R-state we can arrive at by applying the sequence. Next, we check whether the given initial state is also a reset state by traversing a single machine (one factor of a product machine) from the reset state using the implicit method in [12]. Since every state reachable from a reset state is also a reset state, if the initial state is reachable from the R-state, we can say that the initial state is a reset state. Therefore during the traversal, as soon as we detect that the initial state is reached, we exit from the execution by declaring the initial state is resettable. However, if the initial state is not a reset state, we compute all reachable states from the R-state. When multiple states are given as initial state, if any of them is a reset state, we say the given design has a resettable initial state.

The benchmark circuits are from several literatures like [11, 2]. All the experiments were performed on DECstation 5000/200. Our code is built on top of **MIS**[1] whose BDD package is based on [7].

Table 1 can be interpreted as follows. **FF/PI** is the number of latches and primary inputs of a given design. **DR** indicates whether the design is resettable(o)

CKT	FF/PI	DR	IR	LB/SL	TIME
sand	6/11	o	o	19/21	19.7(13.1/4.0)
planet	6/7	o	o	17/29	40.6(26.0/11.9)
mclc	11/11	o	×	1/1	1.7(0.4/0.0)
scf	8/27	o	o	1/1	6.7(2.7/0.1)
sbc	28/40	o	×	2/3	45.8(20.2/1.7)
dsip	224/228	o	o	1/ 1	104.4(34.8/53.4)
pewd	112/117	o	o	1/1	20.2(7.5/6.9)
key	228/258	o	o	1/1	179.2(39.4/110.7)
mM30	90/33	o	o	1/1	95.9(69.7/1.1)
cbp32	32/33	×	-	-/-	3.4(0.0/0.0)
mult32a	32/33	o	o	32/33	338.3(108.4/224.3)
s1196	18/14	o	×	1/1	16.3(1.0/0.2)
s1238	18/14	o	×	1/1	16.2(1.1/0.2)
s1488	6/8	o	o	1/1	2.8(0.6/0.1)
s1494	6/8	o	o	1/1	2.8(0.7/0.1)
s208.1	8/10	×	-	-/-	0.3(0.1/0.0)
s27	3/4	o	o	1/1	0.1(0.0/0.0)
s298	14/3	o	o	2/2	7.5(6.2/0.4)
s344	15/9	o	×	2/2	10.7(6.9/1.0)
s349	15/9	o	×	2/2	10.8(7.0/1.0)
s382	21/3	o	×	1/1	7.9(0.7/0.1)
s386	6/7	o	o	2/2	1.0(0.4/0.1)
s400	21/3	o	×	1/1	7.9(0.8/0.1)
s420.1	16/18	×	-	-/-	0.9(0.2/0.0)
s444	21/3	o	×	1/1	7.8(0.6/0.1)
s510	6/19	o	o	28/75	41.6(32.9/7.2)
s526	21/3	o	o	2/2	22.4(18.7/0.6)
s641	19/35	o	o	1/1	7.2(4.0/0.6)
s713	19/35	o	o	1/1	7.4(4.1/0.6)
s820	5/18	o	o	1/1	1.5(0.4/0.0)
s832	5/18	o	o	1/1	1.5(0.4/0.0)
s838.1	32/34	×	-	-/-	2.5(0.8/0.0)

Table 1: Experimental Results for Resettability

or not(×). **IR** indicates whether the design has a resettable initial state (o) or not (×). For example, **s510** is resettable and the given initial state is a reset state. If we look at the column **LB** and **SL**, the length of the synchronization sequence computed is not much greater than the lower bound of the synchronization sequence except for **s510** and in most cases they are the same which means the synchronization sequence computed is a minimum sequence.

Two time results are given along with the total execution time. They are for computing alignable state pairs and for calculating a synchronization sequence. And the rest of the time is spent preprocessing and checking whether the given initial state(s) is(are) resettable. When a given circuit is well-structured or iterative like **mult32a**, **mM30**, or **key**, we use the structural analysis proposed in [12] for ordering variables.

For **sand**, **planet**, and **s510** which have a small number of latches and a long synchronization sequence, we used the constrained pre-image computation as presented in Sec. 3.2. We had about 2 times faster results for the above circuits using the constrained preimage computation method than without the method. However, as noted previously, for the circuits which have more than 10 latches, the cache hit is important in the preimage computation and therefore the constrained preimage decreased the number of cache hits which resulted in slower results.

5 Conclusions and Future Works

In this paper, we have presented an exact method to compute a synchronization sequence, to compute a tight lower bound for the length of such a sequence, and to check that an initial state given in the specification is a reset state. Also we have shown in the experiments that the proposed method can handle fairly large circuits and the length of the actual synchronization sequence computed is quite close to the lower bound.

As a future study, to handle broader class of circuits, one possible method would be to identify a circuit structure. For example, if a given circuit has a parallel decomposition, then we can apply the proposed algorithm to each sub-machine and by concatenating the synchronization sequences of the sub-machines, we can obtain a synchronization sequence for the whole machine.

References

[1] R. Brayton, R. Rudell, A. Sangiovanni-Vincentelli, and A. Wang, MIS: Multiple-Level Interactive Logic Optimization System, *IEEE Transaction on CAD*, Vol. C6 No. 6, pp1062-1081, November 1987.

[2] O. Coudert, C. Berthet, Jean Christolphe Madre, Verification of Sequential Machines Using Boolean Functional Vectors, *Proceedings of the IMEC-IFIP International Workshop on Applied Formal Methods For Correct VLSI Design*, November 13-16, 1989.

[3] O. Coudert, C. Berthet, and Jean Christolphe Madre, Verifying Temporal Properties of Sequential Machines without Building Their State Diagrams, *DIMACS Technical Report 90-31, volume 2, Workshop on Computer-Aided Verification June 18-21, 1990*, Robert Kurshan and E.M. Clarke, eds.

[4] C. Pixley, A Computational Theory and Implementation of Sequential Hardware Equivalence, *DIMACS Technical Report 90-31, volume 2, Workshop on Computer-Aided Verification June 18-21, 1990*, Robert Kurshan and E.M. Clarke, eds.

[5] C. Pixley and G. Beihl, Calculating Resetability and Reset Sequences, *Proceedings of the IEEE International Conference on Computer Aided Design, pp376-379*, Santa Clara, CA, November 1991.

[6] R.E. Bryant, Graph-Based Algorithms for Boolean Function Manipulation, *IEEE Transactions on Computers*, Vol. C35 No. 8, August 1986.

[7] K. Brace, R. Bryant, and R. Rudell, Efficient Implementation of a BDD Package, *Proceedings of the 27th ACM/IEEE Design Automation Conference*, June 1990, pp. 40-45.

[8] H. Cho, G. Hachtel, S.-W. Jeong, B. Plessier, E. Schwarz, and F.Somenzi, ATPG Aspects of FSM Verification, *Proceedings of the IEEE International Conference on Computer Aided Design, pp. 134-137*, Santa Clara, CA, November 1990.

[9] H.J. Touati, H. Savoj, B. Lin, R.K. Brayton, and A. Sangiovanni-Vincentelli, Implicit State Enumeration of Finite State Machines Using BDDs, *Proceedings of the IEEE International Conference on Computer Aided Design, pp. 130-133*, Santa Clara, CA, November 1990.

[10] F.C. Hennie, Finite-State Models for Logical Machines, *John Wiley & Sons, Inc.*, 1968.

[11] S. Yang, Logic Synthesis and Optimization Benchmarks User GuideVersion 3.0, Microelectronics Center of North Carolina,Research Triangle Park, NC, January, 1991.

[12] G. Hachtel, S.-W. Jeong, B. Plessier, and F.Somenzi, Variable Ordering and Selection for FSM Traversal, to appear in *Proceedings of the IEEE International Conference on Computer Aided Design*, Santa Clara, CA, November 1991.

Functional Approaches to Generating Orderings for Efficient Symbolic Representations*

M. Ray Mercer Rohit Kapur Don E. Ross

Dept. of ECE - ENS 143
The University of Texas
Austin, TX 78712-1084

Dept. of Elec. Eng.
Texas A&M University
College Station, TX 77843

Abstract

Many Boolean functions are efficiently represented as Ordered Binary Decision Diagrams (OBDDs). However, the order in which variables appear in an OBDD can significantly affect their required memory space. In this paper we present a functional approach to generate orderings for representing functions. We develop a cost function which closely mimics the OBDD operations and can be quickly computed. Using the cost as a metric for an ordering we use an annealing procedure to arrive at "good" variable orderings. We compare the results obtained by simulated annealing to orderings generated from heuristics that use circuit topology to arrive at a variable ordering.

1 Introduction

Ordered Binary Decision Diagrams are used to compactly represent logic functions and have found applications in Logic Synthesis, Design Verification and Test Generation. To efficiently manipulate OBDDs, an ordering has to be defined for switching variables which appear in these data structures.

Different variable orderings can result in varied representation costs for the same set of functions. For a single function, the representation cost can grow linearly as variables are added under one ordering, while growing exponentially under another. Because OBDD computational complexity is closely related to its space requirements[BRYA86], it is important to find an ordering that minimizes the memory used during all phases of OBDD manipulation.

There are $n!$ different orderings that could be used to represent a function, where n is the cardinality of the dependency set of the function. The best algorithm so far to deterministically arrive at an optimal ordering for a circuit has a time complexity of $n^2 3^n$ [FRIE90]. Given the search space and the complexity of finding optimal orderings, we must resort to heuristics to obtain "good" orderings with low overheads.

The $n!$ space is similar to that of the traveling salesman problem for which approaches based on simulated annealing have resulted in near optimal solutions. For simulated annealing to be used to select an ordering for OBDDs, an efficient mechanism has to be defined to compute the cost function for a given ordering. Apart from the speed requirements this cost function should be a good estimator for the actual OBDD representation cost. Ordered Partial Decision Diagrams (OPDDs) have been shown to satisfy both the requirements of the cost function. In this paper we show how the orderings generated by simulated annealing using OPDDs compare with other heuristically generated variable orderings [BERM88] [BERM89] [BRAC90] [BUTL91] [FUJI88] [MALI88] [MINA90]. Our simulated annealing method differs from other functional approaches [FUJI91][ISHI91] as it does not require that the OBDD of the initial ordering be able to fit into the memory.

In the next section we discuss Partial Decision Diagrams, and in section 3 we discuss their use as a cost metric for variable orderings. Finally in section 4 we discuss the improvements obtained in the orderings over previously known best results (the orderings are listed in the appendix).

2 OBDDs and OPDDs

The reader is assumed to be familiar with the concept of OBDDs. Their properties have been well documented by Bryant [BRYA86]. Bryant also showed that regardless of ordering, some functions exist which cause an n input function to require memory resources exponential in n [BRYA91]. When memory resources get scarce, OBDDs that require more than available resources fail to provide any information about the function. In such cases, we use Ordered Partial Decision Diagrams (OPDDs [ROSS90]), which bound the

*This work was supported by the National Science Foundation under Grant MIP-8552537, by the Semiconductor Research Corporation under Contract #91-DP-142, and by the Innovative Science and Technology Office of Strategic Defense Initiative Organization, administered through the Office of Naval Research under contract N-00014-86-K-00554.

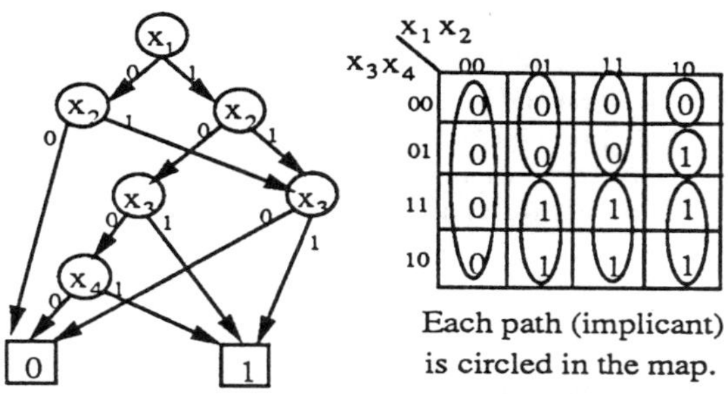

Figure 1: A complete OBDD or an OPDD with the memory limit greater than 6 vertices.

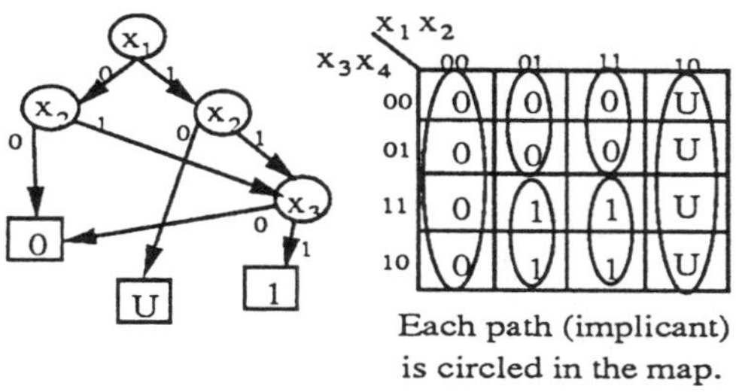

Figure 2: A partial OPDD generated by *Apply* when the memory limit is set to 4 vertices.

memory requirements of an OBDD's non-terminal vertices to a constant k, where a non-terminal vertex is one that does not terminate a path that originates at the root of the OBDD or OPDD. When an expansion algorithm theoretically similar to *Apply* would produce a fully specified graph that requires more than k non-terminal vertices, it terminates the least informative paths (implicants) to a new terminus which signifies an unknown value. The resulting graph, called an OPDD, has partial information about the function. For example, consider the function of Figure 1, where the number of non-terminal (round) vertices in the OBDD representation is six. This would be the functional result produced by *Apply* if the limit for the expansion graph size, k, was set to a value large enough so that the entire expansion could occur. If the expansion limit, k, was set to four, the entire expansion could not fit within the imposed memory limit. Our *Apply* generates four vertices, and then terminates all unexpanded parts of the graph to an unknown terminus, as shown in Figure 2. OPDDs are manipulated in a similar fashion to OBDDs enabling us to generate partial information of functions when OBDDs would fail to provide any information due to excessive memory requirements.

3 Sensitivity of OPDDs to Ordering

Just as the total number of vertices of an OBDD can be used as a metric for the "goodness" of an ordering, a strong correlation exists between the percentage of "known" information calculated using OPDDs and the "goodness" (memory consumption) of the same ordering using OBDDs [ROSS91]. An OPDD can be viewed as a sample of the OBDD for the same ordering. The sampling of the OBDD could be randomly selected, or based on some heuristic. Our algorithm for generating the OPDDs uses a greedy approach to first produce the short paths of the function which hold exponentially more information in linearly less space. The OPDD created by this heuristic will tend to represent the top portion of the OBDD containing the shortest paths to termini. Using OPDDs to select among alternative OBDD orderings is therefore a heuristic selection mechanism, because in the worst case, most of the eventual OBDD is missing. However, empirically we have found that an ordering which explodes the top of the graph dramatically is never better than an ordering that compacts it dramatically, although theoretically this could be the case.

OPDDs provide a very fast and reliable estimate of the "goodness" of the ordering. This enables us to evaluate many variable orderings in a short time to select the best ordering for OBDDs (full functional representation). The sensitivity of OPDDs to orderings is employed in a simulated annealing algorithm.

4 Results

When we run OPDDs using an ordering, one of the following situations may occur:

1. Complete information is obtained for all of the functions represented. In this case we can use the total vertices used to represent all the functions, or the largest graph created as the cost metric for an ordering.

2. Some of the functions are partially represented. In this case, we can use the average amount of information contained in all the functions, or the average information in the functions at the primary outputs of the circuit to be the cost metric for an ordering.

In our implementation we annealed on the largest graph when all functions were completely represented, but we annealed on the average information over all functions when some of the functions were partially represented. Since a move can change the cost from partial functions to complete functions and vice versa we biased our acceptance of the move towards the case when all functions are fully represented. A move is defined by taking a variable out from its present location in the variable ordering and putting it at another

Circuit	Previous best ordering (largest graph/ave. info.)	Ordering results after annealing (largest graph/ave. info.)
C432	524 (nat)	369
C499	4533 (ucb)	4283
C880	2401 (ucb)	1248
C1355	4533 (ucb)	4283
C1908	2279 (uta-b)	1606
C2670	98.889% (ucb)	99.516%
C3540	* (uta-sc)	17747
C5315	1740 (ucb)	1296
C7552	99.729% (ucb)	99.797%

Table 1: Annealing results compared to previously known best orderings. Memory used per OBDD/OPDD is limited to 36000 vertices.

randomly chosen location at most ±5 levels from the current location. We use this instead of a general move to any location because we start our annealing procedure from a heuristically generated ordering. The type of move selected is not of major concern to us, as the thrust of this paper is to show the use of a cost function along with annealing. This allows us to generate variable orderings where full functions can be represented, from seeds that do not allow full functional representations. It also works in situations when orderings for complete functional representations already exist and a more efficient variable ordering is desired.

Table 1 lists the results obtained by simulated annealing as compared to the previous best results generated by any heuristics that traverse the circuit to determine an ordering. The source of each heuristically generated ordering is indicated (*nat* - natural ordering, *ucb* - University of California at Berkeley [MALI88], *uta-b* & *uta-sc* - University of Texas at Austin [BUTL91]). During the annealing process we limited the size of the graphs to 8000 non-terminal vertices. The overall best orderings obtained by annealing were used to generate complete graphs using a memory limit of 36000 vertices per graph. The results show the average information over all functions when graphs exceeding 36000 vertices were created and the largest OBDD created when all functions are fully represented. The results for the smaller circuits indicate the utility of the cost function in simulated annealing when smallest possible representations of the functions are desired. The results for C3540 show a situation where initially no ordering existed which allowed complete functional descriptions for all lines in the circuit and annealing resulted in an ordering that produced full functional information. Moreover, a significant improvement in the orderings generated by simulated annealing over the previous results is observed. Simulated annealing also has the advantage that it does not require an existing circuit topology to find an ordering that results in acceptable resources being used. The scenario (to generate variable orderings) as proposed by us is:

Start with a heuristically selected ordering.
Anneal using OPDDs (with a small limit on the vertices for a quick evaluation of the ordering).
Use the ordering at the end of the annealing procedure for OBDD calculations.

5 Appendix

The following are the orderings that we obtained by simulated annealing on the ISCAS 85 circuits [BRGL85].
Format: The inputs are assumed to be in their natural order. For the input variables the location of the variable in the ordering is specified. Example: Assume a circuit CX with 4 inputs A, B, C, D. If the ordering is defined as CX: 1 0 3 2, it implies an ordering where B is the first variable in the ordering, A the second, D the third and C at the end.

Orderings:
C432: 3 0 4 2 7 1 5 9 6 8 10 13 11 12 14 17 15 16 18 21 19 20 22 25 23 24 26 29 27 28 30 34 31 35 33 32

C499, C1355: 5 15 28 40 1 17 29 36 0 16 30 32 4 14 27 33 2 3 7 6 9 11 8 10 21 22 26 23 18 20 24 25 13 19 31 38 37 34 39 35 12

C880: 0 8 59 16 6 2 43 12 5 14 9 54 56 58 47 13 3 35 51 52 55 21 34 28 22 18 24 30 38 32 37 45 7 17 29 36 20 23 27 10 46 44 19 26 31 42 41 39 11 4 48 50 49 53 57 15 33 40 25 1

C1908: 11 8 7 14 2 6 13 10 9 4 17 18 21 16 15 5 24 25 26 29 12 30 27 28 31 22 20 19 23 3 1 32 0

C2670: 158 145 156 159 133 151 171 110 134 103 164 130 141 125 149 155 129 102 138 147 119 150 114 176 140 124 157 180 21 77 81 146 116 84 93 91 38 45 35 48 62 71 122 96 80 52 69 61 41 72 56 60 139 70 74 83 97 59 57 64 66 76 65 128 63 75 50 82 49 55 39 68 85 178 8 132 25 13 26 28 18 4 7 44 24 115 1 78 40 34 73 31 19 54 6 123 53 58 46 22 15 30 23 12 33 120 10 51 17 32 47 14 20 43 3 37 16 217 218 226 198 213 170 214 36 203 230 193 207 211 0 222 9 228 210 229 216 174 185 196 225 206 220 204 201 194 169 166 202 195 182 232 192 205 212 190 172 189 209 223 163 188 191 199 215 175 186 200 183 187 177 148 184 162 181 161 197 173 179 165 208 168 221 11 219 227 27 231 224 160 137 167 42 136 131 29 135 144 143 154 67 107 79 94 100 98 108 90 89 109 99 126 153 111 113 101 88 105 86 92 2 5 152 118 112 106 142 121 127 117 95 104 87

C3540: 1 7 9 11 4 5 32 28 18 23 35 37 44 49 3 20 0 22 43 41 46 31 14 16 17 15 8 36 29 30 25 12 19 33 38 40 47 13 45 6 42 48 2 34 24 26 39 10 21 27

C5315: 2 96 148 89 151 160 16 172 169 175 166 1 176 158 130 167 115 113 134 124 22 156 98 154 150 162 155 174 164 159 173 0 12 170 147 107 43 78 114 144 119 138 121 105 106 122 117 143 87 23 85 104 111 99

95 136 132 76 38 145 140 141 3 82 163 153 11 128 131 129 116 108 88 93 112 123 97 84 90 165 168 100 101 109 102 118 110 60 66 63 55 56 53 54 47 65 62 51 135 50 52 49 68 69 73 70 81 72 83 74 142 133 40 46 36 35 44 37 31 27 33 29 5 45 24 7 14 8 10 17 26 137 127 4 77 126 67 71 79 48 64 59 61 58 41 30 32 28 18 9 177 15 120 125 152 157 75 86 92 39 57 25 161 171 13 42 21 34 19 146 139 94 103 80 91 6 20 149

C7552: 205 194 33 32 196 4 18 28 20 63 56 95 22 83 37 47 154 144 138 139 199 117 120 137 133 128 109 116 129 49 121 134 130 132 142 124 112 126 131 135 123 115 113 114 101 119 94 97 197 54 46 9 12 92 118 125 122 108 111 105 104 45 39 21 6 11 204 202 68 71 99 57 82 203 80 195 31 24 25 14 66 67 72 87 84 180 0 145 148 160 143 140 146 147 158 187 162 153 149 159 163 152 141 151 136 185 173 190 191 176 166 189 170 169 178 155 164 157 167 161 168 172 192 175 156 165 171 184 183 174 179 177 81 182 90 100 75 86 76 73 35 193 41 61 65 62 44 42 51 53 79 181 19 188 8 13 5 201 15 17 10 29 107 186 198 200 102 93 77 85 96 89 78 91 110 106 98 69 88 74 64 30 27 34 50 127 43 7 40 16 2 206 3 1 23 150 38 48 55 59 36 58 70 60 52 26 103

6 Acknowledgements

The authors would like to thank Sharad Malik of Princeton and Karl Brace of CMU for correspondence and providing us information on their orderings. Also acknowledged are the contributions of Kenneth Butler of Texas Instruments.

References

[BERM88] C.L. Berman, "Circuit width, register allocation, and reduced function graphs," IBM Res. Rep. RC 14127, Nov. 1988.

[BERM89] C.L. Berman, "Ordered Binary Decision Diagrams and Circuit Structure," Extended Abstract, *Proc. International Conference Computer Design*, Oct. 1989, pp. 392-395.

[BRAC90] K.S. Brace, R.E. Bryant and R.L. Rudell, "Efficient implementation of a BDD package," *Proc. ACM/IEEE 27th Design Automation Conf.*, Jun. 1990, pp. 40-45.

[BRGL85] F. Brglez and H. Fujiwara, "A neutral netlist of 10 combinational benchmark circuits and a target translator in FORTRAN," *Proc. IEEE International Symposium on Circuits and Systems (ISCAS)*, Jun. 1985, pp. 695-698.

[BRYA86] R.E. Bryant, "Graph-based algorithms for Boolean function manipulation," *IEEE Transactions on Computers*, Vol C-35, No. 8, Aug. 1986, pp. 677-692.

[BRYA91] R.E. Bryant, "On the complexity of VLSI implementations and graph representations of Boolean functions with application to integer multiplication," *IEEE Transactions on Computers*, Vol. C-40, No. 2, Feb. 1991, pp. 205-213.

[BUTL91] K.M. Butler, D.E. Ross, R. Kapur and M.R. Mercer, "Heuristics to compute variable orderings for efficient manipulation of ordered binary decision diagrams," *Proc. ACM-IEEE 28th Design Automation Conference*, Jun. 1991, pp. 417-420.

[FRIE90] S.J. Friedman and K.J. Supowit, "Finding the optimal variable ordering for binary decision diagrams," *IEEE Trans. on Comput.*, vol C-39, no. 5, May 1990, pp. 710-713. see also *Proc. ACM-IEEE 24th Design Automation Conference*, Jun. 1987, pp. 348-356.

[FUJI88] M. Fujita, H. Fujisawa and N. Kawato, "Evaluation and improvements of Boolean comparison method based on binary decision diagrams," *Dig. Technical Papers, International Conference on CAD*, Nov. 1988, pp. 2-5.

[FUJI91] M. Fujita, Y. Matunaga and T. Kakuda, "On the variable ordering for binary decision diagrams for the application of multilevel logic synthesis," *Proceedings of the European Design Automation Conference*, Feb. 1991, pp. 50-54.

[ISHI91] N. Ishiura, H. Sawada and S. Yajima, "Minimization of binary decision diagrams based on exchange of variables," *Proc. International Conference of Computer-Aided Design*, Nov. 1991, pp. 472-475.

[MALI88] S. Malik, A.R. Wang, R.K. Brayton and A. Sangiovanni-Vincentelli, "Logic verification using binary decision diagrams in a logic synthesis environment," *Dig. Technical Papers, International Conference on CAD*, Nov. 1988, pp. 6-9.

[MINA90] S. Minato, N. Ishiura and S. Yajima, "Shared Binary Decision Diagram with attributed edges for efficient Boolean function manipulation," *Proc. 27th ACM/IEEE Design Automation Conference*, Jun. 1990, pp. 52-57.

[ROSS90] D.E. Ross, "Functional calculations using ordered partial multi decision diagrams," Ph.D. Dissertation, The University of Texas at Austin, Aug. 1990.

[ROSS91] D.E. Ross, K.M. Butler, R. Kapur and M.R. Mercer, "Fast Functional Evaluation of Candidate OBDD Variable Orderings," *Proceedings of the European Design Automation Conference*, Feb. 1991, pp. 4-10.

Inductive Verification of Iterative Systems*

June-Kyung Rho Fabio Somenzi
Department of Electrical and Computer Engineering
University of Colorado at Boulder

Abstract

Recent advances in BDD-based algorithms have brought much larger circuits than before within the reach of verification programs. However, these algorithms only indirectly benefit from the regularity of the circuits being analyzed (or sub-circuits thereof). In this paper we show how inductive proof procedures can derive information on regular circuits in *optimal* time, e.g., they can perform reachability analysis in linear time or check the equivalence of two iterative circuits in time independent of the length of the two arrays. The algorithms presented in this paper rely on the canonicity of BDD's to make the inductive steps efficient. We have implemented them and we briefly discuss their results.

1 Introduction

Sophisticated verification algorithms [12, 9, 23, 13, 11] based on Binary Decision Diagrams (BDD's) [6, 4] have obtained remarkable success in verifying sequential circuits. The numbers of states these algorithms can deal with are exorbitant ($> 10^{120}$), while recent and ongoing work [8, 20, 7, 19, 15, 14] keeps pushing the limits even further. However, those exorbitant numbers are still small when compared to those characterizing complex, real-life systems, for which verification is most important.

Reductions, as proposed in [21], provide a general mechanism to deal with arbitrary complexity. In summary, reduction is the process of deriving a simpler system from a complex one, so that the two systems are related by a homomorphism that preserves the properties being verified. An active area of research is the automatic generation of such homomorphisms.

In this paper we present a procedure for the automatic generation of homomorphisms for iterative systems [18], i.e., systems with a regular structure. The homomorphisms are generated by an inductive procedure and can be used to solve verification problems in optimal time. As an example, we can verify the equivalence of arbitrarily long arrays of combinational circuits by verifying the equivalence of two small finite state machines, whose sizes are independent of the length of the array, or we can build the BDD for the reached states of a sequential circuit in time linear in the number of inputs. Given that the size of the BDD is linear, our result is optimal. For sequential circuits, the homomorphisms relate the sets of reached states to the languages accepted by automata whose sizes are independent of the length of the array.

Browne *et al.* [5] have proposed a restricted form of branching-time temporal logic that can be applied to networks of many identical processes. Kurshan and McMillan [22] discuss an inductive procedure that requires the verifier to produce an invariant process. Theorem provers [3, 16] have been used to assist manual verification of iterative systems. (See, for instance, [2].) Our approach shares the objectives of these works, but takes a different approach, that is fully automated. The results we present are for unilateral iterative systems. Typically, the inductive step of an inductive proof requires the algebraic manipulation of the hypothesis. The canonicity of BDD's is used in our case to simplify and automate this process.

Though few large systems are iterative systems, most do contain iterative subsystems. Hence our research has practical significance. We assume that the high-level synthesis system that generates the input for our procedure passes on the information on repetitive constructs found in the user-provided description. This makes the analysis more efficient, but is not restrictive.

The rest of this paper is organized as follows. Section 2 provides the background material. Section 3 examines the inductive construction of BDD's for combinational systems, whereas Section 4 deals with sequential systems. Section 5 collects the results on the equivalence verification of pairs of iterative implementations. Finally, Section 6 describes the experimental results, presents our conclusions and points to directions for future work.

2 Preliminaries

Binary Decision Diagrams (BDD's) provide a convenient representation for logic functions. A BDD is a directed acyclic graph $(V \cup \Phi \cup \{1\}, E)$ representing a multiple-output logic function F. The variables on which F depends form the *support* of F, indicated by S_F. V is the set of the internal nodes. The outdegree of $v \in V$ is 2. Every node v has a label $l(v) \in S_F$. **1** is the terminal node: Its outdegree is 0. Φ is the set of the function nodes: The outdegree of $\phi \in \Phi$ is 1 and its indegree is 0. The function nodes are in one-to-one correspondence with the components of F. The outgoing arcs of function nodes may have the *complement* attribute. The two outgoing arcs for a node $v \in V$ are labeled T and E, respectively. The E arc may have the *complement* attribute. We use $(l(v), T, E)$ to indicate an internal node and its two outgoing arcs. The variables in S_F are ordered and if v_j is a descendant of v_i $(v_i, v_j \in V)$, then $l(v_i) < l(v_j)$. The function F represented by a BDD is defined as follows:

1. The function of the terminal node is the constant function **1**.

2. The function of an arc is the function of the head node, unless the arc has the complement attribute, in which case the function of the arc is the complement of the function of the node.

3. The function of a node $v \in V$ is given by $l(v) f_T + \overline{l(v)} f_E$, where f_T (f_E) is the function of the T (E) arc.

4. The function of $\phi \in \Phi$ is the function of its outgoing arc.

An arc with (without) the attribute is called a *complement (regular) arc*. Using this mechanism, a function and

*This work was supported in part by NSF/DARPA grant MIP-8719546 and by SRC contract 91-DJ-206.

its complement are represented by the same sub-DAG and complementation takes constant time. In particular, there is only one terminal node, since the constant functions (1 and 0) are the complement of each other. BDD's are canonical (the representation of F is unique for a given variable ordering) if:

1. All internal nodes are descendants of some node in Φ.

2. There are no isomorphic subgraphs.

3. For every node, $f_T \neq f_E$.

Note that the restriction that the T arc may not be complemented is imposed to guarantee canonicity.

In the illustrations, we use a circle to denote the E arc. An empty circle indicates a regular arc, whereas a full circle indicates a complement arc. Also, dangling arcs are meant to point to the constant node. For an example, see Figure 5.

In what follows a (finite state) machine is a 6-tuple $\langle I, S, O, \delta, \lambda, s^0 \rangle$, where I is the input alphabet, S is the set of states, O is the output alphabet, δ and λ are the next-state and output functions, respectively, and s^0 is the initial (reset) state. An automaton is a 5-tuple $\langle I, S, \delta, F, s^0 \rangle$, where I, S, δ, and s^0 are as before. F is a set of final or accepting states.

A state σ is *reachable* if there exist a sequence of states $s_1, \ldots, s_k$ and a sequence of inputs $i_0, \ldots, i_k$ (possibly repeated), such that $\delta(s^0, i_0) = s_1$, $\delta(s_j, i_j) = s_{j+1}$, $j = 1, \ldots, k-1$, and $\delta(s_k, i_k) = \sigma$. We only consider finite strings of inputs in this paper. The set of reachable states of a finite state machine or an automaton is of interest in verification and efficient algorithms have been developed to find it [9, 23, 13, 11].

The language produced by an FSM is the set of all possible output strings, produced by the machine when started in s^0. The language accepted by an automaton is the set of all the input strings that drive the automaton from the initial state to one of the accepting states. For every FSM there exists an automaton that accepts the language produced by the FSM. How to build such an automaton is described in [18, Ch. 2].

An *iterative network* is a circuit obtained by interconnecting multiple instances of the same basic *cell*, as in Figure 1. The signals entering the network from the left are called *boundary signals*. An iterative network is *unilateral* if the *intercell leads* carry signals in only one direction; otherwise it is *bilateral*. If the cell is a combinational circuit, then also the iterative network is combinational. An *iterative system* is the class of all iterative networks with the same cell and boundary signals and different numbers of cells.

Suppose a unilateral system S is built by replicating cell C. The i-th instance of C from the left will be denoted by C_i. The intercell leads will be denoted by W_i, $i \geq 0$, with the boundary signals corresponding to W_0. We shall also use w_i as a synonym of W_{i-1}, for $i > 0$. If C is sequential, the internal present state of cell C_i is denoted by y_i and the next state by Y_i. Finally the primary outputs are denoted by z_i.

Combinational unilateral networks are simply related to sequential machines. A unilateral combinational network can be obtained by unrolling the sequential machine for a given number of time frames.[1] The intercell leads correspond to the state variables at the different time frames. Sequential iterative networks correspond to two-dimensional combinational arrays. More details on iterative networks can be found in [17, 18].

3 Verification of Combinational Iterative Systems

In this section we present methods to generate BDD's for combinational circuits based on the correspondence between automata and BDD's. Though these algorithms can be applied to the verification of combinational circuits, this is not their primary interest, since standard techniques are already efficient in that case. They are presented here because we shall use the concepts developed in this section in dealing with sequential circuits.

Consider the FSM M associated with cell C of a combinational unilateral iterative system S. The present state variables of M correspond to W and the next state variables to w. In the discussion that follows, we shall assume that W consists of binary signals W^j, $j = 1, \ldots, J$. For simplicity we assume that x, the primary input, is a single binary signal. Removing this restriction is straightforward.

Finding the input combinations to S that cause W_n^j to be 1 is easily seen to correspond to finding the input sequences of length n that drive M to states with $w^i = 1$. This defines an automaton A with the same state transition graph as M, where the accepting states are those with $w^j = 1$. The problem can then be solved by unrolling[2] the transition graph TG or A for n steps. This produces an acyclic graph G, where the root corresponds to the initial state (which in turn corresponds to the boundary conditions of S) and the nodes at distance d from the root correspond to the states reachable from the initial states in d steps. By associating a value 1 to the states at distance n from the root (1 for the accepting states and 0 for the others), one obtains a decision diagram for the sequences that lead to the desired states and hence for the inputs of S that produce the desired value at the intercell lead.

The input variables appear naturally ordered from x_1 to x_n. It is therefore sufficient to apply reduction, as described in [6] to obtain an ordered, reduced, BDD. Before we discuss how to obtain a BDD with the ordering $x_n, \ldots, x_1$ and how to take into account the arc attribute described in Section 2, we consider an interesting property of G concerning the regularity of the structure. Since A has a finite number of states, the sets of states reached at each step will eventually repeat regularly. This directly translates into a regular structure for G. It is therefore sufficient to unroll TG in breadth-first manner until the same set of states appears twice. From that point on, G could be expanded by just repeating the segment between the two occurrences of the same set of states. In practice, however, unrolling TG by breadth-first traversal is an efficient way of replicating that pattern.

When the BDD's for the outputs from several cells are desired, it is convenient to order the input variables from x_n to x_1. This allows the sharing of parts of the BDD for W_n^j by the BDD's for W_i^j, $i < n$, so that the overall BDD grows linearly with n. We now illustrate how this can be accomplished. The key idea is to build the *reverse* automaton of A, A^R [18, p. 73]. A^R accepts the reverse language of A, that is, $\sigma_1, \sigma_2, \ldots, \sigma_{p-1}, \sigma_p$ belongs to the language of A if and only if $\sigma_p, \sigma_{p-1}, \ldots, \sigma_2, \sigma_1$ belongs to the language of A^R.

The reverse automaton is built by simply reversing the directions of the arcs in TG and making all the accept-

[1] This fact is used in automatic test pattern generation for sequential circuits. (See, for instance, [1].)

[2] The unrolling can be performed by a breadth-first traversal that does not stop when all nodes are visited.

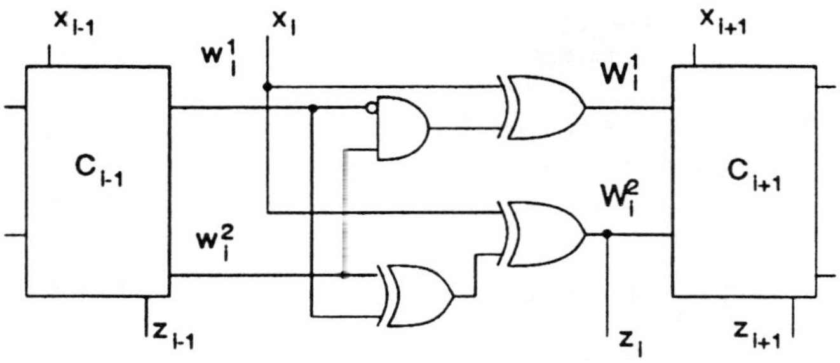

Figure 1: Iterative System Illustrating the Use of the Reverse Automaton.

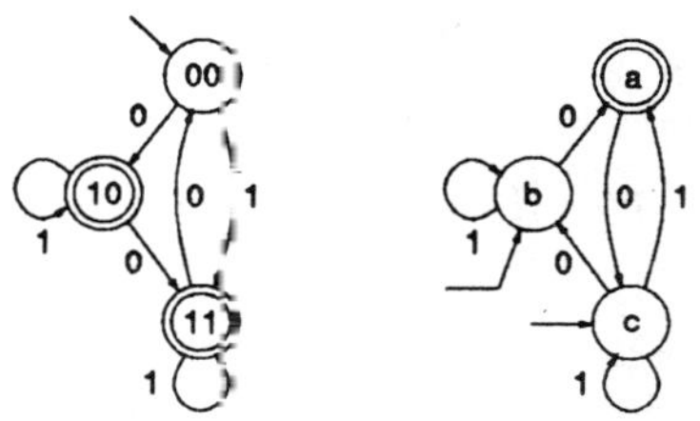

Figure 2: Automata A and N^R for Variable W^1 in the System of Figure 1.

ing states initial and the initial states[3] accepting. The resulting non-deterministic automaton N^R is converted to deterministic form to obtain A^R. As an example, consider the iterative system of Figure 1. Suppose we want to build BDD's for W_i^1, $i = 1, \ldots, n$. We assume the boundary conditions 00. Figure 2 shows the automaton A and the non-deterministic reverse automaton N^R. In N^R the states have been renamed for convenience. Figure 3 shows the deterministic reverse automaton A^R. Once A^R is obtained, the procedure to generate the BDD is exactly as in the case already discussed. The interpretation of the states of A^R is in terms of sets of predecessors states in A.

Observing the deterministic reverse automaton of Figure 3, we see that it has mirror symmetry. This fact is related to the presence of complement arcs in the BDD for W_i^1. Every accepting state can be paired, in this example, with a non-accepting state and vice versa. (This is not the general case, though.)

Exploiting the symmetry, we can form the graph of Figure 4, called the *attributed-edge* graph (AEG), from the graph of Figure 3, by a procedure reminiscent of state min-

[3] We only discuss the case of a single initial state, but the generalization is straightforward.

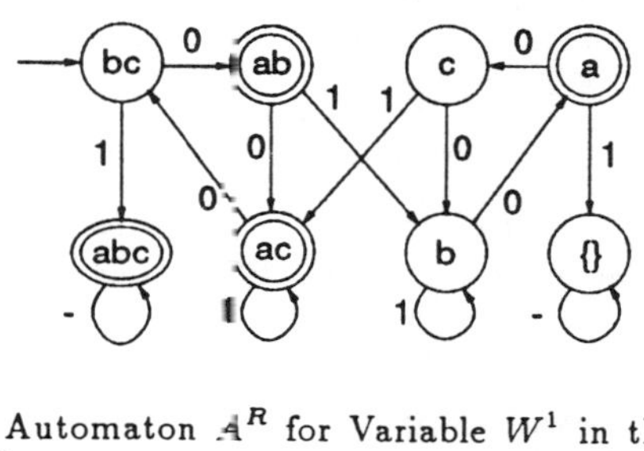

Figure 3: Automaton A^R for Variable W^1 in the System of Figure 1.

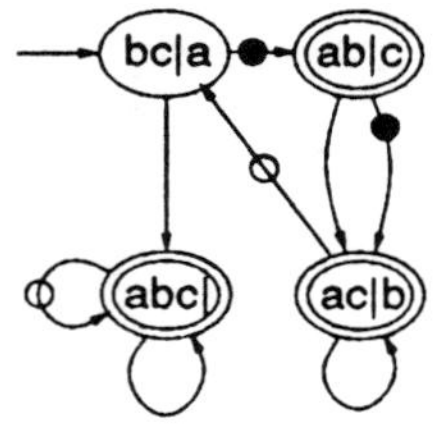

Figure 4: Attributed-Edge Graph for the Reverse Automaton of Figure 3.

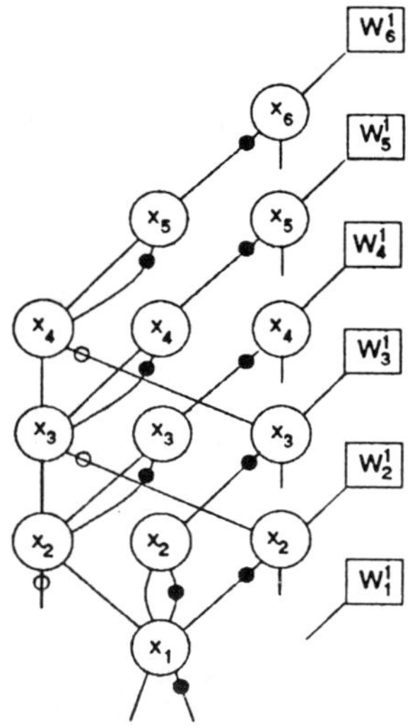

Figure 5: The BDD with Complement Arcs Generated by Unfolding the Graph of Figure 4.

imization.

The unrolling of the AEG directly generates a BDD with complement arcs, as illustrated in Figure 5 for $n = 6$. To complete our discussion, we mention that when C has more than one primary input, the AEG is subjected to the transformation exemplified in Figure 6.

4 Verification of Sequential Iterative Systems

All unilateral combinational systems have linear-size BDD's. In this section we examine the conditions under which the same property is shared by sequential unilateral systems. In particular, we shall be interested in the size of the BDD's representing the set of reached states of the iterative networks. This interest is justified by the central place of

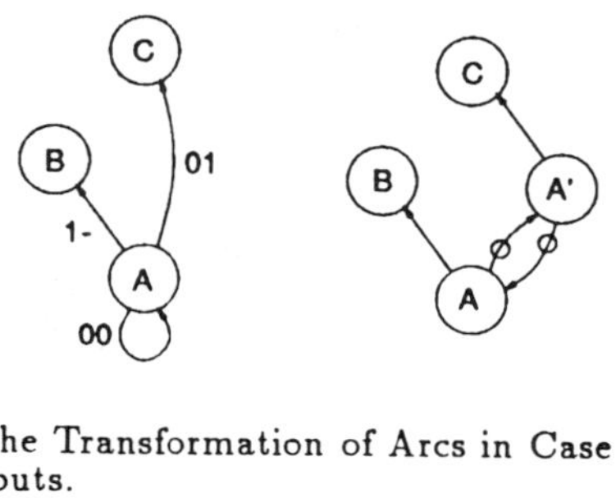

Figure 6: The Transformation of Arcs in Case of Multiple Primary Inputs.

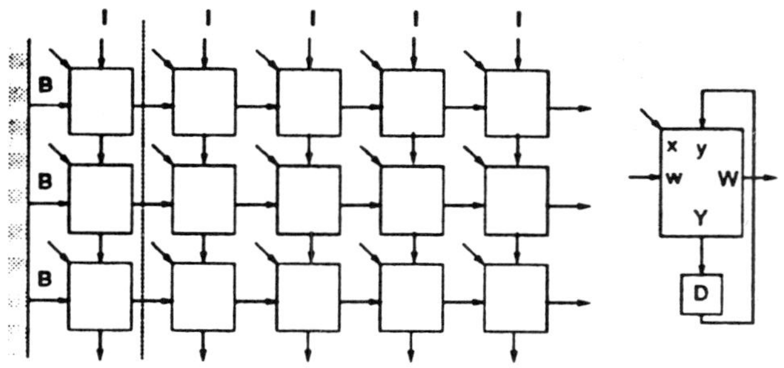

Figure 7: The Two-Dimensional Combinational Equivalent of a Monodimensional Sequential System.

reachability analysis in the verification of finite state machines.

Our main result will be to show that if the basic cell of the iterative system is stable—in a sense to be specified in the following—then the BDD for the reached set is linear in size and regular in structure. We shall also present an approach, based on the result just mentioned, that allows to compare two iterative implementations for equivalence in constant time (i.e., independent of n). The approach, applicable to both combinational and sequential systems is remarkably simple and efficient.

Our analysis considers the two-dimensional combinational correspondent of a sequential (monodimensional) system and regards finite state machines as (regular) language transducers. It was mentioned in Section 2 that the output strings produced by a FSM form a regular language. It was understood that all possible sequences were applicable to the input. If the input strings are restricted to belong to a regular language, the output strings will form another regular language contained in the one produced with unrestricted inputs. We can therefore consider an FSM as a device that, given a regular language, produces a regular language. We shall write $\mathcal{L}_M$ to denote the mapping between languages performed by machine M and we shall call it the *transfer function* of M.

Figure 7 shows the combinational correspondent of a monodimensional, unilateral, sequential system S. It also shows the original sequential cell C and the names of the signals. The vertical dimension of the combinational system represents time, so that the second row in Figure 7 represents the sequential system at the second time of computation.

If we transpose the system of Figure 7, we obtain another two-dimensional combinational system, represented in Figure 8. This new system, can be seen as the expansion in time of a sequential system obtained from the original one by interchanging the roles of state (initial conditions) and intercell leads (boundary conditions), or, more generally, by interchanging space and time.[4] We call this new sequential system S^T the *transpose* of the original system and its basic cell C^T the transpose cell.

We are interested in the transpose cell because its transfer function is strictly related to the sets of reached states of the original sequential system. Consider the lines cut by the dotted line in Figure 7. The image of the three functions is the set of values that may appear on the intercell lead W_1. It is at the same time the set of strings of length 3 produced by the basic cell, when the input language is given by the boundary conditions. This observation can be generalized to say that the characteristic function of a vertical cut in a combinational array with m rows is the same as the characteristic function of the strings of length m in the language produced by the sequential cell in the position immediately to the left of the cut.

[4] For a treatment of space-time transformations, see [18, Ch. 10].

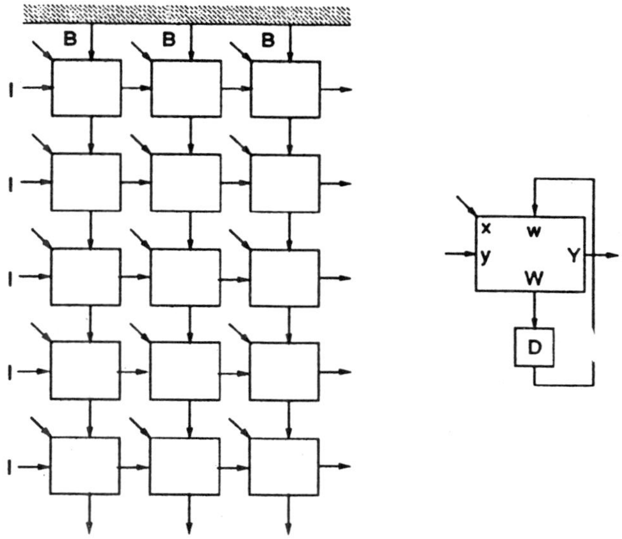

Figure 8: The Transpose of a Two-Dimensional Combinational System.

The same argument can be applied to the transpose array to prove that the strings of length n of the language produced by the transpose cell in column i have the same characteristic function as the image of the intercell leads leaving the same column in a combinational transpose array with n rows. However, by the way we have built the transpose system, the image of the intercell leads leaving column i is the set of states reached by the original system after the i-th step. We are now ready to state our main result.

Lemma 1 *The set of states reached at the i-th step by a sequential unilateral network of n cells can be computed as the characteristic function of the strings of length n produced by the i-th cell of the transpose network (of at least i cells).*

The proof follows from the previous discussion. The importance of this lemma is twofold. On the one hand it tells us that the transpose cells can be unrolled to produce the BDD's for the reached sets, as the basic cell was unrolled to produce the BDD's of combinational functions. On the other hand, Lemma 1 provides us with a sufficient condition for sequential systems to have linear BDD representations for their sets of reached states.

Theorem 2 *Given a sequential unilateral system S with cell C, the BDD's for the sets of states reached at each step by S has size linear in n, the number of cells, if*

$$L_{k+l} \subseteq L_k$$

for some k, l, where L_k is the language produced by the k-th cell in the transpose system.

Proof. If the languages produced by the transpose cells satisfy the condition of the theorem, then all the automata accepting those languages have numbers of states bounded by some constant H. Hence, the widths of the BDD's generated by unrolling those automata are bounded by H and the BDD's themselves grow at most linearly. □

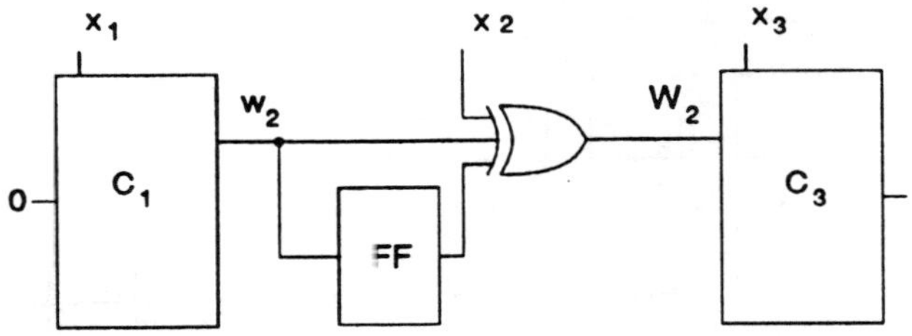

Figure 9: Iterative System Illustrating the Use of Stability.

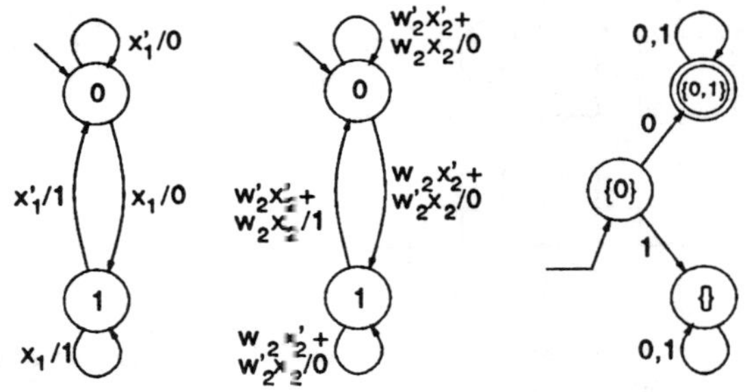

Figure 10: FSM's for Transpose Cells and Automaton Accepting their Language.

If $L_{k+l} \subseteq L_k$, we say that the transpose cell is *stable*. In view of Theorem 2, we have the following procedure to generate the sets of reached states of a sequential unilateral system.

- Test the stability of the transpose cell, with a predetermined limit for $k+l$. If the test fails, fall back to a conventional method of reachability analysis.
- If the test succeeds, generate the BDD's by unrolling the automata accepting the languages produced by the first $k+l$ cells of the transpose array.

As an example, consider the system whose first three cells are depicted in Figure 9. The initial state is assumed to be 0 for all cells. The FSM's for the first two transpose cells, together with the automaton accepting the language produced by both—it is the same—are shown in Figure 10 from left to right. The system is thus stable and the BDD's for the reached states are linear. On the other hand, the system of Figure 11 is unstable. (The initial state is assumed to be 1 for all cells.) One can verify that the i-th transpose cell produces strings with an arbitrary prefix of length i followed by all zeroes. Hence the languages are strictly increasing and convergence does not occur.

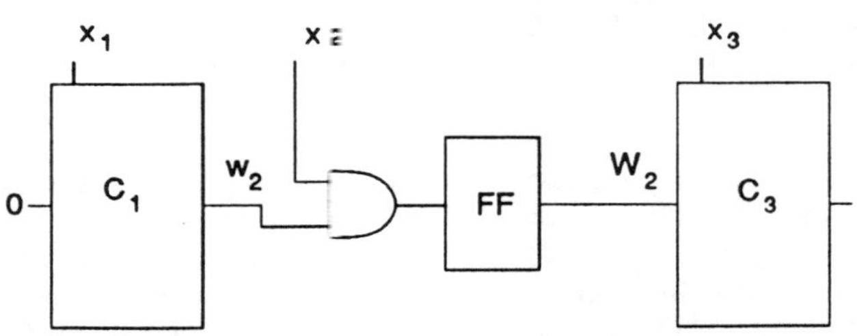

Figure 11: Unstable Sequential Iterative System.

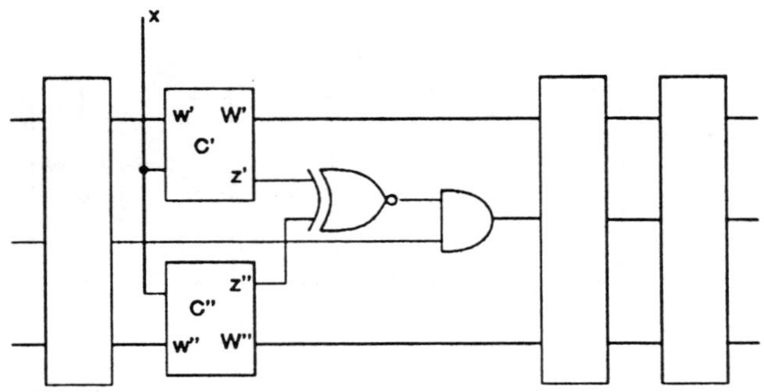

Figure 12: An Iterative Arrangement to Verify Iterative Implementations.

5 Equivalence of Iterative Pairs

One way of checking the equivalence of two sequential machines is to generate the set of reached state of the product of the two machines and checking that the outputs of the two machines are the same in all product states. This amounts to traversing a machine composed of the two machines being compared and a comparator. When checking two iterative implementations for equivalence, the comparator can be apportioned to the cells, so that the product machine to be traversed is also a sequential unilateral network. (See Figure 12.) If the two implementations are combinational, the resulting network is also combinational. In either case, the results of Sections 3 and 4 can then be applied. In particular, if the system is combinational, we have the following result.

Theorem 3 *Let S_1, composed of cells C_1, and S_2, composed of cells C_2, be two combinational unilateral systems. Let M_1' (M_2') be the sequential machine obtained by augmenting M_1 (M_2) with the output logic that produces the observed outputs (those that are checked for equivalence). Then instances of length n of S_1 and S_2 are equivalent for all values of n if and only if M_1' and M_2' are equivalent.*

The proofs follows from the previous discussion. Notice that it is possible to build networks that do not satisfy the condition of the theorem and yet are equivalent for some values of n, but not all.

If the system of Figure 12 is sequential, one checks the stability of the basic cell. If the languages stabilize and the output of the AND gate is one in all strings of the languages, then the two circuits are equivalent. Clearly, for both combinational and sequential systems, the time required to test for equivalence is independent of n. (In the sequential case, the stability condition must be satisfied, though.) A relaxed stability condition can be used: If $L_i \subseteq L_i', \forall i$, then $L_{k+l}' \subseteq L_k'$ implies stability. However, if this condition is used, a failure in the equivalence check does not imply that the two circuits are not equivalent. Hence, this is a heuristic technique that relies on a judicious choice of L_i'.

One can see these verification techniques as inductive proofs. The inductive step is in the proof of stability of the involved automata. The value of the base is determined by trying all values of n until a good value is found or a limit is exceeded. In the case of sequential systems, the proof may not succeed.

6 Results and Conclusions

We have experimented with the techniques described in the previous sections and compared them with the conventional techniques on a few examples. We have been able to verify the expected constant or linear behavior of our procedures and their efficiency in general. For instance, computing the BDD for the reached set for an instance of the *minmax* circuit [12] with 512 cells takes 5.5 seconds. (It takes about 30 seconds to determine the reached set for $n = 30$ with conventional methods.) If the method of Section 5 is applied, two implementations of the *minmax* circuit of the same (arbitrary) length can be verified equivalent in less than 5 seconds. Even more time can be saved by not having to repeat verifications for different values of n.

In this paper we have concentrated on unilateral systems that are the easiest to analyze. We have shown how many problems of interest can be solved in linear or even in constant time by applying inductive arguments. Many aspects, however, are still relatively unexplored: We are currently working on weaker conditions for linearity and on the extension of these results to bilateral networks. We have observed that there are bilateral networks with linear size BDD's, and we are studying the conditions for linearity. Also, we plan to work on the extension of our results to the verification problems that arise when infinite sequences [10], rather than finite strings, are considered. Finally, an area of research is the effective integration of these techniques in a larger synthesis system.

Acknowledgement

The authors wish to thank Randy Bryant, Kurt Keutzer, and Rick Rudell for the interesting discussions on the linearity of BDD's for combinational circuits.

References

[1] M. Abramovici, M. A. Breuer, and A. D. Friedman. *Digital Systems Testing and Testable Design.* Computer Science Press, New York, 1990.

[2] C. M. Angelo, D. Verkest, L. Claesen, and H. De Man. On the comparison of HOL and Boyer-Moore for formal hardware verification. In *Advanced Research Workshop on Correct Hardware Design Methodologies*, pages 419–440, Torino, Italy, June 1991.

[3] R. S. Boyer and J. S. Moore. *A Computational Logic Handbook.* Academic Press, Boston, 1988.

[4] K. S. Brace, R. L. Rudell, and R. E. Bryant., Efficient implementation of a BDD package. In *Proceedings of the 27th Design Automation Conference*, pages 40–45, June 1990.

[5] M. C. Browne, E. M. Clarke, and O. Grumberg. Reasoning about networks with many identical finite state processes. *Information and Computation*, 81(1):13–31, April 1989.

[6] R. E. Bryant. Graph-based algorithms for boolean function manipulation. *IEEE Transactions on Computers*, C-35(8):677–691, August 1986.

[7] J. Burch. Using BDD's to verify multipliers. In *1991 International Workshop on Formal Methods in VLSI Design*, Miami, FL, January 1991.

[8] J. R. Burch, E. M. Clarke, and D. E. Long. Representing circuits more efficiently in symbolic model checking. In *Proceedings of the Design Automation Conference*, pages 403–407, San Francisco, CA, June 1991.

[9] H. Cho, G. D. Hachtel, S.-W. Jeong, B. Plessier, E. Schwarz, and F. Somenzi. ATPG aspects of FSM verification. In *Proceedings of the IEEE International Conference on Computer Aided Design*, pages 134–137, November 1990.

[10] Y. Choueka. Theories of automata on ω-tapes: A simplified approach. *Journal of Computer and System Sciences*, 8:117–141, 1974.

[11] E. M. Clarke, J. R. Burch, and K. L. McMillan. Sequential circuit verification using symbolic model checking. In *Proceedings of the Design Automation Conference*, pages 46–51, June 1990.

[12] O. Coudert, C. Berthet, and J. C. Madre. Verification of sequential machines using boolean functional vectors. In L. Claesen, editor, *Proceedings IFIP International Workshop on Applied Formal Methods for Correct VLSI Design*, pages 111–128, Leuven, Belgium, November 1989.

[13] O. Coudert and J. C. Madre. A unified framework for the formal verification of sequential circuits. In *Proceedings of the IEEE International Conference on Computer Aided Design*, pages 126–129, November 1990.

[14] S. Devadas, K. Keutzer, and A. S. Krishnakumar. Design verification and reachability analysis using algebraic manipulation. In *Proceedings of the International Conference on Computer Design*, pages 250–258, Cambridge, MA, October 1991.

[15] A. Ghosh and S. Devadas. A mixed depth-first/breadth-first technique for sequential logic verification. In *International Workshop on Logic Synthesis*, MCNC, Research Triangle Park, NC, May 1991.

[16] M. J. C. Gordon. HOL: A proof generating system for higher-order logic. In G. Birtwistle and P. A. Subrahmanyam, editors, *VLSI Specification, Verification, and Synthesis.* Kluwer Academic Publishers, Boston, 1988.

[17] F. C. Hennie. *Iterative Arrays of Logical Circuits.* The M.I.T. Press and John Wiley, New York, 1961.

[18] F. C. Hennie. *Finite-State Models for Logical Machines.* John Wiley, New York, 1968.

[19] S.-W. Jeong, B. Plessier, G. D. Hachtel, and F. Somenzi. Extended BDD's: Trading off canonicity for structure in verification algorithms. In *Proceedings of the IEEE International Conference on Computer Aided Design*, pages 464–467, Santa Clara, CA, November 1991.

[20] S.-W. Jeong, B. Plessier, G. D. Hachtel, and F. Somenzi. Variable ordering and selection for FSM traversal. In *Proceedings of the IEEE International Conference on Computer Aided Design*, pages 476–479, Santa Clara, CA, November 1991.

[21] R. P. Kurshan. Analysis of discrete event coordination. *LNCS*, 430:414–453, 1990.

[22] R. P. Kurshan and K. L. McMillan. A structural induction theorem for processes. In *Proceedings of the Eighth Annual ACM Symposium on Principles of Distributed Computing*, pages 239–247, Edmonton, Alberta, Canada, August 1989.

[23] H. Touati, H. Savoj, B. Lin, R. K. Brayton, and A. Sangiovanni-Vincentelli. Implicit enumeration of finite state machines using BDD's. In *Proceedings of the IEEE International Conference on Computer Aided Design*, pages 130–133, November 1990.

The Automatic Generation of Bus-Interface Models*

Yew-Hong Leong William P. Birmingham
Electrical Engineering and Computer Science Department
The University of Michigan
Ann Arbor, MI, USA 48109

Abstract

We describe a system, HIDE, that automatically creates VHDL bus-interface models from timing and bus-state diagrams. HIDE shrinks model generation time to days from months. An Intel iAPX80386 model was developed in less than one week. HIDE can also be used to develop executable specifications, as its inputs are a normal element of the IC design process.

1. Introduction

Simulation of digital systems is common design practice. For effective simulation, accurate models of devices must be available. Model development for highly complex VLSI components is time consuming, sometimes requiring several man-years of effort. This effort is spent in understanding how the device operates from its documentation, and writing the model. Often, models are produced by third parties, not by the component designers. For these groups, understanding the operation of a component is a major task; consider that complete microprocessor descriptions fill several hundred pages.

The modeling process can be hampered by hardware description languages (HDL). While HDLs are intended to facilitate model creation, they are often complex, and require considerable programming skill to be effectively used. Thus, a model writer must be an expert in both hardware and HDLs.

The hardware-modeling task can benefit from automation. In this paper, we describe HIDE (HDL Interface Model Designer), which automatically creates VHDL [7] bus-interface models (BIM) from timing and bus-state diagrams. The information required by HIDE is easily found in component documentation. Thus, HIDE reduces the time for both understanding a component's operation, and in writing the model.

This paper is organized as follows. BIMs are described in Section 2, followed by an overview of HIDE in Section 3. HIDE's algorithms are sketched in Section 4, with experimental results given in Section 5. Section 6 compares HIDE to related tools. Section 7 concludes the paper.

This work was funded by Digital Equipment Corp. and by the National Science Foundation grant MIPS-905781.

2. BIMS

Simulation models for complex components generally are behavioral or BIM, or some combination. Behavioral models implement the full functionality of a component. These models are complex to build, and simulate relatively slowly.

BIMs, also known as hardware-verification [11], bus-functional [4], and chip-level models [2], describe the interface behavior of a device. These models capture a component's signal activity on a bus-cycle[2] by bus-cycle basis. BIMs fits the typical CAE environment, where a schematic is drawn and then simulated. A BIM allows the designer to ensure that a component is communicating properly with other devices. Since a BIM implements a sufficient subset of a component's behavior, it executes more quickly than behavioral models.

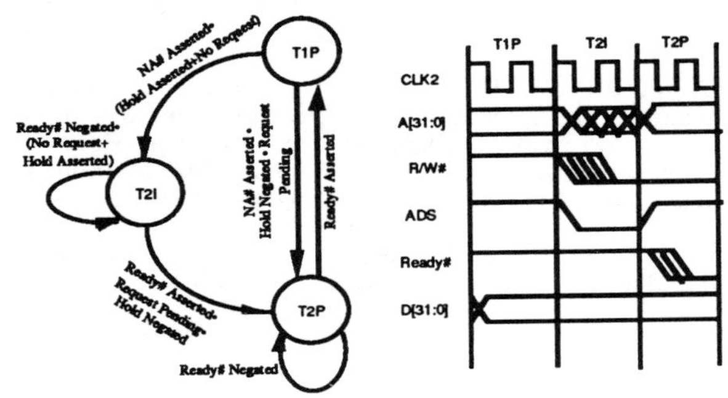

Fig. 1: Bus-state and timing diagram (adopted from [8]).

Bus activity consists of cycles, which are gross bus behavior. Example microprocessor cycles are read and write. Each cycle consists of one or more bus states. For example, the Intel 80386's read cycle is composed of two bus states, T1 and T2 [8]. For some devices, it is possible to map a cycle onto different bus states. Continuing the 80386 example, pipelined read and write allows address and control lines to appear one state earlier than normal read and write. For the pipelined case, the bus cycle has states T1P and T2P [8]. The movement from non-pipelined to pipelined is given by the level of control *signals*. For example, in Figure 1

[2] Bus cycle is used in a general form; a device's interface does not have to be bus-oriented to be amenable to a BIM-style of modeling.

0738-100X/92 $3.00 © 1992 IEEE

going from T1P to T2P occurs when the signals satisfy the following: $\overline{\text{Hold}} \bullet \text{NA\#} \bullet \text{Request Pending}$[3] (1).

Often, this state-transition control information is given as a bus-state diagram or as text. A portion of the 80386 bus-state diagram for pipelined bus cycles is shown in Figure 1. The complete state diagram is more complex in both the numbers of states and transitions. Note that the state diagram is not specific to a cycle type; the state diagram in Figure 1 applies to both read and write.

The bus state in models generated by HIDE is automatically determined by the signal level. Thus, a BIM will evaluate Equation (1) to determine if the next state should be pipelined. It is notable that HIDE can create models with this functionality.

Each signal's assertion level for each state in each cycle is given as (a) timing diagram(s), which also specifies the time between events. An event is a change in the level of a signal.

```
#Command File for read/write of a processor
variable i: integer;
variable start_address: vlbit_vector(31 downto 0);
variable return_data: vlbit_vector(31 downto 0);
start_address := "00";
for i in 1 to 10 loop
  write(2, start_address, "AA");
  read(2, start_address, return_data);
  start_address := increment_address(start_address);
end loop;
```

Fig. 2: Command file for microprocessor read and write.

BIMs are composed of procedures for each cycle. A microprocessor BIM has many procedures that are assembled in a *command file* by a model user to simulate interface behavior. A typical BIM procedure header for a read cycle is:

read(number_of_bytes, start_address, return_data)

which will return *number_of_bytes* of *return_data* when *start_address* is supplied. An example command file is given in Figure 2. The model user must interpret the datum returned.

A BIM is created by HIDE using only timing diagrams and bus state diagrams, as well as information about both the pins of a component and how certain bus activities (data control) are performed. Pin information lists each input and output signal, its corresponding pin number on a schematic symbol, and the pin type (input, output, time multiplexed, and so forth). Data-control information is explained later. Because all this information is readily available from component documentation, creating BIMs is straightforward, and does not require detailed knowledge of component operation.

[3] Req. Pending is not a signal, but an indicator in the 80386 that some bus activity is required.

Since timing and bus-state diagrams are normally produced as part of the component design process, BIMs can be used as executable specifications.

3. System Organization

HIDE is shown in Figure 3. For capturing timing diagrams, the XWave tool is provided. XWave's output is a tabular representation of the timing diagram, which is converted to an event graph (EG) [3] by HIDE. XWave also captures bus-state diagrams[4]. Pin (signal) information is supplied through a file.

Many bus activities involve some form of data transfer. Information regarding how the data is arranged on the data bus is not represented in either the timing or state diagrams. This information is found in the text of a data book. For example, during the 80386 memory-read cycle, the location of the valid data on the bus is given by the BS16# and BE signals. The designer provides this information through the *data-control* file.

4. Generating VHDL for a BIM

HIDE generates VHDL procedures for each bus cycle of a component. Consider Figure 1, the bus cycle begins with state T1P, and depending on some conditions, the bus cycle transits to either T2I (idle state) or T2P. Even though the timing diagram shows all three states in the bus cycle, it does not fully describe this transition between states. Thus, both state and timing diagrams are required.

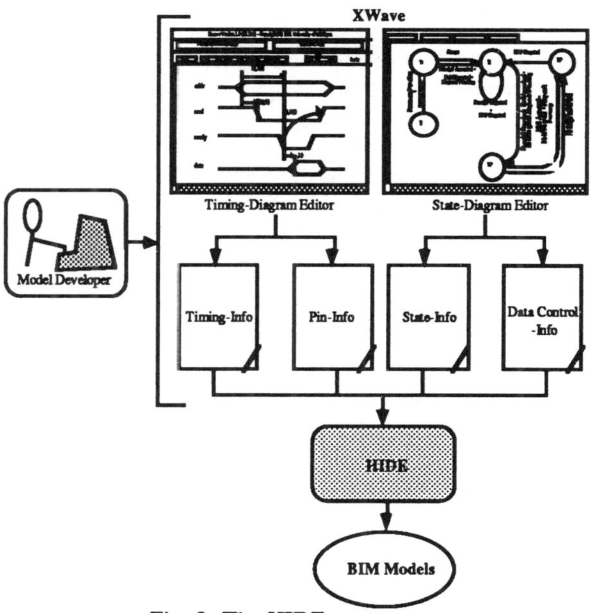

Fig. 3: The HIDE system.

Each cycle's procedure moves the model through bus states, and executes the VHDL code derived from the timing diagrams for signal activity. Figure 4 shows

[4] This version is being constructed; tables are used now.

the general structure of this procedure in VHDL. The procedure initializes the *Next_State* variable to the initial state of the bus cycle. Then, the procedure enters a while loop with case statements for each state of the bus cycle. Within each state, code is generated from the EG, as described in Section 4.3. Code to determine the next state of the bus cycle is easily translated from the state diagram. If necessary, code for proper placement of data on a bus is generated.

```
procedure Bus_Cycle_X
(Input_Parameter1, ... Output_Parameter) is
Next_State = Initial State;
while (NOT (Next_State = EXIT)) loop
  case Next_State is
  when State1 => code for State1 converted
     from EG of State1
     # code for transfer of data if State1
     #involves data transfer
                   ....
     Next_State:=
     Get_Next_State(Next_State, Variable1, ... );
  when State2 =>...
...
end case;
end loop;
end Bus_Cycle_X;
```

Fig. 4. General structure of a VHDL procedure.

Specific input and output parameters for each procedure in a BIM are required. A model builder must specify these parameters to HIDE, and describe how to assign or access them. This information, also input via the *data-control* file, is automatically inserted in the proper place in the VHDL procedure. Figure 5 shows typical parameter access and assignment specification.

```
Input Parameters:      Parameter1, Parameter2;
Output Parameters:     Parameter3;
State:         StateX;
If (Parameter1 = Value1 and Parameter2 = Value2)
     then Parameter3 <= Value_From_Signal1;
     else Parameter3 <= Some_Other_Value;
end if;
```

Figure 5. Typical parameter access and assignment specification.

4.2 Description of the EG

XWave produces a set of timing dependencies from the timing diagrams:

$$T_d = \{ t_d \mid t_d = (e_{from}, e_{to}, t_{min}, t_{max}) \text{ where } e_{from} \text{ occurs before } e_{to} \text{ in time} \},$$

where $t_{min(max)}$ is the minimum(maximum)-timing constraint between the two events, and e_{from} and e_{to} are events E, which is the set of all events. E has a number of attributes: signal name, direction, value, and edge number (an identifier).

Often, the timing dependency between two adjacent clock edges is omitted from the timing diagram. While not shown explicitly, timing dependencies between clock edges do exist. Thus, HIDE appends timing dependencies between every pair of adjacent clock edges, and propagates timing values[1]. For a timing diagram with n clock edges, we have the set

$$T_{clk} = \{ t_{clk} \mid t_{clk} = (e_{from}, e_{to}, t_{clk\_min}, t_{clk\_max}) \text{ where } e_{from} \text{ occurs before } e_{to} \text{ in time} \}$$

where $t_{clk\_min(max)}$ is the minimum (maximum) high or low period of the clock, and

$$e_{from} = (clock, value_i, i), e_{to} = (clock, value_j, j) \text{ for } i = 1, ..., n\text{-}1, j = i + 1.$$

The EG of the timing diagram is $G = (V, E)$, where $V = T_d \, U \, T_{clk}$. Figure 6 shows a portion of the timing diagram, T2I, and its EG representation.

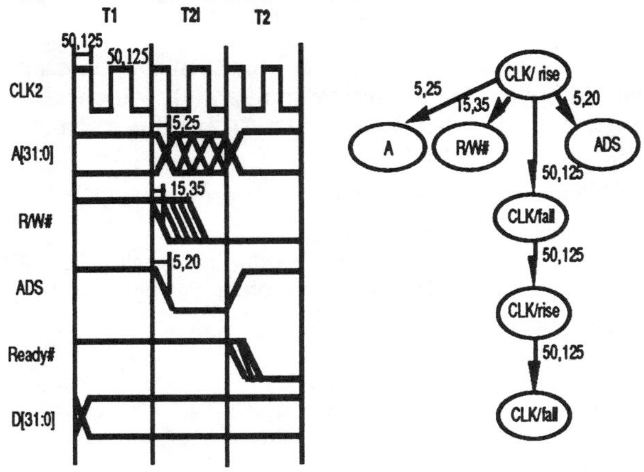

Fig 6. EG for T2I.

4.3 EG Transversal and Code Generation

HIDE builds the EG while reading the input file. The VHDL code for each state is then generated by traversing the EG with the code generation (CG) algorithm sketched in Figure 7.

For each clock event, a wait statement is generated. Then, for each input event that has an EG edge leading directly or indirectly (a path) to a clock event, a timing constraint is propagated (denoted by procedure *Propagate* [1] in CG) from that event to the clock event, and a statement is generated to check the setup time implied by the propagated timing constraint. Similarly, for each event that has an EG edge leading from the clock event to it, the timing constraint is propagated. For an input signal, a statement is generated to check hold times. For an output signal, a VHDL signal assignment statement is generated with the proper delayed timing.

Let d be the average distance of a node from a clock node (*e.g.*, ADS in Fig. 6 has d = 1), and n be the total number of nodes. The average case complexity of CG is $O(nd)$; best case is $O(n)$; and, worst case is $O(n^2)$ [10].

5. Experimental Results

HIDE generated a BIM for the 80386. The model was used to simulate a design containing the 80386 and memory chips in a CAD system. The simulation corresponded exactly to the timing specifications found in the data book. The input to HIDE took less than one man-week to generate. HIDE ran in about two minutes for each bus cycle on an Apollo DN3500 workstation. We estimate a similar model could have taken three to four man-months to write and debug by hand.

```
c = number of clock events;
for y = 1 to c do
  e_clk ∈ E | (e_clk.signal = clock) and
              (e_clk.edgenum = y)
  generate wait statement for e_clk
  forall v_j ∈ V | ∃ ( v_j = (e_j, e_k, min_j, max_j)
              and v_k = (e_k, e_l, min_k, max_k) and
              ... and v_q = (e_q, e_clk, min_q, max_q)
    Propagate(e_j, e_clk)
    if e_j.dir = IN
              generate min, typical or max setup
              time violation check for e_j.signal
      else
              report illegal timing
              constraint specification
      endif
  end forall
  forall v_p ∈ V | ∃ ( v_j = (e_clk, e_j, min_j, max_j) and
              v_k = (e_j, e_k, min_k, max_k) and ... and
              v_p = (e_p, e_q, min_p, max_p) )
      Propagate(e_clk , e_q )
      if e_q.dir = IN
              generate min, typical or max hold
              time violation check for e_q.signal
      else
              generate output statement for signal
              eq.signal with min, typical or max
              timing delay
      endif
  end forall
end for
```

Fig. 7. The CG algorithm.

6. Discussion

Historically, simulation models have been written by hand. Due to the complexity of both components and HDLs, this is often not time effective. A number of approaches have been taken recently to improve the model writer's productivity. In particular, a number of tools produce VHDL code from a variety of input formats. One of the most extensive systems is BIF [6], which is typical of the table-based approach. The behavior of a component, both interface and internal, is entered via sets of state tables, which are converted to efficient VHDL code. The HUM system [12] supports a similar paradigm to BIF, except that the input format is a spreadsheet.

There are a number of important differences between HIDE and table-based approaches. HIDE is dedicated to producing BIMs, not general behavior models, as are the table-based systems. Thus, HIDE's user interface, XWave, is highly efficient when compared to general-purpose-input techniques. XWave does not require specialized training. Thus, the model generation process is much simpler than with general purpose modeling tools. Furthermore, since HIDE only produces BIMs, it takes advantage of language constructs to improve model-simulation performance.

The advantages of using a BIM-style model are discussed by Armstrong [2] and Jennings [9]. Jennings does not, however, provide tools to build these models.

7. Conclusion

The HIDE system provides for the automatic creation of synchronous VHDL bus-interface models without requiring detailed knowledge of a component's behavior. BIMs are an efficient modeling paradigm for hardware verification, and fit naturally into CAE environments. HIDE generates all set-up and hold checks between relative signals, and checks other intra-signal timing. HIDE is efficient ($O(n^2)$ worst case) and has generated a VHDL model for Intel's iAPX80386.

HIDE is currently being extended to generate asynchronous devices, and for devices where internal registers affect bus signals. While HIDE currently generates VHDL, it can be easily be modified to produce models in a variety of HDLs.

Acknowledgments

We would like to thank Dave Kieras for help with XWave, Ajay Daga, and an anonymous reviewer.

References

[1] J. Abramson, W. Birmingham. "An Efficient Method for Representation and Integration of External Timing Constraints", Technical Report CSE-TR-82-91, University of Michigan, Nov. 1990.

[2] J. Armstrong. Chip-Level Modeling with VHDL. Prentice-Hall, Inc., 1989

[3] G. Boriello, "A New Interface Specification Methodology and its Application to Transducer Synthesis", PhD Thesis, University of California, Berkeley, Computer Science Division, 1988.

[4] D. Coelho. The VHDL Handbook. Kluwer Academic Publishers, 1989.

[6] N. Dutt, J. Cho, T. Hadley. "A User Interface for VHDL Behavioral Modeling", CHDL-91, IFIP, 1991

[7] Institute of Electrical and Electronics Engineering, IEEE Standard VHDL Language Reference Manual, IEEE Std 1076-1987, 31 March 1988.

[8] Intel Corporation. Microprocessor and Peripheral Handbook, Vol. I - Microprocessor, 1989.

[9] G. Jennings. "An Exercise in VHDL Timing Back-Annotation", CHDL-91, IFIP April 1991.

[10] Y.H. Leong, W. Birmingham. "Automatically Generating Bus-Interface Models", Technical Report CSE-TR-122-92, University of Michigan, Feb. 1992.

[11] Logic Automation, Inc. Processor Control Language, Users Guide.

[12] Lewis Systems, Inc. HUM User's Manual, VHDL PC Version, July 91.

Superpipelined Control and Data Path Synthesis

Usha Prabhu Barry M. Pangrle

Department of Computer Science,
The Pennsylvania State University,
University Park, PA 16802.

Abstract

This paper describes a superpipelined control and data path synthesis system. The system can 1) handle pipelined modules in the data path, 2) perform functional pipelining of the datapath and 3) schedule the datapath using a pipelined controller. Three control styles – serial, parallel and pipelined, are implemented. The system automatically picks one depending on the data path, the clock frequency, and the functional unit and control path delays. The results show that using a modifiable clock cycle time and a parameterized control style can significantly improve the throughput of high performance systems.

1 Introduction

Much of the earlier work in high-level synthesis has concentrated on scheduling, allocation and binding within basic blocks. However, recent efforts have been directed toward global scheduling in an effort to improve throughput under resource constraints. Throughput is usually defined as the number of clock cycles needed to realize the schedule. Two main directions have been taken in global scheduling – transformational techniques and path-based techniques. Transformational techniques such as trace scheduling [7] and percolation scheduling [18] move operations across basic blocks in order to fully utilize the resources available. Path-based scheduling [5] tries to schedule each possible path optimally, and may thus schedule an operation into different control states depending on the path. The parallelism in the data path is thus exploited at the expense of an increase in control path complexity.

Control path complexity may also increase with the choice of clock frequency in the synthesis of high-performance systems. High throughput designs can sometimes be achieved by choosing the clock cycle time to minimize the total amount of dead-time in a control state. Dead time is the time wasted in a clock cycle because all ready operations need functional units which have finished execution but cannot be scheduled because they have already been scheduled in the current clock cycle. Decreasing the dead time increases the efficiency of the schedule because less time is wasted. Minimizing dead time usually means decreasing the clock cycle time. As the clock cycle time decreases (and the number of states required to realize the schedule increases), the control circuit once again becomes more complex.

Thus efforts to improve throughput may result in the control delay occupying a large percentage of the clock cycle, or even, in some cases, being longer than the clock cycle time. To achieve high throughput, high level synthesis systems must look at ways to execute the data and the control paths in parallel or to pipeline the control path.

We propose that a modifiable clock cycle time and a parameterized control style are necessary when synthesizing high performance systems. The combination of small cycle times and a high degree of pipelining usually improves system throughput. Machines with these characteristics have been called *superpipelined* [9]. Ideally, both the clock cycle time and the pipeline depth should be chosen by the synthesis system. Currently, in this system, the clock cycle time and the level of pipelining in the functional units are supplied by the user. The system automatically picks the pipeline depth in the controller so as to maximize performance. Results show that in many of the benchmark examples, superpipelining improves performance.

This paper describes the control styles incorporated in SandS. Figure 1 shows a block diagram of the system. Slicer [14] and Splicer [15] perform the scheduling and module allocation tasks respectively. Piper [11] performs functional pipelining. The system can thus 1) handle pipelined modules in the data path, 2) perform functional pipelining of the data path, and 3) schedule the data path on a pipelined controller.

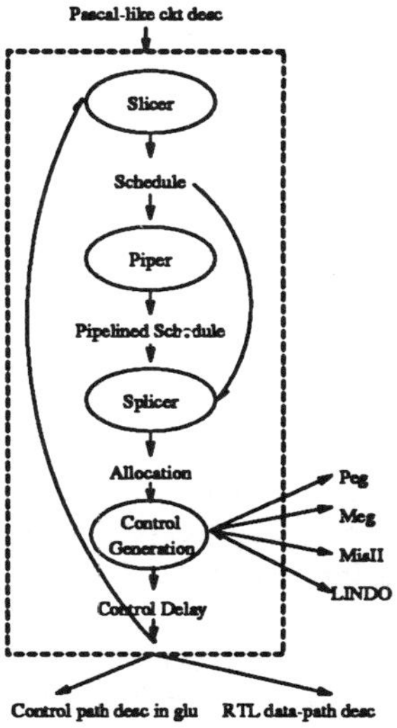

Figure 1: Block diagram of the SandS system

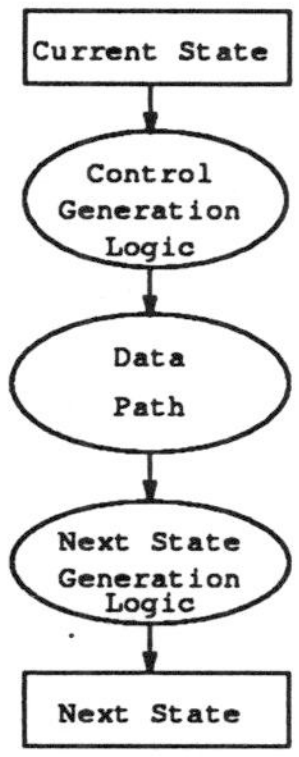

Figure 2: Serial Control Style

2 Previous Work

Most high level synthesis systems which schedule both the data and the control path use an FSM PLA based controller and assume that the clock cycle time is large enough for both control and data path operations. One of the exceptions is Chippe [2]. Chippe calculates the cycle time and picks a control model that minimizes the area required while satisfying the performance criteria. The control styles used are PLA-based control (with or without a condition register), pipelined control (where the control is executed in parallel with the data path), and partitioned pipeline control (where a 2-stage pipelined controller is executed in parallel with the datapath). The main difference in the work described here, is that it allows an arbitrary degree of pipelining in the control path. It can thus handle the combination of long control delays and small cycle times.

Mlinar concentrates on controller area in his thesis [12]. He uses a PLA based controller, and explores the changes in PLA area with different datapath schedules. He does not take controller delay into consideration.

Most systems do not vary the clock cycle time, but take as input functional unit delays in clock cycle time units. An exception is CYOS [6], which tries to generate a schedule using the best possible cycle time. Thus cycle time is one of the variables explored in the design space. It uses simulated annealing to do the scheduling and cycle time optimization.

MAHA [19] also picks a cycle time which allows it to meet its area/time constraints. However, the lower bound on the cycle time it chooses is the delay of the slowest functional unit, i.e., it does not allow multicycle functional units.

3 Generating the Control Path

An iterative algorithm is used to generate both the data and the control path. Assuming a control delay of zero, Slicer and Splicer are used to schedule and allocate operations in the data path. A Moore machine based finite state machine is then generated to represent the control path. This is used as input to PEG [8] which generates the control equations. These are then passed to MisII [3], which is used with a standard script and a library of basic gates. MisII generates the control circuit, maps it into the component library and returns the maximum delay. This delay is fed back to Slicer to generate a new schedule taking into account the control delay and the control style used. (The control styles are described in the next section). This process is iterated until a stable solution is found. The fast response times of Slicer and Splicer allow such an iterative method to be used successfully.

Section 3.1 describes the control styles defined in the system. Section 3.2 gives an overview of the sytem with the help of an example.

3.1 Control styles

Three main control styles are used, serial, parallel, and pipelined, which are depicted in Figures 2-4. The style actually implemented depends on the data path, the clock frequency and the data and control path delays.

Figure 2 depicts a serial control style, where both the data and the control paths are executed in the

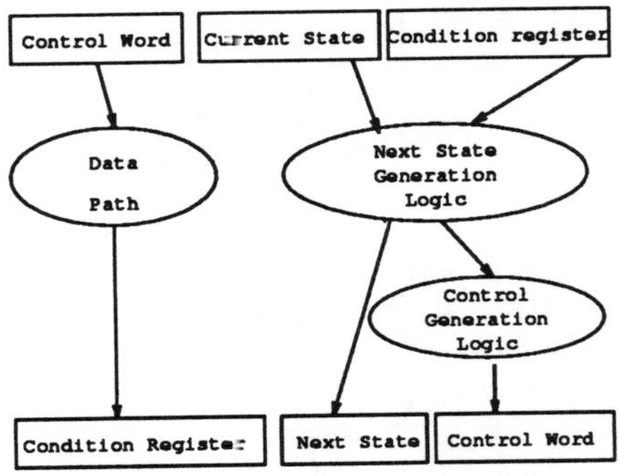

Figure 3: Parallel Control Style

same clock cycle. If this control style is chosen, the data path scheduling algorithm must take into account the smaller time available to execute data path operations. Obviously, this style cannot be used if the control delay is greater than one clock cycle. This control style is usually chosen if the data path contains a large number of conditionals and loops, and there is not too much scope for across-block transformations. The disadvantage of this style is that a significant portion of a clock cycle can be used in generating the control signals.

This problem is overcome in the control style depicted in Figure 3. Here the control path is executed in parallel with the data path, thus allowing operations in the data path to use the complete clock cycle. The penalty paid for this is that registers must be used to store the control signals (marked control word in Figure 3). Also, a NOOP state will have to be inserted in the data path after every conditional. This state may be deleted if the conditional can be scheduled early enough or if code motion can be used to move data-independent code into these states (This is similar to the *delayed branches* used in RISC systems [9]). This control style is usually used if the data path does not contain too many conditionals or loops. A parallel control style may also be viewed as a 2-stage control and datapath pipeline.

If the delay of the control path is greater than the clock cycle time, it may be necessary to break the control path into stages, where the delay of each stage is less than the clock cycle time. A 2-stage pipe is shown in Figure 4. Here registers are needed not only to store the final control signals, but also to store intermediate values between stages. Again, it may be necessary to introduce multiple NOOP states every time there is a branch in the control flow. The number of NOOP states required is the number of stages in the control path.

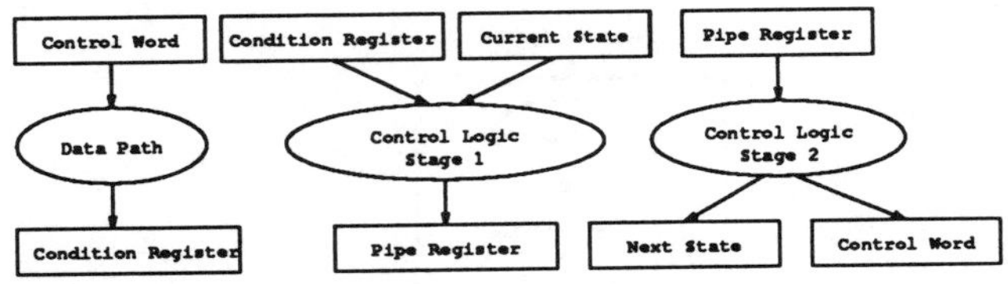

Figure 4: Pipelined Control Style

Note that there is no fixed cycle initialization overhead associated with using parallel or pipelined controllers. All the intermediate registers (control word, pipe registers, condition registers and current state register) can be initialized on reset to values that are possible to precompute.

In this system, the serial control style is represented as a Moore machine. Peg and MisII are then used to generate and time the resulting circuit. The parallel control style is represented as a Mealy machine, hence Meg [21] and MisII are used to generate these designs. The pipelined control style is generated from the parallel design using retiming [10]. Retiming is used to insert registers into the network generated by MisII. The minimum number of registers are inserted such that no stage of the resulting circuit has delay greater than the clock cycle time. This problem is the linear programming dual of a minimum-cost flow problem and can be solved in polynomial time. The linear programming package LINDO [20] is used to solve it.

3.2 System overview

The system is described with the help of an example. The example chosen is from HAL [17], expanded so as to use input and output ports. The available hardware consists of 2 multipliers (each of which takes 60ns to execute), 1 adder, 1 subtractor, 1 comparator (each of which takes 30ns to execute), five input ports and two output ports (each of which takes 5ns to execute). Note that all the execution times given are register-to register times. For this example, the cycle time is set to 70ns.

The system initially sets the control delay to zero. Using this, Slicer generates the 6-state schedule shown in Figure 5. Splicer is then used to do the module allocation and connectivity binding. The system generates a VHDL-like description of the finite state machine for the control.

The finite state machine description is then translated into Peg format. The output of Peg is then fed into MisII to map the description into a library of standard gates. The results in this paper were obtained with the library mcnc.genlib. The maximum

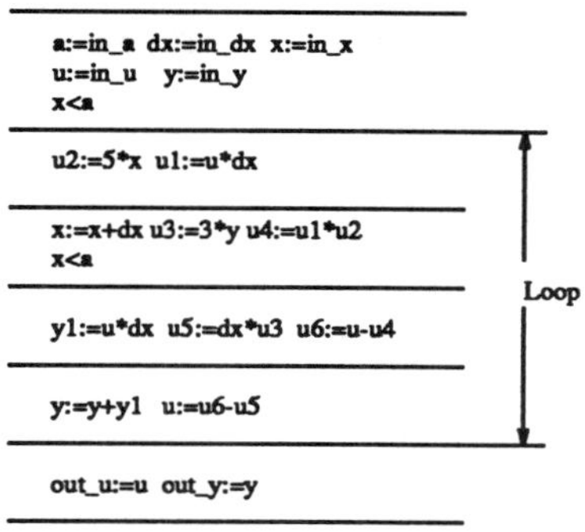

Figure 5: 6 state schedule for diffeq (with a serial control style).

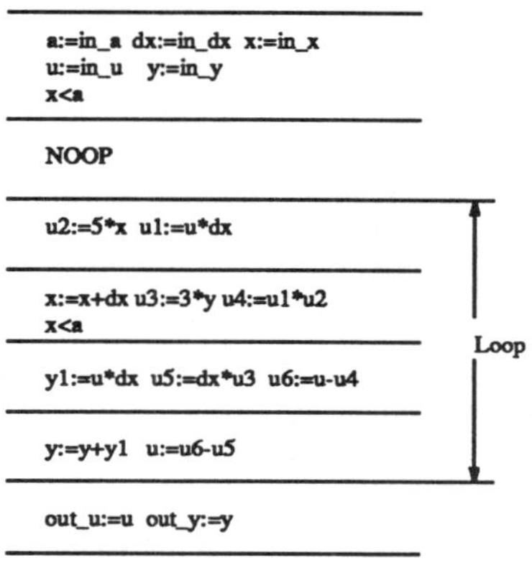

Figure 6: 7 state schedule for diffeq using 5 input ports and a parallel control style.

delay through the network is fed back to the scheduler. The scheduler may then change the schedule based on the new information available. If the control delay is greater than the clock cycle time, retiming [10] is used to introduce registers into the control path until the maximum delay in any stage of this pipelined controller is less than the clock cycle time.

In this example, the maximum delay through the control path is 5ns. Assuming a serial control style, the schedule remains exactly the same. Since there is a dead time of at least 10ns in every clock cycle, the control delay can easily be added to each cycle without modifying the circuit. With a clock cycle time of 70ns, a parallel control style would have given the schedule shown in Figure 6. The number of states has now increased to 7, since an extra NOOP state has to be inserted after the first comparison. Slicer performs in-block transformations that eliminate NOOP states wherever possible. In this case, the comparison within the loop is scheduled early enough so that other operations can be moved into the state after it. Hence a NOOP state does not have to be inserted within the loop. For this clock cycle time, the serial control style gives better throughput and is accepted by the system. The network description is fed to ArtistII [13] which does the layout.

With a cycle time of 60ns, the parallel control style is superior to the serial style and will thus be adopted by the system. The schedule with a serial control style takes 8 states (with the body of the loop taking 6 states), whereas that with the parallel control style takes 7 states (with the body of the loop taking 4 states). Using the parallel controller results in using less states in spite of the fact that a NOOP state has to be added after the comparison. If the loop is repeated only ten times, the system with the parallel control takes 2580ns to execute, whereas that with the serial control takes 3720ns – a savings of over 30%. This saving increases as the number of iterations through the loop increases.

Table 1 shows the results obtained with this example and different clock frequencies. The total time column shows the time taken for ten iterations of the loop. For this example, the best throughput is achieved by setting the clock cycle time to 10ns and the control style to pipelined. Assuming that the loop iterates ten times, the total execution time is now 2170ns – a savings of 1550ns over using the 70ns clock cycle and a serial control style. Thus significant improvements in throughput can be obtained even with very small systems.

Cycle Time	# of States	Control Delay	Control Style	Deg of Piping	Tot Time
70	6	5	Serial	–	2940
60	7	7	Parallel	–	2580
30	11	11	Parallel	–	2220
15	19	12	Parallel	–	2175
10	28	14	Pipelined	1	2170

Table 1: Differential Equation Solver

4 Experiments

This section describes the results obtained with the FIR filter [18] and the Mark1 processor [4]. All the functional unit execution times mentioned in this section are register-to- register delays. Experiments run with other benchmarks from [1] showed similar increases in throughput.

4.1 Mark1 processor

This example was translated from the ISPS description of the Manchester University Mark-1 Computer in [4]. The functional units used were one adder, one subtractor, one comparator and a memory with a read and a write port. Memory operations take 90ns, all other operations take 35ns. Table 2 shows the results obtained with a range of clock frequencies. The numbers in the "Number of States" column are the number of states in the schedule with the chosen control style, and with the "other" control style. For example, with a clock cycle time of 100ns, Mark1 can be scheduled in 13 states with a parallel control style and in 21 states with a serial control style – a savings of 800 ns. When the clock cycle time is reduced to 15ns, the control delay is greater than the cycle time, and a 2 stage pipelined controller is used.

Cycle Time	# of States	Control Delay	Control Style	Deg of Piping	Tot Time
100	13/21	12	Parallel	–	1300
52	23/24	14	Serial	–	1196
20	47/57	15	Parallel	–	940
15	63	16	Pipelined	1	945
10	86	16	Pipelined	1	860

Table 2: Mark1 Processor

4.2 FIR filter

Experiments were run with the FIR filter to illustrate performance improvement with functional pipelining, pipelined functional units and a parallel control path. The register-to-register delay of the adders is 30ns and that of the multipliers is 60ns.

Table 3 compares nonpipelined and (functionally) pipelined designs using a non-pipelined multiplier. The second figure in the total time column for the pipelined case shows the elapsed time for the first output from the pipeline. The first figure in this column is the rate at which results are available after that. In both counts, substantial savings are obtained with the pipelined version as compared to the nonpipelined version. Table 4 compares nonpipelined and pipelined designs using a pipelined multiplier. The multiplier can accept new inputs in every clock cycle. The total time column here has the same meaning as in the previous case.

The best throughput is obtained using a clock cycle time of 30ns, functional pipelining, pipelined multipliers and a parallel control style.

5 Conclusions and Future Work

Increasing throughput in high-performance systems can be achieved by completely utilizing every control step. This can be done either by increasing the hardware available or by changing the clock cycle time and modifying the control style. The results in this paper show that varying the clock cycle time and using pipelined or parallel controllers can lead to significant improvements in the throughput of resource-constrained systems.

Experiments have shown that the right choice of cycle time can decrease the amount of dead time in a control step. Sometimes the improvement can be achieved only by executing the data and the control paths in parallel. Even if the control delay is only a few nanoseconds, decreasing *each* clock cycle by this amount leads to a substantial improvement across the whole schedule. These effects will be much more pronounced in large systems scheduled over hundreds or thousands of clock cycles.

These methods will also be very effective in control dominated systems where the control delay is comparable to functional unit delay. Here the necessity for parallel or pipelined control units is much more obvious.

The system will be expanded to incorporate facilities for specifying external timing constraints on control so as to make the system more useful in control dominated designs. The system should also automatically generate the best cycle time.

References

[1] Benchmarks for the 6th International Workshop on High-Level Synthesis, 1991.

[2] F. Brewer and D. Gajski. Chippe: A System for Constraint Driven Behavioral Synthesis. *IEEE Trans. on CAD*, Vol 9, July 1990, pp 681-695.

[3] R. Brayton, R. Rudell, A. Sangiovanni- Vincentelli, A. Wang. MIS: A Multiple-Level Logic Optimization System. *IEEE Trans. on CAD*, Vol 6, No 6, Nov 1987, pp 1062-1081.

[4] M. R. Barbacci and D. P. Siewiorek. Design and Analysis of Instruction Set Processors. McGraw-Hill, 1982.

[5] R. Camposano. Path-Based Scheduling for Synthesis. *IEEE Trans. on CAD*, Vol 10, No 1, January 1991, pp 85-93.

		Nonpipelined				Functional Pipelining				
Cycle Time (ns)	FUs Used	Num of States	Ctrl Delay (ns)	Control Style	Total Time (ns)	Data Intro Intrvl (cycs)	Comp Time (cycs)	Ctrl Delay (ns)	Control Style	Total Time (ns) DII/CT
100	5+, 3*	5	5	Serial	500	3	9	3	Serial	300/900
70	5+, 3*	6	5	Serial	420	3	9	3	Serial	210/630
60	5+, 3*	6	7	Parallel	360	3	12	4	Parallel	180/720
40	5+, 4*	10	7	Serial	400	4	11	3	Serial	160/440
30	5+, 4*	10	8	Parallel	300	4	11	5	Parallel	120/330

Table 3: FIR Filter: non-pipelined multiplier

		Nonpipelined				Functional Pipelining				
Cycle Time (ns)	FUs Used	Num of States	Ctrl Delay (ns)	Control Style	Total Time (ns)	Data Intro Intrvl (cycs)	Comp Time (cycs)	Ctrl Delay (ns)	Control Style	Total Time (ns) DII/CT
40	5+, 3*	8	5	Serial	320	3	12	3	Serial	120/480
30	5+, 3*	8	7	Parallel	240	3	12	4	Parallel	90/360

Table 4: FIR Filter: pipelined multiplier

[6] J. Cortadella, R. M. Badia, E. Ayguade. Scheduling in a Continuous Area-Time Design Space: A Simulated Annealing Approach. *Proc. of the Fifth International Workshop on High-Level Synthesis*, March 1991, pp 102-117.

[7] J. A. Fisher. Trace Scheduling: A Technique for Global Microcode Compaction. *IEEE Trans. on Computers*, vol. C-30, No. 7, July 1981.

[8] G. Hamachi. Designing Finite State Machines with PEG. UC Berkeley, 1983.

[9] J. L. Hennessy and D. A. Patterson. Computer Architecture: A Quantitative Approach. Morgan Kaufmann Publishers, Inc. 1990.

[10] C. Leiserson, J. Saxe. Retiming Synchronous Circuitry. *Algorithmica*, 6(1), 1991, pp 5-35.

[11] D. Lobo and B. M. Pangrle. Optimization Techniques for Pipelined Scheduling. *Fifth International Conference on VLSI Design*, India, Jan 1992.

[12] M. J. Mlinar. Control Path / Data Path Tradeoffs in VLSI Design. CEng Tech. Report 91-16, University of Southern California.

[13] R. M. Owens and M. J. Irwin. A Comparison of Four 2-Dimensional Gate Matrix Layout Tools. *Proc. 26th Design Automation Conference*, 1989.

[14] B. M. Pangrle and D. D. Gajski. Slicer: A State Synthesizer for Intelligent Compilation. *Proc. of ICCD*, Oct 1987.

[15] B. M. Pangrle. Splicer: A Heuristic Approach to Connectivity Binding. *Proc. 25th Design Automation Conference*, June 1988.

[16] P. G. Paulin and J. P. Knight. Scheduling and Binding Algorithms for High-Level Synthesis. *Proc. 26th Design Automation Conference*, June 1989.

[17] P. G. Paulin, J. P. Knight, E. F. Girczyc. HAL: A multi-paradigm approach to automatic datapath synthesis. *Proc. 23rd Design Automation Conference*, 1986, pp 263-270.

[18] R. Potasman, J. Lis, A. Nicolau, D. Gajski. Percolation Based Synthesis. *Proc. 27th Design Automation Conference*, 1990, pp 444-449.

[19] A. Parker, J. Pizarro, M. Mlinar. MAHA: A Program for Datapath Synthesis. *Proc. 23rd Design Automation Conference*, July 1986, pp 461-466.

[20] L. Schrage. Linear, Integer and Quadratic Programming with LINDO. The Scientific Press, 1984.

[21] D. Wood. MEG. UC Berkeley.

Distributed Design-Space Exploration for High-Level Synthesis Systems *

Rajiv Dutta, Jayanta Roy and **Ranga Vemuri**
Laboratory for Digital Design Environments
Department of Electrical and Computer Engineering
University of Cincinnati, ML 30
Cincinnati, Ohio 45221-0030

Abstract

A parallel algorithm for design-space exploration and trade-off analysis is presented. Coarse-grained parallelism is introduced by generating multiple module bags and performing scheduling and performance analysis of the data flow graph for each module bag in parallel. This algorithm is implemented on a multiple processor machine as part of a distributed high-level synthesis system. Experimental results show reduction in search time, improvement in design quality, and close-to-linear speedup.

1 Introduction

High-level synthesis is the process of generating register-level designs given a behavioral specification, performance goals usually in terms of area and clock period, and a register-level module library. The result of high-level synthesis is a register-level data path which contains modules (registers, functional units etc.) selected from the module library and a finite-state controller to control the sequencing of register transfers in the data path. The high-level synthesis process is usually divided into three tasks: scheduling, allocation and binding. The behavioral specification is typically represented as a data dependency graph which indicates the potential parallelism among the various abstract operations. During the scheduling phase, the abstract operations in this graph are assigned to control steps. Scheduling algorithms take one or more of the following constraints into account: (1) Resource constraint – only the modules available in a module bag can be used; [1] (2) Clock-period constraint – the maximum module delay in any control step should not exceed the user-specified limit on the clock period. (3) Area constraint – the total area occupied by the modules used in the schedule should not violate the user-specified area constraint; (4) Schedule length constraint: the length of the longest path in the scheduled graph should not exceed the user-specified limit. Most scheduling algorithms take resource constraint and clock-period constraint into account. Some scheduling algorithms produce the shortest possible schedule that meets the stated area constraint, whereas others attempt to produce the best area-efficient schedule while meeting the given schedule-length constraint. [7, 10, 11, 9]

Scheduling phase has a profound effect on the overall performance (both area and speed) of the synthesized design. It is important to generate various schedules representing various regions of the design-space and select the best possible schedule. Performance estimation procedures can be used to estimate the area/time characteristics of the designs which would result from a scheduled data flow graph; the best schedule can then be chosen either by the user or by the synthesis system based on a user-supplied cost function. Given the resource, clock-period, area and/or schedule length constraints, very good scheduling algorithms exist to determine near-optimal schedules [12, 2, 1]. Of the four types of constraints, usually the clock-period, area and schedule-length constraints, if any, come from the user. The resource constraint, however, must be generated by the synthesis system based on the module library provided. Clearly, different module bags lead to different schedules which in turn lead to different designs. Given a data flow graph and a module library, a vast number of module bags can be formulated to drive the scheduling algorithm. Arriving at a good schedule for each module bag, estimating the design performance for each schedule, and finally selecting the best among these schedules can be very expensive in computing time. Fortunately, since the schedules under different module bags can be arrived at independently of each other, the computing power of multiple processor systems can be effectively used to reduce this time.

This paper presents an algorithm for module bag selection and parallel design-space exploration, based on this simple idea shown as Algorithm 1. We admit parallelism at the scheduling algorithm level. Each processor in the multiple processor system executes the same scheduling algorithm. Each scheduler has a copy of the same data flow graph and the same area and clock-period constraints. What is different among the schedulers is the resource constraint; each scheduler is driven by a different module bag selected from a pool of valid module bags created a priori. Each scheduler attempts to generate the shortest possible schedule. The scheduling phase in each processor is followed by a performance estimation phase where the area and speed of the design that would result with this schedule are estimated. The performance results are then gathered and presented (to the user or the synthesis system) in the form of tradeoff curves where different tradeoff points correspond to different module bags. The best tradeoff point (and module bag) may then be selected (by the user or the synthesis system) and the corresponding schedule may be passed on to the register allocation and connection optimization phases to complete the design synthesis process.

The number of valid module bags generated (Step 1, in Algorithm 1) decides the amount of parallelism. Clearly, too many module bags (say, more than the number of processors) can slow down design-space exploration process. Too few module bags, on the other hand, may mean under-

*This work is partly supported by the Defense Advanced Research Projects Agency under order no. 7056 monitored by the Federal Bureau of Investigation under contract no. J-FBI-89-094 and by the University of Cincinnati Research Council.

[1] A *bag* is a collection of items with duplicates allowed. A *set* is a collection of items with no duplicates. The module bag is the collection of modules taken from the module library and actually used to synthesize the data path.

Algorithm 1 (Distributed Design-Space Search)
DFG: the data flow graph representing the behavior.
LIB: library of register-level modules available.

1. *Generate the set S of valid module bags.*
2. **for each** *valid module bag* $B \in S$ **do in parallel**
3. **begin**
4. *Schedule the DFG using the modules in the module bag B.*
5. *Estimate the area and speed of the design using the assigned schedule.*
6. **end**
7. *Choose the best schedule and the corresponding module set.*

exploration of the design-space. If a module bag is known to subsume another module bag (not in set theoretic terms, but in terms of the quality of the schedules that these module bags lead to) then the second module bag should not be included in the collection S of module bags. Some synthesis systems assume that there is only one module in the module library that can implement a given function [12, eg.]. Other systems, instead of attempting to find the best modules to use keeping the constraints in view, tend to use all the modules in the module library that can implement a given function [10, eg.]. The former approach is too restrictive and the latter approach could result in the usage of inefficient modules in the design where more efficient modules could have been used. In this paper, we present an efficient algorithm for generating a good set of valid module bags. This algorithm, discussed in detail in Section 3, admits only the most appropriate module bags into the set S and disallows others; it is thus responsible for the amount of the design-space explored and the time take by the tradeoff analysis phase.

We use a variation of force-directed list scheduling, due to Paulin and Knight, [12] as our basic scheduling algorithm. Our performance estimation procedures are extensions of (and are somewhat more rigorous than) those proposed by Jain, Mlinar Kurdahi and Parker [4, 8]. Neither the basic scheduling algorithm nor the performance estimation procedures are discussed in detail in this paper. Interested readers may refer to [5] as well as the the above references. Instead, the theme of this paper is the module bag generation algorithm and its effect on the quality of distributed design-space exploration. Included are some experimental results about the quality of the tradeoff analysis in comparison with other systems and the speedup results due to the use of a multiple processor system.

2 Inputs to the Exploration System

Our distributed design-space exploration system takes four inputs:

1. Data Flow Graph (DFG): The behavior specification is represented by a data flow graph (DFG) [15]. For the purposes of this paper we will not get into the details of this representation except to note that the nodes of the graph can represent either functions (such as +, * etc) or control operations (such as *case*, *call*, etc). The function nodes which are eventually implemented by the functional units available in the library are of interest dur-

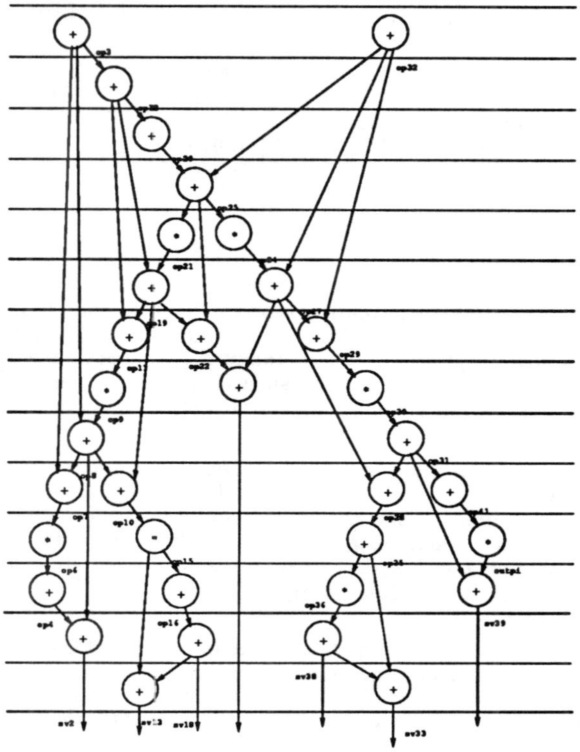

Figure 1: Data Flow Graph of the Elliptic Wave Filter (the edges corresponding to primary inputs are not shown)

ing the module bag generation phase. Figure 1 shows the DFG (data flow graph) representation of an Elliptic Wave Filter. This example was used as a benchmark at the High-Level Synthesis workshop in 1988. All the functions and edges (carriers) in this example have the same bit-width of 16.

Given a DFG, we define the Set of Functions (SF) as the set of all function symbols used in the DFG. For the Elliptic Wave Filter example the Set of Functions is $\{+, *\}$. [2] For each function in SF there should be at least one module that can implement the function in the module library; otherwise, the given DFG cannot be implemented using the given module library.

2. Module Library: The module library contains (all the pertinent information about) the register-level modules that can be used in the design. Specifically, the module library contains several arithmetic-logic units each of which can implement one or more functions such as $+_{16}$, $*_{16}$, $<_{16}$ etc. The set of functions supported by a library module is called the Intrinsic Function Set (IFS). Each

[2]Operations with different bit-widths are considered to be different functions. For example, 16-bit addition is viewed as a different function than 32-bit addition; these are denoted as $+_{16}$ and $+_{32}$ respectively. When there is no ambiguity, we drop the bit-width suffix.

Table 1: Example Information in the Module Library

Module Name	IFS	Area(sq.mm.)	Delay(ns)
ALU	$+_{16}$, $*_{16}$	16.258	90
adder_1	$+_{16}$	0.763	78
adder_2	$+_{16}$	0.920	50
multiplier	$*_{16}$	16.033	90

module also has two performance attributes: delay time and area. [3] Typical information available in the module library is shown in Table 1. Each function node in the data flow graph must be eventually mapped onto a module selected from the library such that the IFS (Intrinsic Function Set) of the module contains that function.

3. Area Constraint (A): Area constraint is a user-specified value. For the purposes of the paper it denotes the upper-limit on the arithmetic sum of areas of the modules in the module bag used to schedule the data flow graph. This constraint is used to delimit the modules selected from the module library. [4]

4. Clock-Period Constraint (C): This user-specified constraint denotes the maximum delay of any module in the module bag used to schedule the DFG. Clock-period constraint is also used to delimit the modules selected from the library.

3 Generation of Valid Module Bags

A <u>*module bag*</u> is a collection of modules, with duplicates allowed, selected from the module library. A <u>*valid module bag*</u> (VMB) satisfies the following properties:

1. *Property 1:* For each function $f \in SF$ (the set of functions in the given DFG), there exists a module M in the bag such that f is supported by M, ie., $f \in IFS(M)$.
2. *Property 2:* The delay of each module in the bag is no more than the user-given constraint on the clock-period (C).
3. *Property 3:* The total area occupied by the modules in the bag should not exceed the area constraint (A) provided by the user. That is, $\sum_{M_i \in VMB} M_i \leq A$.

A <u>*valid module set*</u> is a valid module bag which contains no duplicates.

The module bag generator produces a collection of valid module bags. It produces enough variety of valid module bags so that the most serial design, the most parallel design, and a variety of designs in between are included in the explored design-space. The module bag generator avoids inferior VMB's (VMB's that could lead to inferior schedules) using a heuristic based on an estimate of the maximum number of copies of any library module required to implement the full parallelism available in the DFG. (Note that if there are n nodes of the same function in a control step, then, if we wish to exploit the full parallelism, we need n modules capable of implementing that function.)

The module bag generation algorithm has four major steps: (1) Identification of *useful* modules in the module library; (2) Generation of valid module sets (VMB's with no duplicates); (3) Estimation of the maximum number of copies of each useful module, required to implement the full parallelism available in the DFG; (4) Generation of valid module bags by enlarging the valid module sets generated in Step 2 by introducing duplicates upto the maximum numbers found in Step 3. These four steps are explained in detail in the following sections.

[3] The library may also contain, as it does in our system, parameterized module-generators rather than specific modules. The performance attributes (delay time and area) are then made available in the form of equations in terms of the formal parameters or tables whose entries can be interpolated. In case of compound modules (which support multiple functions) the delay time for each function may be specified separately.

[4] Another type of area constraint such as the overall area taking the registers, wires and controller into account may also be specified as an input to the synthesis system. Such an area constraint can be used to select a tradeoff point *after* the scheduling and performance estimation is done.

Algorithm 2 (Generation of Useful Modules)
SF: the set of function types in the DFG.
SM_f*: the set of modules in the library which support the function* f.
UMS: useful module set.

```
BUILD-USEFUL-MODULE-SET()
begin
   UMS ← NULL
   for each f ∈ SF do
   begin
      SM_f ← Supporting_Modules(f)
      for each module M ∈ SM_f do
         if (delay(M) > C) or (area(M) > A)
         then SM_f ← SM_f − {M}
      for all M_1,M_2 ∈ SM_f do
         if (IFS(M_1) ∩ SF) = (IFS(M_2) ∩ SF)
         then if area(M_1) < area(M_2)
               then SM_f ← SM_f − {M_2}
               else SM_f ← SM_f − {M_1}
      if |SM_f| = 0
      then error ⇒ "no useful modules for f"
      UMS ← UMS ∪ SM_f
   end
   return UMS
end
```

3.1 Identification of Useful Modules

Let the set of function types in the given DFG be SF. For each function type $f \in SF$ the module library is consulted to determine the set of all the modules that support f. From among these modules, those that violate the clock-period constraint (C) or the area-constraint (A) are rejected. Each remaining module is *useful* in the sense that it can be used to implement some function node in the DFG. An additional improvement is possible: if there are two useful modules which support exactly the same set of function types in the DFG, then the one with larger area can be ignored. The resulting module set is called the Useful Module Set (UMS). Algorithm 2 shows the procedure to compute the UMS, given a DFG (actually, the SF of the DFG is enough), a module library, and the area and clock-period constraints. In Algorithm 2, *Supporting_Modules(f)* is a library interface function which returns the set of all the modules in the library that implement f.

For the Elliptic Wave Filter example, assume that the area constraint is 50 sq. mm. and the clock-period constraint is 100 ns. Using the module library information shown in Table 1, ALU, adder_1 and adder_2 can be used to implement the + operation; they all satisfy the clock-period constraint. However, since adder_1 and adder_2 are functionally equivalent, and adder_1 has less area, adder_2 is not included in the set of useful modules.[5] Both the ALU and the multiplier are useful to implement the * operation.

3.2 Valid Module Sets

Recall that the valid module set is the same as a valid module bag except that it is strictly a set (no duplicates allowed). A valid module set contains enough resources to implement all the functions in the DFG; it may not however be able to support the parallelism available in the DFG. Thus a design generated using a valid module set will be relatively slow, but will be small in terms of area. Such designs are usually called 'serial designs'.

Conceptually, the generation of valid module sets involves first deriving the powerset of the UMS (the useful module set computed in the previous step). From this powerset, all the sets that violate Property 1 or Property 3 are discarded to yield the set of all valid module sets. The following technique is used to reduce the time taken to enumerate the elements of the powerset of the UMS:

A *critical module* is a module in the UMS which supports a DFG function that is *not* supported by any other module in the UMS. Hence, all critical modules must be included in any valid module set. Based on this observation, the UMS can be partitioned into two subsets: the set of all critical modules and the set of all non-critical modules. The powerset of the set of non-critical modules is first generated. The set of critical modules is then added to each element of this powerset; each element is then tested for Property 1 and Property 3. Algorithm 3 shows the steps in generating the collection of all the valid module sets.

For the Elliptic Wave Filter example, the following valid module sets are generated: (ALU, adder_1), (ALU, multiplier), (adder_1, multiplier) and (ALU, adder_1, multiplier). All these sets satisfy the area constraint.

3.3 Potential Parallelism for Each Useful Module

A valid module set is sufficient to implement all the functions in the behavior. However, to exploit all the parallelism available in the data flow graph, multiple copies of some of the useful modules may be necessary. For a given module, just how many copies are necessary? The answer depends upon the amount of parallelism in the DFG, in terms of the maximum number of function nodes assignable to the same control step, where each function can be supported by a copy of that module.

Let *SS* denote the set of all possible valid schedules of the DFG. Given a schedule (S) (assignment of function nodes in the DFG to control-steps), a control step (CS), and a library module (M), let $F_S(M, CS)$ denote the bag of function nodes supported by M and assigned to CS. Clearly, for each $f \in F_S(M, CS)$, $f \in IFS(M)$. Let $N_S(M, CS) = |F_S(M, CS)|$ be number of such function nodes. *Maximum parallelism* (MP), with respect to a module M, of the DFG is then defined as the maximum such N. That is, $MP(M) = MAX_{S \in SS, 1 \leq CS \leq K_S}(N_S(M, CS))$, where K_S is the maximum control-step number used in schedule S. The MP(M) (maximum parallelism wrt M) is

Algorithm 3 (Generation of Valid Module Sets)
SCM: Set of critical modules in UMS.
SNCM: Set of non-critical modules in UMS.
SVMS: Set of valid module sets.
SF: Set of function types in the DFG.

```
VALID-MODULE-SETS()
begin
   SVMS ← { }
   P ← powerset(SNCM)
   for each set of non-critical modules S ∈ P do
   begin
      R ← (SCM ∪ S)
      if ((∀f ∈ SF, ∃M ∈ R such that f ∈ IFS(M))
            and CHECK-AREA-CONSTRAINT(R))
      then SVMS ← SVMS ∪ R
   end
return(SVMS)
end
```

```
CHECK-AREA-CONSTRAINT(MB)
MB: A bag of modules.
A:  Area constraint.
area(M): library interface function which returns the area
of module M
begin
   area = 0
   for each M ∈ MB do
      area = area + area(M)
   if (area > A)
   then return false
   else return true
end
```

an upper-limit on the number of copies of the module ever needed in any schedule of the DFG. How to determine MP(M) is the next question.

MP(M) can be rigorously determined by a clique partitioning formulation: First, a graph whose nodes are all the function nodes in the DFG covered by IFS(M) should be constructed. Edges should then be introduceed such that any two nodes that have no direct or indirect data dependence are connected by an edge. Edges thus indicate the potential for assignment of the two nodes to the same control step. The number of nodes in the maximal clique of this graph is the value of MP(M).

Determining the maximal clique in a graph is an NP-complete problem. To avoid this, a good estimate of MP(M) can be obtained by considering the ASAP (As Soon As Possible) schedule of the DFG. Thus, we take MP(M) to be the maximum number of function nodes supported by IFS(M) and assigned to the same control step in the ASAP schedule.[6] Algorithm 4 shows the steps in determining the MP(M) values based on the ASAP schedule. MP values are determined for all useful modules in the set UMS using this algorithm.

On many occasions in practice, we discovered that the estimate of MP(M) obtained using Algorithm 4 works just as well as the rigorous approach based on clique partitioning; and, some times reduces the search-space in a nice

[5] Note that adder_2 should not be deleted from the set of useful modules if a scheduling algorithm that permits chaining is used.

[6] ASAP scheduling is well-known and is not discussed here; see for example, [11].

Algorithm 4 (Maximum Parallelism)
M: a library module
IFS(M): intrinsic function set of M

```
MAXIMUM-PARALLELISM(M)
begin
  max_count ← 0;
  for i ← 1 to Max_ASAP_Cstep do
    begin
      count ← 0;
      for each function node f ∈ DFG do
        if (f ∈ IFS(M) and ASAP_Cstep(f) = i)
        then count ← count + 1;
      max_count ← max(count, max_count);
    end
  return max_count;
end
```

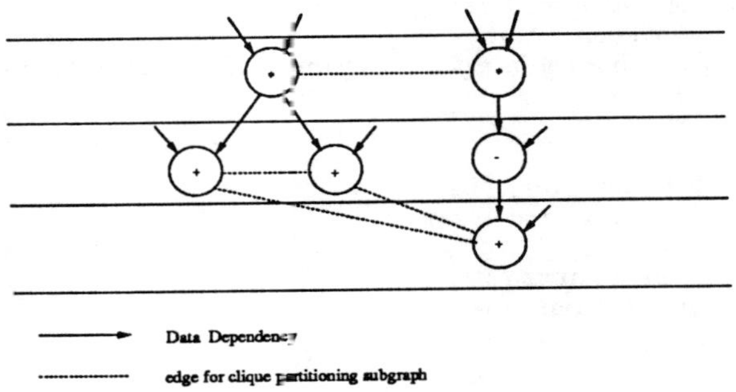

Figure 2: **Illustration for MP(M) Finding Techniques**

Module (M)	IFS(M)	MP(M)
ALU	+, *	4
adder_1	+	2
multiplier	*	4

Table 2: **Maximum Parallelism in EWF DFG**

Algorithm 5 (Generation of Module Bags)
This is a recursive algorithm. The top level call on this algorithm must be of the form MODULE-BAGS(VMS) where VMS is a valid module set.
MS: A module set
SMB: Set of module bags generated using MS (some of these bags may not be valid).
MP(M): Maximum parallelism with respect to M

```
MODULE-BAGS(MS)
begin
  if MS = null then return { }
  SMB ← null
  for each M ∈ MS
    begin
      Q ← MODULE-BAGS(MS − {M})
      for i ← 1 to MP(M)
        if Q = null then
          SMB ← SMB ∪ {M, M, ··· i times}
        else for each Bag ∈ Q

          SMB ← SMB ∪ {Bag ∪ {M, M, ··· i times}}
    end
  return(SMB)
end
```

way. For example, consider the DFG segment in Figure 2. Suppose the modules to be considered are a multiplier ($M1$) and an adder ($M2$). MP(M1) = 2 by both ASAP and clique partitioning techniques. On the other hand, MP(M2) = 2 by the ASAP schedule method and 3 by the clique partitioning method. The critical path length is 3. Highly parallel designs can be generated by a module set consisting of (2 mults, 2 adds, 1 sub), (2 mults, 3 adds, 1 sub) or (1 mult, 3 adds, 1 sub). The ASAP schedule method does not produce the last two module sets saving the search time, but still covers this extreme point (most parallel design) in the design-space by producing the first module set. In practice we found that this happens more often than one would think. If a more rigorous value of MP(M) is needed for a more thorough design-space exploration, then one must use the clique partitioning or similar approach.

Table 2 shows the values of MP(M) for each module in the useful module set for the Elliptic Wave Filter example.

3.4 Valid Module Bags

The final step in the generation of module bags involves duplicating each module (M) in each valid module set as many times as MP(M) and discarding the bags that violate the area constraint. For any given valid module set $VMS = \{M_1, M_2, \cdots M_n\}$, the duplication process creates $\Pi_{i=1,n} MP(M_i)$ module bags. This procedure is shown as Algorithm 5. From the module bags so produced, the ones that violate the area constraint are discarded yielding the collection of valid module bags. This procedure is shown as Algorithm 6.

Each valid module bag so produced represents a point in the design-space. Collectively, the set of valid module bags represents a collection of design points with various area-time trade-offs, including highly serial and highly parallel designs, and many others in-between.

For the Elliptic Wave Filter example, 29 valid module bags are generated. Table 3 lists some of these valid module bags.

Sl	1	2	3	4	5	6	7
adder	1	3	2	4	3	2	0
alu	1	0	0	0	0	0	3
mult	1	2	2	1	1	1	0

Table 3: **Some Sample valid module bags generated for the Elliptic Wave Filter Example.**

Algorithm 6 (Generation of Valid Module Bags)
SVMS: Set of valid module sets.
SVMB: Set of valid module bags.

```
VALID-MODULE-BAGS(SVMS)
begin
SVMB ← null
for each VMS ∈ SVMS do
    for each MB ∈ MODULE-BAGS(VMS) do
        if CHECK-AREA-CONSTRAINT(MB)
        then SVMB ← SVMB ∪ MB
return SVMB
end
```

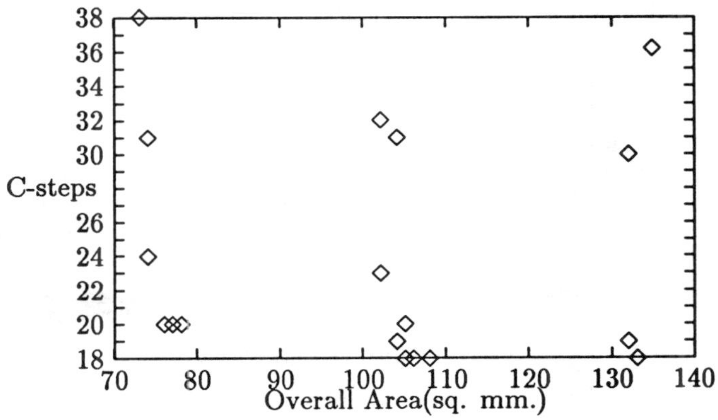

Figure 3: **Performance Plot (Ctrl-steps vs Area) for Elliptic Wave Filter**

4 Distribution of Module Bags and DFG

The valid module bags are then distributed across the processors. Each processor receives a copy of the DFG and one or more valid module bags. Each processor then executes the scheduling and performance estimation phases.

5 Scheduling

Our scheduling algorithm is an extended version of the force directed list scheduling (FDLS) developed by Paulin and Knight [12]. The extensions mostly pertain to the scheduling of control nodes in the data flow graph. [7] Operations in the DFG are scheduled (assigned to control steps) under the resource constraints defined by the module bag.

6 Performance Estimation

For each schedule produced, the overall area and throughput time in terms of control steps are estimated. The area estimate consists of component (registers, wires, controller, ALUs) area estimates. The time estimate is calculated as the number of control steps to execute the longest path in the DFG. Some of our performance estimation techniques are similar to the ones proposed by Jain, Mlinar, Kurdahi and Parker [4, 8, 6]; others are new techniques which yield more rigorous estimates [5].

7 Tradeoff Plots

Finally, the performance estimates produced in the processors are gathered and are presented to the synthesis system or to the user if so desired. The presentation is in the form of tradeoff plots. A variety of plots are produced to give the user a reasonable idea of the performance envelope.

For the Elliptic Wave Filter example, one of these plots is presented in Figure 3. The plot shows the variation in the overall area with respect to the control steps for each of the 29 valid module bags. The most desirable schedule (and module bag) may be chosen based upon the overall performance goals in terms of the number of control steps and/or the overall area.

[7] Our synthesis systems accepts behavior specifications in VHDL. The data flow graph representation we use [15] has nodes to denote various VHDL-specific control constructs such as *wait*. These nodes in turn require special handling during synthesis, especially during the scheduling and controller generation phases. Also, our calculation of deferral forces is somewhat different from that suggested by Paulin and Knight. [12, 5]

Note that the design-space explored by the proposed algorithm subsumes that explored by the HAL synthesis system which also uses the force-directed scheduling algorithm as the basis. (See [12] for the design-space HAL produces for the same Elliptic Wave Filter example.) This, of course, is to be expected due to the rather extensive set of resource bags generated and considered by our algorithms.

8 Implementation and Speedup

The proposed algorithms for distributed design-space exploration have been implemented as part of the DSS (Distributed Synthesis System) [14]. DSS executes on the ESK (Experimental Systems Kit) multiple processor machine developed at the MCC [3]. The version we used has 16 nodes connected in mesh-configuration. There is no shared memory; all communication is through high-speed links. The speed of the message fabric is such that from the the programmer's point of view, the mesh configuration is unimportant; the message communication time between any two processors (regardless of whether the two processors are neighbors) is about the same. DSS is implemented in the extended C++ supported by the ESK system. During the design-space exploration phase, load balancing is attempted by distributing equal number of bags to each processor.

Three examples were considered for experimentation. These examples are described in detail in [17]. Briefly, the first example is the Move Machine (70 lines of VHDL code). Move machine is a control dominated machine. The Move Machine has instructions to move data between memory and registers. The second example is the Elliptic Wave Filter (50 lines in VHDL) which is an arithmetic-dominated machine. The last example is the Viper microprocessor [17] which is a substantially large example (400 lines of VHDL code). Viper is an instruction set processor containing both control and arithmetic statements.

Figure 4 shows the variation in execution time (from parsing the specification to producing the performance plots) with respect to the number of processors. For the Viper and Elliptic Wave Filter examples, close-to-linear speedup is observed upto about 8 processors, after which

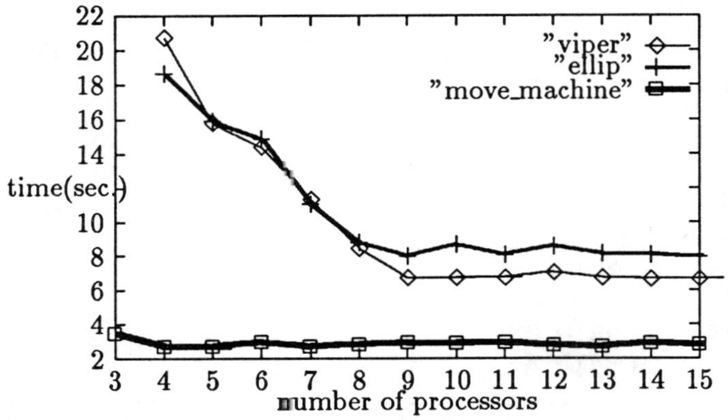

Figure 4: **Execution Time for Various Examples**

speed-up tapers off. This occurs due to the communication overheads which start dominating the overall time. For Move Machine description only 3 valid module bags were generated, which explains why there was no speedup.

9 Discussion

The Elliptic Wave Filter example represents a typical arithmetic-dominated machine and the Move Machine and Viper represent control-dominated machines. The full set of performance trade-off plots for the Elliptic Wave Filter, Move Machine and Viper example can be found in [5].

The DFG representation is not as simple as presented in this paper; it is quite elaborate to accommodate various ideocyncracies of VHDL. These idiosyncrasies have many ramifications on the scheduling algorithms, particularly in force calculations and while making deferral decisions. We have made non-trivial extensions (eg. how to schedule *wait* nodes, how to schedule parallel loop blocks, how to schedule concurrent procedure invocations etc.) to the basic force-directed scheduling algorithm to account for all the features of the DFG. This scheduling algorithm is discussed in detail in [5].

As a part of the ADAM work at USC, Rajiv Jain [4] developed lower-bound area-delay curves which give some indication of the nature of useful module bags. The basic idea (that of generating many module bags) we used is similar but the techniques however are different. The amount of design-space explored in our case is some what more extensive and covers many interesting trade-off points. Also, the performance estimation procedures we have developed (not presented in this paper) are aimed at estimating the actual values instead of determining the lower bounds.

All the algorithms presented in this paper make extensive use of set operations (union, intersection, power-set etc). In our implementation sets are implemented as C++ classes (resulting in a set data type). Within the class, the methods (implementing set operations) are implemented using doubly linked lists and are very efficient in execution time. In our algorithms, we have suggested that all module sets/bags are first generated and then filtered out based on the area constraints. In the implementation, however, the generation and filtering of module sets/bags happens hand-in-hand to improve the space-efficiency of the algorithms. In this paper, for clarity, we have not presented the algorithms at that level of detail. The exact data structures and algorithms used along with their complexity analyses can be found in [5].

Careful analysis of the algorithms presented in this paper will show that there is considerably more parallelism that can be exploited (for example, most of the for-all and for-each loops can be executed in parallel after slight modifications). It remains open to investigate whether it is worth, in terms of communication cost vs. potential speedup, attempting to exploit such fine-grain parallelism.

References

[1] R. J. Cloutier and D. E. Thomas, " The Combination of Scheduling, Allocation and Mapping in a Single Algorithm," 27th Design Automation Conference, 1990.

[2] C. A. Papachristou and H. Konuk, "A Linear Program Driven Scheduling and Allocation Method Followed by an Interconnect Optimization Algorithm," 27th Design Automation Conference, 1990.

[3] "Extensible Software Platform Development Environment Programmer's Guide", MCC Technical Report Number: ACA-ESP-223-58.

[4] Rajiv Jain, "High-Level Area-Delay Prediction with Application to Behavioral Synthesis," Technical Report CENG 89-23, University of Southern California, July 1989.

[5] Rajiv Dutta, "Distributed Design-Space Exploration for High-Level Synthesis Systems," Master's Thesis, Univ. of Cincinnati, 1991.

[6] Fadi Joseph Kurdahi, "Area Estimation of VLSI Circuits," PhD Thesis, Department of Electrical Engineering, University of Southern California, August 1987.

[7] M. C. McFarland, A. Parker, and R. Camposano, "Tutorial on High-Level Synthesis," Proc. Design Automation Conf., pp. 401-439, June 1988.

[8] Mitchell J. Mlinar, "Control Path/Data Path Tradeoffs in VLSI Design," PhD Thesis, Department of Electrical Engineering, University of Southern California, June 1991.

[9] B.M. Pangrle and D.D. Gajski, "Slicer: A State Synthesizer for Intelligent Silicon Compilation," Proceedings of the IEEE International Conference on Computer Design (ICCD), October 1987.

[10] A.C. Parker, J.T. Pizarro and M. Mlinar, "MAHA: A Program for Datapath Synthesis," Proceedings of the 23rd Design Automation Conference, 1986.

[11] R. Camposano, "From Behavior to Structure: High-Level Synthesis", IEEE Design and Test of Computers, pp. 8-19, Oct 1990.

[12] Pierre G. Paulin and John P. Knight, "Force-Directed Scheduling for the Behavioral Synthesis of ASIC's," IEEE Transactions on Computer-Aided Design, June 1989.

[13] Jay Roy and R. Vemuri, "Appropriate Usage of VHDL : The Synthesis Point of View", 53 pp., TM-ECE-DDE-89-08, Dept. of Electrical and Computer Engg., Univ. of Cincinnati, June 1990.

[14] Jay Roy, Nand Kumar, Rajiv Dutta and Ranga Vemuri, "DSS: A Distributed High-Level Synthesis System", IEEE Design and Test of Computers (to appear in June 1992).

[15] Jay Roy,"The VSS Intermediate Format", TM-ECE-DDE-91-16, June 1991.

[16] D.E. Thomas and G.W. Leive, "Automating Technology Relative Logic Synthesis and Module Selection," IEEE Transactions on Computer-Aided Design of Integrated Circuits and Systems, April 1983.

[17] Ranga Vemuri et. al., "Benchmarks for High Level Synthesis", TechnicalMemo-ECE-DDE-91-11, Dept of Electrical and Computer Engg, Univ. of Cincinnati, June 1991.

An Efficient Algorithm for Microword Length Minimization

Ruchir Puri and Jun Gu
Dept. of Electrical Engineering, Univ. of Calgary
Calgary, Canada T2N 1N4
gu@enel.Ucalgary.CA

Abstract

The problem of *microword length minimization* is crucial to the synthesis of microprogrammed controllers in digital systems. The problem is NP-hard. Although various enumerative and heuristics methods have been developed in the past, they usually do not provide fast and efficient solutions to a large size problem. In this paper, we formulate the problem into a graph partitioning problem. We employ a local search approach to further reduce the microword length. This has resulted in an efficient algorithm which outperforms any of the existing techniques available to solve the problem. We have tested our algorithm with practical microcodes and with large size examples generated from random graphs. We will compare our experimental results with other existing methods.

1 Introduction

The optimization of microword length is vital to microprogrammed controller design in a digital system. A microprogram can be implemented into control memory by assigning the microoperations to memory word bits. The control memory size depends on the length of microword it stores. Thus, any reduction in the microword length reduces the control memory's silicon area and increases its real-time performance in a microprocessor.

The problem of microword length minimization is NP-hard [8]. An optimal solution obtained by various enumerative procedures [1, 2, 5, 9] cannot be justified at the expense of exponential time complexity. One way to solve the problem is to employ efficient heuristic procedures which decrease the computing time and provide a near-optimal solution [3, 4, 7, 10] Nagle et al. [7] used an autonomous weight algorithm which takes an inadvertently long time as the parallelism in the microprogram decreases. Researchers have used simulated annealing approach [3]. So far there is little progress. Recently Hong et al. [4] proposed a probabilistic algorithm with an $O(n^3 logn)$ heuristics that uses Kernighan-Lin's graph partitioning method [6]. Their approach, although improving over Nagle et al.'s algorithm, requires many trials before it can find the best solution.

In this paper, we present an efficient heuristic algorithm for microword length minimization problem.

2 Problem Formulation

The microcode of a microprogrammed controller consists of a sequence of microinstructions $\mu_1, \mu_2, \ldots, \mu_M$, where each microinstruction is a set of microoperations $O_1, O_2, \ldots, O_i$. The microoperations in each microinstruction are conflict free in resource accessing. Thus, two microoperations O_i and O_j, belonging to a single instruction, are executed concurrently. They should not be encoded into a single group or *field*. Such pairs of microoperations are ***incompatible***. On the other hand, two microoperations are grouped into a single field, if they do not appear together in any microinstruction. Such a pair of microoperations are *compatible*. A ***compatibility class*** is a set of microoperations, that are pairwise compatible. That is, they are never executed concurrently and thus can be encoded together. Let $\{C_1, C_2, \ldots, C_k\}$ be a set of compatibility classes for a given set of microinstructions such that each microoperation appears in only one of the compatibility classes. Let $|C_i|$ be the number of microoperations in a compatibility class C_i. We can encode the compatibility class C_i into a single field with $\lceil log_2(|C_i| + 1)\rceil$ bits (where one is added to account for the NO-OP condition). The microword length L equals the total number of bits required to encode the compatibility classes $\{C_1, C_2, \ldots, C_k\}$. That is : $L = \sum_{i=1}^{k} \lceil log_2(|C_i| + 1)\rceil$. Microword length optimization problem is to determine a set of compatibility

29th ACM/IEEE Design Automation Conference®

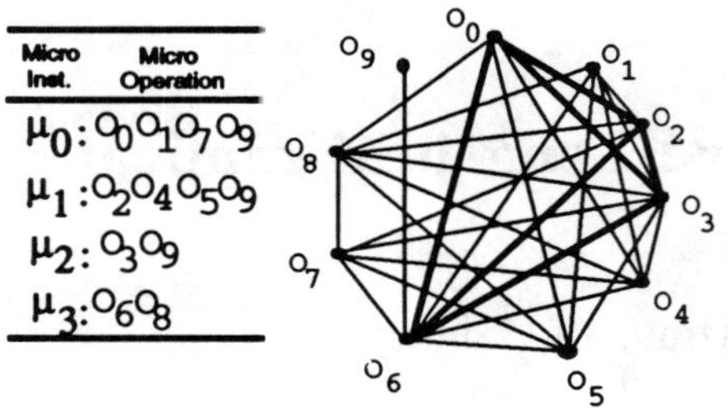

Figure 1: Compatibility Graph of Ex. Microcode.

classes $\{C_1, C_2, \ldots, C_k\}$ such that L is minimum.

We can formulate the problem in a *compatibility graph*. A microoperation O_i is represented by node i in the graph. Two nodes i and j are connected by an undirected arc, if their corresponding microoperations O_i and O_j are compatible. A compatibility class is a complete subgraph such that all its nodes are connected to each other, i.e., they are pairwise compatible. In such a graph model, the microword length optimization problem is equivalent to finding a partitioning of compatibility graph into k disjoint complete subgraphs $\{C_1, C_2, \ldots, C_k\}$ such that the cost function $L = \sum_{i=1}^{k} \lceil log_2(|C_i| + 1) \rceil$ is minimum.

Figure 1 gives a simple microcode example. The microoperations $\{O_0, O_1, \ldots, O_9\}$ are mapped as nodes {0,1,...,9} in the compatibility graph. Microoperations O_0, O_2, O_3 and O_6 do not appear together in any of the microinstructions $\mu_0, \ldots \mu_3$, so they are pairwise compatible. Hence, {0,2,3,6} forms a compatibility class in the compatibility graph (as highlighted in Figure 1).

3 An Efficient Heuristic Algorithm

Most of the techniques [1, 2, 4, 5, 9] proposed to solve the microword length minimization problem build a search space by deriving all the maximal complete subgraphs (i.e., maximal compatibles) in the compatibility graph. Hong et al.'s approach partitions the compatibility graph by recursively extracting a maximal clique and removes it from the graph, until no nodes are left in the graph. We observed that this approach, which is quite time-consuming for graphs with higher densities, does not yield a good initial disjoint partitioning. We now describe our algorithm which consists of two parts: an algorithm for initial graph partitioning and a local search algorithm

```
Input : Compatibility Graph G(V,E).
Output : Compatibility classes {C_1, C_2, ..., C_n},

Procedure initial_graph_partitioning( ) {
  Initialize : number of compat. classes j=1 ;
  while(|V|!=0) {
     for each node v_i in V (i = 1,2,...,|V|) {
        V_i = V ;
        delete nodes in V_i,not connected to v_i;
        for each node v_k in V_i
           delete nodes in V_i, not connected to v_k ;
        if (|V_i| >= |C_j|) then C_j = V_i ;
        /*complete subgraph grown from seed v_i*/
     }
     delete nodes in V, covered by C_j ;
     increment j ;
  }
  n=j ;
  return compat. classes {C_1, C_2, ..., C_n} ;
}
```

Figure 2: Compatibility Graph Partitioning Algo.

for cost minimization.[1]

3.1 A Fast Algorithm for Initial Graph Partitioning

A maximal clique algorithm that partitions a graph into the minimum number of groups is not the best way to solve this problem, since our goal is to minimize cost L. Thus, the absolute maximality of the clique sizes derived from the graph will not contribute to a good engineering solution. Besides it is very costly to generate the maximal cliques (NP-complete). We describe a fast algorithm for an initial partitioning of the compatibility graph, where the graph partitions obtained may not be maximal. The pseudocode of the initial graph partitioning algorithm is shown in Figure 2.

The algorithm initially selects a *seed* node and deletes all the nodes from vertex set V that are not connected to the seed node. Then another seed node is selected from the remaining nodes in the vertex set and the same deletion procedure is performed. This step is repeated until all the nodes remaining in the vertex set are connected to each other. This is called growing a clique from a "seed." A clique is grown from each node of the graph G(V,E) and the one having

[1] A nontrivial case study of local search is given in the Appendix.

maximum number of nodes forms the first compatibility class C_1. The nodes belonging to C_1 and their corresponding arcs are then removed from the compatibility graph. This guarantees that the compatibility classes are disjoint. The process of growing the cliques, deriving the compatibility classes and removing them from the compatibility graph is repeated until the vertex set of the graph is empty. This guarantees that the set of compatibility classes $\{C_1, C_2, \ldots, C_n\}$ covers the compatibility graph.

Applying this procedure to the microcode example shown in Figure 1, we get four compatibility classes: C_1={0,2,3,6}, C_2={1,4,8}, C_3={5,7}, and C_4={9}. Since the number of bits required to encode a compatibility class C_i having $|C_i|$ nodes is $\lceil log_2(|C_i|+1)\rceil$, the number of bits required to encode C_1, C_2, C_3 and C_4 are 3, 2, 2, and 1, respectively. Thus, the initial cost L is : L=3+2+2+1=8. This yields an initial value of cost function L, which can further be reduced using a local search algorithm described in the following section.

3.2 An Efficient Local Search Algorithm for Cost Minimization

The initial partitioning generated four compatibility classes for the compatibility graph shown in Figure 1. Compatibility class C_1 has 4 nodes. Compatibility classes C_2, C_3 and C_4 have 3 nodes, 2 nodes, and 1 node, respectively. If we can move a node from C_1 to C_3, then the cost of encoding C_1 will be reduced by one, whereas that of encoding C_3 will remain the same. The *movement* of a node will reduce the cost L from 8 to 7. When moving a node from C_1 to C_3, we must ensure that the node under consideration is compatible with all the nodes in C_3. Since node 6 in C_1 is compatible with all the nodes in C_3, it is qualified and can be moved to C_3.

In the following discussion, let *overflow*(C_i) be the minimum number of nodes which must be removed from C_i in order to reduce its encoding length by one. That is, *overflow*(C_i) = $(|C_i| + 1) - 2^{\lfloor log_2(|C_i|+1)\rfloor}$. Let *underflow*($C_i$) be the maximum number of nodes which may be added to C_i without increasing its encoding length. That is, *underflow*(C_i) = $2^{\lceil log_2(|C_i|+1)\rceil}$ - $(|C_i|. + 1)$

In the microcode example C_1 has 4 nodes and should give away one node to reduce its encoding length by one. Thus C_1 has an *overflow* of one. Compatibility class C_3 has two nodes and it can accommodate one more node without increasing its encoding length, thus it has an *underflow* of one. A node k can be moved from compatibility class C_i to C_j, if the following four constraints are satisfied :

- *Overflow*(C_i) ≥ 1 (i.e., C_i needs to give away some nodes in order to reduce its cost).
- *Underflow*(C_j) ≥ 1 (i.e., C_j can accept some nodes without increasing its cost).
- Node k in C_i is compatible with all nodes in C_j.
- $\sum_{l=1,l\neq i}^{n} underflow(C_l) \geq overflow(C_i)$. That is, the movement of nodes from C_i can possibly result in a cost reduction.

```
Input: Compatibility Graph G(V,E)
     : Compat. classes {C1,C2,...,Cn}
Output : Final Compatibility classes {C1,C2,...,Cn}.

Procedure cost_minimization( ) {
  while there is a cost reduction {
     for each compatibility class Ci
        if (overflow(Ci) ≥ Σ_{m=1,m≠i}^{n} underflow(Ci) )
          for each node k in Ci
            if (overflow(Ci)>0)
              for each Cj (j > i) {
                if (underflow(Cj)>0
                  if (node k compatible with Cj) {
                    move node k from Ci to Cj;
                    decrement overflow(Ci),
                              underflow(Cj) by one;
                  }
                if qualified nodes ≥ overflow(Ci)
                  move the qualified nodes to Cj;
                else recompute overflow, underflow
                    in each compat. class;
              }
  }
  return modified {C1,C2,...,Cn} ;
}
```

Figure 3: A Local Search Algorithm to Minimize L.

The pseudocode of the local search algorithm is given in Figure 3. It swaps the nodes of compatibility classes under the constraints stated above and minimizes cost L. This algorithm, when applied to microcode shown in Figure 1, reduces the cost L by one. Value 7 of the cost function obtained from our algorithm turns out to be the optimal cost. The final partitioning generated by the procedure is {0,2,3}, {1,4,8}, {5,6,7} and {9}. Thus, the microword for this particular example is programmed as four fields. These fields are : $\{O_0, O_2, O_3\}, \{O_1, O_4, O_8\}, \{O_5, O_6, O_7\}, \{O_9\}$.

Table 1: Results with Example Microcodes. Symbol * stands for optimal solution.

Microcode		*Our Method*		*Hong et al.* [4]			*Nagle et al.* [7]		*Baer et al.* [1]	
Name	Mico-Op.	Cost	Time	Best Cost	Time	Success Rate[2]	Cost	Time	Cost	Time
ex1 [4]	8	6*	0.01	6	0.09	54/60	7	0.07	6	0.25
ex2 [9]	11	9*	0.01	9	0.27	15/60	9	0.03	9	0.11
ex3 [5]	25	16*	0.18	16	3.08	6/60	19	1.38	16	1.93
ex4 [4]	22	7*	0.03	7	0.91	30/30	7	0.47	7	0.4
ex6 [4]	165	33	5.58	32	884.11	4/40	84	>60,000	34	>60,000

4 Experimental Results

1. Results using Practical Microcodes

The algorithm was implemented in C language. To compare our results with other approaches, we ran our algorithm on the examples used by Hong et al. in [4]. In order to maintain a uniform platform for comparison purposes, the examples were tested on a SUN 3/60 system. The results are given in Table 1. By analysing these results, one can draw the following conclusions :

- Baer et al.'s branch and bound algorithm [1] takes exponential time to find a solution. Although it guarantees an optimal solution, it is not of much practical use in optimizing even moderately sized microcodes.
- Nagle et al.'s autonomous weight algorithm [7] may yield a solution in lesser time than Baer's approach but suffers from the similar drawback of exponential time complexity. Also, the solutions obtained are far from the optimal and the non-optimality of the solution increases with the increasing number of nodes and edges.
- Hong et al.'s method [4], depends on random initial starting point. To select the best solution, the approach must perform many trials. Thus, it effectively requires much larger time (more trials) to yield a good solution. Also, the time required for a single trial is quite large.

As evident from results in Table 1, our approach is much faster and gives a near-optimal solutions. Let us consider the large microcode example *ex6* having 165 microoperations and a graph density of 96.64%. Our method yielded a cost of 33 in 5.58 seconds. This is incomparably faster and gives a better cost solution than that obtained from either Nagle'a or Baer's approach. Our algorithm does not depend on any random starting point and thus, is not a test and trial method like Hong's approach. For microcode *ex6*, out of 40 trials, only 4 yielded the best solution (32 bits) with Hong's algorithm and the run time for a single trial was 884.11 CPU seconds on a SUN3/60. Whereas, our method required only 5.58 CPU seconds on the same platform to yield a near-optimal cost of 33 bits.

[2] Ratio of number of trials yielding best solution to total number of trials.

2. Performance Measures on Random Graphs

Table 2: Results with Large Size Examples.

Problem Instances			Results (Non-Opt.)		Results (Optimized)	
Nodes	Edges	Density	Cost	Time	Cost	Time
100	1,306	25%	71	0.45	68	0.68
100	2,015	40%	58	0.39	55	0.54
100	3,242	65%	46	0.41	43	0.48
100	4,497	90%	26	0.42	25	0.45
500	32,494	25%	277	53.48	273	56.86
500	50,977	40%	245	48.53	234	51.99
500	82,226	65%	163	44.19	157	45.21
500	113,426	90%	87	45.28	84	45.53
1,000	129,864	25%	494	429.4	490	440.1
1,000	204,624	40%	447	392.5	434	400.2
1,000	329,675	65%	310	369.4	289	374.0
1,000	454,678	90%	144	376.8	141	377.7
1,500	291,963	25%	728	1,497	717	1,520
1,500	460,386	40%	641	1,370	626	1,386
1,500	742,055	65%	442	1,293	421	1,301
1,500	1,452,908	90%	194	1,277	192	1,279
2,000	519,476	25%	978	2,781	944	2,790
2,000	818,910	40%	821	2,594	809	2,602
2,000	1,319,385	65%	565	2,416	543	2,419
2,000	1,819,312	90%	245	2,410	241	2,416

We tested our algorithm on large size examples generated from random graphs having up to 2000 nodes and more than 1.8 million edges. Experiments were performed with graphs, having graph density varying from 15% to 90% and having the number of nodes varying from 100 to 2000. The experiments were performed on a SPARC 1+ Workstation. The results are given in Table 4. The ***non-optimized results*** indicate that the experiment was performed without using local search cost optimization. The experimental results are plotted in Figure 4. The low slopes of curves (almost horizontal) in Figure 4(a) indicate the independence of our algorithm to variations with graph density. It is observed that the execution time decreases slightly as the graph density increases. This is

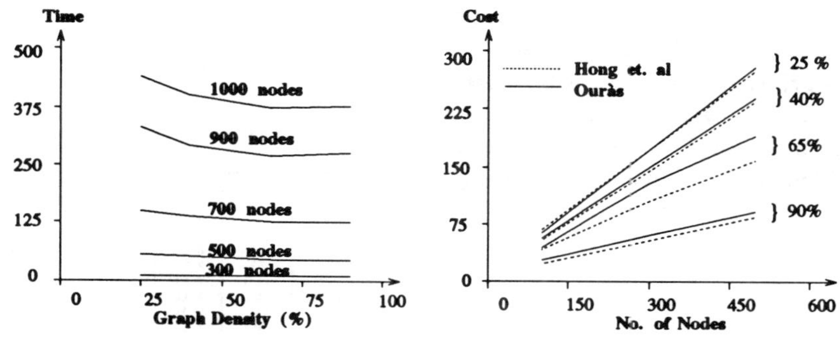

Figure 4: [a] Graph density v/s CPU Time. [b] Cost Comparison with Hong et al.'s Algo.

due to the extraction of larger cliques at higher densities. Thus, the procedure *initial_graph_partitioning* requires less number of iterations as the graph density increases. It was observed that the runtime for a graph having 2000 nodes and 90% density, i.e., having 1.8 million edges was around 2,416 CPU seconds. This is incomparably faster than those of any presently known approaches. Figure 4(b) shows that for almost all the cases, the optimized cost obtained using our algorithm is better than the cost obtained using Hong et al.'s approach.

5 Conclusion

The microword length minimization problem is important to the design of a microprogrammed controller. In this paper, we presented an efficient algorithm to solve the microword length minimization problem. Our algorithm is capable of finding fast and near-optimal solutions for very large size microcodes, efficiently. This performance is achieved because of the application of an efficient initial graph partitioning and the application of an efficient local search algorithm for cost minimization.

6 Acknowledgements

S.K. Hong provided the microcode examples and his results. S.S. Ravi provided their recent work.

References

[1] J. L. Baer and B. Koyama. On the minimization of the width of the control memory of microprogrammed processorors. *IEEE Trans. on Computers*, C-28:310–316, April 1979.

[2] A. Grasselli and U. Montanari. On the minimization of read-only memories in microprogrammed digital computers. *IEEE Trans. on Computers*, C-19:1111–1114, Nov. 1970.

[3] D. Gu, S. S. Ravi, and S. H. O'Leary. Width minimization of microprograms through simulated annealing. Technical Report TR-88-20, CS Dept., SUNY, Albany, Aug. 1988.

[4] S. K. Hong, I. C. Park, and C. M. Kyung. An $o(n^3 logn)$ heuristic for microcode bit optimization. In *ICCAD-90*, pages 180–183, 1990.

[5] T. Jayasri and D. Basu. An approach to organizing microinstructions which minimizes the width of control store words. *IEEE Trans. on Computers*, C-25:514–521, May 1976.

[6] B. W. Kernighan and S. Lin. An efficient heuristic procedure for partitioning graphs. *Bell Sys. Tech. J.*, pages 291–307, Feb. 1970.

[7] A. W. Nagle, R. Cloutier, and A. C. Parker. Synthesis of hardware control of digital systems. *IEEE Trans. on CAD*, pages 201–212, Oct. 1982.

[8] E. L. Robertson. Microcode bit optimization is np-complete. *IEEE Trans. on Computers*, C-28:316–319, April 1979.

[9] S. J. Schwartz. An algorithm for minimizaing read only memories for machine control. In *Proceedings of 10th Annual Symposium on Switching and Automata Theory*, pages 28–33, 1968.

[10] L. F. Sun, J. M. Liaw, and T. M. Parng. Automated synthesis of microprogrammed controllers in digital systems. *IEE Proceedings-Computers and Digital Techniques*, 135, Pt.E(4):231–240, July 1988.

APPENDIX: Local Search for the Satisfiability (SAT) Problem — J. Gu (gu@enel.Ucalgary.CA)

Local search is one of the early techniques proposed during the mid-sixties to cope with the overwhelming computational intractability of NP-hard combinatorial optimization problems. A primitive local search

```
procedure SAT6.0 ()
begin
   get_a_SAT_instance();
   x⃗_0 := select_an_initial_solution();
   f(x⃗_0) := evaluate_object_function();
   k := 0;
   repeat
      for some x_i(k)s ∈ x⃗_k
      begin
         if test_min(f(x_i(k)s)) then
         begin
            x⃗_k+1 := perform_min(f(x_i(k)s));
            f(x⃗_k+1) := evaluate_object_function();
         end
         if |f(x⃗_k+1)| ≤ E then c/d_converter();
      end;
      if local then local_handler();
      k := k + 1;
   until solution_found();
end;
```

Figure 1: **SAT6.0:** A Local Search Algorithm for the Satisfiability (SAT) Problem.

algorithm for the satisfiability (SAT) problem, the $SAT6.0$ algorithm, is given in Figure 1 [1,2]. Procedure *obtain_a_SAT_instance*() initializes a SAT instance with n clauses, m variables and average l literals per clause. An object function of various forms representing the SAT problem was formulated. The SAT problem thus becomes a minimization problem to the object function. To begin, an initial solution point is chosen and the corresponding value of the object function is computed. During a single iterative search step, the test to see if the object function can be minimized is performed by function *test_min*(). If this is true, the minimization operation is performed by procedure *perform_min*(), followed by operation *evaluate_object_function*() that updates the object function. The algorithm terminates when a solution of the SAT instance is found. In practice, during the search, the algorithm could be stuck at a locally optimum point. To improve the convergent performance of the algorithm, a local handler may be added.

Table 1: Computing Performance on a SUN SPARC-1 for practical CSP Problems. Time unit: seconds.

Problems			Backtracks			Execution Times		
n	m	l	Min	Mean	Max	Min	Mean	Max
84	16	2.19	0	3.1	6	0.000	0.015	0.030
170	25	2.18	0	0.0	0	0.010	0.016	0.020
302	36	2.16	0	8.8	38	0.020	0.077	0.230
490	49	2.14	0	3.3	21	0.050	0.081	0.180
744	64	2.13	0	3.2	18	0.090	9.130	0.280
1074	81	2.12	0	4.7	19	0.130	0.199	0.340
1490	100	2.11	0	32.9	113	0.220	0.677	1.730
2002	121	2.10	0	72.2	240	0.340	1.734	4.970
2620	144	2.09	0	25.4	56	0.490	1.081	1.860
3354	169	2.09	0	11.6	53	0.680	1.023	2.220
4214	196	2.08	0	77.9	243	0.940	3.783	9.570
5210	225	2.08	0	35.1	175	1.180	2.784	9.360
6352	256	2.07	0	76.4	269	1.580	5.724	16.80
7650	289	2.07	0	46.6	255	2.130	5.377	21.09
9114	324	2.06	0	9.8	28	2.580	3.351	4.840
10754	361	2.06	0	26.5	76	3.230	5.450	9.820
12580	400	0.26	0	42.2	141	3.900	7.999	16.99
24850	625	2.05	0	47.4	260	9.720	19.62	66.13
43270	900	2.04	0	11.1	45	21.35	25.52	35.52
103560	1600	2.03	0	27.7	131	68.14	87.67	157.5
111602	1681	2.03	0	8.6	59	73.2	81.04	118.2

Algorithms $SAT6.1$, $SAT6.2$, $SAT6.3$ and $SAT6.4$ handle 1-dimensional or multi-dimensional search problems for arbitrary f or $\mathbf{f}$ of the *UniSAT* problem models [1,2]. Some twenty families of such SAT algorithms were developed and experimented, which include, for examples, the discrete local search methods ($SAT1$ family), the steepest descent methods ($SAT7$ family), modified steepest descent methods ($SAT8$ family), newton methods ($SAT10$ family), quasi-newton methods ($SAT11$ family), descent methods ($SAT14$ family), conjugate direction methods ($SAT16$ family), Boolean difference method ($SAT20$ family), and other techniques [1,2,3]. Among hundreds of SAT algorithms developed/experimented, some such SAT algorithms run in $O(knl)$ time and take $O(nl)$ space [1,2,3,4]. Table 1 gives the real execution performance of an early backtracking version of a SAT algorithm for some practical CSP problem instances [2,4]. The ***hardness*** of the SAT problem increases when the number of literals l in each clause decreases or n/m increases. Tested through years of real algorithm executions, our SAT algorithms are capable of computing hard, very large-scale and practical satisfiability problems with excellent convergence performance.

[1] J. Gu, How to Solve Very Large-Scale Satisfiability (VLSS) Problems. 1988.
[2] J. Gu, Benchmarking SAT Algorithms. Technical Report, UCECE-TR-90-002, Oct. 1990.
[3] J. Gu, *Constraint-Based Search.* Cambridge University Press, New York, 1992.
[4] J. Gu and X. Huang, SAT14 Algorithm. Jan. 1991.

Control Optimization in High-Level Synthesis using Behavioral Don't Cares

Reinaldo A. Bergamaschi § Donald Lobo † Andreas Kuehlmann §

§ IBM Research Division
Thomas J. Watson Research Center
Yorktown Heights, N.Y., U.S.A

† University of Washington
Seattle, Washington, U.S.A.

Abstract

This paper presents techniques for optimization of the control part of designs generated by high-level synthesis. The concept of Behavioral Don't Cares is defined and algorithms for extracting behavioral don't care conditions from a high-level description are given. These don't care conditions are used for the optimization of the control logic and the finite-state machine, after high-level synthesis. It is shown that the use of behavioral don't cares combined with finite automata state minimization algorithms, can further optimize the logic and reduce the number of states, after scheduling and allocation. Results from several benchmark examples show that significant area and delay reductions can be obtained with these techniques.

1 Introduction

Most high-level synthesis systems accept descriptions in textual or graph formats, perform high-level synthesis tasks, such as scheduling and allocation, and produce as output a register-transfer level (RTL) description of the synthesized design [1]. This RTL description can be further optimized by means of logic synthesis and finite-state machine (FSM) encoding algorithms [2, 3, 4].

The scheduling and allocation tasks are usually performed on a graph representation (control and data-flow graphs) which is derived from an input description. Generally speaking, an input description specifies the events taking place in the design, in a given order, according to certain input conditions. Hence, it defines the events which should happen under certain conditions, however, it does not specify the behavior if those conditions are not satisfied. By default it is assumed that those events do not happen if the conditions are not met.

In analogy to the logic domain, a high-level hardware description specifies only the ON-set of the behavior. Most existing high-level synthesis systems perform synthesis based only on the ON-set of the specification, which is equivalent to considering the behavioral OFF-set to be the complement of the ON-set, with a null don't care (DC) set. This paper shows that this is a restrictive assumption - don't care information can indeed be extracted during high-level synthesis and used efficiently for control optimization. This paper defines, in a general way, the concept of *Behavioral Don't Cares* and presents algorithms for extracting them during high-level synthesis, as well as for using them for control optimization and state minimization.

Most high-level synthesis systems are not able to extract don't care conditions from the behavioral specification. The BECOME system [5], for example, is only able to derive limited don't care information in specific cases, such as If-Then-Else or Case statements.

Section 2 presents the concept of behavioral don't cares and algorithms for extracting them during high-level synthesis. Section 3 describes state minimization after high-level synthesis using behavioral don't cares. Sections 4 and 5 present results and conclusions, respectively.

2 Behavioral Don't Cares

A high-level description specifies the following relations between a set of operations $\mathcal{O}$ and a set of variables $\mathcal{V}$: (1) Precedence relation $\mathcal{P} = \mathcal{O} \times \mathcal{O}$, which defines the control-flow and (2) Data dependency relation $\mathcal{D} = \mathcal{V} \times \mathcal{O} \times \mathcal{V}$, which defines the data-flow. The ON-set of the behavior is defined by the tuple $\{\mathcal{O}, \mathcal{V}, \mathcal{P}, \mathcal{D}\}$. Most high-level descriptions do not specify $\mathcal{P}$ and $\mathcal{D}$ under the full universe of input conditions. Consequently, they represent incompletely specified behaviors.

In this context, given an operation $o \in \mathcal{O}$, the following definitions apply:

Behavioral ON-set $V^{on}(o)$: is the set of conditions under which operation o is necessarily executed.

Behavioral OFF-set $V^{off}(o)$: is the set of conditions under which operation o must not be executed.

Behavioral DC-set $V^{dc}(o)$: is the set of conditions under which the input/output behavior is unaffected independently of whether operation o is executed or not.

As a simple example, consider a conditional operation on variable v_1 with two branches, and assume that operation o_1 is declared inside branch '$v_1 = 1$'. The condition '$v_1 = 1$' defines the ON-set $V^{on}(o_1)$.

0738-100X/92 $3.00 © 1992 IEEE

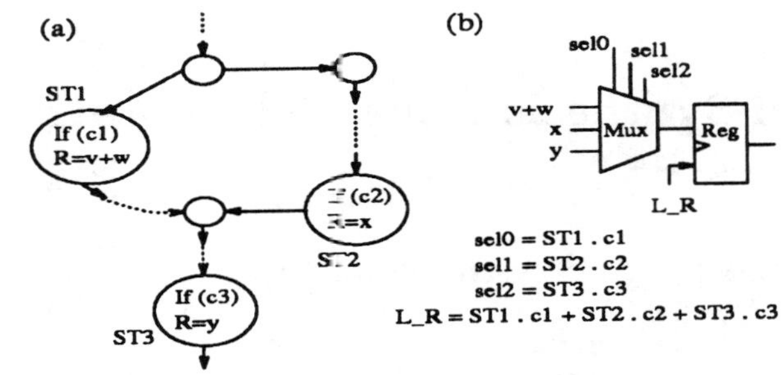

Figure 1: Example of FSM generated by high-level synthesis and corresponding data-path.

Now suppose that operation o_1 could be duplicated and inserted in branch '$v_1 = 0$' without affecting the input-output behavior. In this case '$v_1 = 0$' would be part of the DC-set $V^{dc}(o_1)$.

Scheduling maps incompletely specified precedence relations into defined sequences of states of a FSM, and allocation maps incompletely specified data-dependency relations into hardware elements and interconnection networks. After scheduling and allocation, the behavioral ON-set, OFF-set, and DC-set can then be defined in terms of signals present in the implementation, which involve the state of the FSM (state register), primary inputs and data-path signals. These sets can be used for structural and control optimization at the logic level.

The algorithms presented in this paper have been integrated in the HIS system [6, 7]. HIS output consists of a controller FSM and a data-path. The outputs of the FSM are the control signals for the data-path operations, namely, select signals for multiplexers, select signals for multi-function arithmetic-logic units (ALU) and load-enable signals for registers. Figure 1 shows an example of a FSM and related data-path generated by HIS. This RTL description is subsequently submitted to logic synthesis for further optimizations.

Behavioral Don't Care conditions can be derived by analyzing the relationships between the operations in an input description and the RTL implementation generated by high-level synthesis. Three types of behavioral don't care conditions have been identified, and are described below.

2.1 Don't Cares Associated with Storage Elements

The load enable signal for a register is derived based on the scheduling of the operations which assign a value to the register. This implies that there is a set $S = \{States, Conditions\}$ under which a new value must be loaded into the register (load signal true). Whenever the register is not loaded (load signal false), it is assumed that it should retain its old value.

The rationale for extracting don't care conditions in this case is the following. Based on *lifetime* analysis it can be stated that if a register is *dead* in a state, its value is not going to be used by any operation, before being assigned again. Hence, the input-output behavior is independent of the value of the register in that state. Consequently, it can assume any value, which means that the load signal can assume a don't care value in that state. Furthermore, if the register is *alive* in a state but *dead* in the next state, then the present state and the transition condition can be added to the don't care condition.

The precise formulation is a follows:

Let the ON-set for the load signal *Load_R* of register R be the following set of states and conditions:

$$Load\_R^{on} = \{(s_1, c_1), (s_2, c_2), .., (s_p, c_p)\}$$
$$= (state \equiv s_1)\cdot c_1 + (state \equiv s_2)\cdot c_2 + .. + (state \equiv s_p)\cdot c_p$$

The set of states and conditions under which the load signal may assume a don't care value is the union of two sets: (1) the set of states and conditions in which the register is *dead* and *not loaded*, and (2) the set of states and conditions in which the register is *alive*, *not loaded*, but *dead* in the successor state. As shown below:

$$Load\_R^{dc1} = \{(s_{d1}, c_{d1}), (s_{d2}, c_{d2}), .., (s_{dp}, c_{dp})\} \quad (1)$$

where register R is *dead* is states $s_{d1}, .., s_{dp}$, and $Load\_R^{on} \bigcap Load\_R^{dc1} = \phi$ (not loaded).

$$Load\_R^{dc2} = \{(s_{a1}, c_{a1}), (s_{a2}, c_{a2}), .., (s_{ap}, c_{ap})\} \quad (2)$$

where register R is *alive* is states $s_{a1}, .., s_{ap}$, and *dead* in successor states $\tau(s_{a1}, c_{a1}), .., \tau(s_{ap}, c_{ap})$, and $Load\_R^{on} \bigcap Load\_R^{dc2} = \phi$ (not loaded). $\tau(s, c)$ is the transition function, such that $s_\gamma = \tau(s_\alpha, c_\beta)$ is the successor state from state s_α under condition c_β.

$$Load_R^{dc} = Load\_R^{dc1} \bigcup Load\_R^{dc2}$$
$$= (state \equiv s_{d1}) \cdot c_{d1} + .. + (state \equiv s_{dp}) \cdot c_{dp} +$$
$$(state \equiv s_{a1} \cdot c_{a1} + .. + (state \equiv s_{ap}) \cdot c_{ap}$$

This condition can then be used with $Load\_R^{on}$ for the optimization of *Load_R*.

2.2 Don't Cares Associated with Interconnection Elements and Multi-Function ALUs

Interconnection elements, such as multiplexers, are inserted in the data-path during the allocation phase in order to connect several source signals to one destination.

Select signals for multiplexers are derived according to the scheduling of the operations requiring the interconnections (as illustrated in figure 1). Whenever one of these operations is *active*, a select signal is true and the corresponding input is transferred to the output, to be used somewhere in the circuit.

Alternatively, when none of these operations is active, then all select signals are false. This implies that none of the inputs has to be transferred to the output, which in fact means that the multiplexer output is not being used. These "no operations active" conditions can be used as don't care conditions for the minimization of the multiplexer's select signals.

The rationale above can be expressed more formally in the following way:

Let $S = \{s_1, s_2, .., s_p\}$ be the set of unencoded select signals of a p-to-1 multiplexer connected to the input of register R. The signals in S are dependent on a set $V = \{v_1, v_2, .., v_n\}$ of variables (primary inputs, data-path signals, state register). The ON-set of S, $V^{on}(S)$, is the set of values v such that $s_i(v) = 1$, for any $1 \leq i \leq p$. The OFF-set of S, $V^{off}(S)$, is the set of values v such that $s_i(v) = 0$, for all $1 \leq i \leq p$ (this represents the "no operations active" condition). By definition, $V^{off}(S)$ is the complement of $V^{on}(S)$. Register R is only loaded when one of the select signals in S is true, and it is not loaded otherwise. Hence, the OFF-set, $V^{off}(S)$ can be used as the *Don't Care set*, $V^{dc}(S)$, for the optimization of each select signal individually.

The use of these don't cares is not restricted to cases where multiplexers are followed by registers. They can be applied to multiplexers in general, except in cases where the multiplexer feeds a multi-cycle operation or a primary output directly, where extra conditions must be considered to insure the correct behavior. The use of don't cares might force two select signals to "1" simultaneously and this would be a problem in technologies using pass-transistors. In this cases, extra conditions could be applied to guarantee the mutual exclusiveness of the select signals. By definition, the select signals are mutually exclusive when the ON-set of any of them is active.

Don't cares associated with multi-function ALUs are similar to those for interconnection elements. During allocation, different operators may be merged into multi-function ALUs. Control signals are generated to select the function corresponding to the operation to be executed. When the operation is *active*, an ALU select signal is true, which implies that the ALU is being used. When none of the ALU's operations is active, all select signals are false and the ALU and its output are not used. Hence, the "no operations active" conditions can be used as don't care conditions for the minimization of the ALU's select signals individually.

As an example, consider the control-flow graph shown in figure 2a. Figure 2b.1 shows a FSM generated by the *As-Fast-As-Possible* scheduler [6] which needs three states. The conditions controlling the execution of each operation in a state, as well as the register lifetimes are indicated. The data-path for this example contains three registers (PC, OLDPC, IR) and two 2-to-1 multiplexers. Figure 2c.1 contains the state table for the FSM in figure 2b.1. This State Table was built based on the ON-set of the control signals (load-enable for PC, OLDPC, IR, and unencoded select signals for Mux1 and Mux2). From this table it can be seen that registers IR and OLDPC are not loaded in rows 1, 2, 3, 5, and 7, and in addition they are alive in state 0 but dead in state 1 and state 2, hence their load signals can assume a don't care value in these rows. Similarly, the select signals for multiplexers Mux1 and Mux2 are null in rows 3, 5, 7, and 7, 8, respectively. Therefore, they could also assume don't care values in these rows.

3 State Minimization after High-Level Synthesis

Scheduling binds operations to control steps, or states of a controller. Each state is defined by the set of operations scheduled in it. Irrespective of the scheduling algorithm used, the sets of operations scheduled in any two states are never equivalent (if they were the same the two states could be merged). Moreover, it is assumed that since different states contain different sets of operations, they cannot be combined, or in other words, it is assumed that the FSM produced by high-level synthesis has no ***compatible*** states. This section shows that the above assumption is incorrect. States may be compatible even if the sets of operations scheduled in them are different.

From Finite Automata theory, it is known that two states s_i and s_k are ***compatible*** if they yield ***comparable*** output sequences for every input sequence. Two output sequences are called *comparable* if they are identical *whenever both are specified.*

Hence, in order to determine whether two states are compatible, it is necessary to look at the outputs sequences produced by the two states, and not at operations executed in each state. In the case of FSMs produced by high-level synthesis, the FSM output signals are used as control signals for the data path and/or primary outputs. The important thing to notice here is that two states may contain different operations, but nevertheless they may have comparable output sequences. Furthermore, two output sequences may be made comparable by inserting behavioral don't cares.

In order to check for compatible states, it is necessary to build the *Inputs-Present_State-Next_State-Outputs Table* (also called State or Flow Table). Figure 2c.1 gives and example of such table for the FSM in figure 2b.1, which contains no compatible states. However, if behavioral don't cares are inserted for Mux1 and Mux2 select signals, on rows 3, 5, 7, 8, by converting the "00" entries to "– –", then the output sequences for states 1 and 2 become ***comparable***, which means that states 1 and 2 are compatible and can be merged. The resulting State Table is given in figure 2c.2, while the corresponding FSM is shown in figure 2b.2. The resulting control signals can be significantly optimized due to the don't cares.

The complete state minimization procedure consists of three parts: (1) Generation of the State Table; (2) Insertion of Behavioral Don't cares into the State Table; and (3) Minimization of the state machine (based on the State Table).

Tasks 1 and 3 above can be computationally expensive due to the possibly large number of combinations of input signals in each state, which increases the size of the table. The complexity is reduced by finding a subset of inputs such that, all transition equations from the state and all control signals whose ON-set intersect the state, depend upon. Only combinations of inputs in this subset must be enumerated. This drastically reduces the number of rows in the State Table. Once the State Table has been generated, the insertion of don't cares proceeds as detailed in section 2. The insertion of don't cares is not always a requirement for obtaining compatible states. Task 3 above

(a)

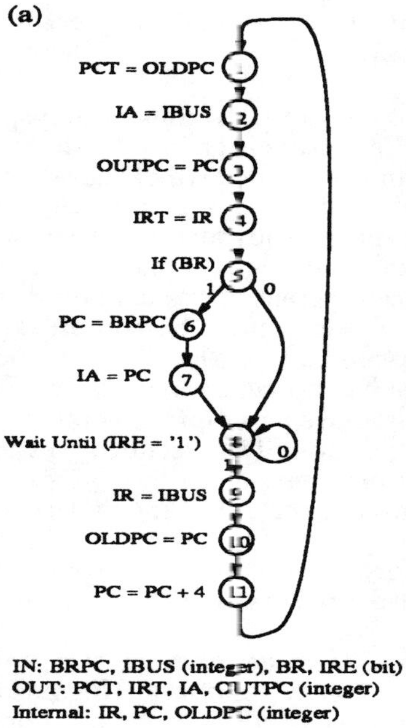

(b.1)

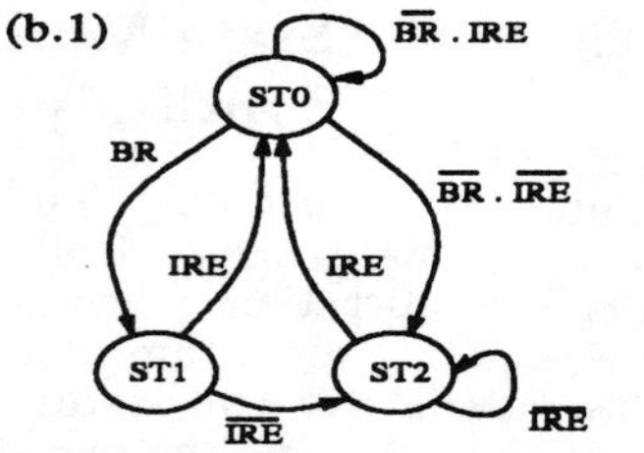

State	Operations	Regs. Alive
ST0	1,2,3,4,5, 6 (BR), 8 (BR), [9,10,11] (BR . IRE)	PC, IR, OLDPC
ST1	7,8, [9,10,11] (IRE)	PC
ST2	8, [9,10,11] (IRE)	PC

(b.2)

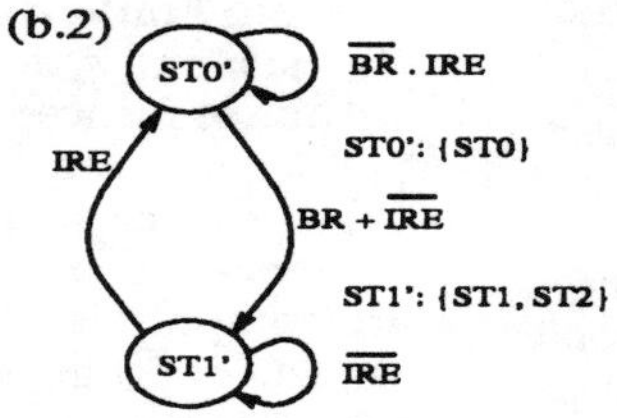

(c.1) State table for initial FSM

	Inputs				Outputs				
	BR	IRE	PS	NS	L_IR	L_PC	L_OLDPC	MUX1	MUX2
1	1	0	ST0	ST1	0	1	0	1 0	1 0
2	1	1	ST0	ST1	0	1	0	1 0	1 0
3	0	0	ST0	ST2	0	0	0	0 0	1 0
4	0	1	ST0	ST0	1	1	1	0 1	1 0
5	-	0	ST1	ST2	0	0	0	0 0	0 1
6	-	1	ST1	ST0	1	1	1	0 1	0 1
7	-	0	ST2	ST2	0	0	0	0 0	0 0
8		1	ST2	ST0	1	1	1	0 1	0 0

(c.2) State table for final FSM

	Inputs				Outputs				
	BR	IRE	PS	NS	L_IR	L_PC	L_OLDPC	MUX1	MUX2
1	1	0	ST0	ST1	-	1	-	1 0	1 0
2	1	1	ST0	ST1	-	1	-	1 0	1 0
3	0	0	ST0	ST1	-	0	-	- -	1 0
4	0	1	ST0	ST0	1	1	1	0 1	1 0
5	-	0	ST1	ST1	-	0	-	- -	0 1
6	-	1	ST1	ST0	1	1	1	0 1	0 1

Figure 2: Design views: (a) control-flow graph, (b) FSM after scheduling (1) and after state minimization (2), (c) State Tables, before (1) and after (2) state minimization.

corresponds to the classical problem of minimization of incompletely specified state machines. A program called STAMINA [8], developed by the University of Colorado, Boulder, is used for this task.

The technique described here can result in state minimization, independently of the scheduling algorithm which generated the FSM, as long as it is able to consider conditional operations and loops. Clearly, a scheduling consisting of a straight sequence of states will not have compatible states.

There is a trade-off between resource sharing in allocation and state minimization after scheduling. Resource sharing may decrease the amount of possible don't cares, which in turn may decrease the number of compatible states.

4 Results

The concepts described in the previous sections were implemented in the HIS system. Several benchmark examples from the 5th International High-Level Synthesis Workshop were automatically synthesized using the steps below:

(1) Run HIS for scheduling and allocation [6]. Optimize control signals (based on their ON-set only).
(2) Generate State Table based on FSM and control/output signals.
(3) Insert behavioral don't cares.
(4) Run STAMINA for state minimization with State Table as input, and update FSM and control signals.
(5) Repeat steps 2 and 3.
(6) Run state encoding program (MUSTANG [4]) with State Table as input, and update FSM and control signals .
(7) Generate RTL description, run LSS [2] for final logic synthesis and technology mapping.

Both STAMINA and MUSTANG use the don't care information inserted in the State Table. The results of these programs are submitted to ESPRESSO [3] for final logic optimization (also using behavioral don't cares). Once the logic has been optimized, an RTL description is generated and input to LSS.

Table 1 presents the number of states, before and after state minimization - only benchmarks which showed a reduction in the number of states are listed (these include almost all non-trivial benchmarks). State minimization was performed using different don't care levels: (1) no don't cares, (2) only Register don't cares, (3) only MUX and ALU don't cares, and (4) MUX, ALU and Register don't cares.

Results in this table show that the number of states can be significantly reduced after scheduling using finite automata techniques. In addition, the use of behavioral don't cares can further reduce the number of states. In examples M6502, HUNT, TX8251 and PREFETCH, state minimization was independent of the use of don't cares. In examples MTLC, FRISC and MEMOP, states could only be minimized with the use of multiplexer and ALU don't cares. Examples RCV8251 and I8251A showed state reduction without don't care and further reduction using multiplexer and ALU don't cares.

Table 2 gives the final cell count and number of levels of logic after running LSS. Three cases are shown: (1) after scheduling and allocation (steps 1, 2, 6 and 7 - no state minimization, no don't cares), (2) state minimization plus logic optimization without don't cares (steps 1, 2, 4, 6 and 7), and (3) state minimization

Design	States / Transitions				
	after Sched.	after State Minimization			
		no DC	r DC	m+a DC	m+a+r DC
m6502	106/204	88/179	88/179	88/179	88/179
hunt	12/24	9/17	9/17	9/17	9/17
tx8251	22/101	20/81	20/81	20/81	20/81
prefetch	4/9	3/7	3/7	3/7	3/7
mtlc	8/14	8/14	8/14	6/10	6/10
frisc	42/88	42/88	42/88	40/86	40/86
memop	6/9	6/9	6/9	5/8	4/6
rcv8251	23/100	20/87	20/87	18/79	18/79
i8251A	15/65	14/62	14/62	11/53	11/53

Table 1: Number of states after scheduling and after state minimization using different don't care levels.

Design	Cells / Logic Levels (after LSS)			Δ%
	Sched. + Alloc.	State Min. + Opt.		
		no DC	with DC (type)	
m6502	7701/27	7580/28	7567/27 (r)	2
hunt	458/6	470/7	407/7 (m+a+r)	11
tx8251	1117/20	1013/23	901/11 (m+a+r)	19
prefetch *	1942/7	1831/5	1561/6 (m+a+r)	20
mtlc	74/6	no chg.	56/4 (m+a)	24
frisc *	3566/19	no chg.	3703/18 (m+a+r)	-4
memop	299/8	no chg.	250/8 (m+a+r)	16
rcv8251	1199/13	1212/13	1111/11 (m+a+r)	7
i8251A	528/8	501/9	375/8 (m+a+r)	29
gcd	417/22	no chg.	372/18 (m+a)	11
addrc	2027/20	no chg.	1904/18 (m+a+r)	6
kalman *	6459/51	no chg.	6351/33 (m+a+r)	2
mahal	621/3	no chg.	677/3 (m+a)	-8
diffeq *	6382/21	no chg.	6317/21 (r)	1

Table 2: Results of synthesis (after LSS): cell count and number of logic levels for original implementation (after scheduling and allocation), and after state minimization and logic optimization with and without behavioral don't cares. (obs.: number of levels do not include some internal modules in examples marked with *).

plus logic optimization with don't cares (all steps). In this last case, the numbers correspond to the best results using don't cares (the type of don't cares used is indicated in parenthesis).

From table 2 it can be seen that almost all benchmarks (except for FRISC and MAHA1) presented area reduction after state minimization and optimization with don't cares. From columns "Sched+Alloc" and "no DC" it can be seen that state minimization alone (without don't cares) does not have a great impact on the final area. However, the use of mux, alu and register don't cares results in a significant area and delay decreases in most cases. In cases where no state reduction was possible, don't cares were used for encoding and logic optimization only. The largest decreases in area were obtained in control dominated circuits, such as I8251A, TX8251 and MTLC. In almost all cases (except HUNT) the number of logic levels decreased or stayed the same, after the optimizations. This decrease was significant in TX8251, KALMAN and GCD, which shows that the number of levels can be affected by the use of don't cares independently of their effect on the area.

5 Conclusions

This paper presented novel techniques for optimization of the control part of designs generated by high-level synthesis systems. These techniques are based on the extraction of Behavioral Don't Care conditions from the high-level specification, and use of these don't cares for logic and state minimization. It was shown that scheduling algorithms do not necessarily produce FSMs with incompatible states, and finite automata state minimization algorithms can be used to minimize the number of states. The use of behavioral don't cares can further improve state minimization. Results for several benchmarks showed that significant improvements in area and delay (logic levels) can be obtained using these techniques.

References

[1] M. C. McFarland, A. C. Parker, and R. Camposano, "The high-level synthesis of digital systems," *Proceedings of the IEEE*, pp. 301–318, February 1990.

[2] J. Darringer, D. Brand, J. V. Gerbi, W. Joyner, and L. Trevillyan, "LSS: A system for production logic synthesis," *IBM Journal of Research and Development*, vol. 28, September 1984.

[3] R. Brayton, G. Hachtel, C. McMullen, and A. Sangiovanni-Vincentelli, *Logic Minimization Algorithms for VLSI Synthesis*. The Netherlands: Kluwer Academic Publishers, 1985.

[4] S. Devadas, H.-K. T. Ma, A. R. Newton, and A. Sangiovanni-Vincentelli, "MUSTANG: State assignment of finite state machines targeting multi-level logic implementation," *IEEE Transactions on Computer-Aided Design*, vol. CAD-7, December 1988.

[5] R. Wei, S. G. Rothweiler, and J. Jou, "BECOME: Behavior level circuit synthesis based on structure mapping," in *Proceedings of the 25th ACM/IEEE Design Automation Conference*, (Anaheim, CA), pp. 409–414, ACM/IEEE, July 1988.

[6] R. Camposano, R. A. Bergamaschi, C. Haynes, M. Payer, and S. M. Wu, "The IBM high-level synthesis system," in *High-Level VLSI Synthesis* (R. Camposano and W. Wolf, eds.), pp. 79–104, Kluwer Academic Publishers, 1991.

[7] R. Bergamaschi, R. Camposano, and M. Payer, "Data-path synthesis using path analysis," in *Proceedings of the 28th ACM/IEEE Design Automation Conference*, pp. 591–596, ACM/IEEE, June 1991.

[8] G. D. Hachtel, J.-K. Rho, F. Somenzi, and R. Jacoby, "Exact and heuristic algorithms for the minimization of incompletely specified state machines," in *Proceedings of The European Conference on Design Automation*, (Amsterdam, The Netherlands), pp. 184–191, IEEE, February 1991.

Transformation-Based High-Level Synthesis of Fault-Tolerant ASICs*

Ramesh Karri and Alex Orailoğlu
Department of Computer Science and Engineering
University of California, San Diego
La Jolla, CA 92093

Abstract *As fault-tolerance becomes common in everyday applications (like computerized automobile control) where cost of design and time-to-market are crucial factors, software mechanisms for the automatic design of fault-tolerant ASICs merit immediate attention. A straightforward replication of a design is expensive. In this paper, we present a transformation-based approach to the high-level synthesis of fault-tolerant ASICs satisfying a given performance constraint but requiring less than proportional increase in hardware over their non-redundant counterparts.*

1 Introduction

With fault-tolerance becoming common in everyday applications like computerized automobile control, issues like cost of design and time-to-market are assuming added importance. Furthermore, interest in automating the design of fault-tolerant ASICs has been stimulated by advances in fault-tolerant computing and VLSI technology.

Straightforward replication of an algorithm will lead to expensive designs. Furthermore, in a typical VLSI design environment this simple strategy renders the final design susceptible to common mode errors[3]. Time redundancy on the other hand can result in inexpensive designs, but at the expense of reduced performance.

An input behavior has several distinct, yet semantically equivalent behavioral representations (each having an associated time and area overhead when implemented in hardware). Behavioral transformations which help explore the rich space of alternative representations are crucial to the realization of affordable on-chip fault-tolerance. Furthermore, a synthesis methodology that exploits such transformations to minimize hardware requirements is important.

There have been many efforts in developing structured design methodologies for fault-tolerant and fault-recovering ASICs [3, 10]. Whereas significant work has been done in the broad area of high-level synthesis[2], high-level synthesis of fault-tolerant ASICs has not received much attention. In general, CAD for fault-tolerance is relatively unexplored. ASSURE [1] is an automated design for dependability tool, [9] is an algorithm for the synthesis of self-recovering ASICs and [5] proposes a synthesis methodology to maximize performance subject to area and fault-tolerance/reliability constraints. Previously, transformations have been used to either improve performance [6], area [7, 8], or both [11], during the synthesis of non-redundant designs.

*This work is supported by NSF I/UC Research Center on ICAS at University of California, San Diego

1.1 Research Contributions

In this paper, we propose a transformation-based approach to high-level synthesis of fault-tolerant ASICs. The cost of fault-tolerance (due to replication) must be minimized, while fault-tolerance and performance have to be maximized.

We propose a synthesis methodology to exploit hardware minimizing transformations. Specifically, we identified and implemented a simple set of transformations that minimize the fault-tolerance overhead. Furthermore, the selected transformations make the final design resilient to common mode failures. These transformations can be composed to form a rich set of complex transformations. Additionally, we present an algorithm to automatically identify structures in a flow graph where transformations can improve hardware utilization, and apply transformations that suite the structure best.

The rest of this paper is organized as follows. In section 2, we present the concepts of fault-tolerance relevant to our work together with our assumptions. In section 3, we present the transformation-based synthesis of fault-tolerant ASICs. The system has been used to schedule several flow graphs. In section 4, we present experimental results, and share our experiences with the transformation-based approach. Finally, section 5 concludes this paper.

2 Definitions and Assumptions

Permanent faults remain in a system indefinitely, while **transient faults** appear and disappear a short time later. **Fault tolerance** is the ability of a system to continue to perform its tasks after the occurrence of faults. **Fault masking** is the process of preventing faults from introducing errors. **N Modular Redundancy (NMR)** uses N modules and majority voting for fault-masking. In figure 1 triple modular redundancy is shown. In this model, the voting is performed at the end of a computation. The voter is a single point of failure. Consequently, care has to be taken while designing the voter to make it fault-tolerant.

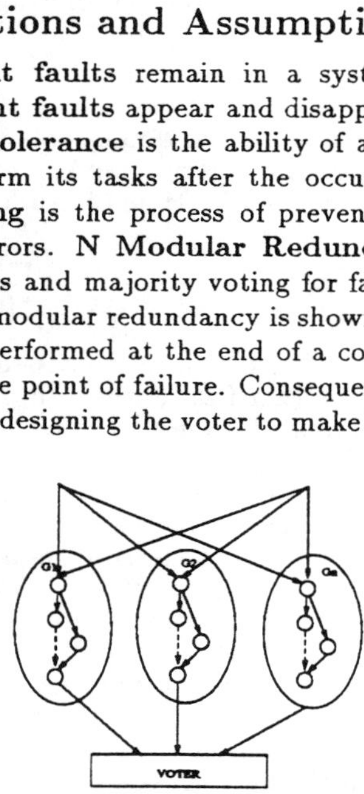

Figure 1: Triple Modular Redundancy

0738-100X/92 $3.00

In the current version of the algorithms, we assume that there are no conditional transfers of control in the input flow graph, that there is no chaining and that the voting circuitry is fault-tolerant.

3 Synthesis of Fault-Tolerant ASICs

Synthesis for fault-tolerance is formulated as follows: Given G, an input flow graph, N, the redundancy ratio, and C, the maximum number of clock cycles, schedule N copies of G into C clock cycles using total hardware less than or equal to N times the minimal hardware required to schedule G. The algorithm is given in figure 2. Initially, time-constrained scheduling is used to compute the schedule of the untransformed flow graph (G^0). As a result of time-constrained scheduling, $H_k^0(j)$, the amount of jth hardware type used in the kth clock, $1 \leq k \leq C$, is determined. For each j, the peak utilization, $\mathcal{H}^0(j)$, is derived as $\mathcal{H}^0(j) = max(H_k^0(j)) \quad 1 \leq k \leq C$.

The $(i-1)^{th}$ copy of the flow graph, G^{i-1} (which is already scheduled) is transformed into G^i. The transformations are driven by hardware usage across clock cycles obtained from scheduling $i-1$ copies. Hardware sharing, the key to minimizing fault-tolerance overhead, is accomplished by utilizing unused hardware $(\Delta H_k^{i-1}(j) = \mathcal{H}^{i-1}(j) - H_k^{i-1}(j))$ in each of the clock cycles. $H_k^{i-1}(j)$ is the sum of the utilizations of the jth hardware type in the kth clock cycle obtained from scheduling i copies (0 to $i-1$), while $\mathcal{H}^{i-1}(j)$ is the maximum of this value over all clock cycles. Subsequently, G^i is scheduled using a time and resource-constrained scheduler. The time constraint is C clock cycles, while the resource constraint in the kth clock cycle is $\mathcal{H}^0(j) + \Delta H_k^{i-1}(j)$.

NMR Synthesis (G, C, N) {
  $G^0 = G$
  $H_k^0(j) =$ Schedule($G^0, C,$)
  for($i = 1; i < N; i = i+1$)
  {
   $G^i =$ Transform (G^{i-1})
   $H_k^i(j) = H_k^{i-1}(j) +$ Schedule(G^i, $C, \mathcal{H}^0(j) + \Delta H_k^{i-1}(j)$))
  } }

Figure 2: Fault-Tolerant ASIC Synthesis

3.1 Transformations

A drawback of straightforward replication is that the hardware to execute N copies of a flow graph is determined by peak hardware usage across clock cycles. This observation motivated the transformation-based approach, and guided the selection of a set of transformations. The selected transformations are simple to apply and can be composed into more complex ones. Also, the flow graph structures on which they are applied are easy to identify.

The transformations currently implemented include associativity, distributivity, duplication, and commutativity. Local transforms like distributivity and associativity applied to non-commutative operators improve utilization of hardware in adjacent clock cycles, while global transforms like associativity impact the hardware utilization across a number of clock cycles. In fact, associativity alters the length of the paths on which it is applied, thereby affecting mobility in the nodes on a path.

The Transform() routine is outlined in figure 3. Critical clock cycles have high utilization of a hardware type (i.e. $\mathcal{H}^{i-1}(j) = H_k^{i-1}(j) \Rightarrow k \in C_{crit}$), while candidate transformations help reduce the hardware. Initially, local transformations are applied on a copy of the flow graph guided by (i) the distribution of utilized hardware due to previously scheduled copies, and (ii) the effects of the preceding transformations on the current copy. The local transforms are easy to apply. The global transformations are applied when no local transformations are applicable. Global transformations improve hardware utilization characteristics across clock cycles and additionally uncover flow graph structures amenable to local transformations. Preconditions for distributivity and associativity will be described. Remaining transformations are elaborated in [4].

Transform (G) {
 while ($\exists$ an useful transformation) {
  while ($\exists$ a Local transformation)
  {
   Identify $c \in C_{crit}$
   Identify candidate transformation
   Apply it to minimize peak utilization in c
  }
  If ($\exists$ a candidate Global transformation)
  Apply it to introduce mobility
 } }

Figure 3: Algorithm to apply transformations

Precondition 1: If $k, k+1$ are adjacent clock cycles with (i) $\Delta H_{k+1}^i(add) > 1$ and $\Delta H_k^i(mult) > 2$, (ii) a flow graph structure of the form $a \times (b+c)$, and (iii) $(b+c)$ is not used elsewhere, then **distributivity** *can be applied.* This can be extended to the case of noncontiguous clock cycles. Additionally, distributivity can be applied to subtract and divide operations.

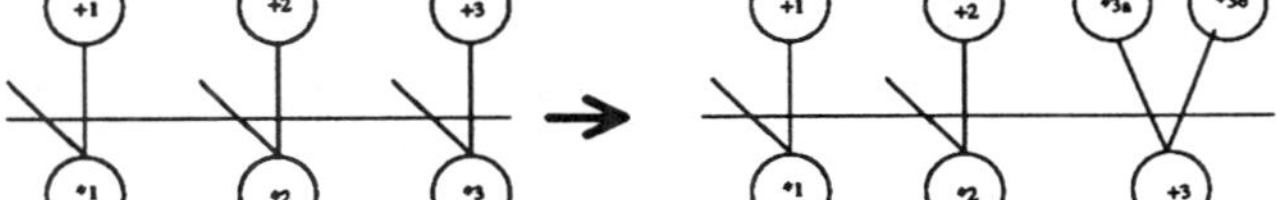

Figure 4: Hardware reduction using distributivity

Example: In figure 4, assuming two clock cycles and no chaining, the $3MR$ schedule on the left hand side of the arrow requires three adders and three multipliers, while the transformed $3MR$ schedule requires only two adders and two multipliers.

Precondition 2: If $P(a), P(b)$, and $P(c)$ are the number of predecessor nodes of a, b, and c respectively, in a flow graph structure of the form $(a+b)+c$, and if $P(a) > max(P(b), P(c))$, or if $P(b) > max(P(a), P(c))$, then **associativity** *can be applied.*

Example: In figure 5, associativity reduces the length of the path on which it is applied. Our motivation in applying associativity is not to improve performance (as the number of clock cycles is given as an input) but to reduce the cost of the final design. Associativity increases the mobility of individual nodes, resulting in less expensive designs via load balancing during scheduling.

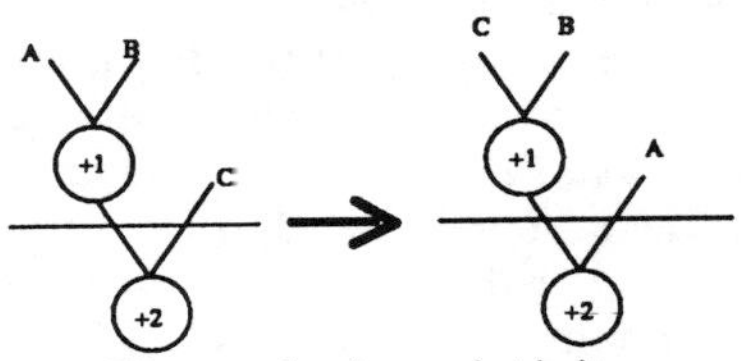

Figure 5: Associativity

3.2 Scheduling Algorithm

In the synthesis methodology, we used a time constrained as well as time and resource constrained scheduler. In figure 6, the time and resource constrained scheduling algorithm is given (of which time constrained scheduling is a special case).

Schedule $(G^i, C, \mathcal{H}^0(j) + H_k^i(j))$ {
  ASAP(G^i, $C, \mathcal{H}^0(j) + H_k^i(j)$)
  ALAP(G^i, $C, \mathcal{H}^0(j) + H_k^i(j)$)
  while ($\exists$ a $Node_l$ with $E(Node_l) \neq L(Node_l)$)
  {
  Select $Node_{crit}$, the critical node
  t = type of $Node_{crit}$
  for ($k = E(Node_{crit}); k \leq L(Node_{crit}); k = k+1$)
   if ($C_t^k < C_t^{Clock_{best}}$) $Clock_{best} = k$
  Assign ($Node_{crit}, Clock_{best}$)
  Propagate
  }
  Return $U_k^i(j)$
}

Figure 6: Scheduling a copy of a flow graph

Initially, a time and resource constrained ASAP and ALAP scheduling is performed to determine the earliest feasible clock cycle (E) and the latest feasible clock cycle (L) for each node ($Node_l$) in the flow graph. A critical node ($Node_{crit}$ of type t) is then selected. Criticality of a node is inversely proportional to the cardinality of its *feasible set.* The feasible set of a node is the clock cycles it can be assigned to, subject to input constraints on clock cycles and available hardware. For $Node_{crit}$, the *least contested* clock cycle (C_{best}) is chosen from its feasible set. The contestation measure C_j^k for an operator type j in clock cycle k, is given as

$$C_j^k = \frac{O_k^i(j)}{\Delta H_z^{i-1}(j) + \mathcal{H}^0(j) - U_k^i(j)}$$

where, the numerator denotes the number of unassigned operators in G^i of the jth hardware type that have clock cycle k in their feasible set, and the denominator is the number of hardware units of the jth type that are available. $U_k^i(j)$ is the number of jth type of hardware units utilized in the kth clock cycle exclusively by the ith copy. Finally, $Node_{crit}$ is assigned to C_{best}. The effects of an assignment include updating hardware utilization. If the hardware in the chosen clock cycle is fully utilized, it is removed from the feasible sets of all unscheduled nodes of the same type. Feasible clock cycles of other nodes are in turn updated automatically by algorithms which propagate the effects of a commitment.

4 Results

We implemented a framework for behavioral synthesis to serve as a vehicle for experimenting with the algorithms outlined in this paper. The hardware platform is a Sun 3/50. The transformations discussed and the transformation-based scheduler for fault-tolerance have been integrated into the synthesis framework with remarkable ease.

In this section, we present the results of our transformation-based scheduler on two examples. The examples are (i) a backsolver for a 4×4 matrix, which can be used as a kernel to solve large simultaneous linear equations using block LU decomposition, and (ii) a 16 point digital filter. The backsolver solves a set of linear equations of the form $SX = B$, where S is an upper triangular matrix and X, B are column vectors.

We assume that: (i) the cost of a multiplier = cost of a divider = 2 units, cost of an adder = cost of a subtractor = 1 unit and (ii) the duration of a control step = the delay through the individual functional units. For simplicity, we do not include the overhead due to voting circuitry, registers, and interconnect in the analyses.

We perform transformation-based scheduling on the backsolver and filter examples. The results obtained without triggering the transformations form a basis for computing the savings afforded by our transformation-based scheduler. The results are summarized in tables 1 and 2. In figure 7, we plot the reduction in hardware cost.

N	Add	Mult	Sub	Div	Cost	% savings
1	1	1	1	1	6	-
2	2	2	2	2	12	0
3	2	3	2	2	14	22.22
4	2	4	2	2	16	33.33
5	3	4	3	3	20	33.33
6	3	5	3	4	24	33.33
7	4	6	3	4	27	35.71

Table 1: Experimental results on backsolver flow graph

Discussion: From the tables 1 and 2, it may be observed that significant savings in hardware can be obtained by using simple transformations on a flow graph. The savings in hardware costs range between 0% and 35.71% for the backsolver example and between 0% and 30% for the filter example. As an illustration, the effect of transformations on the filter flow graph are shown in figure 8, while a

N	Add	Mult	Cost	% savings
1	2	2	6	-
2	4	3	10	16.67
3	6	5	16	11.11
4	8	5	18	25
5	9	6	21	30

Table 2: Experimental results on 16 point filter

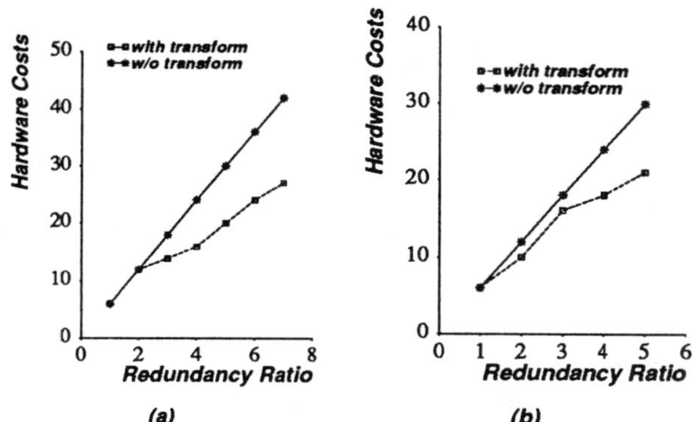

Figure 7: N *vs* Hardware Cost for (a) example 1 and (b) example 2

2MR schedule of the backsolver is shown in figure 9.

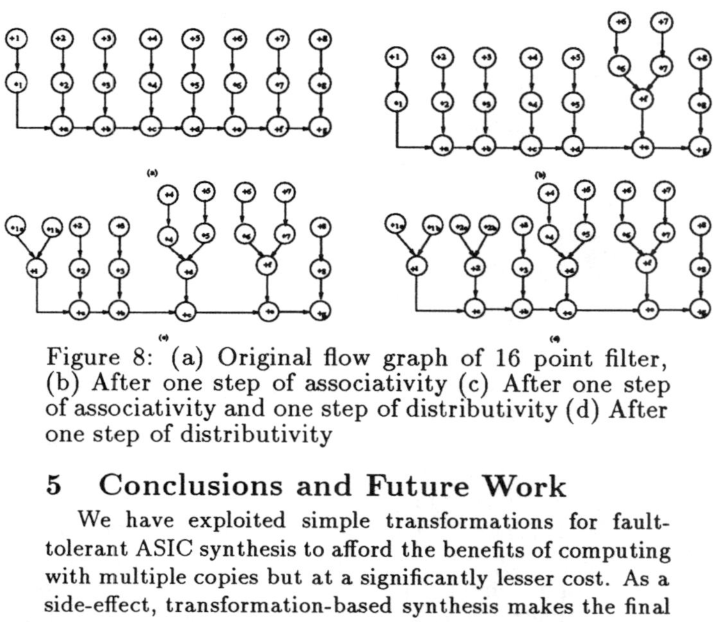

Figure 8: (a) Original flow graph of 16 point filter, (b) After one step of associativity (c) After one step of associativity and one step of distributivity (d) After one step of distributivity

5 Conclusions and Future Work

We have exploited simple transformations for fault-tolerant ASIC synthesis to afford the benefits of computing with multiple copies but at a significantly lesser cost. As a side-effect, transformation-based synthesis makes the final design resilient to common mode failures. The scheduler uses uneven resource constraints across clock cycles and performs both time and resource constrained scheduling.

Hardware sharing across copies may reduce the fault-tolerance capability. This can be overcome by partitioning flow graphs into regions and preventing sharing among copies within a region. On the other hand, this problem does not exist when the target environment has transient faults and self-recovery via duplication is employed. Transformations such as distributivity may introduce errors into the computation. Techniques for identifying and disabling such computations have to be incorporated. As the voter is a vulnerable point, it has to be designed for fault-tolerance. Additionally, the number of voters in a design should be minimized. In future, we will address these issues.

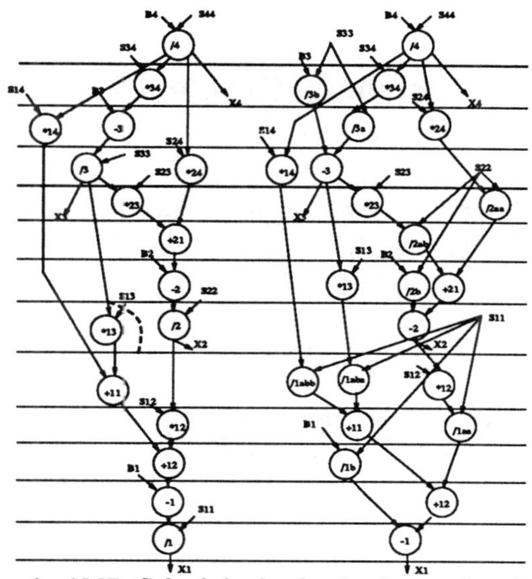

Figure 9: 2MR Schedule for backsolver using 2 multiplies, 2 divides, 1 add and 1 subtract, and 13 clock cycles

References

[1] P. Edmond, *et. al.* ASSURE: Automated design for dependability. In *Proc. of 27th DAC*, 1990.

[2] D.D. Gajski, *et. al. High-Level Synthesis: Introduction to Chip and System Design.* Kluwer, 1992.

[3] B. W. Johnson. *Design and Analysis of Fault-Tolerant Digital Systems.* Addison-Wesley, 1989.

[4] R. Karri and A. Orailoğlu. Transformation-Based High-Level Synthesis of Fault-Tolerant ASICs. TR 215, UCSD, Dept. of CSE, October 1991.

[5] R. Karri and A. Orailoğlu. High-Level Synthesis of Fault-Tolerant ASICs. In *Proc. of ISCAS*, 1992.

[6] D. A. Lobo and B. M. Pangrle. Redundant operator creation: A scheduling optimization technique. In *Proc. of the 28th DAC*, 775–778, 1991.

[7] C. A. Papachristou. Rescheduling transformations for high-level synthesis. In *Proc. of ICCAD*, 1990.

[8] M. Potkonjak and J. Rabaey. Optimizing Resource Utilization using Transformations. In *Proc. of ICCAD*, 88–91, 1991.

[9] V. Raghavendra and C. Lursinsap. Automated Micro-Rollback Self-Recovery Synthesis. In *Proc. of the 28th DAC*, 385–390, 1991.

[10] D. P. Siewiorek and R. S. Swarz. *The Theory and Practice of Reliable System Design.* Digital, 1982.

[11] R. A. Walker and D. E. Thomas. Behavioral transformation for algorithmic level IC design. *IEEE trans on CAD*, 8(10):1115–1128, 1989.

The Electronic Design Interchange Format EDIF: Present and Future

Hilary J. Kahn
Dept. of Computer Science
University of Manchester
Oxford Road, Manchester, M13 9PL UK

Richard F. Goldman
Texas Instruments Inc
P.O.Box 655303
Dallas, TX, USA 75265

Abstract

This paper reviews the present status and current developments of the Electronic Design Interchange Format, EDIF. After a review of the role EDIF has at present, some of the new ideas being developed for inclusion in the new release of EDIF (Version 2 1 0) are introduced.

1: Introduction

Work on the definition of the Electronic Design Interchange Format EDIF started in 1983 and the first balloted version of the EIA standard, EDIF Version 2 0 0, was published in 1987. It became an ANSI standard as well in 1988 [1]. At the time, EDIF offered a way forward for end users and developers alike. For the end user in particular, it gave hope that there really could be a choice of design tools for different design stages (say) by transferring design information between different CAE systems, and the route to silicon or manufacture in general would have choices because a neutral format for transfer was available.

The original enthusiasm for EDIF assumed that it would be feasible - both technically and commercially - for CAD vendors to implement conformant transfer software. The reality has turned out to be rather different in some respects. It is true that the EDIF format is not without flaws; however the format is considerably more powerful and applicable than much current CAD vendor software permits. There have unfortunately been failures to understand the format and somewhat liberal interpretations of its semantics.

Despite problems, EDIF is now the primary mechanism for transfer of connectivity and schematic data. However, real practical problems exist and are being tackled. In this paper some of the difficulties that have arisen are reviewed and some solutions are discussed. The paper also summarises some of the extensions that are to be available as the next release of EDIF.

2: EDIF Version 2 0 0: Technical Status

EDIF Version 2 0 0 is intended primarily as a mechanism for the transfer of design information between CAD systems or from CAD system through to manufacture. At the time when it was first developed, the main interest in the format was to be found among silicon houses and CAD vendors whose products supported design on silicon. It followed therefore that the capabilities of EDIF Version 2 0 0 should have been focussed on meeting the needs of transfers of interest to ASIC designers.

EDIF offers a very structured representation of design data. The information is gathered into a set of libraries, which will typically hold information which is related by use (e.g. a library of schematic symbols) or by technology. Within a library the information is structured into a set of cells which are the main unit of design. Within a cell, the information is further structured into a set of views. The allowable types of view relate closely to the stages of the design process and hence directly to the points in the design process where design transfer may be expected. In EDIF Version 2 0 0, the main design related views are:

- NETLIST. This captures connectivity information together with aspects of timing specification.
- SCHEMATIC. This allows schematic symbols to be defined and provides for the description of multi-page schematic diagrams. The underlying connectivity represented by the schematic may also be captured. Again, timing information may be included.
- MASKLAYOUT. This captures the geometric layout of an IC together with the underlying connectivity.
- SYMBOLIC LAYOUT. This is intended to capture layout information and connectivity in terms of an abstract symbolic representation instead of mask level layout
- LOGICMODEL. This type of view allows the behaviour of a cell to be described using an event driven modelling representation.
- PCBLAYOUT. This view does not have adequate capability in EDIF Version 2 0 0. It is intended to support the definition of printed circuit board geometry and the underlying connectivity it represents.

An important characteristic of the EDIF design representations is that they are all capable of capturing the details of design hierarchy. This means that an EDIF description can accurately capture the instance and definition hierarchies found in databases which support electronic CAD processes.

0738-100X/92 $3.00

The industrial take-up of EDIF Version 2 0 0 has been widespread in the United States, Japan and Europe and active user communities exist throughout the world. The EDIF views which are most widely supported are Netlist and Schematic. These domains allow EDIF to be in regular use not only for the transfer of design data from one CAD system to another, but also to allow 'point tools' to be linked into larger CAD systems to give additional functionality. EDIF is in common use passing design information to or from tools such as simulators, partitioning and placement algorithms, analysis tools and netlist extractors. In addition to its intended role linking CAD systems, EDIF has also been used to provide the nucleus of the inter-tool communication links that are now being addressed by initiatives such as the CAD Frameworks Initiative (CFI).

Practical problems have however been identified in using EDIF. In many cases these are the result of inadequacies in the EDIF Version 2 0 0 and some of these issues are addressed in the rest of this paper. Other problems have arisen as a result of the inadequacy of the translators produced by some CAD vendors. The ultimate goal of EDIF should be to become invisible - to be the unseen link that moves design information across distributed systems or between totally disjoint systems. The reality at present is that there are EDIF translators which fail to conform to the standard or which are selective in the way in which they process EDIF information. It is still possible to find on the market, CAD systems which are even unable to read in the EDIF that they themselves generated!

3: EDIF Methodology

The approach adopted in the EDIF organisation to identifying problem areas in EDIF has involved:

- gathering comments and requirements from active EDIF participants in EDIF Technical Subcommittees
- providing a technical Query Answering Service. This service, in addition to providing answers to individual technical queries (and publishing the questions and official answers for other users [2]), has highlighted ambiguities, misunderstandings and extension needs in EDIF Version 2 0 0.
- gathering review comments and requirements from users and developers of EDIF interfaces through published papers and Users' Group workshops in the US, Europe and Japan.

In order to ensure that problems were correctly understood and to help define solutions unambiguously, the EDIF Technical Committee decided to adopt a formal approach to the specification of proposed extensions to EDIF. Early in 1990, it was agreed that an Information Model written in the information modelling language EXPRESS [3] would be a prerequisite for reviewing or accepting any extensions to EDIF. This decision, which followed a review by the EDIF Technical Committee of alternative formalisms, has meant that before new proposals for EDIF syntax are generated, an agreed information model must be accepted. Appendix 1 outlines briefly what information modelling is and how this is achieved in EXPRESS.

This decision is a formal recognition of the real nature of EDIF. First and foremost EDIF has an inherent understanding of the structure and semantics of electronic design. It is built on a model of the world of electronic design; until now, however, that information model had not been written down in a way which is easy for engineers and tool developers to use. Secondly EDIF, as a transfer format, must have a syntactic representation to allow systems to generate and process files. This syntax complements the underlying information model and happens to use a lisp-like syntax which is easy to parse and to extend.

The decision to adopt information modelling has been successful in a number of ways. Firstly, it has meant that extension proposals come to the EDIF Technical Committee in the form of requirements specifications rather than suggested syntax which, experience showed, often took the form of overloading existing keywords. Secondly, it has meant that proposers of extensions have gone through a careful rationale creation process as they have developed their information models.

Following the decision to adopt information modelling in EXPRESS, there have been various activities involving the modelling of aspects of EDIF. A partial EXPRESS model of the current standard, EDIF Version 2 0 0, has been published [4] and further modelling work is being done by a special EDIF Information Modelling Group. The EDIF Technical Committee has developed an information model for enhanced connectivity representation and schematics [5]. This model was used as the basis for the syntax creation for EDIF Version 2 0 100 [6]. A more complete model covering all aspects of Level 0 EDIF is to be published in mid-1992. This is the model which forms the basis for EDIF Version 2 1 0.

EDIF Technical Subcommittees have also been very active in using information modelling in EXPRESS as the basis for their new proposals. Comprehensive models for EDIF PCB [7] and a new EDIF Test View [8] have been completed. These have been used for the development of draft syntax and interchange trials are under way.

Two additional benefits in the general electronic design domain have arisen from the use of EXPRESS information modelling by the EDIF organisation. Firstly, the availability of advanced models provides major technical input to CFI developments; the CFI Design Representation work is now in a position to draw directly on the work done in EDIF. They can take the EDIF information models and use them as the basis for the syntax crafting of the CFI Programming Interface. This will help to ensure that end-users find a more unified approach to the handling of design information in CAD systems.

A second advantage of using EXPRESS is in the international standards arena. EXPRESS is the information modelling language developed within the ISO Standard for the Exchange of Product Data initiative (STEP). A requirement for a product description mechanism to be considered for inclusion within STEP is that it should be supported by a formal model in EXPRESS. EDIF is at present being considered by the ISO/IEC Working Group on Electronic and Electrical standards within STEP.

4: EDIF Appraisal

An evaluation of the successes and failures of EDIF has indicated a number of technical areas where the EDIF Version 2 0 0 standard is inadequate or ambiguous. Some of these areas are reviewed here, followed by an outline of the approach being adopted to solve the problems raised.

The goals of the changes which have been made to produce EDIF Version 2 1 0, the EDIF release in preparation, include:

- More accurate transfer of design information. This has focussed both on ensuring that the new version of EDIF follows even more closely the concepts needed to represent design and on making it easier for CAD tool developers to provide accurate EDIF translators.
- Correction of known problem areas and removal of ambiguities found in EDIF Version 2 0 0.
- Specific focus on improving the handling of Connectivity and Schematics transfers
- Providing a sound basis for the future extensions which include the proposed PCB representation and test representation

5: Connectivity

The ability to represent connectivity is a key issue in the handling of electronic design data. In EDIF Version 2 0 0, the ability to represent single-bit connectivity is good. The main connectivity object, the net, is used to capture a number of concepts:

- In a net, there is a formal definition of which ports are joined (i.e. made electronically common) by that net
- In the case of views which also carry implementation information (e.g. schematic or maskLayout views), the implementation graphics can be closely bound to the net definition.
- Nested structuring of nets can be captured to allow information such as routing criticality or timing details to be defined for parts of a net or to further bind pieces of implementation geometry to particular sub-nets.

The problem with EDIF connectivity arises in relation to the representation of busses and bundles. These terms, which are themselves ill-defined, typically imply connectivity which is wide or multi-bit rather than single-bit. In EDIF Version 2 0 0, there are two different ways in which wide connectivity can be captured: the arrayed net or the netBundle. These two mechanisms are quite different in nature: the arrayed net is in all respects analogous to a single-bit net, except that the number of 'wires' conceptually involved is greater than 1; the netBundle on the other hand is a wrapper for a set of net definitions. It does not itself establish connectivity nor does it indicate which ports are to become electrically common. The nets defined inside a netBundle (which may themselves be arrayed) establish the detailed connectivity.

Furthermore, although it is true that for simple netlists all the connectivity at a given level of hierarchy is defined in one place, this is not true for more complicated cases. For example, if electrically common information happens to be described on multiple pages in a Schematic view, then in EDIF Version 2 0 0 there is no way of allowing the EDIF writer to indicate the total connectivity in one place. In addition, where information from a wide 'bus' is 'ripped' into other less wide nets or 'busses', a special type of cell is required. This means that a CAD system accepting a design in EDIF Version 2 0 0 must unravel the artificially inserted ripper 'cells' to deduce the underlying connectivity.

5.1: Proposed Solution for EDIF 2 1 0

The problem of representing busses and bundles is resolved in the next release of EDIF by recognising that, in common usage, busses and bundles both represent wide connectivity, but typically one form (the bus) allows constituents to be referenced by index (analogous to membership of an array) whereas members of bundles are commonly referenced by name. In EDIF 2 1 0, both forms of wide connectivity will be coalesced in a *bus*, with support for reference to constituents both by position (index) and name.

The proposed solution to capturing all the connectivity at one level of the design hierarchy is to introduce into EDIF the concept of *signal*. All views that carry connectivity information (e.g. netlist, schematic, pcbLayout, maskLayout and symbolicLayout views) are divided into two sections, the logical connectivity section and the structure section. In the logical connectivity section, the abstract view-wide connectivity is specified in terms of signals which are one bit wide and which join all the ports that are electrically common at the current level of hierarchy. In addition, in the logical connectivity section, signalGroups can be defined. These may be multi-bit and establish a grouping of signals. SignalGroups may themselves be structured in terms of nested signalGroups. These groupings provide a basis for the structured connectivity described by busses.

Figure 1 illustrates a grouping of signal information. In EDIF Version 2 1 0 this grouping would be reflected by having a signalGroup **instruction** which grouped signalGroups **operator, operand** and **address**. Eventually, signalGroup **address**, for example, would itself be defined as a grouping of signals, **ad_0** through **ad_15**. In the structural part of the view, it is likely that signalGroup **address** would itself be represented as an address bus. The members of the bus may be referenced either by position in the bus (e.g. the **0th** element of the bus called **address**) or by reference to their signal names (e.g. **ad_0**).

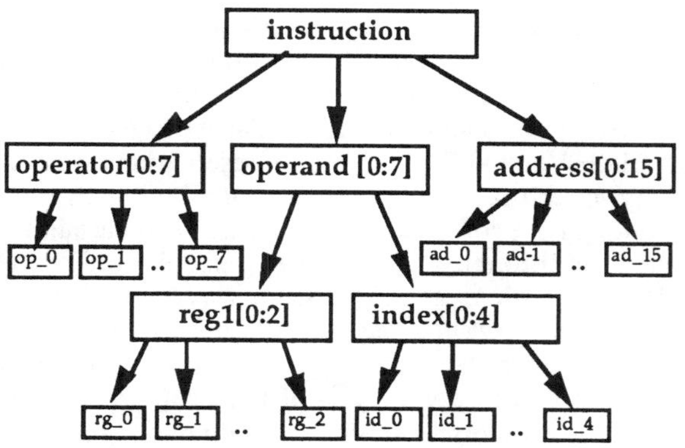

Figure 1: Grouping of signal information

In the structural connectivity section, nets and busses are defined. These allow the ordering of ports to be described. Each net is associated with a single signal and with single-bit ports. Each bus is associated with a signalGroup and with a set of ports of appropriate width.

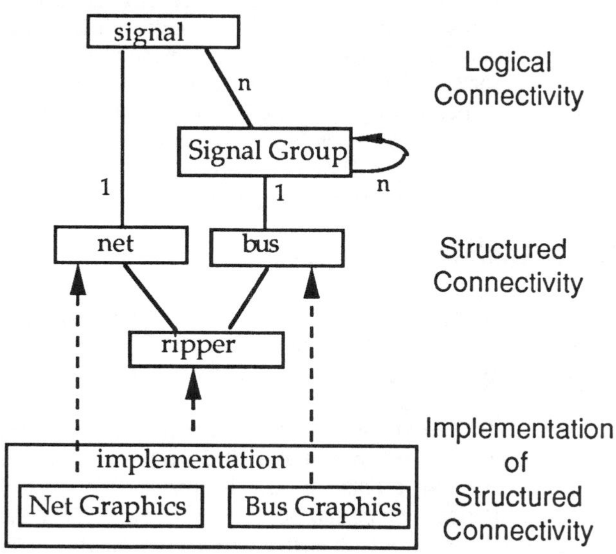

Figure 2: Connectivity relationships

Nets may be further structured into subnets. The structuring into subnets allows specific attributes such as criticality to be associated with particular parts of a connection and supports the net implementation details that follow in views which include graphical implementation details (e.g. The schematic and layout views).

Busses may be subdivided in two orthogonal ways. They may be divided into subBusses or into busSlices. A *subBus* by definition has the same width as its containing structure. It is similar in nature to a subnet in that it allows additional structural information to be given about a part of the bus. A *busSlice* is used to define part of a bus that has a width less than the width of the immediately enclosing structure. A busSlice allows wide structures to be ripped into other less wide structures. Once again, the specific structure of nested busses, subBusses and busSlices used is determined by the need to specify attributes on parts of the structure and the requirements of the graphical implementation that may follow. Figure 2 illustrates the relationships between logical signals and signal groups, structural nets and busses, and implementation information.

6: Characteristics of re-useable objects

In EDIF Version 2 0 0 there is only one way of defining the re-use of an object via the instantiation process. This is the *instance* which may be applied to views of a cell. It followed therefore that the need to allow the re-use of a 'dumb graphics' object led, in EDIF Version 2 0 0, to the definition of a GRAPHIC view type which in all respects other than re-useability bore no resemblance to a 'true' cell view.

The Schematic Technical Subcommittee in developing their requirements for the next release of EDIF identified a number of object types which have the characteristic that they are defined once and re-used many times. Examples of these are off-page connector symbols and on-page connector symbols. These needs could have been met by continued extension of the EDIF view, but this would have resulted in the gross overloading of the view concept. Instead, the concept of 'Reuseable Object' has been recognised and the needs of the Schematic Technical Subcommittee met by defining a sub-class of reuseable objects called 'Instantiatable Objects'.

The goals of the definition of reuseable objects of various types are to:

- differentiate clearly between different classes of re-useable object. The three classes identified are: views, view-related instantiatable objects and view-independent dumb graphics.
- clarify the definition of the semantics of a cell view.
- allow re-useable dumb graphics objects to be defined and used.
- allow objects which have special significance in a given type of view to be defined and re-used. At this time, the examples considered all come from the schematic view, but the concept is being extended to views of other types.

6.1: Characteristics of Cell Views

The instantiation of a cell view implies that a design object has been included in the design at the point of instantiation. A cell view has, or may be given, ports and has the potential for elaboration at a lower level of hierarchy. It is often a functional unit. A cell view may contain within it occurrences of other re-useable objects. Depending on the context (e.g. whether or not the type of view supports graphical information) these re-useable object occurrences may be instances of other cell views or occurrences of instantiatable objects.

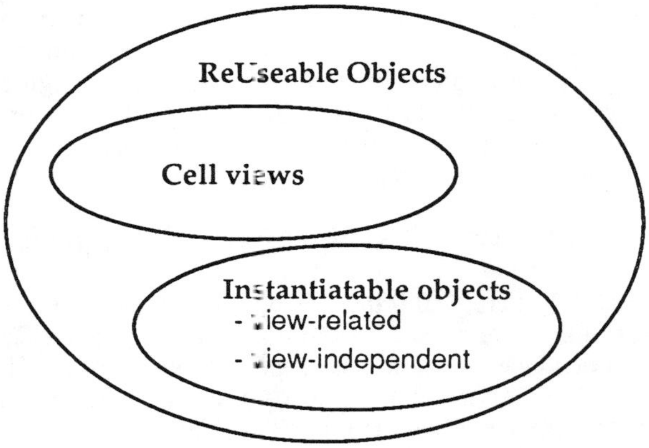

Figure 3: Classes of Re-Useable Objects

6.2: Instantiatable Object Characteristics

An instantiatable object is a re-useable object which has specific semantics relevant to the context in which it is being referenced. Instantiations of such objects typically occur within the implementation parts of cell views. The definition of an instantiatable object may be held in a CAD library. Such definitions do not contain occurrences of cell views or of other view-related instantiatable objects.

Two classes of instantiatable object are currently supported: view-independent instantiatable objects and view-related ones. The view-independent instantiatable objects allow for the definition and re-use of dumb graphics and replace the EDIF Version 2 0 0 view of type GRAPHIC. The view-specific instantiatable objects relate, for example, to the schematic view. They allow for the definition of templates which are 'instanced' on pages. These templates define special symbols to represent such things as on-page and off-page connectors, global ports, master ports and rippers, as well as page borders and title blocks. It is expected that when enhancements to EDIF to offer proper support for PCB representation are made, additional instantiatable objects will be included specifically for the pcbLayout view.

7: Structuring of EDIF Views

In common design practice there are various ways in which a unit of design (EDIF cell) may be represented. One natural way is to use a symbol in a schematic to represent the existence of an instance of a unit of design at some point in the design hierarchy. When thinking about the cell as a 'black box', the interface is the representation considered. At the point at which the design details are explored more fully - say to do a simulation or to complete a layout, then the contents of the design description are elaborated. That information is represented in EDIF as the contents of a particular type of view.

In EDIF Version 2 0 0, these representations were very closely interlinked, so a view contains an interface definition and a contents for elaboration, and certain types of view additionally contain a symbol definition. This model is not close enough to the way in which many CAD systems operate using symbol libraries (and hence to the way in which their users approach the design process). As a result, in the new release of EDIF, the relationships between these representations have been rethought.

In the planned version, views are grouped together into clusters which share a common interface. Within a cluster, there may be different types of view and various symbols which can be used as abstractions of the views. Because it is not realistic to require all views of a cell to share one interface (e.g. a logical connectivity view would not include the power ports that are essential for a layout view), a cell definition may contain multiple clusters.

Furthermore, to help capture the meaning of having multiple views of a given cell, a new viewGroup mechanism has been introduced. This allows views which are related for some user-defined or design configuration reason to be specifically identified. An indication can be given to show that one view is a new version of another, or that one view (say a flattened netlist) is derived from another view (say a hierarchical schematic).

8: Translator Support

The acceptance of EDIF as a truly viable design interchange format has been hampered by the poor quality of much of the translator software available to end users. In part, this has been the result of the fact that EDIF Version 2 0 0 offered considerable flexibility and includes some complex rules. In order to help overcome these problems, the new release of EDIF ensures that where possible a single concept can be defined in only one way. Furthermore, the syntax has been extended so that it is more self explanatory and explicit. More keywords have been introduced into the format - but the vast majority of those keywords are so structured that their use is limited to single contexts. This means that an EDIF reading tool will have the minimum interpretation to do in order to map the objects described in the EDIF file to the CAD system database supporting the design engineer.

9: Conclusions

A new release of EDIF has been built using the experience gained from the widespread use and abuse of EDIF Version 2 0 0. The new version has been designed to minimise problems of understanding for EDIF readers and now includes more comprehensive support for connectivity description and schematic diagram representation. This development has been based on a practical yet formal approach involving the use of information modelling. A new release of EDIF is available for public review. Its supporting documentation includes not only the necessary full Reference Manual and Syntax required for implementation but also a comprehensive Information Model which tool developers and others can use to understand the concepts now included in EDIF.

Acknowledgements

The authors would like to thank all the members of the EDIF Technical Committee for their unstinting efforts over the years. Hilary Kahn would particularly like to thank Alan Williams and Rachel Lau of the University of Manchester for their support and enthusiasm. The contribution made by Hilary Kahn is supported in part by the European Commission through the European CAD Integration Project (ECIP)

References

1. Electronic Industries Association *Electronic Design Interchange Format Version 2 0 0* ANSI/EIA-548-1988, March 1988
2. EDIF Version 2 0 0: *Questions and Answers, Volumes 1 - 5*, University of Manchester, UK
3. ISO STEP ISO/TC184/SC4/WG5 *Express Language Reference Manual* N14. April 1991
4. Rachel Lau *Proposal for an Information Model for EDIF* Technical Report Series: UMCS-91-6-2
5. Electronic Industries Association *An Information Model for EDIF Version 2 0 100*. February 1992
6. Electronic Industries Association, *Electronic Design Interchange Format Version 2 0 100*, February 1992
7. EDIF PCB Technical Subcommittee *Information Model of a PCB (EXPRESS) Version 9*, January 1991
8. EDIF Test Technical Subcommittee *Information Model of the EDIF Test Extension EDIF 2 0 32*, October 1991

Appendix 1: Information Modelling

An information model is a formal way of defining the objects, relationships, attributes and constraints for a given 'Universe of Discourse' (or domain). When the ISO Modelling Language EXPRESS is used, the objects in the domain are defined as entities, and relationships are established between entities. EXPRESS includes powerful ways of defining the constraints that must be adhered to. They may be defined using simple statements or, in complicated cases, using rules and functions.

The EXPRESS language style is relatively easy to learn and to read; experience to date shows that tool developers and engineers learn it readily and rapidly become accustomed to reading information models. The main reason may be that the language looks very much like an ordinary programming language with which most potential model reviewers are familiar. The following example illustrates part of the formal model of an EDIF net showing *inter alia* that a net relates to one signal and has associated ports:

```
ENTITY generic_net
  ABSTRACT SUPERTYPE OF (ONEOF (net, subnet));
   associated_signal          : signal;
   connected_instance_ports   : SET OF instance_port;
   connected_master_ports     : SET OF master_port;
   criticality                : integer_token;
   ......
END_ENTITY;
```

The subtype and supertype specification is used in EXPRESS to define and classify objects in a structured manner and to obtain object-oriented style inheritance in the handling of attributes and constraints.

Appendix 2: EDIF Organisation

The Electronic Design Interchange Format is being developed under the auspices of the EDIF Subdivision of the Electronic Industries Association (EIA). A Steering Committee is responsible for general guidance and a Technical Committee is responsible for the overall technical content of EDIF. Proposals and domain related work are done in Technical Subcommittees which have active participation in the USA, Europe and Asia. Technical advice for potential users or developers of EDIF tools is currently available from an EDIF Technical Centre. General Information may also be obtained from the EIA at the address given below.

Contact Addresses

Electronic Industries Association, Attention Patti Rusher
2001, Pennsylvania Avenue NW
20th & I, Washington DC 20006, USA
Fax: +1-202-457-4985

or

EDIF Technical Centre
University of Manchester
Department of Computer Science, Room IT403
Oxford Road, Manchester, M13 9PL, UK
e-mail: edif_support@cs.man.ac.uk
Fax: +44-61-275-6289

CAD Framework Initiative – A User Perspective

Todd J. Scallan

IBM Corporation, Neighborhood Road, Kingston NY 12401
CAD Framework Initiative, 4030 W. Braker Lane, Austin TX 78759

Abstract

This paper presents an overview of the CAD Framework Initiative (CFI) from a user's perspective. CFI is an industry consortium whose charter is to define interface standards that facilitate integration of design automation tools and design data for the benefit of end users and vendors worldwide. Integration problems faced by CAD users are introduced, followed by a discussion on the concept of a CAD framework. CFI's role in the evolution toward framework standardization is described. Specifically highlighted are key demonstrations of CFI's progress that have been conducted at the Design Automation Conference. The paper concludes with a discussion on end user benefits and expectations with respect to CFI.

1. Introduction

Computer-Aided Design (CAD) is a general term that applies to the use of computer hardware and software to assist with engineering tasks. CAD is employed in a wide variety of disciplines, including electronics, mechanical drafting, software design, and many others.

Manufacturers of technical products rely heavily on CAD tools as an integral part of their design processes. To maintain a competitive edge in areas such as integrated circuit design, many companies have invested considerably in the development of proprietary CAD tools and systems. In addition, commercial vendor products are increasingly utilized for specific tasks, such as the design of printed circuit cards.

As competition becomes more intense and technology cycles shorten, the adaptation of existing design systems to address the requirements of next generation products offers a constant challenge. There exists a myriad of complicating factors, including combinations of purchased and in-house developed software, distributed heterogeneous computing environments, and design methodologies that span multiple disciplines across the manufacturing enterprise.

2. The problem

A tremendous problem for CAD customers is the difficulty encountered when integrating design tools into their production environments. Tools from different vendors, and many times from the same vendor, do not work well together. This is typically due to incompatible data formats, conflicting execution environments, and dissimilar user interfaces. Replacement of a tool requires rewriting data conversion routines and the control mechanisms that manage the tool within a customer's design system. Users must be retrained to interact with the new tool, resulting in lost productivity.

The tool integration problem can be attributed to the nature of the CAD industry, which can be characterized as follows:

- No single source provides a turnkey solution that meets all the needs of an enterprise.
- Many vendors only provide solutions for specific applications, such as simulation.
- Customers require integrated solutions that consist of multi-sourced tools, including in-house developed.
- Existing standards, such as EDIF and VHDL, address only pieces of the integration problem.
- Ongoing consolidation within the CAD industry is forcing vendors to combine tools from other sources with their own products.
- Vendors and customers recognize the need for CAD tool cooperation and ease of replacement.

To answer the tool integration problem, some vendors have introduced proprietary CAD frameworks. However, these products do not provide adequate solutions. In fact, frameworks from two vendors may address completely different sets of problems. For example, some vendor frameworks provide a unified data model and other common services for newly developed or modified tools, while others provide mechanisms for tool invocation and the management of design files without tool modification. In the absence of common guidelines to which CAD vendors can adhere, customers continue to struggle with tool integration in heterogeneous environments. To overcome this dilemma, developers and users agree that standards for CAD frameworks are needed. However, before discussing standardization, the notion of a CAD framework should be explored in further detail.

3. What is a CAD framework?

A *CAD framework* is a software environment that enables the coexistence and cooperation of a variety of design tools. It provides a base on which tools can be developed and integration functions that shield end users from system complexities.

A typical engineering design environment consists of many tools supplied by different vendors, all using various graphics interfaces, design files or data bases, and translators for exchanging data with foreign tools. In addition, many customers develop their own tools to supplement vendor offerings. An engineer is typically a member of a team and often performs several tasks concurrently. Data sharing is required in varying degrees and must be controlled to provide consistency and integrity. The states of designs, libraries, software levels, and other aspects of the design cycle must be managed.

3.1 Framework users

A framework supports activities for three basic classes of users:

1. The *developer* implements CAD tools. Historically, most tools have contained internal frameworks; that is, services for graphics, data manipulation, and access to system resources. The trend today is to externalize those portions of the CAD tool that are not associated with the core algorithm, and to provide them in the form of framework services. This allows developers to focus solely on algorithms and facilitates better integration of tools.
2. The *administrator* customizes the framework. The framework must provide facilities to assist with the administration of the design environment. Tasks include the registration of users and tools, definition of design tasks, and construction of rules that enforce design methodologies.
3. The *CAD user* executes tools within the framework. The framework must insulate the user from the complexities that make up a design environment. Among the many details to be hidden are system and network configurations, establishment of tool environments, data conversions, and methodology rules checking. Activities are centered around tasks, such as "design a chip."

3.2 Framework components

A CAD framework is a modular, extensible structure consisting of several major software components that are tied together by an overall architecture. The primary functional goal of a CAD framework is to facilitate tool *interoperability* and *interchangeability*. Interoperability implies cooperation among tools from multiple sources. Interchangeability refers to ease of tool replacement within a design system, which relates directly to the portability of tools across frameworks and platforms. Framework components include the following:

- A *system environment* provides an operating system independent interface to lower level supporting services and network transparency.
- A *repository* provides a storage mechanism for configuration and control data, as well as actual design data. It is actually the logical aggregation of storage mechanisms that span application domains and which may be distributed across a network.
- *Design data management* refers to the control of information about design data. Versions of designs, interrelationships among components in product configurations, and data states must be tracked to ensure design data integrity and consistency.
- A common *design representation* provides a data model on which multiple tools can operate, hence achieving a significant degree of data integration.
- *Inter-tool communication* mechanisms and protocols allow tools to exchange messages, thereby facilitating cooperative processing. Messaging includes point-to-point, multi-cast, and broadcast modes.
- *Methodology management* provides a mechanism for the definition of rules that comprise a design process and for the enforcement of those rules. For example, a customer may wish to define a policy that stipulates, "Physical design cannot occur until successful completion of a simulation."
- A common *user interface* provides a unified appearance and behavior across design tools.
- A *session manager* serves as an agent that monitors and coordinates all framework activities on behalf of the end user.

4. The CAD industry's response

An industry-wide movement to define framework standards has taken the form of the *CAD Framework Initiative (CFI)*. CFI is a non-profit corporation consisting of approximately 50 member companies. Its charter is to define interface standards that facilitate integration of design automation tools and design data for the benefit of end users and vendors worldwide. CFI is not developing a framework; rather, its end product is a set of standards that includes requirements definitions, an architecture reference, information models, and programming interfaces. CFI is initially focused on electronic design automation. However, by maintaining domain independence for the majority of framework services, CFI will be able to expand its scope to other disciplines, such as mechanical and software design automation.

CFI began meeting on a regular basis after the 1988 Design Automation Conference, and was formally incorporated in February 1989. Since that time, interest has grown and the organization's momentum has continued to rise. Members of CFI include CAD developers, integrators, and users. Representation is international, including companies from North America, Europe, and Japan.

It is CFI's intent to be synergistic with other standards and related activities. It adopts and adapts existing standards, where applicable, and has actively solicited input from numerous other organizations. Where invention is required, CFI attempts not to conflict with existing standards.

4.1 Demonstrations of progress

One of the first major accomplishments for CFI was a project that culminated in a demonstration at the 1990 Design Automation Conference. The DAC-90 Project proved the feasibility of vendor cooperation, which is critical to the success of CFI. In addition, the project illustrated how real problems could be solved through industry acceptance of framework standards. Specifically, the demonstration focused on the classic netlist translation problem.

The DAC-90 Project was based on a programming interface to a design representation of netlist data. The interface and design representation were developed by CFI as a feasibility study and as a preview of standards that would eventually emerge. The programming interface was used to integrate arbitrary pairs of tools that produce or consume netlist data. For example, a schematic editor and a logic simulator would be a producer and a consumer, respectively. It is important to note that this project did not take a data exchange approach like that employed for EDIF. Rather, it illustrated data integration where a consumer tool could directly access design information stored by a producer tool.

Participation in the demonstration exceeded the expectations of many. Twenty vendors contributed 32 tools and five workstations. To continue the momentum achieved through this success, a new expanded project was initiated for demonstration at the 1991 Design Automation Conference.

The '91 Integration Project included four possible roles for company participants: data producer, data consumer, data server, and framework cockpit. In addition to an expanded design representation interface, two additional draft standards were featured: a specification for the encapsulation of CAD tools into frameworks, and an inter-tool communication programming interface that supports notification and request messaging among tools and frameworks. Once again, the project was a tremendous success, with 26 companies contributing CAD and supporting technologies.

4.2 CFI 1.0

The results of the CFI DAC-90 and '91 Integration Projects proved invaluable in testing the feasibility of vendor cooperation and validating key concepts and approaches to framework standards. However, the time has arrived to deliver the standards that will fulfill the first step in CFI's goal of tool and data integration. CFI 1.0 is a set of candidate standards that address requirements for basic "plug and play" of CAD tools. These initial standards include base guidelines for design environment software (system and networking services, error handling, and execution log format), design representation supporting basic netlist manipulation, inter-tool communication providing messaging among tools, and basic techniques for describing tool encapsulation information.

5. What is next?

While CFI has made significant progress in a relatively short time, it must now shift from a primarily technology oriented organization to one that is product oriented and customer driven. CFI's product is its standards specifications. Its customers are the CAD suppliers that implement the standards, and the end users that reap productivity gains from CFI-compliant software. The final measure of CFI's success is not a set of legislated standards, but rather the acceptance of commercial products in the marketplace that implement those standards.

A "pilot program" has been established to facilitate the insertion of CFI 1.0 standards into real products. Its primary goals are to refine the 1.0 specifications, achieve industry acceptance through commercial implementations, and help set the direction for subsequent CFI releases. Several pilot projects have been initiated. Each project involves the implementation of one or more CFI 1.0 standards within commercial products and the deployment of those products in an end user production environment.

Follow-on CFI deliverables are being defined in terms of timely solutions to user problems. The standards that comprise these deliverables fit within the context of an overall architecture and a technical road-map. The CFI technical organization has undergone a restructuring to accelerate the definition of specifications. In addition, the process whereby CFI produces relevant standards is being streamlined to enable rapid progress without sacrificing the attainment of broad industry consensus.

To help support the ongoing activities of CFI, such as the pilot program and 1.0 product release, a sponsor membership class has been established. Several key companies in the CAD industry, representing both major customers and suppliers, have demonstrated their strong commitment to the success of CFI by becoming sponsor members. The extraordinary level of participation and commitment on the part of all CFI member companies is expected to continue.

6. Conclusions

The CAD integration problem must be addressed through vendor cooperation. The CAD Framework Initiative was formed to serve as a vehicle for driving framework standardization. CFI has so far demonstrated a tremendous amount of success. The achievement of rapid progress in a consensus-based environment is evidence of CFI's acceptance by the industry.

CFI was born out of frustration in the end user community. Users have benefited from the establishment of CFI in that CAD suppliers are working together to remove barriers to integration. There are in fact already commercial implementations of CFI draft standards. The CFI 1.0 pilot program will result in final balloting of the 1.0 standards, insertion of the standards into commercial products, and utilization of CFI-compliant software by significant CAD customers.

The CFI 1.0 release offers benefits to both CAD developers and users. For developers, standard interfaces and approaches to framework services provide structural consistency and promote code re-use, thus lowering development and maintenance costs. Once the standards are implemented within commercial products, users will benefit from decreased integration costs and increased productivity. For example, the design representation standard provides a common mechanism for CAD tools to share data and eliminates the need for netlist translation. The user has the flexibility to select the best tool for the task at hand based on the tool's function, and not on whether the tool can be incorporated into the user's existing system. Further, an often repeated step is potentially eliminated from the design process, such as the netlist translation that typically occurs within the design entry-simulation loop.

Beyond CFI 1.0, users can expect solutions to more difficult problems. The next major release, CFI 2.0, has been defined to address the following end user needs:

- Support for mixed level design will allow the representation of design data at various levels of abstraction. A user will be able to select the appropriate level and view for a particular operation.
- A general model for design object management will serve as the basis for library and design information management.
- The ability to allow multiple users and tools to share design data will represent initial support for concurrent design.
- Session and context management will provide each user with a consistent means of managing and preserving the state of the execution environment.

A road-map will provide a clear vision of CFI's future deliverables that will be staged in a timely sequence of releases. The road-map will help CAD customers plan the migration of CFI compliant products into their present design systems. The definition of releases will be more closely tied to end user requirements. Finally, as CFI expands into new areas such as component representation and technology CAD, it is anticipated that CFI's scope will include CAD domains beyond electronic design.

An Efficient Routing Algorithm for SOG Cell Generation on a Dense Gate-Isolated Layout Style

Ryosuke Okuda   Sumio Oguri

ASIC Design Engineering Center
Mistubishi Electric Corporation
Itami City, Hyogo, 664, JAPAN

Abstract

In this paper, we propose an efficient routing algorithm for SOG cell generation on a dense gate-isolated layout style. This layout style is a difficult one for routing, because only one vertical routing track per one P/N pair is allowed. The key point for obtaining a complete routing is how to prevent some routes from blocking others. We propose a routing algorithm based on a simple routing model in which such a blocking can be easily detected and avoided in the global routing phase. About 87% out of 174 cells in our library has been successfully routed, which will reduce a great deal of manual efforts.

1 Introduction

The Sea-Of-Gates(SOG), which enables short design and fabrication time, is widely used in the industrial world. However, the drawback is a great deal of initial efforts which must be paid in order to build a cell library. In this paper, we propose a novel routing algorithm for automatic SOG cell layout generation.

Generally there are two major technologies for SOG to isolate active transistors from one another. Those are the *gate-isolation technology* and the *oxide-isolation technology*. Further, there are two layout styles for the gate-isolation technology. One is the *dense gate-isolated layout style* (DGI) [1], and the other is the *loose gate-isolated layout style* (LGI) [2]. The DGI allows only one vertical routing track per one P/N transistor pair as shown in Figure 1 (a), while the LGI allows more tracks as shown in Figure 1 (b). Since the DGI enables high silicon area utilization and reduced diffusion capacitance, the DGI is used rather than the LGI in the commercial gate array. The proposed routing algorithm is intended for the DGI.

Recently, a comparison of routability among various SOG layout styles has been done, by executing an automatic cell layout generator on each layout style[3]. In the results, the DGI requires the largest number of horizontal routing tracks to get a completely routed result for every cell. This results show that the DGI is most difficult for automatic routing in various layout styles of SOG.

Some routing algorithms for SOG cell layout generation have been proposed so far [3]–[9]. Mostow [4] and Nair [5] proposed single layer routing algorithms. Both algorithms assume that more than two vertical routing tracks per P/N transistor pair are allowed, which means that they are not applicable to the DGI. Noda [6] proposed a new routing algorithm based on the channel routing on an oxide-isolated layout style. Its successful results are due to an additional vertical routing track between the neighbouring P/N transistor pairs. Therefore the algorithm is not applicable to the DGI either.

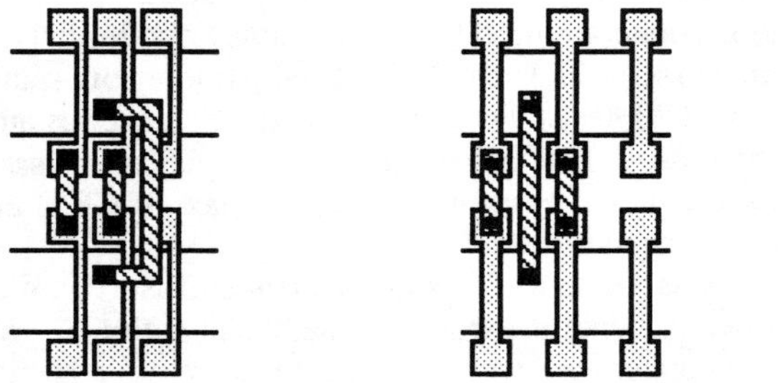

(a) Dense gate-isolated layout style (b) Loose gate-isolated layout style

Figure 1: Two styles of the gate-isolation technology.

The rest of the previous algorithms [3, 7, 8, 9] are all based on the maze routing algorithm, which has few restrictions about layout styles. Therefore they are *in principle* applicable to the DGI.

The most important point of net-by-net routing methods such as the maze routing is how to prevent formerly routed nets from blocking others. For this purpose all of the prior algorithms adopt *rip-up and reroute* techniques. But the rip-up reroute maze routing wastes much run-time, and seldom produces good results. For example, Rowson [8] reported that 126 cells out of 254 had unroutes, although it was on the LGI which is considered to be a fairly easy layout style for automatic routing.

Shin [10] introduced a *wire-pushing* technique in order to avoid blockings. However the wire-pushing operation tends to be limited in a small local area by reason of too much run-time. Codar [3, 7] got successful results by introducing a global routing phase in addition to wire-pushing. The purpose of Codar's global routing is to spread out as evenly as possible the wiring density. This is useful for avoiding blockings, but such an effect on avoiding blockings is indirect, because no blocking can be detected during the global routing phase.

The proposed algorithm is based on a routing model of channel routing. This routing model is a restricted one compared with that of the maze routing. Therefore, blockings on this model are also restricted to some types, and can be detected easily. For example, some blockings are detected by detecting cycles in a *vertical constraint graph* representing the constraints among the vertical positions of horizontal wire segments. We propose a new global routing algorithm, in which most of blockings can be

0738-100X/92 $3.00

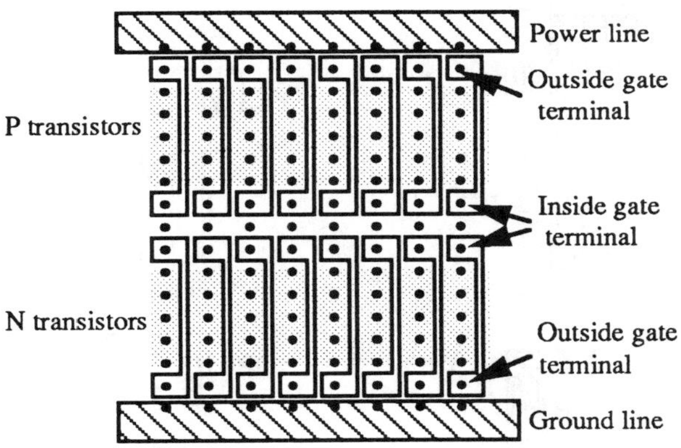

Figure 2: Base array.

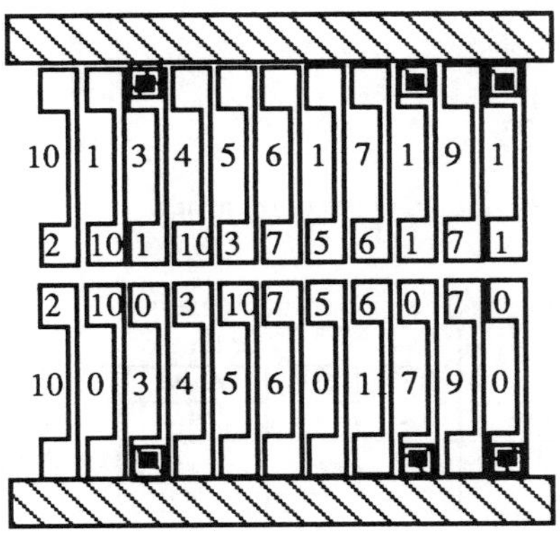

Figure 3: A routing requirement.

detected and be avoided.

2 Problem Formulation

Figure 2 illustrates the base array of transistors. The power line runs horizontally on the top in the first metal layer, and the ground line runs on the bottom. The dots in the figure denote grid points. Each transistor has two grid points on its gate. One next to the power or ground line is called an *outside gate terminal*. And the other is called an *inside gate terminal*. N_P and N_N grid points line up vertically in the P and N diffusion area respectively (in the figure $N_P = N_N = 5$).

There are four kinds of available routing elements, which are:

first metal wire a wire in the first metal layer;

second metal wire a wire in the second metal layer;

contact a via hole which connects a first metal wire with a gate/diffusion of a transistor;

through hole a via hole which connects a first metal wire with a second metal wire.

Each routing element must be laid on grid points. The geometrical design rules such as minimum width or minimum spacing need not be regarded.

A routing requirement is given as shown in Figure 3. In the figure, the numbers shown on inside gate terminals and diffusions of the transistors denote net numbers. The gates and diffusions having a same number must be connected. The net number 0 and 1

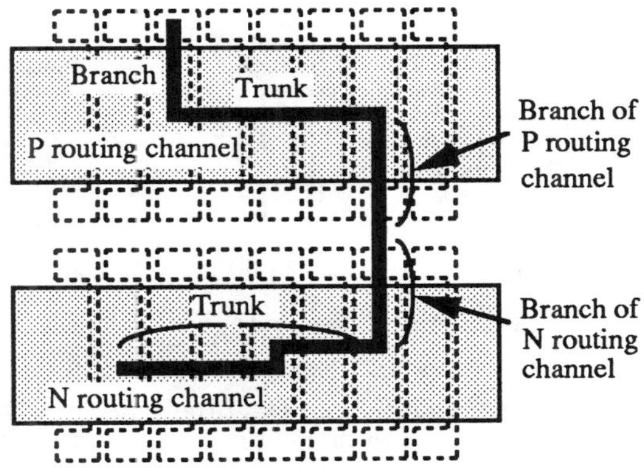

Figure 4: Layout model.

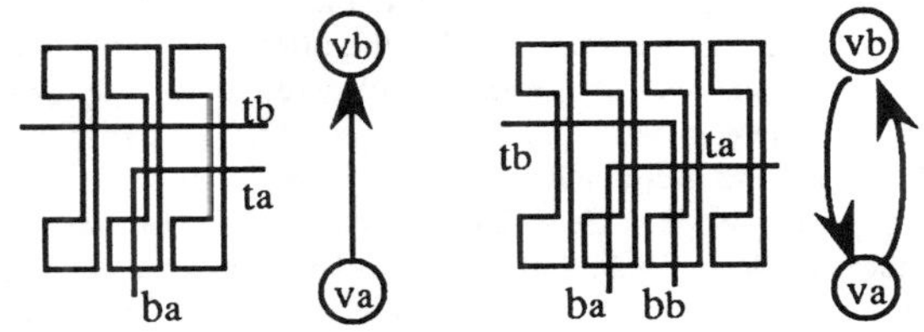

(a) A pliable edge. (b) A pliable cycle.

Figure 5: A pliable edge and a pliable cycle.

correspond to the ground net and the power net respectively. The P/N transistor gates having the net number 1/0 are called *isolation gates*. Every isolation gate is connected to the power/ground line with a first metal wire and a contact beforehand.

The most important goal of the problem is to complete the routing. The objective function is to minimize the number of the columns that have second metal wires, because vertical second metal wires are used as feed-through routings in the chip-level routing phase.

3 Algorithm

3.1 Overview

As shown in Figure 4, we see each P/N diffusion area as a routing channel. The routing channel on the P diffusion is called the *P routing channel*, and the other is called the *N routing channel*.

A *branch* is a vertical wire segment such that its one end is inside of one channel and the other end is outside of the both channels. Note that a vertical wire which connects the P and N routing channels as shown in Figure 4 is considered to be a combination of two branches. A *trunk* is a set of connected wire segments other than branches within a channel. A trunk may have any number of bends. Note that a vertical wire segment which connects two horizontal wire segments within a same channel is not a branch but a part of a trunk. The trunks are always laid in the first metal layer, but the layer of each branch is not decided at first.

Before the vertical positions of trunks are decided, we use a *vertical constraint graph* representing the constraints among the vertical positions of trunks for each routing channel. Each vertex in the graph corresponds to a trunk, and each directed edge denotes a constraint between the vertical positions of two trunks.

Consider Figure 5 (a) for example. There are two trunks t_a

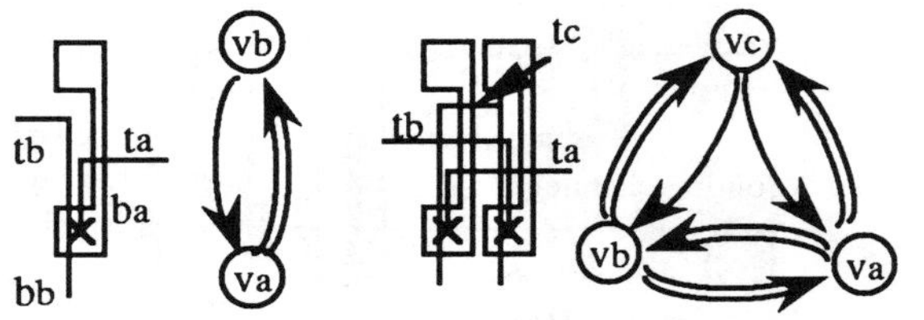

(a) A rigid edge. (b) A rigid cycle.

Figure 6: A rigid edge and a rigid cycle.

and t_b, and a branch b_a connecting to t_a. In order to lay the branch b_a in the first metal layer [1], the trunk t_b must be located above the trunk t_a. The vertices v_a and v_b in the graph correspond to the t_a and t_b respectively, and the directed edge denotes the constraint. However, this constraint is not inevitable, since this constraint would be no longer needed if we lay the branch b_a in the second metal layer. Thus we can eliminate such an edge. An edge which we can eliminate is called a *pliable edge*. And we call a branch laid in the second metal layer a *second metal branch*.

In Figure 5 (b), it is clear that the both of two branchs b_a and b_b cannot be laid in the first metal layer at a same time. The cycle in the graph corresponds to the unsatisfiable set of constraints. We must eliminate such a cycle in order to complete the required routing. In this case, the cycle is eliminated, if we lay either b_a or b_b in the second metal layer. We call this type of cycle a *pliable cycle*. A pliable cycle in the constraint graph denotes that a second metal branch is needed.

Some branchs are forced to be laid in the first metal layer. Consider Figure 6 (a) for example. The "×" in the figure denotes a contact. In this case, the branch b_a must be laid in the first metal layer; otherwise it could not physically connect to the gate terminal. Therefore we can never eliminate the edge directed from v_a to v_b. An edge which we can never eliminate is called a *rigid edge*. A rigid edge is denoted by "$\Longrightarrow$" in the figure.

A cycle made of rigid edges is called a *rigid cycle*. Figure 6 (b) shows an example of such a cycle. It is obvious that the branches cannot be realized in two layers, because the routes block each other. Such a blocking is denoted by the rigid cycle which contains v_a and v_b.

There are three major phases in the proposed algorithm; those are:

- Global Routing
- Layer Assignment
- Detailed Routing.

In the global routing phase, all the nets are routed using trunks and branches. At this time, neither the vertical position of each trunk nor the layer of each branch is not decided yet. The vertical constraint graphs of P/N routing channels are built and maintained during the phase. The main object of this phase are to minimize the number of pliable cycles and to avoid rigid cycles.

After the global routing, the vertical constraint graphs may contain some pliable cycles. The object of the layer assignment phase is to eliminate all the pliable cycles with minimum number of second metal branches.

In the detailed routing phase, the vertical position of each trunk can be decided using a channel routing algorithm [11].

[1] Because we want to reduce the number of second metal wires.

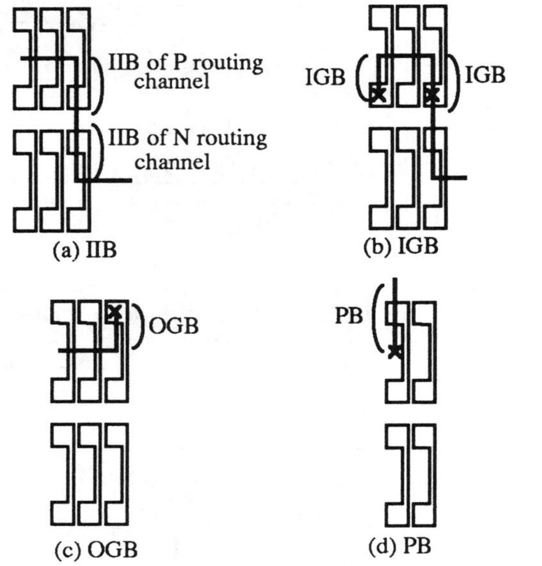

Figure 7: Types of branches.

Table 1: The rule of making edges.

		The type of b_j				
		IGB	IIB	"none"	OGB	PB
The type of b_i	IGB	–	○	⋆	○	○
	IIB	△	△	△	○	○
	"none"	–	–	–	△	○
	OGB	–	–	–	–	○
	PB	–	–	–	–	–

–: make no edge ○: make a rigid edge
△: make a pliable edge
⋆: if an IIB exists at the same column then make a rigid edge, otherwise make a pliable edge.

3.2 Vertical Constraint Graph

In this section, the exact definition of the vertical constraint graph is described. Before that, branches are classified into four types according to what they connect to in the outside of the routing channels. Those types are:

inward I/O branch (IIB) A branch which does not connect to a gate terminal of a transistor, but connects to a branch of the other routing channel. (Figure 7 (a))

inward gate branch (IGB) A branch which connects to an inside gate terminal. An IGB may or may not connect to a branch of the other routing channel. (Figure 7 (b))

outward gate branch (OGB) A branch which connects to the outside gate terminal of a transistor. (Figure 7 (c))

power/ground branch (PB) A branch which connects to the power/ground line. (Figure 7 (d))

Only the PBs are predetermined to be laid in the first metal layer.

Now, how to make a vertical constraint graph is explained. First, vertices are made corresponding to the trunks. Next, on every column that has at least one branch, we must examine a constraint on every combination of two trunks which lie at the

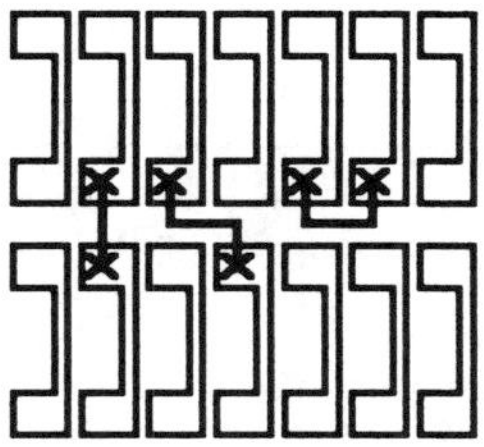

Figure 8: Inter gate routings.

```
procedure: get routes of all the nets ;
  begin
    repeat
      for each net do
        begin
          rip-up the existing route of the net ;
          get a minimum cost route of the net
        end
    until all the routes do not change any more
  end;
```

Figure 9: The procedure of getting routes of all the nets.

```
procedure: global routing ;
  begin
    β ← 0.5 ;
    loop1: for each α of {0, 0.25, 0.5, 0.75, 1} do
      begin
        get routes of all the nets ;
        if Σ_net (A + M) equals 0 then
          return {success}
      end;
    if Σ_net M equals 0 then
      return {success} ;
    α ← 1 ;
    loop2: for each β of {0.6, 0.8, 1.0} do
      begin
        get routes of all the nets ;
        if Σ_net M equals 0 then
          return {success}
      end;
    return {fail}
  end;
```

Figure 10: The procedure of the global routing phase.

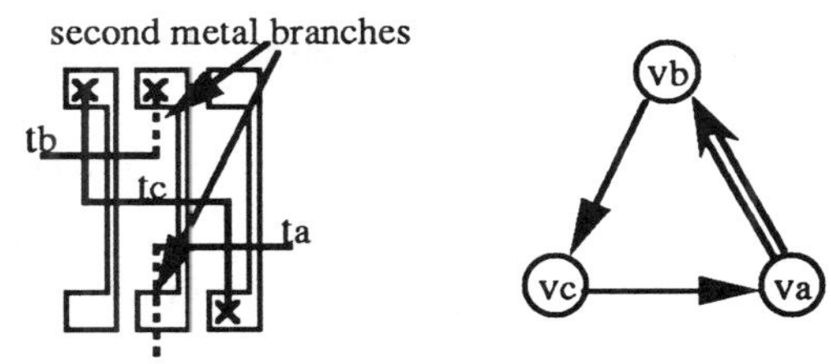

Figure 11: An example of a long cycle.

column. The key point is the rule of making edges on each combination of two trunks. As easily seen from the previous examples, it depends on the types of the branches connected with the trunks at the same column. Now, let t_i and t_j be two trunks which lie at a column. And let b_i and b_j be the branches connected with t_i and t_j respectively at the column. Table 1 summarizes the rule of making a directed edge from the vertex v_i to v_j according to the types of b_i and b_j, where v_i and v_j correspond to t_i and t_j respectively. (Either t_i or t_j may connect to no branch at the column. In such a case, we assume the type of b_i or b_j to be *none*.)

3.3 Global Routing

In addition to trunks and branches, we introduce a new routing element, called an *inter gate routing*, which connects the inside gate terminals of two transistors. For simplicity, the inter gate routings are restricted to connect two transistors that lie within two neighboring columns. Figure 8 shows examples.

In this phase, a minimum cost route of only one net is obtained at a time, in a maze routing manner. Therefore, in order to get routes of all the nets, the iteration of ripping-up and rerouting of each net is continued until it reaches a convergence. Figure 9 shows this procedure.

The cost of a route is calculated by the formula:

$$(1-\alpha)\cdot L + \alpha \cdot \{(1-\beta)\cdot A + \beta \cdot M\}$$

where

L	a subcost on the route length
A	a subcost on second metal branches
M	a subcost on blockings
α, β	coefficients weighting subcosts.

The subcost L is the roughly estimated route length. Currently, the length of a branch and an inter gate routing is assumed to be one, and the length of a trunk is to be the number of columns in which it lies. The subcost A is the number of pliable cycles made of two edges [2]. It denotes the number of second metal branches. The subcost M is the sum of blockings which are detected as the following:

rigid cycle the number of rigid cycles made of two edges in the vertical constraint graphs.

trunk overflow the number of columns where the number of trunks exceeds the *trunk capacity* [3].

branch overflow the number of columns where too many branches exist.

overflow of inter gate routing the number of columns having more than two inter gate routings of different nets.

Figure 10 shows the whole procedure of the global routing phase. In the net-by-net routing method shown in Figure 9, the order of nets tends to greatly affects the final result. To avoid this effect, we adopt the following strategy, in the *loop1* of this procedure. At the beginning, we weight L (the subcost self-determined by one route) much more than A and M (the subcosts on the relationship with the other routes). Then we gradually

[2]Observing only cycles made of two edges is sufficient here, although the graphs may contain longer cycles. The reason will be described in section 3.4.

[3]The trunk capacity is defined on each column of each routing channel. Initially, it is set to the number of horizontal routing tracks, namely N_P or N_N. The trunk capacity will be argued in section 3.6 again.

Table 2: Examples of results.

Circuit	Size (#P/N pairs)	Automatic			Manual		Auto/Manual	
		SZ	CS	CPU (sec)	SZ	CS	SZ_{auto}/SZ_{man}	CS_{auto}/CS_{man}
2 bits decoder	6	10	8	0.73	9	8	1.11	1.00
6 inputs AOI	12	15	8	2.17	13	7	1.15	1.14
4 bits selector	27	36	19	9.77	35	14	1.03	1.36
D-FF	16	19	9	3.52	21	6	0.90	1.50
T-FF	17	21	14	1.63	23	6	0.91	2.33
4 bits D-FF	70	76	50	14.48	89	24	0.85	2.08

SZ: The number of P/N pairs in the cell area. CS: The number of columns which have at least one second metal wire.

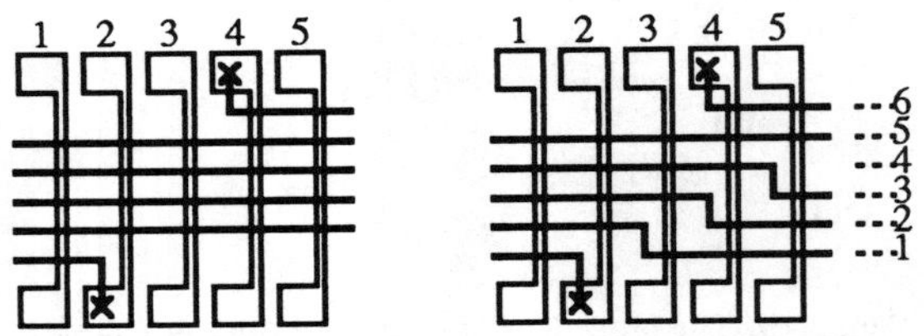

(a) After global routing (b) After detailed routing

Figure 12: Trunk overflow detected in the detailed routing phase.

weight A and M. In the current implementation, α is set to 0 at the beginning, and is increased to 1 by 0.25.

In the *loop2* of the procedure, the subcost A and M are weighted equally at first. Then we gradually weight M, because the sum of M must be 0 in the final result. If the sum of M for all the nets becomes zero during the procedure, a legal result is obtained; otherwise the global routing fails.

3.4 Layer Assignment

The vertical constraint graphs generally contain cycles of any length. In this phase, however, all the cycles made of two edges are eliminated using a greedy method at first, and then, remained longer cycles are eliminated. This strategy is practical, because very few cycles are left after eliminating all the cycles made of two edges, by the following reasons.

Theorem 1 *If all the cycles made of two edges are eliminated without making two second metal branches at any column of a routing channel, the vertical constraint graph of the channel contains no cycle.*

The proof of this theorem is omitted by reason of the space limit. An example of two second metal branches at one column is shown in Figure 11, but such a case has been seldom seen in manually routed layouts.

3.5 Detailed Routing

The trunks are assigned to horizontal tracks using a channel routing algorithm [11]. After that some bends are inserted to trunks in order to lay all the routings in the routing area.

3.6 Overall Algorithm

In this section, the whole procedure of the proposed routing algorithm is described.

First we must explain the *trunk capacities* used for detecting trunk overflow in the global routing phase. The trunk capacity is a positive integer defined on each column of each channel. The

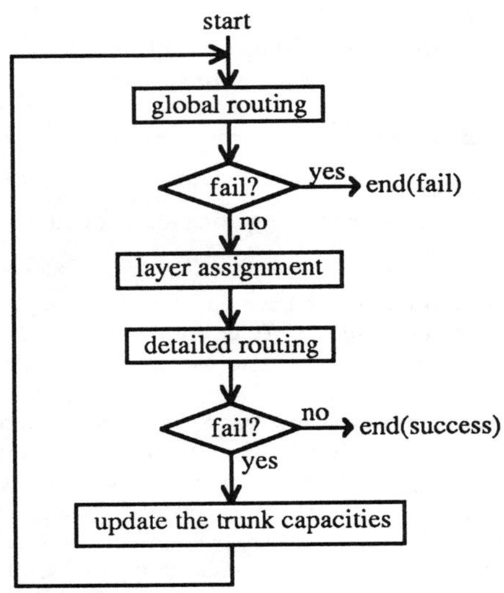

Figure 13: The whole procedure.

trunk capacity of every column is set to the number of horizontal routing tracks, namely N_P or N_N, at the beginning.

Not all the trunk overflow can be detected in the global routing phase. Consider Figure 12 (a), for example. Since the number of trunks does not exceed $N_P = 5$ at any column, no trunk overflow is detected during the global routing phase. However, if we assign the trunks to tracks from bottom to top, one trunk cannot be assigned within five tracks at the column 4 and 5 as shown in Figure 12 (b).

In such a case, we decrease the trunk capacities of the columns where the overflow occurs, and re-do routing from the global routing. In the case of Figure 12 (b), the trunk capacities of the column 4 and 5 are decreased from five to four. The whole procedure of the proposed algorithm is shown in Figure 13.

4 Implementation

A routing program based on the proposed algorithm has been implemented on UNIX machines. Figure 14 shows how a cell is routed by the program. Figure 14 (a) shows the state after the global routing phase. The vertical positions of the trunks displayed here are meaning-less, because they are not decided yet. An arrow directed from one trunk to another denotes an edge in the vertical constraint graphs. Figure 14 (b) shows the state after the layer assignment phase. Four vertical wires are laid in

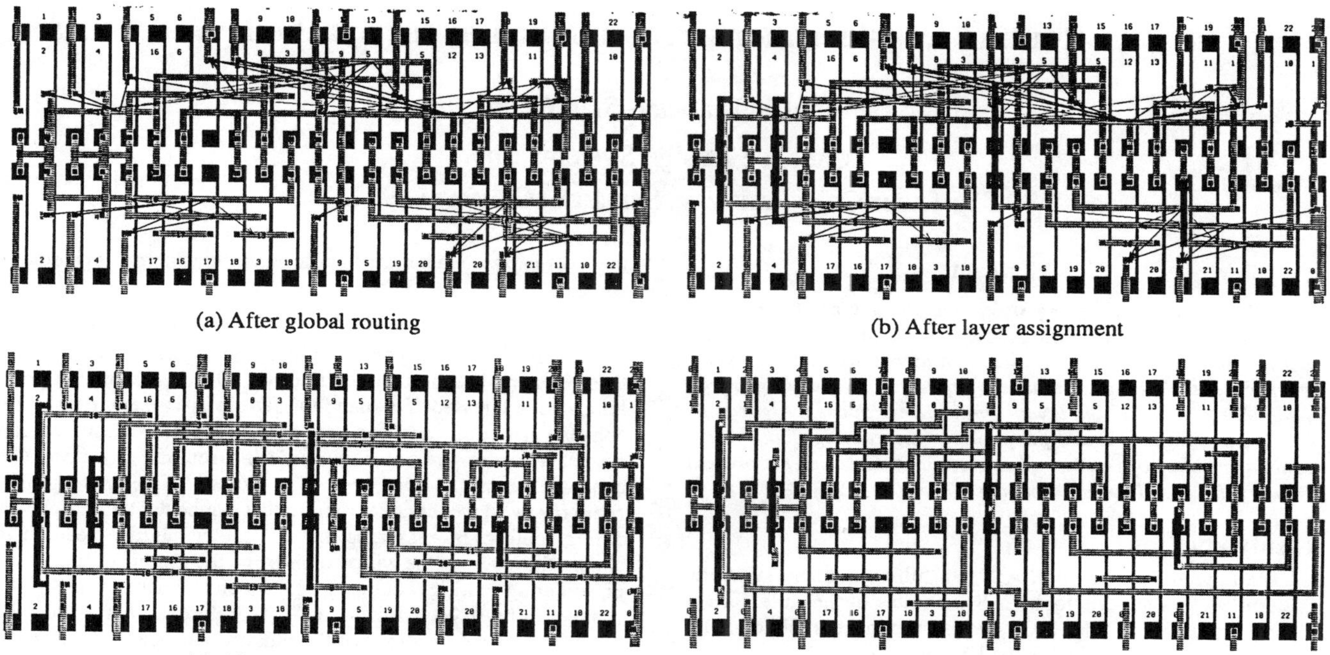

(a) After global routing

(b) After layer assignment

(c) After channel routing

(d) Final result

Figure 14: An example of results.

the second metal layer. Figure 14 (c) shows the state after channel routing. And Figure 14 (d) shows the final result.

A transistor placement program intended for SOG cell generation has been already developed. 152 cells out of 174 in our $0.8\mu m$ CMOS SOG cell library are successfully laid out by those programs. The rest are failed in routing. The routing for them should be re-tried after improving the transistor placement.

Table 2 summerizes the comparison with manually laid out results. In the table, SZ denotes the number of P/N pairs, and CS denotes the number of columns having no second metal wire. Those numbers of automatically laid out results are 1.02 and 1.29 times of those of manually laid out ones in average respectively. The average CPU time is 4.28 second on a SUN–4 work station.

5 Conclusion

In this paper, a novel routing algorithm for the DGI, in which most of the blockings can be detected and avoided in the global routing phase, is proposed. The routing program implemented based on this algorithm successfully routes about 87% out of 174 cells in our SOG cell library.

Acknowledgements

The authors wish to thank Koji Sato, Kaoru Okazaki and Masayuki Terai for their valuable comments and encouragements.

References

[1] K.Sakashita, et.al.: "A 10K-Gates CMOS Gate Array Based on a Gate Isolation Structure," *IEEE J. of Solid-State Circuits*, Vol.SC-20, No.1, pp.413–417, 1985.

[2] G.Boudon, et.al.: "A 20K CMOS Array with 200-ps Gate Delay," *IEEE J. of Solid-State Circuits*, Vol.23, No.5, pp.1176–1181, 1988.

[3] G.D.Adams and C.H.Séquin: "Template Style Considerations for Sea-of-Gates Layout Generation," *Proc. of 26th D.A.C.*, pp.31–36, 1989.

[4] M.A.Mostow: "AGAR: A Single-Layer Router for Gate Array Cell Generation," *Proc. of 1989 ICCAD*, pp.48–51, 1989.

[5] R.Nair: "Single-Level Wireing for Cell Compilers," *Proc. of VLSI 89*, pp.13–22, 1989.

[6] S.Noda, et.al.: "Automatic Layout Algorithms for Function Blocks of CMOS Gate Arrays," *Proc. of 22nd D.A.C.*, pp.46–52, 1985.

[7] P.Tzeng and C.H.Séquin: "Codar: A Congestion-Directed General Area Router," *Proc. of 1988 ICCAD*, pp.30–33, 1988.

[8] J.Rowson and S.Trimberger: "Gate Array Macro Layout Automation," *Proc. of 1986 ICCAD*, pp.448–451, 1986.

[9] A.Chakraverti and M.J.Chung: "Routing Algorithm for Gate Array Macro Cells," *Proc. of 25th D.A.C.*, pp.658–662, 1988.

[10] H.Shin and A.Sangiovanni-Vincentelli: "A Detailed Router Based on Incremental Routing Modifications: Migthy," *IEEE Trans. on Computer-Aided Design*, Vol.CAD-6, No.6, pp.942–955, 1987.

[11] *for example*, T.Yoshimura: "An Efficient Channel Router," *Proc. of 21st D.A.C.*, pp.38–44, 1984.

Routing Considerations in Symbolic Layout Synthesis

Youlin Liao and Stan Chow
Cadence Design Systems, Inc.
555 River Oaks Parkway, San Jose, California 95134

Abstract

This paper discusses routing styles, including a newly proposed hybrid routing style, in basic silicon-level design for a symbolic layout synthesis tool with practical considerations in design rules and process technology. Methods for systematically adding jogs in the layout to achieve high layout quality are proposed. Algorithms for IO-pin assignment and for pin-ordering routing that achieve minimal number of vias are also presented. Experiments on real industry designs showed promising results.

1. Introduction

The *Layout Synthesizer* (LAS) is a fully automatic symbolic custom layout synthesis tool. LAS generates design rule correct layouts from netlists for CMOS designs (not necessarily complementary). As described in [1], the layouts generated by LAS have a row-based structure with each row consists of a pair of complementary N- and P-diffusion strips. The earlier version of LAS [1] could only generate single-row designs and so was limited to leaf cell library development. Many people, e.g. Hwang et al. [2] and many others, have worked in this area. We have further developed LAS so that it also generates layouts with multiple rows. Previous work with this multiple-row capability include those reported in [3],[4],[5],[6],[7],[8]. We call a single-row design a *cell-level* design and a multiple-row design a *block-level* design.

This paper discusses some of the issues in the routing phase of LAS, as listed in the following, that have significant impact on the layout density and quality.

Routing Styles

Layout style, or more specifically, routing style, significantly affects the quality of the final layout. For example, Hwang et al. [2] presented a routing style that is different from that in [1], but only analyzed and compared the routing styles for cell-level designs. In general, consideration in design rules and process technology has a big influence on a designer's choice of routing style. For example, for a given routing style, if stacking contact cuts are used, the layout will be very different from the one that does not use them. Throughout our discussion, we *do not* assume that stacking contact cuts are available. In a symbolic layout environment, where a compactor is used to generate design rule correct and compact layouts, routing style determines, to a large extent, how the compactor will "react" to the layout to be compacted and greatly affects the final layout quality. Since it is not likely that there will be an "universal" routing style that can generate the best possible results for any given design, four different routing styles, including a newly proposed *hybrid routing style*, have been implemented in LAS to enable users to optimize their designs.

Adding Jogs in the Layout

An important issue in symbolic layout generation is adding jogs to routing wires designed for the compactor to help achieve high layout density and quality. By exploiting the structured layout style, we propose methods for layout synthesis tools to add jogs in the *routing phase* in a systematic manner, to generate high quality designs.

IO-Pin Assignment and Pin-Ordering Routing

Hill and Shugard [7] discuss global routing issues in a cell synthesis system, but the input/output-pin (IO-pin) assignment problem is not addressed. We found the IO-pin assignment helpful for congestion control and for user-specified pin-ordering requirements. We also present a pin-ordering routing algorithm that achieves a minimal number of vias. Wang [9] presents an algorithm for two-layer channel routing that can be applied for pin-ordering routing, but assumes that the router has the freedom to choose the layers of the wires coming into the routing channel. Our pin-ordering routing requires the layers on at least one side of the channel to be fixed. Also, Wang's algorithm can not exploit the fact that the final locations of the pins on one side of the channel are flexible for pin-ordering routing.

The rest of this paper is organized according to the flow of the router. Section 2 describes the IO-pin assignment problem and the pin-ordering routing algorithm. Section 3 describes and compares the four different routing styles in LAS. Section 4 describes methods for adding jogs on the routing wires to achieve high layout quality.

2. IO-Pin Assignment & Pin-Ordering Routing

The IO-pin assignment and pin-ordering routing problem can be stated as follows: For a given placement: 1) Assign *the starting* and *ending* locations for each of the IO-nets in the layout so that the cost for routing the IO-pins to the layout boundary will be minimal. Depending on the cost function, the IO-pin assignment can be used to deal with different problems, e.g., congestion control. Our main focus on IO-pin assignment in this paper is to satisfy a set of user-specified pin-ordering constraints; 2) for a given sequence of IO-pins on the layout boundary, route the IO-pins to satisfy the given order such that the routing cost will be minimal.

2.1 IO-Pin Assignment

Because of the row-based layout style used by LAS, the IO-pin assignment problem for IO-pins to be routed to the top and/or bottom side of the layout ("top/bottom IO-pin" for short) and for these to be routed to the left and/or right side of the layout ("left/right IO-pin" for short) are different. The top/bottom IO-pins are routed vertically to the cell boundary through available feed-through slots. So, the problem of IO-pin assignment becomes a matter of selecting starting and ending points for the top/bottom IO-pins so that the routing costs are small, and the IO-pins reach the cell boundary in an order as "close" as possible to the final pin-ordering, if given. On the other hand, for the left/right IO-pins, the IO-pin assignment is a matter of choosing the row in which the IO-pin starting point will be located and the routing region (routing channel inside the row or that in between two adjacent rows) for the IO-pins to be routed to the cell boundary. For both cases, however, the basic algorithm for the IO-pin assignment are the same. Our IO-pin assignment algorithm can be summarized as follows:

Algorithm 2.1

1. Record the locations of possible IO-pin starting points for each of the IO-nets.
2. Estimate and record the costs for routing from these starting points to *the nearest location* on the cell boundary.
3. Choose a combination consisting of one point from each of the sub-nets such that the total IO-pin routing cost is minimal.
4. Calculate *the designated location* on the cell boundary for each IO-pin from the chosen combination.

For the total IO-pin routing cost, we consider routing congestion and the final *designated location* in favor of a possible set of given pin-ordering constraints. In step 3, if the total number of possible combinations is small, we perform an exhaustive search to find the *optimal* combination, and when the number of combinations is large, a simulated annealing algorithm is used to find a solution. During the search, the total costs are updated *dynamically* to help reduce the run time. Step 4 is necessary when there is a set of pin-ordering constraints, and when we need to route the IO-pins toward "favorable locations" on the cell boundary using information we obtain from the calculation of total IO-pin routing costs. We found that the algorithm gives a reasonably good solution with little impact on the total routing time.

2.2 Pin-Ordering Routing

When IO-pins are routed to the cell boundary according to the designated locations obtained in step 4 of Algorithm 2.1 and the pins are not already in the user specified order, it is necessary to perform additional routing to obtain the correct order.

The following definition will serve as the foundation for our pin-ordering routing algorithm, and is one of the key elements in our calculation of the total IO-pin routing cost for IO-pin assignment with pin-ordering requirements.

Definition: The *order* of a numerical sequence **S**, denoted by **O(S)**, is the *length* of *the longest increasing sub-sequence* of **S**. We call a longest increasing sub-sequence of **S** *an ordered set.*

Example 2.1: For sequence **S** = {3, 2, 4, 1, 5, 6}, the order of **S**, **O(S)** = 4 and *one of the ordered set* of **S** is {2,4,5,6}.

Note that the ordered set of a sequence is not unique, for example, {3, 4, 5, 6} is also an ordered set of **S**.

The above definition suggests a possible pin-ordering routing algorithm:

Algorithm 2.2

For a given set of N IO-pins located on cell boundary:

1. Find an ordered set from the sequence consisting of the designated pin-orders extracted from the IO-pins to be routed.
2. Route the pins corresponding to the members in the ordered set directly to the end of the pin-ordering area.
3. Route the rest of the pins, by using a pair of vias, to the location at the end of the pin-ordering area so that the order will be as constrained.

Algorithm 2.2 requires 2(N - **O(S)**) vias. We claim that this is the minimal number of vias if the directions of the routing wire layers are fixed.

Theorem: The minimal number of vias required to route N pins to a given order is 2(N - **O(S)**), where **S** is the initial order of the sequence consisting of designated ordering numbers of the IO-pins to be routed.

Proof: Assume the theorem is not true, i.e., there exists a pin-ordering routing requiring a lesser number of vias. Extract the designated orders from pins that are routed directly to the final location without using vias, this set of order numbers is an increasing subsequence of **S** with length greater than **O(S)**, which contradicts the fact that **O(S)** is the *maximal* length of *all* increasing sub-sequences of **S**.

Having derived our pin-ordering routing algorithm, the problem now becomes finding an *efficient* way to calculate **O(S)**.

A recursive top-down algorithm can be easily derived to calculate **O(S)**; however, the worst case performance of algorithms in that nature is $O(2^N)$. We have derived an $O(N^2)$bottom-up algorithm for finding **O(S)** *and* an ordered set of **S**. Finding not only **O(S)**, but also an ordered set, is important because it enables dynamic maintenance of the total cost function which in turn greatly speeds up the IO-pin assignment process. In the following, for simplicity, we present an algorithm that calculates **O(S)** and *does not* identify an ordered set:

Algorithm 2.3

```
For a numerical sequence S with N elements
ranging from 0 to N - 1:
for (i = 0 to N - 1) {A[i]=N - 1 and B[i] = 1};
for (n = N to 0) {
    i = S(n) - 1;
    if (S(n) <= A[i]) then {B[i] = B[i] + 1;A[i] = S[i]};
    for (i = i - 1;i>=0; i = i - 1) {
        if (S(n) <= A[i]) then B[i] = B[i] + 1;
        if (B[i] <= B[i + 1]) then A[i] = A[i + 1];
    } end for;
} end for;
return (B[0]);
```

Clearly, the worst case run time for Algorithm 2.3 is $O(N^2)$. With minor modifications to Algorithm 2.3, the algorithm can also identify an ordered set of **S** and the run time remains to be $O(N^2)$. The basic idea is to monitor the change of value in A[i]. For brevity, we do not list it here. It should also be pointed out that our pin-ordering algorithm is capable of handling the "same pin-order" constraints, i.e., within a given set of pin-ordering constraints, different pins may have the same order number. Or, translated into the language that Algorithm 2.2 and 2.3 use, the elements of **S** need not be distinct.

Example 2.2: Assume the initial order of a set of IO-pins is {4, 1, 2, 3, 6, 7, 8, 5, 10, 9}. Using Algorithm 2.3 we find that the order is 7 (hence the minimal number of vias = 2 (10 - 7) = 6) and an ordered set can be identified as {1, 2, 3, 6, 7, 8, 10}. The IO-pins may be routed into the required order as shown in Fig. 1.

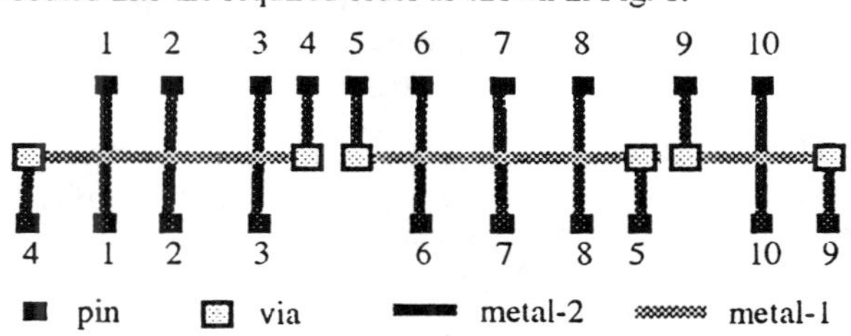

Fig. 2 Pin-Ordering Routing Example

2.3 Experimental Data

We implemented the IO-pin assignment and pin-ordering routing algorithms 2.1 to 2.3 in LAS. We synthesized layouts from a series of benchmark circuits ranging from 20 to over 2000 transistors and 6 to 189 IO-pins with given pin-ordering constraints and observed a consistent improvement in total layout area and total wire length (see Table 1). The last two test cases (cell3 and cell4) are real industry designs. For the comparison, we used a version of the code that routes the IO-pins to the cell boundary from the nearest location in the core of the layout and then routes the IO-pins to the given order around the gravity center of the IO-pins on the cell boundary. It should be pointed out that, pin-ordering routing is usually a small part of the overall layout. The savings percentage numbers listed in Table 1 are savings in terms of the ***total cell area*** and ***total wire length***, not in terms of pin-ordering alone. Hence, the improvement with our algorithm is more dramatic than what is reflected by these numbers.

Table 1. Experimental Data for Pin-Ordering Routing

Cell	# Xtr	# Net	# IO	Ordered Side t / b / l / r	Area%	Wire%
counter	20	14	6	Y Y N N	-16.8	-7.0
reg2	42	22	8	Y Y Y Y	-6.6	-4.9
reg4	80	40	12	Y Y N N	-8.1	+1.7
				Y Y Y Y	-5.7	-8.0
mux16h	88	55	27	Y Y N N	-1.4	0.0
				Y Y Y Y	-8.6	-3.1
cell1	192	96	13	Y Y Y Y	-11.2	-1.6
cell2	544	277	53	Y Y N N	-1.3	-0.4
cell3	1800	880	189	Y Y Y Y	-4.4	0.0
cell4	2181	965	97	Y Y N N	-2.2	+1.1

3. Routing Styles

In practice, design rules and process technology issues, such as the availability of stacking contacts, the track pitch (minimal wire width plus spacing) ratio of different layers of routing wires, timing, and other performance-related issues, should be considered when choosing a routing style. In symbolic layout synthesis, routing style should also be chosen so that the compactor will generate best possible improvement for the final layout. We implemented four routing styles in LAS to meet the users need for quality layouts under a wide range of design specifications. Presently, LAS uses one or two layers of metal plus the polysilicon layer for routing. For all four routing styles, the user can choose to merge the adjacent power supply busses between rows (see Fig. 3.1). The seemingly redundant adjacent power supply busses makes it easier to connect the nets between different rows and makes it possible to control the routing channel height inside a row. On the other hand, merging the adjacent power supply busses will help to achieve a smaller cell area if the savings obtained from merging the power supply busses are enough to compensate the possible routing overhead from eliminating the routing regions between the rows.

3.1 The "Original"

The LAS routing style reported in [1] is depicted in Fig. 3.1, where power and ground busses are on metal-1 as well as all the horizontal routing tracks. Metal-2 is the main vertical routing layer. Polysilicon wires are used for connecting the transistor gates. The main advantages of this scheme are: 1) Easy transition from one routing layer to the other because the metal-1 wires are mostly horizontal and metal-2 and the polysilicon wires are mostly vertical. 2) The metal-2 feed-though can go on top of other layers without difficulty. The main disadvantages of this routing style are: a) It is more difficult to achieve over-the-transistor routing because

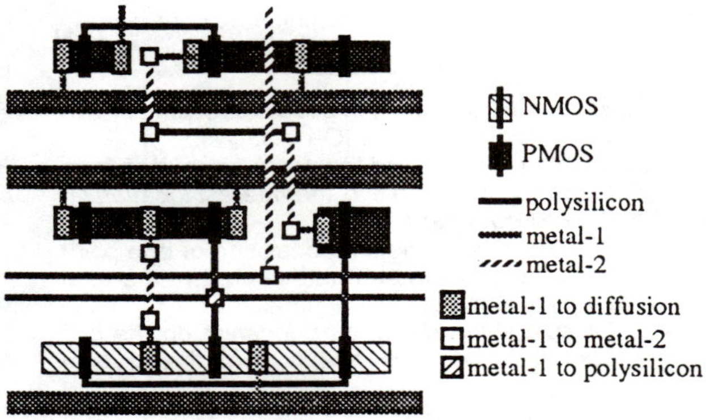

Fig. 3.1 The "Original" Routing Style

of the metal-1 to diffusion contacts. Diffusion jogs, as will be described in Section 4, are necessary for better over-transistor routing; b) The need for metal-1 to metal-2 vias to connect the source/drain to the routing channel. In most cases, this routing style yields a reasonably good final compacted layout.

3.2 Reversed Metal Layer Routing

The "reversed metal layer" routing is a routing style that interchanges the metal-1 and metal-2 direction in the "original" style (see Fig. 3.2). In general, 1) this style has better layer sharing characteristics. For example, over-the-transistor routing is more likely; the metal-2 power supply busses, along with other horizontal metal-2 routing wires, can easily "slide" over transistors during compaction. In addition, 2) the connection between complementary source/drain nodes requires no vias. The "price" one has to pay for using the reversed metal routing style are: a) Metal-1 is used for feed-throughs, which is more restrictive than using metal-2; b) The connection between routing tracks (metal-2) to polysilicon gate pins requires "double contacts" when stacked contact cuts are not available. So, we think reversed metal routing will be helpful for reducing cell height, but one might expect a larger cell width compared to the "original" style.

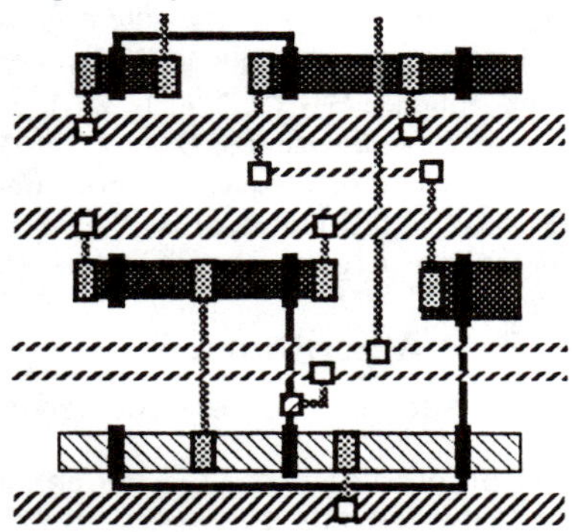

Fig. 3.2 The Reversed Metal Routing Style

3.3 Single Metal Layer Routing

In [1] single metal layer routing is available for cell-level layouts only. We extend the single metal layer routing capability to block-level designs. Single metal layer routing is the "original" routing style with metal-2 replaced by polysilicon. However, for layouts with more than one row, one must solve the problem of using polysilicon wire as feed-through - it is not possible to route polysilicon over diffusion. The possible locations for feed-through are classified into two categories: "opened" and "closed" (see Fig. 3.3 (a) and (b)). If a "closed" feed-through slot is to be used, the diffusion chain will be opened and reconnected by metal-1 wires to accommodate the polysilicon feed-through (Fig. 3.3 (c)). This routing style is useful for generating cell/blocks in two metal layer processes and yields a lower production costs although its die area is larger.

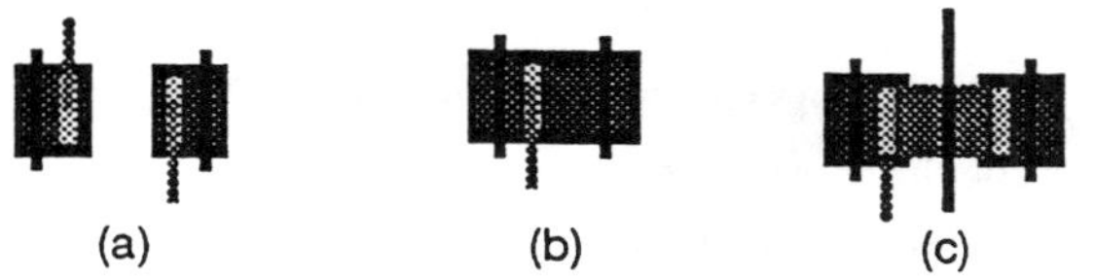

(a) (b) (c)

— polysilicon ▣ metal-1to diffusion — metal-1 ▮ MOS

Fig. 3.3 Polysilicon Feed-Through Slots
(a) Opened (b) Closed (c) Open a closed slot

3.4 Hybrid Routing

In many processes, metal-2 layer has a larger track pitch than that of metal-1 layer. For reversed metal routing, the horizontal metal-2 wires might become the critical path in the vertical direction, restricting the cell height during the compaction process, especially when the widths of the power supply busses are large. We implemented the hybrid routing style to overcome this difficulty. Hybrid routing is a combination of reversed metal routing and single metal routing; single metal routing is used for each of the rows and the reversed metal routing is used for inter-row routing. With this routing style, the metal-2 power supply busses and metal-2 wires for inter-row routing can be "merged" on top of the transistors and even get into the routing region inside the rows. A major concern for hybrid routing is the resistance introduced by the polysilicon wires when the fabrication process has relatively high resistance for the polysilicon layer. To reduce resistance from the polysilicon wires and therefore improve the circuit performance, we need to keep the two complementary transistor rows close to each other. LAS has an option, applicable to all routing styles, that allows fewer tracks in between the complementary transistor rows to help improve the layout quality.

3.5 Experimental Data

We demonstrate the characteristics of different routing styles by comparing layouts generated by LAS.

Example 3.1: We use this small example to demonstrate the layer sharing characteristics of reversed metal routing. For this example, a 0.6 micron technology is used. The power supply busses are relatively wide (five microns, about the same as the average transistor width in the layout) and the metal-1 and metal-2 tracks have about the same pitch. As shown in Fig 3.4, the layout with reversed metal routing (Fig. 3.4(b)) did much better in sharing area occupied by transistors and other routing layers. The cell area is about 26% smaller than that of our "original" routing style (Fig. 3.4(a)).

Example 3.2: For this example, we use an industry design with about 1000 transistors to compare the four routing styles. Also, the effect of merging the adjacent power supply busses is studied. The requirements for this design are: 1) no transistor splitting, 2)use full metal-1 to diffusion contact cut, and 3) wide power supply busses. The transistors in the layout are placed in seven rows. Table 2 shows comparisons between different routing styles.

Again, for this example, reversed metal routing did better than the original style by about 13%, despite the fact that the track pitch ratio for metal-1 and metal-2 is 2/3. However, since the horizontal metal-2 wires have become the limiting factor on the cell height, the hybrid routing further improved the cell area by about an additional 6%, making the layout over 20% smaller than that produced by the original routing style. It is interesting to note that merging

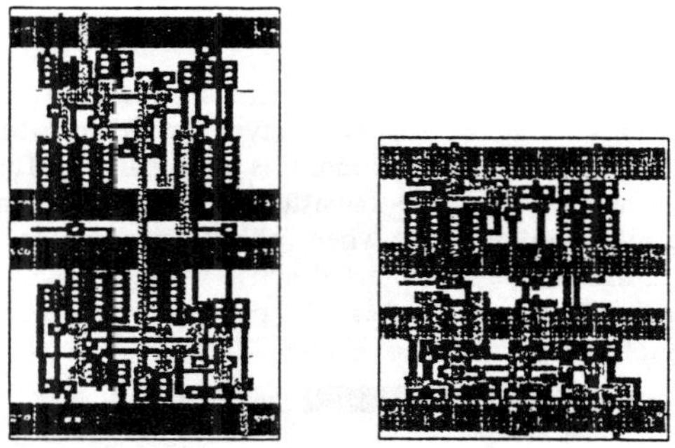

(a) (b)

Fig. 3.4 Layouts Generated by LAS for Example 3.1
(a) Layout with the "original" routing style
(b) Layout with the reversed metal routing

the adjacent power supply busses generates a 13.9% and an 8.0% reduction in the cell area for the original and single metal routing, respectively. However, the savings are not as significant, (or even negative) for the reverse metal and hybrid routing because the power supply busses are already shared with other layers and the routing overhead from merging the supply busses becomes a more significant factor.

Table 2. Cell Area Comparison for Example 3.2

Routing Style	Area Comparison
original	0.0%
original + merge power busses	-13.9%
reverse	-14.5%
reverse + merge power busses	-20.8%
single metal	+ 2.1%
single metal + merge power busses	-5.9%
hybrid	-20.2%
hybrid + merge power busses	-17.1%

We have also some benchmark results showing that the original routing style produces layouts significantly smaller than other routing styles.

Example 3.3: Table 3 shows the cell area comparison results from another industry circuit with 2181 transistors. For this design, all 96 of the IO-pins of the circuit are routed to the top and the bottom of the layouts with a given pin-order constraint. Because of that, many vertical feed-throughs are created in the layouts, and because of the original routing style's better capacity of adapting vertical feed-throughs, it yields a better cell area for this design. The cell area from the original routing style is about 30% smaller than that from the standard cell place and route counterpart.

Table 3. Cell Area Comparison for Example 3.3

Style	original	reverse	single metal	hybrid
Area	0.0%	+6.5%	+1.8%	+22.2%

4. Using Jogs to Improve Layout Quality

For symbolic layouts, adding jogs to the routing wires plays an important role in achieving high layout quality. It is possible exploit the LAS row-based layout style to systematically insert jogs, at certain locations on a carefully chosen group of routing wires during *the routing stage* to help the compactor produce more compacted layouts.

4.1 Diffusion Wire Jogs

We use short vertical wires on diffusion chains to connect diffusion contacts. Fig. 4.1 shows the layouts with the diffusion wire jogs before and after compaction. It is clear that the diffusion jogs promote much better over-transistor routing. It is meaningful to use the diffusion jogs only when full diffusion contact is not required. Diffusion jogs are most effective with our original and single metal routing styles, where the power supply busses and the horizontal routing wires are on metal-1.

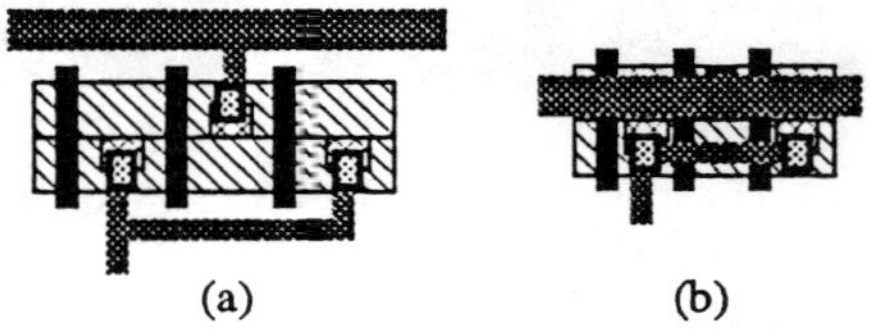

Fig. 4.1 Diffusion Jogs (a) Before compaction
(b) After compaction

4.2 Vertical Wire Jogs

Fig. 4.2 uses two examples to show how the vertical wire jogs are inserted before compaction (drawings on the left) and the results obtained after compaction (drawings on the right) to help improve the cell width and to reduce possible long diffusion wires. Fig. 4.2 (a) is a case where a large number of metal-2 feed-through wires are routed over a diffusion chain and the vertical jogs help the diffusion wire to "shrink" to a shorter length. Fig. 4.2 (b) is a case where the density of the complementary pair of diffusion chains are mismatched and the jogs on the polysilicon wires help reduce the lengths of the bottom diffusion wires.

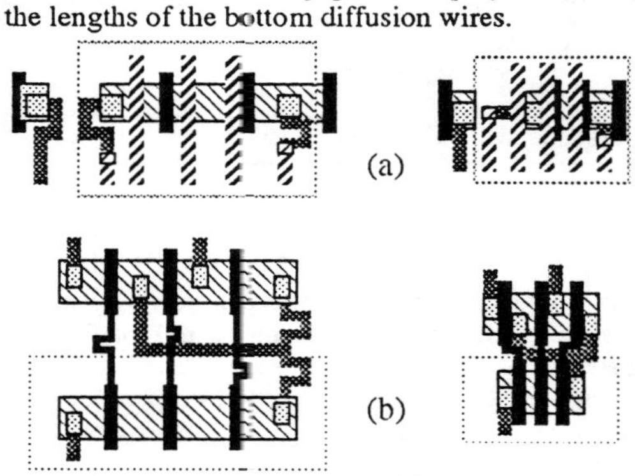

Fig. 4.2 Vertical Wire Jogs (a) Feed-through case
(b) Mismatching diffusion chain density

4.3 Jogs on IOs

For cell-level designs, a *pin-grid* is frequently used to constrain the IO-pins (usually on metal-2) in the layout to a specified grid setting. Often a metal-2 IO-pin is to be attached to a polysilicon wire. For processes where no stacking contacts are available and no contact cuts are allowed on the path of polysilicon wires, two separate contacts have to be used. A possible placement of the two contacts are as shown in the left half of Fig. 4.3 (a). The compacted result of this type of contact placement is often poor due to mismatching between the contact sizes, spacing rules and the given pin-grid spacing (right hand side of Fig. 4.3 (a)). Fig. 4.3 (b) shows a much better contact placement where jogs are added to the short, horizontal metal-1 wires that connect the polysilicon to metal-1 contact and the metal-1 to metal-2 contact. The compacted result shows that the two adjacent IO-pins stayed next to each other without missing a grid spacing.

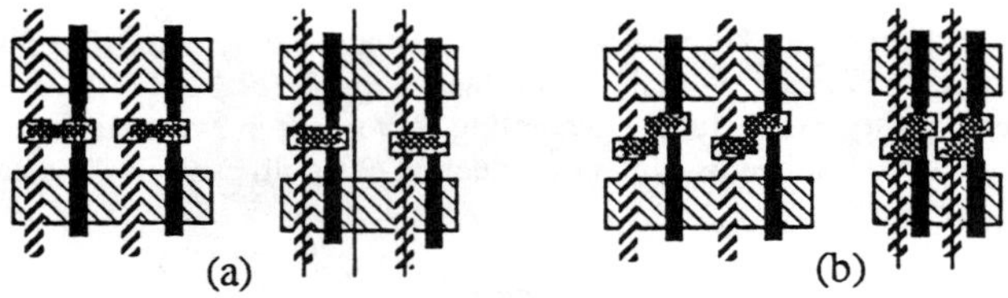

Fig. 4.3 Jogs on IOs (a) Without Jogs (b) With jogs

4.4 Experimental Data

We conclude this section by presenting a set of data we gathered by using LAS on a set of 191 standard cell circuits, with and without pin-grid, on a SUN SPARCstation 1+ with 24 megabytes of memory. The total CPU time for synthesizing the 191 layouts was about one hour. The layouts, without jogs, are about 9.8% larger on the average, than those using jogs if no pin-grid is applied. However, when the pin-grid is applied, the layouts without jogs are, on the average, about 25% larger than those with jogs.

5. Conclusion

Our expansion of the LAS router effectively improved LAS performance in both block and cell-level designs. Based on practical considerations, we have demonstrated that the choice of routing styles and the systematically inserted jogs have dramatic effects on layout quality in symbolic environment. For some test cases, we have observed over 20% reduction in layout area by our newly proposed hybrid routing style. For pin-ordering routing, we set our primary objective to achieving minimal number of vias since the wire lengths can change significantly during compaction. Some consideration in number of tracks might bring further improvement.

Acknowledgments

The authors wish to thank Hungsen Chang and Jimmy Lam for suggestions and help throughout the development of the LAS router. We would also like to thank Dan Schweikert, Stephen Lau and Suzanne Moore for very constructive comments.

References

[1] Chao C. Chen and Shau-Lim (Stan) Chow, "The Layout Synthesizer: An Automatic Netlist-to-Layout System," *Proc. 26th ACM/IEEE DAC*, pp. 232-238, 1989.

[2] C. Hwang, Y. Hsieh, Y. Lin and Y. Hsu, "An Efficient Layout Style for 2-Metal CMOS Leaf Cell and Their Automatic Generation," *Proc. 28th ACM/IEEE DAC*, pp. 481-486, 1991.

[3] D. Baltus and J. Allen, "SOLO: A Generator of Efficient Layouts from Optimized MOS Circuit Schematics," *Proc. 25th ACM/IEEE DAC*, pp. 445-452, 1988.

[4] A. Domic, et. al., "CLEO: a CMOS Layout Generator," *Proc. ICCAD-85*, pp. 172-174, 1985.

[5] P. Kollaritsch and N. Weste, "TOPOLOGIZER: An Expert System Translator of Transistor Connectivity to Symbolic Cell Layout," *IEEE J. Solid-State Circuits*, Vol. Sc-20, No. 3, pp. 799-804, 1985.

[6] Y. Lin and D. Gajski, "LES: A Layout Expert System," *IEEE Trans. on CAD*, Vol. CAD-7, No. 8, pp. 868-876, 1988.

[7] D. Hill and D. Shugard, "Global Routing Considerations in a Cell Synthesis System," *Proc. 27th ACM/IEEE DAC*, pp. 312-316, 1990.

[8] K. Tani et al., "Tow-Dimensional Layout Synthesis for Large-Scale CMOS Circuits," *Proc. IEEE ICCAD-91*, pp. 490-493.

[9] Deborah Wang, "Novel Routing Schemes for IC Layout Part I: Two-Layer Channel Routing," *Proc. 28th ACM/IEEE DAC*, pp. 49-53.

Experiments with a Performance Driven Module Generator

Soohong Kim, Robert M. Owens, and Mary J. Irwin

Department of Computer Science
The Pennsylvania State University
University Park, PA 16802

Abstract

This paper describes a performance-driven module generator for efficient generation of fast static combinational CMOS circuit modules. A new flexible CMOS layout style provides the foundation for implementing fast circuits. Timing optimization is performed via transistor sizing, transistor reordering and the reduction of wiring capacitance on critical paths, all of which are performed in close interaction with the simulated-annealing layout process. The main contributions of this paper are the new layout style that is well-suited for the synthesis of high-speed circuits and the way in which well-known performance optimization techniques are embedded into the layout generation process so as to achieve fine-grain optimization.

1: Introduction

Until recently, reducing the size of layouts was the primary concern of VLSI CAD tool designers. No matter how small a layout is, however, it would be useless if it does not meet certain functionality requirements which almost invariably includes timing characteristics. As design rules scale down and demands for fast and functionally correct circuits grow, timing optimization of VLSI circuits becomes increasingly important.

At the physical design level, the three most important timing optimizations that are extensively exposed in the open literature are: 1) critical net length reduction [2,5,9,12,15]; 2) transistor sizing [3,4,8,10,11,18]; and 3) transistor reordering (also known as input reordering) [6,8,16]. Most often, critical net length optimization is considered independently in the context of placement and routing of pre-designed blocks such as standard cells and gate arrays. Although the cell-based design techniques are widely used today, a chip designed using these techniques, while meeting the timing requirements, can be very inefficient due to the inability to size individual transistors [6]. On the other hand, if transistor sizing and physical layout are treated as two independent steps in the design process, it is unlikely that the final layout will meet the timing specification in an area-efficient fashion, unless whole design process is repeated several times.

We describe, in this paper, an approach to automatically generate small static CMOS circuit modules optimized for their timing performance while keeping the layout area as small as possible. The foundation of our performance-driven module generator, called *perflex*, is a flexible CMOS layout style that supports sizing of individual transistors. In *perflex*, the three timing optimization components for transistor sizing, transistor reordering, and critical net optimization are integrated into the layout process so as to facilitate closer interaction between the timing optimizers and the layout process. This offers nice features such as resizing the transistors as the layout structures become more and more concrete.

The layout process unifies construction of gates, placement of gates, and routing between gates into a single process as in the predecessor tool, *artistII* [14], that is primarily targeted for optimizing layout area. Simulated annealing [13] with two cost functions is used to drive the layout-area and critical-path optimization. One cost function reflects the layout area, and the other the critical path delay. For the transistor sizer, we adopted the posynomial programming approach taken in TILOS [6,8,11]. In *perflex*, however, the transistor sizing is an integral part of the layout process rather than just a preprocessing step.

2: Layout style

In our two-dimensional CMOS layout style, a layout consists of vertically-stacked rows of horizontally-aligned gates. Each row is a pull-up and pull-down plane pair. The number of layout rows is specified by the designer. A static gate consists of a pull-up and a pull-down structure. The pull-up structure is constructed by arranging transistors on one column of P-diffusion in the pull-up plane of the row, and making source-drain connections between the transistors either by abutting adjacent transistors or explicitly connecting them using diffusion contacts and metal one wires. A pull-down structure can contain N-type transistors only, and is constructed in the pull-down plane in the same way.

Figure 1 is a part of a layout in this style showing, from left to right, static gates AOI21, OI2, AOI21, and I1 arranged in one row. The top half of the layout shown is the pull-up plane, and the bottom half the pull-down plane. The two thick horizontal wires running horizontally across the layout in the center are the Vdd and

0738-100X/92 $3.00 © 1992 IEEE

ground rails. The gates are grouped into pairs, and are attached back-to-back so that they can share the power connections as well as poly wires for transistor gate terminals (The latter case is not shown in the figure.) This is the only occasion when poly is used for routing. The rest of the horizontal interconnections are done in the second level metal. Polysilicon and metal-two wires never run vertically until after the layout is compacted. Metal-one wires run vertically when used for inter-row routing and intra-row routing.

Our layout style supports fast circuits in several ways. First, it avoids large parasitic capacitances and resistances through the minimal use of diffusion. The diffusion layer is used only for transistor creation. To our knowledge, our layout style is unique in this respect. Second, it limits the use of polysilicon wires to transistor creation and very short distance routing. Since poly wires are highly resistive compared to metal wires, a poly wire is not allowed to extend over more than a region covered by a back-to-back pair of gates. Allowing short distance poly routing between gate terminals belonging to adjacent gates can greatly help in reducing the size of the layout as well as the capacitance associated with the nets. Third, it uses the two metal layers as primary interconnection media. Metal layers, in general, have the lowest parasitic coefficients of all levels available. Fourth, it allows wires at different levels run overlapped only for a negligible distance; this avoids the creation of unwanted capacitors that can cause extra delay or crosstalk problems. And fifth, it allows individual transistors to be sized. When sizing, the widths of transistors are varied while the lengths are kept at a minimum allowed by the design rules.

3: Timing model and timing analysis

Perflex uses Elmore delays [7] to approximate true gate delays. The model for computing them is similar to those found in [6,8,11,18]. Figure 2 illustrates the RC-tree-based timing model of the pull-down network of a gate when input *b* is triggered. C_g is the sum of fan-out gate capacitances, C_i the interconnection capacitance, $C_{s,x}$ sum of source and drain capacitances, and C_{up} is the sum of the capacitances due to the drains of the transistors in the pull-up network that are connected to the output node. R_a,R_b and R_c are equivalent channel resistances on the pull-down path. The electrical parameters are obtained from the fabrication process parameters (e.g, a BSIM parameter file) supplied by the designer.

In *perflex*, timing analysis consists of two steps: circuit extraction from a given layout description, and stage-oriented event-driven simulation using Elmore delays. First, the input-to-output propagation delays are obtained for all inputs of each gate, considering both rising and falling transitions at the inputs. Then, these values are used to compute signal arrival times at the outputs of the gates in a breadth first search starting from the primary inputs of the circuit.

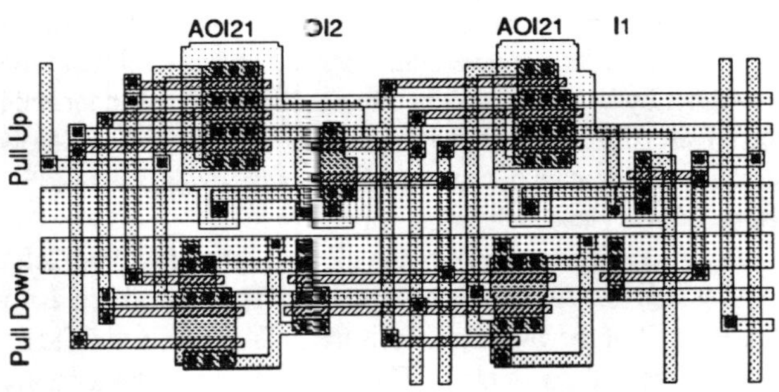

Figure 1: Layout style

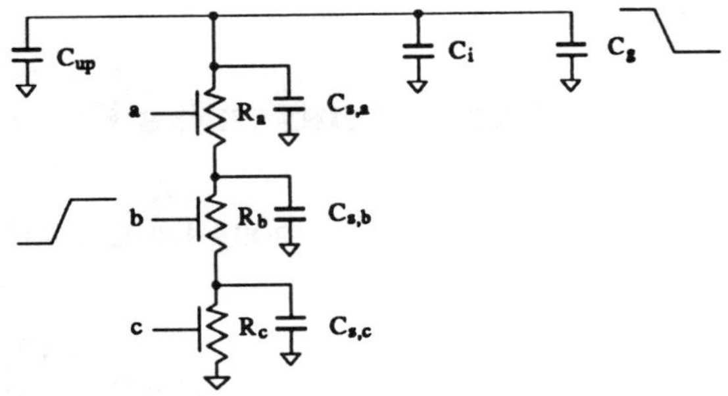

Figure 2: Timing model

The timing analyzer performs a simulation of forward signal propagation from the primary inputs (PI's) to the primary outputs (PO's) to obtain *actual arrival times* at the outputs of the gates. Its other task is assigning net weights that are to be used in the performance cost function during the critical net optimization process. We first define slack, $S_{i,k}$ of net *i* computed from a backward signal propagation from PO *k* to be $S_{i,k} \equiv RAT_{i,k} - AAT_i$ where $RAT_{i,k}$ is the *Required Arrival Time* at node *i* when the required arrival time at PO *k* is set to *max*, the maximum of all actual arrival times at PO's. AAT_i is the actual arrival time at node *i*. To be more precise, each net has two slack values — one for the falling transition and the other for the rising transition; and S_i is the lesser of these two slack values. Then weight W_i for net *i* is

$$W_i \equiv \sum_{k \in \{PO\}} exp\,(K \cdot (max - S_{i,k}))$$

where K is a positive scaling constant. If the slack of a net is smaller (i.e., the net is more critical), the net will be assigned a larger weight. A net through which several critical paths pass will be assigned a correspondingly larger weight.

The timing model described above has the following shortcomings: 1) interaction with side paths (pull-up/pull-down paths that are not the most resistive) are not considered; 2) false paths can be a problem; and 3) slowly changing signals can cause inaccuracy. The techniques presented in [1] or [17] can be applied to cope with the false path problem.

4: Performance optimization

Inputs to *perflex* are a transistor-level net-list description of the circuit, a fabrication process parameter file, and various control options, such as the delay requirements, the number of layout rows to use, the maximum transistor size, optional gate-clustering information, and the number of layouts to be tried during the simulated-annealing-driven timing and layout optimization. Output of the program is a geometrical description of a layout for the circuit.

4.1: Critical net length optimization

We derived *perflex* from *artistII* which uses simulated annealing for optimizing layout area, and the strategy for annealing remains essentially the same. In order to optimize for circuit performance, however, a new cost function that reflects the critical path delay of a layout configuration is incorporated in addition to the layout-area-based cost function. Readers are referred to [14] for details of the simulated annealing and the layout algorithm.

In the early stages of annealing, new configurations are generated by permuting gate placement. During the later, fine-tuning stages, gate placement is fixed and the gate substructures are permuted to see if doing so would create a better configuration. Because the order of series-connected substructures are determined and fixed by the arrival-time-based timing optimizer we describe next, only the parallel substructures are permuted.

The area cost function returns the exact area that is required to map the configuration to a physical layout. The evaluation of this cost function involves construction of gate structures and the implementation of all interconnections. Thus, all mask-level objects that can affect the size of the actual layout are considered. This leads to a somewhat expensive, but exact measurement of the layout area. By-products of this process, accurate geometric figures of gate structures and the interconnections, are used in estimating the timing performance of the configuration.

The timing cost function is a weighted sum of path delays:

$$\text{performance cost} \equiv \sum_{i \in N} W_i R_i (C_{l,i} + C_{w,i} + C_{d,i})$$

where N is the set of all nets, W_i the weight assigned to net i, R_i the resistance of the critical driver (either the pull-up or pull-down driver for the net), $C_{l,i}$ the sum of fan-out gate capacitances, and $C_{w,i}$ the wiring capacitance. $C_{d,i}$ is the sum of capacitances of the drain regions of the transistors connected to net i. $C_{w,i}$ and $C_{d,i}$ are computed from the accurate gate geometry and routing information, and all other values are known through the timing analysis.

Figure 3 displays the consistency of the timing cost function by showing the correlation between the timing cost and the critical path delay reported by HSPICE for 209 different layouts encountered during simulated annealing on an 8-bit ripple carry adder circuit.This shows a high confidence level in our cost function.

The performance cost is used primarily for driving the simulated annealing process. The acceptance decisions for new configurations are made based only on the performance cost. However, the decision as to whether or not to save the new configuration as the best configuration is made based on both costs. The basic strategy is to use the performance cost as the primary determinant, and to apply the area metric when the best-so-far and the current configurations have an equal performance cost.

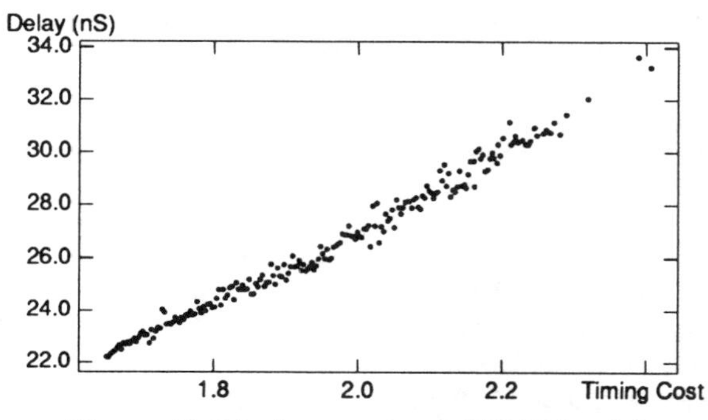

Figure 3: Timing cost vs. HSPICE delay

4.3: Transistor sizing and reordering

The transistor sizing algorithm used in *perflex* is an adoption of the heuristic algorithm used in AT&T's TILOS. In *perflex*, however, transistor sizing is more than just a preprocessing step to the layout process. The transistor sizing procedure can be invoked repeatedly as the simulated-annealing layout optimizer slowly solidifies the gate placement to a final one. Such an ability allows the transistors to be sized more optimally than in a preprocessing approach because more accurate information about the interconnects are available.

In the TILOS heuristic, transistors are sized only to increase their widths. Thus, if the sizer is invoked repeatedly with layouts differing in their interconnection structures, one may end up with some overly-sized transistors. Perflex copes with this problem by invoking a greedy sizer that trims overly-sized transistors.

Transistor reordering is a simple, yet effective transistor-level timing optimization technique. It is achieved by arranging the transistors connected in series in the order of their input arrival times so that the latest-arriving input is assigned to the transistor closest to the output node. This rule can be generalized to include the ordering of series-connected substructures containing one or more transistors, designating the latest of the input arrival times of the transistors in each substructure as the arrival time for the substructure.

In *perflex*, the reordering is carried out once prior to layout, and whenever transistors are resized. However, the ordering between substructures connected in parallel is left undetermined. This gives the simulated-annealing optimizer the flexibility to order them for reducing the layout area.

5: Preliminary results

We have experimented with *perflex* using a 16-bit parity checker (*par16*) and an 8-bit and a 16-bit ripple carry adder circuits. The first experiment shows the effects of transistor reordering and critical-net capacitance reduction. Table 1 summarizes the result. In this experiment, *perflex* is run, for each circuit, followed by two

Circuit (#TR's)	Tool/ Ordering	Area μm^2	Delay (nS)	CPU (Sec)	perflex/artistII Area	perflex/artistII Delay
par16	perflex/Best	87657	7.79-8.23	286	-	-
(152)	artistII/Best	78824	8.13-9.02	234	1.11	0.91
	artistII/Free	75048	8.68-10.73	245	1.17	0.76
add8	perflex/Best	141453	21.61	376	-	-
(224)	artistII/Best	139500	25.22	337	1.01	0.85
	artistII/Free	139500	28.85	322	1.01	0.74
add16	perflex/Best	262539	41.86	778	-	-
(448)	artistII/Best	283943	48.85	650	0.92	0.85
	artistII/Free	262539	52.24	693	1.00	0.80

Table 1: Benchmark results

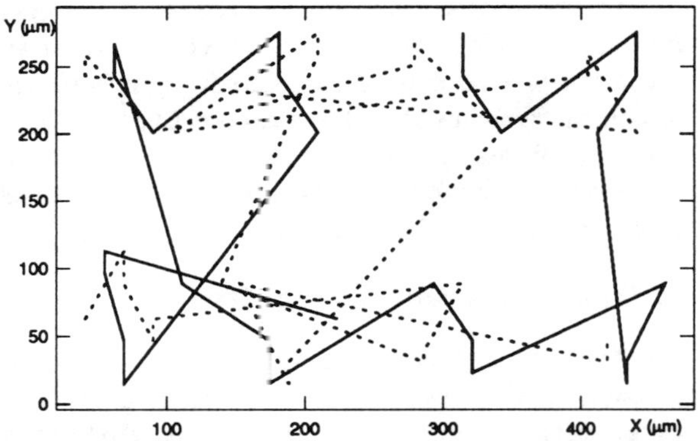

Figure 4: Traces of the critical paths

runs of *artistII*: once using the best transistor orders found by *perflex* (*Best*), and once allowing *artistII* to manipulate the orders freely (*Free*). Our intention by fixing the transistor orders in some *artistII* runs was to make the effect of different annealing strategies visible by suppressing the effect of transistor ordering. However, in the result for the *add16* example, fixing the ordering seems to affect the performance of *artistII* negatively due to the reduced flexibility.

The *par16* circuit consists of XOR gates arranged in a binary tree. The minimum and maximum delays along the sixteen symmetric critical paths are reported. The memory used for the *add16* benchmark circuit was about 900 Kbytes. The *perflex/artistII* column compares the quality of layouts obtained by *perflex* and *artistII*. The widths of the transistors were fixed at 4μm, and 10000 trial configurations were examined during each run. Two layout rows were used for all runs, and a 2μm CMOS technology was used. Run times are in Sun 4/490 CPU seconds.

Figure 4 shows the traces of the most critical paths of the two layouts of the *add8* circuit. The solid and dotted lines represent the critical paths of the layout corresponding to the table entries marked *perflex/Best* and *artistII/Best*, respectively. Results show that typical reduction in delay is 10% to 25%.

In the second experiment, *perflex* was run on the *add8* circuit with transistor sizing enabled. A maximum limit of 15μm on transistor width was imposed. The first run, marked as *schedule A* of Table 2, models a preprocessing approach by sizing transistors only initially. Delays for both the rising and falling output transitions are given. *Schedule B* shows the statistics when the transistor sizer was invoked twice, once initially, and once after final gate placement is determined at the end of the first phase of the annealing process; the fine-tuning sizer was also invoked just before the final layout is to be generated. This resulted in 47.1% decrease in delay and 23.8% increase in area when compared to the entry marked as *artistII/Free*. The final layout is shown in Figure 5.

It is surprising to see that annealing on timing costs works well, because the timing cost function does not usually change smoothly with layout permutations. The ripple carry adder examples behave nicely in this respect. For ill-behaved circuits, hierarchical gate clustering can be used to provide the tool with hints on which gates should be placed close together.

Schedule	Area (μm^2)	Delay (nS)	Sizing Time (Sec)
A	172701	14.48/16.13	47
B	172701	14.66/15.27	88

Table 2: Effect of transistor sizing schedule

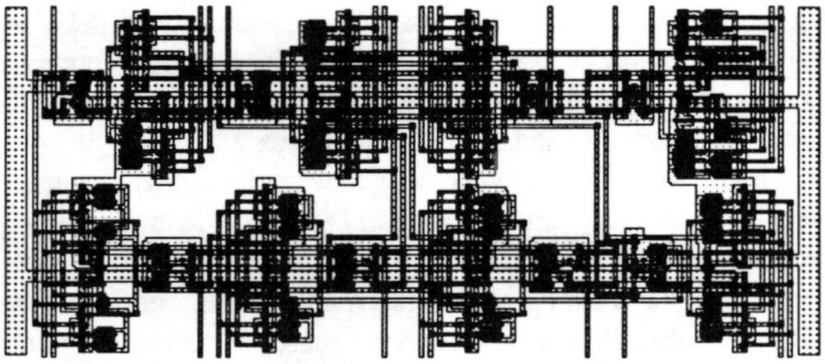

Figure 5: A fast 8-bit ripple carry adder

References

1. J. Benkoski, et al., "Timing Verification Using Statically Sensitizable Paths," *IEEE Trans. CAD*, Vol. 9, No. 10, pp. 1073-1083, Oct. 90.
2. M. Burstein and M. N. Youssef, "Timing Influenced Layout Design," *Proc. 22nd DAC*, pp. 124-130, 1985.
3. H. Y. Chen and S. M. Kang, "A New Circuit Optimization Technique for High Performance CMOS Circuits," *IEEE Trans. CAD*, Vol. 10, No. 5, pp. 670-676, May 1991.
4. M. A. Cirit "Transistor Sizing in CMOS Circuits," *Proc. 24th DAC*, pp. 121-124, 1987.
5. A. E. Dunlop, et. al., "Chip Layout Optimization Using Critical Path Weighting," *Proc. 21st DAC*, pp. 133-136, 1984.
6. A. E. Dunlop, et. al., "Experiments Using Automatic Physical Design Techniques for Optimizing Circuit Performance," *Proc. 1990 IEEE Int'l Symposium on Circuits And Systems*, vol 2, pp. 847-851, 1990.
7. W. C. Elmore, "The Transient Response of Damped Linear Networks with Particular Regard to Wideband Amplifiers," *J. Applied Physics*, vol. 19-1, Jan. 1948.
8. J. P. Fishburn and A. E. Dunlop, "TILOS: A Posynomial Programming Approach to Transistor Sizing," *Proc. ICCAD*, pp. 326-328, 1985.
9. P. S. Hauge, R. Nair, and E. J. Yoffa, "Circuit Placement for Predictable Performance," *Proc. ICCAD*, pp. 88-91, 1987.
10. K. S. Hedlund, "Aesop: A Tool for Automated Transistor Sizing," *Proc. 24th DAC*, pp. 114-120, 1987.
11. D. Hill, et al., *Algorithms and Techniques for VLSI Layout Synthesis*, pp. 153-169, Boston, MA., Kluwer Academic Publishers, 1989.
12. M. A. B. Jackson and E. S. Kuh, "Performance-Driven Placement of Cell Based IC's," *Proc. 26th DAC*, pp. 370-375. 1989.
13. S. Kirkpatrick et. al., "Optimization by Simulated Annealing," *Science*, V220-4598, pp. 671-680, May 1983.
14. S. Kim and R. M. Owens, "A Two Dimensional Gate Matrix Module Generator," *3rd Physical Design Workshop*, May 1991.
15. M. Marek-Sadowska and S. P. Lin, "Timing Driven Placement," *Proc. ICCAD*, pp. 94-97, 1989.
16. M. Marek-Sadowska and S. P. Lin, "Pin Assignment for Improved Performance in Standard Cell Design," *Proc. ICCD*, pp. 339-342, 1990.
17. P. C. McGeer and R. K. Brayton, "Efficient Algorithms for Computing the Longest Viable Path in a Combinational Network," *Proc. 26th DAC*, pp. 561-567, 1989.
18. J-M. Shyu et al., "Optimization-Based Transistor Sizing," *IEEE J. Solid-State Circuits*, Vol. 23, No. 2, pp. 400-409, Apr. 1988.

PLANE PARALLEL A* MAZE ROUTER AND ITS APPLICATION TO FPGAS

Mikael Palczewski

XILINX
2100 Logic Drive, San Jose, CA 95124

ABSTRACT

A plane-parallel maze-routing method for FPGAs is presented. It is demonstrated that a plane-parallel approach is significantly faster than traditional approaches, without compromising quality as measured by routing length. A framework is established for unified code to support traditional wire routing, timing driven routing, plane parallel and global routing.

INTRODUCTION

This new approach is based on A* searching but uses wires (connections) in a parallel fashion. (Nets are processed one at a time.) Although particularly well suited to the FPGA domain, the approach can be applied to other device and design environments. Therefore, the wire-parallel operations are also referred to as plane parallel, where a plane may represent routing layers or tracks. The first objective of this paper is to show that definitions of problem states for wire-parallel expansion can be used for single-wire expansion but not vice versa. The second objective is to show why the parallel A* expansion is so much more efficient than single-wire expansion, timing driven or not, in terms of processing speed and that it does not sacrifice quality as measured by routing length.

CONCEPTS.

Figure 1 shows a section of a typical island–type FPGA

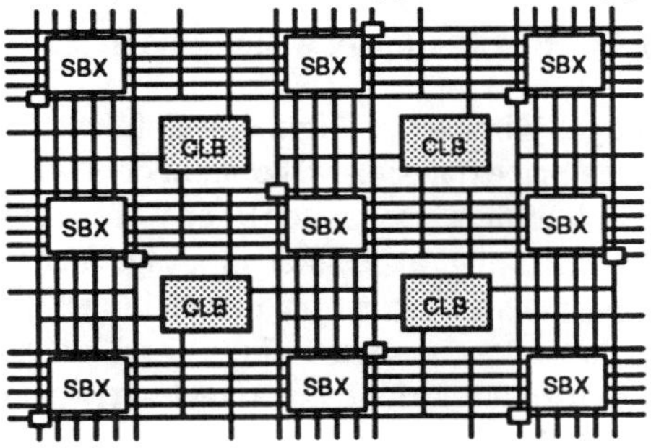

Figure 1 Typical Island-Type FPGA

[1,2,3,4]. Alternative FPGA architectures are described in [5,6,7]. Programmable macro cells (gray boxes), referred to as Configurable Logic Blocks (CLBs), are surrounded by routing channels each carrying a number of wires (six shown). Cell pins (four shown) connect to a subset of wires in adjacent channels via programmable connections. Switchboxes, (two white boxes denoted SBX) allow a predefined set of programmable connections between wires in crossing. The majority of wires within switch boxes connect to only one wire in adjacent channel as illustrated in Figure 2. Such a switchbox makes adaptations of traditional channel routing approaches impossible [11]. Therefore, a general A* maze routing approach [8,9,10,15] was selected, although other approaches such as "family of solutions" [12] are also possible.

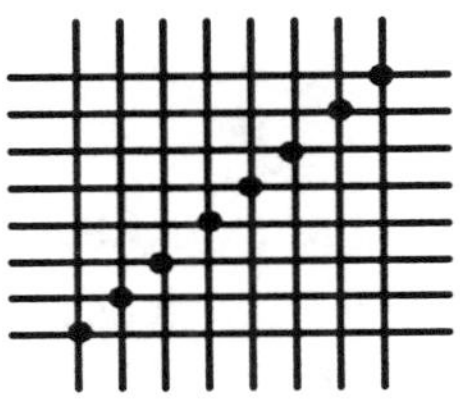

Figure 2 Switchbox Example

The schematic shown in Figure 1 can be viewed as a graph consisting of pins and wires where each pin and wire segment is a graph node and each switchbox connection is an arc. The routing task is an A* graph search [8,9,13] processing one net a time. A traditional search would work on a wire-to-wire basis. The algorithm presented here processes wires in a parallel fashion. Both approaches are illustrated in Figure 3 and Figure 4, using an *ad hoc* switchbox example.

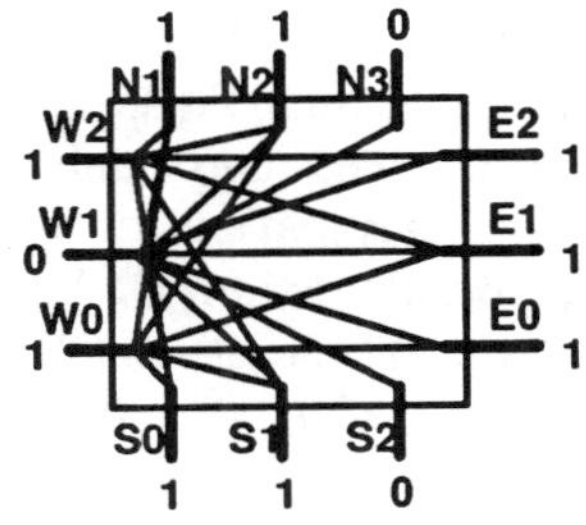

Figure 3 Traditional Switchbox Model

The diagrams show the basic pattern connecting wires in the West channel to wires in the South, East and North channels. During A* expansion, "one" denotes wires being actively expanded and "zero" denotes wires disabled due to blockages generated by resources that have already been

0738-100X/92 $3.00 © 1992 IEEE

Paper 41.4

used. For example, vector [1,0,1] applied on the West is transformed via the switchbox to [1,1,0] in the South and North channels and [1,1,1] in the East channel. These bit vectors are called expansion carriers. If the router works on a wire-to-wire basis, then expansion propagates a carrier with one only wire (one bit set) corresponding to the wire being processed. For example, wire W0 [1,0,0] is propagated to six adjacent wires, one at a time, producing S0 = [1,0,0], S1 = [0,1,0], E0 = [1,0,0], E1 = [0,1,0], N0 = [1,0,0] and N1 = [0,1,0]. Similarly, wire W2 produces six carriers, but only the one at E2 = [0,0,1] will be new. After processing the west channel, a traditional A* routing would produce 7 new problem states for each wire, each of them to be processed separately

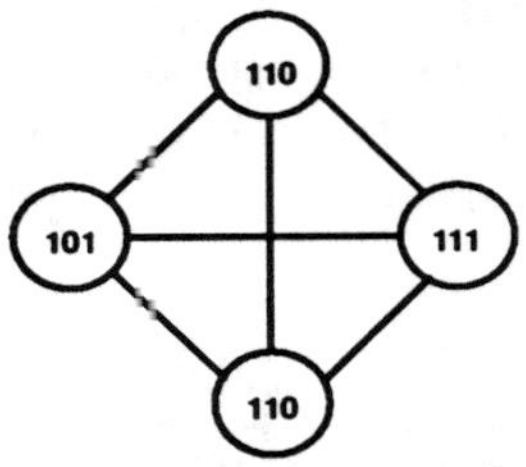

Figure 4 Plane-Parallel Switchbox Model

In plane-parallel routing, operations are performed on bit vectors instead. Carriers may be propagated as above, on a wire-to-wire basis, but the problem states are defined on a channel rather than on a wire basis. Instead of scheduling a new state with each wire, carriers are ORed and accumulated as bit vectors in adjacent channels until all wires in the West channel are processed. The resulting carriers define new problem states. The total number of new states is at most equal to the number of adjacent channels, e.g., three, rather than all possible destinations measured on a wire basis. The graph is also searched on a channel (or subchannel) basis as in global routing [10].

The function that transforms input carriers to output carriers is called a transition function. These functions can be coded as fast table look-up, shown in Figure 5. The figure illustrates computations between two graph nodes. The input carrier is 32 bits wide. It is split into four bytes and each byte is used as an index to a table holding 32-bits wide output carriers. The output carriers from all four tables are then ORed. Experiments with the XC 4000 architecture have shown that this method requires less than 64 KB, if transition tables are saved as library items.

The strength of the plane parallel carrier concept is illustrated with the following examples (Figure 6 and Figure 7). A carrier is emitted from a source and propagated along channels until it dies. In each channel, D out of K planes are occupied. What is the average lifetime of the carrier mea-

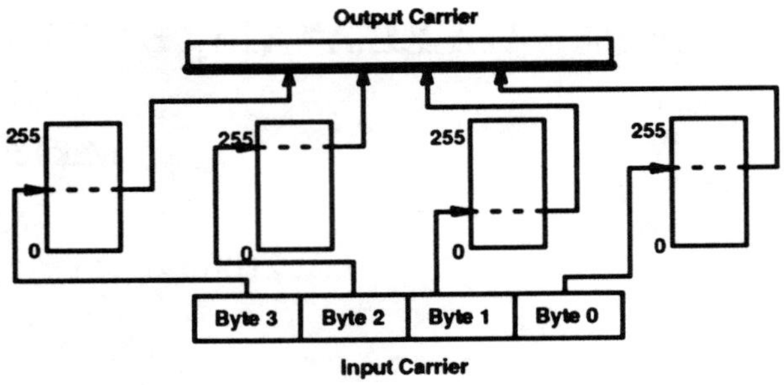

Figure 5 Table Look-Up Transition Function

sured as number of channels that it passes before it dies ? This problem can be represented with this analogy: An urn has K balls, N white and K-N black. D balls are taken per drawing, painted black and returned to the urn. What is the expected number of drawings (L) required to paint all balls black ? In the illustrations below, K=16 and D=3. The plane-parallel carrier starts with N=16 free planes or white balls (Figure 6), vs. N=1, i.e. one free plane or one white

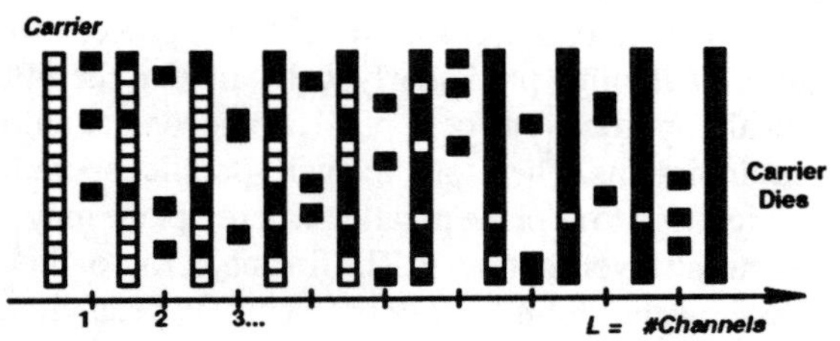

Figure 6 The Lifetime of a Plane-Parallel Carrier

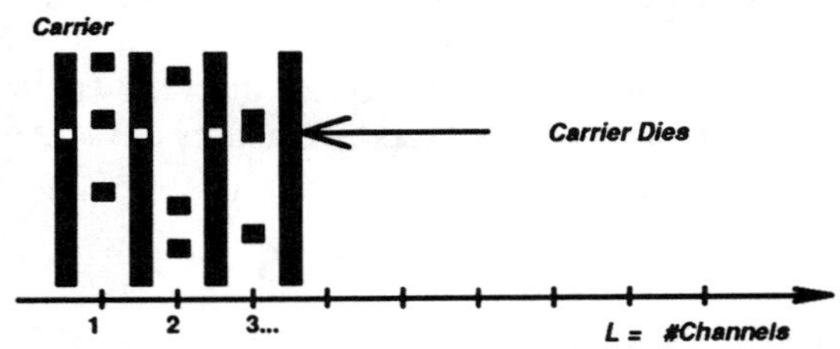

Figure 7 The Lifetime of a Single-Wire Carrier

ball in a single-wire carrier (Figure 7). Columns show states of a carrier along expansion or the number of white balls in the urn. Each column is separated by a set of D=3 black rectangles, representing a channel blockage density of 3. Hence, three balls are taken from the urn in each step. It can be shown that if K increases to infinity, the expected lifetime of the plane parallel carrier (N=K) is at least two times longer (for D=K-1) and at best it is O(logN) times longer (for D=1). These bounds are a specific function of the connection pattern in switchboxes and valid only for the switchbox shown in the Figure 2. In the switchbox shown in Figure 3, vector [1,0,0] on the West side is translated to [1,1,0] on all other sides. Consequently, carriers can be "born" and "die" and the problem changes. Formulating and exploring problems as above is important because they relate density, connectivity patterns, and routability. For example, given a placement and a net list, the average net length can be estimated. If this length is larger than the expected lifetime of a plane-parallel carrier, then the placement can be rejected as very hard to route or unroutable. The problem can also be reversed. What should channel capacity *(K)* and the connection pattern in a switchbox be to secure routability of a net list with an average net length of *L* ?, etc. The concept of the plane-parallel carrier is also very handy for providing theoretical bounds of FPGA hardware. Reference [14] provides alternative methods and insight on switchboxes and the routability of FPGAs.

ALGORITHM

Figure 4 illustrates a wire-parallel model of the wire graph shown in Figure 3. Now, nodes represent channels, and edges are transition functions. Such a construct, referred to as a graph template, uniquely defines routing resources and the rules governing how these resources can be used. Figure 8 shows such a 9 x 9 template. The adjacency lists and tran-

Figure 8 9 * 9 Graph Template

sition functions are coded directly between channels (black circles), and subchannels and macro-cells (grey rectangles). Switchboxes (white circles) are skipped. Such a template has been used to model an XC4010 device as a 41 x 41 array for the examples in this paper. Figure 9 briefly outlines the algorithm operating on this graph template. Shaded items indicate changes when comparing this approach to traditional approaches [10,13,15]. The diagram depicts simple graph searching [11]. PQ denotes a priority queue implemented as a heap [13]. The searching is initialized with one or more sources stored in the priority queue. A unique target is also defined. Multi-pin nets are routed one target pin at a time and PQ is initialized with existing routing and the source pins. The main search loop starts at *Remove Node from PQ;* once a node is removed, it is expanded by an internal loop starting at *Get Successor*. In the traditional A* searching, successors are taken directly from the adjacency list. This is replaced by the vector calculations described earlier. Once the carrier is initiated, bit vector operations compute the successor on all carriers independently of the number of bits (i.e. wires) in a carrier that are set. Hence,

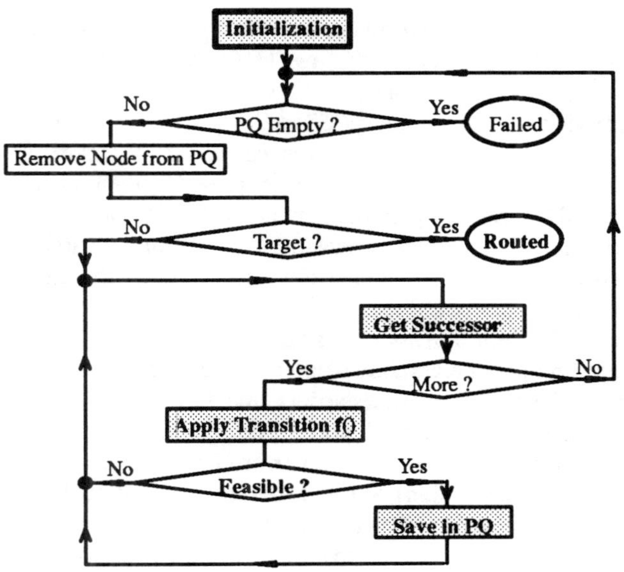

Figure 9 The Algorithm

Initialization and *Get Successor* provide a run-time switch between wire-parallel and wire-by-wire computations.

In the traditional expansion, once a graph node has been expanded, it is never re-introduced into the heap. This fact is a simple consequence of graph searching techniques [13] preventing an infinite loop within the search. In traditional routing approaches a graph node is said to be unique if it has not been visited. The concept of uniqueness is very different in wire-parallel and timing-driven searching. The problem is illustrated below (Figure 10). Expansion starts at node "A" goes to "B" and reaches node "C" at time t2. There is a path from A to C via "E" and "H", and this path is longer. Which path is faster ?

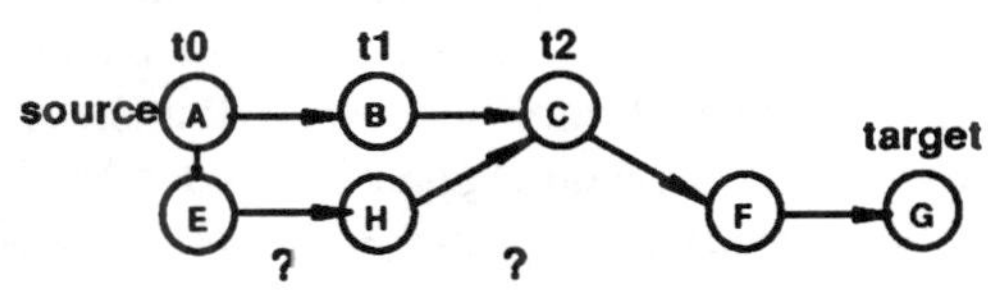

Figure 10 The Fastest Path Problem

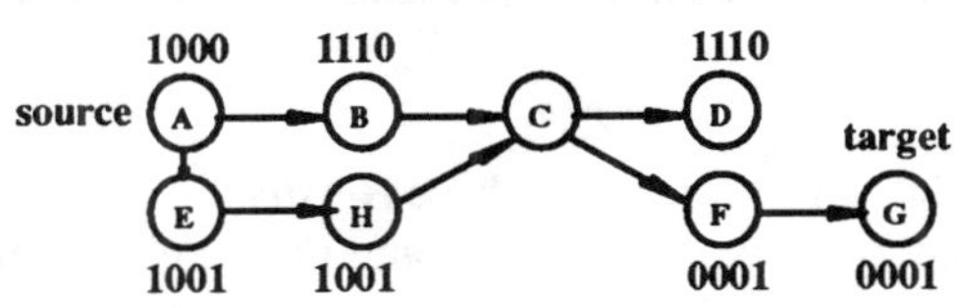

Figure 11 The Feasible Path Problem

To determine the faster route, both paths must be expanded, meaning node "C' could be searched twice. Similarly, with wire parallel expansion, problem states are more closely related to paths than to graph nodes (Figure 11). Expansion starts at node "A" with bit vector [1000], passes through "B" and "C", reaches node "D" as [1110] and then dies. The only path to "G" is via node "F" with carrier [0001]. This carrier has a path via nodes "E" and "H". Again, node "C" is searched twice, but with different carriers. In this case, the uniqueness of the problem states is defined by their bit vectors. Different bit patterns indicate different states. More exactly, wire-parallel expansion accepts unique carriers, while-timing driven expansion accepts only the faster paths. This is possible because with each new state in the heap, the algorithm associates a new record with the *Carrier* and *Time* parameters and a pointer to the current graph node. *Carrier* saves expansion carriers, and *Time* saves the time required to reach a graph node. The essential information about a problem state is stored externally not in the graph. This is because graph nodes may be searched repeatedly and do not represent unique problem states. The problem space is also different. This is the second difference between the traditional A* searches and the presented algorithm.

After applying transition functions, the carrier is tested for feasibility. The test validates the carrier comparing it with free resources and checks its uniqueness. To do so, each graph node keeps two records, *used* and *expanded*. The former saves all used resources and the latter keeps already expanded resources for the net being routed. Computations and the storage associated with transition between two graph nodes N_1 and N_2 are summarized in Figure 12. $F_{12}()$ denotes the transition function from N_1 to N_2. and is used to translate carrier 0001 from a previous state (not shown). Then a new carrier is computed with *used* and *expanded* taken from N_2. Since it is non-zero the state is accepted and added to the priority queue. In addition, timed paths are compared against *time* saved in the graph node N_2. Fastest paths are accepted, slower paths are rejected; in this example, the new carrier is faster. A new record *state 1* is created and saved in PQ. Searched graph node N_2 is OR updated by the new carrier and the new timing data is saved in N_2. The feasibility test reduces to: *feasible* = F(*carrier*) AND NOT(*used*) AND NOT(*expanded*), where all operands are bit vectors. If *feasible* is non-zero, then it also defines a new carrier.

The feasibility test can also be based on *density* and *capacity*, i.e. the total number of wires and the number of used wires in a channel. This simple change to the test can transform the code into a global router. Similarly, the feasibility test is different in timing driven and plane parallel routing. Therefore, the feasibility test should be an external function providing a run–time option. Unifying all these approaches is possible because graph nodes in a plane–parallel routing represent channels (or subchannels), the adjacency lists link

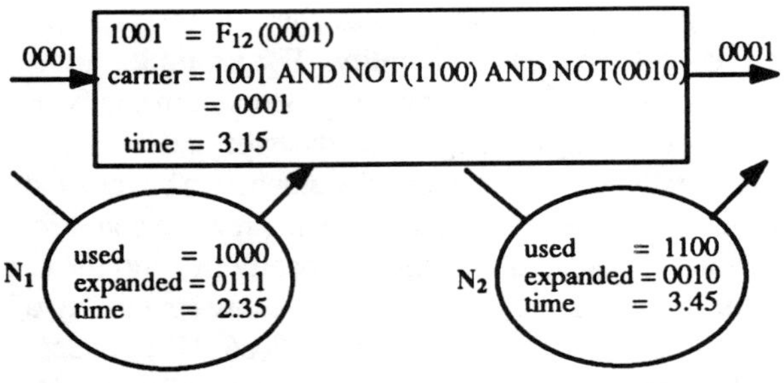

Figure 12 Problem State Transition

channels and wire adjacency is left to transition functions. Together with the previously shown examples, this establishes the first goal of showing that the algorithm can support plane–parallel, timing driven and traditional approaches.

The third difference between wire parallel and wire-by-wire routing is related to how the solution is extracted. Accepted states are linked backward to the state from which they came. Once the loop terminates, reconstructing the path is related to the traversal of linked states as this set of links defines the shortest or the fastest path. The question is which wire in the carrier should be used. Figure 13 illustrates the problem. All three targets can be reached by a vector [1110], but taking decisions a single source to target basis, the program could assign three different wires where one would suffice.

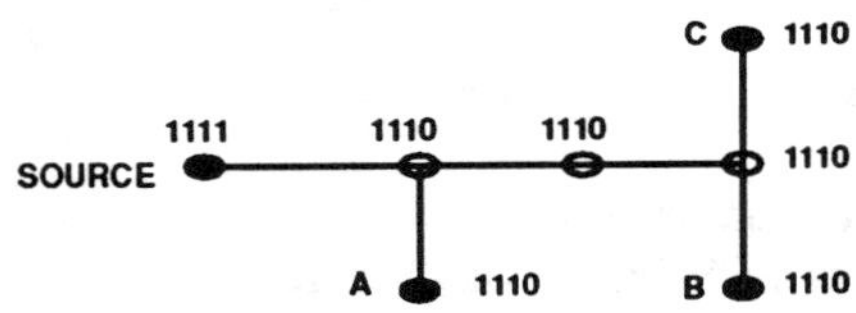

Figure 13 Wire Assignment

Since carriers describe a set of wires, e.g., solutions, it would be sub-optimal to select a wire by looking at only one pair of pins (source and target) at a time. Therefore, the wire assignment is postponed until all pins are routed. Then, paths are backtracked from targets to the source to determine which wires at the source reached each target. The backtracking process is the opposite of search (Figure 11) and must invoke appropriate transition functions. The exact formula to backtrace a carrier is: $C_{N-1} = C_{N-1}$ AND $G(C_N)$, where C_{N-1} and C_N are expansion carriers in two consecutive states S_{N-1} and S_N, and G is a reverse function of the transition function F from S_{N-1} to S_N such that $x = G(F(x))$. After backtracking is completed, each target pin returns appropriate carriers to its source. The set of wires defined in the carrier at the source is pruned to a minimum set. Then, routing is traced again from the source to all targets to assign selected wires and save them as *Used* resources along the path.

A final note about the cost function used to sort records in the heap: it is given by f(n) = g(n) + h(n), where g(n) is the accumulated cost from the source to node "n" and h(n) is a lower bound on the cost from "n" to target [8,9]. Wire-parallel expansion uses Manhattan distance to calculate the cost. It follows that all graph nodes within the rectangle defined by the source and the target pins have the same cost, and the advantage of an A* approach is seemingly lost. To make such a search efficient, a conditional statement has been added to break this tie. If two states have the same cost, the statement selects the node having lower h(n), i.e., closer to the target. This simple fix significantly improves the efficiency of the search without violating the admissibility of A* [8,9]. In the timing-driven approach, the delay is accumulated from the transition tables and h(n) must be user supplied. By default it is zero. This is necessary to guarantee the admissibility of an A* approach in a technology-independent fashion. Far better bounds are possible for selected architectures but this limits the more general applications.

EXPERIMENTAL RESULTS

The experiments presented here use the template described, with some modifications. All sub-channels carried 18 wires implemented as short lines. Within each switchbox, the *N*th wire was connected only to the *N*th wire in all adjacent channels (Figure 2). This decreased the lifetime of the plane-parallel carriers and the influence of wire-parallel expansion, since a carrier having *N* active wires could transforms only to a carrier with *N–I* wires, where *I*=(1,2,...,N) i.e., wires could only "die". At the same time, this model provided the best case for timing driven searches — the same wire cannot be expanded twice. Random net list generators were used, and the number of nets was limited to 1000, the number of pins/net was limited to two, and various average net lengths of 5, 8 and 12 (measured in graph nodes) were used. Output pins were initialized as described earlier. (see *Initialization* and *Get Successor*). Both offer selection of wire-parallel and traditional routing. Input pins were connected to all wires, and output pins were connected to wires in a randomly defined fashion. The number of wires that each output pin could drive was parametrized.

To make qualitative comparison of both methods, efficiency was measured in terms of run-time, duplication factors (DF) and branching factors (BF). Duplication factors measured the number of times each graph node was inserted in the queue. This count was divided by the total number of generated problem states. The importance of this measure is related to modifications allowing the same graph node to be re-expanded. Branching factors were defined as BF = P/E, with P being the number of nodes in the solution path and E being the total number of processed states. In other words, branching factors provide the ratio between "good" and "wasted" problem states.

In the first experiment, the number of planes an output pin could drive was varied from 2 to 18 in increments of four. Figure 14 shows the run times. Traditional operations are indicated by a cross, and wire-parallel operations are indicated by a diamond. The gain measured as a ratio between run times of wire-by-wire and wire-parallel operations is show inFigure 15.

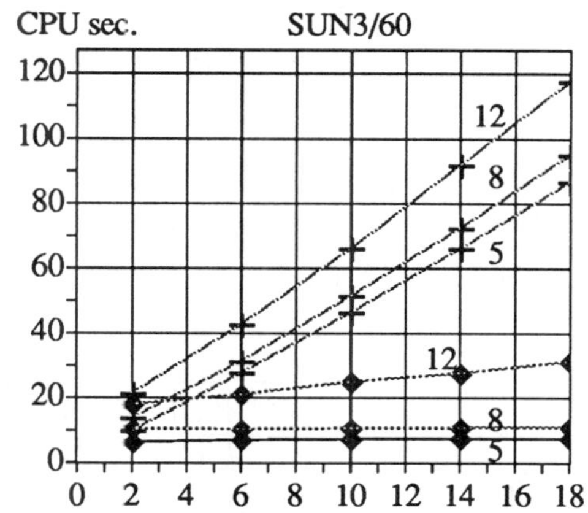

Figure 14 Runtime = f(#planes)

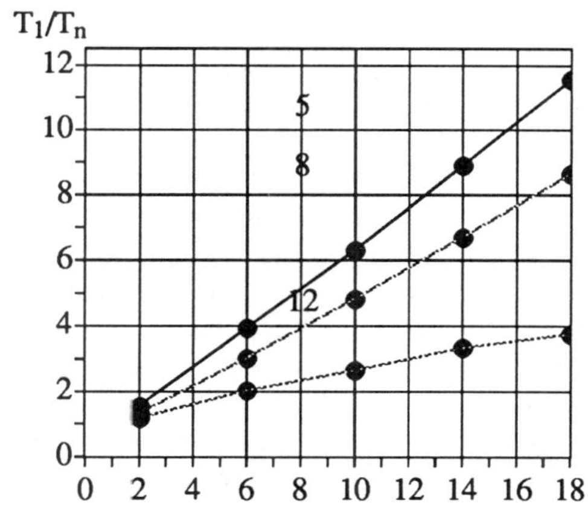

Figure 15 T_1/T_n = **f(#planes)**

Figure 16 shows Duplication Factors and the ratio between the methods, is provided in Figure 17. Figure 18 and Figure 19 show branching factors and the gain. In all these diagrams, each entry has been averaged over 10 runs for the same net list, but with randomized ordering of nets. Figure 14 established the last goal, illustrating that wire-parallel expansion is indeed much faster, but it does not explain why.

The explanation requires a look at the Duplication Factors and the run time. The first item of note is that the data for

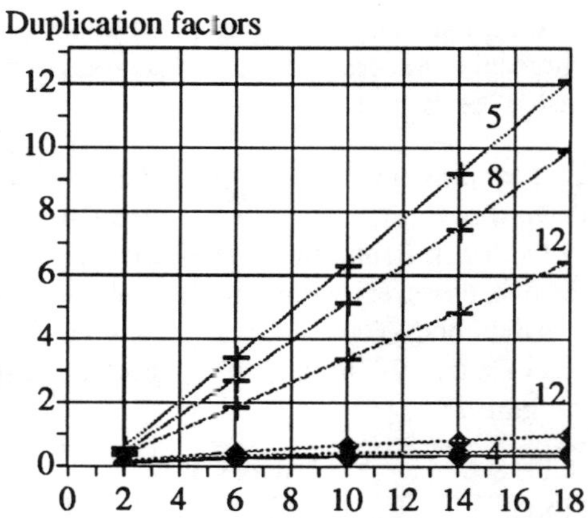

Figure 16 DF = f(#planes)

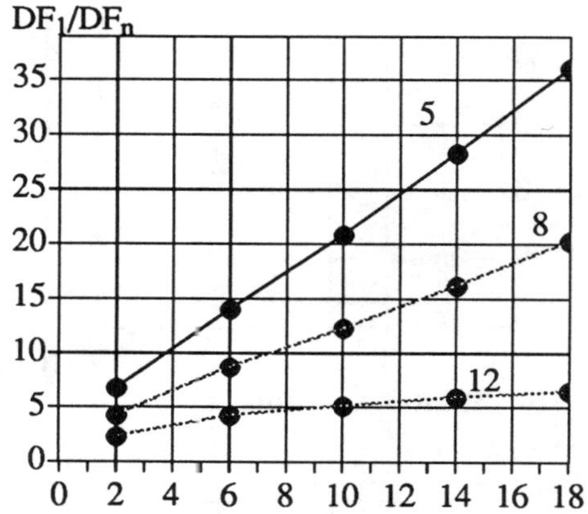

Figure 17 DF_1/DF_n = f(#planes)

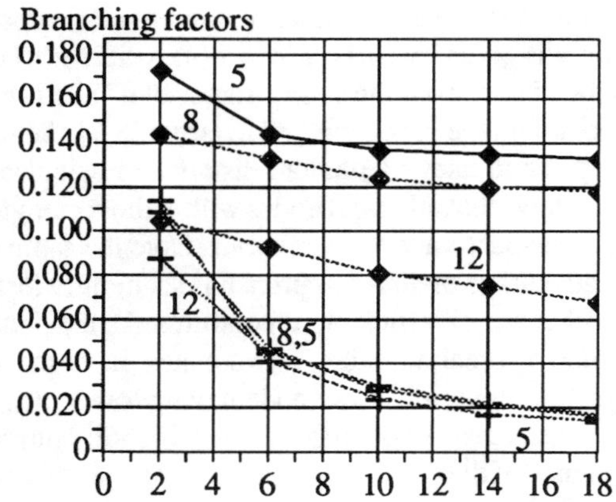

Figure 18 BF = f(#planes)

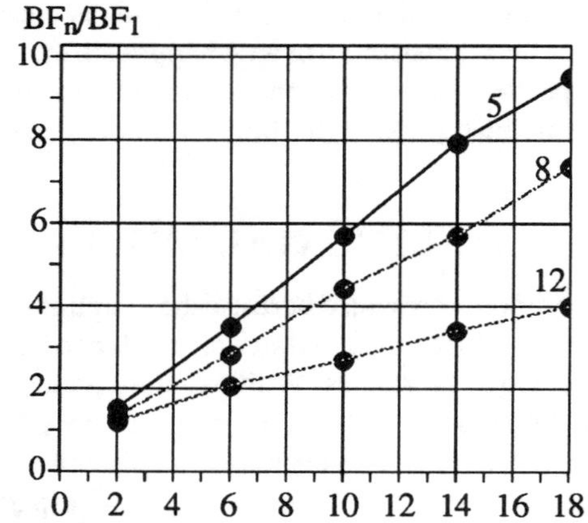

Figure 19 BF_n/BF_1 = f(#planes)

wire-parallel expansion is almost constant. This indicates that routing expanded generally within a rectangle defined by the source and the target, and found solutions in a time proportional to the length of nets. Furthermore, flat Duplication Factors indicate that the state space is approximately equal to the number of graph nodes. A look at the Duplication Factors of traditional expansion shows that number of re-expanded graph nodes increases in proportion to the number of planes. Hence, the only conclusion is that the single-wire carriers die before reaching the target, and the expansion had to restart with a new carrier.

The routing problem is strikingly similar to Bernoulli trials. There are N planes to route, only one has the least cost solution. Considering again an urn with N balls, $N-1$ being black, and one being white, what is the expected number of drawings of one ball at time to get the white ball ? This yields a geometric distribution with $E(X) = 1/p = N$ where p is the probability of success. Obviously, taking more balls at a time increases the probability of getting the white ball. This simple example illustrates the difference between plane-parallel and traditional expansion. The time spent on expanding along "wrong" wires dominates the time devoted to finding the path using "good" wires.

The Branching Factors shown in Figure 18 and Figure 19 illustrate that the efficiency of the search decreases with the utilization of routing resources — as expected. In fact, all net lists with an average net length of 12, due to congestion, were unroutable and never completed, leaving from 5% to 12% of the unrouted nets.

The functionality of this framework is further demonstrated by a rip–up and reroute process. This task starts with removal and re-routing of all nets that failed to complete, e.g., *Unroutes*. Wire-parallel search is used, but with zeroed carriers and without testing feasibility of states. Hence, the output provides a list of subchannels as global routers do. These channels are marked as *Difficult*. After processing all unroutes, *Difficult Channels* are scanned, and nets are ranked by the number of *Difficult Channels* they pass. This list, referred to as *Blockages*, is sorted according to the highest ranking nets. Ties are broken by the number of times each net has already been routed. After sorting, a number of nets is taken from the front of the *Blockages*, unrouted, and appended to the original unrouts. The number of appended *Blockages* is equal to the number of *Unroutes* plus a constant percentage of this number, plus some small number (3 – 10) selected in a random fashion. The list is re-routed, keeping net ordering as described above, *Unroutes* first and then *Blockages*. Each time the router fails to produce better results, the percentage constant is increased. Otherwise, it is restored. If, in a pre-defined number of passes (3+), the router does not produce better results, all *Blockages* are removed and re-routed.

As a final experiment, plane-parallel operations were tested on the standard MCNC benchmark *primary1*. The circuit was converted and placed using currently available production tools, and an XC4009 device. A total of 793 nets and 2158 were selected for routing. The plan-parallel expansion completed this task in 73 CPU seconds with a SPARC 2 after 1 +19 passes of rip-up and re-route. The first pass attempted all, but completed only 86% of the pins in 10 CPU seconds. Total 1028 nets (129%) have been re–routed in the following passes. On the other hand, a timing-driven production router could find the solution in one pass, but it took significantly longer than 1+19 passes . Why ?

In terms of resource assignments, timing oriented evaluations are surprisingly more efficient than manhattan metric. On XC4000, approximately 22% of routing resources belong to the "double lines", and 33% of the resources belong to even longer "long–lines" [3,4]. Hence, more than 50% of routing resources are special purpose hardware dedicated for high–speed routing. Figure 20 illustrates the incompatibility of timing and length–oriented solutions. Programmable switch-points contributing to the delay are dots, cells are rectangles. The dotted line shows the shortest path. It is slower path than the bold route (6 dots vs. 4 dots). Consequently, the strategy of the shortest path fails for assignment of more than 50% of the routing resources. By their nature, wire parallel operations do not differentiate wires within a carrier. Using manhattan metric, the wire–parallel router will very efficiently assign a minority of the resources and then must undo blockages with the rip–up and reroute. On the other hand, timing–driven expansion is significantly slower, but it also explores a wider population of resources. Since most nets are not time critical, plane–parallel operations may dominate routing strategy, provided critical paths are known.

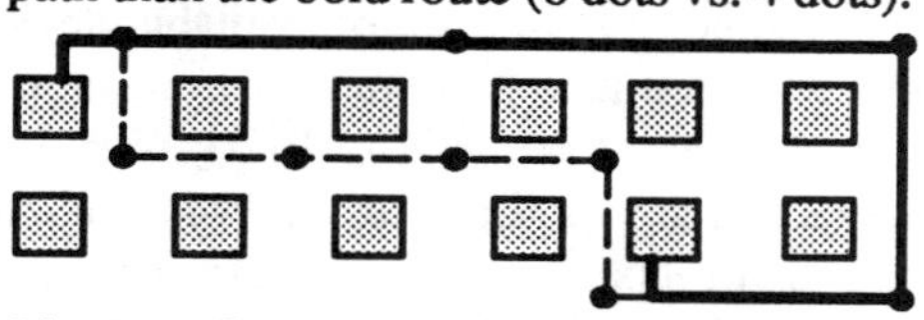

Figure 20 The Shortest and the Fastest Path

CONCLUSIONS

Using a plane-parallel carrier is a viable routing concept with a number of applications beyond A* routing. This paper has addressed its application to routing problems specific to FPGAs. It has been demonstrated that the plane-parallel approach is significantly faster than traditional approaches, without compromising quality as measured by routing length. However, due to the hardware specific implementation of routing resources, the shortest path strategy will very likely fail to produce the fastest paths. These factors explain why this discussion has focused on a runtime compatibility of approaches such as: timing driven routing, plane parallel and global routing.

Many issues discussed are directly related to the timing-oriented hierarchy between routing resources found in FPGAs. This hierarchy not only upsets traditional routing concepts. it changes traditional views of placement problems. Problems associated with management of these resources will undoubtedly dominate algorithmic approaches for both placement and routing and is proposed as a topic for the further research.

REFERENCES

[1] W. Carter, *"A User Programmable Reconfigurable Gate Array"*, Proc. IEEE CICC, 1986

[2] H.C. Hisieh, *"A Second Generation User–Programmable Gate Array"*, Proc. IEEE CICC, 1987

[3] S. Trimberger, *"Beyond Logic – FPGAs for Digital Systems"*, International Workshop on Field Programmable Logic and Applications, University of Oxford, 1991

[4] *"XC 4000 Logic Cell Array Family"*, Technical Data, Xilinx, 1990.

[5] A. El Gamal, J. Greene, J. Reyneri, E. Rogoyski, K. El–Ayat and A. Mohsen, *"An Architecture for Electrically Configurable Gate Arrays"*, IEEE J. Solid–State Circuits", Vol.24, No.2, 1989,

[6] Actel, *"ACT Family Oriented Programmable Gate Array Data"*, 1990

[7] Altera Corporation, *"The Maximalist Handbook"*, 1990

[8] N.J Nilsson, *"Principles of Artificial Intelligence"*, Tioga Publishing Company, 1980.

[9] N,J Nilsson, *"Problem–solving Methods in Artificial Intelligence"*, McGraw–Hill, 1971.

[10] G. W. Clow, *"A Global Routing Algorithm for General Cells"*, Proc. ACM/IEEE 21st Design Automation Conference, 1984,

[11] J. Greene, V. Roychowdhury, S. Kaptanoglu, A. El Gamal, *"Segmented Channel Routing"*, Proc. 27th ACM/IEEE Design Automation Conference, 1990.

[12] S. Brown, J. Rose, Z. Vranesic, *"A Detailed Router for Field Programmable Gate Arrays"*, IEEE Proc. ICCAD–90

[13] R. Sedgewick, *"Algorithms"*, Addison–Wesley Publ., 1988

[14] J. Rose, S. Brown, *"The Effect of Switch Box Flexibility on Routability of Field Programmable Gate Arrays"*, Proc. CICC, 1990.

[15] D.D. Hill, *"A CAD System for the Design of Field Programmable Gate Arrays"*, Proc. 28th ACM/IEEE Design Automation Conference, 1991.

A Mixed-Integer Nonlinear Programming Approach to Analog Circuit Synthesis *

Prabir C. Maulik, L. Richard Carley and Rob A. Rutenbar
Department of Electrical and Computer Engineering
Carnegie Mellon University
Pittsburgh, PA 15213

Abstract

A Mixed-Integer Nonlinear Programming (MINLP) approach to cell-level analog circuit synthesis, allowing simultaneous topology selection and parameter selection, is presented. Topology choices are represented as binary integer variables and design variables (e.g., device sizes and bias voltages) are represented by continuous variables. Examples using a *branch and bound* approach to efficiently solve the MINLP problem for CMOS two-stage op amps are given.

1 Introduction

The development of analog design automation tools has lagged far behind digital tools. Experts rely on years of experience to guide the selection of circuit topologies and parameters for those topologies such as device sizes and bias voltages. Considerable work in the area has been devoted to *parameter selection*, i.e., selecting optimum device sizes and bias points to meet specific performance targets. However, *topology selection* is also crucial in designing high-performance analog circuits; some performance specifications are vastly easier to meet if we are allowed to change not only device sizes, but also the number, type, and interconnection of devices as well. To date, proposed analog synthesis strategies have relied on a limited set of approaches to perform topology selection: user input, heuristics and exhaustive search.

In this paper, we present a new synthesis methodology, based on Mixed-Integer Nonlinear Programming (MINLP) methods[1]. Our problem formulation uses integer variables to model topology choices and continuous variables for the design parameters such as device sizes and bias voltages.

2 Background

Analog synthesis can be broken down into two tasks: (a) selection of a device-level topology, and (b) selection of parameters (e.g., device sizes) for that topology. Various approaches have been taken to accomplish the above tasks. Early work in the area of analog circuit synthesis addressed parameter selection for a fixed analog circuit topology using numerical optimization techniques[2, 3, 4]. In general, these optimization strategies (e.g.[5]), rely on expensive circuit simulations in order to predict performance.

A new generation of circuit synthesis tools have emerged that employ analytical equations to predict the performance of an analog circuit[6]. IDAC[7] requires the human designer to choose one topology from a library, and then follows a procedural routine created by an expert designer to perform coarse parameter selection. OASYS[8] walks down a hierarchical decomposition of an analog cell, alternately selecting topologies and translating performance specifications from each level to the next lower level, and backtracking to handle design failures. Although OASYS can automatically design a wide variety of topologies, creating the heuristics for backtracking is very time consuming. OPASYN[9] uses heuristic rules to choose a topology from a set of available device-level schematics and then uses numerical optimization to perform parameter selection. HECTOR[10] uses a "hierarchical, interactive and expert-system assisted" approach to construct circuit topologies from a library of basic building blocks. In all of these approaches, topology selection employs heuristics which must be provided by an expert analog designer. Unfortunately, it can be very difficult to create heuristics that accurately predict the best topology choice without actually carrying out, at least partially, the parameter selection process. SEAS[11], like OASYS and HECTOR, uses a hierarchical description of a variety of analog circuit topologies. However, it uses simulated evolution to change from one topology to another, based upon a score generated after selecting parameters for the current topology. Although this approach avoids the need for expert input, the methods for scoring subblocks in a topology are not well defined. And, more importantly,

*This work was supported in part by the Semiconductor Research Corporation under Contract 91-DC-068

modifications in topology are made only after complete parameter selection of the current topology.

All of the above approaches separate topology selection and parameter selection. The main contribution in this paper is the breakdown of this separation, i.e., *simultaneous* topology selection and parameter selection.

3 Problem Formulation

Fig. 1 shows a simple common source (CS) amplifier. Equations can be derived for the various performance parameters of this amplifier like gain, bandwidth, etc., in terms of design variables **x**, which, in this case, are device sizes and bias voltages. Using the performance equations, we can formulate the design of this amplifier as a *constrained optimization problem*[12]. The function to be optimized (minimized or maximized) is called the *objective function*, and it is typically a function of area or power. The constraints assure that the performance specifications are met. This formulation has been used to design high performance op amps and regenerative comparators[12] using NPSOL[13].

Fig. 2 shows an alternate topology for the common source amplifier in which cascoding has been used to increase the output impedance and therefore the gain of this amplifier. In order to select between these two topologies, we introduce a binary variable y. When y is 0 the simple topology is selected and when y is 1, the cascoded topology is chosen.

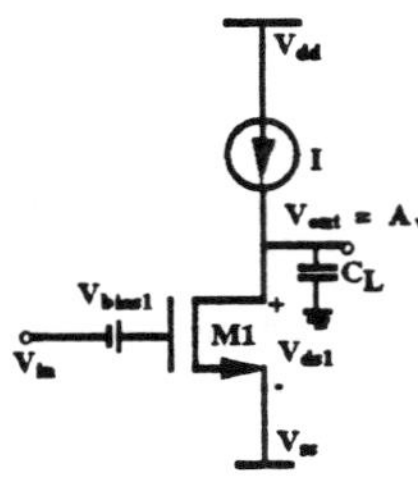

Figure 1:

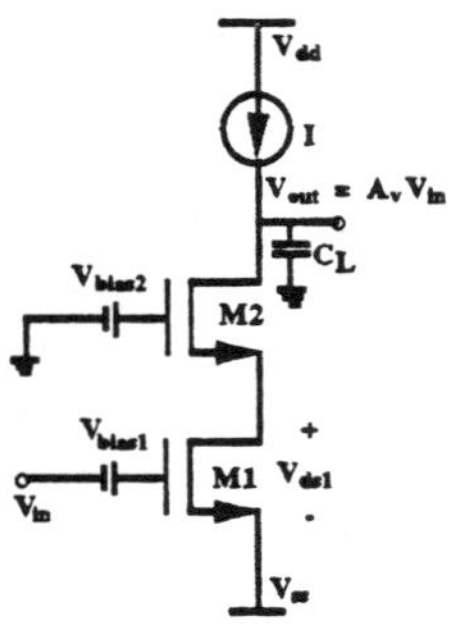

Figure 2:

The complete problem formulation is shown below :

$$\min(W_1^s + W_2^s + L_1^s + L_2^s + y \times \Gamma_0)$$

subject to

$$\left.\begin{array}{lcl} gain_S(\mathbf{x}) + y \times \Gamma_1 & \geq & GAIN \\ gain_C(\mathbf{x}) + (1-y) \times \Gamma_2 & \geq & GAIN \\ \frac{I_{d1}(\mathbf{x})}{C_L} & \geq & SLEW \\ \frac{g_{m1}(\mathbf{x})}{2\pi C_L} & \geq & UGF \end{array}\right\} \text{Perf. constrs.}$$

$$\left.\begin{array}{lcl} I_{d1}(\mathbf{x}) & = & I \\ I_{d1}(\mathbf{x}) - I_{d2}(\mathbf{x}) + (1-y) \times \Gamma_5 & \geq & 0 \\ I_{d1}(\mathbf{x}) - I_{d2}(\mathbf{x}) - (1-y) \times \Gamma_6 & \leq & 0 \end{array}\right\} \text{KCL constrs.}$$

$$\left.\begin{array}{lcl} V_{c1}(x) - V_{ds1}(x) + (1-y) \times \Gamma_7 & \geq & 0 \\ V_{c1}(x) - V_{ds1}(x) - (1-y) \times \Gamma_8 & \leq & 0 \\ V_{c2}(x) - (-V_{ss}) + (1-y) \times \Gamma_9 & \geq & 0 \\ V_{c2}(x) - (-V_{ss}) - (1-y) \times \Gamma_{10} & \leq & 0 \\ V_{ds1}(x) - (-V_{ss}) + y \times \Gamma_{11} & \geq & 0 \\ V_{ds1}(x) - (-V_{ss}) - y \times \Gamma_{12} & \leq & 0 \end{array}\right\} \text{KVL constrs.}$$

$$\left.\begin{array}{lcl} W_{min}^s \leq & W_1^s, W_2^s & \leq W_{max}^s \\ L_{min}^s \leq & L_1^s, L_2^s & \leq L_{max}^s \\ y & \in & \{0, 1\} \end{array}\right\} \text{Variable Range constrs.}$$

where

$$\begin{array}{rcl} gain_S(\mathbf{x}) & = & g_{m1}(\mathbf{x}) r_{o1}(\mathbf{x}) \\ gain_C(\mathbf{x}) & = & g_{m1}(\mathbf{x}) r_{o1}(\mathbf{x}) r_{o2}(\mathbf{x}) \\ V_{c1}(\mathbf{x}) & = & V_{bias2} - V_{gs2}(\mathbf{x}) \\ V_{c2}(\mathbf{x}) & = & V_{ds1}(\mathbf{x}) + V_{ds2}(\mathbf{x}) \\ \Gamma_0 & = & W_{min}^s + L_{min}^s \end{array}$$

The design specifications are in capital letters on the right-hand sides of the performance constraints. The subscripts S and C denote simple and cascode respectively. The Γ_i's are suitable positive constants. The $y \times \Gamma_1$ and $(1-y) \times \Gamma_2$ terms in the first two performance constraints, for example, make sure that when y is zero, the gain of the simple topology is above the gain specification and the gain constraint of the cascoded topology is met trivially, and vice versa. The superscript "s" on the W and L variables signifies that these variables are scaled. For simplicity, parasitic capacitances are ignored in this example. The constraints are not all used in the form they are shown but are transformed and scaled for reasons of numerical efficiency. We use the sum of the W's and L's to be our objective function in order to maintain convexity[14]. The term $y \times \Gamma_0$ in the objective function assures that the simple topology is chosen when the specifications can be met by both topologies with minimum sizes on both transistors.

The second and third KCL constraints assure that M1 and M2 conduct the same current when the cascoded topology is chosen (y = 1). When y = 0, both constraints are met trivially and the two currents need not be equal. The values of the Γ_i constants for the performance constraints can be chosen by simply making Γ_i equal to the right-hand-side of those constraints. The values of the Γ_i constants in the KCL constraints can be chosen by assessing the maximum possible difference in currents between the two transistors. A similar reasoning is applied to the KVL constraints. The KCL and KVL constraints can be extracted automatically given a set of topology choices. All Γ_i values should be chosen as small in magnitude as possible because this directly affects the performance of the algorithms used. Notè that the binary variable occurs only in linear terms in the expressions above, and thus no additional non-convexities are introduced into the problem (non-convexities affect the performance of algorithms in

Slew (V/μs)	10	10	100
Gain	2000	30	30
y (relaxed)	0.976	0	0.674
Obj fn (relaxed)	8.5	0	0.71
Obj fn (y=0)	infeas	0	25.7
Obj fn (y=1)	15	-	16.9
Topology	Cascoded	Simple	Cascoded

Table 1: Topology Selection of CS amp

terms of reduced rate of convergence, possible infeasible subproblems and local minima [14]). The above formulation yields a MINLP problem.

Three amplifiers were designed with the following specifications: UGF = 5 MHz, Load Capacitance = 1 pF and with different values of gain and slew specifications. The results are shown in Table 1. The third row represents the value of y obtained from the problem in which y was a continuous variable (the "relaxed" problem). The fourth, fifth and sixth rows represent the values of the objective function obtained for the relaxed problem, and for y fixed at 0 and 1, respectively.

For the relaxed problem, the first design in Table 1 gave a value of y very close to 1, the second design yielded a value of zero while the third design yielded an intermediate value of y. The cascoded topology turned out to be optimal for the first design because of its high gain and for the last design because of its high slew rate.

In a later section we will address a general method to determine the best topology choice.

4 CMOS two-stage Amplifier

In this section we consider the design of a more complex circuit: a two-stage CMOS op amp. The first stage can either be simple or cascoded for higher gain at the expense of reduced input common mode range and die area. If the first stage is cascoded, a source follower can be added between the first stage and the second to improve the systematic input offset voltage at the expense of increased power dissipation and die area. Similarly, the output stage can be cascoded at the expense of reduced output voltage swing and die area. Although general trends can be ascribed to each topology, the actual numerical tradeoffs among many different performance requirements must be taken into consideration in order to match the circuit topology and parameters to the required task. Fig. 3 illustrates the circuit topology with all the possible additions.

We also allow the type of each transistor to switch (PMOS becomes NMOS and vice versa) along with the power supplies. We use a simple MOS model[15], and use a binary variable to select between the two sets of device

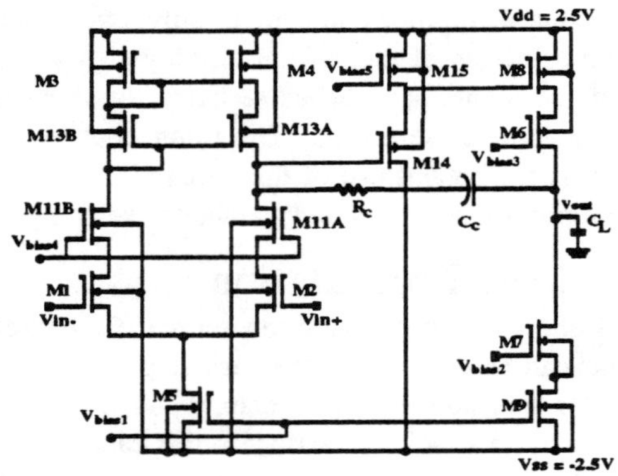

Figure 3: Two-stage CMOS op amp

parameters. Note however, that the same approach can be used with complex device models. Overall, we obtain a suite of twelve device-level circuit topologies for this two-stage CMOS op amp.

We model the topologies by using the following four binary variables :

1. y_{1s} : 1 if first stage simple and 0 if cascoded
2. y_{2s} : 1 if second stage simple and 0 if cascoded
3. y_{sf} : 1 if source follower present and 0 if not
4. y_N : 1 if input transistor N-type and 0 if P-type

For functional correctness, some logical relations between the binary variables are required. For example, we need a relation to assure that a source follower is not used when the first stage is simple i.e. $y_{sf} \Rightarrow \overline{y_{1s}}$. This can be expressed as a simple linear constraint viz. $(1 - y_{sf}) + (1 - y_{1s}) \geq 1$.

The four binary variables above can be taken to be the "primary" variables of our problem. In addition, the performance constraints will require binary variables which signify the topologies of the complete amplifier. For example, if y_{ss} is a binary variable which denotes the topology in which both stages are simple, then the gain constraint corresponding to this topology will be :

$$Gain_{ss}(\mathbf{x}) + (1 - y_{ss}) \times \Gamma_i \geq GAIN$$

Logical operations between the basic variables will yield binary variables denoting the available topologies of the two-stage op amp e.g. $y_{ss} \Leftrightarrow y_{1s} \wedge y_{2s} \wedge \overline{y_{sf}}$. These logical operations can also be expressed as simple linear constraints.

5 Solving the MINLP problem

MINLP problems are difficult to solve and fall into the NP-hard class [16]. The principal solution methods are Branch and Bound[17], Generalized Bender's Decomposition[18] and Outer Approximation[19]. We choose Branch and Bound in this work because it is simple, and for the problem size of our example, performs well enough to bring out the viability of the MINLP formulation. However, it must be remembered, that the other two methods mentioned above are much more sophisticated and may perform significantly better for large problems.

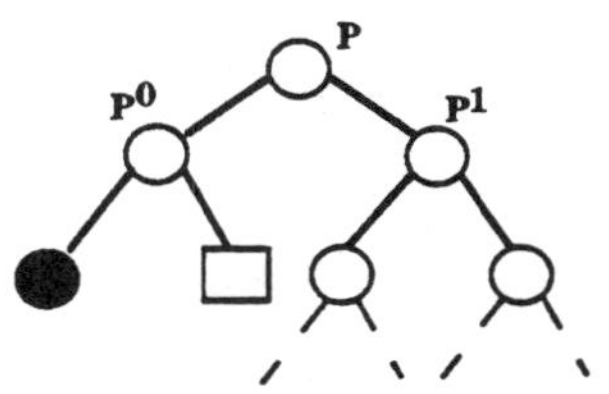

Figure 4: Branch and Bound Tree

5.1 The Branch and Bound Algorithm

We seek the solution x^*, y^* to the problem :

$$P_I : \min_{\mathbf{x},\mathbf{y}} f(\mathbf{x},\mathbf{y})$$

subject to

$$\begin{aligned} g(\mathbf{x},\mathbf{y}) &\leq 0 \\ x_i \quad real \quad & for\ i = 1,\ldots,m \\ y_j \quad &\in \{0,1\}\ for\ j = 1,\ldots,n \end{aligned}$$

We define the problem P from problem P_I by converting the integrality constraints into bound constraints e.g. $0 \leq y_j \leq 1$. This is called a relaxation of the constraint(s), and the corresponding problem is a relaxed problem. If the minimizer x' of P exists and if it is feasible in P_I, then it solves P_I. If not, then there is at least one non-integral y_j. In this case two problems can be defined by branching on one of the variables y_j in problem P, say y_k, fixing it to be 0 in one case (P^0) and 1 in the other (P^1). This process is repeated and a tree structure like the one shown in the Fig. 4 is generated. The □'s are nodes which have integer feasible solutions, and these provide upper bounds on the final solution. The •'s are nodes where the corresponding problem has no feasible point. Assuming that we attain an unique optimum at every node, any node in the tree provides a solution which is a lower bound on the solutions at all nodes below it. Hence, if at any node the solution is greater than the current upper bound, then we need not evaluate any nodes below it. No branching is done at the □ and • nodes also. The current upper bound is updated when a better integer feasible solution is obtained. The method terminates when there are no more nodes to branch on, and the current upper bound provides the final solution x^*. In this way we can find x^* by only a partial evaluation of the tree. The number of nodes evaluated, i.e., the efficiency of the method, is directly dependent on the quality of the lower bounds. In our formulation, the quality of the lower bounds directly depends on how small we can make the values of the Γ_i constants. This is because the larger the values of the Γ_i's, the larger is the feasible region of the relaxed problems.

5.2 Branching Heuristics

The number of nodes evaluated also depends on how quickly we can obtain an upper bound. This in turn depends on how the branching nodes and branching variables are chosen. We used the following heuristics:

- If at any node one of the binary variables turns out to be within a certain specifiable tolerance (say 0.1%) of integer value, then it is assumed to be integer. This can be justified by the fact that there will always be some residual looseness in the formulation.
- The branching node which has the lowest value of lower bound is selected as the next node to be branched on. This is also known as the *Breadth First Strategy* and it leads to minimum number of node evaluations[17].
- The variable which is the farthest from the nearest integer value is selected as the branching variable. This choice is aimed at getting the largest degradation of the objective function when branching is carried out, so that more nodes are "pruned" at an early stage. At present, we decide the selection order once and for all after the solution of the top node.

The Branch and Bound method requires that a unique optimum be reached at each node. This is because of the fact that, only if this condition is met can we guarantee that the solution at a node is a lower bound on all its successors. Preliminary experimentation with fixed topology optimizations has shown that these problems have unique optima given the objective function and class of specifications of our designs. The fact that we are just making linear changes to the original fixed topology optimization problem, and therefore not introducing any additional nonconvexities, suggests that the unique optimum condition will be met.

6 Synthesis Examples

We illustrate the viability of our methodology by some examples. The nonlinear optimizer used was NPSOL [13]. The number of variables was 57, the number of linear constraints was 56 and the number of nonlinear constraints was 48. The load capacitance specification for all the designs was 1 pF.

The results are shown in Figs. 5 to 9. The value of the objective function corresponding to each node and the branching variable and its value corresponding to each edge are shown in the figures. The design in Fig. 6 has a higher slew rate specification than the design in Fig. 5. This required more current in the output stage, and therefore an increase in the value of V_{gs8}. The output stage

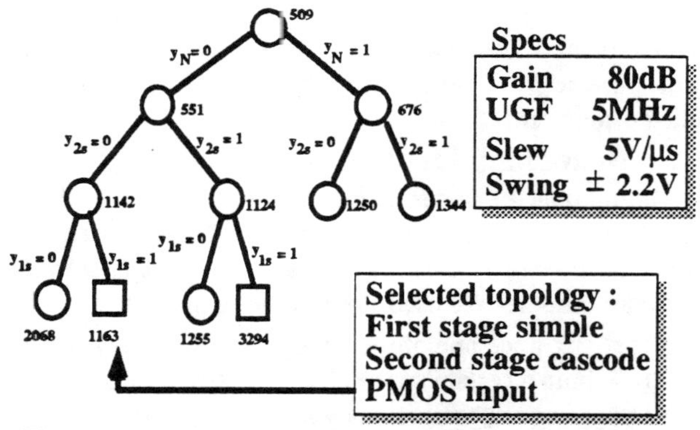

Figure 5: Design 1:- Number of nodes evaluated = 11

could not be cascoded as in the first design, and the first stage had to be cascoded instead in order to meet the gain specification. The design of Fig. 7 has the same specifications as the design of Fig. 5 except for a higher output swing specification. Not only does the output stage have to be simple in this case, but also a source follower has to be placed between the first stage and the second. The fourth design, shown in Fig. 8, has moderate specifications for all performance metrics except the output swing.

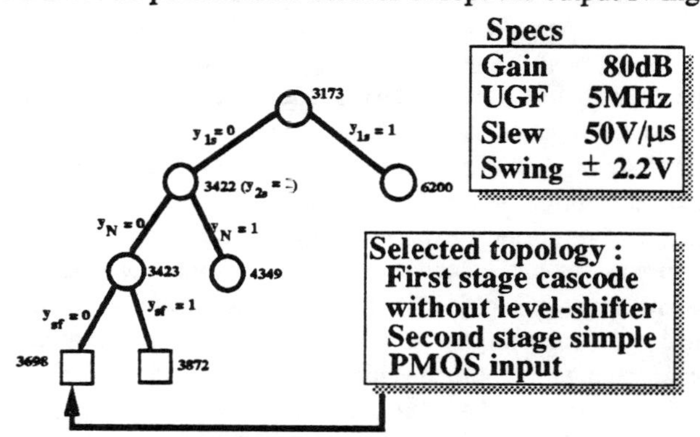

Figure 6: Design 2:- Number of nodes evaluated = 7

The fifth design, shown in Fig. 9, cannot be met by any of the topologies because it demands a very high slew rate and output swing. Note that all the topologies turned out to be PMOS input. This is because M8 turns out to be NMOS in this case, and g_{m8} (higher for NMOS), determines the position of the second pole.

A comparison between the execution times on a DECstation 3100 of our program versus designing all the topologies separately and choosing the best one is shown in Table 2. Optimization runs which could not come up with a feasible point are run multiple times from the previous endpoint for a maximum of 50 seconds. Infeasible points on the branch and bound tree, at nodes other than leaf nodes (nodes at the bottom of the tree), are run for a maximum of 20 secs (less than the 50 seconds because

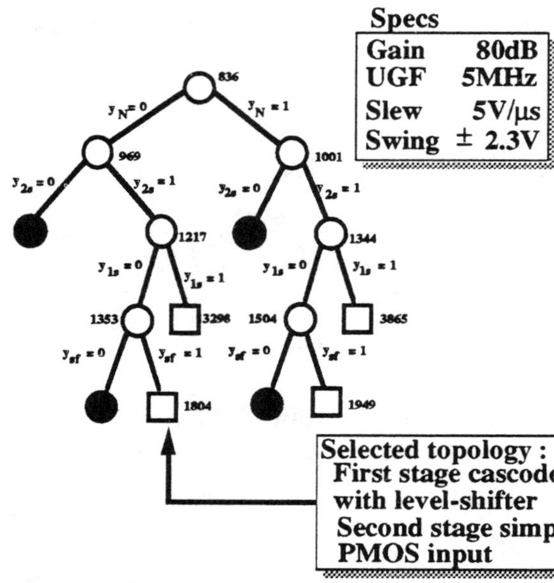

Figure 7: Design 3:- Number of nodes evaluated = 15

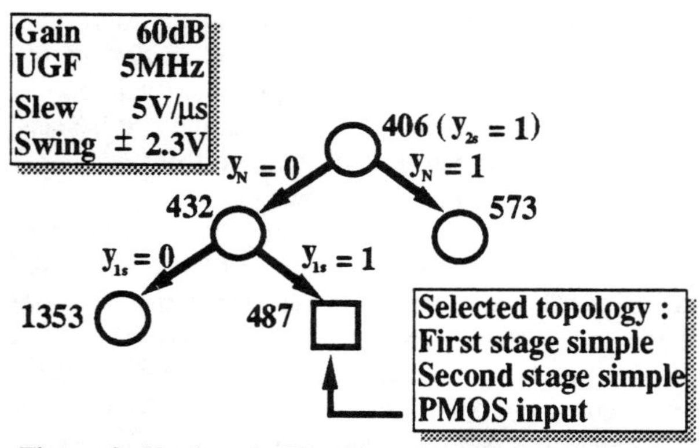

Figure 8: Design 4:- Number of nodes evaluated = 5

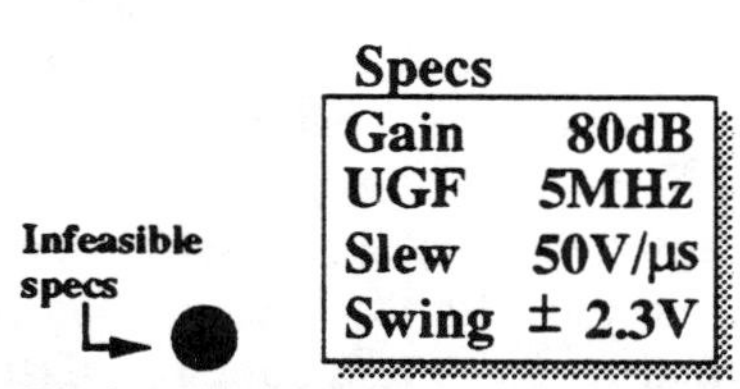

Figure 9: Design 5:- Number of nodes evaluated = 1

Dsn. no.	CPU secs. (BB)	CPU secs. (Exh.)
1	205	216
2	170	424
3	499	440
4	72	432
5	20	600

Table 2: Comparison of CPU times

they are less constrained than each problem in the exhaustive search). The last design brings out the greatest advantage of using this method, in which we have to solve just one optimization problem to discover that the given specifications were not achievable.

It must be remembered that the primary contribution in this work is the MINLP formulation and not the solution method. Branch and bound is a very expensive method for MINLP problems because of the large number of nonlinear programming problems which have to be solved. Branch and bound performed better than exhaustive search in four out of the five designs and much better in two of the designs. Other algorithms [18, 19], may result in much higher speedup. Even in branch and bound there is scope for improvement in terms of better branching variable and node selection strategies. Also, real problems will be much bigger, and then, we believe the advantages over exhaustive search will be more apparent.

Finally, if we desire more accuracy we can take the original design and resynthesize using a more complex device model as in [12].

7 Conclusions

In this paper we have presented a new formulation for analog circuit synthesis that casts it as an MINLP problem. And, using that formulation, we have demonstrated *simultaneous* topology selection and parameter selection of analog circuits for the first time. Our experiments demonstrate a significant overall improvement in design time over exhaustive search. Because adding new design choices exponentially increases the number of possible topologies, we expect that the advantage of the proposed method over exhaustive search will increase dramatically as more realistic and more complex design problems are addressed.

References

[1] I.E.Grossmann. MINLP Optimization Strategies and Algorithms for Process Synthesis. In *Proceedings of FOCAPD Meeting*, pages 105–132, 1989.

[2] Robert K. Brayton et. al. A Survey of Optimization Techniques for Integrated-Circuit Design. *Proceedings of the IEEE*, 69(10):1334–1362, Oct. 1981.

[3] R.A.Rohrer. Fully Automated Network Design by Digital Computer. *Proceedings of the IEEE*, 55(11):1929–1939, Nov. 1967.

[4] S.W.Director and R.A.Rohrer. Automated Network Design: The Frequency Domain Case. *IEEE Transactions on Circuit Theory*, CT-15(2):330–337, June 1968.

[5] William Nye et. al. DELIGHT.SPICE: An optimization-based system for the design of integrated circuits. *IEEE Transactions On Computer-Aided Design*, 7(4):501–519, Apr. 1988.

[6] L.R.Carley and R.A.Rutenbar. How to automate analog IC design. *IEEE Spectrum*, Aug. 1988.

[7] Marc G.R.Degrauwe et. al. IDAC: An Interactive Design Tool for Analog CMOS Circuits. *IEEE Journal of Solid-State Circuits*, sc-22(6):1106–1116, Dec. 1987.

[8] R.Harjani, R.A.Rutenbar and L.R.Carley. OASYS: A Framework for Analog Circuit Synthesis. *IEEE Transactions on Computer-Aided Design*, 8(12):1247–1266, Dec. 1989.

[9] H.Y.Koh, C.H.Sequin and P.R.Gray. OPASYN: A Compiler for CMOS Operational Amplifiers. *IEEE Transactions on Computer-Aided Design*, 9(2):113–125, Feb. 1990.

[10] Koen Swings, Stephane Donnay, and Willy Sansen. HECTOR: A Hierarchical Topology-Construction Program For Analog Circuits Based on a Declarative Approach to Circuit Modeling. In *Proceedings of CICC*. IEEE, 1991.

[11] Zhen-Qiu Ning, Ton Mouthaan, and Hans Wallinga. SEAS: A Simulated Evolution Approach for Analog Circuit Synthesis. In *Proceedings of CICC*. IEEE, 1991.

[12] Prabir C. Maulik and L. Richard Carley. Automating Analog Circuit Design using Constrained Optimization Techniques. In *Proceedings of ICCAD*. IEEE, Nov. 1991.

[13] Philip Gill, Walter Murray, Michael Saunders, and Margaret Wright. User's guide for NPSOL (version 4.0). Technical Report SOL 86-2, Stanford University, Jan. 1986.

[14] Roger Fletcher. *Practical Methods of Optimization*. John Wiley and Sons.

[15] H.Shichman and D.A.Hodges. Modeling and Simulation of Insulated-Gate Field-Effect Transistors. *IEEE Journal of Solid-State Circuits*, SC-3(5):285–289, Sept. 1968.

[16] George L. Nemhauser and Laurence A. Wolsey. *Integer and Combinatorial Optimization*. Wiley Interscience Publications. John Wiley and Sons, Inc., New York.

[17] O.K.Gupta. *Branch and Bound Experiments in Nonlinear Integer Programming*. PhD thesis, Purdue University, Oct. 1980.

[18] A.M.Geoffrion. Generalized Benders Decomposition. *Journal of Optimization Theory and Applications*, 10(4):237–260, 1972.

[19] Marco A. Duran and Ignacio E. Grossmann. An Outer-Approximation Algorithm for a Class of Mixed-Integer Nonlinear Programs. *Mathematical Programming*, 36:307–339, 1986.

An Integrated Approach to Realistic Worst-Case Design Optimization of MOS Analog Circuits

*A. Dharchoudhury and S. M. Kang**
Department of Electrical and Computer Engineering,
Beckman Institute and Coordinated Science Laboratory
University of Illinois at Urbana–Champaign
1101 W. Springfield Ave, Urbana, IL 61801.

Abstract

In this paper we present a new integrated approach for the optimization of MOS analog circuit performances using realistic worst-case device parameter files, each corresponding to a performance measure. Nonlinear response surfaces are constructed for the performance measures of interest, and the worst-case device parameter files are identified by solving a set of suitably cast nonlinear programming problems. The worst-case files are shown to depend on the values of the designable parameters. An efficient method of incorporating this dependence during worst-case design optimization has been developed. This method enables the design of circuits with optimal performances and high parametric yields. Some illustrative analog circuit examples are given to demonstrate the application of the worst-case design optimization procedure.

1 Introduction

The performances of MOS Integrated Circuits (IC), such as voltage gain and speed, are functions of two types of parameters [1]: (a) *designable* parameters, such as transistor channel widths and lengths, which can be set to a desired nominal value by the designer and (b) *noise* parameters which are randomly varying due to manufacturing process fluctuations. For MOS circuits, noise parameters include device model parameters like gate-oxide thickness and threshold voltage.

Worst-case noise parameter files were previously determined using the "corners" technique, but the results have been found to be overly pessimistic. A shortcoming of the "limit parameters" approach proposed by Müller [2] is that it hinges on the assumption that the variation of the performances with the noise parameters is linear and monotonic. This is not always true. Moreover, sensitivities calculated at the nominal noise parameter vector may not be valid at other values of the noise parameters. In our approach, we construct nonlinear response surfaces which accurately model the dependence of the performance on the noise parameters. The worst-case noise files are then obtained as solutions to suitably formulated nonlinear programming problems. This method has been shown to be accurate as well as computationally efficient [3].

An important assumption in statistical circuit design has been that worst-case analysis using noise parameters and design optimization using designable parameters can be done independently. Experimental evidence corroborates to the invalidity of this assumption. In our approach, the interaction between designable and noise parameters during worst-case design optimization is accounted for by a simple yet accurate response surface oriented method.

2 Worst-Case Analysis

Let $\underline{\mathbf{r}}$ denote the n_r-vector of the circuit performance measures of interest. Let the n_d designable parameters be denoted by $\underline{\mathbf{d}}$ and the n_u noise parameters be denoted by $\underline{\mathbf{u}}$. We will assume that $\underline{\mathbf{u}}$ is multivariate Gaussian with mean vector $\underline{\mu}$ and covariance matrix $\mathbf{Q}$. The correlated noise parameters $\underline{\mathbf{u}}$ can be transformed (by diagonalizing the positive definite covariance matrix $\mathbf{Q}$) to a set of *independent zero mean* Gaussian random variables, $\underline{\mathbf{v}}$. This transformation, denoted by $\mathbf{T}$, is linear and invertible. Let

*This research was supported in part by the Joint Services Electronics Program (N00014-90-J-1270), DCCA Illinois Technology Challenge Grant, the Semiconductor Research Corporation (SRC-90-DP-109), and an unrestricted grant from Texas Instruments, Inc.

Ω be the (diagonal) covariance matrix of $\underline{\mathbf{v}}$. The joint probability density function of the transformed noise parameters is given by the following formula :

$$f(\underline{\mathbf{v}}) = C * exp(-\frac{1}{2}\underline{\mathbf{v}}^T \Omega^{-1} \underline{\mathbf{v}}) \qquad (1)$$

where C is a constant term.

At a fixed design point $\underline{\mathbf{d}}$, the circuit simulator implements the following transformation of random variables:

$$r_i = h_i(\underline{\mathbf{v}}), 1 \leq i \leq n_r \qquad (2)$$

where the function $h_i(.)$ is known only implicitly via the circuit equations. The function $h_i(.)$ is approximated by constructing a *response surface* [4] for the performance measure r_i in terms of $v_1, \ldots, v_{n_u}$. The Latin Hypercube Sampling (LHS) [5] strategy is used to select the values of the transformed noise parameters $\underline{\mathbf{v}}$ where the circuit is simulated. The values of r_i are automatically extracted and the response surface $\hat{h}_i$ is constructed using least squares. Such *experimental design* techniques provide an accurate response surface at low computational cost. The mean μ_{r_i} and the standard deviation σ_{r_i} of the performance measure r_i can be easily estimated using the response surface. In general, r_i is non-Gaussian, but it can be well approximated by a Gaussian random variable. Let us define the worst-case direction w_i for the performance r_i as follows: w_i is equal to +1 if increasing the value of r_i makes it worse, and is equal to -1 if decreasing the value makes it worse. Then the worst-case value of r_i, r_i^{wc} can be estimated using

$$r_i^{wc} = \mu_{r_i} + w_i k \sigma_{r_i} \qquad (3)$$

where the constant k is typically equal to 3.

Having estimated r_i^{wc} as above, the worst-case vector of independent transformed noise parameters, $\underline{\mathbf{v}}_i^{wc}$ is the solution to the equation:

$$r_i^{wc} = \hat{h}_i(\underline{\mathbf{v}}) \qquad (4)$$

As in [2], the worst-case vector is chosen to be the one that is "most probable" from among the vectors that satisfy Eq. (4). Thus, $\underline{\mathbf{v}}_i^{wc}$ is the solution to the following equality constrained nonlinear programming problem (NLP)

NP1 :

$$\max f(\underline{\mathbf{v}})$$
$$s.t. \quad r_i^{wc} = \hat{h}_i(\underline{\mathbf{v}}) \qquad (5)$$

From Eq. (1), it can be inferred that the worst-case vector $\underline{\mathbf{v}}_i^{wc}$ is the vector that lies on the hypersurface (4) and has the smallest "probabilistic distance" $\underline{\mathbf{v}}^T \Omega^{-1} \underline{\mathbf{v}}$ from the origin. Therefore, the algorithm for worst-case analysis at a particular design point can be summarized as follows.

procedure $wc(\underline{\mathbf{d}})$
  for each r_i
    begin
      S1. Obtain $\hat{h}_i(\underline{\mathbf{v}})$
      S2. Compute r_i^{wc}
      S3. Solve NP1 for $\underline{\mathbf{v}}_i^{wc}$
      S4. $\underline{\mathbf{u}}_i^{wc} = \mathbf{T}^{-1}\underline{\mathbf{v}}_i^{wc}$
    end

3 Worst-Case Design Optimization

In this section, we describe the procedure of optimizing the circuit performances using the worst-case files derived above. Conventionally, worst-case noise parameter files have been assumed to be independent of the values of the designable parameters. When the noise parameters are at their mean values, the circuit performances are said to be nominal. Since the nominal performances change with the designable parameters, we expect that r_i^{wc} should also be a function of $\underline{\mathbf{d}}$. This in turn implies that for each performance r_i, $\underline{\mathbf{u}}_i^{wc}$ will vary during the optimization. If the procedure wc outlined in the previous section were to be executed at each design point, then the cost of simulating the circuit to construct the response surfaces would quickly become prohibitive.

A computationally efficient way to account for the dependence of r_i^{wc} on $\underline{\mathbf{d}}$ is to construct response surfaces for the performance measures in terms of both the designable and the noise parameters. Let these response surfaces be denoted by $\hat{r}_i(\underline{\mathbf{d}}, \underline{\mathbf{v}})$, $1 \leq i \leq n_r$. At a particular design point $\underline{\mathbf{d}}_k$, $\hat{r}_i(\underline{\mathbf{d}}_k)$ is some function of the transformed noise parameters $\underline{\mathbf{v}}$, say $\hat{t}_i(\underline{\mathbf{v}})$. The mean and the standard deviation of r_i at $\underline{\mathbf{d}}_k$ can be estimated using $\hat{t}_i$. Then for each i, $r_i^{wc}(\underline{\mathbf{d}}_k)$ is given by

$$r_i^{wc}(\underline{\mathbf{d}}_k) = \mu_{r_i}(\underline{\mathbf{d}}_k) + k w_i \sigma_{r_i}(\underline{\mathbf{d}}_k) \qquad (6)$$

$\underline{\mathbf{u}}_i^{wc}(\underline{\mathbf{d}}_k)$ can then be obtained as the solution of NP1 (with h_i replaced by $\hat{t}_i$). This augmented response surface method ensures that the worst-case noise parameter files used at each step in the optimization process are accurate.

Next, we formulate the objective function for the worst-case design optimization problem. In general, a circuit optimization problem involves multiple performances (which are possibly competing against each other) subject to multiple hard and soft constraints on other performance measures. Let $r_1, r_2, \ldots, r_{n_r}$ denote the n_r performance measures in the optimization. The following simple scheme converts the actual performance values r_i into *performance scores* s_i. A "good" value g_i (with a score of 100) and a "bad" value b_i (with a score of 0) is associated with each performance measure. Therefore, for $i = 1, \ldots, n_r$

$$s_i = 100 * (\frac{r_i - b_i}{g_i - b_i}) \tag{7}$$

At each design point $\underline{\mathbf{d}}_k$, define the *circuit score* s as

$$s = \min[s_1, s_2, \ldots, s_{n_r}] \tag{8}$$

This circuit score is maximized during the worst-case design optimization. Notice that this scheme lumps all the performance objectives and constraints into a single circuit score and is similar to the minimax formulation of the multiple-criteria optimization problem. Constraints are handled in this scheme by assigning scores that are high for feasible values but which degrade very quickly with small violations, thus keeping the optimization away from non-feasible regions. The following is the algorithm for worst-case design optimization.

procedure *wcopt*
  begin
    for each performance r_i
      Construct $\hat{r}_i(\underline{\mathbf{d}}, \underline{\mathbf{v}})$
    $k = 1$, $\underline{\mathbf{d}}_k = \underline{\mathbf{d}}_0$
    while (all specs not met) **do**
      begin
        for each performance r_i
          begin
            Obtain $u_i^{wc}(\underline{\mathbf{d}}_k)$
            Simulate circuit at $\underline{\mathbf{d}}_k$, $u_i^{wc}(\underline{\mathbf{d}}_k)$
            Compute $s_i(\underline{\mathbf{d}}_k)$
          end
        Compute $s(vectd_k)$
        Find $\underline{\mathbf{d}}_{k+1}$
        $k = k + 1$, $\underline{\mathbf{d}}_k = \underline{\mathbf{d}}_{k+1}$
      end
  end

In the algorithm presented above, the value $r_i^{wc}(\underline{\mathbf{d}}_k)$ predicted by the response surface can also be used to compute $s_i(\underline{\mathbf{d}}_k)$. However, we prefer to simulate the circuit at the n_r worst-case files, and add the extracted results to the database for continuously updating the response surface. These adaptively fit response surfaces are used to obtain reliable estimates of the worst-case noise vectors. The user is however provided with the option of using the response surfaces to compute the scores.

Two optimization algorithms - the Nelder-Mead Simplex algorithm and the Sequential Quadratic Programming (SQP) algorithm (as implemented in the NPSOL package [6]) are used to compute the next iterate $\underline{\mathbf{d}}_{k+1}$ based on the circuit score. The Nelder-Mead Simplex algorithm is very simplistic and was chosen because it does not require gradients and is robust to the occasional non-convergence problems of the circuit simulator SPICE. However, it has been observed that this method might cause the designable parameters to change such that a sudden large change occurs in the measures of the performances. In such a case, the response surface loses its accuracy. The optimization process will then have to be restarted - such restarts are expensive in terms of circuit simulations. The adaptive fitting strategy used to update the response surfaces was found to be less useful in conjunction with the simplex method due to these sudden movements. Moreover, as the optimization proceeds, the designable parameters move away from the initial region in which the response surfaces are fit (the *sampling box*). This also causes the response surface to lose accuracy. The expensive restarts can be avoided by using a gradient-based algorithm such as SQP. On the other hand, a larger number of SPICE runs would conceivably be required to compute the gradients using finite difference schemes. To reduce this computational cost, we use the current response surfaces to provide the predicted scores for gradient calculation, and periodically use simulations to update the response surfaces. The current sampling box is also specified as bounds on the designable parameters. Unlike the simplex method, adaptive fitting of the response surfaces used with NPSOL provided much more accurate estimates of the worst-case noise vectors at the new design points. The results of these two optimization techniques are compared in Section 4.

4 Applications to MOS Analog Circuits

4.1 Worst-Case Analysis of a Biquad

In this section, we demonstate the application of the worst-case analysis procedure $wc(\underline{\mathbf{d}})$ to the circuit of Figure 1 [7]. The circuit consists of four identical

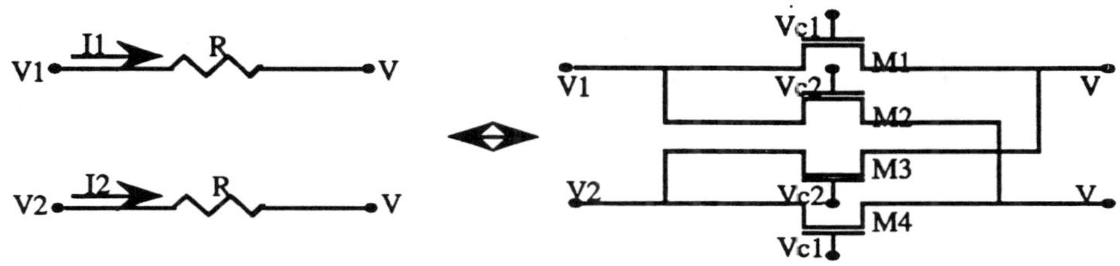

Figure 1: Four MOSFETs and their equivalent resistor pair

	G^{wc}	$\mathbf{u}^{wc}$	prob. dist.
LRS(30-run)	11.52	[373.18, 744.2, 0.026, -0.074]	3.0
QRS(30-run)	11.3	[380.97, 728.1, 0.006, -0.042]	1.97
MC(200-run)	11.4	[386.06, 722.1, 0.045, -0.044]	1.71

Table 1: Worst-case Analysis for Biquad

nMOS transistors simulating a pair of identical resistors R. The noise parameters screened for the worst-case analysis are (i) gate-oxide thickness(A^o), (ii) electron mobility($cm^2/V.s$), (iii) width reduction (μm), and (iv) length reduction (μm). These noise parameters are assumed to be statistically independent. In this example, the performance measure of interest is the conductance $G = 1/R$ (10^{-6}mho). Two response surfaces were used: a linear response surface (LRS) and a quadratic response surface (QRS), both fitted using 30 runs of the circuit simulator. These results were confirmed by simulating the circuit using a 300-run Monte Carlo (MC) sample of the noise parameters, and calculating the worst-case performance. The worst-case noise vector is chosen as the vector whose corresponding performance is closest to the worst-case performance. These results are shown in Table 1.

As seen from Table 1, the results obtained from the QRS are more accurate. Moreover, the QRS gives a worst-case vector with a much smaller probabilistic distance (and hence a larger probability density) than the LRS. This example demonstrates that the nonlinear response surface approach coupled with the NLP formulation for the worst-case analysis problem yields realistic results.

4.2 Worst-Case Design Optimization of CMOS opamp

The circuit of Figure 2 is used to illustrate the worst-case design optimization methodology. The performance measures of interest are the unity-gain frequency f_t (MHz), the low-frequency gain A_v (dB) and the power-dissipation P_d (mW). Five designable parameters are chosen. They are $PW1(\mu m)$, $PW3(\mu m)$, $PW5(\mu m)$, $PW6(\mu m)$ and $PCAP(pF)$. The correlated noise parameters in this example are (i) pMOS and nMOS gate-oxide thickness(A^o), (ii) nMOS threshold voltage(V), (iii) pMOS threshold voltage (V), (iv) nMOS length reduction (μm), (v) pMOS length reduction (μm), (vi) nMOS width reduction (μm), and (vii) pMOS width reduction (μm).

Table 2 shows the results of the worst-case analysis for f_t at two design points, $\underline{\mathbf{d}}_1$ and $\underline{\mathbf{d}}_2$. This illustrates that the worst-case noise parameter vectors do indeed depend on the designable parameters and it is important to use accurate estimates of these worst-case files during the optimization when the design point changes.

We consider the following optimization problem:

$$\begin{aligned} &\max f_t \\ \text{subject to } &A_v \geq 60dB \\ &P_d \leq 0.4mW \end{aligned}$$

Table 3 shows the progress of the worst-case design optimization using the NPSOL program. The limits of the current sampling box are passed as parameter bounds to NPSOL. The response surfaces for the three performance measures are updated as and when the circuit is simulated during the optimization. Each row in the table corresponds to a single pass of the procedure *wcopt* described in Section 3. When the designable parameters reach the edges of the sampling box, and further improvement in the circuit score is possible, a new pass is started. This involves fitting response surfaces in a new sampling box defined around the current point. The number of circuit simulations shown for each pass is the sum of the SPICE trials incurred in fitting the response surfaces and that incurred during the optimization. The adaptively fit response surfaces and the predicted worst-case noise

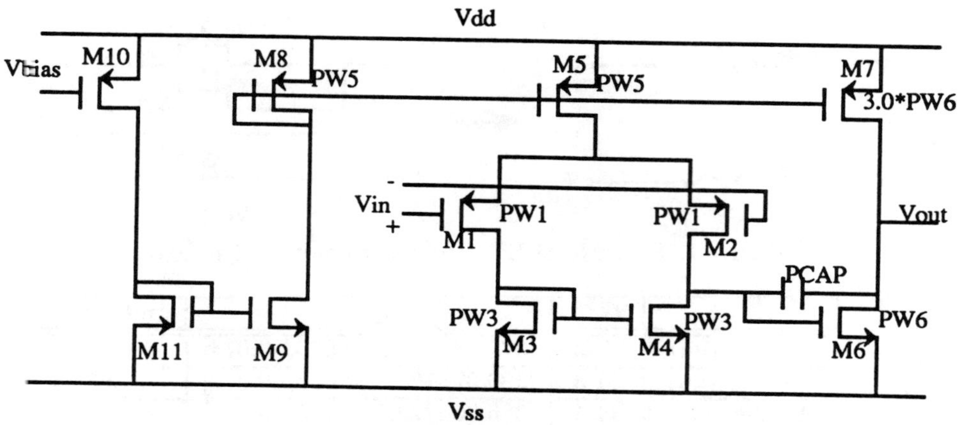

Figure 2: CMOS opamp

	f_t^{wc}	$\underline{u}^{wc}$	prob. dist
$\underline{d}_1$	0.421	[418.6, 0.86, -0.82, -0.14, -0.14, 0.04, 0.05]	2.96
$\underline{d}_2$	1.73	[388.4, 0.77, -0.8, -0.09, -0.03, -0.01, -0.02]	1.73

Table 2: Variation of worst-case files with $\underline{d}$

vectors are seen to be accurate during the course of the optimization. The final design point shown in the table represents a much improved circuit whose worst-case performances satisfy the specifications.

For comparison, the Simplex algorithm was applied to solve the same worst-case design optimization problem. As shown in Table 4, the number of passes required are more, necessitated by the loss in accuracy of the response surfaces as the circuit performances changed suddenly. The more frequent restarts of the optimization process results in a much larger number of circuit simulations in this case. However, the simplex method is much more robust to SPICE nonconvergence problems and can be used with advantage to explore the design space.

5 Summary and Conclusions

In this paper, an integrated approach to the worst-case design optimization of MOS analog integrated circuits is presented. At a particular point in the designable parameter space, the worst-case noise parameter vector is obtained by a combination of response surface methodology and nonlinear programming. This method has been shown to be efficient and accurate in yielding realistic worst-case files. Worst-case design optimization is performed by changing the designable parameters to optimize the worst performances of the circuit. The necessity and existence of different worst-case files for different performance measures and the variation of these worst-case files with the designable parameters have been taken into consideration. The problem of multiple criteria circuit design optimization is formulated as that of maximizing a circuit score. The accuracy and efficiency of the worst-case design optimization procedure developed in this paper have been demonstrated through its application to some analog circuit examples. The methodology described in this paper has been implemented in the iEDISON2 program.

References

[1] T. K. Yu, S. M. Kang, I. N. Hajj, and T. N. Trick, "Statistical Modeling of VLSI Circuit Performances," *IEEE Int. Conf. Computer-Aided Design*, Nov. 1986. Santa Clara, CA.

[2] G. E. Müller-L., "Limit-Parameters: The General Solution of the Worst-Case Problem for the Linearized Case," *Proc. IEEE ISCAS*, 1990.

[3] A. Dharchoudhury and S. M. Kang, "Identification of Correlated Device Model Parameter Values for Worst-Case Circuit Performance Analysis," *VLSI Design '92*, Jan. 1992. Bangalore, India.

[4] G. E. P. Box and N. R. Draper, *Emprical Model Building and Response Surfaces.* New York: Wiley, 1987.

Pass	SPICE runs	$\underline{d}$	f_t^{wc}	A_v^{wc}	P_d^{wc}
0	0	[15, 10, 10, 10, 5]	0.42	27.93	0.37
1	160	[22.5, 5, 15, 5.56, 2.5]	5.28	66.58	0.38
2	88	[23.6, 5.25, 15.75, 5.7, 2.4]	19.77	66.97	0.38

Table 3: Design Optimization using NPSOL

Pass	SPICE runs	$\underline{d}$	f_t^{wc}	A_v^{wc}	P_d^{wc}
0	0	[15, 10, 10, 10, 5]	0.42	27.93	0.37
1	91	[19.87, 10.21, 12.07, 11.42, 1.3]	1.91	30.01	0.39
2	85	[26.5, 8.17, 14.5, 12.6, 0.34]	11.9	35.0	0.42
3	88	[29.7, 7.11, 17.75, 9.27, 0.42]	15.38	41.03	0.42
4	79	[30.4, 7.35, 18.9, 7.31, 0.42]	17.95	40.08	0.37

Table 4: Design Optimization using Simplex Method

[5] M. D. McKay, R. J. Beckman, and W. J. Conover, "A Comparison of Three Methods for Selecting Values of Input Variables in the Analysis of Output from a Computer Code," *Technometrics*, vol. 21, pp. 239–245, May 1979.

[6] P. E. Gill, W. Murray, M. A. Saunders, and M. H. Wright, *User's Guide for NPSOL, version 4.0.* Department of Operations Research, Stanford, California, 1986. Report SOL86-2.

[7] Z. Czarnul, "Modification of the Banu-Tsividis Continuous-time Integrator Structure," *IEEE Trans. on Circ. and Systems*, vol. CAS-33, pp. 714–716, July 1986.

Multipole-Accelerated 3-D Capacitance Extraction Algorithms for Structures with Conformal Dielectrics

K. Nabors J. White

Massachusetts Institute of Technology
Dept. of Electrical Engineering and Computer Science
Cambridge, MA 02139 USA

Abstract

The new three-dimensional capacitance calculation program FASTCAP2 is described. Like the earlier program, FASTCAP, FASTCAP2 is based on a multipole-accelerated algorithm that is efficient enough to allow three-dimensional capacitance calculations to be part of an iterative design process. FASTCAP2 differs from FASTCAP in that it is able to analyze problems with multiple dielectrics, thus extending the applicability of the multipole-accelerated approach to a wider class of integrated circuit interconnect and packaging problems.

1 Introduction

The self and coupling capacitances associated with integrated circuit interconnect and packaging are becoming increasingly important in determining final circuit performance and reliability. However, accurate estimation of these capacitances involves analyzing innately three-dimensional structures with dielectric materials surrounding conductors in a complicated fashion. Integrated circuits, for example, have multiple layers of polysilicon or metal conductors, separated by conformal or space-filling dielectrics. Also, packaging and off-chip interconnection problems often involve connectors passing through several plastic or ceramic dielectrics. The recent development of multipole-accelerated boundary-element methods for three-dimensional capacitance extraction has made accurate analysis of very complex structures in a uniform dielectric computationally inexpensive [5]. This paper extends the original method to problems where the conductors are surrounded by multiple dielectric regions of arbitrary shape, thus allowing the analysis of more realistic integrated circuit interconnect and packaging problems.

The following section describes the equivalent-charge approach to analyzing structures with dielectric interfaces. Section 3 then shows how this formulation can be used as a framework for a multipole-accelerated iterative solution method. The method's utility is demonstrated in Section 4 by applying our implementation of the algorithm, FASTCAP2, to several examples.

This work was supported by the Defense Advanced Research Projects Agency contract N00014-91-J-1698, the National Science Foundation contract (MIP-8858764 A02), and grants from I.B.M. and Digital Equipment Corporation.

2 Equivalent-Charge Formulation

To determine all the self and coupling capacitances of a structure with m conductors, the conductor surface charges must be computed m times, with m different sets of conductor potentials. In particular, if conductor i is raised to unit potential and the rest are grounded, then the total charge on conductor i is numerically equal to conductor i's self capacitance. Furthermore, any other conductor's total charge is numerically equal to the negative of its coupling capacitance to conductor i.

Given the conductor potentials, the conductor surface charges can be computed by replacing the conductor-dielectric and dielectric-dielectric interfaces with a surface charge layer of density $\sigma(x)$ and changing all the dielectric regions to free space. Then in this equivalent free-space problem, $\sigma(x)$ is determined by insisting that it produce a potential which matches conductor potentials at conductor-dielectric interfaces, and satisfies normal electric-field conditions at the dielectric-dielectric interfaces [2, 6].

To numerically compute σ, the conductor surfaces

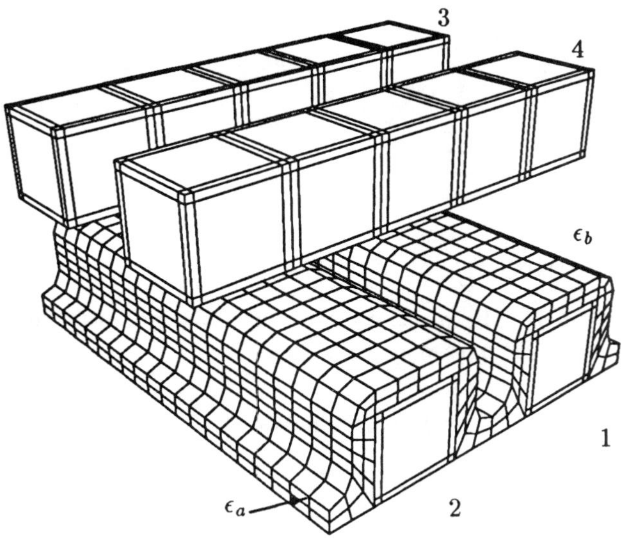

Figure 1: The panels used to discretize a 2×2 dielectric-coated bus-crossing problem. The lower conductors' surfaces are discretized in the same way as the two upper conductors' surfaces.

and dielectric interfaces are discretized into $n = n_p + n_d$ small panels or tiles, with n_p panels on conductor surfaces and n_d panels on dielectric interfaces as in Figure 1. It is then assumed that on each panel i, a charge, q_i, is uniformly distributed. For each conductor surface panel, an equation is written which relates the potential at the center of that i-th panel, denoted p_i, to the sum of the contributions to that potential from the n charge distributions on all n panels. For example, the contribution of the charge on panel j to the potential at the center of panel i is given by the superposition integral

$$\frac{q_j}{a_j} \int_{panel\ j} \frac{1}{4\pi\epsilon_0 \|x_i - x'\|} da', \tag{1}$$

where x_i is the center of panel i, a_j is the area of panel j, ϵ_0 is the permittivity of free space, and the constant charge density q_j/a_j has been factored out of the integral. The total potential at x_i, is the sum of the contributions from all n panels,

$$p_i(x_i) = P_{i1}q_1 + P_{i2}q_2 + \cdots + P_{ij}q_j + \cdots + P_{in}q_n, \tag{2}$$

where

$$P_{ij} \triangleq \frac{1}{a_j} \int_{panel\ j} \frac{1}{4\pi\epsilon_0 \|x_i - x'\|} da'. \tag{3}$$

Similarly, for each dielectric interface panel, an equation is written that relates the normal displacement-field difference at the center of that i-th dielectric interface panel to the sum of the contributions to that displacement field due to the n charge distributions on all n panels. In particular, if panel i lies on the interface between dielectrics with permittivities ϵ_a and ϵ_b, then the Gauss's Law condition

$$\epsilon_a \frac{\partial p_i(x_i)}{\partial n_i} - \epsilon_b \frac{\partial p_i(x_i)}{\partial n_i} = 0 \tag{4}$$

must hold. Here n_i is a normal to panel i. Substituting the expression (2) breaks (4) into a superposition over all the panels,

$$D_{i1}q_1 + D_{i2}q_2 + \cdots + D_{ij}q_j + \cdots + D_{in}q_n = 0, \tag{5}$$

where

$$D_{ij} \triangleq (\epsilon_a - \epsilon_b) \frac{\partial}{\partial n_i} \frac{1}{a_j} \int_{panel\ j} \frac{1}{4\pi\epsilon_0 \|x_i - x'\|} da'. \tag{6}$$

Careful evaluation of the derivative in (6) leads to the important special case [3]

$$D_{ii} \triangleq \frac{(\epsilon_a + \epsilon_b)}{2\epsilon_0}. \tag{7}$$

Collecting all n equations of the form (2) and (5) leads to the dense linear system

$$\begin{bmatrix} P_{11} & \cdots & P_{1n} \\ \vdots & & \vdots \\ P_{n_p 1} & \cdots & P_{n_p n} \\ D_{n_p+1,\,1} & \cdots & D_{n_p+1,\,n} \\ \vdots & & \vdots \\ D_{n1} & \cdots & D_{nn} \end{bmatrix} \begin{bmatrix} q_1 \\ \vdots \\ q_{n_p} \\ q_{n_p+1} \\ \vdots \\ q_n \end{bmatrix} = \begin{bmatrix} p_1 \\ \vdots \\ p_{n_p} \\ 0 \\ \vdots \\ 0 \end{bmatrix}, \tag{8}$$

or

$$\begin{bmatrix} P \\ D \end{bmatrix} \begin{bmatrix} q \end{bmatrix} = \begin{bmatrix} p \\ 0 \end{bmatrix}, \tag{9}$$

where $P \in \mathbf{R}^{n_p \times n}$ is the matrix of potential coefficients, $D \in \mathbf{R}^{n_d \times n}$ is the matrix representing the dielectric interface boundary conditions, $q \in \mathbf{R}^n$ is the vector of panel charges, and $p \in \mathbf{R}^{n_p}$ is the vector of conductor-panel center-point potentials. Using

$$A \triangleq \begin{bmatrix} P \\ D \end{bmatrix}, \quad b \triangleq \begin{bmatrix} p \\ 0 \end{bmatrix}, \tag{10}$$

gives

$$Aq = b \tag{11}$$

as the linear system to solve to for the conductor charge densities. In the standard approach the $n \times n$ linear system (11) is solved using a Gaussian elimination algorithm at a cost of order n^3 operations [8, 6]. Our algorithm uses the multipole-accelerated iterative method of the next section which requires only order mn operations.

3 The Multipole Approach

The dense linear system of (11) can be solved to compute panel charges from a given set of panel potentials, and the capacitances can be derived by summing the panel charges. If Gaussian elimination is used to solve (11), the number of operations is order n^3. Clearly, this approach becomes computationally intractable if the number of panels exceeds several hundred. Instead, consider solving the linear system (11) using a conjugate-residual style iterative method like GMRES [9]. Such methods have the form given below:

Algorithm 1: GMRES algorithm for solving (11)
Make an initial guess to the solution, q^0.
Set $k = 0$.
do {
  Compute the residual, $r^k = b - Aq^k$.
  if $||r|| < tol$, return q^k as the solution.
  else {
    Choose α's and β in
      $q^{k-1} = \sum_{j=0}^{k} \alpha_j q^j + \beta r^k$
    to minimize $||r^{k+1}||$.
    Set $k = k + 1$.
  }
}

Here $||\cdot||$ is the Euclidean norm. The dominant costs of this strategy are calculating the n^2 entries of A using (3) and (6) before the iterations begin, and then performing n^2 operations to compute the matrix-vector product Aq^k on each iteration. It is possible to avoid forming most of A and to substantially reduce the cost of computing Aq^k if an approximation to Aq^k can be tolerated [7].

The approximation used here is the multipole algorithm, an efficient way of calculating the potential due to a charge distribution in free space that reduces the cost of forming Aq^k to order n operations [1]. This does not necessarily imply that each iteration of the GMRES algorithm can be computed with order n operations. If the number of GMRES iterations required to achieve convergence approaches n, then the minimization in each GMRES iteration requires order n^2 operations. This problem is avoided through the use of a preconditioner which reduces the number of GMRES iterations required to achieve convergence to well below n for large problems [4].

The product Aq^k is, using (9),

$$Aq^k = \begin{bmatrix} Pq^k \\ Dq^k \end{bmatrix}. \qquad (12)$$

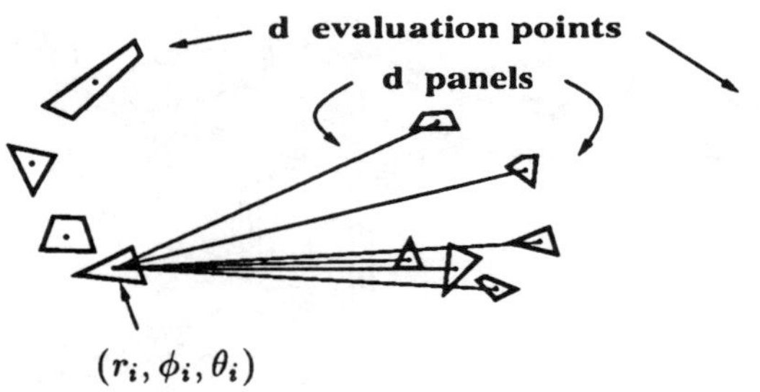

Figure 2: The direct evaluation of the potential due to d panel charges at d points.

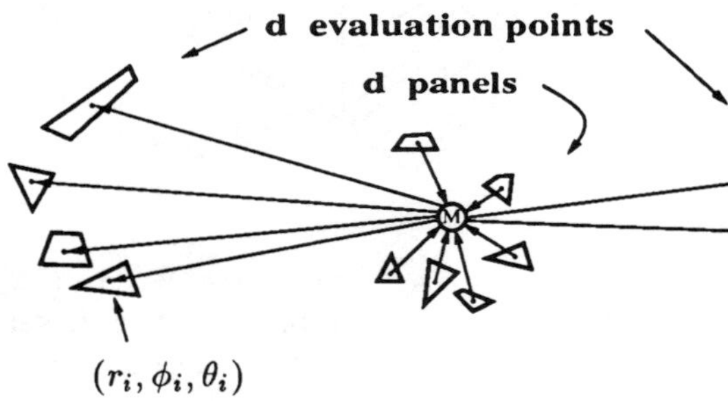

Figure 3: The evaluation of a the potential due to d panel charges at d points using a multipole expansion.

Forming the product Pq^k is equivalent to calculating the potential at all the conductor panel center points, making it possible to approximate it directly with the fast multipole algorithm. Although Dq^k has the dimensions of electric displacement, it can be expressed in terms of potential evaluations, leading to fast evaluation with the multipole algorithm.

3.1 Potential Evaluations

The key approximation employed in the multipole algorithm is the evaluation of potentials using multipole expansions. Consider evaluating the potential at d panel centers due to charges on another d panels as in Figure 2. In a direct evaluation, a single panel's center-point potential is calculated using an explicit equation like (2). In the Figure 2 case, d $P_{ij}q_j$ products are computed and added at a cost of d operations. Repeating the process for all d evaluation points requires d^2 operations.

When the charge panels are well-separated from the evaluation points the potentials can be approximated at much lower cost using a multipole expansion. A multipole expansion is a truncated series expansion of the form

$$p_i(r_i,\phi_i,\theta_i) \approx \sum_{j=0}^{l}\sum_{k=-j}^{j} \frac{M_j^k}{r_i^{j+1}} Y_j^k(\phi_i,\theta_i). \quad (13)$$

where (r_i, ϕ_i, θ_i) are the spherical coordinates of panel i's center point (the evaluation point) measured relative to the origin of the multipole expansion, $Y_k^j(\phi_i, \theta_i)$ are the surface spherical harmonics, l is the expansion order, and M_k^j are the multipole coefficients which are determined from the well-separated panel charges using

$$M_j^k \triangleq \sum_{i=1}^{d} \frac{1}{a_i} \int_{panel\ i} (\rho')^j Y_j^{-k}(\alpha', \beta') da'. \quad (14)$$

The use of the expansion is illustrated in Figure 3. A single multipole expansion for the potential due to the d panel charges is evaluated at the d evaluation points. Since the expansion is used many times, the cost of computing it may be neglected. Thus the multipole evaluation of the same d potentials requires only d operations.

The reduction in complexity resulting from the aggregation of distant panels into multipole expansions which can be used to evaluate potentials at many panel centers is the source of the multipole algorithm's efficiency. Maintaining this efficiency for general distributions of panels while controlling error leads to the hierarchical multipole algorithm used in FASTCAP2. A detailed description of the complete multipole algorithm is given in [1] and its use in the context of capacitance extraction is described in [5, 4].

3.2 Electric Field Evaluations

The evaluation of Dq^k amounts to calculating the left-hand side of n_d equations of the form (5) or, equivalently, (4). The left-hand side of (4) can be approximated by replacing the derivatives by divided differences constructed near panel i as illustrated in Figure 4, yielding

$$\epsilon_a \frac{\partial p_i}{\partial n_i} - \epsilon_b \frac{\partial p_i}{\partial n_i} \approx \epsilon_a \frac{(p_i - p_a)}{h} - \epsilon_b \frac{(p_b - p_i)}{h}. \quad (15)$$

Thus the n multiply-adds required to find one inner product in Dq^k are replaced by three multipole algorithm potential evaluations followed by the divided difference calculation (15). In this way the bulk of the displacement field calculation is performed using the efficient multipole algorithm to compute the potentials $p_i(x_i)$, $p_a(x_a)$ and $p_b(x_b)$.

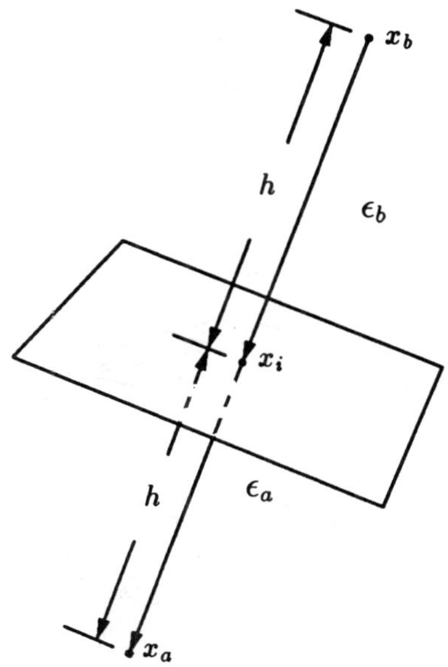

Figure 4: The electric fields on both sides of the dielectric panel are approximated with divided differences.

4 Results

To demonstrate the efficiency and accuracy of the multipole-accelerated capacitance extraction algorithm for problems with multiple dielectrics, the capacitances associated with the easily parameterized bus-crossing structure of Figure 1 are calculated. The Figure 1 problem is called the 2×2 bus-crossing problem and is representative of the 1×1 through 5×5 bus-crossings examined here. In all these problems the lower bus is covered with a layer of conformal dielectric with permittivity $\epsilon_a = 7.5\epsilon_0$ while the surrounding material has permittivity $\epsilon_b = 3.9\epsilon_0$. All the conducting bars have $1\mu m \times 1\mu m$ cross-sections, and all overhang and inter-conductor spacings are $1\mu m$. The conformal dielectric is nominally $0.25\mu m$ thick.

The accuracy attained by our program, FASTCAP2, is investigated using the Figure 1 problem. The smallest coupling and self capacitances in the problem are calculated using FASTCAP2 and by Gaussian elimination applied directly to (11), the standard direct method. The entries in Table 1 represent the capacitances associated with the top, rear conductor in Figure 1. By default FASTCAP2 is configured to produce capacitances within 1% of those calculated using direct factorization, as is clearly the case here. Thus any error in the FASTCAP2 capacitances is dominated by discretization error rather than multipole approximation effects.

	C_{31}	C_{32}	C_{33}	C_{34}
Direct	−0.2112	−0.2112	0.9854	−0.3200
FASTCAP2	−0.2113	−0.2112	0.9886	−0.3212

Table 1: Comparison of capacitances (in fF) for the Figure 1 problem.

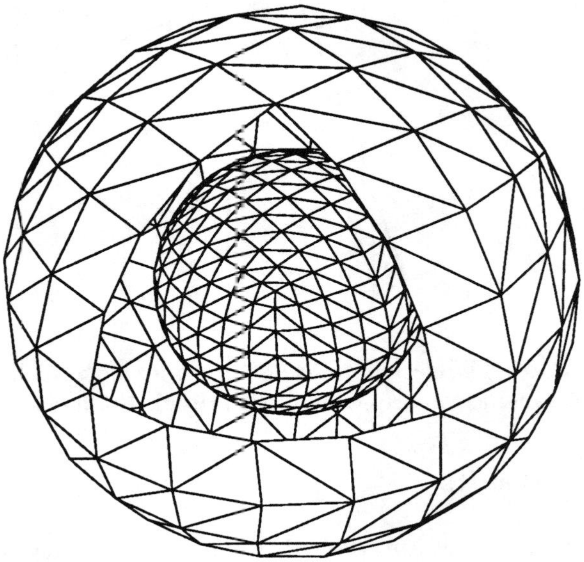

Figure 5: The discretization used to compute the capacitance of a dielectric-coated sphere in free space. Some of the outer dielectric-boundary panels have been removed to show the inner conductor surface panels.

Using the dielectric-coated sphere of Figure 5, FASTCAP2's accuracy is further demonstrated by comparing its result to an analytic value. The inner conducting sphere of radius 1m is coated by a 1m thick dielectric layer with relative permittivity 2. The surrounding region is free space. By Gauss's Law, such a structure has capacitance 148.35pF. The value calculated using FASTCAP2 applied to the discretization of Figure 5 is 148.5pF, well within 1% of the analytic value.

The program's execution speed for the four bus-crossing problems is compared to the speed of the standard direct method in Table 2. The values in parenthesis indicate extrapolated execution times corresponding to problems that could not be solved using the standard method due to excessive memory and time requirements. FASTCAP2's lower complexity leads to much lower execution times for even moderate sized problems like the 2×2 bus crossing. In particular, in the time required to compute the capacitance of the 4×4 bus crossing problem using standard direct methods, FASTCAP2 can perform seventy such calculations.

Problem	1×1	2×2	3×3	4×4
Panels	664	1984	3976	6640
Direct	1.4	41	(320)	(1400)
FASTCAP2	0.44	2.4	8.6	20

Table 2: Comparison of execution times in I.B.M. RS600/540 CPU minutes. Values in parenthesis are extrapolated.

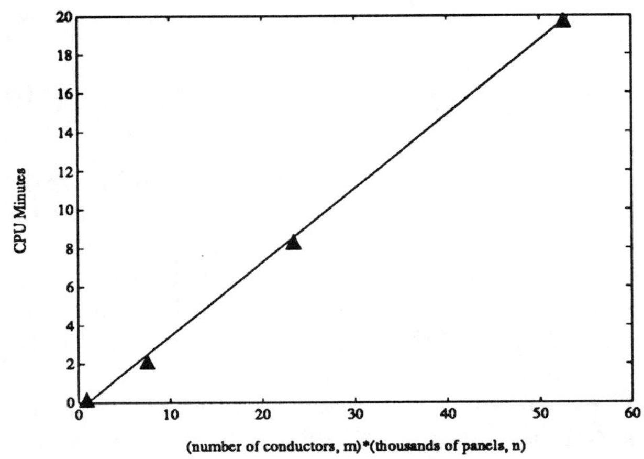

Figure 6: Demonstration of the order mn complexity of FASTCAP2 using the four bus crossing execution times. Times in CPU minutes on an I.B.M. RS6000/540.

Finally, FASTCAP2's execution time is shown to grow roughly linearly with mn, where m is the number of conductors and n is the number of panels. Figure 6 plots the FASTCAP2 execution times in Table 2 verses mn for the bus-crossing problems. As is illustrated, the execution time data points closely follow the best-fit straight line, indicating order mn complexity.

As a final example of the kind of analysis made possible by FASTCAP2, consider the connector in Figure 7. The U-shaped polyester body has a relative permittivity of 3.5; the pins have 0.65mm×0.65mm cross-sections and are spaced 3.25mm center to center. Connectors of this type must be analyzed carefully when used for high-speed bus connections [10]. Using a 9524-panel discretization, FASTCAP2 com-

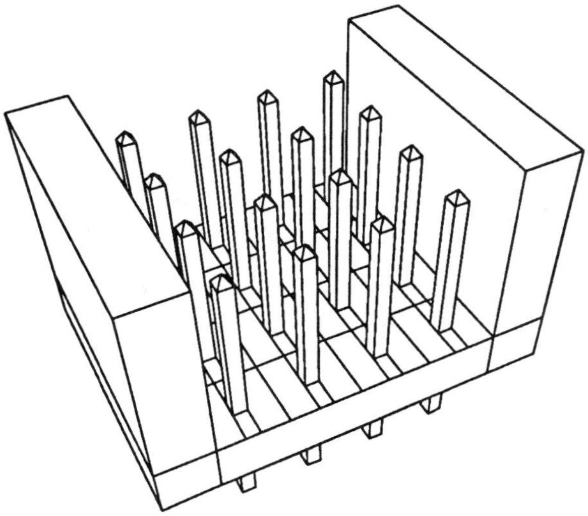

Figure 7: The backplane connector example.

putes all the self and coupling capacitances for the pins in thirty CPU minutes on an IBM RS6000/540. An identical analysis using standard Gaussian elimination algorithms requires roughly eight CPU days on the same machine. The four pins in the the center have the highest self capacitances: 0.547pF. The lowest self capacitance is 0.481pF and is attained by the four corner pins. The strongest coupling capacitances, slightly more than 0.2pF, occur between pin pairs next to the sides of the connector body. By grounding four pins on a diagonal and the four pins on the remaining parallel two-pin diagonals, the maximum signal pin coupling capacitance is reduced to around 0.065pF.

5 Conclusion

The multipole-accelerated capacitance extraction algorithm has been extended to problems with arbitrarily shaped, multiple-dielectric regions. The extended algorithm as implemented in FASTCAP2 has the same 1% accuracy and reduced time and memory requirements of the original algorithm. In particular, FASTCAP2 is fast enough to allow capacitance extraction of complex three-dimensional, multiple-dielectric geometries to be part of an iterative design process.

The authors would like to acknowledge many helpful discussions with Prof. Stephen Senturia and Brian Johnson. We are also grateful to Songmin Kim for his help in implementing FASTCAP2, and to several anonymous reviewers for their careful readings of an earlier draft of this paper.

References

[1] L. Greengard. *The Rapid Evaluation of Potential Fields in Particle Systems.* M.I.T. Press, Cambridge, Massachusetts, 1988.

[2] M. A. Jaswon and G. T. Symm. *Integral Equation Methods in Potential Theory and Elastostatics.* Academic Press, London, 1977.

[3] R. Kress. *Linear Integral Equations.* Springer-Verlag, Berlin, 1989.

[4] K. Nabors, S. Kim, J. White, and S. Senturia. Fast capacitance extraction of general three-dimensional structures. In *Proceedings of the 1991 IEEE International Conference on Computer Design: VLSI in Computers and Processors*, pages 479–484, October 1991.

[5] K. Nabors and J. White. Fastcap: A multipole accelerated 3-D capacitance extraction program. *IEEE Transactions on Computer-Aided Design of Integrated Circuits and Systems*, 10(11):1447–1459, November 1991.

[6] S. M. Rao, T. K. Sarkar, and R. F. Harrington. The electrostatic field of conducting bodies in multiple dielectric media. *IEEE Transactions on Microwave Theory and Techniques*, MTT-32(11):1441–1448, November 1984.

[7] V. Rokhlin. Rapid solution of integral equations of classical potential theory. *Journal of Computational Physics*, 60(2):187–207, September 15, 1985.

[8] A. E. Ruehli and P. A. Brennan. Efficient capacitance calculations for three-dimensional multiconductor systems. *IEEE Transactions on Microwave Theory and Techniques*, 21(2):76–82, February 1973.

[9] Y. Saad and M. H. Schultz. GMRES: A generalized minimal residual algorithm for solving nonsymmetric linear systems. *SIAM Journal on Scientific and Statistical Computing*, 7(3):856–869, July 1986.

[10] W. Whitehead. The role of computer models in connector standardization. In *24th Annual Connector and Interconnection Technology Symposium*, pages 103–110, San Diego, CA, 1991.

SESSION 43

PANEL: THE STATE OF EDA STANDARDS

Chair: John P. Eurich - Engineering DataXpress, Inc., San Jose, CA

During the late 1970's and during the 1980's the Electronic Design Automation industry grew significantly. Many universities and colleges worldwide added Design Automation to their curriculum. The number of companies offering commercial Design Automation products increased by the hundreds. With this growth, many non-standard solutions began to emerge, such as different: operating systems, graphical user interfaces, design description languages, design databases, design editors, data representations formats, frameworks, design, implementation, analysis and verification methodologies, etc. The EDA Tower of Babel was rapidly forming.

To counteract this disparity of solutions, a number of standards groups began to form. Their goal was to provide the EDA industry with standard solutions to specific problem domains, and thus reduce the cost associated with a non-standard approach. Some of the standards that have resulted by these efforts are: CFI, EDIF, IGES, VHDL, Open Verilog, OSF, etc.

But do such standards really work in the EDA industry? This panel addresses just that question. Represented on the panel are people who have helped develop the CFI, EDIF, VHDL, and Open Verilog standards. Also, represented are people who have had years of experience using these standards.

Many questions will be addressed by this panel: "How well have each of these standards met their goal?", "Are they really solving problems or are they creating new problems?", "What changes will be seen in the near future to each of these standards?", "Are these changes addressing the needs of the industry?", "What would the EDA industry be like without these standards?", "What else should be done?", etc.

Panel Members:

Mike Burney - Compaq Computer Corp., Houston, TX
Carlos Dangelo - LSI Logic, Milpitas, CA
Hilary Kahn - Univ. of Manchester, Manchester, UK
Sven Rau - CAE Consulting, Saxau, Germany
Todd Scallan - IBM Corp., Kingston, NY
Moe Shedad - Viewlogic Systems, Inc., Bethesda, MD
K. Tabuchi - Mitsubishi Electric Corp., Itami, Japan

SESSION 44

PANEL: MANUFACTURING INTERFACE

Chair: Charles A. Shaw - Cadence Design Systems, Inc., San Jose, CA
Organizer: Tom Pennino - AT&T Bell Labs., Holmdel, NJ

This panel will focus on the integrated circuit fab and assembly line and the work of its manufacturing engineers. It will explore desirable CAD tool modifications for its products, process, reliability, and test engineers, and brainstorm new workstation-based software to aid them to improve quality, reduce cost and shorten time to market. The panel and the audience will explore alternatives for breaking down the wall between R&D and manufacturing and improving the links to manufacturing process and test computers.

Most of the IC industrial attendees at DAC come from the R&D side. They therefore know that the chip designs produced are breathtaking in their concept. Aided by flawless and magnificent CAD tools, they produce designs that are perfect in construction and performance. But then the product is turned over to manufacturing and sometimes things go wrong!

The panelists will walk us through a day in the life of an IC manufacturing engineer. What are their different specialties? What do they do? What's the impact of their work on the bottom line of the business? How do they work to reduce cost, improve quality and reduce time to market? Panelists will describe tools and facilities developed in their own companies to facilitate this work.

The panel will then invite the audience to join in a brainstorming session on improving the interface and breaking down the wall between R&D and manufacturing. How can ALL the design information be made available to manufacturing efficiently, in a form that it is easy to use? What modifications to CAD tools would adapt them to manufacturing use? What additional Workstation-based facilities would aid the manufacturing engineer?

Panel Members:

Randy Burdick - Advanced Micro Devices, Sunnyvale, CA
Ed Hall - Sematech (Motorola), Austin, TX
James Hogan - National Semiconductor Corp., Santa Clara, CA
Shigeo Ihara - Hitachi Central Labs., Tokyo, Japan
Paul Leonard - Harris Semiconductor Corp., Melbourne, FL

Author Index

NOTES

NOTES

NOTES

NOTES

NOTES